IHS™ Jane's® Land Warfare Platforms

Artillery & Air Defence

2014-2015

Christopher F Foss & James C O'Halloran

ISBN 978 0 7106 3128 2 - Land Warfare Platforms Armoured Fighting Vehicles
ISBN 978 0 7106 3129 9 - Land Warfare Platforms Artillery & Air Defence
ISBN 978 0 7106 3130 5 - Land Warfare Platforms Logistics, Support & Unmanned
ISBN 978 0 7106 3131 2 - Land Warfare Platforms System Upgrades
ISBN 978 0 7106 3121 3 - Land Warfare Platforms Full Set

This book was produced
using FSC® certified paper
Printed and bound in the UK
by Polestar Wheatons

Contents

Front cover montage:

Top left - MBDA SAMP/T surface-to-air missile (SAM) launcher of the French Army deployed in the firing position. This is also referred to as the Mamba by the French Air Force and launches the Aster 30 missile (Christopher F Foss) 1534355

Top right - First production BAE Systems Archer 155 mm/52-calibre self-propelled artillery systems were delivered to Sweden late in 2013 (BAE Systems) 1565418

Bottom left - Launch of Akash missile (DRDO) 1303199

Bottom right - Nexter Systems CAESAR 155 mm/52-calibre (6 × 6) self-propelled artillery system of the French Army deployed in the firing position at Camp Canjuers, France. CAESAR is also deployed by Thailand and the United Arab Emirates and is now in production for Indonesia (Christopher F Foss) 1534354

Alphabetical list of advertisers

Denel Land Systems
PO Box 7710, Pretoria, 0001, South Africa .... *Facing inside front cover*

Roketsan Missiles Industries
PO Box 30, Elmadag, 06780 Ankara, Turkey .... [2]

EDITORIAL AND ADMINISTRATION

Managing Director: Blake Bartlett, e-mail: blake.bartlett@ihs.com

Group Publishing Director: Sean Howe, e-mail: sean.howe@ihs.com

Director IHS Jane's Reference and Data Transformation: Chris Bridge, e-mail: chris.bridge@ihs.com

Director EMEA Editing and Design: Sara Morgan, e-mail: sara.morgan@ihs.com

Product Manager Defence Equipment & Technology: Emma Cussell, e-mail: emma.cussell@ihs.com

Compiler/Editor: Welcomes information and comments from users who should send material to:
Research and Information Services
IHS Jane's, IHS Global Limited, Sentinel House, 163 Brighton Road, Coulsdon, Surrey CR5 2YH
e-mail: yearbook@ihs.com

SALES OFFICES

Europe/Middle East/Africa/Asia Pacific
Tel: (+44 0) 13 44 32 83 00 Fax: (+44 0) 13 44 32 80 05
e-mail: customer.support@ihs.com

North/Central/South America
Tel: Customer care to 1–800–IHS-CARE or 1–800–447–2273
e-mail: customercare@ihs.com

ADVERTISEMENT SALES OFFICES

UNITED KINGDOM
IHS Jane's, IHS Global Limited
Sentinel House, 163 Brighton Road,
Coulsdon, Surrey CR5 2YH, UK
Tel: (+44 20) 32 53 22 89 Fax: (+44 20) 32 53 21 03
e-mail: defadsales@ihs.com

Janine Boxall, Global Advertising Sales Director
Tel: (+44 20) 32 53 22 95 Fax: See UK
e-mail: janine.boxall@ihs.com

Richard West, Senior Key Accounts Manager
Tel: (+44 20) 32 53 22 92 Fax: See UK
e-mail: richard.west@ihs.com

Carly Litchfield, Advertising Sales Manager
Tel: (+44 20) 32 53 22 91 Fax: See UK
e-mail: carly.litchfield@ihs.com

Adam Smith, Advertising Sales Executive
Tel: (+44 20) 32 53 22 93 Fax: See UK
e-mail: adam.smith@ihs.com

UNITED STATES
IHS Jane's, IHS Global Inc.
110 N Royal Street, Suite 200,
Alexandria, Virginia 22314, US
Tel: (+1 703) 683 37 00 Fax: (+1 703) 836 55 37
e-mail: defadsales@ihs.com

Robert Sitch, US Advertising Sales Director,
Tel: (+1 703) 236 24 24 Fax: (+1 703) 836 55 37
e-mail: robert.sitch@ihs.com

Drucie DeVries, South and Southeast USA
Tel: (+1 703) 836 24 46 Fax: (+1 703) 836 55 37
e-mail: drucie.devries@ihs.com

Dave Dreyer, Northeastern USA
Tel: (+1 703) 438 78 38 Fax: (+1 703) 836 55 27
e-mail: dave.dreyer@ihs.com

Janet Murphy, Central USA
Tel: (+1 703) 836 31 39 Fax: (+1 703) 836 55 37
e-mail: janet.murphy@ihs.com

Richard L Ayer, Western USA and National Accounts
127 Avenida del Mar, Suite 2A, San Clemente, California 92672, US
Tel: (+1 949) 366 84 55 Fax: (+1 949) 366 92 89
e-mail: ayercomm@earthlink.com

REST OF THE WORLD
Australia: *Richard West* (UK Office)

Benelux: *Adam Smith* UK Office)

Brazil: *Drucie DeVries* (USA Office)

Canada: *Janet Murphy* (USA Office)

Eastern Europe (excl. Poland): MCW Media & Consulting Wehrstedt
Dr Uwe H Wehrstedt
Hagenbreite 9, D-06463 Ermsleben, Germany
Tel: (+49 03) 47 43/620 90 Fax: (+49 03) 47 43/620 91
e-mail: info@Wehrstedt.org

Germany and Austria: *MCW Media & Consulting Wehrstedt* (see Eastern Europe)

Greece: *Carly Litchfield* (UK Office)

Hong Kong: *Carly Litchfield* (UK Office)

India: *Carly Litchfield* (UK Office)

Israel: *Oreet International Media*
15 Kinneret Street, IL-51201 Bene Berak, Israel
Tel: (+972 3) 570 65 27 Fax: (+972 3) 570 65 27
e-mail: admin@oreet-marcom.com
Defence: Liat Heiblum
e-mail: liat_h@oreet-marcom.com

Italy and Switzerland: *Ediconsult Internazionale Srl*
Piazza Fontane Marose 3, I-16123 Genoa, Italy
Tel: (+39 010) 58 36 84 Fax: (+39 010) 56 65 78
e-mail: genova@ediconsult.com

Japan: *Carly Litchfield* (UK Office)

Middle East: *Adam Smith* (UK Office)

Pakistan: *Adam Smith* (UK Office)

Poland: *Adam Smith* (UK Office)

Russia: *Anatoly Tomashevich*
4-154, Teplichnyi Pereulok, Moscow, Russia, 123298
Tel/Fax: (+7 495) 942 04 65
e-mail: to-anatoly@tochka.ru

Scandinavia: *Falsten Partnership*
23, Walsingham Road, Hove, East Sussex BN41 2XA, UK
Tel: (+44 1273) 77 10 20 Fax: (+ 44 1273) 77 00 70
e-mail: sales@falsten.com

Singapore: *Richard West* (UK Office)

South Africa: *Richard West* (UK Office)

Spain: *Carly Litchfield* (UK Office)

ADVERTISING COPY
Sally Eason (UK Office)
Tel: (+44 20) 32 53 22 69 Fax: (+44 20) 87 00 38 59/37 44
e-mail: sally.eason@ihs.com

For North America, South America and Caribbean only:
Tel: (+1 703) 68 33 700 Fax: (+1 703) 83 65 5 37
e-mail: us.ads@ihs.com

Executive Overview

Artillery

Christopher F Foss

There have been a number of significant developments in the area of artillery systems over the past year, especially around self-propelled (SP).

Early in 2013 the German company Krauss-Maffei Wegmann announced that it had signed a contract with Qatar covering the supply of 62 new Leopard 2A7 main battle tanks and a batch of 24 new PzH 2000 155 mm/52-calibre SP artillery systems together with associated ammunition and training.

These PzH 2000s will replace the currently deployed French 155 mm Mk F3 SP artillery systems that by today's standards are obsolete as they have limited cross-country mobility, no protection for the gun crew, and ammunition has to be carried in separate ammunition re-supply vehicles.

This contract means that the PzH 2000 production line will reopen, which could potentially open the door for additional export sales.

So far PzH 2000 sales have been made to Germany (185), Greece (24), Italy (70), and the Netherlands (57).

As a result of downsizing, quantities of brand new and nearly new PzH 2000s are currently available from Germany and the Netherlands. The German Army PzH 2000 fleet is being reduced from 185 systems to less than 90.

The currently deployed PzH 2000 fires conventional natures of 155 mm ammunition but several users are considering purchasing 155 mm projectiles with a more precision effect; the contenders are the Oto Melara Volcano and the Raytheon Excalibur.

Following the cancellation of the advanced 2S35 twin 152 mm Koalitisya-SV (Coalition), Russia is now working on at least one new SP artillery systems: one tracked and one wheeled.

While China has started to replace its 152 mm towed and tracked SP artillery systems with new weapon systems using the more widely deployed 155 mm calibre, it is by no means clear as to whether Russia will follow suit.

The latest Russian SP artillery system is the 152 mm 2S19 (MSTA-S), which in addition to being in service with the Russian Army has also been exported to a number of countries; the latest being Ethiopia and Venezuela.

In addition to developing a number of upgrades for the 2S19, mainly in the area of fire control, Russia has developed the 2S19M1, which has a modified turret and is armed with the 155 mm/52-calibre NATO gun.

The Russian 152 mm 2S19 ammunition suite consists of the 152 mm projectile and its associated fuze and a cartridge case into which the correct charge is placed. Once fired, the empty cartridge case is ejected through an opening in the lower part of the turret front.

The 2S19M1 can fire standard 155 mm NATO ammunition together with a modular charge system but the exact origins of the latter have not been disclosed. They could have been developed in Russia or been imported.

The currently deployed BAE Systems M109A6 Paladin 155 mm/39-calibre tracked SP artillery system used by the US Army was to have been replaced by the advanced 155 mm Crusader SP artillery system, which had a fully automated ammunition handing system.

Standard Russian 152 mm 2S19M1 SP artillery system in travelling configuration with ordnance in travel lock and roof mounted 12.7 mm machine gun in position (Christopher F Foss) 1527050

If fielded, Crusader would have been the most advanced SP artillery system ever to have been deployed but it was cancelled, however, as it was deemed too heavy at the time.

The Non-Line-Of-Sight Cannon was also cancelled. The 155 mm/38-calibre system was part of the Future Combat System and shared a number of components with the Non-Line-Of-Sight 120 mm mortar (also cancelled).

The US Army then decided to upgrade its current M109A6 Paladin to the M109A6 Paladin Integrated Management (PIM) standard. It placed a low-rate initial production (LRIP) contract with BAE Systems in 2013, following trials with five prototype M109A6 PIM and two PIM Carrier Ammunition Tracked (CAT) vehicles. The LRIP contract covered 19 M109A6 howitzers and 18 PIM CATs.

The latter was originally referred to as the M992 Field Artillery Ammunition Support Vehicle. It carries 155 mm projectiles, charges and fuzes, which are fed to the M109A6 PIM via a conveyor belt when the two systems are back-to-back on the firing point.

Under the terms of the LRIP contract the US Army should take delivery of its first systems in mid-2015 with full rate production expected in 2017.

Two German Army PzH 2000 155 mm/52-calibre self-propelled artillery systems deployed in the firing position (Krauss-Maffei Wegmann) 1043770

One of the five M109A6 PIM prototype systems in travelling configuration with the 155 mm ordnance in travelling lock (BAE Systems) 1340123

It did take delivery of a total of 975 M109A6 Paladins but under current plans it is expected that only 580 M109A6 Paladin PIM systems will be procured, and even this figure could be reduced as the US Army is downsized further.

The trend towards the fielding of wheeled SP artillery systems continues as when compared to their tracked counterparts, they offer the end user a number of significant advantages including lower operating and support costs and great strategic mobility.

The French Nexter Systems CAESAR 155 mm/52-calibre (6 × 6) SP artillery system is already deployed by France, Malaysia and Saudi Arabia and has now been ordered by Indonesia, which has placed a contract for 37 systems.

The French, Indonesian, and Thai CAESAR SP artillery systems use the Renault Trucks Defense Sherpa (6 × 6) chassis, while those for Saudi Arabia use the Mercedes-Benz Unimog (6 × 6) chassis.

The CAESAR has already seen operational service with the French Army in the Lebanon and more recently Mali. The French Army intention is, funding permitting, that all of the currently deployed 155 mm GCT/AUF1 tracked and 155 mm TR1 towed artillery systems will be replaced by CAESAR.

To meet the potential requirements of India, Nexter Systems has worked with Indian company Larsen & Toubro on two 155 mm/52-calibre artillery systems.

For the Indian wheeled SP requirement CAESAR system has recently been integrated on the locally developed Ashok Leyland (6 × 6) truck chassis.

The upper part, for example the mounting and the 155 mm/52-calibre gun is also used on the Trajan towed artillery system, which is being offered to meet the Indian towed gun requirement.

The baseline CAESAR has an unprotected cab to the immediate rear of the engine compartment but armour protected cabs have been developed and deployed; there are also options for different fire-control systems.

The main disadvantage of these wheeled SP artillery systems, with the exception of the BAE Systems Archer, is that the crew have to leave their cab in order to operate the weapon.

To meet the potential requirements of the Italian Army for a wheeled SP artillery system to operate with their Freccia (8 × 8) brigades, Oto Melara has been working on the Centauro 155 mm/39-calibre ultra-lightweight self-propelled wheeled howitzer.

This has a fully protected turret mounted 155 mm/39-calibre ordnance which, like the BAE Systems Archer 155 mm/52-calibre (6 × 6) SP artillery system, is aimed, loaded, and fired with the crew under complete armour protection. This is being developed as a private venture by Oto Melara with some funding from the Italian MoD.

For the Indian market Nexter Systems has teamed with Larsen & Toubro and Ashok Leyland Defense to offer CAESAR, integrated on this local (6 × 6) chassis (Nexter Systems) 1525853

BAE Systems Archer 155 mm/52-calibre SP artillery system carrying out a fire mission (FMV) 1513519

BAE Systems was awarded a contract for 48 Archer systems, 24 for Norway and 24 for Sweden, which took delivery of its first four systems in late 2013.

Norway subsequently cancelled its order whilst production was underway and has stated that it will procure another 155 mm SP system in the future.

Production of the BAE Systems 155 mm/39-calibre M777 has been completed for the time being and this is currently in service with Australia, Canada and the United States (Army and Marines). Production could restart if the long awaited Indian contract for up to 145 systems is ever signed.

Although traditionally operated by the infantry, overall command, control, and operation of mortars is being carried out by the field artillery in a number of countries, especially 120 mm mortars.

Self-propelled mortar systems are therefore included in *IHS Jane's Land Warfare Platforms: Artillery and Air Defence*.

Vehicle-mounted mortar systems are normally mounted in the rear of tracked or wheeled armoured platforms and fire through open roof hatches, leaving the crew exposed to attack from the overhead munitions.

There is also a trend towards turret-mounted mortars that enable the mortar to be aimed and fired with the crew under complete protection not only from small arms fire and shell splinters, but from NBC attacks. They also have a direct fire capability.

The Russian Army has used 120 mm rifled turret-mounted mortar systems for many years including the tracked 120 mm 2S9 system and the 120 mm 2S23 wheeled system, which is based on the BTR-80 (8 × 8) armoured personnel carrier hull.

In addition to having the normal indirect fire capability both of these systems have a direct fire capability which would be useful not only in the self-defence role but also in counter-insurgency type operations.

The latest Russian 120 mm 2S31 Vena SP gun/mortar system, based on a modified BMP-3 infantry fighting vehicle hull, was first shown in public as far back as the mid-1990s but was never produced in production quantities for the Russian Army.

The 2S1 is now in production for the export market with known sales having been made to Azerbaijan and most recently Venezuela.

This is a significant improvement over the earlier 2S9 and 2S23 systems as not only has it a longer 120 mm ordnance but also an on board computerised fire-control system and a defensive aids suite.

It is understood that the Russian Army has taken delivery of a quantity of 2S34 SPMs, which is essentially the older 122 mm 2S1 SP artillery system but with the 122 mm ordnance replaced by the 120 mm 2A80 rifled ordnance of the 2S31.

While the main emphasis of the artillery section is on the actual platform or its associated weapon system, some countries are either upgrading their platforms or procuring more effective ammunition such as projectiles, fuzes and their associated propelling charges.

Production standard 120 mm 2S31 self-propelled gun/mortar system in travelling configuration (Christopher F Foss) 1521951

The range of 155 mm artillery systems can be enhanced by firing rocket-assisted or base-bleed projectiles.

As range increases, however, they often become less accurate meaning that more rounds have to be fired to neutralise the target.

The 155 mm Raytheon Excalibur precision-guided munition (PGM) has been used in combat by Canada and the United States, providing the field artillery with a precision effect. These are expensive projectiles, although their cost could be driven down with higher production quantities.

Artillery systems and mortars have proven to be highly effective indirect fire weapons in recent conflicts and although air power can deliver very accurate munitions, it is only the field artillery that can rapidly provide suppressive fire on a 24-hour basis.

Precision munitions, which are already being fielded by some countries, will enable the field artillery to expand its capability on battlefields in the future.

Air Defence
James C O'Halloran

It only seems like only yesterday that I was sitting here writing my element of the Executive Overview for the Air Defence section of *IHS Jane's Land Warfare Platforms: Artillery and Air Defence* 2013-2014. The year has gone by so quickly, probably because we have all been so busy with air defence matters that we didn't notice the weeks and months fly by; the loss of the Malaysian Airline Boeing 777-200 is at the forefront of our minds.

Air defence in Asia has recently taken a hard knock due to this situation, in particular the use and capabilities of long-range air defence tracking radars.

Questions must be asked and answers found as to why, when so much money is being spent on upgrading older systems, and even more spent on ultra-modern new radar systems deployed within Asia, was this aircraft allowed to fly and/or bypass military radar?

For instance, how can a commercial aircraft (or any other aircraft for that matter), fly without any communication or transpondance anywhere near Diego Garcia?

Have we learnt nothing from New York or Washington? Surely if this is the case, then money invested today on air-defence systems in Asia would be much better spent on improving the lifestyle of the population.

The Southeast Asian area has seen some major changes and upgrades in the last 12 months; India being one country in parrticular.

With the threats in mind from Pakistan and China, India has developed a layered defence that puts the country in the enviable position of being able to defend itself from intercontinental ballistic missiles in the exo-atmospheric area with the Prithvi Defence Vehicle, endo-atmosphere with the Prithvi Air Defence missile, and other air defence systems such as Akash for tactical defence.

It was also reported that Vietnam had taken delivery of the Russian Pantsyr-S1 system and indeed a photograph of such was released. However, this delivery/deployment has not yet been proven but nevertheless remains a possibility as Vietnam could be deploying the Pantsyr in defence of the already purchased S-300PMU-1 systems, much like how Russia is defending the S-400 with the Pantsyr.

China meanwhile allegedly secretly conducted an anti-satellite (ASAT) test in May 2013, the first since the SC-19 test in January 2007. The missile used for this test was reported as the Dong Ning-2 (DN-2) and was described by Chinese sources as a ground-based, high earth-orbit attack missile. The system is also reported to be a kinetic kill vehicle (hit-to-kill). If the assessment of such a system is true, then this is a very significant development in China's ASAT capabilities.

Another coup for China is that Russia's president, Vladimir Putin, in late March 2014, gave authority to sell the S-400 air defence guided missile system to China and provide protection over islands in the East China Sea at the centre of a dispute with Japan.

Beijing has been interested in acquiring the guided missile system since 2011. Export sales of the system may not begin until 2016 or even later, after Russia has filled its quota of the S-400. In light of the Crimea situation, the numbers of these systems will most likely rise.

In Taiwan, further deployments of the US Patriot PAC-3 missile systems to counter the Chinese build up just across the Taiwan Straits continues apace. During April 2013 a PAC-3 system was deployed to the north of the country and it was reported by the then Deputy Minister of National Defense, Andrew Yang, that the next three systems would be deployed in the south, probably in late 2014.

In Seoul, South Korea, the *Yon Hap News* reported on the 12 March 2014 that South Korea will upgrade its present PAC-2 air defence system and buy PAC-3 missiles next year to improve its anti-ballistic missile capability against North Korea. It is assessed that South Korea currently operates 48 Patriot PAC-2 missiles imported from Germany.

In the Middle East there has been some movement in the Turkish desire for a long-range air-defence system; a contract between Turkey and China was initially set up but was then referred back for renegotiations. NATO and the United States came down heavily on Turkey following the announcement that it was going to purchase the Chinese FT-2000 system.

Firstly, questions were asked as to how the system would fit in with the NATO air-defence and the command-and-control. Naturally NATO became increasingly concerned about having Chinese missile experts being given clearance to see and use NATO C3 equipment, including the cryptographic side of things.

Another question in this long drawn out story was if the FT-2000 is so good, why ask NATO and the United States for help in deploying Patriot systems along the border with Syria, and then ask for a six-month extension in time to keep the Patriot systems deployed on the border?

I suppose what can only best be described as a major step forward for peace in the Middle East is the proposal for Israel to make use of the Arrow-3 anti-ballistic missile system to include the defence of its neighbours Jordan and possibly Egypt.

Lockheed Martin photograph showing a Patriot PAC-3 missile being launched (Lockheed Martin) 1525993

IAI Arrow 2 and 3 missiles (IHS/Patrick Allen) 1523776

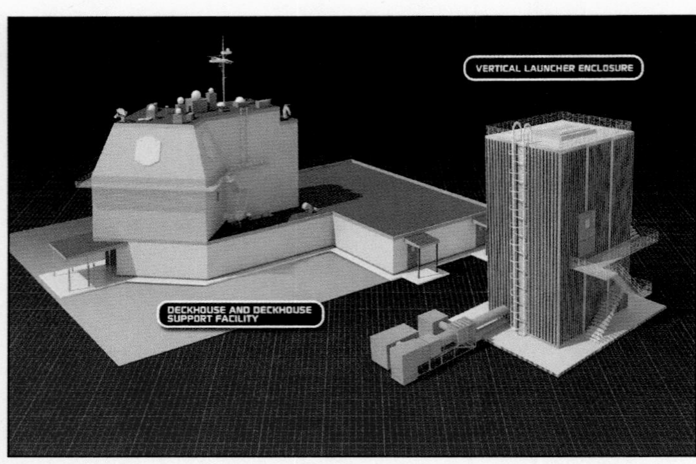

Aegis Ashore, a four-storey deckhouse and launch tower will provide a static home in Romania and Poland for the SPY-1D(V) radar, SM-3 missiles and other systems taken from the US Navy's BMD ships (MDA) 1460804

This suggestion has been cautiously welcomed by Israel but will need a lot of negotiations before such a capability could be extended to include those countries.

The United States will no doubt play its part in any such collaboration probably backing the system and idea with the US Dollar. Israel signed a peace treaty with Egypt in 1979 and with Jordan in 1994.

Meanwhile, in the European Theatre of Operations, Russia continues to dominate the news in particular with the continued development of new and more modern defensive systems.

The S-500 is reported to be nearing the last part of its development with early deployment for some parts of the system probably in late 2015 or early 2016.

This then begs the final question; working on the analysis of past system development within the former USSR and now Russia, is there yet another even longer-range air defence system in the planning stage? There has in the past been mention of the S-1000, maybe this or another air or ballistic missile defensive system will make the open press over the next year.

The other system to make news during early 2014 was the US deployment in Romania of the Aegis Ashore system. It is understood that the first ground clearance and work in the area is already under way.

Acknowledgements

One of the last tasks of the editors is to thank everyone who has been involved, however small the part, in the preparation and production of this title. It is very much a team effort.

The Editors would like to thank the many manufacturers, defence forces and individuals that have and continue to contribute to the Artillery and Air Defence section of the Land Warfare Platforms module.

Special thanks are due to Paul Beaver, Desmond Dempsey, Peter Felstead, Marsh Gelbert, Ben Goodlad, David C Isby, Michael Jerchel, Ian Kemp, Francisco A Marin, David Markov, Barry Marriot, Julio Montes, Rupert Pengelly, E Q Tee, Steven Zaloga and C R Zwart for their most valuable assistance.

The drawings used in *IHS Jane's Land Warfare Platforms: Artillery and Air Defence* have been prepared by a number of people. Steven Zaloga has allowed us to reproduce some of his excellent work, while others have been provided by Vasco Barbic, Sam O'Halloran and Henry Morshead.

Christopher F Foss would like to take this opportunity to thank his wife, Sheila, who has been of invaluable assistance in the preparation of this edition of *IHS Jane's Land Warfare Platforms: Artillery and Air Defence.* Sam Reynolds, Melanie Rovery, and Peter Partridge and team, have also given the editor maximum possible support throughout.

James C O'Halloran would like to give special thanks to his grandson Sam O'Halloran, who has provided some fantastic graphics, and also to Miroslav Gyürösi for all his support.

Thanks are also due to Raymond Cheung, Terry J Gander, Stefan Marx, Cookie Sewell, Pierre Touzin.

Christopher F Foss

Christopher F Foss is the Editor of *IHS Jane's Land Warfare Platforms: Armoured Fighting Vehicles* and Co-editor of *IHS Jane's Land Warfare Platforms: Artillery and Air Defence, IHS Jane's Land Warfare Platforms: Logistics, Support, and Unmanned* and *IHS Jane's Land Warfare Platforms: Systems Upgrades*; he is also the Land Consultant for *IHS Jane's Defence Weekly* and *IHS Jane's International Defence Review.*

Chris began his association with *IHS Jane's* as early as 1970, when he began writing for the then *Jane's Weapon Systems.* Since 1979 he has acted as a self-employed contractor with the company and a number of other organisations and companies throughout the world. His achievements have included conceiving and editing *Jane's Defence Review*, the first magazine ever produced by *Jane's Information Group*, and proposing and launching *Jane's Missiles and Rockets.*

He has written numerous books relating to armoured vehicles and weapons systems both for *IHS Jane's* and for other publishers. He has also lectured in a number of countries and chaired many conferences.

James C O'Halloran

James C O'Halloran is a private consultant whose background includes working for a leading aerospace company as a consultant to the Ministry of Defence on the Defence Intelligence Staff as a weapons systems engineer from June 1984 through to December 1998. This work involved the design and analysis of guided weapon systems and their associated equipment, including related weapon control radars, communications and system simulators.

Prior to joining the Ministry of Defence, Jim served in the Royal Navy for in excess of 20 years within the Communications and Electronic Warfare branch. His naval career included shore time spent at the Government Communications Headquarters in Cheltenham and time at the Ministry of Defence in Naval Intelligence.

Sea time was spent on various ships of the Fleet, including Royal Fleet Auxiliary's, and four years as the operational Chief of the Fleet Electronic Warfare Support Group, a small, smart and select group of individual experts available at a moment's notice to go anywhere in the world where their particular skills could be used to forward the aims of Her Majesty's Government.

Measurements

Military Radar Frequency/Wavelength Values

Military radar frequencies are described by two concurrent systems of 'bands', the first of which has evolved from the system used during the Second World War and is the US IEEE radar band system. During 1969, NATO (the North Atlantic Treaty Organisation) adopted a second system, which originated within the Electronic Warfare (EW) community and is still, somewhat confusingly, referred to as the 'EW' system within some quarters. Both systems are presented here for completeness

Current NATO Radar/EW bands

Designation	*Frequency*	*Wavelength*
A	30-250 MHz	1,000-120 cm
B	250-500 MHz	120-60 cm
C	500 MHz-1 GHz	60-30 cm
D	1-2 GHz	30-15 cm
E	2-3 GHz	15-10 cm
F	3-4 GHz	10-7.5 cm
G	4-6 GHz	7.5-5.0 cm
H	6-8 GHz	5.0-3.75 cm
I	8-10 GHz	3.75-3.0 cm
J	10-20 GHz	3.0-1.5 cm
K	20-40 GHz	1.5-0.75 cm
L	40-60 GHz	0.75-0.50 cm
M	60-100 GHz	0.50-0.30 cm

Historic (IEEE US) radar/radio bands

Designation	*Frequency*	*Wavelength*
VHF	30-300 MHz	1,000-100 cm
UHF	300 MHz-1 GHz	100-30 cm
L	1-2 GHz	30-15 cm
S	2-4 GHz	15-7.5 cm
C	4-8 GHz	7.5-3.75 cm
X	8-12 GHz	3.75-2.5 cm
Ku	12-18 GHz	2.5-1.6 cm
K	18-27 GHz	1.6-1.1 cm
Ka	27-40 GHz	1.1-0.75 cm
MM	40-100 GHz	0.75-0.3 cm

Major Military Radar Applications

Air surveillance	Ground-controlled interception (abbreviated to 'GCI')
	Height-finding
	Target acquisition
	Long-range early warning (including Airborne Early Warning - abbreviated to 'AEW')
	Air traffic management (abbreviated to 'ATM') functions
Tracking and guidance	Fire control for Anti-Aircraft Artillery (abbreviated to 'AAA'), naval gunfire, land artillery and mortars and missile systems
	Missile guidance
	Precision landing and approach systems
	Range instrumentation
Surface search and battlefield surveillance	Sea search and navigation
	Harbour and waterway management
	Ground mapping
	Mortar and artillery location
	Intruder detection
	Mine detection
	Airfield taxiway management
Space and missile surveillance	Ballistic missile warning and acquisition
	Satellite surveillance
Weather surveillance	Observation/prediction of wind and precipitation conditions
	Weather avoidance (for aircraft)
	Clear-air turbulence detection

Other Radar/Radio Frequency/Wavelength Values

In addition and for completeness, the International Telecommunication Union (ITU) and prior old radio communication frequency band designations are covered in the following two tables.

ITU bands

Designation	*Frequency*	*Wavelength*	Meaning
ELF	3-30 Hz	100,000-10,000 km	Extremely Low Frequency
SLF	30-300 Hz	10,000-1,000 km	Super Low Frequency
ULF	300 Hz-3 kHz	1,000-100 km	Ultra Low Frequency
VLF	3-30 kHz	100-10 km	Very Low Frequency
LF	30-300 KHz	10-1 km	Low Frequency
MF	300 kHz-3 MHz	1,000-100 m	Medium Frequency
HF	3-30 MHz	100-10 m	High Frequency
VHF	30-300 MHz	10-1 m	Very High Frequency
UHF	300 MHz-3 GHz	100-10 cm	Ultra High Frequency
SHF	3-30 GHz	10-1 cm	Super High Frequency
EHF	30-300 GHz	10-1 mm	Extremely High Frequency

Old frequency band designations

Designation	*Frequency*	*Wavelength*
I	100-150 MHz	300-200 cm
G	150-225 MHz	200-130 cm
P	225-390 MHz	130-77 cm
L	390 MHz-1.55 GHz	77-20 cm
S	1.55-5.2 GHz	20-5.8 cm
X	5.2-10.9 GHz	5.8-2.8 cm
K	10.9-36 GHz	2.8-0.83 cm
Q	36-46 GHz	0.83-0.65 cm
V	46-56 GHz	0.65-0.53 cm

Russian/NATO Frequency Comparison Table

		Abbreviation				Abbreviaton		
Band No	Designation	Russia	NATO	Frequency	Metric	Russian	NATO	Wavelength
1	extremely low frequencies	KNCH		3-30 Hz	decamegameter waves			100-10 Mm
2	super low frequencies	SNCH	ELF	30-300 Hz	megameter waves			10-2 Mm
3	ultra low frequencies	INCH		0.3-3 kHz	hectokilometer waves			1,000-100 km
4	very low frequencies	ONCH	VLF	3-30 kHz	myriameter waves	SDV		100-10 km
5	low frequencies	NCH	LF	30-300 kHz	kilometer waves	DV	LW	10-1 km
6	medium frequencies	SCH	MF	0.3-3 MHz	hectometer waves	SV	MW	1-0.1 km
7	high frequencies	VCH	HF	3-30 MHz	decameter waves	KV	SW	100-10 m
8	very high frequencies	OVCH	VHF	30-300 MHz	meter waves			10-1 m
9	ultra high frequencies	UVCH	UHF	0.3-3 GHz	decimeter waves			1-0.1 m
10	super high frequencies	SVCH	SHF	3-30 GHz	centimeter waves	UKV	USW	10-1 cm
11	extremely high frequencies	KVCH	EHF	30-300 GHz	millimeter waves			10-1 mm
12	hyper high frequencies	GVCH		0.3-3 THz	decimillimeter waves			1-0.1 mm

The US 'AN' Electronic Equipment Identification System

Starting during the Second World War, the US military established the 'Army-Navy' (abbreviated to 'AN') system for the identification of its electronic equipment. Currently, this system is applied to electronic equipment taken into the inventories of the USAF, US Navy, US Army and USMC. 'AN' designations are also increasingly applied to military electronic systems of American origin but which have not been taken up by the country's military. 'AN' identifiers are presented thus: **AN/APG-66(V)3** in which **APG** identifies the equipment's installation type (first letter), the equipment's type (second letter) and its purpose (third letter); **–66** indicates that the particular equipment is the 66th of its type included in the system; **(V)** indicates that the equipment can be configured to suit a number of platforms and/or system applications and **3** indicates that it is the third such variable configuration produced. The initial installation/type/purpose group can be read as follows:

Installation identifier		*Type identifier*		*Purpose identifier*	
A	Piloted aircraft	**A**	Invisible light/heat radiation	**B**	Bombing
B	Underwater mobile/submarine	**C**	Carrier	**C**	Communications
D	Pilotless carrier	**D**	Radiac	**D**	Direction-finding/surveillance
F	Fixed ground	**G**	Telegraph/teletype	**E**	Release/ejection
G	General ground use	**I**	Interphone/public address	**G**	Fire control
K	Amphibious	**J**	Electromechanical/inertial wire covered	**H**	Recording/reproduction
M	Ground mobile	**K**	Telemetry	**K**	Computing
P	Portable	**L**	Countermeasures	**M**	Test/maintenance
S	Water	**M**	Meteorological	**N**	Navigation
T	Ground Transportable	**N**	Sound in air	**Q**	Special purpose
U	General utility	**P**	Radar	**R**	Receiver
V	Ground vehicular	**Q**	Sonar/underwater sound	**S**	Search/detection/range bearing
W	Water (surface/subsurface applications combined)	**R**	Radio	**T**	Transmitting
Z	Unmanned/piloted air vehicle combination	**S**	Special/combination of purposes	**W**	Automatic flight/remote control
		T	Telephone (wire)	**X**	Identification/recognition
		V	Visible light	**Y**	Surveillance and control
		W	Armament		
		X	Facsimile/TV		
		Y	Data processing		

Accordingly, **AN/APG-66(V)3** can be identified as the third variable configuration subvariant of the 66th airborne fire-control radar identified within the system. **AN/APY-1** identifies the equipment as being the system's first registered airborne surveillance and control radar, while **AN/APX-109(V)** represents its 109th airborne radar identification/recognition system which, like APG-66(V), can be configured for variable applications. The reader should also be aware of the use of a suffix letter to identify succeeding generations of equipment. Thus, **AN/APX-12A** identifies the second-generation of the system's 12th ground mobile radar identification/recognition equipment to be produced.

Units of measurement

Abbreviation	*Unit*	*Measurement*
A	ampere	electric current
Ah	ampere hour	electric charge
atm	standard atmosphere	pressure
B	bel	sound pressure level
Bd	baud	signalling events
bhp	brake horsepower	power
bps	bits/second	data processing
BTU	British Thermal Unit	energy
C	Celsius	temperature
C	coulomb	electric charge
cal	calorie	energy
cd	candela	luminous intensity
cc	cubic centimetre	volume
cl	centilitre	volume
cm	centimetre	length
cp	candelpower	luminous intensity
cps	cycles/second	frequency
cwt	hundredweight	mass
Da	Dalton	mass
daN	decanewton	force
dB	decibel	sound pressure level
dm	decimetre	length
dP	differential pressure	pressure
dwt	deadweight tonnage	mass

Abbreviation	*Unit*	*Measurement*
ehp	equivalent horsepower	power
ekW	equivalent kilowatt	power
emf	electromotive force	electromotive force
eV	electron volt	energy
F	fahrenheit	temperature
F	farad	electrical capacitance
FL	foot lambert	luminance
fl oz	fluid ounce	volume
ft	foot	length
g	gram, gramme	mass
g	gravity	acceleration
gal	gallon	volume
GHz	gigahertz	frequency
gpm	gallons/minute	flow rate
grt	gross registered tonnage	mass
h	hour	time
H	henry	inductance
hp	horsepower	power
Hz	hertz	frequency
ihp	indicated horsepower	power
in	inch	length
ips	instructions/second	data processing
J	Joule	energy
K	kelvin	temperature

Abbreviation	Unit	Measurement
kb	kilobyte	data processing
kg	kilogram	mass
kHz	kilohertz	frequency
kips	thousand instructions/second	data processing
kJ	kilojoule	energy
km	kilometre	length
kmh	kilometres/hour	velocity
kN	kilonewton	force
kops	thousand operations/second	data processing
kph	kilometres/hours	velocity
kt	knot	velocity
kT	kiloton	explosive power
kV	kilovolt	electromotive force
kVA	kilovolt-ampere	electrical power
kW	kilowatt	power
l	litre	volume
lb	pound	mass
lm	lumen	luminous flux
lx	lux	illuminance
m	metre	length
M	mach (speed of sound)	velocity
mb	millibar	pressure
Mbit	megabit	data processing
Mbyte	megabyte	data processing
mg	milligram	mass
MHz	megahertz	frequency
min	minute	time
Mips	million instructions/second	data processing
MJ	megajoule	energy
ml	millilitre	volume
mm	millimetre	length
Mops	million operations/second	data processing
mph	miles/hour	velocity
ms	millisecond	time
mW	milliwatt	power
MW	megawatt	power
N	newton	force (weight)
n mile	nautical mile	length
nm	nanometre	length
ns	nanosecond	time
N s	newton second	impulse (thrust)
oz	ounce	mass
P	poise	viscosity
Pa	pascal	pressure
pF	picofarad	electrical capacitance
pps	pulse per second	frequency
psi	pounds/square inch	pressure
pt	pint	volume
rad	radian	angle
rpm	revolutions/minute	rotation (frequency)
rps	revolutions/second	rotation (frequency)
s	second	time
S	Siemens	conductivity
S	stoke	viscosity
shp	shaft horsepower	power
sr	steradian	angle
st	stone	mass
t	tonne	mass
T	tesla	magnetic flux density
tpi	threads/inch	screw thread
t/c	thickness to chord	ratio
u	atomic mass unit	mass
V	volt	electromotive force
VA	volt-ampere	electrical power
vpm	vibrations/minute	vibration (frequency)
W	watt	power
Wb	weber	magnetic flux
yd	yard	length

Prefixes

Multiples and sub-multiples

Y	yotta	million million million million
Z	zetta	thousand million million million
E	exa	million million million
P	peta	thousand million million
T	tera	million million
G	giga	thousand million
M	mega	million
k	kilo	thousand
h	hecto	hundred
da	deca	ten
d	deci	tenth
c	centi	hundredth
m	milli	thousandth
μ	micro	millionth
n	nano	thousand millionth
p	pico	million millionth
f	femto	thousand million millionth
a	atto	million million millionth
z	zepto	thousand million million millionth
y	yocto	million million million millionth

Conversion factors

Mass

1 grain	=	0.0648 gram
1 gram	=	15.43 grains
1 gram	=	0.03527 ounces
1 kilogram	=	1,000 grams
1 kilogram	=	2.205 pounds
1 tonne	=	1,000 kilograms
1 tonne	=	0.9842 tons (UK)
1 tonne	=	1.102 tons (US)
1 dram	=	1.772 grams
1 ounce	=	16 drams
1 ounce	=	28.35 grams
1 pound	=	453.6 grams
1 pound	=	0.4536 kilograms
1 pound	=	0.00045 tonnes
1 stone	=	14 pounds
1 stone	=	6.35 kilograms
1 quarter	=	2 stones
1 quarter	=	12.7 kilograms
1 hundredweight	=	4 quarters
1 hundredweight	=	50.8 kilograms
1 ton (UK)	=	20 hundredweight
1 ton (UK)	=	1.016 tonnes
1 ton (US)	=	2,000 pounds
1 ton (US)	=	907.18 kilograms

Volume

1 cubic centimetre	=	0.06102 cubic inches
1 cubic metre	=	35.32 cubic feet
1 cubic metre	=	1.306 cubic yards
1 litre	=	61.02 cubic inches
1 litre	=	0.22 imperial gallons
1 litre	=	0.264 US gallons
1 pint	=	20 fluid ounces
1 pint	=	0.568 litre
1 quart	=	2 pints
1 quart	=	1.136 litres
1 imperial gallon	=	4 quarts
1 imperial gallon	=	4.546 litres
1 peck	=	2 gallons
1 peck	=	9.092 litres
1 bushel	=	4 pecks
1 bushel	=	36.4 litres
1 quarter	=	8 bushels
1 quarter	=	291.2 litres
1 US gallon	=	3.785 litres
1 cubic inch	=	16.39 cubic centimetres
1 cubic inch	=	0.01639 litres
1 cubic foot	=	0.028 cubic metres
1 cubic yard	=	0.7646 cubic metres
1 barrel of oil	=	42 US gallons
1 barrel of oil	=	35 imperial gallons

Length

1 millimetre	=	0.03937 inches
1 centimetre	=	10 millimetres
1 centimetre	=	0.3937 inches
1 metre	=	100 centimetres
1 metre	=	3.28 feet
1 metre	=	1.094 yards
1 kilometre	=	1,000 metres
1 kilometre	=	0.6214 miles
1 kilometre	=	0.539 nautical miles
1 inch	=	25.4 millimetres
1 foot	=	12 inches
1 foot	=	0.3048 metres
1 yard	=	3 feet
1 yard	=	0.9144 metres
1 fathom	=	6 feet
1 chain	=	22 yards
1 furlong	=	10 chains
1 mile	=	8 furlongs

1 mile	=	1,760 yards
1 mile	=	1.609 kilometres
1 mile	=	0.868 nautical miles
1 nautical mile	=	1.852 kilometres
1 nautical mile	=	1.152 miles

Angle

1 radian	=	57.3 degrees
1 degree	=	17.45 milliradians
1 degree	=	60 arc minutes
1 milliradian	=	3.438 arc minutes
1 arc minutes	=	0.2909 milliradians
1 arc minutes	=	60 arc seconds

Angular velocity

1 radian/second	=	57.3 degrees/second
1 degree/second	=	0.01745 radian/second

Linear velocity

1 centimetre/second	=	0.033 feet/second
1 metre/second	=	3.281 feet/second
1 kilometre/hour	=	0.621 miles/hour
1 kilometre/hour	=	0.539605 knots
1 foot/second	=	30.48 centimetres/second
1 foot/second	=	0.305 metres/second
1 miles/hour	=	1.609 kilometre/hour
1 mile/hour	=	0.868 knots
1 knot	=	1.852 kilometres/hour
1 knot	=	1.152 miles/hour
1 Mach	=	340.3 metres/second
1 Mach	=	1116.5 feet/second

Mach figure taken at sea level conditions (+15°C, 1.225 kg·m^{-3} and 101,325 Pa).

Area

1 square centimetre	=	0.155 square inches
1 square kilometre	=	0.3861 square miles
1 square metre	=	10.76 square feet
1 square metre	=	1.196 square yards
1 hectare	=	10,000 square metres
1 hectare	=	2.471 acres
1 square inch	=	6.452 square centimetres
1 square foot	=	0.0929 square metres
1 square yard	=	0.8361 square metres
1 square mile	=	2.59 square kilometres
1 acre	=	4,840 square yards
1 acre	=	4,047 square metres
1 acre	=	0.4047 hectares

Power, force and energy

1 kilowatt	=	1.34102 horsepower
1 horsepower	=	0.7457 kilowatt
1 newton	=	0.2248 pound-force
1 pound-force	=	4.448 newtons
1 joule	=	0.2388 calories
1 kilojoule	=	0.9478 British thermal unit
1 British thermal unit	=	1.055 kilojoules
1 calorie	=	4.1868 joules

Temperature

Celsius × ($^{9}/_{5}$ + 32)	=	Fahrenheit
Fahrenheit - (32 × $^{5}/_{9}$)	=	Celsius

°C	°F	°C	°F	°C	°F	°C	°F
1	33.8	26	78.8	51	123.8	76	168.8
2	35.6	27	80.6	52	125.6	77	170.6
3	37.4	28	82.4	53	127.4	78	172.4
4	39.2	29	84.2	54	129.2	79	174.2
5	**41.0**	**30**	**86.0**	**55**	**131.0**	**80**	**176.0**
6	42.8	31	87.8	56	132.8	81	177.8
7	44.6	32	89.6	57	134.6	82	179.6
8	46.4	33	91.4	58	136.4	83	181.4
9	48.2	34	93.2	59	138.2	84	183.2
10	**50.0**	**35**	**95.0**	**60**	**140.0**	**85**	**185.0**
11	51.8	36	96.8	61	141.8	86	186.8
12	53.6	37	98.6	62	143.6	87	188.6
13	55.4	38	100.4	63	145.4	88	190.4
14	57.2	39	102.2	64	147.2	89	192.2
15	**59.0**	**40**	**104.0**	**65**	**149.0**	**90**	**194.0**
16	60.8	41	105.8	66	150.8	91	195.8
17	62.6	42	107.6	67	152.6	92	197.6
18	64.4	43	119.4	68	154.4	93	199.4
18	66.2	44	111.2	69	156.2	94	201.2
20	**68.0**	**45**	**113.0**	**70**	**158.0**	**95**	**203.0**
21	69.8	46	114.8	71	159.8	96	204.8
22	71.6	47	116.6	72	161.6	97	206.6
23	73.4	48	118.4	73	163.4	98	208.4
24	75.2	49	120.2	74	165.2	99	210.2
25	**77.0**	**50**	**122.0**	**75**	**167.0**	**100**	**212.0**

This table reads in both directions, showing equivalent Celsius and Fahrenheit temperatures. It is based on whole number Celsius temperatures.

Where whole number Fahrenheit temperatures occur, they and their Celsius equivalents are shown in bold type. Approximate Celsius equivalent for whole number Fahrenheit temperatures can be worked out for example, 54°F will be between 12°C and 13°C, but nearer 12°C. It is in fact 12.2°C.

Exact Celsius equivalents for Fahrenheit temperatures can be found by using the equation: -32, × 5, ÷ 9. For example 54°F -32 = 22 × 5 = 110 ÷ 9 = 12.2°C.

Periodic table

Symbol	*Element*	*Atomic number*	*Atomic mass*
Ac	Actinium	89	(227)
Ag	Silver	47	107.868
Al	Aluminium	13	26.9815
Am	Americium	95	(243)
Ar	Argon	18	39.948
As	Arsenic	33	74.9216
At	Astatine	85	210
Au	Gold	79	196.9665
B	Boron	5	10.81
Ba	Barium	56	137.34
Be	Beryllium	4	9.0122
Bi	Bismuth	83	208.9806
Bk	Berkelium	97	(249)
Br	Bromine	35	79.904
C	Carbon	6	12.011
Ca	Calcium	20	40.08
Cd	Cadmium	48	112.4
Ce	Cerium	58	140.12
Cf	Californium	98	(251)
Cl	Chlorine	17	35.453
Cm	Curium	96	(247)
Co	Cobalt	27	58.9332
Cr	Chromium	24	51.996
Cs	Caesium	55	132.9055
Cu	Copper	29	63.546
Dy	Dysprosium	66	162.5
Er	Erbium	68	167.26
Eu	Europium	63	151.96
F	Fluorine	9	18.9984
Fe	Iron	26	55.847
Fm	Fermium	100	(253)
Fr	Francium	87	(223)
Ga	Gallium	31	69.72
Gd	Gadolinium	64	157.25
Ge	Germanium	32	72.59
H	Hydrogen	1	1.008
He	Helium	2	4.0026
Hf	Hafnium	72	178.49
Hg	Mercury	80	200.59
Ho	Holmium	67	164.9303
I	Iodine	53	126.9045
In	Indium	49	114.82
Ir	Iridium	77	192.22
K	Potassium	19	39.102
Kr	Krypton	36	83.8
La	Lanthanum	57	138.9055
Li	Lithium	3	6.941
Lr	Lawrencium	103	()
Lu	Lutetium	71	174.97
Md	Mendelevium	101	(256)
Mg	Magnesium	12	24.305
Mn	Manganese	25	54.938
Mo	Molybdenum	42	95.94
N	Nitrogen	7	14.0067
Na	Sodium	11	22.9898
Nb	Niobium	41	92.9064
Nd	Neodymium	60	144.24
Ne	Neon	10	20.179
Ni	Nickel	28	58.71
No	Nobelium	102	(254)
Np	Neptunium	93	(237)
O	Oxygen	8	15.9994
Os	Osmium	76	190.2
P	Phosphorus	15	30.9738

Symbol	*Element*	*Atomic number*	*Atomic mass*
Pa	Protactinium	91	231.0359
Pb	Lead	82	207.2
Pd	Palladium	46	106.4
Pm	Promethium	61	(145)
Po	Polonium	84	(210)
Pr	Praseodymium	59	140.9077
Pt	Platinum	78	195.09
Pu	Plutonium	94	(244)
Ra	Radium	88	226.0254
Rb	Rubidium	37	85.4678
Re	Rhenium	75	186.2
Rh	Rhodium	45	102.9055
Rn	Radon	86	(222)
Ru	Ruthenium	44	101.07
S	Sulphur	16	32.06
Sb	Antimony	51	121.75
Sc	Scandium	21	44.9559
Se	Selenium	34	78.96
Si	Silicon	14	28.086
Sm	Samarium	62	150.4
Sn	Tin	50	118.69
Sr	Strontium	38	87.62
Ta	Tantalum	73	180.9479
Tb	Terbium	65	158.9254
Tc	Technetium	43	(98)
Te	Tellurium	52	127.6
Th	Thorium	90	232.0381
Ti	Titanium	22	47.9
Tl	Thallium	81	204.37
Tm	Thulium	69	168.9342
U	Uranium	92	238.029
V	Vanadium	23	50.9414
W	Tungsten	74	183.85
Xe	Xenon	54	131.3
Y	Yttrium	39	88.9059
Yb	Ytterbium	70	173.04
Zn	Zinc	30	65.37
Zr	Zirconium	40	91.22

Values in brackets indicate mass numbers of most stable isotopes.

Range tables

Surface-to-Air Missile (SAM) range definition

Title	Minimum range	Maximum range	Other data (example of types of SAMs)
VSHORAD	0	5 km	Man-portable SAMs
SHORAD	5 km	20 km	Some man-portable SAM + Infrared homing SAMs or laser beam riders
MRAD	20 km	100 km	Radar controlled SAMs + semi-active
LRAD	100 km	250 km	Radar controlled + active seeker SAMs
VLRAD	250 km	>	Active seeker SAMs + third party targeting

Ballistic missile range definition

Title	Minimum range	Maximum range	Other data
Ballistic missiles			
SRBM	None	1,000 km	
MRBM	1,000 km	3,500 km	3,000 km has also been used
IRBM	3,500 km	5,500 km	
LRBM	3,500 km	5,500 km	Sometimes called Long Range
ICBM	5,500+	No ultimate range	
Tactical ballistic missiles			
TacBM	150 km	300 km	
Theatre ballistic missiles			
TBM	None	3,500 km	

Glossary

AA	Anti-Aircraft
AAA	Anti-Aircraft Artillery
AAAV	Advanced Amphibious Assault Vehicle (today EFV)
AAC	Advanced Armour Concept
AAD	Armoured Amphibious Dozer
AAF	Automatic Ammunition Flow
AAG	Anti-Aircraft Gun
AAK	Appliqué Armor Kit (US)
AAPC	Advanced Armoured Personnel Carrier (Turkey)
AARADCOM	Army Armament Research And Development COMmand
AAS	Appliqué Armor System (US)
AAV	Active Articulation Vehicle
AAV	Assault Amphibian Vehicle
AB	Air Burst
ABIT	Advanced Built-In Test
ABM	Air Bursting Munition
ABM	Anti-Ballistic Missile
ABMS	Air Bursting Munitions System
ABRO	Army Base Repair Organisation (now Defence Support Group)
ABS	Air-Bursting System
ABS	Anti-skid Braking System
ABSV	Armoured Battlegroup Support Vehicle
AC	Alternating Current
ACA	Ammunition Container Assembly
ACCV	Armored Cavalry Cannon Vehicle (US)
ACE	Armored Combat Earthmover (US)
ACLOS	Automatic Command Line-Of-Sight
ACOG	Advanced Combat Optical Gunsight
ACRV	Armoured Command and Reconnaissance Vehicle
ACS	Artillery Communications System
ACTD	Advanced Concept Technology Demonstration
ACV	Armored Cannon Vehicle (US)
ACV	Armoured Combat Vehicle
ACV-S	Armoured Combat Vehicle - Stretched
ACVT	Armored Combat Vehicle Technology program (US)
AD	Air Defence
ADAD	Air Defence Alerting Device
ADAM	Area Denial Artillery Munition
ADAMS	Advanced Deployable Autonomous Mortar System
ADATS	Air Defence Anti-Tank System
ADEA	Army Development and Employment Agency
ADI	Australian Defence Industries
ADMS	Air Defence Mobile System
ADS	Active Defence System
AEA	Advanced Energetic Armour
AESV	Armoured Engineering Squad Vehicle
AEV	Armoured Engineer Vehicle
AF	Air Force
AFARV	Armored, Forward Area, Re-arm Vehicle (US)
AFAS	Advanced Field Artillery System
AFCS	Automatic Fire-Control Systems
AFD	Automatic Feeding Device
AFEDSS	Automatic Fire and Explosion Detection and Suppression System
AFES	Automatic Fire Extinguishing System
AFSV	Armoured Fire Support Vehicle
AFV	Armored Family of Vehicles (US)
AFV	Armoured Fighting Vehicle
AGC	Automatic Gain Control
AGF	Army Ground Forces
AGL	Above Ground Level
AGL	Automatic Grenade Launcher
AGLS	Automatic Gun Laying System
AGM	Artillery Gun Module
AGS	Armored Gun System (US)
AGV	Assault Gun Vehicle
AGVT	Advanced Ground Vehicle Technology
Ah	Ampère hour
AHEAD	Advanced Hit Efficiency and Destruction
AHK	Ammunition Handling Kit
AHRS	Altitude and Heading Reference System
AIFS	Advanced Indirect Fire System
AIFV	Advanced Infantry Fighting Vehicle
AIFV	Armored Infantry Fighting Vehicle (US)
AIFV	Armoured Infantry Fighting Vehicle
AIPS	Advanced Integrated Propulsion System (US)
ALC	Advanced Land Combat (US)
ALS	Advanced Laying System
ALSV	Armoured Logistics Support Vehicle
ALT	Armoured Launching Turret
AMAP	Advanced Modular Armour Protection
AMARTOF	Anti-Missile Ammunition Reduce Time Of Flight
AMC	Advanced Mortar Carrier (Turkey)
AMC	Armoured Mortar Carrier
AMC	Army Matériel Command
AMCCOM	Armament Munitions and Chemical Command
AMDS	Anti-Missile Discarding Sabot
AMF	*Amphibische Mehrzweck-Fahrzeuge* (multipurpose amphibious vehicle)
AML	*Automitrailleuse Légère* (light armoured car)
AMOS	Advanced MOrtar System
AMR	*Automitrailleuse de Reconnaissance*
AMRS	Automatic Muzzle Reference System
AMRWS	Advanced MultiRole Weapon System
AMS	Armoured Mortar System
AMS-H	Advanced Missile System - Heavy
AMV	Armoured Modular Vehicle
AMX	*Atelier de Construction d'Issy-les-Moulineaux*
ANAD	Anniston Army Depot
AOI	Arab Organisation for Industrialisation
AOS	Add-On Stabilisation
AP	Anti-Personnel
AP	Armour-Piercing
APAM	Anti-Personnel, Anti-Matériel
APC	Armoured Personnel Carrier
APC	Armour-Piercing Capped
APC-T	Armour-Piercing Capped - Tracer
APCT-BF	Armour-Piercing Capped Tracer - Base Fuze
APDS	Armour-Piercing Discarding Sabot
APDS-T	Armour-Piercing Discarding Sabot - Tracer
APE	*Amphibisches Pionier-Erkundungsfahrzeug* (amphibious front-line reconnaissance vehicle)
APEP	Armour-Piercing Enhancement Programme
APERS	Anti-Personnel
APERS-T	Anti-Personnel - Tracer
APFIDS	Armour-Piercing Fragmentation Incendiary Discarding Sabot
APFSDS	Armour-Piercing Fin-Stabilised Discarding Sabot
APFSDS(P)	Armour-Piercing Fin-Stabilised Discarding Sabot (Practice)
APFSDS-T	Armour-Piercing Fin-Stabilised Discarding Sabot - Tracer
APG	Aberdeen Proving Grounds
APGM	Autonomous Precision-Guided Munition
APHE	Armour-Piercing High-Explosive
API	Armour-Piercing Incendiary
API-T	Armour-Piercing Incendiary - Tracer
APM	Anti-Personnel Mine
APS	Active Protection System
APS	Advanced Propulsion System
APS	Artillery Pointing System
APSE	Armour-Piercing Secondary Effect
APSE-T	Armour-Piercing Secondary Effect - Tracer
APT	Armour-Piercing - Tracer
APTE	Abrams Power Train Evolution (US)
APU	Auxiliary Power Unit
APV	Armoured Patrol Vehicle
AR/AAV	Armored Reconnaissance/Airborne Assault Vehicle (US)
ARA	Advanced Reactive Armour
ARCS	Acquisition Radar and Control System
ARD	Anti-Reflective Device
ARDEC	Armament Research Development and Engineering Centre
ARDNOT	Automatic Day/Night Optical Tracker
ARE	*Atelier de Construction Roanne*
ARMAD	Armoured and Mechanised unit Air Defence
ARMSCOR	Armament Manufacturing Corporation (South Africa)
ARMVAL	Anti-Armor Vehicle Evaluation (US)
ARP	Anti-Radiation Projectile
ARP	Armoured Repair Plates
ARRADCOM	Armament Research and Development Command
ARRV	Armoured Repair and Recovery Vehicle
ARSV	Armored Reconnaissance Scout Vehicle (US)
ARTEC	Armoured Vehicle Technology
ARV	Armoured Recovery Vehicle
ASA	Advanced Security Agency
ASARC	Army Systems Acquisition Review Council
ASCOD	Austrian Spanish Co-Operative Development
ASEP	Abrams System Enhancement Package
ASLS	Acoustic Sniper Location System
ASM	Armored Systems Modernisation (US)
ASP	Automatic, Self-Powered
ASTROS	Artillery Saturation Rocket System (Brazil)
ASV	Ammunition Supply Vehicle
ASV	Armored Security Vehicle (US)
AT	Anti-Tank
ATCAS	Advanced Towed Cannon System
ATACS	Advanced Tank Cannon System (US)
ATAM	Air-Air Trés Courte Portée
ATAS	Automatic Target Acquisition System
ATBM	Anti-Tactical Ballistic Missile
ATCAS	Advanced Towed Cannon System
ATCOPT	Al Technique Corporation of Pakistan
ATD	Advanced Technology Demonstration

ATD Advanced Technology Demonstrator (US)
ATD Automatic Target Detection
ATDT Automatic Target Detection and Tracking
ATDU Armoured Trials and Development Unit (UK)
ATLAS Affût Terrestre Léger Anti-Saturation
ATD Advanced Technology Demonstrator
ATEC Allison Transmission Electronic Control
ATFCS Advanced Tank Fire-Control System
ATFCS Automatic Targeting and Fire-Control System
ATGM Anti-Tank Guided Missile
ATGW Anti-Tank Guided Weapon
ATG Anti-Tank Gun
ATGM Anti-Tank Guided Missile
ATGW Anti-Tank Guided Weapon
ATIS Advanced Thermal Imaging System
ATLAS Advanced Technology Light Artillery System
ATLV Artillery Target Location Vehicle
ATM Anti-Tank Mine
ATM Anti-Tank Modular
ATMOS Autonomous Truck MOunted howitzer System
ATR Automotive Test Rig
ATR-EDS Automotive Test Rig - Electric Drive System
ATS *Atelier de Construction de Tarbes*
ATT Automatic Target Tracking
ATTC All Terrain Tracked Carrier
ATTS Air-Transportable Towed System
ATV Armoured TOW Vehicle (Turkey)
AVGP Armoured Vehicle General Purpose (Canada)
AVH Armoured Vehicle Heavy
AVL Armoured Vehicle Light
AVLB Armoured Vehicle-Launched Bridge
AVM Armoured Vehicle Medium
AVR Armoured Vehicle Reconnaissance
AVRE Armoured Vehicle Royal Engineers (UK)
AVRE Assault Vehicle Royal Engineers (UK)
AVT Advanced Vehicle Technologies (US)
AWE Advanced Warfighting Experiment
BAFO Best And Final Offer
BARV Beach Armoured Recovery Vehicle
BB Base-Bleed
BB Battery Block
BC Biological Chemical
BCC Battery Control Centre
BCP Battery Command Post
BCS Battery Computer System
BCT Brigade Combat Team (US)
BCV Battle Command Vehicle
BD Base Detonating
BDA Battle Damage Assessment
BE Base Ejection
BELRF Bradley Eyesafe Laser Range-Finder
BFV Bradley Fighting Vehicle
BGT Bodenseewerk Gerattechnik GmbH
BGT Brigade Combat Team
BGTI Battle Group Thermal Imaging
bhp brake horsepower
BIFF Brigade Identification Friend or Foe
BIR Base Inspection and Repair
BIT Built-in Test
BITE Built-In Test Equipment
BL Blank
BL-T Blank - Tracer
BLAAM Bunkers Light Armour and Masonry (for TOW missile)
BLOS Beyond Line Of Sight
BLR *Blindado Ligero de Ruedas*
BMD *Boevaya Machina Desantoya*
BMF Belgian Mechanical Fabrication
BMP *Boevya Machina Pekhota*
BMR *Blindado Medio de Ruedas*
BMS Battlefield Management System
BNS Bill Night Sight
BOCV Battery Operations Centre Vehicle
BOL Bearing Only Launch
BOLTS Bolt On Loading Tray System
BPS Battery Power Source
BRV Bionix Recovery Vehicle
BST Basic Skills Trainer
BTA Best Technical Approach
BUA Bill Under Armour
BW Bacteriological Warfare
CAD Component Advanced Development
CAD Computer-Assisted Design
CADS Close Range Air Defence System
CAL Canadian Arsenals Limited
CAMU Combined Auto Manual Unit
CAP Combustible Augmented Plasma
CARRV Challenger Armoured Repair and Recovery Vehicle
CASW Close Area Support Weapon (Canada)
CAT/LCV Combined Arms Team/Lightweight Combat Vehicle (US)
CATTB Component Advanced Technology TestBed (US)
CAV Composite Armored Vehicle (US)
CAWS Cannon Artillery Weapons Systems (US)
CBC *Companhia Brasileira de Cartuchos*
CCC Combustible Cartridge Case
CCD Charge Coupled Device
CCI Communications Interface
CCM Common Compatible Mount
CCO Close Combat Optic
CCP Computer Control Panel
CCR Crown Copyright Reserved
CCTV Closed-Circuit TV
CCU Central Control Unit
CCV Close-Combat Vehicle
CCV Command-and-Control Vehicle (US)
CCV-L Close-Combat Vehicle - Light
CDM Coastal Defence Missile
CDP Canadian Dry Pin
CDU Command Display Unit
CDU Control Display Unit
CE Chemical Energy
CECOM Communications-Electronics Command
CEIEC China Electronics Import and Export Corporation
CENTCOM Central Command
CEP Circular Error of Probability
CERA Composite Explosive Reactive Armour
CET Combat Engineer Tractor
CEU Computer Electronics Unit
CEV Combat Engineer Vehicle
CF Canadian Forces
CF Controlled Fragmentation
CFE Conventional Forces Europe
CFV Cavalry Fighting Vehicle (US)
CGS Crew Gunnery Simulator
CHARM Challenger Chieftain Armament
CHIP Challenger Improvement Programme
CICM Close-in-CounterMeasure
CIF Customer Interface Unit
CIFV Composite Infantry Fighting Vehicle
CILAS *Compagnie Industrielle des Lasers*
CIS Chartered Industries of Singapore
CIS Commonwealth of Independent States
CITV Commander's Independent Thermal Viewer (US)
CIWS Close-In Weapons System
CKEM Compact Kinetic Energy Missile
CLAMS Clear Lane Marking System (US)
CLASS Computerised Laser Sight System
CLAWS Close Combat Light Armor Weapon System (US)
CLGP Cannon-Launched Guided Projectile (US)
CLOS Command to Line Of Sight
CLU Command Launch Unit
CLV Command and Liaison Vehicle (UK) (Panther)
CM Common Missile
CMC Committee for the Communist Party of China
CMS Compact Modular Sight
CMT Cadmium Mercury Telluride
CMTV Carrier Mortar Tracked Vehicle
CMV Combat Mobility Vehicle (US)
CMWS Common Missile Warner System
CNPMIEC China National Precision Machinery Import and Export Corporation
CNTICM China Thermal Imaging Common Modules
CNVD Clip-on Night Vision Device
COG Course Over Ground
COLDS Common Opto-electronic Laser Detection System
Comp B Composition B
COMVAT Combat Vehicles Armament Technology (US)
COND Clip-on Night Device
COS Chief Of Section
COTAC *Conduite de Tir Automatique pour Char* (tank automatic fire)
COTS Commercial-Off-The-Shelf
COV Counter Obstacle Vehicle (US)
CP Concrete-Piercing
CPMIEC Chinese Precision Machinery Import and Export Corporation
CPS Cardinal Points Specification (UK)
CPS Commanders Periscope Sight
CPV Command Post Vehicle
CR Capability Requirement
C-RAM Counter-Rockets, Artillery and Mortars
CRR *Carro de Reconhecimento Sobre Rodas* (reconnaissance tracking scout car)
CRT Cathode Ray Tube
CRU Cable Reel Unit
CS Communications Subsystem
CSB Combat Support Boat
CSF Combined Service Forces
CSI Computer-Synthesised Image
CSP Capability Sustainment Programme (UK)
CSS Computer Sighting System
CTA Case Telescoped Ammunition

CTCA	Cased Telescoped Cannon and Ammunition
CTI	Central Tyre Inflation
CTIS	Central Tyre Inflation System
CTRA	*Carro de Transporte Sobre Rodas Anfibo* (amphibious tracking scout car)
CTT	Challenger Training Tank
CTWS	Case Telescoped Weapon System
CU	Command Unit
CV90	Combat Vehicle 90
CVAST	Combat Vehicle Armament System Technology (US)
CVR(T)	Combat Vehicle Reconnaissance (Tracked) (UK)
CVR(W)	Combat Vehicle Reconnaissance (Wheeled) (UK)
CVRDE	Combat Vehicle Research and Development Establishment
CVT	Controlled Variable Time
CVTTS	Combat Vehicle Targeting System
CW	Chemical Warfare
CWR	Continuous Wave Radar
CWS	Commander's Weapon Station
CWS	Control Warning System
CWS	Cupola Weapon Station
DA	Direct Attack
DA	Double Action
DACS	Digital Aiming and Stabilisation System
DAHA	Dual Axis Head Assembly
DAO	Double Action Only
DAREOD	Damaged Airfield Reconnaissance Explosive Ordnance Disposal
DARPA	Defense Advanced Research Projects Agency
DAS	Defensive Aids Suite
DAS	Defensive Aid System
DASP	Demountable Artillery Surveillance Pod
DC	Data Couplers
DC	Digital Computer
DC	Direct Current
DCA	*Défense Contre Avions* (anti-aircraft)
DD	Detroit Diesel
DDA	Detroit Diesel Allison
DDEC	Detroit Diesel Electronic Control
DDS	Department of Defense Support
DDU	Digital Display Unit
DECA	Digital Electronic Control Assembly
Def Stan	Defence Standard
DEFA	*Direction des Études et Fabrications d'Armement*
DERA	Defence Evaluation and Research Agency
DESO	Defence Export Sales Organisation
DFCS	Digital Fire-Control System
DFS	Direct Fire Sight
DFSV	Direct Fire Support Vehicle
DFV	Desert Fighting Vehicle
DGI	Directional Gyro Indicator
DHSS	Data Handling SubSystem
DIA	Defense Intelligence Agency (US)
DIP	Driver Instrument Panel
DIRCM	Directable Infra-Red CounterMeasure
DLO	Defence Logistics Organisation
DMC	Digital Magnetic Compass
DMS	Dual Mounted Stinger
DMU	Distance Measurement Unit
DNA	Defence Nuclear Agency
DNRS	Day/Night Range Sight
DNS	Day Night Sight
DNTSS	Day/Night Thermal Sighting System
DoD	Department of Defence
DoD	Department of Defense (US)
DoP	Department of Productivity
DP	Demonstration Purpose
DP	Dual Purpose
DPA	Defence Procurement Agency (UK)
DPICM	Dual-Purpose Improved Conventional Munition (US)
DPTR	Dioptre
DRA	Defence Research Agency
DROPS	Demountable Rack Offloading and Pick-up System
DRS	Dynamic Reference System
DSACS	Direct Support Armored Cannon System (US)
DSETS	Direct Support Electrical Test System
DSG	Defence Support Group (UK)
DS/T	Discarding Sabot/Tracer
DSWS	Division Support Weapon System
DTAT	*Direction Technique des Armements Terrestres*
DTI	Department of Trade and Industry
DTT	Driver Training Tank
DTU	Distance Transmitter Unit
DU	Depleted Uranium
DU	Display Unit
DV	Demining Vehicle
DVE	Driver's Vision Enhancer
DWFK	Deep Water Fording Kit
EA	Electronic Architecture
EAAK	Enhanced Appliqué Armor Kit (US)
EADS	European Aeronautic Defence and Space company
EAOS	Enhanced Artillery Observation System
EAPS	Extended Area Protection and Survivability
EBG	*Engin Blindé Génie* (armoured combat vehicle)
EBR	*Engin Blindé de Reconnaissance* (armoured reconnaissance vehicle)
EBRC	*Engin Blindé a Roues de Contact*
EC	European Community
ECASS	Electronically Controlled Active Suspension System
ECCM	Electronic Counter-Countermeasures
ECM	Electronic Countermeasures
ECOS	Enhanced Combat Optical Sight
ECS	Environmental Control Subsystem
ECTAP	Engine Compartment Test and Alarm Panel
ECU	Electronic Controller Unit
ECV	Enhanced Capacity Vehicle
EDS	Electric Drive System
EEP	Environmental Enhancement Package
EFAB	*Etablissement d'Etudes et de Fabrications d'Armement de Bourges*
EFC	Equivalent Full Charge
EFCR	Equivalent Full Charge Rounds
EFM	Explosives Factory Maribyrnong
EFP	Expanded Feasibility Phase
EFP	Explosively Formed Penetrator
EFP	Explosively Formed Projectile
EFSS	Expeditionary Fire Support System
EFV	Expeditionary Fighting Vehicle (previously Advanced Amphibious Fighting Vehicle - AAAV)
EFVS	Electronic Fighting Vehicle System (US)
EG	External Gun
ELITE	Eye-safe Laser Integrated Telescope Equipment
ELKE	Elevated Kinetic Energy weapon (US)
ELS	Electromagnetic Launcher System
ELSAP	*Elektronische Schiessanlage für Panzer* (electronic fire-control system for tanks)
EM	Electro-Magnetic
EMC	Electro-Magnetic Compatibility
EMC	Executive Management Committee
EMD	Engineering and Manufacturing Development
EMDG	EuroMissile Dynamics Group
EMG	Externally-Mounted Gun
EMP	Electro-Magnetic Pulse
ENGESA	*Engesa Engenheiros Especializados* (Brazil) (no longer trading)
EO	Electro-Optics
EOD	Explosive Ordnance Disposal
EOR	Explosive Ordnance Reconnaissance
EOTS	Electro-Optical Tracking System
EPC	Electronic Plane Conversion
EPC	*Engin Principal de Combat* (future main battle tank) (became the Leclerc MBT)
EPG	European Production Group
EPG	European Programme Group
EPLASS	Enhanced Precision Land Survey System
EPS	Electrical Power Subsystem
EPU	Electronics Processing Unit
ER	Enhanced Radiation
ER	Extended Range
ERA	Explosive Reactive Armour
ERA	Extended Range Ammunition
ERC	*Engin de Reconnaissance Canon*
ERFB	Extended Range Full-Bore
ERFB-BB	Extended Range Full-Bore - Base-Bleed
ERGFCDS	Extended Range Gunnery Fire-Control Demonstration System
ERGP	Extended Range Guided Projectile
ERMIS	Extended Range Modification Integration System
ERO	Extended Range Ordnance
ERP	Extended Range Projectile
ERSC	Extended Range SubCalibre
ERV	Emergency Rescue Vehicle
ERV	Engineer Reconnaissance Vehicle
ES	Extreme Spread
ESD	*Electronique Serge Dassault*
ESLDE	Eye-Safe Laser Daylight Elbow
ESM	Electronic Support Measures
ESPAWS	Enhanced Self-Propelled Artillery Weapon System (US)
ESRS	Electro Slag Refined Steel
ESS	External Suspension System
ESU	External Stabilisation Unit
ETA	Estimated Time of Arrival
ETC	ElectroThermal Cannon
ETM	Electronic Technical Manuals
ETS	Elevated TOW System
ETS	Engineer Tank System
EW	Electronic Warfare
EWK	*Eisenwerke Kaiserslautern Göppner*
EWS	External Weapon Station
FAAD	Forward Area Air Defense (US)
FAAR	Forward Area Alerting Radar (US)

FAASV Field Artillery Ammunition Support Vehicle (US)
FACE Field Artillery Computer Equipment
FAL Fusil Automatique Légère
FAPDS Frangible Armour-Piercing Discarding Sabot
FARS Field Artillery Rocket System
FARV-A Future Armored Resupply Vehicle - Artillery (US)
FAST Forward Area Support Team
FAV Fast Attack Vehicle (US)
FBI Federal Bureau of Investigation
FBRV Future Beach Recovery Vehicle (UK)
FCA Flexible Ceramic Armour
FCC Fire Command Centre
FCC Fire-Control Computer
FCCS Fire-Control Computer System
FCCVS Future Close-Combat Vehicle System (US)
FCE Fire-Control Equipment
FCLV Future Command and Liaison Vehicle (UK)
FCS Fire-Control System
FCS Future Combat System (US) (cancelled programme)
FCU Fire-Control Unit
FDC Fire Direction Centre
FDCV Fire Direction Centre Vehicle
FDSWS Future Direct Support Weapon System
FEBA Forward Edge of Battle Area
FET Future Engineer Tank
FF Fire-and-Forget
FFAR Folding Fin Aerial Rocket
FG Field Gun
FH Field Howitzer
FIBUA Fighting In Built Up Area
FIFV Future Infantry Fighting Vehicle (US)
FINDERS Fast Information, Navigation, Decision and Reporting System
FIS Fuze Interface System
FISTV Fire Support Team Vehicle (US)
FITOW Further Improved TOW (US)
FLEA Frangible Low Energy Ammunition
FLIR Forward-Looking Infra-Red
FLOT Forward Line of Own Troops
FM Fei Ming (Flying Midge)
FM Frequency Modulation
FM Titanium Tetrachloride (code designation)
FMC Food Machinery Corporation
FMS Foreign Military Sales
FMTV Family of Medium Tactical Vehicles
FN *Fabrique Nationale*
FOD Foreign Object Damage
FOG Fibre Optic Gyro
FOM Fibre Optic Missile
FOO Forward Observation Officer
FORTIS Forward Observation and Reconnaissance Thermal Imaging System
FOTT Follow On To TOW
FOV Field of View
FPA Focal Plane Array
FRAG Fragmentation
FRES Future Rapid Effect System (cancelled programme) (UK)
FROG Free Rocket Over Ground
FSCS Future Scout Cavalry System
FSCV Fire Support Combat Vehicle
FSED Full-Scale Engineering Development (US)
FSS Fire Sensing and Suppression
FST Future Soviet Tank/Follow-on Soviet Tank
FSV Fire Support Vehicle
FSV Future Scout Vehicle (US)
FTA Frangible Training Ammunition
FTMA Future Tank Main Armament
FTS Future Tank Study
FTT Field Tactical Trainer
FUE First Unit Equipped
FUG *Felderitö Usó Gépkosci*
FV Fighting Vehicle
FV/GCE Fighting Vehicle/Gun Control Equipment
FVDD Fighting Vehicle Development Division
FVRDE Fighting Vehicle Research and Development Establishment (UK) (now closed)
FVS Fighting Vehicle System (US)
FVSC Fighting Vehicle Systems Carrier (US)
FY Fiscal Year
FYDP Future Years Defence Plan
G Gendarmerie
g Gramme(s)
GAJG GPS Anti-Jam Electronics
GAMA Gun Automatic Multiple Ammunition
GAO General Accounting Office
GAP Gun Aiming Post
GCDU Gunner Control and Display Unit
GCE Gun Control Equipment
GCT *Grande Cadence de Tir* (high rate of fire)
GCU Gun Control Unit
GCV Ground Combat Vehicle (US Army programme)
GCW Gross Combat Weight
GD General Dynamics
GDELS General Dynamics European Land Systems
GDLS General Dynamics Land Systems
GDU Gun Display Unit
GFE Government Furnished Equipment
GH Gun-Howitzer
GIAT *Groupement Industriel des Armements Terrestres* (became Nexter Systems)
GITS GMHG Integrated TOW Sight
GKEM Guided Kinetic Energy Missile
GLATGM Gun Launched Anti-Tank Guided Missile
GLC Gun Lay Computer
GLH-H Ground-Launched Hellfire-Heavy (US)
GLLD Ground Laser Locator Designator (US)
GLPS Gun Laying and Positioning System
GLS *Gesellschaft für Logistischen Service*
GM, MVO General Motors, Military Vehicle Operations
GMC General Motors Corporation
GMG General Machine Gun
GMG Grenade Machine Gun
GMLRS Guided Multiple Launch Rocket System
GMS Gun Management System
GOCO Government-Owned, Contractor-Operated (US)
GONS Guns Orientation and Navigation System
GP General Purpose
GP Guided Projectile
GPMG General Purpose Machine Gun
GPO Gun Position Officer
GPS Global Positioning System
GPS Gunner's Primary Sight
GPSS Gunner's Primary Sight Subsystem
GP-T General Purpose - Tracer
GRP Glass Reinforced Plastic
GS Gunners Sight
GSR General Staff Requirement (US)
GSRS General Support Rocket System
GST General Staff Target
GST *Gesellschaft für System-Technik*
GTCS Gun Test and Control System
GTI German Tank Improvement
GVW Gross Vehicle Weight
GW Guided Weapon
h Hour(s)
HAB Heavy Assault Bridge (US)
HAFCS Howitzer Advanced Fire-Control System
HALO High Activity/Low Observable
HAWK Homing-All-the-Way-Killer (US)
HB Heavy Barrel
HBCT Heavy Brigade Combat Team
HC Hexachlorethane/zinc oxide
HC High Capacity
HC Hollow Charge
HCER High-Capacity Extended-Range
HCHE High-Capacity High-Explosive
HCT HOT Compact Turret
HDTI High Definition Thermal Imaging/Imager
HE High-Explosive
HEAA High Explosive Anti-Armour
HE-APERS High-Explosive Anti-Personnel
HEAP-T High-Explosive Anti-Personnel - Tracer
HEAT High Explosive Anti-Tank
HEAT-FS High-Explosive Anti-Tank Fin-Stabilised
HEAT-MP High-Explosive Anti-Tank MultiPurpose
HEAT-MP(P) High-Explosive Anti-Tank Multipurpose (Practice)
HEAT-T High-Explosive Anti-Tank - Tracer
HEAT-T-HVY High-Explosive Anti-Tank - Tracer - Heavy
HEAT-T-MP High-Explosive Anti-Tank - Tracer - MultiPurpose
HEAT-TP-T High-Explosive Anti-Tank - Target Practice - Tracer
HED-D Hybrid Electric Drive-Demonstrator
HEDP High-Explosive Dual-Purpose
HEER High-Explosive Extended-Range (US)
HEF High-Explosive Fragmentation
HE-FRAG High-Explosive Fragmentation
HE-FRAG-FS High-Explosive Fragmentation - Fin-Stabilised
HE-FS High-Explosive - Fin-Stabilised
HEFT High-Explosive Follow-Through
HEI High-Explosive Incendiary
HEI-T High-Explosive Incendiary - Tracer
HEL High-Energy Laser (US)
HEL Human Engineering Laboratory (US)
HELP Howitzer Extended Life Program (US)
HELS Hellfire Enhanced Laser Seeker
HELTADS High Energy Laser Tactical Air Defence System
HEMAT Heavy Expanded Mobility Ammunition Trailer (US)
HEMP High Explosive Multi-Purpose
HEMTT Heavy Expanded Mobility Tactical Truck (US)
HEP High-Explosive Plastic
HEPD High-Explosive Point-Detonating

HE/PR High-Explosive Practice
HEP-T High-Explosive Practice - Tracer
HERA High-Explosive Rocket-Assisted
HE-S High-Explosive - Spotting
HE-T High-Explosive - Tracer
HET-PF High-Explosive Tracer - Percussion Fuze
HETS Heavy Equipment Transport Semi-Trailer
HE-T-SD High-Explosive - Tracer - Self-Destruct
HESH High-Explosive Squash Head
HESH-T High-Explosive Squash Head - Tracer
HFCC Howitzer Fire-Control Computer (US)
HFHTB Human Factors Howitzer TestBed (US)
HFM Heavy Force Modernisation (US)
HgCdTe Mercury Cadmium Telluride
HIFV Heavy Infantry Fighting Vehicle
HIMAG High-Mobility Agility test vehicle (US) (cancelled programme)
HIMARS High Mobility Artillery Rocket System (US)
HIP Howitzer Improvement Program (US)
HIRE Hughes Infra-Red Equipment
HITP High Ignition Temperature Propellant
HIU Heading Indicator Unit
HMC Howitzer Motor Carriage
HMD Helmet-Mounted Display
HMG Heavy Machine Gun
HMLC High-Mobility Load Carrier
HMMWV High-Mobility Multipurpose Wheeled Vehicle
HN Hong Nu (Red Cherry)
HOE Holographic Optical Element
HOMS Hellfire Optimised Missile System
How Howitzer
HP High Power
HP Hollow Point
hp horsepower
HPFP High-Performance Fragmentation Projectile
HPS Helmet Pointing System (or Sight)
HPT High-Pressure Test
HQ Hong Qi (Red Flag)
HRU Heading Reference Unit
HSDS Hellfire Shore Defence System
HSS Hunter Sensor Suite
HSTV(L) High-Survivability Test Vehicle (Lightweight) (US)
HTTB High-Technology TestBed (US)
HUMS Health Usage Monitoring System
HVAP High-Velocity Armour-Piercing
HVAPDS-T High-Velocity Armour-Piercing Discarding Sabot - Tracer
HVAPFSDS High-Velocity Armour-Piercing Fin-Stabilised Discarding Sabot
HVAP-T High-Velocity Armour-Piercing - Tracer
HVM High Velocity Missile
HVM Hyper Velocity Missile
HVSS Horizontal Volute Spring Suspension
HVSW Hyper Velocity Support Weapon
HVTP-T High-Velocity Target Practice - Tracer
HWSTD High Water Speed Technology Demonstrator
HYPAK Hydraulic Power Assist Kit
I Incendiary
IA Improved Ammunition
IAF Interim Armoured Vehicle
IAF Israeli Air Force
IAFV Infantry Armoured Fighting Vehicle
IAI Israel Aircraft Industries (today Israel Aerospace industries)
IAL Infra-Red Aiming Light
IAS Improved Armour System
IASD Instant Ammunition Selection Device
IBAS Improved Bradley Acquisition System
ICC Information Command-and-Control
ICC Information Co-ordination Centre
ICD Interface Control Document
ICM Improved Conventional Munition
ICM BB Improved Conventional Munition Base-Bleed
ICV Infantry Combat Vehicle
IDAIT Integrated Display and Tactical Terminal
IDF Israel Defense Forces
IED Improvised Explosive Device
IEPG Independent European Program Group
IFCS Improved Fire-Control System
IFCS Integrated Fire-Control System
IFF Identification Friend or Foe
IFV Infantry Fighting Vehicle
IFVwCM Infantry Fighting Vehicle with Integrated CounterMeasures (US)
II Image Intensification/Intensifier
IIR Imaging Infra-Red
ILL Illuminating
ILMS Improved Launcher Mechanical System
ILS Integrated Logistic Support
ILV Improved Low Visibility
IM Insensitive Munition(s)
IMMLC Improved Medium Mobility Load Carrier
IMU Inertial Measurement Units
INS Inertial Navigation System
InSb Indium Antimonide
INU Inertial Navigation Unit
IOC Initial Operational Capability
IOF Indian Ordnance Factory
IPF Initial Production Facility
IPO International Programme Office
IPPD Integrated Product Process Development
IR Infra-Red
IRBM Intermediate-Range Ballistic Missile
IRCCM Infra-Red Counter-Counter Measures
IRCM Infra-Red Counter Measures
IRU Inertial Reference Unit
IS Internal Security
ISD In Service Date
I-SFCS Image - Stabilised Tank Fire-Control System
ISGU Integrated Sight and Guidance Unit
ISU In-arm Suspension Unit
ISU Integrated Sight Unit
ISV Internal Security Vehicle
ITAS Improved Target Acquisition System
ITPIAL Infra-Red Target POINTER/Illuminator/Aiming Laser
ITT Invitation To Tender
ITU Infantry Trials and Development Unit (UK)
ITV Improved TOW Vehicle (US)
ITV Internally Transportable Vehicle
IVPDL Inter-Vehicle Positioning and Data Link (US)
IW Interim Warhead
IWS Improved Weapon System
JBMoU Joint Ballistic Memorandum of Understanding
JGSDF Japanese Ground Self-Defence Force
JLTV Joint Light Tactical Vehicle (US)
JPO Joint Project Office
JSC Joint Steering Committee
JSC Joint Stock Company
JSDFA Japanese Self-Defence Force Agency
KAAV Korean Armoured Amphibious Vehicle
KBM Machine-Building Design Bureau
KBP Instrument-Making Design Bureau
KE Kinetic Energy
KEC Kaman Electro-magnetics Corporation
KEM Kinetic Energy Missile (US)
kg Kilogramme(s)
KIFV Korean Infantry Fighting Vehicle
KPA Korea People's Army
KUR Key User Requirement
LAAG Light Anti-Aircraft Gun
LAD Light Aid Detachment
LADS Light Air Defense System (US)
LAM Laser Aiming Module
LAM Loitering Attack Missile
LAPES Low-Altitude Parachute Extraction System
LASIP Light Artillery System Improvement Plan
LAST Light Appliqué System Technique
LATADS Laser Target Decoy System
LATS Light Armoured Turret System
LAV Light Armored Vehicle (US)
LAV Light Armoured Vehicle
LAV Light Assault Vehicle (US)
LAW Light Anti-armour Weapon
LAW Light Anti-tank Weapon
LBMS Leclerc Battlefield Management System
LC Laser Collimator
LCD Liquid Crystal Display
LCS Loader Control System
LCU Landing Craft Utility
LCV Light Contingency Vehicle (US)
LD Low Drag
LDPS Laser/Daylight Periscopic Sight
LDU Layers Display Unit
LED Light Emitting Diode
LEP Light Emitting Plastic
LEP Life Extension Programme
LEU Launcher Erector Unit
LF Linked Feed
LHR Low Heat Rejection
LIA Linear Induction Accelerator
Li-ion Lithium-ion battery
LIMAWS Lightweight Mobile Artillery Weapon System (cancelled programme)
LION Lightweight Infra-Red Observation Night Sight
LIW Lyttleton Engineering Works
LKP Loader Keyboard Panel
LL Launcher Loader
LLAD Low-Level Air Defence
LLLTV Low Light Level TeleVision
LLM Launch Loader Module
LLR Light Low Recoil
LM Loitering Munition

LMG Light Machine Gun
LML Lightweight Multiple Launcher
LMT Lightweight Medium Turret
LNS Land Navigation System
LNS Laying and Navigation System
LOAL Lock-On After Launch
LOBL Lock-On Before Launch
LOCATS Low-cost aerial target system
LOMO Leningrad Optical Mechanical Association
LOS Line-Of-Sight
LOSAT Line Of Sight Anti-Tank (US programme cancelled)
LOSFH Line of Sight Forward Heavy
LP Liquid Propellant
LPC Launcher Pod Carrier
LPC Launch Pod Container
LPPV Light Protected Patrol Vehicle
LPT Low-Profile Turret
LPTS Lightweight Protected Turret System
LR Long Range
LRAR Long-Range Artillery Rocket
LRASS Long-Range Advanced Scout System
LRAT Long-Range Anti-Tank
LRBB Long-Range Base-Bleed
LRD Long Range Deflagrator
LRF Laser Range-Finder
LRF Low Recoil Force
LRHB Long-Range Hollow Base
LRIP Low-Rate Initial Production
LRM Laser Range-finder Module
LRN Low Recoil NORICUM
LRU Launcher Replenishment Unit
LRU Line Replacement Unit
LRU Line-Replaceable Unit
LSAH Laser Semi-Active Homing
LTA Launch Tube Assembly
LTD Laser Target Designator
LTFCS Laser Tank Fire-Control System
LTP Laser Target Pointer
LUNOS Lightweight Universal Night Observation System
LVA Landing Vehicle Assault (US)
LVS Light Video Sight
LVT Landing Vehicle Tracked (US)
LVTC Landing Vehicle Tracked Command (US)
LVTE Landing Vehicle Tracked Engineer (US)
LVTH Landing Vehicle Tracked Howitzer (US)
LVTP Landing Vehicle Tracked Personnel (US)
LVTR Landing Vehicle Tracked Recovery (US)
LWIR Long Wave Infra-Red
LWL Light Weight Launcher
LWML LightWeight Multiple Launcher
LWMS LightWeight Modular Sight
LWS Laser Warning System
LWT Light Weapon Turret
LY Lei Ying (Falcon)
m Metre(s)
m/s Metres per second
MAC Medium Armored Car (US)
MACS Modular Artillery Charge System
MADLS Mobile Air Defence Launching System
MAMBA Mobile Artillery Monitoring Battlefield Radar
MANPADS Man Portable Air Defence System
MAOV Mobile Artillery Observation Vehicle
MAP Military Aid Package
MAP Military Aid Programme
MAPS Modular Azimuth Position System
MAPS-H Modular Azimuth Positioning System - Hybrid
MARDI Mobile Advanced Robotics Defence Initiative (UK)
MARS Multiple Artillery Rocket System
MAV Maintenance Assist Vehicle
MAV Monitors for Armoured Vehicles
MAVD MLRS Aim Verification Device
MBA Main Battle Area
MBB *Messerschmitt-Bölkow-Blohm*
MBT Main Battle Tank
MCD Missile Countermeasures Device
MCLOS Manual Command to Line Of Sight
MCP Mistral Co-ordination Post
MCP Mobile Command Post
MCRV Mechanised Combat Repair Vehicle
MCS Microclimate Conditioning System
MCS Modular Charge System
MCSK Mine Clearance System Kit (US)
MCT Mercury Cadmium Telluride
MCT MILAN Compact Turret
MCV Mechanised Combat Vehicle
MCWL Marine Corps Warfighting Laboratory
MCWS Minor Calibre Weapons Station (US)
MDU Map Display Unit
MEADS Medium Extended Air Defence System
MEMS Micro-Electro Mechanical Systems
MENA Middle East North Africa
MENS Mission Element Need Statement (US)
MEV Medical Evacuation Vehicle (US)
MEWPP Modular Electronic Warfare Pre-Processor
MEWS Mobile Electronic Warfare System
MEWSS Mobile Electronic Warfare Support System
MEXAS Modular Expandable Armour System
MF MultiFunction
MFC Mortar Fire Controller
MFCS Modified Fire-Control System
MFCS Multi-Fire Control System
MFCV Missile Fire-Control Vehicle
MFF Munition Filling Factory
MFR Multi-Function Radar
MG Machine Gun
MGB Medium Girder Bridge
MGS Missile Guidance Set
MGS Mobile Gun System
MGTS MultiGun Turret System
MGU Mid-course Guidance Unit
MGV Manned Group Vehicles
MICOM Missile Command (US)
MICV Mechanised Infantry Combat Vehicle
MILAN *Missile d'Infantrie Léger Antichar* (light infantry anti-tank missile)
MILES Multiple Integrated Laser Engagement System (US)
MIL-STD MILitary STanDard
MIPS Medium Integrated Propulsion System (US)
MLC Military Load Class
MLC Modular Load Carrier
MLI Mid-Life Improvement
MLOS Miniature Laser Optical Sight
MLRS Multiple Launch Rocket System (US)
mm Millimetre(s)
MMBF Mean Miles Between Failures
MMG Medium Machine Gun
MMS Mast-Mounted Sight
MNVD Monocular Night Vision Device
MoA Memorandum of Agreement
MOA Minute of Angle
Mod Modification
MoD Ministry of Defence
MoDA Ministry of Defence and Aviation
MOLF Modular Laser Fire-control (Germany)
MOTS Military-Off-The-Shelf
MoU Memorandum of Understanding
MOUT Military Operations in Urban Terrain
MPBAV MultiPurpose Base Armoured Vehicle
MPC MultiPurpose Carrier (Netherlands)
MPGS Mobile Protected Gun System (US)
MPIM Multi-Purpose Individual Munition
MPM Metric Precision Munition
MPM Multi-Purpose Munition
MPM Munition à Précision Métrique
MPPV Medium Protected Patrol Vehicle
MPS Maritime Pre-positioning Ships (US)
MPV MultiPurpose Vehicle
MPWS Mobile Protected Weapon System (US)
MR Medium Range
MRAP Mine Resistant Ambush Protected
MRAV MultiRole Armoured Vehicle
MRBF Mean Rounds Before Failure
MRCV Multi-Role Combat Vehicle
MRDEC Missile Research Development and Engineering Centre
MRS Multiple Rocket System
MRS Muzzle Reference System
MRSI Multiple Round Simultaneous Impact
MRV(R) Mechanised Recovery Vehicle (Repair)
MRVR Mechanised Repair and Vehicle Recovery (US)
MSGL Multisalvo Smoke Grenade Launcher
MSR Missile Simulation Round
MSS Multispectral Surveillance System
MSTAR Man-portable Surveillance and Target Acquisition Radar
MSV Manoeuvre Support Vehicle
MSV Modular Support Vehicle
MT Mechanical Time
MTAS Moving Target Acquisition System
MTAS Multisensor Target Acquisition System
MTB Mobility TestBed
MTBF Mean Time Between Failure
MTI Moving Target Indicator/Indication
MTIP Manned Turret Integration Programme
MTL Materials Technology Laboratory (US)
MTR Mobile Test Rig (US)
MTSQ Mechanical-Time and Super-Quick
MTTR Mean Time To Repair
MTU *Motoren und Turbinen-Union*
MTVL Mobile Tactical Vehicle Light
MUGS Multipurpose Universal Gunner Sight
MULE Modular Universal Laser Equipment (US)

MUSS Multifunctional Self-protection System
MV Muzzle Velocity
MVEE Military Vehicles and Engineering Establishment (now closed)
MVRS Muzzle Velocity Radar System
MWIR Mid-Wavelength Infra-Red
MWS Manned Weapon Station (US)
MWS Modular Weapon System
NABK NATO Ballistic Kernel
NALLADS Norwegian Army Low-Level Air Defence System
NATO North Atlantic Treaty Organisation
NBC Nuclear, Biological, Chemical
NBMR NATO Basic Military Requirement
Nd: YAG Neodymium-doped: Yttrium Aluminium Garnet
NDC National Development Complex
NDI Non Development Item
NDU Navigation Display Unit
NEMO New Mortar (Finland)
NEQ Net Explosive Quantity
NFOV Narrow Field of View
NFS North-Finding System
NGFCS New Generation Fire-Control System
Ni/Cd Nickel Cadmium
Ni/MH Nickel Metal Hydride
NLOS Non-Line-of-Sight
NLOS-C Non-Line-of-Sight - Cannon (cancelled programme)
NLOS-LS Non-Line-of-Sight - Launch System (cancelled programme)
NLOS-M Non-Line-of-Sight - Mortar (cancelled programme)
NML Naval Multiple Launcher
NORINCO China North Industries Corporation
NTC National Training Center (US)
NUGP Nominal Unit Ground Pressure
NV Night Vision
NVE Night Vision Equipment
NVESD Night Vision and Electronic Sensors Directorate
NVG Night Vision Goggles
NVS Night Vision Systems
OBR Optical Beam-Riding
OCC *Obus à Charge Creusé* (shaped-charge shell)
OCC Operational Control Centre
OCP Operator's Control Panel
OCSW Objective Crew-Served Weapon
OCU Operational Centre Unit
ODE Ordnance Development and Engineering (Singapore)
OEG Occluded Eye Gunsight
OEM Original Equipment Manufacturer
OEO Optical and Electro-Optical
OFAS Optical Fire Sensor Assembly
OLED Organic Light Emitting Diode
OP Observation Post
OP Operator's Panel
OPV Observation Post Vehicle
OT Operational Test
OTA Overflight Top Attack
OTA Overfly, Top Attack
OTEA Operational Test and Evaluation Agency (US)
OTTC Overhead Turret Technology Carrier
OWS Overhead Weapons Station
P^3I Pre-Planned Product Improvements
P How Pack Howitzer
PADS Position and Azimuth Determining System
PAM Precision Attack Missile
PAT Power-Assisted Traverse
PbS Lead Sulphide
PC Personal Computer
PCB Printed Circuit Board
PCU Power Conditioning Unit
PD Point-Detonating
PD Project Definition
PDFCS Primary Direct Fire-Control System
PDNU Positioning Determiningavigation Unit
PDRR Programme Definition Risk-Reduction
PE Procurement Executive
PEC Printed Electronic Circuits
PELE Penetrator with Enhanced Lateral Effect
PEU Periscope Electronic Unit
PFD Proximity Fuze Disconnecter
PFHE Pre-Fragmented High-Explosive
PFN Pulse Forming Network
PFPX Pre-Fragmented Proximity fuzed
PGM Precision Guided Munition
PIE Pyrotechnically Initiated Explosive
PIM Paladin Integrated Management
PIP Product Improvement Programme
PLA People's Liberation Army
PLARS Position Location And Reporting System (US)
PLC Programmable Logic Control
PLGR Precision Lightweight GPS Receiver
PLO Palestinian Liberation Organisation
PLOS Predicted Line of Sight
PLS Palletised Load System (US)
PM *Porte Mortier* (mortar carrier)
PM Project Manager
PMADS Pedestal Mounted Air Defence Missile System
PMO Program Management Office (US)
PMS Pedestal Mounted Stinger
PNU Position Navigation Unit
POI Point of Intercept
POMALS Pedestal-Operated Multi-Ammunition Launching System
POS *Postes Optiques de Surveillance*
POST Passive Optical Sighting Technique
PPI Plan Position Indicator
PPM Pulsed Power Module
PPS Precise Positioning Service
PPS Pulsed Power Supply
PPV Protected Patrol Vehicle
PRAC Practice
PRAC-T Practice - Tracer
PRC People's Republic of China
PRF Pulse Repetition Frequency
PRI Projector Reticle Image
PSD Propulsion System Demonstrator
PSS Primary Sight System
PTM Power Train Module
PTO Power Take-Off
PVO Voiska Protivovzdushnoy (National Air Defence)
PVP *Petit Véhicule Protégé*
PWI-SR(GR) *Panser Wagen Infanterie-Standaard (Groep)*
PWP Plasticised White Phosphorus
PWS Protected Weapons Station
PWV *Pantser Wiel Voertuig*
QCB Quick Change Barrel
QE Quadrant Elevation
Qian Wei Advanced Guard (Van Guard)
RA Royal Artillery
RAAMS Remote Anti-Armor Mine System (US)
RAC Royal Armoured Corps
RAF Royal Air Force
RAM-D Reliability, Availability, Maintainability and Durability
RAO Rear Area Operations
RAP Rocket-Assisted Projectile
RAPI Reactive Armour Protection
RARDE Royal Armament Research and Development Establishment (UK) (closed facility)
RATAC *Radar de Tir pour L'Artillerie de Campagne* (radar for field artillery fire)
RBL Range and Bearing Launch
RC/MAS Reserve Component/Modified Armament System
RCAAS Remote-Controlled Anti-Armor System (US)
RCDU Remote-Controlled Defence Unit
RCS Radar Cross Section
RCT Royal Corps of Transport
RCU Remote-Control Unit
RCV Robotic Command Vehicle
RCWS Remote-Controlled Weapon Station
RDECOM Research, Development and Engineering Command
RDF Rapid Deployment Forces
RDJTF Rapid Deployment Joint Task Force (US)
RDT&E Research Development Test and Evaluation
RDU Remote Display Unit
REME Royal Electrical and Mechanical Engineers
RF Radio Frequency
RF Rimfire
RFAS Russian Federation and Associated States
RFI Request For Information
RFP Request For Proposals
RFPI Rapid Force Projection Initiative
RFQ Request For Quotations
RHA Rolled Homogeneous Armour
RHA Royal Horse Artillery
RISE Reliability Improved Selected Equipment
RLC Royal Logistic Corps
RLG Ring Laser Gyro
RLNS Ring Laser Gyro Land Navigation System
RMG Ranging Machine Gun
RN Royal Navy
RO Royal Ordnance
ROBAT Robotic Counter-Obstacle Vehicle (US)
RoC Republic of China
ROC Required Operational Characteristics
ROF Rate Of Fire
ROF Royal Ordnance Factory (no longer in existence)
RoK Republic of Korea
RoKIT Republic of Korea Indigenous Tank
ROR Range Only Radar
ROTA Royal Ordnance Training Ammunition
RP Red Phosphorus
RP Rocket-Propelled
RPC Rocket Pod Container
RPG Rocket Propelled Grenade

RPV	Remotely-Piloted Vehicle
RR	Recoilless Rifle
RRPR	Reduced Range Practice Rocket
RRTR	Reduced Range Training Round
RSS	Rosette Scanning Seeker
RSTA	Reconnaissance, Surveillance and Target Acquisition
RTAF	Royal Thai Air Force
RTG	Rheinmetall Towed Gun
RTM	Resin Transfer Moulding
RTT	*Roues Transporteur de Troupes*
RUC	Royal Ulster Constabulary
RWR	Radar Warning Receiver
RWS	Remote Weapon Station
S&T	Science & Technology
SABCA	*Société Anonyme Belge de Constructions Aéronautiques*
SACLOS	Semi-Automatic Command to Line Of Sight
SAD	Safe and Arm Device
SADARM	Sense And Destroy Armor (US)
SADF	South African Defence Force
SADRAL	Système d'AutoDéfense Rapprochée Anti-aérienne Légère
SADS	Small Arms Detection System
SAE	Society of Automotive Engineers
SAF	Singapore Armed Forces
SAHV-IR	Surface-to-Air High Velocity Infra-Red
SAL	Semi-Active Laser
SAM	Surface-to-Air Missile
SAMM	*Société d'Applications des Machines Motrices*
SANDF	South African National Defence Force
SANG	Saudi Arabian National Guard
SAP	Semi-Armour-Piercing
SAPHEI	Semi-Armour-Piercing High Explosive Incendiary
SAPI	Semi-Armour-Piercing Incendiary
SATCP	*Système Anti-aérien à Très Courte Portée* (very short-range anti-aircraft system)
SATES	Stand Alone Thermal Elbow Sighting system
SAVA	Standard Army Vectronics Architecture
SAVAN	Stabilised Aiming, Vertical sensing And Navigation
SAW	Squad Automatic Weapon
SB	Smooth Bore
SBC	Single Board Computer
SCDB	Surface-Coated Double-Base
SCM	Smart Coupling Mechanism
SCORE	Stratified Charge Omnivorous Rotary Engine
SD	Self Destruction
SD	Self-Destruct
SDD	System Design and Development
SDD	System Development and Demonstration
SDV	Swimmer Delivery Vehicle
SEAD	Suppression of Enemy Air Defence
SEME	School of Electrical and Mechanical Engineering
SEN	Shell Extended-range NORICUM
SEOSS	Stabilised Electro-Optical Sighting System
SFCS	Simplified Fire-Control System
SFIM	*Société de Fabrication d'Instrument de Mesure*
SFIRR	Solid Fuel Integral Rocket/Ramjet
SFM	Sensor Fuzed Munitions
SGTS	Second-Generation Tank Sight
SH/PRAC	Squash Head Practice
SHORAD	Short Range Air Defence
shp	shaft horsepower
SICPS	Standard Integrated Command Post Systems
SIP	System Improvement Plan
SIP	System Improvement Programme
SIPS	Small Integrated Propulsion System
SKOT	*Sredni Kolowy Opancerzny Transporter* (armoured personnel carrier)
SLAD	Shell Load Assist Device
SLAP	Sabot Light Armour Penetrator
SLAP	Saboted Light Armour Penetrator
SLEP	Service Life Extension Program (US)
SLID	Small, Low-cost Interceptor Device
SLR	Super Low Recoil
SLWAGL	Super Light Weight Automatic Grenade Launcher
SM	Smoke
SMG	Sub-Machine Gun
Smoke BE	Smoke Base Ejection
Smoke WP	Smoke White Phosphorus
SMP	Surface Mine Plough
SOCOM	Special Operations Command
SOF	Special Operations Force
SOG	Speed Over Ground
SP	Self-Propelled
SPAAG	Self-Propelled Anti-Aircraft Gun
SPAAGM	Self Propelled Anti-Aircraft Gun and Missile
SPAAM	Self-Propelled Anti-Aircraft Missile
SPAG	Self-Propelled Assault Gun
SPARK	Solid Propellant Advanced Ramjet Kinetic energy missile
SPATG	Self-Propelled Anti-Tank Gun
SPAW	Self-Propelled Artillery Weapon
SPDNS	Stabilised Panoramic Dayight Sight
SPG	Self-Propelled Gun
SPH	Self-Propelled Howitzer
SPL	Self-Propelled Launcher
SPLL	Self-Propelled Loader Launcher
SPM	Self-Propelled Mortar
SPS	Standard Positioning Service
SPSM	*Sensorgezundete Panzerabwehr SubMunition*
SPU	Self Propelled Unit
SR	Staff Requirement
SRAMS	Super Rapid Advanced Mortar System (Singapore)
SRC	Space Research Corporation
SRG	Shell Replenishment Gear
SRU	Shop Replaceable Unit
SRU	Slip Ring Unit
SRV	Surrogate Research Vehicle (US)
SS	Shot Sensor
SSA	Special Spaced Armour
SSES	Suite of Survivability Enhancement System
SSG	Single Shot Gun
SSK	Single Shot Kill
SSKP	Single-Shot Kill Potential
ST	Staff Target
STA	Shell Transfer Arm
STAFF	Small Target Activated Fire-and-Forget (US)
STANAG	Standardisation agreement
STARTLE	Surveillance and Target Acquisition Radar for Tank Location and Engagement (US)
STE/ICE	Simplified Test Equipment/Internal Combustion Engine (US)
StFPA	Staring Focal Plane Array
STK	Singapore Technologies Kinetics
STUP	Spinning Tubular Projectile
SV	Specialist Vehicle
SWARM	Stabilised Weapon And Reconnaissance Mount
T	Tracer
TA	Target Acquisition
TACMS	Army Tactical Missile System (US)
TACNAV™	Tactical Navigation
TACOM	Tank Automotive Command
TACOM	Tank-automotive and Armaments Command (US)
TADS	Target Acquisition and Designation System (US)
TAG	Terminal Active Guidance
TAGS	Transparent Armoured Gun Shield
TALIN™	Tactical Advanced Land Inertial Navigator
TAM	*Tanque Argentino Mediano*
TAPV	Tactical Armoured Patrol Vehicle (Canadian requirement)
TAR	Target Acquisition Radar
TAS	Tripod Adapted Stinger
TAS	Target Acquisition System
TAS	Tracking Adjunct System
TAS	Turret Attitude Sensor
TASS	Turret Altitude Sensor
TBAT	TOW/Bushmaster Armored Turret (US)
TBM	Tactical Ballistic Missile
TCC	Turret Control Computer
TCCS	Transmission Command-and-Control System
TCO	Tactical Combat Operations
TCS	Transmission Control System
TCU	Tactical Control Unit
TD	Tank Destroyer
TD	Target Detection
TDCS	Tank Driver Command System
TDD	Target Detection Device
TDP	Technical Demonstrator Programme
TDR	Target Data Receiver
TDS	Traction Drive System
TE	Target Elevation
TE	Target Engagement
TEG	Thermal Elbow Gunner's
TEL	Transporter Erector Launcher
TELAR	Transporter Erector Launcher and Radar
TER	Target Engagement Radar
TES	Target Engagement System
TES	Theatre Entry Standard
TES	Turret Electronics System
TFCS	Tank Fire-Control System
TGMTS	Tank Gunnery Missile Tracking System
TGP	Terminally Guided Projectile
TGS	Tank Gun Sight
TGS	Teplovaya Golovka Samonavedeniya (seeker)
TGSM	Terminally Guided SubMunition
TGTS	Tank Gunnery Training Simulator
TI	Target Illuminator
TI	Thermal Imager
TI	Thermal Imaging
TICCS	Target Information Command and Control System
TICM	Thermal Imaging Common Modules
TIGS	Thermal Imaging Gunner's Sight
TIIPS	Thermal Imaging and Integrated Position System
TILAS	Tank Integrated Laser Sight

TIM	Thermal Imaging Module
TIPU	Thermal Image Processing Units
TIRE	Tank Infra-Red Elbow
TIS	Thermal Imaging System
TLC	Transport Launching Container
TLD	Top Level Demonstrations
TLE	Treaty Limited Equipment
TLR	Tank Laser Range-finder
TLS	Tank Laser Sight
TM	Thermal Module
TM&LS	Textron Marine & Land Systems
TMBC	Turret Management Ballistic Computer
TMS	Turreted Mortar System
TMS	Turret Modernisation System
TMUAS	Turreted Mortar Under Armor System (US)
TNT	Trinitrotoluene
TOC	Tactical Operations Centre
TOC	Time Optimal Control
TOE	Table of Organisation and Equipment
TOGS	Thermal Observation and Gunnery System (UK)
TOI	*Tourelleau d'Observation et d'Intervention*
TOP	Total Obscuring Power
TOPAS	*Transporter Obrneny Pasovy*
TOTE	Tracker, Optical Thermally Enhanced
TOW	Tube-launched Optically tracked Wire-guided (US)
TP	Target Practice
TPCM	Target Practice Colour-Marking
TPDS	Target Practice Discarding Sabot
TP-FL	Target Practice Flash
TPFSDS-T	Target Practice Fin-Stabilised Discarding Sabot - Tracer
TP-S	Target Practice - Signature
TP-S	Target Practice - Spotting
TP-SM	Target Practice - Smoke
TP-SP	Target Practice - Spotting
TP-T	Target Practice - Tracer
TRACER	Tactical Reconnaissance Armoured Combat Equipment Requirement (UK) (cancelled programme)
TRACKSTAR	Tracked Search and Target Acquisition Radar System (US)
TRADOC	Training and Doctrine Command (US)
TS - F	Thermal Sight - Fagot
TSFCS	Tank Simplified Fire-Control System
TS - M	Thermal Sight - Malyutka
TSQ	Time and SuperQuick
TT	*Transport de Troupes* (troop transporter)
TTB	Tank Test Bed (US) (cancelled programme)
TTD	Transformation Technology Demonstrator
TTG	Time To Go
TTM	Tube Temperature Monitor
TTS	Tank Thermal Sight
TU	Terminal Unit
TU	Traversing Unit
TUA	TOW Under Armor (US)
TUR	*Tiefflieger-Überwachungs-Radar* (low-level surveillance radar)
TV	Television
TVC	Thrust Vector Control
TVM	Track via-missile
TWD	Thermal Warning Device
TWMP	Track Width Mine Plough
TWS	Thermal Weapon Sight
TWS	Threat Warning Sensor
TYDP	Ten-Year Defence Programme (Australia)
UAV	Unmanned Aerial Vehicle
UBLE	Universal Bridge Launching Equipment
UDLP	United Defense, Limited Partnership (became BAE Systems)
UDR	Ulster Defence Regiment
UET	Universal Engineer Tractor (US)
UGV	Unmanned Ground Vehicle
UGWS	UpGunned Weapon Station
ULC	Unit Load Container
UOR	Urgent Operational Requirement
UPPV	Urban Protected Patrol Vehicle
USAF	United States Air Force
USMC	United States Marine Corps
USN	United States Navy
UTAAS	Universal Tank and Anti Aircraft System
UTL	Universal Tactical Light
UTM	Universal Transverse Mercator
UTS	Universal Turret System
UV	Ultraviolet
UV	Utility Vehicle
V	Volt(s)
VAB	*Véhicule de l'Avant Blindé* (front armoured car)
VAB	Vickers Armoured Bridgelayer
VADS	Vulcan Air Defense System (US)
VAE	*Vehículo Armado Exploracion*
VAPE	*Vehículo Apoyo y Exploracion*
VARRV	Vickers Armoured Repair and Recovery Vehicle
VARV	Vickers Armoured Recovery Vehicle
VBC	*Véhicule Blindé de Combat* (armoured combat vehicle)
VBCI	*Véhicule Blindé de Combat d'Infanterie*
VBL	*Véhicule Blindé Léger* (light armoured vehicle)
VBM	*Véhicules Blindés Modulaires*
VCA	*Véhicule Chenillé d'Accompagnement* (tracked support vehicle)
VCC	*Veicolo Corazzato de Combattimento*
VCD	Visual Cueing Device
VCG	*Véhicule de Combat du Genie* (armoured engineer vehicle)
VCI	*Véhicule de Combat d'Infanterie; Vehículo combate infanteria* (infantry combat vehicle)
VCR	Variable Compression Ratio
VCR	*Véhicule de Combat à Roues* (wheeled combat vehicle)
VCR/AT	*Véhicule de Combat à Roues/Atelier Véhicule*
VCR/IS	*Véhicule de Combat à Roues/Intervention Sanitaire*
VCR/PC	*Véhicule de Combat à Roues/Poste de Commandement*
VCR/TH	*Véhicule de Combat à Roues/Tourelle HOT*
VCR/TT	*Véhicule de Combat à Roues/Transport de Troupes*
VCTIS	Vehicle Command and Tactical Information System
VCTM	*Vehículo de Combate Transporte de Mortero* (armoured mortar carrier)
VCTP	*Vehículo de Combate Transporte de Personal* (armoured personnel carrier)
VDA	*Véhicule de Défense Anti-aérienne* (anti-aircraft defence vehicle)
VDAA	*Véhicule d'Auto-Défense Anti-Aérienne*
VDC	Voltage Direct Current
VDM	Viscous Damped Mount
VDSL	Vickers Defence Systems Ltd (now BAE Systems Global Combat Systems)
VDU	Visual Display Unit
VEC	*Vehículo de Exploración de Caballerie*
VEDES	Vehicle Exhaust Dust Ejection System (US)
VELR	Vehicular Eye-safe Laser Range-finder
VERDI	Vehicle Electronics Research Defence Initiative (UK) (completed programme)
VHIS	Visual Hit Indicator System
VIB	*Véhicule d'Intervention du Base*
VIDS	Vehicle Integrated Defence System
VINACS	Vehicle Integrated Navigation And Command System
VIRSS	Visual and Infra-Red Smoke Screening system
VITS	Video Image Tracking Systems
VLAP	Velocity-enhanced Long-range Artillery Projectile
VLC	*Véhicule Léger de Combat* (light armoured car)
VLI	Visible Light Illuminator
VLLAD	Very Low Level Air Defence
VLSMS	Vehicle-Launched Scatterable Mine System
VMBT	Vickers Main Battle Tank
VML	Vehicle Mounted Launcher
VMS	Vehicle Motion Sensor
VNAS	Vehicle Navigation Air System
VPLGR	Vehicular Precision Lightweight GPS Receiver
VRL	*Véhicule Reconnaissance Léger* (light reconnaissance vehicle)
VSEL	Vickers Shipbuilding and Engineering Ltd (now BAE Systems Global Combat Systems)
VSHORAD	Very Short Range Air Defence
VT	Variable Time
VTP	*Véhicule Transport de Personnel* (personnel carrier)
VTT	*Véhicule Transport de Troupe* (troop transporter)
VVSS	Vertical Volute Spring Suspension
VXB	*Véhicule Blindé à Vocations Multiples* (multipurpose armoured car)
WAM	Wide Area Mines (US)
WAPC	Wheeled Armoured Personnel Carrier (Canada)
WASAD	Wide Angle Surveillance and Automatic Detection device
WAV	Wide Angle Viewing
WBSV	Warrior Battlegroup Support Vehicle
WCSP	Warrior Capability Sustainment Programme
WEEA	Warrior Enhanced Electronic Architecture
WES	Waterways Experimental Station
WFLIP	Warrior Fightability Lethality Improvement Programme
WFOV	Wide Field of View
WFSV	Wheeled Fire Support Vehicle (Canada) (no longer in service)
WHA	Weapons Head Assembly
WHSA	Weapons Head Support Assembly
WIU	Weapons Interface Unit
WLIP	Warrior Lethality Improvement Programme
WLR	Weapon Locating Radar
WMRV	Wheeled Maintenance and Recovery Vehicle (Canada)
WP	White Phosphorus
WP-T	White Phosphorus - Tracer
WSLI	Warning System or Laser Illumination
WT	Weapons Terminal
X	Experimental
YMRS	Yugoslav Multiple Rocket System
YS	Yield Strength
ZA-PVO	Zenitnaya Artilleriya-PVO, Anti-aircraft artillery
ZNA	Zimbabwe National Army
ZP	Zone of Protection

ZPI Zone Position Indicator radar
ZRB *Zenitnaya Raketnaya Brigada*, Air defence missile brigade
ZRD-SD *Zenitnaya Raketnye Kompleks - Strednoye Deistive*, Anti-aircraft missile system - medium range
ZRV *Zenitnaya Raketnya Voiska*, Zenith Rocket Troops
ZRV-PVO *Zenitnaya Raketnye Voiska*-PVO, Anti-aircraft missile troops

TERMS AND CONDITIONS FOR THE SALE OF HARDCOPY PRODUCTS

All orders for the sale of hardcopy Products are subject to the following terms and conditions:

1. **DEFINITIONS**
   "Client" means the person, firm or company or any other entity that purchases the Products from IHS.
   "Delivery Point" where applicable, means the location as defined in the Order Confirmation where delivery of the Products is deemed to take place.
   "Directory Products" means IHS's proprietary database or any part thereof, including without limitation, details of particular company/organisation, key personnel, financial/statistical information, products/services description, organisational structure and any other information pertaining to such company(s)/organisation(s) operating in various industrial sectors.
   "Fees" means the money due and owing to IHS for Products supplied including any order processing charge and as set forth in the Order Confirmation. Fees are exclusive of taxes, which will be charged separately to the Client.
   "Products" means any publication, database, supplied to the Client in physical or electronic media, more specifically mentioned in the Order Confirmation. Products include Directory Products.
   "Order Confirmation" includes the order form or confirmation email or any other document which IHS sends to the Client to confirm that IHS has accepted the Client's order and which identifies the name of the Client, Product(s) being supplied, period of supply, delivery information, media of supply, Fees and any terms or conditions unique to the particular Product to be supplied hereunder.
2. Client will pay IHS the Fees as set forth in the Order Confirmation within 30 days from the date of the invoice. Any payments not received by IHS when due will be considered past due, and IHS may choose to accrue interest at the rate of five percent (5%) above the European Central Bank "Marginal lending facility" rate. Client has no right of set-off. Client will pay all the value-added, sales, use, import duties, customs or other taxes where applicable to the purchase of Products. IHS may request payment of the Fees before shipping the Products.
3. IHS grants to Client a nonexclusive, nontransferable license to use the Products for its internal business use only. Client may not copy, distribute, republish, transfer, sell, license, lease, give, disseminate in any form (including within its original cover), assign (whether directly or indirectly, by operation of law or otherwise), transmit, scan, publish on a network, or otherwise reproduce, disclose or make available to others, store in any retrieval system of any nature, create a database or create derivative works from the Product or any portion thereof, except as specifically authorized herein. Any information related to third party company and/or personal data included in the Directory Product(s), may be used by Client for the limited purpose of enquiring about the products and services of the companies/organisations listed therein and who have given permission for their data to be used for this purpose only. Client must comply with the UK Data Protection Act and all other applicable data protection and privacy laws and regulations. In particular, Client must not use such data (i) for any unlawful, harmful or offensive purpose; (ii) as a source for any kind of marketing or promotion activity; or (iii) for the purposes of compiling, confirming or amending its own database, directory or mailing list.
4. Client must not remove any proprietary legends or markings, including copyright notices, or any IHS-specific markings on the Products. Client acknowledges that all data, material and information contained in the Products are and will remain the copyright property and confidential information of IHS or any third party and are protected and that no rights in any of the data, material and information are transferred to Client. Client will take any and all actions that may reasonably be required by IHS to protect such proprietary rights as owned by IHS or any third party. Any unauthorised use may give rise to IHS bringing proceedings for copyright and/or database right infringement against the Client claiming an injunction, damages and costs.
5. Any dates specified in the Order Confirmation for delivery of the Products are intended to be an estimated time for delivery only and shall not be of the essence. IHS shall not be liable for any delay in the delivery of the Products. Unless otherwise agreed by the parties, packing and carriage charges are not included in the Fees and will be charged separately. The Products will be despatched and delivered to the Delivery Point as per Client's preferred method of delivery and as agreed by IHS. If special arrangements are required, then IHS reserves the right to additional charges. Except as provided hereunder, for all Products supplied hereunder, delivery is deemed to occur and risk of loss passes upon despatch of Products by IHS.
6. If for any reason IHS is unable to deliver the Products on time due to Client's failure to provide appropriate instructions, documents or authorisations etc; (i) any risk in the Products will pass to the Client; (ii) the Products will be deemed to have been delivered; and (iii) IHS may store the Products until delivery, whereupon the Client will be liable for all related costs and expenses.
7. Except as otherwise required by law, Client will not be entitled to object or to return or reject the Products or any part thereof unless the Products are damaged in transit. IHS's sole obligation and Clients' exclusive remedy for any claim with respect to such damaged Products will be to replace the damaged Products without any charge. No returns will be accepted by IHS without prior agreement and a returns number issued by IHS to accompany the Products to be returned. All return shipments are at the Client's risk and expense.
8. The possession and usage rights of the Products in accordance with clause 3 above will not pass to Client until IHS has received in full all sums due to it in respect of: (i) Fees; and (ii) all other sums which are or which become due to IHS from Client on any account. Until such rights have passed to Client, the Client will: (i) hold the Products in a fiduciary capacity; (ii) store the Products (at no cost to IHS) in such a way that they remain readily identifiable as IHS property; (iii) not destroy, deface or obscure any identifying mark or packaging on or relating to the Products; and (iv) maintain the Products in satisfactory condition and keep them insured on IHS' behalf for their full price against all risks to the reasonable satisfaction of IHS.
9. The quantity of any consignment of Products as recorded by IHS on despatch from IHS' place of business shall be conclusive evidence of the quantity received by the Client on delivery unless Client can provide conclusive evidence proving otherwise. IHS shall not be liable for any non-delivery of the Products (even if caused by IHS' negligence) unless Client provides conformed claims to IHS of the non-delivery. Any such conformed claim for non-receipt of the Products must be made in writing, quoting the account and Order Confirmation number to the IHS' Customer Service Department, within thirty (30) days of the estimated date of delivery as stated in the Order Confirmation.
10. The Products supplied herein are provided "AS IS" and "AS AVAILABLE". IHS does not warrant the completeness or accuracy of the data, material, third party advertisements or information as contained in the Product or that it will satisfy Client's requirements. IHS disclaims all other express or implied warranties, conditions and other terms, whether statutory, arising from course of dealing, or otherwise, including without limitation terms as to quality, merchantability, fitness for a particular purpose and noninfringement. To the extent permitted by law, IHS shall not be liable for any errors or omissions or any loss, damage or expense incurred by reliance on information, third party advertisements or any statement contained in the Products. Client assumes all risk in using the results of the Product(s).
11. If the Products supplied hereunder are subscription based, except as otherwise provided herein the period of supply will run for one calendar year from the start date as specified in the Order Confirmation and the Fees will cover the costs of supply of all issues of the Product published in that year. If Client attempts to cancel the Product subscription anytime during such period; (i) the Fees payable for that year will be invoiced by IHS in full; or (ii) where Client has already paid the Fees in advance, any Fees relating to the remaining period shall be forfeited. In addition to other rights and subject to the provisions of this clause, IHS in its sole discretion may discontinue the supply the Products in the event Client commits breach of any of the provision of these terms and conditions.
12. In the event of breach of any of the provision of these terms and conditions by IHS, IHS' total aggregate liability for any damages/losses incurred by the Client arising out of such breach shall not exceed at any time the Fees paid for the Product which is the subject matter of the claim. In no event shall IHS be liable for any indirect, special or consequential damages of any kind or nature whatsoever suffered by the Client including, without limitation, lost profits or any other economic loss arising out of or related to the subject matter of these terms and conditions. However, nothing in these terms and conditions shall limit or exclude IHS' liability for (i) death or personal injury caused by its negligence; (ii) fraud or fraudulent misrepresentation; or (iii) any breach of compelling consumer protection or other laws.
13. Client represents and warrants that it will not directly or indirectly engage in any acts that would constitute a violation of United States laws or regulations governing the export of United States products and technology.
14. The parties will comply with all applicable country laws relating to anti-corruption and anti-bribery, including the US Foreign Corrupt Practices Act and the UK Bribery Act. The parties represent and affirm that no bribes or corrupt actions have or will be offered, given, received or performed in relation to the procurement or performance of these terms and conditions. For the purposes of this clause, "bribes or corrupt actions" means any payment, gift, or gratuity, whether in cash or kind, intended to obtain or retain an advantage, or any other action deemed to be corrupt under the applicable country laws.
15. All Products supplied herein are subject to these terms and conditions only, to the exclusion of any other terms which would otherwise be implied by trade, custom, practice or course of dealing. Nothing contained in any Client-issued purchase order, Clients' acknowledgement, Clients' terms and conditions or invoice will in any way modify or add any additional terms to these terms and conditions. IHS reserves the right to amend these terms and conditions from time to time.
16. These terms and conditions and any dispute or claim arising out of or in connection with them or their subject matter shall be governed by and construed in accordance with the laws of England and Wales and shall be subject to the exclusive jurisdiction of the English Courts.

How To Use

Content

In order to help our readers we have included reference pages at the beginning of this publication, the first being this page on How to use which gives an overview of what is contained in each section and an explanation of the entry tags we have added.

These pages and the Executive Overview and Contents are printed on coloured paper to make it easier for readers to find the page they need.

IHS Jane's Land Warfare Platforms: Artillery and Air Defence 2014-15 has changes and updates that reflect, on a worldwide basis, the modern ground-based air defence environment.

Structure

For ease of reference, *IHS Jane's Land Warfare Platforms: Artillery and Air Defence* is divided up into the following sections:

Artillery

- Self-propelled guns and howitzers
    - Tracked
    - Wheeled
- Towed anti-tank guns, guns and howitzers
- Self-propelled mortar systems
- Multiple rocket launchers.

Air Defence

- Self-propelled Air Defence
    - Anti-aircraft guns
    - Anti-aircraft guns/surface-to-air missile systems
    - Surface-to-air missiles
- Towed anti-aircraft guns
- Man-portable surface-to-air missiles
- Static and towed surface-to-air missile systems
- Shelter- and container-based surface-to-air missile systems
- Universal (multipurpose: land/sea/air) missiles.

Artillery and Air Defence in service

This section, organised by country, is a list of artillery and air defence systems (respectively) that are assessed to be in country and in service.

Record structure

In order to standardise each record or entry within the publication, a set proforma will be followed.

Title This will always contain the primary contractor for the system followed by the name or designator of the weapon.

Development Where applicable this heading covers the history of the development of a specified item of equipment up to the point where it entered service and, in some instances, details of any major development after entry into service.

Description Under this heading a detailed technical description is given of each system, in so far as military and commercial confidentiality allows. Where equipment integrates with other equipment or systems, this is also noted.

Variant Most missile systems undergo upgrades, modifications and technology changes. This normally leads to what the primary contractor calls a variant of the basic weapon. If the system variant becomes the primary weapon and the basic is removed then a separate entry will be listed.

Specifications Under this heading the main technical parameters of systems are listed, including dimensions and operating parameters (where available) and, in the case of weapons the destructive charge and so on.

Status This lists the current status of the system, whether it is under development, on order, undergoing trials and so on where known. This is further amplified in certain sections by a table listing the operational platforms carrying the equipment.

Contractors Contained in this section will be the primary contractor and principle sub-contractors.

Images

Photographs are provided for equipment wherever possible, however where images are not available, then line drawings and graphics will be substituted or may complement certain entries.

Images appear with a seven-digit barcode to identify them within IHS/Jane's image database.

Other information

Regular readers of what was previously *Jane's Land-Based Air Defence* and *Jane's Armour and Artillery* will have noticed the reluctance of the Editors to remove older systems which remain operational within inventories of those non-countries of origin. Furthermore, as *IHS Jane's Land Warfare Platforms: Artillery and Air Defence* is now the world's primary reference publication for artillery and air defence, most users find it essential to understand those older systems and technologies before attempting to assess modern weapons with newer technologies. Unless a strong case can be put before the Editors then systems like the Chinese HQ-2 (ex-Soviet S-75), which were designed 50 years ago, will continue to be included in *IHS Jane's Land Warfare Platforms: Artillery and Air Defence* until scrapped entirely.

Artillery

Self-Propelled Guns And Howitzers

Tracked

Argentina

TAMSE VCA 155 155 mm self-propelled artillery system

Development

The VCA 155 *(Vehiculo de Combate de Artilleria)* 155 mm full tracked self-propelled artillery system was developed by TAMSE to meet the operational requirements of the Argentine Army.

The VCA 155 is essentially a lengthened locally manufactured TAM MBT hull with seven roadwheels either side, fitted with the same turret as installed on the Italian Oto Melara 155 mm/41-calibre Palmaria self-propelled howitzer, which was developed specifically for the export market.

Oto Melara supplied a total of 20 turrets to TAMSE and it is believed that at least 15 systems were in service by the late 1990s. The remaining five units are now believed to be operational. Previous information was that the TAMSE production line for the TAM had closed.

Details of the Oto Melara Palmaria 155 mm/41-calibre self-propelled artillery system are provided in a separate entry in *IHS Jane's Land Warfare Platforms: Artillery and Air Defence*.

This is in service with Libya (210 units) as well as Nigeria who took delivery of an initial batch of 25 systems ordered in 1982. Since then additional vehicles have been supplied to Nigeria with a further nine systems being delivered as recently as 2007.

In view of the conflict in Libya in 2011, the number of Palmaria systems still operational is unknown.

It is known that a significant number of Palmaria systems were destroyed by air strikes.

Description

The layout of the VCA 155 155 mm self-propelled artillery system is virtually identical to the TAM medium tank, which is only in service with Argentina.

The TAM medium tank was originally developed to meet the requirements of Argentina by the now Rheinmetall Landsysteme (at that time Thyssen Henschel).

The driver front left, power pack to his right and turret in the centre of the hull slightly to the rear. However, an auxiliary generator is mounted in the left side of the hull and, to increase the operational range of the system, two long-range drum type diesel fuel tanks are mounted at the rear of the hull.

TAMSE VCA 155 155 mm self-propelled artillery system with 155 mm/41-calibre ordnance at maximum elevation (T J Gander) 0500768

TAMSE VCA 155 155 mm self-propelled artillery system with 155 mm/41-calibre ordnance in travel lock 0500414

The all-welded aluminium turret is identical to that of the Oto Melara Palmaria 155 mm self-propelled howitzer. Firing standard 155 mm ammunition, a maximum range of 24,000 m can be achieved while a range of 30,000 m can be achieved with special ammunition.

A semi-automatic projectile loading system is installed, with charges being loaded manually. This increases rate of fire and reduces crew fatigue.

Specifications

	VCA 155
Dimensions and weights	
Crew:	5
Length	
overall:	10.30 m
hull:	7.69 m
Width	
overall:	3.30 m
Height	
overall:	3.20 m
Ground clearance	
overall:	0.45 m
Track	
vehicle:	2.62 m
Weight	
combat:	40,000 kg
Ground pressure	
standard track:	0.82 kg/cm^2
Mobility	
Configuration	
running gear:	tracked
Power-to-weight ratio:	18.00 hp/t
Speed	
max speed:	55 km/h
Range	
main with additional fuel tanks:	719 km
Fuel capacity	
main:	873 litres
main, with auxiliary tanks:	1,273 litres
auxiliary tanks:	400 litres
Fording	
without preparation:	1.4 m
Gradient:	60%
Side slope:	30%
Vertical obstacle	
forwards:	1 m
Trench:	2.5 m
Engine	MTU MB 833 Ka-500, 6 cylinders, diesel, 720 hp at 2,400 rpm
Gearbox	
model:	Renk HSWL 204 planetary
forward gears:	4
reverse gears:	4
Steering:	double differential with hydrostatic continuous steering
Suspension:	torsion bar
Electrical system	
vehicle:	24 V
Firepower	
Armament:	1 × turret mounted 155 mm howitzer
Armament:	1 × roof mounted 7.62 mm (0.30) machine gun

	VCA 155
Armament:	8 × turret mounted 76 mm smoke grenade launcher (2 × 4)
Ammunition	
main total:	28
Main weapon	
calibre length:	41 calibres
Main armament traverse	
angle:	360°
Main armament elevation/depression	
armament front:	+70°/-5°
Survivability	
Night vision equipment	
vehicle:	optional
NBC system:	yes
Armour	
hull/body:	steel
turret:	aluminium

Status
Production complete. In service with Argentine Army. This 155 mm self-propelled artillery system was never offered on the export market.

Contractor
TAMSE.

Bulgaria

Bulgarian 122 mm 2S1 Gvozdika self-propelled howitzer

Development
For some years Bulgaria manufactured the Russian 122 mm 2S1 Gvozdika self-propelled howitzer for both the home and export markets. As far as is known this is identical to the Russian system although there may well be local modifications.

The 2S1 has also been built in Poland, while Romania has designed and built a similar system called the Model 1989 which uses a 2S1 turret on a local chassis. Full details of this are provided in a separate entry in *IHS Jane's Land Warfare Platforms: Artillery and Air Defence*.

It is understood that about 42 of these systems were built for the Romanian Army but only 18 of these now remain in front line service together with six of the original Russian 122 mm 2S1 self-propelled artillery systems.

According to the United Nations Arms Transfer Lists, Bulgaria exported the following quantities of 122 mm 2S1 self-propelled howitzers between 1994 and 2010. At this stage it is not known as to whether these were new-build systems or surplus Bulgarian Army 2S1 systems.

It is considered probable that these export systems were surplus Bulgarian Army systems rather than new vehicles.

Country	Quantity	Comment
Chad	2	delivered 2008
Congo	3	delivered 1999
Eritrea	20	delivered 2005
Togo	6	delivered 1997

It has been confirmed that production of the 2S1, Artillery Command and Reconnaissance Vehicle (ACRV) family of vehicles and the various models of the MT-LB full-tracked vehicles were undertaken in Bulgaria by BETA plc of Cherven Briag.

Bulgarian-built 122 mm 2S1 self-propelled artillery system 0085211

Bulgarian-built MTP-1 technical support vehicle in travelling configuration 0500415

Bulgarian-built ACRV without weapons or equipment installed 0126550

Production of the 2S1 commenced in Bulgaria in 1979. Like the original Russian 2S1, the Bulgarian version is fully amphibious being propelled in the water by its tracks. The direct-fire sight has a magnification of ×5.5 while the indirect sight has a magnification of ×3.7.

Standard equipment includes an NBC system, intercom system and a radio system. Production of the 122 mm 2S1 self-propelled howitzer is undertaken in Bulgaria on an as-required basis.

Variants

MTP-1 Engineer Vehicle
This is based on the Armoured Command and Reconnaissance Vehicle (ACRV) hull.

The MTP-1 has been designed to carry out a wide range of roles on the battlefield including the recovery of damaged and disabled vehicles, changing vehicle components, preparing fire positions and vehicle hides and general lifting duties.

The MTP-1 is virtually identical to the ACRV originally developed in Russia. Mounted on the front right side is a one-person turret that is identical to that fitted to the MT-LB multipurpose tracked vehicle, which was manufactured in Bulgaria.

This manually operated turret is armed with a 7.62 mm machine gun, although some Bulgarian turrets have been observed upgunned to 12.7 mm.

Mounted on top of the roof is a crane with a telescopic jib and, when travelling, this is traversed to the front and the hook restrained by an eye at the front of the hull.

The crane is mounted in a small turret-like cupola with the operator being provided with full armour protection. The crane can lift a weight of 3,000 kg with the jib extended out to 3.4 m or 2,000 kg with the jib extended out to 5 m.

Mounted at the rear of the vehicle is an entrenching blade similar to that fitted to the specialist engineer variant of the MT-LB vehicle. This is mounted on hydraulic arms and can prepare a vehicle scrape in about 110 minutes.

The MTP-1 has a combat weight of 14,000 kg and is provided with an onboard winch with a maximum capacity of 30 tonnes with the entrenching device lowered or 10 tonnes with the entrenching device raised.

Like the ACRV (this being its Western designation, its correct designation being the MT-LBus), it is fully amphibious, being propelled in the water by its tracks. An NBC system is fitted as standard.

ACRV
Bulgaria has manufactured the ACRV (MT-LBus) which can be used for a wide range of roles in addition to the version described previously.

KshTMS
This is a Bulgarian ACRV vehicle modified to carry out the role of battlefield command, control and communications and is fitted with an automated equipment complex, life support complex and a power supply complex.

There is an onboard computer which stores, processes and distributes information and controls the input/output devices, to which it is linked by a Bulgarian standard parallel interface, Type IZ.

A shelter is provided which allows the crew to work up to 100 m from the vehicle, to which they are linked by a fibre optic cable.

Communications equipment installed on the vehicle includes the R-17M, R-173M and R-134 and the data transfer is capable of feeding four radio or wire networks. There are two modes of online encryption.

R55 radio system

This is an ACRV vehicle fitted with extensive communications equipment by the Electron company with equipment including R55P radio system, automatic antenna matching device and an operator's control panel.

Specifications

	2S1 Gvozdika
Dimensions and weights	
Crew:	4
Length	
overall:	7.31 m
Width	
overall:	2.88 m
Height	
overall:	2.775 m
Weight	
combat:	15,400 kg
Mobility	
Configuration	
running gear:	tracked
Speed	
max speed:	61 km/h
water:	4.5 km/h
Range	
main fuel supply:	500 km
Engine	YMZ-238H, V-8, water cooled, diesel
Firepower	
Armament:	1 × 122 mm howitzer
Ammunition	
main total:	40 (HE - 35; HEAT - 5)
Main weapon	
maximum range:	15,200 m
Main armament traverse	
angle:	360°
Main armament elevation/depression	
armament front:	+70°/+3°
Survivability	
Armour	
hull/body:	steel
turret:	steel

Status

Production as required. In service with the Bulgarian Army. It is understood that there has been no production of the 2S1 or ACRV in Bulgaria in recent years. Known export sales of the Bulgaria 122 mm 2S1 self-propelled howitzers include Congo (three in 1999), Eritrea (12 in 1999 and 20 in 2005) and Togo (six in 1997).

While the older exports sales could have been brand new production vehicles, it is considered that the 2005 export sales could have been re-conditioned vehicles from Bulgarian Army stocks. It is understood that Bulgaria currently deploys a total of 247 122 mm 2S1 self-propelled howitzers with these being their only self-propelled artillery systems.

Contractor

BETA Joint Stock Company.

China

NORINCO 155 mm self-propelled gun howitzer PLZ-05 (PLZ52)

Development

During a parade held in Beijing in October 2009 to celebrate the 60th Anniversary of the formation of the People's Republic of China (PRC), many new weapons were shown for the first time.

This included a new 155 mm self-propelled artillery system from the Artillery Brigade of the 38th Group Army which is understood to be designated the PLZ-05 in Peoples Liberation Army (PLA) service.

In 2012 China North Industries Corporation (NORINCO) released full information in their latest PLZ52 155 mm/52-calibre full tracked self-propelled (SP) artillery system which they refer to as an SP gun-howitzer. This is used by the People's Liberation Army (PLA) under the designation of the PLZ-05.

This is being marketed alongside the older NORINCO PLZ45 155 mm/45-calibre SP system that is known to have been exported to at least two countries – Kuwait and Saudi Arabia – as well as being deployed by the PLA.

Prototype of the Chinese 155 mm/52-calibre self-propelled howitzer PLZ-05 which is now in service with the PLA 1132159

Scale model of the latest NORINCO PLZ52 155 mm/52-calibre self-propelled artillery system with ordnance partly elevated and showing muzzle velocity radar mounted above rear of ordnance (Christopher F Foss) 1452899

Details of this are provided in a separate entry in *IHS Jane's Land Warfare Platforms: Artillery and Air Defence*.

Development of the PLZ52 started around 2000 with the first prototypes being completed about 2003 and was accepted for service around 2006.

Prototype turrets in the development of the PLZ52 were similar in some respects to those used by the Russian 2S19 152 mm self-propelled artillery system used by the Russian Army.

Description

While similar in appearance to the earlier PLZ45, the latest PLZ52 is a more capable system and is based on a slightly different hull and has a Gross Vehicle Weight (GVW) of 43 tonnes which compares to the 33 tonnes of the earlier PLZ45.

The hull is of all welded steel armour that provides the occupants with protection from small arms fire and shell splinters.

The driver and power pack are at the front of the hull with the fully enclosed turret at the rear.

The driver is provided with a single piece hatch cover that opens to the rear with three integrated day periscopes. The middle one can be replaced by a passive periscope for driving at night.

Suspension is probably of the torsion bar type with six dual rubber tyred road wheels, drive sprocket at the front, idler at the rear and three track return rollers.

The upper part of the suspension is covered by a skirt which helps to keep down dust.

Spacing of the road wheels on the latest PLZ52 is different to those of the earlier PLZ45 which would indicate a new or updated hull.

PLZ52 also features a new power pack which consists of a diesel developing 1,000 hp at 2,300 rpm coupled to a fully automatic transmission which gives the PLZ52 a maximum road speed of up to 65 km/h and a cruising range of up to 450 km.

The PLZ52 has a power-to-weight ratio of 23.25 hp/tonne which is significantly more than the 16.4 hp/tonne of the earlier PLZ45 and would allow for greater battlefield mobility.

Crew of the PLZ52 consists of four rather than five people as in the case of the PLZ45 due to the increased emphasis being placed on the automation of ammunition handing system.

The turret of the PLZ52 is of all welded steel with vertical sides and rear with the front part of the turret sloping backwards to the rear.

The 155 mm/52-calibre ordnance is fitted with a fume extractor and slotted muzzle brake and when travelling is held in a lock installed on the front of the glacis plate.

Turret power traverse is a full 360° with weapon elevation from –3° to +68°.

A muzzle velocity radar is mounted above the rear of the ordnance and this feeds information to the computerised Fire-Control System (FCS).

Maximum range depends on the projectile/charge combination but according to NORINCO a maximum range of 53 km can be achieved firing a locally developed Extended Range Full Bore – Base Bleed – Rocket Assisted (ERFB-BB-RA) projectile, although accuracy at this range has not been revealed.

Latest Chinese PLZ52 155 mm/52-calibre self-propelled artillery system with ordnance elevated and showing muzzle velocity measuring radar at the very rear of the ordnance (NORINCO) 1405959

It can fire the full family of NORINCO 155 mm ammunition that in addition to the previously mentioned ERFB-BB-RA includes ERFB high-explosive, ERFB-BB high-explosive, ERFB white phosphorous, ERFB smoke, ERFB illumination, incendiary and cargo.

It can also fire 155 mm guided artillery projectiles for a more precision effect including the GP1 and GP6 laser homing guided artillery projectiles and the recently revealed GS1 top attack smart projectile.

Like most Chinese artillery systems, the PLZ52 also has a direct fire capability and NORINCO has developed the armour-piercing high explosive projectile BEA1 for this role. This is claimed to be able to penetrate 30 mm of armour at an angle of 30°.

A total of 30 155 mm projectiles plus bi-modular charge system (MCS) propellant are carried with a stated maximum rate of fire of up to eight rounds a minute being possible as well as a four round Multiple Round Simultaneous Impact (MRSI) capability.

The two NORINCO bi-MCS are the BC1 and BC2 which are also claimed to be compatible with NATO 155 mm artillery systems.

A high rate of fire is achieved through the use of an automatic ammunition handing system for the 155 mm projectiles and a semi-automatic handling system for the MCS.

A roof mounted 12.7 mm machine gun is provided for self-defence purposes and mounted either side of the turret is a bank of four grenade launchers.

A digitized computerised on board FCS and a laser/fibre optic land navigation system is fitted which allows the PLZ52 to rapidly carry out autonomous fire missions.

It would normally typically be used in a battery of six weapons or a troop of three with a regiment typically having a total of 18 PLZ52.

The increased range and faster reaction time of the latest PLZ52 would enable it to rapidly come into action, carry out a fire mission and deploy to another fire position to escape any potential counter battery fire.

Variants

PLZ52 artillery system

As well as being marketed as a stand-alone artillery weapon, the PLZ52 is also being offered as part of a complete artillery system at battery and battalion level all integrated into an overall C4ISR system.

Other elements as part of a complete artillery package would include a dedicated ammunition resupply vehicle based on a similar platform (as has been supplied for the PLZ45), command post vehicle, reconnaissance vehicle with advanced day/night sensor pod, Type 704-1 artillery and fire locating radar, Type 702 meteorological radar and Type W653D armoured recovery vehicle.

Future surveillance capability could include an unmanned aerial vehicle to provide a real time target engagement capability.

Command post vehicle

Also seen with the PLZ-05 155 mm self-propelled artillery system during the parade held in Beijing was a command post vehicle.

This appears to be based on a different six wheeled chassis with a raised crew compartment to the rear of the drivers and commanders position at the front left side of the chassis.

Specifications

	PLZ52
Dimensions and weights	
Crew:	4
Length	
overall:	11.6 m
Width	
overall:	3.38 m
Height	
to turret roof:	2.76 m
to top of roof mounted weapon:	3.55 m (machine gun)
Weight	
combat:	43,000 kg
Mobility	
Configuration	
running gear:	tracked
Power-to-weight ratio:	23.25 hp/t
Speed	
max speed:	65 km/h
Range	
main fuel supply:	450 km
Engine	diesel, 1,000 hp at 2,300 rpm
Gearbox	
type:	automatic
forward gears:	6
reverse gears:	1
Suspension:	torsion bar
Firepower	
Armament:	1 × turret mounted 155 mm howitzer
Armament:	1 × roof mounted 12.7 mm (0.50) machine gun
Ammunition	
main total:	30
Turret power control	
type:	powered/manual
Main armament traverse	
angle:	360°
Main armament elevation/depression	
armament front:	-3°/+68°
Survivability	
Night vision equipment	
vehicle:	yes
NBC system:	yes
Armour	
hull/body:	steel
turret:	steel

Status

In production. In service with China.

Contractor

Chinese state factories.

NORINCO 155 mm 45-calibre self-propelled gun-howitzer PLZ45

Development

Late in 1988, the prototype of a new Chinese 155 mm self-propelled gun-howitzer was shown for the first time. In many respects this is very similar to the now BAE Systems (previously United Defense, Ground Systems) M109 series that was first fielded by the US Army in 1962/1963.

The China North Industries Corporation (NORINCO) PLZ45 has a 155 mm/45-calibre ordnance compared with the 155 mm/39-calibre of the final production M109A6 Paladin and is a much heavier system.

It is fitted with a 45-calibre barrel based on that used in the NORINCO towed 155 mm gun-howitzer Type WAC 21 system, which has been in service with the People's Liberation Army for some years.

Late in 1997, Kuwait placed a contract with NORINCO for a complete self-propelled artillery package, which included 27 155 mm PLZ45 155 mm 45-calibre self-propelled artillery systems. The first batch of 18 PLZ45 self-propelled guns and their associated support vehicles were delivered to Kuwait in March 2000.

NORINCO 155 mm GP1 laser-guided projectile as it would appear in part of its flight with rear mounted fins extended and four control fins extended towards the front (NORINCO) 1366480

Production standard PLZ45 155 mm/45-calibre SPGH with ordnance elevated and fitted with roof mounted 12.7 mm machine gun for local defence (NORINCO) 1365001

ZCY45 battalion (ZCL45 battery) command post vehicle showing higher roof level at the rear 0533769

All of these systems have now been delivered and are in service with Kuwait.

Late in 2001 it was stated Kuwait had placed a contract for another batch of PLZ45 self-propelled gun-howitzers, sufficient to equip an additional battalion. This second batch may also consist of 27 systems, plus a complete support package.

It is understood that Saudi Arabia has taken delivery of a batch of NORINCO PLZ45 self-propelled gun-howitzers.

Description

The hull and turret of the PLZ45 system are of all-welded steel armour construction that provides protection from small arms fire and shell splinters.

The layout is conventional, with the driver's and power pack compartments at the front and the fighting compartment and turret at the rear. The PLZ45 has a crew of five consisting of commander, layer, two ammunition loaders and a driver.

The driver is seated at the front of the hull on the left side with the power pack compartment to his right. The driver has a single-piece hatch cover that opens to the rear and in front of this are two day periscopes for use when the hatch is closed. One of these can be replaced by a passive night driving periscope.

The turret is at the rear of the hull with the commander's cupola on the right side. This has five day periscopes for observation, a single-piece hatch cover that opens to the rear and an externally mounted 12.7 mm machine gun for use in the ground defence and anti-aircraft roles.

On either side of the turret is a large forward-opening door with an observation block in its lower part and there is also a large hatch in the left side of the turret towards the rear.

A small door in the turret rear and a larger door in the hull rear both open to the right for 155 mm ammunition resupply. The turret has a diameter of 2.6 m.

The direct fire telescope is mounted in the front of the turret to the left of the 155 mm gun-howitzer while the indirect fire sight is mounted in an armoured cupola on the left side of the turret roof.

The panoramic telescope has a magnification of ×4 and a 9° 30' field of view, while the direct fire telescope has a magnification of ×5.5 and a 1.1° field of view.

The 155 mm 45-calibre barrel is provided with a slotted muzzle brake and a fume extractor and, when in the travelling position, is held in place by a manually released travelling lock mounted on the glacis plate.

The 155 mm/45-calibre ordnance has a semi-automatic horizontal sliding-wedge breech block, a semi-automatic loader and an electrically controlled and hydraulically operated rammer that enables projectile loading to take place at any angle of elevation with the charge being loaded manually.

The actual ammunition handling system consists of the loader, rammer and shell trolley.

Production standard PCZ45 ammunition support vehicle with hydraulic crane traversed to the rear and roof mounted 12.7 mm machine gun fitted for local defence (NORINCO) 1365002

155 mm PLZ45 155 mm/45-calibre SPGH being transported by rail and showing muzzle velocity radar above rear of ordnance (NORINCO) 1133908

The loader and the rammer are driven hydraulically with electric control. The projectile handling process starts with the projectile moving automatically sideways to the take-off position.

When the loader presses the unloading handle the projectile drops into the shell tray. Then the shell loader draws out the shell trolley together with the projectile and moves it to the loading tray.

The rammer automatically swivels the loading tray into the axis of the barrel and rams the projectile into the chamber. The bagged propellant charges are then loaded by hand.

The hydraulically operated rammer allows for 155 mm projectiles to be loaded at any angle of elevation.

Ammunition is stored in the rear compartment of the turret. A total of 30 155 mm projectiles is carried in two positions, 24 rounds in the loader and six rounds on the right side below the loader.

A total of 30 propellant charges are carried in two positions with eight bagged charges placed on the left side of the loader and 22 bagged charges longitudinally placed on the left and right sides of the gun gear.

Until recently China used conventional bag type charges but they have now developed a 155 mm modular charge system consisting of BC1 and BC2 which are claimed to be compatible with NATO 155 mm projectiles.

The 155 mm self-propelled gun-howitzer PLZ45 is fitted with an inertial direction-finder, a gun point display system and a gun display unit, with fire information coming from the battery or battalion command system. Chinese sources state that the system can have either automatic or manual operating and laying.

It is understood that the PLZ45 systems sold to Kuwait have a new fire-control system that includes an automatic laying system, optical sighting system, gun orientation and navigation system and a GPS receiver.

The automatic laying system can automatically carry out azimuth and elevation and tilt correction. The hydraulic equilibrator for elevation/depression is mounted on the roof of the turret to make more space for the crew. Although the turret has full 360° turret traverse, when firing it is normally traversed 30° left and right.

When in the firing position, two spades can be lowered at the rear of the hull, one either side and in line with the track.

Using ERFB-BB (DDBO2) ammunition a range of 39,000 m can be achieved with maximum rate of fire being 4 to 5 rds/min. Sustained rate of fire is quoted as 2 rds/min, with a barrel life of 2,500 EFC (Charge 9).

	ERFB	ERFB-BB	ERFB-WP	ERFB III	ERFB BE
Length	938 mm	938 mm	938 mm	940 mm	940 mm
Weight	45.54 kg	47.6 kg	47.9 kg	45.8 kg	45.7 kg
Muzzle velocity	897 m/s	903 m/s	795.5 m/s[1]	789.2 m/s[1]	974.5 m/s[1]
Max range	30 km	39 km	25 km	24 km	26 km
(The system can also fire NATO 155 mm M107 projectiles)					
[1] with M2 charge 9 system					

NORINCO is now marketing the 155 mm DDBO3 ERFB-BB projectile with rocket assist which is claimed to have a maximum range of 50 km. This is officially designated the 155 mm ERFB-BBRA/HE.

The torsion bar suspension either side consists of six dual rubber-tyred roadwheels with a distinct gap between the first, second and third roadwheels and the drive sprocket at the front, idler at the rear and three track-return rollers. The first and last roadwheel stations are fitted with a hydraulic shock-absorber.

Standard equipment includes an auxiliary power unit, NBC system, an explosion detection and suppression system and a fire suppression system. A muzzle velocity measuring device is fitted over the rear of the ordnance. This feeds information into the onboard fire-control system.

The types of separate loading 155 mm ERFB ammunition that can be fired are shown in the table above. In addition standard NATO 155 mm artillery projectiles and charges can be fired.

NORINCO 155 mm laser-guided projectile

NORINCO is now marketing the GP1 155 mm Laser Homing Artillery Weapon System (LHAWS).

The actual Laser-Guided Projectile (LGP) features an inertial guidance system with Semi-Active Laser (SAL) homing for the final top attack of the target.

In a typical target engagement, the target is first detected by the forward observer who is provided with the tripod mounted Laser Target Designator/Range-finder (LTDR), radio, Fire-Control System (FCS) and the Fire-Control Calculator (FCC).

The GP1 155 mm laser-guided projectile is programmed and when within range is fired and at the appropriate time the observer uses the LTDR to illuminate the target until the target is hit.

The GP1 LGP weighs 51 kg at launch and is 1302 mm long and fitted with a HE warhead.

According to NORINCO the 155 mm GP1 LGP has a hit probability of at least 90 per cent with between two and three targets being engaged by one operator at any time. This depends on a number of factors including range of the firing platform from the target.

The tripod mounted laser designator weighs a total of 23 kg and can designate targets with at a range of 500 to 5,000 m with NORINCO quoting a maximum firing range of 20 km.

In many respects this 155 mm GP1 LGP is almost identical to the Russian 155 and 152 mm Krasnopol LGP developed by the KBP Instrument Design Bureau and has a similar weight.

The 152 mm Krasnapol was developed to meet the requirements of the Russian Army with the export version being in 155 mm calibre with known sales including France and India with the latter county using it in combat against Pakistan.

It should be noted that France only used this for evaluation purposes.

More recently NORINCO has started marketing a 155 mm projectile designated the GS1 which carries two top attack munitions.

PLZ45 artillery system

In addition to marketing the 155 mm 45-calibre PLZ45 self-propelled artillery system, for which the Kuwait Army is the first export customer, NORINCO is now offering this as part of a complete artillery package.

The complete PLZ45 self-propelled artillery system includes:

- PLZ45 155 mm/45-calibre self-propelled gun
- ZCY45 battalion command post/fire-control vehicle that is based on the NORINCO Type 85 (WZ751) armoured ambulance hull
- ZCL45 battery command post vehicle based on the NORINCO Type 85 (WZ751) armoured ambulance
- GCL Type (WZ751) tracked hull or a Chinese-built lightweight 4 × 4 vehicle used for forward observation/reconnaissance vehicles
- PCZ45 ammunition resupply vehicle that uses the same hull as the PLZ45 155 mm self-propelled gun
- Type 84 (modified Type 653A) engineering/recovery vehicle
- optional Type 704-1 fire direction radar mounted on a 6 × 6 truck hull
- optional 702D meteorological radar system mounted on a 6 × 6 hull.

The ammunition resupply vehicle carries 90 155 mm projectiles and associated charges which are fed to the PLZ45 via a conveyor belt at the rate of eight rounds a minute.

Standard equipment includes an auxiliary power unit and a crane for loading pallets of ammunition through the roof.

It is believed that the Kuwait Army ordered the complete PLZ45 artillery system.

Enhanced PLZ45 system

NORINCO are now offering an enhanced version of its PLZ45 155 mm/45-calibre self-propelled gun howitzer system for the export market.

In the original version, the forward observation vehicle was based on a 4 × 4 light cross-country vehicle that had limited cross-country mobility, had no armour or NBC protection for the crew and limited observation equipment.

The latest forward observation vehicle model is called the GCL45 reconnaissance vehicle and is based on a modified NORINCO Type YW534 full-tracked Armoured Personnel Carrier (APC) hull.

To enable it to carry out its specialised role, the YW534 has been modified. The area to the rear of the commander's position, on the left side of the hull, has been raised to provide greater internal volume for the crew and their specialised equipment.

Mounted on the roof at the rear is a sensor package that incorporates a laser range-finder with a maximum range of 10 km, Charged Coupled Device (CCD) with a maximum detection range of 8 km and a maximum recognition range of 5 km and a thermal imager. The latter has a maximum detection range of 6 km and a maximum recognition range of 3.5 km.

Using the information from the roof-mounted sensor package and the onboard north seeker and Global Positioning System (GPS), target data can be rapidly processed and passed to the battalion or battery combat post for subsequent target engagement. The system can also be used to adjust friendly artillery fire.

Other armoured elements of the PLZ45 155 mm/45-calibre artillery system include the ZCY45 battalion and ZCL45 battery command post vehicles, which are also based on a modified NORINCO YW534 APC hull.

Each PLZ45 155 mm/45-calibre weapon has a dedicated PCZ45 ammunition support vehicle that is based on a similar hull and can carry 90 rounds of 155 mm ammunition (projectiles and charges). There is also a Type 653A1 armoured recovery vehicle based on a Type 69 tank hull and this is fitted with a 20-tonne crane, dozer/stabiliser blade and winches.

Finally there are a number of unarmoured systems including the Type 704-1 artillery locating and fire correction radar, 702D meteorological radar station and associated support vehicles. The radar vehicle is based on a Chinese 6 × 6 hull but the other vehicles are based on a Steyr (4 × 4) truck hull.

NORINCO 155 mm self-propelled gun-howitzer PLZ-05

This made its first public appearance during a parade held in Beijing in October 2009 and available details are provided in a separate entry in *IHS Jane's Land Warfare Platforms: Artillery and Air Defence*. As of early 2012 this had not been offered on the export market by NORINCO.

PGZ-07 twin 35 mm self-propelled anti-aircraft gun system

In 2011 it was disclosed that the PLA had taken into service a new full tracked twin 35 mm Self-Propelled Anti-Aircraft Gun (SPAAG) designated the PGZ-07.

PGZ-07 is armed with two 35 mm cannon which are mounted externally one either side of the large turret. Each of these 35 mm cannon is fitted with a muzzle velocity measuring system which feeds information to the on board computerised Fire-Control System (FCS).

Mounted on the turret roof at the rear is the surveillance radar and mounted on the forward part of the turret is the limited traverse tracking radar. Fitted over the tracking radar appears to be a electro-optical package that moves in elevation with the weapons.

This probably includes a day camera, thermal camera and laser range-finder and would enable targets to be tracked under most weather conditions with the radars switched off.

The full tracked hull used for the PLG-07 appears to be very similar to that used as the basis for the PLZ45 155 mm/45-calibre self-propelled artillery system and its associated PCZ45 ammunition support vehicle.

Layout of the PGZ-07 is conventional with the driver front left, diesel power pack to the right, turret in the middle and a compartment at the rear which could contain additional ammunition.

Hull of the PGZ-07 is of all welded steel armour with an access door in the hull rear, like the PLZ45.

Suspension is of the torsion bar type with each side having six dual rubber tyred road wheels with the drive sprocket at the front, idler at the rear and one return roller.

Assuming the PGZ-07 has the same 525 hp diesel power pack as the PLZ45 it would have a maximum road speed of about 55 km/h and a maximum range of up to 450 km.

The 35 mm cannon used on the PGZ-07 are ballistically identical to the Chinese twin 35 mm towed Anti-Aircraft Gun (AAG) system which is based on the now Rheinmetall Air Defence Oerlikon first generation GDF twin 35 mm towed AAG.

This fires ammunition with a muzzle velocity of 1,175 m/s, a maximum range out to 4,000 m and a maximum altitude of up to 3,000 m and are fitted with a self-destruct fuze.

Some 10 years ago China developed to the prototype stage a similar twin 35 mm SPAAG to the latest PGZ-07 but based on a different hull.

This may well have been a prototype system as the more recent PGZ-07 differs in a number of key areas including a different type of surveillance radar.

NORINCO 203 mm (8 in) SPG

This is essentially a US designed and built 175 mm M107/203 mm (8 in) M110 self-propelled hull fitted with a locally developed long-barrelled 203 mm ordnance, with a large double-baffle muzzle brake. As far as it is known, it never entered production or service.

Specifications

	PLZ45
Dimensions and weights	
Crew:	5
Length	
overall:	10.52 m
hull:	6.1 m
Width	
overall:	3.3 m
Height	
to turret roof:	2.6 m
to top of roof mounted weapon:	3.52 m
axis of fire:	2.16 m
Ground clearance	
overall:	0.45 m
Weight	
standard:	29,000 kg
combat:	33,000 kg
Mobility	
Configuration	
running gear:	tracked
Power-to-weight ratio:	16.40 hp/t
Speed	
max speed:	55 km/h
Range	
main fuel supply:	450 km
Fording	
without preparation:	1.2 m
Gradient:	58%
Side slope:	47%
Vertical obstacle	
forwards:	0.7 m
Trench:	2.7 m
Engine	Deutz BF12 L413FC, supercharged, air cooled, diesel, 525 hp
Gearbox	
type:	manual
forward gears:	6
reverse gears:	1
Steering:	hydraulic
Suspension:	torsion bars and hydraulic shock-absorbers
Firepower	
Armament:	1 × turret mounted 155 mm howitzer
Armament:	1 × roof mounted 12.7 mm (0.50) machine gun
Ammunition	
main total:	30
12.7/0.50:	480
Main weapon	
length of barrel:	7.046 m
Rate of fire	
rapid fire:	5 rds/min
Turret power control	
type:	electrohydraulic/manual
Main armament traverse	
angle:	360°
Main armament elevation/depression	
armament front:	+72°/-3°
Survivability	
Night vision equipment	
vehicle:	yes
NBC system:	yes
Armour	
hull/body:	steel
turret:	steel

Status

Production as required. In service with China (several regiments) and Kuwait (two batches each of 27 units plus associated support vehicles).

A quantity of PLZ45 has also been delivered to Saudi Arabia.

Contractor

China North Industries Corporation (NORINCO).

PLZ-07 122 mm self-propelled howitzer

Development

During the major military parade held in Beijing in October 2009, many new weapon systems made their first public appearance.

These included two new self-propelled artillery systems, the 155 mm self-propelled gun howitzer PLZ-05 and the 122 mm self-propelled howitzer PLZ-07.

Both of these self-propelled artillery systems have been in service with the People's Liberation Army (PLA) for a number of years.

Of these two new systems, as of early 2013 NORINCO is only offering the PLZ-05 on the export market under the designation of the PLZ52. It is also still offering the older PLZ45 on the export market.

It is considered that the PLZ-07 122 mm self-propelled howitzer has been developed to supplement and eventually replace some of the older 122 mm self-propelled howitzers in service with the People's Liberation Army.

These include the NORINCO Type 89, Type 85 and Type 70/70-1 systems. The latter two systems have the 122 mm ordnance in an unprotected mount on the hull to the rear with little protection being provided to the crew from small arms fire and shell splinters.

NORINCO is now offering the SH3 122 mm self-propelled howitzer on the export market and details of this are provided in a separate entry in *IHS Jane's Land Warfare Platforms: Artillery and Air Defence*.

It is possible that the SH3 122 mm self-propelled howitzer was developed in competition with the PLZ-07 122 mm self-propelled howitzer which was subsequently adopted by the PLA.

Description

It is probable that the PLZ-07 122 mm self-propelled howitzer has a hull and turret of all welded steel armour that provides the occupants with protection from small arms fire and shell splinters.

The driver is seated at the front of the hull on the left side and is provided with a slightly domed single piece hatch cover that opens to the left.

Forward of the hatch cover are at least three periscopes that provide observation over the frontal arc. The middle drivers periscopes can be replaced by a passive periscope for driving at night.

The diesel power pack is located to the right of the drivers position with the air inlet and out louvres in the roof above the power pack and the exhaust outlet on the left side of the hull above the second road wheel station.

The fully enclosed turret is mounted on the hull to the rear and it is probable that the crew enter and leave the vehicle by a door in the lower part of the hull rear. This is also used for ammunition resupply purposes.

A total of three roof hatches are provided in the turret, one on the left side and two on the right side.

The roof hatch on the left side opens to the rear and in front of this is a 12.7 mm machine gun for anti-aircraft and local defence purposes. The gunner is provided with no protection whilst using this weapon.

The commander is seated on the right forward side of the turret and has a single piece hatch cover that opens to the front with a similar hatch cover to the immediate rear for another crew member.

Sights are provided in the turret roof on the left and right sides and there appears to be a direct fire sight located to the left of the 122 mm ordnance.

It is understood that the main 122 mm armament is ballistically the same as that in other Chinese 122 mm self-propelled and towed artillery systems.

The 122 mm ordnance is provided with a fume extractor and a muzzle brake and whilst travelling is held in position by a travel lock located just to the rear of the fume extractor.

Suspension is probably of the torsion bar type and either side consists of six rubber tyred road wheels with drive sprocket at the front, idler at the rear and three track support rollers. The first, second and last road wheel stations are provided with a hydraulic shock absorber.

The turret has almost vertical sides that slope slightly inwards and mounted on either side of the turret is a bank of six electrically operated smoke grenade launchers that fire towards the front of the vehicle.

A stowage basket is provided at the rear of the turret and there is a vision block in either side of the turret just to the rear of the smoke grenade launchers.

The hull sides are vertical and appear to have attachments points for skirts, additional armour or perhaps even flotation devices for amphibious operations.

Variants

Command post vehicle

Each PLZ-07 self-propelled howitzer battery is normally supported by a command post vehicle.

This is based on a six wheeled tracked hull with the compartment to the rear of the drivers station raised to provide increased space for the command staff.

Specifications

Not available.

Status

In service with the People's Liberation Army and Marines.

Contractor

Chinese state factories.

NORINCO SH3 122 mm self-propelled howitzer

Development

China North Industries Corporation (NORINCO) has completed development of a new 122 mm full-tracked Self-Propelled Howitzer (SPH) called the SH3.

NORINCO SH3 122 mm self-propelled howitzer is based on a new full-tracked chassis (NORINCO) 1168383

Today the PLA uses a number of 122 mm full tracked SPH. These include the Type 89 which is covered in a separate entry in *IHS Jane's Land Warfare Platforms: Artillery and Air Defence* and which has never been offered on the export market by NORINCO.

The latest is the 122 mm PLZ-07 which is now in service with the PLA and was shown for the first time during the major military parade held in Beijing in October 2009. This has yet to be offered on the export market by NORINCO.

Some sources have indicated that the SH3 is already in service with the Peoples Liberation Army as the replacement for the older 122 mm Type 89 SPH. This vehicle may be designated the WMZ322.

Description

The 122 mm SH3 SPH is based on a new full-tracked hull with driver front left and diesel power pack to the right which leaves the remainder of the hull clear for the fighting compartment.

The hull and turret are probably of all-welded steel armour that provides the occupants with protection from small arms fire and shell splinters.

The hull appears to be a new design and from past experience is probably used for a number of other applications.

Suspension is probably of the torsion bar type with six dual rubber-tyred roadwheels either side, drive sprocket at the front, idler at the rear and track return rollers.

The fully-enclosed turret is at the rear of the hull and is armed with a 122 mm ordnance fitted with a fume extractor and muzzle brake.

This ordnance is a development of the Chinese equivalent of the Russian 122 mm D-30 towed howitzer which has been manufactured by NORINCO for the home and export markets for many years.

Turret has traverse through a full 360° with weapon elevation from –3° to +70°. This fires standard 122 mm D-30 separate loading (projectile and charge) ammunition.

Standard 122 mm High-Explosive (HE) projectile has a maximum range of 17,000 m and other natures include smoke and illuminating.

NORINCO is now marketing a new family of 122 mm separate loading ammunition. This includes a HE Base Bleed (BB) rocket assist projectile Type BEE2 with a maximum range of 27,000 m, HE hollow base with a maximum range of 18,000 m and a HE BB projectile with a maximum range of 22,000 m.

Although some countries have banned cargo projectiles carrying submunitions these are still manufactured by a number of countries, including China.

The SH3 can fire a 122 mm HB cargo projectile carrying 33 bomblets with a maximum range of 18,000 m and BB cargo projectile with a maximum range of 22,000 m. Bomblets are 39.2 mm in diameter and are claimed to be capable of penetrating up to 85 mm of conventional steel armour. There is also a 122 mm HB incendiary projectile.

NORINCO is marketing 155 mm laser-guided artillery projectiles but so far has not announced any 122 mm laser-guided artillery projectiles which could be used with this system.

The commander's cupola is on the right side of the turret roof and is armed with an externally-mounted 12.7 mm MG for local and air defence purposes. Mounted either side of the turret is a bank of five electrically-operated grenade launchers.

SH3 has a combat weight of 24.5 tonnes and is powered by a BF8M1015CP water-cooled supercharged diesel engine that develops 440 kW and gives the vehicle a power-to-weight ratio of 17.95 kW/tonne.

SH3 has a crew of five which indicates that it is not fitted with an automatic ammunition handling system. A flick rammer could be fitted to reduce fatigue and increase rate of fire, especially at high elevations. Maximum rate of fire is stated to be six to eight rounds a minute.

Specifications

	SH3
Dimensions and weights	
Crew:	5
Length	
overall:	6.6 m
Width	
overall:	3.28 m
Height	
overall:	2.5 m
Ground clearance	
overall:	0.40 m
Weight	
combat:	24,500 kg
Mobility	
Configuration	
running gear:	tracked
Speed	
max speed:	60 km/h
Range	
main fuel supply:	500 km
Fording	
without preparation:	1 m
Gradient:	60%
Side slope:	47%
Vertical obstacle	
forwards:	0.7 m
Trench:	2.5 m
Engine	BF8M1015CP, supercharged, water cooled, diesel, 590 hp at 2,100 rpm
Firepower	
Armament:	1 × turret mounted 122 mm howitzer
Armament:	1 × roof mounted 12.7 mm (0.50) machine gun
Armament:	10 × turret mounted smoke grenade launcher (2 × 5)
Ammunition	
main total:	40 estimated (122 mm)
Main armament traverse	
angle:	360°
Main armament elevation/depression	
armament front:	+70°/-3°
Survivability	
Night vision equipment	
vehicle:	yes
NBC system:	yes
Armour	
hull/body:	steel
turret:	steel

Status

Development completed. Understood to be in service with the Peoples Liberation Army and already being marketed by NORINCO.

Contractor

China North Industries Corporation (NORINCO).

NORINCO Type 83 152 mm self-propelled gun-howitzer

Development

The China North Industries Corporation (NORINCO) Type 83 152 mm self-propelled gun-howitzer was first seen in public in October 1984, although it is thought it entered service some years before. Recent information has stated that the first prototype was completed in 1981 with the system being type classified in 1983.

The Type 83 uses an entirely new full-tracked hull that is shared by the 122 mm Type 89 (40-round) tracked self-propelled rocket launcher, trench digger, 425 mm mineclearing rocket launcher and the PLZ89 self-propelled anti-tank gun. As far as it is known, production of all these systems is complete and they are no longer being marketed by NORINCO.

Some Chinese sources have also called the Type 83 152 mm self-propelled gun-howitzer the 152 mm PL83 gun-howitzer with its towed equivalent, the Type 66 gun-howitzer being referred to as the PL66.

Description

The overall layout of the Type 83 152 mm self-propelled gun howitzer is conventional with driver seated towards the front of the hull of the left side and the diesel power pack on the right side. The driver is provided with periscopes for forward observation and a single piece hatch cover.

NORINCO Type 83 152 mm self-propelled gun-howitzer showing ordnance with large double-baffle muzzle brake in a slightly elevated position. The roof-mounted 12.7 mm anti-aircraft machine gun is fitted 0003976

NORINCO Type 83 152 mm self-propelled gun-howitzer in travelling configuration with commanders and drivers hatches locked open and 155 mm ordnance in travel lock (INA) 1403897

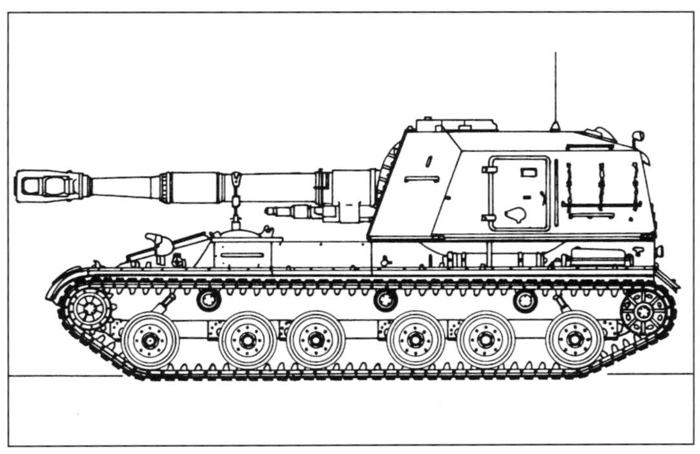

NORINCO Type 83 152 mm self-propelled gun-howitzer in travelling configuration. The roof-mounted 12.7 mm anti-aircraft machine gun is not fitted in this drawing 0500416

The hull and turret of the Type 83 152 mm self-propelled gun-howitzer is constructed of armoured steel that provides the occupants with protection from small arms fire and shell splinters.

The fully enclosed turret is positioned at the rear of the hull and is provided with side doors and roof hatches.

The forward opening door in the left side of the turret is provided with a vision block and a firing port below.

Main armament of the Type 83 152 mm gun-howitzer is based on that used in the NORINCO towed Type 66 152 mm gun-howitzer which is covered in a separate entry in *IHS Jane's Land Warfare Platforms: Artillery and Air Defence* and is fitted with a double baffle muzzle brake and a fume extractor.

NORINCO 120 mm armed PTZ 89 self-propelled anti-tank gun (INA) 1133159

The hull of the NORINCO Type 83 152 mm self-propelled gun-howitzer is used as the basis for a trench digger which is also fitted with a front mounted dozer blade and shown here in the travelling configuration (NORINCO) 0009982

Turret traverse is a full 360° with weapon elevation from –5° to +65°. As far as it is known, no load assist device is fitted for the 152 mm ordnance.

A 12.7 mm Type 54 anti-aircraft machine gun is mounted on the roof and there is also a 7.62 mm tank machine gun fitted next to the main barrel.

It is assumed that the Type 83 152 mm self-propelled gun-howitzer fires the same 152 mm separate loading ammunition as the towed Type 66, HE and smoke. The maximum range is 17,230 m and the muzzle velocity at full charge is 655 m/s.

A Rocket-Assisted Projectile (RAP) known as the MP-152 has been developed and tested. Four rounds can be fired in one minute. Some sources have stated the Type 83 152 mm self-propelled gun-howitzer has a maximum rate of fire of six to eight rounds per minute.

In early 2013 NORINCO was marketing the following types of 152 mm separate loading ammunition for use with this weapon as well as the towed PL66.

Type	HE	HE/PF	Cargo	HE/BB	BB-RA
Length:	703 mm	703 mm	835 mm	874 mm	873 mm
Weight:	45.3 kg	43.6 kg	43.6 kg	43.6 kg	43.5 kg
Muzzle velocity:	655 m/s	655 m/s	655 m/s	660 m/s	674 m/s
Maximum range:	17.2 km	17.2 km	17 km	22 km	28 km

Notes:
- HE pre-fragmented projectile has a lethal radius of 38 m
- Cargo carries 63 bomblets each 39.2 mm in diameter which will penetrate 85 mm of conventional armour and have a lethal radius of 7 m
- HE/BB is high explosive base bleed
- BB-RA is high-explosive base bleed rocket assist

In addition there is also a 152 mm radio jamming projectile

The fire-control system is operated from a panel and all turret movements are power controlled. Time in and out of action is one minute. Panoramic and direct fire sights are provided along with night vision equipment. Communications equipment includes a Type 889D radio set and a Type 803 intercom system.

It also carries a Type 40 unguided rocket launcher.

The suspension system either side uses six roadwheels with the central four wheels arranged in two pairs. There are three track-return rollers, idler at the rear and drive sprocket at the front.

Average road speed is 40/42 km/h while average speed on cross-country roads is claimed to be 30/32 km/h.

Variants

130 mm SPG

This is essentially the Type 83 152 mm SPGH with the 152 mm ordnance removed and replaced by the same ordnance as fitted to the NORINCO 130 mm gun Type 59. It never entered quantity production.

Trench digger

The hull of the Type 83 152 mm self-propelled gun-howitzer is used as the basis for a trench-digging machine capable of digging 250 to 300 m^3 of earth per hour. The digging machine head is carried over the rear of the hull and the crew of three is in a raised cab at the front. A dozer blade is also carried. Combat weight is 30,000 kg.

425 mm mineclearing rocket launcher

The Type 83 is also used as the basis of the carrier/launcher for the Type 762 425 mm mineclearing rocket. As far as it is known, this system has not been exported. Details of this system, which is no longer marketed, are provided in a separate entry in *IHS Jane's Land Warfare Platforms: Armoured Fighting Vehicles*.

120 mm PZT89 self-propelled anti-tank gun

The 120 mm self-propelled anti-tank gun is based on the hull of the NORINCO Type 83 152 mm self-propelled gun-howitzer. This system has a new turret armed with a 120 mm smoothbore gun which fires fixed ammunition with a semi-combustible cartridge case. This system is only used by China and has not been offered on the export market by NORINCO.

It is understood that this was only built in small numbers and is also referred to as the Type 1989.

It should be noted that the 120 mm smoothbore gun installed in the PZT89 was also offered as part of an MBT upgrade programme but as far as it is known this was never sold.

The latest Chinese MBTs, including the Type 99 series, are armed with a 125 mm smoothbore gun that fires separate loading ammunition.

Specifications

	Type 83
Dimensions and weights	
Crew:	5
Length	
overall:	7.005 m
hull:	6.405 m
main armament rear:	9.805 m
Width	
overall:	3.236 m
Height	
overall:	3.502 m
to turret roof:	2.682 m
hull:	1.47 m
to top of roof mounted weapon:	3.19 m
axis of fire:	2.08 m
Ground clearance	
overall:	0.45 m
Track	
vehicle:	2.62 m
Track width	
normal:	480 mm
Length of track on ground:	4.601 m
Weight	
combat:	30,000 kg
Ground pressure	
standard track:	0.68 kg/cm^2
Mobility	
Configuration	
running gear:	tracked
Power-to-weight ratio:	17.33 hp/t
Speed	
max speed:	55 km/h
cross-country:	32 km/h (est.)
Range	
main fuel supply:	450 km
Fuel capacity	
main:	885 litres
Fording	
without preparation:	1.3 m
Gradient:	60%
Side slope:	40%
Vertical obstacle	
forwards:	0.7 m
Trench:	2.5 m
Engine	Deutz Type 12150L, diesel, 520 hp
Electrical system	
vehicle:	24 V
Batteries:	4
Firepower	
Armament:	1 × turret mounted 152 mm howitzer
Armament:	1 × roof mounted 12.7 mm (0.50) Type 54 machine gun
Ammunition	
main total:	30
12.7/0.50:	650
Turret power control	
type:	electric/manual
Main armament traverse	
angle:	360°
Main armament elevation/depression	
armament front:	+63°/-5°
Survivability	
Night vision equipment	
vehicle:	yes
NBC system:	no
Armour	
hull/body:	steel
turret:	steel

Status

Production complete. In service with the People's Liberation Army. Some sources have stated that 400 units are in service with the PLA. There are no known export sales of this system. This system is no longer marketed by NORINCO.

There is an unconfirmed report that Iraq has taken delivery of 18 Type 83 units.

Contractor

China North Industries Corporation (NORINCO).

NORINCO 122 mm Type 89 self-propelled howitzer

Development

The China North Industries Corporation (NORINCO) 122 mm Type 89 self-propelled howitzer was one of the many new weapon systems that were shown for the first time during a large display of land, sea and air systems that took place in China in late 1999.

It is understood that the 122 mm Type 89 system entered service with the People's Liberation Army some time ago.

It is believed to use some automotive components of the Type 77 full-tracked armoured personnel carrier. The Type 77 APC was developed over 30 years ago and, as far as it is known, was never exported. It is no longer marketed by NORINCO.

As of early 2013 there were no known exports of the 122 mm Type 89 self-propelled howitzer.

The NORINCO 122 mm Type 89 SP howitzer has also been referred to as the PLZ89 with each battalion normally having three batteries each with six weapons.

In Peoples Liberation Army service it is now being supplemented by the more recent PLZ-07 122 mm SP howitzer which was first seen in public during late 2009. Details of this are provided in a separate entry in *IHS Jane's Land Warfare Platforms: Artillery and Air Defence*.

Description

The hull and turret of the 122 mm Type 89 self-propelled howitzer is of all-welded aluminium armour construction that provides the occupants with protection from small arms fire and shell splinters.

The driver is seated at the front of the vehicle on the left side with the vehicle commander to his rear and the power pack compartment to his right.

The driver has a single-piece circular hatch cover that opens to the right, in the front of which are three day periscopes, the centre one of which can be replaced by a passive periscope for night driving.

The commander's position to his rear has a single-piece hatch cover that opens to the rear and can be locked in the vertical position.

The engine compartment has air inlet and outlet louvres in the roof with the large rectangular exhaust outlet on the right side.

Suspension is of the torsion bar type with either side having six large single rubber-tyred roadwheels with the drive sprocket at the front and idler at the rear, there are no track-return rollers. It is believed that side skirts are normally fitted to the Type 89 self-propelled howitzer.

The power-operated fully enclosed turret is positioned to the rear and mounted in the forward part is the 122 mm howitzer which is understood to be the Chinese equivalent of the Russian 122 mm D-30 towed howitzer.

The NORINCO 122 mm Type 89 self-propelled howitzer was first seen in public late in 1999
0093548

NORINCO 122 mm Type 89 self-propelled howitzer with roof-mounted 12.7 mm MG covered up 1133909

When travelling, the 122 mm ordnance is held in position by a travelling lock mounted on top of the hull. The 122 mm ordnance is fitted with a multislotted muzzle brake and a fume extractor.

Assuming that the ordnance and ammunition are the same as the NORINCO 122 mm D-30, the following maximum ranges can be achieved:

- 15,300 m with standard ammunition
- 18,000 m with special charge
- 18,000 m with extended range full bore – hollow base
- 22,000 m with extended range full bore – base bleed
- 27,000 m with base bleed rocket assist (BEE2).

It is believed that some form of semi-automatic ammunition handling system is fitted in the turret bustle and that maximum rate of fire is between six and eight rounds per minute.

Additional details of the 122 mm ammunition family are given for the entry on the NORINCO towed 122 mm howitzer D-30 in *IHS Jane's Land Warfare Platforms: Artillery and Air Defence*.

Mounted on either side of the turret is a bank of four electrically operated smoke grenade launchers while mounted on the turret roof on the right side is a 12.7 mm machine gun which is mainly used for anti-aircraft defence. This is mounted on the forward part of a cupola that is provided with a single-piece hatch cover that opens to the side. There are additional hatches in the roof.

Observation devices are provided in either side of the turret and the indirect fire sight is mounted in the roof on the left side.

It is believed standard equipment on the Type 89 self-propelled howitzer includes NBC, night vision equipment, fire detection and suppression system and an onboard, computerised fire-control system.

Some sources have indicated the 122 mm Type 89 self-propelled howitzer has a combat weight of 20.5 tonnes and a maximum road speed of 56 km/h.

Variants

Twin 35 mm SPAAG

China has developed to the prototype stage a twin 35 mm self-propelled anti-aircraft gun system that could be based on a modified 122 mm Type 89 self-propelled howitzer hull.

For this application the hull has been slightly cut down at the rear to accommodate the all-welded steel armour turret armed with twin 35 mm cannon which are mounted externally one either side of the turret.

In many respects this turret is similar to that used on the German Gepard twin 35 mm SPAAG.

Mounted on the front of the turret is the tracking radar with the surveillance radar being mounted on the turret rear. Mounted either side of the lower part of the turret is a bank of electrically operated smoke grenade launchers.

For many years the People's Liberation Army has been using a version of the Swiss twin 35 mm GDF towed anti-aircraft gun system and ammunition used by this would also be compatible with the new twin 35 mm SPAAG. Maximum effective range is about 4,000 m.

As of early 2013 it is understood that production of this twin 35 mm self-propelled air defence system had yet to commenced.

New Chinese 122 mm PLZ-07 self-propelled howitzer

This made its first public appearance during a parade held in Beijing in October 2009; as of early 2013 it had not been offered on the export market by NORINCO.

The latest 122 mm PLZ-07 is similar in overall layout to the 122 mm Type 89 and the 122 mm SH3 but has a different hull and turret design.

Specifications

	Type 89
Dimensions and weights	
Crew:	5
Length	
overall:	11 m
Width	
overall:	3.4 m
Height	
overall:	2 m
Weight	
combat:	20,000 kg
Mobility	
Configuration	
running gear:	tracked
Speed	
max speed:	60 km/h
Range	
main fuel supply:	500 km
Engine	12V150L12, V-12, water cooled, diesel
Gearbox	
type:	manual
Suspension:	torsion bar
Firepower	
Armament:	1 × turret mounted 122 mm howitzer
Armament:	1 × roof mounted 12.7 mm (0.50) machine gun
Ammunition	
main total:	40
Main armament traverse	
angle:	360°
Survivability	
Night vision equipment	
vehicle:	yes
NBC system:	yes
Armour	
hull/body:	steel
turret:	steel

Status

Production probably complete. In service with the PLA. Also used by Marines. Some sources have stated that about 500 Type 89 122 mm systems are in service with PLA. There are no known exports of this system.

Contractor

China North Industries Corporation (NORINCO).

NORINCO 122 mm self-propelled howitzer Type 85

Development

The China North Industries Corporation (NORINCO) 122 mm self-propelled howitzer Type 85 is the self-propelled artillery component of the range of full-tracked vehicles based on the Type 85 armoured personnel carrier, formerly known as the Type YW531H. This system has also been referred to as the Type 85/Type 54-II self-propelled 122 mm howitzer.

Recent information has indicated that the 122 mm self-propelled howitzer Type 85 was not considered to be a success and its design left a lot to be desired. This includes lack of protection for the gun crew from small arms fire, shell splinters and the weather.

In Peoples Liberation Army (PLA) service the 122 mm self-propelled howitzer Type 85 has been supplemented by the 122 mm Type 89 which is covered in detail in a separate entry in *IHS Jane's Land Warfare Platforms: Artillery and Air Defence*.

NORINCO 122 mm self-propelled howitzer Type 85 with ordnance elevated and showing hoops on hull side for pontoons to give system an amphibious capability 0003978

122 mm self-propelled howitzer Type 85 from the front with ordnance covered and in travelling lock 0500325

More recently the PLA has taken into service the 122 mm PLZ-07 self-propelled howitzer.

This made its first public appearance during a parade held in Beijing in October 2009, as of early 2013 it had not been offered on the export market by NORINCO.

The latest 122 mm PLZ-07 is similar in overall layout to the 122 mm Type 89 but the 122 mm SH3 but has a different hull and turret design.

Description

The Type 85 self-propelled howitzer uses the basic all-welded steel hull of the Type 85 APC with a Chinese-produced copy of the Russian 122 mm D-30 towed howitzer mounted in a semi-open superstructure at the rear.

The driver is seated at the front of the hull on the left side and has day periscopes to the front and a single-piece hatch cover that opens to the rear. One of the periscopes can be replaced by a passive periscope for driving at night. Another hatch cover is positioned to the rear of the driver.

The diesel power pack compartment is to the right of the driver with the crew and gun compartment extending to the rear. The semi-open superstructure is equipped with a canvas cover for weather protection. Although the crew compartment has an open top, sides and rear, a shield provides the gun crew with some protection from small arms fire and shell splinters over the frontal arc. No protection is provided against overhead attack.

The 122 mm D-30 howitzer is fitted with a multibaffle muzzle brake and has a maximum elevation of +70°. Traverse is limited to 22.5° either side. The D-30 fires a 122 mm HE projectile weighing 21.76 kg to a maximum range of 15,300 m, but an ERFB projectile has been developed which has a maximum range of 21,000 m. The maximum rate of fire is 6 to 8 rds/min.

NORINCO is now marketing a new family of 122 mm separate loading ammunition that also includes a high-explosive base bleed projectile which has a maximum range of 22,000 m and a high-explosive base bleed rocket assist projectile which has a maximum range of 17,000 m.

The latest long range NORINCO 122 mm projectile is the 122 mm Base Bleed Rocket Assist/High-Explosive (BBRA/HE) which is designated the BEE2 and this has a maximum stated range of 27,000 m.

Additional details of the complete family of NORINCO 122 mm separate loading ammunition are given in the entry for the Chinese NORINCO 122 mm howitzer D-30 in *IHS Jane's Land Warfare Platforms: Artillery and Air Defence*.

When travelling, the 122 mm ordnance is held in a travel lock which folds down on to the hull front when the weapon is deployed in action.

The vehicle is amphibious but requires additional pontoons attached to both sides of the hull to achieve full flotation. When afloat, the vehicle is propelled in the water by its tracks. Maximum water speed is said to be 6 km/h. No secondary armament is carried other than the crew's weapons and the vehicle is equipped with a Type 889 or VRC-83 radio set.

Production of the NORINCO 122 mm self-propelled howitzer Type 85 was completed some time ago and marketing has now ceased.

Variants

NORINCO 122 mm SH3 self-propelled howitzer

This is the latest 122 mm tracked self-propelled howitzer to be marketed by NORINCO and has a combat weight of 24.5 tonnes and a crew of five.

It is fitted with a turret armed with a 122 mm weapon based on the towed D-30 which is also used in the 122 mm Type 89 and 122 mm Type 85 self-propelled artillery systems.

Details of the SH3 are given in a separate entry in *IHS Jane's Land Warfare Platforms: Artillery and Air Defence* and some sources have stated that it is already in production and service with the Peoples Liberation Army as the replacement for older 122 mm self-propelled howitzers such as the Type 89, Type 85 and the oldest Type 70/Type 70-1 systems.

Specifications

	Type 85
Dimensions and weights	
Crew:	6
Length	
overall:	6.663 m
Width	
overall:	3.068 m
Height	
overall:	2.85 m
hull:	1.914 m
Ground clearance	
overall:	0.46 m
Weight	
combat:	16,500 kg
Ground pressure	
standard track:	0.67 kg/cm^2
Mobility	
Configuration	
running gear:	tracked
Power-to-weight ratio:	19.40 hp/t
Speed	
max speed:	60 km/h
Range	
main fuel supply:	500 km
Fuel capacity	
main:	450 litres
Amphibious:	yes
Gradient:	53.1%
Vertical obstacle	
forwards:	0.6 m
Trench:	2.2 m
Engine	Deutz Type BF8L413F, air cooled, diesel, 320 hp
Gearbox	
type:	manual
forward gears:	5
reverse gears:	1
Firepower	
Armament:	1 × hull mounted 122 mm D-30 howitzer
Ammunition	
main total:	40
Main armament traverse	
angle:	45° (22.5° left/22.5° right)
Main armament elevation/depression	
armament front:	+70°/-5°
Survivability	
Night vision equipment	
vehicle:	yes
NBC system:	no
Armour	
hull/body:	steel

Status

Production complete. In service with the People's Liberation Army. No longer marketed. Some sources have indicted that only 200 of these systems were built for the People's Liberation Army.

Contractor

China North Industries Corporation (NORINCO).

NORINCO 122 mm self-propelled howitzer Type 70 and Type 70-1

Development

The first Chinese 122 mm Self-Propelled Howitzer (SPH) was called the Type 70 (or WZ302) and was based on the hull of the Type B531 series armoured personnel carrier (also referred to as the Type 63-1). This retained the standard four single rubber-tyred roadwheels either side.

The Type 70 122 mm SPH is understood to have entered service around 1967 but was only used by the People's Liberation Army.

The second version to enter production and service was the WZ 302A, or Type 70-1. This was followed by an improved model called the WZ 302B or Type 70-2.

Both of these are based on the longer, full-tracked hull of the Type YW531H armoured personnel carrier but retain the older Deutz diesel engine and the commander's position at the front right. These were only used by the People's Liberation Army.

Description

The hull is similar to that of the NORINCO YW531H APC with the crew and engine compartment at the front and with the 122 mm weapon mounted on top of the hull at the rear.

122 mm self-propelled howitzer Type 70-1 with cover in position over the main armament (Christopher F Foss) 0500326

122 mm self-propelled howitzer Type 70-1 showing the shield for the 122 mm howitzer 0500417

The latter is the upper part of the 122 mm Type 54-1 howitzer, which in turn is a copy of the Russian 122 mm M1938 (M30) towed artillery system developed in the late 1930s.

The original howitzer shield is retained and the sides and top have a canvas cover that is stowed around the periphery of the fighting compartment when not in use.

The hull is of all welded steel with a maximum thickness of 12 mm.

The driver sits forward to the left of the engine compartment and has an overhead hatch and two forward-facing day periscopes. There are two white light headlamps, one each side, and on the right is a night driving infra-red lamp with a night vision device for use with one of the driver's periscopes.

The suspension on either side consists of five dual rubber-tyred road wheels, drive sprocket at the front, idler at the rear and three track-return rollers. The first, second and fifth roadwheel arms either side are fitted with shock-absorber cylinders. The track has a cover along its upper length, probably for use during wading when an anti-splash board on the front hull is raised. Rubber-covered track links are used.

The Type 54-1 122 mm howitzer has a maximum range of 11,800 m firing an HE projectile. It fires the variable-charge, case-type separate loading ammunition detailed in the table.

Ammunition type	HE	Smoke	Illuminating[1]
Designation:	Type 54	Type 54	Type 54
Weight:			
(projectile)	21.76 kg	22.55 kg	21.9 kg
(filling)	3.5 kg	3.65 kg	1.515 kg
Type of filling:	TNT	WP	n/a
Muzzle velocity:	515 m/s	509 m/s	496 m/s
Max range:	11,800 m	11,930 m	11,000 m
Type of fuze:	Liu-4	Yan-2	Shi-1

[1] The illuminant burns for 25 seconds with 450,000 candela at a burst height of 450 m

The actual calibre of the weapon is 121.92 mm with normal recoil being 960 to 1,065 mm and maximum recoil being 1,100 mm. Height of the trunnions is 2,304 mm.

The main drawback of the Type 70/Type 70-1 by today's standards is the short range of the 122 mm howitzer and the very small amount of protection provided for the gun crew from small arms fire and shell splinters.

Also carried on the Type 54-1 self-propelled howitzer is a 7.62 mm Type 67 machine gun (Chinese version of the Russian RP-46) with 1,000 rounds of ammunition.

Each vehicle has a Model A-220A radio set with a range of 16,000 m. This has a total of 288 channels. The intercom is a Type A-221A. Two manually operated CO_2 fire extinguishers are also carried.

Variants

There are no known variants of the Type 54-1/Type 70-1.

Specifications

	Type 70
Dimensions and weights	
Crew:	7
Length	
overall:	5.654 m
Width	
overall:	3.06 m
Height	
hull:	1.91 m
axis of fire:	2.729 m
Ground clearance	
overall:	0.456 m
Track	
vehicle:	2.526 m
Length of track on ground:	3.425 m
Weight	
combat:	15,400 kg
Ground pressure	
standard track:	0.592 kg/cm²
Mobility	
Configuration	
running gear:	tracked
Power-to-weight ratio:	16.88 hp/t
Speed	
max speed:	56 km/h
Range	
main fuel supply:	500 km
Fuel capacity	
main:	450 litres
Fording	
without preparation:	1.5 m
Gradient:	46.6%
Vertical obstacle	
forwards:	0.6 m
Trench:	2.0 m
Engine	Deutz Model 6150L, in line-6, water cooled, diesel, 260 hp at 2,000 rpm
Firepower	
Armament:	1 × hull mounted 122 mm howitzer
Ammunition	
main total:	40
Main armament traverse	
angle:	45° (22.5° left/22.5° right)
Main armament elevation/depression	
armament front:	+63°/-2.5°
Survivability	
Night vision equipment	
vehicle:	yes
NBC system:	no
Armour	
hull/body:	steel

Status

Production complete. In service with the People's Liberation Army.

In People's Liberation Army (PLA) service, the 122 mm self-propelled howitzer Type 70 and Type 70-1 has been supplemented by the 122 mm Type 85 and Type 89 self-propelled artillery systems which are covered in separate entries in *IHS Jane's Land Warfare Platforms: Artillery and Air Defence*.

More recently the PLA has taken into service the 122 mm PLZ-07 self-propelled howitzer. This made its first public appearance during a parade held in Beijing in October 2009, as of early 2013 it had not been offered on the export market by NORINCO.

The latest 122 mm PLZ-07 is similar in overall layout to the 122 mm Type 89 and the 122 mm SH3 with the fully enclosed turret mounted at the rear of the chassis but has a new hull and turret design.

Contractor

China North Industries Corporation (NORINCO).

Egypt

SP122 (122 mm) self-propelled howitzer

Development

The SP122 122 mm is based on the hull of the now BAE Systems (at that time United Defense LP, Ground Systems) 155 mm M109A2 self-propelled howitzer/M992 Field Artillery Ammunition Support Vehicle.

The former is no longer in front-line service with the US Army. The M992 FAASV series vehicle is used to support the US Army's M109A6 Paladin self-propelled artillery systems.

Originally there were two competitors for the Egyptian Army requirement for a 122 mm self-propelled artillery system, the now BAE Systems (at that time RO Defence) with the SP 122 based on the RO2000 full-tracked hull and the then BMY Combat Systems SP122, based on the hull of the 155 mm M109/M992 systems.

Initial Egyptian Army trials with these two competing prototypes were completed in 1985, with the final series of trials being completed in 1987.

Following these trials, the BMY Combat Systems (now BAE Systems) system was selected and a total of 76 was built in the US, with the 122 mm weapon and associated elevating mechanism being built in Egypt. Final deliveries were made from US to Egypt in early 1995.

In March 1996, the Arab Republic of Egypt signed a follow-on contract with the company for the supply of an additional batch of 24 SP122 howitzers with a value of USD27.2 million.

The first of these vehicles were delivered to Egypt late in 1997 and final deliveries were made late in 1998. In mid-1999, the company was awarded a contract by Egypt for the supply of another batch of 24 SP122 systems for delivery between August and December 2000. All of these were built at the company facility at York, Pennsylvania and have now been delivered to Egypt.

Description

The now BAE Systems SP122 uses the same basic aluminium armour construction as other vehicles in the M109 range, which provides protection from small arms fire and shell splinters.

The driver sits in the driver's compartment at front left and the diesel power pack is at front right with the raised fighting compartment extending to the rear. The driver has three periscopes for day driving and the centre one of these can be replaced by a passive periscope for driving at night.

The 122 mm D-30 howitzer is mounted in the forward part of the fighting compartment and has manual traverse of 30° left and 30° right with elevation from –5° to +70°. Firing an HE projectile, a maximum range of 15,300 m can be achieved with conventional ammunition.

A dual-purpose sighting system is fitted which comprises an indirect fire sighting device and a direct fire telescopic sight. The former incorporates a range drum that directly computes the required angle of elevation corresponding to the target without additional calculation. The latter is used for engagements of both stationary and moving targets.

The standard towed Russian designed 122 mm D-30 howitzer has been manufactured in Egypt by the Abu Zaabal Engineering Industries Company (qv) which is also known as Factory 100. Details of the Egyptian 122 mm D-30 built howitzer are given in a separate entry in *IHS Jane's Land Warfare Platforms: Artillery and Air Defence*.

The Russian howitzer breech mechanism is a mechanical-type semi-automatic system with automatic case extraction after firing. The breech block is prevented from interfering with the extraction of the fired case and the breech remains open for reloading of the next round.

Hatches are located in the superstructure sides and rear and there is a roof hatch over the gunner's position with accommodation for a .50 (12.7 mm) Browning M2 HB machine gun. The driver's position, engine, transmission and suspension of the M109 vehicle family are retained.

The main fighting compartment houses four of the crew of five and there is space for the stowage of 77 High-Explosive (HE), four High-Explosive Anti-Tank (HEAT) and four illuminating rounds. There is also space for the crew's personal 7.62 mm rifles.

Variants

US supplied M109 SPG

Between 1994 and 2004 Egypt took delivery of 287 M109 SP weapons with 279 delivered in 2000 and eight in 2004.

According to the United Nations Arms Transfer Lists a further 161 overhauled M109A5 155 mm self-propelled howitzers were delivered in 2007 from surplus US Army Stocks.

These were followed by a further 14 M109A5 in 2008 and there were no deliveries in 2009 or 2010 according to the United Nations.

D-30 on T-55 MBT chassis

Egypt has started a programme to fit surplus T-54/T-55 tank hull with a new locally designed turret armed with the locally built 122 mm D-30 howitzer. Factory 100 is responsible for the turret and ordnance and Factory 200 for the hull.

155 mm weapon on T-55 MBT chassis

It is understood that Egypt has test fired a 155 mm/52-calibre weapon in a modified T-55 tank hull. As far as it is known this has yet to enter service with Egypt.

It is understood that the ordnance used was the Finnish Patria 155 mm/52-calibre as used in the towed systems being made under licence in Egypt. A prototype system was built and tested but there are no current plans for this to enter production for the Egyptian Army.

Prototype of the now BAE Systems SP122 for Egypt with 122 mm D-30 ordnance in travel lock (Christopher F Foss) 0500874

Specifications

	SP 122
Dimensions and weights	
Crew:	5
Length	
overall:	6.957 m
Width	
overall:	3.15 m
Height	
to turret roof:	2.821 m
Ground clearance	
overall:	0.46 m
Track	
vehicle:	2.788 m
Track width	
normal:	381 mm
Length of track on ground:	3.962 m
Weight	
standard:	20,909 kg
combat:	23,182 kg
Mobility	
Configuration	
running gear:	tracked
Power-to-weight ratio:	17.47 hp/t
Speed	
max speed:	56.3 km/h
Range	
main fuel supply:	349 km
Fuel capacity	
main:	511 litres
Fording	
without preparation:	1.07 m
Gradient:	60%
Side slope:	40%
Vertical obstacle	
forwards:	0.53 m
Trench:	1.83 m
Engine	Detroit Diesel Model 8V-71T, V-8, water cooled, diesel, 405 hp at 2,350 rpm
Gearbox	
model:	Allison XTG-411-2A cross-drive
forward gears:	4
reverse gears:	2
Suspension:	torsion bar
Electrical system	
vehicle:	24 V
Firepower	
Armament:	1 × hull mounted 122 mm D-30 howitzer
Armament:	1 × roof mounted 12.7 mm (0.50) M2 HB machine gun
Ammunition	
main total:	81
12.7/0.50:	500
Main armament traverse	
angle:	60° (30° left/30° right)
Main armament elevation/depression	
armament front:	+70°/-5°
Survivability	
Armour	
hull/body:	aluminium (superstructure)

Status
Production as required. A total of 124 SP122 systems has been built for Egypt with the last of these being completed in late 2000. If additional orders are placed then production of the SP122 could start again.

Contractor
BAE Systems. (Abu Zaabal Engineering Industries Company of Egypt was also involved in the programme.)

France

Nexter Systems 155 mm GCT self-propelled gun

Development
The 155 mm GCT *(Grande Cadence de Tir)* was developed by the now Nexter Systems (previously Giat Industries) from 1969 to meet a French Army requirement for a self-propelled gun to replace the 105 mm and 155 mm self-propelled weapons, both based on the hull of the AMX-13 light tank, then in service.

The first prototype was completed in 1972 and was shown for the first time at the 1973 Satory Exhibition of Military Equipment.

Between 1974 and 1975 six preproduction vehicles were completed and the vehicle entered production in 1977. First production GCTs were delivered in 1978 to Saudi Arabia.

The 155 mm GCT self-propelled howitzer was officially selected by the French Army in July 1979.

It is understood that a total of 440 GCT were built for the home and export markets as listed in the table below.

Production is now complete and the system is no longer marketed. It is no longer in service with Iraq or Kuwait.

Country	Quantity
France	273
Iraq	86
Kuwait	18
Saudi Arabia	63

The French Army designation for the GCT is the 155 AUF1. From 1988, production switched to the improved 155 AUF1 T with a total of 94 (74 plus 20) delivered to the French Army. This has a number of improvements including a more powerful APU, an improved and more reliable automatic ammunition loading system and an inertial goniometer.

Production of the 155 mm GCT self-propelled gun for the French Army commenced again in 1995 with first production systems being completed in 1996. This batch consisted of 20 vehicles.

The GCT has seen action with the French Army in Bosnia. It also saw extensive service with the Iraqi Army in its war with Iran.

Of the 273 systems built for the French Army a total of 104 were upgraded to the enhanced AUF1-TA standard, details of which are given later on in this entry.

Marketing of this 155 mm system has now ceased and Nexter Systems is now concentrating all marketing on their CAESAR 155mm/52-calibre truck mounted system.

Nexter Systems is no longer marketing the GCT/AUF1 and is concentrating its efforts on its truck mounted CAESAR (CAmion Equipe d'un Systeme d ARtillerie) 155 mm/52-calibre self-propelled (SP) artillery systems.

The French Army has taken delivery of 77 CAESAR of which 72 were brand new production standard systems with the remaining five being the pre-production systems brought up to the latest production standard.

The French artillery still deploy quantities of Nexter Systems 155 mm AUF1 tracked SP artillery systems and the 155 mm TRF1 towed artillery systems.

The standard AUF1 SPG currently in service with the French Army is fitted with a 155 mm/40-calibre barrel (Pierre Touzin) 1164303

French AUF1 fitted with the standard 155 mm/40-calibre barrel (Nexter Systems) 1403898

These are to be replaced by a second batch of 64 CAESAR which would be delivered in the 2015 to 2020 time frame, funding permitting.

This would mean that the CAESAR would be the only conventional 155 mm tube artillery system deployed by the French artillery who also operates the TDA 120 mm rifled mortars and the Multiple Launch Rocket System (MLRS) with the first 15 of the latter already being upgraded to fire the Lockheed Martin Guided MLRS to provide a precision effect.

Nexter Systems has already delivered six CAESAR to the Royal Thai Army while Saudi Arabia has ordered a total of 136 systems, although this customer has never been confirmed by Nexter Systems.

The first 100 (80 + 20) CAESAR has been delivered to the Saudi Arabian National Guard (SANG).

Final assembly and integration of the last batch 36 of CAESAR for Saudi Arabia is taking place in country by CAESAR International using subsystems supplied from the French production line. Final deliveries of this batch of 36 are due to be made in 2014.

Late in 2012 Indonesia placed a contract with Nexter Systems for a batch of 37 CAESAR systems.

Details of the Nexter Systems CAESAR 155 mm/52-calibre truck mounted self-propelled artillery system are provided in a separate entry in *IHS Jane's Land Warfare Platforms: Artillery and Air Defence*.

Description
The 155 mm GCT basically consists of a modified AMX-30 MBT hull fitted with a new turret armed with a 155 mm gun and an automatic loading system. The vehicle has a crew of four that consists of the commander, gunner, loader and driver.

The hull is almost identical to that of the AMX-30 MBT apart from the following modifications which result in the hull being 2,000 kg lighter than the MBT's: the 105 mm ammunition racks in the hull have been removed, and a 5kVA 28 V generator and a ventilator system to supply the turret with cold air have been installed.

The driver is seated at the front of the vehicle on the left side and has a single-piece hatch cover that opens to the left, in front of which are three day periscopes. The centre day periscope can be replaced by an image intensification periscope for night driving or a combined SOPELEM OB-16A day/night periscope.

The turret is in the centre of the vehicle. The power pack is immediately behind the bulkhead that separates the engine compartment from the turret compartment.

The transmission consists of an automatic clutch, a combined gearbox and steering unit, brakes and two final drives. The centrifugal-type clutch is activated electrically by a gearshift lever. The combined gearbox and steering mechanism contains a mechanically operated gearbox giving five speeds both forward and reverse, and a triple-differential steering system.

The brakes are hydraulically operated and are used both as service and parking brakes. Each final drive comprises a spur-type right angle and an epicyclic gear train.

The torsion bar suspension either side consists of five, dual, rubber-tyred roadwheels with the idler at the front, drive sprocket at the rear and five track-return rollers that support the inside of the track only. The first, second, fourth and fifth roadwheels are mounted on bogies and the first and last roadwheel stations have a hydraulic shock-absorber.

The turret is all-welded steel with the commander and gunner seated on the right and the loader on the left. The commander's cupola is equipped with day periscopes for all-round observation and has a single-piece hatch cover that opens to the rear.

The loader, who also operates the roof-mounted 7.62 mm or .50 (12.7 mm) M2 HB anti-aircraft machine gun, has a single-piece rear-opening hatch cover. The crew enters the turret via a door in each side.

The 155 mm 40-calibre barrel has a vertical sliding wedge breech block hermetically gas-sealed by a metal blanking plate. The breech is hydraulically opened with manual controls for emergency use. The barrel has a double-baffle muzzle brake. The 155 mm gun has an elevation of +66°, depression of –4° and the turret can be traversed through 360°. Turret traverse and gun elevation are hydraulic with manual controls for emergency use.

155 mm GCT self-propelled guns of the French Army in the travelling configuration (Pierre Touzin) 0106219

The weapon takes one to two minutes to bring into action and one minute to bring out of action. Average rate of fire is eight rds/min with automatic loading and two to three rds/min with manual loading. For burst fire, six rounds can be fired in 45 seconds.

Forty-two projectiles and 42 cartridge cases are carried in the turret rear arranged in seven racks of six identical projectiles and seven racks of six cartridge cases; in addition, 22 propelling charges for short range are housed near and under the loader's seat. A typical ammunition load would consist of 36 (6 × 6) HE projectiles plus six smoke (1 × 6) or 30 HE (5 × 6), six smoke (1 × 6) and six illuminating (1 × 6); but any combination of HE, illuminating or smoke projectiles is possible in racks of six similar projectiles. The various types of projectile can be selected and fired with automatic loading without any preparation of the loading system.

The gunner can select single shots or bursts of six rounds. The turret is re-supplied with 155 mm ammunition by two large doors in the turret rear which fold down into the horizontal to provide a platform for the crew. The turret can be reloaded by a crew of four in 15 minutes or by a crew of two in 20 minutes. The gun can continue to fire while being reloaded.

The 155 mm gun can fire the following types of separate loading (for example projectile and charge system) 155 mm ammunition:

- 155 mm HE 56/69 with a maximum range of 23,000 m
- 155 mm HE (hollow base) with a maximum range of 23,000 m
- 155 mm HE (base bleed) with a maximum range of 29,000 m
- 155 mm HE US M107 with a maximum range of 18,000 m.

Other types of projectiles including illuminating, smoke and carrier can be fired.

It will also fire the BONUS top-attack anti-tank munition developed jointly by Nexter Munitions of France and the now BAE Systems of Sweden. First production Bonus projectiles were delivered in mid-2002. It is now operational in France and Sweden.

The fire-control system of the standard production GCT consists of the following:

- Gimbal suspension providing for the mounting of the goniometer in the turret and permitting it to move freely by 10° relative to the axis in all directions
- One hermetically sealed conventional optical goniometer. Traverse angles are read on engraved drums and elevation angles appear on an elevation scale
- One azimuth plate that can be manually adjusted and is provided with a spirit level to set the goniometer vertically
- A contrarotation device making it possible to preserve the vertical position of the goniometer during turret rotation
- A direct fire sight for anti-tank firing.

A 7.62 mm or a .50 (12.7 mm) M2 HB machine gun can be mounted at the loader's station for anti-aircraft defence. The 7.62 mm machine gun has an elevation of +45°, a depression of –20°, total traverse of 360° and is provided with 2,050 rounds of ammunition, of which 550 rounds are for ready use. The .50 (12.7 mm) M2 HB machine gun has an elevation of +50°, a depression of –20°, total traverse of 360° and is provided with 800 rounds of ammunition of which 100 are for ready use.

Two electrically operated smoke grenade dischargers are mounted on the forward part of the turret, below the main armament.

The standard optical control system can be replaced by a CITA 20 system which consists of a land navigator coupled with a gyroscope. When the 155 mm GCT comes to a halt the crew knows immediately the co-ordinates of its position and the bearing of its ordnance.

Variants

T-72 hull with GCT turret

This was developed and tested on a number of hulls but is no longer being marketed by Nexter Systems.

AUF1-TA and AUF2 upgrades

In September 1999 the Delegation Generale pour l'Armement (DGA) awarded the then Giat Industries a FFR325 million development contract for the AUF1/AUF2 programme for the French Army.

This covered the upgrade of the first batch of 10 systems that retained the current 155 mm/40-calibre barrel. It also included: installation of the Thales Communications ATLAS FCS; fitting of muzzle velocity radar over 155 mm ordnance; refurbished and upgraded turret and chassis.

The existing power has been replaced by a more fuel efficient and reliable Mack E9 diesel coupled to an ENC 200 transmission.

It was expected that the French Army would have upgraded a total of 174 155 mm AUF 1 weapons, of which 70 would have been fitted with a 155 mm/52-calibre ordnance plus all of the other systems. This would have been designated the AUF 2.

The remaining 104 would have retained the 155 mm/40-calibre barrel plus all of the other systems. These will be designated the AUF1-TA.

The enhanced turrets are mounted on an upgraded AMX-30B2 type hull fitted with a Renault/Mack E9 engine. This engine, with the associated semi-automatic SESM ENC 200 gearbox proves the system with a much-improved power-to-weight ratio.

According to the company, this is greater than the original configuration by 20 per cent. This gives the system mobility and manoeuvrability identical to those of MBTs.

The AUF1-TA turret retains the original ordnance, fully automated loading of the projectiles and charges, hydraulic aiming system and the CITA 20 SAGEM onboard navigation system.

The enhanced 155 mm/52-calibre version provides for the integration of the following new subsystems:

- Ballistic CALP 2G computer from EADS
- Installation of a RDB4 muzzle velocity radar from IN-SNEC
- PRISM electronic fuze programming device from TDA
- Two PR4G type frequency-hopping radio sets.

The first qualification example was delivered by the company to the French DGA late in 2002, who then commenced qualification tests.

Late in 2003 it was revealed that the French Army had changed its policy on the upgrade of the GCT, and the 155 mm/52-calibre version will not be procured. Final deliveries to the French Army were made in 2007.

The French Army has decided to procure an initial batch of 72 CAESAR 155 mm/52 6 × 6 self-propelled artillery systems based on a new Renault Trucks Defense Sherpa (6 × 6) hull.

In addition to the installation of the 155 mm/52-calibre ordnance, the AUF2 upgrade would also have included the installation of a new cradle, a new screw-type breech mechanism, new double baffle muzzle brake and modified recoil system.

ATLAS reduces the engagement cycle as well as increasing accuracy and enables Multiple-Round Simultaneous Impact (MRSI) missions to be accomplished. Part of the fleet has been fitted with a muzzle velocity radar which will feed information to the onboard fire-control system.

The installation of an automatic loading system relays the ordnance after each round has been fired and also gives the upgraded GCT a Multiple Rounds Simultaneous Impact (MRSI) capability.

To come into action, fire a 10-round salvo and come out of action, takes two minutes. It is also fitted with a chamber monitoring device and an automatic feeding system for the primers.

Range depends on projectile and charge combination, but the BONUS top attack anti-tank projectile can be fired to a maximum range of 35 km, the LU 211 base bleed high-explosive shell to a maximum range of 39 km and firing an extended range, full-bore projectile, a range of 42 km can be achieved.

In addition, the AUF2 would have had its automatic loading system adapted to NATO standard modular propellant charges. The now Nexter and SNPE developed the new modular charge system under contract to the DGA, with a maximum of six modules being used with a 155 mm/52-calibre barrel and five modules with a 155 mm/40-calibre barrel. The modular charge consists of two modules, Top Charge System (TCS) and Bottom Charge System (BCS).

The TCM weighs 2.6 kg and uses multibase propellant, while the BCS weighs 2 kg and uses single-base propellant. Both use a black powder ignition system. Maximum muzzle velocity is quoted as 945 m/s.

Development of the new French modular charge system is complete and an industrialisation contract has now been awarded. A typical upgraded GCT will carry 180 TCSs and 20 BCSs.

A total of 42 155 mm projectiles of various types plus modular charges are carried in the turret rear with a maximum stated rate of fire of 10 rds/min.

The AUF is based on a modified AMX-30 Main Battle Tank (MBT) hull that was manufactured by the then Giat Industries at their Roanne facility. This was upgraded at a French Army depot prior to being supplied back to the company where it was integrated with the refurbished and upgraded turret.

Air defence consists of the roof-mounted .50 (12.7 mm) M2 HB machine gun with the option of the GALIX armoured vehicle protection system. Add-on armour is also possible and a collective NBC system is fitted as standard.

In mid-2004 the then Giat Industries facility at Bourges completed the first upgraded 155 mm AUF1 Self-Propelled Gun (SPG) for the French Army under the designation of the AUF1-TA. By the end of 2004 a total of 18 units were delivered, with 33 following in 2005, 33 in 2006 and the remaining 10 in 2007.

Specifications

	155 mm GCT
Dimensions and weights	
Crew:	4
Length	
overall:	10.25 m
hull:	6.7 m
main armament rear:	9.51 m
Width	
overall:	3.15 m
without skirts:	3.115 m
Height	
to turret roof:	3.25 m
axis of fire:	2.16 m

	155 mm GCT
Ground clearance	
overall:	0.42 m
Track	
vehicle:	2.53 m
Track width	
normal:	570 mm
Length of track on ground:	3.9 m
Weight	
standard:	38,000 kg
combat:	42,000 kg
Mobility	
Configuration	
running gear:	tracked
Power-to-weight ratio:	17.14 hp/t
Speed	
max speed:	60 km/h
Range	
main fuel supply:	450 km
Fording	
without preparation:	2.1 m
Gradient:	60%
Side slope:	30%
Vertical obstacle	
forwards:	0.93 m
reverse:	0.48 m
Trench:	1.9 m
Engine	Hispano-Suiza HS 110, 12 cylinders, supercharged, water cooled, multifuel, 720 hp at 2,000 rpm
Auxiliary engine:	Microturbo gas turbine
Gearbox	
forward gears:	5
reverse gears:	5
Steering:	triple differential
Clutch	
type:	centrifugal
Suspension:	torsion bar
Electrical system	
vehicle:	28 V
Batteries:	8 × 12 V, 100 Ah
Firepower	
Armament:	1 × turret mounted 155 mm howitzer
Armament:	1 × roof mounted M2 HB machine gun
Armament:	4 × turret mounted smoke grenade launcher (2 × 2)
Ammunition	
main total:	42
secondary total:	800
Main weapon	
maximum range:	29,000 m
Rate of fire	
rapid fire:	8 rds/min
Turret power control	
type:	hydraulic/manual
Main armament traverse	
angle:	360°
speed:	10°/s
Main armament elevation/depression	
armament front:	+66°/-4°
speed:	5°/s
Survivability	
Night vision equipment	
vehicle:	yes
NBC system:	yes
Armour	
hull/body:	steel
turret:	steel

Status

Production complete. No longer being marketed. In service with:

Country	Quantity	Comment
France	273	104 systems upgraded to AUF1-TA standard (37 are now deployed)
Kuwait	18	in reserve
Saudi Arabia	63	

Contractor

Nexter Systems (previously Giat Industries). (Production was undertaken at its Roanne facility for chassis and Bourges for weapon and turret.)

Nexter Systems 155 mm self-propelled gun Mk F3

Development

The 155 mm self-propelled gun Mk F3, or the Canon de 155 mm Mle F3 Automoteur (Cn-155-F3-Am), was developed in the early 1950s to meet the operational requirements of the French Army and is essentially a modified AMX-13 light tank hull with the rear idler removed and the existing hull modified to accept a 155 mm ordnance and its associated recoil, elevating and traversing mechanism.

The 155 mm gun was designed by the Atelier de Construction de Tarbes (ATS), the hull by the Atelier de Construction Roanne (ARE) while integration of the gun with the chassis and firing trials were undertaken by the Etablissement d'Etudes et de Fabrications d'Armement de Bourges (EFAB).

As the ARE was tooling up for production of the AMX-30 MBT, production of the whole AMX-13 tank family, including the 155 mm self-propelled gun, was transferred to Mécanique Creusot-Loire.

Total production of the Mk F3 amounted to over 600 guns and, in 1993, Mécanique Creusot-Loire became Giat Industries which in 2006 was renamed Nexter. In 1997 France supplied 10 155 mm Mk F3 systems to Morocco.

It is understood that there were ex-French Army systems that were refurbished prior to delivery to Morocco.

Production of this system is complete and it is no longer being marketed.

Description

The hull of the 155 mm self-propelled gun Mk F3 is of all-welded steel armour, which provides the occupants with protection from small arms fire and shell splinters.

The all-welded steel armour hull of the 155 mm self-propelled gun Mk F3 has the following thickness:

Hull front:	155 mm at 55°
Hull sides:	20 mm
Hull top:	10 mm
Hull floor:	
(front)	20 mm
(rear)	10 mm
Hull rear:	10 mm

The driver's compartment at the front on the left, engine compartment to the right and the main 155 mm armament at the rear. A splashboard mounted at the front of the hull stops water rushing up the glacis plate when the vehicle is fording. One replacement road wheel is often carried on the glacis plate.

The driver is provided with a single-piece hatch cover that opens to the left, in front of which are three day periscopes, the centre one of which is replaceable by an image intensification (or thermal) periscope for night driving.

The commander is seated behind the driver and has a two-piece hatch cover that opens either side of his position and three day periscopes in front of him.

The torsion bar suspension either side consists of five single rubber-tyred roadwheels with the drive sprocket at the front and the fifth roadwheel acting as

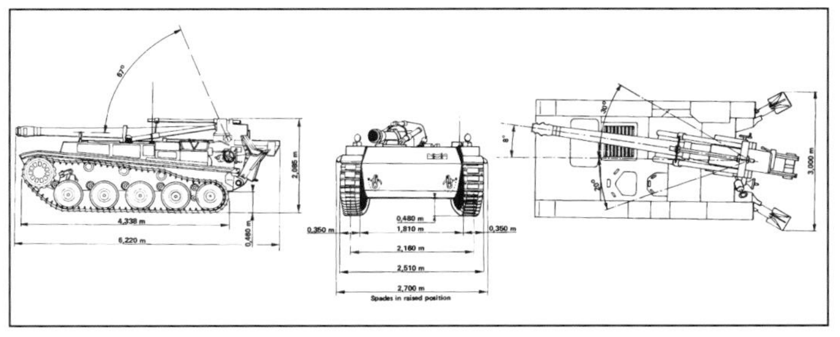

155 mm self-propelled gun Mk F3 0500419

155 mm self-propelled gun Mk F3 deployed in firing position with spades lowered to the ground 0018868

155 mm self-propelled gun Mk F3 in service with Venezuela and shown here in the travelling configuration 0533766

the idler, and three track-return rollers. The first and last roadwheel stations have hydraulic shock-absorbers. The steel tracks can be fitted with rubber pads if required.

There are stowage containers along each side of the upper part of the hull and standard equipment includes a loudspeaker and a cable reel with 400 m of cable.

Each side of the hull at the rear is a large spade which, on arrival at the gun position, is released manually; the vehicle then reverses slightly to bed them in. The vehicle does not have an NBC system and has no amphibious capability.

The other eight members of the gun crew are usually carried in a 6 × 6 truck or an AMX VCA tracked armoured vehicle which also carries 25 × 155 mm projectiles, 25 charges plus fuzes.

For short distances, for example when moving from one gun position to another, four of the gun crew can be carried on the self-propelled gun itself.

When travelling, the 155 mm gun is horizontal and locked 8° to the right of the vehicle's centreline. The gun has a 155 mm 33-calibre barrel, screw breech, double-baffle muzzle brake and a sustained rate of fire of 1 rd/min, which can be increased to 3 rds/min for short periods.

The weapon is provided with an APX L877G direction-finder with a magnification of ×4 and 175° field of view and an APX M321 sighting telescope for direct fire with a magnification of ×4 and 180° field of view.

The 155 mm gun has an elevation of 0° to +67°. When elevated between 0° and +50° it has a traverse of 20° left and 30° right, and when elevated from +51° to +67° it has a traverse of 16° left and 30° right. Elevation and traverse are both manual.

The 155 mm self-propelled gun can fire the following types of separate loading ammunition:

- 155 mm HE projectile Mk 56 weighing 43.75 kg to a maximum range, with charge 9A, of 20,047 m, maximum muzzle velocity 725 m/s
- 155 mm illuminating shell weighing 44 kg to a maximum range of 17,750 m, illumination time of two minutes and intensity of 800,000 candlepower
- 155 mm smoke shell weighing 44.25 kg to a maximum range of 17,750 m (with base bleed projectiles ranges are enhanced, for example to 23,800 m for OE ammunition weighing 42.5 kg to a maximum range of 25,300 m)
- 155 mm HE projectile (type M107 USA) weighing 43.1 kg to a maximum range of 14,700 m (maximum muzzle velocity of 570 m/s) with US charge and 18,000 m (maximum muzzle velocity 690 m/s) with French charge.

Variants

The company did offer the Mk F3 with a Detroit Diesel 6V-53T diesel engine and a kit was available to convert existing petrol-engined versions to this configuration. More recently the Baudouin 6F 11 SRY diesel developing 280 hp at 3,200 rpm and an automatic transmission was offered by the company for this and other members of the AMX-13 family.

RDM Technology upgrade for Qatar

Under a contract awarded late in 1991, RDM Technology of the Netherlands upgraded Qatar's fleet of 155 mm Mk F3 self-propelled artillery systems.

RDM Technology carried out a complete overhaul of these vehicles as well as fitting a new power pack consisting of a Detroit Diesel Model 6V-53T turbocharged diesel developing 280 hp coupled to a Rockford automatic transmission.

In total, 57 vehicles were upgraded plus 22 155 mm self-propelled guns, 33 members of the VCI full-tracked family including ammunition support vehicles and command post vehicles and two armoured recovery vehicles.

RDM Technology of the Netherlands closed down in 2004.

155 mm Mk F3 with 155 mm/39-calibre ordnance

This was developed and tested by Nexter systems but is no longer being marketed.

105 mm self-propelled gun

There was also a 105 mm self-propelled gun on the AMX-13 series hull and this was used by France, Netherlands, Israel and Morocco (which is believed to have five in service).

Specifications

	Mk F3 SPG (petrol)
Dimensions and weights	
Crew:	2
Length	
overall:	6.22 m
Width	
overall:	2.7 m
without skirts:	2.51 m
Height	
transport configuration:	2.085 m
Ground clearance	
overall:	0.48 m
Track	
vehicle:	2.16 m
Track width	
normal:	350 mm
Length of track on ground:	2.997 m
Weight	
combat:	17,400 kg
Ground pressure	
standard track:	0.84 kg/cm^2
Mobility	
Configuration	
running gear:	tracked
Power-to-weight ratio:	14.00 hp/t
Speed	
max speed:	60 km/h
Range	
main fuel supply:	300 km
Fuel capacity	
main:	450 litres
Fording	
without preparation:	1 m
Gradient:	40%
Vertical obstacle	
forwards:	0.6 m
reverse:	0.4 m
Trench:	1.5 m
Turning radius:	4.12 m
Engine	SOFAM Model 8Gxb, 8 cylinders, water cooled, petrol, 250 hp at 3,200 rpm
Gearbox	
type:	manual
forward gears:	5
reverse gears:	1
Steering:	Cleveland-type differential
Clutch	
type:	friction clutch
Suspension:	torsion bar
Electrical system	
vehicle:	24 V
Batteries:	4 × 12 V, 190 Ah
Firepower	
Armament:	1 × hull mounted 155 mm howitzer
Main armament traverse	
angle:	46° (at +51° to +67° elevation is 16° left/30° right)
Main armament elevation/depression	
armament front:	+67°/0°
Survivability	
Night vision equipment	
vehicle:	yes
NBC system:	no
Armour	
hull/body:	steel

Status

Production complete. No longer being marketed. In service with the following countries:

Country	Quantity	Comment
Argentina	24	some sources state 20
Chile	12	some sources state 45/50
Cyprus	12	
Ecuador	12	some sources state 10
Morocco	98	10 delivered in 1997 (Morocco also has five of the older 105 mm systems on an AMX chassis)
Qatar	28	upgraded from 1991
Sudan	10	
UAE	18	
Venezuela	20	some sources state 10/12. Now being replaced by Russian 152 mm 2S19 SPG

Contractor

Nexter Systems (previously Giat Industries).

Germany

Krauss-Maffei Wegmann Panzerhaubitze 2000 (PzH 2000)

Development

Following the cancellation of the international (Germany, Italy and the UK) 155 mm SP-70 programme in July 1986, the German government awarded contracts totalling (at the time of writing) DM183 million to two consortia, for the development and construction of prototypes of a new 155 mm self-propelled howitzer for the German Army called the Panzerhaubitze 2000 (PzH 2000).

The first consortium consisted of Krauss-Maffei, KUKA and Rheinmetall and the second of Wegmann and MaK.

Under Phase I, each of the two consortia built a prototype system which was tested by the German Army and in late 1990 the Wegmann and MaK consortium version was selected.

Under Phase II a development contract was subsequently awarded for an additional four vehicles for further trials. In March 1996, Wegmann and Co GmbH of Kassel was awarded a contract from the German Ministry of Defence for the supply of 185 new 155 mm/52-calibre PzH 2000 self-propelled artillery systems.

On 1 January 1999, Krauss-Maffei Wehrtechnik GmbH and Wegmann and Co GmbH merged to form Krauss-Maffei Wegmann GmbH and Co KG. This company is prime contractor and is responsible for systems integration, with the main subcontractors being:

Company	Subsystem
Rheinmetall Landsysteme	hull
Rheinmetall Waffe Munition	elevating mass
MTU	engine
Renk	transmission
Honeywell	inertial reference unit
ESW	gun laying system
Moog	electric drives
Hensoldt	optical sights
Krauss-Maffei Wegmann Schweisstechnik	hull housing

Production of the 155 mm/52-calibre PzH 2000 self-propelled artillery system for the German Army commenced at the Krauss-Maffei Wegmann facility in Kassel in the second half of 1997.

The first production system was completed early in 1998 and, following company trials, was handed over to the German Army in July 1998 and final deliveries were made on schedule late in 2002.

The German Army placed an order for 185 PzH 2000 systems to replace the 155 mm M109A3G self-propelled howitzers currently in service. Each PzH 2000 battalion has three batteries each with eight guns.

The German Army was the first army in the world to deploy a 155 mm/52-calibre self-propelled artillery system.

Italy has manufactured 68 systems under a licence/co-production programme with Iveco/Oto Melara being prime contractor in Italy.

The Italian contract for the PzH 2000 was finally signed late in 2002 between the Italian Direzione Generale Armamenti Terrestri and Consorzio Iveco-Oto Melara.

Deliveries ran from 2006 through to 2010 and replace the current M109L issued to the Ariete armoured brigade and the Garibaldi and Pinerolo mechanised brigades.

In mid-2001, the Hellenic Army signed a contract worth around EUR190 million with Krauss-Maffei Wegmann for the supply of 24 PzH 2000 155 mm/52-calibre Self-Propelled Howitzers (SPH). Additional details of the Hellenic Army versions are given later on in this entry.

PzH 2000 155 mm/52-calibre self-propelled artillery system of the German Army in travelling configuration with ordnance in travel lock. Note additional armour protection on turret and chassis roof (Krauss-Maffei Wegmann) 1403729

Royal Netherlands Army Krauss-Maffei Wegmann 155 mm/52-calibre PzH 2000 self-propelled artillery system deployed in Afghanistan (Royal Netherlands Army) 1184354

The contract was signed following trials in Germany that demonstrated the PzH 2000 could achieve a maximum range of over 40 km.

These trials used the South African Rheinmetall Denel Munition M2000 Base Bleed (BB) Assegai high-explosive projectile and the Rheinmetall DM72 Modular Charge System (MCS).

The Hellenic Army contract also included Rheinmetall 155 mm ammunition and the DM72 MCS which has been in production for the German Army for some years. It has also been disclosed that the contract included the 155 mm SMArt top attack projectile.

In the first quarter of 2002, the Netherlands finally awarded the German company of Krauss-Maffei Wegmann a contract for the supply of 57 155 mm/52-calibre PzH 2000 self-propelled artillery systems plus associated support equipment for the Royal Netherlands Army (RNLA).

According to the United Nations, the following quantities of PzH 2000 were delivered to the Netherlands between 2004 and 2009:

Year	Quantity
2004	2
2005	2
2006	11
2007	14
2008	11
2009	14

In October 2002, the Netherlands announced another round of defence cuts which included phasing out of service all of the self-propelled M109s and towed M114/39-calibre artillery systems.

It was expected that 18 of the 57 systems ordered for the Netherlands would be passed on to Norway as part of an equipment exchange but this was cancelled in late 2004.

According to the prime contractor, the main features of the PzH 2000 can be summarised as:

- High level of protection – the crew as well as the ammunition are protected by armoured steel in the turret and hull areas as well as by add-on roof protection elements against top attack bomblets and fragments; the hull and turret housings are covered by an additional internal liner. After intensive tests, the add-on roof protection as well as the internal spall liner have passed the qualification programme. The production PzH 2000 are prepared for the adaptation of the roof protection and the internal liner. The retrofit of all series PzH 2000 started in 2001. This protection also allows for the quick adaptation of future armour technologies.

PzH 2000 155 mm/52-calibre self-propelled artillery system of the Royal Netherlands Army in travelling configuration with ordnance in travel lock (Krauss-Maffei Wegmann) 1403730

German Army PzH 2000 155 mm/52-calibre self-propelled artillery system deployed in firing position and with turret traversed right. The turret roof has the mounting points for additional armour protection
(Krauss-Maffei Wegmann) 1403732

- High armament performance – the high rate of fire of three rounds in 10 seconds, up to 10 rounds in less than 60 seconds, is achieved with the assistance of an automatic shell-loading device. Ammunition replenishment of the PzH 2000 with both projectiles and charges (with a total weight of about 3,400 kg) can be carried out in 11 minutes by two crew members using the automatic shell-loading device. Graduated semi-automatic and manual modes are provided in the event of failure of individual parts or the complete automatic system.
- Increased effective range – the required range of 30 km with standard projectiles or almost 40 km with assisted projectiles is achieved with the Rheinmetall 155/52-calibre ordnance as well as by the Rheinmetall Modular Charge System (MTLS), which is already in service with the German Army and has been ordered by Greece, Malaysia, the Netherlands and Norway. The 155 mm/52-calibre ordnance as well as the MTLS are completely compliant with the NATO Joint Ballistic Memorandum of Understanding (MoU). Continued use of the in-service bag charge system is also possible with the minimum range for indirect firing only 2.7 km. The fire-control system with the hybrid navigation system and newly developed and tested flick rammer, with elevation controlled pressure guarantees high accuracy and lowest dispersion at all firing ranges.
- Increased ammunition load – in its shell magazine located at the centre of the hull, the PzH 2000 can carry a total complement of 60 155 mm projectiles. 288 MTLS modules or the corresponding number of bag charges are stowed in the turret bustle. If required, automatic relaying is performed by the system. Loading and firing is possible within the entire gun and turret control range without any restriction.
- Autonomy – an onboard hybrid navigation system permits independent determination of the northern direction, vehicle co-ordinates, including elevation above sea level, as well as spatial positioning of the barrel. This capability and the onboard ballistic computer make each PzH 2000 a stand-alone weapon system and independently combat ready at all times. In addition, the PzH 2000 is linked to a higher-level command and fire-control system via radio data transmission. The system is operated and monitored via menu-controlled displays. The 155 mm/52-calibre main armament of the PzH 2000 howitzer is automatically laid to the required angle and is automatically relaid during firing. In addition, the gunlaying equipment permits semi-automatic or manual back-up operation. The PzH 2000 is equipped with a commander's sight and gunner's direct fire sight, both of which have day and night capabilities.
- Improved mobility – due to the position of the turret at the rear of the hull, minimum barrel overhang is achieved which results in best cross-country mobility and minimum overall vehicle length for good transportability. The PzH 2000 fulfils STANAG 2832B transportation requirements with regard to external dimensions. Maximum interior space by maximum width limited to 3,370 mm is achieved by using Diehl 550 mm wide tracks.
- High endurance and survivability under combat conditions – the onboard fire-control system MICMOS in combination with the hybrid navigation system, the high-mobility performance and the high rate of automation results in a superior shoot-and-scoot capability. After receiving a fire mission on the march it takes the PzH 2000 less than two minutes after stopping to carry out the order, for example fire a 10-round volley and leave its firing position again. All-electric and fully digitised turret, automatic loader and hull with no hydraulic system in the turret, together with the Built-In Test Equipment (BITE) for the crew and maintenance personnel allow for the immediate fault localisation down to circuit board level as line-replaceable units. Due to decentralised vehicle electronics and many back-up systems the crew is able to operate the PzH 2000 in a variety of semi-automatic or manual modes in order to achieve the best performance even in the case of breakdown.
- Three (+2) person crew. The PzH 2000 can be operated in the current configuration by three crew members. This crew is capable of operating the gun, loading the gun and carrying out preventative maintenance. The crew enters the PzH 2000 through a large door in the hull rear independent of the position of the turret. Easy exiting and changing of crew stations even in an emergency situation is possible. The PzH 2000 offers sufficient space for a complete crew of up to five people including storage space for additional equipment and components. In addition to the commander, gunner and driver an additional two loaders can complete the crew. The German Army vehicles have a crew of five.
- Logistic support – In order to minimise time, cost and risk during the fielding phase, for initial training, initial spare parts, special tools and test equipment, as well as for the development of technical documentation, a new logistics concept was developed for the German Army. This concept provides for logistic support to be conducted by industry personnel who train the military and gradually transfer the logistic support activity to the military personnel.
- Force multiplier potential – rationalisation potential – one PzH 2000 is capable of performing the mission of three 155 mm M109A3G howitzers.

During trials that took place at the WTD 91 proving ground in Meppen, Germany, in 1997, one of the four prototypes (PT01) of the 155 mm/52-calibre PzH 2000 self-propelled artillery system, using 48 V technology for some of the autoloader's electric drives attained a rate of fire of 12 rounds in just under 60 seconds.

Other achievements during this trial included: three rounds in 8.4 seconds; eight rounds in 37.1 seconds; 10 rounds in 48.1 seconds; 20 rounds in 1 minute 47 seconds.

In May 1999, the PzH 2000 demonstrated its Multiple Round Simultaneous Impact (MRSI) capability. A production standard system fired five rounds at a range of 17,000 m with 1.2 seconds between the impact of the first and fifth rounds.

Description

The hull and turret of the PzH 2000 are of all-welded steel armour construction, with the driver seated at front right, the power pack to the left and the turret towards the rear.

The driver has a single-piece hatch cover that opens to the rear for driving in the head-out position and mounted forward in this hatch are two day periscopes that give observation to the front of the vehicle, one of which can be replaced by a passive night vision device of the image intensification or thermal type.

The driver, who can also exit to his immediate rear, controls the vehicle using a steering wheel with essential information being provided by means of digital data processing with multifunction colour display and colour displays on the steering wheel.

The MTU diesel engine and transmission can be removed as a complete unit, with the air intakes being located on the top of the hull and the exhaust outlet at the front of the hull on the left side.

The suspension is of the torsion bar type with drive sprocket at the front, idler at the rear, seven double-tyred roadwheels plus track-return rollers. Linear shock-absorbers and hydraulic bump stops using proven components from the Leopard MBT are installed. The upper part of the track is covered by a skirt.

The turret is at the rear of the vehicle and is provided with two roof hatches, one forward on the left side and one on the right side towards the rear. Mounted on the left hatch is a standard German Army 7.62 mm MG3 machine gun for air and self-defence purposes.

The turret has a total of eight 76 mm electronically operated smoke grenade dischargers, four either side of the 155 mm ordnance, firing forwards.

To achieve the long ranges required, Rheinmetall developed the 155 mm 23-litre L52 ordnance for the PzH 2000 and this, together with the matching Rheinmetall Modular Charge System (MTLS), gives the required range and rate of fire.

The L52 ordnance system coupled with the six-zone MTLS has a maximum range of 30 km with the standard L15A2 HE projectile and up to 40 km with extended-range ammunition.

The weapon system includes the cannon assembly with an enhanced slotted muzzle brake, a wedge breech block assembly, recoil and recuperator system, ballistic protection, elevating mechanism interface and a temperature sensor for thermal management.

The 155 mm 52-calibre ordnance is 8.06 m long and is connected to the breech block by an interrupted thread allowing for the quick removal of the gun from the outside of the howitzer without dismantling the weapon system.

Intensive computer simulations of the dynamic firing and recoil process resulted in an optimised barrel having enhanced accuracy with reduced weight. A glass fibre-reinforced plastic bore evacuator ensures an almost gas free crew compartment and also contributes to weight reduction.

The slot-type muzzle brake contributes to an increase in muzzle velocity and also reduces the muzzle flash. Integrated into the breech wedge is an exchangeable primer magazine with an endless conveyor for automatic primer transport, loading and unloading, controlled by the wedge movement. The firing mechanism allows automatic solenoid-actuated firing as well as manual firing. A long wear life is achieved as the barrel is partly chrome-plated.

The Krauss-Maffei Wegmann-developed automatic electrically driven projectile loading system enables loading and firing to take place within the full traverse and elevation range. The 60-round projectile magazine as well as the complete shell handling system is controlled by an automatic data management system.

The PzH 2000 has three ammunition flow modes:

- Automatic/manual ammunition loading and unloading
- Automatic/semi-automatic/manual loading from the magazine to the gun
- Automatic/manual through-loading from the rear of the vehicle directly to the gun.

A burst rate of fire of three rounds in 9.2 seconds has been achieved with the sustained rate of fire being up to 10 rds/min.

Fresh ammunition is loaded into the PzH 2000 from the rear of the vehicle in the lower part of the hull.

The integrated fire-control system and the all-electric gun-control equipment installed in the PzH 2000 enable firing with indirect laying from unprepared firing positions as well as firing with direct laying with a high first round hit probability and short reaction times.

The onboard fire-control equipment, integrated with the gun-control system for indirect target laying, consists of an integrated hybrid navigation system with the appropriate digital computers for automatic operation. This is the primary mode and there is optical/mechanical equipment for the back-up mode.

Geographical north, the vehicle's own position and altitude above sea level are determined automatically with great accuracy. The onboard ballistic computer, in conjunction with the remaining equipment and its datalink to an external fire-control command post enables the gun to conduct a fire mission quickly and independently from an unprepared firing position after receiving target information and data.

The PzH 2000 is also able to lay the 155 mm/52-calibre main armament in accordance with laying and ammunition data radio transmitted by a command post.

The 155 mm/52-calibre armament is automatically laid at high speed and precision, its position checked after every round fired and, if required, relaid automatically.

The commander has direct control as well as monitoring the gunner and the two loaders during system operation and target engagement.

The PzH 2000 is operated from a Graphic User Interface (GUI) terminal using a mission-oriented menu-driven operating system which completely automates the system operation and communication with the fire-control command via data digital radio link, via wire link or via voice radio communications.

A roof-mounted panoramic day/night sight with laser range-finder and a day/night direct fire sight are fitted to the PzH 2000 system.

For improved survivability, especially against top attack weapons, additional armour protection can be fitted to the roof of the vehicle. This can be quickly attached by the crew using onboard equipment. Survivability is also enhanced by compartmentalised charge storage separate from the crew compartment.

Standard equipment includes a fire warning and extinguishing system for the engine compartment and an NBC protection and ventilation system for the crew compartment.

PzH 2000 in Afghanistan

In late 2006 the Royal Netherlands Army deployed three of its PzH 2000 systems to Afghanistan where they worked in conjunction with Canadian Armed Forces 155 mm M777 series towed artillery systems. These PzH 2000 were deployed to Afghanistan by USAF C-17 transport aircraft.

The Royal Netherlands Army used the 155 mm M107C high-explosive projectiles with M557 point detonating fuze and D541 conventional propelling charges.

Late in 2005 the Royal Netherlands Army ordered 10,000 Rheinmetall Rh 40 Base Bleed (BB) projectiles and 69,000 DM92 MTLS.

In 2010 all Royal Netherlands Army PzH 2000 SP artillery systems were withdrawn from Afghanistan.

The German Army has deployed a total of seven PzH 2000 to Afghanistan since 2009 and prior to deployment the only modifications carried out were installation of heat shields on the turret and provision for additional spares.

Under Urgent Operational Requirement (UOR) funding, a number of modifications have been carried out including a cooling system for the propellant charge store and inside of the turret.

It is expected that additional armour will be provided to allow for the PzH 2000 to carry out fire missions whilst deployed away for its fixed firing position.

PzH 2000 demonstrates increased capabilities in South Africa

In 2006 a 155 mm PzH 2000 Self-Propelled (SP) artillery system achieved a maximum firing range of over 56 km during a series of firing trials carried out at the Alkantpan test range in South Africa.

The performance certificate was arranged by the German Bundesamt für Wehrtechnik und Beschaffung (BWB) (Federal Office of Defense Technology and Procurement) on behalf of KMW who are prime contractor for the overall PzH 2000 artillery system.

A maximum range of 56 km was achieved with the 155 mm/52-calibre ordnance at an elevation of 737 mils with a Rheinmetall Denel Munition Velocity-enhanced Long range Artillery Projectile (V-LAP) being fired using six standard Rheinmetall DM72 modular charges.

The maximum elevation of 737 mils was due to range limitations and according to KMW, based on sea level conditions the system has a range of more than 60 km using an increased elevation of 980 mils.

The V-LAP 155 mm HE projectile has been developed by Rheinmetall Denel Munition as a private venture and combines BB and rocket motor assist to give a significant increase in range over existing conventional projectiles.

According to the company, the V-LAP projectile shares an identical external interface with all other projectiles with their existing Assegai and Extended-Range Full-Bore (ERFB) families and uses the same fuzes, charges, packaging, storage and logistics. The V-LAP projectile also includes an Insensitive Munition (IM) main filling as well as Pre-Formed Fragmentation (PFF) HE warhead which provides a significant increase in lethality over currently fielded HE projectiles. It is also compatible with most other 155 mm 39/45/52-calibre artillery systems, towed and self-propelled.

The PzH 2000 has already fired the baseline Assegai 155 mm artillery projectile and the latest test firing of the longer ranger V-LAP was aimed as further enhancing the overall capabilities of the system in the area of extended range.

PzH 2000 155 mm/52-calibre self-propelled artillery system deployed in firing position and fitted with roof mounted 7.62 mm machine gun and showing additional turret roof protection (Krauss-Maffei Wegmann) 1403731

SMArt 155 mm projectile

In addition to firing the standard natures of 155 mm ammunition, the PzH 2000 also fires the German Rheinmetall/Diehl SMArt 155 mm top attack projectile which was originally developed to meet the requirements of the German Army.

When fired from the PzH 2000 this has a maximum range of 27 km and carries two advanced top attack submunitions.

Each of these contains a submunition which is suspended by parachute and rotate over the target zone. A combination of infra-red and mini-radar sensors scans the surface below and when a target is identified the advanced warhead is activated.

Germany placed its first production contract in 1997 and it has also been in service with Australia, Greece and Switzerland. It was also ordered for use by the UK 155 mm/39-calibre AS90 SP artillery systems but this order was subsequently cancelled.

Volcano 155 mm ammunition developments

This was being developed to meet the requirements of Italy and Spain but the latter has now dropped out for financial reasons.

Oto Melara is now continuing development to meet the requirements of the Italian Army and Navy.

The first version of the Volcano will be the unguided Ballistic Extended-Range (BER) version which is expected to be produced in both 127 mm (naval) and 155 mm (land) versions.

These will both have a common 90 mm projectile weight 16 kg which includes fuze, pre-fragmented high-explosive warhead and associated fin unit.

This will be supported by a discarding sabot in either 127 mm or 155 mm depending on its application.

Total weight will be around 30 kg with its programmable multifunction fuze having altimetric, impact, delay, electronic time and self-destruct functions.

Maximum range of the land version is expected to be around 50 km when fired from a 155 mm/39-calibre weapon and about 70 km when fired from a 155 mm/52-calibre weapon such as the PzH 2000.

This will be followed by the Vulcano Guided Long Range (GLR) version which in its 155 mm mode using global positioning/inertial measurement unit (GPS/INS) guidance will have a maximum range of 100 km.

This could be followed by a version with a Semi-Active Laser (SAL) to engage moving targets with a CEP of 5 m.

Hellenic Army PzH 2000

By 2004 the Hellenic Army had taken delivery of the first batch of 24 Krauss-Maffei Wegmann PzH 2000 155 mm/52-calibre self-propelled artillery systems ordered in 2001 under a contract worth EUR190 million.

The first export customer was the Italian Army, but the Hellenic Army was the first export customer to deploy the system. The first PzH 2000s were delivered to the Hellenic Army in mid-2003 and are known as the PzH 2000 GR, which have a number of improvements to meet the specific operational requirements of the Hellenic Army. These include an auxiliary power unit with increased output, cooling system for the charges and crew compartment, aiming and pointing system and the laser range-finder integrated into the gunner's sight has a significant increase in range compared to that installed in German Army systems.

The order also included the Rheinmetall DM72 MTLS Modular Charge System, fuzes, primers and a complete suite of 155 mm projectiles including the new Rh 40 Base Bleed (BB) high-explosive projectile with a maximum range of over 40 km and the 155 mm DM 702 SMArt top-attack munition, which has been developed to neutralise armoured vehicles at long ranges.

Coastal defence role

Although originally developed as a self-propelled field artillery system, in May 1996 two 155 mm/52-calibre PzH 2000 systems carried out a major demonstration in Sweden at the invitation of the Royal Swedish Coastal Artillery.

It should be noted that Sweden has disbanded all of its coastal artillery systems, fixed and mobile.

C-RAM capability for PzH 2000

Krauss-Maffei Wegmann is developing, as a private venture, a C-RAM capability for the current PzH 2000 155 mm/52-calibre and the more recent 155 mm/52-calibre Autonomous Gun Module (AGM).

The latter, when integrated on a new hull developed by General Dynamics European Land Systems - Santa Bárbara Sistemas, is known as the DONAR.

This is being marketed my KMW as "Smart Camp Defence" and is claimed to be the ideal solution to protect a forward operating base against asymmetric rocket and mortar bomb attack.

In a typical engagement, a tower mounted surveillance radar would first detect the incoming target which would then be tracked.

The central fire direction centre would calculate the ballistic trajectory and engagement sequence. This information would be passed onto the remote controlled PzH 2000 that would carry out the actual target engagement with an optimised 155 mm High-Explosive (HE) projectile being fired towards the pre-calculated co-ordinates.

The HE projectile detonates in the immediate vicinity of the threat and the resultant explosion neutralises the incoming threat. PzH 2000 would also have the option of engaging the firing position if this had not been moved.

According to KMW three or four PzH 2000 would be sufficient to provide coverage to a medium size FOB through a full 360°.

As of early 2013 this system remained in its early development stage.

Surplus PzH 2000 systems

The German and Royal Netherlands Army have quantities of PzH 2000 that are now surplus to requirements.

Naval PzH 2000 turret

In 2003 the 155 mm/52-calibre turret installed on the Krauss-Maffei Wegmann PzH 2000 self-propelled artillery system was installed on the F124 class frigate *Hamburg* of the German Navy.

In the end the German Navy did not adopt this proposal and ordered a conventional naval gun turret from Oto Melara of Italy.

Krauss-Maffei Wegmann artillery system

In addition to marketing the PzH 2000 as a standalone artillery system, Krauss-Maffei Wegmann is marketing complete artillery solutions that also include some of their other products.

These include the upgraded MLRS that it is providing to the German Army, GFF4 (6 × 6) protected vehicle for use in the command post role, Dingo 2 (4 × 4) protected vehicle also for the command post role, Fennek for the Forward Observer/Joint Fire Support Team role as supplied to the German Army and the Dingo 2 (4 × 4) protected vehicle fitted with a mast mounted sensor ground based radar system to provide a target acquisition capability.

Variants

The German Army has a small number of PzH 2000 driver training vehicles. This is essentially a PzH 2000 with its turret removed and replaced by a cabin with seats for the instructor and other pupils. A dummy 155 mm/52-calibre gun is also fitted.

Krauss-Maffei Wegmann 155 mm/52-calibre Artillery Gun Module

This has been developed by the company as a private venture and uses the complete 155 mm/52-calibre ordnance of the PzH 2000.

Details of the AGM are provided in a separate entry in *IHS Jane's Land Warfare Platforms: Artillery and Air Defence*, and as of early 2013 this remained at the prototype stage.

DONAR 155 mm/52-calibre system

This has been developed by Krauss-Maffei Wegmann and General Dynamics European Land Systems - Santa Bárbara Sistemas and consists of a new tracked chassis based on components of the latest Spanish ASCOD 2 IFV chassis fitted with the latest version of the Artillery Gun Module. It was first shown in mid-2008.

Details of the DONAR are provided in a separate entry in *IHS Jane's Land Warfare Platforms: Artillery and Air Defence*, and as of early 2013 this remained at the prototype stage.

Specifications

	Panzerhaubitze 2000
Dimensions and weights	
Crew:	5
Length	
overall:	11.669 m
hull:	7.30 m
Width	
overall:	3.58 m
without skirts:	3.37 m
Height	
overall:	3.46 m
to turret roof:	3.06 m
Ground clearance	
overall:	0.44 m
Track width	
normal:	550 mm
Length of track on ground:	4.91 m
Weight	
standard:	49,000 kg
combat:	55,330 kg (est.)
Ground pressure	
standard track:	0.98 kg/cm^2
Mobility	
Configuration	
running gear:	tracked
Power-to-weight ratio:	18.00 hp/t
Speed	
max speed:	60 km/h (est.)
Range	
main fuel supply:	420 km
Gradient:	50%
Side slope:	25%
Vertical obstacle	
forwards:	1 m
Trench:	3.0 m
Engine	MTU 881, diesel, 1,000 hp
Gearbox	
model:	Renk HSWL 284
type:	automatic
forward gears:	4
reverse gears:	2
Suspension:	torsion bar
Electrical system	
vehicle:	24 V
Firepower	
Armament:	1 × turret mounted 155 mm L52 howitzer
Armament:	1 × roof mounted 7.62 mm (0.30) MG3 machine gun
Armament:	8 × turret mounted 76 mm smoke grenade launcher (2 × 4)
Ammunition	
main total:	60
charges:	288
Turret power control	
type:	electric/manual
Main armament traverse	
angle:	360°
Main armament elevation/depression	
armament front:	+65°/-2.5°
Survivability	
Night vision equipment	
vehicle:	yes
NBC system:	yes
Armour	
hull/body:	steel
turret:	steel

Status

Production complete but can be resumed. In service with the following countries:

Country	Quantity	Comment
Germany	185	final deliveries late 2002
Greece	24	delivery 2003 to 2004
Italy	70	2 from Germany rest manufactured in Italy
Netherlands	57	ordered 2002, delivery 2004-2009 (some of these are for sale). Only 18 systems are currently deployed of the 57 purchased

Contractor

Krauss-Maffei Wegmann GmbH and Co KG (production and systems integration is undertaken at its facilities in Kassel).

Krauss-Maffei Wegmann 155 mm/52-calibre Artillery Gun Module (AGM)

Development

Late in 2004, Krauss-Maffei Wegmann revealed that it had built the prototype of a 155 mm self-propelled artillery gun based on the chassis of the US-developed Multiple Launch Rocket System (MLRS).

This system is called the Artillery Gun Module (AGM) and has been developed as a private venture to meet the future potential operational requirements of the German Army.

To reduce development time and risk, wherever possible proven components have been used in the development of the AGM including the MLRS carrier hull and elements of the PzH 2000, which is now in service with four countries.

In 2007 Krauss-Maffei Wegmann completed the verification phase of their 155 mm/52-calibre AGM.

Demonstrator 1 was completed in mid 2004 and used to verify the mechanical stiffness and strength of the gun module and its impact on the reinforced full tracked chassis.

During this period the weapon was fired by remote control with the hull at different angles and the AGM positioned at different azimuth and the gun at different elevations.

Demonstrator 1 was the upgraded with the installation of a modified PzH 2000 projectile loader, new charge loader and a computerised fire-control system based on PzH 2000.

Demonstrator 2 started its firing trials in the first half of 2006 and culminated in the firing of a volley of 10 155 mm rounds in two minutes and 19 seconds with a crew of two being seated in the fully armour protected cab.

So far over 1000 rounds of 155 mm ammunition have been fired from the AGM with many using the maximum of six Rheinmetall DM72 modular charges with its 155 mm/52-calibre ordnance.

It is understood that the German Army has a potential requirement for up to 41 systems but as of March 2013 no orders had been placed by the German Army for the AGM and none were expected in the immediate future.

Description

As previously stated, this 155 mm/52-calibre AGM is based the proven hull used for the US-developed MLRS, which is described in detail in a separate entry in *IHS Jane's Land Warfare Platforms: Artillery and Air Defence*. Prime contractor for the MLRS chassis is the now BAE Systems.

Production of the MLRS chassis has been completed but can be resumed if required for the export market.

A number of countries have reduced the size of their MLRS fleets so significant numbers of MLRS hulls are now available that could be converted into other roles.

The hull and cab of the 155 mm/52-calibre AGM is of all-welded aluminium armour that provides the occupants with protection from small arms fire, shell splinters and NBC attack.

The system has a crew of two people. In the cab is the PzH 2000 ballistic fire-control computer with integrated NATO Artillery Ballistic Kernel and the Krauss-Maffei Wegmann Artillery Command-and-Control System.

Standard equipment includes an inertial reference unit with a Global Positioning System (GPS) connection. It also has a muzzle velocity measuring system that feeds information into the onboard fire-control system. Krauss-Maffei Wegmann claims that the AGM can engage moving as well as stationary targets.

A vehicle independent auxiliary power unit is installed as standard and this allows the complete AGM to be run with the main MLRS carrier diesel engine shut down.

Mounted on the rear of the hull is an unmanned turret weighing 12.5 tonnes, which is armed with the same 155 mm/52-calibre ordnance, elevating mass and some loading subsystems (for example pneumatic rammer) as fitted to the PzH 2000 system used by the German Army.

A total of 30 155 mm projectiles and associated charges are carried and the range and ballistic performance of the AGM is the same as PzH 2000, as is accuracy and time in/out of action. Manual loading is also possible and Multiple Round Simultaneous Impact (MRSI) target engagements can take place.

Maximum range firing conventional ammunition is over 40 km (base bleed/rocket assist) but this can be extended to over 60 km using the Rheinmetall Denel Munitions Velocity-enhanced Long-range Artillery Projectile (V-LAP) which has already been demonstrated fired from the PzH 2000.

The complete system has a combat weight of 30 tonnes and a transport weight of 27 tonnes which means that it can be transported in a new A400M transport aircraft. The actual AGM gun module only weighs 12.5 tonnes.

Prototype of Krauss-Maffei Wegmann 155 mm/52-calibre Artillery Gun Module carrying out a fire mission with turret traversed to the rear (Krauss-Maffei Wegmann) 1403733

The 155 mm/52-calibre Artillery Gun Module, shown here in a stand alone configuration, is a key part of the "Smart Camp Defence" system (Krauss-Maffei Wegmann) 1304180

KMW are continuing to invest in the AGM in a number of key areas including improvements to the automatic loader to increase the rate of fire. This has already been increased to a 10 round volley in one minute 40 seconds in the simulation mode.

The actual AGM concept is fitted with the standard 155 mm/52-calibre barrel and fits the weight limits. Therefore, lighter barrels could be fitted if required by the customer with these including a 155 mm/39-calibre or even a 105 mm weapon.

To take account of the additional weight and firing stresses, the hull has been structurally reinforced and fitted with stiffer torsion bars and extra shock absorbers have been added to assist the stability of the chassis.

Variants

Family concept

Although the original application of the AGM was for installation on the standard tracked MLRS carrier hull, Krauss-Maffei Wegmann is now studying a number of other applications. These include other tracked hull, 6 × 6 and 8 × 8 wheeled hull, a stand alone version and naval applications.

The stand alone version would consist of the complete AGM mounted on a substantial pallet with integrated auxiliary power unit which when deployed in the firing position would be stabilised by four outriggers. This would be a complete autonomous unit with the crew controlling the weapon system from a safe position. This would typically be used for base defence purposes.

AGM with 155 mm/39-calibre ordnance

Although the first example of the AGM is fitted with a 155 mm/52-calibre ordnance it could also be supplied fitted with a shorter 155 mm/39-calibre ordnance. This would have the same ammunition capacity but a shorter range.

DONAR 155 mm/52-calibre self-propelled artillery system

Krauss-Maffei Wegmann has teamed with the Spanish company of General Dynamics European Land Systems - Santa Bárbara Sistemas to develop a new mobile 155 mm/52-calibre Self-Propelled (SP) artillery system called DONAR.

This is a further development of the private venture Krauss-Maffei Wegmann 155 mm/52-calibre Artillery Gun Module (AGM) which was completed several years ago integrated onto a surplus Multiple Launch Rocket System (MLRS) hull for firing trials.

For the DONAR system the latest AGM has been integrated onto a new full tracked hull based on the latest version of the General Dynamics European Land Systems - Santa Bárbara Sistemas Pizaro 2 Infantry Fighting Vehicle (IFV), earlier versions of which are in service with Austria (as the Ulan) and Spain.

155 mm/52-calibre Artillery Gun Module in the travelling configuration. This has the same fire power as the German PzH 2000 system but is lighter and more deployable (Krauss-Maffei Wegmann) 1333629

AGM for C-RAM role

Krauss-Maffei Wegmann is developing, as a private venture, a C-RAM capability for the AGM which is being marketed as "Smart Camp Defence" and is claimed to be the ideal solution to protect a Forward Operating Base (FOB) against asymmetric rocket and mortar bomb attack.

In a typical target engagement, a tower mounted surveillance radar would first detect the incoming target which would then be tracked. The central fire direction centre would calculate the ballistic trajectory and target engagement sequence.

This information would be passed onto the remote controlled AGM that would carry out the actual target engagement with an optimised 155 mm High-Explosive (HE) projectile being fired towards the pre-calculated co-ordinates.

The HE projectile detonates in the immediate vicinity of the threat and the resultant explosion neutralises the incoming threat. The AGM would also have the option of engaging the firing position if this had not moved.

According to KMW three or four AGM would be sufficient to provide coverage to a medium size FOB through a full 360°.

The AGM could be installed on a hull or mounted on a base for use in the static mode. Each AGM weighs about 12 tonnes which makes its easy to deploy to FOB.

As of March 2013 there were no plans for the AGM in the C-RAM role to enter service with the German Army.

Specifications

155 mm/52-calibre Artillery Gun Module

Dimensions and weights	
Length:	10.353 m
Width:	2.89 m
Height:	1.977 m
Weight:	12,000 kg
Mobility	
Configuration:	tracked or wheeled
Firepower	
Armament:	1 × 155 mm
Ammunition:	
(main total)	30
(charges)	145 (modular propellant)
Rate of fire:	
(sustained fire)	12 rds/min
(rapid fire)	6 rds/min
Turret power control:	electric
Main armament traverse:	360°
Main armament elevation/depresion:	
(armament front)	0°/+64°

Status

Prototype.

Contractor

Krauss-Maffei Wegmann GmbH & Co KG.

Rheinmetall 155 mm M109L52 self-propelled howitzer

Development

The US-developed 155 mm M109 Self-Propelled Howitzer (SPH) is the most widely deployed system of its type in the world. Most versions are fitted with a 155 mm/39-calibre barrel, which was optimised for older natures of ammunition.

By today's standards the M109 SPH lacks range, rate of fire and accuracy and is not compatible with many of the latest natures of more effective 155 mm ammunition.

As a private venture Rheinmetall Waffe Munition developed to the prototype stage the new M109L52 SPH, which is fitted with a 155 mm/52-calibre barrel.

This barrel is the same as that used in the more recent PzH 2000 SPH that is already in service with Germany, Greece, Italy and the Netherlands.

Development of the M109L52 was originally started by the Dutch company RDM Defence Technology, which is no longer trading, but the programme has now been taken over by the Rheinmetall DeTec Group with special contribution by Rheinmetall Waffe Munition and development has continued.

For the 155 mm PzH 2000 SPH Rheinmetall DeTec Group supplies over half of the complete system, including the complete elevating mass with the 155 mm/52-calibre barrel supplied by Rheinmetall Waffe Munition which meets the Joint Ballistic Memorandum of Understanding (JBMoU).

As of March 2013, the Rheinmetall 155 mm M109L52 self-propelled howitzer remained at the prototype stage.

Description

The overall layout of the upgraded Rheinmetall M109L52 is virtually identical to the standard production now BAE Systems (previously United Defense, Ground Systems) M109 series of SPH, which is covered in detail in a separate entry.

As different customers have different operational requirements, Rheinmetall Waffe Munition has adopted a three-level approach for its M109L52 upgrade. Level 1 is the baseline model as represented by the current prototype.

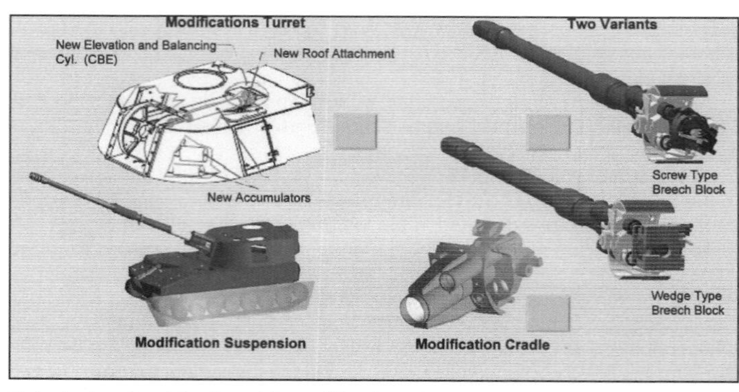

Selected parts of the Level 1 modifications for the M109L52 SPH upgrade (Rheinmetall Waffe Munition) 0590696

The extent of features and equipment contained in Level 2 and Level 3 are typical. If a customer requires another combination of features, provided that it is technically feasible, the upgrade packages can be realised.

The Level 1 M109L52 upgrade includes the 155 mm/52-calibre barrel complete with standard PzH 2000 muzzle brake and bore evacuator, new breech actuator system, modified cradle, new gun mount, modifications to the existing turret, new elevating and balancing cylinder, hydraulic system improvements, new accumulators, and some modifications to the hull and suspension.

The 155 mm/52-calibre barrel is available with a rotating or wedge-type breech mechanism. Maximum range of the M109L52 depends on a number of factors including projectile/charge combination.

Typical ranges using Rheinmetall Waffe Munition 155 mm projectiles are: DM111 30.3 km; DM642 family 27.1 km; DM652 family 34.5 km; and the latest Rh40 out to 40 km. The Rh40 subsequently exceeded this range during company trials.

It can also fire smoke and top-attack projectiles with the latter including the SMArt 155.

As well as using conventional bagged charges the M109L52 can use the more recent Rheinmetall Modular Charge System (MCS). The 52-calibre system can take a maximum of six MCS, while the older 155 mm/39-calibre system can only use a maximum of five MCS.

Studies have shown that as a result of the installation of the 155 mm/52-calibre barrel and turret modifications, trunnion load at recoil is within limits.

The standard M109A3 has a design load of 535 kN and maximum muzzle velocity of 685 m/s. The latest M109L52 with six MCS has a design load of 570 kN and a maximum muzzle velocity of 945 m/s.

Level 2 has the SELEX Galileo combat-proven Laser Inertial Artillery Pointing System (LINAPS). This is already standard equipment on a BAE Systems L118 105 mm Light Guns of the British Army, L119 105 mm Light Guns in service with Thailand and the BAE Systems 155 mm M777 lightweight howitzer of the Canadian Army.

Level 2 also adds a muzzle velocity radar, which is mounted over the rear part of the 155 mm/52-calibre ordnance, Auxiliary Power Unit (APU), onboard fire-control and computing system, rammer, new ammunition stowage boxes, Ammunition Handing Kit (AHK) and a remote travel lock.

The APU allows all of the systems to be run with the main engine switched off while the remote travel lock means that the driver can remain under full armour protection when the weapon comes into action. The rammer increases rate of fire and reduces crew fatigue.

Level 3 is the most advanced model and has all of the features of level 1 and 2 plus a digital communications system, primer magazine, new firing mechanism, AHK and automatic gun laying. This model can carry out onboard ballistic computation and undertake shoot and scoot missions, which makes the platform much more survivable as it is unlikely to be engaged by counterbattery fire.

According to Rheinmetall Waffe Munition, it can come to a halt, fire a six-round burst and then move to a new firing position in under three minutes. It has a burst rate of fire of six rounds per minute and a sustained rate of fire of three rounds per minute.

A driveline upgrade could be added to the Level 3, including a more powerful EURO III compliant 450 hp diesel engine as a replacement for the current Detroit Diesel engine Model 8V-71T developing 405 hp.

In addition to these there are many other improvements that could be installed if required by the customer. These include a laser range-finder for use in the direct-fire role, integrated gun temperature sensor, battle management system, digital crew intercom system, new driver's passive night driving device, air conditioning unit to allow the system to operate in high ambient temperatures, track side skirts to help reduce dust, dust cover/filters, NBC protection, laser warning system, 76 mm electrically operated smoke grenade launchers, fire/explosion detection and suppression system and an automatic fuze setter.

Finally, additional armour could be provided to the hull and turret for a higher level of protection from small arms fire and additional armour could be provided under the hull for increased protection against mines.

Many of these optional extras have already been tested and integrated on vehicles and weapons.

Rheinmetall Waffe Munition could carry out this upgrade work in its own facilities or kits could be provided to enable the customer to upgrade systems in their own facilities.

The German Army has a number of M109A3G SPH, which are now for sale and could be upgraded to the enhanced M109L52 standard before or after delivery.

Rheinmetall Waffe Munition M109L52 SPH with ordnance partly elevated (Rheinmetall Waffe Munition) 1133913

As of March 2013, the only country known to have taken delivery of surplus German Army M109A3GE is Greece, who received 50 units.

These were followed by another batch of 223 ex German Army M109A3GE in 2010 that were surplus to requirements following the introduction of the PzH 2000 and the reduction in the size of the German Army.

Specifications

	M109L52
Dimensions and weights	
Crew:	6
Length	
overall:	11 m
Width	
overall:	3.22 m
Height	
overall:	3.235 m
Weight	
standard:	28,000 kg
Mobility	
Configuration	
running gear:	tracked
Firepower	
Armament:	1 × turret mounted 155 mm howitzer
Main weapon	
calibre length:	52 calibres
Survivability	
Night vision equipment	
vehicle:	yes
NBC system:	optional
Armour	
hull/body:	aluminium
turret:	aluminium

Status

Development. Not yet in service.

Contractor

See text.

Rheinmetall 155 mm M109A3G product improvement programme

Development

In order to update the existing fleet of US-supplied now BAE Systems (previously United Defense) 155 mm M109G self-propelled howitzers of the German Bundeswehr, the now Rheinmetall DeTec AG undertook a programme to modify the vehicles to a new standard.

The main modification is the replacement of the existing 155 mm barrel with a new 39-calibre barrel based on the ordnance of the FH-70 towed howitzer, but other modifications were introduced at the same time.

The M109A3G is no longer in front line service with the German Army, but large quantities are still held in storage.

Description

The installation of the new 39-calibre barrel allows the full family of improved 155 mm ammunition to be fired. This family includes high-explosive, illuminating and smoke, all of which can be fired to a maximum range of 24,700 m. Enhanced range projectiles, such as base bleed, can be fired to a range of 30,000 m. Various 155 mm carrier projectiles such as the Rheinmetall RB 63 and Rh 49 can also be fired.

155 mm M109A3GA1 of the German Army (Michael Jerchel) 1044227

The 155 mm/39-calibre new ordnance is mounted in the cradle and recoil system of the M109G and has a revised obturation system. An external rotation key is located in a cradle extension under the barrel and the barrel dust cover of the original M109G has been modified.

Other innovations to the barrel include a new flexible bellows and a fume extractor, located halfway along the tube. The original M109G muzzle brake is retained but is held in place by new fasteners. Other modifications to the 155 mm ordnance involve machining alterations to the breech ring and buffer.

Basic details of the 155 mm ordnance is as follows:

Calibre:	155 mm
Number of grooves:	48
Twist, clockwise, constant:	8°, 55°, 35°
Lenth:	
(barrell)	6,874 mm
(rifled part)	5,057.7 mm
Chamber volume:	18.845 litres
Effective of muzzle brake:	40%
Weight:	
(complete barrel)	2,380 kg
(liner)	1,420 kg
Maximum range:	
(standard)	24,700 m
(enhanced projectile)	30,000 m plus
Maximum recoil over 910 mils:	465 mm
Maximum gas pressure:	
(charge 8 DM52)	2,300 bar
(charge FH 155-1)	3,628 bar
Muzzle velocity:	
(charge 8 DM52)	685 m/s
(charge FH 155-1)	827 m/s

Modifications to the turret include a new 155 mm projectile magazine at the rear of the turret bustle. This can be reloaded through two outward-opening doors and holds 22 × 155 mm projectiles.

The vehicle can carry a total of 34 155 mm projectiles. New provisions are made for propellant charge stowage. Other changes include door locks, a blast cover for the panoramic sight, and new sealing for the shields. A hand-drive mechanism is provided for the traversing gear and new elevation balancing cylinders are provided.

Changes to the rest of the vehicle include a new 155 mm barrel clamp, revised ventilators, a new air filter system, a new instrument layout for the driver, the location of a driver's control panel in the main compartment, a turbo-supercharger system, reinforced torsion bars on the suspension and a driver's hatch lock. Alterations have been made to the power pack to allow it to be replaced more readily.

Other modifications include the turret hydraulics being altered to suit the metric system and the traversing gear hydraulic motor is now the same as that used on the Krauss-Maffei Wegmann Leopard 1 MBT. An electrical firing circuit has been introduced and various new fire-control data panels have been incorporated.

It is understood that the German Army had a total of 499 M109A3Gs but all of these are in reserve and some have started to be transferred overseas.

According to the United Nations Arms transfer lists for 1994 to 2010 the only export was to Greece who took delivery of 50 M109A3GE systems in 2000. Another 223 followed in 2010.

Rheinmetall Landsysteme upgrade kit for German M109A3G

To improve ammunition stowage and flow inside the 155 mm turret, the German BWB (Federal Office for Defence Technology and Procurement) invited four companies to submit detailed proposals.

The now Rheinmetall Landsysteme (then KUKA) won this competition and was subsequently awarded a contract from the German BWB for two prototypes of the Ammunition Handling Kit (AHK) which the company had originally developed as a private venture in 1994/95.

155 mm M109A3GA1 of the German Army (Michael Jerchel) 1044228

These were installed in two German Army 155 mm M109A3G self-propelled artillery systems for an extensive series of user trials which led to the type classification of the AHK system early in 1997.

It was expected that a production order would soon follow but funding shortages meant that the company was not finally awarded the contract by the BWB for the AHK system until late in 1998.

The contract was for the supply of 262 AHK to the German Army who carried out the installation in its own facilities. These were delivered between July 1999 and June 2000. When fitted with the AHK these are designated the M109A3GA1.

The AHK consists of seven major components. A drawer-type magazine for 24 fuzed 155 mm artillery projectiles is positioned in the turret bustle. Holders for an additional 10 fuzed 155 mm projectiles are mounted one either side of the rear hull entrance door. An electrically driven and electronically controlled 155 mm projectile hoist is installed in the turret.

A projectile chute is also installed, as is an electrically operated rammer which speeds up the loading of 155 mm projectiles at all angles of elevation as well as reducing crew fatigue. Four magazines for propellant charges are provided in the turret and hull. Four large stowage boxes are provided externally at the turret front and rear.

According to the company, the installation of an electrically operated projectile hoist provides for almost weightless projectile transport and handling within the tight confines of the M109A3GA1 turret.

The design of the different magazines for 155 mm fuzed projectiles and their associated propellant charges allows for the entire supply of onboard ammunition, 34 standard NATO 155 mm artillery projectiles and 34 NATO standard artillery charges, to be easily stowed in the M109A3GA1 and be quickly accessible for fire missions.

The M109A3GA1 upgraded with the AHK can also carry 170 of the Rheinmetall Modular Propelling Charge System (MPCS) which is already in volume production for the German Army under the designation of the DM72.

Overall, the installation of the AHK increases the rate of fire of the M109, first fielded over 50 years ago, as well as allowing the system to be operated with no standing personnel behind the weapon during firing missions.

As there is a significant reduction in crew fatigue, the upgraded M109A3GA1 with the AHK installed is more suited to carrying out several 24-hour battlefield days than earlier systems.

The German Army is the launch customer for the AHK, which is already being marketed to other M109 customers. In addition, the company has also developed the Automatic Ammunition Flow (AMF) system, which can be fitted to the M109 and other self-propelled systems.

The AMF system, which has already been successfully tested on an M109A2, allows for 155 mm projectile transfer, ramming and primer loading to be carried out automatically and allows the crew of an M109 to be reduced by one person.

As far as it is known the AMF kit remains at the prototype stage.

Diehl Defence Land Systems upgraded power pack for M109

Whilst most upgrades for the widely deployed M109 SPH concentrate on improving its firepower, accuracy and survivability, little attention is paid to its obsolete power pack.

The now Diehl Defence Land Systems has developed and placed in production an upgraded power pack for the M109 SPH.

This power pack has been developed using proven off-the-shelf components and is claimed to provide a life extension of the M109 SPH of between 10 and 15 years.

The first customer for this new Diehl Defence Land Systems power pack is the United Arab Emirates (UAE) who have taken delivery of 85 units for installation in the M109L47 originally upgraded by the now defunct RDM Technology of the Netherlands.

These Diehl Defence Land Systems power packs were supplied from Germany and fitted in the UAE with final deliveries being made in 2005.

The upgraded M109 SPH diesel power pack consists of a Deutz BF6M 1015 CP six cylinder air cooled diesel developing 442 hp (330 kW) at 2,300 rpm which is coupled to an Allison XTG-411-4 automatic transmission (upgraded from the earlier XTG-411-2A) with four forward and two reverse gears.

A new and more compact cooling system has been developed under the leadership of Diehl Defence Land Systems which includes a new Diehl Defence Land Systems/Behr intercooler, radiator, a fan drive from Krauss-Maffei Wegmann and a new Rocore oil cooler.

A new 24 V 6.6 kW starter has been installed in conjunction with a Bosch 28 V/140 A alternator to provide additional power.

In addition, the upper and lower diesel fuel tanks have been reworked to better interface with the geometry of the new power pack and the existing exhaust pipe.

The new Deutz air cooled diesel engine is not only more powerful but is more fuel efficient. A key feature of this Diehl Defence Land Systems upgraded power pack is that it can be fitted to the M109 SPH without any modifications to the hull.

Specifications

	M109A3G
Mobility	
Configuration	
running gear:	tracked
Firepower	
Armament:	1 × turret mounted 155 mm howitzer
Main weapon	
length of barrel:	6.874 m
rifling:	5.0577 m
maximum range:	24,700 m
Main armament traverse	
angle:	360°
Main armament elevation/depression	
armament front:	+70°/-2.5°

Status

This upgrade programme is complete with the work being carried out by the German Army. The Hellenic Army has taken delivery of 50 M109A3GE systems in 2000 with the first 18 delivered to the Hellenic Army's 32nd Marine Brigade.

Germany transferred an additional 220 M109A3GE to Greece in 2010.

Contractor

Rheinmetall DeTec AG (responsible for upgrade package).

M44T 155 mm self-propelled howitzer

Development

Late in 1986, a German consortium consisting of MTU, GLS and Rheinmetall completed the prototype of an upgraded 155 mm M44 self-propelled howitzer subsequently designated the M44T.

This was built to meet the requirements of the Turkish Land Forces Command and the prototype commenced mobility and fire-power trials at the Turkish School of Artillery in 1987.

Compared to the original US M44 whose design can be traced back to shortly after the end of the Second World War, the more recent 155 mm/39-calibre M44T has increased mobility, fire power and reliability.

In total, 168 M44s of the Turkish Army were upgraded in Turkey with some of the component parts being made under licence in Turkey.

By early 2013, Turkey was the only customer for the M44T upgrade programme.

The 155 mm M44 was originally developed to meet the requirements of the US Army but from 1962 was replaced by the 155 mm M109 series.

Description

The M44T has its original Continental petrol engine replaced by an MTU MB 833 Aa-501 V-6 water-cooled diesel developing 450 hp at 2,300 rpm coupled to the original Allison CD-500-3 transmission via a ZF gearbox.

Increased fuel tankage of 780 litres, together with the MTU diesel engine, provides a range of approximately 620 km compared to the 122 km of the original petrol-engined M44.

On the prototype M44T upgrade, the driver remained in the gun compartment but in production vehicles he is seated in the hull.

Various other components and subsystems have been modernised and a fire warning and extinguishing system has been installed. A new driver's instrument panel has also been fitted.

Replacement of the existing electrical system has also been carried out using new cables, fuse boxes and batteries. New actuators for steering, braking and acceleration have been fitted, and for cold starting of the engine at temperatures below −18°C a preheating system has been fitted. The top engine deck and exhaust system have also been modified.

Modifications to the torsion bar suspension include new hydraulic shock-absorbers and torsion bars and some adaptation of the hydraulic bump stops. A new track (with replaceable rubber pads) has also been installed and snow grips can be fitted if required. New rubber dust skirts have been fitted to help keep dust down.

The original short-barrelled 155 mm ordnance has been replaced by a 155 mm 39-calibre barrel and breech similar to that used on the upgraded German Rheinmetall M109A3G.

This provides a range, firing standard projectiles, of 24,700 m and enhanced range projectiles can now reach 30,000 m. Elevation is from −5° to +65°, with ammunition stowage for 30 155 mm projectiles and charges being provided.

The ordnance has automatic primer feeding as well as automatic opening and closing of the breech mechanism; the recoil system, balancers and ammunition racks have also been modified. The ordnance has a life of 10,000 standard Equivalent Full Charge (EFC) rounds.

M44T showing the new 155 mm ordnance with its double-baffle muzzle brake 0500333

The 155 mm/39-calibre ordnance used by the M44T and M52T has been built in Turkey by MKEK.

Status

Production was completed in 1992. As a follow-on programme, the modification of the M52 (which has a similar chassis to the M44 but with a turret-mounted 105 mm gun) with main components and weapon of the M44 has now been completed. This is a more extensive modernisation programme as it included the removal of the complete turret and rebuild of the hull. It is thought that 425 M52s were upgraded by 1998. Both the upgraded M52 and M44 have seen operational use with the Turkish Army during operations in Northern Iraq.

These upgraded Turkish Land Forces Command M44 and M52 upgraded self-propelled artillery systems are now being supplemented by the new locally manufactured 155 mm/52-calibre Firtina.

Details of this are provided in a separate entry in *IHS Jane's Land Warfare Platforms: Artillery and Air Defence*. It is essentially the South Korean K9 Thunder with minor modifications for Turkey.

Contractor

MTU GmbH.
Rheinmetall DeTec AG.
GLS GmbH.

India

130 mm self-propelled gun (Catapult)

Development

As a result of the effectiveness of the Russian-supplied towed 130 mm Field Gun M-46 in the Indo-Pakistan conflicts, the Indian Army mounted some on Vijayanta tank hull (originally the Vickers MBT).

The conversion is now in Indian Army service, which originally had a fleet of about 2,200 Vijayanta MBTs and about 750 130 mm M-46 guns. The Indian Army did deploy a large fleet of Vijayanta MBT's but these have now all been withdrawn from service.

Description

The standard Russian 130 mm M-46 towed field gun has a range of 27,150 m and fires an HE projectile weighing 33.4 kg with a muzzle velocity of 930 m/s. It can also fire an APHE projectile weighing 33.6 kg, also at a muzzle velocity of 930 m/s, which can penetrate 230 mm of conventional steel armour at an incidence of 0° at a range of 1,000 m.

It would appear that the Vijayanta/130 mm M-46 combination can carry 30 rounds of separate loading ammunition. It is known to the Indian Army as 'Catapult'.

The latest information available indicates that a total of 170 Catapult systems were built of which 100 are operational and the remainder are in reserve.

Some have recently been observed fitted with a horizontal shield over the gun and crew compartments to provide protection against top attack weapons. Some sources have quoted a much lower figure of only 20 Catapult 130 mm self-propelled artillery systems in service. It is also reported that India has a small quantity of Russian 152 mm 2S19 self-propelled artillery systems in service.

Variants

Catapult Mk II 130 mm self-propelled artillery system

The previously mentioned, the Indian Army deployed a number of Catapult Mk I self-propelled artillery systems based on a stretched Vijayanta hull fitted with a Russian 130 mm M-46 field gun firing over the rear arc.

Late in 2012 it was stated that the Indian Defence Research and Development Organisation (DRDO) was to start trials of the Catapult Mk II self-propelled artillery system.

The Indian 130 mm Catapult self-propelled gun consists of a modified Vijayanta MBT hull with the Russian 130 mm M-46 field gun firing to the rear 0500336

This is a modified Arjun Mk I MBT hull also fitted with the Russian 130 mm field gun which is provided with 36 rounds of separate loading ammunition.

If trials are successful it is possible that a series production batch of about 40 Catapult Mk II could be delivered from 2013/2014 onwards and issued to two Indian artillery regiments.

These would replace the early Catapult Mk I systems currently deployed by the Indian Army as their only self-propelled artillery systems.

Status

Production complete. In service with the Indian Army. This is expected to be replaced by a new 155 mm/52-calibre self-propelled gun which has still to be selected. It could well be a mixture of tracked and wheeled systems. As of early 2013, India had not placed any contracts for new self-propelled artillery systems, tracked or wheeled.

Indian 105 mm self-propelled gun

Development

This 105 mm Self-Propelled (SP) artillery system has been developed by the Ordnance Factory Medak, Ordnance Development Center, and is basically a modified version of the Indian Sarath Infantry Combat Vehicle (ICV) hull fitted with a turret armed with the Indian Ordnance Factory Medak 105 mm Light Field Gun ordnance.

The Sarath ICV is the Russian BMP-2 ICV manufactured under licence in India. This has been offered on the export market but there are no known export sales as of early 2013.

Details of the Indian Ordnance Factory 105 mm Light Field Gun are provided in a separate entry in *IHS Jane's Land Warfare Platforms: Artillery and Air Defence*.

As of 2013 it is understood that development of this Indian 105 mm SP had been completed but it had yet to enter production.

Description

For its new role the hull of the Sarath ICV has been modified with the two-person turret removed and the interior modified to enable the 105 mm ammunition to be carried. The turret is provided with ballistic windows for direct side observation.

The driver is seated at the front left with another person being seated to the immediate rear with both crew members provided with a roof hatch.

The diesel power pack is to the right with the remainder on the hull space being taken up by ammunition.

Mounted on top of the hull is a turret that is armed with the same 105 mm ordnance, elevation and traversing system of the standard 105 mm Light Field Gun which has been in service with the Indian Army for many years.

This is ballistically identical to the ordnance installed in the Vickers Defence Systems 105 mm Abbot SP artillery system that was developed to meet the requirements of the British Army who has now phased this weapon out of service. The Indian Army took delivery of the Vickers Defence Systems Abbot Value Engineered.

According to the Indian Ordnance Factory, this 105 mm weapon can be traversed through a full 360° with weapon elevation from –5° to +70°.

When travelling the 105 mm ordnance is held in a travel lock that is mounted towards the front of the well sloped glacis plate. When deployed in the firing position this is manually released.

A total of 42 rounds of 105 mm ammunition can be carried with the High-Explosive (HE) round having a maximum range of 19,000 m.

Other natures of 105 mm ammunition fired include Armour-Piercing Discarding Sabot (APDS), High-Explosive Anti-Tank (HEAT), High-Explosive Squash Head (HESH) and smoke.

Rate of fire is being quoted as four rounds a minute with a maximum rate of fire of six rounds a minute for a short period.

Indian 105 mm self-propelled gun based on a modified BMP-2 ICV hull with ordnance in travel lock (Indian Ordnance Factory Medak) 1365818

The 105 mm SP gun is provided with the same direct (Type 106A) and indirect (Type 104A) fire sights as the standard Indian 105 mm Light Field Gun.

The complete system weighs about 16,000 kg that indicates that unlike the Sarath it is probably not fully amphibious. Maximum road speed is about 65 km/h.

It is operated by a crew of five and the system can brought into and taken out of action in less that one minute.

Variants

Enhanced version

In addition to the base line version of this 105 mm self-propelled artillery system it has been proposed that a version with enhanced capabilities could be manufactured with the following additional features:

- Turret made of composite armour that would provided ballistic protection against attack from small arms fire up to 12.7 mm
- Automatic loading of 105 mm ammunition
- Installation of a computerised Fire-Control System (FCS)
- Installation of a TALIN 500 land navigation system
- Installation of a muzzle velocity radar which could feed information to the computerised FCS
- Ammunition carrying capacity increased from 42 rounds to 92 rounds.

Specifications

Not available.

Status

Prototype completed. Not yet in production or service.

Contractor

Indian Ordnance Factory Medak.

International

DONAR 155 mm/52-calibre self-propelled artillery system

Development

Krauss-Maffei Wegmann of Germany teamed with the Spanish company General Dynamics European Land Systems - Santa Bárbara Sistemas to develop a new mobile 155 mm/52-calibre Self-Propelled (SP) artillery system called DONAR.

This is a further development of the private venture Krauss-Maffei Wegmann 155 mm/52-calibre Artillery Gun Module (AGM) which was completed several years ago integrated onto a surplus Multiple Launch Rocket System (MLRS) hull for firing trials.

For the DONAR system the latest generation AGM has been integrated onto a new full tracked hull based on automotive components from the latest version of the General Dynamics European Land Systems - Santa Bárbara Sistemas Pizaro 2 Infantry Fighting Vehicle (IFV).

This is now in quantity production for the Spanish Army. Earlier versions of the vehicle are in service with Austria (as the Ulan) and Spain Pizaro 1.

While DONAR has been developed as a private venture, it could in the future meet German and Spanish requirements for a highly mobile 155 mm/52-calibre self-propelled artillery system.

DONAR was first shown in mid-2008 and had already carried out extensive firepower and mobility trials in Germany and Spain.

DONAR has the same firepower as the 55-tonne PzH 2000 155 mm/52-calibre self-propelled artillery system, has greater strategic mobility as with an air transport weight of about 31.5 tonnes it can be carried in the A400M transport aircraft.

Description

The new hull for DONAR is of all-welded steel armour that provides the occupants with protection from small arms fire and shell splinters.

As an option a mine protection system and appliqué armour can be fitted for a higher level of protection.

Crew compartment is at the front of the new hull with the power pack located under the floor and to the rear.

Protected crew compartment has a single forward facing door in either side with a bulletproof window in its upper part. To the front of the crew position are three large bulletproof windows and the crew cab is fitted with a fire detection and suppression system.

Suspension is of the torsion bar type with either side having seven dual rubber tyred road wheels with the drive sprocket at the front, idler at the rear and track return rollers. Upper part of the suspension is protected by a rubber skirt.

The AGM is mounted on the rear of the hull and when travelling is traversed to the front. It is loaded, aimed and fired by the crew from within the protected cab and turret traverse and weapon elevation being all electric.

The Rheinmetall 155 mm/52-calibre ordnance is identical to that installed in the Krauss-Maffei Wegmann PzH 2000 self-propelled artillery system that is already deployed by Germany, Greece, Italy and the Netherlands.

Details of the Krauss-Maffei Wegmann PzH 2000 are provided in a separate entry in *IHS Jane's Land Warfare Platforms: Artillery and Air Defence*.

Production of this is complete but it could commence again when additional orders are placed. As of March 2013 the PzH 2000 was in service with Germany (185), Greece (24), Italy (70) and the Netherlands (57). It should be noted that Germany and the Netherlands have quantities of these systems for sale.

Ordnance is fitted with a muzzle brake, thermal sleeve and fume extractor and mounted over the rear of the ordnance is a muzzle velocity radar that feeds information to the on board fire-control computer.

The weapon is aimed, loaded and fired with the crew under complete armour and NBC protection in the forward cab.

The magazine holds 30 155 mm projectiles and 145 associated modular charges are carried with the automatic loader enabling a high rate of fire to be achieved of up to six rounds a minute and it can also carry out MRSI missions.

Maximum range depends on the projectile/charge combination being fired. Firing the South African Rheinmetall Denel Munition enhanced Velocity-enhanced Long-range Artillery Projectile (VLAP) a maximum range of 56 km is obtained depending on ambient conditions.

Additional 155 mm ammunition and modular charges would be carried by a resupply vehicle that could be based on a similar hull or cross-country truck.

A computerised fire-control system is installed which can be linked to a centralised command-and-control system for network centric warfare.

The vehicle also features a hybrid inertial navigation unit and Global Positioning System (GPS) and automatic tracking and correction.

A number of options are available for the DONAR including appliqué armour to the hull and turret as well as a mine protection package.

Variants

DONAR on other platforms

Although the first example of DONAR has been integrated onto a new tracked chassis based on the latest production ASCOD 2 hull, the baseline AGM was originally tested on a MLRS chassis.

DONAR 155 mm/52-calibre self-propelled artillery system deployed in the firing position with unmanned turret traversed left and ordnance elevated (Krauss-Maffei Wegmann) 1403899

DONAR 155 mm/52-calibre self-propelled artillery system with turret traversed to the front (Krauss-Maffei Wegmann) 1403734

It can also be integrated on other hull, tracked and wheeled, with the latter including an 8 × 8 cross-country truck hull with a protected forward control cab.

DONAR for C-RAM role
Krauss-Maffei Wegmann is developing, as a private venture, a C-RAM capability for the AGM/DONAR which is being marketed as "Smart Camp Defence".

This is claimed to be the ideal solution to protect a Forward Operating Base (FOB) against asymmetric rocket and mortar bomb attack.

In a typical target engagement a tower mounted surveillance radar would first detect the incoming target which would then be tracked. The central fire direction centre would calculate the ballistic trajectory and target engagement sequence.

This information would be passed onto the remote controlled DONAR that would carry out the actual target engagement with an optimised 155 mm High-Explosive (HE) projectile being fired towards the pre-calculated co-ordinates.

The HE projectile detonates in the immediate vicinity of the threat and the resultant explosion neutralises the incoming threat. DONAR would also have the option of engaging the firing position if this had not moved.

According to KMW three or four DONAR would be sufficient to provide coverage to a medium size FOB through a full 360°.

Specifications

	DONAR 155 mm/52-calibre self-propelled artillery system
Dimensions and weights	
Crew:	2
Length	
overall:	10.3 m
Width	
overall:	2.8 m
Height	
overall:	3 m
Weight	
standard:	31,500 kg
combat:	35,000 kg
Mobility	
Configuration	
running gear:	tracked
Power-to-weight ratio:	20.57 hp/t
Speed	
max speed:	60 km/h
Range	
main fuel supply:	500 km
Gradient:	60%
Side slope:	30%
Trench:	2.0 m
Engine	MTU, diesel, 720 hp
Gearbox	
model:	Renk
type:	automatic
Suspension:	torsion bar
Electrical system	
vehicle:	24 V
Firepower	
Armament:	1 × turret mounted 155 mm howitzer
Ammunition	
main total:	30
charges:	145
Main weapon	
calibre length:	52 calibres
Turret power control	
type:	electric/manual
Main armament traverse	
angle:	360°
Main armament elevation/depression	
armament front:	+70°/0°
Survivability	
Night vision equipment	
vehicle:	yes
NBC system:	yes
Armour	
hull/body:	steel
turret:	steel

Status
Prototype. Not yet in production or service.

Contractor
General Dynamics European Land Systems - Santa Bárbara Sistemas.
Krauss-Maffei Wegmann GmbH & Co KG.

Iran

Defence Industries Organization Raad-2 155 mm self-propelled gun-howitzer

Description
In the Spring of 2002, the Iranian Defence Industries Organization (DIO), Armour Industries Group, released full details of the Raad-2 155 mm/39-calibre Self-Propelled Gun-Howitzer (SP/GH).

This has been in quantity production and in service with the Iranian Army. It has also been referred to as the Thunder 2 self-propelled artillery system and was first seen in public in 1997.

In developing the Raad-2 series of 155 mm SP/GH, Iran has used components from some of its other armoured vehicle programmes with the obvious logistic and training advantages.

The rubber-tyred roadwheels of the Raad-2 are from the locally produced Russian T-72 series Main Battle Tank (MBT) as is the V-84MS V-12 diesel engine, while the SPAT 1200 transmission is used in the locally upgraded Russian-designed T-54/T-55 MBT.

The DIO have offered the Raad-2 155 mm SP gun howitzer on the export market but as of March 2013 there are no known sales.

Description
In overall appearance, Raad-2 is very similar to the now BAE Systems (previously United Defense LP) 155 mm M109A2 SP/GH, large quantities of which were supplied to Iran many years ago.

Most sources have indicated that a total of 440 M109/M109A2 systems were supplied to Iran by the US. Raad-2 has a crew of five, consisting of the commander, gunner, driver and two ammunition loaders.

The hull and turret of the Raad-2 are of all-welded steel armour with a maximum thickness of 20 mm which provides the occupants with protection from small arms fire and shell splinters. The highest level of protection is over the frontal arc of the Raad-2.

Iranian Raad-2 155 mm/39-calibre self-propelled gun-howitzer deployed in the firing position with spades lowered at the rear of the hull 0095234

Iranian Raad-2 155 mm/39-calibre self-propelled gun-howitzer from the front with ordnance in travelling lock and commanders hatch in open position
0137474

The driver is seated front left and has a single-piece hatch cover above that lifts and swings to the right to open. In front of the driver's position are three day periscopes that give observation over the frontal arc. The middle one can be replaced by a passive periscope for driving at night.

The diesel power pack is located to the right with engine access covers provided in the glacis plate. The air inlet and outlet louvres at located in the forward part of the glacis plate while the exhaust outlet is on the right side of the hull. The exhaust outlet is covered by a distinctive cowl.

The large fully enclosed power-operated turret is at the rear and has a traverse of 360°. There is a single-piece hatch cover in the left side of the turret roof and a cupola with a single-piece hatch cover that opens to the rear on the right side of the roof.

On the forward part on the cupola is a standard pintle-mounted .50 (12.7 mm) M2 HB machine gun. There are also hatches in the turret sides and a stowage basket on the turret rear. There is a door in the hull rear for ammunition resupply purposes.

The indirect fire sight is located in the turret roof on the left side and is protected by an armoured cowl similar to that used on the US-designed M109A2 while the direct fire sight is located parallel to the main 155 mm armament.

It is also fitted with a GPS, gunner's control panel and display, and a commander's control panel and display.

Main armament of the Raad-2 comprises a local version of the 155 mm/39-calibre M185 ordnance which is fitted with a screw breech mechanism, hydropneumatic recuperator, hydraulic recoil system, fume extractor, and a double baffle muzzle brake with elevation limits from –3° to +75°.

Turret traverse is a full 360°, although the Raad-2 would only normally fire over the frontal arc. It is probable that turret traverse and weapon elevation is powered.

This ordnance is supplied by the Hadid facility of the DIO under the designation of the HM44. When travelling the ordnance is held in position by a travel lock located on the glacis plate.

This travel lock is released manually by the driver when the Raad-2 comes into action.

Maximum firing range depends on the type of ammunition (projectile and charge) used but firing a standard M107 High-Explosive (HE) projectile a maximum range of 18,100 m can be achieved while firing a base bleed HE projectile a maximum range of 24,000 m can be obtained. Maximum range with Charge 2 is quoted as 3,000 m.

In late 2006 it was revealed that the Ammunition Industries Group of the Defense Industries Organization (DIO) of Iran had developed and placed in production a number of new artillery projectiles which have a significant increase in range when compared to currently deployed 155 mm conventional artillery projectiles.

Iran has a wide range of towed and SP 155 mm artillery systems including the locally produced 155 mm Raad-2 SP as well as various types of 155 mm/45-calibre towed guns.

These include the South African Denel Land Systems G5 and Austrian NORICUM GH N-45 that were supplied many years ago as well as a small number of more recently acquired China North Industries Corporation (NORINCO) 155 mm/45-calibre Type WA 021 (WAC 21) towed artillery systems which are covered in detail in a separate entry in *IHS Jane's Land Warfare Platforms: Artillery and Air Defence*.

Iran is now marketing a 155 mm HE ERFB-BB artillery projectile which is claimed to have a maximum range of 34 km which is less than some comparable Western-produced projectiles of this type which typically have a maximum range of 39.60 km.

The development and production of these Iranian extended-range artillery projectiles will enable Iran to engage targets at longer ranges without the platform coming under counter battery fire.

For several years Iran has been producing 130 mm and 155 mm cargo artillery projectiles which carry bomblets to attack the vulnerable upper surfaces of tanks and other armoured vehicles.

With the introduction of the ERFB-BB projectiles Iran can now offer a complete family of artillery projectiles as well as their associated fuzes, primers and conventional charge systems. All of these are now being offered on the export market by the Iranian DIO.

Maximum rate of fire is quoted as being four rounds per minute with a total of 30 155 mm projectiles and their associated charges carried. According to the DIO the barrel has a maximum life of 5,000 rounds when firing Charge 8.

According to the DIO, the Raad-2 system takes less than one minute to come to a halt and open fire. Maximum rate of fire is 4 rds/min and minimum rate of fire is quoted as 2 rds/min.

To provide a more stable firing position, two spades are lowered to the ground one either side at the rear of the hull before firing commences.

Suspension is of the torsion bar type with hydraulic shock-absorbers with either side having six rubber-tyred roadwheels with the drive sprocket at the front, idler at the rear and track return rollers.

The track is of the live pad type and the upper part of the suspension is normally covered by a skirt which helps to keep dust down.

The standard Raad-2 is powered by a V-84MS diesel engine developing 840 hp which gives a power to weight ratio of 23 hp/tonne and a maximum road speed of 65 km/h. It is fitted with a 12 kW APU, which allows turret systems to be operated with the main engine off.

Standard equipment includes an NBC system and it can operate in temperatures ranging from –40°C to +55°C. It also has an automatic gunlaying system, automatic and manual fire-extinguishing equipment, radios, digital communications, GPS and an air conditioner.

Variants

Raad-2M

This is powered by a model 5TDF water-cooled diesel engine developing 700 hp that gives a power-to-weight ratio of 19.4 hp/tonne. This engine was developed and manufactured in the Ukraine and was originally fitted in the T-64 series MBT and is a member of a complete family of very compact diesel engines. Maximum road speed of the Raad-2M is quoted as 60 km/h with a maximum road range of 450 km.

Externally the Raad-2M is distinguishable from the Raad-2 as the former has a large oblong exhaust outlet mounted on the right side of the chassis with a distinctive grill above.

155 mm/45-calibre barrel

Iran is also now making a 155 mm/45-calibre barrel (the HM47L) but at this stage it is not known as to whether this is installed in a new Iranian towed or SP artillery system or it is a replacement barrel for its currently deployed 155 mm/45-calibre towed artillery systems which have been in service for many years. Iran is also now producing barrels that are chrome plated to extend their lives.

Specifications

	Raad-2
Dimensions and weights	
Crew:	5
Length	
overall:	9.14 m
Width	
overall:	3.38 m
Height	
to turret roof:	2.60 m
to top of roof mounted weapon:	3.467 m
axis of fire:	2.24 m
Ground clearance	
overall:	0.45 m
Weight	
combat:	36,000 kg
Ground pressure	
standard track:	0.67 kg/cm²
Mobility	
Configuration	
running gear:	tracked
Power-to-weight ratio:	23.00 hp/t
Speed	
max speed:	65 km/h
cross-country:	45 km/h
Range	
main fuel supply:	450 km
Fording	
without preparation:	1.20 m
Gradient:	60%
Side slope:	40%

	Raad-2
Vertical obstacle	
forwards:	0.80 m
Trench:	2.40 m
Engine	V-84MS, diesel, 840 hp
Gearbox	
model:	SPAT 1200
forward gears:	7
reverse gears:	1
Suspension:	torsion bar with hydraulic shock-absorbers
Firepower	
Armament:	1 × turret mounted 155 mm HM44 howitzer
Armament:	1 × roof mounted 12.7 mm (0.50) M2 HB machine gun
Ammunition	
main total:	30
Main weapon	
calibre length:	39 calibres
Turret power control	
type:	manual
Main armament traverse	
angle:	360°
Main armament elevation/depression	
armament front:	+75°/-3°
Survivability	
Night vision equipment	
vehicle:	optional
NBC system:	yes
Armour	
hull/body:	steel
turret:	steel

Status
Production as required. In service with the Iranian Army.

Contractor
Defence Industries Organization (DIO), Shahid Karimi Industrial Complex.

Defence Industries Organization Raad-1 122 mm self-propelled gun-howitzer

Development
Early in 2002, it was revealed by Iran that its 122 mm Raad-1 Self-Propelled Gun-Howitzer (SP/GH) was then in quantity production by the Iranian Defence Industries Organisation (DIO), Armour Industries Group.

This is in service alongside the larger Raad-2 155 mm SP/GH that is covered in a separate entry in *IHS Jane's Land Warfare Platforms: Artillery and Air Defence*. The Raad-1 has also been referred to as the Thunder-1.

The Raad-2 155 mm and Raad-1 122 mm have been offered on the export market but as of March 2013 there were no export sales of these weapons.

Description
In overall concept, the 122 mm Raad-1 is similar in layout to other self-propelled artillery weapon systems of this type with the driver and power pack at the front and the fully enclosed turret and fighting compartment at the rear.

The Raad-1 122 mm SP/GH has an all-welded steel hull that is similar to that used for the Boraq family of Armoured Personnel Carrier (APC) that has been in service with Iran for some years.

In many respects the Boraq is very similar to the China North Industries Corporation (NORINCO) WZ501 IFV which in turn is based on the Russian BMP-1. The Iranian Boraq has M113 type rubber-tyred roadwheels.

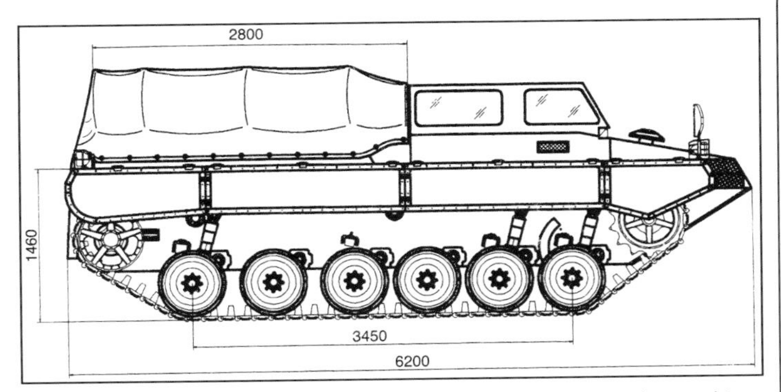

Side drawing of Boraq Armoured Ammunition Carrier which can be used to support the 155 mm Raad-2 and 122 mm Raad-1 self-propelled artillery system (DIO) 1044350

Iranian 122 mm Raad-1 self-propelled gun howitzer in travelling configuration 0127157

According to Iran, maximum armour thickness of the Raad-1 is 17 mm and this provides the occupants with protection from small arms fire and shell splinters.

The driver is seated at the front left and has a single-piece hatch cover above and three day periscopes for observation over the frontal arc. The centre one of these can be replaced by a passive periscope for driving at night.

The power pack compartment is to the immediate right of the driver and is separated from the remainder of the vehicle by a bulkhead. The air inlet and outlet louvres are in the roof with the exhaust outlet on the right side of the hull.

The hull to the immediate rear of the driver's and engine compartment is raised to provide greater internal volume for the gun crew and 122 mm ammunition stowage.

The fully enclosed three-person turret is mounted on the roof at the rear of the hull and this is very similar to that fitted to the Russian 122 mm 2S1 SP/GH which first entered service with the Russian Army as far back as 1972. The 2S1 has also been produced under licence in Bulgaria and Poland for the home and export markets.

Turret traverse is 360° with weapon elevation from –3° to +72° with direct and indirect sights being fitted. Two hatches are provided in the turret roof for crew access and a .50 (12.7 mm) M2 HB MG can be mounted on the turret roof for air defence purposes.

The main armament of the Raad-1 comprises a 122 mm 2A31 39-calibre howitzer that is manufactured in Iran by the Hadid facility of the DIO under the local designation of the HM51. This facility produces all of the artillery and tank barrels for the Iranian Army as well as rocket launchers.

When travelling, the 122 mm ordnance is held in the travelling position by a travel lock mounted on the forward part of the glacis plate which is located forward and right of the driver's position. This is released manually by the driver when the Raad-1 comes into the firing position.

Firing a standard 122 mm High-Explosive (HE) projectile with a ZBK-13 charge the Raad-1 can achieve a maximum range of 15,200 m with a maximum rate of fire of four to five rounds per minute. Muzzle velocity depends on charge used but with ZBK-13 it is 721 m/s.

In late 2006 it was revealed that the Ammunition Industries Group of the Defense Industries Organization (DIO) of Iran had developed and placed in production a number of new artillery projectiles which have a significant increase in range when compared to currently deployed conventional artillery projectiles.

Iran is now producing new 122 mm Base Bleed (BB) and Hollow Base (HB) artillery projectiles with a significant increase in range.

Barrel life is quoted as 3,000 rounds with Charge 8 with a total of 35 122 mm projectiles and charges being carried. Ammunition can be reloaded via a door in the hull rear. When using 122 mm ammunition stockpiled on the ground outside of the vehicle, rate of fire is 1.5 to 2 rds/min.

Some sources have indicated the system is fitted with an automatic loading system of some type. This could well include a flick rammer that would also enable the weapon to be loaded at high angles of elevation as well as increasing its rate of fire.

Ammunition used by the Raad-1 SP/GH is identical to that used by the Russian 122 mm D-30 towed gun-howitzer that has been manufactured in Iran by the Hadid facility under the local designation of the HM40 for many years.

Iran is also self-sufficient in 122 mm ammunition which is of the separate loading type, that is projectile and charge.

The suspension is of the torsion bar type and either side consists of six rubber-tyred roadwheels with the drive sprocket at the front and the idler at the rear. The upper part of the suspension and return rollers are covered by a skirt which helps to keep dust down.

As a trim vane is mounted on the glacis plate of the Raad-1 SP/GH, the system may be fully amphibious, propelled in the water by its tracks. Given the overall weight of the vehicle compared with the Boraq APC it may well only have a fording capability.

The 122 mm Raad-1 is capable of operating in temperatures ranging from –45°C to +45°C. It is fitted with a computer and monitor to display ballistic data.

Variants
There are no known variants of the Raad-1 122 mm self-propelled artillery system although the design of the hull is such that it can be used for a wide range of other applications. The Boraq hull is being used for an increasing number of battlefield roles, including ammunition resupply vehicle and mortar carrier. This could be used to support 122 mm and 155 mm self-propelled artillery systems, as well as towed artillery systems.

Specifications

	Raad-1
Dimensions and weights	
Crew:	4
Length	
overall:	6.50 m
Width	
overall:	2.67 m
Height	
overall:	3.00 m
axis of fire:	2.14 m
Ground clearance	
overall:	0.40 m
Weight	
combat:	17,500 kg
Ground pressure	
standard track:	0.54 kg/cm^2
Mobility	
Configuration	
running gear:	tracked
Power-to-weight ratio:	18.00 hp/t
Speed	
max speed:	65 km/h
cross-country:	45 km/h
Range	
main fuel supply:	400 km
Gradient:	60% (est.)
Side slope:	40% (est.)
Vertical obstacle	
forwards:	0.80 m
Trench:	2.30 m
Engine	8 cylinders, air cooled, 315 hp
Gearbox	
type:	manual
forward gears:	5
reverse gears:	2
Suspension:	torsion bar
Electrical system	
vehicle:	28 V
Firepower	
Armament:	1 × turret mounted 122 mm 2A31 howitzer
Armament:	1 × roof mounted 12.7 mm (0.50) M2 HB machine gun
Ammunition	
main total:	35
Main weapon	
calibre length:	39 calibres
Turret power control	
type:	powered/manual
Main armament traverse	
angle:	360°
Main armament elevation/depression	
armament front:	+72°/-3°
Survivability	
Night vision equipment	
vehicle:	optional
NBC system:	yes
Armour	
hull/body:	steel (not confirmed)
turret:	steel (not confirmed)

Status
Production. In service with Iranian Army.

Contractor
Defence Industries Organization (DIO), Shahid Karimi Industrial Complex.

Israel

Doher 155 mm self-propelled howitzer

Development
The Israel Defense Force has taken delivery of at least 530 BAE Systems (previously United Defense) 155 mm M109 series self-propelled howitzers from the US.

It is believed that Israel placed its first production order for the M109 in 1969 and this consisted of 24 units. Many additional systems were ordered from the US after the 1973 conflict.

In mid-2010 it was stated that the Israel Defense Force was considering replacing its aging fleet of M109 155 mm self-propelled howitzers.

Doher 155 mm self-propelled artillery system which is an M109 with many improvements to meet local operational requirements and shown here laying a smoke screen by injecting diesel fuel into the exhaust outlet (IDF) 0005998

Front runners were said to be German Krauss-Maffei Wegmann 155 mm/52-calibre Artillery Gun Module (AGM) or the Israeli Soltam Systems ATMOS 155 mm truck mounted 6 × 6 self-propelled artillery system.

It is known that the latter has been evaluated by the Israel Defense Force and has already been produced in production quantities for the export market.

Details of both of these systems are provided in separate entries in *IHS Jane's Land Warfare Platforms: Artillery and Air Defence*.

The Israel Defense Force is now moving towards a more balanced artillery arm which includes not only conventional 155 mm tube artillery but also artillery rocket launchers and precision guided munitions, including artillery rockets.

Description
Most of these M109 were upgraded in Israel to a new standard under the local name of the Doher. According to the Israel Defense Force, the upgraded Doher has a number of significant operational advantages including:

- Shorter response time
- High mission availability
- Capability to provide a wide deployment
- Maximum flexibility in operating the artillery battery among others, by leap frogging
- High firing accuracy that results from technological improvements in the systems
- Fighting under NBC conditions and completing the mission under battery fire
- Improving the survivability of the crew by installing a number of new and more modern subsystems including automatic fire detection and suppression system, central NBC filtration system, ability to lay a smoke screen by injecting diesel fuel into the engine exhaust on the right side of the forward part of the hull, installation of another 7.62 mm or .50 (12.7 mm) M2 HB MG on the roof and passive night vision equipment for the driver and commander.

The crew of seven consists of the commander, navigator, gunner, person in charge of parking and refuelling, driver and two soldiers who join the crew from the Alfa ammunition vehicle.

It is understood that all M109s in service with the Israel Defense Force are fitted with a 155 mm/39-calibre ordnance.

Minor modifications carried out to the Doher in IDF service are understood to include additional track links, improved lighting and external bars on the hull sides for additional equipment stowage.

While the Israeli defence industry does not offer completed 155 mm M109s for the export market, it can provide a number of upgrades for this system, especially in the areas of fire-control and turret modernisation. Soltam Systems is the main contractor in Israel for 155 mm artillery systems and their associated ordnance.

Soltam Systems is now part of Elbit which now allows Soltam Systems to provide a complete artillery package including not only the weapon and its ammunition but also a target acquisition system and an artillery fire control system.

Romach 175 mm SPG
This is a slightly modified version of the now BAE Systems 175 mm M107 self-propelled gun.

It is understood that the Israel Defense Forces has 36 remaining M107 which serve alongside a similar number of M110A2 systems.

Specifications

	Doher
Dimensions and weights	
Crew:	7
Length	
overall:	9.014 m
Width	
overall:	3.14 m
Height	
overall:	3.06 m

	Doher
Weight	
standard:	28,200 kg
Mobility	
Configuration	
running gear:	tracked
Speed	
max speed:	50 km/h
Range	
main fuel supply:	350 km
Firepower	
Armament:	1 × turret mounted 155 mm howitzer
Armament:	1 × roof mounted 12.7 mm (0.50) M2 HB machine gun
Turret power control	
type:	hydraulic/manual
Main armament traverse	
angle:	360°
Main armament elevation/depression	
armament front:	+75°/-3°
Survivability	
Night vision equipment	
vehicle:	yes
NBC system:	yes
Armour	
hull/body:	aluminium
turret:	aluminium

Status
Production complete. In service with the Israel Defense Force. Some sources have indicated that the IDF may well upgrade its fleet of M109 systems further in the future.

Contractor
BAE Systems.
Upgrade work has been carried out in Israel by a number of facilities.
Enquiries to SIBAT–Foreign Defence Assistance and Defence Export.

Soltam Systems L33 155 mm self-propelled gun-howitzer

Development
The L33 (the length of the barrel in calibres) 155 mm self-propelled gun-howitzer was developed in the late 1960s by Soltam Systems Limited.

It is basically a Sherman M4A3E8 tank chassis, originally designed and built during the Second World War, fitted with a Cummins diesel engine, a new all-welded steel superstructure and a Soltam Systems 155 mm M-68 gun-howitzer taken from the Soltam Systems 155 mm towed system.

Production of the L33 (which has also been referred to as the 155 mm M68 self-propelled gun) commenced just after the 1967 war and it was used operationally for the first time during the 1973 Middle East conflict.

No production figures for the Soltam Systems L33 155 mm self-propelled gun-howitzer have been released. It is estimated that about 150 units were built for the Israel Defense Forces and most of these are still in service in front line or reserve units.

SIBAT, Foreign Defence Assistance and Defence Export, Land Sales Division, at the Israel Ministry of Defence, is now offering quantities of the L33 systems for sale with the standard Soltam Systems 155 mm/33-calibre barrel or fitted with a new 155 mm/39-calibre barrel. As far as it is known, there have been no export sales of the L33 with a 155 mm/39-calibre barrel.

It is understood that these 150 Soltam Systems L33 self-propelled gun-howitzer are now held in reserve along with the 50 much older 155 mm self-propelled gun howitzer M-50.

Description
The driver is seated at the front of the hull on the left side and has a bulletproof windscreen to the front and sides and a single-piece hatch cover that opens to the rear over the driver.

The steel armour of the L33 hull has a minimum thickness of 12 mm and a maximum thickness of 64 mm.

The commander is seated to the rear and above the driver's position and has a windscreen to the front and sides and a single-piece hatch cover. The anti-aircraft gunner's position is at the front of the hull on the right side. A 7.62 mm (0.30) machine gun is pintle-mounted on the forward part of the cupola which can be traversed through a full 360° and has a two-piece hatch cover that opens either side. The crew enter and leave the vehicle by a single door in each side of the hull that opens to the rear. At the rear of the hull are two large doors for 155 mm ammunition resupply.

The US supplied Cummins VT 8-460-Bi diesel engine is under the gun compartment at the rear of the hull. Power is transferred to the manual transmission in the nose of the vehicle by a two-part propeller shaft.

Soltam Systems L33 155 mm self-propelled gun-howitzer with ordnance in travelling lock and roof-mounted 7.62 mm machine gun in position
(Israel Defense Forces) 1164304

The controlled differential transmits engine power to the final drive unit and also contains a brake system for steering and stopping the vehicle. The suspension each side consists of three two-wheeled bogies, three track-return rollers, the idler at the rear and the drive sprocket at the front.

The Soltam Systems 155 mm M-68 gun-howitzer is mounted in the forward part of the superstructure and has an elevation of +52°, a depression of –3° and a total traverse of 60°. Both elevation and traverse are manual. The weapon has a monobloc barrel and incorporates a horizontal sliding semi-automatic breech block. Mounted on the cradle are the hydropneumatic recuperator, a variable hydraulic damper that varies as a function of elevation (for example 800 mm at 52° elevation and 1,100 mm at 0° elevation) and a pneumatic rammer which permits loading of the gun at all angles of elevation. The barrel has a single-baffle muzzle brake, a bore evacuator and a travelling lock.

The travelling lock has to be released by one of the crew manually by leaving the safety of the vehicle.

The L33 has a panoramic telescope for indirect fire engagements and a telescope for direct fire.

Ammunition is separate loading with obturation achieved by self-sealing obturator discs. Firing is accomplished by a cartridge primer inserted into the breech block and fired by a trip action mechanism. The 155 mm HE M107 projectile weighs 43.7 kg, has a maximum muzzle velocity of 725 m/s and a maximum range of 20,000 m. The following propelling charges are used: M1A2 zone 3 (muzzle velocity 284 m/s), M1A2 zone 4 (muzzle velocity 326 m/s), M1A2 zone 5 (muzzle velocity 390 m/s), M9A2 zone 6 (muzzle velocity 468 m/s), M9A2 zone 7 (muzzle velocity 562 m/s), M9A2 zone 8 (muzzle velocity 647 m/s) and M9A2 zone 9 (muzzle velocity 725 m/s). The M-68 system fires all NATO 155 mm projectiles including illuminating and smoke. Of the 60 rounds of 155 mm ammunition carried, 16 are for ready use.

Variants

L33 with 155 mm/39-calibre ordnance
This is the original L33 with the 155 mm 33-calibre ordnance replaced by the longer 155 mm 39-calibre ordnance but it never entered production or service.

155 mm self-propelled howitzer M-50
An earlier 155 mm self-propelled howitzer was the M-50 which was essentially a modified Sherman tank hull with the French Nexter Systems towed 155 mm M-50 howitzer mounted in an open compartment at the rear. This is no longer in front-line service with the Israel Defence Force but it is believed that up to 50 systems may be held in reserve.

Soltam Systems Slammer 155 mm SPG
This was developed to the prototype stage and is based on a modified Merkava MBT hull. Prototypes were built and tested but it never entered production.

160 mm self-propelled mortar
This was a much modified Sherman hull armed with a Soltam Systems 160 mm mortar. These are no longer deployed by Israel.

Specifications

	L33
Dimensions and weights	
Crew:	8
Length	
overall:	8.47 m
hull:	6.47 m
Width	
overall:	3.5 m
without skirts:	2.86 m
Height	
overall:	3.45 m
Ground clearance	
overall:	0.43 m
Track	
vehicle:	2.28 m

	L33
Track width	
normal:	580 mm
Length of track on ground:	4.6 m
Weight	
combat:	41,500 kg
Mobility	
Configuration	
running gear:	tracked
Power-to-weight ratio:	10.08 hp/t
Speed	
max speed:	38 km/h
Range	
main fuel supply:	260 km
Fuel capacity	
main:	820 litres
Fording	
without preparation:	0.9 m
Gradient:	60%
Side slope:	30%
Vertical obstacle	
forwards:	0.91 m
Trench:	2.3 m
Turning radius:	9.5 m
Engine	Cummins VT 8-460-Bi, diesel, 460 hp at 2,600 rpm
Gearbox	
type:	manual
forward gears:	5
reverse gears:	1
Electrical system	
vehicle:	24 V
Firepower	
Armament:	1 × hull mounted 155 mm L33 howitzer
Armament:	1 × roof mounted 7.62 mm (0.30) machine gun
Ammunition	
main total:	60
Main weapon	
calibre length:	33 calibres
Main armament traverse	
angle:	60° (30° left/30° right)
Main armament elevation/depression	
armament front:	+52°/-4°
Survivability	
Night vision equipment	
vehicle:	optional
NBC system:	no
Armour	
hull/body:	steel

Status
Production complete. In service with the Israel Defense Force reserves. Some are being offered to sale.

Contractor
Soltam Systems Limited (now owned by Soltam Systems).

Italy

Oto Melara Palmaria 155 mm self-propelled howitzer

Development
The Palmaria 155 mm self-propelled howitzer was developed from 1977 by Oto Melara specifically for the export market and first production vehicles followed in 1982.

The first customer to order the Palmaria was Libya, with an order for 210. Nigeria ordered 25 during 1982 and a further 20 Palmaria turrets were ordered by Argentina to be fitted on locally built TAM medium tank hull. These were completed late in 1986 by which time final deliveries had been made to Libya and Nigeria.

Argentina has built lengthened TAM medium tank hull that have been fitted with the turret of the Palmaria. The Argentine system is called the VCA 155. This has been produced in small quantities and is now operational in Argentina. Details are provided in a separate entry in *IHS Jane's Land Warfare Platforms: Artillery and Air Defence*. Production of this system has now been completed.

In 1990, Oto Melara completed a further batch of 25 155 mm Palmaria systems for an undisclosed customer. Some sources indicate that this was a repeat order from Nigeria.

Production standard Oto Melara Palmaria 155 mm self-propelled howitzer with turret traversed right and ordnance at maximum elevation (Oto Melara) 1452900

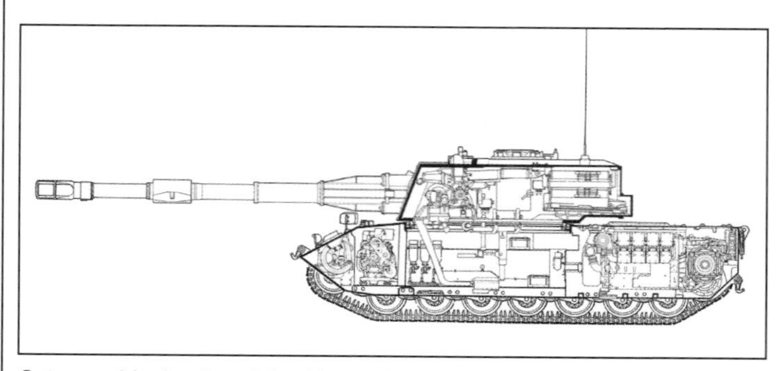

Cutaway side drawing of Oto Melara Palmaria 155 mm self-propelled howitzer with turret traversed to the front (Oto Melara) 1365003

According to the United Nations Arms Transfer List for the period 1992 through to 2012 the only customer for the Palmaria was Nigeria who took delivery of the following complete systems:

Year	Quantity
1992	9
1993	2
1994	2
2001	2
2007	9

Oto Melara built the German Krauss-Maffei Wegmann 155 mm/52-calibre PzH 2000 self-propelled artillery system under licence for the Italian Army.

The first two came from the Krauss-Maffei Wegmann production line in Kassel and the remaining 68 were built under licence in the facilities of Oto Melara in Italy.

Description
The hull of the Palmaria 155 mm self-propelled howitzer is of all welded steel armour which provides the occupants with protection from small arms fire and shell splinters.

Many of the automotive components of the Palmaria 155 mm self-propelled howitzer are also used in the Oto Melara OF-40 Main Battle Tank. A total of 36 of these were built for the United Arab Emirates plus a small number of armoured recovery vehicles on a similar chassis.

These OF-40 MBTs have been phased out of front line service and may be converted into heavy infantry fighting vehicles under the leadership of the Belgian company of Sabiex.

The driver is seated at the front of the hull under armour on the right side and is provided with a single-piece hatch cover and three day periscopes, the centre one of which can be replaced by a passive periscope for night driving. The other four crew members, commander, gunner, charge handler and magazine operator, are seated in the aluminium turret in the centre of the hull.

The commander is seated in the right forward part of the turret which has eight day periscopes for all-round observation and a single-piece hatch cover that opens to the rear. There are two external stowage racks at the rear of the turret. In either side of the turret there is a large rectangular hatch, the left one opening to the rear and the right one opening forwards.

Type	P3 HE	P3 HE LT	P3 HE RAP	P4 Illuminating	P5 Smoke
Weight:					
(projectile)	43.5 kg	43.5 kg	43.5 kg	43.5 kg	43.5 kg
(explosive)	11.7 kg	11.7 kg	8 kg	n/a	n/a
(propellant)	n/avail	1 kg	2.8 kg	n/a	n/a
Length:	933 mm	933 mm	933 mm	933 mm	933 mm
Max range:	24,700 m	27,500 m	30,000 m	24,700 m	24,700 m

A 7.62 or .50 (12.7 mm) M2 HB anti-aircraft machine gun can be mounted on the roof for anti-aircraft defence and four 76 mm smoke grenade dischargers are mounted either side of the turret firing towards the front.

The vehicle has an auxiliary power unit to supply the hydraulic and electric power required to operate the turret, consisting of a diesel engine driving through a gearbox, a generator and a hydraulic pump.

Standard equipment includes an NBC system, an automatic fire-extinguishing system, bilge pumps and an emergency escape hatch in the floor.

The Oto Melara-designed 155 mm barrel is 41-calibres long, is fitted with a fume extractor and a muzzle brake and is loaded semi-automatically. Using the semi-automatic loading system, the howitzer can fire one round every 15 seconds, one round every minute for one hour or one round every three minutes continuously.

The semi-automatic loader is at the rear of the turret and can be loaded via a hatch in the turret side. The 155 mm barrel is loaded at an angle of +2° and the propellant charge is loaded into the chamber manually.

Using semi-automatic loading the turret can accommodate 23 rounds ready for use, with a further seven ready for use elsewhere in the hull. The projectiles are stowed in a linked row of tubes at the rear of the turret ready for use. The firing percussion mechanism is actuated by an electrohydraulic valve.

Firing is by a push-button and using this system an initial burst rate of three rounds in 20 seconds can be achieved. The barrel elevating and balancing mechanism consists of a hydraulically actuated cylinder linked to the cradle and pivoting on the turret structure. The limits of elevation are from –4° to +70°.

There are two modes of operation for the mechanism, hand and power. Using the hand mode the barrel can be moved using a wheel to operate a hand pump linked to the elevation cylinder. In the power mode, with the hydraulic power on, the gunlayer can lay the weapon directly, using a control handle.

After firing, the barrel is returned to the loading position of +2° automatically. Traversing is by a hydraulically operated and controlled gearbox; traverse is a full 360°.

The recoil system consists of two buffers and one recuperator. Firing Charge 8 at an elevation of +70° the recoil force is approximately 45,000 kg.

The Italian company Simmel developed a new family of 155 mm projectiles for the Palmaria self-propelled howitzer, brief details of which are given in the table above.

It can also fire other types of ammunition including that used by the towed FH-70 and BAE Systems FH-77B, as well as US 155 mm ammunition types including the M107, M549A1 and M483A1.

The Palmaria has three sighting systems installed: a P170 direct firing periscope, a P186 panoramic sight for indirect sighting and a dial sight.

The P170 can be used for both observation and sighting and in the latter mode has a magnification of ×8. The optical unit of the sighting system can be replaced by a light intensification system that transforms the day periscope into a low-light device.

The P186 panoramic sight is used for indirect firing and is provided with an illuminated reticle for day or low-light level use. The reticle is fixed and pre-aligned with the optical axis.

The Palmaria can also be provided with an electronic plane converter for rapid and accurate aiming both in direct and indirect fire.

Turrets for the production standard Oto Melara Palmaria 155 mm self-propelled howitzer on the production line at La Spezia (Oto Melara) 1452901

The primary fire-control system of the Palmaria is the P186 telescope that is mounted on a telescope mount. This is used for laying the weapon in azimuth and for indirect fire-control. The sight has a magnification of ×4 with a 10° field of view and weighs 8 kg.

Internal communications are via the normal intercommunication system and an extra external telephone point can be provided. The turret has adequate space for normal radio installations.

The Palmaria is fitted with a 9 kW 3-phase generator with rectifier.

Variants

Oto Melara has proposed that the hull of the Palmaria should be used for other applications, for example fitted with a similar turret to the OTOMATIC 76 mm self-propelled anti-aircraft gun system. All development work on this has now ceased.

Specifications

	Palmaria
Dimensions and weights	
Crew:	5
Length	
overall:	11.474 m
hull:	7.265 m
main armament rear:	9.6 m
Width	
overall:	3.35 m
Height	
to turret roof:	2.874 m
axis of fire:	2.195 m
Ground clearance	
overall:	0.4 m
Track width	
normal:	584 mm
Weight	
standard:	43,000 kg
combat:	46,000 kg
Mobility	
Configuration	
running gear:	tracked
Power-to-weight ratio:	16.30 hp/t
Speed	
max speed:	60 km/h
Range	
main fuel supply:	500 km
Fuel capacity	
main:	800 litres
Fording	
without preparation:	1.2 m
with preparation:	4 m
Gradient:	60%
Side slope:	30%
Vertical obstacle	
forwards:	1 m
Trench:	3.0 m
Engine	MTU MB 837 Ea-500, 8 cylinders, turbocharged, multifuel, 750 hp
Gearbox	
model:	Renk RK 304
forward gears:	4
reverse gears:	2
Suspension:	torsion bar
Electrical system	
vehicle:	24 V
Batteries:	6 × 12 V, 100 Ah
Firepower	
Armament:	1 × turret mounted 155 mm howitzer
Armament:	1 × roof mounted 7.62 mm (0.30) machine gun
Armament:	8 × turret mounted 76 mm smoke grenade launcher (2 × 4)
Ammunition	
main total:	30
Turret power control	
type:	hydraulic/manual
by commander:	no
by gunner:	yes
Main armament traverse	
angle:	360°

	Palmaria
Main armament elevation/depression	
armament front:	+70°/-5°
Survivability	
Night vision equipment	
vehicle:	yes
NBC system:	yes
Armour	
hull/body:	steel
turret:	aluminium

Status

Production as required. In service with the following countries:

Country	Quantity	Comment
Argentina	20	total number of turrets supplied
Libya	210	delivered from 1982. Some sources quote a lower figure of 160 units. (As a result of conflict in Libya in 2011 these numbers have been reduced.)
Nigeria	25	believed that a second batch of 25 systems has been delivered

Contractor

Oto Melara SpA.

Oto Melara 155 mm M109L self-propelled howitzer

Development

Italy purchased 221 now BAE Systems M109s without their 155 mm armament from the US. The armament was manufactured and installed in Italy by Oto Melara and subsequently issued to the Italian Army.

In 1970-71 Oto Melara produced a long-barrelled version of the M109 capable of firing 155 mm FH-70 projectiles to a maximum range of 24,000 m.

At that time there was no Italian Army requirement for this increase in range but, early in 1984, the Italian Ministry of Defence ordered three examples for extensive trials. The trials were successful and a decision was taken in 1986 to upgrade a total of 280 Italian Army M109 series self-propelled weapons to the M109L configuration.

In addition to taking delivery of 221 M109s without armament, the Italian Army also took delivery of 62 M109A1B systems. The first batch of 32 M109Ls was completed in 1986 and final deliveries were made to the Italian Army in 1992.

In Italian Army service the M109L has now been supplemented by the German Krauss-Maffei Wegmann PzH 2000 155 mm/52-calibre self-propelled artillery system. Total Italian Army requirement is 70 systems.

The Italian contract for the PzH 2000 was finally signed late in 2002 between the Italian Direzione Generale Armamenti Terrestri and Consorzio Iveco–Oto Melara.

Deliveries ran from 2006 through to 2011 and replace the current M109L issued to the Ariete armoured brigade and the Garibaldi and Pinerolo mechanised brigades. The first two PzH 2000s were delivered to Italy by Krauss-Maffei Wegmann from the Kassel production line in Germany.

Description

The hull of the M109 has been modified so that it is the equivalent of the BAE Systems M109A3 which is compatible with the new elevating mass and the aiming system of the M109G used by the German Army.

The main change on the M109L is the replacement of the existing 155 mm/23-calibre barrel with a 155 mm/39-calibre barrel. Externally this is configured to the M109 mounting, but internally it is ballistically similar to the towed FH-70 configuration and can thus fire existing 155 mm M107 or FH-70 ammunition. Firing L15 high-explosive projectiles, the M109L has a maximum range of 24,000 m or 30,000 m with Rocket-Assisted Projectiles (RAPs).

In addition to the installation of a new ordnance, the following subsystems have been modified:

Italian Army Oto Melara 155 mm M109L with ordnance in travel lock (Richard Stickland) 1164295

Italian Army Oto Melara 155 mm M109L (Richard Stickland) 1133914

- Recoil buffers and run-out buffer
- Cradle to fit the new ordnance
- Replacement of the equilibrator and elevating cylinder
- Barrel clamp to fit new barrel.

The new 155 mm/39-calibre barrel is fitted with a revised muzzle brake, with the maximum recoil increase on the M109L being only 100 mm. The existing breech and breech ring of the original M109 have been retained.

The elevating mass, as modified, is interchangeable with the one on the basic M109; furthermore, the mechanical interface between the elevating mass and saddle, as well as the mountings of the optical device, do not require any modification.

Specifications

	M109L
Dimensions and weights	
Crew:	6
Length	
overall:	8.133 m
hull:	6.256 m
Width	
overall:	3.295 m
without skirts:	3.149 m
Height	
to top of roof mounted weapon:	3.28 m
Ground clearance	
overall:	0.467 m
Track	
vehicle:	2.768 m
Track width	
normal:	381 mm
Length of track on ground:	3.962 m
Weight	
standard:	20,000 kg
combat:	24,800 kg
Ground pressure	
standard track:	0.68 kg/cm^2
Mobility	
Configuration	
running gear:	tracked
Power-to-weight ratio:	16.33 hp/t
Speed	
max speed:	56 km/h
Range	
main fuel supply:	349 km
Fuel capacity	
main:	511 litres

	M109L
Vertical obstacle	
forwards:	0.533 m
Trench:	1.828 m
Engine	Detroit Diesel Model 8V-71T, V-8, water cooled, 405 hp at 2,300 rpm
Gearbox	
model:	Allison Transmission XTG-411-2A
forward gears:	4
reverse gears:	2
Suspension:	independent torsion bar
Electrical system	
vehicle:	24 V
Batteries:	4 × 12 V
Firepower	
Armament:	1 × turret mounted 155 mm howitzer
Armament:	1 × roof mounted 12.7 mm (0.50) M2 HB machine gun
Armament:	6 × turret mounted 76 mm smoke grenade launcher (2 × 3)
Ammunition	
main total:	28
12.7/0.50:	500
Main weapon	
calibre length:	39 calibres
Turret power control	
type:	hydraulic/manual
by commander:	no
by gunner:	yes
Main armament traverse	
angle:	360°
Main armament elevation/depression	
armament front:	+75°/-3°
Survivability	
Night vision equipment	
vehicle:	yes
NBC system:	no
Armour	
hull/body:	aluminium
turret:	aluminium

Status
In service with the Italian Army (124 units). Production complete. No longer marketed by Oto Melara. The Italian Army has now taken delivery of a total of 70 German Krauss-Maffei Wegmann PzH 2000 155 mm/52-calibre self-propelled artillery systems with the first two production systems coming from the German production line in Kassel and the remainder being manufactured under licence in Italy by Oto Melara at their La Spezia facility.

Contractor
Oto Melara SpA.

Japan

Type 99 155 mm self-propelled howitzer

Development
Following trials with prototype systems, the Type 99 155 mm self-propelled howitzer was type classified by the Japanese Defense Agency. This system is also referred to as the Rongunozu 155 mm self-propelled artillery system.

Prime contractor for the Type 99 is Mitsubishi Heavy Industries (hull, final integration and testing) and the Japanese Steel Works (complete turret and 155 mm armament).

The 155 mm Type 99 is the eventual replacement for the Type 75 155 mm self-propelled howitzer.

Japanese Type 99 155 mm self-propelled howitzer with ordnance in travel lock (Matsuhiro Kadota) 1296079

Production of the more recent Type 99 155 mm self-propelled howitzer is at such a low rate that the older Type75 155 mm self-propelled howitzer will remain in service for many years to come.

By 2013 it is estimated that about 70 systems had been manufactured with production being as follows:

Year	Quantity
2008	8
2007	8
2006	7
2005	7
2004	8
2003	8
2002	7
2001	6
2000	7
1999	4

Description
The overall layout of the Type 99 155 mm self-propelled gun is similar to other recent weapons of this type.

The driver is seated at the front right with a single-piece hatch cover above his position and forward of these are three day periscopes, the centre one of which can be replaced by a passive periscope for driving at night.

The power pack, consisting of diesel engine, transmission and cooling system is mounted to the left of the driver, with the air inlet and outlet louvres in the roof and the exhaust outlet in the left side of the hull.

The fighting compartment is at the rear with normal means of entry and exit for the crew via a door in the lower part of the hull rear.

The fully enclosed turret is at the rear with the commander seated on the right side and provided with a roof-mounted .50 (12.7 mm) M2 HB machine gun mount that is also fitted with a shield.

In addition to the roof hatches there is also a single door in either side of the turret. The one on the left side opens to the rear while the one on the right side opens to the front.

The 155 mm ordnance, which is believed to be a 39-calibre weapon, is provided with a muzzle brake but no fume extractor. A travelling lock is provided at the front of the hull and when not in use, this folds back onto the glacis plate.

It is understood that the ordnance is based on that used in the towed FH-70 155 mm towed artillery system, which has been manufactured under licence in Japan for some years. Maximum stated range of the Type 99 is quoted at 30,000 m.

With extended range ammunition, the range of the Type 99 could be increased to beyond 30,000 m.

It is considered probable that some type of automatic loading system is provided and mounted over the rear part of the 155 mm ordnance in what appears to be a muzzle velocity measuring device. This would feed information into the onboard fire-control system.

Suspension either side consists of seven dual rubber-tyred road wheels with the drive sprocket at the front, idler at the rear and track return rollers.

Standard equipment includes an NBC system and passive night driving aids.

Variants
Ammunition support vehicle

It is understood that the Type 99 155 mm self-propelled howitzer is supported with an armoured ammunition support vehicle based on a different full tracked hull.

Specifications

	Type 99
Dimensions and weights	
Crew:	4
Length	
overall:	11.3 m
Width	
overall:	3.2 m
Height	
overall:	4.3 m
Weight	
combat:	40,000 kg
Mobility	
Configuration	
running gear:	tracked
Speed	
max speed:	49.6 km/h
Engine	air cooled, diesel, 600 hp
Firepower	
Armament:	1 × turret mounted 155 mm howitzer
Armament:	1 × roof mounted 12.7 mm (0.50) M2 HB machine gun

Status
Production. In service with Japanese Ground Self-Defence Force (70+).

Contractor
Mitsubishi Heavy Industries (hull and final assembly) (unconfirmed).
Japan Steel Works/Nihon Seiko Jyo (gun and turret) (unconfirmed).

Type 75 155 mm self-propelled howitzer

Development

In 1969, the Technical Research and Development Headquarters of the Japanese Ground Self-Defence Force started design work on a new 155 mm self-propelled howitzer.

Mitsubishi Heavy Industries built the hull while Nihon Seiko Jyo/Japan Steel Works built the 155 mm gun and turret. The two prototypes, completed in 1971-72, differed only in their ammunition loading systems. Trials with the two prototypes were undertaken between 1973 and 1974 and in October 1975 the vehicle was standardised as the Type 75 155 mm self-propelled howitzer.

By late 1978, the first 20 vehicles had been delivered to the Japanese Ground Self-Defence Force. By the end of 1988, 201 Type 75 self-propelled howitzers were in service with the Japanese Ground Self-Defence Force and production was complete. This system was never offered on the export market. There are no known plans for this system to be upgraded.

The 155 mm Type 75 serves alongside the of locally manufactured US-designed 203 mm M110A2 systems. This has now started to be supplemented by the new 155 mm Type 99 full-tracked self-propelled artillery system, details of which are given in a separate entry in *IHS Jane's Land Warfare Platforms: Artillery and Air Defence*.

As the more recent 155 mm Type 99 self-propelled howitzer is being built in small numbers, typically six or eight per year, the older Type 75 is expected to remain in service for many years.

When compared to the Type 75 155 mm SPH, the more recent Type 99 has improved mobility and firepower, but is also heavier.

While 201 Type 75 155 mm self-propelled howitzers were built it is understood that by early 2013 less than 50 of these remained in front line service.

Type 75 155 mm self-propelled howitzer fitted with a .50 (12.7 mm) M2 HB anti-aircraft machine gun on the turret roof 0126554

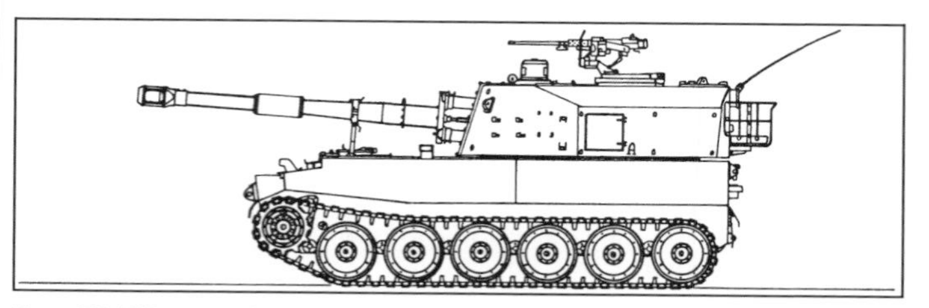

Type 75 155 mm self-propelled howitzer 0500422

Description

The hull and turret of the Type 75 155 mm self-propelled howitzer are made of all-welded aluminium. This provides the crew with protection from small arms fire and shell splinters. The crew of six consists of the commander, layer, two ammunition loaders, radio operator and driver.

The hull is divided into three compartments: driver at the front on the right, power pack to his left and the turret at the rear.

The driver has a single-piece hatch cover, which opens to the right, and in front of this are three day periscopes. The centre one can be replaced by a passive night vision device. The driver can also enter and leave his compartment via the turret.

The turret is fully enclosed, with the commander and gunner seated on the right and the ammunition loaders on the left. The commander has a single-piece hatch cover that opens to the rear and a single day periscope for forward observation. One of the two ammunition loaders is also provided with a single-piece hatch cover that opens to the rear, in front of which is a .50 (12.7 mm) M2 HB anti-aircraft machine gun with a small shield. On each side of the turret is a square hatch that opens to the rear.

At the rear of the hull are two doors that open either side of the vehicle. The torsion bar suspension either side consists of six dual rubber-tyred aluminium road wheels, with the drive sprocket at the front and the sixth road wheel acting as the idler. The first and fifth road wheel stations are provided with a hydraulic shock-absorber. There are no track-return rollers.

The Type 75 is provided with an NBC system, located in the left side of the fighting compartment, a fire extinguishing system, a heater, and night vision equipment.

Main armament consists of a 155 mm 30-calibre weapon with a double-baffle muzzle brake, a fume extractor and a stepped thread interrupted screw-type breech block which opens to the right. A gun travelling lock is fitted on the glacis plate. This has to be released manually by one of the crew.

The weapon fires a Japanese HE projectile (maximum Charge 9) to a maximum range of 19,000 m or an American M107 HE projectile (maximum Charge 7) to a maximum range of 15,000 m. A 155 mm Rocket-Assisted Projectile (RAP) has been developed with a range of around 24,000 m. Some sources have stated that the Type 75 has a maximum range of 19,000 m, firing a projectile of an undisclosed type.

Type 75 155 mm self-propelled howitzer in travelling configuration (Kensuke Ebata) 0003989

The 155 mm/39-calibre gun has an elevation of +65°, a depression of –5° and the turret can be traversed through a full 360°. Gun elevation and turret traverse are hydraulic by means of a single lever.

Movement in the horizontal plane controls the turret position in azimuth and movement in the vertical plane controls the gun in elevation. The traverse of the turret and the elevation of the gun are stopped whenever the layer halts the movement of the lever. Manual controls are provided for emergency use and the turret can also be traversed by the commander.

The gun has a variable recoil mechanism that consists of twin hydraulic cylinders, one on the left above the cradle and the other on the right side under the cradle. The variable recoil system is fitted to prevent the breech hitting the floor when the gun is being fired at a high angle. The recuperator is pneumatic and the recoil damper, which is on the centreline under the cradle, is hydraulic.

When the gun is fired, the 155 mm weapon automatically moves to an elevation of +6° for reloading and returns to its previous firing elevation once reloaded. Mounted in the rear part of the turret are two drum-type rotating magazines, an extensible loading tray and a hydraulic rammer. Each drum holds nine 155 mm projectiles, is rotated electrically or manually and can be reloaded from inside the vehicle or from outside via two small circular doors in the turret rear.

The axis of the drums, two-stage extensible loading tray and the rammer are on the same axis as the gun when it is returned from its firing position for loading. Each of the two magazines holds nine 155 mm projectiles which, coupled with the use of the extensible loading tray and the hydraulic rammer, enable 18 rounds to be fired in three minutes, that is six rds/min. A further 10 projectiles are carried inside the gun compartment plus 56 fuzes and 28 bagged charges.

Fire-control equipment includes a J2 day panoramic sight, with a magnification of ×4 and a 10° field of view, a J3 day direct telescopic sight with a magnification of ×4 and a 10° field of view, an electric elevation sight which includes a cant angle sensor, an angle indicator and an inverter, and finally a collimator. All fire-control equipment is mounted in front of the commander's position and is operated by the layer.

Variants

105 mm Type 74 SPH

A total of 20 of these were built for the Japanese Ground Self Defence Force with the first deliveries taking place in 1975. It is understood that these systems have now been phased out.

Specifications

	Type 75
Dimensions and weights	
Crew:	6
Length	
overall:	7.79 m
hull:	6.64 m
Width	
overall:	3.09 m
Height	
to turret roof:	2.545 m
Ground clearance	
overall:	0.4 m
Track	
vehicle:	2.5 m

	Type 75
Track width	
normal:	500 mm
Length of track on ground:	5.2 m
Weight	
combat:	25,300 kg
Ground pressure	
standard track:	0.64 kg/cm²
Mobility	
Configuration	
running gear:	tracked
Power-to-weight ratio:	17.78 hp/t
Speed	
max speed:	47 km/h
Range	
main fuel supply:	300 km
Fuel capacity	
main:	650 litres
Fording	
without preparation:	1.3 m
Gradient:	60%
Side slope:	30%
Vertical obstacle	
forwards:	0.7 m
Trench:	2.5 m
Engine	Mitsubishi 6ZF, 6 cylinders, air cooled, diesel, 450 hp at 2,200 rpm
Gearbox	
type:	manual
forward gears:	4
reverse gears:	1
Suspension:	torsion bar
Electrical system	
vehicle:	24 V
Batteries:	4 × 12 V
Firepower	
Armament:	1 × turret mounted 155 mm howitzer
Armament:	1 × roof mounted 12.7 mm (0.50) M2 HB machine gun
Ammunition	
main total:	28
12.7/0.50:	1,000
Turret power control	
type:	hydraulic/manual
by commander:	yes
by gunner:	yes
commander's override:	yes
Main armament traverse	
angle:	360°
speed:	12°/s
Main armament elevation/depression	
armament front:	+65°/-5°
Survivability	
Night vision equipment	
vehicle:	yes
NBC system:	yes
Armour	
hull/body:	steel (unconfirmed)
turret:	steel (unconfirmed)

Status

Production complete. In service with the Japanese Ground Self-Defence Force. There have been no exports of this system.

Contractor

Mitsubishi Heavy Industries (hull and final assembly).
Japan Steel Works/Nihon Seiko Jyo (gun and turret).

Korea (North)

North Korean self-propelled artillery

Development/Description

North Korea appears to have followed in the footsteps of countries such as China which have mated Russian-developed artillery barrels and mounts onto locally designed and built full-tracked hull.

North Korean 170 mm M1989 Chuche SPG, which has been withdrawn from UAE front-line service (Christopher F Foss) 1044226

North Korean self-propelled artillery system 170 mm M1989 which appears to use the same ordnance as the 170 mm M1978 0500878

North Korean 122 mm self-propelled artillery system M1977 0106222

North Korean 122 mm self-propelled artillery system M1981 0106223

The US Defense Intelligence Agency has reported the existence of the weapons in the table.

It is probable that there are other North Korean tracked self-propelled artillery systems in addition to those listed below.

During a parade held in North Korea in 2010, a number of new systems were unveiled including a new turret mounted 152 mm self-propelled artillery system.

Designation	Calibre	Weight	Max range	Comment
M1992 SPM	120 mm	16 t	8,700 m	turret-mounted mortar on VT-323 full tracked hull
M1997 SPH	122 mm	18 t	15,400 m	D-30 towed howitzer on modified AT-S hull
M1985 SPG	122 mm	19 t	20,800 m	A-19 on modified AT-S hull
M1981 SPG	122 mm	19 t	24,000 m	D-74 on modified AT-S hull
M1991 SPG	122 mm	22 t	24,000 m	D-74 on modified AT-S hull
M1975 SPG	130 mm	20 t	27,150 m	M-46 on modified AT-S hull
M1991 SPG	130 mm	22 t	27,150 m	M-46 on modified AT-S hull
M1992 SPG	130 mm	20 t	29,500 m	130 mm SM-4-1 coastal gun on modified AT-S hull
M1974 SPH	152 mm	17 t	17,410 m	152 mm D-20 on modified VT-323 APC hull
M1977 SPH	152 mm	15 t	12,400 m	152 mm D-1 on K-63 hull
M1991 SPH	152 mm	19 t	17,265 m	152 mm ML-20 on modified AT-S hull
M1978 SPG	170 mm	40 t	n/known	170 mm gun on T-55 MBT hull
M1989 SPG	170 mm	40 t	n/known	170 mm gun on modified AT-S hull
122 mm Corps gun M1931/37 (A-19), 152 mm howitzer M1943 (D-1), 152 mm gun-howitzer D-20 (see Towed anti-tank guns and howitzers), 122 mm field gun D-74 (see Towed anti-tank guns and howitzers), 122 mm howitzer D-30 (see Towed anti-tank guns and howitzers), 130 mm field gun M-46 (see Towed anti-tank guns and howitzers), 130 mm SM-4-1 is a Russian developed coastal artillery gun system. It should be noted that the original Russian designed and built AT-S hull is not armoured. Most SP models listed above are armoured.				

In recent years, the Korean People's Army has fielded a number of new self-propelled artillery systems including one similar to the Russian 203 mm 2S7.

The UAE took delivery of a quantity of 170 mm M1989 Chuche SPGs, but these are no longer operational.

In addition to the above self-propelled artillery systems it is understood that North Korea has more recently developed and fielded some additional self-propelled artillery weapons with enhanced capability.

Status
Production of all of these systems is considered to be complete. In service with Korean People's Army. It is estimated that North Korea deploys up to 4,200 self-propelled artillery systems.

North Korea has exported some self-propelled artillery systems to the Middle East.

Contractor
North Korean state factories.

170 mm M1978 self-propelled gun Koksan

Development
This 170 mm self-propelled gun was first observed by US sources in the city of Koksan in 1978 and is therefore referred to as the M1978 self-propelled gun Koksan.

A number of these guns were supplied to Iran by North Korea and some were subsequently captured by Iraq in 1987-88.

In North Korean service, it is believed that the 170 mm Koksan was originally employed in regiments with a total of 36 weapons.

It is possible that the 170 mm M1978 self-propelled gun have now been withdrawn from front line service.

Description
The system is based on a modified China North Industries Corporation (NORINCO) Type 59 MBT hull with the existing welded steel turret armed with a 100 mm gun removed and plated over.

A 170 mm gun of a previously unknown type in an open mount has been mounted on top of the hull at the very rear. The driver of the Koksan is seated at the front of the hull on the left side, under full armour protection. As far as it is known, the hull of the Koksan provides the same level of protection as the Type 59 MBT.

At the very rear of the hull are two large spades that are lowered to the ground manually before firing commences; when travelling these are vertical.

A large ordnance travelling lock is pivot-mounted on the glacis plate and this folds forward onto the ground when deployed in the firing position. This has to be released manually.

The long 170 mm ordnance is fitted with a multislotted muzzle brake but has no fume extractor. The screw-type breech mechanism opens to the right with a loading tray mounted on the left side of the weapon. In addition, there is probably a power rammer.

170 mm M1978 Koksan self-propelled gun with ordnance in travelling lock
0501023

Elevation and traverse appear to be powered, although traverse is very limited to avoid overstressing the hull.

It is possible that the 170 mm weapon used may be a Russian naval or coastal weapon supplied to North Korea in the 1950s for use in the coastal defence role. When these were replaced by surface-to-surface missiles, the coastal guns were surplus to requirements and could then have become available for mounting on the tracked hull.

As there is insufficient space for the gun crew and ammunition on the Koksan, another vehicle is required to support the system in action and this would carry the remainder of the crew and some ready use ammunition. It is thought that the ammunition resupply vehicle is based on a full-tracked hull.

It is probable that standard equipment for the 170 mm M1978 includes an NBC system and infra-red night vision equipment for the driver.

Like the Type 59 MBT on which it is based, the 170 mm M1978 self-propelled gun can lay its smoke screen by injecting diesel fuel into the exhaust outlet on the left side of the hull.

Variant
170 mm M1989 SPG

This is understood to use the same weapon and mount as the M1978 but is mounted on a modified AT-S full-tracked carrier hull. The current status of the M1989 is uncertain.

Specifications

	M1978
Dimensions and weights	
Length	
overall:	14.9 m
Width	
overall:	3.27 m
Height	
overall:	3.1 m
Weight	
standard:	40,000 kg
Mobility	
Configuration	
running gear:	tracked
Engine	diesel, 520 hp
Firepower	
Armament:	1 × 170 mm rifled gun
Survivability	
Night vision equipment	
vehicle:	yes
NBC system:	yes
Armour	
hull/body:	steel

170 mm M1978 Koksan self-propelled gun from the rear showing the two large spades
0500771

Status
Production complete. In service with Iran and North Korea. Recent information has indicated that these systems may no longer be deployed in a fully operational role. They have not been seen in any of the recent North Korean military parades.

It is considered probable that Iran no longer has any of these systems operational.

Contractor
North Korean state factories.

Korea (South)

Samsung Techwin 155 mm/52-calibre K9 Thunder self-propelled artillery system

Development
For many years the mainstay of the Republic Of Korea (ROK) self-propelled field artillery units was the now BAE Systems (previously United Defense) 155 mm/39-calibre M109A2 self-propelled howitzer.

Samsung Techwin Defense Program Division, was the ROK prime contractor for the co-production of 1,040 M109A2 SPH, deliveries of which have been completed.

In the late 1980s, the ROK Army drew up its requirements for a new 155 mm/52-calibre tracked SPH to meet the requirements of the 21st Century.

This included deep supporting fire of the corps, qualitatively superior to overcome the numerical inferiority and effective firing support in mountain areas with its longer range. Agency for Defense Development (ADD) started development of a new SPH in 1989.

Key operational requirements included a higher rate of fire, longer range, more accurate fire, faster into and out of action times and greater mobility, all of which lead to a significant increase in battlefield survivability of the weapon system.

Following a competition, Samsung Techwin was selected to be the prime contractor for the K9 Thunder 155 mm/52-calibre SPH.

First prototypes were completed in 1994 under the designation XK9 and put through an extensive series of Development Test/Operational Test (DT/OT) by the ROK Army and the ADD.

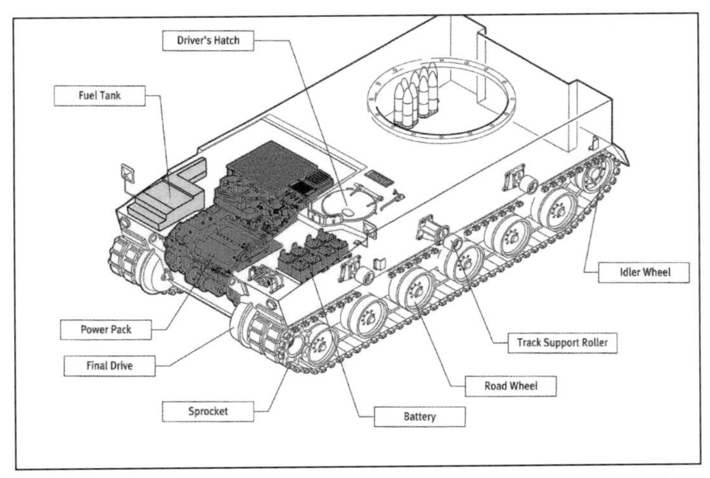

Hull detail of the 155 mm K9 Thunder self-propelled artillery system showing position of main automotive components 0126555

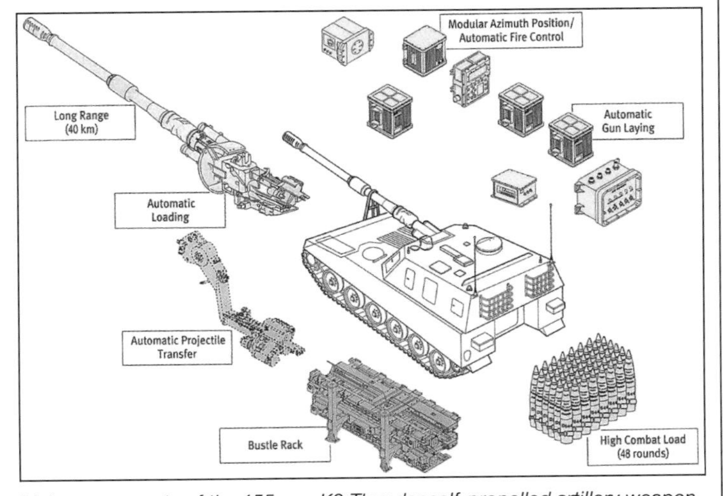

Main components of the 155 mm K9 Thunder self-propelled artillery weapon system showing main ordnance and weapon subsystems 0126556

155 mm/52-calibre K9 Thunder self-propelled artillery system with ordnance in travel lock 0531808

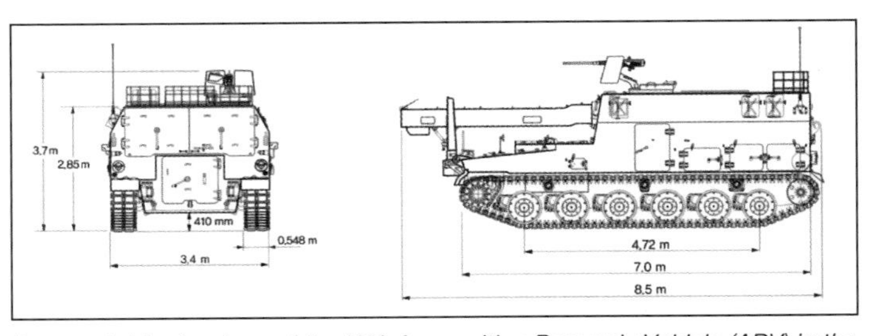

Rear and side drawings of the K10 Ammunition Resupply Vehicle (ARV) in the travelling configuration (Samsung Techwin) 1452902

These were followed by three preproduction vehicles manufactured during the full-scale development phase (Phase II), the last of which was completed in 1998.

The mobility and firing reliability of the K9 Thunder had been verified through the trial of the prototype vehicles in Korea and by the end of 1998 these had travelled 18,000 km and fired 12,000 rounds.

The XK9 Thunder was type classified by the ROK Army in 1998 as the K9, with the first production vehicles being completed earlier in 1999.

No details of production quantities of the 155 mm/52-calibre K9 Thunder self-propelled artillery systems have been released but it is estimated that by early 2013 over 400 units had been produced. Production will continue for some years as in early 2009 MTU stated that it had been awarded a contract for the supply of engine kits and drive components for a total of 428 155 mm/52-calibre Thunder self-propelled artillery systems for delivery over a four-year period.

This brought the total number of engine kits up to 1,206 which includes those for new build vehicles as well as replacement engines. This would indicate that at least 1,000 155 mm/52-calibre K9 Thunder self-propelled artillery systems and their associated K10 Ammunition Resupply Vehicle will be built.

One Thunder battalion has three batteries, each with six weapons. Each battery is normally controlled by one battery fire-direction centre.

The ROK Army was the first army in Asia to deploy a 155 mm/52-calibre SPH and the second in the world.

Late in 2001, it was revealed the Turkish Land Forces Command (TLFC) had signed a contract with the Defense Program Division of Samsung Techwin for the 155 mm K9 self-propelled artillery system.

Under the terms of this deal, South Korea supplied a first batch of K9 systems with progressive licensed production then being undertaken in Turkey.

In 2003 the Turkish Land Forces Command (TLFC) took delivery of the first of its Firtina 155 mm/52-calibre Self-Propelled Guns (SPGs).

Following trials with two prototypes, an order for an undisclosed quantity of systems was placed with the first batch understood to consist of eight units.

Production commenced in 2002 and the total TLFC requirement could be for as many as 300 systems, funding permitting, with each regiment having three batteries each of four systems.

Firtina, or T-155 K/M Obus as it is also referred to, has a combat weight of 47 tonnes and is powered by a MTU 881 series diesel developing 1,000 hp coupled to an Allison X1100-5A3 fully automatic transmission. Maximum road speed is quoted as 65 km/h with a cruising range of 360 to 400 km.

There is a separate entry in *IHS Jane's Land Warfare Platforms: Artillery and Air Defence* for the Turkish 155 mm/52-calibre Firtina self-propelled gun.

This is only in service with the Turkish Land Forces Command and has not been offered on the export market. Additional details are provided later in this entry.

Description
The hull and turret of the K9 are of all-welded steel armour construction with a maximum thickness of 19 mm. This provides the occupants with protection from medium-calibre small arms fire and 155 mm high-explosive projectile fragments.

The K9 has a crew of five; driver and four in the crew compartment, with the latter consisting of commander, gunner, assistant gunner, and ammunition loader.

The driver being seated front left with the power pack to his right and with the remainder of the vehicle being taken up by the turret and crew compartment.

152 mm/52-calibre K9 Thunder self-propelled artillery system in travelling configuration 0085205

The driver enters his compartment by a rear opening roof hatch that can be locked in the horizontal position. The driver has three forward-facing day periscopes, the centre one of which can be replaced by a passive night vision device.

The diesel fuel tank is located at the front right side of the hull with the battery compartment being front left. The air intake and outlet louvres are in the roof, with the exhaust outlet on the right side.

The indirect fire sight is mounted on the left side of the turret roof and is provided with a ballistic cover.

The commander and gunner are seated on the right side of the turret. Mounted on the forward part of the commander's cupola is a .50 (12.7 mm) M2 HB machine gun for air defence/local defence purposes and a single-piece hatch cover is provided which opens to the rear. A door is provided in the left side of the turret and there is also a large door in the hull rear for crew entry and ammunition resupply purposes.

The 155 mm/52-calibre main armament has a chamber volume of 23 litres and is fitted with a slotted muzzle brake and fume extractor.

The large multiple slotted muzzle brake reduces the firing impulse applied to the recoil system and also provides a superior capability of attenuating the blast overpressure and flash.

The vertically sliding breech mechanism automatically opens upward after firing and obturates the propellant gases with the high-strength obturator ring in the barrel assembly.

The primer magazine automatically feeds, inserts and extracts the primers and movement of the breech block allows for the continuous firing capability of 21 rounds.

The gun mount, which consists of two hydraulic recoil brakes with an internal buffer and a pneumatic recuperator, reduces the firing impulse to the hull. The high-rigidity cradle is fitted to the autoloader to the rear, which has enhanced the firing accuracy by minimising the lateral motion of the barrel when firing.

When travelling the 155 m/52-calibre barrel is held in position with a travel lock installed on the forward part of the hull, this is operated by the driver by remote control, so ensuring the NBC integrity of the vehicle.

Mounted over the rear part of the 155 mm/52-calibre ordnance is a muzzle velocity measuring system that feeds information to the onboard computer. This Doppler-type system has a muzzle velocity recording range of from 20 to 2,000 m/s.

A burst rate of fire of three rounds can be fired in 15 seconds, with a maximum rate of fire of six to eight rounds a minute for a period of 3 minutes. Sustained rate of fire is two to three rounds a minute for 1 hour. A thermal warning device is fitted as standard and this feeds the temperature of the barrel to the Automatic Fire-Control System (AFCS).

Firing a standard M107 High-Explosive (HE) projectile a maximum range of 18 km can be achieved. Maximum range firing an M549A1 HE Rocket Assisted Projectile (RAP) with uni-charge of five zones is 30 km. Firing the new locally developed K307 high-explosive base-bleed projectile with a six zone charge, a maximum range of 40 km can be achieved with a chamber pressure of 52,000 psi and a muzzle velocity of 924 m/s. It can also fire the locally developed K310 dual-purpose improved conventional munition base-bleed round to a maximum range of 36 km.

As of early 2013 Samsung Techwin stated that the following 155 mm projectiles and charges could be used by the K9 Thunder:

- M107 high explosive (HE)
- M549A1 high explosive rocket assisted projectile (HE RAP)
- K310 base bleed dual purpose improved conventional munition (BB DP ICM)
- K307 base bleed high explosive (BB HE)
- K315 high explosive rocket assisted projectile (HE RAP)
- Conventional bag charges
- K676 modular charge
- K677 modular charge

Using its onboard computerised fire-control system it can carry out Time On Target (TOT) procedures and put three rounds onto the target at once.

The bustle rack of the turret has been designed to load four types of 155 mm projectile and is fitted with four independently operated electrical drives which can be operated manually in an emergency. A total of 48 projectiles and their associated charges are carried for ready use.

K10 Thunder Ammunition Resupply Vehicle is based on the hull of the K9 Thunder self-propelled artillery system (Samsung Techwin) 1146708

Internally an automatic loading system takes projectiles from the storage position and places them onto the ammunition tray ready for ramming. The projectiles are loaded into the chamber automatically by the flick rammer. The charges are loaded manually. Although it can use a conventional bag type charge system (M3A1, M4A2, M119A1 and M203) the K9 has been optimised for use with a locally developed modular charge system.

The new modular charge system comprises two charges. One is a uni-charge, used for up to five zones (five increments) and another is a module-type charge for zone six only (two increments).

The new K307 projectile is a High-Explosive Extended-Range Full-Bore Base Bleed (HE ER FB BB) type which features an aerodynamic body and base bleed to achieve its enhanced range.

Standard equipment includes a Honeywell Modular Azimuth Position System, Automatic Fire-Control System, powered gun elevation/depression and turret traverse system. The K9 Thunder can open fire within 30 seconds if it is already stationary or within 60 seconds if it is moving.

Information to lay the weapon onto the target can be transmitted from the battery command post to the K9 Thunder via a data digital radio link or voice communications and the system can also calculate laying data using its onboard fire-control equipment.

The computerised fire-control system for the K9 Thunder self-propelled artillery system has been developed by Samsung Thales specifically for this artillery system.

This includes a display unit, system control unit, fire-control unit, communication processor which are all linked via a 1553 databus to the power control unit.

According to Samsung Thales, key features of this computerised fire-control system include automatic/manual data input, common module conception, real time ballistic solution, simple man/machine interface, easy to modify programme through RS422 port, built in test and a high precision ballistic solution.

Suspension system is of the hydropneumatic type, with each side having six dual rubberised roadwheels and with the drive sprocket at the front, idler at the rear and three track-return rollers. Track is of the double-pin type with replaceable rubber pad and designated the GS-2.

According to the prime contractor, the K9 can be towed at a maximum speed of up to 27 km/h and braking distance from a speed of 32 km/h is 12 m.

Subcontractor for the suspension is Tong Myung Heavy Industries who signed a contract with the then Air Log of the UK in 1995 for the supply of an undisclosed quantity of 'Hydrogas' suspension units and licence production of these in South Korea. This is similar to that used on the now BAE Systems (previously RO Defence) 155 mm AS90 self-propelled artillery in service with the Royal Artillery. Air Log is no longer in existence and production is now undertaken by Horstman Defence Limited.

The K9 Thunder is also fitted with an NBC system, heater, internal and external communications system and a manual fire suppression system.

K9 enhancements

The company is already studying a number of enhancements for the K9 Thunder for other export customers in Asia and elsewhere, such as the installation of an air conditioning system.

K10 Ammunition Resupply Vehicle (ARV)

The concept of a fully automated ARV for the ROK Army was conceived during the early stages of development of the K9 Thunder as a firing position partner to intensify the tactical effectiveness of the K9. The practical issue of a fully automated ARV surfaced during the completion of K9 Thunder development in the late 1990s.

In 2001 Samsung Techwin was awarded the full scale development contract for the K10 which was the project name for the ROK ARV.

The ROK Army required a system to substantially increase the amount of ammunition available to the K9 to meet its tactical requirement of higher firing rate, shoot and scoot, a fully automated resupply system to maximise the efficiency of artillery forces. Finally a K9 family vehicle which shares the same chassis, power pack and suspension for maximum efficiency in logistic support and tactical movements.

According to the manufacturer the benefit of using the common system has substantially reduced its development period and the performance of the K10 was completely qualified through development test and evaluation which lasted a year.

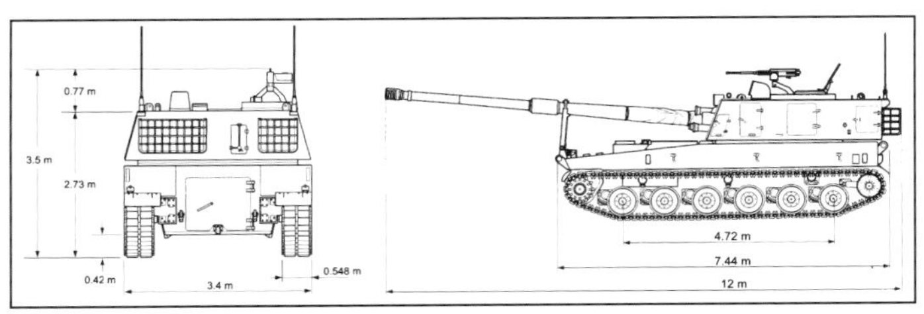

Rear and side drawings of a K9 Thunder self-propelled artillery system in travelling configuration (Samsung Techwin) 1133915

The test included travelling almost 10,000 km with a combination of paved, dirt and off-road conditions for pushing the ammunition handling mechanical systems to endurance limits. In addition it covered operational performance tests for docking with the K9 and other existing 155 mm artillery systems.

K10 is fitted with an electrically driven fully automatic transfer system that consists of magazines for ammunition storage, loading and reloading, manipulators for transferring and a mechanically extendible telescope as well as an automated inventory management system.

This allows any type of ammunition to be transferred at more than 10 rounds a minute. The flow of charges and projectiles is carried out in the same conveyor path while coming from different storage magazines through manipulators and movers.

The armour protected extendible conveyor can be remotely controlled from the driver's seat using a jog switch. This is located at the front of the vehicle to facilitate rapid alignment with the K9 under harsh conditions.

For emergency access to the ammunition, the charge and projectile storage compartment have separate rear access doors for loading or unloading with manual assistance. The ammunition compartment is fitted with a fire detection and suppression system.

As the K10 ARV is based on the same hull, power pack and suspension as the K9 it has the same armour protection level, off-road mobility and survivability.

Development of the K10 ARV is complete and initial production vehicles were deployed in 2006.

Specifications

K10 ARV

Crew:	3
Combat weight:	47,000 kg
Length:	8.43 m
Width:	3.4 m
Height:	3.56 m
Max road speed:	62 km/h
Ammunition stowage:	104 × 155 mm projectiles

K77 Fire Direction Center Vehicle (FDCV)

This is based on a modified M109/M992 hull and uses its onboard Battalion Tactical Computer System (BTCS) in order to calculate firing mission data for K9 Thunder at battery/battalion level.

It has a crew of 10 people when deployed at battalion level and six when deployed at battery level with the complete system weighing 19.8 tonnes.

There have also been proposals for a wheeled (8 × 8) version of the K77 FDCV.

Turkish version of K9

The first export customer for the K9 is Turkey who has selected a modified version known as the TU SpH Storm (or Firtina in Turkey) under a deal signed late in 2001. Details of the Turkish Firtina are provided in a separate entry in *IHS Jane's Land Warfare Platforms: Artillery and Air Defence*. Production of the Firtina is still under way in Turkey.

A Firtina ammunition resupply vehicle has been developed to the prototype stage in Turkey but this is based on a different hull that uses some automotive components of surplus M48/M60 series tank chassis.

Specifications

	K9 Thunder
Dimensions and weights	
Crew:	5
Length	
overall:	12 m
hull:	7.44 m
Width	
overall:	3.40 m
Height	
to turret roof:	2.73 m
to top of roof mounted weapon:	3.5 m
Ground clearance	
overall:	0.41 m
Track width	
normal:	548 mm
Length of track on ground:	4.72 m
Weight	
combat:	46,300 kg
Mobility	
Configuration	
running gear:	tracked
Power-to-weight ratio:	21.60 hp/t
Speed	
max speed:	67 km/h
Acceleration	
from 0 to 32 km/h:	12 s
Range	
main fuel supply:	360 km
Fuel capacity	
main:	850 litres
Fording	
without preparation:	1.5 m
Gradient:	60%
Side slope:	30%
Vertical obstacle	
forwards:	0.75 m
Trench:	2.8 m
Engine	MTU MT 881 Ka-500, water cooled, diesel, 1,000 hp at 2,700 rpm
Gearbox	
model:	Allison Transmission X1100-5A3
type:	automatic
forward gears:	4
reverse gears:	2
Suspension:	hydropneumatic
Batteries:	24 V
Generator:	1 × 28 V
Firepower	
Armament:	1 × turret mounted 155 mm howitzer
Armament:	1 × roof mounted 12.7 mm (0.50) M2 HB machine gun
Ammunition	
main total:	48
12.7/0.50:	500
Main weapon	
calibre length:	52 calibres
Rate of fire	
sustained fire:	3 rds/min
rapid fire:	8 rds/min
Turret power control	
type:	hydraulic/manual
Main armament traverse	
angle:	360°
Main armament elevation/depression	
armament front:	+70°/-2.5°
Survivability	
Night vision equipment	
vehicle:	yes
NBC system:	yes
Armour	
hull/body:	steel
turret:	steel

Status

Production. In service with the Republic of Korea Army. The Turkish version is known as the Firtina and the first production vehicles were delivered in 2004.

Contractor

Samsung Techwin, Defense Program Division.

Netherlands

RDM Technology SPGs

Development

Previous RDM Technology production included the manufacture of parts for Royal Netherlands Army 155 mm M109A2 (supplied by the now BAE Systems (then United Defense) self-propelled howitzers; major assemblies for the Krauss-Maffei Wegmann Leopard 2 MBT for both Dutch and German armies; fitting FMC (now BAE Systems) command and reconnaissance carriers of the Royal Netherlands Army with Oerlikon 25 mm, one-person GBD-A0A turrets; modifying Royal Netherlands Army FMC (BAE Systems) AIFVs to carry the

RDM Technology M109L47 self-propelled artillery systems for the United Arab Emirates 0003990

RDM Technology 155 mm M109A2 in service with Royal Netherlands Army 0106225

Systems and Electronics Inc Improved TOW mount; and production of a large series of AIFVs (together with DAF, Eindhoven) for the Royal Netherlands Army.

For the M109A2 the company manufactured the gun and mount assembly, installing the turret and completing system integration before delivering the complete vehicle.

Description

Canadian M109 upgrade

Late in 1992, RDM Technology received a contract to overhaul 30 M109 series self-propelled howitzers of the Canadian Forces, as well as carrying out a small number of modifications including the installation of an auxiliary power unit. The 155 mm ordnance was retained. This upgrade programme is now complete. Canadian Army M109s have now been withdrawn from front line service.

Royal Netherlands Army M109 upgrade

In early 1991, the Royal Netherlands Army artillery order of battle included 76 M110A2 203 mm self-propelled howitzers, 222 M109 155 mm series self-propelled howitzers, 123 M114 155 mm towed howitzers (of which 82 had been upgraded with the 39-calibre ordnance) and 22 Multiple Launch Rocket Systems.

However, under subsequent plans the artillery strength was then cut to 137 M109 155 mm self-propelled howitzers, 51 155 mm M114/39-calibre towed howitzers and 22 Multiple Launch Rocket Systems. All of the 227 mm (12-round) MLRS of the Royal Netherlands Army have been withdrawn from service and sold to Finland.

It was intended to refurbish all 137 M109 155 mm self-propelled howitzers but late in 1993 it was stated that only 126 would be upgraded, which would be issued to five M109 regiments, each having 24 weapons, with the remaining six to be used as driver training vehicles.

The upgrade work was carried out by RDM Technology under a contract worth USD62.6 million. Of the 126 vehicles upgraded, 88 were M109A2s and the remaining 38 M109A3s.

The upgrade was combined with a programme of basic maintenance and included a better dynamo and voltage regulator, an improved starter motor, improved turret rotation installation and anti-corrosion measures. Modifications were carried out to the aluminium hull to prevent cracks and the 155 mm ammunition stowage racks were replaced. No improvements were carried out to the actual 155 mm weapon. The upgrade has now been completed.

The Royal Netherlands Army 155 mm M109A2/M109A3 self-propelled howitzers have now been replaced by a total of 57 155 mm/52-calibre PzH 2000 self-propelled artillery systems with the first two systems being delivered in 2004.

These are all brand new systems from the Krauss-Maffei Wegmann production line in Germany.

As a result of restructuring, the Royal Netherlands Army no longer has a requirement for all of these 57 PzH 2000 and a number of systems are now surplus to requirements and are being offered for sale.

As of early 2013 the Netherlands had not sold any of its surplus 155 mm PzH 2000 self-propelled artillery systems.

UAE M109 upgrade

In August 1995, Abu Dhabi, one of the states of the United Arab Emirates, purchased 85 155 mm M109A3 self-propelled howitzers that were surplus to the requirements of the Royal Netherlands Army.

These M109A3s were then put through a major upgrade programme by RDM Technology, with the first upgraded systems delivered to Abu Dhabi in 1997.

Early in 1999, the UAE took delivery of the last of the 85 upgraded 155 mm M109L47 self-propelled guns ordered from RDM Technology.

These were upgraded in many areas including the installation of a 155 mm/47-calibre barrel supplied by the now RUAG Defence which is the same as that fitted to the upgraded Panzerhaubitze 88/95 of the Swiss Army.

All of the 85 systems delivered to the UAE have the 155 mm/47-calibre barrel with 60 up to full standard which includes Auxiliary Power Unit (APU), bustle-mounted air conditioning system, new propellant charge racks (NATO and South African), semi-automatic projectile rammer, remote-control travel lock for 155 mm ordnance and revised hydraulic unit.

A total of three UAE artillery regiments are operational with the M109L7 and each of these regiments has three batteries, each with eight guns.

The M109L47 has a maximum range, firing an ERFB-BB projectile, of 38 km.

Diehl Defence Land Systems M109 Repower (Deutz engine)

In 2003 it was stated that the United Arab Emirates (UAE) was to repower its fleet of 85 upgraded 155 mm M109L47 Self-Propelled Guns (SPGs) with a new power pack developed by the German company of Industriewerke Saar (IWS) GmbH which is now known as Diehl Defence Land Systems.

The M109's current power pack consists of a Detroit Diesel 8V-71T 8-cylinder turbocharged diesel developing 405 bhp coupled to an Allison Transmission model XTG-411-4A with four forward and two reverse gears.

In the case of the M109A2, this power pack gives the vehicle a power-to-weight ratio of 16.23 hp/tonne.

Production of the Detroit Diesel 8V-71T engine was completed some years ago and according to Diehl Defence Land Systems the whole power pack is now described as ageing, and there is a question over the long-term support of its subsystems.

According to Diehl Defence Land Systems the new M109 power pack offers a number of significant advantages to the user including long-term support of the engine, greater reliability and reduced life-cycle costs

The new M109L47 power pack consists of the Deutz BF6M 1015 CP 6-cylinder diesel developing 442 hp at 2,100 rpm coupled to the proven Allison XTG-411-4 automatic transmission, which has been modified.

In addition, the power pack is fitted with a new intercooler and radiator developed by Diehl Defence Land Systems/Behr, van drive from Krauss-Maffei Wegmann and a new oil cooler from Rocore. A new Bosch 28 V 140 A generator has also been installed.

The new Diehl Defence Land Systems power packs were assembled in Germany and then installed in the UAE M109L47 by local personnel.

RDM Technology 155 mm/52-calibre M109L52

This is now marketed by the German company Rheinmetall and now uses the complete Rheinmetall 155 mm/52-calibre ordnance of the Krauss-Maffei Wegmann PzH 2000 self-propelled artillery system which is currently in service with Germany, Greece, Italy and the Netherlands.

The M109L52 is being marketed as a modular M109 upgrade that can be tailored to meet the end users specific operational requirements.

Details of the M109L52 are provided in a separate entry in *IHS Jane's Land Warfare Platforms: Artillery and Air Defence* but as of March 2013 this remained at the prototype stage.

Contractor

RDM Technology BV. (This company closed down in 2004.)

Poland

HSW 155 mm/52-calibre Krab self-propelled howitzer

Development

Following an international competition, Poland selected a modified version of the now BAE Systems (previously RO Defence) 155 mm/52-calibre Braveheart self-propelled artillery turret system to meet its future self-propelled artillery requirement.

This 155 mm/52-calibre turret has been installed on a local hull designed and built by OBRUM with HSW being the overall prime contractor for the complete system which is called the Krab.

These two prototype systems were put through an intensive series of trials in Poland.

Although the first two turrets were supplied by the UK, the now BAE Systems has transferred the technical data pack to Poland so that production turrets will be manufactured in Poland.

Qualification trials of the Krab were completed in 2003 and this marked the end of the research and development phase of the programme.

In late 2006 it was stated that the Polish Ministry of National Defence Armament Council had given the go ahead for the procurement of the 155 mm/52-calibre Krab artillery system.

It is expected that up to 48 155 mm/52-calibre Krab systems will be procured which will be issued to four battalions each with 12 systems.

In May 2008 HSW was awarded a contract by the Polish MoD for low rate initial production of the 155 mm/52-calibre Krab self-propelled howitzer.

Under the terms of this PLN223 million (USD101.36 million) contract, two batteries were supplied and tested by 2012.

The two batteries will have a total of eight 155 mm/52-calibre Krab self-propelled howitzers of which six will be brand new and the remaining two the first two units refurbished and upgraded to the latest standard.

In addition there will be three command post vehicles based on the Polish-built 2S1 122 mm self-propelled artillery system chassis, one Honker (4 × 4) all-terrain vehicle and two Jelcz P882 (8 × 8) ammunition resupply/maintenance vehicles.

After trials a further battery of four weapons is expected to be added which will bring the total up to 12 out of a total requirement for 48 weapon systems.

In addition to the above, Poland also has a requirement for a new 155 mm precision artillery projectile with a maximum range of over 70 km.

Description

The hull of the 155 mm/52-calibre Krab self-propelled howitzer is a new design that uses proven parts wherever possible and is of all-welded steel armour construction that provides the occupants with protection from small arms fire and shell splinters.

The driver is seated at the front of the vehicle on the left side and has a single-piece hatch cover in front of which is a single, wide-angle day periscope that can be replaced by a passive periscope for driving at night. Mounted on the forward part of the hull is a gun travel lock for the 155 mm/52-calibre ordnance that is operated by the driver from within the vehicle.

The power pack is located to the right of the driver with the air inlet and outlet louvres in the roof and the exhaust outlet on the right side of the hull so the exhaust gases blow downwards. The diesel fuel tanks are located under the floor of the vehicle.

The suspension is of the torsion bar type with either side consisting of seven dual rubber-tyred road wheels with the drive sprocket at the front and idler at the rear. There are also track return rollers and the upper part of the suspension is covered by a skirt to help reduce dirt.

The turret is positioned at the rear of the vehicle with the crew entering and leaving the vehicle via a large door in the hull rear.

The turret of the HSW 155 mm/52-calibre Krab self-propelled howitzer is almost the same as that installed on the British Army AS90 155 mm/39-calibre system which has also been called the Braveheart. Production of this has been completed by BAE Systems and it us no longer being marketed by the company.

The AS90 is the only conventional tube self-propelled artillery system used by the British Royal Artillery.

The main differences are that it has a locally developed fire-control system, local 12.7 mm machine gun mounted on the left side of the turret roof and a bank of four 81 mm electrically-operated smoke grenades either side of the turret, firing forwards.

According to HSW, the Krab has a burst rate of fire of three rounds in 10 seconds and a maximum rate of fire of 18 rounds in three minutes. A flick rammer is fitted as standard. Minimum range is 4.7 km and maximum range firing an assisted projectile is 40 km. Aiming accuracy in range is 0.50 per cent and in azimuth 0.1 per cent.

The gun commander is provided with a DD-9620 terminal that receives target information from the battery command post. Mounted over the 155 mm/52-calibre ordnance is a muzzle velocity measuring radar that feeds information to the fire-control system. A navigation system is also fitted as standard.

In addition to the internal and external communication system, Krab has a fire-control system that includes a computer to control the turret mechanism, gun layer computer and a ballistic calculation and gun commander computer.

Prototype of the HSW 155 mm/52-calibre Krab self-propelled howitzer during initial firing trials 0096185

155 mm/52-calibre Krab self-propelled howitzer deployed in the firing position (G Holdanowicz) 0530364

There is also a direct fire sight and a laser range-finder mounted coaxial with the 155 mm armament. An auxiliary power unit is provided for the turret that allows all of the Krab systems to be run with the main diesel engine shut down.

Standard equipment on the Krab includes an NBC system, front-mounted dozer blade, fire/explosion detection and suppression system and a laser warning system that is coupled to the two banks of 81 mm electrically operated smoke grenade launchers mounted either side of the turret.

Variants

There are no variants of this system although its chassis could be developed for a wide range of other applications.

Specifications

	Krab
Dimensions and weights	
Crew:	5
Length	
overall:	12.05 m
Width	
overall:	3.58 m
Height	
overall:	3.412 m
to turret roof:	3.13 m
to top of roof mounted weapon:	3.82 m (machine gun)
Ground clearance	
overall:	0.44 m
Weight	
standard:	48,100 kg
combat:	52,140 kg
Ground pressure	
standard track:	0.55 kg/cm^2
Mobility	
Configuration	
running gear:	tracked
Power-to-weight ratio:	16.07 hp/t
Speed	
max speed:	60 km/h
Range	
main fuel supply:	650 km
Fording	
without preparation:	1.0 m
Gradient:	60%
Side slope:	40%
Vertical obstacle	
forwards:	0.8 m
Trench:	2.8 m
Engine	S12-U four-stroke, V-12, turbocharged, water cooled, diesel, 838 hp at 2000 rpm
Gearbox	
model:	transfer box, 2 × gearboxes and coaxial side gears
type:	manual
forward gears:	7
reverse gears:	1
Suspension:	torsion bar
Firepower	
Armament:	1 × turret mounted 155 mm howitzer
Armament:	1 × roof mounted 12.7 mm (0.50) machine gun

	Krab
Armament:	8 × 81 mm smoke grenade launcher (2 × 4)
Ammunition	
main total:	29 (turret, 11 hull)
Main weapon	
calibre length:	52 calibres
Turret power control	
type:	electric/manual
Main armament traverse	
angle:	360°
Main armament elevation/depression	
armament front:	+70°/-3.5°
Survivability	
Night vision equipment	
vehicle:	yes
NBC system:	yes
Armour	
hull/body:	steel
turret:	steel

Status
Low rate initial production for Poland with eight systems to be completed and tested by 2012.

Contractor
Huta Stalowa Wola SA.

Romania

Model 89 122 mm self-propelled howitzer

Development
In the late 1980s, Romania developed a 122 mm full-tracked self-propelled artillery system called the Model 89.

Since then it has been disclosed that the Romanian Army took delivery of about 42 of these systems which are in service alongside six of the Russian equivalent system, the 122 mm 2S1 self-propelled howitzer.

Recent information has indicated that only 18 of these Romanian built Model 89 122 mm self-propelled howitzers remain in service.

Description
The turrets of the two systems are identical and it has been confirmed that the turrets for the Romanian vehicles were supplied by BETA plc of Bulgaria (which also manufactured the Russian MT-LB multirole tracked armoured vehicle), as well as licence production of the 2S1 for the home and export markets for some time.

The hull of the Romanian Model 89 122 mm self-propelled howitzer is of a new design that incorporates parts of the Romanian MLI-84 Infantry Combat Vehicle (ICV), itself a local Romanian equivalent of the older Russian BMP-1.

The turret and hull of the Model 89 are of all-welded steel armour construction that provides protection from small arms fire and shell splinters. The driver is seated at the front of the vehicle on the left side with the diesel power pack to the right and the turret mounted at the very rear. The air inlet and air outlet louvres are in the roof.

There is a single door in the rear of the hull with a firing port and associated vision device.

The turret is armed with a 122 mm 2A31 howitzer which can fire a wide range of separate loading ammunition types, with the standard 122 mm OF-462 high-explosive projectile having a maximum range of 15,300 m; firing an HE rocket-assisted projectile a range of 21,900 m can be achieved. The ordnance has a semi-automatic vertical sliding wedge breech block, a double-baffle muzzle brake and a fume extractor.

Like the 122 mm 2S1, there is a long stowage box on the left side of the turret and carried on the turret rear are slotted covers that cover the forward part of the suspension when afloat to improve water flow.

Romanian 122 mm Model 89 self-propelled howitzer with 122 mm ordnance in travel lock 0500424

Romanian 122 mm Model 89 self-propelled howitzer showing the water-jets at the rear 0500879

The torsion bar suspension either side consists of seven single rubber-tyred road wheels with the drive sprocket at the front, idler at the rear and three track-return rollers.

The major difference between the two systems is the improved amphibious capabilities of the Romanian vehicle, as it is propelled in the water by two water-jets mounted one either side of the entry door in the hull rear. Steering is accomplished by opening and closing the water-jets.

As the Romanian system has a more powerful diesel engine its power-to-weight ratio is also superior to the Russian system.

Romanian Model 89 system is distinguishable from the Russian 2S1 system as the former has seven road wheels and three track-return rollers whereas the 2S1 has seven road wheels and no track-return rollers.

Standard equipment on production Model 89 systems includes an NBC protection system of the overpressure type, infra-red night vision equipment for commander and driver, an R-123 radio and an R-124 intercom system.

Specifications

	Model 89
Dimensions and weights	
Crew:	5
Length	
overall:	7.305 m
Width	
overall:	3.15 m
Height	
overall:	2.665 m
Weight	
combat:	17,500 kg
Mobility	
Configuration	
running gear:	tracked
Power-to-weight ratio:	20.50 hp/t
Speed	
max speed:	65 km/h
water:	7 km/h
Range	
main fuel supply:	500 km
Fuel capacity	
main:	600 litres
Amphibious:	yes
Gradient:	60%
Side slope:	40%
Vertical obstacle	
forwards:	0.80 m
Trench:	2.50 m
Engine	Model 1240 V8DTS, V-8, supercharged, diesel, 360 hp
Steering:	clutch and brake
Suspension:	torsion bar
Firepower	
Armament:	1 × turret mounted 122 mm 2A31 howitzer
Turret power control	
type:	electric/manual
Main armament traverse	
angle:	360°
Main armament elevation/depression	
armament front:	+70°/-3°
Survivability	
Night vision equipment	
vehicle:	yes
NBC system:	yes
Armour	
hull/body:	steel
turret:	steel

Status
Production complete. In service with Romania who had a total of 42 systems in service, plus six of the Russian 122 mm 2S1. There are no known exports of this system. As mentioned in Development, it is understood that only 18 of these now remain in service with Romania.

Contractor
Romanian state factories (chassis and vehicle integration).
Bulgarian state factories (turret).

Russian Federation

203 mm self-propelled gun M1975 (SO-203) (2S7)

Development
In December 1967, the requirement for a new large-calibre longer-range self-propelled artillery weapon for the Russian Army was issued although at this time neither the type of mounting nor the-calibre of the weapon was specified. The key requirement was for a maximum range of 25 km.

Various calibres were studied including 180 mm and 210 mm but, in 1969, 203 mm was selected. Late in 1969, the Leningrad Kirov Factory proposed the hull of the T-64 MBT with a 203 mm gun in an open mount while the Barrikady Factory proposed a model using the Obiekt 429 hull with the 203 mm gun also in an open mount.

The Pion (Peony) chassis was developed by the Kirov Factory under the designation of the Obiekt 216, with the gun and mount being developed by the Barrikady plant.

Following the usual trials and modifications with prototype systems, it was accepted for service as the 203 mm 2S7 Pion (Peony) although it is also referred to as the Malka or the SO-203.

In the West it has been referred to as the M1975 as this is the year when the system was first observed.

The former Soviet ground forces began receiving the 203 mm Model 1975 (M1975) self-propelled gun during 1975. It is estimated that well over 1,000 of these systems were built, although the Russian Federation has not released any production figures.

It is believed that a 2S7 artillery regiment has a total of 24 weapons in three batteries, each with eight weapons. Each battery could be split into two troops, each of four weapons.

Production was completed some time ago and it is no longer being marketed. In 1999 Russia was offering an upgrade to extend the range of the 203 mm projectile. According to the United Nations Arms Transfer Lists, the following quantities of 203 mm 2S7 self-propelled gun were exported between 1994 and 2010.

From	To	Quantity	Comment
Belarus	Azerbaijan	9	delivered in 2009
Belarus	Azerbaijan	3	delivered in 2008
Ukraine	Georgia	5	delivered in 2007

It is assumed that all of these were used weapons as there has been no production of this weapon system for many years.

Description
The hull of the 2S7 is of all-welded steel armour construction that provides the occupants with protection from small arms fire and shell splinters. It is divided into four compartments: driver's, power pack, crew and rear compartment.

It is estimated that the steel armour of the 2S7 has a maximum thickness of 10 mm

When travelling, the commander, gunner and driver/mechanic are seated in the driver's compartment at the front of the vehicle, which is also provided with space to carry a Russian Strela-2 man-portable SAM system.

The commander and driver are each provided with a circular roof hatch and in front of these are day periscopes for forward observation. To their immediate front is a windscreen that, when in a combat area, is covered by an armoured shutter hinged at the top. In addition, there is a single forward-facing day periscope in the forward part of the roof between the commander and driver.

In the left side of the cab towards the rear is an upward-opening hatch cover while in the right side there is a rearward-opening hatch. It is probable that both of these give access to subsystems of the 2S7 rather than the crew compartment.

Mounted under the nose of the vehicle is a dozer blade that can be used to clear obstacles or to provide a firing pit for the weapon.

The Model V-46-I V-12 liquid-cooled diesel engine is located to the rear of the crew compartment cab and is coupled to a manual transmission with eight speeds. The exhaust outlets are either side of the hull top.

203 mm M1975 (SO-203) (2S7) self-propelled gun in travelling configuration (Richard Stickland) 1044225

203 mm M1975 (SO-203) (2S7) self-propelled gun in travelling configuration, from the rear with spade raised (Richard Stickland) 0003993

To the rear of the diesel engine is another crew compartment for the remaining four crew members who enter the vehicle via two circular roof hatches.

On top of the roof compartment is the antenna base for the R-123M radio system, with the later 2S7M probably having the more recent R-173 radio system. A 1V116 crew communication system is fitted as standard.

Along either side of the hull is a series of large stowage boxes with the 203 mm gun, which is not fitted with a fume extractor or a muzzle brake, being mounted at the rear.

The 203 mm ordnance is designated the 2A44 and consists of a composite tube, casing, mount, breech ring and sleeve. The breech mechanism is of the two-cycle screw type with the breech block being opened to the right automatically with the aid of a mechanical drive. In an emergency it can be operated by hand.

The 203 mm barrel has an overall length of 56.2 calibres and weighs a total of 7,800 kg and has a barrel life of about 450 rounds. This does depend on the charge being used. It is not fitted with a fume extractor or muzzle brake.

Recoil braking is hydraulic with recuperation being pneumatic. Total recoil length is a maximum of 1.4 m and, to provide a more stable firing platform, the rear idler wheels are locked, as are the hydraulic shock-absorbers in the suspension.

Mounted at the very rear of the 2S7 is a large recoil spade which, when lowered to the ground, provides stability at all angles and charges. If required, the 2S7 can fire at reduced charges and at low angles of elevation while still in the travel mode and without the spade lowered to the ground.

When travelling, the 203 mm ordnance is held in position by a manually operated travel lock mounted on top of the cab. Elevation is via a sector-type drive, with traverse being accomplished by a screw gear drive. Gun elevation, traverse, loading and operation of the spade are all hydraulic with manual controls for emergency use.

The gun operator is seated at the rear of the vehicle on the left side and for the engagement of targets has a standard PG-1M panoramic day telescope that is used in conjunction with the K-1 collimator. For engaging targets in the direct fire mode he uses the day OP4M-87 telescope.

Ammunition is of the separate loading type, projectile and charge, with a maximum muzzle velocity of 960 m/s. Maximum range, using unassisted ammunition, is 37.5 km.

The standard 203 mm HE round is designated the ZOF 43 and weighs 110 kg, with a total of four projectiles and charges being carried on the 2S7 for immediate use. The remainder of the ammunition load is carried by another vehicle, usually a truck. In addition, there is a rocket-assisted high-explosive projectile that weighs 103 kg and has a maximum range of 47,500 m. Although this has a longer range, it has less high-explosive content.

Full details of the separate loading ammunition are shown in the table below.

It is understood that other natures of ammunition were developed including concrete-piercing, tactical nuclear and chemical but the last two are understood to be no longer in front-line service.

Type	Projectile weight	Filler weight	Charge weight	Muzzle velocity	Max range
HE-FRAG	110 kg	17.8 kg	44 kg	960 m/s	37,500 m
RAP	103 kg	13.8 kg	44 kg	n/avail	47,500 m

Ammunition is brought up to the 2S7 using a two-wheeled hand cart which consists of a frame with two wheels and a carrier. The carrier separates from the frame when loading the projectile from the ground and carries the latter up to the rammer. If the hand cart is not used then an additional six men are required to move the ammunition.

The 2S7 is provided with an ammunition handling system that enables a rate of fire of 2 rds/min to be achieved, this being operated by the loader.

Typical rates of fire can be summarised as: eight rounds in five minutes; 15 rounds in 10 minutes; 24 rounds in 20 minutes; 30 rounds in 30 minutes; 40 rounds in 60 minutes.

Completely combustible bag charges are used in the 2S7 and, as in the case of other Russian weapons, the choice of charges is regulated and these range from the largest fixed charge, weighing almost 45 kg, to the increments, which weigh from 11.8 to 24.9 kg.

The torsion bar suspension has seven dual rubber-tyred road wheels either side, with the drive sprocket at the front and the idler at the rear, and six track-return rollers that support the inside of the track only. Shock-absorbers are provided for the first, second, third, sixth and seventh roadwheel stations. When deployed in the firing position, the hull is lowered at the rear to help provide a more stable firing platform.

The 2S7 takes five to six minutes to come into action and three to five minutes to come out of action.

Standard equipment includes an NBC system of the overpressure type, heater and night vision equipment. A 24 hp 9R4-6U2 auxiliary power unit is provided for the 2S7 so the main turbocharged diesel engine does not have to be kept running as a power supply during firing.

Variants

The final production model of the 2S7 version was designated the 2S7M which has more recent R-173 communications equipment and can carry a total of eight 203 mm projectiles and charges. In addition, rate of fire is increased from 1.5 to 2.5 rds/min, durability of the system has been improved and firing data can be transmitted directly to the gun.

Specifications

	2S7 (M1975)
Dimensions and weights	
Crew:	7
Length	
overall:	13.12 m
Width	
overall:	3.38 m
Height	
overall:	3.0 m
Ground clearance	
overall:	0.4 m
Weight	
combat:	46,500 kg
Ground pressure	
standard track:	0.8 kg/cm²
Mobility	
Configuration	
running gear:	tracked
Power-to-weight ratio:	18.06 hp/t
Speed	
max speed:	50 km/h
Range	
main fuel supply:	650 km
Fording	
without preparation:	1.2 m
Gradient:	40%
Side slope:	20%
Vertical obstacle	
forwards:	0.7 m
Trench:	2.5 m (est.)
Engine	Model V-46-I, V-12, turbocharged, diesel, 840 hp
Firepower	
Armament:	1 × hull mounted 203.2 mm 2A44 howitzer
Ammunition	
main total:	4
Turret power control	
type:	powered/manual
Main armament traverse	
angle:	30° (15° left/15° right)
Main armament elevation/depression	
armament front:	+60°/0°
Survivability	
Night vision equipment	
vehicle:	yes
NBC system:	yes
Armour	
hull/body:	steel 10 mm (est.)

Status

Production complete. In service with:

Country	Quantity
Angola	18
Azerbaijan	12 (delivered 2008/2009 from Belarus)
Belarus	36
Georgia	6
Russia	150
Ukraine	99
Uzbekistan	48

Contractor

Russian state factories.

152 mm/2S35 (twin) Koalitsiya-SV (Coalition) self-propelled artillery system

Development

Russia developed to the technology demonstrator/prototype stage a new twin 152 mm Self-Propelled (SP) artillery system called Koalitsiya-SV (Coalition). This has the industrial designation of the 2S35.

As far as it is known, Russia has not carried out any marketing of this system with main marketing emphasis being on the 152 mm 2S19 (MSTA-S).

This is covered in detail in a separate entry in *IHS Jane's Land Warfare Platforms: Artillery and Air Defence* which is also now being marketed in a 155 mm configuration for the export market.

In 2010 Russia cancelled a large number of Russian Army programmes including the 152 mm (twin) Koalitsiya-SV (Coalition) self-propelled artillery system

There is a possibility that this programme could be restarted in the future. This complete turret system was also being offered for naval applications.

As of March 2013, the latest self-propelled artillery system to enter service with the Russian Army was the 152 mm 2S19 (MSTA).

Description

For trials purposes the first turret is based on a modified Main Battle Tank (MBT) hull but production systems of the new twin 152 mm SP system could be integrated onto a brand new hull.

This would have a crew of two who would be seated in a well protected compartment at the front with the unmanned turret in the centre and the power pack at the rear.

For the first example, the large flat sided turret appears to be based on that of the older 152 mm 2S19 (MSTA-S) SP artillery system which was fitted with a single 152 mm ordnance and entered service with the Russian Army in 1989.

This is armed with a 152 mm weapon fed by an automatic loader which first loads the projectile and then the charge. A total of 50 × 152 mm projectiles and associated charges are carried with the complete system having a crew of five. The driver is in the front of the hull with other four crew members in the turret.

It is considered probable that production Koalitsiya-SV systems would have a new turret that would extend well to the rear to accommodate the 152 mm projectiles which are loaded under remote-control. The charges would be located vertically below the turret and also loaded automatically.

Main armament consists of two long barrelled 152 mm barrels that are each fitted with a pepperpot muzzle brake. No thermal sleeves or fume extractors appear to be fitted on the prototype system but this could change on production systems. When travelling the two barrels are held in a travel lock mounted on the glacis plate.

Russian twin 152 mm Koalitsiya-SV (Coalition) self-propelled artillery system in travelling configuration with weapons in travel lock (INA) 1296083

A muzzle velocity radar does not appear to be fitted at present but again this could be fitted in the future. This would feed information into the on board computerised fire-control system and allow for more accurate target engagements to take place. It has been confirmed that Russia has developed a muzzle velocity radar that could be integrated onto this self-propelled artillery system.

Mounted on the roof is a 12.7 mm machine gun for air defence and self-protection purposes and banks of 81 mm electrically operated smoke grenade launchers are mounted either side of the turret and cover the frontal arc.

It is expected that production Koalitsiya-SV systems would carry a total of 50 projectiles and charges with the system being reloaded via a dedicated armoured transporter loader vehicle.

It is considered probable that Multiple Round Simultaneous Impact (MRSI) target engagements could take place.

When compared to the current 2S19 the new Koalitsiya-SV twin 152 mm SP artillery system has a much longer range and a higher rate of fire.

The prototype system has six road wheels either side but production systems are expected to be based on a brand new full-tracked chassis with seven roadwheels either side, idler at the front, driver sprocket at the rear and track return rollers.

While some countries are now developing or field SP artillery systems for use by rapid deployment forces, systems such as the Russian Koalitsiya-SV have a high rate of fire, large ammunition supply, high cross-country mobility and are in some cases more survivable.

Variants

Single barrel Koalitsiya-SV

Some sources have indicated that Russia is developing a single barrel version of the twin 152 mm Koalitsiya-SV self-propelled artillery system.

Specifications

Not available.

Status

Programme cancelled in 2010.

Contractor

Not known.

152 mm self-propelled artillery system 2S19 (MSTA-S)

Development

The 152 mm 2S19 self-propelled artillery system was developed under the codename MSTA-S from 1985 at the Uraltransmash facility, as the replacement for the older 152 mm 2S3 and 152 mm 2S5 self-propelled artillery systems.

It is understood that whilst under development the 2S19 had the project name of the Ferma with the development designation being the Object 319 (or Obiekt 319 in Russian).

Production of the 152 mm 2S3 and 152 mm 2S5 is complete and these are covered in detail in a separate entry in *IHS Jane's Land Warfare Platforms: Artillery and Air Defence*.

Surplus examples of both of these systems have been offered on the export market by a number of countries.

Following extensive trials, the 2S19 was accepted for service with the former Soviet Army in 1989. Initial production was undertaken by the Uraltransmash facility before being transferred to the Sterlitamak Machine Construction Factory in Bashkiriya.

First production systems were completed in 1989. In the Russian Army the 2S19 is normally deployed in batteries of six guns. Each regiment would normally have three batteries to give a total strength of 18 2S19 systems.

152 mm 2S19 self-propelled gun deployed in firing position with weapon elevated and loading chute at rear deployed 0058380

Close up of the rear of the 152 mm 2S19 self-propelled gun showing unditching beam on rear of hull, snorkel on right side of turret and ammunition resupply chute in stowed position for travelling (Christopher F Foss) 1452903

The 2S19 is also referred to as the MSTA-S with the S standing for *Samokhodnyj*, or self-propelled. The 2S19 uses the same ordnance as the 152 mm 2A65 towed gun-howitzer which is referred to as the MSTA-B with the B standing for *Buksiruemyj*, or towed.

It is estimated that by early 2013 total production of the 2S19 amounted to well over 600 systems.

In the Russian Army service, this system is also sometimes called the Ferma (Farm) because of the large side of the turret.

Description

The hull of the 2S19 is based on MBT assemblies, with the suspension and running gear from the T-80 MBT and the diesel power pack from the T-72 MBT.

The hull and turret is of all-welded steel armour construction which provides protection from small arms fire, shell splinters and mines.

The 2S19 has the driver's compartment at the front, turret in the centre and power pack located at the rear.

The driver has three day periscopes for forward observation, the centre one of which can be replaced by a passive night vision periscope, and a single-piece hatch cover that lifts and opens to the right.

On the glacis plate is a V-shaped splashboard to stop water rushing up the glacis plate and into the driver's compartment when the vehicle is fording.

Mounted under the nose of the vehicle, similar to the T-72/T-80 MBTs, is a dozer blade that can be used to prepare fire positions or clear obstacles without special engineer support.

The suspension is of the torsion bar type with either side having six dual rubber tyred roadwheels, idler at the front, drive sprocket at the rear and five track-return rollers, with the upper part of the track being covered by a skirt. The roadwheels, idler wheels and drive sprocket are the same as those of the T-80 MBT.

The first three road wheel stations either side are locked when the 152 mm 2A64 main armament is fired.

The torsion bar suspension is from the T-80, the first, second and sixth roadwheels have a telescopic shock-absorber and these are locked when the weapon is fired. The track is the same as that used on the T-80 and 2S7.

The gunner is seated in the left side of the turret and operates the roof-mounted panoramic sight and the direct aiming sight, which is mounted in the front left side of the turret.

The vehicle commander is seated on the right and operates the roof-mounted 12.7 mm NSVT machine gun (which can be remotely controlled) and the searchlight (which can be remotely controlled from under full armour protection).

The 12.7 mm NSVT machine gun can be used to engage both ground and air targets. Mounted externally on the right side of the turret are boxes of 12.7 mm NSVT machine gun ammunition for ready use. This weapon can be elevated to +70° with a traverse of 9° left and 255° right.

There are three roof hatches, one for each member of the turret crew.

Main armament of the 2S19 comprises a long-barrelled 152 mm gun, the 2A64, fitted with a fume extractor and a muzzle brake. When the 2S19 is travelling, the ordnance is held in position by a travelling lock mounted on the glacis plate.

The 152 mm 2A64 main armament of the 2S19 is manufactured by the Barrikady State Production Association.

The 152 mm 2A64 gun fires an High-Explosive fragmentation (HE-FRAG) round designated OF-45 to a maximum range of 24.7 km. Other types of projectile include OF-61 base bleed high-explosive with a maximum range of 28.90 km, 30-23 cargo containing 42 HEAT bomblets, 3NS30 jammer projectile with a maximum range of 20 km, smoke, and the Krasnopol laser-guided projectile covered later.

Empty cartridge cases are automatically ejected forwards out of the turret just below the 152 mm ordnance.

Turret traverse is powered through a full 360° with weapon elevation from –3° to +68°. The electric/mechanical turret control system is designated the 2Eh46. Manual elevation and traverse controls are also provided.

Russian 152 mm 2S19 self-propelled artillery system fitted with snorkel for deep fording operations. When not required this is stowed on the rear right side of the turret (INA) 1452905

When using onboard ammunition, the 2S19 has a rate of fire of 7 to 8 rds/min. The 152 mm fuzed projectile is loaded automatically and the charge semi-automatically.

When using ammunition from an external source this is reduced to 5 to 6 rds/min. When using ammunition on the ground, an extra two crew members are required to load the ammunition onto the conveyor, one for the projectile and one for the charge.

Additional details of the ammunition are given in the table.

The 2S19 can also fire all of the ammunition natures used by the older 2S3 (self-propelled) and D-20 (towed) artillery systems.

In addition to selecting the projectile from the magazine, the automatic loading mechanism controls the number of rounds to be fired and has built-in test equipment. Laying in elevation is also automatic with laying in traverse being semi-automatic.

In case of a complete power failure there is a manual ammunition reloading back up system.

In total, 50 155 mm projectiles and charges are carried. Spades are not required by the 2S19 as it is very stable in the firing position.

While the system is firing, the 2S19 can be reloaded by transporters through the turret rear with the help of special mechanisms feeding from the ground. This enables the system to move off to its next fire position with a full load of ammunition on board.

The 152 mm 2S19 is fitted with a 1P22 indirect sight with a magnification of ×3.7 which is stabilised in the horizontal plane and a 1P23 direct fire sight.

Three standard Russian Type 902 81 mm electrically operated smoke grenade dischargers are mounted either side of the turret.

A 16 kW autonomous AP-18D gas-turbine Auxiliary Power Unit (APU) is provided in the turret and this provides power within 30 to 60 seconds of being switched on. This allows the system to be fully operational at temperatures ranging from –50 to +50°C with the main diesel engine being switched off to conserve fuel supplies. The APU can be run for a maximum of eight hours.

Standard equipment for the 2S19 includes the 1V116 intercom system, R-173 radio set and PPO three-way automatic fire extinguishing system, passive night vision equipment for the driver, two filter and ventilation systems and the ability to lay a smoke screen by injecting diesel fuel into the exhaust outlet on the left side of the hull.

Alternative engines

The 2S19 has been manufactured with two different engines. These are the V-12 model V-84A multfuel developing 780 hp which gives a power to weight ratio of 18.57 h/tonne.

152 mm 2S19M

In mid-2000 it was confirmed that two new versions of the 152 mm 2S19 (MSTA-S) self-propelled artillery system had been developed.

The first version, the 152 mm 2S19M, has been developed to meet the requirements of the Russian Army and has the same physical characteristics and is armed with the same 152 mm/47-calibre ordnance as the original 2S19.

The most significant feature of the 2S19M is the installation of the ASUNO Automatic Gun Fire-Control System (AGFCS). This features an onboard ballistic computer which, used in conjunction with the onboard satellite navigation system and automatic gun laying systems gives a faster reaction time and increase in target accuracy.

Data can be stored for up to 10 fire missions. The system is ready for action within three minutes of coming to a halt and can open fire 30 seconds after receipt of the target co-ordinates.

Russian MSTA-S upgrade

In September 2008 it was stated that work was underway to develop an enhanced version of the 2S19 (MSTA-S) for the Russian Army.

Following company trials the upgraded MSTA-S underwent acceptance trials with the Russian Army prior to entering service. The new version of the 2S19 is said to feature a new automated guidance and fire-control system.

The upgrade will allow the fire mission to be fully automated with automatic selection of the required projectile and charge and presumably fuze setting.

At this stage it has not been confirmed as to whether the Russian Army will upgrade in service 2S19 weapons to the new standard, build new weapons to this configuration or go for a mix with the long term aim to have a common build standard.

The existing turret is armed with a 152 mm 2A64 ordnance fed by an automatic loader which first loads the 152 mm projectile and then the charge system is also being upgraded.

As previously stated the original production 2S19 was followed by the 2S19M which is fitted with the ASUNO Automatic Gun Fire-Control System (AGFCS) which is claimed to reduced into action time as well as improving accuracy.

More recently the 152 mm 2S19 fitted with the ASUNO Automatic Firing and Laying Control System has been referred to by the Russians as the upgraded 152 mm self-propelled howitzer 2S19M1.

More recently Russia has also been marketing an Automated Fire-Control System which can be integrated into now only the 2S19 but also other artillery systems including the older 122 mm 2S1 and 152 mm 2S3 SP systems as well as multiple rocket launchers such as the 122 mm BM-21. This may be related to the AGFCS installed in the 2S19M.

155 mm/52-calibre 2S19M1

As an increasing number of eastern countries are now moving away from 152 mm to 155 mm artillery weapons, since 1995 Uraltransmash has been working on 155 mm artillery systems, both towed and self-propelled.

The 2S19M1 has the same ASUNO AGFCS as the 2S19M but is fitted with a NATO standard 155 mm/52-calibre ordnance for which a total of 46 projectiles and charges are carried.

Maximum quoted range, firing a NATO 155 mm L15A1 high-explosive projectile is quoted as 30 km while firing an Extended Range Full Bore - Base Bleed (ERFB-BB) projectile range is 41 km. Firing a US-developed 155 mm M107 high-explosive projectile, a maximum range of 22.6 km can be achieved, while firing Russian 155 mm high-explosive projectiles, a range of 30 km can be achieved.

More advanced, smart projectiles such as the French/Swedish 155 mm BONUS top attack projectile can also be fired, according to the Russians.

Burst rate of fire is three rounds in 15 seconds and seven to eight rounds intense. Sustained rate of fire is being quoted as four rounds per minute.

According to Russian sources, their future 155 mm artillery systems, towed and self-propelled, could have chromed barrels for longer life and advanced cooling systems for enhanced rates of fire.

In addition to ASUNO AGFCS already mentioned, the system can also be used in conjunction with the Falstet-M automated battery/battalion fire-control system.

This includes a number of sub-systems including battery executive officer control vehicle, Kredo-1E portable ground reconnaissance radar, Zoopark-1 reconnaissance and fire-control complex mounted on ACRV hull, battalion commander control vehicle, battery commander control vehicle and an observation helicopter.

Combat weight of this system is being quoted as 43 to 44 tonnes with an overall length of 13.44 m, width of 3.38 m and height including roof mounted 12.7 mm machine gun of 3.35 m.

The 155 mm/52-calibre ordnance can be elevated from –4° to +70° with turret traverse through a 360°.

Projectile designation	Projectile weight	Filler weight	Muzzle velocity	Maximum range
OF-45, HE	43.56 kg	7.65 kg	810 m/s	24,700 m
OF-61, HE-RAP	42.86 kg	7.80 kg	828 m/s	29,060 m
30-23, cargo	42.80 kg	-	-	29,000 m
3NS30, jammer	43.56 kg	-	-	20,000 m
3OF39, LGP Krasnopol	50.00 kg	-	-	20,000 m

Russian 152 mm 2S19 self-propelled artillery system with turret traversed partly right and showing 152 mm ordnance fitted with muzzle brake and fume extractor (Christopher F Foss) 1452904

It is powered by a V-12 diesel engine developing 1,000 hp coupled to a manual transmission with seven forward and one reverse gears.

This gives a power-to-weight ratio of 23.3 hp/tonne and a maximum road speed of 60 km/h and an operating range of up to 450 km on roads.

More recently the 155 mm/52-calibre 2S19M1 has been referred to by the Russians as the upgraded 155 mm MSTA-S self-propelled howitzer 2S19M1-155.

Successful firing trials of the 155 mm/52-calibre 2S19M1 have been carried out, but as of early 2013 there were no known sales of the system and production had yet commence.

152 mm MSTA-K

The 152 mm MSTA-K (Kolyosnyi–wheeled) consists of a modified MAZ (8 × 8) truck on the rear of which is a modified turret with the ordnance of the 2S19. This has so far only reached the prototype stage.

152 mm 2S30 Iset

This is understood to be an improved version of the 2S19 but remains at the prototype stage.

152 mm 2S33 MSTA-SM

This is understood to be an improved version of the 2S19 but remains at the prototype stage.

Krasnopol guided projectile

Designated the 9K25, the 152 mm Krasnopol laser-guided projectile can be fired by the 152 mm 2S19 and 2S3M self-propelled artillery systems as well as the towed 152 mm 2A36, 2A65 and D-20 systems.

During extensive firepower and mobility demonstrations in the Middle East in 1993, the 2S19 fired a total of 40 Krasnopol laser-guided projectiles at targets up to 15 km away. Of these, 38 hit the target.

When the Krasnopol is in flight, four fixed fins unfold to the rear with four movable control surfaces unfolding to the front. It uses inertial guidance for the middle part of the trajectory with semi-active laser homing being used for the terminal phase when the target is illuminated by the tripod-mounted laser designator.

The tripod-mounted laser designator system is called the 1D22 and can be used to designate all of the laser-guided artillery weapons in the Russian inventory and under development.

Late in 1999 it was reported that China had started licensed production of the 152 and 155 mm versions of Krasnopol. It was also stated that India has taken delivery of 1,000 155 mm Krasnopol-M projectiles and 10 laser guidance systems from the KBP Instrument Design Bureau.

Details of the Chinese 155 mm laser-guided projectile, which is currently being offered on the export market by China North Industries Corporation (NORINCO) are provided in the entry for the PLZ45 155 mm 45-calibre self-propelled gun-howitzer covered in detail in a separate entry in *IHS Jane's Land Warfare Platforms: Artillery and Air Defence*.

The complete NORINCO 155 mm laser-guided artillery system is referred to as the GP1 155 mm Laser-Homing Artillery Weapon System.

NORINCO is now marketing another 155 mm laser-guided projectile which is called the GP6 155 mm laser homing artillery weapon system.

This is fitted with an high-explosive warhead and is stated to have a minimum firing range of 6 km and a maximum firing range of 25 km with a first round hit probability of 90 per cent.

Automated Fire-Control System

Russia is now marketing a new Automated Fire-Control System (AFCS) that can be integrated into self-propelled artillery systems such as the full-tracked 152 mm 2S19, 152 mm 2S3 and 122 mm 2S1 as well as multiple rocket launchers such as the 122 mm BM-21 multiple rocket launcher.

The heart of the AFCS is a central computer that receives/sends information from a variety of sources including a gunners indicator display, commander's automated combat station, loaders indicator, digital elevation sensor, Grot-V transportable receiver – indicator, mechanical velocity sensor and a self-orientating system of gyro course and roll indicator.

This AFCS may be related to the ASUNO Automatic Gun Fire-Control System installed in the previously mentioned 155 mm 2S19M developed for the export market.

Basis specifications of the AFCS when installed into a full tracked self-propelled artillery system have been summarised as follows:

Guidance accuracy, deflection angle	0.5
Initial direction accuracy, deflection angle	2
Grid angle storage error, deg/h	0.1
Combat vehicle coordinates detection error with relation to distance covered	0.5% (max)
Firing and launch data generation time	1 s (max)

Chinese 152 mm self-propelled gun

In 2006 it was revealed that China had developed, to the prototype stage, a new 152 mm/52-calibre self-propelled artillery.

This made its first public appearance in Beijing in October 2009 and is designated the 155 mm self-propelled gun-howitzer PLZ-05 and is understood to have been operational with the Peoples Liberation Army for several years.

This is now being offered on the export market by NORINCO under the designation of the PLZ52 and details are provided in a separate entry in *IHS Jane's Land Warfare Platforms: Artillery and Air Defence*. As of March 2013 there are no known export sales of this system.

Specifications

	152 mm Self-Propelled Artillery System 2S19 (MSTA-S)
Dimensions and weights	
Crew:	5
Length	
overall:	11.917 m
hull:	6.04 m
Width	
overall:	3.584 m
without skirts:	3.38 m
Height	
to turret roof:	2.985 m
Ground clearance	
overall:	0.435 m
Track	
vehicle:	2.80 m
Track width	
normal:	580 mm
Length of track on ground:	4.704 m
Weight	
standard:	42,000 kg
Mobility	
Configuration	
running gear:	tracked
Power-to-weight ratio:	18.57 hp/t
Speed	
max speed:	60 km/h
Range	
main fuel supply:	500 km
Fording	
without preparation:	1.5 m
with preparation:	5 m
Gradient:	47%
Side slope:	36%
Vertical obstacle	
forwards:	0.5 m
Trench:	2.8 m
Engine	V-84A, V-12, multifuel, 780 hp, smoke generator in exhaust
Gearbox	
forward gears:	7
reverse gears:	1
Suspension:	torsion bar
Electrical system	
vehicle:	27 V
Batteries:	4
Firepower	
Armament:	1 × turret mounted 152 mm 2A46 howitzer
Armament:	1 × roof mounted 12.7 mm (0.50) NSVT machine gun
Armament:	6 × turret mounted 81 mm smoke grenade launcher (2 × 3)
Ammunition	
main total:	50
12.7/0.50:	300
Turret power control	
type:	electric/manual

	152 mm Self-Propelled Artillery System 2S19 (MSTA-S)
Main armament traverse	
angle:	360°
Main armament elevation/depression	
armament front:	+68°/-3°
Survivability	
Night vision equipment	
vehicle:	yes
NBC system:	yes
Armour	
hull/body:	steel
turret:	steel

Status
Production as required. In service with:

Country	Quantity
Belarus	13
Ethiopia	10
Georgia	1
India	unconfirmed
Russia	600
Ukraine	40
Venezuela	47+

Contractor
Uraltransmash, Ekaterinburg.
Sterlitamak Machine Construction Factory, Bashkiriya.
Design authority is the Uraltransmash at Ekaterinburg (former Sverdlovsk) with production being undertaken at both the Uraltransmash and Sterlitamak Construction Factory in Bashkiriya in parallel.

152 mm self-propelled gun Giatsint (2S5)

Development
Development of the 152 mm 2S5 Giatsint self-propelled artillery system commenced at the Special Design Bureau of the Sverdlovsk (now Ekaterinburg) Machinery Construction Factory late in 1968.

The original intention was to develop two weapons, towed (Giatsint-B) and self-propelled (Giatsint-S), both of which would have the same 152 mm ordnance and fire a new family of 152 mm ammunition.

While the Perm Machinery Construction Plant concentrated on the actual weapon, the Sverdlovsk Transport Machinery Plant concentrated on the hull (which is similar to that used for the 152 mm 2S3 self-propelled howitzer), and NIMI the ammunition.

By late 1969, the user was presented with various designs and the version with an open mount was finally selected. Full-scale design work commenced in mid-1970, and in early 1972 prototypes of the self-propelled and towed versions were presented.

There were at least two gun designs, the 2A37 ordnance used in the Giatsint-S used separate loading ammunition while the 2A43 Giatsint-BK used a bagged charge loading system. In the end it was decided to use a weapon with separate loading ammunition (projectile and charge) and the 2A37 ordnance.

Front view of a Finnish Defence Force 152 mm 2S5 deployed in the firing position (IHS/Peter Felstead) 1044222

Rear view of Finnish Defence Force 152 mm 2S5 deployed in the firing position with spade lowered to the ground (IHS/Peter Felstead) 1133917

Production of the 152 mm 2S5 self-propelled system, which became simply the Giatsint-S, commenced in 1976 and was followed by the 152 mm towed model, the 2A36 which is covered in detail in a separate entry in *IHS Jane's Land Warfare Platforms: Artillery and Air Defence*. Production of this system is complete and it is no longer being marketed.

The 152 mm self-propelled gun 2S5 Giatsint normally deployed in batteries of six weapons, with each battalion having three batteries. The 2S5 is commonly known as the Giatsint (Hyacinth) by the Russian Army.

Production of the 2S5 was completed some time ago and the facility subsequently was reconfigured to build the 152 mm 2S19 (MSTA-S), covered in detail in a separate entry in *IHS Jane's Land Warfare Platforms: Artillery and Air Defence* and by early 2013 it is estimated that over 600 had been built.

According to the United Nations Arms Transfer lists, for the period 1994 through to 2010, Russia exported nine 2S5 weapons to Finland in 1994.

Description
The hull of the 2S5 is of all-welded steel armour construction that is believed to have a maximum thickness of 13 mm, providing the crew with protection from small arms fire and shell splinters.

As with a number of former Russian Army vehicles, including the 152 mm 2S3 self-propelled artillery system and the T-72 MBT, a dozer blade is mounted under the nose of the 2S5. This is used for clearing obstacles and preparing firing positions without specialised engineer support.

The driver is seated at the front of the vehicle on the left and has a single-piece hatch cover that opens to the rear. In front of the hatch cover are two day periscopes, one of which can be replaced by a passive periscope for night driving.

The vehicle commander is seated in a raised superstructure to the rear of the driver and has a cupola that can be traversed through 360°. Mounted externally on the forward part of the cupola is a 7.62 mm PKT machine gun that can be operated by remote control, with an IR white light searchlight mounted on the left side. The commander is provided with day periscopes and other viewing devices.

The diesel power pack is to the right of the driver with the air inlet and air outlet louvres in the roof. The engine will run on diesel or aviation fuel. The diesel engine is coupled to a mechanical stepped transmission with constant-mesh gears, two-power flow, four-gearshift, six forward gears and two reverse.

The other three crew members are seated in the crew compartment at the rear of the hull and enter and leave via a ribbed ramp in the rear. The rear crew compartment is provided with roof hatches and roof-mounted day periscopes to give observation to the sides of the vehicle.

Suspension is the torsion bar type with each side having six dual rubber-tyred roadwheels, the drive sprocket at the front, idler at the rear and four track-return rollers. The first two and last two road stations either side are provided with shock-absorbers.

The 152 mm 2A37 ordnance consists of a monobloc tube, breech ring and the five-part multislotted muzzle brake that is screwed to the ordnance and has an efficiency of 55 per cent. When travelling, the ordnance is held in position by a travelling lock. The breech mechanism is the semi-automatic horizontal loading type.

Recoil braking is hydraulic while recuperation is pneumatic. The recoil component cylinders are attached to the barrel and recoil with it; length of recoil is 730 to 950 mm. Elevation and traverse of the weapon is by sector drives while the balancing system is of the pneumatic rod type with an extra spring mechanism.

The 155 mm 2A37 ordnance installed in the 2S5 is understood to be manufactured by the Motovilikha Plants Corporation.

The gunlayer is seated to the left of the gun and is provided with a shield to his immediate front only. The weapon is aimed using the same sight as installed on the Russian 122 mm BM-21 (40-round) artillery rocket system. This mechanical sight is used with the PG-1M panoramic head sight; direct fire sight is the OP4M-91A. The gun layer operates the loading system and fires the gun.

When deployed in the firing position, a large spade is lowered to the ground to provide a more stable firing platform. Time into and out of action is not more than two minutes.

152 mm 2S5 self-propelled gun deployed in firing position 0058387

Ammunition is of the separate loading type, projectile and charge, and crew fatigue is reduced by means of a semi-automatic loading system. This is operated by remote control with the operator normally being positioned on the left side of the 2S5 on the ground.

The system consists of an electrically driven chain rammer located to the left of the breech which, when not required, folds back through 90° so that it is parallel to the breech, and the projectile and charge loading system.

The charge loading system is pivoted on the right side and has a projectile tray and a charge tray. The projectile and charge are loaded into the trays and then swung upwards through almost 90° where the rammer first rams the projectile and then the charge.

The 2S5 can be supplied with ammunition either from on board via the rear of the vehicle or from ammunition on the ground.

In total, 30 projectiles and charges are carried, with the projectiles being stowed vertically in a carousel device in the left side of the rear compartment with the 30 charges to the right in three rows of 10. Each row of 10 charges is also stowed vertically and they are on a horizontal conveyor belt that returns under the floor of the vehicle.

The charge consists of a conventional cartridge case containing the actual charge and, once fired, the breech automatically opens and the spent cartridge case is ejected.

Maximum range, firing conventional ammunition with a muzzle velocity of 942 m/s, is 28,400 m, which can be extended out to 33,100 m using a Rocket Assisted Projectile (RAP). Minimum range is currently being quoted as 8,600 m. Mention has also been made of a additional high-explosive projectile with an advanced aerodynamic form which has a maximum range of 37,000 m. It has not been confirmed that this projectile has been deployed with the Russian Army.

The normal high-explosive fragmentation projectile weighs 46 kg, with propellant and cartridge weighing a maximum of 34 kg. Other types include concrete-piercing and improved conventional munition. A unit of fire is 30 rounds.

The mechanical ammunition-handling system enables a maximum rate of fire of 5 to 6 rds/min to be achieved and, according to Russian sources, a battery of 2S5s can have 40 projectiles in the air before the first projectile lands on the enemy position.

Standard equipment for the 2S5 includes an NBC system and infra-red night equipment for the driver and commander.

Specifications

	2S5 Giatsint
Dimensions and weights	
Crew:	5
Length	
overall:	8.33 m
Width	
overall:	3.25 m
Height	
overall:	2.76 m
Ground clearance	
overall:	0.45 m (est.)
Weight	
combat:	28,200 kg
Mobility	
Configuration	
running gear:	tracked
Power-to-weight ratio:	18.50 hp/t
Speed	
max speed:	63 km/h
Range	
main fuel supply:	500 km
Fording	
without preparation:	1.05 m
Gradient:	58%
Side slope:	47%
Vertical obstacle	
forwards:	0.7 m
Trench:	2.5 m
Engine	V-59, V-12, supercharged, water cooled, multifuel, 520 hp
Firepower	
Armament:	1 × hull mounted 152 mm 2A37 howitzer
Armament:	1 × roof mounted 7.62 mm (0.30) PKT machine gun
Ammunition	
main total:	30
Turret power control	
type:	electric/manual
Main armament traverse	
angle:	30° (15° left/15° right)
Main armament elevation/depression	
armament front:	+57°/-2°
Survivability	
Night vision equipment	
vehicle:	yes
NBC system:	yes
Armour	
hull/body:	steel

Status

Production complete. Marketing of this system is still undertaken, but there are no recent known sales. In service with:

Country	Quantity	Comment
Belarus	115	
Eritrea	12	
Finland	18	called Telak 91 (delivered in 1994)
Russia	950	estimate
Ukraine	24	
Uzbekistan	n/av	unconfirmed user

Contractor

Uraltransmash.

152 mm self-propelled gun-howitzer 2S3 Akatsiya (M1973)

Development

The requirement for the 152 mm 2S3 Akatsiya (Acacia) self-propelled artillery system was drawn up in July 1967 with Uralmash being responsible for the 152 mm weapon and UZTM in Sverdlovsk responsible for the full-tracked hull.

The 152 mm 2A33 ordnance of the 2S3 was ballistically identical to that used for the 152 mm D-20 towed artillery system and used the same ammunition. The new 152 mm gun had the factory designation of the D-22 and the industrial designation of the 2A33.

The full-tracked hull was developed under the designation of the Obiekt 303 and was a further development of the Obiekt 123 used for the Krug (SA-4 Ganef) surface-to-air missile system. For the 152 mm application the hull was upgraded in many areas including the installation of a more powerful 520 hp diesel engine. The track and torsion bar suspension were also upgraded.

The first two prototypes of the 2S3 were completed by late 1968 and testing continued through to 1969 but problems were encountered with contamination in the turret so an additional four improved systems were built. The system was accepted for service in December 1971 although production had commenced the previous year.

152 mm 2S3 self-propelled gun-howitzer with ordnance in travel lock 1133918

152 mm self-propelled gun-howitzer with 155 mm/28-calibre ordnance in travel lock (Uraltransmash) 1365007

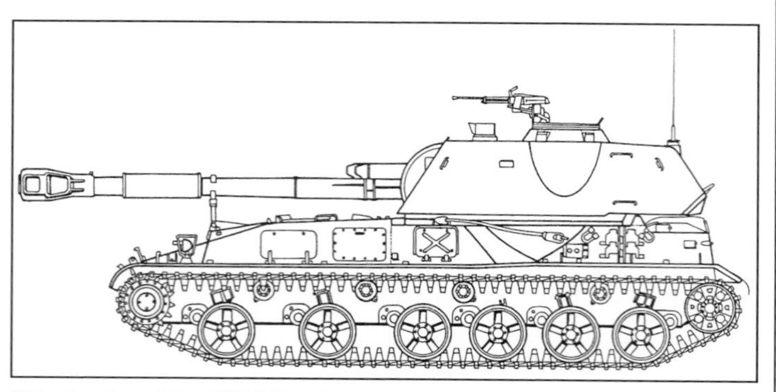

Side drawing of the 2S3M Akatsiya 152 mm self-propelled howitzer (Steven Zaloga) 0500871

Until its correct designation of 2S3 Akatsiya became known it was referred to in the West as the 152 mm M1973. The M1973 indicates the year that it was first observed.

It was originally issued on the scale of 18 per tank division and 18 per motorised rifle division and has now replaced the older towed 130 mm Field Gun M-46 in the former Russian Army artillery brigades and artillery divisions. The 2S3 is normally deployed in batteries of six weapons, with three batteries making up a regiment.

It is estimated that well over 10,000 of these systems were manufactured, although no production figures have been released.

Production of the 2S3 was completed some years ago and its replacement in the Russian Army is the more recent 152 mm self-propelled artillery system 2S19. Due to the reduction in the overall size of the Russian Army, numbers of weapons are becoming available for sale including quantities of artillery systems such as the 2S3.

The hull of the 152 mm 2S5 self-propelled gun shares many common components with the 2S5 hull and is also referred to as the Izdeliye 303.

According to the United Nations Arms Transfer Lists for the period 1994 through to 2010 the following exports of 2S3 systems were recorded. It is of note however that there may be additional exports as in some cases the designation of the system exported is now recorded.

From	To	Quantity	Comment
Bulgaria	Angola	4	supplied 1999
Germany	Finland	447	supplied in 1992, 122 mm and 152 mm systems
Hungary	Czech Republic	3	supplied 2010
Ukraine	Azerbaijan	10	supplied 2010
Ukraine	Congo (Democratic Republic)	12	supplied 2010
Ukraine	Czech Republic	3	supplied 2010
Ukraine	Georgia	6	supplied in 2004 (2S3M version)
Ukraine	Georgia	6	supplied in 2005 (2S3M version)
Ukraine	US	3	supplied in 2000

Description

The all-welded steel hull of the 2S3 self-propelled gun-howitzer is divided into three compartments: driver's at the front on the left, V-12 diesel engine to the right of the driver and the fully enclosed turret at the rear.

It is understood that the welded steel armour hull of the 2S3 has a maximum thickness of 15 mm while the welded steel turret has a maximum thickness of 20 mm.

The driver has a single-piece hatch cover that opens to the rear. In front of this are two day periscopes, the left one of which can be replaced by an infra-red device (or, more recently, a passive device) for driving at night. The driver has an adjustable seat and under this is an emergency hatch. On the forward part of the glacis is a splashboard to stop water rushing up the glacis plate when the vehicle is fording a stream.

The air inlet and outlet louvres are on the top of the hull, with the exhaust outlet on the right side of the hull, just above the track guard.

The large all-welded steel armour turret has a sloped front and well-sloped sides with a vision block in each side. The commander is seated on the left of the turret with the gunner forward and below him and the loader on the right.

The commander has a cupola that can be traversed through a full 360° and a single-piece hatch cover opening to the rear. Vision devices are provided around the lower part of the cupola. Including a TKN-3A device. Mounted on the forward part of the hatch is a 7.62 mm PKT machine gun which can be aimed and fired from inside the turret. There is an OU-3K infra-red/white light searchlight mounted to the left of the 7.62 mm PKT machine gun.

Gunlaying, including firing, is accomplished by the gunner. In addition to the PG-4 sight there is an OP5-38 gunner's telescope. Cants of up to 5° in pitch and roll axes can be offset.

The panoramic telescope has a 10.5° field of view and ×3.7 magnification. The OP5-38 gunner's telescope has an 11° field of view, a magnification of ×5.5 and is normally used for direct fire engagements.

Turret traverse and weapon elevation/depression are powered, with manual controls for emergency use.

The loader feeds the projectiles and charges to the ordnance via a loading tray with a chain-driven rammer used to seat the projectile and then the charge in the chamber.

The loading mechanism is lowered when the gun is fired. The breech opens automatically.

In the rear half of the hull are 33 projectiles, complete with fuzes, which are stowed in three horizontal layers.

On the 2S3M/2S3M1 is a projectile carousel that holds an additional 12 projectiles in the vertical position.

As each round is fired the carousel automatically revolves 30° so positioning the next projectile.

Stowed below the projectiles are 16 charges in two layers of eight with an additional eight being stowed near the loader's position and more being stowed in brackets in the vehicle. An ammunition resupply hatch is provided in the right side of the turret and there is an oval hatch at the rear of the hull that opens downwards.

Wherever possible, 152 mm ammunition would be used from outside the 2S3 with the onboard ammunition supply being retained ready for immediate use.

The two ammunition members are in a 6 × 6 truck and, on arriving at the fire position, one of these plugs into the external connector of the 2S3's intercom system at the rear and listens into the firing commands. The prepared 152 mm projectiles (complete with set fuze) and charge are then passed through the circular openings in the rear of the vehicle. The later 2S3M/2S3M1 has a single opening.

Main armament comprises a 152 mm weapon, designated the 2A33, which is based on the towed 152 mm D-20 gun-howitzer but with a bore evacuator added behind the muzzle brake.

Production of the 152 mm 2A33 ordnance was undertaken by the Motovilkha Plants Corporation. The barrel has a double-baffle muzzle brake, fume extractor and a gun barrel travelling lock. This is operated by the driver, from the driving position, without leaving the vehicle.

The normal projectile fired by the 2S3 is the HE-FRAG OF-540, which is fitted with an RGM-2 fuze, weighs 43.5 kg, contains 5.76 kg of TNT, has a maximum muzzle velocity of 655 m/s and a maximum range of 18,500 m. Other types of projectile fired include BP-540 HEAT-FS (also referred to as HEAT-SS, for spin-stabilised), High-Explosive Rocket-Assisted Projectile (HE/RAP) with a range of 24,000 m, AP-T, illuminating, smoke, incendiary, flechette, scatterable mines (anti-tank and anti-personnel) and the Krasnopol laser-designated projectile.

The 152 mm gun has an elevation of +60° and a depression of -4°. Turret traverse is a full 360°.

Another projectile is the HE-FRAG 3OF25 which with the complete round (projectile and charge) is designated the 3VOF32 (with full propelling charge) and 3VOF33 (with reduced propelling charge).

This projectile weighs 43.56 kg of which 6.8 kg is explosive and is claimed to have 1.5 to 2 times the effectiveness of the OF-540 previously mentioned.

This contains 6.8 kg of explosive and has a maximum range of 17,400 m and can also be fitted with the AR-5 radio fuze.

According to Russian sources, a typical unit of fire of ammunition for the 2S3 is 60 rounds.

155 mm 2S3M3 with a 155 mm/39-calibre ordnance fitted with a fume extractor and a large muzzle brake to reduce recoil forces 0034535

152 mm 2S3 self-propelled gun-howitzer of the Russian Army deployed in a static firing position and with turret traversed slightly to the right (INA) 1403900

The torsion bar suspension consists of six dual rubber-tyred roadwheels each side with the drive sprocket at the front, idler at the rear and four track-return rollers. The first and last return rollers support the inside of the track only. The first, second and sixth road wheel stations have a hydraulic shock-absorber and there is a distinct gap between the first and second and second and third roadwheels.

The 2S3 has infra-red night vision equipment and an NBC system. According to Russian sources, the NBC system also keeps the turret clear of fumes when the gun is fired. An OV-65G heating unit is also fitted. It has no amphibious capability and normally carries an unditching beam at the rear of the hull.

Most 2S3 systems have an entrenching blade mounted at the front of the hull to enable the system to prepare its own fire position without engineer support. This enables the 2S3 to prepare its own firing position in between 20 and 40 minutes depending on the type of terrain.

Command-and-control of the 2S3 series is provided by the 1V12 series of vehicles based on the MT-LBu series full-tracked vehicle. Production of this vehicle was undertaken in the Ukraine and has now been completed.

This is also referred to as the Artillery Command and Reconnaissance Vehicle (ACRV) and is still deployed in large quantities by the Russian Army as well as other countries

Variants

Late production versions of the 2S3 are designated the 2S3M and 2S3M1. The 2S3M and 2S3M1 have an ammunition load of 40 projectiles and charges and also a loading carousel which facilitates loading and therefore increases the rate of fire.

These versions can be distinguished from the rear as they have two small stowage compartment covers positioned to the right and left of the hull rear as opposed to a larger oval cover in the centre.

The 2S3M1 is, in addition, equipped with a data terminal which automatically displays the fire command transmitted to the battery officer's upgraded 1V13M command vehicle. It also has an IP-5 sight.

More recently, Russia has offered a modernised version of the 2S3M1 with a much longer 152 mm barrel. This fires a standard HE projectile to a maximum range of 20,550 m or a HE base bleed projectile to a maximum range of 23,900 m.

Specifications are similar to the standard system but combat weight is 27,800 kg while overall length with ordnance in travelling lock is 9.48 m.

As far as it is known, the 152 mm 2S3 with the longer barrel has yet to enter service.

152 mm 2S3M2

For several years Uraltransmash Federal State Unitary Enterprise in Ekaterinburg, Russia, has been working on a number of enhancements to the 152 mm 2S3 which originally entered service in the early 1970s.

The standard 2S3M has a 152 mm/28-calibre barrel designated the 2A33 but the company is now offering two further enhancements to bring the system up to the 2S3M2 or 2S3M3 standard.

As production of the complete 2S3 series systems was completed some time ago, these upgrades are being offered to existing customers. The upgrades could be carried out by Uraltransmash but it would be most cost effective for the company to provide kits to enable the user to upgrade existing systems in their own facilities.

In the 2S3M2 version the 152 mm/28-calibre barrel is replaced by a new 152 mm/39-calibre barrel which, firing a 3OF61 high-explosive base-bleed projectile enables a maximum range of 25.1 km. It can also fire the KBP Krasnopol laser-guided artillery projectile which is also manufactured in 155 mm-calibre and which has been sold to China, France and India.

155 mm 2S3M3

The 2S3M3 features a western type 155 mm/39-calibre barrel which fires standard NATO types of ammunition. Firing the old 155 mm United States M107 high-explosive projectile a maximum range of 18.2 km is achieved.

Firing the German/Italian/UK L15A1 high-explosive, a range of 24.7 km is quoted while firing the US M549A1 rocket-assisted projectile a maximum range of 30 km is quoted. It can also fire the 155 mm Krasnopol-M laser-guided projectile out to 17 km.

152 mm 2S3 self-propelled gun-howitzer of the Russian Army in travelling configuration from the rear (INA) 1452628

152 mm self-propelled gun howitzer 2S3 (left) compared to the slightly smaller 122 mm self-propelled howitzer 2S1 (right). Both of these self-propelled artillery systems were captured in Iraq (Michael Jerchel) 1452906

As far as it is known, the 155 mm version remains at the prototype stage.

Automated Fire-Control System

Russia is now marketing a new Automated Fire-Control System (AFCS) that can be integrated into self-propelled artillery systems such as the full tracked 152 mm 2S19, 152 mm 2S3 and 122 mm 2S1 as well as multiple rocket launchers such as the 122 mm BM-21 multiple rocket launcher. Full details of the AFCS is provided in the entry for the Russian 152 mm self-propelled artillery system 2S19 (MSTA-S).

Specifications

	2S3 Akatsiya
Dimensions and weights	
Crew:	4
Length	
overall:	8.4 m
hull:	7.765 m
Width	
overall:	3.250 m
Height	
overall:	3.05 m
Ground clearance	
overall:	0.45 m
Track width	
normal:	480 mm
Length of track on ground:	4.94 m
Weight	
combat:	27,500 kg
Ground pressure	
standard track:	0.59 kg/cm²
Mobility	
Configuration	
running gear:	tracked
Power-to-weight ratio:	17.33 hp/t
Speed	
max speed:	60 km/h
Range	
main fuel supply:	500 km
cross-country:	270 km
Fuel capacity	
main:	830 litres

	2S3 Akatsiya
Fording	
without preparation:	1.0 m
Gradient:	60%
Side slope:	30%
Vertical obstacle	
forwards:	0.7 m
Trench:	3.0 m
Engine	V-59, V-12, water cooled, 520 hp
Gearbox	
type:	manual
forward gears:	6
reverse gears:	2
Suspension:	torsion bar
Firepower	
Armament:	1 × turret mounted 152.4 mm 2A33 howitzer
Armament:	1 × roof mounted 7.62 mm (0.30) PKT machine gun
Ammunition	
main total:	46
7.62/0.30:	1,500
Rate of fire	
sustained fire:	1 rds/min
rapid fire:	4 rds/min
Turret power control	
type:	powered/manual
Main armament traverse	
angle:	360°
Main armament elevation/depression	
armament front:	+60°/-4°
Survivability	
Night vision equipment	
vehicle:	yes
NBC system:	yes
Armour	
hull/body:	steel
turret:	steel

Status

Production complete. In service with the following countries:

Country	Quantity	Comment
Algeria	30	
Angola	4	from Bulgaria in 1999
Armenia	28	
Azerbaijan	3/6	plus 10 delivered in 2010 from the Ukraine
Belarus	108	
Congo (Democratic Republic)	12	delivered from Ukraine in 2010
Cuba	20	estimate
Georgia	13	
Kazakhstan	120	
Libya	60	quantity uncertain due to conflict in 2011
Russia	1,600	estimate, also used by Marines (18)
Syria	50	
Turkmenistan	16	
Ukraine	463	
Uzbekistan	17	
Vietnam	30	

Contractor

Russian state factories.

122 mm self-propelled howitzer 2S1 Gvozdika (M1974)

Development

The requirement for the 122 mm 2S1 Gvozdika (Carnation) self-propelled artillery system was drawn up in July 1967 at the same time as the 152 mm 2S3 self-propelled artillery system covered in a separate entry in *IHS Jane's Land Warfare Platforms: Artillery and Air Defence*.

Production of the 152 mm 2S3 self-propelled artillery system is complete but it is still used in significant numbers by the Russian Army and other countries.

Uralmash was responsible for the weapon, based on that used in the towed 122 mm D-30 howitzer, and the Kharkov Tractor Plant (which is now in the Ukraine) for the full-tracked hull which is based on the automotive and running gear of the MT-LB multipurpose tracked armoured chassis.

122 mm 2S1 self-propelled howitzer with 122 mm ordnance partly elevated (Jaroslaw Cislak) 0106215

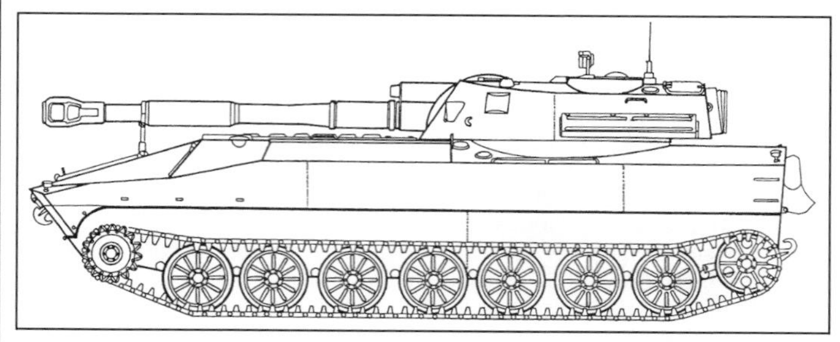

Side drawing of the 2S1 Gvozdika 122 mm self-propelled howitzer (Steven Zaloga) 0500872

122 mm 2S1 self-propelled howitzer in travelling configuration with 122 mm ordnance in travel lock 1044220

Two experimental 122 mm self-propelled howitzers were developed using components of the MT-LB hull and these were designated the D-11 and D-12. Extensive trials of these were carried out between 1967 and 1972.

After comparative testing the D-12 (with the 122 mm ordnance 2A31) was selected for service with the Russian Army. The Russian OKB-9 bureau also investigated a 122 mm self-propelled howitzer based on the BMP-1 infantry fighting vehicle but this was not taken any further.

According to Russian sources, development and production of the 2S1 was carried out by the KhTZ Production Association.

The first four prototypes of the 2S1 were completed in the second half of 1969, it was accepted for service in September 1970 and entered volume production the following year.

The 122 mm 2S1 entered service with Poland and Russia in 1972 and was first seen in public in 1974, hence its western designation of M-1974. It is estimated that over 10,000 vehicles were built with production completed around 1991.

In 2010 Russian sources stated that the Russian Army had a total of 3,050 in service or held in reserve while the Russian Marines had a total of 731 in service or held in reserve.

122 mm 2S1 was issued on the scale of 36 per artillery division, 36 per motorised rifle division and 72 per tank division.

Each 2S1 battalion originally had three batteries each with six guns.

In the past the Russian Army main artillery calibres were 122 mm and 152 mm but in the future will standardise on the latter calibre. There is no direct replacement for the 2S1 apart from the 152 mm 2S19.

The 2S1 Gvozdika had the development designation of the Izdeliye 26.

The 2S1 has also been manufactured under licence in Bulgaria and Poland, while Romania has developed a different version called the Model 89. Details of this are given in a separate entry in *IHS Jane's Land Warfare Platforms: Artillery and Air Defence* and production has now been completed and it is only in service with Romania.

According to the United Nations Arms Transfer Lists the following quantities of 122 mm 2S1 self-propelled howitzers were exported between 1992 and 2010. At this stage it is not known as to whether these were new build systems or surplus Bulgarian Army 2S1 systems.

Scale model of MTK-2 mineclearing system with rocket launcher raised
0018895

RKhM chemical reconnaissance vehicle showing marking pennants at rear of vehicle
0501013

From	To	Quantity	Comments
Bulgaria	Chad	2	delivered in 2008
Bulgaria	Congo	3	delivered 1999
Bulgaria	Eritrea	12	delivered 1999
Bulgaria	Syria	210	delivered 1992
Bulgaria	Togo	6	delivered 1997
Czech Republic	Poland	2	delivered in 2008
Czech Republic	Uruguay	6	delivered 1997
Georgia	Belgium	1	delivered 1992
Germany	Finland	48	delivered 1992
Germany	Finland	27	delivered 1994
Germany	US	6	delivered 1993
Hungary	Czech Republic	10	delivered in 2008
Hungary	Ukraine	12	supplied 2010
Slovakia	Angola	4	delivered 2000
Slovakia	Belarus	34	delivered 2001
Slovakia	Bulgaria	2	delivered 2001
Slovakia	Czech Republic	8	delivered 2001
Ukraine	Azerbaijan	18	delivered in 2008
Ukraine	Azerbaijan	7	supplied 2010
Ukraine	Congo (Democratic Republic)	12	supplied 2010
Ukraine	Hungary	12	supplied 2010
Ukraine	Congo	6	delivered 2000

Description

The all-welded steel hull of the 2S1 is divided into three compartments: driver's at the front on the left, engine behind the driver and the turret at the rear.

It is understood that the welded steel armour hull of the 2S1 has a maximum thickness of 15 mm while the welded steel turret has a maximum thickness of 20 mm.

The driver has a single-piece hatch cover that opens to the rear and, in front of his position, a windscreen covered by a hatch hinged at the top. He has three day periscopes, the middle one of which can be replaced by a TVN-2B infra-red night periscope.

In the top of the hull, to the right of the driver's position, is a large engine access hatch that opens to the right. There is a smaller hatch on the right side of the glacis plate. The exhaust pipe is on the right side of the hull towards the front.

Russian 122 mm 2S1 self-propelled howitzer with ordnance in travel lock
1333631

1V13 artillery deputy battery command vehicle
1405674

122 mm 2S1 self-propelled howitzer with power pack covers over front of hull in open position (INA)
1452629

The all-welded steel armour fully enclosed turret, which does not overhang the sides of the hull, has a well-sloped front and sides with the commander seated on the left, the gunner in front of and below the commander and the loader on the right.

The commander has a cupola that can be manually traversed through a full 360° and a single-piece hatch cover that opens to the front. This cupola has three day periscopes for forward observation and an infra-red searchlight. Next to the commander's position is fitted the R-123M radio.

The indirect sight (PG-2 with a field of view of 10.5° and a magnification of ×5.5) is mounted in the roof of the turret, forward of the commander's cupola with the direct sight (OP5-37 with a field of view of 10.5° and a magnification of ×5.5) mounted to the left of the 122 mm main armament. There is also the 1OP40 periscopic sight. The loader's hatch opens forward, and in front of it is a swivel-type day periscope. A long stowage box is often mounted on the left side of the turret and there is a large door in the rear of the hull, hinged on the left side. The rear door has a single firing port in the centre.

The 2S1's main armament, designated the 2A31, is a modified version of that fitted to the widely deployed 122 mm D-30 towed howitzer. This has a semi-automatic vertical sliding wedge breech block and a firing pin that can be re-cocked in the event of a misfire.

The 122 mm ordnance is fitted with a fume extractor and muzzle brake and is held in position, when travelling, by a lock on the hull glacis plate which is operated by remote control by the driver. A power rammer and extractor are fitted to the folding guard rail to enable a higher rate of fire to be achieved and to permit loading at any angle of elevation. A maximum sustained rate of fire of 5 to 8 rds/min can be obtained with a sustained rate of fire of 70 rounds for the first hour. Of the 40 122 mm projectiles normally carried, 32 are normally HE, six smoke and two HEAT-FS.

Russian Zoopark-1 artillery location radar system with antenna erected (Christopher F Foss) 0121597

Of these, 16 are in standby stowage on the left and right of the sidewalls. When in action, ammunition would normally be used from outside the 2S1 system and fed to the crew inside via a ramp. One of the two ammunition members outside the vehicle is connected to the R-124 vehicle intercom and would fuze the projectiles before they are passed into the vehicle.

The remaining 24 projectiles and charges are stowed near the loader's position with the empty cartridge cases being ejected outside the turret.

The HE projectile has a maximum range of 15,300 m and, in addition to the ammunition listed in the table, leaflet, HE/RAP (range of 21,900 m), armour-piercing high-explosive, flechette, and chemical (no longer used) projectiles are also available.

The 122 mm 2S1 can also fire the KBP Kitolov-2 laser-guided artillery projectile which operates in a similar manner to that of the 152 mm Krasnopol laser-guided artillery projectile covered in the entry for the 152 mm 2S19 self-propelled artillery system in *IHS Jane's Land Warfare Platforms: Armoured Fighting Vehicles*. The Kitolov-2 has a maximum range of 12,000 m, with the complete projectile weighing 25 kg.

Type	HE	HEAT-FS[1]	Illumination	Smoke
Projectile designation:	OF-462	BK-6M	S-463	D4
Weight of projectile:	21.72 kg	21.63 kg	22 kg	21.76 kg
Max muzzle velocity:	690 m/s	740 m/s	690 m/s	690 m/s

[1] Will penetrate 460 mm of conventional steel armour at 1,000 m

According to Russian sources, a typical unit of fire of ammunition for the 2S1 is 80 rounds.

Production of the 122 mm 2A31 ordnance used in the 2S1 Gvozdika self-propelled artillery system was undertaken by the Motovilkha Plants Corporation.

The suspension system is similar to the MT-LB multipurpose tracked vehicle's and either side consists of seven roadwheels with the drive sprocket at the front, idler at the rear, and no track-return rollers.

An unusual feature of the 2S1 Gvozdika is that the suspension can be adjusted to give different heights, which is of particular use when the vehicle is being transported by tactical transport aircraft. The 2S1 is normally fitted with 400 mm wide tracks but when operating in snow or swampy ground 670 mm wide tracks are fitted to give improved traction and a lower ground pressure.

Standard equipment includes infra-red night vision lights and an NBC system. The NBC system for the driver's compartment is designated the FVA-100 while that for the turret compartment is designated the FVA-200. The 2S1 is fully amphibious, being propelled in the water by its tracks.

Before entering the water the bilge pump is switched on, the trim vane is erected at the front of the hull, shrouds are fitted to the hull above the drive sprocket and front road wheels and water deflectors on the rear track covers are lowered.

122 mm 2S1 self-propelled howitzer from the rear. This shows the shrouds on the turret rear that are fitted prior to amphibious operations (Christopher F Foss) 1452630

Any water that enters the hull during amphibious operations is removed via the exhaust outlet using the bilge pump. While afloat only 30 rounds (projectiles and charges) are carried. Covers are also fitted around the engine air intakes to prevent water ingestion into the engine compartment. Russian sources state that the 2S1 takes 20 minutes to be prepared for amphibious operations.

For improved amphibious capability some of the Polish vehicles have been observed fitted with additional buoyancy aids and are powered in the water by two small propellers situated one either side of the hull at the rear.

The standard Russian radio fitted to the 2S1 is a model R-123-M.

Variants

Automated Fire-Control System

Russia is now marketing a new Automated Fire-Control System (AFCS) that can be integrated into self-propelled artillery systems such as the full-tracked 152 mm 2S19, 152 mm 2S3 and 122 mm 2S1 as well as multiple rocket launchers such as the 122 mm BM-21 multiple rocket launcher. Full details of this are provided in the entry for the Russian 152 mm 2S19 self-propelled artillery system in *IHS Jane's Land Warfare Platforms: Artillery and Air Defence*.

120 mm 2S1 with 120 mm 3A80 ordnance

The Motovilka Plants Corporation has proposed that the 122 mm 2A31 ordnance of the 2S1 could be replaced by the complete 120 mm 2A80 rifled ordnance installed in the 120 mm 2S31 Vena self-propelled gun/mortar system.

Development of the 2S31 Vena is complete but as of early 2013 it is understood to remain at the prototype stage. It has already been offered on the export market.

In addition to firing a complete family of standard 120 mm projectiles the 120 mm 2A80 ordnance can also fire a precision guided projectile.

Other features of this upgraded 2S1, called the 2S1M would include the installation of an automated elevation and traverse laying system, automatic survey and orientation system, provision for receiving satellite signals reception and initial orientating equipment.

Huta Stalowa Wola 120 mm Rak turret mortar system

The Polish company of Huta Stalowa Wola (HSW) is developing, as a private venture, a turret mounted 120 mm breech loaded mortar system called Rak to meet the potential requirements of the Polish Army as well as the export market.

This 120 mm Rak turret mortar system can be integrated onto tracked and wheeled chassis with the latter including the Patria Armoured Modular Vehicle (AMV) of which 690 have been ordered by the Polish Army with first deliveries made in 2007.

For trials purposes this turret has been installed on a 2S1 hull as large quantities of these are available in Poland.

Details of this turret system are provided in a separate entry in *IHS Jane's Land Warfare Platforms: Artillery and Air Defence*.

As of early 2013 the 120 mm Rak turret mortar system remained at the prototype stage.

Artillery Command and Reconnaissance Vehicle (ACRV)

The ACRV family of full tracked vehicles are deployed for a wide range of artillery roles including:

Russian designation	Role	NATO designation
1V13	deputy battery commander	M-1974-1
1V14	battery commander	M-1974-1A
1V15	battalion commander	M-1974-2B
1V16	deputy battalion commander	M-1974-3

Other versions include the 1V21 air defence management vehicles, NBC reconnaissance, electronic warfare, radar vehicle to name but a few.

In addition to being manufactured in Russia, the ACRV vehicles have also been manufactured under licence in Bulgaria as has the 122 mm 2S1 self-propelled howitzer.

1V21 series of air defence management vehicles
Russia also deployed a whole family of ACRV-based vehicles for use in the air defence command and control role. These are also referred to as the Rangir family of air defence command vehicles.

These are designated the 1V21,1V22,1V23, IV24 and 1V25 with those originally deployed by Czechoslovakia and Poland being designated the MP21, MP22, MP23, MP24 and MP25.

Zoopark-1 locating radar
The ACRV is for the Zoopark-1 3D multifunctional phased battlefield locating radar which is designated the Zoopark-1 or IL219.

This advanced radar is used to pinpoint threat mortars, artillery and surface-to-surface rocket systems as well as being able to correct friendly artillery fire. It is typically operated by a crew of three.

The system has an instrumented detection range of 45 km and between four and six trajectories can be tracked at any one time.

The multi-functional roof-mounted phased array radar is normally deployed in the horizontal position when travelling and then raised into the vertical position when deployed.

Zoopark-1 is fitted with an auxiliary power unit that allows all of systems to be run with the main engine switched off. It is fitted with an NBC system and like the baseline ACRV it is fully amphibious.

Mineclearing vehicle M-1979 (MTK-2)
This has a turret-like superstructure that contains three rockets on launch ramps. These, together with the upper part of the superstructure, are hydraulically elevated for firing. It is estimated that the range of the rockets is between 200 and 400 m, with each rocket connected, via a towing line, to about 150 m of mine clearance hose that is stowed folded in the uncovered base of the turret on the roof of the vehicle.

The hose is connected by a cable to the vehicle, which allows the crew to position the hose in the optimum breaching position after launching. The hose is then command detonated.

RKhM chemical reconnaissance vehicle
It is now known that the RKhM chemical reconnaissance vehicle is based on the hull of the M-1974 (2S1) self-propelled howitzer and not the MT-LB multipurpose vehicle. It has a raised superstructure on which a 7.62 mm machine gun cupola is mounted and on the rear decking are boxes for the marking pennants.

The vehicle is equipped with a single KZO-2 lane marking system with 19/20 poles on the right side, automatic gas detection apparatus GSP-1M or GSA-12, semi-automatic detection systems PPKhR and VPKhR and radios R-107M, R-123M and R-130M. The RKhM-K is a command version with additional communications systems, large antenna, and navigation system but is not fitted with the KZO-2.

Iranian 122 mm Raad-1
Full details of this system, which uses a 122 mm 2S1 type turret, are given in a separate entry in *IHS Jane's Land Warfare Platforms: Artillery and Air Defence* and this is in service with the Iranian Army but has yet to be exported.

Specifications

	2S1 Gvozdika
Dimensions and weights	
Crew:	4
Length	
overall:	7.26 m
Width	
overall:	2.85 m
Height	
overall:	2.732 m
to turret roof:	2.287 m
Ground clearance	
overall:	0.40 m
adjustable:	0.395 m to 0.415 m
Track width	
normal:	400 mm
Weight	
combat:	15,700 kg
Ground pressure	
standard track:	0.49 kg/cm^2
Mobility	
Configuration	
running gear:	tracked
Power-to-weight ratio:	19.10 hp/t
Speed	
max speed:	61.5 km/h
water:	6 km/h (est.)
Range	
main fuel supply:	500 km
Fuel capacity	
main:	550 litres
Amphibious:	yes
Gradient:	77%
Side slope:	55%
Vertical obstacle	
forwards:	0.7 m
Trench:	2.2 m
Engine	YaMZ-238N, V-8, water cooled, 300 hp
Gearbox	
type:	manual
forward gears:	5
reverse gears:	1
Steering:	clutch and brake
Suspension:	adjustable
Firepower	
Armament:	1 × turret mounted 122 mm 2A31 howitzer
Ammunition	
main total:	40
Turret power control	
type:	electric/manual
Main armament traverse	
angle:	360°
Main armament elevation/depression	
armament front:	+70°/-3°
Survivability	
Night vision equipment	
vehicle:	yes
NBC system:	yes
Armour	
hull/body:	steel
turret:	steel

Status
Production complete. In service with:

Country	Quantity	Comment
Algeria	140	
Angola	16	
Armenia	10	
Azerbaijan	46	
Belarus	198	
Bosnia-Herzegovina	24	
Bulgaria	247	(covered in a separate entry)
Chad	2	
Congo	3	from Bulgaria
Croatia	8	
Cuba	40	
Democratic Republic of Congo	12	
Eritrea	32	from Bulgaria
Ethiopia	12	status uncertain
Finland	36	called PsH 74
Georgia	12	
Iran	60+	also locally developed version called Raad-1 (covered in a separate entry)
Kazakhstan	120	
Kyrgyzstan	18	
Libya	130	quantity uncertain due to conflict in 2011
Poland	373	
Romania	6	plus 18 of local model called Model 89
Russia	2,300	estimate, also used by marines (95)
Serbia	67	
Sudan	10	
Syria	400	include some from Bulgaria
Toga	6	from Bulgaria in 1997
Turkmenistan	40	
Ukraine	580	
Uruguay	6	from Czech Republic in 1997
Uzbekistan	18	
Yemen	25	

Contractor
Bulgarian, Polish and Russian state factories.

(The Kharkov facility was overall prime contractor and built the complete 2S1, while the 122 mm weapon and its associated aiming system was supplied by the Uralmash facility and the diesel engine by the Yaroslavl Avtodizal facility). Polish 2S1 vehicles are fitted with a locally designed and built SW680T diesel engine, new design road wheels and changes to the hydronamics shields for amphibious operations. Bulgarian built vehicles were also used by the Russian Army.

Singapore

Singapore Technologies Kinetics Primus 155 mm/39-calibre self-propelled artillery system

Development

To meet the operational requirements of the Singapore Armed Forces (SAF) for a new self-propelled artillery system to provide fire support for its mechanised units, the SAF, Defence Science and Technology Agency (DSTA) and Singapore Technologies Kinetics (ST Kinetics) developed a new 155 mm/39-calibre Self-Propelled (SP) artillery system called Primus.

This was developed over a six-year period under the codename of the U2. The system is now operational with the Singapore Armed Forces as its first full-tracked SP.

Primus was shown in public for the first time late in 2003 and it is also referred to as the SSPH1 (which stands for Singapore Self-Propelled Howitzer–1).

Primus is a complete self-propelled artillery system that includes not only the self-propelled weapon but also an associated Ammunition Support Vehicle (ASV) and Command Post Vehicle (CPV). The ASV feeds projectiles and charges directly into the weapon under full armour protection.

It is understood that production of the Primus commenced in 2003 with the first battery becoming operational late in 2004. A total of 54 production systems are believed to have been built with the last of these being delivered in 2007.

Although production of the STK Primus 155 mm/39-calibre SP artillery system has been completed, production could commence again if additional orders are placed.

Description

The hull of Primus is based on the proven now BAE Systems M109 155 mm Self-Propelled Howitzer (SPH), which is the most widely used system of its type in the world.

It has been upgraded in a number of key areas and has a new power pack, which is similar to the one fitted to the Singapore Technologies Kinetics Bionix Infantry Fighting Vehicle (IFV) that has been in service with the SAF for 15 years.

The use of common subsystems for Primus and the Bionix IFV offers a number of significant advantages including easier training and reduced logistics.

The driver's compartment is front left with the diesel power pack compartment to the right. This leaves the remainder of the hull clear for the crew compartment, above which is the turret.

The Primus power pack consists of a Detroit Diesel Model 6V-92TIA diesel developing 550 hp coupled to an L3 Combat Propulsion Systems HMPT-500-3EC fully-automatic transmission.

This is more compact and powerful than that fitted in the most recent US M109A6 155 mm/39-calibre Paladin SPH. The engine compartment is fitted with a fire detection and suppression system.

According to ST Kinetics, the armour of the Primus provides protection against 7.62 mm armour-piercing rounds fired from a distance of 30 m and 90 per cent protection against 155 mm high-explosive fragments at 50 m.

A new all-welded aluminium turret has been fitted to the aluminium hull, which is power operated and was developed by ST Kinetics. It is fitted with a locally-developed 155 mm/39 barrel with a muzzle brake and fume extractor. This meets the NATO Joint Ballistics Memorandum of Understanding.

When in the travelling configuration the 155 mm/39-calibre barrel is held in a travelling lock mounted at the front of the hull. The lock folds back onto the glacis plate when not in use.

Firing the old 155 mm M107 High-Explosive (HE) projectile a maximum range of 19,000 m can be achieved.

Firing an HE Extended Range Full Bore - Hollow Base (ERFB-HB) projectile a maximum range of 34,000 m can be achieved which is extended to 30,000 m using an HE Extended Range Full Bore- Base Bleed (ERFB-BB) projectile.

Other types of 155 mm projectile that can be fired by the Primus include illuminating and smoke.

A semi-automatic loading system is provided to increase the rate of fire and reduce crew fatigue. The fuzed projectile is loaded and rammed automatically, with the modular charges loaded manually.

155 mm/39-calibre Primus self-propelled gun in travelling configuration with 155 mm ordnance in travel lock (Singapore Technologies Kinetics) 1044219

155 mm/39-calibre Primus self-propelled gun in travelling configuration (Christopher F Foss) 1365008

It has a burst rate of fire of three rounds in 20 seconds, sustained rate of fire of two rounds for 30 minutes and a maximum rate of fire of six rounds a minute. The bustle-mounted magazine holds a total of 22 155 mm projectiles.

A 7.62 mm machine gun is fitted on the turret roof for local and air defence purposes and two banks of three electrically-operated smoke grenade launchers are installed. These are fitted either side of the turret and cover the frontal arc.

A digital control system automates the complete projectile-loading process and gun laying operation. An ammunition inventory-management system keeps track of all onboard 155 mm ammunition projectiles and accounts for expenditure during firing.

The weapon is laid onto the target using an automatic Fire-Control System (FCS) that includes an onboard positioning and navigation system based on a ring laser gyro system.

This allows Primus to operate autonomously and enables it to react speedily to urgent drills or fire missions. It eliminates time-consuming operations such as alignment, surveying and manual laying, which leads to improved overall system accuracy.

The automatic FCS can receive target information from the battery or regimental command post. It takes 60 seconds to come into action and open fire, and 40 seconds to be re-deployed. A direct-fire sight is fitted that has a maximum range of up to 1.5 km.

Suspension is of the torsion bar type, and either side has seven dual rubber-tyred roadwheels with the drive sprocket at the front, idler at the rear and track return rollers. The upper part of the suspension is covered by a rubber skirt.

Primus has a crew of four, including the commander and driver. A number of enhancements are also underway, including an automatic fuze setter.

Variants

Primus 155 mm/52-calibre system

Singapore Technologies Kinetics has carried out a study to replace the current 155 mm/39-calibre ordnance with a 155 mm/52-calibre ordnance.

As far as it is known there are no plans for the Singapore Armed Forces to take this version into service.

Ammunition Resupply Vehicle

The Singapore Armed Forces use a modified Singapore Technologies Kinetics Bronco All Terrain Tracked Carrier (ATTC) as an ammunition resupply vehicle for the Primus 155 mm/39-calibre self-propelled artillery system.

The rear unit has been modified to carry a total of 48 × 155 mm artillery projectiles plus associated charges and fuzes.

Specifications

	Primus
Dimensions and weights	
Crew:	4
Length	
overall:	10.21 m
hull:	6.6 m
Width	
overall:	3 m
without skirts:	2.8 m
Height	
overall:	3.28 m
Ground clearance	
overall:	0.45 m
Track	
vehicle:	2.8 m
Weight	
combat:	28,300 kg
Mobility	
Configuration	
running gear:	tracked
Power-to-weight ratio:	19.43 hp/t

	Primus
Speed	
max speed:	50 km/h
Range	
main fuel supply:	350 km
Gradient:	60%
Side slope:	30%
Vertical obstacle	
forwards:	0.6 m
Trench:	1.65 m
Engine	Detroit Diesel Model 6V-92TIA, 550 hp
Gearbox	
model:	L3 Combat Propulsion Systems HMPT-500-3EC
type:	automatic
Suspension:	torsion bar
Electrical system	
vehicle:	28 V
Firepower	
Armament:	1 × turret mounted 155 mm howitzer
Armament:	1 × roof mounted 7.62 mm (0.30) machine gun
Armament:	6 × turret mounted 76 mm smoke grenade launcher (2 × 3)
Ammunition	
main total:	26
Main weapon	
calibre length:	39 calibres
Rate of fire	
rapid fire:	6 rds/min
Turret power control	
type:	powered/manual
Main armament traverse	
angle:	32° (16° left/16° right)
Main armament elevation/depression	
armament front:	+63°/-3°
Survivability	
Night vision equipment	
vehicle:	optional
NBC system:	optional
Armour	
hull/body:	aluminium
turret:	aluminium

Status
Production complete. In service with the Singapore Armed Forces.

Contractor
Singapore Technologies Kinetics (STK). It should be noted that while final integration of the hull and turret took place at the facilities of Singapore Technologies Kinetics, the actual hull of the Pegasus was supplied by BAE Systems from their York, Pennsylvania, US, facilities.

Spain

General Dynamics European Land Systems - Santa Bárbara Sistemas M109A5E 155 mm self-propelled howitzer upgrade

Development
The Spanish Army completed extensive user trials of four artillery systems that have been upgraded by the now General Dynamics European Land Systems - Santa Bárbara Sistemas (GDELS-SBS) to enhance their operational capabilities.

Two upgraded BAE Systems towed 105 mm L118/L119 Light Guns and two 155 mm M109A5E Self-Propelled (SP) artillery systems took part in its trials which aim to reduce the time taken to come into action as well as increasing accuracy.

Description
The layout of the Spanish M109A5E 155 mm self-propelled howitzer is identical to the original BAE Systems M109 series of 155 mm self-propelled howitzers.

The M109A5E and the L118/L119 Light Gun have been fitted with a Digital Navigation Aiming and Pointing System (DINAPS).

This is a modular system that combines a hybrid (inertial and global positioning system) navigation system, Muzzle Velocity Radar (MVR) and a navigation and ballistic software which is able to connect to the Spanish Army Command-and-Control system (C2).

The system uses NATO Ballistic Kernel (NABK) as a core for ballistics using a variety of projectiles to enable the weapon to engage stationary and moving targets with a high first round hit probability.

The inertial navigation unit determines the elevation and northing angles of a gun barrel and corrects automatically for any variation in projectile, charge and meteorology.

Automatic Gun Laying System (AGLS) can be used in conjunction with DINAPS in order to automatically lay the weapon onto the target.

In addition to being used with towed and SP weapons it can also be used with rocket launchers. The software installed in DINAPS allows the system to be used with different command-and-control/communications systems.

A M109A5E 155 mm SP artillery systems fitted with DINAPS + AGLS takes only nine minutes to come into action, fire 15 rounds and move to another position to avoid counter battery fire.

According to GDELS-SBS, a standard M109 without this equipment would require about 52 minutes to carry out a similar fire mission (including topographic works) as well as requiring more ammunition to neutralise the target.

If required, to increase the rate of fire of the upgraded M109A5E, it could also be fitted with the GDELS-SBS Full Integrated Ramming System (FIRS) which also reduces crew fatigue.

FIRS integrates four main subsystems which are the hydraulic motion system, breech actuating system, automatic primer magazine and the cannon management system.

It its powered by the onboard hydraulic and electrical systems and only minor modifications are claimed to be required to integrate it onto the platform.

Trials have shown that a M109 with FIRS can fire three rounds in 11 seconds and nine rounds in 56 seconds with the automatic primer magazine holding 10 M82 primers.

FIRS is capable of ramming all NATO 155 mm ammunition with a length of up to 1 m and can be adopted for different types of 155 mm gun.

The Spanish Army has taken delivery of a total of 56 L118 Light Guns which were delivered between 1996 and 1998 which are issued to three regiments each of which has three batteries of six guns.

The weapons can be fitted with the shorter 105 mm L119 barrel for training purposes with 105 mm ammunition being manufactured locally.

The Spanish Army operates a total of 96 M109A5E SP artillery systems it is expected that and all of these will be upgraded to the enhanced configuration.

Towed applications
The Spanish Army also had 34 105 mm M108s which were to have been upgraded to M109 standard but it is understood that this upgrade will no longer move ahead and these are no longer in service.

DINAPS + AGLS has already been type classified for the GDELS-SBS 155 mm/52-calibre artillery systems model V07 and SIAC which has already been ordered by the Spanish Army and these will be fitted with FIRS. Details of this 155 mm/52-calibre towed artillery system are provided in a separate entry in *IHS Jane's Land Warfare Platforms: Artillery and Air Defence*.

For some years the Spanish Army has operated 12 first-generation 155 mm/52-calibre APU SBT in the coastal defence role (V07 standard) and four in the field artillery standard (V06 standard).

These 16 weapons have now been rebuilt to a new common standard called Obus 155/52-calibre standard V07 with an additional 66 being built to the standard SIAC from scratch which will bring the total fleet up to 82 units. Of these 66 will be used by the field artillery and 16 by the coastal artillery

All of the weapons will have DINAPS + AGLS and FIRS and are also referred to as the SIAC (*Sistema Integrado de Artilleria de Campana*) which also includes the command-and-control system and the new Iveco Defence Vehicle (6 × 6) trucks.

The first production standard system was completed late 2006 and handed over to the Spanish Army for trials in December 2009. Production is now complete and final deliveries were made to the Spanish Army early in 2013. Colombia has taken delivery of 13 systems and production can commence again if additional orders are placed.

Specifications
As per standard M109A5.

Contractor
General Dynamics European Land Systems - Santa Bárbara Sistemas.

South Africa

Denel Land Systems 155 mm/52-calibre T6 artillery turret

Development
The Denel Land Systems 155 mm/52-calibre T6 artillery turret has been developed by the company as a private venture for the export market.

It builds on the considerable experience of Denel Land Systems in the design, development and production of the combat proven G6 155 mm/45-calibre 6 × 6 self-propelled artillery system supplied to Oman (24), South Africa (43) and the United Arab Emirates (78).

For trials purposes the T6 artillery turret has been integrated onto a Russian T-72M1 MBT chassis and the Indian Arjun MBT chassis with both of these combinations being tested in India.

155 mm/52-calibre T6 turret system on T-72 chassis with turret traversed to the right 0058391

It is also marketed on an upgraded G6 chassis with the complete combination being known as the G6-52 for which there is a separate entry in *IHS Jane's Land Warfare Platforms: Artillery and Air Defence*.

By early 2013 development of the T6 artillery turret was complete but Denel Land Systems continue to customise its design following its first public unveiling in Abu Dhabi in early 2003.

So far Denel Land Systems have fired over 7,500 155 mm projectiles of all types with various charge systems from the 155 mm/52-calibre T6 artillery turret.

Description

The T6 155 mm/52-calibre turret is a completely autonomous turret system that differs from the G6 in that provision has been made for the storage of 155 mm projectiles and charges in the turret. This required a repackaging of the turret power unit.

The T6 turret system sent to India was fitted with a 155 mm/52-calibre ordnance which, when firing an Extended-Range Full-Bore - Base Bleed (ERFB-BB) projectile, gives a maximum range of 41,000 m.

The chamber volume is 23 litres, similar to the chamber volume of the standard production 155 mm/45-calibre system. The recoil system has been upgraded and now has two diametrically opposed buffers similar in design to the single buffer used on the G6 but with integral oil replenishers. Two diagrammatically opposed, gas-operated recuperators are fitted.

The design of the barrel and cradle has been modified slightly to make it possible to remove and replace the 155 mm/52-calibre barrel from the front. The barrel is fitted with a double-baffle muzzle brake and a steel fume extractor.

The breech mechanism used in the T6 is a newly patented design and is of the slide and swing type with double split rings and an obturator pad. An automatic primer-loader with a 20-round magazine is integrated with the breech.

A barrel cooling fan is fitted on the cradle to reduce recovery time in the event of unsafe barrel temperatures being reached. A barrel temperature measurement system has also been installed, which gives a warning to the commander in case of high barrel temperatures being reached.

The all-welded steel armoured turret is fitted with a fully automated charge and projectile replenishment, handling and loading system.

The T6 turret has been designed to fire all types of NATO ammunition. Using the Rheinmetall Denel Munition Extended-Range Full-Bore (ERFB) boat tail and Extended-Range Full-Bore - Base Bleed (ERFB-BB) projectiles and modular combustible case charge system, muzzle velocity and maximum range (under standard conditions) are as follows:

ERFB projectile	Muzzle velocity	Max range
Charge M51	356 m/s	9,400 m
Charge M52	497 m/s	13,800 m
Charge M51+1 increment	660 m/s	19,800 m
Charge M52+2 increments	815 m/s	26,000 m
Charge M53	925 m/s	31,900 m
ERFB-BB projectile	**Muzzle velocity**	**Max range**
Charge M52+2 increments	808 m/s	31,500 m
Charge M53	922 m/s	42,000 m

The ammunition system (projectile and charge) is identical to that fired by the G5 and G6.

The charge system is the modular combustible case cartridge system developed by Rheinmetall Denel Munitions, which consists of an M51 charge, an M52 charge which can be fitted with one or two charge increments and a stand-alone charge M53 that gives a total of five zones. These charges have special barrel erosion reduction and flash suppression properties.

In firing trials, probable errors of 0.3 to 0.4 per cent of range and 0.35 to 0.4 mil in deflection have been obtained consistently with both boat tail and base bleed ammunition.

Using the M53 A2 top charge, barrel life is claimed to be more than 3,500 rounds.

The 155 mm/52-calibre system has already been qualified with high-explosive, smoke, illuminating and red phosphorus projectiles, of boat tail and base bleed types, to the full range potential. As well as a complete family of South African Rheinmetall Denel Munitions 155 mm projectiles and charges, the T6 artillery turret can also fire more advanced projectiles such as the French/Swedish BONUS top attack smart projectile and the Russian Krasnopol-M laser-guided projectile which has already been successfully test fired.

According to Denel Land Systems, with the appropriate modifications the T6 artillery system could also fire the Raytheon Excalibur 155 mm precision guided munition that has already been used in Afghanistan.

When compared top the earlier G6 155 mm/45-calibre system, the latest G6-52 has a significant number of advantages including a reduction in turret crew requirements, higher rate of fire and an increased firing range which improves its survivability against counter battery fire.

High rate of fire has been achieved due to the installation of a new automatic ammunition handing system that first loads the fuzed projectile and then the charge. It has a burst rate of fire that is three rounds in 15 seconds.

Maximum rate of fire is eight rounds per minute and it can carry out Multiple Round Simultaneous Impact (MRSI) fire missions with five rounds hitting the target at the same time at a range of 25 km.

A sustained rate of five of two rounds a minute can be maintained as long as safe barrel temperatures are indicated on the temperature warning system installed in the turret.

The T6 turret has full power traverse through 360° with the 155 mm/52-calibre ordnance capable of being elevated from –5° to +75°. When travelling, the ordnance is held in position by an 'A' frame that is pivoted at the front of the hull. Weight of the complete turret is 16,000 kg.

The ammunition handling system of the T6 is extremely flexible and can be changed depending on the tactical situation and the customer's exact operational requirements.

The T6 turret carries a total of 40 complete 155 mm rounds and, depending on the hull type selected, an additional 10 complete rounds can be carried.

In the ammunition handling system the projectile and charges are transferred from the carousels to the same position into the charge and projectile loading arms. This facilitates extremely rapid loading of all onboard ammunition. Replenishment of the onboard ammunition can be accomplished either through two doors in the turret rear, or via the turret compartment.

The T6 can also fire ammunition directly from a ground pile, so bypassing the onboard ammunition storage system. For this purpose two openings and trays have been provided in the turret to hull interface, one on the left side and one on the right side.

Loading is by means of two automatic rammer systems, one for the projectiles and one for the charges.

The T6 uses the same laying and navigation system as used in the G6 and full details of this are given in the entry for the G6 155 mm/45-calibre (6 × 6) self-propelled gun howitzer which is currently deployed by Oman, South Africa and the United Arab Emirates.

An autolaying system is coupled to the laying and navigation system as in the G6 and enables rapid relaying between rounds.

As an independent 36 kW turret power unit is fitted in the turret bustle, the main engine of the chassis can be switched off during firing.

The interface ring between the T6 turret and the T-72 chassis allows for the removal of the Russian T-72 power pack with the turret traversed 90° to the side.

The turret is of all-welded steel armour that provides the crew with protection from small arms fire and shell splinters. In total, eight 81 mm smoke grenade launchers are mounted on the front of the turret. A 7.62 mm SS77 machine gun is fitted on the left turret cupola for local and anti-aircraft defence.

Standard G6 systems fitted to the T6 are the crew communication system (intercom and radio) and the automatic fire detection and suppression system. Mounted under the cradle is a muzzle velocity measurement radar. This feeds information to the onboard computerised fire-control system.

VLAP projectile

In early 1997, the Rheinmetall Denel Munition revealed that it had developed a new 155 mm Velocity-enhanced Long-range Artillery Projectile (VLAP) which has a range of more than 50 km.

In addition to being fired from the current 155 mm/45-calibre G5 (towed) and G6/T6 (self-propelled) artillery systems, VLAP can also be fired from other 155 mm/39 and 155 mm/52-calibre systems using standard artillery charges.

No modifications are required to the actual 155 mm weapon in order to fire the VLAP and there would be no changes to drills or logistics.

Firing trials in South Africa are claimed to have shown that the VLAP has excellent reliability, dispersion and range.

The VLAP has the standard elongated shape with welded nubs which is also used for the ERFB and ERFB-BB projectiles.

In addition to the base bleed unit it also has a rocket motor assist which typically extends the range by 10 km depending on the weapon and charge used.

As the rocket motor extends into the VLAP body there is some loss of HE content compared to standard ERFB projectiles.

Rheinmetall Denel Munition stresses that the VLAP is complementary to the company's existing 155 mm ERFB and ERFB-BB projectiles as it would enable friendly artillery systems to engage threat artillery systems while remaining safe from counter battery fire.

In addition, it would enable targets well to the rear to be engaged, such as command centres, air defence systems and logistics, which are now beyond the range of conventional artillery systems.

The design of the VLAP has been completed, first production units were delivered to an undisclosed export customer in 2001.

T-6 with 25-litre chamber

The company was marketing its 155 mm/52-calibre barrels with a choice of a 23- or 25-litre chamber. Late in 2006 it was stated that future efforts would be concentrated on the 23-litre chamber as this meets the JBMOU.

Specifications
(as on T-72 MBT hull)

	T6
Dimensions and weights	
Crew:	4
Length	
overall:	12.49 m
Width	
overall:	4.75 m
Height	
overall:	3.10 m
Ground clearance	
overall:	0.47 m
Weight	
standard:	46,000 kg
Mobility	
Configuration	
running gear:	tracked
Speed	
max speed:	60 km/h
Range	
main fuel supply:	650 km (est.)
Firepower	
Armament:	1 × turret mounted 155 mm howitzer
Armament:	1 × roof mounted 7.62 mm (0.30) S77 machine gun
Armament:	4 × turret mounted 76 mm smoke grenade launcher (2 × 2)
Ammunition	
main total:	40
Main weapon	
calibre length:	52 calibres
maximum range:	42,000 m (for ERFB-BB. VLAP projectile is 52,500 m)
Rate of fire	
sustained fire:	2 rds/min
rapid fire:	6 rds/min
Operation	
stop to fire first round time:	30 s
out of action time:	30 s
Turret power control	
type:	powered/manual
Main armament traverse	
angle:	360° (60° left/60° right)
Main armament elevation/depression	
armament front:	+75°/-5°
Survivability	
Night vision equipment	
vehicle:	optional
NBC system:	optional
Armour	
hull/body:	steel
turret:	steel

Status
Development complete. Ready for production. The 155 mm/52-calibre T6 was one of the four turret systems evaluated in India in 1995 installed on the locally built Arjun MBT chassis. As of early 2013 India had not selected any 155 mm turret system.

Contractor
Denel Land Systems.

Switzerland

RUAG Defence 155 mm M109 self-propelled howitzer upgrade

Development
The Swiss Army was one of the largest users of the 155 mm M109 series of Self-Propelled Howitzers (SPH) and originally took delivery of 146 M109 with the original short barrel and 435 of the improved M109A1 direct from the US.

To enhance the capabilities of the Swiss Army M109 RUAG Defence was awarded a contract by the Swiss Defence Procurement Agency to develop a new version of the M109 with enhanced capabilities, especially increased range.

RUAG Defence upgraded M109 with 155 mm/47-calibre ordnance and showing crew and onboard ammunition stowage (RUAG Defence) 1333632

RUAG Defence built three prototypes for extensive trials and these were followed by a total of five pre-production systems for technical and troop verification trials that were completed in 1994.

All development work was carried out at the main RUAG Defence facility at Thun where a total of 348 M109A1 were upgraded to the enhanced Pz Hb 88/95 standard with deliveries to the Swiss Army running from 1995 through to 2004.

In the Swiss Army, each M109 regiment has a total of three batteries each with six Pz Hb 88/95 systems which also sometimes referred to as the M109L47.

The L47 in the designation refers to he length of the 155 mm ordnance in calibres.

Description
The RUAG Defence M109 SPH upgrade covers the key areas of firepower, survivability and enhancements to reliability/maintainability.

The overall layout of the RUAG Defence M109 upgrade is almost identical to the standard BAE Systems M109 that is covered in a separate entry in *IHS Jane's Land Warfare Platforms: Artillery and Air Defence*.

It should be noted that design of the RUAG Defence M109 upgrade is modular so the end user can select only those items that meet their own specific operational requirements.

The existing 155 mm ordnance is replaced by a new 155 mm/47-calibre ordnance designed and manufactured in Switzerland that has a 60 groove rifling.

This is chrome plated to extend its life and is fitted with a fume extractor and double baffle muzzle brake.

The new ordnance is provided with a 21-litre chamber to provide a significant increase in range and when firing the old US developed 155 mm M107 High-Explosive (HE) projectile a maximum range of 22.4 km is achieved.

When firing an HE base bleed round and the locally developed Charge 10, this is increased to 35.4 km or 25.3 km using an unassisted projectile and Charge 10.

According to RUAG Defence, the fully chrome plated barrel when used in combination with wear additives in recently developed propellants reduces velocity drop, range loss and dispersion due to reduced barrel erosion.

A barrel temperature system is fitted which warns the crew of higher temperatures to avoid a cook off condition when carrying out sustained fire missions.

A total of 85 RUAG Defence 155 mm/47-calibre ordnance were also supplied to the now defunct RDM Technology of the Netherlands for the M109L47 upgraded systems supplied to the United Arab Emirates.

As an alternative to the standard 21-litre chamber, RUAG Defence can supply a 155 mm/47-calibre ordnance with a 23-litre chamber which allows for up to six modular charges to be used.

Using the top zone charge and Extended Range Full-Bore Base Bleed (ERFB-BB) projectiles the maximum firing range is 40.5 km or 31.6 km with the unassisted ERFB round.

The standard M109 carries of a total of 28 projectiles and changes but the Swiss upgrade includes a new bustle mounted storage system which brings the total up to 40 × 155 mm projectiles and up to 64 propellant containers.

Access to this bustle mounted ammunition is via a sliding door from the crew compartment.

Attached to the right of the 155 mm/47-calibre ordnance is a charge standby magazine for the temporary stowage of four pre-prepared charges.

To increase rate of fire and reduce crew fatigue a Semi-Automatic Loader (SAL) is fitted which has been developed by RUAG Defence.

This SAL is hydraulically operated and connected to the standard hydraulic system of the turret and enables all natures of 155 mm artillery projectile to be loaded at any elevation with positive seat ramming.

Installation of the SAL enables a burst rate of fire of three rounds to be fired in 16 seconds, or a sustained rate of fire of six rounds a minute for two minutes. This SAL is also installed on Swiss upgraded UAE M109L47.

To reduce into action times and therefore enhance overall survivability, a Navigation and Positioning System (NAPOS) has been fitted as standard to the Swiss Army Pz Hb 88/95 that is also used by Austria and Chile.

The Ring Laser Gyro (RLG) from different contractors can be used in the Navigation and Positioning System (NAPOS) element of the upgrade.

NAPOS includes a GPS antenna unit, gunner's control and display unit, travel lock sensor, vehicle motion sensor, drivers display unit and a dynamic reference unit.

This allows almost autonomous operation as well as increased platform survivability by using shoot and scoot techniques to avoid counter battery fire. In addition there is a reduction in target acquisition errors as well as eliminating the time consuming requirement for survey.

In the standard M109 the driver has to leave his protected position in order to unlock the 155 mm ordnance prior to coming into action.

For this upgrade a remotely operated Travel Lock (TL) is fitted which is operated by the driver and not only enhances crew survivability to also reduces into/out of action time. This is also fitted to the Austrian, Chilean and UAE systems.

A new 24 V electrical system has been installed in the upgraded M109 that provides sufficient power to carry out sustained operations on a 24 hour basis.

This electrical system includes a new set of shielded cables and associated distribution boxes in hull and turret. The new electrical system meets MIL-STD-1275 as well as Electro Magnetic Compatibility (EMC) and Electro Magnetic Pulse (EMP) requirements.

A new 16-way slip ring ensures 360° data and power transfer between the hull and turret.

All critical components of the system are equipped with the appropriate test connectors for computer supported trouble shooting.

In addition a 1.2 kW air cooled auxiliary power unit is fitted in the rear of the hull which provides power to operate the complete system with the main Detroit Diesel engine shut down.

Other options are also available on the RUAG Defence M109 upgrade include new passive night vision equipment, fire/detection and suppression system for the crew compartment and full on board computation. A new power pack could also be fitted to improve power-to-weight ratio.

Variants

RUAG Defence baseline M109 upgrade

As well as upgrading existing M109s to the M109/47-calibre standard, RUAG Defence is also marketing refurbished standard M109/39-calibre systems.

According to the United Nations Chile has already taken delivery of 24 M109 in 2004 with the United Arab Emirates taking delivery of 40 systems.

These all retained the standard 155 mm/39-calibre barrel but were upgraded in some areas as mentioned in the description.

Specifications

	RUAG Defence 155 mm M109 self-propelled howitzer upgrade
Dimensions and weights	
Crew:	6
Length	
overall:	11.4 m
hull:	7.28 m
Width	
overall:	3.18 m
Height	
overall:	3.20 m
Ground clearance	
overall:	0.45 m
Weight	
combat:	27,000 kg
Mobility	
Configuration	
running gear:	tracked
Speed	
max speed:	60 km/h
Fuel capacity	
main:	500 litres
Firepower	
Armament:	1 × turret mounted 155 mm howitzer
Armament:	1 × roof mounted 12.7 mm (0.50) M2 HB machine gun
Armament:	10 × turret mounted smoke grenade launcher (2 × 5)
Ammunition	
main total:	40 (155 mm projectiles; up to 64 propellant containers)
Main weapon	
calibre length:	47 calibres
Turret power control	
type:	powered/manual
Main armament traverse	
angle:	360°
Main armament elevation/depression	
armament front:	+75°/-3°
Survivability	
Night vision equipment	
vehicle:	yes
NBC system:	no
Armour	
hull/body:	aluminium
turret:	aluminium

Status

Production complete but can be resumed if additional orders are placed. A total of 348 units have been supplied to the Swiss Army. Delivery of modules of this system have also been made for M109 systems used by Austria and the United Arab Emirates. In 2004 Switzerland sold Chile 24 surplus 155 mm M109A3 SPG that had not been upgraded with the 155 mm/47-calibre barrel. In 2004 Switzerland supplied 24 surplus Swiss Army M109s to the United Arab Emirates but these were not upgraded prior to delivery.

Contractor

RUAG Defence.

Taiwan

XT-69 155 mm self-propelled extended range gun

Development

The XT-69 155 mm self-propelled extended range gun was developed by the Combined Service Forces (CSF) and is identical in layout and concept to the basic XT-69.

Description

It is fitted with a 155 mm/45-calibre weapon which may have been developed from a foreign imported 155 mm calibre towed artillery system.

The 155 mm/45-calibre ordnance has a multibaffle muzzle brake, a fume extractor and a travelling lock that folds back onto the glacis plate when not in use.

Maximum range is said to be 30,000 m with the Extended-Range Full-Bore projectile, and over 35,000 m with a base bleed projectile. A maximum rate of fire of 6 rds/min is claimed to be achieved, due to a pneumatically assisted loading device that enables the weapon to be uniformly loaded at any angle of elevation. Elevation is estimated to be from 0° to +70° and traverse 30° left and right. The weapon is used in conjunction with a battery fire-control system developed by the Sun Yat-sen Scientific Research Institute in Taiwan.

The XT-69 system has a crew of five and is fitted with night vision equipment for the driver and a .50 (12.7 mm) M2 HB anti-aircraft machine gun. No NBC system is fitted.

The hull is based on the US now BAE Systems (previously United Defense) M108/M109 self-propelled artillery system covered in a separate entry. Production of the M109 series is undertaken on an as required basis.

For the export market this is normally the M109A5 while the last version to be produced for the US Army was the M109A6 Paladin that was an upgrade of the older system.

Original XT-69 SPH

This has a much shorter 155 mm ordnance and it is thought that this is no longer in front line service.

Taiwanese 155 mm 6 × 6 self-propelled artillery system

Taiwan has completed a prototype of a truck mounted 155 mm (6 × 6) self-propelled artillery system and available details are provided in a separate entry in *IHS Jane's Land Warfare Platforms: Artillery and Air Defence.*

As of early 2013, it is understood that this 6 × 6 system remained at the prototype stage.

Status

In service with Taiwan. Production complete. Some sources have indicated that a total of 20 of these systems are in service under the production designation of the T-69. This system was never offered on the export market. Taiwanese artillery capability has been further strengthened by the recent arrival of 28 now BAE Systems M109A5 systems from the United States.

Contractor

Combined Service Forces, Taiwan.

XT-69 155 mm self-propelled extended range gun in travelling order
(L J Lamb) 0500428

Turkey

155 mm/52-calibre Firtina self-propelled gun

Development

To meet its operational requirements for a new 155 mm/52-calibre full-tracked self-propelled gun to supplement its older upgraded M44 and M52 155 mm/39-calibre systems, the Turkish Land Forces Command (TLFC) selected a version of the Republic of Korea Samsung Techwin K9 Thunder.

This has been operational with the Republic of Korea Army since the late 1990s.

Following trials with two prototypes, an order for an undisclosed quantity of systems was placed in 2001 with the first batch understood to consist of eight units.

Production commenced in 2002 and the total TLFC requirement could be for as many as 300 systems, funding permitting, with each regiment having three batteries each of six systems. By early 2013 it is estimated that about 250 Firtina 155 mm/52-calibre self-propelled guns had been built for the TLFC.

Within the TLFC the K9 Thunder is normally referred to as the Firtina or TUSpH Storm. Programme management of Firtina has been carried out under the Technical and Project Management Department of the TLFC.

Production and assembly of the Firtina is carried out at the TLFC 1010 Ordnance and Maintenance Plant in Arifiye, who have considerable experience in the overhaul and upgrade of self-propelled artillery systems.

Description

The hull and turret of the Firtina are of all-welded steel armour that provides the occupants with protection from small arms fire and shell splinters.

According to the TLFC, the steel armour provides protection against penetration from small arms fire up to 14.5 mm armour-piercing as well as anti-personnel mines. An NBC system is fitted as standard. It is estimated that maximum armour thickness is 19 mm.

The driver is seated at the front of the hull on the left side, with the power pack compartment to his right and the crew/turret compartment extending to the very rear of the hull.

The driver is provided with a single-piece hatch cover in front of which are three day periscopes. The centre periscope can be replaced by a passive periscope for driving at night.

Turret traverse is a full 360° under power control with weapon elevation from –2.5° to +70° also under power control. Manual controls are provided as a back up.

The turret crew enter the vehicle via the large door in the lower half of the hull, which opens to the right. There is also a hatch cover in the left side of the turret roof with the commander's cupola on the right side. On this is mounted the .50 (12.7 mm) M2 HB machine gun for local and air defence purposes.

The 155 mm/52-calibre ordnance meets the NATO Joint Ballistic Memorandum of Understanding (MoU). A standard 155 mm M107 high-explosive projectile can be fired to a maximum range of 18 km or a M549 rocket-assisted high-explosive projectile to a maximum range of 30 km.

Maximum range depends on the projectile/charge combination but firing an Extended-Range Full Bore – Base Bleed (ERFB-BB) projectile a maximum range of over 40 km is claimed. Conventional bag-type charges can be used as well as the more recent modular charges. As far as it is known, Turkey does not currently manufacture the latter charge type.

Firtina 155 mm/52-calibre self-propelled gun (Christopher F Foss) 1164298

Turkish Land Forces Command 155 mm/52-calibre Firtina self-propelled gun with prototype of locally developed Firtina Artillery Ammunition Support Vehicle to the rear (Christopher F Foss) 1403896

The 155 mm/52-calibre barrel has 48 groves with a twist rate of 1 in 20.

It is auto-frettaged to improve its fatigue life and has a chamber volume of 23 litres. When travelling the ordnance is held in a travel lock that is located on the glacis plate. This folds back into the horizontal position when not required.

Breech mechanism is of the sliding type that opens vertically. The barrel is fitted with a multibaffle muzzle brake and fume extractor. The primer mechanism is automatic.

Ammunition loading is automatic and a burst rate of fire of three rounds in 15 seconds can be achieved. For intense fire a rate of six to eight rounds a minute can be achieved, while for sustained fire missions a rate of fire of two rounds a minute is possible. The automatic projectile loading system is electrically driven and electronically controlled.

Direct and indirect optical sights are provided. The Turkish company Aselsan has developed the onboard computerised Fire-Control System (FCS) for Firtina, which allows each system to receive target information from the battery command post.

The FCS includes a computer, gunner's interface unit, power control unit, power conditioning unit, inertial navigation system with embedded Global Positioning System (GPS), assistant gunner's display unit, tactical data/internet communications system and other subsystems. This allows the Firtina to act as an autonomous unit if required.

The hybrid navigation system determines the Firtina's position in UTM co-ordinates and the position of the barrel in the automatic laying mode.

Geographical north, the howitzers real position and altitude above sea level are determined with accuracy and fed into the onboard computer to calculate the ballistic trajectories required to hit the target.

The onboard ballistic computer ensures fully autonomous artillery technical fire-control. It is linked to other onboard equipment and external fire-control command posts. The software has been developed by using the NATO Artillery Ballistic Kernel (NABK) with a Graphical User Interface (GUI) to the gun display unit.

The gunlaying system receives the firing Quadrant Elevation (QE) and Azimuth (AZ) values from the ballistic computer to engage the barrel to the target automatically in the automatic mode. It also has a semi-automatic backup mode and manual laying with hydraulic hand controller.

A muzzle velocity radar is mounted over the 155 mm/52-calibre ordnance and this feeds information to the onboard computer. Firtina can also be integrated into the TLFC Fire Support Automation Project 2000 (FSAP 2000), also developed by Aselsan which is now in production.

After receiving the fire command, Firtina is able to fire on a target within 30 seconds, and complete a fire mission and relocate to another fire position in 90 seconds. With different elevation angles and charge zones, Firtina is able to hit the same target with three projectiles at the same time.

Turkish Firtina Artillery Ammunition Resupply Vehicle clearly showing the extension to the upper superstructure through which 155 mm projectiles and associated charges are fed to the rear of the Firtina 155 mm/52-calibre SP artillery system (Christopher F Foss) 1424122

Firtina 155 mm/52-calibre self-propelled gun with ordnance elevated and travel lock still deployed (Christopher F Foss) 1296080

Suspension is of the hydropneumatic type with each side having a total of six dual rubber-tyred roadwheels with the drive sprocket at the front, idler at the rear and three track return rollers. The tracks are of the double pin type with replaceable rubber pads.

It is fitted with a 5 W and 50 W VHF 9,600 frequency hopping radio system, with the former used for voice communication and the latter for data transfer.

A combined NBC and ventilation system is provided for each crew member. Each is provided with one NBC mask which in individually integrated into an NBC filtering system.

Firtina Artillery Ammunition Resupply Vehicle

To support its growing fleet of Firtina 155 mm/52-calibre Self-Propelled (SP) artillery systems, the Turkish Land Forces Logistics Command (TLFLC) designed and built the first prototype of the Firtina Artillery Ammunition Resupply Vehicle (FAARV).

This will be followed by second pre-production that is expected to be completed in 2013 before the main production run of between 70 and 80 vehicles is built.

Production will be undertaken at the TLFCS facilities at Arifiye as well as undertaking production of the Firtina 155 mm/52-calibre SP artillery system has extensive experience in the upgrading of armoured fighting vehicles and other weapon systems.

It is expected that one Firtina FAARV will be used to support a troop of three Firtina 155 mm SP artillery systems.

Firtina FAARV features a hull of all welded steel armour that provides the occupants with protection from small arms fire and shell sprinters.

The driver is at the front left with the power pack to the right. The latter is from old decommissioned M48 tanks and consists of an AVDS-1790 series V-12 diesel developing 750 hp that is coupled to al Allison CD 850 series automatic transmission.

This gives the FAARV a maximum road speed of 50 km/h and an operational range of up to 360 km.

Combat weight of the FAARV is 47 tonnes and its normally operated by a crew of three.

The suspension is of the hydropneumatic type along similar lines to that fitted to the Firtina 155 mm/52-calibre SP artillery system and provides an improved ride for the crew across rough-terrain.

To the rear of the drivers position is the raised superstructure which contains 155 mm ammunition (projectiles and charges) and its associated ammunition transfer system which has been developed by the local company of Aselsan.

A total of 96 rounds of 155 mm ammunition are carried in the FAARV plus a similar number of charges. These are fed to the Firtina 155 mm/52-calibre SP artillery system by a protected conveyor system that extends over the front of the FAARV and lines up with the rear of the turret of the weapon.

According to the TLFLC a total of 48 × 155 mm artillery projectiles and their associated charges can be transferred to the Firtina 155 mm/52-calibre SP artillery system in 20 minutes.

The 155 mm projectiles would be fitted with the fuze in the Firtina 155 mm/52-calibre SP artillery system.

Today Turkey uses the old bag type 155 mm artillery charges but is expected to move to a modular charge system in the future.

Optional equipment for the FAARV includes a NBC protection system and a fuel transfer system. A roof mounted .50 M2 HB machine gun is provided for air defence and self-protection on the right side of the superstructure roof.

Specifications

	Firtina
Dimensions and weights	
Crew:	5
Length	
overall:	12 m
hull:	7.42 m
Width	
overall:	3.40 m
Height	
to turret roof:	2.85 m
to top of roof mounted weapon:	3.43 m
Ground clearance	
overall:	0.42 m
Track width	
normal:	547 mm
Length of track on ground:	4.72 m
Weight	
combat:	47,000 kg
Mobility	
Configuration	
running gear:	tracked
Power-to-weight ratio:	21.00 hp/t
Speed	
max speed:	65 km/h
Range	
main fuel supply:	400 km (est.)
Fuel capacity	
main:	850 litres
Fording	
without preparation:	1.50 m
Gradient:	60%
Side slope:	30%
Vertical obstacle	
forwards:	0.75 m
Trench:	2.8 m
Engine	MTU 881 Ka-500, diesel, 1,000 hp
Gearbox	
model:	Allison X-1100-5
type:	automatic
forward gears:	4
reverse gears:	2
Suspension:	hydropneumatic
Electrical system	
vehicle:	28 V
Firepower	
Armament:	1 × turret mounted 155 mm howitzer
Armament:	1 × roof mounted 12.7 mm (0.50) M2 HB machine gun
Ammunition	
main total:	48
12.7/0.50:	500
Main weapon	
calibre length:	52 calibres
Turret power control	
type:	hydraulic/manual
Main armament traverse	
angle:	360°
Main armament elevation/depression	
armament front:	+70°/-2.5°
Survivability	
Night vision equipment	
vehicle:	yes
NBC system:	yes
Armour	
hull/body:	steel
turret:	steel

Status

Production. In service with Turkish Land Forces Command (TLFC).

Contractor

TLFC facilities.

155 mm M52T self-propelled artillery system

Development

The Turkish Land Forces Command (TLFC) had a large fleet of US-built 155 mm M44 and 105 mm M52 self-propelled artillery systems which were manufactured in the 1950s.

By today's standards these have long been obsolete as not only do their weapons have a very short firing range but, as they are powered by a petrol engine, their operational range is very short, around 160 km.

The upgraded M52T has a more fuel efficient MTU diesel engine which gives a significant increase in operational range.

The original 105 mm M52 fired a US-developed HE M1 projectile to a maximum range of 11,270 m while the 155 mm M44 fired a US-developed HE M107 projectile to a maximum range of 14,600 m.

In addition to the-calibre of their weapons, the other main difference between the two systems was that the M44 had an open-top gun compartment with the 155 mm weapon mounted in the forward part while the M52 had the 105 mm gun in a fully enclosed welded-steel armour turret.

Turkish 155 mm M52T self-propelled gun with ordnance elevated (Michael Jerchel) 0085191

Turkish 155 mm M52T self-propelled gun from the rear (Christopher F Foss) 1164299

An unusual feature of the M52 was that the driver was seated in the turret, which could be traversed 60° left and right.

In total, 222 M44 were upgraded to the M44T standard. Additional details of the German-developed M44T upgrade are given in a separate entry in *IHS Jane's Land Warfare Platforms: Armour and Artillery*.

In total, 365 M52 have been upgraded to the M52T standard. Trials of the M52T were completed in late 1994 and series production commenced in 1995. The system is also referred to as the M52T K/M Obus L/39 with the 39 being the-calibre length (39 × 155 mm).

In the TLFC the 155 mm M52T and M44T are now being supplemented by the locally built Firtina 155 mm/52-calibre system, which is based on the South Korean Samsung Techwin K9 Thunder system.

Full details of the Turkish Firtina 155 mm/52-calibre self-propelled artillery system are provided in a separate entry in *IHS Jane's Land Warfare Platforms: Artillery and Air Defence*.

By early 2013 it is estimated that about 250 Firtina had been built for the TLFC with production still underway.

Description

The modifications to the original M52 full-tracked hull have been extensive and in the case of the M52 these have included the replacement of the short-barrelled 105 mm howitzer by a 155 mm 39-calibre barrel fitted with a muzzle brake, the same as that fitted to the German-developed 155 mm M109A3, which in turn is based on the NATO standard 155 mm FH-70 towed howitzer.

The 155 mm/39-calibre ordnance has been manufactured under licence in Turkey at the facilities of MKEK which manufactures a wide range of other weapons and ammunition for the home and export markets.

As the 155 mm ammunition is considerably larger than the 105 mm natures fired by the M52, the ammunition racks have also been modified to accommodate the larger-calibre ammunition.

A similar modification was carried out to the M44T and both weapons can fire a standard HE M107 projectile to a maximum range of 18,000 m, a rocket assisted projectile to a maximum range of 24,000 m or a base bleed projectile to a maximum range of 30,000 m.

The main difference between the M44T and the M52T is that the former's 155 mm ordnance has limited traverse while the M52T 155 mm ordnance is mounted in a turret which can be traversed left and right. When deployed in the firing position a spade is lowered at the rear to absorb some of the recoil.

The Turkish company TT–Izmur has developed the computer and command system for the M52T which allows automatic gunlaying. The system consists of the keyboard and display unit, control unit and sensors including cant and ambient air pressure.

The front-mounted power pack has been upgraded, with the original petrol engine having been replaced by a more fuel-efficient German MTU MB 833Aa 501 diesel which develops 450 hp coupled to the Allison CD-500-4 automatic transmission. Operational range has been increased to 400 km with fuel consumption being quoted as 1.28 litres/km.

Other improvements include fire detection and suppression systems, new electrical system, new driver's instrument panel as well as improvements to the steering and a cold starting system for the engine.

In addition the top engine deck and exhaust system have been modified and the suspension upgraded to take into account the increased weight of the vehicle. The original US-type tracks have been replaced by a Turkish T317 track.

A roof-mounted .50 (12.7 mm) M2 HB machine gun is provided for local and anti-aircraft defence and a bank of four 76 mm electrically operated smoke grenade launchers is mounted either side of the turret, firing forwards.

The original 105 mm M52 did not have a spade due to the low recoil force of the 105 mm main armament. The upgraded M52T has a spade that is lowered to the ground before firing commences.

Specifications

	M52T
Dimensions and weights	
Crew:	5
Length	
overall:	8.80 m
Width	
overall:	3.20 m
Height	
to top of roof mounted weapon:	3.30 m
Ground clearance	
overall:	0.49 m
Track	
vehicle:	2.602 m
Track width	
normal:	533 mm
Length of track on ground:	3.793 m
Weight	
combat:	29,500 kg
Mobility	
Configuration	
running gear:	tracked
Power-to-weight ratio:	15.25 hp/t
Speed	
max speed:	60 km/h
Range	
main fuel supply:	420 km
Fuel capacity	
main:	530 litres
Fording	
without preparation:	1.219 m
Gradient:	60%
Vertical obstacle	
forwards:	0.914 m
Trench:	1.828 m
Engine	MTU MB 833Aa 501, V-6, water cooled, diesel, 450 hp at 2,300 rpm
Gearbox	
model:	Allison CD-500-4
forward gears:	4
reverse gears:	2
Suspension:	torsion bar
Electrical system	
vehicle:	24 V
Firepower	
Armament:	1 × turret mounted 155 mm howitzer
Armament:	1 × roof mounted 12.7 mm (0.50) M2 HB machine gun
Armament:	8 × turret mounted 76 mm smoke grenade launcher (2 × 4)
Ammunition	
main total:	30
Main weapon	
calibre length:	39 calibres
maximum range:	18,000 m (M107 projectile)
maximum range:	24,000 m (RAP projectile)
maximum range:	30,000 m (ERFB-BB)
Turret power control	
type:	powered/manual
Main armament traverse	
angle:	120° (60° left/60° right)
Main armament elevation/depression	
armament front:	+65°/-5°
Survivability	
Night vision equipment	
vehicle:	optional
NBC system:	no
Armour	
hull/body:	steel
turret:	steel

Status

Production complete. In service with the Turkish Army. This upgraded system was never offered on the export market. The original M52 may still be used by Jordan who may also have 20 older M44.

Contractor

Upgrade work was carried out in Turkish Army facilities.

United Kingdom

BAE Systems AS90 (Braveheart) 155 mm self-propelled gun

Development

During production of the FH-70 155 mm towed howitzer the now BAE Systems (at that time Vickers Shipbuilding and Engineering Limited) identified a world need for improved self-propelled artillery.

Subsequent concept studies offered a cost-effective solution using the companies private venture GBT 155, a turret based on the 155 mm 39-calibre ordnance of the FH-70 and designed specifically for export.

Conceptual studies commenced in 1982 for the design of a modern artillery hull using a variant of the GBT 155, incorporating all and more of its features but having a larger turret ring to maximise crew space and a frontally positioned power pack.

The project definition for AS90 was completed in March 1985, and the first prototype was shown at the British Army Equipment Exhibition in June 1986.

Following the exhibition, a series of firing trials augmented by mobility trials were undertaken, leading to the award of Live Crew Clearance from the British Ordnance Board in March 1987.

A series of endurance and automotive trials was then initiated, with AS90 travelling some 3,000 km and firing in excess of 1,500 rounds without a mission-relevant failure. These trials, completed in June 1987, covered all aspects of army use.

Following the withdrawal of the UK from the International 155 mm SP70 project late in 1986, the international search for a new self-propelled howitzer started.

This eventually resulted in the selection in June 1989 of the 155 mm AS90 self-propelled howitzer to meet future requirements for the British Army.

The British Army official designation for the AS90 is Howitzer, Self-propelled 155 mm, L131 AS90 155 mm.

The fixed-price contract covered the completion of AS90 development, production of 179 systems and the provision of initial spares. The AS90 artillery system was accepted for service with the British Army in May 1992.

First production AS90s were completed in late 1992, with the first regiment becoming operational in 1993.

AS90s are operational with the Royal Artillery in Germany and the UK and also deployed to Canada for training purposes. They have also served with the Peace Implementation Force (IFOR) operating in the former Yugoslavia. It also saw action in Iraq during 2003.

In 1994, a five-week In Service Reliability Demonstration (ISRD) trial took place in the UK involving a complete battery of eight AS90s. These trials covered mobility and firing reliability equivalent to 80 battlefield days. A total of 6,000 km was travelled and 11,200 rounds fired.

The 155 mm 52-calibre barrel was fitted to one of the guns, proving its performance reliability for future enhancements to the gun.

Final AS90s were delivered to the British Army in the first quarter of 1995, with four systems retained by the company for trials.

In mid-1992 the AS90 (Braveheart) turret system was selected by Poland. For the export market the latest 155 mm/52-calibre AS90 is known as Braveheart. The first two AS90 155 mm/52 turrets were delivered to Poland late in 2000 for installation on a locally designed full-tracked chassis. In late 2006 it was stated that Poland would order a total of 48 systems.

Details of the Polish system, which is called the Krab, and details of this are

British Army AS90 155 mm/39-calibre self-propelled guns being upgraded at the BAE Systems Barrow-in-Furness armament facility (BAE Systems)
1347459

Desert AS90 155 mm/52-calibre ordnance during trials in the Middle East
0073448

provided in a separate entry in *IHS Jane's Land Warfare Platforms: Artillery and Air Defence*.

As a result of a re-organisation of the Royal Artillery, a number of AS90s are now surplus to requirements.

In October 2009 the UK MoD announced that the British Army fleet of AS90 155 mm self-propelled guns would be reduced by 35 per cent. At this time the active AS90 fleet stood at 132 systems.

This is the total number of AS90 upgraded by BAE Systems, out of the 179 systems delivered to the Royal Artillery.

Description

The hull of AS90 is of all-welded steel armour construction with a maximum thickness of 17 mm. Steel rather than aluminium was chosen as it is not only simpler to fabricate but also easier to repair in the field.

The hull and turret of the AS90 provide the occupants with protection from small arms fire and shell splinters.

The driver sits at the front left with the power pack to his right. He normally enters via a single-piece hatch cover over his position that hinges to the left side. This can be locked open for driving in the head-out position. In front of the driver is a single day periscope with a blackout blind and laser filter.

The day periscope can be changed quickly for a passive night vision device by the driver while under armour and without breaking the NBC seal. The driver has a fully adjustable seat and can also enter and leave his position via the fighting compartment at the rear. The driver steers the vehicle using a steering wheel rather than sticks.

The power pack, consisting of the Cummins VTA-903T-660 engine, Renk LSG 2000 transmission and cooling system, can be removed and replaced as a complete unit in well under an hour.

The engine compartment is provided with a fire detection and suppression system that can be operated automatically or manually. Hinged louvres and access covers are provided above the engine compartment for ease of access.

The barrel clamp is located on the front of the vehicle to provide a travelling support for the gun. The barrel clamp is operated from within the vehicle from the driver's position.

AS90 is powered by the Cummins VTA-903T-660 14.8-litre V-8 diesel developing 660 bhp at 2,800 rpm. This is a more powerful version of the engine installed in the US Army's M2/M3 Bradley Infantry Fighting Vehicle/Cavalry Fighting Vehicle and the Multiple Launch Rocket System. The VTA-903T-660 has plenty of stretch potential, having already been run at 750 hp for trial purposes.

The engine is coupled to a fully automatic German Renk Model LSG 2000 automatic transmission, with four forward and two reverse gears.

A 4-stroke diesel-powered auxiliary power unit is installed in the forward part of the fighting compartment, so the main diesel engine does not need to be constantly run to operate the turret systems.

The area to the rear of the engine compartment is the fighting compartment, which extends to the rear of the vehicle. There is a large door opening to the rear of the hull for access and ammunition resupply, and mounted either side of this on the outside are two integral armoured containers which can be used to store supplies and other equipment. Externally mounted on the turret shell are five stowage bins, one for each crew member's use.

The vehicle runs on 12 double roadwheels, six either side, each having hydropneumatic suspension, with a drive sprocket at the front and an idler at the rear. The use of hydropneumatic suspension contributes to the increased headroom in the fighting compartment by eliminating the false floor needed with torsion bar suspension, in addition to giving AS90 a lower profile.

The system has been designed to give the vehicle, with its large wheel travel, progressive rates of springing and powerful damping on every wheel station and a very smooth ride, enabling high average speeds to be maintained across rolling irregular terrain without impairing the safety and comfort of the crew.

The suspension was developed by Air Log with the current design authority being Horstman Defence Systems Limited.

The turret is of all-welded steel construction, with a maximum thickness of 17 mm and it has a large turret ring diameter of 2.7 m. The commander has a cupola on the right side of the turret roof. An air sentry's hatch with pintle mounting for secondary armament is provided on the left-hand side of the turret. There is also a large hatch on the left-hand side of the turret.

The commander is positioned behind the layer on the right-hand side of the turret. Both are provided with seats that have adjustments for height and rotation. The shell and charge loaders stand on the left-hand side of the 155 mm ordnance and are provided with seats, which fold down from the wall of the hull. Turret traverse is through a full continuous 360° and weapon elevation is from –5° to +70°. Elevation and traverse drives are both electric and operate at a rate of 10°/s. Full manual controls are provided as a back-up.

An electric traverse and elevation system, with silent operation from the integral power supply was chosen rather than hydraulic, since it was considered to be safer, more reliable and would require less maintenance. The only hydraulics in the turret are those applicable to the balancing gear, loading tray and shell rammer.

AS90 is fitted with the now BAE Systems 155 mm 39-calibre barrel as standard. The recoil system has two diametrically opposed buffers and one recuperator, each with its own integral reservoir, and a maximum recoil length of 800 mm. The barrel can be drawn out through the front of the turret for replacement in less than one hour.

The removal of the complete elevating mass is simple as the saddle is constructed in two parts and can easily be detached from the trunnions.

The 155 mm ordnance has a double-baffle muzzle brake, fume extractor and a split-block breech. The breech combines the speed of action associated with the sliding block principle and the robustness of the Crossley pad obturating system and has an integral 12-round primer magazine.

AS90 can fire a burst of three rounds in less than 10 seconds, an intense rate of 6 rds/min for three minutes and a sustained rate of fire of 2 rds/min. The ordnance incorporates an electrically initiated percussion firing mechanism using standard DM191A1 or US M282 igniter tubes. There are 48 projectiles and charges carried in the turret and the ordnance will fire all standard NATO projectiles including the US-developed HE M107 and extended range types.

Of the 48 projectiles carried, 31 are stowed in the turret bustle in four magazine modules, each of which has a motor that moves the required projectile to the correct position. This enables it to be pulled forward onto a Shell Transfer Arm (STA). The STA pivots about the left trunnion and aligns with either magazine tray, irrespective of gun elevation. The STA realigns to lock with the elevating mass and the projectile is moved sideways by a motor-driven shell clamp on the loading tray. The loading tray is brought in line with the breech and then rammed by the flick rammer. The charge is then loaded, the breech closed and the ordnance fired by the commander or layer. The loading system has full manual back-up and safety interlocks. The remaining 17 projectiles are statically stowed in the chassis.

Using standard ammunition and charges, the 39-calibre barrel enables a range of 24,700 m to be achieved. Using an assisted projectile, a range of over 30,000 m can be achieved. AS90's area coverage is enhanced even more when the 52-calibre barrel is fitted, with a maximum range of 30,000 m with conventional projectiles and 40,000 m with assisted projectiles.

As of early 2013 there were no plans for the UK to fit its 155 mm AS90 system with a new 155 mm/52-calibre ordnance, although this has been tested.

AS90 is fitted with a full navigation and auto-lay capability using the MAP's standard DRU based on ring laser gyro technology giving the gun complete autonomy. This enables the gun to be laid automatically in bearing as well as elevation.

AS90 155 mm of 26 Regiment Royal Artillery deployed on exercise in the Czech Republic in 2010 (IHS/Nick Brown) 1364223

British Army AS90 155 mm/39-calibre self-propelled artillery system with ordnance in travel lock (Christopher F Foss) 1452907

Muzzle velocity measuring, air conditioning and NBC equipment are provided as standard, the last two being mounted on the rear of the turret bustle.

The then Air Log supplied the complete running gear for AS90. This includes an advanced hydropneumatic suspension system, a Diehl double pin track and sprockets, GLS top rollers and road wheels and the then Air Log designed and manufactured idler gear. The idler gear incorporates a hydraulic track tensioner with an overload protection device, allowing the idler wheel to retract if excessive track tension is built up.

Thales UK designed and developed the day/night direct fire telescope for the AS90 which is designated the DFS90.

A total of eight 12 V batteries are fitted, four in the hull and four in the turret.

Variants

AS90 with 155 mm/52-calibre ordnance

Late in 1998, the UK MoD announced that it was to award the now BAE Systems, Phase Two of a major programme to equip the current in service 155 mm AS90 self-propelled artillery system with a 155 mm/52-calibre Extended Range Ordnance (ERO) and a new Modular Charge System (MCS).

Phase 1, which ran from May 1997 through to October 1998, covered design and development of the ERO and selection of the MCS.

The ERO is a 155 mm 52-calibre barrel manufactured by the now BAE Systems, which could replace the current 155 mm 39-calibre barrel. The new barrel, which is not required to be internally chrome plated, would be fitted at unit level.

The new 52-calibre barrel would use the existing muzzle brake, fume extractor and breech mechanism. The ERO conforms to the Joint Ballistic Memorandum of Understanding signed by France, Germany, Italy, UK and the US.

The current 155 mm 39-calibre barrel achieves a maximum range, firing assisted ammunition, of 30 km, but when fitted with a 155 mm 52-calibre barrel this is raised to 40 km.

Eight AS90 systems arrived at Barrow-in-Furness in March 2001 from 26 Regiment, Royal Artillery. These were expected to be the first to be fitted with the new 155 mm/52-calibre barrel and will be used for system qualification tests and firing trials before rejoining the regiment.

The now Rheinmetall Denel Munition was the winner of the MCS competition and the contract could be worth GBP100 million over a period of 10 years if all options were exercised.

The Rheinmetall Denel Munition MCS for the AS90 is designated the M90 Bi-Modular Charge System (BMCS) and comprises M91A1 low-zone (Z1 and 2) and M92A1 high-zone (Z3 to 6).

In order to fire the existing 155 mm L15 projectile from the 155 mm/52-calibre ordnance, an enlarged nylon obturator band will have to be fitted to each L15 projectile.

The MoD awarded the now BAE Systems the Phase 2 contract in 1999, which was expected to run through to May 2004 to cover full development, trials and production of the ERO and MCS.

The In Service Date (ISD), with first regiment of the Royal Artillery fully converted and trained with ERO/MCS, was expected to be late 2003.

In mid-2003 it was revealed that RO Defence had halted all work on the ERO/MCS programme and presented a number of options to the UK DPA.

As of early 2013 all of the AS90 deployed by the Royal Artillery retained the 155 mm/39-calibre barrel and were still using the older bagged charge system.

British Army AS90 upgrade

In 2009 BAE Systems, completed on time and on budget, the AS90 Capability Enhancement Programme 1 (CEP 1) for the Royal Artillery.

Overall design authority for the AS90 is BAE Systems, Armament Production Facility (APF) at Barrow-in-Furness where the AS90 was originally designed and built.

The key part of the AS90 CEP 1, which got underway in 2003, was to remove obsolescence from the Line Replaceable Units (LRU) which were originally developed over 20 years ago.

Upgrades include a new Turret Control Computer (TCC), Layers Display Unit (LDU) and Data Couplers (DC) with the new generation LRU having growth potential for future upgrades.

British Army AS90 155 mm/39-calibre self-propelled artillery system with ordnance in travel lock (Christopher F Foss) 1452908

In addition, a major upgrade has been carried out on the Turret Electronics Systems (TES) and at the same time an open electronic architecture has been added which will enable any future CEP to be more rapidly carried out on AS90 without re-working the turret.

Prior to the CEP, a total of 32 AS90 were upgraded under a Urgent Operational Requirement (UOR) Environment Enhancement Package (EEP) for Gulf War 2 (Operation Telic).

A number of sub-systems were enhanced under this UOR to enable the AS90 to operate in high ambient temperatures encountered in Iraq including modification of the existing Auxiliary Power Unit (APU).

Most of these were originally deployed to Iraq and according to the National Audit Office achieved a 95 per cent availability rate.

All AS90 have been now returned from Iraq to Germany and the UK. All British Army operational AS90s have also been fitted with the General Dynamics UK Bowman digital communications system.

First AS90 CEP technical demonstrator was completed in 2004 followed by pre-production hardware, assessment phase and trials that were successfully completed in 2006.

Under CEP 1 a total of 132 AS90 have been upgraded of which 68 were upgraded in Barrow-in-Furness, 41 in Germany, 10 in BATUS and 10 at bases in the UK. These were followed by three remaining AS90 which were returned from Iraq.

Given that the projected Out of Service Date (OSD) of AS90 is probably around 2030, a number of future capability enhancements (CEP 2 and CEP 3) have been derived.

This has been carried out in close consultation between BAE Systems and the Defence Equipment & Support (DE&S) organisation Joint Integrated Project Team (JIPT) at Abbey Wood and the Royal Artillery.

This aims to provide a holistic approach in order to introduce further capability in an affordable manner.

These additional capabilities do very much depend on funding, but does provide a flavour of what could be provided in a possible chronological order:

(1) Adding the EEP throughout the British Army AS90 fleet
(2) Barrel thermal monitoring to reduce potential of a cook off when the system is carrying out sustained fire missions in high ambient temperatures
(3) Extending the range and accuracy by the introduction of the 155 mm Excalibur precision guided munition (PGM), early versions of the which have been successfully fired by the M777 artillery system in Afghanistan deployed by Canada and the US. The upgrading of the loading rammer to allow full stroke ramming may be developed to give future guided projectiles a smoother ram and air projectile qualification. In addition, further automation of the Ammunition Handling System could take place to comply with current and future health and safety legislation.
(4) The drivers compartment would be upgraded to include a new instrument panel which could provide AS90 with the option to introduce an enhanced Health Usage Monitoring System (HUMS) and a user friendly diagnostic capability. In addition, live way points could be transmitted to the driver via wireless data couplers
(5) To reduce into and out of action times a barrel automatic stowing system would be introduced that would eliminate the requirement a crew member to leave the gun and physically guide the barrel into the stow position
(6) An Electronic Technical Publication Manual (ETPM) would be introduced to enhance overall system maintainability
(7) The current conventional primer ignition system would be replaced by a laser ignition system which would provide the AS90 firing platform with a much simpler firing chain that would enhance overall reliability and availability
(8) An on board ballistic computation would be provided which is claimed will enhance autonomy and link closely to the C4I UK initiative. The recently installed TCC has this capability but is not currently used
(9) Overall system accuracy may also be enhanced by monitoring muzzle velocity on a round by round basis
(10) On the automotive side a number of key obsolescence issues around the Cummins power pack and improvements to the running gear have been identified. If carried out these would enhance mobility, especially in hot climatic conditions.

AS90 and 155 mm Excalibur PGM

In 2008 it was revealed that an AS90 155 mm/39-calibre system was to be tested firing the US Army/Raytheon 155 mm M982 Excalibur Precision Guided Munition (PGM) which has already been used in combat by Canada and the US.

This trial was successfully carried out in the US but as of early 2013 there were no firm plans for the UK to field the Excalibur PGM.

Upgraded AS90 for Desert Operations

A total of 32 AS90 were upgraded for desert operations in 2003 and some of these were also fitted with an auxiliary power unit.

AS90 support contract

In 2005 BAE Systems announced that it had been awarded a GBP60 million contract by the then UK Defence Logistics Organisation covering a period of five years to support spares for the AS90 self-propelled artillery systems used by the Royal Artillery.

This is known as an Equipment Support Agreement (ESA), under which BAE Systems has taken management control of supplying spare parts and repairs on demand. It has also implemented a help desk to provide co-ordinated technical support to the user.

G6 with AS90 turret

For trials purposes in South Africa, the complete turret of the AS90 with a 155 mm/52-calibre ordnance has been installed on the South African Denel Land Systems 155 mm G6 (6 × 6) self-propelled artillery system chassis.

AS90 Turret on T-72 hull

This was developed to meet the requirements of India but no production orders were placed and it is no longer being marketed.

Desert AS90

As a private venture, the now BAE Systems developed Desert AS90 which has a number of improvements for operations in the high ambient temperatures encountered in the Middle East.

Major improvements incorporated in Desert AS90 include:

- improved engine cooling system
- improved transmission oil cooling system
- enhanced transmission gear range
- upgraded auxiliary power unit cooling system
- solar-reflective paint and thermal cover for turret roof
- enhanced air conditioning
- improved resistance to sand ingress in magazine
- German Diehl Type 940 double pin track.

Desert AS90 has undergone extensive climatic trials in the UK, at Yuma Proving Ground in Arizona and two series of trials in the Middle East during which it operated in temperatures of up to +60°C.

AS90 155 mm barrel for naval applications

Specifications

	AS90 Braveheart
Dimensions and weights	
Crew:	5
Length	
overall:	9.90 m
hull:	7.20 m
Width	
overall:	3.40 m
Height	
overall:	3.00 m
Ground clearance	
overall:	0.41 m
Track width	
normal:	550 mm
Length of track on ground:	4.59 m
Weight	
combat:	45,000 kg
Ground pressure	
standard track:	0.9 kg/cm^2
Mobility	
Configuration	
running gear:	tracked
Power-to-weight ratio:	14.66 hp/t
Speed	
max speed:	55 km/h
Range	
main fuel supply:	370 km (est.)
Fuel capacity	
main:	750 litres (est.)
Fording	
without preparation:	1.50 m
Gradient:	60%
Side slope:	25%
Vertical obstacle	
forwards:	0.88 m
Trench:	2.8 m
Engine	Cummins VTA-903T-660, V-8, diesel, 660 hp at 2,800 rpm

	AS90 Braveheart
Gearbox	
model:	Renk LSG 2000
type:	automatic
forward gears:	4
reverse gears:	2
Suspension:	hydropneumatic
Electrical system	
vehicle:	48 V
Batteries:	8 × 12 V
Firepower	
Armament:	1 × turret mounted 155 mm howitzer
Armament:	1 × roof mounted 7.62 mm (0.30) machine gun
Armament:	10 × turret mounted smoke grenade launcher (2 × 5)
Ammunition	
main total:	48
7.62/0.30:	1,000
Main weapon	
calibre length:	39 calibres
Turret power control	
type:	electric/manual
Main armament traverse	
angle:	360°
Main armament elevation/depression	
armament front:	+70°/-5°
Survivability	
Night vision equipment	
vehicle:	yes
NBC system:	yes
Armour	
hull/body:	steel
turret:	steel

Status
Production complete. A total of 179 were supplied to the British Army of which 132 have been upgraded. A number of surplus British Army AS90 systems are now being offered on the export market.

Contractor
BAE Systems.

United States

BAE Systems M110 series of 8 in (203 mm) self-propelled howitzers

Development
Future US Army requirements for new heavy artillery emphasised air transportability, time into and out of action, common parts and interchangeability area of fire. A feasibility study encompassing these requirements was carried out, presented and approved at a meeting in January 1956.

The Pacific Car and Foundry Company (subsequently PCF Defense Industries) submitted a concept study for a new family of full-tracked armoured self-propelled weapons and was subsequently awarded a contract for the design, development and construction of six prototype vehicles: two 175 mm self-propelled guns designated the T235, three 203 mm (8 in) self-propelled howitzers designated the T236 and one 155 mm self-propelled gun designated the T245.

Major design features included the interchangeability of the 175 mm gun, 8 in howitzer and 155 mm gun in a common mount, on a common hull; use of the 8 in howitzer and 155 mm gun field pieces and portions of the M17 standard mount; and drastic reductions in size and weight over conventional equipment which was made possible through a new hydraulic lockout system.

The hull was also considered suitable for use as a light recovery vehicle and, in 1957, the programme was expanded to include both armoured (T120) and unarmoured (T119 and T121) recovery vehicles.

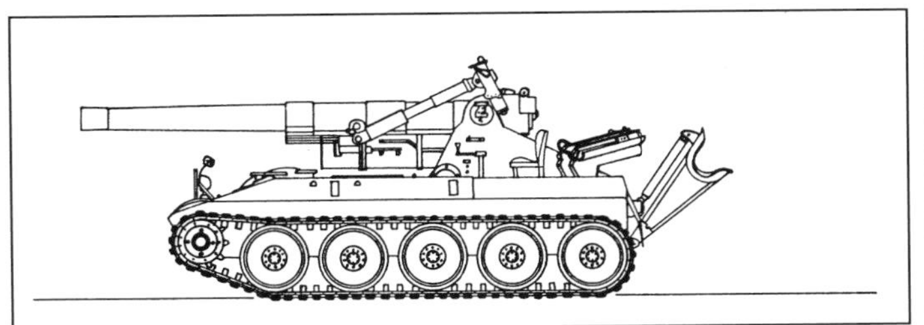

M110 203 mm self-propelled howitzer in travelling order 0500429

M110A2 203 mm self-propelled howitzer in service with Jordan (Paul Beaver) 0533760

Ordnance service tests with the first prototype hull began late in 1958. However, in 1959, a policy was established that diesel rather than petrol engines would be used for future vehicles and three of the prototypes, the T235, T236 and the T120, were retrofitted with Detroit Diesel Model 8V-71T diesel engines and designated the T235E1, T236E1 and the T120E1.

Trials with the T235E1 and the T236E1 were completed early in 1961 and in March both vehicles were standardised, the T235E1 as the M107 and the T236E1 as the M110. The T120E1 was subsequently standardised as the M578 ARV (but the T245, T119 and the T121 were not developed further).

In June 1961, the Pacific Car and Foundry Company (which is no longer involved in defence) was awarded an initial production contract for both the M107 and the M110.

First production vehicles were completed in 1962 and the first M110 battalion was formed at Fort Sill early in 1963. Production of the M110 was also undertaken at a later date by the FMC Corporation of San José and Bowen-McLaughlin-York (BMY) of York, Pennsylvania.

Both these companies merged to become United Defense LP. Today, the latter company is called BAE Systems. Original production of the M110 was completed in the late 1960s, by which time about 750 vehicles had been produced. In FY78, 209 M110A2s were ordered at a cost of USD109.3 million for delivery from 1979.

The first M110A1 was produced by Bowen-McLaughlin-York in May 1980 (it produced its first M110 during 1965). The first M110A2 was produced in February 1980.

In many countries, for example France, Germany, the UK and the US, the M110A2 has been replaced by the Lockheed Martin Missiles and Fire-Control, 227 mm Multiple Launch Rocket System. The main reason the M110 series remained in service for so long was its tactical nuclear capability.

According to United Nations sources, the following quantities of surplus M110 self-propelled howitzers were supplied, often free of charge by Germany, Netherlands and the US between 1992 and 2010:

From	To	Quantity	Delivery dates
Germany	Greece	72	1994
Germany	Turkey	131	1994
Netherlands	Bahrain	13	1994
US	Bahrain	49	1997
US	Greece	84	1992/1993
US	Jordan	18	1997
US	Morocco	60	1997
US	Spain	52	1992/1994
US	Turkey	72	1992/1993

It should be noted that BAE Systems is no longer marketing the M110 8 in (203 mm) self-propelled howitzer or the 175 mm M107 that shares the same hull.

Description
The M110 203 mm (8 in) Self-Propelled Howitzer (SPH) is normally operated by a team of 13, five of whom (commander, driver and three gunners) are carried on the gun, with the rest in the M548 tracked cargo carrier which also carries

M110A2 203 mm self-propelled howitzer in service with Japan 1365009

The M548 tracked carrier is used to support the M110A2 203 mm self-propelled howitzer by some countries (US Army) 1296060

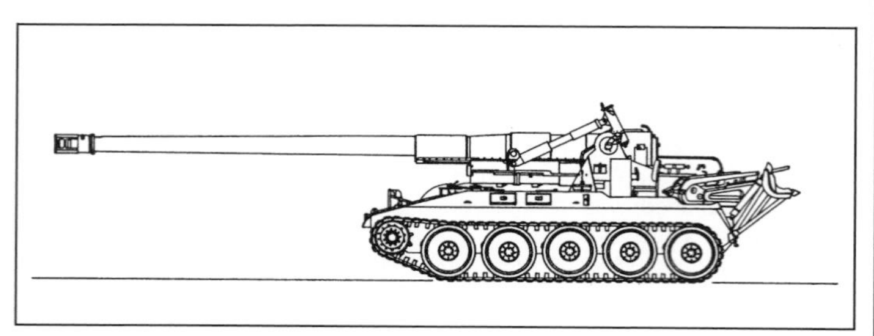

M110A2 203 mm self-propelled howitzer in travelling order 0500430

the ammunition. The M548 unarmoured full-tracked carrier shared many common components with the now BAE Systems M113 series of full-tracked armoured personnel carriers. Production of the M548 was completed some time ago.

The hull of the M110 is identical to that of the M107 and is made of all-welded armour and high-tensile alloy steel with the driver at the front of the hull on the left, the diesel power pack to his right and the main 203 mm (8 in) armament at the rear.

The driver, who is the only member of the crew to be seated under armour, has a single-piece hatch cover in front of which are three M17 day periscopes. The middle M17 day periscope can be replaced by a passive night vision device.

The Detroit Diesel Model 8V-71T engine is coupled to the Allison Transmission XTG-411-2A cross-drive transmission at the front of the hull.

The torsion bar suspension either side consists of five dual rubber-tyred roadwheels with the drive sprocket at the front and the fifth road wheel acting as the idler. There are no track-return rollers.

Attached to each road wheel is a hydraulic cylinder that serves as a shock-absorber, a hydraulic bump stop and a suspension lockout which transmits recoil shock directly to the ground. The tracks are of the single-pin type with removable rubber pads.

The M110 has night vision equipment but no NBC system or amphibious capability.

The M110 is armed with a 203 mm (8 inch) M2A2 howitzer in an M158 mount and has an elevation of +65°, a depression of –2° and a traverse of 30° left and right. Elevation, depression and traverse are hydraulic with manual controls for emergency use. The weapon has a hydropneumatic recoil system, an interrupted screw breech block and a percussion firing mechanism.

Mounted at the rear of the hull on the left side is a rammer and loader assembly which lifts a projectile from the rear or left side of the vehicle, positions it and rams it into the chamber. Mounted at the rear of the hull is a large hydraulically operated spade that is lowered into position before firing begins.

The 8 in (203 mm) M2A2 howitzer fires the following types of conventional ammunition:

HE (M106) with the projectile weighing 92.53 kg, a maximum muzzle velocity of 587 m/s and a maximum range (Charge 7) of 16,800 m;

HE (M404) (carries 104 M43A1 grenades) with a projectile weighing 90.72 kg, a maximum muzzle velocity of 587 m/s and a maximum range (Charge 7) of 16,800 m.

The ammunition is separate loading. Two rounds are carried on the vehicle and the rest in the supporting vehicle. The normal rate of fire is one round every two minutes but two rounds a minute can be fired for short periods.

Fire-control equipment consists of a panoramic sight M115 (magnification of ×4 and 10° field of view) for indirect fire, telescope M116C (magnification of ×3 and 13° field of view) for direct fire, elevation quadrant M15 and gunner's quadrant M1A1.

M110A1 and M110A2

In 1969, the United States Army Armament Command began the development of a new version of the M110, which would have a longer range and fire new types of improved 203 mm ammunition. This was standardised as the M110A1 in March 1976 and entered service in January 1977. The M110A1 replaced both the M110 and M107, which were phased out of service with the US Army in Europe by 1980. It then cost less than USD100,000 to convert the M107/M110 to the M110A1/M110A2 standard.

The M110A1 has a new and much longer 203 mm barrel called the M201, a direct fire elbow telescope, the M139, and a hull identical to the M110's. The M110A2, which was standardised in 1978, is the M110A1 fitted with a double-baffle muzzle brake and can fire Charge 9 of the M118A1 propelling charge whereas the earlier M110A1 can fire only up to Charge 8. The weapon can fire the following projectiles:

- HE (M106) with the projectile weighing 92.53 kg, a maximum muzzle velocity of 711 m/s and a maximum range (Charge 8) of 22,900 m
- HE (M404) with the projectile weighing 90.72 kg, a maximum muzzle velocity of 711 m/s and a maximum range (Charge 8) of 17,200 m. This is also referred to as an Improved Conventional Munition and carries 104 anti-personnel grenades
- HE (M509A1) carrying 180 anti-personnel/anti-materiel grenades with a maximum range (Charge M188A1) of 22,900 m. This is also referred to as an Improved Conventional Munition and carries dual-purpose grenades
- HERA (M650) with a maximum range (Charge 9) of 30,000 m.

Specifications

	M110	M110A2
Dimensions and weights		
Crew:	5	5
Length		
overall:	7.467 m	10.731 m
hull:	5.72 m	5.72 m
Width		
overall:	3.149 m	3.149 m
Height		
overall:	2.93 m (top of barrel)	3.143 m (top of barrel)
hull:	1.475 m	1.475 m
Ground clearance		
overall:	0.44 m	0.393 m
Track		
vehicle:	2.692 m	2.692 m
Track width		
normal:	457 mm	457 mm
Length of track on ground:	3.936 m	3.936 m
Weight		
standard:	24,312 kg	25,492 kg
combat:	26,534 kg	28,350 kg
Ground pressure		
standard track:	0.76 kg/cm²	0.76 kg/cm²
Mobility		
Configuration		
running gear:	tracked	tracked
Power-to-weight ratio:	15.26 hp/t	14.28 hp/t
Speed		
max speed:	56 km/h	54.7 km/h
Range		
main fuel supply:	725 km	523 km
Fuel consumption		
road:	1.568 litres/km	1.88 litres/km
Fuel capacity		
main:	1,137 litres	984 litres
Fording		
without preparation:	1.066 m	1.066 m
Gradient:	60%	60%
Side slope:	30%	30%
Vertical obstacle		
forwards:	1.016 m	1.016 m
Trench:	2.362 m	1.905 m
Engine	Detroit Diesel Model 8V-71T, turbocharged, water cooled, diesel, 405 hp at 2,300 rpm	Detroit Diesel Model 8V-71T, turbocharged, water cooled, diesel, 405 hp at 2,300 rpm
Gearbox		
model:	Allison Transmission XTG-411-2A	Allison Transmission XTG-411-2A
forward gears:	4	4
reverse gears:	2	2
Suspension:	torsion bar	torsion bar
Electrical system		
vehicle:	24 V	24 V
Batteries:	4 × 12 V	4 × 12 V
Firepower		
Armament:	1 × hull mounted 203 mm M2A2 howitzer	1 × hull mounted 203 mm M201 howitzer
Ammunition		
main total:	2	2

	M110	M110A2
Turret power control		
type:	hydraulic/manual	hydraulic/manual
Main armament traverse		
angle:	60° (30° left/30° right)	60° (30° left/30° right)
Main armament elevation/depression		
armament front:	+65°/-2°	+65°/-2°
Survivability		
Night vision equipment		
vehicle:	yes	yes
NBC system:	no	no
Armour		
hull/body:	steel	steel

Status
Production of the M110 has been completed. No longer marketed.

User	M110	M110A1	M110A2
Bahrain	-	-	62 (13 from Netherlands, 49 from US)
Greece	-	-	105
Iran	38[1]	-	about 30 left
Israel	36	-	reserve
Japan	-	-	201 (locally built) (a figure of only 80 was recently quoted)
Jordan	-	-	82 (29 new, rest upgraded)
Korea, South	99	-	estimate (some state only 13, now deployed)
Morocco	-	-	60 (from US)
Pakistan	-	-	40
Taiwan	-	-	60
Turkey	-	-	219

[1] Total delivered, some destroyed or captured in Iran-Iraq War
In some of the above countries these weapons are held in reserve.

Contractor
Pacific Car and Foundry Company, (now PCF Defense Industries) Renton, Washington.
FMC Corporation, San José, California.
Bowen-McLaughlin-York (BMY), York, Pennsylvania.
The last two companies merged to become United Defense. Today this company is called BAE Systems.

BAE Systems M107 175 mm self-propelled gun

Development
Future US Army requirements for new heavy artillery emphasised air transportability, time into and out of action, common parts and interchangeability area of fire.

A feasibility study encompassing these requirements was carried out, presented and approved at a meeting held in January 1956.

The Pacific Car and Foundry Company (which subsequently became PCF Defense Industries) submitted a concept study for a new family of self-propelled weapons and was subsequently awarded a contract for the design, development and construction of six prototype vehicles: two 175 mm self-propelled guns designated the T235, three 203 mm (8 in) self-propelled howitzers designated the T236 and one 155 mm self-propelled gun designated the T245.

Major design features included the interchangeability of the 175 mm gun, 8 in howitzer and 155 mm gun in a common mount, on a common hull; use of 8 in howitzer and 155 mm gun field pieces and portions of the M17 standard mount; and drastic reductions in the weight over conventional equipment which was made possible through a new hydraulic lockout system.

The hull was also considered suitable for use as a light recovery vehicle and, in 1957, the programme was expanded to include both armoured (T120) and unarmoured (T119 and T121) recovery vehicles.

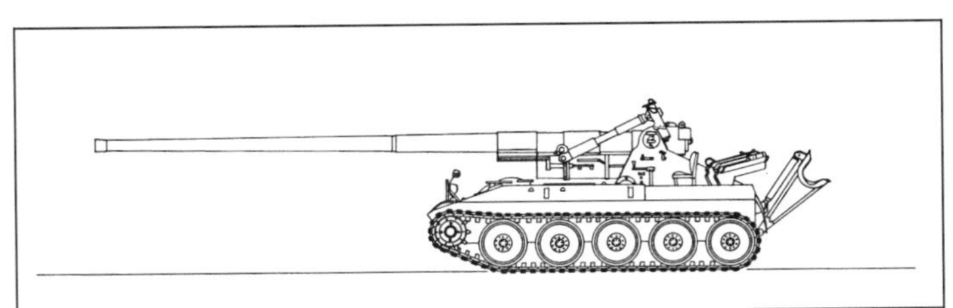

M107 175 mm self-propelled gun in travelling order 0500431

M107 175 mm self-propelled gun in travelling order from rear with spade in raised position (BAE Systems) 1333635

M107 175 mm self-propelled gun in travelling configuration with spade raised (BAE Systems) 1403894

Ordnance service tests with the first prototype hull began late in 1958. In 1959, a policy was established that diesel rather than petrol engines would be used for future vehicles and three of the prototypes, the T235, T236 and T120, were retrofitted with Detroit Diesel diesel engines and designated the T235E1, T236E1 and the T120E1.

Trials of the T235E1 and T236E1 were completed early in 1961 and, in March 1961, both vehicles were standardised, the T235E1 as the M107 and the T236E1 as the M110. The T120E1 was subsequently standardised as the M578 ARV but the T245, T119 and T120 were not developed further.

In June 1961, the Pacific Car and Foundry Company was awarded an initial production contract for both the M107 and M110. First production vehicles were completed in 1962 and the first M107 battalion was formed at Fort Sill in January 1963.

Production of the M107 was also undertaken at a later date by the FMC Corporation of San José and Bowen-McLaughlin-York (BMY) of York, Pennsylvania, (both companies subsequently merged to become United Defense LP) with the first production model coming off the York line in 1965. Production was completed in May 1980, by which time 524 units had been completed by BMY. Today, United Defense is called BAE Systems.

In the US Army M107s were deployed in battalions of 12 guns, held at corps level. All M107s in the US Army and Marine Corps were converted to M110A2s by 1981 and other countries have converted their M107s to the M110A2 configuration.

Most countries have now replaced their M107 and M110 systems with the Lockheed Martin Missiles and Fire-Control, Multiple Launch Rocket System.

A list of arms transfers issued by the United Nations for the period 1992 to 2010 shows that no surplus M107s were exported by any country between these years.

It should be noted that BAE Systems is no longer marketing the 175 mm M107 self-propelled gun or the M110 series of 8 in (203 mm) self-propelled howitzers which shares the same hull.

Description
The M107 Self-Propelled Gun (SPG) is operated by a team of 13, of which the commander, driver and three gunners are carried on the gun, with the rest in the now BAE Systems M548 unarmoured tracked cargo carrier, which also carries the ammunition.

The hull of the M107 is identical to that of the M110 and is made of all-welded cast armour and high-tensile alloy steel, with the driver at the front of the hull on the left, the power pack to his right and the main 203 mm (8 in) armament at the rear.

The driver, who is the only member of the crew seated under armour, has a single-piece hatch cover in front of which are three M17 day periscopes. The middle M17 day periscope can be replaced by a passive periscope for driving at night.

The Detroit Diesel 8V-71T engine is coupled to the Allison Transmission XTG-411-2A cross-drive transmission which is at the front of the hull.

M107 175 mm self-propelled gun of the Israel Defense Force with ordnance elevated and M548 ammunition resupply vehicle in the background
(Israel Defense Force) 1403895

The torsion bar suspension either side consists of five dual rubber-tyred roadwheels with the drive sprocket at the front and the fifth road wheel acting as the idler. There are no track-return rollers.

Attached to each roadwheel arm is a hydraulic cylinder, which serves as a shock-absorber, a hydraulic bump stop and a suspension lockout that transmits recoil shock directly to the ground. The tracks are of the single-pin type with removable rubber pads.

The M107 has infra-red night vision equipment but does not have an NBC system or any amphibious capability. In most countries passive (image intensification or thermal) night vision equipment has replaced the infra-red equipment.

The M107 is armed with a 175 mm M113 gun (development designation T256E3) in mount M158, and has an elevation of +65°, a depression of –2° and a traverse of 30° left and right. Elevation, depression and traverse are all hydraulic, with manual controls available for emergency use.

The gun has a hydropneumatic recoil system, minimum recoil being 0.711 m and maximum recoil 1.778 m. The breech block is of the Welin-step thread type with a percussion firing mechanism. Mounted at the rear of the vehicle on the left side is a rammer and loader assembly which lifts a 175 mm projectile from the rear or left side of the vehicle, positions and rams it into the chamber.

The 175 mm gun M113 only fires an HE projectile M437A2 or M437A1. The projectile weighs 66.78 kg, with the M437A1 containing 13.6 kg of TNT and the M437A2 containing 14.96 kg of Composition B. Range of the 175 mm gun with the standard M437 series US developed 175 mm projectile is as follows:

	Muzzle velocity	Max range
Charge 1	509 m/s	15,100 m
Charge 2	702 m/s	22,100 m
Charge 3	912 m/s	32,700 m

The M107 could fire the former Space Research Corporation ERSC Mk 7 MOD 7 projectile to a maximum range of 40,000 m. This was not, however, adopted by the US Army, although it has been used by Israel. Production of this round was completed by the Space Research Corporation many years ago and the company no longer exists.

The 175 mm ammunition used by the M107 is the separate loading type; two rounds are carried on the vehicle and the rest in the supporting the now BAE Systems M548 unarmoured tracked cargo vehicle. The normal rate of fire is one round every two minutes but two rds/min can be fired for short periods. When travelling, the barrel is retracted slightly to the rear to reduce the overall length of the complete vehicle. Mounted at the rear of the hull is a large hydraulically operated spade, which is positioned before firing begins.

Fire-control equipment includes a panoramic sight M115 on mount M137 (magnification of ×4 and a 10° field of view), elevation quadrant M15, gunner's quadrant M1A1 and direct fire telescope M116C on mount M138 (magnification of ×3 and 13° field of view).

According to the US Army, the maximum number of rounds fired consecutively at maximum rate was 10 rounds. The 175 mm ordnance had an average accuracy life of 400 rounds.

During operations in Vietnam a number of problems became apparent due to the higher rates of fire with new barrels being required after 45 days or less.

There were also failures of the barrel due to the improper storage of the propellant in hot and humid conditions and the high firing rates caused projectiles to become overheated that resulted in in-bore malfunctions.

In 1965 a 175 mm M113 barrel failed in Vietnam after a total of 428 Equivalent Full Charge (EFC) rounds and this was attributed to fatigue.

As a result the fatigue was reset to 400 EFC (300 maximum Zone 3 rounds) fatigue life on the tube.

A programme was initiated to find the cause of this problem and the results were subsequently fed into future US Army artillery barrels including the longer 8 inch (203 mm) M110A2 which had a nuclear capability.

This included a move from a monobloc design to an autofrettaged design.

The M110A2 uses the M201A1 barrel which features:

- An upgraded steel formulation and processing in the barrel
- A higher impulse longer range charge and projectile system
- A muzzle brake to maintain the same impulse level delivered to the carriage.

Variants

There are no variants of the M107 although the M110 8 in (203 mm) self-propelled howitzer and the M578 ARV use the same hull as the M107. The Israel Defense Force calls its 175 mm M107s the Romach.

Specifications

	M107
Dimensions and weights	
Crew:	5
Length	
overall:	11.256 m
hull:	5.72 m
Width	
overall:	3.149 m
Height	
overall:	3.679 m (top of barrel - travelling)
hull:	1.475 m
Ground clearance	
overall:	0.466 m
Track	
vehicle:	2.692 m
Track width	
normal:	457 mm
Length of track on ground:	3.936 m
Weight	
standard:	25,915 kg
combat:	28,168 kg
Ground pressure	
standard track:	0.81 kg/cm²
Mobility	
Configuration	
running gear:	tracked
Power-to-weight ratio:	14.37 hp/t
Speed	
max speed:	56 km/h
Range	
main fuel supply:	725 km
Fuel consumption	
road:	1.568 litres/km
Fuel capacity	
main:	1,137 litres
Fording	
without preparation:	1.066 m
Gradient:	60%
Side slope:	30%
Vertical obstacle	
forwards:	1.016 m
Trench:	2.362 m
Engine	Detroit Diesel Model 8V-71T, turbocharged, water cooled, diesel, 405 hp at 2,300 rpm
Gearbox	
model:	Allison Transmission XTG-411-2A
forward gears:	4
reverse gears:	2
Suspension:	torsion bar
Electrical system	
vehicle:	24 V
Batteries:	4 × 12 V

	M107
Firepower	
Armament:	1 × hull mounted 175 mm M113 howitzer
Ammunition	
main total:	2
Turret power control	
type:	hydraulic/manual
Main armament traverse	
angle:	60° (30° left/30° right)
speed:	5.6°/s
Main armament elevation/depression	
armament front:	+65°/-2°
speed:	5.6°/s
Survivability	
Night vision equipment	
vehicle:	yes
NBC system:	no
Armour	
hull/body:	steel

Status

Production complete. No longer marketed. In service with the following countries:

Country	Quantity	Comments
Iran	25	-
Israel	36	reserve
Korea, South	N/K	-
Turkey	36	-
Vietnam	N/K	-

Contractor

Pacific Car and Foundry Company, (now PCF Defense Industries), Renton, Washington.
FMC Corporation, San José, California.
Bowen-McLaughlin-York (BMY), York, Pennsylvania.
The last two companies merged to become United Defense LP (today BAE Systems).

BAE Systems 155 mm/52-calibre International Howitzer

Development

Based on its experience in the design, development and production of 957 155 mm/39-calibre M109A6 Paladin Self-Propelled Howitzers (SPH) which is covered in a separate entry in *IHS Jane's Land Warfare Platforms: Artillery and Air Defence*, the now BAE Systems developed a new version called the International Howitzer for export only.

Based on the proven M109A6 Paladin system, the International Howitzer improves upon lethality, survivability and tactical mobility, having been designed to meet emerging international artillery requirements.

When compared to the US Army's latest 155 mm M109A6 Paladin, the most significant improvement of the new International Howitzer is the installation of a 155 mm/52-calibre ordnance.

As of March 2013 there had been no orders placed for the BAE Systems 155 mm/52-calibre International Howitzer.

BAE Systems 155 mm/52-calibre International Howitzer, which is now being offered for export 0073766

BAE Systems 155 mm/52-calibre International Howitzer with ordnance in travel lock (BAE Systems) 0568272

Description

In overall layout the 155 mm International Howitzer is identical to the M109A6 Paladin but has major differences at the ordnance and subsystem level.

The basic M109A6 includes an integrated automatic fire-control system with onboard ballistic computation, digital communications, inertial and GPS navigation, automatic 155 mm weapon pointing and an integrated muzzle velocity radar system.

It also includes embedded prognostics/diagnostic capability, improved suspension and drive train as well as a micro-climatic conditioning system that provides either heated or cooled NBC filtered air to the standard crew of four.

Some M109 upgrades use the existing cab/turret but the company believes that a new cab structure is required for the 155 mm/52-calibre application.

The International Howitzer system features a 155 mm/52-calibre ordnance built by Watervliet Arsenal that meets the Joint Ballistic Memorandum of Understanding and is installed in a modified M182 mount.

The 155 mm/52-calibre ordnance is fitted with a fume extractor and a double baffle muzzle brake. This enables targets to be engaged at longer ranges than current M109s.

To increase its rate of fire and reduce crew fatigue, the International Howitzer is fitted with a semi-automatic rammer, with user selectable options for a semi-automatic primer feed or an integrated laser ignition system. Maximum rate of fire is about six rounds a minute. A total of 39 155 mm projectiles and associated charges are carried.

According to the company, the 155 mm International Howitzer receives fire missions whilst on the move, computes firing data, selects and takes up firing position, unlocks and points its 155 mm/52-calibre ordnance, shoots and then moves. It can fire the first round within 45 seconds.

The suspension has been upgraded from the standard M109A6 Paladin to include redesigned suspension with hydropneumatic struts in lieu of standard shock-absorbers.

The power pack of the M109A6 has been retained, this consists of a Detroit Diesel 8V-71T engine developing 440 hp coupled to an Allison XTG-411-4A automatic transmission with quick disconnect.

The all-welded aluminium hull and turret is provided with internal spall liners and supplemental appliqué passive armour for higher battlefield survivability.

Ground pressure with 15 inch track is 0.9892 kg/cm^2 and with the wider 18 inch track it is 0.81 kg/cm^2.

Standard equipment on the International Howitzer includes an Automatic Fire-Control System (AFCS) with modified tactical software to support the new 155 mm/52-calibre main armament and a unique International Howitzer mission capability, with integrated fibre-optic inertial and Global Positioning System (GPS) navigation capability.

Options for the International Howitzer include an auxiliary power unit as well as an integrated direct fire sight with laser range-finder.

Taken overall, the International Howitzer maintains a full 95 per cent component commonality with the combat proven in-service M109A6 Paladin.

Customers for the International Howitzer can purchase either complete brand new systems or new turrets for installation on upgraded chassis.

Variants

NLOS-C 155 mm self-propelled artillery system

The 155 mm/38-calibre Non-Line-Of-Sight – Cannon (NLOS-C) was being developed by the now BAE Systems as part of the US Army's ambitious Future Combat System (FCS).

NLOS-C was a full tracked vehicle fitted with a 155 mm/38-calibre ordnance fed by an automatic loading system that loaded the 155 mm projectile followed by the correct number of modular charges. It had a crew of two.

NLOS-C was part of the Manned Ground Vehicles (MGV) element of the FCS which was cancelled by the US Army in 2009.

As of March 2013 no announcement had been made regarding future new US Army conventional tube artillery system. The M109A6 Paladin 155 mm/39-calibre self-propelled howitzer will now remain in service with part of the fleet being upgraded to the M109A6 Paladin Integrated Management (PIM) standard which is covered in detail in a separate entry in *IHS Jane's Land Warfare Platforms: Artillery and Air Defence*.

The NLOS-C shared many common components with the Non-Line-Of-Sight - Mortar (NLOS-M) armed with a breech loaded 120 mm mortar. But this was also cancelled in 2009 without a direct replacement.

Specifications

	155 mm/52-calibre International Howitzer
Dimensions and weights	
Crew:	4
Length	
overall:	11.78 m
hull:	6.807 m
Width	
overall:	3.922 m
hull:	3.149 m
Height	
to turret roof:	2.794 m
to top of roof mounted weapon:	3.236 m
Ground clearance	
overall:	0.449 m
Track	
vehicle:	2.778 m
Track width	
normal:	381 mm
Length of track on ground:	3.692 m
Weight	
combat:	30,000 kg
Ground pressure	
standard track:	0.9892 kg/cm²
Mobility	
Configuration	
running gear:	tracked
Power-to-weight ratio:	14.66 hp/t
Speed	
max speed:	64 km/h
Range	
main fuel supply:	343 km
Fuel capacity	
main:	504 litres
Fording	
without preparation:	1.05 m
Gradient:	60%
Side slope:	40%
Vertical obstacle	
forwards:	0.53 m
Trench:	1.83 m
Engine	Detroit Diesel Model 8V-71T LHR, turbocharged, water cooled, diesel, 500 hp at 2,300 rpm
Gearbox	
model:	Allison Transmission XTG-411-2A
forward gears:	4
reverse gears:	2
Suspension:	independent torsion bar with high-capacity shock-absorbers and hydropneumatic struts
Electrical system	
vehicle:	24 V
Batteries:	4 × 12 V
Firepower	
Armament:	1 × turret mounted 155 mm howitzer
Armament:	1 × roof mounted 12.7 mm (0.50) M2 HB machine gun
Ammunition	
main total:	39
12.7/0.50:	500
Main weapon	
calibre length:	52 calibres
Turret power control	
type:	hydraulic/manual
Main armament traverse	
angle:	360°
Main armament elevation/depression	
armament front:	+75°/-3°
Survivability	
Night vision equipment	
vehicle:	yes
NBC system:	yes, individual crew member
Armour	
hull/body:	aluminium + appliqué
turret:	aluminium + appliqué

Status

Prototype. Not yet in production or service.

Contractor

BAE Systems.

BAE Systems M109A6 Paladin Integrated Management (PIM) 155 mm self-propelled howitzer

Development

The US Army took delivery of a total of 975 155 mm/39-calibre M109A6 Paladin 155 mm self-propelled howitzers which are all conversions of older systems.

Prime contractor for this upgrade was the now BAE Systems (previously BAE Systems, Ground Systems) and full details of this are provided in a separate entry in *IHS Jane's Land Warfare Platforms: Artillery and Air Defence.* The M109A6 Paladin is only used by the US Army.

Late in 2007 the BAE Systems and the US Army signed a Memorandum of Understanding (MoU) that established a Public Private Partnership (P3) to enhance and maintain the currently deployed M109A6 Paladin 155 mm/39-calibre self-propelled howitzer through to 2050 - 2060.

In consultation with the US Army, BAE Systems designed and built the prototype of the M109A6 Paladin Integrated Management (PIM) SP artillery system in just nine months using company funding.

This was completed in September 2007 and shown for the first time the following month.

In October 2009 it was announced that BAE Systems had been awarded a USD63.9 million contract from the US Army Tank Automotive and Armaments Command for the procurement of five prototype M109A6 PIM and two M992A2 FAASV.

Under the terms of this contract, design and engineering analysis work for the vehicle structure, automotive systems and electronic and vehicle electronics was carried out at BAE Systems facilities in California, Minnesota, Michigan, New York and Pennsylvania.

In addition, US government facilities at the US Army Research and Development Center in Picatinny, New Jersey was also involved.

The first batch is expected to include a total of 177 M109A6 Paladin and a similar quantity of M992A2 FAASV.

Within the US Army, the M109A6 PIM programme is managed by the US Army Project Manager HBCT.

BAE Systems is the prime contractor and system integrator, overseeing system design, development and production.

In early 2010 the US Army officially unveiled its first M109A6 Paladin PIM and as of early 2013 these were being put through US Army trials.

By 2010 these five M109A6 PIM howitzers and the two M992A2 PIM FAASV had been completed and were undergoing trials at various US Army locations.

These included Aberdeen Proving Ground (one M109A6 PIM howitzer and one M992A2 PIM FAASV) and Yuma Proving Ground (three M109A6 PIM howitzers).

BAE Systems have one M109A6 PIM howitzer and one M992A2 PIM FAASV vehicle for development work.

There may be further improvements suggested by the user as a result of these trials and this could include enhanced protection.

Under latest plans it is expected that a total of 440 sets of howitzers and ammunition carriers will be procured for the US Army Heavy Brigade Combat Teams (HBCT) with each unit consisting of one M109A6 PIM howitzer and one associated M992A2 PIM FAASV.

A Milestone C Low Rate Initial Production (LRIP) decision is expected in FY13 to allow for production to commence in FY15 with about 16 to 18 sets being completed a year and Full Rate Production (FRP) in FY17, but this could be changed due to budget considerations.

BAE Systems York facility will manufacture and assemble the hull while Anniston Army Depot will overhaul the 155 mm M284 cannon and its mount and upgrade the turret.

Final assembly and integration of the hull and turret will be undertaken at the BAE Systems facility at the Fort Sill Industrial Park in Elgin, Oklahoma.

According to the US Army, the M109A6 PIM howitzer has been engineered to increase force protection, improve readiness and vehicle survivability and avoid component obsolescence.

Prototype of the BAE Systems, M109A6 PIM with 155 mm/39-calibre ordnance in travel lock and enhanced protection for commander when operating the .50 (12.7 mm) M2 HB machine gun (BAE Systems) 1169366

One of the five prototype 155 mm/39-calibre M109A6 Paladin Integrated Management (PIM) 155 mm self-propelled howitzer in travelling configuration (BAE Systems) 1340123

In January 2012, BAE Systems announced that it had been awarded a USD313 million contract for additional engineering design, logisitics development and test evaluation support to complete the Engineering and Manufacturing Development phase of the M109A6 Paladin PIM programme.

At that time, BAE Systems stated that they had completed all contractor tests as well as Phase One of the US Army Developmental Test.

This additional funding will allow BAE Systems to support the remainder of the US Army's test programme and complete the production planning efforts in support of the LRIP decision. This contract will run from January 2012 through to January 2015.

In December 2012 BAE Systems confirmed that low-rate initial production of the M109A6 Paladin PIM would be undertaken at the Elgin, Oklahoma facility located at the Fort Sill Industrial Park, home of the US Army field artillery.

At the same time it was stated that the LRIP award was expected to take place in the third quarter of 2013 with the first batch to consist of 72 systems.

Key components of the PIM production vehicles, including the hull, will be sent to Elgin facility from BAE Systems manufacturing facilities and suppliers.

As part of final assembly and checkout, BAE Systems will use Fort Sill where they will undergo mobility and firing verification.

Description

M109A6 PIM integrates an upgraded current M109A6 Paladin turret with a brand new chassis developed by BAE Systems.

This programme was developed specifically to address the long term viability and sustainability of the M109 family in the US Army, maximising commonality across the HBCT.

The new all-welded aluminium armour chassis incorporates a Cummins 600 hp diesel and L3 Combat Propulsion Systems HMPT-500 series automatic transmission which are already used in the US Army's Bradley M2 infantry combat vehicle and variants used by the HBCT.

According to BAE Systems, this power pack will improve the power-to-weight ratio as well as the top speed of the M109A6 PIM.

In addition it uses Bradley final drives, torsion bars, roadwheels, road arms and track and the upgraded electrical system includes a 600 V DC 75 kW generator. An HUMS is installed as standard and new in arm rotary dampers will also be fitted.

The driver, seated at the front of the vehicle on the left side, is provided with cameras to provide improved situational awareness front and rear.

Existing M109A6 Paladin turret has been enhanced and now includes rammer and all electric drives that leverage technology from the US Army's cancelled NLOS-C. This was to have been the first element of the US Army's Future Combat System (FCS) to be fielded. The whole of the Manned Ground Vehicles (MGV) was cancelled in 2009.

Crew of the M109A6 PIM will be provided with individual spot cooling as well as the standard cooling vests for use in high ambient temperatures.

The prototype shown in Washington in October 2007 was fitted with an integrated protection system for the commander that provides lateral protection when using the .50 (12.7 mm) M2 HB machine gun or another weapon.

Existing 155 mm/39-calibre M284 ordnance is retained which can fire all natures of ammunition including the recently type classified Raytheon M982 Excalibur long range Precision Guided Munition (PGM).

As the new hull is longer it has greater internal volume and can carry in excess of 43 × 155 mm projectiles including 17 of the M982 Excalibur long range PGM which are already in low rate production and has been fired in combat by Canada (M777) and the US (M198).

It will be able to fire all currently deployed US Army 155 mm artillery projectiles.

In mid-2013 the US Army will start to field the ATK XM1156 Precision Guidance Kit which will provide 155 mm projectiles such as the M549A1 and the M795 with a significant increase in accuracy meaning that fewer rounds are required to neutralise a given target.

The latest digital fire control system will also be installed in M109A6 PIM and the current production Modular Artillery Charge System (MACS) used as a follow on to the old and inefficient bag type charges.

It also features a 600 V on board power system which is designed to accommodate emerging technologies.

Survivability will be enhanced by the use of shoot and scoot tactics to avoid counter battery fire as well as improved ballistic hull design, Nuclear, Biological and Chemical (NBC) protection for both units.

It will be capable of coming to a halt and firing its first 155 mm projectile within 45 seconds due to its on board communications, remote travel lock and automated cannon slew capability.

Specifications

	M109A6 Paladin Integrated Management (M109A6 PIM)
Dimensions and weights	
Crew:	4
Length	
overall:	9.70 m
Width	
overall:	3.91 m
Height	
overall:	3.73 m
Ground clearance	
overall:	0.482 m
Weight	
combat:	31,752 kg
Mobility	
Configuration	
running gear:	tracked
Power-to-weight ratio:	18.89 hp/t
Speed	
max speed:	64.4 km/h
Range	
main fuel supply:	322 km
Fuel capacity	
main:	545 litres
Fording	
without preparation:	0.166 m
Gradient:	60%
Side slope:	40%
Vertical obstacle	
forwards:	0.53 m
Trench:	1.828 m
Engine	Cummins VTA-903T, turbocharged, water cooled, diesel, 600 hp at 2,600 rpm
Gearbox	
model:	L3 Combat Propulsion Systems HMPT-500 series
Suspension:	torsion bar
Electrical system	
vehicle:	24 V
Firepower	
Armament:	1 × turret mounted 155 mm M284 cannon
Armament:	1 × roof mounted 12.7 mm (0.50) M2 HB machine gun
Ammunition	
main total:	39
secondary total:	500
Turret power control	
type:	powered/manual
Main armament traverse	
angle:	360°
Main armament elevation/depression	
armament front:	+75°/-3°
Survivability	
Night vision equipment	
vehicle:	yes
NBC system:	yes
Armour	
hull/body:	aluminium + appliqué
turret:	aluminium + appliqué

Status

First prototype M109A6 PIM was completed late in 2007 and in October 2009 it was announced that the US Army had awarded a contract to BAE Systems covering the supply of five M1096 PIM and two M992A2 FAASV.

Prototypes undergoing US Army trials with a total of 440 units expected to be upgraded in the future.

Contractor

BAE Systems.

BAE Systems M109A6 155 mm Paladin self-propelled howitzer

Development

Following the demise of the Divisional Artillery Support Weapon System (DASWS), several companies were involved with the M109 related project known as the Howitzer Extended Life Programme (HELP).

This programme eventually merged with the M109 Howitzer Improvement Programme (HIP) to bring the M109 up to a new M109A6 standard. Approval for the programme was given in November 1984 and invitations to tender were released in February 1985.

In October 1985, BMY was awarded a development contract to carry out the HIP programme. This involved the conversion of eight prototypes (six US and two for the Israeli Defense Forces) over a three-year period. A Special In-Process review of the programme was carried out in April 1986, with a further In-Process Review carried out during the second quarter of FY87.

The first of the US and Israeli HIP prototypes were rolled out on 30 March 1988 at BMY's York facility and subsequently undertook extensive prototype qualification tests at Aberdeen and Yuma proving grounds.

From September 1988 to February 1989, the US prototypes (the M109A3E2) underwent a technical testing programme, and operational tests were carried out at Fort Sill from May 1989 onwards.

The US Army made a production decision for the M109A6 on 5 September 1990, with the first Low-Rate Initial Production (LRIP) options being awarded in September 1990 for 44 vehicles.

First production M109A6 systems were completed in April 1992. This was followed by two additional LRIP options for 60 vehicles each, which brought the total up to 164 vehicles. The LRIP portion of the M109A6 programme was completed in November 1994.

For the remainder of the M109A6 Paladin requirement there was a competition between BMY Combat Systems, FMC Corporation, Ground Systems and General Dynamics, Land Systems. In April 1993, FMC Corporation, Ground Systems Division was selected.

An initial contract worth USD30.5 million was awarded to FMC on 9 April 1993, to upgrade 60 M109s to the M109A6 Paladin standard, with the whole batch of 630 systems to be completed by October 1998. An additional 83 systems were ordered under options on the basic contract, bringing the total value to USD376 million.

In April 1997, the US Army placed an order with United Defense for the supply of 37 additional M109A6 systems. In November 1997 the US Army exercised an option for a further 36 systems which brought the grand total up to 950 units with final deliveries taking place in June 1999.

The 37 systems ordered in early 1997 marked the beginning of production for the National Guard. It is expected that the National Guard will field a total of 16 battalions with M109A6 Paladins the first three fully operational by mid-1999.

United Defense built the new M109A6 turrets with the upgraded hull supplied by Letterkenny Army Depot. United Defense then integrated the hull and turret and then delivered the complete system to the US Army.

The US Army did deploy the M109A6 in batteries of eight weapons but these have now reverted to six gun batteries.

In mid-2000 the US Army placed an order for an additional seven M109A6 Paladin systems for the National Guard at a total value of USD8.3 million. Deliveries took place from November 2001 through to January 2002. An additional 18 vehicles were ordered under this contract at a total of USD21.2 million. This brought the total number of M109A6 purchased up to 975.

Work was carried out at the United Defense facilities at Aiken and York. Chassis work was carried out at the Anniston Army Depot where they stripped down the chassis and then prepared it for shipment to United Defense.

In 2005 United Defense, Ground Systems became BAE Systems. The company offers the option of RUAG Defence positive ram semi-automatic projectile loader and automatic primer for the M109A6, although this was not adopted by the US Army. This is offered for the export market.

The M284 cannon is a modified M185 cannon with many improvements including a reinforced muzzle brake, advanced bore evacuator, a redesigned chamber and forcing cone, two shallow grooves for torque key replacing one deep groove, a new breech housing assembly, wiring of fasteners, added M49 firing mechanism and improved firing block, improved crank lead springs, strengthened breech handle springs and modified plunger, modification of the breech ring and a longer 155 mm ordnance.

The M182A1 gun mount is a modified M182 with an added temperature sensor, a redesigned recuperator cylinder, a redesigned torque key, improved seals in the recoil cylinders and improved seals and new bearing installed in the elevation/equilibration cylinders.

US Army M109A6 Paladin 155 mm self-propelled howitzer deployed in firing position (right) with M992 Field Artillery Ammunition Support Vehicle (FAASV) to the left (BAE Systems) 1333634

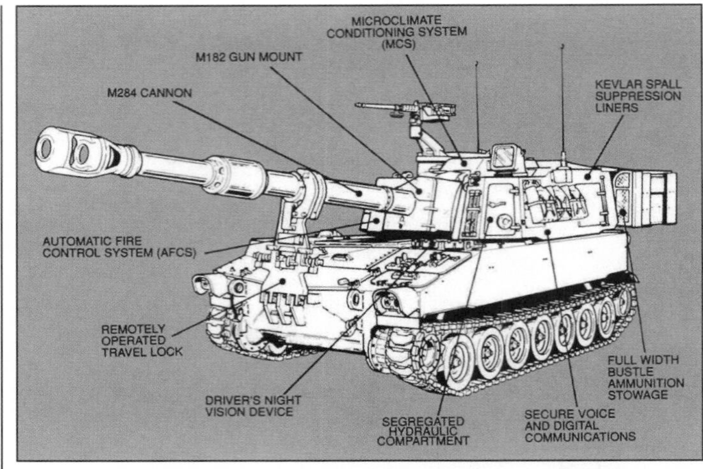

Key elements of the BAE Systems Division M109A6 Paladin (BAE Systems) 1133922

With M203/M203A1 charge a range of 30,000 m can be achieved.

The new Automatic Fire-Control System (AFCS XXI) is being retrofitted into all M109A6 Paladin fielded systems.

This new fire-control system is based on commercial technology with reduced system costs and parts obsolescence issues while providing significant improvements in computation speed, storage capacity and growth capability.

Features of the M109A6 Paladin include:

- New turret with improved armour and Kevlar ballistic lining
- Full-width turret bustle
- Automatic Fire-Control System (AFCS)
- Onboard ballistic fire-control computer and navigation system providing automatic gun pointing
- New gun drive servos
- Muzzle velocity management system
- Modular azimuth positioning System (MAPS)
- Global positioning system
- VIS intercom
- SINCGARS radio (Single Channel Ground/Air Radio System)
- Built in training software
- M182A1 gun mount
- M28 155 mm cannon which allows the M203 charge to be used
- New traversing mechanism
- Upgraded torsion bar suspension system
- Driver's passive night vision system
- Electrical and hydraulic system improvements, including a new 650 A generator
- Increased ammunition stowage
- Microclimate cooling system
- Remotely operated barrel travel lock
- Segregated hydraulic compartment and hydraulic line fuzing
- Engine starter protective circuit
- Air cleaner blower relay
- Engine sensor and connector covers
- Crossover tube protective cover
- New crew compartment water and oil drains
- Slave start circuit
- New sensors to monitor power pack parameters
- Ventilated face piece system for NBC conditions
- Halon fire extinguishing system
- Improved hatch and night vision provisions for the commander
- Mounted water ration heater
- External stowage baskets
- Final drive quick disconnect
- Propellant segregation
- Digital communications link
- Four-man crew
- A flick rammer loader assist was installed on the Israeli version of HIP which yielded burst rates of fire of three rounds in less than 13 seconds.

The M109A6 Paladin can fire all currently deployed US Army 155 mm artillery projectiles including the latest Excalibur 1b precision guided projectile (PGP).

In February 2013 Raytheon announced that it had been awarded a US Army Fiscal Year 2012 contract for the production of the 155 mm Excalibur Increment 1b PGP.

This USD56.6 million contract is for Low Rate Initial Production (LRIP) with first deliveries to commence in the last quarter of 2013 and includes options through to FY2016.

During trials the Excalibur has achieved a radial miss distance of less than 4 m at ranges in excess of 35 km.

In addition, 155 mm artillery projectiles such as the M549A1 and M795 will be able to be fitted with the ATK XM1156 Precision Guidance Kit (PGK) with first deliveries to be made in mid-2013.

This PGK has been procured as an Urgent Material Release and corrects the ballistic trajectory of the projectile using a global positioning system with fuzing functions to improve accuracy so that at least 50 per cent of the time the 155 mm projectiles lands within 50 m of the target.

155 mm M109A6 Paladin self-propelled howitzer (US Army) 1365013

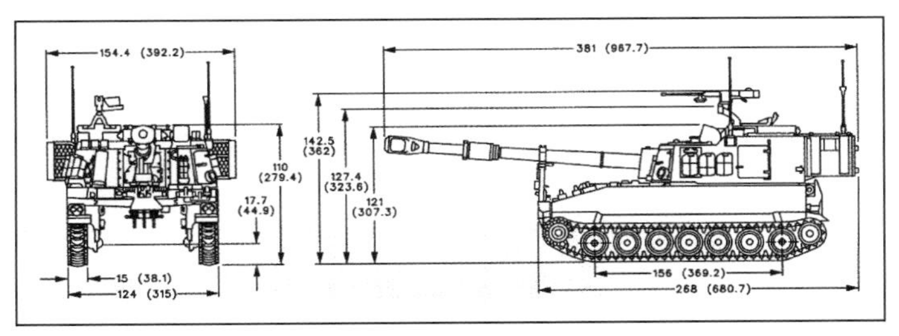

Front and side drawing of the 155 mm M109A6 self-propelled artillery system 0018993

Description

The M109A6 Paladin has a similar layout to the M109 series and details are provided in a separate entry in *IHS Jane's Land Warfare Platforms: Artillery and Air Defence*.

M109A6 HIP

A number of additional potential improvements have been identified by the US Army for the 155 mm M109A6 Paladin self-propelled howitzer and these have been prioritised in a System Improvement Plan (SIP).

The 15 identified subsystem upgrades fall into six major categories: computer growth, improved accuracy, increased rate of fire, survivability, reliability, availability and maintainability and crew safety/comfort.

Additional AFCS upgrades under development include an advanced digital display allowing for future expansion to include items such as Interactive electronic technical manuals and situational awareness screens.

The current microclimate conditioning system that provides NBC protection as well as conditioned air to reduce crew stress is now being redesigned to replace the current R-12 refrigerant with a more environmentally friendly alternative. This is expected to be retrofitted to all current US Army M109A6 Paladin systems.

Improvements to the rate of fire are being studied including addition of a laser ignition system, a semi-automatic loader and an automatic fuze setter. Prototypes of these improvements have already been developed, tested and demonstrated to the US Army for consideration.

Other individual upgrades proposed by the SIP include an upgrade to the digital communications speed, laser range-finder, installation of a driver's thermal viewer to replace the current image intensification device, hull vulnerability reduction and an upgrade to the Prognostic/Diagnostic Interface Unit (PDIU).

M109A6 Paladin Integrated Management

Some of the proposed elements of the M109A6 HIP have been rolled into the new M109A6 Paladin Integrated Management covered in a separate entry in a separate entry in *IHS Jane's Land Warfare Platforms: Artillery and Air Defence*.

In October 2009 it was announced that BAE Systems, had been awarded a USD63.9 million contract from the US Army Tank Automotive and Armaments Command for the procurement of five prototype M109A6 PIM and two M992A2 FAASV.

Under current plans it is expected that 440 M109A6 Paladin will be upgraded to the M109A6 Paladin PIM standard plus associated M992A2 FAASV.

In December 2012 BAE Systems confirmed that LRIP production of the M109A6 PIM would be undertaken at the Elgin, Oklahoma facility which is located at the Fort Sill Industrial Park, home of the US Army field artillery.

Key components of the PIM production vehicles, including the hull, will be sent to Elgin facility from BAE Systems manufacturing facilities and suppliers.

As part of final assembly and checkout, BAE Systems will use Fort Sill where they will undergo mobility and firing verification.

It is expected that the LRIP PIM production contract will be awarded in the third quarter of 2013 and cover an initial 72 systems.

Enhanced Paladin Demonstrator

As a private venture the company has also built a single Enhanced Paladin Demonstrator that successfully underwent US Army trials in 1994.

This is very much a testbed and has a 155 mm 52-calibre barrel, additional armour, semi-automatic ammunition handling, upgraded suspension and additional ammunition capacity. There is no current work being carried out on this vehicle.

Some of the ideas in this system were subsequently used in the 155 mm/52-calibre International Howitzer. As of early 2013, this remained at the prototype stage.

M109 export upgrades

Details of this are given in the entry for the now BAE Systems M109 series of self-propelled howitzers.

International Howitzer

This has been developed from the M109A6 Paladin for the export market and is fitted with a 155 mm/52-calibre ordnance. As of March 2013 this remained at the prototype stage and details are provided in a separate entry.

Specifications

	M109A6 Paladin
Dimensions and weights	
Crew:	4
Length	
overall:	9.677 m
hull:	6.807 m
Width	
overall:	3.922 m
hull:	3.149 m
Height	
overall:	3.62 m
to top of roof mounted weapon:	3.24 m
Ground clearance	
overall:	0.457 m
Track	
vehicle:	2.778 m
Track width	
normal:	381 mm
Length of track on ground:	3.962 m
Weight	
combat:	28,849 kg
Ground pressure	
standard track:	0.95 kg/cm^2
Mobility	
Configuration	
running gear:	tracked
Power-to-weight ratio:	15.25 hp/t
Speed	
max speed:	64.4 km/h
Range	
main fuel supply:	344 km
Fuel capacity	
main:	504 litres
Fording	
without preparation:	1.05 m
Gradient:	60%
Side slope:	40%
Vertical obstacle	
forwards:	0.53 m
Trench:	1.83 m
Engine	Detroit Diesel Model 8V-71T LHR, turbocharged, water cooled, diesel, 440 hp at 2,300 rpm
Gearbox	
model:	Allison Transmission XTG-411-2A
forward gears:	4
reverse gears:	2
Suspension:	independent torsion bar with high-capacity shock absorbers
Electrical system	
vehicle:	24 V
Batteries:	4 × 12 V
Firepower	
Armament:	1 × turret mounted 155 mm M284 howitzer
Armament:	1 × roof mounted 12.7 mm (0.50) M2 HB machine gun
Ammunition	
main total:	39
12.7/0.50:	500
Turret power control	
type:	hydraulic/manual
Main armament traverse	
angle:	360°
Main armament elevation/depression	
armament front:	+75°/-3°

	M109A6 Paladin
Survivability	
Night vision equipment	
vehicle:	yes
NBC system:	yes
Armour	
hull/body:	aluminium
turret:	aluminium

Status
A total of 975 have been delivered to the US Army. Can be placed back in production if further orders are placed.

Contractor
BAE Systems.

155 mm M109A2 self-propelled howitzer in service with Jordan (Paul Beaver)
0533759

BAE Systems M109 Series of 155 mm self-propelled howitzers

Development
In January 1952, a conference was held in Washington DC on the subject of self-propelled artillery which indicated an urgent need for improved self-propelled artillery. Shortly after this conference, preliminary concept studies began for a self-propelled howitzer to replace the 155 mm M44. The first studies for the design of the new vehicle, called the Howitzer 156 mm Self-Propelled T196, were presented to CONARC in August 1952, but were rejected, as were additional studies presented in September 1953. At a conference in May 1954, a concept was finally approved for presentation to CONARC.

In June 1954, a conference was held at Fort Monroe to review the military characteristics and to determine the whole self-propelled programme. It was then decided that future concepts of the T196 would be prepared along the design proposed for the Howitzer 110 mm Self-Propelled T195. Concept studies continued along these lines and, in June 1956, it was finally decided to use the basic hull and turret of the T195 but with a 155 mm howitzer instead of the original 156 mm howitzer.

In October 1956, the mock-up of the T196 was reviewed and verbal authority was given to proceed with development of the first prototype.

The first prototype of the T196 was completed in 1959, about six months later than the 105 mm T195. During preliminary user evaluation at Fort Knox a number of failures occurred in the suspension. The prototype differed from later vehicles in that it had a different shaped hull and turret, the seventh road wheel acted as the idler and it was powered by a Continental petrol engine.

In 1959, a policy was established that diesel rather than petrol engines would be used for future combat vehicles and the prototype of the T196 was then fitted with a Detroit Diesel engine and subsequently re-designated the T196E1.

In February 1961, an order was placed for two T196E1 preproduction vehicles, which were delivered within six months. After further trials the T196E1 was classified as a Limited Production Type in December 1961. Two months before this decision, in October, a letter order was given to the Cadillac Motor Car Division for one year's production of the T196E1 at the Cleveland Army Tank Plant.

The first production vehicles were completed in October 1962. In January 1963, an extension was authorised to continue the classification of Limited Production and, in July the same year, the T196E1 was classified as standard A and designated the Howitzer, Medium, Self-Propelled: 155 mm, M109. Early in 1963, a contract was awarded to Cadillac for the second year of production. The contract for the third year of production, awarded in December 1963, went to the Chrysler Corporation, although production remained at the Cleveland Army Tank Plant. First M109s were issued to the US Army in June 1963.

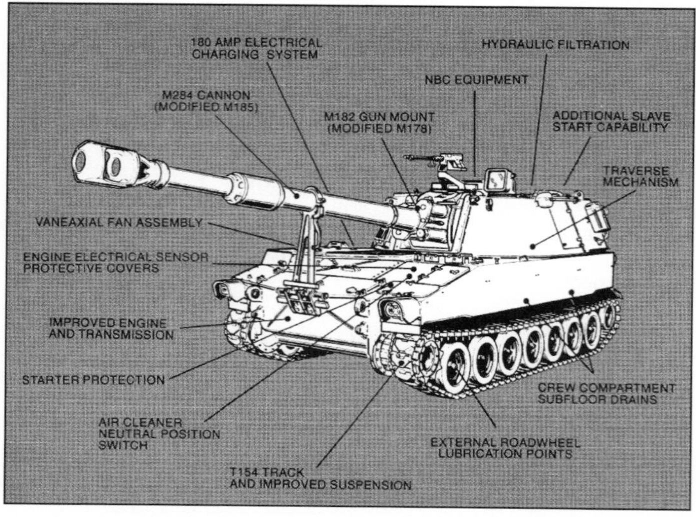

Main features of the 155 mm M109A5 self-propelled howitzer 0500880

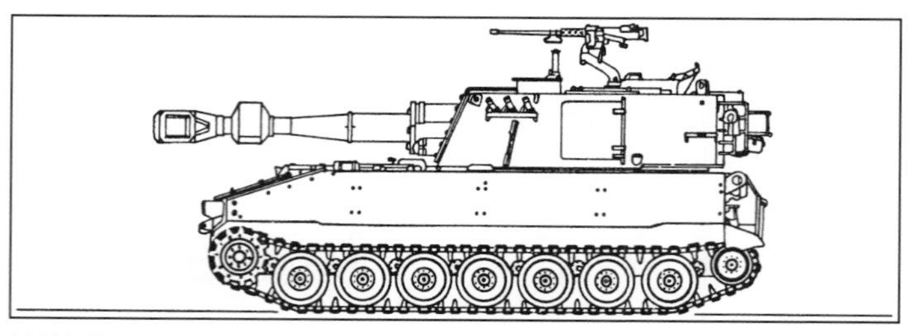

M109 155 mm self-propelled howitzer in travelling order 0500432

Further development of the M109 resulted in the M109A1 and the M109A2 and, by early 1979, production had exceeded 4,000 units. M109 production for the US Armed Forces by the Cleveland Army Tank Plant was as follows:

FY	US Army	US Marine Corps
1962	217	30
1963	150	58
1964	360	-
1965	360	-
1966	454	-
1967	420	36
1968	-	-
1969	-	26

The now BAE Systems began production of the M109 in 1974 and in 1994 released the following production figures on the 155 mm M109 series of self-propelled howitzer:

Year	M109A1B	M109A2	M109A5	Total
1974	0	99	0	99
1975	0	222	0	222
1976	0	169	0	169
1977	0	183	0	183
1978	0	178	0	178
1979	71	45	0	116
1980	231	70	0	301
1981	335	53	0	388
1982	276	260	0	536
1983	199	120	0	319
1984	194	27	0	221
1985	231	75	0	306
1986	209	0	0	209
1987	256	0	0	256
1988	106	0	0	106
1989	119	0	0	119
1990	117	0	0	117
1991	182	0	0	182
1992	105	0	0	105
1993	110	0	0	110
1994	0	0	10	10

There has been no recent new build production for the US Army but production could be undertaken for the export market as and when required. Recent exports sales include Austria, Greece, Taiwan and Thailand. The company is still marketing the M109A5.

M109 HIP (M109A6)
Further development has resulted in the new M109A6 Paladin for which there is a separate entry in *IHS Jane's Land Warfare Platforms: Artillery and Air Defence*. The US Army took delivery of a total of 975 M109A6 Paladin self-

M109A5 155 mm self-propelled artillery system with ordnance in travelling lock and muzzle brake covered up (Michael Jerchel) 0106070

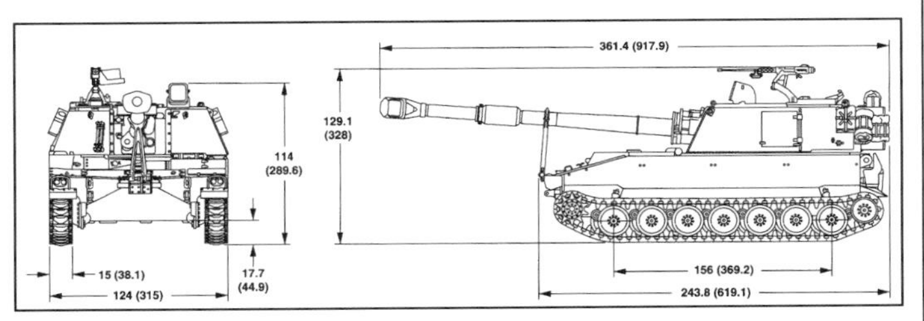

Front and side drawings of the 155 mm M109A5 self-propelled howitzer in travelling configuration 0126563

propelled howitzers and all of these were rebuilds of earlier M109 systems.

Description

(M109)

The M109 has a crew of six, consisting of commander, gunner, three ammunition members and the driver. The hull is made of all-welded aluminium armour, which provides protection from small-arms fire and shell splinters. The driver is at the front of the hull on the left, the Detroit Diesel power pack is to his right and the turret is at the rear.

The driver has a single-piece hatch cover that opens to the left, with three M45 day periscopes in front that can be covered by small metal flaps to prevent damage. The driver can replace one of his M45 day periscopes with a passive night vision periscope from within the vehicle.

The Detroit Diesel Model 8V-71T engine is coupled to an Allison Transmission XTG-411-4A cross-drive transmission, which is at the front of the hull.

The all-welded aluminium armour turret at the rear of the hull has a square hatch in each side that opens to the rear, and twin doors in the turret rear. The commander is seated on the right side of the turret and has a cupola that can be traversed through 360°, a single-piece hatch cover that opens to the rear and an M27 day periscope. Pintle-mounted on the forward part of the commander's cupola is a .50 (12.7 mm) M2 HB local defence machine gun.

The gunner is seated on the left side of the turret and has a square single-piece hatch cover that opens to the right. Twin doors are provided at the rear of the turret for ammunition resupply. Mounted at the rear of the hull, each side of the hull door, is a large spade that is lowered manually to the ground before firing. These spades are only normally deployed firing top charges.

The torsion bar suspension either side consists of seven dual rubber-tyred roadwheels with the drive sprocket at the front and the idler at the rear. There are no track-return rollers. The tracks are of the single-pin, centre guide type with replaceable rubber pads.

The M109 is fitted with night vision equipment but does not have an NBC system. The basic vehicle can ford to a depth of 1.828 m without preparation. It can be fitted with an amphibious kit consisting of nine air bags, four each side of the hull and one at the front.

The bags, which are not carried on the vehicle as part of its normal equipment, are inflated from the vehicle and it can then propel itself across rivers by its tracks at a maximum of 6.43 km/h.

Main armament is an M126 155 mm howitzer in an M127 mount, with a fume extractor and a large muzzle brake. The recoil system is hydropneumatic and the breech block is of the Welin-step thread type. The 155 mm weapon has an elevation of +75°, a depression of –5° and the turret can be traversed through a full 360°.

Gun elevation and depression and turret traverse are hydraulic with manual controls for emergency use. Normal rate of fire is 1 rd/min but 3 rds/min can be fired for a short period. Fire-control equipment includes an elbow telescope M118C for direct fire with a magnification of ×4 and a 10° field of view, panoramic telescope M117 for indirect fire with a magnification of ×4 and a 10° field of view, quadrant fire-control M15 and gunner's quadrant M1A1. The weapon can fire the following types of ammunition:

- Copperhead cannon-launched guided projectile M712 with a range of 16,400 m
- HE (M107) with the projectile weighing 42.91 kg, a maximum muzzle velocity of 562.4 m/s and a maximum range (Charge 7) of 14,600 m
- HE (M795) which is the replacement for the M107; the projectile weighs 46.90 kg
- HE (M449A1) (carries 60 anti-personnel grenades) with the projectile weighing 43.09 kg, a maximum muzzle velocity of 563 m/s and a maximum range of 14,600 m

Upgraded M109A3GN of the Norwegian Army showing different external stowage (Svein Wilger) 0126558

- HE (M483A1 series) (carries 64 M42 and 24 M46 dual-purpose grenades) with the projectile weighing 46.53 kg, a maximum muzzle velocity of 535.2 m/s and a maximum range (Charge 7) of 14,320 m
- HE (M692) (carries 36 anti-tank mines) with the projectile weighing 46.49 kg, a maximum muzzle velocity of 535.2 m/s and a maximum range (Charge 7) of 14,320 m
- M718/M741 (containing nine anti-tank mines) with a maximum range of 14,320 m
- HERA (M549) with the projectile weighing 43.54 kg, with a maximum range (Charge 7) of 19,300 m
- Illuminating (M485) with the projectile weighing 41.73 kg, a maximum muzzle velocity of 576.9 m/s and a maximum range (Charge 7) of 13,586 m
- Illuminating (M818) with the projectile weighing 46.26 kg, a maximum muzzle velocity of 536 m/s and a maximum range of 11,600 m
- Smoke (M825) with the projectile weighing 46.53 kg, a muzzle velocity of 535.2 m/s and a maximum range of 14,320 m
- Smoke BE (M116 series) with the projectile weighing 42.22 kg, a maximum muzzle velocity of 562.4 m/s and a maximum range (Charge 7) of 14,600 m
- Smoke WP (M110 series) with the projectile weighing 44.4 kg, a maximum muzzle velocity of 563.9 m/s and a maximum range (Charge 7) of 14,600 m
- M804 Practice.

Variants

M109A1

The M109A1 is basically the M109 fitted with a new and much longer 155 mm barrel, designated the M185, an improved elevation and traverse system and a strengthened suspension system. The M109A1 fires an HE projectile to a maximum range of 18,100 m compared with the 14,600 m of the M109, while a rocket-assisted projectile extends the range to 24,000 m. The M109A1 was type-classified as Standard A in October 1970 with first conversion kits made available early in 1972 and the first vehicles, which were converted from standard M109s, becoming operational in 1973. The M109A1 weighs 24,070 kg fully loaded and is 9.042 m long including the barrel. The weapon can fire the following US 155 mm projectiles:

- HE (M795) which is the replacement for the M107, the projectile weighs 46.9 kg
- HE (M107) with the projectile weighing 42.91 kg, a maximum muzzle velocity of 684.3 m/s and a maximum range (Charge 8) of 18,100 m
- HE (M483A1) (carries 64 M42 and 24 M46 dual-purpose grenades) with the projectile weighing 46.53 kg, a maximum muzzle velocity of 650 m/s and a maximum range (Charge 8) of 17,500 m
- HE (M692) (carries 36 anti-tank mines) with the projectile weighing 46.49 kg, a maximum muzzle velocity of 650 m/s and a maximum range (Charge 8) of 17,740 m
- HERA (M549A1) with the projectile weighing 43.54 kg and a maximum range of 23,500 m
- M864 Improved Conventional Munition, base bleed, type-classified in December 1987
- Illuminating (M485) with the projectile weighing 41.73 kg, a maximum muzzle velocity of 684.5 m/s and a maximum range (Charge 8) of 17,500 m
- Illuminating (M818) with the projectile weighing 46.26 kg, a maximum muzzle velocity of 684.3 m/s and a maximum range of 18,100 m
- Smoke BE (M116) with the projectile weighing 42.22 kg, a maximum muzzle velocity of 684 m/s and a maximum range (Charge 8) of 18,100 m
- Smoke WP (M110 series) with the projectile weighing 44.4 kg, maximum muzzle velocity of 684.3 m/s and a maximum range of 18,100 m
- Smoke (M825) with the projectile weighing 43.53 kg, a muzzle velocity of 660 m/s and a maximum range of 17,500 m
- HE ICM M449A1 carrying 60 anti-personnel grenades with a maximum range of 18,100 m; weight approximately 43.1 kg
- ADAM M692/M731 Area Denial Artillery Munition containing 36 self-destruct anti-personnel mines, which are ejected in-flight. Maximum range 17,740 m, weight 46.72 kg
- M712 Copperhead, maximum glide range 16,400 m; M823 Trainer is used for crew training and is not fired
- AT RAAMS M718/M741 containing nine anti-tank mines that are ejected during flight. Weight is 46.72 kg, maximum range 17,740 m.

Standard production M992 Field Artillery Ammunition Support Vehicle (FAASV) (BAE Systems) 1365010

With modifications, the more recent M109A6, for example, can fire the longer range M982 Excalibur precision guided munition.

In February 2013 Raytheon announced that is had been awarded a US Army Fiscal Year 2012 contract for the production of the 155 mm Excalibur Increment 1b PGP.

This USD56.6 million contract is for Low Rate Initial Production (LRIP) with first deliveries to commence in the last quarter of 2013 and includes options through to FY2016.

During trials the Excalibur has achieved a radial miss distance of less than 4 m at ranges in excess of 35 km.

In addition the M109A6 Paladin can also fire 155 mm artillery projectiles such as the M549A1 and M795 fitted with the ATK XM1156 Precision Guidance Kit (PGK) with first deliveries to be made in mid-2013.

This PGK has been procured as an Urgent Material Release and corrects the ballistic trajectory of the projectile using a global positioning system with fuzing functions to improve accuracy so that at least 50 per cent of the time the 155 mm projectiles land within 50 m of the target.

M109A2

The M109A2 entered production at BMY Combat Systems in 1978, with first deliveries made early in 1979. The first production quantity of 103 vehicles was with FY77 funding. Major changes from the M109A1 include a redesigned rammer and improved recoil mechanism, engine operation warning devices, a redesigned hatch and door latches, an improved hydraulic system and a bustle designed to carry an additional 22 rounds of 155 mm ammunition.

M109A2 production carried out by BMY for the US Army was as follows:

FY	US Army total
1976	12
1977	103
1978	250
1979	136
1980	96
1981	-
1983	-
1984	123
1985	70

In addition to the above, 36 M109A2s were produced for the US Army National Guard using FY82 funds. There has been no recent procurement of the M109A2/M109A3 by the US Army or Marine Corps. It should be noted that these are no longer used by the US Marine Corps.

M109A3

The M109A3 is a US Army depot converted M109A1. The M109A3 conversion includes:

- boresight alignment driver;
- M127 to M178 gun mount kit;
- selected RAM and safety kits, which included fuel system air purge, driver's instrument panel, bustle/rack, propellant stowage, torsion bar, counter recoil buffer and upper recoil cylinder.

M109A4

The M109A4 is a converted M109A2 or M109A3. The major changes are the incorporation of Nuclear, Biological and Chemical (NBC) requirements and Reliability And Maintainability (RAM) improvements. Selected improvements are found in the following kits: traverse mechanism, NBC equipment, hydraulic power pack filter, starter protection, external slave start, floor drains.

M109A5

The M109A5 is an upgraded M109A4, with the incorporation of the Reserve Component/Modified Armament System (RC/MAS). The RC/MAS is a range capability improvement kit and, firing the M203A1 charge, ranges of 30 km can be achieved with rocket-assisted projectiles. The RC/MAS kit includes: the M284 cannon assembly; and the M182 gun mount.

The M109A5 is available as a new production vehicle from the manufacturer or an M109A2 upgrade kit.

The M109A5 readily accommodates such individual customer requirements as the Global Positioning System (GPS) or the Automatic Fire-Control System (AFCS).

The new higher horsepower Low Heat Rejection (LHR) engine and uprated transmission deliver increased power in all climates due to higher engine torque and more efficient cooling/heat rejection systems.

The engine starter protection device prevents starter burnout. The 180 A alternator extends battery life, even with the increased electrical load from the NBC equipment.

The elevation/equilibration cylinder and the turret traverse clutch assembly are redesigned to prevent failures. Protective covers for engine electrical sensors and conductors rule out accidental damage.

Relocated and easily accessible filters improve hydraulic power pack filtration and easy to replace track pads and external roadwheel lubrication points simplify maintenance in the track area.

An external NATO power receptacle permits the M109A5 to be electrically powered by the M992A1 Field Artillery Ammunition Support Vehicle. The NATO slave start receptacle has been relocated to the driver's compartment for ease of access. The crew compartment sub-floor drains allow easy washdown of NBC contaminant and of other fluids during maintenance.

A ventilated facepiece system gives crew members filtered, conditioned air during NBC operations and as needed.

In addition to the Austrian order for modified M109A5s (covered later), other customers have ordered the M109A5 including Greece (12 delivered from 1998), Taiwan (28 delivered in 1998) and Thailand (20 delivered).

Surplus M109 export sales

As a result of the end of the Cold War, a significant quantity of 155 mm M109 self-propelled artillery systems have become surplus to requirements.

According to the United Nations Arms Transfer Lists for the period 1994 through to 2010, the following quantities of M109 weapons have been transferred overseas but in some cases they are brand new systems and not surplus weapons:

From	To	Quantity	Comment
Belgium	Brazil	6	delivered 1999 via SABIEX (M109A3)
	Brazil	24	delivered 2000 via SABIEX (M109A3)
	Brazil	7	delivered 2001 via SABIEX (M109A3)
	Chile	21	M108 command post and delivered 2008 (ex-Belgium Army)
	Morocco	43	delivered 2008 (M109A2)
Germany	Greece	50	delivered 2000 (M109A3GE)
Germany	Greece	223	delivered 2010
Netherlands	Austria	5	delivered 1997
	UAE	87	delivered 1997 (upgraded by RDM)
Switzerland	Chile	24	delivered 2004
	UAE	40	delivered 2004
	Chile	24	delivered 2006
UK	Austria	51	delivered 1994 (M109A2)
	Austria	32	delivered 1994 (M109A3)
	Austria	18	delivered 1995 (M109A3)
US	Austria	16	delivered 1995 (new M109A50e)
	Austria	27	delivered 1997 (new M109A50e)
	Austria	27	delivered 1998 (new M109A50e)
	Bahrain	20	delivered 2005 (M109A5)
	Egypt	161	delivered 2007 (M109A5)
	Egypt	279	delivered 2000
	Egypt	8	delivered 2004
	Egypt	25	delivered 2005 (M109A5)
	Egypt	1	delivered 2005 (M109A2)
	Egypt	4	delivered 2006 (M109A5)
	Egypt	19	delivered 2009 (M109A5)
	Greece	12	delivered 1998 (M109A5)
	Greece	12	delivered 1999 (M109A5)
	Pakistan	115	delivered 2007 (M109A5)
	Pakistan	115	delivered 2009 (M109A5)
	Pakistan	67	delivered 2008 (M109A5)
	Portugal	14	delivered 2001 (M109A5)
	Saudi Arabia	44	delivered 2002
	Saudi Arabia	4	delivered 2004 (M109A5)
It should be noted that these figures do not always tie up with other published information.			

M109 (Austria)

Austria purchased 38 M109 and 18 M109A2 self-propelled howitzers from the US.

From 1994, Austria purchased 101 M109 series self-propelled artillery systems which were surplus to the requirements of the British Army and these have now been delivered.

Late in 2009 Austria announced that it would phase out of service all of its M109's with the exception of the new build M109A5e (Austria).

M109A50e (Austria)

Early in 1995, through the US Army Armament and Chemical Acquisition and Logistics Activity at Rock Island, Austria awarded the now BAE Systems a contract worth USD48.6 million for the supply of 54 155 mm M109A50e self-propelled howitzers for the Austrian Army.

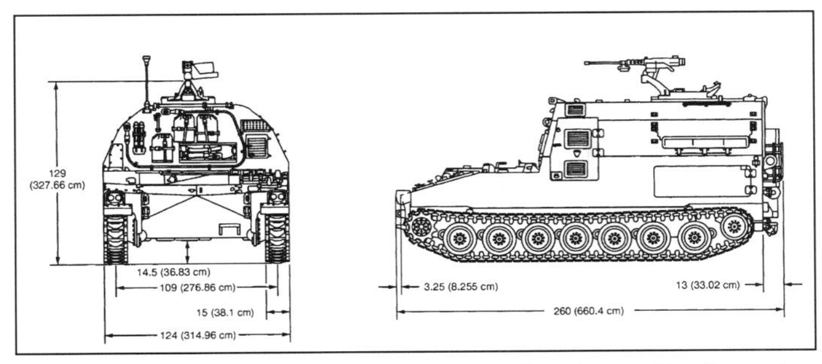

Front and side drawings of the M992A2 Field Artillery Ammunition Support Vehicle (FAASV) 0126562

The M109A50e has a number of unique features that differ from the standard M109A5 including an Austrian semi-automatic flick rammer, inertial positive navigation system, Swiss electrical system and the same 155 mm/39-calibre ordnance as used on the M109A6 Paladin currently in service with the US Army.

The M109A50e is armed with a 155 mm M284 cannon in an M182 mount. Using the US M203A1 charge, ranges of 30 km can be achieved firing a rocket-assisted projectile.

The first production M109A50e for the Austrian Army was rolled out on 6 May 1997 and six vehicles were delivered each month until delivery was complete.

ReStPz M109 (Austrian)

This is an artillery fire-control vehicle used by the Austrian Army under the designation of the Rechenstellenpanzer (ReStPz M109) and has a crew of six.

The turret is retained, although the 155 mm barrel has been removed to provide greater internal volume.

Equipment installed to carry out its specialised role include an Artillery Tactical Computer (ATC) with printer, a KFF-33F-0 radio set and a 28 V SD/43 A generator. Mounted either side of the turret are three 80 mm electrically operated smoke grenade launchers. Mounted on the roof is a .50 (12.7 mm) M2 HB machine gun for local and air defence purposes.

Externally it is very similar to the Belgian VBCL, based on the M108 105 mm self-propelled howitzer chassis which has had its turret removed and the hull sides extended upwards.

The Austrian Army has a total of 19 of these ReStPz M109 artillery fire-control vehicles in service.

M109A2 (Belgium)

In August 1983, the Belgian Minister of Defence signed a contract with BMY for the purchase of 127 M109A2 self-propelled howitzers. The first vehicles were delivered in August 1984 and all were delivered by the end of 1985. These vehicles were used by the six field artillery battalions, the Brasschaat Artillery School and the maintenance reserve. The older M109s were upgraded to M109A3 standard by the Arsenal du Matériel Mécanique et de l'Armement, at Rocourt, near Liège.

From 1998 the operational M109A2s were fitted with a North Finding Module and between 2002 and 2006 a total of 108 M109A2s were upgraded to the M109A4 standard. These are no longer in service.

M109A3 (Brazil)

Brazil has taken delivery of 37 M109A3 SPH from Sabiex International with all delivered by 2001 and issued to three artillery regiments of the Brazilian Army.

The Brazilian Army has a requirement for an additional 36 M109A5 self-propelled artillery systems and would like to bring the currently deployed M109A3 up to the same build standard.

M109 (Canada)

This country took delivery of 50 early M109s, which were upgraded followed by 26 M109A2. Training ceased on these systems in 2005.

M109A3 (Denmark)

Early in 1989, the Danish Army stated that the modernisation of its 76 M109 155 mm self-propelled howitzers to the M109A3 standard had been completed. The modernisation was carried out in Denmark with kits supplied by FMS Corporation (now Marvin Land Systems) of Los Angeles, California.

M109 (Chile)

Late in 2004 it was stated that Chile was to take delivery of 24 ex-Swiss Army M109s. These retain their standard 155 mm barrels.

These were refurbished by RUAG Defence before delivery to Chile.

In 2008 Chile rook delivery from Belgium of 21 M108 that had been converted to the artillery command post role and were no longer required by the Belgian Army.

Egyptian surplus US Army M109A2/A3

In mid-2003 the US Army Tank-automotive and Armaments Command awarded the then United Defense Industries a contract worth USD43.7 million to refurbish 201 surplus US Army 155 mm M109A2/A3 Self-Propelled Howitzers (SPHs) for Egypt. This was funded through the US Foreign Military Sales (FMS) programme.

Over the last few years Egypt has made a number of requests for surplus US Army 155 mm M109A2/M109A3 SPHs, but these letters of offer were not taken up.

Hellenic Army Fire Direction Centre (FDC) vehicle (Christopher F Foss) 1403737

The M109A2/M109A3 SPHs were taken from US Army reserve stocks and prior to delivery to Egypt were returned by the manufacturer to a fully functional state in the areas of drivetrain (engine, transmission and cooling system), fire-control and 155 mm/39-calibre main armament.

The work was carried out at company's facilities at York and Fayette County and Anniston. First vehicles were ready by the end of 2003 and the whole programme completed by November 2005.

According to the US Government Egypt took delivery of 287 systems between 2000 and 2004.

Currently the Egyptian Army operates 164 now BAE Systems 155 mm M109A2 self-propelled howitzers delivered some years ago. These are supported by 51 M992 Field Artillery Ammunition Support Vehicles (FAASV) and 72 Fire Direction Centre (FDC) vehicles also supplied new by the company from its York facility. These are also based on a M109 series chassis.

In 2005 it was stated that the Egyptian Army was seeking to acquire another 200 M109A5 self-propelled howitzers from the US through the FMS programme.

The total value of this contract, when awarded, would be USD181 million and the work would be carried out at the facilities of the now BAE Systems. These will be surplus US Army systems.

First deliveries were made in 2005 and consisted of 26 units with another 161 being delivered in 2007 which brought the total to 165.

Since then additional vehicles have been supplied from US Army stocks and details of these are provided in the table.

M109 (Germany)

The original German Army vehicles were designated the M109G, with a sliding breech developed by Rheinmetall extending the effective range of the weapon to 18,000 m.

In 1983, the German Army purchased from the FMS Corporation (now Marvin Land Systems), Los Angeles, California, 586 kits to convert its M109Gs to the M109A3G. The newly modified M109A3G has an extended range of 24,000 m and uses a recently developed gun tube made by Rheinmetall Weapons & Munitions but still uses the original M109G sliding breech mechanism. The M109A3G also has a different fire-control system and three forward-firing 76 mm smoke grenade dischargers either side of the turret.

The German fleet underwent the AURORA modification at the now Rheinmetall Landsysteme, with the modified vehicles designated the M109A3GEA1.

Germany has a number of M109A3GE surplus to requirements and Greece has already taken delivery of 50 units.

In 2010 Germany supplied Greece with another 223 ex-German Army M109A3GE that were surplus to requirements.

As of early 2013 there were no 155 mm M109 series self-propelled artillery systems in front line service with Germany.

M109A5 (Greece)

In May 1998, Greece placed an order with the company for the supply of 12 M109A5 systems for delivery from late 1998. The contract also included an option for a further 12 M109A5 systems. Total value of the contract, with options, was about USD50 million. These have now been delivered. Germany delivered another 50 surplus M109A3GE systems in 2000. These were followed by another 223 M109A3GE in 2010.

M109 (Israel)

The M109A1s of the Israel Defense Force are designated the M109 AL. They have additional stowage racks on the hull and turret sides, allowing them to carry more ammunition inside. The IDF is planning to upgrade some of its M109s to incorporate selected features of the M109A6. Details of the Israel version of the M109 are given in a separate entry. About 350 remain in service.

M109L (Italy)

This was developed by Oto Melara to meet the requirements of the Italian Army. It is only used by Italy. These are now being supplemented by the locally assembled PzH 2000. Oto Melara is no longer marketing this system. Details of the Italian M109L are provided in a separate entry in *IHS Jane's Land Warfare Platforms: Artillery and Air Defence*.

M109A2 (Korea, South)

A total of 1,040 have been built in South Korea with final deliveries in 1997. These are now being supplemented by the locally designed and built Samsung Techwin K9 Thunder 155 mm/52-calibre self-propelled artillery system. Details

Country	M109	M109A1 kit	PIP kit	M109A1B	M109A2	M109A5
Austria[4]	38	-	-	-	18	54[3]
Belgium	-	-	41	-	127D	-
Canada	50	-	50	-	26D	-
Denmark	76	-	-	-	-	-
Egypt	-	-	-	-	164D	-
Ethiopia	12	-	-	-	-	-
Germany	609[1]	-	2	-	-	-
Greece	-	-	-	51	84	12[5]
Iran	50	50	-	390	-	-
Israel	60	60	-	294 + 75D	-	-
Italy	221[1]	-	-	62	-	-
Jordan	126	-	126	-	108	-
Korea, South	-	-	-	-	1,040	-
Kuwait	-	-	-	6[2]	31[2]	-
Morocco	-	-	-	36	3	-
Netherlands	135	-	135	3	91	-
Norway	127[1]	-	-	-	-	-
Pakistan	-	-	-	-	152	-
Peru	-	-	-	12	-	-
Portugal	-	-	-	-	6	-
Saudi Arabia	-	-	-	87	24	-
Spain	18	18	-	60	6	-
Switzerland	-	146	-	435	-	-
Taiwan	-	-	-	-	197	28
Thailand	-	-	-	-	-	20

D–direct sales
[1] M109G
[2] Status uncertain following Iraqi invasion and Operation Desert Storm
[3] M109A5 plus
[4] 67 M109A2 and 48 M109A3 from UK in 1994/1995 and five from Netherlands in 1997
[5] Ordered 1997 with option on additional 12 (see text)
Note: Some of these users have now phased their M109s out of service (e.g. Canada, Netherlands and Germany) while others have reduced their quantities.

of this system are provided in a separate entry in *IHS Jane's Land Warfare Platforms: Artillery and Air Defence.*

M109 (Netherlands)

The Royal Netherlands Army was the first in Europe to upgrade the M109 to the M109A3. In 1980, the Royal Netherlands Army released its contract to RDM Technology for the conversion of 136 of its M109s to the M109A3 configuration. RDM Technology purchased kits from the now Marvin Land Systems of the US, which was also its contractor for an earlier parts contract for 86 M109A2s co-produced in the Netherlands for the Royal Netherlands Army.

Canada and the Netherlands awarded RDM Technology contracts to refurbish parts of their M109 fleets of self-propelled howitzers. RDM Technology closed down in 2004.

These have now been replaced by the PzH 2000.

M109 (Norway)

These have all been upgraded to the M109A3GN standard, which is almost the same as the German M109A3G configuration.

This are to be replaced by 24 BAE Systems Archer 155 mm/52-calibre (6 × 6) self-propelled artillery systems.

M109A2 (Pakistan)

Pakistan operates a fleet of US-supplied M109A2s, which are overhauled to an as-new condition at the Heavy Industries Taxila facility. The US supplied another 114 surplus M109A5 to Pakistan in 2007.

M109A5 (Portugal)

This country took delivery of 14 M109A5 from the US in 2001 with earlier deliveries including six M109A2.

As of early 2013 Portugal deploys one regiment with a total of 18 M109A5 which are split into three batteries each with six M109A5. The six M109A2 have been canibalised for spare parts.

M109 (Spain)

Today the Spanish Army fleet consists of 96 M109A5E 155 mm self-propelled howitzers which, together with other Spanish artillery weapons are now being upgraded. There is a separate entry on the Spanish 155 mm M109A5E in *IHS Jane's Land Warfare Platforms: Artillery and Air Defence*.

M109 (Switzerland)

In 1968, the Swiss Parliament authorised an order for 140 M109s from the US and these entered service in 1974 under the designation of the Panzerhaubitze 66.

A further 120 M109A1Bs were subsequently purchased and the original M109s were upgraded to the M109A1B standard.

The new M109A1Bs are known as the Panzerhaubitze 74 and the refitted vehicles as the Panzerhaubitze 66/74.

The 1979 armament programme called for the purchase of approximately 207 additional M109A1Bs which entered service as the Panzerhaubitze 79.

The last fleet of 110 M109A1Bs, called Panzerhaubitze 88, entered service as part of the 1988 armament programme.

The Swiss Army M109s have a bank of three forward-firing electrically operated 76 mm smoke grenade launchers mounted either side of the turret front. For improved traction, Swiss M109s are fitted with German Diehl tracks.

Of the original Swiss Army fleet of 577 M109s (some sources have stated 581) at least 348 have now been upgraded in a number of key areas including the installation of a new Swiss designed and built 155 mm/47-calibre barrel. Late in 2004 it was stated that 24 ex-Swiss Army M109 systems would be sold to Chile. These retain the standard 155 mm/39-calibre barrel.

Switzerland has also sold 40 M109 series systems to the UAE and some systems have also been scrapped. Some of the UAE vehicles are understood to have been transferred to a third party.

Details of the RUAG Defence M109/47-calibre upgrade are provided in a separate entry in *IHS Jane's Land Warfare Platforms: Artillery and Air Defence*.

M109 (Taiwan)

Details of the 155 mm SPG based on an M109 chassis are given in a separate entry in *IHS Jane's Land Warfare Platforms: Artillery and Air Defence*. The 155 mm ordnance is mounted in an unprotected position at the rear of the hull with no protection for the weapon or its crew for small arms fire and shell splinters.

Taiwan has also developed its own ammunition support vehicle for the M109A5, called the CM24A1, based on a full-tracked hull. This has an all-welded hull with the crew compartment at the front and the ammunition compartment at the rear.

The driver and commander enter the vehicle via a large door either side that opens to the front and, in the upper part of this, there is a bulletproof window. In the front of the vehicle are three bulletproof windows and a roof-mounted machine gun can be fitted.

The suspension system is of the torsion bar type and either side consists of seven dual rubber-tyred roadwheels with the drive sprocket at the front, idler at the rear and three tracked return rollers.

The CM21A1 backs up to the rear of the M109A5 and feeds ammunition using a conveyor belt. There is a similar vehicle for the 203 mm (8 in) M110A2 self-propelled artillery system called the CM27.

M109A5 (Thailand)

Thailand ordered 20 M109A5 self-propelled artillery systems and 20 Field Artillery Ammunition Support Vehicles through the FMS programme for delivery in 1994. These have been formed into a new self-propelled artillery regiment with three batteries each of six weapons, the remaining two being used for training.

Egyptian SPH122

The hull of the M109/M992 is used as the basis for the Egyptian Army SP122 122 mm self-propelled howitzer, final deliveries of which were made in 1995. By late 2002 Egypt had taken delivery of 124 SP122 systems.

Prime contractor for the SP122 programme was the now BAE Systems and details are provided in a separate entry in *IHS Jane's Land Warfare Platforms: Artillery and Air Defence*.

Denel Land Systems M109 upgrade

In late 2000 it was revealed that the United Arab Emirates (UAE) was upgrading the major part of its 155 mm Self-Propelled (SP) artillery fleet with a new gun laying and navigation/integral gun fire-control system supplied by Denel Land Systems, South Africa.

This has now been installed in the 78 155 mm/45-calibre G6 (6 × 6) SP guns supplied by Denel from 1991 and the 85 M109A3's SP guns upgraded by RDM Technology of the Netherlands to the enhanced M109L47 configuration. The latter were only recently delivered and are fitted with a 155 mm/47-calibre ordnance supplied by the now RUAG Defence.

According to Denel Land Systems, the new system navigates accurately to within a distance of about 16 m using its two integrated sensors. The first of these is inertial and used where the system starts from a known point and then keeps track of the exact movement of the gun and continually calculates the current position. Second is using Global Positioning System (GPS) navigation that calculates its position with signals received from navigation satellites.

The Selex ring laser gyro makes it possible to lay the 155 mm gun accurately to within 1 mil or 0.05° which is considered necessary for long-range artillery fire.

This upgrade enables the weapon to come into action, fire and move away before any counter battery fire is returned so enhancing its speed of response as well as improving its survivability.

In addition, the land navigation computer also receives firing data from the Battery Command Post (BCP) and displays this information on the same screen as that used for navigation. This allows the gun commander to carry out the gun laying roles so saving one crew position.

Finally the system is linked to the projectile muzzle velocity measurement radar mounted over the gun. This measures and stores accurate muzzle velocity information as well as telling the BCP that a round has been fired and updates ammunition information.

This is the first time that Denel Land Systems has supplied an upgrade package for the US-designed M109 self-propelled artillery system. In addition to the navigation and laying system/integral gun fire-control system, the M109 was also fitted with a semi-automatic hydraulic push rammer which is used to ram projectile so increasing rate of fire and reducing crew fatigue.

Other M109 improvements include a remote-operated barrel clamp, auxiliary power unit installed on the turret rear and fitting an air conditioning system to allow the M109 to be operated in high ambient temperatures.

Rheinmetall Denel Munitions provided upgrade kits for both the G6 and M109, which were used in the United Arab Emirates utilising local resources.

M992 Field Artillery Ammunition Support Vehicle (FAASV)

Originally developed as a private venture by Bowen-McLaughlin-York, the XM992 was type-classified M992 in 1983.

The M992 FAASV is basically the hull of a standard M109 155 mm self-propelled howitzer with the turret replaced by a fully enclosed superstructure. Inside this superstructure, which has the same protection as the rest of the vehicle, can be stacked 93 (90 conventional and 3 Copperhead) 155 mm projectiles, 99 propellant charges and 104 fuzes. The ammunition can be loaded into the superstructure by a front-mounted crane and is fed from the vehicle along a conveyor to the recipient self-propelled howitzer; this crane was not fitted to US Army M992 vehicles but is offered for export. All ammunition handling within the M992 is mechanical. In use, the M992 backs up to the user M109 and the conveyor delivers the 155 mm ammunition at the rate of up to 8 rds/min to the M109 bustle or lower hatches. The M992 can have a crew of two plus six passengers and the weight fully loaded is 25,900 kg. Intended primarily for use with M109 field units, the M992 can also be used by units with 175, 203, 120 and 105 mm self-propelled weapons. With 203 mm units the M992 can carry 48 203 mm projectiles, 53 charges and 56 fuzes.

It has been suggested that the M992 could form the basis for a family of battlefield vehicles including a Fire Direction Centre Vehicle (see below), Command Post Vehicle (CPV), Medical Evacuation Vehicle (MEV), Maintenance Assistance Vehicle (MAV) and Armoured Forward Area Rearm Vehicle (AFARV).

US Army testing and decisions relating to the M109A6 Paladin reaffirmed the need for an FAASV in direct, organic support of each self-propelled gun. The US Army has improved the basic M992 through a series of upgrades to M992A1 and M992A2 configurations. The US Army recently type-classified the M992A1. The M992A1 contains a series of improvements based on lessons learned during the extensive operational testing of the M109A6 Paladin self-propelled howitzer.

By 2013, total production of the M992 FAASV for the home and export markets amounted to:

United States	921 (delivered)
Egypt	51 (delivered)
Saudi Arabia	60 (delivered)
Spain	6 (delivered)
Thailand	20 (delivered 1993-94)

In addition 119 Fire Direction Centre vehicles have been built:

Egypt	72
Greece	41
Taiwan	6

Since then the US Army ordered an additional 132 M992A2 FAASV for the National Guard and the first of these was completed in June 1998. Most of these were new build vehicles but the last 36 are rebuilds of M109A2 self-propelled howitzers and these were delivered between May and September 1999. Letterkenny Army depot was also involved in this programme. All 132 vehicles have now been delivered and fielded to the US Army National Guard. In addition, the savings achieved through the conversion programme, saved enough money for the US Army to purchase an additional six more vehicles which were delivered by the end of 2000.

Under a separate programme, some 166 M992 FAASV have been upgraded to the M992A2 standard by Espace Mobile International in Belgium with final deliveries taking place in mid-1999.

Total production/conversion of FAASV for the US Army is now:

M992A0 to A2 conversions	664
M992A1 to A2 conversions	125
New production M992A2	96
M109A2/M109A3 conversions	36
Total	**921**
(To the above figures should be added the six additional vehicles delivered in 2000.)	

M109A6 PIM and M992A2 PIM

In October 2009 it was announced that BAE Systems, had been awarded a USD63.9 million contract from the US Army Tank Automotive and Armaments Command for the procurement of five prototype M109A6 PIM and two M992A2 FAASV. Details of the M109A6 PIM are provided in a separate entry in *IHS Jane's Land Warfare Platforms: Artillery and Air Defence*.

In December 2012 BAE Systems confirmed that LRIP production of the M109A6 PIM would be undertaken at the Elgin, Oklahoma facility which is located at the Fort Sill Industrial Park, home of the US Army field artillery.

Key components of the PIM production vehicles, including the hull, will be sent to Elgin facility from BAE Systems manufacturing facilities and suppliers.

As part of final assembly and checkout, BAE Systems will use Fort Sill where they will undergo mobility and firing verification.

It is expected that the LRIP PIM production contract will be awarded in the third quarter of 2013 and cover an initial 72 systems.

Field Artillery Operation Centre Vehicle (FAOCV)

During TF-XXI, the company converted three FAASV into the FAOCV configuration for use at the National Training Centre (NTC).

Modifications included four compartmentalised workstations, 10 m high telescoping antenna mast, environmental control unit and 400 A auxiliary power unit.

According to the company, results of the exercise demonstrated improved mobility for M109A6 units, FAOCV improved survivability and maintainability for units. Also demonstrated were on-the-move computer operations.

Fire Direction Centre Vehicle (FDCV)

The FDCV, based on the chassis and hull of the M992 FAASV, has been tested at the US Army School of Artillery at Fort Sill to prove the feasibility and function of the concept under field and NBC conditions. The equipment used in the FDCV is a mix of 'off-the-shelf' or prototype products assembled for the concept, for which the internal headroom and armour of the M992 hull is eminently suitable. If required, bolt-on armour can be added to protect the specialist personnel and equipment carried.

Other M109 hull applications

The chassis of the M109, or versions of the M109, have been used for a wide range of other applications. These include the XM-11S threat radar simulator from General Dynamics. This was developed for the US Army Missile and Space Intelligence Centre and is used to simulate the Russian SA-11 ('Gadfly') surface-to-air missile system.

The XM975 Roland hull, based on the M109, was used as the basis for the Robotic Command Center for use in various robotic vehicle programmes.

Specifications

	M109A2
Dimensions and weights	
Crew:	6
Length	
overall:	9.12 m
hull:	6.19 m
Width	
overall:	3.15 m
Height	
to top of roof mounted weapon:	3.28 m
transport configuration:	2.8 m
Ground clearance	
overall:	0.46 m
Track	
vehicle:	2.788 m
Track width	
normal:	381 mm
Length of track on ground:	3.962 m
Weight	
standard:	21,110 kg
combat:	24,948 kg
Mobility	
Configuration	
running gear:	tracked
Speed	
max speed:	56.3 km/h
Range	
main fuel supply:	349 km

	M109A2
Fuel capacity	
main:	511 litres
Fording	
without preparation:	1.07 m
Gradient:	60%
Side slope:	40%
Vertical obstacle	
forwards:	0.53 m
Trench:	1.83 m
Engine	Detroit Diesel Model 8V-71T, turbocharged, water cooled, diesel, 405 hp at 2,300 rpm
Gearbox	
model:	Allison Transmission XTG-411-2A
forward gears:	4
reverse gears:	2
Suspension:	independent torsion bar
Electrical system	
vehicle:	24 V
Batteries:	4 × 12 V
Firepower	
Armament:	1 × turret mounted 155 mm M185 howitzer
Armament:	1 × roof mounted 12.7 mm (0.50) M2 HB machine gun
Ammunition	
main total:	36
12.7/0.50:	500
Turret power control	
type:	hydraulic/manual
by commander:	no
by gunner:	yes
Main armament traverse	
angle:	360°
Main armament elevation/depression	
armament front:	+75°/-3°
Survivability	
Night vision equipment	
vehicle:	yes
NBC system:	no
Armour	
hull/body:	aluminium
turret:	aluminium

Status
The M109 and its derivatives are in service with the US Army and National Guard. Other nations using M109 series weapons are listed (see table).

During the first production phase, carried out initially by the Cleveland Army Tank Plant, US-configured M109s were delivered to the following nations: Austria, Belgium, Canada, Denmark, Ethiopia, Iran, Israel, Jordan, Netherlands, Spain, Switzerland and the UK. Italy, the Netherlands and Norway formed co-production agreements. Canada, Iran, Israel, the Netherlands, Spain, Switzerland and the UK converted their M109s to the M109A1 configuration. M109Gs were delivered to Germany, Italy and Norway.

M109A1B FMS production since 1972 has resulted in systems for Greece, Iran, Israel, Italy, Jordan, Kuwait, Morocco, Netherlands, Pakistan, Portugal, Spain, Taiwan and the UK. M109A2 production since 1980 has resulted in deliveries to Austria, Jordan, Kuwait, Netherlands, Pakistan, Portugal, Spain, Taiwan and the UK.

Known sales of the M109 series of self-propelled howitzers are in the table, which excludes some of the more recent sales by the countries listed under Surplus M109 export sales earlier in this entry.

In addition to the above, there are the following additional M109 series users:

Country	Quantity	Comment
Libya	14	believed to be in service (numbers reduced due to 2011 conflict)
Peru	12	M109A2
US	1,594	includes all models
Venezuela	5	

Contractor
BAE Systems.

Original manufacturer was Cadillac Motor Car Division at Cleveland Army Tank Plant, later Chrysler Corporation at the same plant.

M108 105 mm self-propelled howitzer

Development
In 1958, the first prototype of the T195 self-propelled howitzer was completed but a number of failures occurred in the torsion-bar suspension. The prototype differed from later vehicles in that it had a different shaped hull and turret, the seventh roadwheel acted as the idler and the short-barrelled 105 mm howitzer was fitted with a muzzle brake.

Austrian Army M109 command post vehicle (Michael Jerchel) 1044214

Belgian Army M108 SPH converted into the VBCL command post configuration (Pierre Touzin) 0533758

In 1959, a policy was established that diesel rather than petrol engines would be used for future vehicles and the prototype of the T195, which was powered by a Continental petrol engine, was refitted with a Detroit diesel engine and called the T195E1.

Trials with the latter were carried out in 1960 but there were failures of critical suspension and final drive components. These deficiencies were subsequently corrected and, in December 1961, the T195E1 was classified for limited production as the howitzer, light, self-propelled, 105 mm, M108.

Production of the M108 was undertaken by the Cadillac Motor Car Division of the General Motors Corporation: the first production vehicle was completed in October 1962 and production continued until 1963.

The vehicle was in production for only a short period, as the US Army decided to concentrate on the self-propelled 155 mm M109 rather than the 105 mm M108.

The M108 is almost identical to the M109, and differs only in its armament, loading system, fire-control system, ammunition stowage racks and in that it is not fitted with spades at the rear of the hull. In the case of the M109 these were deployed one either side at the rear to provide a more stable firing platform.

When compared to the 105 mm M108, the 155 mm M109 fired more effective projectiles to a larger range and had additional development potential.

Description
The 105 mm M108 self-propelled howitzer has a crew of five: the commander, gunner, two ammunition numbers and the driver.

The hull is made of all-welded aluminium armour with the driver at the front of the hull on the left, the power pack to the right and the fighting compartment at the rear. The aluminium armour provides the occupants with protection from small arms fire and shell splinters.

The driver has a single-piece hatch cover that opens to the left, in front of which are three M45 day periscopes that can be covered by small metal flaps to prevent damage. The centre day periscope can be replaced by a passive periscope for driving at night.

The Detroit Diesel Model 8V-71T engine is coupled to the Allison Transmission XTG-411-2A cross-drive automatic transmission, which is at the front of the hull.

The all-welded aluminium turret is at the rear of the hull. It has a square hatch in each side that opens to the rear and twin doors in the rear. The commander is seated on the right and has a cupola which can be traversed manually through 360°, a single-piece hatch cover that opens to the rear and an M27 day periscope.

Pintle-mounted on the forward part of the commander's cupola is a .50 (12.7 mm) M2 HB machine gun. The gunner is seated on the left side of the turret and has a square single-piece hatch cover that opens to the right. Twin doors are provided at the rear of the hull for ammunition resupply.

The torsion bar suspension either side consists of seven dual rubber-tyred road wheels with the drive sprocket at the front and the idler at the rear. There are no track-return rollers. The tracks are of the single-pin, centre guide type with replaceable rubber pads.

M108 105 mm self-propelled howitzer with ordnance in travelling lock (C R Zwart) 0500350

The M108 is fitted with night vision equipment but no NBC equipment. The basic vehicle can ford to a depth of 1.828 m without preparation.

Main armament is an M103 105 mm howitzer in an M139 mount. The recoil system is of the hydrospring, constant retarding force type while the breech block is of the sliding wedge drop block type.

The weapon has an elevation of +74°, depression of –4° and 360° turret traverse. Gun elevation/depression and turret traverse are manual. Normal rate of fire is 1 rd/min, but 3 rds/min can be fired for short periods.

Fire-control equipment consists of an M15 elevation quadrant, an M117 panoramic telescope with a magnification of ×4 and a 10° field of view, and an M118 elbow telescope with a magnification of ×4 and a 10° field of view.

The following types of ammunition can be fired:

- HE (M1) with the complete round weighing 18.107 kg, a maximum muzzle velocity of 494 m/s and a maximum range (Charge 7) of 11,500 m
- HE (M444) (carries 18 M39 grenades) with the complete round weighing 19.05 kg, a maximum muzzle velocity of 494 m/s and a maximum range (Charge 7) of 11,500 m
- HE (M413) (carries 18 M35 grenades) with the complete round weighing 19.05 kg, a maximum muzzle velocity of 494 m/s and a maximum range of 11,500 m
- HEP-T (M327) with the complete round weighing 15.17 kg, a maximum muzzle velocity of 559 m/s and a maximum range of 8,685 m
- HERA (M548) with the complete round weighing 17.46 kg, a maximum muzzle velocity of 548 m/s and a maximum range of 15,000 m
- Illuminating (M314 series) with the complete round weighing 21.06 kg, a maximum muzzle velocity of 494 m/s and a maximum range (Charge 7) of 11,500 m
- Illuminating (M314A3) with the complete round weighing 21.06 kg, a maximum muzzle velocity of 494 m/s and a maximum range of 11,500 m
- Leaflet (M84B1) with the complete round weighing 18 kg, a maximum muzzle velocity of 433 m/s and a maximum range (Charge 7) of 9,091 m
- Smoke (M60 series) with the complete round weighing 19.46 kg, a maximum muzzle velocity of 494 m/s and a maximum range (Charge 7) of 11,500 m
- Smoke (M84) with the complete round weighing 19.03 kg, a maximum muzzle velocity of 494 m/s and a maximum range (Charge 7) of 11,500 m

It should be noted that not all of these 105 mm projectiles are still in service.

Variants

Belgian Army M108A2B training vehicles

The Belgian Army modified at least 14 M108 105 mm self-propelled howitzers by mounting these turrets on surplus M109A2 hulls. This is designated the M108A2B and is used for training purposes as 105 mm ammunition is cheaper than 155 mm ammunition.

It is understood that these are no longer deployed.

Belgian Army VBCL fire direction and command post

The Belgian Arsenal du Matériel Mecanique et de l'Armement at Rocourt converted a total of 45 surplus M108 chassis into the VBCL (*Vehicule Blindé de Commandement et de Liaison*) with final deliveries being made in 1994.

It is understood that these are no longer deployed.

The VBCL is a fire direction centre and command post for Belgian Army field artillery battalions and batteries. It is fitted with four BAMS radios, two telescopic antennas PU-8, a 2.7 kW generator DE-CA, tent to increase working area, ABM-1045 field telephone system and a roof mounted .50 (12.7 mm) M2 HB machine gun.

The Fire Direction Centres are fitted with ruggedised portable computers PCI but are otherwise similar to the CP versions.

According to the United Nations, in 2008 Belgium sold 21 of the M108 command post vehicles to Chile.

Austrian Army fire direction and command post vehicles

These are similar to the Belgian vehicles but are based on an M109 155 mm self-propelled system, rather than an M108.

Specifications

	M108
Dimensions and weights	
Crew:	5
Length	
overall:	6.114 m
Width	
overall:	3.295 m
transport configuration:	3.149 m
Height	
to top of roof mounted weapon:	3.155 m
transport configuration:	2.794 m
Ground clearance	
overall:	0.451 m
Track	
vehicle:	2.768 m
Track width	
normal:	381 mm
Length of track on ground:	3.962 m
Weight	
standard:	18,436 kg
combat:	22,452 kg
Ground pressure	
standard track:	0.71 kg/cm^2
Mobility	
Configuration	
running gear:	tracked
Power-to-weight ratio:	18.038 hp/t
Speed	
max speed:	56 km/h
Range	
main fuel supply:	390 km
Fuel capacity	
main:	511 litres
Fording	
without preparation:	1.828 m
Gradient:	60%
Vertical obstacle	
forwards:	0.533 m
Trench:	1.828 m
Engine	Detroit Diesel Model 8V-71T, turbocharged, water cooled, diesel, 405 hp at 2,300 rpm
Gearbox	
model:	Allison Transmission XTG-411-2A
forward gears:	4
reverse gears:	2
Suspension:	torsion bar
Electrical system	
vehicle:	24 V
Batteries:	4 × 12 V
Firepower	
Armament:	1 × turret mounted 105 mm M103 howitzer
Armament:	1 × roof mounted 12.7 mm (0.50) M2 HB machine gun
Ammunition	
main total:	87
12.7/0.50:	500
Turret power control	
by commander:	no
by gunner:	yes
Main armament traverse	
angle:	360°
Main armament elevation/depression	
armament front:	+74°/-4°
Survivability	
Armour	
hull/body:	aluminium
turret:	aluminium

Status

Production complete. In service with:

Country	Quantity	Comment
Brazil	50	estimate. These serve alongside 37 M109A3 acquired from the Belgian company of SABIEX which were delivered between 1999 and 2001 and are being overhauled.
Chile	21	M108 command post and delivered 2008 (ex-Belgium Army)
Taiwan	100	estimate
Turkey	26	M108T

Contractor

Cadillac Motor Car Division of General Motors Corporation. (This company is no longer involved in armoured fighting vehicle development or production.)

Wheeled

China

NORINCO SH1 155 mm/52-calibre self-propelled artillery system

Development

In 2007 China North Industries Corporation (NORINCO) revealed that they had completed development of a new 155 mm/52-calibre Self-Propelled (SP) (6 × 6) artillery system called the SH1.

According to NORINCO, development of this system commenced in 2002 and so far two examples have been completed, one prototype and one pre-production system.

Details of the SH1 155 mm/52-calibre self-propelled artillery system were released at the same time as details of the SH2 122 mm self-propelled artillery system which is mounted on a different 6 × 6 cross-country truck chassis.

During a major parade held in Beijing in October 2009 a number of major new weapon systems were shown for the first time but the SH1 155 mm/52-calibre self-propelled artillery system was not one of these.

The SH1 is integrated on the rear of the WS5252 (6 × 6) cross-country truck chassis which is designed and built by Wanshan Special Vehicle Company who manufacture a wide range of all wheel drive vehicles for a wide range of civil and military applications.

As of March 2013 it had not been confirmed that the NORINCO SH1 155 mm/52-calibre self-propelled artillery systems was in service with the Peoples Liberation Army. It is understood that Myanmar has taken delivery of a quantity of these systems.

The SH1 was shown in public for the first time in the Middle East in early 2011.

Description

SH1 has a combat weight of about 22.5 tonnes and is normally operated by a crew of five who are seated in the armour protected four door cab towards the front of the chassis and to the rear of the diesel power pack. Mounted on the cab roof is a 12.7 mm machine gun for local and air defence purposes.

Mounted at the rear of the cross-country chassis is the 155 mm/52-calibre ordnance that is provided with powered elevation and traverse, muzzle brake and has a maximum elevation of plus 70°.

When travelling, the ordnance is held in a travel lock. Before opening fire, a large spade is lowered to the ground at the rear of the chassis to provide a more stable firing platform. A flick rammer is installed to enable a higher rate of fire to be achieved as well as reducing crew fatigue.

Maximum range of the SH1 depends on projectile/charge combination but when firing the latest high-explosive Extended-Range Full-Bore – Base Bleed Rocket Assist (ERFB-BB-RA) projectile, a maximum range of 53 km can be achieved using charge zone 10 according to NORINCO.

Other types of 155 mm ammunition fired by SH1 include HE ERFB, HE ERFB-BB, ERFB-BB cargo, ERFB illuminating, ERFB smoke base ejection and ERFB white phosphorous.

According to NORINCO, all standard NATO 155 mm artillery projectiles and charges can be fired by the SH1 system.

Firing an ERFB High-Explosive (ERFB HE) a maximum range of 32 km is achieved which increases to a maximum range of 41 km firing an ERFB - Base Bleed - High-Explosive (ERFB BB HE).

A total of 20 × 155 mm projectiles and associated charges are carried. The latter are currently of the bag type but China has developed a modular charge system.

This is a two part Modular Charge System consisting of BC2 and BC1 with up to six of the former being used with 155 mm/52-calibre systems and five with 155 mm/39-calibre systems.

According to NORINCO, this Modular Charge System is compatible with all NATO 155 mm artillery systems and features less wear on the ordnance as well as being well suited to use in automatic loaders.

SH1 155 mm/52-calibre self-propelled artillery system deployed in the firing position with spade lowered at the rear and weapon elevated. A 12.7 mm machine gun in mounted of the roof of the protected cab for self-defence and air defence purposes (NORINCO) 1452631

NORINCO 155 mm/52-calibre SH1 self-propelled artillery system carrying out a fire mission (NORINCO) 1193004

It can also fire a locally manufactured 155 mm laser homing projectile for pinpoint accuracy which is based on the Russian Krasnopol design. This is used in conjunction with a forward observation post that includes a laser designator and synchronizer.

This 155 mm laser-guided projectile is designated the GP1 and is currently being offered on the export market by NORINCO.

More recently NORINCO has started to market the GP6 155 mm Laser-Homing Artillery Weapon System which includes the latest GP6 laser-guided artillery projectile.

This is stated to have a minimum range of 6 km, a maximum range of 25 km and is fitted with a high-explosive warhead.

NORINCO is also now marketing a top attack 155 mm artillery projectile which is called the GS1 Smart Projectile which carries two sub-munitions fitted with sensors to attack the vulnerable upper surfaces of armoured fighting vehicles.

It is claimed that SH1 is very accurate as not only is it fitted with a computerised fire-control system but also fibre optic north seeking positioning and navigation system.

Mounted above the 155 mm/52-calibre ordnance is a muzzle velocity radar that feeds information into the fire-control computer. Gunner, aimer and loader are all provided with their own displays.

The 6 × 6 chassis is claimed to provide the system with not only good strategic mobility but also a high level of cross-country mobility as a central tyre pressure regulation system is fitted as standard.

SH1 weapon system

NORINCO is now marketing a complete SH1 weapon system that includes at the battalion level the following key elements:

- 18 × SH3 (with three batteries each of six weapons)
- 1 × battalion command vehicle
- 3 × battery command vehicles
- 3 × 3 battery reconnaissance vehicles
- 1 × set of artillery locating and fire correction radar (two vehicles)
- 1 × meteorological radar set (two vehicles)
- 1 × mechanical maintenance set (of two vehicles)
- 1 × electrical maintenance set (one vehicle)

It should be noted that a typical SH1 battery has six weapons which can be split into two troops each of three SH1 although another alternative is a battery of eight guns which could be split into two troops each of four guns.

Variants

As far as it is known there are no variants of the SH1 155 mm/52-calibre self-propelled artillery system but there are a number of other recent Chinese self-propelled artillery systems, tracked and wheeled, that have the same SH designator:

NORINCO 155 mm/52-calibre self-propelled artillery system deployed in the firing position with ordnance elevated and spade lowered at the rear of the chassis. On the roof of the cab is the 12.7 mm machine gun (NORINCO) 1452909

- SH2 - 122 mm (6 × 6) self-propelled artillery system.
- SH3 - 122 mm full tracked self-propelled artillery system. The PLA have introduced the 122 mm full tracked self-propelled howitzer called the PLZ-07.This is very similar to the first system but had a different hull and suspension.
- SH5 - 105 mm (6 × 6) self-propelled artillery system.

Note: All systems are covered in a separate entry in *IHS Jane's Land Warfare Platforms: Artillery and Air Defence*.

Specifications

SH1 155 mm self-propelled artillery system

	SH1
Dimensions and weights	
Crew:	6
Length	
overall:	9.68 m
Width	
transport configuration:	2.58 m
Height	
transport configuration:	3.50 m
Ground clearance	
overall:	0.38 m
Weight	
combat:	22,500 kg
Mobility	
Configuration	
running gear:	wheeled
layout:	6 × 6
Speed	
max speed:	90 km/h
Range	
main fuel supply:	600 km
Fording	
without preparation:	1.2 m
Gradient:	58%
Side slope:	40%
Vertical obstacle	
forwards:	0.4 m
Trench:	0.7 m
Turning radius:	11 m
Engine	WD615.44, supercharged, water cooled, diesel, 315 hp at 2,200 rpm
Steering:	powered
Firepower	
Armament:	1 × 155 mm howitzer
Armament:	1 × roof mounted 12.7 mm (0.50) machine gun
Main weapon	
calibre length:	52 calibres
Main armament traverse	
angle:	40° (20° left/20° right)
Main armament elevation/depression	
armament front:	+70°/-3°

Status

Development complete. Production as required. May be in service which China and understood to be in service with Myanmar.

Contractor

Chinese State Factories. Marketing carried out by China North Industries Corporation (NORINCO).

NORINCO SH2 122 mm self-propelled artillery system

Development

In 2007 China North Industries Corporation (NORINCO) revealed that they had completed development of a new 122 mm Self-Propelled (SP) (6 × 6) artillery system called the SH2.

As of March 2013 it is understood that quantity production of the NORINCO SH2 122 mm self-propelled artillery system had yet to commence.

During a major parade held in Beijing in October 2009 a number of new weapon systems were shown for the first time but the SH2 122 mm self-propelled artillery system was not one of these.

Description

The SH2 122 mm self-propelled artillery system is based on a 6 × 6 cross country chassis that appears to be a new design with the front and rear axles, both of which steer, being at the extreme ends of the chassis.

The protected engine compartment is at the front of the chassis with the fully enclosed protected crew compartment to the immediate rear.

NORINCO 122 mm SH2 self-propelled artillery system deployed in the firing position with ordnance elevated and spade lowered at rear (NORINCO) 1296053

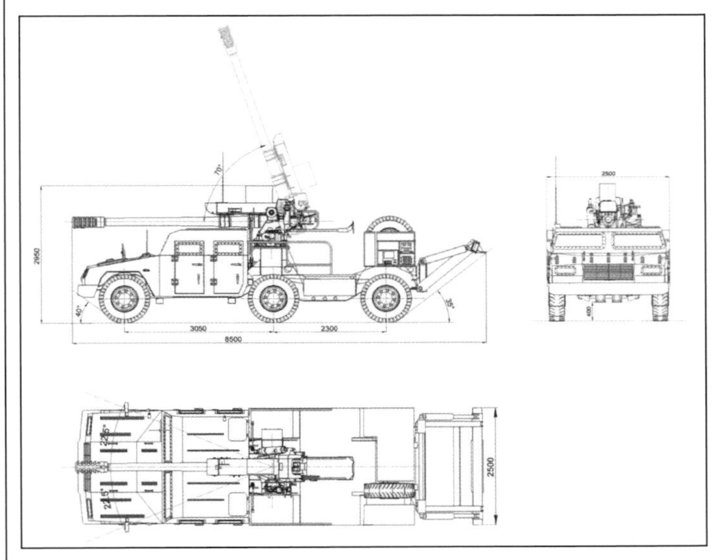

Detailed side and top drawings of NORINCO 122 mm SH2 self-propelled artillery system in travelling configuration (NORINCO) 1296054

This is provided with two large bulletproof windows to the front with two forward opening doors in either side.

Each of these has a large bulletproof window in the upper part. In either side of the roof of the cab towards the rear is a single-piece hatch cover that opens to the rear.

To the rear of the crew cab is a flatbed which is mounted a turntable with the complete 122 mm barrel which is the same as that used in the standard towed 122 mm D-30 which has been manufactured for the home and export markets for some years.

This 122 mm D-30 ordnance is also used in a number of NORINCO full-tracked self-propelled artillery systems including the 122 mm Type 89 and 122 mm Type 85. Both of these systems are in service with the People's Liberation Army but as far as it is known they have not been exported.

These systems are no longer marketed by NORINCO who is now concentrating on this SH2 122 mm (6 × 6) and the SH3 full tracked system which is covered in detail in a separate entry in *IHS Jane's Land Warfare Platforms: Artillery and Air Defence*.

The 122 mm D-30 ordnance fires the same family of ammunition as the towed D-30 with maximum range depending on projectile/charge combination. According to NORINCO the following ranges can be achieved:

- Standard high-explosive (HE) projectile – 15.3 km
- Hollow Base (HB) projectile – 18,000 m
- Base Bleed (BB) projectile – 22,000 m
- Base Bleed Rocket Assist (BB RA) - 27,000 m.

A total of 24 × 122 mm projectiles are carried plus associated charges which are positioned ready for immediate use by the gun crew.

SH2 is fitted with a computerised fire-control system which enables the system to come into action, carry out a fire mission and come out of battery before counterbattery fire arrives.

Optional subsystems include different communications equipment, tactical direction and fire distribution, positioning, orientation and navigation systems and an automatic gun laying ability.

According to NORINCO, the SH2 can operate in battalion level, battery level or single gun configuration.

When deployed in the firing position, a large spade is lowered at the rear of the chassis to provide a more stable firing platform.

Standard equipment includes powered steering and a central tyre pressure regulation system that allows the driver to adjust the tyre pressure to suit the terrain being crossed. A replacement spare wheel is also carried.

Scale model of the NORINCO SH2 122 mm self-propelled artillery system showing the four door protected cab and 122 mm weapon mounted to immediate rear (Christopher F Foss) 1452910

Variants

SH5 105 mm self-propelled artillery system

This is similar to the SH2 122 mm self-propelled artillery system but the 122 mm ordnance has been replaced by a 105 mm ordnance. Details are provided in a separate entry in *IHS Jane's Land Warfare Platforms: Artillery and Air Defence*.

As of March 2013 it is understood that this system remained at the prototype/pre-production phase.

Other NORINCO SH series self-propelled artillery services

There are a number of other recent Chinese self-propelled artillery systems, tracked and wheeled, that have the same SH designator:

- SH1 155 mm/52-calibre self-propelled artillery system.
- SH3 - 122 mm full tracked self-propelled artillery system.

Note: Both systems are covered in a separate entry in *IHS Jane's Land Warfare Platforms: Artillery and Air Defence*.

The SH1 is integrated on the rear of the WS5252 (6 × 6) cross-country truck chassis which is designed and built by Wanshan Special Vehicle Company.

As of March 2013 it had not been confirmed that the NORINCO SH1 155 mm/52-calibre self-propelled artillery systems was in service with the Peoples Liberation Army. It is understood that Myanmar has taken delivery of a quantity of these systems.

Poly Technologies are currently marketing a 122 mm truck mounted self-propelled artillery system which is in service with the Peoples Liberation Army. Details of this are provided in a separate entry in *IHS Jane's Land Warfare Platforms: Artillery and Air Defence* and some sources have also referred to this as the SH3.

A very similar system is also being marketed by NORINCO under the designation of the CS/SH1 122 mm self-propelled howitzer.

122 mm (8 × 8) self-propelled artillery system

A number of 122 mm wheeled self-propelled artillery systems have been developed in China and details of these are provided in separate entries in *IHS Jane's Land Warfare Platforms: Artillery and Air Defence*.

Specifications

	SH2
Dimensions and weights	
Crew:	5
Length	
overall:	8.50 m
Width	
overall:	2.5 m
Height	
overall:	2.95 m
Ground clearance	
overall:	0.4 m
Wheelbase	3.05 m + 2.3 m
Weight	
combat:	11,500 kg
Mobility	
Configuration	
running gear:	wheeled
layout:	6 × 6
Speed	
max speed:	90 km/h
Range	
main fuel supply:	600 km
Angle of approach:	40°
Angle of departure:	35°
Gradient:	60%
Side slope:	40%
Engine	supercharged, water cooled, diesel, 214.5 hp at 4,300 rpm
Steering:	powered
Firepower	
Armament:	1 × 122 mm howitzer
Ammunition	
main total:	24
Main armament traverse	
angle:	45° (22.5° left/22.5° right)
Main armament elevation/depression	
armament front:	+70°/0°

Status

Development complete. Not yet in production or service.

Contractor

Chinese state factories. Marketing carried out by China North Industries Corporation (NORINCO).

Poly Technologies self-propelled Type 86 122 mm howitzer

Development

Poly Technologies of China is now marketing a Self-Propelled (SP) version of the China North Industries Corporation (NORINCO) Type 96 122 mm towed howitzer which in turn is based on the widely deployed Russian 122 mm D-30 howitzer.

Details of the NORINCO 122 mm Type 96 and original Russian 122 mm D-30 towed howitzers are provided in separate entries in *IHS Jane's Land Warfare Platforms: Artillery and Air Defence*.

It should be noted that Poly Technologies is also marketing the 122 mm Type 96 howitzer as the PL96 while NORINCO is marketing this weapon as the D30-2 122 mm howitzer. Older versions of the Type 96 are designated the Type 86.

In Peoples Liberation Army (PLA) service the 122 mm Type 96 howitzer is normally towed by a 6 × 6 cross-country truck that also carries the crew and its ready use ammunition.

By integrating this system onto a 6 × 6 cross-country chassis not only is the mobility of the system enhanced but it can also be brought into action quicker and come out of action quicker which enhances its survivability against counter-battery fire.

This system is in service with the Peoples Liberation Army and some sources have called this system the SH-3 but thus designation has already been applied to a NORINCO SH-3 tracked system.

It should be noted that the Poly Technologies self-propelled Type 86 122 mm howitzer is also being marketed by NORINCO under the designation of the CS/SH1 122 mm self-propelled howitzer.

It should be noted that the ordnance of the Type 96 122 mm towed howitzer is also used in a number of other Chinese self-propelled artillery systems including:

- SH2 122 mm self-propelled howitzer (6 × 6)
- SH3 122 mm self-propelled howitzer (tracked)
- 122 mm PLZ-07 self-propelled howitzer (tracked)
- 122 mm Type 89 self-propelled howitzer (tracked)
- 122 mm Type 85 self-propelled howitzer (tracked)
- 122 mm Type 70 and Type 70-1 self-propelled howitzer (tracked)
- 122 mm self-propelled howitzer (8 × 8) with at least two versions which are understood to remain at the prototype stage.

Of the above it is understood that only the SH2 and SH3 are currently being marketed by NORINCO.

Chinese Type 96 122 mm self-propelled artillery system deployed in the firing position with stabilisers deployed and weapon being used in direct fire role (Poly Technologies) 1365881

The PLZ-07 made its first public appearance in October 2009 during a parade in Beijing but has yet to be marketed by NORINCO.

Description

For the SP application the upper part of the 122 mm Type 96 howitzer has been removed from its standard two wheeled carriage and installed on the rear of a forward control 6 × 6 cross-country truck chassis.

The crew of five are accommodated in the four door cab with a total of 40 rounds of separate loading 122 mm ammunition being carried in lockers to the immediate rear of the cab on the left and right sides.

For this application the upper part of the 122 mm Type 96 howitzer fires to the rear arc and to provide a more stable firing platform stabilisers are deployed either side of the platform before firing commences.

Maximum range depends on the 122 mm projectile/charge combination but it has a stated minimum range of 5,500 m and maximum range firing standard High-Explosive (HE) ammunition of 18,000 m (muzzle velocity of 713 m/s).

This is increased to 22,000 m using a HE Base Bleed (BB) projectile with a muzzle velocity of 725 m/s. A recently developed BB Rocket Assisted (RA) projectile increases the range out to 27 km.

It also has a useful direct fire capability with a maximum rate of fire of up to eight rounds per minute.

According to Poly Technologies the system takes about 1.5 minutes to come into action and it can be rapidly be deployed to a new position to avoid counterbattery fire.

Specifications

	Type 86 122 mm
Dimensions and weights	
Crew:	5
Length	
overall:	7.60 m (travelling)
Width	
overall:	2.65 m (travelling)
Height	
overall:	2.85 m (travelling)
Weight	
combat:	16,500 kg
Mobility	
Configuration	
running gear:	wheeled
layout:	6 × 6
Speed	
max speed:	85 km/h
Range	
main fuel supply:	600 km
Firepower	
Armament:	1 × 122 mm howitzer
Ammunition	
main total:	40
Main weapon	
maximum range:	11,500 m (standard HE projectile)
maximum range:	19,500 m (rocket assisted projectile)
Rate of fire	
sustained fire:	12 rds/min
Turret power control	
type:	powered/manual
Main armament traverse	
angle:	30° (left/right)
Main armament elevation/depression	
armament front:	-3°/+71°

Status

Development complete. Production as required. In service with the Peoples Liberation Army.

Contractor

Poly Technologies.

VN1/ZBD-09 122 mm self-propelled artillery system

Development

During a major military parade held in Beijing in October 2009, a number of new Armoured Fighting Vehicles (AFV) and other weapons were shown for the first time.

These included the ZBD-09 (8 × 8) Infantry Combat Vehicle (ICV) which is already in production and in service with the People's Liberation Army (PLA).

This is being offered on the export market by NORINCO as the VN1 with a 6 × 6 model being called the VN2.

A modified ZBD-09 (8 × 8) ICV chassis is being used for this new 122 mm self-propelled artillery system which as of March 2013 is understood to be still undergoing trials and had yet to enter quantity production. The correct Chinese designation of this system has yet to be revealed.

New Chinese 122 mm self-propelled artillery system based on a VN1/ZBD-09 (8 × 8) ICV chassis with turret armed with 122 mm ordnance partly covered up 1365027

Description

The overall layout of this 122 mm self-propelled artillery system is similar to the ZBD-09 (8 × 8) ICV with the drivers compartment at the front left and the diesel engine compartment to the front right.

Mounted on the roof to the rear of the second road wheel station is what appears to be a fully enclosed turret.

This is armed with the local version of the Russian 122 mm D-30 howitzer which has been in service with the Peoples Liberation Army (PLA) for many years under the designation of the Type 86.

When travelling the 122 mm ordnance is held in a travel lock located to the rear of the drivers position.

Details of this 122 mm weapon and types of ammunition and ranges achieved is provided in a separate entry in *IHS Jane's Land Warfare Platforms: Artillery and Air Defence*.

The prototype of this 122 mm self-propelled artillery system retains the two propellers mounted one either side at the rear to provide an amphibious capability.

This amphibious feature could well be on the prototype vehicles only as the overall weight of this system is expected to be more that the VN1 which is currently being offered on the export market with a combat weight of about 21 tonnes.

Variants

There are no known variants of this 122 mm self-propelled artillery system although a number of other new Chinese self-propelled artillery systems have recently been revealed which include:

SH1 155 mm/52-calibre self-propelled artillery system

This may be in service with the Peoples Liberation Army. It is understood that Myanmar may have taken delivery of a quantity of these systems.

Full details of the SH5 are provided in a separate entry in *IHS Jane's Land Warfare Platforms: Artillery and Air Defence*.

SH2 122 mm (6 × 6) self-propelled artillery system

This is now being marketed by NORINCO and details are provided in a separate entry in *IHS Jane's Land Warfare Platforms: Artillery and Air Defence*. As far as it is known, as of March 2013 this system remains at the prototype stage.

SH3 122 mm full tracked self-propelled artillery system

This is now being marketed by NORINCO and details are provided in a separate entry in *IHS Jane's Land Warfare Platforms: Artillery and Air Defence*. As far as it is known, as of March 2013 this system remains at the prototype stage.

SH5 105 mm (6 × 6) self-propelled artillery system

This has been developed for the export market and is basically a 105 mm version of the SH2. The SH5 is covered in a separate entry in *IHS Jane's Land Warfare Platforms: Artillery and Air Defence*. As far as it is known, as of March 2013 this system remained at the prototype stage.

Specifications

Not available.

Status

Prototypes being tested. Not yet in service with Peoples Liberation Army.

Contractor

Chinese state factories.

NORINCO 155 mm self-propelled gun (6 × 6)

Development

Early in 2002, it was revealed that China had developed a new 6 × 6 self-propelled artillery system that uses the upper part of the standard China North Industries Corporation (NORINCO) 155 mm/45-calibre WA021 towed artillery system.

The WA021 has been in service with the People's Liberation Army (PLA) for some years and has also been offered on the export market with or without its Auxiliary Propulsion System (APU).

The 155 mm/45-calibre WA021 is ballistically identical to the NORINCO PLZ45 155 mm/45-calibre full-tracked armoured self-propelled artillery system, for which the first export customer was Kuwait who took delivery of 27 systems in 2000. Since then an additional 27 PLZ45 have been supplied which brings the total to 54 units.

More recently Saudi Arabia has also purchased the PLZ45 155 mm/45-calibre self-propelled artillery system.

As of March 2013, this 155 mm/45-calibre (6 × 6) self-propelled artillery system had not been offered on the export market and could well have been a development vehicle which led to the SH1 (6 × 6) 155 mm/52-calibre self-propelled artillery system.

In recent years China has also developed a number of other wheeled (6 × 6 and 8 × 8) self-propelled artillery systems which includes:

- SH1 155 mm/52-calibre (6 × 6) self-propelled artillery system which is covered in detail in a separate entry in *IHS Jane's Land Warfare Platforms: Artillery and Air Defence*. This has been produced in production quantities for at least one export customer.
- SH2 122 mm (6 × 6) self-propelled artillery system which is covered in detail in a separate entry in *IHS Jane's Land Warfare Platforms: Artillery and Air Defence*. As far as it is known, as of March 2013 this system remained at the prototype stage.
- SH5 105 mm (6 × 6) self-propelled artillery system which is similar to the SH2 but has a 105 mm ordnance. Details are provided in a separate entry in *IHS Jane's Land Warfare Platforms: Artillery and Air Defence*. As far as it is known, as of March 2013 this system remained at the prototype stage.
- 122 mm (8 × 8) self-propelled artillery system using a 122 mm ordnance installed on at least two chassis. Available details are provided in a separate entry in *IHS Jane's Land Warfare Platforms: Artillery and Air Defence*. As far as it is known, as of March 2013 these systems remained at the prototype stage.
- 120 mm (8 × 8) armoured car/tank destroyer which have been developed to the prototype stage but are understood not to have entered production or service as of March 2013.

As far as it is known, none of these self-propelled artillery systems are currently in production or service with the Peoples Liberation Army.

Of these only the SH1, SH2 and SH5 are currently being offered on the export market by NORINCO.

The SH1 is integrated on the rear of the WS5252 (6 × 6) cross-country truck chassis which is designed and built by Wanshan Special Vehicle Company.

As of March 2013 it had not been confirmed that the NORINCO SH1 155 mm/52-calibre self-propelled artillery systems was in service with the Peoples Liberation Army. It is understood that Myanmar has taken delivery of a quantity of these systems.

Poly Technologies are currently marketing a 122 mm truck mounted self-propelled artillery system which is in service with the Peoples Liberation Army. Details of this are provided in a separate entry in *IHS Jane's Land Warfare Platforms: Artillery and Air Defence*. This is being marketed by Poly Technologies as the self-propelled Type 86 122 mm howitzer.

It should be noted that NORINCO is also marketing this system under their designation of the CS/SH1 122 mm self-propelled howitzer.

In addition it has been confirmed that the Peoples Liberation Army has fielded two systems based on a modified WZ551 (6 × 6) series armoured personnel carrier chassis.

These are the PLL-05 120 mm self-propelled mortar system, which also has a direct fire capability, and the PTL-02 self-propelled anti-tank gun.

The latter is being marketed by NORINCO in 105 mm configuration under the name Assaulter.

Description

This new Chinese self-propelled artillery system, which is understood to be at the prototype stage, is based on a 6 × 6 forward control unarmoured cross-country truck chassis.

It is probable that a central tyre pressure-regulation system is fitted, allowing the driver to adjust the tyre pressure from within the cab to suit the type of ground being crossed.

According to China, the 6 × 6 forward control chassis has the designation of the Type WS2250 and is powered by a diesel developing 236 hp.

On the prototype system, no protection is provided for the crew in the forward control cab from small arms fire and shell splinters.

The chassis however appears identical to the Minsk Wheel Tractor Plant MZKT-8007 series that was originally developed for specialised applications such as being fitted with an excavator planner with a bucket or other attachment.

The Minsk Wheel Tractor Plant is based in the Republic of Belarus and has extensive experience in the design, development and production of cross-country vehicles for civil and military applications.

The latter includes special high-mobility chassis for launching a variety of surface-to-surface missiles, including ballistic types.

One version of this MZKT-8007 chassis is powered by a Russian power pack but it is also offered with a German Daimler-Benz or Deutz diesel engine coupled to a German ZF transmission.

Chinese sources have however stated that the 6 × 6 systems use the elevating mass and ordnance of the 155 mm/45-calibre PLL01 towed artillery system.

This retains the pneumatic rammer that rams the 155 mm projectile with the charge being loaded manually. Maximum rate of fire is stated to be four to six rounds a minute.

A total of 20 155 mm projectiles and charges are carried. Additional ammunition would be carried on a supporting vehicle which could be based on a similar chassis.

The 155 mm/45-calibre weapon has a traverse of 20° left and right with elevation from +20° to +70°. To provide a more stable firing platform, a hydraulically operated spade is lowered to the ground at the rear. The system is said to take one minute to come into action and a similar time to come out of action.

In addition, on the prototype there is an hydraulically operated stabiliser located between the first and second roadwheel stations.

There is also a hydraulically operated travelling lock for the 155 mm ordnance to the rear of the cab which is operated by remote-control.

The 155 mm weapon can be laid on to target manually, semi-automatic or full automatic. A gun display unit is fitted as standard so firing data can be relayed from the battery command post directly to the weapon.

If required, onboard ballistic computation could also be carried out and a land navigation system could be installed. The combination of on board ballistic computation and a land navigation system would allow for fully autonomous fire missions to be carried out

Firing a 155 mm Extended-Range Full-Bore (ERFB) High-Explosive (HE) projectile a maximum range of 30 km can be achieved by the system. Using an ERFB-Base Bleed projectile, this can be extended to 39 km.

More recently, NORINCO has developed a new family of 155 mm ammunition as well as a two part Modular Artillery Charge System.

According to Chinese sources, the system has a crew of six. As the cab cannot seat all of these people, the remainder are carried in a second vehicle which also carries additional ammunition.

Specifications

	155 mm self-propelled gun (6 × 6)
Dimensions and weights	
Crew:	6
Weight	
combat:	22,000 kg
Mobility	
Configuration	
running gear:	wheeled
layout:	6 × 6
Power-to-weight ratio:	10.72 hp/t
Speed	
max speed:	75 km/h
Range	
main fuel supply:	500 km
Engine	diesel, 236 hp
Firepower	
Armament:	1 × 155 mm howitzer
Ammunition	
main total:	20
Main weapon	
calibre length:	45 calibres
Rate of fire	
rapid fire:	6 rds/min
Turret power control	
type:	hydraulic/manual
Main armament traverse	
angle:	40° (20° left/20° right)
Main armament elevation/depression	
armament front:	+72°/-20°

Status

Prototypes of this system have been built and tested. It is possible that this was a tested only.

Contractor

Chinese state factories.

NORINCO 122 mm (8 × 8) self-propelled artillery system

Development

China North Industries Corporation (NORINCO) has developed, to the prototype stage, a new 122 mm 8 × 8 self-propelled artillery system, that is believed to be currently undergoing trials.

At this stage it is not certain as to whether this has been developed for use by the People's Liberation Army (PLA) or for the export market.

Given that the PLA is now placing increased emphasis on expanding its fleet of wheeled Armoured Fighting Vehicles (AFVs), it is considered probable that this system has been initially developed for the home market.

The long-term aim would be to also offer the system on the export market through NORINCO who market virtually all Chinese ground weapon systems.

As of March 2013, this 122 mm (8 × 8) self-propelled artillery system had not been offered on the export market by NORINCO.

It is possible that this 122 mm (8 × 8) self-propelled artillery system was used for developmental purposes.

Description

The chassis used for this as yet unnamed 122 mm self-propelled artillery system appears to be very similar to that recently revealed for the new Chinese 120 mm self-propelled anti-tank gun, which is also understood to be still at the prototype stage.

The hull of the new 122 mm self-propelled artillery system is of all-welded steel armour which provides the occupants with protection from small arms fire and shell splinters.

The driver and one person are seated at the front of the vehicle, each are provided with a window to their immediate front with another to the side. It is probable that there is a hatch cover above each of their positions to enable some of the crew to rapidly enter and leave the vehicle.

In the centre of the vehicle is the fully enclosed all-welded turret with almost vertical sides. It has hatches in the roof and a bank of four electrically operated smoke grenade launchers either side. Pintle mounted on the roof is a 12.7 mm anti-aircraft machine gun that can also be used for self-defence purposes.

The 122 mm ordnance is provided with a slotted muzzle brake and a fume extractor and when travelling is held in position by an small A frame which is pivoted on top of the hull. When not required this folds forwards.

For many years NORINCO has been manufacturing a reverse engineered copy of the Russian 122 mm D-30 towed artillery system. This has a maximum range, firing a standard high-explosive projectile, of 15.30 km.

In Chinese service the Russian 122 mm D-30 is referred to as the Type 96 towed howitzer.

In addition, there is also a special charge with a maximum range of 18 km and an Extended-Range Full-Bore – Base Bleed (ERFB-BB) projectile with a maximum range of 21 km.

This 122 mm ordnance is also fitted to the NORINCO 122 mm full-tracked self-propelled howitzer Type 85, as well as Type 89 which was never offered on the export market by NORINCO.

Details of both of these systems are provided in separate entries in *IHS Jane's Land Warfare Platforms: Artillery and Air Defence*.

Production of both of these tracked 122 mm self-propelled artillery systems is now complete although they remain in service in China.

It is also installed in the full tracked 122 mm PLZ-07 self-propelled howitzer which has been in service with the People's Liberation Army for some years and this was first seen in public in the major parade held in Beijing late in 2009. So far the 122 mm PLZ-07 has not been offered on the export market by NORINCO.

The diesel power pack is at the rear of the vehicle and steering is understood to be power-assisted on the front four wheels. A central tyre pressure-regulation system is probably fitted, allowing the driver to adjust the tyre pressure to suite the terrain being crossed.

There also appears to be an emergency door/hatch in either side of the lower part of the hull, between the second and third roadwheel stations.

There is a possibility that the vehicle is fully amphibious, being propelled in the water by a single propeller mounted one either side at the rear.

Variants

In recent years China has also developed a number of other wheeled (6 × 6 and 8 × 8) self-propelled artillery systems which includes:

- SH1 155 mm/52-calibre (6 × 6) self-propelled artillery system which is covered in detail in a separate entry in *IHS Jane's Land Warfare Platforms: Artillery and Air Defence*. It has not been confirmed that the NORINCO SH1 155 mm/52-calibre self-propelled artillery system was in service with the Peoples Liberation Army. It is understood that Myanmar has taken delivery of some of these systems. This may be in service with the Peoples Liberation Army.
- SH2 122 mm (6 × 6) self-propelled artillery system which is covered in detail in a separate entry in *IHS Jane's Land Warfare Platforms: Artillery and Air Defence*. As of March 2013 it is understood not to have entered production or service.
- SH5 105 mm (6 × 6) self-propelled artillery system which is similar to the SH2 but has a 105 mm ordnance. Details are provided in a separate entry in *IHS Jane's Land Warfare Platforms: Artillery and Air Defence*. As of March 2013 it is understood not to have entered production or service.
- 122 mm (8 × 8) self-propelled artillery system using a 122 mm ordnance installed on at least two chassis. Available details are provided in separate entries in *IHS Jane's Land Warfare Platforms: Artillery and Air Defence*. As of March 2013 it is understood that neither of these systems have entered production or service.
- 120 mm (8 × 8) armoured car/tank destroyer which are understood to remain at the prototype stage as of March 2013.

Of these only the SH1, SH2 and SH5 are currently being offered on the export market by NORINCO.

In addition it has been confirmed that the Peoples Liberation Army has fielded two systems based on a modified WZ551 (6 × 6) series armoured personnel carrier chassis.

These are the PLL-05 120 mm self-propelled mortar system, which also has a direct fire capability, and the PTL-02 self-propelled anti-tank gun.

The latter is being marketed by NORINCO in 105 mm configuration under the name Assaulter.

Specifications

Not available.

Status

Prototype undergoing trials.

Contractor

Chinese state factories.

NORINCO SH5 105 mm self-propelled artillery system

Development

In 2007 China North Industries Corporation (NORINCO) revealed that they had developed and tested the SH2 122 mm (6 × 6) self-propelled artillery system.

Details of this are provided in a separate entry in *IHS Jane's Land Warfare Platforms: Artillery and Air Defence*. As of March 2013 it is understood that production of the SH2 122 mm (6 × 6) self-propelled artillery system had yet to commence and as far as it is known this is not in service with the Peoples Liberation Army.

Further development of the NORINCO SH2 122 mm (6 × 6) self-propelled artillery system has resulted in the SH5 105 mm (6 × 6) self-propelled artillery system.

The major difference between the two systems is that the SH5 has a 105 mm ordnance not a 122 mm-calibre ordnance based on the Russian D-30 which is used in large numbers by the Peoples Liberation Army (PLA) in towed and full tracked self-propelled configurations.

Description

The SH5 is based on a 6 × 6 cross-country chassis that is a new design that is also used for the SH2 122 mm (6 × 6) self-propelled artillery system.

The front and rear axles, which both steer, are at the extreme ends of the chassis.

The protected engine compartment is at the front of the chassis with the fully enclosed protected crew compartment to the immediate rear.

This is provided with two large bulletproof windows to the front with two forward opening doors down either side.

Each of these doors has a large bulletproof window in the upper part. In either side of the roof of the cab towards the rear is a single piece hatch cover that opens to the rear.

To the rear of the cab is a flat bed on which is mounted a turntable with a 105 mm/37-calibre ordnance. This has elevation from 0° to +70° with traverse being 30° left and right.

Maximum range depends on projectile/charge combination but NORINCO have released the following range figures for this 105 mm system:

Projectile Type	Maximum range
M1 HE	12,000 m
HB/HE	15,000 m
BB/HE	18,000 m

A total of 40 × 105 mm projectiles at carried in the ammunition containers located at the rear of the chassis.

To provide a more stable firing platform two stabilisers are lowered one either side at the rear of the chassis.

NORINCO SH5 105 mm self-propelled artillery system with 105 mm ordnance partly elevated (NORINCO) 1452911

NORINCO SH5 105 mm (6 × 6) self-propelled artillery system in travelling configuration (NORINCO) 1353714

The SH5 is fitted with an automatic gun laying system to reduce into action times and make the system more accurate. According to NORINCO this is accurate to one mil in elevation and traverse.

This automatic gun laying system allows the SH5 105 mm self-propelled artillery system to open fire within 40 seconds of coming to a halt.

Specifications

	SH5
Dimensions and weights	
Crew:	4
Length	
overall:	8.35 m
Width	
overall:	2.5 m
Height	
overall:	2.95 m
Ground clearance	
overall:	0.4 m
Wheelbase	3.05 m + 2.3 m
Weight	
combat:	10,000 kg
Mobility	
Configuration	
running gear:	wheeled
layout:	6 × 6
Speed	
max speed:	90 km/h
Range	
main fuel supply:	600 km
Angle of approach:	40°
Angle of departure:	35°
Gradient:	60%
Side slope:	40%
Turning radius:	6.5 m
Engine	supercharged, water cooled, diesel, 160 hp at 4,300 rpm
Steering:	powered
Firepower	
Armament:	1 × 105 mm howitzer
Ammunition	
main total:	40
Turret power control	
type:	powered/manual
Main armament traverse	
angle:	60° (30° left/30° right)
Main armament elevation/depression	
armament front:	+70°/0°

Status

Prototype. Not yet in production or service.

Contractor

Chinese state factories. Marketing carried out by China North Industries Corporation (NORINCO).

Poly Technologies 105 mm wheeled artillery system

Development

It has been revealed that Poly Technologies of China is now marketing a truck mounted 105 mm howitzer which is aimed at the export market as the People's Liberation Army (PLA) does not use this calibre of weapon.

As of March 2013 it is understood that this system remained at the prototype/pre-production phase and could well have been developed to meet the specific requirements of one or more export customers.

Description

The basis of the system is a Chinese 6 × 6 cross-country truck chassis with a four door fully enclosed cab for the driver and crew at the front.

At present this cab is not provided with any protection from small arms fire or shell splinters but this feature could well be incorporated on production systems.

The system has a combat weight of 11.5 tonnes and is operated by a crew of five and a maximum road speed is being quoted as 85 km/h with a range of up to 600 km.

To the rear of the fully enclosed cab is a flatbed on which is mounted a 105 mm howitzer which when travelling is traversed to the front and positioned in a travel lock to the cab rear.

When deployed the 105 mm howitzer fires through the rear arc and can be elevated from –5° to +70° with a traverse of 30° left and right using manual controls.

It is claimed that the system can be deployed in less 60 seconds and to provide a more stable firing platform stabilisers are lowered to the ground either side before opening fire.

Poly Technologies truck mounted 105 mm howitzer deployed in the firing position with stabilisers lowered to the ground and weapon traversed to the rear ready for firing (Christopher F Foss) 1448562

The 105 mm ordnance is provided with a slotted muzzle brake and can fire standard NATO 105 mm ammunition with the old US developed 105 mm M1 High-Explosive (HE) projectile having a maximum range of 11.5 km.

This 105 mm round has been marketed by NORINCO for many years and can also be fired by the NORINCO version of the Italian Oto Melara 105 mm Pack Howitzer as well as the US M101 and M102 and the UK BAE Systems/Royal Ordnance 105 mm L119 Light Gun.

Firing the US M913 High-Explosive Rocket Assist (HERA) projectile maximum range of the Chinese 105 mm self-propelled artillery system can be extended out to 19.5 km but this HERA round has never been manufactured by NORINCO.

A total of 40 rounds of 105 mm ammunition and associated fuzes are carried in stowage lockers arranged around the rear platform.

A 7.62 mm machine gun can be mounted on the roof of the cab for self-defence purposes and a total of 600 rounds of ammunition are provided for this weapon.

At present it is understood that conventional direct and indirect sighting systems are fitted but it the future this could well be supplemented by a more advanced gun laying system for improved accuracy.

In concept, this Poly Technologies 105 mm truck mounted howitzer is very similar to the Dutch RDM Technology 105 mm/33-calibre MOBAT [Mobile Artillery] system.

This was developed by the company as a private venture for the export market and ordered by Jordan but the company ceased trading and these were never delivered.

Poly Technologies are also marketing the 122 mm Type 86 wheeled SP artillery system which is based on a different 6 × 6 cross-country chassis and this is in service with the PLA.

It should be noted that the 122 mm Type 86 wheeled SP artillery system is also being marketed by China North Industries Corporation (NORINCO) under the designation of the CS/SH1 122 mm self-propelled howitzer. Full details of this are provided in a separate entry in *IHS Jane's Land Warfare Platforms: Artillery and Air Defence*.

China North Industries (NORINCO) also market a complete family of wheeled SP artillery systems including the 105 mm SH5 but this 6 × 6 is a different design to the Poly Technologies 105 mm weapon.

Details of the NORINCO 105 mm SH5 (6 × 6) self-propelled artillery system are provided in a separate entry in *IHS Jane's Land Warfare Platforms: Artillery and Air Defence*.

Specifications

	105 mm wheeled artillery system
Dimensions and weights	
Crew:	5
Length	
overall:	7.6 m (travelling)
Width	
overall:	2.65 m (travelling)
Height	
overall:	2.85 m (travelling)
Weight	
combat:	11,500 kg
Mobility	
Configuration	
running gear:	wheeled
layout:	6 × 6
Speed	
max speed:	85 km/h
Range	
main fuel supply:	600 km

	105 mm wheeled artillery system
Gradient:	60%
Firepower	
Armament:	1 × 105 mm howitzer
Armament:	1 × 7.62 mm (0.30) machine gun
Ammunition	
main total:	40
7.62/0.30:	600
Turret power control	
type:	manual
Main armament traverse	
angle:	60° (30° left/right)
Main armament elevation/depression	
armament front:	+70°/-5°
Survivability	
Night vision equipment	
vehicle:	no
NBC system:	no

Status
Development complete. Ready for production on receipt of orders.

Contractor
Poly Technologies.

France

Nexter Systems CAESAR 155 mm self-propelled gun

Development
Giat Industries (which in late 2006 became Nexter Systems) developed a 155 mm self-propelled technology demonstrator called CAESAR (*CAmion Equipé d'un Système d'ARtillerie*, or truck-mounted artillery system) and this was shown in public for the first time in June 1994.

Following trials with this system, a decision was taken by the company to build a pre-production CAESAR and following company trials, this was evaluated by the French Army late in 1998.

In September 2000, the French Délégation Générale pour l'Armement (DGA) awarded the company a contract for the supply of a batch of five 155 mm/52-calibre CAESAR (6 × 6) self-propelled artillery systems.

This contract included the following key aspects:
- Complementary developments featuring the additional French Army requirements
- Type qualification totally relying on the contractor
- Delivery of five CAESAR systems
- Maintenance and support over a period of three years.

These were integrated at the now Nexter Systems facility at Bourges with the cab being supplied by SOFRAME (previously LOHR) which has been installed on the modified Mercedes-Benz UNIMOG (6 × 6) truck chassis.

By this time the company had built two CAESAR systems. The first CAESAR was essentially a technology demonstrator to prove that it was possible to mount a 155 mm/52-calibre artillery system on a modified Mercedes-Benz UNIMOG U 2450 (6 × 6) truck chassis.

This was followed by a second model which has been evaluated by the French Army and late in 1999 went to Asia.

The third CAESAR incorporated all of the lessons learned with the French and Malaysian trials and was a pre-production vehicle.

Latest Nexter Systems CAESAR 155 mm/52-calibre (6 × 6) SPG in travelling configuration (Nexter Systems) 1192691

First example of the Nexter Systems CAESAR 155 mm/52-calibre self-propelled artillery system integrated onto a new Renault Trucks Defence Sherpa 5 (6 × 6) chassis (Nexter Systems) 1132722

This is fitted with a revised fully enclosed crew cab with improved air conditioning and upgraded computer software. This CAESAR was the baseline for the French Army version, which has the Thales ATLAS computerised fire-control system and a land navigation system which is now fitted to the French Army AUF1 upgraded self-propelled artillery system.

The first CAESAR systems were delivered to the French Army late in 2002 with final deliveries taking place in early 2003. These were used by the French Army to form one complete artillery troop of four CAESARs with the 5th being used for training and reserve.

This CAESAR troop was used to evaluate operational procedures for the deployment of the system by the French Army.

Late in 2003, following a detailed analysis, the French Army decided to purchase additional CAESAR SPGs rather that continue with the major upgrade of the AUF1 SP to the enhanced AUF2 standard.

In December 2004, the French DGA awarded the company a contract covering the supply of 72 CAESAR 155 mm/52-calibre self-propelled artillery systems for the French Army.

The first production CAESAR was delivered to the French Army on 16 July 2008 and final deliveries under this first contract were made in March 2011.

The French Army took delivery of 10 CAESAR in 2008, 38 in 2009 25 in 2010 with the remaining four in first quarter of 2011. This includes the original five CAESAR pre-production systems that have been brought up to production standard.

As part of the French Army CAESAR contract Nexter Systems have guaranteed 80 per cent availability with the company holding spare parts for the French Army at 13 locations. In the first year CAESAR achieved 95 per cent availability for nine out of 12 months.

The French Army typically deploys these in batteries of eight CAESAR which are then split into two platoons each of four weapons.

CAESAR has seen combat use by the Royal Thai Army in its border area and in the Lebanon and Afghanistan by the French Army.

A total of eight CAESAR are in Afghanistan, six plus two spare with the latter not required so far.

The French artillery still deploy quantities of Nexter Systems 155 mm AUF1 tracked SP artillery systems (based on an AMX-30 tank hull) and the 155 mm TRF1 towed artillery systems.

These are to be replaced by a second batch of 64 CAESAR which would be delivered in the 2017 to 2020 time frame, funding permitting as a new Socialist government has recently been appointed.

This would mean that the CAESAR would be the only conventional 155 mm tube artillery system deployed by the French artillery.

The French artillery arm also operates the TDA 120 mm rifled mortars and the Multiple Launch Rocket System (MLRS) with the first 15 of the latter already being upgraded to fire the Lockheed Martin Guided MLRS to provide a precision effect.

Standard equipment for French CAESAR includes the Thales PR4G radio, Zodiac muzzle velocity radar mounted over rear part of ordnance, SAGEM SIGMA 30 inertial navigation system, Nexter Systems Top and Ready fire management computers and terminals for the Thales ATLAS battle management system which have the SAGEM CALP2G ballistic computer.

While the weapon system is manufactured at the Bourges facility, integration of the complete weapon with the chassis take place at Roanne.

Production CAESAR systems are based on the new Renault Trucks Defense Sherpa chassis (6 × 6) which is fitted with a fully armour protected cab also supplied by Renault Trucks Defense. The first production verification CAESAR system on the Sherpa chassis (6 × 6) was completed in mid-2006.

Apart from the new chassis and cab this has only minor differences when compared to the first CAESAR systems. The muzzle velocity radar, for example, is now on the right side of the ordnance rather than above the ordnance.

In April 2006 the company announced that Thailand had placed a contract with the company for an initial six systems with the total requirement understood to be for at least 18 units to enable a complete CAESAR regiment to be formed.

Production standard Nexter Systems CAESAR 155 mm/52-calibre self-propelled gun of the French Army in travelling configuration (Christopher F Foss) 1452632

The initial batch of six CAESAR have now been delivered to Thailand and are operational with the Royal Thai Army.

The Saudi Arabian National Guard (SANG) has taken delivery of 100 (80 + 20) CAESAR systems, although this customer has never been confirmed by Nexter Systems.

This was followed by another batch of 36 with final assembly and integration in Saudi Arabia by CAESAR International using subsystems supplied from the French production line. Final deliveries of this batch of 36 CAESAR for SANG are due to be made in 2014.

In addition to the CAESAR weapons, SANG is understood to have ordered a complete suite of Nexter Munitions 155 mm ammunition and associated charges as well as the latest Nexter BACARA [ballistic calculator for artillery].

The SANG CAESAR use a Mercedes-Benz UNIMOG 6 × 6 chassis modified by SOFRAME with a protected cab.

SANG has also ordered 15,000 of the latest Nexter Munitions LU211 Insensitive Munition (IM) compliant 155 mm artillery High-Explosive (HE) projectiles for their CAESAR systems.

SANG has also taken delivery of a complete suite of Nexter 155 mm ammunition including the 155 mm Bonus Mk 2 top attack projectile.

Late in 2012 it was announced that Indonesia had placed a contract with Nexter Systems covering the supply of 37 CAESAR 155 mm/52-calibre self-propelled guns.

These will be issued to two artillery regiments each with 18 weapons plus one for training/reserve. First deliveries are expected in late 2013/early 2014.

Description

It should be noted that prototype CAESAR have been based on a Mercedes-Benz (6 × 6) truck chassis, all production CAESAR systems for France and Thailand are built on a new Renault Trucks Defense Sherpa (6 × 6) truck chassis. CAESAR is based on a 6 × 6 truck chassis which, as well as providing good cross-country mobility, provides good strategic mobility as it can be rapidly be moved around without having to rely on tank transporters and semi-trailers. Additional details of the Renault Trucks Defense Sherpa 6 × 6 chassis are provided later in this entry.

For use in high ambient temperatures the cab is fitted with an air-conditioning system. The fully enclosed crew compartment is at the front with the 155 mm/52-calibre ordnance mounted at the rear. A central tyre pressure regulation system is fitted as standard that allows the driver to adjust the tyre pressure to suit the terrain being crossed. Power steering is also provided.

As CAESAR is mounted on a wheeled chassis its overall life cycle costs are much lower than a comparable tracked vehicle. This is of increasing importance as many users are now looking at the total life cycle costs of weapon systems rather than only the initial procurement cost.

A new fully enclosed cab has been mounted at the front of the vehicle and has individual seats for the crew of five. The cab is of welded steel armour and provides protection from small arms fire up to 7.62 mm in-calibre and shell splinters. The cab windows are 26 mm thick.

Mounted at the rear of the CAESAR is the complete upper part of the 155 mm 52-calibre upgraded version of the TRF1 towed artillery system. The 155 mm 52-calibre ordnance is fitted with a double baffle muzzle brake and when travelling the ordnance is held in position by a clamp, located to the immediate rear of the cab, which is operated by remote-control.

When the system is deployed in the firing position a large spade is hydraulically lowered at the rear to provide a more stable firing platform. The rear four wheels are raised clear of the ground so that the large spade absorbs all the firing stresses.

CAESAR can come into action in less than 1 minute and come out of action in a similar period. According to the company, in less than two minutes this system can fire six 155 mm projectiles, come out of battery and start to move to another position. This means that it would be difficult to engage CAESAR with counterbattery fire.

The 155 mm 52-calibre ordnance, which was developed some years ago for the 155 mm 52-calibre upgraded version of the TRF1, is fitted with a screw breech mechanism that opens upwards automatically with a revolving automatic primer feed mechanism holding 14 primers. It conforms to the latest NATO Joint Ballistic Memorandum of Understanding (JBMoU). An electrical firing device is fitted as standard.

CAESAR is a self-sufficient self-propelled artillery system as no survey team is required due to the installation of a SAGEM SIGMA 30 onboard Reference Package/Position Data System (RP/PDS) which includes a Global Positioning System and which is mounted on the actual gun.

An Intertechnique RDB4 muzzle velocity radar is mounted over the ordnance and this feeds information to the CS 2002-G onboard fire-control computer which is located in the cab together with its printer.

This receives target information from the battery command post with ballistic computation taking place on the system. French Army systems are fitted with the Thales PR4G radios.

The CS 2002-G computer also carries out a number of other functions including 3-D display of friend or foe local situation. With the aid of integral sensors it can also carry out ammunition status management, gun status management and ammunition resupply management.

There is also a temperature control device that advises the crew if the ordnance is becoming too hot and there is a danger of a cook off.

To reduce crew fatigue and increase the rate of fire an automatic projectile loader is mounted to the right side of the breech. The propelling charges, which can be of the conventional type or the more recent modular charge type, are loaded manually.

Maximum rate of fire is 6 rds/min and according to Nexter a burst rate of fire of three rounds in 18 seconds is normally attained.

Elevation and traverse is hydraulic with manual controls being provided as a back-up. For indirect usual firing elevation ranges from +17° to +66° with traverse being 17° left and right. For direct firing elevation ranges from –3° to +10° right, with traverse between 21° left and 27° right. Aiming is carried out automatically via the onboard computer but optical sights are provided as a back-up.

Maximum range depends of the nature of ammunition and charge system but firing a 155 mm Extended-Range Full-Bore Base Bleed (ERFB-BB) a maximum range of 42 km can be achieved. Full range details are given in the ammunition specification in Table 1.

Table 1

Projectile designation	Maximum range	Type
155 OGRE	35 km	Dual Purpose ICM
155 OMI	21.5 km	Cargo, anti-tank mines
155 mm BONUS	35 km	Anti-tank, smart
155 mm LU 211 BB	39 km	HE, base bleed
155 mm LU 211 HB	30 km	HE, hollow base
155 mm LU 214 BB	39 km	Smoke, base bleed
155 mm LU 214 HB	30 km	Smoke, hollow base
155 mm PRAC SR F1	9 km	Short range practice
155 LU 112	29 km	Practice
M107	18.3 km	HE
M485A2	18.3 km	Illuminating
ERFB NR 269 BB	42 km	Dual Purpose ICM, base bleed
ERFB NR 265 BB	42 km	HE, base bleed
ERFB NR 173 BT	33 km	HE, boat tail

In addition to firing the standard French Army LU122 high-explosive Base Bleed (BB) and Hollow Base (HB) projectile CAESAR has also fired the Rheinmetall Denel Munitions Velocity enhanced Long range Artillery Projectile (VLAP) which achieved a maximum range of 58 km.

Trials are also to take place firing the US Raytheon Excalibur precision projectile.

The Spacido course correction fuze has a potential in service date of 2015 but it sufficient funding was available this could be brought forwards.

CAESAR can use conventional bag charges as well as well as a Modular Charge System (MCS) with a maximum of six modules being used to obtain maximum range. Rheinmetall MCS have been used but the French Army will use the MCS developed by Nexter Munitions and EURENCO.

Production standard Nexter Systems CAESAR 155 mm/52-calibre self-propelled guns of the French Army in travelling configuration (DGA) 1363412

Nexter Systems CAESAR 155 mm/52-calibre self-propelled artillery system of the 40th Regiment of Artillery (40 RA) of the French Army deployed in the firing position (Christopher F Foss) 1454074

French Army Nexter Systems CAESAR 155 mm/52-calibre self-propelled artillery system of the 40th Regiment of Artillery (40 RA) of the French Army deploys to a new fire position (Christopher F Foss) 1454075

This consists of the Top Charge System (TCS) consisting of similar Top Charge Modules (TCM) with five being used for a 155 mm/39-calibre system and six for a 155 mm/52-calibre system.

There is also a Bottom Charge System (BCS) consisting of one or two Bottom Charge Modules (BCM) to meet short range and training requirements. Maximum muzzle velocity with six TCM when fired from a 155 mm/52-calibre system is 945 m/s.

The main fire-control computer is located in the cab of the CAESAR but at the rear on the left is the gun display unit for the crew when the system is deployed in the firing position. This provides such information as elevation, traverse, projectile, charge and fuze type.

The main advantages of the CAESAR are that it has similar firepower to existing towed and self-propelled artillery systems with greater strategic mobility and quicker in/out of action times. The last feature also enhances the combat survivability of the system to counterbattery fire. The standard 155 mm TRF1 has a crew of eight while CAESAR has a crew of only five including the commander and driver.

The design of CAESAR is modular so that the customer can select which radio, fire-control system or muzzle velocity measuring system and type of cab required.

Nexter Systems have also developed an ammunition resupply vehicle which carries containers of ammunition (projectiles and charges) which can be rapidly unloaded using an onboard hydraulic crane. A total of six containers are carried which hold a total of 72 rounds (projectiles and charges) of 155 mm ammunition. Conventional bagged charges can be used as well as the more recent modular charge type.

The latter can be of French design or the German Rheinmetall MTLS modular charge system of which over 1.5 million have been built for the home and export markets.

Variants

CAESAR appliqué armour kit

Nexter Systems has also provided the French Army with a total of 32 (10 + 22) appliqué armour kits for the CASEAR which were ordered in September 2007 and all were delivered by March 2009. At Eurosatory in 2010 Nexter Systems showed a new protected cab for CAESAR.

Sherpa chassis for CAESAR

Renault Trucks Defense was awarded a contract by Nexter for 76 examples of its Sherpa 5 (6 × 6) truck chassis for use as the platform for the CAESAR 155 mm/52-calibre self-propelled gun system ordered by the French Army.

Renault Trucks Defense delivered the first development Sherpa 5 truck chassis late in 2005, with delivery of production CAESAR scheduled for 2007 through to 2011.

All previous examples of the CAESAR have utilised a German Mercedes-Benz UNIMOG (6 × 6) truck chassis but within the latest UNIMOG family of cross-country vehicles the 6 × 6 model has been dropped.

The Sherpa 5 (6 × 6) 5 tonne payload tactical truck, along with examples of the Sherpa 10, 15 and 20, were displayed publicly for the first time at Eurosatory in mid-2004.

The Sherpa family of tactical wheeled vehicles was developed following a market analysis that confirmed to Renault Trucks Defense the need for a full range of purpose-designed high-mobility (strategic and tactical) trucks.

With the exception of the Sherpa 20 (8 × 8), all of the Sherpa are in 6 × 6 configuration. The Sherpa 5 has a purpose-designed chassis that has been reinforced for the CAESAR application and the Sherpa 10, 15 and 20 are based on the heavy duty Kerax range commercial chassis.

Some components, including the cab, are manufactured in Turkey. For the CAESAR application an armoured cab has been fitted, which will protect the crew from small arms fire and shell splinters.

FAST-Hit®

Building on their extensive experience in weapon systems development, Nexter has developed the FAST-Hit artillery fire-control system. This is said to combine the flexibility of high performance and easily upgradeable hardware with easier to write specific application programmes.

This aims to break away from the C3I artillery systems, which have a tendency to centralisation. FAST-Hit offers distributed computing that is provided at each level of real management of the parameters and information at its level, while complying with the general rules for the use of artillery.

The modular structure of the system allows for the projection of artillery detachments dimensioned exactly to suite the need defined by the threat and by the mission entrusted to the detachment.

Finally, the interoperability of the system from its concept through to manufacture has been defined to meet all cases from customers wishing to integrate their pieces into C3I artillery of their choice, up to customers requiring a complete system including all the intermediary configurations. FAST-Hit is based on its capability to use any communications and protocol interface level.

Specifications

	CAESAR
Dimensions and weights	
Crew:	4
Length	
overall:	9.94 m
Width	
overall:	2.55 m
Height	
overall:	3.2 m
transport configuration:	2.7 m
Ground clearance	
overall:	0.40 m
Track	
vehicle:	1.87 m
Wheelbase	3.9 m + 1.4 m
Weight	
standard:	15,800 kg
combat:	17,700 kg
Mobility	
Configuration	
running gear:	wheeled
layout:	6 × 6
Power-to-weight ratio:	13.63 hp/t
Speed	
max speed:	100 km/h
Range	
main fuel supply:	600 km (est.)
Fuel capacity	
main:	230 litres
Fording	
without preparation:	1.20 m
Gradient:	40% (est.)
Side slope:	30% (est.)
Engine	Mercedes-Benz LA, 6 cylinders, turbocharged, diesel, 240 hp at 2,600 rpm
Gearbox	
type:	manual
forward gears:	8
reverse gears:	1
Steering:	hydraulic
Clutch	
type:	friction clutch
Brakes	
main:	pneumatic
parking:	spring loaded on rear axles
Tyres:	14.50 × 20
Suspension:	portal axles with hub drive, torque tubes, all-wheel drive and differential locks in each axle, coil springs and telescopic shock-absorbers

	CAESAR
Electrical system	
vehicle:	24 V
Batteries:	2 × 12 V, 125 Ah
Firepower	
Armament:	1 × 155 mm howitzer
Ammunition	
main total:	18
Main weapon	
calibre length:	52 calibres
Main armament traverse	
angle:	34° (17° left/17° right)
Main armament elevation/depression	
armament front:	+66°/-3°
Survivability	
Night vision equipment	
vehicle:	no
NBC system:	no

Status

In production. On order for the following countries:

Country	Quantity	Comment
France	5	ordered September 2000 and delivered 2002/2003
France	72	production systems, delivered between 2008 and 2011
Saudi Arabia	36	final order for SANG, being delivered
Saudi Arabia	100	all delivered for SANG
Thailand	6	ordered April 2007, and all now delivered
Indonesia	37	ordered late 2012

Contractor

Nexter Systems.

India

Indian self-propelled artillery requirement

Development

Following the phase out of the Vickers Defence Systems Abbot 105 mm tracked Self-Propelled (SP) artillery system, the only SP artillery system currently deployed by the Indian Army is the 130 mm Catapult.

This is based on a stretched Vijayanta tank hull with the turret removed and fitted with the Russian 130 mm M-46 field gun firing over the rear arc. Brief details of this are provided in a separate entry in *IHS Jane's Land Warfare Platforms: Artillery and Air Defence*.

Late in 2012 it was stated that the Indian Defence Research and Development Organisation (DRDO) was to start trials of the Catapult Mk II SP artillery system.

This is a modified Arjun Mk I MBT hull also fitted with the Russian 130 mm field gun which is provided with 36 rounds of separate loading ammunition.

Late in 2012 Power Strategic Electronics Division of Tata showed the truck-mounted 155 mm/52-calibre self-propelled artillery system for the first time and shown here with stabilisers deployed and with weapon traversed to the rear 1482295

If trials are successful it is possible that a series production batch of about 40 Catapult Mk II could be delivered from 2013/2014 onwards, issued to two Indian artillery regiments and would replace the early Catapult Mk I systems currently deployed by the Indian Army.

India also has potential requirement for a wheeled SP artillery system which is sometimes referred to as a Gun On A Truck (GOAT) with a potential total requirement for up to 814 systems.

Late in 2012 the Power Strategic Electronics Division of Tata showed the first example of a new 155 mm/52-calibre 8 × 8 SP artillery system to meet the potential requirements of the Indian Army.

It was claimed that some 55 per cent of the system was indigenous with some other elements such as the inertial navigation system being imported.

The system consists of an 8 × 8 cross-country truck chassis fitted with a forward control cab and to the immediate rear of this is the ammunition stowage area.

Mounted at the very rear of the chassis is a turntable-mounted 155 mm/52-calibre weapon with powered elevation and traverse and a flick rammer to increase rate of fire and reduce crew fatigue.

The 155 mm ordnance is fitted with a muzzle brake and fume extractor, and to provide a more stable firing platform, hydraulically operated stabilisers are lowered at the sides and rear of the platform.

Description

Not applicable.

Specifications

Not applicable.

Status

Future Indian Army requirement.

Contractor

Not selected.

International

Denel Land Systems/General Dynamics Land Systems Stryker 105 mm self-propelled gun

Development

Late in 2003 it was revealed that an international consortium headed by General Dynamics Land Systems (Canada and US) and the now Denel Land Systems (South Africa) was developing a new 8 × 8 105 mm Self-Propelled Gun (SPG).

Within the consortium, General Dynamics Land Systems is responsible for the hull, while Denel Land Systems is responsible for the turret and 105 mm weapon system.

General Dynamics Land Systems completed the first prototype of its new 105 mm SPG in the first quarter of 2004.

It is being developed as a private venture to meet emerging user requirements of for an armour-protected SPG that is fully air-portable in a Lockheed Martin C-130 Hercules transport aircraft.

In April 2004 the new General Dynamics Land Systems/Denel Land Systems 105 mm Self-Propelled Gun (SPG) carried out its first series of test firings at Eglin Air Force Base, Florida.

During these trials, the system fired a total of 43 projectiles at ranges varying from 4 km to almost 32 km. The system was then flown in a Lockheed Martin C-130 Hercules transport aircraft to Fort Sill, home of the US Army Field Artillery, where another five rounds were fired.

As usual in early development trials, all of these firings were performed by remote-control and manned crew fire clearance was not obtained at that time.

The complete system then returned to General Dynamics Land Systems facilities in Stirling Heights, Michigan, where all of the information from the two series of firings was collected and evaluated.

Denel Land Systems/General Dynamics Land Systems Stryker 105 mm self-propelled gun with ordnance elevated (Denel Land Systems) 1296055

Denel Land Systems/General Dynamics Land Systems Stryker 105 mm SPG carrying out a fire mission (IHS/Rupert Pengelley) 1405480

After a lull of several years, development of the Stryker 105 mm SPG commenced again in 2010 and in early 2011 the system was demonstrated to an international audience in South Africa at a firepower demonstration organised by Rheinmetall Denel Munitions (RDM).

During these firing trials conducted at the Alkantpan firing range a maximum range of 31 km was achieved using the M0125 base-bleed projectile.

Firing the M1025 in the boat tail configuration a range of about 24 km was achieved, the boat tail is interchangeable with the base-bleed attachment.

RDM is responsible for the complete family of 105 mm ammunition suite and works in close co-operation with General Dynamics Land Systems (the hull provider) and Denel Land Systems (T7 turret designer).

Improvements carried out for this latest series of trials in 2011 included a G6 type swing and slide breech mechanism and a semi-automatic loader that is a scaled down version originally developed for the G6 (6 × 6) self-propelled artillery system.

With this configuration a maximum rate of fire of about six rounds a minute can be achieved but this could be increased using a single tray/single stroke ramming system for the uni-modular charge system.

In July 2011 the Stryker 105 mm SPG was fully man rated in South Africa in mid-2011.

As of March 2013 no production orders had been placed for this 105 mm self-propelled gun.

Description

The Denel Land Systems/General Dynamics Land Systems Stryker 105 mm self-propelled gun is based on a modified chassis of the Stryker infantry carrier vehicle currently in service with the US Army.

It has a hull of all-welded steel armour to which a layer of appliqué armour can be added to provide a higher level of protection. The driver's compartment is front left with the power pack to the right. The remainder of the vehicle is occupied by the fighting compartment with the turret above.

While the baseline vehicle has combat weight of 17,500 kg the weight of production versions would depend on the level of ballistic protection.

The latest production Stryker ICV with the double-V hull and other ballistic enhancements has a combat weight of up to 24,948 kg.

As of March 2013 production of the Stryker DVH was still underway for all variants with the exception of the Mobile Gun System and the NBC versions.

For its new mission, the hull line of the LAV-III has been lowered to the rear of the engine and driver's compartment at the front of the vehicle.

The lower hull height is required to enable the system to be air transportable.

Denel Land Systems/General Dynamics Land Systems Stryker 105 mm SPG with ordnance in travel lock (IHS/Rupert Pengelley) 1405481

Denel Land Systems/General Dynamics Land Systems Stryker 105 mm SPG carrying out a fire mission (Denel Land Systems) 1390384

On the rear is mounted a power operated, fully-enclosed turret, developed by Denel Land Systems, armed with a long-barrelled 105 mm gun.

The gun is the Denel Land Systems 105 mm Light Experimental Ordnance (LEO), which was first revealed several years ago and that fires a new range of enhanced 105 mm ammunition with greater range and lethality. The ordnance is fitted with a muzzle brake and fume extractor.

Firing an unassisted 105 mm boat-tail projectile a range of 30 km can be achieved, while firing an assisted base bleed projectile a range of 40 km is obtained.

The new High-Explosive Pre-Formed Fragmentation (HE – PFF) projectile is claimed to be more lethal than the standard 155 mm M107 high-explosive projectile.

In addition to HE – PFF other 105 mm ammunition types will include bispectral smoke and illumination (visual/infra-red). The new modular charge system has been developed by the now Rheinmetall Denel Munition.

An automated ammunition handling system is fitted, which will first load the projectile and then the modular charges.

The five zone unimodular charge system is essentially a scaled down version of the M90 155 mm charge system that has been produced in production quantities and is Insensitive Munition (IM) compliant.

For the trials carried out in 2011 the system fired RDM's optimised and productionised uni-modular charge system which incorporates up to five XM24A2 charge modules.

This is used in conjunction with standard M82 percussion primers and projectiles from the RDM family of Igala family of extended range 105 mm projectiles.

The number of 105 mm projectiles and associated charges carried depends on the configuration but is expected to be a maximum of 56 in the production version plus a 42 charges.

The crew's personal equipment is carried in pods on either side of the turret, which can rapidly be removed to allow the system to be loaded into a Lockheed Martin C-130 Hercules aircraft.

The system has a crew of two consisting of chief of section and driver, with the weapon aimed and fired by remote-control. An option is a .50 (12.7 mm) remote-control weapon system for air and self-defence purposes.

Standard equipment includes a central tyre-inflation system, hydropneumatic suspension with height control, power brakes and an anti-skid braking system on the rear three axles.

Crew of the system depends on whether a manual or automatic ammunition handing system is installed.

The most recent improvements have been to implement upgrades to the turret management system including the integration of a Honeywell Inertial Navigation Unit (INU) and the General Dynamics Canada Digital Fire-Control System that has already been adopted by the US Army.

A 105 mm projectile rammer system was also integrated to increase rate-of-fire and reduce crew fatigue.

For the firing trials carried out in South Africa in 2011 an ad-hoc fire-control system based on a Denel Land Systems Arachnida type weapon data display operated in combination with a strap-on optical sight.

It is expected that production systems would include either a Honeywell GPS/INU or a Selex FIN3110.

During the firing trials carried out in late 2005/early 2006 the fire mission was controlled by the gun commander using only his display.

The fire-control system automatically received data from the INU and then calculated azimuth and elevation, laid the 105 mm weapon onto the target and provided information on the projectile/charge combination required to neutralise the target.

In the prototype the ammunition is loaded manually onto the loading tray but this process could be fully automated on production systems. A system with a fully automated ammunition handling system would have a crew of three but there are clearly cost implications.

Variants

Turret T7

The Denel Land Systems turret used in this system is designated the T7 and can be integrated with a vehicle or platform than can accommodate a slewing bearing of 2,210 mm and a trunnion recoil force of 25 metric tonnes.

The actual weight of the turret is currently being quoted as 3,750 kg.

Specifications

	Stryker 105 mm SPG
Dimensions and weights	
Crew:	3
Length	
overall:	9.271 m
hull:	6.985 m
Width	
overall:	2.642 m
Height	
overall:	2.667 m
Ground clearance	
overall:	0.47 m
Track	
vehicle:	2.27 m
Wheelbase	3.86 m
Weight	
standard:	17,500 kg
Mobility	
Configuration	
running gear:	wheeled
layout:	8 × 8
Speed	
max speed:	96 km/h
Range	
main fuel supply:	631 km
Fuel capacity	
main:	200 litres
Gradient:	60%
Side slope:	30%
Vertical obstacle	
forwards:	0.635 m
Trench:	1.981 m
Engine	Caterpillar 3126, diesel, 350 hp
Gearbox	
model:	Allison MD 3066
type:	automatic
forward gears:	6
reverse gears:	1
Transfer box:	2-speed
Differentials:	4
Steering:	power assisted, front 2 wheels either side
Brakes	
main:	power
parking:	hydraulic
Tyres:	1200 × R20 XML run flat
Suspension:	hydropneumatic with height adjustment
Electrical system	
vehicle:	24 V
Batteries:	4
Firepower	
Armament:	1 × turret mounted 105 mm howitzer
Armament:	1 × roof mounted 12.7 mm (0.50) M2 HB machine gun
Ammunition	
main total:	56
charges:	42
Turret power control	
type:	hydraulic/manual
Main armament traverse	
angle:	60° (30° left/30° right)
Main armament elevation/depression	
armament front:	0°/+75°
Survivability	
Night vision equipment	
vehicle:	yes
NBC system:	yes
Armour	
hull/body:	steel

Status

Prototype completed in 2004. As of March 2013 there were no plans for this system to enter production or service.

Contractor

Denel Land Systems.
General Dynamics Land Systems - Canada/US.

Iran

Iranian 155 mm wheeled self-propelled artillery system

Development

Late in 2011 it was revealed that the Iranian Defence Industries Organisation (DIO) had developed to the prototype stage a 155 mm/39-calibre 6 × 6 self-propelled (SP) artillery system.

As of March 2013 it is not known as to whether this system was still at the prototype stage or had entered quantity production.

Description

The new system consists of the upper part of the standard Iranian DIO, Hadid Armament Industries Group, 155 mm/39-calibre HM41 towed artillery system (covered in a separate entry in *IHS Jane's Land Warfare Platforms: Artillery and Air Defence*) integrated onto the rear of a forward control MAN type (6 × 6) cross-country truck chassis.

Photographs of the first example show that the three man cab is not fitted with any armour protection but it is considered possible that this would be fitted to production versions.

A 7.62 mm or 12.7 mm machine gun can be mounted on the roof for self-defence purposes.

To the immediate rear of this cab is a small enclosed cabin for additional members of the crew and to the rear of this is a bench seat which is not provided with any protection from the elements.

A large hydraulically operated spade is pivoted at the rear chassis to provide a more stable firing platform when deployed in the firing position.

It does not appear to be fitted with powered elevation or traverse or a flick rammer to enable high rates of fire to be achieved while reducing crew fatigue. These features could however be fitted to production systems.

Firing a standard 155 mm M107 High-Explosive (HE) projectile, a maximum range of 14.6 km can be achieved.

This can be increased to 22 km firing an unassisted HE projectile or 30 km firing a rocket assisted projectile with a slight loss of accuracy.

An analysis of the available photographs indicates that this is probably a prototype system as it does not appear to be fitted with any fire-control system, travelling lock or ammunition stowage.

When in action this 155 mm/39-calibre (6 × 6) self-propelled artillery system would be supported by another vehicle carrying ammunition and crew equipment.

Latest Iranian wheeled 155 mm/39-calibre self-propelled artillery system deployed in firing position with hydraulically operated spade lowered to the ground at the rear (Fars News Agency) 1440862

Latest Iranian wheeled 155 mm/39-calibre self-propelled artillery system in travelling configuration and showing additional crew area to immediate rear of forward control cab (Fars News Agency) 1440864

The main advantage of wheeled SP artillery systems when compared to their heavier tracked counterparts is that they not only have lower procurement costs but also lower operating and support costs.

In addition they have greater strategic mobility as they do not need heavy equipment transporters to be transported to where they are required.

Their main disadvantage is that when deployed the weapon and crew are exposed to counter battery fire but this effect is reduced due to the rapid into and out of action times.

Iran already deploys a number of locally developed tracked armoured SP artillery systems including the Raad-1 122 mm and Raad-2 155 mm and details of these are provided in separate entries in *IHS Jane's Land Warfare Platforms: Artillery and Air Defence*.

Iran has been self-sufficient in 155 mm ammunition including projectiles, changes and fuzes for some time and has offered these as well as artillery systems on the export market.

Specifications

Not available.

Status

Undergoing trials.

Contractor

Defence Industries Organization (DIO).
Hadid Armament Industries Group.

Israel

Soltam Systems 155 mm ATMOS 2000 SPG

Development

Late in 2001, Soltam Systems released details of the latest version of its Autonomous Truck MOunted howitzer System (ATMOS) 2000 155 mm/52-calibre self-propelled artillery system whose existence was first revealed late in 1999. At that time it was also referred to as the 155 mm Self-Propelled Wheeled Gun (SPWG).

ATMOS 2000 has been developed by Soltam Systems as a private venture. It is aimed mainly at the export market, although it has already been demonstrated to the Israel Defense Force.

According to Soltam Systems, wheeled self-propelled guns are not only cheaper to procure than their more common tracked counterparts, but have lower life cycle costs and are easier to operate and maintain.

In addition they also have greater strategic mobility as they do not have to rely on Heavy Equipment Transporters (HETs) to be redeployed over extended distances as do their full-tracked counterparts.

By late 2001 the prototype ATMOS 2000 fired over 1,000 rounds of various natures of 155 mm ammunition, using bagged charges during extensive company trials in Israel.

In mid-2003 it was revealed that an undisclosed export customer had placed a contract with the company worth USD5 million for an undisclosed batch of ATMOS 2000 systems.

From late 2004 the Israel Defense Force (IDF) carried out extensive field tests on the locally developed Soltam Systems ATMOS 155 mm/39-calibre Self-Propelled (SP) artillery system.

Today the mainstay of the IDF field artillery is the US-supplied BAE Systems, M109A2/A3 155 mm SP howitzer, which is called the Doher. This has been modified in a number of areas and has a 155 mm/39-calibre ordnance and an upgraded onboard fire-control system.

Funding permitting, the Israel Defence Force (IDF) is expected to order an initial batch of 18 Soltam Systems ATMOS that will be sufficient to form one artillery regiment with three batteries each with six weapons.

Soltam Systems ATMOS 2000 155 mm/45-calibre (M46S) self-propelled artillery system deployed in the firing position from the rear 0096538

ATMOS 155 mm/39-calibre SP artillery system integrated onto a 6 × 6 chassis with a protected cab and with two stabilisers lowered a the rear of the chassis. Ready use ammunition is carried in the lockers (Elbit Systems) 1364633

While the IDF evaluated an ATMOS 155 mm/39-calibre system, it is expected that production systems may be fitted with a 155 mm/52-calibre ordnance for increased range.

In mid-2011 it was revealed that the IDF was now considering three options to upgrade its 155 mm conventional tube artillery.

This included further upgrades to the current fleet of M109 tracked self-propelled weapons or the procurement of a new system such as the ATMOS or the German Krauss-Maffei Wegmann 155 mm/52-calibre Artillery Gun Module (AGM).

As of March 2013 no announcement had been made on the IDF procurement of any of these new conventional tube artillery systems.

The 155 mm M109A5 is expected to remain in service with Israel for the foreseeable future and in January 2013 it was announced that the IDF had placed a contract with Israel Aerospace Industries for an undisclosed quantity of its TAMAM Modular Azimuth Position System (TMAPS) inertial navigation systems for installation on its M109A5 155 mm self-propelled artillery systems.

TMAPS is based on a Ring Laser Gyro (RLG) and in this application supplies the M109A5 with navigation and laying data and is not dependent on a Global Positioning System.

According to Soltam Systems several undisclosed export customers have already placed orders for the ATMOS 2000 system. According to the United Nations, recent export sales by Soltam Systems of 155 mm towed artillery systems have included Cameroon (18), India (60), Slovenia (18) and Uganda (18).

It has been confirmed that Uganda took delivery of three Soltam Systems ATMOS 155 mm/39-calibre systems in 2005 and that these were mounted on a 6 × 6 cross country chassis with a protected forward control cab.

According to the United Nations Arms transfer lists there were no exports of this system between 2006 and 2008.

In 2009 Uganda took delivery of another three ATMOS systems integrated onto a 6 × 6 cross-country truck chassis.

In April 2011 it was announced that Elbit Systems has been awarded a contract valued at about USD24 million for the supply of a complete ATMOS 155 mm SP artillery system for an undisclosed African country.

Elbit Systems acquired Soltam Systems in 2010 and this was the first contract for a complete artillery system since this acquisition.

Deliveries to the undisclosed customer will start in about 18 months and be completed over a two-year period and the contract also includes training and maintenance support.

In addition to an undisclosed quantity of truck mounted ATMOS 155 mm SP artillery systems the contract also includes the key target acquisition system for use by the forward observer and the computerised Fire-Control System (FCS) which would link the battery to the weapons.

For the latest undisclosed export customer a German MAN cross-country truck chassis will be used.

Description

The Soltam Systems ATMOS, 155 mm/52-calibre self-propelled artillery system essentially consists of a 6 × 6 cross-country truck chassis, on the rear of which is mounted the upper part of a Soltam Systems 155 mm/52-calibre towed TIG 2000 system.

The prototype is fitted with a 155 mm/52-calibre ordnance that conforms to the NATO Joint Ballistic Memorandum of Understanding (JBMoU), although Soltam Systems 155 mm/39 and 155 mm/45-calibre barrels can also be fitted.

The breech mechanism is of the horizontal sliding type and automatically opens to the right with self sealing metal obturating rings.

The buffer is of the one cylinder hydraulic type with the recuperator being of the hydropneumatic type. Recoil length is variable from 850 to 1,100 mm. Two pneumatic equilibrators are fitted.

Weapon elevation and traverse is all hydraulic and computer controlled. The hydraulic power pack operates the howitzer's aiming gears as well as the load assist systems and spades.

When fitted with a 155 mm/52-calibre barrel, a maximum range of 41 km can be obtained firing an Extended-Range Full-Bore – Base Bleed (ERFB-BB) projectile, 30 km firing the NATO L15 High-Explosive (HE) projectile and 24.5 km firing the older M107 HE projectile.

When fitted with a 155 mm/39-calibre ordnance and with a chamber volume of 16.85 litres maximum range is being quoted as 30,000 m firing an ERFB-BB projectile.

Soltam Systems ATMOS 155 mm/39-calibre self-propelled artillery system from the rear showing weapon slightly elevated, stabilisers lowered to the ground and operators platform deployed in horizontal position (IHS/IDR) 1330590

Soltam Systems Semser 122 mm D-30 self-propelled howitzer deployed in the firing position and showing stabilisers lowered (Altair) 1330940

When fitted with a 155 mm/45-calibre ordnance and with a chamber volume of 23.55 litres maximum range is being quoted as 39,000 m firing an ERFB-BB projectile.

ATMOS 2000 carries a total of 32 155 mm projectiles and associated charges and is normally operated by a crew of four, of whom two are loaders positioned one either side at the rear.

A rate of fire of between 4 and 9 rds/min are claimed and to reduce crew fatigue, a flick rammer is fitted as standard. To provide a more stable firing platform, two large spades are lowered to the ground one either side at the rear of the system by remote hydraulic control.

It also has an onboard Advanced Fire-Control System (AFCS). This includes a main computer and a navigation and aiming device and enables ATMOS 2000 to rapidly engage the target after receiving the position of the target from the Forward Observation Officer (FOO).

The AFCS processes the target information, presents this on a display to the gunner and then lays the weapon onto the target in traverse and elevation. Manual back-up is also provided for in case of power failure.

Although the prototype system is installed on a modified Tatra (6 × 6) truck chassis (which has been built under licence in India), Soltam Systems have emphasised that it could be fitted to a wider range of other tracked and wheeled chassis.

As of March 2013 it is understood that all production applications for the Soltam Systems ATMOS 2000 and its variants had been for integration on a wheeled platform.

In October 2012 Soltam Systems announced that it had been awarded a USD40 million contract from an unidentified export customer for artillery and radio systems. This could have included ATMOS 2000 systems.

The forward-control armour-protected cab provides the occupants with protection from small arms fire and shell splinters.

Variants

ATROM 155 mm/52-calibre self-propelled gun

This has been developed by Soltam Systems in co-operation with the Romanian company of Aerostar to meet the requirements of the Romanian Army.

This is based on a local ROMAN (6 × 6) cross-country chassis and details are provided in a separate entry in *IHS Jane' s Land Warfare Platforms: Artillery and Air Defence*.

As of March 2013 it is understood that this system had not entered quantity production for Romania.

ATMOS 2000 155 mm M46S

Soltam Systems have stated that the ATMOS 2000 could be fitted with the saddle and elevating mass of the Russian 130 mm M-46 field gun, with the ordnance replaced by a new 155 mm/45-calibre system.

Soltam Systems has already won a major contract from India to upgrade some 180 Russian-supplied 130 mm M-46 field guns with a 155 mm/39-calibre and 155 mm/45-calibre ordnance in 2000.

This Indian programme is understood to have now been completed.

Soltam Systems Semser 122 mm D-30 self-propelled artillery system

To meet the requirements of Kazakhstan, Soltam Systems developed the Semser which consists of a KaMAZ 63502 (8 × 8) forward control cross-country truck chassis fitted with the upper part of the widely deployed Russian 122 mm D-30 towed howitzer mounted on a turntable at the rear.

To the rear of the forward control cab is space for the crew and ready use ammunition. To provide a more stable firing platform, four hydraulic stabilisers are lowered to the ground, two on either side of the chassis.

Specifications

	ATMOS 2000/52
Dimensions and weights	
Crew:	4+2
Length	
overall:	9.50 m
Width	
transport configuration:	2.50 m
Ground clearance	
overall:	0.40 m
Weight	
standard:	21,000 kg
Mobility	
Configuration	
running gear:	wheeled
layout:	6 × 6
Power-to-weight ratio:	15.00 hp/t
Speed	
max speed:	80 km/h
Range	
main fuel supply:	1,000 km
Fording	
without preparation:	1.40 m
Vertical obstacle	
forwards:	0.60 m
Trench:	0.9 m
Engine	V-12, diesel, 315 hp at 2,200 rpm
Gearbox	
forward gears:	10
Firepower	
Armament:	1 × 155 mm howitzer
Ammunition	
main total:	27
Main weapon	
length of barrel:	8.745 m
calibre length:	52 calibres
rifling:	6,900 m
maximum range:	41,000 m
Main armament traverse	
angle:	50° (25° left/25° right)
Main armament elevation/depression	
armament front:	+70°/0°

Status

Production as required. In service with Cameroon (18 - unconfirmed), Uganda (6 units) and other undisclosed countries.

An undisclosed African country placed a contract for ATMOS 155 mm systems which was announced in April 2011. ATMOS has been evaluated by the IDF.

Contractor

Soltam Systems Limited. This company is now part of Elbit Systems.

Italy

Oto Melara Draco 76 mm multirole remote weapon system

Development

Oto Melara has considerable experience in the design, development and production of not only tracked Armoured Fighting Vehicles (AFV) but also a wide range of weapon systems for land and naval applications.

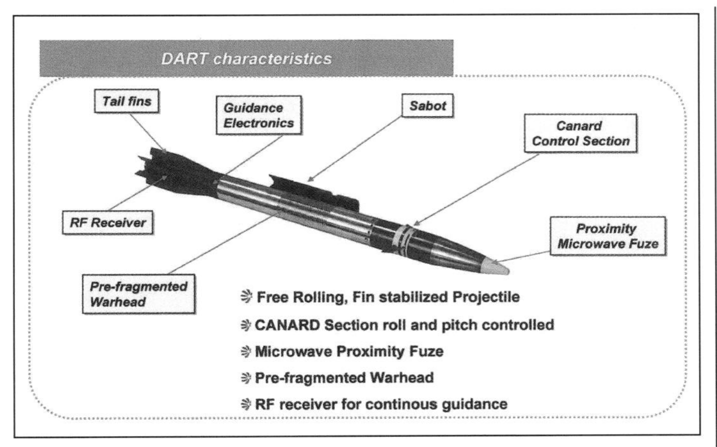

Detailed drawing of Oto Melara 76 mm DART projectile showing main components (Oto Melara) 1347995

In mid-June 2010 Oto Melara unveiled the Draco turret mounted Remote Weapon System (RWS) which the company has developed as a private venture.

Draco has been designed to meet emerging requirements for a mobile anti-air weapon system with a major secondary role capability to protect bases and manoeuvre forces against a variety of current and future threats.

These include the protection of high value targets such as airports, bases and strategic installations against air threats including aircraft, helicopters, air launched weapons and Unmanned Aerial Vehicles (UAV).

In addition there is also a requirement to provide direct and indirect fire support for ground based units and in some countries a coastal defence role as well.

In the longer term Draco will also provide a Counter Rocket and Mortar (C-RAM) capability which is becoming of increasing importance in expeditionary operations.

As of March 2013 initial firing trials of the Draco integrated into an Iveco Defence Vehicles Centauro (8 × 8) chassis had taken place, although design of the Draco is such that it can be integrated onto a wider range of platforms, tracked and wheeled.

Description

At an early stage of development of Draco, Oto Melara carried out a study on the most effective calibre weapon to counter the emerging threat.

In the end the company decided on 76 mm as the optimum calibre as the ammunition has a more effective radius of 10 m when compared to smaller calibre ammunition natures as well as having a much longer range.

Typical target engagement ranges currently being quoted for the 76 mm gun are 8 km for helicopters and 6 km for cruise missiles and UAV.

When being used in the indirect fire role against ground targets a range of 15 km is being quoted and 20 to 22 km for naval targets when being used in the coastal defence role.

It can also be used in the direct fire role against ground targets out to a maximum range of about 3 km with this depending on vehicle type being engaged.

The combat proven Oto Melara 76 mm 62-calibre naval gun was selected for the Draco weapon system.

This has been built in large numbers for the home and export markets and is currently in service with 56 countries with a family of ammunition already in production in Italy as well as numerous overseas countries.

The Draco RWS turret is of all welded steel armour construction that in its basic form provides protection to STANAG 4569 Level III with the option to increase this to Level IV.

The stabilised 76 mm gun has a cyclic rate of fire of between 80 and 100 rounds a minute with 12 rounds of ready use ammunition in the drum type magazine in the rear of the turret and a further 24 rounds stowed in an automatic feeding system.

Oto Melara Draco remote weapon station installed on a Centauro (8 × 8) armoured car hull and clearly showing 76 mm gun with pepperpot muzzle brake (Christopher F Foss) 1347991

Oto Melara Draco remote weapon station installed on a Centauro (8 × 8) armoured car hull showing 76 mm gun, coaxial machine gun and banks of 76 mm grenade launchers (Christopher F Foss) 1347993

For automatic reloading the 76 mm gun is depressed to 0° and 12 rounds can be reloaded in less than 60 seconds. For the successful engagement of most targets a short burst would be sufficient.

The empty 76 mm cartridge cases can be automatically ejected outside of the lower part of the turret.

These could also be kept inside the drum when the external ejection could be dangerous such as when troops are deployed near the system.

Turret traverse is all electric at a maximum speed of 35 °/s with elevation from –10° to +75° at 60°/s.

Draco can fire all current production 76 mm naval rounds including High-Explosive Multirole (HE-MOM) with a pre-formed fragmentation warhead, High-Explosive Semi-Armour Piercing (HE-SAPOM) and High-Explosive, Semi-Armour Piercing Extended-Range (HE-SAPOMER).

There is also an Armour-Piercing Fin-Stabilised Discarding Sabot - Tracer (APFSDS-T) round specifically developed for land applications and is already ready for production.

According to Oto Melara its performance is equal to the current 90 mm APFSDS rounds available on the market.

The latest development by Oto Melara is the 76 mm Driven Ammunition Reduced Time of Flight (DART) guided munition that features a pre-fragmented warhead, microwave proximity fuze and a RF receiver for continuous guidance.

This has undergone successful trials with the Italian Navy and is now in production.

Under study by Oto Melara are two new 76 mm rounds which will further enhance the capability of the Draco and these are special C-RAM and a top attack with Semi-Active Laser (SAL) terminal guidance.

The C-RAM would have a special pre-fragmented warhead that was successfully tested against a variety of threats in 2009.

A 7.62 mm or .50 machine gun can be mounted coaxial to the right of the 76 mm gun. A bank of four 76 mm grenade launchers are mounted on the forward part of the turret either side of the main armament and cover the frontal arc of the Draco.

Mounted on the roof towards the rear is the tracking and guidance antenna that can be folded down under full armour protection when not required.

The two axis monopulse cassegrain antenna is of the fully coherent Doppler type with fast lob switching. This would be modified to provide guidance for the 76 mm DART round.

The commander is provided with a roof mounted stabilised panoramic sight which has a day camera with wide and narrow fields of view and a x10 zoom, thermal imager with wide and narrow fields of view and an eye safe laser range-finder.

In the basic configuration in a typical target engagement the commander would first detect the target by means of the roof mounted panoramic sight and if confirmed as hostile would turn this over to the gunner who carries out the actual target engagement using the same sight.

Optionally, but more expensive, the gunner has a roof mounted stabilised sight that incorporates a day/night image intensification camera, daylight colour TV camera and an eyesafe laser range-finder.

Initial target information would come from an external surveillance radar with the Draco tracking radar then taking over. This is also used as the guidance beam when firing the DART ammunition.

Draco system is operated by a crew of two commander and gunner who are seated in the hull with each provided with a 12 inch LCD display with the commander on the left and gunner on the right.

The Draco RWS weighs about 6.5 tonnes and this allows it to be installed on a wide range of chassis, tracked and wheeled with a minimum weight of about 15 tonnes.

First example of the Draco RWS was shown in mid-2010 integrated onto the hull of the Iveco Defence Vehicles Centauro (8 × 8) armoured car which is already in service with Italy and Spain.

Draco RWS could also be installed on a tracked hull such as the Iveco Defence Vehicles/ Oto Melara Dardo infantry fighting vehicle which is already in service with the Italian Army.

Another alternative would be to install Draco in a shelter that could then be carried on the rear of a cross-country truck and then off loaded using a mechanical handing system.

This would be used in the static role with stabilisers providing a more stable firing platform and could be suited for the C-RAM role.

Draco RWS would also be fitted with a Battle Management System (BMS) and radio suite depending on customer requirements.

Variants

Draco with Scudo ADS

The Draco RWS shown in mid-2010 was fitted with a mock up of the inner layer of the Scudo Active Defence System (ADS) on the right side of the hull.

Scudo has been under development under the leadership of Oto Melara to meet the requirements of the Italian Army who has partly funded the programme since 2002.

Scudo has been designed to provide the platform with a high level of protection against anti-tank weapons such as the RPG-7 rocket propelled grenade launcher, Anti-Tank Guided Weapons (ATGW) and lower velocity gun launched projectiles.

It includes the X-band dual frequency CW detection and tracking sensor, infra-red sensor, management computer, 70 mm fragmentation launcher and a kit of explosive tiles and dedicated optronic trigger sensors.

The Italian Ariete MBT, for example would have two roof mounted launchers each with six munitions in the ready to launch position.

The incoming threat is first detected by the tracking radar and if confirmed as a threat to the platform one or two of the 70 mm interceptor rockets will be launched into the path of the incoming threat.

Each of these has a smart proximity fuze that activates the munitions which showers thousands of tungsten balls into the path of the incoming warhead.

This has a kill probability of over 90 per cent with targets being engaged at a range of between 30 and 100 m.

If the threat is launched at a very short range, for example an RPG-7, this will be engaged by the inner layer which consists of small explosive tiles that also discharge thousands of tungsten balls into the path of the threat.

All components of the Scudo ADS have been tested and Oto Melara is proposing a one year live test programme of the sensor suite connected to a black box to record behaviour of the system in a real environment.

As of March 2013 no time scale for the introduction of the Scudo ADS had been announced.

Specifications

Draco turret

Firepower	
Armament:	1 × turret , punted 76 mm rifled gun
	1 × roof mounted 7.62 mm (0.30) machine gun
	8 × 76 mm smoke grenade launchers (2 × 4)
Ammunition, main total:	36 (× 76 mm)
Turret power control, type:	powered/manual
Main armament traverse, angle:	360°
Main armament elevation/depression:	
(armament front)	+75°/-10°

Status

Under development. Not yet in production or service.

Contractor

Oto Melara.

Oto Melara 155 mm Ultra Light Weight Self-Propelled Wheeled Howitzer (ULWSPWH)

Development

The Italian Army already fields a fleet of Consorzio Iveco Oto (CIO) Centauro 105 mm (8 × 8) armoured cars and is now fielding a fleet of Freccia (8 × 8) Infantry Fighting Vehicles (IFV) plus specialised variants including IFV fitted with Spike anti-tank guided weapons, command post vehicle and 120 mm self-propelled mortar.

To provide artillery fire support for these units, CIO is developing the 155 mm Ultra Light Weight Self-Propelled Wheeled Howitzer (ULWSPWH).

This is being funded by the company as well as the Italian MoD. The latter financed 50% of the study and technological research as well as the complete recoiling mass consisting of the barrel, breech block and breech ring as well as the automatic feed mechanism.

The study also included the installation of the weapon onto a Centauro type 8 × 8 hull which has already proved to be a stable firing platform for turret mounted 105 mm and 120 mm direct fire weapons already in production and service.

A concept vehicle was shown at the Eurosatory defence equipment exhibition held in Paris in June 2012.

The time taken to implement the complete system through to the firing phase depends on funding and it is likely that a phased approach will be adopted to gradually add capacity in a progressive manner.

155 mm Ultra Light Weight Self-Propelled Self-Propelled Wheeled Howitzer concept in travelling configuration and showing ordnance with pepper pot muzzle brake (Oto Melara) 1454104

The first year could see the development of the cradle, turret shell, hull modifications followed by continuation of the development of the automatic loading system and then a critical design review.

This would allow for a total of three years for a fully operational system for firing trials with qualification expected to take another year. Mobility aspects would be complete at an early stage.

The first example of the ULWSPWH is integrated on the original Centauro 105 mm hull of which 400 have been built for the Italian Army and 84 for the Spanish Army.

It is expected that production ULWSPWH howitzers would be based on the latest production Feccia IFV which is currently deployed to Afghanistan with an enhanced protection package against mines and improvised explosive devices.

Description

The ULWSPWH would typically be operated by a crew of three consisting of driver, commander and gun layer and who all be seated in the protected hull.

The 155 mm/39-calibre ordnance is fitted with a bore evacuator and pepper pot muzzle brake and this is mounted in a fully enclosed remote turret mounted towards the rear of the hull.

Turret is of all welded steel armour with traverse of 15° left and 15° right with elevation from –5° to +75°.

When travelling the turret is traversed to the front and the 155 mm ordnance locked by remote control.

A total of 15 × 155 mm projectiles and associated Modular Charge System (MCS) are carried ready for immediate use with additional projectiles and MCS carried in the hull.

The high rate of fire is achieved due to the installation of the automatic ammunition handling system which first loads the projectile and then the charge.

There is also an automatic Primer Feed Mechanism (PFM) and the setting of the projectile fuze is automatic and integrated into the automatic ammunition handling system.

It has a maximum rate of fire up to eight rounds a minute and also features a Multiple Round Simultaneous Impact (MRSI) capability with four 155 mm projectiles landing on the target at the same time.

The 155 mm/39-calibre ordnance meets the NATO Joint Ballistic MoU and in addition to firing all types of standard NATO ammunition can also fire the recently developed Oto Melara Volcano ammunition.

Firing the Volcano Ballistic Extended Range (BER) ammunition a maximum range of more than 40 km is claimed by Oto Melara.

This can be increased to more than 60 km using the Volcano Guided Extended Range (GER) ammunition to provide a precision effect.

155 mm Ultra Light Weight Self-Propelled Wheeled Howitzer concept from rear with an Oto Melara HITROLE remote weapon station installed on turret roof (Oto Melara) 1455516

Close up of the 155 mm Ultra Light Weight Self-Propelled Wheeled Howitzer concept turret showing the 155 mm weapon, banks of 76 mm grenade launchers mounted in the turret and the roof mounted Oto Melara HITROLE remote weapon station (OtoMelara) 1455517

The on board computerised Fire-Control System (FCS) allows the system to come into action and open fire in under three minutes and come out of action in about one minute so avoiding counter battery fire.

A 7.62 mm machine gun (MG) would be mounted for self-protection over the frontal arc and there could also be the option of a roof mounted Oto Melara Hitrole Remote Weapon Station (RWS) that could be armed with a 7.62 mm or 12.7 mm MG or even a 40 mm automatic grenade launcher.

Other options include the installation of a roof mounted panoramic day/thermal sighting system.

Production ULWSPWH would have similar mobility as the latest production Freccia and include a maximum road speed of up to 110 km/h and a typical cruising range of up to 800 km.

It has a fording depth of 1.5 m and can cross a trench of up to 1.2 m and go up a 60 per cent slope and traverse a 30 per cent side slope.

Combat weight would be under 30 tonnes and it would be capable of being transported by rail as per STANAG 2832B and be carried inside the A400M transport aircraft.

Variants

Ammunition resupply vehicle

It is envisaged that the ULWSPWH would be supported by a dedicated ammunition resupply vehicle on a similar hull or a protected wheeled vehicle.

This would be fitted with an automatic ammunition handling system that would rapidly transfer 155 mm projectiles and MCS to the ULWSPWH with a complete ammunition resupply mission taking less than ten minutes.

Specifications

Not released.

Status

Under development for the Italian Army.

Contractor

Oto Melara SpA.

Kazakhstan

152 mm wheeled self-propelled howitzer

Development

In late 2010 it was stated that the South Korean company of Samsung Thales had signed an agreement with the state owned company of Kazakhstan Engineering for the development of a truck mounted wheeled self-propelled howitzer.

It is expected that this will be a 6 × 6 or 8 × 8 truck chassis fitted with the ordnance and elevating mechanism of a 152 mm towed artillery system that is currently deployed by Kazakhstan.

It has been reported that the total value of this contract is about USD200 million and that Samsung Thales would provide the computerised fire-control system and other command-and-control elements.

Soltam Systems of Israel has already developed the Semser 122 mm D-30 self-propelled artillery system to meet the operational requirements of Kazakhstan.

This is essentially a KaMAZ 63502 (8 × 8) forward control truck chassis on the rear of which is mounted a turret armed with the elevating mechanism and ordnance of the Russian 122 mm D-30 howitzer.

Available details of this are provided in the entry for the Soltam Systems 155 mm ATMOS 2000 SPG in *IHS Jane's Land Warfare Platforms: Artillery and Air Defence*.

Description

No firm details have been released but it will probably be a 6 × 6 or 8 × 8 cross-country truck chassis with the 152 mm ordnance mounted at the rear of the chassis. Stabilisers would probably be fitted to provide a more stabilised platform.

It is expected that the weapon to be installed on this chassis will be that of the Russian 152 mm towed howitzer 2A65 (MSTA-B) of which about 90 are still deployed by Kazakhstan.

Another alternative would be the Russian 152 mm towed gun 2A36 of which about 180 are still deployed by Kazakhstan.

Details of both of these towed artillery weapons are provided in separate entries in *IHS Jane's Land Warfare Platforms: Artillery and Air Defence*. Production of these weapons has been completed in Russia.

Specifications

Not released.

Status

Under development.

Contractor

See Development.

Korea (South)

Samsung Techwin EVO-105 105 mm mobile howitzer

Development

The South Korean company of Samsung Techwin, prime contractor for the 155 mm/52-calibre K9 Thunder tracked self-propelled propelled artillery system currently deployed by the Republic of Korea Army, have developed to the prototype stage a 105 mm truck mounted howitzer called the EVO-105.

As of March 2013 it is understood that this system remained at the prototype stage.

It has not been confirmed as to whether this has been developed to meet the potential requirements of the Republic of Korea Army or for the export market.

It could also be a technology demonstrator to meet future home or export requirements for a mobile wheeled artillery system.

In concept, the EVO-105 is similar to the Netherlands RDM Defence Technology MOBAT [Mobile Artillery] system.

This was the upper part of the 105 mm M101 towed artillery system fitted with a new the Royal Ordnance 105 mm/33-calibre barrel, integrated onto the rear of a surplus DAF YA 4440 (4 × 4) truck chassis.

This was developed and tested and 18 production MOBAT systems were ordered by Jordan fitted with the Selex Laser Inertial Artillery Pointing System (LINAPS) but the company closed before any weapons could be delivered.

Description

The EVO-105 consists of a standard South Korean KM500 (6 × 6) 5 ton truck chassis with the rear cargo area modified to accept the upper part of the US 105 mm M101 towed howitzer.

To provide a more stable firing platform a hydraulically operated stabiliser is lowered to the ground to the immediate rear of the cab either side before firing commences.

At present the cab is not provided with any protection from small arms fire and shell splinters but this could be provided on production systems.

The 105 mm howitzer is mounted on a turntable and fires over the rear arc with elevation limits from –5° to +65° and a traverse of 90° left and right.

Elevation and traverse is electric using a joystick which is claimed to provide for more rapid target engagement and there are also manual back up controls.

According to Samsung Techwin, maximum rate of fire is claimed to be 10 rounds a minute with a sustained rate of fire typically being three rounds a minute with a total crew of five.

A total of 60 rounds of 105 mm ammunition are carried and there is a protected area to the immediate rear of the standard cab and on the roof of this is a 7.62 mm or .50 machine gun for local protection which is provided with a shield.

Maximum range of the 105 mm M101 howitzer depends on the projectile and charge combination but firing the standard US M1 High-Explosive (HE) projectile a maximum range of 11.27 km can be achieved.

Firing the locally developed K178 HE assisted projectile the range can be increased out to 18 km.

A computerised fire-control system is fitted which is coupled to an automatic gun laying system for greater accuracy. The system can come into action and fire its first round in just 60 seconds.

Although the primary role is the indirect fire mission, the EVO-105 also has a direct fire capability which could be useful in some tactical situations.

The conventional 105 mm M101 howitzer is normally operated by a crew of eight and is towed by a 6 × 6 truck which also carries some ready use ammunition.

By integrating the 105 mm M101 onto a 6 × 6 cross-country truck chassis the key into and out of action times are dramatically reduced which increases the survivability of the system to counter battery fire. In addition it has a reduced crew requirement.

Although the first example is mounted on the older KM500 (6 × 6) truck chassis it is considered possible that production systems would have a more recent chassis with improved cross-country mobility.

Specifications

	EVO-105
Dimensions and weights	
Crew:	5
Length	
overall:	9 m
Width	
overall:	2.5 m
Height	
overall:	3.2 m
Mobility	
Configuration	
running gear:	wheeled
layout:	6 × 6
Speed	
max speed:	80 km/h
Firepower	
Armament:	1 × 105 mm howitzer
Ammunition	
main total:	60
Main weapon	
maximum range:	11,270 m (HE M1 projectile)
maximum range:	18,000 m (K178 HE projectile)
Rate of fire	
sustained fire:	3 rds/min
rapid fire:	10 rds/min
Turret power control	
type:	powered/manual
Main armament traverse	
angle:	90° (left/right)
Main armament elevation/depression	
turret rear:	+65°/-5°
Survivability	
Night vision equipment	
vehicle:	no
NBC system:	no

Status

Prototype built and tested.

Contractor

Samsung Techwin, Defense Program Division.

Romania

ATROM 155 mm/52-calibre self-propelled gun

Development

During the Expomil 2003 international defence exhibition held late in 2003 in Bucharest, Romania, the ATROM 155 mm/52-calibre self-propelled artillery system was shown for the first time.

This has been developed in Romania under the leadership of the local Aerostar SA aviation company, which has extensive experience in the design, development and production of a wide range of 122 mm truck-based Multiple Rocket Launchers (MRLs) for the home and export markets.

The prototype ATROM 155 mm/52-calibre self-propelled gun has since been fitted with a new fully enclosed armoured cab which is provided with four doors and armoured windows. This was presented during the Expomil 2005 exhibition held in Bucharest.

As of March 2013 it is understood that production of the ATROM had yet to commence.

Description

ATROM essentially consists of the locally developed ROMAN 26.360 DFAEG (6 × 6) cross-country truck chassis fitted with the latest version of the Israeli Soltam Systems Autonomous Truck Mounted-Howitzer System (ATMOS) on the rear.

The vehicle is powered by a MAN 2866 LF 24 diesel engine, which develops 360 hp, coupled to a Steyr VG 1600/300 transmission.

The chassis is fitted with a fully armour-protected forward control cab for its crew of five as well as an auxiliary power unit, which enables the 155 mm/52-calibre gun weapon system to be run with the main diesel engine shut down.

ATMOS was originally developed as a private venture by Soltam Systems based on their considerable experience in the design, development and production of 155 mm 39/45/52 towed systems.

Prototype of the ATROM 155 mm/52-calibre self-propelled gun deployed in the firing position, with stabilisers lowered and 155 mm/52-calibre weapon elevated and showing protected four door cab (Aerostar) 1186174

When deployed in the firing system the 6 × 6 chassis is stabilised by two hydraulically-operated spades that are lowered at the rear. Standard equipment includes a computerised fire-control system coupled to an inertial navigation system. The installation of a power-operated rammer not only increases rate of fire but also reduces crew fatigue.

The complete ATROM 155 mm/52-calibre system was integrated in Romania and fired for the first time late in October 2003 using various types of ammunition, including 155 mm Extended-Range projectile that gives a range of at least 41 km.

As with a number of countries in Eastern Europe, Romania is now moving towards NATO standard-calibres for its weapons.

For the Romanian programme the US-based Valentec Systems, a sister company of Soltam Systems, is heading up the programme from the weapon side.

Available details of the ATMOS 155 mm self-propelled artillery, which entered production for an undisclosed export customer in 2003, are given in a separate entry in *IHS Jane's Land Warfare Platforms: Artillery and Air Defence*. This is known to be in service with Cameroon and Uganda with an undisclosed African customer placing a USD24 million contract on 2011.

Since then additional export orders have been placed for the ATMOS 2000 but Soltam Systems have not released any specific details.

The computerised fire-control system installed on the ATROM is the same as that used in the Israel Military Industries/Aerostar LAROM multiple surface-to-surface rocket system covered in a separate entry. This has been in service with the Romanian Army for several years.

Specifications

	ATROM
Dimensions and weights	
Crew:	6
Weight	
combat:	26,000 kg
Mobility	
Configuration	
running gear:	wheeled
layout:	6 × 6
Firepower	
Armament:	1 × 155 mm howitzer
Ammunition	
main total:	24
Main weapon	
calibre length:	52 calibres
maximum range:	41,000 m
Turret power control	
type:	powered/manual
Main armament traverse	
angle:	50° (25° left/25° right)
Main armament elevation/depression	
armament front:	+70°/-8°

Status

Prototype undergoing trials. Not yet in production or service.

Contractor

Aerostar SA, Group Industrial Aeronautic.

Serbia

Yugoimport NORA B-52 155 mm/52-calibre self-propelled gun

Development

Following extensive trials over several years with a testbed of a 155 mm/45-calibre 8 × 8 Self-Propelled Gun (SPG), Yugoimport then commenced work on a new 155 mm/52-calibre system with the first prototype completed in 2003.

155 mm M03 (NORA K-1) truck mounted artillery system in travelling configuration (Yugoimport) 1403739

The first testbed is designated the self-propelled 155/45 mm NORA B M96 and is fitted with a 155 mm/45-calibre barrel that, firing an Extended-Range Full-Bore Base Bleed (ERFB-BB) projectile M98, enables a maximum range of 39 km to be achieved.

Late in 2004 it was disclosed that development of the NORA B-52 155 mm/52-calibre self-propelled gun had been successfully completed and an undisclosed customer had placed the first export order for an undisclosed quantity of systems.

While first examples of the NORA B-52 155 mm/52-calibre (8 × 8) self-propelled artillery system were based on a locally developed chassis the first production systems, for an unidentified export customer, were based on a KamAZ 63501 (8 × 8) cross-country chassis.

Unconfirmed reports have indicated that the first export customer is Myanmar with first batch of vehicles delivered.

The total quantity could be as high as 30 units which as of early 2013 are understood to all be in service.

In late 2011 it was stated that Bangladesh had placed a contract for a total of 18 NORA B-52 155 mm/52-calibre self-propelled guns and that these would be fitted with a SAGEM Sigma 30 inertial navigation platform system for autonomous navigation and pointing.

Description

The platform used in the original system is based on a locally designed and manufactured FAP 2882 (8 × 8) cross-country chassis with a forward control cab that was originally developed as a cargo truck but was subsequently used for a wide range of other applications. This includes the 262 mm (12-round) LRSV M-87 'Orkan' multiple rocket launcher, which is covered in a separate entry.

This 8 × 8 chassis has good cross-country mobility and is fitted with a central tyre-pressure-regulation system that allows the driver to adjust the tyre pressure to suit terrain being crossed.

The fully enclosed forward control two-door cab can be tilted forwards to allow access to the power pack for maintenance purposes.

Some of the SPG crew are carried in the cab with the remainder being seated in a fully enclosed pod to the immediate rear of the cab. To the rear of this is ammunition stowage that carries 36 155 mm projectiles, combustible charges and fuzes.

The 155 mm/45 mm-calibre ordnance is mounted on a turntable at the very rear of the chassis with a shield, and is normally traversed to the front to reduce the overall length of the system for travelling.

When deployed in action the weapon is normally fired over the rear arc. The prototype chamber has a capacity of 22.5 litres with the ordnance having 48 grooves and maximum pressure of powder gases being 3,000 bars.

To provide a more stable firing platform, four hydraulic stabilisers are lowered to the ground by remote-control, two either side of the chassis. These are positioned to the rear of the second and forth axles.

Elevation is from –3° to +51° at a maximum rate of 20 mils/s, while traverse is 30° left and right at a maximum speed of 100 mils/s. Elevation and traverse are electro-hydraulic with manual controls provided as a back up.

According to Yugoimport, the system has a maximum rate of fire of up to five rounds per minute and it takes three minutes to come into action. A burst round of three 155 mm projectiles can be fired in 20 seconds due to the installation of a semi-automatic gun hydraulic rammer.

The vehicle can be fitted with a 7.62 mm or 12.7 mm machine gun on the roof for local protection and two banks of 82 mm electrically-operated smoke grenade launchers are integrated into the platform.

A camouflage net is carried, which provides protection in a number of radiation bands including visible, near infra-red, thermal and radar.

The system is provided with power elevation and traverse and is fitted with a flick rammer to reduce crew fatigue. Information from the battery fire-control system is sent to the gun display unit on the weapon.

The new 155 mm/52-calibre NORA B-52 system is similar to the original testbed but incorporates a number of improvements as a result of the field trials and will be almost to production standard. This includes a new breech block with a new self-sealing system.

Yugoimport also markets this system with a flexible battery level computerised fire-control system.

For many years the former Yugoslav Army fielded the 152 mm gun-howitzer M84/M84B1/M84B1 which in turn was based on the Russian 152 mm D-20 towed system. There was also a conversion of the Russian 130 mm M-46 to a 155 mm 45-calibre system, but this never entered production.

SOKO SP RR 122/105/122 mm truck mounted artillery system in firing position with outriggers extended. This model is armed with a 122 mm D-30 type ordnance (Yugoimport) 1403740

Variants

KamAZ 63501 (8 × 8) based 155 mm/52-calibre NORA B-52 self-propelled artillery system

In 2008 it was revealed that Yugoimport had supplied a quantity of its latest generation 155 mm/52-calibre NORA B-52 series Self-Propelled (SP) artillery systems to an undisclosed export customer.

Latest production system has many improvements and is based on a more recent Russian KamAZ 63501(8 × 8) cross-country chassis that is already used for a wide range of applications.

This truck is fitted with a conventional semi-elliptical leaf spring suspension system, powered steering and is fitted with a central tyre inflation system that allows the driver to adjust the tyre pressure to suit the terrain being crossed.

For this application the chassis has been fitted with an additional welded steel frame to the rear of the forward control cab on which the weapon system and its associated equipment is mounted.

First production systems are designated the M03 version K0 with the enhanced K1 system said to be in the final development phase.

The first version includes an Artillery Fire Command-and-Control System (AFCCS) which includes two forward observation officers with their target acquisition systems.

This consists of a tripod mounted electronic goniometer and laser range-finder, GPS device and communications system. This provides information to the battery command post.

This in turn provides target information to the gun display units installed in the commanders cabin and has a built in GPS unit and VHF radio data modem.

The 155 mm/52-calibre ordnance is mounted on a turntable at the very rear of the chassis and when travelling is traversed to the front and held in position by a travel lock mounted to the immediate rear of the fully enclosed forward control cab. The barrel is unlocked by the gunner pressing a button with the ordnance then being elevated using the joystick.

The 155 mm/52-calibre ordnance is fitted with a double baffle muzzle brake but does not have a fume extractor or a thermal sleeve.

Traverse is 30° left and right with elevation limits from –5° to +65°. Elevation and traverse is hydraulic.

When deployed in the firing position the ordnance normally fires over the rear arc with four telescopic electrohydraulic stabilisers being lowered to the ground to provide a more stable firing platform.

In an emergency the weapon can be fired over the frontal arc but in this application the area covered by the weapon is reduced.

According to Yugoimport, the system takes 1.5 minutes to come into action and a similar time to come out of action. Maximum rate of fire is claimed to be five rounds a minute and a total of 36 × 155 mm projectiles and charges are carried.

M09 105 mm truck mounted artillery system deployed in the firing position with turret traversed to rear and stabilisers lowered to the ground (Yugoimport) 1403741

Serbian NORA B-52 M03 version K0 155 mm/52-calibre self-propelled artillery system in travelling configuration with ordnance in travel lock (Yugoimport) 1330538

Latest production standard NORA B-52 155 mm/52-calibre self-propelled artillery system deployed in the firing position with stabilisers lowered and turntable mounted ordnance traversed to the rear (Yugoimport) 1403742

Maximum range depends of the projectile/charge combination but firing a standard locally produced High-Explosive Extended-Range Full-Bore Base Bleed (HE ERFB BB) projectile a maximum range of 44 km is obtained.

Under development is a High-Explosive Rocket-Assist Base Bleed (HE RA BB) projectile which is claimed will have a maximum range of 65 km.

Other types of projectile that can be fired include the old US M107 HE projectile but this has a maximum range of only 18.4 km, as well as smoke and illuminating projectiles.

In addition, it can also fire the Russian KBP Instrument Design Bureau Krasnopol-M laser-guided artillery projectile out to a maximum range of 25 km when used in conjunction with the associated ID20 laser designator and range-finder and communications equipment.

The 155 mm Krasnopol-M has already been produced in production quantities for the export market and has seen action with the Indian Army with its now BAE Systems (previously Bofors) 155 mm Field Howitzer 77B (FH-77B).

Charges are of the locally developed monoblock type although the system can also use conventional bag type charges according to Yugoimport.

The 155 mm/52-calibre ordnance is fitted with a semi-automatic horizontal sliding breech mechanism which remains open after the first round is fired. The six round primer magazine holds standard M82 priming caps.

The latest version has a automatic loader which has two rotary type magazines located on the rear of the mount with each of which hold 12 projectiles or monoblock charges.

Installation of an automatic loader not only increases the rate of fire but also reduces crew fatigue. The projectile is loaded first followed by the charge.

The system has a crew of five which includes the gunner and commander who are seated in fully enclosed cabs positioned one either side of the mount. The other three crew members are the driver and two loaders.

The commander, seated in the right cabin, is responsible for monitoring the overall combat situation and is the link with higher command such as battery or battalion headquarters.

The gunner, seated in the left cabin, is responsible for the automatic loading function, laying the weapon onto the target and firing the weapon.

While the first version is designated the M03 version K0, the next generation K1 has a number of additional features including automatic target laying and the installation of an auxiliary power unit which enables all of the functions to be carried out with the main truck engine switched off.

The K1 also features a commanders integrated Battle Management System (BMS) which can carry out a number of functions including collection and processing of information and enemy and friendly forces and displaying this information on a digitised map. It is also used for mission planning, decision making, transfer of information and data encryption.

Other Serbian wheeled self-propelled artillery systems

155 mm M03 (NORA K-1)

This has been developed to the prototype stage and is an 8 × 8 protected chassis with the crew compartment at the front and the turret mounted 155 mm ordnance mounted at the rear.

Standard equipment includes an auxiliary power unit, ammunition handing system, powered elevation and traverse, on board navigation and fire-control system.

When deployed in the firing position two hydraulic stabilisers are lowered to the ground either side of the rear unit.

The complete system weighs 34 tonnes and is powered by a turbocharged diesel developing 410 kW which gives a maximum road speed of 90 km/h. Maximum range at a speed of 80 km/h is being quoted as 1,000 km.

SOKO SP RR 122/105/122 mm

This family of self-propelled artillery systems has been developed by Yugoimport as a private venture and is aimed mainly at the export market.

It consists of a locally developed FAP 2228 (6 × 6) cross-country truck chassis with a protected cab at the front.

At the rear of the vehicle can is fitted a protected turret which can be armed with various 100 mm, 105 mm and 122 mm weapons.

Additional details are provided in a separate entry in *IHS Jane's Land Warfare Platforms: Artillery and Air Defence*.

As of March 2013 it is understood that production of this system has yet to commence.

M09 105 mm armoured truck mounted howitzer

This is a 6 × 6 cross-country truck chassis with a protected crew compartment at the front and a turret with an open roof and back mounted on the rear of the platform.

This is armed with a locally manufactured 105 mm howitzer M56 which is covered in a separate entry in *IHS Jane's Land Warfare Platforms: Artillery and Air Defence.*

A total of 60 rounds of 105 mm ammunition are carried. Maximum range firing a HE ER projectile is 15,000 m and when firing an HE ER-BB this increases to 18,000 m.

Gross vehicle weight is being quoted as 18 tonnes maximum road speed 90 km/h and a range of up to 450 km. It is normally operated by a crew of five.

As of March 2013 it is understood that production of this system had yet to commence.

Specifications

	NORA B-52
Dimensions and weights	
Crew:	5
Length	
overall:	10.645 m
main armament rear:	14.00 m
Width	
transport configuration:	2.665 m
Height	
transport configuration:	3.2 m
axis of fire:	3.72 m
Weight	
standard:	29,400 kg
Mobility	
Configuration	
running gear:	wheeled
layout:	8 × 8
Speed	
max speed:	80 km/h
cross-country:	25 km/h (est.)
Range	
main fuel supply:	500 km
Engine	diesel, 315 hp
Firepower	
Armament:	1 × 155 mm howitzer
Armament:	1 × roof mounted 12.7 mm (0.50) machine gun
Armament:	2 × 82 mm smoke grenade launcher
Ammunition	
main total:	36
Main weapon	
calibre length:	52 calibres
Turret power control	
type:	powered/manual
Main armament traverse	
angle:	60° (30° left/30° right)
Main armament elevation/depression	
armament front:	+65°/-5°

Status

Production as required. In service with at least one undisclosed export customer. This is understood to be Myanmar with up to 30 systems now delivered.

Late in 2011 Bangladesh placed a contract for 18 NORA B-52 155 mm/52-calibre self-propelled guns.

Contractor

Yugoimport - SDPR.

Yugoimport SOKO SP RR 122 mm self-propelled artillery system

Development

Yugoimport are currently marketing a number of upgrade packages that convert existing 100 mm/105 mm/122 mm towed artillery systems into Self-Propelled (SP) artillery systems.

A typical example is the SOKO Self-Propelled Rapid Response (SP RR) 122 mm truck mounted gun-howitzer that is now being offered for export but as of March 2013 there are no known sales.

Description

The standard Russian 122 mm D-30 towed howitzer (or its Serbian equivalent the D-30J) is normally towed by a cross-country truck that as well as its crew also carries a quantity of ready use 122 mm separate-loading ammunition.

The towed 122 mm D-30 howitzer takes time to come into and come out of action and is also exposed to counter battery fire and lacks cross-country mobility.

The SOKO SP RR 122 mm system has been developed and tested by Yugoimport to provide the required increased levels of mobility and protection.

The first example is based on the locally produced FAP 2228 (6 × 6) cross-country truck chassis but the concept could be applied to other types of wheeled chassis.

The armour protected cab is at the front for the crew of four and is provided with bullet/splinterproof windows and side entry doors.

Mounted on the chassis at the rear is a protected turret armed with the complete elevating mechanism and ordnance of the towed 122 mm D-30 artillery system.

The protected turret is fitted with a forward opening door in either side of the turret with the door in the left side having a small observation window to the immediate front upper part. The door in the right side has a bullet/splinter window in its upper part.

When travelling the turret is traversed to the front and locked in position with a travel lock located at the very rear of the cab on the roof.

It is claimed that the cab of the SOKO SP RR 122 mm system provides the occupants with protection from small arms fire and shell splinters to STANAG 4569 Level 1.

The turret has a traverse of 35° left and right with the weapon elevation from -7° to +65° which is accomplished using electric power controls with manual back up.

To provide a more stable firing platform two hydraulically operated stabilisers are lowered one either side at the rear of the chassis.

A total of 51 rounds of 122 mm ammunition are carried of which 46 would typically be High-Explosive Fragmentation (HE-FRAG) and the remaining five High-Explosive Anti-Tank (HEAT).

This ammunition is carried in a protected compartment to the immediate rear of the crew cab.

According to Yugoimport, the system has a maximum rate of fire of up to six rounds a minute with this being achieved using a semi-automatic loading system and a pneumatic rammer.

Firing a standard 122 mm HE-FRAG projectile maximum range is being quoted as 17.3 km which can be extended to 21 km with a HE Base Bleed (HE BB) projectile.

With the correct fire-control system it could also be used to fire the Russian 122 mm Kitolov laser artillery to provide a precision effect.

Although the main role of the SOKO SP RR 122 mm SP howitzer is to carry out indirect fire missions it also has a useful direct fire capability with the 122 mm HEAT rounds capable of neutralising most light armoured fighting vehicles and inflicting damaged to heavier vehicles.

For increased accuracy and reduced time into action the SOKO SP RR 122 mm system is fitted with a computerised fire-control system which concludes a ballistic computer.

Banks of three electrically operated grenade launchers can be mounted either side of the cab with a 7.62 mm machine gun (MG) mounted on the turret roof on the right side in a protected mount.

Suspension is of traditional semi-elliptical leaf springs and hydraulic shock absorbers with powered steering on the front wheels as standard.

A hydraulic winch with a capacity of 10 tonnes is fitted for self-recovery.

SOKO SP RR 122 mm truck mounted artillery system in firing position with stabilisers lowered and turret traversed to the right (Yugoimport) 1452708

SOKO SP RR 122 mm truck mounted artillery system in firing position with outriggers extended and turret traversed to the front (Yugoimport) 1452707

Variants

In addition to being marketed with the 122 mm D-30 ordnance it is also being marketed fitted with a 100 mm or 105 mm ordnance.

Details of the Yugoimport upgraded 105 mm M56 howitzer and the current 4 × 4 and 6 × 6 self-propelled versions are provided in a separate entry in *IHS Jane's Land Warfare Platforms: Artillery and Air Defence*.

Specifications

	SOKO SP RR 122 mm self-propelled artillery system
Dimensions and weights	
Crew:	4
Length	
overall:	8.385 m
Width	
overall:	3.09 m
Height	
overall:	3.176 m
axis of fire:	2.533 m
Ground clearance	
overall:	0.45 m
Track	
front:	2.030 m
rear:	2.020 m
Wheelbase	3.8 m + 1.40 m
Weight	
standard:	17,000 kg
combat:	22,000 kg
Mobility	
Configuration	
running gear:	wheeled
layout:	6 × 6
Speed	
max speed:	100 km/h
Range	
main fuel supply:	600 km
Fording	
with preparation:	1.2 m
Angle of approach:	33°
Angle of departure:	33°
Gradient:	60%
Side slope:	30%
Vertical obstacle	
forwards:	0.63 m
Engine	Daimler Chrysler OM-906 6.7-litre, diesel, 279 hp
Gearbox	
model:	ZF 9S
type:	manual
forward gears:	6
reverse gears:	1
Steering:	hydraulic, ball and nut
Brakes	
main:	air
Tyres:	15.00-21 PR 12
Suspension:	leaf springs and hydraulic shock absorbers
Electrical system	
vehicle:	24 V
Batteries:	2 × 143 Ah
Firepower	
Armament:	1 × turret mounted 122 mm D-30J howitzer
Armament:	1 × 7.62 mm (0.30) machine gun
Armament:	6 × smoke grenade launcher (2 × 3)
Ammunition	
main total:	51
Main weapon	
calibre length:	35 calibres

	SOKO SP RR 122 mm self-propelled artillery system
Turret power control	
type:	electric/manual
by commander:	no
by gunner:	yes
Main armament traverse	
angle:	35° (left/right)
Main armament elevation/depression	
armament front:	+65°/-7°
Survivability	
Night vision equipment	
vehicle:	optional
NBC system:	optional
Armour	
hull/body:	steel
turret:	steel

Status
Development complete. Ready for production on receipt of orders.

Contractor
Yugoimport - SDPR.

Singapore

ST Kinetics 155 mm Advanced Mobile Gun System (AMGS)

Development
Based in their extensive experience in the design, development and production of a wide range of 155 mm towed and self-propelled artillery systems for the home and export markets, Singapore Technologies Kinetics (STK) is carrying out feasibility studies on the 155 mm Advanced Mobile Gun System (AMGS) as a private venture.

Description
The AMGS is based on an 8 × 8 cross-country chassis for greater strategic mobility than a conventional tracked SP artillery system as well as having lower operating and support costs.

Maximum road speed is being quoted as 80 km/h and cross-country speed is 30 km/h but the latter depends on the type of terrain being crossed.

Gross system weight is being quoted as 28 tonnes which includes crew and ammunition.

The crew of three is seated in the fully protected four door cab at the front of the chassis with the remote controlled turret armed with a 155 mm/52 calibre ordnance to the rear.

When travelling the ordnance is traversed to the front and locked in position over the cab. To the immediate rear of the cab is additional ammunition stowage.

According to STK, the 155 mm AMGS can come into action and open fire in about 30 seconds and come out of action in a similar time frame.

To provide a more stable firing platform two hydraulically operated stabilisers are lowered to the ground by remote control either side of the platform before firing commences.

The 155 mm/52-calibre barrel is fitted with a muzzle brake and fume extractor and firing a 155 mm Extended Range Full Bore - Base Bleed (ERFB-BB) projectile a maximum range of 40 km can be obtained.

All standard 155 mm projectiles can be fired from the system as well as more advanced projectiles.

Firing the older 155 mm M107 HE projectile a maximum range of 19,000 m can be achieved and this increases to 30,000 m with an Extended Range Full Bore – Hollow Base (ERFB-HB) projectile.

The 155 mm/52-calibre ordnance used in the STK AMGS is the same as that used in the FH-2000 towed artillery system fitted with an auxiliary power unit that is in service with the Singapore Armed Forces.

Details of this system are provided in a separate entry in *IHS Jane's Land Warfare Platforms: Artillery and Air Defence*.

The on board computerised Fire-Control system (FCS) and automatic ammunition handling system allows the system to be controlled by the crew from within the protected crew compartment at the front of the chassis.

A muzzle velocity measuring system is fitted above the rear of the 155 mm weapon and this feeds information to the FCS.

According to STK a burst rate of fire of three rounds in 20 seconds can be achieved with a maximum rate of fire of six rounds a minute for a period of three minutes.

This is due to the installation of an automatic ammunition handing system which first loads the fuzed 155 mm projectile and then the required number of modular charge systems (MCS) required.

To enable the maximum range of 40 km to be achieved firing an ERFB-BB projectile a total of six MCS would be required.

Scale model of the 155 mm/52-calibre Advanced Mobile Gun System deployed in the firing position with stabilisers lowered (STK) 1448578

Side view of the scale model of the 155 mm/52-calibre Advanced Mobile Gun System stabilisers lowered and showing additional ammunition stowage to rear of protected cab (STK) 1448580

Powered traverse is a 30° left and right with weapon elevation from –3° to +70° with a total of 26 × 155 mm projectiles of ready use ammunition plus associated MCS systems being carried.

In concept the STK AMGS is similar to the 155 mm/52-calibre German Krauss-Maffei Wegmann Artillery Gun Module (AGM) which has been designed as a private venture and remains at the prototype stage.

When integrated onto a new tracked chassis developed by General Dynamics European Land Systems this is known as the Donar and this also remains at the prototype stage.

Specifications

	155 mm Advanced Mobile Gun System (AMGS)
Dimensions and weights	
Crew:	3
Length	
overall:	12.3 m
Width	
overall:	3 m
Height	
overall:	3.8 m
Weight	
combat:	28,000 kg
Mobility	
Configuration	
running gear:	wheeled
layout:	8 × 8
Speed	
max speed:	80 km/h
Firepower	
Armament:	1 × 155 mm howitzer
Ammunition	
main total:	26 (ready use)
Main weapon	
calibre length:	52 calibres
Main armament traverse	
angle:	360°
Main armament elevation/depression	
armament front:	-3°/+70°

Status
Feasibility study.

Contractor
Singapore Technologies Kinetics (STK).

Slovakia

ZTS 155 mm/52-calibre Zuzana A1

Development

The 155 mm/52-calibre Zuzana A1 (8 × 8) self-propelled artillery system was developed as a private venture by ZTS.

This is a further development of the earlier ZTS 155 mm/45-calibre (8 × 8) self-propelled artillery system that has been built for the home (Slovakia) and export markets (Cyprus).

Details of this system, which are still being marketed are provided in a separate entry in *IHS Jane's Land Warfare Platforms: Artillery and Air Defence*.

The more recent ZTS 155 mm/52-calibre Zuzana A1 has many improvements over the earlier system including a brand new chassis and an upgraded turret fitted with a 155 mm/52-calibre ordnance which provides a significant increases in range.

As of March 2013 it is understood that trials with the first prototype system were still underway and the system had yet to enter production.

Description

The latest ZTS 155 mm/52-calibre Zuzana A1 self-propelled artillery is based on a modified Tatra 815 8 × 8 chassis, with the driver seated in the fully armoured compartment at the front. The fully enclosed turret is in the centre with the new armour protected power pack at the rear.

Main armament comprises a 155 mm/52-calibre ordnance fitted with a double-baffle muzzle brake, which is externally-mounted between the two elements of the turret system.

According to ZTS, the 155 mm/52-calibre ordnance with a 23-litre chamber meets the NATO Joint Ballistic Memorandum of Understanding (JBMoU).

Maximum rate of fire is quoted as eight rounds in the first minute and 22 rounds in three minutes. The right part of the turret accommodates a conveyor for 40 155 mm projectiles in the vertical position with the left part of the turret containing the conveyor for a similar number of charges.

A 155 mm modular charge system comprising three different modules is said to already be in production.

The weapon can be loaded with 155 mm projectiles and charges at all angles of elevation and the fuzes can also be set automatically.

Maximum range depends on charge/projectile combination but when firing Extended-Range Full-Bore - Base Bleed (ERFB-BB) projectiles it is 41.5 km.

Both parts of the turret have air conditioning and an NBC system and the complete turret has its own Auxiliary Power Unit (APU), which allows it to be operated with the main diesel engine switched off. Turret traverse is 360°, with the firing envelope 60° left and right.

It has a direct-fire sight with laser range-finder and mounted above the ordnance is a muzzle-velocity radar that feeds information into the fire-control system, which is connected to the digital map system.

The Zuzana A1 can carry out Multiple Round Simultaneous Impact (MRSI) fire missions in which all of the projectiles land on the turret at the same time.

Zuzana A1 155 mm/52-calibre self-propelled artillery system with ordnance elevated (ZTS) 1333655

First example of the ZTS 155 mm/52-calibre Zuzana A1 in the travelling configuration and clearly showing protected drivers position at front of the chassis (ZTS) 1365014

For this application the Tatra 815 chassis has been upgraded and the power pack now consists of a MAN D28 76 LF diesel engine developing 338 kW coupled to an Allison HD 4560 PR automatic transmission. The 8 × 8 chassis has been developed by Tatra Sipox.

To provide a more stable firing platform, hydraulic stabilisers can be lowered to the ground to provide a more stable firing platform. These are positioned one either side between the second and third road wheel stations, and one either side at the rear.

Standard equipment includes a central tyre-pressure system, powered steering on front four wheels and a roof-mounted 12.7 mm NSV machine gun.

Status

Prototype. Not yet in production or service.

Contractor

ZTS Dubnica nad Váhom (Slovak Republic).

ZTS 155 mm/45-calibre self-propelled gun-howitzer Zuzana (8 × 8)

Development

In December 1992, ZTS completed the first of two prototypes of a 155 mm 45-calibre version of the Dana 152 mm (8 × 8) self-propelled gun-howitzer, which is fully described in a separate entry in *IHS Jane's Land Warfare Platforms: Artillery and Air Defence*.

It should be noted that production of the older Dana 152 mm (8 × 8) self-propelled gun-howitzer is complete and it is no longer being marketed. A number of countries may have surplus quantities of these weapons for sale.

According to the manufacturer, an initial production batch of 10 155 mm/45-calibre Zuzana self-propelled guns has been completed and delivered to Slovakia.

This turret has also been installed on a modified T-72M1 MBT hull for the Indian competition which took place in 1995. As of March 2013 India had not made any purchases of any new 155 mm self-propelled artillery systems, tracked or wheeled.

Early in 1997, the Slovak Defence Minister confirmed that the Slovak Army had formally accepted the 155 mm 45-calibre Zuzana (8 × 8) self-propelled gun/howitzer under the designation Howitzer Model 2000.

The Slovak Army was the first member of the former Warsaw Pact to adopt the NATO 155 mm-calibre for its self-propelled artillery systems.

The first Slovak order was for eight Zuzana systems but further systems were purchased in the future as additional funding becomes available.

Unit cost of the 155 mm 45-calibre Zuzana is between USD3.4 and USD3.6 million.

Ammunition resupply vehicle on Tatra (8 × 8) chassis for use with 155 mm/45-calibre Zuzana showing fully armoured cab (PB) 0130601

Slovakian 155 mm/45-calibre Zuzana (8 × 8) self-propelled artillery system in service with Cyprus 1133153

Compared with tracked self-propelled artillery system, Zuzana has greater strategic mobility (maximum road speed 80 km/h) and lower operating and maintenance costs.

It has been reported that the Slovakian Army requirement is for a total of 78 systems, funding permitting.

In May 1998 a further order was placed for 40 systems plus a quantity of Tatrapan (6 × 6) command vehicles.

Cyprus has taken delivery of a total of 12 systems, which are believed to be designated the Zuzana 2000G. These have a number of modifications to meet local requirements including 76 mm smoke-grenade launchers, Thales radios and intercom system and no 12.7 mm anti-aircraft/local defence machine gun. These arrived in Greece in 2001 and were then shipped to Cyprus.

According to the United Nations Cyprus took delivery of a further batch of 12 Zuzana 2000G systems in 2007 which brought the fleet up to a total of 24 units.

The Tatra 8 × 8 chassis used in this system has the Tatra designation of the 815 VP 31 29.265 8 × 8.1R and weighs 16,400 kg in running configuration (without armour or payload). Maximum system weight quoted by Tatra is 29,000 kg.

Description

The Zuzana is essentially the earlier 152 mm Ondava self-propelled gun-howitzer fitted with a new 155 mm 45-calibre ordnance capable of firing all Western types of 155 mm ammunition.

Firing a 155 mm ERFB-BB projectile, a maximum range of 39.6 km can be achieved. Minimum range is quoted as 5.9 km.

The 155 mm 45-calibre ordnance has a new breech mechanism and is fitted with a double-baffle muzzle brake but no fume extractor. The automatic loading system enables a maximum of 30 rounds to be fired in six minutes with a total of 40 projectiles and charges being carried.

Of the 40 155 mm projectiles carried, 36 are in the automatic magazine for ready use with the remaining four in a fixed magazine. Of the 40 charges carried, 30 are in the automatic magazine and 10 in the fixed magazine. Ammunition handling is fully automatic with ramming and loading of projectiles and charges. The primer magazine holds a total of 30 primers. According to the manufacturer, the automatic loading system enables Multiple Round Simultaneous Impact (MRSI) engagements to take place.

Turret traverse is 60° left and right with weapon elevation from –3.5° to +70°. When in the firing position, the system is stabilised by three hydraulic jacks. These are positioned one at the rear and one either side between the second and third axles. The driver is responsible for unlocking the turret and lowering the stabilisers before the weapon starts its fire mission.

Maximum sustained rate of fire is claimed to be 90 rounds per hour. Maximum rate of fire with manual reloading is two rounds per minute.

The Zuzana has a crew of four and standard equipment includes an NBC system, air conditioning system, a computerised fire-control system, a land navigation system and a sight with automatic levelling.

The chassis is the Tatra 815 VP31 29265 8 × 8.1R. This features full 8 × 8 drive, powered steering and a central tyre pressure regulation system which allows the driver to adjust the tyre pressure to suit the terrain being crossed.

Variants

ZTS 155 mm/52-calibre Zuzana

Details of this system, which is fitted with a 155 mm/52-calibre ordnance, are given in a separate entry. It remains at the prototype stage and will be marketed alongside the current Zuzana 155 mm/45-calibre system.

ZTS 155 mm self-propelled gun-howitzer Zuzana T-72M1 A 40

This is essentially a 155 mm/45-calibre Zuzana turret system integrated onto a locally manufactured T-72M1 MBT chassis with the complete system also being called the Self-propelled Howitzer Model 2000. It has not entered production or service.

Tatrapan (6 × 6) command post

The Tatra T-815 6 × 6 chassis has also been used for a number of applications including use as an artillery fire-control system for use at battery, regiment and brigade command levels.

Other versions include an air defence command post vehicle that is used by Slovakia and the MOKYS communications system that is also used by Slovakia.

ZTS 155 mm/45-calibre Zuzana (8 × 8) self-propelled artillery system from the rear deployed in firing position and showing stabilisers lowered (ZTS) 1403743

Tatrapan (6 × 6) armoured all-terrain vehicle being used in the command post role with the 155 mm/45-calibre self-propelled gun-howitzer Zuzana 0121598

Sipox 8 × 8 chassis

More recently, Tatra Sipox has developed a new 8 × 8 chassis based on their earlier 8 × 8 series. This is powered by a MAN D 2876 LF 02 liquid-cooled diesel engine developing 453 hp with an Allison TC-551 clutch with hydraulic torque converter and an Allison HD 4560 PR transmission.

The driver is seated in a fully enclosed and armour protected compartment at the front of the vehicle. Bulletproof windows are provided to the front and sides and there is an entry hatch above his position.

The basic chassis dimensions are:

Weight: 17,000 kg
Length: 10.4 m
Width: 2.5 m
Height: 2.05 m
Ground clearance: 0.4 m

Ammunition resupply vehicle

ZTS have also completed the prototype of an ammunition resupply vehicle based on a Tatra (8 × 8) chassis that has a fully armoured cab at the front. This carries a total of 120 155 mm projectiles and charges and has a maximum weight of 30,562 kg. This also features a power pack consisting of a MAN diesel coupled to an Allison automatic transmission.

Skoda 35 mm SPAAG

This was only a concept and never reached the prototype stage.

Specifications

	ZTS 155 mm Zuzana
Dimensions and weights	
Crew:	4
Length	
overall:	12.97 m
hull:	9.89 m
Width	
overall:	3.015 m
Height	
overall:	3.530 m
to turret roof:	3.30 m
Ground clearance	
overall:	0.41 m
Track	
front:	2.044 m
rear:	1.988 m
Wheelbase	1.65 m + 3.07 m + 1.45 m

	ZTS 155 mm Zuzana
Weight	
combat:	28,600 kg
Mobility	
Configuration	
running gear:	wheeled
layout:	8 × 8
Speed	
max speed:	80 km/h
Range	
main fuel supply:	750 km
Fuel capacity	
main:	500 litres
Fording	
without preparation:	1.4 m
Gradient:	58%
Side slope:	27%
Vertical obstacle	
forwards:	0.6 m
Trench:	2.00 m
Turning radius:	14.0 m
Engine	Tatra 3-930-52, V-12, diesel, 361 hp
Gearbox	
forward gears:	20
reverse gears:	4
Steering:	power assisted, front four wheels
Tyres:	15.00 - 21 T03
Electrical system	
vehicle:	26 V
Firepower	
Armament:	1 × turret mounted 155 mm howitzer
Armament:	1 × roof mounted 12.7 mm (0.50) NSV machine gun
Armament:	6 × turret mounted 81 mm smoke grenade launcher (2 × 3)
Ammunition	
main total:	40
charges:	40 estimated
Main weapon	
calibre length:	45 calibres
Rate of fire	
sustained fire:	1 rds/min
rapid fire:	6 rds/min
Turret power control	
type:	hydraulic/manual
Main armament traverse	
angle:	120° (60° left/60° right)
Main armament elevation/depression	
armament front:	+70°/-3.5°
Survivability	
Night vision equipment	
vehicle:	yes
NBC system:	yes
Armour	
hull/body:	steel
turret:	steel

Status
Production as required. In service with Cyprus (12 + 12) and Slovakia (48).

Contractor
ZTS Dubnica nad Váhom (Slovak Republic). KERAMETAL Joint-Stock Company (Slovak Republic) (marketing).

ZTS 152 mm self-propelled gun-howitzer Dana

Development
The Dana 152 mm self-propelled gun-howitzer was developed in the late 1970s by ZTS to meet the operational requirements of the Czech Army for a highly mobile self-propelled artillery system.

Dana is based on automotive components of the well-known Czech Tatra 815 VP 31 29265 8 × 8.1R truck which has a good demonstrated cross-country capability.

Following trials with prototype systems, the Dana was accepted for service with the Czech Army and first production models were issued to the Kiev Artillery Battalion Jan Zizka of Trocnov based at Tabor, Southern Bohemia, in 1981.

The Czech Army designation for the system is the vzor 77 self-propelled howitzer Dana.

152 mm Dana self-propelled gun-howitzer deployed in the travelling system with ordnance in travel lock (Michael Jerchel) 0106068

By early 1994, total production of this system, for the home and export markets, had reached over 750 units. As far as is known there has been no recent production of the 152 mm Dana system.

This system, and further developments of this system, are now manufactured and marketed by ZTS Dubnica nad Váhom of the Slovak Republic.

Production of the 152 mm self-propelled gun-howitzer Dana has now been completed and ZTS is now concentrating all of its marketing activities on upgraded versions with a 155 mm/45-calibre ordnance. This is already is service with Cyprus and Slovakia.

According to the United Nations, Slovakia delivered a batch of 12 152 mm self-propelled gun-howitzer Dana to Georgia in 2006 and these took part in combat operations in mid-2008.

According to the United Nations Arms transfer lists, in 2009 Slovakia transferred 26 Dana systems to the Czech Republic.

It is possible that these were for re-export in the future. This contract is not reflected the table at the end of this entry.

In late 2009 it was stated that Poland would deploy a battery of five 152 mm self-propelled gun-howitzer Dana with the International Security Assistance Force (ISAF).

These have been upgraded with a recently developed battery level variant of the WB Electronics Topaz battalion level command-and-control system.

Description
The Dana 152 mm SPH is based on automotive components of the Tatra 815 (8 × 8) cross-country truck chassis with the driver's compartment at the front, fully enclosed turret in centre and the diesel engine compartment at the rear.

All three compartments are armoured and provide protection against small arms fire and shell splinters.

The combat weight of the 152 mm Dana SPH depends on a number of factors.

Weight with a full load of fuel is quoted as 25,100 kg. When loaded with a total of 40 × 152 mm rounds of ammunition it has a combat weight of 28,100 kg and when loaded with a total of 60 × 152 mm rounds of ammunition it has a combat weight of 29,250 kg.

The maximum road speed also depends on the number of rounds of ammunition being carried.

For example when 40 rounds of 155 mm ammunition are carried maximum road speed 80 km/h and when the maximum load of 60 rounds of 155 mm ammunition are carried maximum road speed is reduce to 70 km/h.

A central tyre pressure regulation system is fitted which allows the driver to change the tyre pressure to suit the type of terrain being crossed. The tyre pressure can be changed when the vehicle is on the move. Power steering is provided for the front four wheels.

The driver is seated at the front of the vehicle on the left with the vehicle commander to his right. Both are provided with a windscreen to the front and when in a combat area this is protected with two shutters, vision to the front then being maintained through two vision blocks in each shutter.

The commander and driver are each provided with a forward-opening roof hatch and a vision block to give observation to the sides of the vehicle. Four circular firing ports are provided, one in each side of the compartment and one either side of the shutters.

In addition to operating the communications system, the commander also has access to the night sight.

The driver operates the turret unlocking system as well as lowering or raising the stabilisers for firing.

The 152 mm ordnance is fitted with a muzzle brake and fires the same ammunition of the separate loading type, projectile and charge, as the Russian 152 mm 2S3 tracked self-propelled artillery system, as well as locally produced ammunition.

The breech mechanism is of horizontal wedge type with the empty cartridge cases being ejected as the breech opens. The cartridge cases are then removed by means of a moving belt.

The actual 152 mm ordnance has a overall length of 5.58 m with the recoil length being a maximum of 1,300 mm and a minimum of 750 mm.

The Dana is fitted with a fully automatic loading system, which loads the projectile and then the charge using an automatic rammer. The weapon can be loaded at all angles of elevation.

Dana 152 mm self-propelled gun-howitzer modernised by Excalibur Army to the latest Dana-M1 CZ Modernisation standard with ordnance at maximum elevation and showing steering on front wheels (Excalibur Army) 1452912

The automatic loading system is hydraulic and is controlled by electrical elements and the gunner can select either single shots or fully automatic fire.

The system is ready to fire within two minutes of coming to a halt and can move off again one minute after the last round is fired.

To provide a more stable firing platform, three hydraulic jacks are lowered to the ground, one at the rear and one either side between the second and third roadwheels.

Optical systems fitted include a ZZ-73 rocking bar sight collimated with a PG1 M-D telescope for indirect fire and an OP 5-38-D sight for direct fire. All the sights are in the left side of the turret with two openings being provided in the turret front as well as a roof-mounted periscope.

Turret traverse is limited to 225° to each side due to cables, so no full traverse through 360° is possible.

Firing is normally carried out over the frontal arc of 90°, 45° left and right.

The turret is in two parts with the 152 mm ordnance being mounted in the aisle between them. Each half of the turret is fully enclosed and is provided with access doors, roof hatches and vision devices. As the 152 mm weapon is outside the turret, no fumes can enter the cabin.

The left cabin houses the gunner and loader operator while the ammunition handler is in the right cabin and sets the fuzes of the 155 mm projectiles manually.

Before the weapon is fired for the first time, the breech has to be opened manually. This is achieved by one of the crew members in the left cabin using his hand which is encased in a rubber glove attached to the side of the turret.

On the right side of the turret roof is mounted the 12.7 mm NSV machine gun which can be used in the direct fire or air defence roles.

The automatic loading system enables 30 rounds of ammunition to be fired in 30 minutes or 90 rounds in one hour with the projectiles and charges stowed in magazines.

Maximum rate of fire with manual loading is two rounds a minute.

The projectiles and charges are rammed using a chain-driven rammer, which is located above the ordnance. Firing is accomplished using a foot pedal.

A total of 60 155 mm projectiles and charges is carried in the chassis and inside the turret. This limits the maximum road speed to 70 km/h. Approximately half the ammunition is stowed in the turret vertically, with the charges in the left cabin and the projectiles in the right cabin.

The standard HE projectile (EOF D-20) weighs 43.56 kg and has a maximum velocity of 693 m/s. This has a maximum range of 18,700 m. Using a HEAT projectile maximum muzzle velocity is 717 m/s. There is also an enhanced projectile, the Czech EOFd, with a maximum range of 20,080 m.

More recently the EKK projectile has been developed which is of the base bleed type and contains 42 fragmentation bomblets, each of which are claimed will penetrate 100 mm of conventional steel armour. This projectile has a maximum range of 28,230 m.

Minimum range, at an elevation angle of +70° and using charge 6, is 4,600 m.

Standard equipment includes an NBC system which is provided with filter ventilation and an air conditioning system which distributes fresh air to both the cab and the turret.

Pressure for the hydraulic system of the self-loading system as well as for weapon elevation and turret traverse is supplied by a pump driven from the main diesel engine, although manual back-up systems are provided for all powered elements.

Variants

152 mm ShKH Ondava

This is a further development of the Dana and has a 152 mm 47-calibre ordnance which fires a new high-explosive projectile with a base bleed unit which enables a maximum range of 32,000 m to be achieved.

N-22 low-level radar system based on armoured Tatra 815 (8 × 8) chassis in the travelling configuration (Michael Jerchel) 0106067

Onboard ammunition supply is only 40 projectiles and charges. As far as is known, this system remains at the prototype stage. It is no longer being marketed.

Excalibur Army Dana-M1 CZ Modernisation

Excalibur Army of the Czech Republic has developed and tested an upgrade package for the 152 mm self-propelled gun-howitzer Dana which is referred to as the Dana-M1 CZ Modernisation.

This features a new digital computerised Fire-Control System (FCS) which also allows the platform to be integrated into an overall C4I system.

The new computerised FCS includes an inertial navigation system, gun commander's ballistic computer, gun commander's smart terminal and communications system.

The overall lay out of the crew compartment at the front of the vehicle has been redesigned and a vehicle blackout lighting system and reversing camera have been installed.

The cab is also provided with an independent air conditioning system and heating system and new crew seats have been fitted.

The steering system has been upgraded and the T3-930 diesel engine has been upgraded with new turbochargers and an intercooler have been fitted which is claimed to provide enhanced torque and power curve.

The company is also marketing a new 152 mm long range projectile and charge system called the Long Distance Shell DN1 CZ.

This can be fired from existing 152 mm Dana systems without any modifications and consists of the projectile 152 mm OFdDV (High Explosive Extended range – Base Bleed) and the associated cartridge case P740 which is stated to have a maximum range of up to 25.50 km.

A total of 34 rounds of ready use ammunition are carried plus 24 additional rounds in reserve.

The 12.7 mm machine gun is retained on the turret roof and this is provided with a total of 300 rounds of ready use ammunition.

155 mm/45-calibre Zuzana

This is being marketed by ZTS and is fitted with a 155 mm/45-calibre ordnance and is currently in service with Greece and Slovakia. Details are provided in a separate entry in *IHS Jane's Land Warfare Platforms: Artillery and Air Defence*.

Export marketing is now being concentrated on this new model.

155 mm/52-calibre Zuzana A1

This is the latest development and is a new design with a turret armed with a 155 mm/52-calibre ordnance integrated onto a new 8 × 8 chassis. Details of this system are provided in a separate entry in *IHS Jane's Land Warfare Platforms: Artillery and Air Defence*. As of March 2013 this remained at the prototype stage.

Tatrapan ACV

In 1993, the prototype of an armoured command post based on the Tatra T 815 (6 × 6) chassis was completed. Full details of this model, a small quantity of which have been built to support the new ZTS 155 mm self-propelled gun-howitzer Zuzana (8 × 8) systems now in service with Slovakia, are given in a separate entry.

MUR-20 EW system

The MUR-20 is a mobile system for the identification of radar systems in the forward battlefield area, for example ELectronic INTelligence/Electronic Support/Surveillance Measures (ELINT/ESM). Typically three or four MUR-20 systems would be used together. It is based on an armoured Tatra 8 × 8 chassis with automatic levelling and an antenna mast.

RADWAR N-22 low-level surveillance radar

This is mounted on a similar Tatra 815 8 × 8 chassis and has a fully enclosed armoured cab in the centre and an elevatable mast at the rear. It is used to provide target information to low-level air defence systems such as the Russian 9K33 Osa (NATO SA-8 'Gecko') self-propelled surface-to-air missile system. Prime contractor for the N-22 is the RADWAR company of Poland.

Specifications

	Dana
Dimensions and weights	
Crew:	5
Length	
overall:	11.156 m
Width	
overall:	3.00 m
Height	
to turret roof:	2.85 m
axis of fire:	2.41 m
Ground clearance	
overall:	0.41 m
Track	
front:	2.044 m
rear:	1.988 m
Wheelbase	1.65 m + 3.07 m + 1.45 m
Weight	
combat:	29,250 kg
Mobility	
Configuration	
running gear:	wheeled
layout:	8 × 8
Power-to-weight ratio:	13.74 hp/t
Speed	
max speed:	80 km/h
Range	
main fuel supply:	740 km
dirt road:	650 km
Fording	
without preparation:	1.4 m
Gradient:	60%
Side slope:	30%
Vertical obstacle	
forwards:	0.6 m
Trench:	2.00 m
Turning radius:	12.8 m
Engine	Tatra 2-939-34, air cooled, diesel, 345 hp
Gearbox	
forward gears:	20
reverse gears:	4
Steering:	power assisted, front four wheels
Tyres:	15.00 - 21 T03
Firepower	
Armament:	1 × turret mounted 152.4 mm howitzer
Armament:	1 × roof mounted 12.7 mm (0.50) NSV machine gun
Ammunition	
main total:	60 (× 155 mm)
Rate of fire	
sustained fire:	4.3 rds/min
rapid fire:	5 rds/min
Turret power control	
type:	hydraulic/manual
Main armament traverse	
angle:	450° (225° left/225° right)
Main armament elevation/ depression	
armament front:	+70°/-4°
Survivability	
Night vision equipment	
vehicle:	yes
NBC system:	yes
Armour	
hull/body:	steel
turret:	steel

Status

Production complete. In service with the countries listed in the table.

Country	Quantity
Czech Republic	57
Georgia	12 (from Slovakia in 2006)
Libya	80 (status uncertain due to conflict in 2011)
Poland	86
Slovakia	131

Contractor

ZTS Dubnica nad Váhom (Slovak Republic).

South Africa

Rheinmetall Defence RWG-52 155 mm/ 52-calibre self-propelled artillery system

Development

In 2010 Rheinmetall Defence released some details of their Rheinmetall Wheeled Gun 52 (RWG-52) which it has developed since 2008 for the international market using company funding.

The RWG-52 is based on the same design concept as the G6 Denel Land Systems 155 mm/45-calibre Self-Propelled (SP) artillery system which is in service with the South African National Defence Forces.

The G6 (6 × 6) with a 155 mm/45-calibre ordnance was developed 30 years ago. Details of this system, which is still being marketed, are provided in a separate entry in *IHS Jane's Land Warfare Platforms: Artillery and Air Defence*.

The first prototype of the RWG-52 was completed in South Africa in 2009 followed by mobility and firing trials.

The RWG-52 is also referred to as the Nashorn but as of March 2013 remained at the prototype stage.

Description

The Rheinmetall Defence RWG-52 features a brand new 6 × 6 chassis that has been re-designed and built by Industrial Automotive Design (IAD) in South Africa.

In February 2013 it was announced that the South African Paramount Group had taken over the Industrial Automotive Design company.

This has been fitted with a new fully autonomous turret called RT-52 that is a further development of the original turret with a number of major modifications.

The most significant of these is the installation of the complete 155 mm/52-calibre ordnance used in the combat proven PzH 2000 SP artillery system currently in service with Germany (185), Greece (24), Italy (70) and the Netherlands (57).

PzH 2000 first saw operational use in Afghanistan with the Royal Netherlands Army (RNLA and these were withdrawn in 2010. The RNLA PzH 2000 were replaced by seven German Army PzH 2000 systems in 2010.

The 155 mm/52-calibre ordnance has a 23-litre chamber. A total of 40 × 155 mm artillery projectiles and associated modular charges are carried in two magazines in the rear of the turret. The 155 mm projectiles and charges are loaded automatically with a manual back up mode provided.

Maximum range depends on projectile/charge combination but it is 42 km when firing a High-Explosive Extended-Range Full-Bore Base Bleed (HE ERFB BB) projectile with a sustained rate of fire of six rounds a minute.

RWG-52 can carry out multiple round simultaneous impact (MRSI) fire missions and takes only 30 seconds to come into or come out of action which makes it highly survivable from counterbattery fire.

Four independent hydraulic stabilisers are lowered to the ground, lifting the vehicle up when firing, thus providing for a very stable firing platform.

RWG-52 is fitted with an automatic gun laying and navigation system featuring a muzzle velocity radar and Inertial Navigation Unit (INU) mounted on top of the rear end of the ordnance, feeding information to the on board fire-control system.

Full combat weight is currently being quoted a 48 tonnes with the Deutz 523 hp six-cylinder diesel coupled to a ZF 6-speed automatic transmission giving it a maximum road speed of up to 80 km/h and a road range of up to 700 km.

RWG-52 has permanent 6 × 6 drive with a Central Tyre Inflation System (CTIS) fitted as standard as well as Run-Flat Inserts (RFI) on each wheel.

It normally has a crew of three or four people plus the driver who is seated in a well protected position at the front of the vehicle between the first wheels.

The all welded steel armour hull and turret provides ballistic protection against 7.62 mm armour-piercing attack through a full 360° and protection against 14.5 mm ammunition to 30° either side of the front.

An Environmental Control System (ECS) consisting of an air conditioning system and NBC system are standard for the driver and crew compartments. Mine protection is provided against the Russian TM-46 anti-tank mine or equivalent.

RWG-52 (6 × 6) SP artillery system in travelling configuration clearly showing the 155 mm/52-calibre ordnance in travel lock (Rheinmetall) 1401342

RWG-52 (6 × 6) SP artillery system deployed in the firing position and carrying out a fire mission (Rheinmetall) 1401343

A total of 8 × 76 mm smoke grenade launchers are fitted to the front of the turret, four either side, while a bracket is provided on the left hand top hatch to mount a machine gun as secondary armament.

Excalibur 155 mm PGM

The standard G6 155 mm/45-calibre SP artillery system has successfully fired the US Raytheon 155 mm Excalibur precision guided munition which could also potentially be fired from the RWG-52 155 mm/52-calibre SP artillery system.

Variants

Although the turret is installed on a wheeled chassis for the Indian trials, it could also be integrated onto a full tracked chassis such as a stretched T-72, T-90 or Arjun MBTs used by the Indian Army. This version is referred to as the RTG-52 (Rheinmetall Tracked Gun 52-calibre).

Specifications

Not available.

Status

Prototype undergoing trials.

Contractor

Rheinmetall Defence.

Denel Land Systems G6-52 155 mm self-propelled artillery system

Development

During the IDEX 2003 defence exhibition held in the United Arab Emirates in March 2003, the South African company of Denel Land Systems finally revealed its new 155 mm/52-calibre G6-52 self-propelled artillery system.

The 155 mm/52-calibre G6-52 self-propelled artillery system was under development for over 10 years and will now replace the current combat proven 155 mm/45-calibre G6 self-propelled artillery system offered on the export market.

G6-52 is a further development of the original combat proven 155 mm G6, which has a 155 mm/45-calibre ordnance and is in service with Oman (24), South Africa (43) and United Arab Emirates (78).

Denel Land Systems continue to customise design of its latest generation G6-52 155 mm self-propelled artillery system.

Denel Land Systems has now fired over 8,000 155 mm projectiles of all types with various charge systems. Maximum firing range of the G6-52 depends on the ambient temperature and the projectile/charge combination.

The system was originally marketed with a 23- or 25-litre chamber but all marketing is now concentrated on the 23-litre chamber which meets the NATO Joint Ballistic Memorandum of Understanding (JBMoU).

As well as a complete family of South African Rheinmetall Denel Munitions 155 mm projectiles and charges, the G6-52 can also fire more advanced projectiles such as the French/Swedish BONUS top attack smart projectile and the Russian Krasnopol-M laser-guided projectile which has already been successfully test fired.

According to Denel Land Systems, with the appropriate modifications the G6-52 could also fire the recently deployed Raytheon Excalibur 155 mm precision-guided munition that has already been used in Afghanistan.

When compared to the earlier G6 155 mm/45-calibre, the latest G6-52 has a significant number of advantages including a reduction in turret crew requirements, higher rate of fire and an increased firing range which improves its survivability against counterbattery fire.

New production G6-52 would have a similar layout to earlier systems but feature a new and lighter diesel power pack and larger tyres which would be fitted with a central tyre pressure regulation system to allow the driver to adjust the tyre pressure to the terrain being crossed.

Description

The overall layout of the latest G6-52 is similar to the current in-service G6 with the driver in the fully enclosed compartment at the front, diesel power pack in the centre and the crew compartment and turret at the very rear of the hull.

The hull and turret is of all-welded steel armour. This protects the occupants from 7.62 mm armour-piercing attack through a full 360° fired at a range of 30 m. Over the frontal arc, protection is provided against attack from 14.5 mm armour piercing fired from a range of 1,000 m.

The G6-52 can withstand the detonation of a Russian TN-46 anti-tank mine or an equivalent landmine under any wheel station.

Denel Land Systems are quoting the following maximum range for the 155 mm/52-calibre ordnance used with the Rheinmetall Denel Munitions M64 Modular Charge System:

Projectile	Maximum range
M9 Extended-Range Full Bore (Boat Tail):	32.7 km
M9 Extended-Range Full Bore (Base Bleed):	44.2 km
Velocity enhanced Long range Artillery Projectile:	58 km

Late in 2003 the following ammunition was qualified for use with M64 BMCS: pyrotechnic carrier (red phosphorous, illuminating, screening smoke and bi-spectral screening smoke); submunition; field exchangeable base bleed up to zone 6; and High-Explosive and HE-VLAP up to zone 5.

It was decided in 2003 not to qualify HE and HE-VLAP with a TNT filling, and instead an Insensitive Munition (IM) filling will be used.

The M9000 series projectiles have a number of other improvements for use with the G6-52 including increased strength, increased diameter and reduced tolerance on nubs and a new double driving band.

The new cluster projectile carries 42 bomblets each of which can penetrate 120 mm of conventional steel armour and fitted with a self-destruct fuze.

A new base bleed unit has also been developed and type classified that features an improved body, male thread for improved propellant volume and a six-hole nozzle for improved bleed.

In earlier G6 systems ammunition was manually loaded, but the new G6-52 has a fully automatic ammunition handling system that increases the rate of fire to 8 rds/min. This is accomplished by means of computer programmed ammunition carousels located inside the rear of the turret. The left carousel holds the projectiles while the right carousel holds the modular charges.

As well as increasing the rate of fire this feature enables the crew of the G6-52 to be reduced. According to the manufacturer, the system takes 10 minutes to be reloaded. There are 80 percussion tubes in four primer magazines.

For continuous bombardment from fixed firing positions, ammunition may be fed from the rear of the gun using two replenishment arms, one for 155 mm projectiles and one for charges.

The turret is fitted with an integrated diesel power unit that is located between the projectile and charge magazines. This allows the complete turret system to be operated with the main diesel engine shut down.

There is also an automated ammunition inventory system that records and updates the status of the inventory as each projectile is fired. A total of 48 rounds of 155 mm ammunition are carried, 40 in the turret and 10 in the vehicle. Fuze setting is also automated.

Using Multiple Rounds Simultaneous Impact (MRSI) techniques the G6-52 can deliver up to six rounds at a range of 25 km (with a 10 second impact tolerance) using the proven African Defence Systems AS2000 artillery target engagement system.

Highly accurate laying and navigation is provided by the onboard Ring Laser Gyro and with a GPS secondary navigation aid. The fully automatic laying system achieves accuracy better than 1 mil RMS.

According to the company, the G6-52 can come into action and fire its first round within 45 seconds and come out of action in 30 seconds. This improves its survivability against counterbattery fire.

The G6-52 also has a direct fire range that is quoted as being a minimum of 500 m and a maximum of 3,000 m.

G6-52 self-propelled artillery system with 155 mm/52-calibre ordnance at maximum elevation (Denel) 0572179

Denel Land Systems G6-52 155 mm self-propelled artillery system redeploying to a new firing position (Denel Land Systems) 1296044

Prime contractor for the original G6 155 mm/45-calibre self-propelled artillery system chassis was the now BAE Systems Land Systems South Africa.

This has permanent 6 × 6 drive, maximum road speed being up to 80 km/h with an operating range of 700 km. Standard equipment includes an automatic tyre-inflation system, run-flat inserts and an NBC system.

The new G6-52 is also being marketed as a complete artillery system that also includes the ammunition suite (projectile, charge and fuze), Seeker observation drone developed by the Denel Aerospace and African Defence Systems AS2000 fire-control system.

Variants

The turret fitted to the G6-52 is identical to that of the T6-52 turret system that has already been successfully demonstrated on a T-72 and Indian Arjun MBT hull. The turret has an integrated APU to allow the main platform engine to be shut down.

The towed G5-52 gun is ballistically identical to the G6-52 (6 × 6) as is the T5 (8 × 8) truck-mounted system that was originally called Condor.

RWG-52 155 mm/52-calibre self-propelled artillery system

In 2010 Rheinmetall Defence released some details of their Rheinmetall Wheeled Gun 52 (RWG-52) which it has developed since 2008 for the international market using company funding. Details of this are provided in a separate entry in *IHS Jane's Land Warfare Platforms: Artillery and Air Defence.*

This is also referred to as the RWG-52 Nashorn and as of March 2013 remained at the prototype stage.

Specifications

	G6-52
Dimensions and weights	
Weight	
combat:	49,000 kg
Mobility	
Configuration	
running gear:	wheeled
layout:	6 × 6
Speed	
max speed:	80 km/h
Range	
main fuel supply:	700 km
Suspension:	fully independent swing arm with torsion bars
Firepower	
Armament:	1 × turret mounted 155 mm howitzer
Ammunition	
main total:	50
Main weapon	
calibre length:	52 calibres
maximum range:	32,700 m (ERFB-BT)
maximum range:	44,200 m (ERFB-BB)
maximum range:	58,000 m (VLAP)
Rate of fire	
rapid fire:	8 rds/min
Turret power control	
type:	powered/manual
Main armament traverse	
angle:	80° (40° left/40° right)
Main armament elevation/depression	
armament front:	+70°/-2°
Survivability	
Night vision equipment	
vehicle:	yes
NBC system:	yes
Armour	
hull/body:	steel
turret:	steel

Status

Development complete. Ready for production.

Contractor

Denel Land Systems.

Denel Land Systems G6 155 mm self-propelled gun-howitzer

Development

In the 1960s, the South African Artillery deployed three main artillery weapons, the UK 25-pounder Field Gun (G1), the 5.5-inch Medium Gun (G2) and the Canadian-built Sexton 25-pounder self-propelled gun.

During early South African operations in Angola the poor range of South African artillery compared to opposing Russian-supplied 122 mm D-30 and 130 mm M-46 towed artillery weapons and BM-21 122 mm (40-round) multiple rocket launchers led to the fielding of the Denel Land Systems towed 155 mm G5 gun-howitzer and the 127 mm (24-round) Valkiri multiple rocket system.

While the towed G5 is perfect for many South African Artillery requirements, it does not have the cross-country mobility of the now BAE Systems Land Systems South Africa Ratel (6 × 6) infantry fighting vehicle.

The South Africans weighed the options of wheels or tracks for their new self-propelled artillery system and came down firmly in favour of wheels. The possible areas of operation of the G6 are likely to be some distance from their main bases, perhaps up to 1,000 km, and tracked vehicles have distinct disadvantages over long distances.

Fuel consumption of a wheeled vehicle is a lot less than a tracked vehicle so less logistic support is required. The decision was therefore taken to choose a wheeled artillery system.

The G6 prototype was completed in 1981 and shown for the first time in public in September 1982. By late 1986, four complete G6s had been built plus one additional turret. Of these four, two were considered as prototypes and two advanced development vehicles, for they incorporated many improvements as the result of user trials and changes in operational requirements.

The first advanced development vehicle was completed early in 1984 and the second during 1986.

There were four engineering development models completed during 1987 for final troop trials and confirmation of the modifications carried out on earlier vehicles.

The first production G6s were completed during 1988, with Denel Land Systems being the prime contractor. BAE Systems Land Systems South Africa delivered the hull to Denel Land Systems who integrates the turret and delivers the complete system to the user.

The production G6s were the first self-propelled howitzers with a full autonomous laying and navigation capability.

As with the earlier G5, the G6 is a complete artillery system and includes not only the vehicle but also the ammunition system, meteorological station, muzzle velocity analyser, fire-control system and a special artillery helmet radio communication system.

Denel Land Systems G6 155 mm 45-calibre (6 × 6) self-propelled artillery system deployed in the firing position (Denel Land Systems) 1133154

Using the new VLAP projectile and the M64 bi-modular charge system, the Denel Land Systems G6 155 mm 45-calibre G6 (6 × 6) self-propelled artillery system has a maximum range of 53.6 km 0096018

Denel Land Systems 155 mm/45-calibre G6 self-propelled artillery system in travelling configuration with ordnance in travel lock (IHS/Patrick Allen) 1441139

The G6 is deployed in batteries of six with the South African Artillery, each regiment having three batteries.

In 1990, Abu Dhabi, part of the United Arab Emirates, placed an order for the supply of 78 G6 155 mm self-propelled howitzers with the first of these being delivered early in 1991. All have now been delivered and are organised into three regiments each of 24 weapons, with the remaining vehicles being used for training and war reserve. A comprehensive long-term product support package was also successfully delivered and implemented.

Early in 1994, Oman placed an order for 24 G6 155 mm self-propelled howitzers, spares, logistic support, ammunition and a training package. All have now been delivered to Oman.

Late in 1996, it was disclosed that the South African National Defence Force had a total of 43 G6 systems in service under the designation GV6.

There has been no production of the G6 155 mm/45-calibre system for more than 15 years.

Marketing of G6 continues and it is expected that hulls for future production systems would be manufactured by the South African company of Industrial Automotive Design.

In February 2013 it was announced that the South African Paramount Group had taken over the Industrial Automotive Design company.

Description

The hull of the G6 is all-welded steel armour, with protection against a variety of threats including small arms fire and shell splinters. The G6 has a double armoured floor for increased protection against anti-tank mines.

The hull is essentially divided into three compartments, driver at the front, power pack in the centre and fighting compartment at the rear.

The driver sits at the front between the front two roadwheels and, as there is no hull armour above these, any blast resulting from the vehicle running over a mine will be vented upwards and away from the driver.

To the front and sides of the driver are large bulletproof windows which afford excellent visibility through the front in excess of 180°. If required, the central front window can be covered by an armoured shutter, hinged at the bottom, simply by activating a lever. When this is raised the driver observes through a periscope.

The driver enters and leaves the vehicle via a roof hatch that opens to the left. Communication between the driver and the gun crew is via the gun's intercom system.

Steering is power assisted and the G6 now has permanent 6 × 6 drive (prototypes had the option of 6 × 4 with the front axle being disengaged). The driver can select either fully automatic or semi-automatic drive.

Forward of the driver's compartment is a lateral wedge-shaped box that doubles as a container for 155 mm projectiles and as a bush-clearing device capable of cutting through trees and shrubs.

The diesel engine compartment is to the rear of the driver's position with air inlet louvres in the front of the engine venting through the roof and the air outlet louvres through the sides, all of which can be opened to allow access to the diesel engine for maintenance. The engine compartment is fitted with a fire detection and suppression system.

The G6 is powered by an air-cooled diesel developing 518 hp, it is coupled to an automatic transmission with a torque converter giving a total of six speeds forward and one in reverse.

Drive shafts take the power from the transmission via differentials to each roadwheel station.

The crew compartment, with the turret above, has seats for the commander, gun layer/navigator, breech operator, loader and an ammunition handler. There is an escape hatch in the floor. The turret crew enter via the main door in the rear right-hand side of the turret or via two roof hatches in the turret, one either side.

The fully enclosed armoured steel turret has firing ports either side, with vision blocks above for close-in defence. Standard rifles and grenades are carried for close defence.

Mounted either side of the forward part of the turret is a bank of four 81 mm electrically operated grenade launchers which fire smoke grenades.

The gun commander sits at the right front of the crew compartment. A Gun Management System (GMS) is provided to assist the commander with command, control, communication and information functions in and around the G6. The GMS contains a Gun Display Unit (GDU) situated at the commander station which is used for sorting and displaying navigation data, gunlaying data and fire orders received via radio data communication link to each G6 from a Battery Command Post (BCP). Data transmission is also performed when the gun is mobile, ensuring an in action time of one minute and out of action time of 30 seconds.

Navigation and gunlaying data is transmitted via serial communication to the Laying and Navigation System (LNS). Navigation data contains UTM map datum points, routes containing waypoints and danger points with selectable warning radii.

The GMS interfaces with a Muzzle Velocity Radar System (MVRS). The MVRS, which is mounted above the rear part of the 155 mm ordnance, calculates a predicted mean normalised muzzle velocity. This calculated value is obtained utilising a moving average of muzzle velocities of an already fired group of ammunition. This value is used during target laying data calculations.

The GMS also interfaces with a Thermal Warning Device (TWD) warning the crew of unsafe barrel and recoil oil temperatures. Digital and analogue values are displayed on a separate TWD display by the commander. The TWD is connected to the intercom system providing an audible warning. A barrel-cooling fan is provided to reduce recovery time in the event of unsafe temperatures being reached.

The GMS contains an Ammunition Display Unit (ADU) mounted on the left rear of the vehicle providing the ammunition replenishment crew outside the gun with ammunition preparation and gunlaying data ensuring silent operation on the deployment site.

Data transmission makes use of the UHF/VHF frequency-hopping system of up to 50 km in range or back-up voice UHF radio link of up to 10 km in range. The commander in turn is in constant communication with the crew of the gun by using an internal intercom system or external local VHF band system of up to 100 m in range with the crew carrying belt-mounted transceivers.

Safety during firing is ensured with the aid of firing safety interlocks. The gun can only be fired once the breech operator, gun leader and gun layer have depressed their safety/ready switches and the gun has been laid within 1 mil onto target.

The gun layer/navigator sits at the left front of the crew compartment.

A Laying and Navigation System (LNS) assists the gun layer with laying and navigation functions. The LNS, using a Ring Laser Gyro (RLG) inertial unit, provides an autonomous all-weather day and night fighting capability from unsurveyed sites with all time ready state navigation and gun orientation ability. Navigation and gun orientation is performed without the use of external reference points or signals.

An additional Fibre Optic Gyro (FOG) inertial unit can be provided as an option giving a full laying and navigation back-up capability ensuring continuous availability of the gun.

Both the RLG and FOG inertial units receive inputs from a Distance Transmitting Unit (DTU) mounted on the gearbox of the vehicle. An additional embedded Global Positioning System (GPS) input complements the navigation accuracy. Advanced filtering techniques are used to optimise gunlaying and navigation performance.

Navigation information and centre of arc is displayed for the driver ensuring orientation of the gun during deployment.

The LNS requires no scheduled service or recalibration during operation and no scheduled maintenance during storage of the gun. Cold start is reduced to 12 minutes for both inertial systems. The Mean Time Between Failures (MTBF) of the RLG inertial sensor is 7,000 hours compared to 1,000 hours of the previous mechanical gyro system used in previous models.

A fully automatic gunlaying capability in both traverse and elevation directions is provided. This automatic gunlaying system enables rapid and accurate relaying of the 155 mm ordnance between rounds at high rates of fire. Accurate gunlaying is activated by a single touch button on the laying and navigation touchscreen. Target data is downloaded with the data communication link from the BCP via radio to the GMS in the gun. The GMS in turn downloads the target data to the LNS ensuring ease of operation. Target data can also be keyed into the LNS via the touchscreen display.

Direct firing is done by the gun layer under the commander's command. A telescopic direct sight is provided for initial aiming onto a direct target by the gun layer. Automatic onboard calculations are done by the LNS utilising direct firing range tables. A laser range-finder is integrated with the LNS to provide

Table 1

Type	HE	HE BB	Smoke	Illum	RP	Leaflet	Submunition
Length (fuzed)	938 mm	958 mm	938 mm	938 mm	938 mm	938 mm	938 mm
Weight (nominal)	45.3 kg	47.7 kg	44.3 kg	44.3 kg	44.3 kg	44.3 kg	44.3 kg
Payload	8.7 kg	8.7 kg	10.9 kg	11.8 kg	±120 pellets	±3,500 message discs	13.72 kg

instant target range. The ordnance is aimed automatically on target utilising a single touch button on the laying and navigation touchscreen.

The 155 mm 45-calibre autofrettaged ordnance of the G6 is a modified version of that used in the Denel Land Systems towed G5. It has a single-baffle muzzle brake, a semi-automatic screw and swing breech mechanism and an electrical firing system. A steel fume extractor is fitted. The recoil system consists of a single buffer and single recuperator.

The turret can be traversed through 180° (90° left and right) under electrohydraulic control with manual controls provided for maintenance, although for practical purposes the firing arc of the turret is 40° left and right. The 155 mm ordnance can be elevated from –5° to +75° under electrohydraulic servo control.

Although a single-touch auto layer provides the main laying function, a digital hand controller is provided for manual elevating and traversing the ordnance. A Hydraulic Control Panel (HCP) provides hydraulic system status. The HCP also has a joystick providing additional back-up for traversing and elevating the ordnance during emergency conditions or during maintenance.

A semi-automatic hydraulic flick rammer is mounted on the left of the breech with the projectile being placed on the rammer tray by hand. This is then aligned with the breech and rammed. The rammer can be used at all elevation angles and is operated by the ammunition loader. Propelling charges are loaded directly into the breech by hand by the breech operator.

A rate of 4 rds/min is achieved when firing with the maximum charge and can be continued for a period of 15 minutes, after which chamber temperatures have to be monitored. A rate of 5 rds/min is achieved when firing charge 2. A burst capability of three rounds in 25 seconds can be obtained.

The 155 mm ammunition used with the G5 and G6 is of the ERFB type. Rounds available include HE, Smoke, Illuminating, RP, Submunition and Leaflet. The HE projectiles are forged from A151 9260 high-fragmentation steel with welded nubs. Filling may be TNH at the choice of the customer. The cargo rounds mentioned above are of the base ejection type. All rounds can be field-fitted with base-bleed. Full details of the projectiles are contained in Table 1. above

Prime contractor for the 155 mm family is the now Rheinmetall Denel Munition.

The Illuminating projectile has 1,000,000 candela for 120 seconds while the Smoke (SCR SMK) produces a smoke screen for 120 seconds from four canisters. The submunition round contains 42 bomblets and has an anti-personnel as well as a 120 mm armour-piercing capability. This round is sometimes referred to as the Cluster.

The charge system comprises five zones made up of three charges (combustible cartridge cases) and is optimised for use with 155 mm 45-calibre ballistic systems. Details are in Table 2.

Table 2

Charge	Muzzle velocity	Range (sea level)
1 M51A1	350 m/s	4.5 to 9.1 km
2 M52A1	483 m/s	8.7 to 13.4 km
2 + 1 increment	645 m/s	12.6 to 19.0 km
2 + 2 increment	795 m/s	17.7 to 25.4 km
2 + 2 increment (BB)	789 m/s	22.0 to 31.0 km
3 M53A2	897 m/s	21.6 to 30.0 km
3 M53A2 (BB)	895 m/s	27.5 to 39.3 km
3 M53A2 VLAP	895 m/s	50 km
Area coverage: ERFB BB 1,062 km^2, VLAP 1,700 km^2		

The fuzes used with the projectiles include the PD M841, the M9220 electronic time fuze and the M8513 proximity. The PD M841 and M8513 are used with HE. The M9220 is used with Smoke, Illuminating, RP, Submunition and Leaflet. The PD M841 fuze incorporates a rain/cloud desensitising device to replace the US PD M572/PD M739 fuze and is compatible with base bleed projectiles.

At 75 per cent of maximum range, the 50 per cent zone for dispersion (2PE) is 0.96 per cent of range in length and less than 2 mils in deflection.

All 155 mm projectiles and charges are stowed below the turret bearing ring. It takes three men 15 minutes to reload the G6 with 45 projectiles and 50 charges. Fuzed projectiles and charges are transferred from an ammunition support vehicle through the rear of the G6 on conveyor chutes.

The G6 can be fired from the wheels, but is more stable when the four outriggers are lowered hydraulically to the ground by the driver or commander, two at the rear and one between the first and second wheels either side.

The crew compartment is provided with an explosion detection and suppression system with four outlets for the suppressant around the turret. Blowout panels are provided in the hull rear and the stowage of all ammunition is below the turret to improve the survivability.

The driver's compartment is provided with forced ventilation with the option of air conditioning, while the crew compartment has full air conditioning. Temperatures of 25°C lower than ambient can be maintained at ambient temperatures of up to 55°C.

A 38 kW Turret Power Unit (TPU) is mounted in the turret bustle to charge batteries and drive the air conditioning system. The latter is fitted with an NBC filtration system functioning on an overpressure principle.

The production G6 has permanent 6 × 6 drive and diff-locks on all differentials for improved traction. It is fitted with a central tyre-pressure regulation system that enables the driver to adjust the tyre pressure to suit the type of terrain while on the move.

The G6 has an independent suspension with torsion bars, hydraulic shock-absorbers and bump stops all-round. The 21.00 × 25 tyres are fitted with run-flat elements.

The G6 is supported in action by an ammunition supply vehicle carrying 156 rounds in four ammunition units and support equipment.

The turret and driver air conditioning systems of the G6 have been upgraded to accommodate ozone friendly R134a refrigerant gas. Halon gas in the fire extinguishing system has been replaced by NAF SIII gas, these being ozone friendly products.

The G6 has a 24 V electrical system which includes 175 A batteries in the hull and 370 A batteries in the turret.

155 mm T6 turret

This was developed for the Indian Army competition for a 155 mm turret system that could be fitted onto the hull of the locally built T-72M1 chassis. It forms the basis for the latest Denel Land Systems G6-52 155 mm self-propelled artillery system that was launched in 2003. Details of this are provided in a separate entry in *IHS Jane's Land Warfare Platforms: Artillery and Air Defence*. As of March 2013 the T6 turret system had yet to enter production or service.

More range for G6

In 2001, Denel Land Systems announced that their 155 mm G6/45-calibre self-propelled artillery system could now fire accurately out to a range of 53.6 km.

According to the company, this makes the G6 the first fully qualified and in-service 155 mm/45 G5 gun and ammunition system breaching the 40 km range barrier by a considerable margin.

The 53.6 km range was achieved using the new Velocity enhanced Long range Artillery Projectile VLAP, which combines base bleed and rocket motor technology and the new M64 bi-modular charge system. The latter has also been fully qualified and is in quantity production for export. The new charge gives a maximum muzzle velocity of 910 m/s.

During extensive qualification firings in South Africa from a G6, probable errors in ranges were constantly better than 0.38 per cent. To make space for the base bleed and rocket assist elements, the VLAP carries less high-explosive. Its lethality is however still considerably higher than the older 155 mm M107 high-explosive projectile.

The VLAP can be used with the 155 mm G5 towed and G6 self-propelled artillery systems and has already entered production for use with an undisclosed export customer equipped with the G6 system. It is probable that the country concerned is the United Arab Emirates (UAE).

PRO-RAM projectile

In 2001, it was disclosed that South Africa was working on yet another new 155 mm long-range artillery projectile called the PROjectile RAM (PRO-RAM). This has an integrated ramjet propulsion system and will allow ranges of at least 70 km to be achieved. It will also be very accurate when fitted with a low-cost course correction module.

Extended-range G5 and G6

Late in 1992, it was revealed that the company had built the prototype of a 155 mm ordnance with a-calibre length of 52. It was used in development work which could be used to upgrade the G5 and G6 in the future, enabling their range advantage to be retained.

A number of developments is also under way on 155 mm ammunition in South Africa to give the user greater operational flexibility. Now available is the new M90 bi-modular charge system comprising identical modules of around 2.2 kg which can be used in 39, 45 and 52-calibre systems. A new 155 mm high-explosive Extended-Range Full-Bore – Base Bleed (ERFB-BB) projectile has been developed, containing 8.4 kg of high-explosive which gives a range of 41 km. When fired from a 155 mm 52-calibre ordnance a range of 43.5 km is achieved.

The base bleed unit is of the screw-on type and can be removed under field conditions if not required.

More recently, the Velocity-enhanced Long-range Artillery Projectile (VLAP) has been introduced which is a rocket motor-assisted 155 mm projectile combined with the base bleed unit. This extends the range of the 45-calibre G5, G6 and T6 systems to 50 km and the 52-calibre gun to 53.6 km.

This maximum range does depend on a number of factors including height when the fire mission is carried out and the ambient temperature.

UAE G6/M109L47 upgrade

The United Arab Emirates awarded Denel Land Systems a contract to upgrade their G6 and RDM Technology M109L47 to a common gunlaying and navigation integral fire-control system. It should be noted that RDM Technology ceased trading in 2004.

155 mm GV6

In 1996, it was stated that the first upgraded and retrofitted GV6 had been delivered to the South African Army. It is not known at present what this upgrade includes.

G6 M1A3

This is the latest model of the G6, which is being marketed in the Middle East. This retains the 155 mm/45-calibre barrel with a 1:20 twist rifling and 48 grooves 1.27 mm deep, which are common to 155 mm 39 and 52-calibre tubes that meet the NATO Joint Ballistic Memorandum of Understanding, as well as a 23-litre chamber.

The original glass fibre fume extractor has been replaced by a new steel fume extractor. Additional safety interlocks have been installed, as has an improved rammer, to increase rate of fire from 3 to 5 rds/min.

The Rheinmetall Denel Munition M64 bi-modular charge system and the ERFB-BB give a maximum range of 40.5 km. Firing the latest VLAP, a range of 53.6 km can be achieved.

It also fitted with the latest AS2000 fire-control system to give a four round MRSI capability. The commander has an African Defence Systems launcher management computer with a Windows based touch screen, while the gunner has a Denel WM205 gun data display and a new digital hand controller. The Selex Galileo FIN3110 RLG and integrated GPS sensor is retained.

The hull has been fitted with a three-level central tyre pressure regulation system. Other enhancements include a more powerful turret APU, improved insulation and a cool water dispenser.

As well as featuring improved reliability, these upgraded systems were also fitted with barrel temperature measurement, a barrel cooling fan on the cradle, improved air conditioning system and a backup laying system.

Excalibur 155 mm PGM

In 2012 it was announced that the standard G6 155 mm/45-calibre SP artillery system had successfully fired the US Raytheon 155 mm Excalibur precision guided munition (PGM) during trials in an undisclosed country in the Middle East.

A total of four standard production 155 mm Excalibur PGM were successfully fired from the 155 mm/45-calibre ordnance with a 23 litre chamber out to a maximum range of 38 km with all projectiles landing within 5 m of the target.

The range quoted suggests that it was probably the longer range M982 Block 1a-2 version of the Excalibur. The earlier Block 1a-1 being restricted to a maximum range of 28 km using a MACS Zone 4 charge.

Raytheon would not release details of where the trial took place but it is probable that this was the United Arab Emirates who deploys a fleet of 78 G6.

Rheinmetall Defence RWG-52

Details of this system developed by Rheinmetall Defence are provided in a separate entry in *IHS Jane's Land Warfare Platforms: Artillery and Air Defence*.

The Rheinmetall Defence RWG-52 features a brand new 6 × 6 hull that has been re-designed and built by Industrial Automotive Design (IAD) in South Africa.

This has been fitted with a new fully autonomous turret called RT-52 that is a further development of the original turret with a number of major modifications.

Specifications

	G6
Dimensions and weights	
Crew:	6
Length	
overall:	10.335 m
hull:	9.20 m
Width	
overall:	3.4 m
Height	
to turret roof:	3.3 m
to top of roof mounted weapon:	3.8 m
Ground clearance	
overall:	0.45 m
Track	
vehicle:	2.8 m
Weight	
standard:	42,500 kg
combat:	47,000 kg
Mobility	
Configuration	
running gear:	wheeled
layout:	6 × 6
Power-to-weight ratio:	11.17 hp/t
Speed	
max speed:	90 km/h
Range	
main fuel supply:	700 km
Fuel capacity	
main:	700 litres
Fording	
without preparation:	1 m
Gradient:	40% (est.)
Side slope:	30%
Vertical obstacle	
forwards:	0.50 m
Trench:	1.00 m
Turning radius:	12.5 m
Engine	air cooled, diesel, 518 hp
Gearbox	
type:	automatic
forward gears:	6
reverse gears:	1
Clutch	
type:	torque converter
Tyres:	21.00 × 25 with run flat inserts and CTIS
Suspension:	independent with torsion bars and hydropneumatic damping system
Electrical system	
vehicle:	24 V
Firepower	
Armament:	1 × turret mounted 155 mm howitzer
Armament:	1 × roof mounted 12.7 mm (0.50) M2 HB machine gun
Armament:	8 × turret mounted 81 mm smoke grenade launcher (2 × 4)
Ammunition	
main total:	45
charges:	50
Main weapon	
calibre length:	45 calibres
Turret power control	
type:	powered/manual
Main armament traverse	
angle:	90° (50° left/40° right)
Main armament elevation/depression	
armament front:	+75°/-5°
Survivability	
Night vision equipment	
vehicle:	optional
NBC system:	yes
Armour	
hull/body:	steel
turret:	steel

Status

Production as required. In service with the countries listed in the table.

Country	Quantity	Comment
Oman	24	delivery from 1996
South Africa	43	deliveries complete
United Arab Emirates	78	delivery from 1991

Contractor

Denel Land Systems.

Denel Land Systems 155 mm T5 Condor self-propelled artillery system

Development

The Denel Land Systems 155 mm/52-calibre Condor T5 self-propelled artillery system has been developed to meet the potential requirements of India.

Teamed with Denel Land Systems is the Indian company of Bharat Earth Movers Limited who are the local manufacturer for the Czech Republic Tatra 8 × 8 chassis on which the system is based.

Development of the Condor T5 is understood to have commenced in 2001, with the first prototype being completed in 2002. The system made its first public appearance at the Africa Aerospace & Defence exhibition held in Pretoria in September 2002.

Development of the baseline Denel Land Systems T5 Condor self-propelled artillery system is considered to be complete but as of March 2013 no sales of the system had been made but marketing continued.

Description

Condor essentially consists of a modified Tatra 8 × 8 truck chassis on the rear of which is mounted the upper part of the latest Denel Land Systems 155 mm G5-2000 towed artillery system.

The first model had a 155 mm/45-calibre ordnance while the second model had a 155 mm/52-calibre ordnance.

In addition to the complete ordnance of the turret used in the Denel Land Systems G5-52 155 mm/52-calibre self-propelled artillery system, some of the subsystems are identical on the Condor T5, including the breech mechanism, recoil and counter re-coil system, the cradle, crew communications, automatic laying system, fire-control system, primer loading system and the complete rammer system.

To provide a more stable firing platform, the Tatra 8 × 8 truck has three hydraulically operated stabiliser blades, one at the rear and one either side. The latter also incorporate steps to allow the crew to obtain access to the platform.

The standard two-door fully enclosed unarmoured Tatra cab is retained and this has seats for three people and to the rear of the cab is equipment stowage, crew and a quantity of ready use 155 mm ammunition.

The system has powered elevation and traverse with the elevation arc from –3° to +70° and with a traverse of 40° left and right. The system is also capable of engaging targets in the direct fire role.

The installation of laser gyro based automatic laying system reduces the time into action and also ensures quick and accurate re-laying to improve dispersion at high rates of fire. Maximum rate of fire is six rounds per minute.

When deployed in the firing position the system normally fires to the rear. In an emergency it can fire towards the front and in this case elevation is limited from +15° to +70° with a total traverse of 20°, e.g. 10° left and 10° right.

When fitted with the 155 mm/45-calibre barrel it has a push rammer and when fitted with a 155 mm/45-calibre barrel it has semi-automatic loading system.

The 155 mm/52-calibre system has an autonomous laying and navigation system as standard while the 155 mm/45-calibre system this is an option.

Denel Land Systems Condor self-propelled artillery system based on Tatra (8 × 8) chassis and fitted with 155 mm/52-calibre barrel and stabilisers deployed (Denel Land Systems) 1296045

Denel Land Systems Condor self-propelled artillery system on Tatra (8 × 8) chassis with 155 mm/45-calibre ordnance in travel lock and showing hydraulically operated stabilisers lowered at side and rear (Christopher F Foss) 1452913

Denel Land Systems Condor self-propelled artillery system based on Tatra (8 × 8) chassis and fitted with 155 mm/52-calibre barrel and stabilisers deployed (Denel Land Systems) 1296046

Ammunition is the same as that used on the in-service G5 155 mm/45-calibre self-propelled artillery system with Rheinmetall Denel Munition being the prime contractor for projectiles and charges.

Denel Land Systems are quoting the following maximum range for the 155 mm/52-calibre ordnance fitted to the T5 Condor when used with the Rheinmetall Denel Munitions M64 Modular Charge System:

Projectile	Maximum Range
M9 Extended-Range Full-Bore (Boat Tail):	32.7 km
M9 Extended-Range Full-Bore (Base bleed):	44.2 km
Velocity enhanced Long range Artillery Projectile:	58 km

Condor T5 with 155 mm/45-calibre system

Denel Land Systems have shown the truck mounted Condor T5 fitted with a 155 mm/45 calibre ordnance.

Specifications

	T5 Condor	T5/45
Mobility		
Configuration		
running gear:	wheeled	wheeled
layout:	8 × 8	8 × 8
Speed		
max speed:	80 km/h	80 km/h
Range		
main fuel supply:	600 km	600 km
Firepower		
Armament:	1 × 155 mm howitzer	1 × 155 mm howitzer
Ammunition		
main total:	22	22
charges:	26	26
Main weapon		
calibre length:	52 calibres	45 calibres
maximum range:	44,200 m (Base Bleed (BB))	39,200 m (Base Bleed (BB))
maximum range:	58,000 m (VLAP)	54,000 m (VLAP)
Rate of fire		
rapid fire:	8 rds/min	4 rds/min
Operation		
stop to fire first round time:	90 s	90 s
out of action time:	90 s	90 s
Main armament traverse		
angle:	80° (40° left/40° right)	80° (40° left/40° right)
Main armament elevation/depression		
armament front:	+70°/-3°	+70°/-3°

Status

Prototype. Not yet in production or service.

Contractor

Denel Land Systems.

Sudan

Military Industry Corporation self-propelled 122 mm D-30 howitzer Khalifa

Development

Early in 2013 it was revealed that the Sudanese Army had deployed a self-propelled version of the widely deployed Russian towed 122 mm D-30 howitzer under the local name of the Khalifa.

This has been developed by the Military Industry Corporation (MIC) of the Sudan and the upper part of the refurbished towed 122 mm D-30 howitzer is mounted on the rear of a modified Russian Kamaz (6 × 6) cross-country truck chassis.

This system has been offered on the export market by the MIC of the Sudan but as of March 2013 there are no know sales.

The system has also been designated the GHY02 by the MIC, but this may be just for marketing purposes.

While the first application of the self-propelled D-30 Khalifa is on a Russian 6 × 6 cross-country truck chassis it could also be integrated on other 6 × 6 chassis or even a larger 8 × 8 chassis which would have greater payload and mobility.

Description

The 6 × 6 chassis has been fitted with a locally designed forward control four door protected cab for the crew with provides some protection to the crew from small arms fire and shell splinters.

Scale model of the Sudanese Khalifa 122 mm D-30 howitzer integrated on the rear of a Russian (6 × 6) truck chassis and showing protected cab at front of vehicle (Christopher F Foss) 1455569

Rear view of the Sudanese Khalifa 122 mm D-30 howitzer integrated on the rear of a Russian (6 × 6) truck chassis and showing stabilisers deployed at rear (MIC Sudan) 1484356

The Khalifa system is operated by a crew of five people compared to the crew of seven normally used with the standard 122 mm D-30 howitzer.

The vehicle is provided with a central tyre inflation system for improved cross-country which allows the driver to adjust the tyre pressure from the cab to meet the terrain conditions encountered. This is of particular use in the desert conditions encountered in the Sudan.

To provide a more stable firing platform two stabilisers are lowered to the ground one on either side at the rear of the vehicle by remote control.

The operator's controls for deploying the stabilisers are located at the rear of the chassis on the right side.

To the immediate rear of the weapon when deployed in the firing position is a flat folding panel that is deployed horizontally to allow the crew to load and fire the weapon.

The standard direct and indirect fire sights are mounted in the normal position on the left side and as the weapon is higher off the ground a seat is provided for the gun layer.

A Fire-Control System (FCS) is said to be fitted for increased accuracy but no details of this have been released by the MIC.

The 122 mm ordnance does not appear to be fitted with a muzzle velocity measuring radar which is fitted would feed information to the FCS if fitted.

Traverse is 40° left and 40° right with the elevation arc being from -5° to +70° outside the vehicle cabin and from +15° to +70° when firing above the cabin.

When travelling the 122 mm D-30 barrel is held in a travel lock which is located on top of the ammunition stowage area.

A replacement wheel and tyre is carried to the immediate rear of the cab on the right side.

A quantity of 122 mm (projectile and charge) ammunition is carried to the immediate rear of the cab and the sides fold down to provide increased working space and allow rapid access to the projectiles and charges.

The sides are lowered using a hydraulically operated jack which is located to the forward part of the sides.

According to the MIC, the complete Khalifa has a combat weight of around 20.5 tonnes which includes the crew and a total of 45 × 122 mm projectiles and associated charges.

Maximum stated rate of fire is up to eight rounds a minute and maximum range using standard ammunition is quoted as 17 km.

This is slightly different to the Russian 122 mm D-30 which has a maximum range of 15.4 km firing a High-Explosive Fragmentation (HE-FRAG) projectile and up to 21.9 km for an assisted HE-FRAG projectile.

Maximum recoil length is stated to be 930 mm and minimum recoil length is 740 mm.

When compared to the standard 122 mm D-30 towed howitzer the Khalifa can be brought into action in just 90 seconds and come out of action in a similar time which makes the system more survivable against counter battery fire.

The MIC is also now manufacturing a wide range of artillery ammunition including the projectiles, charges and brass cases not only for the 122 mm D-30 but also for the Russian 130 mm M-46 field gun and the US 105 mm M101 howitzer.

Specifications

	122 mm D-30 howitzer Khalifa
Dimensions and weights	
Crew:	5
Length	
overall:	9.055 m (travelling)
Width	
overall:	2.665 m (travelling)
Height	
overall:	3.49 m (travelling)
axis of fire:	2.20 m
Ground clearance	
overall:	0.385 m
Track	
vehicle:	2.05 m
Weight	
standard:	18,900 kg
combat:	20,500 kg
Mobility	
Configuration	
running gear:	wheeled
layout:	6 × 6
Firepower	
Armament:	1 × 122 mm D-30 howitzer
Ammunition	
main total:	45
Main armament traverse	
angle:	80° (40° left/right)
Main armament elevation/depression	
armament front:	-5°/+70° (outside cab; +15°/+70° above cab)
Survivability	
Night vision equipment	
vehicle:	no
NBC system:	no
Armour	
hull/body:	steel (cab only)

Status

Production as required. In service with the Sudan.

Contractor

Military Industrial Corporation (MIC).

Sweden

BAE Systems FH-77 BW L52 Archer (6 × 6) self-propelled artillery system

Development

Under contract to the Swedish Defence Materiel Administration (FMV), the now BAE Systems (previously BAE Systems Bofors) fitted a 155 mm FH-77 artillery system on the rear of a modified VME A25C (6 × 6) all-terrain chassis.

The first driving and firing rig was completed late in 1995 and, following company trials, was delivered to the Swedish Army for six months of troop trials.

This chassis is widely used in the construction industry and has proved to be very reliable and to have good cross-country mobility.

Under a GBP135 million (USD200 million) contract awarded to BAE Systems in March 2010 the company will build a total of 48 of the latest FH-77 BW L52 155 mm Archer Self-Propelled (SP) artillery systems.

Of these 24, of these will be supplied to the Norwegian Army and 24 to the Swedish Army and in each case these will be the only artillery systems to be deployed in the future.

In Norwegian service Archer is to replace the currently deployed M109A3GN full tracked SP artillery systems while in the Swedish Army they are to replace the currently deployed towed 155 mm FH-77B.

It is expected that each battalion will have a total of 12 Archer that will be deployed in three batteries each of four weapons.

There will also be an ammunition resupply vehicle based on an 8 × 8 cross-country truck.

Under the terms of this contract two pre-production Archer were supplied in 2010 followed by the first production batch of six systems in 2011 of which Sweden will get four and Norway two.

Full rate production will be underway by 2012 and all systems will be delivered by 2014.

As part of the deal, all of the Archer will be fitted with the Norwegian Kongsberg Protector Remote Controlled Weapon Station (RCWS) armed with a .50 M2 HB MG.

Description

For its new mission the Volvo chassis has been modified with a fully armoured cab fitted at the front which would also cover the engine compartment. The cab has bulletproof windows and can accommodate a six-man crew.

155 mm FH77 BW L52 Archer 155 mm/52-calibre self-propelled artillery system deployed in the firing position (BAE Systems) 1116747

BAE Systems 155 mm FH-77 BW L52 Archer (6 × 6) self-propelled artillery system deployed in the firing position (BAE Systems) 1363990

A machine gun is mounted on the roof for anti-aircraft/local defence. In addition to the circular hatch in the roof there is also a single door on either side at the rear.

By mounting the system on a 6 × 6 chassis this combination not only provides greater mobility but, when travelling, the crew will be under armour protection and the system will also be able to come into and out of action faster and therefore be less vulnerable to counterbattery fire.

Following extensive Swedish Army trials with the driving and firing test rig, the user suggested a number of improvements that were carried out by the company.

These included the requirement for the crew to stay in the armoured cabin when carrying out the fire mission with a magazine for 24 155 mm projectiles being mounted above the weapon and protected by armour.

The first prototype driving and firing rig was then rebuilt and tested in 1999 by the Trials Wing of the Swedish Artillery School. The main features of this enhanced model can be summarised as:

- Possibility of deploying and taking the gun out of action from inside the cabin
- Facilities that make it possible to lay and fire the gun from inside the cabin
- The onboard gun computer is also capable of taking care of the meteorological information as well as acting as a ballistic computer
- The same computer will in future be capable of storing information about the enemy, friendly and also different target co-ordinates
- A magazine for 24 155 mm projectiles
- The charge box has been removed to the right side of the vehicle
- The crew has been reduced to five but seating is still provided for six people
- When filling up the magazine the computer will be told which type of projectile is where in the magazine.

Communication between the crew members is performed with a short-range radio set. Time to go into and come out of action is about 30 seconds.

The latest FH-77BD is a further development of the original prototype system based on a commercial VME A25C (6 × 6) articulated all-terrain chassis evaluated by the Swedish Army.

For the Indian trials the FH-77BD was fitted with a BAE Systems designed and built 155 mm/45-calibre barrel with a 25-litre chamber. This, firing an High-Explosive Extended-Range Full-Bore - Base Bleed (HE ER FB BB) projectile, will achieve a range of around 40 km.

Late in 2003 the company was awarded a contract from the Swedish Defence Material Administration (FMV) to build two system demonstrators of the FH-77BD L52 6 × 6 Self-Propelled (SP) artillery system. The total value of this contract, which was announced in early 2004, is SEK200 million.

The first system demonstrator of the FH77 BW L52 (6 × 6) Self-Propelled (SP) artillery system was completed in early 2005 and by October 2005 was well into its extensive firing and mobility trials in Sweden.

In 2006, a second system demonstrator joined the trials programme and has a number of improvements including new software and a roof-mounted LEMUR day/night device armed with a 7.62 mm machine gun for local defence which can be aimed and fired from within the cab.

155 mm FH-77 BW L52 Archer self-propelled artillery system deployed in the firing position from the rear (IHS/IDR) 1184763

Late in 2006 the FMV awarded the company a contract worth SEK40 million for additional development work on the system.

In the Summer of 2007 a government Memorandum of Understanding (MoU) was signed between Norway and Sweden where the two countries expressed their willingness to jointly develop and introduce the Archer 155 mm/52-calibre self-propelled artillery system into service with both countries.

In January 2009 it was announced that the FMV had awarded a USD70 million contract to complete development of the 155 mm FH-77 BW L52 (6 × 6) Self-Propelled (SP) artillery system for future deployment by Norway and Sweden.

The FH77 BW L52 is based on the latest Volvo A30E 6 × 6 all-terrain articulated carrier chassis, which has been modified to accept the 155 mm/52-calibre weapon with a 25.4-litre chamber.

To reduce costs, the cradle and recoil system are from the current in service FH-77B towed artillery system.

An automatic loading system is installed, which has enabled the crew to be reduced to just three people. This loads the 155 mm projectile and then the charge and enables a burst rate of fire of three rounds in 13 seconds to be achieved.

According to the company, a full magazine of 21 rounds can be fired in 2.5 minutes. A Multiple Rounds Simultaneous Impact (MRSI) of six rounds can be achieved for greater weapon effectiveness.

Mounted over the rear part of the ordnance is a MVS-470 muzzle-velocity radar, which feeds information into the onboard fire-control system.

The crew are seated in an armour- and NBC-protected cab in the front unit. The loading, laying and firing of the weapon is undertaken from within the cab via remote-control.

When deployed in the firing position a large hydraulic stabiliser is lowered at the rear of the chassis to provide a more stable firing platform.

In addition to the computerised fire-control system, standard equipment includes an inertial navigation and laying system that not only reduces the time taken to bring the weapon into action but also increases its accuracy.

Archer can fire a wide range of current and future 155 mm artillery projectiles including the locally developed 155 High-Explosive Extended-Range (HEER) base bleed, 155 High-Explosive 77, and 155 TR 54/55 training projectile.

More advanced and effective 155 mm projectiles include the M982 Excalibur precision guided munition (already used in combat by Canada and the United States) and the now BAE Systems/Nexter Munitions 155 Bonus top attack projectile which is already in service with France and Sweden. When fired from a 155 mm/52-calibre weapon Bonus has a maximum range of 34 km.

A direct fire day/night sight is fitted which enables targets to be engaged out to 2,000 m.

Maximum range depends on the 155 mm projectile/charge combination but is typically 40 km with conventional projectiles and 60 km with the 155 mm M982 Excalibur.

Firing the French/Swedish Bonus top attack projectile maximum range is stated to be 35,000 m.

A direct-fire day/night sight is fitted which enables targets to be engaged out to 2,000 m.

The total weight of the weapon system is being quoted as 33.5 tones, which will enable it to be carried in the A400M transport aircraft. Maximum road speed is quoted as 70 km/h.

FH77 BW L52 is also being marketed as part of a complete system called Archer, which also includes an ammunition resupply vehicle and support vehicle based on a similar 6 × 6 chassis, artillery C4I system, a locally developed modular charge system and a 155 mm family of projectiles.

In 2009 the Swedish company of Åkers Krutbruk was awarded a contract by BAE Systems for the development of the complete passive armour package for the 155 mm FH-77 BW L52 Archer (6 × 6) self propelled artillery system.

Under the terms of this contract, for an undisclosed amount, Åkers Krutbruk has developed a new passive protection system for the 155 mm FH-77 BW L52 Archer which will provide a high level of protection against small arms fire, fragments and mines.

Åkers Krutbruk has considerable experience in the design, development and production of armour systems for tanks, infantry fighting vehicles and support vehicles.

Variants

BAE Systems is already offering Archer on the export market with another version, called the Mounted Gun System (MGS) being aimed at the potential Indian Army requirement for a 155 mm/52-calibre wheeled system.

This has the same chassis as the Archer but is fitted with the ordnance and elevating mechanism of the FH-77B05 towed weapon that is loaded manually.

This weapon is a contender for the Indian towed gun requirement. For the Indian market BAE Systems is teamed with the local company of Mahindra.

Specifications

	FH-77 BW L52 Archer
Dimensions and weights	
Crew:	3
Length	
overall:	14.55 m
Width	
overall:	3.0 m
Height	
to top of roof mounted weapon:	4.007 m
transport configuration:	3.3 m
Ground clearance	
overall:	0.45 m
Weight	
standard:	33,500 kg
Mobility	
Configuration	
running gear:	wheeled
layout:	6 × 6
Speed	
max speed:	70 km/h
Range	
main fuel supply:	500 km (est.)
Fuel capacity	
main:	400 litres
Fording	
without preparation:	0.75 m
Gradient:	40%
Side slope:	30%
Engine	Volvo model D9AACE2, diesel, 325 hp
Firepower	
Armament:	1 × 155 mm howitzer
Armament:	1 × roof mounted 12.7 mm (0.50) M2 HB machine gun
Ammunition	
main total:	21
charges:	18
Main weapon	
calibre length:	52 calibres
maximum range:	40,000 m
Rate of fire	
sustained fire:	1.25 rds/min
rapid fire:	8 rds/min
Main armament traverse	
angle:	150° (75° left/75° right)
Main armament elevation/depression	
armament front:	+70°/0°
Survivability	
Night vision equipment	
vehicle:	optional
NBC system:	yes
Armour	
hull/body:	steel + appliqué

Status

In production for Norway (24 units) and Sweden (24 units) with production running from 2011 through 2014.

Contractor

BAE Systems.

Taiwan

Taiwanese 155 mm (6 × 6) self-propelled artillery system

Development

Since 2005 the Taiwanese 202nd Plant Material Production Center Armaments Bureau has been developing a new wheeled 155 mm Self-Propelled (SP) artillery system to meet the potential requirements of the Taiwanese Army.

Today Taiwan deploys a mix of towed and self-propelled artillery systems with most of the latter being US supplied BAE Systems, 155 mm/39-calibre M109 series.

As of March 2013 it is understood that production of this Taiwanese 155 mm/45-calibre (6 × 6) self-propelled artillery system had yet to commence.

It is considered that the first example of the Taiwanese 155 mm/45-calibre (6 × 6) self-propelled artillery system was a technology demonstrator and that any production systems would be based on a chassis with at least a protected crew cab and a chassis with more cross-country capability.

Description

This new system is based a US conventional commercial bonneted truck on the rear of which is mounted the complete elevating mechanism of a 155 mm/45-calibre towed artillery system.

This is understood to have been taken from an existing towed artillery system currently deployed in Taiwan.

This ordnance is fitted with a double baffle muzzle brake and a fume extractor and when travelling the 155 mm ordnance is held in a travel lock to the immediate rear of the cab.

To provide a more stable firing platform hydraulically operated stabilisers are lowered to the ground. These are positioned one either side between the first and second road wheels with an additional stabiliser positioned one either side at the chassis rear.

Maximum range of this 155 mm/45-calibre SP artillery system depends on the projectile/charge combination but is currently being claimed as 38 km.

Natures of 155 mm ammunition fired include conventional High-Explosive (HE) types as well as Extended-Range Full-Bore – Base Bleed (ERFB-BB).

Maximum rate of fire is claimed to be six rounds a minute with a total of 34 projectiles and associated charges being carried in two ammunition lockers which are positioned one either side of the platform to the cab rear.

Propelling charges are understood to be of the conventional bag type rather than being of modular charge type that is being rapidly introduced by many armies with significant logistical advantages.

A flick rammer is installed to reduce crew fatigue as well as increasing rate of fire. Standard equipment includes an Inertial Navigation System (INS) coupled with a Global Positioning System (GPS).

This, when used with the on board computerised fire-control system enables the system to come into action, carry out its fire mission and then rapidly deploy to another fire position before counterbattery fire is returned.

Mounted over the rear part of the 155 mm/45-calibre ordnance is a muzzle velocity measuring radar that feeds information to the on board fire-control computer.

Elevation and traverse is believed to be powered with manual back up controls being provided for the reversionary role. Maximum elevation is claimed to be +65° with a traverse of 15° left and right.

It is claimed that the system is operated by a crew of four but another vehicle would probably accompany the system with additional ammunition and crew equipment as there is insufficient space on the vehicle to carry all of their individual equipment and rations.

First example has a slightly modified commercial type cab with some blast protection provided for the crew but production systems could well have a more robust cross-country truck chassis. Maximum combat weight is about 18.5 tonnes and maximum road speed is about 100 km/h.

Specifications

	Taiwanese 155 mm self-propelled artillery system
Dimensions and weights	
Crew:	4
Length	
overall:	9.00 m
Width	
overall:	2.5 m
Height	
overall:	3.2 m
Weight	
standard:	18,500 kg
Mobility	
Configuration	
running gear:	wheeled
layout:	6 × 6
Speed	
max speed:	100 km/h
cross-country:	50 km/h
Firepower	
Armament:	1 × 155 mm howitzer
Ammunition	
main total:	34
Main weapon	
calibre length:	45 calibres
maximum range:	38,000 m (ERFB-BB projectile)
Main armament traverse	
angle:	30° (15° left/15° right)
Main armament elevation/depression	
armament front:	+65°/0°

	Taiwanese 155 mm self-propelled artillery system
Survivability	
Night vision equipment	
vehicle:	no
NBC system:	no

Status
Prototype system. Not yet in production or service.

Contractor
See development.

Ukraine

Tasko 155 mm self-propelled howitzer

Development
The Tasko 155 mm Self-Propelled Howitzer (SPH) has been developed as a private venture by the Tasko corporation for the home and export markets.

As of March 2013 it is understood that production of the Tasko 155 mm self-propelled artillery system had yet to commence.

It is considered that the first example of the this 155 mm (6 × 6) self-propelled artillery system was a technology demonstrator and that any production systems would be based on a 6 × 6 cross-country chassis with at least a protected crew cab.

Description
Prototype of the Tasko 155 mm SPH consists of a standard production KrAZ-6322 (6 × 6) cross-country truck chassis that retains its standard bonneted unarmoured cab.

Mounted at the very rear of the chassis is the mount that can be armed with a 155 mm/45- or 155 mm/52-calibre barrel which is fitted with a double baffle muzzle brake.

It is considered that the complete mount, elevating mass and some other subsystems used on the Tasko 155 mm SPH are taken from an existing towed artillery system rather than being designed from scratch.

According to the Tasko Corporation, maximum range of the 155 mm/45-calibre version is 39.6 km while the 155 mm/52-calibre system has a maximum range of 42 km. It is assumed that this is achieved using a high-explosive base bleed projectile.

At this stage it is not certain as to whether the system has powered or manual elevation and traverse. It is possible that a flick rammer is fitted to enable a higher rate of fire to be achieved and to reduce crew fatigue.

The first example does not appear to be fitted with a muzzle velocity measuring radar. It is considered probable that production systems could well have this feature as well as an onboard computerised fire-control system that could be integrated with a land navigation system.

The system has a typical crew of six of whom three are seated in the cab with the remainder being seated to the rear of the fully enclosed cab. The prototype carries a total of 16 × 155 mm projectiles and associated charges.

To provide a more stable firing platform, a large stabiliser is lowered to the ground at the rear before firing commences.

The system takes one minute to come into or be taken out of action and a burst rate of fire of three rounds can be achieved in 15 seconds.

Chassis is fitted with a central tyre pressure regulation system that allows the driver to adjust the tyre pressure to suit the terrain being crossed. Tyres are 13.00 × 530 – 533.

The vehicle is powered by a V-8 turbocharged diesel developing 265 hp at 1,200 rpm which is coupled to a manual transmission.

Variants
While the prototype is integrated onto a KrAZ-6322 (6 × 6) chassis the system could be integrated on other 8 × 8 or even 6 × 6 cross-country chassis.

Tasko 155 mm self-propelled howitzer integrated onto a KrAZ-6322 6 × 6 truck chassis (Tasko Corporation) 1040802

Specifications
Not available.

Status
Prototype. Not yet in production or service.

Contractor
Tasko Corporation.

United States

Mandus Group Hawkeye 105 mm soft-recoil self-propelled artillery system

Development
In October 2011, the US Mandus Group, who has extensive experience in the upgrade of field artillery systems, showed in public for the first time their Hawkeye soft recoil 105 mm Self-Propelled (SP) artillery system.

This has been developed as a technology demonstrator with the first example being integrated onto the rear of a Mack Trucks/Renault Trucks Defense Sherpa (4 × 4) two door protected truck for trials purposes.

In the 1970s the US Army Rock Island Arsenal developed and tested the 105 mm XM204 soft-recoil howitzer which weighed 1,814 kg. This was type classified as the 105 mm M204 but never entered production or service.

The Mandus Group have developed this soft-recoil technology further using the latest engineering and technology with the first test bed being completed in the third quarter of 2011 and based on the 105 mm M137A1 ordnance used in the 105 mm M102 towed howitzer.

Prior to being shown for the first time in public, a 25-round test of the Hawkeye soft-recoil system was carried out using an hydraulic recoil simulator at Rock Island Arsenal.

This was followed by the first firing trials on 16th September 2011 with the firing of 25 'sand' rounds at Zone Five and another 24 at Zone Seven.

The first high-explosive ballistic rounds were fired on 1 October 2011 of which 21 were at Zone Six and a further four at Zone Four, which was the maximum permissible at high elevation because of range constraints.

Traverse limits for the trials were 18° left and right with a maximum elevation of +72° and a depression of –5°.

The Mandus Group is based in Rock Island, near to the US Government owned Rock Island Arsenal (RIA) who has many years of experience in artillery weapons.

While the Hawkeye has been developed using internal research and development funding, some parts have been made by RIA.

The Mandus Group and RIA have entered into a public/private partnership arrangement whereby the finished product is to be co-marketed by the two companies. This also gives the Mandus Group access to the RIA soft-recoil technology database.

Description
The soft-recoil principle aims to reduce the howitzers recoil impulse by imparting a forward velocity to its recoiling part immediately prior to charge ignition. This is claimed to reduce the recoil force by 50 to 70 per cent.

This is turn reduces the loads passed to the platform through the gun trunnions and therefore allows the host platform to be substantially lighter than howitzers that fire from the conventional ("battery") position with the barrel/ordnance fully run out.

To start the soft-recoil firing cycle, the recoiling parts of the weapon are initially held at the latch (mid-recoil) position by a catch, against the pressure of a nitrogen-filled recuperator.

Close up of the elevating mass of the Hawkeye 105 mm soft-recoil demonstrator. Visible above the cradle is Selex FIN3110 inertial navigation unit with its integral GPS and below the Weibel muzzle velocity radar and direct fire camera (IHS/Rupert Pengelley) 1364829

First example of the Hawkeye 105 mm soft-recoil demonstrator carrying out a firing by remote control. On the right is the tripod-mounted GLU for the Selex LINAPS fire control system (Mandus) 1364828

Hawkeye 105 mm soft-recoil demonstrator in travelling configuration and integrated onto rear of Sherpa (4 × 4) cross-country truck (Mandus) 1364807

When the firing handle or switch is tripped, the catch is released and the recoiling mass moves forwards.

A sensor monitors its progress and velocity, the propelling charge only igniting once the recoiling mass reaches a pre-determined velocity.

The recoil forces generated on firing have to first arrest and then reverse the initial forward motion of the 105 mm barrel, which reduces the forces transmitted to the mount by up to 70 per cent.

The residual energy is used to automatically reset the recoiling mass to its pre-firing position and ready for the next round to be loaded.

The Hawkeye as such is not a true soft-recoil system as it uses the Fire-Out-Of-Battery (FOOB) technique only to mitigate a portion of the recoil forces.

Hawkeye mitigates two-thirds of the recoil with a FOOB run-up and recompression cycle, with the remainder using a conventional recoil system.

In the event of ignition variability or hang fire incidents, it is therefore afforded a failsafe mechanism to take care of the remaining impulse through a throttling process that relies on conventional techniques.

The Hawkeye system has no need for magnetorheological fluids or complex computer controlled recoil technologies.

The Hawkeye weighs around 998 kg including the elevating mass, top carriage and traverse bearing. Weight of the complete unit, e.g. Sherpa and weapon is about 7,246 kg.

Although the first example uses the 105 mm M137A1/27-calibre barrel, this could be replaced by alternative weapons.

The first example has a maximum range of 11.5 km firing a standard 105 mm M1 high-explosive projectile or 15.1 km firing a 105 mm M927 high-explosive rocket-assisted projectile.

Traverse is through a full 360° with elevation from –5° to +72° and it can come into action in 15 to 20 seconds.

It would normally be operated by a crew of three and have a maximum rate of fire of 10 to 12 rounds a minute and a sustained rate of fire of three rounds a minute.

It can be fitted with various fire-control options with the first example being fitted with the combat proven Selex LINAPS artillery pointing system.

This incorporates an FIN3110 laser-gyro inertial navigation unit with integral Global Positioning System (GPS).

For this application, the commander has a demountable LINAPS Gun Layer's Unit (GLU) which is used for vehicle navigation when the system is travelling and on a tripod for weapon aiming when carrying out a fire mission.

Mack Defense Sherpa (4 × 4) vehicle fitted with protected cab and Hawkeye 105 mm soft-recoil weapon on the rear platform with ordnance elevated and traversed slightly to the right (Mandus) 1452914

This GLU for the first example of Hawkeye is based on a DRS MRT handheld unit but this could change in the future.

This is connected to the Selex FIN3110 laser-gyro inertial platform with its integral Garmin GPS receiver which is mounted on the cradle as is the Weibel muzzle velocity radar and the direct fire video camera.

Elevation and traverse is all electric with manual back up controls and laying can be carried out semi-automatically using a joystick mounted on the GLU or potentially via touchscreen buttons on the display itself.

For the direct fire role it has a computerised aiming mark which is driven by the ballistic kernel within the layers display unit.

The first example of the Hawkeye is manually aimed and loaded but the Mandus Group has studied a fully digital solution which would be able to automatically activate the servos and lay the weapon onto the target.

The elevation and traverse systems are being designed to stabilise the weapon on its mount during firing.

Although the first example of the Hawkeye 105 mm weapon system has been demonstrated integrated onto the rear of a Sherpa (4 × 4) chassis, its design is such that it can be integrated onto a wide range of other platforms, track and wheeled, according to customer requirements.

The first example of the Hawkeye 105 mm system is regarded as a technology demonstrator and the same principles could be applied to other weapons such as a 155 mm howitzer.

According to the Mandus Group, the soft-recoil principle is well suited to replace 120 mm mortar systems and would offer potential customers a number of significant operational advantages including improved minimum and maximum ranges as well as a useful direct fire capability.

Variants

It is expected that development of a soft-recoil 155 mm system with optional 39-, 45- or 52-calibre barrel sizes will commence in the future.

Specifications

	Hawkeye
Mobility	
Configuration	
running gear:	wheeled
Firepower	
Armament:	1 × 105 mm M137A1 howitzer
Main weapon	
maximum range:	11,500 m (HE M1 projectile)
maximum range:	15,100 m (HE M927 projectile)
Rate of fire	
sustained fire:	12 rds/min
Main armament traverse	
angle:	360°
Main armament elevation/depression	
armament front:	-5°/+72°

Status

Under development. Not yet in production or service.

Contractor

Mandus Group Limited.

Towed Anti-Tank Guns, Guns And Howitzers

Argentina

155 mm L 45 CALA 30/2 Cannon

Development

The 155 mm L 45 CALA 30/2 Cannon towed artillery system was developed by the Argentine Armed Forces Research and Technology Institute (CITEFA) with the first prototypes being built by the local military manufacturing establishment (DGFM) which was also subsequently responsible for production of the weapon.

Final trials of the second prototype, which incorporated a number of improvements over the first prototype, were carried out in November 1996 during which it is claimed the L 45 CALA 30/2 Cannon achieved a maximum range of 39,000 m firing a 155 mm Extended Range Full Bore - Base Bleed (ERFB-BB) high-explosive projectile.

The Argentine Armed Forces placed an initial production order for this system that comprised not only the 155 mm L 45 CALA 30/2 Cannon but also a new family of 155 mm artillery projectiles including the PACU (HE) with hollow base and the PALA 37 (Extended Range Full Bore) in two versions, hollow base and base bleed. When fired these longer range 155 mm artillery projectiles increase the survivability of the weapon to counter battery fire.

In addition, it can fire all standard types of 155 mm HE projectiles such as the US-designed high-explosive M107 projectile.

This 155 mm M107 HE projectile was designed many years ago but by todays standards it has a short range and is of limited effectiveness.

As far as it is known, Argentina uses conventional bag-type charges with this weapon, rather than the more recent modular charge system.

Funding permitting, it is probable that in the longer term, the 155 mm L 45 CALA 30/2 Cannon will supplement and perhaps eventually replace the older L 33 X1415 CITEFA Models 77 and 88 which are covered in detail in a separate entry in *IHS Jane's Land Warfare Platforms: Artillery and Air Defence*. It should be noted that production of both of the latter weapons in now complete.

Description

The 155 mm L 45 CALA 30/2 Cannon artillery system is mounted on a conventional split trail carriage and when deployed in the firing position rests on the two spades and a hydraulic firing jack under the carriage.

The 155 mm 45-calibre ordnance is fitted with a double-baffle muzzle brake, screw breech with automatic opening mechanism and has a fixed recoil system - length does not vary with elevation.

Elevation limits are from –3° to +67° while traverse is 30° left and right. The equilibrators are gas-operated. The aiming unit consists of a panoramic optical periscope. The sighting system is mounted on the left side and includes direct and indirect fire sights.

155 mm L 45 CALA 30/2 Cannon deployed in the firing position with road wheels raised clear of the ground and trail wheels raised 0006030

155 mm L 45 CALA 30/2 Cannon in travelling position with ordnance traversed over closed trails and spades in vertical position 0006031

When being towed, the upper carriage and 155 mm ordnance is traversed through 180° and locked in position over the closed trails. The trails are provided with wheels to assist in bringing the weapon into action.

The system is fitted with an auxiliary power unit with a Deutz air-cooled diesel engine that provides power to the four large roadwheels.

When in the self-propelled mode the ordnance is traversed to the rear with the driver being seated at the front left of the carriage.

The suspension of the carriage is of the walking-beam type with each side having two rubber-tyred roadwheels for improved mobility in rough terrain. The wheels are fitted with air brakes and the carriage is fitted with an anti-unhitch safety device.

As previously stated, the L 45 CALA 30/2 Cannon can fire all standard NATO 155 mm projectiles as well as locally developed projectiles with ranges of the latter being as follows:

Projectile	Muzzle velocity	Range
M56 or CX4 (HE)	820 m/s (max)	23,000 m (max)
PACU (HE)	900 m/s (max)	27,000 m (max)
PALA 37/CH (HE)	897 m/s (max)	31,000 m (max)
PALA 37/BB	897 m/s (max)	39,000 m (max)

The actual 155 mm ordnance has a barrel length of 6.975 m with rifling of 48 right turn, variable pitch grooves, 20/30 calibre.

While the maximum quoted towing speed is up to 70 km/h on the road whilst being towed on a dirt road this is reduced to 40 km/h and when cross-country to 20 km/h.

These towing speed figures do depend on a number of factors such as actual terrain conditions.

Specifications

	155 mm L 45 CALA 30/2
Mobility	
Carriage:	split trail
Speed	
towed:	70 km/h
Firepower	
Armament:	1 × carriage mounted 155 mm cannon
Main weapon	
length of barrel:	6.975 m
muzzle brake:	double baffle
Rate of fire	
sustained fire:	2 rds/min
rapid fire:	6 rds/min
Main armament traverse	
angle:	60° (30° left/30° right)
Main armament elevation/depression	
armament front:	+67°/-3°
Survivability	
Armour	
shield:	no

Status

Production as required. Believed to be in service with Argentina in small numbers (25). United Nations sources stated that there were no exports of 155 mm artillery systems from Argentina between 1992 and 2010. It is understood that Argentina deploys a total of 110 155 mm towed weapons. These are a mixture of 155 mm 33 and 45-calibre systems, with the majority of the former type. Argentina also deploys a small batch of about 15 TAMSE VCA 155 155 mm self-propelled artillery systems based on a stretched TAM tank chassis manufactured in Argentina.

Details of this system are provided in a separate entry in *IHS Jane's Land Warfare Platforms: Artillery and Air Defence*. The 155 mm/31-calibre turret fitted to this chassis is the same as that used in the Oto Melara Palmaria 155 mm/41-calibre self-propelled artillery system currently in service with Libya and Nigeria.

Contractor

Instituto de Investigaciones Cientificas y Tecnicas de la Fuerzas Armadas (CITEFA) (design authority).

Direccion General de Fabricaciones Militares (DGFM) (production).

155 mm howitzer L33 X1415 CITEFA Models 77 and 81

Development

These 155 mm/33-calibre towed weapons are basically the top carriage (barrel, cradle, recoil system and equilibrators) of the French now Nexter Systems (Giat Industries and before that Creusot-Loire) 155 mm self-propelled gun Mk F3 (of which 24 are in service with Argentina) mounted on a new bottom carriage designed by the Scientific and Technical Research Institute of the armed forces (CITEFA).

155 mm howitzer L33 X1415 CITEFA Model 77 in firing position with barrel at maximum elevation and with spades bedded in 0500352

A small number of these 155 mm weapons believed to be 12, has been observed in service in Croatia.

Total production of this system is believed to have amounted to around 120 systems. A small quantity of these were exported. It is understood that Argentina deploys a total of 110 155 mm towed weapons.

In the future, these systems will be supplemented by the more recent 155 mm L 45 CALA 30/2 Cannon also developed in Argentina. This features a 155 mm 45-calibre ordnance and an auxiliary propulsion unit. Full details of this system, which has yet to be exported, are given in a separate entry in *IHS Jane's Land Warfare Platforms: Artillery and Air Defence* with production being undertaken on an as required basis.

Description

The top carriage of both versions is a welded steel structure, which contains the elevating brackets and traverse mechanism, and forms the connection with the cradle trunnion attachment and the bottom carriage. The traversing mechanism is mounted inside the front and bottom of the top carriage and, to obtain acceptable handwheel loads, two pneumatic equilibrators are connected to the top carriage and cradle.

The bottom carriage is a welded and machined steel unit. For firing, the wheels are raised clear of the ground and the weapon rests on a firing base at the front and on the trails at the rear. The firing base is a circular steel structure mounted on a ball socket, suspended underneath the bottom carriage, which compensates for uneven ground contour.

The firing base is normally carried on the bottom carriage when travelling, as the ground clearance of 0.3 m makes removal unnecessary during movement. The suspension is of the spring type.

The L33 X1415 fires a locally manufactured US-developed M107 HE projectile weighing 43 kg with a maximum muzzle velocity of 765 m/s, illuminating and smoke ammunition.

The Model 81 is a slightly revised version of the Model 77, fitted with a locally produced copy of the French 155 mm Mk F3 barrel. Some changes have been made to the trail legs and to the recuperator and equilibrators. The Model 81 has exactly the same specifications as the Model 77 and fires the same types of separate loading 155 mm ammunition.

105 mm exports

According to United Nations sources, in 1995 Argentina exported 18 105 mm guns of an undisclosed type to Venezuela.

Specifications

	L33 X1415 CITEFA 155 mm howitzer
Dimensions and weights	
Length	
overall:	10.15 m
Width	
overall:	2.67 m
Height	
overall:	2.2 m
Ground clearance	
overall:	0.3 m
Track	
vehicle:	2.31 m
Weight	
standard:	8,000 kg
Mobility	
Carriage:	split trail
Firepower	
Armament:	1 × carriage mounted 155 mm howitzer
Main weapon	
length of barrel:	5.115 m
muzzle brake:	double baffle
maximum range:	22,000 m
maximum range:	25,300 m (RAP)
Rate of fire	
sustained fire:	1 rds/min
rapid fire:	4 rds/min
Main armament traverse	
angle:	70° (35° left/30° right)
Main armament elevation/depression	
armament front:	+67°/-0°
Survivability	
Armour	
shield:	no

Status

Production complete. In service with Argentina (84). A quantity of these 155 mm/33-calibre weapons has been observed in Croatia. United Nations sources stated that there were no exports of 155 mm artillery systems from Argentina between 1992 and 2010.

Contractor

Instituto de Investigaciones Cientificas y Tecnicas de la Fuerzas Armadas (CITEFA), (design authority).

Direccion General de Fabricaciones Militares (DGFM), (production).

Belgium

SRC International GC 45 155 mm gun-howitzer

Development

The GC 45 155 mm gun-howitzer was developed from 1975 as a private venture by SRC International of Belgium. This was a company established jointly by the now defunct Space Research Corporation of Canada and PRB of Belgium (which subsequently went bankrupt).

The first prototype of the GC 45 was completed in 1977 and 12 weapons were subsequently purchased by the Royal Thai Marines. Of these, two were produced and assembled in Canada, the remaining 10 being produced in Canada and assembled in Austria by Voest-Alpine.

Licence for the production of a weapon based on this design was transferred to Voest-Alpine which manufactured this, with a number of improvements, as the GH N-45.

The Swiss company T&T Technology Trading purchased the complete manufacturing rights of this weapon as well as all of the remaining weapons and spare parts. Additional details of this weapon, now marketed by the NORICUM Division of T&T Technology Trading, are given in a separate entry in *IHS Jane's Land Warfare Platforms: Artillery and Air Defence*. As far as it is known, recent production of this weapon, called the GH N-45, has been for the Royal Thai Marine Corps.

According to United Nations sources, Thailand took delivery of 18 weapons in 1992, 18 in 1996 and 36 in 1998.

According to the United Nations, there have been no recorded exports of the 155 mm GH N-45 towed artillery system since then.

SRC International GC 45 gun-howitzer in travelling mode with ordnance over closed trails 0501026

SRC International GC 45 155 mm gun-howitzer in firing position. Note that the trails are not bedded in 0500354

It is understood that the Royal Thai Marines still use the original SRC International GC45 155 mm gun howitzer while the Royal Thai Army uses the more recent 54 GH N-45 supplied by the now NORICUM Division of T&T Technology Trading.

GC 45 stands for Gun-calibre 45 with the figure indicating the length of the 155 mm gun in-calibres.

This weapon was the baseline for many other 155 mm 45-calibre weapon systems built in other countries. The NORINCO 155 mm gun-howitzer Type WA 021 is similar in many respects to this weapon.

The origins of the South African Denel Land Systems 155 mm G5 (towed) and G6 (6 × 6 self-propelled) artillery systems can be traced back to the SRC International GC 45. Details of these systems are provided in separate entries in *IHS Jane's Land Warfare Platforms: Artillery and Air Defence*.

Production of the G5 and G6 is now complete but they could be placed back in production if any additional orders are placed by overseas customers. Further development has resulted in 155 mm/52-calibre models and main sales effort is now being concentrated on these versions.

Description

The GC 45 is of the split trail carriage configuration. It has a 155 mm 45-calibre barrel designed to fire all standard and new 155 mm NATO ammunition and charges, with the same muzzle velocity and range as would be achieved with a 155 mm/39-calibre barrel of the International FH-70 or US M198 type.

This performance is achieved at a lower breech pressure owing to a larger chamber, matched to the 45-calibre length. A significant feature of the GC 45 is that it meets the NATO requirement for a range of 30,000 m with a completely ballistic solution, without rocket booster or other external devices. When fired with the SRC 155 mm ERFB projectile and the top zone M11 charge, the maximum range of 30,000 m is achieved. At high angle and low charge (M3A1 zone 2) the GC 45 achieves a minimum range of 2,800 m with the same projectile.

Using an ERFB/Base Bleed projectile with the M11 charge, the maximum range is 39,000 m.

All structural components are built of a high-strength alloy steel, the 155 mm barrel is made of high-yield steel and autofrettaged throughout. The breech assembly incorporates an automatic breech-opening cam system: the breech opens automatically when the barrel assembly returns to battery and closes automatically when the cam is swung out of engagement.

The elevating system incorporates pneumatic equilibrators and results in small handwheel loads during elevation. Adjustment capability in the equilibrator system geometry, together with different nitrogen pressures, allows for ease of elevation at all gun system operating ambient temperatures.

A low-friction traverse bearing eases traversing. A friction device system is also incorporated to minimise feedback loads into the traversing system from any gun barrel whip.

A swing-aside loading tray mounted on the rear cradle of the recoil system moves in elevation with the barrel assembly. A pneumatically operated telescoping ramming cylinder is rotated into place behind the projectile for loading at high-elevation angles, obviating the need to lower the weapon between shots for ramming. The pneumatic rammer also provides a high degree of uniformity from shot to shot. A pneumatic system consisting of a bottle, mount, valving, regulator, accumulator, hoses and fittings is installed to power the ramming cylinder. An added feature of this system is that the bottle can be recharged in the field.

The suspension of the GC 45 incorporates a walking beam (two wheels each side) enabling the gun to be towed at speeds of up to 90 km/h. The freely pivoting walking beams also allow greater mobility during towing in rough terrain.

The design of the screw jack and float system, which is mounted under the forward part of the carriage, incorporates a telescoping system and ball screw jack. One crew member can easily jack the gun in 90 seconds. Disengagement of the ball jack will result in the immediate settling of the gun on its four undercarriage wheels.

The trails and spades, with the built-in towbar system, allow for efficient handling during emplacement and disemplacement. The spades are pivoted, allowing four angular positions, so that handling of these heavy components is minimised. It is possible to spread the trails to very large angles permitting large traverses of the barrel assembly at all angles of elevation.

The trail lifting assemblies, at the outboard rear of each trail, are designed to assist the gun crew during the emplacement and disemplacement operations. They are fully swivelling and fitted with rubber tyres, and are raised and lowered by handwheels at the forward end of each assembly. The force at the handwheel is transmitted to the wheel through a ball screw actuator connected to the wheel forks by a ball and crank arrangement. The ball screw is reversible and the trails can be simply dropped to the ground during emplacement. An additional feature of these assemblies is that they enable the gun crew to manhandle the gun through 360° in traverse.

The GC 45 fires the following separate loading ammunition: an SRC ERFB Mk 10 Mod 2 projectile weighing 45.4 kg and containing 8.8 kg of HE, illuminating projectile weighing 45.2 kg, smoke projectile weighing 45.7 kg and WP projectile weighing 47.7 kg. An ERFB/Base Bleed projectile weighing 47.5 kg fired with an M11 charge achieves a maximum range of 39,000 m. Firing the ERFB Mk 10 Mod 2 projectile, the weapon has the following range capabilities:

Charge	Muzzle velocity	Range
M3A1 zone 3	266 m/s	5,900 m (max)
M3A1 zone 4	302.1 m/s	7,300 m (max)
M3A1 zone 5	359.9 m/s	9,300 m (max)
M4A2 zone 6	458.1 m/s	12,400 m (max)
M4A2 zone 7	541.9 m/s	15,100 m (max)
M119 zone 8	658 m/s	19,200 m (max)
M2 zone 9	795.5 m/s	24,700 m (max)
M11	897 m/s	30,000 m (max)

In addition to the above natures of ammunition it will also fire NATO/US natures of ammunition. The SRC family of ammunition for the GC 45 was originally manufactured in Canada. Production ceased there many years ago. Many other countries now manufacture ERFB and ERFB-BB for the GC 45 and other 155 mm towed and self-propelled artillery systems.

The 155 mm GC 45 gun-howitzer is normally towed by a five-tonne (6 × 6) cross-country truck which carries the crew plus a supply of ready use 155 mm ammunition (projectiles, charges and fuzes).

The weight at the lunette in the towed mode with muzzle aft is being quoted as 1,815 kg and weight at the lunette in the towed mode with the muzzle forwards is being quoted as 1,134 kg.

Length travelling with the ordnance traversed to the rear is 13.614 m and this would be the normal towing configuration.

Specifications

	GC 45
Dimensions and weights	
Crew:	8
Length	
overall:	10.82 m (firing)
transport configuration:	9.144 m
Width	
overall:	10.364 m (firing)
transport configuration:	2.692 m
Height	
transport configuration:	3.28 m
Ground clearance	
overall:	0.355 m
Weight	
standard:	8,222 kg
combat:	8,222 kg
Mobility	
Carriage:	split trail
Firepower	
Armament:	1 × carriage mounted 155 mm howitzer
Main weapon	
length of barrel:	6.975 m
muzzle brake:	multibaffle
maximum range:	30,000 m
maximum range:	39,000 m (base bleed)

	GC 45
Rate of fire	
sustained fire:	2 rds/min
rapid fire:	4 rds/min
Main armament traverse	
angle:	80° (40° left/40° right)
Main armament elevation/depression	
armament front:	+69°/-5°
Survivability	
Armour	
shield:	no

Status

Production complete. In service with the Royal Thai Marines (12). No longer marketed.

Contractor

SRC International SA (no longer trading).

China

NORINCO 155 mm gun-howitzer Type WA 021 (WAC 21) PLL01

Development

The China North Industries Corporation (NORINCO) 155 mm gun-howitzer Type WA 021 (also referred to as the WAC 21) is very similar in many respects to the Austrian NORICUM GH N-45 gun-howitzer, to the extent that the carriage of the Type WA 021 can accommodate the ordnance of the NORICUM weapon.

The Type WA 021 is produced by the Heping Machinery Factory of NORINCO. It would appear that an interim 155 mm weapon known as the MF 45 was produced in China, probably as part of the development programme that led to the Type WA 021.

By the end of 1986, 10 Type WA 021s had been produced, two as prototypes and the remainder as pre-production examples.

The WA 021 was first seen in public during a defence equipment exhibition held in Beijing in 1986. The WA 021 system is based on a design supplied by the now defunct SRC International of Belgium headed by Dr Gerald Bull who also designed the GC 45. Details of the GC 45 are provided in a separate entry in *IHS Jane's Land Warfare Platforms: Artillery and Air Defence*.

Production of this was completed many years ago and it is only used by the Royal Thai Marines who took delivery of a total of 12 weapons.

The Type WA 021 was observed during the major military parade held in Beijing late in 1999 towed by a 6 × 6 truck. The PLA is understood to call this weapon the Type 88. Iran has taken delivery of a batch of 15 systems.

In 1991, officials of NORINCO stated that the WA 021 was in service with the People's Liberation Army (PLA).

Until recently this system was referred to as the NORINCO 155 mm gun-howitzer Type WA 021 (WAC 21), but more recently NORINCO has referred to this weapon as the PLL01 155 mm towed gun-howitzer.

NORINCO has also developed and placed in production the PLZ45 155 mm 45-calibre self-propelled artillery system. This uses the same 155 mm/45-calibre ordnance as the WA 021 and details are given in a separate entry.

The first known export customer for the 155 mm 45-calibre PLZ45 was Kuwait, which placed an order for 27 systems and all of these have now been delivered to Kuwait together with associated fire control and support equipment. It is understood that these were followed by a second batch of 27 systems.

More recently it is understood that Saudi Arabia has placed a contract with NORINCO which covers the supply of a total of 54 (27 × 2) 155 mm 45-calibre PLZ45 self-propelled artillery systems.

NORINCO 155 mm gun-howitzer Type WA 021 deployed in firing position 0006032

AH1 155 mm/45-calibre gun howitzer deployed in the firing position with ordnance elevated (NORINCO) 1366485

During a parade held in Beijing, China in October 2009 to celebrate the 60th Anniverary of the formation of the People's Republic of China (PRC), many new weapons were shown for the first time.

This included a new 155 mm self-propelled artillery system from the Artillery Brigade of the 38th Group Army which is understood to be designated the PLZ-05 in Peoples Liberation Army (PLA) service.

This is now being marketed by NORINCO under the designation of the PLZ52 and is based on a new hull and turret armed with a 155 mm/52 calibre ordnance.

This is currently being offered on the export market by NORINCO alongside the older PLZ45 but as of March 2013 there are no known export sales.

Description

The NORINCO Type WA 021 has an auto-frettaged 155 mm barrel 45-calibres long, fitted with a multibaffle muzzle brake, which has an efficiency of 30 per cent.

A screw-type breech mechanism opening to the right is employed and the chamber volume is 22.95 litres. The rifling employs 48 grooves with a twist of 1 in 20-calibres, and the grooves are understood to be three times deeper than the rifling depths found on comparable Western 155 mm designs.

The 155 mm 45-calibre ordnance has maximum recoil of 1.524 m and a minimum recoil length of 1.041 m. EFC has been quoted as 1,000 rounds.

The recoil system consists of the recuperator cylinder (top) and the recoil cylinder (below). The recoil mechanism can change the recoil distance with the variation of elevation angle so there is no need to dig a firing pit when the weapon fires at a high elevation.

The upper carriage includes the elevating mechanism, traversing mechanism, equilibrator and gear for traverse. The equilibrator is a pneumatic-hydraulic balancer. When the temperature changes and the unbalanced torque increases or decreases, the torque of the handwheel in the elevating mechanism will change. In this case the hydraulic pressure can be adjusted to obtain suitable torque of the handwheel in the elevating mechanism.

The all-welded steel carriage of the Type WA 021 is the split trail type and employs a screw jack firing platform under the carriage pivot. For travelling, the barrel is swung through 180° and held in a clamp on the left-hand trail leg.

A walking beam suspension is used and pivoting dolly wheels are provided on each trail leg for ease of handling. A pneumatic flick rammer is located on the left-hand trail leg.

Elevating and traversing systems are to the left of the barrel. Sighting equipment includes the M115 panoramic sight and the M137 telescopic sight.

Batteries of up to six WA 021s can be controlled by a fire-control system mounted on a command vehicle. The system includes a 16-bit fire-control computer, a command control console, a tape driver/printer, a company command relay display unit, gun display units, automatic observation equipment and a laser range-finder.

The Type WA 021 fires a complete family of NORINCO developed Extended-Range Full-Bore (ERFB) projectiles. These include ERFB HE, ERFB BB HE, ERFB BB WP, ERFB illuminating, ERFB smoke, ERFB BB illuminating, ERFB improved conventional munition and ERFB improved conventional munition BB. The improved carries 72 bomblets. These are of the High-Explosive Anti-Tank (HEAT) type and are designed to attack the vulnerable upper surfaces of armoured vehicles and are also highly effective against trucks and troops. The ERFB projectiles have welded nubs. Range with ERFB is 30,000 m and 39,000 m with ERFB BB. Dispersion at maximum range is stated to be 0.45 per cent longitudinal and 0.7 mil in traverse. Barrel service life is claimed to be 1,000 EFCs.

More recently NORINCO has developed a new family of 155 mm artillery projectiles which, in addition to being fired from locally developed 155 mm artillery systems, are also claimed to be compatible with NATO 155 mm towed and self-propelled (SP) artillery systems.

This family of 155 mm artillery projectiles is of the Extended-Range Full-Bore (ERFB) type with welded nubs which was originally pioneered by the late Dr Gerald Bull of the now defunct Canadian Space Research Corporation (SRC).

The complete family currently consists of Hollow Base/High Explosive (HB/HE), Base Bleed/High Explosive (BB/HE), Base Bleed Rocket Assist High Explosive (BB-RA/HE), Base Bleed Cargo (BB Cargo), Illuminating, White Phosphorous and Smoke.

NORINCO 155 mm gun-howitzer Type WA 021 fitted with auxiliary propulsion unit and in travelling configuration 0006033

AH2 155 mm/52-calibre gun howitzer with trails closed and showing muzzle brake on ordnance and APU on front of carriage (NORINCO) 1366487

Brief details of this family of 155 mm ERFB ammunition is as follows:

Type	HB/HE	BB/HE	BB-RA/HE	BB/cargo III		WP
Number:	DDB01	DDB02	DDB03	DDM02	M92	M92
Length:	940 mm	960 mm	960 mm	960 mm	941 mm	941 mm
Weight:	45.54 kg	48 kg	47.5 kg	48 kg	46.2 kg	47.3 kg
Range:	30 km	39 km	50 km	39 km	25 km	25 km

Notes:
(1) Maximum range quoted is when fired from a 155 mm/45-calibre system
(2) BB cargo round carries 56 top attack High-Explosive Anti-Tank (HEAT) bomblets each of which are claimed to penetrate 100 mm of conventional steel armour at 0°
(3) Illumination time is quoted as 90 seconds
(4) WP lasts for 55 seconds depending on ambient conditions
(5) There is also ERFB Smoke Type M92 which is 940 mm long, has a maximum range of 25 km with smoke lasting 120 seconds

NORINCO still manufactures conventional 155 mm bagged changes but has now developed a 155 mm Modular Charge System (MCS) which consists of BC1 and BC2 which are claimed to be compatible with NATO 155 mm artillery projectiles and can be fired from 155 mm 39/45 and 52-calibre systems.

The normal sustained rate of fire is 2 rds/min. This can be increased to between 4 and 5 rds/min for short periods.

According to NORINCO, maximum muzzle velocity when firing an ERFB projectile is 897 m/s and when firing an ERFB-BB projectile is 903 m/s.

It is normally towed by a 7- to 10-tonne (6 × 6) cross-country truck that in addition to its crew carries a quantity of ready use ammunition (projectiles, changes and fuzes).

Maximum towing speed on roads is up to 90 km/h but cross-country this is reduced to between 15 and 50 km/h with this figure depending on terrain conditions.

NORINCO 155 mm laser-guided projectile GP1

NORINCO is now marketing a 155 mm laser-guided artillery projectile that is similar in concept and appearance to the Russian KBP Instrument Design Bureau 152 mm Krasnapol laser-guided artillery projectile developed some years ago.

Although fitted with a high-explosive warhead, NORINCO claims its 155 mm system can engage tanks, infantry combat vehicles and self-propelled artillery systems moving up to a maximum speed of 36 km/h.

In addition, this laser-guided artillery projectile can also be used to engage static targets with a higher level of accuracy. It also has a coastal defence application.

The projectiles can be fired from all types of 155 mm towed and self-propelled artillery weapons when fitted with the correct communications equipment. This links the firing platform with the tripod-mounted laser designator.

In a typical target engagement the target is first detected by the forward observer and information passed to the firing platform. The 155 mm laser-guided projectile is then fired and the initial part of the trajectory is carried out using a normal ballistic trajectory.

The forward observer illuminates the target using the tripod-mounted laser target designator that has a minimum range of 500 m and a maximum range of 5,000 m.

Synchronisation of information between the gun position and the forward observer equipped with the laser designator is critical to the efficient operation of the 155 mm laser homing artillery weapon system.

The complete laser-guided 155 mm system is called the GP1 laser-homing artillery weapon system.

Specifications

NORINCO 155 mm guided projectile GP1

Calibre:	155 mm
Weight of projectile:	50 to 51 kg
Length of projectile:	1,302 mm
Warhead type:	HE
Max range:	20,000 m (approx.)
Hit probability:	90%
Firing speed:	2 to 3 rds/min
Accuracy of designator:	10 m
Weight of designator:	10 kg

NORINCO 155 mm laser-guided projectile GP6

The more recently developed 155 mm laser guided projectile GP6 is being marketed by NORINCO alongside the earlier GP1.

This is claimed to have a minimum range of 6 km and a maximum range of 25 km with a first round hit probability of 90 per cent and is capable of engaging stationary and moving targets.

The projectile weighs 52 kg and is fitted with a high-explosive warhead and then compared to the earlier GP1 is claimed to have a higher resistance to jamming and its multiple laser coding technology enables co-operative multi-target engagement.

NORINCO 155 mm top attack projectile GS1

NORINCO is now marketing a 155 mm top attack projectile that is similar to the German 155 mm SMART artillery projectile.

The Chinese 155 mm top attack projectile GS1 contains two sub-munitions which are dispensed over the target to attack the vulnerable upper surfaces of armoured vehicles. It is claimed that this is already in production and service with the PLA.

155 mm gun-howitzer with APU

The 155 mm gun-howitzer Type WA 021 is also available fitted with an APU. This enables the gun to move in to or out of the firing position as well as travel in the self-propelled mode for a short distance.

The APU can also be mounted on other towed artillery systems such as the NORINCO 130 and 152 mm types.

As far as it is known, as of March 2013 there were no known sales of this APU for other towed artillery applications.

The APU consists of a four-cylinder diesel developing 110 hp. It is mounted on the forward part of the carriage giving the system a maximum road speed of 20 km/h. When fitted with the APU the complete system has a turning radius of 7 m and a maximum range without refuelling of 100 km. It can also be towed on roads up to a maximum speed of 90 km/h.

The four main wheels are hydraulically powered with the front wheels being used for differential steering.

In addition giving the system a self-propelled capability, the APU also assists in the automatic opening and closing of the trails, automatic lowering and raising the spades and automatic traversing of the 155 mm/45-calibre ordnance through 360°. The version is also referred to as the 155 mm Auto-propulsion Gun Howitzer.

NORINCO has more recently started to refer to this system as the AH1 155 mm auto-propulsion gun howitzer AH1 is marketed in two versions. AH1 with a conventional screw breechblock and AH1A with a laterally sliding wedge breechblock.

The latest AH1APU develops 110 hp and gives the system a maximum speed of up to 20 km/h and a range of up to 100 km.

The installation of the APU allows the system to be deployed away from its prime mover if required and also makes it easier to deploy in rough terrain. It also assists in bringing the weapon into action more quickly.

Rate of fire is being quoted as 4 rpm and a pneumatic rammer is fitted to reduce crew fatigue and maintain rate of fire. This rammer is powered by a cylinder located on the left trail leg.

When travelling the 155 mm/45-calibre ordnance is traversed to the rear and locked in position over the closed trails. Combat weight of the AH1 is currently being quoted as 12.5 tonnes.

155 mm Auto-propulsion Gun Howitzer System

This consists of the self-propelled model integrated into a complete artillery battalion/regiment. These would have three batteries each with six weapons, plus battery/battalion command post vehicles, Type 704-1 locating radars, Type 702-D meteorological radars and reconnaissance vehicles based on a special version of the WZ551 series of 6 × 6 armoured personnel carrier.

Specifications

155 mm Auto-propulsion Gun Howitzer

Calibre:	155 mm/45
Length:	
(firing)	11.54 m
(travelling)	9.85 m
Width:	
(firing)	9.93 m
(travelling)	2.7 m
Height:	
(firing)	1.97 m
(travelling)	2.25 m
Weight:	12,500 kg

155 mm coastal defence system

This is being offered on the export market by NORINCO and additional details are provided in a separate entry in *IHS Jane's Land Warfare Platforms: Artillery and Air Defence*.

As of March 2013 there are no known sales of this coastal defence system.

NORINCO AH2 155 mm APU gun-howitzer

The AH2 155 mm/52-calibre is a further development of the AH1 155 mm/45-calibre system and is also fitted with an APU but as it has a longer barrel is slightly heavier at 14 tonnes.

Maximum range firing a standard ERFB HB HE is being quoted as 32 km and this can be extended to 41 km with a BB projectile and 53 km with a BB RA HE projectile.

AH4 155/39-calibre lightweight howitzer

This was first revealed in 2010 and is very similar in appearance to the combat proven BAE Systems 155 mm/39-calibre M777 and is claimed to use advanced materials in order to keep weight down to 4000 kg.

Available details are provided in a separate entry in *IHS Jane's Land Warfare Platforms: Artillery and Air Defence*.

As of March 2013 it is understood that production of this system had yet to commence and as far as it is known this system was not in service with the PLA.

NORINCO PLZ45 155 mm/45-calibre self-propelled gun-howitzer

This uses the same ordnance and ammunition system as the Type WA 021 and is covered in detail in a separate entry in *IHS Jane's Land Warfare Platforms: Artillery and Air Defence* In addition to being in service with China it has also been exported to Kuwait and Saudi Arabia.

NORINCO 155 mm gun XP52

This is understood to have been an experimental 155 mm 52-calibre gun but never entered production or service.

NORINCO 203 mm (8 in) howitzer

This system was developed to the prototype stage but is no longer being marketed by NORINCO.

It was mounted on a split trail carriage and had a maximum range of 40 km firing ERFB ammunition or up to 50 km firing ERFB-BB ammunition.

Specifications

	Type WA 021
Dimensions and weights	
Crew:	9
Length	
overall:	11.4 m (firing)
transport configuration:	9.068 m
Width	
overall:	9.931 m (firing)
transport configuration:	2.67 m
Height	
overall:	2.187 m
transport configuration:	2.23 m
axis of fire:	1.659 m
Ground clearance	
overall:	0.28 m
Weight	
standard:	9,500 kg
Mobility	
Carriage:	split trail
Speed	
towed:	90 km/h
Firepower	
Armament:	1 × carriage mounted 155 mm howitzer
Main weapon	
length of barrel:	7.045 m
rifling:	5.83 m
muzzle brake:	multibaffle
Rate of fire	
sustained fire:	2 rds/min
rapid fire:	5 rds/min
Main armament traverse	
angle:	70° (30° left/40° right)
Main armament elevation/depression	
armament front:	+72°/-5°
Survivability	
Armour	
shield:	no

Status

Production as required. In service with the PLA (150 estimate). Recent information has indicated that Iran has taken delivery of at least 15 of these systems from China. According to the United Nations Arms Transfer Lists for the period 1992 through to 2010, China did not export any towed 155 mm artillery systems. It should be noted however that there are a number of exports of Chinese towed artillery systems where the actual-calibre of the weapon has not been disclosed.

Contractor

Heping Machinery Factory.

China North Industries Corporation (NORINCO).

NORINCO 155 mm coastal defence gun system

Development

Although most countries have now phased out of service all of their gun and missile coastal defence systems there is still a potential market in some parts of the world for a highly mobile gun coastal defence system that can be rapidly deployed where they are required in times of tension.

China North Industries Corporation (NORINCO) are now marketing an integrated 155 mm coastal defence gun system that brings together a number of systems that are already in production and service.

Artists impression of an integrated NORINCO coastal defence system engaging a sea and air threat with the former being engaged by 155 mm guns (NORINCO)

1461675

NORINCO has been offering this modular coastal artillery system on the export market for several years but as of March 2013 there are no known sales.

Description

A typical NORINCO 155 mm coastal defence gun regiment would consist of three batteries of 155 mm towed artillery weapons with each of these having a Battery Command Post (BCP) and six 155 mm towed artillery systems.

In addition there would be a battalion command post, a Type 702D truck based meteorological radar system, truck mounted field logistic support system and Type 905 coast defence radar system.

In addition there would be between three and six portable observation posts equipped with a tripod mounted day/night surveillance system which includes a laser range-finder. The latter could if required be integrated on a cross-country platform.

The 155 mm artillery system would be the NORINCO 155 mm/45-calibre AH1 gun howitzer which is fitted with an Auxiliary Power Unit (APU) to allow the weapon to come into and be taken out of action quicker as well as being able to be manoeuvred more easily in a confined space or over rough terrain.

Full details of the AH1 are provided in a separate entry in *IHS Jane's Land Warfare Platforms: Artillery and Air Defence*.

AH1 weighs 12.5 tonnes and maximum range firing a NORINCO Extended Range Full Bore - Base Bleed (ERFB-BB) High-Explosive (HE) projectile is claimed to be 39 km which can be extended to 50 km firing an ERFB-BB Rocket Assist (ERFB-BB-RA) projectile.

AH1 is normally towed by a 6 × 6 cross-country truck that also carries the crew and a quantity of ready use 155 mm ammunition (projectiles, charges and fuzes).

The coastal artillery system could also use the more recent AH2 155 mm/52-calibre howitzer which also has an APU and weighs 14 tonnes. This has a maximum range of 41 km firing an ERFB-BB projectile which increases to 53 km firing the ERFB-BB-RA projectile.

A highly accurate computerised Fire-Control System (FCS) is required to enable a surface craft to be successfully hit at a long range due to the target engagement cycle and the extended flight time of the 155 mm projectile at longer ranges.

At this stage it has not been confirmed as to whether target information is supplied to each weapon automatically or not.

The optimum projectile to engage naval targets would probably be HE with cargo projectiles using sub-munitions to neutralise small landing craft or damage sensors and radars on larger ships.

Laser-guided projectiles could also be used but these typically have a range of around 20 km with NORINCO currently marketing the GP1 and GP6 Laser Homing Artillery Weapon Systems (LHAWS) both of which feature a HE warhead.

In addition to employing 155 mm towed artillery systems, unguided surface-to-surface rockets could also be used for area effect with the sub-munitions being highly effective against landing craft.

The Type 702D meteorological system is the same as that used with standard field artillery systems and takes about 20 minutes to deploy and is used to carry out meteorological detection in the upper atmosphere using a balloon carrying radio sonde.

According to NORINCO, it can provide accurate data of wind speed, wind direction, atmospheric temperature, humidity and pressure below the altitude of 25,000 m. This information is fed to the FCS and enhances the accuracy of the 155 mm artillery projectiles.

The Type 702D meteorological system is operated by a crew of four and consists of two 4 × 4 trucks, one of which carries the radar and associated operating equipment and the other which carries the 10 balloons, five hydrogen bottles and 10 radio sondes.

The radar is claimed to be able to track radio sondes up to a maximum range of at least 200 km.

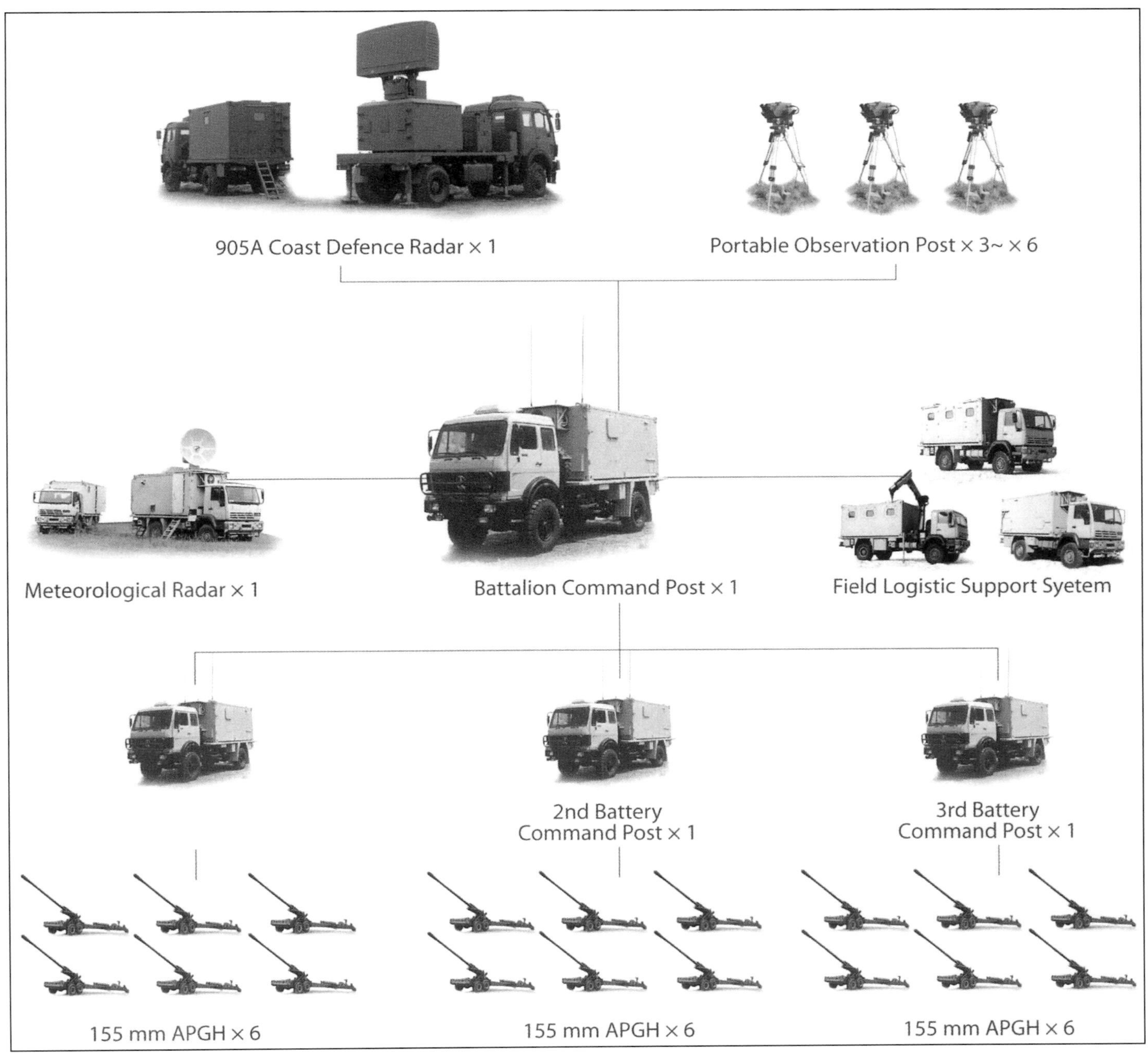

A typical 155 mm coast defence weapon system would include three batteries as well as a complete target acquisition system (NORINCO) 1461676

The truck mounted Type 905 coast defence radar is mounted in two containers each of which is carried on the rear of a 4 × 4 truck.

One of these has the operator's station and the other X-band radar the antenna of which is hinged forwards through 90° to lower the overall height for travelling.

When deployed four stabilisers are lowered to the ground and a generator is mounted to the immediate rear of the cab to power the radar system.

The radar is claimed to be able to detect surface craft out to a range of up to 150 km which would probably be the larger surface ships and can track up to 20 targets at once.

It can correct the fire of the 155 mm weapons by using splash spotting techniques and can also detect low flying aircraft and helicopters.

As well as marketing this 155 mm coastal defence gun system, NORINCO are also marketing a small mobile RC3 medium/short range coastal defence X-band radar.

This is claimed to be able to detect a 10,000 tonne surface craft at a range of 60 km and a smaller 1,000 tonne surface craft at a range of 40 km.

This is installed on a light 4 × 4 vehicle to allow for rapid deployment to where it is required and if required can also be fitted with an Automatic Identification System (AIS).

NORINCO is also marketing the CY-1011 fixed long/medium range 2D costal defence radar which could provide early warning for the 155 mm coastal defence gun system or surface-to-surface coastal defence missiles which are marketed by China National Precision Machine Import and Export Corporation (CPMIEC).

NORINCO claim that this mast mounted radar can detect a 10,000 tonne surface ship at a range of up to 140 km with a smaller 1,000 tonne surface ship being detected at a range of up to 100 km with a maximum of 200 potential targets being tracked at any one time.

As well as acquiring and tracking targets using radar and the tripod mounted day/night observation systems, targets could also be tracked using an unmanned aerial vehicle which could provide a real time target acquisition capability against the moving targets.

Specifications

Not applicable.

Status

Development complete. Ready for production when orders are placed. Individual elements of this 155 mm coastal artillery system are having already been produced in production quantities.

Contractor

China North Industries Corporation (NORINCO).

NORINCO AH4 155 mm/39-calibre Lightweight Gun Howitzer

Development

NORINCO is now marketing at least three towed 155 mm artillery systems, the AH1 155 mm/45-calibre, AH2 155 mm/52-calibre and the AH4 155 mm/39-calibre and details of the AH1 and AH2 are provided in separate entries in *IHS Jane's Land Warfare Platforms: Artillery and Air Defence.*

The recently revealed AH4 155 mm/39-calibre lightweight gun howitzer is very similar in appearance to the combat proven BAE Systems 155 mm/39-calibre M777 and is claimed to use advanced materials in order to keep weight down to 4,000 kg.

AH1 155 mm/39-calibre lightweight gun howitzer deployed in firing position with ordnance elevated (NORINCO) 1366477

AH4 155 mm/39-calibre lightweight gun howitzer deployed in firing position and with ordnance at 0° (NORINCO) 1366483

Details of the NORINCO AH4 155 mm/39-calibre Lightweight Gun Howitzer were released in 2010 but as of March 2013 it is understood that production had yet to commence.

Description

The system is based on a new carriage and it is understood that when in the travelling configuration it is towed by its muzzle.

When deployed in the firing position two spades are deployed either side of the platform to provide a more stable firing platform.

It is normally operated by a crew of seven and takes about three minutes to come into action and two minutes to come out of action. This is claimed to be due to the installation of a hydro-pneumatic mechanism.

Elevation limits are from –3 to + 72° with a traverse of 22.5° left and right which is very similar to the M777 and the 155 mm/39-calibre ordnance is fitted with a multibaffle muzzle brake.

Maximum range firing a HE ERFB-HB projectile is being quoted as 25 km which increases to 30 km firing a HE ERFB-BB round and 40 km firing the HE ERFB-BB RA projectile.

Rate of fire is claimed to be between four and five rounds a minute but the system does not appear to be fitted with a rammer or load assist device. If fitted this would not only increase rate of fire but also reduce crew fatigue.

As well as the previously mentioned HE ERFB-HB, HE ERFB-BB and HE ERFB-BB RA projectiles, these weapons can also fire a number of other NORINCO 155 mm NORINCO ERFB projectiles.

These include cargo (with 56 bomblets), illuminating, white phosphorous, smoke and incendiary.

155 mm conventional projectiles (e.g. not ERFB) include anti-personnel HE and standard HE with the latter having a maximum range of only 18 km when fired from a 155 mm/39-calibre system.

These projectiles can be used with a conventional bag charges as well as the BC1/BC2 modular propellant charge which are claimed to be compatible with NATO 155 mm artillery systems.

As well as these conventional natures of ammunition, these can also fire a 155 mm laser guided projectile which forms part of the G1 155 mm laser homing artillery weapon system. The actual projectile has a maximum range of 20 km and is fitted with an high explosive warhead.

NORINCO 155 mm laser-guided projectile GP6

The more recently developed 155 mm laser guided projectile GP6 is being marketed by NORINCO alongside the earlier GP1.

This is claimed to have a minimum range of 6 km and a maximum range of 25 km with a first round hit probability of 90 per cent and is capable of engaging stationary and moving targets.

The projectile weighs 52 kg and is fitted with a high-explosive warhead and then compared to the earlier GP1 is claimed to have a higher resistance to jamming and its multiple laser coding technology enables co-operative multi-target engagement.

NORINCO 155 mm top attack projectile GS1

NORINCO is now marketing a 155 mm top attack projectile that is similar to the German 155 mm SMART artillery projectile.

The Chinese 155 mm top attack projectile GS1 contains two sub-munitions which are dispensed over the target to attack the vulnerable upper surfaces of armoured vehicles. It is claimed that this is already in production and service with the PLA.

Specifications

	AH4 155 mm/39-calibre
Dimensions and weights	
Crew:	7
Length	
transport configuration:	8.764 m
Width	
transport configuration:	2.57 m
Height	
transport configuration:	2.049 m
Weight	
standard:	4,000 kg
Firepower	
Armament:	1 × 155 mm howitzer

	AH4 155 mm/39-calibre
Main weapon	
length of barrel:	6.045 m
calibre length:	39 calibres
muzzle brake:	multibaffle
maximum range:	40,000 m (HE ERFB-BB RA)
maximum range:	30,000 m (HE ERFB-BB)
maximum range:	25,000 m (HE ERFB-HB)
minimum range:	4,400 m (HE ERFB-HB)
Rate of fire	
sustained fire:	5 rds/min
Main armament traverse	
angle:	45° (22.5° left/22.5° right)
Main armament elevation/depression	
armament front:	+72°/-5°
Survivability	
Armour	
shield:	no

Status
Prototypes undergoing trials. Not yet in production or service.

Contractor
China North Industries Corporation (NORINCO).

NORINCO 152 mm gun Type 83

Development
The China North Industries Corporation (NORINCO) 152 mm Type 83 gun is believed to now equip reserve artillery units. It uses a long 152 mm barrel, believed to be about 45-calibres long, on a much revised carriage and recoil system based on the NORINCO 152 mm gun-howitzer Type 66 which is covered in a separate entry in *IHS Jane's Land Warfare Platforms: Artillery and Air Defence*. Production of the 152 mm gun-howitzer Type 66 is complete and it is no longer marketed by NORINCO. Quantities of these were supplied by China to Iraq and some of these were captured by Allied forces in the 1991 campaign to retake Kuwait. Some are also understood to have taken part in the Iraq war of early 2003.

Production of the NORINCO 152 mm gun Type 83 was completed some time ago and it is no longer marketed.

Description
The 152 mm barrel of the NORINCO 152 mm gun Type 83 is fitted with a single-baffle muzzle brake and the right-hand pitch rifling uses a uniform twist.

A shield is fitted and the split trail legs are equipped with hydraulically operated wheels to assist handling. The carriage rests on a firing platform when emplaced, with the main wheels raised off the ground.

When deployed in the firing position, a large spade is fitted to each trail. When not required these are carried one on each trail. When travelling, the barrel is swung through 180° and clamped over the closed trail legs. The trails are then attached to a steerable limber. Short moves can be made without having to stow the barrel in the reverse position.

The Type 83 gun fires a standard NORINCO 152 mm HE projectile also designated the Type 83. This projectile weighs 48 kg, is 830 mm long and uses a propellant case 960 mm long weighing 37 kg with a full charge and maximum muzzle velocity of 955 m/s.

NORINCO 152 mm Gun Type 83 in travelling configuration and showing limber attached (Steven Zaloga) 0501027

NORINCO 152 mm Gun Type 83 in travelling configuration with ordnance over closed trails 0500356

NORINCO is now marketing a new family of ammunition which can be fired from the NORINCO 152 mm gun Type 83 as well as the NORINCO 152 mm gun-howitzer Type 66 (D-20).

This family of 152 mm projectiles includes High-Explosive (HE), High-Explosive Pre-Fragmented (HE-FRAG), cargo, Base Bleed High-Explosive (BB-HE), Base Bleed- Rocket Assist High Explosive (BB-RA/HE), and radio jamming.

Brief details of this family of Chinese 152 mm projectiles are as follows:

Type:	HE	HE-FRAG	Cargo	BB-HE	BB-RA/HE	Jamming
Length:	703 mm	703 mm	835 mm	874 mm	873 mm	713 mm
Weight:	43.5 kg	43.6 kg	43.6 kg	43.6 kg	43.5 kg	43.6 kg
Muzzle Velocity:	655 m/s	655 m/s	655 m/s	660 m/s	674 m/s	655 m/s
Range:	17.2 km	17.2 km	17 km	22 km	28 km	17.2 km

Notes:
(1) Cargo round carries 63 attack High-Explosive Anti-Tank (HEAT) bomblets each of which are claimed to penetrate 85 mm of conventional steel armour at 0°
(2) Radio jamming projectile has a 60 minute jamming time at a range of 700 m

The optical fire-control system installed on the Type 83 includes a Type 58 panoramic sight weighing 1.942 kg with a magnification of ×3.7, Type 58 collimator sight weighing 1.3 kg and a Type CS-43 direct fire sight weighing 4.6 kg with a magnification of ×5.5.

According to NORINCO, the 152 mm gun Type 83 has a maximum effective direct fire range of 1,170 m.

The 155 mm ordnance has normal recoil of 910 mm and maximum recoil of 950 mm.

A number of these weapons was supplied to Iraq and captured during Desert Storm in 1991. Some of these were provided with vertical attachments on their long barrels so that camouflage netting could be quickly installed.

Variant
NORINCO 155 mm gun XP52

This developmental 155 mm/52-calibre weapon based on the 152 mm gun Type 83 but never entered production or service.

Specifications

	Type 83
Dimensions and weights	
Length	
transport configuration:	9.4 m
Width	
transport configuration:	2.67 m
Height	
transport configuration:	1.865 m
axis of fire:	1.4 m
Weight	
standard:	10,500 kg
combat:	9,700 kg
Mobility	
Carriage:	split trail
Speed	
towed:	60 km/h
Firepower	
Armament:	1 × carriage mounted 152 mm howitzer
Main weapon	
muzzle brake:	single baffle
maximum range:	30,000 m
Rate of fire	
rapid fire:	4 rds/min
Main armament traverse	
angle:	50° (24° left/26° right)

	Type 83
Main armament elevation/depression	
armament front:	+45°/-2.5°
Survivability	
Armour	
shield:	yes

Status

Production complete. In service with China. They were also used by the Iraq Army that has now ceased to exist. Some may also have been sold to some other countries in Asia and the Middle East. As far as it is known the NORINCO 152 mm gun Type 83 is no longer marketed. United Nations sources state that China did not declare any exports of these systems between 1992 and 2010.

Contractor

China North Industries Corporation (NORINCO).

NORINCO 152 mm gun-howitzer Type 66 (D-20)

Development

The China North Industries Corporation (NORINCO) 152 mm gun-howitzer Type 66 is the Chinese version of the Russian 152 mm gun-howitzer D-20. It differs in appearance from the original in few details.

According to United Nations sources, between 1992 and 2010 there was only one export sale of the NORINCO 152 mm gun-howitzer Type 66 and that was to Angola. This country took delivery of four systems from China in 1999.

It is known that production of the NORINCO 152 mm gun-howitzer Type 66 (D-20) was completed some time ago and it is no longer marketed by the company. It is possible that as newer 155 mm towed artillery systems enter service with the People's Liberation Army quantities of 152 mm gun Type 66 will become available for export.

Description

The 152 mm Type 66 may or may not be fitted with a shield. It is normally towed by a 6 × 6 truck at speeds of up to 60 km/h. The carriage is equipped with a Type 58 panoramic sight weighing 1.942 kg with a magnification of ×3.7 and a 10° field of view, a Type 58 collimator weighing 1.3 kg and a Type 66 telescopic sight weighing 4.6 kg with a magnification of ×5.5 and an 11° field of view.

The weapon is mounted on a split trail carriage with each trail being provided with a castor wheel to assist in bringing the weapon into action. When deployed in the firing position these castor wheels rest on the top of each trail.

When deployed in the firing position, the weapon is supported on a circular firing jack and the spade on each trail leg.

The 152 mm ammunition (of the separate loading type, e.g. projectile and charge) fired by this system is also fired by the NORINCO 152 mm Type 83 self-propelled gun-howitzer and both systems may use a similar ordnance. Full details of the 152 mm Type 83, which is understood to be only in service with the People's Liberation Army, are given in a separate entry in *IHS Jane's Land Warfare Platforms: Artillery and Air Defence*. Production of the Type 83 SP system was completed some time ago and it is no longer being marketed.

The hull of the 152 mm Type 83 is used for a wide range of other applications. These include a rocket propelled mineclearing system and a 122 mm self-propelled rocket launcher called the Type 89 (40-round) system. This is no longer being offered on the export market by NORINCO.

The 152 mm gun-howitzer Type 66 has an effective direct fire range of 800 m and a minimum range in the indirect fire role of 7,800 m.

It is normally towed by a 6 × 6 cross-country truck that in addition to its crew carries a quantity of ready use ammunition,

The Type 66 is used to fire at least three types of 152 mm projectile using a variable six-charge bag propellant system. Two types are HE and smoke. Details are as follows:

Type	HE	Smoke
Designation:	Type 66	Type 66
Weight:		
(projectile)	43.56 kg	43.56 kg
(filling)	5.86 kg	6.626 kg
Length:		
(projectile)	702 mm	689.7 mm
Type of filling:	TNT	WP
Muzzle velocity:	655 m/s	655 m/s
Max range:	17,230 m	17,230 m
Fuze:	Liu-4	Yan-2

The third type is an RAP projectile known as the Type MP-152. This weighs 43.56 kg and provides a range increase of 27 per cent to 21,880 m. Accuracy is claimed to be the same as that for a conventional projectile.

NORINCO is now marketing a new family of ammunition which can be fired from the NORINCO 152 mm gun Type 83 as well as the NORINCO 152 mm gun-howitzer Type 66 (D-20).

NORINCO 152 mm gun-howitzer Type 66 in firing position 0500357

Type	Incendiary	Smoke (red)	Illuminating
Weight: (projectile)	43.56 kg	43.5 kg	40.6 kg
Length: (projectile)	704.48/ 707.3 mm	699.81/ 707.14 mm	699.25/ 701 mm
Max muzzle velocity:	650 m/s	650 m/s	651 m/s
Max range:	17,300 m	17,200 m	16,828 m

More recently three additional 152 mm projectiles have been developed by NORINCO:

This family of 152 mm projectiles includes High Explosive (HE), High Explosive Pre-Fragmented (HE-FRAG), cargo, Base Bleed High Explosive (BB-HE), Base Bleed - Rocket Assist High Explosive (BB-RA/HE), and radio jamming.

Brief details of this NORINCO family of 152 mm projectiles are as follows:

Type	HE	HE-FRAG	Cargo	BB-HE	BB-RA/HE	Jamming
Length:	703 mm	703 mm	835 mm	874 mm	873 mm	713 mm
Weight:	43.5 kg	43.6 kg	43.6 kg	43.6 kg	43.5 kg	43.6 kg
Muzzle Velocity:	655 m/s	655 m/s	655 m/s	660 m/s	674 m/s	655 m/s
Range:	17.2 km	17.2 km	17 km	22 km	28 km	17.2 km

Notes:
(1) Cargo round carries 63 attack high explosive anti-tank (HEAT) bomblets each of which are claimed to penetrate 85 mm of conventional steel armour at 0°
(2) Radio jamming projectile has a 60 minute jamming time at a range of 700 m

Specifications

	Type 66
Dimensions and weights	
Crew:	12
Length	
transport configuration:	8.69 m
Width	
transport configuration:	2.42 m
Height	
transport configuration:	2.52 m
axis of fire:	1.22 m
Weight	
standard:	5,720 kg
Mobility	
Carriage:	split trail
Speed	
towed:	60 km/h
Firepower	
Armament:	1 × carriage mounted 152.4 mm howitzer
Main weapon	
length of barrel:	5.195 m
breech mechanism:	vertical sliding block (semi-automatic)
muzzle brake:	double baffle
maximum range:	17,230 m
Rate of fire	
rapid fire:	8 rds/min
Main armament traverse	
angle:	58°
Main armament elevation/epression	
armament front:	+45°/-5°
Survivability	
Armour	
shield:	optional

Status

Production complete. In service with the Chinese armed forces. It is believed that Albania has 18 of these systems in service. Angola took delivery of four systems from China in 1999, Sri Lanka (46). No longer marketed.

Contractor

China North Industries Corporation (NORINCO).

NORINCO 130 mm field gun Type 59-1

Development

The first towed 130 mm field gun manufactured by China North Industries Corporation (NORINCO) was the Type 59, a virtually direct copy of the Russian 130 mm M-46.

At around the same time, China was also producing its 122 mm field gun Type 60, a virtual copy of the Russian D-74. Some of the features of the Type 60 were carried over to the Type 59 to produce the Type 59-1.

Much of the development work on the Type 59-1 weapon was carried out by the No.127 Factory of the Fifth Ministry of Machine Building.

Production of the NORINCO 130 mm field gun Type 59-1 has been completed and all marketing of new systems has now ceased.

Surplus weapons have been sold on the export market following the introduction into service by the People's Liberation Army (PLA) of more capable artillery systems.

Description

The NORINCO Type 59-1 uses a scaled-up version of the Type 60's muzzle brake, recoil mechanism and breech system, all mounted on an all-welded steel box section carriage featuring some components from the Type 60.

This amalgamation of Type 59 and Type 60 features has produced a weapon that is more manoeuvrable than the original Type 59 (M-46). The carriage no longer requires a limber for towing and weight has been saved by using a smaller shield.

According to the United Nations Arms Transfer Lists for the period 1992 through to 2010 the following quantities of 130 mm field gun Type 59-1 were exported by China. It is considered probable that all of these were ex-People's Liberation Army (PLA) stocks and not new-build weapons.

From	To	Quantity	Year
China	Gabon	10	2004
China	Iran	106	1992
China	Pakistan	27	1998
China	Pakistan	60	2000

When the 130 mm field gun Type 59-1 is deployed in the firing position, it rests on the jack under the carriage and the two spades at the rear. To assist bringing the weapon into action each spade is provided with a castor wheel. When travelling or deployed in the firing position this rests on top of the trail.

The Type 59-1 fires two NORINCO types of projectile, HE and illuminating, although at one stage of the gun's development an AP-T projectile was proposed, capable of penetrating 250 mm of conventional steel armour set at 0° at a range of 1,500 m. Details of the two original projectiles are as follows:

Type	HE	Illum[1]
Designation:	Type 59	Type 59
Weight:		
(projectile)	33.4 kg	29.1 kg
(filling)	3.6 kg	1.96 kg
Length:		
(projectile)	668 mm	663 mm
Muzzle velocity:	930 m/s	950 m/s
Max range:	27,490 m	24,500 m[2]
Fuze:	Liu-5	Shi-1

Notes:
[1] burns for 40 seconds producing 500,000 candela
[2] air burst

Both projectiles are fired using a two-charge full-variable charge or a three-charge reduced variable charge.

NORINCO 130 mm field gun Type 59-1 fitted with auxiliary propulsion unit and shown in travelling configuration 0006036

NORINCO 130 mm field gun Type 59-1 in firing position and showing castor wheel on top of each carriage trail 0006035

There have been two enhanced-range Chinese 130 mm projectiles developed for the Type 59-1 gun. One is a Rocket Assisted Projectile (RAP) known as the MP-130. It weighs 33.4 kg and provides a maximum range of 34,360 m, an increase of 25 per cent. A RAP known as the MP-132 has been developed and tested. The second projectile is an ERFB HE, which is fired using a multisection charge system employing four bags, an increment and 2.5 kg of loose propellant. Details of this ERFB round are as follows:

Weight:	
(complete round, with full charge)	58 kg
(complete round, with reduced charge)	50.7 kg
(complete round, packed)	79 kg
(projectile)	32.41 kg
(charge)	3.3 kg
Length:	
(projectile)	799 mm
Type of charge:	TNT
Muzzle velocity:	944 m/s
Max chamber pressure:	3,150 kg/cm^2
Max range:	32,000 m
Fuze:	Liu-5

Early in 1993, NORINCO announced that a new family of ammunition had been developed for the 130 mm field gun Type 59-1 and the Russian 130 mm M-46, brief details of these are as follows:

- HE, weight of projectile 33.4 kg with maximum range of 27.49 km
- HE incendiary, weight of projectile 33.4 kg with maximum range of 27.49 km
- ERFB-BB, weight of projectile 33.4 kg with maximum range of 38 km
- ERFB (boat tail), weight of projectile 32.7 kg with maximum range of 30 km
- ERFB-A (Nubb), weight of projectile 32.41 kg with maximum range of 32 km
- ERFB-B, weight of projectile 32.7 kg with maximum range of 30.7 km
- Illuminating, weight of projectile 29.1 kg with maximum range of 25 km
- Shrapnel (10,000 flechettes), weight of projectile 33 kg with maximum range of 25 km
- Cargo (35 HEAT top attack bomblets), weight of projectile 33 kg with maximum range of 25 km
- Smoke, weight of projectile 32.47 kg with maximum range of 27 km.

NORINCO is now marketing a new family of 130 mm projectiles which can be fired from the NORINCO 130 mm field gun Type 59-1 (or its Russian) equivalent the 130 mm M-46 field gun.

This family includes Base Bleed - Rocket Assist High-Explosive (BB-RA/HE), cargo and Base Bleed High-Explosive (BB-HE).

Brief details of this family of 130 mm projectiles are as follows:

Type	BB-RA-HE	Cargo	BB-HE
Length:	815 mm	687 mm	814 mm
Weight:	31.3 kg	33.3 kg	32.2 kg
Muzzle velocity:	947 m/s	926 m/s	947 m/s
Maximum Range:	44 km	25 km	37.5 km

Note:
Cargo round carries 35 top attack High-Explosive Anti-Tank (HEAT) bomblets each of which are claimed to penetrate 85 mm of conventional steel armour at 0°.

The 130 mm field gun Type 59-1 is fitted with a panoramic sight Type 58 with a magnification of ×3.7 and a 10° field of view which weighs 1.942 kg, collimator sight Type 58 with a 10° 40' field of view and weighing 1.3 kg and a direct fire sight with a magnification of ×5.5° and an 11° field of view and weighing 4.6 kg.

Country	Quantity	Comment
Bangladesh	62	
Cambodia	n/avail	
Cameroon	12	
China	n/avail	
Democratic Republic of Congo	42	
Egypt	420	local production (see separate entry)
Gabon	10	from China in 2004
Iran	985	estimate, these may be Russian weapons
Korea, North	n/avail	local production, also three self-propelled models (available details are given in a separate entry)
Oman	12	understood to be held in reserve
Pakistan	400	including 27 in 1998 and 60 in 2000
Somalia	n/avail	may be non-operational
Sri Lanka	40	
Sudan	30	includes Russian M-46
Tanzania	30	
Thailand	12-15	gift from China
UAE	20	reserve
Vietnam	n/avail	includes Russian M-46

Maximum range in the indirect fire role is 7,800 m and effective range in the direct fire mode is 1,170 m.

The 130 mm field gun Type 59-1 is normally towed by a 6 × 6 truck that in addition to carrying its crew carries a quantity of ready use ammunition up to a maximum speed of up to 60 km/h which depends on the terrain conditions.

It can also be towed by a tracked vehicle up to a maximum speed of 35 km/h but this does depend on the terrain conditions.

Variants

NORINCO has developed an Auxiliary Propulsion Unit (APU) which can be fitted to 155, 152 and 130 mm towed artillery weapons such as the Type 59-1 field gun to increase their battlefield mobility.

This APU consists of one diesel-driven engine, which is mounted on the front of the carriage, with the driver being seated on the left, and the hydraulic transmission systems.

The total weight of the APU is 1,800 kg with the engine having an output of 75 hp and a maximum road speed of 15 km/h.

Sufficient fuel is carried for three hours of continuous operations and time taken to go from travelling mode into firing mode is two minutes. Steering is accomplished by using the two smaller roadwheels at the rear of the carriage.

NORINCO 155 mm gun-howitzer Type GM-45

This was developed by NORINCO for the export market and is a NORINCO 130 mm field gun Type 59-1 (or its Russian equivalent the 130 mm M-46 upgrade) with a 155 mm 45 calibre barrel. As far as it is known this system never entered service and is no longer being marketed by NORINCO.

Specifications

	Type 59-1
Dimensions and weights	
Crew:	10
Length	
transport configuration:	10.8 m
Width	
transport configuration:	2.42 m
Height	
transport configuration:	2.75 m
axis of fire:	1.214 m
Weight	
standard:	6,300 kg
Mobility	
Carriage:	split trail
Speed	
towed:	60 km/h (15/60 km/h (truck); 35 km/h (tracked vehicle))
Firepower	
Armament:	1 × carriage mounted 130 mm howitzer
Main weapon	
breech mechanism:	vertical sliding block (semi-automatic)
muzzle brake:	double baffle
maximum range:	27,150 m
Rate of fire	
rapid fire:	10 rds/min
Main armament traverse	
angle:	58°
Main armament elevation/depression	
armament front:	+45°/-2.5°
Survivability	
Armour	
shield:	yes

Status

Production complete. No longer marketed by NORINCO. As new 155 mm NORINCO marketed artillery systems are introduced into service with the People's Liberation Army, some of the older systems such as the 130 mm field gun Type 59-1 have now become available for sale. In service with the countries listed in the table in the opposite column.

Contractor

China North Industries Corporation (NORINCO).

NORINCO 122 mm Type 96 towed howitzer

Development

For many years China North Industries Corporation (NORINCO) has been manufacturing and marketing a locally manufactured version of the Russian 122 mm D-30 howitzer which is in service with the Peoples Liberation Army as the 122 mm Type 96 howitzer. NORINCO marketing material normally refers to this simply as the "122 mm D-30 howitzer".

According to the United Nations Arms Transfer list for the period 1992 through to 2010 the following quantities of 122 mm Type 96 towed howitzers were exported by China:

From	To	Quantity	Year
China	Bangladesh	18	2006
China	Pakistan	80	2003
China	Pakistan	63	2004

It should be noted that there were additional export sales of Chinese towed artillery systems in this period which were not identified by system type.

These include 65 of an unidentified type to Bangladesh in 2006, a further 36 of an unidentified type to Bangladesh in 2007 and six weapons to Rwanda in 2007.

Description

The Chinese Type 96 version of the 122 mm howitzer D-30 is almost identical to the original Russian version apart from a probably minor changes to meet local manufacturing requirements.

The Chinese Type 96 fires an HE projectile weighing 21.76 kg that is identical to the Russian original. However, the Chinese Type 96 also fires a locally developed long-range HE projectile using a special charge that produces a maximum range of 18,000 m (standard range is 15,300 m).

The long-range HE projectile is longer and uses a more aerodynamic body. It has a wide knurl at the bottom of the body that serves to increase friction between the driving band and the body surface to prevent sliding and shedding of the drive band under high chamber pressures. The knurl on the driving band edge also improves the chamber gas sealing.

NORINCO has also developed two 122 mm cargo projectiles that can be fired from all versions of the D-30. The standard 122 mm HB cargo projectile contains 33 top attack bomblets and has a maximum range of 18 km, while the 122 mm base bleed cargo projectile has a maximum range of 22 km.

Both of these cargo rounds carry 33 bomblets that are designed to attack the vulnerable upper surfaces of armoured vehicles and are also highly effective against troops in the open.

The projectiles are fitted with a MS15 fuze and the High-Explosive Anti-Tank (HEAT) bomblets and can penetrate up to 80 mm of conventional steel armour

Type 96 is the NORINCO copy of the Russian 122 mm howitzer D-30 in firing position 0500358

under ideal static conditions. These new 122 mm projectiles are being offered on the export market.

In 2009 NORINCO was marketing the following family of 122 mm projectiles for use with the locally manufactured Type 96 and its self-propelled versions.

This family includes Base Bleed - Rocket Assist High-Explosive (BB-RA/HE), Hollow Base High-Explosive (HB/HE), Base Bleed High-Explosive (BB/HE), Hollow Base Cargo (HB Cargo), Base Bleed Cargo (BB Cargo) and Hollow Base Incendiary (HB Incendiary).

Brief details of this family of 122 mm projectiles are as follows:

Type	BB-RA/HE	HB/HE	BB/HE	HB Cargo	BB Cargo	HB Incen-diary
Length:	666 mm	660 mm	660 mm	704 mm	718 mm	704 mm
Weight:	21.4 kg	21.76 kg	22.25 kg	21.76 kg	22.25 kg	21.76 kg
Muzzle velocity:	744 m/s	725 m/s	730 m/s	719 m/s	730 m/s	719 m/s
Maximum range:	27 km	18 km	22 km	18 km	22 km	18 km
Notes: Cargo projectiles carry 33 top attack High-Explosive Anti-Tank (HEAT) bomblets each of which are claimed to penetrate 85 mm of conventional steel armour at 0°						

On-carriage fire-control instruments mounted on the Chinese Type 96 include a Type PG-1M panoramic sight weighing 1.9 kg (with a magnification of ×4 and a 10° field of view), a Type OP4M-45 telescopic sight weighing 5 kg (with a magnification of ×5.5 and an 11° field of view) and a Type K-1 indicator. The latter weighs 1.3 kg and has a 10° field of view.

The elevation of the weapon is 360° at elevation –7° to +22°. At elevation +22° to +70°, the elevation is 66° between movable trails and 124° between fixed mount and left and right trail

NORINCO 122 mm howitzer Type D-30-2

In 2004 NORINCO was marketing the 122 mm howitzer Type D-30-2, which appears to be very similar to the original system that has been marketed by the company for some years. As of March 2013 there are no known sales of this version of the 122 mm D-30-2 towed howitzer. Specifications do, however, show some differences between the earlier system and the more recent Type D-30-2.

Specifications

Type D-30-2 howitzer

Calibre:	122 mm
Length: (travelling)	5.4 m
Width: (travelling)	1.95 m
Height: (travelling)	1.66 m
Weight:	3,370 kg
Max range:	
(HE-HB projectile)	18,000 m
(HE-BB projectile)	22,000 m
Elevation/depression:	+70°/–7°
Traverse:	
(elevation range –5 to +22°)	360°
(elevation range +22 to +70°)	66° (between moveable trails); 29° (between front fixed mount and left and right trails)
Rate of fire:	6-8 rds/min
Max towing speed:	60 km/h

NORINCO 122 mm howitzer Type D-30-3

In 2004 NORINCO stated that it was marketing another version of the D-30 called the 122 mm howitzer Type D-30-3. This is essentially an older towed artillery system, believed to be the NORINCO 85 mm Field Gun Type 56, upgraded with the addition of the 122 mm ordnance and recoil system of the D-30. As far as it is known, this has yet to enter quantity production.

Specifications

Type D-30-3 howitzer

Calibre:	122 mm
Length: (travelling)	8.4 m
Width: (travelling)	2.15 m
Height: (travelling)	2.1 m
Weight:	3,200 kg
Maximum range:	
(HE projectile)	15,300 m
(HE-HB projectile)	18,000 m
(HE-BB projectile)	22,000 m
Elevation/depression:	+70°/–3°
Traverse:	54°
Maximum towing speed:	60 km/h

Variants

PLZ-07 122 mm self-propelled howitzer

This full tracked 122 mm SP howitzer was seen for the first time in public in late 2009 and is in service with the Peoples Liberation Army under the designation of the PLZ-07 122 mm SP howitzer. This uses the same 122 mm ordnance as the Type 96 towed howitzer. As of March 2013 this had not been offered on the export market by NORINCO. Details are provided in a separate entry in *IHS Jane's Land Warfare Platforms: Artillery and Air Defence*.

NORINCO 122 mm Type 89 self-propelled howitzer

This has been in service with the Peoples Liberation Army for many years but was never offered on the export market. This uses the same 122 mm ordnance as the Type 96 towed howitzer. Details are provided in a separate entry in *IHS Jane's Land Warfare Platforms: Artillery and Air Defence*.

Production of the NORINCO 122 mm Type 89 self-propelled howitzer is now complete.

NORINCO 122 mm Type 85 self-propelled howitzer

This has been in service with the Peoples Liberation Army for many years and has a 122 mm Type 96 ordnance in an unprotected mount installed at the rear of the tracked chassis.

Production was completed many years ago and it is no longer marketed. Details are provided in a separate entry in *IHS Jane's Land Warfare Platforms: Artillery and Air Defence*.

NORINCO 122 mm SH3 self-propelled howitzer

This full tracked system is currently being offered on the export market by NORINCO and is provided with a turret armed with a 122 mm ordnance based on the Type 96 towed howitzer. Details are provided in a separate entry in *IHS Jane's Land Warfare Platforms: Artillery and Air Defence*.

NORINCO 122 mm SH2 self-propelled howitzer

NORINCO is now offering the SH2 122 mm self-propelled howitzer on the export market and details of this are provided in a separate entry in *IHS Jane's Land Warfare Platforms: Artillery and Air Defence*.

This 6 × 6 system has a 122 mm Type 96 ordnance mounted at the rear of the cross-country chassis.

122 mm (8 × 8) self-propelled howitzer

A number of prototypes of 8 × 8 self-propelled artillery systems that use a 122 mm Type 96 ordnance have been developed to the prototype stage but as far as it is known none have entered production.

Poly Technologies 122 mm (6 × 6) self-propelled howitzer

This is a 6 × 6 forward control cross-country truck with the complete upper part of the 122 mm type 96 towed howitzer mounted on the upper part.

Details of this system are provided in a separate entry in*IHS Jane's Land Warfare Platforms: Artillery and Air Defence*. This is understood to be in service with the Peoples Liberation Army under the designation of the SH-3.

This is also being marketed by NORINCO as the CS/SH1 122 mm self-propelled howitzer.

NORINCO is quoting a combat weight of 16.5 tonnes, a crew of five and a total of 40 rounds of 122 mm ammunition carried.

Specifications

	Type 96
Dimensions and weights	
Length	
transport configuration:	5.4 m
Width	
transport configuration:	1.95 m
Height	
transport configuration:	1.66 m
Weight	
combat:	3,200 kg
Mobility	
Speed	
towed:	60 km/h
Firepower	
Armament:	1 × carriage mounted 122 mm howitzer
Main weapon	
muzzle brake:	multibaffle
maximum range:	15,300 m (with standard ammunition)
minimum range:	4,070 m (with standard ammunition)
maximum range:	18,000 m (with special charge)
maximum range:	18,000 m (with ERFB-HB)
maximum range:	21,000 m (with ERFB-BB)
maximum range:	27,000 m (with BB-RA)
Rate of fire	
rapid fire:	8 rds/min
Main armament traverse	
angle:	360° (at elevation -7° to +22°)

	Type 96
Main armament elevation/depression	
armament front:	+70°/-7°
Survivability	
Armour	
shield:	yes

Status
Production as required. In service with a number of countries including Bangladesh (at least 40), China and Pakistan (at least 143).

Contractor
China North Industries Corporation (NORINCO).

NORINCO 122 mm howitzer Type 54-1

Development
The China North Industries Corporation (NORINCO) 122 mm howitzer Type 54-1 is the Chinese version of the Russian 122 mm howitzer M1938 (M-30). It is virtually identical to the original apart from some small manufacturing expedients.

This was the first large-calibre artillery weapon to be manufactured in China and may have been produced as early as 1953.

According to United Nations sources, between 1992 and 2010 the only export sales of the NORINCO Type 54-1 were to Bolivia, which took delivery of 36 systems in 1992. These are understood to be refurbished rather than brand new weapons.

Production of the NORINCO 122 mm howitzer Type 54-1 is completed and in People's Liberation Army service it is being replaced by more recent weapons with a longer firing range.

Description
The NORINCO 122 mm Type 54-1 howitzer is fitted with a Type 58 panoramic sight weighing 1.3 kg (with a magnification of ×3.7 and a 10° field of view) and a collimator sight Type 58 (with a 10° 40' field of view).

The Type 54-1 uses a variable nine-charge propellant system and can fire at least four types of projectile: HE, smoke, illuminating and incendiary. No details are available regarding the incendiary projectile and it is thought to be little used. Details of the other three NORINCO 122 mm standard artillery projectiles are as follows:

Type	HE	Smoke	Illum[1]
Weight:			
(projectile)	21.76 kg	22.55 kg	21.9 kg
(filling)	3.5 kg	3.65 kg	1.515 kg
Length:			
(projectile)	559 mm	551 mm	531 mm
Muzzle velocity:	515 m/s	509 m/s	496 m/s
Maximum range:	11,800 m	11,930 m	11,000 m[2]
Fuze:	Liu-4	Yan-2	Shi-1

Notes:
[1] burns for up to 30 seconds producing 450,000 candela
[2] † air burst

A leaflet shell containing 1.1 kg of leaflets is also used. A 122 mm cargo round with six layers of HEAT bomblets has been developed to the prototype stage. These HEAT bomblets are dispersed over the target area and are highly effective against the upper surfaces of tanks and other armoured vehicles, as well as soft skinned vehicles and troops in the open.

The recoil system of the 122 mm howitzer Type 54-1 consists of a hydraulic buffer and a hydropneumatic recuperator.

The 122 mm howitzer Type 54-1 is normally towed by a 6 × 6 cross-country truck that in addition to its crew carries a supply of ready use ammunition.

NORINCO 122 mm howitzer Type 54-1 with trails closed for travelling 0500359

The upper part, mount and 122 mm weapon are also used in the NORINCO 122 mm Type 70 and Type 70-1 self-propelled artillery system that are based on a full-tracked chassis. These are only used by the People's Liberation Army (PLA). Production was completed many years ago and they were never exported.

Specifications

	Type 54-1
Dimensions and weights	
Crew:	8
Length	
transport configuration:	5.9 m
Width	
transport configuration:	1.975 m
Height	
transport configuration:	1.82 m
axis of fire:	1.2 m
Weight	
standard:	2,500 kg
Mobility	
Carriage:	split trail
Speed	
towed:	25 km/h
Firepower	
Armament:	1 × carriage mounted 121.92 mm howitzer
Main weapon	
length of barrel:	2.8 m
breech mechanism:	interrupted screw
muzzle brake:	none
maximum range:	11,800 m
Rate of fire	
rapid fire:	6 rds/min
Main armament traverse	
angle:	49°
Main armament elevation/depression	
armament front:	+60.5°/-3°
Survivability	
Armour	
shield:	yes

Status
Production complete. The NORINCO 122 mm howitzer Type 54-1 is known to be in service with the following countries. As the weapon is virtually identical to the Russian 122 mm M1938 (M-30) it is possible that there are additional users.

Country	Quantity	Comment
Bangladesh	57	
Bolivia	18	delivered in 1992
China	n/avail	
Iran	100	estimate
Pakistan	400	
Rwanda	20	estimate
Sudan	50+	
Sri Lanka	n/avail	
Tanzania	80	
Vietnam	n/avail	

Contractor
China North Industries Corporation (NORINCO).

NORINCO 105 mm light howitzer

Development
Early in 1997, it was revealed that China North Industries Corporation (NORINCO) was marketing a 105 mm portable light howitzer which is almost identical to the widely deployed Italian Oto Melara 105 mm Model 56 pack howitzer which is covered in detail in a separate entry in *IHS Jane's Land Warfare Platforms: Artillery and Air Defence*.

As of March 2013 it had not been confirmed that the NORINCO 105 mm light howitzer had entered production for the People's Liberation Army or for the export market. It is however still being offered on the export market by NORINCO.

Oto Melara confirmed that it supplied two examples of the 105 mm Pack Howitzer to China some time ago, although no follow-up orders were placed by the People's Liberation Army (PLA). According to United Nations sources there were no declared exports of the NORINCO 105 mm light howitzer between 1992 and 2010.

NORINCO is now marketing this weapon as the AH3 105 mm Light Howitzer.

Description
Like the Italian Oto Melara 105 mm Pack Howitzer, the NORINCO 105 mm weapon can be used as a howitzer or a low-profile anti-tank weapon firing High Explosive Anti-Tank (HEAT) ammunition.

NORINCO-built 105 mm portable light howitzer deployed in anti-tank firing position and showing three-part trails 0021005

This HEAT projectile has adequate performance against conventional steel armour but is considered ineffective against the latest advanced armour systems.

Firing a standard US High-Explosive (HE) M1 projectile, a maximum range of 10,222 m can be achieved. NORINCO now has the capability to manufacture 105 mm ammunition for this and other Western weapons of this-calibre including the standard US developed HE M1 round.

The NORINCO 105 mm light howitzer can be quickly disassembled in 11 key parts which can be carried by men or horses and a crew of seven can disassemble the weapon in three or four minutes and assemble the weapon again in four to five minutes.

The NORINCO 105 mm light howitzer can be used in two configurations, high configuration as a normal howitzer and in the low-profile position for use as an anti-tank weapon. When used in the anti-tank configuration it has a lower profile and greater traverse.

Today NORINCO is marketing two types of 105 mm projectile for use with the locally manufactured 105 mm light howitzer as well as other towed and self-propelled artillery systems. These are High-Explosive and Hollow Base High-Explosive Incendiary (HB/HEI).

Brief details of these 105 mm projectiles are as follows:

Type	**HE**	**HB/HEI**
Length:	492 mm	599 mm
Weight:	15 kg	15 kg
Muzzle velocity:		
(from M56 PH)	420 m/s	n/avail
(from M101)	472 m/s	n/avail
Maximum range:		
(from M56 PH)	10.2 km	11 km
(from M101)	11.27 km	12 km
Notes:		
(1) HE projectile is available with a Composition B or TNT filling		
(2) HB/HEI has a claimed lethal radius of 35 m		

The recoil system of the NORINCO 105 mm light howitzer consists of a hydraulic buffer and helical recuperator.

While the maximum range of the weapon is 10,222 m in the indirect fire role the minimum range in the indirect fire role is being quoted as 2,200 m.

If required the shield of the NORINCO 105 mm light howitzer can be removed to save weight.

It should be noted that the physical characteristics of the NORINCO 105 mm light howitzer depend on as to whether it is being used in the conventional howitzer role or the lower profile anti-tank role:

	Howitzer Role	**Anti-Tank Role**
Length: (firing)	4.2 m	4.7 m
Width: (firing)	2.53 m	4 m
Height: (firing)	1.93 m	1.55 m
Axis of fire:	1.07 m	690 mm

Variants

NORINCO 105 mm SH5 (6 × 6) self-propelled howitzer

NORINCO is also marketing the 105 mm (6 × 6) self-propelled howitzer which is not related to the NORINCO 105 mm light howitzer covered in this entry.

Details of this system, which as of March 2013 is understood not to have entered production are provided in a separate entry in *IHS Jane's Land Warfare Platforms: Artillery and Air Defence*.

Specifications

	105 mm Light Howitzer
Dimensions and weights	
Crew:	7
Length	
overall:	4.70 m
transport configuration:	1.90 m
Width	
overall:	4.00 m
transport configuration:	1.50 m
Height	
overall:	1.93 m
transport configuration:	3.65 m
axis of fire:	0.69 m
Weight	
standard:	1,250 kg
Mobility	
Carriage:	split trail
Firepower	
Armament:	1 × carriage mounted 105 mm howitzer
Main weapon	
length of barrel:	1.478 m
breech mechanism:	vertical sliding block
muzzle brake:	multibaffle
maximum range:	10,222 m
Rate of fire	
sustained fire:	4 rds/min
rapid fire:	8 rds/min
Main armament traverse	
angle:	36° (18° left/18° right howitzer role; 56° (28°left/28°right) anti-tank role)
Main armament elevation/depression	
armament front:	+65°/-5° (howitzer role; +25°/-5° anti-tank role)
Survivability	
Armour	
shield:	yes

Status

Production as required. There are no known export sales of this system. As far as it is known this system is not used by the PLA. In March 2013 this system was still being marketed by NORINCO.

Contractor

China North Industries Corporation (NORINCO).

NORINCO 100 mm anti-tank gun Type 86

Development

The China North Industries Corporation (NORINCO) 100 mm Type 86 anti-tank gun is very similar to the NORINCO 85 mm field gun Type 56, which is covered in a separate entry in*IHS Jane's Land Warfare Platforms: Artillery and Air Defence*. This has not been offered on the export market by NORINCO for many years.

It incorporates a new 100 mm ordnance and recoil system, but it may be a development of the NORINCO Type 73 anti-tank gun, which is believed to be a copy of the Russian 100 mm T-12 towed anti-tank gun which has a smooth barrel.

There are no known export sales of the Type 73 and no mention of the weapon has been made in any recent NORINCO sales literature.

Description

The Type 86 100 mm anti-tank gun ordnance is smoothbored and features a distinctive slotted muzzle brake. The breech mechanism has a vertical sliding block and the sighting mechanism is mounted to the left of the gun. The carriage uses split trails with a castor wheel on the left trail leg to assist in bringing the gun into action. A shield with sloping sides is fitted with an opening for the sight on the left side.

Recoil is normally 1.05 m with a maximum recoil of 1.18 m which depends on the projectile being fired.

It is normally towed by a 4 × 4 or 6 × 6 cross-country truck that also carries a quantity of ready use ammunition.

Although the Type 86 100 mm anti-tank gun is primarily an anti-tank weapon it can be used for the indirect fire role, even though the potential maximum range is limited by the maximum gun elevation of +38°. Maximum rate of fire is quoted as 8 to 10 rds/min.

A night vision device for sighting purposes is available.

NORINCO 100 mm anti-tank gun Type 86 in firing position with trails spread but not embedded 0500396

NORINCO 100 mm anti-tank gun Type 86 in travelling mode with trails locked together 0500397

The Type 86 100 mm anti-tank gun fires the following types of 100 mm fixed ammunition:

- APFSDS Type 86 with a tungsten penetrator at a muzzle velocity of 1,610 m/s with an effective range of 1,800 m; when firing this round the maximum rate of fire is 4 to 5 rds/min
- APDS Type 73 with a tungsten penetrator at a muzzle velocity of 1,500 m/s with an effective range of 1,700 m
- HEAT Type 73 fin-stabilised projectile at a muzzle velocity of 1,010 m/s.
- APFSDS Type 73 with tungsten alloy penetrator with a maximum direct fire range of 1,700 m
- APFSDS Type 73 with the projectile having a length-to-diameter ratio of 13.3:1 and a maximum range in the direct fire role of 1,730 m
- HE Type 73 fin-stabilised projectile at a muzzle velocity of 900 m/s with a maximum range of 13,700 m.
- When firing these projectiles, the usual recoil length is between 1 and 1.14 m; maximum recoil length is 1.18 m.

NORINCO is now marketing a 155 mm laser-guided artillery projectile and laser-guided (for example 105 mm) tank-launched projectiles. As far as it is known, no 100 mm laser-guided projectiles have been developed for this weapon. It is understood that these Chinese laser-guided anti-armour projectiles are based on Russian technology transfer.

Specifications

	Type 86
Dimensions and weights	
Length	
transport configuration:	9.52 m
Width	
transport configuration:	2.12 m
Height	
transport configuration:	1.838 m
axis of fire:	1.088 m
Weight	
standard:	3,660 kg
Mobility	
Carriage:	split trail
Speed	
towed:	50 km/h
Firepower	
Armament:	1 × carriage mounted 100 mm smoothbore gun
Main weapon	
muzzle brake:	slotted
maximum range:	13,654 m
Rate of fire	
rapid fire:	10 rds/min
Main armament traverse	
angle:	100° (50°left/50°right)
Main armament elevation/depression	
armament front:	+38°/-4°
Survivability	
Armour	
shield:	yes

Status

Production complete. In service with the People's Liberation Army. It is understood that Albania has 50 of these weapons in service. There has been no recent marketing of this weapon by NORINCO.

Contractor

China North Industries Corporation (NORINCO).

NORINCO 85 mm field gun Type 56

Development

The Russian 85 mm Divisional gun D-44 was first supplied to China during the Korean War. China began production of this gun during the early 1960s under the designation Type 56.

NORINCO 85 mm field gun Type 56 in firing position 0500362

Some slight changes were made to the original Russian D-44 design to suit Chinese production methods, one of them being the location of a push-rod firing device in the centre of the elevating handwheel.

Description

The China North Industries Corporation (NORINCO) Type 56 field gun does not appear to use the Russian infra-red searchlight and sight system. It has been observed that some NORINCO Type 85 field guns have a castor wheel under the left trail leg to assist in bringing the weapon quickly into action as well as coming out of action.

Ammunition currently available for the Type 56 field gun includes HE, HEAT-FS and HESH. In the past, AP-T and HE-FRAG have been used. The only current projectile for which definite information is available is the NORINCO Type 56 HEAT-FS. This one-piece round weighs 12.5 kg and is 990 mm long. The direct fire maximum range is 970 m at which it will penetrate 100 mm of conventional steel armour set at an angle of 65°. A Dian-1A fuze is used and the muzzle velocity is 845 m/s.

Optical equipment fitted includes a Type 58 panoramic sight with a magnification of ×3.7, Type 85 direct fire sight with a magnification of ×5.5 and a collimator Type 58.

Recoil system consists of a hydraulic buffer and hydropneumatic recuperator with a maximum recoil of 1.18 m.

While the maximum range in the indirect fire role is quoted as 15,650 m its maximum effective range in the anti-tank role firing a HEAT round is 970 m.

It is normally towed by a 4 × 4 or 6 × 6 cross-country truck with carries the crew and a quantity of ready use ammunition.

Maximum towing speed when on roads of 60 km/h and maximum cross-country towing speed of 15 km/h.

Both of these figures depend on terrain conditions and type of towing vehicle.

Variants

Russia developed and fielded an auxiliary propelled version of the D-44 called the SD-44 but as far as it is known this was never built in China for the People's Liberation Army or offered on the export market by NORINCO. This was developed for use by the Russian airborne forces.

NORINCO 122 mm howitzer Type D-30-3

This is understood to be the NORINCO 85 mm field gun Type 56 fitted with the ordnance and recoil system of the Chinese D-30. Details of this system are given in the entry for the NORINCO 122 mm Type 96 towed howitzer. As far as it is known the NORINCO 122 mm howitzer Type D30-3 has not entered production or service as of March 2013.

Specifications

	Type 56
Dimensions and weights	
Crew:	8
Length	
transport configuration:	8.34 m
Width	
transport configuration:	1.73 m
Height	
transport configuration:	1.42 m
Weight	
standard:	1,750 kg
Mobility	
Carriage:	split trail
Speed	
towed:	60 km/h
Firepower	
Armament:	1 × carriage mounted 85 mm howitzer

	Type 56
Main weapon	
length of barrel:	4.693 m
breech mechanism:	vertical sliding block
muzzle brake:	double baffle
maximum range:	15,650 m
Rate of fire	
rapid fire:	20 rds/min
Main armament traverse	
angle:	54°
Main armament elevation/depression	
armament front:	+35°/-7°
Survivability	
Armour	
shield:	yes

Status

Production complete. As far as it is known this system is no longer marketed by NORINCO. The NORINCO 85 mm field gun Type 56 is used by the following countries:

Country	Quantity
Cameroon	n/avail
China	n/avail
Congo, Democratic Republic	10
Korea, North	n/avail
Mozambique	12
Pakistan	200
Sri Lanka	8
Tanzania	75
Vietnam	200

Contractor

China North Industries Corporation (NORINCO).

NORINCO 82 mm Type W99 towed mortar

Development

China North Industries Corporation (NORINCO) is now offering on the export market an 82 mm towed rapid-firing mortar system called the Type W99.

This was first revealed early in 2001, but it is not known as to whether this system is currently deployed by the People's Liberation Army (PLA).

In appearance the NORINCO Type W99 appears to be almost identical to the Russian 82 mm AM 2B9 Vasilek automatic mortar system. This entered service with the Russian Army around 1971 and has also been manufactured under licence in Hungary.

There are a number of self-propelled versions of the Russian 82 mm AM 2B9 Vasilek automatic mortar system.

As far as it is known, none of these self-propelled models ever entered production.

Description

The NORINCO 82 mm Type W99 mortar weighs 650 kg and is normally operated by a crew of four or five. It has a minimum range of 800 m and a maximum range of 4,270 m. When deployed in the firing position it rests on a circular screw jack lowered under the forward part of the split trail carriage and

NORINCO 81 mm mortar system from the front clearly showing firing jack deployed and trails extended but not embedded (NORINCO) 1365029

Rear view of Chinese NORINCO 82 mm Type W99 mortar deployed in the firing position and showing clips of 82 mm ammunition between the spread trails 0095233

the two trails, each of which are provided with a spade. Each trail is also provided with a handle to assist in bringing the weapon into action.

Compared with conventional muzzle-loaded 81 mm and 82 mm mortars the 82 mm Type W99 offers a number of operational advantages including a higher rate of fire and the gunner can select single shots or burst firing. It can be also used in direct fire role as well as the normal indirect fire role. A conventional mortar can only be used in the indirect fire role.

According to NORINCO, their 82 mm Type W99 mortar is fitted with a standard sighting system and a firing calculator, although no details of the latter have been released.

It can fire a W99 82 mm high-explosive mortar bomb or modified versions of existing 82 mm mortar bombs with maximum muzzle velocity being 272 m/s. Clips of four 82 mm mortar bombs are loaded from the right and four rounds can be fired in 1.5 seconds.

The 82 mm ordnance has a traverse of 30° left and right with elevation from −1° to +85°. Breech loading is normally used at the lower angles of elevation but it can also be muzzle loaded.

The complete Russian 2K21 Vasilek mortar system comprises the actual 82 mm AM 2B9 Vasilek mortar, the 2F54 transport vehicle based on the GAZ-66 (4 × 4) 2,000 kg truck and 226 82 mm mortar bombs (of which 96 are fitted with fuzes ready for use in 24 clips).

The 2B9 Vasilek is normally carried in the rear of the GAZ-66 and unloaded when required for action, using onboard handling gear. NORINCO has stated that it is also offering a complete system built around the 82 mm Type W99 mortar.

Maximum range is 4,270 m with a minimum quoted range of 800 m.

While the maximum rate of fire is up to 170 rds/minute its practical rate of fire is around 120 rds/minute.

In addition to carrying the crew, the towing vehicle also carries a quantity of ready used ammunition.

While the maximum towing speed is up to 60 km/h on hard roads, this is reduced to 20 km/h on dirt roads, but this depends on a number of factors.

According to NORINCO, the Type W99 has a mean time between failure of around 360 rounds and can be deployed in an operational ambient temperature range from −40 to +50°C.

In addition to being marketed by NORINCO, this system in its 81 mm and 82 mm versions is also being marketed by Poly Technologies under the designation of the 81/82 mm Rapid Fire Mortar Type PP02.

Poly Technologies is marketing this is as a complete system that includes the mortar, its ammunition, platform "lighting device" and firing data calculator.

Variants

NORINCO 81 mm Automatic Mortar

More recently, NORINCO has marketed an 81 mm version, which weighs 700 kg, has a minimum range of 800 m and a maximum range of 4,000 m. This is chambered for Western-calibre 81 mm ammunition but as far as it is known this remains at the prototype stage.

NORINCO 81 mm mobile mortar system

In 2008 it was revealed that China North Industries Corporation (NORINCO) had completed development of a new 81 mm self-propelled mortar system which is already being offered on the export market.

The 81 mm mortar installed on a turntable at the rear of a locally produced 4 × 4 cross-country truck chassis that allows it to be rapidly traversed through 360° onto a new target. The weapon can be elevated from −1° through to +85°.

It is understood that this weapon is the NORINCO 81 mm Automatic Mortar covered above which is a further development of the NORINCO 82 mm Type W99 towed mortar system.

Details of the NORINCO 81 mm mobile motar system are provided in a separate entry in *IHS Jane's Land Warfare Platforms: Artillery and Air Defence*. As of March 2013 this is understood to remain at the prototype stage.

The 81 mm mobile mortar system is being marketed by NORINCO under the designation of the CS/SM1 81 mm self-propelled rapid mortar system with a combat weight of 4.8 tonnes and carrying a total of 64 rounds of 81 mm ammunition in clips of four rounds.

Specifications

	Type W99
Dimensions and weights	
Crew:	5
Length	
transport configuration:	4.115 m
Width	
transport configuration:	1.576 m
Height	
overall:	1.18 m
Ground clearance	
overall:	0.26 m
Weight	
standard:	650 kg
Mobility	
Speed	
towed:	60 km/h
Firepower	
Armament:	1 × carriage mounted 82 mm mortar
Main weapon	
maximum range:	4,270 m
Rate of fire	
rapid fire:	120 rds/min
Main armament traverse	
angle:	60° (30° left/30° right)
Main armament elevation/depression	
armament front:	+85°/-1°
Survivability	
Armour	
shield:	no

Status
Development complete. Production as required. It is not known as to whether this system is currently in service with the PLA.

Contractor
China North Industries Corporation (NORINCO).

Other Chinese towed artillery systems

Description
In addition to the China North Industries Corporation (NORINCO) towed anti-tank guns, guns and howitzers covered in separate entries, NORINCO has also designed and built a number of other towed artillery systems. Brief details of these are given below. Recent information has indicated that none of the systems listed below are currently being marketed by NORINCO.

It is however possible that some of these weapons that are now surplus to the requirements of the People's Liberation Army could be offered on the export market by NORINCO or Poly Technologies.

NORINCO 203 mm (8 in howitzer)
This was developed to the prototype stage but never entered production. It is no longer being marketed by NORINCO.

NORINCO 155 mm gun XP52
This was developed to the prototype stage but never entered production. It is no longer being marketed by NORINCO.

NORINCO 152 mm Type 86 cannon
Chinese sources have stated that there is a Type 86 152 mm cannon system which was successfully developed in 1986 and has a maximum range of 30,000 m. Development work on this system started in the late 1950s but was suspended several times.

Final development work was carried out by No 123 and No 127 Factories. This weapon is of the towed type with a welded steel carriage and a 152 mm ordnance with a highly efficient muzzle brake. Types of 152 mm separate loading (projectile and charge) ammunition fired include a hollow base extended range.

NORINCO 152 mm Type 54 howitzer
This is the Chinese-built version of the Russian D-1. It has a combat weight of 3,600 kg, rate of fire of 4 rds/min and a maximum range of 12,400 m. Quantities of surplus PLA weapons are available for export.

Rwanda has about 30 of these weapons in service.

NORINCO 122 mm Type 83 howitzer
This new light howitzer was first seen in public in 1984 during the National Day Parade and the following outline specifications are available:

Calibre:	121.92 mm
Length:	
Weight: (travelling)	2.157 m
Width:	2.157 m
Elevation/depression:	+65/–3°
Traverse: (total)	65°
Rate of fire:	7-8 rds/min
Range:	
(maximum)	18,000 m (with extended-range ammunition)
(minimum)	3,700 m

Recent Chinese sources have indicated that this system had a protracted development period and started again, in 1978, for the fourth time.

Development was carried out by the Number 247 and 724 Factories and features subsequently incorporated include a movable equilibrator, integrated counter recoil system, semi-automatic breech block plus ammunition including a new single-base propellant system and new fuze. Also involved in the ammunition development was No 202 Research Institute.

NORINCO 122 mm Type 60 field gun
This is the Chinese-built version of the Russian D-74, it has a combat weight of 5,500 kg and a maximum range of 24,000 m.

Known export customers for the 122 mm Type 83 include:

Country	Quantity
Iran	100
Pakistan	200
Sudan	n/avail
Vietnam	n/avail
Zimbabwe	18

NORINCO 100 mm Type 73 anti-tank gun
This is possibly a copy of the Russian 100 mm T-12 anti-tank gun and has a maximum direct-fire range of 1,700 m.

This is no longer being marketed by NORINCO.

Chinese towed artillery exports
According to the United Nations, China exported to following quantities of towed artillery weapons between 1992 and 2010 but there could well be other unidentified export orders:

Weapon type	Supplied to	Quantity	Date
Not disclosed	Bangladesh	42	1999
Not disclosed	Bangladesh	18	1996
Not disclosed	Bangladesh	209	1998
Not disclosed	Bangladesh	65	2006
122 mm Type 96 field gun	Bangladesh	18	2006
122 mm Type 54-1 howitzer	Bolivia	36	1992
Type 65 guns	Bolivia	18	1992
130 mm Type 59-1 guns	Gabon	10	2004
105 mm HM2 howitzer	Gabon	4	2004
130 mm Type 59-1 guns	Iran	100	1992
Not disclosed	Jordan	150	2006
Not disclosed	Niger	6	2006
130 mm Type 59-1 guns	Pakistan	27	1998
130 mm Type 59-1 guns	Pakistan	60	2000
122 mm D-30 howitzer	Pakistan	80	2003
122 mm D-30 howitzer	Pakistan	63	2004
Not disclosed	Sudan	18	1992

It should be noted that some of these exports are of re-conditioned artillery systems while others, probably the 122 mm D-30 howitzers, or Type 96 as it is referred to by China, were brand new weapons.

Status
In service with the People's Liberation Army. It is understood that no marketing of these weapons is being carried out and that none of these weapons are currently in production.

Contractor
China North Industries Corporation (NORINCO).

Croatia

RH ALAN 122 mm howitzer D-30 HR M94

Development

This RH ALAN 122 mm howitzer D-30 HR M94 is almost identical to the Russian 122 mm D-30 towed weapon and is used for general fire support. It should be noted that RH ALAN is an export marketing organisation and that it does not manufacture the weapons.

The 122 mm howitzer D-30 HR M94 has been offered on the export market but as of March 2013 there are no known sales an it is understood that there has been no recent production of this weapon in Croatia.

Description

According to RH ALAN, its 122 mm howitzer D-30 HR M94 has the following modifications when compared to the original Russian D-30 towed weapon:

- The original slotted muzzle brake has been replaced by a new double-baffle muzzle brake
- The fixed trail leg has been redesigned and parts of the braking system and electrical installations are fitted onto it
- On the prototype a sighting device with a scale division 360 (=6,000 mils) was retained, whereas on the serial production weapons a sighting device with a scale division 360 (=6,400 mils) has been fitted together with the mechanisms that enable simple use of firing tables according to NATO standards
- For handling of the weapon in the firing position, a manual hydraulic firing jack is fitted which enables the turning of the weapon with deployed trail legs
- The hydraulic recoil brake is of a single-chamber design with a hydraulic compensator placed inside the hydraulic brake cylinder. The standard production Russian 122 mm D-30 weapon had a compensator outside the hydraulic brake which RH ALAN claims to be unreliable.

The 122 mm ordnance of this weapon consists of a monobloc barrel, with a new high-efficiency double-baffle muzzle brake and collars are used for suspension on slides on the cradle. A semi-automatic vertical wedge-type breech block enables a high rate of fire to be achieved.

The counter-recoil system, which consists of a hydraulic brake, compensator and hydropneumatic recuperator, is accommodated in the cradle of the weapon.

The transition from travelling into the firing position and vice versa is readily and quickly made by means of a hydraulic jack mounted under the carriage. For manoeuvring in the firing position, an auxiliary wheel is attached to the towing lunette mounted under the muzzle brake.

A well trained crew can bring the 122 mm howitzer D-30 HR M94 into action in 1.5 to 2.5 minutes. The 122 mm howitzer D-30 HR M94 is normally manned by a six-man crew plus the section commander.

For towing, a vehicle fitted with a standard hook is used. This vehicle must have a payload of at least 3 tonnes and tow a trailer with a loaded weight of at least 3.30 tonnes. Maximum towing speed of the 122 mm howitzer D-30 M94 is quoted as 60 km/h.

When the weapon is elevated from −5° to +18° it can be traversed through a full 360°, but when elevated from +18° to +70° traverse is limited to between two of the trails.

The 122 mm D-30 HR M94 fires ammunition of the separate loading type, for example projectile and charge with reduced, variable and full-type propellant charges. Maximum operating pressure of the ammunition must not exceed 2,600 bars and stated rate of fire is between 7 and 8 rds/min.

As well as the normal natures of 122 mm ammunition, a locally developed family of separate loading (projectile and charge) ammunition has been developed for this weapon and its foreign equivalent weapons.

Two standard natures of 122 mm projectile are manufactured in Croatia, the TF462 and the M76. The former has a maximum range of 15,300 m and the latter a maximum range of 17,133 m. In addition there is the M95 HE round that has a maximum range of 16,000 m with a full charge and 17,133 m with super charge.

122 mm howitzer D-30 HR M94 deployed in the firing position with stakes ready to be driven into ground 0021006

In addition, there are two extended-range high-explosive locally developed 122 mm projectiles, the ER 122 and the ER BB 122. The former is of the Extended Range Full Bore (ERFB) type with nubs and has a maximum range of 16,530 m with full charge and 17,630 m with super charge.

The ERFB ER BB 122 is similar and has a Base Bleed (BB) unit attached, this has a maximum range of 18,850 m with full charge and 20,050 m with super charge.

Type	Muzzle velocity
HE projectile M76	735 m/s (full charge)
HE projectile TF462	690 m/s (full charge)
HE projectile TF462	276/600 m/s (reduced charge)

Variants

According to RH ALAN, the carriage can also be fitted with 100 or 105 mm barrels and the former is in service in Croatia in the coastal defence role. As far as it is known, this has not been produced in quantity.

The recoil of the 122 mm ordnance depends on the projectile/charge combination but is between 740 to 930 mm.

Muzzle velocity for the various types of projectile and charges is as follows:

Specifications

	D-30 HR M94
Dimensions and weights	
Length	
transport configuration:	5.330 m
Width	
transport configuration:	1.95 m
Height	
transport configuration:	1.66 m
axis of fire:	0.9 m
Weight	
standard:	3,440 kg
combat:	3,350 kg
Firepower	
Armament:	1 × carriage mounted 122 mm howitzer
Main weapon	
length of barrel:	4.785 m
breech mechanism:	vertical sliding block
muzzle brake:	double baffle
maximum range:	15,300 m
Rate of fire	
rapid fire:	8 rds/min
Main armament traverse	
angle:	360° (at elevation -5°/+18°; 66° at elevation +18°/+70°; 29° left/29° right between fixed and mobile trails)
Main armament elevation/depression	
armament front:	+70°/-7°
Survivability	
Armour	
shield:	yes

Status

Production as required. In service with Croatia. There are no known exports of the weapon. According to the United Nations the only exports by Croatia between 1992 and 2010 were 40 120 mm mortars to Guinea in 2000.

Contractor

RH ALAN d.o.o.

Czech Republic

100 mm field gun M53

Development

The 100 mm field gun M53 was originally developed in Czechoslovakia in the early 1950s and fulfils a similar role to the Russian 100 mm M1944 (BS-3) field gun and fires the same 100 mm ammunition. Its ammunition is not interchangeable with the more recent Russian smoothbore 100 mm T-12 anti-tank gun.

However, the M53 can be used in both the field and anti-tank roles and has a much longer range than the 100 mm M1944, 15,400 m, as its barrel can be elevated to +42° compared with the +20° of the M1944.

Description

The 100 mm field gun M53 can be distinguished from the 100 mm M1944 (BS-3) by single roadwheels, a shield with a straight top and sides that slope to the rear rather than the M1944's dual roadwheels and a shield with a wave shaped top.

100 mm field gun M53 in firing position (via Franz Kosar) 0500376

The split trail carriage is provided with a castor wheel on each trail to assist in bringing the weapon into action.

The M53 can be fitted with the large diameter APN-3-5 infra-red night device over the rear part of the 100 mm barrel.

The recoil system of the 100 mm field gun M-53 consists of a hydraulic buffer and hydropneumatic recuperator.

While the maximum range of the 100 mm field gun M53 in the indirect fire role is 21,000 m, its effective range when being used in the anti-tank role is only 1,000 m.

The 100 mm HE projectile has a maximum muzzle velocity of 900 m/s while the HVAPDS-T round has a maximum muzzle velocity of 1,415 m/s.

The 100 mm field gun M53 was typically towed by a Tatra T-138 truck (6 × 6) truck which in addition to the gun crew also carried a quantity of ready use ammunition.

There are no known plans to upgrade these weapons to extend their operational lives. The 100 mm field gun M53 is also referred to as the 100 mm field gun K53.

85 mm field gun M52

As far as it is known, this is no longer in front line service with any country.

Specifications

	M53
Dimensions and weights	
Crew:	6
Length	
transport configuration:	9.1 m
Width	
transport configuration:	2.36 m
Height	
transport configuration:	2.606 m
axis of fire:	1.25 m
Ground clearance	
overall:	0.35 m
Track	
vehicle:	1.98 m
Weight	
standard:	4,280 kg
combat:	4,210 kg
Mobility	
Carriage:	split trail (with castor wheel on each)
Tyres:	11.00 × 22
Firepower	
Armament:	1 × carriage mounted 100 mm
Main weapon	
length of barrel:	6.735 m
breech mechanism:	horizontal sliding block
muzzle brake:	double baffle
maximum range:	21,000 m
Rate of fire	
rapid fire:	10 rds/min
Main armament traverse	
angle:	60°
Main armament elevation/depression	
armament front:	+42°/-6°
Survivability	
Armour	
shield:	yes

Status

Production complete. No longer marketed. This weapon is no longer in front line service with the Czech Republic or Slovak Republic having been replaced by ATGM. Czech Republic supplied 26 of these weapons to Latvia in 1995 under the designation K53. There has been no marketing of this weapon for many years. It is possible that additional quantities of these weapons are now surplus to requirements and available for export.

As far as it is known, Latvia is the only user of the 100 mm field gun M53 but in the long term, spare parts could be a problem.

Contractor

Former Czech state factories.

Egypt

Factory 100 artillery systems

Description

Adu Zaabal Engineering Industries Company, also known as Factory 100, has reverse-engineered the Russian 122 mm D-30 towed howitzer, Chinese 130 mm field gun Type 59 and the Russian twin 23 mm ZU-23-2 towed anti-aircraft gun.

The first 122 mm D-30 was completed in Egypt early in 1984 with the now BAE Systems of the US providing some part-machined components which were completed in Egypt.

Factory 100 has also been closely involved with BAE Systems in the upgunning of the Russian-supplied T-55 MBT with the BAE Systems 105 mm L7A3 rifled gun.

The Egyptian defence industry also manufactures ammunition for these weapons.

For the D-30-M 122 mm system natures include HE, HEAT, illumination and smoke while for the M59-1M 130 mm natures include HE, armour-piercing, smoke and illuminating.

Recent information has stated that the Egyptian Army has a fleet of about 190 D-30 and 420 Type 59 towed artillery weapons.

The Egyptian designations for these weapons are the D-30-M (D-30) and M59-1M (Type 59).

130 mm Type 59-1M

The maximum pressure of the 130 mm ordnance of the Egyptian Type 59-1M is quoted as being 3,250 kg/cm^2 and maximum muzzle velocity is 939 m/s.

It is normally towed by a 6 × 6 cross-country truck that in additional to carrying its crew also carries a quantity of ready use ammunition.

122 mm D-30M

The maximum pressure of the 122 mm ordnance of the Egyptian D-30M is being quoted as 2,300 kg/cm^2 and maximum muzzle velocity is being quoted as 690 m/s.

Egyptian 130 mm M59-1M guns in firing position (Christopher F Foss) 0500377

Egyptian 122 mm D-30-M towed howitzer in travelling position (Christopher F Foss) 0500378

It is normally towed by a 6 × 6 cross-country truck that in addition to its crew also carries a quantity of ready use ammunition.

SP122

The complete ordnance and elevating system of the Egyptian-built 122 mm D-30 is fitted to the SP122 self-propelled artillery system. This was developed by the now BAE Systems under contract to the Egyptian Army. A total of 124 systems have been purchased with final deliveries taking place late in 2000. Production can commence again if additional orders are placed.

It should be noted that the SP122 was never offered on the export market by BAE Systems.

The hull used for the SP122 is based on that used for the M109/M992 also used by Egypt.

T-54/T-55 with D-30

Early in 1999, it was revealed that a new 122 mm self-propelled artillery system was entering production in Egypt. This consisted of a modified T-54/T-55 tank hull fitted with a new all-welded steel turret armed with the locally produced 122 mm D-30. Prime contractor for the chassis was to be Factory 200 (Egyptian Tank Plant) while prime contractor for the turret and D-30 was to be Factory 100.

Egyptian production of 155 mm 155 GH 52

In August 1999, the now Patria (at that time Patria Weapon Systems) of Finland signed an agreement with the Arab Republic of Egypt under which Finland would transfer technology for the local production of the 155 mm 155 GH 52 weapon system. Some reports have stated that 16/18 are in service with Egypt. This model is not fitted with an APU and has the designation of the 155 mm EH52. Finland delivered the first complete system to Egypt in 2000. Unconfirmed reports have also mentioned a self-propelled version based on a modified T-54/T-55 tank chassis. It is understood that this system remains at the prototype stage.

Specifications

	M59-1M	D-30-M
Dimensions and weights		
Crew:	9	7
Length		
transport configuration:	10.8 m	5.4 m
Width		
transport configuration:	2.42 m	1.95 m
Height		
transport configuration:	2.75 m	1.66 m
Ground clearance		
overall:	0.3 m	0.325 m
Weight		
standard:	6,300 kg	3,210 kg
combat:	6,100 kg	3,150 kg
Mobility		
Speed		
towed:	60 km/h	60 km/h
Firepower		
Armament:	1 × carriage mounted 130 mm howitzer	1 × carriage mounted 122 mm howitzer
Main weapon		
length of barrel:	7.15 m	4.27 m
breech mechanism:	horizontal sliding block	vertical sliding block
muzzle brake:	double baffle	multibaffle
maximum range:	27,150 m	15,300 m
Rate of fire		
rapid fire:	8 rds/min	8 rds/min
Operation		
stop to fire first round time:	180 s	90 s (est.)
out of action time:	180 s	90 s (est.)
Main armament traverse		
angle:	58°	360°
Main armament elevation/ depression		
armament front:	+45°/-2.5°	+70°/-7°
Survivability		
Armour		
shield:	yes	yes

Status

Production as required. In service with the Egyptian Army: Type 59-1M (400); 122 mm D-30M (190). Egypt has supplied Bosnia-Herzegovina with 12 D-30-M and 12 M59-1M weapons. According to United Nations sources Egypt delivered six D-30 weapons to Rwanda in 1992.

Contractor

Abu Zaabal Engineering Industries Company.

Finland

Patria 155 mm 155 GH 52 APU

Development

In early 1994, it was revealed that the Finnish company VAMMAS, previously known as Tampella and more recently Patria, had completed the prototype of a new 155 mm 52-calibre towed artillery system fitted with an Auxiliary Propulsion Unit (APU).

This weapon is known as the 155 GH 52 APU and was originally developed as a private venture together with the Finnish Defence Forces (FDF).

It is a further development of the earlier 155 mm M-83 towed howitzer which was produced in two versions, neither of which has an APU.

Development of the 155 GH 52 APU commenced in 1990 with the first prototype being completed the following year. Firing trials were successfully completed in 1993.

It would be possible to upgrade earlier weapons to the new 155 mm 52-calibre version.

Following trials with prototype weapons under the designation of the KX, the Finnish Army placed an order for a quantity of weapons under the designation of the 155K98 with first deliveries taking place in 1998.

Late in 2000, another batch of 155 mm 155 GH 52 APU systems was ordered for the Finnish Defence Force, which brought the total number of systems up to 36 units.

This figure of 36 consisted of 27 + 9 with final deliveries under this contract made late in 2002. The first regiment was fully operational by 2001 and has a total of 18 guns in three batteries each of six guns, which can be further split into two troops each of three guns.

In the FDF the 155K98 is deployed as the Corps level artillery system with a total of three regiments formed with a total of 54 guns.

Final production models of the 155K98 for the FDF were fitted with a Tactical Advanced Land Inertial Navigation (TALIN) system, a modern version of the Honeywell Modular Azimuth Position System (MAPS).

MAPS was originally developed for the US Army M109A6 Paladin with over 2,000 built for the home and export markets. The installation of TALIN on the 155K98 reduces the time taken to come into action time as well as increasing the accuracy of the weapon.

In August 1999, Patria signed an agreement with the Arab Republic of Egypt under which Finland would transfer technology for the local production of the 155 mm 155 GH 52 which has the local designation of the 155 mm EH52 towed howitzer.

The first complete Patria 155 mm 155 GH 52 system was delivered to Finland in 2000. Prime contractor in Egypt is Factory 100 (Abu Zaabal Engineering Industries Company). Recent information is that this programme has come to a halt.

Description

The 155 GH 52 APU weapon consists of five key components, barrel assembly, top carriage, bottom carriage, sighting devices and the APU.

The barrel assembly consists of the barrel, muzzle brake, fume extractor, breech piece, primer housing lock, firing safety lever, breech block safety lever and flick rammer safety lever.

The top carriage consists of the saddle, elevating mechanism, traversing mechanism gearbox, equilibrators, semi-automatic mechanism cocking device, layers pressure trigger, cradle, buffer, buffer replenisher, recuperator and the flick rammer.

The bottom carriage consists of the gun jack, firing platform, central structure, trail, trail support cylinder, walking beam, wheel, dolly wheel, trail connecting lock, trail support plate, towing eye, cabin, APU, mechanical control of hydraulic units, electric control of hydraulic units, external output for accessories and manual hydraulic pump.

Sighting devices covers indirect and direct firing. The sighting devices for indirect firing include elevation-setting mechanism, pitch levelling mechanism, cross-levelling mechanism, telescope mount holder and panoramic telescope illuminating equipment.

Close-up of Patria 155K98 field gun of the Finnish Defence Force fitted with the TALIN (Christopher F Foss) 0130007

Patria 155K98 field gun of the Finnish Defence Force deployed in the firing position (Christopher F Foss) 0130008

Patria 155K98 field gun of the Finnish Defence Force in travelling configuration with ordnance locked in position over closed trails (Christopher F Foss) 0130009

The panoramic telescope can also be used for direct firing but a special telescope, an elbow telescope sight, is normally used for direct firing. There is a mount for the elbow telescope on the gun layer's side of the gun.

The APU includes the control devices, steering panel, gauge panel, electric control of hydraulic units, mechanical control of hydraulic units, steering joystick, external output for accessories and manual hydraulic pump.

The 155 mm 52-calibre ordnance has a 23-litre chamber, semi-automatic horizontal sliding wedge breech mechanism, pneumatic rammer located on the left side to ram projectiles and charges at all angles of elevation and a single-baffle muzzle brake. A fume extractor is fitted as standard. The 23-litre chamber conforms to the NATO Joint Ballistic Memorandum Understanding (JBMoU).

The recoil brake is hydraulic while the recuperator is pneumatic. Recoil length depends on the charge and elevation but maximum recoil stroke is 1,500 mm.

The manual elevating and traverse mechanism is the same as that fitted to the earlier Patria M-83 towed howitzer and is fitted on the left side of the carriage together with the direct and indirect sights. The split trail carriage is of a new design.

When deployed in the firing position, the forward part of the carriage rests on a hydraulically operated firing platform under the carriage, with each of the trails being provided with a spade, each of which is fitted with pegs to avoid the need to dig them in.

The weapon takes two minutes to come into action and is normally operated by a nine-man crew .

When travelling, the 155 mm 52-calibre ordnance is traversed through a full 180° and locked in position over the closed trails. The travelling lock for the 155 mm ordnance is located on the right trail.

The carriage has four rubber-tyred roadwheels, two either side, for high-speed towing with the suspension being of the walking beam type for improved cross-country mobility. In addition, there is a single dolly wheel on each trail leg to assist in bringing the weapon into action.

Mounted on the forward part of the carriage is the APU, a Deutz air-cooled turbocharged diesel developing 104.6 hp, coupled to a hydrostatic transmission which in turn powers the four motors located one at each wheel station.

Maximum stated rate of fire is 10 rds/min but normal rate of fire is being quoted as eight rounds a minute with a burst rate of fire of three rounds in 12 seconds.

While a typical crew is nine it can be operated by a crew of two for a short period and with a reduced rate of fire.

Maximum speed in two wheel drive configuration is 15 km/h and in four wheel drive configuration is 7.5 km/h. Weight at the towing eye is being quoted as 1,000 kg.

Maximum gradient using APU is 40 per cent and maximum side slope using APU is 30 per cent.

The driver is seated on the forward part of the carriage on the left side and steers the system using a small joystick. The fuel tank holds five litres of diesel fuel.

The installation of the APU enables the 155 GH 52 APU to be taken quickly out of action and deployed into positions that are not normally accessible by the prime mover.

The FDF uses a local version of the French Nexter Systems (previously Giat Industries) 155 mm LU111 high-explosive projectile under the designation 155 tkr88 and this has a maximum range of 23.8 km without a Base Bleed (BB) unit.

The FDF has also adopted and taken into service the German Rheinmetall 155 mm DM 642 cargo projectile that has 63 bomblets each with a self-destruct mechanism.

Patria also market two of their own 155 mm projectiles, the 155 ERFB-N (Extended Range Full Bore - Nubbs) which has a maximum range with base bleed unit fitted of 42.5 km and the 155 mm ERFB-S (streamlined) which has a maximum range of 40.5 km with BB.

Variants

There is also a 155 mm 45-calibre version, without an APU, which is called the 155 GH 45. This was developed and tested but never entered production and is no longer marketed.

Specifications

	155 GH 52 APU
Dimensions and weights	
Crew:	9
Length	
overall:	11.0 m
transport configuration:	10.90 m
Width	
transport configuration:	2.82 m
Height	
transport configuration:	2.25 m
Ground clearance	
overall:	0.40 m
Track	
vehicle:	2.40 m
Wheelbase	1.45 m
Mobility	
Carriage:	split trail
Speed	
max speed:	15 km/h
towed:	100 km/h
Fording	
without preparation:	0.7 m
Gradient:	40%
Side slope:	30%
Firepower	
Armament:	1 × carriage mounted 155 mm howitzer
Main weapon	
length of barrel:	8.06 m
calibre length:	52 calibres
breech mechanism:	horizontal sliding block
muzzle brake:	single baffle
maximum range:	3,000 m (direct fire)
maximum range:	27,000 m (standard ammunition)
maximum range:	41,300 m (BB)
Rate of fire	
rapid fire:	10 rds/min
Main armament traverse	
angle:	70° (35° left/35° right)
Main armament elevation/depression	
armament front:	+70°/-5°
Survivability	
Armour	
shield:	no

Status

Production complete. No longer marketed. In service with the Finnish Army. May now be manufactured under licence in Egypt. Sixteen/eighteen may be in service with Egypt.

Prime contractor in Egypt is the Abu Zaabal Engineering Company who has also manufactured other weapons including the Russian 122 mm D-30 and the Chinese 130 mm Type 59.

Contractor

Patria.

Patria 155 mm M-83 howitzer

Development

The 155 mm M-83 howitzer was originally developed and produced by Tampella, which then became known as VAMMAS Ordnance and more recently Patria.

The Finnish Army designation for this system is the 155 K 83. United Nations sources state that Finland did not export any 155 mm artillery systems between 1992 and 1998. Between 1999 and 2000 Finland exported only one 155 mm artillery system. This was a 155 mm/52-calibre system to Egypt based on the Patria 155 mm 155 GH 52 APU. This is called the EH52 as it is not fitted with an APU.

Patria 155 mm M-83 howitzer deployed in firing position 0501034

Patria 155 mm M-83 howitzer in travelling configuration with ordnance locked in position over closed trails 0501035

Patria has completed production of their 155 mm 155 GH 52 APU and is now concentrating on its 120 mm turret-based mortar systems.

These include the win 120 mm AMOS in production for the Finnish Defence Force on the Armoured Modular Vehicle (AMV) hull and the single 120 mm NEMO Turret Mortar System, in production for Saudi Arabian National Guard (on LAV hull) and the United Arab Emirates (naval application).

Description

The 155 mm M-83 howitzer has a split trail carriage and can be fitted with a 39 or 45-calibre ordnance which has a distinctive T-type muzzle brake. As an alternative to the normal 155 mm ordnance, the gun can be supplied with a 122, 130 or 152 mm barrel allowing Chinese/Russian types of ammunition to be fired rather than 155 mm NATO-type ammunition.

The semi-automatic horizontal sliding breech mechanism and a special loading mechanism enable a high rate of fire to be achieved with obturation being obtained by a steel ring.

Maximum rate of fire is being quoted as 10 rds/minute while a burst rate of fire of three rounds in 12 seconds can be achieved with a well trained crew.

The special loading system is operated by compressed air, with the cylinder for the latter located on the right side of the mount. The sights, elevation and traverse wheels are located on the left side of the mount.

According to Patria, the weapon is very easy to handle with only a few men required to bring the weapon into firing or travelling configuration.

When in the travelling configuration the ordnance is traversed through 180° and locked in position over the closed trail legs. When in the firing position the weapon is supported on a firing jack under the forward part of the carriage and on the two trails, each of which is fitted with a spade.

Steel stakes are provided for the spades and these are carried in racks on the inside of each trail.

The 155 mm M-83 fires the standard NATO M107 HE projectile as well as hollow base and two types of base bleed projectile, one type with nubs and the other without.

The design of the charge system is optimised so that range areas cover each other in low- and high-elevation angles. Compared to the standard NATO charge system (at least four different charges), the Patria system consists of three different charges, incremental charge and two types of full charges.

The make-up of the incremental charge is similar to NATO charges M4A2 and M3A1, including four increments, zones 5, 6, 7 and 8. Full and top charges are similar to NATO M2 and M11, zones 9 and 10. The latter is only used in the 155 mm 45-calibre gun.

Some 155 mm M-83 systems have been observed fitted with a muzzle velocity radar measuring system over the ordnance. This feeds information to the fire-control computer.

Variants

There are no variants of the Patria 155 mm M-83 howitzer and the company has ceased all production and marketing of 155 mm artillery systems to concentrate on 120 mm motar systems.

Specifications

	M-83/45	M-83/39
Dimensions and weights		
Weight		
combat:	9,500 kg	9,500 kg
Mobility		
Carriage:	split trail	split trail
Speed		
towed:	80 km/h	80 km/h
Firepower		
Armament:	1 × carriage mounted 155 mm howitzer	1 × carriage mounted 155 mm howitzer
Main weapon		
length of barrel:	7.1 m	5.99 m
calibre length:	45 calibres	39 calibres
breech mechanism:	horizontal sliding block	horizontal sliding block
muzzle brake:	single baffle	single baffle
maximum range:	39,600 m	30,000 m
Rate of fire		
rapid fire:	10 rds/min	10 rds/min
Main armament traverse		
angle:	90° (45° left/ 45° right)	90° (45° left/ 45° right)
Main armament elevation/ depression		
armament front:	+70°/-5°	+70°/-5°
Survivability		
Armour		
shield:	no	no

Status

Production complete. In service with the Finnish Army. No longer being marketed.

Contractor

Patria.

France

Nexter Systems Trajan 155 mm/52-calibre towed gun system

Development

Late in 2011 Nexter Systems revealed that to meet the potential requirements of the export market, especially India, the company was developing the Trajan 155 mm/52-calibre towed artillery system.

Nexter Systems have already signed an agreement with Larsen & Toubro of India under which Trajan will be offered to meet the Indian Army requirement for a 155 mm Towed Artillery Gun Programme (TAGP).

The first two examples of the Trajan were completed early in 2013 with the carriages being manufactured in India. One of these was shown in public for the first time at the IDEX exhibition held in the United Arab Emirates in February 2013.

One of these Trajan systems will be retained in France while the second will be available for overseas trials in India when requested.

Description

Trajan essentially consists of the complete elevating mass and 155 mm/52-calibre ordnance, loading and aiming system of the Nexter Systems CAESAR (6 × 6) self-propelled artillery system integrated on the carriage of the 155 mm/39-calibre TR1 towed artillery system.

The latter was developed to meet the requirements of the French Army and is also deployed by Cyprus and Saudi Arabia.

The two carriages for the first Trajan were manufactured in India at the facilities of Larsen & Toubro and sent to the Nexter Systems facility in Bourges for final integration.

The CAESAR (6 × 6) 155 mm/52-calibre self-propelled artillery system is currently in service with France (5 + 72), Saudi Arabia (136) and Thailand (6) and has recently been ordered by Indonesia (37).

CAESAR has seen operational service with the French Army in Afghanistan, Lebanon and Mali.

When deployed in the firing position, the forward part of the Trajan carriage is raised off the ground and supported by a firing jack with each trail being fitted with a spade.

First example of the Nexter Systems Trajan 155 mm/52-calibre towed gun system with ordnance at maximum elevation and showing APU mounted on forward part of carriage (IHS/Patrick Allen) 1510155

When travelling the 155 mm/52-calibre ordnance is traversed to the rear and locked in position over the closed trails.

The 155 mm/52-calibre ordnance has a 23-litre chamber and is fully compliant with the Joint Ballistic MoU and is fitted with a double baffle muzzle brake and an upward opening screw breech mechanism.

Trajan can be brought into action in less than 90 seconds and is claimed to have a maximum rate of fire of three rounds in 30 seconds, up to six rounds a minute and a sustained rate of fire of 12 rounds in less than three minutes.

Firing a 155 mm Extended Range Full Bore projectile the Trajan can achieve a maximum range of 42 km or more than 55 km using a rocket assisted projectile.

The split trail carriage of the Trajan is fitted with an 80 hp diesel Auxiliary Power Unit (APU) that speeds up deployment of the weapon as well as reducing crew fatigue.

A flick loader is also installed and automatic aiming of the weapon is carried out via the on board ballistic computer which also enables automatic re-aiming after each round is fired for increased accuracy.

A day/night sight is fitted on the carriage for direct target engagements against stationary and moving targets out to a range of 2,000 m.

The embedded fire-control system provides for increased accuracy as well as being able to be connected to any artillery field management system.

It also features automatic positioning through the highly accurate on board 3D inertial navigation unit with a muzzle velocity radar feeding information to the FCS.

Specifications

	Trajan 155 mm/52-calibre towed gun system
Dimensions and weights	
Crew:	6
Length	
overall:	11 m (travelling)
Width	
overall:	3 m (travelling)
Height	
overall:	2.5 m
Ground clearance	
overall:	0.3 m
Weight	
standard:	14,000 kg
Mobility	
Carriage:	split trail
Speed	
max speed:	10 km/h (APU road)
cross-country:	30 km/h (APU 5 km/h)
towed:	80 km/h
Fording	
without preparation:	1 m
Firepower	
Armament:	1 × carriage mounted 155 mm howitzer
Main weapon	
calibre length:	52 calibres
muzzle brake:	double baffle
maximum range:	42,000 m (ERFB-BB)
minimum range:	4,500 m (ERFB-BB)
maximum range:	55,000 m (Rocket assisted projectile)
Rate of fire	
sustained fire:	6 rds/min
Main armament traverse	
angle:	34° (17° left/right)
Main armament elevation/ depression	
armament front:	+73°/-2°
Survivability	
Armour	
shield:	no

Status

Development complete. Ready for production on receipt of orders.

Contractor

Nexter Systems (but see development).

Nexter Systems 155 mm towed gun TR

Development

This 155 mm towed gun TR (*Le Canon de 155 mm Tracte*) was developed by the now Nexter Systems (previously Giat Industries) to meet the requirements of the French motorised infantry divisions and rapid action forces as the replacement for the older French-built 155 mm howitzer Model 50.

The weapon was shown for the first time at the 1979 Satory Exhibition of Military Equipment. Following trials with prototype systems, a pre-production batch of six weapons was delivered by the company facility at Bourges for troop trials from December 1987.

The main production run commenced in 1989 and a total of 105 weapons were delivered to the French Army.

Export sales of the 155 mm towed gun have been made to at least two countries, Cyprus (12) and Saudi Arabia (28).

The 155 mm towed gun TR was deployed to Saudi Arabia and took part in Operation Desert Storm, the liberation of Kuwait, in the first quarter of 1991.

Production of components for the 155 mm towed gun TR was undertaken by various Nexter Systems facilities but system's integration was undertaken at Bourges where all French artillery and tank barrels are produced.

In French Army service the Nexter Systems 155 mm towed gun TR has already started to be replaced by the Nexter Systems CAESAR truck mounted 155 mm/52-calibre self-propelled artillery system.

Following trials with five pre-production CAESAR a total of 72 production systems were supplied by Nexter Systems to the French Army from 2008 and deliveries were completed in 2011.

These have seen operational service with the French Army in Afghanistan, Lebanon and Mali.

The French artillery still deploys quantities of Nexter Systems 155 mm AUF1 tracked SP artillery systems (based on a modified AMX-30 tank hull) and the 155 mm TRF1 towed artillery systems.

Nexter Systems 155 mm TR gun with 52-calibre barrel deployed in firing position 0501036

These are expected to be replaced by a second batch of 64 CAESAR which would be delivered in the 2017 to 2020 time frame, funding permitting.

This would mean that the CAESAR would be the only conventional 155 mm tube artillery system deployed by the French artillery.

The French artillery also operates the TDA 120 mm rifled mortars and the Multiple Launch Rocket System (MLRS) with the first 15 of the latter already being upgraded to fire the Lockheed Martin Guided MLRS to provide a precision effect.

Description

A crew of eight is required for the tractor and gun: commander; layer/gunner who also drives the gun in the self-propelled mode; loader who operates the loading device during firing, controls the raising and lowering of the gun and operation of the trails when moving into and out of action; two men who are responsible for the preparation of the propelling charges and supplying the gun; two loaders responsible for the preparation of the projectiles and supplying the loading mechanism; and finally the tractor driver. If the tasks are divided differently, the crew can be reduced to seven, six of whom serve the gun. If required, the gun can be brought into and out of action by a crew of three.

The 155 mm 39-calibre barrel has a double-baffle muzzle brake which when travelling is swung through 180° and secured in position over the closed trails.

The 155 mm 39-calibre barrel has a hydropneumatic recoil system and is claimed to have a tube life of 3,000 rounds.

Mounted on the front of the carriage is a 39 hp engine which drives three hydraulic pumps, one for each of the two road wheels and one to provide power for functions such as elevation, traverse, raising the suspension, trail wheel jacks and the loading mechanism.

Travelling under its own power the gun has a maximum road speed of 9 km/h, can climb a gradient of 60 per cent and ford to a depth of 1 m in the self-propelled mode and 1.5 m in the towed mode.

In French Army service the 155 mm TR is towed by a Renault Trucks Defense TRM 10 000 (6 × 6) truck which carries the crew of eight and 48 rounds ready use of 155 mm ammunition. This combination has a maximum road speed of 80 km/h and a road range of 600 km, but it can be towed by various other types of prime mover.

On arriving at the battery position, the weapon is first uncoupled from the prime mover and then the barrel is swung from its travelling position through 180° into its firing position. The wheels are raised off the ground so that the weapon rests on its firing jack under the carriage. The trails are then swung apart, which is a simple operation as castor wheels are fitted on each trail. When opened out, the small wheels are raised clear of the ground and the trails staked into position. The maximum time taken to bring the weapon into battery is 90 seconds and it can be brought out of battery in the same time due to the hydraulic system.

The layer is seated on the left of the mount and has hydraulic aiming controls. The sights, which can be used by day and night, are the GA 81 goniometer, indirect sight and the direct fire telescopic sight.

After indicating the elevation angle on a graduated drum, the barrel position is checked using the level shown on the elevating mechanism's elevation meter.

The ranging (GR) and firing (GT) angles can be read directly through the goniometric sight. The direct fire telescope can be used to engage targets up to a maximum range of 2,000 m. In an emergency the gun can be fired without the trails being deployed.

The automated hydraulic system for ramming the projectile into the ordnance operates at all angles of elevation, reduces crew fatigue and ensures all projectiles are rammed at the same pressure, giving a stable muzzle velocity. The ramming system permits a high rate of fire of three rounds in the first 18 seconds and 6 rds/min thereafter.

Due to its total interoperability, the 155 mm towed gun TR can fire projectiles with the French or NATO propelling system to the following ranges:

- HE hollow base to 24,000 m
- HE base bleed to 30,000 m
- HE 56/59 to 23,000 m
- HE M107 to 18,500 m
- HE ERFB BB to 32,900 m
- 155 mm HE hollow base shell Type F1 (HE 155 F1) or Type F2 (LU 111 HB) with the projectile weighing 43.25 kg (containing 8.83 kg of HT 50/50), a muzzle velocity of 830 m/s and a maximum range of 24,000 m

Nexter Systems 155 mm towed gun TR of the French Army being prepared for towing (Nexter Systems) 1044601

Nexter Systems 155 mm towed gun TR of the French Army (Nexter Systems) 0073704

- 155 mm phosphorus smoke shell HB Type F2A (OFUM 155 F2A) with the projectile weighing 43.15 kg (containing 8.7 kg of white phosphorus), a muzzle velocity of 830 m/s and a maximum range of 24,000 m
- 155 mm training shell Type F1 (OX 155 F1) with the projectile weighing 43.5 kg (containing 1.2 kg of black powder), a maximum muzzle velocity of 830 m/s and a maximum range of 24,000 m
- 155 mm illuminating shell Type F1 (OECL 155 F1) with the projectile weighing 43.5 kg, a maximum muzzle velocity of 760 m/s and a maximum range of 19,250 m
- 155 mm anti-tank mine shell Type H1 (OMI 155 H1) with the projectile weighing 46 kg. This shell contains six anti-tank mines each weighing 0.55 kg and although fired at a maximum of charge 5, no maximum range has yet been announced
- 155 mm HE BB shell Type H2 (BBE 155 H2) with the projectile weighing 43.5 kg (containing 10 kg of HT 50/50), a maximum muzzle velocity of 830 m/s and a maximum range of 29,000 m
- Nexter Munitions cargo round projectile OGRE 155 G1 containing 63 bomblets with a maximum range of 28,500 m.

The TR will also fire the more recent Nexter Munitions/BAE Systems BONUS 155 mm artillery projectile, which contains two top attack sub-munitions. This is now in service with France and Sweden.

Nexter are now delivering to the French Army the latest version of their well established LU211 high-explosive projectile manufactured to the Insensitive Munition (IM) standard.

This is designated the 155 LU211-IM which meets the STANAG 4224 and STANAG 4439 requirements. The French Army has ordered an initial batch of 5,000 projectiles.

The TR has been provided with a number of emergency systems to allow continued use in the event of damage. These are a battery for the hydraulics which makes it possible to fire a further six rounds, or take the weapon out of action if the thermal engine stops, connection of the gun to the tractor's hydraulic system powering all of the gun's firing operations and coming into and out of action (or two guns can be connected together), manual pumps for elevation and traverse and manual mechanisms for traversing, opening and closing the breech and firing.

According to Nexter Systems, upgrades settled during 2003 on 155 mm TR weapons in service with the French Army included the means to improve reliability, security, ergonomics and interoperation of the fire-control system. These could also be offered to export customers of the TR. The 155 mm barrel has a maximum life of 3,000 rounds.

Variants

155/45 gun

A prototype was completed and tested but marketing is now being concentrated on the 155 mm/52-calibre system.

155/52 gun

In 1990, the now Nexter Systems announced that it had completed the prototype of a 155 mm towed gun TR with a 52 calibre ordnance that enables a range of 30,000 m to be achieved with basic projectiles and 42,000 m with ERFB-BB projectiles. This conforms to the new NATO ballistic MoU and was fired for the first time early in 1990.

The main modifications include replacement of the barrel by the new 52 calibre ordnance, modified towbar, modified rammer and some modifications to the recoil system.

The 155/52 has similar specifications to the 155/39 except that it is slightly longer in the firing position and weighs 11,000 kg.

The complete upper part of the 155/52 calibre weapon forms the basis of the Nexter Systems CAESAR 155 mm (6 × 6) self-propelled artillery system which was first shown in 1994.

The first 72 production CAESARs for the French Army were delivered on 16 July 2008 and under the terms of the contract all systems.

Trajan 155 mm towed gun

In September 2011, to meet the potential requirements of the export market, especially India, Nexter Systems of France has developed the Trajan 155 mm/52-calibre towed artillery system.

Trajan essentially consists of the complete elevating mass and 155 mm/52-calibre ordnance, loading and aiming system of the Nexter Systems CAESAR (6 × 6) self-propelled artillery system integrated on the carriage of the 155 mm/TR1 towed artillery system.

Trajan has a maximum rate of fire of up to six rounds a minute and firing Extended Range Full Bore - Base Bleed (ERFB-BB) ammunition can achieve a range of 42 km or more than 55 km using a rocket assisted projectile.

The carriage of the Trajan is fitted with an auxiliary power unit that speeds up deployment of the weapon as well as reducing crew fatigue.

A flick loader is also installed and automatic aiming of the weapon is carried out via the on board ballistic computer which also enables automatic re-aiming after each round is fired for increased accuracy.

Nexter Systems of France have already signed an agreement with Larsen & Toubro of India under which the Trajan will be offered to meet the Indian Army requirement for a 155 mm Towed Artillery Gun Programme (TAGP).

Under this agreement, local production of the gun carriage would be carried out in India as well as serial integration with other Indian contractors also involved. By February 2013 two pre-production Trajan 155 mm/52-calibre weapons had been completed and additional details are provided in a separate entry in *IHS Jane's Land Warfare Platforms: Artillery and Air Defence*.

Nexter Systems 155 mm M114 upgrade

This was developed to the prototype stage and fitted with a 155 mm/39-calibre barrel but never entered production and is no longer being marketed by Nexter Systems.

Specifications

	TR 155 mm
Dimensions and weights	
Crew:	7
Length	
overall:	10 m (firing)
transport configuration:	8.75 m
Width	
overall:	8.4 m (firing)
transport configuration:	3.09 m
Height	
axis of fire:	1.65 m
Ground clearance	
overall:	0.4 m
Weight	
standard:	10,750 kg
Mobility	
Carriage:	split trail
Speed	
max speed:	9 km/h
towed:	80 km/h
Fording	
without preparation:	1.5 m
Gradient:	60%
Firepower	
Armament:	1 × carriage mounted 155 mm howitzer
Main weapon	
length of barrel:	6.2 m
breech mechanism:	horizontal sliding block
muzzle brake:	double baffle
maximum range:	24,000 m (HE hollow base)
maximum range:	18,500 m (HE M107)
maximum range:	30,000 m (HE base bleed)
maximum range:	32,900 m (HE ERFB BB)
Rate of fire	
rapid fire:	6 rds/min
Main armament traverse	
angle:	65° (27° left/38° right)
Main armament elevation/depression	
armament front:	+66°/-6°
Survivability	
Armour	
shield:	no

Status

Production complete. No longer marketed. In service with:

Country	Quantity	Comment
Cyprus	12	
France	105	43 now in service
Saudi Arabia	28	Saudi Arabia National Guard (SANG)

Contractor

Nexter Systems, Centre de Bourges.

Nexter Systems 155 mm howitzer Model 50

Development

The 155 mm howitzer Model 50 was developed in France immediately after the Second World War and was also manufactured in Sweden by the now BAE Systems (at that time Bofors) for the Swedish Army, which called it the 15.5 cm Field Howitzer Fr. The French call the weapon the OB 155-50 BF.

The 155 mm Model 50 was used widely by the Israel Defense Force but has been withdrawn in favour of self-propelled artillery pieces such as the US-supplied 155 mm M109. The Israel Defense Force also used a self-propelled version of the Model 50, which is based on a modified Sherman chassis. It is possible that up to 120 of these may still be held in reserve.

In the French Army the Model 50 has now been replaced by the Nexter Systems 155 mm towed gun TR and replaced in the Swedish Army by the now BAE Systems FH-77B.

A number of these weapons were captured by the Israel Defense Force from the PLO during the fighting in Lebanon in the Summer of 1982.

More recently, some of these weapons have been observed to be operational in Ecuador, although their exact source is not known.

According to the United Nations, between 1992 and 2010 France exported 18 of these systems to Morocco, all of these were delivered in 1997. These were from French Army stocks rather than new-build weapons.

Description

The 155 mm barrel, with the breech ring, slides in the cradle during recoil and the breech ring is secured to the piston rod of the recoil buffer and to the recuperating rod.

The recoil length varies automatically with elevation. The barrel, breech ring and cradle, which together form the elevating mass, are secured to the upper mounting by cradle trunnions.

The elevating mass moves vertically in relation to the upper mounting when the elevating device is turned. The upper mounting rests on the lower carriage and turns horizontally in relation to it when the traversing device is turned. On the bottom of the lower mounting there is a pivot plate. The trails are pivoted on the sides of the lower carriage. Secured to each trail is a wheel bar, which in turn holds two pneumatic puncture-proof tyres.

The 155 mm howitzer Model 50 has a split trail carriage and, on the right trail, there is an unlimbering eye which during transport is hooked to the corresponding hook on the towing vehicle. During transport, the elevating mass is locked to the trails by a device in the rear part of the cradle. The weapon is fitted with a compressed air braking system, which is operated from the towing vehicle.

In the firing position the 155 mm howitzer Model 50 rests on the pivot plate and the outer end of the trails. The trails, when opened out, are secured to the ground by pickets. The pivot plate is pivoted in two directions at right angles to each other and thus can be placed so that the carriage remains horizontal regardless of the ground angle.

It is fitted with a hydropneumatic recoil system and the overall length of the barrel from the front of the muzzle brake to the rear of the breech ring is 4.41 m.

The 155 mm howitzer Model 50 fires an HE projectile weighing 43 kg with a maximum muzzle velocity of 650 m/s.

Specifications

	155 mm howitzer Model 50
Dimensions and weights	
Crew:	11
Length	
overall:	7.15 m (firing)
transport configuration:	7.8 m
Width	
overall:	6.8 m (firing)
transport configuration:	2.75 m
Height	
transport configuration:	2.5 m
axis of fire:	1.65 m

155 mm howitzer Model 50 (Israeli Ministry of Defence) 0500407

	155 mm howitzer Model 50
Ground clearance	
overall:	0.25 m
Track	
vehicle:	2.31 m
Weight	
standard:	9,000 kg
combat:	8,100 kg
Mobility	
Carriage:	split trail
Firepower	
Armament:	1 × carriage mounted 155 mm howitzer
Main weapon	
length of barrel:	4.41 m
rifling:	2.727 m
muzzle brake:	multibaffle
maximum range:	18,000 m (standard ammunition)
maximum range:	23,300 m (with rocket assisted projectile)
Rate of fire	
rapid fire:	4 rds/min
Main armament traverse	
angle:	80°
Main armament elevation/depression	
armament front:	+69°/-4°
Survivability	
Armour	
shield:	no

Status
Production complete. In service with:

Country	Quantity	Comment
Lebanon	14	estimate
Morocco	40	including 18 delivered in 1997
Peru	30	
Senegal	6	

Contractor
Nexter Systems, Centre de Bourges.

Nexter Systems 105 mm LG1 Mk II Light Gun

Development
The 105 mm LG1 Light Gun has been developed by Nexter Systems (previously Giat Industries), as a private venture, specifically for the export market.

By 1987, three prototypes had been built by the Etablissement d'Etudes et de Fabrications d'Armement de Bourges for extensive trials. These were followed by three pre-production weapons.

Late in 1990, following a competition between Giat Industries (now Nexter Systems) with the 105 mm LG1 Light Gun and the now BAE Systems of the UK with its 105 mm Light Gun, Singapore placed an order for 37 105 mm LG1 Light Guns. In Singapore the weapons were issued to two artillery battalions each of which has three batteries with six guns. These were delivered between 1992 and 1993.

Nexter Systems 105 mm LG1 Mk II deployed in position with firing jack deployed (Christopher F Foss) 1333674

Belgian Army 105 mm LG II Light Gun deployed in firing position (Nexter Systems) 0556088

Belgian Army 105 mm LG II Light Gun coupled to a UNIMOG (4 × 4) truck (BP) 1044602

Early in 1994, Indonesia placed a firm order with Giat Industries (now Nexter Systems) for 20 105 mm LG1 Mk II Light Guns plus a significant quantity of ammunition and a training package. These were delivered in 1996.

In June 1994, the Public Works and Government Services Canada, on behalf of the Canadian Department of National Defense, placed an order for 28 105 mm LG1 Mk II Light Guns worth CAD18 million. These were delivered between 1996 and 1997.

In early 2000, the Canadian Army deployed its LG1 Mk II Light Guns for the first time when A Battery, 1st Regiment Royal Canadian Horse Artillery deployed six weapons to Bosnia.

Late in 1995, the company was awarded a contract from the Belgian Army for the supply of 14 105 mm LG1 Mk II guns valued at BFR330 million (USD11.5 million).

The LG1 Mk II Light Guns were delivered from late 1996 through to early 1997 and have been issued to two batteries of the Belgian Army with each battery having six guns. The remaining two are used for training/reserve.

Early in 1996, it was announced that Nexter Systems had been awarded a contract from Thailand for the supply of 24 LG1 Mk II guns, which were delivered from late 1996 through to early 1997.

The original LG1 Mk I gun was subsequently replaced in production by the LG1 Mk II weapon which differs from the earlier weapon in having a new autofrettaged barrel to enable it, in the future, to fire ammunition with a higher pressure.

The layout of the recoil mechanism has been improved to allow for easier maintenance and the shield has also been removed to reduce weight.

Nexter Systems has developed a kit which enables existing users of the LG1 Mk I to upgrade these to the latest production LG1 Mk II standard in their own facilities.

Production of the 105 mm LG1 Mk II Light Gun is undertaken at the Nexter Systems facility in Bourges where all French artillery and tank barrels are produced.

The 105 mm LG1 Mk II is now proposed with a 3-D inertial aiming and positioning system and a muzzle velocity radar system. The latter is installed above the ordnance.

When so equipped, the weapon can also be connected up to any type of fire-control system such as the FAST-Hit. This system is claimed to significantly reduce the time required to bring the weapon into action.

Late in 2009 it was stated that Nexter Systems had been awarded a contract from an undisclosed country, believed to be Colombia, covering the supply of 20 of the LG1 Mk III weapons for delivery in 2010.

These are fitted with a digitalised fire control system that is also fitted the General Dynamics European Land Systems - Santa Bárbara Sistemas 155 mm/52-calibre APU SBT systems already in service with Colombia.

Nexter Systems 105 mm LG Mk III fitted with Kearfott inertial reference unit and Astronautics layers display (IHS/Rupert Pengelley) 1347487

As of March 2013 marketing of the Nexter Systems 105 mm LG1 light gun was still underway as a complimentary artillery system to the Nexter Systems CAESAR 155 mm/52-calibre truck mounted self-propelled artillery system and the Trajan 155 mm/52-calibre towed artillery system with auxiliary power unit which has been developed by Nexter Systems for the Indian market.

Description

The 105 mm LG1 Mk II Light Gun has a split trail carriage and a 30-calibre ordnance that is fitted with a double-baffle muzzle brake. It can be towed with the 105 mm ordnance over the closed trails or with the ordnance deployed in the normal firing position. The operational life of the barrel is stated to be greater than 7,700 full load rounds.

The weapon is fitted with a semi-automatic breech and once the weapon is fired the breech opens automatically and ejects the spent cartridge case. It is then ready for immediate reloading.

The hydropneumatic recoil length of the weapon varies as a function of the boresight angle of the gun and according to Nexter Systems a recoil pit does not have to be dug at steep boresight angles.

These three anchoring points not only provide stability for the weapon when it is being fired but also allow quick orientation by pivoting the gun around the platform through a full 360°.

When deployed in the firing position, it is supported on its two trails and a circular firing platform that is lowered under the forward part of the carriage.

The 105 mm LG1 Light Gun Mk II can be brought into or taken out of action within 30 seconds. A simple hydraulic system operated by a hand pump assists in opening out the trails. Nexter stress that when bringing the weapon into action there is no part to be assembled or disassembled.

The weapon can be loaded and fired at any elevation from –3° to +70° and also has a direct-fire capability out to 2,000 m.

It does not require a pit to be dug to allow firing at steep angles thanks to an adjustable recoil travel according to barrel elevation. It has a semi-automatically operated vertical breech and automatic ejection of the shell case after firing.

According to Nexter Systems, only two types of ammunition are required to cover ranges between 1.4 and 18.5 km, while the new Nexter Munitions base bleed projectiles cover the ranges from 7.3 to 18.5 km. Nexter Munitions has also proposed a hollow base projectile covering the ranges 2.5 to 15 km.

The Nexter Systems 105 mm LG1 Light Gun can be towed by a wide range of light vehicles including an ACMAT (4 × 4) truck, Peugeot P4 (4 × 4), Renault TRM 2000 (4 × 4), Land Rover (4 × 4) and Toyota (4 × 4). It can also be towed by the widely deployed AM General HMMWV (4 × 4). The weapon can be towed in two configurations, either with the barrel folded back over the closed trails or with the barrel deployed in the firing position.

It is fully air-portable slung under a helicopter and four can be transported inside of a Lockheed Martin C-130 Hercules transport aircraft.

Nexter is now offering a number of options to the LG1 Mk II light gun to further improve its operational capabilities.

These include the installation of a positioning and north seeking system to reduce the into-action time and increase accuracy, ballistic computer and a muzzle velocity measuring radar system which is mounted above the ordnance.

Future LG1 developments

Nexter Systems has carried out outline studies to further enhance the capabilities of the LG1.

The current LG1 Mk II would be fitted with an inertial 3-D aiming and positioning system and muzzle velocity radar. It can also be connected to the EADS STAREM C3I system. This would be the Mk III.

Further in the future are the Mk IV with automatic aiming and crew reduced to three and the Mk V with all of the earlier improvements plus automatic loading.

It would also be possible the install the weapon on the rear of a truck such as the US AM General High Mobility Multipurpose Wheeled Vehicle (HMMWV) (4 × 4) used in large numbers by the US Armed Forces.

105 mm M-1950 howitzer

This was phased out of service many years ago but four are still in service with Côte d'Ivoire.

Specifications

	LG1 Mk II
Dimensions and weights	
Crew:	5
Length	
overall:	6.6 m (firing)
transport configuration:	5.32 m
Width	
transport configuration:	1.97 m
Height	
transport configuration:	1.34 m
Ground clearance	
overall:	0.3 m
Track	
vehicle:	1.71 m
Weight	
standard:	1,600 kg
Mobility	
Carriage:	split trail
Firepower	
Armament:	1 × carriage mounted 105 mm howitzer
Main weapon	
length of barrel:	3.15 m
calibre length:	30 calibres
breech mechanism:	vertical sliding block
muzzle brake:	double baffle
maximum range:	11,400 m (HE M1 ammunition)
maximum range:	15,000 m (HE HB)
maximum range:	18,500 m (HE BB)
Rate of fire	
rapid fire:	12 rds/min
Operation	
stop to fire first round time:	30 s (est.)
out of action time:	30 s (est.)
Main armament traverse	
angle:	360° (on a turntable; 18° left/right)
Main armament elevation/depression	
armament front:	+70°/-3°
Survivability	
Armour	
shield:	no

Status

Production as required. In service with the following countries (see table below):

Contractor

Nexter Systems, Centre de Bourges.

Country	Order placed	Quantity	Model	Comment
Belgium	1996	14	Mk II	all delivered in 1997
Canada	1994	28	Mk II	8 delivered in 1996 and 20 in 1997
Colombia	2008	20	Mk III	delivery 2010
Indonesia	1994	20	Mk II	all delivered in 1996, Marines
Singapore	1990	37	Mk I	26 delivered in 1992 and 7 in 1993 which completed deliveries*
Thailand	1996	24	Mk II	24, all delivered in 1996

* - Phased out of service and replaced by Singapore Technologies Kinetics 155 mm/39-calibre Pegasus

Nexter Systems 105 mm M101A1 modernisation package

Development
Nexter Systems has developed, as a private venture, a modernisation package for the US 105 mm M101A1 towed howitzer, which is used by many countries.

The kit has been developed by Nexter Systems (formerly Giat Industries) at facilities in Bourges where all production of artillery and tank barrels in France is undertaken.

The modifications can be carried out by Nexter Systems or the company can supply kits to enable the user to upgrade in his own facilities using his own personnel.

As the French Army does not deploy any 105 mm artillery systems, this 105 mm M101 conversion was developed by Nexter Systems specifically for the export market.

Description
The modernisation includes replacing the original US-made 105 mm barrel with the more recent Nexter Systems LG1 Mk I/II Light Gun already in service with Belgium, Canada, Indonesia, Singapore (now phased out of service and replaced by locally developed Singapore Technologies Kinetics 155 mm/39-calibre Pegasus Light Weight Howitzer) and Thailand. The Nexter Systems M101A1 upgrade package involves:

- Replacing the existing 105 mm/23-calibre barrel with a new 105 mm/30-calibre barrel whose ballistics are identical to those of the 105 mm LG1 Mk II weapon
- Fitting this barrel with a double-baffle muzzle brake so as not to increase the forces on the recoil mechanism, which is retained in this upgrade
- Replacing the existing equilibrator
- Adding two ballasts on the trails
- Installing a breech that is identical to the original but made from a new steel that has been designed to withstand the pressures of longer range 105 mm projectiles.

Nexter Systems have stressed that there is no modification of the gun structure, or of the recoil mechanism in their upgrade package.

The upgraded M101A1 is 233 kg heavier than the standard weapon but can fire all standard NATO ammunition including the old US M1 high-explosive projectile to a maximum range of 11,400 m. It can also fire the more recent Nexter Munitions 105 BB base bleed projectile to a range of 18,500 m or a Nexter Munitions HB HE projectile to a maximum range of 15,000 m. It can also fire the new Nexter HE ER G2 round to a maximum range of 19,200 m after changing the breech ring.

In mid-1996, Nexter Systems announced that it had been awarded a contract from the Philippines to upgrade a first batch of 12 old US-supplied 105 mm M101 towed artillery systems.

The Philippines is the second customer for the upgraded M101, the first being Thailand who ordered a total of 285 kits, all of which were delivered to Thailand by early 1997.

Thailand has also taken delivery of significant quantities of 105 mm High-Explosive Base Bleed (HE-BB) rounds from Nexter for use with these upgraded M101 weapons and the more recently acquired LG1 Mk II Light Guns covered in detail in a separate entry in *IHS Jane's Land Warfare Platforms: Artillery and Air Defence.* Production of these upgrade kits is undertaken on an as required basis for the export market.

Status
Production as required. In service with the Philippines (12) and Thailand (285). As far as it is known, there has been no recent production of this upgrade package.

Contractor
Nexter Systems, Centre de Bourges.

105 mm M101A1 towed howitzer modernised by Nexter Systems and deployed in the firing position (IHS/Rupert Pengelley) 0105750

India

Upgraded 155 mm/45-calibre Metamorphosis artillery system

Development
Early in 2003, it was revealed that the Indian Ordnance Factory Board had developed an upgrade package for the widely deployed Russian designed 130 mm M-46 field gun which is used in large numbers by the Indian Army.

The upgraded system is also referred to as the Metamorphosis and was demonstrated during the IDEX 2003 exhibition held in Abu Dhabi in March 2003.

The Indian Army has already placed an order with the Israeli company of Soltam Systems for their 130 mm M-46 upgrade with a 155 mm/45-calibre barrel.

The first contract covered the upgrade of 180 systems with follow on contracts expected to cover an additional batch of 200 to 250 units.

Recent information has indicated that no follow on orders were placed with Soltam Systems. Details of the Soltam Systems 130 mm M-46 field gun upgrade are provided in a separate entry in *IHS Jane's Land Warfare Platforms: Artillery and Air Defence.*

It is not known as to whether the recently revealed Indian Ordnance Factory Board Metamorphosis M-46 155 mm/45-calibre upgraded weapon was developed in competition to the Soltam Systems design or as a possible alternative.

The Indian Ordnance Factory Board upgraded weapon is also referred to as the IOB M46 Field Gun (FG).

This upgraded weapon is being offered on the export market by the Indian Ordnance Factory Board.

As of March 2013 it is understood that the Metamorphosis upgrade remained at the prototype stage.

It is known to have been demonstrated overseas, including the Middle East.

Description
The existing 130 mm ordnance of the M-46 field gun has been replaced by a new 155 mm/45-calibre ordnance which was developed and manufactured at the Indian Ordnance Factory at Kanpur.

This ordnance 155 mm/45-calibre is fitted with a large double baffle muzzle brake. The ordnance has a 23 litre chamber with the ordnance having 48 grooves and a 1 in 20-calibre rifling rate (right hand constant twist).

The actual weight of the recoil mass has been quoted as 3,390 kg with the ground clearance at the bottom of the carriage being 0.40 m and that of the limber of 0.375 m.

According to the Indian Ordnance Factory Board, no modifications are made to the existing breech block. Failure of self-sealing systems during combat has been overcome by the use of a stub cartridge case obturator similar to the obturating system of the original Russian 130 mm M-46 field gun.

Conversion lead time is said to be minimal since there is no modification to the breech block mechanism and no change in the travel lock, cradle and recoil systems of the original Russian 130 mm M-46 field gun.

The only modification to the horizontal sliding breech block is a widening to allow for the insertion of the larger-calibre 155 mm projectiles.

The split trail carriage, elevating mechanism, shield and two-wheeled limber of the existing M-46 are also retained on the upgraded weapon.

When travelling the 155 mm/45-calibre ordnance is withdrawn to the rear by the standard 130 mm M-46 field gun chain mechanism located on the right side. The existing two-wheeled limber is retained.

To reduce operator fatigue, a three cylinder telescopic rammer with eight bar nitrogen gas pressure and pneumatic circuit is fitted. There is also an in-built safety mechanism that ensures that the rammer is withdrawn to the rear before the breech is closed and the weapon fired.

The upgraded system fires standard NATO types of 155 mm ammunition (projectile and charge) including the old US developed USM107 high-explosive projectile as well as the more recent Extended Range Full Bore (ERFB) and Extended Range Full Bore - Base Bleed (ERFB-BB) types.

When firing at the higher altitudes, which are encountered in many parts of India, the ranges quoted in the specifications tables can be exceeded.

The upgraded M-46 155 mm/45-calibre system can employ either bagged or modular charges with the intention being that the latter will be produced in India.

Russian 130 mm M-46 field gun upgraded with a 155 mm/45-calibre barrel by the Indian Ordnance Factory Board (T J Gander) 0543044

During loading, the 155 mm projectile is introduced first, followed by the propellant charge and then the stub case into which the primer is inserted.

Specifications

	Metamorphosis
Dimensions and weights	
Length	
overall:	11.30 m
transport configuration:	11.90 m
Width	
transport configuration:	2.45 m
Height	
transport configuration:	2.55 m
Ground clearance	
overall:	0.375 m (imber)
Weight	
standard:	8,700 kg
combat:	8,000 kg
Firepower	
Armament:	1 × carriage mounted 155 mm howitzer
Main weapon	
breech mechanism:	vertical sliding block
maximum range:	18,000 m (HE M107 ammunition)
maximum range:	30,000 m (ERFB ammunition)
maximum range:	39,000 m (ERFB-BB ammunition)

Status
Prototype. Not yet in production or service.

Contractor
Indian Ordnance Factory Board.

Indian Ordnance Factories 105 mm Light Field Gun

Development
The Indian Ordnance Factories' 105 mm Light Field Gun (LFG) bears a resemblance to the now BAE Systems 105 mm Light Gun. This is partly due to the fact that both were derived from the same 105 mm L13 gun used on the Vickers FV433 Abbot self-propelled gun.

The Indian Army took delivery of a total of 68 Value Engineered Abbot 105 mm self-propelled guns. The Abbot is understood to have been phased out of front line service with the Indian Army. Some sources have stated that these are still held in reserve.

The first 105 mm barrel, developed in India from the British original, was used on the 105 mm Field Gun Mk 1 (see separate entry). This uses tubular bow trails similar to those of the British now BAE Systems 105 mm Light Gun.

By early 1996, a total of 533 105 mm LFGs had been manufactured for the Indian Army. There are no known export sales of this weapon. It is believed that these weapons are mainly deployed with artillery units deployed in mountainous terrain.

As far as it is known there has been no recent production of the Indian Ordnance Factories 105 mm Light Field Gun. This has been confirmed by the United Nations who have stated that India did not export any artillery systems between 1992 through to the year 2010.

Description
To reduce weight the weapon has been constructed using light, high-strength alloy steels giving a weight in action of 2,380 kg.

The 105 mm Light Field Gun ordnance has a maximum recoil length of 1.06 m with the actual overall length of the ordnance being 4.4 m with the rifling being uniform 1 in 19 calibres right handed.

The 105 mm 37-calibre ordnance consists of a monobloc autofrettaged barrel with a vertical, sliding, hand-operated breech block.

Firing is electrical. The cradle, saddle and undercarriage are lightweight fabricated structures. Other features include spring-type balancing gears, a telescopic type of hydropneumatic recuperator, a suspension system and bow-shaped tubular trails.

The normal towing position is with the barrel folded back across the trails. A 360° traversing firing platform is provided. The LFG can be parachute dropped and carried by helicopter.

The Indian Ordnance Factories 105 mm Light Gun can be towed by a variety of cross-country trucks including the Shaktiman Model 4200 (4 × 4) or a more recent 4 × 4 or 6 × 6 vehicle.

The 105 mm Light Field Gun fires separate ammunition produced by the Indian Ordnance Factories using a seven-charge system (1, 2, 3, 4, 4.5, 5 and Super). Projectiles include HE, smoke, base ejection, star and High-Explosive Squash Head (HESH).

Maximum range is 17,200 m. The normal rate of fire is 4 rds/min and the intense rate of fire is 6 rds/min for up to 10 minutes (charge 5). The sustained fire rate is one round every two minutes for up to two hours (charge Super).

The 105 mm LFG is fitted with a dial sight with a magnification of ×3 and a 13° field of view for use in the indirect fire role and a day telescope with a magnification of ×2 and a 16° field of view for use in the direct fire role. The sighting system is situated on the left side of the mount.

Indian Ordnance Factories 105 mm Light Field Gun deployed in firing position and showing wheels resting on turntable 0526375

Lighter Version of 105 mm LFG
Early in 1993, the Indian Ordnance Factories stated that they were developing a lighter version of their 105 mm LFG with the following features:

- Subassemblies design optimised to reduce weight from 2,380 kg down to 2,100 kg;
- Time into and out of action reduced by first under-slinging the firing platform and second by modifying the wheel and saddle design to reverse the barrel over the ordnance without removing the right wheel.

As far as it is known this version has yet to enter production.

Indian 105 mm self-propelled gun
There is a separate entry in *IHS Jane's Land Warfare Platforms: Artillery and Air Defence* for the Indian 105 mm self-propelled artillery system which has been developed to the prototype stage.

This consists of a modified Sarath chassis (essentially the Russian BMP-2 infantry combat vehicle) fitted with the ordnance of the Indian Ordnance Factory 105 mm Light Field Gun.

As of March 2013 it is understood that this self-propelled version of the 105 mm Light Field Gun has not entered service.

Specifications

	105 mm Light Field Gun
Dimensions and weights	
Length	
overall:	7.216 m (firing)
transport configuration:	5.054 m
Width	
transport configuration:	1.82 m
Height	
overall:	1.494 m
Ground clearance	
overall:	0.33 m
Track	
vehicle:	1.465 m
Weight	
combat:	2,380 kg
Mobility	
Carriage:	bow
Firepower	
Armament:	1 × carriage mounted 105 mm howitzer
Main weapon	
length of barrel:	3.8905 m
calibre length:	37 calibres
muzzle brake:	double baffle
maximum range:	17,200 m
Rate of fire	
sustained fire:	0.5 rds/min
rapid fire:	6 rds/min
Main armament traverse	
angle:	360° (on turntable; 5° left/right)
Main armament elevation/depression	
armament front:	+73°/-5°
Survivability	
Armour	
shield:	no

Status
Production complete. In service with the Indian Army. The total Indian Army fleet of 105 mm weapons is estimated to be about 1,300 units.

Contractor
Gun Carriage Factory.
Enquiries to: The Secretary, Ordnance Factory Board.

Other Indian artillery developments

Description

In addition to the Indian Ordnance Factories 105 mm Light Field Gun (covered separate entry in *IHS Jane's Land Warfare Platforms: Artillery and Air Defence*), at least two other Indian towed artillery equipments were developed and placed in production.

The lightest of these is the 75 mm Pack Gun-Howitzer Mk 1 which has a 24-calibre barrel and was designed for use in mountainous terrain.

The carriage is of the split trail type and, if required, the wheels can be removed to allow the weapon to pivot on the trails and a firing platform under the carriage. The barrel has a single- or double-baffle muzzle brake. At least two slightly different models have been developed.

The Mk 1A has a two-part flat shield with the top part of each half folding forwards. The Mk 1B has a one-piece shield with a flat lower half and a top half that slopes to the rear. This 75 mm Pack Gun-Howitzer is in service with the Indian Army.

Recent information has indicated that the requirement for a 75 mm gun to replace the old British 3.7 in howitzers was first identified in 1959.

This was followed by limited production of a Canadian 75 mm mountain gun pending the purchase of the Yugoslav (which is now two separate countries) 76 mm Mountain Gun M48. It is believed that India has a total of 900 locally built 75 mm guns which serve alongside the 215 76 mm M48 weapons. There is a possibility that the latter are now held in reserve.

The second weapon developed in India is the 105 mm Indian Field Gun. Development of this weapon began in the mid-1960s with a gun development team under the control of the Director General Ordnance Factories.

The 105 mm barrel was manufactured at the Ordnance Factory, Kanpur, the elevation and traversing mechanism at Machine Tool Prototype Factory, Ambarnath and the carriage and final equipment was produced by the Gun Carriage Factory, Jabalpur.

The 105 mm barrel of the Indian Field Gun was not autofrettaged. The carriage is of the split trail type with hydropneumatic fabricated recoil system and pneumatic balancing gear. The weapon has a maximum range of 17,400 m.

The 105 mm Indian Field Guns and the more recent 105 mm Light Field Gun have now replaced the British-supplied 25-pounder field guns. The 105 mm Indian Field Gun was replaced in production by the 105 mm Light Field Gun covered in detail in a separate entry in *IHS Jane's Land Warfare Platforms: Artillery and Air Defence*.

It is believed that the total Indian fleet of 105 mm guns amounts to around 1,350 weapons of all types.

130 mm M-46 field gun

It is believed that the Indian Army has as many as 900 Russian-supplied 130 mm M-46 field guns in service with ammunition now being manufactured in India. A total of 36 regiments are equipped with the M-46.

130 mm M-46 upgrade

Early in 2000, India awarded Soltam Systems of Israel a contract to upgrade 180 130 mm M-46 field guns to a 155 mm/45-calibre standard. The first 180 weapons are understood to have been upgraded in Israel with the follow-on batch, expected to consist of 220 to 250 guns, being upgraded at the Indian Ordnance Factory Board facilities with kits provided by Soltam Systems. The work in India was carried out at Kanpur and Jabalpur. It is understood that India is the first customer for this Soltam Systems 130 mm M-46 upgrade package.

For details of the Soltam Systems 130 mm M-46 upgrade to the enhanced 155 mm configuration are provided in a separate entry.

Recent information has indicated that as of March 2013 India had not placed any follow on orders for the Soltam Systems 130 mm M-46 upgrade to a 155 mm/45-calibre system.

Indian Ordnance Factories 130 mm M-46 upgrade

In 2003 the Indian Ordnance Factories demonstrated an upgrade for the Russian 130 mm M-46 field gun that includes the replacement of the existing barrel by a new 155 mm/45-calibre barrel. This remains at the prototype stage.

105 mm Indian Field Gun (Indian Army) 0500409

Indian 75 mm Pack Gun-Howitzer Mk 1B (Indian Army) 0500410

Details of this are provided in a separate entry in *IHS Jane's Land Warfare Platforms: Artillery and Air Defence*.

155 mm FH-77B

India has taken delivery of all the 410 former Bofors 155 mm FH-77B field howitzers together with associated ammunition, fire-control systems, tractors and training equipment. Under the second and third phases of the programme it was expected that local assembly and then local production would be undertaken in India.

From July 1999 India started to place contracts with the former Bofors for spare parts for Indian 155 mm FH-77B.

In September 1999 India awarded a contract to Russia for the supply of 1,000 155 mm Krasnopol-M laser-guided artillery projectiles plus 10 associated tripod-mounted laser target designators and support equipment.

Late in 2011 it was reported that the Indian Ordnance Factory Board may build six prototypes of the Swedish 155 mm FH-77B by the end of 2013.

Of these two would be the standard 155 mm FH-77B, two FH-77B with upgraded computers and two FH-77B with 155 mm/45-calibre barrels.

130 mm self-propelled gun

The Indian Army has modified a number of Vijayanta (Vickers Mk 1 MBTs) to mount the Russian 130 mm M-46 field gun. The modifications to the hull have been extensive and include an additional roadwheel either side. Details of this system are given in a separate entry in *IHS Jane's Land Warfare Platforms: Artillery and Air Defence*. Only a small number of these were built to provide an interim self-propelled artillery capability to the Indian Army.

Catapult Mk II 130 mm self-propelled artillery system

The previously mentioned, the Indian Amy deployed a number of Catapult Mk I self-propelled artillery system based on a stretched Vijayanta hull fitted with a Russian 130 mm M-46 field gun firing over the rear arc.

Late in 2012 it was stated that the Indian Defence Research and Development Organisation (DRDO) was to start trials of the Catapult Mk II self-propelled artillery system.

This is a modified Arjun Mk I MBT hull also fitted with the Russian 130 mm M-46 field gun which is provided with 36 rounds of separate loading ammunition.

If trials are successful it is possible that a series production batch of about 40 Catapult Mk II could be delivered from 2013/2014 onwards and issued to two Indian artillery regiments.

These would replace the early Catapult Mk I systems currently deployed by the Indian Army as their only self-propelled artillery systems.

155 mm self-propelled gun

In 1995, four 155 mm turret systems were evaluated in India on a locally manufactured T-72M1 MBT hull to meet an Indian Army requirement for a 155 mm self-propelled artillery system.

These systems were the French GCT, Slovakian Zuzana, South African T6 and the UK AS90.

As of March 2013 India had not ordered any new self-propelled artillery systems.

New 155 mm towed gun

The Indian Army also has a requirement for a new 155 mm/52-calibre towed artillery system with 400 expected to be procured, funding permitting.

In mid-2002, India tested 155 mm systems supplied by the then Bofors (FH-77 B05L 52), Denel (G5-52) and Soltam Systems, but none have been ordered.

In April 2011 BAE Systems announced that it had withdrawn the FH-77 B05L 52 from the competition due to changes in the Indian requirement.

Indian 155 mm lightweight artillery requirement

Early in 2008 India stated that it was to acquire 145 ultra-lightweight 155 mm/39-calibre artillery systems.

The two contenders for this are the BAE Systems 155 mm M777 (already deployed by Australia, Canada, US Army and US Marine Corps) and the Singapore Technologies Kinetics 155 mm Pegasus (already deployed by the Singapore Armed Forces.

In the end only the 155 mm M777 was tested and this was fitted with the Selex Laser Inertial Artillery Navigation and Pointing System (LINAPS) which is already in service with Canada, Thailand and the UK.

Two 155 mm M777 underwent extensive firepower and transport trials in India in 2011 but as of March 2013 no production contracts had been awarded.

Through the US Foreign Military Sales programme India has requested a total of 145 M777 fitted with the Selex Laser Inertial Artillery Pointing Systems worth a total of USD647 million.

Status
Both the 75 mm Gun-Howitzer and 105 mm Field Gun are in service with the Indian Army.

Contractor
See Development.

155 mm FH-70 in self-propelled mode with 155 mm/39-calibre ordnance over closed trails (BAE Systems) 1365026

International

155 mm field howitzer 70 (FH-70)

Development
In the early 1960s, Germany, the UK and the US agreed that they had a requirement for a new 155 mm towed field howitzer. Germany and the US wanted to replace their 155 mm M114 howitzers (covered in detail in a separate entry in *IHS Jane's Land Warfare Platforms: Artillery and Air Defence* and no longer deployed by the US Army) and the UK wanted to replace its 5.5 in guns. In the end, the US went on to develop the 155 mm M198 which has no APU.

Germany and the UK agreed to develop a new weapon, which became known as the FH-70.

The FH-70 MoU was signed in 1968 and the requirement was for a 155 mm howitzer to meet the following military and tactical requirements:

- Continuous high rate of fire with a burst-fire capability
- High mobility with minimum effort for deployment
- Increased range and lethality and a new family of 155 mm separate loading ammunition, for example 155 mm projectiles and bagged charges.

The UK was the project leader for the towed FH-70 and Germany for the self-propelled version called the SP-70, development of which was suspended in 1986.

The first six FH-70 prototypes were completed between 1969-70 and, after accepting the Agreed Operational Characteristics, Italy joined the collaborative project in 1970 as a full partner. The second batch of eight prototypes was completed between 1971 and 1973 and the first international trials battery of six FH-70s was formed in 1975.

A total of 19 prototypes of the FH-70 were completed and after extensive trials the weapon was accepted for service in 1976, with the first production weapons being completed in 1978. Germany ordered 216, Italy 164 and the UK 71 FH-70s.

In May 1982, Saudi Arabia placed an order with International Military Services of the UK for the supply of 72 FH-70 howitzers plus ammunition, muzzle-velocity measuring equipment, fire-control systems and training equipment. By early 1983, the first 18 weapons had been delivered.

The British designation for the FH-70 was the L121 while the Germans call it the FH 155-1. Production for all three countries was completed in 1982. Malaysia took delivery of nine FH-70 systems in 1989 with a further three being delivered in 1993. The FH-70 has also been manufactured under licence in Japan together with its ammunition.

The trilateral FH-70 responsibilities for production were as follows:

- United Kingdom: carriage, traversing gear, high-explosive projectile (L15) and propelling charge (cartridges numbers 1 and 2)

155 mm FH-70 deployed in the firing position and showing APU mounted on forward part of carriage (BAE Systems) 1365025

- Germany: ordnance, loading system, auxiliary propulsion unit, suspension, sighting equipment, smoke and illuminating ammunition and propelling charges (cartridge number 2)
- Italy: cradle, recoil system, sights bracket, elevating gear and arc, high-explosive projectile, smoke and illuminating ammunition and propelling charges (cartridges numbers 2 and 3).

FH-70s have been withdrawn from service by Germany and the UK. The now BAE Systems did take a batch of 24 FH-70s for possible export sales, but these have now been scrapped.

According to the United Nations Arms Transfer Lists, the following quantities of 155 mm field howitzer FH-70 weapons were transferred between 1992 and 2010. In each case these were from existing stocks rather than new build weapons.

From	To	Quantity	Date
Germany	Estonia	4	2003
Germany	Estonia	16	2004
Germany	Estonia	4	2005
Germany	Norway	15	1995
UAE	Netherlands	15	1995
UK	Malaysia	3	1993
UK	Malaysia	3	1994

It should be noted that production of the FH-70 has been completed and that only surplus weapons would now be available from Germany and Italy.

Description
The elevating mass of the FH-70 comprises the ordnance, cradle with integrated recoil system and the loading mechanism. The ordnance consists of a monobloc barrel fitted with a double-baffle muzzle brake, a breech mechanism incorporating a semi-automatic vertically sliding block and an automatic tube loader with a capacity of nine tubes, spent tube ejection and an emergency firing mechanism.

The cradle, which carries the 155 mm/39-calibre ordnance on slides, includes the recoil system, which consists of a buffer with high-angle cut-off gear and a recuperator.

The mechanical main firing system, which is independent of elevation, is carried through one of the cradle trunnions.

The semi-automatic loading system operates at all elevations and traverses and consists of a loading tray, which presents the projectile to the chamber. Together with an automatic tube loader, this system permits a normal rate of fire of 6 rds/min and a burst rate of three rounds in 13 seconds.

An automatic loading (flick ramming) system was available as an optional extra and provides a burst fire rate of three rounds in eight seconds.

The carriage is of lightweight construction with saddle, split trails, self-digging spades, main and trail wheels that are operated hydraulically, a soleplate and a detachable APU. When travelling, the 155 mm barrel is rotated through 180° and clamped between the trails.

The APU is housed in a space frame attached to the forward part of the carriage and consists of a commercial Volkswagen 1,800 cc engine with a gearbox, differential, batteries and a hydraulic pump.

Country	Quantity	Comment
Estonia	24	from Germany in 2003/05
Germany	216	quantity delivered new, no longer in front-line service
Italy	164	quantity delivered new
Japan	420	local production, which has now been completed
Malaysia	12	from UK, including three in 1993
Morocco	35	from Saudi Arabia
Netherlands	15	from UAE in 1995/96, not deployed
Norway	15	from Germany in 1995 (not deployed)
Oman	12	
Saudi Arabia	72	quantity delivered new, 35 transferred to Morocco

In the APU mode the FH-70 can attain speeds up to 16 km/h and is capable of negotiating slopes up to 34 per cent. When being towed, it can ford to a depth of 1.5 m and when using the APU to 0.75 m.

The functions of the APU include driving the main wheels for moving the howitzer, providing hydraulic power for steering and raising and lowering both the main and trail wheels. This facility allows for rapid unhooking, lowering the howitzer onto its soleplate, easy opening and lowering of the trails when going into action, as well as extracting the spades from the ground, lifting the equipment, rapid closing of the trails and limbering up. In an emergency, or when the APU has been removed, power is provided by a hand pump.

The main wheels are hydraulically suspended on swinging arms attached to the carriage body and are cushioned by hydraulic accumulators. Similar suspension elements are used for raising and lowering the trail wheels. Braking of both main wheels is possible from the towing vehicle, or from the APU driver's position on the left side of the carriage.

The trail wheels are attached to the trail end and can be steered, raised and lowered. The hydraulic units providing these functions are controlled from the driver's position and can operate with the trails open or closed. The steering is capable of 60° of movement either side of the centre and for spreading the trails the steering wheels can be turned through 90°.

The sighting system mounted on the left of the saddle consists of a support bracket, a dial sight carrier incorporating a quadrant elevation scale and levelling bubbles, a periscopic dial sight and a direct fire telescope. A direct fire night sight can also be fitted. All scales, graticules and bubbles are self-illuminating. The sight carrier incorporates transducers for azimuth and elevation and in the German and Italian armies these are connected to a Digital Display Unit (DDU) which also receives firing data (azimuth and elevation) from a fire-control computer. The layer lays the gun by nullifying the differences on the DDU and then laying the sight on the gun aiming point. The DDU was not used in UK service.

The FH-70 can be towed on a road at a maximum speed of 100 km/h. Germany used the MAN (6 × 6) 7,000 kg truck to tow the FH-70, Italy used the FIAT 6605 TM (TM69) and the UK used the Foden (6 × 6) medium-mobility vehicle. The FH-70 is air-portable in a Lockheed Martin C-130 aircraft or underslung under the Boeing CH-47D Chinook helicopter.

A family of 155 mm ammunition was developed for the FH-70 which gives the weapon a significant increase in range and lethality for both direct and indirect firing against armoured targets. The FH-70 can also fire NATO standard ammunition consisting of the following projectiles:

- HE projectile (L15A1) which weighs 43.5 kg and is thin walled, with high fragmentation effect and HE capacity of 11.3 kg
- Base Ejection Smoke projectile (DM105), where four smoke canisters fall to the ground giving a dense cloud of smoke in 30 seconds
- Illuminating (DM106) which provides illumination of 2.1 million candelas for 60 seconds.

The Extended Range Projectile (ERP) which will increase the range of the FH-70 to over 30,000 m is US M549A1 rocket-assisted projectile and RO30 Base Bleed (BB).

The eight-zone charge system uses triple-based propellant and is divided into three separate cartridges, containing zones 1 and 2, 3 to 7 and 8 respectively. The charge system is ballistically balanced, has a clean combustion and is initiated by a percussion inner tube (DM 191A1). Muzzle velocity ranges from 213 to 827 m/s.

Using the now BAE Systems BIS 14 propelling charge the FH-70 can fire an ERFB BB projectile to a range of 31,000 m.

Using the APU, the FH-70 can be unhooked, manoeuvred into position and put into action in less than two minutes and taken out of action in a similar period. If a target moves outside the traverse limits the trails can be raised hydraulically, thereby pulling the spades from the ground. The gun can then be wheeled about its soleplate to a new target. With the trails lowered the first shot will embed the spades.

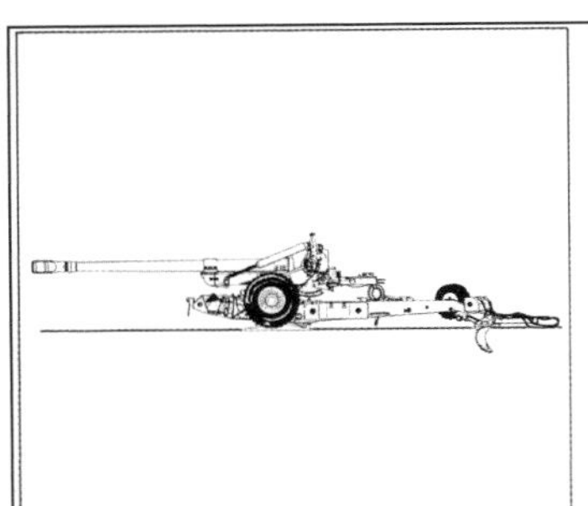
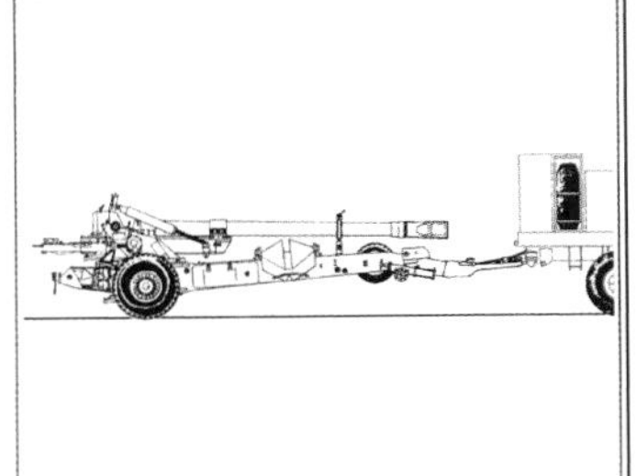
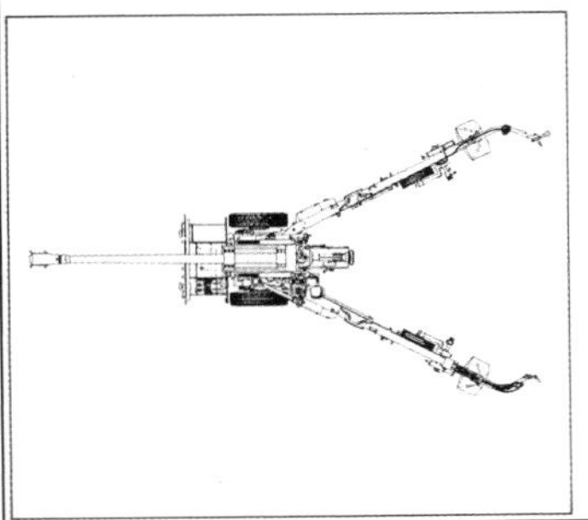
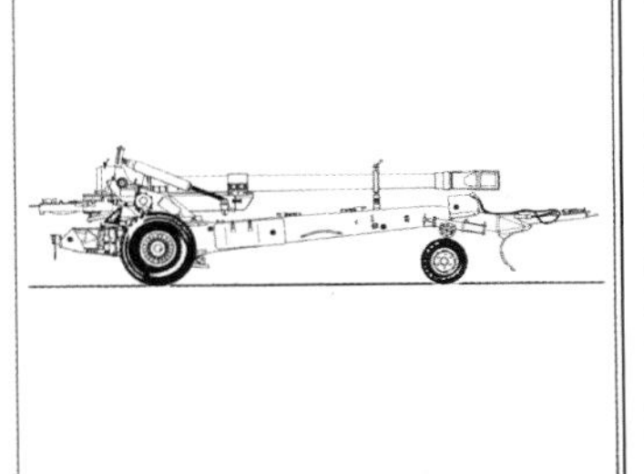

Four view general arrangement drawing of FH-70 155 mm/39-calibre artillery system in travelling and firing configurations (Oto Melara) 1452921

Italian Army FH-70 155 mm/39-calibre artillery system deployed in the firing position from the rear (Oto Melara) 1452920

Apart from the usual preliminary checking of the equipment, the gun detachment has only to open the breech by hand, load the first projectile and propelling charge and place the second projectile on the loading tray. The breech is closed and an ignition tube is automatically loaded from the nine-tube magazine. After this sequence, which takes less than five seconds, the FH-70 is ready for action. When the FH-70 is ready and laid correctly, all that is required for firing is for the layer to pull the firing handle. Thereafter, the sequence of operation becomes semi-automatic. The recoil system and muzzle brake jointly absorb the recoil forces as the gun recoils on the cradle slides to a maximum of 1.4 m. On running out to the firing position, the following operations take place simultaneously: the breech is opened automatically by a cam, the spent tube is ejected and the next 155 mm projectile is lifted to the breech and aligned in the chamber before final hand ramming. The detachment then rams the projectile and loads the propelling charge, places the next projectile in the loading tray, closes the breech, if necessary relays the gun and fires again. A flick ramming device is available as an optional extra giving a burst rate of three rounds in eight seconds and reducing crew fatigue.

Other options included a direct fire image intensification night sight, deflector boxes and mainwheel mudguards.

Variants

Japanese FH-70

The Japanese Ground Self-Defence Force selected the FH-70 to replace its 155 mm M114 towed howitzers and production was undertaken under licence in Japan. Prime contractor in Japan was the Japan Steel Works. Rheinmetall was responsible for the Japanese contract.

Export version

For the world market a simplified sighting system was available, which utilised the now BAE Systems 105 mm Light Gun sighting system with a modified graticule in the direct fire telescope. It can be fitted to the standard gun via an adaptor on the left-hand trunnion.

FH-70 with 52-calibre ordnance

For trials purposes Rheinmetall fitted an FH-70 with a 52-calibre ordnance that conforms with the new quadrilateral ballistic agreement. This enables a range of 30,000 m to be achieved with the L15A1 projectile, or 40,000 m with BB projectiles. This is no longer marketed.

ARIS APU for FH-70

The Italian company Applicazioni Rielaborazioni Impianti Speciali (ARIS) have completed development of a new diesel Auxiliary Power Unit (APU) for installation on the 155 mm FH-70 towed artillery system.

The original production FH-70 has the APU installed on the front of the carriage. This consists of a Volkswagen 1,800 cc engine with a gearbox, differential, batteries and hydraulic pump.

The installation of the APU allows the FH-70 to attain a maximum road speed of 16 km/h as well as climbing slopes of up to 34 per cent.

In addition to providing the power to move the FH-70 system over short distances, especially over terrain that cannot be reached by the 6 × 6 prime mover, the APU also provides hydraulic power for steering and raising and lowering the main and trail wheels. The installation of the APU reduces crew fatigue as well as speeding into and out of action times.

The new ARIS diesel powered APU can be installed on the 155 mm FH-70 in only two hours and uses standard proven Commercial-Off-The-Shelf (COTS) components to reduce overall life cycle costs.

The system has been evaluated by the Italian Army and is almost a direct replacement for the current APU and takes less than two hours to install. It gives the FH-70 a maximum road speed of 15 km/h and it can climb gradients of 40 per cent.

Specifications

	FH-70
Dimensions and weights	
Crew:	8
Length	
overall:	12.43 m (firing)
transport configuration:	9.8 m
Width	
overall:	7.5 m (firing)
transport configuration:	2.58 m
Height	
overall:	2.192 m (firing)
transport configuration:	2.45 m
axis of fire:	1.525 m
Ground clearance	
overall:	0.3 m
Track	
vehicle:	2.195 m
Wheelbase	5.15 m
Weight	
standard:	9,300 kg
combat:	9,300 kg
Mobility	
Carriage:	split trail with APU
Fording	
without preparation:	1.5 m
Turning radius:	9.25 m
Tyres:	14.00 × 20
Firepower	
Armament:	1 × carriage mounted 155 mm howitzer
Main weapon	
length of barrel:	6.022 m
muzzle brake:	double baffle
maximum range:	24,700 m (standard projectile)
maximum range:	31,500 m (base-bleed projectile)
Rate of fire	
sustained fire:	2 rds/min
rapid fire:	6 rds/min
Main armament traverse	
angle:	56°
Main armament elevation/depression	
armament front:	+70°/-4.5°
Survivability	
Armour	
shield:	no

Status
Production complete. No longer marketed.

Contractor
BAE Systems.
Rheinmetall DeTec AG.
Oto Melara SpA.

NORICUM GH N-45 155 mm gun-howitzer

Development
From 1989, NORICUM manufactured an improved version of the 155 mm SRC International GC 45 155 mm gun-howitzer, under the designation GH N-45 (Gun-Howitzer, NORICUM, 45-calibre). A total of 12 of the original GC 45 systems were provided to the Royal Thai Marines.

Details of the original GC 45 155 mm/45-calibre system are provided in a separate entry in *IHS Jane's Land Warfare Platforms: Artillery and Air Defence*.

NORICUM assembled all but two of the 12 guns (two six-gun batteries) supplied to the Royal Thai Marines and then began deliveries of over 600 guns to various customers.

In June 1987, the Thai Cabinet approved a purchase of six GH N-45 weapons at a cost of USD7 million, all of which have now been delivered. These models are fitted with an auxiliary power unit.

In 1994, the Swiss-based company T&T Technology Trading Ltd purchased the design rights, weapons and spare parts of the Austrian NORICUM 155 mm GH N-45 towed artillery system.

No figures for Austrian production of the GH N-45 have been released but it is estimated that between 500 and 600 weapons were produced with production reaching 10 weapons per month.

Production by Voest-Alpine stopped in the late 1980s when it became apparent that the weapons were reaching the Middle East.

Known customers include Iran, Iraq and Thailand, but in some of these cases they were not supplied direct from Austria.

In mid-1997, the Royal Thai Army placed an order for a second batch of 36 155 mm GH N-45 howitzers in the latest A1 towed configuration and these were delivered in 1998.

NORICUM GH N-45 155 mm gun-howitzer captured by Iran with ordnance over rear and trail partly open 0501040

The NORICUM 155 mm GH N-45 howitzer, firing an Extended Range Full Bore - Base Bleed (ERFB-BB) projectile, has a maximum range of 39.6 km so making it the longest range artillery system in service with the Royal Thai Army.

According to the United Nations Arms Transfer Lists, the following quantities of GH N-45 were exported by Austria to Thailand between the years 1992 and 2010:

Year	Quantity
1992	18
1996	18
1997	36

It is understood that small batches of GH N-45 systems could be made from existing stocks of spare parts. In the future new-build weapons will be supplied. As of March 2013 it is understood that no brand new GH N-45 155 mm gun-howitzers had been built.

Description
The GH N-45 is, in most respects, similar in appearance to the Belgian GC 45 but has been redesigned in certain key areas to improve both handling in the field and reliability and to make the weapon easier to manufacture.

Production of the GC 45 155 mm/45-calibre system was completed many years ago and it is no longer marketed as the company has ceased to exist.

Details of this are provided in a separate entry in *IHS Jane's Land Warfare Platforms: Artillery and Air Defence*.

The carriage uses a walking beam suspension for improved cross-country mobility and a secondary float is fitted for easy emplacement, displacement and general weapon handling. There is a quick release in the traversing gear that brings the barrel from the travelling jack-knife position to the firing position in a few seconds and there is a steering device that allows the weapon to travel in the tracks of the towing vehicle.

The GH N-45 can be delivered in a basic towed form or with an APU; an APU kit can be supplied for fitting to towed models by the user. The main power source is a Porsche 236 6.9 four-cylinder, four-cycle air-cooled 2.3-litre engine delivering 120 hp.

The APU is mounted on the forward part of the carriage with the driver on the right. The APU enables the weapon to propel itself at a maximum speed of 30 km/h and two 30-litre fuel tanks provide sufficient fuel for ranges of up to 130 km. With the APU the weapon can climb 40 per cent gradients at a speed of 5 km/h.

The GH N-45 has a 45-calibre barrel with a 22.94 m^3 chamber volume to enable base bleed ammunition to be fired to a maximum range of 39,600 m. The barrel has a semi-automatic breech mechanism with a pneumatic rammer.

NORICUM 155 mm GH N-45 artillery systems in the travelling configuration. The weapon on the left has no APU while the weapon on the right is fitted with an APU (NORICUM) 1159841

The overall length of the barrel, including muzzle brake and breech mechanism is 7,720 mm with the length of rifling being 5,830 mm and the length of chamber being 1,173 mm. The rifling has 48 grooves.

It has a burst rate of fire of three rounds in 16 seconds. Length travelling with the barrel forwards is 13.97 m and height with the barrel traversed to the rear for travelling is 2.26 m.

The 155 mm GH N-45 gun-howitzer is normally towed by a 5- or 10-tonne 6 × 6 cross-country truck that also carries a supply of ready use ammunition (projectiles, charges and fuzes).

While the maximum towing speed on roads is up to 105 km/h, its maximum towing speed across-county is up to 50 km/h, but this does depend on terrain conditions.

Extras available for the GH N-45 included a power elevating and traverse system, and an emergency ammunition bracket holding six standby 155 mm projectiles, both for the APU-powered version. A further option for the APU version was an ammunition-handling device.

The jib and power unit for this device are mounted on the top right of the barrel and are used to raise the ammunition tray from ground level to the loading position. Tracks could be fitted to the APU-powered version to increase traction over soft ground and enable gradients of up to 45 per cent to be climbed. Also available for the APU version was a remote-control system that enabled a 6 × 6 tractor to use the traction of the powered carriage over adverse terrain giving a 10 × 10 configuration.

In addition to standard NATO ammunition, the 155 mm GH N-45 can fire a complete family of 155 mm ammunition, including:

- HE/BT, maximum range 30,300 m (charge N10)
- HE/BB, maximum range 39,600 m (charge N10)
- HC/smoke, maximum range 39,600 m (base bleed), 30,300 m (boat tail) (charge N10)
- WP/smoke, maximum range 39,600 m (base bleed), 30,300 m (boat tail) (charge N10)
- Illuminating, maximum range 39,600 m (charge N10)
- HE, 24,700 m (charge N10)
- HE (M107), 17,800 m (charge M119A2).

In addition, various types of artillery fuze can be supplied by NORICUM as well as various types of bagged and modular charge systems.

Variants

NORICUM 155 mm/39-calibre M114 upgrade

This was developed to the prototype stage and was also referred to as the GH N-39 but it is no longer being marketed.

NORICUM 105 mm anti-tank gun ATG N 105

This was armed with a modified 105 mm L7/M68 tank gun and was developed to the prototype stage but never entered production or service.

NORICUM 155 mm GH N-45 artillery system without an APU being towed by a 6 × 6 forward-control truck (NORICUM) 1159840

Specifications

	GH N-45 APU	GH N-45 Towed
Dimensions and weights		
Crew:	6	6
Length		
overall:	11.4 m (firing)	11.4 m (firing)
transport configuration:	9.725 m	9.72 m
Width		
overall:	9.93 m (firing)	9.93 m (firing)
transport configuration:	2.75 m	2.5 m
Height		
transport configuration:	2.05 m	2.05 m
Ground clearance		
overall:	0.315 m	0.31 m
Track		
vehicle:	2.47 m	2.22 m
Weight		
standard:	12,400 kg	10,100 kg
Mobility		
Carriage:	split trail	split trail
Speed		
towed:	105 km/h	105 km/h
Tyres:	11.00 × 20	9.00 × 20
Firepower		
Armament:	1 × carriage mounted 155 mm howitzer	1 × carriage mounted 155 mm howitzer
Main weapon		
length of barrel:	7.045 m	7.045 m
rifling:	5.82 m	5.82 m
muzzle brake:	multibaffle	multibaffle
maximum range:	30,300 m (ERFB ammunition)	30,300 m (ERFB ammunition)
maximum range:	39,600 m (ERFB-BB ammunition)	39,600 m (ERFB-BB ammunition)
Rate of fire		
sustained fire:	2 rds/min	2 rds/min
rapid fire:	7 rds/min	7 rds/min
Main armament traverse		
angle:	70° (est.) (30° left/40° right)	70° (est.) (30° left/40° right)
Main armament elevation/depression		
armament front:	+72°/-4°	+72°/-5°
Survivability		
Armour		
shield:	no	no

Status

Production as required. In service with a number of countries including Iran (about 120) and Thailand (army and marines).

Contractor

T&T Technology and Trading GmbH.

Iran

DIO 155 mm 39-calibre HM41 howitzer

Development

The 155 mm HM41 howitzer was developed to meet the requirements of the Iranian ground forces and in appearance is very similar to the Republic of Korea upgraded 155 mm/39-calibre M114, which is called the KH179.

A number of facilities in Iran are involved in the manufacture of this weapon including the Hadid Armament Industries Group of the Iranian Defence Industries Organization (DIO). This facility supplies the 155 mm/39-calibre barrel.

The 155 mm HM41 howitzer has been offered on the export market but as far as it is known there have been no export sales of this weapon.

Description

The 155 mm towed artillery system is designated the HM41 and is understood to be an upgraded version of the US 155 mm M114 towed howitzer, which was originally developed before the Second World War.

This fired a standard 155 mm M107 high-explosive projectile, which has been manufactured in Iran for many years by the Ammunition Division of the DIO, to a maximum range of 14,600 m.

DIO 155 mm HM41 howitzer with 39-calibre barrel deployed in the firing position 0526378

The original M114 towed howitzer has been in service with the Iranian Army for many years. By upgrading these with a 155 mm/39-calibre barrel the range of the weapon is increased, allowing it to be deployed more to the rear, so increasing its survivability against counter battery fire.

According to Iranian sources, the 155 mm/39-calibre HM41 weighs 6,890 kg and firing a rocket-assisted High-Explosive (HE) projectile, a maximum range of 30,000 m can be achieved. Firing an unassisted HE projectile, a maximum range of 22,000 m can be obtained.

In 2006 it was revealed that the Ammunition Industries Group of the Defense Industries Organization (DIO) of Iran had developed and placed in production a number of new artillery projectiles which have a significant increase in range when compared to currently deployed conventional artillery projectiles.

This includes a 155 mm HE ERFB-BB artillery projectile which is claimed to have a maximum range of 34 km which is less than some comparable Western-produced 155 mm projectiles of this type which typically have a maximum range of 39.6 km.

At this time there is no indication that Iran has developed a 155 mm Modular Charge System (MCS) which is now being increasingly deployed by a number of western countries.

A maximum rate of fire of 4 rds/min can be achieved but there is no load assist device fitted. Elevation range is from 0° to plus 66° with traverse being 25° right and 23.5° left, all under manual control. Direct and indirect fire sights are fitted.

The new 155 mm/39-calibre ordnance is fitted with a double-baffle muzzle brake to reduce recoil forces, with the breech mechanism being of the screw type. The recoil mechanism of the 155 mm HM41 howitzer is of the variable hydropneumatic type. It can be fired using a hand lever or a lanyard. The 155 mm/39-calibre barrel has a total of 48 grooves.

When deployed in the firing position the HM41 rests on the firing jack under the forward part of the carriage and on the spades attached to the two trails.

The 155 mm 39-calibre HM41 howitzer is normally towed by a 5- or 10-tonne 6 × 6 cross-country truck.

In addition to the gun crew this also carries a quantity of ready use ammunition (projectiles, charges and fuzes).

Development of this system, which is based on the old US M114 towed artillery system, commenced in 1979, with deliveries to the ROK starting in 1983. Full details of the KH179 are given in a separate entry.

As far as is known, there have been no exports of the KH179.

Variants

155 mm/45-calibre barrel HM47L

Iran is also now making a 155 mm/45-calibre barrel (the HM47L) but at this stage it is not known as to whether this is installed in a new Iranian towed or SP artillery system or it is a replacement barrel for its currently deployed 155 mm/45 towed artillery systems which have been in service for many years. Iran is also now producing barrels that are chrome plated to extend their lives.

DIO 155 mm/39-calibre self-propelled howitzer

In 2011 it was revealed that the DIO had developed to at least the prototype stage a 6 × 6 155 mm/39-calibre self-propelled howitzer.

This consists of a two door forward control 6 × 6 cross-country chassis with the complete upper part of the DIO 155 mm/39-calibre HM41 towed howitzer mounted on the rear.

To provide a more stable firing platform hydraulic spades are lowered to the ground before firing commences.

Available details of this are provided in a separate entry in *IHS Jane's Land Warfare Platforms: Artillery and Air Defence*.

As of March 2013 it is not known as to whether this remains at the prototype stage or it has entered service with Iran.

Other Iranian towed artillery systems

It is now known that Iran has produced a version of the Russian 122 mm towed howitzer D-30, under the local designation of the HM40. This fires the same family of 122 mm separate loading ammunition as the Thunder 1 122 mm self-propelled artillery system. This is called the 122 mm D-301 howitzer and available details are given in a separate entry in *IHS Jane's Land Warfare Platforms: Artillery and Air Defence*. This has been offered on the export market.

Specifications

	HM41
Dimensions and weights	
Weight	
standard:	6,890 kg
Mobility	
Carriage:	split trail
Firepower	
Armament:	1 × carriage mounted 155 mm howitzer
Main weapon	
calibre length:	39 calibres
muzzle brake:	double baffle
maximum range:	22,000 m (unassisted projectile)
maximum range:	30,000 m (rocket assisted projectile)
Rate of fire	
rapid fire:	4 rds/min
Main armament traverse	
angle:	48.5° (23.5° left/25° right)
Main armament elevation/depression	
armament front:	+66°/-0°
Survivability	
Armour	
shield:	no

Status

Production as required. In service with Iran. This has been offered on the export market but there are no known sales.

Contractor

Defence Industries Organization (DIO).
Hadid Armament Industries Group.

DIO 122 mm Shafie D-301 howitzer

Development

For some years the Hadid Industries Group of the Defence Industries Organization, Armament Industries Group, has been manufacturing a version of the Russian 122 mm D-30 towed howitzer under the local name of the 122 mm Shafie D-30l howitzer. This weapon has been offered on the export market but as of March 2013 there are no known declared export sales. .

Iran has exported weapons and equipment to a number of countries, including Sudan who is known to deploy the standard 122 mm D-30 towed howitzer as well as a recently developed self-propelled model.

The latter is integrated onto the rear of a 6 × 6 cross-country truck chassis with a protected cab and is called the Khalifa and available details are provided in a separate entry in *IHS Jane's Land Warfare Platforms: Artillery and Air Defence*.

Description

This is understood to be identical to the Russian D-30 covered in detail in a separate entry in *IHS Jane's Land Warfare Platforms: Artillery and Air Defence*. This is still used in large numbers by the Russian Army and many other countries all over the world.

The 122 mm Shafie D-30 howitzer is normally towed by a 4 × 4 or 6 × 6 cross-country truck that in addition to carrying the gun crew also carries a quantity of ready use ammunition (projectiles, charges and fuzes).

Some of the technical specifications supplied by Iran show that there are some physical differences between the Iranian and Russian weapons.

Iranian 122 mm Shafie D-301 howitzer in travelling position from the rear (DIO) 0567815

For example, the Iranian weapon is slightly heavier. This increase in weight could well be due to different manufacturing processes, rather than a different design.

The 122 mm barrel is monobloc with a muzzle brake and draw bar which is located under the muzzle. The barrel is connected with the breech ring through the breech ring nut.

The actual weight of the 122 mm barrel used the weapon is being quoted as 1,050 kg and length of barrel with muzzle brake is 4.785 m and without out muzzle brake (on which is fitted the towing eye) length of the barrel is 4.27 m.

The length of the powder chamber is 594 mm with recoil varying between 740 and 930 mm. The rifling has a total of 36 grooves.

The weapon fires 122 mm ammunition of the semi-fixed type including high explosive, illuminating and smoke projectiles with various charge combinations. A maximum range of 17,000 m can be achieved using standard ammunition.

In late 2006 it was revealed that the Ammunition Industries Group of the Defense Industries Organization (DIO) of Iran had developed, and placed in production, a number of new artillery projectiles which have a significant increase in range when compared to currently deployed 122 mm and 155 mm conventional artillery projectiles.

This includes a new 122 mm Base Bleed (BB) and Hollow Base (HB) artillery projectile with a significant increase in range.

The semi-automatic breech block is of the vertical wedge type. The recuperator is hydropneumatic, whereas the recoil brake is hydraulically controlled.

The cylinders of the hydraulic recoil brake and recuperator are connected to the projections on the breech ring and recoil together with the barrel.

The carriage shield serves for protection of the crew from the effect of small arms fire, shell fragments and muzzle blast.

The equilibrator is of a pneumatic pushing type. It is intended for the equilibration of the weapon parts moveable in elevation throughout the field of action of the cradle with the barrel.

This also facilitates the lifting of the wheels when changing from the travelling to the firing position of the weapon.

The firing jack, located under the carriage, is intended for lifting the howitzer during changing from travelling to the firing position and vice versa.

The travelling parts are connected to the top carriage and the wheels are provided with a pneumatic travelling brake, which is actuated from the vehicle.

For the direct engagement of stationary and moving targets, the optical sight is used. The panoramic telescope, with the range quadrant can be used for engaging indirect targets as well as for the engagement of direct targets. The sights, elevating and traverse mechanisms are mounted on the left side of the carriage.

During darkness, the sighting devices are illuminated by means of the instrument light and autonomous tritium illumination.

Specifications

	Shafie D-301
Dimensions and weights	
Crew:	7
Length	
overall:	7.80 m (firing)
transport configuration:	5.33 m
Width	
transport configuration:	1.95 m
Height	
transport configuration:	1.66 m
axis of fire:	1.42 m
Weight	
standard:	3,440 kg
combat:	3,350 kg
Mobility	
Speed	
towed:	60 km/h
Tyres:	9.00 × 20
Firepower	
Armament:	1 × carriage mounted 121.92 mm howitzer
Main weapon	
length of barrel:	4.785 m
rifling:	3.4 m
breech mechanism:	vertical sliding block
muzzle brake:	multibaffle
maximum range:	17,000 m
Rate of fire	
rapid fire:	8 rds/min
Main armament traverse	
angle:	360°
Main armament elevation/depression	
armament front:	+70°/-7°
Survivability	
Armour	
shield:	yes

Status

Production as required. In service with Iran (550 all sources). Has been offered on the export market but there are no known sales.

Contractor

Defence Industries Organization (DIO).
Hadid Armament Industries Group.

Israel

Soltam Systems 155 mm TIG gun-howitzer family

Development

The 155 mm TIG 2000 gun-howitzer family was originally developed by Soltam Systems as a private venture and is a follow-on to the earlier Soltam Systems Model 839/839P and Model 945/945P towed gun-howitzers. The P in this designation means that it is fitted with an Auxiliary Propulsion Unit (APU).

In the designation TIG 2000, the TIG stands for Towed Independent Gun. According to Soltam Systems, the TIG 2000 is unique in that it can be fitted with a 155 mm 39, 45 or 52-calibre ordnance by the replacement of the barrel assembly and towing rod only. This can be accomplished at field level.

The TIG 2000 can be supplied as a conventional towed artillery system or fitted with an APU. In both cases the system can be operated by a small crew of four people.

It is fully air-transportable in aircraft such as the Lockheed Martin C-130 and the model with an APU can enter and leave the aircraft under its own power.

According to the United Nations Arms Transfer Lists, the following quantities of Soltam 155 mm artillery systems were transferred between 1992 and 2010.

To	Quantity	Year	Comment
Cameroon	4	1997	
Cameroon	4	1998	
Cameroon	4	1999	
Cameroon	6	2001	
India	10	2001	possibly upgraded M-46
India	56	2004	possibly upgraded M-46
Slovenia	10	1996	confirmed as M845
Slovenia	8	1997	confirmed as M845
Uganda	18	2002	
Uganda	3	2005	confirmed as ATMOS 155 mm SP weapons
Uganda	3	2009	confirmed as ATMOS 155 mm SP weapons

The Slovenian company RAVNE was involved in local production of the 155 mm Soltam Systems towed artillery systems. The Slovenian 155 mm 155/45-calibre model is not fitted with an APU. These are referred to as the TN-90.

Description

The TIG 2000 has a conventional split carriage and when in the travelling position the 155 mm/52-calibre barrel is traversed through 180° and locked in position over the closed trails.

The 155 mm ordnance conforms to the Joint Ballistic Memorandum of Understanding (JBMoU) and is of the monobloc autofrettaged type forged from high-strength steel. The ordnance is fitted with a muzzle brake and a fume extractor.

Soltam Systems TIG 2000 artillery system with APU deployed in travelling configuration 0058419

Soltam Systems TIG 2000 artillery system with APU deployed in firing position 0526374

Barrel specifications

	39-calibre	45-calibre	52-calibre
Barrel length:			
(with breech)	6,670 mm	7,700 mm	8,745 mm
(without breech)	6,026 mm	7,035 mm	8,080 mm
Chamber entrance:	164 mm	167 mm	167 mm
Number of grooves:	48	48	48
Rifling:	progressive	constant	constant
Rifling inclination:	ended 1:20	1:20	1:20
Obturation:	self-sealing rings	self-sealing rings	self-sealing rings
Firing mechanism:	trip action	trip action	trip action

The breech ring mechanism consists of the breech ring assembly, breech block assembly and the automatic breech block operating mechanism.

The cradle, pivoted to the saddle by the trunnions, carries all parts of the gun which participate in recoil as well as in the vertical movement during elevation and depression.

The central part of the cradle carries the barrel assembly. On top are the recuperator cylinders and at the bottom, the buffer cylinder and elevating arc. At its lower ends are the trunnions, left and right and at its front end, the hinges for the equilibrators. The recoil length cut-off gear is located at the lower right front end of the cradle.

The recoil mechanism consists of a hydropneumatic recuperator, hydraulic buffer and recuperator and a mechanical recoil length cut-off gear.

Two equilibrators, energised by compressed gas, act as self-adjusting force, balancing the barrel, to allow for each laying and adjusting in elevation and depression.

The mount includes the saddle, which carries the 155 mm ordnance and enables laying in elevation and azimuth. The mounts main assemblies are saddle, elevating and traversing mechanisms.

The saddle is a welded steel construction that is bolted to the bearings rotating part in the bottom and carrying the cradle. The elevating and traversing gears are mounted on the saddle with the gunner's controls being on the left side.

The elevating mechanism comprises a number of gears incorporating worm and worm wheel with the complete mechanism being fitted on the left side of the saddle.

The traverse mechanism is mounted on the forward part of the saddle and serves to lay the gun in azimuth. A total deflection of 75° (1,334 mm) is possible without shifting the gun.

For fast deflection or when rotating the 155 mm barrel from the firing position to the travelling position, or vice versa, the mechanism can be disengaged, using a quick release clutch allowing free lateral movement of the barrel.

The carriage serves as the weapon's main support in action and transport. It also carries all of the auxiliary systems that are required for operating the Soltam Systems 155 mm TIG gun-howitzer.

The main carriage assemblies are the central structure and main bearing, firing platform and jack, trails and towing rod assembly, running gears and walking beam, braking system, spade assemblies, support wheels and the electrical system and installation.

The central structure serves as the backbone of the whole weapon and is of all-welded steel construction carrying on its top the main bearing. On the underside is the firing jack with firing platform and on each side is a trail with running gear. On the front part of the structure is the electric battery and hydraulic controls compartment.

The main bearing is of the slewing wire, heavy duty type. The lower and upper rings of the bearing are stationary and bolted to the central structure.

The central bearing ring carries the saddle and by means of a traversing gearbox, mounted on the saddle and being engaged with the bearing ring, allows traversing of the weapon in relation to the carriage.

The firing jack with its associated platform serves two purposes. First, to assure stability of the gun when it is firing with the gun resting on the platform at the front and spades at the rear. Second, to allow for the spreading and folding of the trails when the gun is brought into and out of action.

The jack is operated hydraulically by the APU or by a back-up electric pump powered by the batteries. Should the electric system fail then the jack will be operated by the manually operated hydraulic pump.

The trails also carry the APU (left trail) and associated systems including braking system, support wheels, spades, towing rod, travelling clamp and the rammer's pressure bottle.

Both trails are fitted with fixtures and mounts to carry the different accessories required to operate the gun. Handles are provided at the rear of each trail to facilitate movement of the weapon, this is in addition to sockets in each trail in which lifting rods can be inserted.

Standard size, four (two either side) running wheels that incorporate hydraulic motors in their hubs are mounted on walking beams for road and cross-country use. Each individual wheel incorporates a disc brake with two braking callipers.

The system is fitted with a hydropneumatic braking booster, which is energised by NATO standard dual-line pneumatic connections. The braking system, built into one of the howitzer's trails, incorporates an emergency pressure tank.

In ordinary operation, the howitzer's brakes are automatically applied simultaneously with the towing vehicle's brakes by the driver's pedal. In case the howitzer and pneumatic lines become disconnected from the towing vehicle, the howitzer's brakes are automatically activated, stopping the gun.

To support the gun against horizontal firing loads two self-digging spades are used. When travelling, the spades are carried on each trail at its rear and when required for action fold down. In this position, a special dual-purpose mechanism locks and secures the blade. The same lock is used for holding the blade in the travelling position.

One support wheel is located on each trail on the outboard sidewall. The wheels serve for supporting the gun while travelling as well as spreading and folding the trails when coming into and out of action.

The electrical system comprises two 24 V, 120 Ah batteries as a power source, 24 V DC series motor driving a hydraulic pump, electronic controller for operating the pump motor and monitoring the battery charge state, lighting and signalling system and an electrical harness connection to the towing vehicle.

The TIG 2000 is fitted with a loading tray which incorporates a pneumatic ramming piston enabling uniform loading of the projectile into the barrel chamber at all angles of elevation without the need to lower the barrel.

According to Soltam Systems, the weapon imposes no limitations on the towing vehicles speed when being towed. Its low centre of gravity, overall length and the tandem wheels mounted on the walking beams assure smooth riding at all speeds, bends and rough terrain. The APU fitted to the towed independent version does not influence the towing properties in any way.

The automatic horizontal sliding breech mechanism, fume extractor and wide angle of traverse, together with the loader assist device, enable effective high rates of fire to be achieved. The load-assist device transfers 155 mm projectiles from the supply truck, or from the ground, to the loading tray of the howitzer. A manual back up is provided in case there is a power failure.

In the TIG 2000 fitted with the APU, the latter provides power not only for driving the howitzer but also for bringing it into and out of action. It also assists in preparing the howitzer for the transport mode and for ammunition hoisting.

The driver of the APU has a control lever of the joystick type, which controls direction and speed, other controls include starter, speed range selector and parking brake.

By pushing the lever to the desired direction the system moves in that direction, as soon as it is released it returns to the neutral position and the weapon comes to a halt.

As this version can move in any direction up to 17 km/h, it can enter firing positions that could not be reached if the APU was not fitted.

The driver's seat and console are located on the left side of the howitzer and when deployed in the firing position, the driver's console is folded downwards. The system can be manoeuvred from the driver's seat or from the ground with the operator walking alongside the gun.

The hydraulic power supplied from the APU powers the howitzer through hydrostatic motors incorporated into the four, (two either side) driving wheels. In addition, it lifts the howitzer off the wheels onto its firing platform, operates the auxiliary jack to hook or unhook the howitzer from the towing vehicle, spreads the trails and afterwards undigs the spades.

In the conventional towed version of the TIG 2000, an assistance hydroelectric power unit is installed to ease crew member's tasks.

Automotive specifications

	39-calibre	45-calibre	52-calibre
Wheels:			
(driving)	4 × 11 × 20	4 × 11 × 20	4 × 11 × 20
(support)	2 × 8.5 × 10	2 × 8.5 × 10	2 × 8.5 × 10
Low speed range with APU:	0 to 8 km/h	0 to 8 km/h	0 to 8 km/h
High speed range with APU:	0 to 17 km/h	0 to 17 km/h	0 to 17 km/h
Towed hook load:			
(towed configuration)	750 kg	800 kg	900 kg
(APU configuration)	1,000 kg	1,200 kg	1,400 kg

The TIG 2000 family can fire all natures of 155 mm ammunition, with the range depending on the type of projectile and charge system used, details of which can be found in the table.

Maximum range

Projectile family	39-calibre	45-calibre	52-calibre
M107/M107A1/M101	21,900 m	24,000 m	24,500 m
M56/M56A1	23,500 m	25,800 m	26,000 m
L15	24,700 m	27,500 m	30,000 m
M483A1	21,500 m	23,500 m	25,000 m
M397/401, LR-BB	30,000 m	33,000 m	34,500 m
ERFB-BB	28,000 m	39,000 m	41,000 m

The TIG 2000 can be used in the direct and indirect fire roles and to achieve these two objectives, separate sighting systems are installed. The main sighting system is intended for long-range fire with the secondary system serving for the direct fire role, especially self-defence. The sights are mounted on the left side of the TIG 2000.

The TIG 2000 can also be linked to any artillery command, control and communication system.

In place of the conventional indirect laying system, an Advanced Laying Instrument (ALI), based on ring laser gyro techniques, can be fitted. This also provides options for North Finder, GPS and navigation modules.

The Soltam Systems 155 mm TIG gun-howitzer is normally towed by a 5- or 10-tonne (6 × 6) cross-country truck that in addition to its crew carries a quantity of ready use ammunition (projectiles, charges and fuzes).

ATHOS 2052

Further development of the 155 mm TIG gun-howitzer family by Soltam Systems has resulted in the 155 mm Autonomous Towed Gun ATHOS 2052 (the last two digits [52] stand for the length of the barrel in-calibres).

In addition to being fitted with an APU, ATHOS 2052 is also fitted with an automatic laying mode which is integrated with a fully computerised system to enable automatic control, accurate navigation and target acquisition.

Standard equipment on ATHOS 2052 includes a flick rammer, GPS and onboard firing computer. It can receive digital radio target data directly from the forward observer or from the target acquisition system and fully automatic laying is possible. A muzzle velocity radar can be installed on the barrel if required. This could feed information into the fire-control system.

Two 12 V batteries are provided and these are charged when the gun is towed or by the diesel engine of the APU. These batteries provide the power necessary for the various electronic devices and also allow the use of an electro-hydraulic pump for silent operation.

According to Soltam Systems, the hydraulic system installed on ATHOS 2052 enables easy deployment of the weapon using hydraulic jacks and auxiliary wheels. It requires a crew of four to six people to fully operate the system.

To reduce crew fatigue, an ammunition loader assist is installed. The weight of ATHOS 2052 is quoted as 13,000 kg.

In addition to being marketed in 155 mm/52-calibre it is also being marketed in other-calibres such as 155 mm/39-calibre and 155 mm/45-calibre.

Autonomous Truck-MOunted System (ATMOS)

This is a self-propelled version of ATHOS and is normally installed on a 6 × 6 or 8 × 8 truck chassis. The Israel Defense Force has evaluated this system, but the first known customer is Uganda who took delivery of a batch of three 155 mm/39-calibre systems on a forward control 6 × 6 chassis in 2005. Details of this are provided in a separate entry in *IHS Jane's Land Warfare Platforms: Artillery and Air Defence.*

It has been confirmed that Uganda took delivery of an additional three ATMOS systems in 2009 which brought their total fleet to six units.

Cameroon is understood to have taken delivery of 18 ATMOS 155 mm systems.

ATROM

This is the Romanian version of ATMOS and was first shown late in 2003 and is installed on a Romanian 6 × 6 truck chassis. As far as it is known this system remains at the prototype stage and is covered in detail in a separate entry in *IHS Jane's Land Warfare Platforms: Artillery and Air Defence*.

Specifications – *see table below*

Status

Production as required. In service with undisclosed countries. The earlier Soltam Systems 155 mm Model 845 is known to be in service with Slovenia, which took delivery of 18 weapons between 1996 and 1998 (called TN-90). Cameroon has taken delivery of 12 Soltam Systems 155 mm weapons of an undisclosed type. Botswana has taken delivery of 12 Soltam Systems 155 mm system of an unknown type. Uganda has taken delivery of 18 × 155 mm M-839 artillery systems from Israel. Some reports have indicated that Israel may have taken delivery of 80 systems.

Contractor

Soltam Systems Limited (now part of the Elbit Group).

Soltam Systems 155 mm M-71 gun-howitzer

Development

The Soltam Systems towed 155 mm M-71 gun-howitzer is a further development of the Soltam Systems towed 155 mm M-68 gun-howitzer and uses the same recoil system, breech and carriage. The first prototype was completed in 1974 and first production units the following year.

Production of this system is undertaken for the export market on an as required basis.

Soltam Systems is now concentrating its export marketing on the more recent TIG gun-howitzer which is available with a 39-/45-/52-calibre barrel according to customer requirements. Details of this are provided in a separate entry in *IHS Jane's Land Warfare Platforms: Artillery and Air Defence*.

Description

The main differences between the M-71 and the M-68 are that the former has a longer 155 mm/39-calibre barrel and is fitted with a rammer driven by compressed air. This permits rapid loading at all angles of elevation. The rechargeable cylinder for the M-71 is mounted on the right trail.

	TIG 2000 52 cal	TIG 2000 45 cal	TIG 2000 39 cal
Dimensions and weights			
Length			
overall:	10.0 m	9.00 m	9.00 m
Width			
transport configuration:	2.60 m	2.60 m	2.60 m
Height			
overall:	2.10 m	2.10 m	2.10 m
Ground clearance			
overall:	0.38 m	0.38 m	0.38 m
Weight			
standard:	13,000 kg (fitted with APU; 9200 kg towed model)	12,200 kg (fitted with APU; 9400 kg towed model)	12,000 kg (fitted with APU; 9800 kg towed model)
Mobility			
Speed			
max speed:	17 km/h	17 km/h	17 km/h
towed:	100 km/h	100 km/h	100 km/h
Gradient:	35%	35%	35%
Auxiliary engine:	Deutz 106 hp air cooled diesel	Deutz 106 hp air cooled diesel	Deutz 106 hp air cooled diesel
Firepower			
Armament:	1 × carriage mounted 155 mm howitzer	1 × carriage mounted 155 mm howitzer	1 × carriage mounted 155 mm howitzer
Main weapon			
breech mechanism:	horizontal sliding block	horizontal sliding block	horizontal sliding block
maximum range:	24,500 m (M107 projectile)	24,000 m (M107 projectile)	21,900 m (M107 projectile)
maximum range:	41,000 m (ERFB-BB)	39,000 m (ERFB-BB)	28,000 m (ERFB-BB)
Main armament traverse			
angle:	75°	75°	75°
Main armament elevation/depression			
armament front:	+70°/-3°	+70°/-3°	+70°/-3°
Survivability			
Armour			
shield:	no	no	no

Soltam Systems 155 mm M-71 gun-howitzer in firing position and showing ordnance lock on left trail 0500385

Soltam Systems 155 mm M-71 gun-howitzer in travelling position with barrel swung through 180° and locked over trails (Israeli Ministry of Defence) 0500411

The barrel has 48 grooves, the length of the rifling is 5.038 m, the width of the groove is 6.6 mm, the width of the land is 3.5 mm and the depth of the groove is 1.2 mm. The recoil system is of the hydropneumatic type.

Soltam Systems is now concentrating its marketing on the latest 155 mm TIG 2000 artillery system which is available with a 39-, 45- or 52-calibre ordnance. Details of this system, which is understood to have been developed for the export market are given in a separate entry in *IHS Jane's Land Warfare Platforms: Artillery and Air Defence.* This is fitted with an auxiliary propulsion unit.

This 155 mm M-71 gun-howitzer is normally towed by a 5- or 10-tonne (6 × 6) cross-country truck that in addition to its crew also carries a quantity of ready use ammunition (projectiles, charges and fuzes).

Variants

Soltam Systems developed to the prototype stage a self-propelled model of the M-71 gun-howitzer based on a Centurion MBT hull. This never entered service and is no longer marketed.

Specifications

	M-71
Dimensions and weights	
Crew:	8
Length	
transport configuration:	7.5 m
Width	
transport configuration:	2.58 m
Height	
transport configuration:	2.115 m
Ground clearance	
overall:	0.38 m
Track	
vehicle:	2.2 m
Weight	
standard:	9,200 kg
Mobility	
Carriage:	split trail
Speed	
towed:	100 km/h
Firepower	
Armament:	1 × carriage mounted 155 mm howitzer
Main weapon	
length of barrel:	6.045 m
calibre length:	39 calibres
breech mechanism:	horizontal sliding block
muzzle brake:	single baffle
maximum range:	23,500 m (M107 projectile)
maximum range:	30,000 m (BB projectile)
Rate of fire	
sustained fire:	2 rds/min
rapid fire:	5 rds/min
Main armament traverse	
angle:	84°
Main armament elevation/depression	
armament front:	+52°/-3°
Survivability	
Armour	
shield:	no

Status

Production complete but can be resumed if additional orders are placed. In service with a number of countries including:

Country	Quantity	Model
Botswana	12	probable more recent TIG 155 mm
Chile	24	from South Africa (called G4) (coastal defence role as well)
Chile	8	M-71 original order
Chile	11	M-68 original order
Israel	100	M-68/M-71
Myanmar	-	unconfirmed
Philippines	50	M-68/M-71
Singapore	45	M68, now in reserve
Singapore	38	M71S, now in reserve
Thailand	32	M-71

Contractor

Soltam Systems Limited (now part of the Elbit Group).

Soltam Systems 155 mm/45-calibre M-46S field gun

Development

Based on its extensive experience in the design, development and production of towed and self-propelled artillery systems, Soltam Systems has completed and qualified the prototype of a Russian designed 130 mm M-46 field gun upgraded with a 155 mm/45-calibre ordnance.

Soltam Systems can either provide the kit, enabling the user to upgrade existing 130 mm M-46 weapons in its own facilities, or the company can undertake this work in its facilities in Israel.

The Israeli Defense Force is estimated to have 100 captured 130 mm M-46 guns in service but it is not known if this upgrade kit was aimed at the local requirement, for the export market or for both the home and export markets.

Early in 2000, India awarded Soltam Systems a contract to upgrade 180 130 mm M-46 field guns to a 155 mm/39-calibre and 155 mm/45-calibre standard.

The first 180 weapons are understood to have been upgraded in Israel with the follow-on batch, expected to consist of 220 to 250 guns, being upgraded at the Indian Ordnance Factory Board facilities with kits provided by Israel.

Recent information has stated that the total value of the first contract to upgrade 180 weapons in Israel is about INR2.07 billion (USD48.13 million).

Recent information has indicated that no follow on contracts have been placed by the Indian MoD with Soltam Systems for additional 130 mm M-46 field gun upgrade kits. India has a requirement for additional new 155 mm towed artillery systems but none had been ordered as of March 2013.

Soltam Systems 155 mm/45-calibre M-46S field gun in travelling configuration and with limber attached 0500883

Soltam Systems 155 mm/45-calibre M-46S field gun in firing configuration
0500884

This upgrade is also applicable to the China North Industries Corporation (NORINCO) equivalent of the Russian M-46, the Type 59-1.

According to the United Nations Arms Transfer Lists, the following quantities of Soltam 155 mm artillery systems were transferred between 1992 and 2010.

To	Quantity	Date	Comment
Cameroon	4	1997	
Cameroon	4	1998	
Cameroon	4	1999	
Cameroon	6	2001	
India	10	2001	possibly upgraded M-46
India	56	2004	possibly upgraded M-46
Slovenia	10	1996	confirmed as M845
Slovenia	8	1997	confirmed as M845
Uganda	18	2002	
Uganda	3	2005	confirmed as ATMOS 155 mm SP version
Uganda	3	2009	confirmed as ATMOS 155 mm SP version

The above figures indicate that the first batch consisted of 66 weapons. In addition Israel supplied India with five 155 mm artillery barrels in 2000 which are assumed to be for the 155 mm/45-calibre M-46S field gun upgrade programme.

Description

The Soltam Systems upgraded M-46 is fitted with a monobloc autofrettaged 155 mm barrel 45-calibres long that incorporates a horizontal sliding breech block which is automatically operated. It uses standard M82 primers that are automatically ejected after firing.

The complete weapon system can be divided into two major assemblies, the ordnance and carriage.

Whereas the carriage, understructure and recoil mechanism remain unchanged, the ordnance is completely replaced by the new 155 mm/45-calibre barrel assembly that, together with the pneumatic ramming device, comprises the conversion kit.

The aiming and sighting systems can be brought up to Western standards as an option and a modern inertial aiming device can also be fitted.

The standard sighting system of the Russian 130 mm M-46 consists of optical sights for long-range indirect fire and direct fire. The sight for indirect laying includes a sight instrument and a panoramic periscope. The sight for direct aiming includes a telescope mount and an elbow telescope.

The sight instrument is mounted on the left side of the saddle while the panoramic telescope, mounted on the top of the sight is used for azimuth laying. The elbow telescope is used for direct fire in case of defence against enemy armour or when flat trajectory fire is required.

The 155 mm M-46 conversion kit comprises the 45-calibre ordnance consisting of the barrel, breech ring assembly, breech block assembly, automatic breech block operating mechanism, new single-baffle muzzle brake, fume extractor and cradle adaptors with the loader assist device and its associated installation.

Specifications of the 155 mm ordnance include a chamber diameter of 172 mm, a volume of 25.7 litres, a rifling of 48 grooves and a rifling inclination of 1:20 constant right-hand twist.

According to Soltam Systems, the upgraded weapon has a number of unique features:

- Firing cannot commence unless the breech block is completely closed and locked
- The horizontal breech mechanism, operated together with the pneumatic ramming device, enables a high rate of accurate fire to be achieved without the need to lower the ordnance for loading
- The gun crew is clear of the overpressure effects and clean of residual post-firing gases due to the efficient muzzle brake and bore evacuator.

The Soltam Systems upgraded M-46 gun can fire all the standard 155 mm types of ammunition in worldwide use starting with the US-developed M107 HE family of ammunition and ending with the latest 155 mm ERFB-BB family of long-range ammunition.

With unassisted ammunition, ranges of up to 30,000 m can be achieved while, with assisted ammunition, ranges of up to 39,000 m can be obtained.

According to Soltam Systems, the upgraded 155 mm/45-calibre M-46 field gun can be deployed using its original sighting system with a direct fire telescope, or can be upgraded by adding 6400 artillery mils.

The sighting system, as well as inertial navigation and aiming system, can be incorporated with an onboard firing computer if required by the customer.

Variants

Self-propelled 155 mm M-46S

Soltam Systems have developed a new 155 mm/52-calibre self-propelled artillery system on a Tatra (6 × 6) truck chassis called the ATMOS 2000. This system could also be fitted with the Soltam Systems 155 mm/45-calibre M-46S upgrade. ATMOS is already in production for an undisclosed country and is marketed on a variety of 6 × 6 and 8 × 8 chassis. Recent information is that the first production batch of ATMOS systems used the upper part of the 130 mm M-46 fitted with a 155 mm ordnance. According to the United Nations, Uganda took delivery of a batch of three ATMOS from Soltam Systems in 2005 in the 155 mm/39-calibre configuration.

It has been confirmed that Uganda took delivery of an additional three ATMOS systems in 2009 which brought their total fleet to six units.

Specifications

	M-46S
Dimensions and weights	
Length	
transport configuration:	11.73 m
Width	
transport configuration:	2.45 m
Height	
transport configuration:	2.55 m
Weight	
standard:	8,850 kg
Mobility	
Carriage:	split trail
Firepower	
Armament:	1 × carriage mounted 155 mm howitzer
Main weapon	
calibre length:	45 calibres
rifling:	5.851 m
breech mechanism:	horizontal sliding block
muzzle brake:	single baffle
maximum range:	25,800 m (standard projectile)
maximum range:	39,500 m (ERFB-BB projectil)
Main armament traverse	
angle:	50° (25° left/25° right)
Main armament elevation/depression	
armament front:	+45°/-2.5°
Survivability	
Armour	
shield:	yes

Status

Production as required. In early 2000 India ordered this upgrade package in 155 mm/39-calibre and 155 mm/45-calibre configurations. As far as it is known India is the only customer for the upgraded 155 mm M-46S artillery system.

Contractor

Soltam Systems Limited (now part of the Elbit Group).

Soltam Systems 155 mm M114S howitzer

Development

A number of manufacturers have developed modernisation packages that utilise a 155 mm/39-calibre ordnance, for the old US 155 mm M114 towed howitzer which was developed before the Second World War and is now obsolete.

This is still used by a number of countries and details are provided in a separate entry in *IHS Jane's Land Warfare Platforms: Artillery and Air Defence.*

The upgraded Soltam Systems M114S utilises a 155 mm/33-calibre barrel enabling a standard 155 mm M107 HE projectile to be fired to a maximum range of 18,300 m compared to the 14,600 m of a standard US-developed M114 and 18,100 m when fired from a 39-calibre barrel.

The conversion can be carried out in the user's own facilities using a kit supplied by Soltam Systems or Soltam Systems can carry out the upgrade work.

The Israel Defense Force has about 50 155 mm M114 towed howitzers held in reserve, as far as it is known these have not been modified by Soltam Systems to the M114S standard. United Nations sources have shown that none of these weapons were exported by Israel between 1992 and 2010.

Description

The 155 mm/33-calibre M114 upgrade version does not require such extensive modifications to the existing weapon when compared to the other 155 mm/39-calibre modernisation packages and, according to Soltam Systems, it is a more cost-effective solution.

Soltam Systems 155 mm M114S upgraded howitzer in firing position. It should be noted that the spades have not been attached in this photograph (Soltam Systems) 1044604

The new 155 mm/33-calibre barrel has a high-efficiency muzzle brake and is fitted with the original breech ring and breech mechanism of the M114. The new ordnance has a constant 1:20 twist rifling with 48 lands and grooves. The muzzle brake is the same as that fitted to other Soltam Systems 155 mm towed artillery systems.

While 155 mm/33-calibre barrel has an overall length of 4,990 mm it has a chamber volume of 17.5 litres and a chamber diameter of 172 mm.

The rifling has 48 grooves with an inclination of 1:20 constant with a right hand twist.

The cradle remains unchanged although a front bushing has to be added by inserting a new bushing into the original one without the need for extensive machining. The existing spring-loaded equilibrators remain untouched but the upper location is changed, using new brackets bolted to the original points on the saddle.

All other assemblies and components remain untouched, as well as all the integrated logistics aspects.

A loader assist device based on a pneumatic ramming piston is provided as an add-on kit. The loader assist device enhances the rate of fire and loading accuracy as well as easing crew workload by reducing the need to lower the gun for loading.

The conversion kit includes the following:

- 155 mm/33-calibre tube and muzzle brake
- Pneumatic ramming system and its installation (the pneumatic cylinder is located on the right trail leg)
- Mounts and fixtures kit
- Direct firing telescope as an option
- Upgrading instruction and technical data pack.

The Soltam Systems upgraded M114S can fire all the standard types of 155 mm ammunition in worldwide use, starting with the M107 family of ammunition and ending with the latest ERFB–BB long-range ammunition family.

The following maximum ranges are obtained:	
M107 projectile	18,300 m
M56 projectile	19,100 m
ERFB-BB projectile	22,500 m

Due to improved gun balance, six crew members are enough for the deployment of the M114S instead of eight with the original M114.

The upgraded Soltam Systems 155 mm M114S howitzer is normally towed by a 5- to 10-tonne (6 × 6) cross-country truck which also carries its crew and a quantity of ready use ammunition (projectiles, charges and fuzes).

Specifications

	155 mm M114S Howitzer
Dimensions and weights	
Crew:	6
Length	
transport configuration:	9.20 m
Width	
transport configuration:	2.40 m
Height	
overall:	1.80 m
Mobility	
Carriage:	split trail
Firepower	
Armament:	1 × carriage mounted 155 mm howitzer
Main weapon	
calibre length:	33 calibres
rifling:	4.2 m
muzzle brake:	single baffle
maximum range:	18,300 m (M107 projectile)
maximum range:	22,500 m (ERFB-BB projectile)
Main armament traverse	
angle:	49° (25° left/24° right)
Main armament elevation/depression	
armament front:	+63°/-2°
Survivability	
Armour	
shield:	optional

Status

Development complete. Ready for production on receipt of orders.

Contractor

Soltam Systems Limited (now part of the Elbit Group).

Italy

Oto Melara 105 mm Model 56 Pack Howitzer

Development

The 105 mm Model 56 Pack Howitzer was developed by the Italian Army and produced by Oto Melara in the mid-1950s and entered production in 1957. Since then over 2,500 have been built and sales have been made to more than 30 countries.

The advantage of the Oto Melara Model 56 is that it can be dismantled into 11 subassemblies for transport across rough terrain, lifted by helicopters such as the Bell UH-1 and its profile can be lowered for anti-tank use.

According to the United Nations Arms Transfer List for the period 1992 through to 2010 the following quantities of 105 mm Model 56 Pack Howitzer were exported:

From	To	Quantity	Date	Comment
Canada	Brazil	8	1999	confirmed M-56
Canada	Brazil	8	2000	confirmed M-56
Italy	Bangladesh	115	1998	assumed M-56
Italy	Brazil	18	1995	confirmed M-56
Italy	Brazil	18	1997	assumed M-56
Italy	Nigeria	18	2003	confirmed M-56
Italy	Thailand	12	2002	confirmed M-56

Early in 1997, it was revealed that China North Industries Corporation (NORINCO) had built a weapon almost identical to the Oto Melara 105 mm Model 56 Pack Howitzer and details of this are available in a separate entry in *IHS Jane's Land Warfare Platforms: Artillery and Air Defence.* It is understood that most recent sales have been of surplus rather than new build weapons. Some sources have stated that Mexico has taken delivery of some NORINCO versions.

Ammunition for the Oto Melara 105 mm Model 56 Pack Howitzer is available from many sources and is the same as that used in the US M101 towed howitzer.

Oto Melara 105 mm Model 56 Pack Howitzer 0500386

In March 2013 Oto Melara confirmed that they still had the capability to manufacture new 105 mm Model 56 Pack Howitzer's but most export requests were for the resale of surplus weapons which could be upgraded prior to delivery.

The firing range of the weapon can also be improved with the fielding of ammunition with an enhanced range.

Description

The 105 mm barrel assembly of the Model 56 Pack Howitzer is composed of the barrel, breech ring, slipper and muzzle brake. The weapon has a multibaffle muzzle brake.

The vertical sliding breech block has a percussion firing mechanism which can be cocked automatically or by hand. Firing is either by hand lever or by lanyard.

The gun carriage comprises the undercarriage, saddle, cradle, upper and lower recoil systems, two balancing springs, two sights and the shield.

The undercarriage consists of a cross head housing the saddle centre pivot pin, two stub-axles, spring equalising rocker arm and two articulated arms. Each of the stub-axle supporting arms has two bearings into which the stub-axle can be fitted, this carries the wheel assembly and its handbrake. Each of the articulating brackets carries a trail leg of two or three sections, which can be folded up and interlocked. The end section of the trail may be anchored to the ground by a knife or rock spade.

The saddle, with its centre pivot mounting, is provided with two side members with rear-mounted lugs that carry the elevating and traversing gear. The recoil system is contained in two cradles, upper and lower, which slide on top of each other.

The upper cradle contains the two recuperator-buffers and the lower cradle contains one buffer and two recuperators. The two balancing springs have helical springs operating in sliding sleeves and the cradle is mounted on two trunnion bearings.

The recoil system of the 105 mm Model 56 Pack Howitzer consists of a hydraulic buffer and a helical recuperator.

The sighting equipment consists of an indirect sight on the left and an anti-tank direct sight on the right. The splinter-proof shield mounted on the saddle is in two halves, both of which can be removed to lighten the weapon.

As a field weapon the 105 mm Pack Howitzer wheels are overslung and the barrel has an elevation of +65°, a depression of –5° and a traverse of 18° left and right. As an anti-tank weapon the wheels are underslung and the barrel has an elevation of +25°, a depression of –5° and a traverse of 28° left and right. When deployed in the anti-tank role it has a lower profile and increased traverse arc.

It should be noted that the specifications in the table are for the 105 mm Model 56 Pack Howitzer in the normal indirect fire role.

When being used in the lower profile anti-tank role it has the following specifications:

Length firing:	4.7 m
Width firing:	4.0 m
Height firing:	1.55 m
Axis of bore firing:	0.693 m
Track:	1.32 m

When being used in the anti-tank role firing High-Explosive Anti-Tank (HEAT) ammunition it has a maximum rate of fire of 8 rds/min.

The weapon can be dismantled into 11 subassemblies in three minutes (the heaviest of which weighs 122 kg) and can be reassembled in four minutes.

The 105 mm Pack Howitzer fires the same range of ammunition as the US 105 mm M101; HE (M1) with the complete round weighing 21.06 kg, a muzzle velocity of 472 m/s and the fixed HEAT (M67) round, which weighs 16.7 kg, has a muzzle velocity of 387 m/s and will penetrate 102 mm of conventional steel armour.

NORINCO is now marketing two types of 105 mm projectile for use with the locally manufactured 105 mm light howitzer which is a copy of the Oto Melara 105 mm Model 56 Pack Howitzer.

This 105 mm ammunition can also be fired from other towed and self-propelled artillery systems.

Oto Melara 105 mm Pack Howitzer in travelling configuration with three-part trail stowed one above the other 0500387

The two types of 105 mm ammunition marketed by NORINCO are High-Explosive (HE) and Hollow Base High-Explosive Incendiary (HB/HEI).

Brief details of these 105 mm projectiles are as follows:

Type:	**HE**	**HB/HEI**
Length:	492 mm	599 mm
Weight:	15 kg	15 kg
Muzzle velocity:		
(from M56 PH)	420 m/s	n/avail
(from M101)	472 m/s	n/avail
Maximum range:		
(from M56 PH)	10.2 km	11 km
(from M101)	11.27 km	12 km
Notes:		
(1) HE projectile is available with a Composition B or TNT filling		
(2) HB/HEI has a claimed lethal radius of 35 m		

It is normally towed by a long wheelbase Land Rover or similar 4 × 4 which carries some of its crew and some of its ready use 150 mm ammunition with a similar vehicle carrying remainder of crew and ammunition.

If a larger vehicle is used such as a 2.5- or 5-tonne (4 × 4) cross-country truck then all of the gun crew and ready use ammunition can be carried.

Specifications

	105 mm Model 56 Pack Howitzer
Dimensions and weights	
Crew:	7
Length	
overall:	4.8 m (firing)
transport configuration:	3.65 m
Width	
overall:	2.9 m (firing)
transport configuration:	1.5 m
Height	
overall:	1.93 m (firing)
transport configuration:	1.93 m
axis of fire:	1.07 m
Track	
vehicle:	1.14 m
Weight	
standard:	1,290 kg
Mobility	
Carriage:	split trail
Firepower	
Armament:	1 × carriage mounted 105 mm howitzer
Main weapon	
length of barrel:	1.478 m
breech mechanism:	vertical sliding block
muzzle brake:	multibaffle
maximum range:	10,575 m
Rate of fire	
sustained fire:	4 rds/min
Main armament traverse	
angle:	36° (56° when in anti-tank configuration)
Main armament elevation/depression	
armament front:	+65°/-5° (+25°/-6° when in anti-tank configuration)
Survivability	
Armour	
shield:	optional

Status

Production as required. In service with the following countries:

Country	Quantity	Comment
Argentina	82	used by army and marines (12)
Bangladesh	56	
Botswana	6	
Brazil	100	including 36 supplied from Italy and 16 from Canada
Chile	54	
Ecuador	24	
Greece	18	
India	50	
Kenya	8	
Malaysia	100-130	
Mexico	24 plus	believed to be Chinese NORINCO model
Nigeria	50	from Italy in 2003
Nepal	14	
Pakistan	110	
Peru	40	

Country	Quantity	Comment
Philippines	n/avail	
Portugal	24	
Sierra Leone	n/avail	unconfirmed user
Somalia	90	
Spain	170 + 12	army + marines
Thailand	12	from Italy in 2002
Venezuela	40	marines also have 12
Zambia	18	

Contractor
Oto Melara SpA.

Korea (South)

WIA Corporation 155 mm KH179 howitzer

Development
Development of the KH179 155 mm howitzer by the Kia Machine Tool Company (which is today known as the WIA Corporation) began in 1979 and was completed in 1982. It has been in service with the Republic of Korea Army since early 1983.

The conversion programme is believed to have been completed some years ago.

It is estimated that South Korea has about 1,800 155 mm towed artillery systems but it is not known as to how many of these are M114s upgraded to the KH179 standard.

It is understood that there has been no recent production of the 155 mm KH197 howitzer.

It has been offered on the export market but as of March 2013 there are no known export sales.

Description
The 155 mm KH179 howitzer is a conversion of the American M114A1 howitzer and carriage upgraded to mount a new 155 mm/39-calibre barrel. Certain other parts of the carriage have been altered to suit the new 155 mm barrel and new fire-control equipment has been fitted.

The KH179 is light enough to be airlifted by a Boeing CH-47C/D Chinook helicopter and can also be carried inside a Lockheed Martin C-130 Hercules transport aircraft.

The new 155 mm/39-calibre barrel is of monobloc construction with an interrupted screw breech mechanism. The rifling has a constant 1:20 twist and there are 48 grooves. High-strength alloy steel is used throughout the barrel construction and heat treatment and autofrettaging are employed to provide maximum barrel life. Separate recoil and recuperator systems are used in what is stated to be a variable length, hydropneumatic, independent design.

Maximum recoil length is quoted as 1.524 m and minimum recoil length is quoted as 1.041 m.

On the carriage, the layer is on the left of the breech to operate the traverse handwheel on the left of the top carriage. Another member of the gun crew turns the elevating handwheel on the right of the breech. The layer is provided with a dial sight and telescope for aiming and another telescope for direct fire. All fire controls and some of the accessory equipment on the KH179 are equipped with radio-activated light sources. Pneumatic equilibrators are connected between the yoke of the recoil mechanism and the top carriage, to keep the handwheel loads within acceptable limits.

For indirect firing, a panoramic telescope with a ×4 magnification is fitted on the left side. For direct firing, the elbow telescope with a ×3.5 magnification and a 13.5° field of view is mounted on the right and has an effective performance up to 1,500 m.

The KH179 can fire existing NATO 155 mm ammunition and can also accommodate enhanced performance ammunition such as the ERFB projectiles. Range with RAP is stated to be 30,000 m and for normal HE projectiles the range is 22,000 m. Maximum rate of fire is 4 rds/min, dropping to 2 rds/min for sustained fire.

155 mm KH179 howitzer in travelling configuration 0130003

To reduce weight, the shield of the M114 has been removed on the upgraded 155 mm/39-calibre KH179 weapon.

It is normally towed by a 6 × 6 cross-country truck that in addition to its crew carries a quantity of ready use ammunition (projectiles, changes and fuzes).

Towing speed on good roads is a maximum of 70 km/h with a maximum towing speed of 16 km/h cross-country, but this depends of the type of terrain.

Variants
Iranian 155 mm howitzer

The Iranian DIO has produced a very similar weapon to the K179 howitzer called the HM41. Details are of this are given earlier in a separate entry in *IHS Jane's Land Warfare Platforms: Artillery and Air Defence.* There are no known exports of the Iranian HM41. Illustrations of the Iranian 155 mm HM41 howitzer are identical to those of the KH179.

Specifications

	155 mm KH179 Howitzer
Dimensions and weights	
Length	
transport configuration:	10.389 m
Width	
overall:	2.438 m
Height	
transport configuration:	2.77 m
Ground clearance	
overall:	0.28 m
Weight	
standard:	6,890 kg
Mobility	
Carriage:	split trail
Speed	
towed:	70 km/h
Firepower	
Armament:	1 × carriage mounted 155 mm howitzer
Main weapon	
length of barrel:	7.013 m
calibre length:	39 calibres
breech mechanism:	interrupted screw
muzzle brake:	double baffle
maximum range:	22,000 m (M107 projectile)
maximum range:	30,000 m (RAP projectile)
Rate of fire	
sustained fire:	2 rds/min
rapid fire:	4 rds/min
Main armament traverse	
angle:	48.7° (23.5°left/25.2° right)
Main armament elevation/depression	
armament front:	+68°/-0°
Survivability	
Armour	
shield:	no

Status
Production complete. In service with the Republic of Korea Army. As far as it is known this weapon has not been exported.

Contractor
WIA Corporation.

WIA Corporation 105 mm KH178 light howitzer

Development
In the preliminary stages of the design of a new 105 mm light howitzer for use by the Republic of Korea Army, the Kia Machine Tool Company (today the WIA Corporation) procured examples of the now BAE Systems 105 mm Light Gun and the German Rheinmetall conversion of the US 105 mm M101 howitzer.

Aspects of both 105 mm weapons were subsequently incorporated in the KH178 105 mm light howitzer. Production of the KH178 started in 1984 and it is now in service with the Republic of Korea Army. This conversion programme is believed to have been completed some years ago.

It is believed that South Korea took delivery of about 1,700 M101s from the United States and that a significant portion of these were upgraded to the 105 mm KH178 configuration.

Description
The KH178 comprises the CN78 105 mm cannon, the RM78 recoil mechanism, the CG78 carriage and a new fire-control system. The barrel is 34-calibres long and is fitted with a double-baffle muzzle brake.

The overall barrel length is being quoted as 4.48 m with the actual tube length being quoted as 3.922 m.

A horizontal sliding breech block is used with a percussion firing mechanism. The barrel is autofrettaged and has a life of 7,500 EFC. Progressive twist rifling is used, changing from 1:35 to 1:18 at the muzzle.

105 mm KH178 howitzer in firing position. Note that the spades are not bedded into the ground 0500388

The recoil mechanism is of the hydropneumatic type with a constant recoil length. The recoil and recuperator cylinders are mounted over and under the barrel. In order to reduce elevation handwheel loads, the equilibrators are of a new type using a hydropneumatic and spring system.

On the split trail carriage the trail legs are reinforced and a shield may be fitted. Run-flat tyres are used. All the fire-control equipment and some accessory devices are fitted with self-luminous sources.

The KH178 can fire all current standard 105 mm ammunition plus a new single-increment super propellant M200. This gives the weapon a range of 14,700 m. Firing a Rocket-Assisted Projectile (RAP) the range is 18,000 m.

It is normally towed by a 2.5-tonne 6 × 6 cross-country truck that in addition to its crew carries a quantity of ready use ammunition (projectiles, charges and fuzes).

Towing speeds on a good road is up to 56 km/h but cross-country this is reduced to 16 km/h, but this depends on the type of terrain.

Specifications

	105 mm KH178 Light Howitzer
Dimensions and weights	
Width	
transport configuration:	2.146 m
Height	
overall:	1.575 m
Ground clearance	
overall:	0.386 m
Weight	
standard:	2,650 kg
Mobility	
Carriage:	split trail
Speed	
towed:	56 km/h
Firepower	
Armament:	1 × carriage mounted 105 mm howitzer
Main weapon	
length of barrel:	4.48 m
calibre length:	34 calibres
muzzle brake:	double baffle
maximum range:	14,700 m (standard ammunition)
maximum range:	18,000 m (RAP projectile)
Rate of fire	
sustained fire:	5 rds/min
rapid fire:	15 rds/min
Main armament traverse	
angle:	45.5°
Main armament elevation/depression	
armament front:	+65°/-5°
Survivability	
Armour	
shield:	optional

Status

Production complete. In service with the Republic of Korea Army. It is understood that the Chilean Marines have taken delivery of 16 105 mm KH178 weapons. It has been estimated that the Republic of Korea has about 1,700 105 mm towed artillery systems of all types in service. These are being supplemented by an increasing number of 155 mm K9 Thunder self-propelled weapons.

Contractor

WIA Corporation.

Romania

152 mm Model M1985 gun-howitzer

Development

This 152 mm Model M1985 gun-howitzer was developed to meet the requirements of the Romanian Army as a follow on weapon to the Romanian 152 mm M1981 howitzer which is covered in a separate entry in *IHS Jane's Land Warfare Platforms: Artillery and Air Defence.*

Romania has completed production of the 152 mm Model M1985 gun-howitzer but quantities of these weapons are now surplus to requirements and are available for export.

The only known export customer as of March 2013 was Nigeria who took delivery of four 152 mm M1985/M1981 systems in 1993.

Romanian 152 mm M1985 gun-howitzer deployed in the firing position and with ordnance at maximum elevation. Note that the spades have not been bedded in 0130022

Romanian 152 mm M1985 gun-howitzer in travelling configuration 0500606

Description

The weapon is mounted on a split trail two-wheeled carriage, with each of the trails provided with a castor wheel to assist in rapidly bringing the weapon into and taking it out of action.

The barrel of the 152 mm M1985 gun-howitzer has an overall length of 7.195 m without its muzzle brake with the barrel having 48 grooves . Maximum recoil is being quoted as 910 mm.

The carriage appears to be very similar to that used for the Russian 152 mm D-20 gun-howitzer although the Romanian weapon has a much longer 152 mm barrel and therefore a longer range. Ammunition for this weapon is made in Romania.

The Romanian 152 mm Model 1985 fires an HE projectile weighing 43.56 kg, maximum muzzle velocity of 825 m/s, with the propelling charge weighing 11.95 kg to a maximum range of 24,000 m.

According to the manufacturer, the 152 mm Model M1985 gun-howitzer fires a number of separate loading (projectile and charge) 155 mm ammunition types including:

- OF-540 high-explosive incendiary projectile (maximum range of 24,000 m)
- OF-550 high-explosive incendiary projectile (maximum range of 22,000 m)
- BR-540B armour-piercing projectile.

When deployed in the firing position, the 152 mm M1985 gun-howitzer is supported at three points, two spades at the rear and a circular firing jack which is mounted under the forward part of the carriage. In the firing position the castor wheels are swung upwards and rest on top of the trails.

Variants

Romanian sources have stated that there is another 152 mm gun-howitzer called the Model 1984. This has an identical range and barrel length as the Model M1985 but elevation limits are from –5° to +57° with weight in firing position being 7,500 kg.

Croatia is understood to have taken delivery of the M1985 under the local designation of the M84H1. This could well be the export Model 1984.

Types of 152 mm projectile fired are HE-FRAG PF-540 and OF-540 RM with a maximum muzzle velocity of 822 m/s and APC-T BR-540B.

Specifications

	152 mm Model M1985 Gun-Howitzer
Dimensions and weights	
Length	
transport configuration:	11.70 m
Width	
transport configuration:	2.53 m
Ground clearance	
overall:	0.35 m
Weight	
standard:	7,550 kg
combat:	7,500 kg
Mobility	
Carriage:	split trail
Firepower	
Armament:	1 × carriage mounted 152.4 mm howitzer
Main weapon	
length of barrel:	8.025 m
rifling:	6.102 m
muzzle brake:	double baffle
maximum range:	22,000 m (OF-550 HEI projectile)
maximum range:	24,000 m (OF-540 HEI projectile)
Rate of fire	
rapid fire:	4 rds/min
Main armament traverse	
angle:	50°
Main armament elevation/depression	
armament front:	+57°/-5°
Survivability	
Armour	
shield:	yes

Status

Production complete. In service with the Romanian Army. It is understood that Romania has a total of 103 of these weapons in service. According to United Nations sources, Nigeria took delivery from Romania of four 152 mm M1985/M1981 systems in 1993.

Contractor

Arsenalul Armatei.

152 mm M1981 howitzer

Development

The 152 mm M1981 howitzer was developed in Romania to meet the operational requirements of the Romanian Army. Production of this weapon was completed many years ago and it is no longer marketed. Surplus weapons could be offered on the export market.

Romanian 152 mm M1981 howitzer in travelling configuration (PB) 0085644

Description

It has split box-type trails, with a castor wheel being provided on each trail which assists the crew in bringing the weapon into action more quickly.

When not required the 152 mm barrel is swung back and rests over the closed trails. The carriage of this weapon is similar in some respects to the Russian 152 mm gun-howitzer D-20 and 122 mm field gun D-74, as a circular firing pedestal is fitted as standard and is inverted and secured just forward of the shield when travelling.

When in the firing position, the circular firing pedestal is lowered to the ground allowing the weapon to be traversed fairly quickly through 360° using the castor wheels mounted on each trail leg.

The 152 mm ordnance is fitted with a large double-baffle muzzle brake with a distinctive circular ring to its rear.

The carriage used by this weapon is very similar to that used by the Romanian 130 mm gun Model 1982, which is covered in a separate entry. This is very similar to the Russian 130 mm M-46 field gun.

Types of separate loading 152 mm ammunition (for example, projectile and charge) fired by the 152 mm M1981 howitzer include the following:

- HE full variable charge, projectile weight 43.56 kg, maximum range 17,000 m
- HE full variable charge OF-550, projectile weight 43.56 kg, maximum range 24,000 m
- HE reduced variable charge, projectile weight 43.56 kg, maximum range 13,000 m
- HEAT, projectile weight 36.5 kg, maximum range 13,000 m
- HE full variable charge (APC-550), projectile weight 43.56 kg, maximum range 17,000 m
- HE full variable charge (CG-540), projectile weight 44.6 kg, maximum range 17,000 m
- Leaflet, projectile weight 43.56 kg, maximum range 17,000 m
- Illuminating, projectile weight 43.56 kg, maximum range 17,000 m.

Status

Production complete. In service with the Romanian Army. It is understood that the Romanian Army has 241 of these weapons in service. Recent information has indicated that Romania may be considering converting some of its 152 mm weapons into the NATO standard 155 mm. According to United Nations sources, in 1992 Nigeria took delivery from Romania of four 152 mm M1981/M1985 towed howitzers. Romania has about 41 of the Russian 122 mm M-1938 (M-30) howitzers held in reserve.

Contractor

Arsenalul Armatei.

130 mm gun Model 1982

Development

The Romanian 130 mm Model 1982 field gun was developed to meet the requirements of the Romanian Army but as far as it is known none of these weapons remain in front line service. Some of these surplus weapons have been exported and available details of these are provided in the Status section of this entry.

It is possible that this weapon was developed with Russian assistance or based on Russian technology transfer.

In many respects the Romanian 130 mm gun Model 1982 is equivalent in range capabilities to the widely deployed Russian 130 mm field gun M-46 which is no longer in front line service with the Russian Army but is still widely deployed all over the world.

The main difference is that the Romanian 130 mm gun Model 1982 does not have a two wheeled limber which is attached to the closed trails.

Romanian 130 mm gun Model 1982 in travelling configuration 0501045

Description

Some Romanian sources have also referred to this as the 130 mm M-46 field gun as its ballistic characteristics are almost the same. They also fire the same family of separate loading 130 mm ammunition.

There are however differences in weight and physical characteristics between the Romanian Model 1982 and Russian M-46 weapons.

It has a long barrel with a double-baffle muzzle brake and no fume extractor is fitted. When deployed in the firing position, it rests on the two spades and a circular firing jack under the forward part of the carriage. When not required for use, the firing jack is swung upwards through 180° and is locked in position under the barrel.

The carriage and ordnance appear to have no relationship with the Russian 130 mm M-46 field gun. Each trail leg has a castor wheel to assist in bringing the weapon into action. According to the manufacturer, the 130 mm gun Model 1982 takes between two and three minutes to bring into action with a well-trained crew.

When being used in the direct fire mode it has an effective quoted range of 1,100 m.

Types of separate loading ammunition (for example, projectile and charge), fired by the 130 mm gun Model 1982 include HE full charge and HE reduced charge, both of which can also be fired from the Russian 130 mm M-46 field gun.

While its maximum quoted towed speed on a good road surface is 50 km/h but this is reduced to around 20 to 25 km/h when towed cross-country but this figure depends on the nature of the terrain.

Type	HE	HE
	(full charge)	(reduced charge)
Fuze:	V-429	V-429
Total weight:	59 kg	52 kg
(cartridge)	25.6 kg	18.6 kg
(projectile)	33.4 kg	33.4 kg
Muzzle velocity:	930 m/s	705 m/s
Max range:	27,150 m	22,000 m

The manufacturer states that types of projectile fired by this system include the following:

- OF-482M high-explosive incendiary
- BR-482B armour piercing
- and newly created dedicated ammunition.

According to the United Nations Arms Transfer Lists for the period 1992 through to 2010, the following quantities of 130 mm Model 1982 weapons were exported by Romania. As far as it is known all of these were surplus weapons rather than new build:

From	To	Quantity	Date
Romania	Bosnia	8	1998
Romania	Cameroon	12	1992
Romania	Guinea	12	2000
Romania	Nigeria	4	1992
Romania	Nigeria	7	1993

Specifications

	130 mm Gun Model 1982
Dimensions and weights	
Crew:	7
Length	
transport configuration:	10.80 m
Width	
transport configuration:	2.59 m
Height	
transport configuration:	2.65 m
Weight	
standard:	6,200 kg
combat:	6,150 kg
Mobility	
Carriage:	split trail
Speed	
towed:	50 km/h
Firepower	
Armament:	1 × carriage mounted 130 mm howitzer
Main weapon	
length of barrel:	6.85 m
muzzle brake:	double baffle
maximum range:	22,000 m (HE reduced charge)
maximum range:	27,150 m (HE reduced charge)
Rate of fire	
rapid fire:	7 rds/min
Main armament traverse	
angle:	49° (27° left/22° right)
Main armament elevation/depression	
armament front:	+45°/-2°
Survivability	
Armour	
shield:	yes

Status

Production complete. Recent information is that these are no longer in front line service with Romania. Known export sales of surplus Romanian 130 mm gun Model 1962 artillery weapons include Bosnia-Herzegovina (8), Cameroon (12), Guinea (12) and Nigeria (4 + 7).

It is possible that Romania still retains quantities of surplus 130 mm gun Model 1982 for sale overseas.

Contractor

Arsenalul Armatei.

100 mm towed anti-tank gun M1977

Development

The 100 mm towed anti-tank gun M1977 was specifically developed by the Romanian defence industry to meet the requirements of the Romanian Army for a towed weapon to engage tanks and other armoured fighting vehicles under both day and night conditions, as well as being used in the fire support role.

The Romanian Army nickname for the 100 mm towed anti-tank gun M1977 is the Misdrache. The weapon is also referred to by the manufacturer as the 100 mm anti-tank gun md.1977.

It is understood that the Romanian Army has a total of 208 M1977 (or Gun 77 as they are also referred to) and 72 M1975 (or Gun 75 as they are referred to).

It is believed that the M1975 is an earlier model of the M1977 with similar capabilities. For some reason Israel took delivery of one 100 mm anti-tank gun from Romania in 1993. This could have been for trials work.

Romania also has about 25 Russian supplied SU-100 assault guns armed with a 100 mm gun.

Description

The carriage of the 100 mm towed anti-tank gun is of the split trail type with one castor wheel being provided to assist in bringing the ordnance into action.

The 100 mm rifled ordnance of the M1977 towed anti-tank gun has an overall length of 5,350 mm of which 4,630 mm is rifling with 40 grooves.

The width of the lands is 3.5 mm ±0.3 mm with the grooves being 4.7 mm ±0.33 mm.

The normal recoil is between 750 and 830 mm with a maximum recoil of 840 mm with a typical recoil cycle being between one and two seconds.

The fixed one-piece 100 mm ammunition fired by the M1977 is identical to that fired by the 100 mm D-10 series guns installed in the Russian T-54/T-55 MBTs, so the ordnance may well be related.

Romanian sources have stated that this ammunition can also be fired from the Russian 100 mm field gun M1944 (BS-3) which was first fielded in 1944.

Romanian sources give the M1977 a maximum range of 4,000 m against tanks and armoured vehicles in the indirect fire mode and 1,100 m in the direct fire mode firing an armour-piercing projectile against a target height of 2 m. In the indirect fire mode using HE ammunition it has a maximum range of 20,700 m.

To bring the weapon into action from its travelling configuration takes 60 seconds. Each trail is provided with a lifting device to assist in bringing the weapon into action.

Romanian sources provide the 100 mm fixed ammunition data in the table.

Romanian sources also state that it also fires a BM-412Sg APFSDS projectile with a muzzle velocity of 1,400 m/s that will penetrate 444 mm of conventional steel armour at an inclination of 90° or at a range of 500 m or 328 mm at a range of 4,000 m.

In addition to the previously mentioned Russian rounds, the 100 mm M1977 can also fire these locally manufactured 100 mm projectiles (see table at the bottom of the next page):

Optical equipment installed on the 100 mm towed anti-tank gun M1977 includes a mechanical sight, a panoramic sight with a magnification of ×4 and a 10° field of view, an optical sight with magnifications of ×4 and ×8 and a collimator.

Romanian 100 mm towed anti-tank gun M1977 in travelling configuration with spades closed and castor wheel in position under rear of closed trails
0130023

Specifications

	100 mm Towed Anti-Tank Gun M1977
Dimensions and weights	
Crew:	8
Length	
transport configuration:	9.37 m
Width	
transport configuration:	2.24 m
Height	
transport configuration:	1.9 m
axis of fire:	0.9 m
Ground clearance	
overall:	0.36 m
Firepower	
Armament:	1 × carriage mounted 100 mm rifled gun
Main weapon	
calibre length:	30 calibres
rifling:	4.63 m
muzzle brake:	multibaffle
maximum range:	20,700 m (OF-412 HE-FRAG)
maximum range:	4,000 m (APC-T)
maximum range:	2,200 m (HEAT)
Main armament traverse	
angle:	54° (27° left/27° right)
Main armament elevation/depression	
armament front:	+37°/-5°
speed:	0.76°/s
Survivability	
Armour	
shield:	yes

Status
Production complete. In service with Romanian Army. Iraq is understood to have taken delivery of some of these systems and these were used in the war against Iran.

It is possible that quantities of surplus 100 mm towed antitank gun M1977 are available for export.

Contractor
Arsenalul Armatei.

98 mm Model 93 mountain howitzer

Development
For some years it has been known that the Romanian Army deployed a 76.2 mm mountain gun and a 100 mm mountain howitzer.

Under the Conventional Forces Europe Treaty each country has had to declare its tracked and self-propelled artillery systems, and mortars, with a calibre of more than 100 mm.

The 76.2 mm mountain gun did not have to be declared and the 100 mm mountain howitzer was never declared. The reason for this weapon not being disclosed is that its actual calibre is 98 mm and therefore it did not have to be declared in any case. Recent information has revealed that it is called the 98 mm Model 93 mountain howitzer.

Photographs now reveal that the 76.2 mm mountain gun, which has no known designation, and the 100 mm Model 93 weapons are similar in some respects and it is possible that the 98 mm Model 93 weapon is a further development of the older 76.2 mm weapon. It is understood that production of these weapons was completed some time ago.

Description
The Romanian 98 mm Model 93 mountain howitzer is mounted on a conventional split trail carriage with the rear half of each carriage folding upwards and forwards through about 180° to rest on the front part of the trail when travelling.

The trails are then closed together and a locking pin with towing eye attached is fitted. Under this, a small rubber-tyred roadwheel is mounted to make deployment of the weapon easier.

When deployed in the firing position, spades help to absorb the recoil although stakes can be driven into the ground through the rear part of each trail if required.

The carriage has conventional rubber tyres, the wheels of which are provided with towing eyes. Positioned on the forward part of the carriage are two handbrakes, which are used when deploying the 98 mm Model 93 mountain howitzer.

Elevation is manual from −5° to +65° with traverse being manual from 25° left and right. The elevation and traverse controls are located on the left side of the carriage where the direct and indirect sights are located. The weapon is provided with a small shield.

The 98 mm ordnance is fitted with a multibaffle muzzle brake and typical quoted rate of fire is 6 rds/min.

This weapon fires ammunition of the separate loading type, for example projectile and charge which is in a brass cartridge case. Two natures of 98 mm high-explosive ammunition are fired, the OF-402 and the OF-403.

Both of these are packed three rounds to a case and are fitted with a nose-mounted fuze that can be set for impact, short delay and delayed functioning.

The only other known nature of ammunition is a fin-stabilised High-Explosive Anti-Tank (HEAT) type, which is packed three to a case and fitted with a piezoelectric fuze. This HEAT projectile has limited effectiveness against the latest compound and explosive reactive armour systems.

If required, the weapon can be quickly disassembled into three loads, each of which is carried on a carriage towed by a horse. Its normal towing vehicle is a DAC 444 (4 × 4) 2.5-tonne truck, which carries the crew and a supply of ready-use 98 mm ammunition.

Type	HE-FRAG	APC-T	HEAT	HEAT
Designation:	OF-412	BR-412D	3BK-5M	BK-412R
Muzzle velocity:				
(max)	900 m/s	887 m/s	900 m/s	950 m/s
(reduced)	600 m/s	n/a	n/a	n/a
Weight:				
(propellant)	5.5 kg	5.5 kg	4.5 kg	4.5 kg
(projectile)	15.6 kg	15.88 kg	12.2 kg	10.625 kg
Maximum range:	20,700 m	4,000 m	2,200 m	2,200 m

Notes:
(1) Also BR-412 and BR-412B with a muzzle velocity of 895 m/s
(2) Weight of reduced propellant charge is 4.5 kg

Type	HE (full charge)	HE (reduced charge)	AP-T	HEAT
Fuze:	V-429	V-429	MD-8	GPV-2R
Total weight:	30.27 kg	26.74 kg	30.40 kg	21 kg
(cartridge)	11.5 kg	8.29 kg	11.4 kg	10.47 kg
(projectile)	15.6 kg	16.66 kg	15.88 kg	10.625 kg
Maximum range:	16,000 m	11,000 m	4,000 m	3,000 m
Direct fire range:	–	–	1,000 m	1,000 m
Muzzle velocity:	900 m/s	600 m/s	887 m/s	950 m/s
Armour penetration:	–	–	175 mm	80 mm

Ammunition specifications

Type	HE	HE	HEAT
Designation:	OF-402	OF-403	nil
Fuze designation:	UP-2	UP-2	GPV-2
Length:			
(total)	756 mm	759 mm	758 mm
(cartridge case)	241 mm	241 mm	241 mm
(projectile)	515 mm	518 mm	517 mm
Weight:			
(total)	18 kg	18 kg	16 kg
(cartridge)	2.8 kg	2.8 kg	2.8 kg
(projectile)	13.2 kg	14.6 kg	12.2 kg
Muzzle velocity:	235/470 m/s	225/430 m/s	470 m/s
Maximum range:	10,800 m	10,500 m	10,800 m
Notes: (1) Depends on charge			

Specifications

	98 mm Model 93 Mountain Howitzer
Dimensions and weights	
Crew:	8
Width	
transport configuration:	1.65 m
Height	
overall:	1.70 m
Weight	
standard:	1,500 kg
combat:	1,500 kg
Mobility	
Carriage:	split trail
Firepower	
Armament:	1 × carriage mounted 98 mm howitzer
Main weapon	
muzzle brake:	multibaffle
maximum range:	10,800 m (indirect)
maximum range:	600 m (direct)
Rate of fire	
rapid fire:	6 rds/min
Main armament traverse	
angle:	40° (17° left/23° right)
Main armament elevation/depression	
armament front:	+70°/-5°
Survivability	
Armour	
shield:	yes

Romanian 98 mm Model 93 mountain howitzer in travelling configuration 0500604

Romanian 98 mm Model 93 mountain howitzer in firing configuration 0500605

Status
Production complete. No longer marketed. In service with Romanian forces. There are no known exports of this weapon.

It is possible that surplus quantities of the 98 mm Model 93 mountain howitzer are available for export.

Contractor
Arsenalul Armatei.

76.2 mm mountain gun Model 1984

Development
The Romanian-developed 76.2 mm Model 1984 mountain gun was specifically developed to meet the requirements of Romanian mountain units and can be quickly disassembled without specialised tools into eight pack-animal loads. Production is now complete.

It has been offered on the export market but there has been no known sales.

It is possible that surplus weapons could be offered on the export market.

Description
The crew of seven consists of commander, aimer, breech mechanism operator, loader, fuze handler, projectile handler and charge handler.

When being towed by a vehicle or horse team, the weapon can be brought into action in about one minute.

When being carried by pack horse, the weapon takes between six and eight minutes to come into action. This takes longer as the individual sections have to be assembled. Each trail leg consists of two parts that can be rapidly disassembled.

Types of 76.2 mm semi-fixed ammunition fired include high explosive (maximum muzzle velocity 398 m/s), smoke and High-Explosive Anti-Tank (HEAT), which has limited capabilities against the more recent armour systems.

The ordnance has a maximum recoil length of 830 mm.

The elevation limits depend on its configuration and are +45° to –15° with lengthened trails and +60° to 0° with shortened trails.

Rates of fire quoted are:

- First minute - 25 rounds
- First three minutes - 40 rounds
- First five minutes - 50 rounds
- First 10 minutes - 60 rounds
- First 15 minutes - 70 rounds
- First 30 minutes - 90 rounds
- First 60 minutes - 120 rounds
- First 120 minutes - 210 rounds
- Each additional 60 minutes - 80 rounds
- Life of barrel - 15,000 rounds.

As stated in the specifications the maximum effective range when firing an HEAT projectile is 1,000 m.

Effective range against a target with a height of 2 m is 460 m and effective range against a target with a height of 3 m is quoted as 550 m.

According to the manufacturer, the 76 mm mountain gun (or howitzer as it is also referred to), fires the following ammunition types as listed below. It should be noted that some of these with the M series designation were also used by the former Yugoslavia 76 mm mountain gun M48 which is closely related to the Romanian 76.2 mm mountain gun/howitzer Model 1984.

Romanian 76.2 mm Model 1984 mountain howitzer deployed in the firing position with stakes ready to be driven into the ground 0130024

Romanian 76.2 mm Model 1984 mountain howitzer with one trail leg either side, and showing maximum elevation (Arsenalul Armatei) 0567809

- Charge round - HE OF-350M, fuze UP3
- Charge round - M55, fuze M51A5
- Charge round - smoke WP M60, fuze PD M48A3
- Charge round - smoke D-350 TM, fuze UP3
- Charge round - HEAT BK 354, piezoelectric fuze CVP-312
- Charge round - shell M50, fuze PD MK-451.

Optical equipment fitted to the weapon consists of a day telescope with a magnification of ×3. This is mounted on the left side of the carriage.

It has recently been revealed that the 76.2 mm mountain gun is related to the 98 mm Model 93 mountain howitzer covered in a separate entry in *IHS Jane's Land Warfare Platforms: Artillery and Air Defence.* Production of this weapon has been completed.

As far as is known the 76.2 mm and 98 mm Model 93 mountain howitzers have never been exported although they have been offered on the export market.

Specifications

	76.2 mm Model 1984
Dimensions and weights	
Crew:	7
Length	
overall:	3.10 m (firing)
transport configuration:	2.45 m
Width	
overall:	2.65 m (firing)
transport configuration:	1.33 m
Height	
overall:	1.66 m
Ground clearance	
overall:	0.18 m
Track	
vehicle:	1.28 m
Weight	
standard:	730 kg
combat:	722 kg
Firepower	
Armament:	1 × carriage mounted 76.2 mm howitzer
Main weapon	
length of barrel:	1.178 m
muzzle brake:	multibaffle
maximum range:	8,600 m (indirect HE)
maximum range:	1,000 m (direct HEAT)
Rate of fire	
sustained fire:	2 rds/min
rapid fire:	25 rds/min
Main armament traverse	
angle:	50°
Main armament elevation/depression	
armament front:	+45°/-15°
Survivability	
Armour	
shield:	yes

Status

Production complete. In service with Romanian Army.

It is considered possible that there are quantities of surplus 76.2 mm Model 1984 mountain howitzers for sale.

Contractor

Arsenalul Armatei.

Russian Federation

155 mm 2A45M-155 artillery system

Development

The 155 mm 2A45M-155 artillery system has been developed by Artillery Plant No 9 at Yekaterinburg who also developed the widely deployed Russian 122 mm D-30 howitzer which is covered in a separate entry in *IHS Jane's Land Warfare Platforms: Artillery and Air Defence.*

Today Artillery Plant No 9 is known as the Joint Stock Company Spetstehnika and more recently has developed and tested the 125 mm 2A45M (Sprut-B) towed anti-tank gun.

This is covered in a separate entry in *IHS Jane's Land Warfare Platforms: Artillery and Air Defence* and as March 2013 this is understood to remain at the prototype stage but has been offered on the export market.

It is understood that firing trials of the 155 mm 2A45M-155 system have been carried out but as of March 2013 no sales had been made.

Russian 155 mm 2A45M-155 artillery system deployed in the firing position with ordnance at maximum elevation, three trails spread out and showing APU on front of carriage (JSC Spetstehnika) 1452646

Today 152 mm and 122 mm are the main calibres of towed and self-propelled artillery system deployed by the Russian Army and the 155 mm 2A45M-155 is aimed at the export market.

Description

The 155 mm 2A45M-155 artillery system is essentially the 2A45M (Sprut-B) towed anti-tank gun with the 122 mm ordnance replaced by a 155 mm/39-calibre ordnance to allow it to fire NATO and Western types of ammunition.

It is also possible that this system uses elements of the Russian 152 mm 2A61 towed howitzer which is covered in a separate entry in *IHS Jane's Land Warfare Platforms: Artillery and Air Defence* which is understood to remain at the prototype stage.

It should be noted that the 155 mm Light Howitzer M-389 is also based on the 2A61 and this is understood to also remain at the prototype stage.

The 155 mm 2A45M-155 artillery system is fitted with an Auxiliary Power Unit (APU) which provides for a maximum speed of up to 14 km/h.

When travelling the three trails are closed together and the weapon is towed by its muzzle with the 6 × 6 cross-country truck also carrying the crew and a quantity of ready use ammunition.

The 155 mm/39-calibre barrel ordnance is fitted with a large double battle muzzle brake and when firing a NATO 155 mm L15A1 high-explosive projectile has a maximum range of 24,000 m which can be increased to 30,000 m using a high-explosive base bleed projectile.

The weapon can be traversed through a full 360° and elevated from –6° to +70° and in addition to the indirect fire role also has a direct fire capability.

Specifications

Not available.

Status

Prototype complete. Not yet in production or service.

Contractor

Joint Stock Company Spetstehnika.

152 mm howitzer 2A65 (M1987) (MSTA-B)

Development

As well as the 152 mm gun 2A36 (original NATO designation M1976 as this was the first year that it was identified by western sources), which is covered in detail in a separate entry in *IHS Jane's Land Warfare Platforms: Artillery and Air Defence* Russia also deployed a new 152 mm towed howitzer, designated the 2A65, which was allocated the NATO designation of the M1987.

This was the year that the weapon was first identified by Western intelligence. According to United Nations sources there were no exports of this weapon between 1992 and 2010.

The 152 mm 2A65 is also referred to as the MSTA-B and the weapon forms the main armament of the 152 mm self-propelled artillery system, the 2S19, which is also referred to as the MSTA-S. Full details of the latter are given in a separate entry in *IHS Jane's Land Warfare Platforms: Artillery and Air Defence*. This has been in service with the Russian Army since 1989/1990 and it is estimated that about 600 have been built.

In the designation MSTA-B the latter stands for Buksiruemyi, or towed.

Description

The 152 mm howitzer 2A65 (M1987) is mounted on a conventional split trail carriage and, when deployed in the firing position, rests on three points, the hydraulic circular firing jack under the forward part of the carriage and the two spades at the rear.

Each of the box section trails has a castor wheel to assist the gun crew in bringing the weapon into action. When deployed in the firing position, these swing upwards through 180° and rest on top of each trail.

152 mm towed howitzer 2A65 (M1987) deployed in the firing position 0500366

The 152 mm ordnance is fitted with a muzzle brake and a semi-automatic breech mechanism, spring-operated ramming system, hydraulic counter-recoil device and a liquid-cooled recoil brake.

Elevation and traverse is manual, two-speed, with the direct and indirect sighting devices being located on the left side of the weapon. Pneumatic brakes are fitted as standard.

The 152 mm 2A65 fires the same 152 mm ammunition types as the 152 mm 2S19 self-propelled artillery system and, more recently, a new family of separate loading (for example projectile and charge) 152 mm has been introduced.

The OF45 high-explosive projectile weighs 43.56 kg, has a maximum muzzle velocity of 810 m/s and a maximum range of 24.7 km. The charges include OF72 (long range), OF58 (full charge) and OF73 (reduced charge).

The OF45 projectile can be fitted with different rear ends, for example various types of screw-on boat tails, or the OF61 base bleed projectile which weighs 42.86 kg, has a maximum muzzle velocity of 828 m/s and a maximum range of 29 km.

The OF23 cargo projectile weighs 42.8 kg, has a maximum range of 26,000 m and contains 42 High-Explosive Anti-Tank (HEAT) bomblets, each of which can penetrate a maximum of 100 mm of conventional steel armour.

Other types of projectile include the HS30 jamming round and the Russian 152 mm Krasnopol laser-guided projectile. Details of this are provided in the entry for the 152 mm 2S19 MSTA-S self-propelled artillery system in *IHS Jane's Land Warfare Platforms: Artillery and Air Defence.*

The Russian KBP Instrument Design Bureau has also developed a 155 mm version of this laser guided projectile and China North Industries Corporation is also marketing a similar round which is based on Russian technology.

In addition to these new 152 mm projectiles the 2A65 can fire all standard types of 152 mm ammunition fired by the older Russian D-20 towed gun-howitzer and the 2S3 self-propelled gun-howitzer.

A unit of fire is being quoted as 80 rounds of ammunition (projectiles, charges and fuzes).

In Russian Army service the 152 mm howitzer 2A65 (M1987) (MSTA-B) is normally towed by a KrAZ-260 (6 × 6) or Ural 4320 (6 × 6) cross-country truck that in addition to its crew also carries a quantity of ready use ammunition (projectiles, charges and fuzes).

Maximum towing speed on roads is 80 km/h and maximum towing speed cross-country is 30 km/h, but this does depend on terrain factors.

Variants

It is understood that a longer barrel version of the 2A65 is currently under development. It is known that there is a 155 mm version of the 2S19 self-propelled artillery system. Russia has also developed to the prototype stage a 152 mm self-propelled artillery system based on a 8 × 8 forward control truck chassis, this is called the MSTA-K.

As of March 2013 this is understood to remain at the prototype stage.

Kazakhstan 152 mm self-propelled howitzer

Late in 2010 it was revealed that Samsung Thales of South Korea had signed a Memorandum of Understanding (MoU) with Kazakhstan for the joint development of a 152 mm self-propelled howitzer.

It is understood that this will be based on an existing 152 mm towed artillery system currently in service with Kazakhstan.

This could be the 152 mm 2A65 or the 152 mm gun 2A36, both of which are deployed by Kazakhstan in significant numbers.

It is estimated that total value of this contract will be about USD200 million and this will be carried out in conjunction with Kazakhstan Engineering.

This self-propelled artillery system will probably be based on a cross-country truck chassis, either a 6 × 6 or an 8 × 8.

Specifications

	152 mm Howitzer 2A65 (M1987) (MSTA-B)
Dimensions and weights	
Crew:	8
Weight	
combat:	7,000 kg
Mobility	
Carriage:	split trail
Speed	
towed:	80 km/h
Firepower	
Armament:	1 × carriage mounted 152.4 mm howitzer
Main weapon	
muzzle brake:	multibaffle
maximum range:	24,700 m (OF45 HE projectile)
maximum range:	29,000 m (OF45 HE projectile with boat tail)
Rate of fire	
rapid fire:	7 rds/min
Main armament traverse	
angle:	54° (27° left/27° right)
Main armament elevation/depression	
armament front:	+70°/-3.5°
Survivability	
Armour	
shield:	yes

Status

Production as required. Known to be in service with the following countries:

Country	Quantity
Belarus	132
Georgia	10
Kazakhstan	90
Russia	720 (600 in reserve)
Ukraine	185

Contractor

Motovilikha Plants Corporation.

152 mm gun 2A36 (M1976)

Development

The towed 152 mm gun 2A36 was developed from December 1968 at the Perm Machine Works following a requirement issued by the Russian Army for a new 152 mm weapon to replace the old towed 130 mm M-46 field gun. The main role of the 152 mm 2A36 system is counter-battery fire.

Two different versions were developed, the towed 2A36 called the Giatsint-B (or Hyacinth) and the self-propelled Giatsint-S which has a 152 mm ordnance designated the 2A37.

The latter was eventually type classified as the 2S5, which is covered in a separate entry in *IHS Jane's Land Warfare Platforms: Artillery and Air Defence.* Production of this is complete but it remains in service with the Russian Army as well as a number of other countries.

Both towed systems shared identical ballistics but fired different types of ammunition.

Full development of the towed 152 mm 2A36 commenced in 1970 with the first two prototypes being completed in early 1971.

Full scale production commenced in 1976 and it was first observed by Western intelligence the same year and subsequently designated the 152 mm gun M1976 until its correct Russian designation became known.

The 152 mm 2A36 was first seen in public during a parade held in Moscow in May 1985.

The 152 mm 2A36 is normally deployed in batteries of six or eight guns, with three batteries making up a battalion.

Although the 2A36 was first observed being towed by the KrAZ-260 (6 × 6) truck, it can also be towed by the KrAZ-255B (6 × 6) or lighter Ural 4320 (6 × 6) truck.

According to United Nations sources, there were no exports of this weapon outside of the Russian Federation between 1992 and 2010.

Production of this system is understood to have ceased in the 1980s. Total production is understood to have amounted to at least 1,500 systems. It has been supplemented by the more recent Russian 152 mm 2A65 (M1987) (MSTA-B) weapon.

Description

The 152 mm 49-calibre barrel of the 2A36 is fitted with a multislotted muzzle brake weighing 141 kg. The recoil system consists of a buffer and a recuperator which are mounted above the ordnance towards the rear.

The horizontal sliding breech mechanism opens to the right automatically.

The ordnance pivots on the cradle, which is of cast and welded steel construction. Elevation and traverse is manual with the former being from –2° 30' to +57° and traverse 25° left and right.

The direct and indirect sights are mounted on the left side of the carriage as are the elevation and traverse mechanisms.

Some of the eight-person crew are provided with protection from shell splinters by the armoured shield that slopes to the rear and extends over the wheels.

Mounted below and to the rear of the breech is the load assist system, which is referred to as a quick-firing loading system by Russia. This includes a hydraulic rammer operated from the hydropneumatic accumulator with controls provided for ramming and return.

A fuzed 155 mm projectile is placed on the loading tray that is mounted to the right of the breech. When the breech is opened, this slides to the left until it lines up with the breech. The projectile is then rammed into the ordnance and the loading tray slides back to the right. The cartridge-cased charge is then located in a similar manner and, when the breech is closed, the rammer automatically returns to the side and the weapon is ready to fire. If the hydraulic rammer fails, loading takes place manually, although rate of fire is reduced.

152 mm gun 2A36 (M1976) from the rear in travelling configuration 0500400

Ukraine has developed and fielded the Kvitnyk 152 mm SAL guided projectile (top) shown with is prior to launch configuration with standard 152 mm high explosive projectile below (Ukraine MoD) 1482301

Russia quotes a rate of fire of 6 rds/min for the 2A36 and that a battery can put almost one tonne of ammunition onto a target in one minute.

Types of separate-loading ammunition fired include High-Explosive Fragmentation (HE-FRAG) and Armour-Piercing Tracer (AP-T). The former weighs 46 kg and has a maximum muzzle velocity, using the top charge, of approximately 800 m/s. Before loading the projectile the fuzing mode is selected either for fragmentation or high-explosive action.

The AP-T round is used in the direct fire mode against tanks and other armoured fighting vehicles. The propelling charge, which weighs a maximum of 11 kg, is placed in a conventional cartridge case.

The standard HE-FRAG projectile has a maximum range of 27,000 m although a rocket-assisted HE-FRAG projectile, called the active-reactive projectile by Russia, can also be fired. This has a maximum range of 40,000 m.

Other types of projectile that can probably be fired by the 152 mm 2A36 include smoke, concrete-piercing and incendiary.

The ammunition system used by the 152 mm 2A36 is of a new design and is not interoperable with that of earlier artillery systems, such as the 152 mm 2S3 and may well be of a new streamlined design.

The walking beam suspension gives the 152 mm gun 2A36 improved cross-country mobility and has a total of four rubber-tyred braked roadwheels for high-speed towing and improved cross-country mobility. The tyres have a normal pressure of 4.8 kg/cm^2. Total wheel travel is 150 mm. Each of the front roadwheels has a hydraulic shock-absorber and a manual handbrake.

When in the firing position, the 152 mm 2A36 is supported on a circular jack located under the forward part of the carriage and on each of the box-type welded trails which is provided with a spade.

The winter spade is standard, but when firing in soft terrain much larger summer spades are used.

When coming out of action, the jack is raised clear of the ground and secured under the front, the spades removed and stowed on the trails, the trails brought together and clamped with a tie mechanism. The hinged towbar is fitted to the right trail and the weapon is then hitched up to the prime mover.

A unit of fire is being quoted as 60 rounds of ammunition (projectiles, charges and fuzes).

In Russian Army service the 152 mm 2A36 (M1976) weapon is normally towed by a KrAZ-260 (6 × 6) or Ural 4320 (6 × 6) cross-country truck that in addition to its crew also carries a quantity of ready use ammunition (projectiles, charges and fuzes).

Maximum towing speed on roads is 80 km/h and maximum towing speed cross-country is 30 km/h, but this does depend on terrain factors.

Variants

Improved Giatsint-B

Mention has also been made of an improved version of the 2A36 with a 53.8-calibre 152 mm ordnance compared to the 49-calibre of the standard production weapon. The longer 152 mm barrel would give a slight increase in range.

152 mm gun 2A36 (M1976) (PB) 0085645

Giatsint-BK
This did not pass the prototype stage.

Kazakhstan 152 mm self-propelled howitzer
Late in 2010 it was revealed that Samsung Thales of South Korea had signed a Memorandum of Understanding (MoU) with Kazakhstan for the joint development of a 152 mm self-propelled howitzer.

It is understood that this will be based on an existing 152 mm towed artillery system currently in service with Kazakhstan.

This could be the 152 mm 2A65 or the 152 mm 2A36, both of which are deployed by Kazakhstan in significant numbers.

It is estimated that total value of this contract will be about USD200 million and this will be carried out in conjunction with Kazakhstan Engineering.

This 152 mm self-propelled artillery system will probably be based on an 6 × 6 or 8 × 8 cross-country truck chassis.

New Kvitnyk 152 mm guided artillery projectile
Late in 2012 the Ukraine stated that it had taken into service the 152 mm Kvitnyk fitted with a semi-active laser (SAL) guidance system that operates along similar lines to the Russian Krasnopol SAL guided projectile.

This has been manufactured by the Central Design and Project Technological Office Tocknost and is claimed to have a maximum range of 20 kg and is fitted with a high explosive warhead.

This 152 mm SAL guided projectile can be fired by the 152 mm 2A36 (M1976) weapon as well as other Russian weapons with this calibre.

Specifications

	152 mm Gun 2A36 (M1976)
Dimensions and weights	
Crew:	8
Length	
overall:	12.3 m (firing)
transport configuration:	12.92 m
Width	
transport configuration:	2.788 m
Height	
overall:	2.76 m
Ground clearance	
overall:	0.475 m
Track	
vehicle:	2.34 m
Weight	
standard:	9,800 kg
combat:	9,760 kg
Mobility	
Carriage:	split trail
Speed	
towed:	80 km/h
Firepower	
Armament:	1 × carriage mounted 152.4 mm howitzer
Main weapon	
length of barrel:	8.197 m
breech mechanism:	horizontal sliding block
muzzle brake:	multibaffle
maximum range:	27,000 m (standard HE-FRAG projectile)
maximum range:	40,000 m (rocket assisted HE-FRAG projectile)
Rate of fire	
rapid fire:	6 rds/min
Main armament traverse	
angle:	50° (25° left/25° right)
Main armament elevation/depression	
armament front:	+57°/-2.5°
Survivability	
Armour	
shield:	yes

Status
Production complete. In service with the following countries:

Country	Quantity	Comment
Armenia	26	
Azerbaijan	18	
Belarus	48	
Finland	24	local designation 152 K 79
Georgia	3	
Kazakhstan	180	
Moldova	21	
Russia	1,100	Reserve
Uzbekistan	140	
Ukraine	287	

Contractor
Perm Machine Works.

152 mm gun-howitzer D-20

Development
The towed 152 mm gun-howitzer D-20 was developed shortly after the end of the Second World War by the F F Petrov Artillery Design Bureau at Artillery Plant No 9 located in the town of Yekaterinburg.

The 152 mm gun-howitzer D-20 was first seen in public during the May Day parade in Moscow in 1955. It is understood that this weapon is no longer in front-line service with the Russian Army.

Significant numbers of the 152 mm gun-howitzer D-20 are held in reserve and for potential sale overseas.

According to United Nations sources, Russia did not expect any surplus 152 mm gun-howitzer D-20 weapons between 1992 and 2010.

Bulgaria did however export four D-20s to Angola in 1999, while China exported 18 of its equivalent China North Industries Corporation (NORINCO) Type 66 to Bolivia in 1992.

Description
The D-20 uses the same carriage and recoil system as the 122 mm field gun D-74 which was also shown for the first time in 1955. The D-20 can be distinguished from the D-74 by a much shorter and fatter stepped barrel and a larger double-baffle muzzle brake.

The shield has an irregular top with a sliding centre section. The top of the shield can be folded down to reduce the overall height of the weapon.

A circular firing pedestal is fitted as standard on all D-20s and is inverted and secured just forward of the shield when travelling. In the firing position the firing pedestal is lowered to the ground under the carriage, allowing the weapon to be traversed quickly through a full 360°. On each of the split box section trails is a castor wheel (on top) and a spade (underneath).

Targets are engaged by the gunner seated on the left side of the weapon. The gunner's sights are the indirect sight PG1M and the telescope OP4M for direct fire.

The ordnance of the 152 mm self-propelled gun-howitzer 2S3 (covered in a separate entry) is a development of that used on the towed D-20. Production of the 152 mm 2S3 was completed some time ago and it is no longer marketed. Surplus quantities of 152 mm 2S3 weapons are available from a number of sources.

The 152 mm gun-howitzer D-20 fires the case-type, variable-charge, separate-loading ammunition listed in the table on the next page.

Other types of ammunition include HE/RAP (range of 24,000 m), HEAT, illuminating (S-540), smoke (D-540) HEAT-SS (Spin-Stabilised), flechette, scatterable mines (anti-tank and anti-personnel) and semi-active laser guided. The last of these is called the Krasnopol by the RFAS and details of this are given in the entry for the 2S19 152 mm self-propelled gun in *IHS Jane's Land Warfare Platforms: Artillery and Air Defence*. It is estimated that about 600 2S19 systems have been built.

152 mm gun-howitzer D-20 in travelling configuration (Christopher F Foss)
0500774

152 mm gun-howitzer D-20 of the Sandinista Army (Julio A Montes) 0021013

Ammunition type	FRAG-HE	CP	AP-T
Projectile designation:	OF-540	G-545	BR-540
Fuze model:	RGM-2	KTD	MD-7
Weight of projectile:	43.51 kg	56 kg	48.78 kg
Weight of bursting charge:	6.25 kg	4.22 kg	1.2 kg
Type of bursting charge:	TNT	TNT	n/a
Muzzle velocity:	655 m/s	670 m/s	600 m/s
Armour penetration at 0°:	n/a	n/a	124 mm/1,000 m
Notes: (1) Other FRAG-HE projectiles also available (2) Other CP projectiles available			

A unit of fire is being quoted as 65 rounds of 152 mm ammunition (projectiles, charges and fuzes).

In Russian Army service the 152 mm howitzer D-20 gun howitzer was normally towed by a Ural 375 (6 × 6) cross-country truck that in addition to its crew also carries a quantity of ready use ammunition (projectiles, charges and fuzes). Other countries use different 4 × 4 or 6 × 6 cross-country chassis.

Maximum towing speed on roads is 60 km/h and maximum towing speed cross-country is 15 km/h, but this does depend on terrain factors.

Yugoslavia (today Serbia) produced a long-barrelled version of the D-20 known as the 152 mm howitzer M84. As far as is known there has been no production of this weapon in recent years. Details are provided in a separate entry in *IHS Jane's Land Warfare Platforms: Artillery and Air Defence*.

Variants

Other Russian 152 mm and 122 mm towed artillery systems

A number of older Russian 152 mm and 122 mm towed artillery systems are still in service outside of Russia with some of these dating back to before the Second World War.

152 mm Howitzer M1943 (D-1)

This was developed to replace the earlier 152 mm howitzer M1938 (M-10) and is based on the carriage of the 122 mm M1938 (M-30).

It has a split trail carriage and a maximum range of 12,400 m firing a HE-FRAG projectile.

This remains in service with the following countries:

Country	Quantity
Armenia	2
Belarus	6
Costa Rica	n/avail
Korea, North	n/avail
Kyrgystan	16
Mozambique	12
Russia	700 (reserve)
Rwanda	29 (Chinese model)
Turkmenistan	17

152 mm Gun-Howitzer M1937 (ML-20)

This was developed shortly before the Second World War and has a split trail carriage to which a two wheeled dolly is attached prior to being connected the prime mover. It has a maximum range of 17,265 m firing a HE-FRAG projectile.

This remains in service with the following countries:

Country	Quantity
Algeria	20
Korea, North	n/avail
Libya	25 (status uncertain due to 2011 conflict)
Mongolia	n/avail
Russia	100 (reserve)

These weapons must now be considered obsolete and obtaining spare parts is now a major problem.

122 mm Corps Gun M1931/37 (A-19)

This was developed shortly before the Second World War and has a split trail carriage to which a two wheeled dolly is attached prior to being connected the prime mover.

It has a maximum range of 20,800 m firing a HE-FRAG projectile and is still in service with the following countries:

Country	Quantity
Algeria	100
Egypt	36
Guinea	12
Korea, North	n/avail
Yemen	30

These weapons must now be considered obsolete and obtaining spare parts is now a major problem.

Specifications

	152 mm Gun-Howitzer D-20
Dimensions and weights	
Crew:	10
Length	
transport configuration:	8.69 m
Width	
transport configuration:	2.40 m
Height	
transport configuration:	1.925 m
axis of fire:	1.22 m
Ground clearance	
overall:	0.38 m
Track	
vehicle:	2.03 m
Weight	
standard:	5,700 kg
combat:	5,650 kg
Mobility	
Carriage:	split trail
Speed	
towed:	60 km/h
Tyres:	12.00 × 20
Firepower	
Armament:	1 × carriage mounted 152.4 mm howitzer
Main weapon	
length of barrel:	5.195 m
breech mechanism:	vertical sliding block
muzzle brake:	double baffle
maximum range:	17,410 m (standard HE-FRAG projectile)
maximum range:	24,000 m (rocket assisted HE-FRAG projectile)
Rate of fire	
rapid fire:	6 rds/min
Main armament traverse	
angle:	58°
Main armament elevation/ depression	
armament front:	+63°/-5°
Survivability	
Armour	
shield:	yes

Status

Production complete. In service with the countries listed in the table.

Country	Quantity	Comment
Albania	18	Chinese Type 66
Angola	4	from Bulgaria in 1999
Armenia	34	
Azerbaijan	24	
Bosnia	13	
Bulgaria	132	
China	1,400	Chinese Type 66
Congo	5	
Croatia	20	
Cuba	n/k	
Hungary	18	
Iran	30	
Korea, North	n/k	
Moldova	31	
Mozambique	12	D-1
Nicaragua	30	reserve
Russian Federation	1,075	estimate (held in reserve) + naval infantry
Serbia/Montenegro	18	plus 52 of local M84 variant
Sri Lanka	46	Chinese Type 66
Syria	20	

Country	Quantity	Comment
Turkmenistan	72	+ 17 D-1
Ukraine	215	
Vietnam	n/k	
Yemen	10	

Contractor
Joint Stock Company Spetstehnika.
China North Industries Corporation (NORINCO).

152 mm 2A61 howitzer

Development
Early in 1993, the RFAS released outline information of a new lightweight 152 mm towed howitzer designated the 2A61.

This was developed by the famous F F Petrov artillery design bureau at Artillery Plant No 9, Yekaterinburg, which has designed many previous artillery weapons, including the Russian 122 mm D-30 towed howitzer.

It has been confirmed that this passed all of its official tests in 1991 but was never placed in production for the Russian Army. It is still being marketed but as of March 2013 no production orders are known to have been placed.

Description
The 152 mm 2A61 light towed howitzer is based on the modified carriage of the late production 122 mm D-30A (2A18M) howitzer, which has been modified to accept a larger calibre 152 mm ordnance.

The latter is fitted with a large distinctive double 'T' muzzle brake with the towing eye located below.

A semi-automatic projectile rammer is provided with the manual charge loading. A rate of fire of six to eight rounds a minute is claimed with the D-30 carriage allowing traverse through a full 360°.

When deployed in the firing position, the two pneumatic roadwheels are raised clear of the ground and the three trails are staked to the ground to provide a more stable firing platform. The weapon can be quickly traversed through a full 360° to be laid onto a new target.

When in the travelling position, the ordnance is over the closed trails. A small shield is fitted to provide some protection to the gun crew.

Recoil length of the 152 mm ordnance of the 2A61 howitzer depends on elevation.

With an elevation of up to +20° it is 0.4 m and at an elevation of over +20° it is 0.98 m.

Typical towing vehicles for the 122 mm 2A61 howitzer are the MT-LB multipurpose full tracked armoured vehicle or the ZIL-131 (6 × 6) cross-country truck, or more recent 6 × 6 cross-country vehicles.

It is possible that this weapon was developed as a replacement for the 122 mm D-30 used by the RFAS air assault divisions. It is expected that if and when placed in production, the 152 mm 2A61 will utilise existing D-30 carriages rather than brand new carriages.

155 mm/30-calibre Light Howitzer M-389 deployed in firing position with ordnance at maximum elevation (Artillery Plant No 9) 1184811

152 mm 2A61 towed howitzer with ordnance horizontal 0501029

Variants
155 mm Light Howitzer M-389

This is a further development of the 152 mm 2A61 howitzer but fitted with a 155 mm/30-calibre barrel that fires standard NATO ammunition. In addition the recoil system has been improved and existing carriage trunnions have been strengthened.

The system weighs 4,350 kg with elevation from –5° to + 70° and traverse through a full 360°. The breech mechanism is of the sliding wedge type with the projectile being loaded using a semi-automatic rammer and the charge being loaded manually.

Maximum range using a NATO standard 155 mm M107 high-explosive projectile is being quoted as 15,000 m with a maximum rate of fire of between six and eight rounds per minute.

This is understood to have been developed to meet potential export requirements rather than the requirements of the Russian Army.

Specifications

	152 mm 2A61 Howitzer
Dimensions and weights	
Crew:	7
Length	
transport configuration:	6.36 m
Width	
transport configuration:	2.2 m
Height	
transport configuration:	1.97 m
axis of fire:	0.93 m
Ground clearance	
overall:	0.35 m
Track	
vehicle:	1.84 m
Weight	
standard:	4,350 kg
Mobility	
Speed	
towed:	80 km/h
Firepower	
Armament:	1 × carriage mounted 152.4 mm howitzer
Main weapon	
muzzle brake:	multibaffle
maximum range:	15,000 m
Rate of fire	
rapid fire:	8 rds/min
Main armament traverse	
angle:	360°
Main armament elevation/depression	
armament front:	+70°/-5°
Survivability	
Armour	
shield:	yes

Status
Development complete. Ready for production.

Contractor
Former state factories.

130 mm field gun M-46

Development
In April 1946, the Soviet Artillery Committee issued the technical requirements for towed 130 and 152 mm artillery systems, which were to use the same split trail carriage.

130 mm M-46 field gun complete with two wheeled dolly and clearly showing large pepper pot muzzle brake (Israel Defence Force) 1452922

130 mm M-46 deployed in firing position 0500401

Development was carried out by Factory No 172 and the Scientific Research Institute for Artillery Armaments with the 130 mm weapon being designated the M-46 and the 152 mm weapon the M-47.

The first prototypes of the 130 mm M-46 and 152 mm M-47 were completed in mid-1948. After trials, further work was carried out which resulted in the building of four additional 130 mm M-46 and 152 mm M-47 guns for troop trials that took place late in November 1950 after which they were accepted for service.

During 1951, production of both weapons commenced at Factory No 172 with 20 M-46 and four M-47 being built in that year. In the 1950s the following quantities of weapons were built:

Year	130 mm M-46	152 mm M-47
1951	20	4
1952	26	nil
1953	nil	nil
1954	16	16
1955	50	20
1956	62	32
1957	79	50

Production of both weapons continued after this date although the 152 mm M-47 was soon phased out of production. As far as it is known the 152 mm M-47 weapon is no longer deployed by the Russian Army.

The 130 mm M-46 was first seen in public during a parade held in Moscow in 1954 and was the replacement for the 122 mm M1931/37 (A-19) field gun.

The 130 mm M-46 has seen combat in Asia and the Middle East and has also been built in China as the China North Industries Corporation (NORINCO) Type 59-1 and in Egypt. The weapon has now been phased out of front line service with the Russian Army although it is still used in many parts of the world.

Although production of the 130 mm field gun M-46 was completed many years ago, significant quantities of this weapon have been sold on the export market by various other countries in recent years.

According to the United Nations, between 1992 and 2010 the following quantities of weapons were exported (see table at the top of the following page):

Description

When travelling, the 130 mm barrel of the M-46 field gun is withdrawn to the rear by a mechanism on the right trail from battery to the rear to reduce the overall length of the weapon.

The carriage is of the split trail type and is provided with a two-wheeled limber. When travelling, the spades are removed and carried on top of the two trails. The recoil system is mounted under the barrel and in front of the shield, which has been removed on some models, is an inverted U-shaped collar.

The recoil system consists of a hydraulic buffer and hydropneumatic recuperator and when the weapon has an elevation between −2° 30 minutes and +20°, this is between 1,150 mm and 1,320 mm.

130 mm field gun M-46 deployed in the firing position from the rear with ordnance elevated 0130002

130 mm M-46 field gun complete with two wheeled dolly in travelling configuration (Christopher F Foss) 1452923

With elevation from +34° to +45° (which is the maximum elevation) recoil is between 734 mm to 815 mm.

While the carriage has a ground clearance of 400 mm the ground clearance under the associated limber is 350 mm.

The 130 mm M-46 field gun has the OP4M-35 direct fire sight with a field of view of 11° and a magnification of ×5.5, and an APN-3 active/passive night sight.

The 130 mm M-46 fires case-type, variable-charge, separate loading (projectile and charge) ammunition as shown in the table:

Ammunition type	RAG-HE	APC-T
Projectile designation:	OF-482M/OF-33	BR-482B
Fuze model:	RGM-2	DBR
Weight:	33.4 kg	33.6 kg
(of projectile)		
(of bursting charge)	4.63 kg	0.127 kg
Type of bursting charge:	TNT	RDX/alum
Muzzle velocity:	930 m/s	930 m/s
Armour penetration at 0°:	n/a	230 mm/1,000 m

Other types of 130 mm projectile include:

- PB-42, SAP-HE, weight 33.4 kg
- SP-46, illuminating, weight 25.80 kg, muzzle velocity 687 m/s
- OF-3S-42, semi-armour-piercing, weight 42 kg
- DTs-1, target marker smoke, weight 32.80 kg, muzzle velocity 930 m/s.

Israel Military Industries has marketed a 130 mm HE projectile identical to the original but fitted with a PD M739 fuze. This projectile is used by the Israel Defense Force.

China North Industries Corporation (NORINCO) of China has developed a new generation of 10 130 mm projectiles and brief details of these are given in the entry for the NORINCO 130 mm field gun Type 59-1.

Details of this are provided in a separate entry in *IHS Jane's Land Warfare Platforms: Artillery and Air Defence*. Production of the Type 59-1 has been completed and it is no longer being marketed by NORINCO as a brand new system although surplus weapons are available from China.

From	To	Quantity	Date
Bulgaria	Eritrea	30	1999
China	Iran	106	1992
China	Pakistan	87	27 in 1998 and 60 in 2000
China	Gabon	10	2004
Germany	Finland	166	1994
Kazakhstan	Ethiopia	6	2000
Kazakhstan	Russia	24	1999
Romania	Bosnia-Herzegovina	8	1998
Romania	Cameroon	12	1992
Romania	Guinea	12	2000
Romania	Nigeria	13	4 in 1992 and 7 in 1993
Russia	India	120	1994

The now Rheinmetall Denel Munition of South Africa also developed new generation types of 130 mm ammunition for this weapon. As far as it is known, South Africa has not placed any of these 130 mm rounds in quantity production.

When in service with the Russian Army in large numbers the 130 mm field gun M-46 was typically towed by a 6 × 6 cross-country truck or a full tracked tractor.

Typical examples of the latter include the Russian AT-S, ATS-59 and M1972, but these are no longer in service with the Russian Army.

In addition to the crew these prime movers also carried a quantity of ready use ammunition (projectiles, changes and fuzes).

Iranian 130 mm BB ammunition

The Iranian Army currently employs an estimated 1,000 130 mm M-46 field guns and its Chinese equivalent the China North Industries Corporation (NORINCO) Type 59-1.

The Ammunition Group of the Iranian Defence Industries Organization (DIO) has revealed that they have developed and placed in production a new 130 mm HE fragmentation projectile with a composite Base Bleed (BB) unit.

This new Iranian 130 mm HE BB projectile weighs 35.5 kg and contains 3.7 kg of TNT, this TNT being smaller than the standard OF-482M due to the added weight of the base bleed unit.

The new round is fired like a conventional 130 mm HE projectile with a maximum muzzle velocity of 930 m/s. When in flight the base bleed unit is activated and enables the projectile to achieve a maximum range of 37,000 m, a significant increase over the 27,150 m of the standard OF-482M projectile.

This allows the 130 mm M-46 field gun to engage targets at a much longer range and also reduces the number of enemy weapons that can engage the system with counter-battery fire.

Variants

Modified 130 mm M-46s with a longer barrel, recuperator and cradle appeared in Russian Army service in the mid-1970s. India has deployed a self-propelled model based on a lengthened Vijayanta MBT chassis. This is called the Catapult by the Indian Army. Brief details of this are given in a separate entry in *IHS Jane's Land Warfare Platforms: Artillery and Air Defence*. This is only deployed by the Indian Army and is currently the only self-propelled artillery system deployed.

155 mm conversions

The Chinese NORINCO concern has marketed a 155 mm conversion known as the GM-45 and Serbia has offered conversion. Soltam Systems of Israel is also offering a conversion pack for the 130 mm M-46 to 155 mm/45-calibre standard. This was ordered by India early in 2000.

In 2002 the Indian Ordnance Factories completed the first prototype of a 130 mm M-46 fitted with a 155 mm/45-calibre ordnance. Details of the Soltam Systems and Indian Ordnance Factory 155 mm conversion kits for the 130 mm field gun M-46 are provided in separate entries in *IHS Jane's Land Warfare Platforms: Artillery and Air Defence*.

Serbia upgraded Russian 130 mm

Yugoimport of Serbia is marketing its type classified upgrade package for the Russian 130 mm M-46 field gun which, firing standard Russian 130 mm ammunition has a maximum range of 27.15 km.

The Yugoimport upgrade includes replacing the existing barrel with a 155 mm/52 calibre barrel with the complete upgraded system being designated the M46/10 cal 155 mm.

The upgrade includes a new barrel, breechlock and muzzle brake plus modifications to the original hydraulic brake and recuperator.

According to Yugoimport it can fire all NATO 155 mm projectiles and charges and firing a 155 mm Extended Range Full Bore - Base Bleed projectile a maximum range of 42 km is claimed.

As an option the upgraded system can be fitted with a flick rammer which reduces crew fatigue and increases rate of fire.

It would typically be integrated with a computerised fire control system with a gun display unit being installed on the right side of the carriage near the shield.

With the significant number of Russian 130 mm M-46 field guns still deployed around the world, as well as its Chinese equivalent, the 130 mm Type 59-1, this upgrade is regarded as a cost effective alternative to the procurement of brand new 155 mm artillery systems.

Specifications

	130 mm Field Gun M-46
Dimensions and weights	
Crew:	8
Length	
transport configuration:	11.73 m
Width	
transport configuration:	2.45 m
Height	
transport configuration:	2.55 m
axis of fire:	1.38 m
Ground clearance	
overall:	0.375 m
Track	
vehicle:	2.06 m
Weight	
standard:	8,450 kg
combat:	7,700 kg
Mobility	
Carriage:	split trail with limber
Speed	
towed:	50 km/h
Tyres:	13.50 × 38
Firepower	
Armament:	1 × carriage mounted 130 mm howitzer
Main weapon	
length of barrel:	7.6 m
breech mechanism:	horizontal sliding block
muzzle brake:	pepperpot
maximum range:	27,150 m
Rate of fire	
rapid fire:	6 rds/min
Main armament traverse	
angle:	50° (25° left/25° right)
Main armament elevation/depression	
armament front:	+45°/-2.5°
Survivability	
Armour	
shield:	optional

Status

Production complete. In service with the countries listed in the table:

Country	Quantity	Comment
Afghanistan	n/avail	
Algeria	10	
Angola	48	
Azerbaijan	36	
Bangladesh	62	Chinese Type 59-1
Bosnia-Herzegovina	61	including 12 from Egypt and 8 from Romania
Cambodia	n/avail	Chinese Type 59-1
Cameroon	24	12 from China (Type 59-1) and 12 from Romania
China	250	Chinese Type 59-1
Congo	5	
Croatia	79	
Cuba	n/avail	
Democratic Republic of Congo	42	Chinese Type 59-1
Egypt	420	local production
Eritrea	30	from Bulgaria in 1999
Ethiopia	6	
Finland	36	called K 54
Gabon	10	from China in 2004 (Type 59-1)
Guinea	12	from Romania in 2000

Country	Quantity	Comment
Guyana	6	status uncertain
India	1,200	also some self-propelled. Including 120 from Russia in 1994. Some being upgraded to 155 mm
Iran	990	Russian M-46 and Chinese Type 59-1
Israel	20	reserve
Korea, North	n/avail	also some self-propelled versions
Lebanon	16	
Libya	330	status uncertain owing to conflict in 2011
Laos	10	
Mali	6	reserve
Mongolia	64 (estimate)	
Morocco	18	
Mozambique	6	
Myanmar	16	from North Korea
Nigeria	7	
Oman	24	12 Russian M-46 and 12 Chinese Type 59-1
Pakistan	410	Chinese Type 59-1 (all of these are held in reserve)
Peru	36	
Russia	650	reserve
Serbia/Montenegro	18	
Somalia	n/avail	status uncertain
Sri Lanka	40	Chinese Type 59-1
Sudan	40	M-46
	35	Type 59-1
Syria	700/800	
Tanzania	30	Chinese Type 59-1
Thailand	15	Chinese Type 59-1
UAE	54	Chinese Type 59-1 (reserve)
Uganda	12	
Vietnam	n/avail	
Yemen	60	
Zambia	18	

Contractor's

Motovilikha Plants Corporation.
China North Industries Corporation (NORINCO).
Abu Zaabal Engineering Industries Company (Egypt).

125 mm 2A45M (Sprut-B) towed anti-tank gun

Development

The 125 mm 2A45M (Sprut-B) auxiliary-propelled towed anti-tank gun is also known as the Sprut-B (Russian for Octopus) and was developed as the potential replacement for the 100 mm MT-12/T-12 towed anti-tank gun which entered service in 1955.

The weapon was developed in the late 1980s by the F F Petrov design bureau at the Artillery Plant No 9 at Yekaterinburg, which also designed the famous 122 mm D-30 artillery system and numerous tank guns.

According to Russian sources, the development of the 2A45M (Sprut-B) towed anti-tank gun can be traced back to 1967, when it became apparent that existing in-service smoothbore 100 mm T-12/MT-12 series towed ant-tank gun could not defeat new generation tanks such as the UK's Chieftain and the MBT-70 (which in the end did not enter service).

In January 1968 OKB-9, which subsequently became part of the now Joint Stock Company Spetstehnika, was selected to develop a new 125 mm anti-tank gun. It had the same ballistics as the 125 mm D-81 smoothbore tank which subsequently became the standard Russian tank gun.

The 125 mm smoothbore gun is installed in many MBTs including the Russian T-64, T-72, T-80 and T-90 as well as in the T-84 MBT developed in Ukraine.

Originally two versions of the 125 mm anti-tank gun were developed to the prototype stage, the D-13 towed and the SD-13 self-propelled. Further development of the latter resulted in the 2A45M (Sprut-B).

As of march 2013 this weapon remained at the prototype stage and the older Russian 100 mm T-12 towed anti-tank gun remained in service with the Russian Army in declining numbers.

Description

The 125 mm 2A45M (Sprit-B) system is based on the well-known 122 mm D-30 towed howitzer carriage which has been in service with RFAS and many other countries for well over 40 years.

The 122 mm rifled ordnance of the D-30 has been replaced by a modified version of the 125 mm D-81 smoothbore ordnance used in the T-64/T-72/T-80/T-84/T-90 MBTs and features a thermal sleeve and a large single 'T'-type baffle muzzle brake with a towing eye mounted underneath like the earlier D-30.

125 mm 2A45M (Sprut-B) towed anti-tank gun deployed in the firing position
0501030

Ballistically, the 125 mm 2A45M and the 125 mm tank guns are identical. The cradle is also taken from the 125 mm D-81 tank gun.

The barrel of the 2A45M consists of a tube with a muzzle brake attached by a casing to the chamber part and a breech ring. The breech mechanism is of the semi-automatic mechanical vertical sliding wedge type. This is opened manually before the first round is loaded and after that opens automatically after the weapon has fired.

The recoil system is a counteracting piston-type hydraulic buffer with the counter recoil system being pneumatic, with both mounted above the ordnance.

The 2A45M towed anti-tank gun has a mechanised system to convert from the firing position to the travelling position and this consists of a hydraulic jack and hydraulic cylinders. With the aid of the jack the carriage is raised to the required height for folding or unfolding the trails, then lowered onto the ground. The hydraulic cylinders raise the gun to the maximum clearance and also raise and lower the wheels.

When deployed in the firing position, the weapon rests on three trails that are normally staked to the ground to provide a more stable firing platform.

The system takes between 90 and 120 seconds to come into action and between 120 and 160 seconds to come out.

The system can be brought into action manually or using hydraulic assistance with the Auxiliary Propulsion Unit (APU) running.

The design of the carriage is such that the upper part of the weapon can be quickly traversed through a full 360° to be laid onto new targets.

A well-sloped shield provides the crew with limited protection from small arms fire and shell splinters. The 2A45M is fitted with an auxiliary propulsion unit consisting of an MeMZ-967A four-stroke petrol engine with hydraulic drive which allows it to propel itself around the battlefield at a maximum speed of 10-14 km/h.

The engine is mounted on the right side with power transmitted to the road wheels via cardan shafts. The driver's seat is mounted on the left side and is provided with the necessary controls for controlling the system when the APU is being used.

Maximum range when being used in the APU mode is up to 50 km but this does depend on the terrain conditions. A small wheel attached to one of the three trail legs is then lowered and the 2A45M moves under its own power to the firing position.

The driver is seated in front of the shield on the left side and steers using a conventional steering wheel which folds forward when not required.

The two large rubber-tyred roadwheels of the carriage are powered and claimed to be bulletproof with steering accomplished by altering the power to either road wheel.

The 2A45M weighs 6,575 kg when deployed in the firing system compared to the standard 122 mm D-30 towed howitzer which weighs 3,210 kg.

The sights, elevation and traverse equipment are positioned on the left side. Optical systems fitted include the OP4M-48A direct fire day telescope and the 1PN53-1 night sight. The iron sight 2Ts33 is used together with the panoramic sight PG-1M for indirect firing.

The 125 mm 2A45M fires the same family of separate-loading ammunition (projectile and semi-combustible cartridge case) as the T-64/T-72/T-80/T-84/T-90 MBTs and maximum rate of fire is quoted as 6 to 8 rds/min.

A total of six ready rounds (projectile and charge) can be carried on the system for ready use with the remainder being carried in the towing vehicle.

The APFSDS-T projectile is used to engage MBTs and has an effective range of 2,100 m while the HE-FRAG projectile is used to engage general battlefield targets such as personnel. It contains 3.4 kg of A-IX-2 explosive and is fitted with a V-429E impact fuze.

Maximum range in the indirect fire role is 12,200 m compared to the 15,400 m (without using a rocket-assisted projectile) of the 122 mm D-30, as the latter has a maximum elevation of +70°.

The HEAT projectile has a V15 nose-mounted base activated fuze and contains 1.76 kg of A-IX-1 or OCFOL explosive which will penetrate the conventional passive first-generation MBTs such as the M60, M48 and Leopard 1 unless they are fitted with explosive reactive armour.

All projectiles use the same propelling charge, which is 408 mm long, and all that remains after firing is the stub case. Basic details of the key 125 mm ammunition types are given in the accompanying table:

Designation	Type	Muzzle velocity	Round weight	Projectile weight	Projectile length	Weight of filler
BM-17	APFSDS	1,700 m/s	20.7 kg	7.05 kg	620 mm	nil
BK-10	HEAT	905 m/s	29.58 kg	19.08 kg	675 mm	1.76 kg
OF-36	HE	850 m/s	32.5 kg	23.00 kg	675 mm	3.4 kg

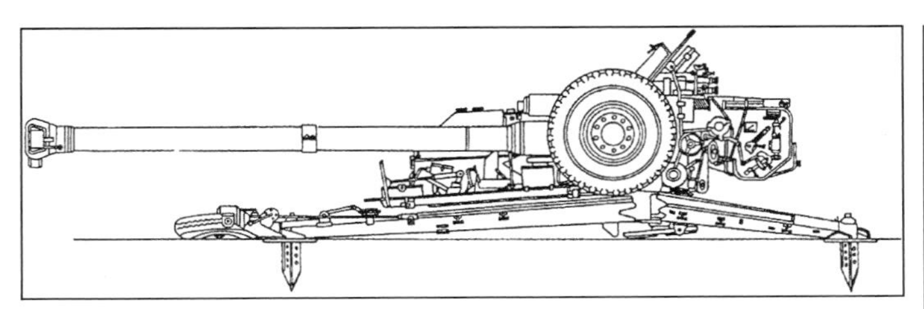

Provisional drawing of the 125 mm 2A45M (Sprut-B) towed anti-tank gun deployed in the firing position 0500776

The 125 mm 2A54M anti-tank gun system can fire the 125 mm laser-guided projectiles fired by the latest RFAS MBTs, the 9K120 Refleks or Sniper. Both these projectiles have a range of 5,000 m and a single HEAT warhead, which will penetrate 700 mm of conventional steel armour. More recent versions have a tandem HEAT warhead to defeat Explosive Reactive Armour (ERA).

In order to fire these 125 mm laser-guided projectiles a 9S53 laser fire-control system has to be used. This would be integrated onto the weapon.

Towing vehicle of the 125 mm 2A45M (Sprut-B) towed anti-tank gun would typically be a Ural-4320 (6 × 6) cross-country truck or the MT-LB armoured multipurpose vehicle.

In addition to the crew these would also carry a quantity of ready use 100 mm ammunition.

155 mm howitzer 2A45M-155

This is based on the carriage of a 125 mm 2A45M (Sprut) towed anti-tank gun and is covered in a separate entry in *IHS Jane's Land Warfare Platforms: Artillery and Air Defence*. As of March 2013 this weapon remained at the prototype stage.

Specifications

	125 mm 2A45M (Sprut-B) Towed Anti-Tank Gun
Dimensions and weights	
Crew:	7
Length	
overall:	6.790 m
transport configuration:	7.120 m
Width	
transport configuration:	2.66 m
Height	
transport configuration:	2.09 m
axis of fire:	0.925 m
Ground clearance	
overall:	0.36 m
Track	
vehicle:	2.20 m
Weight	
standard:	6,800 kg
combat:	6,575 kg
Mobility	
Speed	
max speed:	14 km/h
towed:	80 km/h
Fording	
without preparation:	0.9 m
Firepower	
Armament:	1 × carriage mounted 125 mm smoothbore gun
Main weapon	
breech mechanism:	vertical sliding block
muzzle brake:	double T type
maximum range:	2,000 m (BM-17APFSDS projectile)
maximum range:	5,000 m (missile)
maximum range:	12,000 m (OF-46 HE projectile)
Rate of fire	
rapid fire:	8 rds/min
Main armament traverse	
angle:	360°
Main armament elevation/depression	
armament front:	+25°/-6°
Survivability	
Armour	
shield:	yes

Status

Development complete. As far as is known the 125 mm 2A45M anti-tank gun has not entered volume production. It is being offered for export. It is considered that if placed in volume production then existing 122 mm D-30 carriages may probably be used rather than brand new carriages.

Contractor

Joint Stock Company Spetstehnika.

122 mm field gun D-74

Development

The 122 mm field gun D-74 was developed in the late 1940s by the F F Petrov design bureau and was seen for the first time in public during the 1955 May Day parade in Moscow.

It appears that the 122 mm field gun D-74 and the 130 mm field gun M-46 were both designed to meet the Russian Army requirement for a gun with a longer range to replace the 122 mm corps gun M1931/37 (A-19).

After extensive troop testing the 130 mm field gun M-46 was selected in preference to the D-74 as it fired a heavier high-explosive projectile to 27,150 m compared with the 24,000 m of the D-74.

The D-74 was, however, produced in some numbers, most of which were subsequently exported. It is now considered obsolete by the RFAS. The 122 mm field gun D-74 uses the same carriage as the 152 mm gun-howitzer D-20 but the D-20 can easily be distinguished from it by a much thicker barrel and a larger muzzle brake.

According to United Nations sources, there were no exports of the 122 mm field gun D-74 by any country between 1992 and 2010.

The 122 mm field gun D-74 has also been manufactured for the home and export market by China North Industries Corporation (NORINCO) under the Chinese designation Type 60. This has not been built or marketed by the Chinese for some years.

The 122 mm field gun Type 60 is still in service with the People's Liberation Army.

The 122 mm field gun D-74 is no longer used by the Russian Army and none are understood to be held in reserve by the Russian Army.

Description

The shield, which slopes slightly to the rear and extends over the roadwheels, has an irregular top with a centre section which slides upwards so the ordnance can be elevated. A circular firing pedestal fitted as standard on all D-74s is inverted and secured just forward of the shield when travelling. In the firing position the circular firing pedestal is lowered to the ground, allowing the D-74 to be quickly traversed through a full 360°.

The recoil system of the 122 mm field gun D-74 consists of a hydraulic buffer and a hydropneumatic recuperator.

On each of the split box section trails is a castor which folds up on to the top of the trail when travelling and a spade which folds up underneath the trail when travelling.

Maximum effective range when firing an 122 mm APHE projectile is being quoted as 1,070 m. Sustained rate of fire is quoted as 75 rounds for the first hour.

A unit of fire of the 122 mm field gun D-74 is a total of 90 rounds of ammunition.

It is typically towed by a Ural-375D (6 x 6) cross-country truck that in addition to its crew also carries a quantity of ready use ammunition (projectiles, charges and fuzes).

The D-74 fires variable charge, case-type, and separate loading ammunition (see Table 1 on next page).

Variants

North Korea

This country is understood to have removed the upper part of the towed 122 mm D-74 field gun and integrated this onto a full-tracked armoured chassis.

US sources have given the following designations to these locally developed systems:

Designation	Weight of system
M1981:	19 t
M1991:	22 t

122 mm field gun D-74 on firing pedestal with spades lowered and support wheels raised (US Army) 0500368

Table 1

Ammunition type	FRAG-HE	APC-T
Projectile designation:	OF-472	BR-472
Fuze model:	V-429	DBR
Weight:		
(projectile)	27.3 kg	25 kg
(bursting charge)	2.95 kg	0.091 kg
Type of bursting charge:	TNT	RDX/alum
Maximum muzzle velocity:	885 m/s	885 m/s
Armour penetration at 0°:	n/a	185 mm/1,000 m

Notes:
Other types of projectile include illuminating (m/v 885 m/s) and smoke (m/v 885 m/s with projectile weighing 21.7 kg)

Specifications

	122 mm Field Gun D-74
Dimensions and weights	
Crew:	10
Length	
transport configuration:	9.875 m
Width	
transport configuration:	2.35 m
Height	
transport configuration:	2.745 m
axis of fire:	1.668 m
Ground clearance	
overall:	0.4 m
Track	
vehicle:	2.03 m
Weight	
standard:	5,550 kg
combat:	5,500 kg
Mobility	
Carriage:	split trail (with castor wheel on each)
Tyres:	12.00 × 20
Firepower	
Armament:	1 × carriage mounted 121.92 mm howitzer
Main weapon	
length of barrel:	6.45 m
breech mechanism:	vertical sliding block
muzzle brake:	double baffle
maximum range:	24,000 m (OF-472 HE-FRAG projectile)
maximum range:	1,070 m (BR-472 APC-T projectile)
Rate of fire	
sustained fire:	1.25 rds/min
rapid fire:	7 rds/min
Main armament traverse	
angle:	58°
Main armament elevation/depression	
armament front:	+45°/-5°
Survivability	
Armour	
shield:	yes

Status

Production complete. In service with the countries listed in the table:

Country	Quantity	Comment
Algeria	25	
Cambodia	n/avail	NORINCO Type 60
China	n/avail	local designation Type 60
Congo, Democratic Republic	14	NORINCO Type 60
Iran	40	NORINCO Type 60
Korea, North	n/avail	possible local production
Libya	60	
Mauritania	20	
Nigeria	50	estimated figure
Pakistan	200	NORINCO Type 60
Somalia	n/avail	status uncertain
Sudan	n/avail	
Syria	n/avail	
Vietnam	n/avail	NORINCO Type 60
Zimbabwe	18/20	NORINCO Type 60

Contractor

Joint Stock Company Spetstehnika.
China North Industries Corporation (NORINCO).

122 mm howitzer D-30

Development

The towed 122 mm howitzer D-30 entered service with the former Soviet Army in the early 1960s as the replacement for the towed 122 mm howitzer M1938 (M-30).

The 122 mm howitzer D-30 was designed by the F F Petrov design bureau at Artillery Plant No 9 in Sverdlovsk. The main improvements over the earlier weapon are increased range and the ability to rapidly traverse the weapon through 360°. This allows the D-30 to engage new targets much quicker than a normal towed artillery system.

Late production models of the 122 mm D-30 are designated the D-30M and have a number of modifications including a new double-baffle muzzle brake, a new central baseplate which is square rather than round and a towing lunette assembly. It also features modifications to its cradle, carriage and recoil system. As it is ballistically the same as the earlier 122 mm D-30 it has the same range and capabilities.

According to the UN, between 1992 and 2010 the following quantities of D-30 122 mm howitzer weapons were exported:

Description

The 122 mm howitzer D-30 is towed muzzle first and on arrival at the battery position the crew first unlock the barrel travelling lock which then folds back onto the central trail. The firing jack under the carriage is then lowered, raising the wheels clear of the ground. The two outer trails are each spread through 120° and the firing jack is raised until all three trail ends are on the ground, when they are staked into position. The lunette is mounted either under the muzzle brake or on the barrel just to the rear of the muzzle brake. In either case it is normally swung through 180° so that it lies under the barrel before firing begins.

Fire-control instruments include a Type PG-1M panoramic sight, Type OP4M-45 telescopic sight and a Type K-1 collimator.

The 122 mm D-30 is typically towed by an Ural-375D (6 × 6) cross-country truck or the MT-LB multipurpose tracked vehicle.

In addition to the crew the prime mover also carries a quantity of ready use ammunition (projectiles, charges and fuzes).

An unusual feature of the D-30 is that the recoil system is mounted over the barrel, which is designated the 2A18. Some of the projectiles fired by the D-30 are interchangeable with those fired by the 122 mm howitzer M1938 (M-30) but the D-30 also fires a High Explosive Anti-Tank (HEAT) projectile of the non-rotating fin-stabilised type.

122 mm D-30 howitzer of the Finnish Defence Forces deployed in the firing position (IHS/Peter Felstead) 1133934

122 mm D-30 in travelling configuration 0085648

Afghan National Army (ANA) firing a Russian 122 mm howitzer D-30 during a fire mission carried out late in 2013 (US Army) 1455344

The D-30 fires the following variable charge, case-type, separate loading ammunition:

Ammunition type	FRAG-HE	HEAT-FS
Projectile designation:	OF-462[1]	BK-6M[2]
Fuze model:	RGM-2	GPV-2
Weight:		
(projectile)	21.76 kg	21.63 kg
(bursting)	3.675 kg	n/a
Type of bursting charge:	TNT	RDX
Muzzle velocity:	690 m/s	740 m/s
Armour penetration at 0°: (conventional armour)	n/a	460 mm/any range

Notes:
[1] There are variants of projectile design giving different weights. Various other fuzes may also be used. The HEAT projectile is of limited effectiveness against the more recent advanced armour systems
[2] Rotating BP-463 may also be used

Other types of projectiles include illuminating (S-463 weighing 22.4 kg), smoke (D-462 weighing 22.3 kg), leaflet, flechette and incendiary. Recently a Rocket-Assisted Projectile (RAP) has been introduced with a maximum range of 21,900 m.

The main armament of the Russian 122 mm self-propelled howitzer 2S1 is based on the ordnance of the D-30.

Details of this are provided in a separate entry in *IHS Jane's Land Warfare Platforms: Artillery and Air Defence*.

Production of the 2S1 was completed some years ago but it is still used in large numbers with a significant number of surplus vehicles being sold on the international market.

Upgrade package

Russia now has available an upgrade package for the 122 mm D-30 which enables it to be towed at higher speeds, improving its reliability and enabling it to come into action more quickly.

122 mm D-30A

This is also referred to as the 2A18M and has a number of minor improvements including a new cradle and modified recoil system. It also features a new muzzle brake.

Brief specifications are:

Calibre:	121.92 mm
Maximum range:	15,300 m
Rate of fire:	6-8 rds/min
Axis of fire:	900 mm
Traverse:	360°
Elevation/depression:	+70/–7°
Length: (travelling)	5.4 m
Width: (travelling)	2.2 m
Height: (travelling)	1.8 m
Ground clearance:	0.325 m
Track:	1.85 m
Maximum towing speed:	80 km/h

Variants

Chinese 122 mm D-30 howitzer

The Chinese version of the D-30 is designated the Type 86 122 mm towed howitzer although it has also been referred to as the W86.

This is being offered on the export market by China North Industries Corporation (NORINCO) and is covered in detail in a separate entry in *IHS Jane's Land Warfare Platforms: Artillery and Air Defence*.

NORINCO has also developed and placed in production a new family of 122 mm ammunition for the Type 86 and D-30 122 mm howitzers.

The ordnance of the Type 86 122 mm towed howitzer is also used in a number of Chinese self-propelled artillery systems including:

- SH2 122 mm self-propelled howitzer (6 × 6)
- SH3 122 mm self-propelled howitzer (tracked)
- 122 mm PLZ-07 122 mm self-propelled howitzer (tracked)
- 122 mm Type 89 self-propelled howitzer (tracked)
- 122 mm Type 85 self-propelled howitzer (tracked)
- 122 mm self-propelled howitzer (8 × 8) (at least two versions which as of March 2013 remain at the prototype stage).

Details of all of the above are provided in separate entries in *IHS Jane's Land Warfare Platforms: Artillery and Air Defence*. Of the above, only the SH2 and SH3 122 mm systems are currently being marketed by NORINCO.

The 122 mm PLZ-07. Type 89, Type 85, Type 70 and Type 70-1 are all deployed by the Peoples Liberation Army with the PLZ-07 making its first public appearance in October 2009 during a parade in Beijing.

Croatian 122 mm D-30 howitzer

Croatia has developed and placed in production an upgraded version of the D-30 under the designation of the D-30 HR M94 and details are provided in a separate entry in *IHS Jane's Land Warfare Platforms: Artillery and Air Defence*.

Egyptian 122 mm D-30 howitzer

The 122 mm D-30 howitzer has been manufactured in Egypt by Factory 100 (Abu Zaabal Engineering Industries Company) for the home and export markets are details are provided in a separate entry in *IHS Jane's Land Warfare Platforms: Artillery and Air Defence*.

Iranian 122 mm D-30 howitzer

The Defence Industries Organization. Hadid Armament Industries Group, builds the 122 mm D-30 under the designation of the 122 mm Shafie D-301 howitzer which is covered in a separate entry in *IHS Jane's Land Warfare Platforms: Artillery and Air Defence*. The 122 mm D-30 ordnance is also used in the Iranian Raad-1 self-propelled artillery system. The latter is currently

From	To	Quantity	Date	Comment
Belarus	Azerbaijan	30	2010	
China	Pakistan	80	2003	Chinese built D-30
China	Pakistan	63	2004	Chinese built D-30
Czech Republic	Peru	6	1998	
Czech Republic	Slovakia	24	2001	
Czech Republic	Slovakia	9	2002	
Czech Republic	Georgia	30	2006	
Czech Republic	Ukraine	44	2010	
Egypt	Rwanda	6	1992	
Germany	Finland	218	1992	
Germany	Belgium	1	1992	
Kazakhstan	Angola	4	1998	
Kazakhstan	Ethiopia	100	2000	
Kazakhstan	Angola	24	1999	
Romania	Moldova	18	1992	not confirmed as D-30
Slovakia	Bulgaria	7	1994	
Slovakia	Uganda	9	2002	
Slovakia	Afghanistan	24	2006	
Slovakia	Czech Republic	2	2010	
Ukraine	Congo, Democratic Republic	36	2010	

deployed by the Iranian Army and is essentially the turret similar to the Russian 122 mm 2S1 fitted to a locally designated Iranian hull.

Iraqi 122 mm D-30 howitzer

The 122 mm D-30 howitzer was manufactured in Iraq and called the Saddam.

Serbian 122 mm D-30

Serbia has developed and placed in production an upgraded version of the 122 mm D-30 under the local name of the D-30J and details of this are provided in a separate entry in *IHS Jane's Land Warfare Platforms: Artillery and Air Defence*.

Country	Quantity	Comment
Afghanistan	120	
Algeria	160	
Angola	500	estimate
Armenia	69	
Azerbaijan	129	
Bangladesh	54	Type 56
Belarus	48	
Benin	n/avail	
Bosnia-Herzegovina	268	
Burundi	18	
Cambodia	n/avail	
China	n/avail	local production
Congo	10	
Congo, Democratic Republic of (Zaïre)	36	estimate
Costa Rica	n/avail	
Croatia	53	local production
Cuba	n/avail	
Djibouti	6	
Estonia	42	
Egypt	190	local production (also SP122 self-propelled model)
Eritrea	10/20	
Ethiopia	460	includes 122 mm M-30
Finland	84	local designation H-63
Georgia	53	
Ghana	6	
Guinea Bissau	9	estimate
India	550	
Iran	550	estimate, local production
Israel	5	reserve
Korea, North	n/avail	also deploys a self-propelled tracked model called the M1997 by the US
Kazakhstan	400	including truck mounted (8 × 8) version
Kyrgyzstan	72	
Laos	20	
Lebanon	24	
Libya	190	status uncertain as result of conflict in 2011
Macedonia	142+	
Madagascar	12	
Mali	8	
Mauritania	20	
Mongolia	120	
Montenegro	12	
Morocco	50	
Mozambique	24	includes 122 mm M-30
Myanmar	100	
Nicaragua	12	
Nigeria	30	estimate
Oman	30	
Pakistan	143	from China
Peru	42	including 6 in 1998 (36 Army; 6 Marines)
Russia	4,600	estimate
Rwanda	6	from Egypt in 1998
Serbia/Montenegro	78	local production
Seychelles	3	
Slovakia	48	
Somalia	12+	
Sudan	25	also 6 × 6 self-propelled model
Syria	500	some were fitted onto T-34 chassis
Tajikistan	12	
Tanzania	20	from China
Turkmenistan	180	
Ukraine	369	
Uzbekistan	60	
Vietnam	n/avail	
Yemen	130	
Zambia	25	
Zimbabwe	4	

A self-propelled (4 × 4) model of the D-30J has been developed to the prototype stage.

Serbian SOKO SP RR 122/105/122 mm self-propelled artillery system

This is FAP 2228 (6 × 6) cross country chassis fitted with an armour protected cab at the front with the ordnance mounted in a protected turret at the rear of the chassis.

When deployed in the firing position two stabilisers are lowered at the rear of the chassis to provide a more stable firing platform.

The turret can be fitted with a 100 mm, 105 mm or 122 mm ordnance with the latter being based on the widely deployed Russian designed D-30 towed howitzer or its Serbian equivalent.

The complete system has a maximum gross combat weight of 22 tonnes and is powered by a OM 906 diesel engine developing 205 kW (279 hp) coupled to a manual transmission. This gives a maximum road speed of 100 km/h with a typical operating range of 1,000 km.

As of March 2013 it is understood that trials with the first prototype system were still underway and the system had yet to enter production.

Sudanese self-propelled 122 mm D-30 howitzer Khalifa

Early in 2013 it was revealed that the Sudanese Army had deployed a self-propelled version of the widely deployed Russian towed 122 mm D-30 howitzer under the local name of Khalifa.

This has been developed by the Military Industry Corporation (MIC) of the Sudan and the upper part of the refurbished towed 122 mm D-30 howitzer is mounted on the rear of a modified Russian Kamaz (6 × 6) cross-country truck chassis.

This system has been offered on the export market by the MIC of the Sudan but as of March 2013 there are no known sales.

The system has also been designated the GHY02 by the MIC, but this may be just for marketing purposes.

Details of this are provided in a separate entry in *IHS Jane's Land Warfare Platforms: Artillery and Air Defence*.

SP122

This was developed by the now BAE Systems of the US and consists of a modified 155 mm M109 self-propelled howitzer/M992 Field Artillery Ammunition Support Vehicle (FAASV) hull with a fully enclosed superstructure at the rear in which is mounted the complete upper carriage of the Egyptian-built D-30 towed howitzer.

A total of 124 SP122 have been delivered to the Egyptian Army and details are provided in a separate entry in *IHS Jane's Land Warfare Platforms: Artillery and Air Defence*.

Soltam Systems 122 mm D-30 (8 × 8) self-propelled howitzer

To meet the requirements of Kazakhstan Soltam Systems of Israel has developed a self-propelled version of the Russian 122 mm howitzer D-30 which is now in service.

This is based on a KamAZ-63502 (8 × 8) forward control truck chassis. To the immediate rear of the cab is space for the crew and ammunition.

Mounted at the very rear of the chassis is a turntable on which is mounted a 122 mm howitzer D-30 with powered elevation and traverse.

To provide a more stable firing position two hydraulic stabilisers are lowered either side before firing commences.

Specifications

	122 mm Howitzer D-30
Dimensions and weights	
Crew:	7
Length	
transport configuration:	5.4 m
Width	
transport configuration:	1.95 m
Height	
transport configuration:	1.66 m
axis of fire:	0.9 m
Ground clearance	
overall:	0.325 m (est.)
Track	
vehicle:	1.85 m
Weight	
standard:	3,210 kg
combat:	3,150 kg
Mobility	
Tyres:	9.00 × 20
Firepower	
Armament:	1 × carriage mounted 121.92 mm howitzer
Main weapon	
length of barrel:	4.875 m
breech mechanism:	vertical sliding block
muzzle brake:	multibaffle
maximum range:	15,400 m (OF-462 HE-FRAG projectile)
maximum range:	21,900 m (HE-FRAG rocket assisted projectile)
Rate of fire	
rapid fire:	8 rds/min

	122 mm Howitzer D-30
Main armament traverse	
angle:	360°
Main armament elevation/depression	
armament front:	+70°/-7°
Survivability	
Armour	
shield:	yes

Status
Production complete. In service with the following countries (see table on previous page):

Contractor
Joint Stock Company Spetstehnika.

122 mm howitzer M1938 (M-30)

122 mm howitzer M1938 (M-30) in travelling configuration (IDF Spokesman)
0501032

Development
The towed 122 mm howitzer M1938 (M-30) entered service with the Russian Army in 1939 after standardisation in September 1938. It was designed by the F F Petrov design bureau at Artillery Plant No 172 at Perm.

Until the introduction of the 122 mm howitzer D-30 it was the standard division towed howitzer of the Warsaw Pact and was issued on the scale of 36 per motorised rifle division (two battalions each consisting of 18 M1938s with three batteries each with six howitzers) and 54 per tank division (three battalions each consisting of 18 M1938s with three batteries each with six howitzers).

It is no longer in front-line service with the RFAS, although it can still be found in service in various parts of the world. By today's standards the 122 mm howitzer M1938 (M-30) is obsolete as it has poor mobility and a very short range.

According to the United Nations Arms Transfer Lists for the period 1992 through to 2010, there were the following declared exports of the old 122 mm howitzer M1938 (M-30).

It should be noted however that there could well be more as some countries only provide information as 'artillery' without specific details.

From	To	Quantity	Date	Comment
Bulgaria	Macedonia	108	1999	
Bulgaria	UN	1	1994	
China	Bolivia	36	1992	Chinese Type 54-1
Slovakia	Bulgaria	7	1994	probably for resale

Description
The carriage of the 122 mm howitzer M1938 (M-30) is of the riveted box section split trail type and is identical to that used for the 152 mm howitzer M1943 (D-1), which can easily be distinguished from the 122 mm weapon by its longer and fatter barrel with a double-baffle muzzle brake.

The top half of the shield slopes to the rear and in the centre of the shield is a section that slides upwards so the weapon can be elevated. The recoil system is under the 122 mm barrel and the counter-recoil system above. The recoil system of the 122 mm howitzer M1938 (M-30) consists of a hydraulic buffer and a hydropneumatic recuperator.

Typical rate of fire is between five and six rounds a minute with a quoted sustained rate of fire of up to 75 rounds in 60 minutes.

The elevating mechanism is on the right side of the carriage and the traverse mechanism is on the left. The 122 mm howitzer M1938 (M-30) can be fired without spreading its trails, in which case traverse is limited to 1.5°. Maximum towing speed is 48 km/h. Each trail has two spades, one fixed for use on hard surfaces and one hinged for softer ground.

122 mm howitzer M1938 (M-30) with standard sponge rubber-filled tyres showing sliding vertical section in centre of shield 0501033

The 122 mm howitzer M1938 (M-30) fires the following variable charge, case-type, separate-loading ammunition which was developed many years ago and can also be fired by the more recent Russian 122 mm D-30 howitzer covered in a separate in *IHS Jane's Land Warfare Platforms: Artillery and Air Defence.*

Unit of fire for the 122 mm howitzer M1938 (M30) is typically 80 rounds of ammunition.

It is normally towed by an Ural-375 (6 × 6) cross-country truck that in addition to the gun crew also carries a quantity of ready use ammunition.

Projectile type	FRAG-HE	HEAT
Projectile designation:	OF-462[1]	BP-463[2]
Fuze model:	RGM-2	GKB
Weight:		
(projectile)	21.76 kg	14.8 kg
(bursting)	3.675 kg	n/a
Type of bursting charge:	TNT	n/a
Muzzle velocity:	515 m/s	570 m/s
Armour penetration at 0°:	n/a	180 mm

Notes:
[1] Other FRAG and FRAG-HE projectiles available
[2] Fired at full charge. Better design than earlier BP-460A (weight 14.8 kg) which was fired at charge 4 giving only 325 m/s. The old HEAT projectile is of limited effectiveness against the more recent armour systems

Other projectiles include illuminating (S-463) weighing 21 kg, smoke (D-462) weighing 22.4 kg and leaflet (A-462) weighing 21.5 kg. China North Industries Corporation (NORINCO) also manufacture 122 mm ammunition for this weapon.

Variants
The 122 mm howitzer M1938 (M-30) used by Bulgaria has new disc wheels with spoke-type impressions but retains the sponge rubber-filled tyres of the original model.

The 122 mm howitzer M1938 (M-30) has also been manufactured by China North Industries Corporation (NORINCO) as the Type 54 and Type 54-1 and was originally issued on the scale of 12 per infantry division and 12 per armoured division. A self-propelled model of the Type 54-1 is essentially a Type 531 fully-tracked APC with the Type 54-1 mounted on the rear of the vehicle firing forwards. This is only used by the People's Liberation Army (PLA). This is still used in small numbers by the PLA but has been replaced by more recent self-propelled artillery systems armed with a 122 mm D-30 series ordnance.

Details of the Chinese versions of the Russian 122 mm howitzer M1938 (M-30), towed and self-propelled, are provided in separate entries in *IHS Jane's Land Warfare Platforms: Artillery and Air Defence.*

Polish upgraded 122 mm M1938 (M-30)
In the 1980s Poland upgraded some of their 122 mm howitzer M1938 (M-30) weapons to improve their direct fire capability. These are known as the wz 1938/1985.

It should be noted that Poland no longer has any 122 mm M1938 (M-30) howitzers in service but they may be found in service overseas.

Specifications

	122 mm Howitzer M1938 (M-30)
Dimensions and weights	
Crew:	8
Length	
transport configuration:	5.9 m
Width	
transport configuration:	1.975 m

	122 mm Howitzer M1938 (M-30)
Height	
transport configuration:	1.71 m
axis of fire:	1.2 m
Ground clearance	
overall:	0.33 m
Track	
vehicle:	1.6 m
Weight	
standard:	2,450 kg
combat:	2,450 kg
Mobility	
Carriage:	split trail
Tyres:	11.80 × 15
Firepower	
Armament:	1 × carriage mounted 121.92 mm howitzer
Main weapon	
length of barrel:	2.8 m
breech mechanism:	interrupted screw
muzzle brake:	none
maximum range:	11,800 m (OF-462 HE-FRAG projectile)
maximum range:	630 m (BP-463 HEAT effective)
Rate of fire	
sustained fire:	1.25 rds/min
rapid fire:	6 rds/min
Main armament traverse	
angle:	49°
Main armament elevation/depression	
armament front:	+63.5°/-3°
Survivability	
Armour	
shield:	yes

Status
Production complete. In service with countries listed below. In many cases these weapons are held in reserve or used for training purposes.

Country	Quantity	Comment
Algeria	60	
Bangladesh	57	Chinese Type 54
Bolivia	36	from China in 1992
Bosnia-Herzegovina	n/avail	
Bulgaria	20	108 sold to Macedonia in 1999
Cambodia	n/avail	Chinese Type 54
China	n/avail	Chinese Type 54
Congo (Democratic Republic)	10	estimated figure
Costa Rica	n/avail	phasing out
Croatia	29	phasing out
Cuba	n/avail	
Egypt	300	
Ethiopia	460	D-30 and M-30
Guinea-Bissau	9	
Indonesia	28	Marines
Iran	100	Chinese Type 54
Korea, North	n/avail	
Kyrgyzstan	35	
Laos	20	estimate, includes D-30
Lebanon	32	
Macedonia	108	from Bulgaria in 1999
Moldova	17	
Mongolia	n/avail	
Mozambique	24	D-30 and M-30
Pakistan	490	Chinese Type 54
Russia	3,000	reserve
Somalia	n/avail	
Sudan	24	M-30
	76	Chinese Type 54
Syria	150	Estimate
Tanzania	80	including Chinese Type 54
Uganda	18	
Ukraine	2	reserve
Vietnam	n/avail	including Type 54
Yemen	40	

Contractor
Joint Stock Company Spetstehnika.
China North Industries Corporation (NORINCO).

122 mm assault gun M-392

Development
The 122 mm Assault Gun M-392 was developed by Artillery Plant No 9 to provide infantry units with a lightweight system capable of being towed by a 4 × 4 light vehicle such as the GAZ-66 or more recent UAZ-469 or slung under a helicopter such as the Mi-24.

As of March 2013 it is understood that the 122 mm assault gun M-392 has yet to enter quantity production.

Description
The 122 mm Assault Gun M-392 is based on a standard split trail carriage with each trail being provided with a spade. A two part shield is provided for the crew with elevation, traverse and sights being mounted on the left side of the carriage.

According to the manufacturer the system takes one minute to be brought into action.

The short barrel is not fitted with a fume extractor or a muzzle brake with the recoil system being mounted above.

It fires the same family of ammunition as fired by the old 122 mm howitzer M1938 (M-30) which was developed before the Second World War.

Full details of this and its family of ammunition are provided in a separate entry. The high-explosive fragmentation projectile weighs 21.76 kg and when fired from the 122 mm Assault Gun M-392 has a maximum stated range of 8.6 km.

The 122 mm assault gun M-392 is normally towed by a GAZ-66 or UAZ-469 (4 × 4) cross-country vehicles but two vehicles would be required to tow the gun, carry the crew and a quantity of ready use ammunition.

Specifications

	122 mm Assault Gun M-392
Dimensions and weights	
Weight	
standard:	1,300 kg
Mobility	
Carriage:	split trail
Firepower	
Armament:	1 × carriage mounted 122 mm
Main weapon	
muzzle brake:	none
maximum range:	8,600 m
Rate of fire	
sustained fire:	8 rds/min
Main armament traverse	
angle:	50° (25° left/25° right)
Main armament elevation/depression	
armament front:	+70°/-7°
Survivability	
Armour	
shield:	yes

Status
Development complete. Ready for production.

Contractor
Artillery Plant No 9.

122 mm Assault Gun M-392 being towed by a light (4 × 4) vehicle (Artillery Plant No 9) 1040792

120 mm 2B16 (NONA-K) combination gun

Development
For some years the Russian Army deployed two 120 mm gun/mortar systems, the 2B16 (NONA-K) towed system and the 2S9 full-tracked self-propelled system.

The 2S9 is based on the BMD-1 airborne combat vehicle hull which was developed at the Izhevsk plant with production of the latter BMD-2 being undertaken at the Volgograd Tractor Plant.

Type	HE-FRAG	HE-RAP	HEAT
Calibre:	120 mm	120 mm	120 mm
Round weight:	19.8 kg	19.8 kg	13.17 kg
HE content:	4.9 kg	3.25 kg	n/a
Round length:	820 mm	825 mm	960 mm
Muzzle velocity:	367 m/s	367 m/s	560 m/s
Maximum pressure:	100 MpA	100 MpA	110 MpA
Maximum range:	8,850 m	12,850 m	1,000 m
Minimum range:	1000 m	1000 m	40 m

It is now known that these two weapons share the same 120 mm rifled ordnance and fire the same family of ammunition with the obvious logistic and training advantages to the user.

These 120 mm weapons combine the features of a howitzer and mortar in one system. In many respects they are unique weapons as there is no similar equivalent weapon system in the West.

The 120 mm 2B16 (NONA-K) towed system was developed in the 1970s at the famous Perm artillery facility and the Central Scientific Research Institute for Precision Machinery Construction.

It was accepted for service with the Russian Army in 1986 with the main user being the Russian air assault forces.

More recently, Russia has developed a new full-tracked self-propelled artillery system called the 2S31 which fires the same family of ammunition as the 2B16, 2S9 and 2S23 as well as a complete new family of 120 mm ammunition.

The 120 mm 2S31 is based on a much-modified BMP-3 infantry combat vehicle hull. This remains at the prototype stage and details are given in a separate entry in *IHS Jane's Land Warfare Platforms: Artillery and Air Defence*. As of March 2013 it is understood that the 120 mm 2S31 had yet to enter quantity production.

UN sources have stated that Russia did not declare any export of the 120 mm 2B16 (NONA-K) for the years 1992 through to 2010.

Description

Within the designation NONA-K the K (or Kolesnaya in Russian) stands for wheeled, with the weapon being mounted on a split trail carriage and each trail being provided with a castor wheel to assist in bringing the weapon quickly into action.

When travelling, the trails are closed together and locked and the complete upper part of the weapon is traversed through 180° and locked in position over the closed trails. This reduces the overall length of the system for travelling, which is considered important for air transport.

The 2B16 is normally towed by the GAZ-66 (4 × 4) 2,000 kg light truck which also carries the five-man crew and a quantity of ready to use 120 mm ammunition. Maximum towed speed on a good road is quoted as 80 km/h. For short distances it can be towed by the smaller UAZ-469 (4 × 4) 695 kg light vehicle.

When deployed in the firing position, the NONA-K is supported on a small circular baseplate that is located under the forward part of the carriage. The rubber-tyred road wheels are raised clear of the ground and the equipment rests on the two box-type spread trails, each of which is fitted with a spade.

The 120 mm/24.2 calibre ordnance, designated the 2A51, has 40 constant twist rifling grooves and is breech loaded. It can fire both Russian and Western natures of 120 mm rifled ammunition (for example that used by the French TDA 120 RT-61 towed mortar).

To reduce recoil forces on the barrel, a large box-type multibaffle muzzle brake is fitted which absorbs about 30 per cent of the recoil.

The 120 mmordnance/barrel has a maximum recoil distance of 400 mm.

The hydraulic recoil system is mounted above the ordnance and extends about halfway along the barrel with the recuperator being of the hydropneumatic type. The breech mechanism is of the vertical sliding type with plastic gas obturator and a chamber indent device is fitted to retain the 120 mm projectile in place when the weapon is fired at high elevation.

The thin shield either side of the ordnance provides the crew of the 2B16 with some protection from small arms fire and shell splinters, with the left side of the shield being slightly higher than the right to protect the direct and indirect sighting system. The manual elevation and traverse controls are also on the left side of the weapon.

120 mm 2B16 (NONA-K) combination gun deployed in the firing system and showing large muzzle brake and castor wheel on trail to assist in bringing the weapon into action 0130025

Types of Russian 120 mm special projectiles fired by the 2B16 include High-Explosive Fragmentation (HE-FRAG), High-Explosive Anti-Tank (HEAT) and smoke as well as 120 mm mortar bombs.

Details of the pre-rifled projectiles fired by the 120 mm 2B16 (NONA-K) combination gun are given in the table in the oppositie column.

The HE-FRAG projectile is called the OF-49, has a steel body and contains A-IX-2 explosive. There is also the OF-51 HE-FRAG projectile which also contains A-IX-2 explosive.

The rocket-assisted projectile is called the OF-50 with the motor cutting in 10 to 13 seconds after the projectile leaves the barrel.

The HEAT projectile is fin-stabilised and will penetrate 650 mm of conventional steel armour at 90° to the vertical.

Standard 120 mm mortar bombs fired include the OF-843B, OF-34 and OF-38 high-explosive fragmentation, S-843 and 3S9 illumination, 3Z2 incendiary and 3D5 smoke. These have muzzle velocities of between 119 and 331 m/s and ranges of between 430 and 7,150 m.

The 120 mm ammunition for this as well as the 2S9, 2S23 and latest 2S31 has been designed and manufactured by the Bazalt State Research and Production Enterprise.

It has also been confirmed that the 2B16 will fire the Kitolov-2 laser-guided projectile out to a maximum range of 9 km. According to the manufacturer this has a 90 per cent hit probability on a stationary target and a 80 per cent hit probability on a moving target. Cyclic rate of fire is being quoted as three laser-guided rounds a minute.

In order to fire this laser guided missile a fire control system would have to be added as well as communications system.

The 2S9, also referred to as the NONA-S, has a similar ordnance to that used in the NONA-K but this is fitted with a system to purge the ordnance after the weapon has been fired. A pneumatically assisted rammer is fitted to increase the rate of fire and load the weapon at higher elevations. The ordnance is not fitted with a muzzle brake.

The latest 120 mm gun/mortar system to enter service is the 2S23, or NONA-SVK which is based on the hull of the Russian BTR-80 (8 × 8) APC which has been in service with the Russian Army and Marines.

This uses a modified version of the 120 mm 2A51 ordnance, which is called the 2A60, but ballistics and characteristics remain the same. The 120 mm 2S23 is in service with the Russian Army and Marines and is covered in detail in a separate entry in *IHS Jane's Land Warfare Platforms: Artillery and Air Defence*.

The Chinese have a similar 120 mm self-propelled system based on a modified WZ551/WMZ551 (6 × 6) amphibious APC chassis called the PLL-05.

Details of this system, which is currently being offered on the export market by NORINCO, are provided in a separate entry in *IHS Jane's Land Warfare Platforms: Artillery and Air Defence*.

Specifications

	120 mm 2B16 (NONA-K) Combination Gun
Dimensions and weights	
Crew:	5
Width	
transport configuration:	1.79 m
Weight	
standard:	1,200 kg
Mobility	
Carriage:	split trail
Firepower	
Armament:	1 × carriage mounted 120 mm rifled gun
Main weapon	
calibre length:	24.2 calibres
muzzle brake:	multibaffle
maximum range:	8,700 m (HE-FRAG projectile)
maximum range:	12,850 m (rocket assisted projectile)
maximum range:	7,100 m (HE-FRAG mortar bomb)
Rate of fire	
rapid fire:	10 rds/min
Main armament traverse	
angle:	60° (30° left/30° right)
Main armament elevation/ depression	
armament front:	+80°/-10°
Survivability	
Armour	
shield:	yes

Status

Production complete. In service with the following countries:

Country	Quantity
Azerbaijan	12
Russian Federation	150
	18 (Naval Infantry)
Ukraine	2

Contractor

Perm Artillery Factory.

100 mm anti-tank gun T-12 (2A19) and MT-12 (2A29)

Development

The towed 100 mm T-12 anti-tank gun was developed as the replacement for the older 85 mm D-48 towed anti-tank gun. This may still be found in service in some parts of the world.

The 100 mm T-12 anti-tank gun has the industrial designation 2A19, was accepted for service in 1961 and was followed by the improved MT-12 which has the industrial designation 2A29. The MT-12 anti-tank gun entered production in 1970. The 100 mm T-12 was the first smoothbore anti-tank gun in the world to enter production.

According to Russian sources, the designers of the 100 mm T-12 smoothbore anti-tank gun at Factory No 75 decided on a smoothbore gun rather than a rifled gun for a number of reasons, this included:

- A spinning projectile tends to lose armour penetration from the jet of gases and hot metal that is produced when the High Explosive Anti-Tank (HEAT) projectile hits the target.
- As the smoothbore projectile is a tighter fit than a rifled projectile, the smoothbore gun can contain gas pressures that are higher when compared to a rifled tank gun. This means that the smoothbore gun has a higher muzzle velocity and therefore greater armour penetration characteristics than a comparable rifled gun.
- A rifled gun has lands and grooves that wear out when firing projectiles at high-muzzle velocities. Smoothbore guns have no lands or grooves and therefore have a longer life. This is especially true when rifled guns fire APFSDS projectiles.

The T-12 anti-tank gun can be distinguished from the earlier 85 mm D-48 anti-tank gun by its large breech ring, slightly differently shaped shield with an opening in the left side for the sighting systems and its smaller diameter pepperpot muzzle brake which is also parallel to the muzzle.

Although the T-12/MT-12 has been designed for use in the direct fire mode, both weapons also have indirect fire sights. Their range in this role is, however, limited as maximum elevation is only +20°. For elevations above +15°, a pit has to be dug under and to the rear of the breech or the breech will hit the ground during recoil.

According to the United Nations, the following quantities of 100 mm T-12 towed anti-tank guns were exported between 1992 and 2010. There could however be other exports of these weapons as some countries simply state 'artillery' without giving specific details of the calibre and type of weapon involved.

From	To	Quantity	Date
Ukraine	Azerbaijan	72	2002

Description

The 100 mm anti-tank gun T-12/MT-12 is operated by a six-man crew consisting of detachment commander, driver of the towing vehicle, gunlayer on the left, loader on the right and two ammunition numbers.

The weapon comprises four main components, 100 mm barrel and breech ring, semi-automatic breech, top carriage with sight and lower carriage. The actual mount used in the T-12 is the same as that used in the 85 mm D-48 rifled anti-tank gun.

The 100 mm smoothbore barrel is 6.30 m long with the front part strengthened and fitted with a perforated muzzle brake. When in the travelling position, the 100 mm smoothbore barrel is clamped tightly to the trails with a lug on the breech ring.

The breech ring contains the vertical breech block and some parts of the semi-automatic loading system. The breech block is manually opened and closed by the loader.

Manually opening the breech is only required before the first round is fired and after that the semi-automatic loading system to the right of the breech ring on the upper carriage opens and closes the breech block. Consequently, the loader only has to reload.

The sights, elevating and traverse system are on the left. The aiming system of the 100 mm anti-tank gun consists of the S71-40 mount with the PG-1M panoramic sight for indirect aiming, the OP4M-4OU direct sight and the APN-6-40 (or older APN-5-40) for direct aiming at night. A range drum is fitted as standard.

The OP4M-4OU direct sight has a magnification of ×5.5 and an 11° field of view. The image intensification night sight has a magnification of ×6.8 and a 7° field of view.

100 mm anti-tank gun MT-12 being towed by an MT-LB multipurpose armoured tracked vehicle 0500403

The hydraulic buffer and the hydropneumatic recuperator are mounted on top of the ordnance forward of the vertical breech mechanism. Recoil is between 680 and 760 mm and the equilibrator of the MT-12 is mounted parallel and to the right of the recoil system.

The carriage of the MT-12 is of the split trail type with a castor wheel mounted on the left trail to assist bringing the weapon into action. The carriage is provided with a small shield with sides that slope to the rear. This provides protection for some of the crew from small arms fire and shell splinters.

The wheels of the T-12 can be removed and replaced by LO-7 type skis, which allow the weapon to be fired on ice or snow at elevations of up to 16°.

The main difference between the T-12 and the MT-12 is the new carriage. The former was normally towed by a truck whereas the MT-12 is towed by a MT-LB multipurpose tracked armoured vehicle. Production of the MT-LB was completed some time ago.

The original T-12 often turned over when being towed causing the MT-12 with the improved carriage to be introduced. The wheels are the same as those on the ZIL-150 (6 × 6) truck and have sponge rubber-filled tyres.

The suspension of the MT-12 consists of one torsion bar and one shock-absorber per wheel. When the weapon is deployed in the firing position, the suspension system is disengaged to increase gun stability.

There are three basic types of fixed 100 mm ammunition fired by the T-12/MT-12 as detailed in the table:

Type	Muzzle velocity	Range	Weight of round
APFSDS	1,575 m/s	3,000 m (direct)	19.34 kg
HEAT	975 m/s	5,955 m (direct)	23.00 kg
HE	700 m/s	8,200 m (indirect)	18.6 kg

The APFSDS round will penetrate the following Rolled Homogenous Armour (RHA) at 90°:

Penetration	Range
230 mm	500 m
215 mm	1,000 m
190 mm	1,500 m
180 mm	2,000 m
165 mm	2,500 m
140 mm	3,000 m

The HEAT round will penetrate 350 mm of RHA at 90°. The T-12/MT-12 can also fire the 9M117 Kastet (Knuckleduster) (NATO designation Stabber) laser-guided projectile, which has a range of 100 to 4,000 m with its single HEAT warhead penetrating between 550 and 600 mm of conventional steel armour.

The laser designator is mounted on a tripod, normally to the left of the weapon and all the operator has to do is keep the cross-hairs of his sights on the target to ensure a hit.

The original version of the 9M117 Kastet missile only had a single HEAT warhead but more recently a version with a tandem HEAT warhead has been developed which will penetrate 550 mm of RHA protected by explosive reactive armour.

The laser-guided projectile version has been offered on the export market but there are no known sales as of March 2013.

The 100 mm anti-tank gun T-12 (2A19) and the more recent MT-12 (2A29) are normally towed by a Ural-375D (6 × 6) cross-country truck that also carries a quantity of ready use ammunition.

The T-12 has a maximum road towing speed of 60 km/h while the more recent MT-12 has a higher road towing speed of 70 km/h.

The T-12 has a maximum quoted cross-country towing speed of 15 km/h while the more recent MT-12 has a maximum towing speed of 25 km/h, but both of these figures depend on terrain conditions.

100 mm anti-tank gun MT-12 with trails closed but spades still fitted and castor wheel lowered to ease field handling 0500402

100 mm M1944 (BS-3) anti-tank gun
This was fielded during the Second World War and is still used by:

Bulgaria	16
Cyprus	20
Democratic Republic of Congo	10
Ethiopia	n/avail
Kyrgyzstan	18
Mali	6
Mongolia	100
Mozambique	20
Nicaragua	24
Sudan	n/avail
Vietnam	n/avail
Yemen	20

Specifications

	100 mm T-12 (2A19)	MT-12 (2A29)
Dimensions and weights		
Crew:	6	6
Length		
transport configuration:	9.48 m	9.65 m
Width		
transport configuration:	1.795 m	2.31 m
Height		
overall:	1.565 m	1.60 m
transport configuration:	1.565 m	1.60 m
Ground clearance		
overall:	0.380 m	0.380 m
Track		
vehicle:	1.465 m	1.92 m
Weight		
standard:	2,750 kg	3,050 kg
Mobility		
Carriage:	split trail	split trail
Tyres:	9.00 × 20	9.00 × 20
Firepower		
Armament:	1 × carriage mounted 100 mm smoothbore gun	1 × carriage mounted 100 mm smoothbore gun
Main weapon		
breech mechanism:	vertical sliding block	vertical sliding block
muzzle brake:	pepperpot	pepperpot
Rate of fire		
rapid fire:	14 rds/min	14 rds/min
Main armament traverse		
angle:	27°	27°
Main armament elevation/ depression		
armament front:	+20°/-6°	+20°/-6°
Survivability		
Armour		
shield:	yes	yes

Variants
Serbia placed the ordnance of the T-12 (designated the 2A19M) on a modified D-30 122 mm howitzer carriage. The complete system is designated the 100 mm anti-tank gun TOPAZ. As far as is known this never entered quantity production or service.

Status
Production complete. In service with the following countries:

Country	Quantity	Comment
Algeria	12	
Armenia	36	status uncertain
Azerbaijan	30	
Belarus	40	
Bosnia	27	
Bulgaria	126	
Cambodia	n/avail	
China	120	local designation Type 73
Croatia	133	
Cuba	n/avail	
Georgia	40	
Hungary	106	
Kazakhstan	68	
Kyrgyzstan	18	
Moldova	36	
Mongolia	24	
Russia	500-600	including marines
Serbia/Montenegro	283	
Turkmenistan	72	
Ukraine	450	
Uzbekistan	36	
Vietnam	n/avail	

Contractor
Factory Number 75.

85 mm divisional gun D-44

Development
The towed 85 mm divisional gun D-44 was developed by the F F Petrov design bureau during 1943-44 as the replacement for the 76 mm division gun M1942 (ZIS-3) but was not issued until after the war.

The barrel is a development of the 85 mm gun mounted in the T-34/85 medium tank and is basically identical to both this and the 85 mm M1939 (KS-12) anti-aircraft gun.

Ammunition, which is of the fixed type, is fully interchangeable with that used in the Czech 85 mm M52 field gun and the auxiliary-propelled version of the D-44 called the SD-44.

Production of the 85 mm divisional gun commenced in 1945 and large numbers were built:

Year	Quantity
1945	7
1946	474
1947	1,803
1948	2,606
1949	2,500
1950	2,510
1951	1,508
1952	n/avail
1953	110

As far as is known, this weapon is no longer in front-line use with the RFAS, but may still be used for training.

There is a separate entry under China for its version, which is called the Type 56. As far as it is known, there has been no recent production of the Type 56 in China.

NORINCO is now offering an upgrade for the 85 mm Type 56 in which the existing 85 mm weapon and recoil system is replaced by the complete ordnance and recoil system of the local version of the howitzer 122 mm D-30. The Chinese version of the Russian 122 mm D-30 howitzer is called the Type 86 and is currently being offered on the export market by NORINCO.

Poland modified its D-44 and D-44ND with an electric system and these are now designated the D-44M and D-44MN.

It should be noted that they are no longer deployed by Poland.

The 85 mm divisional gun is normally towed by a 4 × 4 or 6 × 6 cross-country truck that also carries the crew and a quantity of ready use ammunition.

Description
The shield of the 85 mm divisional gun D-44 has a wavy top and a small sliding panel in front of the main gun shield, moving in elevation with the barrel.

The recoil system is mounted behind the shield and above the cradle and breech ring. The recoil system consists of a hydraulic buffer and a hydropneumatic recuperator. A single castor mounted towards the spade on one of the trails assists in manoeuvring the weapon into or out of action. When fitted with an APN3-7 infra-red night device, which is mounted to the rear of the shield, the D-44 is known as the D-44-N.

85 mm divisional gun D-44 of the Polish Army with trails spread (Jaroslaw Cislak) 0021015

The weapon is also fitted with gunner's telescope model OP1-7 and periscope PG-1M.

Ammunition is of the fixed type and the types available for this weapon include:

Type	Designation	Penetration[1]
AP-T	BR-357	120 mm
AT-T	BR-365	100 mm
AP-T	BR-365K	95 mm
HVAP-T	BR-367P	180 mm
HVAP-T	BR-365P	100 mm

Notes:
[1] At a range of 1,000 m with conventional steel armour at an angle of 90°.

85 mm Auxiliary Propelled Field Gun SD-44
As far as is known this weapon is no longer in front-line service. This was developed mainly for use by Russian air assault divisions.

First production SD-44s were in fact conversions of earlier production D-44 weapons. A total of 88 were converted in 1954 with a further 250 the following year.

Specifications

	85 mm Divisional Gun D-44
Dimensions and weights	
Crew:	8
Length	
transport configuration:	8.34 m
Width	
transport configuration:	1.78 m
Height	
transport configuration:	1.42 m
axis of fire:	0.825 m
Ground clearance	
overall:	0.35 m
Track	
vehicle:	1.434 m
Weight	
standard:	1,725 kg
combat:	1,703 kg
Mobility	
Carriage:	split trail
Tyres:	6.50 × 20
Firepower	
Armament:	1 × carriage mounted 85 mm rifled gun
Main weapon	
length of barrel:	4.693 m
breech mechanism:	vertical sliding block (semi-automatic)
muzzle brake:	double baffle
maximum range:	15,650 m (indirect fire; with trails dug in to provide 45° elevation, maximum range of 18,870 m can be achieved)
maximum range:	1,150 m (effective HVAP)
Rate of fire	
rapid fire:	20 rds/min
Main armament traverse	
angle:	54°
Main armament elevation/depression	
armament front:	+35°/-7°
Survivability	
Armour	
shield:	yes

Status
Production complete. It is considered possible that Russia and some of the other users of the 85 mm divisional gun may now have weapons surplus to requirements. In service with the following countries:

Country	Quantity	Comment
Afghanistan	n/avail	Chinese Type 56 (current status uncertain)
Albania	61	Chinese Type 56
Algeria	80	
Angola	n/avail	
Armenia	35	
Azerbaijan	20	
Bulgaria	85	reserve
Cameroon	n/avail	Chinese Type 56
China	n/avail	Chinese Type 56
Congo	n/avail	Chinese Type 56
Croatia	50	
Cuba	n/avail	
Democratic Republic of Congo	10	Chinese Type 56
Egypt	48	
Eritrea	12	
Ethiopia	n/avail	
Georgia	40	estimate
Guinea	6	
Guinea Bissau	8	
Iran	100	
Korea, North	n/avail	
Mali	6	
Mongolia	n/avail	
Morocco	36	
Mozambique	24	Russian Type D-44
	12	Chinese Type 56
Pakistan	200	Chinese Type 56
Russia	n/avail	reserve
Tanzania	75/80	Chinese Type 56
Serbia	45	
Somalia	100	
Sri Lanka	6	Chinese Type 56
Sudan	100	estimate
Syria	n/avail	
Vietnam	n/avail	
Yemen	100	

Contractor
Joint Stock Company Spetstehnika.
China North Industries Corporation (NORINCO).

82 mm Vasilyek (2B9) automatic mortar

Development
The 82 mm Vasilyek (Cornflower) (2B9) towed automatic mortar was introduced into the former Soviet Army in the early 1970s and saw extensive combat use with the Russian Army in Afghanistan.

It has now been replaced in front-line service by the conventional 120 mm 2B11 mortar system, but remains in service with RFAS air assault units. The 120 mm 2B11 is normally transported in the rear of a Russian GAZ-66 truck (4 × 4) 2,000 kg.

The latest 120 mm Russian mortar system is designated the 2S12 which consists of the 120 mm 2B11 mortar, 2L81 two wheeled mortar carriage and the 120 mm B11 mortar with the complete system being operated by a crew of five. This is also referred to as Sani.

The 120 mm 2B11 mortar is carried on the rear of an Ural 43206-0651 (4 × 4) cross-country truck fitted with a four door cab.

The rear cargo area is provided to assist in loading and unloading the mortar on its 2L81 two wheel mortar carriage.

Firing a 120 mm high explosive mortar bomb a maximum range of 7.1 km is claimed with a maximum rate of fire of up to 10 rounds a minute.

Traverse on the bipod is 5° left and right which elevation limits are from +45° to +80°.

Description
The complete system is called the 2K21 and consists of the 82 mm 2B9 automatic mortar and the 2F54 transport vehicle based on the GAZ-66 (4 × 4) truck.

Compared to existing mortar systems the 82 mm 2B9 has the added advantage that it can be used in the direct and indirect fire roles.

Vasilyek (2B9) 82 mm automatic mortar installed in the rear of an HMMWV (4 × 4) (US Army) 0577772

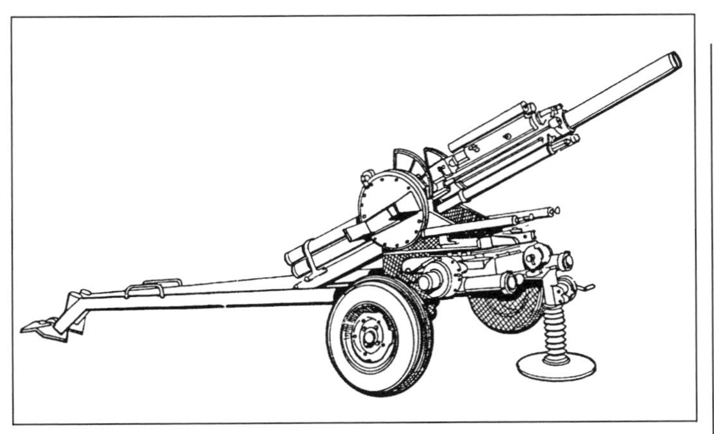

82 mm Vasilyek automatic mortar (2B9) in firing position (Steven Zaloga) 0500405

This flexibility allows a variable charge system to be used. Up to three charges can be used when the weapon is firing at high angles to provide the indirect fire support normally associated with conventional mortars.

A separate fixed charge can be used with the mortar when it is used to provide direct fire, similar to an anti-tank gun. According to Eastern sources, the weapon can produce consistently accurate fire in single shots, bursts, or at a sustained rate of fire of 120 rds/min. Its cyclic rate of fire is 170 rds/min.

When firing in the automatic mode the 82 mm mortar bomb has a maximum muzzle velocity of 270 m/s and when being used in the single-shot mode (muzzle loaded) it has a maximum muzzle velocity of 210 m/s. The primary charge weighs 7.5 g while the automatic loading charge weighs 75 g.

The 82 mm HE-FRAG mortar bomb weighs 3.1 kg and is fitted with a direct action or graze fuze which is armed after the projectile has travelled 10 m. There is also an anti-armour projectile of the High-Explosive Anti-Tank (HEAT) type that will penetrate 100 mm of conventional steel armour. This is not effective against the more advanced armours.

Before loading, the moving parts are pulled back by the charger until they come to rest on the automatic sear. The lever of the disconnector mechanism is set at a position corresponding to the selected type of fire, automatic or single shot.

The mortar is then loaded and a clip holding four 82 mm mortar bombs inserted into the weapon through a cover on the right trunnion. The weapon is then fired by pressing the trigger lever.

As stated, the 82 mm Vasilyek (2B9) mortar can be fired singly or automatically and converts the energy of powder gases generated in firing the weapon to operate the charger.

The weapon fires during counter-recoil and as a result the recoil energy develops and is spent to decelerate and stop the moving parts and impart them with recoil energy.

The recoil energy is damped by the return springs with its excess being absorbed by the ring shock-absorbers of the piston rods.

After each 82 mm mortar bomb is fired, the feed mechanism automatically moves the clip by one pitch bringing the next mortar shell to the ramming line. When being used in the sustained fire role the loader links up during firing the first clip with the following clips so as to form a continuous belt. This enables a high rate-of-fire to be achieved.

Provision is made for a hydraulic buffer to absorb the counter-recoil energy of the moving parts in the event of a misfire or hangfire.

To enable sustained fire missions to be undertaken, the barrel is water cooled. This consists of a welded structure consisting of a jacket, steam collector and two valves, relief and check.

The screw breech block rams a mortar shell into the bore, indents the primer of the shell primary cartridge, obturates the powder gases and cocks the moving parts via the pressure of the powder gases.

To ensure an equal rate of counter-recoil at all angles of elevation, the mortar is provided with a booster mechanism. Interlocks exclude the triggering of the moving parts when the charger is engaged and there is no clip in the mortar as well as during the clip movement to bring the next round to the ramming line.

82 mm Vasilyek automatic mortar (2B9) mounted on the rear of an MT-LB multipurpose armoured vehicle 0500882

Polish 82 mm Vasilyek automatic mortar (2B9) in travelling configuration (Richard Stickland) 0130017

The 82 mm barrel is mounted on a towed, split trail, wheeled carriage and in action the weapon rests on the trail legs, each of which is provided with a spade and a forward extendable leg that is lowered by a handwheel, with the main carriage raised off the ground.

The 82 mm ordnance then rests on a turning plate allowing 30° of traverse but the saddle is not fixed directly to the plate. Instead, it is supported by hydraulic elevators that can extend the ordnance into a high angle for indirect fire.

For this the saddle, cradle, ammunition feed tray and ordnance are elevated together and moved forward so that at full elevation the trunnions are above the carriage axle.

Total traverse of the weapon is 60°, 30° left and right, while elevation is from −1° to +85°. The optical sight has magnifications of ×2.5 and ×3. The sight is designated the PAM-1 with the Lutch-PM2M illuminator enabling fire missions to be carried out in the dark. The sighting system is mounted on the left side of the mount.

The 82 mm Vasilyek (2B9) can be towed but is normally transported under canvas covers on the cargo area of a modified GAZ-66 (4 × 4) 2,000 kg truck. For the export markets other types of 4 × 4 chassis could be used.

The vehicle is fitted with two ramps to assist in loading and offloading the 82 mm 2B9 automatic mortar.

In addition to being towed by a 4 × 4 light vehicle it can also be towed a larger vehicle such as the Ural-375D (6 × 6) cross-country truck that can also carry its crew and a significant quantity of ammunition.

Variants

Vasilyek on MT-LB

During the Russian fighting in Afghanistan a number of MT-LB multipurpose tracked armoured vehicles were observed fitted with the 82 mm Vasilyek mortar with the wheels removed on their rear decking. It was originally understood that this was a battlefield expedient and was first observed in Afghanistan. It has been retained in front line service with the Russian Army.

Vasilyek on ACRV

For trials purposes, Hungary mounted the 82 mm Vasilyek automatic mortar system on a modified ACRV chassis. There are no known plans for this to enter production.

Vasilyek on BMP-1

Hungary has built the prototype of a self-propelled version of the Vasilyek mounted on a modified BMP-1 infantry combat vehicle hull. There are no known plans for this to enter production.

Chinese Vasilyek

China North Industries Corporation (NORINCO) have developed a similar system called the 82 mm Type 82 mm Type W99 mortar and details of this are provided in a separate entry in *IHS Jane's Land Warfare Platforms: Artillery and Air Defence.* There are no known export sales of the Chinese version of the Vasilyek. More recently NORINCO has developed an 81 mm version for the export market. This is aimed at those countries who use Western natures of 81 mm ammunition rather than Eastern 82 mm types.

NORINCO 81 mm mobile mortar system

In 2008 it was revealed that China North Industries Corporation (NORINCO) had completed development of a new 81 mm self-propelled mortar system which is already being offered on the export market.

The 81 mm mortar installed on a turntable at the rear of a locally produced 4 × 4 cross-country truck chassis that allows it to be rapidly traversed through 360° onto a new target. The weapon can be elevated from −1° through to +85°.

It is understood that this weapon is the NORINCO 81 mm Automatic Mortar covered above, which is a further development of the NORINCO 82 mm Type W99 towed mortar system.

There is a separate entry on this system in *IHS Jane's Land Warfare Platforms: Artillery and Air Defence* but as of March 2013 this self-propelled version is understood to remain at the prototype stage.

HMMWV with 2B9

In 2004, for trials purposes, the US Army fitted an AM General High Mobility Multipurpose Wheeled Vehicle (HMMWV) with the 82 mm Vasilyek (2B9) automatic mortar in the rear. This work was carried out by engineers from Picatinny Arsenal using a Hungarian-supplied 82 mm Vasilyek (2B9) automatic mortar with the complete system called Scorpion. This was demonstrated to the US Army and Marine Corps.

According to the US Army Scorpion is a proof of principle demonstrator with the following features:

- 191-round ammunition storage box
- Maximum cyclic rate of fire of 120 rounds per minute
- Round-to-round: 0.3 s
- 4-round burst in 1.03 s
- Direct-fire range: 1,000 m
- Maximum range: 4,273 m at 748 mils
- Minimum range: 800 m at 1,387 mils
- Azimuth of fire: 360°
- Maximum elevation: +85°
- Minimum elevation: –1.25°
- Chassis M1097 HMMWV with a combat weight of 4,218 kg
- Armoured cab provides ballistic and blast overpressure protection.

It should be noted that there are some differences in the above specifications when compared to those supplied from Russian sources. As of March 2013 there were no plans for this to enter service with the US Army.

Conventional Russian 82 mm mortars

Russia is currently marketing the 82 mm 2B24 and the 82 mm Podnos conventional mortar systems.

The 82 mm 2B24 mortar has a maximum range of 4 km using standard ammunition which can be increased to 6 km using the latest 3-0-26 mortar bomb.

The older 82 mm Podnos mortar has a maximum range of 4.12 km.

Specifications

	82 mm Vasilyek (2B9) Automatic Mortar
Dimensions and weights	
Crew:	6
Length	
transport configuration:	4.115 m
Width	
transport configuration:	1.576 m
Height	
overall:	1.18 m
Ground clearance	
overall:	0.26 m
Weight	
standard:	645 kg
Mobility	
Speed	
towed:	60 km/h
Firepower	
Armament:	1 × carriage mounted 82 mm mortar
Main weapon	
maximum range:	4,270 m
Rate of fire	
rapid fire:	170 rds/min (maximum; 120 rds/min practical)
Main armament traverse	
angle:	60° (30° left/30° right)
Main armament elevation/depression	
armament front:	+85°/-1°
Survivability	
Armour	
shield:	no

Status

Production complete. In service with the following countries:

Country	Quantity	Comment
Bosnia-Herzegovina	n/avail	
Bosnian Serbs	n/avail	
Croatia and HVO	n/avail	
Hungary	n/avail	local production
Montenegro	n/avail	
Poland	n/avail	
Russia	n/avail	
Serbia	n/avail	

Contractor

Russian state factories. Has also been manufactured under licence in Hungary. NORINCO builds a similar system called the Type W99.

76 mm mountain gun GP (M1966)

Development

The 76 mm mountain gun GP is known in the West by the reporting designation M1966, but has been referred to as the M1969. The gun was first reported in 1966 and was used by Russian troops operating in Afghanistan.

Production of the 76 mm mountain gun GP (M1966) is now complete. Its potential replacement is the 122 mm assault gun (M-392) which is covered in a separate entry in *IHS Jane's Land Warfare Platforms: Artillery and Air Defence*. As of March 2013 it is understood that this still remained at the prototype stage.

Description

The 76 mm mountain gun GP has a split trail and a carriage arrangement similar to that used by the Italian Oto Melara Model 56 pack howitzer, including a split shield.

The short barrel has no muzzle brake and reciprocates in a square cross-section slipper. The breech is a horizontal sliding wedge opening to the right. Panoramic and direct fire optical sights are located to the left of the barrel. The split trails can fold sideways and the right-hand trail leg is provided with a metal dolly wheel for ease of handling with the trails folded.

The rubber-tyred wheels are on stub axles that can be raised and lowered to alter the height of the 76 mm mountain gun GP (M1966) in action and to level the carriage on rough ground. It is assumed that the gun can be disassembled into loads for pack transport in a manner similar to the Oto Melara Model 56 Pack Howitzer.

It has been reported that the gun can fire any type of ammunition used by the 76 mm divisional gun M1942 (ZIS-3) including HE/FRAG, HEAT-FS, HEAT, HVAP and AP-T. The HE/FRAG projectile is fired at a muzzle velocity of 600 m/s, while the HEAT projectile can penetrate 300 mm of conventional steel armour set at right angles at any range.

Maximum rate of fire is 15 rds/min with sustained fire being 100 rds/h. As the primary role of this system is fire support the effectiveness of the anti-tank projectiles, for example HEAT, against the latest armour systems is limited.

It is typically towed by a GAZ-66 (4 × 4) 2,000 kg light vehicle or the larger Ural-375D (6 × 6) 4,000 kg truck with the latter being able to carry additional ammunition due to its greater payload.

Specifications

	76 mm Mountain Gun GP (M1966)
Dimensions and weights	
Crew:	7
Length	
transport configuration:	4.8 m
Width	
transport configuration:	1.5 m
Height	
transport configuration:	1.4 m
Weight	
standard:	780 kg
combat:	780 kg
Mobility	
Carriage:	split trail
Firepower	
Armament:	1 × carriage mounted 76.2 mm rifled gun
Main weapon	
breech mechanism:	horizontal sliding block
muzzle brake:	none
maximum range:	11,500 m (est.)
Rate of fire	
sustained fire:	2 rds/min
rapid fire:	15 rds/min
Main armament traverse	
angle:	50°
Main armament elevation/depression	
armament front:	+65°/-5°
Survivability	
Armour	
shield:	yes

76 mm mountain gun GP (M1966) being towed by a GAZ-66 (4 × 4) truck

0500406

Status
Production complete. In service with the Russian Army in limited quantities. This weapon has not so far been offered on the export market.

Contractor
Former Soviet state factories.

Serbia

Yugoimport 155 mm converted gun M46/84

Development
The now Serbian 155 mm Gun M46/84 is a conversion of the Russian 130 mm Gun M-46 (or the Chinese North Industries Corporation (NORINCO) Type 59-1) to accommodate a new 155 mm 45-calibre barrel capable of firing Extended Range Full Bore – Base Bleed ammunition of the separate loading type (projectile and charge).

This conversion gives a range increase of 45 per cent over the original 130 mm M-46 when firing ERFB-BB ammunition. The efficiency of the 155 mm ERFB projectile is claimed to be about three times that of the older Russian-designed 130 mm OF-482 high-explosive projectile.

It is believed that this upgrade package was originally designed by the former Space Research Corporation headed by the late Dr Gerald Bull.

The converted gun M46/84 was also sold to Iraq, but it is understood Iraq did not carry out any upgrades on its existing fleet of weapons prior to the Gulf conflict of 1990/1991.

Description
The now Serbian M46/84 conversion involved replacing the existing 130 mm barrel with a new 155 mm barrel 45 calibres long. This is designed to make use of the original 130 mm sliding breech and cradle parts.

The barrel is fitted with a high-efficiency double-baffle muzzle brake of a new design. An updated equilibrator sealing system is introduced, involving a nitrogen pressure adjustment and an updated sealing system for the recuperator mechanism is also required.

According to Yugoimport the recoiling parts weigh 2,930 kg while the elevating parts weigh 3,860 kg.

Recoil length depends on elevation but is being quoted as 1.28 m with elevation from –2.5° to +20° and 0.775 m with elevation from +36° to +45° with the latter being the maximum elevation limit.

Rates of fire are being quoted as:
- Normal - 4 rds/min
- In five minutes - 15 rds
- In 10 minutes - 25 rds
- In 30 minutes - 50 rds.

The existing original 130 mm M-46 sights are also revised to accommodate the new 155 mm ballistics with the optical sight graticule being changed. A new range quadrant drum with respective range scales is mounted.

The conversion allows existing Russian 130 mm M-46 and the China North Industries Corporation (NORINCO) Type 59-1 users to fire standard 155 mm M107, ERFB and ERFB-BB ammunition to the ranges shown in the table below:

Variants
The converted gun could also be delivered in 152 mm calibre with this model being designated the cal. 152 M46/86. It uses the same system of propelling charges and Serbian and Montenegrin projectiles M84 and M84-GG.

155 mm gun M46/84 in travelling configuration with two wheeled limber attached 0500620

155 mm gun M46/84 in firing position, but note that the spades are not embedded 0500621

Maximum range with the M84 projectile is 27,500 m and with the M84-GG base bleed it is 34,000 m. The 152 mm model also has a 45-calibre barrel.

At the request of customers, both versions of the converted gun could be supplied for the use of ammunition with separate loading propelling charges.

Yugoimport NORA B-52 155 mm/52-calibre SP gun

Details of this weapon, based on a 8 × 8 truck chassis, are provided in a separate entry in *IHS Jane's Land Warfare Platforms: Artillery and Air Defence.*

While first prototype systems were integrated onto a locally developed FAP 2832 (8 × 8) cross-country chassis production systems are integrated onto the rear of a Russian KamAZ 63501 (8 × 8) cross-country chassis.

It is understood that the latest production version is called the K1 and features a fully protected cab at the front and the 155 mm/52-calibre weapon is installed in a protected gun compartment at the rear.

The 152 mm/52-calibre weapon normally fires over the rear arc and to provide a more stable firing platform two hydraulically operated stabilisers are lowered to the ground either side.

The latest K1 version is currently deployed by at least two countries, Bangladesh (18 systems) and Myanmar (30 systems).

Specifications

	155 mm Converted Gun M46/84
Dimensions and weights	
Crew:	9
Length	
transport configuration:	11.17 m
Width	
transport configuration:	2.4 m
Height	
transport configuration:	2.65 m
axis of fire:	1.38 m
Ground clearance	
overall:	0.35 m
Weight	
standard:	8,428 kg
combat:	7,680 kg
Mobility	
Carriage:	split trail with limber
Firepower	
Armament:	1 × carriage mounted 155 mm rifled gun
Main weapon	
length of barrel:	6.975 m
breech mechanism:	horizontal sliding block
muzzle brake:	double baffle
maximum range:	17,850 m (M107 HE projectile)
maximum range:	30,300 m (ERFB projectile)
maximum range:	39,000 m (ERFB-BB projectile)
Rate of fire	
sustained fire:	4 rds/min
Main armament traverse	
angle:	50°
Main armament elevation/depression	
armament front:	+45°/-2.5°
Survivability	
Armour	
shield:	yes

Projectiles						
Charge zone	**ERFB-BB**		**ERFB**		**M107**	
	M/V	**Range**	**M/V**	**Range**	**M/V**	**Range**
3	-	-	266 m/s	5,900 m	292 m/s	7,100 m
4	-	-	302 m/s	7,300 m	334 m/s	8,800 m
M4A2 5	-	-	360 m/s	9,400 m	390 m/s	10,200 m
6	-	-	458 m/s	12,500 m	469 m/s	12,200 m
7	-	-	542 m/s	15,200 m	555 m/s	14,500 m
M2 8	-	-	658 m/s	19,300 m	675 m/s	17,850 m
9	-	-	796 m/s	25,100 m	-	-
M11 10	897 m/s	38,600 m	897 m/s	30,300 m	-	-

Status

Development complete. Ready for production. As far as it is known, no production/upgrade weapons were made before the start of the Yugoslav civil war. This is still being marketed by Yugoimport - SDPR.

Contractor

Yugoimport - SDPR.

Yugoimport 152 mm gun-howitzer M84, M84B1 and M84B2

Description

The Serbian 152 mm gun-howitzer M84 is based on a similar carriage to the Russian 152 mm gun-howitzer D-20 and can fire D-20 ammunition. The barrel is much longer than the D-20 and has a semi-automatic vertical wedge-type breech mechanism.

The 152 mm ordnance of the M84 is 39.73-calibre and is mounted on a carriage similar to the Russian design. It is modified to improve ballistic performance and to enable easier and more rapid handling in the firing position.

The 152 mm gun-howitzer M84 features a hydropneumatic recoil system.

It should be noted that while the M84 has a deployed firing weight of 7,080 kg the M84B1/M84B2 has a deployed firing weight of 6,880 kg.

It is possible to fire at elevations of +45° to +63°. A counter-recoil system, consisting of a hydraulic brake, compensator and hydropneumatic recuperator, is located over the barrel and, when combined with the high-efficiency (approximately 50 per cent) multibaffle muzzle brake, the resultant recoil length is short even at high angles of elevation. The maximum rate of fire is 6 rds/min.

The welded-steel carriage has a shield and split box section trails fitted with castor wheels to assist in handling.

There are two hydraulic pumps provided to assist the nine-man crew in bringing the weapon from the travelling to the firing position. These set the weapon on a circular firing platform mounted on the forward part of the carriage and raise the wheels and suspension clear of the ground.

Sighting equipment includes a range quadrant, a direct fire ×5.5 optical sight, a ×3.7 panoramic sight, a gunner's quadrant, a collimator and a sight illuminating set.

The 152 mm gun-howitzer M84 takes between three and five minutes to be brought into and out of action. It is provided with a single and double line braking installation and a handbrake as well as a lighting system.

The 152 mm M84 can fire the complete range of Russian D-20 ammunition, including the OF-540 FRAG-HE, to a range of 17,190 m with a muzzle velocity of 647 m/s.

The M84 can also fire an M84 HE round weighing 43.56 kg, 7.677 kg of which is the cast TNT:RDX (50:50) explosive payload. When firing the M84 HE at a muzzle velocity of 810 m/s the maximum range is 24,160 m (minimum range 5,000 m).

The propellant charge system uses one full charge and six increments. The M84B1 and M84B2 versions of the artillery weapon can fire the Russian OF-540 projectiles with propellant charges.

Maximum range depends of the projectile/charge but includes:

- M84 HE projectile - 24,160 m (M84 weapon)
- M84 HE projectile - 24,000 m (M84B1 and M84B2)
- M84-GG base bleed projectile - 27,000 m.

It should be noted that the minimum range with the M84 HE projectile is 5,400 m.

The M84 B2 version is fitted with a pneumatic loader which is operational at all gun elevations, with the capacity of over 120 work cycles from one standard compressed air tank mounted on one of the trails.

The Serbian illuminating projectile is called the M88 and falls to the ground at a speed of 5 m/s. It gives a maximum intensity of 1.3 million candela for a maximum period of one minute. There is also an High-Explosive/Improved Conventional Munition (HE/ICM) projectile which carries 63 KB-2 bomblets with a maximum range of 22,500 m. Each bomblet is 40 mm in diameter, 85 mm long and weighs 250 g, 35 g of which is FH-5 explosive. The bomblets are designed to attack the vulnerable upper surfaces of armoured vehicles.

The M84-GG base bleed projectile has been developed, the performance of which has already been proven with long-barrel systems. Maximum range of the M84B1 and M84B2 weapons is 27,000 m.

152 mm gun-howitzer M84 in travelling configuration 0500623

152 mm gun-howitzer M84 in firing position 0500622

It is understood that the 152 mm M84 series is also referred to in Serbia as the Nona.

In original Yugoslav service this weapon was towed by the locally produced FAP 2026 BS/AV (6 × 6) cross-country truck that has also been used for a wide range of other roles.

When towing this artillery weapon the FAP 2026 BS/AV (6 × 6) cross-country truck can also carry the gun crew and a quantity of ready use ammunition (projectiles, charges and fuzes).

Variants

MB-160

This is used by the Bosnian Serbs (Minobaca 160 mm) and has the original 152 mm ordnance replaced by a heavy 160 mm mortar barrel without a muzzle break.

New 8 × 8 155 mm SPG

Following extensive trials over several years with a testbed of a 155 mm/45-calibre 8 × 8 Self-Propelled Gun (SPG), Yugoimport has built the prototype of the NORA-B 155 mm/52-calibre self-propelled gun based on an FAP 2832 (8 × 8) truck chassis. This has been produced in production quantities based on the Russian KamAZ 63501 (8 × 8) cross-country chassis for an export customer. This is understood to be Myanmar.

In 2011 Bangladesh ordered 18 of these systems. Additional details are provided in a separate entry in *IHS Jane's Land Warfare Platforms: Artillery and Air Defence.*

155 mm M03 (NORA K1)

This has been developed to the prototype stage and is an 8 × 8 protected chassis with the crew compartment at the front and the turret mounted 155 mm ordnance mounted at the rear.

Standard equipment includes an auxiliary power unit, ammunition handing system, powered elevation and traverse, on board navigation and fire-control system.

When deployed in the firing position two hydraulic stabilisers are lowered to the ground either side of the rear unit.

It is understood that the K1 is the production system and this is the version deployed by Bangladesh (18 systems) and Myanmar (30 systems).

The K1 version features a fully protected cab at the front and the 155 mm/52-calibre weapon is installed in a protected gun compartment at the rear.

The 152 mm/52-calibre weapon normally fires over the rear arc and to provide a more stable firing platform two hydraulically operated stabilisers are lowered to the ground either side.

Specifications

	152 mm Gun-Howitzer M84
Dimensions and weights	
Crew:	9
Length	
overall:	9.67 m (firing)
transport configuration:	11.21 m
Width	
overall:	5.73 m (firing)
transport configuration:	2.415 m
Height	
transport configuration:	2.16 m
axis of fire:	1.385 m
Ground clearance	
overall:	0.37 m
Weight	
combat:	7,080 kg
Mobility	
Carriage:	split trail

	152 mm Gun-Howitzer M84
Speed	
towed:	70 km/h
Tyres:	12.00 × 20
Firepower	
Armament:	1 × carriage mounted 152 mm howitzer
Main weapon	
length of barrel:	6.056 m
muzzle brake:	multibaffle
maximum range:	24,160 m
Rate of fire	
sustained fire:	6 rds/min
Main armament traverse	
angle:	50°
Main armament elevation/depression	
armament front:	+63°/-5°
Survivability	
Armour	
shield:	yes

Status
Production complete. No longer marketed. In service with:

Country	Quantity	Comment
Bosnia-Herzegovina	17	-
Croatia	23	-
Serbia/Montenegro	36	-

Contractor
Yugoimport - SDPR.

Yugoimport 122 mm howitzer D-30J

Development
The 122 mm howitzer D-30J is a Serbian version of the Russian 122 mm howitzer D-30 with minor differences. Full details of the Russian 122 mm D-30 are given in a separate entry in *IHS Jane's Land Warfare Platforms: Artillery and Air Defence*. The Serbian 122 mm howitzer D-30J has been upgraded in a number of key areas to meet local requirements.

The D-30J is fitted with a more simple double-baffle muzzle brake.

Description
Installation of a hydraulic firing jack (in place of the normal mechanical jack) is claimed to facilitate and speed up handling of the weapon in the firing position.

The 122 mm howitzer D-30J has a recoil of between 740 and 930 mm and the barrel length, without the muzzle brake is being quoted as 4,270 mm.

Maximum muzzle velocity is being quoted as 735 m/s and maximum powder charge pressure is being quoted as 2,600 bar.

The maximum range when firing a standard HE projectile with a full charge is given as 17,300 m. The D-30J can fire the full family of ammunition developed for the Russian D-30.

The former Yugoslavia had under development an M82 High-Explosive Anti-Tank (HEAT) projectile fitted with a UT PE M82 fuze and an M82 illuminating projectile fitted with an ETU 102 time fuze. The Serbian M76 projectile has a maximum range of 17,300 m. The M76 projectile with full charge has an m/v of 730 m/s, while the TF-462 projectile has variable charge from 276 to 565 m/s.

Iraq manufactured a version of the 122 mm D-30J under the local name of the Saddam. Production of this is complete and as far as it is known it was not exported.

The 122 mm howitzer D-30J is normally towed muzzle first by a 6 × 6 cross-country truck that in addition to its crew also carries a quantity of ready use ammunition (projectiles, charges and fuzes).

ANA 122 mm howitzer D-30 deployed in the firing position (ISAF) 1440738

122 mm D-30J deployed in firing position with trails staked to the ground 0500624

Variants

D-30 RH 94

RH ALAN of Croatia has built a further development of the D-30 called the 122 mm howitzer D-30 RH M94 which, firing an Extended Range Full Bore - Base Bleed projectile, has a maximum range of 20,050 m. Full details of this system are given in a separate entry in *IHS Jane's Land Warfare Platforms: Artillery and Air Defence*. Serbia has been offered the D-30J on the export market but as of March 2013 there are no known export sales.

122 mm D-30 howitzers for Afghanistan

In mid-2011 Bosnia and Herzegovina (BiH) delivered the first of a total of 60 refurbished 122 mm D-30 howitzers to the Afghan National Army.

Thirty 122 mm D-30 howitzers were taken from the forces of HiB Serb Republic and 30 from the Bosnia-Croat Federation of Bosnia and Herzegovina.

Prior to delivery by air to the Afghan National Army, the 60 122 mm D-30 howitzers were overhauled by the BiH companies BNT-TMiH of Novi Travnik and TRZ of Hadzici with this cost, and the cost of air transport, being paid by the US Government.

100 mm coastal gun

This uses a D-30J carriage and a 100 mm ordnance. It is used by Croatia and Serbia and Montenegro.

100 mm M87 TOPAZ anti-tank gun

This is a 122 mm D-30J carriage fitted with a 100 mm ordnance for use in the anti-tank role. It never entered production or service as far as it is known.

Serbian 122 mm D-30J self-propelled howitzer

Yugoimport has studied a wheeled self-propelled version of the Serbian 122 mm howitzer D-30J under the designation of the SORA (*Samohodno ORude Artiljerije*). As of March 2013 this is understood not to have entered production.

SOKO SP RR 122/105/122 mm

This is FAP 2228 (6 × 6) cross country chassis fitted with an armour protected cab at the front with the ordnance mounted in a protected turret at the rear of the chassis.

When deployed in the firing position two stabilisers are lowered at the rear of the chassis to provide a more stable firing platform.

Then turret can be fitted with a 100 mm, 105 mm or 122 mm ordnance with the latter being based on the widely deployed Russian designed D-30 towed howitzer or its Serbian equivalent.

Full details of this system are provided in a separate entry in *IHS Jane's Land Warfare Platforms: Artillery and Air Defence*.

Specifications

	122 mm Howitzer D-30J
Dimensions and weights	
Length	
transport configuration:	5.33 m
Width	
transport configuration:	1.95 m
Height	
transport configuration:	1.66 m
axis of fire:	0.9 m
Ground clearance	
overall:	0.325 m (est.)
Track	
vehicle:	1.855 m
Weight	
standard:	3,440 kg
combat:	3,350 kg
Mobility	
Speed	
towed:	60 km/h
Tyres:	9.00 × 20
Firepower	
Armament:	1 × carriage mounted 122 mm howitzer

	122 mm Howitzer D-30J
Main weapon	
length of barrel:	4.785 m
rifling:	3.4 m
breech mechanism:	vertical sliding block
muzzle brake:	double baffle
maximum range:	17,300 m (M76 HE projectile)
maximum range:	15,300 m (TF-462 HE projectile)
Rate of fire	
rapid fire:	8 rds/min
Operation	
stop to fire first round time:	80 s
Main armament traverse	
angle:	360°
Main armament elevation/depression	
armament front:	+70°/-7°
Survivability	
Armour	
shield:	yes

Status
Production complete. In service with the following countries:

Country	Quantity	Comment
Bosnia-Herzegovina	116	of these 60 were supplied to Afghanistan in 2011
Croatia	42	local production
Serbia/Montenegro	304	including 70 D-30

Contractor
Yugoimport - SDPR.

Yugoimport 105 mm howitzer M56

Development
In addition to using the US 105 mm M101 (M2A1) howitzer and the German 105 mm M18, M18M and M18/40, the former Yugoslavia also deployed a 105 mm howitzer of local design and manufacture called the M56.

According to the United Nations Arms Transfer Lists for the period 1992 through to 2010, the former Yugoslavia exported the following artillery systems during this period:

Customer	Quantity	Year	Comment
Bangladesh	108	1998	type not disclosed
Burma	36	2001	105 mm M2A1
Burma	18	2001	confirmed as 105 mm M56
Burma	36	2004	type not disclosed

Description
In many respects the M56 is similar to the German M18/40 but it fires semi-fixed ammunition of the US M1 High-Explosive pattern, rather than the separate loading type used by the German 105 mm howitzers.

The semi-fixed ammunition fired by the M56 consists of three types: HE shell M1, weighing 15 kg and filled with 2.2 kg of explosive, fired at a maximum muzzle velocity of 570 m/s and with a fuze that can be set for graze or 0.05 second delay; smoke shell WP M60 filled with white phosphorous and weighing 15.8 kg (this shell has the same fuze as the HE M1); the AP-T shell M67 High-Explosive Squash Head - Tracer (HESH-T) weighing 10 kg and fired at a muzzle velocity of 670 m/s to a maximum range of 12,000 m.

Close-up of rear of 105 mm M56 howitzer in service with El Salvador showing breech detail (Julio A Montes) 0105792

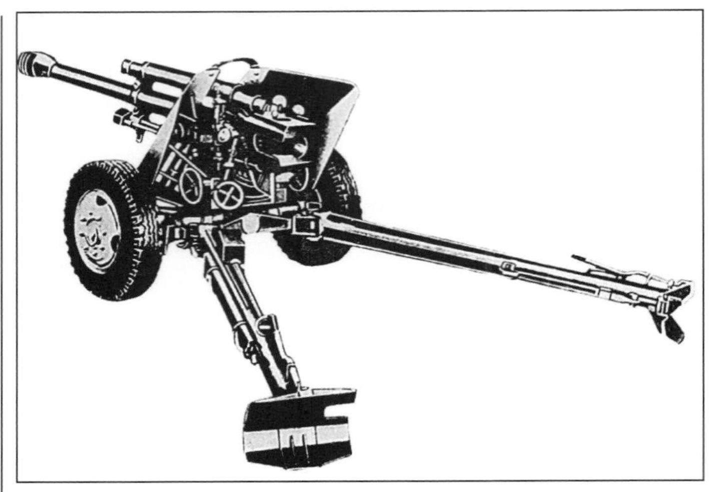

105 mm M56 howitzer in firing position. This model has US-type wheels with pneumatic tyres 0500625

In use the M67 HESH range is normally limited to 870 m. The 2.2 kg of explosive filling can penetrate a conventional steel armour up to 100 mm thick, set at an angle of 30°.

On such a plate, the HESH projectile will scab off a film of material opposite the strike of the projectile that is 25 mm thick, with the fragments weighing up to 5 kg and travelling at a velocity of 300 m/s. The 105 mm howitzer M56 may be fitted with a 20 mm subcalibre training device.

The 105 mm barrel is mounted on a tubular split trail carriage with a split shield sloping to the sides and rear. The most common model has pressed alloy wheels with solid rubber tyres, similar to those on the German M18 series weapons. However, some M56s are fitted with pneumatic-tyred wheels similar to those fitted to the US 105 mm M101 (M2A1). With the pneumatic wheels the M56 can be towed at speeds up to 70 km/h.

The recoil system of the Yugoimport 105 mm howitzer M56 consists of a hydraulic recoil buffer and hydropneumatic recuperator.

When originally deployed it was towed by a locally produced TAM 1500 (4 × 4) cross-country truck that in addition to its crew also carried a quantity of ready use ammunition (projectiles, charges and fuzes).

Normally the M56 is supplied with a range quadrant, panoramic telescope (dial sight), direct firing anti-tank telescope and a gunner's quadrant, all equipped with lights for night firing. The direct fire telescope has a magnification of ×2.8. The panoramic telescope has a magnification of ×4 and weighs 6 kg. All controls and sights are mounted on the left side of the carriage.

While the carriage of the M56 is normally used for firing with the trail legs spread, in an emergency it can be fired from the travelling position.

In this configuration the maximum elevation is limited to +16° and the total traverse to 16°.

Variants

105 mm howitzer M56 upgrade
This is currently being offered by Yugoimport - SDPR and details are given in a separate entry in *IHS Jane's Land Warfare Platforms: Artillery and Air Defence.* As of March 2013 this remained at the prototype stage.

Self-propelled 105 mm howitzer M56
This has been developed to the prototype stage but as of March 2013 no production orders had been announced.

105 mm armoured truck mounted howitzer
This is a 6 × 6 cross-country truck chassis with a protected crew compartment at the front and a turret with an open roof and back mounted on the rear of the platform.

105 mm M56 howitzer in service with Cyprus 0130011

This is armed with a locally manufactured 105 mm howitzer M56 for which a total of 60 rounds of 105 mm ammunition are carried. Maximum range firing a HE ER projectile is 15,000 m and when firing an HE ER-BB this increases to 18,000 m.

Gross vehicle weight is being quoted as 18 tonnes maximum road speed 90 km/h and a range of up to 450 km. It is normally operated by a crew of five.

As of March 2013 this 6 × 6 105 mm truck mounted howitzer system remained at the prototype stage.

Specifications

	105 mm Howitzer M56
Dimensions and weights	
Crew:	7
Length	
overall:	5.46 m (firing)
transport configuration:	6.17 m
Width	
transport configuration:	2.15 m
Height	
transport configuration:	1.56 m
axis of fire:	1.13 m
Ground clearance	
overall:	0.335 m
Track	
vehicle:	1.8 m
Weight	
standard:	2,100 kg
combat:	2,060 kg
Mobility	
Carriage:	split trail
Firepower	
Armament:	1 × carriage mounted 105 mm howitzer
Main weapon	
length of barrel:	3.48 m
breech mechanism:	horizontal sliding block
muzzle brake:	multibaffle
maximum range:	13,000 m
Rate of fire	
rapid fire:	16 rds/min
Main armament traverse	
angle:	52° (see Description)
Main armament elevation/depression	
armament front:	+68°/-12°
Survivability	
Armour	
shield:	yes

Status

Production complete. Still being marketed by Yugoimport - SDPR. In service with the following countries:

Country	Quantity	Comment
Bosnia-Herzegovina	101	
Burma	36	delivered 2001
Croatia	56	
Cyprus	72	still operational
El Salvador	18	reserve
Guatemala	56	
Indonesia	10	
Macedonia	14	
Mexico	40	army
	16	marines
Serbia/Montenegro	162	

Some sources have indicated that the Mexican guns are in fact Chinese versions of Italian 105 mm Pack Howitzer.

Contractor

Yugoimport - SDPR.

Yugoimport Upgraded 105 mm howitzer M56-2 (M56A1)

Development

The former Yugoslavia built around 800 105 mm M56 towed howitzers for the home and export markets with known earlier export users including Burma, Cyprus, El Salvador, Guatemala, Indonesia and Mexico.

Yugoimport of Serbia has completed development and testing of an upgrade package for the M56 under the designation of the M56A1 which increases the range of the weapon. This has also been referred to as the upgraded 105 mm howitzer M56-2.

Self-propelled model of the upgraded 105 mm M56 artillery system with stabilisers lowered (Yugoimport - SDPR) 0577518

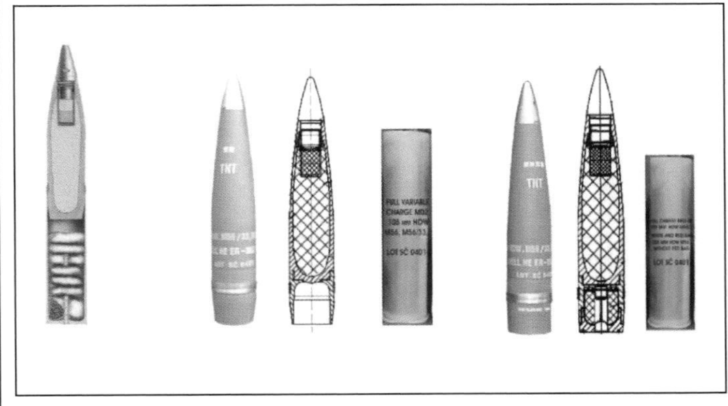

Increased range of the upgraded M56A1 howitzer is achieved by the installation of a new 105 mm/33-calibre barrel and new family of ammunition (Yugoimport) 1459341

According to Yugoimport, development of this upgrade package is complete and production can commence when orders are placed but as of March 2013 there are no known export customers.

This upgrade work could be carried out using kits provided by Yugoimport or the weapons could be returned to Serbia where the work could be carried out, weapons tested and then returned to the customer.

Description

The standard production 105 mm M56 howitzer is fitted with a 105 mm/28-calibre ordnance. The upgraded M56-2 model has a longer 105 mm/33-calibre barrel, which is also made of stronger steel and therefore allows 105 mm projectiles to be fired with a higher charge and therefore a longer range.

The original barrel assembly weighed 526 kg while the new 105 mm/33-calibre barrel weighs 616 kg with the barrel on its own weighing 410 kg. Other modifications include the installation of a new double-baffle muzzle brake while the existing horizontal sliding breech and breech block are replaced by identical ones made of stronger steel.

The recoil system is of the hydropneumatic type with the ordnance having 36 grooves with a uniform right hand twist angle. Chamber volume is 2.5 litres.

Typical rate of fire is up to 8 rds/minute with 16 rounds fired in four minutes and 30 rounds in 10 minutes.

In addition there have been minor modifications of the existing pneumatic equilibrator with the pressure being increased in order to compensate for the added weight of the new 105 mm barrel assembly. The standard split trails have been retained but additional mass has been fitted to each trail.

The new barrel has a claimed life of at least 18,000 Equivalent Full Charge (EFC) rounds.

Upgraded M56A1 can fire the standard 105 mm M1 High-Explosive (HE) projectile with a special new locally developed S1 charge to a maximum range of 14.5 km and minimum range of 2 km.

All of the ammunition fired by the 105 mm M56/M56A1 is of the semi-fixed type, e.g. projectile and cartridge case.

The HE projectile weighs 15 kg of which 2.2 kg is the Composition B TNT explosive with maximum muzzle velocity being 530 m/s.

A new generation of 105 mm ammunition has been developed which is not only more effective but also increases the range of the weapon.

The High-Explosive Boat Tail (HE-BT) gives a maximum muzzle velocity of 680 m/s and increased maximum range to 15 km with minimum range being 4.6 km.

The projectile weighs 13.1 kg of which 2.85 kg is the Composition B TNT explosive with a maximum m/v of 680 m/s.

The High-Explosive Extended-Range (HE-ER) projectile is fitted with a Base Bleed (BB) unit which give a maximum muzzle velocity of 675 m/s and range increased to 18 km with minimum range being 7 km.

Projectile weighs 18.5 kg of which 2.85 kg is the Composition B TNT explosive with maximum muzzle velocity being 680 m/s.

Upgraded 105 mm M56A1 howitzer deployed in the firing position with spades well dug in and muzzle velocity radar installed to the left of the weapon for trials purposes (Yugoimport) 1459340

The weapon is normally operated by a crew of 8 and is towed by a 4 × 4 or 6 × 6 truck that also carries a quantity of ready use ammunition.

Variants

Complete 105 mm artillery system

As well as marketing the upgraded 105 mm M56 howitzer and its family of enhanced ammunition, Yugoimport is also marketing the Artillery Fire Command and Control System (AFCCS) as part of a complete artillery system package, e.g. weapon, ammunition and fire control system.

This includes elements at the regiment command post as well as at the battery and forward observation post with each weapon in the typical battery of six weapons being provided with a gun display unit.

105 mm (6 × 6) truck mounted howitzer

This has been followed by a similar concept based on a forward control 6 × 6 cross-country chassis that has increased cross-country mobility and payload.

This also has the complete upper part of the upgraded 105 mm M56 howitzer mounted on the rear which can be rapidly covered by the rail mounted tarpaulin system.

This system is designated the 105 mm/33-calibre M56A2 TM/TMP and to provide a more stable firing platform two hydraulically operated stabilisers are lowered to the ground at the rear.

105 mm (6 × 6) Crossbow truck mounted howitzer

Yugoimport is also studying a lower profile 6 × 6 105 mm SP artillery system called Crossbow for the export market which would have a combat weight of around 15 tonnes.

On the rear would be fitted a new turntable mounted remote controlled 105 mm/60-calibre weapon with elevation limits from −10° to +66°.

This would have a fully automatic gun laying system coupled to a computerised fire-control system and an automatic ammunition loading system.

According to Yugoimport this would have a maximum range firing the HE-ER-BB round of around 30 km which a maximum rate of fire of up to 12 rounds a minute.

To provide a more stable firing platform two hydraulically operated stabilisers are mounted two either side at the rear.

This would be able to come into action in just 60 seconds and come out of action in around 45 seconds which makes it highly survivable against counter battery fire.

105 mm M56-2 self-propelled howitzer

Yugoimport has also developed a new Self-Propelled (SP) model of the upgraded M56 towed howitzer mounted on the rear of a modified locally produced FAP 1117BS/AV (4 × 4) forward cross-country truck chassis.

To the immediate rear of the fully-enclosed cab is a supply of ready use 105 mm ammunition and mounted on the flatbed at the very rear of the chassis is the complete upper part of the upgraded M56 weapon.

When travelling the weapon is depressed, traversed to the front and covered by a sliding bows and tarpaulin cover that is normally retracted to the cab rear.

This makes the weapon system very difficult to detect from a standard cross-country vehicle. Mounted on the roof of the cab is a 12.7 mm machine gun for local and air defence purposes.

To provide a more stable firing platform two hydraulically-operated stabilisers are lowered to the ground either side before the weapon is fired. Initial firing trials of this configuration have been completed but as of March 2013 it is understood that this remains at the prototype stage.

In concept and appearance this is very similar to the Dutch RDM Technology and Defense Systems MOBAT 105 mm SP artillery system which was developed and tested but never entered production. RDM Technology and Defense Systems is no longer trading.

SOKO SP RR 122/105/122 mm

This is FAP 2228 (6 × 6) cross country chassis fitted with an armour protected cab at the front with the ordnance mounted in a protected turret at the rear of the chassis.

The projected future Crossbow 6 × 6 self-propelled model would feature a 105 mm/60-calibre barrel which firing a HE-ER-BB projectile would enable a range of around 30 km to be achieved (Yugoimport) 1459907

When deployed in the firing position two stabilisers are lowered at the rear of the chassis to provide a more stable firing platform.

Then turret can be fitted with a 100 mm, 105 mm or 122 mm ordnance with the latter being based on the widely deployed Russian designed D-30 towed howitzer or its Serbian equivalent.

105 mm armoured truck mounted howitzer

This is a 6 × 6 cross-country truck chassis with a protected crew compartment at the front and a turret with an open roof and back mounted on the rear of the platform.

This is armed with a locally manufactured 105 mm howitzer M56 for which a total of 60 rounds of 105 mm ammunition are carried. Maximum range firing an HE-ER projectile is 15,000 m and when firing an HE ER-BB this increases to 18,000 m.

Gross vehicle weight is being quoted as 18 tonnes , with a maximum road speed of 90 km/h and a range of up to 450 km. It is normally operated by a crew of five.

Specifications

	105 mm Howitzer M56-2
Dimensions and weights	
Crew:	8
Length	
overall:	6.77 m
Width	
transport configuration:	2.15 m
Height	
overall:	1.56 m
axis of fire:	1.107 m
Ground clearance	
overall:	0.335 m
Track	
vehicle:	1.8 m
Weight	
standard:	2,220 kg
Mobility	
Carriage:	split trail
Speed	
towed:	70 km/h
Firepower	
Armament:	1 × carriage mounted 105 mm howitzer
Main weapon	
length of barrel:	3.5 m
rifling:	3.119 m
breech mechanism:	horizontal sliding block
muzzle brake:	double baffle

	105 mm Howitzer M56-2
maximum range:	11,800 m (standard ammunition)
maximum range:	18,500 m (new ammunition)
Rate of fire	
sustained fire:	6 rds/min
rapid fire:	8 rds/min
Main armament traverse	
angle:	52° (open trails; 16° closed trails)
Main armament elevation/depression	
armament front:	+65°/-9° (spread trails; +16°/-12° closed trails)
Survivability	
Armour	
shield:	yes

Status
Development complete. Ready for production.

Contractor
Yugoimport - SDPR.

Yugoimport Upgraded 105 mm M101 howitzer

Development
Yugoimport - SDPR is now marketing a number of artillery upgrades, including one for the US-developed 105 mm M101 howitzer.

The main objectives of this upgrade are to extend the range and firepower of the M101 towed howitzer, whose original design can be traced back to well before the Second World War.

The company has developed, tested and qualified the upgraded version of the 105 mm M101 US-developed towed howitzer, which is now being offered on the export market.

As of March 2013 there are no known sales of this upgrade package by Yugoimport.

The M101 weapon can be upgraded in the facilities of Yugoimport or the company can provide kits, which enable the user to upgrade the weapons at their own facility. The total weight of the upgrade kit is stated to be only 233 kg.

According to the United Nations Arms Transfer Lists for the period 1992 through to 2010, the former Yugoslavia exported the following artillery systems during this period:

Customer	Quantity	Date	Comments
Bangladesh	108	1998	type not disclosed
Burma	36	2001	105 mm M2A1
Burma	18	2001	confirmed as 105 mm M56
Burma	36	2004	type not disclosed

Notes:
At this stage it is not known as to whether 36 105 mm M2A1 (or M101 as it is also referred to) weapons supplied to Burma were upgraded prior to delivery

Description
The M101 105 mm upgrade includes replacing the current 105 mm/23-calibre barrel with a new 105 mm/30-calibre barrel, which is capable of withstanding higher pressures.

The new barrel is fitted with a new double-baffle muzzle brake. This is claimed to have an efficiency of 40 per cent.

According to Yugoimport the barrel of this upgraded M101 howitzer weighs 450 kg.

Upgraded 105 mm M101 howitzer deployed in firing position (Yugoimport - SDPR) 1044606

The existing breech mechanism, including breech ring and breech block, is replaced by an identical one that has stronger steel, which can withstand the pressures of firing extended-range projectiles.

The existing equilibrator spring has been replaced by a new one that will compensate for the added weight of the new barrel assembly. Finally, additional mass is added to the trails.

The upgraded 105 mm M101 howitzer can fire all existing types of 105 mm ammunition as well as the more recent locally developed 105 mm High-Explosive Extended Range - Base Bleed (HE-ER-BB), which enables a maximum range of 18.5 km to be achieved.

Additional details of the 105 mm ammunition fired by this upgraded 105 mm howitzer M56-2 are given in the table.

Type	M1	HE-BT	HE-ER-BB
Weight:	15 kg	13.1 kg	13 kg
Muzzle velocity:	502 m/s	675 m/s	685 m/s
Maximum range:	11,800 m	15,000 m	18,500 m

The upgraded 105 mm M101 howitzer is normally towed by a 6 × 6 or 4 × 4 cross-country truck that in addition to the crew also carries a quantity of ready use ammunition.

Specifications

	Upgraded 105 mm M101 howitzer
Dimensions and weights	
Crew:	8
Length	
overall:	7 m
Width	
transport configuration:	2.146 m
Height	
overall:	1.56 m
axis of fire:	0.97 m
Weight	
standard:	2,263 kg
Firepower	
Armament:	1 × carriage mounted 105 mm howitzer
Main weapon	
length of barrel:	3.175 m
rifling:	2.789 m
muzzle brake:	double baffle
maximum range:	11,800 m (M1 HE projectile)
maximum range:	15,000 m (HE boat tail projectile)
maximum range:	18,500 m (HE ERFB-BB)
Rate of fire	
rapid fire:	8 rds/min
Main armament traverse	
angle:	46°
Main armament elevation/depression	
armament front:	+66°/-5°
Survivability	
Armour	
shield:	yes

Status
Development complete. Ready for production.

Contractor
Yugoimport - SDPR.

Yugoimport 76 mm mountain gun M48

Development
The 76 mm mountain gun M48, which is often called the 'Tito Gun', was developed after the Second World War specifically to meet the requirements of Yugoslav mountain units, although it can also be used as a field gun.

A recent list of weapons currently being marketed by Yugoimport still listed the 76 mm mountain gun M48. At this stage it is not clear as to whether these are surplus weapons or new production weapons.

It is considered probable that these would be surplus weapons and not new build weapons.

It is possible that quantities of these weapons have been exported but as they have a calibre of 76.2 mm they do not fall under the United Nations Arms transfer lists.

Description
There have been at least four variants of the M48. The M48 (B-1) has pneumatic tyres and a maximum towing speed of 60 km/h. It can also be towed by animals in tandem or disassembled into eight pack loads.

76 mm Mountain Gun M48 with rectangular cross-section trail legs 0500627

The recoil system of the 76 mm mountain gun M48 is of the hydraulic type.

The M48 (B-1A1-I) has the pneumatic tyres and wheels as fitted to the M48 (B-1), plus some of the features of the suspension of the M48 (B-1A2). The M48 (B-1A2) can also be used as a field piece but cannot be towed by animals or disassembled for pack transport.

The M48 (B-1A2) has light alloy wheels with solid rubber tyres and modified suspension, its maximum towing speed is 30 km/h. The final production model of the 76 mm mountain gun M48 was called the B-2 about which little is known at present.

Ammunition is of the semi-fixed type with four charges. It is based on that used for the obsolete Russian 76 mm Regimental Gun M1927 which fired fixed ammunition: HE M55 projectile weighing 6.2 kg with a muzzle velocity between 222 and 398 m/s; High-Explosive Anti-Tank (HEAT) projectile weighing 5.1 kg which will penetrate 100 mm of conventional steel armour at a range of 450 m; and a smoke shell WP M60 weighing 6.2 kg.

The 76 mm mountain gun M48 is related to the Romanian 76 mm mountain howitzer M1984 (or Model 1984 as it is also referred to).

According to the Romanian manufacturer, the 76 mm mountain howitzer fires the following ammunition types as listed below. It should be noted that some of these with the M series designation are also used by the Yugoslav 76 mm mountain gun M48:

- Charge round – HE OF-350M, fuze UP3
- Charge round – M55, fuze M51A5
- Charge round – smoke WP M60, fuze PD M48A3
- Charge round – smoke D-350 TM, fuze UP3
- Charge round – HEAT BK 354, piezoelectric fuze CVP-312
- Charge round – shell M50, fuze PD MK-451.

All models of the M48 have split trails with each trail hinged in the centre and folding forwards through 180° for towing. At the end of each trail is a slot for the stake. The shield has winged sides and the centre of the shield has to be raised to obtain maximum elevation.

Normal rate of fire is typically 25 rds/minute with 40 rounds being fired in three minutes, 50 rounds in five minutes, 70 rounds in 15 minutes and 90 rounds in 30 minutes.

When originally deployed by the former Yugoslavia it was normally towed by a Zastava (4 × 4) vehicle but at least two of these would be required to carry the complete gun crew and adequate ammunition.

While maximum towing speed on a good road has been quoted as up to 60 km/h, maximum towing speed on a lower quality road has been quoted as 40 km/h and off road as low as 15 km/h, but this does depend on terrain conditions.

Variants

In 2004 Yugoimport was marketing the 76 mm mountain gun M48 B1 A5, which has an almost identical specification to that given below apart from the fact that it has a crew of seven.

As of March 2013, it is understood that there had been no known export sales of this version by Yugoimport.

Specifications

	M48 B-1
Dimensions and weights	
Crew:	6
Length	
transport configuration:	2.42 m
Width	
overall:	2.65 m (firing)
transport configuration:	1.46 m
Height	
transport configuration:	1.22 m
axis of fire:	0.73 m
Ground clearance	
overall:	0.18 m
Track	
vehicle:	1.3 m
Weight	
standard:	720 kg
combat:	705 kg
Mobility	
Carriage:	split trail (folding)
Speed	
towed:	60 km/h
Tyres:	6.00 × 16
Firepower	
Armament:	1 × carriage mounted 76.2 mm howitzer
Main weapon	
length of barrel:	1.178 m
muzzle brake:	multibaffle
maximum range:	8,750 m
maximum range:	8,600 m (M55 HE projectile)
Rate of fire	
sustained fire:	25 rds/min
Main armament traverse	
angle:	50°
Main armament elevation/depression	
armament front:	+45°/-15°
Survivability	
Armour	
shield:	yes

Status

Production complete. This is still being marketed by Yugoimport SDPR. In service with the following countries:

Country	Quantity	Comment
Bosnia-Herzegovina	n/avail	unconfirmed user
Croatia	57	unconfirmed user
India	215	
Indonesia	50	
Macedonia	55	
Myanmar	100	
Serbia/Montenegro	97	
Sri Lanka	14	

Contractor

Yugoimport - SDPR.

Singapore

Singapore Technologies Kinetics 155 mm/52-calibre FH2000 artillery system

Development

The Singapore 155 mm FH2000 project was first conceived in January 1990 with the first prototype being completed the following year, the prime contractor being Ordnance Development & Engineering (ODE), a company of Chartered Industries of Singapore. Early in 2000 this company became part of Singapore Technologies Kinetics.

This was followed by a pre-production example with a number of improvements. Final acceptance tests were completed in December 1993.

By mid-1995, a total of 18 155 mm FH2000 towed artillery systems had been deployed with the 23rd Battalion Singapore Artillery (23SA) in three batteries each of six guns.

The 155 mm FH2000 towed artillery system is a further development of the Singapore Technologies Kinetics 155 mm FH-88 gun howitzer which has a 39-calibre ordnance and entered service in 1987.

155 mm/52-calibre FH2000 gun-howitzer deployed in the firing position 0021019

155 mm/52-calibre FH2000 gun-howitzer in self-propelled configuration
0021020

Some key parts of the earlier FH-88 are used in the FH2000 including the sighting system and the 155 mm ammunition.

The 155 mm/52-calibre FH2000 artillery system is now being offered for export. As of March 2013 there are no known exports of the FH2000 artillery system.

Description

The most significant change compared to the earlier Singapore Technologies Kinetics 155 mm FH-88 is the installation of the longer 155 mm/52-calibre ordnance with a 23-litre chamber which conforms to the NATO Joint Ballistics Memorandum of Understanding.

The 155 mm/52-calibre ordnance is autofrettaged for durability and has a welded double-baffle muzzle brake.

The interrupted screw breech mechanism is semi-automatic with the breech opening automatically during counter-recoil. To allow space for a future automatic loading device and to enable easy charge loading, the breech opens upwards. The breech can also be operated remotely in the case of a misfire.

The primer feeding mechanism has been relocated on the cradle for built-in reliability. A new recoil system has been incorporated to tap the recoil energy for operation of the flick rammer, breech opening, primer feeding and thermal compensating equilibrators.

The FH2000 is fitted with a temperature indicator on the gun barrel that warns the gun crew of the temperature condition of the barrel before and during firing. This avoids cook-offs.

The electronically controlled and hydraulically operated flick rammer has a safety device to ensure the safe loading of 155 mm projectiles into the ordnance.

For example, if there is no 155 mm projectile in the ramming tray, it will not ram. If a projectile has been loaded into the chamber but not fired, it will not ram another round into the barrel.

The electronically controlled rammer has four modes of operation: automatic, automatic back up, electric manual and hydraulic manual. Therefore, if there is a failure, one of these alternatives is available to ensure successful accomplishment of the mission.

Traverse, elevation and firing of the 155 mm/52-calibre ordnance are purely mechanical. As a result, the weapon can still function when APU or hydraulic lines are damaged.

The four-wheeled split trail carriage is provided with an APU, which is mounted on the forward part of the carriage. Each of the two trail legs has a retractable dolly wheel to assist in bringing the weapon into action. When in the travelling configuration, the 155 mm/52-calibre ordnance is swung through 180° and secured by the barrel clamps on the trail legs.

The APU consists of a 75 hp air-cooled turbo-charged diesel engine with a hydrostatic drive system, which, in addition to providing the howitzer with a self-propelled capability, also powers the firing platform and the trail wheels. Steering is through the trail wheels.

If required the APU can be activated from the cab of a towing vehicle to provide additional traction in difficult terrain.

Maximum speeds of the FH-2000 depend on as to its configuration, for example in the towed configuration or in the APU configuration. These speeds are:

- Towed configuration on paved roads - 80 km/h
- Towed configuration on unpaved roads - 50 km/h
- Self-propelled mode with APU on paved roads - 16 km/h
- Self-propelled mode with APU cross-country - 8 km/h.

It should be noted that the above speeds depend on a number of factors including terrain conditions.

The weapon can be bought into action in two minutes in the following sequence:

- The weapon is unlocked from the tractor (normally a 6 × 6 cross-country chassis) by lowering the trail wheels to the ground hydraulically which in turn lifts the trail end
- The trails are opened hydraulically
- Trails are lowered to the ground
- Spades are self-embedded
- The circular firing platform is lowered beneath the front end of the carriage.

The direction of fire can be changed in less than two minutes by raising the firing platform, lowering and steering the trail wheels, executing a pivot turn, steering and raising the trail wheels, embedding the spades and lowering the firing platform.

When deployed in the firing position the FH2000 rests on the firing platform under the carriage and the two trails, each of which is provided with a spade.

For improved reliability and for superior maintainability, all electrical relays in the APU have been replaced by a Programmable Logic Control (PLC) system.

The standard sighting system comprises a cant compensating mount, a peri-telescopic sight, a display unit and a direct aiming sight. The exact configuration selected can be tailored to meet particular battlefield conditions and to match contemporary military needs.

The technology of the sighting system ranges from a simple optical sight to a sophisticated electronic sight.

The display unit provides a visual aid for both layer and commander to fine-tune elevation and traverse (azimuth) settings for pinpoint accuracy. It is connected to the sighting system and receives firing data from the fire-control computer.

An onboard independent battery unit acts as a back-up system to ensure mission performance. Alternatively, the hydraulic system can be operated by a hand pump, which gives the advantage of silent operation at night.

When designing the FH2000, the manufacturer incorporated a number of features for ease of maintenance and increased durability, especially when operating in a jungle environment.

All cables, wires and pipes are hidden away in the two trail legs while the batteries which power the gun's engine and electronic systems are housed in one of the two trail legs.

The number of parts has also been reduced where possible. The driver's instrument panel is made up of three modules so that the faulty module can be quickly replaced.

It has a burst rate of fire of three 155 mm rounds in 20 seconds with a sustained rate of fire of six rounds a minute and a maximum of six rounds a minute for three minutes.

In addition to firing the standard US developed 155 mm M107 HE projectile and the locally manufactured 155 mm Extended Range Full Bore - Hollow Base (ERFB-HB) and Extended Range Full Bore - Base Bleed (ERFB-BB) projectiles, Singapore has developed a new 155 mm cargo for the Singapore Army.

Two versions of this have been developed, High Explosive - Base Bleed (HE-BB) and High Explosive - Hollow Base (HE-HB). Both of these carry dual purpose bomblets that have been designed to attack the vulnerable upper surfaces of armoured vehicles. Both carry 64 bomblets that are manufactured in Singapore and are fitted with a self-destruct fuze mechanism.

It is normally towed by a 6 × 6 cross-country truck that in addition to its crew carries a quantity of ready use ammunition (projectiles, charges and fuzes).

Specifications

	155 mm/52-calibre FH2000 Artillery System
Dimensions and weights	
Crew:	6
Length	
overall:	12.90 m (firing)
transport configuration:	11.05 m
Width	
overall:	9.726 m (firing)
transport configuration:	2.80 m
Height	
overall:	3.88 m (with display unit; 2.286 m without display unit)
Ground clearance	
overall:	0.35 m
Weight	
standard:	13,200 kg
Mobility	
Carriage:	split trail
Speed	
max speed:	16 km/h
cross-country:	8 km/h
towed:	80 km/h
Fording	
without preparation:	0.76 m
Gradient:	45%
Firepower	
Armament:	1 × carriage mounted 155 mm howitzer
Main weapon	
breech mechanism:	interrupted screw
muzzle brake:	double baffle
maximum range:	19,000 m (M107 HE projectile)
maximum range:	30,000 m (ERFB-HB projectile)
maximum range:	40,000 m (ERFB-BB projectile)
Rate of fire	
sustained fire:	2 rds/min
Main armament traverse	
angle:	360° (on carriage; 30° left/30° right on carriage)
Main armament elevation/depression	
armament front:	+70°/-3°
Survivability	
Armour	
shield:	no

Status

Production as required. In service with Singapore (18). As of March 2013 there were no export sales of the FH2000 artillery system, although marketing is still being carried out.

Contractor

Singapore Technologies Kinetics (STK).

Singapore Technologies Kinetics 155 mm/39-calibre Pegasus Light Weight Howitzer (LWH)

Development

The 155 mm/39-calibre Pegasus Light Weight Howitzer (LWH) was developed by Singapore Technologies Kinetics (STK), Singapore Defence Science and Technology Agency and the Singapore Armed Forces (SAF).

The Pegasus LWH was first revealed in late 2005 by which time it was fully operational with the SAF. It is understood that the SAF has formed at least two regiments equipped with this weapon.

This is the replacement for the French Nexter Systems 105 mm LG1 light guns which have now been withdrawn from service.

A total of 37 105 mm LG1 light guns were delivered by the now Nexter Systems (at that time Giat Industries) between 1992 and 1994. It is understood that these are now up for sale.

This 155 mm/39-calibre artillery system has been marketed overseas but as of March 2013 there are no known sales.

Description

To reduce weight, the structure of the 155 mm/39-calibre LWH is made of lightweight materials, such as titanium and high alloy aluminium. It is claimed that an innovative recoil management design serves to reduce recoil by one-third.

Through a simple see-saw action that shifts the gun's centre of gravity to suit different missions, it is claimed that the Pegasus LWH can be deployed with a crew of eight people in less that 2.5 minutes.

In addition to providing power to enable the Pegasus LWH to act as an autonomous system over short distances, the Auxiliary Power Unit (APU) also powers the ammunition loading system, which reduces crew fatigue during sustained firing missions.

The hydrostatic drive and in hub radial piston motor power the road wheels of the 155 mm Pegasus LWH. The brakes are disc and steering is power assisted.

It is claimed that a burst of three rounds can be fired in 24 seconds, with a maximum rate of fire of four rounds per minute for three minutes and two rounds per minute for 30 minutes.

All types of 155 mm artillery projectile and associated charges can be fired by the 155 mm Pegasus LWH. The Pegasus 155 mm/39-calibre LWH weighs 5.4 tonnes and is fitted with an APU that is fitted on one of its trail arms.

The APU is a 21 kW Lombardini 9LD625-2 diesel engine that provides power to the two main roadwheels located at the front of the carriage, with steering via the two smaller wheels at the rear. Using the APU, a maximum road speed of 12 km/h is claimed.

Mounted on the left side of the weapon is a conventional mechanical sight that is claimed to be able to withstand firing shocks of up to 90 g.

The system can be carried internally of a Lockheed Martin C-130 Hercules transport aircraft and by the CH-47D Chinook helicopter, both of which are used by the SAF.

The 155 mm/39-calibre Pegasus LWH is normally towed by a 5- or 7-tonne 6 × 6 cross-country truck that also carries the crew and a quantity of ready use ammunition.

Singapore Technologies Kinetics Pegasus 155 mm/39-calibre Light Weight Howitzer (LWH) deployed in the firing position
(Singapore Technologies Kinetics) 1116793

Singapore Technologies Kinetics Pegasus 155 mm/39-calibre Light Weight Howitzer (LWH) deployed in the firing position from the front
(Singapore Technologies Kinetics) 1040793

Maximum towing speed on roads is being quoted as 70 km/h, maximum towing speed cross-country is being quoted as 50 km/h and maximum speed when being used in the APU mode is being quoted as 12 km/h but these figures do depend on terrain conditions.

Turning radius is as follows:

- Towing (kerb to kerb) - 21 m
- Towing (wall to wall) - 25 m
- Self-propelled (kerb to kerb) - 21 m
- Self-propelled (wall to wall) - 22 m.

Specifications

	155 mm/39-calibre Pegasus Light Weight Howitzer (LWH)
Dimensions and weights	
Crew:	8
Weight	
standard:	5,400 kg
Mobility	
Gradient:	30%
Turning radius:	21 m
Auxiliary engine:	air-cooled diesel engine developing 28 hp
Brakes	
main:	disc on both wheels
Suspension:	hydropneumatic
Firepower	
Armament:	1 × carriage mounted 155 mm howitzer
Main weapon	
muzzle brake:	double baffle
maximum range:	19,000 m (M107 HE projectile)
maximum range:	24,000 m (ERFB-HB projectile)
maximum range:	30,000 m (ERFB-BB projectile)
Rate of fire	
sustained fire:	2 rds/min
rapid fire:	4 rds/min
Survivability	
Armour	
shield:	no

Status

Production complete. Can be resumed if additional orders are placed. In service with the Singapore Armed Forces.

Contractor

Singapore Technologies Kinetics (STK).

Singapore Technologies Kinetics 155 mm/39-calibre FH-88 gun-howitzer

Development

Ordnance Development and Engineering of Singapore (ODE), a company of Chartered Industries of Singapore, developed a 155 mm gun-howitzer with a 39-calibre barrel known as the FH-88. Early in 2000 the company became part of Singapore Technologies Kinetics.

There were five FH-88 prototypes produced over a period of four years beginning in 1983. These were followed by a pre-production batch of six 155 mm/39-calibre FH-88 gun-howitzers that incorporated a number of improvements as a result of trials with the prototype weapons.

First production FH-88s were completed in 1987, with the weapon becoming operational with the Singapore Army in 1988. It is understood that a total of 52 155 mm 155 mm/39-calibre FH-88 gun-howitzers were delivered.

These replaced Israeli older Soltam Systems 155 mm M71S (with the S standing for Singapore, as they have been modified) systems which have been placed in reserve. Indonesia took delivery of five 155 mm FH-88 systems in 1997. It is understood that these were brand new weapons.

The 155 mm/39-calibre FH-88 has been followed in production by the Singapore Technologies Kinetics 155 mm/52 calibre FH2000 which is covered in a separate in *IHS Jane's Land Warfare Platforms: Artillery and Air Defence.*

The more recent FH-2000 artillery system uses some components of the FH-88 including the sighting system, APU and the complete family of 155 mm ammunition.

More recently Singapore Technologies Kinetics has developed the new 155 mm/39 calibre Pegasus Light Weight Howitzer (LWH) which is already in service with the Singapore Armed Forces (SAF). Details of this system are provided in a separate entry in *IHS Jane's Land Warfare Platforms: Artillery and Air Defence.*

The Pegasus LWH has been offered on the export market but as of March 2013 there were no known sales.

Description

The monobloc 155 mm barrel is 39 calibres long and is autofrettaged. A double-baffle muzzle brake is fitted. The breech is of the hinged, interrupted screw type and automatically opens during the latter half of the return stroke of the recoil movement.

The camplate lifting mechanism incorporates a safety lock to prevent accidental closing of the breech screw and the firing mechanism cannot be activated if the breech screw is not fully closed.

Obturation is achieved using a resilient pad. The firing mechanism is a mechanical 'trip action' type which is actuated either by pulling the lanyard or operating the firing handle at the layer's seat. Primer feeding and extraction is automatic using a 12-round magazine. A flick rammer is used with power derived from an independent power pack.

The hydropneumatic recoil mechanism is of the independent-type single-recoil system. The recoil mechanism, connected to the recoiling and static parts of the gun, consists of a recoil brake and a recuperator/counter-recoil mechanism. The governor rod in the recoil brake varies the orifice for hydraulic oil to flow throughout the recoiling stroke, absorbs and ensures minimum and uniform recoil on the carriage. The pressurised nitrogen gas in the recuperator/counter-recoil mechanism is to return the barrel after recoil and also holds the barrel in its fully run-out position at all elevations.

There are two independent pneumatic equilibrators that balance the weight of the elevated mass throughout the full range of elevation. The equilibrator is made up of a cylinder, piston rod and a receiver using nitrogen gas as the working medium.

The cradle, which is made of high-strength alloy steel, supports and guides the recoiling parts. It houses the recoil mechanism and provides attachments for the equilibrators, camplate arm, elevating arc, flick rammer and the sighting system.

The saddle, which is made of high-strength alloy steel, comprises two side frames welded to a circular baseplate. It supports the elevating mass through the trunnion bearings and houses the elevating and traversing mechanism. It also provides attachments for the equilibrators and sighting equipment.

The elevating mechanism, with the aid of the equilibrators, enables the 155 mm/39-calibre ordnance to be elevated or depressed with accuracy and ease. The traversing mechanism, mounted in front of the saddle, enables the ordnance to be swivelled horizontally.

The flick rammer is an electronically controlled hydraulic system, which in turn operates the mechanical system. The latter basically consists of the swivel arm and the ramming unit. The swivel arm, with the loading tray attached, is pivoted at the rear end of the cradle.

155 mm FH-88 gun-howitzer in travelling configuration with ordnance in travel lock over closed spades 0500609

155 mm FH-88 gun-howitzer deployed in the firing position 0501046

The chassis serves as the base frame to join the upper carriage to the lower carriage. During firing it transmits all of the load on the upper carriage to the ground through the firing platform and trail legs.

The chassis is a welded structure of high-strength steel on a box section design, with provision for the mounting of the main bearing, rocker beams, firing platform, trail legs and the APU.

The firing platform is connected to the chassis assembly through the pad arrangements and lifts the weapon off the ground for firing. It is operated by a telescopic cylinder, which is completely isolated from all firing loads when the platform is fully deployed. An indicator is incorporated to show when the platform is fully deployed.

The trails are made of high-strength steel and, together with the firing platform, serve as the main support for the weapon. In the closed position they serve as a base for locking the barrel in a fixed position for travelling. Hydraulically powered operations permit the trails to be spread or closed within 10 seconds.

The two trail wheels, attached to the trail ends, can be steered, raised or lowered by hydraulically powered operations controlled by the driver.

The spades, an integral part of the trails, transmit the firing loads to the ground. As the spades are permanently hinged at the end of the trails, no carrying and mounting of spades is required during deployment.

The carriage is of the split trail type and is fitted with an APU. The 155 mm FH-88 can be towed at speeds up to 80 km/h and is self-propelled at speeds up to 16 km/h.

Maximum towing speed on secondary roads is being quoted as 50 km/h. In the self-propelled mode maximum cross-country speed is being quoted as 8 km/h while being used in the remote drive configuration maximum speed is being quoted as 4 km/h.

It is normally towed by a 6 × 6 cross-country truck that in addition to the gun crew also carries a quantity of ready use ammunition (projectiles, charges and fuzes).

The APU is a 96 hp Deutz air-cooled turbocharged diesel which powers a hydrostatic drive system on the front wheels. Back-up systems include a battery-operated pump and a manual pump. The APU can be operated in tandem with the towing vehicle for maximum traction and can be remote controlled. The APU has a 50-litre fuel tank providing an operating range of 60 km.

The diesel-powered APU consists of a hydrostatic drive system, braking system, operation hydraulic system and a back-up system. The basic functions of the APU can be summarised as follows:

- Self-propelled capability of the howitzer
- Mechanised hydraulic operation during deployment
- Ability to assist the prime mover in manoeuvring over rough terrain.

The rapid change in the zone of fire by the FH-88 to engage targets outside the primary zone is achieved by:

- Raising the firing platform
- Lowering and steering trail wheels
- Executing pivot turn
- Steering and raising trail wheels
- Embedding spades
- Lowering firing platform.

The basic sighting system of the 155 mm FH-88 consists of a cant compensating mount, peri-telescopic sight (×4 magnification) and a direct aiming sight (×6 magnification). However, this can be upgraded to a sophisticated electronic sight and linked to battery fire-control computers.

The trail legs have wheels for ease of handling. They can be opened and closed by hydraulic power and the firing platform is raised and lowered by a double-stage hydraulic cylinder.

The trail spades are self-embedding. With a crew of six the gun can be brought in or out of action in less than 1 minute.

The 155 mm FH-88 gun-howitzer can fire standard NATO ammunition such as the M107 HE projectile to a maximum range of 19,000 m. The ERFB HB projectile can be fired to 24,000 m while an ERFB BB can be fired to 30,000 m. The manufacturer claims a range error of less than 0.5 per cent, with deflection error of less than 0.1 per cent at two-thirds of maximum range. The flick rammer can be used to achieve a burst rate of 3 rds/15 s. The sustained fire rate is 2 rds/min for 1 hour.

In addition to firing the standard 155 mm M107 HE projectile and the locally manufactured 155 mm Extended Range Full Bore - Hollow Base (ERFB-HB) and Extended Range Full Bore - Base Bleed (ERFB-BB) projectiles, Singapore has developed a new 155 mm cargo projectile that is now in service.

Two versions of this have been developed, High Explosive - Base Bleed (HE-BB) and High-Explosive Hollow Base (HE HB). Both of these carry dual purpose bomblets that have been designed to attack the vulnerable upper surfaces of armoured vehicles. Both carry 64 bomblets that are manufactured in Singapore and are fitted with a self-destruct fuze mechanism.

When fired using the C30 combustible propelling charge a maximum muzzle velocity of 800 m/s is obtained together with a maximum range of 27.6 km.

Variants

Pegasus 155 mm/39-calibre Light Weight Howitzer (LWH)

This is now operational with the SAF. Details of the LWH are given in a separate entry in *IHS Jane's Land Warfare Platforms: Artillery and Air Defence.* This is in service with two artillery regiments of the Singapore Armed Forces as the replacement for the French Nexter Systems 105 mm LG1 light guns that have now been withdrawn from service.

LWSPH

As a private venture, Singapore Technologies Kinetics developed to the prototype stage the 155 mm Light Weight Self-Propelled Howitzer (LWSPH), which is ballistically identical to the towed 155 mm FH-88 gun-howitzer. This was a test bed only and used in development of the 155 mm/39-calibre Pegasus LWH now in service with the SAF.

Details of the 155 mm/39-calibre Pegasus LWH are provided in a separate entry in *IHS Jane's Land Warfare Platforms: Artillery and Air Defence.*

Primus SPH

Now in service with the SAF is the Singapore Self-Propelled Howitzer 1 Primus (SSPH1).

This full-tracked artillery system was developed by the Singapore Defence Science and Technology Agency (DSTA) and Singapore Technologies Kinetics.

Details of the Primus 155 mm/39 calibre self-propelled artillery system are provided in a separate entry in *IHS Jane's Land Warfare Platforms: Artillery and Air Defence.* Production of this has now been completed and it is understood that a total of 54 production systems have been supplied.

Specifications

	155 mm FH-88 Gun-Howitzer
Dimensions and weights	
Crew:	6
Length	
overall:	9.88 m (firing)
transport configuration:	9.16 m
Width	
overall:	8.20 m (firing)
transport configuration:	2.8 m
Ground clearance	
overall:	0.35 m
Weight	
standard:	12,800 kg
Mobility	
Carriage:	split trail
Speed	
max speed:	16 km/h (APU)
cross-country:	8 km/h (APU)
towed:	80 km/h
Fording	
without preparation:	0.76 m
Gradient:	45%
Firepower	
Armament:	1 × carriage mounted 155 mm howitzer
Main weapon	
length of barrel:	6.1 m
breech mechanism:	interrupted screw
muzzle brake:	double baffle
maximum range:	19,000 m (M107 HE projectile)
maximum range:	24,000 m (ERFB HB projectile)
maximum range:	30,000 m (ERFB-BB projectile)
Rate of fire	
sustained fire:	2 rds/min
rapid fire:	6 rds/min
Main armament traverse	
angle:	60°
Main armament elevation/depression	
armament front:	+70°/-3°
Survivability	
Armour	
shield:	no

Status

Production complete. In service with Indonesia (5) and Singapore (54). Production can be resumed if further orders are placed.

Contractor

Singapore Technologies Kinetics (STK).

South Africa

Denel Land Systems G5-52 155 mm/52 calibre gun howitzer

Development

Based on their experience in the design, development and production of the 155 mm/45-calibre G5 towed gun-howitzer and the G6 155 mm/45-calibre (6 × 6) self-propelled gun howitzer, Denel Land Systems developed a number of enhanced 155 mm artillery systems.

These include the 155 mm/52-calibre T6 artillery system which has undergone extensive trials on a T-72 tank hull as well as installed on the Indian Arjun tank hull. The latter combination is known as the Bhim.

In 2002, Denel Land Systems unveiled two new systems, the G5-52 155 mm/52-calibre towed artillery system and the Condor 155 mm/52-calibre 8 × 8 self-propelled artillery system.

The Denel Land Systems G5-52 was originally referred to as the G5-2000. As of March 2013 no production orders for the G5-52 155 mm/52-calibre gun howitzer had been placed.

Description

According to Denel Land Systems, the G5-52 has been developed keeping in mind the flexibility required for modern warfare. The modular design of the system allows for ballistic upgrades as well as the upper part (for example saddle, elevating mass and ordnance) being mounted on other platforms, tracked and wheeled.

The latest G5-52 155 mm/52-calibre system is similar in design to the older 155 mm/45-calibre G5 towed artillery system, although it has a new and longer barrel and many significant improvements at the subsystem level.

The 155 mm/52-calibre barrel has a 23 litre chamber and is similar to that used in the T6 turret system and this meets the NATO Joint Ballistics Memorandum of Understanding (JBMoU).

The ordnance has a higher pressure limit than the normal JBMoU. This allows it to fire both compliant and heavier non-compliant projectiles using standard NATO six-zone bimodular charge systems such as Rheinmetall Denel Munition M91/M92. In addition it can fire Rheinmetall Denel Munition five zone M64 bimodular charge system which is optimised for Extended Range Full-Bore projectiles.

The following subsystems are fully interchangeable with the T6 turret:

- 155 mm Rheinmetall Denel Munition ammunition and charge system
- breech mechanism
- counter recoil system
- recoil system
- barrel

Denel Systems G5-52 155 mm/52-calibre howitzer deployed in the firing position and clearly showing large muzzle brake (Denel Land Systems)
1133935

Denel Land Systems G5-52 155 mm/52-calibre howitzer deployed in the firing position 0567818

- cradle
- crew communication system
- automatic laying system
- navigation system
- fire-control system
- launcher management system
- primer loading mechanism
- ramming system.

The 155 mm/52 calibre barrel is autofrettaged and is fitted with an automatic swing and slide type breech. It has an integrated double recoil system and an electrical firing system. To reduce crew fatigue and increase rate of fire it is fitted with a hydraulically-operated automatic primer loader.

The loading system includes an automatic loading tray that loads the 155 mm projectile and charge and an electrically-operated rammer. Ammunition can be loaded at all elevation angles.

While this system has a burst rate of fire of three rounds in 10 seconds, an intense rate of fire of up to 8 rds/min can be attained with a sustained rate of fire being 3 rds/min.

The gun control equipment includes combined hydraulic elevation and balancing cylinders, hydraulic traverse gearbox with clutch for rapid traverse of the barrel from the travelling to the firing position.

The gun can cover an area of 1,000 m^2 and with the Rheinmetall Denel Munition base-bleed ammunition range at sea level is in access of 40 km while the range at higher altitudes exceeds 50 km.

Range at sea level when using the Rheinmetall Denel Munition long-range Velocity-enhanced Long-range Artillery Projectile (VLAP) projectiles is more than 50 km. According to Denel, the probable error specification at 75 per cent maximum range in the lower trajectory is a proven 0.48 per cent of range and 1 min in deflection.

The launcher management system includes a launcher management computer, muzzle velocity radar system, radio link to battery command post and a link to the automatic laying and navigation system on the weapon.

The bottom carriage is of high strength steel with split trails and has four driving wheels on a walking beam suspension. There are also two trail wheels that steer when the gun moves under its own power. Each trail is provided with a self dig-in spade.

Mounted under the carriage is a large hydraulically operated stabilising platform that is lowered to the ground when the weapon comes into action. When travelling the barrel is traversed through 180° and locked in position over the closed trails. There is also a rammer clamp for the travelling position.

Mounted on the forward part of the carriage is the Auxiliary Power Unit (APU) which enables the weapon to propel itself across the battlefield as well as reducing into and out of action times.

The APU includes a 70 kW air-cooled four-cylinder diesel engine with a main hydraulic pump for propulsion and a separate hydraulic pump for deployment. A diesel tank with a capacity of 102 litres is provided.

This feature not only makes the gun independent of the serviceability of a prime mover, but also allows the towing vehicle to be used for other logistic tasks after the gun has been deployed in a defensive position. Using the APU, four people can bring the G5-52 into action in four minutes.

In the event of a hydraulic failure, the emergency hydraulic power is supplied to the G5-52 by the towing vehicle. In extreme terrain conditions, the APU can provide additional tractive effort forming an all-wheel combination with the towing vehicle.

According to Denel Land Systems, when being used in the APU mode it has a maximum self-propelled speed on a hard surface of 16 km/h which reduces to 8 km/h whilst in the APU mode on sand. This figure does however depend on terrain conditions.

A telescope and sight for direct fire up to a range of 3,000 m is fitted as standard. The G5-52 is fitted with an accurate inertial navigation and positioning system that eliminates survey and orientation procedures during employment.

The automatic laying system enables the gun to be relayed between rounds, even at the maximum rate of fire.

The artillery communication system installed on the G5-52 includes a battery position intercommunication system with a gun control net and a crew net.

The G5-52 can be provided either alone or as part of a complete system than includes, projectiles, charges, fuzes, primers, command and fire-control system, meteorological systems, observation system, crew communication system, communication system (voice and data), logistic support vehicles, ammunition resupply vehicle and the gun tractor. The latter is usually a 6 × 6 type which can be specified by the customer.

The prototype G5-52 has the African Defence Systems WM205 weapon management system, which can be integrated with a higher-level artillery command and control system, such as the African Defence Systems AS2000.

It is normally towed by a 6 × 6 cross-country truck that in addition to its crew also carries a quantity of ready use ammunition.

Specifications

	G5-52 155 mm/52 Calibre Gun Howitzer
Dimensions and weights	
Crew:	5
Mobility	
Carriage:	split trail with APU
Speed	
max speed:	16 km/h
towed:	90 km/h
Firepower	
Armament:	1 × carriage mounted 155 mm howitzer
Main weapon	
calibre length:	52 calibres
muzzle brake:	double baffle
maximum range:	44,200 m (ERFB-BB projectile)
maximum range:	58,000 m (VLAP projectile)
Rate of fire	
sustained fire:	3 rds/min
rapid fire:	8 rds/min
Main armament traverse	
angle:	82° (41° left/41° right)
Main armament elevation/depression	
armament front:	+72°/-2°
Survivability	
Armour	
shield:	no

Status
Prototype. Not yet in production or service.

Contractor
Denel Land Systems.

Denel Land Systems G5 155 mm/45-calibre gun-howitzer

Development
During operations carried out in 1975, the South African Defence Force (SADF) found its own artillery out-ranged and out-gunned by various forms of Russian artillery used by the Angolan forces.

Conventional tubed artillery used by Angola included the Russian 122 mm D-30 howitzer and the 130 mm M-46 field gun.

Following this experience, the South African artillery initiated a staff requirement for a new artillery system that would encompass not only a new artillery piece but also an ammunition system, a new gun tractor, fire-control equipment, a fire-control computer system and other accessories.

The G5 became operational with the SADF during 1983. In mid-1996, it was stated that 72 155 mm G5 systems were in service with the South African Artillery.

The G5 155 mm gun-howitzer has seen combat service with the SADF in Southern Angola and South West Africa (now Namibia) and with the Iraqi Army during its war with Iran. Quantities of these weapons were captured by the Allied Coalition forces during Operation Desert Storm, the liberation of Kuwait, which took place early in 1991.

In November 2000, it was announced that an agreement worth about USD50 million had been signed between Denel Land Systems and Malaysia for the supply of an artillery system.

Under the terms of this contract, Denel Land Systems supplied 22 G5 155 mm Mark 3 gun-howitzers to Malaysia together with ammunition, modular charges, fire-control systems, logistics support and training. The fire-control system deployed is the African Defence Systems AS2000, which was originally developed to meet the operational requirements of the South African Defence Force.

First deliveries were made late in 2001 and all systems were delivered by the end of 2002.

More recently an order has been placed for an additional batch of G5s, and these have now been delivered. It is understood that the second order was for a further six guns, which brings the total Malaysian G5 fleet up to 28 units. The Malaysian G5s are towed by MAN (6 × 6) trucks, which have been modified by local company DEFTECH.

Denel Land Systems 155 mm G5 gun-howitzer deployed in the firing position with ordnance elevated 0105799

Description

The 155 mm G5 howitzer uses a 45-calibre barrel designed around the performance of the ERFB round originally developed by the now defunct Space Research Corporation of Canada.

The 155 mm G5 howitzer is produced and manufactured by Denel Land Systems at Centurion, near Pretoria. The monobloc autofrettaged barrel has a single-baffle muzzle brake and uses a semi-automatic interrupted screw-type breech.

The 155 mm G5 incorporates a manually controlled, telescopic pneumatic cylinder rammer mechanism for ramming the projectiles in the chamber, at a uniform position whatever the elevation. Charges (modular or bag type) are loaded by hand.

The carriage is fitted with an APU to facilitate hydraulic powered deployment and self-propulsion. This APU comprises a 79 hp air-cooled diesel engine.

Hydraulic powered deployment includes the following actions:

- Aligning the trail wheels
- Lifting and lowering of the trail wheels
- Spreading and closing of the trail wheels
- Lifting and lowering of the firing platform.

The G5 can be brought into action or taken out of action in two minutes. In case of APU failure, slave sockets and pipes are provided to couple the hydraulic system to that of another gun or tractor.

The G5 gun crew consists of five people, although in an emergency two can operate the weapon. Elevation and traverse controls are manual.

The G5 can be driven at a maximum speed of at least 16 km/h and has a gradient capability of 40 per cent. With the diesel fuel tank capacity of 100 litres the G5 has an operating range under its own power of more than 100 km.

When being towed, the 155 mm ordnance can either be clamped over the trail or in the firing position. When travelling the ordnance will be clamped over the trail, while clamping in the firing position is intended for tactical movement.

In SANDF service the G5 is towed by a 6 × 6 tractor, based on a locally built SAMIL 100 (6 × 6) chassis. The gun tractor not only tows the G5 but also has accommodation for a crew of eight (normally five serve the gun in action). Storage is provided on the gun tractor for 32 155 mm projectiles and 48 charges.

To facilitate towing of the G5 system with virtually any prime mover, an extended towbar is fitted.

Denel Land Systems 155 mm G5 gun howitzer in travelling configuration with ordnance locked to front and APU being used (Denel Land Systems) 0567812

Denel Land Systems G5 155 mm gun-howitzer deployed in the firing position (Denel Land Systems) 1333675

The ammunition system of the G5, which comprises the projectile, charges and fuzes, is identical to that of the Denel Land Systems G6 (6 × 6) self-propelled artillery system covered in detail in a separate entry in *IHS Jane's Land Warfare Platforms: Artillery and Air Defence.* Production is undertaken on an as required basis but there has been no production of the G6 for at least 10 years. This is currently in service with Oman (24), South Africa (43) and United Arab Emirates (78).

The layer's position on the left of the breech is provided with an upgraded metricated panoramic sight and a direct fire telescope for use up to 5,000 m.

The G5 features a gun management system that provides the layer with all relevant laying data.

A temperature warning device is available as an option. This warns the gun crew of high barrel-temperature conditions. Thus a high rate of fire can be maintained while avoiding a possible 'cook-off'.

The AS2000 Artillery Target Engagement System has been designed specifically for use with the G5 and G6 (6 × 6) artillery systems. This also includes a muzzle velocity meteorological ground station and a crew communication system.

Additional technical details, not provided in the specifications table are:
Length on tow and barrel to the rear - 12.2 m
Length with extended tow bar - 13.1 m
Height with barrel camped forwards - 2.85 m
Height of towing eye travelling - 0.95 m.

Variants

Improved G5 Mk 3 155 mm towed gun

In co-operation with the South African National Defence Force, Denel Land Systems has further improved the G5. Designated the G5 Mk 3, a total of 35 modifications and improvements were carried out to enhance the already proven reliability and operational efficiency of the gun.

User feedback based on many years of maintenance and operational experience has been incorporated in the G5 Mk 3, as well as results obtained from Denel Land Systems research and development centre.

On the main weapon the recoil/buffer cylinder and replenisher have been improved and a gas trap has been added which greatly improves the efficiency of the recoil system. The improved G6 breech and the G6 camplate have been adopted in the G5 Mk 3. A barrel temperature indicator and warning light has been installed.

Electrical power distribution to the top carriage has been improved to accommodate the state-of-the-art Artillery Communication System (ACS), muzzle velocity analyser and fire-control computer. The ACS has two dome antennas fitted on the gun which allow closed-circuit communication between the gun crew and gun commander and external communications between the gun commander and the battery control post.

Accessibility to the traverse bearing bolts has been improved. Improvements have been made to the traverse drive clutch, traverse drive shaft and elevation gearbox for easier maintenance and more efficient functioning.

The advantages of these changes are improved gun alignment and greater durability. A greatly improved brake system has been introduced which automatically applies the brakes should the hydraulic drive system fail to stop the gun. When the APU is not running the brakes are engaged and only released if the air pressure provided by the newly fitted air compressor, is sufficient.

Attention to detail and to the users' requirements has led to minor modifications such as an easier to read diesel tank dipstick, additional bottom carriage drainholes, improved trail leg stops that improve reliability and ease of replacement, dog-clutch disengagement indication light, improved muzzle brake screw lock and improved hydraulic tank indicator, which all give the system better performance and reliability.

Trials with the upgraded G5, subsequently called the G5 Mk 3, were completed in 1996 and troop trials with a battery of six weapons took place in 1997 with conversions running through to 2000.

155 mm G5-52

In 2002, Denel Land Systems unveiled the G5-52, which has many improvements, including a 155 mm/52-calibre barrel. Details of this system are

Denel Land Systems G5 155 mm gun-howitzer deployed in the firing position (Denel Land System) 1365028

provided in a separate entry in *IHS Jane's Land Warfare Platforms: Artillery and Air Defence.*

As of March 2013 the G5-52 remained at the prototype stage. The towed G5-52 has the same ordnance as the G6-62 (6 × 6) self-propelled artillery system which as of March 2013 remained at the prototype stage.

This turret is marketed on its own as the T6-52 turret system and for trials purposes this has been integrated onto a number of hulls including the Russian T-72 and Indian Arjun.

Velocity enhanced Long-range Artillery Projectile (VLAP)

The new Velocity enhanced Long-range Artillery Projectile (VLAP), first revealed early in 1997, is a rocket motor-assisted 155 mm projectile combined with the base bleed unit which extends the range of the 45-calibre gun to 50 km.

First production VLAP were completed early in 2001.

India is known to have recently ordered large quantities of 155 mm ammunition (projectiles and charges) from South Africa and could be the launch customer for the VLAP.

The VLAP was originally designed as a private venture by Rheinmetall Denel Munition to complement their existing family of 155 mm artillery projectiles which includes High Explosive - Extended Range Full Bore - Base Bleed (HE ERFB BB).

The current VLAP has a high explosive payload although other payloads are now under investigation by Rheinmetall Denel Munition.

Rheinmetall Denel Munition stress that the VLAP is complimentary to their existing family of projectiles and allows friendly artillery units to engage threat targets while positioned well to the rear and beyond range of enemy counter-battery artillery fire. This will lead to higher survivability of friendly artillery systems.

To achieve this significant increase in range the VLAP incorporates base bleed and rocket assist. It can be fired from most existing towed and self-propelled 155 mm artillery systems including the South African Denel Land Systems 45-calibre G5 (towed), 45-calibre G6 (6 × 6) self-propelled and the more recent 52-calibre T6 turret systems which has undergone extensive tests by the Indian Army.

Ranges of the VLAP compared to other types of 155 mm projectiles are as follows:

Weapon type	Boat tail	Base Bleed	VLAP
155 mm/39-calibre	23 km	30 km	39 km
155 mm/45-calibre	30 km	39 km	50 km
155 mm/52-calibre	32.5 km	41 km	52.5 km

The VLAP can be used in conjunction with a number of charge systems including the Rheinmetall Denel Munition M53, M90 Bi-Modular Charge System (BMCS).

PRO-RAM 155 mm projectile

Currently under development is a new 155 mm projectile called the PRO-RAM. This could have a range of up to 80 km, depending on which it is fired from.

140 mm GV2 gun

This is the South African designation for the old British 5.5 inch Medium Gun. This was used in large numbers by South Africa but has now been phased out of service. In 1998 South Africa transferred 24 of these weapons to Namibia.

Specifications

	G5 155 mm Gun-Howitzer
Dimensions and weights	
Crew:	5
Length	
overall:	11 m (firing)
transport configuration:	9.5 m
Width	
overall:	8.7 m (firing)
transport configuration:	2.53 m
Height	
transport configuration:	2.3 m
axis of fire:	1.77 m
Ground clearance	
overall:	0.31 m
Weight	
standard:	13,750 kg
Mobility	
Carriage:	split trail with APU
Speed	
max speed:	16 km/h (APU)
Fording	
without preparation:	0.6 m
Gradient:	40%
Tyres:	14.00 × 20
Firepower	
Armament:	1 × carriage mounted 155 mm howitzer
Main weapon	
length of barrel:	7.045 m
breech mechanism:	interrupted screw
muzzle brake:	single baffle
maximum range:	30,000 m (standard HE projectile)
minimum range:	4,500 m (standard HE projectile)
maximum range:	39,000 m (ERFB-BB projectile)
maximum range:	50,000 m (VLAP)
Rate of fire	
sustained fire:	3 rds/min
Main armament traverse	
angle:	82° (total up to +15° elevation; 65° total above +15° elevation)
Main armament elevation/depression	
armament front:	+75°/-3°
Survivability	
Armour	
shield:	no

Status

In production. In service with the following countries:

Country	Quantity	Comment
Chile	28	marines
Iran	100	estimate, believed 200 delivered
Malaysia	28	Mark 3s, delivered in two batches
South Africa	72	
Uganda	4	

Contractor

Denel Land Systems, a division of Denel (Pty) Ltd.

Denel Land Systems 105 mm Light Experimental Ordnance (LEO)

Development

The origins of the Denel Land Systems 105 mm Light Experimental Ordnance, sometimes referred to as the LEO, can be traced back to 1995. This was when the South African Defence Force set the following design goal to Denel Land Systems to 'develop the technology for a light gun with a light logistic load that will equal or better the range and equal or better the terminal effect of current 155 mm light guns'.

A ballistic testbed was completed and test fired in 1997, with the first complete example of the 105 mm Light Experimental Ordnance completed in 2001. This system has undergone extensive firing trials at the Alkantpan range in South Africa.

Denel Land Systems stress that this is a technology demonstrator and is not representative of a type classified and ready for production 105 mm artillery system.

If adopted for service it will probably be given the designation of the GV7 or G7 for international marketing purposes. Its primary role is close support.

According to Denel Land Systems, there is no other artillery system in the world with the capabilities of the 105 mm Light Experimental Weapon which includes a maximum range of 30,000 m.

This is similar to that of 155 mm/39-calibre artillery systems with the new 105 mm high-explosive artillery projectile, having a lethal area of 1,900 m^2 against targets. This is claimed to be better than a first-generation 155 mm M107 HE artillery projectile.

A 105 mm Light Experimental Weapon together with 100 rounds of 105 mm ammunition currently weighs 6,000 kg which is being reduced still further. According to Denel Land Systems this should be compared to more than 10,000 kg for a 155 mm system plus a similar number of rounds. Because of the lighter 105 mm ammunition, the crew requirement is reduced and it is also suitable for the 'lighter built' soldier.

Denel Land Systems 105 mm Light Experimental Ordnance (LEO) deployed in the firing position (Denel Land Systems) 0526370

The first example of the 105 mm Light Experimental Ordnance weighs 3,800 kg and the development is expected to reduce this to 2,500 kg making it easier to deploy.

According to Denel Land Systems the reduction in weight is especially important for fielding in peacekeeping operations or for early entry forces/light forces for rapid support of allies and for close fire support.

This weapon system has also been referred to as the 105 mm Artillery Ballistic System. In addition to the conventional split trail carriage used for the prototype system, a number of other carriages are under consideration.

LEO will form the basis of the new 105 mm artillery system which will also include its family of ammunition, on board fire control systems plus logistics and training.

This system is called the Advanced Multirole Light Artillery Gun Capability (AMLAGC) and Phase I and is now expected to include man-rating of the existing LEO 105 mm ordnance and its ammunition as well as starting development of the new gun.

This is expected to be followed by Phase II which will include the construction and test of a complete 105 mm weapon, interface for mounting this on other chassis and final qualification of the weapon and its ammunition.

As of early 2013 it is understood that Denel Land Systems had ceased all work on their 105 mm Light Experimental Ordnance towed artillery system for the time being.

The company is heavily committed to supply a significant number of two person turrets for the Turkish FNSS Pars AV8 (8 × 8) vehicles to be supplied to Malaysia and in 2013 they expect to be awarded the contract for the Badger (8 × 8) infantry fighting vehicles and variants for the SANDF.

Denel Land Systems expect to start work on the 105 mm Light Experimental Ordnance again in the future and when fully type classified it is expected to attract considerable overseas interest.

Description

The 105 mm Light Experimental Ordnance has a split trail carriage of high-strength steel and when deployed in the firing position is supported by a jack lowered under the forward part of the carriage and by the spade on the end of each trail.

The carriage is provided with an adjustable towing hook for towing by 4 × 4 vehicles and lifting hooks for transportation slung under a helicopter.

The 105 mm autofrettaged ordnance is 52 calibres long, or 57 calibres when fitted with the advanced muzzle brake, and has a rifling twist of 1:22. The high-efficiency pepperpot muzzle brake has special Laval-shaped nozzles and is rifled on the inside, which effectively adds five calibres to the barrel. The 105 mm ordnance has a chamber volume of 12 litres.

According to Denel Land Systems the 105 mm Light Experimental Ordnance takes two minutes to be brought into action by five men and three minutes to come out of action by five men.

The system is fitted with a braking system with the inertia brake activated by the tow hook coupled to the gun tractor with drums on the main wheels. The parking brake form part of the brake system.

The ordnance has a fixed recoil length of 1 m with a gas type counter system being fitted. Counter recoil is also used to retract the 105 mm ordnance using the hydraulics of the traverse and elevation system.

The breech mechanism is of the semi-automatic swing and slide type.

The traverse and elevation system of the 105 mm Light Experimental Ordnance consists of the following key parts:

- Hydraulic system fitted with proportional control valves
- Hydraulic elevation cylinders with integrated pneumatic equilibrators
- Traverse gearbox with hydraulic drive motor
- Hydraulic hand pump to supply hydraulic pressure manually
- Electrical driven hydraulic pump to supply the hydraulic pressure required
- Hydraulic hand pumps for manual elevation and traverse of the gun
- Traverse gearbox fitted with a clutch for rapid traverse of barrel from travelling to the firing position
- Joystick control fitted for controlling elevation and traverse of the gun.

The bottom carriage is of the split trail type made of high-strength steel. The two main wheels are mounted on a trailing arm suspension and are adjustable for height. The trails are provided with dig in spades designed for use in hard ground.

Mounted on the forward part of the carriage is a large manually operated stabilising platform that is lowered to the ground before firing commences. There is also a barrel clamp on the trail legs to clamp the barrel when the gun is towed in the travelling position.

The high-strength steel saddle is of the trough type with self-adjustable type trunnion bearings. The gun control equipment is also mounted on the saddle on the left side.

An adjustable towing hook is provided to allow towing by a 4 × 4 or any five-tonne type of vehicle and lifting hooks can be fitted to allow transport of the gun by helicopter. The weapon can be towed on roads up to a maximum speed of 100 km/h with the ordnance in the firing position or locked in position over the closed trails.

To power the onboard electrical equipment, two 12 V maintenance-free batteries are fitted together with electrical harnesses, control boxes, tail lights, indicator lights and brake lights.

The basic model is fitted with telescopic sight for direct fire up to a range of 3,000 m and a trunnion-mounted optical panoramic sight with compensating mechanism for trunnion cant for indirect fire. This is installed on the left side of the weapon. Aiming posts and beta lights are stored on the gun.

The advanced model will retain the telescopic sight for direct fire up to a range of 3,000 m but will also have a ring laser gyro, display unit and an automatic laying system. This has already been proven by Denel Land Systems, on other towed and self-propelled artillery systems.

The basic model is fitted with a manual loading system (hand ram) which enables the weapon to be loaded at all angles of elevation. The improved model will be fitted with a positive chain-push rammer, which will be hydraulically operated and controlled. This will be able to ram both the projectile and charge at any elevation.

Firing Rheinmetall Denel Munition standard ammunition a maximum range of 24.6 km will be achieve which increases to 29.3 km with base bleed and 36 km with a Velocity-enhanced Long-range Artillery Projectile (V-LAP).

The new family of Igala 105 mm ammunition consists of the projectile which can be fitted with an interchangeable boat tail or base bleed unit and a unimodular charge system which gives a maximum pressure of 427 Mpa.

The ammunition suite includes HE (natural fragmentation), HE (pre-formed fragmentation), practice, smoke (visual and IR screening) and illumination (visual or infra-red). In addition, a Armour Piercing Fin Stabilised Discarding Sabot (APFSDS) round has been tested with a m/v of 1,700 m/s.

In fact, development of the baseline 105 mm family has already been completed with the last round, the V-LAP still under development. The later leverages off of a 155 mm V-LAP which has already been produced in production quantities for an export customer.

It should be noted that all of this 105 mm ammunition is IM compliant and according to Denel is backwards compatible with existing 105 mm weapons.

Trials have already shown that the 105 mm PFF has a greater lethal radius that the older US 155 mm M107 HE projectile but at the same time a 76 per cent smaller danger area.

A locally developed Denel Arachnida (or Selex LINAPS) will be fitted to reduce into action times and improve accuracy with the former already standard on South African G5/G6 guns and 127 mm (40-round) Valkiri Mk II artillery rocket launcher.

Ammunition is of the separate loading type, projectile and charge and the system has a maximum rate of fire of six rounds a minute. Various fuzes, including the new Fuchs electronics multi-option fuze have been tested with the 105 mm Light Experimental Ordnance. Fuze options include impact fuze M841, proximity fuze M8513 and time fuze M9220.

According to Denel Land Systems, at 75 per cent of maximum range, dispersion is 0.3 per cent of range and 0.5 mil in deflection.

During the full scale development phase, Denel Land Systems stated that the following improvements will be addressed:

- Development of a base bleed unit to increase the range of the system to more than 30 km
- Further weight reduction making use of high-strength materials
- Methods of further improving accuracy and dispersion
- Adding an automatic laying system on models to be fitted with a laser inertial laying system
- A load-assist system
- Mobility and deployability improvements
- Changes to make it possible to fire other types of ammunition for training purposes.

In addition to the standard split trails of the technology demonstrator 105 mm LEO, other carriages are also under consideration.

It would typically be towed by a 4 × 4 cross country truck that in addition to its crew would also carry a quantity of ready use 105 mm ammunition.

Variants

GDLS 105 mm SPG

In addition to the development of the 105 mm LEO, Denel Land Systems teamed with General Dynamics Land Systems of the USA have developed a Self-Propelled (SP) version.

This consists of a LAV (8 × 8) chassis fitted with a new turret armed with the same 105 mm ordnance as fitted to the LEO.

Details of this system are provided in a separate entry in *IHS Jane's Land Warfare Platforms: Artillery and Air Defence* with additional trials and demonstrations taking place in 2010 and 2011.

As of March 2013 the 105 mm SP version remained at the prototype stage.

Ultra lightweigt 105 mm LEO

Denel Land Systems have also carried out some studies on a ultra lightweight version of LEO 105 mm using advanced materials which would have a new carriage and weigh about 2,000 kg.

Close up of gunners display and controls on first example of LEO 105 mm artillery system and also showing spread trails and turntable
(Denel Land Systems) 1296091

Truck mounted 105 mm LEO
Using the upper part of LEO 105 mm a truck mounted version (6 × 6) could be developed which would also have space for the crew and some ready use 105 mm ammunition.

Specifications

	105 mm Light Experimental Ordnance
Dimensions and weights	
Crew:	5
Length	
transport configuration:	6.9 m
Width	
transport configuration:	2.02 m
Height	
transport configuration:	2.1 m
Ground clearance	
overall:	0.3 m
Track	
vehicle:	1.82 m
Weight	
standard:	3,800 kg
Mobility	
Carriage:	split trail
Firepower	
Armament:	1 × carriage mounted 105 mm howitzer
Main weapon	
muzzle brake:	pepperpot (with special Laval-shaped nozzles)
maximum range:	24,600 m (standard ammunition)
minimum range:	500 m (standard ammunition)
maximum range:	29,300 m (BB)
maximum range:	36,000 m (VLAP)
Rate of fire	
rapid fire:	6 rds/min
Main armament traverse	
angle:	80° (40° left/40° right)
Main armament elevation/ depression	
armament front:	+75°/-5°
Survivability	
Armour	
shield:	no

Status
Prototype. Not yet in production or service.

Contractor
Denel Land Systems.

Spain

General Dynamics European Land Systems - Santa Bárbara Sistemas 155 mm 155/52 APU SBT howitzer

Development
Based on the experience in the design and development of the 155 mm SB 155 mm/39-calibre towed howitzer, which never entered production, the now General Dynamics European Land Systems - Santa Bárbara Sistemas, in association with other Spanish defence equipment manufacturers, developed a new 155 mm/52-calibre system called the 155/52 APU SBT howitzer.

The first prototype was completed in December 1997 and by May 1998 had fired over 170 rounds of 155 mm ammunition including Extended Range Full Bore and Extended Range Full Bore - Base Bleed.

Early in 2001, it was stated that the Spanish Army had begun forming a Mobile Coastal Artillery Group with the first batch of six 155/52 APU SBT howitzers.

The 155/52 APU SBT Coastal Howitzer Command-and-Control (C2) system will co-ordinate data from coastal defence tracking radars to enhance target engagement capability.

In mid-2005 the Spanish Ministry of Defence (MoD) signed a contract with General Dynamics European Land Systems - Santa Bárbara Sistemas worth EUR189.5 million for the supply of 70 155 mm 155/52 APU SBT howitzers for the Spanish Army.

The Spanish Army originally operated 12 General Dynamics European Land Systems - Santa Bárbara Sistemas 155 mm/52-calibre APU SBT. Of these, four (V06) are in the field artillery role and eight (VO7) are integrated into a coastal artillery group.

In the designation '52' represents the calibre length of the 155 mm barrel and 'APU' stands for Auxiliary Power Unit. The latter allows the system to travel on roads up to a maximum speed of 18 km/h, as well as moving around the battery position without its prime mover.

These weapons are a further development of the Spanish SITECSA 155 mm 45-calibre weapon developed some years ago in two configurations, standard towed and fitted with an Auxiliary Power Unit (APU).

Under the terms of this new contract the 12 existing weapons have been overhauled and rebuilt to a new common enhanced standard with the complete weapon system being called the Obus 155/52 version V07.

The contract included the development of a new version of the howitzer named SIAC (*Sistema Integrado de Artilleria de Campaña* or Field Artillery Integrated System) with improved mobility and new capabilities including C2 integration and a more complete truck integration.

The 70 brand-new weapons comprised four units delivered late in 2006 which correspond to the V07 configuration and the remaining 66 were built to the new SIAC standard. This will bring the total Spanish Army fleet up to a total of 82 systems (16 V07 and 66 SIAC).

Of these, 66 are deployed with the field artillery which will be organised into four groups each of 16 weapons with eight weapons being allocated to a battery. Each battery consists of two sections each of four weapons. The other two units will be used for training purposes.

The 16 V07 is used by the coastal artillery that will have two groups each of eight which was operated by two batteries each of four weapons. These are deployed in Southern Spain to defend the Straits of Gibraltar.

Spanish Army 155 mm/52 APU SBT howitzer used in the coastal defence role by the Spanish Army
(General Dynamics European Land Systems - Santa Bárbara Sistemas) 0548783

Spanish Army General Dynamics European Land Systems - Santa Bárbara Sistemas 155 mm/52-calibre APU towed howitzer carrying out a fire mission (General Dynamics European Land Systems - Santa Bárbara Sistemas)
1460766

General Dynamics European Land - Systems Santa Bárbara Sistemas 155 mm 155/52 APU SBT howitzer deployed in firing position (General Dynamics European Land Systems - Santa Bárbara Sistemas)
1333677

The first prototype of the complete weapon to the latest SIAC V07 production build standard was completed late in 2006 and was subsequently evaluated.

This was followed by pre-production systems with the main production run of SIAC V07 starting in 2009 and running through 2013 at the rate of 13 to 14 weapons a year.

Final deliveries were made to the Spanish Army early in 2013 but production can commence again if additional orders are placed.

In January 2012 Spain deployed a number of these 155 mm/52-calibre artillery systems to its Melilla enclave off of North Africa to replace older coastal artillery weapons.

The major contract also covered the supply of the onboard digital navigation and automatic laying system, Thales PR4G communications system (which will be supplied by the Spanish company of Amper) and Italian Iveco Defence Vehicles (6 × 6) cross-country trucks. In addition the contract covers the complete logistical in-service support. The Iveco Defence Vehicles (6 × 6) prime mover is referred to as the Vehiculo Tractor (VET) and these were delivered at the rate of 16 a year from late 2009.

VET is provided with a forward control cab and an additional fully-enclosed cabin to the rear for the remainder of the gun crew. Space is also provided for a quantity of ready use 155 mm ammunition (projectiles, charges and fuzes).

According to the company the onboard computerised fire-control system coupled to an inertial navigation system will provide for greater accuracy as well as reducing the time taken to come in and be taken out of action. This will enhance its survivability to counter-battery fire.

According to the United Nations Arms Transfer Lists, Colombia took delivery of a total of 13 General Dynamics Santa European Land Systems Bárbara Sistemas 155 mm 155/52 APU SBT howitzers in 2007.

According to General Dynamics European Land Systems - Santa Bárbara Sistemas, the Colombian weapons are in the V07 configuration and have a fire-control system virtually identical to that used by Spain except for the software ballistic module.

Description

The 155 mm 155/52 APU SBT howitzer has a conventional split trail carriage and when deployed in the firing position is supported on a circular firing platform mounted under the carriage and on the two trails, each of which is provided with a spade and a small wheel. The latter is used to assist in bringing the weapon into action.

The carriage has four wheels and mounted on the front of the carriage is the 106 hp (78 kW) diesel Auxiliary Power Unit (APU) with the driver being seated on the left side. The alternator and batteries are also located on the APU.

The suspension is of the walking beam type and the diesel APU supplies hydraulic power to the four drive wheels to enable the weapon to be driven across country under its own power.

The 155 mm 155/52 APU SBT can be towed on roads up to a maximum speed of 90 km/h which drops to 50 km/h on trails and 15 km/h cross-country, but these figures depend on terrain conditions.

Maximum speed when being used in the APU mode is 18 km/h on roads which drops to 15 km/h on trails and 8 km/h cross-country, but again these figures depend on terrain conditions.

In addition to providing power for the four main road wheels the APU also provides power for raising and lowering of the auxiliary wheels, opening of the trails and operation of the firing platform and aiming systems.

When travelling, the upper part of the system, complete with 155 mm/52-calibre ordnance, is traversed through 180° and locked in position over the closed trails.

The 155 mm/52-calibre autofrettaged ordnance has a 23-litre chamber and is fitted with a three part slotted muzzle brake with an efficiency of 35 per cent. The screw breech mechanism is of the screw type with a plastic pad obturator and upon recoil opens automatically. The ordnance meets the NATO Joint Ballistic MoU.

The firing system includes a detonation and primer magazine system with 10 M82 type primers in a drum type magazine.

The counter recoil system is of the oleo-pneumatic type. Recoil system is of the hydraulic rod type with an absorber. The maximum recoil of the first stage is 1,525 mm and the maximum recoil of the second stage is 1,041 mm. The counter recoil system is of the oleo pneumatic type and the recoil brake consists of recoil and counter recoil system and a hydraulic rod and absorber. Elevation and traverse are hydraulic with the layer being seated on the left side of the carriage.

To enable a high rate of fire to be achieved and load the weapon at high angles of elevation, a loading tray and hydraulic projectile rammer are fitted as standard.

According to the company, the 155 mm 155/52 APU SBT howitzer has a maximum range of 31.7 km firing standard projectiles and 41 km firing base bleed projectiles.

Standard bagged charges or modular artillery charge systems can be used with this equipment. Standard equipment on the weapon includes a flick rammer, muzzle velocity measuring radar, ten round primer magazine, round counter, recoil meter and chamber temperature detector.

The software used in these new systems will be the same as that used in the fire-control systems used with the Spanish Army's 105 mm Light Guns and 155 mm M109E5 self-propelled artillery systems.

Under a separate contract the company has also been awarded a contract for the design and development of a new family of 155 mm ammunition for use with these 155 mm/52-calibre artillery systems.

Optional equipment includes a hydraulic rammer, automatic breech opening system, recoil meter, navigation system and various types of laying system.

The following aiming devices are fitted:

- Telescope M137 (magnification ×4)
- Telescope mount M171
- Telescope square M17
- Direct aiming sight M138 (magnification ×4)
- Sight mounting M172
- Sight square M18
- Muzzle-centring device SKS155.

The latest Spanish Army systems are fitted with a Digital Navigation Aiming and Pointing System (DINAPS) which is also being installed in Spanish Army 105 mm L118/L119 Light Guns (56 units) and M109A5E 155 mm self-propelled artillery system (96 units).

DINAPS is a modular system that combines a hybrid (inertial and global positioning system) navigation system, Muzzle Velocity Radar (MVR) and a navigation and ballistic software which is able to connect to the Spanish Army Command-and-Control system (C2).

The system uses NABK (NATO Ballistic Kernel) as a core for ballistics using a variety of projectiles to enable the weapon to engage stationary and moving targets with a high first round hit probability.

The inertial navigation unit determines the elevation and northing angles of a gun barrel and corrects automatically for any variation in projectile, charge and meteorology.

The 155 mm 155/52 APU SBT is also fitted with an Automatic Gun Laying System (AGLS) which is used in conjunction with DINAPS to automatically lay the weapon onto the target.

In addition to being used with towed and SP weapons it can also be used with rocket launchers. The software installed in DINAPS allows the system to be used with different command and control/communications systems.

DINAPS + AGLS has already been type classified for the GDSBS 155 mm/52-calibre artillery systems model V07 and SIAC.

Specifications

	155 mm 155/52 APU SBT Howitzer
Dimensions and weights	
Length	
overall:	12.70 m (firing)
transport configuration:	10.35 m
Width	
transport configuration:	2.82 m
Height	
overall:	2.20 m
Ground clearance	
overall:	0.30 m
Weight	
standard:	13,500 kg

	155 mm 155/52 APU SBT Howitzer
Mobility	
Carriage:	split trail with APU
Fording	
without preparation:	0.70 m
Gradient:	30%
Auxiliary engine:	diesel developing 78 kW
Tyres:	11 × 20
Firepower	
Armament:	1 × carriage mounted 155 mm howitzer
Main weapon	
length of barrel:	8.12 m
rifling:	6.890 m
breech mechanism:	interrupted screw
muzzle brake:	slotted
maximum range:	31,700 m (ERFB projectile)
maximum range:	18,000 m (standard M107 HE projectile)
maximum range:	41,000 m (ERFB-BB projectile)
Main armament traverse	
angle:	80° (40° left/40° right)
Main armament elevation/depression	
armament front:	+72°/-3°
Survivability	
Armour	
shield:	no

Status

Production complete but can be restarted again if additional orders are placed. The Spanish Army has taken delivery of a total of 12 units, four in the field artillery role and eight in the coastal defence role. In 2005 a further order was placed for an additional 70 units. This will bring the total fleet up to 82 units. In 2007 Colombia took delivery of a batch of at least 13 weapons.

Contractor

General Dynamics European Land Systems-Santa Bárbara Sistemas.

Sweden

BAE Systems 155 mm Field Howitzer 77B (FH-77B)

Development

The now BAE Systems 155 mm FH-77B (now frequently referred to just as the FH-77) was originally developed by Bofors as an export model of the FH-77A.

The main change was the ability to fire NATO standard 155 mm ammunition, other improvements being made to its general handling, cross-country mobility and the introduction of a mechanised ammunition handling system.

Following an FH-77B order from Nigeria, the largest export order in Swedish history came when India placed an initial order for 410 FH-77B weapons and equipment. AB Bofors (today BAE Systems) acted not only as manufacturer but as the major supply contractor for items such as Saab-Scania trucks, Barracuda camouflage netting, Marconi Command and Control Systems Quickfire fire-control systems, Fairey Australia muzzle velocity radars, BEAB sighting equipment and surveying equipment from Wild of Switzerland and navigation equipment from the then Ferranti UK. First production FH-77Bs for India left the Bofors factory in August 1986.

The first Indian order for the FH-77B was for a total of 410 weapons with final systems delivered in 1990. The company delivered FH-77Bs to India in complete batteries to enable Indian artillery units to become fully operational as complete units.

The ammunition included smoke, illuminating (Bofors) and new High-Explosive Extended-Range (HEER) projectiles for which India was the first customer. There were two versions of the HEER supplied, one with boat tail and the other with base bleed.

When the original order was placed by India, it was envisaged that the total Indian requirement would consist of some 1,500 weapons.

The second batch of FH-77Bs would have been assembled in India from components supplied by the company, with the third batch being made under licence in India. For a number of reasons, India did not place any further orders for the FH-77B.

In July 1999, the company announced that it had received a small but important contract from India for spare parts for the 410 155 mm FH-77B field howitzers used by the Indian Army. This was the first order placed by India with the company since 1989.

The order covered the continuance of spare parts within the framework of the original FH-77B contract placed in 1986.

A second contract was placed in India with the company in September 1999 for further spare parts worth a total of SKR190 million.

BAE Systems 155 mm FH-77B deployed in firing position. This weapon has the new muzzle brake 0500612

In March 1990, the company announced that the Swedish Matériel Defence Administration had placed a USD162 million contract with the then Bofors for the supply of 48 FH-77B artillery systems, spares, software, ammunition and associated charges. The actual quantity was subsequently increased and a total of 51 are currently in service in Sweden.

The Swedish Army FH-77Bs are similar to those produced for India but with some modifications to enable them to be operated at lower temperatures.

The Nigerian FH-77Bs and the original Indian FH-77Bs had a parallel pepperpot muzzle brake but, following trials with the more recent ammunition types including HEER used with Charge 9, a new and more efficient muzzle brake of the double-baffle type was developed. This was fitted to late production FH-77Bs and has now been retrofitted to earlier weapons.

Ammunition in the Swedish Army order included the locally-developed HEER projectile, developed to meet the requirements of the Indian Army which, when used with Charge 9, has a range of 30,000 m. High-Explosive Extended-Range - Base Bleed (HEER-BB) was included in the order and the FH-77B could also use ammunition already stockpiled by the Swedish Army.

As of March 2013, the FH-77B is the only artillery system in service with the Swedish Army and is deployed with A 9 Artillery Regiment in Boden.

In Swedish Army service this has already started to be replaced by the Archer 155 mm/52-calibre 6 × 6 self-propelled artillery system which is covered in detail in a separate entry in *IHS Jane's Land Warfare Platforms: Artillery and Air Defence*.

Under the terms of a contract placed by BAE Systems a total of 24 Archer are being delivered to Sweden with a similar number being supplied to Norway.

In both countries the Archer will be the only conventional tube artillery system deployed.

Description

The FH-77B differs from the earlier FH-77A in three main areas; a slightly longer barrel, the introduction of a breech screw mechanism using bag charges, and a new hydraulic loading system.

The FH-77B can be towed at speeds of up to 70 km/h, normally by the Saab-Scania SBAT 111S (6 × 6). For cross-country travel the driver of the tractor vehicle can remotely engage the gun roadwheels by starting the gun hydraulic motor making an 8 × 8 combination. Maximum speed in this configuration is 7 km/h and when this speed is exceeded the gun wheels are disengaged.

Steering is by regulating the drive to the independent roadwheels of the gun. The same hydraulic motor can also be used for several other functions. To get the gun into action the driver on the gun uses the hydraulic power to raise the gun castor wheels on the trail to disengage from the tractor towbar.

In the self-propelled mode the trails could be locked together or split. The gun driver then reverses the gun to dig the trail spades into the ground. For making changes of fire direction the same process can be repeated. Using this system the driver on the gun can make rapid fire direction changes, get the gun in and out of action and move the gun from one firing location to another alone.

The ammunition loading system uses a hydraulic crane driven from the hydraulic system. The crane grab is designed to lift any 155 mm projectile in groups of three at a time, offered to the grab either horizontally or vertically and almost any projectile length can be handled, including the longer enhanced-range projectiles.

From the crane the projectiles are loaded onto a loading table and are then guided onto the loading tray, while another loader places the propellant charge onto a separate loading trough/rammer. In two operations the projectile loading tray is swung over to line up with the chamber and the charge tray swings over behind it. The projectile is then hydraulically rammed forward into the chamber with the charge pushed right forward behind the projectile. Once the projectile is seated in the driving bands the loading trough/rammer is withdrawn until the charge is stripped from the tray by a retaining catch. The rammer mechanism is fully withdrawn, the breech screw is closed and a primer cartridge automatically loaded into the primer seat from a magazine holding eight tubes.

Using this system, projectiles and charges can be loaded into the gun at angles of elevation of up to +70°. The rate of fire, with three people on the gun, can be three rounds in 12 seconds. The subsequent rate of fire is governed only by the ammunition supply. An added bonus of the hydraulic ramming is that uniform ramming reduces the resultant projectile dispersion.

The full crew of the FH-77B is six: the gun commander, a layer and four loaders. One loader operates the mechanism for loading the projectiles from the loading table to the loading tray, one prepares the projectiles by inserting the fuzes, setting them and loading the projectiles in groups of three into the loading crane, one places the charges onto the loading tray and another loader operates the ammunition crane. If necessary, only the layer, crane operator and projectile loader combined, and a charge loader can operate the gun in action.

There is a special internal communication system built into each gun, separate from the usual fire-control communication systems, with a 10 m range based on an induction loop with receivers built into the crew's ear protectors.

Gunlaying is fully hydraulic. Two levers, one on the right for elevation and depression and one on the left for traverse allow barrel movement. Several types of sight may be fitted but the usual model fitted is the electronic BEAB BAAB RIA.

The layer sits high on the left of the breech with a protective open cage over the seat and sight (the loading crane is on the right of the breech). A ×4 telescope sight is fitted for direct fire. The BAAB NK24 night sight fitted in front of the ordinary telescopic sight permits direct fire under night conditions.

As the FH-77B depends on hydraulic power for many of its systems, an electric motor for all operations using an internal or external power supply or two hydraulic hand pumps are provided in case the main system fails. One, situated just forward of the layer's position and in the power compartment itself, can provide manual hydraulic power for traverse, elevation and ramming. The other pump is on the other side of the gun and provides manual hydraulic power for operating the support wheels.

Unlike the FH-77A the ammunition system of the FH-77B is based on a bagged charge system which enables a wide range of 155 mm ammunition to be fired, including the following main types of projectile: the now BAE Systems HE M/77 B, which is a high-capacity shell weighing 42.6 kg and containing 8 kg of TNT or Composition B. This has a maximum range of 24,000 m when fired from the FH-77B howitzer. The round is fully compatible with all Western-manufactured guns and internationally adopted fuzes.

HE M107, the standard NATO 155 mm projectile weighing 42.91 kg. It can be fired to a range of approximately 19,000 m.

Development of a base bleed round designated High-Explosive Extended-Range (HEER) has been completed by the now BAE Systems. HEER is compatible with all Western-manufactured 155 mm guns and international fuzes. Maximum range is 30,000 m and the weight is the same as conventional 155 mm rounds. This involves less stress on the barrel than firing other types of projectiles such as base bleed with a greater weight. Production is also made easier by omitting ERFB nubs.

Other rounds available are different types of smoke (including infra-red smoke) and the MIRA illuminating round in two versions, with maximum ranges of 17,000 m and 24,000 m respectively. For both, the time of illumination is 60 seconds, descending velocity 4 m/s and the mean illuminating intensity is 2.2 Mcd.

Standard NATO and other propellant charges can be used with the FH-77B but the company has developed its own charges especially for the FH-77B. Charges available include M3 which provides a maximum muzzle velocity of 375 m/s to give a range of 10,300 m and M4 which provides a maximum muzzle velocity of 563 m/s to give a range of 15,500 m.

Charges include the Charge 8 which is roughly equivalent to M119 but gives more favourable pressure and burning properties at extreme temperatures. Charge 8 provides a muzzle velocity of 685 m/s to give a range of 19,000 m. Charge 9 is a new development to provide a muzzle velocity of 827 m/s and a range with standard projectiles of 24,000 m and with base bleed projectiles of 30,000 m.

The FH-77B can also fire the 155 mm BONUS top-attack artillery projectile developed to meet the operational requirements of the French and Swedish armies by the now BAE Systemss of Sweden and Nexter of France. First production BONUS projectiles were handed over to France and Sweden in mid-2002.

The Bonus 155 mm top attack artillery projectile has also been supplied to Saudi Arabia by Nexter Systems as part of a complete CAESAR 155 mm/52-calibre 6 × 6 artillery system.

BAE Systems 155 mm FH-77B from the front clearly showing new muzzle brake (BAE Systems) 1044608

BAE Systems 155 mm FH-77 BO5 L52 deployed in the firing position from the front (BAE Systems) 1133938

In Swedish and Indian Army service this weapon was originally towed by a Saab Scania SBAT IIIS (6 × 6) cross-country truck that in addition to the gun crew carries a quantity of ready use ammunition. It could however be towed by other vehicles.

Variants

Original 155 mm Field Howitzer 77A

The FH-77A was developed to meet the requirements of the Swedish Army and a total of 200 were built between 1978 and 1984. All of these have now been phased out of service and scrapped. Further development of the FH-77A resulted in the FH-77B.

FH-77B with MAPS

The 51 155 mm FH-77Bs delivered to the Swedish Army by late 1992 were fitted with the Honeywell Modular Azimuth Position System (MAPS) land navigation system. The Honeywell MAPS system was fitted to the small number of 155 mm Bandkanon 1A self-propelled artillery systems used by the Swedish Army. These have now been withdrawn from service and scrapped.

Swedish FH-77B upgrade

The now BAE Systems built 200 of the earlier FH-77 towed systems for the Swedish Army of which 100 are in reserve and 100 have been sold.

A total of 51 of the improved FH-77B were built and these are now the only front line artillery systems deployed by Sweden.

Archer 155 mm self-propelled system

Under a GBP135 million (USD200 million) contract awarded to BAE Systems in March 2010 the company will build a total of 48 of the latest FH-77 BW L52 155 mm Archer Self-Propelled (SP) artillery systems.

Of these, 24 are being supplied to the Norwegian Army and 24 to the Swedish Army and in each case these will be the only artillery systems to be deployed in the future.

Under the terms of this contract two pre-production Archer were supplied in 2010 followed by the first production batch of six systems in 2011 of which Sweden will get four and Norway two.

Full rate production was underway by 2012 and all systems will be delivered by 2013.

As part of the deal, all of the Archer will be fitted with the Norwegian Kongsberg Protector Remote Controlled Weapon Station (RCWS) armed with a .50 M2 HB MG.

Archer will use elements of the 155 mm FH-77B but will have a 155 mm/52-calibre ordnance and be integrated onto the rear of a Volvo (6 × 6) all terrain chassis.

Details of the Archer are provided in a separate entry in *IHS Jane's Land Warfare Platforms: Artillery and Air Defence*.

Mounted Gun System

BAE Systems is already offering Archer on the export market with another version, called the Mounted Gun System (MGS) being aimed at the potential Indian Army requirement for a 155 mm/52-calibre wheeled system.

This has the same chassis as the Archer but is fitted with the ordnance and elevating mechanism of the FH-77B05 towed weapon that is loaded manually.

This weapon is a contender for the Indian towed gun requirement. For the Indian market BAE Systems is teamed with the local company of Mahindra.

BAE Systems FH-77 B05 L52 L52

The now BAE Systems has upgraded two examples of the FH-77B 39-calibre with a 155 mm/52-calibre ordnance and this is designated the FH-77 B05 L52. They were successfully demonstrated in India in 2002, 2003, 2004 and 2006. As of March 2013 no production contracts had been awarded by India for any 155 mm artillery systems.

The FH-77 B05 L52 enables a range of 41+ km to be achieved with standard base bleed projectiles. It is fitted with an integrated navigation, positioning and aligning system and onboard ballistic computation.

The new gun computer has a man-machine interface similar to standard commercial PCs, which facilitates the training and operation of the computer, according to the company.

Many procedures are automatic including indirect laying and execution of MRSI firings. According to the now BAE Systems this automation improves speed, accuracy and safety when firing. In addition, the operator will also have access to manuals stored in the computer.

The onboard gun computer and navigation system give the system an autonomous capability. It is also fitted with a new turbocharged diesel engine. The communications equipment installed allow it to communicate with not only with other weapons but also other batteries up to a range of 5 km.

It can fire all standard types of 155 mm artillery projectiles using modular- or bag-type charges. It also has a direct-fire capability and a MRSI with up to five projectiles hitting the target within three seconds from one weapon. It has a direct-fired range out to 2,000 m.

It has a burst rate of fire of three rounds in 13 seconds and it is claimed to be able to come into action in 90 seconds and come out of action in a similar time.

Many of the components of the latest FH-77 B05 L52 are identical to the existing FH-77B, which allows existing customers to upgrade their weapons with a 155 mm/52-calibre barrel.

This systems weighs about 13.1 tonnes and the 155 mm/52-calibre ordnance gives a maximum range of 41,000 m firing a base bleed projectile.

Specifications

	155 mm Field Howitzer 77B
Dimensions and weights	
Crew:	6
Length	
overall:	11.16 m (firing)
transport configuration:	11.6 m
Width	
overall:	7.18 m (firing)
transport configuration:	2.65 m
Height	
transport configuration:	2.82 m
axis of fire:	1.7 m
Ground clearance	
overall:	0.42 m
Track	
vehicle:	2.2 m
Weight	
standard:	12,000 kg
Mobility	
Carriage:	split trail with APU
Speed	
max speed:	8 km/h (APU)
towed:	70 km/h
Fording	
without preparation:	0.75 m
Auxiliary engine:	petrol or diesel
Firepower	
Armament:	1 × carriage mounted 155 mm howitzer
Main weapon	
length of barrel:	6.045 m
breech mechanism:	interrupted screw
muzzle brake:	double baffle
maximum range:	24,000 m (standard)
maximum range:	30,000 m (BB)
Rate of fire	
rapid fire:	10 rds/min
Main armament traverse	
angle:	60°
Main armament elevation/depression	
armament front:	+70°/-3°
Survivability	
Armour	
shield:	no

Status
Production complete. In service with India (410), Nigeria (48) and Sweden (51).

Contractor
BAE Systems (previously Bofors).

Taiwan

155 mm extended range gun

Development
In the late 1970s, the Combined Service Forces (CSF) developed manufacturing capability for a version of the Israeli Soltam Systems 155 mm M-68/M-71 family of towed howitzers.

It is believed that two prototypes were built in Taiwan. From the late 1970s, all effort was concentrated on the 155 mm towed extended range gun.

It is estimated that Taiwan has about 350 155 mm towed artillery systems of all types.

Description
The 155 mm towed extended range gun uses the same ordnance and recoil system as the XT-69 self-propelled weapon. This is based on the hull of the now BAE Systems M108 (105 mm)/M109 (155 mm) full-tracked self-propelled artillery systems which were supplied from the United States. Details of this system, which has not been exported, are given in a separate entry in *IHS Jane's Land Warfare Platforms: Artillery and Air Defence*.

The ordnance is mounted on a two-wheeled split trail carriage similar to that of the US 155 mm M114 howitzer and incorporates some of the features found on the SRC International modernised M114/39 howitzer.

A handwheel is fitted at the elevation controls and the pneumatic equilibrators are of the high-capacity type. Mounted over the left trail are a pneumatic rammer, pressurised air bottle, swing-aside loading tray and a telescopic rammer mounted near the rear cradle.

This enables a high rate of fire to be achieved and the weapon to be loaded at all angles of elevation. It fires standard ammunition of the separate loading type (for example projectile and charge).

The 155 mm 45-calibre ordnance is fitted with a multibaffle muzzle brake and a fume extractor. The weapon can fire all standard US/NATO ammunition as well as ERFB (with a maximum range of 30,000 m) which is now being manufactured in Taiwan. It is believed that 155 mm base bleed ammunition has now been developed which has extended the range of the weapon to around 39,000 m.

The original US M114 was also built in Taiwan under the designation of the T-65 and as far as it is known, this is identical to the US model covered in a separate entry in *IHS Jane's Land Warfare Platforms: Artillery and Air Defence* It is estimated that about 250 of these are in service with Taiwan, as well as about 90 older M59 155 mm Long Tom towed artillery weapons.

Variants
155 mm self-propelled artillery system

Taiwan has developed to the prototype stage a new truck mounted (6 × 6) 155 mm self-propelled artillery system whose upper mount and ordnance may be related to the 155 mm extended range towed artillery system which is believed to have a 39-calibre barrel.

As of March 2013 this 155 mm self-propelled artillery system remained at the prototype stage. It is considered that this is probably a technology demonstrator and would be followed by a system based on a chassis with more cross-country mobility.

Status
Production complete. In service with Taiwanese Army. The system was never offered on the export market.

Contractor
Combined Service Forces, Taiwan.

Thailand

105 mm M425 howitzer

Description
The Royal Thai Army has deployed a 105 mm towed howitzer designated the M425. This is believed to be the US-supplied 105 mm M101 upgraded in a number of key areas including the installation of a new and longer 105 mm barrel which has a life of 3,000 rounds.

105 mm M425 towed howitzer with trails spread 0501048

In 1997, it was confirmed that Thailand had taken delivery from Nexter Systems, France of a total of 285 kits to upgrade its old US-supplied M101s with the complete barrel of the Nexter Systems LG1 Light Gun.

As well as the locally developed 105 mm M425, other sources have indicated that Thailand may have in service 32 locally built 105 mm guns under the designation of the M-618A2. At this stage it is not known if the M425 and the M-618A2 are in fact the same weapon.

The M425 has a conventional split carriage with a shield that extends either side over the wheels. The barrel is provided with a double baffle muzzle brake to reduce recoil forces when the weapon is fired.

The 105 mm M425 howitzer has a hydropneumatic recoil system.

Imports of towed artillery to Thailand

According to the United Nations Arms Transfer Lists the following quantities of towed artillery weapons were imported by Thailand between 1992 and 2010:

From	Quantity	Date	Comment
Austria	18	1992	155 mm GH N-45
Austria	18	1996	155 mm GH N-45
Austria	36	1998	155 mm GH N-45
France	24	1995	105 mm LG1
Italy	12	2002	105 mm Pack Howitzers
US	20	1994	artillery
US	20	1995	artillery

BAE Systems has more recently supplied Thailand with a batch of 24 L119 Light Guns.

Specifications

	105 mm M425 Howitzer
Mobility	
Carriage:	split trail
Firepower	
Armament:	1 × carriage mounted 105 mm howitzer
Main weapon	
breech mechanism:	horizontal sliding block
muzzle brake:	double baffle
maximum range:	14,500 m
Rate of fire	
sustained fire:	4 rds/min
rapid fire:	10 rds/min
Main armament traverse	
angle:	46°
Main armament elevation/depression	
armament front:	+61.87°/-5.34°
Survivability	
Armour	
shield:	yes

Status

Believed to be in service with the Royal Thai Army. Production complete. Not offered on the export market. In 2004 Thailand placed an order with BAE Systems for 20 L119 Light Guns and deliveries of these took place in 2006.

Contractor

Upgrade carried out in facilities of the Royal Thai Army.

Turkey

155 mm/52-calibre Panter howitzer

Development

The 155 mm/52-calibre Panter towed howitzer was developed in the 1990s to meet the operational requirements of the Turkish Land Forces Command (TLFC).

Following trials and modifications with a number of prototype systems, the first production order was placed and the first batch of six 155 mm/52-calibre Panter systems were completed in mid-2002.

These were built at the Cankiri facilities of MKEK CANSAS and handed over to the 105th Artillery Regiment in Corlu with a ceremony held at the 1011th Ordnance factories in Ankara.

The first production batch of 155 mm/52-calibre Panter consisted of 18 units, which was sufficient to equip one artillery regiment, which has three batteries each of six weapons.

While Turkey has carried out extensive upgrades on old US supplied M44 (155 mm) and M52 (105 mm) self-propelled weapons, which have been fitted with a 155 mm/39-calibre barrel, the 155 mm/52-calibre Panter is the first complete artillery system to have been developed in Turkey.

The 155 mm/52-calibre Panter has started to supplement the current M114 towed howitzer, which only has a range of 14.6 km and fires standard ammunition.

155 mm/52-calibre Panter howitzer deployed in the firing position without spades embedded (MKEK) 1133939

The TLFC have a total of 517 M114s in service but only part of this fleet will be replaced by the much more effective 155 mm/52-calibre Panter.

The TLFC is now taking delivery of the locally manufactured Firtina 155 mm/52-calibre self-propelled artillery system. This is essentially the Republic of Korea 155 mm/52-calibre K9 Thunder. Details of both self-propelled systems are given in separate entries in *IHS Jane's Land Warfare Platforms: Artillery and Air Defence*.

In is understood that local production and/or assembly of the 155 mm/52-calibre Panter howitzer is now being undertaken in Pakistan, probably at the Heavy Industries Taxila (HIT) facility.

Description

The 155 mm/52-calibre Panter howitzer is mounted on a conventional split-trail carriage. When in the travelling position the 155 mm/52-calibre ordnance is traversed through 180° and locked in position over the closed trails. The travel lock is mounted on the right trail.

Mounted on the forward part of the carriage is an Auxiliary Power Unit (APU) that enables the 155 mm/52-calibre Panter to propel itself at a maximum speed of 18 km/h. The APU includes a German DEUTZ air-cooled diesel developing 160 hp.

According to MKEK the APU gives this system a cruising range of up to 200 km.

A burst rate of fire of three rounds in 15 seconds can be achieved.

When deployed in the firing position the weapon is supported on a circular baseplate mounted under the carriage and the two trails each of which is provided with a spade. When deployed in the firing position, the four main roadwheels are raised clear of the ground.

Each trail leg has a small wheel to assist in bringing the weapon into the firing position. These are also used in conjunction with the four main wheels when the weapon is being used in its self-propelled mode.

The sighting system, as well as the laying equipment, is mounted on the left side where the seat for the layer is provided.

The 155 mm/52-calibre Panter is used in conjunction with the new locally developed Aselsan BAIKS-2000 Field Artillery Battery Fire Direction system. This will improve response times as well as improving first round hit probability under a variety of weather conditions at all ranges.

Specifications

	155 mm/52-calibre Panter Howitzer
Dimensions and weights	
Crew:	5
Length	
transport configuration:	11.6 m
Width	
transport configuration:	3.3 m
Height	
transport configuration:	2.6 m
Weight	
standard:	15,000 kg
Mobility	
Carriage:	split trail with APU
Speed	
max speed:	18 km/h (APU)
towed:	80 km/h
Gradient:	45%
Vertical obstacle	
forwards:	0.3 m
Firepower	
Armament:	1 × carriage mounted 155 mm howitzer
Main weapon	
length of barrel:	8.060 m
muzzle brake:	double baffle
maximum range:	24,000 m (M107 HE projectile)
maximum range:	30,000 m (M549A1 RAP projectile)
maximum range:	40,000 m (est.) (ERFB-BB projectile)

	155 mm/52-calibre Panter Howitzer
Rate of fire	
sustained fire:	2 rds/min
rapid fire:	6 rds/min
Main armament traverse	
angle:	34° (17° left/17° right)
Main armament elevation/ depression	
armament front:	+69°/-3°
Survivability	
Armour	
shield:	no

Status
Production. In service with Turkish Land Forces Command (TLFC). It is understood that the weapon is also entering service with Pakistan under an agreement signed with MKEK in mid-2007. At least 30+ have been built.

Contractor
See development.

United Kingdom

BAE Systems 155 mm Lightweight Howitzer (M777)

Development
The 155 mm Lightweight Howitzer was originally developed as a private venture. Its origins can be traced back to the early 1980s when Vickers Shipbuilding and Engineering Limited (VSEL - which today is BAE Systems) originally perceived a potential market for a lightweight 155 mm towed howitzer.

In the Spring of 1987 the project definition was completed. Its objective was to have a weapon with the same range as the US Army's M198 155 mm towed howitzer but weighing no more than 4,000 kg.

The earlier M198 weighs 7,163 kg which limits its air mobility; it can only be carried by two US helicopters, the US Army Boeing CH-47 or the US Marine Corps Sikorsky CH-53.

The US Army was fully briefed on the system and agreed that if the company built a prototype of the system with its own money it would carry out a complete evaluation of the system.

In September 1987, the main board gave approval to build two prototypes of the system, which is today called the 155 mm Lightweight Howitzer. Both were completed in late 1989.

The complete upper part of the weapon was test fired at Eskmeals in June 1989 with a total of 50 rounds being fired at all elevations, 12 of which were zone 8S (top charge).

Although the weapon was originally targeted at the US Army, the US Marine Corps took the initiative as it was looking for a lightweight 155 mm system to replace all current 105 mm and 155 mm towed artillery systems.

Following its unveiling at the 1989 Association of the United States Army Exhibition in Washington DC, one of the two prototypes went to the US for early evaluation.

This evaluation, under the supervision of the US Army Armament Research and Development Command on behalf of the US Marine Corps, took place in three phases through to 1990.

Canadian BAE Systems 155 mm M777 Lightweight Howitzer deployed in Afghanistan (Canadian DND) 1169586

155 mm Lightweight Howitzer fitted with Selex Galileo LINAPS (Christopher F Foss) 1040795

It also completed limited land mobility trials and airlift certification in single and split mode. At the end of Phase 1 the system was awarded limited live crew clearance for the US.

Phase 2 was conducted at the US Marine Corps Base at Camp Lejeune, North Carolina and at the Naval Base at Little Creek, Virginia. During Phase 2 the system achieved a single lift with the UH-60L Black Hawk helicopter. Amphibious trials were carried out successfully at Little Creek.

The final phase took place at Aberdeen Proving Ground, Maryland where the system carried out successful climatic chamber firings at temperatures ranging from −25°C to +145°C. These climatic firings were followed by air transportability (split lift) trials and 622 km of land mobility trials on test tracks ranging from trails to Belgian blocks and included wading to a depth of 1.5 m.

The US then had a competition which involved extensive tests with the 155 mm Lightweight Howitzer and the Light Towed Howitzer developed at the then Royal Ordnance facility at Nottingham. In the end the former was selected.

For the US programme, Textron Marine & Land Systems was selected to be the prime contractor with the then Vickers Shipbuilding and Engineering Limited being the main subcontractor.

By 1998 it was clear that the US programme was running into problems and early in 1999 the now BAE Systems assumed the role of prime contractor of the troubled XM777 towed artillery system from its team member Textron Marine & Land Systems. This company no longer has any involvement with the programme.

In September 2000, following an extensive competition, BAE Systems finally selected its core industrial supplier base for US production of the XM777 155 mm weapon.

The body assembly is manufactured by HydroMill Inc of Chatsworth, California, stabilisers, spades and trails are supplied by Major Tool and Machining Inc of Indianapolis, Indiana, the breech operating load tray system is provided by Rock Island Arsenal, Rock Island, Illinois, with titanium being supplied by RTI International Metals Inc of Niles, Ohio.

In late 2002, BAE Systems was awarded a USD135 million contract by the US DoD for the Low Rate Initial Production (LRIP) of the M777 following its type classification.

Under the initial phase of the LRIP contract, BAE Systems has built 94 M777s for the US Marine Corps, with first weapons delivered in February 2003 from the company's Hattiesburg, Mississippi facility.

The M777, which while under development was called the XM777, has now replaced all front line M198 155 mm towed artillery systems in service with the US Army and Marine Corps.

Under the five-year Engineering and Manufacturing Development (EMD) contract a total of nine systems were built at the BAE Systems facility at Barrow-in-Furness. These underwent an extensive series of tests in the US during which more than 10,000 rounds of ammunition have been fired.

The nine EMD guns were followed by two pre-production (PP1 and PP2) guns from the US production line to test and validate the US production base.

According to BAE Systems, about 70 per cent of the M777 is made in the US, including the 155 mm/39-calibre barrel, which is provided by Watervliet Arsenal. Barrow-in-Furness manufacture the upper cradle as well as the suspension and running gear.

In March 2005, BAE Systems was awarded a contract worth USD834 million covering the supply of 495 M777A1 155 mm/39-calibre lightweight howitzers for the US Army and Marine Corps.

The 495 M777A1 were delivered over a four-year period starting in July 2006 and running through to late 2009.

Marines from the 3rd Battalion, 11th Marine Regiment stationed at Twentynine Palms, California, were the first unit to receive the M777.

Early in 2002, BAE Systems was awarded a USD41 million contract by the DoD for the development of the M777E1 over a three-year period.

Under this contract, BAE Systems integrated the General Dynamics Armament and Technical Products Towed Artillery Digitization/Digital Fire-Control System (TAD/DFCS) onto the weapon.

BAE Systems is prime contractor, with General Dynamics being main subcontractor. The M777E1 was Type Classified as the M777A1 in January 2007. The M777A2 is the M777A1 modified to fire the Raytheon 155 mm Excalibur precision guided munition. This has a maximum range of 40 km and is accurate to 10 m.

155 mm M777 deployed in the firing position (BAE Systems) 1340082

The M777A2 also has a hard wired fuze setter which is located near the assistant gunner's position and used to programme the Excalibur fuze prior to loading. Excalibur can also be fired by other towed and self-propelled artillery systems with a modified fire-control system.

US Army M777A1s are fitted with TAD/DFCS and this has been back fitted to the US Marine Corps LRIP M777, which are designated the M777A1.

Installation of the TAD/DFCS enables the M777A1 to fire the Raytheon Excalibur precision guided projectile out to a range of 40 km with an accuracy of 10 m.

Late in 2005 Canada took delivery of the six M777 155 mm lightweight towed artillery weapons and these were issued to the 1st Regiment Royal Canadian Horse Artillery.

These were supplied to Canada by the US Marine Corps through the US Foreign Military Sales (FMS) programme from the BAE Systems production line at Hattiesburg in the US. In early 2006 these were deployed by Canada to Afghanistan in support of operation Archer. This was the first operational deployment of the M777.

Since then an additional six weapons have been supplied which brings the total up to 12.

The Canadian M777 are fitted with the Selex Galileo Laser Inertial Artillery Pointing System (LINAPS).

LINAPS has been installed on all operational BAE Systems 105 mm L118 Light Guns of the Royal Artillery (RA) which proved highly effect in the Second Gulf campaign.

In early 2008 another contract was placed by the US DoD for an additional 87 weapons of which 25 were for the US Marine Corps and 62 for this US Army.

This brought the total up to 676 units of which 381 are for the US Marine Corps and 295 for the US Army.

In August 2008 BAE Systems was awarded another contract for 47 weapons worth GBP42.8 million (USD85.6 million) for an additional 43 weapons. At this time the company had delivered over 400 weapons to Canada, US Army and US Marine Corps.

In April 2009 BAE Systems delivered the 500th M777 155 mm/39-calibre towed lightweight howitzer to the US Armed Forces.

At this time the M777 order book stood at 737 units which included the original eight Engineering, Manufacturing and Development (EMD) weapons and 12 weapons supplied to Canada to meet an Urgent Operational Requirement (UOR).

Current US production model is the M777A2 and all earlier weapons have now been brought up to this standard which includes the Digital Fire Control System (DFCS) which is the primary element of the Towed Artillery Digitisation (TAD).

Block 1a software allows the M777A2 to fire the M982 Excalibur precision guided munition which gives the field artillery pinpoint accuracy at ranges out to 40 km.

Production of the M777 series peaked at 14/16 weapons per month with production now being lowered to about 10 weapons a month as required by the customer. In 2009 the company delivered 150 M777A2 to the US Armed Forces.

In May 2009 BAE Systems announced and order worth USD118 million for an additional 63 weapons of which 38 are for the US Army and Marines Corps and 25 to Canada through the Foreign Military Sales programme to add to the 12 it has already received.

In July 2009 the US ordered another 62 additional M777A2 under a contract worth USD117 million.

The UK did put some funding in the early days of the M777 programme but for the time being at least has decided to retain its 105 mm Light Guns as its only towed artillery weapon.

In July 2010 BAE Systems announced another order for an additional 93 M777 howitzer which took the total order book to 955 systems and total sales of GBP1 billion.

Of these 93 systems, 58 were for the US Army and Marine Corps and 35 for Australia which are being purchased through the Foreign Military Sales (FMS) programme.

In February 2011 BAE Systems announced another US Army contract for a further 46 M777 weapons which brought the grand total up 1,001.

In October 2011 it was announced that the US Department of Defense had awarded BAE Systems a contract worth USD134 million (GBP87 million) for a further batch of 70 M777A2 155 mm/39-calibre lightweight howitzers which will start to equip the US Army Infantry Brigade Combat Teams (IBCT).

This latest contract will mean that the M777A2 production line will now keep running through to at least October 2013 and brings the total order book up to 1071 weapons for the US Army, Marines Corps and export customers.

All export sales of the M777 series are through the Foreign Military Sales (FMS) programme and by early 2013 export orders have been placed by Australia (35) and Canada (37) with the latter country being the first to fire the weapon in action in Afghanistan.

During combat operations there have been problems with the power supply for the on board electronic equipment and in combat the weapon has had to be slaved to an external power source to enable the computerised Fire-Control System (FCS) to work.

This source is typically a High Mobility Multi-Purpose Wheeled Vehicle (HMMWV) or seven-tonne (6 × 6) truck.

BAE Systems has been awarded a contract to design, develop, qualify and place in production an improved Power Conditioning Control Module (PCCM) to supply power to the FCS.

When development is completed a total of 1,049 units will be produced for installation on current and future M777A2 weapons.

In addition the PCCM will be able to accommodate future accessories to enhance the capabilities of the M777 series such as electronic thermal management and laser ignition.

Current system uses lead-acid batteries but the new PCCM will also be able to use lithium ion batteries.

While the US uses the Advanced Field Artillery Data Systems (AFADS) computerised FCS, Canada has opted for the Selex Galileo Laser Inertial Pointing System (LINAPS) for all of its M777 and this is also fitted to all UK 105 mm Light Guns and well as a number of other countries.

In 2011 Saudi Arabia made a request for a major artillery package that included 36 M777 weapons but this was not taken up.

In October 2012 Australia stated that it would order an additional 19 M777A2 weapons from the US which would bring the total Australian M777 fleet up to 54 weapons.

In November 2012 the Indian MoD sent a letter of request to the US government for the acquisition of 145 M777 fitted with the Selex LINAPS system already fitted to the M777 used by Canada and all British Army 105 mm M118 Light Guns.

Description

Although the M777 uses advanced materials in its construction, it is claimed to be simple to operate and maintain under field conditions.

The 155 mm/39-calibre ordnance (M776E2) is essentially that of the M284 barrel used by the US Army's M109A6 Paladin self-propelled howitzer which is fitted with the M199 muzzle brake as used by the current towed M198 howitzer but modified to take a towing eye.

The conventional screw breech is hydraulically operated and opens vertically. For this application the breech of the 155 mm M776A2 cannon has the screw breech turned 90° to allow vertical operation between the cradle tubes.

A derivative of the M49 firing mechanism is mounted to the primer feed mechanism. The magazine has the capacity for 10 M82 primers.

The 155 mm/39-calibre ordnance can be elevated to 1,275 mils (71°) and depressed to −43 mils (3°). Top traverse is 400 mils (23°) left and right.

To keep the weight down, extensive use has been made of extruded titanium alloy fabrications, designed by BAE Systems but subcontracted to other companies, as well as efficient structural design.

Some of the parts perform two functions, for example the Horstman Defence Systems designed hydropneumatic suspension system also operates as a hydraulic jack.

The elevating mass comprises two subassemblies, the cradle and the 155 mm/39-calibre cannon tube assembly.

The recoil system of the M777 is of the hydropneumatic type with a maximum recoil of 1.39 m.

Trunnion height is being quoted as 0.65 m when the M777 is deployed in the firing position.

It should be noted that the weight quoted in the specifications table of 4,128 kg includes section equipment.

Australian 155 mm Lightweight Howitzer M777 deployed in the firing position (Australian MoD) 1459971

The cradle includes four extruded titanium tubes, an accumulator, two recoil cylinders and balancing gear. The cannon tube assembly includes the cannon tube, muzzle brake, towing eye, primer feed mechanism and screw breech.

The carriage comprises two subassemblies, the body and the saddle. The body includes two forward stabilisers and two split trails fitted with self-digging spades and dampers.

Mounted on either side of the body is a hydrogas suspension unit fitted with stub axle and aluminium rimmed road wheels. A small hydraulic hand pump is installed at each wheel station to raise and lower the system into and out of action.

A centrepost on the body receives the saddle, and a rack segment attached to the rear of the body engages with the saddle-mounted traverse gearbox enabling the saddle to be rotated through 400 mils left and right.

The saddle is fitted with a yoke-mounted roll screw and nut elevation gear actuated by handwheels on either side of the saddle.

In the baseline model, a universal sight mount is fitted on the left side but provision is made for a second sight on the right side for customers requiring a two-sight system.

A loading tray is provided to the rear of the breech, which is opened automatically. BAE Systems is considering future options, which could include a flick rammer, but the aim for the present is to keep the weapon as simple as possible.

When in the travelling configuration it is towed by its muzzle in a similar manner to the Russian 122 mm D-30 howitzer by any truck with a two-tonne capacity (or greater) fitted with air-operated brakes.

While the M777 is normally operated by a seven person detachment but can be operated by a five person detachment.

The US Marine Corps uses the Oshkosh Medium Tactical Vehicle Replacement (MTVR) (6 × 6) to tow the M777 while the US Army uses the BAE Systems/Oshkosh Defence Family of Medium Tactical Vehicle (MTVR) to tow their M777. It can however be towed by many types of 6 × 6 and 4 × 4 tactical vehicle.

Towing speed on roads is being quoted as up to 80 km/h while towing speed across-country is being quoted as up to 50 km/h, but these figures depend on the type of terrain.

It can be internally transported in a number of transport aircraft including the C-130, C-141, C-5 and C-160 Transall. It can also be transported slung under the UH-60L/UH-60M Black Hawk and CH-53E/CH-53D helicopters and the MV-22 Osprey.

Variants

155 mm M777 Tilt Bed Carrier

This was developed to the prototype stage but is no longer marketed.

155 mm Lightweight Howitzer (M777) with APS

For trials purposes the system has been fitted and tested with the now Selex Galileo Laser Inertial Artillery Pointing System (LINAPS). This is already installed on all L118 Light Guns of the UK Royal Artillery. LINAPS is also installed in the M777 supplied to Canada.

M777 portee

This was developed to meet the British Army requirements for a Lightweight Mobile Artillery Weapon System (Gun) and based on an 8 × 6 chassis.

In the end, the UK MoD cancelled the LIMAWS (G) requirement and all development and marketing work on this weapon has now ceased by BAE Systems.

In 2008 the UK MoD also cancelled the LIMAWS (R) programme for which the prime contractor was Lockheed Martin UK.

Specifications

	155 mm Lightweight Howitzer (M777)
Dimensions and weights	
Crew:	7
Length	
overall:	10.236 m
transport configuration:	10.584 m
Width	
overall:	3.721 m
transport configuration:	2.589 m
Height	
transport configuration:	2.336 m
axis of fire:	1.219 m
Ground clearance	
overall:	0.66 m
Track	
vehicle:	2.30 m
Weight	
standard:	4,128 kg
Mobility	
Speed	
towed:	88 km/h
Firepower	
Armament:	1 × carriage mounted 155 mm howitzer
Main weapon	
length of barrel:	6.096 m
calibre length:	39 calibres
breech mechanism:	interrupted screw
muzzle brake:	double baffle
maximum range:	24,700 m (unassisted projectile)
maximum range:	30,000 m (assisted projectile)
maximum range:	40,000 m (M982 Excalibur)
Rate of fire	
sustained fire:	2 rds/min
rapid fire:	4 rds/min
Operation	
stop to fire first round time:	180 s (est.)
out of action time:	120 s (est.)
Main armament traverse	
angle:	46° (23° left/23° right)
Main armament elevation/depression	
armament front:	+71°/-3°
Survivability	
Armour	
shield:	no

Status

Production on order with the countries listed in the table:

Country	Quantity	Comment
Australia	19	Australia stated that it would order weapons in October 2012
Canada	12 + 25	with LINAPS - delivered
US	1,071	Army and Marines

Contractor

BAE Systems.

BAE Systems 105 mm Light Gun

Development

In 1965, the British Army issued a requirement for a new 105 mm Light Gun to replace the Italian Oto Melara 105 mm Pack Howitzer then in use with the light regiments of the Royal Artillery.

105 mm L118 Light Gun of the Royal Artillery deployed in the firing position with Pinzgauer (6 × 6) vehicle to the right (Pinzgauer) 1044605

The requirement was for a gun with a longer range, a more stable firing platform, improved reliability and the ability to be towed across rough terrain at high speed. Design work began at the Royal Armament Research and Development Establishment at Fort Halstead the following year.

Trials with the prototypes were successful and in 1973 the 105 mm Light Gun (L118) was accepted for service with the British Army.

First production 105 mm Light Gun in the L118 version was handed over to the British Army at the Royal Ordnance Factory Nottingham in 1974

This major production facility subsequently became part of BAE Systems and eventually closed.

The 105 mm Light Gun, which uses the same family of 105 mm ammunition as the Abbot self-propelled gun, replaced all 105 mm Pack Howitzers in regular units of the Royal Artillery and has been issued to TA units. Each 105 mm Light Gun Regiment typically has three batteries, each with six guns.

Late in 1981, Australia selected the 105 mm Light Gun, where it is known as the Hamel Gun, to meet its future requirements.

In early 1986, following an extensive series of trials and demonstrations, the Light Gun was type-classified by the US Army as the M119. The Light Gun version involved is the L119 with a barrel suitable for firing US 105 mm M1 ammunition. The type classification followed the evaluation of six Light Guns leased to the US Army in late 1984, followed in mid-1985 by a further 14 guns. These were evaluated by the 9th Infantry Division at Fort Lewis, Washington, as the XM119.

The US Army is the largest user of the 105 mm Light Gun having taken delivery of 147 direct from Nottingham with another 280 manufactured under licence in the USA by Watervliet Arsenal (ordnance) and Rock Island Arsenal (carriage). Licensed production of the original 105 mm Light Gun has been completed in the US. Production of the more recent L119A2 has now started.

In mid-1991, the Brazilian Army placed an order for four 105 mm Light Guns in the L118 configuration. These were delivered early in 1992.

In December 1994, Brazil placed an order for a further 36 105 mm L118 Light Guns and the first of these were delivered early in 1995. This was followed by an order for 18 L118 Light Guns for the Brazilian Marines.

Early in 1995, the Royal Netherlands Marine Corps received eight 105 mm Light Guns from the United Arab Emirates.

It should be noted that these were not deployed by the Netherlands.

In July 1995, the Spanish Cabinet approved the purchase of 56 105 mm L118 Light Guns for use by the Spanish Army Rapid Deployment Forces (FAR).

These 105 mm Light Guns are used by three battalions each consisting of three batteries, each with six guns. The remaining two were for the artillery school.

Total value of the contract, which covered the supply of the guns, spares and ammunition was estimated to be PTA8,850 million (USD48 million) over a five-year period ending in 1999.

EXPAL and the then General Dynamics - Santa Barbara Sistemas were involved in the production of 105 mm ammunition in Spain. In addition, the contract included 38 kits to convert the L118 to the L119 standard to allow standard 105 mm HE M1 ammunition to be fired for training purposes.

The first batch of L118 105 mm Light Guns, together with a quantity of ammunition, spare parts and kits to convert the weapons to the L119 configuration for training purposes, were shipped in January 1996.

By early 2013, over 1,100 105 mm Light Guns had been built, including those produced under licence in Australia and the US.

Late in 1996, following an extensive series of user trials between the French Nexter Systems 105 mm LG1 Mk II Light Gun and the UK's 105 mm Light Gun, the latter was selected to meet the requirements of the Portuguese Army.

The contract for 21 105 mm L119 Light Guns was signed late in December 1997 and the first batch of eight guns was shipped in February 1998.

Portugal selected the 105 mm L119 gun which fires standard 105 mm HE M1 semi-fixed high-explosive ammunition. The Portuguese order included a 100 per cent offset.

L118 105 mm Light Gun fitted with Selex Galileo LINAPS 0036868

In 2004 it was confirmed that Thailand had placed a contract with BAE Systems covering an initial batch of 20 105 mm Light Guns.

The initial batch of weapons was for the Royal Thai Army but in the future it is expected that additional weapons will be procured for the Royal Thai Marine Corps, with the total requirement for both users between 60 and 80 weapons in the long term.

Thailand selected the L119 version, which uses the standard US 105 mm M1 HE projectile with a maximum range of 11.2 km as this is already used by other 105 mm weapons currently in service.

The first batch of 105 mm L119 Light Guns were delivered to Thailand late in 2005.

The Thai 105 mm L119 weapons are fitted with the now Selex Galileo Laser Inertial Pointing System (LINAPS), which is standard on all British Army 105 mm Light Guns. This will reduce into-action time, provide enhanced survivability against counter-battery fire and provide greater accuracy.

Production of the 105 mm Light Gun was originally undertaken at BAE Systems facility at Nottingham which is now closed.

The company had to find new suppliers for almost every part of the weapon and these have now been qualified during an extensive series of firing and towing trials with two L119 weapons in the UK.

The 105 mm ordnance was supplied by the BAE Systems facility at Barrow-in-Furness, while HSW of Poland supplied a number of elements including fabrications, trail and cradle. The saddle was provided by Paul Fabrications of the UK.

Description

The 105 mm Light Gun consists of the following main components: elevating mass, saddle and controls, trail and spades, platform, suspension and sighting system.

The elevating mass consists of the 105 mm ordnance, recoil system, cradle including trunnions, balancing gear and electric firing gear. The gun is fitted with a multibaffle muzzle brake, which can be easily removed for cleaning.

The vertical sliding breech block is actuated by a lever mounted on the top. The hydropneumatic recoil system has a separate recuperator and the buffer is fitted with a cut-off gear to shorten the recoil length as elevation is increased. The recoil system is mounted in a lightweight fabricated cradle, which carries the elevating arc and a simple helical compression spring-balancing gear.

The recoil system of the 105 mm L118 Light Gun is of the hydropneumatic type and at 0° quadrant elevation this is a maximum of 1.07 m while at 70° quadrant elevation it is 0.33 m.

The weight on the towing eye of the 105 mm L118 Light Gun is 168 kg while the weight of the elevating mass on its own is 1,066 kg and the weight of the lower carriage and stores is 794 kg.

While the length in the firing position is 7.01 m the length in the travelling position with 105 mm ordnance forwards is being quoted as 6.629 m, length travelling with 105 mm ordnance traversed over the rear trails is 4.876 m. It should be noted that the latter is the normal towing configuration.

Height travelling with the ordnance forwards is 2.63 m while height travelling with ordnance traversed over the closed trails is 1.371 m.

The firing gear, which is mounted on the cradle, is fully waterproofed and is of the electric magneto type, specially designed for use with the Cap Conducting Composition primer of the 105 mm Abbot ammunition system.

The saddle is a lightweight fabrication on which the complete elevating mass is carried. It provides a top traverse of 11°. The elevating gear is a simple mechanical system providing constant handwheel effort throughout the range of elevation. A clutch is provided to reduce abnormal firing loads on the equipment.

Differences between 105 mm Light Gun Ordnances

Weapon designation	Ordnance designation	Rifling length	Number of grooves	Twist	Firing mechanism	Muzzle brake	Max range
L118A1	L19A1	3.21 m	28	1 in 19 constant	electric	double baffle	17,200 m
L119A1	L20A1	2.779 m	36	1 in 18 to 1 in 35	percussion	single baffle	11,500 m
L127A1	L27A1	3.052 m	32	1 in 25.6	percussion	double baffle	n/av

105 mm Light Gun L118 deployed in firing position and with wheels on turntable (BAE Systems) 1333678

The trail is fabricated in high-strength corrosion-resistant steel. It is bow shaped, enabling the breech operator and loader to remain within the trail and maintain a high rate of fire at all elevations. The platform is carried on top of the trail during travelling.

The forward end carries the saddle, traversing gear, suspension and the layer's seat. At the rear are the non-rotating towing eye, overrun brake, barrel clamp and spades. The spade system, of which there are three types, caters for various firing and ground conditions. The first type is a combined rock and digger spade for use, normally with the platform, on rock or firm ground; the second is a field spade for use on very soft ground or when firing without a platform and the third is a snow spade.

The platform is a circular lightweight fabrication that gives a firm base and permanent gun stability under all conditions. The tyres run on the outer edge of the platform, which is connected to the underside of the gun by four wire stays.

The wheels, which have large section 9.00 × 16 tyres, incorporate special hydraulically operated overrun brakes. This ensures safe towing at a high speed with a light vehicle such as a Land Rover (4 × 4). The brakes can also be operated during firing by a lever at the rear end of the trail. For manhandling purposes, an individual handbrake is fitted at each wheel. The trailing arm suspension has laminated torsion springs and shock-absorbers. The suspension system remains in operation during firing and assists in maintaining the stability of the equipment and reducing carriage stresses.

Indirect and direct sight system and controls are operated by the layer while seated. Indirect laying has been simplified by the abolition of a separate setting for angle of sight. The quadrant elevation is set on the elevation scale. A direct fire telescope is fitted which incorporates a moving illuminated graticule, which is adjusted for lead to allow for target movement and range.

A direct fire night sight is also available. Trilux activated light sources are fitted to illuminate all scales and graticules, eliminating the use of electric batteries on the equipment.

The L35 charge contains the first five charge zones in a single brass cartridge case. The propellant charge is ignited by the L10 electrical primer, which is more reliable than a percussion primer. The Supercharge, L36, for maximum range, has a separate brass case and also has an L10 electrical primer. Polycarbonate spoiler rings can be clipped onto the nose of the projectile enabling targets to be engaged between 2,400 m and 3,600 m; these reduce the range by causing aerodynamic drag.

In mid-1986, it was revealed that the now BAE Systems was working on extended-range (base bleed) 105 mm round as a private venture. The former has been fired with standard Supercharge L36 and has achieved a range of over 20,000 m compared to the 17,200 m of the standard projectile. The base bleed round is currently designated the X/RO 0381 ERHE which during trials in 1998 achieved a range of around 20.60 km. This extended-range projectile has yet to enter production.

The 105 mm Light Gun fires the following BAE Systems-manufactured separate loading ammunition: HE (L31) with the projectile weighing 16.1 kg; HESH (L42) with the projectile weighing 10.49 kg (no longer produced); illuminating (L43); SH/practice (L41– no longer produced); smoke (L45) with the projectile weighing 15.89 kg; smoke, coloured red (L37) and smoke, coloured orange (L38).

The charge system for the L118 Light Gun is as follows:

Charge 1	2,500 to 5,700 m (minimum range using shell spoilers)
Charge 2	2,700 to 7,200 m (minimum range using shell spoilers)
Charge 3	5,900 to 9,500 m
Charge 4	7,900 to 12,200 m
Charge 4½	8,700 to 13,600 m
Charge 5	15,300 m (maximum)
Charge Super	17,200 m (maximum)

It can also fire US 105 mm M1 semi-fixed ammunition by using a complete interchangeable percussion-fired ordnance (L119) which is available. The charge coverage is as follows:

105 mm Light Gun being towed by Pinzgauer (4 × 4) vehicle (Michael Jerchel) 0105797

Charge 1	1,800 to 3,400 m
Charge 2	2,100 to 4,100 m
Charge 3	2,600 to 4,800 m
Charge 4	4,100 to 6,300 m
Charge 5	4,200 to 7,900 m
Charge 6	5,000 to 9,500 m
Charge 7	6,200 to 11,500 m

A range of 14,300 m can be achieved by the L119 Light Gun firing the M760, which is basically an M1 projectile with an M200 charge system. A range of 19,500 m can be achieved with the M913 HERA projectile and 19,000 m with the new ERM1 base bleed projectile.

Rates of fire of the 105 mm L118 Light Gun are as follows:

- For one minute - 12 rounds a minute
- For three minutes - six rounds a minute
- Sustained - three rounds a minute.

The 105 mm Light Gun can be carried slung under a Puma helicopter as one load. The elevating mass can be removed to give two light helicopter (for example, Wessex) loads and the complete weapon can be reassembled with one simple tool in less than 30 minutes.

The 105 mm Light Gun is normally towed with the gun in the forward (firing) position. For long distances the gun is towed with the barrel rotated through 180° and clamped to the trail.

This attitude is achieved by jacking up the equipment with a jack stowed on the trail, removing the quick-release wheel and traverse gear pin, clamping the barrel and replacing the wheel. The complete conversion takes less than one minute.

Firing US M1 105 mm ammunition and using the M200 charge system originally developed for the M204, a range of 14,000 m is achieved. Firing the M913 HERA cartridge, with the M229 charge 8 propelling charge, a range of 19,500 m is achieved.

US changes include: modifications to the sight to permit use of the deflection method of laying, revision of the graticule pattern of the sight to incorporate 5 mil graduations instead of 10 mil, addition of attachment brackets for the battery computer, provision of attachment brackets for M90 muzzle velocity measurement system, labelling of all items containing radioactive elements (tritium sights), modification of tie down brackets, use of standard US paint type and US camouflage pattern, use of nitrogen in the recoil system instead of air, use of US tail-light patterns, incorporation of recent UK MoD minor modifications, elevation level guidance plate for M200 charge system, modified engraving and limited US thread forms in some areas of the weapon.

It should be noted that the US Army has introduced a new family of enhanced 105 mm ammunition for its M119 Light Guns.

Variants

To meet the requirements of the Swiss Army the L127A1 Light Gun was developed by BAE Systems, with the ordnance being designated the L27A1. This is very similar to the British Army version but can fire Swiss ammunition. There were two guns delivered to Switzerland in 1979-80 with a further four following in 1981, but no production order was placed.

105 mm L119 Light Gun complete with turntable above trails but without sights fitted (Christopher F Foss) 1365030

Enhanced ammunition for 105 mm Light Gun

The then UK Defence Procurement Agency in 2001 awarded a contract to the now BAE Systems to manage the Assessment Phase for enhanced ammunition for the 105 mm Light Gun used by the Royal Artillery.

In 2005 following an international competition, the then UK Defence Procurement Agency (DPA) awarded BAE Systems a contract to supply a new generation of High-Explosive (HE) ammunition for the 105 mm L118 Light Gun used by the Royal Artillery.

Under the terms of the contract, awarded by the DPA's Future Artillery Weapons Systems (FAWS) Integrated Product Team (IPT), an initial batch of 105 mm HE ammunition were supplied for qualification.

The new Improved Ammunition (IA) for the 105 mm L118 Light Gun is of the Insensitive Munition (IM) type and replace the in-service 105 mm L31 HE projectile, which has a maximum range of 17.2 km.

The new 105 mm HE IM round will also have improved lethality, at least 20 per cent, and will be the first IM artillery round to enter service with the UK. It will be ballistic match to the in-service L31 HE round.

Production of the new IA 105 mm ammunition is undertaken at the BAE Systems main ammunition production facility at Glascoed in South Wales.

In recent years the company has invested some GBP15 million in this new IM facility at Gloscoed, which is now the ammunition HQ of BAE Systems.

BAE Systems has also developed, in association with German company Buck Neue Technologien, a new 105 mm smoke projectile with a red phosphorous payload. This offers bi-spectral screening in the vision and infra-red spectra.

105 mm Light Gun upgrade
(Mid-Life Upgrade)

BAE Systems has developed an upgrade package for the L118 and L119 Light Gun which can be incorporated into new production weapons or retrofitted to existing weapons.

The main improvements can be summarised as:

- Currently one of the wheels has to be removed when rotating the gun to its stowed position. This is time consuming and adds to the burden of the gun crew. Instead of removing the wheel slide it partly out of the axle and let it hinge down. When the gun is rotated simply lift the wheel, slide it back into position and lock it in place
- The current platform must be handled by the crew when bringing the weapon into action. The platform is relatively heavy and weight saving would reduce crew fatigue and improve overall gun performance. A new platform has been constructed from aluminium and thus weight is significantly reduced
- The maintenance burden of the existing recoil system is relatively high and a maintenance free recoil system has been introduced with fewer parts and modern sealing technology
- When hosing the gun down with high-pressure hose jets water can ingress into the gearbox. A thicker cover plate has been introduced with a modern sealing capacity
- Under some circumstances there is difficulty in attaching the 105 mm Light Gun to the towing vehicle and there is limited crew assistance for a large traverse movement. Handgrips have been introduced to the gun trails to assist the user
- The existing brake system is maintenance intensive and the handbrake lever can be difficult to operate. An updated brake system is introduced for easier user operation
- The existing 'A' frame under the forward part of the weapon can become distorted and has difficult to engage locking pins. Into and out of action times are slower due to pin engagement difficulties. The existing A frame is replaced with a single strut and centre traverse lock
- The clearance between the gun and towing vehicle can cause user problems with towing and manoeuvring. A new extended towing eye is introduced.

105 mm Light Gun with LINAPS

Early in 1999, following an international competition, the UK MoD selected the now Selex Galileo Laser Inertial Pointing System (LINAPS).

The contract, valued at over GBP10 million, was for the supply of 137 APS, or Laser Inertial Automatic Pointing Systems (LINAPS) as it is called by the company, for installation on the BAE Systems 105 mm L118 Light Guns used by the Royal Artillery.

First production LINAPS were delivered late in 1999 and deliveries ran to the year 2002.

The LINAPS consists of the proven FIN3110L integrated Inertial/Global Positioning System (GPS), a large screen Layers Display and Control Unit (LDCU), a wheel hub mounted odometer and a Power Management System (PMS). In addition there is a Navigation Display Unit (NDU) for use in the cab of the Pinzgauer Turbo D 4 × 4 gun-towing vehicle.

The mounting of the INS/GPS directly above the barrel permits the direct measurement of all elevation and azimuth and hence the rapid and accurate pointing of the gun onto the target. In addition, LINAPS provides the precise location of the 105 mm gun platform with a route navigation capability.

It provides the crew with orientation and location data without the requirement for much of the existing sighting and survey equipment and its associated training.

The installation of the APS on the BAE Systems 105 mm Light Gun L118 has made for a much-reduced into-action time as well as greater accuracy. LINAPS is also installed on the Royal Thai Army 105 mm L119 Light Guns and on the Canadian 155 mm M777.

UK upgrade to 105 mm Light Gun with new LDCU

Late in 2011 it was stated that the British Royal Artillery (RA) was to upgrade its fleet of 105 mm L118 Light Guns with a new Layers Display and Control Unit (LDCU).

This was carried out under a GBP4.3 million (EUR5 million) contract awarded to Selex Galileo by the Defence Equipment & Support, Artillery Systems IPT.

Deliveries started late in 2012 with the LDCU being fitted by the Defence Support Group with the assistance of Selex Galileo.

The LDCU forms a key part of the Selex Galileo Laser Inertial Artillery Pointing System (LINAPS) which was first installed on all RA 105 mm L118 Light Guns and accepted for service in July 2001.

The LDCU is the man-machine interface with LINAPS which is a self-contained weapon mounted navigation, pointing and weapon management system that allows the weapon to come into action more quickly as well as increasing accuracy.

The existing LCDU, supplied by the South African company Denel Land Systems, is to be replaced by a new DRS Technologies (also a Finmeccanica company) 10 inch display with enhanced capability.

The LCDU, installed on the left side of the L118 Light Gun, includes a touch sensitive, electro-luminescent display, powerful processing for the on-board fire control and ballistic calculations and the ability to interface with a range of additional sensors and equipment such as a muzzle velocity radar.

According to Selex Galileo, LINAPS provides the weapon with a navigation and fire-control capability with minimal hardware content within a unique single box solution.

It provides continues and accurate 3D location with and without global positioning and continuous determination and display of gun barrel direction and elevation.

Under the earlier LINAPS Capability Enhancement Programme (CEP) carried out by Selex Galileo the original Power Management System (PMS) was changed to Battery Power Management (BPM). The installation of the new DRS Technologies LCDU is called the LCDU Obsolescence Programme.

105 mm Light Gun in saluting role

For many years the British Army used the 25 pounder as its standard saluting gun. These have now been replaced by a modified 105 mm Light Gun.

This work was carried out by the now Defence Support Group (previously ABRO). The guns were stripped of their components and all paint was chemically removed. All components were given a thorough safety check and the firing mechanism was altered to allow only blanks to be fired.

Finally the weapon was reassembled and specially painted, although some parts were chromed in order to improve its appearance.

Defence Support Group 105 mm Light Gun base overhaul

Following a competition, the Defence Support Group was selected by the UK MoD to carry out vase overhaul of the BAE Systems 105 mm Light Gun. This is carried out at the Defence Support Group facility in Stirling.

Specifications

	105 mm Light Gun (L118)
Dimensions and weights	
Length	
overall:	7.01 m (firing)
transport configuration:	4.876 m
Width	
overall:	1.778 m
Height	
transport configuration:	1.371 m
Ground clearance	
overall:	0.5 m
Track	
vehicle:	1.4 m
Weight	
standard:	1,860 kg
combat:	1,860 kg
Mobility	
Carriage:	bow
Tyres:	9.00 × 16
Firepower	
Armament:	1 × carriage mounted 105 mm howitzer
Main weapon	
breech mechanism:	vertical sliding block
muzzle brake:	double baffle
maximum range:	17,200 m
Rate of fire	
sustained fire:	3 rds/min
rapid fire:	12 rds/min
Main armament traverse	
angle:	360° (on turntable)
Main armament elevation/depression	
armament front:	+70°/-5.5°
Survivability	
Armour	
shield:	no

Status

Production as required. In service as follows:

Country	Quantity	Comment
Abu Dhabi	70	less eight supplied to Netherlands in 1995 and 18 to Bosnia-Herzegovina
Australia	111	of which 105 were built under licence
Bahrain	8	
Benin	12	
Bosnia-Herzegovina	18	L118 from UAE (some sources state 36)
Botswana	36	Including 12 delivered in 1996
Brazil	58	40 army, 18 marines
Dubai	18	
Ireland	24	including 12 delivered in 1998
Kenya	58	
Malawi	9	
Morocco	36	refurbished
Nepal	8	delivered 1992
Netherlands	8	from Abu Dhabi in 1995 (not deployed)
New Zealand	24	built in Australia under licence
Oman	42	
Portugal	21	deliveries from 1998
Spain	56	ordered 1995, first deliveries 1996 and all delivered by 1998
Thailand	20	order placed in 2004, first deliveries late 2005
UK	166	including reserve and training
US	147	manufactured by BAE Systems
	280	manufactured under licence in the US
	377	current new build requirement

Contractor

BAE Systems.

United States

8 in (203.2 mm) howitzer M115

Development

The Westervelt Board, which was established early in 1919, recommended the development of a howitzer with a calibre of approximately 8 in (203 mm) mounted on a carriage with an elevation of 0 to +65°, a traverse of 360° and a maximum range of 18,000 yards (16,800 m). It was also stated that it would be desirable for a carriage to be developed which could mount either an 8 in howitzer or a 155 mm gun.

There were two 8 in howitzers subsequently developed, the M1920 and M1920M1, plus a carriage which could also be used for a 155 mm gun.

Owing to a shortage of funds the project was suspended until 1927. Between 1927 and 1929 the carriage was redesigned for high-speed transport and the breech ring of the 8 in howitzer modified. The howitzer was known as the T2 and was to have a centrifugally cast barrel, but as facilities for centrifugal casting were not then available, it was decided to design a barrel of cold-worked forged steel under the designation T3. Tests of the T3 howitzer began at Aberdeen Proving Ground, Maryland in 1939 and the following year the T3 howitzer was standardised as the 8 in Howitzer M1 and the 155 mm Gun/8 in Howitzer Carriage T2E1 with minor modifications was standardised as the M1.

Production commenced in 1941, with first weapons being completed the following year and by the end of the Second World War, over 1,000 had been completed.

After the Second World War the complete weapon was redesignated the Howitzer, Heavy, Towed: 8 in: M115. This consisted of Cannon Howitzer M2 or M2A1, Recoil Mechanisms M4 series (M4, M4A1, M4A2 or M4A3) and Carriage M1.

By today's standards the 8 in howitzer M115 is obsolete as it is heavy, has a low rate of fire, a short range and takes too long to come into and to be taken out of action. It also has limited cross-country.

As far as it is known, the US has not exported any of its surplus 8 in howitzer M115 systems in the last 10 years.

Description

The M2/M2A1 8 in cannon consists of the barrel, breech ring and breech mechanism. The M2A1 cannon is ballistically and physically interchangeable with the cannon M2, the major difference being that the M2A1 was constructed with stronger steels. Recoil system is of the hydropneumatic type.

The carriage consists of equilibrator assemblies, elevating and traversing mechanisms, two single-wheel, single-axle heavy limber M5, two-axle bogie with eight tyres and two trails. Four spades, carried on the trails, are used to emplace the weapon. In recent years some armies have towed the weapon without the limber.

The 8 in howitzer M115 fires the following separate loading (projectile and bagged charge) 8 in ammunition:

8 in howitzer M115 in travelling configuration and with limber attached (Richard Stickland) 0526365

- HE (M106) with the projectile weighing 92.53 kg, maximum muzzle velocity of 587 m/s to a maximum range (Charge 7) of 16,800 m
- HE (M404) (carries 104 M43A1 grenades) with the projectile weighing 90.72 kg, maximum muzzle velocity of 587 m/s to a maximum range (Charge 7) of 16,800 m
- HE (M509) (carries 195 M42 grenades) with the projectile weighing 93.66 kg, maximum muzzle velocity of 594.4 m/s to a maximum range of 16,000 m.

It should be noted that not all of these 8 in projectiles are currently used, especially by the remaining export customers.

In many countries the 8 in M115 was subsequently replaced by a self-propelled version called the M110.

When deployed by the US Army, the 8 in (203.2 mm) howitzer was normally towed by a M125 (6 × 6) cross-country truck that also carried a quantity of ready use ammunition.

Versions of the M110 with the longer barrel are the M110A1 (no muzzle brake) and the M110A2 (fitted with muzzle brake). The 8 in M110 series of self-propelled howitzers were finally phased out of service with the US Army (Regular) in March 1995.

The last unit was the 5th Battalion, 18th Field Artillery, based at Fort Sill, Oklahoma. There was no direct replacement as such of the 8 in howitzer in US Army service but its role was taken over by the Multiple Launch Rocket System (MLRS).

In most countries the M115 and the M110 have been phased out of service due to short range. In NATO service their prime mission was to fire tactical nuclear projectiles, which have now been phased out of service.

Prime contractor for the M110 series of self-propelled tracked howitzers is the now BAE Systems.

Variants

155 mm M59 (Long Tom)

This uses the same carriage as the 8 inch (203.2 mm) M115 howitzer and is still used by Jordan (18), Bosnia (three) and Taiwan (about 70), but in each case it is understood that these are held in reserve.

It should be noted that this was originally designated the M1, then the M2 and after the Second World War the M59.

It is referred to as the Long Tom because of its long range and accuracy.

It could fire a 155 mm High-Explosive (HE) projectile to a maximum range of 22,000 m with other projectiles include armour piercing (M112) and smoke (M104).

Specifications

	8 in Howitzer M115
Dimensions and weights	
Crew:	14
Length	
transport configuration:	10.972 m
Width	
overall:	6.857 m
transport configuration:	2.844 m
Height	
transport configuration:	2.743 m
Ground clearance	
overall:	0.318 m
Track	
vehicle:	1.98 m
Weight	
standard:	14,515 kg
combat:	13,471 kg
Mobility	
Carriage:	split trail with limber
Tyres:	11.00 × 20
Firepower	
Armament:	1 × carriage mounted 203.2 mm howitzer
Main weapon	
length of barrel:	5.142 m
breech mechanism:	interrupted screw
muzzle brake:	none
maximum range:	16,800 m
Rate of fire	
sustained fire:	0.5 rds/min
rapid fire:	1 rds/min

	8 in Howitzer M115
Main armament traverse	
angle:	60°
Main armament elevation/depression	
armament front:	+65°/-2°
Survivability	
Armour	
shield:	no

Status

Production complete. In service with the following countries, in some cases these are held in reserve:

Country	Quantity	Comment
Croatia	19	phasing out
Iran	20	
Jordan	4	reserve (also has 18 155 mm M59 guns in reserve)
Pakistan	28	reserve
Saudi Arabia	8	reserve
Taiwan	70	
Turkey	162	reserve

Status

US Government facilities.

155 mm howitzer M198

Development

Development of a new 155 mm towed howitzer to replace the standard 155 mm towed howitzer M114/M114A1 began in 1968 under the direction of the Project Manager for Cannon Artillery Weapon Systems (CAWS) of the US Army Armament Command at Rock Island, Illinois.

A test rig was built to demonstrate how a lightweight towed 155 mm howitzer could fire the increased range of ammunition, with reasonable probability of stable dynamics and structural soundness.

Design and fabrication of an advanced development prototype began in 1969 and firing trials commenced in 1970.

Rock Island was responsible for the carriage and recoil mechanism, Frankford Arsenal for the fire-control equipment, Watervliet Arsenal for the 155 mm ordnance and Picatinny Arsenal and the Harry Diamond Laboratory for the 155 mm ammunition.

The first two prototypes were delivered in April and May 1972 and were followed by a further eight prototypes.

After trials, the XM198 was standardised as the M198 and production commenced at Rock Island Arsenal in 1978 with first production weapons completed in July 1978.

The M198 replaced M114A1s in separate field artillery batteries assigned to corps/army and the general support field artillery battalions of the light divisions.

The first battalion of M198s became operational at Fort Bragg in April 1979. The first Marine Corps battalion to be equipped with the M198 was the 2nd Battalion, 10th Marines, at Camp Lejeune.

Production of the M198 was carried out at Rock Island Arsenal which also manufactured all components with the exception of the 155 mm ordnance (which came from Watervliet Arsenal), fire-control system and tyres.

In February 1992, the 1,672nd M198 rolled off the production line at Rock Island Arsenal. During FY91, a total of 120 were manufactured followed by 120 more in FY92.

155 mm howitzer M198 in firing position 0500614

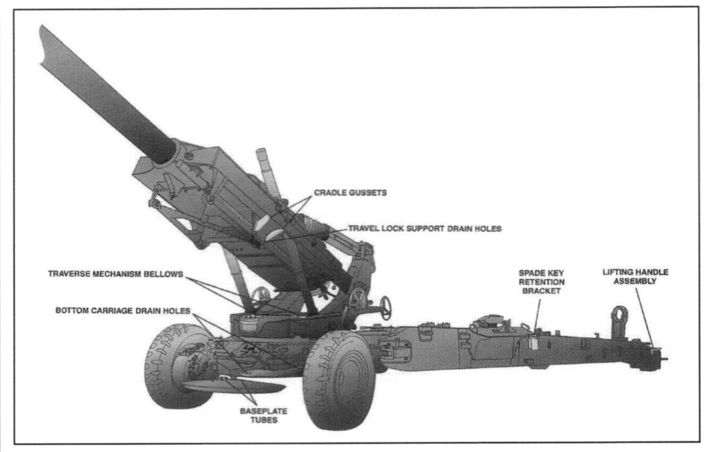

The Mandus Group is offering a number of improvements for the M198 155 mm towed artillery system including the elements shown in this rear view drawing (Mndus Group) 1452924

In June 1992, Watervliet Arsenal was awarded a USD5.402 million increment as part of a USD5.509 million firm fixed-price contract for 292 product improvement kits for the M198 towed howitzer with all work completed by April 1995.

In December 1993, Watervliet Arsenal was awarded a USD5.525 million increment as part of a USD10.730 million firm fixed-price contract for 602 product improvement kits for the M198 towed howitzer (quantity increased from the original 292) with the work completed by April 1995.

According to United Nations sources the following quantities of 155 mm M198 towed howitzers were exported by the United States between 1992 and 2010:

Country	Quantity	Comment
Bahrain	19	delivered in 1992
Lebanon	41	delivered in 2009
Lebanon	30	delivered in 2010
Pakistan	24	delivered in 1996
Saudi Arabia	18	delivered in 2010

The 155 mm howitzer has now been replaced in front line US Army and Marine Corps by the BAE Systems 155 mm/39-calibre M777 lightweight howitzer which was covered in a separate entry in *IHS Jane's Land Warfare Platforms: Artillery and Air Defence*.

In addition to being in service with the US Army and Marine Corps the M777 has also been sold to Australia and Canada with the latter country being the first to use the weapon in combat in Afghanistan.

The 155 mm M198 saw extensive combat with the US Army and Marine Corps during the invasion of Iraq in 2003. In this campaign it was used by the US Army and Marine Corps.

While no longer in front line service with the US Army or Marine Corps significant numbers of surplus 155 mm M198 howitzers have been exported in recent years.

The Mandus Group in co-operation with Rock Island Arsenal are now offering a number of upgrades for the 155 mm M198 howitzer to improve its reliability and details of this are provided later in this entry.

Description

The 155 mm howitzer M198 is mounted on a split trail carriage, fitted with a two-position rigid suspension system which can be rotated upwards to raise the wheels. This allows the weapon to rest on a non-anchored firing platform. When towed, the cannon is in the forward position but for storage it can be traversed through 180° and locked in position over the closed trails.

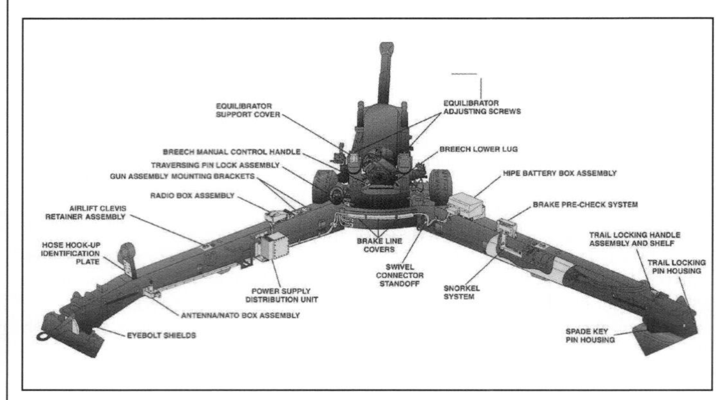

The Mandus Group is offering a number of improvements for the M198 155 mm towed artillery system including the elements shown in this front view drawing (Mandus Group) 1452925

155 mm howitzer M198 in service with Ecuador in travelling position 0130012

The carriage consists of the top carriage, elevating mechanism, equilibrators, traversing mechanism, bottom carriage, wheel suspension assembly, speed shift, trails and firing base.

The recoil mechanism is of the hydropneumatic type with a variable recoil length. The autofrettaged ordnance has a screw-on double-baffle muzzle brake, and a screw breech mechanism with a conventional obturator pad and split ring seal obturator.

When in travelling configuration with the ordnance traversed to the normal front firing position, the M198 has an overall length of 12.34 m but with the ordnance traversed through 180° and locked in position over the closed trails for travelling this is reduced to 7.44 m.

The fire-control equipment includes an M137 panoramic telescope with a magnification of ×4 and a 10° field of view, two elevation quadrants (M17 and M18) and an M138 elbow telescope with a magnification of ×8 and a 6° field of view.

The US M939 (6 × 6) five-tonne (US) truck was originally used by the US Army.

This was subsequently replaced in the US Army by the Family of Medium Tactical Vehicles (FMTV) (6 × 6) truck while the US Marine Corps used the Medium Tactical Vehicle Replacement (MTVR) (6 × 6) truck. Both of these are now used to tow the 155 mm/39-calibre M777.

While the maximum towing speed of the 155 mm M198 is quoted as 72 km/h on firm roads this is reduced to 40 to 48 km/h on secondary roads and as low as 8 km/h cross-country, but this depends on terrain conditions.

The internal configuration of the Cannon Assembly M199 is in accordance with the German/British/American/Italian quadrilateral agreement and will fire all current standard and developmental 155 mm ammunition of the anti-material and anti-personnel types. The weapon can fire the following types of 155 mm ammunition:

- Anti-tank (M718) (containing nine anti-tank mines) with the projectile weighing 46.72 kg, a muzzle velocity of 660 m/s and a maximum range of 17,740 m
- Anti-tank (M741) (containing nine anti-tank mines) with the projectile weighing 46.72 kg, muzzle velocity of 660 m/s and a maximum range of 17,740 m
- Copperhead Cannon Launch Guided Projectile (M712) to a range of 16,400 m
- HE (M107) with the projectile weighing 42.91 kg, muzzle velocity of 684.3 m/s and maximum range of 18,100 m
- HE (M449) (containing 60 anti-personnel grenades) with the projectile weighing 43.09 kg and a maximum range of 18,100 m
- HE (M483) (containing 88 dual-purpose grenades) with the projectile weighing 46.53 kg, muzzle velocity of 660 m/s and a maximum range of 17,500 m
- HE (M692) (containing 36 anti-personnel mines) with the projectile weighing 46.5 kg, muzzle velocity of 660 m/s and a maximum range of 17,740 m
- HE (M731) (containing 36 anti-personnel mines) with the projectile weighing 46.5 kg, muzzle velocity of 660 m/s and a maximum range of 17,740 m
- HE (M795) registration projectile for M483 and maximum range of 22,400 m
- HERA (M549 series) (RAP) with the projectile weighing 43.54 kg, maximum muzzle velocity of 826 m/s and a maximum range of 30,000 m
- Dual-Purpose Improved Conventional Munition (DPICM) M864 (type-classified late in 1987)
- Illuminating (M118 series) with the projectile weighing 46.26 kg, muzzle velocity of 684.3 m/s and a maximum range of 18,100 m
- Illuminating (M485 series) with the projectile weighing 41.73 kg, muzzle velocity of 684 m/s and a maximum range of 17,500 m
- Smoke WP (M110 series) with the projectile weighing 44.4 kg, muzzle velocity of 684.3 m/s and a maximum range of 18,100 m
- Smoke BE (M116 series) with the projectile weighing 42.22 kg, muzzle velocity of 684.3 m/s and a maximum range of 18,100 m
- Smoke (M825), base ejection, with a muzzle velocity of 797 m/s and a maximum range of 22,600 m
- M982 Excalibur precision guided munition.

It should be noted that not all of these 155 mm projectiles are currently and some were only issued to US artillery units.

Variants

Improved M198

Since it was first introduced the M198 has been upgraded on an almost continuous basis. According to the US Army the following upgrades have been incorporated to date:

155 mm howitzer M198 of the Royal Thai Army in the travelling configuration towed by a 6 × 6 truck (INA) 1452635

1982 - Add Gun Assembly Bracket

This modification revises the right and left trail assemblies providing a suitable mounting area for the installation of the Battery Computer System (BCS) gun assembly bracket.

1987 - Modification of the Breech Mechanism Assembly

This modification had two purposes. First a change in the release lever of the breech block assembly was to add a 'click' to the latch so that the user could be sure that the breech is closed. This feature improved safety during night time operations. Second machining the lower lug of the breech ring assembly to allow high angle firing (over 1,200 mils).

1992 - Modification of the Howitzer, Medium Towed: 155 mm, M198

This modification improved problem areas of the M198 and increased user satisfaction. Problem areas to be improved were equilibrator adjusting screw, traversing mechanism and locking devices, strengthening of the cradle travel lock strut, firing baseplate, retaining devices and addition of a brake pre-check system.

1993 - Additional Equilibrator Support Cover Screws

This modification addressed a problem with the top of the top carriages towers. The interim fix was adding additional screws and the finalised fix was a stronger cover plate and additional attachment screws for the equilibrator support cover.

1994 - Cradle Drain Holes

This modification provided drain holes in the cradle assembly.

1999 - Hydraulic Power Assist Kit

This modification provided an automated lifting system to the howitzer. It tied into the existing system and added a hydraulic pump, filter, solenoid box to the bottom carriage. A NATO receptacle for powering the system was located on the rear of the left trail. An on/off switch was located by the hydraulic selector valves. This saves time and physical effort during operations of the howitzer.

2002 - Drain holes in bottom carriage for ring gear

This modification added two to four 12.7 mm drain holes in the bottom carriage in the area where the traversing ring gear is attached.

2004 - Howitzer Improvement Power Enhancement Program (HIPE)

This modification added four new components to the M198. They were an antenna/NATO box, Power Supply Distribution Unit (PSDU), radio box and the battery box. These four components allow the howitzer to operate independently from the prime mover for the period of time by providing 24 V power supply to operate the HyPak, the Gunner Display Units (GDUs), the Muzzle Velocity System (MVS) and the SINCGARS radio. The modification commenced in 2006.

2005 - Breech modification for MACS

This modification upgraded the breech mechanism assembly to allow the firing of the M232 Modular Artillery Charge System (MACS) zone five. The original breech block housing assembly, firing mechanism block assembly and firing mechanism were replaced.

Mandus Group 155 mm M198 upgrades

The Mandus Group in co-operation with Rock Island Arsenal, is now offering a number of upgrades to the still widely deployed 155 mm M198 howitzer.

These include all of the upgrades previously mentioned under variants.

In addition, the Mandus Group/Rock Island Arsenal team can offer:

(1) Total life cycle support including parts, upgrades, modifications, and repair
(2) Technical skills to assist in training personnel to maintain customers current fleet of 155 mm M198 howitzers.
(3) technical manuals and updates
(4) Special tools and equipment that have been designed to efficiently support the howitzer
(5) Advice and recommendations about design changes tailored to meet customers specific user requirements.

Specifications

	155 mm Howitzer M198
Dimensions and weights	
Crew:	11
Length	
overall:	11 m (firing)
transport configuration:	12.34 m
Width	
overall:	8.534 m (firing)
transport configuration:	2.794 m
Height	
transport configuration:	2.9 m
axis of fire:	1.803 m
Ground clearance	
overall:	0.33 m
Track	
vehicle:	2.362 m
Weight	
standard:	7,163 kg
combat:	7,163 kg
Mobility	
Carriage:	split trail
Speed	
towed:	72 km/h
Tyres:	16.5 × 19.5
Firepower	
Armament:	1 × carriage mounted 155 mm howitzer
Main weapon	
length of barrel:	6.096 m
calibre length:	39 calibres
breech mechanism:	interrupted screw
muzzle brake:	double baffle
maximum range:	22,000 m (M483A1 projectile, Z8C charge)
maximum range:	18,150 m (M107 projectile, Z8 charge)
maximum range:	30,000 m (M549A1 RAP)
Main armament traverse	
angle:	45°
Main armament elevation/depression	
armament front:	+72°/-5°
Survivability	
Armour	
shield:	no

Status
Production complete but can be resumed if required. In service with following countries:

Country	Quantity	Comment
Australia	35	replaced by 155 mm M777
Bahrain	28	including 19 delivered in 1992
Ecuador	12	
Honduras	4	
Lebanon	77 + 30 (in 2010)	
Morocco	26	
Pakistan	124	including 24 delivered in 1996
Saudi Arabia	66 + 8 (in 2010)	40 National Guard
Somalia	18	status uncertain
Thailand	62	
Tunisia	57	
US (Army)	656	Replaced by M777A1/M777A2
US (Marines)	595	Replaced by M777A1/M777A2

Contractor
Rock Island Arsenal, Illinois.

155 mm howitzer M114

Development
In 1934, the development of a new split trail carriage was authorised under the designation T3. Development of the T3 was subsequently cancelled in favour of a new 155 mm howitzer (the T3) and a new carriage (the T2), which were standardised as the 155 mm Howitzer M1 and Carriage M1 in April 1941. The carriage was the same as that used for the 4.5 in field gun. By the time production of the M1 had been completed, just over 6,000 had been built.

After the Second World War the complete 155 mm weapon was redesignated the Howitzer, Medium, Towed: 155 mm, M114, consisting of the Cannon (M1 or M1A1), Carriage (M1A1) and Recoil System (M6, M6A1, M6B1 or M6B2). The M114A1 is almost identical to the M114 except that it has an M1A2 carriage.

US Army M114 155 mm howitzer in firing position (US Army) 0500615

US Marine Corps M114 155 mm howitzer in travelling configuration without shield (Ray Young) 0500616

According to United Nations' sources, between 1992 and 2010 the United States exported surplus M114s to Bosnia, which took delivery of 126 systems in 1997.

Description
The 155 mm howitzer M114 howitzer consists of the breech ring, breech block, breech mechanism and 155 mm barrel. The carriage M1A1 or M1A2 consists of top and bottom carriages, wheels and brake mechanisms, trails, equilibrators and the firing jack. The carriage is of the unsprung two-wheel split trail type with the wheels fitted with combat tyres for high-speed towing.

The carriage is equipped with airbrakes, which are operated from the prime mover, and there are handbrakes on each wheel for parking.

The barrel is balanced by spring-type equilibrators. The recoil mechanism, elevating and traversing mechanism, left and right shields, telescope mount and panoramic telescope are attached to the top carriage. The bottom carriage supports the weight of the top carriage and the barrel, and transmits operating stresses to the firing jack and trails.

The counter-recoil system consists of the recuperator cylinder, counter-recoil cylinder, and the counter-recoil and recuperator cylinder head stuffing box that links the two cylinders hydraulically.

The recoil system consists of the recoil cylinder, replenisher and variable recoil cam assembly. Recoil system is of the hydropneumatic type.

The M1 155 mm barrel is similar to the M1A1 but the M1A1 uses steels with increased physical properties. The M1A1 carriage has a rack-and-pinion-type firing jack while the M1A2 has a screw-type firing-jack. The travelling lock of the M1A2 is equipped with a firing jack hanger whereas that of the M1A1 is not.

Stated US Army rates of fire of the 155 mm howitzer M114 are:
- First 30 seconds - two rounds
- First four minutes - eight rounds
- First 10 minutes - 16 rounds
- First hour - 40 rounds (sustained).

The M114 fires the following types of US 155 mm separate-loading (for example projectile and bagged charge) ammunition:
- HE (M107) with the projectile weighing 42.91 kg, maximum muzzle velocity of 563.9 m/s and a maximum range (charge 7) of 14,600 m
- HE (M449) (carries 60 anti-personnel grenades) with the projectile weighing 43.09 kg, maximum muzzle velocity of 563.9 m/s and a maximum range (charge 7) of 14,600 m
- Illuminating (M485) with the projectile weighing 41.73 kg, maximum muzzle velocity of 576 m/s and a maximum range (charge 7) of 13,600 m
- Smoke (M116 series) with the projectile weighing 42.22 kg, maximum muzzle velocity of 563.9 m/s and a maximum range of 14,600 m
- Smoke WP (M110 series) with the projectile weighing 44.4 kg, maximum muzzle velocity of 563.9 m/s to a maximum range (charge 7) of 14,600 m
- M804 practice with a maximum range of 14,600 m.

Note: Not all these types of 155 mm ammunition are in current use.

In US Army service the 155 mm howitzer M114 was normally towed by a five-tonne (US) (6 × 6) cross-country truck that in addition to its crew carried a quantity of ready use ammunition (projectiles, charges and fuzes).

Variants

M114A2
It was originally understood that the M114A2 had the complete ordnance of the 155 mm M198.

It has now been confirmed that the barrel of the 155 mm M114A2 is only slightly longer than that of the M114A1 with the key identification feature being a groove that is cut around the outside of the barrel about 50 to 70 mm from the end.

The main tube difference is internal with the M114A2 tube possessing a 1 in 20 twist as opposed to the 1 in 12 twist used in earlier models. The greater twist extends the range of existing projectiles and other than the tube there are no additional changes in the M114A2 design.

In mid-1997, the US government announced that it was to supply the Bosnian Federation with 116 ex-US Army 155 mm M114A2 towed howitzers held in reserve and all of these were delivered by the end of 1997. An additional 145 M114A2 were earmarked for spare parts.

155 mm M114 howitzer of the Hellenic Army in travelling configuration (INA)
1452636

155 mm M114A2 Towed Howitzer (US Army) 0008270

Modified M114 (Netherlands)
Under the leadership of RDM Technology of the Netherlands, which has now ceased trading, the M114 155 mm towed artillery system was upgraded in a number of key areas including the installation of a 155 mm/39-calibre ordnance. The latter was supplied by the now BAE Systems (originally Bofors) of Sweden.

This was produced for Denmark (96), Netherlands (82) and Norway (48), but these have all now been phased out of front-line service. RDM Technology also offered new-build M114/39-calibre systems but these never entered production or service.

Modified M114 (Israel)
Full details of this conversion, developed as a private venture by Soltam Systems of Israel, are given in a separate entry in *IHS Jane's Land Warfare Platforms: Artillery and Air Defence*. As far as it is known, this remains at the prototype stage.

155 mm howitzer M65
This was the Serbian version of the M114 but was built only in small numbers. This is no longer marketed by Serbia.

M114 (France)
Nexter System (previously Giat Industries) of France developed an upgrade package for the M114 which included installation of a new 155 mm/39-calibre barrel. This was never sold and is no longer marketed.

KH179 (South Korea)
Details of the South Korean KH179, a conversion of the M114, are given in a separate entry in *IHS Jane's Land Warfare Platforms: Artillery and Air Defence*. This has been offered on the export market by South Korea but there are no known sales as of March 2013.

M114 (Iran)
For some years Iran has deployed a 155 mm/39-calibre towed artillery system, which is believed to be based on an M114 carriage. This is called the HM41, details of which are given in a separate entry in *IHS Jane's Land Warfare Platforms: Artillery and Air Defence*. A self-propelled version of this was revealed in 2011 and details of this system are provided in a separate entry in *IHS Jane's Land Warfare Platforms: Artillery and Air Defence*.

Specifications

	155 mm Howitzer M114
Dimensions and weights	
Crew:	11
Length	
transport configuration:	7.315 m
Width	
transport configuration:	2.438 m
Height	
transport configuration:	1.803 m
axis of fire:	1.676 m
Ground clearance	
overall:	0.229 m
Track	
vehicle:	2.07 m
Weight	
standard:	5,800 kg
combat:	5,760 kg
Mobility	
Carriage:	split trail
Tyres:	14.00 × 20
Firepower	
Armament:	1 × carriage mounted 155 mm howitzer
Main weapon	
breech mechanism:	interrupted screw
muzzle brake:	none
maximum range:	14,600 m
Rate of fire	
sustained fire:	40 rds/min
rapid fire:	4 rds/min
Main armament traverse	
angle:	49° (25° left/24° right)
Main armament elevation/depression	
armament front:	+63°/-2°
Survivability	
Armour	
shield:	optional

Status
Production complete. In service with the following countries:

Country	Quantity	Comment
Bosnian Federation	119	delivered 1997 (now 116)
Brazil	95	army
	8	marines
Ecuador	12	
Greece	129	
Iran	70	estimate, plus upgraded version that is called the HM41
Israel	50	reserve
Jordan	18	
Korea, South	1,800	also local M114 versions called KH179
Laos	12	
Lebanon	18	
Morocco	20	
Pakistan	144	
Peru	36	
Philippines	12	
Saudi Arabia	50	reserve
Serbia/Montenegro	66	including small number of M65
Somalia	6	status uncertain
Spain	44	
Sudan	12	
Taiwan	250	also local self-propelled model
Thailand	50	
Tunisia	12	
Turkey	517	
Uruguay	8	
Vietnam	n/avail	
Yemen	15	
Venezuela	12	

Contractor
Rock Island Arsenal, Illinois.

105 mm towed howitzer M119A1/M119A2

Development
Following the success of the now BAE Systems 105 mm L118 Light Gun during the Falkland Islands campaign, the US Army purchased operational testing weapons to determine the feasibility of using this weapon in the US Army's new light division concept.

The key in the US Army's evaluation was the ability to airlift an entire division artillery within the light division air transport constraints.

Following extensive trials, the US Army negotiated both a licence agreement, for production within the US and a production contract with the now BAE Systems in 1987.

Deliveries to the US Army commenced in December 1989 with the First Unit Equipped (FUE) being the 7th Infantry Division, Fort Ord, California.

Since then, production of the weapon has started again in the US and details of this are provided later in this entry.

105 mm M119A1 howitzer deployed by the US Army (US DoD) 0577750

In addition to being manufactured for the US Army, production of the 105 mm towed howitzer M119 can also be undertaken for the export market through the Foreign Military Sales (FMS) programme.

Description

The M119A1 105 mm towed howitzer is essentially the BAE Systems (originally Royal Ordnance Factories) 105 mm L119 Light Gun, covered in detail in a separate entry in *IHS Jane's Land Warfare Platforms: Artillery and Air Defence* with some modifications. For example, the installation of a US-type sighting system, to meet the requirements of the US Army.

Production of the 105 mm towed howitzer M119A1 was undertaken at Rock Island in the US with the 105 mm barrels being provided by Watervliet Arsenal.

The US Army describes the M119A1 as a lightweight, air-mobile, air-droppable, low-altitude air extraction weapon system that provides direct and indirect fire support for use in selected field artillery battalions in light infantry, airborne and air assault divisions.

Using standard 105 mm ammunition, the M119A1 can reach targets at a range of 14,300 m. The M119A1 replaced the older US developed M101A1 and M102 towed howitzers in front-line service.

The 105 mm M102 remains in service with some reserve units but these are being replaced by new build 105 mm M119A2 weapons.

There have recently been a number of new 105 mm artillery projectiles developed in the US which give the M119/M119A1 a significant range advantage.

The M915 rocket assisted projectile significantly increases lethality and improves range by 40 per cent over the standard US M1 HE projectile.

In contrast to the older M548 RAP, which uses the low end M176 propellant, the M913 uses the M229 maximum propelling charge (Charge 8, high end). The increased range is obtained by using the M229 propellant charge coupled with a modern rocket motor, which will function about 1.5 seconds after leaving the barrel. With the rocket on the M913 it provides a range of more than 19.5 km. The 105 mm towed howitzer M119A1 can also fire the more advanced natures of ammunition from South Africa.

The M916 is the same projectile as the M915 but will be fired with the M67 propellant (low end charge).

Both the M915 and M916 projectiles have a payload consisting of 42 M80 submunitions which incorporate a self-destruct mechanism.

The onboard fire-control system consists of an M137 panoramic telescope with a magnification of ×4, M90A2 direct fire telescope with a magnification of ×3, M186 direct fire mount and the M187 telescope mount.

In US Army service the M119A1 is normally towed by the AM General M1097 (4 × 4) Heavy HMMWV and can be quickly moved and employed to provide maximum fire power with minimum of combat loaded weight.

Light Artillery System Improvement Programme (LASIP)

As the M119A1 will remain in service well into the 21st Century the US Army embarked on a major modernisation and improvement programme for the M119A1 weapon to enhance its Reliability, Availability, Maintainability and Durability (RAM-D) characteristics under the Light Artillery System Improvement Programme (LASIP).

LASIP Block I modifications included:

- Low Temperature Recuperator (LTR)
- Improved 12 inch brakes
- Modified trunnion adapter
- Travelling and firing stays
- Improved cam follower
- Lift handles
- Trail end step.

LASIP Block II modifications includes:

- LED/battery illuminated fire control (replaced tritium illumination)
- Improved elevation gearbox
- Improved buffer and LTR interconnection
- Improved rammer extraction
- Roll bar.

After completion of LASIP Block I and Block II, the howitzer model designation was changed to M119A2. All M119A2 in the US Army inventory have now been converted to M119A2 standard.

M119A1 105 mm towed howitzer in travelling configuration being towed by an M1097 (4 × 4) Heavy HMMWV (Steven Zaloga) 0021136

Additional 105 mm M119A2 weapons

Achieving US Army modularity requirements requires the acquisition of new M119A2 howitzer. Currently the plan is to acquire an additional 377 howitzers.

Initial funding for a portion of this requirement was provided in 2005 and in subsequent years. Current plans are to replace all M102 howitzers in Army National Guard Units of Action (UoA) with M119A2 Howitzers.

The first purchase was for 35 weapons through the FY05 supplemental budget for Low Rate Initial Production (LRIP).

M119A2 with digital fire-control system

Late in 2011 the US Army took delivery of a batch of four 105 mm M119A2 upgraded howitzers for extensive trials.

This upgraded 105 mm M119A2 howitzer is fitted with an integrated digital fire control system that includes Global Positioning System (GPS) technology and an inertial navigation system to reduce the time into action and increase accuracy.

According to the US Army, this allows the weapon to come into action and fire the first round within two or three minutes.

The software development and integration of the digital fire-control system was carried out in house by the US Army Armament Research Development and Engineering Centre (ARDEC) and uses 90 per cent of the software derived from the 155 mm M777A2 already deployed.

Following trials it was expected to be type classified late in 2012 and a total of 603 M119A2 are expected to be upgraded from early 2013.

New M20A1 breech for M119A2

In March 2013 it was stated that the US Army had taken delivery of an initial batch of 19 redesignated M20A1 breech blocks from Watervliet Arsenal as part of the M119A2 towed howitzer upgrade programme.

More than 650 upgrade kits were ordered by the US Army under a USD22.6 million contract in an effort to improve crew safety and reduce logistical support. Under the terms of this contract final deliveries are due in August 2015.

Manufactured in collaboration with The US Army Benet Laboratory, the new kits are designed to rectify a safety flay in the M20A1.

This is claimed to require the crew to verify and measure the firing pin protrusion prior to each live firing manoeuvre.

In addition to addressing the firing pin protrusion issue, the upgraded breech block kits are claimed to enhance soldier safety during misfire as the re-cock mechanism function has been removed from behind the weapon and incorporated into the breech ring side.

In addition, the kit improvement lowers logistical footprint and maintenance time by reducing the number of breech block assembly parts by about 30 per cent.

US 105 mm artillery developments

In 2012 the US Army is expected to field the improved 105 mm M1130 High-Explosive Fragmentation - Base Bleed (HE FRAG-BB) round for use with its 105 mm M119 series towed howitzers.

The 105 mm M1130 HE FRAG BB couples the South African Rheinmetall Denel Munitions projectile with the legacy M67 propelling charge.

Trials are also underway of the 105 mm M1130E1 round which could be fielded in 2017.

This combines the 105 mm M1130 HE FRAG BB projectile with an optimized six-zone XM350 propelling charge.

When fielded the 105 mm M1130E1 would replace four types of in service 105 mm HE projectiles and two models of propelling charge with significant logistic savings. Maximum range would be just over 17 km.

It is envisioned that if and when fielded the 105 mm M1130E1 would replace the M1, M760, M927 and M1130 projectiles as well as the M67 charge and M200 top charge.

Today Rheinmetall Denel Munitions is working with General Dynamics Ordnance and Tactical Systems under the US Foreign Comparative Test and Market Survey on this programme.

Saudi Arabian M119A2 requirement

In September 2011 a FMS deal to Saudi Arabia included the potential sale of 36 155 mm M777A2 under a USD886 million deal also included radars, AFADS, ammunition, HMMWV, radios and 54 105 mm M119A2 howitzers.

As of March 2013 this potential sale had not gone ahead.

Specifications

	105 mm M119A1
Dimensions and weights	
Width	
overall:	1.78 m
Height	
overall:	1.37 m
Ground clearance	
overall:	0.50 m
Track	
vehicle:	1.40 m
Weight	
combat:	1,936 kg
Mobility	
Carriage:	bow
Tyres:	9.00 × 16
Firepower	
Armament:	1 × carriage mounted 105 mm howitzer
Main weapon	
breech mechanism:	vertical sliding block
muzzle brake:	single baffle
maximum range:	11,500 m (M1 HE projectile)
maximum range:	14,000 m (unassisted projectile zone 8)
maximum range:	19,000 m (M913 rocket assisted projectile)
Rate of fire	
sustained fire:	3 rds/min
rapid fire:	8 rds/min
Main armament traverse	
angle:	360° (on turntable)
Main armament elevation/depression	
armament front:	+70°/-5.5°
Survivability	
Armour	
shield:	no

Status

Production started again in 2007. In service with US Army. A total of 147 were supplied from the UK production line, with another 280 built under licence in the US. All US Army systems have been upgraded to the M119A2 standard. Production of new-built M119A2s has started. The M119A2 has seen combat use in Operation Iraq Freedom and Operation Enduring Freedom.

In late 2011 the potential sale of 54 M119A2 howitzers to Saudi Arabia was announced as part of an FMS USD886 million contract.

Contractor

Rock Island Arsenal, Illinois.
BAE Systems.

105 mm howitzer M102

Development

The requirement for a light towed 105 mm howitzer to replace the 105 mm M101 (which is covered in detail in a separate entry in *IHS Jane's Land Warfare Platforms: Artillery and Air Defence*) in selected units was issued in 1960.

The first prototype was completed at Rock Island Arsenal in 1962 and was type-classified as Standard A in December 1963. First production M102s were completed at Rock Island in December 1965 and the howitzer was first used in combat in South Vietnam the following month. The weapon subsequently underwent major modifications to solve design problems.

105 mm howitzer M102 deployed in firing position (Mike Green) 0500617

105 mm M102 towed howitzer in service with El Salvador in travelling configuration (Julio A Montes) 0105795

Over 1,200 M102s were built by Rock Island Arsenal before production was finally completed in the early 1970s. A further 22 were built for foreign military sales in FY80.

The main advantage of the M102 over the earlier M101 is that it only weighs 1,496 kg compared to 2,258 kg. It can also be traversed rapidly through 360° to engage targets in other sectors.

In 1985, the US Army had 526 M102 howitzers in its inventory. The M102 has been replaced in front-line service by the BAE Systems 105 mm Light Gun, designated the M119 (L119 in British Army service) in US Army service. The US Army holds quantities of the M102 for overseas sales to authorised countries.

Figures released by the United Nations for the years 1992 to 2010 show that there were no exports of the M102 in this period.

In mid-2004 it was stated that a total of 11 US Army National Guard field artillery battalions deploy the 105 mm M102. It is now expected that these are now being replaced by an additional order of the 105 mm M119A2.

Through a US FMS programme valued at USD886 million, in late 2011 Saudi Arabia requested a major artillery package that included 54 M119A1 howitzers.

It is understood that these are a replacement for the current 105 mm M102 howitzers deployed by the Saudi Arabian National Guard (SANG).

As of March 2013 the Saudi arabian FMS contract for a complete artillery package that included the 155 mm M777A2 and the 105 mm M119A1 had now been placed with the US.

Description

The 105 mm howitzer M102 consists of the M137 105 mm cannon, M37 recoil mechanism, M31 carriage and fire-control system.

The two-wheeled box carriage is constructed of aluminium. In the firing position, a circular baseplate is lowered to the ground under the forward part of the carriage and the complete carriage can be quickly swung through 360° by means of a roller located at the rear of the trail assembly. The baseplate is staked to the ground before firing begins.

The elevating mechanism utilises a pair of elevating screw assemblies and permits an elevation of +75° and depression of −5°. The variable-length recoil system used on this weapon eliminates the need for a recoil pit. The recoil system is of the hydropneumatic type.

The fire-control system consists of the M1A2 quadrant, M14A1 quadrant, M113A1 panoramic telescope and an M114A1 elbow telescope. Radioactive light sources are used to illuminate the fire-control system. The tritium light sources are a heavy radioactive isotope of hydrogen gas, sealed in a Pyrex glass container, coated with phosphor.

Weapons Command initiated a self-illumination development programme at Frankford Arsenal, Delaware in January 1969, in response to numerous reports of poor instrument light performance in Vietnam. The equipment successfully completed testing in February 1970. Production contracts for the modification kits were awarded in September 1973 and the first kits were issued in 1975.

The 105 mm howitzer M102 is normally towed by a 6 × 6 cross-country truck that in addition to its crew also carries a quantity of ready use ammunition. Some countries tow the weapon with a 4 × 4 cross-country truck.

The M102 fires the same family of ammunition as the M101 but with a slightly higher muzzle velocity and, therefore, a slightly longer range:

- APERS-T (M546) with the complete round weighing 17.35 kg, maximum muzzle velocity of 549 m/s and a maximum range (Charge 7) of 12,400 m
- HE (M1) with the complete round weighing 21.06 kg, maximum muzzle velocity of 494 m/s and a maximum range (Charge 7) of 11,500 m
- HE (M413) (carries 18 M35 grenades) with the complete round weighing 19.06 kg, maximum muzzle velocity of 494 m/s and a maximum range (Charge 7) of 11,500 m
- HE (M444) (carries 18 M39 grenades) with the complete round weighing 19.06 kg, maximum muzzle velocity of 494 m/s and a maximum range (Charge 7) of 11,500 m
- HEP (M327) with the complete round weighing 15.27 kg, maximum muzzle velocity of 559 m/s and a maximum range of 8,685 m
- HERA (M548) (RAP) with the complete round weighing 17.46 kg, maximum muzzle velocity of 548 m/s and a maximum range of 15,100 m
- Illuminating (M314 series) with the complete round weighing 21.06 kg, maximum muzzle velocity of 494 m/s and a maximum range (Charge 7) of 11,500 m
- Smoke (M60 series) with the complete round weighing 19.46 kg, maximum muzzle velocity of 494 m/s and a maximum range (Charge 7) of 11,500 m
- Smoke (M84 series) with the complete round weighing 19.03 kg, maximum muzzle velocity of 494 m/s and a maximum range (Charge 7) of 11,500 m
- HE (M760) to a maximum range of 14,000 m.

First example of the Mandus Hawkeye soft recoil howitzer integrated on the rear of a Renault Trucks Defense Sherpa (4 × 4) chassis in travelling configuration with 105 mm M137 ordnance traversed to the rear (Mandus) 1364807

It should be noted that all of these 105 mm projectile were supplied to export customers for the M102 howitzer.

Variants

Hawkeye soft-recoil howitzer

The private venture Mandus Hawkeye 105 mm soft-recoil howitzer, first shown in October 2011 integrated onto the rear of a Renault Sherpa (4 × 4) cross-country chassis, uses the 105 mm ordnance of the M102 howitzer.

Details of this system are provided in a separate entry in *IHS Jane's Land Warfare Platforms: Artillery and Air Defence*.

Specifications

	105 mm M102
Dimensions and weights	
Crew:	8
Length	
transport configuration:	5.182 m
Width	
transport configuration:	1.964 m
Height	
transport configuration:	1.594 m
Ground clearance	
overall:	0.33 m
Weight	
standard:	1,496 kg
Mobility	
Carriage:	box trail
Tyres:	7.00 × 16
Firepower	
Armament:	1 × carriage mounted 105 mm howitzer
Main weapon	
length of barrel:	3.382 m
breech mechanism:	vertical sliding block
muzzle brake:	none
maximum range:	11,500 m (M1 HE projectile)
maximum range:	14,000 m (M760 HE projectile)
maximum range:	15,100 m (M548 rocket assisted projectile)
Rate of fire	
rapid fire:	10 rds/min
Main armament traverse	
angle:	360°
Main armament elevation/depression	
armament front:	+75°/-5°
Survivability	
Armour	
shield:	no

Status

Production complete. In service with the following countries:

Country	Quantity	Comment
Brazil	19	
El Salvador	36	
Guatemala	8	
Honduras	24	
Jordan	54	18 used by special forces
Lebanon	10	
Philippines	230	M101/M102
Thailand	12	
Tunisia	48	M101/M102
Saudi Arabia	50	National Guard
Uruguay	8	
US	434	still used by National Guards but being replaced by new build M119A2 weapons
Vietnam	n/available	

Contractor

Rock Island Arsenal, Illinois.

105 mm howitzer M101

Development

The development of the 105 mm M101 towed howitzer can be traced back to 1919. It entered service with the US Army in 1940. Shortly afterwards the 105 mm Howitzer Carriage M2A1 was standardised.

By the end of the Second World War, 8,536 105 mm towed howitzers had been built and post-war production continued at Rock Island Arsenal until 1953, by which time 10,202 had been built.

The M101 105 mm howitzer was replaced in front-line US Army service by the M102 howitzer which in turn has now been replaced by the now BAE Systems 105 mm Light Gun.

According to UN sources, between 1992 and 2010 there were only five declared exports of the 105 mm howitzer M101 and these were as follows:

From	To	Quantity	Comment
China	Gabon	4	HM2, delivered 2004
Denmark	Lithuania	72	delivered 2002
Netherlands	Brazil	6	3 in 1994 and 3 in 1997
US	Colombia	78	delivered 1992
US	Macedonia	18	delivered 1999

Description

After the Second World War the complete weapon was redesignated the Howitzer, Light, Towed: 105 mm, M101 and M101A1. The M101 consists of 105 mm cannon M2A1 or M2A2, carriage M2A1 and recoil system (M2A1, M2A2, M2A3, M2A4 or M2A5).

The M101A1 is identical apart from an M2A2 carriage. The difference between the 105 mm cannon M2A1 and the M2A2, is that the M2A1's muzzle end is straight and the M2A2's is bell-shaped.

The M2A2 carriage is fitted with a main shield group composed of right and left upper and lower shields, right and left top flaps, a bottom flap, and left and right auxiliary shields; while the earlier M2A1 is equipped with left and right main shields and a top shield.

105 mm M101 series towed howitzer in travelling configuration (PB) 0085640

105 mm M101 series towed howitzer of the Canadian Forces deployed in the firing position (Canadian Forces) 0500619

RDM Technology 105 mm M101/33-calibre upgraded howitzer deployed in the firing position 0006046

RDM Technology 105 mm M101/33-calibre upgraded howitzer in service with the Chilean Army (Julio A Montes) 1365031

The recoil system of the 105 mm howitzer is of the hydropneumatic type. While the barrel bore length is 2.363 m and the length from the front of the muzzle to the rear of the breech ring is 2.574 m.

In addition there are slightly different weighs between the different models with the M101 in travelling and firing configuration weighs 2,030 kg while the M101A1 in the travelling and firing position weighs 2,258 kg.

The barrel consists of the barrel assembly, breech ring and locking ring. The carriage is of the single axle split trail type with a drawbar for securing to the prime mover. The carriage consists of the equilibrator, shield, elevating mechanism, cradle gears, elevating arcs, traversing mechanisms, top carriage, wheels and trails.

Fire-control equipment consists of a Telescope Elbow M16A1D with a magnification of ×3 and a 13° field of view, a Telescope Panoramic M12A7S with a magnification of ×4 and a 10° field of view and quadrant M4A1.

The M101 fires the following semi-fixed ammunition:

- APERS-T (M546) with the complete round weighing 17.35 kg, maximum muzzle velocity of 504 m/s and a maximum range (Charge 7) of 11,600 m
- HE (M1) with the complete round weighing 21.06 kg, maximum muzzle velocity of 472.4 m/s and a maximum range (Charge 7) of 11,270 m
- HE (M444) (carries 18 M39 grenades) with the complete round weighing 19.05 kg, maximum muzzle velocity of 472.4 m/s and a maximum range (Charge 7) of 11,270 m
- HEP (M327) with the complete round weighing 15.17 kg, maximum muzzle velocity of 559 m/s and a maximum range of 8,685 m
- HERA (M548) (RAP) with the complete round weighing 17.46 kg, maximum muzzle velocity of 548 m/s and a maximum range of 14,600 m
- Illuminating (M314 series) with the complete round weighing 21.06 kg, maximum muzzle velocity of 472.4 m/s and a maximum range (Charge 7) of 11,270 m
- Smoke (M60 series) with the complete round weighing 19.46 kg, maximum muzzle velocity of 472.4 m/s and a maximum range (Charge 7) of 11,270 m
- Smoke (M84 series) with the projectile weighing 19.03 kg, maximum muzzle velocity of 472.4 m/s and a maximum range of 11,270 m
- APERS-T (M546) with a maximum range of 11,600 m.

It should be noted that not all of these 105 mm projectiles were supplied to export customers.

It is normally towed by a 6 × 6 cross-country truck that in addition to carrying the gun crew also carries a quantity of ready use ammunition (projectiles, charges and fuzes).

Variants

Upgraded M101 (France)

Details of the Nexter Systems (previously Giat Industries) upgraded M101 are given in a separate entry in *IHS Jane's Land Warfare Platforms: Artillery and Air Defence*. The first customer for this was Thailand. In 1996 the Philippines became the second customer and ordered 12 upgrade packages from Giat Industries (now Nexter Systems). The total number of M101 systems upgraded by Thailand was 285. There is a separate entry under Thailand for the upgraded M101, which has the local designation of the M425.

Upgraded M101 (Korea, South)

Details of the M101 upgrade package from the Kia Machine Tool Company are provided in a separate entry in *IHS Jane's Land Warfare Platforms: Artillery and Air Defence*. In addition to being in service with the South Korean Army, it is believed that Chile (Marines) have taken delivery of 16 of these which have the designation of the K-178.

Upgraded M101 (Netherlands)

The now defunct RDM Technology of the Netherlands upgraded a number of 105 mm M101 towed howitzers for the export market and these remain in service.

A total of 96 Canadian M101 were upgraded under a contract signed March 1994 and these were fitted with a 105 mm/33-calibre barrel with all work completed by 1998. This has a maximum range of 11,900 firing an HE M1 projectile.

Chile upgraded a total of 66 M101 units with the 105 mm/33-calibre barrel with most of the work being carried out in Chile with kits. This programme was completed in 2002.

Self-propelled M101 (South Korea)

The South Korean company of the Samsung Techwin has developed a self-propelled version of the M101 under the designation of the EVO-105.

This is based on a 6 × 6 cross-country truck with a 105 mm M101 type weapon mounted on a modified rear platform.

Available details of the EVO-105, which may well be a technology demonstrator, are provided in a separate entry of *IHS Jane's Land Warfare Platforms: Artillery and Air Defence*.

M7 105 mm self-propelled howitzer

This is armed with an M2 or M2A1 howitzer with an elevation of +35°, a depression of –5°, and a traverse of 30° left and 15° right.

Specifications

	105 mm M101
Dimensions and weights	
Crew:	8
Length	
transport configuration:	5.991 m
Width	
overall:	3.657 m (firing)
transport configuration:	2.159 m
Height	
overall:	3.124 m (firing)
transport configuration:	1.524 m
axis of fire:	1.3 m
Ground clearance	
overall:	0.356 m
Track	
vehicle:	1.778 m
Weight	
standard:	2,030 kg
Mobility	
Carriage:	split trail
Tyres:	9.00 × 16
Firepower	
Armament:	1 × carriage mounted 105 mm howitzer
Main weapon	
length of barrel:	2.363 m
breech mechanism:	horizontal sliding block
muzzle brake:	none
maximum range:	11,270 m (M1 HE projectile)
maximum range:	14,600 m (M548 rocket assisted projectile)
Rate of fire	
sustained fire:	3 rds/min
rapid fire:	10 rds/min
Main armament traverse	
angle:	46°
Main armament elevation/depression	
armament front:	+66°/-5°
Survivability	
Armour	
shield:	yes

Status

Production complete. In service with the countries listed in the table:

Country	Quantity	Comment
Argentina	6	marines
Benin	4	
Bolivia	25	
Bosnia	25	
Brazil	205	estimate, army and marines, some replaced by 105 mm Light Gun
Burkina Faso	8	
Cameroon	20	
Canada	125	96 upgraded to C3 configuration
Chile	90	army and marines, most upgraded, including 78 from the US in 1992
Colombia	86	including 78 delivered in 1992
Croatia	61	
Chad	5	
Dominican Republic	4	
Ecuador	54	
El Salvador	8	reserve
Gabon	4	
Germany	10	reserve
Greece	263	10 to be upgraded
Guatemala	12	
Indonesia	60	
Iran	130	estimate
Israel	70	reserve
Korea, South	1,700	including local K-178 model
Laos	20	
Lebanon	13	
Libya	42	status uncertain due to 2011 conflict
Lithuania	72	
Madagascar	5	
Mauritania	35	
Mexico	60	
Morocco	20	
Mozambique	12	
Myanmar	96	
Pakistan	216	
Paraguay	15	
Peru	68	
Philippines	150	
Portugal	9	
Saudi Arabia	100	estimate, M101/M102
Senegal	6	
Serbia/Montenegro	54	
Sudan	20	status uncertain
Taiwan	650	including local T64
Thailand	280+	including local models, some upgraded by Nexter
Togo	4	
Tunisia	48	M101/M102
Turkey	680/700	
Uruguay	28	
US	331	USMC
Venezuela	40	
Vietnam	n/avail	
Yemen	25	

Contractor

Rock Island Arsenal, Illinois.

Self-Propelled Mortar Systems

China

NORINCO 120 mm SP mortar PLL-05

Development

In mid-2001, it was revealed that China North Industries Corporation (NORINCO) had developed a 120 mm self-propelled mortar system.

It is based on a modified hull of the WMZ551 (6 × 6) series Armoured Personnel Carrier (APC). This is in turn a further development of the WZ551 (6 × 6) APC which has been in service with the People's Liberation Army (PLA) for many years.

It has been confirmed that this NORINCO 120 mm self-propelled mortar system is now in service with the Chinese Army under the designation of the PLL-05 120 mm self-propelled mortar howitzer.

This was shown in public for the first time during a major parade held in Beijing in October 2009.

The 120 mm SP mortar PLL-05 has been offered on the export market by NORINCO but as of March 2013 there are no known export sales of this system.

The 120 mm SP mortar PLL-5 has also been referred to on the export market by NORINCO as the "WMA029 120 mm self-propelled mortar howitzer".

Description

Mounted on the roof in the centre of the WMZ551 hull is the fully enclosed turret. This is very similar to that used by the Russian 120 mm 2S23 NONA-SVK self-propelled mortar system mounted on a modified BTR-80 (8 × 8) APC hull.

Details of the Russian 2S23 system, which is still being marketed, are provided in a separate entry in *IHS Jane's Land Warfare Platforms: Artillery and Air Defence*.

In 1997, it was reported that Russia was to supply China with 100 120 mm 2S23 NONA-SVK self-propelled mortar systems, whose existence was first revealed in 1990.

The United Nations Arms transfer lists for the period 1992 through to 2009 show no transfers of complete 2S23 systems from Russia to China. It does show the transfer of three 120 mm 2S9 systems from Ukraine in 2000.

Russian military sources however never confirmed this deal or the transfer of technology to China allowing local production of the 2S23 system, ever took place.

While the three-person turret resembles the Russian 2S23 NONA-SVK armed with a 120 mm 2A60 rifled ordnance, the actual ordnance used for the Chinese system is rather longer.

It is similar to the Russian 120 mm 2A80 installed in the 2S31 Vena self-propelled gun/mortar system. This is based on a considerably modified BMP-3 hull and has yet to enter volume production.

The fully enclosed power-operated turret is mounted on the roof of the vehicle at the rear and is armed with a 120 mm rifled barrel fitted with vertical sliding breech mechanism. Although its primary role would be indirect fire, it also has a secondary direct-fire role.

The 120 mm ordnance does not currently feature a fume extractor, thermal sleeve or muzzle brake. Turret traverse is through a full 360° with weapon elevation from –4° to +80°.

This fires three main types of 120 mm ammunition, mortar bomb, High-Explosive (HE) and High-Explosive Anti-Tank (HEAT) projectiles with a total of 36 rounds being carried.

Details of the 120 mm ammunition types fired are given in Table 1. The 120 mm HEAT projectile is used in the direct-fire role and all of the projectiles have the propelling increments attached to the rear of the projectile.

The High-Explosive Rocket-Assisted Projectile (HE-RAP) is still under development. It has a muzzle velocity of 265 m/s, a maximum range of about 13,000 to 13,500 m and contains 3.2 kg of explosive.

Like the HEAT and HE-RAP, the 120 mm cargo projectile has a casing that is pre-engraved, which means that it has a tighter fit in the rifled ordnance.

A total of 30 dual-purpose anti-armour/anti-personnel submunitions are carried, which are claimed to be able to penetrate 90 mm of conventional steel armour. This projectile has a muzzle velocity of 425 m/s and a maximum range of 9,300 m.

These two new natures of 120 mm ammunition offer a significant enhancement capability to the NORINCO 120 mm SPM.

According to NORINCO, rate of fire depends on the type of ammunition used and the aiming and operating mode with maximum rates of fire being 6 to 8 rds/min for HE, 10 rds/min for mortar bomb and 4 to 6 rds/min for a HEAT round.

Latest Chinese Type 07PA (8 × 8) fitted with turret armed with 120 mm breech loaded mortar/howitzer and roof mounted 12.7 mm anti-aircraft machine gun (Poly Technologies) 1446011

NORINCO 120 mm self-propelled mortar PLL-05 with 120 mm ordnance partly elevated. This system is also referred to as the WMA029 120 mm self-propelled howitzer-mortar (NORINCO) 1452915

A semi-automatic loading system is fitted with direct and indirect fire sights provided. There are three aiming modes, automatic, semi-automatic and manual.

The complete system has a combat weight of 16.5 tonnes and a normal crew of four, consisting of the commander, gunner, loader and driver.

It is also fully amphibious, being propelled in the water by two propellers mounted one either side under the hull at the rear. Before entering the water a trim vane is erected at the front of the hull and the bilge pumps activated. Maximum water speed is 8 km/h.

Standard equipment includes powered steering on the front four roadwheels, NBC system and a central tyre pressure-regulation system.

A 12.7 mm machine gun is installed on the roof for local and air defence purposes and a bank of three electrically operated grenade launchers is mounted either side of the turret firing forwards.

This system has been designed to carry out direct and indirect fire engagements against a variety of targets including 'participating in landing and anti-landing operations'.

Variants

Chinese Type 07A 120 mm mortar system

In 2012 it was revealed that Poly Technologies of China was now marketing another version of the Type 07P (8 × 8) Armoured Personnel Carrier (APC) fitted with a turret armed with a 120 mm breech loaded mortar/howitzer under the designation of the Type 07PA.

Table 1

Type	HE (WMA 029)	HE (WMA 029)	HEAT (WMA 029)	Cargo (BM3)	RAP
Length:	709 mm	600 mm	625 mm	621 mm	600 mm
Weight:	13.9 kg	17.3 kg	10 kg	17.3 kg	17.3 kg
Payload:	1.8 kg TNT	5 kg TNT	HEAT	30 bomblet	3.25 kg
Range:					
maximum)	8,500 m	9,500 m	1,200 m[1]	9,000 m	13,500 m
(minimum)	400 m	1,000 m	600 m	n/avail	n/avail

[1] Used in direct-fire missions and is fin stabilised

This is aimed at the export market and is being marketed alongside the original Type 07P APC which is normally fitted with a one-person power operated turret armed with a 30 mm cannon and 7.62 mm coaxial machine gun (MG).

The 120 mm turret appears to be identical to that fitted to the NORINCO PLL-05 self-propelled mortar currently deployed by the PLA.

Mounted on the forward part of the turret roof is a manually operated 12.7 mm MG which can be traversed through 360° with weapon elevation from –5° to +85° with a total of 500 rounds of ammunition being carried.

On the PLL-05 the unprotected 12.7 mm MG is mounted on the forward part of the commander's cupola but on the Type 07PA this is pintle mounted on the forward part of the turret roof.

The Type 07PA is quoted as having a combat weight of 17 tonnes and has a crew of four consisting of commander, gunner, loader and driver.

It is powered by a Dong Feng Cummins 6LTAA8.9-C340 diesel engine which develops 250 kW and gives the vehicle a maximum road speed of up to 100 km/h.

The baseline Type 07P APC is fully amphibious being propelled in the water at a speed of up to 4 km/h by its wheels. Due to its increased weight it has not been confirmed that the Type 07PA is fully amphibious.

Standard equipment on the Type 07PA includes powered steering on the front four wheels and a central tyre inflation system that allows the driver to adjust the tyre pressure to suit the terrain conditions without leaving his position.

Specifications

	PLL-05
Dimensions and weights	
Crew:	4
Length	
overall:	6.695 m
Width	
overall:	2.8 m
Height	
overall:	2.82 m
Ground clearance	
overall:	0.41 m
Track	
vehicle:	2.44 m
Wheelbase	1.9 m + 1.9 m
Weight	
combat:	16,500 kg
Mobility	
Configuration	
running gear:	wheeled
layout:	6 × 6
Power-to-weight ratio:	19.39 hp/t
Speed	
max speed:	85 km/h
water:	8 km/h
Range	
main fuel supply:	800 km
Fuel capacity	
main:	400 litres
Amphibious:	yes
Gradient:	60%
Side slope:	47%
Vertical obstacle	
forwards:	0.55 m
Trench:	1.2 m
Turning radius:	9.5 m
Engine	Deutz BF8L413FV, V-8, air cooled, diesel, 320 hp at 2,500 rpm
Gearbox	
model:	ZF 6HP500
type:	automatic
forward gears:	5
reverse gears:	1
Steering:	hydraulic, first and second axles
Clutch	
type:	torque converter
Brakes	
main:	dual circuit, air/hydraulic
Tyres:	14.00 × 20 run flat
Suspension:	coil springs and hydraulic shock absorbers
Electrical system	
vehicle:	24 V
Batteries:	2 × 24 V, 180 Ah
Firepower	
Armament:	1 × turret mounted 120 mm mortar
Armament:	1 × roof mounted 12.7 mm (0.50) machine gun
Ammunition	
main total:	36
Turret power control	
type:	powered/manual
by commander:	no
by gunner:	yes
Main armament traverse	
angle:	360°
Main armament elevation/depression	
armament front:	+80°/-4°
Gun stabiliser	
vertical:	no
horizontal:	no
Survivability	
Night vision equipment	
vehicle:	yes
NBC system:	yes
Armour	
hull/body:	steel
turret:	steel

Status

Production. In service with China under the designation of the PLL-05 (50+ units).

Contractor

Chinese state factories. Enquiries to China North Industries Corporation (NORINCO).

NORINCO 81 mm mobile mortar system

Development

In 2008 it was revealed that China North Industries Corporation (NORINCO) had completed development of a new 81 mm self-propelled mortar system which is already being offered on the export market.

As far as it is known this 81 mm mortar system has not entered service with the People's Liberation Army.

This 81 mm mobile mortar system has been offered on the export market by NORINCO but as of March 2013 there are no known export sales of this system.

Description

The NORINCO 81 mm mortar system has been integrated onto a modified Dong Feng high mobility utility vehicle (4 × 4) chassis.

This is very similar to the US AM General High Mobility Multi-purpose Wheeled Vehicle (HMMWV) which is used in large numbers by the US Army and many other countries.

The Chinese Dong Feng vehicle has a typical Gross Vehicle Weight (GVW) of 4,700 kg with the baseline vehicle having a four door configuration with a hinged tailgate at the rear.

It is powered by a Dong Feng/Cummins supercharged diesel developing 110 kW coupled to a five speed transmission and transfer case.

Maximum road speed is being quoted as 135 km/h with fuel consumption being 18 litres per 100 km.

The 81 mm mortar installed on a turntable at the rear that allows it to be rapidly traversed through 360° onto a new target. The weapon can be elevated from –1° through to +85°.

While the primary role is that of indirect fire, the system has a very useful direct fire capability. The vehicle would normally come to a halt before opening fire.

NORINCO 81 mm mobile mortar system deployed in the firing position (NORINCO)

1303185

Ammunition is fed to the breech loaded 81 mm mortar in clips of four rounds which enables a very high rate of fire of four rounds to be achieved in just two seconds.

The 81 mm mortar has a maximum range of 6,200 m with a minimum range of 800 m being quoted. As well as firing the standard High-Explosive (HE) mortar bombs with a maximum muzzle velocity of 316 m/s, it can also fire other natures including illuminating, incendiary and smoke.

By integrating this 81 mm mortar system onto a highly mobile cross-country chassis the system can be brought into action much quicker than a conventional towed system which would make it more survivable against counter battery fire.

The new mortar is very similar in appearance to the earlier NORINCO 81 mm/82 mm towed systems but has a much longer barrel for increased range.

For some years NORINCO has been marketing the standard towed 82 mm W99 rapid mortar system which is based on the Russian 82 mm Vasilyek (2B9) automatic mortar.

This uses a conventional split trail carriage and weighs about 650 kg. Maximum range is quoted as 4,270 m and minimum range is 800 m. Full details of this are provided in a separate entry in *IHS Jane's Land Warfare Platforms: Artillery and Air Defence* and is still being offered on the export market by NORINCO.

They have also developed an 81 mm version of the W99 which is optimised for western 81 mm natures of ammunition.

This version weighs 700 kg and has a maximum quoted range of 4,000 m. As far as it is known this system has yet to enter quantity production.

NORINCO is also marketing at least two types of conventional 81 mm mortars; the Type W91 long range mortar (maximum range of 8,000 m) and the Type W87 (maximum range of 5,600 m). They are also marketing a complete family of 81 mm mortar ammunition for use with these mortars.

Specifications

Calibre:	81 mm
Range:	
(maximum)	6,200 m
(minimum)	800 m
Max rate of fire:	4 rounds in 2 seconds
Feed:	4-round clip
Elevation/depression:	–1°/+85°
Traverse:	360°

Status

Development complete. Ready for production.

Contractor

Chinese state factories. Enquiries to China North Industries Corporation (NORINCO).

Finland

Patria NEMO 120 mm mortar system

Development

Patria have considerable experience in the design, development and production of a variety of mortar systems and their associated families of ammunition.

While Patria is teamed with Hägglunds for the AMOS twin 120 mm mortar system which is now in production for the Finnish Defence Force (FDF) and covered in a separate entry in *IHS Jane's Land Warfare Platforms: Artillery and Air Defence*, the company believed that there was a significant market for a lighter and more compact 120 mm mortar system with similar range capabilities.

Patria 120 mm NEMO Turreted Mortar System integrated onto a Patria Armoured Modular Vehicle (8 × 8) platform which has also been fitted with appliqué armour (Patria) 1460884

Production standard 120 mm NEMO turreted mortar system integrated on a Patria Armoured Modular Vehicle (Patria) 1347594

This led to the development, using company funding, of the Patria NEMO [NEw MOrtar] Turreted Mortar System (TMS) which was first shown in 2005 integrated onto the Patria Armoured Modular Vehicle (AMV) (8 × 8) hull and this was first shown in 2006.

Launch customer for the 120 mm NEMO was Slovenia who ordered 135 AMV in various versions including 10 fitted with the NEMO TMS, but this part was subsequently cancelled, through no fault with the actual mortar system.

The Saudi Arabian National Guard (SANG) placed a contract with General Dynamics Land Systems – Canada for an additional 724 Light Armoured Vehicle (LAV) in 10 versions including a batch of 36 NEMO TMS.

Patria delivers the NEMO TMS to General Dynamics Land Systems Canada who integrates these onto the LAV hull and delivers the complete system to SANG.

Previous SANG contracts included one for 73 LAV fitted with the Delco/Royal Ordnance, 120 mm Armoured Mortar System (AMS) but all marketing for this has ceased.

Patria can supply not only the NEMO integrated onto a variety of platforms put also as part of a complete indirect fire system that includes sensor to shooter a complete mortar system that also includes a FCS, communications assets, forward observers, unmanned aerial vehicles and a meteorological system.

Description

The main advantage of NEMO 120 mm mortar turret is its light weight of around 1,650 kg which allows it to be integrated onto a much wider range of hull, tracked and wheeled, than AMOS which typically weighs around four tonnes.

The baseline NEMO turret provides protection from small arms fire and shell splinters but a modular ballistic protection system is available for higher levels of ballistic protection. NEMO turret also incorporates a number of signature reduction measures.

From an early stage in the development programme it was decided to make NEMO as simple as possible in the base line configuration to drive down costs but with a clear growth path to the future.

The 120 mm smooth bore breech loaded mortar barrel used in NEMO is ballistically identical to that installed in the twin 120 mm AMOS.

NEMO was successfully fired for the first time in February 2006 with the full up system being fired in May the same year.

A typical NEMO self-propelled system would have a crew of four consisting of driver, commander and two loaders who would prepare/load ammunition trays located below the turret ring.

NEMO is fitted with a semi-automatic loading system with the fuze being set and the required number of increments being selected prior to being loaded into the loading trays which in turn feeds these to the breech loaded mortar.

The amount of ammunition being carried will depend on the hull that NEMO is integrated onto but in the case of the AMV this is expected to be 60 rounds of 120 mm ammunition. All standard natures of smooth bore mortar ammunition can be fired as well as smart projectiles.

Time to fire the first projectiles upon coming to a halt is 30 seconds and time taken to come out of action is 10 seconds.

According to Patria a burst rate of fire of three rounds can be fired in 15 seconds with the system also capable of six Multiple Round Simultaneous Impact (MRSI) missions.

Maximum range depends on projectile/charge combination but is typically over 10 km. Turret traverse and weapon elevation is all electric with manual back up controls.

NEMO has a day/night sight for direct fire as well as the normal indirect fire capability with all of the crew in the hull under full armour protection.

The turret could also be fitted with a 7.62 mm machine gun and electrically operated smoke grenade launchers if required.

The first example of NEMO has the same Fire-Control System (FCS) as that installed in the AMOS currently being supplied to the Finnish Defence Forces (4 + 20).

Other FCS could be also be installed according to customer requirements as could a muzzle velocity radar system which would feed information to the FCS to improve accuracy.

Although all NEMO trials to date have been on an AMV (8 × 8) hull, its design is such that it can be integrated on a wide range of other hull, tracked and wheeled (8 × 8 and 6 × 6).

The low profile of NEMO allows the system to be integrated onto a wheeled hull at the same time as retaining its air transport requirement in a Lockheed Martin C-130 Hercules aircraft.

Variants

NEMO with remote weapon station

In mid-2012 the NEMO TMS was shown fitted with a Norwegian Kongsberg Protector Super Lite remote weapon station.

Naval application

Late in 2007 Patria completed a Phase One study contract awarded by the Finnish Defence Force (FDF) for the integration of a modified version of 120 mm NEMO TMS onto a Watercat M12 surface craft developed by Marine Alutech Oy for potential use by the Finnish Coastal Jaeger Battalions.

This was followed by a FDF Phase Two contract for a complete prototype of a naval version of NEMO for extensive user trials. This has now been delivered.

For the marine application the current 120 mm NEMO TMS has been modified for its new role including measures to stop salt water corrosion in the naval environment.

In 2008 the United Arab Emirates ordered eight NEMO systems for installation on their surface craft and these have all now been delivered with first systems already fitted.

For the naval application the NEMO TMS has been modified to allow it to operate in the harsh sea environment and has a different FCS that is optimised to allow a fire on the move capability.

NEMO indirect fire support system

In addition to marketing the 120 mm NEMO TMS as a stand alone turret for integration on tracked and wheeled armoured vehicles, Patria have recently started to market a complete mortar indirect fire support system.

This includes not only the 120 mm NEMO TMS installed on the customers own platform but also fire control system, target acquisition, sounding system and an Patria mini-unmanned aerial vehicle to locate potential targets. A typical 120 mm NEMO TMS would consist of two platoons each with three mortars.

Specifications

NEMO turret

Weight:	1,700 kg
Main armament:	1 × 120 mm smoothbore mortar
Barrel length:	3,000 mm
Recoil system:	hydropneumatic
Loading system:	electric, semi-automatic
Laying system:	electric with manual back up
Traverse:	360°
Elevation:	–3° to +85°
Range:	10,000 m (maximum)

Status

Production. In service with Saudi Arabia (36 being integrated onto a General Dynamics Land Systems – Canada Light Armoured Vehicle) and the United Arab Emirates (eight for naval application).

Contractor

Patria.

France

TDA 120 mm 120R 2M self-propelled mortar system

Development

The 120 mm 120R 2M recoiling rifled mortar system was developed as a private venture for the export market by TDA (previously Thomson Brandt Armements) and can be fitted to a wide range of vehicles, tracked and wheeled, in the 10 to 15 tonnes class.

The 120 mm 120R 2M recoiling rifled mortar was developed from 1992 and was first revealed in 1993 with the first complete prototype being shown in 1994.

The system has been demonstrated in France and Saudi Arabia installed in the MOWAG Piranha (8 × 8) APC and has also been installed in a Turkish FNSS Savunma Sistemleri AS Armoured Infantry Fighting Vehicle (AIFV) for trials in Turkey.

One TDA 120 mm 120R 2M installed in a VAB series APC has been ordered by the French Army and this was delivered in December 2003 for extensive trials.

French Army units equipped with the VAB currently use the towed TDA 120 mm rifled mortar which takes time to come into and be taken out of action.

The Italian Army has selected the system for installation in specialised versions of its Freccia infantry fighting vehicle (8 × 8) and Dardo (tracked vehicles) and prototypes of these are now undergoing trials in Italy.

Renault Trucks Defense VAB (6 × 6) with TDA 120 mm Recoiling Rifled Mounted Mortar (2R2M) in rear of vehicle (Christopher F Foss) 1403746

Renault Trucks Defense VAB (6 × 6) with TDA 120 mm Recoiling Rifled Mounted Mortar (2R2M) in rear of vehicle (Christopher F Foss) 1403747

In early 2008 TDA Armements SAS was awarded a contract from Oman for an undisclosed quantity of 120 mm 2R2M systems.

These 120 mm 2R2M mortars were integrated into a French Renault Trucks Defense VAB (*Vehiclule de l'Avant Blinde*) Light Armoured Vehicles (LAV) which are already in service with Oman and are being upgraded.

In addition to supplying the 120 mm mortar system, the contract also included a command-and-control system, target acquisition system and a comprehensive logistics support package, training, spares and accessories.

These new Omani 120 mm 2R2M mortar systems are probably earmarked for the Omani Royal Guard Brigade who is known to operate a fleet of VAB light armoured vehicles supplied some time ago.

In April 2008 the Royal Malaysian Army paced a contract for eight Turkish FNSS Armoured Combat Vehicle – Stretched (ACV-S) armed with the 120 mm 2R2M system which were delivered by 2010.

The actual TDA 120 mm 120R 2M mortar systems for these ACV-S were not however ordered until April 2011 under an MYR60 million contract placed with TDA.

It is understood that Saudi Arabia has taken delivery of a quantity of these systems integrated into the rear of upgraded M113 series Armoured Personnel Carriers (APC).

The Turkish company of FNSS has been awarded a contract from Malaysia covering the supply of a total of 257 Pars (8 × 8) vehicles for the Malaysian Army with these to be supplied via the local company of DEFTECH.

Within this contract are 12 variants which include one fitted with the TDA 120 mm 120R 2M rifled mortar which will fire through the open roof hatches.

A total of eight will be provided fitted with the TDA 120 mm 120R 2M rifled mortar.

Description

The 120 mm mortar and its associated hydraulic recoiling system is mounted on a 1 m circular turntable with ball-bearing enabling it to be quickly traversed under power through 220°. Elevation and traverse is controlled using a joystick with elevation from +42° to +85°.

When travelling, the 120 mm mortar barrel is lowered into the horizontal position and the two roof hatches are closed making the vehicle difficult to distinguish as a mortar carrier.

The complete 120 mm mortar system weighs around 1,500 kg and comprises the rifled mortar barrel, two hydraulic recoil brakes and a bearing-mounted base plate. Recoil force on the platform is about 35 tonnes and maximum recoil travel is about 300 mm.

The 122 mm 120R 2M mortar is ballistically identical to the well-known TDA 120 mm MO-120-RT rifled towed mortar which is used by more than 20 countries.

Scale model of the FNSS Pars (8 × 8) vehicle for Malaysia fitted with the TDA 120 mm 120R 2M recoiling rifled mortar shown firing through the open roof hatches (FNSS) 1460886

Maximum range using rocket-assisted 120 mm ammunition is 13 km and maximum rate of fire is 6 to 10 rds/min for three minutes and sustained rate of fire is 2 to 3 rds/min. When fitted with a semi-automatic loading system a rate of fire of 18 rds/min can be achieved.

The system fires the same family of 120 mm mortar bombs as the standard 120 mm towed mortar including high-explosive (with impact/delay fuze or with proximity fuze), smoke (white phosphorus), anti-armour personnel carrier (with pre-fragmented body), illumination, cargo (with 20 grenades) and practice (black powder).

It will also fire 120 mm smart mortar projectiles such as the Swedish Strix precision mortar bomb.

It should be noted that this Strix 120 mm guided mortar bomb is only in service with Sweden and Switzerland. Production of this has been completed and it is no longer marketed.

Standard French TDA 120 mm mortar bombs have a maximum range of 8,135 m. The rocket-assisted projectile can be fitted with an impact/delay fuze or a proximity fuze. The rocket-assisted projectile has a maximum range of 13,000 m and an accuracy of 26 × 18 m.

About 40 mortar bombs, charges and associated fuzes are carried with the actual number depending on the hull used.

A semi-automatic loading device driven by a programmable automatic controller is fitted as standard as is a weapon aiming and pointing system developed by Allied Signal.

This comprises a north finder (ring laser gyro), chief or section control display and gunners display unit. A manual back-up mode system is fitted for traverse, elevation and loading.

As an option, a modern location and positioning system (GPS and land navigation system) and fire-control system (ballistic computer and automatic data transmission system) can be fitted.

The 120 mm 120R 2M mortar system does not have a direct fire capability as TDA believes that there is little requirement for this.

Variants

A smoothbore version of this system is available under the designation of the 120 RS2M system. TDA also developed the Dragon Fire remotely controlled 120 mm rifled mortar system. This was developed to meet the requirements of the US Marine Corps and it uses the 120R 2M ordnance. The TDA MO 120 RT 120 mm towed mortar is the basis for the US Marine Corps Expeditionary Fire Support System (EFSS) covered in a separate entry in *IHS Jane's Land Warfare Platforms: Artillery Air Defence*. This is only used by the US Marine Corps. It should be noted that the US Army deploys 120 mm smoothbore mortar systems.

This is deployed by the US Marine Corps field artillery units, some of which are issued with the BAE Systems, Global Combat Systems M777 series towed artillery system and the 120 mm EFSS.

Specifications

120 mm 120R 2M

Calibre:	120 mm
Total weight:	1,500 kg
Barrel weight:	117 kg
Recoil brake:	350 kg
Baseplate:	400 kg
Recoil brake:	strain reduced from 150 t to less than 35 t
Firing from horizontal platform:	
(elevation)	+42/+85°
(traverse)	220° (left and right)
Maximum range:	8,135 m (standard bombs)

Status

In production. In service with a number countries including Malaysia (integrated in a FNSS Armoured Combat Vehicle - Stretched) and Oman (integrated into a VAB 6 × 6).

Undergoing trials in the Dardo (tracked) and Freccia (wheeled) IFVs of the Italian Army. A total of eight systems are to be supplied for integration into Pars (8 × 8) vehicles of the Malaysian Army.

Also understood to be in service with Saudi Arabia integrated onto an upgraded M113 series vehicle.

Contractor

TDA.

Germany

Rheinmetall Landsysteme Wiesel 2 120 mm mortar system

Development

Like some other countries, the German Army is placing increased emphasis on its rapid reaction forces. This has led to the development and fielding of much lighter vehicles and weapon platforms that can more readily be transported by aircraft and helicopter. The latter includes the Sikorsky CH-53G that is used in large numbers by Germany

Some years ago, as a private venture, MaK System GmbH of Kiel (today known as Rheinmetall Landsysteme), started work as a private venture on an enhanced version of the Wiesel 1 light armoured vehicle.

This enhanced Wiesel finally emerged as the Wiesel 2 and has a significant increase in internal volume and load carrying capacity, enabling it to undertake a much wider range of battlefield missions.

The first customer for the Wiesel 2 is the German Army for its new low-level air defence system, for which the prime contractor is Rheinmetall Defence Electronics.

A total of 67 Wiesel 2 platforms have been supplied, 50 for the Ozelot weapon platform, 10 for the platoon command post and fitted with the HARD radar and seven battery command post vehicles.

The Wiesel 2 air defence system was operated by the German Army but these have all now been handed over to the German Air Force.

The Wiesel 2 has since been developed by Rheinmetall Landsysteme as the baseline for a complete family of other vehicles including the 120 mm self-propelled mortar system.

Rheinmetall Landsysteme is responsible for the complete system including the vehicle, ammunition handing, fire-control concept, computer system and the electric system.

Rheinmetall Weapons & Munitions is responsible for the actual 120 mm mortar system and its associated family of 120 mm ammunition. Other members of the Wiesel 2 120 mm mortar team are Honeywell who are responsible for the navigation and laying system and rear support drive, ESG for the fire-control software and Wittenstein Motion Control for the actual gear drives.

The first prototype of the Wiesel 2 120 mm mortar system was completed in 1997 and based on a surplus Wiesel 1 hull. This proved that the concept was viable and during company trials, over 200 rounds of 120 mm ammunition were fired.

Late in 2002, the German BWB awarded Rheinmetall Landsysteme a contract for the supply of two Wiesel 2 120 mm mortar systems, which were delivered for extensive trials in 2004.

The German Army designation for the vehicle is the Leichter Panzermorser 120 mm (Light Armoured Mortar 120 mm).

Following firepower and mobility trials in Germany, these two vehicles were put through extensive user trials in Yuma, Arizona for hot weather trials and in Sweden for cold weather trials. Almost 2,000 120 mm mortar bombs were fired with various charges.

In mid-2009 the German Army placed its first production contract with Rheinmetall Landsysteme for the Wiesel 120 mm mortar system.

This EUR54 million contract covered the supply of eight Wiesel 2 120 mm mortar systems.

Production standard Rheinmetall Wiesel 2 120 mm self-propelled mortar in travelling configuration showing the mortar integrated into the hull rear (Christopher F Foss) 1455562

Rear view of one of the two pre-production Wiesel 2 120 mm self-propelled mortar systems deployed in the firing position and with stabilisers deployed and mortar ready to fire (Rheinmetall) 1132724

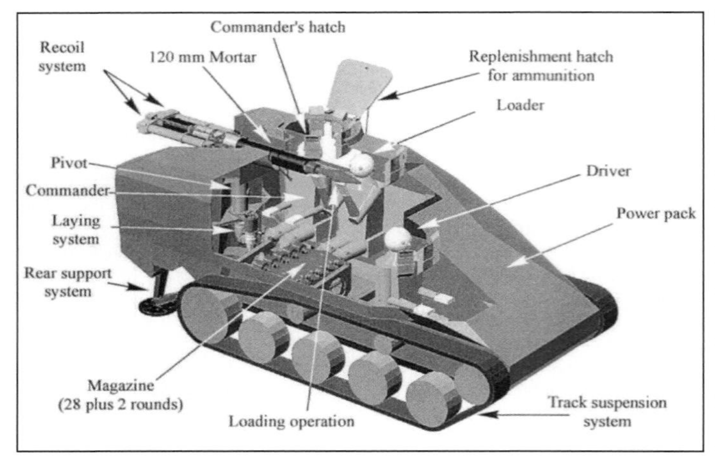

General arrangement drawing of Wiesel 2 120 mm self-propelled mortar system showing position of main components 0526377

It should be noted that the German Army has now phased out of service all of its 120 mm M113 based self-propelled mortar systems.

Description

The hull is almost identical to that used for the Wiesel 2 vehicle. The driver is seated at the front right with the power pack, consisting of the diesel engine, transmission and cooling system, front left. The commander and ammunition loader are seated on the left and face inwards while the commander is also provided with a hatch and periscopes for all-round observation.

The actual 120 mm mortar is pivoted at the rear and when being loaded, is lowered into the horizontal position with the mortar bombs being fed into the muzzle manually.

The 120 mm mortar then returns to its designated firing position, fired and then returns to the loading position again. Loading the 120 mm mortar is carried out under complete NBC protection.

The 120 mm muzzle loaded smoothbore mortar has been fitted with a special recoil system and firing a standard 120 mm mortar bomb, a maximum range of 8,000 m is obtained.

It can also fire smart 120 mm rounds, for example the Swedish terminally guided Strix up to a maximum length of 1,000 m. This is already in service with Sweden and Switzerland.

It should be noted that production of the Strix has been completed and it is no longer being marketed.

A total of 25 rounds of 120 mm ammunition are carried plus two smart rounds. It is normally operated by a crew of three: commander, loader and driver.

Under a separate contract, Rheinmetall Weapons and Munitions has developed a new generation of more effective ammunition for use with the 120 mm smoothbore mortar.

This will include an improved high-explosive mortar bomb with a multifunctional fuze with a maximum state range of 8 km. In addition there is a new illuminating mortar bomb and smoke/obscurant. These 120 mm mortar bombs are all insensitive munition compliant.

The Wiesel 120 mm SP mortar system has a onboard computerised fire-control system to allow for autonomous fire missions to be carried out.

Before firing commences, a stabiliser mounted to the immediate rear of each of the last road wheel station is lowered to the ground under electric power from within the vehicle. The mortar can be traversed 30° left and right and elevated from +35° up to +85° without moving the vehicle.

Using its onboard computer and hybrid navigation and laying system, the platform can automatically determine, not only its heading angle, but also its exact position and height.

The 120 mm mortar can be brought into action much quicker than a conventional mortar system and can also quickly redeploy to a new fire position before counter battery fire arrives.

The main computer sends and receives information from the laying system, rear support system, safety electronics and navigation systems. Information is displayed on the commander's MRT 86 Central Operating and Display Unit (CODU) that is also linked to the communications system. The 120 mm mortar is laid onto the target and fired from the CODU.

According to Rheinmetall Landsysteme, it can open fire in under 60 seconds of coming to a halt. After each 120 mm mortar bomb is fired, the computer automatically adjusts the mortar so that firing accuracy is maintained. A rapid rate of fire of three rounds in 20 seconds can be achieved with a sustained rate of fire of six rounds a minute for three minutes.

The 120 mm mortar can be aimed, loaded and fired by the crew of three, with the commander, loader and driver being under complete protection from small arms fire and NBC attack.

The Wiesel 2 120 mm self-propelled mortar system can be carried internally in a CH-53G helicopter or, as an alternative, it can be slung underneath of the helicopter.

Standard equipment includes an NBC system but as an option an air conditioning system can be fitted. Other options include onboard ballistic computation and automatic fuze selection.

The quoted range of 400 km comprises 40 per cent of cross-country travel and 60 per cent of road travel.

Wiesel 2 MultiPurpose Carrier

The German Army has so far taken delivery of the following production quantities of the Wiesel 2 multipurpose carrier:

- Air defence system - 69 (deliveries 1999 through to 2004)
- Ambulance - 33 (deliveries 2001 through to 2008)
- Battalion command post - 32 deliveries (2004 through to 2009)
- Combat engineer reconnaissance - 9 (deliveries 2001 through to 2006).

Specifications

	Wiesel 2 120 mm self-propelled mortar
Dimensions and weights	
Crew:	3
Length	
overall:	4.52 m
Width	
overall:	1.820 m
Height	
overall:	1.81 m
Ground clearance	
overall:	0.30 m
Track	
vehicle:	1.62 m
Track width	
normal:	200 mm
Length of track on ground:	2.427 m
Weight	
combat:	4,300 kg
Ground pressure	
standard track:	0.49 kg/cm^2
Mobility	
Configuration	
running gear:	tracked
Power-to-weight ratio:	24.20 hp/t
Speed	
max speed:	70 km/h
Range	
main fuel supply:	400 km (est.)
Fuel capacity	
main:	95 litres
Fording	
without preparation:	0.5 m
Gradient:	60%
Side slope:	30%
Trench:	1.2 m
Engine	Audi TDI, 4 cylinders, diesel, 110 hp
Gearbox	
model:	ZF LSG 300/4
type:	automatic
Steering:	hydrostatic cross-drive
Electrical system	
vehicle:	24 V
Firepower	
Armament:	1 × hull mounted 120 mm mortar (smoothbore)
Ammunition	
main total:	30
Main armament traverse	
angle:	60° (30° left/30° right)
Main armament elevation/depression	
armament front:	+35°/+85°

	Wiesel 2 120 mm self-propelled mortar
Survivability	
Night vision equipment	
vehicle:	yes
NBC system:	yes
Armour	
hull/body:	steel

Status
Private venture prototype completed in 1997. In 2002, a contract was placed for two systems, which were delivered for German Army trials in 2004.

In mid-2009 the production contract was placed for eight 120 mm Wiesel systems.

Production of the Wiesel 2 120 mm self-propelled mortar is now underway for the German Army.

Contractor
Rheinmetall Landsysteme GmbH.

India

Indian 81 mm Carrier Mortar Tracked Vehicle

Development
The 81 mm Carrier Mortar Tracked Vehicle (CMTV) was developed under the direction of the Indian Ordnance Factory (IOF) to meet the operational requirements of the Indian Army to provide indirect fire support for its mechanised infantry.

The Indian Ordnance Factory Medak has manufactured the Russian Kurgan Machine Construction Plant BMP-2 Infantry Combat Vehicle (ICV) for many years under the local name of Sarath.

Since the first Sarath rolled off the production line as far back as 1987, it is estimated that about 1,500 Sarath ICVs have been built. As far as it is known there have been no exports of the Sarath or its variants by India.

The first prototype of the 81 mm CMTV was completed in 1997 and following trials this was accepted for service by the Indian Army, with first production vehicles completed in 2000. So far more than 100 vehicles have been delivered according to the IOF.

Production of the Indian BMP-2 Sarath has been completed although more specialised versions of the vehicle are manufactured as required.

The Indian Army will replace its BMP-2 series vehicle by the Future Infantry Fighting Vehicle (FIFV) but as of early 2013 no development contracts had been awarded.

Description
The new 81 mm CMTV is based on the hull of the Sarath ICV, which has had its two-person 30 mm turret removed and the hull modified for its new role.

The 81 mm mortar is mounted in the rear of the hull and fires through a two-part roof hatch that opens left and right.

The 81 mm mortar has a traverse of 24° left and right with elevation from +40° to +85°. A total of 108 × 81 mm mortar bombs and associated propelling charges are carried.

A mortar baseplate is carried externally, which allows the mortar to be dismounted and deployed away from the vehicle, if required by the tactical situation.

Mounted on the forward part of the hull is a 7.62 mm machine gun for local and air defence purposes and for this a total of 2,350 rounds of ammunition are carried.

81 mm Carrier Mortar Tracked Vehicle based on the Sarath ICV is now in service with the Indian Army and shown here with roof hatches open (IOF)
0556066

As well as the small arms of the crew, an 84 mm Carl Gustaf anti-tank weapon is carried plus 12 rounds of ammunition. As it is used in high ambient temperatures 320 litres of drinking water are carried.

The 81 mm CMTV is operated by a crew of six, which includes the driver and commander, and has a combat weigh of 14.3 tonnes and a power-to-weight ratio of 21 hp/t.

It is fully amphibious and is propelled in the water by its tracks at a maximum speed of 7 km/h. Before entering the water a trim vane is erected at the front of the hull and the electrically-operated bilge pumps are switched on.

The 81 mm CMTV is the latest in a long line of Sarath variants developed by the IOF for the Indian Army with earlier models including an ambulance, engineer reconnaissance and armoured dozer.

Stretched and modified versions of the Sarath hull have been developed for a number of missile applications including the Nag anti-tank guided weapon and the Akash and Trishul surface-to-air missile systems. As far as it is known, there have been no export sales of this 81 mm Carrier Mortar Tracked Vehicle.

It is understood that production of the specialised versions of the Sarath ICV are undertaken on an as required basis.

Status
Production as required. In service with the Indian Army.

Contractor
Indian state factories.

International

AMOS 120 mm self-propelled mortar system

Development
In June 1996, Hägglunds Vehicle (which today is BAE Systems and Patria) of Finland signed an agreement to develop a new turret-mounted twin 120 mm smoothbore mortar system called AMOS (which stands for Advanced MOrtar System).

The now BAE Systems has considerable experience in the design, development and production of tracked vehicles and turrets, while Patria has developed and produced a complete range of mortar systems and their associated ammunition.

Under the terms of the agreement, BAE Systems, Global Combat Systems is responsible for the turret with Patria Defence responsible for the twin 120 mm smoothbore mortar system.

Two versions of the AMOS 120 mm self-propelled mortar system were originally developed: Model A muzzle loaded and Model B, breech loaded. A decision was subsequently taken to concentrate all work on the breech-loaded version.

It should be noted that AMOS was originally being developed to meet the requirements of Finland and Sweden but the latter country has now cancelled its involvement in this programme.

In December 2010 Finnish Defence Forces (FDF) awarded Patria Hägglunds a contract for an undisclosed amount for the supply of 18 twin 120 mm Advanced Mortar Systems (AMOS) for delivery from 2013.

Of these 14 will be integrated onto the chassis of new build Patria Armoured Modular Vehicle (AMV) (8 × 8) which is already in service with the FDF.

Under the original contract, the FDF took delivery of an initial four AMOS in 2006 integrated onto the AMV (8 × 8) hull that have been put through an intensive series of user trials.

The latest contract for 18 units consists of 14 brand new build twin 120 mm AMOS turrets plus the original four AMOS turret which will be brought up to the same production standard including a Computerised Fire-Control System (CFCS). There is also an option for additional AMOS systems.

One of the first four AMOS twin 120 mm mortar systems for the Finnish Defence Force integrated onto a Patria Armoured Modular Vehicle (AMV) (8 × 8) (Christopher F Foss)
1296047

AMOS twin 120 mm mortar system integrated onto a BAE Systems CV9040 tracked hull for trials (Patria) 1365024

Description

AMOS has two breech-loaded mortars, each of which is 3 m long. The barrels are mounted in a common cradle but in this case both of them have a hydraulic recoil brake and pneumatic recuperator of their own, thus enabling each barrel to recoil independently.

Each 120 mm smoothbore barrel is fitted with a fume extractor to keep toxic gases inside the vehicle within acceptable levels.

The breech system is a vertical downward-opening wedge type which is operated by a toggle link-type semi-automatic mechanism.

Breech obturation and positive shell positioning is achieved by means of a special reusable stub case adapted to each projectile.

Each barrel is provided with a revolver-type projectile feeder, although for the prototype stage loading is manual by means of a loading tray.

The first breech-loaded example was completed in 1997 and was shown for the first time in June of that year.

The turret is of all-welded steel armour construction which provides protection from small arms fire and shell splinters. It also provides muzzle blast and NBC protection for the turret crew of two.

Turret traverse and weapon elevation is all electric, with manual controls being provided as a back-up for emergency use.

A wide range of fire-control options is available, depending on user requirements.

The prototype system had an optical laying system but growth potential could include an automatic loading system and an electronic fire-control system.

Ammunition load would depend on the hull but typically would be 50 conventional mortar bombs (for example HE, smoke and illuminating) and six guided mortar bombs and resupply would take 10 minutes.

As well as the BAE Systems CV90 full-tracked chassis, the AMOS 120 mm mortar system could be fitted to a wide range of other hull including the Russian MT-LB multipurpose tracked armoured vehicle (used in some numbers by Finland and Sweden), the now BAE Systems M113 series tracked vehicle and the Swiss MOWAG Piranha (8 × 8) light armoured vehicle.

In mid-1999 the Finnish Defence Force awarded the company a contract worth EUR4.5 million for the twin 120 mm AMOS. This contract covered the delivery to the Finnish Army of a study of some aspects of AMOS, including the loading system.

It also covered the supply of one complete AMOS installed on a Patria XA-203 (6 × 6) chassis.

In the first quarter of 2000, the Swedish Defence Materiel Administration (FMV) and its Finnish counterpart signed a project agreement for the development of the vehicle-based twin 120 mm Advanced Mortar System (AMOS).

This agreement is based on a 1994 agreement signed between Denmark, Finland, Norway and Sweden for the joint development of material to meet the operational requirement of all four countries. For the AMOS project Denmark and Norway had observer status.

The new agreement covered a study and prototype phase that ran through to December 2002 with a total value of SEK100 million. Prime contractor is Patria Hägglunds.

In mid-2001, the Finnish Defence Forces completed extensive firepower and mobility trials of the AMOS installed on a Patria XA-203 (6 × 6) Armoured Personnel Carrier (APC) chassis.

These trials started in November 2000 and were carried out by a FDF crew under a wide range of operational conditions all over Finland. Over 1,300 120 mm mortar bombs were fired and 3,000 km of road and cross-country running were successfully carried out.

This was the second AMOS turret and was the one ordered in June 1999 and installed on the latest production Patria Defence XA-203 (6 × 6) chassis. This was delivered late in 2000 and is referred to as the PT1 and incorporates all of the improvements suggested with the earlier AMOS.

This AMOS turret is of all-welded steel construction providing the occupants with protection from 7.62 mm ball small arms fire and shell splinters.

To this can be added a layer of additional passive armour to provide a higher level of protection. A 7.62 mm or 12.7 mm machine gun can be fitted on the roof for local protection and electrically operated smoke grenade launchers can be mounted either side of the turret. In addition, PT1 has a new and more streamlined shape and incorporates stealth characteristics in its design.

The PT1 turret weighs 5,800 kg but this does depend on armour protection and if the bustle-mounted ammunition handling system is fitted.

Without the bustle-mounted ammunition handling system, turret weight is reduced to 3,640 kg. The turret crew consists of three, commander (left), gunner (right) and loader.

Also conventional ammunition (high-explosive, illumination, smoke and practice) from MECAR SA has been qualified for use in AMOS with extensive trials being carried out in 2004 to 2005.

Turret traverse is through a full 360° with weapon elevation from –3° to +85°. Elevation and traverse is electric with manual back-up in case of power failure.

AMOS is fitted with a computerised fire-control system developed by BAE Systems, which enables the system to use shoot-and-scoot techniques. This was called the Hägglunds Control System (HCS) and also includes training, built in test and battle management functions.

According to the company, AMOS can open fire within 30 seconds of coming to a halt and move off again within 10 seconds of completing its fire mission. A total of four rounds can be fired in about five seconds and with a well-trained crew, a maximum rate of fire is stated to be 16 rds/min with a semi-automatic loading system.

A key feature of AMOS is its ability to employ Multiple Rounds Simultaneous Impact (MRSI) techniques where all rounds hit the target at once for maximum effect.

Following extensive trials with the FDF the AMOS turret was removed from the XA-203 (6 × 6) hull and sent to Sweden where it was integrated into a BAE Systems CV 90 full-tracked hull.

The first four AMOS turrets integrated onto an AMV chassis have been put through a series of extensive trials.

Late in 2010 the FDF placed a contract for 18 AMS, 14 of these were brand new AMOS turret system plus the original four which are being brought up to production standard.

Turret structures will continue to be supplied by BAE Systems in Sweden with final hull integration taking place in Finland.

The FDF version of AMOS features a semi-automatic loading system that will enable a cyclic rate of fire of 16 rounds per minute. It has a direct-fire range of up to 1,000 m with an indirect-fire range of 10 km but this depends on the type of ammunition/charge combination.

It is claimed that AMOS is the only mortar system on the market that enables Multiple Round Simultaneous Impact (MRSI) target engagements to take place, with up to 10 mortar projectiles impacting the target at the same time.

It has also been integrated into a BAE Systems CV9040 infantry fighting vehicle chassis. The Swedish Army has already taken delivery of 40 CV90 chassis, which were stored pending delivery of production-standard AMOS turrets.

In June 2006, the Swedish Defence Material Administration (FMV) placed a SEK500 million (USD70 million) contract covering the supply of two twin 120 mm AMOS for the Swedish Army.

In 2009 the Swedish FMV cancelled the AMOS programme in order to fund other programmes.

The 40 BAE Systems CV9040 chassis for these turrets have already been built and have now been stored.

Under a contract signed in December 2004, Patria is supplying new artillery and 120 mm mortar ammunition to the FDF. The latter includes illumination, infra-red smoke, cargo and extended-range high-explosive. The high-explosive round is being offered on the export market as the 120 Mortar Extended-Range High-Explosive (MEHRE). This features a new charge system utilising modern double base propellant hermetically sealed in an aluminium tube.

The main charge is an NGL-impregnated NC propellant in combustible containers, with a maximum muzzle velocity of 480 m/s. It contains 3.1 kg of Composition B high-explosive.

Although this new 120 mm MERHE has been optimised for use with the 3 m long barrel of AMOS, it can be fired from conventional mortars using a reduced charge system. The standard bomb is fitted with a proximity fuze but this can be replaced by a conventional point-detonating fuze.

Variants

Naval AMOS

In 2003 sea trials were undertaken of the twin 120 mm Advanced MOrtar System (AMOS) installed on a Combat Boat 90 of the Swedish Navy.

In 1999 one of the early twin 120 mm AMOS turret systems was installed on a Combat Boat 90 for form and fit trials. The Swedish Defence Material Administration (FMV) subsequently awarded Patria Hägglunds a contract for an AMOS Naval Application Demonstrator (NAD).

For the AMOS naval trials carried out in the Swedish archipelago the prototype AMOS turret was used, which has already undergone extensive trials in the Finnish and Swedish armed forces on tracked (CV 9040) and wheeled (XA 6 × 6 series) hull.

It is envisioned that production AMOS turrets for a naval application would be smaller than the army version and have a turret of all-welded aluminium rather than steel, which would reduce the overall weight of the turret to about 2,500 kg.

Typical AMOS hull applications

Type	Combat weight
AMV (8 × 8)	24.00 tonnes
BMP-3 tracked	18.7 tonnes
CV90 tracked	27.6 tonnes
LAV-III (8 × 8)	16.5 tonnes
M113S tracked	18.5 tonnes
XA-203 (6 × 6)	22.5 tonnes

Specifications

AMOS turret

Crew:	2
Turret weight:	3,600 kg
Armament:	2 × 120 mm smoothbore mortar
Barrel length:	3,000 mm
Loading:	semi-automatic
Secondary armament:	1 × 7.62 mm MG, 8 smoke grenade launchers (2 × 4)
Gun control equipment	
Turret power:	electric/manual
Turret traverse:	360°
Elevation/depression:	+85°/–3°
Ammunition:	50+ × 120 mm
Time to fire:	30 s
Time to scoot:	10 s
Maximum rate of fire:	16 rds/min
First six rounds:	10 s
Maximum range:	16 km
Simultaneous impact on target:	16 rounds
Armour:	steel (turret)

Status

Four systems supplied to Finland for extensive trials with the turret integrated onto an Armoured Modular Vehicle chassis. Contract placed late in 2010 for 18 production systems (14 new plus original four brought up to common production standard).

Production of this batch is now underway for the Finnish Defence Force.

Contractor

BAE Systems, Sweden.
Patria, Finland.

General Dynamics Land Systems - California Technical Center/RO Defence 120 mm Armoured Mortar System (AMS)

Development

The 120 mm Armoured Mortar System (AMS) comprises a 120 mm smoothbore breech loading, recoiling mortar integrated with a full solution fire-control system and mounted in a lightweight steel turret that can be fitted to a wide variety of light armoured vehicle hull.

The now BAE Systems first developed the 120 mm breech-loading mortar as a private venture in 1985. The first prototype of the mortar, complete with its elevating mass, was completed and test fired in mid-1986. During 1987, the turret system was integrated onto a now BAE Systems (previously United Defense) M113A2 hull and test fired.

In the Autumn of 1991, the now General Dynamics Land Systems – Canada Light Armoured Vehicle (LAV) (8 × 8) was fitted with the system and successfully completed initial trials at a UK weapons range.

During these trials around 150 rounds of 120 mm mortar bomb were fired at high and low angles in the indirect fire mode as well as in the direct fire mode. For the trials, the turret was also fitted with elevation and traverse drives and a mock-up of a fire-control system.

The then BAE Systems, RO Defence was responsible for the turret and 120 mm ordnance, General Dynamics Land Systems – California Technical Center of the US was responsible for the electronics and sighting system and General Dynamics Land Systems – Canada was responsible for the LAV (8 × 8) hull and turret integration.

During 1992, General Dynamics Land Systems – Canada/RO Defence undertook the design and construction of a complete turret system with integrated fire-control that they then demonstrated.

120 mm Armoured Mortar System installed on a General Dynamics Land Systems - Canada LAV (8 × 8) hull 1133229

The system comprised a new all-welded steel turret, the 120 mm ordnance and a complete fire-control/sighting system integrated on an 8 × 8 LAV hull.

In late 1995, another 120 mm AMS turret was produced and then fitted to the private venture the now BAE Systems Mobile Tactical Vehicle Light hull, a further development of the widely deployed M113 series. This system was then subjected to extensive trials and a customer demonstration in a country in the Middle East.

Early in 1996, the then General Dynamics Land Systems – California Technical Center awarded the then BAE Systems, RO Defence a contract worth GBP37 million for the supply of 73 120 mm AMS and associated ammunition.

Under the terms of the contract, the now BAE Systems supplied the turret, including the 120 mm mortar, General Dynamics Land Systems – California Technical Center supplied the computerised fire-control system and sights.

The complete turret was then fitted to the General Dynamics Land Systems – Canada Light Armoured Vehicle (LAV) (8 × 8) hull.

The now BAE Systems order included a substantial quantity of 120 mm ammunition that was developed by MECAR of Belgium and includes high-explosive (Composition B), illuminating and smoke (white phosphorus) natures.

Under a Foreign Military Sales (FMS) agreement, General Dynamics Land Systems – Canada has supplied a total of 1,117 LAVs in 10 versions to the Saudi Arabian National Guard (SANG). The 73 120 mm AMS variants are within this figure of 1,117 vehicles.

General Dynamics Land Systems - California Technical Center closed down in late 2004.

As a result of mergers and acquisitions, RO Defence no longer exists as a separate company and has been consumed within BAE Systems.

The 120 mm turret shell and 120 mm ordnance was designed and built at Nottingham which has now closed.

In late 2009 it was announced that the Saudi Arabian National Guard had placed a contract with General Dynamics Land Systems - Canada for the supply of another 724 LAV (8 × 8) vehicles for delivery from April 2011.

Within this contract there was an order for 36 LAV fitted with the Finnish Defence NEMO 120 mm mortar turret.

Description

The 120 mm mortar has a semi-automatic conical screw, swinging breech mechanism with obturation being achieved with a Crossley pad.

Firing is percussion mechanical activated by a solenoid. The recoil mechanism consists of two buffers and a pneumatic recuperator.

Sustained rate of fire is 4 rds/min, rapid rate is 8 rds/min (sustained for three minutes maximum). The ability to fire in direct mode at targets in excess of 1,000 m provides the ground force commander with enhanced operational capabilities while also improving the speed of response to conventional indirect fire roles.

The mortar can fire all standard 120 mm conventional smoothbore ammunition including the new generation smart bombs currently under development. With the 120 mm high-performance mortar ammunition, ranges in excess of 9,000 m were achieved. Also, as the weapon can be fired in the indirect mode of fire at much lower angles of elevation than a conventional mortar, it is much more difficult to detect with mortar locating radar.

The turret is fabricated from armour steel and has stations for two crew. The commander is to the right and the loader to the left. Each crewman has a hatch for access and egress, and an array of vision periscopes.

Mounted either side of the turret is a bank of four electrically operated smoke grenade launchers which fire over the frontal arc and mounted on the turret roof is a 7.62 mm machine gun for the commander.

The complete system is all-weather capable and can revert to manual back-ups should a power failure occur. The turret traverse and elevation drives are hydraulically operated.

The full solution fire-control system employs a differential Global Positioning System (GPS) aided Turret Attitude Sensor System (TASS). The latter continuously and automatically updates location, turret bearing and tilt and cant parameters to the fire-control system.

This allows the AMS quickly to engage the enemy with into action times of under a minute being possible. With the information and target location data stored in the computer, re-engagement of any target from any location within range limitations requires only the selection of that target from the computer menu. A number of targets can be entered into the system to allow rapid switching between them.

For the direct fire role a thermal sight assembly with integral laser range-finder is fitted.

The 120 mm ammunition for the Saudi Arabian National Guard requirement was supplied by the Belgian company of MECAR.

120 mm Turreted Mortar System

This is similar in concept to the 120 mm Armoured Mortar System and has been developed by BAE Systems. This is no longer being marketed by BAE Systems.

Specifications

AMS turret

Crew:	2 (commander and loader)
Turret length:	
(with barrel)	4.68 m
(without barrel)	2.81 m
Turret width:	2.20 m
Turret height:	0.89 m (above hull top)
Barrel length:	3.00 m

Recoil length:	
(maximum)	600 mm
(maximum load)	170 kN
Turret weight:	2,800 kg
Maximum range:	9,200 m (with HE M530A1)
Rate of fire:	
(rapid)	8 rds/min
(sustained)	4 rds/min
(MRSI)	4 rds/min
Armament:	
(main)	1 × 120 mm smoothbore breech-loading mortar
(secondary)	1 × 7.62 mm MG
(smoke grenade dischargers)	2 × 4
Control:	
(traverse)	360° electrohydraulic with manual back-up
(elevation)	–5° to +80° electrohydraulic with manual back-up
Fire-control system:	multitarget input capability with automatic weapon positioning, gun angle calculations and weapon compensation for vehicle attitude and meteorological conditions
Navigation:	integrated GPS system for vehicle position, heading and attitude
Optics:	integrated all-weather day/night system with thermal imager and integrated laser range-finder for automatic range FCS input
Armour:	steel (turret)

Status
Production complete. No longer marketed. A total of 73 AMS have been supplied to the Saudi Arabian National Guard on General Dynamics Land Systems Land Systems - Canada (8 × 8) LAV hull.

Contractor
General Dynamics Land Systems - Canada (LAV).
General Dynamics Land Systems - California Technical Center (turret integration - closed down late 2004).
BAE Systems (originally RO Defence).

Israel

Soltam Systems 120 mm Advanced Deployable Autonomous Mortar System (ADAMS)

Development
To meet emerging requirements for a highly mobile lightweight mortar system for use by rapid deployment forces, Soltam Systems of Israel is now developing a 120 mm mortar system mounted on a modified US AM General High Mobility Multipurpose Wheeled Vehicle (HMMWV) chassis. This is called the Advanced Deployable Autonomous Mortar System (ADAMS). The first example was completed in 2005. It is being developed as a private venture using company funding.

As of March 2013 it is understood that no production orders had been placed for the Soltam Systems 120 mm ADAMS.

M113 APC armed with a Soltam Systems CARDOM 120 mm mortar carrying out a fire mission (Soltam Systems/Elbit Systems) 1440865

Artist's impression of ADAMS, which is a US AM General HMMWV (4 × 4) fitted with turntable-mounted Soltam Systems 120 mm Computerised Autonomous Recoil rapid Deployed Outrange Mortar (CARDOM) system (Soltam Systems) 0590199

Description
For this application, the rear of the AM General HMMWV has been modified to take the Soltam Systems turntable-mounted 120 mm Computerised Autonomous Recoil rapid Deployed Outrange Mortar (CARDOM) system.

For this application the CARDOM 120 mm mortar is fitted with a special recoil system to reduce strain on the firing platform.

The 120 mm CARDOM is currently in production for the Israel Defense Force (IDF) installed in the rear of a now BAE Systems (previously United Defense) M113 series full-tracked Armoured Personnel Carrier (APC). It is also installed in the General Dynamics Land Systems - Canada Stryker (8 × 8) Mortar Carrier for the US Army.

To provide a more stable firing platform, when deployed in the firing position two outriggers would be deployed at the rear with a firing jack lowered under either side of the HMMWV chassis between the first and second wheel stations.

The Soltam Systems CARDOM 120 mm mortar system weighs 670 kg. Traverse is a full 360° and elevation is from +40° to +85°.

Maximum range using conventional ammunition is 7,200 m while rate of fire is between 12 and 15 rounds per minute. With extended range ammunition range can be increased to 9,500 m.

The first prototype of the ADAMS was completed in 2005 and for this application the system has a limited traverse of 30° left and right. A small quantity of ready use ammunition is carried with another vehicle carrying additional ammunition.

The IDF 120 mm self-propelled mortar system is used with a computerised fire-control and powered laying system, while the US Army model has a conventional manual laying system.

The US Army also uses an earlier Soltam Systems 120 mm smoothbore mortar in the towed (M120) and tracked (M121) versions (with the vehicle being designated the M1064A3) as a replacement for the older 4.2 in mortars.

In addition to being in service with the IDF and US Army, the Soltam Systems 120 mm mortar has also been tested in an M113 APC of the Turkish Land Forces Command and the RN-94 (6 × 6) APC. Neither of these versions have entered service with the Turkish Land Forces Command.

IDF M113 with CARDOM 120 mm mortar
The Israel Defense Force deploys the Soltam Systems CARDOM 120 mm mortar integrated into the rear of its M113A3 series full tracked APCs to provide rapid fire support to its infantry battalions.

For this application the CARDOM 120 mm mortar is coupled to a computerised inertial navigation and aiming system.

This computes the target co-ordinates and lays the CARDOM 120 mm mortar onto the target.

In September 2011 Soltam Systems announced that it had been awarded a contract worth USD40 million for the supply of additional CARDOM 120 mm mortars to be delivered to the Israel Defense Force over a four-year period.

Within the Israel Defense Forces these systems are referred to as the Keshet (or Bow).

Mercedes-Benz chassis with CARDOM 120 mm mortar
Soltam Systems developed, tested and placed in production an additional version of its combat proven 120 mm CARDOM. Integrated onto the rear of a Mercedes-Benz (4 × 4) cross-country truck, the turntable 120/81 mm recoil mortar is intended for an undisclosed customer in Africa. This is understood to be Cameroon with the first batch consisting of eight units.

The first battery has already been delivered with additional batteries to follow. The CARDOM vehicle is provided with space for the crew and ready to use onboard ammunition. To provide a more stable firing platform, two hydraulic spades are lowered to the ground one either side at the rear of the chassis before firing commences.

BMP-1 with CARDOM 120 mm mortar
Soltam Systems has also integrated a 120 mm CARDOM mortar into a Russian BMP-1 Infantry Fighting Vehicle (IFV) for another undisclosed export customer with first units already in delivery.

For this application the complete turret of the BMP-1 IFV has been removed and a slightly raised superstructure added. The 120 mm CARDOM mortar is mounted in the centre of the hull and fired through a two part hatch cover that opens left and right. No modifications have been required for the torsion bar suspension of the BMP-1 IFV.

Soltam Systems CARDOM 120 mm mortar system integrated onto a 4 × 4 chassis for the export market (Soltam Systems) 1193006

Soltam Systems CARDOM 120 mm mortar system integrated into a Russian BMP-1 IFV chassis (Soltam Systems) 1169373

Trailer mounted 120 mm mortar

Late in 2006 Soltam Systems revealed that it had developed a new trailer-based version of its combat proven CARDOM 120 mm smoothbore mortar system designed the CARDOM-T.

The standard turntable-mounted CARDOM 120 mm smoothbore muzzle-loaded mortar is normally mounted in the rear of an Armoured Fighting Vehicle (AFV) such as the now BAE Systems M113 or the General Dynamics Land Systems – Canada Stryker.

For use by rapid deployment type forces, the company has developed and tested the trailer-mounted CARDOM-T. This system weighs about 2,000 kg and can be towed by a light vehicle such as an AM General High Mobility Multipurpose Wheeled Vehicle (HMMWV) which can also carry its crew of four plus a quantity of ready use 120 mm ammunition. Additional 120 mm ammunition would be carried by another vehicle.

According to Soltam Systems, the trailer-mounted CARDOM-T 120 mm mortar can come into action in less that two minutes and be taken out of action in a similar time.

This is much less than a standard ground-based mortar which takes much more time to come into action and even more to come out of action. The system can be split into two, for example, HMMWV and CARDOM-T to allow it to be transported under a helicopter.

Trailer-mounted CARDOM-T 120 mm mortar system has full power operation using an onboard Auxiliary Power Unit (APU) which allows it to be deployed away from its towing platform. Elevation of the mortar is from +43° to +85° with traverse 30° left and right. When in the firing position the wheels are raised clear of the ground so that the specially reinforced floor of the trailer absorbs the full recoil. The 120 mm smoothbore barrel can be fitted with a blast attenuation device if required.

The CARDOM-T 120 mm smoothbore mortar system can fire all NATO-standard natures of ammunition with a maximum range of 7,000 m being claimed.

It is laid onto the target using a Soltam Systems computer-controlled automatic laying system but an optical sight and manual controls are provided as a back up. Other computerised fire-control systems could also be integrated with this CARDOM-T 120 mm system if required.

Although the first version is fitted with a proven turntable mounted 120 mm smoothbore mortar it could also be fitted with a standard 81 mm mortar in which case traverse would be through a full 360°.

Spanish 81 mm Soltam Systems contract

Following a competition Spain opted for a CARDOM 81 mm version integrated onto the rear of its VAMTAC (4 × 4) cross-country vehicles under a USD8.5 million contract placed in late 2011 with first six units delivered in mid-2012.

These were integrated in Spain by General Dynamics European Land Systems - Santa Bárbara Sistemas.

Soltam Systems mortar exports

According to the United Nations Arms transfer lists, the following quantities of mortars were exported by Soltam Systems in 2008 and 2010:

Country	Quantity	Comment
Botswana	18	120 mm mortars
Colombia	18	81 mm mortars
Portugal	30	120 mm mortars CARDOM (for Pandur II (8 × 8) APC)
	33	81 mm mortars
US	57	120 mm mortars

Soltam Systems 160 mm self-propelled mortar system

This was developed to meet the requirements of the IDF and integrated onto a much modified Sherman tank hull. It is understood that this has now been withdrawn from service with the IDF.

Specifications

120 mm Advanced Deployable Autonomous Mortar System (ADAMS)

Calibre:	120 mm
Weight (mortar):	670 kg
Maximum range:	
(conventional ammunition)	7,200 m
(extended-range ammunition)	9,500 m
Elevation:	+40° to +85°
Traverse:	360°
Rate of fire:	12-15 rds/min

Status

Development. Not yet in production or service. Baseline 120 mm CARDOM mortar system is already in service with a number of countries including Israel and the US. The former has the system integrated into a modified M1064 series tracked vehicle named Keshet and is operational with IDF Infantry Brigades.

The US Army deploys the system with its Stryker (8 × 8) Brigade Combat Teams. Portugal has taken delivery of 33 systems integrated into a Steyr-Daimler-Puch Pandur II (8 × 8) armoured personnel carrier chassis with first deliveries made in 2008.

Kazakhstan is understood to have 18 × CARDOM 120 mm mortar systems integrated onto an armoured platform. Spain deploys the CARDOM 81 mm version integrated onto the rear of a Spanish VAMTAC (4 × 4) cross-country platform.

Contractor

Soltam Systems Limited. This company was taken over by Elbit Systems late in 2010.

Japan

Type 96 120 mm self-propelled mortar

Development

The Type 96 120 mm self-propelled mortar system was developed in the early 1990s to meet the operational requirements of the Japanese Ground Self-Defence Force for a mobile and protected mortar system to provide fire support for its mechanised infantry units.

Following trials with prototype systems it was type classified as the Type 96 120 mm self-propelled mortar and placed in production. It is believed that production has been at a very low rate with just three being procured in 1999.

By March 2013 it is estimated that about 40 of the Type 96 self-propelled mortar systems were in service.

The Japanese Ground Self-Defence Force did deploy two self-propelled mortar systems based on the SU 60 tracked Armoured Personnel Carrier (APC). One was armed with a 81 mm mortar and the other a 107 mm mortar but both of these systems have now been phased out of service as has the SU 60 APC.

Description

The Type 96 120 mm self-propelled mortar is based on a full-tracked armoured hull that provides the occupants with protection from small arms fire and shell splinters. It is probably fitted with an NBC system and passive night vision devices for some of the crew.

The front of the vehicle is well sloped with the hull sides vertical and with a slight chamfer to their upper part. The hull roof slopes slightly down at the rear with the very rear of the vehicle being vertical.

The diesel power pack is situated front left of the hull with the air inlet louvres in the roof and the air outlet louvres and exhaust outlet in the left side of the hull.

The driver is seated at the front of the hull on the right side and is provided with a single-piece hatch cover that opens to the left. In front of his position are day periscopes, the centre one of which can be replaced by a passive periscope for driving at night.

Above and to the rear of the driver's position, in the front of the vehicle, is another crew station, probably that of the vehicle commander. This is provided with roof hatches and day vision devices below.

Mounted in the forward part of the roof is a third crew station which is provided with twin opening roof hatch covers. Mounted in front of this is a .50 (12.7 mm) M2 HB machine gun that is provided with a shield. It should be noted that no side, rear or overhead protection is provided for the gunner.

Mounted in the rear of the hull firing to the rear is the locally built French TDA 120 mm RT rifled mortar which has been manufactured under licence in Japan for many years. It is understood that the Japanese Ground Self-Defence Force now deploys about 410 of these TDA 120 mm RT towed mortars.

This 120 mm mortar has a maximum rate of fire of 15 to 20 rds/min with maximum range using standard ammunition of 8,100 m.

Using a rocket assisted projectile a maximum range of 13,000 m can be obtained. Various types of 120 mm ammunition can be fired including high explosive, smoke and illuminating.

The suspension either side is understood to be of the torsion bar type with each side having the drive sprocket at the front, six dual rubber-tyred roadwheels and the drive sprocket at the rear. There are no track-return rollers.

Specifications

	Type 96
Dimensions and weights	
Crew:	5
Length	
overall:	6.70 m
Width	
overall:	2.99 m
Height	
overall:	2.95 m
Weight	
combat:	23,500 kg
Mobility	
Configuration	
running gear:	tracked
Speed	
max speed:	50 km/h
Firepower	
Armament:	1 × 120 mm mortar (rifled)
Armament:	1 × roof mounted 12.7 mm (0.50) M2 HB machine gun

Status

Production as required. In service with the Japanese Ground Self-Defence Force. Like all Japanese armoured fighting vehicles, and other defence-related systems, the Type 96 120 mm self-propelled mortar has not been offered on the export market.

Contractor

Not known.

Poland

Huta Stalowa Wola (HSW) 120 mm Rak turret mortar system

Development

Huta Stalowa Wola (HSW) is developing a turret mounted 120 mm breech loaded mortar system called Rak to meet the potential requirements of the Polish Army as well as export markets.

This 120 mm Rak turret mortar system can be integrated onto tracked and wheeled hull with the latter including the Patria Armoured Modular Vehicle (AMV) of which 690 have been ordered by the Polish Army with first deliveries made in 2007.

Within the Polish Army the AMV is known as the Rosomak with the first vehicles coming from the Finnish production line with subsequent production being undertaken in Poland.

For trials purposes, the first example of the HSW 120 mm turret mortar system has been installed on the hull of the 2S1 122 mm Self-Propelled Howitzer (SPH). This Russian-designed system was built in large numbers by HSW for the home and export markets.

Late in 2010 the 120 mm Rak turret mortar system was shown integrated onto the locally manufactured Armour Modular Vehicle (8 × 8) hull.

While originally called the Rak this is now usually referred to as the MSK-HSW (*Mozdrzierz Samobiezny Kolowyz* HSW) turret system.

The Polish MoD awarded the company a development contract for the system in 2007 that runs through to 2012 under which two turrets have been provided, one for a tracked vehicle and one for a wheeled vehicle.

The contract also covers a command post vehicle, 4 × 4 reconnaissance vehicle and ammunition and maintenance vehicles based on a locally developed Jelcz 6 × 6 cross-country truck chassis.

HSW 120 mm Rak turret mortar system integrated onto a Rosomak (8 × 8) armoured personnel carrier which is the locally built Armoured Modular Vehicle (AMV) (Grzegorz Holdanowicz) 1364188

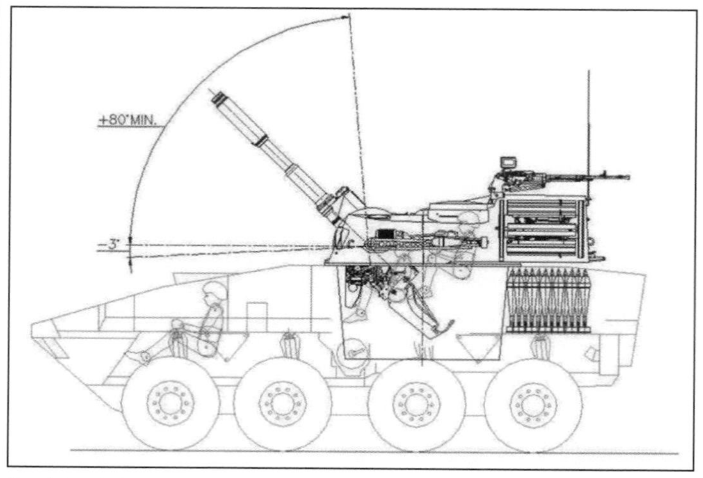

Provisional drawing of the HSW Rak 120 mm turret mortar installed on a Polish Rosomak (8 × 8) armoured personnel carrier (HSW) 1169039

When deployed each unit of this system would consist of eight mortars in two platoons, a reconnaissance group with thee vehicles, three command vehicles, two ammunition trucks and a maintenance truck.

Poland has a requirement for 12-16 units of which 64 would be mounted on the Rosomak and 32 on a tracked hull with deliveries due to commence, funding permitting, in mid-2014.

Description

The turret is of all-welded steel armour protection that provides protection from small arms fire and shell splinters.

For improved accuracy the 120 mm breech loaded mortar is laid onto the target using a computerised fire-control system that is already used by the Polish Army. The system will be operated by a crew of four people.

As the system is turret mounted it will be capable of carrying out direct and indirect fire missions with maximum range of the latter depending on the projectile/propellant combination. Maximum stated rate of fire is between 10 and 12 rounds a minute.

It is understood that when being used in the direct fire role it will have a minimum range of 500 m and a maximum range of 12,000 m when being used in the indirect fire role.

Total number of projectiles carried will depend on the hull but will probably be about 60 of which about 20 would be in the turret bustle for ready use and loaded using a semi-automatic loading system.

While HSW is developing the Rak 120 mm turret mortar system, the Polish Ministry of National Defence Armament Technology Institute is developing the family of 120 mmammunition.

The 120 mm Rak turret mortar system will be used in conjunction with the currently deployed WB Electronics computerised fire-control system.

The aiming system is called the ZIG-T-2 and includes a Charge Coupled Device (CCD) camera, TV camera, thermal camera and eye-safe laser range-finder and has a range of between 50 and 10,000 m.

A forced air ventilation system is fitted which is activated after each round is fired to clear fumes from the turret.

Specifications

	Rak
Dimensions and weights	
Crew:	4
Weight	
standard:	21,000 kg (est.)

	Rak
Mobility	
Configuration	
running gear:	tracked
Engine	diesel
Firepower	
Armament:	1 × 120 mm mortar
Main weapon	
maximum range:	8,000 m (standard ammunition)
maximum range:	12,000 m (extended range ammunition)
maximum range:	-
minimum range:	500 m
Rate of fire	
sustained fire:	12 rds/min (est.)
Turret power control	
type:	powered/manual
Main armament traverse	
angle:	360°
Main armament elevation/depression	
armament front:	+85°/-3°
Survivability	
Night vision equipment	
vehicle:	yes
NBC system:	yes
Armour	
turret:	steel

Status

Under development under contract to the Polish MoD.

As of March 2013 it is understood that no production orders has been placed for the 120 mm Rak turret mortar system.

Contractor

Huta Stalowa Wola (HSW).

Russian Federation

240 mm self-propelled mortar M-1975 (SM-240) (2S4)

Development

The 240 mm self-propelled mortar, originally known to the West as the M-1975, is known as the SM-240 (2S4) by the Russian Army, although its more common name is the Tyul'pan, or Tulip Tree. This system reached full operational capability with the former Soviet Army in 1975.

It is estimated that about 450 of these systems were manufactured, although the Russian Federation has not released any production figures.

Recent information has indicated that very few, if any, of these systems now remain in front line service as the weapon has such a short range and low rate of fire.

The 2S4 was developed in the early 1970s by the Uraltransmash Works, with the 240 mm 2B8 mortar system being developed by the Perm Machine Construction Works. The development designation of the system was Obiekt 305.

The hull used for this system is a modified version of that used in the 2K11 Krug (SA-4 Ganef) surface-to-air missile system.

240 mm 2S4 self-propelled mortar in travelling configuration with 240 mm ordnance in travel lock 1365016

240 mm 2S4 self-propelled mortar in travelling configuration from the rear and clearly showing large spade in raised position 1365017

The 2S4 has also been used by the Czech Republic and Slovakia, Iraq and Lebanon but as far as it is known it is no longer in front line service with these countries.

Recent information has indicated that the 2S4 did see operational service in Chechnya in conjunction with the Smel'chak laser-guided projectile.

The 240 mm 2S24 self-propelled mortar system is no longer in front line service with any nation, but Russia may have up to 400 of these systems held in reserve.

Description

The hull of the 240 mm self-propelled mortar M-1975 (SM-240) 2S4 is of all-welded steel armour construction that provides the crew with protection from small arms fire and shell splinters.

The 2S4 has a crew of four with the other five crew members being carried in the ammunition support vehicle. This could typically be a 6 × 6 truck or the MT-LB armoured full-tracked multirole vehicle.

The driver and vehicle commander are seated front left with the diesel power pack to the right and the crew/ammunition compartment extending to the rear.

The driver has a single-piece hatch cover that opens to the rear with day periscopes for forward observation. The centre one can be replaced by a passive night vision device. Behind this is the raised commander's position. The commander's cupola can be rotated through a full 360° and is provided with an externally mounted 7.62 mm PKT machine gun and an infra-red searchlight, with additional hatches being provided in the roof.

The 240 mm smoothbore mortar, designated the 2B8, is carried complete with its baseplate in the horizontal position on top of the hull and is pivoted at the rear.

When in the firing position, the mortar is hydraulically swung to the rear under remote control so that the mortar barrel faces rearwards.

The 240 mm mortar has powered elevation from +50° to +80° with powered traverse of 10° left and right.

Inside the hull of the 2S4 are two drum-type magazines, each of which holds 20 240 mm mortar bombs. When the drums rotate, they line up with a hatch in the roof of the vehicle and a specialised telescoping mechanism flips the muzzle down and loads the barrel first with the projectile and then with the bagged charge.

The mortar is fired by either electrical or mechanical means and two basic types of 240 mm mortar bomb are fired.

A conventional high-explosive fragmentation bomb (3OF-864), which weighs 130 kg and has a maximum range of 9,650 m, and a high-explosive fragmentation rocket-assisted projectile weighing 228 kg with a maximum range of 18,000 m. Minimum range is 800 m.

The mortar system is aimed with a hydroelectric drive system by remote control. At an elevation of +60° rate of fire is one round in 62 seconds and with an elevation of +80° a maximum rate of fire of one round every 77 seconds can be achieved.

Other types of 240 mm projectile are understood to have originally included chemical, concrete-piercing and tactical nuclear. Russian sources have stated that the tactical nuclear and chemical projectiles have now been phased out of service.

Suspension is of the torsion bar type with each side having six dual rubber tyred roadwheels and the drive sprocket at the front, idler at the rear and four track-return rollers.

Standard equipment on the 2S4 includes a NBC system and night vision equipment. Mounted at the front of the hull is a dozer blade that is used to prepare firing positions.

Smel'chak guided projectile

In late 1993, it was revealed that the 240 mm 2S4 self-propelled mortar system and the 240 mm M-240 towed mortar could both fire the Smel'chak (Daredevil) laser-guided mortar projectile.

The Smel'chak was developed in the 1980s and was successfully trialled in Afghanistan, being fired from the 240 mm M-240 towed mortar.

In addition to the actual projectile and mortar, other elements of the system include the 1D15 laser target indicator/laser range-finder, the 1A35K/1A35J synchronisation system and communications equipment.

Typically, the forward observer first locates the target and this information is relayed back to the command post which calculates the relevant target-laying information. The 240 mm Smel'chak guided projectile is then loaded, aimed at the target and fired.

As the projectile gets to within 400 to 800 m of the target the self-guidance system is activated. This reacts to the illumination of the target by the laser designator and the projectile then homes onto the target. If the projectile is off course, a rocket motor is engaged to correct its flight.

Compared to the US Copperhead 155 mm Cannon-Launched Guided Projectile, the Smel'chak is very difficult to jam as the target is illuminated for such a short period of time.

The 1D15 laser target indicator/designator has a ×10 monocular sight and can designate targets from 200 to 10,000 m to an accuracy of 10 m.

The Smel'chak has a minimum range of 3.6 km and a maximum range of 9.2 km with laser target designation range being 0.2 to 5.0 km. The actual projectile, the 3F5, weighs 134.2 kg of which 32 kg is the warhead.

Specifications

	M-1975 2S4 self-propelled mortar
Dimensions and weights	
Crew:	4
Length	
overall:	7.94 m
Width	
transport configuration:	3.25 m
Height	
transport configuration:	3.225 m
Ground clearance	
overall:	0.40 m
Weight	
combat:	27,500 kg
Ground pressure	
standard track:	0.6 kg/cm²
Mobility	
Configuration	
running gear:	tracked
Power-to-weight ratio:	18.90 hp/t
Speed	
max speed:	60 km/h
Range	
main fuel supply:	500 km
Fuel capacity	
main:	850 litres
Fording	
without preparation:	1 m
Gradient:	60%
Side slope:	30%
Vertical obstacle	
forwards:	1.10 m
Trench:	2.79 m
Engine	V-59, V-12, diesel, 520 hp
Firepower	
Armament:	1 × hull mounted 240 mm 2B8 mortar
Armament:	1 × roof mounted 7.62 mm (0.30) PKT machine gun
Ammunition	
main total:	40
7.62/0.30:	1,500
Turret power control	
type:	powered/manual
Main armament traverse	
angle:	20° (10° left/10° right)
Main armament elevation/depression	
armament front:	+80°/+50°
Survivability	
Night vision equipment	
vehicle:	yes
NBC system:	yes
Armour	
hull/body:	steel

Status

Production complete. In service with the Russian Army (reserve).

Contractor

Russian state factories.

120 mm 2S31 Vena self-propelled gun/mortar system

Development

In early 1997 Russia finally revealed the 120 mm 2S31 Vena self-propelled gun/mortar system which had been under development for some years.

120 mm 2S31 Vena self-propelled gun/mortar system with ordnance at maximum elevation (Christopher F Foss) 0073007

Early development examples of the 2S31 were based on the Obiekt 934 light tank hull developed by the Volgograd Tractor Plant but the first complete system, shown for the first time in the Middle East, is based on a much modified BMP-3 Infantry Fighting Vehicle (IFV) hull.

The BMP-3 IFV entered service with the Russian Army in 1990 and is also being used for an increasing number of more specialised supporting roles of which the 2S31 is the latest.

While the prime contractor for the BMP-3 chassis is the Kurgan Machine Construction Plant, the prime contractor for the 2S31 Vena system is the Motovilikha Plants Corporation based in Perm which is involved in a number of other Russian artillery systems as well as multiple launch rocket systems.

Development of the 120 mm 2S31 Vena self-propelled gun/mortar system is complete.

The 120 mm 2S31 Vena system has been offered on the export market but as of March 2013 no export sales had been made of this system.

Description

For the 2S31 application, the hull of the BMP-3 IFV has been modified in a number of areas. The crew position either side of the driver has been removed as have the two bow-mounted forward-firing 7.62 mm PKT machine guns.

The existing two-person turret has been removed and replaced by a much larger turret with vertical sides and rear and a curved front. The hull rear of the vehicle has been modified and all roof hatches and rear doors have been removed.

A new and much smaller downward-opening hatch is provided in the rear of the hull on the left side. There are two ammunition resupply hatches in the right side, one in the hull and one in the turret.

The 2S31 Vena has a four-man crew consisting of driver seated at the front in the centre, commander in the turret on the right, gunner in the turret on the left and the ammunition loader who sits to the rear of the commander when travelling.

Turret traverse is powered through a full 360° with weapon elevation from −4° to +80°; manual controls are provided for emergency use.

Each vehicle is also fitted with an onboard automatic survey and orientation system. The advanced fire-control system includes a computer that calculates the information required to lay the 120 mm main armament onto the target and ensure a high first-round hit probability.

The commander's cupola is mounted on the right side of the turret roof and has a large electro-optical sensor package on the right side, which is covered by a door when not required. This cupola can be traversed through a 90° arc.

This sensor package includes 1P51 day/image intensification sights and a 1D22S laser range-finder/designator with some components believed to be used in the BMP-3K reconnaissance vehicle.

A 7.62 mm PKT machine gun is mounted on top of the cupola which can be laid and fired under complete armour protection and can be used to engage ground and air targets. The gunner has both direct and indirect fire sights.

The 2S31 Vena is fitted with a bank of six 81 mm electrically operated smoke/decoy grenade dischargers either side of the turret.

A laser detector is mounted on the lower part of the turret front, either side of the 120 mm 2A80 ordnance and on either side of the turret roof at the rear which is used in conjunction with the smoke/decoy grenade system.

The long 120 mm 2A80 barrel is rifled and like other Russian 120 mm systems is not fitted with a fume extractor or a muzzle brake as it is of the low pressure type. The 120 mm ordnance is fitted with a combined breech mechanism, pneumatic rammer and a bore scavenging system.

This weapon has a direct and indirect fire capability and fires a new family of 120 mm ammunition.

It can also fire the same family of ammunition as the current 120 mm 2S9 (tracked) and 2S23 (8 × 8) self-propelled gun mortar systems. According to Russian sources the high-explosive projectile has the equivalent destructive power of a 122 mm/152 mm artillery projectile.

The Russian BAZALT State Research and Production Enterprise has developed two new 120 mm projectiles for the 2S31 Vena self-propelled gun/mortar system, both of which weigh 26 kg.

The high-explosive fragmentation projectile has a minimum range of 2,000 m and a maximum range of 18,000 m with each projectile having a damage area of 2,200 m^2.

The second is a cargo round which contains 35 high-explosive/fragmentation submunitions each of which will penetrate about 100 mm of conventional steel armour. These are designed to penetrate the vulnerable upper surfaces of armoured vehicles. This projectile has a minimum range of 1,500 m and a maximum range of 11,000 m. Each projectile has a damage area of about 10,000 m^2 and the 2S31 Vena can also fire western natures of 120 mm rifled mortar ammunition.

More recently the Motovilikha Plants Corporation has stated that the maximum range when firing normal HE projectiles is 13,000 m and when firing 'mines' projectiles is 7,200 m.

A total of 70 120 mm projectiles is carried of which 22 (11 + 11) are in ready use magazines for rapid loading using a ramming system. Maximum rate of fire is quoted as 8 to 10 rds/min.

Also carried are 10 KBP Instrument Design Bureau 120 mm Kitolov-2M laser-guided projectiles which have a maximum range of 13 to 14 km with a 0.8 to 0.9 hit probability. Targets can be designated from the 2S31 Vena or from a normal tripod-mounted designator. The Kitolov-2M laser-guided projectiles are loaded manually.

The Kitolov was first revealed some time ago in 122 mm calibre for the 2S1 self-propelled artillery system.

Maximum quoted range for HE projectile is 13 km while for a standard mortar projectile it is 7.2 km.

The 2S31 has a combat weight of 19,500 kg and a power-to-weight ratio of 26.31 hp/tonne, maximum road speed is 70 km/h and cruising range on roads is 600 km. Other hull details and specifications are almost identical to those of the BMP-3 ICV that is currently in production.

Like other members of the BMP-3 family, Vena is fully amphibious being propelled in the water by two water-jets mounted low down one either side of the hull at the rear.

Before entering the water the trim vane is erected at the front of the vehicle by the driver without leaving his seat and the bilge pumps are switched on.

Standard equipment for the 2S31 Vena includes an NBC system of the overpressure type, land navigation system, night vision equipment, vehicle washdown capability and a toilet.

Although the prototype of the 2S31 Vena is based on the hull of the BMP-3 infantry combat vehicle, the turret system can be installed on a wide range of other chassis, tracked and wheeled.

Variants

Russian 120 mm 2S34 SP mortar system

Russia has developed to the prototype stage the 120 mm 2S34 SP mortar which for trials purposes has been integrated onto the hull of the 2S1 122 mm self-propelled howitzer and the widely deployed and MT-LB multipurpose tracked vehicle.

The 120 mm 2S34 SP mortar has yet to enter production and if and when it does it is expected to be integrated onto the hull of the new Russian Kurganets tracked vehicle now under development. Kurganets is the potential follow on to the BMP-3 IFV.

Status

Prototype. Not yet in production or service.

Contractor

Motovilikha Plants Corporation (turret and prime contractor).
Kurgan Machine Construction Plant (chassis).

120 mm NONA-SVK 2S23 self-propelled gun-mortar system

Development

The 120 mm 2S23 self-propelled gun-mortar system, which is also referred to as the NONA-SVK, is now in service with the Russian Army in small numbers.

Compared with conventional towed mortar systems, self-propelled systems such as the full-tracked 2S9 (tracked) and the NONA-SVK 2S23 (wheeled) have many advantages.

These include the ability to keep up with the units for which they are to provide fire support; rapid deployment times; the crew is provided with protection from small arms fire, shell splinters and NBC attack and they also have a direct fire capability, which can be very useful in some tactical situations.

120 mm 2S23 self-propelled gun-mortar system 0572184

The actual 120 mm weapon is also referred to as the NONA with the 2S9 being the NONA-S, the 2S23 the NONA-SVK and the towed 120 mm 2B16 combination gun the NONA-K, all of which use the same family of 120 mm ammunition.

The existence of the 2S23 was first revealed at a defence exhibition in Asia in late 1990 when a complete system was displayed for the first time. According to Russian sources the 2S23 entered service with its ground forces in 1990.

It is believed that hull integration of the 2S9 and more recent 2S23 was undertaken at the IV Gavalov Design Bureau which is located at the Volgograd Tractor Plant where production of the BMD-1, BMD-2 and BMD-3 airborne assault vehicles was undertaken.

More recently a new 120 mm self-propelled gun-mortar system, based on a modified BMP-3 hull, has been developed called the 2S31. This has a longer barrel than the 2S23 but fires the same family of ammunition. Details of this system, which as of early March 2013 remains at the prototype stage, are provided in a separate entry in *IHS Jane's Land Warfare Platforms: Artillery and Air Defence*.

As of March 2013 the only confirmed export customer for the 120 mm 2S23 was Venezuela with the first batch of 23 being delivered by 2011/2012 with additional vehicles probably following.

Recent information has indicated that China purchased a technology package, rather than complete 2S23 systems. This technology was subsequently incorporated into the NORINCO 120 mm SPM based on a modified WZ/WMZ551 (6 × 6) series armoured personnel carrier chassis, are provided in a separate entry in *IHS Jane's Land Warfare Platforms: Artillery and Air Defence*.

China North Industries Corporation (NORINCO) have been offering this on the export market for many years.

This is now in service with the Peoples Liberation Army under the designation of the PLL-05 120 mm wheeled self-propelled mortar-howitzer group. This was first shown in public during a parade held in Beijing late in 2009.

This system has been offered on the export market by NORINCO but as of March 2013 there are no known export sales of this system.

Description

The 2S23 is based on a modified BTR-80 (8 × 8) Armoured Personnel Carrier (APC) hull with the turret-mounted mortar over the forward part of the hull.

The BTR-80 is widely used by Russian motorised rifle divisions while the tank divisions use the BMP-1/BMP-2 and, in small numbers, the latest BMP-3.

Compared to the BMPs and the 2S9, the BTR-80 and 2S23 have much greater strategic mobility and also lower life cycle costs as they are wheeled rather than tracked. The 2S23 may also be used by Russian naval infantry units.

The hull of the 2S23 is of all-welded steel armour construction that provides protection over the frontal arc from 12.7 mm armour-piercing rounds at a distance of 1,000 m with the remainder of the vehicle providing protection against 7.62 mm armour-piercing rounds.

The driver is seated at the front left with the commander to his right, turret and ammunition compartment in the centre and the power pack compartment at the rear. The latter is provided with an automatic fire detection and extinguishing system, a preheater, fuel primer, heat exchangers for the radiator and crank case, diesel fuel tanks and a bilge pump.

Unlike the BTR-80 APC, the 2S23 does not have any firing ports and apart from the emergency hatch between the second and third roadwheels the only means of entry are the hatches in the turret roof and above the commander's and driver's station.

The turret of the 2S23 is of a slightly different design from that of the 2S9 but has a later 120 mm 2A60 rifled gun-mortar and fires the same family of new ammunition. The direct fire sight is to the left of the 120 mm ordnance and the indirect sight is in the turret roof on the left side.

The commander has a raised cupola which can be traversed through a full 360° and also has vision devices, a 7.62 mm PKT externally mounted machine gun and a TKN-3A infra-red searchlight which is mounted to the left of the machine gun. The machine gun be aimed and fired from within the turret by remote control as it is connected to the TKN-3A sighting device.

The 120 mm 2A60 rifled ordnance, which has a calibre length of 24.2, has an elevation from –4° to +80° with turret traverse being 35° left and right.

A total of 30 120 mm mortar projectiles and their associated charges is carried and the automatic loading system enables a rate of fire of 8 to 10 rds/min to be achieved.

120 mm 2S23 self-propelled gun-mortar system with thermal shielding (Christopher F Foss) 0018998

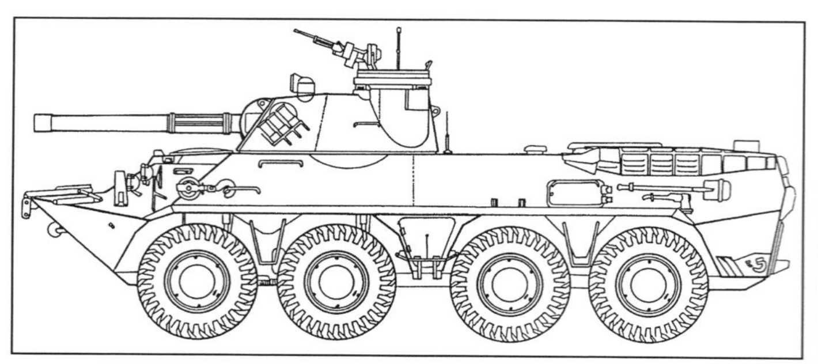

Side drawing of the 120 mm 2S23 self-propelled gun-mortar system NONA-SVK (Steven Zaloga) 0500873

New ammunition is loaded into the 2S23 via a small door in the lower part of the hull between the second and third roadwheels.

Ammunition for the 120 mm 2S23 is manufactured by the Bazalt State Research and Production Enterprise and brief details are as follows:

Type	**HE-FRAG**	**HE-RAP**	**HEAT**
Calibre:	120 mm	120 mm	120 mm
Round weight:	19.8 kg	19.8 kg	13.17 kg
HE content:	4.9 kg	3.25 kg	n/a
Round length:	820 mm	825 mm	960 mm
Muzzle velocity:	367 m/s	367 m/s	560 m/s
Maximum pressure:	100 MPa	100 MPa	110 MPa
Maximum range:	8,850 m	12,850 m	800 m
Minimum range:	1,000 m	1,000 m	40 m

The High-Explosive Fragmentation (HE-FRAG) round is designated the OF-54, with the actual projectile being called the OF-49, and has a maximum range of 8,850 m while the OF-50 High-Explosive Rocket-Assisted Projectile (HE-RAP) has a maximum range of 12,800 m and has less explosive content. Both of these mortar bombs have a body of C60 steel and a nose-mounted point-detonated B35 fuze.

The fin-stabilised High-Explosive Anti-Tank (HEAT) round has a point-blank range of 500 m and will penetrate 650 mm of rolled homogeneous conventional armour. When in flight the fins unfold to provide flight stabilisation. This bomb has a point-initiated base-detonated fuze.

While this 120 mm HEAT round will not penetrate the latest MBTs over their frontal arc, it will inflict damage to their sides and rear as well as being capable of destroying most light armoured fighting vehicles and self-propelled artillery systems likely to be encountered on the battlefield.

All of these three 120 mm mortar bombs feature a pre-engraved copper driving band to the rear with the propellant charge being secured to the base.

In addition to the Russian types of 120 mm ammunition, the weapon can fire ammunition designed for the French TDA 120 RT-61 towed rifled mortar system which is used by many countries around the world.

Russia has disclosed that it has developed a 120 mm laser-guided mortar projectile called the Gran which has been designed to be launched from muzzle-loaded 120 mm mortar systems, including vehicle-carried systems. Basic specifications of the Gran are:

Calibre:	**120 mm**
Warhead type:	**HE-FRAG**
Weight:	
(projectile)	25 kg
(warhead)	11 kg
(HE)	5.1 kg
Length:	
(projectile)	**1,225 mm**
Maximum range:	**7.5 km**

It is probable that the Gran laser-guided mortar projectile has been designed to be fired only from 120 mm smoothbore mortar tubes. Russia has, however, developed to the prototype stage the 122 mm Kitolov-2 laser-guided artillery projectile which can be fired from the 122 mm 2S1 self-propelled artillery system.

Late in 1993, it was revealed that this could also be produced in 120 mm version and an artist's impression showed this being fired from the 120 mm 2S23 system. For the 2S23 application a maximum range of 13 to 14 km would be achieved with a 0.8 to 0.9 hit probability. The latest Kitolov-2M has a tandem High-Explosive Anti-Tank (HEAT) warhead which will defeat armoured vehicles fitted with Explosive Reactive Armour (ERA).

A bank of three standard 81 mm electrically operated smoke grenade dischargers is mounted either side of the turret firing forwards.

Like the BTR-80 APC, the 2S23 is fully amphibious, being propelled in the water by a single water-jet which is located in the lower part of the hull rear. Before entering the water a trim board is erected at the front of the hull by the driver, without leaving his seat and the bilge pumps are activated.

The 2S23 is transportable by transport aircraft, can be dropped by parachute and has been designed to operate in temperatures ranging from –50 to +50°C.

Standard equipment includes an NBC system of the overpressure type, a central tyre-pressure regulation system that allows the driver to adjust the tyre pressure to suit the ground being crossed without leaving his seat, and night vision equipment for the driver.

During a defence equipment exhibition held in Nizhny Novgorod in September 1995, a 2S23 was observed with what appeared to be a thermal shield on its turret and chassis.

Variants

Russian 120 mm 2S34 SP mortar system

Russia has developed to the prototype stage the 120 mm 2S34 SP mortar which for trials purposes has been integrated onto the hull of the 2S1 122 mm self-propelled howitzer and the widely deployed and MT-LB multipurpose tracked vehicle.

The 120 mm 2S34 SP mortar has yet to enter production and if and when it does it is expected to be integrated onto the hull of the new Russian Kurganets tracked vehicle now under development. Kurganets is the potential follow on to the BMP-3 IFV.

As of March 2013 Russia was still marketing the 2S31 (tracked) and 2S23 (wheeled) systems and this is expected to continue for the immediate future.

Specifications

	NONA-SVK 2S23
Dimensions and weights	
Crew:	4
Length	
overall:	7.40 m
Width	
overall:	2.90 m
Height	
overall:	2.495 m
Ground clearance	
overall:	0.475 m
Track	
vehicle:	2.41 m
Weight	
combat:	14,500 kg
Mobility	
Configuration	
running gear:	wheeled
layout:	8 × 8
Power-to-weight ratio:	17.93 hp/t
Speed	
max speed:	80 km/h
water:	10 km/h
Range	
main fuel supply:	600 km
Fuel capacity	
main:	290 litres
Amphibious:	yes
Gradient:	60%
Side slope:	30%
Vertical obstacle	
forwards:	0.5 m
Trench:	2.00 m
Turning radius:	7.0 m
Engine	V-8, water cooled, diesel, 260 hp
Gearbox	
type:	manual
forward gears:	5
reverse gears:	1
Transfer box:	2 speed
Steering:	power-assisted
Brakes	
main:	hydraulic on all wheels
Tyres:	13.00 × 18

	NONA-SVK 2S23
Electrical system	
vehicle:	24 V
Firepower	
Armament:	1 × turret mounted 120 mm 2A60 mortar (rifled)
Armament:	1 × roof mounted 7.62 mm (0.30) PKT machine gun
Armament:	6 × smoke grenade launcher (2 × 3)
Ammunition	
main total:	30
secondary total:	500
Main armament traverse	
angle:	70° (35° left/35° right)
Main armament elevation/depression	
armament front:	+80°/-4°
Survivability	
Night vision equipment	
vehicle:	yes
NBC system:	yes
Armour	
hull/body:	steel
turret:	steel

Status
Production as required. In service with Russia (Army (30+) and Naval Infantry (20+)). The People's Liberation Army has a similar system in service on a modified WZ/WMZ551 (6 × 6) hull with the complete system designated the PLL-05 120 mm wheeled self-propelled mortar howitzer.

The first batch of 13 2S23 was delivered to Venezuela in 2011 followed by additional vehicles.

Contractor
Motovilikha Plants Corporation (prime contractor).

120 mm Anona 2S9 self-propelled gun/mortar

Development
The 120 mm SO-120 (2S9) Anona (Anemone) self-propelled gun/mortar was first seen in public in May 1985 and is an airborne artillery assault vehicle that was developed to carry out two tactical functions.

One role is as a conventional artillery equipment to replace existing mortars and howitzers in the Russian Army's air assault divisions and the second role could be as a direct fire anti-tank weapon system firing HEAT projectiles.

The SO-120 is used by the Russian Army *Vozdushno-desantnyye Voyska* (or VDV) air assault divisions. The 2S9 entered service in 1981 and was deployed in Afghanistan.

Although it was originally developed to meet the requirements of former Soviet air assault divisions, it has also been identified in service with regular ground forces and naval infantry. A quantity was left behind in Afghanistan when the Soviets withdrew. During the fighting in Afghanistan in 2001 to 2002 none of these systems were observed in action.

The lead developer for the complete 2S9 system was TsNII Tochmash (Klimovsk). The 120 mm weapon components were developed by the Motovilikha Factory in Perm, while the ammunition suite was developed by GNPP Bazalt.

The hull used as the basis for the 2S9 system is based on that of the BMD airborne combat vehicle designed by the I V Gavalov Design Bureau located at the Volgograd Tractor Plant, where production of the BMD-1, 2, 3 and 4 has been undertaken.

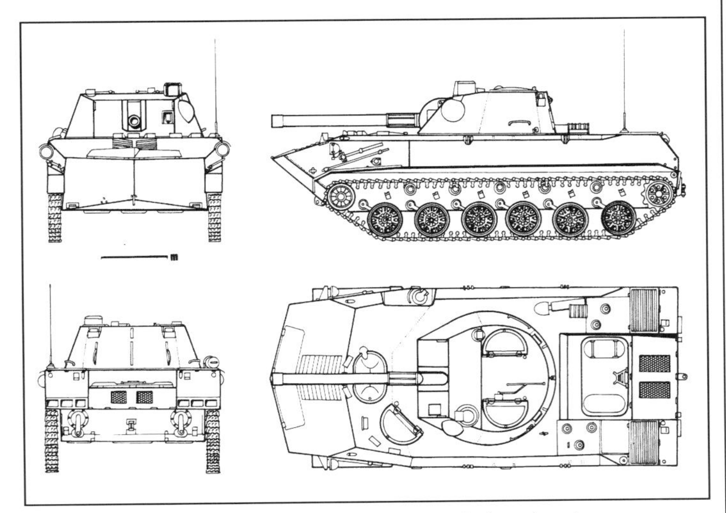

Four-view drawing of the 120 mm 2S9 self-propelled gun/mortar (Steven Zaloga) 0500418

Russian Army 120 mm Anona 2S9 self-propelled gun/mortar with ordnance slightly elevated (Michael Jerchel) 1365020

It is estimated that well over 1,000 of these 2S9 systems have been completed, although Russia has not released any production figures.

The 2S9 has also been observed being transported by the Pomornik-class air cushion vehicle used by the Russian Navy.

The 120 mm 2S9 Anona is also referred to as the Nona-S with the S standing for Samokhodnaya, or self-propelled.

According to the United Nations Arms Transfer Lists for the period 1992 through to 2010 there was only one export of the 2S9. Ukraine supplied three to China in 2000. It is possible that these systems were for development work.

More recently, Russia has developed the 120 mm 2S23 self-propelled gun/mortar system. This is essentially the BTR-80 (8 × 8) armoured personnel carrier, fitted with a modified version of the turret fitted to the 2S9.

Details of the 2S23 are provided in a separate entry in *IHS Jane's Land Warfare Platforms: Artillery and Air Defence* and this is still being marketed by Russia with production being undertaken on an as required basis.

As of March 2013 the only confirmed export customer for the 120 mm 2S23 was Venezuela with the first batch of 23 being delivered by 2011/2012 with additional vehicles probably following.

Description
The SO-120 (2S9) has a crew of four: commander, driver/mechanic, gunner and loader.

The driver/mechanic is seated in the front centre of the hull and is provided with three day periscopes with the centre block replaceable by an infra-red night driving device. The driver has a single-piece hatch cover that opens to the right.

The commander is located at the front left of the hull. He is provided with three day vision devices and the communication sets, a gyrocompass and land navigation devices. The commander has a single-piece hatch cover that opens towards the rear.

The gunner is on the left of the turret. His position has a panoramic day sight fitted to the turret roof and a telescopic sight for direct fire. The loader is to the right of the gun. He is provided with a single vision block and a seat that folds away while he is serving the gun.

The SO-120 (2S9)'s aluminium hull is a version of the BTR-D airborne APC. It is divided into three compartments: the command compartment, the fighting compartment and the power pack compartment.

The command compartment is in the front of the hull, forward of the turret mount. It accommodates the commander and the driver/mechanic. The fighting compartment has stowage for 25 rounds of 120 mm ammunition behind the turret and is in the middle of the hull.

The fully enclosed all-welded turret is located above the fighting compartment. At the rear of the hull is the diesel power pack compartment.

The engine is a 5D20 diesel developing 240 hp, coupled to a five-speed manual transmission. Diesel fuel is stored in the engine compartment.

The two-man turret is of welded aluminium construction with 16 mm thick frontal armour. The turret roof has two hatches, one for the gunner and the other for the loader. Turret traverse is limited to 35° either side. Mounted in either side of the turret rear is a TNPO-170A day observation device.

The track is the same as that used on the BMD-1 airborne combat vehicle and the suspension is hydropneumatic with an adjustable ground clearance of between 100 and 450 mm. The ground clearance is adjustable by the driver/mechanic from within the vehicle. There are six road wheels each side and five track-return rollers. Unlike many Russian vehicles, there are no track skirts fitted to the 2S9.

The SO-120 (2S9) is amphibious and is powered in the water by two water-jets at the rear. Power can be delivered to the water-jets and the tracks at the same time. When in the water, splash vanes can be erected in front of the driver's position. Electrically operated bilge pumps are activated to expel any water that enters the vehicle during amphibious operations.

The SO-120 (2S9) is armed with a 120 mm breech-loaded gun 2A51 with a barrel approximately 1.8 m long. The gun is probably provided with an interrupted-screw breech mechanism and a chamber detent to retain a round in place when the barrel is elevated.

Ammunition is fixed and loading is manual, although ramming is automatic. After the loader has selected a round from a ready rack it is placed in a feed tray. After an electrical button has been pressed a rammer automatically seats

Russian Army 120 mm 2S9 self-propelled gun/mortar in travelling configuration with commanders and drivers hatches open (INA) 1452633

120 mm 2S9 self-propelled gun/mortar (left) with 1V119 artillery fire direction vehicle (right) 0501017

the round in the chamber and closes the breech. The rammer is a pneumatic device operating at a pressure of 150 kg/cm². As the breech opens after firing, the rammer bleeds compressed air into the chamber to force firing fumes from the muzzle.

The 120 mm mortar has a rate of fire of 6 to 8 rds/min. Muzzle velocity is 560 m/s for the HEAT round and 367 m/s for the artillery rounds. The direct fire HEAT projectile is estimated to defeat up to 650 mm of conventional steel armour. In the indirect fire role the mortar fires high-explosive, white phosphorus and smoke rounds. The standard HE mortar bomb has a maximum range of 8,855 m while the 120 mm rocket-assisted mortar bomb has a maximum range of 13,000 m. Additional details of the ammunition family are provided in the entry for the 120 mm 2B16 (NOKA-K) combination gun which is mounted on a split trail carriage.

Mounted below the turret rear is an ammunition loading hatch and mounted on top of the hull is a device for loading ammunition from the ground directly into the fighting compartment. This allows sustained fire missions to be carried out while preserving onboard ammunition supply.

The SO-120 (2S9) takes 30 seconds to come into action and a similar time to come out of action. When in the firing position the suspension is raised to provide a more stable firing platform.

When deployed in the firing position the overall height of the 2S9 is 1.9 m.

The SO-120 (2S9) has no coaxial machine gun in the turret and there are no bow or turret-mounted machine guns. Standard equipment includes an NBC system, night vision equipment and an electric bilge pump.

For air transport the SO-120 (2S9) can be carried by any Russian Military Air Transport (VTA) medium or heavy cargo aircraft. It can be paradropped from the An-32 'Cline', Il-76 'Candid' or An-22 'Cock' aircraft using the PRSM-915 heavy drop system, from heights of between 300 and 1,500 m.

Variants

There are no known variants of the 2S9 although also used with these units are the 1V118 artillery observation post vehicle and the 1V119 artillery fire direction vehicle, both of which are based on the chassis of the BMD-1 airborne combat vehicle.

2S9-1

Russian sources have recently mentioned the 2S9-1. This is understood to be almost identical to the standard production 2S9 but has a combat weight of 8.7 tonnes and a 120 mm ammunition load increased from 25 to 40 rounds.

Russian 120 mm 2S31 SP mortar system

This is currently being marketed by Russia and is based on a modified BMP-3 infantry fighting vehicle hull.

As of March 2013 it is understood that the 2S31 has not entered quantity production with no export sales having been made.

Details of this system are provided in a separate entry in *IHS Jane's Land Warfare Platforms: Artillery and Air Defence*.

Russian 120 mm 2S34 SP mortar system

Russia has developed to the prototype stage the 120 mm 2S34 SP mortar which for trials purposes has been integrated onto the hull of the 2S1 122 mm self-propelled howitzer and the widely deployed and MT-LB multipurpose tracked vehicle.

The 120 mm 2S34 SP mortar has yet to enter production and if and when it does it is expected to be integrated onto the hull of the new Russian Kurganets tracked vehicle now under development. Kurganets is the potential follow on to the BMP-3 IFV.

As of March 2013 Russia was still marketing the 2S31 (tracked) and 2S23 (wheeled) systems and this is expected to continue for the immediate future.

Specifications

	Anona 2S9
Dimensions and weights	
Crew:	4
Length	
overall:	6.02 m
Width	
overall:	2.63 m
Height	
overall:	2.3 m
Ground clearance	
adjustable:	0.10 m to 0.45 m
Track	
vehicle:	2.38 m
Track width	
normal:	250 mm
Length of track on ground:	3.23 m
Weight	
combat:	8,700 kg
Ground pressure	
standard track:	0.5 kg/cm²
Mobility	
Configuration	
running gear:	tracked
Power-to-weight ratio:	27.58 hp/t
Speed	
max speed:	60 km/h
water:	9 km/h
Range	
main fuel supply:	500 km
water:	90 km (est.)
Amphibious:	yes
Gradient:	60%
Side slope:	33%
Vertical obstacle	
forwards:	0.8 m
Trench:	1.8 m
Engine	5D20, diesel, 240 hp
Gearbox	
type:	manual
forward gears:	5
reverse gears:	1
Suspension:	hydropneumatic, adjustable
Firepower	
Armament:	1 × turret mounted 120 mm 2A60 mortar (rifled)
Ammunition	
main total:	25
Main armament traverse	
angle:	70° (35° left/35° right)
Main armament elevation/depression	
armament front:	+80°/-4°
Survivability	
Night vision equipment	
vehicle:	yes
NBC system:	yes
Armour	
hull/body:	aluminium
turret:	aluminium

Status

Production complete. No longer marketed. In service with:

Country	Quantity	Comment
Azerbaijan	18	
Belarus	48	
China	3	from Ukraine in 2003 (trials only)
Kazakhstan	25	
Kyrgyzstan	12	
Moldova	9	
Russia	790	Army
	75	Marines
Turkmenistan	17	
Ukraine	67	
Uzbekistan	54	
Vietnam	n/avail	unconfirmed user

Contractor

Russian state factories.

Singapore

Singapore Technologies Kinetics 120 mm Super Rapid Advanced Mortar System (SRAMS)

Development

The 120 mm smoothbore Super Rapid Advanced Mortar System (SRAMS) was developed as a private venture by Singapore Technologies Kinetics (STK). It was first revealed in late 2001 and development was completed late in 2006.

To allow it to be installed on lighter platforms it incorporates an advanced recoil system that has reduced the peak recoil to 10 tonnes when firing a 120 mm standard mortar bomb with maximum charge 9.

When firing a 120 mm standard Extended-Range (ER) mortar bomb with maximum charge 8, peak recoil is below 18 tonnes.

According to STK, this feature allows SRAMS to be installed on a much wider range of tracked and wheeled vehicles such as the AM General High Mobility Multipurpose Wheeled Vehicle (HMMWV).

Although the 120 mm SRAMS was first revealed as far back as 2001 and development was completed in 2006, it is understood that by the latter date it had already been ordered by the Singapore Armed Forces (SAF).

The 120 mm SRAMS deployed by the SAF has been integrated onto the rear of the STK Bronco All Terrain Tracked Carrier (ATTC) hull which is used in large numbers by the SAF for a wide range of battlefield roles and missions.

The 120 mm SRAMS also forms part of the Agrab (Scorpion) 120 mm Mobile Mortar System (MMS) developed from early 2006 by the International Golden Group of the United Arab Emirates.

Following trials with the latest configuration of the Agrab (Scorpion) Mk 1 MMS, in 2011 the International Golden Group (IGG) was awarded an AED786 million contract for the delivery of 72 Agrab Mk 2 production systems to the UAE Armed Forces.

IGG is the main contractor for the Agrab programme and will have systems responsibility and will facilitate and manage integration of the subsystems and the acceptance of all systems to the UAE Armed Forces.

The original Agrab Mk 1 consisted of a BAE Systems Land Systems South Africa RG31 Mk 5 (4 × 4) Mine Protected Vehicle (MPV) with a fully protected cab.

On the rear of this is a Singapore Technologies Kinetics 120 mm Super Rapid Advanced Mortar System (SRAMS) controlled by an integrated Fire-Control System (FCS).

Following extensive trials with the first four Agrab Mk 1 MMS the system was accepted for service with the UAE Armed Forces.

A number of improvements to enhance the overall capabilities of the system were successfully made.

This included using the latest BAE Systems Land Systems South Africa RG31 Mk 6E MPV and an enhanced computerised FCS. Full details of the Agrab Mk 2 are provided in a separate entry in *IHS Jane's Land Warfare Platforms: Artillery and Air Defence*.

The first 10 of these are now being integrated in South Africa with the remaining 62 to be integrated at the facilities of the IGG in the UAE.

Full details of this system are provided in a separate entry in *IHS Jane's Land Warfare Platforms: Artillery and Air Defence.*

Description

According to STK, SRAMS can be customised at a high rate of fire of up to 10 rounds per minute. This is achieved due to installation of a semi-automatic loading device that takes the 120 mm projectile from the lower part of the mortar up to the muzzle where it is automatically loaded.

The forged-steel smoothbore barrel is 1.8 m long and is fitted with a unique blast diffuser to enhance user safety.

This allows a significant amount of high-pressure gas to escape through intermediate chambers of the diffuser before the mortar bomb leaves the barrel. This reduces blast overpressure by a considerable amount.

The mortar system has powered elevation limits from +40° to +80°. The weapon is laid onto the target using a joystick. Elevation is electrohydraulic and traverse electric. Traverse is 45° left and right and the mortar is normally fired to the front of the vehicle.

SRAMS integrated on to the rear platform of an AM General HMMWV (4 × 4) chassis (Christopher F Foss) 0590857

Singapore Technologies Kinetics 120 mm SRAMS integrated onto the rear unit of a Bronco All-Terrain Tracked Carrier (ATTC) as used by the Singapore Armed Forces (STK) 1460885

Due to the installation of a navigation and positioning system based on a Ring Laser Gyro (RLG) and an Automatic Fire-Control System (AFCS), SRAMS can rapidly come into and be taken out of action. The AFCS includes a flat panel display.

Future growth potential could include a Global Positioning System (GPS), which will enable a 'shoot and scoot' capability. This would increase the survivability of the platform to counterbattery fire.

STK has tested and can provide data to show that the first 120 mm mortar bombs can be fired within 60 seconds of the vehicle coming to a halt, also helping to increase its survivability against counterbattery fire. Target accuracy is increased due to the installation of the navigation and positioning system and the AFCS.

Maximum range of the SRAMS depends of the type of 120 mm mortar bomb, barrel length and the charge combination.

Maximum range of the 120 mm SRAMS depends on the projectile/charge combination but is currently being quoted as 6.7 km using a standard high-explosive mortar bomb which can be increased using a rocket assisted projectile.

A Bronco ATTC would be able to carry about 50 × 120 mm mortar bombs for the SRAMS system, with a crew of three required to operate the system.

Variants

SRAMS on Bronco ATTC

The Singapore Armed Forces deploy the 120 mm SRAM on the rear unit of the STK Bronco ATTC.

SRAMS on HMMWV

In February 2005 SRAMS was shown at the IDEX 2005 defence equipment exhibition integrated into the rear of an AM General HMMWV chassis.

For this application the HMMWV chassis has been modified and a larger hydraulic stabiliser blade has been mounted under the rear that is lowered to the ground by remote control before firing commences. When the stabiliser blade is lowered to the ground, the rear wheels of the HMMWV are raised clear of the ground so that the rear suspension is not over-stressed. This provides a more stable firing platform.

It is envisioned that production versions would carry about 12 rounds of 120 mm ammunition, with another HMMWV carrying additional ammunition. The HMMWV is used by many countries in the Middle East for a wide range of roles and missions.

As of March 2013 this version of the SRAMS remained at the prototype stage.

SRAMS on Spider (4 × 4) chassis

The 120 mm SRAMS has been developed with such a low recoil force that enables the mounting of the mortar onto very light platforms such as the Spider Light Strike Vehicle (LSV) that was shown for the first time in 2006.

In this configuration the 120 mm SRAMS can be deployed by small units operating independently in dismounted or in heliborne operations.

The latest Spider version of 120 mm SRAMS is seen by the company as a major step in improving the mobility of the system with particular emphasis on its use by rapid deployment type forces.

To provide a more stable firing platform two stabiliser options are under consideration, first a suspension lock out system and second a stabiliser that is lowered to the ground under the chassis.

Given the very high mobility of the Spider and its high rate of fire, it will be able to come into action, engage the target and move to another fire position before the system can be detected and engaged by counterbattery fire.

Each 120 mm SRAMS firing platform can be designed to carry a quantity of ready use ammunition with another platform carrying additional ammunition.

It has also been revealed that STK have completed development of a new 120 mm Dual Purpose Improved Conventional Munition (DPICM) mortar bomb designated the P138. This is already in quantity production and is now being offered on the export market.

This 120 mm mortar bomb, fitted with a Junghans DM93 Mod 505.54 mechanical time fuze, carries a total of 25 dual-purpose anti-personnel and anti-armour bomblets each of which will penetrate around 75 mm of Rolled Homogenous Armour (RHA). This is sufficient to penetrate the top armour of almost all armoured fighting vehicles currently deployed.

This cargo bomb weighs 15.7 kg and is 916 mm long. Maximum range, when fired from a standard 120 mm mortar, is claimed to be 6,600 m with a 120 mm light mortar achieving a maximum range of 6,350 m. Minimum range in both cases is 980 m.

As of March 2013 the Spider fitted with the 120 mm SRAMS remained at the prototype stage with no known plans for production.

SRAMS on RG31 Mk 5 MPV

This is the version deployed by the UAE and details are provided in a separate entry in *IHS Jane's Land Warfare Platforms: Artillery and Air Defence*. A total of 72 production Agrab Mk 2 are being supplied integrated on the RG31 Mk 6E MPV.

Specifications

	Super Rapid Advanced Mortar System (SRAMS)
Dimensions and weights	
Weight	
combat:	1,200 kg
Mobility	
Configuration	
running gear:	tracked
Firepower	
Armament:	1 × 120 mm smoothbore gun
Main weapon	
length of barrel:	1.8 m
maximum range:	6,700 m (standard ammunition)
maximum range:	13,000 m (assisted ammunition)
Rate of fire	
sustained fire:	10 rds/min
Main armament traverse	
angle:	45° (left/right)
Main armament elevation/depression	
armament front:	+40°/+80°

Status

Production. In service with Singapore (integrated onto a Bronco All Terrain Tracked Carrier). A total of 72 SRAMS integrated onto the rear of a BAE Systems, Global Combat Systems South Africa RG31 Mk 6E MPV were ordered in early 2011 by the UAE. Production is now underway for the UAE who are taking delivery of a total of 72 units.

Contractor

Singapore Technologies Kinetics (STK).

Slovakia

ZTS PRAM-S 120 mm self-propelled mortar system

Development

In 1990, ZTS completed the prototype of a 120 mm self-propelled mortar system called the PRAM-S.

The Russian-designed BMP-1 and the more recent BMP-2 Infantry Fighting Vehicle (IFV) were manufactured under licence in Czechoslovakia for some years. As far as it is known there has been no recent production of the BMP-2 IFV in Slovakia.

ZTS PRAM-S 120 mm self-propelled mortar with trim vane extended at the front of the vehicle for amphibious operations 0501021

ZTS PRAM-S 120 mm self-propelled mortar 0501020

In early 1992, ZTS stated that the PRAM-S was in production and would be deployed in batteries of six systems. With the break-up of the former Czechoslovakia, production and marketing of this system is now carried out by ZTS in the Slovak Republic.

It is understood that of the 12 systems built, eight are in service with the Czech Republic and the other four with the Slovak Republic.

Although production of the PRAM-S 120 mm self-propelled mortar system was completed some years ago, it is still being offered on the export market by ZTS of Slovakia and production can resume if further significant orders are placed.

It is considered that if production did start again it would probably be integrated on a more recent hull. This would probably be of Western design.

Description

The PRAM-S 120 mm self-propelled mortar system is based on a modified Russian designed BMP-2 IFV hull, which has had its turret removed and roof raised to accept a new 120 mm mortar system that can be aimed, loaded and fired under complete armour protection.

The hull of the original BMP-2 IFV has been lengthened and raised at the rear and now has a total of seven roadwheels compared to the standard vehicle which has six.

The drive sprocket is at the front, idler at the rear and there are three track-return rollers. Hydraulic shock-absorbers are provided for the first and last road wheel stations.

The driver is seated at the front left with the engine compartment to his right, with the area to the rear of the vehicle being raised to provide sufficient space for the 120 mm mortar, ammunition and crew.

Mounted in the forward part of the roof is the 120 mm breech-loaded mortar which has a traverse of 15° left and right and can be elevated from +40° to +80°. Elevation and traverse of the 120 mm mortar is manual.

Maximum range of the 120 mm mortar is 8,036 m with minimum range being 504 m, which in both cases depends on the type of 120 mm mortar ammunition (for example projectile and propelling charge) being used.

Maximum rate of fire using the Czech D charge is 18 to 20 rds/min, 40 rounds in five minutes and 70 rounds in 10 minutes. The mortar can be fired on longitudinal slopes of up to 5° and transverse slopes of up to 5°.

On coming to a halt, the first rounds can be fired in one minute and the system can come out of action one minute after the last round has been fired.

A total of 80 rounds of 120 mm ammunition are carried of which 21 are for ready use. After these have been expended the magazine has to be reloaded manually.

While the height to the top of the hull is 2.25 m, the overall height with the 120 mm mortar at an elevation of +40° is 2.5 m and when elevated to +80° is 3.38 m.

When travelling, the 120 mm mortar is at the minimum elevation angle and when required for action armoured covers unfold either side of the mortar to allow for greater elevation.

The commander is seated to the right of the 120 mm mortar and is provided with a roof hatch, three day periscopes and an infra-red searchlight.

The gunner is seated opposite the commander, has a roof-mounted day laying system and can enter the vehicle via a forward-opening door in the left side of the hull that is also provided with a vision block.

The vehicle commander operates the R-123 UHF/VHF radio with the antenna for this being located forward and to the left of the driver's hatch.

In addition, there is a hatch in the roof of the vehicle to the rear of the commander on which an externally mounted 12.7 mm NVS machine gun is fitted for local and anti-aircraft defence.

The loader is seated towards the rear of the mortar and is responsible for preparing the 120 mm mortar bombs and filling the automatic loader from the ammunition containers to the left and right rear of the combat compartment.

The 120 mm mortar bombs are stowed vertically, the loading system being chain driven with a rammer provided to ram the mortar bombs into the mortar.

In the rear of the vehicle is a door for ammunition resupply and 80 rounds of 120 mm mortar ammunition are carried, of which 21 rounds are in the conveyor and 59 rounds in the magazines.

The PRAM-S is fully amphibious, being propelled in the water by its tracks, and is fitted with an NBC system and a full range of night vision devices.

The vehicle can lay its own smoke screen by injecting diesel fuel into the exhaust outlet located on the right side of the hull which, when compared to Russian BMP-1/BMP-2 type vehicles, has a distinctive cowl-type shroud.

Variants

In late 2001, it was reported that a number of enhancements were under active consideration for this system, including integrating on a wheeled platform. As of March 2013 the wheeled version remained at the concept stage.

Specifications

	PRAM-S
Dimensions and weights	
Crew:	4
Length	
overall:	7.47 m
Width	
overall:	2.94 m
Height	
hull:	2.25 m
Ground clearance	
overall:	0.42 m
Track	
vehicle:	2.55 m
Track width	
normal:	300 mm
Length of track on ground:	3.60 m
Weight	
combat:	16,970 kg
Mobility	
Configuration	
running gear:	tracked
Power-to-weight ratio:	17.67 hp/t
Speed	
max speed:	63 km/h
water:	7.4 km/h
Range	
main fuel supply:	550 km
Fuel capacity	
main:	462 litres
Gradient:	60%
Side slope:	30%
Vertical obstacle	
forwards:	0.7 m
Trench:	2.5 m
Engine	Model UTD-20, 6 cylinders, diesel, 300 hp at 2,600 rpm
Gearbox	
type:	manual
forward gears:	5
reverse gears:	1
Clutch	
type:	friction clutch
Suspension:	torsion bar, hydraulic shock-absorbers on 1st, 2nd and 6th road wheels
Electrical system	
vehicle:	22 V
Batteries:	2 × 120 Ah
Firepower	
Armament:	1 × hull mounted 120 mm mortar
Armament:	1 × roof mounted 12.7 mm (0.50) NSV machine gun
Ammunition	
main total:	80
Main armament traverse	
angle:	30° (15° left/15° right)
Main armament elevation/depression	
armament front:	+80°/+40°
Survivability	
Night vision equipment	
vehicle:	yes
NBC system:	yes
Armour	
hull/body:	steel

Status

Production complete but can be resumed if further orders are placed. In service with the Czech Republic (8 and called SPM-85) and Slovak Republic (4).

Contractor

ZTS Dubnica nad Váhom (Slovak Republic).

Spain

EXPAL Integrated Mortar System 60 mm/81 mm (EIMOS)

Development

The Spanish company of EXPAL (*Explosivos Alaveses*), using company funding, developed the EXPAL Integrated Mortar System (EIMOS) for use by rapid deployment forces.

It is understood that this EIMOS in its 81 mm configuration has been evaluated by the Spanish Army integrated into the rear of a locally built URO VAMTAC (4 × 4) high mobility tactical vehicle which is used in significant numbers by the Spanish Army.

As of March 2013 there were no known sales of the EXPAL Integrated Mortar System 60/81 mm.

It should be noted that following a competition, in October 2011 Soltam Systems of Israel and now part of the Elbit Group, was awarded an USD8.5 million contract by the Spanish Army to supply 81 mm CARDOM mortars within 12 months.

Soltam Systems is the lead contractor for the project that includes the installation of 81 mm CARDOM autonomous mortars in the rear of Spanish VAMTAC (4 × 4) cross-country chassis.

The first six units were delivered in mid-2012 and integrated on the VAMTAC (4 × 4) chassis in Spain by General Dynamics European Land Systems - Santa Bárbara Sistemas.

Description

EIMOS essentially consists of a turntable mounted 60 mm or 81 mm EXPAL mortar integrated onto the rear of a 4 × 4 light tactical vehicle with powered elevation and traverse.

The mortar is provided with all electric traverse through a full 360° and electric elevation with the hydropneumatic recoil system having a maximum recoil of 300 mm.

According to EXPAL, this recoil system reduces the recoil force by more that 90 per cent and allows it to be integrated onto light tactical vehicles with no modification to the chassis.

A quantity of ready use 81 mm mortar bombs are carried on the platform with additional ammunition carried by a support vehicle on a similar chassis.

The first concept demonstrator of the EIMOS is hand loaded and laid onto the target using manual elevation and azimuth controls, but production systems could have both autolay and autoloader systems incorporated.

The first concept demonstrator has a laptop-based fire-control system loaded with EXPAL developed STANAG compliant ballistic software and a differential GPS navigation and pointing system.

Concept demonstrator EXPAL EIMOS armed with an 81 mm muzzle loaded mortar integrated onto the rear of a Spanish URO VAMTAC (4 × 4) high mobility tactical vehicle (IHS/Rupert Pengelly) 1347484

Close up of the EXPAL EIMOS integrated onto the rear of a 4 × 4 cross-country chassis and showing ammunition stowage (left) and turntable mounted mortar (right) complete with controls and ready used ammunition (EXPAL) 1452634

It is envisaged that production EIMOS systems would be integrated with the EXPAL developed TECHFIRE fire-control system and data management.

This links the forward observers with the unit commander who can control up to 16 mortars.

This enables the EIMOS to come to a halt, carry out a fire mission and move to another firing position before counter battery fire is returned.

The system corrects changes to the vehicle caused by firing the mortar or slight displacement of the vehicle position.

According to EXPAL, the system comes into action within 10 to 20 seconds of coming to a halt.

When fitted with an EXPAL 81 mm mortar with a barrel length of 1.45 m, the system has a maximum range of 6,900 m using Charge 6.

When fitted with a 60 mm mortar with a barrel length of 1 m, maximum range is 4,900 m using Charge 5.

It takes only three minutes to change the mortar barrel.

Types of 81 mm ammunition fired include high-explosive, smoke WP, Smoke HC and illuminating.

It is stated that if required, the mortar system could be dismounted and fired from the ground.

Specifications

Calibre:	60 mm	81 mm
Range:	4,900 m (maximum)	6,900 m (maximum)
Height:	1.75 m	2.05 m
Diameter of base:	780 mm	780 mm
Weight:	500 kg	500 kg

Status

Under development.

Contractor

EXPAL.

Switzerland

RUAG Defence Bighorn 120 mm recoiling mortar system

Development

Since the 1990s, RUAG Defence has been developing the vehicle-mounted Bighorn 120 mm smoothbore recoiling mortar system as a private venture.

A total of two systems have been built, one prototype and one pre-production, and during the development period over 1,700 120 mm mortar bombs of various types were fired.

RUAG Defence Bighorn 120 mm recoiling mortar integrated into an FNSS Armoured Combat Vehicle - Stretched showing operators controls on right side of mount (RUAG Defence) 1296049

RUAG Defence 120 mm recoiling mortar integrated onto a FNSS Armoured Combat Vehicle - Stretched (RUAG Defence) 1365021

For trials purposes the Bighorn 120 mm recoiling mortar system has already been tested on a number of tracked and wheeled platforms, with the latter including a Swiss General Dynamics European Land Systems - MOWAG Piranha (8 × 8) light armoured vehicle chassis.

More recently it has been installed on the Turkish FNSS Savunma Sistemleri Armoured Combat Vehicle - Stretched.

The Swiss Army is understood to have a requirement for a new 120 mm self-propelled mortar system to replace its current, older ground-based mortar systems.

In 2007 nine foreign delegations attended a demonstration of the RUAG Defence Bighorn 120 mm recoiling mortar system.

During this demonstration the Bighorn 120 mm smoothbore recoiling mortar system fired a wide range of mortar bombs including High-Explosive, cargo, smoke, illumination, training and Strix precision guided mortar munition. It also demonstrated shoot and scoot tactics.

Development of the RUAG Defence Bighorn 120 mm recoiling mortar system is complete but as of March 2013 no production orders had been placed although it was still being marketed.

Description

The Bighorn 120 mm recoiling mortar system features an automatic muzzle loading device and an integrated laser gyro inertial navigation and pointing system, vehicle motion sensors and commander and gunner controls.

The commander and gunner controls are the same as those used on the RUAG Defence upgraded 155 mm/47-calibre upgraded M109s of the Swiss Army. Details of this are provided in a separate entry in *IHS Jane's Land Warfare Platforms: Artillery and Air Defence*.

The installation of the advanced navigation and pointing system allows the system to carry out 'shoot and scoot' fire missions. An integrated hydraulic power pack feeds the gun-laying system and the semi-automatic loader.

According to RUAG Defence, the system takes 30 seconds to prepare for firing on coming to a halt, 60 seconds for a typical fire mission and 30 seconds to come out of the firing position.

When deployed in the travelling position the 120 mm mortar is under complete armour protection and the appearance of the vehicle is very similar to that of a standard armoured personnel carrier.

The standard system has a traverse of 190° left and right but, as an option, this system can be delivered with a traverse through a full 360°.

Maximum range depends on a number of factors. During trials a range of 9.2 km was achieved when firing at a temperature of +21°C with a Swiss zone 8 charge. Maximum range achieved during hot weather trials in Kuwait in 2001 was 9.4 km at a temperature of +53°C ambient.

The installation of an automatic loader is claimed to reduce the workload on the crew, as well as increasing crew survivability.

The weapon fires all standard natures of 120 mm ammunition, as well as smart (for example the Swedish Strix top-attack projectile) and cargo types. It should be noted that production of the Strix is complete and it was only sold to Sweden and Switzerland.

The amount of 120 mm ammunition carried depends on the type of hull but is typically 40 standard plus eight larger rounds. For some larger chassis this can be increased to between 45 and 72 rounds.

The 81 mm barrel insert system allows training with standard 81 mm smoothbore ammunition. It is easy to remove and all operations and procedures are identical to the 120 mm system.

In case of total hydraulic failure a manually operated hydraulic pump is provided for the Bighorn 120 mm recoiling mortar system.

Specifications

Bighorn

Dimensions and weights	
Weight:	
(combat)	1,425 kg
Firepower	
Armament:	1 × 120 mm
Main weapon:	
(length of barrel)	2 m
(maximum range)	9,000 m (est.)
Rate of fire:	
(sustained fire)	6 rds/min (est.) (in 20 s, or 8 rds/min)
Operation:	
(stop to fire first round time)	30 s
Main armament elevation/depression:	
(armament front)	+40°/+85° (option for 360°)

Status

Development complete. Ready for production on receipt of orders.

Contractor

RUAG Defence.

United Arab Emirates

AGRAB (Scorpion) 120 mm Mobile Mortar System

Development

Development of the AGRAB (Scorpion) 120 mm Mobile Mortar System (MMS) commenced early in 2006 by the International Golden Group with the first prototype system completed in mid-2006.

Funding for development was provided by the International Golden Group in anticipation of future operational requirements for a weapon system of this type in the Gulf Area.

By early 2007 AGRAB 120 mm MMS had undergone extensive firepower and mobility trials with over 800 120 mm mortar bombs being fired.

In early 2012 following trials with the latest configuration of the Agrab (Scorpion) Mk 1 120 mm MMS, the International Golden Group (IGG) was awarded an AED786 million contract for the delivery of 72 Agrab Mk 2 production systems to the UAE Armed Forces.

Following trials with the first prototype in 2007, the following year the UAE Armed Forces awarded a contract to the IGG covering the supply of four units for operational evaluation.

Latest Agrab Mk 2 120 mm Mobile Mortar System ordered by the UAE Armed Forces (IGG) 1424096

ST Kinetics 120 mm SRAMS is fitted to the Agrab Mk 2 120 mm Mobile Mortar System and fires to the rear arc (IGG) 1424098

Rear view of the Agrab Mk 2 120 mm Mobile Mortar System with rear folded down and ST Kinetics SRAMS unlocked (IGG) 1424101

These four systems were subsequently delivered for extensive operational trials in 2008 through to 2010.

The original Agrab Mk 1 consisted of a BAE Systems Land Systems South Africa RG31 Mk 5 (4 × 4) Mine Protected Vehicle (MPV) with a fully protected cab.

On the rear of this is a Singapore Technologies Kinetics 120 mm Super Rapid Advanced Mortar System (SRAMS) controlled by an integrated Fire-Control System (FCS).

Following extensive trials with the first four Agrab Mk 1 MMS the system was accepted for service with the UAE Armed Forces.

A number of improvements to enhance the overall capabilities of the system were successfully made.

This included using the latest BAE Systems Land Systems South Africa RG31 Mk 6E MPV and an enhanced computerised FCS.

The first 10 of these are being integrated in South Africa starting early in 2012 with the remaining 62 being integrated at the facilities of the IGG in the UAE with a potential for additional follow on orders.

IGG is the prime contractor for the Agrab programme and has systems responsibility and will also facilitate and manage integration of the subsystems and the acceptance of all complete Agrab Mk 2 MMS to the UAE Armed Forces.

Description

The Agrab Mk 2 is based on the latest BAE Systems Land Systems South Africa RG31 Mk 6E MPV and provides the occupants with a high level of protection against a variety of battlefield threats including small arms fire, shell splinters, Anti-Tank Mines (ATM) and Improvised Explosive Devices (IED).

When compared to the earlier Mk 5, the Mk 6E has a higher payload capacity, wider track and larger tyres and gross vehicle weight is 19.5 tonnes combat ready.

It is powered by a Cummins turbocharged diesel developing 300 hp coupled to an Allison 6-speed automatic transmission which gives a maximum road speed of up to 90 km/h and a road range of more than 800 km.

The crew can enter and leave the vehicle via front doors in either side of the crew compartment, a door in the rear or a roof hatch.

All of the doors are provided with a bullet/splinterproof window in their upper part.

Mounted on the roof is a protected weapon station that would typically be armed with a .50 M2 HB Machine Gun (MG) or 40 mm Automatic Grenade Launcher (AGL) for self-defence purposes.

In January 2013 it was announced that W & E Platt had been awarded a contract to supply its manually operated protected ring mount weapon stations for the Agrab Mk 2 MMS with deliveries starting early in 2013.

This contract was placed with the IGG and is for the MR550 Bi-Metal Round Mount which can be armed with a 5.56 mm, 7.62 mm or 12.7 mm machine gun or a 40 mm automatic grenade launcher.

Traverse is through a full 360° with elevation from –35° to +65° depending on the platform.

It weighs around 400 kg and is provided with STANAG 4569 Level 3 ballistic protection.

Customer selected grenade launchers such as the 81 mm or the recently developed Rheinmetall ROSI grenade launchers can be installed to provide screening coverage through 360°.

Standard equipment for the vehicle includes an air conditioning system, central tyre inflation system and a front mounted winch for self-recovery operations or to recover other vehicles.

A 40-litre water system is also provided and additional stowage is provided in the external bins which blow off in the event of a ATM or IED attack.

The Singapore Technologies Kinetics 120 mm SRAMS is already in service with the Singapore Armed Forces (SAF) and fires over the rear arc and has a maximum stated rate of fire of up to 10 rounds a minute. A total of 52 ×120 mm mortar bombs are carried.

The 120 mm SRAMS has a low recoil force of 26 tonnes and powered traverse is 40° left and right with powered elevation limits are from +40° to +80°. Additional details of the 120 mm SRAM are provided in a separate entry in *IHS Jane's Land Warfare Platforms: Artillery and Air Defence*.

The 120 mm ammunition is supplied by Rheinmetall Denel Munitions and natures include high-explosive, bi-spectral smoke and illuminating.

According to the IGG the ranges of the 120 mm mortar bombs are as follows:

Type	Maximum Range
High-Explosive	8.2 km
Screening smoke	7.54 km
Illumination	7.25 km
Laser guided (future)	7 km
Rocket assisted (future)	13 km

Agrab Mk 2 would normally be deployed in units of four but are also capable of being used as autonomous units.

Production standard Agrab Mk 2 will feature a number of improvements and are based on the latest RG31 Mk 6E MPV and a computerised FCS supplied by Thales South Africa Systems coupled to a Selex inertial navigation systems.

The IGG is quoting an into action time of 30 seconds and a similar out of action time. When fitted with a GPS navigation accuracy is around 10 m.

The FCS provides for a full automatic laying capability, ballistic computation including met, weapon management system and a high command connectivity.

When compared to the RG31 Mk 5 the Mk 6E has a greater payload, wider track and larger tyres with gross vehicle weight being 20 tonnes combat ready.

It is fitted with a full air conditioning system, central tyre inflation system, winch for self-recovery operations and a 40-litre water system.

Specifications

	AGRAB
Dimensions and weights	
Crew:	4
Length	
overall:	7.95 m
Width	
overall:	2.95 m
Height	
overall:	3.27 m (including gun mount)
Weight	
standard:	18,000 kg
combat:	20,000 kg
Mobility	
Configuration	
running gear:	wheeled
layout:	4 × 4
Power-to-weight ratio:	18 hp/t
Speed	
max speed:	90 km/h
Range	
main fuel supply:	800 km
Engine	Cummins, turbocharged, diesel, 360 hp at 2,600 rpm
Gearbox	
model:	Allison
type:	automatic
forward gears:	6
reverse gears:	1
Firepower	
Armament:	1 × 120 mm mortar (muzzle loaded)
Ammunition	
main total:	52
Turret power control	
type:	powered/manual
Main armament traverse	
angle:	40° (left/right)
Main armament elevation/depression	
armament front:	+40°/+80°
Survivability	
Night vision equipment	
vehicle:	optional
NBC system:	optional
Armour	
hull/body:	steel

Status

Development complete. In production for the United Arab Emirates Armed Forces who placed a contract for 72 units based on the BAE Systems Land Systems South Africa RG31 Mk 6E (4 × 4) mine protected vehicle.

The first 10 have been integrated in South Africa and the remaining 62 are being integrated in the United Arab Emirates.

Contractor

International Golden Group.

United States

BAE Systems M1064A3 120 mm self-propelled mortar

Development

For many years the US Army deployed two Self-Propelled Mortars (SPM) based on a modified BAE Systems M113 series full-tracked Armoured Personnel Carrier (APC).

These were the 4.2 inch (107 mm) M106/M106A1/M106A2 and the 81 mm M125/M125A1/M125A2.

These have now been replaced in US Army service by the BAE Systems M1064A3 120 mm self-propelled mortar system.

The hull used for the M1064A3 120 mm self-propelled mortar is essentially the late production M113A3 APC modified for its specific role.

It was expected that the BAE Systems M1064A3 120 mm self-propelled mortar would have been replaced by the Non-Line-Of-Sight Mortar (NLOS-M) which was part of the manned ground vehicle element of the US Army's Future Combat System (FCS).

This was cancelled in 2009 and as of March 2013 the US Army had not announced any plans to replace the BAE Systems 120 mm M1064A3 self-propelled mortar system.

It should be noted that the US Army Stryker Brigade Combat (BGT) deploy the General Dynamics Land Systems - Canada Stryker (8 × 8) vehicles in many configurations including the M1129 armed with a 120 mm muzzle loaded mortar.

As a private venture, BAE Systems has proposed that the 120 mm mortar system integrated into this M1064A3 self-propelled mortar carrier could also be integrated into a reconfigured member of the Bradley family of vehicles.

M1064A3 120 mm self-propelled mortar from the rear with ramp lowered and three-part roof hatch in open position
0572185

The US has supplied quantities of the M1064A3 self-propelled mortar carrier vehicles to Israel in recent years and it is considered probable that these have been delivered without mortars. This are provided by Soltam Systems of Israel.

According to the United Nations Arms Transfer lists the USA provided 25 M1064A3 in 2009 and 15 in 2008 with an undisclosed number being supplied in 2007.

In mid-1995 Thailand ordered 82 brand new M113 series vehicles including 12 120 mm M1064A3 mortar carriers which were delivered by 1997.

Description

The hull is of all-welded aluminium armour, which provides the occupants with protection from small arms fire and shell splinters with the highest level of protection over the frontal arc.

The driver is seated at the front of the vehicle on the left side with the diesel power pack to the right.

The driver is provided with a single-piece hatch cover that opens to the rear, and to his front and side are four M17 day periscopes with an AN/VVS-2 image intensification or AN/VAS-5 thermal viewer provided in the roof.

The diesel power pack compartment is to the right of the driver's position and is fitted with a fire-extinguishing system that can be operated by the driver or from outside the vehicle.

The air inlet and outlet louvres and the exhaust pipe outlet are in the roof and there is an engine access door in the front of the hull that hinges forwards.

The power train consists of the Detroit Diesel power pack (engine, transfer gear case and transmission) steering control differential, pivot steer, final drive and associated drive shafts and universal joints. The diesel fuel tanks are mounted externally either side of the hull at the rear and are armour protected.

In the middle of the roof, to the rear of the driver's position, is the commander's cupola, which is provided with vision blocks and a single-piece hatch cover that opens to the rear. On this is mounted a .50 (12.7 mm) M2 HB machine gun that is used for local or air defence purposes.

Mounted either side at the front of the hull is a bank of electrically-operated smoke grenade launchers that cover the frontal arc.

The crew enter the vehicle through the large power-operated ramp at the rear, which is also provided with a door for emergency use.

The torsion bar suspension either side consists of five dual rubber-tyred roadwheels with the drive sprocket at the front and the idler at the rear. The first and last road wheel stations are provided with a hydraulic shock-absorber. There are no track return rollers.

A rubber track shroud on either side of the hull controls the flow of water over the tracks when the vehicle is afloat. The M1063A3 is fully amphibious and is propelled in the water by its tracks. Steering when afloat is the same as on land.

Before entering the water the two bilge pumps are switched on and the trim vane, which folds back onto the glacis plate when not in use, is extended at the front of the hull.

For its specialised role the floor features a welded-in crossbeam and additional floor-support structure to withstand mortar reaction forces.

Mounted in the rear is the 120 mm muzzle-loaded mortar, which fires through a three-part roof hatch. The mortar is normally fired to the rear through a 90° arc.

A baseplate is carried on the left side of the hull, which allows the mortar to be dismounted for ground use if required by the tactical situation.

Types of 120 mm ammunition fired by the M1064A3 self-propelled mortar include the following:

- 120 mm high-explosive M933
- 120 mm high-explosive M934 (as M933 but with different fuze)
- 120 mm smoke WP M929
- 120 mm M930 illuminating
- 120 mm M983 infra-red illuminating
- 120 mm practice M931.

The M120 mm M933 has a maximum range of 7,240 m and a minimum range of 300 m. Maximum rate of fire is 15 rounds per minute and sustained rate of fire is four rounds per minute.

BAE Systems is offering a wide range of optional equipment for this vehicle including a more powerful Detroit Diesel engine, more powerful generator, automatic fire detection and suppression system in the engine compartment, enhanced protection against direct-fire weapons and mines, air conditioning system, various types of land navigation system and mortar fire-control systems.

Additional specifications of the M11064A3 self-propelled mortar include width over tracks of 2.54 m and height to top of commanders cupola of 2.22 m excluding weapon.

Variants

BAE Systems can supply kits to convert M113, M106 and M125 vehicles to the latest M1064A3 configuration.

Specifications

	M1064A3
Dimensions and weights	
Crew:	6
Length	
overall:	5.30 m
Width	
overall:	2.69 m
Height	
hull:	1.85 m
to top of roof mounted weapon:	2.53 m
Ground clearance	
overall:	0.43 m
Track	
vehicle:	2.159 m
Track width	
normal:	381 mm
Length of track on ground:	2.67 m
Weight	
standard:	10,596 kg
combat:	12,809 kg
Ground pressure	
standard track:	0.63 kg/cm^2
Mobility	
Configuration	
running gear:	tracked
Power-to-weight ratio:	21.46 hp/t
Speed	
max speed:	66 km/h
water:	5.8 km/h
Range	
main fuel supply:	483 km
Fuel capacity	
main:	360 litres
Amphibious:	yes
Gradient:	60%
Side slope:	40%
Vertical obstacle	
forwards:	0.61 m
Trench:	1.68 m
Turning radius:	2.97 m
Engine	Detroit Diesel Model 6V-53T, V-6, turbocharged, water cooled, diesel, 275 hp
Gearbox	
model:	Allison X200-4B crossdrive
forward gears:	3
reverse gears:	1
Suspension:	torsion bar
Electrical system	
vehicle:	24 V
Batteries:	4 × 12 V
Firepower	
Armament:	1 × hull mounted 120 mm mortar
Armament:	1 × roof mounted 12.7 mm (0.50) M2 HB machine gun
Ammunition	
main total:	69
12.7/0.50:	2,000
Survivability	
Night vision equipment	
vehicle:	yes, passive
NBC system:	yes
Armour	
hull/body:	aluminium (12-44 mm)

Status

Production complete but could be resumed. In service with the US Army. It should be noted that the US Army continues to upgrade early M1064A2 120 mm self-propelled mortars to the enhanced M1064A3 standard. The US has also provided quantities of these M1064A3 self-propelled mortar systems to Israel as mentioned in the development section. Thailand has taken delivery of 12 of these M1064A3 120 mm self-propelled mortar systems.

Contractor

BAE Systems.

Expeditionary Fire Support System (EFSS)

equipmentDictionary

Development

Following a competition, late in 2004 a team led by General Dynamics Ordnance and Tactical Systems was selected to develop two new systems for the US Marine Corps.

These are the Expeditionary Fire Support System (EFSS) and the Internally Transportable Vehicle (ITV). Both of these systems are required to be carried inside the V-22 Osprey tilt rotor aircraft.

According to the manufacturer, EFSS provides the US Marines with short-range indirect fire support capability with a range of 7 to 20 km with the system being transported by the ITV. It should be noted that the 20 km range is a long term objective as of early 2013 this is not in the US Marine Corps inventory.

EFSS system includes the French TDA MO 120 RT 120 mm mortar
(General Dynamics Ordnance and Tactical Systems) 0578762

EFSS ammunition supply system
(General Dynamics Ordnance and Tactical Systems) 1039092

One of the ITVs carries some crew and tow the 120 mm mortar while the second unit carries additional crew and tow a trailer carrying up to 36 rounds of 120 mm ammunition. It is expected that Special Operations Forces (SOF) will also use the ITV as a light strike vehicle.

General Dynamics Ordnance and Tactical Systems was awarded an initial USD18.2 million contract to carry out the System Development and Demonstration (SDD) of the EFSS/ITV programme.

In mid-2009 the US Marine Corps fielded the EFSS with Bravo Battery, 1st Battalion, 10th Marine Artillery Regiment located at Camp Lejeune, North Carolina.

Each battalion of the 10th Marine Regiment has one battery of six EFSS with each of these comprising a pair of prime movers (one for the weapon and one for the ammunition trailer), one 120 mm M327 mortar, 120 mm ammunition and a trailer.

Description

The weapon system is the French TDA MO 120 RT 120 mm rifled towed mortar with the ITV being based on the American Growler's commercially available UVDB 1000.

Growler embodies some components of the M151A2 (4 × 4) vehicle and is powered by a 4-litre 4-cylinder turbocharged diesel engine which is coupled to a four-speed fully automatic transmission. Operational range is being quoted as 580 km with an unarmoured curb weight of 1,678 kg.

The two-wheeled ammunition trailer has a maximum payload of 1,588 kg and a maximum capacity of 36 × 120 mm mortar bombs. It is fitted with rigged recyclable insensitive munition compliant ammunition packaging, surge brakes and a landing leg.

The TDA MO 120 RT 120 mm rifled mortar is already in service with the armed forces of 24 countries, four of which are in NATO. In addition the 120 mm rifled mortar is also capable of firing 120 mm smoothbore ammunition.

It should be noted that the US Army uses a 120 mm smoothbore mortar system rather than a 120 mm rifled mortar.

General Dynamics Ordnance and Tactical Systems produces ammunition for the EFSS in the United States. Natures to be procured are:

- High-Explosive (maximum range 8,100 m)
- High-Explosive Rocket-Assisted Projectile (HE RAP) (maximum range 12,500 m)
- Illuminating (maximum range 7,000 m)
- Practice
- Smoke
- Submunition (maximum range 8,170 m)
- Anti-personnel carrier (maximum range 8,135 m).

Other subcontractors for the EFSS programme include Tec-Masters of Huntsville (responsible for integrated and contractor logistics support) and General Dynamics Canada (who are developing the EFSS Ballistic Computer (EBC)).

Variants

120 mm Precision Extended Range Munition

With the increased emphasis on urban operations there is a requirement for more precision artillery and mortar projectiles to reduce collateral damage.

The US Marine Corps has a requirement for a 120 mm Precision Extended Range Munition (PERM) for use with its currently deployed Expeditionary Fire Support System.

Late in 2012, a 24 month PERM Engineering and Manufacturing Development (EMD) contracts were awarded to ATK teamed with GDOTS and Raytheon teamed with Israel Military Industries for a competitive shoot off.

It is expected that the new PERM will have an effective range of up to 17 km compared to 8.1 km of the currently deployed 120 mm ammunition.

The fielding of more accurate 120 mm mortar projectiles will not only improve accuracy and reduce collateral damage but also mean that fewer rounds are required which will have a dramatic effect on reducing ammunition supply problems.

Specifications

MO 120 RT (baseline towed mortar)

Calibre:	120 mm (rifled mortar)
Barrel length:	2,060 mm
Range:	
(maximum, HE RAP)	13,000 m
(minimum)	100 m
Maximum rate of fire:	1 rd/min
Deployment time:	1 min
Total weight:	643 kg
Weight:	
(base plate)	200 kg
(carriage)	310 kg
(barrel)	133 kg

	Light Strike Vehicle
Dimensions and weights	
Crew:	1+1
Length	
overall:	4.14 m
Width	
overall:	1.524 m
Weight	
standard:	2,964 kg
combat:	2,057 kg
Mobility	
Configuration	
running gear:	wheeled
layout:	4 × 4
Power-to-weight ratio:	44.53 hp/t
Speed	
max speed:	104.58 km/h
Fuel capacity	
main:	90.85 litres
Fording	
without preparation:	0.762 m
Engine	Navistar Defense, in line-4, turbocharged, diesel, 132 hp at 2,000 rpm
Gearbox	
model:	General Motors 4L70E
type:	automatic
Transfer box:	2 speed manual
Steering:	four wheel powered
Suspension:	gas bladder
Electrical system	
vehicle:	24 V
Batteries:	2 × 12 V

Status

Production. Entered service with the US Marine Corps in 2009. It should be noted that the US Marine Corps artillery can deploy with the 120 mm EFSS or 155 mm M777 towed artillery.

Contractor

See text.

Multiple Rocket Launchers

Argentina

TAMSE VCLC series armoured rocket launchers

Development

During the early 1980s, the Argentine TAMSE (*Tanque Argentino Mediano Sociedad del Estado*) concern, which was originally formed to manufacture the German Rheinmetall TAM medium tank, began to investigate a joint venture with Israel Military Industries Ltd to fit the company's modular rocket systems covered in detail in a separate entry in *IHS Jane's Land Warfare Platforms: Artillery and Air Defence*.

In 1986, the first prototype vehicle, called the VCLC (*Vehiculo de Combate Lanza Cohetes*) underwent firing trials with the Israel Military Industries 160 mm Light Artillery Rocket System (LARS 160) integrated on to the roof of the vehicle.

Two years later a second version was tested fitted with the Israel Military Industries 350 mm Medium Artillery Rocket System (MARS 350) integrated on to the roof of the vehicle.

It is understood that a small quantity of VCLC-CAL 160 are now in service with the Argentine Army. Production of the TAM hull on which the VCLC is based was completed some years ago.

The TAM tank is only used by Argentina, although it was offered on the export market by Argentina along with the original German version, called the TH 301. Rheinmetall has ceased all marketing of this family of tracked vehicles.

Description

The major differences between the VCLC and the TAM medium tank hull (originally designed and developed in Germany by the now Rheinmetall (at that time Thyssen Henschel) is the modification of the full-tracked hull to incorporate additional hydraulic suspension, movement suppression systems to absorb the recoil forces during rocket launching and fitting of a new two-person fully enclosed armoured turret.

The actual rocket launcher is mounted on top of this turret towards the rear. This allows the launcher to be traversed left and right. It is provided with bulletproof windows for crew observation purposes.

Support arms for the rocket system modules are fitted at the rear of the turret on either side. They are traversed and elevated using the turret's electrohydraulic power unit. No modifications are needed if a change between the rocket types is required.

For self-protection and limited air defence, a pintle-mounted 7.62 mm MAG or 0.30 in Browning MG is mounted above the front of the turret.

When the LARS 160 rocket system is carried the vehicle is designated the VCLC-CAL. This has two 18-round 160 CAL pod container-launchers that, once fired, are unloaded and replaced in a secure area by a supporting vehicle.

In Argentine Army trials, the vehicle used was a US supplied M809 Series (6 × 6) five-tonne flatbed truck fitted with a 12,000 kg capacity hydraulically operated extendable crane and carrying two of the 160 CAL pods. The whole reloading operation takes approximately 10 minutes.

If fitted with the MAR 350 system the vehicle is designated the VCLC-CAM and has two pairs of side-by-side cylindrical container-launchers for CAM 350 unguided rockets. The resupply operation is similar to that used for the VCLC-CAL vehicle.

In combat the basic six-launcher VCLC battery will have a truck-mounted fire-control radar system. Its job is to monitor the first ranging shot for between two-thirds to three-quarters of its flight. The round is deliberately aimed offset from the target so as not to alert it.

The tracking data obtained is then fed into the vehicle's fire-control computer and used to calculate the rocket launch information, which is then transmitted to the battery launchers.

Variants

TAMSE VCA 155 155 mm self-propelled artillery system

This is a modified TAMSE MBT hull fitted with the turret that is also used for the Oto Melara Palmaria 155 mm/41-calibre self-propelled artillery system which is covered in a separate entry in *IHS Jane's Land Warfare Platforms: Artillery and Air Defence*.

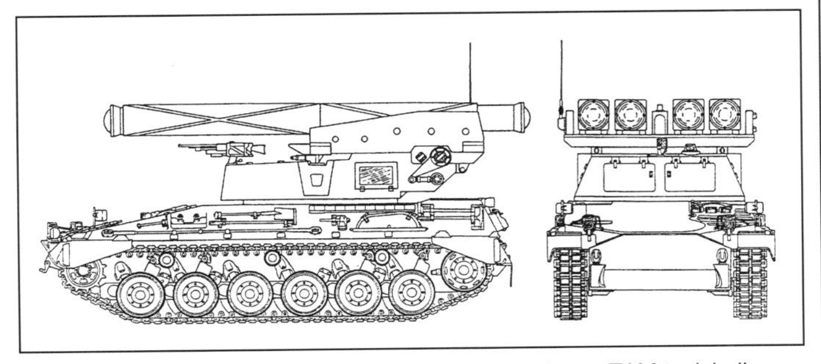

Argentine CAM 350 multiple rocket system mounted on a TAM tank hull 0500628

VCLC-CAL 160 multiple rocket launcher based on a modified TAM tank hull 0500630

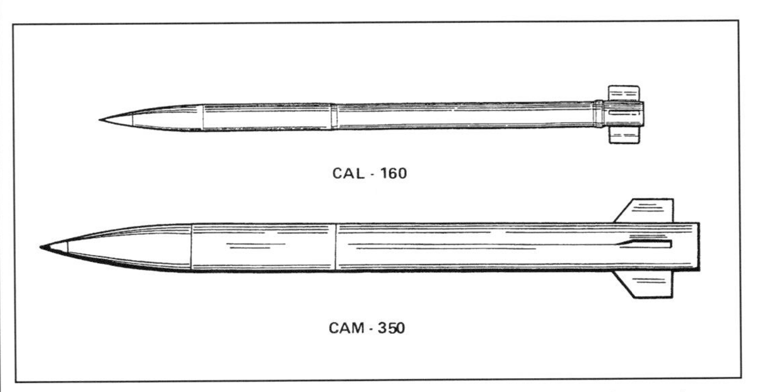

General arrangement drawings of CAL 160 (top) and CAM 350 (bottom) rockets with fins unfolded as in flight 0500629

This is in service with Libya (210 units) as well as Nigeria who took delivery of an initial batch of 25 systems ordered in 1982.

Since then additional vehicles have been supplied to Nigeria with a further nine systems being delivered as recently as 2007.

It should be noted that as a result of operations in Libya in 2011 the exact numbers of Palmaria now in operation in Libya is uncertain.

Specifications

	VCLC rocket launcher
Dimensions and weights	
Crew:	3
Length	
hull:	6.75 m
Width	
hull:	3.12 m
Height	
overall:	3.05 m
Ground clearance	
overall:	0.45 m
Track	
vehicle:	2.62 m
Weight	
combat:	32,582 kg
Ground pressure	
standard track:	0.84 kg/cm^2
Mobility	
Configuration	
running gear:	tracked
Power-to-weight ratio:	22.1 hp/t
Fuel capacity	
main:	680 litres
main, with auxiliary tanks:	1,080 litres
Engine	MTU MB 833 Ka-500, 6 cylinders, supercharged, diesel, 720 hp at 2,400 rpm
Gearbox	
model:	Renk HSWL 204 planetary
forward gears:	4
reverse gears:	4
Steering:	double differential with hydrostatic steering control

	VCLC rocket launcher
Firepower	
Armament:	36 × turret mounted 160 CAL rocket system (2 × 18)
Armament:	alternate main 4 × turret mounted 350 CAM rocket system
Survivability	
Armour	
hull/body:	steel
turret:	steel

Status

A small quantity of the VCLC-CAL 160 are in service with the Argentine Army but the VCLC-CAM with the larger rockets is understood to be at the prototype stage. Production complete. This was not offered on the export market.

It should be noted that solid propellant rockets have a typical life of about 10-15 years after which time the rocket motor is time expired and the rockets are not safe to fire.

Contractor

TAMSE (hull).

DGFM 127 mm (28-round) multiple rocket launcher

Development

This 127 mm (16-round) multiple rocket launcher was developed to meet the requirements of the Argentina by the DGFM.

It is understood that a total of 30 systems were built under the local designation of the SLAM Pampero but only four systems remain in service.

It has recently been reported that the DGFM has developed a new rocket launcher which can fire either the 105 mm Pampero unguided rockets or the more recent 127 mm SS30/CP30 unguided rockets.

This is stated to have three layers each of nine launcher tubes to give a total of 27 launcher tubes.

DGFM 127 mm (28-round) CP-30 rocket showing wraparound fins at rear in unfolded position as they would appear in flight 0008304

Original DGFM 127 mm (28-round) rocket launcher mounted on rear of Mercedes-Benz UNIMOG 416 (4 × 4) truck chassis 0008303

FIAT 697 BN (6 × 6) chassis with one pod of nine 127 mm rockets with system in travelling configuration (DGFM) 1172810

FIAT 697 BN (6 × 6) chassis with one pod of nine 127 mm rockets with stabilisers lowered and launcher traversed right. Note hydraulic crane to cab rear for reloading purposes (DGFM) 1172809

A complete salvo of 27 rockets, which can be launched in about 15 seconds with a maximum range of 30 km, is claimed for the 127 mm SS30/CP30 rocket.

Description

To provide a more stable platform, four hydraulically-operated stabilisers are lowered to the ground. Traverse and elevation are also hydraulically operated systems. Sights are provided for direct and indirect fire.

Single-round and ripple firings up to the full salvo are possible. The 127 mm steel-bodied rocket uses a PHE-2 solid propellant filled rocket motor casing with four wraparound pop-up tail-mounted guidance fins.

The warhead types available include: training, drill, HE anti-personnel, HE incendiary and anti-tank/anti-personnel submunition.

Maximum rocket range is stated to be 16,300 m. The warshots have either contact or delayed action fuzing options.

When originally introduced into service with Argentina the system was integrated onto the rear of an unprotected German Mercedes-Benz UNIMOG 416 (4 × 4) cross country chassis.

It should be noted that production of this chassis has been completed and it is no longer being marketed.

Mounted to the rear of the fully enclosed cab was a manually operated launcher with two layers each of eight 127 mm tubular rocket launcher tubes.

Variants

More recently the DGFM has developed a new system called the SS30 which is based on a FIAT 697 BN (6 × 6) forward control cross-country chassis with a forward control cab with a powered launcher to the rear which has a total of 36 launcher tubes. These 127 mm rockets have a maximum range of 30,000 m and a launch velocity of 1,000 m/s.

Specifications

	127 mm (28-Round) CP-30 rocket launcher
Dimensions and weights	
Length	
overall:	7.8 m
Width	
overall:	2.9 m
Height	
overall:	3.57 m
Weight	
standard:	18,870 kg (unloaded)
combat:	20,940 kg

	127 mm (28-Round) CP-30 rocket launcher
Mobility	
Configuration	
running gear:	wheeled
layout:	6 × 6
Engine	diesel, 254 hp
Gearbox	
forward gears:	8
reverse gears:	2
Firepower	
Armament:	36 × 127 mm rocket system (2 × 18)

Rocket Specifications

	SAPBA SS	CP30 rocket
Accuracy:	Probable CEP less than 1%	Probable CEP 1.5%
Initiation:	electrical	electrical
Rocket motor case:	steel	steel
Propellant:	extruded PHE-2 double base	composite
Fuze:	nose-mounted point detonating with optional delay	nose-mounted point detonating with optional delay
Stabilisation:	wraparound tail fins	wraparound tail fins

Status
Production complete. In service with the Argentine Army. Offered for export. As far as it is known, there has been no recent production of the complete system. Some sources have indicated that only four units are now operational with the Argentinian Army.

Contractor
Dirección General de Fabricaciones Militares (DGFM).

DGFM 105 mm (16-round) Pampero multiple rocket launcher

Development
Developed by the Research and Development Institute of the Argentine Armed Forces (the DGFM), the 105 mm (16-round) Pampero multiple rocket launcher entered operational service in the mid-1980s.

Production of this system has been completed and it is no longer marketed.

It has recently been reported that the DGFM has developed a new rocket launcher that can fire with the 105 mm Pampero unguided rockets or the more recent 127 mm SS30/CP30 unguided rockets.

This is stated to have three layers each of nine launcher tubes to give a total of 27 launcher tubes.

A complete salvo of 27 rockets which can be launched in about 15 seconds with a maximum range of 30 km, is claimed for the 127 mm SS30/CP30 rocket.

Description
The 105 mm 16-round launcher assembly, with two rows of eight tubes, is mounted on the rear bed of a modified Mercedes-Benz UNIMOG 416 (4 × 4) 1,500 kg light cross-country truck chassis.

If required, the launcher can be fitted to other chassis types with differing numbers of tubes. Production of the Mercedes-Benz UNIMOG 416 (4 × 4) series chassis was completed many years ago.

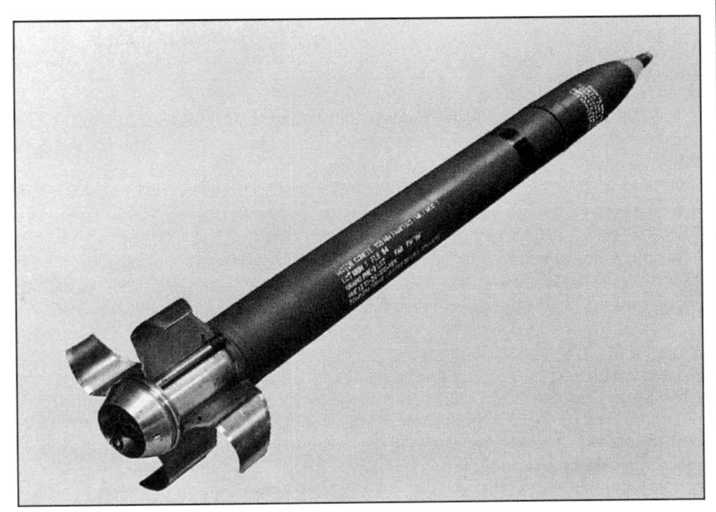

DGFM 105 mm Pampero rocket showing four unfolded fins at rear of rocket as it would appear in flight 0008306

DGFM 105 mm (16-round) Pampero rocket launcher deployed in the firing position on Mercedes-Benz UNIMOG 416 (4 × 4) truck chassis 0008305

This system is also referred to as the SLAM Pampero and about 40 systems are understood to have been built, all based on the Mercedes-Benz UNIMOG 416 1,500 kg (4 × 4) chassis with most being delivered to the army.

To provide a more stable firing platform four hydraulically-operated stabilisers are lowered, two at the front of the vehicle and two at the rear. Traverse and elevation are manually set via handwheels. Sights are provided for both indirect and direct fire modes.

Single round, partial ripple or full salvo firings are possible. A full salvo takes 7.5 seconds and will fall within a 300 × 200 m area with a maximum claimed round dispersion CEP of less than one per cent. The system has to be manually reloaded.

The actual weight of the 105 mm rocket is 28.5 kg of which 10 kg is the warhead with the overall length of the rocket being 1.476 m.

The single-stage 105 mm rocket utilises a PHE-2 solid propellant fuel filled rocket motor casing with four pop-up tail-mounted guidance fins. The warhead types include training, HE fragmentation, HE incendiary and HE anti-personnel. Maximum range is stated to be 10,100 m. The warshots have either contact or delayed action warhead fuzing options.

There is also an air-to-ground version of the rocket. These rockets are launched from a six round aluminium launcher that is normally installed on Bell UH-1 helicopters.

Specifications

	105 mm (16-Round) Pampero rocket launcher
Dimensions and weights	
Crew:	3
Length	
overall:	5.22 m
Width	
transport configuration:	2.12 m
Height	
transport configuration:	2.38 m
Weight	
standard:	6,035 kg
combat:	6,490 kg
Mobility	
Configuration	
running gear:	wheeled
layout:	6 × 6
Engine	4 cylinders, petrol
Firepower	
Armament:	16 × 105 mm rocket system
Main weapon	
maximum range:	10,100 m
muzzle velocity:	530 m/s
Rate of fire	
rapid fire:	16/7.5 s
Main armament traverse	
angle:	180° (90° left/90° right)
Main armament elevation/depression	
armament front:	+70°/+10°

Status
Production complete. In service with the Argentine Army (36) and Marine Corps. There has been no recent production of this system. No longer marketed. Some sources have indicated that the Pampero is only now in service with the army, who deploy only four units.

Contractor
Dirección General de Fabricaciones Militares (DGFM).

Belgium

Forges de Zeebrugge FZ 70 mm (40-round) LAU97 multiple rocket launcher system

Development

Developed specifically for the export market by Forges de Zeebrugge SA, the 70 mm (40-round) LAU97 is a highly mobile light artillery rocket launcher system designed to provide saturation and holding fire at the regimental level.

Recent information has stated that the Army of Shajah in the United Arab Emirates (UAE) took delivery of 18 of these systems, fitted on the rear of a French ACMAT VLRS 4.15 LRM (4 × 4) cross-country truck chassis.

The company has now ceased marketing of this land based multiple rocket launcher system.

The company continues to manufacture and market rockets for air-to-ground applications.

Description

The LAU97 system of a launcher box with five horizontal rows of eight 70 mm tubes and a 360° traversable base that allows elevation arcs of 0° to +55°.

The system could be fitted on any military vehicle or trailer capable of carrying 1,200 kg payloads. The launcher's azimuth and elevation laying is performed by either electric motors or handwheels.

A BR2 mortar sight or artillery panoramic sight is fitted for aiming the system with an electrically operated remote-control unit for firing.

The 70 mm FZ series of unguided rockets can be launched either singly or in ripple sequences up to the full 40-round salvo, which takes less than six seconds. Using the FZ90 solid propellant rocket motor and a 4.3 kg warhead, the maximum range is 9,000 m at a firing angle of 40° elevation.

The rockets, using the standard FZ-68 70 mm rocket motor and a 4.3 kg warhead, have a maximum range of 7,900 m at a launcher elevation of +40°; if the elevation is set at +15° then the range is reduced to 6,000 m. If the Mk 40 rocket motor is used, the maximum range at +40° elevation is reduced to 7,500 m.

A full complement of 70 mm unguided rockets can saturate an area measuring 200 × 300 m at maximum range. The FZ-100 6.2 kg cargo warhead with a payload of nine 0.48 kg anti-personnel/anti-tank bomblets was also available.

The bomblets have a lethal radius of 10.5 m and are capable of penetrating up to 105 mm of conventional steel armour plate. The submunitions are released by a delay-adjustable time fuze over the target area.

Other warheads available include: the 4.3 kg FZ210 practice impact marking practice; the 3 kg FZ-49 anti-armour capable of piercing more than 350 mm of conventional steel armour plate; the 4.3 kg FZ-71 anti-personnel which forms more than 8,000 fragments on detonation and is lethal to between 18 and 21 m radius; and the 4.9 kg M257 illuminating type with a 120 second duration and one million candle lighting capacity. The launcher is operated by a crew of three, a layer, a commander and a gunner, who also reload the system by hand.

Additional rockets would be carried by a resupply vehicle.

Early in 2006, the manufacturer stated that production of this system was complete and it was no longer being marketed.

LAU97 launcher installed on rear of ACMAT VLRS 4.15 LRM (4 × 4) truck chassis with stabilisers lowered and launcher traversed right 0130029

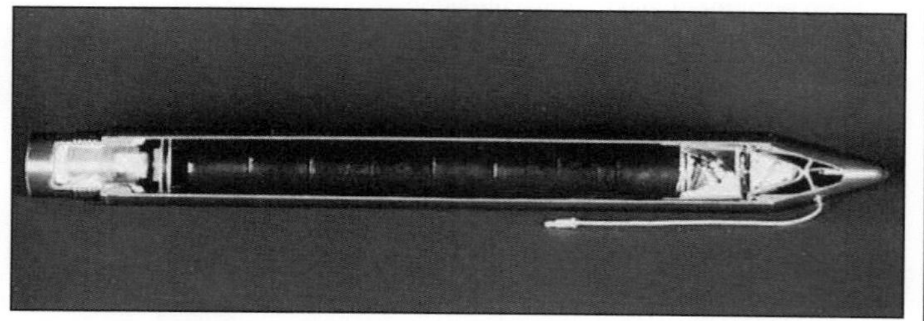

FZ-100 submunition warhead 0500889

Variants

There is also a towed version.

The FZ90 rocket motor with the flash warhead FZ181 is used in the Roxel (previously the Rocket Motors Division of RO Defence) Reduced Range Practice Rocket (RRPR), which was developed to meet the requirements of the British Army. This is now used by the British Army.

South Korean 70 mm (40-round)

This is very similar to the Belgian system and has been developed by Hanwa. Details are given in a separate entry in *IHS Jane's Land Warfare Platforms: Artillery and Air Defence.*

As of March 2013 it is understood that this system had yet to enter production.

Specifications

	70 mm (40-Round) LAU97 rocket launcher
Mobility	
Configuration	
running gear:	wheeled
Firepower	
Armament:	40 × 70 mm rocket system
Ammunition	
main total:	160
Main weapon	
maximum range:	9,000 m (depends on rocket type)
Main armament elevation/depression	
armament front:	+55°/0°

Status

Production complete. No longer marketed. In service with Sharjah in the United Arab Emirates (UAE) (18 units) on ACMAT chassis. From 1985 IPTN of Indonesia started producing the 70 mm rocket system FFAR under licence from Forges de Zeebrugge SA. Indonesia has developed its own 70 mm launcher system (which is covered in detail in a separate entry) and also now manufactures the rockets under licence.

Contractor

Forges de Zeebrugge SA, a TDA subsidiary.

Brazil

AVIBRAS ASTROS II rocket system

Development

AVIBRAS developed the Artillery SaTuration ROcket System (ASTROS) for the Brazilian Army and the export market. The first production units were completed in late 1983.

According to the United Nations Arms Transfer Lists for the period 1992 through to 2010, Brazil exported the following quantities of ASTROS II rocket systems and associated rockets:

Country	Quantity	Date	Comment
Malaysia	12	2002	rocket launchers plus associated vehicles and equipment
	3,232	2009	2,304 SS30 and 928 SS40 rockets
	Not disclosed	2009	rocket launchers
Qatar	384	1992	SS30 and SS60 rockets
Saudi Arabia	50,328	1992	SS30. SS40 and SS60 rockets

In 2002 the Malaysian Army took delivery of an initial batch of 18 ASTROS II rocket systems plus support vehicles which are used by the 51 Artillery Regiment which supports the 3rd Division.

A second contract was placed in 2007 for an additional batch of 18 ASTROS rocket systems which were delivered from 2009 through to 2010. These are used by the 52 Artillery Regiment which supports the 2nd Division.

Each of these regiments has three batteries each with six launchers plus associated support vehicles including the fire-control unit with its associated radar system. In Malaysian service this system is known as the Keris SPRM.

In June 2012 the Brazilian Army ordered three batteries of ASTROS 2020 systems for delivery by 2015.

The framework contract was awarded in December 2011 and became effective in January 2012 followed by discussions and subsequent award.

The ASTROS 2020 has had a number if improvements over the earlier systems including a new digitised command-and-control system and new navigation and communications system.

It will also have the ability to fire the AV-TM long-range guided tactical cruise missile which is claimed to have a maximum range of 300 km. Available details of this are provided later in this entry.

Complete battery of six ASTROS II launchers awaiting delivery to the Brazilian Army (AVIBRAS) 1452649

It is understood that these systems will be based on the Tatra T815-790PR39 (6 × 6) cross-country truck chassis with the battery command post vehicles and meteorological systems being based on the Tatra T815-7A0R59 (4 × 4) cross-country truck chassis.

Late in 2012 it was announced that AVIBRAS was to supply Indonesia with two batteries/regiments of ASTROS II systems plus the transfer of technology to allow for local industry to carry out maintenance work on the systems.

It is understood that a total of 36 launchers plus a similar number of ammunition resupply vehicles will be supplied to Indonesia as well as fire-control vehicles, battery command vehicles and associated support vehicles.

Late in 2013/early 2014, the Brazilian Marine Corps (Corpo de Fuzileiros Navais) are due to take delivery of one battery of ASTROS CFN 2020 systems which is expected to comprise six launchers.

Production of the ASTROS II rocket system and its associated unguided rockets is undertaken on an as required basis.

Description

The ASTROS II system consists of the launching vehicles (Universal Multiple Launcher: AV-LMU), ammunition resupply vehicles (AV-RMD with two complete reload sets for an AV-LMU) with onboard hydraulic cranes, a command-and-control vehicle (AV-VCC), a vehicle with radar for fire-control (AV-UCF - optional) and several mobile workshop vehicles for electronic and mechanical field maintenance.

It should be noted that the actual scale of issue of the ASTROS II rocket launcher and its associated support vehicles depends on the end user.

All the vehicles are based on the same 6 × 6 10,000 kg cross-country truck (AV-VBA) designed and built by the AVIBRAS subsidiary company TECTRAN.

There is a command-and-control vehicle at battery level (AV-PCC), a command-and-control vehicle at the battalion level (AV-VCC) and a meteorological mobile station (AV-MET) as part of the system.

All of these are parts of the ASTROS II rocket system and are based on a UNIMOG (4 × 4) chassis, also designed and built by the AVIBRAS subsidiary company TECTRAN.

According to AVIBRAS, a typical ASTROS II rocket system battery would comprise one AV-PCC, one AV-MET, one AV-UCF, six AV-LMU and six AV-RMD. For a battalion level headquarters there would be one AV-VCC and three mobile workshops (AV-OFVE).

The AV-VCC provides co-ordination and fire-control direction for up to three ASTROS II batteries but other battery configurations are possible depending on user requirements. Other battery configurations possible include four or eight AV-LMUs.

The launcher can fire four different types of unguided surface-to-surface rocket: the AVIBRAS SS-30 (32 rounds per launcher), SS-40 (16 rounds per launcher), SS-60 (four rounds per launcher) and SS-80 (four rounds per launcher). The larger SS-40, SS-60 and SS-80 rounds have cluster munition warheads. The SS-40 warhead has 20 and the SS-60 warhead 65 dual-effect anti-armour/anti-personnel bomblets, which improve the area-saturation effectiveness. The weight of the SS-60 warhead is 212 kg.

The SS-80 rocket has an increased range that has been achieved by carrying more solid propellant and reducing the weight of the warhead.

The complete family of AVIBRAS ASTROS II rockets, from top to bottom: SS-80, SS-60, SS-40 and SS-30 0500890

TECTRAN (6 × 6) AV-RMD resupply vehicle, left, loading new launcher pods onto TECTRAN (6 × 6) AV-LMU launcher, right 0500631

The Multiple Warhead Rockets (MW) of the ASTROS II rocket system use anti-armour/anti-personnel submunitions only. Each submunition has a self-destruct mechanism that acts a few seconds after impact on the terrain, in case of failure of the impact fuze. In this way, no dangerous submunition remains on the terrain.

There are two calibre tyres of submunition, 55 mm or 70 mm with the following specifications:

According to the company, ASTROS II is a conventional artillery rocket system that employs only authorised and controlled materials such as propellants and explosives.

In the context of the United Nations, the system fulfils the Protocol on Explosive Remnants of War (Protocol V to the 1980 Convention) referred to cluster munitions, as well as the Conventional on the Prohibition of anti-personnel mines (Protocol II to the 1980 convention).

The range of the unguided rockets vary from 9 to 90 km according to type. They all have wraparound fins for improved accuracy.

The command-and-control vehicle is equipped with military computers for fire-control calculations and a radio set for data transfer and voice communication with the launchers, fire-control radar vehicle, ammunition resupply vehicles, forward observers and higher command echelons.

The ASTROS II rocket systems of the Iraqi Army saw extensive combat service in both the 1980-88 and 1990-91 Gulf wars. Saudi Arabian systems saw extensive service in the 1990-91 Gulf War against Iraq.

Training

In 1997 AVIBRAS introduced a new concept for ASTROS II training means, which is the use of AVIBRAS Sub Calibre Training Ammunition 70 mm (AV-SS 09-TS).

This has been designed to help trainees to assimilate the operational, handling and storage procedures of the AVIBRAS ASTROS II rockets.

The AVIBRAS AV-SS 09-TS is an unguided rocket 70 mm (2.75 in) in diameter using composite propellant, wraparound fins and spin stabilisation with high kinetic energy and superior accuracy.

The rockets are fired from their own reusable AVIBRAS Sub Calibre Training Container Launcher (AV-CL/55-09 TS) and are fully compatible with the AV-LMU Universal Multiple Launcher.

The AVIBRAS (AV-CL/55-09 TS) with eight tubes has been specifically designed to cope with high-energy rockets.

The ASTROS is also being marketed as the ASTROS II Coastal Defence System.

As far as it is known, Brazil is the only user of the ASTROS II rocket system in its coastal defence application.

These are actually based on Mercedes-Benz (6 × 6) heavy duty truck chassis model 2028A/38 which has been modified by TECTRAN and designated the AV-VBA.

AVIBRAS AV-UCF radar fire-control vehicle deployed in operational configuration 0501054

Scale model of the AVIBRAS TM Tactical Missile as it would appear in flight (Christopher F Foss) 1333679

This model features a protected crew cab, banks of smoke grenade launchers, communications, automatic aiming processor, rear mounted manual aiming system and a ring mount for a .50 M2 HB machine gun.

The Brazilian Army SS-40 ASTROS system were delivered and filled with the Metsman meteorological sensor produced in the UK by the now Thales UK.

ASTROS II Mk 5 rocket system

So far all army production ASTROS II rocket systems have been based on the locally produced TECTRAN (6 × 6) cross-country truck chassis.

The ASTROS II Mk 5 is based on the more widely deployed Tatra T-815-7 cross-country truck chassis with a fully protected crew compartment at the front of the chassis.

SS-150 guided rocket

This guided rocket has a calibre of 450 mm, range of between 40 to 150 km with each launcher having two rockets in separate launch containers. This is understood to have not yet entered production.

FOG 60 rocket

The Fiber Optic Guided (FOG) missile version of the AVIBRAS II family has a calibre of 300 mm and a stated rage of up to 60 km and is still at the development stage.

Tactical Missile

Late in 2000, AVIBRAS confirmed that it had carried out the first launch of its latest Tactical Missile from their Artillery SaTuration ROcket System (ASTROS II).

The Tactical Missile is being developed as a private venture by the company to enhance the capability of the ASTROS II system to engage targets at a much longer range and with a significant increase in terminal accuracy over the current unguided rocket system.

AVIBRAS Climatised Mobile Ammunition Depot with doors open to show pods of rockets inside and handling rails in position (AVIBRAS) 1452650

ASTROS II rocket system of the Malaysian Army launching unguided rockets (AVIBRAS) 1401374

The first test, conducted at a Brazilian Army range, was to test the launch of the Tactical Missile from the ASTROS II launcher and the unfolding of the wings.

This was the first in a number of anticipated test launches, although AVIBRAS is seeking external funding to complete full development.

An ASTROS II launcher would typically carry two Tactical Missiles, which would be launched from the 450 mm launcher tubes. The solid propellant motor would boost the rocket out of the launcher after which the wings would unfold and the turbojet motor would cut in.

Typical warheads could range from a high-explosive warhead to one that would dispense submunitions out of either side of the rocket's nose.

The ASTROS TM missile has a range of 300 km and uses an inertial guidance system, which presumably incorporates a Global Positioning System (GPS) data for mid-course and possible terminal guidance.

No details of the payload of ASTROS TM have been released, although AVIBRAS have stated that its range and payload complies with the range and payload constraints of the Missile Technology Control Regime.

This suggests a payload of less than 500 kg and probably includes high explosive and submunition types.

ASTROS 2020 project

In 2011 the Brazilian government released an initial BRL45 million (USD28 million) in funding for the Brazilian Army's ASTROS 2020 project.

This covers the supply of three batteries as mentioned under development with a total of 49 vehicles including launchers, ammunition resupply vehicles and command-and-control vehicles.

ASTROS 2020 will feature a fully digitised command-and-control system and in addition to firing the currently deployed rockets will also be able to fire the AVIBRAS Tactical Missile (or AV-TM as it is also referred to) previously mentioned.

AV-VBL 4 × 4 vehicle

To support their ASTROS II ARS, AVIBRAS have also now developed the Light Armoured Vehicle (LAV). This uses a Mercedes-Benz UNIMOG (4 × 4) truck chassis fitted with an armoured body.

It provides the occupants with protection from small arms fire and shell splinters. This has already been manufactured in production quantities, with the first customer being Malaysia, who took delivery of a batch of ASTROS II (6 × 6) systems in 2002. This is also referred to as the Avibras Viatura Blindada Leve (AV-VBL) and can also be used as an APC.

It is based on a German Mercedes-Benz UNIMOG U2150L (4 × 4) truck chassis and is powered by an OM 366LA six-cylinder diesel engine developing 240 hp. This gives the vehicle a maximum road speed of 102 km/h. It has a maximum combat weight of 11.6 tonnes and is normally armed with a .50 (12.7 mm) M2 HB machine gun, which is mounted on the roof.

Climatised Mobile Ammunition Depot

This has been supplied to the Brazilian Army and typically holds pods of different unguided rockets that can be rapidly unloaded using rails carried in the container. This is called the AV-DMMC by AVIBRAS.

AVIBRAS ASTROS III

Development of this has ceased and it is no longer being marketed by the company.

Specifications

	SS-30
Dimensions and weights	
Length	
overall:	2.9 m
Diameter	
body:	127 mm
Weight	
launch:	68 kg
Performance	
Range	
min:	9 km
max:	40 km

	SS-40
Dimensions and weights	
Length	
overall:	4.2 m
Diameter	
body:	180 mm
Weight	
launch:	152 kg
Performance	
Range	
min:	15 km
max:	35 km

	SS-60	SS-80
Dimensions and weights		
Length		
overall:	5.5 m	5.5 m
Diameter		
body:	300 mm	300 mm
Weight		
launch:	578 kg	591 kg
Performance		
Range		
min:	20 km	20 km
max:	60 km	890 km

Status

Production of the ASTROS II system is undertaken on an as required basis. An ASTROS II battery was ordered in early 1994 for the Brazilian Army. The first contract was for the supply of four launchers, two ammunition vehicles and one fire-control vehicle.

Early in 1998 it was announced that the Brazilian Army had placed another order with AVIBRAS for the supply of a further four batteries of the ASTROS II system.

As part of this contract, in May 1998 a batch of four launchers and two ammunition supply vehicles were delivered to the Brazilian Army. In 1999 another eight launchers, five ammunition vehicles, two fire-control vehicles, two maintenance vehicles and five meteorological post vehicles were supplied to the Brazilian Army.

The Brazilian Army ordered a further three batteries of the latest generation ASTROS 2020 systems for delivery by 2015.

The Brazilian Marines will take delivery of a battery of ASTROS II in 2013/2014. Indonesia ordered 36 ASTROS II systems late in 2012.

Other users of the ASTROS II system are known to include Qatar (three batteries), Malaysia (two regiment) and Saudi Arabia (believed to have a total of 10 batteries each with six launchers).

Late in 2012 it was announced that AVIBRAS was to supply Indonesia with two batteries/regiments of ASTROS II systems. It is understood that a total of 36 launchers plus associated command vehicles and resupply vehicles will be supplied.

The former Iraqi Army used the ASTROS II system, which was also manufactured locally under the designation of the Sijell. The Sijell 30 used the SS-30 rocket, the Sijell 40 the SS-40 rocket and the Sijell 60 the SS-60 rocket.

As far as it is known none of these rocket launchers are now operational in Iraq.

Contractor

Avibras Indústria Aeroespacial S/A.

AVIBRAS SBAT-127 127 mm rocket system

Development

This 127 mm unguided surface-to-surface rocket, formerly manufactured by AVIBRAS, is an adaptation of the unguided 127 mm SBAT-127 High Velocity Aircraft Rocket (HVAR).

As of March 2013 there are no known exports of this system and the company concentrates its marketing on the longer range ASTROS II rocket system which is covered in a separate entry in *IHS Jane's Land Warfare Platforms: Artillery and Air Defence.*

Description

The 12-rail launcher was designed to be mounted either on a trailer or a vehicle.

There are two types of high-explosive warhead used and these differ in weight.

The heavier one produces a flight time of 70 seconds and the lighter one a flight time of 68 seconds for the rockets. Maximum ranges are achieved at an elevation of +47°. The system was developed for export. As far as it is known, there have been no sales of this system in the surface-to-surface role.

Specifications

AVIBRAS SBAT-127 rocket system

Calibre:	127 mm
Weight:	
(complete round)	45/58 kg (rocket)
(warhead)	19/32 kg (rocket)
Maximum range:	15,400/13,700 m (depend on rocket type)
Maximum speed:	690/680 m/s
Number of rails:	12
Launcher:	trailer or vehicle

Status

Production as required. No known land sales.

Contractor

Avibras Indústria Aeroespacial S/A.

AVIBRAS 108 mm 108-R rocket system

Development

This 108 mm unguided surface-to-surface rocket, which is manufactured on an as required basis by AVIBRAS, has the Brazilian Army designation FGT108-RA1 and is fitted with a single-stage solid propellant motor with a thrust of 1,250 kg.

It is spin stabilised in flight by one central and six canted rocket nozzles.

The solid propellant motor burns for half a second to attain a maximum velocity of 409 m/s at burnout.

This 108 mm unguided surface-to-surface rocket system has been offered on the export market but there are no known exports as of March 2013.

Description

The 108 mm rocket is launched from the X2A1 trailer launcher, which has 16 × 1.1 m long tubes. The launcher, also manufactured by AVIBRAS, weighs 530 kg empty, 802 kg loaded and is 2.95 m long, 1.6 m wide and 1.37 m high.

The wheelbase is 1.4 m and the tyre size 5.00 × 16. The launcher is normally towed by a ¾-tonne 4 × 4 vehicle. It can also be airlifted by helicopter. The maximum rate of fire is 2 rds/s.

The elevation is –1° to +50° and the traverse is ±12°. A 50 m cable is attached to the launcher for firing purposes. The maximum height of the rocket's trajectory is 4,000 m.

The crew consists of four and reloads are carried in 16-round glass fibre containers. A battery will normally have four launchers, each having at least 64 rockets available on its towing vehicle.

Specifications

	AVIBRAS 108-R
Dimensions and weights	
Length	
overall:	0.931 m
Diameter	
body:	108 mm
Weight	
launch:	16.8 kg
Performance	
Speed	
max speed:	1,584 km/h
Range	
max:	9.10 km
Ordnance components	
Warhead	7.8 kg

X2A1 trailer launcher for 108-R rocket (Ronaldo S Olive) 0569047

Status
Production as required. In service with Brazil. There are no known exports of this rocket system.

Contractor
Avibras Indústria Aeroespacial S/A.

AVIBRAS 70 mm SBAT rocket system

Development
This unguided 70 mm surface-to-surface rocket system was developed by AVIBRAS to meet user requirements for a highly mobile system that can be rapidly transported by land, sea and air.

Production of this trailer mounted rocket system is undertaken for the export market on an as required basis.

As of March 2013 there are no known exports of this system and the company concentrates its marketing on the longer range ASTROS II rocket system which is covered in a separate entry in *IHS Jane's Land Warfare Platforms: Artillery and Air Defence.*

Description
This unguided surface-to-surface rocket is essentially a standard AVIBRAS 70 mm SBAT-70 folding-fin air-to-ground rocket.

For land application a 36-tube LM-07/36 (also referred to as the SS-07/36 system) launcher is used, which requires a crew of four.

The maximum range is about 8,000 m. The following warhead types can be fitted: AVC-70/AC high-explosive anti-tank; AVC-70/AP high-explosive fragmentation (anti-personnel); AVC-70/AC/AP high-explosive anti-tank/anti-personnel; AVC-70/F flechette; AVC-70/FB white phosphorus smoke; AVC-70/TS practice smoke and AVC-70/E inert practice.

The launcher weighs 700 kg empty and approximately 1,000 kg loaded. The length is 3.17 m, width 1.67 m, height 1.19 m and wheelbase 1.47 m. The elevation is 0° to +50° and the traverse ±12°. A 50 m cable is attached to the launcher for firing purposes.

Specifications
AVIBRAS LM-07/36 rocket

Calibre:	70 mm
Weight:	
(complete round)	8.5 kg
(warhead)	3.2 kg
Maximum range:	8,000 m
Launcher:	LM-07/36 trailer

Variants
AVIBRAS Skyfire-70 Weapon System
Further development of the original AVIBRAS 70 mm SBAT rocket system has resulted in the new generation Skyfire-70 weapon system that includes the air-to-ground AV-SF-70 unguided rockets, the AV-LM 70-SF airborne multiple launchers and the land based LM-12/36 rocket launcher.

This has been developed by AVIBRAS as a private venture and the system comprises unguided rockets with a 70 mm (2.75 in) calibre using composite technology, wrap around fins and spin stabilisation in flight with higher kinetic energy and superior accuracy.

Each 70 mm rocket is composed by a rocket motor with a warhead and fuze. The combination among the types of rocket motors and warheads allows the assembly of specific rockets for the determined operational employment.

The rockets can be launched from airborne multiple tube launchers that are compatible with American (NATO) airborne 70 mm standard rockets may be utilised with a variety of aircraft.

The AVIBRAS AV-LM 70-SF airborne multiple launchers are reusable with seven or 19 tubes.

These rockets can also be used for surface-to-surface applications using the LM-12/36 rocket launcher.

AVIBRAS Skyfire family, LM 70/7 airborne SF launcher (background), Skyfire 70 M8/M9 and M10 folding fin rockets (left) and warheads (right) 0021035

LM-07/36 70 mm rocket launcher being carried slung under a helicopter 0058452

Skyfire rockets
The AV-SF-70 Skyfire rocket system includes different types of warheads for several tactical applications such as the high-explosive anti-personal/anti-material (HE) warhead, the hollow charge anti-armour/anti-personal (AC/AP) warhead, the anti-personnel flechette (F) warhead, incendiary/smoke white phosphorous (FB) warhead and the anti-runway/anti-material PE warhead.

In addition, the system has two types of training warhead, one inert and the other for impact signalling.

The AV-SF-70 Skyfire rockets can be powered by three models of solid propellant rocket motors, AV-SF-70 M8, AV-SF-70 M9 and AV-SF-70 M10 with an effective air-to-ground range of up to 4,000 m, 4,700 m and 6,000 m respectively. Surface-to-surface ranges are up to 10.5 km, 10.5 km and 12 km.

Skyfire rocket motors
The AV-SF-70 M8 rocket motor, which uses composite propellant, utilise four folded fins for stabilisation. The Skyfire M8 rocket motor has electrical contact and retention system compatible with the US MK4/MK40 and the Brazilian SBAT-70 rockets.

The AV-SF-70 M9 and M10 rocket motors, which use composite propellant, utilise four wraparound fins for stabilisation. The Skyfire M9 and M10 rocket motors have electrical contact and retention system compatible with the US MK66 rockets.

Skyfire warheads
There are essentially two classes of warheads referred to the weight. The 3.8 kgf class warheads are to be used with the Skyfire M8 or M9 rocket motor. The 6.0 kgf warheads are to be used with the Skyfire M10 rocket motor.

The AVC-70 HE high-explosive warheads are fitted with an impact fuze and used as anti-personnel/anti-material weapons for area saturation.

The AVC-70 AC/AP anti-armour/anti-personnel warheads are equipped with an impact fuze and also used as anti-armour/anti-personnel weapons for area saturation.

The AVC-70 F anti-personnel flechette warheads have hardened metallic flechettes and are used as an anti-personnel area coverage weapon. It is intended to be used with the Skyfire M8 or M9 rocket motor.

The AVC-70 FB white phosphorous warheads are fitted with an impact fuze and are used as incendiary and smoke generator weapons for area saturation.

The AVC-70 PE penetration warhead has a special shape and contains high explosive for use as an anti-runway/anti-matériel weapon.

Standard AVIBRAS 36-tube multiple launcher AV-LM-12/36 launching the folding fin rocket 0021036

The AVC-70 E inert practice warhead and the AVC-70 TS training/signalling warheads are used from training purposes.

AVIBRAS 70 Rocket Multiple Launcher 12/36
For land applications, AVIBRAS has developed the 36-tube AV-LM-12/36 towed launcher. This weighs 650 kg empty, 1,250 kg loaded and can either be towed or installed on a dedicated vehicle chassis.

This launcher integrates the AV-SS 12/36 artillery rocket saturation system, a mobile 70 mm rocket system for field artillery employed for area saturation up to 12,000 m.

The AV-SS 12/36 includes a set of AV-LM 12/36 launchers, each with 36 tubes, 70 mm unguided ground-to-ground rockets and equipment for firing computation and control.

The ammunition system is based on Skyfire 70 ammunition with the standard ammunition being the AV-SS-10 HE rocket comprising a AV-SS-70 M10 rocket motor and an AVC-70 HE M2 high explosive warhead of 6 kg.

AV-SS-10 HE rocket and LM 12-36 launcher Specifications

Calibre:	70 mm
Weight:	
(complete round)	14.5 kg (rocket)
(warhead)	6 kg (rocket)
Maximum range:	12,000 m (depends on rocket type)
Number of tubes:	36
Launcher:	AV-LM-12/36

Status
Production as required. In service with undisclosed countries.

Contractor
Avibras Indústria Aeroespacial S/A.

China

Beijing Bao-Long Science & Technology 300 mm (8-round) ANGEL-120 multiple rocket launcher

Development
The 300 mm calibre (8-round) ANGEL-120 multiple rocket launcher, understood to have been developed by the Beijing Bao-Long Science & Technology Developing Incorporation Ltd, is based on a heavyweight 8 × 8 truck chassis fitted with a launcher assembly of two by four rocket tubes on its rear decking.

The 8 × 8 cross-country chassis is fitted with a fully enclosed two-door cab.

To the rear of this is an additional cabin for further crew members.

The 300 mm ANGEL-120 MRLS would be supported by an associated rocket resupply vehicle.

When deployed in the firing position a number of stabilisers are lowered to the ground to provide a more stable firing platform.

Description
The 300 mm guided rocket is stabilised in flight by a spin imparted to it at launch by the rocket tube and is fitted with a composite propellant rocket motor.

The warhead fitted is a cluster munition type and can carry either 40 anti-personnel bomblets or 40 dual anti-personnel/anti-armour bomblets.

Other warhead types such as a minelet carrying one could be developed according to customer requirements.

The guided 300 mm rocket comprises five major subsystems, warhead, simplified guidance control system, propellant, stabilising system and the actual rocket structure. The rocket has wraparound fins, which unfold as the rocket leaves the launcher.

The lethal radius of one rocket round is stated to be 100 m while the full 8-round salvo can cover a 300 m radius.

A typical ANGEL-120 battery would comprise four launcher vehicles, two rocket resupply vehicles with 32 reloads in total, a fire direction (control) vehicle, technical support facilities and a truck-mounted crane.

The manufacturer claims the following key advantages for the ANGEL-120 artillery rocket system: high mobility, long range, high accuracy, great lethality and high volume of fire.

It is also claimed that the ANGEL-120 has a number of advanced features including a simplified guidance system, cluster type warhead, composite solid propellant motor and rocket spin imparted by the launch tube when the rocket is launched.

There has been no recent information on the ANGEL-120 multiple rocket launcher system and it is possible that development of the system has been halted. As far as it is known this system is not currently deployed by the Peoples Liberation Army.

Specifications
ANGEL-120 multiple rocket launcher

Calibre:	300 mm
Number of rounds:	8
Weight:	
(complete round)	740 kg
(warhead)	155 kg
Length:	
(complete round)	6.9 m
Rate of fire:	8 rds/30 s
Range:	
(minimum)	50,000 m
(maximum)	120,000 m
Time into action:	10 min
Launcher:	heavyweight (8 × 8) cross-country truck chassis

Status
As of March 2013, the exact status of the ANGEL-120 system is uncertain. Offered for export. As far as it is known, the ANGEL-120 is not used by the People's Liberation Army.

Contractor
Beijing Bao-Long Science & Technology Developing Incorporation Ltd.

SCAIC 400 mm WS-2 multiple rocket weapon system

Development
The Sichuan Aerospace Industry Corporation (SCAIC) has developed the 400 mm WS-2 (Wei Shi 2) Multiple Rocket Weapon System (MRWS).

The complete system is also being marketed by Poly Technologies integrated onto the rear of a TA5450 8 × 8 cross-country truck chassis.

This is fitted with a forward control four door fully air conditioned cab with an additional fully enclosed crew compartment to the rear.

Information released by Poly Technologies on the actual 400 mm unguided rockets is identical to that of released by SCAIC.

Description
This system is to be carried on the rear of an 8 × 8 cross-country truck chassis, with the powered launcher with two layers each of three launch tubes mounted at the rear.

To provide a more stable firing platform, stabilisers are lowered to the ground by remote-control.

Some sources have indicated that the overall system has been developed by the China National Precision Machinery Import and Export Corporation (CNPMIEC).

It is considered possible that CNPMIEC is the overall contractor, with SCAIC responsible for the actual guided rockets that can be launched from the cab or away from it with the aid of a remote-control system. CEP is being quoted as 600 m and a maximum flight speed of Mach 5.6.

The 400 mm guided rockets can be fitted with the following types of warhead to meet different user requirements:

- Submunition carrying 540 bomblets with a High-Explosive Anti-Tank (HEAT) warhead that is also highly effective against troops. The HEAT warhead will penetrate up to 85 mm of conventional steel armour and has a lethal radius of 7 m.
- Comprehensive effect cluster warhead carrying 61 submunitions with each submunition warhead containing approximately 200 preformed fragments and a shaped charge. It is claimed that the shaped charge will penetrate 180 mm of conventional steel armour. In addition the warhead also has a incendiary area effect of about 50 m^2 (4 m radius).

400 mm (6-round) WS-2 multiple rocket weapon system integrated onto a TA5450 (8 × 8) cross country truck chassis with stabilisers lowered and launcher elevated into the firing position (Poly Technologies) 1365050

400 mm rocket out of its launch tube with fins and control surfaces extended (SCAIC) 1154665

- Fuel Air Explosive (FAE) called 'Cloud blasting warhead' by the company. This contains 120 kg of FAE and produces a high blast overpressure upon functioning (both duration and peak pressure). The over pressure is quoted as 0.1 MPa (29 psi) at 25 m and 0.02 MPa (2.9 psi) at 50 m. No data on pulse duration is provided.
- High-explosive fragmentation incendiary which is called 'Blast burning warhead' by the company. This contains an undisclosed quantity of high explosive and incendiary pellets in addition to approximately 40,000 spherical steel preformed fragments. The lethal radius is claimed to be 85 m.
- High-explosive fragmentation called 'Lethal blasting warhead' by the company. It contains about 38,000 spherical preformed fragments and is claimed to have a lethal radius of 85 m.

Variant

400 mm WS-3 multiple rocket weapon system

This is currently being marketed by Poly Technologies and appears to be the same as the standard 400 mm WS-2 multiple rocket weapon system but the rockets are much more accurate as they are guided.

Using an inertial navigation system the 400 mm rockets are claimed to have a CEP of 300 m which can be reduced to 50 m using an inertial navigation and GPS guidance system.

The WS-3 rockets are about 7.15 m long and can be fitted with various types of warhead weighing up to 200 kg. The rockets have four wraparound fins at the rear and control surfaces towards the front.

The earlier unguided WS-2 system is still being marketed by Poly Technologies with a number of different warhead types.

These originally included submunition carrying 540 bomblets, comprehensive effect cluster, fuel air explosive, high-explosive fragmentation incendiary and high explosive fragmentation. The rockets have a shelf-life of 10 years.

WS-2C

This is understood to have a range of up to 300 km and may have a new terminal guidance system for use against ships as well as land-based targets.

WS-2D

Late in 2010 it was reported that the SCAIC had developed a new version of this system called the WS-2D. This has also been referred to as the Wei-Shi/Guardian-2D and may have entered service as far back as 2004.

This is larger than the standard 400 mm of the WS-2 with a diameter of 425 mm and an overall length 8.1 m with range increased to 400 km and a CEP of less than 600 m due to the installation of a low cost inertial guidance system.

In addition to sub-munition warheads the WS-2D is said to be able to carry three new unidentified "killer unmanned aerial vehicles".

It is understood that the launcher would carry six of nine WS-2D rounds in individual launch tubes that would be rapidly replaced by new rockets.

Specifications

	WS-2
Dimensions and weights	
Length	
overall:	7.15 m
Diameter	
body:	400 mm
Weight	
launch:	1,275 kg
Performance	
Range	
min:	70 km
max:	200 km
Ordnance components	
Warhead:	200 kg

Status

Development complete. Ready for production on receipt of orders.

Contractor

Sichuan Aerospace Industry Corporation (SCAIC) (but see text).

SCAIC 252 mm WS-1D Short Range Multiple Launcher Rocket Weapon System (SRMLRWS)

Development

The Sichuan Aerospace Industry Corporation (SCAIC) has marketed the 252 mm WS-1D Short Range Multiple Launcher Rocket Weapon System (SRMLRWS).

The photographs released by SCAIC show these 252 mm unguided surface-to-surface rockets being launched from the current China North Industries Corporation (NORINCO) 253 mm (10-round) Type 81 minesweeping rocket system based on a 6 × 6 truck chassis.

This system, covered in detail in a separate entry in *IHS Jane's Land Warfare Platforms: Artillery and Air Defence*, fires short-range 252 mm rockets with a long nose probe to clear minefields. It should be noted that NORINCO no longer markets the 252 mm minesweeping system.

Description

The recent WS-1D rockets are a new design and are different in appearance. They are much shorter and are fitted with four fins that unfold after they have left the launcher and a short nose probe, which may well extend in flight.

When compared to the older 253 mm rockets, the WS-1D rockets have a longer range, are heavier and the launcher has eight to 24 launch tubes.

In addition to the normal 6 × 6 wheeled chassis, a new launcher with three rows of eight rockets has also been observed mounted on an armoured superstructure, installed on the roof of a standard PLA NORINCO Type 59 tank hull that has had its turret removed. The armoured launcher mechanism has powered traversed left and right.

As production of the Type 59 MBT was completed some time ago, it is possible that this hull was just used for trials purposes.

SCAIC claims that the 252 mm unguided surface-to-surface rockets can be fitted with various types of warhead including blasting, blast burning and cloud blasting. It is probable that the last of these is of the fuel air explosive type.

It is assumed that these can be launched from the existing NORINCO 253 mm (10-round) Type 81 minesweeping rocket system, although it may have to be modified.

These rockets would clear a mine field by creating an overpressure to active the mine fuze.

252 mm WS-1D launcher on modified Type 59 tank chassis (SCAIC) 1154667

252 mm WS-1D rockets awaiting packing (SCAIC) 1154666

Specifications

WS-1D rocket

Calibre:	252 mm
Number of launch tubes:	8 to 24
Length of rocket:	1,600 mm
Weight:	
(rocket)	150 kg
(warhead)	90 kg
Range:	
(maximum)	3,500 m
(minimum)	400 m
Dispersion:	1/300 at longitudinal; 1/300 at traversal

Status

Production as required. See Description.

Contractor

Sichuan Aerospace Industry Corporation (SCAIC).

SCAIC 122 mm WS-6 Multiple Launcher Rocket Weapon System (MLRWS)

Development

The Sichuan Aerospace Industry Corporation (SCAIC) is currently marketing a 122 mm (40-round) Multiple Launcher Rocket Weapon System (MLRWS).

As of March 2013 there are no known direct export sales by SCAIC of this 122 mm WS-6 Multiple Launcher Rocket Weapon System.

It is considered probable that the SCAIC only have the capability to supply the 122 mm rockets and not the complete system (e.g. rockets and launcher).

Description

The photographs released by SCAIC show the China North Industries Corporation (NORINCO) 122 mm (40-round) Type 89 system based on a tracked chassis and the NORINCO 122 mm (40-round) Type 81 system on a 6 × 6 chassis.

It should be noted that NORINCO is no longer marketing either of these 122 mm (40-round) rocket launchers and is now concentrating all of its marketing on the more recent 122 mm (40-round) Type 90B based on a locally-built Mercedes-Benz (6 × 6) cross-country chassis.

It is considered possible that SCAIC is offering a new generation of 122 mm unguided surface-to-surface rockets as well as a new computerised fire-control system.

According to SCAIC, the 122 mm rockets can be fitted with various types of warhead including blasting, submunition and cloud blasting. It is probable that the last of these is of the fuel air explosive type.

All these 122 mm rockets are claimed to be capable of being launched from any 122 mm rocket launcher of Chinese or foreign manufacture.

It should be noted that NORINCO also manufactures and markets a 122 mm rocket with a maximum range of 40 km. This is slightly different in specifications to the SCAIC 122 mm rockets.

It is considered possible that while NORINCO is currently marketing this new family of 122 mm extended range rockets, they have actually been developed and manufactured by SCAIC.

A new command vehicle based on a forward control German MAN (6 × 6) cross-country chassis has been mentioned, which contains a new digital firing command system.

Specifications

WS-6 multiple launcher rocket system

Calibre:	122 mm
Number of launcher tubes:	40
Length of rocket:	2.863 m
Weight of rocket:	67 kg (with warhead)
Weight of warhead:	19 kg
Range:	
(maximum)	40,000 m
(minimum)	20,000 m

Status

Production as required. See Description.

Contractor

Sichuan Aerospace Industry Corporation (SCAIC).

China Precision Machinery Import & Export Corporation (CPMIEC) SY400 400 mm (8-round) Guided Rocket Weapon System (GRWS)

Development

The SY400 400 mm (8-round) GRWS is understood to have been developed by the China Precision Machinery Import & Export Corporation (CPMIEC) from the mid-1990's and was first shown in public around 2008.

SY400 Guided Rocket Weapon System deployed in the firing position and with launcher elevated at the rear of the chassis and stabilisers deployed (CPMIEC) 1403748

Scale model of SY400 Guided Rocket Weapon System in travelling configuration with 8-round launcher lowered into horizontal position (Christopher F Foss) 1365045

Scale model of SY400 Guided Rocket Weapon System deployed in firing position with 8-round launcher raised in vertical position (Christopher F Foss) 1365046

Development of the system is now said to be complete and production can commence when orders are placed.

This is already being offered on the export market by CPMIEC but as of March 2013 there have been no known export orders.

Description

SY400 GRWS is a brand new design and consists of an 8 × 8 cross-country chassis with a four door fully enclosed cab at the front.

Pivoted at the rear is an 8-round launcher with two pods each of four 400 mm guided rockets.

When deployed in the firing position, four hydraulically operated stabilisers are lowered to the ground, two either side to provide a more stable firing platform.

While most ARS/MLRS have the launcher mounted on a turntable with limited elevation, SY400 launcher pivots at the rear and has no traverse and is raised into the vertical position by an hydraulic ram.

The rocket is a new design with an overall length of 6.451 m and is front part is 400 mm in diameter and can be fitted with various warheads with weights of 300 kg or 200 kg.

When fitted with the former range is being quoted 80 to 150 km and when fitted with the latter a range of 80 to 200 km is being quoted.

When fitted with a 300 kg warhead rocket launch weight is quoted as 1,276 kg and when fitted with the lighter 200 kg warhead rocket launch weight is reduced to 1,176 kg.

Poly Technologies A200 301 mm (8-round) Guided Multiple Rocket Launcher deployed in the firing position with launcher elevated into the vertical position (Poly Technologies) 1452647

Poly Technologies A200 301 mm rocket out if its launcher and with four tail fins extended and four control surfaces extended towards the front as they would appear in flight (Poly Technologies) 1452648

The rocket is guided to enable targets to be engaged at long range with a precision effect.

When fitted with an Inertial Navigation System (INS) a Circular Error or Probability (CEP) of 160 m is being quoted. When fitted with INS and Global Positioning System (GPS) CEP is reduced to 50 m.

At least four types of warhead can be fitted to the SY400 rocket and these include:

- anti-armour (carrying 560 or 660 bomblets which will penetrate between 80 and 120 mm of conventional steel armour). The 200 kg warhead carries not less than 560 bomblets with a lethal radius of not less than 7 m. The 300 kg warhead carries not less than 560 bomblets with a lethal radius of not less than 7 m.
- integral blast fragmentation with two versions with fragments, tungsten balls and steel balls. The 200 kg version contains a total mass of 110 kg while the 300 kg version contains a total mass of 180 kg with both having a lethal radius of not less than 150 m.
- integral blast fragmentation combustion - similar to the above but contains incendiary with same mass and lethal radius as above.
- fuel air explosive with the 200 kg warhead containing not less than 120 kg of explosive and the 300 kg warhead containing not less than 180 kg of explosive. Both have a lethal radius of not less than 50 m.

According to the CPMIEC the SY400 GRWS takes 15 minutes to come into action and rockets can be launched at eight second intervals.

Once the rockets have been launched the system would normally move to another position and new pods of rockets loaded from an associated load transfer vehicle.

In addition to the SY400 GRWS, other elements include a load transfer vehicle, probably based on a similar chassis and fitted with a crane to load pods of four rockets and associated command-and-control unit.

The SY400 GRWS is based on the WS series of heavy duty cross-country vehicles which ranges from 6 × 6 configuration up to 14 × 14 configurations with some of these having a fully protected cab.

This can be used to transport and launch a variety of missile types as well as being used for a wide range of support roles.

The latest SY400 GRWS is being marketed alongside older CPMIEC artillery rocket systems including the 320 mm (4-round) WS-1, 302 mm WS-1B (4-round) and 300 mm (10-round) A100 system. They are also marketing a number of twin round tactical missile systems such as the P611M with a maximum range of up to 260 km.

Poly Technologies A200 301 mm (8-round) Guided Multiple Rocket Launcher (GMLR)

Poly Technologies is currently marketing the A200 301 mm (8-round) GMLR which appears to be almost identical to the CPMIEC 400 mm SY 400 mm system apart from the calibre of the rockets.

According to Poly Technologies the 301 mm rockets have been designed to engage targets out to a range of 50 through to 200 km.

The rockets have an overall length of 7,264 mm, a diameter of 301 mm, launch weight of 750 kg and a wingspan of 615 mm.

During the course of the flight, the warhead front part of the rocket is separated from the solid propellant rocket and guided to its target using combined Inertial Navigation System/Global Positioning System (INS/GPS) guidance which provides a CEP of about 50 m.

A salvo of eight × 301 mm rockets can be fired in 50 seconds and it takes about eight minutes to be prepared for firing.

Once a fire mission has been completed it takes two minutes to come out of action and move to another position for re-loading to avoid counterbattery fire.

Specifications

	SY400 (300 kg warhead)	SY400 (200 kg warhead)
Dimensions and weights		
Length		
overall:	6.451 m	6.451 m
Diameter		
body:	696 mm (aft motor)	696 mm (aft motor)
Flight control surfaces		
span:	0.69 m (wing)	0.69 m (wing)
tail span:	0.40 m	0.40 m
Weight		
launch:	1,276 kg	1,176 kg
Performance		
Range		
min:	80 km	80 km
max:	150 km	200 km
Ordnance components		
Warhead:	300 kg	200 kg

Status

Development complete. Ready for production on receipt of orders.

Contractor

China Precision Machinery Import & Export Corporation (CPMIEC).

China Precision Machinery Import & Export Corporation (CPMIEC) 400 mm/600 mm guided rocket system

Development

The China Precision Machinery Import & Export Corporation (CPMIEC) is now marketing a rocket system that allows an 8 × 8 mobile launcher system to transport and launch a pod of four SY400 400 mm guided rockets and a pod of containing one BP-12A 600 mm long range guided missile.

The standard SY400 system has two pods each of four rockets while the standard BP-12A system has two pods each of one rocket.

Full details of the SY400 400 mm (8-round) launcher are provided in a separate entry in *IHS Jane's Land Warfare Platforms: Artillery and Air Defence*.

The combination of two rocket systems onto a single platform provides the user with increased mission flexibility as a much wider range of targets can be engaged from a single platform with precision effect.

Both systems are based on the Wanshan Special Vehicle Company (WSVC) WS 2400 (8 x 8) cross-country chassis with a four door fully enclosed cab at the front.

This chassis is already used by China for a number of missile applications and has a gross vehicle weight of 41 tonnes depending on its application.

As of March 2013 it is understood that development of the system was complete but that production had yet to start.

Description

Mounted at the very rear of the chassis is the launcher that is horizontal for travelling and rapidly raised in the vertical position for launching the guided rockets and missiles.

Stabilisers are lowered to the ground to provide a more stable firing platform.

The SY400 rocket is fitted with four rear-mounted stabilising fins, four mid-body strakes approximately 2 m in length and an ogival nose.

It has an overall length of 6.5 m, a body diameter of 400 mm and a rear fin span of 696 mm.

It can be fitted with four different types of warhead types:

- unitary blast/fragmentation;
- unitary blast/fragmentation/incendiary;
- unitary fuel air explosive;
- and anti-armour and anti-personnel blast fragmentation submunition.

Each of these warheads is available in 200 or 300 kg versions.

A 300 kg warhead the SY400 rocket has a maximum range of 80 to 150 km and a stated launch weight of 1,276 kg. Maximum range increases to 250 km with the lighter 200 kg warhead.

Scale model of the combined BP-12A/SY400 guided rocket system deployed in the firing position and showing one pod of four SY400 rockets and one pod containing one BP-12A long range rocket (Christopher F Foss) 1423206

When an Inertial Navigation System (INS) based guidance system is fitted, the SY400 is claimed to have a CEP of 160 m but the addition of a Global Navigation Satellite System (GNSS)-based guidance system reduces this to 50 m.

Guidance corrections from either systems are actioned by four vanes in the rocket nozzle which direct the motor's efflux.

The guided BP-12A is fitted with four rear-mounted stabilising fins, a long ogival nose section and a smooth mid-body with no strakes.

It has an overall length of 6.3 m which is slightly less than the SY400 rocket but its increased diameter provides far greater internal volume.

The increased volume gives a higher payload and propellant carrying ability which provides a longer range.

The launch weight is 2,293 kg of which 480 kg is the warhead. The latter can be one of five types that are not interchangeable.

These are High-Explosive (HE), two types of high-explosive fragmentation and two submunition variants.

One of the latter carries 24 large submunitions each with an effective radius of 80 m and the other 850 submunitions each with an effective radius of 7 m.

The BP-12A missile has a minimum range of 80 km and a maximum range of 280 km and can be fitted with an INS guidance system providing a CEP 100 m or an INS/GNSS guidance system that is more accurate and provides a CEP of 50 m.

As for the SY400, the BP-12A uses four control vanes in the rocket efflux to make guidance corrections.

For the combined system both of these rocket systems use the same launch control and ground test equipment.

Both rockets are hot launched and targets can be engaged through 360° without moving the launcher.

Current quoted preparation launch time for the SY400 system is less than 15 minutes with an out of action time of three minutes while the BP-12A system has a launch quoted preparation time of less than 12 minutes and a similar out of action time.

Once a fire mission has been carried out the system would normally withdraw to a safe area for missile resupply with new pods of rockets being rapidly reloaded.

Specifications
Not available.

Status
Development complete. Ready for production.

Contractor
China Precision Machinery Import & Export Corporation (CPMIEC).

China Precision Machinery Import & Export Corporation (CPMIEC) 320 mm (4-round) WS-1 artillery rocket system

Development
China Precision Machinery Import & Export Corporation (CPMIEC) has developed a long-range unguided surface-to-surface artillery rocket weapon system called the WS-1 (alternative designation believed to be M-1).

According to CPMIEC, the WS-1 system has been designed to bridge the gap between conventional towed and self-propelled artillery systems and tactical missile systems.

It is understood that all export marketing is now being concentrated on the more recent CPMIEC 302 mm WS-1B (4-round) artillery rocket systems.

As of March 2013 there are no known exports of this artillery rocket system.

CPMIEC 320 mm (4-round) WS-1 artillery rocket system (right) with rockets in foreground with 6 × 6 forward control cargo truck to the immediate rear 0500632

Description
Mounted on the rear of a forward control Mercedes-Benz type 6 × 6 cross-country truck (which is manufactured under licence in China by China North Industries Corporation (NORINCO) are four cylindrical launcher tubes for the WS-1 rocket.

The WS-1 rocket has four fixed fins and reaches a maximum flight altitude of 30,000 m with a maximum speed of Mach 3.6.

Maximum range is quoted as 80,000 m with a minimum range of 20,000 to 30,000 m with probable deviation of 1 to 1.25 per cent of range.

According to CPMIEC, who are carrying out marketing of the WS-1 ARS, the unguided rocket is fitted with a conventional warhead that weighs 150 kg, which includes 70 kg of explosive.

The warhead detonation contains about 26,000 pre-formed fragments plus an undisclosed amount of natural fragments. It is claimed that the fragments have an effective lethal radius of 70 m.

The unguided rockets have a maximum velocity of Mach 4.2, with an initial rotation of 360 revolutions per minute. Flight time is quoted as 170 seconds.

Recent information has indicated that the rockets are manufactured by the Sichuan Aerospace Industry Corporation and that at least four types of warhead are available, which are also used for the more recent WS-1B rocket system covered in a separate entry in *IHS Jane's Land Warfare Platforms: Artillery and Air Defence*. These four warheads are:

- High-explosive fragmentation, named 'Blasting warhead' by the manufacturer, weighs 150 kg, as mentioned in the text above.
- Submunition warhead with a total weight of 152 kg. It contains an undisclosed quantity of sub-munitions. Each sub-munition projects approximately 475 high velocity pre-formed fragments upon detonation, these having a lethal radius of 7 m. It is assumed that the submunitions are of the combined effects type (fragmentation/blast/shaped charge) and are designed to attack the vulnerable upper surfaces of armoured vehicles and/or troops in the open.
- Fuel Air Explosive (FAE), named 'Cloud Blasting' by the company, total filling weight of 70 kg. This has an effective casualty radius of 70 m and a lethal over pressure radius of 50 m.
- High-explosive fragmentation incendiary, named 'Blast burning' by the company. It has a high-pressure explosive filling weight of 70 kg and an effective lethal radius of 70 m. There is no disclosed information on the incendiary effects of this warhead.

The 4-round launcher is mounted on a turntable at the rear of the chassis and has powered traverse and elevation. Hydraulically-operated stabilisers have to be lowered at the rear of the vehicle to provide a more stable firing platform.

Once the four rockets have been launched, new rockets are reloaded manually into the launcher tubes from a truck backed up to the rear of the firing platform.

It is stated that an onboard digital computer provides assistance to increase the accuracy of the rockets. Accuracy is stated to be within one per cent. As these rockets are unguided, their main drawback is lack of accuracy. With this type of rocket, as the range increases there is a loss of accuracy.

Specifications
WS-1 artillery rocket system

Calibre:	320 mm
Number of rockets:	4
Range:	
(maximum)	80,000 m
(minimum)	40,000 m
Length:	4.732 m
Weights:	
(complete round)	524 kg
(warhead)	150 kg
Maximum speed:	M 3.6

Variants
WS-1B rocket system

The CPMIEC is now understood to be concentrating marketing on the more effective 302 mm WS-1B (4-round) artillery rocket systems. Details of this are given in a separate entry in *IHS Jane's Land Warfare Platforms: Artillery and Air Defence*.

Status

Production as required. Understood to be in service with the People's Liberation Army. Early in 1997 it was reported that China and Turkey signed a contract to produce, in Turkey, five batteries of CPMIEC 320 mm (4-round) WS-1 artillery rocket systems for the Turkish Land Forces Command.

Recent information indicates that these were in fact WS-1B systems with a Roketsan producing the 300 mm (4-round) system for the Turkish Land Systems Command.

Details of the Turkish Rokestan 300 mm (4-round) Artillery Rocket System are provided in a separate entry in *IHS Jane's Land Warfare Platforms: Artillery and Air Defence*. These rockets have a maximum range of 100 km.

It is understood that production of these systems for the Turkish Land Systems Command is now complete. The rockets are designated the TR-300 with the launcher designated the T-300.

It is understood that the United Arab Emirates have taken delivery of some 300 mm rocket systems which are launched from the Jobaria Defense Systems Multiple Cradle Launcher covered in a separate entry.

Contractor

China Precision Machinery Import & Export Corporation (CPMIEC).

China Precision Machinery Import & Export Corporation (CPMIEC) 302 mm WS-1B (4-round) artillery rocket system

Development

The China Precision Machinery Import & Export Corporation (CPMIEC) 302 mm WS-1B (4-round) Artillery Rocket System (ARS), is a further development of the 320 mm (4-round) WS-1 that is described in detail in an separate entry in *IHS Jane's Land Warfare Platforms: Artillery and Air Defence.*

Production of the WS-1 has now been completed and marketing is now concentrated on the WS-1B.

The latest 302 mm WS-1B ARS is based on the German Mercedes-Benz (6 × 6) forward control cross-country chassis that is manufactured under licence by China North Industries Corporation (NORINCO).

This is used for a wide range of military applications including the NORINCO 122 mm (40-round) Type 90 series of ARS, covered in a separate entry in *IHS Jane's Land Warfare Platforms: Artillery and Air Defence*, as well as towing a variety of artillery systems. The latter include the 155 mm gun-howitzer Type WA 021.

The major difference between the WS-1 and WS-1B is that the latter has smaller diameter rockets, which have a significant increase in range.

The latest WS-1B ARS has a new high-performance solid-propellant rocket motor and can be supplied fitted with different warheads to meet different users operational requirements.

To meet the requirements of the Turkish Land Forces Command (TLFC) the local company of Roketsan has developed and placed in quantity production a 300 mm (4-round) Artillery Rocket System which was first revealed several years ago.

This is understood to be based on the Chinese CPMIEC 302 mm WS-1B (4-round) Artillery Rocket System but customised to meet the operational requirements of the TLFC.

The Roketsan 30 mm (4-round) Artillery Rocket System is based on a MAN forward control (6 × 6) cross-country truck chassis and includes the T-300 MultiBarrel Rocket Launcher (MBRL) and the TR-300 artillery rocket which has a maximum range of 100,000 m.

The United Arab Emirates have taken into service the Jobaria Defense Systems Multiple Cradle Launcher (MCL) which can launch Roketsan 122 mm or 300 mm surface-to-surface rockets. Details of this system are provided in a separate entry in *IHS Jane's Land Warfare Platforms: Artillery and Air Defence.*

This was first shown in public in February 2013 and is understood to have entered service with the United Arab Emirates in 2011/2012.

The 302 mm WS-1B (4-round) rocket launcher platform is designated the HF-4 and is shown here with launcher elevated 0526850

QY-88B transport and loading truck showing hydraulic crane mounted to rear of fully enclosed cab and rockets in foreground 0526851

Although the WS-1 series is marketed by the CPMIEC, it is understood that the actual rockets are manufactured by the Sichuan Aerospace Industry Corporation (SCAIC).

In addition to being marketed by the CPMIEC this system has also been marketed overseas by Poly Technologies.

Thailand and China are co-operating on further development of the 302 mm WS-1 surface-to-surface rocket and additional details are provided under variants later in this entry.

Description

The WS-1B ARS is based on a Mercedes-Benz (6 × 6) cross-country truck chassis that has been manufactured under licence by China North Industries Corporation (NORINCO).

The layout of the 302 mm WS-1B ARS is similar to the larger calibre 320 mm (4-round) WS-1 system with the two-door fully enclosed forward control cab at the front and the powered launcher, which is offered with four or eight launcher tubes, mounted at the very rear. To the immediate rear of the cab is a fully enclosed cabin for additional crew members.

When deployed in the firing position, four hydraulically-operated stabilisers are lowered to the ground hydraulically to provide a more suitable firing platform.

One of these is mounted either side to the immediate rear of the cab with the other two stabilisers mounted at the very rear of the chassis.

Launcher elevation is powered from 0° to +60° at a speed of between 0.1° and 3° per second while traverse is powered at a speed of 30° left and right at 0.1 to 4° per second. Manual controls are also provided.

The 302 mm unguided surface-to-surface rocket consists of the warhead and its associated fuze, FG-43 solid-propellant rocket motor and the tail section with its four fixed fins.

The FG-43 is a single-chamber solid-propellant rocket motor with an advanced Hydroxy-Terminated PolyButadine (HTPB) composition rocket propellant.

The rocket can be fitted with two types of warhead, the ZDB-2 blasting warhead or an SZB-1 submunition. The ZDB-2 is a blast type warhead high-explosive, surrounded by steel balls and prefabricated fragments. The warhead weighs 150 kg of which 70 kg is high-explosive. It explodes with 26,000 fragments and is said to have an effective lethal radius 70 m.

The SZB-1 submunition warhead is optimised for use against area targets such as concentrations of tanks and other armoured vehicles. When over the target area the warhead is activated and attacks the vulnerable upper surfaces of the target. This warhead is said to be 1,380 mm long and contains 466 'bullets' with a dispersion area of 28,000 m^2. Each of the 'bullets' will penetrate 70 mm of conventional steel armour. It is assumed that these bullets are bomblets fitted with a single High-Explosive Anti-Tank (HEAT) warhead. It is not known if these bomblets are fitted with a self-destruct fuzing system.

According to the manufacture, the WS-1B takes about 20 minutes to prepare for action with information being transmitted from the DZ-88B firing command truck to the HF-4 rocket launcher.

The DZ-88B firing command truck contains the fire-control computer and simulation trajectory system, global positioning system, gyro theodolite direction system, infra-red ranging system, radio communication and data transmission system. A field meteorological detection system is also provided as in a communication control unit.

A typical WS-1B rocket battery consists of:

- 1 × DZ-88B firing command truck (6 × 6);
- 6 to 9 HF-4 rocket launcher truck (6 × 6);
- 6 to 9 QY-88B transport and loading truck.

The QY-88B transport and loading truck has a similar chassis and carries rockets which can be loaded using mechanical means to the launcher when the two elements are back to back.

The rockets are packed in individual boxes and are then removed and loaded onto the QY-88B transport and loading truck by a crane mounted on the vehicle. This hydraulic crane is mounted to the immediate rear of the enclosed cab.

The system can operate up to a maximum height of 3,000 m above sea level and in ambient temperatures from −30°C to +45°C and with a relative humidity of up to 98 per cent (at 15°C). It can also operate with a wind velocity of up to 8 m/s and in rain or snow of 3 mm/h.

WS-1B surface-to-surface rockets with warheads fitted 0526852

WS-1/WS-1B warheads

According to the Sichuan Aerospace Industry Corporation, the WS-1 and WS-1B have the same family of warheads, which consists of four different types to meet different user requirements. These are:

- Blasting warhead, with a total weight of 150 kg, containing 70 kg of explosive, which has 26,000 fragments and an effective lethal radius of 70 m. This is believed to be the ZDB-2 warhead mentioned in the text.
- Submunition warhead, weighing 152 kg, which contains 475 submunitions, each with an effective lethal radius of 7 m. This is assumed to be the SZB-1 warhead mentioned in the text.
- Cloud-blasting warhead, with a total weight of 70 kg, which creates a damage radius of 70 m with a lethal over-pressure radius of 50 m. This is assumed to be of the fuel air explosive type.
- Blast burning warhead, with a charge mass of 70 kg and an effective lethal radius of 70 m.

The Sichuan Aerospace Industry Corporation give the WS-1B rocket an overall length of 6.276 m and a maximum launch weight of 725 kg. The rocket has a maximum speed of Mach 5.2.

Chinese/Thailand 302 mm rocket development

In March 2013 Thailand's Defence Technology Institute (DTi) signed an memorandum of understanding with the Royal Thai Army to develop multiple launch rocket systems under the designations of the DTi-1 and DTi-1G which are based on the Chinese 302 mm WS1 series.

It is understood that Thailand purchased the technology for the DTi-1 in 2009 followed by technologies for the DTi-1G late in 2012.

It is expected that production of the DTi-1 will commenced in Thailand in 2013/2014 followed by eventual production of the guided DTi-1G which has increased accuracy for precision effect.

Specifications

WS-1B artillery rocket system

Calibre:	302 mm
Number of rockets:	4 or 8
Range:	
(maximum)	180,000 m
(minimum)	60,000 m
Maximum flight altitude:	60,000 m
Length:	
(complete rocket)	6.376 m
(rocket motor)	4.709 m
Weights:	
(complete rocket)	708 kg (at launch)
(rocket motor)	538 kg
(propellant)	370 kg
(warhead)	150 kg
Total impulse:	888 kN/s
Maximum thrust:	275 kN
Motor burn time:	5.3 s (maximum)
Maximum speed:	M 5.00 approx
Error of probability:	1 to 1.5%
Maximum acceleration:	40 *g*
Elevation:	0° to +60°
Traverse:	30° left and right

Notes:
Earlier sources have stated that the 302 mm rocket had a minimum range of 80,000 m and was 6.182 m long. Launch weight has also been quoted as 735 kg

CPMIEC 302 mm WS-1B (4-round) artillery rocket system launching a rocket (CPMIEC) 1151504

Status

Production as required. In service with China and Turkey (see Development). Production has been competed in Turkey for the Turkish Land Forces Command.

As mentioned under development it is understood that the Roketsan 300 mm rocket is also deployed by the United Arab Emirates with a different trailer mounted launcher. Thailand will manufacture the 302 mm rocket in the future in unguided DTi-1 and guided DTi-1G versions.

Contractor

China Precision Machinery Import & Export Corporation (CPMIEC).

China Precision Machinery Import & Export Corporation (CPMIEC) 300 mm (10-round) A100 multiple rocket system

Development

Early in 2000 it was revealed that China had developed a new 300 mm (10-round) Multiple Rocket System (MRS) to meet the operational requirements of the People's Liberation Army (PLA) called the A100.

As early as 1995, sources from Moscow stated that China was seeking a technology transfer of the latest Splav Russian 300 mm BM 9A52 (12-round) Smerch MRS which has the longest range of all of the Russian unguided surface-to-surface rocket systems.

The 300 mm BM 9A52 (12-round) Smerch MRS was developed by the Splav Scientific Production Concern to meet the operational requirements of the Russian Army and was first fielded in 1987.

The prime contractor for the new 300 mm (10-round) A100 MRS is now known to be China Precision Machinery Import & Export Corporation (CPMIEC).

According to the CPMIEC, their new A100 MRS is now in quantity production and in service with the People's Liberation Army (PLA). It is also being offered on the export market. As of March 2013 there are no known exports of the A100 system.

In 2008 it was stated that NORINCO was marketing a possible further development of this system under the name of the AR2 300 mm Multiple Launch Rocket System.

This system is also based on an 8 × 8 chassis and supported by a resupply vehicle on a similar chassis.

AR2 fires at least three types of surface-to-surface rocket which are designated the BRC3 (maximum range 70 km and fitted with a cargo type warhead), BRE2 (maximum range 130 km and fitted with a High-Explosive (HE) warhead) and BRC4 (maximum range of 130 km and fitted with a cargo warhead). Combat weight of the system is quoted as about 42 tonnes.

Details of the AR2, which is in service with the PLA, are provided in a separate entry in *IHS Jane's Land Warfare Platforms: Artillery and Air Defence.*

Chinese A100 300 mm (10-round) MRS on a new 8 × 8 cross-country truck launching a rocket (CPMIEC) 1151505

300 mm (10-round) A100 multiple rocket system deployed in the firing position from the rear and showing stabilisers lowered to the ground (Poly Technologies) 1365049

Description

The new Chinese 300 mm (10-round) A100 system is very similar to the Russian 300 mm BM 9A52 (12-round) Smerch in rocket diameter, range, rate of fire and overall appearance.

The standard production Russian 300 mm BM 9A52 (12-round) Smerch is based on the older MAZ-543M (8 × 8) chassis but the Chinese version is based on a different locally developed cross-country 8 × 8 chassis with a different stabiliser design.

A100 fires unguided 300 mm rockets with a minimum range of 40 to 50 km and a maximum range of 85 to 120 km. The current production warhead weighs 200 kg and carries an undisclosed quantity of anti-armour/anti-personnel submunitions. It is understood that these are fitted with a High-Explosive Anti-Tank (HEAT) warhead. Other warheads are under development.

The solid propellant rockets are 7.276 m long, have a total weight of 780 kg and a projected shelf-life of 10 years. A new family of unguided rockets are under development by CPMIEC. These will have a minimum range of 70 km and a maximum range of 180 km, which can be fired from the A100 launcher.

The launcher and its associated resupply vehicle share the same 8 × 8 WS 2400 cross-country chassis that features a fully enclosed forward control cab. This has a combat weight of about 22 tonnes and a maximum road speed of about 60 km/h.

The 8 × 8 chassis features steering on the front four wheels and a central tyre pressure regulation system that allows the driver to adjust the tyre pressure to suit the terrain being crossed.

When deployed in the firing position, the A100 (8 × 8) launcher is stabilised by four hydraulically operated stabilisers that are lowered to the ground, two on either side.

On coming to a halt, the system takes six minutes to be prepared for action. Target information is provided to the launcher by the battery fire-control system installed on a 6 × 6 chassis.

Rocket accuracy is enhanced as the platform is fitted with a Global Positioning System (GPS) which also reduces target response times.

A complete salvo of 10 unguided surface-to-surface rockets can be launched in 60 seconds.

The powered launcher, mounted at the rear of the chassis, is automatically laid on to the target from within the cab.

Once these rockets have been launched the A100 would normally redeploy to a reloading area where the rocket resupply vehicle would be waiting.

According to the manufacturer, the system takes two minutes to come out of action. This is based on the same WS 2400 8 × 8 chassis and new rockets in their launch tubes are then reloaded using the onboard hydraulic crane. According to the CPMIEC, a complete system takes 20 minutes to be reloaded with 10 new rockets.

The latest A100 is being marketed alongside the older CPMIEC WS-1 and WS-1B (4-round) launchers that are installed on a Mercedes-Benz 6 × 6 truck chassis manufactured under licence by NORINCO.

WS-1 launches rockets with a maximum range of 80 km, while the more recent WS-1B launches rockets with a maximum range of 180 km. At least two types of warhead are available for the WS-1B, the ZDB-2B blast fragmentation and the SZB-1 that carries 466 submunitions.

Chinese sources indicated that the new A100 MRS has been developed to bolster the PLA's surface-to-surface firepower capability against the outlying islands of Taiwan and the Penghu Archipelago.

Poly Technologies A100

This system is also being marketed by Poly Technologies of China but they are quoting a maximum range of 80 km, a minimum range of 40 km with a warhead weight of 235 kg.

The system is claimed to come into action in eight minutes, fire a complete salvo of 10 rockets in about 60 seconds and come out of action in about three minutes. The system takes about 20 minutes to reload.

Specifications

A100 multiple rocket system

Calibre:	300 mm
Number of launch tubes:	10
Length of rocket:	7.276 m
Weight:	
(rocket total)	840 kg
(warhead)	235 kg
(system loaded)	22,000 kg
Range:	
(maximum)	85-120 km
(minimum)	40-50 km
Maximum road speed:	80 km/h

Status

Production as required. In service with the People's Liberation Army. There are no known exports of this system.

It is possible that marketing of this system has ceased in favour of the NORINCO AR2 300 mm (12-round) multiple rocket system.

Contractor

China Precision Machinery Import & Export Corporation (CPMIEC).

It should be noted that this system has also been marketed by Poly Technologies on the export market.

NORINCO AR3 370 mm/300 mm multiple launch rocket system

Development

China North Industries Corporation (NORINCO) is now marketing the AR3 370 mm/300 mm Multiple Launch Rocket System (MLRS) based on an 8 × 8 cross-country truck chassis.

It is understood that development of the AR3 MLRS is complete and production can commence when orders are placed.

As of March 2013 there are no known sales of the NORINCO AR3 370 mm/300 mm MLRS.

The chassis appears identical to that used in the NORINCO AR1A 300 mm (10-round) MLRS that is covered in a separate entry in *IHS Jane's Land Warfare Platforms: Artillery and Air Defence.*

The main difference is that the AR3 can be fitted with eight (4 + 4) 370 mm rocket tubes or 10 (5 + 5) 300 mm rockets with the latter being the same as that used in the NORINCO AR1A and AR2 systems.

These rockets are in pods which enable the system to be reloaded with replacement pods much more rapidly that loading the rockets one by one.

As the AR3 MLRS system can be deployed with two types of rocket system this provides the user with considerable operational flexibility.

Description

NORINCO AR3 launcher has a four door fully enclosed air conditioned cab at the front with an additional fully enclosed crew cab to the immediate rear of the main cab with the system normally being operated by a crew of four.

Mounted at the very rear of the chassis is the turntable mounted launcher with powered elevation and traverse. When travelling this is traversed to the front and lowered into the horizontal position.

The launcher is normally laid onto the target from within the cab but a manual backup sight is fitted as standard and mounted externally on the left side of the launcher.

The on board navigation and fire control system is claimed to provide a laying accuracy of 1 mil.

To provide a more stable firing platform two stabilisers are lowered to the ground by remote control with these being situated one either side between the third and fourth road wheels.

AR3 MLRS with 2 × 4 370 mm launcher tubes deployed in firing system with launcher traversed left and hydraulic stabilisers lowered to the ground between the last two road stations either side (NORINCO) 1452642

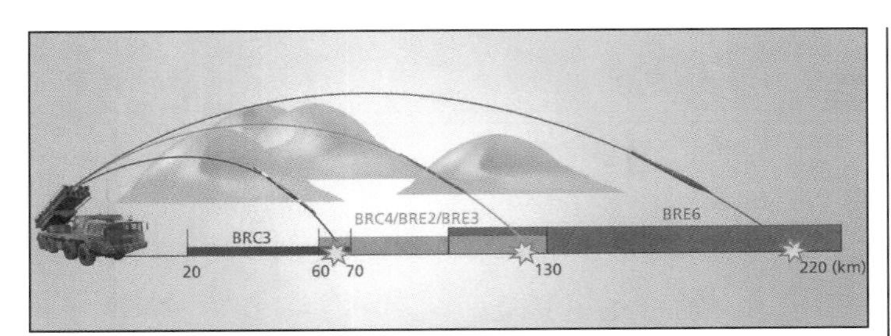

Range table showing capability of the AR3 MLRS which can fire rockets with a minimum range of 20 km out to a maximum range of 220 km (NORINCO) 1452643

The system can be fitted with pods that can launch solid propellant rockets and once these have been fired the system would normally deploy to another position for reloading to avoid counter-battery fire.

According to NORINCO this system can fire the following rockets:

Designation	Calibre	Range	Type
BRC3	300 mm	70 km	cargo (unguided)
BRC4	300 mm	130 km	cargo (unguided)
BRE2	300 mm	130 km	high-explosive (unguided)
BRE3	300 mm	130 km	guided rocket
BRE6	370 mm	220 km	guided rocket

BRC3 rocket is fitted with a cargo type warhead that carries 623 dual purpose anti-personnel/anti-tank bomblets and has a maximum range of 70 km and a minimum range of 20 km which is the shortest range of all of these 300 mm rockets.

BRE2 rocket is fitted with a 180 kg high explosive warhead and has a maximum range of 130 km and is understood to have a minimum range of 60 km.

BRC4 rocket carries 414 dual purpose anti-personnel/anti-tank bomblets which has a maximum range of 130 km and a minimum range of 60 km.

The BRE3 and longer range BRE6 guided rockets are claimed to have a circular error of probability of 50 m.

According to NORINCO, reaction time of the system at battalion level is being quoted as about 20 seconds and at battery level about 15 seconds.

It can be deployed up to a maximum altitude of 3,000 m above sea level.

AR3 would typically be deployed a battalion or battery level for maximum effect but it could also be deployed at platoon or individual level if required by the tactical situation.

It would also be integrated into a fire-control system linked with a meteorological survey system, unmanned aerial vehicle and other target acquisition systems.

Specifications

	AR3 300 mm MLRS
Dimensions and weights	
Crew:	4
Weight	
combat:	45,000 kg
Mobility	
Configuration	
running gear:	wheeled
layout:	8 × 8
Speed	
max speed:	60 km/h
Range	
main fuel supply:	650 km
Firepower	
Armament:	10 × 300 mm rocket system (5 + 5)

	AR3 370 mm MLRS
Dimensions and weights	
Crew:	4
Weight	
combat:	45,000 kg
Mobility	
Configuration	
running gear:	wheeled
layout:	8 × 8
Speed	
max speed:	60 km/h
Range	
main fuel supply:	650 km
Firepower	
Armament:	8 × 370 mm rocket system (4 + 4)

Status

Development complete.

Contractor

Chinese state factories; China North Industries Corporation (NORINCO).

NORINCO AR2 300 mm (12-round) multiple launch rocket system

Development

China North Industries Corporation (NORINCO) is now marketing the AR2 300 mm (12-round) Multiple Launch Rocket System (MLRS) based on an 8 × 8 cross-country truck chassis.

This is understood to already be in service with the People's Liberation Army (PLA) under the local designation of the PHL03 (Type 03) and may entered front line service in 2004/2005.

In appearance the AR2 300 mm (12-round) MLRS is very similar to the earlier China Precision Machinery Import-Export Corporation A-100 300 mm (10-round) MLRS which is covered in detail in a separate entry in *IHS Jane's Land Warfare Platforms: Artillery and Air Defence.*

This in turn is very similar to the Russian Spav 300 mm BM 9A52 (12-round) Smerch which was supplied to China in small numbers. The A100 fires 300 mm rockets with a shorter range than the more recent AR2.

Unconfirmed reports have indicated that Pakistan has ordered sufficient of these NORINCO AR2 300 mm (12-round) Multiple Launch Rocket Systems for deployment by one regiment.

These would be a counter to the Indian Army deployment of the locally-produced 214 mm (12-round) Pinacha and has recently taken delivery of the Russian Spav 300 mm BM 9A52 (12-round) Smerch systems.

Description

NORINCO AR2 launcher has a four door fully-enclosed air conditioned cab at the front with an additional fully-enclosed crew cab to the immediate rear with the system normally being operated by a crew of four. This cab is fitted with a roof-mounted air conditioning system.

Mounted at the very rear of the chassis is the 300 mm (12-round) launcher with powered elevation and traverse. When travelling this is traversed to the front and lowered into the horizontal position.

The launcher is normally laid onto the target from within the cab but a manual backup sight is fitted as standard and mounted externally on the left side of the launcher.

To provide a more stable firing platform two stabilisers are lowered to the ground by remote-control with these being situated one either side between the third and fourth roadwheels.

AR2 fires at least three types of unguided solid propellant surface-to-surface rocket that are designated BRC3, BRE2 and BRC4.

BRC3 rocket is fitted with a cargo type warhead that carries 623 dual purpose anti-personnel/anti-tank bomblets and has a maximum range of 70 km.

BRE2 rocket is fitted with a 190 kg high explosive warhead and has a maximum range of 130 km.

BRC4 rocket carries 414 dual purpose anti-personnel/anti-tank bomblets and has a maximum range of 130 km.

NORINCO has also stated that this system can fire the BRE3 300 mm guided rocket out to a maximum range of 130 km, although the launcher would have to have the fire-control system modified.

According to NORINCO, reaction time of the system at battalion level is being quoted as about 20 seconds and at battery level about 15 seconds. It can be deployed up to a maximum altitude of 3,000 m above sea level.

Once the launcher has carried out a fire mission, it would normally rapidly move away to avoid counterbattery fire where reloading would take place.

AR2 300 mm (12-round) MLRS being reloaded by its dedicated resupply vehicle based on a similar chassis (NORINCO) 1332599

AR2 300 mm (12-round) MLRS deployed in the firing position with launcher elevated and hydraulic stabilisers lowered to the ground (NORINCO) 1365037

According to NORINCO, this system takes less than five minutes to come into action and launch the first rocket. A complete salvo of 12 rockets can be launched in less than 60 seconds and time out of action is being quoted as one minute.

The AR2 is supported by a dedicated resupply vehicle on a similar 8 × 8 cross-country chassis which is provided with two hydraulically-operated stabilisers.

For reloading, the resupply vehicle backs up to the AR2 launcher and replacement rockets are reloaded one at a time using a hydraulic crane and rammer.

AR2 would typically be deployed a battalion or battery level for maximum effect but it could also be deployed at platoon or individual level if required by the tactical situation.

It would also be integrated into a fire-control system linked with a meteorological survey system, unmanned aerial vehicle and other target acquisition systems.

Variant

NORINCO AR1A 300 mm (10-round) multiple launch rocket system

This is similar to the NORINCO AR2 300 mm (12-round) multiple launch rocket system but is mounted on a different 8 × 8 chassis fitted with a forward control cab.

The AR1A (10-round) multiple launch rocket system fires the same family of rockets as the NORINCO AR2 300 mm (12-round) system and details are provided in a separate entry in *IHS Jane's Land Warfare Platforms: Artillery and Air Defence.*

NORINCO AR3 MLRS

This is similar to the AR2 but can launch 370 mm or 300 mm rockets, and details are provided in a separate entry in *IHS Jane's Land Warfare Platforms: Artillery and Air Defence.*

Specifications

	AR2 300 mm (12-round) MLRS
Dimensions and weights	
Crew:	4
Weight	
combat:	44,500 kg
Mobility	
Configuration	
running gear:	wheeled
layout:	8 × 8
Speed	
max speed:	60 km/h
Range	
main fuel supply:	650 km
Firepower	
Armament:	12 × 300 mm rocket system
Main weapon	
maximum range:	70,000 m (BRC3)
minimum range:	20,000 m (BRC3)
maximum range:	130,000 m (BRE2)
minimum range:	60,000 m (BRE2)
maximum range:	130,000 m (BRC4)
minimum range:	60,000 m (BRC4)

Status

Production as required. In service with China (96 estimated) and understood to be on order for Pakistan. Some sources have stated 36 systems have been ordered/delivered, but may be 300 mm AR1.

Contractor

Chinese state factories. China North Industries Corporation (NORINCO).

NORINCO AR1A 300 mm (10-round) multiple launch rocket system

Development

China North Industries Corporation (NORINCO) is now marketing the AR1A 300 mm (10-round) Multiple Launch Rocket System (MLRS) that was first shown in 2008/2009.

This is being marketed alongside the NORINCO AR2 300 mm (12-round) MLRS that is based on a slightly different 8 × 8 cross-country truck chassis which is covered in a separate entry in *IHS Jane's Land Warfare Platforms: Artillery and Air Defence.*

This is already be in service with the Peoples Liberation Army (PLA) under the local designation of the PHL-03 (Type 03) and may have entered front line service in 2004/2005.

Development of the AR1A 300 mm (10-round) MLRS is complete but as of March 2013 there are no known export sales. As far as it is known this system is not in service with the Peoples Liberation Army (PLA).

The main advantage of the AR1A over the earlier AR2 is that the former has two pods each of five rockets and when these are launched each pod is removed and replaced by a new pod of rockets. The AR2 has 12 × 300 mm rocket tubes which have to be reloaded individually.

AR1A 300 mm (10-round) MLRS with launcher traversed right. Note that stabilisers are not lowered to the ground (NORINCO) 1365048

AR1A 300 mm (10-round) MLRS with launcher elevated. Note that stabilisers are not lowered to the ground (NORINCO) 1365047

Description

The NORINCO AR1A launcher has a four door fully enclosed air conditioned cab at the front with an additional fully enclosed crew cab to the immediate rear with the system normally being operated by a crew of four.

Mounted at the very rear of the chassis is the 300 mm (10-round) launcher with powered elevation and traverse. When travelling this is traversed to the front and lowered into the horizontal position.

The launcher is normally laid onto the target from within the cab but a manual backup sight is fitted as standard and mounted externally on the left side of the launcher.

To provide a more stable firing platform two stabilisers are lowered to the ground by remote-control with these being situated one either side between the third and fourth road wheels.

AR2 fires at least three types of unguided solid propellant surface-to-surface rocket that are designated BRC3, BRE2 and BRC4.

BRC3 rocket is fitted with a cargo type warhead that carries 623 dual purpose anti-personnel/anti-tank bomblets and has a maximum range of 70 km. Each of these bomblets will penetrate 50 mm of armour.

BRE2 rocket is fitted with a 180 kg high explosive warhead and has a maximum range of 130 km.

BRC4 rocket carries 414 dual purpose anti-personnel/anti-tank bomblets which has a maximum range of 130 km. Each of these bomblets will penetrate 50 mm of armour.

It is considered that other types of warhead are under development as well as ways to make the system more accurate and it is considered possible that the PLA uses 300 mm rockets that have not so far been released for export.

According to NORINCO, reaction time of the system at battalion level is being quoted as about 20 seconds and at battery level about 15 seconds. It can be deployed up to a maximum altitude of 3,000 m above sea level.

The AR1A s supported by a dedicated resupply vehicle on a similar 8 × 8 cross-country chassis which is provided with two hydraulically operated stabilisers.

Replacement pods of five rockets are rapidly reloaded from the dedicated resupply vehicle which is fitted with a hydraulic crane.

The AR1A would typically be deployed a battalion or battery level for maximum effect but it could also be deployed at platoon or individual level if required by the tactical situation.

It would also typically integrated into a fire control system linked with a meteorological survey system, unmanned aerial vehicle and other target acquisition systems.

NORINCO AR3 370 mm/300 mm Multiple Launch Rocket System

Details of this are provided in a separate entry in *IHS Jane's Land Warfare Platforms: Artillery and Air Defence* and this can fire either 300 mm or 370 mm rockets.

This is being offered on the export market by NORINCO but as of March 2013 it is understood not to have entered production or service.

Specifications

	AR1A 300 mm (10-round) multiple launch rocket system
Dimensions and weights	
Crew:	4
Weight	
combat:	42,000 kg
Mobility	
Configuration	
running gear:	wheeled
layout:	8 × 8
Speed	
max speed:	60 km/h
Range	
main fuel supply:	650 km
Firepower	
Armament:	12 × 300 mm rocket system
Main weapon	
maximum range:	70,000 m (BRC3)
minimum range:	20,000 m (BRC3)
maximum range:	130,000 m (BRE2)
minimum range:	60,000 m (BRE2)
maximum range:	130,000 m (BRC4)
minimum range:	60,000 m (BRC4)

Status
Development complete. Ready for production.

Contractor
Chinese state factories. China North Industries Corporation (NORINCO).

NORINCO AR1 300 mm (8-round) multiple launch rocket system

Development
China North Industries Corporation (NORINCO) has been marketing Multiple Launch Rocket Systems (MLRS) for many years including the AR3 370 mm/300 mm, AR2 300 mm (12-round system) and the AR1A 300 mm (10-round system).

Details of these systems are provided in a separate entry in *IHS Jane's Land Warfare Platforms: Artillery and Air Defence*.

One of the latest NORINCO MLRS is the AR1 300 mm (8-round) based on a Mercedes-Benz (8 × 8) cross-country truck chassis which has a combat weight of 32.5 tonnes and a cruising range of up to 650 km.

As the AR1 300 mm (8-round) system is lighter than the AR3, AR2 and AR1A it is claimed to have greater strategic mobility and increased mission flexibility.

Development of the AR1 300 mm (8-round) system is complete but as of March 2013 there are no known sales.

Description
As previously stated the AR1 300 mm (8-round) system is based on a Mercedes-Benz (8 × 8) chassis fitted with a two-door forward control cab and the turntable mounted launcher at the rear.

Instead of reloading the individual launcher tubes after a fire mission, new pods of four 300 mm rockets are reloaded from the resupply vehicle using a crane.

This is similar in concept to the latest Russian Splav 300 mm CV 9A52-4 (6-round) system.

The AR1 300 mm (8-round) MLRS has an onboard computerised fire-control system and land navigation system.

System reaction time at the battalion level is quoted as 20 seconds and reaction time at the battery level is quoted as 15 seconds.

The AR1 300 mm (8-round) system can fire the following types of rockets which have a minimum range of 20 km and a claimed CEP of about 50 m.

It is considered that the latter relates to the guided BRE3 rather than the unguided rockets.

AR1 300 mm (8-round) multiple launch rocket system integrated onto the rear of a Mercedes-Benz (8 × 8) cross-country chassis and shown here in the travelling configuration with launcher in the horizontal position (NORINCO)
1365820

Designation	Maximum range	Warhead type
BRC3	70 km	cargo
BRC4	130 km	cargo
BRE2	130 km	HE
BRE3	130 km	guided

BRC3 cargo rocket carries 623 bomblets each of which is claimed to be able to penetrate 50 mm of conventional steel armour and is designed to neutralise light armoured vehicles as well as being effective against soft skin vehicles and troops.

The BRC4 cargo rocket has more propellant for increased range but carries only 480 bomblets.

BRE2 has a pre-fragmented high-explosive warhead weighing 190 kg that is claimed to have a lethal radius of 100 m.

BRE3 is the guided version to provide a precision strike capability.

Specifications

	AR1 300 mm (8-round) MLRS
Dimensions and weights	
Crew:	4
Weight	
combat:	32,500 kg
Mobility	
Configuration	
running gear:	wheeled
layout:	8 × 8
Power-to-weight ratio:	laden
Speed	
max speed:	130 km/h
Range	
main fuel supply:	650 km
Gearbox	
type:	manual
Clutch	
type:	centrifugal
Firepower	
Armament:	8 × 300 mm rocket system (2 × 4)
Main weapon	
maximum range:	70,000 m (BRC3)
maximum range:	130,000 m (BRC4)
maximum range:	130,000 m (BRE2)
maximum range:	130,000 m (BRE3)

Status
Development complete. Ready for production on receipt of orders.

Contractor
Chinese state factories; China North Industries Corporation (NORINCO).

NORINCO 425 mm Type 762 (2-round) mineclearing rocket system

Development
This is the largest of the China North Industries Corporation (NORINCO) heavy mineclearing rocket systems. It is mounted on a cradle assembly carried on the rear of a tracked vehicle hull which is very similar to that of the NORINCO Type 83 152 mm self-propelled gun-howitzer. This hull is also used for the Type 1989 120 mm tank destroyer.

As far as it is known, neither of these two systems nor the NORINCO 425 mm Type 762 (2-round) mineclearing rocket system have been exported.

It is understood that there has been no recent production of these full-tracked hull by NORINCO.

It is considered possible that these launchers could be removed from their original hull and be integrated on a more recent hull.

Description
The hull of the NORINCO 425 mm Type 762 (2-round) mineclearing rocket system is of all welded steel armour that provides the occupants with protection from small arms fire and shell splinters. Highest level of protection is probable over the frontal arc.

The driver is seated at the front of the hull on the left side with the power pack, consisting of the diesel engine, transmission and cooling system, to the right.

The remainder of the crew are seated in the fully enclosed and protected compartment that is located at the rear.

The warhead section of the rocket weighs 621 kg and probably contains a sectionalised payload of fuel-air explosive containers, in order to create the blast overpressure required to clear a minefield. The range is from 800 to 1,000 m using a 129 kg, fin-equipped, solid propellant rocket motor section, to attain a maximum flight velocity of 98 m/s.

Each rocket can clear a passage through a minefield 130 m long and between 12 and 22 m wide. The vehicle is equipped with a 7.62 mm machine gun for self-defence purposes which can be dismounted for use on a special anti-aircraft tripod.

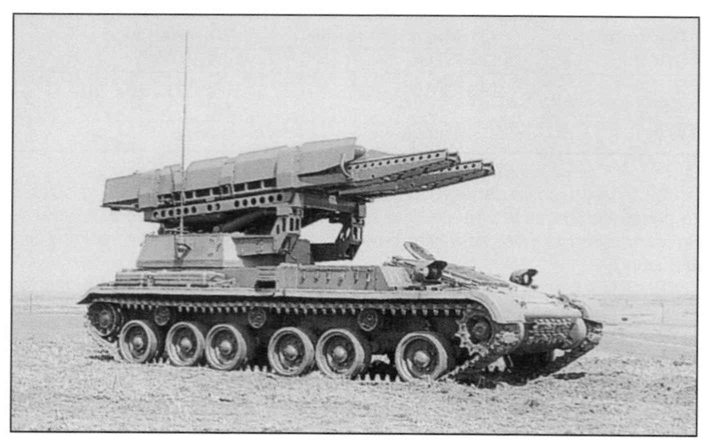

NORINCO 425 mm Type 762 mineclearing rocket system deployed in firing position 0058454

Firing ports for rifles are located on either side of the crew compartment and at the rear where a door is provided for access. The launcher system is hydraulically controlled.

Other systems available for the crew include Type 803 intercom system, Type 889 radio set, panoramic sight Type WG002, 0.5 m stereo-optical rangefinder system and a Type TYG-1 night vision set for the driver.

Specifications

	Type 762 mine clearing rocket system
Dimensions and weights	
Crew:	4
Length	
overall:	7.162 m
hull:	6.405 m
Width	
overall:	3.236 m
Height	
overall:	3.2 m
Ground clearance	
overall:	0.45 m
Track	
vehicle:	2.62 m
Track width	
normal:	480 mm
Length of track on ground:	4.601 m
Weight	
combat:	27,000 kg
Ground pressure	
standard track:	0.62 kg/cm²
Mobility	
Configuration	
running gear:	tracked
Power-to-weight ratio:	19.3 hp/t
Speed	
max speed:	55 km/h
cross-country:	32 km/h (est.)
Range	
main fuel supply:	450 km
Fording	
without preparation:	1.3 m
Gradient:	60%
Side slope:	40%
Vertical obstacle	
forwards:	0.7 m
Trench:	2.7 m
Engine	Type 12150L, diesel, 520 hp
Firepower	
Armament:	2 × turret mounted 425 mm rocket system
Armament:	1 × 7.62 mm (0.30) machine gun
Ammunition	
7.62/0.30:	500
Main weapon	
maximum range:	1,000 m (est.)
Main armament traverse	
angle:	180° (90° left, 90° right)
Main armament elevation/depression	
armament front:	+45°/+5°
Range-finding device:	laser
Survivability	
Armour	
hull/body:	steel
turret:	steel

Status

Production complete. There are no known export sales of this system. In service with the People's Liberation Army. No longer marketed by NORINCO.

Contractor

Chinese state arsenals. China North Industries Corporation (NORINCO).

NORINCO 305 mm (8-round) Type 79 minelaying rocket system

Description

Mounted on the same CA-30A (6 × 6) truck chassis as the China North Industries Corporation (NORINCO) 305 mm fuel air explosive minesweeping rocket system and NORINCO 284 mm Type 74 minelaying rocket system, the NORINCO 305 mm (8-round) Type 79 minelaying rocket system uses the same welded tubular steel launcher assembly frame of the former system instead of the rails of the Type 74.

The elevation and depression of the assembly is +4° to +50°, the traverse is 95° left and 45° right. The sighting system is mounted on the left side of the launcher.

The Type 79 minelaying rocket shell carries a payload of 10 Type 69 plastic anti-tank mines. A ring stabiliser is fitted to the tail fins. Maximum range is 2,600 m. Four vehicles can lay a 600 × 350 m minefield. Firing interval between the unguided solid propellant rockets is 1.75 seconds.

Ready for launch each Type 79 minelaying rocket weighs 156 kg to which the rocket motor adds an additional 77 kg to bring the total up to 233 kg.

The actual rocket has an overall length of 2.927 m and a maximum chamber pressure is being quoted as 220/260 kg/cm².

Launch velocity is quoted as 45 m/s with maximum velocity being 185 m/s. It can be used in ambient temperature conditions from –40°C to +40°C.

Production of this system and its associated old design 6 × 6 cross-country truck chassis has been completed and it is no longer marketed.

It is understood that the actual 305 mm rockets used in this system were manufactured by another contractor rather than NORINCO.

Specifications

	NORINCO 305 mm (8-round) Type 79 minelaying rocket system
Dimensions and weights	
Length	
overall:	6.918 m
Width	
overall:	2.20 m
Height	
overall:	3.00 m
Weight	
combat:	8,730 kg
Mobility	
Configuration	
running gear:	wheeled
layout:	6 × 6
Speed	
max speed:	65 km/h
Fording	
without preparation:	0.85 m
Firepower	
Armament:	8 × 305 mm rocket system (rails)

NORINCO 305 mm (8-round) Type 79 minelaying rocket system with launcher elevated 0021039

	NORINCO 305 mm (8-round) Type 79 minelaying rocket system
Main weapon	
maximum range:	2,600 m
muzzle velocity:	45 m/s
Rate of fire	
rapid fire:	1/1.75 s
Main armament elevation/depression	
armament front:	+50°/+4°

Status
Production complete. In service with the People's Liberation Army. NORINCO has ceased marketing of this system.

Contractor
Chinese state factories. China North Industries Corporation (NORINCO).

NORINCO 305 mm (8-round) Fuel Air Explosive minesweeping rocket system

Development
This system was developed to meet the requirements of the Peoples Liberation Army to clear minefields ahead of advancing troops. Production has ceased and is no longer marketed by NORINCO.

It is assumed that this system is still in service with the People's Liberation Army and may now have been integrated onto a new 6 × 6 chassis.

It is considered possible that this system could have been replaced in PLA service by a more capable mine clearing system.

Description
Mounted on the same CA-30A (6 × 6) chassis as the China North Industries Corporation (NORINCO) 284 mm Type 74 and NORINCO 305 mm Type 79 minelaying rocket systems, the NORINCO 305 mm (8-round) Fuel Air Explosive (FAE) minesweeping rocket system uses the same welded tubular steel launcher assembly frame of the latter system instead of the rails of the NORINCO Type 74.

When deployed in the firing position no stabilisers are used. The sighting system is fitted on the left side of the launcher.

The elevation and depression of the assembly is +4° to +50° with the traverse from 95° left to 45° right. Elevation and traverse are manual.

The unguided solid propellant FAE rocket is 2.927 m long, weighs 156 kg at launch and has a flight velocity of 185 m/s. A ring stabiliser is fitted to the tail fins. A single round can clear an area 48 m in diameter of anti-personnel mines and thick grass/undergrowth.

When used as an anti-personnel weapon the round can kill and/or wound all exposed personnel within an area of 24 to 60 m in diameter.

The FAE warhead is ignited over the target to create a significant overpressure to neutralise the mines. Maximum range is 2,700 m. For transportation, the rocket and its warhead are carried in separate boxes with the rocket box weighing 71 kg and warhead box weighing 152 kg. The system can be used in ambient temperatures of from –15°C to +50°C.

Specifications

	NORINCO 305 mm (8-round) FAE minesweeping rocket system
Dimensions and weights	
Length	
overall:	2.927 m
Diameter	
body:	305 mm
Weight	
less warhead:	156 kg
Performance	
Speed	
max speed:	666 km/h
Range	
max:	2.7 km

Status
Production complete. In service with People's Liberation Army. There are no known export sales of this system. NORINCO is no longer marketing this system. As the chassis is now very old it is possible that the launcher could in the future be fitted to a more recent 6 × 6 truck chassis.

Contractor
Chinese state factories. China North Industries Corporation (NORINCO).

NORINCO 284 mm (10-round) Type 74 multiple rocket minelaying system

Development
The People's Liberation Army has several minelaying heavy rocket systems in service including the China North Industries Corporation (NORINCO) 284 mm (10-round) Type 74 multiple rocket minelaying system, which is mounted on the rear platform of a CA-30A (6 × 6) truck.

NORINCO 284 mm Type 74 (10-round) rocket minelayer on Chinese (6 × 6) CA-30A truck chassis with stabilisers lowered to the ground 0500634

Production of the NORINCO 284 mm (10-round) Type 74 multiple rocket minelaying system is now complete and it is no longer marketed by NORINCO. It is considered possible that it has been replaced by a new generation minelaying system.

It is understood that the actual 284 mm unguided rockets used by this system were supplied by another contract rather than NORINCO.

As far as it is known this system is still in service with the People's Liberation Army but these may have been removed from their older chassis and integrated on a new 6 × 6 or 8 × 8 chassis.

Description
This launcher comprises a 4 m long five-rail I-shaped frame system for 10 rockets with a team of four required to load each unguided rocket manually. The fin-stabilised Type 74 rockets each carry 10 plastic-cased Type 69 or Type 72 anti-tank mines in a bulbous warhead.

The Type 74 launchers are normally deployed in batteries of four, which can lay a single 400 × 400 m minefield with one salvo. The maximum range of the rocket is 1,500 m and all 10 284 mm unguided rockets can be fired within 15 seconds.

The usual crew totals six, four of whom sit behind the driver's cab on an open bench seat. A replacement wheel and tyre is stowed in the vertical position to the immediate rear of the cab on the right side.

For firing, the launcher has two stabiliser legs that are lowered by the vehicle crew using the manual controls at the rear of the chassis and steel shutters which are used to protect the cab windows.

A later solid propellant rocket type used is the NORINCO 284 mm Type 85 mine-laying rocket shell. This has a maximum range of 3,500 m and contains 10 anti-tank mines.

A complete battery salvo has a laying zone of 700 × 500 m. The unguided solid propellant rockets are 2.6 m long and weigh 156 kg at launch.

The rockets are claimed to have a maximum velocity of 120 m/s.

It is understood that there has been no production of the CA-30A (6 × 6) truck for some years. If production was started again it is considered that a more recent 6 × 6 cross-country chassis would be used for this system.

Specifications

	NORINCO 284 mm (10-round) Type 74 multiple rocket minelaying system
Dimensions and weights	
Crew:	6
Length	
overall:	6.44 m
Width	
transport configuration:	2.27 m
Height	
transport configuration:	2.93 m
Weight	
standard:	6,954 kg
combat:	8,780 kg
Mobility	
Configuration	
running gear:	wheeled
layout:	6 × 6
Speed	
max speed:	40 km/h
Firepower	
Armament:	10 × 284 mm rocket system
Main weapon	
maximum range:	1,500 m
muzzle velocity:	34.6 m/s

	NORINCO 284 mm (10-round) Type 74 multiple rocket minelaying system
Main armament traverse	
angle:	135° (90° left, 45° right)
Main armament elevation/depression	
armament front:	+48°/+7°

Status

Production complete. In service with the People's Liberation Army. There are no known export sales of this system. Production of this system is no longer being marketed by NORINCO.

Contractor

Chinese state factories. China North Industries Corporation (NORINCO).

NORINCO 273 mm (8-round) WM-80 multiple rocket system

Development

This China North Industries Corporation (NORINCO) 273 mm (8-round) WM-80 multiple rocket system has been designed for use in brigade (regiment), battalion and battery units.

According to the United Nations Report on Conventional Arms for 2001, between 1992 and 2010 there was only one export sale of the WM-80. This was to Armenia who took delivery of four units in 1999.

Given that the typical life of a solid propellant rocket when stored in ideal conditions is between 10 and 15 years, if Armenia has not taken delivery of replacement rockets then these original WM-80 systems may now be non-operational.

According to NORINCO, a complete WM-80 multiple rocket system would typically consist of four subsystems, the Firepower system, Fire Command-and-Control System, Reconnaissance System and Technology Service System.

These are further broken down. The Firepower System covers the rocket launcher, ammunition carrier and the actual rockets. The Fire Command-and-Control System includes the battalion command vehicle, battery command vehicle and meteorological radar. The Reconnaissance System includes reconnaissance vehicle and unmanned aerial vehicle. The Technology Service System includes the mechanical maintenance vehicle and the electronics maintenance vehicle.

Each brigade would typically have four battalions, each with three batteries. A total of six launcher vehicles and six resupply vehicles would be attached to each battery.

The standard combat unit is considered to be the battalion with the customer choosing his own Table Of Equipment (TOE) if he so wishes.

Fire-control and command duties throughout the brigade are accomplished by use of an automatic distribution network. This consists of main computers located at the brigade, battalion and battery HQ command levels with various terminals and radio/wire communications links to automatically distribute the data as directed.

To undertake the command-and-control functions from brigade level downwards the following assets are attached:

- At the brigade headquarters level – a Brigade Fire-Control and Command Vehicle (to act as the Brigade Command Station) and a meteorological radar unit;
- At each battalion headquarters level – a Battalion Fire-Control and Command Vehicle (to act as the Battalion Command Station) and a surveillance/spotting radar set;
- At each battery headquarters level – a Battery Fire-Control and Command vehicle (to act as the Battery Command Station).

NORINCO 273 mm WM-80 multiple rocket system deployed in the firing position and launching a rocket. Note hydraulic stabilisers lowered to the ground to provide a more stable firing platform (NORINCO) 1365039

NORINCO 273 mm rocket out of its box-type launcher and showing four fins at the rear 0500636

A 4 × 4 light vehicle is used as the forward observation vehicle for the unit's command and control system. This is fitted with a laser range-finder, a location and orientation navigation system, radio and target co-ordinate information transmission equipment.

This forward observation vehicle would be deployed at battery or battalion level and provide target information in real time.

NORINCO is NOW marketing a new version of its earlier WM-80 273 mm (8-round) Multiple Launch Rocket System (MLRS) called the WM-120. Details of this are given later in this entry.

NORINCO is no longer marketing the 273 mm (8-round) WM-80 multiple rocket system and is concentrating its efforts on the more capable WM-120 system.

The original WM-80 has a combat weight of 36,500 kg while the latest WM-120 is being quoted with a combat weight of 38,000 kg.

The 8 × 8 chassis has been improved and it is now stated to have a maximum road speed of up to 80 km/h and a maximum cruising range of 500 km.

Description

The WM-80 launcher is mounted on the rear of a TA-580 (8 × 8) cross-country truck chassis. An eight-round launcher assembly (comprising two four-round rail-type container-launcher box banks) is fitted on a hydraulically raised rear deck turntable platform.

Attached to the assembly is the aiming system, which comprises the elevating mechanism, traversing mechanism and sight.

The actual 273 mm rocket launch tubes are 4.885 m long.

Inside the five-person fully-enclosed forward control vehicle cab is the launcher firing display (to display any command vehicle data inputs and the launch vehicle status), the operating and aiming system to lay the launcher assembly and the system operating console and control panel.

Optional wire-link or radio communications systems are also available to receive fire-control and targeting data and allow automatic weapon aiming and firing.

The operator can also fire the launcher himself or, if required, perform remote off-vehicle firing. Single round and salvo launches can be performed.

For firing, four hydraulic jacks (two at the rear and one on either side section) are lowered to stabilise the vehicle. To protect the front windscreen when the rockets are fired, special shutters are activated from inside of the vehicle.

The resupply vehicle is based on a TA-580 8 × 8 cross-country truck chassis and carries two four-round launcher-container boxes. These are transferred to the launch vehicle by means of a 10,000 kg capacity hydraulic crane fitted just behind the truck cab and carried over the centre line of the truck bed. It is locked in place on a raised frame at the truck rear for travel purposes.

The resupply vehicle is called the WM-80DY and is fitted with four stabilisers that are lowered to the ground when the rocket launcher is being resupplied.

The rocket itself is 4.582 m long, 273 mm in diameter and weighs 505 kg in total. A small rotary motor is carried to impart rotational spin stabilisation to the rocket in flight. There are also four rear-mounted fins fitted. The rockets are normally launched at five-second intervals.

Maximum velocity is being quoted as 1,140 m/s.

According to NORINCO, the following types of warhead can be fitted:

NORINCO 273 mm WM-80 multiple rocket system deployed in the firing position with launcher elevated (NORINCO) 1365038

- 150 kg HE blast fragmentation containing 34 kg of TH (40/60) high-explosive. The detonation of the filling shatters the casing, creating approximately 17,000 fragments. The resulting fragments having an effective casualty radius of approximately 35 m.
- 150 kg submunition dispensing warhead with a payload of 380 dual-purpose bomblets, each of which has a High-Explosive Anti-Tank (HEAT) warhead which can pierce 80 to 100 mm of conventional steel armour plate.

There are two types of fuzes fitted: the WJ-6A mechanical impact fuze for direct contact; and the MD-23A electronic proximity fuze for airburst or submunition dispensing, height depending upon warhead type used. Both fuzes can be selected and preprogrammed with the appropriate parameters at any time during the engagement sequence by the computerised fire-control system.

More recently NORINCO stated that the WM-80 rocket system can fire rockets fitted with three different warheads, high explosive, high explosive incendiary and cargo carrying sub-munitions.

The rocket is launched by the explosive charge at a velocity of 40 m/s. After the rocket has travelled a safe distance from the vehicle the main solid propellant rocket motor fires, rapidly building the velocity up to the burnout velocity of 1,140 m/s. The rocket attains an apogee of 31,000 m at maximum range launch elevation. Flight duration is 165 seconds with a maximum range of 80,000 m plus. Minimum range is stated to be 34,000 m. The rocket motor section uses 205.5 kg of case-bonded central channel HPTB solid propellant fuel.

According to NORINCO, the dispersion Circular Error of Probability (CEP) is between one and two per cent, but this depends on a number of factors including range. The system has an operating temperature range of between −40°C and +50°C and claimed to have an effective storage period of 10 years.

Variants

273 mm (8-round) WM-120 multiple rocket system

This fires guided 273 mm solid propellant rockets out to a maximum claimed range of 120 km with a claimed Circular Error of Probability (CEP) of about 50 m. Minimum range is being quoted as 34 km.

The powered launcher is mounted on the rear of the TA-580 (8 × 8) cross-country truck chassis and has a traverse of 20° left and right with elevation limits from +20° to +60°.

No details of the actual guidance system of the WM-120 rockets have been revealed but it is probably of the Global Position System plus Inertial Navigation System (GPS+INS) which has also been used on some other recently developed Chinese surface-to-surface rockets.

The significant increase in accuracy would allow targets to be engaged with a more precision effect as well as requiring less rounds to neutralise a given target. This in turn would lead to a significant reduction in logistic support.

The increased range over the earlier system is probably obtained by increasing the amount of propellant.

The 273 mm rockets are in pods of four rockets and once fired the platform would normally deploy to a re-load area to avoid counter battery fire.

According to NORINCO, it takes only five minutes to come into action and launch the rockets and eight minutes to load the launcher with two pods of new rockets as the transloader vehicle is fitted with a hydraulic crane.

Specifications

	273 mm (8-Round) WM-80
Dimensions and weights	
Crew:	5
Length	
overall:	9.55 m
Width	
overall:	3.06 m
Height	
overall:	3.3 m
Weight	
combat:	36,500 kg
Mobility	
Configuration	
running gear:	wheeled
layout:	8 × 8
Speed	
max speed:	70 km/h
Range	
main fuel supply:	400 km
Gradient:	35%
Engine	air cooled, diesel, 525 hp
Firepower	
Armament:	8 × 273 mm rocket system
Main weapon	
maximum range:	80,000 m (est.)
Rate of fire	
rapid fire:	8/5 s
Operation	
stop to fire first round time:	300 s (est.)
out of action time:	300 s
reload time:	480 s (est.)
Main armament traverse	
angle:	20° (left and right)
Main armament elevation/depression	
armament front:	+60°/+20°

Status

Production as required. In service with Armenia (4). As far as it is known this system is not currently in service with People's Liberation Army.

Contractor

Chinese state factories. China North Industries Corporation (NORINCO).

NORINCO 273 mm (4-round) Type 83 multiple rocket system

Development

Developed by the No 123 and No 127 factories, the China North Industries Corporation (NORINCO) Type 83 273 mm (4-round) multiple rocket system had a protracted development period that started in the 1960s.

It was redesigned several times before it gained its type classification in 1983. It is understood that there has been no recent production of this system by NORINCO.

NORINCO is now concentrating its marketing efforts on other types of multiple rocket systems. Production of the NORINCO Type 60-1 full-tracked chassis was completed some time ago.

It is understood that while this was deployed by the People's Liberation Army it could be now have been replaced by more recent systems firing rockets with a longer range.

The People's Liberation Army has now moved to wheeled chassis for their multiple rocket systems as these have lower operating and support costs and have greater strategic mobility.

Description

The system is based on the unarmoured Type 60-1 full-tracked artillery tractor with a five-man, fully enclosed cab at the front and a turntable with four 273 mm unguided rockets at the very rear.

NORINCO 273 mm Type 83 MRS deployed in firing position with launcher elevated (NORINCO) 1452950

NORINCO 273 mm Type 83 MRS in firing position from rear with stabilisers lowered 0500637

It should be noted that the chassis is not provided with any armour protection for the crew.

Before launching, the launch platform's two stabilisers are lowered manually at the rear of the chassis to provide a more stable firing platform.

Each unguided solid propellant rocket has a high-explosive warhead, although other types of warhead could easily be accommodated in a rocket of this calibre. The rocket has a maximum velocity of 811 m/s and all four rockets can be salvo-fired in 7.5 seconds with a time into action of one minute. Once the four rockets have been fired, the system is reloaded manually.

The actual solid propellant rockets weigh 484 kg at launch and have an overall length of 4.753 m.

To avoid counterbattery fire it would probably deploy to another position to enable reloading to take place.

The launcher is traversed and elevated manually, with the controls mounted on its left side. Fire-control equipment consists of one panoramic sight Type 85 and one indicator Type 85.

Specifications

	273 mm (4-Round) Type 83 multiple rocket system
Dimensions and weights	
Crew:	5
Length	
overall:	6.19 m
Width	
transport configuration:	2.60 m
Height	
overall:	3.18 m
hull:	2.81 m
Ground clearance	
overall:	0.40 m
Track	
vehicle:	2.15 m
Length of track on ground:	3.248 m
Weight	
standard:	15,134 kg
combat:	17,542 kg
Mobility	
Configuration	
running gear:	tracked
Speed	
max speed:	45 km/h
Range	
main fuel supply:	400 km
Gradient:	60%
Side slope:	30%
Vertical obstacle	
forwards:	1 m
Turning radius:	2.15 m
Engine	12150L-1, V-12, direct injection, 4 stroke, diesel, 300 hp at 1,600 rpm
Firepower	
Armament:	4 × 273 mm rocket system
Main weapon	
maximum range:	40,000 m
Main armament traverse	
angle:	20°
Main armament elevation/depression	
armament front:	+56°/+5°

Status

Production complete. In service with People's Liberation Army. There are no known export sales of this system. No longer being marketed. It should be noted that there has been no production of the Type 60-1 full-tracked artillery tractor for some years by NORINCO.

Contractor

Chinese state factories. China North Industries Corporation (NORINCO).

NORINCO 253 mm (24-round) Type 87 minesweeping rocket system

Development

This system was developed by China North Industries Corporation (NORINCO) to provide a rapid means of clearing anti-personnel minefields. Production has been completed and it is no longer marketed by NORINCO.

It is understood that this system is still in service with People's Liberation Army.

It is considered possible that this is now being supplemented by more recent mine clearing systems.

These could be of the fuel air explosive type or of another type.

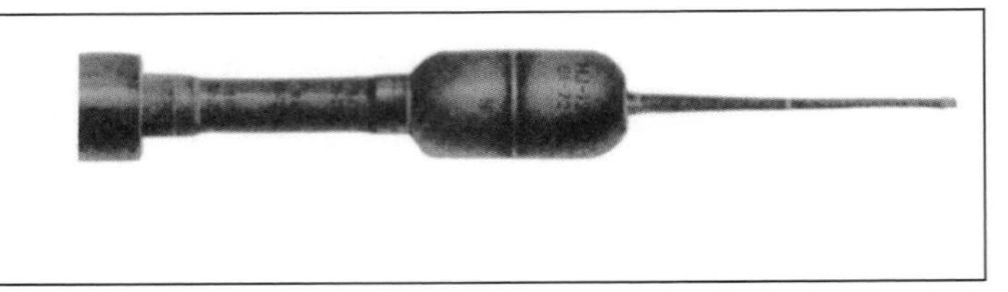

253 mm minesweeping rocket shell Type 81-II 0058455

Description

Mounted on the hull of a modified China North Industries Corporation (NORINCO) Type 59 series MBT, the NORINCO 253 mm (24-round) Type 87 minesweeping rocket system uses a rectangular turret assembly fitted with 24 253 mm rocket launcher tubes in three layers of eight.

It is also equipped with a laser range-finder to accurately measure minefield target distances. The system is used to clear anti-personnel minefields.

The system uses the following unguided minesweeping solid propellant rocket types, which are also used in the NORINCO 253 mm (10-round) Type 81 minesweeping rocket system covered in a separate entry in *IHS Jane's Land Warfare Platforms: Artillery and Air Defence.*

Model	Type 81	Type 81-II
Calibre:	253 mm	253 mm
Muzzle velocity:	24 m/s	n/avail
Weight: (rocket)	n/avail	74.5 kg
Maximum range:	1,400 m	2,000 m (with jet efflux cone); 3,000 m (without jet efflux cone)
Mineclearing zone: (salvo fire)	n/avail	10,000 m^2

The Type 81 unguided surface-to-surface rocket can be used in mountainous and jungle regions and apart from clearing anti-personnel mines, both rocket types may also be used for anti-personnel and fortification demolition roles.

The major difference between the two rounds is that the Type 81-II has a rigid-spike piezoelectric fuze fitted to its nose. This controls the height of the warhead's airburst, optimising its explosive effectiveness against semi-buried or ground-laid anti-personnel mines.

A single Type 81-II rocket can also clear a 35 m diameter circle of anti-personnel mines. This type of system is also highly effective against infantry positions and other battlefield targets at close range. The rockets may be of the Fuel Air Explosive (FAE) type.

Specifications

	Type 81	Type 81-II
Dimensions and weights		
Diameter		
body:	253 mm	253 mm
Weight		
less warhead:	-	75.4 kg
Performance		
Range		
max:	1.4 km	2 km (with jet efflux cone)
	-	3 km (without jet efflux cone)

	Type 87 Minesweeping Rocket System
Dimensions and weights	
width	
calibre:	253 mm
Number of tubes:	24
Performance	
Traverse:	
angle:	360°
Elevation	
angle:	45°/0°

Status

Production complete. In service with the People's Liberation Army. There are no known exports of this system. No longer marketed.

Contractor

Chinese state factories. China North Industries Corporation (NORINCO).

NORINCO 253 mm (10-round) Type 81 minesweeping rocket system

Development

The China North Industries Corporation (NORINCO) 253 mm (10-round) Type 81 is a truck-mounted system for clearing anti-personnel mines for infantry assaults.

NORINCO 253 mm (10-round) Type 81 minesweeping rocket system launching a single rocket 0021041

NORINCO 253 mm (10-round) Type 81 minesweeping rocket system in travelling configuration 0021040

It is assumed that this system is still in service with the Peoples Liberation Army but it is no longer being marketed by NORINCO.

It is considered possible that this system has been replaced by a more modern mine clearing rocket system.

Description

It comprises 10 rocket launcher tubes (in two banks of five), fitted on a manually-operated turntable assembly at the very rear of the truck's aft decking.

Immediately to the fore of the assembly is a housing for 10 reload 253 mm rounds and a fuze container. The truck used is a 6 × 6 model and is fitted with an extended four door fully-enclosed cab. When deployed in the firing position, shutters are raised to protect the windscreen and the windows in the upper part of the side door.

When not required, the side blast shields fold down through 180°, while the front shields fold down 90°.

To provide a more stabilised firing platform when the rockets are being fired, two stabilisers are manually lowered to the ground one either side at the rear of the chassis. A replacement wheel and tyre are carried in the horizontal position to the cab rear.

The NORINCO 253 mm (10-round) Type 81 minesweeping rocket system fires the same Type 81 and Type 81-II unguided rockets as the Type 87 system covered in a separate entry in *IHS Jane's Land Warfare Platforms: Artillery and Air Defence.* Production of this system has been completed and it is no longer marketed by NORINCO.

Rate of fire is one round every 1.2 to 1.4 seconds. Once the rockets have been fired, the system would normally deploy to another position where the rockets would be reloaded. This rapid move is required to reduce counterbattery fire.

It is understood that a small quantity of additional rockets are carried on the launcher.

A full 10 round ripple of 253 mm rockets is claimed to be able to clear at area of 60 × 10 m, although these figures could be an underestimate.

Variants

Details of the Sichuan Aerospace Industry Corporation (SCAIC) WS-1D 252 mm series of rocket are given in a separate entry in *IHS Jane's Land Warfare Platforms: Artillery and Air Defence.* These are understood to be launched from the NORINCO 253 mm (10-round) Type 81 minesweeping rocket system.

Specifications

	NORINCO 253 mm (10-round) Type 81 minesweeping rocket system
Dimensions and weights	
Length	
overall:	6.80 m
Width	
transport configuration:	2.20 m
Height	
transport configuration:	2.45 m
Weight	
combat:	8,300 kg
Mobility	
Configuration	
running gear:	wheeled
layout:	6 × 6
Speed	
max speed:	80 km/h
Fording	
without preparation:	0.850 m
Firepower	
Armament:	10 × 253 mm rocket system
Ammunition	
main total:	20
Main weapon	
maximum range:	1,400 m
Rate of fire	
rapid fire:	10/14 s (est.)
Main armament traverse	
angle:	135°
Main armament elevation/depression	
armament front:	+50°/+20°

Status

Production complete. In service with People's Liberation Army. There are no known exports of this system. No longer marketed.

Contractor

Chinese state factories. China North Industries Corporation (NORINCO).

NORINCO SR5 Universal Artillery Rocket Launcher

Development

In mid-2012 China North Industries Corporation (NORINCO) revealed that they had developed and tested the SR5 Universal Artillery Rocket Launcher (UARL).

It is understood that this has been developed to meet the potential requirements of the People's Liberation Army (PLA) and export customers.

As of April 2013 development of the SR5 Universal Artillery Rocket Launcher (UARL) is understood to have been completed but production had yet to start.

Description

The SR5 UARL is based on the rear of a Chinese TA 5310 (6 × 6) cross-country chassis fitted with a four door unprotected forward control cab.

The 6 × 6 vehicle has powered steering on the front wheels and is fitted with a central tyre pressure regulation system that allows the driver to adjust the tyre pressure to suit the terrain being crossed.

Mounted on the rear of this chassis is a power operated turntable rocket launcher which can accept two pods of surface-to-surface rockets, one with 20 × 122 mm rockets and one with six longer 220 mm rockets, or two pods of either rockets.

This allows the user greater operational flexibility and allows the type of rocket launched to be matched to the target to be engaged out to a maximum range of 70 km for the larger calibre 220 mm rockets.

These pods of 122 mm or 220 mm rockets can be rapidly loaded using a mechanical handing system operated by remote control. This is very similar in concept to that used by the widely deployed US Lockheed Martin Missiles and Fire Control tracked 227 mm (12-round) MLRS which is covered in detail in a

Chinese SR5 (6 × 6) Universal Artillery Rocket Launcher with stabilisers lowered and launcher traversed to the left. This has one pod of 20 × 122 mm rockets and one pod of longer 6 × 220 mm rockets in the ready to launch position (NORINCO) 1452688

separate entry in *IHS Jane's Land Warfare Platforms: Artillery and Air Defence*.

According to NORINCO the SR5 UARL takes five minutes to come into action and fire its first rockets and only one minute to come out of action.

Once the rockets have been fired the SR5 UARL would normally redeploy to another position for rapid reloading and to avoid counter battery fire.

The empty rocket pod is then slid out on rails and lowered to the ground and another pod of new rockets is then fitted, raised until it locks into the launcher arm and is then retracted into the launcher ready for the next fire mission.

To reload the SR5 system with two pods of new rockets takes about five minutes and is carried out by remote control.

The launcher has powered traverse of 70° left and right and powered elevation from 0 to +60°.

Prior to launching the surface-to-surface rockets two hydraulically operated stabilisers are lowered to the ground one either side at the rear of the vehicle by remote control.

The SR5 can launch the complete family of NORINCO 122 mm unguided rockets which have ranges of 30, 40 and 50 km depending on the type. Details of this family of rockets are provided in a separate entry in *IHS Jane's Land Warfare Platforms: Artillery and Air Defence*.

For the first time NORINCO is now offering a guided 122 mm rocket which is claimed to have a maximum range of 40 km and is fitted with an high explosive warhead with a lethal radius of 60 m.

Chinese SR5 (6 × 6) Universal Artillery Rocket Launcher in travelling configuration with launcher lowered and traversed to the front (NORINCO)
1452689

Chinese SR5 (6 × 6) Universal Artillery Rocket Launcher from the rear in deployed position with stabilisers lowered and launcher traversed to the left containing one pod of 122 mm rockets and one pod of 220 mm rockets (NORINCO)
1452690

Chinese SR5 (6 × 6) Universal Artillery Rocket Launcher loaded with one pod of 6 × 220 mm rockets and one pod of 20 × 122 mm rockets being lifted in position using mechanical handing system which is operated by remote control (NORINCO)
1452691

It is fitted with a Global Positioning System/Inertial Navigation (GPS/INS) guidance system which is claimed to provide a Circular Error of Probability (CEP) of 25 m.

The new generation 220 mm rockets have a maximum range of up to 70 km and can also have a GPS/INS guidance system with a CEP of just 3 m. This is the first time that NORINCO has developed and marketed a 220 mm rocket system.

Combat weight of the SR5 is being quoted as 25 tonnes with a maximum road speed of 85 km/h and a cruising range of 600 km.

SR5 is fitted with an on-board computerised fire-control system which includes a very accurate land navigation system which provides for increased accuracy.

Variants

There are no known variants of the SR5 UARL.

Specifications

	SR5 Universal Artillery Rocket Launcher
Dimensions and weights	
Crew:	3
Weight	
combat:	25,000 kg (depends on rockets carried)
Mobility	
Configuration	
running gear:	wheeled
layout:	6 × 6
Power-to-weight ratio:	laden
Speed	
max speed:	85 km/h
Range	
main fuel supply:	600 km
Firepower	
Armament:	40 × 122 mm rocket system (2 × 20 optional)
Armament:	12 × 220 mm rocket system (2 × 6 optional)
Main armament traverse	
angle:	70° (left/right)
Main armament elevation/ depression	
armament front:	0°/+60°

Status

Development complete. Ready for production on receipt of orders.

Contractor

China North Industries Corporation (NORINCO).

NORINCO 130 mm (30-round) Type 82 and Type 85 multiple rocket systems

Development

The China North Industries Corporation (NORINCO) 130 mm Type 82 multiple rocket system is an improved version of the NORINCO 130 mm Type 63/70 family which is covered in detail in a separate entry in *IHS Jane's Land Warfare Platforms: Artillery and Air Defence*.

All marketing of these 130 mm multiple rocket launchers has ceased by NORINCO. It is possible that surplus launchers are available for export and new-build 130 mm rockets could be provided.

It is assumed that this system is still in service with the Peoples Liberation Army. It is possible that these launchers could be integrated onto a new chassis.

When compared to the artillery rocket systems currently being marketed by NORINCO, these older 130 mm systems have a short range and their launch platforms would be vulnerable to threat artillery systems.

Description

The Type 82 consists of a 6 × 6 cross-country truck with a four-door, fully-enclosed forward control cab. To the rear of this are 30 reserve rockets for immediate use and, at the very rear of the chassis is a 130 mm rocket launcher with three layers of 10 tubes each.

Two salvos, each of 30 unguided rockets, can be launched in five minutes. The system would then rapidly redeploy to avoid counterbattery fire.

As the rocket launcher is mounted at the very rear of the chassis the forward control cab is not fitted with any armoured shutters for its windows.

The launcher is also mounted on the top of the NORINCO Type 85 tracked APC to improve its cross-country capability.

The Type 85 full-tracked APC is also referred to as the NORINCO YW531H and production has now been completed.

The 130 mm unguided rocket types available are shown in the table on the following page.

There is also a 1.05 m long single man-portable tripod-mounted launcher tube for a 130 mm rocket which is designed for use by commando, guerrilla or special forces units.

NORINCO Type 85 full-tracked APC fitted with 130 mm (30-round) Type 85 multiple rocket system 0085760

NORINCO 130 mm (30-round) Type 82 multiple rocket system in firing configuration and with 30-round launcher traversed to the left 0085761

This can be disassembled into small loads for carriage-launch tube 10.5 kg, tripod 11.9 kg, sight 1.2 kg and remote firing device 1.25 kg. Its elevation range is +5° to +40° and its traverse range 7° to the left or right. Time to bring the weapon into action is about 1¾ minutes and to disengage between 1½ and two minutes.

Specifications

	130 mm (30-Round) Type 82	130 mm (30-Round) Type 85
Dimensions and weights		
Crew:	7+6	7+6
Length		
overall:	6.438 m	6.125 m
Width		
transport configuration:	2.250 m	3.060 m
Height		
transport configuration:	2.525 m	3.014 m
Weight		
combat:	9,000 kg	14,500 kg
Mobility		
Configuration		
running gear:	wheeled	tracked
layout:	6 × 6	
Speed		
max speed:	80 km/h	60 km/h
Amphibious:	no	yes
Fording		
without preparation:	0.85 m	n/a
Gradient:	33% (est.)	33% (est.)
Firepower		
Armament:	30 × 130.65 mm rocket system	30 × 130.65 mm rocket system
Ammunition		
main total:	60	60
Main weapon		
length of barrel:	1.050 m	1.050 m
Rate of fire		
rapid fire:	30/17.4 s (est.)	30/17.4 s (est.)
Operation		
stop to fire first round time:	60 s	60 s
out of action time:	120 s (est.)	120 s (est.)
Main armament traverse		
angle:	170°	30° (15° left, 15° right)
Main armament elevation/depression		
armament front:	+50°/0°	+50°/0°
Survivability		
Armour		
hull/body:	-	steel

Status
Production complete. No longer marketed. In service with the People's Liberation Army (Type 82 and Type 85), Democratic Republic of Congo (3) and Thailand (Type 85). According to the United Nations Report on Conventional Arms issued in 2001, between 1992 and 2010 China exported the following multiple rocket launcher systems:

Country	System	Quantity
Armenia	WM-80	4
Gabon	107 mm	16 (delivered 2004)
Gabon	122 mm	4 (delivered 2004)
Sudan	not released	18
Bangladesh	not released	42

It is not known as to whether the latter two countries took delivery of 130 mm or 122 mm calibre systems. In recent years NORINCO has recently placed its main sales emphasis on the 122 mm and larger calibre artillery rocket systems.

Contractor
Chinese state factories. China North Industries Corporation (NORINCO).

NORINCO 130 mm (19-round) Type 63 and Type 70 multiple rocket systems

Development
The China North Industries Corporation (NORINCO) designed and manufactured 19-tube 130 mm calibre Type 63 (19-round) rocket launcher was developed by the No 743 and No 247 factories and type classified in 1963.

All marketing of these 130 mm multiple rocket launchers has ceased by NORINCO. It is possible that surplus 130 mm launchers are available for export and new-build 130 mm rockets could be provided.

It is assumed that this system is still in service with the Peoples Liberation Army (PLA).

Description
It entered service with the People's Liberation Army mounted on the rear platform of the NJ-230 4 × 4 truck. There are two variants of the Type 63 mounted on this truck, the major difference being that the second variant, Type 63-1 has a covered crew cabin.

Designation	**Type 63 HE rocket**	**Type 82 HE fragmentation rocket**	**Type 82 HE-I rocket**	**Extended Range HE rocket**
Diameter:	130.43 mm	130.43 mm	130.43 mm	130.43 mm
Length:	1.051 m	1.063 m	1.043 m	1.056 m
Weight: (complete)	33 kg	32 kg	32.8 kg	32.5 kg
Maximum range:	10,100 m	10,217 m	10,100 m	15,000 m
Warhead type:	HE fragmentation. Produces approx. 1,900 fragments upon detonation. Lethal radius of 24 m	HE preformed fragmentation with 2,600 6.35 mm diameter steel spheres. Produces approx. 3,800 fragments upon detonation. Lethal radius 42.1 m. Minimum range is 3,000 m	HE incendiary fragmentation. Approx. 5,000 incendiary pellets. Produces over 2,300 fragments upon detonation. Incendiary effect radius of 30 m	HE fragmentation with lethal radius of 32 m
Fuze type and settings:	proximity or MJ-1 impact with instantaneous and delay actions			

Latest NORINCO EQ2060E (4 × 4) cross-country truck chassis fitted with 130 mm (19-round) rocket launcher on the rear (NORINCO) 1365044

NORINCO 130 mm Type 70 MRS on Type YW531 APC with launcher traversed left 0085762

NORINCO 130 mm Type 63-1 MRS with covered crew cab on NJ-230 (4 × 4) truck chassis (Andrew Li) 0500638

In both cases the firing units are normally grouped into batteries of six vehicles.

The Type 70 version was developed from the Type 63 rocket launcher and mounted on the NORINCO Type 63 tracked APC chassis. It was originally assigned to rocket launcher units attached to the armoured and mechanised infantry divisions of the People's Liberation Army.

The launch tubes are arranged in two rows, the top one of 10 over another of nine. Once the rockets are launched, the system would normally redeploy to another position where new rockets would be manually loaded. A rapid move is required to avoid counterbattery fire.

The 130 mm unguided rocket types available are given in the entry for the NORINCO 130 mm (30-round) Type 82 wheeled and Type 85 tracked multiple rocket launchers in a separate entry in *IHS Jane's Land Warfare Platforms: Artillery and Air Defence.* Production and marketing of these systems has also ceased by NORINCO.. These were replaced in production by the 130 mm (30-round) Type 82 and Type 85 multiple rocket systems based on more modern tracked and wheeled chassis.

The Type 85 system is mounted on a full-tracked chassis with five roadwheels, while the older Type 70 is based on a smaller chassis with four larger roadwheels.

There is also a 1.05 m long single man-portable tripod-mounted launcher tube for 130 mm rockets which is designed for use by commando, guerrilla or special forces units. This can be disassembled into small loads for carriage - launch tube 10.5 kg, tripod 11.9 kg, sight 1.2 kg and remote firing device 1.25 kg. Its elevation range is +5° to +45° and its traverse range 7° to the left or right. Time to bring the weapon into action is about 1¾ minutes and to disengage between 1½ to two minutes.

Variant

NORINCO EQ2060E (4 × 4) chassis with 130 mm (19-round) launcher

NORINCO has fitted their EQ2060E (4 × 4) cross-country truck chassis with the 130 mm (19-round) rocket launcher removed from the original Type 63-1 system.

To the immediate rear of the fully enclosed cab are seats that face outwards for the additional crew members with the 130 mm (19-round) launcher mounted at the very rear of the chassis.

To provide a more stable firing platform two jacks are lowered manually at the rear of the chassis for the unguided 130 mm rockets are launched.

This vehicle is powered by a Cummins 6BT 5.9-litre six-cylinder water cooled diesel developing 155 hp which gives a maximum road speed of 90 km/h. The vehicle has a gross weight of 6,350 kg of which 1,500 kg is payload.

The baseline NORINCO EQ2060E is fitted with a cargo type body with drop sides and rear, bows and a tarpaulin cover.

Specifications

	130 mm Type 63-1	130 mm Type 70
Dimensions and weights		
Crew:	7	6
Length		
overall:	6.659 m	5.746 m
Width		
transport configuration:	2.140 m	2.798 m
Height		
overall:	2.435 m	2.625 m
Weight		
combat:	6,887 kg	14,000 kg
Mobility		
Configuration		
running gear:	wheeled	tracked
layout:	4 × 4	
Speed		
max speed:	50 km/h (est.)	60 km/h
water:	n/a	5.2 km/h
Fording		
without preparation:	0.85 m	1.232 m
Gradient:	33% (est.)	34% (est.)
Firepower		
Armament:	19 × 130.65 mm rocket system	19 × 130.65 mm rocket system
Ammunition		
main total:	38	38
Main weapon		
length of barrel:	1.050 m	1.050 m
Rate of fire		
rapid fire:	19/11.5 s (est.)	19/11.5 s (est.)
Operation		
stop to fire first round time:	120 s	60 s
out of action time:	120 s	120 s
Main armament traverse		
angle:	180° (-32° left/ +148° right)	180°
Main armament elevation/depression		
armament front:	+50°/0°	+50°/0°

Status

Production complete. In service with China, North Korea and Vietnam. This system is no longer marketed by NORINCO. According to the United Nations Report on Conventional Arms issued in 2001, between 1992 and 2010 China exported the following multiple rocket launcher systems:

Country	Systems	Quantity
Armenia	WM-80	4
Gabon	107 mm	16 (delivered 2004)
Gabon	122 mm	4 (delivered 2004)
Sudan	not released	18
Bangladesh	not released	42

It is not known as to whether the latter two countries took delivery of 130 mm or 122 mm calibre systems.

NORINCO has now ceased all marketing of these 130 mm multiple rocket launcher systems.

It is possible that quantities of these 130 mm multiple rocket systems – that are now surplus to the requirement of the People's Liberation Army following the introduction of systems with a longer range – could be available for export.

Contractor

Chinese state factories. China North Industries Corporation (NORINCO).

Aerospace Long-march International Trade Company WS-15 122 mm (40-round) multiple launch rocket system

Development

The WS-15 122 mm (40-round) Multiple Launch Rocket System (MLRS) is currently being offered on the export market by the Chinese Aerospace Long-march International Trade Company (ALIT) as a supplement to conventional tube artillery systems.

In many respects the WS-15 is similar to other types of 122 mm MLRS but has the added advantage that it carries an additional 40 × 122 mm rockets in two pods of 20 for rapid reloading.

The WS-15 has been offered of the export market for several years but as of March 2012 there are no known export sales.

As far as it is known this system is not in service with the People's Liberation Army (PLA).

Description

The WS-15 122 mm (40-round) MLRS consists of a 6 × 6 cross-country truck chassis with a fully enclosed forward control cab.

This is typically the Chinese Sinotruk Howo 371 (6 × 6) chassis made by CNHTC, although other similar chassis could be used if required by the end user.

Mounted on the rear of the chassis is the power operated launcher which has two pods each of 20 × 122 mm rocket launcher tubes with elevation and traverse being powered and with manual back up controls.

Once these 40 × 122 mm rockets have been fired the launcher would then be rapidly re-loaded with two new pods of 122 mm × 20 rounds which are carried to the immediate rear of the cab.

This would be carried out using the onboard hydraulic crane mounted to the rear of the cab.

To provide a more stable firing platform, two stabilisers are lowered to the ground at the very rear of the chassis prior to a fire mission being carried out.

The rear of the vehicle is provided with a tarpaulin cover which can be rapidly slid right to the rear of the chassis to enable the complete launcher and its associated resupply rockets to be covered.

This would make the vehicle difficult to detect as being a dedicated MLRS platform and would therefore increase its survivability.

The 122 mm solid propellant rockets are claimed to have a maximum range of between 10 and 45 km with each rocket having a launch weight of 71 kg and an overall length of 3 m.

The 122 mm rocket is fitted with a 22.5 kg blast fragmentation warhead which is claimed to have a lethal radius of 40 m.

It can also fire other suppliers 122 mm rockets which are available from many sources.

According to ALIT, the 122 mm rockets can be fired when the launcher is deployed up to a maximum altitude of about 3,000 m above sea level and in an environment temperature of 0°C to +60°C and a ground wind speed of up to 8 m/s.

The 122 mm rockets have a typical storage life of ten years at a storage temperature of 0°C to +30°C.

According to ALIT, a typical WS-15 battery would consist of nine launchers, nine ammunition resupply vehicles on a similar 6 × 6 cross-country chassis, one command and communications vehicle on a 4 × 4 cross-country chassis and one meteorological survey vehicle to provide accurate weather information for increased accuracy.

Variants

ALIT WS-22 122 mm (40-round) MLRS

This is the same as the WS-15 but launches a new family of 122 mm guided rockets to provide a precision effect.

This would mean that less 122 mm rockets would be required to neutralise a given target.

These 122 mm guided rockets are 3.03 m long and have a total launch weight of 73 kg and are fitted with a 21 kg warhead with maximum range of between 20 and 45 km.

The increased accuracy is achieved by a combination of an Inertial Navigation System (INS) and a Global Positioning System (GPS) with the antennas for the latter being located on either side of the rocket's nose section.

To the rear of this antenna are six rows of small sideways firing rocket thrusters with each row having seven to provide a total of 42 thrusters which are used to make course corrections while the rocket is in-flight.

Information from the INS and GPS is used by the guidance computer to initiate the thrusters at the appropriate time.

The combined INS/GPS guidance system gives the 122 mm rocket a claimed CEP of less than 100 m at maximum range.

The 122 mm WS-22 guided rockets out of their launch tubes and showing the GPS antenna (the small, white rectangle near the nose) immediately ahead of the rows of course correcting thrusters) (ALIT) 1331989

WS-22 122 mm (40-round) MLRS deployed in the firing position with stabiliser jacks deployed at the rear and launcher elevated and traversed to the right (ALIT) 1331987

Other Aerospace Long-march International surface-to-surface rockets

WS-32 300 mm (4-round) guided MLRS rocket

The WS-32 MRLS fires 300 mm guided rockets which use a combined INS/GPS guidance system.

Guidance commands are sent to actuators attached to four small nose mounted canards. The rockets are said to have a range of 60 to 150 km and a CEP of less than 40 m at maximum range.

The 300 mm rockets are available with a least seven types of warheads including blast fragmentation, incendiary blast-fragmentation, fuel air explosive, cargo, shaped charge/fragmentation submunition, Kinetic Energy (KE) submunitions and a so called "terminal sensitive".

The KE warhead is most likely similar to one of the warheads proposed for the US GMLRS Alternative Warhead Programe (AWP) which contained multiple tiers of flechettes which are dispensed over the target and used to attack personnel, matériel and light armour.

The terminal sensitive warhead could be of the US Sense And Destroy Armor (SADARM) type which is optimised to attack the more vulnerable upper surfaces of armoured vehicles.

It has been marketed on a Baotou North Benz ND2629A (6 × 6) chassis with four 300 mm launchers or the larger TA5450 (8 × 8) chassis with eight 300 mm launcher tubes.

The former version can come into action in seven minutes and come out of action in three minutes respectively.

WS-33 200 mm (8-round) guided MLRS

This is based on the TA5310 (6 × 6) cross-country truck chassis which launches 200 mm guided rockets with each rocket having four folding cruciform wings positioned three quarters of the way down the body and four tail mounted cruciform control surfaces.

A large umbilical conduit runs from the nose/seeker guidance sections to the rear control surface section.

The 200 mm rockets are 3,300 mm long, 200 mm in diameter and weigh 200 kg at launch with the warhead said to weigh 23 kg.

Minimum range is stated to be to 10 km with a maximum range of at least 70 km.

The rockets use a combination of INS/GPS guidance during the initial portions of the flight, switching to a nose mounted optronic seeker in the terminal flight phase which according to ALIT enables moving targets to be engaged with a CEP of 10 m or less.

WS-1/WS-1B 302 mm (8-round) artillery rocket system

This is also marketed by CPMIEC and details are this is provided in a separate entry in *IHS Jane's Land Warfare Platforms: Artillery and Air Defence*.

ALIT state that their system can fire the WS-1 or WS-1B rocket both of which are fitted with a 150 kg warhead which can be of the blast-fragmentation, incendiary blast-fragmentation, fuel air explosive or anti-personnel or armour-piercing dual-purpose warhead type.

Artists impression of WS-15 122 mm (40-round) MLRS with crane deployed to rear and launcher traversed slightly to the right. Note that the stabilisers are not deployed at the rear of the chassis (ALIT) 1452656

Computer generated image of the WS-33 200 mm (8-round) launcher vehicle with launcher in elevated position (ALIT) 1331992

The WS-1 rocket is being quoted as having a range of between 40 and 100 km with the rocket being 4,737 mm long and having a launch weight of 524 kg. The rocket motor is designated the FG-42 with a claimed accuracy of 1 per cent of range.

The WS-1B rocket is being quoted as having a range of between 60 and 180 km with the rocket being 6,376 mm long and having a launch weight of 725 kg and a claimed accuracy of 1 to 1.25 per cent of range.

WS-2 400 mm (6-round) Guided MLRS

This is also marketed by Poly Technologies and SAIC and this is covered in a separate entry in *IHS Jane's Land Warfare Platforms: Artillery and Air Defence*.

According to ALIT the 400 mm guided rockets are claimed to have a minimum range of 70 km and a maximum range of 200 km and each rocket is 7,302 mm long and has a launch weight of 1,285 kg of which 200 kg is the warhead.

The latter can be of the blast-fragmentation, incendiary blast-fragmentation, fuel air explosive or anti-personnel and armour-piercing dual-purpose type.

Time taken to launch the rockets is 12 minutes from when the vehicle comes to a halt.

WS-3 400 mm (6-round) Guided MLRS

This is also marketed by Poly Technologies and SAIC and this is covered in a separate entry in *IHS Jane's Land Warfare Platforms: Artillery and Air Defence*.

According to ALIT the guided 400 mm rockets have a minimum range of 70 km and a maximum range of 200 km and each rocket is 7,150 mm long and has a launch weight of 1,282 kg.

Types of warhead that can be fitted include a basic blast-fragmentation, cluster (bomblet submunition), fuel air explosive and incendiary blast-fragmentation.

The guided rockets feature a GNSS/SINS integrated navigation system and whole course control technology to provide attitude control and guidance control and to provide a precision effect.

The control system comprises a navigation controller, GNSS antenna, low noise amplifier, actuator system, on board battery, power distributer, onboard cable net and flight control software.

This is claimed to provide a CEP of 50 m, or 300 m when inertial navigation only is being used and takes about seven minutes to come into action after coming to a halt.

A100 MLRS 310 mm (10-round) artillery rocket system

This has 10 × 310 mm diameter tubes and can launch the A100-311 or the A100-111 solid propellant rockets.

The former has a launch weight of 835 kg and a range of 60 to 120 km while the A100-111 has a launch weight of 845 kg and a range of 40 to 80 km.

A variety of warhead types can be fitted including blast-fragmentation, unitary blast and shaped charge/fragmentation submunition.

It should be noted that there is a separate entry in *IHS Jane's Land Warfare Platforms: Artillery and Air Defence* for the CPMIEC A100 300 mm (10-round) multiple rocket system.

Upon coming to a halt it takes eight minutes to launch its first rocket and reloading time using a rocket resupply vehicle on a similar chassis fitted with a crane is being quoted as about 20 minutes.

A200 301 mm (8-round) Guided MLRS

This is based on an 8 × 8 cross-country truck chassis that is also claimed to be used for the A100 system.

According to ALIT the A200 system is able to come into action and launch its first rockets in eight minutes with a salvo of eight rockets being launched in about 50 seconds.

M20 Tactical Missile Weapon System deployed in firing position with stabilisers deployed and launcher elevated (ALIT) 1452658

A200 301 mm (8-round) Guided MLRS with launcher elevated (ALIT) 1452657

Mounted at the rear of the chassis are two box type launchers each containing four × 301 mm rockets which have a maximum range of between 50 and 200 km.

The rockets are 7,264 mm long and have a launch weight of 736 kg and can be fitted with various type of warhead including blast fragmentation, unitary blast and shaped charge/fragmentation submunition warhead.

The rockets are fitted with an integrated INS/GPS guidance system and towards the end of the trajectory the warhead separates from the solid propellant rocket and is guided to the target.

According to the ALIT the system can launch eight rockets and eight different targets covering an area of 5,000 × 5,000 m^2.

Once the rockets have been expended the launcher would deploy to another area to be reloaded with the sequence taking about 20 minutes.

M20 Tactical Missile Weapon System

This is an 8 × 8 cross-country chassis fitted with a two round launcher for two rockets which have a range of between 100 and 280 km with a claimed CEP of 30 m. The solid propellant rockets are fitted with a warhead weighing 480 kg.

According to ALIT the systems rakes 12 minutes to come into action and launch its first rocket and about three minutes to come out of action.

Specifications

	WS-15 122 mm (40-round)
Mobility	
Configuration	
running gear:	wheeled
layout:	6 × 6
Firepower	
Armament:	40 × 122 mm rocket system (2 × 20)
Ammunition	
main total:	80 (20 × 2 × 2)

Status

Development complete. Ready for production.

Contractor

Long-march International Trade Company.

NORINCO 122 mm (40-round) Type 89 self-propelled multiple rocket system

Development

The China North Industries Corporation (NORINCO) 122 mm Type 89 (40-round) self-propelled rocket launcher was developed by No 674 factory, No 5137 factory and No 207 Research Institute and type classified for use by the People's Liberation Army in 1989.

The system uses a full-tracked armoured hull that is also used in a modified form for the NORINCO Type 83 152 mm self-propelled gun-howitzer, NORINCO Type 762 (2-round) mineclearing surface-to-surface rocket system and the NORINCO Type 1989 120 mm self-propelled tank destroyer.

Production of all of these systems was completed some time ago. Although all of these systems were offered on the export market, as far as it is known none of these were ever sold overseas.

The 122 mm Type 89 (40-round) system has been designed to accompany mechanised forces and, unlike many other self-propelled multiple rocket systems, has a fully armoured hull. This provides the occupants with protection from small arms fire and shell splinters.

It is assumed that this system is still in service with the Peoples Liberation Army but could have been replaced in front line units by the more recent 122 mm (40-round) Type 90 series system which is based on a Mercedes-Benz (6 × 6) truck chassis.

This is a more mobile system and has lower operating and maintenance costs and is currently being offered on the export market by NORINCO.

Details of the NORINCO 122 mm Type 90 and the latest 122 mm Type 90B are provided in separate entries in *IHS Jane's Land Warfare Platforms: Artillery and Air Defence.*

Description

The engine compartment is front right with the driver on the left-hand side. He is provided with a single-piece hatch cover that opens to the rear, a heated day periscope WG501 and a TYG-1 night driving periscope.

The other four crew members are seated in the fully enclosed armoured compartment at the back with bulletproof observation windows on either side and an entry door at the rear.

It has torsion bar suspension and each side has six dual rubber-tyred roadwheels, a drive sprocket at the front, an idler at the rear and three track-return rollers. The first and last roadwheel stations either side have a hydraulic shock-absorber.

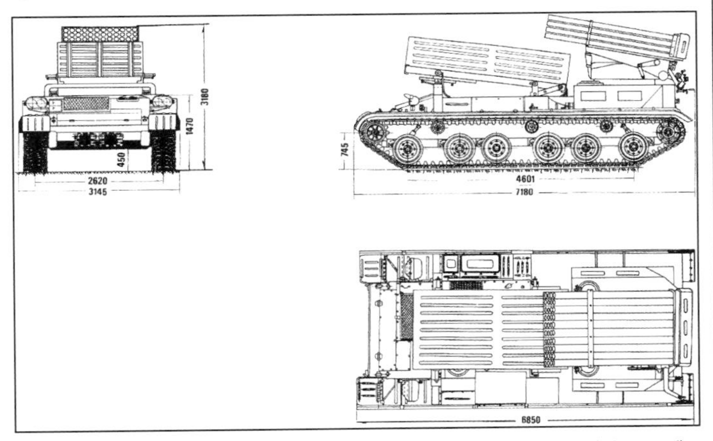

General arrangement drawing of the NORINCO 122 mm Type 89 (40-round) self-propelled multiple rocket system in travelling configuration 0500891

Close-up of late production NORINCO 122 mm Type 89 (40-round) self-propelled multiple rocket system showing new armoured layer's position and different cover arrangement for forward pack of 40 rockets 0500892

While the top road speed is being quoted as 55 km/h the average road speed has been quoted as 40 to 42 km/h and the average cross-country speed 30 to 32 km/h.

The vehicle commander has a Type WG501 heated periscope while the gunner has a Type 59 tank periscope, Type 84 panoramic telescope and a Type 66-1 direct fire sight.

The 122 mm (40-round) launcher is mounted on a turntable over the rear of the vehicle. It is almost identical to the 122 mm Russian BM-21 (40-round) system introduced in the early 1960s.

The launcher can be traversed and elevated under full electric power and manual controls are provided for emergency use. The rockets can be fired singly or in a salvo with all 40 being launched in 20 seconds.

Mounted over the forward part of the vehicle is a reload pack with a further 40 122 mm rockets which can be reloaded into the launcher automatically.

This reload feature is similar to that used on the Czech 122 mm RM-70 (40-round) multiple rocket system which made its first appearance in 1972.

The NORINCO 122 mm rocket types available are shown in the table below. It can also fire the new NORINCO family of 30-40 km range unguided rockets covered in a separate entry.

NORINCO has also recently developed a new family of 40 km and 50 km unguided rockets.

China has also developed a combined Inertial Navigation System/Global Positioning System (INS/GPS) 122 mm rocket to provide a precision effect at longer ranged.

This is claimed to provide the 122 mm rocket with a CEP of less than 100 m at maximum range.

Late production NORINCO 122 mm Type 89 (40-round) self-propelled multiple rocket systems have a number of modifications including a different cover arrangement for the forward pack of 40 rockets and a fully enclosed armoured compartment for the layer on the left side of the launcher at the rear.

Type	HE rocket	HE fragmentation rocket	HE incendiary rocket	HE cargo rocket	HE fragmentation incendiary rocket	Minelaying rocket	Type 84 minelaying rocket
Diameter:	122 mm	122 mm	122 mm	122 mm	122 mm	122 mm	122 mm
Length:	2.87 m	2.87 m	2.87 m	3.03 m	2.87 m	2.95 m	2.83 m
Weight:							
(complete)	66.83 ± 0.66 kg	66.88 ± 0.66 kg	66.83 ± 0.66 kg	67 kg	66.8 ± 0.66 kg	58.2 kg	63 kg
(warhead)	18.3 kg	18.3 kg	18.3 kg	19.25 kg	18.3 kg	25.7 kg	33 kg
Max range:	20,580 m	20,580 m	20,580 m	20,580 m	20,580 m	15,000 m	7,000 m
Warhead type:	HE	6 kg HE with 4,100 steel balls	HE with 6,000 incendiary pellets and burning radius of 30 m	39 bomblets with 110 mm armour penetration and 7 m lethal radius	HE with 3,400 steel balls and burning radius of 30 m	6 AT mines or 96 anti-personnel mines (900 × 800 m full salvo laying zone)	8 AT mines or 128 anti-personnel mines (650 × 650 m full salvo laying zone)
Fuze type and settings:	impact with instantaneous, short or long delay	proximity or impact with instantaneous, short or long delay	impact with instantaneous, short or long delay	proximity	impact with instantaneous, short or long delay	proximity	proximity

Notes:
Extended-range (30,000 m) versions of the HE (lethal radius 28.2 m, 2.7 m long, 60 kg total round weight), HE cargo, HE fragmentation, HE incendiary and HE fragmentation incendiary (lethal radius 44.2 m and burning radius of about 30 m) rocket rounds are available. Details of these new rockets are given in a separate entry in *IHS Jane's Land Warfare Platforms: Artillery and Air Defence.* It should be noted that at least two Chinese companies are marketing unguided 122 mm rockets for use by the NORINCO 122 mm self-propelled multiple rocket systems, NORINCO and the Sichuan Aerospace Industry Corporation (SCAIC). The latest NORINCO family of 122 mm unguided rockets cover ranges of 20 km, 30 km and 40 km with all of these being of the unguided type.

NORINCO 122 mm Type 89 (40-round) self-propelled multiple rocket system deployed in firing position with launcher elevated 0501055

WS-6 rockets
Sichuan Aerospace Industry Corporation is marketing the WS-6 MRS, which uses 122 mm rockets with a maximum range of 40,000 m and that can be fired by the 122 mm (40-round) Type 89 system.

Specifications

	122 mm (40-Round) Type 89
Dimensions and weights	
Crew:	5
Length	
overall:	7.18 m
hull:	6.85 m
Width	
overall:	2.62 m
Height	
overall:	3.18 m
hull:	1.47 m
Ground clearance	
overall:	0.45 m
Track	
vehicle:	2.62 m
Length of track on ground:	4.601 m
Weight	
combat:	29,900 kg
Ground pressure	
standard track:	0.68 kg/cm^2
Mobility	
Configuration	
running gear:	tracked
Power-to-weight ratio:	17.39 hp/t
Speed	
max speed:	55 km/h
Range	
main fuel supply:	450 km
Fording	
without preparation:	1.3 m
Gradient:	60%
Side slope:	30%
Vertical obstacle	
forwards:	0.7 m
Trench:	2.6 m
Firepower	
Armament:	40 × 122.4 mm rocket system
Ammunition	
main total:	80
Main weapon	
maximum range:	20,580 m
Rate of fire	
rapid fire:	40/20 s (est.)
Main armament traverse	
angle:	168° (102° left, 66° right)
Main armament elevation/depression	
armament front:	+55°/0°
Survivability	
Night vision equipment	
vehicle:	yes
NBC system:	yes
Armour	
hull/body:	steel
turret:	steel

Status
Production complete. No longer marketed. In service with the People's Liberation Army. According to the United Nations Report on Conventional Arms issued in 2001, between 1992 and 2010 China exported the following multiple rocket launcher systems:

Country	Systems	Quantity
Armenia	WM-80	4
Gabon	107 mm	16 (delivered 2004)
Gabon	122 mm	4 (delivered 2004)
Sudan	not released	18
Bangladesh	not released	42

It is not known as to whether the latter two countries took delivery of 130 mm or 122 mm calibre systems.

Contractor
Chinese state factories. China North Industries Corporation (NORINCO).

NORINCO 122 mm (40-round) Type 90B multiple rocket system

Development
For many years China North Industries Corporation (NORINCO) has been marketing the 122 mm (40-round) Type 90 multiple rocket launcher based on a locally manufactured Mercedes-Benz (6 × 6) truck chassis. Full details of this are given in a separate entry in *IHS Jane's Land Warfare Platforms: Artillery and Air Defence.*

Marketing of the original 122 mm (40 round) Type 90 system is complete and main export emphasis is now on the NORINCO 122 mm (40-round) Type 90B. This has been in service with the People's Liberation Army for some years and has also been exported.

Further development of the 122 mm (40-round) Type 90 by NORINCO has resulted in the latest Type 90B, which is a complete weapon system.

According to NORINCO, the latest Type 90B differs from the earlier Type 90A as a reconnaissance vehicle has been added to the system. In addition, the overall mobility of the system is said to have been improved as has the reliability of the system.

A typical Type 90B battalion would consist of the following:
- Three reconnaissance vehicles based on the NORINCO WZ551 (6 × 6) armoured personnel carrier chassis.
- The baseline WZ551 has been in service with the PLA for many years with the latest model being the enhanced WMZ551 which for the export market has been referred to as the VN2A.
- Battalion command vehicle based on a Mercedes-Benz (4 × 4) truck chassis.
- Meteorological radar based on two Steyr (4 × 4) truck chassis.
- Maintenance vehicles based on Steyr (4 × 4) truck chassis, one would be for mechanical maintenance and the other for electronic maintenance.

Three batteries with each consisting of:
- Battery command vehicle on Mercedes-Benz (4 × 4) truck chassis.
- Six 122 mm rocket launchers on Mercedes-Benz (6 × 6) truck chassis.
- Six 122 mm rocket resupply vehicles based on Mercedes-Benz (6 × 6) truck chassis.

Development of the Type 90B is complete but there are no known export orders for the system as of March 2013.

The actual 122 mm unguided rockets used in the NORINCO 122 mm (40-round) Type 90B multiple rocket system are understood to be manufactured by another Chinese contractor.

Description
The 122 mm (40-round) launcher is based on a Beifang Benchi 2629 series 6 × 6 truck chassis, which in turn is a German Mercedes-Benz design.

Modified NORINCO WZ551 (6 × 6) reconnaissance vehicle with sensor pod raised 0526855

NORINCO 122 mm (40-round) Type 90B rocket supply vehicle with insert showing crane deployed in position (NORINCO) 1043229

The vehicle is fitted with a forward control type cab. This chassis is used for a number of applications by the People's Liberation Army including towing artillery systems. The Mercedes-Benz 6 × 6 cross-country truck chassis is made by NORINCO.

Equipment installed on the chassis includes the launching tubes, automatic operational and laying device, display terminal, radio unit, differential GPS, north-finding gyro, automatic loading device and a semi-automatic collapsible awning device.

The 122 mm launcher can launch all NORINCO 122 mm unguided solid propellant rockets with a minimum range of 20,000 m and a maximum range of 40,000 m.

NORINCO has also recently developed a new family of 40 km and 50 km unguided rockets.

China has also developed a combined Inertial Navigation System/Global Positioning System (INS/GPS) 122 mm rocket to provide a precision effect at longer ranged.

This is claimed to provide the 122 mm rocket with a CEP of less than 100 m at maximum range.

The Type 90B command vehicle has a fully-enclosed rear cabin in which is installed a message processing system, commander's computer and operating console, operator's computer and operating console, status display, communication controller, differential GPS, radio set and intercom. The unit is fully air conditioned and is normally operated by a crew of five.

The reconnaissance system is based on a modified WZ551 (6 × 6) amphibious APC with a raised superstructure to the rear of the commander and driver positions at the front of the vehicle. The WZ551 is used in significant numbers by the People's Liberation Army and has also been exported to a number of countries. These are known to include Oman and Sri Lanka.

Mounted on the roof is a reconnaissance pod that includes a laser range-finder and an optical device. This can be retracted into the vehicle under full armour protection if required.

The laser range-finder has a minimum range of 150 m and a maximum range of 10,000 m and can measure distances to an accuracy of 5 m.

The thermal viewer has two fields of view, narrow and wide, with detection range being quoted as 5,000 m and recognition range being 2,500 m. The CCD camera has two fields of view, wide and narrow and has a detection range of 8,000 m and an identification range of 4,000 m.

Mounted inside of the reconnaissance vehicle is a message processor, display, differential GPS as well as a radio and intercom system.

Armament consists of a roof-mounted, pintle-mounted 12.7 mm machine gun and a bank of four electrically operated smoke grenade dischargers either side of the full firing forwards. The reconnaissance vehicle has a crew of four.

The WZ551 is fully amphibious, being propelled in the water by two propellers mounted one either side, at the rear.

The Type 702D meteorological radar is based on two Steyr 4 × 4 trucks, with one having the meteorological radar and the other the support equipment.

The meteorological radar shelter consists of the following main subsystems. Antenna assembly, elevating ball screw mechanism, main control console, terminal computer system, communications, power distribution box and an air conditioning unit.

NORINCO 122 mm (40-round) Type 90B launcher based on a Mercedes-Benz truck chassis with launcher traversed left but stabilisers not deployed (NORINCO) 1043228

The meteorological support shelter consists of the following main subsystems. A 12 kW diesel generator, hydrogen bottles, water bucket, automatic balloon releasing device, ground meteorological instrument, meteorological equipment cabinet, radiosondes, balloons, reflectors, filling pipe and a weight set.

Detection range of the system is quoted as 200 km, while detection altitude is quoted as 25 km. Meteorological data measured includes windspeed, wind direction, air temperature, air humidity and air pressure.

The mechanical maintenance vehicle is mainly used to perform replaceable repair of mechanical and hydraulic components for the Type 90B launcher.

The electronic maintenance vehicle is mainly used to perform failure diagnostic and location as well as emergency trouble shooting of electronic, electric and optical electronic equipment of the system.

Specifications

	122 mm (40-Round) Type 90B
Dimensions and weights	
Crew:	5
Length	
overall:	9.315 m
Width	
transport configuration:	2.485 m
Height	
overall:	3.028 m
Mobility	
Configuration	
running gear:	wheeled
layout:	6 × 6
Speed	
max speed:	85 km/h
Range	
main fuel supply:	800 km
Gradient:	60%
Side slope:	30%
Firepower	
Armament:	40 × 122.4 mm rocket system
Main weapon	
maximum range:	40,000 m
Main armament traverse	
angle:	102° (left and right)
Main armament elevation/depression	
armament front:	+55°/0°

Status

Development complete. Ready for production. In 2004 China supplied Gabon with four 122 mm multiple rocket systems but it is not known which version these were. The original NORINCO 122 mm (40-round) Type 90 multiple rocket system is known to be in service with China and Oman (6) and possibly some other countries.

Contractor

Chinese state factories. China North Industries Corporation (NORINCO).

NORINCO 122 mm (40-round) Type 90 multiple rocket system

Development

The China North Industries Corporation (NORINCO) Type 90 122 mm 40-round multiple rocket system is an indigenously designed and built system equipped with an automatic operating and laying system, an electric firing system and an automatically reloadable pack of 40 122 mm unguided surface-to-surface rockets.

Production of the NORINCO 122 mm (40-round) Type 90 multiple rocket system has ceased and NORINCO is now concentrating its marketing on the similar 122 mm (40-round) Type 90B. This has enhanced capabilities and is covered in detail in a separate entry in *IHS Jane's Land Warfare Platforms: Artillery and Air Defence* but as of March 2013 there are no known exports of this version.

NORINCO Type 90A (6 × 6) multiple rocket system showing method of reloading 40 new 122 mm rockets to immediate rear of cab 0058456

NORINCO Type 90A (6 × 6) multiple rocket system deployed in the firing position 0058453

Description

The chassis used is the Tiema SC2030 6 × 6 truck model with four rows of 10 tubes mounted above each other on a rotating cradle assembly. Elevation and traverse are performed by electrical systems. A manual back-up system is also provided.

The chassis is a German Mercedes-Benz design that has been manufactured under licence by NORINCO for a variety of civil and military roles.

A hydraulically-operated collapsible awning is mounted over the rear deck for use as protection and camouflage. Conversion to the combat mode takes between 1½ and two minutes following arrival at the launch site.

The automatic operating and laying system consists of five main subsystems: an HJ-1 mini-computer; a launcher direction-finder (which takes less than three seconds to display the launcher's magnetic azimuth angle); longitudinal and transversal slope sensors; launcher assembly elevation and traverse position sensors; and a special vehicle power supply interface power pack unit.

The launcher is controlled by the computer through man/machine interaction. An interface is provided for connection to a battalion central command system, instructions from which can either be input directly via the interface or manually input by push-buttons.

There are three modes of operating and laying the system: manual, electrical or automatic. Hydraulic jacks at the rear and sides of the vehicle are used to minimise the effect of uneven ground or vibration during firing. The computer automatically corrects the firing solution for any deviation of the firing angle caused by ground slope.

The electrical firing system fires the unguided rockets either singly or in a salvo with 0.5 seconds between rounds. The system comprises an electric igniter, remote firing device and ignition contacts and cables. The unguided 122 mm rocket can be launched from the driver's cab or remotely from outside the vehicle.

The fin-stabilised rockets used are of the 122 mm calibre extended-range composite solid propellant type and full details of these are given in a separate entry in *IHS Jane's Land Warfare Platforms: Artillery and Air Defence.*

These rockets can be launched by other countries' 122 mm multiple rocket systems such as those manufactured by a number of countries including Croatia, Egypt, India, Iran, Pakistan, Poland, Russia, Serbia, Sudan, Slovak Republic and Turkey and probably other countries as well.

NORINCO has also recently developed a new family of 122 mm calibre 30/40 km range unguided rockets and a new family of 40 km and 50 km unguided rockets. Details of these are provided in separate entries in *IHS Jane's Land Warfare Platforms: Artillery and Air Defence.*

China has also developed a combined Inertial Navigation System/Global Positioning System (INS/GPS) 122 mm rocket to provide a precision effect at longer ranged.

This is claimed to provide the 122 mm rocket with a CEP of less than 100 m at maximum range.

The actual length of the 122 mm launch tubes used by this system is 3 m.

The automatic loading system consists of elevator, rack and feeder. Loading of the rockets can be performed either automatically, by push-button operation in the driver's cab, or manually by external operations. The loading mechanism can be rotated 90°. Automatic reloading of the launcher takes less than three minutes. A rocket-in-position and after-fire chamber display system in the driver's cab monitors the weapon status after loading or firing.

Variants

122 mm (40-round) Type 90A multiple rocket system

This was the second version of the Type 90 to enter production and the four main improvements can be summarised as:

- ability to launch all types of unguided 122 mm rocket out to a maximum range of 40 km including high explosive and high explosive incendiary.
- installation of a new computerised fire-control system which includes a Global Positioning System (GPS) and north finding device for greater accuracy. The panoramic sight is retained on the left side of the launcher for manual laying if required.
- high level of automation for the operator.
- command post vehicle can lay and control a number of Type 90A systems by remote control for maximum firepower.

The Type 90A has a modified Mercedes-Benz 6 × 6 chassis which is designated the Tienna XC 2200 (6 × 6) and this has a similar specification to the above apart from an overall length of 9.7 m, height of 3.20 m, fording depth of 0.7 m and a turning radius of 11 m. Prime contractor for the chassis in China is NORINCO.

A typical Type 90A multiple rocket launcher battalion would consist of a battalion command vehicle based on a 6 × 6 chassis, three forward observation post vehicles mounted on a 4 × 4 light all terrain vehicle chassis and three batteries. Each of the latter has one battery command vehicle, six rocket launchers, six rocket resupply vehicles and one maintenance vehicle.

122 mm (40-round) Type 90B multiple rocket system

This is the latest version of the Type 90 to be marketed and details are provided in a separate entry in *IHS Jane's Land Warfare Platforms: Artillery and Air Defence.* There are no known exports of this system. As far as it is known, all export marketing is now being concentrated on the NORINCO 122 mm (40-round) Type 90B.

Specifications

	122 mm (40-Round) Type 90
Dimensions and weights	
Length	
overall:	9.84 m
Width	
transport configuration:	2.5 m
Height	
transport configuration:	3.245 m
Ground clearance	
overall:	0.412 m
Weight	
combat:	20,000 kg (est.)
Mobility	
Configuration	
running gear:	wheeled
layout:	6 × 6
Speed	
max speed:	85 km/h
Range	
main fuel supply:	600 km
Fording	
without preparation:	0.9 m
Turning radius:	12 m
Engine	air cooled, diesel, 300 hp
Firepower	
Armament:	40 × 122 mm rocket system
Ammunition	
main total:	80
Main weapon	
length of barrel:	3.0 m
maximum range:	30,000 m (maximum 40,000 m with long range rocket)
Rate of fire	
rapid fire:	40/20 s (est.)
Operation	
reload time:	180 s
Main armament traverse	
angle:	102° (left and right)
Main armament elevation/depression	
armament front:	+55°/0°

Status

Production as required. In service with the People's Liberation Army. According to the United Nations Report on Conventional Arms issued in 2001, between 1992 and 2010 China exported the following multiple rocket launcher systems; (some sources have stated that the UAE has taken delivery of a quantity of 122 mm Type 90 systems):

Country	Systems	Quantity
Armenia	WM-80	4
Gabon	107 mm	16 (delivered 2004)
Gabon	122 mm	4 (delivered 2004)
Sudan	not released	18
Bangladesh	not released	42

It is not known as to whether the latter two countries took delivery of 130 mm or 122 mm calibre systems. Late in 2001, it was revealed that Oman had taken delivery of a quantity of Chinese 122 mm Type 90 series multiple rocket systems, which have been issued to at least one battalion.

Contractor

Chinese state factories. China North Industries Corporation (NORINCO).

NORINCO 122 mm calibre 40 km and 50 km range rockets

Development

In addition to marketing the 122 mm calibre 30 km/40 km range family of rockets, which is covered in a separate entry in *IHS Jane's Land Warfare*

Platforms: Artillery and Air Defence, China North Industries Corporation (NORINCO) is now marketing a new family of 40/50 km range of 122 mm rockets.

It is possible that this family of 122 mm rockets is manufactured by the Sichuan Aerospace Industry Corporation (SCAIC) and with NORINCO carrying out additional marketing.

Production of these unguided 122 mm rockets is undertaken on an as required basis and it is assumed that these 122 mm rockets can be fired from any 122 mm rocket launcher of Chinese or foreign manufacture.

Description

There are at least five unguided solid propellant 122 mm rockets in the latest NORINCO 122 mm 40 km and 50 km family of rockets.

All of these 122 mm unguided rockets feature a solid composite rocket motor with wraparound fins at the rear.

The SHE-40 122 mm rocket is fitted with a warhead that contains a smaller amount of high-explosive surrounded by 5,710 steel balls and is claimed to have a lethal radius of 47.4 m.

The C-40 122 mm cargo rocket contains 49 bomblets each with a diameter of 42.5 mm and each of these bomblets is claimed to have a lethal radius of 7 m.

The SHE-50 is the longest range rocket and is slightly heavier and is fitted with a warhead that contains a smaller amount of high explosive surrounded by 7,400 steel balls and is claimed to have a lethal radius of 50 m.

It addition there is a 122 mm fuel air explosive rocket and SHEI rocket which contains steel balls, high-explosive and an incendiary element.

Specifications

	SHE-40
Dimensions and weights	
Length	
overall:	2.903 m
Weight	
launch:	67.5 kg
Performance	
Range	
max:	40 km
Ordnance components	
Warhead:	22 kg

	C-40
Dimensions and weights	
Length	
overall:	3.023 m
Weight	
launch:	68.1 kg
Performance	
Range	
max:	40 km
Ordnance components	
Warhead:	22.6 kg

	SHE-50
Dimensions and weights	
Length	
overall:	3.037 m
Weight	
launch:	73.9 kg
Performance	
Range	
max:	50 km
Ordnance components	
Warhead:	24.3 kg

Status

Development complete. Production as required.

Contractor

China North Industries Corporation (NORINCO).

NORINCO 122 mm calibre 30 km/40 km range rocket family

Development

China North Industries Corporation (NORINCO) has developed a new family of 122 mm unguided calibre rockets for use with its range of wheeled and tracked 122 mm multiple rocket launchers and similar foreign made systems.

It is possible that these rockets are manufactured by the Sichuan Aerospace Industry Corporation (SCAIC) with NORINCO carrying out additional marketing.

Production of these 122 mm unguided rockets is undertaken on an as required basis for the home and export markets.

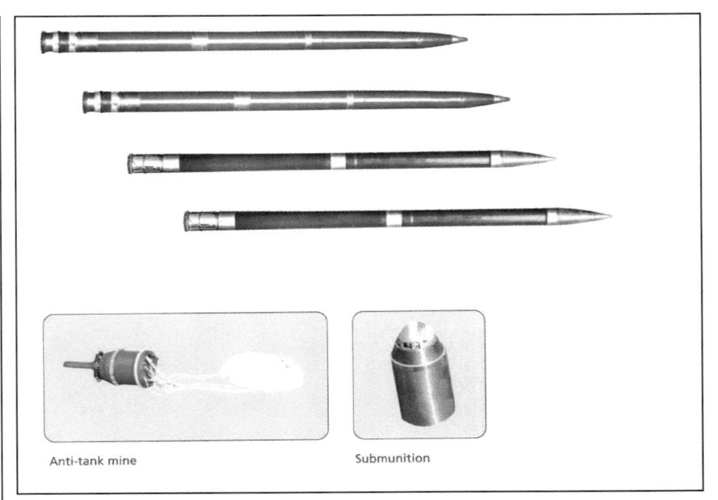

Some members of the NORINCO 122 mm family of unguided surface-to-surface rockets with anti-tank mine and submunition below (NORINCO)
1043230

The main advantage of this family of NORINCO 122 mm unguided rockets is not only their increased range, which increases the survivability of the launcher to counter battery fire, but also a wide range of warhead types to meet different operational requirements.

Description

The solid propellant unguided rockets are called the 30 km range series with a further development being a 40 km range series. They all feature a composite solid propellant motor with wrapround fins at the rear.

In addition to being launched from Chinese North Industries Corporation 122 mm multiple rocket launchers, these new 122 mm rockets can also be launched by 122 mm systems manufactured by other countries including Croatia, Egypt, India, Iran, Pakistan, Poland, Romania, Russia, Serbia, Sudan, Slovak Republic and Turkey.

NORINCO 122 mm calibre 40 km/50 km range rocket family

This family of longer range 122 mm rockets is now being marketed by NORINCO and available details are provided in a separate entry in *IHS Jane's Land Warfare Platforms: Artillery and Air Defence*.

Specifications

Type	Length	Weight	Warhead weight	Max. range	Min. range
HE	2.87 m	67 kg	18.3 kg	20,000 m	9,600 m
HE[1]	2.87 m	67 kg	18.3 kg	20,000 m	9,600 m
HE[1]	2.87 m	67 kg	18.3 kg	20,000 m	12,000 m
HE	2.757 m	61 kg	18.3 kg	33,000 m	12,700 m
HE[1]	2.757 m	61 kg	18.3 kg	33,000 m	12,400 m
HE[1]	2.757 m	61 kg	18.3 kg	33,000 m	12,700 m
HE[1]	2.900 m	67 kg	22 kg	40,000 m	20,000 m
Cargo[2]	3.037 m	66 kg	18.3 kg	20,000 m	9,600 m
Cargo[3]	3.037 m	68 kg	28 kg	26,000 m	13,000 m
Cargo[2]	2.927 m	61 kg	18.3 kg	32,000 m	15,000 m
Cargo[4]	3.008 m	67 kg	22 kg	40,000 m	20,000 m
Mine[5]	2.950 m	58 kg	26 kg	15,000 m	6,000 m

[1] contains high-explosive and steel balls for greater fragmentation effect
[2] contains 39 bomblets
[3] contains 74 bomblets
[4] contains 44 bomblets
[5] contains six anti-tank mines
Note: In addition there are also steel high-explosive incendiary rockets that are designated SHEI.

Status

Production as required. In service with Chinese People's Liberation Army. Offered for export. It is considered probable that some of these 122 mm unguided rockets have been exported to undisclosed countries.

Contractor

Chinese state factories. China North Industries Corporation (NORINCO).

NORINCO 122 mm (40-round) Type 81 multiple rocket system

Development

The China North Industries Corporation (NORINCO) 122 mm (40-round) Type 81 multiple rocket launcher system was developed by the No 743 and No 5137 factories and type classified in 1981.

NORINCO 122 mm (40-round) Type 81 multiple rocket system with launcher elevated 0500893

Production of the NORINCO 122 mm (40-round) Type 81 multiple rocket launcher system has been completed and it is no longer marketed by NORINCO.

Currently NORINCO is marketing the 122 mm (40-round) Type 90B system based on a Mercedes-Benz (6 × 6) cross-country truck chassis. Details of this are provided in a separate entry in *IHS Jane's Land Warfare Platforms: Artillery and Air Defence.* This may be in service with the People's Liberation Army.

Description

The 40-tube launcher is carried on a CQ261 Hongyan 6 × 6 truck chassis. The chassis is fitted with a four-door fully-enclosed steel cab. To the immediate rear of the cab is a spare wheel and tyre, and a bench seat for additional crew members.

There is also a short 122 mm HE rocket which is used in conjunction with a 2.5 m long single-tube animal or man-portable tripod-mounted 122 mm rocket launcher. This is designed for use by guerrilla, commando or special forces units.

The standard 122 mm unguided rocket weighs 46.3 kg complete and is 1.927 m long. Its maximum range is 13,400 m with the launcher having an elevation range of –3° to +50° and a traverse of +15°.

The unguided solid propellant rockets are fitted with a nose-mounted fuze with pre-selected instantaneous, short or long delay actions.

Total weight of the launcher when assembled for firing is 59 kg (which is made of the 25 kg tripod, the 30 kg launch tube and 4 kg for the sight and remote firing device with its 25 to 35 m long connector cable).

The launcher can also be fixed on a special mount for use on a light cross-country vehicle. In this application the launcher can either be fired from the vehicle or dismounted for ground use. A total of 10 to 15 reload rounds would normally be carried.

The Type 81 multiple rocket launcher system fires the same family of rockets as the 122 mm Type 83 already described. It can also fire the latest generation NORINCO family of 122 mm long range rockets covered in a separate entry in *IHS Jane's Land Warfare Platforms: Artillery and Air Defence.*

One part of this 122 mm solid propellant unguided rocket family has a maximum range of 30 km with the other having a maximum range of 40 km. There is also an anti-tank mine laying rocket with a maximum range of 15 km.

The standard indirect fire sighting system is on the left side of the launcher at the rear.

New rockets

The Sichuan Aerospace Industry Corporation is marketing the WS-6 rocket systems, which fire 122 mm rockets with a maximum range of 40 km. These can also be fired by the 122 mm (40-round) Type 81 MRS.

Details of these 122 mm rockets are provided in a separate entry in *IHS Jane's Land Warfare Platforms: Artillery and Air Defence.*

Production of these 122 mm rockets is undertaken for the home and export markets on an as required basis.

Variant

Upgraded 122 mm (40-round) Type 81 multiple rocket system

This is currently being marketed by Poly Technologies together with an associated 122 mm rocket re-supply vehicle. Details are provided in a separate entry in *IHS Jane's Land Warfare Platforms: Artillery and Air Defence.*

Specifications

	122 mm (40-Round) Type 81
Dimensions and weights	
Crew:	7
Length	
overall:	7.12 m
Width	
transport configuration:	2.5 m
Height	
transport configuration:	3.082 m
Weight	
combat:	15,160 kg
Mobility	
Configuration	
running gear:	wheeled
layout:	6 × 6
Firepower	
Armament:	40 × 122 mm rocket system
Ammunition	
main total:	40
Main weapon	
length of barrel:	3.0 m
maximum range:	30,000 m (see text)
Rate of fire	
rapid fire:	40/20 s (est.)
Operation	
reload time:	600 s (est.)
Main armament traverse	
angle:	172° (102° left, 70° right)
Main armament elevation/depression	
armament front:	+55°/0°

Status

Production complete. No longer marketed. In service with Chad (18), Ghana (3), the People's Liberation Army and Sudan (18).

Contractor

Chinese state factories. China North Industries Corporation (NORINCO).

NORINCO 122 mm (24-round) Type 83 multiple rocket system and Type 84 minelaying rocket system

Development

Development of the China North Industries Corporation (NORINCO) Type 84 minelaying rocket system began in 1977 and was jointly carried out by the People's Liberation Army Engineer Corps Technical Equipment Research Institute and factories Nos 743 and 5137.

Production and marketing of these rocket launchers and rockets has now ceased by NORINCO.

Currently NORINCO is marketing the 122 mm (40-round) Type 90B system based on a Mercedes-Benz (6 × 6) forward control cab truck chassis. Details of this system are provided in a separate entry in *IHS Jane's Land Warfare Platforms: Artillery and Air Defence.*

It is understood that development of the latter system is complete but as of March 2013 there are no known export sales of this system.

Description

The Type 84 consists of a 24-round 122 mm multiple rocket system mounted on the rear of a 6 × 6 cross-country chassis. The engine is at the front, the four-door, fully-enclosed cab in the centre and the rocket launcher at the rear.

The sighting system is mounted on the left side of the launcher and consists of a panoramic sight Type 58 and a dependent yoke type sight collimator Type 58.

Speed of electro-operated laying is 5.7° in elevation and 9.3° in traverse. Speed of manual operated laying is 1.1 mil per turn of handle in elevation and 1.8 mil per turn of handle in traverse. The manual controls are mounted near the sighting system.

Reload rockets are carried by a dedicated resupply vehicle. The 122 mm minelaying cargo round carries eight small anti-tank mines or 128 anti-personnel mines and is stated to have a maximum range of 7,000 m.

A single vehicle salvo can lay a 650 m long by 400 m wide minefield at 6,000 m range. The rocket containing anti-tank mines weighs 62 kg and has a launch velocity of 350 m/s. The rocket is 2,770 mm long.

In addition to these there is also a short 122 mm HE rocket which is used in conjunction with a 2.5 m long single-tube, animal- or man-portable, tripod-mounted 122 mm rocket launcher. This is designed for use by guerrilla, commando or special forces units.

The unguided solid propellant 122 mm rocket weighs 46.3 kg complete and is 1.927 m long. Its maximum range is 13,400 m with the launcher having an elevation range of –3° to +50° and a traverse of +15°. A contact fuze with pre-selected instantaneous, short or long delay actions is fitted.

Total weight of the complete 122 mm launcher when assembled for firing is 59 kg (which is made up of the 25 kg tripod, the 30 kg launch tube and 4 kg for the sight and remote firing device with its 25 to 35 m long connector cable).

The launcher can also be fixed on a special mount for use on a light cross-country vehicle. In this application the launcher can either be fired from the vehicle or dismounted for ground use. A total of 10 to 15 reload rounds would normally be carried.

The elevation/depression of the launcher is +55° to 0° and the traverse ±90° of the central axis. A full salvo of 122 mm rockets takes between 15 and 20 seconds to fire.

The designation of the standard 122 mm (24-round) multiple rocket launcher system is Type 83.

The multiple rocket launcher system fires the same family of rockets as the 122 mm Type 83 system already described.

The Type 83 will also fire the latest generation of NORINCO 30 km and 40 km 122 mm unguided rockets, which are covered in a separate entry in *IHS Jane's Land Warfare Platforms: Artillery and Air Defence*. These are fitted with a wide range of warheads enabling different battlefield targets to be engaged.

Production of these 122 mm unguided rockets is undertaken on an as required basis for the home and export markets.

New rockets

The Sichuan Aerospace Industry Corporation is now marketing the WS-6 rocket system, which fires 122 mm rockets with a maximum range of 40 km. It is understood that these can also be fired by the 122 mm (24-round) Type 83 system.

There is a separate entry in *IHS Jane's Land Warfare Platforms: Artillery and Air Defence* for these 122 mm rockets.

Production of these 122 mm unguided rockets is undertaken on an as required basis for the home and export markets.

Specifications

	122 mm (24-Round) Type 83
Dimensions and weights	
Crew:	6
Length	
overall:	6.51 m
Width	
transport configuration:	2.2 m
Height	
transport configuration:	2.7 m
Ground clearance	
overall:	0.285 m
Weight	
combat:	8,700 kg
Mobility	
Configuration	
running gear:	wheeled
layout:	6 × 6
Speed	
max speed:	80 km/h
Range	
main fuel supply:	550 km
Fording	
without preparation:	0.85 m
Gradient:	60%
Engine	diesel, 135 hp
Firepower	
Armament:	24 × 122 mm rocket system
Main weapon	
length of barrel:	3.0 m
maximum range:	20,580 m (HE rocket)
maximum range:	7,000 m (minelaying rocket)
Main armament traverse	
angle:	180° (90° left, 90° right)
Main armament elevation/depression	
armament front:	+55°/0°

Status

Production complete. In service with People's Liberation Army. This system is no longer marketed and, as far as it is known, it was not exported. It is believed that the only 122 mm multiple rocket system currently being marketed by NORINCO is the 122 mm (40-round) Type 90B based on a locally manufactured Mercedes-Benz (6 × 6) cross-country truck chassis.

Contractor

Chinese state factories. China North Industries Corporation (NORINCO).

NORINCO 107 mm (12-round) Type 63 and Type 81 multiple rocket systems

Development

The China North Industries Corporation (NORINCO) 107 mm (12-round) Type 63 MRS was developed in the late 1950s as the replacement for the 102 mm (6-round) Type 427 and Type 488 multiple rocket systems which were withdrawn from front-line service with the People's Liberation Army many years ago.

The Type 63 was developed by the No 803 and No 847 factories and type-classified in 1963. It was issued on the scale of 18 per Chinese infantry division.

It is considered probable that this system has been replaced in front line units of the People's Liberation Army by more recent NORINCO rocket launchers (for example 122 mm calibre and larger) with a longer range and firing more effective rockets.

Pedestal-mounted 107 mm (12-round) Type 63 multiple rocket launcher captured in Lebanon (Israeli Ministry of Defence) 0501056

NORINCO 107 mm (12-round) Type 81 multiple rocket launcher on rear of 4 × 4 truck 0021044

Large numbers of trailer and truck (for example GAZ-66) mounted 107 mm (12-round) multiple rocket launchers were observed during the fighting in Afghanistan late in 2001.

At this stage it is not known as to the source of these systems, as the 107 mm rocket system has been made by many countries. These include Iran, Iraq, North Korea, South Africa, Sudan and Turkey.

All of these can in theory launch the unguided surface-to-surface rockets.

In a number of recent operations these unguided 107 mm surface-to-surface rockets have also been launched on an individual basis from improvised launchers or even pointing in the general direction of the target.

NORINCO has now ceased all marketing of these 107 mm launchers but production of the 107 mm rockets may well be carried out on an as required basis.

Description

The 107 mm (12-round) Type 63 towed rocket launcher has three banks of four 107 mm barrels and is mounted on a rubber-tyred split pole-type carriage.

In the firing position the wheels are removed and the launcher is supported by the two trails at the rear which are fitted with spades, which are carried on the top of the trails when travelling, and two short legs in the front of the carriage. The unguided 107 mm solid propellant rockets are spin stabilised.

An improved version of the Type 63 is mounted on a 4 × 4 truck, designated Type 81, with an enlarged fully-enclosed cab to accommodate the launcher crew of four and 12 reload rounds. In transit the launcher is covered by a tarpaulin.

There is a seat for one crew member outside the cab and to the rear. When firing, the vehicle suspension is locked and the entire system can be removed from the vehicle if required and placed on a towing carriage.

Firing rate is 12 rockets in 10 to 12 seconds from the vehicle or 12 rockets is six to nine seconds when deployed on the ground.

According to NORINCO, reload time for both systems is about three minutes.

In the specifications, the weight of the 107 mm (12-round) Type 63 trailer mounted system without the rockets is 385 kg.

In the specifications, the weight of the 107 mm (12-round) Type 81 4 × 4 system with one set of 107 mm rockets and a crew of four is quoted as 3,709 kg.

The launcher can be fired from inside the cab or using a cable from a remote position. Maximum road speed is 100 km/h.

A pack model developed for use by airborne and mountain units weighs 281 kg in the firing position and can be dismantled into manpack loads.

A lighter version of the Type 63 has been given the Western designation of Type 63-I. It is 136 kg lighter than the basic model and is distinguishable from it by its smaller spoked roadwheels and its four banks of three rockets.

107 mm (12-round) Type 63 multiple rocket launcher 0021045

The 107 mm unguided rocket types available are given in the table below.

There is also a 0.9 m long Type 85 single man-portable tripod-mounted launcher tube for 107 mm rockets which is designed for use by commando, guerrilla or special forces units.

This weighs 22.5 kg in total and can be disassembled into small loads for carriage. Its elevation range is +5° to +45° and its traverse range 7° to the left or right. Time to bring the weapon into action and to disengage is about 30 seconds. Typical rate of fire is 3 rds/min. A remote firing device is used to ignite the rocket.

The 107 mm unguided rocket can also be fired by laying it on a suitable piece of sloped or raised terrain and pointing it in the direction of the target. In this application a set of small batteries with an electrical cable can be used as the firing mechanism.

The accuracy is not very good against point targets unless the launch occurs very close to them but is more than sufficient for area targets such as population centres or large military installations.

All of the 107 mm rockets have a muzzle velocity of 34 m/s and can be used in ambient temperatures from −40°C to +50°C.

Variants

During the 1982 invasion of Lebanon, a number of locally modified Type 63 MRSs were captured from the Palestinian guerrillas by the Israelis. These consisted of the basic 12-round launcher fitted to a crudely built pedestal and mounted on the rear of a light 4 × 4 (or similar chassis) truck chassis to increase mobility.

Iranian 107 mm MRS

The Defence Industries Organisation (DIO) of Iran is currently manufacturing a number of 107 mm (12-round) multiple artillery rocket systems including the FADGR and HASEB. Details of these are given in a separate entry in *IHS Jane's Land Warfare Platforms: Artillery and Air Defence.* It is considered possible that quantities of these rocket launchers and associated unguided rockets have been exported.

Iraqi 107 mm MRS

Iraq manufactured a copy of the Chinese Type 63 107 mm (12-round) Type 63 multiple artillery rocket launcher. There has been no production of this for some years.

Jordanian 107 mm MRS

This country has a small quantity of US supplied AM General HMMWV (4 × 4) vehicles with a modified rear, on which is mounted a 107 mm (12-round) rocket launcher.

NORINCO 130 mm single-tube rocket launcher deployed in the firing position 0021046

Details of this are provided in a separate entry in *IHS Jane's Land Warfare Platforms: Artillery and Air Defence* and is only used by Jordan.

North Korean 107 mm MRS

These are known to include a 107 mm (18-round) launcher on the VTT 323 (M1992) full tracked armoured personnel carrier and the 107 mm (24-round) M1992/2 self-propelled multiple rocket launcher.

South African 107 mm MRS

The South African company Mechem Developments reverse engineered the 107 mm (12-round) Type 63 system under the local designation of the RO 107 (12-round) Towed Multiple Rocket Launcher System. There is also a single 107 mm launcher called Inflict. Details are given in a separate entry in *IHS Jane's Land Warfare Platforms: Artillery and Air Defence.*

Sudanese 107 mm MRS

The Military Industrial Corporation of the Sudan is currently marketing an indigenous 107 mm (12-round) multiple rocket launcher and associated ammunition under the local name of the Taka. Available details are provided in a separate entry in *IHS Jane's Land Warfare Platforms: Artillery and Air Defence.*

Turkish 107 mm MRS

MKEK of Turkey has manufactured a 107 mm (12-round) multiple rocket launcher and details of this are given in a separate entry in *IHS Jane's Land Warfare Platforms: Artillery and Air Defence.* Roketsan make the rockets for this system. Production of the launcher has been completed although Roketsan continues to manufacture 107 mm rockets for the home and export markets as and when required.

Specifications

	107 mm (12-Round) Type 63	107 mm (12-Round) Type 81
Dimensions and weights		
Crew:	5	4
Length		
overall:	2.6 m	4.6 m
Width		
transport configuration:	1.4 m	1.964 m

Designation	Type 63-II HE rocket	Type 63-III HE rocket	Type 63-I HE incendiary rocket	HE fragmentation rocket	Type 63 HE Incendiary rocket	Jamming rocket
Calibre:	107 mm	107 mm	107 mm	107 mm	107 mm	107 mm
Length:	0.841 m	0.897 m (with mechanical fuze)	0.915 m	n/avail	0.840 m	0.920 m
		0.947 m (with electronic fuze)				
Weight: (complete)	18.84 kg	n/avail	18.74 kg	18.9 kg	18.8 kg	18.3 kg
Maximum range:	8,500 m	10,000 m	8,000 m	7,800 m	8,500 m	7,800 m
Warhead type:	HE fragmentation. Produces approx. 1,200 fragments. Lethal radius of 18.5 m	HE fragmentation. Lethal radius 12.5 m	HE fragmentation incendiary. 40 s incendiary burning time.	HE preformed fragmentation 1,600 6.35 mm diameter steel spheres, in addition to casing fragments	HE fragmentation incendiary. 1,600 incendiary pellets. Incendiary effect radius of about 21 m	parachute-deployed jammer unit with duration of up to 15 mins
Fuze type and settings:	radio frequency proximity or MJ-1 contact with instantaneous and delay actions					

	107 mm (12-Round) Type 63	107 mm (12-Round) Type 81
Height		
transport configuration:	1.2 m	2.185 m
Weight		
combat:	613 kg	3,615 kg
Mobility		
Configuration		
running gear:	wheeled	wheeled
layout:	-	4 × 4
Firepower		
Armament:	12 × 107 mm rocket system	12 × 107 mm rocket system
Ammunition		
main total:	12	24
Rate of fire		
rapid fire:	12/7 s (est.)	12/10 s (est.)
Operation		
reload time:	180 s	180 s
Main armament traverse		
angle:	30°	105° (+45°/-60°)
Main armament elevation/depression		
armament front:	+60°/0°	+60°/0°

Status
Production as required. In service with the following countries:

Country	Quantity	Comment
Bosnia-Herzegovina	28	
Burkina Faso	4-6	and Marines
Cambodia	25-30	
China	n/avail	Army and Marines
Democratic Republic of Congo	12	
Gabon	16	delivered by China in 2004
Iran	600-700	local production
Jordan	n/avail	including some installed on HMMWV chassis
Korea, North	n/avail	local production
Libya	250-300	
Mongolia	30	
Myanmar	30	
Nicaragua	33	
Pakistan	50	
Sudan	400	estimate
Syria	200	
Uganda	6-12	
Vietnam	360	
Zimbabwe	18	

They are also widely used by various guerrilla forces in many parts of the world.

Contractor
Chinese state factories. China North Industries Corporation (NORINCO).

Poly Technologies upgraded 122 mm (40-round) Type 81 multiple rocket system

Development
Poly Technologies are now marketing an upgraded version of China North Industries Corporation (NORINCO) 122 mm (40-round) Type 81 Multiple Rocket System (MRS).

Details of the latter system, which is no longer manufactured or marketed by NORINCO, are provided in a separate entry in *IHS Jane's Land Warfare Platforms: Artillery and Air Defence.*

At this stage it is not known as to whether this upgraded system is in service with the Peoples Liberation Army (PLA).

This has been offered on the export market by Poly Technologies for several years but as of March 2013 there are no known exports of this system.

Description
On the original NORINCO 122 mm (40-round) Type 81 MRS, once the 40 122 mm unguided rockets had been launched the vehicle would normally deploy to another position where new rockets would be reloaded manually one at a time.

A resupply vehicle has now been developed on a similar 6 × 6 cross-country chassis with a four door fully enclosed forward control cab.

Mounted to the rear of the chassis is a pod of 40 × 122 mm rockets on a mechanically operated reloading platform with the noses pointing to the rear.

Standard NORINCO 122 mm (40-round) Type 81 multiple rocket system with launcher elevated and traversed to the left and shutters in position over forward control cab windows (Poly Technologies) 1365051

On the left is NORINCO 122 mm (40-round) Type 81 multiple rocket system with launcher traversed to the left and on the right is the rocket resupply vehicle ready to reload the launcher with another 40 122 mm rockets (Poly Technologies) 1365052

Once the 122 mm (40-round) Type 81 MRS has carried out a fire mission it would now move to a pre-determined area where the resupply vehicle would be waiting.

The 122 mm (40-round) Type 81 MRS comes to a halt and the empty launcher is traversed 90° to the left at 0° elevation.

The 122 mm rocket resupply vehicle then backs up to the launcher and the pod of rockets is aligned with the launcher by the operator using a remote-control panel to the rear of the cab.

Once aligned the 122 mm rockets are pushed into the launcher mechanically and once reloaded the resupply vehicle moves away with the complete sequence taking less than five minutes as well as requiring less manpower.

To allow the rocket resupply platform to be correctly aligned with the launcher the platform can be moved in the vertical plane to ±100 mm and in length 2,520 mm. Tilting of the rocket resupply platform is ±5° on traverse and ±5° in longitude.

The 122 mm (40-round) Type 81 MRS can fire a complete salvo of 40 × 122 mm rockets in 18 to 22 seconds with the launcher having a total traverse of 172° and with elevation from 0° to +55°.

A new computerised fire-control system has been developed for the 122 mm (40-round) Type 81 MRS which not only improves accuracy but also reduces the time taken for the weapon system to be deployed.

The standard 122 mm unguided surface-to-surface rockets has a maximum range of 20.50 km and in addition to being fitted with a High-Explosive (HE) warhead can also be fitted with other types including HE fragmentation, HE incendiary, HE cargo, HE fragmentation incendiary and minelaying.

China is now marketing two new families of unguided 122 mm rockets called the 30 km family and the 40 km family.

These longer range rockets enable the launcher to be deployed more towards the rear to avoid potential counter battery fire.

The 30 km family have a minimum range of 12.7 km and a maximum range of 32.7 km with four warheads being marketed with the 40 km family having a maximum range of about 40 km and with three warhead types.

Details of these NORINCO 122 mm unguided rockets are provided in a separate entry in *IHS Jane's Land Warfare Platforms: Artillery and Air Defence.*

NORINCO has now developed and is offering on the export market two new families of 122 mm rockets with a longer range.

Details of these are provided in a separate entry in *IHS Jane's Land Warfare Platforms: Artillery and Air Defence* and are called:

122 mm rocket, 40 km and 50 km range series

122 mm rocket, 30 km range series

In addition to being launched from the numerous Chinese 122 mm rocket systems, these can also be used from foreign rocket launchers of a similar calibre which are the most widely used type in the world.

The 122 mm rocket was originally developed and manufactured in Russia but has been subsequently copied or manufactured under license by many countries including Croatia, Egypt, India, Korea (North), Pakistan, Poland, Romania, Serbia, Sudan, Slovakia and Turkey.

It should be noted that not all of these countries are currently manufacturing these 122 mm unguided rockets.

In overall concept this reload system is very similar to the former Czechoslovakian RM-70 122 mm (40-round) system concept based on Tatra (8 × 8) truck chassis.

In addition to the launcher mounted at the very rear of the chassis this has a reload pod of 40 × 122 mm rockets to the immediate rear of the cab.

NORINCO is no longer marketing the 122 mm (40-round) Type 81 launcher and instead is concentrating its efforts on the latest 122 mm (40-round) Type 90B which is covered in a separate entry in *IHS Jane's Land Warfare Platforms: Artillery and Air Defence.*

Specifications

	122 mm (40-round) Type 81
Dimensions and weights	
Length	
overall:	7.12 m
Width	
overall:	2.50 m
Height	
overall:	2.90 m
Ground clearance	
overall:	0.355 m
Weight	
standard:	11,800 kg
combat:	14,500 kg
Mobility	
Configuration	
running gear:	wheeled
layout:	6 × 6
Speed	
max speed:	60 km/h
Fording	
without preparation:	1.2 m
Turning radius:	9 m

Status

Development complete. Production as required.

Contractor

Poly Technologies.

Other Chinese MRL systems

Description

The People's Liberation Army is understood to have a number of other multiple rocket systems in service on which only limited information is available at present. It is understood that all of these use unguided rockets supplied by China North Industries Corporation (NORINCO).

As far as it is currently known, none of these rocket systems are currently in production or are being marketed.

Variants

180 mm (10-round) Type 71 multiple rocket launcher system

This is a truck-mounted system and has 10 cage-type launch rails for unguided 180 mm calibre rockets. These rockets are 2.737 m long, weigh 134.8 kg at launch and carry a 45 kg HE warhead. Movable tail fins are fitted for stabilisation. Maximum velocity of the rocket is around 620 m/s, which gives a maximum range of 20,000 m.

20-round minelaying rocket system

This is believed to be truck-mounted. No other details are available.

10-round mineclearing rocket system

This is of unknown calibre and is carried on the rear of a 4 × 4 truck chassis. No other details are available.

Sichuan rockets

In addition to the NORINCO and China Precision Machinery Import and Export Corporation (CPMIEC) series of MRL, it has been revealed that the Sichuan Aerospace Industry Corporation (SCAIC) manufactures rockets. These systems include:

- WS-2 (400 mm calibre)
- WS-1D (252 mm calibre)
- WS-6 (122 mm calibre).

Available details of these are given in separate entries in *IHS Jane's Land Warfare Platforms: Artillery and Air Defence.* These are being marketed overseas but there are no known sales to date. Some of these rockets appear to be in direct competition to those marketed by NORINCO. It is however possible that NORINCO provides the launcher and its systems integrator while SCAIC provides some of the rockets. This especially applies to the NORINCO 122 mm rockets.

In addition to Aerospace Long-march International Trade Company, NORINCO, CPMEIC, SCAIC other contractors in China also market multiple rocket launchers including Poly Technologies.

Status

As far as it is known none of these multiple launch rocket systems has been exported outside of China. It is understood that there has been no recent production of any of these complete systems. None of the recent marketing material issued by NORINCO has listed any of these systems.

Sichuan Aerospace Industry Corporation (SCAIC) is also marketing complete rocket systems as well as individual rockets.

Contractor

Chinese state factories. It is probable that most of these systems have been developed by China North Industries Corporation (NORINCO).

Croatia

RH ALAN 128 mm (24-round) LOV RAK 24/128 self-propelled rocket launcher

Development

The LOV RAK 24/128 rocket launcher is based on the 4 × 4 LOV light armoured vehicle hull which in its basic form is used as an armoured personnel carrier.

Production of the LOV-OP and its variants is now complete and it is no longer being marketed by Croatia.

It is understood that there has been no recent marketing of this rocket system and it is probable that production of the complete system is complete.

Description

The baseline LOV hull is almost identical for this application with the 128 mm (24-round) launcher mounted on the roof.

The all welded steel armour of the LOV is claimed to provide protection from 7.62 × 51 mm armour piercing ammunition and high explosive shell fragments.

For this application the lower part of the hull has been cut away and is now flat above the last road wheel station to facilitate the loading of new 128 mm rockets when a fire mission has been carried out.

The sighting system consists of an PC-1 panoramic sight with a TT-2 hand held computer being provided to enhance accuracy.

The 128 mm (24-round) rocket launcher can fire two types of locally manufactured unguided 128 mm rocket, designated the M91 and M93, both fitted with a nose-mounted impact fuze.

It should be noted that the specifications provided in the table are different from the Serbian 128 mm rockets covered in detail in a separate entry in *IHS Jane's Land Warfare Platforms: Artillery and Air Defence.*

The launcher is fitted with an electronic elevation system of –5° to +45° and can be traversed a full 360°. An additional 24 rockets are stored in the rear section of the hull. The launcher system can be automatically levelled when at the firing position.

Training the system and controlling the firing sequence is performed either on the vehicle or remote from it using a hand-held computer. In a normal fire mission the launcher would redeploy to another position to avoid countrybattery fire.

The official army designation for the system is the SVLR 128 M91 RAK-24 with SVLR standing for Samovozni Visecijevni Lanser Raketa.

RH ALAN 128 mm (24-round) LOV RAK 24/128 self-propelled rocket launcher in firing position, with launcher traversed to the right 0008308

Specifications

	128 mm (24-Round) LOV RAK 24/128
Dimensions and weights	
Crew:	4
Length	
overall:	5.89 m
Width	
overall:	2.36 m
Ground clearance	
overall:	0.315 m
Weight	
combat:	8,500 kg
Mobility	
Configuration	
running gear:	wheeled
layout:	4 × 4
Speed	
max speed:	100 km/h
Range	
main fuel supply:	700 km (est.)
Fuel capacity	
main:	170 litres
Fording	
without preparation:	1 m
Gradient:	65%
Side slope:	30%
Vertical obstacle	
forwards:	0.5 m
Engine	Torpedo BF6L912S, turbocharged, diesel, 130 hp at 2,650 rpm
Gearbox	
type:	manual
forward gears:	5
Transfer box:	2 ranges
Steering:	hydraulic, power-assisted
Axles	
front:	rigid with differential locks
rear:	rigid with differential locks
Tyres:	14.5R20 with run-flat inserts
Electrical system	
vehicle:	24 V
Firepower	
Armament:	24 × 128 mm rocket system
Ammunition	
main total:	48
Main weapon	
length of barrel:	1.3 m
maximum range:	8,550 m (standard rocket)
maximum range:	13,000 m (extended range)
Rate of fire	
rapid fire:	24/16.8 s
Main armament traverse	
angle:	360°
Main armament elevation/depression	
armament front:	+45°/-5°

Rockets

Designation	M91	M93
Calibre:	128 mm	128 mm
Weight:		
(complete rocket)	23.5 kg	26 kg
(warhead)	8.5 kg	9 kg
(propellant)	4.5 kg	7.5 kg
Spinning speed:	16,000 rpm	26,000 rpm
Maximum range:	8,550 m	13,000 m

Status
Production complete. No longer marketed. In service with the Croatian and BiH armies. There are no known exports of this system.

Contractor
RH ALAN d.o.o.

RH ALAN 128 mm (12-round) VLR 128 M91A3 RAK 12 multiple rocket launcher

Development
The RH ALAN 128 mm (12-round) VLR 128 M91A3 RAK 12 is a mobile trailer-mounted rocket system which fires the same family of 128 mm unguided rockets as the RH ALAN 128 mm (24-round) LOV RAK 24/128 unguided

RH ALAN 128 mm (12-round) VLR 128 M91A3 RAK 12 multiple rocket launcher deployed in firing position, with rocket leaving right launcher 0008309

128 mm self-propelled rocket launcher covered in a separate entry in *IHS Jane's Land Warfare Platforms: Artillery and Air Defence.*

It should be noted that RH ALAN is a marketing organisation. All key parts of this system have been manufactured by other contractors.

It is understood that production of this system has now been completed and it is no longer being marketed.

Description
The 128 mm (12-round) rocket launcher can fire two types of locally manufactured 128 mm solid propellant rocket, both fitted with a nose mounted impact fuze and details of these rockets are listed in the table:

Specifications
Rocket

Designation	M91	M93
Calibre:	128 mm	128 mm
Weight:		
(complete rocket)	23.5 kg	26 kg
(warhead)	8.5 kg	9 kg
(propellant)	4.5 kg	7.5 kg
Spinning speed:	16,000 rpm	26,000 rpm
Maximum range:	8,550 m	13,000 m

Note:
It should be noted that the specifications provided here are different from the Serbian 128 mm rockets covered in detail in a separate entry in *IHS Jane's Land Warfare Platforms: Artillery and Air Defence.*

VLR 128 M91A3 RAK 12 rocket launcher

Calibre:	128 mm
Number of barrels:	12
Length of barrel:	1.25 m
Weight:	
(loaded)	850 kg
(unloaded)	580 kg
Maximum range:	
(M91 rocket)	8,500 m
(M93 rocket)	13,000 m
Rate of fire:	single round fire with 1 rocket/s, ripple fire with 1 rocket/0.75 s
(traverse)	±15°
(elevation)	−1° to +46°

Status
Production complete. No longer marketed. In service with the Croatian Army and Croatian Defence Council. There are no known exports of this rocket system.

Contractor
RH ALAN d.o.o.

RH ALAN SVLR 122 mm (32-round) M96 Typhoon self-propelled multiple rocket launcher

Development
The RH ALAN 122 mm (32-round) M96 Typhoon self-propelled multiple rocket launcher system is based on the former Czechoslovakia Tatra T813 (8 × 8) cross-country truck chassis. In the designation SLVR stands for *Samovozni Visecijevni Lanser Raketa*.

It should be noted that RH ALAN is a marketing organisation. The actual launchers and rockets for the system are made by other contractors.

This system was offered on the export market by Croatia but as far as it is known no export sales were made. Production of the system is now complete and as far as it is known it is no longer marketed.

RH ALAN SVLR 122 mm (32-round) M96 Typhoon multiple rocket launcher showing pack of replacement 122 mm rockets to rear of cab 0008310

Description

The Tatra T813 (8 × 8) cross-country truck chassis is fitted with the standard unarmoured fully-enclosed two-door cab of the forward-control type with a pod of 32 rockets to the rear of the cab ready for rapid loading into the 32-round launcher assembly.

The launcher assembly is controlled by a microprocessor-based control panel system in the cab and mounted on the aft end of the truck's rear decking. Loading, under full power control, takes two minutes.

When travelling, a sliding tarpaulin cover extends right to the rear of the vehicle covering the launcher and the rockets so making the system difficult to distinguish from a conventional covered Tatra (8 × 8) truck.

With minor modifications the launcher system could be fitted to other cross-country truck types such as a MAN or Mercedes.

When deployed in the firing position four stabilisers are lowered to the ground. Standard equipment for the chassis includes a central tyre pressure-regulation system that allows the driver to adjust the tyre pressure to suit the ground being crossed and a front-mounted hydraulically-operated winch.

The original chassis could also be fitted with a front-mounted dozer blade, which could also be used to clear obstacles and prepare a fire position.

The suspension system can also be locked when the truck is deployed in the firing system. The suspension locking system is automatic/manual. The launcher has a powered traverse of 105° left and right with powered elevation from 0° to +50°. Manual back-up systems are fitted.

Levelling is ±6° longitudinal and ±10° in traverse.

The 122 mm unguided solid propellant surface-to-surface rockets are normally fired in salvos with an interval of 0.6 seconds between each rocket being launched. If the operator selects single shots then one round is fired every second. The methods of firing are electronic and can be either from inside the cab itself or from a remote position up to approximately 30 m from the truck.

The latter also has a manual triggering option. The Typhoon is fitted with a Global Positioning System (GPS) and a gyroscope for accurate firing position and targeting purposes. Time to fire two complete salvos from the same launch position is five minutes.

The locally produced 122 mm unguided RAK 122 M93 rocket has a maximum range of 20,400 m and is 2.878 m long. The rocket weighs 66 kg in total of which the warhead accounts for 18.5 kg and the propellant 20.5 kg. The maximum speed at motor burnout is 699 m/s. A point detonating fuzing system is fitted.

The full range of Russian Federation 122 mm calibre rockets and its clones can also be used. The Croatians are also using the 122 mm M21-OF 9M22U rocket with a point detonating fuze and similar 20,400 m maximum range to the M95.

In many respects, the RH ALAN 122 mm (32-round) M96 Typhoon self-propelled rocket launcher is a 122 mm version of the Yugoslav (today Serbian) 128 mm (32-round) M-77 Oganj multiple rocket launcher developed in the early 1970s on a 6 × 6 chassis. As far as it is known there has been no recent production of this system. As production of the Tatra T813 truck chassis was completed some time ago, any new production would have to be on the more recent Tatra T815 (8 × 8) truck chassis. This is in production for a wide range of civil and military applications and is manufactured under licence in India.

Specification

	SVLR 122 mm (32-Round) M96 Typhoon
Dimensions and weights	
Length	
overall:	9.4 m
Width	
overall:	2.5 m
Height	
transport configuration:	3.4 m
deployed:	4.26 m (at max elevation)
Track	
vehicle:	2.03 m
Wheelbase	1.65 m + 2.2 m + 1.45 m
Weight	
standard:	19,000 kg
combat:	23,500 kg
Mobility	
Configuration	
running gear:	wheeled
layout:	8 × 8
Speed	
max speed:	80 km/h
cross-country:	25 km/h
Range	
main fuel supply:	500 km
Fuel capacity	
main:	520 litres
Fording	
without preparation:	1.4 m
Gradient:	65%
Vertical obstacle	
forwards:	0.6 m
Engine	Tatra T-930-3, 12 cylinders, air cooled, diesel, 250 hp at 2,000 rpm
Gearbox	
forward gears:	20
reverse gears:	20
Brakes	
main:	air
parking:	mechanical
Tyres:	15.00 × 21
Electrical system	
vehicle:	24 V
Firepower	
Armament:	32 × 122 mm rocket system
Ammunition	
main total:	64
Main weapon	
length of barrel:	3.00 m
maximum range:	20,400 m
Main armament traverse	
angle:	210° (105° left/105° right)
Main armament elevation/depression	
armament front:	+50°/0°

Status

Production complete. In service with the Croatian Army. There are no known exports of this system although it is has been offered for export. It is probable that Croatia only has one of these systems deployed.

Contractor

RH ALAN d.o.o.

RH ALAN 70 mm (40-round) VLR 70 M93A3 Heron multiple rocket launcher

Development

The RH ALAN 70 mm (40-round) VLR 70 M93A3 Heron multiple rocket launcher is a single-axle trailer-mounted all-weather cross-country assembly designed for the area fire support artillery mission.

It should be noted that RH ALAN is a marketing organisation. The actual 70 mm rockets, launcher and other subsystems are manufactured by a number of subcontractors.

It is understood that there has been no recent marketing of this system and production of the system has now been completed. It was offered on the export market but there are no known export sales.

Description

The modular system has 40 70-mm launch tubes arranged in banks in two 20-round boxes. Each 1.6 m long tube is smooth, with a single stainless steel guide tube and a locking device to hold the rocket secure once loaded.

When fully loaded the trailer is capable of being towed at speeds up to 15 km/h across country; on asphalt roads at speeds up to 80 km/h; and on tarmac roads at speeds up to 50 km/h. It is also air-portable as an underslung helicopter load.

The Heron has the ability to launch unguided 70 mm calibre solid propellant surface-to-surface rockets either singularly at one per second or in ripples of two rockets per second up to the full capacity of 40 rounds. A remote-control firing device can be used at a distance up to 25 m from the launcher.

When deployed in the firing position, the launcher is supported by a jack lowered under the tow bar and one either side on outriggers. The latter are fitted with screw type jacks that can be operated by hand to stabilise the platform before firing commences.

RH ALAN 70 mm (40-round) VLR M93A3 Heron multiple rocket launcher deployed in firing configuration, with stabilisers lowered to the ground
0008311

The current unguided 70 mm rocket used is the RAK 70 mm TF M95, which has a launch weight of 8.6 kg with 1.7 kg of double base propellant and can be fitted with the following types of warhead:

- 3.7 kg M95 HE-fragmentation (with 1.45 kg HE and instantaneous action impact fuze)
- 3.7 kg M95-O illumination
- 3.7 kg M95-D smoke
- 3.7 kg M95-I incendiary
- 3.7 kg M95-T training.

The actual 70 mm rocket has an overall length of 1.244 m and maximum speed at burnout is being quoted as 440 m/s. The unguided rockets can be launched one at a time or in salvo ripple.

An improved 70 mm range rocket, which increases the maximum range to 10,000 m from 8,100 m, is in development. The rocket launch weight is increased to 10 kg with the additional weight accounted for by a larger solid propellant rocket motor filled with 2.4 kg of double base propellant rather than the current 1.7 kg. Maximum velocity of the rocket is also increased to 550 m/s from 440 m/s.

The sighting system consists of a CN-2 mortar sight and an M2HR quadrant.

Specifications

	M93A3
Dimensions and weights	
Length	
overall:	3.15 m
Width	
overall:	1.85 m
Height	
overall:	1.4 m
Weight	
standard:	920 kg
combat:	1,250 kg
Mobility	
Configuration	
running gear:	wheeled
Firepower	
Armament:	40 × 70 mm rocket system
Ammunition	
main total:	40
Main weapon	
length of barrel:	1.6 m
maximum range:	8,100 m (M95 rocket system)
Operation	
stop to fire first round time:	180 s
out of action time:	90 s
Main armament traverse	
angle:	30° (±15°, 360° optional)
Main armament elevation/depression	
armament front:	+46°/-1°

Status

Production as required. In service with the Croatian Army. It is understood that there has been no recent production of this system.

Contractor

RH ALAN d.o.o.

RH ALAN 60 mm (24-round) VLR 60 M91 multiple rocket launcher

Development

RH ALAN has developed and manufactured small quantities of a mobile 24-round 60 mm calibre rocket launcher.

It is understood that there has been no recent marketing of this system and production of the system has now been completed. It was offered on the export market but there are no known sales.

Description

The 24 rocket 60 mm launcher tube assembly is fitted to the trailer system used in the 12-round 128 mm RAK-12 system. The crew is three to five, with the 60 mm HE unguided 60 mm rocket having a maximum range of 7,200 m.

No other details are known at present. It should be noted that RH ALAN is a marketing organisation. The actual rockets, launcher and other subsystems are made by various subcontractors.

Status

Production complete. No longer marketed. In service with the Croatian Army.

Contractor

RH ALAN d.o.o.

RH ALAN 60 mm (4-round) 4RL 60 M93A1 'Obad' self-propelled multiple rocket launcher

Development

RH ALAN has developed and manufactured a quantity of man-portable 4-round 60 mm calibre rocket launcher. It should be noted that RH ALAN is a marketing organisation. The actual rockets and launcher are made by a number of Croatian contractors.

It is understood that there has been no recent marketing of this system and production of the system has now been completed. It was offered on the export market but there are no known sales.

Description

The launcher has four 60 mm rocket launcher barrels in a rectangular formation and is disassembled into two single-man loads for carriage. It weighs 49 kg in total. The crew is two to three with the 60 mm unguided HE rocket having a maximum range of 7,200 m. No other details are known at present.

Variant

The RH ALAN LOV-IZV reconnaissance vehicle version of the standard Torpedo LOV-OP 4 × 4 APC is equipped with a pedestal rocket launcher on the roof. The pedestal is equipped with a four-round M92A1 launcher assembly on either side.

Production of the LOV family of 4 × 4 light armoured vehicle was undertaken in Croatia. Production has been completed and there are no known exports.

The LOV-OP (4 × 4) is also used as the basis for the RH ALAN 128 mm (24-round) LOV RAKL 24/128 self-propelled rocket launcher which is covered in a separate entry in *IHS Jane's Land Warfare Platforms: Artillery and Air Defence.*

Production of this is complete and only a small number were built.

Status

Production complete. In service with the Croatian Army and Croatian Defence Council. It is understood that there has been no recent production of this system.

Contractor

RH ALAN d.o.o.

LOV-IZV reconnaissance vehicle fitted with two four-round 60 mm M93A1 rocket launcher systems
0501059

Egypt

SAKR 122 mm series multiple rocket launchers

Development

Following development work in the late 1970s, the SAKR Factory for Developed Industries, which comes under the control of the National Organisation of Military Production, successfully test-fired the SAKR-36 (or Hawk-36) 122 mm calibre unguided multiple rocket system in May 1981.

Production of these rocket launchers and associated rockets is undertaken on an as required basis. As far as it is known there has been no recent production of these rocket systems for the Egyptian Army.

While Egypt has not declared any exports of these multiple rocket launchers and their associated unguided rockets, it is considered possible that there have been some undeclared exports.

Description

SAKR-36 system

Based on the Russian 122 mm Grad BM-21 rocket, the weapon is fitted with interface systems allowing it to be fired from a number of rocket launcher types including the Helwan Machine Tools Company's 40-round launcher on a ZIL (6 × 6) truck chassis, the 30-round RL-21 launcher on a 3,500 kg Russian ZIL-131 6 × 6 or Japanese Isuzu 6 × 6 truck chassis, the 21-round launcher on a 4,000 kg modified Romanian 4 × 4 Bucegi SR-114 truck chassis, the lightweight 4/8-round light vehicle systems and the Russian BM-21 system.

The RL 21 can also be mounted on other cross-country chassis such as the more recent Russian Ural-4320 (6 × 6) and German Mercedes-Benz 1222A/36 (4 × 4).

Another alternative is installing the system on the rear of the Russian supplied ATS-59G full tracked prime carrier as illustrated in this entry.

It should be noted that production of the ATS-59G has been completed many years ago and it is no longer in front line service with the Russian Army.

The trend in many countries is to base rocket launchers on wheeled chassis rather than tracked chassis as the former have greater strategic mobility and lower operating and support costs than their tracked counterparts.

It should be noted that the 122 mm rockets listed in this entry are regarded as the first-generation 122 mm rockets. The SAKR Factory for Developed Industries is now marketing a complete family of 122 mm rockets called SAKR-10/SAKR-18/SAKR-36 and SAKR-45, details of which are given in a separate entry in *IHS Jane's Land Warfare Platforms: Artillery and Air Defence.*

PRL-113 (3-round) and PRL-111 (single-round) rocket launchers which have a maximum range of 10,800 m (Christopher F Foss) 0500647

SAKR-36 rocket being launched during trials 0500646

Model RL 21 (30-round) 122 mm rocket launcher mounted on rear of ATS-59G full-tracked carrier 0500648

As far as it is known, export marketing is now being concentrated on these new-generation 122 mm rockets. The United Nations Arms Transfer Lists for the period 1992 through to 2010 show no exports by Egypt of rocket launchers or rockets. In 2004 however, Slovakia supplied Egypt with 4,000 122 mm rockets.

Model RC 21 rocket carrier system

Associated with the 122 mm MRL launchers is the Model RC 21 rocket carrier system. This is designed to hold 40 122 mm calibre 3 m long rockets in two 20-round protected stacks 3.5 m long, 1 m wide and 1 m high. There are also two protected boxes provided, one for rocket brake rings and the other for the rocket fuzes. Total weight of the carrier loaded is 3,150 kg and 450 kg empty.

The carrier is designed for mounting on any suitable military vehicle, typically a chassis of similar type to the launcher platform.

SAKR-18 system (bomblets)

To complement the SAKR-36 Egypt has also fielded the 122 mm calibre SAKR-18 system. This uses a two-second burn rocket motor with 20.45 kg of double base propellant grain as used in the Egyptian copy of the Russian 122 mm Grad rocket. This gives a maximum range of 18,000 to 20,000 m at +48° launcher elevation. Rocket length is 3.25 m and weight complete is 67 kg.

The warhead contains 72 dual-purpose High-Explosive Anti-Tank (HEAT) bomblets that can penetrate up to 110 mm of conventional steel armour plate and has an effective anti-personnel lethal radius of 15 m. The fuze fitted is an electromechanical type which allows for dispensing altitudes between 2,500 and 3,000 m.

Russian-built Ural-4320-10 (6 × 6) truck chassis fitted with Egyptian-built RL-21 122 m (30-round) multiple rocket system 0008312

Model RC 21 rocket carrier on ZIL (6 × 6) chassis which is designed to carry 40 122 mm rockets. In this photograph the actual 122 mm rockets are not actually being carried 0008313

Model PRL-111 122 mm single-tube rocket launcher deployed in firing position 0500649

Model PRL-81 107 mm single-tube rocket launcher deployed in firing position 0500650

122 mm (4-round) SAKR launcher

Details of this are given in a separate entry in *IHS Jane's Land Warfare Platforms: Artillery and Air Defence.* Production of this system is undertaken on an as required basis.

This has been offered on the export market but there are no known exports as of March 2013.

Lightweight man-portable and vehicle-mounted systems

In 1987, the Helwan Machine Tools Company Factory No 999, which manufactures the 30-round 122 mm rocket launcher system, exhibited two types of man-portable tripod-mounted 122 mm launchers for use by special forces and other mobile units.

The single-round launcher is known as the PRL-111 while the three-round system has the designation PRL-113. Both have an optical mortar-type sight on their left sides which can be used both for direct and indirect fire roles. They are capable of being traversed 7° left and right and elevated from –10° to +40°. The rocket type used is the 122 mm SAKR-10 which gives the system a minimum range of 3,000 m and a maximum range of 10,800 m. There is also a quadruple launcher, which can be fitted on a 4 × 4 light vehicle.

A 107 mm man-portable single-tube rocket launcher, the PRL-81 has also been manufactured for the Egyptian Armed Forces and export. This is similar in many respects to the Chinese NORINCO 107 mm Type 85 single-round man-portable system, which is covered in a separate entry in *IHS Jane's Land Warfare Platforms: Artillery and Air Defence.* NORINCO is no longer marketing this 107 mm system.

More recently the Helwan Machine Tools Company has developed the Model RL812/TLC rocket launcher which is mounted on the rear of a 4 × 4 light vehicle. Details of this system are provided in a separate entry in *IHS Jane's Land Warfare Platforms: Artillery and Air Defence.*

This is understood to be in service with the Egyptian Army and has been offered on the export market but there are no known export sales.

Specifications

	ZIL-131 (6 × 6) Launch Platform
Dimensions and weights	
Length	
overall:	7.25 m
deployed:	7.25 m
Width	
transport configuration:	2.6 m
deployed:	3.00 m
Height	
transport configuration:	2.9 m
deployed:	4.2 m
Weight	
standard:	8,500 kg
combat:	11,250 kg
Mobility	
Configuration	
running gear:	wheeled
layout:	6 × 6
Speed	
max speed:	60 km/h

30-tube RL-21 launcher

Calibre:	122.4 mm
Number of tubes:	30
Length: (tube)	3 m
Rate of fire:	30 rds/15 s
Elevation:	0° to +55°
Traverse:	102° left, 72° right

RL-21 - rockets

Type	Long	Short
Calibre:	122 mm	122 mm
Weight:		
(complete rocket)	66 kg	45.25 kg
(HE content)	6.4 kg	6.4 kg
Maximum range:	20,400 m	11,000 m
Maximum velocity:	690 m/s	450 m/s
Note: The above 122 mm unguided rockets can be fitted with HE fragmentation, smoke and illuminating warheads		

Portable rocket launchers

Designation	PRL-111	PRL-113	PRL-81
Calibre:	122.4 mm	122.4 mm	107 mm
Rocket types:	SAKR-10, RFAS 9M22M	SAKR-10, RFAS 9M22M	107 mm models
Warhead types:	HE fragmentation, illumination, smoke	HE fragmentation, illumination, smoke	HE fragmentation, illumination, smoke
Number of tubes:	1	3	1
Time into combat:	2.5 min	3 min	2 min
Rate of fire:	1 rd/min	3 rds/min	2 rds/min
Elevation:	+10° to +40°	+10° to +40°	0° to 45°
Traverse:	7° left and right	7° left and right	7° left and right
Range:			
(minimum)	3,000 m	3,000 m	3,500 m
(maximum)	10,800 m	10,800 m	8,000 m
Length:			
(deployed)	2.5 m	2.5 m	0.9 m
(rocket tube)	2.5 m	2.5 m	0.9 m
Width: (deployed)	1.5 m	1.5 m	0.78 m
Height: (maximum, deployed)	2.5 m	1.79 m	0.97 m
Weight:			
(tube)	22 kg	22 kg (single)	8.8 kg
(tripod)	32 kg	74 kg	13.3 kg
(fully deployed)	56 ±2 kg	142 ±2 kg	23 kg

PRL-81 - rockets

Calibre:	107 mm
Weight:	
(complete rocket)	18.7 kg
(warhead)	8.33 kg
(HE content)	1.26 kg
Fuze:	M491

Status

SAKR-36, production as required. In service with Egypt (with 21-round SAKR, 30-round SAKR and 40-round SAKR and BM-21 heavyweight rocket launchers) and Sudan (50).

SAKR-18, production as required. In service with Egypt (with 21-round SAKR, 30-round SAKR and 40-round SAKR and BM-21 heavyweight rocket launchers).

SAKR-10, production as required. In service with Egypt (with single-round PRL-111, three-round PRL-113).

PRL-81 and 107 mm rocket. Production as required. In service with Egypt and other countries.

It is estimated that Egypt has about 500 MRL of all types.

Egypt has taken delivery for the US of 26 Lockheed Martin Missiles and Fire Control Systems 227 mm (12-round) Multiple Launch Rocket Systems based on a full tracked chassis.

In addition to the standard rockets Egypt has also taken delivery of 227 mm extended range rocket pods that were delivered between 2002 and 2003.

Contractor

SAKR rocket systems and four/eight-round lightweight launchers: SAKR Factory for Developed Industries.

PRL-81/111/113, 21-, 30- and 40-round launchers: Helwan Machine Tools Company.

Enquiries to Helwan Machine Tools Company.

SAKR 122 mm unguided rocket family

Development

The Egyptian SAKR Factory for Developed Industries has developed and placed in production a complete family of 122 mm unguided surface-to-surface rockets, which can be fired not only by the locally built 122 mm rocket launchers but also foreign systems of this calibre.

The latter include the widely deployed Russian BM-21 122 mm (40-round) as well as many similar and reverse engineered examples.

Production of these 122 mm rockets is undertaken on an as required basis.

It is considered possible that quantities of these 122 mm unguided rockets could have been export by Egypt.

According to the United Nations Arms Transfer Lists, in 2004 Slovakia supplied Egypt with 4,000 122 mm rockets.

Description

The SAKR-10 and SAKR-18 are the older 122 mm rockets and have a double-base solid propellant and an S-shaped fin. The more recent SAKR-36 and SAKR-45 use composite propellant and have a straight fin arrangement.

The high-explosive solid propellant 122 mm rockets are fitted with an impact/proximity fuze, while the cargo rockets - or submunitions as they are referred to - are fitted with a time fuze that operates when the rocket is over the target area.

The submunitions are understood to be of the High-Explosive Anti-Tank (HEAT) type, which are highly effective against troops in the open, as well as the vulnerable upper surfaces of armoured vehicles.

The SAKR-36 leaflet rocket contains 4,000 leaflets. This leaflet rocket has the same physical characteristics as the SAKR-36 SM rocket.

There are also training versions of all of these 122 mm rockets, which have their own high-explosive warheads replaced by a pyrotechnic charge used to detect the point of impact.

According to the manufacturer, each 122 mm rocket saturates a target area of between 35 and 90 m long and 35 to 50 m wide, but this depends on the range and type of warhead fitted.

Accuracy is currently being quoted as one per cent of range. As with all unguided surface-to-surface rockets, as the range is increased accuracy is reduced.

While a number of other companies are currently marketing 122 mm rockets with increased ranges, when compared to the original Russian 122 mm rockets the SAKR Factory for Developed Industries claims that it is the only company producing 122 mm rockets with a range of 45 km.

It should however be noted that the China North Industries Corporation (NORINCO) has also recently developed a new family of 122 mm calibre 30/40 km longer range unguided rockets and a new family of 40 km and 50 km unguided rockets.

Details of these are provided in separate entries in *IHS Jane's Land Warfare Platforms: Artillery and Air Defence*.

Specifications

Type	**SAKR-10**	**SAKR-18**	**SAKR-36**	**SAKR-45**
Warhead type:	HE	HE & SM[1]	HE & SM	HE & SM
Weight:				
(rocket)	44 kg	66.5/71.5 kg	58.5/67.5 kg	63.5/67.5 kg
(warhead)	19.5 kg	19.5/24.5 kg	19.5/28.5 kg	20.5/24.5 kg
Length of rocket:	1.933 m	2.928/3.34 m	2.6/3.125 m	2.9/3.31 m
Maximum range:	11 km	20/17 km	36/31 km	45/42 km
Number of bomblets:	nil	nil/72	nil/98	nil/72
Fuze:	I/P	I/P/time	I/P/time	I/P/time
Propellant:	DB	DB/DB	C	C
Fin assembly:	S	S/S	SF	SF

Notes: [1] plus leaflet, which has a maximum range of 31 km

Key:
I/P - Impact/Proximity
DB - Double Base
C - Composite
S - S-shaped fin
SF - Straight Fin

Status

Production as required. In service with Egypt as well as undisclosed countries.

Contractor

SAKR Factory for Developed Industries.

SAKR 122 mm (4-round) rocket launcher

Development

The Egyptian SAKR Factory for Developed Industries has developed and placed in production a 122 mm (4-round) rocket launcher, which can launch locally produced SAKR 122 mm unguided surface-to-surface rockets.

In addition there is a single portable tripod-mounted 122 mm (1-round) rocket launcher.

This has been designed for use by guerrilla and special forces units where light weight and ease of handling are required.

Production of these systems is undertaken by the SAKR Factory for developed industries on an as required basis.

This system has been offered on the export market but as of March 2013 there are no known sales.

Description

There are two versions of the 122 mm (4-round) rocket launcher: one mounted on the rear of a 4 × 4 cross-country truck chassis and the other on a rapidly erected stand.

The specifications of the SAKR 122 mm (4-round) rocket launcher are for the system installed on the rear of a US supplied M720 (4 × 4) cross-country chassis, although similar or other types of launcher chassis could also be used.

Both of these have an optical aiming device for direct and indirect target engagement and a remote firing device.

In the case of the former version, the manually mounted 122 mm (4-round) launcher is mounted on the rear deck. To provide a more stable firing platform four stabilisers are lowered to the ground: two are positioned at the front of the vehicle and two at the rear.

The 122 mm (4-round) stand-mounted version has four adjustable legs, which can be adjusted to level the system prior to the rockets being launched.

While the 4-round launcher has a total traverse of 120° left and right it has a fine adjustment of 7° left and right.

The single round 122 mm rocket launcher has a mortar type sight with a magnification of 2.5.

SAKR Factory for Developed Industries 122 mm (4-round) rocket launcher mounted on rear of a 4 × 4 cross-country vehicle
(SAKR Factory for Developed Industries) 1151506

Variants
Other SAKR land systems products that are beyond the scope of *IHS Jane's Land Warfare Platforms: Artillery and Air Defence* include the following:
Fateh-1 anti-personnel mineclearing system (ground based)
Fateh-2 anti-tank mineclearing system (trailer mounted)
Fateh-3 anti-tank mineclearing system (mounted on T-54/T-55 MBT hull)
Fateh-4 anti-tank mineclearing system
Fateh-5 anti-personnel mineclearing system
Nather 1 and Nather 2 anti-tank mine dispensing systems
Magnetic mine activation system SMAS-II
Anti-tank mineclearing system (for T-54/T-55 MBT).

Specifications

	M720 122 mm (4-Round)
Dimensions and weights	
Length	
overall:	4.347 m
Width	
overall:	2.044 m
Height	
overall:	2.58 m
Weight	
combat:	3,000 kg
Mobility	
Configuration	
running gear:	wheeled
layout:	4 × 4
Speed	
max speed:	75 km/h
Firepower	
Armament:	4 × 122 mm rocket system
Main armament traverse	
angle:	240° (120° left and right)
Main armament elevation/depression	
armament front:	+55°/0°

	122 mm (1-round) Rocket Launcher
Dimensions and weights	
Length	
overall	3 m (9 ft 10 in) (tube)
Diameter	
body	122 mm (4.80 in)
Performance	
Elevation	
angle	17°/55°

Status
Production as required. Understood to be in service with the Egyptian Army.

Contractor
SAKR Factory for Developed Industries.

107 mm (12-round) RL812/TLC multiple rocket launcher

Development
For some years the Helwan Machinery and Equipment Factory marketed a 107 mm man-portable single-round type rocket launcher for the home and export markets called the PRL-81.

Details of this are given in the entry for the SAKR 122 mm series multiple rocket launchers, which also includes details of the other 107 mm systems.

107 mm (12-round) RL812/TLC rocket launcher mounted on 4 × 4 cross-country chassis, with launcher traversed to the left 0130018

More recently the Helwan Machinery and Equipment Factory (or Factory 999 as it is also referred to) has developed the Model RL812/TLC rocket launcher system.

This is mounted on the rear of a 4 × 4 light cross-country vehicle, although it can be fitted to various other types of chassis, tracked and wheeled.

It is not known as to whether this system is currently deployed by the Egyptian Army.

As far as it is known, as of March 2013, there have been no export sales of this rocket system.

Description
The layout of the 107 mm (12-round) RL812/TLC multiple rocket launcher system is conventional with the power pack at the front, fully-enclosed two door cab in the centre and a flatbed type area at the rear.

A replacement wheel and tyre are carried to the immediate rear of the cab and there is also additional unprotected seating at the rear.

Mounted on the rear of the chassis is a 12-round 107 mm rocket launcher with sighting system, manual traverse and elevation controls on the left side. This fires 107 mm unguided solid propellant surface-to-surface rockets to a maximum range of 8,200 m.

These unguided 107 mm rockets, which have been manufactured by the SAKR Factory for Developed Industries, can be fitted with various warheads including high explosive, smoke and illuminating.

The 107 mm unguided rocket has a maximum speed of 375 m/s with the complete rocket weighing 18.7 kg of which 8.33 kg is the warhead and of this 1.26 kg is the high-explosive content. It is fitted with a nose mounted M491 fuze.

Specifications

	107 mm (12-Round) RL812/TLC
Dimensions and weights	
Crew:	3
Height	
transport configuration:	2.08 m
deployed:	2.48 m (launcher at maximum elevation)
Weight	
standard:	2,900 kg
combat:	3,108 kg
Mobility	
Configuration	
running gear:	wheeled
layout:	4 × 4
Firepower	
Armament:	12 × 107 mm rocket system
Main weapon	
length of barrel:	0.900 m
maximum range:	8,200 m
Rate of fire	
rapid fire:	12/7 s (est.)
Main armament elevation/depression	
armament front:	+45°/0°

Status
Development complete. Production as required. There are no known export sales of this system.

Contractor
Helwan Machine Tools Company.

India

DRDO 214 mm (12-round) Pinaka multiple rocket launcher system

Development
The Indian Defence Research and Development Organisation (DRDO) has developed a 214 mm calibre (12-round) multiple rocket launcher system.

It is estimated that a typical battery of six launchers firing 72 unguided 214 mm rockets can neutralise a target area of 700 × 500 m.

The system has been designed for shoot-and-scoot fire missions with each launcher being provided with its own computerised fire-control system and automatic positioning system.

A typical Pinaka multiple rocket launcher battery would consist of six launchers, rocket resupply vehicles, command post vehicle and a vehicle fitted with the Digicora met radar system for improved target accuracy.

According to the United Nations Arms Transfer Lists, India has recently taken delivery of the following quantities of Russian Splav BM 9A52 (12-round) Smerch multiple rocket systems:

DRDO 214 mm (12-round) Pinaka multiple rocket launcher system launching a rocket (DRDO) 0008314

Year	Quantity	Type
2007	16	9A52-2T Smerch launchers
2007	6	Smerch transloaders
2008	18	9A52-T Smerch launchers
2008	6	Smerch transloaders

It is understood that these are based on a Tatra (10 × 10) chassis rather than the normal Russian 8 × 8 cross-country truck chassis.

This is due to the fact that the Indian Army deploys a large number of Tatra (8 × 8) trucks for a wide range of roles including missile launcher platform.

In early 2013 it was stated that the DRDO had conducted development trials of an advanced version of the Pinaka rocket system at the Proof and Experimental Establishment at Chandipur in Orrisa, India.

During these trials, carried out by the Pune-based Armament Research and Development Establishment, three 214 mm rockets were successfully fired out to a range of 30 km, although its maximum range is 38 to 40 km.

According to Bharat Earth Movers, the actual 8 × 8 locally manufactured chassis used for this system is designated the BEML - Tatra 27ER96 28 300 8 × 8 1R/50T.

Description

Mounted on the chassis of the Bharat Earth Movers Limited licence-built former Czechoslovakian Tatra 815 (8 × 8) Kolos cross-country truck chassis, the system comprises a fully enclosed five-man cab at the front with two pods each containing six 214 mm rockets mounted at the rear of the chassis.

Once the 214 mm rockets have been launched, replacement rockets have to be loaded one at a time. To avoid counterbattery fire, reloading would normally be carried out away from the original launch/firing position.

The fully enclosed forward-control cab is fitted with an NBC system and standard equipment fit includes a central tyre regulation system that allows the driver to adjust the tyre pressure to suit the type of ground being crossed and passive night vision driving equipment.

The maximum road speed of the vehicle is 80 km/h. When firing the unguided rockets, the cab windows are protected by shutters.

When deployed in the firing position, the system is stabilised by four hydraulically operated stabilisers which are lowered to the ground, two either side.

According to the DRDO the system can come into action and launch its first 214 mm rockets within three minutes with a salvo of 12 rockets being launched in 44 seconds.

CEP is currently being quoted as between one and two per-cent of range.

While the DRDO is quoting a maximum range of 38 km other sources have quoted a longer range of up to 45 km.

The launcher assembly has electromechanical elevation and traverse, with traverse being 90° left and right of the centreline and elevation up to 55°.

There is also a resupply vehicle for use with the 214 mm (12-round) Pinaka multiple rocket launcher system.

Status

In low rate production for Indian Army. As of March 2013 no export sales of the system had been announced.

Specifications

Rocket

Calibre:	214 mm
Range	
(maximum)	38 km
(minimum)	10 km
Weight at launch:	276 kg
Length:	4.95 m
Warhead weight:	100 kg
Warhead types:	high-explosive blast fragmentation, high explosive, anti-personnel minelet and incendiary

Contractor

Designed by the DRDO. Produced by Indian state factories.

122 mm (40-round) multiple rocket launcher system

Development

India has built an indigenously developed 40-round rocket launcher based on the Russian Splav Scientific Production Concern 122 mm BM-21 (40-round) (6 × 6) design.

Production of this is undertaken on an as required basis and it is understood that there has been no recent production of this system in India.

It is understood that some of these launchers have now been integrated on the rear of an Ashok Leyland (6 × 6) cross-country chassis.

Description

Mounted on the locally produced Shaktiman (6 × 6) 5,000 kg truck chassis, the launcher fires a new longer-range unguided 122 mm rocket called the Long Range Artillery Rocket (LRAR).

This has been developed by the Armament Research and Development Establishment in Pune. The warhead contains 18 kg of TNT explosive and the rocket is fitted with a double-base solid propellant motor.

The LRAR can also be fired from the Russian-designed and built 122 mm BM-21 (40-round) multiple rocket launcher. A full 40-round salvo can be fired within 20 seconds and reloading is thought to be manual.

Recent information has indicated that India has 150 122 mm systems in service. At this time it is not clear as to whether these are all Russian supplied BM-21 systems or a mixture of BM-21 and locally developed versions. These 122 mm (40-round) multiple rocket systems have seen widespread use during the Indian/Pakistani border dispute.

Status

Production complete. In service with the Indian Army. As far as it is known, India has not exported any of these locally developed 122 mm (40-round) MRLS.

This 122 mm (40-round) multiple rocket launcher system has now started to be supplemented by the Russian Splav 300 mm BM 9A52 series (12-round) Smerch multiple rocket system and the locally developed DRDO 214 mm (12-round) Pinacha multiple rocket launcher system. Both of these have a much longer range that the older 122 mm (40-round) system and are covered in separate entries in *IHS Jane's Land Warfare Platforms: Artillery and Air Defence.*

Contractor

Indian Ordnance Factories.

122 mm (40-round) BM-21 in service with the Indian Army on an original Russian truck chassis (Indian Army) 0130034

122 mm (40-round) BM-21 in service with the Indian Army on an original Russian truck chassis (Indian Army) 0130035

Indonesia

IPTN 70 mm NDL-40 (20-round) multiple rocket launcher

Development

IPTN has designed the 70 mm NDL-40 multiple rocket launcher for use in the ground support role. In the designation NDL-40, the former stands for Nusantara Dual Function Launcher.

As mentioned in the entry, the 70 mm rockets and possibly the launcher is related to the Belgian FZ 70 mm (40-round) LAU97 multiple rocket launcher system which is covered in a separate entry in *IHS Jane's Land Warfare Platforms: Artillery and Air Defence.*

It should be noted that this launcher is no longer manufactured or marketed by the Belgian company although the rockets continue to be manufactured for the air-to-ground application.

Indonesia has offered this system on the export market but as of March 2013 there are no known sales.

Description

It comprises a launch assembly unit containing 20 70 mm launch tubes (in four rows of five) in a box mounted on a traversable base that is attached to a two-wheeled trailer fitted with four stabilisers, two at the front and two at the rear.

The complete unit weighs 754 kg and is aimed using a clip-on sight. The 70 mm unguided rockets may be fired singly or in salvos using a small remote-control device connected to the launcher by cable.

The standard production launcher has 20 launch tubes but according to the manufacturer this could be increased to 40 thus doubling the firepower of the system.

A vehicle as small as a jeep can tow the trailer. The 20-round launcher can also be mounted on a truck chassis or a fast patrol boat. If required, individual rocket tubes can be removed from the box and mounted on a tripod to give a fully man-portable system.

As far as it is known, this version has not been deployed by Indonesia.

Close-up of 70 mm NDL-40 (20-round) multiple rocket launcher being loaded with a 70 mm unguided rocket
0130028

IPTN 70 mm NDL-40 (20-round) multiple rocket launcher with stabilisers raised for travelling
0021050

The 70 mm rocket used is the 2.75 in Folding Fin Aerial Rocket (FFAR) round fitted with the FZ-68 solid propellant motor and FZ-71 warhead. Maximum range in the ground role is 8,500 m, with a complete salvo covering a 200 × 300 m area. These rockets are the same as those used in the Belgian Forges de Zeebrugge FZ 70 mm (40-round) LAU97 multiple rocket launcher covered in detail in a separate entry in *IHS Jane's Land Warfare Platforms: Artillery and Air Defence.*

The sighting system can be rotated over 6,500 mils and has a covered view of 200 mils and can be used for direct and indirect target engagement.

A typical battery would consist of six NDL-40 (20-round) rocket launchers, which would be connected to a central fire-control computer with forward observers providing target information for the computer prior to the targets being engaged.

The firing system is 265 mm long, 140 mm wide and 150 mm high, while the sighting system is 195 mm long, 145 mm wide and 200 mm high.

While the launcher has a combat weight of 754 kg, the individual firing system weighs 10 kg, the sighting system weighs 4.5 kg while the commanders firing control system weighs 2 kg.

The standard launcher has a total of 20 × 70 mm launcher tubes, it could also be supplied in 40 × 70 mm version. The actual 70 mm launcher tubes have an overall length of 1.860 m.

According to the manufacturer, the 70 mm launcher tube has a life of 400 rounds before requiring replacement, while the detainer and connectors have a life of 4,000 rounds before requiring replacement.

The main advantage of the IPTN 70 mm NDL-40 (20-round) multiple rocket launcher is its weight and small size that allows it to be easily transported in rough terrain and lifted by helicopter.

Specifications

	70 mm NDL-40 (20-Round)
Dimensions and weights	
Length	
overall:	3.595 m
Width	
overall:	1.995 m
Height	
overall:	1.600 m
Weight	
combat:	754 kg
Mobility	
Configuration	
running gear:	wheeled
Firepower	
Armament:	20 × 70 mm rocket system
Main weapon	
length of barrel:	1.860 m
maximum range:	8,750 m
Main armament traverse	
angle:	360°
Main armament elevation/depression	
armament front:	+65°/-3°

Status

Production as required. In service with Indonesian Army (Special Forces). There are no known exports of this system.

Contractor

Industri Pesawat Terbang Nusantara (IPTN).
Enquiries to Divisi Sistem Hankam.

International

Chinese and Thailand 302 mm Multiple Rocket Launcher

Development

Under an agreement signed in 2008/2009 the Defence Technology Institute (DTi) of Thailand is building a version of the China Precision Machinery Import & Export Corporation (CPMIEC) 302 mm WS-1B (4-round) unguided artillery rocket system.

Full details of this system are provided in a separate entry in *IHS Jane's Land Warfare Platforms: Artillery and Air Defence*.

A version of this is also in service with the Turkish Land Forces Command (TLFC) integrated onto a MAN (6 × 6) cross-country chassis.

Prime contractor for this system is the Turkish company of Roketsan and details of the system are provided in a separate entry in *IHS Jane's Land Warfare Platforms: Artillery and Air Defence*.

The Roketsan launcher is designated the T-300 while the unguided rocket is designated the TR-300.

The Thai version is understood to be called the DTi-1 Multiple Rocket Launcher (MRL) with a guided version under development under the designation of the DTi-1G with the latter letter standing for guidance.

According to Thai sources the original DTi-1 has the same range as the original Chinese 302 mm WS-1B has a maximum range of 180 km and a minimum range of 60 km. It is possible that the DTi-1G will have a longer range.

The agreement for the DTi-1G was finally signed in early 2013 and is expected to take three years and include joint testing and evaluation and will be supported by logistics and training packages.

It is not expected that production of the DTi-1G will commence before 2016/2017.

Some sources have indicated that Thailand has already taken delivery of a batch of 18 Chinese 302 mm WS-1B (4-round) systems.

Description
This is understood to be identical to the original Chinese 302 mm WS-1B (4-round) system.

Specifications
Not released but probably similar to the original Chinese 302 mm WS-1B (4-round) system.

Status
In is expected that production of the DTi-1 will commence in 2013/2014. Some sources have indicated that Thailand has already taken delivery of a batch of 18 302 mm WS-1B (4-round) systems from China.

Contractor
See development.

King Abdullah II Design and Development Bureau/Hanwha Corporation 70 mm (50-round) multiple rocket launcher

Development
The Jordanian King Abdullah II Design and Development Bureau (KADDB) and the Hanwha Corporation of South Korea have jointly developed this self-propelled 70 mm (50-round) Multiple Rocket Launcher (MRL) to meet the potential requirements of the Jordanian Armed Forces (JAF).

The two contractors started development work on this system in 2008 and it was first shown in public at the Special Operations Forum and Exhibition (SOFEX) held in Jordan in 2010.

KADDB has provided the chassis and is responsible for system integration while the Hanwha Corporation have provided the launcher, unguided 70 mm rockets and the computerised fire-control system.

It should be noted that prior to this KADDB developed and built the AB19 107 mm (12-round) MRS which is covered in a separate entry in *IHS Jane's Land Warfare Platforms: Artillery and Air Defence.*

This is essentially the US AM General High Mobility MultiPurpose Wheeled Vehicle (HMMWV) (4 × 4) fitted with the China North Industries Corporation 107 mm (12-round) MRL on the rear of the chassis. This is in service with the Jordanian Armed Forces.

As of March 2013 there are no known sales of this KADDB/Hanwha Corporation 70 mm (50-round) MRL which is aimed at the Special Forces market where platform weight and speed is of considerable importance.

King Abdullah II Design and Development Bureau/Hanwha 70 mm (50-round) MRL integrated on the rear of a Dodge Ram 2500 (4 × 4) cross-country truck chassis (Robert Hewson) 1401072

Description
This system essentially consists of a Dodge Ram 2500 (4 × 4) unprotected cross-country pick up chassis on the rear of which has been integrated the latest Hanwha Corporation 70 mm (50-round) MRL.

It should however be noted that the system can be integrated onto a variety of other chassis.

Details of the Hanwha Corporation 70 mm rocket launcher and their associated unguided surface-to-surface rockets are provided in a separate entry in *IHS Jane's Land Warfare Platforms: Artillery and Air Defence.*

This MRL can deliver a salvo of 50 × 70 mm rockets in 13 seconds which would cover an area of 250 × 250 m.

Types of 70 mm unguided rockets fired include high explosive, high explosive dual purpose and high explosive multipurpose submunition.

Individual rockets and smaller salvoes can also be fired by the system with a maximum stated range of 8,000 m.

While the current family of 70 mm rockets are unguided, the Hanwha Corporation are also developing fire-and-forget infra-red rocket for air launched applications.

Specifications
Not available.

Status
Prototype undergoing trials. Not yet in production or service.

Contractor
King Abdullah II Design and Development Bureau (KADDB) (Jordan) and Hanwha Corporation (South Korea).

LAROM 160 multiple rocket launcher system

Development
The LAROM 160 Multiple Artillery Rocket System is a joint development between Aerostar of Romania and Israel Military Industries.

The system essentially consists of the old Romanian Aerostar 122 mm (40-round) APRA multiple rocket launcher system, based on a DAC 26.410 (6 × 6) cross-country chassis upgraded and modified to accept the Israel Military Industries LAR 160 mm rockets. Full details of the LAR 160 rockets are given in a separate entry in *IHS Jane's Land Warfare Platforms: Artillery and Air Defence.*

Israel Military Industries provides the 160 mm rockets and their associated pods and the Fire-Control System (FCS) computers while Aerostar have integrated the new subsystems of the existing platform.

It is understood that the Romanian Army hopes to upgrade its complete fleet of existing 122 mm (40-round) APRA systems to the enhanced LAROM 160 configuration, with the first batch consisting of 18 units (three batches each with six launchers).

According to UN sources, the first three LAROM 160 systems were sent to Romania in 1999. All subsequent conversions have been undertaken in Romania.

As of March 2013 it is understood that Romania deployed 133 of the older 122 mm (40-round) rocket launcher systems and 54 of the upgraded and more capable LAROM systems.

Marketing of the LAROM 160 multiple rocket launcher system is undertaken for the export market on a case-by-case basis. As previously stated the first customer for this system was Romania but according to the United Nations Arms Transfer Lists for 2007 Israel transferred a total of four 122 mm/160 mm multiple rocket launcher systems to Georgia and it is assumed that these are the LAROM 160.

According to the United Nations Arms Transfer Lists there were no declared exports of these systems from Israel in 2009 or 2010.

Description
The main improvement is that the system is fitted with two Israel Military Industries 13-round pods of 160 mm LAR unguided rockets in place of the two

160 mm rocket being launched from LAROM 160 artillery rocket system 0073645

LAROM 160 artillery rocket system in travelling configuration with stabilisers raised and launcher traversed to the front 0105967

20-round pods that launch 122 mm Russian Grad type 9M22/9M22U rockets. The pods are loaded on to the launcher using an ammunition resupply vehicle fitted with a hydraulic crane.

To provide a more stable firing platform, four hydraulic stabilisers are lowered to the ground. These are positioned two either side. The first is positioned to the rear of the cab and the second at the very rear of the chassis.

By launching 160 mm LAR 160 rockets the range of the system is increased from just over 20 km to over 45 km. The 26 Mk 4 160 mm rockets can be launched in just over 46 seconds.

The solid propellant LAR 160 cargo rockets each have a payload of 104 Israel Military Industries top attack bomblets each of which has a self-destruct fuze.

The warhead canister is opened at the appropriate height on trajectory by an electronically remote set time fuze, automatically regulated by the computerised fire-control system before launch.

Two versions of the LAR 160 rockets are currently available, the standard unguided Mk 4 and the other fitted with the ACCULAR guidance system and both have a maximum quoted range of 45 km.

For training and fire-adjustment missions LAR practice rockets can be used. These have a smoke signature charge that can be clearly seen by the forward observer.

In addition there is a training pod that has the same shape, dimensions and weight as the tactical pod launch container and is loaded with 13 inert 160 mm rockets that can be used in all operational modes. It is used to train the LAROM crew in launcher positioning, aiming, firing and reloading.

When deployed in the firing position the launcher is stabilised by four hydraulic jacks that are lowered to the ground by remote-control.

One of these is positioned either side of the chassis between the first and second roadwheels with the other two at the rear of the chassis.

The LAROM 160 upgrade also adds a modern, automated FCS with Elbit computers being integrated into the launcher and in the Fire-Control Unit (FCU) which is installed in the fully enclosed forward control cab.

Before the rockets are launched, shutters/protection shields are positioned over the windows of the cab to protect its windows from the blast.

The FCU provides an improved man machine interface and is operated from the cab of the vehicle or from a nearby shelter by a remote-control 100 m long extension cable.

This indicates the number, type and status of the rockets in the pod, the physical and technical condition of the rockets and launchers.

It also allows for the input of all aiming sighting details, including meteorological conditions. The firing modes are manual or automatic and the operator can select either single shots, partial or complete salvos at four firing intervals, ranging from 0.5 seconds though to 1.8 seconds between rounds. At the slowest firing rate the complete load of 26 160 mm rockets can be launched in 46.8 seconds.

The new communication system is based on the Thales (previously Racal) Panther - EDR frequency hopping digital radio which is made in Romania by ELPROF S.A. - Bucuresti company that is part of the AEROSTAR's industrial group, this information to be transmitted back to higher command elements and firing requests to be sent direct to the fire units.

This information flow enables the LAROM 160 system to come into action much quicker than the normal launcher and therefore gives a quicker target response time.

Variants

According to Aerostar, this upgrade is also applicable to the older Romanian developed 122 mm (12-round) jeep mounted light rocket launcher called Aurora (Awora), which is covered in detail in a separate entry in *IHS Jane's Land Warfare Platforms: Artillery and Air Defence.* As of March 2013 it is understood that this version had yet to reach the prototype stage.

LAROM with ACCULAR

Late in 2004 Romania signed a contract with Israel Military Industries covering the supply of USD40 million worth of ACCULAR 16 rockets for use with the LAROM 160 system. These are fitted with a Trajectory Correction System for greater accuracy, and can carry 104 top attack M85 bomblets. Additional details of ACCULAR are given in the entry Israel Military Industries LAR 160 rocket system in a separate entry in *IHS Jane's Land Warfare Platforms: Artillery and Air Defence.*

Status

Production. In service with the Romanian Army. In 2007 Georgia took delivery of four 122 mm/160 mm multiple rocket launcher systems and it is assumed that these are the LAROM 160.

Contractor

Aerostar SA, Group Industrial Aeronautic.
Israel Military Industries (IMI).

MORAK 122 mm (40-round) rocket launcher

Development

The German company Diehl Munitionssysteme was selected by the Slovakian Ministry of Defence (MoD) as system house for modernising the RM-70 122 mm (40-round) Multiple Rocket System (MRS) of former Czech-Slovakian origin, based on a Tatra 813 series (8 × 8) cross-country truck chassis.

This enterprise, part of Diehl VA Systeme (Überlingen, Germany), signed an accord with Slovakian MoD officials. There were two phases to the programme, development and production.

The second phase covered modernised items for the RM-70 122 mm (40-round) MRS, especially 122 mm rockets.

Besides Slovakia, many other countries have the RM-70 system in service. The RM-70, or its more recent Mod 70/85, are also used by many other countries including Angola, Czech Republic, Ecuador, Finland, Georgia, Greece, Indonesia, Libya, Poland, Rwanda, Sri Lanka, Uganda, Uruguay and Zimbabwe.

Production of the original RM-70 system is undertaken on an as-required basis but it is understood that most recent sales have been from surplus stocks.

A full list of known export sales of the Slovakian 122 mm RM-70 (40-round) multiple rocket systems are provided in a separate entry in *IHS Jane's Land Warfare Platforms: Artillery and Air Defence.*

This co-operation is regarded as having a future high-value potential, since all former Warsaw Pact countries have 122 mm MRS rocket systems in service, especially the Russian BM-21. The RM-70 upgrade is also applicable to these systems.

In mid-2005 the Slovak Army took delivery of the first of 26 production RM-70 MORAK at the facilities of VOP 28 (Military Repair Factory 28) at Presov.

The first seven were delivered by the end of 2005, with a further 10 in 2006 and the remaining systems in 2007.

Diehl Munitionssysteme is teamed for the RM-70 project with the Slovakian company Konstrukta Defence and Honeywell Deutschland.

According to the development team, the upgraded RM-70 system increases combat capabilities, safety and comfort of the crew.

While the first prototype of the MORAK 122 mm (40-round) rocket launcher upgrade retained the complete armour protected cab of the original RM-70 system, the pre-production model has a different cab.

The new cab is of all-welded steel that provides the occupants with protection from small arms fire and shell splinters. Bulletproof windows are provided in the front and sides of the cab.

When the rockets are being launched the front windows are protected by armoured shutters that are operated from inside the cab.

The chassis is provided with four stabilisers that are lowered to the ground to provide a more stable firing platform. These are positioned two either side, one between the first and second roadwheel station and another to the rear of the last roadwheel station.

Mounted to the rear of the cab is a hydraulically-operated crane that can be used to load and unload fresh pods of six 227 mm rockets.

Description

Modifications include improvements to the 122 mm rocket, including a new solid propellant motor and fuze as well as improvements to fire-control and launcher. Two versions were developed:

- RM-70/85 modernised rocket launcher which is then designated the RM-70/85M (this has a standard two-door unarmoured truck cab)

Prototype of MORAK on RM-70 launching a 227 mm rocket (Diehl) 0569057

Prototype of MORAK on RM-70 chassis with launcher with pod of six 227 mm rockets traversed left and stabilisers lowered at the rear of the chassis (Diehl) 0569056

- RM-70 modernised rocket launcher which is then designated the MODULAR (this has the fully enclosed armoured cab).

The main difference between the two systems is that the RM-70/85M retains the 122 mm (40-round) launcher and its resupply pack of 40 122 mm rockets.

MODULAR has a new turntable mounted launcher that takes a six pack from the US Lockheed Martin Missile and Fire Control Multiple Launch Rocket System (MLRS). First successful trials of this version were made in late 2001. Full details of the complete family of rockets fired by the upgraded six round pod are provided in the entry for the Lockheed Martin Missiles and Fire Control Multiple Launch Rocket System (MLRS) in *IHS Jane's Land Warfare Platforms: Artillery and Air Defence.*

It should be noted that many countries are now phasing out the original 277 mm Phase I M77 unguided rocket as this contains 644 sub-munitions that are not fitted with a self-destruct mechanism.

The two launchers are interchangeable so that the shorter range 122 mm rockets could be used for training with the longer range 227 mm rockets being used for combat.

The RM-70/85M is fitted with a computerised fire system that enables onboard navigation, autonomous computation and setting firing data. The TALIN-3000 three-axis navigation system is provided by Honeywell Deutchland.

Target data is received from the battery headquarters. The launcher is aligned onto the target from the safety of the cab. If required it can also be fitted with a Global Positioning System (GPS).

According to the consortium, the system can launch its rockets within 60 seconds of coming to a halt and leave the position in a similar time frame. A full salvo of 40 122 mm rockets can be fired within a maximum of 22 seconds.

Specifications

	MORAK 122 mm (40-Round)
Dimensions and weights	
Crew:	3
Length	
overall:	9.26 m
Width	
overall:	2.65 m
Height	
overall:	2.655 m
Weight	
standard:	20,300 kg
combat:	23,400 kg (122 mm pod)
Mobility	
Configuration	
running gear:	wheeled
layout:	8 × 8
Firepower	
Main weapon	
maximum range:	20,380 m (122 mm rockets)
maximum range:	30,000 m (227 mm rockets)
Operation	
reload time:	240 s
Survivability	
Armour	
hull/body:	steel (cab)

Status

Production as required. A total of 26 production systems have been developed to the Slovakian Army.

As of April 2013 there are no known export sales of this upgraded rocket system.

Contractor

Diehl Munitionssysteme GmbH & Co KG (Germany).
Konstrukta Defense as (Slovak Republic).

Iran

Aerospace Industries Organization Fadjr-5 333 mm (4-round) artillery rocket system

Development

The Fadjr-5 333 mm (4-round) Artillery Rocket System (ARS) was developed by the Shahid Bagheri Industries division of the Aerospace Industries Organization of Iran and uses the same 6 × 6 chassis as the Fadjr-3 240 mm (12-round) artillery rocket system, which is covered in detail in a separate entry in *IHS Jane's Land Warfare Platforms: Artillery and Air Defence.*

The Iranian Fadjr-5 and Fadjr-3 unguided surface-to-surface rocket systems have been used by Hezbollah during operations against Israel conducted from Southern Lebanon.

Hezbollah also deploys a wide range of other rockets including the M600 which is understood to be a Syrian clone of the Iranian Fateh-100 (which is also referred to as the Fateh A-110) solid propellant Short Range Ballistic Missile (SRBM) which is beyond the scope of *IHS Jane's Land Warfare Platforms: Artillery and Air Defence.*

This is understood to have a range of 250 km and carry a 500 kg conventional high-explosive warhead.

This rocket has increased accuracy as it is fitted with an inertial navigation system that provides a Circular Error of Probability (CEP) of 500 m.

Description

The Fadjr-5 artillery rocket system is based on a modified older generation Mercedes-Benz 2624 series (6 × 6) truck chassis with the diesel engine and fully enclosed two-door cab at the front. To the immediate rear of the cab is a small crew compartment. When launching rockets the cab windscreen would normally be protected by shutters.

Mounted on the rear of the chassis is the turntable with the manually operated elevating and traversing mechanism on the left side.

The Fadjr-5 rocket system has a total of four 333 mm diameter rocket launcher tubes in one layer with launcher elevation being from 0° to +57° and with a traverse of 45° left and right.

It launches a solid propellant 333 mm unguided rocket with fins that unfold after launch and the rockets can be launched with an ambient temperature of between –40°C to +50°C.

The unguided 333 mm rocket has an overall length of 6,485 mm and a launch weight of 915 kg of which 175 kg is the warhead which contains 90 kg of high explosive. Maximum range is being quoted as 30,000 m.

It should be noted that earlier versions of this rocket had quoted an overall length of 6.46 m, a launch weight of 860 kg and a maximum range of 68,000 m.

The Fadjr-5 system has a rate of fire of one rocket every four or eight seconds and replacement 333 mm Fadjr-5 rockets are packed one to a crate which weighs 1.21 tonnes.

To provide a more stable firing platform, four stabilisers are lowered to the ground before the rockets are launched, two of these are positioned at the rear of the vehicle with another one between the second and third axles on either side.

The rockets can be launched singly or in ripple fire and once the launcher has expended its rockets it would normally deploy to another position when new rockets would be loaded using a crane.

To help avoid counterbattery fire, the Fadjr-5 would normally redeploy to another position for reloading once the four 333 mm rockets have been launched.

It is possible that some of the technology of these rocket systems could have come from North Korea.

It is assumed that these rockets have a high-explosive warhead, although other types of warhead are considered possible.

Variants

Upgraded Fadjr-5 artillery rocket system

The Iranian Aerospace Industries Organization (AIO) have developed an upgraded version of the Fadjr-5 333 mm (4-round) unguided surface-to-surface Artillery Rocket System (ARS).

Fadjr-5 333 mm (4-round) artillery rocket system in travelling configuration 0067779

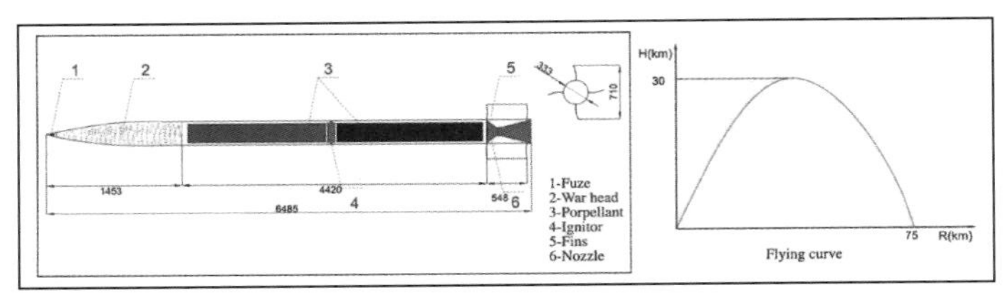

General arrangement drawing of Fadjr-5 rocket (left) and flying curve (right) 0526857

This is understood to be in production and service with Iran but at this stage it is not clear as to whether these are brand new production systems or upgraded versions of the original production Fadjr-5 333 mm (4-round) ARS.

The baseline Fadjr-5 333 mm (4-round) ARS has been in service with Iran for some years installed on an older generation Mercedes-Benz (6 × 6) cross-country truck chassis.

The latest production of the Fadjr-5 ARS is installed on a new Mercedes-Benz (6 × 6) forward control chassis with the platform now integrated into a complete weapon system rather than an individual launcher.

The new chassis has improved cross-country mobility with the forward control fully enclosed cab providing space for the driver and two passengers.

To the immediate rear of the cab is another fully enclosed cabin for the remainder of the crew. To provide a more stable firing platform four hydraulically operated stabilisers are lowered to the ground before firing commences.

This Mercedes-Benz chassis is similar to that manufactured in China and used as the basis for the China North Industries Corporation (NORINCO) 122 mm (40-round) Type 90 ARS which is covered in a separate entry in *IHS Jane's Land Warfare Platforms: Artillery and Air Defence* and is also used for a wide range of other Chinese military roles including towing 155 mm artillery systems.

In the original version of the Fadjr-5 333 mm (4-round) ARS, each launcher had to be laid onto the target on an individual basis. This took time and reduced the overall effectiveness of the system.

The new upgraded system is fully computerised with a typical troop having four launchers with a command post vehicle. Two troops would make up a battery with each battalion typically having two batteries to give a total of 16 launchers.

At the battery level there is a command post vehicle based on a 4 × 4 truck chassis with a meteorological unit and battalion command vehicle at battalion level.

At the troop level there is also a repair facility, two transportation and loading vehicles and two loading vehicles.

Fire missions would normally be carried out through the chain of command with each launcher being provided its own computer which would receive information via a data link.

The GPS would inform the platform of its exact position with the computer providing information to level the launcher and then lay the launcher onto the target.

In addition to launching the rockets from inside of the launch platform, there is also a remote fire capability from a distance of 1,000 m as well as from the command post vehicle that could be up to 20 km away.

It is understood that the 333 mm unguided rocket is the same as that deployed by the earlier launcher although it is considered probable that an upgraded rocket with enhanced capabilities could be under development.

This baseline rocket has a maximum range of 75 km with a claimed Circular Error of Probability (CEP) of four per cent of range. The rocket is 6.485 m long and has a launch weight of 915 kg with the High-Explosive (HE) warhead being fitted with an impact or proximity fuze but other types of warhead could be fitted including sub-munition and HE incendiary.

The earlier German Mercedes-Benz (6 × 6) chassis is also used as the basis for a number of other Iranian AIO ARS including the Fadjr-3 240 mm (12-round) ARS.

According to the AOI it would be possible to field a mixed unit equipped with upgraded versions of the Fadjr 333 mm (4-round) systems and the Fadjr 240 mm (12-round) ARS.

Specifications

	Fadjr-5 333 mm (4-Round)
Dimensions and weights	
Length	
overall:	10.45 m
Width	
transport configuration:	2.54 m
Height	
transport configuration:	3.34 m
Weight	
combat:	15,000 kg
Mobility	
Configuration	
running gear:	wheeled
layout:	6 × 6
Speed	
max speed:	60 km/h
Firepower	
Armament:	4 × 333 mm rocket system

Status

Production as required. In service with Iranian Army. It is understood that quantities of these rockets and/or launchers have been supplied to other users including Hezbollah.

Contractor

Aerospace Industries Organization (AIO), Shahid Bagheri Industries.

Aerospace Industries Organization Fadjr-3 240 mm artillery rocket system

Development

The Fadjr-3 240 mm (12-round) Artillery Rocket System (ARS) was developed by the Shahid Bagheri Industries division of the Aerospace Industries Organization of Iran and uses the same 6 × 6 cross-country chassis as the Fadjr-5 333 mm (4-round) artillery rocket system.

It is understood that quantities of these rockets and/or launchers have been supplied to other users including Hezbollah.

Description

The Fadjr-3 artillery rocket system is based on a modified Mercedes-Benz 2624 series (6 × 6) truck chassis with the diesel engine and fully enclosed cab at the front.

To the immediate rear of the cab is a small crew compartment. When launching rockets the cab windscreen would normally be protected by shutters.

Mounted on the rear of the chassis is the turntable with the manually operated elevating and traversing mechanism on the left side. The Fadjr-3 has a total of 12 240 mm diameter rocket launcher tubes in two layers of six with launcher elevation being from 0° to +57° and with a traverse of 90° left and 100° right.

It launches a solid propellant 240 mm rocket with fins that unfold after launch and the rockets can be launched with an ambient temperature of between −40°C to +50°C.

The solid propellant rocket has an overall length of 5.20 m and has a launch weight of 407 kg of which 90 kg is the warhead that contains 45 kg of high-explosive.

While maximum range is being quoted as 43,000 m its maximum altitude being quoted is 17,000 m.

The Fadjr-3 unguided artillery rocket system has a rate of fire of one rocket every four or eight seconds and replacement Fadjr-3 rockets are packed three to a crate which weighs 1.65 tonnes.

To provide a more stable firing platform four stabilisers are lowered to the ground before the rockets are launched, two of these are positioned at the rear of the vehicle with another one between the second and third axles on either side.

The 240 mm rockets can be launched singly or in ripple fire and once the launcher has expended its rockets it would normally deploy to another position when new rockets would be loaded using a crane.

To help avoid counterbattery fire the Fadjr-3 would redeploy to another area for reloading once the 12 rockets had been fired.

It is possible that some of the technology of these rocket systems could have come from North Korea.

When first observed in 1996, this system was installed on a Japanese Isuzu (6 × 6) truck chassis and these were deployed with Hezbollah units operating in Southern Lebanon.

It is assumed that these unguided rockets have a high-explosive warhead, although other types of warhead are considered possible.

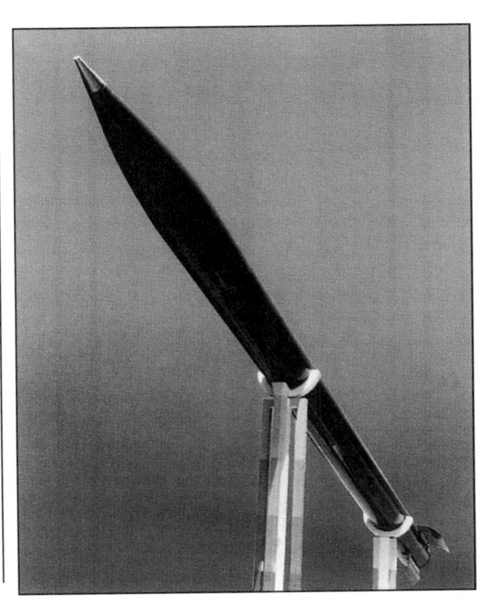

240 mm Fadjr-3 rocket out of launcher with fins unfolded 0008317

Fadjr-3 240 mm (12-round) artillery rocket system with launcher elevated
0067780

Variant

It is considered probable that these systems will be installed on a new forward-control chassis which is already used for the latest Fadjir-5 system.

Details of the upgraded Fadjr-5 333 mm (4-round) artillery rocket system are provided in a separate entry in *IHS Jane's Land Warfare Platforms: Artillery and Air Defence* and this is already in service with the Iranian Army.

This is a more recent Mercedes-Benz (6 × 6) chassis which is used for a wide range of military roles including towing 155 mm artillery systems.

Iran has also developed a 155 mm/39-calibre self-propelled artillery system, that also uses the latest generation MAN (6 × 6) cross-country chassis and details of this are provided in a separate entry in *IHS Jane's Land Warfare Platforms: Artillery and Air Defence.*

Specifications

	Fadjr-3 240 mm
Dimensions and weights	
Length	
overall:	10.45 m
Width	
transport configuration:	2.54 m
Height	
transport configuration:	3.34 m
Weight	
combat:	15,000 kg
Mobility	
Configuration	
running gear:	wheeled
layout:	6 × 6
Speed	
max speed:	60 km/h
Firepower	
Armament:	12 × 240 mm rocket system

Status

Production as required. In service with the Iranian Army. Also used by Hezbollah.

Contractor

Aerospace Industries Organization (AIO), Shahid Bagheri Industries.

Aerospace Industries Organization 610 mm Zelzal 2 heavy Artillery Rocket System (ARS)

Development

The Zelzal 2 unguided heavy artillery rocket is launched from a single rail launcher and is spin stabilised in flight by small solid-propellant rocket motor nozzles located just behind the nose-mounted HE warhead and electromechanical fuzing section. All available details are given in the accompanying table.

It is understood that quantities of the rockets and/or launchers have been supplied to other users including Hezbollah.

These have been launched against Israel from positions in Southern Lebanon.

Description

Like all other Iranian rockets the 610 mm Zelzal 2 heavy artillery rocket system is produced by the Aerospace Industries Organization, Shahid Bagheri Industries facility.

According to the manufacturer, CEP of the Zelzal 2 rocket at maximum range of 210 km is five per cent. The solid propellant rocket motor has a specific impulse of 235 seconds with the solid composite (PPG base) propellant weighing 1,840 kg.

Zelzal 2 610 mm unguided artillery rocket on its launcher, based on a German Mercedes-Benz truck. Note stabilisers lowered to ground to the immediate rear of first axle and at the rear of chassis
0105902

It should be noted that earlier details of the Zelzal 2 rocket, released by DIO, quoted a maximum range of 200 km with the rocket having an overall length of 8.46 m and a launch weight of 3,545 kg.

The launcher is based on a modified Mercedes-Benz three-axle truck. When deployed in the firing position, four stabilisers are lowered to the ground. Of these four stabilisers, one is mounted on either side to the immediate rear of the cab and the other two are mounted at the very rear of the chassis.

Variants

Fateh A-110 ballistic missile

Hezbollah also deploys a wide range of other rockets including the M600 which is understood to be a Syrian clone of the Iranian Fateh-100 (which is also referred to as the Fateh A-110) solid propellant Short Range Ballistic Missile (SRBM) which is beyond the scope of *IHS Jane's Land Warfare Platforms: Artillery and Air Defence.*

This is understood to have a range of 250 km and carry a 500 kg conventional high-explosive warhead.

This rocket has increased accuracy as it is fitted with an inertial navigation system that provides a Circular Error of Probability (CEP) of 500 m.

In this past it has been reported that the M600 was based on the Chinese CSS-8 SRMB which is a development of the Chinese built Russian SA-2 'Guideline' Surface-to-Air Missile (SAM) optimised for the surface-to-surface role.

Shahab 3A ballistic missile

This liquid-fuel ballistic missile is also referred to as the Ghadr 101 and is beyond the scope of *IHS Jane's Land Warfare Platforms: Artillery and Air Defence.* Mention has also been made of a Shahab 4.

Specifications

	Zelzal 2 rocket
Dimensions and weights	
Length	
overall:	8.325 m
Diameter	
body:	610 mm
Weight	
launch:	3,400 kg
Performance	
Range	
max:	210 km
Ordnance components	
Warhead:	600 kg HE

Status

Production as required. In service with Iranian Army. Some have been supplied to Hezbollah.

Contractor

Aerospace Industries Organization (AIO), Shahid Bagheri Industries.

Aerospace Industries Organization NAZEAT Artillery Rocket System (ARS) family

Development

The NAZEAT unguided heavy artillery rocket system family is an indigenous development by the Aerospace Industries Organization, Shahid Bagheri Industries facility.

NAZEAT 6 355.6 mm unguided artillery rocket being launched 0021059

The NAZEAT family of surface-to-surface rockets is known to comprise the following versions: N5, N6, N7, N8, N9 and N10 (with a maximum range of 80,000 up to 150,000 m depending upon model type).

All of these rockets use a solid propellant and all available model details are given below.

It is considered that in the future the remaining systems will be integrated onto a new forward-control chassis which is also used for some other Iranian surface-to-surface rocket systems.

The launcher currently uses an old Mercedes-Benz (6 × 6) bonneted cross-country truck chassis.

It is expected that these chassis will be replaced by the more recent Mercedes-Benz forward control chassis which is also used for the upgraded version of the Fadjr-5 333 mm (4-round) artillery rocket system.

Details of this system are provided in a separate entry in *IHS Jane's Land Warfare Platforms: Artillery and Air Defence* and this is already in service with the Iranian Army.

Description

355.6 mm calibre NAZEAT N4 artillery rocket system

The unguided NAZEAT N4 carries an HE warhead fitted with an electromechanical-type fuze.

355.6 mm calibre NAZEAT N5 artillery rocket system

The unguided NAZEAT N5 heavy unguided artillery rocket system is carried either on a single rail elevating launcher fitted to the rear of a modified Mercedes-Benz 2624 (6 × 6) truck chassis to act as a Tractor-Erector-Launcher (TEL) or a single rail semi-trailer launcher. This Mercedes-Benz (6 × 6) chassis is also used for a number of Iranian rocket systems including the 610 mm Zelzal 2 heavy artillery rocket system.

Different warhead types can be fitted. The preparation time for a launch is approximately one hour. Maximum range is 120,000 m with a flight time of 180 seconds and maximum velocity of 1,800 m/s.

355.6 mm calibre NAZEAT N6 artillery rocket system

The unguided NAZEAT N6 carries an HE warhead fitted with an electromechanical-type fuze. According to Iranian sources, the rocket has a CEP of five per cent of range with the solid composite (PPG base) rocket motor having a maximum burn time of nine seconds and specific impulse of 237 seconds. Total weight of propellant is stated to be 420 kg.

NAZEAT 10 450 mm unguided artillery rocket being launched 0023375

NAZEAT 6 355.6 mm unguided artillery rocket ready to be launched from the back of its Mercedes-Benz (6 × 6) truck chassis 0130036

The NAZEAT 6 rocket is carried and launched from a Mercedes-Benz (6 × 6) truck chassis. When deployed in the firing position four stabilisers are lowered to the ground to provide a more stable firing platform.

NAZEAT 6-H artillery rocket

Further development of the NAZEAT 6 has resulted in the NAZEAT 6-H artillery rocket that is launched from the same Mercedes-Benz 6 × 6 chassis. This has more propellant and a longer range.

According to the manufacturer, the CEP at maximum range is five per cent. Average action time is nine seconds, while specific impulse is 240 seconds. The HTPB solid propellant weighs 420 kg and has a quoted shelf-life of seven years.

450 mm calibre NAZEAT N10 artillery rocket system

The unguided NAZEAT N10 heavy unguided artillery rocket system comprises the following systems:

- Mercedes-Benz 911 Command-and-Control (4 × 4) truck with a rear-mounted cabin containing a Model 380DX computer processing system, communications equipment and other necessary apparatus.
- Mercedes-Benz 911 (4 × 4) meteorological truck with a rear-mounted cabin, containing the necessary equipment to measure the required meteorological conditions such as ambient air temperature, pressure and humidity, wind velocity and direction that can affect the rocket performance.
- Model WM 101 Global Positioning System console for surveying the launch platform site.
- Mercedes-Benz 2624 (6 × 6) transporter vehicle with four NAZEAT 10 reloads within its rear loading bed. It is fitted with a remote cable-operated hydraulic crane just behind the cab and two hydraulically-operated stabilising jacks to facilitate loading of the launcher.
- A launcher platform which can be one of the following types:

- A version of the Mercedes-Benz 2624 (6 × 6) truck chassis fitted with a single rail elevatable launcher assembly on its rear decking to act as a TEL vehicle. Maximum elevation angle is 80° and azimuth angle ±7°. Time to reload is 30 minutes. This is designated the NAZEAT 10 L1 Model Launcher. Maximum road speed is quoted as 60 km/h with a maximum dirt road speed of 20 km/h.

- A four-wheel semi-trailer, with a different style single rail hydraulically operated elevatable launcher assembly from that fitted on the Mercedes-Benz truck version. The maximum elevation and traverse angles remain the same, as does the reload time. This is known as the NAZEAT 10 L2 Model Launcher.

- The standard Russian BAZ-135L4 (8 × 8) Luna-M (FROG-7) artillery TEL vehicle. This is known as the NAZEAT 10 L3 Model Launcher. Production of the Russian FROG-7 system was completed some time ago, but it remains in service with a declining number of countries. Details are given in a separate

Iranian 355.6 mm NAZEAT rocket 0500660

	NAZEAT N4	NAZEAT N5	NAZEAT N6	NAZEAT N6-H	NAZEAT N10	NAZEAT N10-H
Dimensions and weights						
Length						
overall:	5.8 m	6 m	6.29 m	6.290 m	8.03 m	8.030 m
Diameter						
body:	355.6 mm	355.6 mm	355.6 mm	355.6 mm	450 mm	455 mm
Weight						
launch:	650 kg	850 kg	900 kg	970 kg	1,850 kg	1,830 kg
Performance						
Speed						
max speed:	-	6,480 km/h	-	-	-	-
Range						
max:	80 km	120 km	90 km	100 km	125 km	140 km
Ordnance components						
Warhead:	150 kg	150 kg HE	130 kg HE	150 kg HE	240 kg	230 kg HE
Propulsion						
type:	solid propellant	solid propellant	solid propellant	solid propellant	solid propellant	solid propellant

entry. 'FROG' stands for 'Free Rocket Over Ground' and was originally a NATO-applied designation. It should be noted that in front line Russian service the FROG-7 was replaced by the more recent SS-21 missile system.
The NAZEAT 10 is fitted with a single-stage solid propellant rocket motor with a maximum thrust of 280 kN, a propellant weight of 840 kg and specific impulse of 240 seconds. The fuze is of an electro-mechanical type.

NAZEAT 10-H artillery rocket

Further development of the NAZEAT 10 has resulted in the NAZEAT 10-H artillery rocket, which is launched from the same 6 × 6 chassis. According to the manufacturer, CEP at maximum range is five per cent. Average action time is quoted as 10 seconds, while specific impulse is quoted as 240 seconds. The rocket uses an HTPB base solid-composite propellant with a maximum shelf-life of seven years.

Specifications – *see table above*

Status

Production as required. In service with the Iranian Army and the IRGC. It is understood that some of these Iranian rockets have been supplied to Hezbollah in Southern Lebanon.

Contractor

Aerospace Industries Organization (AIO), Shahid Bagheri Industries.

Aerospace Industries Organization 333 mm SHAHIN 1 artillery rocket system

Development

The 333 mm SHAHIN 1 unguided heavy artillery rocket system is an indigenous development of the Aerospace Industries Organization.

It is understood that production of this 333 mm rocket system has been completed.

Trailer-mounted Iranian SHAHIN rocket system with 333 mm SHAHIN 1 in centre position and 333 mm SHAHIN 2 on two outer rails 0500659

It is considered possible that the 333 mm SHAHIN 1 artillery rocket system has been replaced in front line Iranian service by more modern rocket systems.

By today's standards the 13 km range of the Aerospace Industries Organization 333 mm SHAHIN 1 ARS is very short and the launcher could be neutralised by counterbattery fire unless it uses shoot-and-scoot procedures.

Description

The SHAHIN 1 has a solid propellant rocket motor with seven exhaust nozzles and is fitted with a 190 kg HE blast warhead. Other types of warhead may have been developed for this rocket. Maximum range is stated to be 13,000 m. It has four fixed fins at the rear.

The launcher is a four-wheel trailer-mounted elevating triple-rail assembly. Before firing, two pairs of stabiliser jacks have to be lowered to the ground manually that provide a stable firing platform. The launcher platform is assigned to heavy artillery units.

According to the manufacturer, this rocket has a storage temperature range from 30°C to +60°C.

Specifications

	SHANIN 1 rocket
Dimensions and weights	
Length	
overall:	2.9 m
Diameter	
body:	333 mm
Weight	
launch:	384 kg
Performance	
Speed	
max speed:	1,620 km/h
impact velocity:	1,620 km/h
Range	
max:	13 km
Ordnance components	
Warhead:	190 kg HE blast
Propulsion	
type:	solid propellant

Status

Production complete. In service with the Iranian Army and the IRGC. Not thought to have been exported.

Contractor

Aerospace Industries Organization (AIO), Shahid Bagheri Industries.

Aerospace Industries Organization 333 mm SHAHIN 2 Artillery Rocket System (ARS)

Development

The 333 mm SHAHIN 2 unguided heavy artillery rocket system is a further indigenous development of the 333 mm SHAHIN 1 system by the Aerospace Industries Organization, Shahid Bagheri Industries.

It is understood that production of the 333 mm SHAHIN 2 artillery rocket system is complete.

It is considered possible that the 333 mm SHAHIN 2 artillery rocket system has been replaced in front line service by more modern rocket systems.

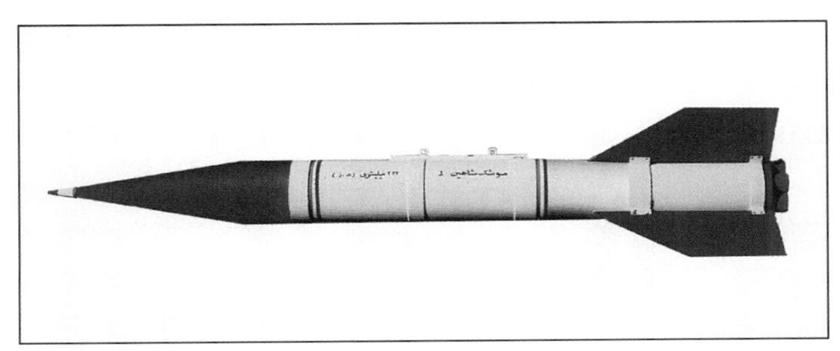

333 mm SHAHIN 2 artillery rocket 0021055

By today's standards, the 20 km range of the Aerospace Industries Organization 333 mm SHAHIN ARS is very short and the launcher could be neutralised by counterbattery fire unless it uses shoot-and-scoot procedures.

Description

The 333 mm SHAHIN 2 has a lengthened version of the solid propellant rocket motor that is fitted to the SHAHIN 1. It has the same seven-nozzle exhaust system and 190 kg HE blast warhead. Maximum range is stated to be 20,000 m. The SHAHIN 2 has four fixed fins at the rear of the rocket.

The launcher is normally a four-wheel trailer-mounted elevating triple-rail assembly. Before firing, two pairs of stabiliser jacks have to be lowered manually from the trailer to provide a stable firing platform. The launcher platform is assigned to heavy artillery units.

The SHAHIN 2 can also be used as an unguided air-to-ground rocket. The SHAHIN 2 artillery rocket system has a storage temperature of –30°C to +60°C.

As far as it is known, the only warhead fitted to the SHAHIN 2 unguided rocket is of the conventional high-explosive type. As with all rockets of this type, SHAHIN is an area weapon rather than a precision attack weapon.

Specifications

	SHANIN 2 rocket
Dimensions and weights	
Length	
overall:	3.9 m
Diameter	
body:	333 mm
Weight	
launch:	530 kg
Performance	
Speed	
max speed:	2,376 km/h
Range	
max:	20 km
Ordnance components	
Warhead:	190 kg HE blast

Status

In service with the Iranian Army and the IRGC. It is probable that production of the 333 mm SHAHIN unguided ARS has been completed. It is understood that marketing of the SHAHIN 2 is no longer undertaken.

Contractor

Aerospace Industries Organization (AIO), Shahid Bagheri Industries.

Aerospace Industries Organization 333 mm (1-round) Falaq-2 Artillery Rocket System (ARS)

Development

It is understood that this system was developed in the 1990s by the Shahid Bagheri Industries, which is part of the Aerospace Industries Organization.

It is considered possible that the 333 mm unguided surface-to-surface artillery rocket used in the Falaq-2 artillery rocket system is related to the longer-range SHAHIN 1 and SHAHIN 2 systems.

It is considered possible that production of the earlier SHAHIN 1 and SHAHIN 2 333 mm unguided rocket systems in now complete.

The Falaq-2 however, is much shorter and therefore carries less solid propellant and has a shorter range.

According to the Aerospace Industries Organization, the 333 mm rockets are 'designed for strike, attack and defence conditions'.

In addition it can be used for the 'destruction of artillery cannons, tank groups, gathering trenches, combat engineering equipment and damaging the command stations'.

The 333 mm (1-round) Falaq-2 system has been offered on the export market.

As of April 2013 there were no known export sales of this unguided rocket system. Iran is however known to have provided unguided surface-to-surface rockets to various factions in the Middle East and elsewhere.

333 mm Falaq-2 rocket during test firing from a static mount (Aerospace Industries Organization) 0559360

333 mm (1-round) Falaq-2 ARS mounted on a 4 × 4 light cross-country chassis with stabilisers raised in travelling configuration (Aerospace Industries Organization) 0559359

Description

The only known 333 mm Falaq-2 system fielded to date by Iran is based on a 4 × 4 jeep light cross-country vehicle with the engine at the front and two door fully-enclosed cab in the centre.

Mounted on the rear of the chassis is a single tube-type launcher for the 333 mm rockets. Before the rockets are launched, two stabilisers are manually lowered to the ground one at each side at the rear of the vehicle, to provide a more stable firing platform.

The 333 mm spin-stabilised rockets have a maximum range of 10,800 m and are fitted with a nose-mounted fuze. Propellant used is of the double base solid propellant type and there are eight nozzles at the rear of the rocket.

Variants

The same chassis is used as the launch platform for the Iranian Falaq-1, which has six 240 mm rockets in the ready-to-launch position in a frame type launcher on the rear. This launches 240 mm unguided surface-to-surface rockets to a maximum range of 10,000 m. Details of this are given in a separate entry in *IHS Jane's Land Warfare Platforms: Artillery and Air Defence.*

Specifications

	Falaq-2 rocket
Dimensions and weights	
Length	
overall:	1.82 m
Diameter	
body:	333 mm
Weight	
launch:	255 kg
Performance	
Range	
max:	10.8 km
Altitude	
max:	3,200 m
Ordnance components	
Warhead:	120 kg HE

Status

Production as required. Probably in service with Iran.

Contractor

Aerospace Industries Organization (AIO), Shahid Bagheri Industries.

Aerospace Industries Organization 240 mm (6-round) Falaq-1 Artillery Rocket System (ARS)

Development

It is understood that this mobile artillery rocket system system was developed in the 1990s by the Shahid Bagheri Industries, which is part of the Aerospace Industries Organization.

The 240 mm unguided surface-to-surface rocket is very similar to the rocket used with the Russian 240 mm (12-round) BM-24 system based on a ZIL-157 (6 × 6) truck chassis. This was first fielded by the Russian Army in the 1950s but has long since been replaced in Russian service by more modern systems.

The Russian 240 mm (12-round) BM-24 system is understood to have been deployed by a number of other countries including Algeria, Egypt, former East Germany, Israel, North Korea, Poland and Syria. Production of the BM-24 was completed some time ago and it has been phased out of service with the Russia Army.

The Russian BM-24 fired two types of 240 mm unguided surface-to-surface rockets. The so called long rocket had a length of 1.29 m, weighed 109 kg and had a maximum range of 10,200 m. The so called short rocket had a length of 1.225 m, weighed 112 kg and had a maximum range of 6,575 m.

Israel captured a number of the BM-24 systems and these entered service with the Israel Defense Force (IDF) with Israel Military Industries manufacturing the rockets. It is understood that these systems have now been phased out of front line service with the IDF but these systems may well be held in reserve. Some BM-24 rocket launchers are still deployed by North Korea.

Description

The only known 240 mm system fielded to date by Iran is based on a 4 × 4 jeep light cross-country vehicle with the engine at the front and two door fully enclosed cab in the centre.

Mounted on the rear of the chassis is what appears to be a six-round frame-type launcher with two layers each of three for the 240 mm rockets. Before the rockets are launched, two stabilisers are manually lowered to the ground, one at each side at the rear of the vehicle. These stabilisers provide stability to the platform when the rockets are launched. Once fired the launcher would rapidly be moved to a concealed area where it would be reloaded manually.

As a cab is not fitted with blast protection shutters, its is considered that the launcher is traversed left or right prior to rockets being launched.

Six-round 240 mm Falaq-1 ARS mounted on a 4 × 4 light cross-country chassis (Aerospace Industries Organization) 0559357

240 mm Falaq-1 rocket being tested from a static mount (Aerospace Industries Organization) 0559358

The 240 mm spin stabilised rockets have a maximum range of 10,000 m and are fitted with a nose-mounted fuze. Propellant used is of the solid double-base type.

Variants

The same chassis is used as the launch platform for the Iranian Falaq-2, which has a single tube-type launcher mounted on the rear. This launches a single 333 mm unguided surface-to-surface rocket to a maximum range of 10,800 m. Details of this are given in a separate entry in *IHS Jane's Land Warfare Platforms: Artillery and Air Defence.*

Specifications

Falaq-1 rocket

Calibre:	240 mm
Weight:	
(total)	111 kg
(warhead)	50 kg
Length:	1,320 mm
Maximum range: (at sea level)	10,000 m
Maximum flight altitude:	3,500 m

Status

Production as required. Probably in service with Iran.

Contractor

Aerospace Industries Organization (AIO), Shahid Bagheri Industries.

Aerospace Industries Organization 230 mm OGHAB Artillery Rocket System (ARS)

Development

The 230 mm OGHAB (Eagle) unguided Artillery Rocket System (ARS) was developed indigenously during the 1980-88 Gulf War. The initial combat use of the rocket was in December 1986 when several rounds were fired at Basra.

From then to the end of the 1980-88 Gulf War approximately 330 rounds were launched by rocket units of both the regular army and the ground forces of the Iranian Revolutionary Guard Corps (IRGC).

It is understood that production of this rocket and associated launcher is now complete.

There are no known exports of this rocket but Iran is known to have provided unguided surface-to-surface rockets to various factions in the Middle East and elsewhere.

It has now been replaced in production by new systems with a longer range.

Description

The 230 mm calibre OGHAB rocket has four fixed tail-mounted fins and is spin stabilised in flight. It carries a 70 kg HE fragmentation warhead. A contact fuze with instantaneous and delay actions is fitted.

Propulsion is by a solid propellant rocket motor that develops a maximum thrust in the order of 76,480 kN.

The launcher is an elevatable triple-rail launcher assembly fitted to the rear of a Mercedes-Benz LA911B (4 × 4) 4,500 kg truck chassis. The truck is also known as the LA1113 and before 1963 as the LA328.

Before firing, two pairs of stabiliser jacks have to be lowered to provide a stable firing platform. The launcher vehicle is assigned to heavy artillery units.

Like all other Iranian unguided artillery rockets, the 230 mm OGHAB has been developed and produced by the Aerospace Industries Organization, Shahid Bagheri Industries.

Earlier sources gave this as giving the 230 mm OGHAB artillery rocket a maximum range of 34,000 m. It is possible that this related to a first generation 230 mm OGHAB rocket, with the later production model having a range increased to 45,000 m.

Iranian 230 mm OGHAB rocket 0500661

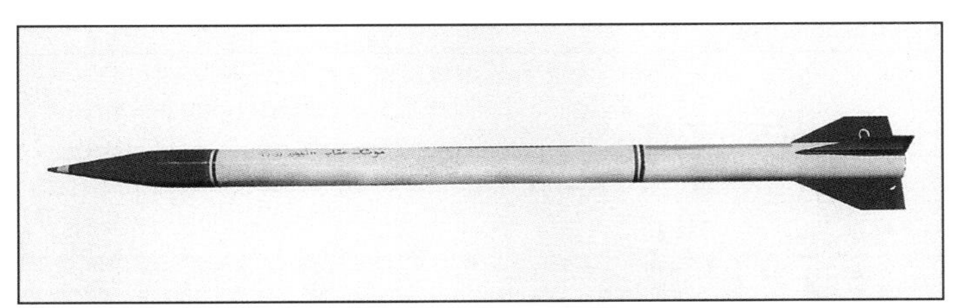

230 mm OGHAB unguided artillery rocket 0021057

Specifications

	OGHAB rocket
Dimensions and weights	
Length	
overall:	4.82 m
Diameter	
body:	230 mm
Weight	
launch:	360 kg
Performance	
Speed	
max speed:	2,700 km/h
Range	
max:	45 km
Ordnance components	
Warhead:	70 kg HE fragmentation
Propulsion	
type:	solid propellant

Status
Production complete. In service with the Iranian Army and the IRGC. Not thought to have been exported. Believed to have been supplemented by 240 mm Fadjr-3 ARS. This rocket system is no longer being marketed by Iran.

Contractor
Aerospace Industries Organization (AIO), Shahid Bagheri Industries.

Aerospace Industries Organization 122 mm (40-round) HADID multiple rocket launcher system

Development
The 122 mm HADID rocket launcher system is a 40-round launcher assembly system mounted on the rear of a Mercedes-Benz 2624 (6 × 6) truck chassis.

These chassis are also used for the Fadjr 333 mm (4-round) and Fadjr 240 mm (12-round) artillery rocket systems that are covered in a separate entry.

The technology has certain operational characteristics that are similar to the China North Industries Corporation (NORINCO) 122 mm (40-round) Type 81 MRS and the Russian 122 mm (40-round) BM-21 systems, both of which are covered in detail in a separate entry in *IHS Jane's Land Warfare Platforms: Artillery and Air Defence.*

It should be noted that some of these systems have been upgraded and installed on a new Mercedes-Benz (6 × 6) forward-control truck chassis. It is considered possible that the 122 mm (40-round) HADID multiple rocket launcher system could also be installed on this chassis in the future.

It is understood that quantities of these rockets and/or launchers have been supplied to other users including Hamas and Hezbollah.

Iranian HADID 122 mm (30-round) MRS mounted on Mercedes-Benz LA911B (4 × 4) 4,500 kg truck chassis with stabilisers lowered to the ground 0500662

HADID 122 mm (40-round) HM20 multiple rocket launcher system on Mercedes-Benz 6 × 6 chassis with launcher elevated and traversed to the right 0526859

HADID 122 mm (40-round) HM20 multiple rocket launcher system on Mercedes-Benz 6 × 6 chassis in travelling configuration 0526858

Description
The launcher assembly has two banks of 20 122 mm barrels in a side-by-side 4 × 5 row arrangement and is mounted on a rotating cradle fitted to the rear of the truck.

The unguided rockets are spin stabilised and are of the 122 mm type. The full 40-round load can be salvo-fired in 20 seconds, with the firing operations conducted either from the vehicle cab or from a ground position situated at a safe distance from the launcher. Reloading is performed manually by the launcher crew. The sighting system is mounted on the left side of the launcher at the rear.

The actual 122 mm launcher tubes used in this system weigh 23.4 kg.

These rocket systems have been developed by Aerospace Industries Organization, Shahid Bagheri Industries. Details of the Iranian 122 mm ARASH artillery rockets are given in a separate entry in *IHS Jane's Land Warfare Platforms: Artillery and Air Defence.* The 122 mm unguided surface-to-surface rockets can be launched from any similar 122 mm rocket launcher system.

The actual launchers have been developed by the Armament Industries Division of the Defence Industries Organization. The launcher has also been referred to as the Hadid 122 mm (40-round) HM20 multiple rocket launcher system.

This facility also manufactures tank (for example 125 mm) and artillery (for example 122 mm, 152 mm and 155 mm) barrels.

These are used in locally developed Iranian weapon systems now in service with Iran.

The chassis used for this system is the Mercedes-Benz 2624 (6 × 6) cross-country chassis but it is possible that this has now been replaced by a more recent Mercedes-Benz (6 × 6) chassis provided with a forward control cab as this is used for a number of other recent Iranian weapon systems.

Variants
122 mm (30-round) multiple rocket launcher system
A copy of the North Korean 30-round 122 mm calibre BM-11 rocket launcher system has also been produced. This comprises a single bank of three rows of 10 tubes mounted on a rotating cradle assembly. The chassis used is the Mercedes-Benz LA911B (4 × 4) truck chassis.

Four stabilising jacks are lowered for firing and a hydraulic crane unit is also fitted immediately aft of the cab for use in reloading the rocket tubes.

The rear stabilisers are lowered to the ground manually while those to the immediate rear of the cab are lowered to the ground using an hydraulic system. It is possible that the launcher is fitted with an auxiliary power unit.

122 mm (16-round) multiple rocket launcher system
This has been designed for installation on boats and rafts and has two banks, each of eight 122 mm launcher tubes. This is designated the HM23.

122 mm naval rocket launcher
An 8-round launcher assembly has been produced for use on naval vessels. The total launcher weight is 450 kg with the system having a maximum range of 14,000 m. Elevation range is 0° to 20°. The standard 122 mm rocket types are fired from the 3 m long, 23.4 kg weight barrels.

122 mm single barrel rocket launcher

A man-portable single-round 122 mm calibre rocket launcher has been produced. Mounted on a tripod the system has a maximum range of 20,400 m, an elevation range of 5° to 50° and a traverse of ±12°. The total assembly weighs 85 to 90 kg and uses the standard 3 m long, 23.4 kg weight barrel. This is designated the HM21.

Specifications

	122 mm (40-Round) HADID
Dimensions and weights	
Crew:	7
Weight	
combat:	13,150 kg
Mobility	
Configuration	
running gear:	wheeled
layout:	6 × 6
Fording	
without preparation:	1.5 m
Firepower	
Armament:	40 × 122.4 mm rocket system
Ammunition	
main total:	40
Main weapon	
length of barrel:	3 m
maximum range:	20,400 m
Rate of fire	
rapid fire:	40/20 s
Operation	
reload time:	480 s (est.)
Main armament traverse	
angle:	174° (102° left/72° right)
Main armament elevation/depression	
armament front:	+55°/+1°

Status

Production as required. In service with the Iranian Army and the IRGC. It should be noted that Iran also uses the Russian 122 mm BM-21 system and the North Korean 122 mm BM-11 systems mounted on a 6 × 6 chassis.

It is probable that Hezbollah has some of these rockets deployed in Southern Lebanon and the Gaza Strip.

Contractor

Aerospace Industries Organization (AIO), Shahid Bagheri Industries.

Aerospace Industries Organization 122 mm ARASH artillery rockets

Development

The unguided 122 mm ARASH artillery rockets are similar in appearance to the Russian and Chinese 122 mm calibre unguided artillery rockets in widespread use. The unguided 122 mm rocket can be used in all the Aerospace Industries Organization, Shahid Bagheri Industries 122 mm rocket launchers.

In addition to being used in Chinese and Iranian 122 mm rocket launchers these Iranian 122 mm unguided rockets can also be fired from 122 mm systems built or marketed by many other countries including Croatia, Czechoslovakia (today Czech Republic and Slovakia), Egypt, India, Italy, North Korea, Pakistan, Romania, Russia, Serbia, Sudan and Turkey.

These include the 122 mm (40-round) HADID multiple rocket launcher system and foreign-supplied 122 mm rocket launcher versions currently in Iranian use. The standard 122 mm rocket has a maximum range of 21,500 m.

It is understood that these rockets have been supplied to other users including Hamas and Hezbollah.

These have been used against Israel from positions in Southern Lebanon.

Description

The 122 mm calibre ARASH unguided rocket has four wraparound tail-mounted fins and is spin stabilised in flight. It carries an HE fragmentation warhead. A contact fuze with instantaneous and delay actions is fitted. It is probable that individual 122 mm rockets have been exported by Iran. The standard Russian Federation type 122 mm rocket is also manufactured. All of these rockets are of the solid-propellant type with wrap round fins at the rear. There is a possibility that other types of warhead have been developed for these 122 mm Iranian rockets in addition to the HE fragmentation warhead.

The availability of different types of 122 mm rocket warheads would provide the launcher with a significant capability enhancement and enable a wide target array to be engaged.

Variant

Extended-range 122 mm rocket

An extended-range 122 mm rocket is available. This rocket has a maximum range of 26,000 m. It has an operational temperature range of 30°C to +50°C.

Specifications

	122 mm ARASH	122 mm ARASH Extended Range
Dimensions and weights		
Length		
overall:	2.815 m	3.20 m
Diameter		
body:	122 mm	122 mm
Weight		
launch:	65 kg	72 kg
Performance		
Speed		
max speed:	2,556 km/h	3,780 km/h
Range		
max:	21.5 km	29 km
Ordnance components		
Warhead:	18.38 kg	18 kg

Status

Production as required. In service with the Iranian Army and the IRGC.

It is probable that quantities of these rockets have been used by Hezbollah in Southern Lebanon.

Contractor

Aerospace Industries Organization (AIO), Shahid Bagheri Industries.

Aerospace Industries Organization 122 mm NOOR artillery rockets

Development

The 122 mm NOOR unguided artillery rocket system is similar in appearance to the Russian and Chinese 122 mm calibre short version artillery rockets in widespread use.

It can be used in all of the Iranian Aerospace Industries Organization, Shahid Bagheri Industries facility, and foreign-supplied 122 mm rocket launcher versions currently in Iranian use.

Description

The 122 mm calibre NOOR solid-propellant rocket has four wraparound tail-mounted fins and is spin stabilised in flight. It carries an HE fragmentation warhead. A contact fuze with instantaneous and delay actions is fitted. It is possible that individual 122 mm rockets have been exported by Iran. It is considered possible that other types of 122 mm warhead have been developed apart from the 122 mm Fadjr-6 minelaying rocket covered in a separate entry in *IHS Jane's Land Warfare Platforms: Artillery and Air Defence.*

Like other Iranian 122 mm unguided artillery rockets the 122 mm NOOR rocket can be stored in the temperature range of −30°C to +60°C. Details of Iranian 122 mm multiple rocket launcher systems are given in a separate entry in *IHS Jane's Land Warfare Platforms: Artillery and Air Defence.*

Production of these is undertaken on an as required basis and quantities of these launchers and associated 122 mm rockets have been exported.

Specifications

	NOOR
Dimensions and weights	
Length	
overall:	2.050 m
Diameter	
body:	122 mm
Weight	
launch:	45 kg
Performance	
Speed	
max speed:	2,592 km/h
Range	
max:	18 km
Ordnance components	
Warhead:	18.35 kg HE fragmentation

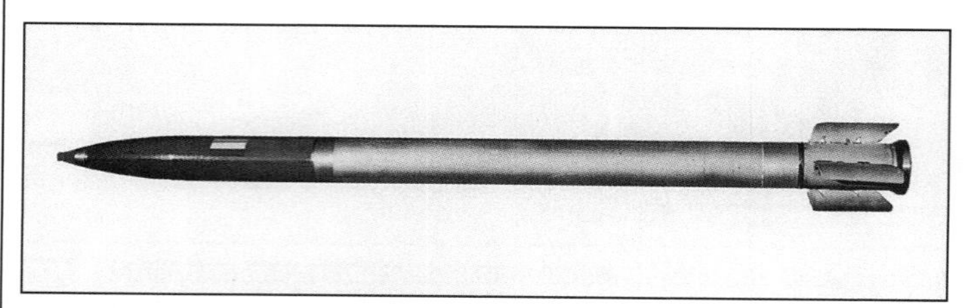

122 mm NOOR artillery rocket with fins unfolded 0021058

Status

Production as required. In service with the Iranian Army and the IRGC.

It is understood that these rockets have been supplied to Hamas and Hezbollah.

Contractor

Aerospace Industries Organization (AIO), Shahid Bagheri Industries.

Aerospace Industries Organization 122 mm Fadjr-6 minelaying rocket

Development

The Aerospace Industries Organization, Shahid Bagheri Industries, unguided 122 mm calibre Fadjr-6 minelaying rocket can be launched from standard 122 mm calibre multiple rocket launcher systems. Details of the Iranian developed and manufactured 122 mm (40-round) HADID multiple rocket launcher are given in a separate entry in *IHS Jane's Land Warfare Platforms: Artillery and Air Defence.* While Iran has supplied quantities of rockets and launchers to Hezbollah it is considered unlikely that they have been supplied with minelaying rockets.

As far as it is known, none of the 122 mm Fadjr-6 minelaying rockets have been exported by Iran.

Description

Fadjr-6 carries a total of eight anti-tank/anti-personnel mines, which are stabilised after release by parachute. All available details are given in the accompanying table.

The pressure-activated mines are manufactured by the Ammunition Industries Division of the Defence Industries Organization.

Specifications

	Fadjr-6
Dimensions and weights	
Length	
overall:	2.83 m
Diameter	
body:	122 mm
Weight	
launch:	63 kg
Performance	
Range	
min:	3.5 km
max:	6 km
Ordnance components	
Warhead:	32 kg 8 anti Personnel, armour piercing mines

Status

Production as required. In service with Iranian Army.

Contractor

Aerospace Industries Organization (AIO), Shahid Bagheri Industries.

Aerospace Industries Organization 107 mm (12-round) Fadjr-1 multiple rocket launcher system

Development

The 107 mm Fadjr-1 multiple rocket launcher system is similar in appearance to the China North Industries Corporation (NORINCO) 107 mm Type 63 MRS which is covered in a separate entry in *IHS Jane's Land Warfare Platforms: Artillery and Air Defence.* NORINCO is no longer marketing this system or its rockets. In its standard two-wheeled trailer form it can be moved by vehicle, animals or personnel.

107 mm (12-round) Fadjr-1 multiple rocket launcher being towed by a light 4 × 4 vehicle which has another system mounted in rear 0105900

107 mm (12-round) Fadjr-1 multiple rocket launcher deployed in firing position from the rear 0526860

All of these launchers are manufactured by the Armament Industries Division of the Defence Industries Organisation.

It is understood that these 107 mm rockets have been exported to various countries in Asia and the Middle East, especially guerrilla units.

These 107 mm unguided rockets have been used against Israel from Gaza and Southern Lebanon and have also been encountered in Afghanistan and Iraq.

In addition to being fired from the standard trailer mounted 107 mm (12-round) multiple rocket launcher system single 107 mm rockets are often fired by remote-control in the general direction of the target.

Description

The launcher has three banks of four 107 mm barrels and in the trailer form is mounted on a rubber-tyred split pole-type carriage.

In the firing position the wheels are normally removed and the launcher is supported by the two trails at the rear which are fitted with spades which are carried on the top of the trails when travelling, and two short legs in front of the carriage. It can also be fired with the two rubber-tyred roadwheels still in position on their stub axles.

The optical sight is fitted on the left side of the launcher together with the elevation and traverse handwheels.

The unguided spin stabilised 107 mm rockets used by this launcher are manufactured by the Aerospace Industries Organization, Shahid Bagheri Industries.

It is assumed that these Iranian-manufactured rockets can also be fired by foreign 107 mm systems produced by a number of countries, including China, Iran, North Korea, South Africa, Sudan and Turkey.

A full salvo of 12 107 mm rockets can cover an area of 200 × 80 m. The full load of 12 rockets can be fired in seven to eight seconds and once fired, the launcher is manually reloaded. The shelf-life of the rockets is quoted as 13 years when stored in ideal conditions.

Variants

107 mm (12-round) vehicle mount

The standard 107 mm (12-round) launcher assembly is available with a 50 kg weight, 1.4 × 0.585 × 0.22 m base plate for fitting to light 4 × 4 vehicles or 6 × 6 truck chassis. The optical sight is fitted on the left side of the launcher.

107 mm single-round rocket launcher

A single-round rocket launcher is available. Mounted on a tripod it weighs 23 kg in total and uses a generator or battery as the trigger system. Elevation range is 0° to 50° and traverse ±7°. The rocket tube is 0.8 m long. The optical sight is fitted on the left side of the launcher.

107 mm twin-round rocket launcher

This is similar to the single-round launcher but has two barrels. Total weight is 32 kg, elevation range is 0° to 50° and traverse ±10°.

107 mm (12-round) naval mounts

Two naval mounts exist, the Type ML-2 and Type ML-4. The former has 10 barrels in a 4:4: offset 2 arrangement and the latter the standard 12 barrel configuration.

Specifications

	107 mm (12-Round) Fadjr-1
Dimensions and weights	
Crew:	5
Length	
overall:	2.6 m
Width	
transport configuration:	1.4 m
Height	
transport configuration:	1.2 m
Weight	
standard:	195 kg (carriage)
combat:	380 kg

	107 mm (12-Round) Fadjr-1
Mobility	
Configuration	
running gear:	wheeled
Firepower	
Armament:	12 × carriage mounted 107 mm rocket system
Ammunition	
main total:	12
Main weapon	
maximum range:	8,500 m
Rate of fire	
rapid fire:	12/7 s (est.)
Operation	
reload time:	180 s
Main armament traverse	
angle:	30° (15° left/15°right)
Main armament elevation/depression	
armament front:	+60°/0°

Fadjr-6 minelaying rocket

Calibre:	122 mm
Weight:	
(complete round)	63 kg
(warhead)	32 kg
Length:	2.83 m
Payload:	8 anti-personnel/anti-tank mines
Range:	3,500-6,000 m

Status

Production as required. In service with the Hezbollah, Iranian Army, Iranian Navy and the Iranian Revolutionary Guard Corps (IRGC). It is estimated that Iran has about 900 107 mm rocket launchers from a variety of sources.

It is considered that quantities of these have been supplied to other countries and guerilla units in the Middle East.

Contractor

Defence Industries Organization (DIO), Armament Industries Division.

Israel

Israel Aerospace Industries/Israel Military Industries 306 mm EXTRA artillery rocket system

Development

In 2005 Israel Military Industries, Rocket Systems Division and Israel Aerospace Industries, MLM Division of its Systems, Missiles & Space Group, stated that they were developing the EXTended Range Artillery (EXTRA) rocket.

The new-generation EXTRA rocket which has now completed its development phase and according to the contractors provides a much greater range and greater accuracy. This will mean that less rockets will have to be fired to neutralise a given target and this, in turn, will reduce the logistical burden.

Although EXTRA is being developed as a private venture it could well meet future Israel Defence Force (IDF) requirements for a rocket with greater range and effect than the current in-service MLRS system. These are already being upgraded with the Israel Military Industries trajectory correction system.

Description

Few details of the EXTRA solid propellant surface-to-surface rocket have been released but it is expected to have a maximum range of at least 135 to 150 km and have a Circular Error of Probability (CEP) of about 10 m.

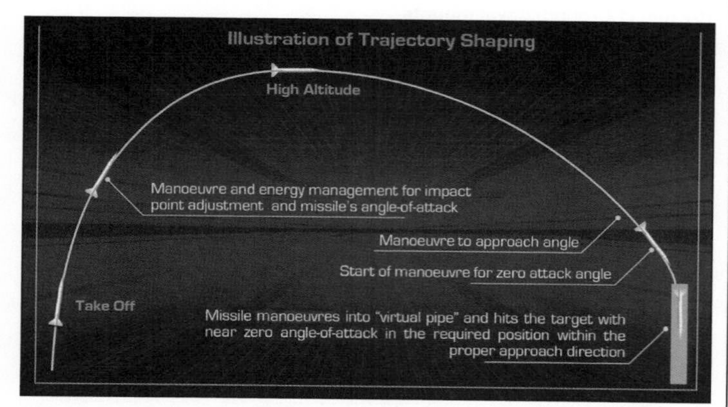

LORA long range artillery rocket launching sequence showing illustration of trajectory shaping (Israel Aircraft Industries) 1403749

Long Range Artillery (LORA) surface-to-surface guided missile during a test flight (IAI) 1327844

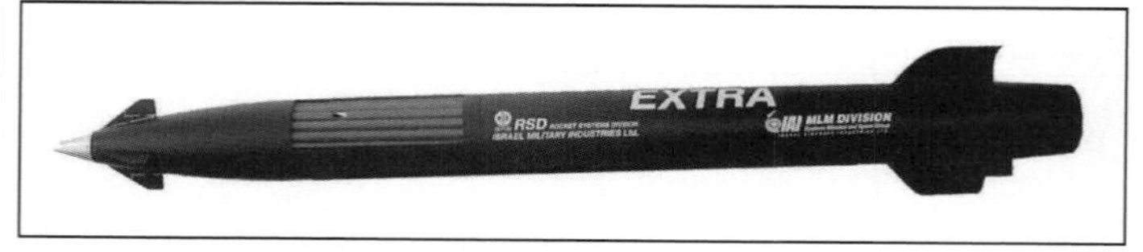

First drawing released of EXTRA surface-to-surface rocket as it would appear in flight with tail fins extended (IAI) 0590912

The 306 mm EXTRA artillery rocket system the Israel Aerospace Industries GPS-based inertial navigation system, with a gas generator thruster correcting the trajectory in flight. Target information is fed into the rocket prior to launch from the cab of the launch platform.

Various types of warhead weighing up to 120 kg could be fitted, including high-explosive or cargo-carrying submunitions. It is estimated that about 500 Israel Military Industries Bantam submunitions could be carried.

Bantams are fitted with a self-destruct fuze mechanism and are claimed to be highly effective against a variety of battlefield targets. Large numbers of these have been built for a variety of applications, such as artillery projectiles, mortar bombs, rockets and air-launched weapons.

According to Israel Military Industries, EXTRA can be launched from the in-service US supplied Lockheed Martin Missiles and Fire Control 227 mm Multiple Launch Rocket System (MLRS) that is used by the IDF and many other countries.

As the EXTRA rocket has a larger calibre, it is assumed that some modifications would be required to the launcher. It is understood that it would be provided in pods of four rockets. The current 227 mm MLRS rockets come in pods of six rockets. EXTRA rockets are supplied in pods of four rockets.

Variants

Long Range Artillery (LORA)

LORA has been developed by Israel Aerospace Industries, has a range from 30 to 300 km and can carry a warhead weighing 440 to 600 kg. Accuracy is being quoted as less than 10 m CEP. Up to four missiles would be carried in a single launcher. According to Israel Aircraft Industries main advantages of LORA can be summarised as:

- Pinpoint accuracy due to installation of GPS/INS navigation system
- Seekerless, shaped trajectory missile, homing in on target coordinates
- Super fast sensor-to-shooter cycle
- Various warhead types can be fitted to suit mission requirements
- Very short response time
- Supersonic velocity and minimal activation - choose target and fire
- Various types of target can be engaged including permanent and time critical
- Day and night operation
- Can be deployed in all weather and/or low visibility conditions
- Can be used in heavily defended areas.

In addition to being marketed for land based applications, LORA is also being marketed for surface-to-surface naval applications.

Land based version of LORA would typically be transported by a quadruple cross-country truck based system which according to IAI, would be able to launch its first LORA within ten seconds of coming to a halt.

It is understood that the first test flights of LORA took place in March 2004 and it has been offered to a number of potential customers including India, Israel and Turkey but as of April 2013 no firm sales had been announced.

Specifications

	EXTRA rocket
Dimensions and weights	
Length	
overall:	3.97 m
Diameter	
body:	306 mm
Weight	
launch:	450 kg
Performance	
Range	
min:	30 km
max:	150 km
Ordnance components	
Guidance:	GPS, INS

Status
Development. Not yet in production or service.

Contractor
Israel Military Industries (IMI) and Israel Aerospace Industries (IAI).

Israel Military Industries MAR-350 350 mm Medium Artillery Rocket System

Development
Israel Military Industries developed a 350 mm calibre artillery rocket system called the MAR-350 for installation on a similar range of tracked and wheeled chassis as the widely deployed LAR 160 mm system.

It is understood that the first firing trials took place in 1986 with a further series of trials following in 1988. As far as it is known the only customer for this is Argentina who has a small number of systems installed on a TAM medium tank hull. Available details of the system on a TAM chassis are provided in a separate entry in *IHS Jane's Land Warfare Platforms: Artillery and Air Defence.* It should be noted that Argentina has completed production of this hull.

It is understood that production of these rockets was completed some time ago and they are no longer marketed by Israel Military Industries.

In the designation MAR-350, MAR stands for Medium Artillery Rocket and 350 for the calibre of the rocket in mm.

Given the shelf-life of a solid propellant rocket is typically 10 to 15 years it is considered possible that these rockets may no longer be operational.

Description
The launcher is almost identical to that used in the Israel Military Industries LAR 160 mm rocket system fully covered in a separate entry in *IHS Jane's Land Warfare Platforms: Artillery and Air Defence.* This is fitted with two launcher pod containers/launchers in a side-by-side arrangement.

The weight of a pod of two rockets is 2,000 kg. The unguided surface-to-surface rockets can be launched one at a time or all four can be launched in 30 seconds.

The fuzed rockets are factory seated in their launch pod containers, two per container, which are loaded onto the launcher using a transport vehicle fitted with a hydraulic crane.

The 350 mm unguided surface-to-surface rockets are fitted with a cluster-munition-type warhead that contains 770 CL-3022-S4 bomblets and a remotely set time fuze. These bomblets are fitted with a self-destruct mechanism.

Variants
Israel Military Industries is no longer marketing the MAR-350 Medium Artillery Rocket System.

Today the company is marketing the following rocket systems:
- LAR 160 mm multiple rocket launcher system
- Strikes guided 122 mm rocket
- Jumper autonomous artillery system for ground forces.

Details of all of these systems are provided in separate entries in *IHS Jane's Land Warfare Platforms: Artillery and Air Defence.*

Specifications

	MAR-350
Dimensions and weights	
Length	
overall:	6.2 m
Diameter	
body:	350 mm
width	
body:	970 mm
height	
body:	450 mm

Status
Production complete. The only known user is Argentina. Some sources have stated that Israel has 20 units.

No longer marketed.

Contractor
Israel Military Industries (IMI).

Israel Military Industries LAR 160 mm multiple rocket launcher system

Description
The Light Artillery Rocket (LAR) system consists of a tracked or wheeled (typically 6 × 6) all-terrain vehicle or trailer on which is mounted a multiple rocket launcher holding two factory-sealed, expendable Launch Pod Containers (LPCs) with fuzed 160 mm diameter rockets in their launch tubes.

The LPCs, which also serve as storage and transportation containers for the rockets, can be rapidly replaced in the field after the rockets have been fired. When the rockets are fired, they burst frangible covers over the forward and aft ends of the launch tubes.

The Mk I system was first deployed on a French Nexter Systems (previously Giat Industries) AMX-13 light tank hull by the Venezuelan Army.

The Mk I was followed in production by the Mk II and more recently by the Mk IV systems. The Mk IV system, having the same dimensions as the other rockets but using a modified propellant, has a range of 45 km.

The solid propellant rockets are 160 mm in diameter and 3.4 m long, employing composite solid propellants. Wraparound stabilising fins deploy when the rocket exits the launcher.

The Mk I version of the LAR rocket, weighing 100 kg, carries a 40 kg HE-COFRAM warhead activated by a super-quick impact fuze or proximity fuze. Maximum effective range is 30 km. Its launch pod containers hold 13 or 18 rockets each, with two pods on the launcher.

The production Mk II version rocket weighs 110 kg. Its 46 kg warhead can be an HE-COFRAM type or a cluster warhead containing 104 CL-3022-S4 AP/AM submunitions. A remotely set electronic time fuze opens the canister at the appropriate height to give area coverage of about 31,400 m^2 for each cluster warhead. The submunitions are fitted with a self-destruct fuze.

Thirteen rockets are loaded in each launch pod container. Like the other versions of the LAR, the Mk II has a minimum range of 12 km. Its maximum range is 35 km.

The launchers are being mounted on heavy truck chassis, for example a German Mercedes-Benz (6 × 6). Suitable chassis for mounting the modular launcher are light tanks (such as AMX-13), medium tanks (such as M47 and M48), heavy trucks (such as M809, MAN, Steyr and Mercedes), self-propelled howitzer (M109) and a trailer towed by a heavy truck.

In its standard configuration, each launcher accommodates two 13-rocket launch pod containers. A light version with one launch pod is available for towing behind jeeps.

The LAR incorporates a modern C3I system called ACCS, which has a total interface capability to all common artillery elements including meteorological unit, forward observers as well as sophisticated mapping, GPS and other items.

Elevation and traverse of the launchers are performed by an electrohydraulic system, which is backed up by a manual system.

When the system is fitted on a wheeled chassis, two hydraulically-operated stabilisers are lowered to the ground to provide a more stable firing platform.

The launcher is operated by a computerised fire-control system, which is fully integrated with the ACCS, so enabling each one to operate as a one-gun battery.

All 26 160 mm unguided rockets can be fired in less than 60 seconds and the launchers are reloaded in less than five minutes by the truck driver and launcher crew from conventional army trucks equipped with a 15 t/m hydraulic crane.

160 mm ACCULAR being fired from Mercedes-Benz (6 × 6) truck-mounted LAR system 0021063

Pod of 18 160 mm rockets being loaded onto a launcher mounted on a French AMX-13 series full tracked hull (IMI) 0569053

LAROM upgrade

Israel Military Industries, teamed with Aerostar of Romania, is upgrading the Romanian Aerostar 122 mm (40-round) APRA multiple rocket launcher system to a new configuration called LAROM. This integrates the Israel Military Industries 160 mm rocket system into the existing platform. Full details of the LAROM 160 mm upgrade are given in a separate entry in *IHS Jane's Land Warfare Platforms: Artillery and Air Defence.* In addition to being in service with Romania it is understood that Georgia has also taken delivery of a batch of four systems.

Light LAR

A light launcher, which can be tailored for specific transport modes, has been specially supplied for several customers. It can be carried by helicopters and was designed to provide long-range, heavy fire support to light, highly mobile airborne units. This is an independent system that includes a launcher towed by a light vehicle such as a jeep or US AM General HMMWV. The towing vehicle is equipped with a light crane for loading the LPCs. A second vehicle, such as a light truck, carries the initial ammunition, which includes at least two LPCs of eight or more rockets each. The launcher system is equipped with an optical aiming system, communications equipment and a land navigation system. The trailer-mounted launcher is stabilised when deployed in the firing position by four hydraulically operated stabiliser legs.

The LAR launcher is not restricted to firing 160 mm rockets. Any launch pod container with a suitable interface can be mounted on the launcher. The pods are fully interchangeable with pods containing other size rockets, so a commander has a wide range of options without having to field completely separate systems. As new types of submunitions and mines are developed, they are being incorporated into submunition warheads.

ACCULAR – trajectory corrected artillery rocket system

ACCULAR, a unique Israel Military Industries development, is an accurate artillery rocket system, based on the LAR family. With its unique trajectory correction system, accuracy of the rockets is said to be equal to that of conventional tube artillery.

According to Israel Military Industries, it was only natural that this artillery rocket system will be significantly upgraded to better meet the requirements of the modern battlefield.

As with other members of Israel Military Industries family of artillery rockets, the system can be used on various ground platforms including tanks, standard trucks, trailers and others.

In addition, for rapid deployment forces and their requirement for air transportability, special towed models are also available.

The modular design of the system allows varying the quantities of rockets (eight, 10 or 13) in sealed launch pod containers on the launcher to meet the load capacity.

The ACCULAR concept can be adapted to different types and sizes of Israel Military Industries rockets and is valid through all artillery ranges.

The ACCULAR is claimed to represent only a moderate increase in cost. The major investment (technological and economic) is made on the existing command post while the cost increase in the rocket is relatively marginal.

By providing artillery rockets with increased accuracy to a level of that obtained by conventional tubed artillery systems, ACCULAR enables the optimal deployment of smart munitions (such as top attack type or scatterable anti-tank mines) which require deployment in close vicinity to the target.

The Israel Military Industries ACCULAR concept may be adapted to different models and types of rockets and claimed to be a major breakthrough in this field as the upgrading of a free-flight rocket has now become a reality.

This system has now been applied to the MLRS used by the Israel Defense Force under a contract awarded in mid-1998 to Israel Military Industries. Lockheed Martin Missiles and Fire Control, prime contractor for the 227 mm MLRS system, is also involved in this programme.

Late in 2003, Israel Military Industries announced that it had successfully completed flight tests for its trajectory correction system for the Lockheed Martin Missiles and Fire Control 227 mm MLRS rocket.

The design verification, conducted late in September 2003, for a range of 35 km, marked the end of a 10-year development contract to enhance the accuracy of the 227 mm MLRS rockets used by the Israel Defense Force. The system is now in production for the 227 mm MLRS used by the IDF.

US 6 × 6 truck towing LAR 160 mm trailer-mounted rocket system with two pods each carrying 13 rockets (IMI) 0569054

According to Israel Military Industries, the trajectory correction system will reduce by 95 per cent the number of rockets required to neutralise a given target when compared to standard production unguided rockets.

The 160 mm ACCULAR is fitted with a Trajectory Correction System (TCS) which includes a steering unit mounted towards the nose that includes an electronic unit, gas generator and valve.

After launch, the ground control unit establishes a datalink with the rocket and calculates the current trajectory and corrections necessary to reach the target.

The actual course correction phase is carried out as the rocket is accumulating environmental data that affects its flight course. Through this process, course correction is performed in a closed loop as the ground control unit commands the rocket steering to correct its trajectory.

The type of warhead depends on the user's operational requirement, for example a cargo warhead containing 104 High-Explosive Anti-Tank (HEAT) bomblets fitted with a self-destruct fuze mechanism.

Specifications of the 160 mm ACCULAR rocket are given as these are slightly different to those listed:

Rocket specifications

Calibre:	160 mm
Length:	3.70 m
Rocket weight:	120 kg
Payload weight:	60 kg
Payload:	104 bomblets with self-destruct fuze
Maximum range:	45 km
Pod	
Number of rockets:	13
Total weight loaded:	2,050 kg
Length:	4,000 mm
Width:	980 mm
Height:	685 mm

Late in 2004 Romania became the first export customer for ACCULAR when it placed a USD40 million contract with Israel Military Industries for an undisclosed quantity of 160 mm ACCULAR rockets. This is referred to as the ACCULAR-160.

Israel Military Industries is also understood to be marketing a laser-guided 160 mm rocket.

CAL 160 MRS

Argentina uses the LAR 160 mm on a TAM hull under the designation of the CAL 160. Details of this and the 350 mm CAL 350 are given in a separate entry. Production of this is understood to be complete.

NAVLAR MRS

This is a fully stabilised Naval Artillery Rocket (NAVLAR) system based on the ground-based LAR 160 mm system but that has been modified for naval applications. According to Israel Military Industries, it can fire all 160 mm rockets and the system can be installed on surface craft as small as 1,000 tonnes.

306 mm EXTRA rocket

In 2005 it was revealed that Israel Aerospace Industries and Israel Military Industries were developing a new surface-to-surface rocket called EXTRA (EXTended Range Artillery). This has a maximum range of 150 km and available details are provided in a separate entry.

Long Range Artillery (LORA) rocket system

This has been developed by Israel Aerospace Industries and details are provided in a separate entry in *IHS Jane's Land Warfare Platforms: Artillery and Air Defence.*

It is understood that LORA, which has a maximum range of up to 300 km and has pinpoint accuracy, has yet to enter service.

Artillery rocket specifications

Model	Mk I	Mk II	Mk IV	350
Rocket calibre:	160 mm	160 mm	160 mm	350 mm
Rocket length:	3.4 m	3.4 m	3.4 m	5.8 m
Rocket weight:	100 kg	110 kg	117 kg	850 kg
Warhead weight:	40 kg	46 kg	46 kg	300 kg
Range:				
(minimum)	12 km	12 km	12 km	30 km
(maximum)	30 km	35 km	45 km	100 km

Status
Production as required. In service with a number of countries including Argentina (on TAM chassis), Chile (8/12), Israel (about 50) and Venezuela (20 on modified AMX-13 light tank chassis – phasing out of service).

Contractor
Israel Military Industries (IMI).

Israel Aerospace Industries Strikes 122 mm guided rocket

Development
Based on their experience in the design, development and production of rocket systems Israel Aerospace Industries, MLM Division, has developed the Strikes 122 mm guided rocket.

Development of this is understood to have been completed and production can commence when export orders are placed.

The baseline Russian designed 122 mm unguided rocket typically has a range of around 20 km but according to Israel Aerospace Industries, with the addition of the Strikes 122 mm guidance system range is increased to 40 km.

As the Israel Defense Force does not currently deploy any 122 mm rocket launchers, the Strikes 122 mm guided rocket is aimed at the export market.

As of April 2013 there are no known sales of the Strikes 122 mm guided rocket.

Description
The Israel Aerospace Industries Strikes 122 mm guided rocket has been developed to enhance the capabilities of Russian BM-21 122 mm (40-round) rocket system which is used by many countries, or the many similar systems.

The standard 122 mm unguided rocket is fitted with a steering and avionics section which is located between the solid propellant rocket motor and the warhead.

The warhead with the attached steering section, including the manoeuvring section, separates towards the end of the flight.

The steering section contains the Global Positioning System/Inertial Navigation System (GPS/INS) and at rocket motor burnout the latter is separated from the steering section.

The four steering fins are deployed and the manoeuvring section navigates to the target location coordinates.

This provides a pinpoint accuracy with a claimed Circular Error of Probability (CEP) of 10 m at any range.

According to Israel Aerospace Industries, Strikes reduces collateral damage as well as the number of rockets required to neutralise a given target.

Specifications
Not available.

Status
Development complete. Ready for production.

Contractor
Israel Aerospace Industries (IAI), MLM Division.

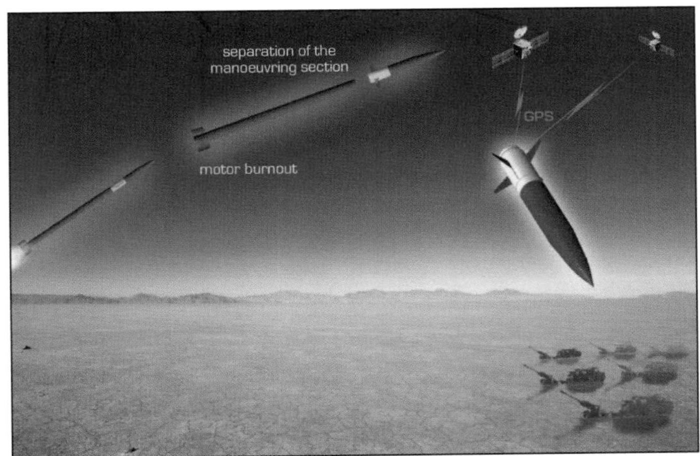

Artists impression of mode of operation of Strikes 122 mm guided rocket with steering/warhead section deployed on right (IAI) 1403750

Israel Aerospace Industries Jumper guided rocket system

Development
Based on their experience in the design, development and production of rocket systems, Israel Aerospace Industries, MLM Division, has developed the Jumper "autonomous artillery for the ground forces".

Artists impression of mode of operation of Jumper autonomous artillery for the ground forces with Jumper missile starting to exit its vertical launcher (IAI) 1403754

This has been developed as a private venture by Israel Aerospace Industries but could meet the potential requirements of the Israel Defense Force (IDF).

In many respects Jumper is similar to the US Army Non-Line-Of-Sight Launch System (NLOS-LS) which was part of the US Army Future Combat System (FCS) that was consequently cancelled.

Prime contractor for NLOS-LS was Netfires and this consisted of a box type vertical launcher that could launch the Precision Attack Missile (PAM) or a Loitering Attack Missile (LAM).

The Jumper system was announced for the first time in late 2009 and as of April 2013 it is understood this was still under development but was already being marketed on an international basis.

Description
The Israel Aerospace Industries Jumper consists of a Vertical Launcher Hive (VLH) with a 3 × 3 vertical elements with eight of these for the actual Jumper missiles and the ninth for the integrated command-and-control element of the system.

It is estimated that the VLH is 1.4 m long, 1.4 m wide and has an overall height of 2 m.

Each Jumper missile has a overall length of 1,800 mm, diameter of 150 mm and a launch weight of 63 kg.

The Jumper missile uses a Global Positioning System/Inertial Navigation System (GPS/INS) guidance system for pinpoint accuracy with a maximum stated range of up to 50 km.

A semi-active laser homing system can be used for terminal homing.

Jumper will be offered with different types of warhead to meet different target patterns with the baseline missile probably having a high-explosive fragmentation warhead.

Main advantages of the Jumper claimed by Israel Aerospace Industries are the provision of autonomous precision fire for manoeuvring echelons, rapid fire response, can be deployed and engage targets under all weather conditions and minimal collateral damage.

If deployed by the IDF it would supplement their currently deployed 155 mm artillery systems and rocket systems such as the US supplied 227 mm (12-round) Multiple Launch Rocket System (MLRS).

Specifications

	Jumper guided missile
Dimensions and weights	
Length	
overall:	1.8 m
Diameter	
body:	150 mm
Weight	
launch:	63 kg
Performance	
Range	
max:	50 km

Status
Development. Not yet in production or service.

Contractor
Israel Aerospace Industries (IAI), MLM Division.

Italy

BPD FIROS 30 122 mm (40-round) multiple rocket launcher

Development

The FIROS 30 Field Rocket System is an area-saturation weapon system and is based on the 122 mm BPD Difesa E Spazio SpA unguided rockets. The system is intended for use either in battery or isolated missions and can cover large areas in the 8 to 34 km zone.

The FIROS 30 122 mm (40-round) multiple rocket launcher and its associated family of 122 mm unguided rockets was originally developed as a private venture by BPD Difesa E Spazio for the export market.

It was not adopted by the Italian Army who currently deploy the US Lockheed Martin Missiles and Fire Control 227 mm (12-round) MLRS.

All production and marketing of the FIROS 30 122 mm (40-round) multiple rocket launcher has now ceased.

As far as it is known the only current user of this system is the United Arab Emirates who have upgraded most of their systems under a contract awarded to the Turkish company of Roketsan.

Available details of this upgrade programme, which has now been completed, are provided later in this entry under variants.

Description

A typical battery would consist of a command post equipped with a fire-direction system, six firing units and six to 12 escort units. The firing unit is a modular rocket launcher installed on a heavy (6 × 6) truck, for example Iveco (Italy) or Mercedes-Benz (Germany).

It was available in two different versions: a standard version, with a motorised/manual movement system simple to operate; and an automatic version, equipped with an inertial navigator, a servo-control system and a ballistic computer.

This firing unit configuration enables the crew to perform the topographic and ballistic calculations and the aiming procedures through an Automatic Control Unit installed in the vehicle cab.

The launcher consists of two removable modules with 20 launching tubes each. The escort unit can carry four launching modules. The loading operations are carried out by removing the empty 122 mm (20-round) module with a jib crane fitted on the support vehicle and installing the filled one ready to fire.

The firing unit has a fire-control system and a link with the command post to acquire the necessary firing data in real time. The fire-control system has been designed to enable selection of the number (from 1 to 40) and type of rockets to be fired.

The FIROS 30 122 mm unguided rockets differ in propellant types and maximum range and can carry a variety of warheads.

The United Arab Emirates use the FIROS system on Mercedes-Benz (6 × 6) truck chassis.

The 122 mm rockets launched by the 122 mm FIROS 30 system can also be launched by the widely used 122 mm Russian, Chinese or similar BM-21 systems.

Production of this 122 mm (40-round) rocket and its associated rockets has been completed and they are no longer marketed.

Variants

Upgraded United Arab Emirates systems

The United Arab Emirates (UAE) has now taken into service an upgraded version of its Italian supplied BPD FIROS 122 mm (40-round) multiple rocket launcher system.

These were originally delivered to the UAE many years ago and each launcher system originally consisted of a Mercedes-Benz (6 × 6) cross-country chassis on the rear of which is mounted a turntable with two pods of 20 122 mm unguided surface-to-surface rockets.

FIROS 30 multiple rocket launcher in action 0130021

United Arab Emirates Mercedes-Benz (6 × 6) chassis upgraded by Al Jaber prior to being fitted with the Turkish Roketsan launcher (Al Jaber) 1452637

These 122 mm rockets have now run out of shelf life and the Turkish company of Roketsan was awarded a contract to provide a new launcher system together with a new family of 122 mm rockets and a modified fire-control system.

First deliveries were made in 2005 and continued through to 2008. The upgrade work to the launcher was carried out in the UAE with rockets provided direct from Turkey to UAE.

It is understood that the UAE has about 36/48 launchers but at this stage it is not known how many were upgraded.

Existing launcher system has been removed and replaced by a new turntable with powered launcher which carries two pods each of 20 × 122 mm rockets.

Once these rockets have been fired the launcher would normally redeploy to another position to avoid counterbattery fire where the pods would be rapidly removed and replaced by two new pods.

So far the UAE has taken delivery of at least two types of Roketsan 122 mm rockets which have already been produced in production quantities for the Turkish Land Forces Command.

First UAE batch consisted of 3,040 rockets in two models, TRB-122 and TRK-122. The former is fitted with a high explosive warhead and has a stated maximum range of over 40 km with a minimum range of 21 km without drag ring and 10 km with drag ring.

TRK-122 has a maximum range of 30 km and a minimum range of 16 km and is fitted with a warhead containing 50 anti-personnel/anti-material munitions and six incendiary bomblets. Both rockets have a composite solid propellant rocket motor. These were delivered in 2005 with a similar number being delivered in 2006.

It is understood that this upgrade work was carried out in the United Arab Emirates by a number of contractors including the Al Jaber Group.

The Mercedes-Benz 6 × 6 chassis now features a protected cab with the new launcher mounted on the rear of the chassis.

A central tyre inflation system is fitted for improved cross-country mobility in sand terrain conditions.

Details of the Roketsan 122 mm rockets used in the upgraded system are provided in a separate entry in *IHS Jane's Land Warfare Platforms: Artillery and Air Defence.*

The 122 mm unguided rockets can be launched from the cab or away from the cab using a remote control device.

Specifications

BPD 122 mm rocket

Calibre:	122 mm
Length of rocket:	
(conventional warhead)	2.815 m
(submunition warhead)	3.321 m
Weight of rocket:	
(conventional warhead)	65 kg
(submunition warhead)	71 kg
Maximum range:	
(conventional warhead)	34,000 m
(submunition warhead)	30,000 m
Propellant weight:	24 kg
Total impulse:	6,000 kg/s
Chamber pressure:	210 kg/cm^2
Average thrust:	3,000 kg
Combustion time:	2.05 s (at 50° maximum pressure)
Propellant type:	composite
Conventional warheads	
Weight: (total)	26 kg
Trajectory safety:	150 m
Types of warhead:	High-Explosive (HE), Target Practice (TP)
Fuze:	Point Detonating (PD)
Submunition warheads	
Weight: (total)	32 kg
Characteristics	
Time into action:	5-10 min (1 min using automatic firing unit)
Ripple time:	16 s (40 rockets)
Disengagement time:	1 min (30 s using automatic firing unit)
Launcher Unit	
Number of tubes:	40 (20 per module)
Elevation:	0° to +60°
Azimuth:	±105°

Crew:	3 (2 using automatic firing unit)
Loading Module	
Length:	3.7 m
Width:	0.82 m
Height:	0.69 m
Weight:	
(empty)	400 kg
(loaded)	
(conventional warhead)	1,710 kg
(submunition warhead)	1,810 kg

Status
Production complete. No longer marketed. In service with the United Arab Emirates (36/48 units). It is understood that the systems used by the UAE have now been upgraded by Roketsan of Turkey to fire its latest generation 122 mm rockets. It has been confirmed that Roketsan has also supplied rockets to the UAE. In 2005 Turkey supplied the UAE with 3,040 122 mm rockets for the upgraded launchers. A similar number were delivered in 2006 by Roketsan.

Contractor
BPD Difesa E Spazio SpA. (No longer involved in the design, development, or production of land-based rocket systems.)

BPD FIROS 51 mm multiple rocket system

Development
The FIROS 51 (48-round) multiple rocket system was developed by BPD Difesa E Spezio in the 1970s as a private venture specifically for the export market.

As far as it is know the only customer for this short range unguided system were the Mexican Marines who purchased a small batch of systems which were integrated onto the rear of a French ACMAT (4 × 4) cross-country truck chassis.

Production and marketing of this 51 mm (40-round) multiple rocket launcher has ceased.

Given the shelf-life of these 51 mm unguided surface-to-surface rockets, it is considered that these systems could be non-operational in the future.

Description
The FIROS 6 is a 48-round 51 mm MRS mounted on the rear of light 4 × 4 vehicles. The launcher is moved and aimed in traverse and elevation by handwheels and partial or full ripples can be fired at the rate of 10 rockets a second.

The crew can fire the rockets standing alongside the vehicle or from a remote position up to 30 m from the launcher via a cable. The 51 mm 40-round launcher can be reloaded by hand within five minutes.

The 51 mm unguided solid propellant rockets used in this system are the same solid propellant, with four folding fins, BPD Difesa E Spazio SpA 2 in rockets originally developed for the air-to-ground role.

51 mm rocket Specifications

Calibre:	51 mm
Length:	
(without warhead)	744 mm
(with warhead)	1.05 m
Weight:	
(without warhead)	2.6 kg
(with warhead)	4.8 kg
(propellant)	1.06 kg
Maximum range:	6,550 m
Time of flight to maximum range:	39 s
Propellant type:	extruded double base
Total impulse:	210 kg/s
Maximum chamber pressure:	100 kg/cm^2
Average thrust:	200 kg
Acceleration:	60 g
Speed at burnout:	515 m/s
Motor case:	steel, cold extruded
Nozzle fin:	with thrust deflectors and folding free rotating fins
Warhead types:	Anti-Tank/Anti-Personnel (AT/AP), High-Explosive Incendiary (HEI), Illuminating (Ill), Pre-Formed Fragmentation (PFF), Smoke (SM), Target Practice (TP), Target Practice Flash (TP-FL), Target Practice Smoke (TP-SM), Target Practice Spotting (TP-SP)

The anti-tank/anti-personnel rocket can penetrate over 250 mm Rolled Homogenous Armour (RHA) steel in addition to the secondary anti-personnel effect.

These 51 mm rockets have a typical shelf-life of 10 to 15 years depending on their storage conditions. Against this background it is considered that the shelf-life of these rockets has now expired.

FIROS 6 multiple rocket system on an ACMAT 4 × 4 chassis in service with Mexican Marines (Julio A Montes) 0500902

Specifications
FIROS 6 launcher module

Number of tubes:	48
Length: (module)	2.08 m
Width: (module)	0.67 m
Height: (module)	0.93 m
Weight:	
(empty)	250 kg
(loaded)	480 kg
Elevation/traverse:	manual
Elevation limits:	–5° to +45°
Traverse:	360°
Power supply:	24 V DC
Rate of fire:	10 rockets/s

Status
Production complete. No longer marketed. The only known user is the Mexican Marines, who have installed the launcher on the rear of a 4 × 4 cross-country truck chassis.

Contractor
BPD Difesa E Spazio SpA. (No longer involved in the design, development or production of land-based rocket systems.)

Jordan

AB19 HMMWV 107 mm (12-round) multiple rocket system

Development
The HMMWV AB19 107 mm (12-round) multiple rocket system was developed from late 2001 in Jordan to meet the operational requirements of the Jordanian Special Operations Command.

The system was designed and built by the Jordanian King Abdullah II Design and Development Bureau (KADDB) and a total of six systems were built and delivered by 2002.

These AB19 HMMWV systems, together with other equipment, were deployed by the Jordanian Special Operations Command to provide protection for Jordanian Armed Forces deployed to Afghanistan as part of the International Security Assistance Force.

Production of this system has been completed and it has not been offered on the export market by Jordan.

It should be noted that these 107 mm unguided rockets can be supplied from a number of countries.

Description
The system consists of a US supplied AM General High Mobility Multipurpose Wheeled Vehicle (HMMWV) (4 × 4) unarmoured M998 chassis with the rear modified to accept the complete upper part (including mount, launcher and elevating/traverse mechanism) of a China North Industries Corporation (NORINCO) 107 mm (12-round) Type 63 multiple rocket launcher.

This is normally mounted on a trailer and towed by a light vehicle, although there are also self-propelled versions Full details of the NORINCO 107 mm (12-round) Type 63 system are provided in a separate entry in *IHS Jane's Land Warfare Platforms: Artillery and Air Defence.* It should be noted that NORINCO has ceased marketing this 107 mm rocket system.

Maximum range of the unguided rockets is 8,500 m or 10,000 m depending on the version.

The HMMWV (4 × 4) is normally armed with a roof-mounted .50 (12.7 mm) M2 HB Browning machine gun, which is mounted on an unprotected pintle mount on the roof of the vehicle.

It is understood that the HMMWV chassis carries a number of 107 mm rockets for manual reloading.

Specifications
As far as it is known, no detailed specifications of the complete AB19 system have been released. The 107 mm (12-round) launcher is understood to be identical to that of the China North Industries Corporation (NORINCO) Type 63 system.

KADDB AB19 HMMWV (4 × 4) rocket launcher with 107 mm (12-round) launcher traversed to the left (IHS/Rupert Pengelly) 0532526

Status
Production as required. In service with Jordan.

Contractor
King Abdullah II Design and Development Bureau (KADDB).

Korea (North)

North Korean 107 mm (12-round) rocket launcher

Development
Initial production of these multiple rocket launchers began in North Korea during the early 1960s, with a locally-built version of the very old 107 mm China North Industries Corporation (NORINCO) Type 63 system. Well over 5,000 were produced for the home and export markets.

It should be noted that NORINCO is no longer marketing this unguided surface-to-surface rocket system.

With the Korean People's Army (KPA) the trailer 107 mm system was originally found in the infantry division and brigade multiple rocket artillery battalions and batteries.

In the mechanised and motorised division MRL battalions and in the specialised combined arms brigades battery the trailer version was replaced by either a tracked version or a truck-mounted variant.

Considerable numbers have been exported and these have seen combat service with the Palestinian Liberation Organisation (PLO) and Syria in Lebanon and with Iran in the 1980-88 Gulf War.

It should be noted that these 107 mm unguided rockets are manufactured in many countries.

While production of these 107 mm (12-round) rocket launchers has been completed for the North Korean Army, it is considered probable that additional rocket launchers could be manufactured for the export market if required.

An alternative would be to provide surplus 107 mm (12-round) rocket launchers from North Korean Army stocks as in many units these have been replaced by more mobile systems with a longer range which increases their survivability from counter battery fire.

Description
The former has the launcher mounted on the licence-built Chinese full-tracked M1973 (Modified NORINCO Type YW531 full-tracked armoured personnel carrier). Apart from the 12-round type, the self-propelled variant also exists in two other versions, the 18-round VTT-323 (M1992) system (which uses the launcher, in three layers of six tubes, mounted on the rear upper decking of the VTT-323 tracked APC) and the 24-round M1992/2 system (which uses the launcher, in three layers of eight tubes, mounted on the rear upper decking of the indigenously developed 4 × 4 armoured personnel carrier).

Details of the VTT-323 tracked launcher are given in a separate entry. It is understood that production of this system was completed some time ago.

Specifications

	M1992/2
Dimensions and weights	
Length	
overall:	6.10 m
Width	
transport configuration:	2.80 m
Height	
transport configuration:	2.60 m
Weight	
combat:	12,000 kg
Mobility	
Configuration	
running gear:	tracked
Engine	diesel, 250 hp
Firepower	
Armament:	24 × 107 mm rocket system
Survivability	
Armour	
hull/body:	steel

Status
107 mm towed system. Production as required. In service with Iran, Libya, North Korea (towed and self-propelled versions), Hezbollah the PLO and Syria.

As the 107 mm towed system is made by a number of countries, it is possible that some of these countries could have been supplied by other sources, rather than North Korea.

Contractor
North Korean state factories.

240 mm (22-round) M1991 multiple rocket launcher

Development
The 240 mm (22-round) M1991 Multiple Rocket Launcher (MRL) was first seen in 1991 and is mounted on a heavy 6 × 6 truck chassis with the cab located over the engine.

Like the North Korean 240 mm M1985 system the M1991 is assigned to MRL battalions comprising three batteries, each of five 240 mm MRLs.

It is considered that the production of the unguided surface-to-surface rocket system has now been completed.

The Iranian Aerospace Industries Organization Fadjr-3 240 mm artillery rocket system, covered in detail in a separate entry in *IHS Jane's Land Warfare Platforms: Artillery and Air Defence,* may be related to this North Korean System.

Description
The system fires the same 240 mm calibre unguided surface-to-surface rockets as the M1985 and has similar HE, smoke, incendiary and chemical warhead capabilities.

The designation M1991 was originally allocated to this system by the US Department of Defense. In the designation M1991 the 1991 indicates the year that it was first observed by US intelligence.

The 6 × 6 forward control cross-country truck chassis, on which this system is based, is understood to be based on an imported design.

Status
Production as required. In service with the North Korean Army. It is possible that technology to build these 240 mm rockets has been transferred to Iran. Recent information has indicated that at least one country in the Middle East, not Iran, has taken delivery of complete 240 mm multiple rocket launchers of an undisclosed type from North Korea.

Myanmar may have taken delivery of a batch of M1991 MRL.

Contractor
North Korean state factories.

240 mm (12-round) M1985 Multiple Rocket Launcher

Development
The 240 mm (12-round) M1985 Multiple Rocket Launcher (MRL) was developed to meet the requirements of the North Korean Army.

The designation of the M1985 was given by the US Army as this was the first year that the system was identified by Western sources.

As far as it is known production of this multiple rocket launcher has been completed.

The 240 mm (12-round) M1985 Multiple Rocket Launcher (MRL) is mounted on a Japanese Isuzu 6 × 6 cross-country truck chassis and comprises 12 × 240 mm rocket tubes mounted in two banks of six on the rear decking. Each bank has two rows of three tubes.

It is considered possible that North Korea has developed and deployed additional 240 mm truck mounted multiple rocket launchers.

Description
Prior to the 240 mm rockets being launched, four manually-operated stabilising jacks, two mounted at the rear of the vehicle and the other pair located just behind and either side of the cab, have to be lowered. All the traverse and elevation controls are manually operated.

The 240 mm calibre unguided solid-propellant surface-to-surface rocket is spin-stabilised in flight and fitted with a contact detonation fuzing system. Maximum range is said to be 43,000 m. The standard warhead used is a 90 kg HE fragmentation type although other types such as smoke, incendiary and chemical agent are believed to exist.

The launcher is supported by a transloader vehicle, which carries at least one complete reload and a crane to assist in the manual reloading operation. The unguided 240 mm rockets are loaded one by one.

Reloading would normally be carried out well away from where the rockets were launched to avoid counterbattery fire.

It is understood that the Isuzu (6 × 6) chassis used for the 240 mm (12-round) M1985 multiple rocket launcher is the same as that used for the North Korea BM-11 122 mm (30-round) multiple rocket launcher.

It is understood that technology to build these 240 mm rockets could have been transferred to Iran. The Iranian Defence Industries Organisation 240 mm Fadjr-3 (12-round) system is very similar. Iran is believed to have taken delivery of nine 240 mm (12-round) M1985 MRLs. Recent information gives this system a maximum range of 43,000 m with the 90 kg warhead containing 45 kg of high-explosive.

Details of the Iranian 240 mm Fadjr-3 artillery rocket system which is integrated onto a Mercedes-Benz (6 × 6) chassis are provided in a separate entry in *IHS Jane's Land Warfare Platforms: Artillery and Air Defence.*

Variant

240 mm M1989 multiple rocket launcher

A 240 mm multiple rocket launcher, designated M1989 by the US, is known to exist. No other details are known at present.

US sources have indicated that there is also a 240 mm MRS designated the M1999 which is probably a new system.

Status

Production is probably complete. Both the M1985 and M1989 are in service with the North Korean Army. The designations M1985 and M1989 has been allocated by the US DoD as these were the years that they were first sighted. It is understood that technology to build these 240 mm rockets has been transferred to Iran. Recent information has indicated that at least one Middle Eastern country, not Iran, has taken delivery of complete 240 mm multiple rocket launchers of an undisclosed type from North Korea.

Contractor

North Korean state factories.

North Korean 122 mm BM-11 (30-round) rocket launcher

Development

Following delivery during the late 1960s of numbers of Russian 122 mm BM-21 (40-round) MRL systems, the North Koreans began production of their own derivative in the mid-1970s.

Large numbers of these systems were built for the export market. Some countries have installed this 122 mm (30-round) rocket launcher on other chassis. The Lebanon, for example, has integrated these onto the rear of US supplied M35 series (6 × 6) 2.5-tonne cross-country truck chassis.

Description

Known as the BM-11, this has two side-by-side 15-round banks of launcher tubes and is usually mounted on locally-produced versions of the Russian Ural-375D (6 × 6) 4,000 kg truck, although for export purposes it can be mounted on other chassis.

An example of this is the PLO's Japanese-built Isuzu (6 × 6) 2,500 kg systems which were captured by Israel during the 1982 war in Lebanon. Some of these were taken into service with the Israel Defense Force.

In KPA service the BM-11 is used at army and corps level within MRL brigades (up to 72 systems). One of the roles assigned to these units is to provide chemical warfare delivery capability at the strategic level using chemical warheads that have been indigenously designed, manufactured and filled with agents including nerve gases.

The BM-11 may also be found in the MRL battery of some of the combined arms brigades replacing the six towed 107 mm systems found in that unit.

It is assumed that these North Korean BM-11 (30-round) rocket launchers fire the standard 122 mm unguided rockets which have a maximum range of 20,100 m.

A number of countries such as China, Egypt, Poland and Russia have developed and placed in production 122 mm rockets with much longer ranges, in some cases out to 40 km.

North Korea may deploy some of these Russian 240 mm (12-round) BM-24 multiple rocket systems shown here with loading device rail attached at rear of the launcher (INA) 1452926

North Korean BM-11 version of Russian 122 mm MRS with two banks of 15 launcher tubes instead of single-round bank, mounted on rear of Japanese Isuzu (6 × 6) 2,500 kg truck chassis (Israeli Ministry of Defence) 0500669

North Korea may still deploy some of these Russian supplied 140 mm (16-round) RPU-14 towed multiple rocket systems (INA) 1452927

It is considered possible that North Korea has developed and placed in production enhanced 122 mm rockets.

Variants

There are two other 122 mm systems known. The 40-round M1977 version, which may be a direct copy of the Russian BM-21 on a locally built chassis and the follow-on 40-round M1985, which is mounted on a Japanese Isuzu (6 × 6) truck chassis. The latter is believed to have a 40-round reload pack as part of the rear deck-mounted system.

The older rocket launcher systems are known to include varying numbers of truck-mounted Russian-designed 132 mm BM-13-32, 140 mm BM-14-16, 200 mm BMD-20 and 240 mm BM-24 weapons plus small numbers of the trailer 140 mm RPU-14 for airborne forces use.

These rocket launchers were supplied to North Korea by Russia many years ago and unless North Korea has produced replacement rockets, it is probable that these are now held in reserve or could have well been phased out of service. China is also known to have provided a number of its 130 mm Type 63 systems. Both can be produced as required and are in service with North Korea.

Status

Production as required. In service with Algeria, Egypt (50), Iran (20), Israel, North Korea, Lebanon, Libya (200), Nicaragua, Pakistan, PLO, Polisario Front, Syria, Uganda and Zimbabwe.

This system is also used by a number of insurgent elements in the Middle East and elsewhere.

Contractor

North Korean state factories.

Korea (South)

South Korean Long Range Multiple Rocket Launcher

Development

In 2011 the South Korean Agency for Defence Development confirmed that it is developing a new Multiple Rocket Launcher (MRL) with the first prototype expected to be rolled out in 2013.

According to officials of the South Korean Defense Acquisition Program Administration (DAPA) the new locally developed MRL will have a range of 80 km and be used to booster the artillery capability of the South Korean Army.

No details of the calibre, type of rocket or warhead fitted, or whether it will be based on a tracked or wheeled chassis, have been released. Some sources however have indicated that this system will have a calibre of 230 mm.

MRL are complimentary to conventional tube towed and self-propelled artillery and are capable of rapidly providing suppressive fire over a much wider area and more rapidly than conventional tube artillery.

The rockets can be fitted with various types of warhead including high-explosive or carrying submunitions as the latter are highly effective against mass attack by armoured vehicles or infantry.

Neither North or South Korea have signed up to the Ottawa Convention on cluster munitions/submunitions as there has yet to be developed a direct and compliant replacement.

The new ROK MRL may well have a guided rocket with a blast type warhead to provide a precision fire capability against high value targets such as air defence weapons, command posts as well as key infrastructure such as bridges.

Today the ROK Army deploys a number of shorter range unguided MRS including about 156 of the locally developed 6 × 6 truck mounted Kooryong and 29 US supplied Lockheed Martin Missiles and Fire Control 227 mm (12-round) Multiple Launcher Rocket Systems (MLRS) based on a full tracked chassis.

Details of both of these systems are provided in *Jane's Artillery and Air Defence* and production of these has been completed.

The Kooryong was developed in the ROK by Doosan Infracore Defense Products and is a 6 × 6 chassis on the rear of this is a 130 mm (36-round) launcher with the unguided rockets being developed by the Hanwha Corporation. The current system is not provided with a protected cab for the crew.

The K30 unguided rocket has a maximum range of 23 km while the Improved K33 has its range increased to 33 km but carries less high-explosive.

The MLRS fires an unguided rocket carry 644 submunitions out to a maximum range of 31.6 km but the launchers can also fire the larger Army Tactical Missile System (Army TACMS) with a range of over 165 km. The US supplied MLRS is provided with a fully protected cab for the crew.

As Doosan Infracore Defense Products and Hanwha Corporation have been involved in the currently deployed Kooryong MRS it is possible that they and other ROK contractors are involved in the new MRL system.

When deployed the new South Korean MRS would help counter the MRS and conventional towed and self-propelled artillery weapons used by North Korea.

Description

As of early 2012, no details of the calibre, type of rocket or warhead fitted to the new ROK MRL have been released although some sources have mentioned a calibre of 230 mm.

This would enable a large warhead, perhaps incorporating a guided system which would enable high value targets to be engaged at long range with a precision effect.

It is considered probable that the future ROK MRL would be based on a protected wheeled platform, probably 6 × 6 or 8 × 8, to provide greater strategic mobility as well as lower operating and support costs when compared to their tracked counterparts.

Specifications

Not available.

Status

Under development to meet the requirements of the Republic of Korea Army.

Contractor

Not revealed.

Doosan Kooryong 130 mm (36-round) multiple rocket system

Development

The Doosan Infracore Defense Products (previously Daewoo Heavy Industries & Machinery Ltd) Kooryong 130 mm 36-round, multiple rocket launcher system was originally designed and built in South Korea for corps and divisional level MRL units.

Although fitted on the rear of the locally built five-tonne (6 × 6) KM809A1 truck chassis in South Korean Army service, it can be fitted to any wheeled or tracked vehicle with a 5 ton class load capacity.

As far as it is known, all production applications of the multiple rocket system have been on the locally manufactured five-tonne (6 × 6) KM809A1 truck chassis.

Production of this chassis was completed some years ago and it is considered possible that in the future the system would be removed from this chassis and integrated on a more recent cross-country truck chassis.

In the longer term this Doosan Kooryong 130 mm (36-round) multiple rocket system will be replaced by the new Hanwha Corporation Chun-Mu Long Range Multiple Rocket System which is covered in a separate entry in *IHS Jane's Land Warfare Platforms: Artillery and Air Defence*.

This is currently under development and is based on an 8 × 8 cross-country vehicle with protected forward control cab.

On the rear of this a launcher with two pods each of six rockets which probably have a maximum range of up to 80 km.

Description

The 130 mm (36-round) launcher uses a hydraulic motor to drive the azimuth, elevation and stabilising jack systems. If required all can be operated manually as well. The fire-control system consists of three solid-state components, the firing control box, the ignition distribution box and a circuit tester unit.

A resupply KM813A1 five-tonne (6 × 6) truck is used to carry 72 reload rounds and to act as the loading platform. Reloading takes about 10 minutes. To avoid counterbattery fire, reloading would normally take place well away from the original launch position.

The Kooryong 130 mm (36-round) multiple rocket launcher system fires at least two types of 130 mm unguided solid propellant rockets which are manufactured by the Hanwha Corporation, these are called the standard rocket (K30) and the improved model (K33) and details are given in the Table.

These unguided rockets have the same warhead but the Improved Kooryong II has more solid propellant for increased range.

Single round, partial or full ripple salvo firings can be undertaken using the fire-control box from either the driver's seat, or off-vehicle from a site several metres away. A firing table or the fire direction calculator is used to provide the ballistic data. The latter provides a faster and more accurate method of calculation than the former. Claimed CEP is 387 m.

The K30 and K33 solid propellant rocket motors are packed one round per wooden box with 12 boxes packed on a pallet for ease of transport in the field. The K37 and K38 warheads are packed two per wooden box and 16 of these boxes are then packed on a pallet.

While the solid propellant unguided rockets have a maximum range of 36,000 m they have a minimum range of 12,000 m and a maximum velocity of 1,200 m/s.

Surface wind measurements are made using a 10.25 m mast installation mounted on the back of a 1 1/4-tonne 4 × 4 KM450 light truck and connected to a microprocessor. This provides both digital average and instant windspeed data for the fire-control calculations.

It is believed that the South Korean Army has a total of 156 of these Kooryong 130 mm (36-round) multiple rocket launchers in service.

They have now been supplemented by the US Lockheed Martin Missiles and Fire Control 227 mm (12-round) Multiple Launch Rocket Systems.

These came from the US production line with at least 29 delivered to date. In addition to the conventional MLRS rockets, the ROK has also taken delivery of the longer range ATACMS.

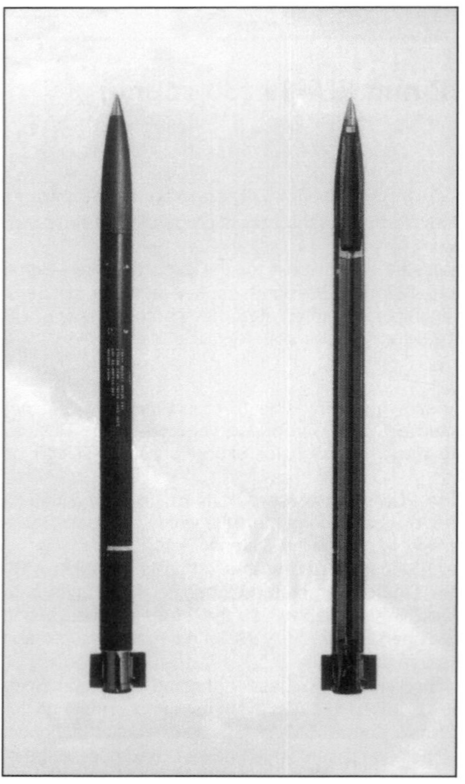

The two 130 mm rockets fired by the 130 mm (36-round) Kooryong multiple rocket system are the K30 (left) and the K33 (right) with fins unfolded as they would in flight 0021064

Kooryong 130 mm (36-round) Multiple Rocket Launcher System mounted on a locally produced KM809A1 (6 × 6) truck chassis, with launcher traversed to the left (Doosan Infracore Defense Products) 1043231

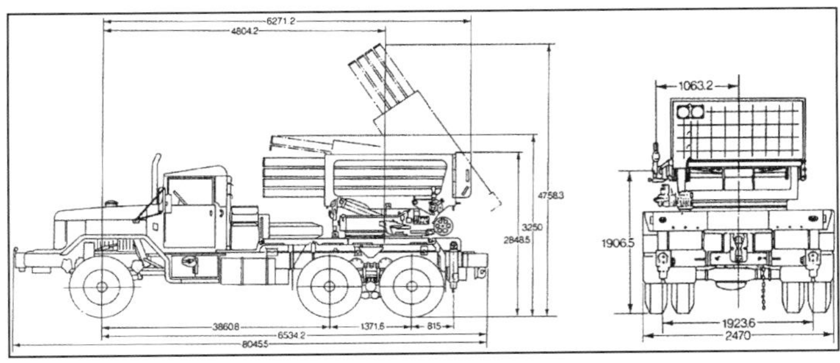

General configuration Kooryong 130 mm (36-round) Multiple Rocket Launcher System mounted on KM809A1 (6 × 6) truck chassis 0500903

It is understood that all of the ROK MLRS 227 mm (12-round) are now ATACMS capable systems.

Variants

According to Doosan, this 130 mm (36-round) multiple rocket system can be integrated onto a wide range of other chassis, tracked and wheeled.

As far as it is known the only production application so far is to be integrated on the rear of the locally manufactured KM809A1 (6 × 6) cross-country truck chassis.

The specifications of the Kooryong 130 mm (36-round) multiple rocket system are for this system integrated on the rear of the locally manufactured KM809A1 (6 × 6) chassis.

Specifications

	Kooryong 130 mm (36-Round)
Dimensions and weights	
Crew:	3
Length	
overall:	8.0455 m
Width	
transport configuration:	2.47 m
Height	
transport configuration:	2.8485 m
Weight	
combat:	17,100 kg
Mobility	
Configuration	
running gear:	wheeled
layout:	6 × 6
Speed	
max speed:	80 km/h
Gradient:	55%
Engine	6 cylinders, diesel, 236 hp
Gearbox	
type:	manual
forward gears:	5
reverse gears:	1
Electrical system	
vehicle:	24 V
Firepower	
Armament:	36 × 130 mm rocket system
Main weapon	
length of barrel:	3.4 m
maximum range:	36,000 m (est.)
Rate of fire	
rapid fire:	2/1 s
Operation	
reload time:	600 s (est.)
Main armament traverse	
angle:	180° (90° left and right)
Main armament elevation/depression	
armament front:	+55°/0°

	K30	K33
Dimensions and weights		
Length		
overall:	2.390 m	2.528 m
motor:	1.79 m	1.924 m
Diameter		
body:	130 mm	130 mm
Weight		
launch:	55 kg	64 kg
Performance		
Speed		
max speed:	3,276 km/h	4,320 km/h
Range		
max:	23.0 km	36 km
Ordnance components		
Warhead:	20.37 kg HE K37, Composition B fragmentation	20.37 kg HE K38, Composition B 16,000 fragmentation steel balls
Fuze:	impact, K505	impact, K505
Propulsion		
type:	solid propellant (Kooryong I, composite (CTPB))	solid propellant (Kooryong I, composite (HTPB))

Status

Production of the launcher is complete but could be resumed if additional orders are placed. It is considered that the system could be installed on a more recent wheeled chassis in the future. Production of the 130 mm rocket is undertaken on an as-required basis. The only known user is the Republic of Korea Army. It is understood that the Republic of Korea has 156 of these in service alongside a smaller number of 227 mm MLRS. Some sources have stated that Egypt has taken delivery of 36 of these systems.

Contractor

Doosan Infracore Defense Products BG (launcher).

Hanwha Corporation, Defense Business Division (130 mm rockets).

Hanwha 70 mm (40-round) multiple rocket launcher

Development

In 2000, the Hanwha Corporation started development, as a private venture, of a lightweight self-propelled 70 mm (40-round) multiple rocket launcher.

The company has considerable experience in the design, development and production of a wide range of munitions including the 130 mm Kooryong series of unguided surface-to-surface rockets. These are launched by the 130 mm (36-round) multiple rocket system covered in detail in a separate entry in *IHS Jane's Land Warfare Platforms: Artillery and Air Defence.*

The 70 mm (40-round) system has already undergone trials with the Republic of Korea Army but, as of April 2013, it is understood that quantity production of the system had yet to commence.

The Republic of Korea Army already uses the locally developed Kooryong series of 130 mm (36-round) multiple rocket launcher, as well as the US Lockheed Martin Missiles and Fire Control 227 mm (12-round) Multiple Launch Rocket System (MLRS).

Description

The Hanwha 70 mm (40-round) multiple rocket launcher system consists of a tracked or wheeled chassis fitted with a launcher that contains a total of 40 launch tubes (five rows each of eight rockets) in the ready-to-launch position.

For some years, the company has been manufacturing a variety of 2.75 in (70 mm) solid propellant unguided rockets for use in the air-to-ground role and these can be fired from a variety of fixed-wing aircraft and helicopters.

In addition to the unguided 70 mm rockets listed below, under development is a guided munition that will feature GPS/laser/imaging IR guidance and a claimed CEP accuracy of 1 to 2 m.

Details of these 2.75 in (70 mm) rocket motors and associated warhead are given in the table.

The K223 solid propellant rocket motors are typically used by high-performance aircraft, as well as helicopters and other low-speed aircraft, and have a substantially higher thrust and longer range than the standard MK40 rocket. They have wrap-around fins at the rear.

Hanwha 70 mm (40-round) multiple rocket launcher on a locally produced Kia Motors KM45 (4 × 4) truck chassis (R K Karniol) 0590161

Standard Hanwha 70 mm (40-round) multiple rocket launcher deployed in the firing position and with launcher traversed right and firing an unguided surface-to-surface rocket. Note that the stabilisers at the rear of the chassis are deployed on the ground (Hanwha) 1333682

The MK rocket motor is used for higher speed aircraft, while the MK40 is used by low-speed aircraft and helicopters.

Rocket specifications

Model	**K223**	**MK4 and MK40**
Diameter:	70 mm	70 mm
Rocket length:	1.06 m	0.997 m
Rocket weight:	6.2 kg	5.1 kg
Propellant type:	K674	MK43 MOD 1
Burn time:	1.05 to 1.10 s	1.55 to 1.69 s
Average thrust:	1,300/1,415 lb	720 lb
Impulse:	1,500 lb/s	1,150 lb/s
Max range:	10,426 m	8,080 m

Various types of warhead can be attached to these rockets. The M151 is a High-Explosive (HE) warhead that can be assembled in five different combinations of motors and fuzes.

The High-Explosive Dual-Purpose (HE DP) M247 warhead is designed to simultaneously meet the requirements of defeating enemy armour and personnel.

The High-Explosive MultiPurpose Sub-Munition (HE MP SM) warhead contains nine submunitions for use against personnel, matériel and light armour.

Warhead type	**M151**	**M247**	**K224**
Type:	HE	HE DP	HE MP SM
Weight:	3.95 kg	3.99 kg	6.17 kg
Length: (without fuze)	0.328 m	0.474 m	0.682 m
Filler:			
(type)	Comp B4	Comp B4	M73 HE MP SM
(weight)	1.04 kg	0.91 kg	4.9 kg

The first example of the system shown in 2004 was mounted onto the rear of a locally produced Kia Motors KM45 1.25-tonne (4 × 4) cross-country truck chassis. This is stated to have a total weight of 4,200 kg without the rockets and is operated by a crew of four. The KM45 is used for a wide range of roles and missions by the Republic of Korea Army.

The latest version features a fully automatic computerised fire-control system that includes a fire-control unit, laser range-finder, Precision Lightweight GPS Receiver (PLGR), vehicle motion sensor, system control unit/actuation control unit and an inertial navigation system.

Variants

It should be noted that when originally revealed the 70 mm system had a total of 40 launch tubes. The latest version, fitted with the computerised fire-control system, has a total of 32 launcher tubes arranged in four rows each of eight launcher tubes.

This is being marketed with the high-explosive, multipurpose sub-munition and flechette type warheads.

In addition to the previously stated 70 mm unguided rockets, a 70 mm guided munition is claimed to be under development with a number of guidance methods being considered. These include GPS, laser and imaging infra-red sensor.

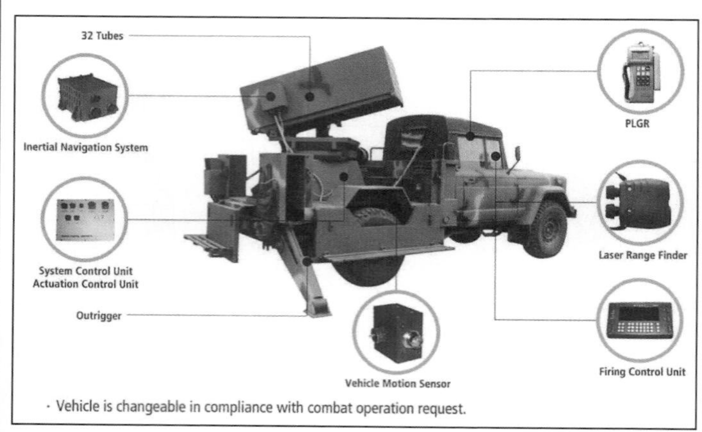

Key parts of the fire-control system of the Hanwha 70 mm (40-round) multiple rocket launcher (Hanwha) 1040810

This latest system with the 32-round launcher is also being marketed as the "Fast Attack" version with a firing rate of four rounds per second.

As an automatic aiming system is fitted, this version can rapidly come into action, carry out a fire mission and move to another position for reloading before counterbattery fire arrives.

Specifications

Not available.

Status

Prototype. Not yet in production or service.

Contractor

Hanwha Corporation, Defense Business Division.

Hanwha Corporation Chun-Mu Long Range Multiple Rocket System

Development

In May 2009 the Hanwha Corporation, Defense Business Division, was selected as the supervisory contractor for the research and development of the Chun-Mu Long Range Multiple Rocket System (LRMRS).

In 2011 the South Korean Agency for Defence Development confirmed that existence of the Chun-Mu LRMRS and it is understood that the prototype systems were completed in 2012 and are now undergoing company trials.

When deployed this would supplement the other artillery rocket systems deployed by the Republic of Korea Army which include the locally developed Doosan Kooryong 130 mm (36-round) multiple rocket system and the US Lockheed Martin Missile and Fire Control 227 mm (12-round) Multiple Launch Rocket System (MLRS).

Description

The Chun-Mu LRMRS is based on a Doosan DST 8 × 8 cross-country chassis fitted with a fully protected forward control cab for the crew and the computerised fire control system.

The cab is provided with an NBC system and a circular roof hatch is provided on the right side. At present this does not appear to be fitted with a weapon station.

On the rear of the chassis is a launcher with powered elevation and traverse which contains two pods of rockets which in appearance are very similar to the

Chun-Mu (12-round) Long Range Multiple Rocket Launcher deployed in firing position with launcher traversed right and showing protected cab at the front and hydraulic stabilisers lowered to the ground (Doosan ST) 1452928

Scale model of the Chun-Mu Long Range Multiple Rocket Launcher deployed in the firing position with launcher elevated and traversed to the right (Christopher F Foss) 1452930

The Chun-Mu Long Range Multiple Rocket Launcher is supported by a dedicated rocket resupply vehicle based on a similar 8 × 8 cross-country chassis. Pods of new rockets are placed on the ground and then loaded onto the Chun-Mu Long Range Multiple Rocket Launcher using an integrated self-loading system (Doosan ST) 1452929

US Lockheed Martin Missiles and Fire Control 227 mm (12-round) MLRS which is already deployed by South Korea.

To provide a more stable firing platform two hydraulic stabilisers are lowered to the ground either side.

These are positioned one to the immediate rear of the first road wheel station and another to the immediate rear of the last road wheel.

The 8 × 8 chassis features an independent suspension system for a high level of cross-country mobility and the tyres are provided with a central tyre inflation system and run-flat inserts.

No details of the calibre of the rockets have so far been released but a calibre of 230 mm has been mentioned and a maximum range of up to 80 km.

In addition to firing unguided rockets the system is also expected to fire guided rockets in the future for increased accuracy.

The computerised fire-control system includes the Fire Control Console whose security is stated to be guaranteed by Variable Message Format (VMF) and data with the radio network, Weapon Control Unit (WCU) which carries out missile/rocket function management and missile/rocket ballistic computation, Launcher Control Unit (LCU) which provides hydraulic servo control and isolated electrical power control and the Position Navigation Unit (PNU) which is a Global Positioning System/Inertial Navigation System (GPS/INS).

Once the rockets have been fired, the Chun-Mu would typically move to another position to avoid counter battery fire when the system would be rapidly reloaded using its Self-Loading System (SLS).

New pods of six rockets are carried on a similar 8 × 8 vehicle with a protected cab which is fitted with a hydraulic crane at the ear of the chassis.

A stabiliser is lowered at the rear and the pods are unloaded onto the ground where they are picked up by the telescopic arms of the SLS of the Chun-Mu launcher.

Specifications

Not released.

Status

Under development to meet the requirements of the Republic of Korea Army.

Contractor

Hanwha Corporation, Defense Business Division.

Myanmar

Defence Products Industries 81 mm BA84 artillery rocket system

Development

The Burmese Defence Products Industries has developed and placed in production a mobile 81 mm calibre unguided surface-to-surface rocket system known as the BA84. No other details are known at present. It is possible that this is an unguided 81 mm air-to-ground rocket modified for ground-to-ground use.

Description

At this time it is not known as to whether this system is a single-round launcher or a truck or trailer-mounted multiple rocket launcher.

It is thought that the warheads are the same as those used on unguided air-to-ground rockets.

Status

Production as required. In service with the Burmese Army. There are no known exports of this artillery rocket system.

Myanmar is also reported to have deployed about 30 NORINCO 107 mm (12-round) Type 63 trailer-mounted multiple rocket launchers and an undisclosed quantity of BM-21 122 mm (40-round) multiple rocket launchers which are manufactured in a number of countries.

It has also been reported that Myanmar has some 240 mm (22-round) multiple rocket launchers supplied from North Korea.

Myanmar is known to have taken delivery of 155 mm/52-calibre self-propelled (wheeled) artillery systems and may have taken delivery of longer range rocket launchers.

Contractor

Defence Products Industries.

Pakistan

Institute of Industrial Control Systems 122 mm (30-round) multiple rocket launcher system

Development

In many respects the actual 122 mm (30-round) launcher, which is designed and built by the now Institute of Industrial Control Systems (previously Dr A Q Khan Research Laboratories), is very similar to the widely deployed Russian 122 mm BM-21 (40-round) multiple rocket launcher.

Some sources have indicated that this system is based on the China North Industries Corporation 122 mm (30-round) Type 83 system, but based on a US chassis rather than the original Chinese (6 × 6) truck chassis. It is understood that Pakistan has about 50 of these 122 mm (30-round) rocket launchers in service.

This system has also been referred to as the 122 mm (30-round) Azar multiple rocket launcher system.

Some sources have indicated that this system is based on a Romanian rocket design rather than a 122 mm NORINCO design.

Pakistan also deploys a number of Chinese supplied 107 mm (12-round) rocket launchers and a number of more recently supplied Chinese 300 mm rocket launchers on an 8 × 8 cross-country chassis.

Pakistan Army 122 mm (30-round) Institute of Industrial Control Systems MRS mounted on a US-supplied M35 series (6 × 6) truck 0500671

Pakistan Army 122 mm (30-round) Institute of Industrial Control Systems MRS mounted on a Japanese 6 × 6 truck chassis 0021065

Description

The Pakistan 122 mm (40-round) system however consists of two banks of two launch tubes, each of which consists of 15 launch tubes in three layers each of five tubes. The stainless steel tubes are 3.1 m long and have an actual calibre of 122.4 mm with the unguided rockets being manually loaded from the rear.

Each launch tube consists of a guide tube, projectile stopper and a contact device bracket. The guide tube has a spiral groove that has a uniform twist of 2-2.5° and at the rear end of the tube is a clamp, which is fitted with a projectile stopper. The latter prevents the loaded 122 mm rocket from slipping out of the firing tube backwards and when activated, releases the guide pin when the propelling force of the rocket reaches 600 to 800 kg. This ensures the optimum muzzle velocity when fired.

The solid propellant unguided rockets, which are manufactured by the Pakistan Ordnance Factories (POF) have wraparound fins at the rear and a nose mounted high-explosive warhead and fuze system. The actual rockets are referred to as the Yarmuk by POF.

The 122 mm rocket warhead weighs 19.4 kg with fuze or 18.4 kg without fuze. The rocket is 2,875 mm long and weighs 66 kg of which 20.4 kg is the double base solid propellant and 6 kg the composition B bursting charge.

The 122 mm rockets can be launched by an electric firing mechanism which is located in the vehicle cab, or, more usually, from a safe position up to 50 or 60 m from the launcher using a remote-control device which is connected to the launcher by a cable.

The rockets can be launched one at a time, or in salvo, with maximum stated range of the unguided rockets being 20,580 m. The launcher can be traversed through 180° (90° left and 90° right), with elevation being from 0° to +55°. Launcher traverse and elevation is electric with manual controls being provided for emergency use. The latter are mounted at the rear of the launcher on the left side where the sighting system is also located.

In Pakistan Army service the system is mounted on the rear of a US supplied M35 series 6 × 6 truck or a Japanese 6 × 6 truck, although it can be installed on a wide range of other chassis, tracked and wheeled.

To provide a more stable firing platform two stabilisers are lowered to the ground, one either side at the rear.

In addition to manufacturing the 122 mm (30-round) rocket launcher, the Institute of Industrial Control Systems is also involved in a number of other rocket and missile projects for the Pakistan armed forces. According to the United Nations, in 2000 Pakistan exported six 122 mm rocket launchers to Sri Lanka.

As a counter to the Indian 214 mm (12-round) Pinacha multiple rocket launcher system and the more recently acquired Russian Splav 300 mm BM 9A52 (12-round) Smerch multiple rocket system, Pakistan is understood to have placed an order late in 2008 for one regiment of Chinese multiple rocket launchers.

It is understood that these are the CPMIEC 300 mm (10-round) A100 multiple rocket systems or the NORINCO AR2 300 mm (10-round) multiple rocket system with all of these based on an 8 × 8 cross-country chassis with an associated resupply vehicle on a similar chassis.

It is understood that these systems have now been delivered to Pakistan.

Variant

New 122 mm Roxel extended range rocket

Late in 2009 it was announced that the European company of Roxel had signed a memorandum of understanding with Pakistan Ordnance Factories to upgrade 122 mm unguided surface-to-surface rockets to increase the range of the system from 20 km out to 40 km.

Specifications

122 mm (30-round) MRL rocket

Calibre:	122 mm
Weight of warhead:	
(without fuze)	18.4 kg
(with fuze)	19.4 kg
Weight of complete rocket:	66 kg
Weight of propellant:	20.4 kg
Weight of bursting charge:	6 kg
Length of rocket:	2,875 mm
Muzzle velocity:	400 m/s
Maximum time of flight:	78 s
Maximum range:	
(without brake ring)	20,580 m
(with small brake ring)	15,890 m
Accuracy:	
(PE in range and deflection)	80 and 168 m respectively when fired at an elevation of +50°; 49 and 90 m respectively when fired at an angle of +45°
(back blast to rear)	50 m

Status

Production as required. In service with Pakistan. A total of six 122 mm launchers were supplied to Sri Lanka in 2000. It is assumed that these were 122 mm (30-round) systems.

Contractor

Institute of Industrial Control Systems.

Poland

Polish upgraded 122 mm BM-21 (40-round) multiple rocket launcher WR-40 Langusta

Development

The most widely used Multiple Rocket Launcher (MRL) in the world is the Russian 122 mm BM-21 (40-round) system, which is in service with well over 50 countries.

The original Russian 122 mm BM-21 system was developed from the late 1950s, with the Splav Scientific Production Concern being the overall system prime contractor and responsible for the 122 mm unguided rockets and the Motovilikha Plants Corporation being responsible for the launcher.

The original system entered service with the Russian Army in 1963 and has since seen combat in many parts of the world, especially Asia and the Middle East.

The baseline 122 mm BM-21 (40-round) system has also been copied or manufactured by many other countries, some of which use their local chassis, normally wheeled but some times tracked.

These systems some times incorporate a number of improvements to meet local operational requirements.

The Polish Army has a large number of 122 mm BM-21 (40-round) systems installed on Russian Ural-375D (6 × 6) truck chassis.

Some years ago a decision was taken to retain some of these systems in service by carrying out a major upgrade covering the chassis, 122 mm unguided rockets and fire-control system.

Following trials with the prototype BM-21M (40-round) upgraded system, late in 2006 the Polish Ministry of National Defence awarded an initial production contract for the system.

Following extensive trials with prototype systems, in March 2007 the Polish Army took delivery of its first upgraded BM-21 systems.

These were originally expected to be called the BM-21M but are now designated the WR-40 Langusta with first units delivered to the Torun based artillery training centre.

The first operational Polish Army unit, 2nd Artillery Regiment, was issued with the WR-40 Langusta systems in early 2008.

It is expected that two regiments will be equipped with the upgraded systems per year with current requirement being for six regiments each with 18 launchers plus additional systems for training and reserve.

Today Poland has about 236 122 mm BM-21 (40-round) and RM-70 (40-round) MRL in service but it is not expected that the latter, based on an armoured Tatra (8 × 8) chassis, will be upgraded.

As of April 2013 it is understood that the Polish Army had placed orders for a total of 62 WR-40 Langusta multiple rocket launchers with additional orders expected to be placed in the future.

Polish WR-40 Langusta (40-round) multiple rocket launcher deployed in the firing position and launching an unguided 122 mm surface-to-surface rocket (HSW) 1169163

The new 122 mm rockets for the upgraded Polish system have been developed by the Polish company Pressta SA and Roxel of France (Christopher F Foss) 0096295

Description

The prototype of the Polish upgraded BM-21 (40-round) multiple rocket launcher was based on a Polish Star 1466 (6 × 6) truck chassis as a replacement for the original Russian Ural-375D (6 × 6) chassis.

As the Star 14666 is no longer in production, the BM-21M system is now based on the 10-tonne Jelcz P622D.35 (6 × 6) cross-country chassis which has a fully enclosed six-person cab.

This forward-control cab features STANAG 4569 Level 1 ballistic protection, NBC system and an air conditioning system. The vehicle is powered by a Cursor 8, EURO III diesel engine developing 259 kW which gives a maximum road speed of 85 km/h.

This chassis has excellent cross-country mobility and is fitted with a central tyre-pressure regulation system that allows the driver to adjust the tyre pressure from within the cab to suit the terrain being crossed.

Mounted on the rear of the cab is the 122 mm (40-round) launcher, although the actual calibre of the launcher tubes is 122.4 mm. The launcher has elevation from 0° to +55° with traverse being 70° right and 102° left.

Elevation is powered at the rate of up to 7° a second with traverse being powered at the rate of up to 5° a second. In addition, there are manual back-up controls mounted on the left side of the launcher.

Prime contractor HSW Stalowa Wola who is also prime contractor for the new Polish 155 mm/52-calibre Krab self-propelled artillery system.

The standard Russian 122 mm BM-21 (40-round) MRL fires a 122 mm 9M22U unguided rocket to a maximum range of 20,380 m. By today's standards this was considered to be too short.

For many years the Polish company of Pressta Spotka Akcyjna (SA) has been engaged in the manufacture of a wide range of metal parts for ammunition as well as 122 mm rockets.

In the early 1990s, development of a new and more effective 122 mm unguided artillery rocket commenced by the company with three new and more effective warheads being developed which could be retrofitted to existing rockets. These were:

- High-explosive type with increased fragmentation effect called Spall which is designated the F-M-21 OB and weighs 18.4 kg.
- Cargo type carrying five Polish developed scatterable anti-tank mines called Platan and which is designated the F-M-21 MK and weighs 25.6 kg. The mines are designated the MN-121 and are made by BZU Belma.
- Cargo type carrying 42 Polish developed High-Explosive Anti-Tank Fragmentation (HEAT-FRAG) bomblets to attack the vulnerable upper surfaces of armoured vehicles. These will penetrate 120 mm of conventional steel armour. This is designated the F-M-21 K1 with the warhead weighing 21.7 kg.

Polish WR-40 Langusta (40-round) multiple rocket launcher integrated on a Jelcz P622D.35 (6 × 6) truck chassis (Grzegorz Holdanowicz) 1165106

Recent information has stated that these three warheads are actually manufactured by Fabryka Produkcji Specjainej, with export marketing carried out by the BUMAR Group.

The next step in the upgrade was to increase the range of the rocket and this part has been developed by the French company Roxel who have extensive experience in rocket motor development.

One of the key requirements was to increase the range of the system to 30 km. According to Pressta SA, not only has the new family of 122 mm rockets have increased range, but they are also more accurate and have longer shelf-life.

Roxel developed a new composite solid-propellant rocket motor for the 122 mm rocket called Phenix (or Feniks). Modification of the airframe nozzle system and stabiliser unit was also carried out. The nose-mounted fuzes on all 40 rockets can be set in three minutes.

During acceptance trials in Poland in 1999, a maximum range of 41,000 m was achieved with the new rocket, with all rockets being launched in 20 seconds.

These unguided rockets are spin and fin stabilised with the latter unfolding after they have left the launch tube.

Specifications of the Spall and Platan type unguided rockets are as follows:

Type	HE	Cargo
Maximum range:	41 km	36 km
Length:	2.67 m	2.82 m
Weight:	61 kg	65 kg
Weight of warhead:	19 kg	23 kg

According to Roxel, the main advantage of this Phenix rocket motor are:

- Mature composite solid-propellant motor technology
- Previous total impulse increased by 50 per cent
- High strength and lightweight steel case
- Single nozzle assembly.

In addition to being launched from the Russian 122 mm BM-21 (40-round) MRL and its equivalent systems, the new rockets can also be launched from the RM-70 122 mm (40-round) system on a Tatra (8 × 8) series cross-country truck, which is used by many countries.

In Polish Army service, the effectiveness of the upgraded BM-21 system has also been improved as it is provided with a new Command, Control, Communications, Computer and Intelligence (C4I) system developed by the local company of W B Electronics.

For the Polish upgraded BM-21 system this includes a Battle Field Computer (BFC) 201 as a fire-control system terminal, RRC 9500 digital radio, FONET digital intercom and a navigation system.

Variants of this C4I system are also used on other Polish artillery mortar systems, including the Russian supplied 122 mm 2S1 self-propelled artillery system, Polish 155 mm/52-calibre Krab self-propelled artillery system and the 98 mm M-98 mortar.

Specifications

	Upgraded BM-21
Dimensions and weights	
Crew:	3+3
Weight	
combat:	17,000 kg
Mobility	
Configuration	
running gear:	wheeled
layout:	6 × 6
Firepower	
Armament:	40 × 122 mm rocket system
Main weapon	
maximum range:	41,000 m
Main armament traverse	
angle:	172° (102° left, 70° right)
Main armament elevation/depression	
armament front:	+55°/0°

	9M22U HE	9M22U Cargo
Dimensions and weights		
Length		
overall:	2.67 m	2.82 m
Weight		
launch:	61 kg	65 kg
Performance		
Range		
max:	41 km	36 km
Ordnance components		
Warhead:	19 kg	23 kg

Status

Production. In service with Polish Army.

This has been offered to the export market but as of April 2013 there are no known export sales.

Contractor

Huta Stalowa Wola SA (HSW).

Romania

Aerostar 122 mm (40-round) APRA multiple rocket launcher system

Development

There are two versions of the Romanian Aerostar 122 mm (40-round) APRA-40 multiple rocket launcher system known to exist.

The basic model, covered below, is based on the locally built DAC 665T (6 × 6) truck chassis, has the locally developed APRA 122 mm (40-round) launcher produced by Aerostar as shown in the photographs and this features handrails mounted on the rear mudguards. This version is also referred to as the LRSV BM-21 (Croatia) and the SVLR 122 'Grad' (Bosnia).

The second and more recent model has a launcher with two removable modules (Launch Pod Containers - LPC) as mentioned later in the text below, and the handrails have been replaced with sheet plate to protect the tyres.

This is based on the Romanian DAC 26140 DFAEG (6 × 6) diesel powered cross-country truck chassis. In Croatian service, this is referred to as the LRSV M-91 'Vulcan', while the Bosnian Army calls the system the SVLR 122 APR-40.

The first model to enter service is called the APRA-40 (*Aruncator de Proiectile Reactive*) while the model with the two launcher modules (LPC) is called 40 APRA 122 FMC.

Production of this 122 mm rocket launcher system is undertaken on an as required basis but it is understood that there has been no recent production.

Emphasis is now concentrated on the Aerostar BM-21 upgrade covered later in this entry.

Description

The 6 × 6 cross-country chassis is of the forward control armoured type with the two-door cab providing space for the complete crew of the system. A central tyre-pressure regulation system is fitted as standard and allows the driver to adjust the tyre pressure to suit the type of terrain being crossed.

The launcher consists of two removable modules (LPC) each with 20 × 122 mm launching tubes that are banked in four rows of five.

An ammunition re-supply vehicle called the MITC is based on a Romanian DAC 26 360 (6 × 6) truck chassis. This weighs 16,440 kg when fully loaded and carries two LPC.

The chassis is fitted with a rear-mounted MH 2.2 × 5.3 t/m hydraulically operated crane arm with a maximum lifting load of 2.2 tonnes at 6 m lifting height from ground level and 1.8 m below ground level.

Replacing the expended rocket modules takes about 10 minutes. The resupply vehicle can also tow a simple four-wheeled two-axle, Type RM7 5/2-S, trailer that carries two LPC of 122 mm rockets that can be used to reload the launcher by crane. Total weight of a fully loaded trailer is 7,500 kg.

The fire-control unit is mounted in the forward-control cab but if required the rockets can be launched by remote-control via a cable up to 100 m from the cab.

The operator can select either single shots or a complete salvo at variable firing rates of up one, 0.7 second or 0.5 second.

The sighting system is mounted on the left side of the actual launcher where the elevation and traverse controls are located while the firing circuits are on the left side of the chassis, just above the diesel fuel tank.

Standard equipment includes a radio that is installed in the cab, this is used to communicate with other launchers or the next chain of command. When fitted with an antenna 1.2 m high the radio has a range of 10 to 12 km. The radio can also be operated up to 100 m away from the vehicle using a remote-control handset.

At maximum range the area covered by the 40-round salvo is 210,000 m^2.

The Romanian companies of CN ROMARM and CN ROMTEHNICA are now marketing 122 mm unguided rockets for the locally developed APRA system as well as the more widely deployed Russian 122 mm BM-21 series of truck mounted (6 × 6) multiple rocket systems.

Two 122 mm rockets are currently being marketed, the Type M21-OF-FP and the shorter range Type M21-OF-S and brief details of these 122 mm rockets are given in the table:

Type	M21-OF-FP rocket	M21-OF-S rocket
Calibre:	122 mm	122 mm
Length of rocket:	2,870 mm	1,927 mm
Weight of rocket:	65.4 kg	46.6 kg
Range:		
(maximum)	20,400 m	12,700 m
(minimum)	5,000/6,000 m	1,000 m
Weight of explosive in warhead:	6.35 kg	6.35 kg
Weight of propellant:	20.65 kg	11.10 kg

Both of these rockets have a nose-mounted fuze which is set before launch to give instantaneous, short or long delay functions.

Variants

Resupply vehicle

The ammunition resupply vehicle, called the MITC, is based on a DAC 26 360 (6 × 6) truck chassis and weighs 16,440 kg fully loaded. It carries two LPC and a rear mounted MH 2.2 × 5.3 t/m hydraulically operated crane.

122 mm (40-round) APR 40 MRS based on DAC 665T (6 × 6) chassis (BA)
0085764

This crane has a maximum lifting load of 2.2 tonnes a 6 m lifting height from ground level and 1.8 m below ground level.

Replacing the expended rocket modules takes about 10 minutes. The resupply vehicle can also tow a four-wheel two axle Type RN 5/S-S trailer that carries four LPCs a maximum lifting load of 122 mm rockets that can be used to reload the LPC. Total weight of the loaded trailer is 7,500 kg.

LAROM upgrade

This upgrade has been carried out by Israel Military Industries and Aerostar and is covered in detail in a separate entry in *IHS Jane's Land Warfare Platforms: Artillery and Air Defence*. The initial batch of 18 systems are now in service with Romania.

Georgia has taken delivery of six rocket systems and it is assumed that these are of the LAROM type.

Complete 122 mm (40-round) APRA multiple rocket launcher system

In addition to marketing the existing 122 mm (40-round) APRA MRLS, the company is also marketing a complete system that includes the following elements:

- Launchers - 122 mm/APRA MRLS equipped with 40 Grad rounds being containerised in two LPC
- Associated container loading and transport truck (or resupply truck) mounted on a DAC 26 360 (6 × 6) truck chassis
- Brigade/battalion/battery command post (firing management posts)
- Upper air sounding station with radiotheodolite mounted on DAC 22 310 (6 × 6) truck chassis
- Artillery forward observer (reconnaissance post) based on the TABC-79 (4 × 4) light armoured fighting vehicle and associated target acquisition system equipped with thermal camera, command and communications equipment. The TABC-79 is still used by the Romanian Army but production of this vehicle has been completed although it remains in service with the Romanian Army.

All of these components of the programme are being integrated in a modern C3I artillery command and control system as used in the LAROM and ATROM programmes.

Aerostar BM-21 upgrade

Aerostar is also marketing a kit to upgrade the widely deployed Russian BM-21 (family) system in the following key areas:

- Launcher: the existing system can be modified to accommodate other types of Launch Pod Containers (LPCs)
- Cradle: increasing the length, reinforcement of the main beam, incorporating lockers into the LPC and an additional electrical adapter

122 mm (21-round) A226-21 based on Bucegi SR-114 (4 × 4) chassis. This is no longer used by the Romanian Army (BA)
0085765

- Traverse system: locking devices will be added to prevent azimuth drift (especially when launching the longer-range rockets)
- Hydraulic system: four hydraulic stabilisers will be added, two on either side
- Cabin window protection shield: if required, a detachable shield will be added for the window. This is positioned when the rockets are being launched, in order to protect the windows
- Fire-control system: the existing system will be replaced by a more up-to-date system and, where applicable, modified to fire the latest rockets.

Coastal defence model

Special coastal and border defence versions of the 122 mm (40-round) rocket systems have also been proposed by Aerostar.

Typically these would be located underground together with their associated reloading mechanisms and raised into the operating position when required. As far as it is known, this remains at the concept stage.

122 mm APRN

This is a 122 mm system for naval applications and consists of 2 × 40 122 mm rocket launchers which retract under the deck of the ship for rapid reloading purposes. The system is intended to be integrated into river monitor type surface craft.

As far as it is known this system has not entered production.

A226-21 122 mm (21-round) multiple rocket launcher system

This old system is no longer in service with Romania. It is based on a locally developed Bucegi SR-114 (4 × 4) chassis/platform and has three layers each with seven 122 mm launch tubes. It was exported to some countries, for example Morocco, with Aerostar carrying out upgrade work.

Specifications

	122 mm (40-Round) APRA
Dimensions and weights	
Crew:	5
Length	
overall:	7.39 m
Width	
transport configuration:	2.51 m
Height	
transport configuration:	3.25 m
Ground clearance	
overall:	0.39 m
Weight	
standard:	15,010 kg
combat:	17,650 kg
Mobility	
Configuration	
running gear:	wheeled
layout:	6 × 6
Speed	
max speed:	80 km/h
cross-country:	25 km/h
Range	
main fuel supply:	1,000 km
Fording	
without preparation:	1.2 m
Gradient:	60%
Side slope:	30%
Engine	D2156 HMN8, 6 cylinders, water cooled, diesel, 215 hp at 2,200 rpm
Firepower	
Armament:	40 × 122 mm rocket system
Main weapon	
length of barrel:	3 m
maximum range:	20,380 m
Main armament traverse	
angle:	210° (110° left/100° right)
Main armament elevation/depression	
armament front:	+55°/0°

Status

Production as required. In service with:

Country	Quantity	Comment
Bosnia-Herzegovina	36	delivered 1997/98
Botswana	20	delivered 1996
Cameroon	20	delivered 1996
Croatian Army	n/avail	possibly BM-21
Iranian National Liberation Army	n/avail	-
Liberia	n/avail	-
Nigeria	18/25	APR-21 model
Romania	133	first batch of 18 upgraded to LAROM configuration.

Contractor

Aerostar SA, Group Industrial Aeronautic.

Aerostar 122 mm (12-round) Aurora multiple rocket launcher system

Development

The Romanian 122 mm (12-round) Aurora multiple rocket launcher system was originally developed as a private venture by Aerostar SA which is also prime contractor for the 122 mm (40-round) APRA system on a 6 × 6 chassis and, the 122 mm (single-round) rocket launcher covered in detail in a separate entry in *IHS Jane's Land Warfare Platforms: Artillery and Air Defence.*

This has also been referred to as the 12 ARO 122 mm system. According to United Nations sources there were no exports of this system between the years 1992 to 2010.

It is understood that there has been no recent marketing of this system.

Description

The 122 mm (12-round) Aurora is essentially a locally built ARO (4 × 4) light vehicle with the diesel engine at the front, two door, two person fully-enclosed cab centre and the 122 mm (12-round) rocket launcher carried on the rear decking.

The launcher has a Launch Pod Container (LPC) consisting of two banks of six 122 mm rocket launcher short tubes within the restraining metal framework.

This comprises a complete pod that can be rapidly removed from the vehicle for reloading purposes. To provide a more stable firing platform, a hydraulically-operated stabiliser is lowered to the ground either side at the rear of the vehicle.

The ADFM-01 fire-control unit is mounted in the cab but if required the rockets can be launched by remote-control via a cable up to 100 m away from the cab. This is the same fire unit as used in the 122 mm (40-round) APRA system.

The operator can select either single 122 mm shots of a complete salvo at variable firing rates of up one, 0.7 second or 0.5 second. When firing at the rate of 0.5 rockets per second a complete salvo can be fired in six seconds and, when firing at the rate of one rocket per second, a complete fire sequence takes 12 seconds. Once the rockets have been launched, the launcher would normally redeploy to another position where the LPC would be reloaded.

The sighting system is mounted on the left side of the actual launcher where the manual elevation and traverse controls are located. This is also used on the APR 40 122 mm (40-round) multiple rocket launcher system.

Standard equipment includes a radio that is installed in the cab, this is used to communicate with other launchers or the next chain of command. When fitted with an antenna 1.2 m high the radio has a range of 10 to 12 km. The radio can also be operated up to 50 m away from the vehicle using a remote-control handset.

The 122 mm unguided rocket types normally used are the short version Russian DKZ-B and equivalent 122 mm Chinese round.

122 mm unguided rockets for this and other Romanian rocket launchers is now manufactured by the Romanian company of C.N. ROMARM S.A.

It should be noted the overall height of the Aerostar 122 mm (12-round) Aurora multiple rocket launcher system when the launcher is elevated to +35° is 2.815 m.

While the maximum range of the 122 mm unguided rockets is being quoted as 13,000 m, minimum range is being quoted as 1,500 m.

Variants

LAROM 160 multiple rocket launcher system

According to Israel Military Industries, parts of the LAROM 160 upgrade, covered in detail in a separate entry in *IHS Jane's Land Warfare Platforms: Artillery and Air Defence* and now in service with Romania, are applicable to this 122 mm (12-round) Awora (Aurora) system.

Aurora 122 mm system

This is similar to the previous described system, but is mounted on a new chassis. It is provided with four hydraulic stabilisers that are lowered to the ground to provide a more stable firing platform. The platform is available powered by a Peugot or Toyota diesel engine and with a chassis with a wheelbase of 3.2 m.

122 mm (12-round) Aurora MRS mounted in rear of ARO (4 × 4) light chassis deployed in firing position with launcher traversed left 0058462

Trailer system
A 122 mm (18-round) trailer-type multiple rocket launcher system was developed to the prototype stage.

Specifications

	122 mm (12-Round) Awora
Dimensions and weights	
Crew:	2
Length	
overall:	4.830 m
Width	
transport configuration:	1.775 m
Height	
transport configuration:	1.850 m
Ground clearance	
overall:	0.196 m
Wheelbase	3.2 m
Mobility	
Configuration	
running gear:	wheeled
layout:	4 × 4
Speed	
max speed:	80 km/h
cross-country:	25 km/h
Range	
main fuel supply:	1,000 km
Engine	diesel
Firepower	
Armament:	12 × 122 mm rocket system
Main weapon	
length of barrel:	2.17 m
maximum range:	13,000 m
Rate of fire	
rapid fire:	1/0.5 s
Main armament traverse	
angle:	200° (90° left, 110° right)
Main armament elevation/depression	
armament front:	+55°/0°

Status
Prototype. Not known to be in service. Offered on the export market.

Contractor
Aerostar SA, Group Industrial Aeronautic.

Aerostar 122 mm (single-round) rocket launcher

Development
As well as the Romanian 122 mm multiple rocket launchers mounted on the rear of 6 × 6 and 4 × 4 cross-country truck chassis, Romania developed a short 122 mm HE rocket which is used in conjunction with a 2.5 m long single-tube animal or man-portable tripod-mounted 122 mm rocket launcher.

This is designed for use by guerrilla, commando and special forces units.

Prime contractor for all of these Romanian 122 mm unguided rocket systems is Aerostar SA. This has also been referred to as the 122 mm Short-Type Rocket Launcher for Infantry and Paratroops.

Production of this single 122 mm rocket launcher is undertaken on an as required basis.

Description
The solid propellant unguided 122 mm rocket weighs 46.3 kg complete and is 1.927 m long. Its maximum range is 11,400 m, the launcher has an elevation range of –3° to +50° and a traverse of +15°. A fuze with pre-selected instantaneous, short or long delay actions is fitted to the unguided 122 mm rocket.

Total weight of the launcher when assembled for firing is 63.2 kg (which is made up of the 25 kg tripod, the 30 kg launch tube and 4.2 kg for the sight and remote firing device with 25 to 35 m connector cable).

The launcher can also be fixed on a special mount for use on a light cross-country vehicle. In this application the launcher can either be fired from the vehicle or dismounted for ground use.

A total of 10 to 15 reload 122 mm rounds would normally be carried. The number of reload rockets carried depends on the carrier platform.

122 mm unguided rockets for this and other Romanian rocket launchers is now manufactured by the Romanian company C.N. ROMARM S.A.

These rockets can also be used in other countries' 122 mm artillery rocket systems. These include systems that have been manufactured or developed by China, Croatia, Czech Republic, Egypt, Iran, Russia, Serbia, South Africa, Sudan, and Turkey.

Variant
This system can also be installed and fired from a variety of platforms, such as trucks and light cross-country vehicles.

Romanian 122 mm (single round) rocket launcher in firing position with remote-control system on right 0500672

Specifications

	Aerostar 122 mm (single-round) rocket launcher
Dimensions and weights	
Length	
overall:	2.54 m
width	
calibre:	122 mm
Performance	
Traverse	
angle:	360°
Elevation	
angle:	1°/50°

Status
Production as required. In service with the Romanian Army.

This has been exported to a number of countries and according to the United Nations Arms Transfer List for 2010 a total of 12 of these launchers were sold to the United States, probably for transfer to another country.

Contractor
Aerostar SA, Group Industrial Aeronautic.

Russian Federation

Free Rocket Over Ground (FROG) artillery rocket system (Luna-M)

Development
The first Russian tactical rocket system was the 250 mm Korshun (Kite). It used the liquid propellant 3R7 tactical rocket, which had a high-explosive warhead. This was launched from an SM-44 launcher based on a YAZ-214/KrAZ-214 (6 × 6) cross-country truck chassis.

This entered production in 1957 and had six 250 mm tubes with the unguided rocket achieving a maximum range of 55 km. This was phased out of service many years ago.

This was followed by the first solid-propellant tactical rockets, the 3R1 Mars and the 3R2 Filin (Eagle Owl) with the latter entering service in 1955 and fitted with a tactical nuclear warhead.

The 3R1 Mars was a 324 mm diameter unguided rocket with a maximum range of 17.5 km while the 3R2 Filin was a 612 mm diameter unguided rocket with a maximum range of 25.7 km.

The single-rail Filin (NATO designation FROG-1) was carried on a modified IS-2 heavy tank hull designated the 2P2 and the shorter-range single-rail Mars (NATO designation FROG-2) was carried on the 2P4 non-amphibious variant of the PT-76 light amphibious tank chassis.

By 1960, the cumbersome Filin (Eagle Owl) system was being replaced with the similar performance 2K6 Luna family (3R-8 Luna, 3R-9 Luna-1 and 3R-10 Luna-2, respective NATO designations FROG-3, FROG-4 and FROG-5) based on a further modified version of the Mars non-amphibious PT-76 light tank hull which was designated the 2P16.

The FROG-7 resupply vehicle is designated the 9T29 and is fitted with a hydraulic crane to transport replacement FROG-7 rockets onto the launcher 0569046

The major differences between the family variants that warranted the individual NATO designations were in the shape of the warheads.

It is now known that the 3R-9 was a high-explosive fragmentation warhead whereas the 3R-10 was a tactical nuclear warhead.

More recently it has been stated that the 3R9 Luna rocket could be fitted with a tactical nuclear or high-explosive warhead. A later high-explosive fragmentation warhead was designated the 3N15 while a new rocket, the 3R10, was developed for the tactical nuclear warhead designated the 3N14.

Complete with its rocket, the 2P16 system weighed a total of 18.8 tonnes and had a maximum road speed of 40 km/h. It took 11 minutes to come into action and launch an unguided rocket and was normally operated by a crew of 11 men.

The launcher complex also included a ZIL-157 (6 × 6) tractor truck which towed a trailer carrying two replacement rockets and a crane truck which was required to lift the rockets from the trailer and load them onto the launcher.

Development of the new Luna-M (9M21) commenced in 1961 with the emphasis on mobility of the launcher and greater range and accuracy of the unguided rocket.

The first test launch took place in December 1961 and it was accepted for service with the Russian Army in 1964 with production of the system being undertaken at the Barrikady facility.

The Luna-M (9K52) system is also referred to in the West as the FROG-7a with the BAZ-135L4 (8 × 8) chassis being the Transporter-Erecter-Launch (TEL) platform.

The Luna-M used the 9M21F rocket with the 9N18F warhead filled with 200 kg of TGA-40/60 high-explosive. Other rockets were the 9M21B with AA-22 nuclear warhead using a radio fuze and the 9M21B1 warhead with an AA-38 nuclear warhead.

The 9M21G rocket was equipped with the 9N18G chemical warhead. The 9M21A rocket was equipped with a 9N18A leaflet warhead. The 9M210F rocket has the 9N180F submunition warhead which carries 42 submunitions.

Each of these submunitions weighs 7.5 kg and contains 1.7 kg of high-explosive. The warhead is fitted with a radio fuze that disperses the submunitions over the target at a height of 1,000 to 1,400 m. This was accepted for service in 1969.

While the designation of the complete system is the 9K52 Luna-M, the actual 8 × 8 wheeled launcher is designated the 9P113, which could launch up to 200 FROG-7 rockets before being overhauled.

The 8 × 8 transporter/resupply vehicle is the 9T29 and has a crew of two. It is fitted with a hydraulic crane with a lifting capacity of 2.6 tonnes and can lift complete rockets or replace warheads.

Training rockets were the 9M21Ye and 9M21Yel. All of these used the 3Kh18 solid propellant rocket motor.

Rear view of FROG-7 rocket clearly showing the small exhaust nozzles around the large exhaust venturi 0500783

FROG-7 artillery rocket system deployed in the firing position and with launcher elevated 0501057

There was also a special helicopter version called the 9K53 Luna-MV with the trailer launcher called the Br-257 (9P114) which could also propel itself up to a distance of 40 to 45 km. Prototypes of this system were built in 1964 and 1965. These were never placed in quantity production.

The FROG-7a was joined in service during 1968 by the FROG-7b variant (Russian designation R-70) which has a longer warhead section (total rocket length extended to 9.4 m from 8.95 m).

This is fitted with a pair of air brakes on the rear of the rocket, which can be locked open. The purpose of this is to lower the range of the rocket.

The Luna-M rocket had a CEP of 1,200 to 2,000 m while the new Luna-Z (or R-70) was required to be much more accurate. This was launched from a new complex called the 9K52M, which could fire either the Luna-M or Luna-Z rockets. Trials showed that the Luna-Z had even a larger CEP so this never entered quantity production.

The FROG-1 and FROG-2 systems have been considered obsolete for many years but the FROG-3, FROG-4 and FROG-5 weapons may be in limited service in some parts of the world. As their shelf-life has long expired, it is thought that these no longer have any military value.

The NATO-designated FROG-6 weapon is a non-operational dummy training round, which is designated the PV-65 *(Prakticheskaya Vystrel-65).*

Early FROG types were exported to the former Warsaw Pact, Algeria, Cuba, Egypt, North Korea and Syria while the FROG-7 was originally supplied to the former Warsaw Pact, Algeria, Cuba, Egypt, Iraq, North Korea, Kuwait, Libya, Syria and the Yemen.

Iraq did manufacture its own variant of the FROG-7 called the Laith 90. This had a cluster munition warhead and an increased maximum range of 90,000 m. Iraq and North Korea are also known to have produced their own chemical warheads for FROG rockets.

Iraq subsequently used its FROG-7 systems in the 1990-91 Gulf War against northern Saudi Arabia and against rebel positions in the ensuing civil war.

In 1985, the Soviet Army started using its FROG-7b systems in Afghanistan. Fitted with HE and submunition warheads the rockets were used against village targets as part of the effort to deny basic agricultural food supplies to the guerrilla forces.

In 1989, it was revealed that the Afghan Army had taken delivery of a number of FROG-7 launchers. These used life-expired FROG-7a/b rounds from Soviet stocks and took the place of the 1,000-odd Scud-B rounds delivered previously and fired against various guerrilla targets. The Serbian Army has also used FROG-7 weapons in the Yugoslav troubles.

In Russian Army service, the FROG-7 was replaced by the inertially guided US/NATO designated SS-21 Scarab (Russian name Tochka, meaning 'point') missile system. First deployed operationally in 1976, the weapon had, by 1989, replaced the majority of the FROG-7 systems in Eastern Europe and had been exported to Czechoslovakia, East Germany and Syria. Apart from various conventional warheads the SS-21 can carry the AA-60 nuclear warhead with 5, 10, 20 or 200 kT selectable yields available.

Description

(FROG-7)

A typical FROG-7 battalion consists of an headquarters battery and two firing batteries, each with two FROG-7 9P113 TELs, a GAZ-66 (4 × 4) fire direction vehicle and a trailer-mounted D-band RMS-1 (NATO designation End Tray) long-range meteorological radar. The TELs have no NBC protection for their crews.

Amongst its vehicles the headquarters battery contains four BAZ-135L4 9T29 TZMs *(Transportna-zaryazhyushcha Machina:* transporter-loader vehicle) which carry three reload FROG-7 rounds each. These are placed on the TEL 9P113 by using the vehicle's own onboard hydraulically-operated goose-neck crane which has a capacity of 2.6 tonnes which is positioned on the right side of the launcher rail just behind and above the wheel arch of the centre pair of wheels. The reloading operation takes around 20 minutes in total.

Preparation time for firing can take 15 to 30 minutes depending upon what operations need to be performed and whether the unit has only just occupied the position. The Luna-M itself is aimed by raising the TEL's rail to the desired elevation and adjusting the weapon's speed brakes so as to operate during the flight to give the correct range. If needed, a limited traverse angle can also be set on the launch rail assembly to compensate for any last minute offset required.

FROG-7 artillery rocket system in travelling configuration and showing rocket reloading crane 0058463

For optimum usage the associated RMS-1 radar has to be used to establish the meteorological parameters that might affect the rocket's flight trajectory.

Offensive operations with FROG-7s require them to be positioned well forward, around 8 to 18 km from the line of contact in order to engage tactical targets in the enemy's rear. In defensive operations these distances may be up to 25 to 50 per cent further back.

The FROG-7 unguided rocket is of conventional single-stage design with a cylindrical warhead of the same diameter as the rocket body. The combined booster-sustainer rocket motor uses solid propellant fuel and has a large exhaust venturi surrounded by 16 small booster nozzles. The four tail-mounted fins act to stabilise the rocket in flight.

Flight time to maximum range is about 160 seconds. Burn time of the engine is between 6.8 and 10.8 seconds depending upon the propellant quality. Airbrakes can be fixed to the rear of the rocket to achieve ranges between 15 and 29 km.

The FROG-7 warheads available include nuclear, conventional HE, chemical and (in the case of the FROG-7b) cluster munition. Known rocket designations include 9M21 (Luna-M rocket), 9M21B (Luna-M rocket with one of three tactical nuclear warhead types), 9M21E (Luna-M rocket fitted with 9N18E dispenser warhead) and 9M21F (Luna-M rocket with 450 kg 9N18F unitary HE-fragmentation warhead). There is also a leaflet type warhead which is designated the 9M21D.

The 400 kg cluster munition warhead has the designation 9N18E and carries a payload of 42 9N22 HE bomblets. The 9N22 weighs 7.5 kg, is 540 mm long and 74 mm in diameter. A 9E237 impact fuze is fitted with a pyrotechnic back-up self-destruction function. Upon exploding, 690 pre-fragmented splinters are formed by the bomblet, each weighing from 1 to 4 g each. This results in a pure anti-personnel effect.

The three nuclear warhead types are the AA-22 initial version with 3, 10 and 20 kTs selectable yields, the AA-38, an improved type with same selectable yields as the AA-22, and the AA-52 with 5, 10, 20 and 200 kTs selectable yields.

The chemical warheads have now been phased out of service.

The launch weight of a FROG-7 unguided rocket depends on the actual rocket but is typically between 2,432 and 2,450 kg which depends on the actual warhead type with a typical high-explosive warhead weighing 450 kg.

The FROG-7a rocket has an overall length of 8.96 m while the FROG-7b has an overall length of 9.4 m with the rocket having a diameter of 544 mm.

Maximum range is between 67,000 and 68,000 m with minimum range between 12,000 and 15,000 m with a CEP of 400 m.

Variants

Export 9K52TS

In the late 1960s a special version of the 9K52 Luna-M was developed for use in hot climates, this was designated the 9K52TS. This used the 9P113TS launcher and 9T29TS transporter and was used only for rockets with conventional warheads.

LRSV M-1996 Orkan-2

This is based on the 9P113 TEL but is believed to be equipped with four barrels from the Yugoslav (Serbian) 262 mm M-87 Orkan. Available details of this system are provided in the entry for the 262 mm M-87 Orkan. This system is designated the LRSV M-1996 Orkan-2.

This system is also referred to as the LRSV M-1996 Orkan-2 with the M-1996 being the year of introduction into service. This was also built in very small numbers and available details are provided in the entry for the baseline 262 mm M-87 Orkan in *IHS Jane's Land Warfare Platforms: Artillery and Air Defence*.

Specifications

	FROG 7 Artillery Rocket System
Dimensions and weights	
Crew:	7
Length	
overall:	10.69 m
Width	
transport configuration:	2.8 m
Height	
hull:	2.86 m
transport configuration:	3.35 m
Ground clearance	
overall:	0.475 m
Track	
vehicle:	2.30 m
Weight	
combat:	17,560 kg
Mobility	
Configuration	
running gear:	wheeled
layout:	8 × 8
Speed	
max speed:	65 km/h
Range	
main fuel supply:	650 km
Fording	
without preparation:	1.2 m
Vertical obstacle	
forwards:	0.685 m
Trench:	2.63 m
Turning radius:	12.5 m
Engine	diesel, 360 hp
Firepower	
Main armament traverse	
angle:	14° (–7° to +7°)
Main armament elevation/depression	
armament front:	+65°/+15°

Status

Production complete. In service with the following countries, this includes launchers held in reserve:

Country	Version	Number of launchers
Egypt	FROG-7	9
Korea, North	FROG-5	9
	FROG-7	18
Libya	FROG-7	40 probably non-operational owing to conflict in 2011
Russian Federation	FROG-7	n/avail (reserve)
Syria	FROG-7	18
Ukraine	FROG-7	50
Yemen	FROG-7	12, status uncertain

Contractor

Former Soviet state factories.

GUP TOS-1 220 mm (30-round) rocket system

Development

It is understood that the TOS-1 'Fighting Vehicle' 220 mm (30-round) rocket system was developed by the GUP Design Bureau of Transport Machine Building based in Omsk in the late 1980s and subsequently fielded in small numbers in the early 1990s. It is reported that this system has the nickname of the Buratino.

Some sources have indicated that as few as 24 TOS-1 'Fighting Vehicle' systems were manufactured with a small number of these having been deployed to Chechnya in 1999/2000 where they are claimed to have been very useful in urban fighting.

The exact operational role of the TOS-1 'Fighting Vehicle' is still not clear but it is possible that the system could have been originally developed for use in Afghanistan where it would be highly effective against guerrillas in mountainous terrain.

Another possible role could have been against dug in anti-armour teams in key defended areas, especially in urban terrain, which would be difficult to dislodge with conventional weapons.

The TOS-1 'Fighting Vehicle' may have been deployed at front level by the Russian Army.

More recently a modified version of the TOS-1 has been observed and this has been offered on the export market.

In September 2011 it was revealed that the Republic of Kazakhstan had introduced into service some of the latest Russian army equipment and this was shown for the first time during a parade held in the capital Astana (previously Akmola).

This included the second generation BMPT Tank Support Fighting Vehicle and the latest version of the TOS-1 220 mm (24-round) rocket launcher.

As far as it is known Kazakhstan is the first export customer for these weapons which are rarely seen in public and have so far been built in small numbers.

TOS-1 220 mm (30-round) rocket launcher based on a T-72 MBT hull which has probably been designed for use in the defence suppression role. Note the dozer blade lowered under front of chassis 0048633

Kazakhstan has the latest generation TOS-1 system which is also referred to as the Buratino and this has a total of 24 fuel air explosive rockets which a longer range of 6,000 m.

The first model of the TOS-1 had a total of 30 rocket tubes with the rockets having a maximum range of only 3,500 m.

During the SOFEX 2012 defence and security show held in Amman, Jordan, in 2012 it was stated that Russia had offered to sell the TOS-1 system integrated onto the hull of the US supplied M60 MBT as used by the Jordanian Armed Forces.

As of April 2013 Jordan had not purchased the TOS-1 system but when integrated onto the M60 hull this has been referred to by Russian sources as the Obyekt 634 - M60.

While the overall system is referred to as the TOS, with the latest version being the TOS-1A, the actual launcher is referred to as the BM-1 while the support vehicle, which is fitted with a hydraulic crane with a maximum lifting payload of 1,000 kg, is referred to as the TZM-1.

Description

The TOS-1 'Fighting Vehicle' is essentially a T-72 MBT with its original 125 mm two person turret removed and replaced by a low profile two person armoured turret superstructure with armoured sights in its upper part.

On top of the turret is mounted a large box type structure with powered traverse and elevation.

Within this structure are 30 × 220 mm rocket tubes in four layers, the top layer has six launcher tubes while the lower three layers have eight launcher tubes.

The unguided 220 mm rockets have a maximum range of 3,500 m and a minimum range of 400 m with a full salvo of 30 rockets being launched in 7.5 seconds, although the operator can also select salvoes of two or more rockets.

The unguided rockets have a fuel air type explosive (or thermobaric) warhead which is highly effective against troops in the open or dug in, as well as being effective against soft-skinned and light armoured vehicles.

The fuel air explosive warhead creates a massive overpressure in the target area for which there is no practical defence.

In recent years Russia has developed thermobaric warheads for its anti-tank guided weapons such as the 9M133 Kornet (NATO AT-14) and 9M114 Kokon (AT-6).

Russia has also developed thermobaric type warheads for a wide range of man-portable anti-tank weapons, including the widely deployed RPG-7 Rocket Propelled Grenade (RPG) system.

The latest TOS-1 'Fighting Vehicle' has a 220 mm (24-round) rocket launcher on the roof that launches rockets with a thermobaric warhead out to a maximum range of 6,000 m (KBTM Design Bureau GUP) 0531670

TOS-1 220 mm resupply vehicle based on T-72 series MBT, in travelling configuration from the rear 1151749

Combat experience in Afghanistan and elsewhere showed that the normal High-Explosive Anti-Tank (HEAT) warhead, used to defeat armoured targets, was not very effective against troops.

Standard equipment for the TOS-1 includes an NBC system, ability to lay a smoke screen by injecting diesel fuel into the exhaust outlet on the left side of the hull, unditching beam that is normally carried on the right side of the hull and a dozer blade which is mounted at the front of the hull. This is used to clear obstacles and prepare firing positions.

The first production batch of TOS-1 were installed on a modified T-72 series MBT hull but the system can also be installed on other MBT chassis, such as the Russian T-80 or T-90 and foreign equivalent vehicles.

To enable the rockets to be positioned on the target with greater accuracy, the system is fitted with a computerised fire-control system that includes an optical sight, laser range-finder, cant sensor and an electronic ballistic computer.

The laser range-finder allows the target to be measured to an accuracy of 10 m. Target information is fed into the computer that calculates the required elevation and traverse of the launcher, which is then laid onto the target. The unguided rockets are then launched with the number of 220 mm rockets launched depending on targets to be neutralised.

Variants

220 mm (24-round) TOS-1

It has been disclosed that the Russian Omsk Design Bureau for Transport Machine Building has developed a new version of the TOS-1.

The new system is similar in concept to the earlier TOS-1 but the launcher now has only 24 rocket tubes, not 30. The unguided rockets have a minimum range of 400/600 m but maximum range has now been extended to 6,000 m.

As the rockets are longer and heavier, the overall weight of the system is being quoted as 44.3 tonnes.

The first version of TOS-1 had an upper layer of six rockets and three layers of eight rockets while the latest generation system has three layers of eight rockets in a rectangular box-type launcher with no chamfered sides.

This is a significant range improvement as the TOS-1 system can now be deployed out of the range of most ground based anti-tank weapons, which typically have a range extending out to 4,000 m.

This is also referred to as the TOS-1A and has a maximum road speed of 60 km/h and a quoted road range of 500 km.

Each rocket has a launch weight of 216 kg and has an overall length of 3,723 mm with a typical interval between launch of 0.5 seconds.

A full salvo being launched in 12 seconds or two rockets every six seconds.

Time to fire against a line of sight target is 90 seconds after the vehicle has come to a halt.

122 mm TOS

The prototype of a 122 mm version of TOS has been completed based on a T-80, rather than a T-72, MBT hull as used in the first-generation TOS-1. This has a powered launcher with a top layer of 10 rocket tubes and below this are another five layers of 14 rocket tubes, giving a total of 70 tubes.

TOS-1 220 mm (30-round) rocket launcher based on a T-72 MBT hull in travelling configuration 1043234

TOS-1 resupply vehicle

This is based on a T-72 MBT hull with its turret removed and replaced by two pods of new 220 mm rockets, which are loaded into TOS-1 one by one using a hydraulic crane mounted between the two pods.

The re-supply vehicle has a crew of three and a combat weight of 39 tonnes with a maximum road speed of 60 km/h and a road range of between 550 and 600 km.

The system takes about 45 minutes to be re-loaded with 24 rockets and 24 minutes to reload the actual TOS-1A launcher.

Specifications

	TOS-1 220 mm (30-Round)
Dimensions and weights	
Crew:	3
Length	
overall:	6.86 m
Width	
overall:	3.46 m
Height	
overall:	2.60 m
Weight	
combat:	46,000 kg
Mobility	
Configuration	
running gear:	tracked
Speed	
max speed:	60 km/h
Range	
main fuel supply:	550 km
Firepower	
Armament:	30 × turret mounted 220 mm rocket system
Main weapon	
maximum range:	3,500 m
Survivability	
Night vision equipment	
vehicle:	yes
NBC system:	yes
Armour	
hull/body:	steel + advanced + explosive reactive armour

Status

Production complete. In service with the Russian Army in small numbers.

Late in 2011 the TOS-1A was observed to be in service with Kazakhstan.

Contractor

GUP Design Bureau of Transport Machine Building.

Splav 300 mm BM 9A52 (12-round) Smerch multiple rocket system

Development

The development of the 300 mm Smerch (Tornado) Front (Military District) level RSZO (*Reaktivaya Sistema Zalpovogo Ognya:* Rocket System for Salvo Fire) took place in the late 1970s/early 1980s under the guidance of the Splav Government Scientific Production Concern at Tula.

It was first seen by the West in 1983 and was given the STANAG designation M1983 by NATO. It was accepted for service with the Russian Army in 1987 and entered front line service the same year.

While the overall prime contractor is the Splav Scientific Production Concern the actual launcher and transloader vehicles and associated equipment was designed and built by the Motovilikha Plants Corporation based in Perm.

The main role of the Smerch system is to suppress missile, artillery and mortar batteries, destroy strongpoints and eliminate enemy nodes of resistance. It is assigned to Military Districts and was deployed in brigades of four battalions each with 12 9A52 launchers.

The Russian industrial index number for the original launcher is 9A52 and the complete system has the designation 9K58. The latest launcher variant has the designator 9A52-2 and is known as Smerch-M. This is based on the MAZ-543M (8 × 8) chassis.

In 1993, at the Idex-93 international weapons exhibition, this version was being marketed under the export name Smerch (Sandstorm). The system has also been referred to as the RS30 (Reaktivnaya Systema - Missile System).

The first export customer is Kuwait which took delivery of nine launchers in 1995 and 18 in 1996.

In July 1996, it was revealed that the United Arab Emirates had placed an order for a Smerch battery of six 9A52-2 launchers, six 9T234-2 loader-transporters and a Vivariy automated fire-control system.

In mid-2002, it was revealed that the Indian Army had carried out a series of test firings of the Splav 300 mm BM 9A52 (12-round) Smerch system off the east coast of India.

In 2005 the Ukraine sold 11 BM 9A52 (12-round) Smerch systems to Azerbaijan.

In 2006 India finalised a USD500 million contract with Russia covering the supply of two regiments of 300 mm Smerch system based on the Tatra T816 (10 × 10) truck chassis.

Rocket used in Splav 300 mm BM 9A52-2 (12-round) MRS the with tail fins folded (Christopher F Foss) 0085766

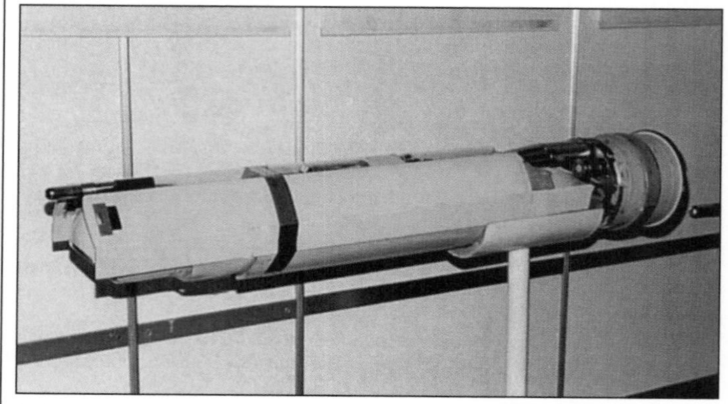

R-90 expendable UAV with wings and tails in the folded position 0008318

A total of 28 launchers and associated resupply vehicles and rockets have been ordered. It is understood that India has ordered a complete set of long-range (90 km) rockets for use with these systems.

According to the United Nations Arms Transfer Lists, India has recently taken delivery of the following quantities of Russian Splav BM 9A52 (12-round) Smerch multiple rocket systems:

Year	Quantity	Type
2007	16	9A52-2T Smerch launchers
2007	6	Smerch launchers
2008	18	9A52-2T Smerch launchers
2008	6	Smerch launchers

Although India placed its contract for Smerch system as far back as 2006, in 2012 it was stated that India and Russia had agreed to form an industrial joint venture to manufacture the Smerch system in India.

This will be via joint venture between the state owned Indian Ordnance Board and Russia on a 50/50 basis with five versions to be manufactured in India.

The latest export customer for the Smerch system is Venezuela who ordered 12 systems with training being carried out in Russia.

These have now been delivered and are in service with the 435th Multiple Rocket Field Artillery Group as the replacement for the Israel Military Industries LAR-160 systems which have now been phased out of service.

In addition to the Smerch systems Venezuela has also ordered and taken delivery of 24 122 mm BM-21 systems.

Description

The 9A52-2 RSZO battalion comprises the following components:

- Battalion and battery 1K 123 Vivary fire-control system command-and-control units contained in KamAZ-4310 (6 × 6) truck chassis-based command vehicles fitted with a K4310 fully enclosed box body, each of which has its own power generator and a secure data and radio communications system that will support VHF communications to a range of 50 km and HF communications to a range of 350 km. The vehicles also contain the elements of the Vivariy automated fire-control system that can undertake the automated or non-automated command and control of a Smerch-equipped multiple rocket launcher brigade. The actual Command and Staff Vehicle (CSV) used by the brigade commander and brigade staff is also the KamAZ-4310 (6 × 6) vehicle described previously. All command element vehicles carry onboard digital computers and associated displays and printers, in addition to the communications equipment mentioned. The latest version is designated the MP32M1 command and staff vehicle and includes workstations based on IBM-compatible computers with i80486 microprocessors interconnected via the Ethernet network with SVGA monitors for digital map visualisation and LCSs for alphanumeric data presentation, as well as a printer for information logging. It also includes a set of data and voice HF/VHF radio communications sets, channel encryption and switching equipment, power supply, NBC and air conditioning system.

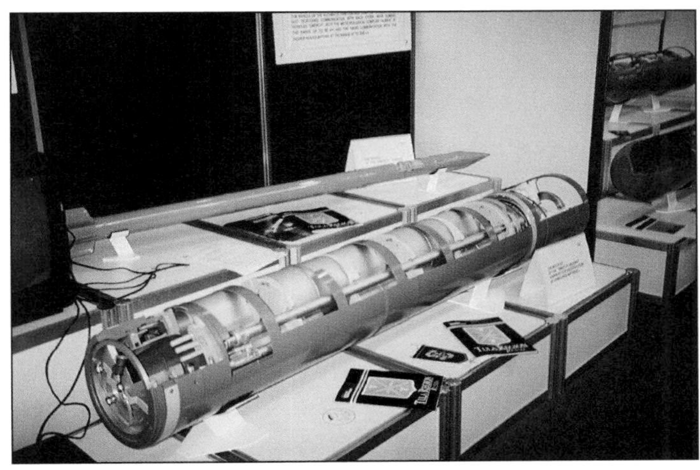

9M55K1 warhead for the Splav 300 mm BM 9A52-2 (12-round) MRS cutaway to show interior and MOTIV-3F submunitions (Christopher F Foss) 0500895

All brigade-level CSV are equipped with GLONASS/NAVSTAR receivers that indicate the precise time and location co-ordinates as well as training facilities. Other elements include 1B44 meteorological station on ZIL-131 (6 × 6) unarmoured truck chassis, PM-2-70 MTO-V maintenance and repair unit also on ZIL-131 (6 × 6) truck chassis and the IT12M-2M topography vehicle on GAZ-66 (4 × 4) unarmoured truck chassis. Mention has also been made of the Slepok-M1 automated fire-control system.

- Three batteries each of four launchers. The 9A52-2 launcher is based on a modified MAZ-543M (8 × 8) cross-country truck chassis and is fitted with a 12-round elevating launcher assembly on its rear decking. The 300 mm rocket tube arrangement is two separate banks of four with a connecting roof of the remaining four tubes overlying the inner tubes of the banks. Before firing, two stabilisers are lowered to the ground. These are positioned one either side between the rear two road wheels. The MAZ-543M chassis is fitted with powered steering on the front two axles and a central tyre-pressure regulation system that allows the driver to adjust the tyre pressure to suit the terrain being crossed from within the cab. The launch vehicle's cabin accommodates the launch preparation and firing equipment. Single-round or salvo firing is possible. The driver is seated at the front of the vehicle on the left side with the vehicle commander to his immediate rear. The fully enclosed cab is provided with an NBC system and two entry doors in the left side that open to the front. Recent developments include an installation of an automated launch preparation equipment that has allowed for autonomous execution of fire missions. This allows for a sharp reduction in fire preparation time from unprepared positions. This equipment provides for the topographical surveying and navigation of the launch vehicle, computation in the launch vehicle of fire settings and flight mission data of the basis of target and meteorological data provided by the Automated Fire-Control System, ramp laying on the target by the crew staying in the cab and using no aiming points, visual monitoring of rocket ramp laying, presentation on the display of the launch vehicle location, pre-launch testing on rocket projectile equipment, flight mission data input in rocket projectile onboard equipment and finally rocket launch with take-off checks. A complete salvo of rockets can be fired by the system in 38 seconds. It would then redeploy to another position for reloading.
- Accompanying each 9A52-2 launcher is a 9T234-2 transloader vehicle based on the MAZ-543A (8 × 8) chassis. This carries 12 reload rounds and a hydraulic crane that can lift a maximum load of 850 kg on its rear decking to load and unload the 9A52-2. This vehicle is also referred to as the Oplot. The system takes a well trained crew 20 minutes to reload as the 300 mm rockets have to be loaded one at a time.
- The flight corrected rocket used is of 300 mm calibre, fitted with a 'self-contained powered-flight phase control system which provides an increase in fire precision by a factor of two and in accuracy by a factor of three as compared to unguided rockets'. The accuracy is claimed as being 0.23 per cent of range. Minimum range is 20,000 m and maximum range 70,000 m.

Available rocket types

The flight corrected solid propellant rocket types available are:

- 9M55K which carries 72 unguided fin-stabilised high-explosive fragmentation submunitions which are highly effective against soft skin and lightly armoured targets. The submunitions weigh 1.75 kg and are fitted with a self-destruct mechanism that operates after a period of 110-160 seconds. The rocket weighs 800 kg and has a minimum range of 20 km and a maximum range of 70 km. This 300 mm rocket has an overall length of 7.6 m and a launch weight of 800 kg and has a maximum range of 70,000 m and a minimum range of 20,000 m.
- 9M55K1, which carries five Bazalt State Research and Production Enterprise parachute-retarded MOTIV-3F top-attack anti-armour submunitions. Each of the latter weigh 15 kg each, and are fitted with a two-spectrum infra-red seeker. The warhead will penetrate 70 mm of conventional steel armour. The submunition has a self-destruct time of 50 seconds. The seeker has a 30° field of view and when it is within some 150 m of the target the warhead is activated. The same munitions are used in the air-delivered RBK-500 SPBE-D 500 kg cluster bomb. The submunition in this application is designated the MOTIV-3F. It has a minimum range of 20 km and a maximum range of 70 km.

Splav 300 mm BA 9A52 (Smerch) multiple rocket system with launcher traversed right and stabilisers not lowered (Christopher F Foss) 1296602

- 9M55K4 carries 25 anti-tank mines each of which weighs 4.85 kg and carries 1.85 kg of high-explosive. The mines are fitted with a self-destruct mechanism that operates after between 16 and 24 hours. The rocket weighs 800 kg and has a minimum range of 20 km and a maximum range of 70 km.
- 9M55F with a unitary warhead carrying 95.5 kg of high-explosive which separates over the target and is slowed down by a parachute. This has a self-destruct time of 100-160 seconds. This rocket weighs 810 kg and has a minimum range of 25 km and a maximum range of 70 km.
- 9M55S is the fuel air explosive (thermobaric) member of the family. The rocket weighs 800 kg and has a minimum range of 25 km and a maximum range of 70 km. The actual rocket weighs 800 kg and contains 100 kg of fuel air explosive, which is fitted with a self-destruct mechanism that operates after a period of 110 to 160 seconds. The rocket is launched over the target where the FAE is dispensed to form a thermal cloud some 25 m in diameter. This is then detonated and the resultant overpressure is highly effective against, not only troops in the open, but also personnel who are sheltering in fortifications as well as trucks and light armoured vehicles.
- In addition there is a special RPV called the R-90 (Typchak) that can be launched from a 300 mm rocket tube. This was developed by the NITs ENIKS design bureau and could be used to acquire targets for the Smerch launcher or carry out battlefield reconnaissance after the target has been engaged. As far as it is known this remains at the prototype stage. This rocket is designated the 9M534 and has a launch weight of 815 kg with the rocket having a minimum range of 25 km and a maximum range of 90 km. The UAV weighs 40 kg, has a 20 minute patrol time and an information sending range of 70 km.
- The recent 9M55K5 rocket carries 646 shaped charge fragmentation submunitions that are dispensed over the target. The submunitions have a self-destruct time of 130 to 260 seconds and the HEAT warhead will penetrate 120 mm of conventional steel armour. The rockets have a minimum range of 25 km and a maximum range of 70 km. The rockets weigh 800 kg at launch.
- 9M55K6 carries five self-targeting top attack munitions. The rocket weighs 800 kg and has a minimum range of 25 km and a maximum range of 70 km. Each submunition will penetrate 120 mm of conventional steel armour.
- 9M55K7 carries 20 small self-targeting submunitions fitted with an advanced seeker with the HEAT warhead capable of penetrating 70 mm of conventional steel armour. The rocket weighs 815 kg and has a minimum range of 25 km and a maximum range of 70 km.
- 9M55UT is a training rocket and weighs 800 kg. This is used to training crews in handling the war shot rocket.

New long-range rockets

More recently a new family of longer-range solid propellant rockets have been developed by Splav. These are all numbered in the 9M500 series and known rockets include:

- The 9M525 is an enhanced version of the 9M55K and carries 72 submunitions each of which weigh 1.75 kg and are now fitted with a self-destruct fuze mechanism which operates after a period of 110 seconds. This rocket weighs 815 kg and has a minimum range of 25 km and a maximum range of 90 km.

300 mm 9T234-2 Splav-M transloader vehicle 0500642

Splav 300 mm BA 9A52 (Smerch) multiple rocket launcher from the rear with launcher traversed right and stabilisers not lowered to the ground (Christoper F Foss) 1296603

- The 9M526 is an enhanced version of the 9M55K1 and carries five parachute-retarded, self-targeting submunitions that weigh 15 kg and will penetrate 70 mm of conventional steel armour. This rocket weighs 815 kg and has a minimum range of 25 km and a maximum range of 90 km.
- The 9M527 carries 25 anti-tank mines which are scattered over the target area. Each mine weighs 4.85 kg and carries 1.85 kg of high-explosive and is fitted with a self-destruct fuze that operates after a period of 16 to 24 hours. The rocket weighs 815 kg and has a minimum range of 25 km and a maximum range of 90 km.
- The 9M528 is a new round that features a single, separable, high-explosive fragmentation warhead which is parachute retarded. This rocket weighs 815 kg and has a minimum range of 25 km and a maximum range of 90 km. The warhead weighs 95 kg and has a self-destruct fuze time of 100-160 seconds.
- The 9M529 is another new round and has a fuel air explosive-type warhead that carries 100 kg of explosive and is fitted with a self-destruct fuze mechanism that operates after a period of between 110 and 160 seconds. The fuel air explosive is claimed to have an effective area of at least 25 m. This rocket weighs 815 kg and has a minimum range of 25 km and a maximum range of 90 km.
- The 90M530 is yet another new round that has been designed to penetrate key military infrastructure facilities such as airfields and fuel depots. It has a hardened high-explosive warhead, which is designed to penetrate the target before it explodes. The warhead contains 75 kg of high-explosive. The rocket weighs 815 kg and has a maximum range of 90 km.
- The new 9M531 is of the high-explosive fragmentation type and contains 646 submunitions that are designed to attack troops in the open, light armoured vehicles or trucks. It carries a total of 646 submunitions, each of which is fitted with a self-destruct fuze mechanism that operates after a period of 130 to 260 seconds. Each submunition is claimed to be able to penetrate 120 mm of conventional steel armour. The rocket weighs 815 kg and has a minimum range of 25 km and a maximum range of 90 km.
- The 9M532 carries 20 self-targeting top attack munitions with each of these weighing 6.7 kg and capable of penetrating 50 mm of conventional armour. The rocket weighs 815 kg at launch, has a minimum range of 25 km and a maximum range of 90 km.
- The 9M533 carries five 9H268 self-targeting top attack munitions with each of these being fitted with a dual mode seeker. Each submunition weighs 17.25 kg and has a warhead that will penetrate 120 mm of armour. These have a minimum range of 25 km and a maximum range of 90 km.
- The 9M537 carries 32 submunitions each of which weigh 4.5 kg. The rocket weighs 815 kg and has a minimum range of 25 km and a maximum range of 90 km.
- The 9M536 carries 70 submunitions each of which weighs 2 kg and are fitted with a HEAT warhead that will penetrate 20 mm of steel armour. The rocket weighs 815 kg and has a minimum range of 25 km and a maximum range of 120 km.
- In addition, there is an improved version of the original 9M55K designated the 9M55K5. This carries 646 submunitions as carried by the 9M531, which are fitted with a self-destruct fuze mechanism that operates after a period of 130 to 260 seconds. The rocket weighs 800 kg and has a minimum range of 25 km and a maximum range of 90 km. Some Russian sources have stated that this has a maximum range of 70 km rather than 90 km.

300 mm 9T234-2 Splav-M transloader vehicle from the front (Christopher F Foss) 1365041

Variants

120 km range rockets

The new long-range rockets mentioned in the text have a maximum range of 90 km but some Russian sources have mentioned a new family of Smerch rockets that have a maximum range of 120 km. No further details are available.

Modernised combat vehicle Smerch with CCCE and AGFCS

An upgrade package is now being marketed by Splav to further enhance the capabilities of the Smerch system.

The upgrade includes the introduction of the Combat Control and Communication Equipment (CCCE) and the Automated Guidance and Fire-Control System (AGFCS). This provides for the automated rapid reception and transmission of data, autonomous survey, navigation and terrain orientation displayed on the electronic map of the display.

Additionally, it provides for automated calculation of fire mission settings and the flight programme and automated tube cluster guidance, without the use of a standard sight and with crew remaining in the cab of the launcher. An air-conditioning system has also been installed.

Other optional equipment for the Smerch system, in addition to the automated fire-control system previously mentioned, include an air conditioning system for the launch vehicle, collective protector for the launch vehicle and a rocket flight path correction system.

Smerch 9T234-T transloader

This is used to resupply the launcher with new rockets and is based on the same 8 × 8 chassis and has the same power pack.

It carries a total of 12 × 300 mm rockets that are loaded one at a time using an on board hydraulic crane with a maximum lifting capacity of 850 kg.

It has similar automotive performance to the launcher and is operated by a crew of three with a loaded weight of 41,500 kg and an empty weight of 30,000 kg

9A52-2T

This is the latest version of the Splav 300 mm BM 9A52 (12-round) Smerch multiple rocket system and is based on a new Tatra T816 10 × 10 cross-country truck chassis. Between the fourth and fifth road wheel station is a hydraulic stabiliser that is lowered to the ground before firing commences, providing a more stable firing platform. This is the model ordered by India.

This version has been developed specifically for the export market and launches the same family of rockets as the standard Splav 300 mm system.

This has a crew of three and a typical combat weight of 39,500 kg and is fitted with powered steering which provides a turning radius of 15 m.

A central tyre pressure inflation system that allows the driver to adjust the tyre pressure from his position to suit the terrain being crossed. Tyres are Michelin 16.00 × R20 XZL.

The vehicle can ford to a depth of 1.25 m with a trench crossing capability of 2 m.

Basic specifications include an overall length of 12.4 m, width of 3.035 m and an overall height of 3.425 m when in the travelling configuration. Ground clearance is being quoted as 450 mm.

It is powered by a Deutz BF8M1015C diesel developing 544 hp which is coupled to a Twin Disc TD-61-1175 automatic transmission which gives a maximum road speed of 90 km/h and a maximum range of 850 km.

The fully enclosed cab is provided with an Italian Diavia air conditioning system to allow the system to operate in high ambient temperatures.

This version is also referred to as the 9K58 Smerch MRLS consisting of the 9A52-2 combat vehicle/launcher, the 9T234-2 transporter-loader, the MP32M1 unified command and staff vehicle, radio direction finding and meteorological support complex (Index Number 1B44) plus the associated family of 300 mm rockets.

300 mm Tornado-G

The Russian Army is to receive the 300 mm Tornado-G as the replacement for the currently deployed 122 mm BM-21 series of truck mounted rocket launchers.

Spav 300 mm (6-round) CV 9A52-4 launcher

This is the latest development and consists of a KamAZ-6350 (8 × 8) cross-country chassis fitted with pod of six 300 mm rockets which are the same as that used in the standard system.

Development of this system is said to be complete and details are provided in a separate entry in *IHS Jane's Land Warfare Platforms: Artillery and Air Defence.* As of April 2013 there are no known sales of this system.

Chinese 300 mm (10-round) A100

China has revealed a 300 mm (10-round) A 100 multiple rocket system that is probably related to the Russian 300 mm BM 9A52 (12-round) Smerch system. This is being marketed by China National Precision Machinery Import-Export Corporation (CNPMIEC).

More recently, China North Industries Corporation (NORINCO) has developed a similar system on an 8 × 8 cross-country chassis which is called the AR2 300 mm (12-round) Multiple Launch Rocket System. This launches various rockets with a maximum range of up to 130 km.

It is understood that Pakistan has now taken delivery of of sufficient 300 mm Multiple Launch Rocket Systems which will be sufficient for one regiment. Today Pakistan only has shorter range 122 mm (40-round) truck-mounted systems in service.

Details of both of these Chinese multiple rocket systems are provided in separate entries in *IHS Jane's Land Warfare Platforms: Artillery and Air Defence.*

Specifications

	9A52-2
Dimensions and weights	
Crew:	4
Length	
overall:	12.1 m
Width	
transport configuration:	3.05 m
Height	
transport configuration:	3.05 m
Ground clearance	
overall:	0.44 m
Weight	
standard:	33,700 kg
combat:	43,700 kg
Mobility	
Configuration	
running gear:	wheeled
layout:	8 × 8
Speed	
max speed:	60 km/h
Range	
main fuel supply:	650 km
Fording	
without preparation:	1.1 m
Vertical obstacle	
forwards:	0.78 m
Trench:	2.50 m
Turning radius:	15 m
Engine	V-12, 4 stroke, diesel, 518 hp at 2,200 rpm
Firepower	
Armament:	12 × 300 mm rocket system
Ammunition	
main total:	12
Rate of fire	
rapid fire:	12/40 s (est.)
Operation	
stop to fire first round time:	180 s
out of action time:	180 s
Main armament traverse	
angle:	60° (30° left and right)
Main armament elevation/depression	
armament front:	+55°/0°

Status

Production as required. In service with following countries:

Country	Quantity	Comment
Algeria	18	
Azerbaijan	11	from the Ukraine in 2006
Belarus	36	
India	28	delivered from 2007
Kuwait	27	delivered between 1995 and 1996
Russia	106	
Turkmenistan	6	
Ukraine	80	
United Arab Emirates	6	
Venezuela	12	delivered 2012/2013

Contractor

Splav Scientific Production Concern (prime contractor and rockets).
Motovilikha Plants Corporation (launcher and transloader).

Splav 300 mm CV 9A52-4 (6-round) multiple rocket system

Development

The original Splav 300 mm BM 9A52 (12-round) Smerch Multiple Rocket System (MRS) has 12 launch tubes that have to be reloaded one at a time. Details of this system are provided in a separate entry in *IHS Jane's Land Warfare Platforms: Artillery and Air Defence.*

This is still being offered on the export market in the original 8 × 8 configuration as well as a more recent 10 × 10 configuration based on a Tatra chassis which is aimed at the export market.

The latest 300 mm CV 9A52-4 MRS is based on a KAMAZ-6350 (8 × 8) series cross-country chassis and has launcher which carries a pod of six standard 300 mm Smerch rockets.

The main advantage of this system over the original BM 9A52-4 is that it is much lighter and easier to transport by air and can be reloaded much quicker.

This system is understood to have been developed from about 2005 and was shown in public in 2007.

Development of this system is understood to be complete but as of March 2013 this is understood not to have entered quantity production.

The overall prime contractor of this system is the Splav Scientific Production Concern while the launcher and transloader has been developed by the Motovilikha Plants Corporation and the 8 × 8 chassis is provided by KAMAZ Incorporated.

Description

The system is based on a standard production KAMAZ-6350 (8 × 8) cross-country truck chassis that has a maximum stated payload of 10,500 kg.

The three-person fully enclosed forward control cab is at the front and mounted at the rear of the chassis is a new launcher with has a pod of six 300 mm rockets in two rows each of six rockets.

As far as it is known the 300 mm rockets launched by the system are identical to those fired by the original Splav 300 mm BM 9A52 (12-round) system covered in a separate entry in *IHS Jane's Land Warfare Platforms: Artillery and Air Defence.*

The 300 mm rockets can be launched from within the fully enclosed cab or by remote control depending on the tactical situation.

A pod of six 300 mm rockets weighs 6,220 kg and is designated the 9Ya295 is of composite material and is discarded after use. The actual container has an overall length of 7,512 mm and is 1,130 mm wide and 996 mm high and has an unloaded weight of 1,420 kg. The reusable six round metal pod is designated the MZ-196 and has the same launch tubes as the standard Smerch 300 mm launcher.

The actual pod has an overall length of 7,400 mm and is 1,310 mm wide and 954 mm high. Loaded weight is 7,300 kg and unloaded weight is 2,500 kg.

The launcher has an elevation from +15° up to +55° with a traverse of 30° left and right. Launcher elevation and traverse is powered.

Splav 9A52-4 300 mm (6-round) multiple rocket system 9T234-4 transloader in travelling configuration complete with pod of rockets (Splav) 1452638

Splav 9A52-4 300 mm (6-round) multiple rocket system with MZ-196 container with six tubes (S V Gurov) 1331391

Splav 9A52-4 300 mm (6-round) multiple rocket system 9T234-4 transloader with 9Ya295 single use composite container in foreground (S V Gurov)
1331392

The system is fitted with a Automated Guidance and Fire-Control System (AGFCS) which not only includes a fire-control function but also an autonomous survey and navigation capability to reduce the time taken to come into action. The system takes about three minutes to come into action and launch its first rockets.

Information can be exchanged with other launch platforms as well as a higher chain of command using the onboard command-and-control system.

The vehicle is supported by a reloading vehicle which is also based on a KAMAZ-6350 which is designated the 9T234-4 transloader.

Specifications

	Splav 300 mm CV 9A52-4 (6-round) multiple rocket system
Dimensions and weights	
Crew:	2
Length	
overall:	11.2 m
Width	
overall:	2.5 m
Height	
overall:	3.15 m
Ground clearance	
overall:	0.39 m
Track	
vehicle:	2.01 m
Wheelbase	1.94 m + 3.34 m + 1.32 m
Weight	
standard:	19,650 kg
combat:	24,650 kg
Mobility	
Configuration	
running gear:	wheeled
layout:	8 × 8
Speed	
max speed:	90 km/h
Range	
main fuel supply:	1,000 km
Fuel capacity	
main:	375 litres
Fording	
without preparation:	1.5 m
Gradient:	68%
Side slope:	44%
Turning radius:	13.9 m
Engine	KAMAZ-740.50–360, 11.75 litre, V-8, water cooled, diesel, 360 hp at 2,200 rpm
Gearbox	
model:	KAMAZ
type:	manual
forward gears:	9
reverse gears:	1
Tyres:	425/85R 21
Suspension:	semi-elliptic leaf springs with telescopic shock absorbers, front;inverted leaf springs rear
Electrical system	
vehicle:	24 V
Firepower	
Armament:	6 × 300 mm rocket system

Status

Development complete. Ready for production.

Contractor

Splav Scientific Production Concern (prime contractor and rockets).
Motovilikha Plants Corporation (launcher and transloader).

Splav 220 mm BM 9P140 (16-round) Uragan multiple rocket system

Development

Like all Russian Multiple Rocket Launcher (MRL) systems, the 220 mm BM 9P140 (16-round) Uragan (Hurricane) was developed under the guidance of the Splav Government Scientific Production Concern at Tula.

While the overall prime contractor for the complete Uragan system is the Splav Scientific Production Concern, the actual launcher and transloader vehicles and associated equipment has been designed and built by the Motovilikha Plants Corporation based in Perm.

The system was designed in the early 1970s, with development being completed in 1975. It entered service with the Soviet Army in the same year and was allocated the STANAG designation of the M1977 by NATO, this being the year it was first identified.

The 220 mm BM 9P140 (16-round) replaced the older 240 mm (12-round) BM-24 and 200 mm (4-round) BMD-20 systems.

The complete BM 9P140 system has the industrial complex number of 9K57, which comprises the launcher, solid propellant unguided 220 mm rockets and the essential rocket resupply vehicle. In the past the system has also sometimes been referred to as the BM-22 or BM-27.

Until the Splav 300 mm BM 9A52 (12-round) Smerch rocket system entered service with the Soviet Army in the early 1980s, the 220 mm 9P140 (16-round) was the largest system of its type in service.

The main advantage of the 220 mm BM 9P140 (16-round) launcher system is that it is mounted on a ZIL-135LM (8 × 8) chassis with improved tactical and strategic mobility.

Its 220 mm rockets have a much longer range and carry a variety of warheads, enabling a wider range of battlefield missions to be undertaken using the same launch platform.

Until the introduction of the larger 300 mm BM 9A52 (12-round) Smerch system, the 220 mm BM 9P140 (40-round) was normally deployed in rocket artillery regiment and brigade formations at Army or Front (Military District) level.

In the case of the former, the regiment Table of Organisation and Equipment (TOE) normally comprises three battalions of 12 launchers, which is normally increased to 18 launchers in time of war.

220 mm BM 9P140 (16-round) Uragan (Hurricane) multiple rocket launcher with cab shutters lowered for travelling and launcher traversed to the right (Christopher F Foss)
1296604

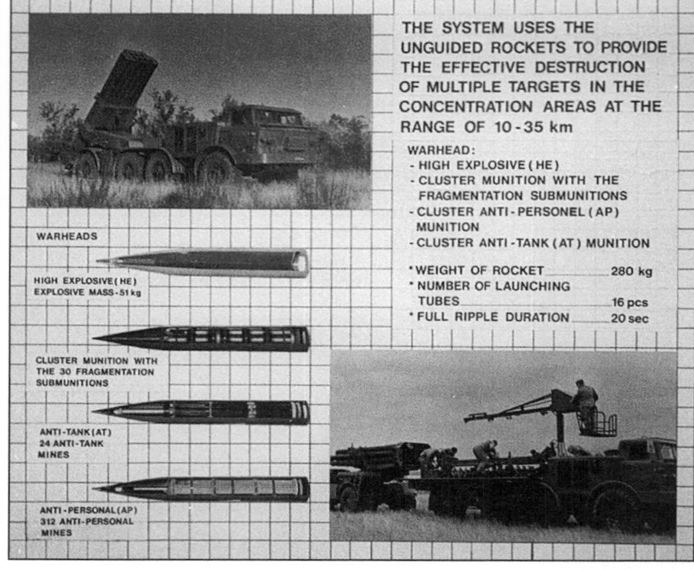

This photograph shows the four main warhead types (lower left) with the system being resupplied (lower right) (Christopher F Foss)
0500899

In a Russian Military District, an independent Uragan rocket artillery regiment has the same TOE. But if the Uragan unit is assigned to an artillery division within the Military District, then the unit may be either a regiment or brigade with four battalions each with 12 launchers (which would be increased to 18 in time of war). If the unit were already a brigade then it would take on the extra six launchers per battalion to achieve its wartime TOE.

Typical roles of the Uragan system would be to suppress enemy missile, artillery and mortar batteries as well as destroying strong points and other areas of resistance. It would normally use 'shoot and scoot' techniques to avoid counterbattery fire.

In some units it has replaced the shorter-ranged 122 mm BM-21 (40-round) rocket system mounted on a 6 × 6 truck chassis. The Uragan has been designed to operate in temperatures ranging from –40°C to +50°C.

The Uragan battalion is controlled by the Kapustnik-B automated fire-control system. This has a Battalion commander's vehicle based on the BTR-80 (8 × 8) APC chassis, a Battalion Chief of Staff vehicle based on the Ural-4320 (6 × 6) truck chassis, three Battery Commander's vehicles based on the BTR-80 APC chassis and three Battery Senior Officer's vehicles based on the Ural-4320 (6 × 6) truck chassis.

The Kapustnik-B comprises subsystems for reconnaissance, initial battalion orientation and location fixing, weather reconnaissance and ballistic tracking information, inter-battalion and higher echelon communication and data transfer. All the elements are connected to a single data processing computer.

If required, the system can automatically accept external reconnaissance data transfers and be connected to higher echelon automated control systems.

According to the United Nations Arms Transfer Lists for 1992 through to 2010 the following systems were exported:

From	To	Date	Quantity
Belarus	Eritrea	2007	9
Moldova	Guinea	2000	3 plus 3 transloaders
Moldova	Yemen	1994	13

Notes:
Guinea also took delivery of 430 9M27F and 860 9M27K rockets
It is understood that there has been no recent production of this system although marketing continues and surplus systems are also available for export.

Description

The system is based on a modified ZIL-135LM 8 × 8 truck chassis, which has excellent cross-country mobility and is used for a wide range of other battlefield missions. These include Transporter, Erector, Launcher (TEL) for the FROG-7 surface-to-surface unguided rocket system (and its associated reload vehicle), Sepal cruise missile carrier, cargo truck and tractor truck, to name but a few.

The chassis was developed at the Likhachev Motor Vehicle Plant near Moscow but production was undertaken at the Bryansk Automobile Works, for which reason the family is sometimes referred to as the BAZ-135L4.

The fully enclosed unarmoured crew compartment is at the front with the two diesel engines to the rear. The turntable-mounted rocket launcher is mounted at the very rear of the chassis and is traversed to the front for travelling.

The rocket launcher pod consists of an upper layer of four tubes and two lower layers of six tubes so giving a total of 16 launch tubes. The sighting system is mounted on the left side of the launcher and access to this is via a folding ladder.

The crew compartment accommodates the launch preparation and firing equipment and the operator can select single rockets or salvo firing. It is not fitted with an NBC system and as far as it is known does not have a land navigation system.

Steering is power assisted on the front and rear axles and a central tyre-pressure system is fitted as standard. This allows the driver to adjust the tyre pressure to suit the terrain being crossed without leaving his seat.

The ZIL-135LM cross-country vehicle is powered by two diesel engines each of which drive the four wheels on one side of the vehicle.

When deployed in the firing position, steel shutters are normally raised over the front windscreens and two stabilisers are lowered at the very rear to provide a more stable firing platform.

9T452 transloader vehicle for 220 mm BM 9P140 (16-round) Uragan (Hurricane) multiple rocket system showing centrally mounted crane. The rockets are not being carried 0500897

300 mm 9T234-2 Splav-M transloader vehicle from the front (Christopher F Foss) 1365042

The launcher is mounted on a turntable at the rear with powered elevation from 0° to +55° and powered traverse of 30° left and right.

There are at least five types of unguided 220 mm artillery rocket launched by the BM 9P140 system, see Table 1.

Table 1

Weight at launch	Min range	Max range
280 kg	10 km	35 km
280 kg	10 km	35 km[1]
270 kg	10 km	35 km
280 kg	10 km	35 km[1]
270 kg	10 km	35 km

Notes:
[1] No longer marketed by Splav

There is also a 9M27K fragmentation submunition cargo warhead rocket that weighs 270 kg and has a minimum range of 10 km and a maximum range of 35 km. It has 30 fragmentation submunitions.

The PGMDM scatterable anti-tank mine is also used in a number of other applications and consists of a liquid explosive charge contained in a thin flexible cover. It uses a pressure-activated fuze, which is also used in the PFM-1 anti-personnel mine.

Unconfirmed reports indicate that there was a chemical rocket for this system and RFAS reports state that a Fuel-Air Explosive (FAE) submunition rocket is also available.

The 9M59 220 mm rocket carries nine anti-tank mines and is 5.178 m long and weighs 270 kg. It is fitted with a warhead that weighs 89.5 kg with the nine anti-tank mines capable of covering an area of 250 hectares.

The anti-tank mines weigh 5 kg and are fitted with a proximity fuze. They carry a 1.85 kg explosive direction charge which has a high armour-piercing capability. The mines have a self-destruct time of 16 to 24 hours.

The 9M27K2 rocket has a 89.5 kg cluster warhead carrying 24 anti-tank mines which cover an area of 150 hectares. The PTM-1 mines have been designed to destroy the tracks of tanks and other armoured vehicles, rather than to pierce the armour of the vehicle.

These mines weigh 1.5 kg and are fitted with a 1.1 kg PW-12S high-explosive charge. The self-destruct fuze time of the PTM-1 anti-tank mine is 3 to 40 hours.

The rockets are fin- and spin-stabilised and have wraparound fins at the rear that unfold after the rockets are launched.

According to RFAS sources, each salvo has a destructive area of 20,000 to 460,000 m^2. If the rockets are launched singly then one rocket is launched every 8.8 seconds. A complete salvo of 16 rockets takes 20 seconds to launch.

Maximum time quoted for the system to be prepared for action from the travelling configuration to ready to fire is three minutes with a similar time to come out of action.

Each BM 9P140 launcher is normally accompanied by at least one and sometimes two, 9T452 transloader vehicles based on the same ZIL-135LM 8 × 8 chassis.

The transloader vehicle carries 16 reload rounds arranged in two stacks positioned one either side of the vehicle. Mounted to the immediate rear of the cab is a hydraulic crane operated by one person, which is used to resupply the rocket launcher. This crane has a maximum lifting weight of 300 kg with the operator being seated to the rear.

For reloading, the launcher has to be in the horizontal position and traversed to one side. The new rocket is then loaded into the tube via the crane and rammed home by the rammer. It is estimated that the reload time is 20 to 30 minutes, a long time when compared to the 227 mm US Lockheed Martin Missiles and Fire Control Multiple Launch Rocket System (MLRS).

Ukraine mobility upgrades for Russian rocket launchers

MRL developed by the Spav Scientific Production Concern have been built in large quantities, not only for the Russian Army but also numerous export customers as well.

The KrAZ company of the Ukraine manufactures a wide range of cross-country trucks for the civilian and military markets and some of the latter have now been modified to take the complete Russian MRL.

The 220 mm BM 9P140 (16-round) Uragan (Hurricane) is still in service with a number of countries and the complete system is also referred to by the industrial designation of the 9K57.

220 mm BM 9P140 (16-round) Uragan (Hurricane) multiple rocket launcher with launcher extended and rocket out of launcher in the foreground with fins extended (Christopher F Foss) 1333683

Production of this 8 × 8 chassis was completed many years ago and KrAZ has integrated this launcher onto the rear of a KrAZ-6322PA Bastion-03 (6 × 6) cross-country truck chassis.

Details of these upgrades for Russian rocket launchers are provided in a separate entry in *IHS Jane's Land Warfare Platforms: Artillery and Air Defence.*

Specifications

	220 mm BM 9P140 (16-Round) Uragan
Dimensions and weights	
Crew:	3
Length	
overall:	9.63 m
deployed:	10.83 m
Width	
transport configuration:	2.8 m
deployed:	5.34 m
Height	
transport configuration:	3.225 m
deployed:	5.24 m (max elevation)
Ground clearance	
overall:	0.58 m
Track	
vehicle:	2.3 m
Wheelbase	2.41 m + 1.5 m + 2.4 m
Weight	
standard:	15,100 kg
combat:	20,000 kg
Mobility	
Configuration	
running gear:	wheeled
layout:	8 × 8
Speed	
max speed:	65 km/h
Range	
main fuel supply:	570 km
Fuel capacity	
main:	768 litres
Fording	
without preparation:	1.2 m
Gradient:	57%
Vertical obstacle	
forwards:	0.6 m
Trench:	2.00 m
Turning radius:	12.5 m
Engine	2 ×, 8 cylinders, 4 stroke, petrol, 177 hp at 3,200 rpm
Gearbox	
model:	hydromechanical
Steering:	power-assisted, 1st and 4th axles
Firepower	
Armament:	16 × 220 mm rocket system
Main weapon	
maximum range:	35,000 m (est.)
Rate of fire	
rapid fire:	1/8.8 s (constant)
Operation	
stop to fire first round time:	180 s
out of action time:	180 s
Main armament traverse	
angle:	60° (±30°)
Main armament elevation/depression	
armament front:	+55°/0°

Status

Production complete. No longer marketed. It is possible that a number of surplus Russian Army systems could be available for export. In service with the following countries:

Country	Quantity	Comment
Afghanistan	n/avail	
Belarus	72	
Eritrea	9	delivered 2007
Guinea	3	delivered 1999
Kazakhstan	180	
Kyrgyzstan	6	
Moldova	11	delivered 2000
Russia	900	including 600/700 reserve
Syria	n/avail	unconfirmed user
Turkmenistan	60	
Ukraine	139	
Uzbekistan	48	
Yemen	13	delivered 1994

Contractor

Splav Scientific Production Concern (prime contractor and rocket manufacturer).

Motovilikha Plants Corporation (launcher and transloader).

Splav 122 mm 9K51 BM-21 (40-round) multiple rocket launcher upgrade

Development

The Splav State Unitary Enterprise, prime contractor for all Russian Multiple Rocket Launcher Systems (MRLSs) has developed a package of enhancements for the 122 mm 9K51 BM-21 (40-round) 'Grad' system.

The system is used by many countries around the world, with an estimated 3,000 systems built by Russia alone.

Development of this is complete but as of April 2013 there are no known exports of this system.

Description

The improvements to the 122 mm BM-21 (40-round) system include the development of a computerised fire-control system for more rapid and effective target engagement and the development of a new family of unguided 122 mm solid propellant surface-to-surface rockets.

Not only do these 122 mm rockets have a longer range, due to the use of a new solid propellant, but also more effective warheads enabling the BM-21 to engage a wider range of battlefield targets at longer ranges with increased effectiveness.

The automated fire-control system is called the Kapustnik-BM. This includes the 1V153M fire-control post mounted on a Ural-375 series (6 × 6) truck chassis, or the more recent Ural-4320 (6 × 6) truck.

This receives target information from reconnaissance means such as aircraft, helicopters and battlefield radars or a command and observation post based on a BTR-80 (8 × 8) APC designated the 1V52M.

Target information is calculated and sent to each launcher that is fitted with a Automated Targeting and Fire Control System (ATFCS). The BM-21 is then rapidly laid onto the target and the rockets launched, with the crew remaining in the cab.

According to Splav, the time taken for a 122 mm BM-21 to open fire from an unprepared firing position is between 25 and 35 minutes, with the ATFCS installed this is reduced to six minutes.

The crew of the BM-21 normally consists of three people, with ATFCS it has been reduced to two.

The standard 122 mm unguided rocket is designated the 9M22U and is fitted with a high-explosive fragmentation warhead and has a maximum range of 20,380 m. The rocket weighs 66.6 kg and contains 18.4 kg of high-explosive.

For engaging tanks and other armoured vehicles the 122 mm 9M217 rocket has been developed which has a maximum range of 30, 000 m and carries two self-targeting submunitions. Each of these has a special self-forging fragment warhead that will penetrate 60 to 70 mm of conventional steel armour. This would be used to attack the vulnerable upper surfaces of armoured vehicles.

The 122 mm 9M218 rocket contains 45 shaped charged (e.g. High-Explosive Anti-Tank) submunitions that are highly effective against light armoured vehicles. This rocket has a maximum range of 30,000 m.

The 122 mm 9M521 high-explosive fragmentation rocket is said to be twice as effective as the standard 9M22U rocket and has a maximum range of 40,000 m and carries 21 kg of high-explosive plus spherical fragments.

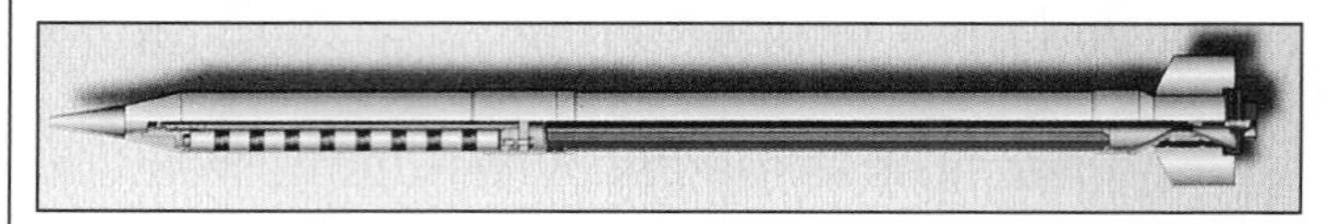

The 122 mm 9M218 rocket carries 45 shaped charge submunitions to attack the vulnerable upper surfaces of armoured vehicles 0067870

Upgraded Splav 122 mm BM-21 (40-round) multiple rocket launcher firing a surface-to-surface rocket 0569055

The 122 mm 9M522 rocket has a maximum range of 37,500 m and is fitted with a high-explosive warhead that separates from the rocket over the target and is stabilised by a parachute. This explodes over the target for maximum effectiveness.

Other types of warhead are probably available including one carrying jammers, that is believed to be designated the 9M519 and possibly a fuel air explosive rocket.

These improvements will be offered on new production 122 mm BM-21 (40-round) MRLS as well as being capable of being back fitted to older systems.

The upgraded BM-21 Combat Vehicle (CV) is called the BM-21-1 and this is supported in action by the 9T254 Transport Vehicle (TV) with the 9F37M unified set of racks.

The CV is quoted as having a combat weight of 13.7 tonnes while the TV is quoted as having a loaded combat weight of 13.6 tonnes. It takes seven minutes to manually reload the CV with 40 new 122 mm rockets.

It should be noted that production of the Ural 375 (6 × 6) cross-country truck that was used as the basis for most production BM-21 (40-round) rocket launcher has been completed.

The Ural 375 was replaced in production by the Ural 4320 (6 × 6) which have a number of improvements including increased payload.

Baseline Ural 4320-31 (6 × 6) cross-country truck has a payload of 6,500 kg and is powered by a YaMZ-238M2 diesel developing 240 hp coupled to a five speed manual transmission and a two speed transfer case. Maximum road speed is 85 km/h with a cruising range of 1,080 km at a constant speed of 60 km/h.

Variants

Upgraded BM-21 Launch Vehicle (LV)

In addition to the new generation of 122 mm rockets, the Splav Scientific Production Concern is has also developed and upgrade Launch Vehicle (LV) package. Elements of this are also included in the ATFCS.

This LV upgrade consists of the following key sub-elements:

- Layers panel in cab
- Odometer
- Satellite navigation system and associated aerial
- North seeing course and roll gyro system
- Inductive fuze setter
- Laptop computer
- Remote data input equipment.

Israel Aerospace Industries 122 mm rocket upgrade

Israel Aerospace Industries is marketing an upgrade package for the 122 mm rocket under the name of Strikes.

Details of this are provided in a separate entry in *IHS Jane's Land Warfare Platforms: Artillery and Air Defence.*

This is claimed to have a maximum range of 40 km with a circular error of probability of 10 m at any range.

As of April 2013 it is understood that this 122 mm upgrade had yet to enter production.

Ukraine mobility upgrades BM-21

The KrAZ company of the Ukraine manufactures a wide range of cross-country trucks for the civilian and military markets.

Some of the latter have now been modified to take the complete Russian 122 mm BM-21 (40-round) launcher and details of these systems are provided in a separate entry in *IHS Jane's Land Warfare Platforms: Artillery and Air Defence.*

Specifications

Table 1

Rocket	9M22U	9M217	9M218	9M521	9M522
Maximum range:	20.1 km	30 km	30 km	40 km	37.5 km
Weight of rocket:	66.6 kg	70 kg	70 kg	66 kg	70 kg
Weight of explosive:	18.4 kg	25 kg	25 kg	21 kg	25 kg

Status

Development complete. Ready for production.

Contractor

Splav Scientific Production Concern.

Splav 122 mm BM-21 multiple rocket launcher family

Development

122 mm (40-round) BM-21 Grad

The development of the 122 mm BM-21 Grad (Hail) divisional-level RSZO (*Reaktivaya Sistema Zalpovogo Ognya:* Rocket System for Salvo Fire) took place in the mid-1950s under the guidance of the Splav Scientific Production Concern at Tula.

The development phase was completed in 1958 and the BM-21 Grad entered operational service in 1963. The Grad is also known in the West by the US designation BM-21a.

BM-21 was accepted for service with the Russian Army in 1963 and is typically deployed at the divisional level.

The BM-21 was first used in combat during the 1969 Sino/Soviet border skirmishes on Damanskiy Island. It was also used in action by the Soviet Army in Afghanistan and by the Russian Army and Chechen fighters during the First and Second Chechen Wars.

The main role of the BM-21 Grad system is to support the division with suppressive fire to counter anti-tank missile, artillery and mortar positions, destroy strong points and eliminate enemy nodes of resistance on the immediate battlefield.

The BM-21 Grad is assigned to divisions as part of the divisional artillery structure in a battalion of 12 launchers (expanding to a Table of Organisation & Equipment [TOE] of 18 launchers in wartime). It may also be found at army or front (military district) level in independent regiments of three battalions each of 12 launchers (these expanding to a TOE of 18 launchers in wartime).

The BM-21 Grad system comprises the BM-21 launcher vehicle, a 9F37 ammunition resupply vehicle and the 9M22U index rocket (also designated M-21-OF). Scale of issue of the associated resupply vehicle depends on a number of factors. In time of war it would be one per launcher.

122 mm BM-21 (40-round) MRS of the Polish Army deployed in the firing position with gunner using sight (Michael Jerchel) 0105895

Russian 122 mm BM-21 (40-round) multiple rocket launchers deployed in the firing position and showing part of the 122 mm launcher which has four rows each of ten 122 mm rocket launcher tubes (INA) 1452942

Ural-4320 (6 × 6) truck used for the 122 mm BM-21 (40-round) MRS of the Russian Army (Stefan Marx) 1124713

While the prime contractor for the overall 122 mm BM-21 multiple rocket system is the Splav Scientific Production Concern based in Tula, the Motovilkha Plants Corporation based in Perm is responsible for the actual launcher mechanism and in 1998 stated that it had manufactured over 2,000 BM-21 series systems for the home and export markets.

In addition to being manufactured in Russia, a number of other countries have designed and manufactured systems with similar capability, usually on local chassis.

According to UN Report on Conventional Arms for 2001, during the period 1992 through to 2010, Russia did not export any of these systems, although parts of the former USSR did:

From	To	Quantity	Delivery
Belarus	Congo (DR)	6	1997
Belarus	Côte d'Ivoire	6	2003
Belarus	Sudan	2	2003
Czech Republic	Ukraine	21	2009
Czech Republic	US	5	2008
Hungary	Czech Republic	20	2006
Hungary	Czech Republic	5	2009
Kazakhstan	Angola	4	1998
Kazakhstan	Congo (DR)	2	1999
Kazakhstan	Russia	2	1999
Kazakhstan	Congo (DR)	1	2000
Ukraine	Azerbaijan	12	2010
Ukraine	Czech Republic	12	2010
Ukraine	Macedonia	6	2001
Ukraine	Israel	1	2008
Ukraine	Kenya	6	2008
Ukraine	Kenya	5	2007
Ukraine	US	4	2010

In addition, there were a number of unidentified transfers of undisclosed rocket systems, some of which could be 122 mm BM-21 type systems:

From	To	Quantity	Delivery
Russia	Angola	14	1993
Turkmenistan	Russia	54	1996

As of April 2013 Russia was still marketing the 122 mm (40-round) BM-21 multiple rocket launcher with the latest customer being Venezuela who took delivery of 24 systems in 2012/2013 together with a batch of 12 300 mm (12-round) Smerch systems.

Description

The first production BM-21 Grad RSZO launcher was originally based on a Ural-375D (6 × 6) 4,000 kg cross-country truck chassis and is fitted with a 40-round elevatable launcher assembly on its rear decking. The rocket tube arrangement is a single rectangular bank of four layers each with 10 tubes. Before firing, two stabilisers are lowered to the ground; these are positioned one either side at the rear of the vehicle. The launch vehicle's cabin accommodates the launch preparation and firing equipment. Individual round, selective ripple or salvo firing is possible from either the cab or from a remote-control unit connected to the vehicle via a 60 m cable.

Launcher has powered elevation and traverse with a manual backup mode as a reserve. Aiming speed in elevation under power control is not less than 4 °/s and in traverse is not less than 7 °/s.

The resupply vehicle is based on the same chassis and is designated the 9F3700 rack set.

Variants

Splav 122 mm BM-21 (36-round) Grad-1 multiple rocket launcher

At the same time as the BM-21 Grad was deployed as a divisional system, a requirement was drawn up for a regimental MRL artillery system for issue to the independent six-launcher MRL batteries assigned to selected tank and motorised rifle regiments. The BM-21 Grad-1 launcher entered service in 1976 with the Russian military index number 9P138. It is also known by the US designation BM-21b and the NATO STANAG designation BM-21 M1976. The BM-21 Grad-1 system was subsequently produced on the MT-LB tracked chassis.

122 mm BM-21b (36-round) multiple rocket launcher on ZIL-131 (6 × 6) 3,500 kg truck chassis from the rear in travelling position 0130020

The BM-21 Grad-1 system comprises the 9P138 launcher, the 9T450 transport vehicle, 9F380 ammunition resupply vehicle and the 122 mm 9F28F rocket with HE warhead.

The associated 122 mm rockets were also modernised. The standard HE fragmentation round was produced with a pre-fragmented warhead, while the incendiary round was modified to carry 180 incendiary elements as part of its payload.

This is also mounted on a 6 × 6 chassis with the launcher mounted at the rear with a traverse of 60° left and 60° right with a maximum elevation of +55°.

Combat weight of the system is 13,845 kg and has an overall length of 7.34 9 m, width of 2.43 m and a height of 2.68 m. All of these figures relate the system in the travelling configuration.

Maximum road speed is 85 km/h and maximum road range is being quoted as 1,040 km.

This version is known to be in service with the following countries:

Country	Quantity	Comment
Belarus	5	
Russia	420	Army
	96	Naval infantry
Turkmenistan	9	
Ukraine	20	
Uzbekistan	24	

122 mm BM-21M or BM-21b

This is a modified version of the 122 mm BM-21 (40-round) multiple rocket launcher family based on a ZIL-131 (6 × 6) 3,500 kg truck chassis, with the complete system also being referred to as the BM-21b.

The main difference is that the BM-21b has had the two centre tubes in each of the lower banks removed to give a total of 36 rather than the normal 40 tubes.

This has allowed the installation of a new elevation mechanism. A quantity of these was supplied to the Iraqi Army and some of these were subsequently captured in 1991. These vehicles were fitted with a thermal shield over the rear part of the rocket launcher to prevent excessive solar heating of the rocket tubes.

This has a total of 36 × 122 mm launcher tubes and the rear mounted launcher has a traverse of 75° left and 106° right with a maximum elevation of +55°.

A full salvo of 36 rockets can be fired in 18 seconds after which the launcher would re-deploy to another firing/reloading area to avoid counterbattery fire.

Time into action is being quoted as 2.5 minutes, time out of action 30 seconds and to reload the launcher seven to eight minutes using the five person crew.

Weight fully loaded is 10,500 kg with the complete system having an overall length of 6.9 m, width of 2.5 m and a height of 2.48 m.

It is powered by a ZIL-131 V-8 water-cooled diesel developing 150 hp at 3,200 rpm coupled to a manual transmission with gives a maximum road speed of 80 km/h. The 340-litre fuel tank gives a maximum range of up to 525 km.

Variants

A number of countries have produced direct copies or modified versions of the BM-21 family and its 122 mm rockets. These include China, Croatia, Egypt, India, Iran, North Korea, Pakistan, Romania, Slovakia, and Sudan (see relevant entries in *IHS Jane's Land Warfare Platforms: Artillery and Air Defence*). All of these countries claim that their 122 mm rockets are fully compatible with the original Russian system.

Most of these foreign 122 mm rockets have the same ballistics as the original Russian rockets rather than the more recent 122 mm rockets available from Russia which have a longer range and are available with different types of warhead.

Specifications

	Grad BM-21 Ural 375D
Dimensions and weights	
Crew:	6
Length	
overall:	7.35 m
Width	
transport configuration:	2.4 m
deployed:	3.1 m
Height	
transport configuration:	3.09 m
deployed:	4.35 m (max elevation)
Weight	
standard:	10,870 kg
combat:	13,800 kg
Mobility	
Configuration	
running gear:	wheeled
layout:	6 × 6
Speed	
max speed:	75 km/h
Range	
main fuel supply:	1,000 km
Fuel capacity	
main:	360 litres
Fording	
without preparation:	1.5 m
Gradient:	60%
Vertical obstacle	
forwards:	0.65 m
Trench:	0.875 m
Turning radius:	10.5 m
Engine	ZIL-375, 7 litre, V-8, water cooled, petrol, 180 hp at 3,200 rpm
Gearbox	
type:	manual
forward gears:	5
reverse gears:	1
Transfer box:	2-speed
Steering:	double-thread worm, hydraulic booster
Clutch	
model:	twin dry disc
type:	friction clutch
Brakes	
main:	air/hydraulic
parking:	mechanical
Tyres:	14.00 × 20
Suspension:	(front) longitudinal semi-elliptical springs with hydraulic shock-absorbers; (rear) bogie with longitudinal semi-elliptical springs
Electrical system	
vehicle:	12 V
Firepower	
Armament:	40 × 122.4 mm rocket system
Rate of fire	
rapid fire:	40/20 s
Operation	
stop to fire first round time:	150 s
out of action time:	30 s
reload time:	420 s
Main armament traverse	
angle:	172° (102° left, 70° right)
Main armament elevation/depression	
armament front:	+55°/0°

Status

Production as required. In service with following countries:

Country	Quantity	Comment
Afghanistan	n/avail	current status uncertain
Algeria	48	
Angola	50	including 4 delivered in 1998
Armenia	47	
Azerbaijan	43	
Belarus	126	
Bosnia	4	
Bulgaria	124	
Burundi	12	
Cameroon	20	
Cambodia	8	
Congo	10	
Côte d'Ivoire	6	from Belarus in 2003
Croatia	31	
Cuba	150	estimate
Cyprus	4	
Democratic Republic of Congo	30	includes 6 delivered in 1997
Ecuador	18	
Egypt	300	including local models
Eritrea	50	
Ethiopia	50	
Georgia	16	
India	150	
Iran	100	estimate
Israel	50	
Kazakhstan	100	
Kenya	11	all in 2007/2008
Korea, North	n/avail	
Kyrgyzstan	15	
Lebanon	25	
Libya	236	owing to the conflict in 2011 the quantities and status of these launchers in uncertain
Macedonia	6	from Ukraine in 2001
Mali	2	
Mongolia	130	
Montenegro	n/a	
Morocco	35	
Mozambique	20	
Myanmar	n/avail	
Namibia	5	
Nicaragua	18	
Pakistan	50	
Peru	22	
Poland	75	
Russia	2,500	also coastal defence
Serbia	n/avail	
Seychelles	2	
Somalia	n/avail	status uncertain
Sudan	90	
Sudan, South	15	
Syria	280	
Tajikistan	3	
Tanzania	58	
Turkmenistan	56 + 9	9P138
Uganda	12/30	
Ukraine	315 + 20	9P138
Uzbekistan	36 + 24	9P138
Venezuela	24	delivery 2012/2013
Vietnam	350	
Yemen	280	
Zambia	30	

Contractor

Splav Scientific Production Concern (prime contractor and rockets).
Motovilikha Plants Corporation (rocket launcher).

122 mm (12-round) Grad-V multiple rocket launcher

Development

At the same time as the BM-21 Grad 122 mm (40-round) multiple rocket launcher received its baptism of fire during the 1969 Sino-Soviet border skirmishes, the Splav Concern was given a requirement from the VDV headquarters for a lightweight MRL system that could be transported by aircraft and dropped with troops via a parachute and carriage platform to make a soft landing.

The result was the Grad-V or BM-21V (with the V standing for *Vozdushnodesantii*, or airborne) system.

This first entered service with the Russian army in 1967. As far as it is known, this 122 mm (12-round) Grad-V system has never been offered on the export market.

There has been no production of this 122 mm (12-round) Grad-V multiple rocket launcher for many years.

It is possible that small numbers of this system may still be used by Russia or held in reserve.

Description

The Grad-V has also been referred to in the West as the BM-21 M-1975, as this was the year that it was first observed by Western sources. It is fitted with two tiers of six 122 mm tubes mounted on the lower chassis of the reduced weight airborne forces GAZ-66B (4 × 4) 2,000 kg truck.

122 mm (12-round) Grad-V multiple rocket launcher on GAZ-66 chassis (Splav) 1328079

The truck has a collapsible canvas cab, removable doors and windscreen, telescopic steering wheel and tie-down points for aiding parachute dropping and landing.

The launcher assembly is traversed towards the cab for travelling and for firing requires the two stabilising jacks at the rear of the vehicle to be lowered.

The rockets can be either fired singly or in a salvo lasting six seconds. Manual reloading by the two-man crew takes approximately five minutes. Launcher traverse is 70° left and right. The system is limited to firing HE fragmentation-type rockets only.

The BM-21V assembly has also been mounted on a tracked chassis, believed to be a variant of the BMD series airborne combat vehicle.

As production of the original GAZ-66 (4 × 4) vehicle was completed many years ago it is considered possible that the remaining 122 mm (12-round) Grad-V multiple rocket launchers could have been installed on a more recent 4 × 4 cross-country chassis.

Specifications

	122 mm (12-Round) Grad-V
Dimensions and weights	
Crew:	3
Length	
overall:	5.655 m
Width	
transport configuration:	2.4 m
Height	
transport configuration:	2.44 m
Weight	
combat:	6,000 kg
Mobility	
Configuration	
running gear:	wheeled
layout:	4 × 4
Speed	
max speed:	85 km/h
Range	
main fuel supply:	875 km
Fuel capacity	
main:	210 litres
Engine	ZMZ-66, 4.254 litre, V-8, water cooled, petrol, 115 hp at 3,200 rpm
Firepower	
Armament:	12 × 122.4 mm rocket system
Rate of fire	
rapid fire:	12/6 s
Operation	
stop to fire first round time:	210 s
out of action time:	30 s (est.)
reload time:	540 s

Status

Production complete. In service with Russia and possibly other countries.

Contractor

Splav Scientific Production Concern.

Serbia

Yugoimport Kosava (Whirlwind) rocket system

Development

In mid-2007, Yugoimport – SDPR revealed that it had developed to the prototype stage a rocket system called Kosava (Whirlwind).

TAM-110 (6 × 6) chassis fitted with three rail launch for Kosava (Whirlwind) laser-guided rocket system (Miroslav Gyurosi) 1290438

Artists impression of the Kosava (Whirlwind) rocket which mates the LVB-250F laser-guided bomb with 262 mm calibre Orkan rocket (Yugoimport) 1290437

This has been developed as a private venture for the export market. As of April 2013 it is understood that this system remained at the prototype stage.

While some elements of the Kosava (Whirlwind) were shown in mid-2007 it is not certain if an all-up system (launcher and complete rocket) has been tested.

Description

Kosava (Whirlwind) consists of a modified TAM 110 (6 × 6) forward control truck chassis on the rear of which are fitted a three rail beam-type launcher.

Each rail can launch a new tactical surface-to-surface guided rocket which consists of a solid propellant rocket motor mated with a semi-active laser-guided bomb.

The latter is the LVB-250F laser-guided bomb developed by Military Technical Institute VTI in Belgrade, Serbia. This is 3.1225 m long, 325 mm in diameter, has a wingspan if 1.315 m and has a launch weight of 310 kg.

This laser guided bomb is mated to the front end of the solid propellant rocket motor already developed for the 262 mm (12-round) LRSV M-87 'Orkan' multiple rocket launcher system.

This was originally developed by the former Yugoslavia (now Serbia) in conjunction with Iraq. Details of this system are provided in a separate entry in *IHS Jane's Land Warfare Platforms: Artillery and Air Defence*.

It is understood that the laser guided bomb separates from the 262 mm rocket motor towards the end of its flight. It is claimed to have a maximum range of 26 km, 120 second time of flight along an apogee of 4 km.

Targets for the Kosava (Whirlwind) would have to be laser designated and then acquired by the nose mounted guidance system of the laser guided bomb at a range of up to 10 km from target impact. Terminal velocity is being quoted as 180 m/s and a Circular Error of Probability (CEP) of 5 m is being claimed.

A number of improvements are being studied in order the increase the operational range of the system. These include using a new solid propellant rocket motor with a higher specific impulse and a weight reduction programme for the original laser guided bomb.

In addition to the twin round launcher, other elements of the Kosava (Whirlwind) rocket system include a command post vehicle and a logistic support vehicle carrying additional rockets. The latter is based on the FAP 2026 (6 × 6) cross-country chassis and has rails for five Kosava (Whirlwind) rockets.

The FAP 2026 (6 × 6) country truck chassis is also used as the basis for the 128 mm (32-round) M-77 'Oganj' multiple rocket system covered in detail in a separate entry in *IHS Jane's Land Warfare Platforms: Artillery and Air Defence.* Production of this system is complete.

The TAM 110 (6 × 6) cross-country chassis is used for a number of other applications including cargo truck, towing artillery as well as the launch platform for the 128 mm (32-round) Plamen-S multiple rocket launcher system which is covered in a separate entry in *IHS Jane's Land Warfare Platforms: Artillery and Air Defence.*

It is understood that production and marketing of the 128 mm (32-round) Plamen-S multiple rocket launcher system is now complete.

Specifications

	Kosava
Dimensions and weights	
Length	
overall:	6.145 m
Weight	
launch:	610 kg
Performance	
Range	
max:	26 km

Status
Prototype system. Not yet in production or service.

Contractor
Yugoimport - SDPR.

Yugoimport 262 mm (12-round) LRSV M-87 'Orkan' multiple rocket launcher system

Development
Late in 1988, at a military exhibition in Iraq, the Iraqi army displayed a 262 mm 12-round multiple rocket launcher system known as the Ababeel 50.

Examination of available evidence indicated that the system was Yugoslav in origin and mounted on the rear of the Yugoslav FAP 3235 heavy-duty (8 × 8) cross-country truck.

Since then further information, gained at the Iraqi's First Baghdad International Exhibition for Military Production in 1989, has shown that it was originally a joint Iraqi-Yugoslav project with Iraq contributing to the development and producing some of the parts locally.

Recent information is that a small production batch of about 10 systems was completed, with most of these systems remaining in the former Yugoslavia. As far as it is known the system was not used in an operational form by Iraq.

Description
A battery of M-87 262 mm (12-round) launchers would typically consist of:
- four 8 × 8 launchers
- four 8 × 8 resupply vehicles (each with 24 reload rockets)
- one 8 × 8 command post vehicle
- two 4 × 4 topographic survey light vehicles
- two 4 × 4 observation post light vehicles
- one 4 × 4 meteorological survey vehicle.

The command post vehicle is fitted with a German Teldix land navigation system. The resupply and command post vehicles are both on the same 8 × 8 chassis as the launcher vehicle and have a similar tarpaulin system to make identification of the launcher difficult.

Loading of the launcher is performed by a semi-automatic system with the preparations to fire taking two minutes. Firing of the rockets can be either singly or in a ripple mode by means of an electronic trigger in the vehicle cab. If needed, firing can also be done up to 20 to 50 m off-vehicle by means of a remote-control box with attached cable. All the traverse and elevation functions are automatically performed via control units with manual systems available as a back-up.

In Serbian Army service the launcher is fitted with a 12.7 mm heavy AA machine gun on the cab roof and four 82 mm smoke dischargers integrated either side in the front bumper.

The five-man crew has two 7.62 mm fully automatic rifles, two 7.62 mm semi-automatic rifles and a 90 mm RBR-90 hand-held anti-tank unguided rocket launcher for self-defence.

Russian FROG-7 TEL fitted with four 262 mm launcher tubes (Yugoimport) 1123830

262 mm (12-round) LRSV M-87 Orkan MRLS deployed in firing position 0500679

The 262 mm unguided rocket, known as the M-87, is 4.656 m long and uses a two-stage solid-propellant rocket motor. The booster stage contains 10 kg of fuel and burns for 200 ms to generate 8,000 kg of thrust to lift the rocket clear of the launcher tube. Once this happens the main sustainer motor cuts in with its 130 kg fuel load to give 18,000 kg of thrust during its 5 second burn time. This gives the rocket a maximum velocity of 1,200 m/s.

Flight time to the maximum range of 50,000 m is 110 seconds with the highest possible altitude reached, depending upon launch angle, being 22,000 m. By opening and closing the four available aerodynamic brake assemblies one of four different ballistic range trajectories can be chosen: 24,000, 28,000, 37,000 or 50,000 m.

A TV camera made by Bosch of Germany is mounted on top of the left side of the launcher. It is used in a similar manner to the German 110 mm LARS (which has now been phased out of service) in that a single rocket is launched and at some point in its flight is detonated so as not to alert the target area. The camera monitors this and the data collected is used to calculate the correct firing solution so that the remainder of the battery or vehicle salvo can be fired.

The APHE rocket has a launch weight of 389 kg with the warhead having a weight of 91 kg.

The anti-tank cluster warhead contains 24 KB-2 anti-tank mines and has a launch weight of 381 kg with the actual warhead weighing 83 kg.

Rocket velocity is stated to be 1,200 m/s and dispersion at the maximum range is quoted as 200 m in range and 1 m to 5 m in azimuth.

Typical rate of fire is one rocket every 2.3 to 4 seconds.

Several types of warhead can be fitted:
- 91 kg unitary APHE. An inertial fuze with instantaneous or delay action is fitted
- 91 kg cluster munition with a payload of 288 HEAT fragmentation bomblets. The shaped charge equipped bomblet also has some 420 steel spheres to enhance the fragmentation effect. Lethal radius is approximately 10 m and conventional armour plate penetration is claimed to be 60 mm plus. A typical dispersion pattern for a warhead at payload release height would be an ellipse 180 × 165 m. An alternative warhead with 420 anti-personnel submunitions has also been mentioned
- 83 kg cluster munition with a payload of 24 small cylindrical shaped-charge anti-tank KB-2 mines, each fitted with four flip-out curved vanes to stabilise their flight to the ground and ensure correct orientation. The mine can penetrate up to 40 mm plus of conventional steel armour plate.

Both cluster munition warheads are fitted with an electronic time fuze for bomblet/mine delivery.

The vehicle is fitted with a central tyre pressure regulation system which is operated by the driver from within the cab and to provide a more stable firing platform four hydraulically operated stabilisers are lowered to the ground, by remote control, one either side to the rear of the second roadwheel and two at the very rear. When travelling, the launcher is traversed to the front and the whole launcher is covered by a tarpaulin cover with integral bows.

The FAP 3235 (8 × 8) heavy-duty chassis has also been used for a number of other applications.

262 mm (12-round) LRSV M-87 Orkan MRLS in travelling configuration with bows and tarpaulin cover stowed to immediate rear of cab 0500678

Variants

Resupply vehicle

There was also an 8 × 8 resupply vehicle on a similar chassis that carried a total of 24 262 mm rockets. This is understood to have reached the prototype stage.

It is understood that production of the FAP 3235 (8 × 8) chassis is now complete.

If any future production of this system was undertaken then it is expected that it would be integrated onto a more recent 8 × 8 cross-country chassis.

Serbia has also developed and placed in production the NORA B-52 155 mm/52-calibre self-propelled artillery system which uses a Russian KamAZ-63501 (8 × 8) chassis and details of this are provided in a separate entry in *IHS Jane's Land Warfare Platforms: Artillery and Air Defence.*

This, or a similar chassis, could be used as the basis for any new production of the 262 mm (12-round) LRSV M-87 'Orkan' multiple rocket launcher system.

LRSV M-1996 Orkan-2

Yugoimport SDPR, the marketing organisation for all defence equipment being manufactured in Serbia, has released some details of another application for some elements of the 262 mm (12-round) LRSV M-87 'Orkan' multiple rocket launcher system.

This is designated the LRSV M-1996 Orkan-2 and a small batch is currently in service with Serbia. It consists of the Russian FROG-7 (8 × 8) launcher, with the original elevating launch rail fitted with four launch tubes from the in-service 262 mm (12-round) LRSV M-87 'Orkan' system. It is assumed that the rockets launched by this system are the same as those used in the original system.

At this stage it is not certain as to whether or not the system could be switch back to transporting and launching the FROG-7 unguided surface-to-surface rocket.

Specifications

	262 mm (12-Round) LRSV M-87 Orkan
Dimensions and weights	
Crew:	5
Length	
overall:	9.00 m
Width	
overall:	2.64 m
Height	
overall:	3.84 m
Ground clearance	
overall:	0.38 m
Weight	
combat:	32,000 kg
Mobility	
Configuration	
running gear:	wheeled
layout:	8 × 8
Speed	
max speed:	80 km/h
Range	
main fuel supply:	600 km
Fording	
without preparation:	1.2 m
Gradient:	60%
Side slope:	25%
Trench:	1.8 m
Turning radius:	13 m
Engine	OM422, turbocharged, diesel, 354 hp
Electrical system	
vehicle:	24 V
Firepower	
Armament:	12 × 262 mm rocket system
Main weapon	
maximum range:	50,000 m
Rate of fire	
rapid fire:	1/2.3 s (est.)
Operation	
stop to fire first round time:	120 s
out of action time:	60 s
Main armament traverse	
angle:	220° (110° left and right)
Main armament elevation/depression	
armament front:	+56°/0°

Status

Production complete. In service with Bosnia-Herzegovina (5), Croatia (2) and Serbia. Recent information has indicated that limited marketing of this system has started again under the leadership of Bosnia-Herzegovina. If this system was placed back in production it would probably be on a different chassis.

According to the United Nations, in 2006 Bosnia-Herzegovina transferred 500 of these 262 mm rockets to Georgia.

Contractor

Yugoimport - SDPR.

Yugoimport Multitube Modular Rocket Launcher

Development

Yugoimport of Serbia has added to is weapons portfolio of unguided surface-to-surface rocket launchers with the development of the self-propelled Multitube Modular Rocket Launcher (MMRL).

The main advantage of this system is that the pods of unguided rockets fired can be tailored to meet the target set to be engaged.

It should be noted that the MMRL has also been referred to as the LRSVM Self-propelled MMRL.

Description

The first example of this is integrated onto the rear of a forward control 4 × 4 cross-country truck chassis that is also provided with central tyre pressure regulation system that allows the driver to adjust the tyre pressure to suit the ground being crossed.

To the immediate rear of the cab are stowed are bows and a tarpaulin cover mounted on rails which can be rapidly deployed to the very rear of the vehicle to cover the launcher and this makes the system indistinguishable from a conventional cargo truck.

Mounted on the rear of the chassis is the turntable mounted rocket launcher that can be fitted with pods of various types of rocket according to the range and type of target to be engaged.

The powered launcher can be fitted with two pods of rockets with the 107 mm pods having 24 tubes per pod, 128 mm pods having 16 tubes per pod and the 122 mm Grad/128 mm Oganj M77 having 12 tubes per pod.

It should be noted that the 122 mm and 128 mm rockets are being marketed by Yugoimport with the 122 mm rockets having enhanced ranges when compared to their original Russian versions.

The launcher has a computerised Fire Control System (FCS) which includes an Inertial Navigation System (INS) for increased accuracy backed up by a Global Positioning System (GPS) and a standard optical sighting system as a reversionary mode.

The onboard FCS allows the system to be deployed as an autonomous unit or as part of a battery with four or six launchers.

When deployed in the firing position a total of four hydraulically operated stabilisers are lowered to the ground to provide a more stable firing platform and when being deployed these also compensate for the slope of the terrain.

Two of these stabilisers are positioned at the very rear of the chassis and the other two one either side of the front wheels.

According to Yugoimport, the system takes onto 45 seconds to come into action and launch its first rockets and once the rockets have been launched it takes about 30 seconds to come out of action.

After a target has been engaged the MMRL would normally rapidly redeploy to another fire position to avoid counter battery fire.

According to Yugoimport, the MMRL would normally be deployed in a battery of four or six launchers plus associated command-and-control and resupply vehicles.

Serbian Multitube Modular Rocket Launcher deployed in the firing position with stabilisers lowered to the ground. The launcher is traversed to the right and has two pods each of 12 × 122 mm unguided rockets (Yugoimport)
1452678

Rear view of Serbian Multitibe Modular Rocket Launcher with stabilisers raised and two pods of 12 × 122 mm unguided rockets traversed to the right (Yugoimport)
1452679

	M07	Grad-M	Grad-2000	Plamen A	Plamen D	Ojanj M77
Dimensions and weights						
Diameter						
body:	107 mm	122 mm	122 mm	128 mm	128 mm	128 mm
Performance						
Range						
max:	11.5 km	28 km	40 km	8.6 km	12.6 km	20.6 km

Serbian Multitube Rocket Launcher in travelling configurations and with rocket launcher covered by bows and a tarpaulin cover to make it indistinguishable from a conventional cargo truck (Yugoimport) 1452680

The MMRL would be supported by a dedicated rocket resupply vehicle fitted with a hydraulic loading crane which would carry pods of new rockets which could be rapidly loaded.

Although the first example of the MMRL integrated onto the rear of a 4 × 4 chassis its design is such that it can be integrated onto the other chassis, tracked and wheeled.

Specifications

Due to the modular nature of the Multitube Modular Rocket Launcher, detailed specifications are not available.

Status

Development complete. Ready for production on receipt of orders.

Contractor

Yugoimport - SDPR.

Yugoimport 128 mm (32-round) M-77 'Oganj' multiple rocket system

Development

The M-77 128 mm (32-round) 'Oganj' (Fire) multiple rocket system was developed in the early 1970s to meet the operational requirements of the former Yugoslav Army. It is normally found in batteries of six launchers. It performs a similar role to the former Czech 122 mm RM-70 (40-round) MRS.

It does not, however, have an armoured cab or the excellent cross-country performance of the Czech vehicle, which is based on the Tatra 813 (8 × 8) truck chassis.

First major production version of M-77 Oganj 128 mm MRL on rear of FAP 2026 BDS/AV (6 × 6) truck chassis, tarpaulin lowered showing bank of 32 128 mm rockets behind cab and rocket launcher at the rear. The stabilisers are mounted at the very rear of the vehicle and to the immediate rear of the front roadwheels 0500680

Production of this is now completed and it is not expected that it will enter production again in the future.

According to the United Nations Arms Transfer List for 2010, Bosnia transferred 10 × 128 mm launchers and 20,000 × 128 mm rockets to Azerbaijan. It is not known what version these were.

Description

The M-77 128 mm MRS basically consists of a modified FAP 2026 BDS/AV (6 × 6) series cross-country truck with a 32-barrel 128 mm rocket launcher mounted at the immediate rear of the chassis and an additional pack of 32 rockets to the rear of the cab for rapid reloading.

The launcher fires a 128 mm calibre, 2.6 m long unguided solid-propellant rocket to a maximum range of 20,600 m.

Weight of the rocket is 67 kg. The 19.5 kg warhead contains 2,562 steel spheres 6.35 mm in diameter and 930 steel spheres of 10.32 mm diameter that are lethal out to a radius of over 30 m. The explosive content is 3.8 kg and a fuze with pre-selected super quick, inertia and delay action options is fitted as standard.

The rockets are either launched singly or in a ripple salvo, the latter taking a total of 25.6 seconds. The rockets can be launched either by manual, semi-automatic or fully automatic control with an optional 25 m off-vehicle remote-control box cable system. Once the 128 mm rockets have been launched, the launcher is traversed to the rear and depressed to the horizontal, the pack of 32 rockets to the rear of the cab is raised until it lines up with the launcher and the rockets are pushed into it. The reloading system then goes back onto the floor of the truck and the launcher is traversed forwards ready for firing again.

The complete reloading sequence takes two minutes with the second salvo capable of launch within five minutes of the first. Reloading can also be performed manually. When travelling the launcher is lowered and the launcher and reloading pack are covered by a tarpaulin that is folded up under the rear in the firing position. The time required to occupy or leave a firing position is less than 20 seconds, with the time taken for occupying a position, firing both rocket loads and leaving being four to five minutes in total.

It is believed that the original FAP-2220 BDS (6 × 4) chassis was used only for early production vehicles as it is not suitable for travelling across very rough country because of its low ground clearance.

The chassis was then changed to the FAP 2020 BS (6 × 6) and FAP-2026 BDS/AV (6 × 6) cross-country trucks. These vehicles have large tyres with single rather than dual rear wheels and are fitted with a central tyre pressure-regulation system that allows the driver to adjust the ground pressure to suit the type of ground being crossed. Recent information has indicated that production of this system and its chassis was completed some time ago.

Variants

M-77 Oganj Mod

This is a Bosnian version which, instead of having 32 × 128 mm rocket tubes, has 2 × 4 128 mm barrels and 2 × 2 122 mm barrels. The latter can launch standard 122 mm unguided rockets of Russian or equivalent types which are widely available.

128 mm single-tube launcher

To complement the M-77 there is a 128 mm lightweight single-tube rocket launcher for use by commando, special forces and territorial defence units.

In addition to being used by Serbia, this 128 mm single-round launcher is known to be used by Macedonia (60) and Slovenia (56).

M-77 Oganj 128 mm MRL on rear of FAP 2026 BDS/AV (6 × 6) truck chassis with tarpaulin covering launcher 0500682

128 mm lightweight single-tube rocket launcher in firing position, showing bipod and baseplate 0500681

Basic specifications of the M-77 Oganj 128 mm single-round launcher include a weight of 50 kg, traverse of 10° left and right and elevation limits from +7° to +48°.

Maximum range of the 122 mm rockets is being quoted as 20,600 m and firing of the launcher is performed using a remote hand held controller and cable at a maximum distance of 25 m.

RH ALAN 122 mm MRS

The Croatian company of RH ALAN has marketed a very similar system to the M-77 Oganj but based on a more mobile Tatra (8 × 8) chassis called the 122 mm (32-round) M96 Typhoon. In addition to having greater mobility it has the added advantage of firing the more widely used Russian family of 122 mm unguided artillery rockets. As far as it is known the Croatian system has not been exported. Full details are provided in a separate entry in *IHS Jane's Land Warfare Platforms: Artillery and Air Defence.*

LRSVM Modular rocket launcher system

This can be fitted with pods of 128 mm, 122 mm or 107 mm rockets and details are provided in a separate entry in *IHS Jane's Land Warfare Platforms: Artillery and Air Defence*.

LRSV M-91 Vulkan

This appears to be a modified version of the M-77 Oganj 128 mm system. It is believed that seven systems are in service with the HV.

Specifications

	128 mm (32-Round) M-77 Oganj
Dimensions and weights	
Crew:	5
Length	
overall:	8.4 m
Width	
transport configuration:	2.49 m
Height	
transport configuration:	3.1 m
Track	
vehicle:	2.02 m
Wheelbase	3.4 m + 1.4 m
Weight	
combat:	22,400 kg
Mobility	
Configuration	
running gear:	wheeled
layout:	6 × 6
Speed	
max speed:	80 km/h
Range	
main fuel supply:	600 km
Engine	V-8, direct injection, water cooled, diesel, 256 hp
Tyres:	15.00 × 21
Firepower	
Armament:	32 × 128 mm rocket system
Main weapon	
maximum range:	20,600 m
Operation	
reload time:	120 s
Main armament traverse	
angle:	185° (left and right)
Main armament elevation/depression	
armament front:	+50°/0°

Status

Production complete. In service with Bosnia-Herzegovina (12), Croatia (190) and Serbia (60).

Contractor

Yugoimport - SDPR.

Yugoimport 128 mm (32-round) Plamen-S multiple rocket launcher

Development

The 128 mm (32-round) Multiple Rocket Launcher (MRL) has in the past also been referred to as the M-85.

This system is an upgraded version of the trailer-mounted 128 mm (32-round) M-63 multiple rocket system integrated onto the rear of a 6 × 6 chassis for greater tactical mobility. It also carries 16 reserve rockets which are reloaded manually.

Production of this system is now complete and it is no longer being marketed.

Description

It consists of a TAM (6 × 6) unarmoured cross-country chassis with a forward control cab, which is fitted with a canvas roof that can be folded to the rear.

Mounted at the very rear of the chassis is the 128 mm (32-round) launcher that has four layers each of eight rockets, which is mounted on a turntable. The elevation and traverse controls are mounted on the left side of the mount, as is the sighting system.

In addition to the 32 × 128 mm rockets in the ready-to-launch position there is a rack of 16 reserve 128 mm rockets to the immediate rear of the cab, which can be covered by a hinged tarpaulin that folds up horizontally when not required.

To provide a more stable firing platform two hydraulic stabilisers are deployed to the ground, positioned one either side to the rear of the last road wheel.

Although the launcher has a total traverse of 360°, it would normally fire to the left or right as the cab is not protected.

When compared to the standard 128 mm (32-round) M-63 towed Plamen MRS, covered in detail in a separate entry in *IHS Jane's Land Warfare Platforms: Artillery and Air Defence*, this 6 × 6 self-propelled system has a greater survivability as it can come into action and start firing its rockets within 30 seconds. Once a fire mission has been carried out, it would rapidly redeploy to another position to help avoid counterbattery fire.

According to Yugoimport, the 32 128 mm rockets can be fired in 6.4, 12.8 or 19.2 seconds with the range depending on the type of unguided rocket.

The rockets can be launched from within the cab or up to 25 m from the vehicle using a remote-control firing device.

Ranges of 8.6 to 13 km are currently being quoted. The standard unguided solid-propellant rocket weighs 23.1 kg and is fitted with a high-explosive warhead. This rocket is the same as that used with the towed 128 mm (32-round) M-63 Plamen system and has a maximum range of 8,545 m.

The more recent M-87 improved rocket weighs 25.5 kg and is fitted with a high-explosive warhead and has a maximum range of 13,000 m.

Variants

The TAM (6 × 6) unarmoured chassis is also used as the basis for the Serbian Kosava rocket system covered in a separate entry.

Available details of this are provided in a separate entry in *IHS Jane's Land Warfare Platforms: Artillery and Air Defence*. As of April 2013 it is understood that this system remained at the prototype/development stage.

Self-propelled Multi-tube Modular Rocket Launcher

This is a recent development by Serbia and is a 4 × 4 cross-country truck chassis with a forward control cab.

128 mm (32-round) Plamen-S multiple rocket launcher deployed in the firing position (Yugoimport) 1043237

On the rear of this is a powered launcher which can be fitted with pods of 107 mm, 122 mm or 128 mm rockets. Details of this system are provided in a separate entry in *IHS Jane's Land Warfare Platforms: Artillery and Air Defence*.

Specifications

	128 mm (32-Round) Plamen-S
Dimensions and weights	
Crew:	3
Weight	
combat:	9,600 kg
Mobility	
Configuration	
running gear:	wheeled
layout:	6 × 6
Firepower	
Armament:	32 × 128 mm rocket system
Main weapon	
maximum range:	8,545 m (standard rocket)
maximum range:	13,000 m (long range rocket)
Main armament traverse	
angle:	360°
Main armament elevation/depression	
armament front:	+48°/0°

Status
It is understood that production of this system is now complete and it is no longer being marketed by Yugoimport. This system is understood to be in service with Bosnia-Herzegovina, Croatia and Serbia. Marketing of this system is still underway.

Contractor
Yugoimport - SDPR.

Yugoimport 128 mm (32-round) M-63 'Plamen' multiple rocket system

Development
The 128 mm (32-round) M-63 'Plamen' (Flame) MRS was developed in the former Yugoslavia (today Serbia) to meet the requirements of the then Yugoslav Army. It is normally found in MRL battalions of three batteries, each with four launchers. It is estimated that total production of this system amounted to about 500 units.

Production of this trailer-mounted system has been completed and there has been no recent marketing of this system.

According to the United Nations Arms Transfer List for 2010, Bosnia transferred 10 × 128 mm launchers and 20,000 × 128 mm rockets to Azerbaijan. It is not known what these versions were.

Description
The 128 mm (32-round) M-63 'Plamen' launcher is mounted on a split trail carriage with elevation, traverse, launch controls and sight on the left side of the launcher.

The spin-stabilised solid propellant 122 mm M-63 unguided solid propellant rockets fired by the M-63 are 0.814 m long, weigh 23.1 kg, have a maximum velocity of 444 m/s and are fitted with a 7.55 kg warhead that is filled with 2.3 kg of high explosive.

A UTI M-63 (OV) point detonating fuze with super quick inertia action is used. A practice round of similar dimensions and weight for loading and unloading training is also used. The rockets are either fired singly or in a ripple salvo with set intervals of 0.2, 0.4 or 0.6 seconds.

Close-up of rear of 128 mm (32-round) M-63 rocket launcher showing sights on the left and trails deployed 0500684

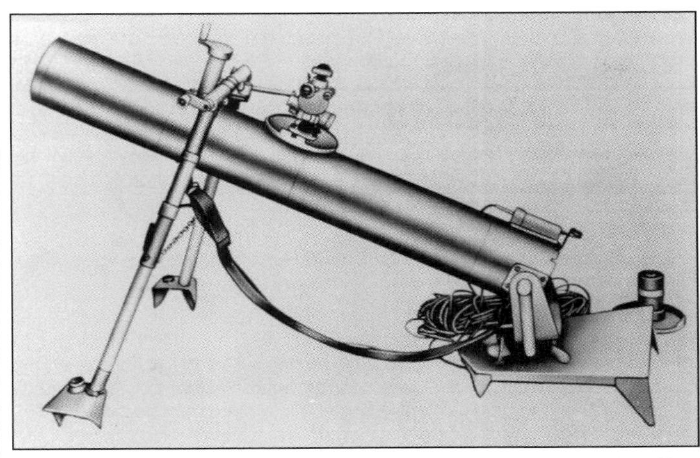

128 mm M-71 'Partizan' lightweight single-tube rocket launcher deployed in firing position and showing bipod and baseplate 0500685

128 mm (32-round) M-63 rocket launcher in firing position with trails open 0500683

An improved version of the M-63 with a wider carriage (2.212 m in travelling position) was also manufactured, and both 8- and 16-round launchers are available for export. If required, these can be mounted on a tracked or wheeled self-propelled system.

All versions of the M-63 launcher can be adapted for towing by animals.

When originally introduced into service with the former Yugoslavia this system was towed by a TAM 1500 (4 × 4) cross-country truck or a similar vehicle.

Variants
128 mm (32-round) 'Plamen' (6 × 6)

This is based on a locally manufactured chassis and is covered in detail in a separate entry in *IHS Jane's Land Warfare Platforms: Artillery and Air Defence*. It is also referred to as the 128 mm Plamen-S.

It is understood that production of this system is complete and it is no longer being marketed by Yugoimport.

128 mm M-71 'Partizan' single-tube rocket launcher

A 128 mm M-71 lightweight single-tube rocket launcher known as the 'Partizan' was also produced for use by commando, special forces and territorial defence units. It is known to be in service with Syria.

Available details of the three-man bipod-mounted weapon are outlined in the table.

M-71 Partizan rocket launcher Specifications

Calibre:	128 mm
Length:	
(in travelling configuration)	1.133 m
(in combat configuration)	1.13 m
Weight:	
(in travelling configuration)	21.5 kg
(in combat configuration, unloaded)	18.5 kg
(in combat configuration, loaded)	45 kg
Elevation:	+5 to +45°
Range:	
(minimum)	800 m
(maximum)	8,564 m
Sight unit:	NBS-4A

Time required for emplacement is five minutes and for displacement (in fired state) three minutes. The 128 mm unguided rocket type used is the M-63 type previously described.

It is understood that the correct designation of this system is the LRL 128 M-71 'Partizan'.

128 mm M-87 Improved Rocket

In 1980, the former Yugoslavia introduced the improved 128 mm solid propellant M-87 HE unguided rocket for use with both the M-63 and M-71 launchers. This weighs 25.5 kg in total, is 0.96 m long and uses an M-84 point detonating fuze with super quick inertia action. The warhead contains 3.3 kg of HE filler.

Maximum velocity is increased to 740 m/s with 7.8 kg of solid-propellant fuel carried giving a minimum range of 800 m and a maximum range of 12,800 m.

128 mm M-92 Improved version

This is an improved model used by Bosnia-Herzegovina and is fitted with a carrying handle.

128 mm (24-round) LOV RAK 24/128

This system was developed in Croatia and marketed by RH ALAN. Details are given in a separate entry in *IHS Jane's Land Warfare Platforms: Artillery and Air Defence.* It is understood that marketing of this system has ceased.

128 mm (12-round) VLR 128 M91A3 RAK 12

This system was also developed in Croatia and marketed by RH ALAN. Details are given in a separate entry in *IHS Jane's Land Warfare Platforms: Artillery and Air Defence.* It is understood that marketing of this system has ceased.

Self-propelled Multi-tube Modular Rocket Launcher

This is a recent development by Serbia and is a 4 × 4 cross-country truck chassis with a forward control cab.

On the rear of this is powered a launcher which can be fitted with pods of 107 mm, 122 mm or 128 mm rockets according to mission requirements.

When travelling, these pods of rockets are covered by a sliding tarpaulin cover which makes it difficult to distinguish from a normal carry truck rather than a rocket launcher.

Specifications

	128 mm (32-Round) M-63 Plamen
Dimensions and weights	
Crew:	7
Length	
overall:	3.682 m
deployed:	3.65 m
Width	
transport configuration:	2.212 m
Height	
transport configuration:	1.278 m
Ground clearance	
overall:	0.268 m
Track	
vehicle:	1.892 m
Weight	
standard:	1,395 kg
combat:	2,134 kg
Mobility	
Configuration	
running gear:	wheeled
Tyres:	8.25 × 20
Firepower	
Armament:	32 × carriage mounted 128 mm rocket system
Main weapon	
length of barrel:	1.03 m
maximum range:	8,545 m
Operation	
reload time:	300 s
Main armament traverse	
angle:	30° (15° left, 15° right)
Main armament elevation/depression	
armament front:	+48°/0°

	LRL 128 M-71
Dimensions and weights	
Length	
overall	1.133 m (3 ft 8½ in)
width	
calibre	128 mm (5.04 in)
Weight	
loaded	45 kg (99 lb)
empty	18.5 kg (40 lb)
Performance	
Elevation	
angle	5°/45°

Status

Production complete. No longer marketed. In service with the following countries:

Country	Quantity
Bosnia	21
Croatia	2
Cyprus	18
Macedonia	11
Serbia/Montenegro	20
Slovenia	4

Contractor

Yugoimport - SDPR.

Yugoimport 122 mm Grad Extended Range Rocket Family

Development

Serbia has completed development of an upgrade programme to enhance the capabilities of the Russian 122 mm Grad unguided surface-to-surface rocket which is fired by the widely deployed BM-21 (40-round) multiple rocket launcher and similar systems.

These new 122 mm Grad Extended range rockets are being offered on the export market but there are no known sales as of April 2013.

Description

The Serbian Grad Extended Range rocket family covers two types of 122 mm Extended Range rockets, GRAD POB and G-2000. The first of these, GRAD POB is an upgraded Russian 122 mm rocket while the G-2000 is a new rocket.

GRAD POB Extended Range Rocket takes the existing warhead and associated fuze and exhaust nozzle with vanes and fits these to a new solid propellant rocket motor developed in Serbia.

This solid propellant rocket motor uses an advanced thermo-plastic composite propellant which is claimed to be based on original technology.

This propellant has a similar combustion temperature to the original one but has a higher specific impulse which means that the original exhaust nozzle assembly can be used.

As well as the new propellant charge, new parts include the actual motor chamber, motor ignition assembly, motor bottom and parts for thermal insulation.

Parts of the old rocket used include the warhead with fuze, exhaust nozzle assembly with vanes, rocket guide, contact cover and rocket packing.

The standard Russian rocket has a maximum range 20,300 m which is increased to 27, 800 m with the new and more powerful rocket motors.

For range decrease, the upgraded POB rocket can also use braking rings which are positioned between the fuze and the warhead. In addition, modified POB rockets can be supplied as completely new rockets.

According to Yugoimport, the upgrade can be carried out at the customers own facilities using kits provided by the company.

Variants

The G-2000 rocket is a new design rocket motor but uses the same warhead and fuze of the PROB rocket. It is also possible to make use of braking rings (between the fuze and the warhead) in order to reduce range of the rocket. Maximum range of the G-2000 is claimed to be 38.4 km.

It has two types of propellant that differ in combustion speed and this consists of thermo-plastic composite that contains a high percentage of aluminium.

This is claimed to provide a low smoke capability with high specific impulse. The steel exhaust nozzle has an impressed thermal insulation and a single graphite neck.

Specifications

Model	GRAD-POB	G-2000
Calibre:	122 mm	122 mm
Length:	2.940 m	3.040 m
Weight:	69.3 kg	68.3 kg
Weight:		
(warhead with fuze)	19.1 kg	19.1 kg
(propellant)	25.6 kg	27 kg
Rocket motor burn time:	2.3 s	2.3 s
Total rocket motor impulse:	51,400 Ns	62,250 Ns
Specific rocket motor impulse:	2,010 Ns/kg	2,280 Ns/kg
Maximum speed:	892 m/s	1,100 m/s
Peak trajectory:	11,200 m	15,300 m
Time of flight:	96 s	114 s
Elevation:	50°	50°
Range at above elevation:	27,800 m	38,400 m

Status

Development complete. Production as required.

It is considered possible that this enhanced rocket has been produced in production quantities for the home and export markets.

Contractor

Yugoimport - SDPR.

Slovakia

122 mm RM-70 (40-round) multiple rocket system

Development

The 122 mm RM-70 (40-round) MRS, which has the NATO designation M1972, was first seen in public during a parade in the former Czechoslovakia following the SHIELD-72 manoeuvres.

It is understood that there has been no recent production of the 122 mm RM-70 (40-round) multiple rocket launcher. It is believed that all of the recent sales declared through the United Nations have been for surplus systems.

While the range of the original 122 mm rockets is short by today's standard, the range of the system can be increased by firing new generation rockets with a longer range.

Description

The 122 mm RM-70 (40-round) multiple rocket system is essentially an armoured version of the Czech Tatra 813 (8 × 8) truck chassis fitted with the same launcher as the Russian 122 mm BM-21 (40-round) MRS at the rear of the hull and an additional pack of 40 122 mm rockets to the immediate rear of the cab for rapid loading.

Main advantages of the RM-70 over the Russian BM-21 are its rapid reloading capability, complete armour protection for the crew from small arms fire, NBC attack and shell splinters and much improved cross-country mobility.

The Tatra 813 chassis is fitted with a central tyre-pressure regulation system, which enables the driver to adjust the tyre pressure to suit the type of ground being crossed.

Some vehicles were fitted with a hydraulically-operated BZ-T dozer blade at the front of the hull for preparing fire positions and clearing obstacles. It is also fitted with a winch with a capacity of 22,000 kg.

According to United Nations sources the following quantities of 122 mm RM-70 (40-round) series multiple rocket system were exported between 1992 and 2010:

Country	Quantity	Comment
Angola	40	from Slovakia in 1994
Angola	12	from Slovakia in 1999
Angola	6	from Slovakia in 2000
Bulgaria	10	from Czech Republic in 2009
Czech Republic	29	from Slovakia in 2008
Czech Republic	10	from Slovakia in 2009
Czech Republic	7	from Slovakia in 2010
Ecuador	6	from Slovakia in 1995
Georgia	4	from Czech Republic in 2003
Georgia	2	from Czech Republic in 2004
Georgia	12	from Bulgaria in 2009
Greece	150	from Germany in 1994
Indonesia	4	from Czech Republic in 2003
Indonesia	2	from Czech Republic in 2004
Indonesia	3	from Czech Republic in 2007
Indonesia	4	from Poland (assumed RM-70) in 2007
Rwanda	5	from Slovakia in 1997
Slovakia	12	from Czech Republic in 1999
Slovakia	12	from Czech Republic in 2000
Slovakia	6	from Czech Republic in 2001
Slovakia	4	from Czech Republic in 2002
Sri Lanka	8	from Czech Republic in 2000
Sri Lanka	8	from Czech Republic in 2001
Uganda	6	from Slovakia in 2002
Uruguay	2	from Czech Republic in 1996
Uruguay	1	from Czech Republic in 1998
Uruguay	1	from Czech Republic in 1999
Uruguay	1	from Czech Republic in 2000
Zimbabwe	20	from Czech Republic in 1992
Zimbabwe	6	from Czech Republic in 2000

The above figures do not always correspond exactly with those listed in the Status table as in many cases the RM-70 were delivered before the period 1992-2008.

At this stage it is not known as to whether the RM-70 systems listed above were brand new systems or surplus to requirements. It is considered that most were surplus or reconditioned vehicles.

In addition to the above 122 mm RM-70 (40-round) multiple rocket systems, the United Nations Arms Transfer lists for 2007 and 2010 show the following export of the actual 122 mm unguided rockets:

From	To	Quantity
Slovakia	Bulgaria	16,576
Slovakia	Indonesia	764
Slovakia	Turkey	380

Details of the 122 mm solid propellant unguided surface-to-surface rockets are provided in the table.

The RM-70 122 mm (40-round) multiple rocket launcher has a fully armour and NBC protected cab. This example belongs to the Czech Republic (Stefan Marx) 1151746

Original RM-70 122 mm (40-round) multiple rocket launcher with fully protected cab and with launcher traversed to the rear (Paul Beaver) 1452639

Original RM-70 122 mm (40-round) multiple rocket launcher with fully protected cab and with launcher traversed to rear and re-loading sequence already underway (Paul Beaver) 1452640

It should be noted that the standard 122 mm rocket HE fragmentation warhead contains a mixture of 86 per cent TNT and 14 per cent Al with the HE content being 6.6 kg and with the rocket fuze weighing 0.9 kg.

Variants

122 mm Mod 70/85

This was introduced into service in the mid-1980s. It is the more recent Tatra T815 VNN (8 × 8) truck chassis fitted with the same 122 mm (40-round) multiple rocket system as the original RM-70 system.

The major difference is that the Mod 70/85 is not fitted with a fully enclosed and armoured cab.

Standard equipment includes a front-mounted winch with a capacity of 12,000 kg and 85 m of cable and the cab is fitted with an NBC system of the collective and individual type.

The chassis has a central tyre-pressure regulation system that allows the ground pressure at each wheel station to be adjusted to suit the type of terrain being crossed. At the front of the vehicle is an SSP 1000 arrow-type snowplough. A BZ-T815 dozer blade can also be quickly installed.

Specifications of the more recent 122 mm Mod 70/85 are almost identical to the original RM-70 with the fully protected cab, the former is however slightly lighter.

Russian 122 mm rocket designations	M-21-OF, DB-1B	M-21-OF	DKZ-B
Industrial index number:	9M22U	9M22M	9M28
Diameter:	122 mm	122 mm	122 mm
Length:	3.226 m	2.870 m	1.905 m
Weight:			
(complete)	77.5 kg	66 kg	45.8 kg
(HE fragmentation warhead)	19.4 kg	18.4 kg	19.4 kg
Range:			
(maximum)	20,380 m	20,000 m	10,800 m
(minimum indirect)	1,500 m	1,500 m	2,500 m
(minimum direct)	500 m	500 m	400 m
Warhead types:	HE fragmentation, smoke, incendiary		
Fuze settings:	instant/single time delay/safe		
Fuze arming distance:	150-400 m	150-400 m	150-400 m
Burnout velocity:	690 m/s	700 m/s	450 m/s
Launchers:	Grad	Grad	Grad
	RM-70	Grad-V	Grad-V
	BM-21/SR-114	RM-70	RM-70
	122 mm/DAC 665T	BM-21/SR-114	9P132
	Grad-1	122 mm/DAC 665T	BM-21/SR-114
		Grad-1	122 mm/ARO
			122 mm/Trailer
			122 mm/DAC 665T
			Grad-1

According to Czech Republic sources, the RM-70 122 mm launcher, when elevated at an angle of 50°, has a dispersion of 1/180 in range and 1/110 in traverse.

A complete salvo of 40 122 mm rockets can be fired in 18 to 22 seconds. The system takes 150 seconds to come into action and 180 seconds to come out of action.

RM-70 upgrade

This system has been developed by the German company Diehl and Konstrukta Defence of the Slovak Republic.

Details of this system are provided in a separate entry in *IHS Jane's Land Warfare Platforms: Artillery and Air Defence* and it is only in service with the Slovak Republic.

122 mm M96 (32-round) Typhoon

Croatia has developed and placed in production the 122 mm M96 (32-round) Typhoon system on an unarmoured Tatra 8 × 8 chassis and details are provided in a separate entry in *IHS Jane's Land Warfare Platforms: Artillery and Air Defence*. As far as it is known this has only been built in small numbers and has not been exported.

Original RM-70 122 mm (40-round) multiple rocket launcher with launcher and rockets covered for transport (INA) 1452943

Latest version of Mod 70/85 122 mm (40-round) multiple rocket launcher fitted with front-mounted dozer blade 0021048

Specifications

	122 mm RM-70 (40-Round)
Dimensions and weights	
Crew:	6
Length	
overall:	8.8 m
Width	
transport configuration:	2.55 m
Height	
transport configuration:	2.96 m
Ground clearance	
overall:	0.4 m
Track	
vehicle:	2.03 m
Wheelbase	1.65 m + 2.2 m + 1.45 m
Weight	
combat:	25,300 kg
Mobility	
Configuration	
running gear:	wheeled
layout:	8 × 8
Speed	
max speed:	75 km/h
Range	
main fuel supply:	1,100 km
Fuel capacity	
main:	400 litres
Fording	
without preparation:	1.4 m
Gradient:	60%
Vertical obstacle	
forwards:	0.6 m
Trench:	1.6 m
Engine	Tatra T-930-3, V-12, air cooled, diesel, 270 hp at 2,700 rpm
Gearbox	
forward gears:	20
reverse gears:	4
Tyres:	15.00 × 20
Firepower	
Armament:	40 × 122.4 mm rocket system
Ammunition	
main total:	80
Operation	
reload time:	35 s
Main armament traverse	
angle:	195° (125° (left), 70° (right))
Main armament elevation/depression	
armament front:	+55°/0°

Status

Production as required. In service with:

Country	Quantity	Comment
Angola	40	from Slovakia
Czech Republic	60	from Slovakia
Ecuador	6	
Finland	36	from Germany, called RaK H 89
Georgia	6	from Czech Republic
Georgia	12	from Bulgaria in 2009
Greece	150	from Germany (112 operational)
Indonesia	13	for marines
Libya	100	status uncertain
Poland	30	
Rwanda	5	from Slovakia
Slovakia	59	some upgraded to MORAK standard (25)
Sri Lanka	8	from Czech Republic in 2000 (additional batch delivered, making 22)
Uganda	6	from Slovakia in 2002
Uruguay	5	from Czech Republic
Zimbabwe	52	some sources state 60

Contractor
ZTS Dubnica nad Váhom (Slovak Republic).

Military and Police Group, Technopol International 122 mm (40/80-round) VP 14 Krizan multiple rocket launcher system

Development
The Slovak company Military and Police Group, Technopol International has developed the 122 mm calibre 40-round VP 14 Krizan multiple rocket system for use with its 122 mm 40 Kus anti-personnel and Krizna minelet-carrying cargo rockets.

It should be noted that under the Ottawa Convention, anti-personnel mines have been banned by most countries.

It is understood that there has been no recent production of these systems.

Description
The 122 mm launcher is mounted on the rear chassis of a Tatra T815VNN 8 × 8 truck as part of a combat engineer's minelaying system which comprises two vehicles, the minelaying vehicle itself and the ammunition resupply vehicle.

The minelaying vehicle has a variable superstructure onto which can be mounted either a mechanical minelaying system, or two 122 mm 40-round rocket minelaying systems or the mechanical minelaying system and a 122 mm 40-round rocket minelaying system.

The actual 122 mm rocket launcher system is understood to be the same as that used in the Slovakian 122 mm Mod 70/85 122 mm (40-round) system covered in detail in a separate entry in *IHS Jane's Land Warfare Platforms: Artillery and Air Defence* and which is also based on a Tatra T815 series 8 × 8 cross-country truck.

The Kus rocket payloads of PPMi-S1 minelets can either be used alone to produce an anti-personnel minefield or be overlaid on a Krizna deployed anti-tank minelet field to hinder clearance.

This system has a fully enclosed and armoured cab to protect the crew from small arms fire and shell splinters. This is of a different design to that used on the original Czechoslovakian RM-70 (8 × 8) 122 mm (40-round) multiple rocket system.

Tatra T815 (8 × 8) chassis with two VP 14 Krizan launchers of 122 mm (40-round) rockets for laying anti-tank/anti-personnel minefields 0500644

Tatra T815 (8 × 8) chassis with VP 14 Krizan 122 mm (40-round) launcher in centre and mechanical minelaying system at the rear 0500645

Status
Production as required. In service with armed forces (engineer units) of the Czech Republic and Slovakia. There are no other known exports of this system.

It is understood that there has been no recent production of this multiple rocket launcher system.

Contractor
Military and Police Group, Technopol International, Slovak Republic.

Military and Police Group, Technopol International 122 mm cargo rocket system

Development
The Slovak company Military and Police Group, Technopol International has developed a series of 122 mm calibre cargo rockets capable of laying remote anti-tank or anti-personnel mines. A mixture of the two can also be laid.

As far as it is known there are has been no recent production of this 122 mm cargo rocket system.

Under the Ottawa Convention, these anti-personnel mines and bomblet type sub-munitions have been banned and most countries in Europe have now phased these out of service and disposed of their stock of anti-personnel mines and munitions carrying submunitions.

Description
The 122 mm unguided rockets are:

- Krizna-R – used for long-range engagements (up to 17 km) and carrying four PTMi-D anti-tank minelets. It is fitted with a timing fuze to burst the casing open and deploy the minelets. The normal launching platforms are the heavier 122 mm calibre systems such as the Russian BM-21 (40 or 36 round) (6 × 6) series and the former Czech RM-70/85 (8 × 8). Full details of both of these systems, which remain in service in large quantities in many parts of the world, are given in separate entries in *IHS Jane's Land Warfare Platforms: Artillery and Air Defence.* It should be noted that production of both of these systems is complete but surplus systems have been exported in significant quantities by the Czech Republic and Slovakia in recent years.The PTMi-D minelet is 116 mm in diameter, 124 mm high in armed position and weighs 2.7 kg. A 1.073 kg HE charge is used which enables it to destroy tank tracks and penetrate up to 110 mm of conventional steel rolled homogenous belly armour. When released from the cargo round's casing the minelet deploys a small parachute to orientate its downward trajectory and slow its descent. Upon impact with the ground, nine spring-loaded legs are released to ensure that it sits upright with the warhead uppermost.The contact also primes the fuzing system and releases a short tilt rod, which detonates the mine when it is driven over by a heavy vehicle. The mine is also detonated if it is directly struck by a vehicle track and will self-destruct at one of three delay times that have been selected and set prior to launch. The warhead weighs 19.3 kg and the solid propellant rocket motor with fuel 46.9 kg.
- Krizna-S – used for short-range engagements (up to 3 km) and carrying the same fuzing system and number of PTMi-D anti-tank minelets as the Krizna-R. It is normally used with the MV-3 launcher assembly. The rocket motor weighs 14.7 kg and the warhead 19 kg. The solid propellant fuel load is 3.45 kg. Details of the MV-3 launcher are given in a separate entry in *IHS Jane's Land Warfare Platforms: Artillery and Air Defence.* As far as it is known, there have been no exports of this system.
- Kus – used with either the MV-3 launcher assembly (covered in a separate entry in *IHS Jane's Land Warfare Platforms: Artillery and Air Defence*) or the larger 40-round VP 14 Krizan system (covered in a separate entry in *IHS Jane's Land Warfare Platforms: Artillery and Air Defence*) mounted on the rear of a Tatra T815 (8 × 8) cross-country truck chassis. The Kus carries five PPMi-S1 anti-personnel minelets. The minelets can either be used to form a separate anti-personnel minefield or be overlaid on a PTMi-D minefield to hamper clearance. Production of the PPMi-S1 anti-personnel mine has ceased and it is no longer marketed.

The PPMi-S1 minelet is 115.5 mm in diameter, 96 mm high and weighs 1.75 kg (with a 0.17 kg explosive charge). It contains some 1,900 fragments that are dispersed on detonation. The fragments have a minimum effective range of 8 m. The minelet also carries four decoys to mislead engineers trying to clear the minefield.

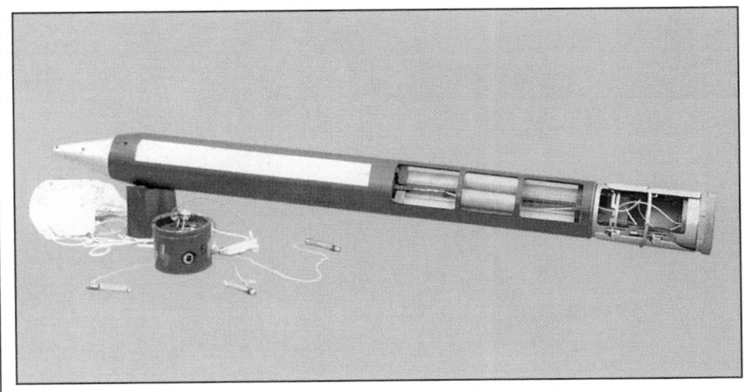

122 mm Kus cargo rocket which carries five PPMi-S1 anti-personnel mines 0500901

Specifications

Rocket	Krizna-R	Krizna-S	Kus
Calibre:	122 mm	122 mm	122 mm
Length:	3.187 m	1.56 m	1.4 m
Weight:	67.5 kg	34.6 kg	32 kg
Cargo:	4 × PTMi-D anti-tank mines	4 × PTMi-D anti-tank mines	5 × PTMi-S1 anti-personnel mines
Range:	8-17 km	0.5-3 km	0.5-3 km
Launcher:	RM-70/85	MV-3	MV-3
Type:	wheeled	tripod	tripod

Status

Production as required. In service with armed forces (engineer units) of the Czech Republic and Slovak Republic. There are no other known exports of these rockets.

It is understood that there has been no recent production of this 122 mm cargo rocket system.

Contractor

Military and Police Group, Technopol International, Slovak Republic.

Military and Police Group, Technopol International 122 mm MV-3 (3-round) rocket launcher system

Development

The Slovak company Military and Police Group, Technopol International has developed the 122 mm calibre MV-3 (3-round) launcher for use with its 122 mm unguided Krizna-S and Kus minelet cargo-carrying rockets.

This has been offered on the export market but there were no known sales as of early 2013.

It should be noted that while anti-tank mines can still be deployed most countries have now banned the manufacture and deployment of anti-personnel mines.

These have been banned under the Ottawa Convention and most countries in Europe have now phased these out of service and disposed of their stock of anti-personnel mines.

Description

The MV-3 consists of a lightweight wheeled trailer, which can either be pulled by two men or mounted on the rear of a light vehicle. The launcher is set up by extending the three legs (one of which also serves as the towing bar in the dismounted role), placing the unit on the ground pointing in the required direction and connecting the cable from the remote firing box.

Two 1.5 V batteries are then inserted in each minelet through openings in the rocket casing, following which the 122 mm unguided projectile is placed in a canister on the launch tray.

The only adjustments are in launch angle, which is achieved using a small handwheel on the left of the launcher and in divergence (up to a maximum of 8° of the three projectiles).

The remote firing box is also used to check the 122 mm unguided projectiles before launch, set the release timing fuzes according to the desired range and then carry out the actual launch. The maximum width of the minefield laid by one salvo is 100 × 100 m.

122 mm MV-3 (three-round) rocket launcher deployed in the firing position with stabilisers in position, but with control box not attached 0130033

An optical sight is mounted on the left side of the launcher and the two rear tripod legs have adjustable feet.

The 122 mm container rocket carrying anti-tank mines type PT Mi-D 1 M is now supplied by Policske Strojirny as of the Czech Republic. Each 122 mm unguided rocket projectile weighs 54 kg and carries four anti-tank mines which are ejected over the target area and then float to the ground on a parachute.

Range of mine laying is quoted as from 500 to 3,000 m. The fuze can be set from 1 to 22 seconds with an accuracy of 0.5 seconds. The anti-tank mine can be activated by a magnetic sensor or pressure.

If the mine is not activated it can be set detonated at a preset time. This can be 3, 12, 24 or 48 hours. The warhead of the mine will penetrate 110 mm of conventional steel armour.

Specifications

MV-3 rocket launcher system

Calibre:	122 mm
Number of tubes:	3
Length of tube:	1.35 m
Rocket types:	Krizna-S, Kus
Crew:	2
Elevation:	+15° to +40°
Traverse:	15° (total)
Length: (travelling)	1.60 m
Height: (travelling)	0.90 m
Diameter of wheels:	310 mm
Total weight:	103 kg
Weight of bag with launching apparatus:	12 kg
Launching facility:	2.55 kg
Weight: (total)	130 kg (without rockets)
Dispersion:	up to 8°
Rate of fire:	3 rounds in 4 s

Status

Production as required. In service with armed forces (engineer units) of the Czech Republic and Slovakia. There are no other known exports of this system.

It is understood that there has been no recent production of the 122 mm MV-3 (three-round) rocket launcher.

Contractor

Military and Police Group, Technopol International, Slovak Republic.

Military and Police Group, Technopol International AGAT cargo warhead for 122 mm rockets

Development

The Military and Police Group, Technopol International of the Slovak Republic is now marketing an improved cargo warhead, the AGAT, for use with 122 mm calibre unguided surface-to-surface rockets of the type used with the RM-70, Mod 70/85, Russian BM-21 and equivalent 122 mm multiple rocket systems.

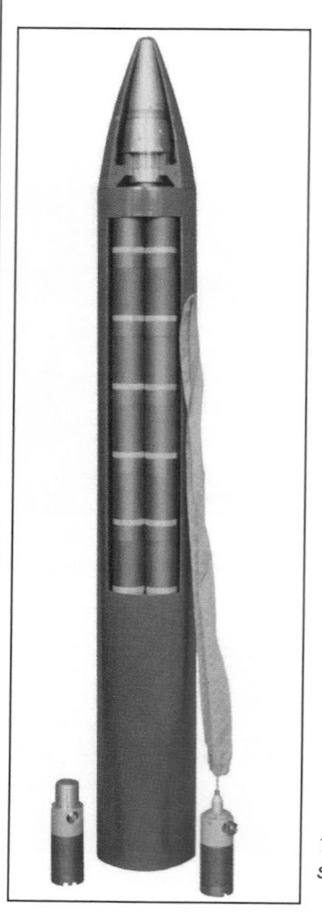

122 mm AGAT cargo projectile cutaway to show interior with two submunitions below 0058458

This has been offered on the export market but there are no known sales as of early 2013.

It should be noted that most countries have now banned the manufacture and deployment of munitions carrying submunitions.

Description

The AGAT can quickly replace the existing 122 mm warhead or, alternatively, completely new rockets fitted with the warhead can be supplied. In the latter case the rear part of the rocket is designated JRM 122 mm thruster Type 70 with the complete rocket known as the JRKK-G. The weight of the complete round is 68 kg with range limits of 5,300 m minimum to 16,500 m maximum.

The AGAT warhead contains 56 dual-purpose anti-armour/anti-personnel bomblets, which are ejected over the target area using a nose-mounted TM-120 time fuze. The bomblets are stabilised in flight by a rear-mounted ribbon. Each of the bomblets weighs 0.28 kg as is 38 mm in diameter. The cylindrical case is made from steel and contains a single High Explosive Anti-Tank (HEAT) warhead that can penetrate between 110 and 130 mm of conventional steel armour. The casing also allows for a secondary fragmentation effect. An impact fuze and a self-destruct system is also fitted. The fuzing is designed to allow attacks on the vulnerable upper surfaces of armoured vehicles.

Specifications

AGAT 122 mm rocket warhead

Calibre:	122 mm
Number of bomblets:	56
Bomblet diameter:	38 mm
Weight of bomblet:	0.28 kg
Warhead type:	HEAT with steel bomblet body for fragmentation effect against personnel and matériel
Armour penetration:	110-130 mm conventional steel plate

Status

Production as required. Believed to be in service with the armies of the Czech Republic and Slovak Republic. There are no other known customers of these rockets so far.

Contractor

Military and Police Group, Technopol International, Slovak Republic.

South Africa

Rheinmetall Denel Munition 127 mm (40-round) Valkiri Mk II Multiple Artillery Rocket System (MARS)

Development

The 127 mm (40-round) Valkiri MARS is a second-generation version of the Rheinmetall Denel Munition Valkiri Multiple Artillery Rocket System (MARS) mounted on an all-terrain Samil 100 (6 × 6) truck chassis fitted with a five-man armoured (against small arms and shell splinters) and mine-protected crew cabin.

Development of the system began in 1985 under the leadership of Rheinmetall Denel Munition (at that time SOMCHEM, a Division of Denel), with the first production units completed in 1989.

Known by the name Bateleur to the South African National Defence Force, the first battery of eight launchers was delivered in March 1990 as the replacement for the 24-round Valkiri Mk I systems used by the mechanised brigades.

127 mm (40-round) Valkiri Mk II MRS deployed in the field with launcher traversed left and stabilisers at rear lowered into position (T J Gander) 0500905

127 mm rockets showing folded fins and nose-mounted fuze 0500788

Details of this older 127 mm rocket system, only used by the South African National Defence Force, are given in a separate entry in *IHS Jane's Land Warfare Platforms: Artillery and Air Defence.* As far as it is known these Mk 1 systems have now been placed in reserve.

Description

The larger Samil 100 (6 × 6) chassis has sufficient space for all the launcher system equipment and rations, stores and water required for a 14 day operational cycle. The modular launch pack unit contains five rows of eight 127 mm rocket launch tubes and is fitted with electrohydraulically operated elevation and traverse gear. A manual back-up system is also fitted. The elevation and traverse controls are located on the left side of the launcher.

The Samil 100 chassis and suspension have been strengthened with the front axle upgraded from 6,500 to 7,500 kg. Improvements to the V-10 air-cooled diesel engine and transmission give the vehicle 45 per cent better traction and 90 per cent better gradient ability than a standard Samil 100.

A new option is a five-speed automatic gearbox, which has improved acceleration and top speed and reduced fuel consumption. Additional diesel fuel tanks extend the range to 800 km at a speed of 80 km/h.

The complete launcher unit can be mounted on other suitable 10,000 kg vehicle types if required. All production Valkiri II systems supplied to the now South African National Defence Force were fitted onto the locally built Samil 100 (6 × 6) chassis.

Sophisticated electronic navigation and north seeking equipment as well as computer-controlled semi-automatic laying equipment has been developed for the weapon system. Together with a data communication radio, this brings the fire-control system of the weapon up to most modern standards.

Firing is usually carried out with the two hydraulically-operated stabiliser jacks lowered at the rear and is performed from within the armoured cab or via an optional remote hand-held firing console.

The 127 mm unguided rockets may be fired either singly or in ripples with a separation interval of 0.5 second. Total salvo time for all 40 × 127 mm rockets is 20 seconds. Reload time for 40 × 127 mm rockets is between seven and 10 minutes.

The 2.95 m long, 127 mm extended-range rockets weigh 63 kg overall and use a rocket motor with a high-energy double-base propellant. They can be fitted with a number of warhead types, the predominant one being the very effective HE pre-fragmented anti-personnel and light vehicle warhead with a thin-walled cast epoxy resin cylinder containing 9,700 steel balls. A proximity fuze assembly with an adjustable burst height between three and eight metres increases the kill probability. The proximity action is backed up by a super-quick point-detonation action.

While the 127 mm unguided rockets have a maximum range of 36,000 m they have a minimum range of 8,000 m.

The system can also fire the standard 54 kg, 22,000 m range 127 mm rocket.

Manual reloading in the field is undertaken by using an ergonomically designed loading platform. The option of mechanical reloading by replacement of the complete launch tube pack is also available.

Specifications

	127 mm (40-Round) Valkiri Mk II Multiple Artillery Rocket System
Dimensions and weights	
Crew:	3
Length	
overall:	9.3 m
Width	
transport configuration:	2.35 m
Height	
transport configuration:	3.4 m
Ground clearance	
overall:	0.35 m
Weight	
combat:	21,500 kg
Mobility	
Configuration	
running gear:	wheeled
layout:	6 × 6
Speed	
max speed:	93 km/h
Range	
main fuel supply:	800 km (at 80 km/h)
Fuel capacity	
main:	400 litres
Fording	
without preparation:	1.2 m
Gradient:	36% (est.)
Vertical obstacle	
forwards:	0.5 m
Engine	V-10, 4 stroke, diesel, 315 hp at 2,500 rpm
Gearbox	
type:	automatic
forward gears:	5
Tyres:	16.00 × R20 radial ply (fitted with run flat)
Electrical system	
vehicle:	24 V
Firepower	
Armament:	40 × 127 mm rocket system
Main weapon	
length of barrel:	3.0 m
maximum range:	36,000 m
Rate of fire	
rapid fire:	2/1 s
Operation	
reload time:	600 s
Main armament traverse	
angle:	110°
Main armament elevation/depression	
armament front:	+50°/0°
Survivability	
Armour	
hull/body:	steel (welded (cab))

Status

Production as required. In service with the South African National Defence Force. It is believed that a total of 16 production systems were built.

Valkiri Mk II system has been demonstrated in a number of countries including the Middle East, but as of May 2013, no export sales had been made. As the Samil 100 (6 × 6) chassis went out of production many years ago, any new production would have to be on a different chassis.

Contractor

Rheinmetall Denel Munition.

Rheinmetall Denel Munition 127 mm (24-round) Valkiri Mk I 22 multiple rocket launcher system

Development

Development of the 127 mm (24-round) Valkiri Mk 1 22 multiple rocket launcher was undertaken by Rheinmetall Denel Munition (at that time SOMCHEM Division of Denel) from November 1977 as a counter to the Russian BM-21 (40-round) 122 mm MRS in service with neighbouring African countries.

Development was completed in March 1981, with the system entering series production in April and first deliveries beginning later in 1982. It is understood that a total of 25 production systems were built. The system is no longer being marketed.

127 mm (24-round) Valkiri Mk I 22 MRLS in firing position 1151745

The 127 mm (24-round) Valkiri Mk I 22 is designed to be deployed either on its own in batteries of two troops with three launchers each, or in conjunction with more conventional tubed artillery against area targets such as guerrilla camps, troop concentrations and soft-skinned vehicle convoys.

The 127 mm (24-round) Valkiri Mk I 22 saw extensive combat service in southern Angola after it entered South African National Defence Force inventory.

The system was given the name *Chindungu* (the red pepper that bites) by UNITA. Within the South African National Defence Force this system has also referred to as the Visared.

Production of this UNIMOG-based system was completed some time ago and it was replaced in production by the more mobile and more well protected Rheinmetall Denel Munition 127 mm (40-round) Valkiri Mk II system on a Samil 100 (6 × 6) chassis.

Details are provided in a separate entry in *IHS Jane's Land Warfare Platforms: Artillery and Air Defence.* It should be noted that the 127 mm (40-round) Valkiti Mk II is no longer marketed.

Description

The launcher unit consists of a 24-round pack of three rows of eight 127 mm launcher tubes mounted on the rear hull of an unarmoured Mercedes-Benz UNIMOG (4 × 4) truck chassis. Overhead canopy railings are fitted to disguise the vehicle as a normal military cross-country vehicle. When travelling, the canopy is rolled down either side to cover the rocket launcher.

The vehicle carries hydraulically-operated elevation and traversing gear for the launcher and is stabilised during firing by a pair of hydraulic supports mounted at the vehicle rear.

A conventional artillery sight is mounted on the left side of the launcher and is stored in a special container on the vehicle when not in use.

Firing is either by a firing panel within the cab or via a 50 m plug-in cable connected to a remote firing unit. A test circuit is incorporated to check the rocket status and circuitry prior to firing.

The unguided 127 mm rockets are launched either singly or in ripples of 2 to 24. The firing rate is one rocket per second.

The 127 mm standard 2.68 m, 53 kg weight rocket has a double-base propellant motor and a pre-fragmented anti-personnel warhead, with a thin-walled cast epoxy resin cylinder containing 8,500 steel balls to give a lethal area of 1,500 m^2.

A contact or proximity fuze is screwed to the front of the warhead. The range varies from 8,000 to 22,000 m depending on whether spoiler rings are used on the rockets to slow them.

Reloading is manual and takes about 10 minutes. Upper atmosphere and surface meteorological data required for firing in order to achieve a high first hit probability is obtained by means of weather balloons, a vehicle-mounted telescopic mast with wind vane and anemometer close to the launchers and a wind gun that fires small calibre projectiles to estimate the prevailing ground wind conditions.

The time taken for the launcher to deploy in a pre-surveyed site is five minutes; to come out of action takes only two minutes. A radio is fitted within the vehicle cab as standard with an extension to the remote fire unit.

Unit of fire is typically 96 × 127 mm rockets which is sufficient for four fire missions.

Specifications

	127 mm (24-Round) Valkiri Mk I 22
Dimensions and weights	
Crew:	2
Length	
overall:	5.35 m
Width	
transport configuration:	2.3 m
Height	
transport configuration:	2.32 m
Ground clearance	
overall:	0.3 m

	127 mm (24-Round) Valkiri Mk I 22
Weight	
combat:	6,400 kg
Mobility	
Configuration	
running gear:	wheeled
layout:	4 × 4
Speed	
max speed:	90 km/h
Range	
main fuel supply:	400 km
Fuel capacity	
main:	90 litres
Fording	
without preparation:	0.6 m
Gradient:	33% (est.)
Vertical obstacle	
forwards:	0.35 m
Engine	6 cylinders, water cooled, 4 stroke, diesel, 99 hp at 2,800 rpm
Gearbox	
type:	manual
forward gears:	4
reverse gears:	1
Tyres:	12.50 × R20
Electrical system	
vehicle:	24 V
Firepower	
Armament:	24 × 127 mm rocket system
Ammunition	
main total:	24
Main weapon	
length of barrel:	3.0 m
maximum range:	22,000 m
Rate of fire	
rapid fire:	1/1 s
Operation	
reload time:	600 s
Main armament traverse	
angle:	110°
Main armament elevation/depression	
armament front:	+50°/0°

Status
Production complete. It is understood that these systems are no longer in front line service with the South African National Defence Force, but are held in reserve. This system is no longer being marketed.

Contractor
Rheinmetall Denel Munition.

Mechem Developments 122 mm RO 122 artillery rocket system

Development
Mechem Developments (a division of Denel) developed a 122 mm unguided calibre rocket, the RO 122, for use with 122 mm multiple rocket launcher systems which are now manufactured by many countries including China, Croatia, Egypt, India, Iran, North Korea, Pakistan, Romania, Russia, Serbia, Slovakia, Sudan, and Turkey.

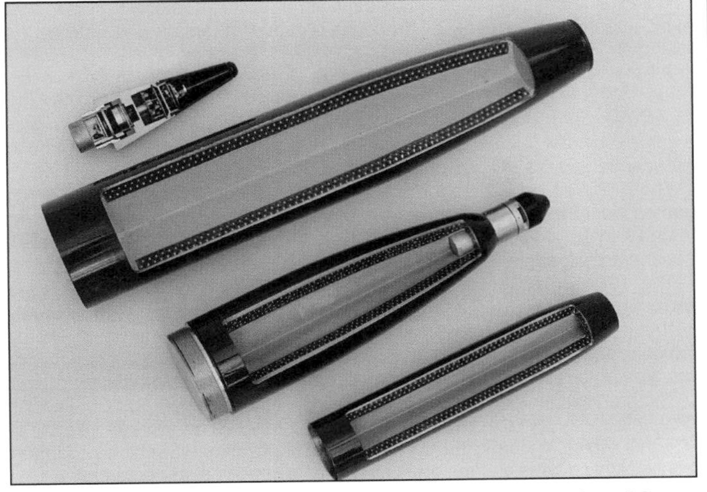

Mechem pre-fragmented rocket warheads, from top to bottom, fuze, RO 122, RO 107 and RO 68. The RO 107 is shown fitted with a nose-mounted Fuchs Electronics fuze 0500908

These solid propellant 122 mm unguided rockets have been offered on the export market but there are no known export sales.

Description
The 122 mm unguided rocket has a new HE fragmentation warhead packed with some 9,800 steel balls and is fitted with a radio proximity fuze. It is understood that the latter has been developed by Fuchs Electronics.

Mechem has developed a complete family of pre-fragmented warheads for the RO 122, RO 107 and RO 68 rockets covered in a separate entry in *IHS Jane's Land Warfare Platforms: Artillery and Air Defence.* In this designation the number after the RO indicates the actual calibre of the rocket in millimetres.

Status
Development complete. Production as required. This family of 122 mm rockets has been developed for the export market.

Contractor
Mechem Developments a member of the Denel Group.

Mechem Developments 107 mm RO 107 (12-round) towed multiple rocket launcher system

Development
The Mechem Developments 107 mm RO 107 (12-round) towed multiple rocket launcher MRS is designed for use as an anti-personnel weapon by light artillery, infantry and special forces units.

It is basically a reverse engineered Chinese China North Industries Corporation (NORINCO) Type 63 107 mm trailer-mounted MRS, which is covered in a separate entry in *IHS Jane's Land Warfare Platforms: Artillery and Air Defence.*

It should be noted that NORINCO is no longer marketing the original 107 mm (12-round) Type 63 trailer mounted or the 107 mm (12-round) Type 81 truck mounted systems.

In addition to supplying its own version of the complete launcher, Mechem can supply its own version of the 107 mm unguided rocket.

It is understood that there has been no recent production or marketing of this system.

A quantity of these systems have been deployed by the South African National Defence Force (SANDF).

As far as it is known there has been no recent marketing of the Mechem Developments 107 mm RO 107 (12-round) towed multiple rocket launcher or its associated 107 mm unguided rockets.

Description
The towed launcher has 12 barrels in three rows of four, weighs 380 kg and can be dismantled into units of 30 kg to allow manhandling across rough terrain.

The carriage is of the split trail type and can be towed by a variety of light wheeled vehicles which, depending on their size, can also carry a quantity of ready use ammunition.

Maximum towing speed on paved roads is being quoted as 60 km/h whilst on unpaved roads this is reduced to 15 km/h and cross-country is 7 km/h.

It takes about five minutes to dismantle and can be reassembled in 10 minutes. Elevation and traverse are set with handwheels mounted on the left. A simple mortar day sight is used for aiming.

All the standard Chinese 107 mm unguided rockets can be fired and because the upgraded Mechem RO 107 rounds use the same range tables as those prescribed for the standard ammunition, mixed salvos of standard and upgraded rockets can be fired using the same launch setting.

Mechem is prime contractor for the upgraded RO 107 rocket: the rocket motor and rocket motor filling are supplied by Rheinmetall Denel Munition, the warhead by Aserma and the RO 107 proximity fuze by Fuchs.

The improved 107 mm rocket is spin stabilised by seven canted venturis, and rotates at approximately 22,000 rpm. It develops an impulse of 6,000 N, which gives it a maximum range of 8,500 m at sea level. Minimum firing angle can be as low as 4°.

Mechem Developments 107 mm RO 107 (12-round) towed multiple rocket launcher system (Christopher F Foss) 0058467

Mechem Developments 107 mm RO 107 (12-round) towed multiple rocket launcher system deployed with a 107 mm unguided rocket being launched (Mechem Developments) 1452944

Mechem Developments 107 mm (12-round) palletised multiple rocket launcher system carrying out a fire mission integrated onto the rear of a 4 × 4 cross-country protected platform (Mechem Developments) 1452945

The 107 mm solid propellant rocket has a maximum velocity of 370 m/s and a launch weight of 19 kg of which 7.6 kg is the high explosive pre-fragmented warhead and the fuze weighs 0.28 kg.

Approximately 80 per cent of the rounds fired to maximum range will fall within an area 150 × 200 m. The rocket can also be fired by equivalent China North Industries Corporation (NORINCO) 107 mm MRS types.

The rocket is fitted with a Fuchs Electronics UHF radio band proximity fuze which arms after four seconds of flight time (approximately 1,400 m from the launcher). Frequency agility provides immunity to enemy jamming action. The proximity fuze causes a predictable airburst irrespective of the type of terrain. A point detonation back-up system is also provided.

The pre-fragmented warhead is packed with approximately 5,200 steel balls cast into epoxy resin. The shrapnel produced reaches an average velocity of 900 m/s which ensures the penetration of soft- skinned vehicles, support trucks, vehicle tyres and radiators as well as being lethal to personnel.

Typical kill and maim areas are 120 m^2 and 504 m^2 respectively for an upgraded round, compared to 34 m^2 and 140 m^2 for a standard cast-iron high-explosive warhead fitted with a point impact fuze.

Variants

The RO 107 (12-round) 107 mm rocket system can be mounted on a variety of chassis, tracked and wheeled. For trials purposes it has already been installed on the rear of the Mamba light armoured vehicle but as far as it is known this combination never entered service.

As far as it is known, there have been no known customers for the vehicle-mounted version of this system. As well as being installed on the Mamba armoured vehicle, it could also be fitted onto unarmoured vehicles.

For vehicle applications, the 107 mm RO 107 (12-round) multiple rocket launcher system is palletised.

The upper part contains the complete rocket launcher with its elevation and traverse mechanism, while under the lower part are tubes that hold reserve rockets.

Specifications

	107 mm RO 107 (12-Round) Towed
Dimensions and weights	
Length	
overall:	2.6 m
Width	
transport configuration:	1.4 m
deployed:	1.7 m
Height	
overall:	1.2 m
Weight	
standard:	380 kg
combat:	610 kg
Mobility	
Configuration	
running gear:	wheeled
Firepower	
Armament:	12 × 107 mm rocket system
Main weapon	
maximum range:	8,500 m
muzzle velocity:	30 m/s
Rate of fire	
rapid fire:	12/9 s
Main armament traverse	
angle:	30° (±15°)
Main armament elevation/depression	
armament front:	+60°/0°

Status

Production as required. In service with several undisclosed countries. It is believed that this system is in service with at least one country in Africa. A small quantity of these systems may be used by the South African airborne forces. In 1998 it was reported that Peru had taken delivery of four RO 107 rocket launchers and 5,000 rounds of 107 mm rockets from South Africa.

Contractor

Mechem Developments, a member of the Denel Group.

Mechem Developments 107 mm Inflict (single-tube) rocket launcher

Development

The 107 mm Inflict single-tube lightweight rocket launcher was developed for the export market by Mechem Developments, a division of Denel, for use as an anti-personnel weapon by commando, airborne or other specialised units.

It is fully compatible with the Mechem Developments 107 mm RO 107 12-round trailer-mounted MRS, which is covered in a separate entry in *IHS Jane's Land Warfare Platforms: Artillery and Air Defence.*

As far as it is known there has been no recent production or marketing of the Mechem Developments 107 mm Inflict (single-tube) rocket launcher.

Description

The complete Inflict 107 mm launcher weighs 26 kg empty and can easily be carried by a two-man team into inaccessible terrain. The rocket tube is made of steel and is fitted inside a rectangular aluminium alloy holding frame with three foldable aluminium alloy legs. The fire-control unit consists of a hand-operated generator and a control box with two push buttons, connected to the launcher by 15 m of cable.

The frame has a screw to adjust the traverse between –8° and +8°, a geared handwheel to change the elevation between 0° and +45° and a simple mortar sight attachment for aiming. The sight is mounted on the left side of the launcher.

All the standard China North Industries Corporation (NORINCO) 107 mm unguided rocket types can be launched as well as the upgraded RO 107 round. Mechem is prime contractor for the upgraded RO 107 rocket: the rocket motor and rocket motor filling is supplied by Rheinmetall Denel Munition, the warhead by Aserma and the RO 107 proximity fuze by Fuchs Electronics.

The improved unguided surface-to-surface rocket is spin stabilised by seven canted venturis and rotates at approximately 22,000 rpm. It develops an impulse of 6,000 N, which gives it a maximum range of 8,500 m at sea level. Minimum firing angle can be as low as 4° for a direct-fire capability.

The 107 mm rocket is fitted with a Fuchs Electronics UHF radio band proximity fuze which arms after four seconds of flight time (approximately 1,400 m from the launcher). Frequency agility provides immunity to enemy jamming action. The proximity fuze causes a predictable airburst irrespective of the type of terrain. A point detonation back-up system is also provided.

The pre-fragmented warhead is packed with approximately 5,200 steel balls cast into epoxy resin. The shrapnel produced reaches an average velocity of 900 m/s which ensures the penetration of soft- skinned vehicles, support trucks, vehicle tyres and radiators as well as being lethal to personnel.

Typical kill and maim areas are 120 m^2 and 504 m^2 respectively for an upgraded round, compared to 34 m^2 and 140 m^2 for a standard cast-iron warhead fitted with a point impact fuze.

Mechem Developments 107 mm Inflict single-tube rocket launcher deployed and with remote launcher control box below 0500673

An extended-range 107 mm calibre rocket with a range of 11,000 m is under development by Mechem.

Specifications

	RO 107 mm Inflict (single-tube) rocket launcher
Dimensions and weights	
Length	
overall:	0.92 m
Weight	
launch:	19 kg
loaded:	45 kg
empty:	26 kg
Number of tubes	1
Performance	
Speed	
max speed:	1,332 km/h
Range	
min:	1.5 km
max:	8.5 km
Traverse	
angle:	8°
Elevation	
angle:	45°/0°
Ordnance components	
Warhead:	7.6 kg HE fragmentation
Fuze:	proximity (radio band with contact back-up)

Status

Production as required. In service with undisclosed countries. As far as is known the Mechem Developments 107 mm Inflict (single-tube) rocket launcher is not used within South Africa.

As far as it is known there has been no recent marketing of the Mecham Developments 107 mm Inflict (single-tube) rocket launcher.

Contractor

Mechem Developments, a member of the Denel Group.

Mechem Developments 68 mm RO 68 (6-round) portable multiple rocket launcher system

Development

The 68 mm RO 68 (6-round) portable rocket launcher was developed and built by Mechem Developments (a division of Denel) as a private venture for use as an anti-personnel weapon by special forces units.

As far as it is known there has been no recent production or marketing of the Mechem Developments 68 mm RO 68 (6-round) portable multiple rocket launcher system.

Mechem Developments 68 mm RO 68 rocket with warhead cutaway and fins folded as they would be in the launcher 0500912

Mechem Developments 68 mm RO 68 (6-round) portable multiple rocket launcher system (Christopher F Foss) 0500674

It has never confirmed to be in service with the South African National Defence Force although it would be well suited to special forces deployment.

Description

In the designation RO 68, the RO stands for rocket and 68 is the calibre of the rocket in millimetres. It is basically a 68 mm air-to-surface unguided rocket, which is normally fired from a pod carried under the wings of the fixed-wing aircraft or on the sides of helicopters and fitted with an upgraded HE fragmentation warhead for the surface-to-surface role and mated to a low-cost six-round portable lightweight launcher assembly.

The launcher has a single bank of six extruded aluminium rocket tubes in two vertical rows of three. The bank is mounted on a collapsible tripod unit and is fitted with a standard mortar sight for aiming purposes.

The launcher may be used in two modes:

- Conventionally, with the crew present and the rockets fired singly or in a rippled salvo by the control unit.
- Unconventionally, in the stand-off role when the six rockets can be fired in a salvo, without the crew being present, by use of a preset timer. The low cost of the launcher permits abandonment of it if necessary. The control unit may also be fitted with a self-destruct device as an option.

The 68 mm calibre RO 68 rocket has a pre-fragmented warhead containing over 3,000 steel balls.

A turbine-powered UHF radio band proximity fuze is fitted causing the warhead to airburst at a consistent height of 3 m above ground. This creates a lethal radius in excess of 15 m against personnel and is also highly effective against soft- skinned vehicles. A back-up point detonation system is also fitted. Maximum rocket range is 6,500 m. The rocket has folding fins which unfold when the rocket leaves the launcher.

Specifications

	RO 68 (6-round) portable multiple rocket launcher system
Dimensions and weights	
Length	
overall:	1.4 m
Diameter	
body:	68 mm
Number of tubes:	6
Weight	
empty:	45 kg
Performance	
Traverse	
angle:	8°
Elevation	
angle:	55°/0°

	RO 68 (6-round) portable multiple rocket launcher system
Range	
max:	6.5 km
Ordnance components	
Warhead:	HE fragmentation
Fuze:	proximity (3 m airburst setting and back-up contact)

Status
Production as required. In service with undisclosed countries. As far as is known the Mechem Developments 68 mm RO 68 (6-round) portable multiple rocket launcher system is not used within South Africa. It is understood that there has been no recent marketing of this system, or its associated 68 mm unguided rockets, by South Africa.

Contractor
Mechem Developments, a member of the Denel Group.

Spain

General Dynamics European Land Systems - Santa Bárbara Sistemas 140 mm (40-round) Teruel Multiple Rocket Launcher System

Development
The 140 mm (40-round) Teruel Multiple Rocket Launcher System (MRLS) was developed by the Spanish Council for Rocket Research and Development of the Spanish Ministry of Defence to replace the older MRLS in Spanish Army service.

Production of this system is complete and it is not expected that production will start again. As far as is known, the system was sold to two countries, Spain (18 systems) and Gabon (eight systems).

In November 2011 it was stated that this system was to be phased out of service with the Spanish Army.

The 140 mm (40-round) Teruel MLRS was only deployed by the Regimiento de Artilleria Lanzacohetes de Campana No 62 which was based in Astorga in Leon province in north-central Spain.

This regiment has now been re-equipped by the General Dynamics European Land Systems - Santa Bárbara Sistemas 155 mm 155/52 APU SBT howitzer which is covered in detail in a separate entry in *IHS Jane's Land Warfare Platforms: Artillery and Air Defence.*

As Spain has not produced any 140 mm unguided rockets for many years, the effectiveness of the eight systems stated to be deployed by Gabon must by now be in doubt.

Description
The launcher has two 20-round packs arranged in five rows of four 140 mm launcher tubes each, mounted on the rear of a Pegaso 6 × 6 truck chassis. The vehicle is fitted with an armoured, fully enclosed forward control six-man crew cab that can mount a light 7.62 mm machine gun on the roof for local defence and anti-aircraft purposes.

General Dynamics European Land Systems - Santa Bárbara Sistemas 140 mm (40-round) Teruel multiple rocket launcher system in the travelling configuration 0021075

General Dynamics European Land Systems - Santa Bárbara Sistemas 140 mm (40-round) Teruel multiple rocket launcher deployed in the firing position with stabilisers lowered at the rear 0569052

A longer 140 mm rocket has also been developed and is fitted with a double-grain solid-fuel motor that, when fitted with aerodynamic airbrakes, allows three distinct trajectories to be flown using the same firing elevation.

The unguided 140 mm rockets also have pop-out fins for stabilisation in flight. The warhead can be either a 18.6 kg HE fragmentation or, in the case of the long rocket, a 21 kg submunition cargo warhead.

This is available in four versions carrying 42 anti-personnel grenades each, with either 950 3.2 mm diameter steel balls, 60 g of Composition B explosive and a PO model percussion fuze, or 28 High-Explosive Anti-Tank (HEAT) grenades with contact fuzes and capable of penetrating 110 mm of conventional steel armour, six pressure-activated anti-tank mines fitted with anti-disturbance features or 14 smoke grenades each providing four minutes of smoke.

The smoke warhead weighs 21 kg and maximum speed of the unguided 140 mm rockets is 687 m/s.

The fuzes for the 140 mm rockets can either be the PD M34A2 contact type, proximity or timed. All six rockets can be mixed within the launcher and can be selected according to the target to be engaged.

A Pegaso 6 × 6 resupply vehicle with four blocks of 20 long-rocket reload rounds or six blocks of 20 short-rocket reload rounds is assigned to each vehicle as part of the battery. Reloading is manual and takes five minutes while the time taken to emplace a launcher is two minutes, to fire a complete salvo takes 45 seconds and to come out of action takes another two minutes.

The short HE rocket is 2.044 m in length, weighs 56.3 kg and has a maximum range of 18,000 m. The burn time of the solid propellent motor is 1.6 seconds. The long rocket is 3.23 m in length, weighs 76 kg and has a burn time of 2.7 seconds, giving a maximum range of 28,000 m. Minimum range for both rockets is 6,000 m.

Specifications

	140 mm (40-Round) Teruel
Dimensions and weights	
Crew:	5
Mobility	
Configuration	
running gear:	wheeled
layout:	6 × 6
Firepower	
Armament:	40 × 140.5 mm rocket system
Ammunition	
main total:	40
Main weapon	
maximum range:	28,000 m
Operation	
reload time:	300 s
Main armament traverse	
angle:	240°
Main armament elevation/depression	
armament front:	+55°/0°

Status
Production complete. No longer marketed, Phased out of service with Spain who deployed 14 launchers. Still deployed by Gabon who took delivery of a total of eight Teruel units but given the age of these 140 mm rockets the long term effectiveness of this systems is in doubt.

Contractor
General Dynamics European Land Systems - Santa Bárbara Sistemas.

Sudan

Military Industry Corporation Sudan Taka-2 122 mm (8-round) multiple rocket launcher

Development

In addition to manufacturing and marketing the Taka (BRY01) 107 mm (12-round) Multiple Rocket Launcher (MRL) in various configurations, which is covered in a separate entry in *IHS Jane's Land Warfare Platforms: Artillery and Air Defence*, the Military Industry Corporation (MIC) of the Sudan has placed in production at least two 122 mm rocket launchers.

These are the Taka-2 122 mm (8-round) MRL, which is also referred to as the GRL01, and the tripod mounted Taka-1 122 mm rocket launcher (GRL02).

Both of these rocket launchers are already deployed by the Sudanese Army and are already being marketed overseas but as of May 2013 there are no known sales.

The exact origin of these Sudanese 122 mm rocket launcher has not been disclosed but in many respects they are very similar to those manufactured by the Iranian Aerospace Industries Organization, Shahid Bagheri Industries.

This company is known to be marketing a variety of 122 mm launcher in single round, eight-round, 16-round, 30-round and 40-round configurations.

The Iran eight-round and 16-round MRL are primarily naval weapon systems with the latter having two banks each of 8 × 122 mm rocket launcher tubes.

Description

The first of these is the Taka-2 (GRL01) which is a manually operated pedestal type launcher that has two rows each of four 122 mm launcher tubes one on top of the other.

Traverse is 45° left and 45° right with elevation limits from –6° to +52° with both of these being via manual controls.

The complete Taka-2 122 mm (8-round launcher) weighs 920 kg and fires standard 122 mm unguided rockets out to a maximum range, using the standard rocket of 20 km.

These 122 mm unguided rockets are very much area effect weapons and are not very accurate with Circular Error of Probability (CEP) increasing as the range is increased.

Taka-2 122 mm (8-round) launcher can be installed on the rear of the tracked and wheeled platforms or even on naval craft and it is understood that the latter may be the first Sudanese application.

Variants

Taka-1 (GRL02) 122 mm tripod mounted single rocket launcher

The MIC of the Sudan is also manufacturing and marketing the Taka-1 (GRL02) 122 mm tripod mounted single barrel rocket launcher.

Details of this are provided in a separate entry in *IHS Jane's Land Warfare Platforms: Artillery and Air Defence*.

Specifications

	Taka-2 122 mm (8-round) multiple rocket launcher
Dimensions and weights	
Length	
tubes:	3 m
Number of tubes	8 × 122 mm
Weight	
empty:	920 kg
Performance	
Traverse	
angle:	45° (left/right)
Elevation	
angle:	52°/-6°

Sudanese 122 mm (8-round) Taka-2 pedestal type rocket launcher (MIC) 1430807

Status

Production as required. In service with the Sudan.

Contractor

Military Industry Corporation (MIC), Sudan.

Military Industry Corporation Sudan Taka-3 122 mm single barrel rocket launcher

Development

In addition to manufacturing and marketing the Taka (BRY01) 107 mm (12-round) Multiple Rocket Launcher (MRL) in various configurations, which is covered in a separate entry in *IHS Jane's Land Warfare Platforms: Artillery and Air Defence*, the Military Industry Corporation (MIC) of the Sudan has placed in production at least two 122 mm rocket launchers.

These are the Taka-2 122 mm (8-round) MRL, which is also referred to as the GRL01, and the tripod mounted Taka-3 122 mm single barrel rocket launcher (GRL02).

Both of these rocket launchers are already deployed by the Sudanese Army and are already being marketed overseas but as of May 2013 there are no known sales.

The exact origin of these Sudanese 122 mm rocket launchers has not been disclosed but in many respects they are very similar to some of those manufactured by the Iranian Aerospace Industries Organization, Shahid Bagheri Industries.

This company is known to be marketing a variety of 122 mm launcher in single-round, 8-round, 16-round, 30-round, and 40-round configurations.

The Iran 8-round and 16-round MRL are primarily naval weapon systems with the latter having two banks each of 8 × 122 mm rocket launcher tubes.

Description

MIC Taka-3 122 mm single barrel rocket launcher is ideal for use in difficult terrain by special and irregular forces and fires the same family of 122 mm unguided rockets out to a maximum range of 20 km as the Taka-2 122 mm (8-round) MRL previously mentioned. Typical rate of fire is about two 122 mm rockets a minute.

This single barrel launcher can be easily disassembled for transport and has manual traverse of 10° left and right with manual elevation from –5° to +55°.

The launcher complete with its tripod weighs only 70 kg and can be disassembled for ease of transportation.

Specifications

	Taka-3 122 mm single barrel rocket launcher
Dimensions and weights	
Length	
overall:	3 m
Number of tubes:	1 × 122 mm
Weight	
empty:	70 kg
Performance	
Traverse	
angle:	10° (left/right)
Elevation	
angle:	55°/-5°

The Sudanese 122 mm (single barrel) Taka-3 rocket launcher is well suited for use by irregular forces due to its low weight and compact dimensions and is shown here in its deployed configuration (MIC) 1430803

Status
Production as required. In service with Sudan.

Contractor
Military Industry Corporation (MIC), Sudan.

Military Industrial Corporation Sudan 107 mm (12-round) Taka multiple rocket launcher

Development
In early 2013 it was revealed that the Military Industry Corporation (MIC) of the Sudan had placed in production a 107 mm (12-round) multiple rocket launcher and its associated 107 mm unguided surface-to-surface rocket.

These are already in production and service with Sudan and are now being marketed overseas.

As of May 2013 there are no known sales of this 107 mm (12-round) rocket launcher or its associated 107 mm rocket.

Description
The MIC 107 mm (12-round) multiple rocket launcher is also referred to as the Taka (BRY01) and is mounted on a welded steel frame that can be attached to a light vehicle or a trailer.

The baseline launcher weighs 384 kg with a total weight of 611 kg including the rockets.

The launcher has manual traverse of 15° left and 15° right and has elevation from 0 to +60° with the controls being mounted on the left side of the launcher.

The 107 mm unguided rockets can be fired one at a time or in a salvo of 12 rockets in just six seconds.

The 107 mm solid propellant unguided rocket is also referred to as the Taka MRY01 and is fitted with a nose mounted impact fuze.

According to the MIC the 107 mm rocket has a launch weight of 18.85 kg of which 1.26 kg is the High-Explosive (HE) content.

Launch velocity is currently being quoted as 28.66 m/s with a maximum range of 8,500 m.

In addition to the 107 mm high explosive war shot rocket there is also a 107 mm inert training rocket and it is considered probable that additional types of rocket could be fielded with different warheads.

In appearance this Sudanese 107 mm (12-round) rocket launcher system is very similar to that manufactured by the Aerospace Industries Organization of Iran 107 mm (12-round) Fadjr-1 rocket launcher system which has a similar range.

Military Industrial Corporation of the Sudan 107 mm solid propellant unguided surface-to-surface rocket fitted with nose mounted fuze (Christopher F Foss) 1455632

Military Industrial Corporation of the Sudan 107 mm (12-round) rocket launcher mounted on the rear of a Karaba (4 × 4) cross-country vehicle (Christopher F Foss) 1455633

Iran has also manufactured a trailer mounted version of this rocket launcher as well as naval versions.

Details of these Iranian systems are provided in a separate entry in *IHS Jane's Land Warfare Platforms: Artillery and Air Defence*.

It should be noted that many other countries also manufacture similar 107 mm rocket launchers as well as 107 mm rockets for this system.

Variants
Self-propelled 107 mm (12-round) multiple rocket launcher

The Sudanese 107 mm (12-round) Taka rocket launcher has been shown integrated on the locally produced Karaba (4 × 4) tactical vehicle which can also be fitted with a wide range of weapons to the rear of the commander and driver position.

These include a 7.62 mm or 12.7 mm machine gun, Chinese Red Arrow series Anti-Tank Guided Weapon (ATGW), SPG-9 73 mm recoilless rifle and the US M40 series 106 mm recoilless rifle.

The vehicle can also be used as a command vehicle fitted with communications equipment.

The Karaba (4 × 4) tactical vehicle has a gross vehicle weight of up to 2,350 kg and is powered by a Nissan diesel engine developing 105 hp which gives a maximum road speed of up to 130 km/h.

It can also tow a trailer or weapon up to a maximum weight on up to 250 kg off road.

Sudanese 122 mm rocket launchers

The Military Industrial Corporation of the Sudan has also manufactured and is marketing the following 122 mm rocket launchers:

122 mm (single barrel) launcher

This is also referred to as the Taka-3 and is claimed to be able to fire a single 122 mm rocket to a maximum range of 20,000 m.

122 mm (8 barrel launcher)

This pedestal mounted launcher has two layers each of four rows of 122 mm rockets with manual elevation and traverse and can fire 122 mm rockets out to a maximum range of 20,000 m.

Specifications

	107 mm (12-round) Taka multiple rocket launcher
Dimensions and weights	
Length	
overall:	0.836 m
Diameter	
body:	106.7 mm
Number of tubes:	12 × 107 mm
Weight	
loaded:	611 kg
empty:	384 kg
launch:	18.85 kg
Performance	
Speed	
launch:	103 km/h
Traverse	
angle:	15° (left/right)
Elevation	
angle:	60°/0°
Range	
max:	8.5 km
Ordnance components	
Warhead:	1.26 kg HE fragmentation

Status
Production as required. In service with Sudan.

Contractor
Military Industry Corporation (MIC), Sudan.

Taiwan

RT2000 Artillery Multiple Launch Rocket System (AMLRS)

Development
Early in 2000, the Missile & Rocket Systems Research Division of the Chung-Shan Institute of Science and Technology revealed that it had developed a new Artillery Multiple Launch Rocket System (AMLRS) called the RT2000 to meet the operational requirements of the Republic of China (ROC) Army.

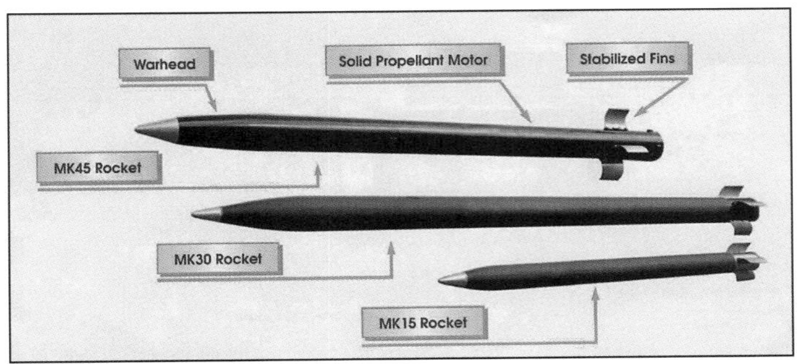

The RT2000 Artillery Multiple Launch Rocket System can launch three types of unguided rocket, the MK 15, MK 30 and MK 45, all have wraparound fins at the rear 0085769

This is now in production and in service with and is also referred to as the Ray Ting ('Thunder').

When compared to earlier systems used by the ROC Army the RT2000 has a significant increase in range and lethality. In addition to using its rocket systems in the conventional fire-support roles for ground forces, Taiwan also uses these in the anti-invasion craft role.

When originally unveiled the system was integrated onto the rear of the US Oshkosh Defense M977 series Heavy Expanded Mobility Tactical Truck (HEMTT) as stated in the description.

More recently this system has been demonstrated into a different 8 × 8 cross-country truck chassis fitted with a forward control cab.

When deployed in the firing position the forward cab windows are protected by shutters.

It is understood that at least 40 systems have now been deployed by Taiwan with production still underway.

They are normally deployed in battalions of 18 launchers with each battery having six launchers plus associated resupply vehicles, command vehicles and maintenance vehicles.

Description

RT2000 is based on the chassis of the US Oshkosh Defense M977 series Heavy Expanded Mobility Tactical Truck (HEMTT) (8 × 8) which is used in large numbers by the ROC and has a nominal payload of 10 tons (US).

Mounted on the rear of the chassis is the powered launcher that can be fitted with pods of unguided surface-to-surface rockets that are aimed and fired with the crew remaining in the forward-control cab.

To provide a more stable firing platform two hydraulic stabilisers are lowered to the ground on either side of the truck.

A computerised fire-control system is fitted which is coupled to global positioning system.

This reduces into action time to less than eight minutes and fire its first unguided surface-to-surface rockets.

The RT2000 fires three types of solid propellant, unguided rocket all of which have four wrap round fins at the rear that unfold after launch.

These rockets are designated the MK 15 (117 mm), MK 30 (180 mm) and MK 45 (230 mm) with the figure of 15, 30 and 45 indicating the range of the rocket in km.

The rockets require no maintenance and are treated as 'wooden rounds'.

The rockets can be fitted with two types of warhead, high-explosive and submunition carrying M77 dual-purpose bomblets that are highly effective against a variety of battlefield targets. The rocket warhead fuzes are set by remote control before the rockets are launched.

The rockets are transported and launched from pods that are rapidly loaded onto the launcher using a crane mounted above and to the rear of the second axle.

The MK 15 rockets come in pods of 20 rockets, MK 30 in pods of nine and MK 45 in pods of six. When using the MK 30 rockets the system carries a total of 27 rockets in the ready to launch position.

Latest version of the RT2000 Artillery Multiple Launch Rocket System (AMLRS) on latest 8 × 8 cross-country chassis with stabilisers deployed and launcher with pod of nine MK 30 180 mm rockets traversed slightly to the right (Michael Cole) 1487951

Scale model of the RT2000 Artillery Multiple Rocket System (AMLRS) which is based on the US HEMETT (8 × 8) truck chassis (Craig Hoyle) 0067603

Mounted in the fully enclosed forward-control cab is the fire-control panel which allows the crew to aim and fire the rockets with key elements of the system including the electronic unit, position and attitude determining system, controller box, servo motors, amplifier and the communications processor.

Primary role of the RT2000 system is the rapid neutralisation of amphibious craft before they are able to deposit troops and equipment on the beach.

Current Taiwanese unguided rocket systems, covered in separate entries in *IHS Jane's Land Warfare Platforms: Artillery and Air Defence*, include the 126 mm (40-round) Kung Feng III/IV and the 117 mm (45-round) Kung Feng VI mounted on a US-supplied, five-tonne (6 × 6) series truck. The latter rockets have a maximum range of 15 km and are similar to those used in the latest RT2000 AMLRS.

Specifications

RT2000 Artillery Multiple Launch Rocket System

Chassis:	8 × 8
Calibre:	
(MK 15 rocket)	117 mm
(MK 30 rocket)	180 mm
(MK 45 rocket)	230 mm
Maximum range:	
(MK 15 rocket)	15,000 m
(MK 30 rocket)	30,000 m
(MK 45 rocket)	45,000 m

Status

Production. In service with the Republic of China Army.

Contractor

Chung-Shan Institute of Science and Technology, Missiles & Rocket Systems Research Division.

126 mm (40-round) Kung Feng III/IV Multiple Rocket Systems

Development

The Kung Feng 126 mm (40-round) MRSs were designed in Taiwan by Sun Yat-sen Scientific Research Institute in the early 1970s.

Production of this system was completed some years ago and they were never offered on the export market.

This has now been supplemented by the more effective RT2000 Artillery Multiple Launch Rocket System (AMLRS).

Details of this more recent system are provided in a separate entry in *IHS Jane's Land Warfare Platforms: Artillery and Air Defence.* As this is mounted on an 8 × 8 cross-country chassis it has greater mobility than the older systems and the unguided rockets have a much longer range and more effective warheads.

Description

The basic Kung Feng ('Worker Bee') consists of two banks of 20 126 mm calibre rocket launcher tubes arranged in five rows of four. The tubes are over 1.6 m long and fire spin-stabilised unguided solid-propellant surface-to-surface rockets.

The rocket length is about 0.78 m and the weight approximately 24.94 kg. The maximum range is estimated to be 9,000 m, with the 40 electrically fired 126 mm rockets ripple fired in 16 seconds.

The launcher can be mounted on a variety of platforms. In Taiwanese service the following are known.

It cannot be confirmed that all of these platforms are currently in service with Taiwan.

Kung Feng IV multiple rocket launcher on modified BAE Systems M113A1 APC chassis (L J Lamb) 0058470

Towed version of Kung Feng IV multiple rocket launcher on parade being towed by an M37 (4 × 4) truck 0500676

Variants

Trailer-mounted version

A towed version for use by the infantry. The launcher is mounted on a two-wheeled carriage with a tubular frame and towed by a US supplied 3/4 ton M37 (4 × 4) truck. This system is believed to be designated Kung Feng III.

M113A1 rocket launcher

A self-propelled version on the United States supplied the now BAE Systems (previously United Defense) M113A1 APC for mechanised and armoured units. The launcher is mounted just behind the troop compartment hatch pointing forward. Elevation and depression are hydraulically operated but there is no facility for traversing.

A blast shield is fitted over the rear hull top to protect the dome ventilator. In order to fire, the crew must disembark. Use of the troop compartment is limited by the launcher's angle of elevation. This system is known as Kung Feng IV.

LVTP5 rocket launcher

A self-propelled version on the old US supplied Landing Vehicle Tracked Personnel 5 (LVTP5) amphibious assault personnel carrier for the marine corps. The launcher is fitted just to the rear of the commander's hatch and can be fired while the vehicle is afloat. This is also known as the Kung Feng IV.

An ECM chaff-dispensing system known as the Kung Feng Sea is in use on Taiwanese naval destroyers, as far as it is known this rocket is not fired by the land-based multiple rocket launchers.

Status

Production of all these 126 mm multiple rocket systems has been completed, although replacement rockets are built as and when required. In service with the Taiwanese armed forces. It is estimated that the ROC has about 300 MRSs currently in service. There are no known exports of this system.

These rocket systems are now being supplemented by the latest locally developed RT2000 Artillery Multiple Launch Rocket System (AMLRS) based on an 8 × 8 cross-country chassis which fires rockets with a greater range and more effective warheads.

It is understood that by early 2013 at least 40 of the latest RT2000 AMLRS systems had been deployed by Taiwan.

Contractor

Unknown.

117 mm (45-round) Kung Feng VI Multiple Rocket Launcher System

Development

The 117 mm Kung Feng VI (45-round) Multiple Rocket Launcher System was designed by the Sun Yat-sen Scientific Research Institute in the late 1970s.

This is now being supplemented by the more effective RT2000 Artillery Multiple Rocket System (AMLRS).

Details of this more recent system are provided in a separate entry in *IHS Jane's Land Warfare Platforms: Artillery and Air Defence.* As this is mounted on an 8 × 8 cross-country chassis it has greater mobility than the older systems and the unguided rockets have a much monger range and more effective warheads.

It is estimated tat by early 2013 at least 40 of thE latest RT2000 AMLRS systems had been deployed by Taiwan.

117 mm Kung Feng VI 45-round multiple rocket launcher mounted on rear of US supplied 5 ton 6 × 6 cross-country truck (L J Lamb) 0058471

Description

Intended only as a self-propelled MRS, the present model is designated Kung Feng VIA and is a 117 mm (45-round) launcher with five layers of nine tubes mounted on a traversable pedestal with hydraulic elevation equipment.

The US supplied carrier vehicle normally used is the M52A1 (6 × 6) five-tonne truck chassis with the launcher mounted on the rear. The cab is unarmoured so that the traverse is normally beyond 90° when firing. A support vehicle carrying reload 117 mm rocket rounds is required. Reloading is manual and takes up to 15 minutes for all 45 barrels.

The unguided solid propellant 117 mm rocket is 1.8 m long, weighs approximately 60 kg and is spin stabilised in flight by four fold-out fins at the tail. Maximum range is estimated to be 15,000 m, with a full salvo having a lethal area of some 30,000 m^2.

The rockets are electrically fired with a 0.5 second delay between each, giving a total ripple firing time of 22 seconds. The warheads are of the high explosive blast fragmentation type.

A computer-based fire-control capability is being developed, which is expected to further increase the accuracy of the system.

When deployed in the firing position, stabilisers are lowered to the ground to provide a more stable firing platform. The sighting system is mounted on the left side of the launcher.

The 117 mm (40-round) Kung Feng VI is now being supplemented by the new RT2000 Artillery Multiple Launch Rocket System based on an Oshkosh M977 (8 × 8) Heavy Expanded Mobility Tactical Truck (HEMTT) series truck.

This system, which fires MK 15 (117 mm), MK 30 (180 mm) and MK 45 (230 mm) rockets. The MK 15 rocket is similar to that used with the 117 mm (45-round) Kung Feng VI system.

It is possible that these launchers could be removed from the older M52A1 (6 × 6) five-tonne truck chassis and placed on a more recent 6 × 6 cross-country truck chassis.

Variant

An experimental version has been identified. A simple non-traversable launcher with hydraulic elevation controls was mounted in place of the twin 40 mm turret on an M42 tracked self-propelled anti-aircraft gun chassis. The current status is not known. The M42 twin 40 mm system was deployed in the 1950s and was phased out of service with the US Army many years ago.

Status

Production of the complete system has been completed but production of the rockets is undertaken as and when required. In service with the Republic of China Army. It is estimated that the ROC Army deploys about 300 multiple rocket launchers of all types. There are no known exports of any of the multiple rocket launcher systems developed by Taiwan.

These rocket systems are now being supplemented by the latest locally developed RT2000 Artillery Multiple Launch Rocket System (AMLRS) based on an 8 × 8 cross-country chassis which fires rockets with a greater range and more effective warheads.

Contractor

Unknown.

Turkey

Roketsan 300 mm (4-round) Artillery Rocket System

Development

In 2007 Roketsan finally released some details of a 300 mm (4-round) Artillery Rocket System (ARS) which it has now delivered to the Turkish Land Forces Command (TLFC).

This system is based on technology transfer but modified and customised to meet the requirements of the TLFC. It is understood that China supplied much of the technology used in the 300 mm rocket used with this system.

In many respects the launcher is very similar to the China Precision Machinery Import and Export Corporation 302 mm WS-1B (4-round) artillery rocket system. This is covered in detail in a separate entry in *IHS Jane's Land Warfare Platforms Artillery and Air Defence* which is still being marketed.

Roketsan are the overall prime contractor of the system and at the subsystem level it leverages off of its considerable experience in the design, development and production of unguided surface-to-surface rocket systems for the home and export markets.

In TLFC, the 300 mm system provides long-range fire support with the Roketsan 122 mm (40-round) T-122 ARS covering shorter ranges out to 40 km with the latest extended range rockets.

The system is understood to be typically deployed in batteries of six or nine launchers with a similar number of resupply vehicles on the same Rheinmetall MAN Military Vehicles (6 × 6) cross-country chassis plus a battery command post vehicle.

According to the United Nations Arms Transfer List for 2010 Turkey exported one 122 mm/300 mm launcher together with 48 x 122 mm rockets and 32 T-300 rockets to the United Arab Emirates.

The United Arab Emirates is known to have upgraded some of its Italian supplied truck mounted (6 × 6) systems with the Roketsan 122 mm (40-round) T-122 Multibarrel Rocket Launcher System which is covered in a separate entry in *IHS Jane's Land Warfare Platforms Artillery and Air Defence.*

Early in 2013 Jobaria Defence Systems (JDS) based in the United Arab Emirates (UAE) unveiled the Multiple Cradle Launcher which has remote controlled pods which can launch 122 mm or 300 mm rockets.

This trailer mounted system is now in service with the United Arab Emirates and is covered in a separate entry in *IHS Jane's Land Warfare Platforms Artillery and Air Defence*.

Description

The Roketsan 300 mm (4-round) ARS comprises two key parts; the T-300 MultiBarrel Rocket Launcher (MBRL) and TR-300 Artillery Rocket (AR).

The T-300 MBRL is based on the German Rheinmetall MAN Military Vehicles 26.372 (6 × 6) 10 ton-tonne cross-country truck chassis which is used by the TLFC for a number of other applications.

The cab windows are provided with shutters which are lowered before the rockets are launched.

This has a fully enclosed forward control cab to the immediate rear of which is the Auxiliary Power Unit (APU) and additional fully enclosed crew space. A .50 (12.7 mm) M2 HB machine gun is mounted on the cab roof for air defence and local defence purposes.

Mounted at the very rear of the chassis is a power operated turntable on which is mounted the four tube 300 mm rocket launcher with a traverse of 30° left and right with elevation limits from 0° to +60°.

To provide a more stable firing platform, four hydraulic stabilisers are lowered to the ground. One of these is positioned to the rear of the cab on either side with the other two at the very rear of the chassis.

Combat weight, complete with four rockets, is being quoted as 23 tonnes and the rockets can be fired in a complete salvo of four rockets with an interval of six seconds between launch or one rocket can be fired at a time.

The rockets can be launched from within the cab of the launcher by a battery command post.

Each launcher has its own computerised fire-control system and land navigation system for reduced into action times and more accurate target engagement.

The 300 mm unguided rockets use a composite solid propellant (HTPB) with a maximum burn time of 4.5 seconds and are fitted with a warhead developed by Roketsan.

This warhead contains about 80 kg of high explosive and 26,000 steel balls are packed for maximum effect against targets in the open.

The rocket is fitted with a nose-mounted proximity fuze with a lethal range of 70 m being quoted although fragments go out well beyond this range.

The rockets have a minimum range of 40 km with a maximum range with drag rings of 80 km and 100 km without drag rings. The rockets have an overall length of 4.75 m and have four fixed fins at the rear with a quoted launch weight being given as 525 kg of which 150 kg is the warhead. Maximum flight time is quoted as 170 seconds.

Roketsan T-300 MBRL based on Rheinmetall MAN Military Vehicles (6 × 6) truck chassis and with launcher elevated. Note that the stabilisers have not been deployed (Roketsan) 1403751

Roketsan T-300 MBRL based on Rheinmetall MAN Military Vehicles (6 × 6) truck chassis with launcher in horizontal position for travelling (Roketsan) 1403752

In a typical target engagement, once the battery has launched its four rockets it would normally redeploy to another position for reloading to avoid counterbattery fire.

Replacement rockets are carried in individual containers and then unpacked and loaded onto the Rheinmetall MAN Military Vehicles (6 × 6) resupply truck using its onboard hydraulic crane. The rockets are then loaded into the launcher tubes using a hydraulic rammer.

Roketsan are understood to be studying alternative warheads for the TR-300 system, as well as perhaps having the rockets in pods of two or four rockets for more rapid loading in the field.

Project J Missile

The TLFC has for several years deployed a tactical ballistic missile system with a range of at least 150 km with a precision strike capability.

This has been referred to a Project J (J-600T Yildirum - Thunderbolt) and is understood to have been developed by Roketsan. The system has also been referred to as Project Kasirga.

The solid propellant rocket is transported and launched from a Rheinmetall MAN Military Vehicles 26.372 (6 × 6) cross-country truck chassis with a similar vehicle carrying replacement missiles which are loaded onto the single rail launcher using a crane.

The missile has an Inertial Guidance System (INS) and can be fitted with various types of conventional warhead that would enable high value targets such as command posts and air defence systems to be neutralised at long range with precision effect.

Baseline 480 kg warhead is probably High-Explosive (HE) and the missile is about 6.2 m in length, 600 mm in diameter and has a launch weight of about 2,100 kg.

Maximum range of the first version to enter service is 150 km but a second version, the Yildirum II is understood to have range extended to 300 km.

It is believed that this Project-J is based on the China Precision Machinery Import-Export Corporation B-611 series missile system. This launches an unguided rocket with a minimum range of 40 km and a maximum range of 100 km fitted with a HE fragmentation warhead.

Variants

Inertial Guidance System

Roketsan has developed an inertial guidance system which includes an active control actuation unit that includes an electric motor with gears, a ball screw assembly and associated electronics.

The avionics package includes an inertial measurement unit, global positioning system and a computer.

As of May 2013 it is understood that this remains at the prototype/development phase and it has not been confirmed as to whether this is applicable to Roketsan 122 mm or 300 mm surface-to-surface rocket systems.

Specifications

	T-300 300 mm (4-round) launcher
Dimensions and weights	
Weight	
standard:	20,500 kg
combat:	23,000 kg
Mobility	
Configuration	
running gear:	wheeled
layout:	6 × 6
Firepower	
Armament:	4 × 300 mm rocket system
Main armament traverse	
angle:	60° (30° (left and right))
Main armament elevation/depression	
armament front:	+60°/0°

TR-300 Artillery Rocket Specifications

Diameter:	300 mm
Length:	4.750 m
Span:	600 mm
Weight:	525 kg
Warhead weight:	150 kg
Warhead type:	TNT/RDX
Warhead explosive weight:	80 kg
Number of lethal fragments:	26,000 steel balls plus fragments
Range:	
(minimum)	40,000 m
(maximum with drag rings)	80,000 m
(maximum range without drag rings)	100,000 m
Warhead fuze type:	proximity
Lethal radius:	70 m
Fragment effective radius:	1,000 m
Rocket burnout time:	4.5 m
Rocket propellant:	composite solid propellant (HTPB)
Maximum flight time:	170 seconds

Status
Production as required. In service with the Turkish Land Forces Command and the United Arab Emirates (one launcher and 32 rockets delivered in 2010).

As stated in the development the United Emirates has deployed a trailer mounted system with remote controlled launchers that can be fitted with pods of 122 mm or 300 mm unguided surface-to-surface rockets supplied by Roketsan.

Contractor
Roketsan.

MKEK 107 mm (12-round) multiple rocket launcher system

Development
The Turkish company MKEK-Çansas, reverse-engineered the China North Industries Corporation (NORINCO) 107 mm (12-round) Type 63 multiple rocket launcher (details of which are given in a separate entry in *IHS Jane's Land Warfare Platforms: Artillery and Air Defence.* It should be noted that NORINCO no longer markets this system which has also been copied by a number of other countries.

Production of the MKEK 107 mm (12-round) multiple rocket launcher system is undertaken on an as required basis although it is understood that there has been no recent production of this system. It has been offered on the export market but as of May 2013 there are no known sales.

Description
Details of the system are believed to be similar to the NORINCO Type 63 specifications except where manufacturing techniques reflect local variations. It is understood that Turkish Land Forces Command (TLFC) facilities were also involved in production of this launcher.

The full range of 107 mm calibre rockets can be fired including the Roketsan TR-107 family. Details of this are given in a separate entry.

This rocket is now in production on an as required basis for the TLFC and has a maximum range of just over 11,000 m and is available with two types of warhead, high-explosive and steel ball. This has been offered on the export market but there are no known sales to date.

A hand-held fire-control system is used to calculate firing data automatically for 107 mm rockets by using precise flight simulation software. The panoramic day sighting telescope is on the left side of the launcher, while the firing cable is on the right.

MKEK 107 mm (12-round) multiple rocket launcher system deployed in the firing position (Christopher F Foss) 0501065

Basic details of the Roketsan T-107 launcher are given below.

Variants
Cobra with 107 mm (12-round) multiple rocket launcher system

For trials purposes the Turkish Otokar Cobra (4 × 4) light armoured vehicle has been fitted with a roof mounted 107 mm (12-round) multiple rocket launcher system towards the rear.

Mizrak

MKEK also manufacture a 122 mm unguided surface-to-surface rocket called Mizrak, which has a maximum range of 22,000 m and weighs 67.5 kg.

Specifications

	107 mm	107 mm (12-round)
Dimensions and weights		
Length		
overall:	0.84 m	2.525 m
width		
body:	-	1,525 mm
calibre:	107 mm	
height		
body:	-	1,600 mm
Weight		
launch:	18.80 kg	
loaded:	663 kg	
empty:	435 kg	
Number of tubes:	12	
Performance		
Speed		
max speed:	1,332 km/h	
Range		
min:	8.5 km	
max:	11 km	
Elevation		
angle:	0°/55°	
Ordnance components		
Warhead:	8.5 kg (18.00 lb) HE TNT	
Propulsion		
type:	single stage (N5 double base)	

Status
Recent information has indicated that the Turkish Army has deployed 48 of these MKEK 107 mm (12-round) multiple rocket launcher systems. Production complete. No longer marketed.

Contractor
Makina ve Kimya Endüstrisi Kurumu (MKEK-Çansas).

MKEK 70 mm (40-round) MAKSAM RA-7040 multiple rocket launcher

Development
As part of the Turkish Land Forces Command weapons diversification programme, MKEK-Çansas developed and placed in production a 70 mm (2.75 in) multiple rocket launcher known as the RA-7040.

In the latter designation, the 70 is for the calibre of the rockets in millimetres and 40 is the number of launcher tubes.

A total of 24 production systems were delivered to the Turkish Land Forces Command. Production has now been completed and it is no longer marketed by MKEK.

It should be noted that the 70 mm solid propellant unguided rockets used in this system were originally developed for the air-to-ground application.

Description
The MKEK RA-7040 multiple rocket launcher system is of a modular design and is 40-round 70 mm mounted on a two-wheeled trailer with four hydraulically-operated jacks for stabilisation purposes during firing.

Any vehicle with a 600 kg towing capacity can tow the launcher. It can be adapted to fit on both land and amphibious vehicles if required. As far as it is known all examples built to date have been in the trailer-mounted configuration.

The trailer-launcher weighs 1,320 kg when empty and can fire all 40 70 mm rockets in either single or ripple launch modes within 10 to 40 seconds. All types of 70 mm unguided solid-propellant rockets can be fired with a 40-round salvo typically covering an area of 200 × 300 m at a range of 7,400 m.

MKEK-Çansas 70 mm (40-round) MAKSAM RA-7040 MRS deployed in firing position 0500677

The actual solid propellant unguided rocket is also used in the air-to-surface role (fired from fixed-wing aircraft and helicopters) with the double-base propellant giving a burn time of 1.42 seconds.

Types of warhead that can be fitted include M151 (anti-personnel), M229 (anti-personnel), M427 (anti-personnel), flechette, Mk 5 and Mk 427 (anti-armour), M156 (smoke) and M245 (chemical). The warheads are understood to be identical to those originally developed for air-to-ground applications.

The launcher has a full 360° traverse capability and is controlled by a remote-control device with 25 m of cable.

Specifications

	RA-7040
Dimensions and weights	
Length	
overall:	3.470 m
Width	
transport configuration:	2.060 m
Height	
transport configuration:	2.135 m
Weight	
standard:	1,320 kg
combat:	1,690 kg
Mobility	
Configuration	
running gear:	wheeled
Electrical system	
vehicle:	24 V (trailer)
Firepower	
Armament:	40 × 70 mm rocket system
Main armament traverse	
angle:	360°
Main armament elevation/depression	
armament front:	+55°/0°

Status

Production complete. To date, 24 have been produced for the Turkish Land Forces Command. No longer marketed. There are no known exports of this system.

Contractor

Makina ve Kimya Endüstrisi Kurumu (MKEK-Çansas).

Roketsan 122 mm Extended Range (+40 km) MultiBarrel Rocket Launcher Ammunition, TR-122

Development

The Turkish company Roketsan has type classified a 122 mm calibre Extended Range MultiBarrel Launcher Ammunition designated the TR-122. This new unguided rocket extends the 122 mm calibre rocket launcher system's range to over 40,000 m.

Roketsan 122 mm (40-round) T-122 MultiBarrel Launcher System firing a Roketsan 122 mm TR-122 extended-range rocket (Roketsan) 1178866

The United Arab Emirates has upgraded some of its old Italian FIROS 122 mm (40-round) systems with a new launcher system which fires Roketsan 122 mm rockets.

Details of this upgrade are provided in the Roketsan 122 mm (40-round) T-122 MultiBarrel Rocket Launcher System in *IHS Jane's Land Warfare Platforms: Artillery and Air Defence.*

The United Arab Emirates may have taken delivery of some of these Roketsan extended range 122 mm rockets but these have not been declared to the United Nations.

According to recent United Nations Arms Transfer lists the following quantities of 122 mm unguided rocket have been supplied to the United Arab Emirates by Turkey:

Year	Quantity	Type
2005	3,040	TRB-122 and TRK-122
2006	3,040	type not stated
2007	2,000	TRB-122 and TRK-122
2008	740	TRB-122
2008	740	TRB-122 rockets

According to the United Nations Arms Transfer Lists, in 2010 Turkey exported to the United Arab Emirates one 122 mm rocket launcher and 480 122 mm rockets as well as one 300 mm rocket launcher and 32 TR-300 rockets.

Early in 2013 Jobaria Defence Systems (JDS) based in the United Arab Emirates (UAE) unveiled the Multiple Cradle Launcher which has remote controlled pods which can launch 122 mm or 300 mm rockets.

This trailer mounted system is now in service with the United Arab Emirates and is covered in a separate entry in *IHS Jane's Land Warfare Platforms: Artillery and Air Defence*.

Description

The TR-122 unguided rocket can be fired from a Roketsan made 122 mm T-122 MultiBarrel Rocket Launcher and its foreign equivalents, such as the widely deployed Russian 122 mm BM-21 Grad.

The T-122 rocket can also be fired by other 122 mm systems, such as the Czech RM-70 and the Romanian APRA 40. The latest production T-122 rocket system is based on a Rheinmetall MAN Military Vehicles (6 × 6) cross-country truck chassis and has been in service with Turkish Land Forces Command since 1996.

The complete Roketsan 122 mm rocket family consists of the following types:

- TR-122 with HE warhead and maximum range 40 km, minimum range without drag ring is 21 km while with drag ring it is 10 km. Propellant is composite (HTB/AP/Al) and a point detonating fuze is fitted.
- TRB-122 with steel ball warhead and maximum range of 40 km, minimum range without drag ring is 21 km while with drag ring it is 10 km. Propellant is composite (HTB/AP/Al) and a proximity fuze is fitted. The warhead contains 5,500 steel balls and 4 kg of TNT + RDX and has a lethal radius of 40 m.
- TRK-122 with submunition warhead and maximum range of 30 km and a minimum range of 16 km. Propellant is composite (HTB/AP/Al). An electronic time fuze is fitted. The warhead contains 50 anti-personnel/anti-armour munitions and six incendiary munitions. The former will penetrate between 120 and 150 mm of conventional steel armour. A self-destruct fuze system is fitted.

Variants

Inertial Guidance System

Roketsan has developed an inertial guidance system which includes an active control actuation unit that includes an electric motor with gears, a ball screw assembly and associated electronics.

The avionics package includes an inertial measurement unit, global positioning system and a computer.

	TR-122	TRB-122	TRK-122
Dimensions and weights			
Length			
overall:	2.93 m	2.93 m	3.24 m
Diameter			
body:	122 mm	122 mm	122 mm
Weight			
launch:	65.9 kg	65.9 kg	71.6 kg
Performance			
Range			
min:	21 km (without drag ring)	21 km; (without drag ring)	16 km (without drag ring)
	10 km (with drag ring)	10 km (with drag ring)	
max:	40 km (est.)	40 km (est.)	30 km
Ordnance components			
Warhead:	18.4 kg HE	18.4 kg 5,500 steel ball	22.9 kg HE 50 APAM
			incendiary 6
Fuze:	impact	proximity	time delay, self destruct

As of May 2013 it is understood that this remains at the prototype/development phase and it has not been confirmed as to whether this is applicable to Roketsan 122 mm or 300 mm surface-to-surface rocket systems.

Specifications – *see table above*

Status

Production as required. Multiyear contract placed by Turkish Land Forces Command. Offered for export. The United Arab Emirates has now taken delivery of an upgrade package from Roketsan for its current FIROS 30 122 mm rocket systems. This is understood to include some 122 mm extended-range rockets.

As stated in the development the United Emirates has deployed a trailer mounted system with remote controlled launchers that can be fitted with pods of 122 mm or 300 mm unguided surface-to-surface rockets supplied by Roketsan.

Contractor

Roketsan.

Roketsan 122 mm (40-round) T-122 MultiBarrel Rocket Launcher System

Development

The Turkish company, Roketsan, is the prime contractor for a 122 mm calibre multibarrel rocket launcher system designated the T-122 which entered service with the Turkish Land Forces Command in 1996 and is now being offered on the export market.

According to Roketsan, the T-122 multibarrel rocket launcher is an autonomous, artillery weapons system that is used for indirect fire-support missions under day and night conditions and in all weather conditions against area targets.

Early in 2013 Jobaria Defence Systems (JDS) based in the United Arab Emirates (UAE) unveiled the Multiple Cradle Launcher which has remote controlled pods which can launch 122 mm or 300 mm rockets.

This trailer mounted system is now in service with the UAE and is covered in a separate entry in *IHS Jane's Land Warfare Platforms: Artillery and Air Defence*.

Description

The Rocketsan 122 mm (40-round) T-122 multi-barrel rocket launcher is based on a Rheinmetall MAN Military Vehicles (6 × 6) 26.281 cross-country truck fitted with a forward-control cab. To the rear of this is another fully enclosed cab for the additional crew members. On the rear of this chassis are two pods each of 20 × 122 mm rocket tubes.

Rocketsan 107 mm/122 mm multibarrel rocket launcher cradle fitted with three pods each of 20 × 107 mm rocket tubes (Roketsan) 1403753

There have been two types of 122 mm pods developed for this T-122 launcher. The original one was the steel (multi-use) pods followed by the recent composite sealed (single use) pods.

These composite pods can withstand the adverse effects of high temperature, humidity, sand and salty conditions.

According to Roketsan, the latest technology has been used in the design of the T-122 multibarrel rocket launcher, including an advanced fire-control computer, digital secure communication system, armour protected cab, NBC protection, composite pods, remote fuze and an automatic stabilisation system.

The launcher is hydraulically traversed and elevated via FCC with manual handwheels as a back up. The unguided rockets can be fired singly or in salvo. A full salvo of 40 rockets can be launched in 80 seconds. Reloading is via a hydraulic loading crane, which is mounted on the launcher.

When deployed in the firing position, four stabilisers are lowered to the ground under hydraulic control to provide a more stable firing platform.

The launcher can fire the full range of 122 mm unguided rockets available including Russian 122 mm GRAD rockets and Roketsan's 122 mm rockets including the 122 mm TR-122, TRB-122 and TRK-122 which are covered in detail in the entry for the Roketsan 122 mm Extended Range (+40 km) MultiBarrel Rocket Launcher Ammunition, TR-122.

Upgraded T-122 launcher

The company is currently marketing an enhanced version of the T-122 MBRLS based on the latest generation Rheinmetall MAN Military Vehicles (6 × 6) cross-country truck chassis.

To the rear of the forward-control cab is a fully enclosed cab for the remainder of the crew, which normally comprises five people. It is envisioned that production systems will have full armour-protected cabs, which would also have a nuclear, biological and chemical defence system of the overpressure type, as well as an air-conditioning system.

Installed in the front cab is a new computerised Fire-Control System (FCS), which carries out a number of functions including built-in test before and during firing. It can store up to 20 target locations and take into account meteorological information in various formats.

The installation of a global positioning system/inertial navigation system improves the system's overall accuracy. A secure digital communications system has been installed to interface with the battery command centre, which typically controls six launchers plus support elements.

Close-up of the latest T-122 launcher showing two disposable pods of 20 × 122 mm rockets and hydraulic crane for reloading purposes (Christopher F Foss) 0563326

Roketsan 122 mm (40-round) T-122 multiple rocket launcher on latest Rheinmetall MAN Military Vehicles (6 × 6) cross-country chassis (Roketsan)
1178868

The FCS calculates the firing data automatically for the 122 mm rockets using precise flight simulation software. It automatically lays the rocket launcher onto the target and fires the 122 mm rockets either one at a time or in intervals of two seconds. The fuzes can be set by remote control.

The installation of the new FCS allows the system to come into action, fire a salvo of 40 122 mm rockets and depart in under five minutes. This reduces the exposure of the rocket system to counterbattery fire.

Although the first-generation launcher had to be reloaded manually, the latest system has two pods of 20 rockets that can be rapidly replaced by the resupply vehicle on a similar chassis using its onboard hydraulic crane. The pods are delivered in sealed units and are discarded once fired.

UAE upgraded FIROS 122 mm (40-round) system

The United Arab Emirates (UAE) has now taken into service an upgraded version of its Italian supplied BPD FIROS 30 122 mm (40-round) multiple rocket launcher system.

These were originally delivered to the UAE many years ago and each launcher system consists of a Mercedes-Benz (6 × 6) cross-country chassis on the rear of which is mounted a turntable with two pods of 20 122 mm unguided surface-to-surface rockets.

As the 122 mm rockets have now run out of shelf-life, the Turkish company of Roketsan was awarded a contract to provide a new launcher system together with a new family of 122 mm rockets and a modified fire-control system.

For contractual reasons, Roketsan has never released any details of this major export contract but first deliveries were made in 2005 and are understood to have been completed in 2008.

The upgrade work to the launcher was carried out in the UAE with rockets provided direct from Turkey. It is understood that the UAE has about 48 launchers but at this stage it is not known how many have now been upgraded.

Existing launcher system has been removed and replaced by a new turntable with powered launcher which carries two pods each of 20 × 122 mm rockets.

Once these rockets have been fired the launcher would normally redeploy to another position to avoid counterbattery fire where the pods would be rapidly removed and replaced by two new pods.

So far the UAE has taken delivery of at least two types of Roketsan 122 mm rockets which have already been produced in production quantities for the Turkish Land Forces Command.

The first UAE batch consisted of 3,040 rockets in two models, TRB-122 and TRK-122. The former is fitted with a high-explosive warhead and has a stated maximum range of over 40 km with a minimum range of 21 km without drag ring and 10 km with drag ring.

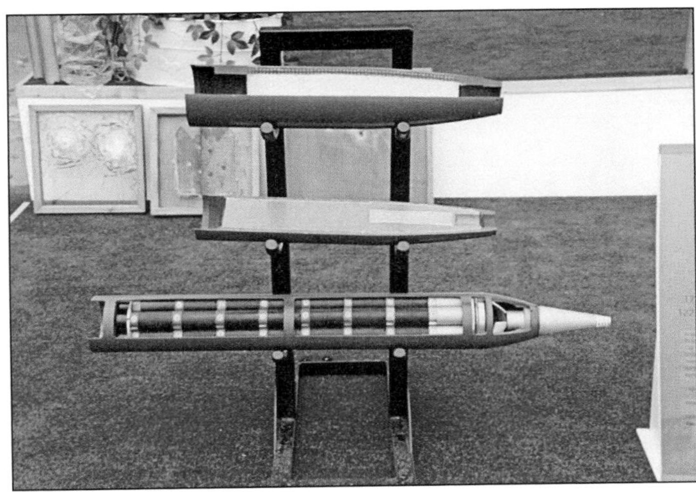

Close-up of three different types of Roketsan 122 mm rocket warheads. From top to bottom: fragmentation with steel balls; high-explosive; and submunition (Christopher F Foss)
0563325

Roketsan 122 mm (40-round) T-122 multiple rocket launcher on latest Rheinmetall MAN Military Vehicles (6 × 6) cross-country truck deployed in firing position with launcher traversed left and stabilisers lowered (Roketsan)
1178869

TRK-122 has a maximum range of 30 km and a minimum range of 16 km and is fitted with a warhead containing 50 anti-personnel/anti-material munitions and six incendiary bomblets. Both rockets have a composite solid propellant rocket motor.

Year	Quantity	Type
2005	3,040	TRB-122 and TRK-122
2006	3,040	type not stated
2007	2,000	TRB-122 and TRK-122
2008	740	TRB-122
2008	740	TRB-122 rockets

According to the United Nations Arms Transfer Lists, in 2010 Turkey exported to the United Arab Emirates one 122 mm rocket launcher and 480 122 rockets as well as one 300 mm rocket launcher and 32 TR-300 rockets.

New multirole launcher

To meet the potential requirements of some export customers, Roketsan has developed to the prototype stage the 107 mm/122 mm MultiBarrel Rocket Launcher Cradle (MBRLC).

This can launch 122 mm or 107 mm unguided surface-to-surface rockets from the same platform.

This can carry one pod of 20 × 122 mm rockets or three pods of 20 × 107 mm rockets.

The latter are the same as those uses in the Roketsan HMMWV 107 mm (24-round) rocket launcher that is covered in a separate entry in *IHS Jane's Land Warfare Platforms: Artillery and Air Defence.*

This is understood to be for the United Arab Emirates and was shown in February 2013 by Jobaria Defence Systems integrated onto the rear of a 6 × 6 cross-country chassis.

Available details are provided in a separate entry in *IHS Jane's Land Warfare Platforms: Artillery and Air Defence.*

This new launcher is unusual in that it can be elevated to launch 122 mm rockets and elevated and tilted when used to launch the shorter range 107 mm rockets.

Roketsan has been making 107 mm rockets for many years and these are normally fired from a 107 mm (12-round) towed launcher based on a Chinese design.

The original 107 mm rockets have a maximum range of 8.5 km but Roketsan has developed and placed in production a new family with enhanced warheads and a maximum range of over 11 km.

Details of these 107 mm unguided rockets and their launchers are provided in separate entries in *IHS Jane's Land Warfare Platforms: Artillery and Air Defence.*

Inertial Guidance System

Roketsan has developed an inertial guidance system which includes an active control actuation unit that includes an electric motor with gears, a ball screw assembly and associated electronics.

The avionics package includes an inertial measurement unit, global positioning system and a computer.

As of May 2013 it is understood that this remains at the prototype/development phase and it has not been confirmed as to whether this is applicable to Roketsan 122 mm or 300 mm surface-to-surface rocket systems.

Specifications

	122 mm (40-Round) T-122
Dimensions and weights	
Length	
overall:	9.2 m
Width	
overall:	2.5 m
Height	
overall:	3.10 m
Weight	
combat:	22,200 kg

	122 mm (40-Round) T-122
Mobility	
Configuration	
running gear:	wheeled
layout:	6 × 6
Speed	
max speed:	75 km/h
Range	
main fuel supply:	970 km
Gradient:	60%
Firepower	
Armament:	40 × 122 mm rocket system
Main weapon	
maximum range:	40,000 m (est.) (TR-122 and TRB-122 rockets with drag ring)
minimum range:	10,000 m (TR-122 and TRB-122 rockets with drag ring)
minimum range:	21,000 m (TR-122 and TRB-122 rockets without drag ring)
maximum range:	30,000 m (est.) (TRK-122 rocket with 56 submunitions)
minimum range:	16,000 m (TRK-122 rocket with 56 submunitions)
Rate of fire	
rapid fire:	40/80 s
Main armament traverse	
angle:	220° (±110°)
Main armament elevation/ depression	
armament front:	+55°/0°

Status
In serial production. Multiyear contract placed by Turkish Land Forces Command. Upgraded launcher in service with the United Arab Emirates.

As stated in the development the United Emirates has deployed a trailer mounted system with remote controlled launchers that can be fitted with pods of 122 mm or 300 mm unguided surface-to-surface rockets supplied by Roketsan. In addition the UAE has shown a 6 × 6 cross-country vehicle fitted with the Multirole Launcher.

Contractor
Roketsan.

Roketsan HMMWV 107 mm (24-round) rocket launcher

Development
During the IDEX Defence Exhibition held in Abu Dhabi in 2005, Roketsan exhibited a US AM General High Mobility Multipurpose Wheeled Vehicle (HMMWV) fitted with a full-scale mock-up of a 107 mm (24-round) rocket launcher. It is being developed as a private venture to meet the possible requirements of an undisclosed customer in the Middle East.

The AM General HMMWV is used for a wide range of roles in the Middle East and this 107 mm (24-round) rocket launcher could be used to provide rapid fire support to special forces units.

In early 2013 a NIMR (6 × 6) vehicle was shown in the United Arab Emirates with two pods each of 24 × 107 mm unguided rockets but it is understood that these were not supplied by Roketsan.

Description
The system consists of a standard production unarmoured AM General HMMWV (4 × 4) with a modified rear, on which is mounted a turntable with two pods of 12 launcher tubes for the standard Roketsan 107 mm unguided surface-to-surface rockets that are covered in detail in a separate entry in *IHS Jane's Land Warfare Platforms: Artillery and Air Defence*.

The turntable has powered traverse and elevation. The rockets would normally be launched with the operator seated in the cab, which is provided with a computerised fire-control system coupled to an inertial navigation system.

This would enable the system to come to a halt, lay the launcher onto the target, launch the rockets and then redeploy before counterbattery fire was returned. According to Roketsan, the system would take less than one minute to come into action.

Currently, two types of 107 mm calibre spin-stabilised extended-range rockets can be fired: TR-107 high-explosive and TRB-107 steel ball, which have a maximum range of over 11,000 m. These are manufactured by Roketsan for the home and export markets.

It is envisioned that no chassis stabiliser would be required at the rear of the HMMWV as the onboard computer would ensure that the rockets are launched at the correct sequence to avoid overstretching the chassis.

Roketsan has also developed the 107 mm/122 mm MultiBarrel Rocket Launcher Cradle (MBRLC) which can take one pod of 20 × 107 mm rockets or three pods of 20 × 107 mm rockets.

This is understood to be for the United Arab Emirates and was shown in February 2013 by Jobaria Defence Systems integrated onto the rear of a 6 × 6 cross-country chassis.

AM General High Mobility Multipurpose Wheeled Vehicle (4 × 4) with mock-up of 107 mm (24-round) rocket launcher on rear (Christopher F Foss) 1040806

Available details are provided in a separate entry in *IHS Jane's Land Warfare Platforms: Artillery and Air Defence*.

Specifications
Not available.

Status
Development. Not yet in production or service.

Contractor
Roketsan.

Roketsan 107 mm TR-107 Extended Range MultiBarrel Rocket Launcher Ammunition

Description
The Turkish company, Roketsan, has type classified a 107 mm calibre unguided rocket designated the TR-107 which extends the trailer-mounted 107 mm calibre rocket launcher system range to over 11,000 m.

The TR-107 can be fired from systems such as the China North Industries Corporation (NORINCO) trailer-mounted Type 63 and its foreign equivalents such as the South African RO 107 and the Turkish MKEK 12-round launcher.

It is understood that the Turkish Land Forces Command has 48 of these 107 mm TR-107 systems in service. Details of the MKEK 107 mm (12-round) rocket launcher are given in a separate entry in *IHS Jane's Land Warfare Platforms: Artillery and Air Defence*. It is understood that there has been no recent production of this launcher.

The solid propellant TR-107 rocket is spin-stabilised and has a reduced smoke solid propellant to provide the extended range. It can be fitted with one of two types of warhead: HE, or SB (containing steel balls, which are highly effective against soft targets). The warhead is activated by a nose-mounted impact fuze.

The TR-107 contains 1.25 kg of HE (TNT + RDX) warhead has a maximum of 500 lethal fragments. A point detonating fuze is fitted.

The TRB-107 contains 1.25 kg of HE (TNT + RDX), 2,800 steel balls. A proximity fuze is fitted.

Variants
In addition to being fired from the standard MKEK 107 mm (12-round) rocket launcher system these unguided rockets can also be fired from a number of other similar rocket launcher systems.

Roketsan 107 mm calibre TR-107 extended-range solid propellant rocket being launched from an MKEK 107 mm (12-round) launcher 0021081

Cobra (4 × 4) armoured personnel carrier with Roketsan 107 mm (12-round) rocket launcher mounted on roof towards the rear (Roketsan) 1452644

Cobra with 107 mm launcher

For trials purposes a Turkish Otokar Cobra (4 × 4) armoured personnel carrier has had its roof-rear modified to accept the upper part of the 107 mm (12-round) trailer-mounted mortar. It is laid onto the target by remote control from within the vehicle.

As of May 2013 this combination remained at the prototype stage.

HMMWV with 107 mm launcher

As a private venture, Roketsan has developed a variant of the US AM General High Mobility Multipurpose Wheeled Vehicle (HMMWV) (4 × 4) fitted with a turntable on the rear deck. Two pods are mounted on this, each with 12 × 107 mm rockets. Elevation and traverse is powered. Additional details of this are given in a separate entry in *IHS Jane's Land Warfare Platforms: Artillery and Air Defence.*

It is understood that this was originally developed to meet the potential requirements of at least one customer in the Middle East.

As of May 2013 this system remains at the prototype stage and it also been integrated onto at least one other chassis.

Specifications

	TR-107	TRB-107
Dimensions and weights		
Length		
overall:	0.840 m	0.840 m
width		
body:	107 mm	107 mm
Weight		
launch:	19.5 kg	19.5 kg
Performance		
Range		
min:	3 km	3 km
max:	11 km (est.)	11 km (est.)
Ordnance components		
Warhead:	8.4 kg HE	8.4 kg HE 2,800 steel balls
Fuze:	impact	proximity

Status

Production as required. Multiyear contract with Turkish Land Forces Command.

Contractor

Roketsan.

Ukraine

Ukraine mobility upgrades for Russian rocket launchers

Development

Multiple Rocket Launchers (MRL) developed by the Russian Spav Scientific Production Concern have been built in large quantities, not only for the Russian Army but also numerous export customers as well.

Details of these systems are provided in separate entries in *IHS Jane's Land Warfare Platforms: Artillery and Air Defence.*

The Ukraine is now marketing upgrades for these Russian designed and built MRL by taking the launchers from the older Russian chassis and installing these onto locally manufactured KrAZ (6 × 6) cross-country truck chassis.

KrAZ-6322PA Bastion-02 (6 × 6) chassis with standard BM-21 122 mm (40-round) rocket launcher mounted at very rear of chassis and additional rockets to rear of cab for manual reloading (KrAZ) 1421673

KrAZ-6322PA Bastion-03 (6 × 6) chassis modified to carry the Russian 220 mm (16-round) Uragan launcher system shown here in the deployed position with stabilisers deployed at the rear (KrAZ) 1421676

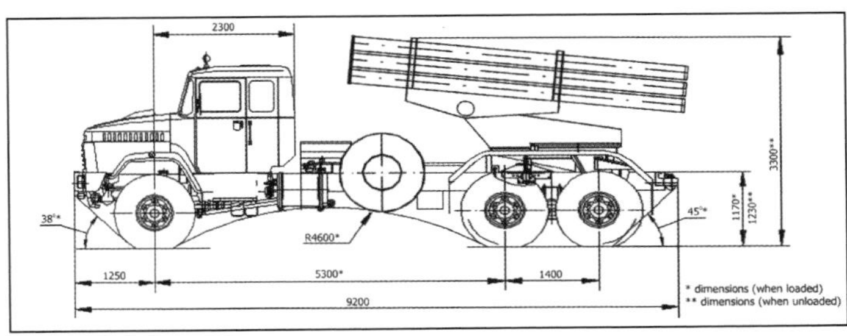

Detailed drawing of the KrAZ-6322PA Bastion-3 (6 × 6) modified to carry the Russian 220 mm (16-round) Uragan rocket launcher shown here in travelling configuration with launcher traversed to the rear (KrAZ) 1421674

It is understood that the Ukraine has supplied a number of countries with KrAZ (6 × 6) chassis for rocket launcher upgrades.

Description

The most widely deployed system is the Russian 122 mm 9K51 BM-21 (40-round) MRL with the original production system based on the Russian Ural-375D (6 × 6) cross-country truck chassis that was followed by a version using the Ural-4320 (6 × 6) chassis.

The Russian 122 mm BM-21 (40-round) MRL has been fitted to the KrAZ-6322PA Bastion-01 (6 × 6) truck chassis for improved mobility and increased fuel efficiency. The baseline cargo model has a typical cross-country payload of 10,000 kg.

Standard equipment for all of these KrAZ trucks includes powered steering and a central tyre pressure inflation system allowing the driver to adjust the tyre pressure to suit the terrain being crossed.

These trucks are also offered in Left Hand Drive (LHD) or Right Hand Drive (RHD) configuration with the option of a winch for self recovery operations or to recover other vehicles.

There is also the option of an armoured cab to protect the occupants from small arms fire and shell splinters up to STANAG 4569 Level 2a.

This also includes engine sump armouring as well as armouring the fuel and air tanks mounted under the chassis of the vehicle.

Unspecified electronic devices to provide protection against Improvised Explosive Devices (IED) could also be fitted and these would be probably Government Furnished Equipment (GFE) by the end user, as would any communications equipment required.

While the normal production vehicle is fitted with a locally produced KrAZ diesel developing up to 400 hp which gives a maximum road speed of up to 85 km/h it is also being marketed with a Western Cummins or Deutz diesel engine.

A long wheelbase (5.3 m + 1.4 m) chassis version has also been developed for the 122 mm (40-round) BM-21 MRL designated the KrAZ-6322PA Bastion-02.

This also has the MRL mounted at the very rear of the chassis but an additional 40 × 122 mm rockets are carried to the immediate cab rear on the horizontal position for manual reloading.

This is powered by a locally produced turbocharged diesel developing 330 to 400 hp depending on the version which gives a maximum road speed of 85 km/h.

It is also being marketed fitted with a Cummins or Deutz diesel engine.

KrAZ-6322PA Bastion-01 (6 × 6) chassis with standard BM-21 122 mm (40-round) rocket launcher covered up on rear of chassis (KrAZ) 1421671

The Russian Spalv Scientific Production Concern 220 mm BM 9P140 (16-round) Uragan (Hurricane) MRL entered service with the Russian Army as far back as 1975 and is still in service with a number of countries.

The complete system is also referred to by the industrial designation of the 9K57 and is based on the modified ZIL-135LM (8 × 8) cross-country truck chassis.

Production of this 8 × 8 chassis was completed many years ago and KrAZ has integrated this launcher onto the rear of a KrAZ-6322PA Bastion-03 (6 × 6) cross-country truck chassis.

When travelling the launcher is traversed to the front to reduce the overall length of the platform for travelling.

When deployed in the firing position, two stabilisers are lowered to the ground one either side at the rear and the launcher traversed through 180° to fire over the frontal arc. The sighting system is mounted on the left side of the launcher.

The complete system has a combat weight of about 21.6 tonnes complete with rockets and crew and is normally powered by a locally produced diesel engine developing up to 400 hp that gives a maximum road speed of up to 85 km/h.

As with the other platforms, alternative diesel engines can be fitted provided by western contractors such as Cummins and Deutz.

As far as it is known the Ukraine does not manufacture any 122 mm or 220 mm rockets for these systems.

Specifications
Not available. Ranges of the unguided rockets are the same as the original Russian rockets.

Status
Development complete. Production as required. The BM-21 version is in service with the Ukraine and is understood to have been exported.

Contractor
Various. See Development.

United Arab Emirates

Jobaria Defense Systems 122 mm Multi Cradle Launcher (MCL) Dinosaur

Development
Unveiled for the first time at IDEX held in Abu Dhabi late in February 2013 were three new Multiple Rocket Launcher Systems (MLRS) which were developed in the United Arab Emirates (UAE) with the assistance of some overseas contractors to meet the operational requirements of the UAE Armed Forces.

Jobaria Defense Systems Multiple Cradle Launcher (MCL) in the travelling configuration showing Oshkosh Defense HET (6 × 6) and semi-trailer carrying three rather than the normal four 122 mm rocket launchers (IHS/Patrick Allen) 1510085

Jobaria Defense Systems Multiple Cradle Launcher (MCL) showing Oshkosh HET (6 × 6) and semi-trailer with four launchers of 122 mm rockets traversed slightly to the right (Christopher F Foss) 1452947

The largest system on show at IDEX 2013 was the Multi Cradle Launcher (MCL) developed by the local company of Jobaria Defense Systems which is also called the Dinosaur due to its larger size.

This system is understood to have entered service with the UAE Armed Forces several years ago.

The Turkish company of Roketsan has been a key player in the development of the MCL and has supplied a number of key parts including the actual power operated cradle, pods of rockets, rockets and the advanced computerised Fire Control System (FCS).

Description
The MCL is the largest system of its type in the world and consists of a US-supplied Oshkosh Defense Heavy Equipment Transporter (HET) (6 × 6) cross-country vehicle towing a huge 10 wheeled semi-trailer with four launchers carrying pods of 122 mm unguided surface-to-surface rockets.

The complete system is stated to weigh 105 tonnes with the 940-litre fuel tank giving a range of not less than 450 km.

It is understood that early in the development of the MCL, a German Rheinmetall MAN Military Vehicles (8 × 8) HET was used with a protected cab but this was subsequently changed for the Oshkosh Defense (6 × 6) HET.

The latter has been fitted with a three-person fully protected cab which provides protection to STANAG 4569 Level II and is provided with a full air conditioning system and an NBC system.

The HET is fitted with large desert tyres with a Central Tyre Inflation System (CTIS) which allows the driver to adjust the terrain to suit the terrain being crossed.

Crew consists of commander, navigator and driver who can deploy and launch the 122 mm rockets from within the cab or away from the cab with the aid of a remote control device.

The semi-trailer is provided with an Auxiliary Power Unit (APU) which is used to provide power to extend the five hydraulically operated stabilisers both sides and the electrically operated launchers.

The APU is provided with 200 litres of fuel which is sufficient for 24 hours of operation with the main HET diesel engine switched off.

The four rocket launchers have three pods each of 20 × 122 mm launcher tubes with the launchers having powered traverse of 50° left and right which gives a total of 240 122 mm rocket launcher tubes.

The system fires the Turkish Roketsan TR-122 and TRB-122 rockets with a maximum stated range of up to 40 km with a Circular Error of Probability (CEP) of 1.3 per cent of range, but this does depend on a number of factors. Minimum range of the unguided rockets is being quoted as 16 km.

The TR-122 rocket is fitted with a High-Explosive (HE) warhead which is activated by a point detonating fuze.

The TRB-122 is fitted with an HE-steel ball warhead that is activated by a proximity fuze.

A Roketsan computerised FCS is fitted to reduce the into action time as well as making for increased accuracy.

The fire mission is controlled from the protected cab and the number of pods fired depends on the target to be engaged and a maximum of 240 × 122 mm rockets can be fired before it needs to be resupplied.

Accuracy is further enhanced by the installation of a Global Positioning System/Inertial Navigation System (GPS/INS) and its own meteorology system.

The INS is fitted to each cradle as well as in the cab and is the Selex ES FIN 3110 Inertial Navigation Unit (INU).

The one in the cab is coupled to the drivers display unit with a moving map display and the Roketsan System Management computerised FCS.

The MCL can be deployed as an autonomous unit, as part of a troop or battery, or integrated into an artillery system that also includes conventional tube artillery.

Once a fire mission has been carried out, the system would normally rapidly re-deploy to avoid counter battery fire.

The system is supported by a similar HET and semi-trailer which carry new pods of 122 mm rockets which are rapidly loaded using a hydraulically operated crane.

Jobaria Defense Systems Multiple Cradle Launcher (MCL) showing Oshkosh HET (6 × 6) and semi-trailer with four launchers of 122 mm rockets and with hydraulically operated trailer outriggers deployed (Christopher F Foss) 1452948

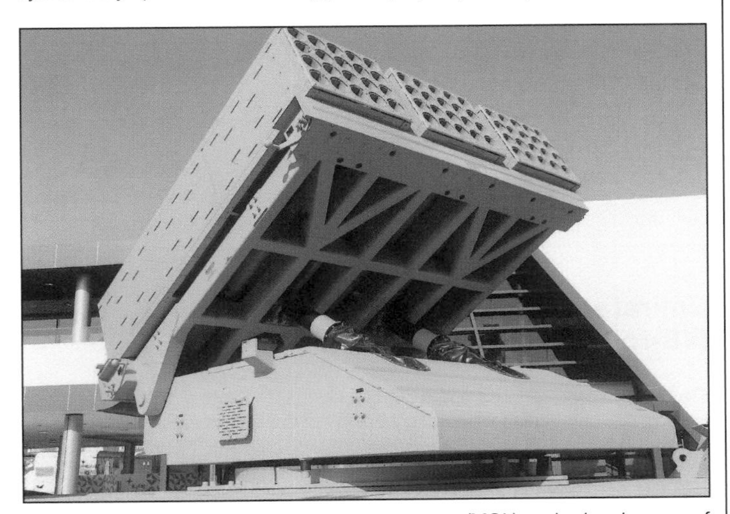

Close up of one of the Multiple Cradle Launchers (MCL) pods showing one of the launchers with three pods each of 20 × 122 mm rocket launcher tubes (Christopher F Foss) 1452946

As this is an MCL, it is considered probable that the 122 mm rocket pods could well be replaced by larger calibre rockets with a significant increase in range.

This is supported by the United Nations Arms Transfer List for 2010 which states that Turkey transferred to the UAE a 122 mm/300 mm rocket launcher plus 480 × 122 mm rockets and 32 300 mm TR-300 rockets.

The latter are already used by the Turkish Land Forces Command (TLFC) with four launcher tubes integrated onto the rear of a Rheinmetall MAN Military Vehicles 6 × 6 truck.

Details of this system are provided in a separate entry in *IHS Jane's Land Warfare Platforms: Artillery and Air Defence*.

The 300 mm unguided rockets have a HE fragmentation warhead with steel balls activated by a proximity fuze.

Maximum range of the 300 mm rockets without drag rings is stated to be 100 km and with drag rings, is reduced to 80 km.

107 mm/122 mm launcher based on 6 × 6 FMTV chassis

Jobaria Defense Systems have also developed a smaller system which they call the JDS - 107/122 System and this has been shown integrated onto the rear of a US Family of Medium Tactical Vehicle (FMTV) (6 × 6) cross-country vehicle fitted with a protected two-person cab and a CTIS.

This can be fitted with pods of 107 mm or 122 mm unguided rockets and details of this are provided in a separate entry in *IHS Jane's Land Warfare Platforms: Artillery and Air Defence*.

As of May 2013 it is understood that this was still under development but is expected to enter service with the UAE in the future.

Nimr (6 × 6) 107 mm multiple rocket launcher

In February 2013 Emirates Defense Technology showed a 107 mm MLRS integrated onto the locally produced Nimr (6 × 6) high mobility tactical vehicle which has a protected two-door cab fitted with an NBC system.

On the flatbed to the rear of the cab is a power operated launcher with two composite pods each of 24 × 107 mm rockets.

Before launching the 107 mm rockets, two stabilisers are lowered to the ground either side of the platform.

It is understood that this is now in production for the UAE Armed Forces and available details are provided in a separate entry in *IHS Jane's Land Warfare Platforms: Artillery and Air Defence*.

Specifications

	122 mm Multi Cradle Launcher
Dimensions and weights	
Crew:	3
Length	
overall:	29 m
Width	
overall:	4 m
Height	
overall:	3.8 m (cab)
Weight	
combat:	105,000 kg
Mobility	
Configuration	
running gear:	wheeled
layout:	6 × 6
Firepower	
Armament:	240 × 122 mm rocket system (4 × 60)
Main weapon	
maximum range:	40,000 m
Main armament traverse	
angle:	50° (left/right)
Survivability	
Night vision equipment	
vehicle:	optional
NBC system:	yes
Armour	
hull/body:	steel

Status

Production as required. In service with the United Arab Emirates.

Contractor

Jobaria Defense Systems.

Jobaria Defense Systems 107 mm/122 mm multiple rocket launcher

Development

Unveiled for the first time at IDEX held in Abu Dhabi late in February 2013 were three new wheeled Multiple Rocket Launcher Systems (MLRS).

These have been developed in the United Arab Emirates (UAE) with the assistance of some overseas contractors to meet the operational requirements of the UAE Armed Forces.

The largest system on show at IDEX 2013 was the Multi Cradle Launcher (MCL) developed by the local company of Jobaria Defense Systems that is also called the Dinosaur due to its large size.

This is already in service with the UAE Armed Forces and details are provided in a separate entry in *IHS Jane's Land Warfare Platforms: Artillery and Air Defence*.

Also shown was the Jobaria Defense Systems 107 mm/122 mm multiple rocket launcher system which is also referred to as the JDS MRL-107/122 mm system or the 221/701 system.

Jobaria Defense Systems Multiple Rocket Launcher with power operated launcher on rear fitted with a mockup of a pod of 20 × 122 mm unguided rockets with the side showing 60 × 107 mm rocket launcher tubes (Christopher F Foss) 1452949

FMTV (6 × 6) cross-country vehicle fitted with power operated launcher on the rear fitted with mockup of one composite pod containing 20 × 122 mm launcher tubes (Christopher F Foss) 1481032

While Jobaria Defense Systems is the overall prime contractor, a number of foreign contractors have also been involved in this programme including the Turkish company of Roketsan.

Description

The Jobaria Defense Systems' 107 mm/122 mm MRL consists of a US supplied Family of Medium Tactical Vehicles (6 × 6) cross-country with a protected forward control cab.

The cab is fitted with an air conditioning system and a central tyre inflation system which enables the driver to adjust the tyre pressure to meet the terrain conditions being crossed.

The FMTV is powered by a 330 hp Euro III compliant diesel coupled to an automatic transmission.

Mounted on the rear platform is a powered launcher which can be fitted with one pod of 20 × 122 mm unguided rockets or three pods of 20 × 107 mm unguided rockets.

To provide a more stable firing platform, hydraulically operated stabilisers are lowered to the ground before the rockets are launched.

All of the rockets are in sealed and thermally insulated composite pods and are normally fired at the rate of one rocket per second.

It is fitted with a computerised Fire Control System (FCS) that includes Global Positioning System/Inertial Navigation System (GPS/INS) with an integrated meteorological system.

The INS is fitted to the cradle as well as in the cab and is a Selex ES FIN 3110 Inertial Navigation Unit (INU).

The one in the cab is couple to the drivers display unit with a moving map display and the Roketsan System Management computerised FCS.

This allows the system to quickly come into action, carry out a fire mission and deploy to a new fire position before counter battery fire is returned.

The crew consists of two, commander/navigator and driver who can deploy and launch the 107 mm or 122 mm rockets from within the cab or away from the cab with the aid of a remote control device.

The system fires the Turkish Roketsan 122 mm TR-122 and TRB-122 rockets with a maximum stated range of up to 40 km with a Circular Error of Probability (CEP) of 1.3 per cent of range, but this does depend on a number of factors. Minimum range of the unguided rockets is being quoted as 16 km.

The TR-122 rocket is fitted with a High-Explosive (HE) warhead which is activated by a point detonating fuze.

The TRB-122 is fitted with a HE-steel ball warhead that is activated by a proximity fuze.

The Roketsan 107 mm TR-107 extended range rocket features a reduced smoke solid-fuelled rocket with a maximum range of 11 km and a minimum range of 3 km, and there is also a direct fire capability.

The rocket is fitted with an 8.4 kg HE warhead activated by a point detonating fuze and is claimed to have an effective radius of 24 m.

The system can be deployed as an autonomous unit, as part of a troop or battery or integrated into an artillery system that also includes conventional tube artillery.

Once a fire mission has been carried out the system would normally rapidly re-deploy to avoid counter battery fire.

Nimr (6 × 6) 107 mm multiple rocket launcher

In February 2013 Emirates Defense Technology showed a 107 mm MLRS integrated onto the locally produced Nimr (6 × 6) high mobility tactical vehicle which has a protected two-door cab fitted with an NBC system.

On the flatbed to the rear of the cab is power operated launcher with two composite pods each of 24 × 107 mm rockets.

Before launching the 107 mm rockets, two stabilisers are lowered to the ground either side of the platform.

It is understood that this is now in production for the UAE Armed Forces and available details are provided in a separate entry in *IHS Jane's Land Warfare Platforms: Artillery and Air Defence*.

Specifications

	107 mm/122 mm multiple rocket launcher
Dimensions and weights	
Crew:	2
Mobility	
Configuration	
running gear:	wheeled
layout:	6 × 6
Clutch	
type:	centrifugal
Firepower	
Armament:	20 × 122 mm rocket system (optional)
Armament:	60 × 107 mm rocket system (optional)
Main weapon	
maximum range:	40,000 m (122 mm rockets)
minimum range:	16,000 m (122 mm rockets)
maximum range:	11,000 m (107 mm rockets)
minimum range:	3,000 m (107 mm rockets)
Survivability	
Night vision equipment	
vehicle:	optional
NBC system:	yes
Armour	
hull/body:	steel

Status

Under development.

Contractor

Jobaria Defense Systems.

Emirates Defense Technology 107 mm (48-round) rocket launcher

Development

In early 2013 Emirates Defense Technology of the United Arab Emirates (UAE) showed a 107 mm (48-round) multiple rocket launcher integrated onto the locally produced Nimr (6 × 6) high mobility tactical vehicle.

It is understood that this is now in quantity production for the UAE Armed Forces as a complimentary system to the larger calibre 122 mm and 300 mm systems developed by Jobaria Defense Systems and covered in separate entries in *IHS Jane's Land Warfare Platforms: Artillery and Air Defence*.

Description

The system consists of the locally manufactured Nimr (6 × 6) cross-country vehicle with a fully enclosed protected two-door cab which is fitted with an NBC system.

The diesel powerpack is at the front and consists of a Cummins diesel coupled to an Allison SP 3000 automatic transmission and two speed transfer case.

Standard equipment includes a central tyre inflation system that allows the driver to adjust the tyre pressure to suite the terrain being crossed and run flat tyres.

On the flatbed to the rear of the cab is a power operated launcher with two composite pods each of 24 × 107 mm rockets.

Before launching the 107 mm unguided rockets two stabilisers are lowered to the ground either side of the platform by remote control and in addition the front suspension can be locked out.

Nimr (6 × 6) vehicle with protected cab and fitted with a powered launcher on the rear with two composite pods each containing 24 × 107 mm rocket tubes (Christopher F Foss) 1481034

According to Emirates Defence Technology it takes only 90 seconds to be prepared for its first fire mission when in the automatic mode and 180 seconds when in the manual mode.

Coming out of action in the automatic mode is 60 seconds and in the manual mode it is 120 seconds.

When travelling the rocket pods are shielded by a protective cover that extends to the rear so making it difficult to distinguish the system from a conventional Nimr protected cargo vehicle.

The system is fitted with a Battle Management System (BMS) and a computerised Fire Control System (FCS) that includes GPS/INS for increased accuracy.

The rocket launcher has a maximum powered elevation of +55° with a powered traverse of 110° left and right with this being controlled from the cab but with manual back up.

The firing module consists of two composite modules each containing 24 107 mm rocket tubes or as an option two steel modules each containing 24 107 mm rocket tubes.

The solid propellant unguided 107 mm rockets have a maximum range of over 11 km and are fitted with a high-explosive warhead. The rockets can be fired every 0.5, 1 or 5 seconds.

According to Emirates Defense Technology the system has a Gross Vehicle Weight (GVW) of around 13 tonnes and a maximum road speed of up to 80 km/h, although maximum speed of the base vehicle is currently being quoted as up to 130 km/h.

Production systems have banks of electrically operated grenade launchers for self-screening purposes. Standard equipment includes an air conditioning system.

The rocket launcher is supported by a resupply vehicle also based on a Nimr (6 × 6) vehicle which is fitted with a hydraulic crane for re-loading purposes.

Specifications

	107 mm (48-round) rocket launcher
Dimensions and weights	
Crew:	2
Length	
overall:	6.45 m
Width	
overall:	2.37 m
Height	
overall:	2.05 m
Ground clearance	
overall:	0.5 m
Track	
front:	1.89 m
rear:	1.89 m
Wheelbase	3.30 m
Weight	
standard:	8,000 kg
combat:	13,000 kg
Payload	
land:	5,000 kg
Mobility	
Configuration	
running gear:	wheeled
layout:	6 × 6
Speed	
max speed:	80 km/h
Range	
main fuel supply:	700 km
Fording	
without preparation:	1.2 m
Angle of approach:	51°
Angle of departure:	40°
Gradient:	60%
Side slope:	30%
Vertical obstacle	
forwards:	0.40 m
Engine	Cummins ISBe 300, in line-6, water cooled
Gearbox	
model:	Allison SP 300
type:	automatic
forward gears:	6
reverse gears:	1
Transfer box:	2-speed
Steering:	hydraulic
Brakes	
main:	hydraulic disc all round
parking:	spring applied air over hydraulic released sliding calliper
Tyres:	335/80 R20
Suspension:	double wishbone, independent over coil spring, hydraulic shock absorbers
Electrical system	
vehicle:	24 V
Firepower	
Armament:	48 × 107 mm rocket system (24 × 2)
Main armament traverse	
angle:	110° (left/right)
Main armament elevation/depression	
armament front:	0°/+55°
Survivability	
Night vision equipment	
vehicle:	optional
NBC system:	yes
Armour	
hull/body:	steel

Status

In production for UAE Armed Forces.

Contractor

Emirates Defence Technology.

United Kingdom

MBDA Fire Shadow Loitering Munition

Development

Following a competition, early in 2005 SD (UK) Limited was selected by the UK MoD to carry out concept demonstration of the technologies for a system that was then called the Low Cost Loitering Carrier (LCLC).

This ran for about two years and during this phase a number of areas were studied including range, number of hours the system would loiter and types of payload.

Another programme was the Loitering Munitions Concept Demonstration (LMCD) based on the Sparrow Unmanned Aerial Vehicle (UAV) designed and built by EMIT Aviation Consult for which prime contractor was Ultra Electronics.

Experience from these activities was fed into the Fire Shadow which is a key part of the UK "Team Complex Weapons" Portfolio.

Team Complex Weapons started in 2007 and aims to provide the British Army, Royal Air Force and Royal Navy with a new range of land, sea and air based precision missiles for the 21st Century.

Fire Shadow was developed by MBDA as a low cost, all weather weapon able to carry out precision attacks under day or night conditions against time sensitive and/or hard-to-engage surface targets that cannot be rapidly engaged by other weapons.

It will be of particular use against moving targets in complex and urban environments that cannot currently be neutralised by other ground based elements.

The system was funded by the UK MoD and forms part of the UK MoD Indirect Fire Precision Attack (IFPA) programme.

When deployed it will be operated by the Royal Artillery and compliment their other assets such as the Lockheed Martin Missiles and Fire Control Guided MLRS (GMLRS) and conventional tube artillery.

Rapid development of the Fire Shadow demonstrator started in January 2007 under private venture funding.

The first of six prototype airframes were delivered by Marshall Specialist Vehicles within 12 days of contract award.

To reduce costs and development time, the Fire Shadow demonstrator used Commercial Off-The-Shelf/Military Off-The-Shelf (COTS/MOTS) components wherever it was possible.

British troops carrying out Fire Shadow Loitering Munition launch procedure using the initial single round launch container designed to provide an early operational capability (MBDA) 1364282

Fire Shadow munition being launched from its trailer launcher during trials at the Vidsel Test Range in Sweden in November 2010 (Vidsel Test Range)
1168591

These included IMU and actuators for an existing system, weapon computer based on Xylinx board, engine from the Banshee target drone, clutch taken from a go-kart and a COTS fuel tank.

Some 80 per cent of the technical de-risking objectives of this programme were carried out within 12 months of the start of the programme.

The first test launch was successfully carried out at the Aberporth test range in Wales on 30 April 2008.

This was made from a single rail launcher with the engine started before launch and the wings pre-deployed. Drive to the propeller was engaged after jettisoning the Roxel booster rocket.

The Fire Shadow demonstrator munition climbed to a pre-determined altitude, followed by a waypoint trajectory and then executed a terminal dive to represent a simulated target engagement.

Onboard video imagery and position data was transmitted from the Fire Shadow demonstrator via a data link to the ground control station.

The Demonstrator Phase was followed by the Assessment Phase that was awarded by the UK MoD in June 2008 and was successfully completed in under two years. By mid-2009 more that 15 prototype Fire Shadow had been completed.

The project then moved into the Demonstration and Manufacture (D&M) phases as part of the MoD/MBDA Bilateral Interim Portfolio Management Agreement contract awarded to MBDA in March 2010.

In September 2011 MBDA announced that the Fire Shadow had successfully completed a series of demonstration trials.

These included a more complex scenario where the man-in-the-loop functionality was fully exercised.

In this trial, which took place in Sweden, the operator was able to successfully engage a representative target.

In addition, various "hand on trials" at the MBDA integration facilities at Filton and Bedford have enabled the British Army personnel to tailor the "look and feel" of the system, to refine the Tactics, Techniques and Procedures (TTPs).

By early 2012 the Fire Shadow weapon system was in production for the British Army and all contractor firing trials had been successfully completed.

During a Fire Shadow trial at the Vidsel range in Sweden late in 2011 the complete man-in-the-loop flexibility of the system was demonstrated which included a complex mission profile over an extended duration.

Under the original plan Fire Shadow weapon system was to be used by 39 Regiment, Royal Artillery who currently deploy the Lockheed Martin Missiles and Fire Control Guided Multiple Launch Rocket System (GMLRS) and the Rafael Advanced Defense Systems Spike Non-Line-Of-Sight (NLOS) Exactor precision guided missile.

It was expected that Fire Shadow would have been deployed to Afghanistan but in the end this did not happen. MLRS used by 39 Regiment have now been withdrawn from Afghanistan.

The longer term future of Fire Shadow is still not clear as a number of key weapon systems that were expected to be deployed by the Riyal Artillery have been cancelled as part of changing requirements and budget cutbacks.

These include the Lightweight Mobile Artillery Weapon System – Gun (LIMAWS-G), Lightweight Mobile Artillery Weapon System – Rocket (LIMAWS-R), 155 mm Ballistic Sensor Fuzed Munition (SFM) and the Large Long Range Rocket (LLLR).

Description

Fire Shadow munition is capable of loitering for several hours and to be ready to strike at any suitable target that emerges.

It would be especially suitable for engaging manoeuvring fleeting targets that cannot currently be neutralised by other battlefield assets.

The Fire Shadow munition is a one way system and will have a "man in the loop" at all times. This person will take the ultimate decision as to whether to engage the target or not.

Potentially when the Fire Shadow munition is loitering and waiting for a potential target it could also provide a real time Intelligence, Surveillance, Target Acquisition and Reconnaissance (ISTAR) capability to ground based units.

Fire Shadow weapon system will be interoperable with other battlefield assets and will operate within normal battlespace management process.

Integration within a modern battlespace headquarters was successfully achieved during participation in Coalition Warrior Interoperability Demonstration in June 2010.

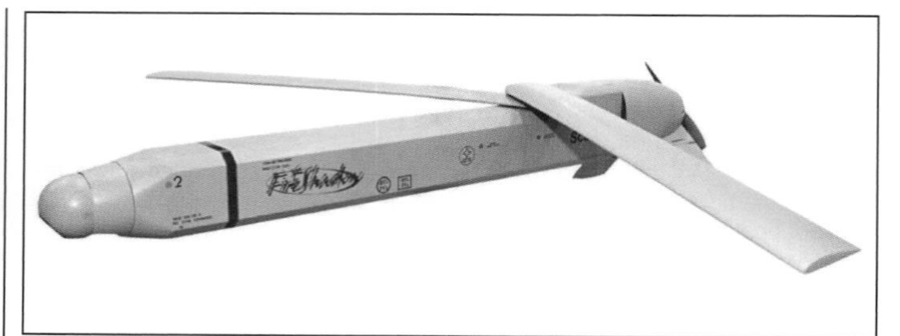

Fire Shadow loitering munition as it would appear in flight with wings unfolded (IHS/Patrick Allen)
1451599

First version of Fire Shadow munition to be fielded will be in a single launcher that can be rapidly deployed by tactical helicopter and can be quickly readied for launch in the field.

The munition is programmed on the launcher via an umbilical cable and can also be re-targeted whilst in flight.

The munition features a blunt ended box-section fuselage. The rear mounted Wankel rotary sustain engine drives a two blade fixed pitch propeller.

Lift is provided by two high mounted forward swept flip out wings with the two tailfins arranged in an inverted-V configuration.

Total launch weight is estimated to be around 200 kg. The rocket boost motor is automatically dispensed shortly after launch.

With a length of 4 m and a wing span of 4 m, the weapon system has a range of approximately 100 km and the munition is capable of about six hours endurance.

Transit height would typically be up to 15,000 ft and depending on operational scenario, with a variable speed in the range of 150 to 300 km/h.

No landing gear is fitted as the Fire Shadow munition is a one way system.

The Selex ES day/night sensor is mounted in the nose with a probable circular error of probability of 1 m and this has already been tested on an Islander aircraft as part of sensor gathering trials.

No details of the warhead have been released but is probably high explosive or high explosive fragmentation with a weight of about 22 kg with a claimed low collateral damage.

The data link between the operator and the Fire Shadow LM has been adapted from the High Integrity DataLink (HIDL) adopted from that originally deployed for the Thales Watchkeeper UAV.

Variants

To provide an early operational capability the first version of the Fire Shadow weapon system used singleton launchers.

Alternative launch platforms would be available subject to customer requirements.

For example, this could include a vehicle launched system which could include a removable pod with six Fire Shadow munitions in canisters which would be elevated prior to launch.

The range and loitering time of the Fire Shadow munition could be extended with the latter having a potential of up to 10 hours.

There is also the potential of alternative warheads and seekers with the latter including semi-active laser homing.

Naval application of Fire Shadow

Late in 2011 it was revealed that MBDA had commenced a concept study to assess the feasibility of applying the Fire Shadow loitering munition concept to a maritime fire support role.

According to MBDA it has identified a potential capability window for a system to provide long range fire support ashore and engage mobile targets sets with a high level of precision.

Specifications

See Description.

Status

One launcher and about 20 Fire Shadow munitions have been delivered to the British Army but they have now been deployed on operations.

Contractor

MBDA Limited.

United States

Lockheed Martin Missiles and Fire Control 227 mm Multiple Launch Rocket System (MLRS)

Development

Early in 1976, the United States Army Missile Command (today US Army Aviation and Missile Command) at Redstone Arsenal, Alabama, initiated feasibility and concept formulation of a General Support Rocket System using

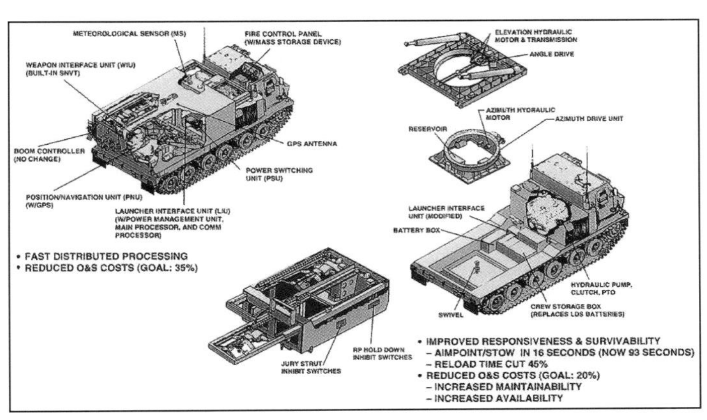

Main components of the IFCS/ILMS for the Multiple Launch Rocket System (MLRS) 0008322

a low-cost unguided rocket that would be handled like a round of conventional ammunition and have a high rate of fire. This system would be used against troops and light equipment, air defence systems and command centres.

In March 1976, five companies, Boeing, Emerson Electric, Martin Marietta, Northrop and Vought were awarded contracts for the concept definition of the General Support Rocket System (GSRS).

After evaluating all five concept-definition studies, in September 1977, Boeing Aerospace and the Vought Corporation (now Lockheed Martin Missiles and Fire Control) were each awarded contracts for the 29-month validation phase of the GSRS competition.

Under the terms of the contract each company delivered three prototype Self-Propelled Launcher Loaders (SPLLs) and rockets for competitive testing at White Sands Missile Range, New Mexico.

Early in 1978, the US Army Missile Research and Development Command redirected the GSRS towards a standard NATO weapon that could be developed and produced in both the US and Europe. The programme was renamed Multiple Launch Rocket System or MLRS.

In 1980, the then Ling Tempco Vought of Dallas, Texas (now Lockheed Martin Missiles and Fire Control) was selected as prime US contractor for MLRS and was awarded a full-scale development contract.

Teamed with the company were the Atlantic Research Corporation for the solid-propellant rocket motor, Bendix Guidance Systems Division for the stabilisation reference package/position determining system, the Brunswick Corporation for launch tubes, Norden Systems for the fire-control system and Vickers for the launcher drive system.

European participation was introduced by an MoU between the governments of the US, Germany, the UK, France and Italy, allowing for the joint development, production and subsequent deployment of a multiple-launch, free flight rocket system.

In 1983, the first MLRS units entered service with the US Army. In the same year MLRS-Europaische Produktions Gesellschaft GmbH (MLRS-EPG), was established in Munich to undertake European production. Diehl of Germany, Hunting Engineering (now Lockheed Martin UK) of the UK, Aerospatiale (now MBDA) of France and SNIA BPD of Italy were invited to form this Prime Contract Consortium and sharing of manufacture was established in accordance with national offtake ratios.

The main European subcontractors to EPG were Aerospatiale (now MBDA) for launcher integration, BPD for the rocket motor, Diehl for rocket integration and Thyssen Henschel (which has now become Rheinmetall Landsysteme) for the full-tracked carrier vehicle.

The first European-built MLRS systems were handed over late in 1989 and production peaked at nine to 10 launchers per month. The European MLRS programme involved the manufacture of:

- 284 launchers
- 128,000 M77 rockets and practice rockets
- 73,000 AT2 rockets (for Germany and UK only).

During Operation Desert Storm the US deployed over 230 MLRS launchers while the British Army deployed 16 launchers. The US Army deployed MLRS to the Gulf in 2003 for the invasion of Iraq. The British Army did not deploy any MLRS systems for this campaign.

Description

The development of the MLRS was planned in three distinct phases. Phase I covered the vehicle, launcher and an M26 anti-matériel/anti-personnel warhead containing M77 submunitions. Phase II addressed the AT2 scatterable anti-tank mine warhead. Phase III was to be the development of a Terminally Guided anti-armour Warhead (TGW). Development of this was eventually cancelled.

Phase I - M270 launcher

The lightly armoured M270 self-propelled launcher is a stretched version of the now BAE Systems M2 Bradley infantry fighting vehicle. The Launcher Loader Module, mounted on the rear of the vehicle hull consists of a base, turret and cage. It contains the computerised fire-control system, a stabilisation reference package/position determining system, a launcher drive system and a twin boom crane unit for self-loading and unloading.

The three-man crew consists of the driver, gunner and section chief, all of whom are seated in the fully enclosed cab at the front of the vehicle. This is protected against small arms fire and shell fragments by aluminium armour and louvred windows.

British Army M270B1 MLRS upgraded for operations in Afghanistan showing additional bar armour fitted to three-person cab (Christopher F Foss) 1308385

The US Army has fitted some of its M270 launchers with a roof-mounted 40 mm MK 19 series automatic grenade launcher.

An overpressure ventilation system prevents rocket fumes from entering the cab and the vehicle is also equipped with a filtration unit to protect the crew from chemical, biological and radiological particles.

Equipment capabilities permit a reduced crew, or even one person, to accomplish a complete fire mission including the loading and unloading operations.

US upgraded M270A1 launcher

The lightweight armoured M270 self-propelled launcher was manufactured by the now BAE Systems which then sends it to Lockheed Martin Missiles and Fire Control for systems integration and final delivery to the customer.

The US Army has now upgraded some of their launchers to a new and enhanced configuration called the M270A1 with the first batch consisting of 35 units.

Of these, the now BAE Systems has remanufactured 18 launchers while the US Army Red River Army Depot has remanufactured the remaining 17.

In mid-1998, Lockheed Martin Missiles and Fire Control was awarded another contract worth USD63 million to upgrade an additional 21 MLRS systems to the M270A1 standard with deliveries starting in April 2000.

As well as generally re-manufacturing the launcher so that it is returned to the US Army in an almost new condition, the launcher was also upgraded with the Improved Fire-Control System/Improved Launcher Mechanical System (IFCS/ILMS), which offers a number of significant advantages to the user including quicker into action times, faster reloading and reduced operating and support costs.

The IFCS involves a new fire-control panel with a mass storage device and the addition of a GPS augmented Positioning Determining/Navigation Unit (PDNU).

The ILMS upgrade involves electronic and hydraulic system modifications to decrease system response time. With current software improvements to the response time, this has been reduced from about five minutes to 2.5 minutes but with the ILMS the figure is further reduced to 1.5 minutes.

Under the US Army programme, the M270A1 is a re-capitalisation of the M270 legacy launcher designed to meet the US Army vision. Unlike the older M270 launcher, the M270A1 is capable of launching the Guided MLRS (GMLRS) rocket and Army TACMS (ATACMS) Block IA munitions variants.

In May 1999, the M270A1 launcher successfully fired an ATACMS Block I missile and in November it conducted a successful launching of a rocket with six AT2 munitions.

According to the prime contractor, the conversion to the enhanced M270A1 standard extends the operational life of the MLRS system by a further 15 to 20 years. First systems were fielded at Fort Hood in 2002 followed by South Korea.

A total of 226 US Army M270 have been upgraded to the M270A1 standard with final deliveries in 2005.

First upgraded MARS II 227 mm (12 round) rocket system has been handed over to the German Army (Krauss-Maffei Wegmann) 1401955

External view of the ATK warhead selected for the US Army's GMLRS alternative warhead programme (ATK) 1364818

According to Lockheed Martin, the latest M270A1 has an operational readiness rate of 98 per cent.

The M270 v1 will be feature a refurbished and improved cab, modified FCS and long range communications.

Future improvements to the M270A1 are expected to include the installation of a 600 hp Cummins diesel engine which will be more fuel efficient, improved weapons interface unit and a new Environmental Control Unit/Auxiliary Power Unit (ECU/APU).

It should be noted that the US Army no longer has any of the original M270 MLRS deployed.

New European FCS

The four members of the European Multiple Launch Rocket System (MLRS) consortium, France, Germany, Italy and the United Kingdom, were considering a programme to replace the current US designed computerised fire-control system installed in their European built 227 mm (12-round) MLRS systems.

The current system was designed in the US but built under licence in Europe under a build to print contract awarded to the then Marconi Command and Control Systems of the UK.

The components of the current European-built FCS are now obsolete and the US Army has now upgraded its (now) Lockheed Martin Missiles and Fire Control MLRS with a new fire-control system.

Europe decided to develop and place in production its own fire-control system with the French Service des Programmes d'armement Terrestre (SPART) acting as the executive agency for the four European countries.

In the end the UK decided to purchase the US Lockheed Martin Missile and Fire Control Improved Fire Control System (IFCS) which is already in quantity production for the US Army while France, Germany and Italy opted for a European solution.

In mid-2003, EADS/Dornier was awarded a EUR41.5 million contract to supply a European Fire Control System (EFCS) for Germany (150 launchers), France (67 launchers) and Italy (20 launchers). The contract ran for 36 months.

New components of the EFCS are the Power Supply Unit (PSU), Main Control Unit (MCU), Position & Navigation Unit (PNU) with a GPS receiver and a Launcher Control Unit (LCU).

New fire-control system for UK MLRS

In 2005 Lockheed Martin Missiles and Fire Control was awarded a contract worth GBP10 million by the then UK Defence Procurement Agency (DPA), Future Artillery Weapon Systems (FAWS) Integrated Project Team (IPT), via the Foreign Military Sales (FMS) programme for the supply of 15 of the latest US Future Fire-Control Systems (FFCS).

By April 2007 the Future Fire-Control System (FFCS) had been integrated into the first batch of upgraded MLRS M270B1 tracked launchers and this met its In Service Date (ISD) with the British Army's Royal Artillery (RA).

The original MLRS M270 launcher cannot fire GMLRS and the UK also ordered the FFCS under a separate contract. Under Phase 1 a total of 12 in service M270 launchers have been upgraded to the M270B1 standard with an additional batch of 15 following. These are not being fitted with the Electric Drive Launcher System (EDLS).

The first system was upgraded in the Lockheed Martin Missiles and Fire Control facility in Camden, Arkansas. Follow on work was carried our in the UK at the facilities of the now Defence Support Group (previously ABRO) in Donnington in two batches, the first of six (2-7) and the second of five (8-12). The upgrade also includes an upgraded Cummins diesel power pack and other enhancements.

These were issued to 39 Regiment, RA who deployed a troop of M270B1 systems to Afghanistan in the second half of 2007.

Until recently the MLRS used by the RA fired Phase 1 and Phase II rockets. The first carries a total of 644 sub-munitions that are not fitted with a self-destruct fuze mechanism out to a maximum range of 31.6 km. The Phase II rocket carries 28 anti-tank mines.

The Phase I rockets have now been phased out of service with the UK and are now being decommissioned.

Following the cancellation of the Lightweight Mobile Artillery Weapon System (Rocket) programme in 2008, the UK MoD decided to upgrade an additional 12 M270 launchers to the latest M270B1 standard which will give the Royal Artillery a total of 36 M270B1 capable of firing the precision strike GMLRS.

For operations in Afghanistan a small batch of M270B1 have been upgraded and fitted with bar armour, mine protection, new energy absorbing seats, additional situational awareness, wire cutters, roof mounted machine gun, environmental control unit and an upgraded suspension system to take into account the additional weight of the vehicle.

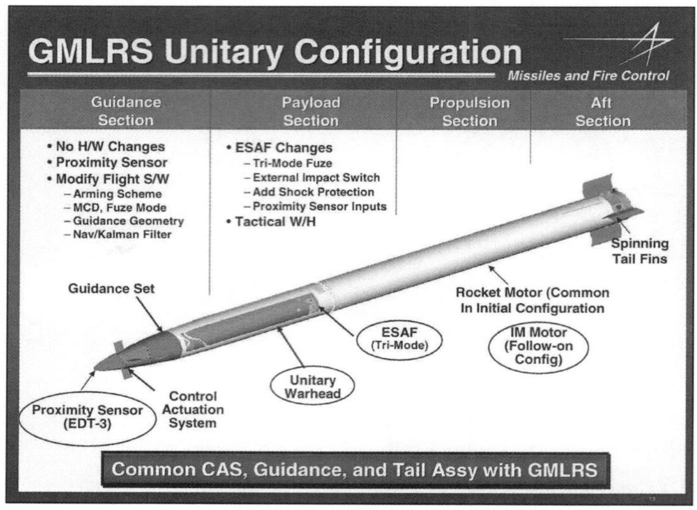

Key components of the Guided MLRS unitary 227 mm rocket (Lockheed Martin Missiles and Fire Control) 1043239

Early in 2013 it was revealed that the Royal Artillery had withdrawn its M270B1 systems from Afghanistan.

Bahrain M270 Upgrade

Bahrain is upgrading its systems to the M270C1 standard which is the export equivalent of the M270A1 used by the US Army.

Finnish upgraded MLRS

In May 2011 it was announced that Lockheed Martin Missiles and Fire Control had been awarded a USD45.3 million contract to upgrade the complete fleet of Finnish MLRS to the M270D1 standard.

The company will provide kits to enable the existing M270 launchers to be upgraded in Finland under a technical assistance agreement with Millog Oy with modifications including the installation of the Universal Fire-Control System.

Finland is the third international customer for the UFCS upgrade with the others being Bahrain (M270C1) and the UK (M270B1).

The M270D1 upgrade is similar to that carried out on the US Army MLRS but without the Improved Launcher Mechanical System (ILMS).

When upgraded the Finnish M270D1 will be able to fire the latest GMLRS and Army Tactical Missile System Unitary to provide the Finnish Army with a precision effect at long range.

Finland purchased the complete Royal Netherlands Army fleet of MLRS as well as pods of Phase 1 unguided rockets in 1997.

Germany has provided Finland with Phase II rockets containing anti-tank mines as well as training rockets.

In mid-2012 it was stated that Finland had made a formal request to the US Defense Security Cooperation Agency for the acquisition of 70 Block 1A ATACMS Block 1A unitary missiles at a total value of USD132 million including associated parts, equipment, logistics and training.

German upgraded MLRS

In April 2011 Krauss-Maffei Wegmann handed over the first MARS II rocket launcher at the German Army Artillery School in Idar Oberstein, Germany.

MARS is the German name for the US developed Multiple Launch Rocket System (MLRS).

Under a contract awarded in December 2008 by the Bundesamt fur Wehrtechnik und Beshchaffung (BWB); the Krauss-Maffei Wegmann facility in Kassel is upgrading the first four units of the German Army fleet to the enhanced MARS II standard. Discussions are underway for the second batch of 16 units.

The upgrade includes a new fire-control system and upgraded launcher that allows it to fire the current production Lockheed Martin Missiles and Fire Control Guided MLRS (GMLRS) rocket.

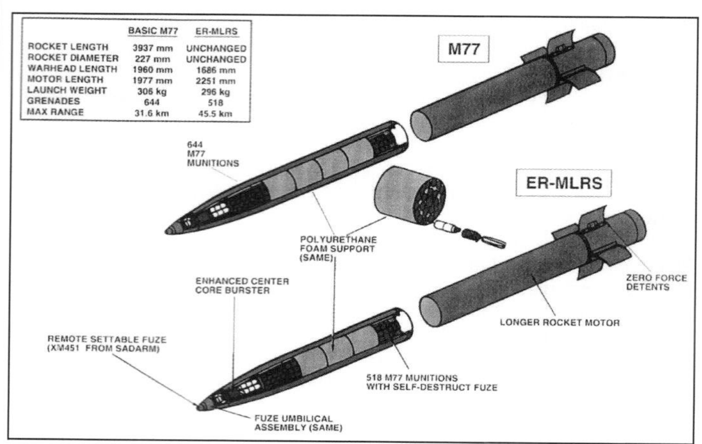

Lockheed Martin Missiles and Fire Control standard M77 rocket (top) with extended-range rocket (below) 0021089

M270 AVMRL with full-scale mock-up of Army TACMS missile and standard rocket pod 0500658

This has a conventional unitary high-explosive warhead and enables targets to be engaged with a precision effect out to ranges of over 70 km.

When originally introduced into service the German Army MLRS fired the standard Phase 1 rocket carrying 644 M77 submunitions out to a range of 31.60 km. This type of munition is now banned under the Ottawa Convention on Cluster Munitions.

MLRS rocket pod

The standard rocket pod, which is also the transportation and storage container, weighs 2,308 kg when loaded. It comprises six glass fibre tubes held rigidly in an aluminium case structure. Blow-off covers seal each rocket in its tube, so the rocket requires no maintenance, inspection or crew servicing from the factory to the launch site.

The rocket pods are loaded by the integral twin-boom loader system of the launcher.

M26 warhead

The Phase I rocket has a minimum 20 year shelf-life, weighs 306 kg, is 3.937 m long and has a diameter of 0.227 m. The rocket is centred in the launch tube by a set of four forward sabots, two aft riders and two spin lugs that ride along helical rails.

When the rocket is discharged, the sabots are discarded and four folding fins at the rear of the rocket motor are released by a cable-release initiated pyrotechnic delay device, attached to the launch tube.

The warhead weighs 156 kg, has an aluminium skin and carries 644 M77 Dual-Purpose Improved Conventional Munition (DPICM) shaped-charge blast fragmentation bomblets weighing 213 g each, nestled with eight polyurethane foam supports. Conventional steel armour penetration is 76 to 102 mm. An M445 remotely set electronic time fuze in the warhead initiates dispersal of the ribbon-stabilised bomblets in mid-air over the target. The range of this version is 31,600 m. At mid-range each warhead of a salvo will saturate a 200 × 100 m sized target.

The actual ripple pattern size varies with the range and aimpoints plotted within the target zone. The Phase I rocket was used operationally with great success during Operation Desert Storm in 1991. The main drawback of this warhead is the submunitions are not fitted with a self-destruct fuze mechanism.

A number of countries are phasing out this rocket as it contains submunitions.

This rocket has now been phased out of production.

Lockheed Martin extended-range rocket (XR-M77)

As a result of the Iraqi use of superior ranged artillery, Lockheed Martin Missiles and Fire Control began development work on an extended-range rocket for the MLRS in February 1991. The first prototype was tested in November 1991 and attained a range of 45,000 m.

In April 1992, the US Army Missile Command announced its intention to buy the weapon from Lockheed Martin Missiles and Fire Control. The warhead is shortened and contains only 518 bomblets instead of the standard 644. However, the reduced payload is compensated for by an increase in accuracy and a 274 mm stretch in rocket motor length to increase the range to 45,500 m. The extended-range rocket uses the current rocket pod container-launcher. A high-altitude fuze is fitted. Firing trials have also taken place at the French *Centre D'Essais de la Mediterranée* (CEM) range. The overall rocket weight is reduced slightly to 296 kg. The first export customer for this was Greece who ordered them in July 1999.

This rocket has now been phased out of production.

Guided MLRS

The five members (France, Germany, Italy, UK and US) that signed the MLRS MoU, formulated their requirements for a new Guided MLRS (GMLRS) which will have a greater range and accuracy.

In November 1998, Lockheed Martin Missiles and Fire Control was awarded a 48-month Engineering and Manufacturing Development (EMD) contract worth USD121 million for the development of the Guided MLRS (GMLRS) rocket.

The GMLRS is an international co-operative programme with France, Germany, Italy, UK and the US who worked together throughout the EMD phase of the rocket's life cycle.

M270A1 Multiple Launch Rocker System (MLRS) upgraded with a new BAE Systems redesigned cab for enhanced crew survivability
(IHS/Rupert Pengelley) 1347674

The GMLRS rocket has an Inertial Measurement Unit (IMU) aided by a Global Positioning System (GPS) integrated on a GMLRS rocket body. In addition, canards on the GMLRS nose add basic manoeuvrability to further enhance the accuracy of the system.

Under the terms of this contract, the four European countries contributed 12.5 per cent each to the EMD phase while the US contributed 50 per cent.

In its basic form, the GMLRS has a warhead containing 404 Dual-Purpose Improved Conventional Munitions (DPICM), each of which will have self-destruct fuze.

The GMLRS is 3.937 m long, have a diameter of 227 mm and weigh 302 kg.

The European companies involved in the programme are Diehl, MBDA and Fiat Avio.

Under the US programme, the GMLRS rocket is a development effort to replace the M26 free flight tactical MLRS rocket and is designed to meet the US Army vision. It will increase accuracy and range while reducing the overall logistic burden to the force.

It also provides MLRS with a guided rocket bus capability for the delivery of various future payloads. It can be launched from the upgraded M270A1 MLRS launcher and the HIMARS (6 × 6) wheeled variant of the MLRS.

The flight test programme for Guided MLRS was completed late in 2002. Late in 2002 GMLRS completed a 48-month System Development and Demonstration (SDD) phase on schedule after successfully completing Production Qualification Flight Testing.

Late in April 2003, the US Army approved the GMLRS to enter Low Rate Initial Production.

In June 2003 Lockheed Martin Missiles and Fire Control was awarded the Low Rate Initial Production I (LRIP I) GMLRS contract, under which it supplied 145 pods.

In March 2004 Lockheed Martin was awarded a USD85 million contract to manufacture 840 (140 pods) GMLRS rockets under a Low Rate Initial Production II (LRIP II) contract.

By October 2011 over 1,685 GMLRS had been fired with a 99 per-cent reliability rating and over 11,500 had been manufactured in Camden for the US and export customers.

In August 2012 Lockheed Martin Missiles and Fire Control stated that they had delivered the 20,000 GMLRS unitary rockets to the US Army.

By that time, over 2,300 had been fired in combat with a combat reliability rate of 98 per cent.

GMLRS - unitary warhead

Late in 2000 it was stated that the US would develop a unitary warhead variant of the GMLRS to provide commanders with a precision strike capability in urban and complex terrain.

This has an improved guidance system to allow for precision strikes. It also has a new differential global positioning system with an antenna that has greater resistance to jamming.

These upgraded systems will feed mid-course updates to the separated warhead, allowing its four canards to make steering corrections that will improve the CEP from 10 m on the standard GMLRS to about 2 m on the unitary warhead variant.

In addition, the unitary warhead will also have the capability to selectively create an airburst, point detonation or delayed detonation for penetrating buildings and bunkers protected by 0.90 to 2.4 m of concrete.

The guided unitary MLRS has a range of over 70 km and be fitted with a 90 kg plus unitary warhead, which would be insensitive munitions compliant. It would feature IMU guidance and be GPS aided.

The Guided Unitary version of GMLRS is also moving into the SDD phase. This version will utilise spiral development to improve the GMLRS by installing a 90 kg plus high-explosive warhead into the rocket.

It will also be insensitive munitions compliant and have a multirole fuze that could be set for airburst, point detonation or delayed.

The programme successfully completed an initial flight demonstration in August 2002, which was funded by Lockheed Martin's internal research and development.

British Army MLRS Repair and Recovery Vehicle towing a MLRS (Lockheed Martin UK) 1333684

It is envisioned that first production models would have point detonate, delay and airburst capabilities. The follow-on model would have initial plus insensitive munition and trajectory shaping.

Phase II - AT2 warhead

The 1.959 m long AT2 anti-tank mine was developed by Germany and has been in service in the ground-scatterable version for some years. Production of this has been completed. The mine warhead is 236.6 mm in diameter, weighs 108 kg and contains 28 AT2 DM1399 anti-tank mines. Due to its lighter 258 kg weight, the AT2 rocket has a range of 40 km, somewhat longer than the M77 rocket with the M26 warhead. The warhead is activated by a DM42 adjustable timer electronic fuze and is opened in flight by a small explosive charge. The seven mine dispensers, each with four mines, are pneumatically ejected and then retarded to release the mines, which are parachute-stabilised in flight. The dispensers measure 105.7 mm in diameter, 548 mm in length and weigh 10.8 kg.

The 3.935 m long rocket usually dispenses the mines when overflying the target area at an altitude of about 1,200 m. The 128.2 mm high AT2 anti-tank mine has a diameter of 103.5 mm and a deployed height of 165 mm with its ground orientation springs open it weighs 2.25 kg. Detonation of the shaped-charge full-width attack mine occurs either when an armoured vehicle passes over it or when it reaches 97 per cent of its useful lifetime, determined before launch from one of six available programmable settings.

Conventional armour penetration is in excess of 140 mm. An anti-handling/removal device is incorporated to disrupt mineclearing operations. The two launcher loads of AT2 rockets can create a minefield of 2,000 × 115 m.

The Phase II AT2 warhead version was only supplied to Germany and the UK.

According to the United Nations Arms Transfer List for 2009 a total of 167 of these rockets were transferred from Germany to Finland. These are for use with the 22 M270 MLRS systems supplied by Netherlands who has phased this system out of service.

Enhanced blast warhead

Early in 2007 it was stated that Lockheed Martin, teamed with Aerojet and MBDA, was developing a thermobaric Enhanced Blast Warhead (EBW) for GMLRS. The first industry funded live fire testing of the warhead took place in mid-2006 at White Sands Missile Range.

The EBW could be fitted in place of the current unitary warhead and would have the capability to be programmed before launch to produce a variable effect.

GMLRS alternative warhead

Following a competition, Alliant Techsystems was selected in 2011 as the prime contractor for the Alternative Warhead (AW) for the GMLRS. Also competing for this contract was Aerojet and General Dynamics Ordnance and Tactical Systems.

It is understood that the ATK AW design uses a combination of tungsten fragments and explosively formed penetrators to achieve the desired anti-armour/anti-material effects against the specified target array.

The ATK warhead uses a PBXN-109 insensitive high-explosive fill and is initiated with the current proximity sensor and electronic safe and arm associated with the current GMLRS.

Variants

US Army Tactical Missile System (Army TACMS)

Late in March 1986, the US Army announced that the now Lockheed Martin Missiles and Fire Control had been selected to design and develop the Army Tactical Missile System (Army TACMS or ATACMS) and was subsequently awarded a USD180.3 million contract.

Army TACMS is fired from the MLRS M270 launcher using off-axis guidance techniques to prevent enemy radars from plotting the trajectory and initiating counterbattery fire against its launch point.

Each Block 1 missile is approximately 3.962 m long, 0.61 m in diameter and is powered by an advanced solid-propellant fuel to a range of over 165 km.

Each 591 kg warhead contains 950 M74 anti-personnel/anti-matériel bomblets which weigh 0.59 kg each, are 0.059 m in diameter and are fitted with a tungsten fragmenting wall with steel overlay. They are dispersed over the target area to detonate upon impact.

Teamed with Lockheed Martin Missiles and Fire Control are Atlantic Research Corporation for the solid rocket fuel propulsion and Honeywell Incorporated for the onboard inertial guidance system.

The first Army TACMS missile was fired in 1988 and the same year Lockheed Martin Missiles and Fire Control was awarded an initial contract for 66 production missiles worth USD44 million.

Early in 1990, the company was awarded a USD59.6 million fixed price contract for low-rate initial production of 104 missiles was delivered by late 1991. In December 1991, a USD114 million contract was awarded to the company for full-scale production. The first year procurement was for 300 missiles and launching assemblies. Further contracts have since been placed. First missiles were deployed to support Operation Desert Storm. A total of 32 out of 105 TACMS rounds taken were successfully used by the US Army in the 1991 Gulf War. Production of the Block I has been completed.

The US Army also awarded Lockheed Martin Missiles and Fire Control a USD83 million contract to integrate Army TACMS into the existing M270 AVMRL. The changes are principally in the software of the system's fire-control computer and will allow the system to fire standard MLRS rockets or Army TACMS missiles.

Army TACMS allows battlefield commanders to strike well behind enemy lines, beyond the range of existing artillery, rockets or missiles.

In early 1994, Lockheed Martin Missiles and Fire Control was awarded an USD18.7 million contract to design and test a long-range variant of ATACMS for the US Navy.

Following this, the US Army approved plans for the development of the Block IA extended-range ATACMS with considerably improved accuracy. The Block IA round is the same size as the current weapon but has twice its range. The same solid fuel rocket motor is used but the range increase is due to a reduction in payload to 275 submunitions. An in-flight receiver for GPS satellite updates is fitted so that the missile ring laser gyro guidance system can make flight corrections. The first Block IA flight test was successfully performed in January 1995.

The US Army ordered 573 Block IA missiles, which were delivered by February 2003. Minimum range is 70 km and maximum range is 300 km.

Greece (71), South Korea and Turkey (72) placed orders for ATACMS rockets. Greece has since placed additional orders for ATACMS with the last one being in July 1999 worth USD39 million for ATACMS and ER rockets. Late in 2001, Korea placed a contract for 111 ATACMS Block IA missiles, with deliveries completed by April 2004.

The latest ATACMS is the Block II and, in June 1999, the company was awarded a Low Rate Initial Production contract from the US Army for a total of 24 missiles at a cost of USD138.5 million. This has a range of more than 128 km.

The Block IA Unitary features a new internal structure and a navy warhead, which is the same as that installed in the SLAM/Harpoon missile system. This warhead is designated the WAU-23B with the missile having a maximum range of 300 km.

A follow-on ATACMS Unitary Missile programme (SDD) will convert 500 Block I missiles from the current US Army inventory to Unitary Missiles. It has a range of 300 km. Block upgrades will include:

- Improved guidance set
- Washed out and re-grained solid rocket motors or re-certified motors
- Improved warhead
- Elimination of demilitarisation costs
- Multifunction fuze.

In Operation Iraqi Freedom, the US fired the following ATACMS:

- Unitary: 16
- Block 1: 373
- Block 1A: 67.

The ATACMS is also out of production by but continues to be deployed by the US and other countries with the former to go through a SLEP.

GMLRS+

Under development as a private venture by the company is GMLRS+ which will extend range to about 130 km and feature GMLRS unitary guidance and fuze system and a terminal seeker.

The control surfaces will be to the rear with strakes added towards the front for improved manoeuvrability.

The HE warhead will be scaleable and the missile will also have the ability to engage fleeting targets.

This was first successfully test fired late in 2010 at the White Sands Missile Range which it was fitted with a Semi-Active Laser (SAL) and fired from a HIMARS and reached a maximum range of 40 km.

Since then trials have continued in 2011 the GMLRS+ achieved a maximum range of 120 km.

Diehl GMLRS with SMArt warhead

Diehl BGT Defence under contract to the German MoD has developed a GMLRS carrying four SMArt top-attack munitions which are already installed in the 155 mm SMArt artillery projectile already in service with the German Army and export customers.

Tactical Missile System Penetrator (TACMS-P)

This version, still at the development phase, features a government furnished equipment re-entry body, earth penetrator and a warhead fin/ACS control system. During trials at White Sands Missile Range, the missile range was between 140 and 170 km with total weight being 1,332 kg. Motor currently used is the Block I/IA with adapted T2K guidance.

Range of production systems will be 140 to 220 km but a follow-on programme funded by the US DoD includes a new motor that will increase the range to about 500 km.

US Army ATACMS Brilliant Submunition
This programme has been cancelled and it never entered production or service.

P44 missile
This was being developed as a private venture by Lockheed Martin Missiles and Fire Control but all development has now ceased.

M270C1 launcher
This is the official US designation for non-US or non-British MLRS (international users) that upgrade their original M270 launchers.
The C1 in the designation means that it has been upgraded with an Improved Fire Control System (IFCS) that allows the launcher to fire all of the current MLRS missile and rocket munitions including GMLRS.

Upgraded M270A1 launcher
In late 2009 it was revealed that BAE Systems, had developed using Independent Research and Development (IR&D) funding, a redesigned cab for the M270A1 MLRS currently deployed by the US Army.
This has been designed in conjunction with the US Army who intends to retain a fleet of 225 full tracked M270A1 227 mm (12-round) launchers through to 2050 alongside its fleet of M142 227 mm (6-round) High Mobility Artillery Rocket System (HIMARS).
It should be noted that deliveries to the US Army of the M270A1 and M142 have now been completed and the earlier M270 MLRS has now been phased out of service.
Development of the redesigned cab commenced in January 2009 at the BAE Systems, facility in Santa Clara, California.
When compared to the existing cab, the redesigned cab for the latest M270A1 MLRS launcher not only has an additional 40 per cent internal volume but also an integrated armour and mine protection package for enhanced crew survivability which meets the latest armour protection requirements.
The wider and higher welded aluminium armour cab features three new mine blast resistant seats that are now attached to the sides rather than being attached to the floor for enhanced crew survivability.
This cab also features a Blue Force Tracker (BFT), Driver Vision Enhancement (DVE) with two flat panel displays (for driver and commander) and long range communications equipment.
In the future the US Army expects to upgrade the Fire-Control System (FCS) of the M270A1.
Some of the subsystems used in the redesigned cab also used in other currently deployed US Army platforms including HIMARS which is based on a BAE Systems Family of Medium Tactical Vehicle (FMTV) (6 × 6) five-tonne cross-country chassis with a protected cab and Mine Resistant Ambush Protected (MRAP) vehicles.
The M270A1 is an improved version of the original M270 MLRS and features an Improved FCS, improved launcher mechanical system and 600 hp Cummins diesel engine.
The power pack of the M270A1 is retained which consists of a Cummins VTA-903 series diesel developing 600 hp coupled to an L3 Combat Propulsion Systems HMPT-500 series automatic transmission.
With a combat weight of about 30 tonnes this gives a power to weight ratio of 20 hp/tonne. The new generation cab also features a power assisted device that tilts the cab forwards to allow access to the power pack located below the cab.
While production of the M270 MLRS launcher was completed so time ago, BAE Systems still has the capability to manufacture new systems for the export market if required.
In addition, the US Army has a number of M270 MLRS launchers that could be refurbished for the US Army or for the export market.

Resupply
The resupply vehicles used by the US Army are a Oshkosh Heavy Expanded Mobility Tactical Truck (HEMTT) (8 × 8), which carries four fully loaded rocket pods and its Heavy Expanded Mobility Ammunition Trailer (HEMAT), which carries a further four fully loaded rocket pods. The truck is air-transportable by C-130 or C-141 aircraft.

Multiple Launch Rocket System (MLRS) of the Norwegian Army with one pod of six rockets and launcher traversed to left (Svein Wiiger Olsen) 0130019

Upgraded British Army M270B1 227 mm (12-round) launcher fires a GMLRS during trials at White Sands Missile Range (Lockheed Martin) 1401590

The European armies selected their own national resupply systems. The UK, for example uses its Demountable Rack Offloading and Pick-up System (DROPS). This is also a truck-trailer system consisting of the Leyland Medium Mobility Load Carrier (MMLC) and a trailer, both of which carry four rocket pods. Germany uses its MULTI system, which is very similar to DROPS.

High-Mobility Artillery Rocket System (HIMARS)
Details of this systems are provided in a separate entry in *IHS Jane's Land Warfare Platforms: Artillery and Air Defence* and is currently in production for the home and export markets.

M28 practice rocket
The practice warhead was used as a training device for personnel. It uses steel pipe in place of the tactical munitions. Additionally, three smoke canisters are located in bay number 4. These canisters are made from plexiglass and contain 100 mg of titanium tetrachloride ($TiCl_4$).
On detonation, the canisters are broken by the central core burster and the $TiCl_4$ reacts with moisture in the air to produce smoke. This cloud is visible for a minimum of 10 seconds from 6,000 m. The M28 has been replaced by the M28A1 rocket.

Lockheed Martin Missiles and Fire Control M28A1 Reduced Range Practice Rocket (RRPR)
This value engineered replacement for the M28 went into production in 1993. A blunt nose limits range to between 7.5 and 14.3 km so allowing its use on short artillery ranges. The M28A1 also deletes the time fuze and centre core burster in preference for a point detonating fuze and smoke charge on impact. The round uses a standard tactical rocket motor for training realism and launcher validation. Like the tactical rocket, no maintenance is required.

MORAK 122 mm (40-round) rocket launcher
Details of this system are given in a separate entry. It is an RM-70 or RM-70/85 system based on a Tatra 813 series 8 × 8 chassis. It has a modified launcher on which is fitted a six pack of 227 mm rockets as used in the MLRS system. The first of 26 production MORAK systems were delivered to Slovakia in mid-2005.

Israel upgraded MLRS
Late in 1998, Israel Military Industries awarded the now Lockheed Martin Missiles and Fire Control of the US a contract worth USD60.2 million for development and production of a portion of the Multiple Launch Rocket System (MLRS) to be integrated with the Israel Military Industries-developed Trajectory Correction System (TCS).
Under the terms of this contract, Lockheed Martin Missiles and Fire Control, prime contractor for the MLRS, built the rockets and supported flight testing in Israel during the development stage. The production phase runs concurrently. Development of this system was completed in 2003.
Most of the work in the US has been carried out at the Lockheed Martin Missiles and Fire Control headquarters in Dallas and at the MLRS rocket production facility at Camden, Arkansas.

The installation of the TCS onto the currently unguided MLRS rocket, provides in-flight control of the rocket's flight path and gives a much greater level of accuracy. This in turn means that fewer rockets are required to neutralise a given target.

Overall prime contractor for the MLRS TCS upgrade is Israel Military Industries who were awarded a USD170 million contract by the Israeli MoD. In addition to Lockheed Martin Missiles and Fire Control there are a number of other sub-contractors including Elisra Electronic Systems in Israel.

The TCS is an add-on upgrading kit and can also be fitted to the Israeli Military Industries LAR 160 rocket which has a range of about 45 km and the MAR350 rocket with a range of about 80 km. Details of these are given in a separate entry in *IHS Jane's Land Warfare Platforms: Artillery and Air Defence.*

MLRS RRV

Following a competition, in 2004 the now Lockheed Martin UK was awarded a GBP2.4 million contract from the UK's then Defence Logistics Organisation (DLO) for the design, development and manufacture of a Repair and Recovery Vehicle (RRV) to support MLRS in service with the Royal Artillery.

Following trials with a prototype vehicle, a total of four units were supplied, with final deliveries taking place in 2006. This work was carried out at the Lockheed Martin UK facility at Ampthill in Bedfordshire.

For the RRV role, the complete 227 mm (12-round) rocket launcher mounted to the rear of the cab has been removed and replaced by a repair and recovery modification kit developed by the company.

The kit includes a crane and a winch. The former can be used to lift subsystems such as a complete MLRS power pack. The winch will enable the MLRS RRV to rapidly recover damaged and disabled MLRS vehicles under almost all situations. When the winch is being used, an earth anchor is lowered at the rear of the vehicle to provide a more stable platform for recovery operations.

The RRV is also equipped with other repair equipment, including an air compressor, battery charger and stowage boxes for tools and other specialised equipment. For travelling, the crew are seated in the armour and NBC protected cab at the front of the vehicle, as in the standard in-service MLRS vehicle.

155 mm Artillery Gun Module

A surplus German Army MLRS chassis is used as the basis for the German Krauss-Maffei Wegmann 155 mm/52-calibre Artillery Gun Module (AGM) which has been developed by the company as a private venture. Full details of this system, which currently remains at the prototype stage, are provided in a separate entry in *IHS Jane's Land Warfare Platforms: Artillery and Air Defence.* As of May 2013 this remained at the prototype stage. Further development by General Dynamics European Land Systems - Santa Bàrbara Sistemas of Spain and Krauss-Maffei Wegmann has resulted in a new system called Thor. This is based on a new hull fitted with the latest version of the AGM and details are provided in a separate entry in *IHS Jane's Land Warfare Platforms: Artillery and Air Defence.* As of May 2013 this remained at the prototype stage.

Dynamit Nobel subcalibre training system

This training system, developed by Dynamit Nobel to meet the requirements of the German Army, is based on the use of 110 mm Light Artillery Rockets (LARs). The LAR has now been phased out of service with the German Army.

These rockets meet the requirements concerning danger zones and low unit cost. They have a maximum range of 14,000 m and a minimum range of 6,000 m.

The use of the Dynamit Nobel subcalibre training system allows all MLRS training activities to be carried out including all tasks from the preparation of loaded Rocket Pods (RP) to rocket launching itself under realistic conditions even with the following constraints:

- Limitations/restrictions on national ranges (area, danger zones)
- Minimising the cost of practice fire missions.

Sensor pod on MLRS

Under contract to the German MoD, Krauss-Maffei Wegmann has fitted an MLRS carrier chassis with a mast-mounted sensor pod that can be elevated to 11 m.

This was only an experimental system and never entered service with the German Army.

Specifications

	M270 Multiple Launch Rocket System
Dimensions and weights	
Crew:	3
Length	
overall:	6.972 m
Width	
overall:	2.972 m
Height	
transport configuration:	2.617 m
deployed:	5.925 m
Ground clearance	
overall:	0.43 m
Track width	
normal:	533 mm
Length of track on ground:	4.33 m
Weight	
standard:	20,189 kg
combat:	25,191 kg
Mobility	
Configuration	
running gear:	tracked
Speed	
max speed:	64 km/h
Range	
main fuel supply:	483 km
Fuel capacity	
main:	617 litres
Fording	
without preparation:	1.1 m
Gradient:	60%
Side slope:	40%
Vertical obstacle	
forwards:	1 m
Trench:	2.29 m
Engine	Cummins VTA-903, 8 cylinders, turbocharged, diesel, 500 hp at 2,400 rpm
Gearbox	
model:	L3 Combat Propulsion Systems HMPT-500 hydromechanical
Steering:	hydrostatic
Suspension:	torsion-bar-in-tube
Electrical system	
vehicle:	24 V
Survivability	
Night vision equipment	
vehicle:	optional
NBC system:	yes
Armour	
hull/body:	aluminium

Status

Production complete of the M270/M270A1 launcher but could be resumed. In service with the following countries:

Bahrain - initial order for seven launchers increased to nine. All delivered in 1992.

Denmark - order for eight launchers, 300 ER rocket pods, 25 practice rocket pods and spares placed. Total cost USD146 million. The first four MLRS launchers were delivered to Denmark in 1999. These are now for sale.

Egypt - late in 2001, Egypt placed a contract for the supply of 26 MLRS launchers under the Foreign Military Sales (FMS) programme. Egypt also placed a contract worth USD72.2 million for the supply of more than 400 Extended Range MLRS rocket pods. These were delivered to Egypt between late 2002 and 2003.

Finland - took delivery of all of the 22 Royal Netherlands Army 227 mm MLRS plus AT2 rockets and training rockets from the German Army. According to the United Nations Arms Transfer Lists in 2007 Finland took delivery of a total of 22 MLRS and 522 rocket pods. These are all being upgraded and additional details are provided in the entry.

France - first launcher from US with remainder built in Europe. Total of 55 launchers with final deliveries made in 1995. Two regiments, 12th and 74th were originally formed with 24 MLRS each.

Germany - six launchers from the US with remainder built in Europe. Total of 150 launchers required. First unit formed in 1991 with eight battalions originally formed, the 12th, 42nd, 52nd, 62nd, 122nd, 150th, 702nd, and 802nd, fielded. Batteries have nine launchers in platoons of four and five launchers respectively. Two battalions form part of Germany's Rapid Reaction Force. Part of this fleet is being upgraded and details are provided in the entry with the first four being delivered in 2011.

Greece - order for nine launchers and 132 reduced range practice rockets placed in 1994. Greece took delivery of its first nine MLRS launchers in 1995 followed by 14 in 1999 and 18 in 2000. Greece has also ordered standard rockets, ER rockets and three lots of ATACMS. A total of 36 are now in service.

Israel - initial order for six launchers, 726 tactical and 720 reduced range practice rockets placed in 1994 for delivery by December 1996. A further USD103.5 million package for 42 launchers and 1,500 rockets was placed in late 1995. According to the United Nations MLRS deliveries to Israel were as follows:

Year	Quantity
1995	6
1996	6
1997	16
1998	20
1999	16

Italy - one launcher from the US with remainder built in Europe. A total of 24 launchers acquired. Single regiment, 3rd Volturno, equipped with 20 launchers in service.

Japan - fourth of four annual contracts each for nine launchers, with associated rockets and support equipment, signed in 1995 with Lockheed Martin Missiles and Fire Control. The first contract was signed in late 1992. The first three launchers were shipped in May 1994, followed by three more in June and the final three of the first contract in August 1994. Upon delivery the now IHI

Aerospace and the official systems integrator for MLRS in Japan, makes a number of modifications primarily to the fire-control system before the launchers are handed over to the Japanese Ground Self-Defence Force. The last nine launchers have now been delivered. The initial order included 1,300 tactical rockets and 47 launcher trainer pods. By the end of 1999 Japan had taken delivery of a total of 45 MLRS systems. According to the UN, MLRS deliveries were as follows:

Year	Quantity
1994	9
1995	9
1996	nil
1997	14
1998	12
1999	33
2000	nil

South Korea - As part of a USD624 million arms package from the US, South Korea has received 29 MLRS launchers, 271 MLRS rocket pods, 168 reduced range practice rocket pods and nine MLRS fire-control proficiency trainers. Also included in the sale were 111 ATACMS rounds to counter hostile long-range artillery and rocket systems. 10 MLRS systems were delivered in 1998 and a further 19 in 1999.

Netherlands - 22 launchers and 16,000 rockets delivered by Lockheed Martin Missiles and Fire Control. First European Army to deploy MLRS. These have been sold to Finland and are no longer operational with the Royal Netherlands Army.

Norway - order placed for 12 launchers, 360 ER tactical rocket pods, 564 practice rocket pods and spares. Total cost USD199 million. Six systems were delivered to Norway in 1997 and a further six in 1998. These are expected to be phased out of service in the future.

Turkey - initial nine launchers delivered in 1989 from the US with a further six in 1992. The latter order included more than 2,100 tactical rockets, 144 practice rockets and 12 training pods. Has taken delivery of ATACMS.

UK - four from US with remainder built in Europe. Total of 63 launchers required. Part of this fleet is being upgraded and details are provided in the entry. The UK has also converted four of its surplus MLRS into Repair and Recovery Vehicles.

US - 830 launchers and 477,378 tactical rockets have been delivered to the US Army (including National Guard units). The US Army has now upgraded 226 MLRSs to the enhanced M270A1 standard.

Contractor

Lockheed Martin Missiles and Fire Control.

Lockheed Martin Missiles and Fire Control M142 227 mm (6-round) High-Mobility Artillery Rocket System (HIMARS)

Development

The High-Mobility Artillery Rocket System (HIMARS) was initially developed as a private venture by Lockheed Martin Missiles and Fire Control to meet a US Army requirement for a lighter and more mobile version of MLRS that is capable of being carried by the Lockheed Martin C-130 transport aircraft.

According to Lockheed Martin Missiles and Fire Control, the smaller HIMARS requires 30 per cent fewer airlifts to transport a battery.

Crew training is the same as the current MLRS system with the lightweight chassis offering higher road speeds and lower operating costs than its full-tracked counterpart.

The system was first unveiled in the US late in 1993 and was shown for the first time in Europe in September 1994.

Early in 1996, the US Army Missile Command (MICOM) awarded Lockheed Martin Missiles and Fire Control a USD23.2 million contract to build four prototypes of the High-Mobility Artillery Rocket System (HIMARS) launcher.

Under the terms of the 53-month contract, the Lockheed Martin Missiles and Fire Control built four systems, which were subjected to a two year extended user evaluation. The fourth system was retained by the company for its testing and evaluation.

The contract awarded in 1996 was under the US Army Rapid Deployment Force Initiative (RFPI) Advanced Concept Technology Demonstration (ACTD) programme.

In July 1998 the US Army successfully conducted the first firing of an Army Tactical Missile Systems (ATACMS) from HIMARS.

Three of the four prototype systems built were delivered to the 18th Airborne Corps Artillery for a two year extended user evaluation while the fourth has been retained by Lockheed Martin Missiles and Fire Control for testing and evaluation.

In December 1999, the US Army Aviation & Missile Command awarded Lockheed Martin Missiles and Fire Control, a USD65 million 36-month maturation/Engineering and Manufacturing Development (EMD) contract for the High Mobility Artillery Rocket System (HIMARS).

Under the terms of this contract, the company's Dallas, Texas and Camden, Arkansas facilities built six HIMARS launchers for extensive US Army trials and evaluation before the award of a Low Rate Initial Production (LRIP) contract.

The first of six HIMARS launchers were delivered in the 4th quarter of 2001 with the sixth being delivered in the second quarter of 2002.

Latest version of HIMARS on FMTV (6 × 6) chassis with fully protected cab and launcher with pod of six rockets traversed to the left
(Lockheed Martin Missiles and Fire Control) 1452645

During an exercise at Fort Bragg, North Carolina, a US Army platoon equipped with HIMARS was delivered to the base by a Lockheed Martin C-130 and conducted a live fire missions within 10 minutes of arrival.

In mid-2000, it was announced that an HIMARS had successfully fired a six-round ripple of standard MLRS rockets.

Under the US programme, HIMARS supports the US Army vision of a more deployable, lethal, survivable and tactically mobile force. It will launch all MLRS Family of Munitions (MFOM) rockets and Army TACMS (ATACMS) missiles.

The United States Marine Corps placed an order for two HIMARS systems for trial purposes. Total requirement is 45 systems.

In April 2003, the company was awarded a USD96.4 million contract by the US Army to commence Low Rate Initial Production (LRIP) of HIMARS.

The first units equipped with HIMARS were battalions 3-321 and 3-27 Field Artillery at Fort Bragg, North Carolina in FY05.

A small quantity of HIMARS were deployed to the Gulf and used operationally for the first time in early 2003.

As of late 2004 a total of 55 HIMARS systems had been ordered in two batches (30 + 25). Of the 55, 52 are for the army and three for the marine corps.

The third tranche was for a total of 38 systems, with a value of USD109 million. Of these, 37 are for the army and one is for the marine corps.

Late in 2006 it was stated that under a potential USD752 million acquisition under the US governments Foreign Military Sales (FMS) programme, the United Arab Emirates was to acquire HIMARS.

This package comprises 20 Lockheed Martin Missiles and Fire Control 227 mm (6-round) HIMARS that are already in quantity production for the US Army.

The UAE was the first customer in the Middle East to have GMLRS and requires two versions, the dual-purpose model which carries top attack sub-munitions and the unitary High-Explosive (HE) warhead version.

They also require the longer range Army Tactical Missile System (ATACMS) fitted with the standard anti-personnel/anti-material warhead and a unitary HE warhead.

In addition to the 20 HIMARS the contract also covers the supply of 20 M1084A1 Family of Medium Tactical Vehicles (FMTV) (6 × 6) trucks for resupply purposes, three M1089A1 FMTV (6 × 6) wrecker vehicles, 101 ATACMS Block 1A dual purpose rocket pods, 101 ATACMS Block 1A unitary rocket pods, 130 Guided MLRS dual-purpose improved conventional munition rocket pods. 130 unitary high-explosive GMLRS rocket pods, 20 MLRS practice rocket pods and 104 standard M26 unguided rocket pods.

In January 2011 Lockheed Martin Missiles and Fire Control was awarded a US Army contract for an additional 44 HIMARS for delivery by January 2013. This will bring the US Army HIMARS fleet up to a total of 375 units.

This was the final HIMARS order to be placed by the US Army as their objective fleet is 375 units.

These are fitted with the latest Increased Crew Protection (ICP) cab which provides a higher level of protection against a variety of battlefield threats including small arms fire, mines, improvised explosive devices and toxic fumes associated with rocket launchers.

In early 2010 it was confirmed that Jordan had placed a contract valued at USD26.9 million covering the supply of 12 HIMARS with first deliveries made in 2011.

This contract covered the supply of not only 12 HIMARS but also 73 MLRS pods each containing six rockets, 36 practice rockets, Raytheon Advanced Field Artillery Data System (AFATDS) with a total value of USD220 million.

It also included three recovery vehicles, three HMMWV and 45 Single Channel Ground and Airborne Radio Systems (SINGARS) plus support and training.

The rockets are of the GMLRS type but as with some other customers it is not clear if the Jordanians have access to the target location mensuration system that matches the precision of the GMLRS.

In 2009 the Singapore Armed Forces (SAF) took delivery of their first Lockheed Martin Missiles and Fire Control 227 mm (6-round) HIMARS.

The 18 HIMARS launchers have been issued to one battalion of the SAF which will have three batteries, each with six launchers.

HIMARS launching a Raytheon Surface-Launched Advanced Medium-Range Air-to-Air Missile (SLAMRAAM) (Raytheon) 1331668

The SAF HIMARS are integrated onto a BAE Systems Global Tactical Vehicles Family of 6 × 6 Medium Tactical Vehicles (FMTV) cross-country truck chassis.

As well as the HIMARS, the SAF also has 32 pods of Guided Multiple Launch Rocket Systems (MLRS) each of which has six global positioning system guided MLRS rockets fitted with a unitary high-explosive warhead.

Other elements of the package include 30 M28A1 MLRS practice rocket pods, nine FMTV M1084A1 FMTV trucks, one M1089A1 FMTV recovery vehicle, radios and associated support equipment, plus contractor support.

The introduction of HIMARS provide the SAF with a precision strike capability out to a range of 70 km and in addition to targeting high value land targets, it will also be used in the air-defence suppression role.

All of these systems have now been delivered to Singapore and are issued to one battalion which has 18 launchers in three batteries of six launchers.

Description

HIMARS consists of the Medium Tactical Vehicles (FMTV) five-tonne (6 × 6) truck chassis, which is in full-scale production and service for the US Army, on the rear of which has been mounted a launcher which can accept a single pod of six MLRS rockets or a single ATACMS missile pod.

For its new role the suspension of the FMTV has been enhanced and fish plates added to stabilise the vehicle when the rockets are fired.

The HIMARS can be fired from the side of the vehicle although it must wait a small amount of time between rounds are fired to return to its aiming point.

To reduce the overall height of HIMARS transport in a Lockheed Martin C-130 aircraft, the tyres are slightly deflated and re-inflated once it has left the aircraft using the onboard central tyre-inflation system.

The system has the same operating procedures as the upgraded M270A1 launcher including fire-control, electronics and communications sub-units.

It also retains the self-loading and autonomous features that have made the MLRS so successful. Although it has a three-man crew it can be operated by just one man.

HIMARS can fire to the same ranges as all the family of MLRS rockets in service and under development.

By late 1994, the HIMARS system had successfully fired the standard Phase I rocket, the Reduced Range Practice Rocket and the Army Tactical Missile System (ATACMS).

Although the prototype system and all production systems use the 6 × 6 version of the FMTV chassis, the HIMARS launcher can be fitted to other 6 × 6 tactical truck chassis types.

The forward control standard cab of the HIMARS provides the crew with no ballistic protection. Late in 2000, the now BAE Systems, revealed that they were developing two protected cabs for FMTV, Level 1 and Level 2.

Level 1 has been designed to protect the occupants against both missile blast and Foreign Object Debris (FOD) during firings and to maintain cab overpressure.

For Level 2, the company has further expanded FMTV armouring possibilities. This would include increasing the ballistic integrity to protect the occupants against 7.62 mm M80 ball ammunition throughout the cab.

In addition, mine blast protection could also be offered with a combination of special wheels and rims, deflective armour under the cab and special energy absorbing seats.

Utilising the modular systems and bulkhead connections of the HIMARS cab, a fully armoured cab could be installed on existing vehicles in about four hours.

Production HIMARS systems include an onboard munitions loading and reloading system.

In addition it would also include an Improved Position Determining System (IPDS) to allow operations with the Army Tactical Missile System Block 1A with global positioning update.

Production of the FMTV (6 × 6) on which HIMARS is based was originally manufactured by BAE Systems but as a result of a competitive re-buy production is now undertaken by Oshkosh Defense.

Variants

HIMARS for common launcher role

In 2009 the US Army successfully fired two air-defence Surface-Launched Advanced Medium-Range Air-to-Air Missiles (SLAMRAAMS).

This was regard as a proof-of-principle test demonstration to test the concept of a common launcher that could launch GMLRS (and other rockets) or the SLAMRAAM.

For this application the missiles were integrated into an ATACMS launch pod and the fire-control system was updated.

This concept is also applicable for the standard US Army M270 tracked launcher.

HIMARS with P44 missile

The HIMARS launcher has been used for some of the trials with the private venture Lockheed Martin Missiles and Fire Control P44 missile. This programme has now been cancelled.

Specifications

	High-Mobility Artillery Rocket System
Dimensions and weights	
Crew:	1+2
Length	
overall:	7.765 m
Width	
overall:	2.4 m
Height	
overall:	2.91 m
Ground clearance	
overall:	0.564 m
Wheelbase	4.1 m
Weight	
standard:	13,592 kg (carrier vehicle)
combat:	15,873 kg (complete system)
Mobility	
Configuration	
running gear:	wheeled
layout:	6 × 6
Speed	
max speed:	94 km/h
Range	
main fuel supply:	484 km (loaded)
Fording	
without preparation:	0.923 m
with preparation:	1.538 m
Angle of approach:	40°
Angle of departure:	63°
Gradient:	60%
Engine	Caterpillar 3116 ATAAC, 6.6 litre, 6 cylinders, turbocharged, diesel, 330 hp at 2,600 rpm
Gearbox	
model:	Allison MD-D7
type:	automatic
forward gears:	7
Transfer box:	Allison single-speed
Steering:	power assisted
Tyres:	395 R 20XML
Suspension:	parabolic leaf springs
Electrical system	
vehicle:	24 V
Alternator:	100 A
Survivability	
Night vision equipment	
vehicle:	optional
NBC system:	yes
Armour	
hull/body:	steel + appliqué

Status

Production. In service or on order for:

Country	Quantity	Comment
Jordan	12	deliveries complete
Singapore	18	deliveries complete
United Arab Emirates	20	deliveries complete
US Army	375	final deliveries 2014
US Marine Corp	45	deliveries complete

Contractor

Lockheed Martin Missiles and Fire Control.

Air Defence

Self-Propelled Anti-Aircraft Guns

Belarus

ZSU-23-4M5 Shilka

Type
Self-Propelled Anti-Aircraft Gun (SPAAG)

Development
The Russian-designed ZSU-23-4 quad 23 mm Self-Propelled Anti-Aircraft Gun (SPAAG) system is one of the most popular and widely deployed weapons of its type in the world.

In the designation, ZSU is the Russian acronym for self-propelled anti-aircraft mount, 23 is the calibre of the cannon in millimetres and 4 is for the number of weapons. Within Russia it is also referred to as the Shilka (Awl).

It is estimated that some 30 countries still deploy the ZSU-23-4 SPAAG. With the increasing pressures on many countries' defence budgets, including Europe, Middle East and Asia, where most ZSU-23-4 operators are found, an increasing number of countries are now looking at upgrading their existing ZSU-23-4 systems instead of purchasing new SPAAG systems.

The existence of the ZSU-23-4M5 upgrade was first revealed in early 1999 when one of the six prototypes was exhibited in Abu Dhabi for the first time.

The main export thrust is the Middle East, with Egypt known to have a requirement to upgrade its fleet of 117 systems.

The major ZSU-23-4M5 upgrade has been carried out by the following:
- Minotor Service Enterprise of the Republic of Belarus
- Peleng Joint Stock Company of the Republic of Belarus
- Ulyanovsk Mechanical Plant of Russia
- Zakłady Mechaniczne Tarnow (ZM Tarnow) of Poland.

It was originally intended that the Ulyanovsk Mechanical Plant would develop two versions of modifications; the ZSU-23-4M4 and ZSU-23-4M5 upgrade packages.

Within the overall consortium, Minotor Service Enterprise is responsible for systems integration while the Peleng Joint Stock Company has developed the optics. Some elements of this upgrade are also used in the Ulyanovsk ZSU-23-4 upgrade package, details of which are given later in this section.

ZSU-23-4M5 SPAAG upgraded showing one of the four laser detectors on the front right side of hull with bank of 81 mm smoke/decoy grenade launchers (Christopher F Foss) 0069191

ZSU-23-4M5 SPAAG upgraded in travelling configuration showing radar retracted (Christian Dumont) 0100278

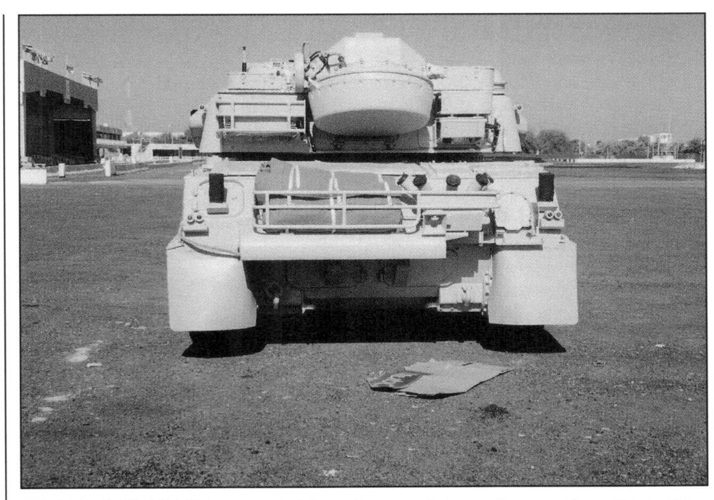

ZSU-23-4M5 SPAAG upgraded from the rear in travelling configuration with radar retracted. Note new stowage box on hull rear and smoke/decoy grenade launchers (Christopher F Foss) 0069192

Description
The modernisation involved the improvement of search capabilities, automation of battle performance, increase in accuracy of fire with particular emphasis on manoeuvring targets as well as equipping the gun mount with advanced fire-control systems.

In the basic ZSU-23-4M5 upgrade, the four water-cooled 23 mm AZP-23M cannon are retained. These have an effective slant range of 2,500 m and a similar range when used in the ground/ground role. A total of 2,000 rounds of 23 mm ammunition is carried. These are fired in maximum bursts of 150 rounds per barrel, which are water-cooled.

One of the options, however, includes the installation of pods of the Igla family of man-portable surface-to-air missiles fire-and-forget type. In a typical target engagement the SAM would be used to engage targets at ranges of over 5,000 m with the 23 mm cannon being used to engage close-in targets. It is believed that this capability has not so far been fitted to any of the prototype ZSU-23-4M5 upgraded systems. The ZM Tranow upgrade is to replace the Igla with Polish produced Grom or Grom-2 missiles.

To improve the overall effectiveness of the ZSU-23-4 system, extensive modifications have been made to the turret. These include:
- Upgraded radar
- New digital computing system
- Installation of a Digital Instrumental Follow up System (DIFS)
- New three-channel Optical Locating System (OLS)
- Automation of some onboard systems, laser warning system
- TV system for the driver (colour front, black and white rear)
- Modernised land navigation system, new AC generator
- Upgraded transmission and mine protection for the driver's compartment.

The integration of new electronics and other subsystems has enabled the crew of the upgraded ZSU-23-4M5 system to be reduced from four to three people. The radar, which carries out both tracking and surveillance functions, has a detection range of 12 km and a tracking range of 10 km. The radar is now solid state and utilises a digital range-finding method. It also has enhanced jam resistance.

ZSU-23-4M5 SPAAG upgraded with radar retracted (Christian Dumont) 0100279

The OLS has three data channels: day with TV camera, passive night with TV camera and laser range-finder. This system allows targets to be tracked under day and night conditions with the radar switched off. The day channel operates in the 0.5 to 0.8 micron range and can detect targets at 8,000 m and track targets from 7,500 m.

The front TV camera is full colour, whilst the after TV camera is black and white only.

The night channel operates in the 8 to 12 micron band. This will detect an F-16 fighter at a range of 20,000 m and identify it at a range of 10,000 m. The 1.06 micron laser range-finder has a maximum range of 7,000 m and feeds target information into the fire-control computer. The data updating rate of the day and night channels and the laser range-finder is 25 Hz.

This feature increases the overall combat survivability of the system as it cannot be detected by radar warning receivers and is therefore unlikely to be attacked by air-launched Anti-Radiation Missiles (ARM).

The laser warning system includes sensors mounted at each corner of the chassis, which are connected to a central warning system in the turret. Banks of 81 mm electrically operated smoke/decoy launchers are mounted at the front right and rear right sides of the chassis.

Side skirts have been added either side, as has a cover below the glacis plate, to assist in keeping the dust down. A new stowage box has been added at the rear of the hull.

The installation of a Multifunctional Training Device allows for crew training to be carried out using simulated targets in clear and jammed environment under surveillance, detection, lock on and tracking modes.

The current ZSU-23-4 has a maximum road speed of 50 km/h. The installation of a differential turning mechanism with hydraulic gears and a hydraulically assisted engine drive train has increased maximum speed to 60 km/h. The driver's seat is now suspended from the roof and a motor cycle-type steering wheel is installed as a replacement for the normal driver's tillers. By suspending the driver's seat from the roof, the driver is provided with a higher level of protection against anti-tank mines.

Other options include a new Auxiliary Power Unit (APU) as a replacement for the existing system. The installation of an APU enables the turret systems to operate with the main engine shut down to save fuel. It can also be fitted with passive night driving equipment.

Standard equipment includes an NBC system, fire detection and suppression system, radio, land navigation system and an air conditioning unit, which is considered essential for operations in the Middle East. The R-163-50V radio system is installed, as is a TNA-4-1 land navigation system.

The system could also be delivered without the radar fitted; this system would use the OLS only and also be fitted with eight fire-and-forget SAMs on the turret rear.

In addition to the ZSU-23-4M5 upgrade and the one already being marketed by the Ulyanovsk Mechanical Plant, there are known to be a number of other ZSU-23-4 upgrades being offered on the world market.

These include one from the State Scientific and Technical Centre of Artillery and Rifle Arms (Ukraine) and the ZSU-30-2 (Russia). The latter has the four 23 mm cannon removed and replaced by the twin 2A38 30 mm cannon as used in the Russian 2S6M Tunguska gun/missiles air defence system. It also has a stabilised Kedr electro-optical package mounted on the right side of the turret, which has day, thermal, and laser range-finding channels.

Specifications

	ZSU-23-4M5
Dimensions and weights	
Crew:	3
Length	
overall:	6.535 m
Width	
overall:	3.120 m
Height	
transport configuration:	2.60 m
Ground clearance	
overall:	0.40 m
Track	
vehicle:	2.88 m
Length of track on ground:	3.70 m
Weight	
combat:	19,200 kg
Ground pressure	
standard track:	0.65 kg/cm²
Mobility	
Configuration	
running gear:	tracked
Power-to-weight ratio:	14.58 hp/t
Speed	
max speed:	60 km/h
Range	
main fuel supply:	400 km
Fuel capacity	
main:	250 litres
Gradient:	60%
Side slope:	30%
Vertical obstacle	
forwards:	1.1 m
Trench:	2.8 m
Engine	Model V-6R, in line-6, water cooled, diesel, 209 hp at 2,000 rpm
Gearbox	
forward gears:	5
reverse gears:	1
Suspension:	torsion bar
Electrical system	
vehicle:	24 V
Firepower	
Armament:	4 × turret mounted 23 mm AZP-23M cannon
Armament:	8 (est.) estimated × hull mounted 81 mm smoke grenade launcher (12 can be fitted)
Ammunition	
main total:	2,000
Main weapon	
cartridge:	23 mm × 152 mm B
Turret power control	
type:	powered/manual
by commander:	yes
by gunner:	yes
Main armament traverse	
angle:	360°
speed:	70°/s
Main armament elevation/ depression	
armament front:	+85°/-4°
speed:	60°/s

Status

Development complete, the system and its improvements are offered for sale on the open arms market.

Contractor

Minotor Service Enterprise.
Peleng Joint Stock Company.
Ulyanovsk Mechanical Plant.

China

LD2000

Type

Self-Propelled Anti-Aircraft Gun (SPAAG)

Development

In response to the ever-changing threat and the increasing use on the battlefield of cruise missiles and Unmanned Aerial Vehicles (UAV), China North Industries Corporation (NORINCO) has developed a Ground-Based Close-In Weapon System (GB-CIWS) based on an already proven Type 730 CIWS naval design gun that is in itself is a copy of the General Electric GAU-8 30 mm seven-barrel Avenger gun used on the A-10. There is also some question by certain western analysts as to the origin of the system being based on the Thales Nederland Goalkeeper. Information on the system, known as LuDun (Land Shield) 2000 (LD2000, also referred to as LD 2000 and LD-2000) was first released in 2005. Later in 2006, the system was shown complemented with the Yitian self-propelled anti-aircraft missiles system that employs the TY-90 SAM to provide a multi-layer gun-missile integrated air-defence system.

Reports from Chinese sources have mentioned that at least one of the systems was transferred by a heavy air-transporter (an AN-124) in June 2009 to the United Arab Emirates for trials and evaluation purposes.

Ground vehicle element of the NORINCO LD2000 Ground-Based CIWS deployed as part of a firing system and with its 30 mm cannon firing over the rear arc (NORINCO)
1303249

LD2000 ground-based CIWS (NORINCO) 1303250

Description

The LD2000 combines the proven seven-barrel naval 30 mm Type 730B gun, firing both Armour-Piercing Discarding Sabot (APDS) and High-Explosive Incendiary (HEI) projectiles at selectable rates of fire to a maximum of 4,200 rounds per minute. Complemented with advanced H-band search radar, J-band tracking radar and also TV/IR, the system has been designed to defend against Mach 2 and below low-flying targets possessing Radar Cross Sections (RCS) in the 0.1 m^2 range, such as cruise missiles, UAVs, helicopters and very short-range precision-guided munitions.

The LD2000 also features closed-loop calibration technology, to guarantee fully automatic all-weather target detection and engagement.

A fire unit that consists of an Intelligence and Command Vehicle (ICV) and up to eight 8 × 8 Combat Vehicles (CVs) of the type Wanshan WS-2400 (Chinese copy of the Russian MAZ 543 missile TEL) supported by ammunition resupply vehicles, power supply vehicles, warning radar and maintenance vehicles.

Intelligence and Command Vehicle

The Intelligence and Command Vehicle (ICV) combines search radar with intelligence consoles, and provides increased situations awareness, automatic target search, acquisition, identification, threat evaluation, target indication and optimal control of subordinate combat vehicles.

Radar

A J-Band Type 347G (1) tracking radar with a 12 km search and 9 km tracking range depending on the size of the target mounted on top of the gun mount with a day/thermal sighting system incorporating a laser range-finder. The cannon is operated by a gunner in the fully enclosed module behind the cab.

The Type 347G (1) tracking radar is the same as the LR66 (Western name) that is produced by CETC International Company Ltd, Beijing, China. The type number is the Chinese designation.

Variants

PLAN Type 730 CIWS

A second variant of the system has mounted on the gun cupola two banks of three low-level, short-range, TY-90 air-defence missiles introduced in 2006. This new/upgraded system is also mounted on a fully mobile cross country 8 × 8 vehicle that is much heavier than its predecessor. The gun/missile system has a new target acquisition radar mounted just aft of the lorry cab and what appears to be a much larger power supply unit. Photographic interpreters will also notice that the hydraulic jacks used to stabilise the vehicle are of a much heavier construction, thus giving the impression that this variant is of a much greater overall weight. Reports emanating from within China have confirmed that the ammunition for the Type 730 gun include HEAT and Armoured-Piercing Discarding Sabot (APDS).

PLAN Type 730 CIWS showing the two of side-mounted missile pod groups and three of the four increased-capacity hydraulic stabilisers 1334301

Specifications

LD2000 gun

Type:	Type 730B 30 mm seven-barrel Gatling gun
Maximum range:	2,500-3,000 m
Maximum effective range:	1,000-1,500 m
Maximum ammunition capacity:	1,000 rds
Rate of fire:	≤4,200 rds/min (adjustable)
Ammunition natures:	APDS, HE, HEI and TP
Muzzle velocity:	
(APDS)	1,150 m/s
(HEI and TP)	930 m/s
Maximum target speed:	780 m/s
Kill probability:	n/avail
Reaction time:	9.8 s
Deployment time:	n/avail
Reload time:	n/avail

LD2000 detection systems

Frequency:	
(search radar)	H-band (6-8 GHz)
(tracking radar)	Type 347G J-band (10-20 GHz)
Frequency agile bandwidth:	>2 GHz
Detection range:	
Tracking radar:	
(2 m^2 RCS)	>18 km
(0.1 m^2 RCS)	>9 km
Search radar:	
(2 m^2 RCS)	n/avail
(0.1 m^2 RCS)	12 km
TV/IR:	
(good visibility)	18 km
(normal)	5 km
Tracking target altitude:	>15 m
Accuracy: (RMSS)	
(azimuth)	1.2 mrad
(elevation)	1.2 mrad
(range)	5 m
Maximum targets:	48

Status

The weapon system has completed its state firing test and is offered by the China North Industry Corporation (NORINCO) to the international market. A relatively small number of pre-series production vehicles will almost certainly have been produced in order to carry out trials and concept operations. The system has been offered for export during arms exhibitions primarily to Middle Eastern countries. The People's Liberation Army Navy (PLAN) has already deployed the Type 730 CIWS, on which the LD2000 is based, onboard its Type 052B and 052C class of DDGs.

Contractor

China North Industries Corporation (NORINCO).

Xi'an Kunlun Industrial Group, a subsidiary of the China North Industries Group are the manufacturers.

Type 80

Type

Self-Propelled Anti-Aircraft Gun (SPAAG)

Development

The China North Industries Corporation (NORINCO) Type 80, twin 57 mm Self-Propelled Anti-Aircraft Gun (SPAAG) was primarily an export model developed for use by armoured and mechanised infantry units to provide defence against air attack at slant ranges of up to 5,500 m by targets flying at speeds of up to 350 m/s.

If required, the vehicle can also be used in a limited ground role to support conventional anti-tank weapons.

The system consists of a modified NORINCO Type 69-II Main Battle Tank (MBT) chassis. This is fitted with a Chinese version of the turret installed on the Russian ZSU-57-2 twin 57 mm self-propelled anti-aircraft tank, but using a twin barrel variant of the locally produced Type 59 which is a copy of the Russian single S-60 57 mm towed anti-aircraft gun. Full details of the 57 mm S-60 are given in the Towed anti-aircraft gun section.

NORINCO Type 80 twin 57 mm self-propelled anti-aircraft gun system with turret traversed to front 0509844

This weapon is a clear weather system and has limited capabilities against high-speed aircraft.

Production of the Type 80 SPAAG is complete and it is no longer marketed by NORINCO.

Description

The hull of the Type 80 twin 57 mm self-propelled anti-aircraft gun system is divided into three compartments:

- Driver's at the front
- Fighting in the centre
- Power pack the rear.

The driver sits at the front of the hull on the right and has a single-piece hatch cover that opens to the left. An infrared periscope is fitted for night driving together with a normal optical day viewing periscope.

The open-topped welded steel turret is mounted in the centre of the chassis. A wire cage, mounted externally at the turret rear, is provided for the empty 57 mm cartridge cases and links.

The main armament consists of two Type 59, 57 mm cannon with their control system firing the 57 × 347 SR ammunition in either in a semi-automatic, electro-hydraulically powered manner or manually. In the former mode the elevation is from –1 to +81°, in the latter it is –5 to +85°. Fire-control is by an automatic optical vector mode sight. A crew of five is carried in the turret: commander, gunner, fuze setter and two loaders. For internal crew communication, a Type 803 intercom system is fitted. A 20 to 25 km range Type 889 radio is standard for external communications.

The 57 mm guns are fully automatic and recoil operated with each barrel having a maximum cyclic rate of fire of 105 to 120 rds/min. The practical rate of fire is 70 rds/min. Muzzle velocity is 1,000 m/s. The complete system carries up to 300 rounds of ammunition.

The weapons can be fired while on the move or on slopes of up to 15° but the best results are obtained when it is stationary on the ground with a slope of 3° or less.

Maximum horizontal range is 12,000 m, maximum vertical range 8,800 m and effective anti-aircraft range 5,500 m. The guns fire two types of fixed ammunition; a 6.47 kg High-Explosive Tracer (HE-T) round with a percussion fuze and a 6.45 kg Armour-Piercing Capped Tracer (APC-T) round with a base fuze. A self-destruct device is incorporated into the HE-T shell to ensure automatic detonation after 6,000 m of flight. A total of 300 rounds are carried.

The Type 80 has torsion bar-type suspension and on either side of the vehicle; five road wheels (the Russian equivalent ZSU-57-2 has only four), an idler at the front, a drive sprocket at the rear and no track return rollers. Unlike the ZSU-57-2, the Type 80 is fitted with track skirts.

The Type 80 twin 57 mm can lay its own smoke screen by injecting diesel fuel into the exhaust outlet on the left side of the hull.

Specifications

	Type 80
Dimensions and weights	
Crew:	6
Length	
overall:	8.42 m
hull:	6.243 m
main armament rear:	8.24 m
Width	
overall:	3.307 m
transport configuration:	3.27 m (without side skirts)
Height	
overall:	2.748 m
axis of fire:	1.94 m
Ground clearance	
overall:	0.425 m
Track width	
normal:	580 mm
Length of track on ground:	3.485 m
Weight	
combat:	31,000 kg
Mobility	
Configuration	
running gear:	tracked
Power-to-weight ratio:	18.7 hp/t
Speed	
max speed:	50 km/h (est.)
Range	
main fuel supply:	440 km (est.)
Fording	
without preparation:	1.4 m
Gradient:	60%
Side slope:	30%
Vertical obstacle	
forwards:	0.8 m
Trench:	2.7 m
Engine	12150-7BW, diesel, 580 hp at 2,000 rpm
Gearbox	
type:	manual
forward gears:	5
reverse gears:	1
Steering:	clutch and brake
Suspension:	torsion bar
Electrical system	
vehicle:	28 V
Batteries:	4 × 12 V, 280 Ah
Firepower	
Armament:	2 × turret mounted 57 mm cannon
Ammunition	
main total:	300
Main weapon	
cartridge:	57 mm × 347 mm SR
Turret power control	
type:	powered/manual
Main armament traverse	
angle:	360°
speed:	36°/s (semi-automatic)
Main armament elevation/depression	
armament front:	+85°/-5°
speed:	20°/s (semi-automatic)
Survivability	
Night vision equipment	
vehicle:	no
commander:	no
gunner:	no
driver:	no
NBC system:	yes

Status

Production complete, no longer marketed.

Contractor

Enquiries to China North Industries Corporation (NORINCO).

Type 95

Type

Self-Propelled Anti-Aircraft Gun (SPAAG)

Development

China North Industries Corporation (NORINCO) has been promoting since at least 1999, a quadruple 25 mm Self-Propelled Anti-Aircraft Gun (SPAAG) system. At the development stage the system was known as the Type 90-II and Type 90-III, this system was first shown publicly at the Beijing Military Parade in 1999. The Chinese industrial number for the system is the PLZ95 (also known as the PGZ95) the Pinyin name is 95 Shi Zixing Gaoshepao. The system was designed and developed by the Northwest Institute of Mechanical and Electrical Engineering based in Xianyang, Shaanxi province.

Description

The Type 95 SPAAG is a self-propelled anti-aircraft gun system for the air defence role in armoured troops at brigade and regiment level. The system is fitted with an independent day and night combat capability, with the option of adding four SAMs. The system consists of a new full-tracked armoured chassis, on the roof of which has been installed a new one-person, power-operated turret, armed with four 25 mm cannons, two on each side of the turret, with the optional four fire-and-forget SAMs mounted two to each side. The cannons are assessed to be based on those used Type 87 towed 25 mm AAA gun introduced into the PLA in the mid-1980s. The cartridge used by the cannons being a type unique to China, the 25 × 183 B. The system makes use of an electro-optic fire-control system, industrial number PGZ95, which is fed from the electro-optical package mounted on the forward part of the turret, which includes a TV tracking camera (range 6,000 m), infrared tracking camera (range 5,000 m), and a Laser Range-Finder (LRF) with a range 500 to 5,500 m.

Battery Command Vehicle 0116544

25 mm self-propelled anti-aircraft gun 0116590

The PGZ95 also has the functions of simulated training and malfunction diagnosis. The malfunction diagnosis covers the components only to locate the failure to PCBs. For the simulated training, the system itself can perform the static and dynamic tests and simulation training for the operators. It can also provide many typical simulated flight paths to the trainees.

A CLC-1 pulse Doppler search radar is also fitted and has a maximum range of 11 km against targets with an RCS of 2 m^2. The CLC-1 is optimised to detect low-flying aircraft and attack helicopters. A bank of four electrically operated forward-firing smoke grenade launchers is mounted either side on the lower part of the turret for self-protection.

Another radar mounted on the Command Vehicle is the CLC-2 S-band low-altitude surveillance radar, employing Pulse Doppler technology. The detection range of the CLC-2 is less than 45 km.

Engagement scenario

In a typical engagement, the China Electric National Import and Export Corporation (CEIEC) CLC-1 pulse Doppler radar would detect and track targets at ranges out to 11 km. The targets are then handed off to the fire-and-forget SAMs. In turn, the missiles would then be used to engage the targets at the longer ranges within the firing envelope of the missiles, which, in the case of the QW-2, is from 500 to 6,000 m, at altitudes of 10 to 3,500 m. The 25 mm cannon is used to engage aircraft and helicopters with the aid of the electro-optic fire-control system and ranges of up to 2,500 m and at altitudes of 2,000 m.

The Type 95 SPAAG/SAM is part of a complete mobile air-defence system that also includes a battery command post. The battery command post is mounted on a similar chassis to the SPAAG/SAM, this can typically control up to six or eight systems simultaneously.

The battery command post is fitted with CEIEC CLC-2 surveillance radar with a maximum range of 45 km and a maximum altitude of up to 4.5 km. Mounted on the front is a 12.7 mm heavy machine gun for self defence.

Specifications

	Type 95	Type 95 Battery Command Vehicle
Dimensions and weights		
Crew:	3	5
Length		
overall:	6.71 m	6.8 m
Width		
overall:	3.2 m	3.2 m
Height		
overall:	4.82 m	4.58 m
transport configuration:	3.40 m	3.38 m
deployed:	4.82 m	4.58 m
Weight		
combat:	22,500 kg	22,500 kg
Mobility		
Configuration		
running gear:	tracked	tracked
Speed		
max speed:	53 km/h	53 km/h
Range		
main fuel supply:	450 km	450 km
Firepower		
Armament:	4 × turret mounted 25 mm cannon	1 × roof mounted 12.7 mm (0.50) machine gun
Armament:	4 × turret mounted QW-2 SAM	-
Main weapon		
maximum range:	2,500 m (gun)	-
maximum range:	6,000 m (QW-2)	-
Rate of fire		
cyclic:	3,200 rds/min	-

Status

Production as required. In service with People's Liberation Army (PLA). As of November 2013 there have been no known exports.

Contractor

China North Industries Corporation (NORINCO).
Northwest Institute of Mechanical and Electrical Engineering, based in Xianyang.

Czech Republic

M53/59

Type

Self-Propelled Anti-Aircraft Gun (SPAAG) system

Development

The M53/59 self-propelled anti-aircraft gun was introduced into the Czech Army in the late 1950s. It consists of an armoured version of the standard Praga V3S (6 × 6) truck chassis produced by the Prague Automobile Factory Ltd, of Prague, with a removable modified version of the standard towed M53 twin 30 mm anti-aircraft guns (details of which are given in the Towed anti-aircraft guns section) mounted on the rear.

The guns are often removed for ground use. The mounting has also been seen on the PTS unarmoured tracked amphibious vehicle. This system has seen widespread combat use in the surface-to-surface role in the conflict in the former Yugoslavia.

Description

The M53/59 (also known in Czech as Praga PLDvKvz 53/59 Jesterka (lit. lizard)), has an all-welded steel armour hull with the engine at the front, crew compartment in the centre and the armament at the rear. The hull armour is well sloped for maximum possible protection from small arms fire and shell splinters within the weight limit available.

The driver is seated on the left side and the commander on the right, with a windscreen in front of them which is covered in action by an armoured cover, hinged at the top, with a vision slit. Both the driver and commander also have a door in the side of the hull, the top half of which has a vision slit and hinges

M53/59 twin 30 mm self-propelled anti-aircraft gun system on Praga (6 × 6) chassis (BP) 0100280

M53/59 twin 30 mm self-propelled anti-aircraft gun system on Praga (6 × 6) chassis (BP) 0100281

downwards on the outside. The two loaders are seated one either side at the back of the crew compartment facing the rear and have a vision slit in the side, immediately behind the side door. At the rear of the crew compartment is a two-piece hatch cover, which folds down into the horizontal to act as a platform for the loaders. It has a single vision slit in either side for the loaders. The commander has a hemispherical plexi glass cupola in the roof for observation.

The weapons have an elevation of +85° and a depression of −10°, except over the crew compartments where depression is limited to +2° and over the commander's cupola where there is none. A steel plate at the rear of the crew compartment stops the gun barrels hitting the roof. The turret can be traversed a full 360° and elevation, depression and traverse are all hydraulic with manual controls available for emergency use. The gunner is seated on the left side of the weapons and has frontal, side and rear armour protection. When originally introduced, the barrels had multibaffle muzzle brakes but more recently the weapons have been observed fitted with conical flash hiders. The barrels can be changed quickly when they become overheated and spare barrels are kept at regimental level as part of basic equipment.

The twin 30 mm cannon are fully automatic, gas-operated weapons and have a cyclic rate of fire of 420-450 rds/barrel/min and a practical rate of fire of 150 rds/barrel/min. Maximum horizontal range is 9,700 m, maximum vertical range is 6,300 m and effective anti-aircraft range is 3,000 m. Each of the vertical magazines holds 50 rounds of ammunition, which are fed to the magazines in clips of 10 rounds. The towed M53 twin 30 mm guns have a lower rate of fire as they have no magazines and the ammunition has to be fed to each weapon in clips of 10 rounds. The 30 mm guns can fire the following types of fixed 30 × 210 ammunition:

- API projectile weighing 0.45 kg, with a muzzle velocity of 1,000 m/s, which will penetrate 55 mm of conventional steel armour at an incidence of 0° at a range of 500 m;
- HEI projectile weighing 0.45 kg with a muzzle velocity of 1,000 m/s.

Yugoimport - SDPR (this was previously known as the Yugoslav Federal Directorate of Supply and Procurement) manufactures ammunition for the M53/59 twin 30 mm SPAAG system, which it calls the anti-aircraft gun 30/2 mm M53, 53/59 (CS). Two types of ammunition are manufactured: HE-T designated the M69, and an action-operating round designated M78. Details of these are as follows:

Designation	M69	M78
Type:	HE-T	break-up shot
Length:	331 mm (total)	331 mm (total)
Weight:		
(complete round)	1.14 kg	0.96 kg
(projectile)	435 g	250 g
(propellant charge)	195 g	195 g
Muzzle velocity:	997 m/s	n/avail
Chamber pressure:	314 MPa	176 MPa
Cartridge case:	steel	steel
Propelling charge:	NC powder	NC powder

The M69 has a tracer, which burns for four seconds and is fitted with an super-quick action impact fuze with mechanical self-destruct. The M78 action-testing round uses a 250 g iron-powder-filled plastic projectile which breaks up after leaving the barrel, but is still capable of operating the cannon's mechanism.

Filled magazines are carried in the rear of the crew compartment fastened to the floor by quick-release catches, a further three filled magazines are carried either side of the platform.

Stowed under the rear of the vehicle are two ramps, which assist the crew to deploy the system when it is being used in the ground role. When in the latter role the mount is stabilised on four jacks. A winch is provided to haul the mount back on to the vehicle via the ramps.

The main drawback to this system is its lack of all-weather capability. The weapon is optically aimed, and can only be used during daylight with good weather conditions. The vehicle has no NBC system, no night vision equipment, no amphibious capability and no central tyre pressure regulation system.

Variants

The former Czechoslovakia (now the Czech Republic) offered an export version of the M53/59 called the M53/70, which appears to be almost identical to the earlier version but with an improved fire-control system. Neither the M53/59 nor the M53/70 have been marketed for some years.

Specifications

	M53/59
Dimensions and weights	
Crew:	4
Length	
overall:	6.92 m
Width	
overall:	2.35 m
Height	
overall:	2.95 m
transport configuration:	2.585 m (excluding magazines)
axis of fire:	2.41 m
Ground clearance	
overall:	0.4 m
Wheelbase	3.58 m + 1.12 m
Weight	
combat:	10,300 kg
Mobility	
Configuration	
running gear:	wheeled
layout:	6 × 6
Power-to-weight ratio:	11.67 hp/t
Speed	
max speed:	60 km/h
Range	
main fuel supply:	500 km
Fuel consumption	
road:	0.24 litres/km
Fuel capacity	
main:	120 litres
Fording	
without preparation:	0.8 m
Gradient:	60%
Side slope:	30%
Vertical obstacle	
forwards:	0.46 m
Trench:	0.69 m
Engine	Tatra T 912-2, in line-6, air cooled, diesel, 110 hp at 2,200 rpm
Gearbox	
type:	manual
forward gears:	4
reverse gears:	2
Transfer box:	2-speed
Tyres:	8.25 × 20
Firepower	
Armament:	2 × turret mounted 30 mm cannon
Ammunition	
main total:	800 (est.) estimated
Main weapon	
cartridge:	30 mm × 210 mm
Turret power control	
type:	hydraulic/manual
Main armament traverse	
angle:	360°
Main armament elevation/ depression	
armament front:	+85°/-10°
Survivability	
Night vision equipment	
vehicle:	no
commander:	no
gunner:	no
driver:	no
NBC system:	no
Armour	
hull/body:	10 mm (est.)

Status

Production complete, the vehicle is now obsolete.

Contractor

Avia Závody.

Prague Automobile Factory Ltd, (Praga) of Prague, Praga V3S chassis.

Egypt

Sinai 23

Type
Self-Propelled Anti-Aircraft Gun (SPAAG)

Development
In January 1984, the Egyptian government awarded competitive contracts to the then Thomson-CSF and Dassault Electronique (that in January 1999 became Thomson-CSF DETEXIS and subsequently Thales) for the development and construction of prototypes of a twin 23 mm self-propelled anti-aircraft gun system. This basic idea was to use two proven subsystems, the M113A2 chassis and the ZU-23-2 together with other system components to develop an enhanced mobile air-defence system.

Both systems were completed in 1984 and displayed for the first time at the 1984 Cairo Defence Equipment Exhibition. They are based on the former FMC (now United Defence LP) M113A2 Armoured Personnel Carrier (APC) chassis and use the Russian-designed 23 mm ZU-23 light towed anti-aircraft gun, manufactured in Egypt by Military Egyptian Factory Abu Zaabal and fully described in the Towed anti-aircraft guns section.

A typical Sinai 23 platoon would consist of one leader vehicle fitted with the RA-20S E-band radar and three or four satellite fire units each with twin 23 mm cannon. The Egyptian Air Defence Command conducted extensive trials of the weapon system in Egypt over a six-month period in 1986.

These trials, which included operating the system in a hostile jamming environment (ground-based and airborne jammers), demonstrated that the system was easy to operate, targets could be detected quickly and the system had very short reaction, acquisition, designation and turret tracking times. Following these trials, the system was adopted by the Egyptian Air Defence Command and first units were delivered to Egypt in 1989.

It is assessed that between 40 and 50 of these systems have been in service in Egypt. Thales is no longer marketing this turret system.

Description
The Sinai 23 system is a twin 23 mm self-propelled gun that mounts two launchers each containing three VSHORAD man-portable SAMs such as the Sakr Eye (or Stinger) either side of the turret. The system is contained in a modified M113A2 chassis along with the RA-205 surveillance radar. The turret is designated TA-23E which in itself is a further development of the TA-20.

The acquisition/surveillance vehicle has the Thales Defence Systems DETEXIS RA-20S radar operating in E-band which can detect aircraft and moving helicopters at a range of 12 km or helicopters hovering at a range of 5 km. Targets can be localised to 1°. This vehicle provides target information to the firing unit.

The firing unit consists of an M113A2 armoured personnel carrier chassis fitted with a one-man TA-23E turret. This is a further development of the TA-20 turret, armed with twin 20 mm cannon which is also fitted on the Panhard M3 (4 × 4) air defence system and VDAA Renault covered later in this section.

The turret has full hydraulic traverse and weapon elevation and is armed with two 23 mm cannon with three Egyptian-built Sakr Eye (the locally produced variant of the Russian 9M32 (SA-7/Grail) or similar man-portable passive infrared homing missiles) fire-and-forget surface-to-air missile launchers mounted either side of the turret. Details of the Sakr Eye missile are given in the Man-portable surface-to-air missile systems section under Egypt.

When travelling, the pod of three Sakr Eye missiles is normally swung down to rest alongside the 23 mm cannon and is raised into the horizontal position when required. The gunner has an optical sight with a magnification of ×6 and a computer. Information can be received from the acquisition/surveillance vehicle positioned up to 2,000 m away. Each cannon has 300 rounds of ready-use 23 ×152 B ammunition. The vehicle commander with control console sits at the rear of the vehicle on the left side, to the rear of the driver.

The Sinai 23 system is of modular construction and can be coupled to thermal night vision systems. The firing unit can also be fitted with the RA-20S radar system enabling two targets to be engaged simultaneously by the four vehicles (one leader and three satellites).

Sinai 23 firing unit showing one-man turret armed with two 23 mm cannon and six Sakr Eye SAMs 0509747

Sinai 23 firing unit showing power-operated turret armed with twin 23 mm cannon, three Sakr Eye SAMs either side of turret and RA-20S radar on turret rear 0509412

Acquisition/surveillance vehicle base on M113A2 chassis and fitted with RA-20S radar system 0509748

Variants
Thales was studying the installation of the Sinai 23 turret system on the Fahd (4 × 4) APC that was in production in Egypt for both the home and export markets.

Specifications
Turret

Designation:	TA-23E
Armament:	
(cannon)	2 × 23 mm ZU cannon
(missile)	6 × Sakr Eye SAM (2 rows of 3 missiles)
Ammunition:	
(ready to use)	300 rounds per gun
(reserve)	600 rounds total
Cartridge:	23 × 152 B
Turret power control:	hydraulic
Sight:	day, magnification ×6
Reaction time:	5 s (total)
Data transmission range:	≤2 km
Radar type:	pulse Doppler
Detection range:	
(moving targets)	12 km
(hovering helicopters)	5 km
Target localisation:	1°
Surveillance radar:	yes
Tracking radar:	no
NBC system:	yes
Night vision equipment:	no

Status
Production complete. Forty-five estimated to be in service with Egyptian Air Defence Command. As far as is known, Egypt remains the only user of this system and the system is no longer being marketed.

Contractor
Thales.

France

AMX-30SA

Type
Self-Propelled Anti-Aircraft Gun (SPAAG)

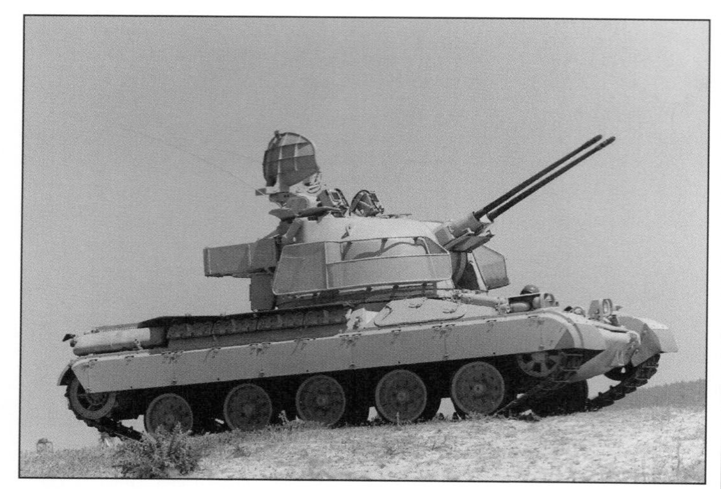

AMX-30 twin 30 mm (Thomson-CSF) 0018608

Development

In the late 1960s, a then Giat Industries (now Nexter Systems) began development of the air-defence variant of the AMX-30 Main Battle Tank (MBT) utilising the standard chassis which was designated the AMX-30 SA. At the same time, the AMX-30 MBT was fitted with the SAMM S401A turret, this is the same turret as fitted on the AMX-13 DCA. A total of 60 systems were supplied to the French Army but these have been phased out of service.

In 1975, Saudi Arabia placed an order with the then Thomson-CSF for the development of the Shahine low-altitude surface-to-air missile system, a further development of the Crotale but mounted on an AMX-30 MBT chassis rather than a 4 × 4 wheeled vehicle.

At the same time Saudi Arabia placed an order for 53 vehicles of an improved version of the AMX-30 DCA called the AMX-30 SA.

Description

The turret, designated the TG 230A, is a further development of the SAMM S401A and is armed with two Hispano-Suiza 30 mm 831 A automatic guns, each of which has 300 rounds of 30 × 170 mm ammunition in the ready racks. Natures of 30 mm ammunition fired are given in the table:

30 × 170 mm ammunition

Designation (NATO)	**HEI**	**HEI-T**	**SAPHEI**	**TP**	**TP-T**
Weight:					
(projectile)	360 g	360 g	360 g	360 g	360 g
(explosive)	40 g	25 g	20 g	n/avail	n/avail
(propellant)	160 g	160 g	160 g	160 g	160 g
(complete round)	870 g	870 g	870 g	870 g	870 g
Muzzle velocity:	1,080 m/s	1,080 m/s	1,080 m/s	1,080 m/s	1,080 m/s

In addition, a further 900 rounds are carried inside the hull in reserve. Both the commander and gunner have a day periscope for using the weapons in the ground-to-ground role and SAGEM sights with a magnification of ×1 and ×6 (12° field of view) respectively for anti-aircraft use. The SAGEM sights are used at a range of 3,000 to 3,500 m's to identify the target.

The AMX-30 SA has the improved Thales Defence Systems Oeil Vert (Green Eye) D-band pulse Doppler radar with a maximum range of over 15 km for remote surveillance and over 7.5 km for close surveillance of pop-up targets. The radar can be used against targets between zero and 3,000 m in altitude with radial velocities from 30 to 300 m/s, as well as hovering helicopters.

The then Thomson-CSF developed a number of other self-propelled and trailer-mounted air-defence gun systems but none of these ever entered production and the company is now concentrating its efforts on air-defence systems such as radars and the Crotale New Generation.

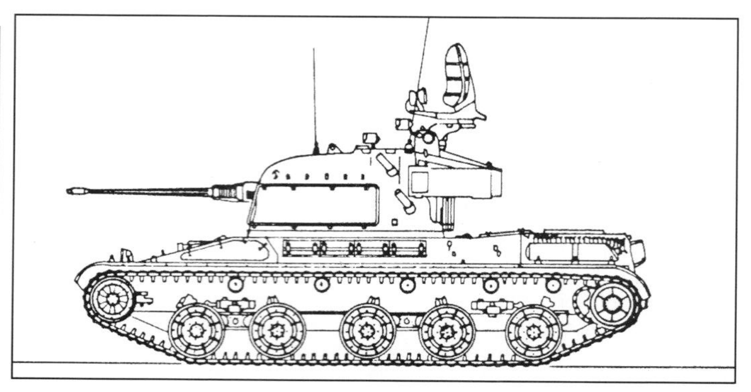

AMX-30 SA 30 mm SPAAG with radar erected 0509414

Specifications

	AMX-30 DCA	**AMX-30 SA**
Dimensions and weights		
Crew:	3	3
Length		
hull:	6.59 m	6.59 m
Width		
overall:	3.1 m	3.1 m
Height		
overall:	3.80 m	3.80 m
transport configuration:	3.00 m	3.00 m
deployed:	3.80 m	3.80 m
Ground clearance		
overall:	0.44 m	0.44 m
Track		
vehicle:	2.53 m	2.53 m
Track width		
normal:	570 mm	570 mm
Length of track on ground:	4.12 m	4.12 m
Weight		
standard:	34,000 kg	34,000 kg
combat:	36,000 kg	36,000 kg
Ground pressure		
standard track:	0.77 kg/cm^2	0.77 kg/cm^2
Mobility		
Configuration		
running gear:	tracked	tracked
Power-to-weight ratio:	20 hp/t	14.9 hp/t
Speed		
max speed:	65 km/h	65 km/h
Range		
main fuel supply:	600 km (est.)	600 km (est.)
Fuel capacity		
main:	970 litres	970 litres
Fording		
without preparation:	1.3 m	1.3 m
Angle of approach:	-	60°
Angle of departure:	-	30°
Gradient:	60%	-
Side slope:	30%	-
Vertical obstacle		
forwards:	0.93 m	0.93 m
Trench:	2.9 m	2.9 m
Engine	Hispano-Suiza HS 110, 12 cylinders, supercharged, water cooled, diesel, 720 hp at 2,000 rpm	Hispano-Suiza HS 110, 12 cylinders, supercharged, diesel, 537 hp at 2,000 rpm
Gearbox		
forward gears:	5	5
reverse gears:	5	5
Steering:	triple differential	triple differential
Clutch		
type:	centrifugal	centrifugal
Suspension:	torsion bar	torsion bar
Electrical system		
vehicle:	28 V	28 V
Batteries:	8 × 12 V, 100 Ah	8 × 12 V, 100 Ah
Firepower		
Armament:	2 × turret mounted 30 mm HS 831 A cannon	2 × turret mounted 30 mm HS 831 A cannon
Armament:	4 × turret mounted smoke grenade launcher	4 × turret mounted smoke grenade launcher
Ammunition		
main total:	1,500 (600 ready use)	1,500 (600 ready use)
Turret power control		
type:	electro hydraulic/ manual	electro hydraulic/ manual
by commander:	yes	yes
by gunner:	yes	yes
commander's override:	yes	yes
Main armament traverse		
angle:	360°	360°
speed:	80°/s	80°/s
Main armament elevation/depression		
armament front:	+85°/-8°	+85°/-8°
speed:	45°/s	45°/s

	AMX-30 DCA	AMX-30 SA
Survivability		
Night vision equipment		
vehicle:	yes	yes
commander:	no	no
gunner:	no	no
driver:	yes	yes
NBC system:	yes	yes

Status
AMX-30 SA, production complete. An unspecified number are in service only with Saudi Arabia.

Saudi Arabia also uses the Thales Defence Systems Shahine self-propelled surface-to-air missile system, also based on an AMX-30 MBT chassis supplied by the then Giat Industries (now Nexter Systems).

Contractor
The then Thomson-CSF, Division Systèmes de Missiles, which is now part of the Thales Group (prime contractor and systems integrator).
The then Giat Industries now Nexter (chassis).

M3 twin 20 mm VDAA SPAAG in firing position with outriggers deployed. Note that the system in the centre is fitted with a radar and the ones either side are not 0509678

M3 VDAA

Type
Self-Propelled Anti-Aircraft Gun (SPAAG)

Development
The M3 Véhicule de Défense Anti-Aérienne (VDAA) 20 mm self-propelled anti-aircraft gun was a joint development between Panhard and Dassault Electronique (which in January 1999 became Thomson-CSF and is today Thales) and consisted of a Panhard M3 (4 × 4) APC fitted with a new turret. Major subcontractors were Hispano-Suiza (turret), Oerlikon Contraves (guns) and Galileo (sight).

Design work on the M3 VDAA began in 1972 and the first prototype was completed in December 1973.

Manufacturers' trials took place between January and March 1974 and qualification trials in May 1974.

Production began in April 1975 and since then sales have been made to a number of countries. Production of this system was completed many years ago and it is no longer being marketed.

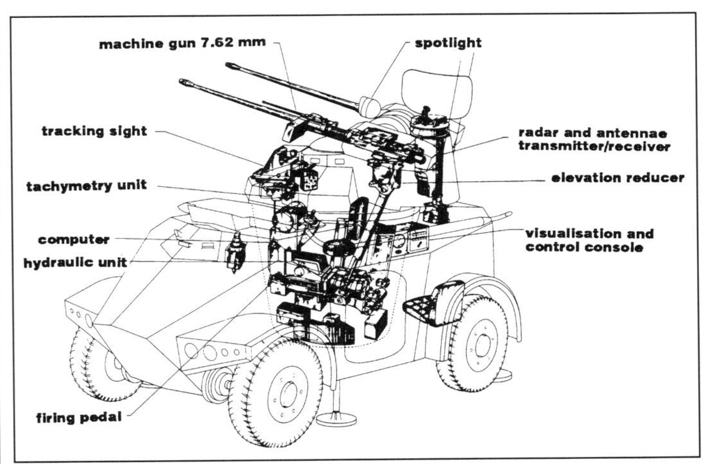

M3 VDAA SPAAG fitted with radar on turret rear 0509415

Description
The hull of the M3 VDAA is almost identical to that of the M3 Armoured Personnel Carrier (APC) and is of all-welded steel armour construction, which provides the crew with protection from small arms fire and shell splinters.

The driver is seated at the very front and has a single-piece hatch that opens to the left with three integral day periscopes. The middle day periscope can be quickly replaced by a passive night driving periscope.

The petrol engine is to the immediate rear of the driver.

In either side of the hull is a door, which opens to the front and there are two doors in the hull rear that open outwards. The only difference between the two is that the M3 VDA has four hydraulically operated outriggers that are lowered to the ground before firing to provide a more stable firing platform. In an emergency, the weapons can be fired without the outriggers in position at the lower rate of fire of 200 rds/min/gun.

The gunner is seated in the TA-20 turret and the commander in the hull to the rear of the turret. The gunner has a single-piece hatch cover, six day periscopes and the Galileo P56T sight, which has a magnification of ×5 and a 12° field of view for engaging aircraft, a sight for engaging ground targets and a wiper and washer.

Main armament consists of two Oerlikon (now Rheinmetall Air Defence) 20 mm cannon mounted externally on the rear of the turret. Reloading and cocking are carried out from inside the turret and the empty cartridge cases and links are ejected externally. The gunner can select single shots, bursts or full automatic and the weapons have two cyclic rates of fire; 200 or 1,000 rds/barrel/min. The gunner can select either left or right weapons or both together. Each barrel is provided with 300 rounds of ready-use ammunition.

Close-up of turret of M3 VDAA twin 20 mm SPAAG with weapons elevated and radar erected 0509846

The weapons have an elevation of +85°, depression of –5° and turret traverse of 360°. Gun elevation and turret traverse are hydraulic with an electric back-up system and manual controls for emergency use. Maximum effective altitude of the 20 mm cannon is 1,500 m and maximum horizontal range is 2,500 m.

The full range of 20 mm ammunition can be fired, including:

- High-Explosive Incendiary (HEI)
- High-Explosive Incendiary Tracer (HEI-T)
- Armour-Piercing High-Explosive Incendiary (APHEI)
- Armour-Piercing Incendiary Capped Tracer (APIC-T)
- Target Practice (TP).

A 7.5 mm or 7.62 mm machine gun with 200 rounds is provided for local protection and two electrically operated smoke grenade dischargers are mounted either side of the lower part of the turret and fire forwards.

Mounted on the rear of the turret is a Thales Defence Systems RA-20 1 to 8 km range fully coherent pulse Doppler radar, which operates in E-band and rotates at 40 rev/min (0.67 Hz), carries out both surveillance and target acquisition (track-while-scan) and can track four targets simultaneously.

The commander is provided with a radar implementation control, Plan Position Indicator, target assignment and acquisition controls, sound alarm that warns when a target has entered the surveillance area and a warning light that indicates the presence of jamming. Once the commander has decided that the target is to be engaged the radar feeds the target bearing to the optical sight and the range and target speed to the computer. The gunner keeps the optical sight on the target with the aid of a joystick, the computer calculates the lead angle and when the target is in range firing begins.

Optional equipment for the M3 VDAA included an Identification Friend-or-Foe (IFF) system, laser range-finder and a TV tracking system.

The M3 VDAA can provide target information by radio to up to four other vehicles without radar. Vehicles without radar can also be fitted with a satellite device on the commander's console, which allows the M3 VDAA with radar to pass data directly to the vehicle. Vehicles without the radar system were delivered with all the mountings and cabling to enable the radar to be retrofitted.

The vehicles without the radar system have two Postes Optiques de Surveillance (POS), one for the driver and one for the commander, each with a magnification of ×1.5 and a 40° field of view and covering an arc of 190°. When the vehicle is stationary, both the commander and driver search for targets. As soon as one is detected, they line up the sight and transfer the target to the turret, which automatically swings in azimuth, the guns are simultaneously elevated on to the target. When the target is within range firing begins.

Production of the M3 armoured personnel carrier has been completed and the vehicle is no longer marketed.

Variants

There are no variants other than those mentioned in the description. Algeria has bought a quantity, known as TA-23, which use Russian 23 mm cannon in place of the 20 mm guns. It is not known on which vehicle Algeria has mounted the turrets.

Specifications

	M3 VDAA
Dimensions and weights	
Crew:	3
Length	
overall:	4.45 m
Width	
overall:	2.4 m
Height	
transport configuration:	2.995 m
Ground clearance	
overall:	0.35 m
Track	
vehicle:	2.05 m
Wheelbase	2.7 m
Weight	
combat:	7,200 kg
Mobility	
Configuration	
running gear:	wheeled
layout:	4 × 4
Power-to-weight ratio:	12.5 hp/t
Speed	
max speed:	90 km/h
Range	
main fuel supply:	1,000 km
Fuel capacity	
main:	165 litres
Amphibious:	yes
Gradient:	60%
Side slope:	30%
Vertical obstacle	
forwards:	0.3 m
Trench:	0.8 m
Turning radius:	5.25 m
Engine	Panhard Model 4 HD, 4 cylinders, air cooled, petrol, 90 hp at 4,700 rpm
Gearbox	
type:	manual
forward gears:	6
reverse gears:	1
Clutch	
type:	centrifugal
Brakes	
main:	hydraulic, dual circuit
parking:	handbrake operating on gearbox output shaft
Tyres:	11.00 × 16
Suspension:	independent, coil spring and hydropneumatic shock-absorbers acting on suspension trailing arms of wheel mechanism
Electrical system	
vehicle:	24 V
Batteries:	2 × 12 V
Firepower	
Armament:	2 × turret mounted 20 mm cannon
Armament:	1 × turret mounted 7.5 mm machine gun
Armament:	alternate secondary 1 × turret mounted 7.62 mm (0.30) machine gun
Armament:	4 × turret mounted smoke grenade launcher (2 × 2)
Ammunition	
main total:	600
secondary total:	200
Turret power control	
type:	electrohydraulic/manual
by commander:	no
by gunner:	yes
Main armament traverse	
angle:	360°
speed:	60°/s
Main armament elevation/depression	
armament front:	+85°/-5°
speed:	90°/s
Survivability	
Night vision equipment	
vehicle:	no
commander:	no
gunner:	no
driver:	no
NBC system:	no
Armour	
hull/body:	12 mm
turret:	10 mm

Status

Production complete. No longer being marketed.

Contractor

Panhard General Defense (chassis).
Thales Defence Systems (turret integration).

S530

Type

Self-Propelled Anti-Aircraft Gun (SPAAG)

Development

This self-propelled anti-aircraft gun system was developed specifically for the export market and consists of a Panhard General Defense AML 60 [Automitrailleuse Leger] (4 × 4) light armoured car chassis fitted with the Société d'Applications des Machines Motrices (SAMM) S530 twin 20 mm powered anti-aircraft turret. It was first shown at the Satory exhibition of military equipment in 1971.

A total of 12 systems were delivered to Venezuela in 1973, of these, an uncertain number possibly remain in service today, although an ever increasing number are probably being used as spares in order to keep a limited quantity serviceable.

SAMM and Panhard General Defense are no longer marketing this system.

Description

The basic chassis is virtually identical to the standard AML. Over 4,800 AMLs were built in France and under licence in South Africa by the now BAE Systems. Details of the baseline Panhard AML (4 × 4) armoured car are given in *IHS Jane's Land Warfare Platforms: Armoured Fighting Vehicles*.

The all-welded steel hull of the AML is divided into three compartments:

- Driver's at the front;
- Fighting in the centre;
- Power pack at the rear.

Variants

The final version marketed was the S530 F, also known as the TAB 220.

The S530 F has the M411 sight and a moving prism episcopic channel for observation with an elevation from –10° to +70°, a magnification of ×1, a horizontal field of view of 77°, a vertical field of view of 32° plus seven observation periscopes. This turret could also be fitted with a now MBDA Mk 3

Panhard General Defense AML S530 twin 20 mm SPAAG with turret traversed partly left and showing space for spare wheel and tyre on side door
0509416

Panhard General Defense ERC Sagaie (6 × 6) Kriss SPAAG, a further development of the S530 turret with slightly different front and sides 0509417

self-contained gyroscopic sight and receive target information from a vehicle with a radar system. Typically, one radar vehicle would control 6 × twin M621 20 mm cannon systems.

When fitted on to the Panhard General Defense Sagaie (Kriss 6 × 6) armoured car the system is known as the Kriss.

Four of these have been sold to Gabon and installed on the Panhard General Defense ERC Sagaie (Kriss 6 × 6) chassis.

Specifications

	S530
Dimensions and weights	
Crew:	3
Length	
overall:	3.90 m
hull:	3.79 m
Width	
overall:	1.97 m
Height	
overall:	2.24 m
hull:	1.385 m
Ground clearance	
overall:	0.33 m
Track	
vehicle:	1.62 m
Wheelbase	2.5 m
Weight	
combat:	5,500 kg
Mobility	
Configuration	
running gear:	wheeled
layout:	4 × 4
Power-to-weight ratio:	16.36 hp/t
Speed	
max speed:	90 km/h
reverse:	5.5 km/h
Range	
main fuel supply:	600 km
Fuel capacity	
main:	156 litres
Fording	
without preparation:	1.1 m
Gradient:	60%
Side slope:	30%
Vertical obstacle	
forwards:	0.3 m
Trench:	0.8 m
Turning radius:	6 m
Engine	Panhard Model 4HD, 4 cylinders, air cooled, diesel, 90 hp at 4,700 rpm
Gearbox	
type:	manual
forward gears:	6
reverse gears:	1
Clutch	
type:	centrifugal
Brakes	
main:	hydraulic, dual circuit
parking:	handbrake operating on gearbox output shaft
Tyres:	11.00 × 16
Suspension:	independent spring and hydropneumatic shock absorbers acting on suspension trailing arms of wheel mechanism
Electrical system	
vehicle:	24 V
Batteries:	2 × 12 V
Firepower	
Armament:	2 × turret mounted 20 mm M621 cannon
Armament:	4 × turret mounted smoke grenade launcher (2 × 2)
Ammunition	
main total:	20
Main weapon	
cartridge:	20 mm × 102 mm
Turret power control	
type:	electrohydraulic/manual
Main armament traverse	
angle:	360°
Main armament elevation/ depression	
armament front:	+75°/-10°
Survivability	
Night vision equipment	
vehicle:	yes, passive IR, WL searchlight
NBC system:	no
Armour	
hull/body:	8 mm
turret:	14 mm

Status

Production complete. No longer being marketed. An unspecified number may still be in service with Venezuela.

As of November 2013, the Panhard General Defense ERC Sagaie (Kriss 6 × 6) is assessed to still be in service with Gabon (four), but as to there operational capability, this may be in question.

Contractor

Panhard General Defense (chassis).
Société d'Applications des Machines Motrices (SAMM) (turret).

VDAA

Type

Self-Propelled Anti-Aircraft Gun (SPAAG)

Development

The then Renault Véhicules Industriels and Dassault Electronique (which, in January 1999, became Thomson-CSF DETEXIS and today Thales) developed the Véhicule d'Auto-Défense Anti-Aérienne (VDAA) as a private venture specifically for the export market.

It consists of either the 4 × 4 or 6 × 6 version of the VAB fitted with a Thales TA-20 twin 20 mm turret which can also be fitted with a Thales RA-20S E-band one to 12 km range fully coherent pulse Doppler radar system.

Description

A typical VDAA unit will comprise two vehicles: a leader vehicle with the TA-20 turret and the RA-20S radar which carries out both tracking and surveillance functions and a satellite vehicle with the TA-20 turret only. A crew of three, comprising the driver, vehicle commander and gunner, is carried.

VDAA twin 20 mm SPAAG system based on Renault VAB (6 × 6) chassis 0509677

This turret is also fitted to the Panhard General Defense M3 VDAA (4 × 4) twin 20 mm self-propelled anti-aircraft gun system that is fully described in a separate entry.

Specifications

	VDAA
Dimensions and weights	
Crew:	3
Length	
overall:	5.98 m
Width	
overall:	2.49 m
Height	
hull:	2.06 m
Ground clearance	
overall:	0.4 m
hull:	0.5 m
Track	
vehicle:	2.035 m
Wheelbase	1.5 m + 1.5 m
Weight	
combat:	14,200 kg
Mobility	
Configuration	
running gear:	wheeled
layout:	6 × 6
Power-to-weight ratio:	16.54 hp/t
Speed	
max speed:	92 km/h
water:	7.5 km/h
Range	
main fuel supply:	1,000 km
Amphibious:	yes
Gradient:	50%
Side slope:	30%
Vertical obstacle	
forwards:	0.4 m
Turning radius:	9 m
Engine	MAN D 2356 HM72, in line-6, water cooled, diesel, 235 hp at 2,200 rpm (original)
Engine	Renault VI MIDS 06.20.45, in line-6, turbocharged, water cooled, diesel, 235 hp at 2,200 rpm (from 1983)
Gearbox	
model:	transfluid
forward gears:	5
reverse gears:	1
Steering:	re-circulating ball, hydraulically assisted
Brakes	
main:	disc
Tyres:	14.00 × 20
Suspension:	independent (torsion bar and telescopic shock-absorbers)
Firepower	
Armament:	2 × turret mounted 20 mm cannon
Armament:	4 × turret mounted smoke grenade launcher (2 × 2)
Turret power control	
type:	powered/manual
by commander:	no
by gunner:	yes
Main armament traverse	
angle:	360°
Main armament elevation/ depression	
armament front:	+85°/-5°
Survivability	
Night vision equipment	
vehicle:	no
commander:	no
gunner:	no
driver:	no
NBC system:	no

Status

Production complete.

Known to have been in service with the Oman National Guard (total of nine systems; three with radar and six without).

This system is no longer being marketed.

As of November 2013, no further information is available on this obsolete system.

Contractor

Thales, (turret).

Renault Trucks Defense, (current 6 × 6 chassis).

International

Gepard and CA 1

Type

Self-Propelled Anti-Aircraft Gun (SPAAG)

Development

In 1965, a decision was taken by Krauss-Maffei Wegmann and the then Oerlikon Contraves to develop a new self-propelled twin 35 mm anti-aircraft gun for the then West German Army as a replacement for the clear-weather, United States M42. It would have an all-weather fire-control system and be based on the chassis of the Leopard 1 Main Battle Tank (MBT). In June 1966, a contract was placed for the development of a twin 30 mm vehicle that later became known as the Matador. Prime contractors for this were Rheinmetall (armament and turret), AEG-Telefunken (target tracking radar and computer), Siemens (search radar and IFF) and Krauss-Maffei/Porsche (chassis and power supply system).

At the same time, a contract for two vehicles armed with twin 35 mm cannon was awarded to a consortium consisting of Oerlikon (armament and turret) Contraves (computer and systems integration), Siemens-Albis (tracking radar), Hollandse Signaalapparaten (search radar) and Krauss-Maffei/Porsche (chassis and power supply system). These two prototypes were called the 5PFZ-A and were delivered in 1968.

Following comparative trials with the Matador and the 5PFZ-A, a decision was taken in 1970 to concentrate further development on the twin 35 mm version developed in Switzerland. Before this decision, an order had been placed for a further four twin 35 mm versions called the 5PFZ-B, which were delivered in 1971 and had their original Dutch search radar replaced by a German Siemens MPDR 12 radar with a Siemens MSR-400 IFF system. A pre-production batch of 12 vehicles was subsequently ordered and delivered by 1973. In September 1973, an order for 420 Gepards was placed for the German Army and the first production vehicles were delivered late in 1976; production was finally completed late in 1980. The first 195 vehicles were delivered as 5PZF-B2 and designated the Flakpanzer 1 Gepard and the remaining 225 as the 5PZF-B2L and fitted with a Siemens laser range-finder. These were also designated the Flakpanzer 1 Gepard A1. The laser range-finder is installed on top of the tracking radome installed on to the turret front. The 5PZF-B2s were fitted with the laser range-finder to bring them up to the 5PZF-B2L standard. Belgium ordered 55 vehicles that were delivered between 1977 and 1980.

In 1969, the Netherlands ordered a version with the same chassis and turret as the 5PFZ-B. These were fitted with Hollandse Signaalapparaten integrated X-band Doppler surveillance radar with moving target indication and peak mission power output of 160 kW and also X-band monopulse tracking radar with an integrated K_a band pencil beam tracker. This model was called the 5PFZ-C and was followed by five pre-production vehicles, which were delivered in 1971/72. The Netherlands subsequently ordered 95 production vehicles, under the designation CA 1, which were delivered between 1977 and 1979. The Netherlands Army designation for the vehicle is the Pantser Rups Tegen Luchtdoelen (PRTL).

Late in 1998 the German MoD stated that it was to donate 43 Gepard 35 mm SPAAG to the Romanian Army as well as maintenance and training equipment. The German Army delivered first systems to Romania in 1999.

During early 2000, an upgrade to the Gepard SPAAG was carried out for the German Army. This upgrade included:

- The installation of a new digital computer
- New 35 mm high velocity Frangible Armour Piercing Discarding Sabot (FAPDS) ammunition
- New cooling system
- Built-in test equipment
- The ability to link Gepard with other air-defence assets.

The upgrade was completed in 2002.

Latest German Army Gepard 1A2 SPAAG with tracking and surveillance radars deployed for action
0045026

Table 1

Ammunition type	HEI	HEI-T	HEI (BF)	SAPHEI-T	FAPDS	APDS-T	APFSDS-T	TP/TP-T	AHEAD
Weight of projectile:	0.55 kg	0.535 kg	0.55 kg	0.55 kg	0.375 kg	0.38 kg	0.38 kg	0.55 kg	0.75 kg
Weight of propellant:	0.33 kg	0.33 kg	0.33 kg	0.33 kg	0.33 kg	0.33 kg	0.33 kg	0.33 kg	0.33 kg
Weight of explosive:	0.112 kg	0.098 kg	0.085 kg	0.012 kg	n/avail	n/avail	n/avail	n/avail	0.001 kg
Weight of complete round:	1.58 kg	1.565 kg	1.58 kg	1.552 kg	1.44 kg	1.46 kg	1.46 kg	1.58 kg	1.78 kg
Muzzle velocity:	1,175 m/s	1,175 m/s	1,175 m/s	1,175 m/s	1,440 m/s	1,440 m/s	1,400 m/s	1,175 m/s	1,050 m/s

Description

The all-welded steel hull of the Gepard is slightly longer than the standard production Leopard 1 MBTs and has slightly thinner armour.

The driver is seated at the front of the hull on the right side and has a single-piece hatch cover that opens to the left of the position and three-day periscopes in front of them. To the left of the driver is the Daimler-Benz OM314 95 hp auxiliary power unit, for which the exhaust pipe runs along the left side of the hull to the rear.

The two man all-welded steel turret is in the centre of the hull with the commander seated on the left and the gunner on the right. Over their position is a single-piece hatch cover that opens to the rear and day periscopes arranged around it give all-round observation with the hatch closed.

In front of both the commander's and gunner's position is a roof-mounted fully stabilised panoramic telescope with magnifications of ×1.5 (50° field of view) and ×6 (12.5° field of view), each of which has a swing-in sun filter, screen washer and wiper and a de-icing and de-misting heater. The sights can be slaved automatically to the tracking radar and therefore have an elevation of +85°, a depression of −10° and a total traverse of 360°. The sights are used for optical target acquisition, battlefield surveillance and the lying of the guns against ground targets. The commander standing in the open hatch operates an optical target indicator mounted on the commander's panoramic telescope.

The Gepard has a vehicle navigation system and the screen of the radar is always north oriented. Information from the radar system can be transmitted by radio in digital form and displayed on a similar monitor at headquarters. The vehicle is also provided with an NBC system and four 76 mm smoke grenade dischargers are mounted on either side of the turret.

Mounted on the rear of the turret is the fully coherent pulse Doppler search radar, which operates in the E/F bands and has a range of 15 km. It rotates at 60 rpm and provides continuous airspace surveillance with an IFF capability. The radar can be operated when the vehicle is moving and, when not required it can be folded down behind the rear of the turret to reduce overall height. As soon as an aircraft appears on the scope the crew is alerted. The target is displayed in terms of azimuth, angle and range and is identified as friend or foe. If the aircraft is hostile, information is passed to the coherent Siemens-Albis pulse Doppler-tracking radar. This radar is mounted on the front of the turret, which has a range of 15 km, operates in the Ku-band and when not in use can be traversed 180° so that the antenna is facing the front of the turret. The tracking radar tracks the target automatically in terms of azimuth, elevation and range. At the same time the search radar is still maintaining a search for other targets. The acquisition range of the tracking radar allows target acquisitions within an angle of about 200° without rotating the turret.

The analogue computer calculates the lead angles taking into account weather, continuously measured muzzle velocity and the cant of the vehicle. Wind speed and direction, ballistic air pressure and ballistic air temperatures are fed in manually once a day. The guns normally open fire when the aircraft is between 3,000 and 4,000 m away, with the rounds reaching the target at a range of 2,000 to 3,000 m. The duration of the burst is a function of the range with a normal burst consisting of 20 to 40 rounds.

Main armament consists of two Oerlikon Contraves 35 mm KDA cannon with a cyclic rate of fire of 550 rds/barrel/min. The guns are mounted externally one either side of the turret and the anti-aircraft ammunition is fed via fixed and moving chutes which are hermetically sealed from the fighting compartment. Each cannon is provided with 310 rounds of anti-aircraft and 20 rounds of armour-piercing 35 × 228 mm ammunition. Nine types of ammunition can be fired (see Table 1 at the top of the page). The FAPDS can be fired by the Gepard after it has been upgraded while further modifications would be required in order to fire the AHEAD ammunition which is already in production for three countries for use with their twin 35 mm GDF series towed anti-aircraft guns. Full details of the AHEAD ammunition are given under Switzerland in the Towed anti-aircraft gun section.

The computer is provided with an automatic unit that can check the total serviceability of the complete fire-control system. If the main computer fails, the crew can switch to an independent standby computer.

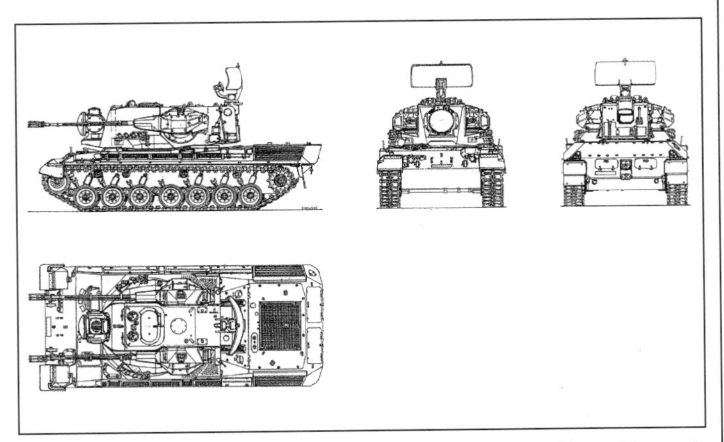

Four-view drawing of Gepard SPAAG of the German Army with tracking and surveillance radars erected (Henry Morshead) 0509418

Variants

Gepard upgrade

In mid-1996, Krauss-Maffei Wegmann was awarded a contract worth DM500 million to upgrade 147 Gepard twin 35 mm self-propelled air defence systems for the German Army and 60 similar CA 1 systems for the Royal Netherlands Army.

Key parts of the Gepard upgrade can be summarised as follows:

- Replacement of the current analogue fire-control computer by a new digital computer plus the adaptation of new computer control panels and interface electronics.
- Procurement of a new generation of 35 mm Frangible Armour-Piercing Discarding Sabot (FAPDS) ammunition with a muzzle velocity of 1,440 m/s. A new muzzle velocity measuring device is also being installed.
- Adapting both German and Netherlands systems for data linking with their respective air defence command-and-control systems by means of new radios (SEM 93, FM 9000) and system interfaces. The former system is the FlaAFuSys for the German Army while the Netherlands system is called the Target Information Command-and-Control System (TICCS).
- Installation of a new cooling system for the crew compartment and electronic components.
- Integration of menu-controlled system Built-In Test Equipment (BITE).
- Modifications of the 76 mm self-protection system installed on the Gepard to allow the new infrared smoke screening ammunition to be launched. This has a very short reaction time.
- A general overhaul of the complete Gepard System.
- Procurement of new training equipment.

When upgraded, the German system has the designation of Gepard Flak Pz 1A2 while the Netherlands version will be the PRTL 35 mm GWI.

Romanian Gepard twin 35 mm self-propelled anti-aircraft gun system overhauled by Krauss-Maffei Wegmann 0067959

Latest Royal Netherlands Army upgraded PRTL 35 mm GWI Cheetah SPAAG with tracking and surveillance radars deployed for action 0100282

This upgrade is expected to extend the operational life of this system to the year 2015. First pre-production vehicles were delivered late in 1997 and first upgraded versions were delivered in the second half of 1998. Deliveries of systems to the German Army will continue until 2002.

The initial requirement is for 147 upgraded vehicles for the German Army and 60 upgraded vehicles for the Royal Netherlands Army.

The German company, Krauss-Maffei Wegmann in Munich is the overall main contractor for the bilateral programme in which the German company, European Aeronautic Defence and Space Company (EADS - previously DASA), is responsible for the complete fire-control system of both vehicles. Signaal is a major subcontractor to DASA.

Signaal will deliver the TICCS interfacing units and modify the Signaal radars used in the PRTL which are different from those installed in the German Gepard systems.

The TICCS interfacing provision includes a full colour Plan Position Indicator (PPI) screen that displays the complete tactical air picture. This includes data from the PRTL own sensors (for example, tracking and surveillance radars) and inputs from the TICCS.

Gepard with Stinger SAM

In late 1999 Krauss-Maffei Wegmann revealed the Gepard 1A2 SPAAG hybrid configuration. This has four fire-and-forget surface-to-air missiles in the ready-to-launch position. These can be of the Stinger or Igla type with two missiles being positioned externally on each 35 mm cannon mount. Additional details of this version, developed as a private venture by Krauss-Maffei Wegmann, are given in the Self-propelled anti-aircraft guns/surface-to-air missile section.

Romanian Gepards

Romanian Army formally received its first operational Gepard twin 35 mm self-propelled anti-aircraft gun on 3 November 2004.

The German government donated 43 surplus Gepards to Romania under a 1998 agreement between the two countries defence ministries. The package also includes logistics support, the training of 25 Romanian personnel in the operation of the Gepard and a concept study on logistics integration.

Krauss-Maffei Wegmann, the Gepard manufacture, is refurbishing 36 of the weapons at its Munich facility while the other seven will be used to provide spare parts.

Romania tested the first of the refurbished weapons on 4 September 2004 at the army's Capu Midia range.

This is the first system to enter the Romanian service that is totally compatible with NATO systems.

The Romanian Gepards are not being upgraded to the latest enhanced standard, as are the 147 Gepards for the German Army and 60 for the Royal Netherlands Army.

Specifications

	Gepard A1	CA 1
Dimensions and weights		
Crew:	3	3
Length		
overall:	7.73 m	7.73 m
hull:	6.85 m	6.85 m
Width		
overall:	3.37 m	3.37 m
Height		
overall:	4.03 m	3.7 m
transport configuration:	3.01 m	3.01 m
axis of fire:	2.37 m	2.37 m
deployed:	4.03 m	3.7 m
Ground clearance		
overall:	0.44 m	0.44 m
Track		
vehicle:	2.7 m	2.7 m
Track width		
normal:	550 mm	550 mm
Weight		
standard:	44,800 kg	44,800 kg
combat:	47,300 kg	47,300 kg
Ground pressure		
standard track:	0.95 kg/cm²	0.95 kg/cm²
Mobility		
Configuration		
running gear:	tracked	tracked
Power-to-weight ratio:	17.54 hp/t	17.54 hp/t
Speed		
max speed:	65 km/h	65 km/h
Range		
main fuel supply:	550 km	550 km
cross-country:	400 km	400 km
Fuel capacity		
main:	985 litres	985 litres
Fording		
without preparation:	2.5 m	2.5 m
Gradient:	60%	60%
Vertical obstacle		
forwards:	1.15 m	1.15 m
Trench:	3 m	3 m
Engine	MTU MB 838 Ca M500, 10 cylinders, multifuel, 830 hp at 2,200 rpm	MTU MB 838 Ca M500, 10 cylinders, multifuel, 830 hp at 2,200 rpm
Gearbox		
model:	ZF 4 HP 250 planetary gear shift	ZF 4 HP 250 planetary gear shift
forward gears:	4	4
reverse gears:	2	2
Steering:	regenerative double differential	regenerative double differential
Clutch		
type:	torque converter	torque converter
Suspension:	torsion bar	torsion bar
Electrical system		
vehicle:	24 V	24 V
Firepower		
Armament:	2 × turret mounted 35 mm Oerlikon Contraves KDA cannon	2 × turret mounted 35 mm Oerlikon Contraves KDA cannon
Armament:	8 × turret mounted 76 mm smoke grenade launcher (2 × 4)	12 × turret mounted 76 mm smoke grenade launcher (2 × 6)
Ammunition		
main total:	660 (620 AA, 40 AP)	660 (620 AA, 40 AP)
Turret power control		
type:	electric/manual	electric/manual
by commander:	yes	yes
by gunner:	yes	yes
Main armament traverse		
angle:	360°	360°
speed:	90°/s	90°/s
Main armament elevation/depression		
armament front:	+85°/-10°	+85°/-10°
speed:	45°/s	45°/s
Range-finding device:	ND Yag, laser	ND Yag, laser
Survivability		
Night vision equipment		
vehicle:	yes	yes
NBC system:	yes	yes

Status

Production of the original system is complete.

Contractor

Krauss Maffei Wegmann GmbH & Co KG.

Italy

SIDAM 25

Type

Self-Propelled Anti-Aircraft Gun (SPAAG)

Development

The Quad 25 mm self-propelled anti-aircraft gun system, also known as the SIDAM 25, is a joint development between the Italian Army and Oto Melara SpA with the assistance of a number of other Italian manufacturers, to meet an Italian Army requirement.

Technical and operational evaluations of the two prototypes SIDAM 25 systems commenced in 1983 with the final tests being carried out in 1986. Production of the turret systems commenced at La Spezia facility in 1988 with first production systems completed in 1989. The original Italian Army requirement was for a total of 350 SIDAM 25 systems, but this was subsequently reduced to 280.

Production of the system has now been completed but marketing still continues for the export market.

The SIDAM 25 is mounted on an upgraded United Defence M113 series Armoured Personnel Carrier (APC) which is used in large numbers by the Italian Army. To take account of the increased weight of the complete system, which is around 14,500 kg combat loaded, the M113 chassis has been upgraded to the improved M113A2 configuration.

Close-up of gunner's position in rear of Oto Melara SIDAM 25 SPAAG 0509419

The SIDAM 25 turret system can be fitted easily on many other types of armoured vehicle, both tracked and wheeled, which include the Santa Bárbara Sistemas BMR-3560 (6 × 6) APC and the Dardo HITFIST developed by Oto Melara under contract to the Italian Army. This is now in volume production for the Italian Army under a contract awarded late in 1998.

Other major subcontractors to Otobreda include Galileo Avionica Division for the MADIS primary stabilised sighting system and Oerlikon Italiana for the four 25 mm KBA cannon.

The M113 chassis is modified by Astra with the internal fuel tank being removed and replaced by two diesel fuel tanks mounted one either side of the rear power-operated ramp, installation of the more powerful Detroit Diesel 6V-53T engine developing 265 hp coupled to the existing Allison TX-100 fully automatic transmission, fitting a reverse flow cooling system and modifying the existing torsion bar suspension to allow for a gross vehicle weight of up to 15 tonnes.

Description

The overall layout of the chassis is almost identical to the basic United Defence LP (manufactured under licence some years ago by Oto Melara for the Italian Army) M113 series armoured personnel carrier with the driver being seated at the front left and provided with a single-piece hatch cover and the power pack to the right. The turret is mounted on the roof just to the centre of the vehicle

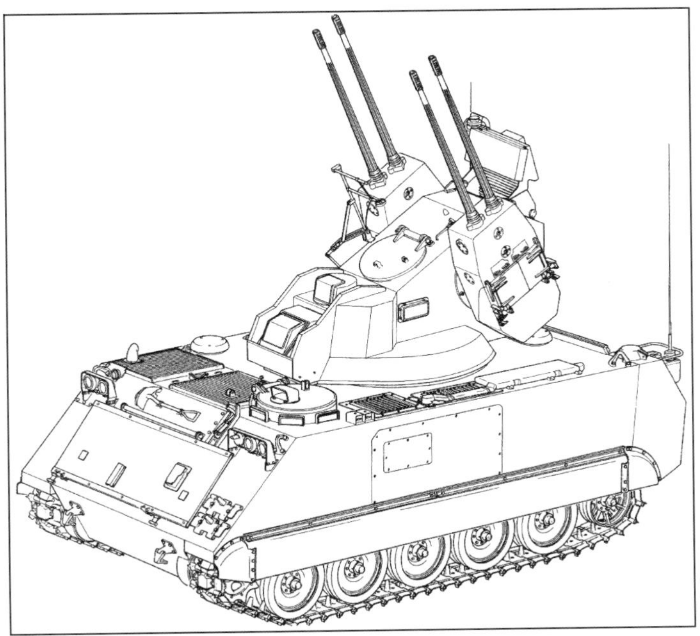

Detailed drawing of the Oto Melara SIDAM 25 turret on M113A2 chassis showing APU in left side of hull 0509679

Oto Melara Quad 25 mm SPAAG on M113A2 APC chassis 0525738

with the power-operated ramp being retained at the rear. The torsion bar suspension either side consists of five rubber-tyre road wheels, drive sprocket at the front, and idler at the rear. There are no track return rollers and a rubber skirt covers the upper part of the suspension.

To the rear of the driver on the left side of the hull is the auxiliary power unit, while on the right side of the hull are the fire-control computer, TV tracker electronics, stable element, power supply unit, radio sets and the gunner's console.

The one-person, all-welded, aluminium alloy turret has four externally mounted Oerlikon-Italiana 25 mm KBA cannon which fire NATO standard 25 × 137 mm ammunition. The maximum cyclic rate of fire for each gun is 600 rounds per minute, although in operational use this is usually kept to the firing of short bursts of 15 to 25 rounds per gun. The turret is fitted with a day clear weather and low light level TV camera sighting system capable of tracking targets automatically. The SIDAM carries 600 HE-FRAG and 30 AP rounds.

The fire-control systems use the Galileo Avionica Madis sighting and drive system which was developed for the Otobreda SIDAM 25 self-propelled anti-aircraft gun system. Madis comprises three modular units and the system can also be installed in other anti-aircraft vehicles either in total or in part.

The three modules of the Madis are:

- The daylight optical head contains the visual telescope. This is self-stabilised by a gyro directly coupled to the line-of-sight to allow for sighting operations on the move. The unit can also operate without the gyro as a servo-optical sight. Provision is made for compensation of image rotation due to the azimuth angle. The head can also accommodate a laser range-finder and a TV camera with integration of the visual, laser and TV channels into a single output path to ensure the required parallelism.
- Night optical head with an integrated low-light TV camera. The line-of-sight can be servo-stabilised to enable sight operations to be conducted on the move.
- Power servo system for the turret and weapons. This consists of an axial pump and hydraulic motor with an asynchronous motor as the prime mover. The unit also contains the first stage on the reduction gearing and, on its upper part, the fire-control computer and operations panel.

The fire-control system also uses an Alenia laser range-finder, FIAR TV components and displays and an Italtel IFF system. A laser range-finder, connected to a digital computer, provides the fire-control data. The system has, however, no all-weather capability. Two operators control the fire-control system, with the commander being seated in the turret and the gunner in the vehicle. An IFF subsystem can be installed as an option. An electronic tracking unit connected to the electro-optic sight performs the angular tracking of the target. Direct target detection and the commander in the turret can carry out acquisition by sight or via an optical sight. From the commander's acquisition, the target is assigned to the gunner who performs automatic tracking or manual tracking using a joystick. As an accurate stabilisation device is installed, all the operations, with the exception of firing, can be carried out while the vehicle is moving. Target designation from an external source can also be accepted through a digital datalink.

For this purpose, the system is furnished with the threat calculator module, which provides the operators with the following main information: first, threat priorities of each detected target (up to 10) are displayed on the monitors by means of a histogram; second, engagement priority order is shown.

Extensive and successful integration tests have been carried out with the Oto Melara OTOMATIC 76 mm as a search and early warning system as well as other radar systems such as the Contraves trailer-mounted LPD-20.

For all-weather and night operations, a video-compatible thermal imager unit can be connected to the electro-optic sight for use in haze and smoke conditions together with a passive infra-red night and day sight. An inertial navigation system for improved attitude sensing may also be fitted. The

Oto Melara SIDAM 25 SPAAG with three MBDA Mistral SAMs mounted above each bank of two 25 mm cannon 0509848

practical rate-of-fire of the four guns is 2,400 rds/min, with the 640 HEI-T ready-use rounds carried providing sufficient ammunition for eight two-second bursts or full-automatic fire. A further 30 APDS-T ready-use rounds are carried for ground defence. The maximum range of the guns is 2,500 m although their effective range is 2,000 m. The maximum traverse rate of the turret is 120°/s, and the maximum elevation rate 100°/s. The 25 mm guns can be elevated to +87° and depressed to –5°. The mode of fire is single shot 1 × 4, 9 × 4 or 18 × 4 bursts or continuous fire. Two of the guns have a dual feed arrangement for firing both the APDS-T and HEI-T rounds. Other round types that can be fired is SAPHEI-T and TP-T (the latter for practice).

SIDAM 25 with SAM

Oto Melara has completed studies of a SIDAM 25 system fitted with two packs of three MBDA Mistral fire-and-forget missiles that are already in volume production and selected by over 20 countries. Full details of the MBDA Mistral SAM are given in the Man-portable surface-to-air missile systems section. The 25 mm cannon would be used to engage close-in targets, with the Mistral missiles being used to engage targets at longer ranges.

The Gulf War of 1991 demonstrated that air defence systems, relying on radar for surveillance and tracking, are highly vulnerable to a variety of defence suppression methods including anti-radiation missiles and electronic jamming. The fire-control system of the SIDAM 25 includes an automatic Electro-Optical (EO) TV tracking system that cannot be jammed.

The EO TV tracking system can be integrated with other search sensors via a datalink that can accept data related to a number of targets. The automatic threat evaluation system indicates which target is the most dangerous as well as that most likely to be effectively engaged.

Specifications

	VCC-1	Dardo	BMR-3560 with SIDAM 25
Dimensions and weights			
Length			
overall:	5.041 m	6.705 m	6.15 m
Width			
overall:	2.686 m	2.98 m	2.50 m
Height			
hull:	1.828 m	1.75 m	2.00 m
Ground clearance			
overall:	0.406 m	0.40 m	0.40 m
Weight			
combat:	15,100 kg	20,000 kg	16,300 kg
Mobility			
Configuration			
running gear:	tracked	tracked	tracked
Speed			
max speed:	64.4 km/h	70 km/h	100 km/h
Range			
main fuel supply:	550 km	600 km	850 km
Gradient:	60%	60%	40%
Side slope:	30%	40%	30%
Vertical obstacle			
forwards:	0.61 m	0.85 m	0.6 m

Status

Production complete. In service only with Italian Army. Has also been installed, for trials purposes, on the Spanish General Dynamics European Land Systems - Santa Bárbara Sistemas BMR-3560 (6 × 6) armoured personnel carrier chassis.

Contractor

Oto Melara SpA.

Japan

Type 87

Type

Self-Propelled Anti-Aircraft Gun (SPAAG)

Development

During the early and mid-1980's, the Japanese Ground Self-Defence Force had a requirement for a new self-propelled anti-aircraft gun system to replace the ageing M42 that had been previously supplied by the US.

Engineering of the system first began in 1982 with the first prototype manufactured in 1983 by the Mitsubishi Heavy Industries (MHI).

The system now known as the Type 87 (87 *hati-nana-shiki-jisou-kousya-kikan-hou*) has been in production for the Japanese Armed Forces since at least 1987.

The new Self-Propelled Anti-Aircraft Gun (SPAAG) was allocated the provisional designation of AW-X. The new system consists of the modified chassis of the MHI Type 74 Main Battle Tank (MBT) fitted with a new turret armed with twin 35 mm Oerlikon Contraves KDA cannon and all-weather surveillance and tracking radar mounted on the roof at the turret rear.

Description

The chassis is divided into three compartments:

- Driver's at the front
- Turret in the centre
- Power pack at the rear.

Mounted externally on the left side of the tracking radar is a flat box which is believed to contain a laser range-finder, optical tracker and possibly a Low-Light Level TV (LLLTV). The commander and gunner sit in the forward part of the turret and have a joint single-piece hatch cover that opens to the rear with gunnery sights to the front and two fixed day observation periscopes either side.

The prime contractor for the fire-control system is believed to be Mitsubishi-Denki with the Nippon Seiko-Jyo Company responsible for the armament. It was first delivered in 1980 with modified 35 mm KDA cannon delivered initially in late 1979.

Prototype turret, turret stabilisation and drive systems were produced in 1981 and full scale engineering of the total gun system began in 1982.

The first complete prototype AW-X was assembled at the Sagamihara works of Mitsubishi Heavy Industries in early 1984. Testing started later in the year.

Following extensive trials with two prototypes of the AW-X twin 35 mm self-propelled anti-aircraft gun system, it was type classified as the Type 87 self-propelled anti-aircraft gun system with the first production batch consisting of 12 vehicles.

Type 87 twin 35 mm SPAAG with tracking and surveillance radars erected 0509749

Close-up of the Type 87 twin 35 mm SPAAG system (Kensuke Ebata) 0509750

Standard equipment on the Type 87 self-propelled anti-aircraft gun system includes a turret-mounted laser warning device, NBC system, auxiliary power unit and a bank of three electrically operated smoke grenade dischargers either side of the turret.

Specifications

	Type 87
Dimensions and weights	
Crew:	3
Length	
overall:	7.62 m
hull:	6.7 m
Width	
overall:	3.18 m
Height	
to turret roof:	2.812 m
deployed:	4.10 m
Ground clearance	
adjustable:	0.2 m to 0.65 m
Weight	
combat:	44,000 kg
Mobility	
Configuration	
running gear:	tracked
Power-to-weight ratio:	17 hp/t
Speed	
max speed:	53 km/h
Range	
main fuel supply:	300 km
Fuel capacity	
main:	950 litres
Fording	
without preparation:	1 m
Gradient:	60%
Side slope:	30%
Vertical obstacle	
forwards:	1 m
Trench:	2.7 m
Engine	Mitsubishi 10ZF Type 22 WT, 10 cylinders, air cooled, diesel, 750 hp at 2,200 rpm
Suspension:	hydropneumatic
Firepower	
Armament:	2 × turret mounted 35 mm KDA cannon
Armament:	6 × smoke grenade launcher
Main weapon	
maximum range:	4,000 m
cartridge:	35 mm × 228 mm
Rate of fire	
cyclic:	550 rds/min
Survivability	
Night vision equipment	
vehicle:	yes
NBC system:	yes

Status

In service with the Japanese Ground Self-Defence Force.

Like all Japanese armoured fighting vehicles, the Type 87 SPAAG has not been offered on the export market.

In 1997 it was stated that the Japanese Ground Self-Defence Force had a total of 41 of these systems in service since this date, however, no numbers have been confirmed. Then later in 2010, reports coming from within Japan quoted the number as 52, neither of these numbers have been confirmed.

Contractor

Main: Mitsubishi Heavy Industries (MHI).
35 mm KDA cannon: The then Oerlikon Contraves, now Rheinmetall Air Defence.
Fire-control system: Mitsubishi-Denki.
Other: Nippon Seiko-Jyo Company.

Korea (North)

M1984

Type

Self-Propelled Anti-Aircraft Gun (SPAAG)

Development

The M1984 was designed to meet the requirements of the North Korean Army and is a North Korean-built version of the RFAS T-62 Main Battle Tank (MBT) chassis fitted with a new-welded steel armoured turret for use in the anti-aircraft role.

Description

The driver is seated at the front of the vehicle on the left with the turret in the centre and the power pack at the rear.

In many respects the overall layout of the turret is very similar to the Russian ZSU-23-4 self-propelled anti-aircraft gun system except that it is armed with four 14.5 mm machine guns instead of four 23 mm cannon.

Mounted on top of the turret at the rear is a tall dome beneath which is probably a surveillance only radar. Some sources have assessed that this is the target tracking radar.

The turret is armed with the North Korean-built version of the Russian ZPU-4 14.5 mm towed anti-aircraft gun system, which is covered in detail later in this book.

The turret can be traversed through a full 360° with weapon elevation from −10° to +90°. Elevation and traverse is probably powered. Effective anti-aircraft range is 1,400 m.

Suspension is of the torsion bar type with each side having five road wheels with the drive sprocket at the rear and the idler at the front; there are no track return rollers.

As on the Russian T-62 MBT there is an exhaust outlet above the 4th/5th road wheel station on the left side and diesel fuel can be injected to allow the vehicle to lay its own smoke screen. Long-range fuel tanks can be carried at the rear of the hull as can an unhitching beam.

Full details of the Russian T-62 MBT chassis, on which the M1984 14.5 mm (quad) self-propelled anti-aircraft gun system is carried, are given in *IHS Jane's Land Warfare Platforms: Armoured Fighting Vehicles*.

Specifications

	M1984
Dimensions and weights	
Length	
overall:	6.60 m
Width	
overall:	3.30 m
Height	
overall:	3.10 m
Weight	
combat:	35,000 kg
Mobility	
Configuration	
running gear:	tracked
Power-to-weight ratio:	16.57 hp/t
Engine	diesel, 580 hp
Firepower	
Armament:	4 × 14.5 mm KPV machine gun
Main weapon	
cartridge:	14.5 mm × 114 mm
Survivability	
Night vision equipment	
vehicle:	yes
NBC system:	yes

Status

Production probably complete, only in service with North Korea. Open source information on North Korean weapon systems is rarely available, hence the exact status of this system is unknown.

Contractor

North Korean state factories.

M1985

Type

Self-Propelled Anti-Aircraft Gun (SPAAG)

Development

The M1985 self-propelled anti-aircraft gun system consists of a modified version of a North Korean full-tracked armoured chassis fitted with an open-topped turret armed with a single 57 mm gun.

The 57 mm gun of the M1985 is probably the Russian 57 mm S-60, which has also been manufactured in China under the local designation of the Type 59. The self-propelled version does not appear to have a shield, although a gun travel lock is provided at the front of the hull.

The hull is of all-welded steel armour construction, which provides the occupants with protection from small arms fire and shell splinters.

Description

The M1985 is a twin 57 mm Self-Propelled Anti-Aircraft Gun (SPAAG) that is mounted on a modified YW 531 Armoured Personnel Carrier (APC). The driver is seated front left with another crewman to his right with each being provided with a single-piece hatch cover that opens to the rear. The turret is in the centre of the hull with the engine compartment at the very rear.

The 57 mm anti-aircraft gun has a maximum effective anti-aircraft range of 6,000 m (maximum range 8,800 m in altitude) with the ammunition being fed to the weapon in four round clips from the left. It is understood that elevation and traverse are powered with manual controls as backup.

As far as is known, the 57 mm M1985 self-propelled anti-aircraft gun system does not have any onboard radar fire-control system so is limited to clear weather operations. However, it is understood that the system can and often is seen with the Fire Can and Flap wheel fire-control radars in close proximity.

A mounting for the use of a manportable SAM such as the Strela-2/Strela-2M has also been observed on the chassis.

Suspension is of the torsion bar type, with each side having six road wheels with the drive sprocket at the rear and idler at the front; there are no track return rollers.

North Korea uses this chassis for a number of other applications including self-propelled field artillery systems.

Specifications

	M1985
Dimensions and weights	
Length	
overall:	10 m
Width	
overall:	2.78 m
Height	
overall:	3.2 m
Weight	
combat:	22,000 kg
Mobility	
Configuration	
running gear:	tracked
Power-to-weight ratio:	13.63 hp/t
Engine	diesel, 300 hp
Firepower	
Armament:	1 × 57 mm machine gun
Main weapon	
maximum range:	4,000 m
cartridge:	57 mm × 347 mm SR
Rate of fire	
cyclic:	180 rds/min (est.)
Main armament traverse	
angle:	360°
Main armament elevation/depression	
armament front:	+87°/0°
Survivability	
Night vision equipment	
vehicle:	yes
NBC system:	yes

Status
Production probably complete. In service with North Korean forces only.

Contractor
North Korean state factories.

M1992

Type
Self-Propelled Anti-Aircraft Gun (SPAAG)

Development
The M1992 self-propelled anti-aircraft gun system consists of a modified version of the North Korean AT-S full-tracked chassis fitted with an open-topped turret armed with twin 37 mm cannon.

North Korean M1992 37 mm (twin) self-propelled anti-aircraft gun systems
0509752

The hull is of all-welded steel armour construction, which provides the occupants with protection from small arms fire and shell splinters.

Description
The driver is seated front left with another crewman to the right with each being provided with a single piece hatch cover that opens to the rear. The power pack compartment is on the right side of the hull with the turret towards the rear.

It is probable that the twin 37 mm guns are based on the Russian 37 mm automatic anti-aircraft gun M1939 that has been manufactured in both single and twin towed configurations. Details of these guns are given in the Towed anti-aircraft guns section.

The 37 mm automatic anti-aircraft gun has an effective slant range of 2,499 m. Effective altitude limit with an elevation of +45° is 1,768 m and with an elevation of +65° is 2,865 m. Self-destruct range is 4,389 m with the ammunition being fed to each weapon in five round clips.

The twin 37 mm automatic cannon have also been manufactured more recently in China by China North Industries Corporation (NORINCO) as the Type 56, Type 65, Type SD and Type P793. Full details of these are given under China in the Towed anti-aircraft gun section.

As far as is known the M1992 37 mm (twin) self-propelled anti-aircraft gun system does not have any onboard radar fire-control system so is limited to clear weather operations.

It is understood that turret traverse and weapon elevation is powered with manual controls as a back up.

The suspension is a torsion bar type with each side having five road wheels with the drive sprocket at the front and idler at the rear and there are no track return rollers.

Specifications

	M1992
Dimensions and weights	
Length	
overall:	6.5 m
Width	
overall:	2.78 m
Height	
overall:	3 m
Weight	
combat:	19,000 kg
Mobility	
Configuration	
running gear:	tracked
Power-to-weight ratio:	15.78 hp/t
Engine	diesel, 300 hp
Firepower	
Armament:	2 × turret mounted 37 mm cannon
Main weapon	
maximum range:	4,750 m (est.)
Survivability	
Night vision equipment	
vehicle:	yes (probable)
NBC system:	yes (probable)

Status
Production probably complete. In service with Armoured Brigades and Mechanised Infantry Brigades within North Korea. As of January 2013, no further data was available on this system.

Contractor
North Korean state factories.

M1992 (37 mm cannon)

Type
Self-Propelled Anti-Aircraft Gun (SPAAG)

Development
The M1992 self-propelled anti-aircraft gun system consists of a modified version of the North Korean AT-S full-tracked chassis fitted with a turret similar to that mounted on the Russian ZSU-23-4 self-propelled anti-aircraft gun system but armed with two rather than four 23 mm cannon. Effective anti-aircraft range is 2,500 m. There has been and still is some discussions in the west about the guns mounted on the chassis, some reports have suggested these are 30 mm whilst others continue with the 23 mm. What is most likely is that there are several versions of the M1992 with a mix of either 23 or 30 mm guns.

Description
The chassis is somewhat higher than that of the Russian ZSU-23-4 with the driver at front left fully enclosed turret in the centre and the power pack at the rear.

Mounted on the turret roof at the rear is radar similar to the 'Gun Dish' installed on the Russian ZSU-23-4 although it is not certain whether this has tracking and surveillance functions as the original Russian radar system has.

North Korean M1992 23 mm (twin) self-propelled anti-aircraft gun systems
0509751

Turret traverse and weapon elevations are powered with manual controls for emergency use.

Suspension is of the torsion bar type with each side having six road wheels with the drive sprocket at the rear and idler at the front; there are no track return rollers.

Specifications

	M1992
Dimensions and weights	
Length	
overall:	7 m
Width	
overall:	2.78 m
Height	
overall:	4.35 m
Weight	
combat:	19,000 kg
Mobility	
Configuration	
running gear:	tracked
Power-to-weight ratio:	15.78 hp/t
Speed	
max speed:	50 km/h
Engine	diesel, 300 hp
Firepower	
Armament:	2 × turret mounted 23 mm cannon
Main weapon	
maximum range:	4,500 m (est.)
cartridge:	23 mm × 152 mm B
Survivability	
Night vision equipment	
vehicle:	yes (probable)
NBC system:	yes (probable)

Status
Production probably complete. In service only with North Korea. No further information on this system was available to date.

Contractor
North Korean state factories.

VTT 323

Type
Self-Propelled Anti-Aircraft Gun (SPAAG)

Development
The VTT 323 (M-1973 Sinhung) is a North Korean-built version of the Chinese China North Industries Corporation (NORINCO) Type 63/YW 531 full-tracked armoured, amphibious personnel carrier modified for use in the anti-aircraft role.

Description
The VTT 323 is operated by a crew of four personnel made up of the driver, commander, gunner and loader. The driver is seated at the front of the vehicle on the left side with the commander to the rear and the engine compartment to the right.

Mounted in an open topped compartment at the rear is the North Korean version of the Russian ZPU-4 14.5 mm towed anti-aircraft gun system, which is covered in detail separately.

The Armoured Personnel Carrier (APC) may also carry the North Korean version of the Strela-2 or Igla-1 man-portable surface-to-air missiles and/or the Russian 9K11 Malyutka (AT-3 'Sagger') Anti-Tank Guided Missile (ATGM) system.

M1992 self-propelled anti-aircraft system
0509752

The mount can be traversed through a full 360° with elevation from −10° to +90°. Elevation and traverse are both manual and effective anti-aircraft range is 1,400 m. This is a clear weather system only.

Suspension is of the torsion bar type with each side having five road wheels with the drive sprocket at the front and the idler at the rear; there are no track return rollers.

Variant
M-2010 is a longer variant with upgraded and improved optics.

Specifications

	VTT 323
Dimensions and weights	
Length	
overall:	6.476 m
Width	
overall:	2.978 m
Height	
overall:	2.58 m
Weight	
combat:	12,600 kg
Mobility	
Configuration	
running gear:	tracked
Power-to-weight ratio:	22.85 hp/t
Speed	
max speed:	80 km/h
Range	
main fuel supply:	500 km (diesel)
Engine	diesel, 320 hp
Firepower	
Armament:	4 × turret mounted 14.5 mm KPV machine gun
Armament:	1 × AT-3 ATGM ATGW
Main weapon	
cartridge:	14.5 mm × 114 mm
Survivability	
Night vision equipment	
vehicle:	yes (probable)
NBC system:	yes (probable)

Status
Production probably complete. In service only with North Korea.

Contractor
North Korean state factories.

Korea (South)

20 mm Vulcan Self-Propelled Anti-Aircraft Gun (SPAAG)

Type
Self-Propelled Anti-Aircraft Gun (SPAAG)

Development
The South Korean Army has taken into service a 20 mm clear weather self-propelled anti-aircraft gun system to provide air defence cover for its mechanised units. No details of total quantities of this system built have been released although some sources quote a figure of 150 units.

Air-defence version of Korean Infantry Fighting Vehicle armed with 20 mm Vulcan cannon 0509422

The then Special Project Division, Daewoo Defence Products, now known as Doosan Infracore, was the primary contractor. From late 2000, the only known export customer for the Daewoo Korean Infantry Fighting Vehicle is Malaysia, which has taken delivery of a total of 111 vehicles in three batches. The 20 mm self-propelled anti-aircraft gun system was not among the variants sold to Malaysia.

In the Republic of Korea (RoK) Army, the 20 mm self-propelled anti-aircraft gun system is now starting to be supplemented by the more effective Daewoo Flying Tiger (Bi Ho) twin 30 mm self-propelled anti-aircraft gun system.

Description

The 20 mm Vulcan carrier has been designed to provide an effective suppression against low flying aircraft and ground targets. Vulcan is deployed for the protection of armour and mechanised infantry troops. This system consists essentially of a modified Korean Infantry Fighting Vehicle (KIFV) K200A1. The chassis is fitted with a 20 mm Vulcan air defence turret system, which is identical to that fitted to the now General Dynamics Armament and Technical Products (originally General Electric) M163 20 mm Vulcan self-propelled anti-aircraft gun system and radar with tracking capability of 5 km. This gun provides effective suppression against not only aircraft at 1.5 km low altitude but also ground targets at 3 km.

The hull is of an all-welded aluminium armour construction to the front and sides of which an extra layer of spaced laminated armour has been added for improved battlefield survivability. This composite armour provides better protection for less weight and the closed polyurethane foam in the armour system gives increased buoyancy for amphibious operations.

The driver is seated at the front left with the power pack to the right. The driver has a single-piece hatch cover that opens to the right and there are four M27 day periscopes to the front and side of this position, a passive periscope for driving at night can replace the centre one.

Mounted on the front of the hull above the trim vane is a bank of six forward-firing, electrically operated smoke grenade dischargers. Day and night lights are fitted as standard.

Mounted on the roof of the hull in the centre is a one-man power-operated turret armed with a six-barrelled 20 mm M168 cannon with range-only radar mounted to the right with target detection capability of 5,200 m and frequency hopping ECCM measures. This is integrated with a digital fire control system. A total of 1,850 rounds of 20 × 102 ammunition are carried.

There is a roof hatch over the rear compartment but unlike the KIFV there are no vision blocks in the side of the compartment. The crew can also enter the vehicle via the power-operated ramp in the hull rear.

The suspension is of the torsion bar-type with dual, rubber tyre road wheels, drive sprocket at the front and idler at the rear; there are no track return rollers. The first, second and last road wheel stations have hydraulic shock-absorbers. The suspension system is locked when the 20 mm cannon is being fired to provide a more effective platform. The track is a steel block, rubber bushed type with removable rubber pads.

Air-defence version of Korean Infantry Fighting Vehicle armed with 20 mm Vulcan cannon 0509421

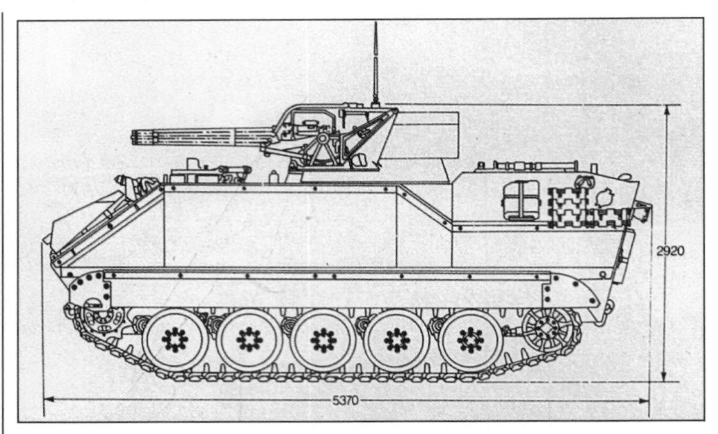

Detailed side drawing of air-defence version of Korean Infantry Fighting Vehicle armed with 20 mm Vulcan cannon 0509680

The vehicle is fully amphibious, propelled in the water by its tracks. Before entering the water the bilge pumps are switched on and a trim vane, which is stowed on the glacis plate when not required, is erected at the front of the hull. An NBC system is fitted as standard.

A fixed CO_2 fire extinguisher is provided in the engine compartment with a portable CO_2 fire extinguisher being provided in the crew compartment.

Specifications

	20 mm Vulcan
Dimensions and weights	
Crew:	4
Length	
overall:	5.37 m
Width	
overall:	2.846 m
hull:	2.545 m
Height	
to turret roof:	2.92 m
Ground clearance	
overall:	0.41 m
Track width	
normal:	381 mm
Weight	
standard:	11,000 kg
combat:	13,200 kg
Ground pressure	
standard track:	0.64 kg/cm^2
Mobility	
Configuration	
running gear:	tracked
Power-to-weight ratio:	26 hp/t (other sources quote 26.5 hp/t)
Speed	
max speed:	70 km/h
water:	6 km/h
Range	
main fuel supply:	480 km
Fuel capacity	
main:	400 litres
Amphibious:	yes
Gradient:	60%
Side slope:	30%
Vertical obstacle	
forwards:	0.63 m
Trench:	1.68 m
Engine	Doosan Model D2848T, V-8, 4 stroke, diesel, 350 hp at 2,300 rpm
Engine	Doosan D2848M, turbocharged, 4 stroke, diesel, 280 hp
Gearbox	
model:	Allison X200-5K
type:	automatic
Suspension:	torsion bar in tube
Electrical system	
vehicle:	28 V
Batteries:	2 × 12 V, 100 Ah
Alternator:	100 A
Firepower	
Armament:	1 × turret mounted 20 mm M168 with 6 barrels cannon
Armament:	6 × smoke grenade launcher
Ammunition	
main total:	1,850
Main weapon	
maximum range:	2,000 m
maximum altitude:	1,500 m
cartridge:	20 mm × 102 mm

	20 mm Vulcan
Rate of fire	
cyclic:	3,000 rds/min
Turret power control	
type:	powered
Main armament traverse	
angle:	360°
Main armament elevation/ depression	
armament front:	+85°/-5°
Survivability	
Night vision equipment	
vehicle:	yes
NBC system:	yes

Status
Production complete and in service with the South Korean Army. There are no known exports of this system.

Contractor
Doosan Infracore Co Ltd (chassis).

Flying Tiger (K30 Bi Ho)

Type
Self-Propelled Anti-Aircraft Gun (SPAAG)

Development
Since 1983, the then Daewoo Defence Products, now known as Doosan Infracore have been developing the K30 Bi Ho (Flying Tiger) twin 30 mm self-propelled anti-aircraft gun system. This development is to meet the specific operational requirements of the Republic of Korea (RoK) Army for a highly mobile self-propelled air-defence system suited to the operational conditions and terrain of South Korea.

Key system requirements include:
- Good mobility and agility
- Day and night combat capability

Four modes of operation:
- radar
- semi-automatic
- manual
- ground

Three man crew consisting of:
- commander
- gunner
- driver

Reaction time of between six and seven seconds.

Late in 1996, following extensive trials with several prototypes of the Flying Tiger twin 30 mm self-propelled anti-aircraft gun system, the first production order was placed by the South Korean Ministry of Defence (MoD) and it is understood that first production systems were completed in 1999.

The prime contractor for the Flying Tiger SPAAG is Doosan, with major South Korean subcontractors being Samsung Precision Instruments (electro-optical devices) and the LG Precision Company (surveillance radar).

An unknown number of vehicles have been developed on the new chassis, the Flying Tiger twin 30-mm self-propelled anti-aircraft gun and a field artillery ammunition support vehicle.

Description
The Bi Ho has been designed for the protection of troops and main facilities against both day and night air attack from low altitude aircraft and helicopters.

A secondary capability on offer, to destroy armoured vehicles in close combat with its 30 mm guns, is also available.

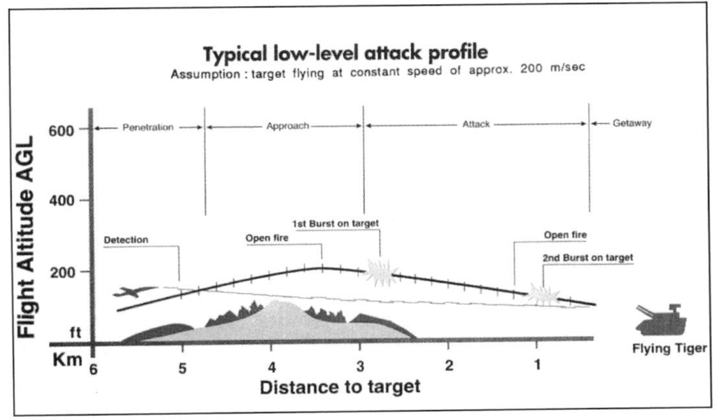

Low-level attack profile 0509852

Flying Tiger (Bi Ho) twin 30 mm SPAAG system with radar surveillance erected 0536505

According to the prime contractor, this system has the following key features:
- All-weather day/night operation
- High kill probability and firing rate, each barrel can fire up to 600 rounds/min which equates to a turret rate of 1200 rounds/min
- Fast reaction time (6 to 13 seconds)
- High ECCM capability
- High mobility with 520 hp diesel engine
- Built-in test equipment (BITE) for self-test and fault isolation
- High survivability with both ballistic and NBC protection.

Key system elements are:
- Chassis (520PS type)
- Turret
- Roof-mounted panoramic periscope
- Surveillance radar
- Target detecting capability on the move
- Fire-control system
- Electro-optical tracking system
- Two 30 mm cannon.

The hull of the twin 30 mm self-propelled anti-aircraft gun system is of all-welded construction with the driver seated at the left and the power pack to the right with the turret being mounted on the hull roof in the centre.

The driver has three periscopes for forward observation and the centre one can be replaced by a passive night vision device. In addition to the single-piece hatch cover, the driver is also provided with a forward opening door in the left side of the hull.

Suspension is of the torsion bar-type and either side consists of six, dual rubber tyre road wheels with the drive sprocket at the front and the idler at the rear, plus return rollers. A rubber skirt covers the upper part of the suspension.

The two 30 mm cannon fitted are Oerlikon Contraves (now Rheinmetall Air Defence) KCB-B models. The cannon fire the 30 × 170 ammunition type, with a cyclic rate of fire of 600 rds/gun/min, each one being provided with 250 rounds of ready-use ammunition. The weapons have an effective anti-aircraft range of about 3,000 m with each barrel being fitted with a muzzle velocity measuring system which feeds information to the system's onboard computer.

Turret traverse and weapon elevation is electric with manual controls being provided for emergency use. Turret traverse is 360° with elevation from –10° to +85°. Traverse speed is 90°/s.

Mounted on the turret rear is the LG GSTAR-30X, X-band search radar, which has an effective range of about 30 km and detection range of 17 km. The system also has a secondary Identification Friend or Foe (IFF) radar capable of operating in four modes: 1, 2, 3/A and 4. When not required, this folds down to the rear of the turret.

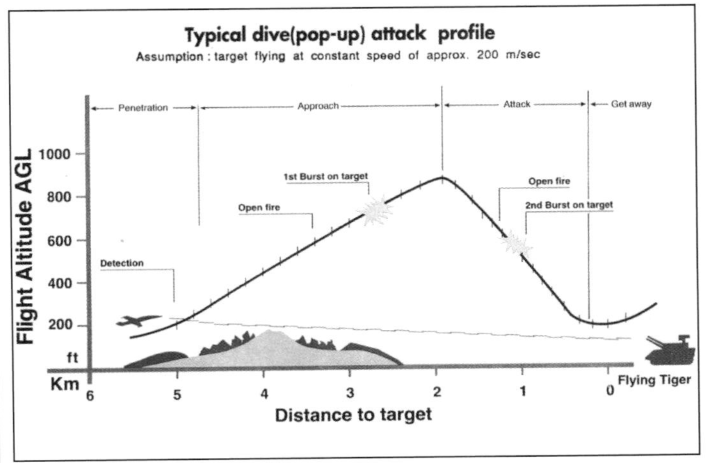

Dive (pop-up) attack profile 0509851

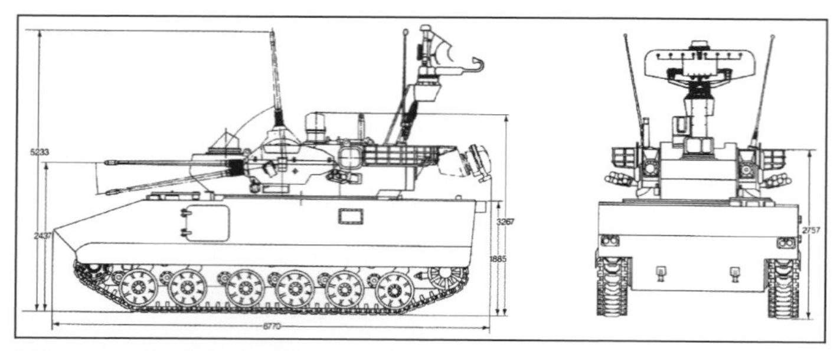

Side and front views of the Doosan Flying Tiger (Bi Ho) twin 30 mm self-propelled anti-aircraft gun system 0509850

The commander has a roof-mounted sight and a computerised fire-control system is fitted. Between the two 30 mm cannon are the now Raytheon Systems Company, Sensors and Electronic Systems, Electro-Optical Tracking System (EOTS).

The EOTS was originally developed as a private venture by the then Hughes Aircraft Company which was awarded a USD5 million production contract by South Korea for delivery from late 1998.

EOTS combines thermal imaging, TV sensors and a high repetition rate, eye-safe laser range-finder to track aerial targets such as aircraft and helicopters under all weather conditions. It also has an automatic dual-mode tracker, which can lock onto targets at low level as they fly through clutter.

The system has four modes of operation, automatic, semi-automatic, periscope and ground. The electro-optical tracking system has a traverse of 360° and elevation limits from –10° to +75° and two fields of view, narrow and wide. Surveillance range is 7 km and tracking range is 6 km with the laser range-finder having a range of 5 km. The tracking speed is 1 rad/sec.

The roof-mounted SAGEM periscopic sight has a traverse of 360° and elevation limits from –10° to +60°. It has two day fields of view (wide and narrow) and one night field of view.

Mounted either side of the turret is a bank of five forward-firing electrically operated smoke grenade dischargers.

Standard equipment on the Flying Tiger includes built-in test equipment, an IFF system and an ECCM capability.

Variants

The chassis of the Flying Tiger twin 30 mm self-propelled anti-aircraft gun system is also used as the basis for the Pegasus surface-to-air missile. This system entered low rate production in 1998. The chassis is also used for the Military Logistic Support Vehicle that is used to supply artillery systems with ammunition.

Specifications

	K30 Bi Ho
Dimensions and weights	
Crew:	4
Length	
overall:	6.95 m
Width	
overall:	3.2 m
Height	
hull:	1.885 m
deployed:	4.5 m
Weight	
combat:	25,000 kg
Mobility	
Configuration	
running gear:	tracked
Power-to-weight ratio:	21 hp/t
Speed	
max speed:	60 km/h
Range	
main fuel supply:	500 km
Fording	
without preparation:	1.2 m
Gradient:	60%
Side slope:	30%
Engine	Doosan D2840L 10V, 4 stroke, diesel, 520 hp
Gearbox	
model:	General Dynamics Land Systems HMPT-500-3EK
type:	automatic
forward gears:	3
reverse gears:	1
Steering:	hydrostatic
Brakes	
main:	multiple wet plate
Firepower	
Armament:	2 × turret mounted 30 mm Oerlikon Contraves KCB-B cannon
Armament:	10 × smoke grenade launcher
Ammunition	
main total:	600
Main weapon	
maximum range:	3,000 m
cartridge:	30 mm × 170 mm
Turret power control	
type:	powered
Main armament traverse	
angle:	360°
Main armament elevation/ depression	
armament front:	+85°/-10°
Survivability	
NBC system:	yes

Status

In production for the South Korean Army.

Prime contractor

Doosan Infracore Co Ltd, Changwon Plant Number 2.

KW2 Anti-Aircraft Gun Vehicle (AAGV)

Type

Self-Propelled Anti-Aircraft Gun (SPAAG)

Development

The KW2 30 mm Anti-Aircraft Gun Vehicle (AAGV) has been mounted on a Rotem developed and produced KW2 chassis in order to produce a Self-Propelled Anti-Aircraft Gun (SPAAG) for the South Korean forces.

Rotem is a South Korean company that manufactures rolling stock, defence products and plant equipment. It is also part of the Hyundai Motor Group. The Company is a relatively new organisation that was founded in 1999 and employs approximately 3,500 personnel. Presumably then the chassis that has been developed to include the KW2 also started life at or around this time.

Description

KW2 is fitted with two 30 mm automatic cannon. The cannon use an optical sighting system for the laying of fire on to the target. The system is also fitted with a Thermal Imaging system, TV tracking, Laser Rangefinder and advanced Fire Control system.

Tracking range for the system is over 7,000 m with an effective fire range of 3,000 m.

Variants

None are known of at present.

Specifications

	KW2
Dimensions and weights	
Crew:	1+2
Length	
overall:	7.2 m
Width	
overall:	2.7 m
Height	
overall:	3 m
Weight	
combat:	18,500 kg
Mobility	
Configuration	
running gear:	wheeled
layout:	8 × 8
Speed	
max speed:	100 km/h
Range	
main fuel supply:	800 km
Fording	
with preparation:	1.2 m
Gradient:	60%
Side slope:	30%
Trench:	2 m
Engine	420 hp
Gearbox	
type:	automatic
Firepower	
Armament:	2 × 30 mm cannon

	KW2
Main weapon	
length of barrel:	2 m
maximum range:	8,800 m
maximum range:	3,000 m (effective)
Turret power control	
type:	electric
Main armament traverse	
speed:	75°/s
Main armament elevation/ depression	
armament front:	+85°/-10°
speed:	50°/s

Status
Development complete, offered for sale at the Seoul Defence Exhibition, as of November 2013 no known exports.

Contractor
Hyundai Rotem Company.

Viper

Type
Self-Propelled Anti-Aircraft Gun (SPAAG)

Development
The Viper mobile air defence gun system has been developed by NEX1 Future Co. Ltd and is part of the improvement programme for the Vulcan with Enhanced Radar and Fire-Control Systems (FCSs). The aim of the development programme was:
- Significantly improved hit probability and footprint
- Easy and effective operation and maintenance
- Cost reduction.

Description
Viper is a tracked mobile air-defence gun system that may also be found mounted on a towed-wheeled chassis. The system offers defence against low altitude aircraft and soft skinned surface targets. Associated with the system is the Low Altitude Surveillance Radar TPS-830KE, that is also mounted on a wheeled truck. The TPS-830KE has been designed with air defence in mind, and as such may be integrated with:

Vulcan/Viper improvements (NEX1 Future Co Ltd) 0554720

Viper mounted on track chassis (NEX1 Future Co Ltd) 0554719

- AA Guns
- Vulcan
- Oerlikon
- Mistral
- Stinger
- KP-SAM.

The radar has been designed specifically to detect low flying and hovering helicopters, fixed-wing aircraft and other types of targets, including cruise missiles.

A new digital Fire-Control Radar (FCR) is normally found mounted to the right of the gun mounting and boasts a new digital processor, built-in test equipment, increased MTBF and decreased MTTR and much reduced maintenance costs.

The complete Viper system, (including radar) may be found in defence of the following:
- Fixed high value assets
- Command posts
- Radar sites
- Airfields.

The system will also be found deployed with infantry at the divisional level.

Variants
The Viper (M167A3) is an improved Vulcan (M167A1).

Specifications
Viper

Lethal range:	>1.8 km
Detection range:	>5.2 km
Signal processing:	digital
Transmit power:	40 W
Target aiming:	direct
ECCM:	automatic frequency changes (11 channels)
Bit function:	module checking
MTTR:	less than 30 minutes
Low Altitude Surveillance Radar Link:	auto-link

Status
Available for export. As of December 2013, no known orders have yet been addressed.

Contractor
NEX1 Future Co. Ltd, Seoul, South Korea.

Poland

Loara-G

Type
Self-Propelled Anti-Aircraft Gun (SPAAG)

Development
The Loara-G integrated self-propelled anti-aircraft gun system was designed and produced in the late 1990s by Radwar and Bumar Sp zoo, Poland, for destroying attack aircraft (fixed and rotary wing), cruise missiles and UAVs flying at very low, low and medium altitudes. The system can also engage light armoured ground and sea-surface targets.

Loara with Mistral 2 SAM (Radwar) 1331179

Loara-G 0594336

Three variants of the basic system were known to exist:
- Loara-G gun-mounted version;
- Loara-M missile-mounted version;
- Loara G/M gun/missile system.

As of January 2014 only the Loara-G is still know to be in existence.

Description

The Loara-G is an autonomous weapons system based on the Polish PT-91 Main Battle Tank (MBT), itself based on a modernised T-72 tank chassis, that ensures high automotive performances when travelling through any terrain conditions. Loara-G is capable of performing its tasks independently or when inter-operating within an air defence system. The vehicle is fitted with two Rheinmetall Air Defence (then Oerlikon Contraves) KDA 35 mm cannon and incorporates modern land navigation systems, using laser gyros and integrated with a Global Positioning System (GPS) receiver.

The system includes:
- Two state-of-the-art KDA 35 mm, license produced cannon;
- A three-dimensional target acquisition and search radar with IFF (ISZ-01);
- Integrated electro-optical target tracking system;
- Fire-control computer;
- Armoured turret;
- Cable and wireless communications systems.

35 mm KDA Cannon

License produced 35 mm KDA cannon, each with:
- A high rate of fire of 550 rounds/min;
- Low dispersion and high accuracy;
- The capability of using different types of ammunition, including AHEAD, APDS and FAPDS (the latter two both sub-calibre);
- A quick-change, dual-feed ammunition system for easy change-over of ammunition types;
- High durability of the barrel and reliability of gun mechanisms;
- The ability able to engage targets with speed up to 500 m/s;
- The ability able to acquire and engage air targets at a range of 4,000 m (6,000 m FAPSD);
- The ability able to acquire and engage air targets up to a ceiling of 3,000 m (4,000 m FAPSD).
- The ability able to effectively engage light armoured ground targets.

Target Acquisition and Search Radar

The Loara Target Acquisition and Search Radar is the short-range 3D Mubler [MultiBeam Littoral Environment Radar] search radar, featuring high average power to provide sufficient coverage on small Radar Cross Section (RCS) targets and high ECCM performances. Typically, it takes three to four antenna scans to automatically initiate tracking a new target. A symbol is displayed on the commander's or operator's monitor, the symbol shape indicating whether a friendly or hostile aircraft has been detected. Following the computer aided threat evaluation on his display, the commander designates the target that has to be engaged. The target coordinates and velocity vector are sent to the fire-control computer. At the same time, the tracking-head operator chooses which sensor shall be used for engaging the target, and the driver is given a command to stop the vehicle. This target acquisition phase can be performed while the vehicle is on the move. The following are features of the Target Acquisition and Search Radar:
- Frequency coverage 3.2-3.4 GHz;
- Instrumented/detection range 6-26 km;
- Elevation coverage 55°, accuracy is 3°;
- Azimuth accuracy 0.5°;
- Data renewal time 1 second;
- Automatic target detection and track-while-scan of up to 64 manoeuvring targets;
- Pop-up helicopter detection;
- Highly efficient ECCM;
- Built-in IFF interrogator (ISZ-01);

Loara-M 0111475

- Transmitter peak power output of 10 kW;
- Average power output is approximately 400 W;
- Pulse width 10 µs.

Target Tracking System

The Fire Control Computer controls the turret drive and the turret is slewed to an azimuth, which is within the tracking head horizontal scanning limits. The tracking head is then driven to intercept the target. In this mode, the tracking head scans small azimuth and elevation sectors around the designated target direction and searches for the target within a narrow range gate at the designated target range. Once the active tracking sensor has detected the target, the whole tracking system locks on the target and switches over to the tracking mode. The target is automatically tracked, its position in azimuth, elevation and range precisely calculated. In this phase, the turret and guns are continuously aimed at the lead direction predicted from the target track parameters and ballistic calculations, aided by the following features:
- TV camera;
- FLIR camera;
- Optical tracking system (video tracker) using images of both cameras;
- Laser range-finder.

NBC Protection

The Loara-G is equipped with NBC protection system against contamination with filter and ventilation system.

Variants

Loara Mod India/Mistral 2, this system has been offered to India to meet the Indian requirement for a SHOrt-Range Air-Defence System (SHORADS). As of February 2014 no such contract has been issued.

Status

The Loara-G has completed development and has been offered for sale.

Late in February 2008, information released by Bumar, suggested that up to 100 Loara (SPAAG) had been sold to India, delivery dates have yet to be specified.

As of October 2004, the Loara-M is no longer in development, this has been stopped due to lack of interest in the system by the Polish Armed Forces. Currently, there is no intention to combine the gun and missile system either.

The first Loara-G was handed over to the Polish Armed Forces on 9 December 2004.

Contractor

CNPEP Radwar SA.
Bumar Sp zoo.

ZSU-23-4MP BIALA

Type

Self-Propelled Anti-Aircraft Gun (SPAAG)

Development

As a cheaper alternative to the locally produced Loara self-propelled anti-aircraft gun system, and in order to extend the life of the Russian-produced ZSU-23-4 system, the then OBR SM in Poland has developed the ZSU-23-4MP BIALA Self-Propelled Anti-Aircraft Gun and Missile system (SPAAGM).

Development of the system began in 1999 and was completed with successful trials in June to July 2003. This is an improved version of the 23 mm ZSU-23-4, Russian Shilka anti-aircraft gun system in which the following objectives have been attained:
- Increased range and effectiveness in combating air targets out to 12 km and land targets at 8 km
- Reduced detection of the system on the battlefield by means of a digital passive system operation utilising a Zeiss Optronik Charge-Coupled Device (CCD) TV day colour camera

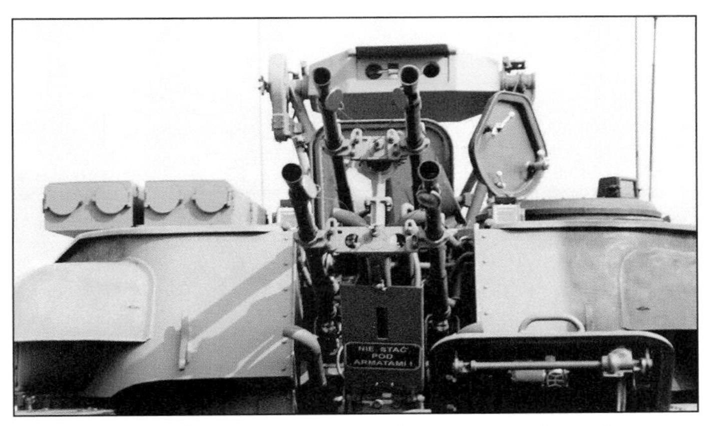

Close-up of turret showing new sensor pod installed above four water-cooled 23 mm cannon (Christopher F Foss) 0522875

Upgraded ZSU-23-4MP Biala SPAAG, which retains its four 23 mm AZP-23M cannon with four Grom missiles in lowered position (Christopher F Foss) 0522874

- Automatic operation capability in the anti-aircraft command system, with full autonomy
- Gradual improvement of tactical, technical and operational parameters, with optional cost reduction depending on the future user's requirements.

Description

In order to attain specific results, the ZSU-23-4 Shilka was subjected to the following improvements:

- Two double containers for the Polish produced man-portable Grom and Grom-2 anti-aircraft missiles, with electronic and mechanical components
- A digital fire-control system to handle the currently used OFZT (HE-T) and BZT (AP-T) and modern APDS-T and FAPDS-T projectiles as well as Grom missiles
- A system of passive sensors for target detection, identification and tracking (day TV camera, thermal camera, video tracker and laser range-finder), replacing the previous radar that was easy to detect followed by attack or stand-off
- New computing navigation system utilising GPS
- Previous UZE power system replaced by a new stand-alone power generator, without a turbo engine that used to radiate, providing a relief to the main drive engine
- Digital internal and external communication systems based on RRC-9500-3 radio station and a terminal-ensuring co-operation with an automatic anti-aircraft command system
- It is probable that the ISZ-01 or IKZ-02 Identification Friend or Foe (IFF) will also be available with the system.

Specifications

ZSU-23-4MP BIALA

Target detection range:	8-10 km
Target tracking range:	7-8 km
Armament:	
(gun)	4 × 23 mm AZP-23M
(missile)	4 × Grom
Calibre:	23 mm
Cartridge:	23 × 152 B
Maximum range:	
Gun, slant:	
(AP-T, HE-T)	1,500 m
(APDS-T, FAPDS-T)	2,000 m
Gun, effective:	
(AP-T, HE-T)	2,000 m
(APDS-T, FAPDS-T)	2,500 m
Grom missile:	5,200 m
Maximum target ceiling:	3,000 m (at missile interception)
Minimum target ceiling:	10 m (at missile interception)
Maximum target speed:	400 m/s (for missiles)

Status

According to the prime contractor, firing trials of the upgraded ZSU-23-4MP were completed in October 2002 using the 23 mm cannon and new sensor package. The missile Grom and Grom-2 firings were successfully completed during June to July 2003. It has been reported that up to 70 of these vehicles were produced for the Polish Army. Reports from Poland have suggested the Biala is cheaper than the Loara and will therefore complement each other.

Funding

Funding for the Polish Army to upgrade 18 to 24 of its existing ZSU-23-4 systems between 2003 and 2008 was approved. The upgraded systems were a stop gap for the Polish Army until the locally developed Loara twin 35 mm self-propelled air-defence gun system came on line.

Contractor

Development was undertaken by Osrodek Badawczo-Rozwojowy Sprzetu Mechanicznego (OBR SM). The company is now known as Zakłady Mechaniczne Tarnow.

Modernisation will be performed by the 4th Regional Technical Depot (4 OWT) in Zurawica.

Russian Federation

ZSU-23-4

Type

Self-propelled anti-aircraft gun (SPAAG)

Development

In 1957, authorisation was given to develop two new full-tracked self-propelled anti-aircraft gun systems to engage aircraft flying at low level, these were the Shilka (ZSU-23-4) and the Enisei (ZSU-37-2).

The main drawback of the twin 57 mm ZSU-57-2 self-propelled anti-aircraft gun was its slow rate of fire and lack of any radar or fire-control system.

The Shilka was armed with four water-cooled 23 mm cannon (developed from the weapon of the same calibre used in the ZU-23-2 towed weapon) while the Enisei used twin 37 mm cannon.

These two systems were developed to meet different operational requirements. The Shilka being developed to protect motorised rifle regiments against targets flying below 1,500 m and the Enisei to protect tank regiments against targets flying below 3,000 m.

The chassis was designated as GSh 575 and was developed by the OKB 40 bureau with parts taken from the SU 85 self-propelled gun. Many of these automotive parts are also used in the PT 76 light amphibious tank.

Prototypes of both the ZSU-23-4 and ZSU-37-2 were completed in 1960 and these underwent extensive trials from late 1960 through to late 1961 when it was demonstrated that both systems were superior to any systems in service at that time.

Although the ZSU-37-2 was the more effective weapon system of the two, it was also the more expensive. Thus, the ZSU-23-4 was adopted.

The ZSU is the Russian designation *Zenitnaia Samokhodnaia Ustanovka* (self-propelled anti-aircraft mount), 23 is the calibre of the weapons (23 mm) and 4 represents the number of weapons in the system. The common name for the ZSU-23-4 is the *Shilka* (Awl).

Production commenced in 1965 with first units being equipped in 1966. However, its first public appearance was in late 1965 when a batch of prototype ZSU-23-4 systems was shown in the annual parade held in Red Square, Moscow. Production continued until 1983 and it is estimated that total production amounted to between 7,000 and 8,000 systems. A quantity of systems was also built in the former Czechoslovakia.

The ZSU-23-4 was first used operationally by Egypt and Syria during the 1973 Middle East conflict where it accounted for about 30 per cent of the aircraft lost by the Israeli Air Force.

ZSU-23-4V1 SPAAG with 'Gun Dish' radar erected. This vehicle is in service with the Polish Army (Jaroslaw Cislak) 0045020

ZSU-23-4 SPAAG with radar in position and weapons elevated (C R Zwart)
0100272

On its own the ZSU-23-4 can be overcome but, used in conjunction with other Russian air defence systems such as the Kub (NATO SA-6 Gainful), it is highly effective.

It has replaced the clear weather ZSU-57-2 self-propelled anti-aircraft gun in front line units and is issued on the scale of four ZSU-23-4s per Russian Motorised Rifle and Tank Regiment anti-aircraft battery with four Strela-1 (NATO SA-9 Gaskin) or Strela-10 (NATO SA-13 Gopher) vehicles) to give a total of 16 ZSUs per Motorised Rifle and Tank Division. They usually operate in pairs with approximately 150 to 200 m between the individual vehicles.

Description

The all-welded steel hull of the ZSU-23-4 is divided into three compartments: driver's at the front, fighting in the centre and engine at the rear.

The four-man crew of the ZSU-23-4 consists of the commander, search and aiming operator, ranging operator and driver/mechanic.

The driver is seated at the front of the vehicle on the left and has a single-piece hatch cover to his front that opens upwards on the outside. When it is raised, a windscreen and wiper can be positioned in front of the driver. When the driver's hatch is closed, forward observation is maintained by a BM 130 day periscope, which can be replaced by an infra-red TVN 2 periscope for night driving. Either side of the driver's position is a vision block.

Mounted on the lower part of the glacis plate is a splashboard to stop water rushing up the front of the vehicle when it is fording a stream.

The other three crew members, commander, search radar operator/gunner and range operator, are all seated in the large square turret. The guns and ammunition are in the forward part of the turret and separated from the crew by a gas tight and armoured bulkhead. Access to the guns and ammunition is by two large hatches, one either side of the turret roof, which are hinged in the centre and open vertically. The commander is seated on the left side of the turret and has a cupola, which can be traversed 360°. The cupola has a single-piece hatch cover that opens to the rear and, in the forward part, three day periscopes, the centre one a TPKU 2 which can be replaced by a TKN 1T infrared periscope for night use. This has a range of 200 to 250 m. The commander can have an infrared searchlight mounted on the forward part of the cupola.

To the right of the commander is a large single-piece hatch cover that opens to the rear in front of which are two BM 190 day periscopes.

The engine and transmission are at the rear of the hull as is the DG4M 180 hp gas turbine. The torsion bar suspension system consists of six single rubber-tyre road wheels with the idler at the front and the drive sprocket at the rear. There are no track return rollers. Hydraulic shock-absorbers are provided for the first and last road wheel stations.

Standard equipment on all vehicles are an air filtration and overpressure NBC system, FG 125 infrared driving lights and a vehicle navigation system for both the driver and commander which allows them to plot their exact position at any given time. An R 123 radio is used for communications.

Main armament comprises four AZP 23M 23 mm cannon (the same guns used on the towed ZU-23) with an elevation of +85°, depression of –4° and 360° turret traverse. The 23 mm cannon is gas operated with a vertically moving breech block locking system, which drops to unlock and has a cyclic rate-of-fire of 800 to 1,000 rds/barrel/min. The ZSU-23-4 can engage targets using only one or two of the four cannon. Normally, bursts of three to five, five to 10 or a maximum of 30 rds/barrel are fired. The 23 mm barrels are water-cooled and are provided with flash hiders. The weapons have a maximum effective slant anti-aircraft range of 2,500 m and a maximum effective ground range of 2,500 m. A total of 2,000 rounds of 23 mm ammunition is carried in 40 box magazines containing 50 belted rounds each. The following types of fixed ammunition are fired:

- API-T with the projectile weighing 0.189 kg and a muzzle velocity of 970 m/s, which will penetrate 25 mm of conventional steel armour at a range of 500 m and 19.3 mm at 1,000 m
- HEI-T with the projectile weighing 0.19 kg and a muzzle velocity of 970 m/s.

Each ammunition belt of 500 rounds contains one API-T and three HEI-T rounds in sequence. Supply trucks, which follow the ZSU-23-4 at a distance of 1.5 to 2.5 km, carry additional 3,000 rounds for each of the vehicles.

A number of companies have developed, or are developing, new 23 mm ammunition for the ZSU-23-4 and the towed ZU-23. These include Pretoria Metal Pressings (PMP) of South Africa (part of the Denel Group) and Oerlikon Contraves Pyrotech (now RWM Schweiz AG). The latter has developed a Frangible Armour-Piercing Discarding Sabot Tracer (FAPDS-T) round with a longer range. This is designated the PMA 276.

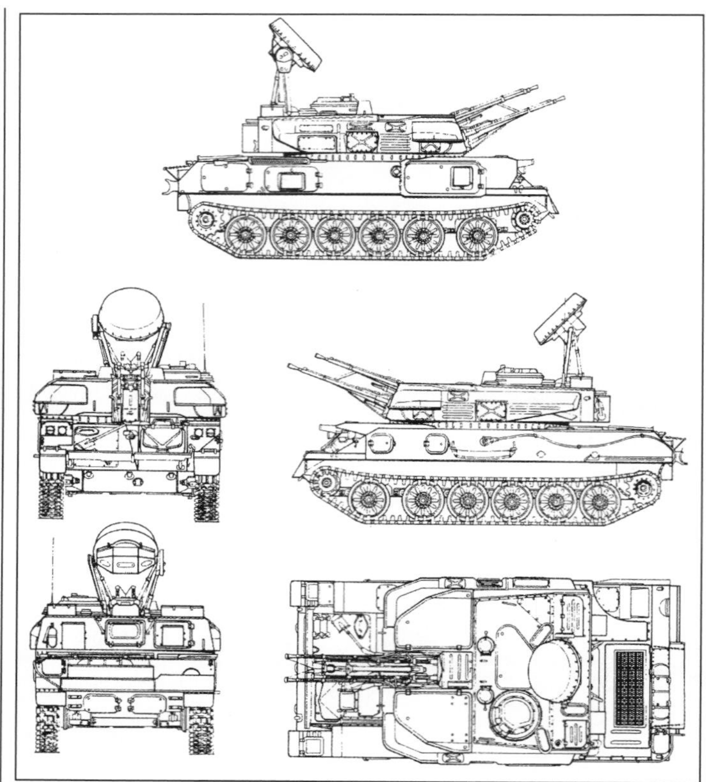

ZSU-23-4V1 Model 1972 SPAAG (Steven Zaloga) 0509411

The fire-control system consists of the radar, sighting device, computer, line-of-sight and line-of-elevation stabilisation units.

The radar designated RPK 2 (Tobol), (NATO 'Gun Dish'), operates in the J band and is mounted at the rear of the turret and the 1 m diameter antenna can be folded down to the turret rear to reduce the overall height of the vehicle for air transport. The radar performs search, detection, automatic tracking, and range to target and angular position. Range of the radar in a panoramic search mode is said to be up to 20 km and in the target tracking mode up to 18 km. An optical sight enables the weapons to be used in an ECM environment. The ZSU-23-4 can fire while stationary, on the move at speeds up to 25 km/h or on slopes with inclines up to 10°. However, gunfire accuracy is reduced by up to half when firing on the move. The on board fire-control radar is subject to ground clutter interference when employed against targets flying below 200 m or so.

The ZSU-23-4 is fitted with a complicated on board stabilisation system that stabilises both the lines of sight to the target as well as the firing direction.

A typical target engagement is believed to take place as follows: the search operator/gunner and range operators first observe the target on their scope when the radar is being used for surveillance or sector scan. If required, the target data can also be accepted from other target acquisition or tracking radars of the division. If the target is confirmed as hostile, the radar is switched to automatic tracking and target data are fed into the computer to determine the gun lead angle. When the target is in effective range of the weapons the computer advises the commander or search operator and the guns engage. The weapons can be aimed when the vehicle is travelling across country and

ZSU-23-4 SPAAG from the rear with radar antenna erected (C R Zwart)
0100273

Russian-supplied Iranian ZS-23-4 self-propelled anti-aircraft gun with mast-mounted laser sensor pod and inset display module that is installed near the commander's position 0092700

can also be laid without the use of the radar, computer or stabilisation system. Acquisition, lock on and firing take 20 to 30 seconds. The ZSU-23-4 is credited with being 50 per cent more accurate than the American Vulcan 20 mm anti aircraft system and having a 66 per cent greater effective range than the same system.

If the radar of the ZSU-23-4 is jammed or senses an incoming missile, the radar automatically shuts itself down, the gunner then engages the target using the optical sighting system.

Variants

At least nine identifiable separate versions of the ZSU-23-4 have been seen these include:

- ZSU-23-4 model 1965 (original version);
- ZSU-23-4 model 1965 (initial production version);
- ZSU-23-4V model 1968;
- ZSU-23-4V1 model 1972;
- ZSU-23-4M model 1977;
- ZSU-23-4M1;
- ZSU 23-4M2;
- ZSU-23-4M4;
- ZSU-23-4M5.

Most differ only in stowage, external fittings or cooling vents. Large ammunition panniers, mounted on the turret sides, were introduced in an intermediate production model. The latest variant, the ZSU-23-4M features these panniers, three (instead of two) access ports on each side of the hull and an armoured cover for the guns. It also has a digital computer, improved RPK 2 radar and can be linked to off-carriage radar and fire-control equipment if required. The RPK 2 radar on the ZSU-23-4M is capable of being used independently in the search mode whereas on previous versions it had been slaved to the gun tubes.

In 1985, a modified ZSU-23-4M was seen with protrusions on the right and left sides of the RPK 2 radar dome and vanes down its centre. The vanes are side lobe clutter reducing devices and the protrusions are IFF receivers.

The most significant changes in late production versions of the ZSU-23-4 have included a major change to the air cooling supply system as well as the radio and electronic systems of the vehicle. These changes have improved the overall reliability of the system. An improved ventilation system for the fighting and crew compartments has been installed.

ZSU-23-4 without radar

For use in Afghanistan, a number of ZSU-23-4 systems had their radars removed which enabled ammunition supply to be increased from 2,000 to 4,000 rounds. In addition, night vision equipment was fitted to aim the weapons. These were used in the ground to ground role, for example convoy escort in high threat areas.

ZSU-23-4 upgrade

In 1996, the original manufacturer of the ZSU-23-4, the Ulyanovsk Mechanical Plant, announced its development of an upgrade package for this system.

ZSU-23-4 upgrade (Ukraine)

Details of this extensive upgrade, which includes the addition of fire-and-forget missiles, are given in the Self-propelled Anti-aircraft Guns/Surface-to-air

ZSU-23-4 SPAAG stowed for rail transport with guns covered up and radar lowered to rear and covered up 0045022

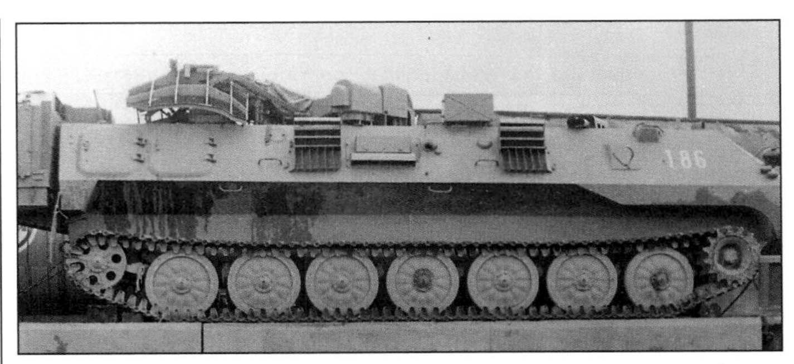

'Dog Ear' radar vehicle in travelling configuration with antenna retracted on to hull roof 0509744

Missile Systems section under Ukraine. As far as is known, this remains at the prototype stage.

ZSU-23-4 (Ukraine) Donets

Details of this system based on a T-80U MBT chassis with a modified ZSU-23-4 turret system armed with SA-13 Gopher SAMs are given in the Self-propelled-Anti aircraft Guns/Surface-to-air Missile Systems section under Ukraine. This is understood to remain at the prototype stage.

ZSU-23-4 upgrade (Belarus)

Details of this upgrade, which was first revealed in 1999, are given in this section under Belarus.

ZSU-23-4 upgrade (Russian)

Details of this are given in the previous entry as it has 30 mm cannon. This understood to remain at the prototype stage.

ZSU-23-4M1 (Russian)

This upgrade is understood to contain modernisation of the radar system with modern solid-state components.

ZSU-23-4M2 (Russian)

In this modernisation, engagements are carried out using an optical sight or night vision device.

New 23 mm cannon

The Tulamashzavod Joint Stock Company has proposed that the current 2A7 23 mm cannon could be replaced by the latest 23 mm cannon, which has a longer barrel life.

Iranian warning system for AFVs

Early in 2000, it was revealed that the Research Institute for Iranian Defence Industries had developed a laser detection, tracking and warning sensor to enhance the battlefield survivability of its armoured fighting vehicles and Self-Propelled Anti-Aircraft Gun (SPAAG) systems.

The mast-mounted laser sensor pod is typically mounted on the turret of the AFV or SPAAG to give coverage through a full 360° in azimuth and up to 60° in elevation. The display and control module is mounted near to the commander's position in the turret.

The system is capable of distinguishing between laser range finders used by Main Battle Tanks (MBTs) and forward artillery observers and laser beams used for guiding air and ground launched anti tank guided missiles onto a target.

In addition, the display module informs the commander of the direction of the laser as well as to its type so that immediate action can be taken.

In the case of a hostile lock on, the system immediately launches smoke grenades in a predetermined pattern that is claimed to be capable of disrupting the illumination of a target by hostile laser guidance systems.

The new Iranian laser warning system has been installed on a Russian supplied ZSU-23-4 quad 23 mm SPAAG and is also said to be integrated into the Iranian developed Zulfiqar MBT, which has recently entered production.

'Dog Ear' radar vehicle

A version of the Artillery Command and Reconnaissance Vehicle (ACRV) is used as the basis for the PPRU (Mobile Reconnaissance and Control Post). This is fitted with a radar scanner and provides target information not only to the ZSU-23-4 and 2S6M but also to other air defence systems such as the SA-9 and SA-13 surface-to-air missile systems.

9S482M7 (PU-12M7) Command Post

The 9S482M7 (PU-12M7), which can be used as a Command Post for the ZSU system was upgraded and modernised during 2007/8. The original version was based on a BTR-60 8 × 8 Armoured Personnel Carrier (APC) and relied on data from external radar sources. The latest variant uses a K1Sh1 wheeled vehicle based on the BTR-80 (GAZ 5903) 8 × 8 APC and incorporates a telescopic sensor mast. Within the telescopic mast are located the SRTR (*Stantsiya Radiotekhnicseskoy Razvedki*), passive ESM system and an electro-optical unit that is located at the masthead.

The mast also boasts four static antennas mounted halfway up the mast for short-range radar. Each antenna covers a sector, which when combined provides all-round coverage.

A trainable electro-optical unit mounted on top of the turret contains day and night cameras, providing hemispherical coverage. The PU-12M7 can accept target data from external surveillance radar, other fire control systems and radar equipped air-defence weapons. Communications with assets up to

15 km away can be achieved by wire, with radio being used for ranges out to 40 km.

This then allows the PU-12M7 tactical control of an area out to 100 km, handling information for as many as 99 airborne targets. The operator can simultaneously track up to nine targets.

Specifications

	ZSU-23-4
Dimensions and weights	
Crew:	4
Length	
overall:	6.535 m
Width	
overall:	3.125 m
Height	
overall:	2.576 m
axis of fire:	1.83 m
deployed:	3.572 m
Ground clearance	
overall:	0.4 m
Track	
vehicle:	2.67 m
Track width	
normal:	360 mm
Length of track on ground:	3.8 m
Weight	
combat:	20,500 kg
Ground pressure	
standard track:	0.69 kg/cm^2
Mobility	
Configuration	
running gear:	tracked
Power-to-weight ratio:	13.7 hp/t
Speed	
max speed:	50 km/h
Range	
main fuel supply:	450 km
Fuel consumption	
road:	0.96 litres/km
Fuel capacity	
main:	250 litres
Fording	
without preparation:	1.07 m
Gradient:	60%
Side slope:	30%
Vertical obstacle	
forwards:	1.1 m
Trench:	2.8 m
Engine	Model V 6R, in line-6, water cooled, diesel, 280 hp
Gearbox	
type:	manual
forward gears:	5
reverse gears:	1
Suspension:	torsion bar
Electrical system	
vehicle:	24 V
Firepower	
Armament:	4 × turret mounted 23 mm cannon
Ammunition	
main total:	2,000
Main weapon	
cartridge:	23 mm × 152 mm B
Turret power control	
type:	powered
by commander:	yes
by gunner:	yes
Main armament traverse	
angle:	360°
speed:	70°/s
Main armament elevation/depression	
armament front:	+85°/-4°
speed:	60°/s
Survivability	
Night vision equipment	
vehicle:	yes
NBC system:	yes

Status
Production complete.

Contractor
Ulyanovsk Mechanical Plant, Russian Federation.

ZSU-23-4M4

Type
Self-Propelled Anti-Aircraft Gun (SPAAG)

Development
Joint Stock Company (JSC) Ulyanovsk Mechanical Plant, (also known as Ulyanovskiy Mekhanicheskiy Zavod JSC) is the original manufacturer of the widely deployed ZSU-23-4 Self-Propelled Anti-Aircraft Gun (SPAAG), has developed an upgrade package concept aimed to launch a new family of both advanced technology and design for this system which is now being marketed to all users.

The main part of the ZSU-23-4M4 upgrade package is to replace old and obsolete components with new and more efficient units and therefore extend the operational capabilities of the system as well as enhancing its overall operational capabilities.

Those subsystems that have not been upgraded are overhauled before being installed on the upgraded ZSU-23-4M4 SPAAG.

According to the manufacturer, the upgraded ZSU-23-4M4 features improved target tracking, raising the kill probability from 0.3 to 0.6 for one fly over and increases the overall responsibility by 2 to 2.5 times.

Description
The ZSU-23-4 Shilka AA mobile gun system was designed to provide air defence of land forces in all kinds of combat theatre of operations, as well as that of fixed installations. The system can detect and engage fixed, rotary-wing aircraft, and other low-altitude air targets from stationary position, short halts or on the move as well as ground/surface targets.

Modernisation of the Shilka system known as the ZSU-23-4M4 was to improve against one air target passing through its engagement envelope with an increased hit probability of 0.6. A special 23 mm round with composite projectiles has been developed as a modernisation of the ZSU-23-4M4, this is to be used primarily against cruise missiles. The main parts of the Ulyanovsk Mechanical Plant ZSU-23-4 upgrade can be summarised as follows:

- Integration of upgraded ZSU-23-4 SPAAG systems into the unified army information system of surveillance and air target designation by the introduction of Mobile Reconnaissance and Fire-Control Post (MRCP) Sborka-M1 into a field battery as the command post. One command post would typically control up to six upgraded ZSU-23-4 SPAAG.
- Inclusion of two pairs of man-portable SAMs.
- 81 mm smoke grenade launchers.
- Laser emission sensors.
- Electro-optical vision devices including TV system for the driver.
- Replacement of the existing 'Gun Dish' radar with a new improved specification and system. Semi-conductors and microcircuits of high integration replace the majority of the tubes.
- Replacement of the current analogue computer with a new and enhanced digital computer.
- Improvement of the fire-control system with the introduction of moving target indication system in the automatic angle channels.
- Installation of equipment of receiving the telecode channel of data communication and external data on target designation from the command post.
- Mobility improved to the level of main battle tanks including the modernisation of the full-tracked chassis and improvements in the crew compartment. This includes improvements to the controllability and manoeuvrability of the system, increased reliability of the engine, which is now easier to start. The crew compartment has been enhanced by the installation of an air conditioning system. A built-in test system has also been installed.

Two versions of the upgraded ZSU-23-4 SPAAG are currently being marketed and these are designated the ZSU-23-4M4 with radar fire-control system and the ZSU-23-4M5 with radar and optic fire-control system.

According to the manufacturer, the upgraded ZSU-23-4M4 can engage aircraft and helicopters flying at an altitude from 25 to 1,500 m. While the ZSU-23-4 M5 can engage similar targets flying at an altitude from 0 to 1,500 m.

ZSU-23-4 self-propelled anti-aircraft gun system showing two pods of fire-and-forget surface-to-air missiles (IHS/Peter Felstead) 0100275

ZSU-23-4 self-propelled anti-aircraft gun system showing two pods of fire-and-forget surface-to-air missiles (IHS/Peter Felstead) 0100276

In both cases, aerial targets can be engaged out to a range of 2,500 m and with a speed of up to 500 m/s. The probability of air targets being successfully engaged in a single flyover is also significantly increased.

Along with the upgrading of technical parameters, the following operational features are also improved:

- Routine monitoring of Radio Device Complex (RDC) and its unit's operational ability are provided;
- The installation of a training system enables the radar operator to be trained in an intensive ECM environment without a real aircraft being used for the first time. Up to five targets can be tracked with the training system also simulating operation in passive and ECM environments.

Listed in the following table is a comparative feature list covering the before and after capabilities of the ZSU system.

Upgrade comparison

Main Features	Before Upgrade	After Upgrade
Detection envelope within unified air-defence system:	12 km	34 km
Target detection time by radar in fast circular search scan:	18 s	≤6 s (external target designation from Command Post)
Main computer processing time:	10 s	≤3.5 s
Radio channel target tracking lowest altitude:	100 m	20 m
Antenna targets engagement probability, using radio channel:	0.07-0.12	0.3-0.6
Mean value targets engaged by ammunition (2,000 rds):	0.8	≤4
Mean value targets engaged by barrel (4,500 per barrel):	4-7	36-40
Round consumption per target:	3,300-5,700	300-600
Number of salvos to engage target (300 rds per salvo):	11-19	1-2
Radio facilities operation check:	n/avail	available
Spare parts full package delivery:	n/avail	available

Variants

Belarus ZSU-23-4 SPAAG upgrade

The Belarus ZSU-23-4 SPAAG upgrade package also contains some of the features of the Russian ZSU-23-4 upgrade package.

Poland ZSU-23-4MP upgrade

The Polish OBR SM w Tarnowie has improved the system with:

- Increased range and effectiveness in combating targets;
- Reduced detection of the system at the battle field (passive system operation);
- Automatic operation capability in the anti-aircraft command system, with full autonomy;
- Gradual improvement of tactical, technical and operational parameters, with optional cost reduction depending on the future user's requirements.

Upgraded ZSU-23-4 with SAMs

At a customers request, at least one upgraded ZSU-23-4 has been observed fitted with two Strelets-23 launch modules. These are designed to carry and launch four Igla-type missiles, as well as an IFF transponder complying with the customer's existing standards.

Each pod, which has been designed by Kolomna KBM, prime contractor for all Russian man-portable SAMs, contains an integral power supply to activate the missiles up to four times. The SAMs used in the system are the Igla (9K38) or Igla 1 (9K310).

In a typical target engagement, the SAMs would be used to engage targets at long range with the 23 mm cannon being used to engage close-in aerial targets as well as having a secondary ground role.

Status

Development complete. Ready for production and offered for sale.

Contractor

JSC Ulyanovsk Mechanical Plant.
Rosoboronexport (for sales).
Almaz/Antei Concern of Air Defence.

ZSU-30-2

Type

Self-Propelled Anti-Aircraft Gun (SPAAG)

Development

In late 1998, the then Russian Promexport Federal State Unitary Enterprise, based in Moscow, (now part of Rosoboronexport) revealed that it was developing the ZSU-30-2, which is an upgrade package for the widely deployed Russian ZSU-23-4 quad 23 mm self-propelled anti-aircraft gun system which is covered in detail in a separate entry.

It is understood that this system has been developed mainly for the export market with the first complete prototype being completed some time in 1998.

When upgraded, the ZSU-23-4 is called the ZSU-30-2 with the 30 indicating calibre of the cannon, 30 mm, and 2 the actual number of weapons installed.

When compared to the ZSU-23-4, developed over 30 years ago, the upgraded ZSU-30-2 is a more effective system against air and ground targets as its weapons have a longer range and a more advanced electro-optical fire-control system has been fitted.

The manufacturer has also proposed that the user could adopt a phased upgrade of the ZSU-23-4 to the latest ZSU-30-2 standard.

Phase one would include:

- installation of twin 30 mm 2A38 cannon
- OESS electro-optical tracker
- large width armoured glass
- TVNE-4PA and TNPO-163 viewers and new vehicle paint.

Phase two would include:

- installation of the APD receiving equipment and R-162 radio set
- installation of Baget central computer system
- installation of anti-smart weapon system
- installation of servo-mechanical devices
- ZETs11-2 fire detection and suppression system.

The upgraded ZSU-30-2 can engage air targets out to 4,000 m in range and 3,000 m in altitude and surface targets out to 2,000 m. According to the manufacturer, the upgraded system has a 0.56 hit probability against a US A-10 ground attack aircraft on one over flight.

Description

In the ZSU-30-2 upgrade, the baseline ZSU-23-4 chassis and turret are retained, but both have been upgraded in a number of areas to extend their operational lives as well as increasing operational effectiveness. It is also claimed to be easier to operate and maintain. The modernisation has also enabled the crew to be reduced from four to three people.

The four 23 mm cannon are replaced by the Tulamashzavod JSC 2A38 or 2A38M twin cannon, two of which are installed in the latest Russian 2S6M self-propelled anti-aircraft gun/missile system.

Mounted on the forward right side of the turret is the Kedr-type stabilised electro-optical package (also referred to as the OESS) which includes a day sight, thermal sight and laser range-finder. This is already used for a number of other applications.

For improved target tracking and lead angle information the Baget central digital computer system has been fitted to the ZSU-30-2.

APD type equipment for receiving and transmitting data has been fitted, as has an R-163 radio set for communications with the PPRU-M1 centre of air target reconnaissance and control.

To improve the commander's view a large width glass armour panel has been installed in the turret roof on the left side. The crew is also provided with

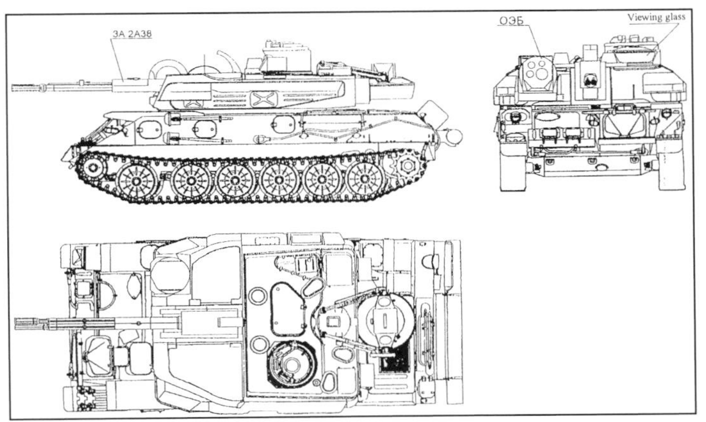

General arrangement drawing of the ZSU-30-2 twin 30 mm self-propelled anti-aircraft gun system 0069184

ZSU-30-2 self-propelled anti-aircraft gun system 0045033

the latest TVNE-4PA and TNPO-163 observation devices, improved systems of cleaning viewers and headlights.

An air conditioning system has been fitted and special paint has been provided for the outside of the hull to reduce the thermal signature of the vehicle. The ZETs11-2 fire detection and suppression system has been fitted and additional armour provided under the driver's seat for a higher level of protection against anti-tank mines.

The chassis has also been upgraded with new track, which has replaceable rubber pads.

As an option, an air-starting device for the main diesel engine can be provided and the original gas turbine can be replaced by a diesel engine with fuel being supplied from external armoured fuel tanks.

Status

Development is complete and the system has been reported to be in service with Russia and Serbia. However, conflicting reports from several sources have suggested that this system has never entered series production. It is further reported that although produced mainly for export, the system has been trialled with Russian forces and this may have led to the supposition that the system was deployed within Russia.

As of November 2013, no further information on the system is available.

Contractor

Promexport Federal State Unitary Enterprise.

ZSU-57-2

Type

Self-Propelled Anti-Aircraft Gun (SPAAG)

Development

The ZSU-57-2 SPAAG was designed and developed in early 1951 at the Design Bureau of Omsk Works Number 174 and Research Institute Number 58 in Kaliningrad, Moscow Region.

The system was first identified in public during a parade in Moscow in November 1957.

ZSU is the Russian designation for self-propelled anti-aircraft guns (ZSU stands for *Zenitnaia Samokhodnaia Ustanovka*), 57 for the calibre of the guns (57 mm) and 2 is for the number of guns.

The system consists of a chassis based on T-54 Main Battle Tank (MBT). The components within the chassis are the same, but with much thinner armour. Four rather than five road wheels and a large open topped turret armed with twin 57 mm S-68 guns which have the same performance and use the same ammunition as the towed single S-60 anti-aircraft gun which is fully described in the Towed anti-aircraft guns section.

The weapon was originally deployed in Russian tank and motorised rifle divisions but has now been replaced by the much more effective ZSU-23-4 system. The ZSU-57-2 entered front-line service with the Soviet Army in 1955 and is usually referred to by them as the Pair (or Sparka).

Description

The all-welded hull of the ZSU-57-2 SPAAG is divided into three compartments: driver's at the front, fighting in the centre and power pack at the rear.

The driver is seated at the front of the hull on the left and has a single-piece hatch cover that opens to the left, in front of which is two day periscopes. One of these is replaceable with an infrared periscope, which is used in conjunction with the infrared headlamp mounted on the right side of the glacis plate. Mounted at right angles to the glacis plate is a narrow board to stop water rushing up when the vehicle is fording shallow rivers. A link of replacement track is usually carried below this narrow board.

The large turret has slightly sloping sides, well curved corners and is fitted with external grab rails on either side. The engine, mounted transversely at the rear of the hull, is provided with a compressed air system for cold weather starting; for normal use there is an electric starter. The air inlet and air outlet louvres are in the roof of the engine compartment at the rear with the exhaust outlet on the left track guard.

ZSU-57-2 SPAAG 0045023

ZSU-57-2 SPAAG 0509745

The torsion bar suspension consists of four dual rubber tyre road wheels with the drive sprocket at the rear, idler at the front but no track return rollers. The first and last road wheel stations are provided with a hydraulic shock absorber. The all steel track has steel pins that are not secured at the outer end and are free to travel towards the hull. A raised piece of metal welded to the hull just forward of the drive sprocket drives the track pins into position each time they pass.

The ZSU-57-2 has no CBRN system and no amphibious capability. It is believed that the vehicle is not fitted with a smoke-laying system similar to that on T-54/T-55 and T-62 tanks. Long-range fuel drums can be fitted to the rear of the hull to increase the operational range of the ZSU-57-2. An unditching beam is often carried at the rear of the hull.

Main armament of the ZSU-57-2 consists of twin 57 mm S-68 cannons with 24 lands and grooves, an elevation of +85°, depression of –5° and 360° turret traverse. Elevation, depression and turret traverse are powered, with manual controls available for emergency use. The ammunition, in clips of five rounds, is fed to the magazines each side of the weapon by a loader seated in the forward part of each side of the turret. The right hand gun is modified to be loaded from the right to avoid loading problems. There are two gun aimer/pointers (vertical and horizontal) with each having an automatic anti-aircraft sight. In order that there would not be a lag or delay in engaging a target, the designers provided for hydroelectric drives for laying with hand wheels as a back up for both elevation and traverse. In addition, in case the automatic sight was disabled, a simplified mechanical sight was also provided.

The guns are fully automatic, recoil operated and each gun has a cyclic rate of fire of 105 to 120 rds/min with a practical rate of fire of 70 rds/gun/min. Maximum horizontal range is 12,000 m, maximum vertical range 8,000 m, although effective ranges are less than this. Effective slant range is 3,993 m, effective altitude limit with weapons elevated at +45° is 2,835 m and effective altitude limit with weapons elevated at +65° is 4,237 m. The weapon does not have the same effective anti-aircraft range as the towed 57 mm S-60, as the latter can be used in conjunction with off carriage fire-control equipment. Fire-control for the ZSU-57-2 is achieved by an optical mechanical computing reflex sight. The maximum traverse rate is 30°/s and the maximum elevation rate 20°/s. The weapons can fire the types of fixed ammunition listed separately in Table 1.

Yugoimport-SDPR (previously the Yugoslav Federal Directorate of Supply and Procurement) produced an HE-T round for the ZSU-57-2 and the towed S-68 and details of this are Table 2.

Table 1

Ammunition type	HEI-T (Frag-T)	HEI-T (Frag-T)	APHE-T
Designation:	OR-281	OR-281U	BR-281 and BR-281U
Fuze:			
(model)	MG-57	MGZ-57	MD-10
(type)	PD SD	PD SD	BD
Weight:			
(projectile)	2.81 kg	2.85 kg	2.82 kg
(bursting charge)	0.168 kg	0.153 kg	0.013 kg
Material:			
(bursting charge)	A-IX-2 (RDX/Al/wax)	A-IX-2 (RDX/Al/wax)	A-IX-2 (RDX/Al/wax)
Muzzle velocity:	1,000 m/s	1,000 m/s	1,000 m/s
Armour penetration:	N/A	N/A	96 mm/1,000 m/0° obliquity

Table 2

Ammunition type	HE-T
Designation:	M66
Fuze type:	impact, super quick action with pyrotechnical self-destruction
Weight:	
(complete round)	6.386 kg
(projectile)	2.85 kg
Length:	
(complete round)	536 mm
(projectile)	258 mm
Material:	
(fuze)	steel
(projectile body)	steel
(cartridge case)	brass
(bursting charge)	RDX/Al/wax
(propelling charge)	NC powder
(primer)	fulminate based
Muzzle velocity:	1,000 m/s

The ammunition is not interchangeable with that used in the 57 mm ASU-57 air-portable self-propelled anti-tank gun or 57 mm towed anti-tank guns. The empty cartridge cases and clips are deposited on a conveyor belt, which runs under the weapons. This takes the cases and clips to the rear of the turret where they are deposited in the wire cage mounted externally on the turret rear. A total of 300 rounds of 57 mm ammunition is carried in clips of five rounds.

As the barrels were air cooled, the practical rate of fire is said to be low. Typically, after 50 rounds were fired, firing stopped to allow the barrels to cool down again.

The main drawback of the ZSU-57-2 is its lack of an all-weather fire-control system. It is however, highly effective in the ground role and is capable of destroying most AFVs on the battlefield with the exception of MBTs, and even those would be very vulnerable on their sides and rear to penetration by the 57 mm APHE projectile.

Late production ZSU-57-2 SPAAGs were fitted with a more sophisticated sighting system. Two small ports in the forward upper portion of the turret front identify this model.

The combat vehicle is also fitted with the following equipment:

- 10RT-26 or R-113 radio equipments
- TPU-47 or R-120 intercom equipments
- TWN-1 driver night vision equipment.

Variants

China has manufactured the China North Industries Corporation (NORINCO) Type 80 self-propelled anti-aircraft tank, which uses a Type 69-II MBT chassis fitted with a Chinese-built copy of the turret of the ZSU-57-2. Additional details of this are given in this section under China. As far as it is known, China has not exported any of these self-propelled anti-aircraft gun systems.

Specifications

	ZSU-57-2
Dimensions and weights	
Crew:	6
Length	
overall:	8.48 m
hull:	6.22 m
main armament rear:	7.82 m
Width	
overall:	3.27 m
Height	
overall:	2.75 m
to top of roof mounted weapon:	1.9 m
axis of fire:	2.05 m
Ground clearance	
overall:	0.425 m
Track	
vehicle:	2.64 m
Track width	
normal:	580 mm
Length of track on ground:	3.84 m
Weight	
combat:	28,000 kg
Ground pressure	
standard track:	0.63 kg/cm²
Mobility	
Configuration	
running gear:	tracked
Power-to-weight ratio:	18.50 hp/t
Speed	
max speed:	50 km/h (est.)
cross-country:	30 km/h (est.)
Range	
main fuel supply:	420 km (est.)
dirt road:	320 km (est.)
Fuel consumption	
road:	1.9 litres/km
Fuel capacity	
main:	812 litres
Fording	
without preparation:	1.4 m
Angle of approach:	30°
Angle of departure:	30°
Gradient:	60%
Side slope:	30%
Vertical obstacle	
forwards:	0.8 m
Trench:	2.7 m
Engine	Model V-54, V-12, water cooled, diesel, 520 hp at 2,000 rpm
Gearbox	
type:	manual
forward gears:	5
reverse gears:	1
Steering:	clutch and brake
Clutch	
model:	multiplate
Suspension:	torsion bar
Electrical system	
vehicle:	28 V
Batteries:	4 × 12 V, 280 Ah
Firepower	
Armament:	2 × 57 mm S-68 cannon
Ammunition	
main total:	300
Main weapon	
maximum range:	5,500 m
maximum vertical range:	4,000 m
minimum vertical range:	100 m
cartridge:	57 mm × 347 mm SR
Rate of fire	
practical:	50 rds/min (est.)
Turret power control	
type:	powered
Main armament traverse	
angle:	360°
speed:	30°/s
Main armament elevation/depression	
armament front:	+85°/-5°
Survivability	
Night vision equipment	
vehicle:	yes
commander:	no
gunner:	no
driver:	yes
NBC system:	no

Status

Production complete. The system has seen service with several countries. The operational capability of the system in some of those countries cannot be guaranteed.

Contractor

Muromskiy Mashinostroitelnyy Zavod.

OAO/Murommashzavod is a known producer of the T-54 and T-55 MBTs and is therefore likely to be involved in the production of this system.

Slovenia

BOV-3

Type

Self-Propelled Anti-Aircraft Gun (SPAAG)

Development

The BOV-3 (*Borbeno Oklopno Vozilo* - 3, literal translation, Combat Armoured Vehicle) triple M-55 20 mm self-propelled anti-aircraft gun system is part of an indigenously designed family of wheeled armoured vehicles by MPP Vorzila doo that was shown officially for the first time at the 1984 Cairo Defence Exhibition.

Since 1991 this system has been used in action during fighting in the former Yugoslavia, in the ground-to-ground role, in addition to the anti-aircraft role for which it was originally designed. In earlier editions of *IHS Jane's Land-Based Air Defence*, the BOV-3 system appeared under Yugoslavia. The manufacturer is now located in Slovenia and continues to provide support for those vehicles in country. Full details of the complete BOV family of 4 × 4 light armoured vehicles are given in *IHS Jane's Land Warfare Platforms: Armoured Fighting Vehicles* 2002-2003, under Slovenia. Production of the BOV family of 4 × 4 armoured personnel carriers and variants has now been completed and it is no longer being marketed by MPP.

Information dated September 2006 indicated that former Yugoslavia has the capability to manufacture the complete turret, armed with three 20 mm cannon that is fitted to the BOV-3 SPAAG.

Description

The hull of the BOV-3 is of all-welded steel construction. The driver sits at the front left and the commander to the right, each with a rear opening single-piece hatch cover.

The driver has three day periscopes for forward observation and a single vision block and firing port in the left side of the hull; the commander has a single forward-facing day periscope. There is a crew entry hatch in the left side of the hull and a single roof hatch to the rear of the driver. The other two crew members, the gunner and loader, normally sit inside the central combat compartment of the vehicle when not manning the anti-aircraft weapons.

The engine compartment is at the rear of the hull with air inlet and air outlet louvres on the top and an engine access door at the rear. It also contains the air filters, heating device, and fuel piping and control mechanism. The vehicle is powered by a German Deutz Type F6L 413 F six-cylinder diesel developing 110 kW (148 hp) at 2,650 rpm. This is coupled to a manual gearbox with five forwards and one reverse gear and a two-speed transfer case. Two hatches in the roof and the rear access door allow for servicing of the engine and other systems.

Steering is power assisted to reduce driver fatigue and a central tyre pressure system is fitted as standard. Pressure can be adjusted from 0.7 to 3.5 bars to suit the ground being crossed. The main brakes are air hydraulic dual circuit with a hand-operated parking brake.

Suspension consists of leaf-type springs with telescopic shock absorbers with 1300 - 18 PR10 cross-country tyres fitted as standard. The differential locks are controlled electro-pneumatically.

M09/BOV-3 with added Strela-2M launcher boxes (Miroslav Gyürösi) 1331380

BOV-3 20 mm SPAAG in travelling configuration 0081999

Standard equipment includes a Jugo-Webasto 7.5 kW heater, day and night vision equipment, intercom and radios.

Mounted in the centre of the roof is the open-topped turret with an external turret basket at the rear. Access to the turret is only from outside the vehicle with extension pieces provided on the hull top either side of the turret for reloading.

The turret is based on the standard towed 20/3 mm gun M55 A4 B1 that has been in service with the former Yugoslav Army for some years. The turret has full 360° hydraulic traverse at a speed of 80°/s with an acceleration of 120°/s^2. Weapon elevation is from –4.5° to +83° at 50°/s with an acceleration of 60°/s^2.

Each of the three 20 mm cannons has a drum type magazine holding 60 rounds of 20 × 110 mm ammunition with an external indicator showing how many rounds are available (10, 30, 40, 50, 60). The weapons have a cyclic rate-of-fire of 750 rds/barrel/min and the gunner, seated at the turret rear, can select either single shots, bursts of 10 rounds, bursts of 10 to 20 rounds or sustained fire. A total of 1,500 rounds of ammunition is carried, which, apart from the three ready use drums in the turret, is stowed internally on special racks for quick access. Ammunition types that can be fired are shown in Table 1.

Additional rounds include:

- TP-T (M57)
- TP (M57)
- Blank (M77)
- Drill.

All of the rounds have brass cartridge cases 110 mm long with propellant being NitroCellulose (NC)-based.

Maximum anti-aircraft engagement altitude is quoted as 2,000 m, although, effective anti-aircraft range is between 1,000 and 1,500 m. The 20 mm weapons can also be used in the direct fire support role against ground targets when the maximum range is approximately 2,000 m.

The triple 20 mm cannon are aimed via a J-171 sight mounted to the rear of the gun shield. The gunner has a joystick to control elevation and traverse and a foot pedal for firing the cannon. The BOV-3 is a clear weather system only with no provision for external fire-control, although general warning of targets approaching could be given over the radio net.

Variants

The BOV-3 20 mm self-propelled anti-aircraft gun system is a member of the BOV 4 × 4 series of wheeled light armoured vehicles. These include the BOV-1 anti-tank vehicle armed with the locally built Sagger ATGW and the BOV-M

BOV-30 twin 30 mm SPAAG in travelling configuration with hatches open 0509755

Table 1

Type	HEI-T	HE-T	HEI	HE	API	API-T	AP-T
Designation:	M57	M57	M57	M57	M60	M60	M60
Weight of projectile:	137 g	137 g	132 g	132 g	142 g	142 g	142 g
Type of bursting charge:	TNT or RDX/Al	TNT or RDX/Al	TNT + incendiary	TNT or RDX/Al	nil	nil	nil
Total weight of round:	261 g	261 g	257 g	257 g	274 g	274 g	274 g
Muzzle velocity:	850 m/s	850 m/s	850 m/s	850 m/s	840 m/s	840 m/s	840 m/s

BOV-3 triple 20 mm SPAAG in travelling configuration with all hatches open
0509856

armoured personnel carrier. In 1985, the BOV-30 self-propelled anti-aircraft gun system was revealed. This uses the chassis of the BOV-3 but with a smaller, upright turret with the gunner seated in a raised cupola at the rear and is armed with two externally mounted 30 mm cannon. The turret has a rate-of-fire of 1,200 rds/min. Three smoke grenade launchers are attached to each side of the turret for smoke rounds. The BOV-3 mm SPAAG has been built in production quantities.

M09

Yugoimport SDPR has created a variant of the BOV-3 known as the M09, by adding two four-round packs of 9K32M Strela-2M man-portable surface-to-air missiles and an improved sighting system, this has effectively turned the BOV-03 into an effective hybrid gun-missile system with an extended engagement range.

Model 30/2 twin 30 mm SPAAG

This system is based on a BVP M80A full tracked mechanised infantry vehicle did not enter production. A small number of prototypes of this system were built and it is understood that these were deployed in Kosovo. The Model 30/2 twin 30 mm turret system is now being marketed again by Yugoslavia.

This turret weighs 2,700 kg and is armed with two 30 mm M86 cannon, each of which has a cyclic rate-of-fire of 550 to 600 rounds per minute. Each barrel is provided with 250 rounds of ready-to-use ammunition and muzzle velocity is 1,100 m/s with maximum effective range being 3,000 m.

For improved target engagement the turret is fitted with the Gunking sighting and fire-control system and the former SAMM TE-06 electric servo drive system.

Specifications

	BOV-3
Dimensions and weights	
Crew:	4+10
Length	
overall:	5.970 m
hull:	5.775 m
Width	
overall:	2.525 m
Height	
overall:	3.285 m
Ground clearance	
overall:	0.325 m
Track	
vehicle:	1.9 m
Wheelbase	2.75 m
Weight	
combat:	9,400 kg
Mobility	
Configuration	
running gear:	wheeled
layout:	4 × 4
Power-to-weight ratio:	15.74 hp/t
Speed	
max speed:	93.4 km/h
Range	
main fuel supply:	500 km
Fording	
without preparation:	1.1 m
Gradient:	55%
Side slope:	30%
Vertical obstacle	
forwards:	0.54 m
Trench:	0.64 m
Turning radius:	7.75 m
Engine	Deutz type F 6L 413 F, 6 cylinders, diesel, 148 hp at 2,650 rpm
Gearbox	
type:	manual
forward gears:	5
reverse gears:	1
Suspension:	leaf springs and hydraulic shock absorbers
Electrical system	
vehicle:	24 V
Batteries:	2 × 12 V, 143 Ah
Firepower	
Armament:	3 × turret mounted 20 mm M55 A4 B1 cannon
Ammunition	
main total:	1,500
Turret power control	
type:	powered
Main armament traverse	
angle:	360°
Main armament elevation/depression	
armament front:	+83°/-4.5°

Status

Production complete. In service with Montenegro, Serbia and with other parts of the former Yugoslavia including Croatia and Slovenia. To date September 2013, there have been no known exports of this system.

Contractor

MPP Vorzila dooand VTI (designer).

South Africa

Ystervark

Type

Self-Propelled Anti-Aircraft Gun (SPAAG)

Development

The Ystervark 20 mm self-propelled anti-aircraft gun system was developed by the African Defence Systems (Pty) Ltd (now Thales) to meet an urgent South African requirement for an armoured self-propelled anti-aircraft gun system which could also be used to escort convoys.

It was first deployed in the Operational Area (Namibia) in the mid 1980s and during fighting in 1988 successfully engaged and shot down a number of high-performance jet aircraft of the Angolan Air Force.

Before the introduction of the Ystervark, the South African Army used a Buffel (4 × 4) tractor fitted with a similar weapon, but these lacked armour protection when compared to the Ystervark.

Description

The Ystervark 20 mm self-propelled anti-aircraft gun system has a crew of two and is essentially a SAMIL 20 (4 × 4) two-tonne truck chassis that is provided with full length protection for mines and an armoured engine compartment at

Ystervark 20 mm SPAAG on SAMIL 20 truck chassis 0509423

the front. In the driver's cab to the right, the Swiss designed and built Oerlikon Contraves (now Rheinmetall Air Defence) 20 mm GAI-C01 cannon, fully described in the Towed anti-aircraft guns section, is mounted on top at the rear. An armoured cabin is provided to protect the gun detachment from shell splinters and small arms fire. The position of the 20 mm cannon is such that its field of fire is about 200° through the rear arc.

The Ystervark has a combat weight of 7.7 tonnes and is powered by a diesel engine giving a maximum road speed of 93 km/h. The 200-litre fuel tank gives an operating range of 950 km.

The South African Army has now introduced a much more capable self-propelled anti-aircraft gun system into service called the Zumlac. This is based on an armoured SAMIL 100 (6 × 6) and has a significant increase in armour, mobility and firepower over the older Ystervark system.

Status

Production complete. A total of 84 of these systems have been in service with the South African Army. This system was not being offered on the export market. The official government defence site quotes the system as being in reserve.

If no further data comes to light during 2014, the file will be removed to archive.

Contractor

Lead contractor for the system
Thales.

20 mm cannon
Rheinmetall Air Defence AG.

Zumlac

Type

Self-Propelled Anti-Aircraft Gun (SPAAG)

Development

During the early to mid-1980's, it was decided to develop a new self-propelled anti-aircraft gun system, later called the Zumlac, based around the captured twin ZU-23-2 LAAG.

During the Angolan conflict, significant numbers of Russian twin 23 mm ZU-23-2 Light Anti-Aircraft Gun (LAAG) were captured and taken back to South Africa. Armscor selected Altech Defence Systems to be prime contractor. In 1994, the Company then became known as the African Defence Systems Ltd, but is today known as Thales.

Zumlac twin 23 mm SPAAG clearly showing weapons with boxes of 23 mm ammunition alongside for rapid reloading (Christopher F Foss) 0045025

Zumlac twin 23 mm SPAAG in the firing position 0525737

Prior to this, the only self-propelled anti-aircraft gun system in service with the South African Army was the 20 mm Ystervark, based on an armour protected 4 × 4 truck, which is covered in a separate entry.

The Ystervark was used during the fighting in southern Angola in the late 1980s and successfully engaged a number of Angolan high-performance aircraft.

This system, however, had a number of disadvantages including lack of cross-country mobility, lack of firepower and limited space for crew, ammunition and stores.

The first prototype Zumlac was completed in 1989 and, following trials and a number of modifications was accepted for service. The South African Army ordered 36 systems, with first deliveries made in 1992, and deliveries being completed in 1993. Two regiments, each of which has 18 weapons in three batteries of six, use the 36 systems.

Compared with the earlier Ystervark, the Zumlac has a significant increase in armour, mobility and firepower and is able to keep up with other South African 6 × 6 cross-country vehicles. The South African National Defence Force name for this system is the Bosvark.

Description

Zumlac is based on a Truckmakers SAMIL 100 (6 × 6) mine-protected truck, already used for a wide range of missions by the South African National Defence Forces (SANDF). Production of the SAMIL 100 (6 × 6) truck was completed some time ago.

The mine-protected engine compartment is at the front with the mine-protected cab to its immediate rear. The cab has seats for the five-man crew including the commander and driver, two entry doors in each side, bulletproof windows to the front and sides and two firing ports each side. The roof hatch can mount a 7.62 mm or 5.56 mm machine gun for local protection.

To the rear of the cab is the main flatbed that extends to the rear of the vehicle. On this is mounted the logistics bin which contains stores, camouflage nets and ammunition.

The twin 23 mm ZU-23-2 LAAG is mounted at the very rear of the flatbed and when travelling, the weapon, ammunition and on board equipment are protected by armoured side panels. When required, these are swung outwards through 90° where they are locked into the horizontal position to form a platform for the crew.

The whole rear unit is known as the interface sub-frame as this integrates the ZU-23-2 LAAG, called the GA6 in South African service, with the vehicle chassis to provide a rigid mounting surface.

Three gun mounts to the interface subsystems mount the ZU-23-2 system on the rear of the vehicle. The weapon can be elevated from –10° to +90° and traversed through 360°. Cyclic rate-of-fire is per cannon is 800-1,000-rds/min. Each 2A14-derived cannon has 50 rounds of ready-use ammunition with a further 1,000 rounds carried in reserve.

Pretoria Metal Pressings (PMP), part of the Denel Group, has developed a new family of 23 × 152 B ammunition for use with the Zumlac and standard 23 mm ZU-23-2 towed anti-aircraft gun systems. The family includes improved High-Explosive Incendiary (HEI) and Armour-Piercing Capped Incendiary

Zumlac twin 23 mm SPAAG in travelling configuration with hull sides raised and weapons covered (Christopher F Foss) 0045024

Zumlac twin 23 mm SPAAG in the firing position (Christopher F Foss) 0525736

Tracer (APCI-T) natures, with a base-fuzed Semi-Armour-Piercing High-Explosive Incendiary (SAPHEI) nature under development.

Maximum effective slant range in the anti-aircraft role is 2,500 m and the system can be ready for action within 30 seconds of the vehicle stopping. If required, the weapon can be removed from the Zumlac and used in the conventional ground role.

Optional equipment includes communications and navigation systems and an integrated air sentry surveillance post.

Although the SANDF Zumlac is mounted on a SAMIL 100 (6 × 6) truck it is possible to use a lighter 4 × 4 truck in which case one vehicle would carry the gun and the other would carry the stores and ammunition.

For users of the ZU-23-2, a system upgrade and a data transmission interface Mk II can be provided. This work can be undertaken in South Africa although normally kits would be provided to enable the user to upgrade systems in their own country.

The inexpensive servo upgrade includes the installation of new azimuth and elevation servo motors and controls electronics and allows the gunner to control the complete weapon using a hand controller.

The normal hand wheel controls are retained as a back up and the standard sight is also retained. The installation of these new servos enables the weapon to engage a target more quickly. To enable the ZU-23-2 to be slaved to an external designation source, for example a fire-control unit, acquisition radar or an optical director (which could include a laser range finder), the Data Transmission Device Mk 11 can be fitted. The guns can be fired remotely from the Data Transmission Device Mk 11 or locally from the 23 mm gun while it is being directed.

The guns can also be linked to a fire-control unit as part of a fire-control system and additional interfaces for training purposes include a mirror firing system and data encryption unit for possible radio link to a remote evaluation system.

Specifications

	Zumlac
Dimensions and weights	
Crew:	5
Length	
overall:	10.273 m
Width	
overall:	2.425 m
Height	
overall:	2.5 m
Ground clearance	
overall:	0.355 m
Wheelbase	5.25 m + 1.38 m
Weight	
combat:	17,000 kg
Payload	
land:	2,000 kg
Mobility	
Configuration	
running gear:	wheeled
layout:	6 × 6
Speed	
max speed:	100 km/h
Range	
main fuel supply:	800 km
Fuel capacity	
main:	500 litres
Angle of approach:	32°
Angle of departure:	27°
Vertical obstacle	
forwards:	0.5 m
Trench:	0.5 m
Turning radius:	11.9 m
Engine	V-10, air cooled, diesel, 268 hp at 2,650 rpm
Gearbox	
type:	manual
forward gears:	6
reverse gears:	1
Transfer box:	2-speed
Steering:	power-assisted, ball and nut
Brakes	
main:	duplex, compressed air
Tyres:	14.00 × 20 PR 18
Suspension:	telescopic hydraulic shock absorbers, leaf springs
Electrical system	
vehicle:	24 V
Batteries:	2 × 12 V, 120 Ah
Firepower	
Armament:	2 × 23 mm 2A14 cannon
Ammunition	
main total:	1,100
Main weapon	
cartridge:	23 mm × 152 mm B
Rate of fire	
cyclic:	2,000 rds/min (est.)
practical:	400 rds/min (est.)
Turret power control	
type:	manual
Main armament traverse	
angle:	360°
Main armament elevation/depression	
armament front:	+90°/-10°
Survivability	
Night vision equipment	
vehicle:	no
NBC system:	no

Status

Production of the system is complete. The Zumlac replaced the Ystervark and 36 are believed to be in service with the South African Army under the name Bosvark, most of which are in use with 10 Air Defence Artillery Regiment at Kimberley. It is very likely that some of these systems will already be broken up and used for spares.

Contractor

Thales, formerly Altech Defence Systems and African Defence Systems Ltd.
Pretoria Metal Pressings (PMP), part of the Denel (Pty) Ltd Group.

Sweden

CV 9040 (CV90AD/AA)

Type

Self-Propelled Anti-Aircraft Gun (SPAAG)

Development

HB Utveckling AB, a company then jointly owned by Bofors Defence (now part of BAE Systems), developed the Combat Vehicle 90 (CV 90 (Stridsfordon 90)) family of fully tracked armoured vehicles for the Swedish Army.

The then Alvis Hägglunds, was responsible for the design of the vehicle chassis and the then Bofors Defence, was responsible for the complete turret with weapons, sensors and fire-control computer.

The first version to enter service with the Swedish Army was the CV 9040 infantry fighting vehicle and the first production example of this was completed in October 1993.

Following a competition in 1994, Norway selected the CV 9030 to meet its future requirements for an infantry fighting vehicle. In 1999, the Swiss Army selected the CV 9030CH to meet its future operational requirements. Late in 2000, Finland selected the CV 9030 to meet its future operational requirements too.

The air defence member of this family is designated as the Lvkv 90 with the first prototype being handed over to the Swedish Defence Matériel Administration (FMV) in December 1991 for extensive trials which were undertaken between 1992 and 1993.

In June 1993, HB Utveckling AB received a production contract for the first batch of Lvkv 90. This is similar to the CV 9040 infantry fighting vehicle but has a slightly modified two man turret armed with a Bofors Defence 40 mm L/70

Production version of Combat Vehicle Anti-Aircraft Vehicle (CV 90 AAV) with muzzle velocity radar over 40 mm barrel 0045018

gun, improved sensors, advanced fire-control computer and search radar which is mounted on the turret rear.

The contract, placed in June 1993, was worth SKR1,300 million for a further four versions of the CV 90:

- The Lvkv 90 anti-aircraft vehicle (also known as the CV 9040 AAV and CV90AD by local Air Defence troops)
- Forward Observation Vehicle (CV 90 FOV)
- Forward Command Vehicle (CV 90 FCV)
- The Armoured Recovery Vehicle (CV 90 ARV) with deliveries running from 1995 onwards.

In October 1993, the then Thales Defence Systems radar was selected and a contract worth over FFR100 million (at the time of writing) was placed for the radars to be delivered from 1995 through to 1999.

While the Swedish Army calls this system the Lvkv 90 anti-aircraft system, it is also known as the Combat Vehicle 90 Anti-Aircraft Vehicle (CV 9040 AAV). As of late 2012, there were no export customers for the CV 9040 AAV.

Description

The vehicle has a six-man crew consisting of the driver in the front, vehicle commander in the turret on the left side and gunner in the turret on the right side.

The remaining three crew members are seated in the rear of the vehicle and consist of the radar operator, air defence combat commander and the plotting operator.

The chassis of the vehicle is virtually identical to other members of the Combat Vehicle 90 family of armoured vehicles.

The hull of the CV 90 AAV is of all-welded steel armour construction with an additional layer of advanced armour for added protection.

The engine compartment is at the front of the hull on the right side with an access plate in the glacis plate. The driver sits at the front on the left with a left opening (on the prototypes the hatch opened upwards to the rear) single-piece hatch cover which has three integral day periscopes for forward observation. A passive periscope for night driving can replace the centre periscope.

The vehicle is powered by a Scania DS 14 diesel engine developing 410 kW (550 hp) coupled to a Perkins Engines Company X-300-5 fully automatic cross-drive transmission with four forward and two reverse gears.

Suspension is of the torsion bar type with each side consisting of seven dual rubber-tyre road wheels with the drive sprocket at the front and idler at the rear.

Turret is of all-welded steel construction and is in the centre of the hull with the vehicle commander and gunner being provided with a single piece hatch cover that opens to the rear. In addition, there is a large door in the hull rear. The CV 9040 infantry fighting vehicle has roof hatches but these are not fitted on the CV 9040 AAV.

The Bofors Defence 40 mm L/70 gun, is an inverted version of the well known anti-aircraft gun equipped with a hydraulically operated 24 round magazine. The latter is divided into three compartments permitting a quick change of ammunition depending on the tactical situation.

Muzzle velocity radar can be mounted above the 40 mm gun and feed information to the fire-control computer.

The gun can fire all types of Bofors 40 mm ammunition with the cyclic rate of fire being 300 rds/min, although single shots or bursts can also be fired. The gun has an elevation arc of –8° to +50° with turret traverse through a full 360°. Turret traverse and weapon elevation is powered, with manual controls being provided for emergency use.

Although the 40 mm L/70 gun can fire all varieties of 40 mm L/70 ammunition, the CV 9040 AAV normally fires only two natures of ammunition. The six-mode programmable all-purpose 3P round for air, soft and lightly-armoured targets and the APFSDS-T round for engagement of heavily-armoured targets. The programming device for the 3P round is included in the production batch of CV 9040 AAV vehicles ordered for the Swedish Army in 1993.

The gunner's sight comprises an optical day channel, a thermal imaging system and a laser range-finder. All sensors have a common optical line-of-sight, via a servo-head mirror, controlled by the fire-control computer.

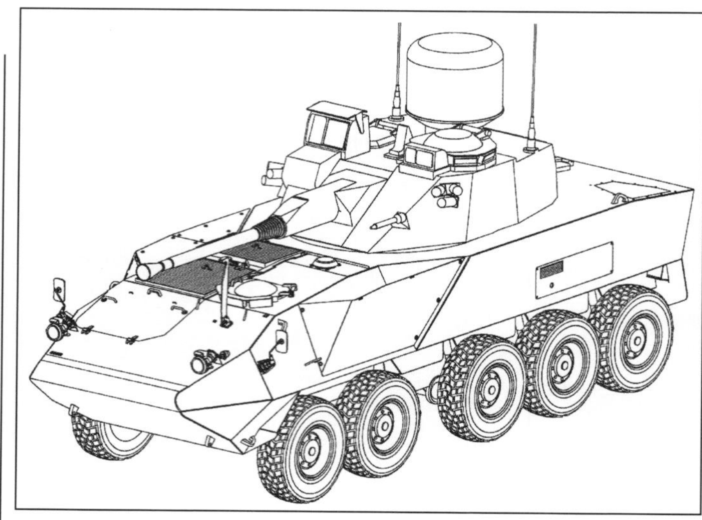

Detailed drawing of Swiss MOWAG Piranha (10 × 10) armoured combat vehicle fitted with CV 9040 AAV turret system 0509853

The day sight, with a magnification of ×8, is provided with a graticule for combating air and ground targets and also includes a reticule that corresponds to the width of the laser beam.

The thermal sight provides two fields of view (wide and narrow) and focuses automatically to optimise surveillance and weapon aiming tasks.

Normally the narrow field of view, which is used when combating aerial targets, provides a magnification of ×9. The thermal sight uses a CCIR standard 625 line display for both commander and gunner.

The laser rangefinder is of the Nd-YAG type and has a pulse repetition rate of 8 Hz and beam width of 1.5 milliradians with maximum range being over 9,000 m. If required, the gunner's sight can be equipped with an automatic tracking unit. The gunner also has a number of standard day observation sights.

The commander's sight is traversable and adjustable in elevation. For quick checks, the commander may use the direct outlook or make more detailed observations via a periscope with a magnification of ×6.

This is equipped with a graticule for firing at ground targets. The commander can command the turret, including the gunner's sight to the bearing of his sight or alternatively fire direct. The commander also has a number of standard day periscopes for all round observation.

The fire-control computer controls the sight-servo and the laying system of the gun/turret while providing automatic threat evaluation of detected targets.

Mounted on either side of the turret firing forwards is a bank of six electrically operated grenade dischargers.

The Thales Defence Systems Gerfaut TRS 2620 search radar can detect aircraft at a range of about 14 km. This permits the positive discrimination of helicopters from aircraft by the signature of its blade, or rotor hub at a range of about 8 km.

The radar antenna is housed in a radome mounted on the rear of the turret with all electronics being mounted inside the vehicle for added protection. Six targets can be tracked simultaneously. An IFF system can be mounted back on the antenna, if required.

Standard equipment includes an NBC system and a fire detection and suppression system.

A typical target engagement takes place as follows. The search and acquisition radar is constantly scanning for targets with the Plan Position Indicator (PPI) display-and-control unit being mounted in the rear of the hull, where the radar operator is positioned.

The PPI is mounted on an adjustable sliding mechanism that can be ergonomically adopted by the operator.

Automatic tracking of a detected target is initiated by marking the target with a joystick. Each target is assigned a number and a symbol indicating the type of target.

The same number with corresponding target data (range, bearing and velocity) is indicated on the PPI which can be set for 15 km or 7.5 km range.

The radar automatically makes a threat evaluation and provides acquisition information to the fire-control computer about the most dangerous target. The commander and radar operator can manually override the threat evaluation.

On command, the computer slew's the turret and gunner's sight to the indicated bearing and searches in elevation, until the gunner detects the target in the sight.

The gunner manually starts tracking and when the laser range-finder has provided the computer with accurate range information, the computer provides tracking support and predicts the intercept point.

The lead angle is given to the 40 mm gun, maintaining the line-of-sight for the gunner. When in range, the gunner opens fire.

Variants

The complete turret of the CV 90 AAV is now also being offered for installation on other vehicles, tracked and wheeled, including the Swiss MOWAG Piranha (10 × 10) armoured combat vehicle.

Successful firing and mobility tests of the standard CV9040 turret were carried out on this 10 × 10 chassis early in 1998.

Specifications

	CV 9040 AAV
Dimensions and weights	
Crew:	6
Length	
overall:	6.55 m
hull:	6.471 m (some sources suggest 6.55 m)
Width	
overall:	3.192 m
Height	
overall:	2.7 m
to turret roof:	2.5 m
hull:	1.73 m
Ground clearance	
overall:	0.45 m
Track width	
normal:	533 mm
Length of track on ground:	3.98 m
Weight	
combat:	23,000 kg
Ground pressure	
standard track:	0.49 kg/cm²
Mobility	
Configuration	
running gear:	tracked
Power-to-weight ratio:	24.1 hp/t
Speed	
max speed:	70 km/h
reverse:	43 km/h
Range	
main fuel supply:	600 km
Fuel capacity	
main:	525 litres
Gradient:	60%
Side slope:	30%
Vertical obstacle	
forwards:	1 m
Trench:	2.4 m
Engine	Scania DS 14, diesel, 550 hp
Gearbox	
model:	Perkins Engines Company X-300-5N
type:	automatic
forward gears:	4
reverse gears:	2
Steering:	crossdrive
Clutch	
type:	torque converter
Suspension:	torsion bar
Electrical system	
vehicle:	24 V
Firepower	
Armament:	1 × 40 mm L/70 cannon
Armament:	1 × coaxial mounted 7.62 mm (0.30) Browning machine gun
Armament:	6 × smoke grenade launcher
Ammunition	
main total:	24 (ready to use)
Main armament traverse	
angle:	360°
Main armament elevation/depression	
armament front:	+50°/-8°
Survivability	
Night vision equipment	
vehicle:	yes
NBC system:	yes

Status
Production of the system is now complete.

Contractor
BAE Systems AB.
BAE Systems, Global Combat Systems.

Retaliator

Type
Developmental Self-Propelled Anti-Aircraft Gun (SPAAG)

Development
In mid 2000, the former Bofors Defence (now BAE Systems, Global Combat Systems) revealed that for some years it had been studying a new modular air defence concept under contract to the Swedish Defence Material Administration (FMV). This system has the commercial name of Retaliator.

Artist's impression of Retaliator air defence system installed on an 8 × 8 armoured chassis shown launching a 120 mm guided round 0067902

Description
Changes in the air-defence threat with an increasing number of smaller targets like the cruise missile, UAV's, RPV's, and so on, have been recognised by Bofors Defence with the development of the Retaliator system.

The Retaliator is based around a rapidly rotating in trajectory projectile with a calibre of about 120 mm and fitted with a smart laser radar (LADAR) sensor and an advanced warhead that would be optimised for air threat. It would also have a secondary capability against ground targets such as light armoured vehicles.

The projectile would be of the fire-and-forget type with the sensor being activated towards the end of its flight to discriminate, classify and determine the position of the target before the warhead is released. Once the sensor detects the target, the smart munition is released engaging the target at stand-off ranges between 50 and 100 m.

According to Bofors, the 120 mm missile type fined projectile has a stand-off capacity with an effective area 10 times that of standard proximity air defence ammunition.

To ensure a high kill, it is probable that several 120 mm projectiles would be fired at one target and these would possibly be in a pattern for maximum target effectiveness.

The provisional artist's impression shows a twin multibarrel launcher mounted on a pedestal which can be rapidly reloaded under armour protection. Precise details have not been released but the system appears to combine features of a gun and missile system.

In addition to being installed on tracked and wheeled land platforms, Retaliator could also be installed on surface craft in a naval application.

Bofors Defence demonstrated the principle of the Retaliator in 2003/4 and in 2005 showed the function, using a round in trajectory, equipped with a smart sensor and focusing warhead. Some subsystem hardware has already been completed.

In recent years, Bofors has established a considerable amount of experience in the design, development and production of advanced munitions including the 40 mm and 57 mm Pre-fragmented Programmable Proximity (3P) fuzed and the 155 mm Bonus smart artillery projectile developed with the then Giat Industries (now Nexter) of France.

Status
Uncertain, but believed to be still in development. The Retaliator concept is a system designed to meet the threat of precision guided weapons. As of December 2013, no further information has become available on this system.

Contractor
The then Bofors Defence AB, now BAE Systems, Global Combat Systems.

Turkey

Korkut

Type
Self-Propelled Anti-Aircraft Gun (SPAAG)

Development
The primary contractor Turkish Defence and Electronics Company Aselsan has been working since the first half of 2010 on the new self-propelled air defence system Korkut with other Turkish organisations FNSS who provide the amphibious tracked engineering vehicle chassis and MKEK (sometimes referred to as MKA) who provided the twin 35 mm guns.

The system is currently in the qualification phase and is expected to complete this by late 2014. Customer qualification is not expected until at least 2015. The Primary Contractor Aselsan is under contract to deliver two Korkut 35 mm Gun Systems and a Command and Control System to the Turkish Undersecretariat for Defence Industries (SSM).

Aselsan displayed a prototype 35 mm gun-based very short range air defence system at IDEF 2013 (IHS/Nick de Larrinaga) 1513279

Description

The Korkut Very Short-Range Air Defence (VSHORAD) system consists of:

- 3 × self-propelled gun vehicles
- 1 × command-and-control vehicle.

The turret itself is unmanned but, the chassis has three crew members:

- Pilot
- Gunner
- Commander.

Aselsan sources have reported that the twin 35 mm cannon vehicles are equipped with 3-D Ku-band fire control radar that has an elevation coverage of –5° to +85° and an instrumentation range of 30 km. Also mounted on the chassis is a thermal imaging system with a day TV camera.

The command-and-control vehicle features a 3-D I-band mobile search radar with an elevation coverage of -5° to +70° and an instrumented range in excess of 70 km. The communications link between the C2 chassis and the cannon vehicles is provided by a VHF wide-band radio, whilst each of the vehicles also is capable of interfacing with the Turkish C4I system.

The 35 mm cannon are based on the Rheinmetall Air Defence KBD weapons and are capable of a rate of fire of 1,100 rounds per minute. The cannon also have the capability of using the 35 mm Air Burst Munition (ABM) currently in development by Aselsan.

Specifications

Korkut cannon chassis

Crew:	3
Combat weight:	30,000 kg for cannon chassis 25,500 kg for C2 vehicle
Power-to-weight ratio:	20 hp/t
Ground pressure:	Not available
Length:	7.07 m
Width:	3.9 m
Height:	3.62 m
Ground clearance:	0.39 m
Length of track on ground:	Not available
Maximum road speed:	65 km/h
Maximum water speed:	6 km/h
Fuel capacity:	Not available
Maximum road range:	500 km
Fording:	1.0 m
Gradient:	60%
Vertical obstacle:	0.9 m
Trench:	2.0 m
Electrical system:	24 V
Armament:	2 × 35 mm cannon with an optional extra of Igla type MANPADs
Ammunition:	Air Burst Munition (ABM)

Status

The system is currently in the qualification phase and is expected to complete by late 2014. Customer qualification is not expected until at least 2015.

Contractor

Aselsan, Turkey.
FNSS, Turkey.
MKEK, Turkey.

M113 M55 AT2S

Type

Self-Propelled Anti-Aircraft Gun (SPAAG)

Development

AT2S ZPT 12.7 mm self-propelled anti-aircraft gun system. In the 1990s, the Turkish Land Forces Command facility at Kayseri modified a United Defense M113 series armoured personnel carrier into the air defence role.

Following trials with prototype systems, it was accepted by the Turkish Land Forces Command and designated as the M113 M55 AT2S. The system entered service in 1999 with the conversion work being carried out at the facilities of the Turkish Land Forces Command.

Description

The M113AT2S ZPT system consists of a United Defense LP M113 series fully tracked armoured personnel carrier, used in large numbers by the Turkish Land Forces, upgraded to the latest M113AT2S standard which includes the replacement of the standard petrol engine by a more fuel efficient Detroit Diesel, diesel engine.

Mounted on the roof of the vehicle, to the rear of the driver's position, is the upper part of the United States 12.7 mm (quad) anti-aircraft machine gun M55 which was developed during the early part of the Second World War. Full details of this system are given in the Towed anti-aircraft gun section.

The M55 system is armed with four .50 (12.7 mm) M2 HB machine guns, two each side, with each being provided with 210 rounds of ready-to-use ammunition. Additional ammunition is carried in reserve in the hull. Four-ammunition cans each holding 210 rounds of 12.7 mm ammunition are also carried externally on either side of the hull towards the rear. A bank of eight electrically operated smoke grenade launchers is mounted on the forward part of the hull.

Turret traverse and weapon elevation is powered with manual controls for emergency use. This is a clear-weather system only.

The M113AT2S is fitted with a NBC system and night vision equipment and is fully amphibious while being propelled in the water by its tracks.

Muzzle Velocity Radar (MVR)

The 7941 MVR System is designed to measure a wide variety of field, anti-aircraft and naval artillery shells and provides accurate muzzle velocity data to enhance first round hit probability.

Reports have suggested that the gun types using the 7941 are all standard weapons. Functionally, the system measures muzzle velocity and calibrates its host weapon by measuring the velocities of multiple rounds. The 7941 MVR operates in the old X-band, now I-band (8-12.5 GHz) and covers the muzzle velocity range of 50 to 2,000 m/s.

Status

The M113 M55 AT2S system production is now complete. As far as is known, the M113 M55 AT2S is still in service with the Turkish Land Forces Command.

Contractor

Turkish government facilities.
Turkish Land Forces Command facility at Kayseri.

Turkish M113 M55 A2TS ZPT 12.7 mm self-propelled anti-aircraft gun system 0045032

STAbilised Machine gun Platform (STAMP)

Type
Self-Propelled Anti-Aircraft Gun (SPAAG)

Development
Aselsan has developed a remote-controlled stabilised machine gun known as STAMP for defending against asymmetric threats. The system is envisaged for use on land and sea-based platforms including coastal and border defence. Currently, the system is deployed with the Turkish Coast Guard MRTP33 Fast Patrol/Attack Boats.

Description
STAMP utilises a stabilised turret that enables the line-of-sight of the gun to be aimed at the target at all times. The system is reported capable of engaging stationary or moving targets while the platform is on the move. For land use, STAMP can be mounted on both tracked and wheeled vehicles and used for close-in air-defence roles of static sites and the protection of troops while on the move. The modular structure enables easy installation on various platforms, including naval, land vehicles or for use at stationary control and surveillance locations while serving the purpose of asymmetric warfare, coastal and land border defence and protection of strategic facilities. The system has the following sub-units:

- Turret
- Gun control unit
- Fire-control computer
- Commander control unit
- Thermal camera
- TV camera
- Laser range-finder (optional)
- Gun re-cocking unit (optional)
- Target tracker (optional)
- Video recording unit (optional)
- Machine gun interface unit (optional)
- Batteries and charging unit (optional).

STAMP has a stabilised turret that enables the line-of-sight of the gun to be aimed at the target at all times. Due to the stabilizing feature, the system can perform precise firings against stationary or moving targets whist the platform is on-the-move.

The system can be operated remotely by using the remote gun control unit and hence, it enables gunner protection against counter fire. Additionally, the system has features of defining the firing zones both in azimuth and elevation

Stamp on armoured 4 × 4 vehicle (Aselsan) 1181651

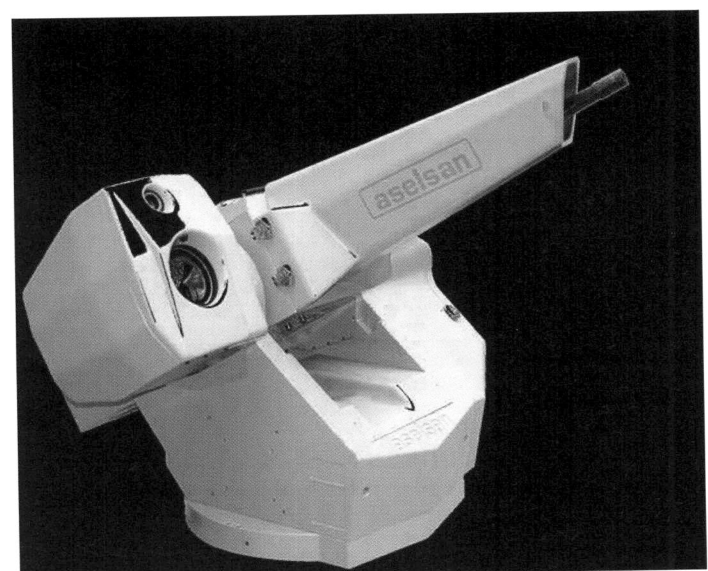

Stamp mounting (Aselsan) 1181647

Stamp on tracked chassis (Aselsan) 1181652

and automatic charging of the gun remotely. The Remote Gun Control Unit (RGCU) has an ergonomic design with a ruggedised LCD monitor that displays reticle, target range scale, turret azimuth and elevation axis position in numerical and graphical form, Field-Of-View (FOV) of the sight system, number of remaining rounds, system operation mode, fault warning symbols and other information for the gunner.

In July 2012, Aselsan revealed a new and further advanced version of the STAMP utilising a General Dynamics GAU-19 Gatling gun. The GAU-19 fires the NATO standard .50 calibre BMG rounds with a firing rate of between 1,000 and 2,000 rounds per minute, depending on the target size and distance.

Muzzle Velocity Radar (MVR)
The 7941 MVR System is designed to measure a wide variety of field, anti-aircraft and naval artillery shells while providing accurate muzzle velocity data to enhance first round hit probability.

Reports have suggested that the gun types using the 7941 are all standard weapons. Functionally, the system measures muzzle velocity and calibrates its host weapon by measuring velocities of multiple rounds. The 7941 MVR operates in the old X-band (8-12.5 GHz) and covers the (muzzle velocity) range of 50 to 2,000 m/s.

Variants
Naval variant in use with Turkish Coast Guard on Patrol Boats.

Specifications
STAMP

Type:	remote-controlled stabilised machine gun
Weight:	220-225 kg with ammunition
Gun type:	12.7 mm MG with an option of 7.62 mm MG or 40 mm automatic grenade launcher
Elevation angle:	–15° to +55°
Traverse angle:	*n* × 360°
Maximum elevation speed:	60°/s
Maximum traverse speed:	60°/s
Ready-to-fire rounds (12.7 mm):	100 standard, 200 optional
Power supply:	24 V DC

Status
The STAMP land system is believed to be in development at the performance and firing trials stage.

Contractor
Aselsan A.S.

United States

M42

Type
Self-Propelled Anti-Aircraft Gun (SPAAG)

Development
In August 1951, authorisation was given to design, develop and build prototypes of a twin 40 mm self-propelled anti-aircraft gun, designated the T141 interim vehicle, a twin 40 mm self-propelled anti-aircraft gun, designated the T141E1 ultimate vehicle and a carrier fire-control vehicle, designated the

M42 twin 40 mm SPAAGs showing ready-use ammunition stowed on turret rear (US Army) 0509428

T53. In May 1952, the T141E1 and its associated T53 fire-control vehicle were cancelled. The Cadillac Motor Car Division of the General Motors Corporation designed the T141 with the first prototype being completed late in 1951.

The T141 had a very short development period as it was based on components of the M41 light tank which was also being produced by the Cadillac Motor Car Division of the General Motors Corporation at the Cleveland Tank Plant and used the same turret with twin 40 mm guns as the earlier M19A1.

The T141 was standardised as the Gun, Twin 40 mm Self-propelled M42, in October 1953 after the vehicle had already been in production for more than a year.

The M42 was in production at Cleveland Tank Plant from late 1951 to June 1956 and at ACF Industries Incorporated of Berwick, Pennsylvania, from early 1952 to December 1953. Production of the M42 amounted to 3,700 units.

To improve the fuel economy and increase operational range, a fuel injection system was designed for the AOS-895-3 engine. The fuel injection model became known as the AOSI-895-5 and trials showed fuel savings of 20 per cent. In February 1956, the basic M42 was reclassified as limited standard and the M42A1 as standard. Most M42s were subsequently brought up to M42A1 standard.

The M42 was replaced in front-line service with the US Army by the M163 20 mm Vulcan Air Defence System from 1969, but it remained in service with the US Army National Guard until 1990-91 when it was finally phased out of service. In US Army service the M42 SPAAG was never upgraded.

Description

The all-welded steel hull of the M42 is divided into three compartments: drivers and commanders at the front, turret in the centre and power pack at the rear.

The driver is seated at the front of the hull on the left with the commander/radio operator to the right. Both have a single-piece hatch cover that opens to the outside of the vehicle, with a single M13 day periscope, which can be traversed through 360°. The other four-crew members are seated in the open-topped turret in the centre of the vehicle.

The engine compartment at the rear of the hull is separated from the fighting compartment by a fireproof bulkhead and has a fire extinguisher operated by the driver. The engine is mounted towards the front of the engine compartment and the transmission at the rear.

The torsion bar suspension consists of five dual rubber tyre road wheels with the idler at the front, drive sprocket at the rear and three track return rollers. The first, second and fifth road wheel stations have a hydraulic shock absorber. The steel tracks have replaceable rubber pads.

The M42 has no amphibious capability or NBC system but most vehicles were originally fitted with infra-red driving lights.

Main armament comprises twin 40 mm cannon M2A1 in a turret with a traverse of 360°. The weapons have hydraulic elevation from –3° to +85° and manual operation from –5° to +85°. The weapon is recoil operated, having a vertical sliding breechblock. The gunner can select either single shots or full automatic.

Early vehicles had flash hiders but they were replaced on later production vehicles by flash suppressers, which were subsequently retrofitted to the earlier vehicles. Practical rate of fire is 120 rds/barrel/min, maximum anti-aircraft range is 5,000 m and maximum ground-to-ground range is 9,475 m. The following types of US fixed 40 mm L/60 (40 × 311 R) ammunition can be fired:

- AP-T (M81-series) with the complete round weighing 2.077 kg, muzzle velocity of 872 m/s;
- HE-T (Mk 2) with the complete round weighing 2.15 kg, muzzle velocity of 880 m/s;
- TP-T (M91) with the complete round weighing 2.14 kg, muzzle velocity of 872 m/s.

Of 480 rounds of 40 mm ammunition carried, most are stored in the ammunition containers along the tops of the track guard either side of the turret. Three sighting devices are incorporated into the fire-control system of the M42: computing sight M38, reflex sight M24C and the speed ring sight. The computing sight M38 is designed to provide an effective means of controlling fire of the 40 mm cannon against both air and ground targets. The reflex sight M24C is designed to superimpose a graticule pattern in the gunner's line of sight and is used in conjunction with the computing sight M38 during power operation. The speed ring sight is used during manual operation if a power failure or local control system malfunction occurs.

M42 twin 40 mm SPAAG used by Taiwan with locally produced 32-round magazine above 40 mm weapon (DTM) 0509426

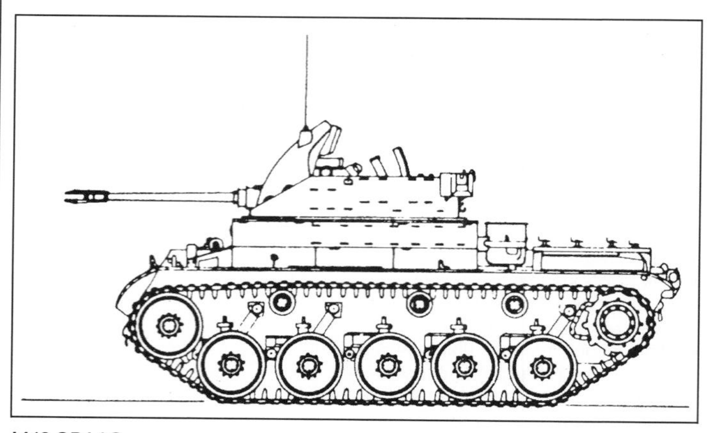

M42 SPAAG 0509427

Mounted on the left rear of the turret is a 7.62 mm (0.30 inch) M1919A4 machine gun, which was replaced by an M60 MG in the US Army. The machine gun has a traverse of 360°, elevation of +76° to the front and an elevation of +60° at the rear.

Variants

In 1982, an M42 was fitted with the same NAPCO power pack as the M41 light tank but with a modified cooling system. The vehicle was also fitted with a then Cadillac Gage Textron weapon control system to improve target tracking. Firing trials were undertaken at the US Army Air Defence School at Fort Bliss, Texas. During these trials, the upgraded vehicle successfully shot down one of the target drones engaging it. By the year 2000, no country had adopted this improvement package.

Taiwan is known to have refitted several of the M42A2s in service with its army as Raytheon Electronic Systems TOW anti-tank guided weapons vehicles.

Taiwan has produced a 32-round magazine for the 40 mm gun of the M42, which is mounted above the weapon.

Specifications

	M42
Dimensions and weights	
Crew:	6
Length	
overall:	6.356 m
hull:	5.819 m
Width	
overall:	3.225 m
Height	
overall:	2.847 m
Ground clearance	
overall:	0.438 m
Track	
vehicle:	2.602 m
Track width	
normal:	533 mm
Weight	
standard:	20,094 kg
combat:	22,452 kg
Ground pressure	
standard track:	0.65 kg/cm^2
Mobility	
Configuration	
running gear:	tracked
Power-to-weight ratio:	22.26 hp/t
Speed	
max speed:	72.4 km/h

	M42
Range	
main fuel supply:	161 km
Fuel consumption	
road:	3.29 litres/km
Fuel capacity	
main:	530 litres
Fording	
without preparation:	1.016 m
Gradient:	60%
Side slope:	30%
Vertical obstacle	
forwards:	0.711 m
Trench:	1.828 m
Engine	Continental AOS-895-3, 6 cylinders, supercharged, air cooled, petrol, 500 hp at 2,800 rpm
Engine	Lycoming AOS-895-3, 6 cylinders, supercharged, air cooled, petrol, 500 hp at 2,800 rpm
Auxiliary engine:	GMC A41-1 or A41-2
Gearbox	
model:	General Motors Allison Division crossdrive model CD-500-3
forward gears:	1
reverse gears:	1
Suspension:	torsion bar
Electrical system	
vehicle:	24 V
Batteries:	4 × 12 V
Firepower	
Armament:	2 × 40 mm M2A1 cannon
Armament:	1 × 7.62 mm (0.30) M1919A4 machine gun
Ammunition	
main total:	480
secondary total:	1,750
Main weapon	
cartridge:	40 mm × 311 mm R
Turret power control	
type:	hydraulic/manual
by commander:	no
by gunner:	yes
Main armament traverse	
angle:	360°
speed:	40°/s
Main armament elevation/depression	
armament front:	+87°/-5° (manual)
speed:	25°/s
Survivability	
Night vision equipment	
driver:	yes
NBC system:	no
Armour	
hull/body:	12.7 mm (maximum)
turret:	15.87 mm (maximum)

M163A1 20 mm SPAAG fitted with Product Improved Vulcan Air Defense System (PIVADS) 0509429

M163 20 mm SPAAG system (Michael Jerchel) 0069183

Status

Production complete.

As of October 2010, no further data is available on this system. If no further information on this system is available during 2014 it will be removed from the files.

Contractor

Cadillac Motor Car Division of General Motors Corporation, Cleveland Tank Plant.

ACF Industries Incorporated.

Neither of these companies are now involved in armoured vehicle/self-propelled anti-aircraft gun activities.

M163 Vulcan

Type

Self-Propelled Anti-Aircraft Gun (SPAAG)

Development

Development of the 20 mm Vulcan Air Defence System began under the direction of the US Army Weapons Command at Rock Island Arsenal, Illinois, in 1964. Two versions of the Vulcan were subsequently developed, a self-propelled model called the M163 (development designation XM163) and a towed model called the M167 (development designation XM167). The prime contractor for both models is the now General Dynamics Armament and Technical Products of Burlington, Vermont. At that time, the company was part of the General Electric (GE) Corporation. The company ceased marketing of the 20 mm Vulcan series of towed and self-propelled anti-aircraft guns some years ago.

After trials carried out by the US Army Air Defence Board at Fort Bliss, Texas and at Aberdeen Proving Ground, Maryland, in 1965, the system was accepted for service in that year as the replacement for the twin 40 mm M42 self-propelled anti-aircraft gun.

First production M163s were delivered to the US Army in August 1968 and final deliveries were made in 1970. Since then the system has been placed back in production for export during 1975 to 1979 and briefly again in 1982 before finally closing down.

Israel used the system during the 1982 invasion of Lebanon, when it destroyed several Syrian aircraft, including one Sukhoi Su-7 fighter-bomber and several Gazelle HOT Anti-Tank Guided Weapons (ATGW) helicopters. The Israelis have also developed the M163 for use in the ground defence role in support of the infantry.

By early 1987, 671 M163 systems had been built, of which 601 were for the US Army and 70 for export. However, a considerable number of US Army vehicles have been diverted to foreign users.

The M163 system was used in the ground fire support role during Operation Desert Storm in 1991 but by 1994 it had been phased out of service with the US Army without a direct replacement being fielded.

20 mm M163A1 SPAAG from the rear showing extensive external stowage (C R Zwart) 0509431

Reasons cited for the withdrawal included high costs of operating and maintaining the system, limited effectiveness of its 20 mm Vulcan cannon, lack of armour and lack of cross-country mobility with the systems that it is designed to protect.

Description

The M163 basically consists of an M113A1 Armoured Personnel Carrier (APC) fitted with a one-person, electrically driven turret which has a 20 mm Vulcan cannon, Navy Mk 20 Mod 0 gyro lead computing sight and an EMTECH AN/VPS-2 range-only radar mounted on the right side of the turret. The chassis, which is designated as M741, differs from the M113A1's in minor details only, including:

- The provision of a suspension lockout system to provide a more stable firing platform when the weapon is being fired;
- The installation of buoyant pods on either side of the hull;
- A buoyant trim vane at the front of the hull to improve its amphibious characteristics and an additional circular hatch cover in the hull roof on the right side.

The 20 mm six-barrelled M168 cannon is a development of the weapon originally developed for aircraft and has two rates of fire, 1,000 and 3,000 rds/min. The original version fired up to 6,000 rds/min. The 1,000 rds/min rate is normally used in the ground role and the 3,000 rds/min rate in anti aircraft defence:

- Maximum effective anti-aircraft range is 1,200 m;
- Ground range 3,000 m;
- Indirect fire range 4,500 m;
- The kill probability per engagement of the basic system is quoted as 0.35 against targets with velocities between 0 and 450 kts.

The weapon has an elevation of +80°, depression of –5° and 360° turret traverse. The gunner can select 10-, 30-, 60- or 100-round bursts. The dispersion pattern can be increased by fitting a special muzzle adapter, which causes the pattern to be spread, resulting in an increased hit probability. The link-less ammunition feed system in the M163 carries 1,180 rounds ready to fire and additional 1,100 rounds in reserve inside the hull. Rows of 20 mm ammunition are held in the feed drum in lateral tracks. A helix mounted in the centre of the drum, moves 20 mm ammunition along the tracks to the exit port of the drum. The 20 mm ammunition is carried from the feed drum to the gun feeder through conveyor shutting. Solid-state, rate servo-amplifiers, one for elevation and two for azimuth, control the turret drive. These servo-amplifiers are interchangeable in function, as are the drive motors they control. Power for the system is provided by three 24 V nickel/cadmium batteries, two of which drive the 20 mm M168 cannon and the third drives the turret.

The batteries are charged either by the vehicle generator or an Auxiliary Power Unit (APU). Fire-control for the system consists of:

- A gyro lead computing gun sight;
- A sight current generator;
- An EMTECH AN/VPS-2 fully coherent I-band pulse Doppler range-only radar with target track range limits of 250 to 5,000 m.

The gunner visually acquires and tracks the target with the gyro lead computing gun sight. The antenna axis of the radar is servo-controlled to the optical line of sight. The radar supplies target range and range rate data to the sight generator. With range, range rate and angular tracking of the optical line of sight (measured by a freely gimballed gyro), the sight automatically computes the future target position and adds the required super elevation to hit the target. The lead angle is equal to the angular rate of the target multiplied by the time of flight of the projectile to the future target position. Turret fire-control is a disturbed line of sight system. The sight case and gun bore is physically fixed in alignment. The sight graticule, which defines the optical line of sight, is positioned by the gyro and is displaced from the gun bore as the gunner tracks the target, thereby establishing the proper lead angle. The amount of optical line of sight displacement is dependent on the range and range rate inputs to the sight. The required tracking time to establish the lead angle is about one second. A green light appears in the sight optics, signalling that the radar has acquired the target and that the target is within the effective range of the turret system. In the manual mode the gunner must estimate target range and speed and set the estimates on indicator dials on the control panel. The gyro lead computing gunsight then computes the lead angle based on these estimates.

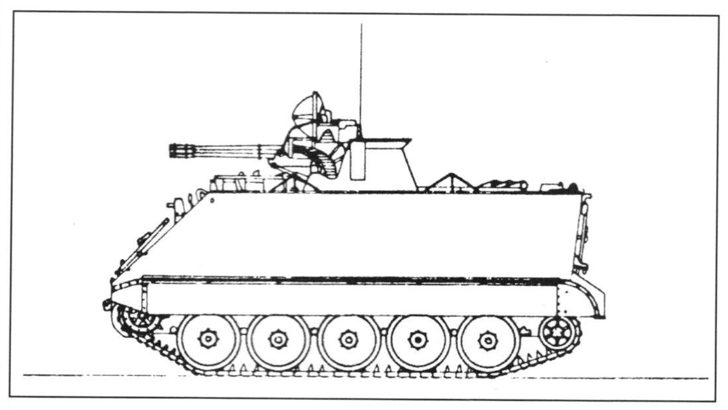

M163 20 mm SPAAG 0509430

The 20 mm M168 cannon fires the 20 × 102 ammunition series (reload time is five minutes) and can fire the following types of fixed ammunition:

Designation	M53	M54	M55A2	M56A3	M220	M242
Nature:	Armour-Piercing Tracer (AP-T)	High Pressure Test (HPT)	Target Practice (TP)	High-Explosive Incendiary (HEI)	Target Practice Tracer (TP-T)	High-Explosive Incendiary Tracer (HEI-T)
Weight:						
(round)	258 g	285 g	255 g	257 g	255 g	245 g
(projectile)	100 g	127 g	98 g	103 g	97 g	94 g
Muzzle velocity:	1,030 m/s	n/avail	1,030 m/s	1,030 m/s	1,030 m/s	1,030 m/s

Variants

Product Improved Vulcan Air Defence System (PIVADS)

The now General Dynamics Armament and Technical Products has modified the fire-control system of the current Vulcan air-defence system family and incorporated the ability to fire the Mk 149 Armour-Piercing Discarding Sabot (APDS) round, which increases the effective range of the system to 2,600 m. The fire-control improvements include replacing the disturbed line of sight currently used with a director sight to give a rate-aided tracking capability and linking a digital fire-control computer to the range-only radar for more accurate lead and superelevation commands to the cannon. The net result of the programme is to increase effectiveness while greatly simplifying the operation. The PIVADS programme is available in the form of modification kits for both the towed and self-propelled Vulcan air-defence systems. In September 1982, the US Army awarded the then Lockheed Electronics a contract for 285 kits for both the towed and self-propelled versions of the 20 mm Vulcan air defence systems, with first deliveries made in June 1984. These were installed on towed and self-propelled Vulcan air-defence systems, with final deliveries in 1988.

Israeli Upgraded Vulcan

A number of Israeli M163 Vulcan systems have been upgraded in key areas, including installation of a pod of four Stinger SAMs. This pod is the same as that fitted to the Avenger (4 × 4) and details of this system, called the Machbet, are given in a separate entry.

ROK 20 mm Vulcan

The Republic of Korea (RoK) has for some years deployed a similar system based on the locally developed Korean Infantry Fighting Vehicle (KIFV). Prime contractor for the complete system is Daewoo Defence Products and full details are given in a separate entry.

ROK 20 mm Vulcan Viper upgrade

The Republic of Korea (RoK) has upgraded its towed and self-propelled 20 mm Vulcan air-defence systems to the enhanced Viper standard. The system configuration has been simplified and stabilised through digitisation, combined numbers of units, reduced cables and direct sight for improved performance, easy operation and maintenance. The local company NEX1 Future Co Ltd developed this upgraded package and details are given in a separate entry.

Specifications

	M163
Dimensions and weights	
Crew:	4
Length	
overall:	4.86 m
Width	
overall:	2.85 m
transport configuration:	2.54 m
Height	
overall:	2.736 m
hull:	1.83 m
Ground clearance	
overall:	0.406 m

	M163
Track	
vehicle:	2.159 m
Track width	
normal:	381 mm
Length of track on ground:	2.667 m
Weight	
combat:	12,310 kg
Ground pressure	
standard track:	0.61 kg/cm^2
Mobility	
Configuration	
running gear:	tracked
Power-to-weight ratio:	17.47 hp/t
Speed	
max speed:	67.59 km/h
water:	5.6 km/h
Range	
main fuel supply:	483 km
Fuel capacity	
main:	360 litres
Amphibious:	yes
Gradient:	60%
Side slope:	30%
Vertical obstacle	
forwards:	0.61 m
Trench:	1.68 m
Engine	Detroit Diesel model 6V-53, 6 cylinders, water cooled, diesel, 215 hp at 2,800 rpm
Gearbox	
model:	Allison TX-100
forward gears:	6
reverse gears:	2
Clutch	
type:	torque converter
Suspension:	torsion bar
Electrical system	
vehicle:	24 V
Firepower	
Armament:	1 × turret mounted 20 mm M168 cannon
Ammunition	
main total:	2,280
Main weapon	
cartridge:	20 mm × 102 mm
Turret power control	
type:	electric/manual
by commander:	no
by gunner:	yes
Main armament traverse	
angle:	360°
speed:	60°/s
Main armament elevation/depression	
armament front:	+80°/-5°
speed:	45°/s
Survivability	
Night vision equipment	
driver:	yes
NBC system:	yes
Armour	
hull/body:	38 mm (maximum)

Status
Production complete. The system was replaced in the US Army service by the M1097 Avenger and the M6 Linebacker before finally going out of commission in 1994.

Contractor
General Dynamics Armament and Technical Products.

Vulcan-Commando

Type
Self-Propelled Anti-Aircraft Gun (SPAAG)

Development
Development of the Vulcan-Commando air defence system was conceived to meet the operational requirements of the Saudi Arabian National Guard (SANG) who are the only users of this system. The 20 mm Vulcan turret used is designed as a 'drop-in' installation, requiring only correct location of mounting holes in the Cadillac Gage (now Textron Marine & Land Systems) LAV-150 vehicle hull, while remaining operationally independent of the vehicle.

Vulcan-Commando SPAAG with turret traversed to left and 20 mm cannon firing 0509432

These LAV-150 vehicles are now being replaced by 1,117 Diesel Division, General Dynamics Land Systems - Canada, LAV (8 × 8) vehicles in numerous models. As of late 2000, SANG had not placed an order for the LAV-AD version, covered in detail in the Self-propelled Anti-aircraft Guns/Surface-to-air Missile Systems section.

Description
The Vulcan-Commando system uses the same turret, weapon and associated feed, power and range only radar equipment as the M163 Vulcan air defence system. The LAV-150 vehicle is fitted with three hydraulic stabilising jacks controlled from inside the vehicle. The system has four operational modes:

(1) **Radar** - the most accurate of the three anti-aircraft engagement modes. The radar supplies continuous range and range rate information to the system's analogue computer so that computations for the gyro lead sight can be made

(2) **Manual** - the gunner estimates the engagement range and target speed and then manually enters them on the control panel

(3) **External** - a second person, off-mount, estimates the target's range and enters the data on a hand-held potentiometer connected to the fire-control system by a cable

(4) **Ground** - the gyro sight is not operated and the lead angle not computed. The sight is mechanically caged at zero lead angles and 7 mils of super elevation.

The vehicle has a crew of four: driver, gunner, commander and radio operator.

Specifications

	Vulcan-Commando
Dimensions and weights	
Crew:	4
Length	
overall:	5.689 m
Width	
overall:	2.26 m
Height	
hull:	1.981 m
Wheelbase	2.667 m
Weight	
combat:	10,206 kg
Mobility	
Configuration	
running gear:	wheeled
layout:	4 × 4
Speed	
max speed:	89 km/h
water:	5 km/h
Amphibious:	yes
Gradient:	60%
Side slope:	30%
Vertical obstacle	
forwards:	0.914 m
Turning radius:	8.382 m
Engine	V-8, 202 hp at 3,300 rpm
Gearbox	
type:	automatic
Firepower	
Armament:	1 × 20 mm M168 cannon

	Vulcan-Commando
Ammunition	
main total:	1,300
Main weapon	
cartridge:	20 mm × 102 mm
Survivability	
Night vision equipment	
vehicle:	no
NBC system:	no

Status

Production complete. No longer marketed. A total of 20 systems were delivered to the Saudi Arabian National Guard (SANG). No further information is available on this system, therefore this file will be removed from the publication in next years edition.

Contractor

General Dynamics Armament and Technical Products (gun).
Textron Marine & Land Systems (chassis).

Anti-Aircraft Guns/Surface-To-Air Missile Systems

Belarus

A3

Type
Self-Propelled Anti-Aircraft Gun and Missile (SPAAGM) system

Development
Tetraedr, a Belarusian company is developing a new multipurpose towed wheeled gun-missile system designated A3 [Anti-air, Anti-terrorism and Anti-armour] for an undisclosed Middle Eastern customer. The system is expected to come in at least three variants:
- Towed wheeled chassis
- Tracked vehicle
- Naval warships.

The system will make use of the Igla man-portable surface-to-air missile, the Konkurs and/or Shturm anti-armour missiles and a GSh-6-23 six barrel 23 mm cannon. Prototypes of the command post and unmanned combat module have been assembled, although development has not yet been completed. Combat algorithm software was developed during early 2009, which resulted in initial trials of the system in the fourth quarter of 2009.

Description
The system will consist of a single command post and up to six fire units that have been described by the developer as 'Combat Modules'. The A3 is intended to defend high-priority fixed targets such as command posts, airports and nuclear power stations from attacks mounted by fixed and rotary wing aircraft, UAVs, missiles and guided bombs. The system can also engage ground targets such as Main Battle Tanks (MBT), Infantry Fighting Vehicles (IFV) and Armoured Personnel Carriers (APC) with either or the Konkurs and Shturm Anti-Tank Guided Missiles (ATGMs).

The towed wheeled version will be mounted on an unmanned remotely controlled installation that is itself mounted on a two-axle trailer produced by Minskiy Traktorniy Zavod (MTZ). The command post for the wheeled version has been visually identified on a Russian-produced Kamaz 43118 6 × 6 truck.

The tracked variant is based on a modified GM-352M1E chassis originally developed for the Russian Pantsyr-S1E. The command post will be manned by a commander/operator and driver, it is expected to be equipped with:
- Passive Electro-Optic (EO) system
- Target-tracking system
- Communications sub-systems
- Automated work stations
- Test facilities
- Built-in crew trainer
- Power-generation system.

Communications
Between the Command Post (CP) and combat modules will be via radio or wire links. The system is expected to use the Asterix protocol thereby enabling it to connect to other Asterix compatible surveillance sensors and weapon systems.

Command Post
The CP has work stations for the commander and two operators. The commander will have information regarding the tactical situation and the location of each of the combat modules, while each operator will be able to control up three combat modules.

Variants
A towed variant is known to be in development. A naval variant has also been proposed.

Computer generated firing unit A3 (Tetraedr) 1331046

Computer generated command post for A3 system (Tetraedr) 1331045

Specifications

Maximum target detection range:	20 km
Maximum range:	
(missile)	5 km
(gun)	1.5 km
Maximum altitude:	
(missile)	5 km
(gun)	1.5 km
Maximum target velocity:	900 m/s

Status
At the prototype stage of development in January 2009 but has progressed considerably through to 2013 where at least several hardware components were in trials.

Contractor
Tetraedr.

China

FB-6A

Type
Self-Propelled Anti-Aircraft Gun and Missile (SPAAGM) system

Development
Poly Technologies of China in association with Aerospace Long March International have developed a mobile air-defence system, the FB-6A Missile Launching Vehicle (MLV). The system was unveiled to the public for the first time in 2005. The FB-6A MLV was developed as a high cross-country-manoeuvrability, Self-Propelled, Anti-Aircraft Gun and Missile (SPAAGM) system.

The FB-6A MLV is almost identical to the US Boeing Avenger air defence system in appearance and, like the Avenger, is based on an non-armoured 4 × 4 cross-country chassis that bears a striking resemblance to the US HMMWV (Humvee). China has in the past been supplied with the HMMWV for civilian applications.

Description
FB-6A vehicle
The FB-6A system consists of missile launching vehicles, command vehicle, planar array radar for search and engagement and other support vehicles. It is a short-range air defence weapon with integrated capability of information, command-and-control, fire-control and launching systems.

The FB-6A is equipped with:
- Electro-optic detection assembly
- Servo system
- Locating and direction finding system
- Communication system
- Fire-control system for both missiles and machine gun.

Mounted on the rear of the FB-6A vehicle is a one-person power-operated turret, either side of which is a pod of four Surface-to-Air Missiles (SAMs) in the ready-to-launch position. The MLV is also equipped with a 12.7 mm machine

FB-6A Missile Launching Vehicle (MLV) (Poly Technologies Inc) 1116070

gun for local and close defence purposes. An electro-optic package that can track small targets such as Unmanned Aerial Vehicles (UAVs) at ranges of over 10 km is mounted under the left missile launcher. Also included is a TV camera, a thermal camera and a laser range-finder.

A typical configuration of the FB-6A battery is:

- 1 × command vehicle
- 6 × missile launch vehicles
- power supply vehicle
- missile supply vehicle
- system test vehicle.

The customer can also be supplied with a loading and aiming training missile, fire-control tester and a launching control tester.

In combat, each FB-6A MLV would normally be supported by a missile re-supply vehicle, which would carry up to 24 additional missiles that could be loaded at the rate of one per minute. The FB-6A can also be integrated into an overall air defence system with target information provided to the FB-6A fire unit from a command vehicle. The will allocate targets to a particular FB-6A fire unit and avoid the possibility of two fire units engaging the same target.

The overall system is designed to deal with fighter aircraft, armed helicopters, UAVs and subsonic cruise missiles.

Missile

The missiles so far associated with the FB-6A system are the FN-6 fire-and-forget man-portable SAMs that can be independently supplied by Poly Technologies.

12.7 mm machine gun

The 12.7 mm machine gun is fitted for close in defence and local engagements of ground forces.

IFF system

The FB-6A MLV has been shown to be fitted for but not with Identification Friend or Foe (IFF). It is therefore logical that in the future, when the system is accepted either into the People's Liberation Army (PLA) forces or exported abroad, a suitable IFF system will be available.

Subsystems

Standard equipment includes communications components and a land navigation and alignment system. Also included is an auxiliary unit to power dedicated storage batteries that can provide sufficient power for up to eight hours of continuous operations and built-in test equipment as standard.

Specifications

	FB-6A MLV
Dimensions and weights	
Crew:	2
Mobility	
Configuration	
running gear:	wheeled
Firepower	
Armament:	12.7 mm (0.50)
Ammunition	
main total:	8
missile:	8
Main weapon	
maximum range:	5,500 m
maximum vertical range:	3,800 m
minimum vertical range:	15 m
Fire-control	
system:	IR/TV/Laser
Main armament traverse	
angle:	360°
Main armament elevation/depression	
armament front:	+70°/-10°

	FN-6
Dimensions and weights	
Length	
overall	1.495 m (4 ft 10¾ in)
Diameter	
body	72 mm (2.83 in)
Weight	
launch	10 kg (22 lb) (est.)
Performance	
Range	
min	0.3 n miles (0.50 km; 0.3 miles)
max	3.0 n miles (5.50 km; 3.4 miles)
Altitude	
min	15 m (49 ft)
max	3,800 m (12,467 ft)
Ordnance components	
Warhead	HE
Guidance	passive, IR (homing seeker)
Propulsion	
type	solid propellant (boost, sustain stage)

Status

Development of the system is complete and it has been with the end-user for some time.

Contractor

Poly Technologies Inc of China (primary contractor and system enquiries).

FLG-1

Type

Self-Propelled Anti-Aircraft Gun and Missile (SPAAGM) system

Development

The China Aerospace Science and Industry Corporation (CASIC), sometimes known as COSIC, have been developing a new family of Ground-Based Air-Defence (GBAD) systems with the export title of FL (Feilong/Flying Dragon). The FLG-1 is a mobile air-defence gun/missile system that has been in development since at least the late 1990s. The system is reported to be in use with China's People's Liberation Army (PLA). The FLG-1 is a combination of the FLV-1 short-range light air-defence missile and the 23 mm (twin) Giant Bow (export name) anti-aircraft gun.

As of September 2005, it has become apparent that the FL (Flying Dragon) is indeed a family of weapon systems with the theme of small vehicular highly mobile surface-to-air missile systems, some vehicles having radar and electro-optical target tracking facilities, others with a mix and match type. The missiles associated (where appropriate) with this FL family are based on the QW series of man-portable SAMs.

FLG-1 (CASIC) 0549672

Description

The system configuration, first displayed at the Zhuhai Air Show in November 2002, showed a mobile air-defence system with twin-quadruple missile launchers mounted on top of a wheeled vehicle. Three twin-barrelled, small-calibre anti-aircraft guns supplement this. The QW-1/2 missile is known to be produced by the CASIC and may be the weapon associated with this system. The guns seen in the publicity brochures appear to be the NORINCO twin 23 mm (Giant Bow) upgraded system that was originally based on the Russian ZU-23-2.

The vehicle appears to be based on a 4 × 4 version of the China North Industries Corporation (NORINCO) WZ 550/551-series chassis, a 6 × 6 variant of which has been in service with the PLA for some years. The rear-engined vehicle weighs 8.5 tonnes, and has a maximum road speed of 90 km/h from an engine that produces 132 kW (177 hp).

The platform features an unmanned roof-mounted turret, on either side of which are mounted four missiles in the ready-to-fire position. Between the banks of missiles is an electro-optical package, while mounted on top is what appears to be a surveillance radar or IFF system. Integrals within the vehicle are command-and-control and communications. In concept, this fire unit is very similar to the German STN Atlas Elektronik ASRAD system that has seen service with the German Army which is installed on a Wiesel 2 tracked armoured chassis.

The primary purpose reported for the system is the protection of high value targets from attack helicopters, cruise missiles and low flying aircraft.

CASIC says the FLG-1 has been in production for more than two years and can launch missiles using either infrared or laser semi-active homing guidance systems.

Variants

During the Paris Air Show 2003, the Chinese Precision Machinery Import and Export Corporation (CPMIEC) displayed the same publicity photographs of the FLG-1 firing vehicle as the FLV-1. There would appear to be a new family of FL Chinese systems. These are: FL-2000, FL-2000N, FL-2000B, FLG-1, FL-2000V and FLS-1.

Specifications

	FLG-1 Launch Vehicle
Dimensions and weights	
Weight	
standard:	8,500 kg
Mobility	
Configuration	
running gear:	wheeled
layout:	4 × 4
Speed	
max speed:	90 km/h
Range	
main fuel supply:	600 km

Missile
(Probably the QW-2)

Type:	2-stage low-altitude SAM
Length:	2.1 m
Diameter:	unknown
Weight:	23 kg
Propulsion:	2-stage solid propellant boost and sustain
Guidance:	semi-active laser homing
Warhead:	
(type)	HE fragmentation (probably)
(weight)	3.2 kg
Average missile cruise speed:	750 m/s
Range:	
(maximum)	8 km
(minimum)	600 m
Altitude:	
(maximum)	5,000 m
(minimum)	4 m
Kill radius:	3 m
Kill probability:	85%

Status

Reported to be in service with the PLA.

Contractor

China Aerospace Science and Industry Corporation (CASIC) (weapon systems).
China North Industries Corporation (NORINCO) (vehicle launcher).

Overseas Sales

China Precision Machinery Import and Export Corporation (CPMIEC).

Type 95

Type

Self-Propelled Anti-Aircraft Gun and Missile (SPAAGM) system

Development

For some years now, China North Industries Corporation (NORINCO) has been developing a new, quadruple 25 mm self-propelled anti-aircraft gun system. Originally known as the Type 90-II and Type 90-III, this system was first shown publicly at the Beijing Military Parade in 1999.

Description

The Type 95 SPAAG is a self-propelled anti-aircraft gun system with the option of added 4 × Very Short-Range Air Defence Missiles (VSHORAD), with an independent day and night combat capability.

These missiles are most likely the QW-2 (or at least another member of the QW family of missiles) although other types of short-range man portable type may also be employed.

The system consists of a new full-tracked armoured chassis, on the roof of which has been installed a new one-person, power-operated turret armed with four 25 mm cannon and optional is four fire-and-forget SAMs.

Also fitted is an electro-optic fire-control system, which is fed from the electro-optic package mounted on the forward part of the turret, which includes a TV tracking camera (range 6,000 m), infrared tracking camera (range 5,000 m), and a laser range-finder (range 500 m to 5,500 m).

A CLC-1 pulse-Doppler search radar is also fitted and has a maximum range of 11 km. The CLC-1 is allegedly an 'S' Band radar optimised to detect low-flying aircraft and attack helicopters.

A bank of four electrically operated forward-firing smoke grenade launchers is mounted either side on the lower part of the turret for self-protection.

Operational

In a typical engagement, CLC-1 pulse-Doppler radar would detect and track targets at ranges out to 11 km. These would be handed over to the fire-and-forget SAMs, which in turn would then be used to engage airborne targets at the longer ranges within the firing envelope of the missiles. In the case of the QW-2, is from 500 to 6,000 m at altitudes of 10 to 3,500 m.

The 25-mm cannon is used to engage aircraft and helicopters with the aid of the optronic fire-control system at ranges of up to 2,500 m and at altitudes of 2,000 m. The 25-mm cannon that is already in service with the PLA has a cyclic rate-of-fire between 600 and 800 rounds per minute, per barrel. The vehicle is expected to carry in the region of 1,000 rounds of ammunition for the gun.

The Type 95 SPAAG/SAM is part of a complete mobile air-defence battery that includes:

- 6 × 25 mm self-propelled anti-aircraft guns
- 1 × battery acquisition and command vehicle with a CLC-2 search and acquisition radar
- 1 × battery testing vehicle
- 3 × ammunition vehicles
- 1 × power supply vehicle
- 1 × simulator.

The battery command post is mounted on a similar chassis to the SPAAG/SAM.

The battery CLC-2 surveillance radar has a maximum range of 45 km and a maximum altitude of up to 4.5 km. Mounted on the front is a 12.7 mm heavy machine gun for self-defence.

Specifications

	Type 95 SPAAG/SAM	Type 95 Battery command vehicle
Dimensions and weights		
Crew:	3	5
Length		
overall:	6.71 m	6.8 m
Width		
overall:	3.2 m	3.2 m

NORINCO Type 95 self-propelled anti-aircraft gun/surface-to-air missile system with guns elevated 0100739

	Type 95 SPAAG/SAM	Type 95 Battery command vehicle
Height		
overall:	3.4 m	-
transport configuration:	3.40 m	3.38 m
deployed:	4.82 m	4.58 m
Mobility		
Configuration		
running gear:	tracked	tracked
Speed		
max speed:	53 km/h	53 km/h
Range		
main fuel supply:	450 km	450 km
Engine	diesel	
Firepower		
Armament:	4 × turret mounted 25 mm cannon	1 × 12.7 mm (0.50) machine gun
Armament:	4 × turret mounted QW-2 SAM	-
Ammunition		
main total:	1,000	-
Main weapon		
operation:	gas automatic	gas automatic
cartridge:	B	B
Rate of fire		
cyclic:	3,200 rds/min	-

Status
The Type 95 is in production and is in use with the PLA in small numbers. As of November 2013, no export sales have yet been reported.

Contractor
China North Industries Corporation (NORINCO).

Germany

Gepard

Type
Self-Propelled Anti-Aircraft Gun and Missile (SPAAGM) system

Development
The basic Flugabwehrkanonenpanzer (Flakpanzer) Gepard (Cheetah) is an autonomous all-weather-capable, twin 35 mm cannon armed, Self-Propelled Anti-Aircraft Gun (SPAAG) system which was developed in the 1960s and fielded in the 1970s. The basic system has been upgraded several times over the years since the Initial Operational Capability (IOC) model in order to keep the system fully front-line capable.

In 1999 it was disclosed that Krauss-Maffei Wegmann, the prime contractor for the Gepard SPAAG system, was developing an add on fire-and-forget Surface-to-Air Missile (SAM) system to enhance the operational effectiveness of the system. This was developed as a private venture and was aimed at existing and future customers for the Gepard SPAAG, which has seen service with Germany, The Netherlands, Belgium and Romania.

Description
The current Gepard SPAAG, which is based on a stretched Leopard-1 chassis with an 830 hp engine, is armed with two fully-stabilised Oerlikon Contraves twin-belt-feed 35 mm KDA cannon. These have a cyclic rate-of-fire of 550 rounds, per cannon, per minute and can elevate up to +85° to give total overhead protection. The effective air defence range of the cannon is about 3,500 m and more than 4,500 m with the new Frangible Armour-Piercing Discarding-Sabot (FAPDS) ammunition.

On the upgraded Gepard, mounted externally either side of the 35 mm mount is the LFK Standard Stinger Launching System (SSLS) that was originally developed for helicopter operations but is now being offered in the air defence role.

This has two Raytheon Systems Company Stingers in the ready-to-launch position, together with a rechargeable coolant bottle suitable for 40 engagement operations. An adapter provides the mechanical interface between the launcher and the platform.

There is also an interface electronics unit that provides the interface between the launchers and the launch platform fire-control system.

The crew of the Gepard would use the surveillance radar, which has a range of about 18 km, with integrated Identification Friend or Foe (IFF) system and tracking radar for lock on and identification purposes. Actual target engagement is carried out using the roof-mounted optical sights.

In a typical target engagement, the Stinger SAM would be used to engage targets at longer ranges, with the 35 mm cannon being used to engage close in and pop-up targets.

Although the US designed Stinger is being proposed, Krauss-Maffei Wegmann have stressed that other fire-and-forget SAMs could also be used. SAMs such as the Russian KBM Instrument Design Bureau Igla and Igla-1 family, both of which are more widely deployed in certain areas of the world, especially Eastern Europe, the Middle East and Asia, could also be used.

Close up of the twin Stinger SAM system integrated onto the Gepard twin 35 mm self-propelled anti-aircraft gun system 0067215

Krauss-Maffei Wegmann at Munich undertook production of the GEPARD SPAAG between 1976 and 1980. The German Army took delivery of 420 systems with export sales being made to Belgium (55 units) and The Netherlands (95 units with different tracking and surveillance radars). The Belgian vehicles have been phased out of service and have been offered for sale.

Krauss-Maffei Wegmann have completed a major modernisation programme on 147 German and 60 Netherlands Gepards. The first upgraded vehicles were completed in 1998, with final deliveries made in 2003.

Krauss-Maffei Wegmann are marketing surplus Gepards to select customers with the authority of the German government. Romania has already take delivery of 43 unmodified Gepards in a government to government deal. Of these, 18 will be fully operational weapons with the remainder being used for spare parts and training purposes.

Variants
The three main variants to the Gepard SPAAGM are:
Gepard 1A2 upgrade variant with new fire-control system
Gepard CA1 Dutch variant
Upgraded Dutch variant PRTL with new radios.

Status
The Gepard system has seen service in Belgium, Chile, Germany, the Netherlands and Romania.

Contractor
Krauss-Maffei Wegmann GmbH & Co KG.

International

Blazer

Type
Self-Propelled Anti-Aircraft Gun and Missile (SPAAGM) system

Development
As a private venture, General Dynamics Armament and Technical Products of the US and the then Thomson-CSF Airsys of France (now Thales) consolidated to develop the Blazer air defence turret.

Thales General Dynamics Armament and Technical Products Blazer Air Defence Turret installed on a General Dynamics Land Systems - Canada LAV (8 × 8) chassis 0100858

Thales General Dynamics Armament Systems Blazer Air Defence Turret installed on General Motors Defence LAV (8 × 8) chassis 0100857

The complete prototype was shown for the first time in June 1994 and again in late 1994. Firing trials with both the 25 mm cannon and the MBDA Mistral Surface-to-Air Missiles (SAMs) have been undertaken in France.

Description

Blazer essentially consists of a 25 mm, five-barrel, externally powered, GAU-12/U Gatling gun and four Raytheon Systems Company, Stinger or MBDA Mistral fire-and-forget surface-to-air missiles on a two-man turret.

The Blazer turret is power operated and of all-welded construction with light armour. It can be mounted on any tracked or wheeled vehicle that could accommodate a 1.625 m diameter turret ring. This all-electric turret has been installed on the US United Defence LP M2 Bradley chassis and the MOWAG Piranha (8 × 8). The Piranha 8 × 8 has also been manufactured under licence by FAMAE, in Chile, and the then Alvis Vickers (now BAE Systems, Land Systems UK, Weapons and Vehicles), as have the Textron Marine and Land Systems Cadillac Gage LAV-300 (6 × 6) and LAV-150S (4 × 4), and similar wheeled and tracked vehicles.

The basic Blazer turret features a digital fire-control system, eye-safe laser range-finder, FLIR/TV stabilised sight and a two-man crew consisting of the commander and gunner.

The system is fitted with the Thales TRS 2630P 2-D digital radar, the MBDA Mistral or Raytheon Systems Company, Stinger missiles or, as a growth option, Command-to-Line-Of-Sight (CLOS) missiles.

The two-man turret houses both gunner and commander, each capable of full system operation including acquisition, tracking, weapon selection and firing. Vision is through armoured windows on the front and sides.

Main armament of the Blazer turret comprises the General Dynamics Armament and Technical Products 25 mm GAU-12/U Gatling gun, which fires the Bushmaster family of ammunition at a cyclic rate of fire of 1,800 rounds/min. The Gatling gun is especially effective in the air defence role because of its high rate of fire. It places a short, dense pattern of projectiles in the path of the target for high hit probability.

According to General Dynamics, this approach effectively masks the effects on random target motion owing to buffeting and jinking. In addition, four or eight infrared seeker missiles are mounted above the gun cradle and integrated into the Blazer's fire-control system.

A FLIR/TV sight is included for viewing and auto tracking. The system has demonstrated day/night capability and the ability to track and fire while the vehicle is moving at up to 50 km/h over uneven terrain.

The Thales TRS 2630P 2-D radar has a range of 17 km, IFF, automatic track-while-scan and data exchange for netting capability. The last feature allows one system to act as a command unit for several units without operating acquisition sensors. The 25 mm GAU-12/U five-barrel cannon can engage targets up to 2,500 m and the missiles at 6,000 m. A total of 400 ready rounds is in the magazine, with up to 600 stowed rounds in the vehicle. Internal loading of the 25 mm gun can be accomplished in 15 minutes. Depending on the chassis, additional missiles can be stored internally. Electronically operated smoke dischargers are located in banks of four on either side of the turret.

Variants

A different version of this system, without the radar and armed with Stinger SAMs was built for the US Marine Corps.

Specifications

	Blazer
Dimensions and weights	
Crew:	2
Mobility	
Configuration	
running gear:	wheeled
layout:	8
Firepower	
Armament:	1 × turret mounted 25 mm GAU-12/U cannon
Armament:	8 × turret mounted Mistral SAM
Rate of fire	
cyclic:	4,200 rds/min
Main armament traverse	
angle:	360°
Main armament elevation/depression	
armament front:	+65°/-8°

Status

The LAV-AD is reported to be in service with the US Marine Corps (USMC).

Contractor

General Dynamics Armament and Technical Products.
Thales.

NIMRAD

Type

Self-Propelled Anti-Aircraft Gun and Missile (SPAAGM) system

Development

The NIMR Air-Defence (NIMRAD) system is an international joint venture between:

- Advanced Industries of Arabia
- MBDA
- Rheinmetall Defence Electronics.

The Vehicle was first shown at the IDEX 2007 in Abu Dhabi where the system was offered for sale. The NIMR family of vehicles, of which the NIMRAD is a variant of, have been in development since 2005.

Advanced Industries of Arabia is part of the multinational Bin Jabr Group of Companies. It is the exclusive distributor of the NIMR armoured vehicle, an off-road extreme-utility-family of military vehicles. NIMR is the Arabic equivalent of Tiger.

Description

The NIMRAD is a self-propelled anti-aircraft gun and missile system variant that mounts two banks of two Mistral 2 missiles either side of the turret with four reload rounds stored inside the vehicle. Also associated with the turret is a heavy machine gun for self-defence of the vehicle.

The system uses the Rheinmetall Defence Electronics Electro-Optical Sensor Suite (EOSS) that has both day and night capability. The EOSS has the ability to monitor the local airspace at a range of up to 10 km depending on weather conditions. NIMRAD can fight autonomously or be networked with other air-defence assets making use of modern communication systems mounted internally.

Variants

A second combat vehicle, based on the same architecture and subsystems, is named NIMRAT and this uses the MILAN anti-tank missile.

NIMRAD front view (IHS/Patrick Allen) 1322052

NIMRAD Turret with Gun, EOSS and Missiles (IHS/Patrick Allen) 1322051

The NIMRAT and NIMRAD are based on a common architecture comprising the following:

- Armoured wheeled off-road capable vehicle
- Automated turret
- Fire-control system with electro-optic sensor suite
- Self-defence machine gun mounted on the turret
- Communication systems.

There are also many other variants of the NIMR vehicle, the NIMR II is an armoured remote weapon station vehicle, there are also multi-rocket launcher systems, troop carriers and a multipurpose cargo vehicles.

Specifications

NIMRAD

Type:	SPAAGM
Missile:	Mistral 2, four in total, mounted two on either side of the turret
Gun:	heavy machine gun
FCS system:	electro-optical sensor suite with day and night capability

Status

Presumably development is complete because the system has been offered for sale at IDEX.

Contractor

Advanced Industries of Arabia LLC (NIMR 4 × 4 wheeled off road-armoured vehicle).
MBDA (Mistral 2 surface-to-air missiles).
Rheinmetall Defence Electronics GmbH (RDE) (fire-control system with EOSS).

Israel

Machbet

Type

Self-Propelled Anti-Aircraft Gun and Missile (SPAAGM) system

Development

The Machbet self-propelled combined gun/missile short range air defence system was developed to meet the operational requirements of the Israel Air Force (IAF) by a joint team established by the IAF and the then Israel Aircraft Industries, MBT Division (now Israeli Aerospace Industries - IAI). The existence of the Machbet was first revealed by the IAF in December 1995, by which time it was already in service.

Description

The Machbet is a further development of the US designed and built United Defense/General Dynamics Armament Systems M163 Vulcan Air Defence System (VADS) 20 mm Self-Propelled Anti-Aircraft Gun (SPAAG) system, fully covered in the Self-propelled anti-aircraft gun system section and based on an M113 series full-tracked armoured personnel carrier chassis.

The main drawback of the M163 is that its 20 mm Vulcan cannon lacks range and lethality to deal with the latest air threats.

To enable Machbet to engage a wider range of targets at longer battlefield ranges, mounted on the left side of the turret is a four-round launcher for the Raytheon Systems Company Stinger fire-and-forget missile launcher.

This Stinger launcher is the same as fitted to a number of US air defences systems including the Avenger, Bradley Linebacker and Light Armoured Vehicle - Air Defense.

Machbet has also been fitted with MBT's Eagle Eye II passive electro-optical unit level fire-coordination-and-control system with enhanced electro-optical sensors. The system extends the engagement envelope of the M-163 Vulcan SPAAG with the addition of Stinger FIM-92 missiles. A monitor, joystick and operating push buttons are used to detect the target where lead angles are calculated by the computer and are integral with the firing sequence.

IAI Machbet combined gun/missile SHORADS assembly in Stinger configuration mounted on tracked M113 APC 0009326

The prime contractor states that the system has been designed to exploit the maximum performance from the ammunition, with enhanced accurate firing capability, especially against attacking helicopters, ultra light and other aerial targets.

In developing the Machbet, special attention has been given to human engineering aspects, both for operation and maintenance. Every malfunctioned unit is identified and viewed on the operations screen.

The commander is seated in the hull of the vehicle and views remote radar pictures adjusted to his position on a control unit. By means of a joystick, the commander designates a selected target. The turret automatically turns towards the selected target and control is then passed to the operator seated in the turret.

At this stage the commander receives the target TV track image on the screen of the control unit to identify and decide whether to engage the target by the gun or missile.

In a typical target engagement the missiles would be used to engage aerial targets at long range with the 20 mm Vulcan cannon being used to engage close-in aerial targets as well as having a secondary ground role. Machbet considerably expands the Vulcan's engagement envelope and can kill enemy helicopters and aircraft at ranges of 500 to 6,000 and 8,000 m. Similar configurations are also available for other low altitude man portable SAMs, including the Russian 9K310 (NATO SA-16 'Gimlet'), 9K39 (NATO SA-18 'Grouse') and the French Mistral.

The baseline Machbet system has a day-only camera but this can be quickly replaced by a plug-in modular FLIR camera.

The ELTA EL/M-2106 point defence radar is also offered with the system for enhanced self-deployment radar capabilities.

A typical engagement would involve the unit commander viewing a remote radar picture adjusted to his position on the control unit. With the joystick he can designate targets and select those with the highest priority for engagement. The weapon's turret is slewed automatically to place the weapon at the designated position and signals the gunner located in the turret with ready signals, waiting for firing command. At this stage, the gunner can engage the target with either weapon.

Variants

In IAF service Machbet is armed with Raytheon Systems Company Stinger FIM-92 fire-and-forget missiles but, according to IAI, other types of fire-and-forget missile can be carried, such as the MBDA Mistral or the Russian KBM Kolomna Igla (NATO SA-18 'Grouse') or Igla-I (NATO SA-16 'Gimlet') family. As far as it is known, the only production application to date is with the Stinger SAM.

Status

Production as required. In service with the Israeli Air Force.

Contractor

Israel Aerospace Industries (IAI) Ltd, MBT Systems Missiles and Space Group.

Russian Federation

2K22/2K22M Tunguska

Type

Self-Propelled Anti-Aircraft Gun and Missile (SPAAGM) system

Development

The 2K22M Tunguska anti-aircraft gun/missile system produced at the Ulyanovsk Mechanical Plant was designed to provide Air Defence (AD) for ground forces in all kinds of combat operations, as well as AD cover for installations.

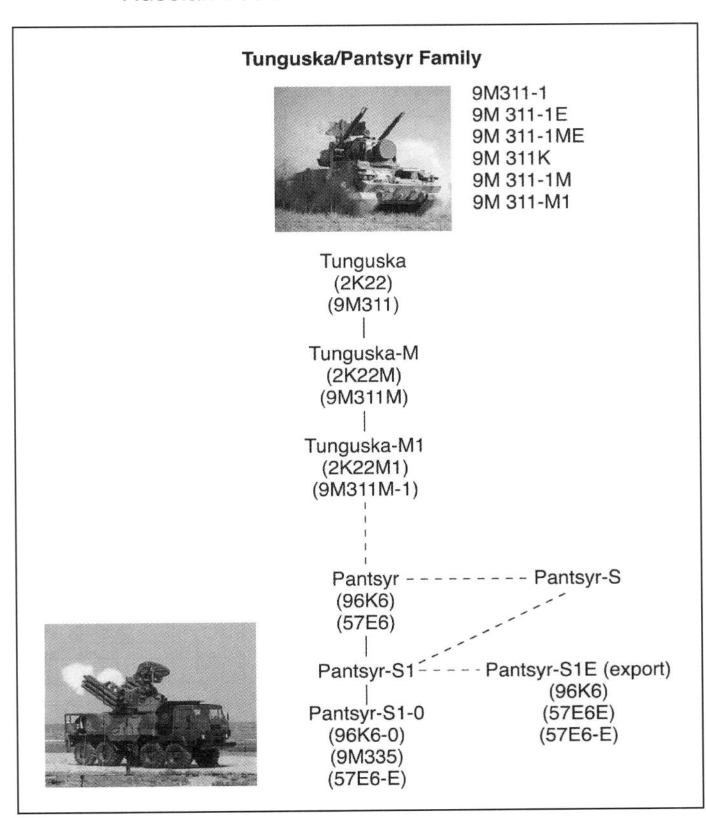

Tunguska family flow diagram (James C O'Halloran) 1154768

Prototype of an improved Tunguska (Miroslav Gyürösi) 1331426

Development of the 2K22M Tunguska self-propelled gun/missile system can be traced back to 1970, when the KBP Instrument Design Bureau based in Tula was authorised to start design work on a new self-propelled air defence gun/missile system called the ZSU 2K22 Tunguska and based on the GM-352 combat vehicle tracked chassis.

As the KBP Instrument Design Bureau was fully committed to the design of the 2K22 gun/missile system of the ZSU Tunguska, planning for the radar system was passed over to the Ulyanovsk Mechanical Factory that came under the then Ministry of Radio Production.

Subsequently, the Ulyanovsk Mechanical Factory took over as the prime contractor and they produce and market the complete 2K22M Tunguska system.

Other major subcontractors include the Electromechanical Scientific Research Institute (NIMI) for the computing sight and the Minsk Tractor Factory for the full-tracked chassis.

The 2K22M Tunguska system was developed as the replacement for the older ZSU-23-4 23 mm self-propelled anti-aircraft gun. The Tunguska is a more effective system as it utilises the 9M311 missiles to engage targets at long range, and the 30 mm 2A38M guns to engage targets at shorter ranges. The system is gradually being replaced with the Pantsyr weapon, a natural follow-on system to the Tunguska.

The first version entered service in 1986 with the self-propelled vehicle 2K22, and had four missiles, two either side, in the ready-to-launch position. It is probable that this was a pre-production system and not built in very large numbers.

The second version to enter service, and the current production model vehicle, is designated the 2K22M. This carries a total of eight missiles in the ready-to-launch position; four either side.

The complete vehicle (missiles, guns, platform and associated support equipment) is known as the 2K22/2K22M system.

The surveillance radar is mounted on the rear of the 2K22M turret and, when not in action, can be folded down to the rear (Christopher F Foss) 0100741

2K22M self-propelled air defence system (Christopher F Foss) 0100740

The missiles used on the 2K22M self-propelled air-defence vehicle are also used in a naval version known as the Kashtan-M. The Kashtan-M (previously known as Kortik (Dirk) uses the 6K30GSh automatic gun that is a derivative of the GSh-6-30K automatic anti-aircraft gun. The system was designed to equip naval missile/gun systems for close-in air-defence and has a maximum rate of fire of 9,000 rounds per minute, 4,500 rounds per minute per gun. The Kashtan system has the NATO designation SA-N-11.

As the 2K22M is mounted on a full-tracked chassis, it has the required mobility characteristics to keep up with mechanised units.

According to the manufacturer, the system can engage targets such as aircraft and helicopters with its 30 mm cannon when stationary or moving, while targets can only be engaged with missiles, when the 2S6M vehicle is static.

A typical 2K22/2K22M battery would consist of six 2S6/2S6M self-propelled air defence vehicles, six reload vehicles and associated support and maintenance vehicles.

This system fills the gap between man-portable surface-to-air missile systems such as the Strela-3, Igla, Igla-1 and the Tor self-propelled surface-to-air missile system. In addition to replacing the ZSU-23-4, the 2K22/2K22M also replaces the older Strela-1 and Strela-10 self-propelled surface-to-air missile systems. The Strela-10 is also being replaced by the Luchnik-E very short-range air defence missile system that has replaced the older 9M35 missiles with a Strelets package of eight (four either side of the cupola) Igla-S missiles.

In late 2000, the only known export customer for this system was India, which is believed to have taken delivery of over 50 systems, which are used by three regiments. The system has now additionally been sold to Algeria, China, Morocco and Myanmar (Burma).

2K22M air defence system with 30 mm weapons at maximum elevation but without SA-19 missiles fitted (Christopher F Foss) 0069232

The 2K22M has no equivalent in the West as it has a combination of both guns and missiles to engage targets at extended ranges (Christopher F Foss) 0069230

Early warning and cueing for Indian Air defence systems, including the recently introduced 2K22M, will be provided by the Dutch Reporter radar system. This is already in service with India and will be made under licence by Bharat Electronics.

Late in October 2009, Russia's KBP design Bureau announced that a further upgrade to the Tunguska would finish trials by the end of 2009. This upgrade was first identified at the Moscow Air Show (MAKS 2009) where the prototype was shown publicly for the first time.

Description

The 2K22 Tunguska is a tracked, self-propelled gun-missile system that has been designed to take-on low flying targets in addition to infantry and vehicles. The 2K22 system has four ready-to-fire SAM rounds, whereas a modified version with the designator 2K22M has eight ready-to-fire rounds. The army designator 9K22 has sometimes been used as a replacement for the 2K22. The basic missile is the 9M311 Treugolnik (triangle).

The first upgrade to the 2K22 the 2K22M is very similar to the German Krauss-Maffei Wegmann Gepard twin 35 mm self-propelled anti-aircraft gun system, covered under the Self-propelled anti-aircraft guns section, with the cannon mounted externally of the turret.

The 2K22M is based on the chassis of the GM-352M which is also used as the platform for the Buk-1M (NATO reporting name 'Gadfly') and 'Tor' SAM systems.

The hull and turret are of all-welded steel armour construction that provides protection from small arms fire and shell splinters.

The driver is seated at the front of the hull to the left side, with the gas-turbine auxiliary power unit to the right, turret, which is designated as the 2A40M, is in the centre and the power pack at the very rear of the hull.

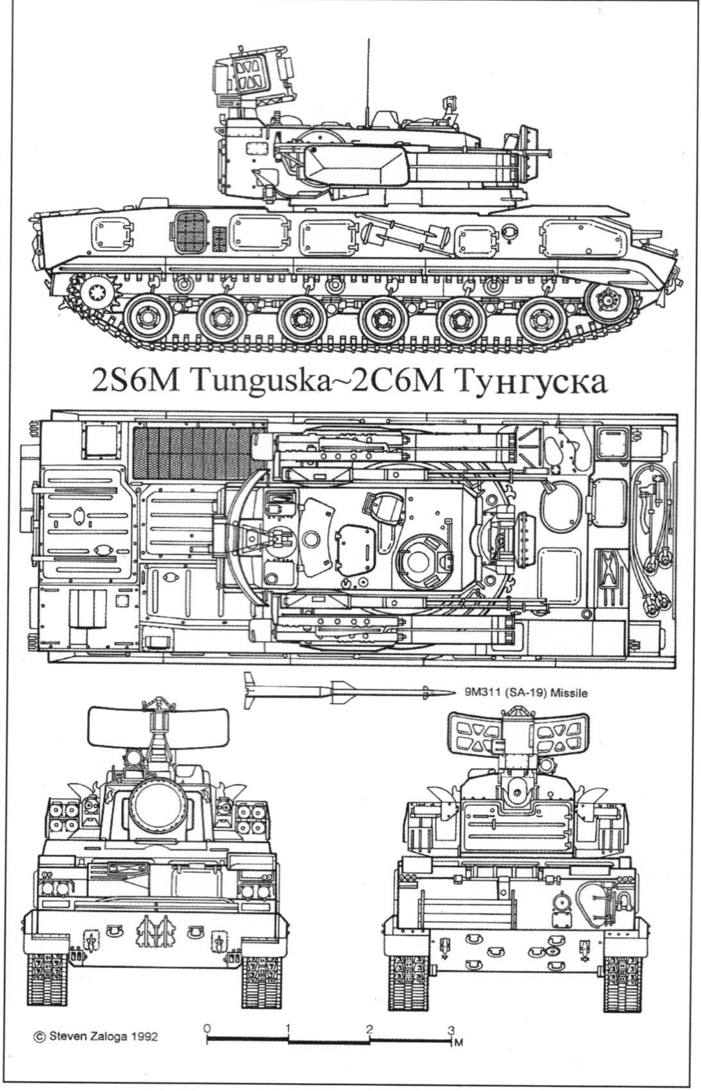

Four-view drawing of the latest production 2K22M (Tunguska) self-propelled air defence system (Steven Zaloga) 0509409

The radar operator is seated in the left front side of the turret with the commander on the right side and gunner to the rear in the centre; the gunner tracks the target in the optical mode.

The commander has a circular hatch cover that opens forward and this also has an integral hatch that opens to the rear. In the former are three forward-facing day periscopes, the centre one of which is similar to that fitted to RFAS Armoured Fighting Vehicles (AFVs) with two eyepieces and two control handles. Some 2K22s have been observed with an infrared searchlight mounted on the commander's cupola.

The complete radar system is designated the 1RL144M. The system includes the E-band surveillance radar, which is mounted on the turret rear. This has a maximum detection range of 18 km. The J-band tracking radar is mounted on the front of the turret. This has a maximum tracking range of 16 km.

The radar system features a special search mode to detect low-flying aircraft (typically 15 m) with the scanning being carried out with an elevation of only +1° along with a strong ground return signal suppression mode.

If a target is detected at a range less than 8 km, this can be transferred automatically to the tracking radar on the front of the turret.

The commander uses the PK console to his right to select type of weapon (gun or missile) and to then engage the target; either the commander or gunner can open fire.

The radar operator has three consoles which he uses to control the tracking and fire-control radar which sends information such as range, bearing and elevation to the fire-control computer which in turn computes the laying commands for the weapons.

The NATO code name for the complete radar system mounted on the 2K22/2K22M vehicle is 'Hot Shot'. When not required, the surveillance radar swings back to the turret rear to reduce the overall height of the system. The Identification Friend-or-Foe (IFF) system fitted has the Russian designation 1RL138.

The roof-mounted optical sight is designated the 1A29M while the on board computer system is designated the 1A26M. The cooling system for the radar equipment and the 1K28 air conditioning system is also in the turret.

The optical sight has a magnification of × 8 and a field of view of 8°. In addition to missile coordinate discrimination equipment, which automatically generates angular coordinates of the missile relative to the line-of-sight, it also ensures the changeover to semi-automatic target tracking at a range of up to 16 km and missile guidance at a range of up to 10 km.

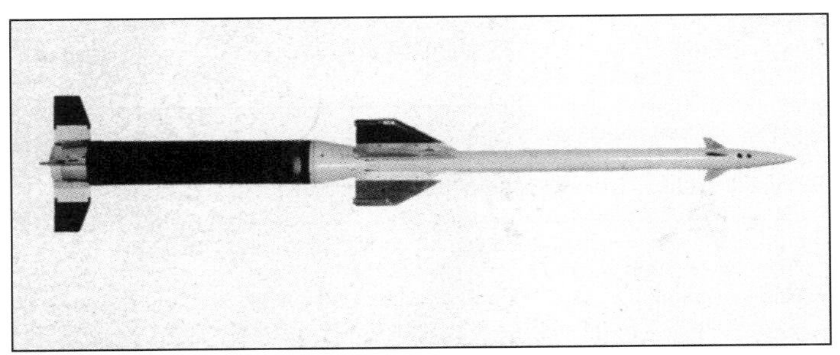

SA-19 'Grison' surface-to-air missile out of its launcher and showing the larger booster towards the rear of the missile 0509410

In the forward part of the turret is the Plan Position Indicator (PPI) screen of the search radar, with the PUIM console installed above the PPI. The latter is used for operating the 1A26M onboard computer system and selecting the operating modes.

In addition, it also incorporates warning signals that indicate to the commander that an aerial target will enter the fire zone of the selected weapon system within five seconds and that it is within range of the weapons.

To the right of the PPI is the OK1M console that is used by the commander to operate the search radar and the 1RL138 IFF system.

The gunner uses the roof-mounted 1A29M optical sight and also has two engagement consoles, NBC detection systems and the manual controls for emergency use. The stabilised sight can also be used as a back-up channel and relay data to the onboard fire-control computer. In addition, it is used to calculate the deviation of a missile's flight path from the line of sight, with this information being fed to the computer which in turn generates the correction signals to the missile.

Two twin-barrelled 30 mm 2A38M cannon are mounted on either side of the power-operated turret, these having an effective range of up to 3,000 m in altitude and 4,000 m in slant range. Russian sources quote the crossover ranges as being up to 2,000 m. The cannon have a total cyclic rate of fire of 4,000 to 5,000 rds/min with the empty cartridge cases and links being ejected externally of the turret. Ammunition is reloaded in containers through the turret roof using a crane. Maximum range against ground targets is quoted as 4,000 m.

The cannon are water-cooled, gas-operated, electrically fired weapons and each cannon has an automatic muzzle velocity measuring device that feeds data to the fire-control computer. The cannon are built by the Tulamashzavod Joint Stock Company; each is fired in turn to give a cyclic rate of fire of between 1,950 and 2,500 rds/min.

The 30 mm ammunition feed system of each cannons is independent, so that if one jams the other weapon continues to fire. In the case of a misfire, three pyrotechnic charges per cannon are provided.

Two types of 30 × 165 mm ammunition are fired, HEI and HEI-T, which are normally belted together both fitted with the A-670M Point-Detonating (PD) fuze. Muzzle velocity is 960 m/s plus or minus 30 m/s.

Typically, bursts of one (83 rounds) or three (250 rounds) seconds would be used, depending on the type of target to be engaged. In addition, the commander can switch to full-automatic control in which the computer decides the length of burst.

The gunner has two modes of operation, radar and optical, for engaging aircraft targets with the anti-aircraft guns.

The radar mode is fully automatic with the guns being laid in elevation and traverse by the tracking radar, which also supplies range data. This mode can be used when the system is stationary or moving.

The optical mode is used when the system is stationary, the guns being laid onto the target using the optical sight with the tracking radar providing range data.

The missiles are launched only when the system is stationary, the target being tracked in elevation and traverse by the gunner using the optical sight. The tracking radar that also transmits commands to the missile, with the receiving antenna for the latter being in one of the missile's fins provides target range.

Close up of turret front of 2K22M air defence system showing tracking radar (left) and missiles (right) (Christopher F Foss) 0045013

2K22M from the rear with radar elevated and two of the missiles and cannon elevated (Christopher F Foss) 0100859

Mounted outwards and below the 30 mm cannon is a bank of four 9M311 SAMs in two blocks of two, which can elevate vertically independently of each other.

At least five variants of the missile are known to exist, these are:

9M311 also known as the naval 3M87
9M311-1 the export variant
9M311K probably the original naval missile, but later the letter K was dropped from use
9M311M also known as the naval variant 3M88
9M311-M1 the second modification of the missile for use with the Tunguska-M1.

The 9M311/3M87 missiles are used with the 2K22M system and the naval equivalent Kortik. The 9M311 naval missile originally had the designator 9M311K but the letter K was subsequently dropped and the missile is now known as the 3M87.

The 9M311M are the first true variant of the missile and are also used with the 2K22M. However, this missile is believed to be the naval variant 3M88.

Typically, two missiles would be fired at each target. Just before launch, the turret is turned slightly off axis so that the smoke of the missiles being launched does not obstruct the gunner's sight. Once the missiles have been launched, the launcher is lowered into the –6° position and locked.

The 9M311 can engage aerial targets with altitudes between 15 and 3,500 m and from 2,400 to 8,000 m in slant range with the target having a maximum speed of 500 m/s. Russian sources quote the crossover range as being up to 4,000 m.

The 9M311 missile has two stages, with the large jettisoned booster having four wraparound fins which boost it to a maximum speed of 900 m/s before the front part of the missile separates and carries on to the target. The latter has four fixed fins and four control surfaces mounted towards the front of the missile. The 9M311 missile uses the so called 'duck' (canard) type aerodynamic control system and is claimed to be very manoeuvrable and can pull a maximum manoeuvre load of 32*g*.

The missile has a 9 kg high-explosive fragmentation warhead which is actuated by a contact fuze or an active laser fuze; thus, if the missile passes within 5 m of the target, the warhead is activated.

The warhead is derivative of the continuous-rod class. The warhead's rods are nearly 600 mm long and are between 4 to 9 mm in diameter. When the warhead is detonated the rods are accelerated outwards eventually forming a continuous ring with a maximum diameter of 10 m, this ring perpendicular to the axis of the missile. After reaching 10 m in diameter the extreme deformation and most likely some form of stress riser (notching, embrittlement, etc) causes the rods to break-up, producing fragments weighing 2 to 3 g each. Due to the difficult nature of launching the rods without unwanted breakage or deformation, the maximum rod (fragment) velocity is normally limited to around 1,600 m/s. Latest production missiles have the designation of the 9M311M and are slightly heavier and longer.

Export models (2004/5) of the missile have the Russian designator 9M311-1M.

Method of target engagement is semi-automatic, radar to command line-of-sight, with the gunner tracking the target, using the roof-mounted stabilised optical sight. When not required, the sight is covered by an armoured hatch, which swings backwards through about 150°.

It is claimed that the 30 mm 2A38M cannon have a 0.6 kill probability with the 9M311 missiles having a 0.65 kill probability.

Suspension is of the hydropneumatic type, with each side having six dual rubber-tyres road wheels, idler at the front, drive sprocket at the rear and three track-return rollers. A hydraulic track tensioning system is fitted as standard and this allows the driver to adjust the track tension without leaving his seat.

Standard equipment includes a heating system, 1V16 internal communications system, fire detection and suppression system, land navigation system, NBC system, night vision aids and internal and external communications equipment.

The 2K22M was being offered for export at around USD8-10 million and India has already taken delivery of a quantity of systems as the replacement for its older ZSU-23-4s.

KamAZ-43101 (6 × 6) 2F77M 2K22M resupply vehicle (Christopher F Foss)
0069233

In addition to developing the actual 2K22M air defence systems, other supporting systems have also been developed and these are offered as part of an overall package.

The 9F810 and 9F810M simulators/training devices produced and offered by Tulatocsmash, other key elements in the support package include:

- Missile resupply vehicle 2F77M
- IR10-1M repair and maintenance truck based on a Ural-43203 (6 × 6) truck with box body
- 2F55-1M repair and maintenance truck based on Ural-43203 (6 × 6) with box body
- 2V-110-1 maintenance truck on Ural-43203 (6 × 6) truck with box body
- MTO-ATG-M1 maintenance shop on ZIL-131 (6 × 6) chassis and the 9V921 automatic mobile test station on a GAZ-66 (4 × 4) chassis with K-66 body.

The 2F77M-resupply vehicle is based on the KamAZ-43101 (6 × 6) truck chassis that has a forward control cab. The flatbed to the rear of the cab has been re-configured to carry out its new role as a forward ammunition re-supply vehicle for the 2K22M.

The KamAZ-43101 (6 × 6) is similar to the KamAZ-4310 (6 × 6) series of vehicles which is used in some numbers by the Russian Army in the cargo role. In its normal version, the KamAZ-43101 (6 × 6) has drop sides, drop tailgate, removable bows and a tarpaulin cover. Cross-country payload is 5,000 kg and towed load is 7,000 kg.

Mounted at the very rear of the flatbed is a hydraulically operated crane, which can lift a maximum load of 280 kg to a maximum height of 3.475 m. The crane normally points to the front while travelling and can be traversed through 200° so, if required, could reload a 2K22M positioned either side.

The vehicle carries a total of eight missiles and 32 containers of 30 mm ammunition. The missiles are carried four either side of the vehicle in vertical racks. This ammunition load is sufficient to re-supply a 2K22M with one complete reload of missiles and two complete reloads of 30 mm ammunition.

Total time taken to re-supply the 2K22M with its full load of eight missiles and 1,904 rounds of 30 mm ammunition is quoted as 16 minutes.

The vehicle is also fitted with a 30 mm cartridge belt loading and unloading machine and a radio communications system to enable it to deploy exactly as required.

Standard equipment for the (6 × 6) chassis includes a central tyre pressure regulation system that allows the driver to adjust from the cab the tyre pressure to suit the ground being crossed.

For maintenance purposes, the forward control cab can be tilted forwards. Standard features of the KamAZ-43101 (6 × 6) include:

- An air filter working from the exhaust gases
- Air intake above the cab roof on the left side
- Centrifugal air cleaner
- Chemical and heat-treated crankshaft
- Hydraulic power steering on the front wheels
- Pneumatic clutch control
- Pneumatic pre-selector
- Heater and driver's seat with suspension.

Command Post

The 9S482M7 (PU-12M7) was upgraded and modernised during 2007/8. The original version was based on a BTR-60 8 × 8 Armoured Personnel Carrier (APC) and relied on data from external radar sources. The latest variant uses a K1Sh1 wheeled vehicle based on the BTR-80 (GAZ 5903) 8 × 8 APC and incorporates a telescopic sensor mast. Within the telescopic mast are located the SRTR (*Stantsiya Radiotekhnicseskoy Razvedki*), passive ESM system and an electro-optical unit that is located at the masthead.

The mast also boasts four static antennas mounted halfway up the mast for short-range radar. Each antenna covers a sector, which when combined provides all-round coverage.

A trainable electro-optical unit mounted on top of the turret contains day and night cameras, providing hemispherical coverage. The PU-12M7 can accept target data from external surveillance radar, other fire-control systems and radar equipped air-defence weapons. Communications with assets up to 15 km away can be achieved by wire, with radio being used for ranges out to 40 km.

This then allows the PU-12M7 tactical control of an area out to 100 km, handling information for as many as 99 airborne targets. The operator can simultaneously track up to nine targets.

Variants

Tunguska-M1

The KBP Tula, Russia and the Izhevsk Electromechanical Plant (Kupol, Russia) announced in April 2003 that a modernised Tunguska M1 system has successfully completed state trials.

Tunguska M1 is an updated and modernised version of the basic 2K22 Tunguska system. The main improvements include:

- Installation of a digital interface for the Ranzhir automated command-and-control system, which enables remote, automated target designation for any Tunguska launch vehicle from higher command level
- Modernisation of the fire-control radar, which includes employing travelling-wave tubes in the transmitter and modernisation of some radar components
- Improved fire-control software, which increases the engagement capabilities against fast-moving targets in electro-optical mode
- A new 9M311M or 9M311-1M missile (replacing the 9M311) with a range increased from 8 to 10 km and possessing a new radar proximity fuse in place of a laser proximity fuse, which increases the kill probability against cruise missiles
- Modernisation of target-angular-speed-measuring devices, which increases accuracy of fire from the both missiles and guns.

The Tunguska M1 will not be a newly produced system, but existing Tunguska combat vehicles will be gradually upgraded to the new standard. It will increase their capabilities and their service life, until new systems will be available after 2010-2015.

Other improvements include:

- Automatic external target, designation reception and transmission and processing equipment by radio with the battery command post. This allows automatic target distribution among the launchers to be performed by the battery command post
- A gunners assistance system has been introduced which performs automatic high-speed two-coordinate optical tracking
- Due to the use of the new missile, the coordination calculation equipment has been upgraded and this has led to a significant increase in jamming resistance
- The heading and roll/pitch control system has been modified which has provided a decrease in disturbances exerted on the gyros whilst the system is moving. This reduces line of sight stabilisation errors and enhances the stability of the 30 mm cannon.

Upgraded Tunguska

The KBP Instrument Design Bureau has proposed another version of the Tunguska fitted with a new sensor package.

The dm wavelength search radar mounted on the turret rear is used to detect aerial targets and to automatically measure azimuth elevation, range and radial velocity of 20 targets simultaneously.

The dual-band (mm/cm) tracking radar mounted on the turret front provides automatic target acquisition and tracking by three points, missile localisation without minimum altitude restrictions, encoding and transmission of guidance commands on board the missile.

Mounted on the forward part of the turret roof is the new infrared tracker that ensures detection, automatic acquisition and tracking of aerial as well as ground based targets plus missile tracking in a wide range of illumination and weather dependent visibility conditions.

Maximum detection range against aircraft is quoted as 36 to 38 km while tracking range is quoted as 24 to 30 km. Maximum range of missiles is from 1,000 to 18,000 m in range and from 5 to 10,000 m in altitude.

While the 2A38M guns are the same as that used in the standard Tunguska 2K22M, this version uses the 57E6 SAM and brief details of this are listed in the 57E6 Missile specifications box.

An upgraded version of the system was shown at the Moscow Air Show (MAKS 2009) that displayed a new surveillance radar and the addition of a thermal camera sight to supplement the original day only optical gunner's sight. This new camera is mounted above the optical sight and both sensors are protected by a single housing with a folding cover. The gunner also has a TTA (*Teleteplovizniy Avtomat*) automatic target tracker that can use television or thermal camera data to track air or ground targets.

Specifications

	2K22M
Dimensions and weights	
Crew:	4
Length	
overall:	7.93 m
Width	
overall:	3.236 m
Height	
transport configuration:	3.356 m
deployed:	4.021 m
Weight	
combat:	34,000 kg
Mobility	
Configuration	
running gear:	tracked
Speed	
max speed:	65 km/h
Range	
main fuel supply:	500 km

	2K22M
Fording	
without preparation:	0.8 m
Gradient:	60%
Side slope:	30%
Vertical obstacle	
forwards:	1 m
Trench:	2 m
Engine	V-46–4, V-12, turbocharged, water cooled, 4 stroke, diesel, 780 hp at 2,000 rpm
Gearbox	
model:	hydro-mechanical
Suspension:	hydropneumatic
Firepower	
Armament:	2 × turret mounted 30 mm 2A38M cannon
Armament:	8 × turret mounted 9M311/9M311M/9M311–1M SAM
Ammunition	
main total:	1,904
missile:	8
Rate of fire	
cyclic:	5,000 rds/min (est.)
Turret power control	
type:	hydraulic
Main armament traverse	
angle:	360°
Main armament elevation/depression	
armament front:	+87°/-10°
Survivability	
Night vision equipment	
vehicle:	yes
NBC system:	yes

Status
In production with an emphasis on export and overseas sales, the system in use in Russia will be replaced with the Pantsyr-S over a period of time. In service with China, India (60+), Morocco (delivered 2005/2006), Algeria (delivered 2006) and the Russian Federation. Of particular interest is the fact that the 2K22M was the only air-defence gun/missile taken into Georgia during the 2008 Russian police action.

Contractor
Primary contractor: Ulyanovsk Mechanical Plant.
Production and systems integrator: KBP Instrument Design Bureau.
State Scientific Research Institute of Machine Building, FGUP/GosNIIMASH (warhead design for the 9M311 missile family).
Signal, All-Russian Scientific and Research Institute, GUP/VNII Signal (developed the AC and DC electric and hydraulic electric system).
Tulamashzavod Joint Stock Company (produces the twin-barrelled automatic AA 2A38M cannon).

2K22 Tunguska System (Christopher F Foss) 0088376

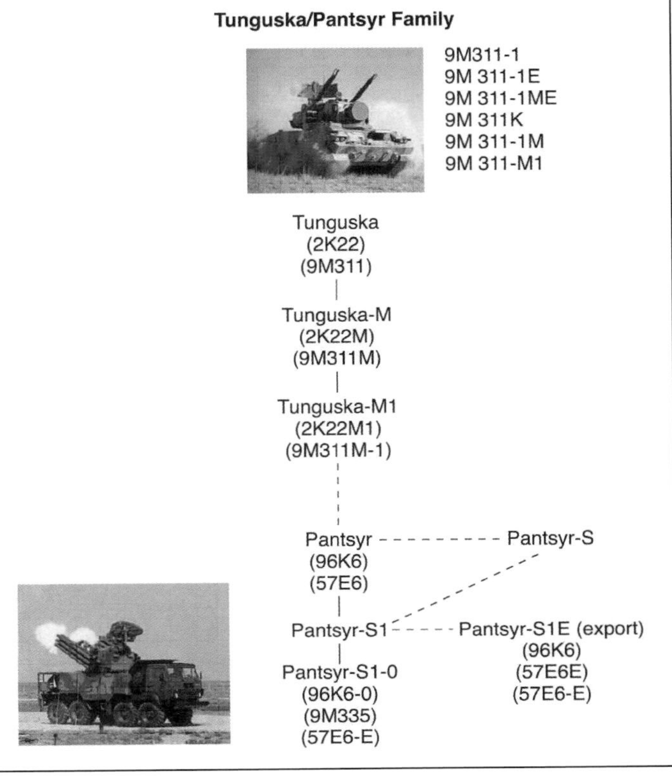

Tunguska family flow diagram (James C O'Halloran) 1154768

2K22/9M311 Tunguska

Type
Self-propelled Surface-to-Air Missile (SAM) and gun system

Development
The Tula based KBP design bureau has developed a tube launched, two stage, supersonic, low-altitude Surface-to-Air Missile System (SAM) for use on the Tula KBP designated 2K22 (Tunguska) system, which utilises the GM-352 combat vehicle tracked chassis.

Description
The 2K22 Tunguska is a tracked, self-propelled gun-missile system that has been designed to take-on low flying targets in addition to infantry and vehicles. The 2K22 system has four ready-to-fire SAM rounds, whereas a modified version with the designator 2K22M has eight ready-to-fire rounds. The army designator 9K22 has sometimes been used as a replacement for the 2K22. The basic missile is the 9M311 Treugolnik (triangle).

Missile variants
At least five variants of the missile are known to exist, these are:
- 9M311 also known as the naval 3M87
- 9M311-1 the export variant
- 9M311K probably the original naval missile, but later the letter K was dropped from use
- 9M311M also known as the naval variant 3M88
- 9M311-M1 the second modification of the missile for use with the Tunguska-M1.

The 9M311/3M87 missiles are used with the 2K22M system and the naval equivalent Kortik. The 9M311 naval missile originally had the designator 9M311K but the letter K was subsequently dropped and the missile is now known as the 3M87.

The 9M311M are the first true variant of the missile and are also used with the 2K22M. However, this missile is believed to be the naval variant 3M88.

The 9M311M/3M88 missile is 2.652 m long and has diameters of 0.17 m and 0.076 m at its widest and narrowest points respectively; the booster section being the larger in diameter. The missile is mounted in two elevatable launcher tube pairs on either side of the turret and is designed primarily for use against anti-tank helicopters.

A 9 kg HE-fragmentation and continuous rod warhead is fitted; either the contact fuze or active laser proximity fuze activates this. The lethal radius of the warhead is up to 5 m. In this warhead a set of rods are laid over the HE charge and are constrained by a rod ring. The rods are about 500 mm long, range from four to 9 mm in diameter and have a layer of cube-shaped prefabricated fragments weighing 2-3 g arranged over them. These fragmentation effects are combined to ensure that the targets' skin and framework is ripped, thus destroying the airframe structure and causing an instantaneous break-up in mid air.

The 9M311M missile has a minimum range of 1,500 m, while maximum is 8,000 m. Altitude limits are from 15 to 4,000 m. Missile launch weight is 42 km. Maximum missile speed is 900 m/s. The maximum target speed is 500 m/s. A second improved missile the 9M311-M1 is slightly heavier and longer, and was shown for the first time in 1998. This missile has a longer-range capability and is associated with the Tunguska-M1.

Command Post
The 9S482M7 (PU-12M7) was upgraded and modernised during 2007/8. The original version was based on a BTR-60 8 × 8 Armoured Personnel Carrier (APC) and relied on data from external radar sources. The latest variant uses a K1Sh1 wheeled vehicle based on the BTR-80 (GAZ 5903) 8 × 8 APC and incorporates a telescopic sensor mast. Within the telescopic mast are located the SRTR (*Stantsiya Radiotekhnicseskoy Razvedki*), passive ESM system and an electro-optical unit that is located at the masthead.

Tunguska Resupply Vehicle (Christopher F Foss) 0069233

The mast also boasts four static antennas mounted halfway up the mast for short-range radar. Each antenna covers a sector, which when combined provides all-round coverage.

A trainable electro-optical unit mounted on top of the turret contains day and night cameras, providing hemispherical coverage. The PU-12M7 can accept target data from external surveillance radar, other fire-control systems and radar equipped air-defence weapons. Communications with assets up to 15 km away can be achieved by wire, with radio being used for ranges out to 40 km.

This then allows the PU-12M7 tactical control of an area out to 100 km, handling information for as many as 99 airborne targets. The operator can simultaneously track up to nine targets.

Status

In production and in service with the Russian Federation, India and possibly China, furthermore it is reported that Peru has ordered the 2K22M system from Belarus and that the Russian Federation is to supply the Tunguska-M1's to Morocco.

The Tunguska is gradually being replaced within the Russian Army by the Pantsyr system.

Contractor

Designer: The KBP Instrument Design Bureau, Tula, Russian Federation.
Production: Ulyanovsk Mechanical Plant, Federal State Unitary Enterprise.
Warhead design: State Scientific Research Institute of Machine-Building FGUP/GosNIIMASH.
Command-and-control complexes: Signal, All Russian Scientific Research Institute GUP/VNII Signal.
Guns: Tulamashzavod, Joint Stock Company.
Mobile training simulators: Tulatocsmash.
Thermal Sight: The thermal sight for the Tunguska-M1 is designated 1TPP1-T and operates in the 3.0-5.0 μm-band with 256 × 320 pixels and is part of the TTA (*Teleteplovizionniy Avtomat Soprovozhdenia Tselej*) thermal and tracking device. The system has two Fields of View (FoV) wide 7.3 × 6.9 and narrow 2.0°-1.5°. The overall weight is reported as 6.5 kg. The TTA was developed by VKT Scientific-Designing Centre for Video and Computing Technologies that is part of the Ryazan Instrument Factory GRPZ. The thermal sight is associated with the Applied Optics Institute NPO GIPO in Kazan.

96K6 Pantsyr-S1

Type

Self-Propelled Anti-Aircraft Gun and Missile (SPAAGM) system

Development

The KBP Instrument Design Bureau, in Tula, has developed a self-propelled air defence weapon family known as the Pantsyr and Pantsyr-S1 (Russian for 'armour') (NATO SA-22 Greyhound) as a replacement for the Tunguska gun/missile system. The system was originally intended for strategic defence of high-value targets (hence the designation) such as ballistic missile silos, airfields and communications nodes against airborne, ground or sea borne threats. The very early missile used in this system was the 9M311 but later the 57E6.

Since development commenced in 1989, the strategic threat has diminished and the Pantsyr system has matured to meet new tactical threats. A tactical missile, the 9M335 which has replaced the 9M311 is now available with the system to meet this new threat, and the weapon is advertised for the defence against fixed-wing aircraft, hovering helicopters, cruise missiles, UAVs, precision-guided weapons and moving ground targets.

The system is designated ZRPK BD (*Zenitnyi Raketno-Pushechnyi Kompleks Blizhnego Deistiviya*; Russian for 'short-range air defence gun and missile system). The Russian industrial index number for the system is 96K6.

A typical battery could comprise two to three, four to six, or eight to 10 fire units (designated BM combat vehicles) with the larger battery sizes having a centralised command-and-control vehicle on the same chassis.

ZU-23-2 Axel with missile (James C O'Halloran) 1480199

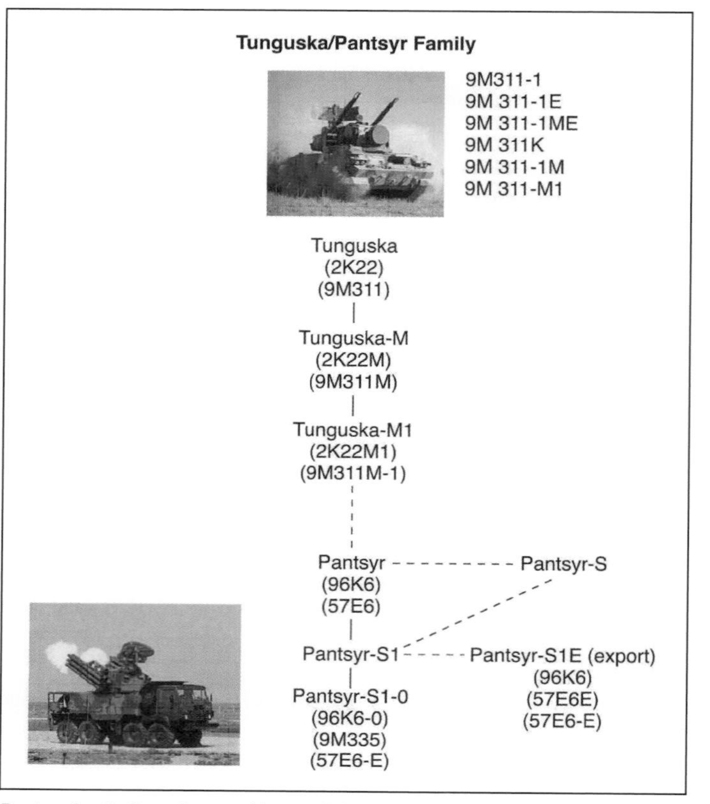

Pantsyr family flow diagram (James C O'Halloran) 1154768

In mid-2000 the United Arab Emirates signed a contract worth USD734 million with the KBP Instrument Design Bureau for final development and subsequent delivery of 50 96K6 Pantsyr-S1 self-propelled air defence systems.

KBP Instrument Design Bureau completed development of the system in 2003. Under this contract, 12 systems were delivered in 2003, 24 were scheduled for delivery during 2004 and 14 for 2005. Of these 50 systems, 26 will be on a wheeled chassis and 24 on a tracked chassis.

During 2007 the system was extensively tested and tuned at the Ashuluk Missile Test Range in the Astrakhan Region of Southern Russia. Testing was completed late in 2007.

In December 2013, there was the first indication that Russia is developing a modernised variant of the system known as the Pantsyr-SM. The modernisation will take into account increasing the effectiveness against UAVs and also the system will be adapted for deployment in Arctic conditions. The system is due to be deployed sometime in 2015.

Further live firings have taken place this time at the Kapustin Yar Missile Test Range (KYMTR) of the Pantsyr-S system. Video evidence of these firings confirmed at least two types of missile being used and the appearance of a new an novel way of launching a single missile from an old ZU-23-2 axle with the wheels folded down.

Description

The Pantsyr (NATO SA-22 Greyhound) is a short-range mobile air-defence system combining two 30 mm 2A38M automatic cannon anti-aircraft guns and 12 , 57E6 or 9M335 surface-to-air missiles. It can simultaneously engage two separate targets. The early systems were mounted on the rear of a Ural-5323.4 (8 × 8) 20,000 kg forward control cross-country truck chassis, however, later systems designated as Pantsyr-S1E and destined for the United Arab Emirates (UAE), utilise the Kamaz-6560 four-axle truck. The Kamaz-6560 weighs 34 tonnes and has a maximum road speed of 70 km/hour. A further variant is mounted on a tracked chassis.

New Phased Array Radar (James C O'Halloran) 1459414

Pantsyr-S1E for UAE (Miroslav Gyürösi) 1331019

The system components consist of:

- Combat vehicle (up to six CVs in a battery)
- Battery command post
- SAMs
- Radar and EO tracking devices
- 30 mm rounds
- Transloader vehicles (1 vehicle for 2 CVs)
- Training facilities
- Maintenance facilities
- Common spare parts (SPTA) kit.

A large turret is mounted on the top of a shelter-type compartment, which has two banks of six or eight ready-to-fire surface-to-air missile container-launchers on either side. Also mounted with the missiles are two model 2A38M 30 mm cannon with up to 1,400 rounds of ready-use 30 × 165 mm ammunition. Once the missiles have been fired, the system has to be reloaded from an external source.

Typically, the missiles would be used first to engage targets at long range and then the 30 mm cannon to engage any leakers at shorter ranges.

Mounted on top of the turret is the 2RL80E S-band surveillance radar, which can be folded flat to reduce the overall system height when travelling. Mounted below the surveillance radar and between the two banks of SAMs are the Phazotron 1RS2-1E Schlem multimode adaptive target tracking and missile guidance radar and electro-optic low-light/passive infra-red sensor package 1TPP1. Target engagement can be fully automatic with a capability to engage two targets simultaneously, if required. The Schlem radar was designed and is produced by Phaztron-NIIR on the basis of airborne radars. Schlem is a two-band radar controlling the operation of gun and missile against classes of targets; aircraft, helicopters, remote controlled aircraft, precision-guided weapons, and mobile ground targets. The target acquisition range (RCS equals 2 m^2), is 24 to 28 km with angular coordinates 0.2 to 0.3 milliradians.

The new-generation attack radar employs phased array technology for detecting targets and missiles and is built as a multi-range coherent station, adapted to the target environment. It will be able to ensure operation modes in accordance with priorities in tactical situations:

- Path tracking of up to 20 targets
- Detection of classes of targets in accordance with given criteria
- Simultaneous tracking of up to 10 objects (target plus missile in different combinations)
- Autonomous sector observation over the environment in tactical situations, when target detection assets cannot detect the target or measure its coordinates with sufficient accuracy (for example, in the sector of intense jamming or of targets flying by a dense group)
- Assisting observation over optical guidance devices of air defence systems to specify information on ranges
- Forming of control commands for missiles at all stages of their guidance.

Pantsyr new Phased Array Radar (James C O'Halloran) 1459415

Close-up of the Pantsyr-S1 turret showing 12 missiles carried in the ready-to-launch position (Steven Zaloga) 0509863

The two-stage 9M335 or 57E6 missile are essentially similar to the 9M311 weapon used on the Tunguska 2S6M but has a thicker fuselage cross-section in the engine area and a longer solid fuel booster stage. It has a maximum altitude capability of 6,000 m, a minimum altitude capability of 5 m and a maximum range of 12,000 m.

In addition to this 57E6 missile, there is also the 57E6E which has a larger warhead and longer range that takes the missile to a range of 20 km.

The 57E6 designator represents a PVO missile, with the 'E' variant an export missile.

A contact and radar proximity fuzing system is fitted to the 16 kg HE-fragmentation continuous-rod warhead. The warhead's rods and fragments maximum attainable velocity is said to be 1,600 m/s. The booster section is a larger calibre than the sustainer/main missile body.

The 8 × 8 chassis used as the basis for the Pantsyr-S1 system features powered steering on the front two axles and a central tyre pressure system that allows the driver to adjust the tyre pressure to suit the terrain being crossed.

When deployed in the firing position, the Pantsyr-S1 is normally stabilised by four jacks, which are mounted two either side.

During October 2009, BAZ (*Bryanskiy Avtomobilniy Zavod*) the Russian truck developer and producer of the Pantsyr-S1 released a photograph claiming it to be of the Russian Army version of the system with the designator Pantsyr-S. Although the basic turret, cabin and power supply are nearly identical to those on the earlier export version, the most obvious change is that the system has only eight ready-to-fire 57E6 missiles rather than the 12 fitted to export systems.

The 9M335 missile used in the Pantsyr-S1 is a further development of the 9M311 missile used in the 2S6M self-propelled gun/missile system (Steven Zaloga) 0044127

This Russian Army version is mounted on a BAZ-6909 8 × 8 four axle truck chassis with a kerb weight of 19,000 kg and a load-carrying capacity of 22,000 kg. It was also reported that trials of this particular system were conducted in mid-2008.

Variants

The following variants have been developed:

- A shelter version with the crew accommodated in a fully equipped and protected shelter. The system is operated remotely using control panels.
- A naval version known as Pantsyr-M for use on all new classes of Russian combat ships, from corvettes to cruisers. There is also an export variant known as Pantsyr-ME that has a response time of 3-5 seconds and can track and destroy simultaneously up to four targets. The export version has a range of 20 km and can hit targets at altitudes from 2 m to 15 km whilst its guns have a range of 4 km and can hit targets at altitudes up to 3 km.
- A simplified version using only an electro-optic fire-control system. This is ideally suited for countries where there are no limitations imposed on operational use in adverse weather conditions, for example the Middle East.
- Pantsyr-S1-0, a mobile self-propelled system that employs the 57E6 missile and no guns.
- Pantsyr-S1E for Middle Eastern customers.
- A new and upgraded version with a new two faced phased array radar has been developed for Algeria.

Pantsyr-SM

In December 2013, there was the first indication that Russia is developing a modernised variant of the system known as the Pantsyr-SM. The modernisation will take into account increasing the effectiveness against UAVs and also the system will be adapted for deployment in Arctic conditions. The system is due to be deployed sometime in 2015.

Pantsyr-S1E

In early 1999, details were released of an export missile designated the 57E6E, which is faster than the current missile used by the Pantsyr-S1 and therefore has a shorter time of flight.

The new advanced fuze has a 9 m radius of action and is of the contact plus proximity, radar and adaptive type.

The new missile has a maximum speed of 1,300 m/s and an average speed at a range of 18 km of 780 m/s. The missile and its associated launch container weighs 85 kg with the actual missile on its own weighing 71 kg.

Overall length of the missile and its launch container is 3.20 m, with the sustainer having a diameter of 90 mm and the booster a diameter of 170 mm. The warhead weighs 20 kg with the explosive content being 5.5 kg.

According to Russian sources, the new missile has a maximum engagement range from 1,200 to 20,000 m and from 5 to 10,000 m in altitude.

In December 2005, Phazotron-NIIR released information on the 1RS2-E combined target and missile tracking radar that has been developed specifically for the Pantsyr-S1E system. In the Russian designation architecture, the addition of the capital "E" normally refers to export variants. This would then tie up with the export to the United Arab Emirates of the Pantsyr-S1 and the 57E6E missiles. The 1RS2-E radar was revealed to be a dual-band radar operating at centimetric and millimetric wavelengths, and based on an open architecture. The radar is designed to detect and track fixed-wing aircraft, hovering helicopters, cruise missiles, UAVs, precision-guided weapons and moving ground targets. The radar has a maximum lock-on range against a 2 m^2 cross section air target of 30 km. The system will track only one target while providing guidance for two missiles simultaneously. A second target may be tracked using the electro-optic low-light/passive infra-red sensor package that is mounted on the turret close to the radar and between the two banks of missiles.

By mid-2006, a further update to the Pantsyr-S1E had taken place, this time the 1RS2-E radar was replaced with the MRLS (*Mnogofunktsionainiy Radiolokator Soprovazhdeniya*) 1RS2-1E phased-array fire-control radar and the new pattern chassis for the wheeled variant of the system.

The new chassis was displayed by KBP in model form showing the MZKT-7930 Astrolog wheeled chassis developed and produced in Belarus, while a promotional leaflet showed the system installed on the Russian Kamaz-6350 Mustang chassis. Both are four-axle heavy trucks, the tracked chassis will be a GM-352M1E vehicle developed and produced by Minskiy Traktorniy Zavod from Belarus.

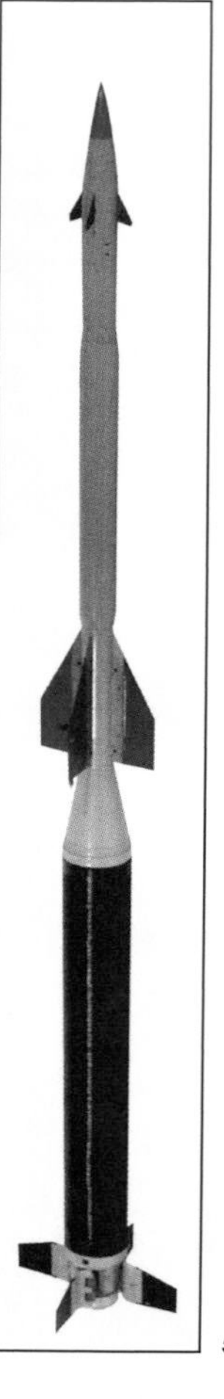

57E6 missile 0010178

A further new chassis has also been identified, this time the Pantsyr-S1E was observed mounted on a Kamaz-6560 four-axle heavy truck. This vehicle is powered by a Kamaz 740.35 engine with a nominal power output of 294 kW and a maximum exploitable power output of 272 kW at 2,200 revolutions per minute.

The new MRLS radar has reportedly given the Pantsyr system a major boost in multitarget capability. Originally the Pantsyr with the 1RS2 could only engage two simultaneous targets, one using the radar while the other used the electro-optic sensor unit. This new MRLS radar however has a multirole tracking and fire-control capability that uses a passive phased-array antenna (*prokhodnaya fazirovanaya antennaya reschotka*) and operates at a wavelength of eight mm, which corresponds to a frequency of about 40 GHz at the upper end of K-band. The antenna has a coverage of 45° in both horizontal and vertical planes, can simultaneously track up to four targets while guiding missiles against three. A fourth missile in flight is possible using the electro-optic sensor that operates in the 0.8 to 9.0 microns wavelength range.

The maximum tracking range is 24 to 28 km. The surveillance radar is left unchanged as are the guns and missiles.

Tracked system

In 1999, a model of a full tracked chassis was shown fitted with a new turret with eight missiles in the ready-to-launch system, with the surveillance radar on the turret roof at the rear and the tracking radar on the turret front. By 2009, this variant was shown for the first time at the IDEX Defence Exhibition in Abu Dhabi.

Algerian variant

During February 2012 the Deputy Chairman of the Russian Government, Dmitry Rogozin was shown the new Algerian version of the Pantsyr-S1 with a new radar detection (two panel antennas with phased array) mounted on a KAMAZ-6560 vehicle. This new radar has been designed for the detection and tracking of targets including electronic counter-counter measures including active and passive jamming.

Radar performance

Target detection and tracking with RCS 1m.kv:	40 km
Number of simultaneously tracked targets:	>40
Altitude of detected targets:	5 m to 20 km
Radial velocity:	30-1200 m/s
Overload during manoeuvre:	10 g
Coverage in azimuth:	
(mechanical)	0-360°
(electronic)	90°
Coverage in elevation (electronic):	0-60°
Weight of radar:	432 kg

Comparative table of old 1RS1 (VNIIRT) radar and new Algerian RLM SOC (TsKBA) radar

	RLM SOC	1RS1
Field of view azimuth:	0-360°	0-360°
Coverage (electronic scanning):	0-60°	0-60° and 40-80°
Number of processed targets:	40	8-10 and 20
Detection range (2 m²):	more than 40 km	36 km
Target velocities:	30-1,200 m/s	0-1,000 m/s

Specifications

	Pantsyr-S1
Dimensions and weights	
Crew:	3
Weight	
combat:	20,000 kg
Mobility	
Configuration	
running gear:	wheeled
layout:	8 × 8
Firepower	
Armament:	2 × turret mounted 30 mm 2A38M cannon
Armament:	12 × turret mounted 9M335 SAM
Armament:	alternate main 12 × turret mounted 57E6 SAM
Ammunition	
main total:	1,400
Main weapon	
maximum range:	4,000 m (gun)
maximum altitude:	3,000 m (gun)
minimum altitude:	0 m (gun)
Operation	
stop to fire first round time:	180 s

	9M335	57E6-E
Dimensions and weights		
Length		
overall	3.2 m (10 ft 6 in)	3.2 m (10 ft 6 in)
Diameter		
body	170 mm (6.69 in)	90 mm (3.54 in)
	-	170 mm (6.69 in) (booster)
	-	76 mm (2.99 in) (warhead)
Performance		
Speed		
max speed	2,138 kt (3,960 km/h; 2,461 mph; 1,100 m/s)	2,527 kt (4,680 km/h; 2,908 mph; 1,300 m/s)
Range		
min	0.5 n miles (1.0 km; 0.6 miles)	0.6 n miles (1.2 km; 0.7 miles)
max	6.5 n miles (12.0 km; 7.5 miles)	10.8 n miles (20.0 km; 12.4 miles)
Altitude		
min	5 m (16 ft)	15 m (49 ft)
max	6,000 m (19,685 ft)	15,000 m (49,213 ft)
Ordnance components		
Warhead	16 kg (35 lb) HE fragmentation	20 kg (44 lb) HE fragmentation
Fuze	impact, proximity	impact, proximity
Guidance	IR (radio command)	IR (radio command)
Propulsion		
type	solid propellant, two stage	solid propellant, two stage

Status

Development of the system has been completed. The order for United Arab Emirates that was expected to be completed by the end of 2005 had been delayed over the Fazotron Shlem tracking and guidance radar. KBP had to find another developer for the system thus delaying the trials that have now completed in Russia. Series production for the UAE order has now begun with deliveries underway.

Talks between Algeria and Russia have reached the contract stage for the supply of the Pantsyr-S1, however, Algeria had been informed by Rosoboronexport that until the UAE contract is complete, work on an Algerian order cannot start. However, the system for Algeria includes a new two faced S-band phased array radar known in Russia as the RLM SOC (TsKBA). Trials/acceptance tests of the system were completed between May and July 2011.

As of April 2007, Jordan and Syria were reported as customers for the Pantsyr-S1E variant although deliveries will not commence until after the trials at Ashuluk and deliveries have been made to the UAE. Early August 2007, deliveries to Syria have commenced and will continue until all 36 have been completed. In mid-May 2013, Hassan Nasrallah the then Hizbullah group leader revealed in a press conference that Syria had supplied the Pantsyr-S1E and the S-200 air defence systems to the group for impending use against Israel. It is easy to understand the Pantsyr system with Hizbullah but a lot more difficult to accept the S-200.

Saudi Arabia (Riyadh) is planning to create a comprehensive air defence system, which promises major contracts for the Russian defence industry. It is ready to buy nearly everything from Russia, such as the Pantsyr-S1 truck mounted system.

It was reported in March 2008 by the KPB in Tula, that so far, orders for 64 systems had been forthcoming and that up to 24 of these systems were delivered during 2008. A Syrian delegation arrived in Tula during mid-April 2008 on a pre-delivery inspection visit.

In March/April 2010, 10 of the Pantsyr-S1 systems were delivered to the Russian Air Force and operationally deployed in the Moscow Air-Defence ring as point defence of the S-400 system. Again in late November 2013, a further estimated 100 Pantsyr systems were on order for the next batch of S-400 systems to be deployed in the defence of Russia late in 2013 and during 2014.

Early October 2012, a contract between Russia and Iraq for 42 × Pantsyr-S1 was agreed and signed. No date for delivery has been set, but this is part of an overall contract valued at USD4 billion.

In early February 2013, it was announced that Brazil has approached the KBP in Russia for the purchase of three batteries of the Pantsyr-S1 and two batteries of Igla missiles. No contract has yet been signed.

Early in February 2014, a photograph allegedly taken in Vietnam of a container-based shelter mounted Pantsyr-S1 was identified. The system was on a low-loader trailer with two Vietnamese military personnel alongside, the military personnel were further identified from their uniforms and shoulder flashes. It is not yet known if this is a genuine picture.

Contractor

System development: KBP Instrument Design Bureau, Tula.
State Institute of Applied Optics, Science and Production Association, FNPTs, GUP/NPO, GIPO (infra-red system for the Pantsyr SAM).
ShchEGLOVSKIJ VAL is the name of the production facility.

IMADS - Integrated Mobile Air Defence System

Type

Self-propelled Surface-to-Air Missile (SAM) system

Development

A consortium of Russian defence organisations has developed the Integrated Mobile Air Defence System (IMADS) in response to the multi-layered defence market. In the past, layered defence has been relatively static, used for the protection of high-value targets, military and politically sensitive areas and, where available, densely populated civilian areas. However, this new concept is based on highly mobile assets for the defence of the army while on the move.

Description

The IMADS system will include an air defence missile and anti-aircraft artillery system of various classes, as well as other weapons. Data and information exchange is based on various radar and electronic intelligence assets, command-and-control and data exchange components.

The IMADS system is offered on the basis of:

- Buk-M1-2
- Tor-M1
- Tunguska-M
- Pantsyr-S1
- Shilka
- 85V6-A Vega (ELINT) system
- SPN-2 and SPN-4 (jammers).

Status

The components of the IMADS system have all been developed and are known to be operational at this time, however, integration and operability presumably will be carried out at the customer's request and requirements.

As of July 2012, there have been no official customers of the IMADS system as a whole unit, however, individual units continue to sell through Rosoboronexport.

A system very similar in appearance and capabilities to the IMADS has been developed in South Korea with Russian advisers. This system is known to use a variant of the 9M96 missile commonly used with the S-400 air defence system. In South Korea the system is known as the Iron Hawk II (also known as Cheongung, Chun Koong (this is translated as The Heaven's Arrow).

Contractor

A Russian consortium consisting of:
Tikhmirov NIIP Instrument-Making Research Institute.
Dolgoprudny Scientific Production Enterprise.
Ulyanovsk Mechanical Plant.
Kupol Electromechanical Plant in Izhevsk.
Kalinin Machine-Building Plant in Yekaterinburg.
Almaz/Antei Concern of Air Defence.
Instrument-Making Design Bureau KBP in Tula.

MT-LBM1 (Variant 6M1B5)

Type
Self-Propelled Anti-Aircraft Gun and Missile (SPAAGM) system

Development
The Muromteplovoz MT-LBM1 carrier is intended to transport infantry squads to the battlefield, for reconnaissance and for protecting military columns and objects, defeating enemy manpower and lightly armoured vehicles as well as low flying air targets, including conditions of natural background and man-made thermal interference.

Description
Based on the now familiar MT-LBM chassis that was first introduced by the Soviets in the late 1960s, this gun and missile system is heavily armed for its size. The weapons suite consists of a 30 mm twin-barrelled GSh-30K automatic cannon, a Kord 12.7 mm heavy machine gun and a 30 mm AG-17 automatic grenade launcher. Also featured is a dual launcher for the Igla (family) of air-defence missiles.

Variants
Many variants of the MT-LBM has been constructed. Variants include those with air defence and ground-to-ground roles.

Those with a purely ground-to-ground role are as follows:

- **MT-LBM 6MA** - with 14.5 mm BTR-80 type turret
- **MT-LBM 6MA1** - with 14.5 mm BTR-80 type turret plus AG-17 30 mm automatic grenade launcher
- **MT-LBM 6MA2** - with 14.5 mm BTR-80 type turret with KPVB 23 mm instead of KPVT
- **MT-LBM 6MA3** - with 4 ATGM Kornet, PKTM and AG-30
- **MT-LBM 6MA4** - with 14.5 mm BTR-80 type turret with KPVB 23 mm instead of KPVT
- **MT-LBM 6MB** - with 30 mm gun 2A72 with PKTM
- **MT-LBM 6MB2** - with 2A42 30 mm gun PKTM and AG-17.

Those with an air defence and a subsidiary ground-to-ground are as follows:

MT-LBM 6MB3
The MT-LBM 6MB3 variant is fitted with a twin-barrelled GSh-23V 23 mm (23 × 152 B) automatic cannon, an AG30 30 mm (30 × 29 B) Automatic Grenade Launcher (AGL) and a Kord 12.7 mm (12.7 × 108) Heavy Machine Gun (HMG).

MT-LBM 6MB4
The MT-LBM 6MB4 variant is fitted with a twin-barrelled GSh-30K 30 mm (30 × 165) automatic cannon, an AG-30 30 mm AGL and Kord 12.7 mm HMG.

MT-LBM 6MB5
The MT-LBM 6MB5 variant is fitted with either a twin 23 or 30 mm automatic cannon, in addition to an Igla SAM and improved sighting equipment.

MT-LBM1 6M1
The MT-LBM1 6M1 variant is fitted with either a twin-barrelled 23 or 30 mm automatic cannon, in addition to an Igla SAM, improved sighting equipment and an improved 300-310 hp engine.

MT-LBM1 (Variant 6M1B5) (Miroslav Gyürösi) 1123835

Specifications

	MT-LBM1 (Variant 6M1B5)
Dimensions and weights	
Length	
overall:	6.45 m
Width	
overall:	2.86 m
Height	
overall:	2.63 m
Ground clearance	
overall:	0.400 m
Track	
vehicle:	2.50 m
Weight	
combat:	11,950 kg
Payload	
land:	1,000 kg
Ground pressure	
standard track:	0.51 kg/cm^2
Mobility	
Configuration	
running gear:	tracked
Speed	
max speed:	66 km/h
water:	6 km/h (est.)
Amphibious:	yes
Gradient:	33% (est.)
Side slope:	28% (est.)
Engine	YaMZ-238BL, 4 stroke, diesel, 228 hp
Gearbox	
model:	mechanical with hydraulic drives of slewing mechanism

Status
Development is complete, the system is reportedly in use with the Russian Armed Forces.

Contractor
Muromteplovoz JSC.

FGUP PO UOMZ (Uralskiy Optiko-Mekhanicseskiy Zavod) from Yekaterinburg (electro-optics).

Yug

Type
Self-Propelled Anti-Aircraft Gun and Missile (SPAAGM) system

Development
The Yug system has been designed and developed for the protection of troop units in combat and on the march against low-flying threats such as fixed wing assets, helicopters, UAVs and cruise missiles.

Description
The Yug system is mounted on the MT-LBM tractor chassis for complete mobility. For air-defence targets the Yug is armed with the GSh-30K twin-barrelled 30 mm cannon and a module carrying two Igla MANPADs.

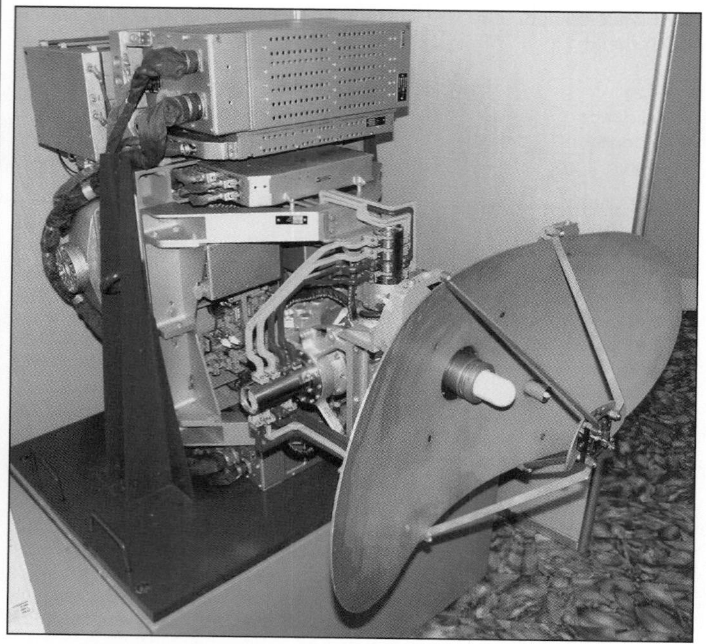

Arbalet Radar used with Yug (Piotr Butowski) 0566966

For self-defence against ground targets the Yug uses a 12.7 mm 6P51 Kord Heavy Machine Gun (HMG) and the either the AG-17 or AG-30 30 mm Automatic Grenade Launcher (AGL).

For sighting the vehicle is fitted with TOES-523 turret-mounted electro-optic system (a moveable turret with TV and thermal imaging devices) and a TKN-4GA reserve sight.

Variants

None known.

Specifications

Yug

Type:	self-propelled anti-aircraft gun and missile system
Target engagement range:	6,000 m
Target engagement altitude:	≤3,500 m
Target speed:	≤450 m/s
Armament:	
(cannon)	30 mm GSh-30K twin-barrel
(missile)	2 × Igla man-portable SAM
(HMG)	12.7 mm 6P51 Kord
(AGL)	30 mm AG-17 or AG-30
Sights:	
(main)	TOES-523
(reserved/backup)	TKN-4GA
Deflection angle:	
(azimuth)	270°
(elevation)	–10° to +80°
Detection range:	
(at night in PMU, ALCM target)	≥6 to 8 km
(at night in PMU, AN-64 target)	8-10 km
(maximum)	12 km
Target lock on and tracking device:	Okhotnik-V
Radar coverage:	
(azimuth)	360°
(elevation)	0° to +80°

Status

Offered for sale as of 2006/07. No known customers at present.

Contractor

Muromteplovoz JSC (Murom, Vladimir Region, Russia).

ZU-23/ZOUM1

Type

Static and towed gun and missile system

Development

The ZU-23/ZOUM1 has been developed from a modernised 23 mm ZU-23 (2A14) twin-barrel air-defence gun combined with either the 9M342 Igla-S and/or the 9M333 Strela-10M3 missiles. This anti-aircraft gun-missile complex is designed for effective protection of stationary and slow-moving objects against strikes of tactical and attack aviation, helicopters, cruise missiles and UAVs. The gun may also be used in the surface role for the protection of troops and above water lightly armoured boats in both day and night conditions.

Description

The system requires a crew of two, mainly for tasks such as deployment of the unit or preparing it for travel; for loading and reloading of the missiles. The system consists of the following subsystems.

ZU-23/ZOM1-SM (Strelbovoy modul)

Gun firing unit, this is an adapted ZU-23 gun unit modernised by the installation of:

- New vertical and horizontal servo drives
- Control and display consoles for the operator
- Automatic target tracker
- Computing unit
- Collimator sight
- Rechargeable batteries.

The gun has 100 rounds of ready use ammunition and may also be fitted with two 9M342 Igla-S or 9M39 missiles installed above the gun mount utilising the PM-23M1-1 launch module.

ZU-23/ZOM1-PM (Puskovoy modul)

Missile firing/launching unit is essentially a ZU-23 gun carriage but without guns, fitted instead with new servo drives and a frame able to carry four transporting/launching containers for the 9M333 Strela-10M3 missiles. The system may also make use of the older but still very capable 9M37M, 9M37M1 and 9M37MD missiles used by the older Strela-10.

The launcher may be slewed/trained through 360° in the azimuth and –5° to +70° in the vertical.

ZU-23/ZOM1-MU (Modul upravleniya)

Electro-optic remote-control unit is used to search for, detect and track targets, then pass the relevant data to the missile launching units, it consists of:

- Daylight TV camera working in the 0.6-1.1 micron waveband
- Thermal camera working in the 8-14 micron waveband
- Laser rangefinder working in the 1.54 micron waveband
- Computing unit
- Display monitor.

The electro-optic unit is mounted on a tripod that provides azimuth and elevation coverage the same as for the missile launcher unit.

Autonomous power supply

The ZU-23/ZOM1 can operate in three modes:

- Automatic
- Semi-automatic
- Manual.

Targets can be engaged at slant ranges of 200 to 7,000 m, 4,500 m in altitude and at crossing ranges out to 4,500 m. The gun can engage targets with speeds up to 415 m/s. The kill probability is:

- 0.2 with the gun
- 0.3 to 0.6 with the Igla-S
- 0.5 to 0.7 using the Strela-10M3 (9M333).

No figures have been given when using the older Strela-10 missiles. Given the use of all three weapons within the system it is assessed that the kill probability will be around 80 per cent.

The time required to move off road and prepare for combat is 10 minutes maximum, although this will probably be reduced with a well worked up crew.

Status

Partial development of the system is complete, the following units have been offered for sale in five configurations:

(1) ZU-23ZOM1-1 consisting of (a) ZU-23/ZOM1-MU electro-optic remote control unit and (b) ZU-23/ZOM1-PM missile launching unit
(2) ZU-23/Zom1-2 consisting of (a) ZU-23/ZOM1-SM gun firing unit instead of a missile launcher
(3) ZU-23/ZOM1-3 consisting of (a) ZU-23/ZOM1-SM gun firing unit and (b) ZU-23/ZOM1-PM missile launching unit
(4) ZU-23/ZOM1-4 consisting of (a) upgraded gun firing unit
(5) ZU-23/ZOM1-4U (U = *Uproschonniy* = simplified) consisting of (a) less comprehensively modernised ZU-23/ZOM1-SM gun firing unit.

As of August 2008, only the ZU-23/ZOM1-4 and ZU-23/ZOM1-4U were available to potential customers.

Other variants are still in development and firing trials of the ZU-23/ZOM1-PM missile launching were planned for 2009. It is not known if these launches took place in Russia, no further information on this system was available as of August 2012.

Contractor

Pemz Spetsmash (Podolskiy elektromekhanicseskiy zavod) also known in the west as Open Joint Stock Company Podolsky Electromechanical Plant of Special Engineering (development and production).

Rosoboronexport (enquiries).

NTC ELINS Scientific Technical Centre, also known as the Closed Joint Stock Company NTC Elins (remote-control module and computing unit).

Switzerland

Skyranger

Type

Self-Propelled Anti-Aircraft Gun and Missile (SPAAGM) system

Development

Oerlikon Contraves, now a subsidiary of the German Rheinmetall Air Defence, developed the Skyranger gun and missile system in order to meet emerging requirements for a highly mobile air defence system that is capable of engaging a wide range of future air defence threats. In addition, elements of the system have a ground defence mission.

Skyranger is a highly effective, self-propelled multi-mission system for the protection of mobile units and stationary assets. Although Skyranger was developed by Oerlikon Contraves, the system is at present built by Lockheed Martin Mission Systems & Sensors.

Description

Skyranger is a modular multi-mission mobile air defence system that makes use of an unmanned turret that consists of three key items, each of which can be installed on an armoured personnel carrier chassis, for example the widely deployed Swiss 8 × 8 Piranha III. Other types of chassis both tracked and wheeled can also be used. Depending on the chassis, the system is fully air transportable in a Lockheed Martin C-130 transport aircraft. The mobility of the Skyranger relies on the platform vehicle, that can be either wheeled or tracked.

The three elements of the Skyranger are:

(1) Skyranger Gun System
(2) Skyranger Missile Launcher System
(3) Skyranger Radar System with integrated Control Centre.

Skyranger 35 mm self propelled air defence system mounted on Piranha armoured fighting vehicle (IHS/Patrick Allen) 1066275

Skyranger radar control section (IHS/Patrick Allen) 1066317

The combination would depend on the mission but could typically be one Skyranger Radar System that would coordinate two Skyranger Gun Systems and one Skyranger Missile System.

Skyranger Gun System

The Skyranger Gun System is a remote controlled turret that is armed with the Rheinmetall Air Defence 35/1000 revolver cannon that has already been produced in production quantities for the Skyshield air defence system.

The weapon which has a dual ammunition feeding system that allows the operator a choice of two types of shell primarily fires the 35 mm Advanced Hit Efficiency and Destruction (AHEAD) ammunition, which although optimised for the air defence role is highly effective against trucks as well as being able to neutralise the optics of armoured vehicles. Typical air defence targets would be; cruise missiles, air-to-ground missiles, attack drones and Unmanned Aerial Vehicles (UAVs).

Each AHEAD shell contains 152 sub-projectiles that are ejected just ahead of the incoming target, this release triggered by a precision programmable time fuze. Targets are destroyed or neutralised by multiple impacts of tungsten-alloy, spin-stabilized sub-projectiles. The AHEAD ammunition has been in production for several years and has already been adopted by at least six countries.

The 35/1000-revolver gun has a cyclic rate of fire of 1,000 rounds a minute with a typical aerial target being engaged by a burst of 20 to 24 rounds. A total of 220 rounds are carried and there is also the option of a dual-feed system with the secondary nature being Frangible Armour Piercing Discarding Sabot (FAPDS) ammunition.

Mounted on the roof of the turret is the electro-optical tracking system that automatically tracks the target with initial information being provided by the Skyranger Radar System. The passive tracking system includes an InfraRed (IR) camera, TV camera and a laser range-finder.

Fire-Control Unit

The Skyranger 3-D search and acquisition I-band fire-control unit is capable of detecting, locking onto, and tracking small, fast and steeply-attacking targets. The radar classifies detected targets and carries out a threat analysis. This shortens system reaction times and plays a key role in assuring that targets are assigned to the right effector. The Skyranger radar assures continuous detection of targets at ranges of up to 25 km and is extremely resistant to electronic countermeasures.

Skyranger Missile System

To engage targets at longer range the Skyranger Missile Launcher System is used. This is the Rheinmetall Defence Electronics ASRAD that has already been sold to three countries; Germany, Greece and Poland. For this application the launcher has four Saab Bofors Dynamics Bolide Surface-to-Air Missiles (SAM) in the ready-to-fire position.

Bolide is a laser-guided missile that is said to be almost impossible to jam and is a follow on to the earlier RBS 70 series. Other missiles could however be used including fire-and-forget types such as Stinger, Mistral or Igla.

Target information is provided by the Skyranger Radar System, which has an Rheinmetall Air Defence advanced 3-D I-band search and acquisition radar that is capable of detecting, locking onto and tracking a wide range of aerial targets.

The radar, which has a maximum range of 25 km, classifies detected targets, carries out threat analysis and then allocates the target to a weapon platform that carries out the actual target engagement. In the travelling configuration the radar is retracted.

Specifications

Skyranger turret/gun

Crew:	two - remotely located (has also been reported as 3)
Calibre:	35 mm
Cartridge:	35 × 228
Muzzle velocity:	1,050-1,175 m/s
Rate of fire:	1,000 rounds/min
Main gun load:	220 rounds
Maximum slant range:	4 km
Barrel length:	2,766 mm
Traverse:	n × 360°
Elevation:	−15° to +85°
Length:	5.75 m (approx.)
Height:	1.50 m (approx.)
Weight:	3,500 kg (approx.)
Power requirements:	8 × 12 V batteries, petrol/diesel-powered generator unit, 2.5kVA built-in battery charger

Skyranger ammunition

Cartridge:	35 × 228
Muzzle velocity:	
(HEI/TP-T)	1,175 m/s
(AHEAD)	1,050 m/s
Weight:	
(round - HEI/TP-T)	1,580 g
(projectile - HEI/TP-T)	550 g
(round - AHEAD)	1,770 g
(projectile - AHEAD)	750 g

Status

The system has undergone an extensive trial programme. First shown at Eurosatory 2004 and is currently offered for sale.

Contractor

Rheinmetall Air Defence, original developer of the gun and radar systems.
Lockheed Martin Mission Systems & Sensors, current manufacturer.
Saab Bofors Dynamics, Bolide missile manufacturer.

Surface-To-Air Missiles

Belarus

Buk-MB

Type
Self-propelled Surface-to-Air Missile (SAM) system

Development
The Ahat Research and Development Enterprise - a Minsk-based state research and manufacturing organisation - has developed and tested the Buk-MB an upgrade/modification to the basic Russian Buk air-defence system. The decision to modernise the Buk system in Belarus was taken in 2004 as the then Russian variant was considered unable to meet the Belarussian high demands.

As of November 2007, the then Belarussian President Alexander Lukashenko ordered the upgrade to be made available to Belarus air-defence forces by 2008.

The Buk system was first accepted into the Soviet Army in 1979, and the upgrade Buk M-1 in 1983.

Description
The Buk Surface-to-Air Missile (SAM) system consists of:
- A circular surveillance radar for detection and target designation
- The control post
- Six self-propelled firing units
- Three launching - loading units.

The main requirements for the new complex are:
- Simultaneous engagement of up to six targets
- Target detection range of not less than 100 km
- A kill zone range from 3 to 30 km
- The altitude from 25 m to 18,000 m.

The Buk-MB is a digital systems retrofit carried out on the older 9K37-1 and 9K37M1 systems. The retrofit involves the replacement of the 1980s based technology systems in the 9S470 command post, the 9A39 TEL and 9A310 TELAR. The ex-Soviet Cathode Ray Tube (CRT) based displays have also been replaced with modern Liquid Crystal Displays (LCDs). A more modern Electro-Optic (EO) system, the 9Sh38-3 that now includes a laser range-finder has also been added to provide range firing solutions. The new digital command post has also been integrated with Polyana DM (9S52M1RB) sector command post.

The upgraded system has now been designated:
- 9S470MB
- 9A39MB
- 9A310MB
- 9Sh38-3.

Ahat has revealed that it has upgraded the command post vehicle and the self-propelled launcher by 80 per cent and the reloading vehicle by 20 per cent.

The new Buk-MB is capable of tracking 255 targets simultaneously with a radar detection and tracking range increased from 80 to 100 km. It is also reported that the missiles now have an increased range to complement the tracking radar's increased range.

Other upgrades include the ability to use the system against surface targets such as aircraft at airfields and surface warships at ranges up to 25 km.

In addition to the upgrading of the technology systems, the tracked chassis of the vehicle has undergone a complete overhaul.

Variants
The SRPA Ahat organisation in Belarus is also marketing an upgrade/modification to the BUK SAM system. Photographic literature indicates the use of both missiles: 9M38 and 9M317 or variants thereof. The details so far released include the fact that the aim of the modernisation is to increase the combat efficiency of the BUK SAM system by:
- Informational and technical interface of the combat control post and the Polyana automated control system

9M38 missile 0517566

Buk-M1 0044960

- Increased reliability of the system due to replacement of its elements with modern ones and implementation of automatic troubleshooting system
- Prolongation of the system life by 15 years due to overhaul of the equipment which cannot be modernised
- Improved noise protection and survivability of the system
- Improved maintenance and ergonomic characteristics
- Improved quality of crew training due to implementation of new high-performance training and simulation facilities in the combat control centre and the self-propelled firing unit.

Status
Development is complete and the system is believed to be in service with the Belarussian forces.

It is understood that the Azerbaijan Ministry of Defence ordered from Belarussia the Buk-MB upgrade but for some unknown reason this has been referred to as the Buk-RB.

Contractor
Ahat Research Development and Manufacturing Enterprise.

Osa-1T

Type
Self-propelled Surface-to-Air Missile (SAM) system

Development
An upgrade/modernisation to the original Russian Osa-AK/AKM system has been undertaken by the Tetraedr Scientific Industrial Unitary Enterprise in order to increase the combat effectiveness of the Surface-to-Air Missile (SAM) system in all types of jamming. This upgrade/modernisation can be carried out on site during the systems standard mid-life renewal and repair time. Testing of the upgraded Osa at the Tetraedr enterprises test ground, in the Brest Region of Belarus have been completed. This self contained mobile air defence missile system is designed to defend Army units, industrial and military objectives from all types of modern and advanced aerial attack assets with the RCS of 0.03 m square and above flying nap of the earth.

A further upgrade, dated May 2007, also covers the 9M33M2 missiles which will not only modernise the system but will extend the life of the missiles into the next decade. This upgrade involves the 9V242-1 automated mobile test station, 9V914 adjusting vehicles and 9V210M3 technical service vehicles.

By December 2008, these vehicles had been further updated yet again, this time the 9V242-1 becomes the 9V242-1T; the 9V210-1 becomes the 9V210-1T; and the 9V914-1 becomes the 9V914-1T.

A totally rebuilt version of the 9Sh33 Karat daylight TV camera used by the Osa-1T (and the S-125-2T Pechora-2T) with a passive thermal imagery system 9Sh38-2 are also part of this modernisation package.

In June 2008, Tetraedr began work on a new 9A6922-1T three axle chassis launching vehicle. This vehicle is developed from the MZKT-6922 chassis design but incorporates changes requested by Tetraedr in order that the system should be air-transportable by Il-76 transport aircraft. All the electronics within the radar system have been digitised to increase Mean Time Between Failure (MTBF) and reliability, whilst also making use of the space saved. This new launcher will be supported in the field by a KAMAZ-43118 loading/reloading vehicle that is also boasts three axles.

Description
The upgrade/modernisation to the Osa-AK/AKM is primarily designed to increase the system's survivability against Homing Anti-Radiation Missiles (HARM) whilst also upgrading the SAM system power supply and installing air

Osa-1T (Tetraedr Scientific Industrial Unitary Enterprise) 1036513

Osa-1T (Tetraedr Scientific Industrial Unitary Enterprise) 1036511

conditioning in the combat vehicle operators position. The modernised system has been reported to be equipped with a more efficient homing capability and has been optimised to engage stealth aircraft and small size targets, such as Unmanned Aerial Vehicles (UAVs).

The following sub-units make up the Osa-1T (9K33-1T) system:

- SKO-1T system air conditioning
- Command encoder
- SRP-1T computer instrumentation unit
- ARM-1T automated work station
- R-173M radio station
- 9F632T training simulator
- VSO error signal discrimination system
- Solid state amplifiers
- EOS-1T (also referred to OES-1T) electro-optical system that also includes a new automatic target-tracking system.

The EOS-1T (OES-1T), is mounted on top of the system, and consists of a Thermal Imaging (TI) camera, laser transmitter and two daylight television cameras. The upper daylight camera is for general use, whilst the lower camera is used by the fire-control system. Both cameras operate at wavelengths of 380-960 nanometres. The general view camera has a Field-of-View (FoV) of 14.2 × 10.6°, while the fire control covers 4.9-3.7°. Using these sensors a typical target of MiG-29 size can be tracked out to 25 km.

The thermal channel operates in the 3.0-5.0 µm-bands, its zoom lens can provide a 9.1 × 6.9° FoV for general observation, or a narrower 2.3 × 1.7° mode for fire-control. The TI sensor is reported to be able to detect a MiG-28 target at ranges of at least 35 km.

The laser range-finder operates at the 1,064 µm-band up to 17 km with an accuracy of ±2.5 m.

Increasing combat effectiveness of the SAM system

The improved Osa-1T system now provides for a two-channel target engagement capacity, an automated crew commander's work station in the combat vehicle and an increase in the effectiveness in defeating jamming of the system.

In the Osa-1T combat vehicle, the new SRP-1T instrumentation unit replaces the computing device and the command encoder. The SRP-1T system combines the function of:

- the launch device;
- the commands generating system;
- the commands encoder;
- the combat operation documenting system.

The SRP-1T is designed as one unit in order to replace the eleven units in the SP-2M1 and SP-1M1 instrument racks of the Osa-AK/AKM.

Loader/Transloader for Osa-1T (Tetraedr) 1331207

The SRP-1T system makes use of two universal guidance methods:

- Kinematic Differential Control (KDC)
- Modified Three Point (MTP) method.

These guidance methods make it possible for the system to:

- increase the probability of the target kill by one missile;
- increase the distant and upper limits of the engagement zone;
- allow for greater speeds of the targets engaged;
- decrease the per-target expenditure of the missiles.

With the Osa-1T the aim of the engagement of the target is with one-missile salvos.

Installation of the 9F632T trainer simulator

The 9F632T trainer-simulator has been incorporated in the Osa-1T SAM system instrumentation array. The simulator represents a single unit installed in the SL-1M rack in lieu of the OO04-12M functional control unit. This has made it possible to completely discard the 9F632T trainer-simulator hitherto mounted on a separate ZIL-131 truck chassis. With the 9F632T now installed into the firing vehicle, the simulator is able to:

- Perform comprehensive check out of all the systems of the combat vehicle in the electronic firing mode;
- Stage independent training sessions for the combat vehicles crews, where simulated environments are superimposed on real situations;
- Simulate the complete combat operation cycle from the crews assigned mission through estimating and grading the simulated missile launches results.

During May 2008, users of the 9K33M2 Osa-AK and 9K33M3 Osa-AKM (SA-8 'Gecko') were offered a modernised version of the associated 9F632/9F632M/9F632T simulators. Developed by the Belarus Company Tetraedr, the 9F632-1T incorporates new software and hardware that take advantage of modern signal processing and other digital and electronic technologies. These changes are intended to improve the performance and mean time between failures while reducing the time needed for maintenance and repairs. The modernised simulator 9F632-1T is intended to train the crews of 9A33BM2 and 9A33BM3 in the task of engaging targets and handling complex air-defence scenarios. It is mounted in a cabin with air conditioning, heating and ventilation system that is carried by a MAZ-6517/6317 three-axle truck. An SEP-1T electrical generator powers the simulator.

The cabin contains two subsystems:

(1) The IVO-1T (air situation simulator);
(2) IBM (combat vehicle simulator), plus communications equipment.

The main modules of the IVO-1T include:

- TO12-1T video signals simulator for the IBM and combat vehicle;
- TO14-1T echo signals simulator for the IBM;
- TO14-1TBM echo signals simulator for the combat vehicle;
- TO19-1-1T training operator console;
- TO19-2-1T air-situation display and crew work examination;
- SRP-1T computing unit;
- VBU-55 video receiving device.

The modernised simulator can operate in three modes:

(1) BR (combat operation);
(2) RR (normal operations);
(3) FK (functional check-up).

In BR mode, the simulator can be used autonomously, with the crew trained inside the simulator, coupled to a combat vehicle to allow one crew to be trained in the simulator and a second in the vehicle, or in an 'inserted working' mode in which a crew in the combat vehicle can monitor the real air situation, into which radar targets can be inserted by the simulator. Crews can be trained in scenarios starting from a single target flying a simple flight path in the absence of enemy jamming. Trainees can be introduced to the effects of target manoeuvres and different types of jamming and instructed as to how to engage

New launcher vehicle for Osa-1T (Miroslav Gyürösi) 1331206

groups of up to four targets flying at different flight levels and approaching from several directions. Low-flying targets, including helicopters, can also be simulated and engaged. Crews can also practise engagements against an enemy armed with anti-radiation missiles. The RR mode is used to set up the IVO and IBM, while the FK mode allows the IVO and IBM to be given a full functional test.

Upgrading the UHF receiver channel
Modernising the UHF receiver channel consists of replacing the standard Travelling-Wave Tube (TWT) HF amplifiers with solid-state HFA. As a result, sensitivity of the target acquisition and target tracking radars of the upgraded system surpasses the Osa-AK/AKM by up to five times. As a consequence, the Osa-1T is better placed for the detection and tracking of small and stealthy sized targets.

Increased reliability of the system
The Error Signal Extraction (ESE) system and the commands encoder in the upgraded Osa-1T SAM have been converted to the new technology that provides for their greater reliability, a longer service life and shorter time spent on maintenance. The command encoder is designed as a single board mounted in the SRP-1T system. The ESE also represents a single unit that replaces the former OO51-7M2, OK51-6M2 and OS51-8M2 units.

Combat assets
9A33-1T combat vehicle
9M33M2(3) missiles
Note: if the T382 missiles are included in the composition of the 9K33-1T (Osa-1T), it is designated as the T38 Stilet ADMS.

Technical support means
9T217-1T transporation and loading vehicle
9V210-1T maintenance vehicle
9V214-1T type alignment vehicle
9V242-1T automatic mobile check up and testing station (AKIPS)
9F16M2 ground equipment kit (GEK).

Specifications

	Osa-AK/AKM	Osa-1T
Travel mode to deployment time:	4 min	4 min
Target channel capacity:	1	1 (2)
Missile channel capacity:	2	2
Maximum target speed:		
(head-on)	500 m/s	700 m/s
(receding)	300 m/s	350 m/s
Altitude:		
(minimum)	25 m	15 m
(maximum)	5,000 m	8,000 m
Minimum range:	1.5 km	1.5 km
Maximum range:		
(tactical aircraft)	10.3 km	12 km (22 km[1])
(helicopters)	6.5 km	10 km (17 km[1])
(crossing target)	6 km	12 km
Missile guidance method:	TP, H, Phi High Trajectory	KDC, MTP
Kill probability (1 × missile):		
(tactical fighter)	0.5-0.7	0.6-0.8
(helicopter)	0.4-0.7	0.6-0.8
(manoeuvring target)	0.2-0.5	0.4-0.7

[1] Valid when firing 9M33M3-1 missile

Status
Development complete. Offered for export.

Contractor
Tetraedr Scientific Industrial Unitary Enterprise, Minsk, Belarus.

Overhaul and Maintenance of Air Defence Vehicles in Belarus

Type
Overhaul and maintenance services of anti-aircraft systems

Description
Since its formation in 1991 the Minotor Service Enterprise of Belarus has built a business based on the overhaul, maintenance and modernisation of military tracked vehicles originating from the Russian Federation (CIS) manufacturers. Minotor also designs, manufactures and supports land and amphibious tracked combat vehicle designed for use within ground forces and Marine Corps units. Included in the vehicle inventory the firm is capable of handling are the following specialised air defence vehicle chassis:

- Tunguska - GM-352 chassis
- Buk family - GM-569, GM-577 and GM-579 chassis
- Kvadrat family - GM-568 and GM-578 chassis
- Shilka - GM-575 chassis (Minotor also offers an upgrade package for the system)
- Tor family - GM-355 chassis.
-

Upon special request, the firm can also provide overhaul services to the above air defence systems.

The main activity areas of Minotor Services are:

- Research and development, design work, and experimental development
- Modernisation of the military equipment that is currently in service
- Repair and servicing of special vehicles
- Spare parts supplies, export and import.

Status
Production as required. Services offered on the export market and to the Belarussian MoD. Minotor is contracted to the Russian MoD to provide specialist repair and overhaul of armoured and air defence vehicles.

Contractor
Minotor Service Enterprise, Minsk, Belarus.

Canada

ADATS

Type
Self-propelled Surface-to-Air Missile (SAM) system

Development
In 1979, Oerlikon-Bührle of Switzerland, with the then prime subcontractor Martin Marietta (now Lockheed Martin Missiles and Fire Control - Orlando) of the US, commenced development of a low-level missile system to defeat both air and ground threats.

Extensive analysis of current and projected threats determined that an Air Defence Anti-Tank System (ADATS) must be an all-weather missile system, designed to protect field troops and Vital Point (VP) targets against low-flying fixed-wing aircraft, attack helicopters, Remotely Piloted Vehicles (RPVs) and cruise missiles, as well as armoured vehicles.

In June 1986, after an exhaustive evaluation, ADATS was selected to fulfil the Canadian Forces Low-Level Air Defence System (CF LLADS) requirements. Oerlikon Aerospace Inc (now Rheinmetall Canada Inc., part of the Rheinmetall group in Germany) of St-Jean-sur-Richelieu in Quebec was formed in 1986 to carry out the CAD1.09 billion CF LLADS contract and to take technical, and subsequently commercial, responsibility for selling the ADATS missile system worldwide.

Canadian Forces ADATS on modified M113A2 chassis during trials in the Middle East 0044102

ADATS missile being launched from M113-series APC during trials in Thailand (Rheinmetall Canada) 1419786

In 1987 ADATS was one of the four weapon systems that competed in the US Army's Forward Area Air Defense - Line Of Sight - Forward-Heavy (FAAD-LOS-F-H) competition. In November 1987, ADATS was selected as the winner after a side-by-side system evaluation exercise that included acquisition and tracking tests in conditions representative of typical combat scenarios. After only eight ADATS with MIM-146 missiles were delivered to the US Army, the programme was halted in 1992 because of US DoD budgetary cuts.

For the CF LLADS programme, the ADATS is installed on a three-man (driver, commander/radar operator, electro-optical operator) modified United Defense LP (now part of BAE systems) M113A2 APC platform. For the FAAD-LOS-F-H programme it was installed on the United Defense LP M3 Bradley. Both versions provide the cross-country mobility and armour protection required for the defence of field forces. The ADATS in its M113A2 configuration is also readily suitable for drive-on/drive-off air transportability by a Lockheed Martin C-130 Hercules tactical transport aircraft. A support vehicle with a further three-man crew is also used to support the M113A2 ADATS vehicle in service with the Canadians.

In its various test programmes, the ADATS has undergone highly successful qualification trials in all types of conditions and environments, from the very cold (–40°C) to the upper extremes of heat (+70°C).

In 1994, ADATS fire units conducted successful live firings against both air and ground targets, in adverse weather and battlefield conditions, in the Middle East and Southeast Asia. In 1996, Canadian Forces ADATS units participated in the US Army's annual Air Defence Exercise 'Roving Sands' at Fort Bliss, Texas, where ADATS' tactical flexibility and effectiveness (including against low-flying cruise missiles) was demonstrated. The built-in capability to interface ADATS with theatre level air defence networks and air defence assets such as Patriot batteries for the purpose of early warning and fire coordination was also demonstrated. A live firing shoot conducted at the end of the exercise gave a system performance of 95 per cent (21 out of 22 missiles fired).

The Royal Thailand Air Force has conducted operational firings of ADATS against Snipe target drones in a coastal environment on an annual basis and scored a 100 per cent success rate.

In June 2002, ADATS units were deployed to protect the G8 summit site at Kananaskis, Alberta, Canada.

Well over 400 missiles have been fired; the high success rate of over 80 per cent, by first-shot operators, against small low-flying and ground targets at extended ranges, demonstrates ADATS ease of operation and system performance.

ADATS can be integrated with other armoured vehicles both tracked and wheeled and shelter systems (see Shelter- and container-based surface-to-air missile systems section).

Description

ADATS is a multipurpose, all-weather low-level missile system designed to defeat both air and ground targets. The air targets, including attack helicopters flying nap of the earth at ranges of up to 10 km, can be engaged at very low altitudes.

The X-band 25 km plus range, Pulse-Doppler frequency and pulse-agile surveillance radar has enhanced ECM resistance features. It is fitted with search-on-the-move, track-while-scan on up to 20 targets and automatic threat prioritisation for up to 10 target capabilities. Targets are automatically interrogated by the onboard AN/MPX-12 (Mk XII) IFF system.

The Electro-Optic (EO) system used comprises a Forward-Looking Infra-Red (FLIR) and high-resolution TV to provide a completely passive acquisition and tracking capability. Each of the eight prepared to fire missiles are enclosed in a protective canister that functions as the launch tube. The missile, which is stored in its canister, requires no maintenance in storage or in preparation for launch. A complete launch canister with its missile weighs 67 kg. The complete set of eight missiles can be loaded by two men, in less than 10 minutes without the use of special equipment.

The missile is a wingless cruciform design, Mach 3 plus weapon with a 10,000 m range. It is fitted with a 12.5 kg HE fragmentation/shaped charge warhead that has both impact and proximity fuzing systems. In its anti-armour role it can penetrate over 900 mm of rolled homogeneous armour. The missile is guided to the target by a CO_2 laser, which is detected by two receivers

ADATS series production model for the Canadian Forces 0044103

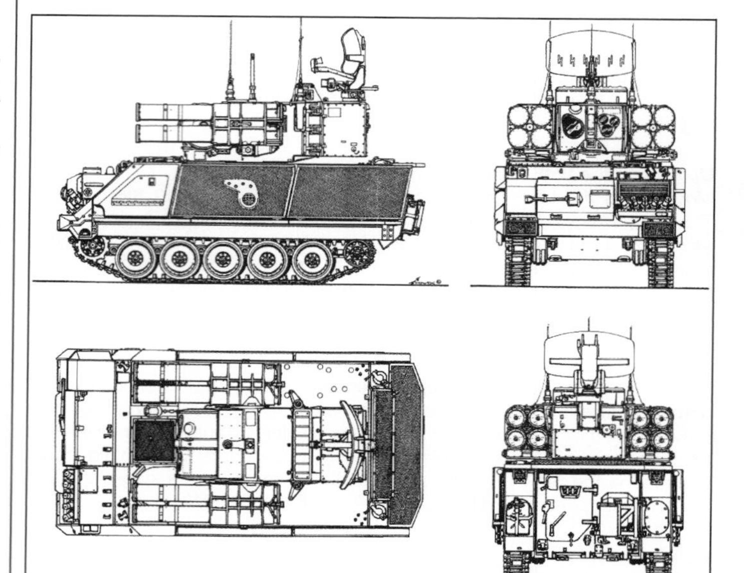

Four-view drawing of ADATS as used by the Canadian Forces on a modified M113A2 series APC chassis (Henry Morshead) 0509756

located in the guidance fins at the rear of the weapon. This passive guidance system combined with the Hercules smokeless solid fuel rocket propellant further enhances the system's survivability on the battlefield.

The ADATS integral Command, Control and Communications (C3) system allows the netting of up to six ADATS into one tactical unit, under the control of the commander in the designated control ADATS. If required, any of the ADATS can assume control with the reconfiguration only taking seconds. Secure jam-resistant communications are achieved by using high capacity data radios. The C3 net also allows the ADATS unit to be linked with higher echelon defences and command centres, other radar and weapon systems, including BM/C4I networks, for automatic response to saturation and time-compressed coordinated attacks. A network of six ADATS can prioritise and engage up to 20 air and ground targets at any one time.

A two-man crew, a radar operator (who is also the system commander) and an EO operator (who is also the gunner) operate the ADATS. The EO operator can also operate both console stations during periods of low activity.

Any targets that are detected using the surveillance radar are passed to the EO target tracking assembly for engagement. The FLIR and TV sensors, both online and operating simultaneously, are available to the EO operator at all times. These imaging sensors provide the operator with complete situational awareness thus allowing him to detect instantly and react to any target attempt to use countermeasures. The EO sensors are optimised for the target tracking role rather than being simple back-up devices to the radar system. The large diameter optics also makes the EO system highly effective in all weather conditions.

The commander in the designated control ADATS has a complete tactical picture of the surrounding area with inputs from the entire subordinate ADATS including:

- Status information of each ADATS
- Accurate geographical position from each unit's Vehicle Navigation and Attitude Reference Systems (VNARS).

The commander transmits the weapon control orders, airspace management information and target assignments to the individual ADATS units over a datalink. One or more surveillance radar of the ADATS network can provide necessary target information to the other ADATS, which can then track and engage them in the radar-silent mode using their EO systems.

Specifications

ADATS missile system

Type:	single-stage, multipurpose anti-armour and low-altitude air defence
Length:	
(missile at launch)	2.05 m
(canister with missile)	2.20 m
Diameter:	
(missile)	0.125 m
(canister)	0.273 m
Weight:	
(missile at launch)	51 kg
(canister with missile)	67 kg
Propulsion:	smokeless solid propellant rocket motor
Guidance:	laser beam-rider using digitally coded CW CO_2 laser
Warhead:	
(weight)	12.5 kg
(type)	dual-purpose HE fragmentation/shaped charge
Fuze:	range-gated laser proximity with variable delay and a nose-mounted mechanical impact/crush
Maximum speed:	Mach 3 plus
Maximum effective range:	
(air and ground targets)	10,000 m
(high-speed manoeuvring aerial targets)	8,000 m
Engagement altitude:	
(maximum)	7,000 m
(minimum)	ground level
Radar type:	X-band pulse-Doppler: 25 km range at 38 rpm scan rate, 17 km range at 57 rpm scan rate
Passive target tracking:	8-12 μm waveband FLIR and 0.7-0.9 μm waveband TV, 1.06 μm Nd-YAG laser range-finder

Status

Production as required. The system has seen service with the Canadian Forces - a total of 36 ADATS were delivered from October 1988 to early 1994 (as of February 2009, 34 remained in the inventory) - and Thailand (shelter-mounted version with Skyguard fire-control unit as target information source). A total of eight ADATS were delivered to the US Army before the US DoD cut funding in 1992 due to overall budget cuts.

The ADATS 401 configuration will include a new display for the electro-optical/gunner operator and a new vehicle navigation system in addition to the new radar/commander console and BMC4I system using high capacity data radios that were recently integrated on the -400 series.

The ADATS system is to be phased out of the Canadian inventory as of 2012 due to cutbacks. This was announced by the Canadian government in 2012.

Contractor

Rheinmetall Canada Inc.

China

FM-90

Type

Self-propelled Surface-to-Air Missile (SAM) system

Development

At the end of 1998, the China National Precision Machinery Import and Export Corporation (CNPMIEC) first offered for sale the FM-90 (Chinese Fei Meng-90, English Flying Midge-90) an enhanced version of the earlier FM-80 (HQ-7/Crotale) all-weather surface-to-air missile system (see entry in Shelter-based SAM section).

By the mid-to-late 2006 and early 2007 the system had again been modified and upgraded to the standard of the technology that had improved since its initial conception. The HQ version now finds itself deployed with the PLA, PLA(N) and PLAAF.

A typical FM-90 mobile fire unit would consist of a search-and-command system on a 4 × 4 armoured chassis and two or three launcher and guidance systems.

FM-90 launcher arrangement and guidance system 0045079

FM-90 search-and-command system 0045080

FM-90 SAM launching a missile 0044104

Description

The FM-90 system uses all digital electronics with the missile being tracked in flight by the launcher. Each of the firing units is based on the same 4 × 4 armoured chassis, each is fitted with hydraulic stabilisers which are lowered to the ground when the launcher is deployed to provide a more stable firing platform. Each launcher carries four ready-to-fire missiles in their cylindrical containers on a pedestal mounting.

The mounting also has a new dual-band monopulse tracking radar with both frequency diversity and agility abilities. The radar can monitor 30 targets and track 12 of them simultaneously. An improved day/night electro-optic television based sighting system is also fitted to the pedestal. Target information between the launcher and the search-and-command system is transmitted via a pole-like antenna datalink system similar to that fitted to the late production of the then Thomson-CSF Crotale 4000. The whole system has been redesigned by the 706 Institute using digital technology, replacing the older analogue design.

The system uses the Chinese missile variant of the RM-90 that has a maximum velocity of 900 m/s and is designed to engage a wide range of aircraft, helicopters, Remotely Piloted Vehicles (RPVs), and anti-radiation, cruise and other missiles. In a typical engagement the target is required to be located by the search-and-command system using its pulse Doppler surveillance radar with low side lobe antenna and constant false alarm receiver. If a target is confirmed as hostile, it is then handed over to the appropriate fire and guidance unit to carry out the missile engagement.

Variants

FM-90(N)

The FM-90(N) is a towed and static version of the system and was developed as a naval variant.

Specifications

RM-90 missile

Length:	3.12 m
Diameter:	0.156 m
Wingspan:	0.55 m
Weight:	84.5 kg
Propulsion:	solid propellant rocket motor
Velocity:	900 m/s
Guidance:	command and electro-optical tracking
Warhead:	HE fragmentation
Warhead weight:	15 kg
Fuze	proximity
Operational range:	
(maximum)	15,000 m (all target types)
(minimum)	700 m
Engagement altitude:	
(maximum)	6,000 m
(minimum)	15 m
Single shot kill probability:	0.8
System reaction time:	6-10 s
Maximum target detection range:	25,000 m
Maximum target tracking range:	20,000 m

Status

It is possible that Iran may have taken delivery of several systems for trial purposes and to upgrade its FM-80 network. It is also likely that both FM-80 and FM-90 technology has been incorporated into Iran's Ya-zahra Project.

The FM-90 is deployed with the PLA's 38th Group Army Air Defence Brigade (second battalion), garrisoned in South West of Beijing for the defence of Beijing.

Contractor

Enquiries

China National Precision Machinery Import and Export Corporation (CNPMIEC).

Manufacturer

2nd Aerospace Academy (now The China Academy of Defence Technology) is the primary contractor.

23rd Institute is responsible for the development of the radar fire-control system.

206 Institute is responsible for the ground equipment.

706 Institute is responsible for the analogue to digital transition and communications update.

HN-5C

Type

Self-propelled Surface-to-Air Missile (SAM) system

Development

The CNPMIEC Hongying-5C (or HN-5C) is the designation used to describe a vehicle-mounted HN-5 (Hongying-5) system (see entry in Man-portable surface-to-air missile systems section). Evaluation trials were completed in June 1986 and it is understood that this system entered low-rate production. As far as is possible to ascertain, there were no known exports.

The Hongying-5C (HN-5C) is essentially an HRB-230 truck chassis, on the rear of which is mounted a pedestal with four Hongying-5A (HN-5A) or -5B (HN-5B) SAMs either side 0509383

Model of NORINCO WZ 551 (6 × 6) APC fitted with the Hongying-5C (HN-5C) SAM system with eight Hongying-5A (HN-5A) or -5B (HN-5B) missiles in the ready-to-launch position 0509384

The main roles of the HN-5C were stated to be:

- Protection of mechanised units
- Air defence for forward units
- Air defence for high-value targets such as airfields.

Description

The first HN-5C version shown was an HRB-230 (Harbin) 4 × 4 cross-country vehicle with a forward fully enclosed control cab, which contains the fire-control equipment. To the rear of the cab is a pedestal upon which a bank of four HN-5, HN-5A or HN-5B missiles is mounted either side in the ready-to-fire position. These missiles may be replaced with the latest QW series of man-portable SAMs.

Between the two banks of missiles is the fire-control electro-optical package which consist of:

- An infra-red tracker
- A range-finder
- A TV camera.

Detecting and tracking the target and launching the missiles can be accomplished either manually or automatically.

The weapon assembly can also be mounted on other chassis, as it only weighs approximately 2,000 kg. During the 1987 Paris Air Show, a model of the WZ 551 (6 × 6) armoured personnel carrier chassis was exhibited with the HN-5C system mounted on the vehicle's roof. The WZ 551 version had turret elevation limits of –2° to +80°, a traverse of 360° and total turret system weight of 2,100 kg. The HN-5C is operated by a two-man crew, the maximum effective engagement range was stated as being 4,200 m between altitude limits of 50 and 2,300 m. Typically, eight reserve missiles can be carried within the launch platform. A further chassis this time, the ZSL93 (Type 93) which is the APC variant of the WZ523, was also suggested as a carrier of the system. However, nothing further has been seen of this variant.

Variants

It is possible that more modern versions of this system have been developed using the later QW-1 and QW-2 man-portable surface-to-air missile systems.

Specifications

	HN-5C
Dimensions and weights	
Crew:	2
Weight	
combat:	2,100 kg (est.)
Mobility	
Configuration	
running gear:	wheeled
layout:	4 × 4
Firepower	
Armament:	8 × HN-5 missiles SAM
Main armament traverse	
angle:	360°
Main armament elevation/depression	
armament front:	+80°/-2°

Status

Production complete. Has seen very limited service with the Chinese People's Liberation Army and is most likely obsolete. As of June 2013, there are no known exports of this system.

Contractor

Manufacturer: Chinese state factories.

Export marketing: China National Precision Machinery Import-Export Corporation (CNPMIEC).

HQ-9/FT-2000

Type

Self-propelled Surface-to-Air Missile (SAM) system

Development

The FT-2000 and FT-2000A are export variants of the Chinese in country HQ-9 system that has been developed with the best of the US Patriot system and the Russian S-300P family.

The China National Precision Machinery Import and Export Corporation (CNPMIEC) is offering the FT-2000 (Chinese; *Fei Tung* = FT), a Chinese-developed enhancement to the Russian S-300PMU/PMU1 missile system. The system uses an indigenously designed surface-to-air Anti-Radiation Missile (ARM) to engage either single or multiple Electro-Magnetic (EM) radiating airborne targets. These targets such as an Airborne Early Warning And Command System (AWACS) or Suppression of Enemy Air Defence (SEAD) jamming aircraft, for example the EA-6B Prowler family radiate in the 2-18 GHz frequency band region.

The FT-2000 is the anti-radiation variant of the HQ-9 whereas the FD-2000 is the air-defence version. FD is believed to be transliterated as Fang Dun meaning defensive shield.

During trials dating from 13th September 1999, according to an official quote in the Beijing Evening News, the missiles have struck their target every time.

Development of the HQ-9 system at the China Academy of Defence Technology (CADT), based in Beijing probably began in the early 1980s; CADT is subordinate to the China Aerospace Science and Industry Corporation (COSIC) (also known as CASIC 2nd Academy), and is also known as China Changfeng Mechanics and Electronics Technology Academy.

A naval variant (HHQ-9) is known to exist and is fitted to the PLA Navy's Type 052C air defence missile destroyers.

Description

FT-2000

The FT-2000 system includes a wideband passive radar detecting station, the specially developed ARM vertically launched missile and a four-round launcher platform similar to that used with the S-300PMU/PMU1. The missiles are launched from an 8 × 8 Transport Erector Launcher (TEL) vehicle carrying four missile launch tubes.

The missiles are also capable of detecting and locking on to random electronic interference (jamming), targeting in this case is provided by up to four ESM stations (some reports have suggested Ukrainian Kolchuga type) deployed at distances of up to 30 km in a triangle with a central fourth unit acting as command-and-control.

HQ-9

The HQ-9 however uses Track-Via-Missile (TVM) for terminal guidance with an active fuze that goes active at approximately 5 km from the target and has an effective range of 35 m.

All missiles are vertically cold launched from a Transporter Erector Launcher (TEL) Taian TAS5380 that is also an 8 × 8 transport vehicle. The HQ-9 missiles use inertial guidance in the initial stages with radio command mid-course correction and TVM for end game.

CNPMIEC FT-2000 surface-to-air anti-radiation missile system on 8 × 8 chassis in travelling configuration 0044105

HQ-9 TEL on MAZ 543 (Christopher F Foss) 0569695

Variants

FT-2000A is a further development of the basic FT-2000. This system is believed to be still in the developmental stages but is expected to be a static weapon system. The missile is believed to be of the ARM type operating in the 2-6 GHz frequency band.

A naval variant of the HQ-9 is known to exist and is fitted to the PLA Navy's Type 052C class of air defence missile destroyer.

FD-2000 is believed to be the air-defence export variant of the system.

Specifications

FT-2000 ARM/HQ-9 missiles

	FT-2000	**HQ-9**
Weight:	1,300 kg	unknown
Length:	6.8 m	unknown
Diameter:	0.466 m	unknown
Guidance:	2-18 GHz passive homing operating band seeker with memory function	track-via-missile
Operational range:		
(maximum)	100 km	200 km
(minimum)	12 km	unknown
Engagement altitude:		
(maximum)	20,000 m	30,000 m
(minimum)	3,000 m	unknown
Launcher:	TAS5380 8 × 8	TAS5380 8 × 8
Radar:	n/avail	HT-233 3D fire-control radar
Maximum loading:	14 *g*	unknown

Status

Latest reports indicate that production could have started as early as the 2005 time frame. The system has been offered for export with the potential first customer being either Pakistan or Iran. During discussions between China and Pakistan in February 2004, the offer was made by China to supply the FT-2000/FT-2000A to counter the Indian threat to Pakistan of the Indian Agni missile systems.

As of October 2011, the HQ-9 is reported to be operational with the PLA and PLA(N).

An announcement was made in late September 2013 that Turkey has procured the FD-2000 variant of the system for its long-range air-defence requirements. The FD-2000 beat off rivals from Russia (S-400), the US (Patriot PAC-3) and Europe (SAMPT).

Contractor

Primary contractor: China Aerospace Science and Industry Corporation (COSIC/CASIC).
Development: China Academy of Defence Technology (CADT).
Enquires: China National Precision Machinery Import and Export Corporation (CNPMIEC)

HQ-16/HQ-16A

Type

Self-propelled Surface-to-Air Missile (SAM) system

Development

Official Chinese press reported in late September/early October 2011 that an alleged co-development between Russia and China of the HQ-16 (Red Flag-16) self-propelled surface-to-air missile system is complete and has reached operational capability within the People's Liberation Army (PLA).

The HHQ-16 is the naval shipborne variant mounted on the Type 956EM Sovremenny-class DDGs, whereas the HQ-16A is the land-based.

The system has been delivered to the Shenyang Military Region. Other reports have suggested that during development the system had a project number "Project 0910" whilst the export market will know the system as LY-80. It is understood that development within China commenced as early as 2005.

Description

The HQ-16 is a third generation, medium-range high-altitude system that has been designed to fill the gap between the HQ-7 and HQ-9 systems currently deployed within China. Reports coming from China indicate that the system has been identified to counter in particular the cruise missile threat.

The basic HQ-16/LY-80 unit consists of a Command-and-Control (C2) cell made up of a surveillance radar vehicle and a C2 vehicle and three firing batteries. Each firing battery consists of a tracking and guidance radar vehicle and four vertical launchers each armed with a missile pack containing six launchers.

Technical support is provided by missile transportation and loading vehicles, a single power supply vehicle, maintenance vehicle and a missile test vehicle.

Launcher

The missile is cold launched from the vertical in a canister that is one of a six pack mounted on the rear of a 3 × axle lorry. A standard battery will consist of four launchers. One tracking and guidance radar can control firing against up to four targets simultaneously.

LY-80 Launcher (Aerospace Long March International) 1331878

Missile

The missile closely resembles the Russian designed and produced 9M38 family of missiles used in the Russian Buk (NATO SA-11 Gadfly) system. The missile uses inertial mid-course guidance, followed by semi-active radar terminal homing with intermittent illumination of the target.

Fire Control Radar

The Fire Control (FC) radar is an all digital electronic system that is mounted on the launcher vehicle. The radar is lifted hydraulically from the vehicle on a pneumatic two length arm.

Variants

HHQ-16

A naval variant known as the HHQ-16 (Hai Hong Qiu) is known to exist and deployed on board the Type 054A Jiangkai II-class guided missile frigates (aka Sovremenny). The naval system was first identified in January 2008.

HQ-17

Reported to be an improved variant of the HQ-16 with an increased range and altitude.

Specifications

	HQ-16
Performance	
Speed	
max speed	2,644 kt (4,896 km/h; 3,042 mph; 1,360 m/s)
Range	
min	1.9 n miles (3.5 km; 2.2 miles)
max	27.0 n miles (50 km; 31.1 miles) (est.)
Altitude	
min	10 m (33 ft) (est.)
max	30,000 m (98,425 ft)
Ordnance components	
Warhead	HE fragmentation
Guidance	semi-active, radar
Propulsion	
type	solid propellant

Status

Initial Operation Capability (IOC) declared for land use (PLA) in late September and early October 2011 but for PLA(N) in January 2008.

Contractor

Shanghai Academy of Spaceflight Technology (SAST).
Aerospace Long March International (ALIT).

LY-80

Type

Self-propelled Surface-to-Air Missile (SAM) system

Development

The Chinese Company Aerospace Long March International (ALIT) has offered a new surface-to-air missile system to the open market designated LY-80. ALIT is authorized by the Chinese government, as a company devoted to exporting aerospace-related equipment and technology.

Description

The system is based on a Semi-Active-Radar (SAR) guided single-stage missile cold-launched from a vertical launcher. In external appearance the missile closely resembles the 9M38 family of missiles used by the Russian Buk (SA-11 'Gadfly') system. A basic LY-80 unit consists of:

- A Command-and-Control (C2) cell made up of a surveillance radar vehicle
- A C2 vehicle
- Three firing batteries.

Each firing battery consists of:

- A tracking and guidance radar vehicle
- Four launchers each armed with six containerised missiles.

LY-80 vertical launch (ALIT) 1331879

LY-80 Launch Vehicle (ALIT) 1331878

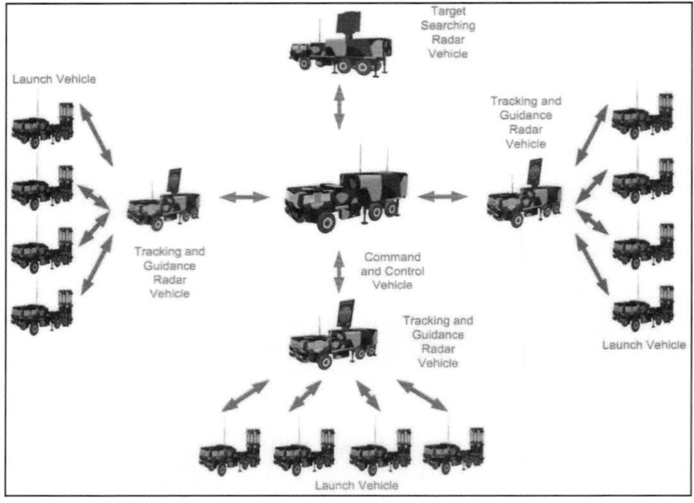

LY-80 battery layout (ALIT) 1331877

The associated technical support equipment includes missile transportation and loading vehicles, a power supply vehicle, a maintenance vehicle and a missile test vehicle.

Chinese sources report that the radars illuminate their targets intermittently and have a good Electronic Counter-CounterMeasures (ECCM) capability.

According to ALIT, aircraft targets can be engaged at ranges of 3.5-40 km, figures that fall to 3.5-12 km for cruise missiles. Maximum and minimum target altitudes are 15,000 m and 15 m respectively. A Single Shot Kill Probability (SSKP) of >0.85 is claimed against aircraft and >0.60 against cruise missiles.

One tracking and guidance radar can control firings against up to four targets simultaneously.

The missile uses inertial mid-course guidance, followed by SAR terminal homing with intermittent illumination of the target. The system can be deployed from the march in 12 minutes. Once it is operational at the chosen location, it has a reaction time of only 12 seconds.

Following one or more engagements, the system can be ready to move within 6 minutes.

LY-80 systems can operate from sites ranging from sea level to an altitude of 3,000 m. It can cope with temperatures from –40°C up to +50°C and humidity levels of up to 95 (±3) per cent at +35°C. Storage temperatures can be from –50°C to +70°C.

Variants
HQ-16A.
HHQ-16 (HQ-16).

Specifications

	LY-80
Performance	
Range	
min	1.9 n miles (3.5 km; 2.2 miles)
max	21.6 n miles (40 km; 24.9 miles)
Altitude	
min	15 m (49 ft)
max	15,000 m (49,213 ft)

Status
The LY-80 has been offered for sale on the open arms market, as far as is known the system has been allocated a PLA Type number HQ-16A and HQ-16 and is operational within China; all LY designations are for export only.

Contractor
Aerospace Long March International (ALIT), China.

PL-9C

Type
Self-propelled Surface-to-Air Missile (SAM) system

Development
The China North Industries Corporation (NORINCO) is promoting the low-altitude SAM system PL-9C that was first displayed in model form at the 1989 Paris Air Show. The PL-9C is allegedly a Chinese copy of the US developed and produced AIM-9 Sidewinder.

This is the latest version of the basic PL-9 air-to-air missile that also has a surface-to-air variant. The system launch unit that is available in both towed and self-propelled arrangements, comprises a four-rail launcher assembly cupola for third-generation PL-9C missiles. The Chinese term for the system is either 'Pili', meaning 'Thunderbolt' or 'Pen Lung', meaning 'Air Dragon'. The PL missiles are normally used in the air-to-air role.

The associated target acquisition radar and electro-optical instruments are mounted on the rear decking of a NORINCO WZ 551D (6 × 6) air-defence vehicle chassis, a derivative of the WZ551 APC design.

The launcher can also be refitted to fire other types of surface-to-air or modified air-to-air missiles of a similar configuration to the PL-9C, according to the customer's requirements.

Each WZ 551D carries:
- A vehicle commander
- A driver
- Three missile operators
- Four ready-to-fire PL-9C missiles.

The target acquisition radar has a detection range of around 18,000 m and an altitude capability up to 6,000 m. The passive electro-optic target sighting/tracking system with TV monitor has an operational range of 15,000 m and a laser range-finder unit with a 10,000 m range capability.

The vehicle troop compartment is fitted with the system fire-control computer, the operator's launch and fire-control console with visual displays for the radar and electro-optical sensors and the servo-mechanisms for controlling the launch cupola in azimuth and elevation.

A WZ 551D unit requires the following technical service facilities:
- Missile test equipment on launch vehicles
- Missile test equipment at technical battery site
- Launcher test equipment
- Missile transloader vehicles for re-supply
- A cryogenic gas recharge carrier vehicle
- A test and maintenance vehicle for the fire unit launcher assemblies.

Description
The PL-9C self-propelled variant is mounted on a 6 × 6 WZ551 Armoured Personnel Carrier (APC).

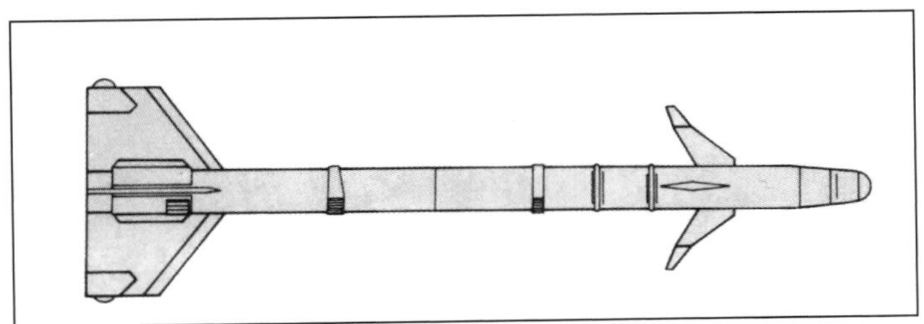

Scale drawing of the NORINCO PL-9C surface-to-air missile 0080399

Scale model of NORINCO PL-9C low-altitude surface-to-air missile system on WZ 551D (6 × 6) air defence vehicle chassis (Christopher F Foss) 0080400

The missile used is the PL-9C surface-to-air passive infrared (IR) terminal homing version of the PL-9 air-to-air weapon which appears to be roughly similar in physical appearance to the Israeli Python 3 air-to-air missile. The PL-9C is capable of ±40° off bore sight angles and uses an all-aspects cryogenic liquid nitrogen gas-cooled seeker head unit utilising proportional navigation guidance techniques. Flight control is by long span pointed delta fins at the front of the missile with Sidewinder-type slipstream driven 'rollerons' on the aft tail fin surfaces to prevent roll and so enhance the operation of the guidance system.

Maximum effective range is 10,000 m at targets up to a 4,500 m altitude limit. The Single-Shot Kill Probability (SSKP) for a single missile launch at an approaching target is 0.8.

The PL-9C surface-to-air missile is also used in the Brigade (Regiment) Level 390 Integrated Gun Missile Air Defence System under the launcher designation DK-9. The Pakistan Air Force has apparently bought 24 DK-9 for air base defence. A naval surface-to-air variant is designated the JK-9.

Fire-control computer system and test units
The Chinese North Automatic Control Technology Institute, which is a subsidiary of NORINCO, has developed the LLP12 fire-control computer system and test unit for use with the PL-9C and the associated gun systems. The LLP12 is used to receive target information and compute firing data from the radar, electro-optic or system or C3I system. It is configured to different fire units such as more than two different guns, missiles and gun-missile-combined systems. The LLP12 is of a modular build with high accuracy for different demands and consists of:
- Simulated co-ordinates input unit
- Accuracy test unit
- Console test unit
- Servo system test unit
- Integral test unit
- Radar simulator
- Fire-control computer system.

The test unit for the fire-control system adopts advanced microprocessor techniques to realise automatic fault diagnosis. It can locate the malfunction to PCB levels.

A second system, also designed by the Chinese North Automatic Control Technology Institute, is the Sky Shield Short Range Optronic Fire-Control System for Air Defence. This system integrates battalion-scaled command posts, search radar, optical director, thermal imaging, laser range-finder, fire-control computer, automatic guns and missiles together into a combat system. The system allows the commander to direct each combat unit via automatic control and communication devices. Four types of Sky Shields have been reported, these are: Expanded Configuration 1 - to control three new double-barrel 37 mm air defence guns and one PL-9C missile launcher. Expanded-Configuration 2 - to control two new double-barrel 35 mm air defence guns and one PL-9C launcher. Extended Configuration 1 - controls six 74-Type double-barrel 37 mm and three new double-barrel 37 mm air defence guns and one PL-9C missile launcher. Extended Configuration 2 - controls six 74-Type double-barrel 37 mm and two double-barrel 35 mm air defence guns and one PL-9C missile launcher.

The types of gun systems, which can be used with the LLP12, include:
- Type 74 double barrel 37 mm
- Type 59, 57 mm
- Type 59, 100 mm
- Type 35 double barrel 35 mm.

Variants
It is possible that the system has been further developed to incorporate later generation missiles other than the PL-9.

The DK-9 uses a towed, hydraulically stabilised launcher with four ready-to-fire missiles on missile rails. It is probable that the DK-9 can also be used for point defence of airfields, ports and other strategic locations.

AF902 Radar/Twin 35 mm AA Gun/PL-9C Missile Integrated Air Defence System. The AF902 Fire-Control Radar has a number of new features such as:
- An additional K^a-band tracking radar
- X-band search radar with Thermal Weapon Sight (TWS) that can track up to 20 targets at a time

- Multiple operation mode under severe Electronic CounterMeasures (ECM) environment and Anti-Radar Missile (ARM) attacks.

The 35 mm towed twin-barrel AA gun is new to NORINCO's arsenal of AA gun family. Its features include:

- Fully automatic operation controlled by the AF902 FCR
- High muzzle velocity up to 1,175 m/s
- High rate of fire up to 2 × 550 rounds/minute.

Ammunition that may be used with the 35 mm gun includes:

- HEI
- HEI-T
- SAPHEI-T
- TP-T.

The PL-9C SAM is a passive infra-red homing weapon that uses an all-aspect seeker head with multi-elements in the detector. When using the AF902 radar, it has a maximum effective range of 8,000 m against targets up to 5,500 m altitude.

The PL-9C SAM system can be employed on different platforms both on the ground and at sea, and integrated as one part of the whole air defence system.

An upgraded variant known as the PL-9D (Type 390) is also known to exist.

Specifications

	WZ551
Dimensions and weights	
Crew:	5
Length	
overall:	6.65 m
Width	
overall:	2.8 m
Height	
transport configuration:	3.4 m
deployed:	5 m
Ground clearance	
overall:	0.41 m
Track	
vehicle:	2.44 m
Wheelbase	1.9 m + 1.9 m
Weight	
combat:	16,000 kg (est.)
Mobility	
Configuration	
running gear:	wheeled
layout:	6 × 6
Engine	Deutz F8L 413F, 4 cylinders, air cooled, diesel, 256 hp at 2,500 rpm
Gearbox	
type:	manual
forward gears:	9
reverse gears:	1
Steering:	hydraulic first and second axles
Brakes	
main:	dual circuit, air/hydraulic
Tyres:	14.00 × 20 (run-flat)
Suspension:	coil springs and hydraulic shock-absorbers
Electrical system	
vehicle:	27 V (DC)
Firepower	
Armament:	× SAM
Main armament traverse	
angle:	360°
speed:	60°/s
Main armament elevation/depression	
armament front:	+80°/-5°
speed:	45°/s

	PL-9C	PL-9D
Dimensions and weights		
Length		
overall	2.90 m (9 ft 6¼ in)	2.99 m (9 ft 9¾ in)
Diameter		
body	157 mm (6.18 in)	167 mm (6.57 in)
Flight control surfaces		
span	0.856 m (2 ft 9¾ in) (wing)	0.81 m (2 ft 8 in) (wing)
Weight		
launch	120 kg (264 lb)	120 kg (264 lb)
Performance		
Speed		
max Mach number	2	2 (est.)
Range		
min	0.5 n miles (1.0 km; 0.6 miles)	0.5 n miles (1.0 km; 0.6 miles)
max	5.4 n miles (10.0 km; 6.2 miles)	8.1 n miles (15.0 km; 9.3 miles)
Altitude		
min	30 m (98 ft)	10 m (33 ft)
max	4,500 m (14,764 ft)	4,500 m (14,764 ft)
Ordnance components		
Warhead	11.8 kg (26 lb) HE fragmentation, continuous rod	11.8 kg (26 lb) HE
Fuze	laser, proximity	
Guidance	passive, IR (proportional navigation)	passive, IR (proportional navigation)
Propulsion		
type	solid propellant (rocket motor)	single stage, solid propellant (rocket motor)

Status

Production as required (since at least 1991). In service with the People's Liberation Army and Pakistan where it is known as the DK-9.

Contractor

Enquiries

China North Industries Corporation (NORINCO).

Front-end technology

Luoyang Optoelectro Technology Development Centre (also known as Institute 612).

Yitian

Type

Self-propelled Surface-to-Air Missile (SAM) system

Development

China has developed a modular-designed Self-Propelled Surface-to-Air Missile (SPSAM) system known as the Yitian. First promoted by the China North Industry Corporation (NORINCO) in 2004/05, the system is based on an earlier developed short-range air-to-air missile Tian Yan (Heavenly Swallow)-90 (TY-90).

The China Aviation Industries Corporation (AVIC I) with its subsidiaries, China National Aero Technology Import and Export Corporation (CATIC), The Luoyang Electro-Optical Equipment Research Institute (LEOER) and China Air-to-Air Missile Research Institute (AAMRI) developed the TY-90 missile during the early to mid-1990s. The missile, which was first publicly displayed in 1998, will normally be found mounted on either side of a Z9G helicopter in pods of four, the same configuration being used in the Yitian SAM system.

Description

As the main carrier of the system, Yitian utilises the WZ 550/1 (or WZ 551A) 6 × 6 wheeled Armoured Personnel Carrier (APC) that has been in service with the People's Liberation Army (PLA) for some years and has also been exported in small numbers. The system has also been noted on tracked and other wheeled type chassis.

Yitian Electro-Optics (IHS/Patrick Allen)
1376724

TY-90 missile (IHS/Patrick Allen) 1376710

Eight missiles are mounted in two banks of four either side of the turret, each missile is contained within its individual launcher container that also serves for transport purposes. Above them is a 3-D X-band search radar that may be folded down into the horizontal position for travelling.

An electro-optical sensor package consisting of a day/thermal sighting system has also been observed with the system, this package also includes a laser range-finder and an automatic target tracker. The electro-optics can detect a typical target out to a range of about 12 km and start tracking at a range of about 10 km depending on the ambient weather conditions.

The primary role of the Yitian is to provide a mobile air-defence capability for mechanised units against a variety of air threats such as aircraft, Unmanned Aerial Vehicles (UAVs), helicopters and cruise missiles. In addition, it may also provide close-in protection against high-value targets, such as command centres and river-crossing points.

The system requires at least two operators who are seated in the rear of the vehicle; each has a display and associated controls that include Command, Control, Communications, Computers and Intelligence (C4I) and communications systems.

A typical Yitian SHORAD battery would probably consist of:

- Headquarters section with a Command Post (CP) vehicle and IBIS-80 radar
- 1 × Battery command vehicle
- 6 × Yitian systems
- 2 × Missile resupply vehicles
- 1 × Missile testing and maintenance vehicle
- 1 × Mechanical/electronic maintenance vehicle.

Radar
According to Chinese source information provided by NORINCO, the 3-D X-band radar has a detection range of 18 km and a tracking range of 10 km. Targets can be tracked either in the radar mode or in the electro-optic mode, with the latter being especially useful when there is a threat of high electronic counter measures. Electro-optical passive tracking has a capability to a maximum range of 12 km. A typical reaction time is six to eight seconds.

Optional equipment associated with the radar is an Identification-Friend-or-Foe (IFF) system that would typically be integrated into an overall C4 air-defence system that could allocate targets to a particular weapon.

Missile
The TY-90 missile is a fire-and-forget weapon that is transported and launched from a box-type container. The missile has four fins at the rear and four control surfaces at the front. TY-90 utilizes a multi-element infrared detector, three-channel control and laser fuze, and is also fitted with the latest in new technologies for counter-counter-measures. Once the missiles have been expended, new missiles are reloaded manually using a support vehicle.

The missile can be equipped on helicopters for air-to-air combat, it can also be combined with anti-aircraft guns to form an integrated gun/missile system for short-range ground and shipboard air defence.

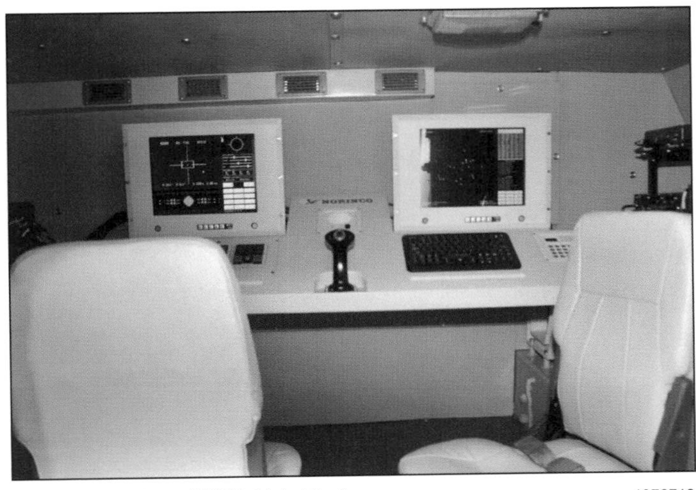

Yitian Workstations (IHS/Patrick Allen) 1376713

Yitian system (NORINCO) 0590753

WZ 550/1 Chassis
The NORINCO standard version of the WZ 550/551A chassis is a 6 × 6 APC vehicle that has been in service with the PLA for some years. Standard equipment carried within the vehicle includes a nuclear, biological and chemical warfare protection system and a central tyre-pressure-regulation system that allows the driver to adjust the tyre pressure to suit the terrain being crossed.

A 12.7 mm machine gun is mounted at the front right side of the vehicle for local defence, with a bank of three electrically operated smoke grenade launchers mounted either side of the turret.

Although the Yitian is being marketed on a WZ 550/1 series 6 × 6 APC chassis, it can also be integrated onto other chassis tracked or wheeled.

Variants
The Yitian has also been shown carried by a Dongfeng Motor EQ2050 high-mobility vehicle and a Chinese version of the US Hummer (Lie Shou) Hunter. The Hunter has not been observed with a radar but does have a YY-1 electro-optic search and track system. It also has TV and Forward Looking InfraRed (FLIR) channels with tracking ranges of 5 km and 4 km respectively.

The TY-90 missiles have also been observed mounted on a two-wheeled chassis that was previously used with the Type 80 (23 mm) Light Anti-Aircraft Gun (LAAG). This configuration was first seen in 2004 and was known at that time as the Shen Gong- (Divine Bow) 2.

Specifications

	TY-90
Dimensions and weights	
Length	
overall	1.862 m (6 ft 1¼ in)
Diameter	
body	90 mm (3.54 in)
Weight	
launch	20 kg (44 lb)
Performance	
Speed	
max speed	1,322 kt (2,448 km/h; 1,521 mph; 680 m/s)
Range	
min	0.3 n miles (0.5 km; 0.3 miles) (est.)
max	3.2 n miles (6.0 km; 3.7 miles)
Altitude	
min	15 m (49 ft)
max	4,000 m (13,123 ft)
Ordnance components	
Warhead	3 kg (6.00 lb) continuous rod
Fuze	laser, proximity
Propulsion	
type	solid propellant (rocket)

Status
Development is complete. Reports dating from IDEX 2009 suggest the system is in limited use with the PLA.

Contractor
China North Industries Corporation (NORINCO) (publicity and sales).
China Air-to-Air Missile Research Institute (missile).
China Aviation Industries Corporation (primary contractor).
Luoyang Electro-Optical Equipment Research Institute (missile front-end technology).
China National Aero Technology Import and Export Corporation (CATIC) (missile).

Croatia

Strijela 10 CRO

Type
Self-propelled Surface-to-Air Missile (SAM) system

Development
As part of the drive for self-sufficiency following the 1991 war with Serbia, Croatia developed an indigenous arms industry.

One of the systems that has been produced is the RH-Alan Torpedo (now Agencija Alan doo) Strijela 10 CRO air defence vehicle. The vehicle is in low-rate production and has seen service with the Croatian Air Force Air Defence Force. The system can fire its missiles either when stationary or when the vehicle is in motion at a speed of up to 30 km/h.

Description
Strijela 10 has an armoured APC type hull mounted on the TAM-150 6 × 6 general truck chassis. On the hull above the rear axles, a standard Russian Strela turret assembly with four elevating ready-to-fire missile container launchers for the 9M31, 9M37, 9M37M or 9M333 missiles used in the Strela-1 (NATO SA-9 'Gaskin') Strela-10M 3 (NATO SA-13 'Gopher') missile systems.

A search radar and an electro-optic TV with a thermal imaging system target detection and acquisition assembly is fitted with ranging and radial/angular speed measurement capabilities. Locally produced computer software architecture has been incorporated into the system to enhance the capabilities of the launch, target tracking and missile guidance subsystems.

A number of the missile components are also locally produced. Full descriptions of the missile types used are given in the Russian Federation and Associated States (CIS) entries for the Strela-1 and Strela-10 in this section.

Specifications
Strijela 10 CRO

Length:	2.19 m
Diameter:	0.12 m
Guidance	IR passive homing with photo contrast
Velocity:	5500 m/s
Warhead:	
(type)	HE fragmentation
(weight)	3-5 kg (missile dependant)
Fuze:	active laser proximity and impact
Armament:	4 × ready-to-fire missiles and 4 reloads on vehicle (9M37, 9M37M, 9M31 or 9M333 missiles)
Altitude:	
(maximum)	3,500 m (some sources suggest 3,000 m)
(minimum)	25 m
Maximum range:	5,000 m
Maximum target speed:	
(head-on)	417 m/s
(crossing)	306 m/s

Croatian Air Force Strijela 10 CRO Air Defence Vehicle from the rear 0080357

Croatian Air Force Strijela 10 CRO Air Defence Vehicle (Roderick de Normann) 0080401

Status
In low-rate production and believed to be in service with the Croatian Army and Air Force Air Defence Force units. As of February 2013 no further information is available on this system.

Contractor
Agencija Alan doo.

Czech Republic

Strela-S10M

Type
Self-propelled Surface-to-Air Missile (SAM) system

Development
The Czech Company of Retia has developed a modernisation package for the Russian 9K35 Strela-10M (US/NATO designation SA-13 'Gopher', local Czech designation Strela-S-10M) self-propelled missile system to improve its overall response capabilities against low-flying targets. In all other respects the Strela-10M remains the same. The package was subsequently chosen in the mid-nineties by the Czech MoD to modernise the Czech Army's Strela-10M force. First delivery of a modernised Strela-S-10M system was made in 1996.

A further upgrade to the system has now been completed to bring the S-10M into the 21st Century and also up to NATO standard where necessary.

Description
The Czech Strela-10M system is based on the light tracked MT-LB that has been gradually upgraded, especially in electronic control equipment and software installed in the command compartment. The upgraded software enables the processing of data from the air situation, identification of a target with NATO IFF, homing missiles using information feedback on overall combat activities.

The original Strela-10 had no automated or digitised connections to higher echelon anti-aircraft command centres so Retia has integrated the following items to overcome these limitations.

- **Data communication equipment** - information from a radar or higher echelon air defence command network is received and processed in the relevant battery command centre and relayed to the appropriate weapons. The new equipment's role as fitted to receiving SA-13 is to present a real-time picture of the air threat situation for the crew and provide a data channel in addition to the original battery-launcher voice communication channel. The data channel equipment comprises a special communications interface and a VHF radio compatible with the system fitted to the battery command post. The vehicle's original whip antenna is employed for both the data and voice radio communications channels. The main role of the special interface is to receive encoded data information and then act as a filter and to demodulate and decode the incoming data. The processed data is then transferred to the vehicle's Combat Control Terminal. The interface also receives the heading data from the gyrocompass and the odometer. From all the information received an audio signal is then generated which is proportional to the target position.
- **Combat Control Terminal (CCT)** - the CCT is the cornerstone of the modernisation package and is based around an ultra-rugged metal-cased notebook PC. The man/machine interface utilises a multiwindow operating system to provide easy-to-use control. The PC is used for real-time warning and cueing of the launcher and other ground-based air-defence systems.
- **TNA-3 Inertial Navigation System (INS)** - in order to calculate the target co-ordinates and the turret slew position, the weapon platforms own geographical position and orientation to the North must be known. The original TNA-3 inertial platform provides the latter data. By linking and integrating the TNA-3 to the CCT through the special communications interface, a continuous indication of the turret position is displayed, together with the route of the vehicle and any surrounding 'objects' that may have an

Modernised S-10M 0099540

influence on the missile engagement. At the same time, an acoustic signal corresponding to the turret bearing relative to the target is generated in the operator's headset. Both functions can be activated or inhibited by the software menu on the CCT. As an option, a Global Positioning System (GPS) can be integrated into the system.
- **Automatic turret slew-to-target** - this is controlled by the CCT and involves the fitting, behind the gunner, of an electronics turret control unit that controls a servo loop. This unit also receives data from the turret encoder, which defines the position of the turret relative to the hull. After all the received data is processed the turret's actual position is then displayed on the CCT.
- **Identify Friend or Foe (IFF)** - optional IFF system integration.

Strela-10M Upgrade Package 2006

The latest upgrade package for the Strela-10M (S-10M) includes the following:
- Installation of the Commander terminal PVK-10;
- Integration of IFF Mark 12 interrogator;
- Automatic turret turn to target;
- Graphic and acoustic indication of the turret position;
- Integration of the TNA-3 INS, or hybrid GPS/INS system;
- Communication with superior command-and-control systems (ASPD/ASPD-U, REDIL).

Specifications

	9M37
Dimensions and weights	
Length	
overall	2.2 m (7 ft 2½ in)
Diameter	
body	120 mm (4.72 in)
Weight	
launch	55 kg (121 lb)
Performance	
Range	
min	0.8 n miles (1.5 km; 0.9 miles)
max	5.4 n miles (10 km; 6.2 miles)
Altitude	
min	10 m (33 ft)
max	5,000 m (16,404 ft)
Ordnance components	
Warhead	6 kg (13.00 lb) HE fragmentation
Propulsion	
type	solid propellant

Status

Production as required. In service in the Czech Republic (modernised 9K35 Strela-S10M (NATO SA-13 'Gopher') systems). The system has been offered for export, as of January 2014 there have been no known customers outside the Czech Republic.

Contractor

Retia AS (responsible for IFF interrogator antennae).

France

Aspic

Type

Self-propelled Surface-to-Air Missile (SAM) system

Development

The then Thales Defence Systems Ltd (now Thales) Aspic system is a fully automated Fire Unit (FU) for very short-range Surface-To-Air Missiles (SAMs). It is designed to defend Vital Point (VP) sites or troops on the move and operates with automated target tracking and engagement facilities to shorten reaction times. If required, the automated functions can be switched to manual control at any time. All the gunner has to do, in either case, is trigger the missile launch.

The two-axis Aspic servo-controlled turret assembly has automated target tracking and engagement facilities to ensure easy operation and maximise target kill probability.

In September 1995 it was revealed that firing trials had taken place in the UK with the SMS Starburst missile. A total of 10 rounds were launched, all of which were successful. Integration of Starburst with the Aspic vehicle took place less than nine months after the project had been launched.

Description

Aspic is a four missile-capable launch-unit on a pedestal mount fitted with a television sight and tracking unit. It can be mounted on a variety of chassis and requires a minimum payload capability of 1,500 kg. Examples of vehicles suitable for Aspic mounting includes 4 × 4 light vehicles such as the Eagle II and Peugeot P4. In the 4 × 4 application, a total of four prepared-to-fire missiles are carried on the launcher and four reloads on the vehicle bed. Up to eight missiles can be carried (depending upon missile type).

Thales Aspic firing post mounted on rear of Peugeot P4 (4 × 4) light cross-country vehicle with four MBDA Mistral SAMs in ready to launch position in service with French Air Force (Pierre Touzin) 0069498

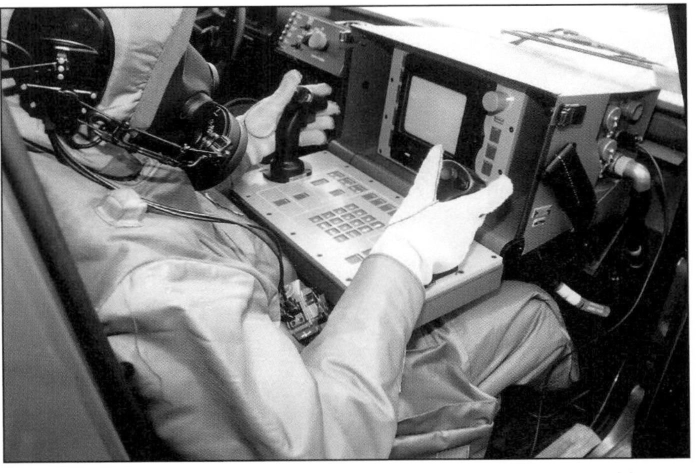

Combat control display mounted inside the Thales Aspic firing post; this can be deployed away from the vehicle if required and is shown here being operated under full NBC protection 0080360

Close-up of the Thales Aspic firing post showing the sensor package in the centre and the two MBDA Mistral fire-and-forget missiles either side 0509455

The missiles may be of the:
- Mistral
- Stinger
- Starburst
- Starstreak
- RBS 70
- or any other suitable VSHORAD type.

The vehicle is fitted as standard with a land navigation and north-seeking system.

The crew of the Aspic is usually two, the vehicle driver and operator-gunner. However, for the coordinated operational role this can be reduced to one and for the stand-alone operating mode the driver should be an operator-driver so that he can perform the target 'pointer' role.

The FU assembly is fitted with a fire-control system based on a TV tracking package. This includes a TV camera, automatic tracker and digital computer. This allows fast and automatic target acquisition, accurate target tracking, optimisation of lead angle calculation plus validation of assigned target lock on when using a system such as Mistral and accurate missile guidance when used with a Semi-Active Command-To-Line-Of-Sight (SACLOS) system such as Starburst.

MOWAG Eagle II (4 × 4) fitted with Thales Aspic 0099524

French Air Force Aspic system from rearward direction 0099525

An InfraRed (IR) camera provides for nighttime and reduced visibility engagement capability. Also available are a thermal imager, laser range-finder and an IFF transponder.

In the coordinated operational mode, data are fed to and from the Aspic FU via an automatic radio communications link to an early warning command-and-control centre such as the then Thomson-CSF (now Thales) Samantha or Clara systems. Once a target is confirmed as hostile by the C2 post, the target designation information is provided to the selected Aspic FU and engaged. Full details of the Samantha and Clara systems can be found under the Anti-aircraft control systems section.

If the Aspic is operated in the stand-alone mode, the target designation is provided by an IR search and tracking device on an optical target designator called ARES. The unit uses the operator-driver as the target 'pointer'. In order to do this he wears the ARES as a helmet-mounted optical target designator system. This allows for fast deployment and almost instantaneous reaction time from the operator-gunner at the deployment site.

Turret traverse is 360° and the assembly may be remotely controlled by the operator from up to 50 m away using a remote firing control console, thus giving the crew additional protection.

Status

Production as required.

No longer in service with the French Air Force, all systems were decommissioned in September 2010. French Air Force systems were previously equipped with thermal imaging cameras supplied by BAE Systems.

The system is still in use by the Chilean Armed Forces.

Contractor

Thales.

Crotale

Type

Self-propelled Surface-to-Air Missile (SAM) system

Development

In 1964, South Africa placed a development contract with the French company, Thomson-Houston (later Thomson-CSF, Thomson-CSF Airsys, which then became Thales Defence Systems) for a mobile, all-weather, low-altitude surface-to-air missile system. The Electronic Systems Division of the then Thomson-CSF was prime contractor for the complete system including the radar and electronics, and Matra was responsible for the missile.

Improved Crotale P4R firing unit with fully automatic electro-optic target and missile tracking mode and four ready-to-launch VT-1 missiles 0080366

Crotale P4R series 4000 firing unit with radio link system mast raised and launcher traversed to right 0080406

The South African government paid 85 per cent of the development costs for the system, which it calls the Cactus, and the remaining 15 per cent were paid by France. After trials in 1971 the first of seven platoons was delivered to South Africa with the final one delivered in 1973.

In February 1971, the French Air Force placed an order for one acquisition vehicle and two firing units which were delivered in 1972. After extensive trials with these units the French Air Force ordered the Crotale (Rattlesnake) system for airfield defence and by late 1978, 20 batteries had been delivered.

Lebanon ordered the Crotale in the late 1960s but the order was cancelled before the systems were delivered. In 1975, Saudi Arabia ordered a new version of the Crotale; mounted on the chassis of the Giat Industries (now Nexter) AMX-30 MBT, it is known as the Shahine, for which there is a separate entry in this section, as the system has a number of improvements over the standard Crotale. This entry title is 'Thales Defence Systems Shahine 1/Shahine 2 low-altitude surface-to-air missile systems' The Saudis also ordered the standard Crotale in late 1978 and an upgraded version in 1990 for their air force.

As produced, Crotale is normally mounted on a P4R (4 × 4) vehicle and can also be shelter-mounted for use in static defence (see Thales defence systems Ltd Crotale shelter-mounted SAM system entry under the Shelter-and Container-based surface-to-air missile systems section). The first Crotale, produced in 1969, was called the 1000 series. This was followed by the 2000 series in 1973 with IFF and TV camera, the 3000 series (originally designed for the French Air Force with automatic TV tracking) in 1978, 4000 series with radio datalink in 1983, the 5000 series in 1985 and the improved Crotale in 1994.

Crotale 3000 fire and acquisition units are not ready for action as soon as they come to the halt but have to be connected together by cables at a maximum distance of 800 m apart. The 4000 series has the LIVH (*Liaison Inter Véhicule Hertzienne*) radio link and mast which not only allows them to come into action faster but also to have up to 10,000 m between the acquisition units and 3,000 m between the acquisition unit and its firing units. The evolved system includes a better ECM performance and the passive tracking (by FLIR) of targets and missiles in both day and night conditions.

Improved Crotale P4R acquisition unit with new planar antenna, associated data processing and ECCM devices, in service with the French Air Force
0080367

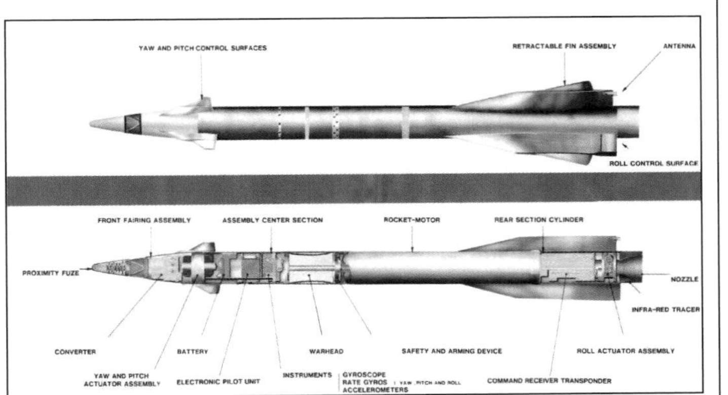

Main components of Crotale R440 missile
0080365

The 5000 series was designed for a French Air Force modernisation programme and included:

- The addition of an optical tracker.
- A new antenna that extended the surveillance range to 18,000 m.

The system was also modified to accept two Matra BAe Dynamics (now MBDA) Mistral missiles on either side of the two container-launcher canisters to help defeat saturation attacks but these were never fielded in practice.

In November 1988, at the second ASIANDEX exhibition in Beijing, the China Precision Machinery Import and Export Corporation (CPMIEC) revealed the FM-80 (HQ-7) land-mobile shelter-mounted surface-to-air missile system on two-axle trailers. The FM-80 is similar in concept to the Crotale shelter-mounted version and its Chinese naval equivalent HHQ-7. Details of FM-80 are given in a separate entry.

In 1998, China revealed a new system, the FM-90, which appears to be an enhanced version of the FM-80, full details are given in this section under China.

In mid-1999 Iran revealed its Ya-zahra Project, a reverse-engineered (locally produced subsystems for modernising and updating) version of Crotale, using technology captured from Iraq during the Iran/Iraq War. It is possible that some Chinese technology from their FM-80/FM-90 programmes may also have been incorporated into the project.

Pakistan has undertaken an indigenous overhaul and upgrade package for its Crotale systems. The package effectively modifies the existing Crotale 2000 systems to the Crotale 4000 standard, at a cost of approximately one half that if the systems were sent to the then Thomson-CSF for the modification. The update has also increased the engagement capability from 20 to 30 km. The indigenous package may have been undertaken with help from Thomson-CSF.

Description

The basic Crotale has an all-weather capability. A typical platoon consists of one Acquisition and Co-ordination Unit (ACU) and two to three firing units, with a battery having two platoons. All the operators have one ACU vehicle to two firing units with the exception of Libya and Bahrain, which have one ACU vehicle to three firing units. The system cannot operate on the move but takes less than five minutes to become operational once it has stopped. Once the target has been detected the missile can be launched within about 6.5 seconds. The system has been designed to combat targets flying at a speed of M1.2 at an altitude of 50 to 3,000 m and an equivalent radar cross-sectional area of 1 m^2 fluctuating. Data is transmitted from the Crotale 4000 ACU to the Crotale 4000 firing units via a cable that allows operations up to 800 m away or via a radio link.

Both vehicles have an all-welded steel hull with the driver at the front, electronics and operators in the centre and the thermal motor at the rear. There is a door in the right side of the hull, which opens to the rear. The thermal motor provides energy. An alternator, driven by the thermal motor, produces power, the output of which is rectified and then fed to a series of DC motors which in turn drive each of the four road wheels by epicyclic reduction gears. Sufficient

Close-up of Crotale P4R series 4000 acquisition and co-ordination unit with radio link system mast raised
0044106

electric power is provided for all the vehicle's electrical systems including the electronics, air conditioning system and the hydraulic circuit, which operates the three levelling jacks, steering, suspension and brakes. Each road wheel station has a hydraulic and pneumatic suspension system designed by Messier. This acts as a pneumatic spring, suspension spring and shock absorber simultaneously. A selector valve connected to differential gear controls the position of each jack and the driver has a lever, which enables him to select one of five positions.

The ACU carries out target surveillance, identification and designation. Mounted on the top of the vehicle is a Thales Defence Systems Ltd E-band Mirador IV pulse-Doppler radar. The radar has a fixed-echo suppression which rotates at 60 rpm and has a maximum detection range of 18.5 km against low-level targets with speeds of between 35 and 440 m/s and altitude limits between zero and 4,500 m. The system also has:

- An IFF interrogator-decoder;
- A non-saturable extractor;
- Real-time digital computer;
- Display console;
- A digital datalink for transmitting information to the firing units.

The computer, which is the same as that installed in the firing unit, is used to generate accurate data for confirmation of threat evaluation. Thirty targets can be processed per antenna revolution with the 12 most dangerous targets automatically evaluated and tracked by the system.

Once the target has been detected, the computer triggers the IFF interrogator and the final threat information is displayed. The target is then allocated to one of the firing units and target designation data and operational orders are transmitted by the datalink, which also supplies information from the firing unit on operational status, for example, the number of missiles available.

The firing unit has a J-band monopulse 17 km range single target tracking radar mounted concentrically with the launcher turret, which carries four ready-to-launch missiles, two each side. The system also has:

- An I-band 10° antenna beamwidth command transmitter;
- Differential angle-error measurement infra-red tracking and gathering system with a ±5° wide field of view (and in French Air Force systems a further narrow field of view mode for passive operations);
- An integrated TV tracking mode as a low-elevation back-up;
- An optical designation tripod-mounted binocular device (which is controlled manually by a handlebar arrangement and used primarily in a heavy ECM environment or whenever passive operation is required);
- Digital computer;
- Operating console and a digital datalink.

All the vehicles are fitted with an inter-vehicle link network to transmit data and orders by cable and for radio communication by a VHF radio link.

The radar can track one target and guide one or two missiles simultaneously. The missiles, fired 2.5 seconds apart, are acquired immediately after launch by the 1.1 degree tracking beam of the radar with the help of infra-red detection and radar transponders during the gathering phase. Initially, the transponder was the 8,000 m range Thomson-CSF Stresa but this was replaced in 1990 by the then Thomson CSF RTKu M Ku-band which has an extended operational range of 10,000 m or greater and uses a solid-state transmitter with integrated processing. There is also a TV tracking mode possibility. Guidance signals are transmitted to the missiles by a remote-control system.

No spare missiles are carried on the vehicle and fresh missiles are brought up by a truck and loaded with a light crane. A well-trained crew of three can load four missiles in about two minutes.

The missile is designated the R440 and weighs 84 kg, has an overall length of 2.89 m, span of 0.54 m and a diameter of 0.15 m. The missile complete with its transport/launch container weighs 100 kg. The HE high-energy focused fragmentation warhead in the centre of the missile weighs 15 kg. The warhead has a lethal radius of 8 m for the 2,300 m/s initial velocity fragments and is activated in the original R440 missiles by either the infra-red proximity fuze (the fuze is commanded to activate 350 m before interception) or back-up contact fuze. The missile has an SNPE Lens III rocket motor with 25.45 kg of solid propellant powder. The missile reaches a maximum speed of 750 m/s in 2.8 seconds. The Naval Crotale fires a slightly modified missile, the R440N fitted with a then Thomson-CSF FPE pulse-Doppler I/J-band proximity fuze.

For 1 m^2 radar fluctuating cross-section targets, with velocities of 50 and 250 m/s respectively, the engagement parameters table.

Engagement parameters

Target velocity	50 m/s	250 m/s
Head-on target:		
(maximum operational intercept range)	>10,000 m	9,500 m
(minimum operational intercept range)	500 m	500 m
Crossing target:		
(maximum operational intercept range)	9,700 m	5,500 m
(minimum operational intercept range)	500 m	2,000 m
Target altitude:		
(maximum)	>5,000 m	4,500 m
(minimum)	15 m	15 m

The Single-Shot Kill Probability (SSKP) for a single missile is 0.8 and for a salvo of two 0.96.

The missile is itself capable of the following performance:

Missile performance

Range	Manoeuvrability	Flight time
5,000 m	27 *g*	10 s
6,000 m	18 *g*	13 s
10,000 m	8 *g*	28 s
13,000 m	3 *g*	46 s

The manoeuvrability (or load factor) of the missile in terms of time is the maximum number of *g* which can be applied to the weapon in pitch and/or yaw when under guidance.

The maximum range to which Crotale has been guided against a slow moving target (for example, helicopter) is 14,600 m.

Minimum flight time is 2.2 seconds (the time required to arm warhead section).

In early 1987, TDA tested a new HE fragmentation warhead for Crotale. This uses a time-space convergence technique to ensure that the warhead fragments arrive coincidentally within a 40 cm band at a distance of 5 to 8 m irrespective of the missile/target miss distance. The fragments are capable of penetrating up to 10 mm of steel plate within this range or severing the aluminium alloy body of a missile.

Variants

Crotale NG

The Crotale NG (new generation) uses the VT1 missile. Covered in greater detail in a separate entry.

Crotale Mk 3

The Crotale Mk 3 uses the VT1 missile. Covered in the Crotale NG entry.

Shahine

Shahine is a tracked version of the Crotale using either the R440 or R460 missile made specifically for Saudi Arabia. Covered in greater detail in a separate entry.

Naval Crotale

The Naval Crotale variant uses the R440N missile. Sold to the People's Republic of China, France and Saudi Arabia.

Thaqeb and Ya-zahra Project

The Thaqeb and Ya-zahra Project is a reverse engineered version Crotale manufactured in Iran. Covered in greater detail in a separate entry.

Specifications

	Crotale Acquisition Vehicle	Crotale Launch Vehicle
Dimensions and weights		
Crew:	2	2
Length		
overall:	6.22 m	6.22 m
Width		
overall:	2.72 m	2.72 m
Height		
transport configuration:	3.05 m	3.41 m
Ground clearance		
overall:	0.45 m	0.45 m
Wheelbase	3.6 m	3.6 m
Weight		
combat:	12,620 kg	14,950 kg
Mobility		
Configuration		
running gear:	wheeled	wheeled
layout:	4	4
Power-to-weight ratio:	laden	laden
Speed		
max speed:	70 km/h	70 km/h
Range		
main fuel supply:	600 km	600 km
Fording		
without preparation:	0.68 m	0.68 m
Gradient:	40%	40%
Vertical obstacle		
forwards:	0.3 m	0.3 m
Survivability		
Armour		
hull/body:	5 mm	5 mm

	R440 missile
Dimensions and weights	
Length	
overall	2.89 m (9 ft 5¾ in)
Diameter	
body	0.15 mm (0.01 in)
Flight control surfaces	
span	0.54 m (1 ft 9¼ in) (wing)
Weight	
launch	100 kg (220 lb)
Performance	
Speed	
max speed	1,458 kt (2,700 km/h; 1,678 mph; 750 m/s)
Range	
min	0.4 n miles (0.7 km; 0.4 miles)
max	5.4 n miles (10.0 km; 6.2 miles)
Altitude	
min	15 m (49 ft)
max	5,500 m (18,045 ft) (est.) (dependent on target velocity)
Ordnance components	
Warhead	15 kg (33 lb) HE fragmentation
Fuze	IR, proximity, impact
Propulsion	
type	solid propellant (rocket motor)

Status

Production as required (over 250 ground systems, 25 naval systems and some 7,000 missiles by 2000). The system has seen service in many countries.

Contractor

Thales (France).

Crotale NG

Type

Self-propelled Surface-to-Air Missile (SAM) system

Development

The Crotale New Generation (NG), is a low-altitude SAM system that was launched in 1988 as an improvement of the LIBERTY system proposed for inclusion in the US Army's FAADS-LOS-FH competition.

The modernised Crotale NG entered production in 1990.

Crotale New Generation for Finland mounted on modified XA-180 (6 × 6) APC

0080361

Operator's console of Crotale New Generation 0080362

NG differs from the original Crotale system in having all the acquisition, tracking, firing and computer units mounted in a single vehicle with only one system operator.

The modular all-weather system is divided into six main subsystems, which can be carried on a variety of platforms. These range from a simple shelter, to armoured vehicles, such as the Korean Daewoo Heavy Industries Infantry Fighting Vehicle (KIFV), the Finnish wheeled Patria Vehicles XA-180 (6 × 6) APC or the MOWAG Piranha (10 × 10).

The shelter version is available on the United Defense, LP designed MLRS carrier FVS 987.

The NG system's VT-1 missile was developed for the then Thomson-CSF in 1986 by the then Missiles Division of the American Company LTV Missiles and Electronics Group (subsequently Lockheed Martin Vought Systems). The VT-1 successfully completed a series of test firings in March 1989. The VT-1 missile is a land- and sea-based missile designed specially for short-range surface-to-air defence systems. The VT-1 equips both the land-based and naval versions of the Crotale Short Range Air Defence Systems.

The first prototype shelter system was used in 1988 for tracking trials, while the first complete system trials, using the shelter unit, took place in 1990 with the VT-1 missile. System integration was undertaken in France with the initial batch of 1,000 VT-1 missiles and 42 reusable launch pod containers built in the US. Production of these began in mid-1989 with the series being shipped in 1992, under a contract signed by LTV with Thomson-CSF Airsys, in 1986.

The initial American missile production was followed by a progressive production shift to Europe. In September 1991, the then Thomson-CSF Airsys based in Bagneux in France signed an agreement with the Euromissile consortium (Aérospatiale and the then Daimler-Benz Aerospace) for the production of the VT-1 missile. Apart from its use with Crotale NG, the weapon has been integrated into the Euromissile Roland 3 programme (see entry in this section).

Work shares are split between the French and German partners with Daimler-Benz Aerospace and Thales Defence Systems Ltd, concentrating on the sub-assemblies and Aérospatiale undertaking final assembly.

Elements of the Crotale NG system such as the VT-1 missile, electro-optical module and acquisition unit can also be introduced into the Crotale 4000/Crotale 5000/Improved Crotale system, either as new-build items or retrofitted into existing Crotale systems.

In 1988, the Crotale NG was selected by the Finnish Army, as its low-level air defence weapon, to fill the gap between the man-portable Strela-2 'Grail'/Strela-3 'Gremlin'/Igla-1 'Gimlet'/Igla 'Grouse' weapons and the static S-125 'Pechora' systems it operates. Initial deployment of the 20 fire units, with the then Thomson-CSF TRS 2630 radars, ordered and mounted on locally designed and built Patria Vehicles XA-180 (6 × 6) APCs, occurred in 1992. The APC hull, complete with its APU and NBC systems, was sent to France for fitting out.

Following this Finnish order, the Royal Netherlands Air Force originally selected the Crotale NG system to protect its air bases with the shelter version fitted on a trailer. However, financial problems have resulted in the programme being combined with a Dutch Army requirement and a reappraisal of available systems. The Crotale NG is being offered as a contender.

VT-1 missile showing position of main components 0080363

Crotale New Generation installed on US Multiple Launch Rocket System carrier chassis launching a missile 0080364

By mid-1991, the French Air Force had selected the shelter form of Crotale NG to protect its air bases. This version is fitted on a completely self-contained air-transportable shelter suitable for rapid overseas deployment operations. The French Air Force system includes a commander's console. An initial batch of 12 units has been ordered, with operational deployment by the French Air Force now complete.

South Korea has developed the Pegasus all-weather surface-to-air missile system based on the Crotale NG (see entry in this section).

At the 1998 Defendory International exhibition in Athens the Greek Defence Ministry revealed a requirement for 11 Crotale NG systems estimated to be worth more than USD200 million. Nine systems are for the Greek Air Force and two for the Greek Navy. The missile used is the VT-1. In June 1999, the Greek Defence Ministry signed the order with the then Thomson-CSF for the 11 systems as well as associated services and support. The contract cost was stated to be in excess of FRF1 billion (at the time of writing) by the then Thomson-CSF.

In February 2007, the new long-range version of the Crotale NG known simply as the Mk 3 was test fired at the *Les Landes* Test Range in Southwest France. The missile successfully intercepted its target at a range in excess of 14 km. This new version of the system further enhances the performance of the Crotale NG combining the shortest reaction time and full control throughout the engagement.

Description

The standard Crotale NG entered production in 1990 and provides air-situation and threat assessment, extended detection range, IFF, multi-target detection plus automated acquisition, tracking and engagement, and all weather operation.

The Crotale NG electrically driven turret weighs around 4,800 kg and includes:

- A Thales TRS 2630 E-band surveillance radar with associated IFF subsystem
- A cupola housing a tracking radar, electrical optical equipment including a day and night Forward-Looking InfraRed (FLIR) camera
- A daylight only TV camera
- Video tracking
- An IR localiser
- Eight ready-to-fire missiles in two packs of four container-launcher tubes.

The standard option surveillance radar is the Thales 2630 S-band frequency-agile pulse-compression Doppler model with a 40 rpm planar-array antenna, improved ECCM features (including strobe-on-jam, low side lobes, wideband frequency agility and constant false alarm rate) and search-on-the-move capability. Detection range against high-performance aircraft is 20,000 m and around 11,000 m on a hovering helicopter. Altitude coverage is from ground level to 5,000 m. An automatic track-while-scan capability provides track details on up to eight targets while simultaneously evaluating the threat.

The tracking radar fitted is of the frequency-agile monopulse Doppler type with improved ECCM features (including low sidelobes, wideband frequency agility, multimode (burst-to-burst or pulse-to-pulse frequency agility), constant false alarm rate and jammer tracking). Operating in the Ku-band with a range of up to 30,000 m on target types from hovering helicopters to M2 plus aircraft. Beam width is 1.2°.

The passive operation electro-optical systems comprise:

- Castor double-field-of-view, electronic magnification (8.1 × 5.4° wide and 2.7 × 1.8° narrow), thermal imaging, TV camera with a maximum acquisition range of up to 19,000 m which reduces to around 10,000 m in optical visibility conditions;

Crotale New Generation on MOWAG Piranha (10 × 10) chassis 0080405

- Mascot day-use single 2.4 × 1.8° field of view CCD TV camera (with a range of up to 15,000 m);
- Video tracker for automatic tracking of the target and missile and a large field of view IR localiser to track the missile in its initial few seconds of flight.

The VT-1 missile itself weighs 75 kg at launch (with its container-launcher tube the total weight is raised to 95 kg) and is 2.29 m long and 0.165 m in diameter. Four folding steel fins open out after launch to stabilise it. A TDA 13 kg focused fragmentation HE warhead is carried which uses the then Thomson-CSF FPNG *(Fusee de Proximity Nouvelle Generation)* pseudo-randomly pulse-modulated I/J-band broadband electromagnetic proximity fuze. This is activated by the missile's onboard firing circuit processor using a time delay set to operate at a time between 0.2 and 0.5 seconds before the projected target interception point is reached. The lethal radius of the warhead is 8 m.

The basic missile has a maximum range of 12,000 m, a minimum range of around 500 m and an altitude engagement limit of very low to more than 6,000 m. The Mk 3 NG missile has a range in excess of 15 km.

The maximum speed of the missile is 1,190 m/s. This is achieved by using an improved TX883 reduced smoke version (of the Sidewinder air-to-air missile) solid propellant rocket motor, with HTPB propellant encased in a graphite-epoxy composite canister, developed by Thiokol specifically for the VT-1 project. The weapon has a flight time of 10.3 seconds to 8,000 m range and is capable of manoeuvring under load factors of up to 35 *g* at this distance. The airframe can withstand up to 50 *g*. The motor weighs 37.9 kg and contains 31.4 kg of propellant.

The VT-1 missile that equips the Crotale Mk3 is currently produced by Thales in Belfast, Northern Ireland.

Missile guidance is by an IR deviation measurement system or narrow radar beam using the multisensor guidance principle which the then Thomson-CSF has incorporated into its naval Crotale system.

This principle involves using all the sensors to send their data to the on board computer which then processes it, after filtering out such interferences as clutter, decoys or jamming in a few milliseconds, to determine the guidance control commands to be up linked to the missile.

The Crotale Mk 3 has two operating modes:

(1) Stand-alone mode – minimises time to deploy in roles requiring a fast response, such as protection of forces on the move;
(2) Cooperative mode – enhances the system's interception capability within a radius of 15 km of each firing unit and provides a more robust defence system because of inherent redundancies.

In a normal engagement mode, both the tracking radar and the electro-optical systems operate together and constantly check each other.

Any control orders to the missiles are passed through the narrow beam, frequency-agile remote-control guidance radio uplink channel of the radar system.

A colour console displays the alphanumeric data on the targets, the TV and thermal images and the video images from the tracking radar, the surveillance radar scope and information on the available missile. All the operator has to do is follow a computer-generated menu displayed on the console and select the desired functions by pressing buttons.

The target engagement cycle from detection to interception is entirely automatic with the gunner only pressing buttons twice to ensure the safety of friendly aircraft. Reaction time is very short at five seconds or less with the total engagement time for target detection and final interception at 8,000 m range being estimated at approximately 15 seconds. Re-engagement time is one to two seconds, depending upon whether the target is isolated or in a group. It is theoretically possible for a single firing unit to engage successively two separate groups of four aircraft each and destroy all of them at a distance of between 500 to 11,000 m. Reloading the two missile packs takes around 10 minutes.

Variants

Thales has teamed with Russia's Fakel missile design bureau to produce a Vertical Launched (VL) version of the VT-1 missile. Intended for use with the Crotale Naval New Generation fire-control system, the VL-VT-1 configuration was achieved without any modification of the VT-1 missile beyond the addition of a pitch over module, thus preserving its performance. Ejection firings have been successfully performed.

The missile and its pitch-over module are repackaged in a Fakel designed launch canister for a cold-gas vertical launch. This projects the missile to 40 m, where upon the pitch-over module's side thrust gas jets sets the missile on course. The VT-1 computer, through a mechanical linkage with the missile's fins, controls the module. Only after the pitch-over is completed and the module jettisoned is the main rocket motor ignited. The missile is then gathered to the target line-of-sight and is commanded in the normal way to the interception point. The VL-VT-1 may have possibilities for a ground-based version.

The South Korean Pegasus Programme (Chun Ma), is almost certainly developed on the basis of those systems that were passed to Seoul in the mid-late 1990s.

China has developed its own variant of the system known as the Yitian. The chassis is based on a NORINCO WZ 551 series, the missiles are designated by the Chinese as the TY90 but the radar looks very similar to that used in the Crotale NG. China has had the basic Crotale for many years, both the land and navel systems.

The VT-1 missile is also used with the Roland 3 self-propelled SAM system.

A shelter mounted version is also available on the open arms market.

Specifications

	VT1	VT1 Mk 3
Dimensions and weights		
Length		
overall	2.29 m (7 ft 6¼ in)	2.29 m (7 ft 6¼ in)
Diameter		
body	165 mm (6.50 in)	165 mm (6.50 in)
Weight		
launch	76 kg (167 lb)	76 kg (167 lb)
Performance		
Speed		
max speed	2,313 kt (4,284 km/h; 2,662 mph; 1,190 m/s)	2,313 kt (4,284 km/h; 2,662 mph; 1,190 m/s)
Range		
min	0.3 n miles (0.5 km; 0.3 miles)	-
max	6.5 n miles (12.0 km; 7.5 miles)	8.1 n miles (15.0 km; 9.3 miles)
Altitude		
max	6,000 m (19,685 ft)	9,000 m (29,528 ft)
Ordnance components		
Warhead	13 kg (28 lb) HE fragmentation	13 kg (28 lb) HE fragmentation
Fuze	proximity, impact	
Guidance	CLOS, radar (electro-optical sensors)	CLOS, radar (electro-optical sensors)
Propulsion		
type	solid propellant (rocket motor)	solid propellant (rocket motor)

Status

In production. The naval version, Naval Crotale NG, has been sold to several countries including Oman and France.

Contractor

Prime contractor and radar
Thales.

VT-1 and Mk 3 Missiles
Thales.

Integrated tactical communications
Intracom Defense Electronics.

DHV

Type

Developmental Surface-to-Air Missile (SAM) system

Development

The five-year DHV (*Demonstrateur Hyper Veloce* or Hyper Velocity Demonstrator) programme that started in 1994 is under the control of the French MoD's Armament Directorate Department of Tactical Missile Programmes. Its main role is to show the feasibility of new technologies that are required for hyper velocity missiles in the 21st Century.

Description

These technologies include:

- New propulsion concepts such as the *Propulseur Hautes Performances* (high-performance motors)
- New component/system manufacturing processes, including resin transfer moulding (as used for the four composite fins of the DHV test vehicle)
- Thermal protection used on the ogive of the DHV, which was not fitted with an actual warhead.

Close-up of the MBDA DHV on its test launch rail 0080359

First test launch of the MBDA DHV in 1997 0080358

Future applications of the hyper velocity missile concept include the NATO VSHORAD/SHORAD feasibility study (see separate entry in this section) and the European EUCLID definition study carried out by France, Germany, Spain and the UK on multirole anti-tank and anti-helicopter systems.

The prime contractor for the DHV programme is MBDA, the main subcontractors are:

- CELERG (for the single-stage solid propellant rocket motor)
- Aerospatiale Batteries (for the thermal batteries)
- Matra BAe Dynamics (MBDA) (for the onboard telemetry equipment).

The first launch of a DHV test vehicle took place successfully from a static single-rail launcher at the Centre d'Essais des Landes (CEL) firing range in Southwest France.

The launch verified the link between the missile and the ground, which, because of the rocket efflux and debris kicked up at launch, can be difficult. Further trials are taking place to test other aspects of the DHV vehicle design.

The German company of BGT is developing the new HFK/KV hypersonic missile concept (see Germany) and it is possible that France and Germany may merge their two programmes to reduce development time and cost.

Status

Advanced technology demonstrator. First firing successfully concluded July 1997. Additional firing are taking place to prove the concept.

As of December 2013 no further information concerning this missile is available.

Contractor

Prime contractor

MBDA.

Shahine 1/Shahine 2

Type

Self-propelled Surface-to-Air Missile (SAM) system

Development

In 1975, the then Thomson-CSF was awarded a contract from Saudi Arabia for the Saudi Army for the design, development and production of a mobile, all-weather, low-altitude, self-propelled surface-to-air missile system based on a modified Crotale 2000, this was to be formerly known as Shahine 1. The contract covered the supply of several Shahine 1 batteries, each comprising two acquisition units and four firing units mounted on modified Giat Industries (now Nexter Systems) AMX-30 MBT chassis. The batteries were delivered to the Saudi Arabian Army from January 1980 through to 1982. The Saudi government also ordered the standard Crotale in 1978 and an upgraded version in 1990 for the Saudi Air Force.

In addition to the Crotale SAM orders Saudi Arabia also ordered the Nexter Systems AMX-30SA twin 30 mm self-propelled anti-aircraft gun system to operate with the Shahine batteries. Details of the AMX-30SA are given in the Self-propelled anti-aircraft gun section.

Shahine firing unit on Nexter AMX-30 MBT chassis 0080402

In January 1984, Saudi Arabia announced the placing of the USD4 billion 'Al-Thakeb' contract with the then Thomson-CSF. This mainly involved the supply of an improved Shahine version based on the Crotale 4000 missile, known as Shahine 2, mounted either on the AMX-30 chassis or on a towed shelter, the latter being known as the Air Transportable Towed System (ATTS). The modular design of the acquisition and fire unit systems allows them to be installed on either platform system. The surveillance radar range is increased to 19,500 m, the addition of a Shahine Data Link (SHADL) to allow data from a Litton TSQ-73 command-and-control centre to be passed to Shahine and the addition of an optional radar-frequency proximity fuze in place of the original InfraRed (IR) one.

Another contract was awarded to the then Thomson-CSF in 1990 to upgrade several Shahine 1 batteries to the Shahine 2 standard. This upgrading was performed from 1991 to 1993 and had the programme name Dattier. The value of the contract was then FFr2.5 billion (USD492.7 million).

In 1994, a FFr3.4 billion (USD670 million) contract was awarded to the then Thomson-CSF Airsys for the upkeep of its Shahine and self-propelled anti-aircraft gun systems.

Thales acts as the prime contractor for the radar and electronics of the weapon systems. MBDA is responsible for the missile.

Description

Shahine 1

The basic Shahine version is mounted on the chassis of the then Giat Industries AMX-30 MBT which has improved cross-country mobility in the desert climate compared to wheeled vehicles. An added advantage is that Saudi Arabia also operates the AMX-30 MBT.

The 32,700 kg acquisition unit has a pulse Doppler E/F-band surveillance radar (18.5 km range and 6,000 m altitude target detection capability) and a digital receiver for the MTI function. A computer that can record up to 40 targets and automatically initiate tracks on the 12 most threatening carries out the automatic data processing, including threat evaluation. A TV system featuring a TV turret, which is both concentric with and independent of the radar turret, provides ground monitoring of moving fire units and optical target reconnaissance.

The 38,800 kg firing unit comprises a monopulse Doppler J-band 17 km range fire-control radar that simultaneously tracks the target and localises one or two missiles. The radar has a digital receiver and a circularly polarised antenna. The fire unit also includes a remote-controller system which sends the missile guidance commands.

During the first part of the flight the missile is gathered in the radar beam by an infra-red sensor. In case of jamming, a TV system is integrated into the firing turret to ensure a back-up mode for the target and missile tracking functions. The firing unit has six ready to fire missiles. When expended, reload missile container-launchers are brought up and loaded by a Missile Transport and Loading Vehicle (MTLV).

Shahine acquisition unit on Nexter AMX-30 MBT chassis 0080404

Shahine firing unit on Nexter AMX-30 MBT chassis with missile leaving upper container
0080403

The R460 missile of the Shahine system is an improvement of the R440 Crotale weapon. It is 3.12 m long, with a diameter of 0.156 m (sometimes incorrectly reported as 0.16 m) and a wing span of 0.59 m. Total weight is 100 kg of which 15 kg is the high-energy focused fragmentation (splinter) warhead. This is triggered by either a passive IR or optional Thomson-CSF FPE pulse Doppler I/J-band proximity fuze. The lethal radius of the 2,300 m/s velocity fragments is about 8 m.

Behind the missile nosecone, which contains the proximity fuzing circuitry, is the section that houses the converter and pitch and yaw actuators. The latter actuate control fins mounted on canard surfaces. The next section along accommodates the battery, electronic control box and interconnections, various support instruments, the warhead and the safety and arming device. The next section contains the rocket motor that boosts the missile to its maximum speed of 933 m/s (M2.8). The solid propellant used was manufactured by the then Protac (now Roxel) and is of the Epictète Cast Double Base (CDB) type. The combustion period lasts about 4.5 seconds.

The final section of the missile comprises the remote-control receiver and transponder, the roll actuator, the infra-red tracer and the nozzle. Large cruciform surfaces around this section are equipped with antennas or roll fins driven by the roll actuator.

For a typical target such as a fighter with 250 m/s (M0.75) velocity and 1 m^2 fluctuating radar cross-section the following engagement parameters apply:

Head-on target:	
(maximum operational intercept range)	11,800 m
(minimum operational intercept range)	500 m
Crossing target:	
(maximum operational intercept range)	8,000 m
(minimum operational intercept range)	2,000 m
Target altitude:	
(maximum)	6,800 m
(minimum)	15 m

The Single-Shot Kill Probability (SSKP) for a single missile is 0.9 and for a salvo of two, 0.99.

The performance of the R460 missile is listed in Table 1:

Table 1

Range	Manoeuvrability[1]	Flight time
6,000 m	35 *g*	11 s
10,000 m	15 *g*	23 s
14,000 m[2]	8 *g*[2]	45 s[2]

[1] The manoeuvrability (or load factor) of the missile is the maximum number of *g* which can be applied to the weapon in pitch and/or yaw when under guidance.
[2] Approximate values.

The Inter-Vehicle Positioning and Data Link (IVPDL) allows complete Shahine system deployment to occur over a large area via J-band microwave data links and reciprocal location information to be exchanged between acquisition and firing units. The IVPDL capability covers 500 to 4,000 m between firing and acquisition units and 7,000 m between acquisition units. Other datalinks can be connected to the acquisition unit so that it can communicate with a higher command echelon, enabling the system to be integrated into an overall air defence network.

Specifications

	R460
Dimensions and weights	
Length	
overall	3.12 m (10 ft 2¾ in)
Diameter	
body	156 mm (6.14 in)
Flight control surfaces	
span	0.59 m (1 ft 11¼ in) (wing)
Weight	
launch	100 kg (220 lb)
Performance	
Speed	
max Mach number	2.8
Range	
min	0.3 n miles (0.5 km; 0.3 miles)
max	5.4 n miles (10.0 km; 6.2 miles)
Altitude	
min	15 m (49 ft)
max	6,800 m (22,310 ft)
Ordnance components	
Warhead	15 kg (33 lb) HE fragmentation
Fuze	IR, proximity, impact
	radar, proximity, impact
Propulsion	
type	solid propellant (Protac rocket motor)

Status
Production complete - a total of 36 acquisition and 73 firing units were built on AMX-30 chassis, and 10 acquisition and 19 firing units in the ATTS shelter configuration. A limited number have seen service or are still in service only with the Royal Saudi Land and Air Defence Forces. Details and illustrations of the ATTS are given in the Shelter- and container-based surface-to-air missile systems section. This system is no longer being marketed.

Contractor
Thales (radar and weapon systems).
MBDA (missile).

VL MICA

Type
Self-propelled Surface-to-Air Missile (SAM) system

Development
The VL MICA (*Missile d'Interception, de Combat et d'Autodefense*) SHORAD system was first unveiled at the Asian Aerospace 2000 Show, held in Singapore. It is designed for both ground defence and naval use and has now been offered in a land-based variant, truck-mounted on a five-tonne class vehicle. The system can be produced without any development risks at a cost that is affordable for the defence budgets of most countries.

VL MICA in launch canister
0110056

Artist's impression of a VL MICA battery in action 0099544

The MICA programme began in the early 1970s as an air-to-air missile replacement for the Magic and Super 530 systems. Development and testing continued through the 1980s and into the early to mid-1990s.

On 22 February 2005, MBDA's VL MICA was test fired using an InfraRed (IR) seeker against a small drone flying at a low-altitude. The test was a success with the drone being hit and destroyed at around 10 km. This test was a demonstration for the Indian Army and took place at the Centre d'Essais de Lancement de Missiles (CELM) missile test centre near Biscarosse on the South West Coast of France. Earlier tests at the same site during 2001 were designed to validate the vertical launch capability of the missile, which is expected to be available for land and naval systems.

On 22nd December 2005, the French DGA (*Délégation Générale pour l'Armement*) announced a contract as a first step towards the introduction of the VL MICA into the three services of the French armed forces. Expected deployment within the French armed forces are the protection of air bases and highly sensitive sites for the Air Force, the protection of deployed forces for the Army and fort the point defence of major vessels (support and command ships, aircraft carriers) for the Navy.

On 24 April 2006, a further test of the vertical launch system was carried out at DGA's CELM test facility. This time the launch was from a naval vertical launcher against a target flying at very low altitude simulating a sea skimming anti-ship missile. The target was intercepted at 10 km and the test was considered a complete success by MBDA.

On 23 October 2008, the final firing of 14 validation trials was completed when the VL MICA was launched against a Banshee drone flying at low level over the sea at a range of 15 km. This missile utilised its onboard J-band seeker and scored a direct hit thus proving the system's precision of the missile guidance system.

During mid-October 2013, the Royal Navy of Oman conducted an operational missile firing from the Al Shamikh OPV. This live firing took place at the French MoD's test range off the coast of the Lie du Levant.

Production of the first VL MICA for an export customer is underway as of November 2011, this is the first of four overseas customers for the system. Deliveries were expected to start in 2012. Meanwhile, the French Air Force has already taken part in the VL MICA land programme in 2009 which carried out the technical and operational assessment of the system demonstrator. The equipment produced and assembled in MBDAs facilities in France has been deployed at an Air Force airbase where the various vehicles have been networked to validate a standard operational configuration.

Description

Vertical Launch MICA is a short-range air-defence system which utilises the MICA missile that is available with two state-of-the-art seekers, IR or Radio Frequency (RF). Both versions of the MICA use a Tactical Operations Centre (TOC) and between three and six multi-round launchers with the missiles in clusters of four rounds. The truck-mounted system typically has a round launch assembly that elevates mounted on its rear decking. The system uses a distributed architecture and is capable of processing information collected by various sensors. This allows for easy integration into a larger air defence network and a greater inherent degree of survivability.

Missile

The VL MICA offers a real multiple target engagement capability regardless of weather or electronic warfare environment and can provide 360° coverage against aircraft, helicopters and air-to-ground missile threats. The MICA missile itself uses a butalite solid propellant booster and sustainer motor that is supplied by Protac, also a Thrust Vector Control (TVC) system that allows for the vertical-launch capability and a very short reaction time. A launch rate of less than two seconds between two missile launches is attainable. The rounds are completely autonomous after launch, flying initially under inertial control and then homing in on the target by means of the fitted seeker system.

The 112 kg VL MICA missile is available in two versions, both of which are identical to the original air-to-air MICA variants. The active radar MICA RF, is fitted with the same programmable J-band (10-20 GHz) pulse-Doppler AD4A radar seeker design that was developed for the Aster missile family; whereas the passive imaging InfraRed (IIR) MICA IR is fitted with a Sagem passive dual-band IIR seeker head with a rounded IR transparent glass dome. The seeker also uses a closed cycle cooling. The missiles' aerodynamic configuration combines four long, narrow-chord fixed fins on the missile main body with four movable L-shaped guidance fins at the rear end. This combined with the TVC that uses four slotted vanes on a circular plate that covers the exhaust area to direct the exhaust flux, to produce an extremely manoeuvrable weapon (reportedly capable of up to 50 *g*). Both versions use the same active radar proximity and contact fusing assembly to detonate a 12 kg HE blast focused-fragmentation warhead. The launchers can be loaded with any combination of the two missile types.

Radar and Optronics

An off-the-shelf radar or optronic surveillance device can be used for pre-launch designation of a target – the only information that the missile must have before it is launched.

System Structure

The system consists of the following:

- Tactical Operations Centre
- 3-D radar
- Launchers, (3-6) mounted on separate vehicles.

All of the vehicles are interconnected via a fibre optic and UHF link. The structure of the system has been designed for ease of deployment, integration into other global air-defence networks and gives the system a high level of survival.

One VL MICA launcher has the capability to protect a 360° area of 700 m^2 against subsonic threats.

Variants

VL Naval

VL Naval is designed to offer a highly effective, rapid-reaction all-weather air defence against a wide range of threats. As with the air-to-air weapon, the VL Naval MICA variant is a fire-and-forget missile featuring TVC system and two seeker variants, active RF or IIR.

Specifications

	MICA RF	MICA IR
Dimensions and weights		
Length		
overall	3.1 m (10 ft 2 in)	3.1 m (10 ft 2 in)
Diameter		
body	160 mm (6.30 in)	160 mm (6.30 in)
Flight control surfaces		
span	0.48 m (1 ft 7 in) (wing)	0.48 m (1 ft 7 in) (wing)
Weight		
launch	112 kg (246 lb)	112 kg (246 lb)
Performance		
Speed		
max Mach number	3 (est.)	3 (est.)
Range		
max	10.8 n miles (20.0 km; 12.4 miles) (est.) (against manoeuvring target)	10.8 n miles (20.0 km; 12.4 miles) (est.) (against manoeuvring target)
Altitude		
max	10,000 m (32,808 ft) (est.)	10,000 m (32,808 ft) (est.)
Ordnance components		
Warhead	12 kg (26 lb) HE blast fragmentation	12 kg (26 lb) HE blast fragmentation
Fuze	proximity, impact	proximity, impact
Guidance	INS, mid-course update	INS, mid-course update
	active, radar (programmable strap-down, monopulse-Doppler J-band AD4A, terminal homing 10 - 20 GHz)	passive, IR (programmable strap-down)
Propulsion		
type	solid propellant (composite Roxel, rocket motor)	solid propellant (composite Roxel, rocket motor)

Status

At least four overseas customers have been identified for the VL MICA. The system was demonstrated with firings from both land and naval launchers to India as a possible replacement for the defunct Trishul, however, the alternative Israeli system Spyder is known to have won the contract. Military sources have confirmed that evaluations of the report on trial firings by France's MBDA's VL MICA and Israel's Rafael Armament Development Authority's Spyder took place in February 2005.

Reports from Oman late in November 2008 have confirmed that the first confirmed naval customer for the VL MICA Naval will be Oman for its new Khareef-class corvettes. With contracts between MBDA and Oman having been signed.

The Romanian Air Force in July 2009 at the Paris Air Show contracted MBDA to supply VL MICA and the Mistral for land-based air-defence. No further details were released at that time concerning deliveries.

Contractor

MBDA.

Germany

ASRAD

Type
Self-propelled Surface-to-Air Missile (SAM) system

Development
The then STN ATLAS Elektronik GmbH Atlas now Rheinmetall Air Defence, based in Bremen, Germany, developed the Advanced Short-Range Air Defence (ASRAD) weapon platform as a private venture. Based on modular construction, it can be mounted on a wide variety of tracked and wheeled platforms, for example, the Mercedes-Benz Wie-Wagen (4 × 4) light vehicle, the Rheinmetall 2 tracked vehicle, the HMMWV, the Fennek etc. It comprises a fully powered pedestal mount with four short-range SAM missiles in the ready-to-fire position on the vehicle's platform. Four reload rounds are normally carried below the mount for manual reloading.

As part of an air defence network, ASRAD can be linked to surveillance radar for target acquisition. Single radar can provide target information for up to eight ASRAD vehicles. The ASRAD platform can be carried inside a CH-53 helicopter.

Following work with Bodenseewerk Gerätetechnik, the system was designed to be compatible with a variety of missile types such as:
- The Raytheon Systems Company Stinger (Basic, Post and RMP)
- Kolomna KBM Igla-1
- MBDA Mistral
- Saab Bofors Dynamics RBS 70 and Bolide.

The German Army selected the ASRAD on a Wiesel 2 chassis in 1995 for its LeFlaSys air-transportable mobile air defence system.

The total requirement was for three batteries, with each battery comprising an HQ unit and three platoons (each of which has a command-post and five missile firing units). A total of 50 firing platforms, 10 platoon command-post vehicles and seven battery command vehicles are being purchased.

The 20 km plus range Ericsson Microwave Systems of Sweden Improved HARD 3-D air defence radar has been procured as the target designation and surveillance radar for the system. The first units, including a trials platoon command-post vehicle with Improved HARD radar, were delivered to the German Army in 1996 with the firing unit given the name Ozelot. The reconnaissance vehicle is based on the Mercedes-Benz Wolf (4 × 4) light vehicle.

The first export order was placed by Greece in August 2000, the second by Finland in July 2002. The Finnish systems are mounted on Sisu Nasus and Mercedes-Benz Unimog 5000 vehicles.

Description
The ASRAD system consists of the following:
- Pedestal with traverse and elevation drive; combined sensor assembly and system electronics including missile interface electronics; multipurpose launcher assembly

ASRAD (Rheinmetall Defence Electronics (RDE) GmbH) 1035278

Weapon platform Ozelot (left) and platform command-post (right) of the German LeFlaSys. Both are based on the Rheinmetall Landsysteme Wiesel 2 chassis 1112398

- Pedestal electronics
- Control-and-display unit with remote-control capability using the same unit up to 100 m from the launch platform.

In the normal operational mode, the ASRAD weapon platform receives its target cueing data from a platoon command-post equipped with an Improved HARD search radar (with integral IFF) and/or integral Thales Optronics ADAD. The Improved HARD I/J-band search radar features Low Probability of Intercept (LPI) capability and is a frequency-agile track-while-scan system. More than 20 targets and five jammers can be tracked automatically. Maximum height capability is 10,000 m. Low level aircraft can be tracked out to 20,000 m and helicopters to 10,000 m.

If the ASRAD weapon platform is equipped with its own search radar or IRST, target cueing can be done by these systems as well. The ASRAD operator always has an air picture display as an overlay to the IR or TV monitor picture. The target to be engaged is presented as a triangle on the display.

The operator presses the 'Target Allocate' button on the right joystick, then the pedestal automatically moves to the target's azimuth bearing. If the cueing data is sent by a 3-D radar or an IRST, the pedestal moves in both traverse and elevation and the target is immediately visible in the IR or TV monitor picture. The cursor of the PPI presentation now points to the triangle.

If the cueing data is sent by 2-D radar, the pedestal moves only in traverse to the bearing of the target and the operator has to perform a manual elevation search until the target is discovered on the monitor.

When the operator recognises the target on the monitor, the track gate is moved over the target by means of the joystick and the autotrack mode is then initiated. Once the target is tracked automatically and the operator presses the 'Engagement' button, the ASRAD system commences to measure automatically and calculate all the necessary fire-control solution data on the chosen target while activating one of the onboard missiles. The target data are displayed on the monitor.

Missile lock-on as well as superelevation and lead angle setting are automatically performed and displayed on the monitor. The lock-on is also announced acoustically to the operator.

When the target enters the missile engagement envelope, this is also presented on the monitor and the gunner can then press the 'Fire' button.

All the above mentioned automatic functions can be manually overridden if needed. After the launch of the missile the gunner can either watch the engagement until the missile hits or he can initiate a second engagement against the same target if it is deemed necessary, or against a different one.

Variants
ASRAD 2

Also known as Skyarcher, the ASRAD 2 is a mobile air defence system that is compatible with various VSHORAD missiles. The system is capable of using various fire and forget missiles including MANPADS.

ASRAD 2 is capable of autonomous and networked mobile operation making use of its electro-optical target tracking sensor.

A 12.7 mm machine gun for self-defence is an optional extra.

Ozelot weapon platform based on latest Rheinmetall Landsysteme Wiesel 2 chassis launches a Stinger SAM during trials 0080372

HMMWV with ASRAD platform of Rheinmetall Air Defence (Rheinmetall Defence Electronics (RDE) GmbH) 1112399

Platoon command-post of the German LeFlaSys on Rheinmetall Landsysteme Wiesel 2 with the Ericsson Improved HARD 3-D radar elevated 0080371

ASRAD-R RBS 70 Version

Rheinmetall Defence Electronics is working with Saab Bofors Dynamics AB of Sweden on a version designated ASRAD-R for the export market. This involves the installation of a four-round RBS 70 fitted ASRAD pedestal on an M113 APC with an Ericsson Microwave Systems HARD 3-D search and acquisition radar on top of the assembly. The pedestal is fully stabilised and contains the electro-optic package of the RBS 70 system. Although fitted to the M113 for prototype trials, the installation can be fitted to a wide range of wheeled and tracked chassis. The Finnish Defence Forces have awarded the first series production contract for ASRAD-R to RDF in July 2002. The ASRAD-R platform therefore will be mounted on a Mercedes-Benz Unimog 5000.

Specifications

ASRAD system

Weight:	320 kg (approx.)
Pedestal:	
(traverse)	360°
(elevation)	–10° to +70°
Sensor LOS: (relative to the platform)	
(traverse)	≥±15°
(elevation)	≤+4 to ≥–16°
Stabilisation accuracy: (of the LOS Sensor)	≤0.05 mrad
LOS accuracy:	
(pedestal)	≤0.2°
(pedestal aiming velocity)	≥56°/s
Sensors:	integrated and stabilised LOS 8-12 µm waveband IR, TV camera and eye-safe laser range-finder
Control box:	fitted with monitor, keyboard and joystick; remote-control up to 100 m
Interfaces:	for data transmission; GPS, inertial and north-finding navigation system; radio telephone; and vehicle intercommunication
Weapons:	
(main)	2-4 ready-to-fire Stinger, Mistral, Igla-1, Starburst or RBS 70 SAMs
(secondary)	7.62 mm machine gun
Power supply:	18-32 V DC vehicle supply

Status

In production and in service with the German Army (Wiesel 2 tracked vehicle).

Authority for final production received mid-1998, with the first platoon becoming fully operational in 2000. The final platoon entered service in 2003. Offered for export to approved countries.

In August 2000, the Hellenic Army awarded a USD116.6 million contract to supply 54 ASRAD (Stinger) systems. The system known locally as ASRAD-HELLAS is related to the German Army's LeFlaSys light mechanised air defence system, and especially to its Ozelot fire unit.

Contractor

Rheinmetall Defence Electronics GmbH, a subsidiary of Rheinmetall Defence.

European vehicle-mounted Low-Level Air Defence System (LLADS)

Type

Self-propelled Surface-to-Air Missile (SAM) system

Development

The LFK-Lenkflugkörpersysteme (now MBDA Deutschland LFK GmbH) European private venture Low-Level Air Defence System (LLADS) has been designed as a cost-effective low-weight modular air defence system to fill the gap between man-portable missile weapons and complex heavy wheeled or tracked air defence missile systems.

Computerised operation of the missile launcher, sensors and an interface to a battlefield command, control, communications and intelligence system allows night and all-weather quick reaction capability and function in a multi-target environment. Typical system employment would be for point defence or escort (using its optional shoot whilst moving capability).

The system is air-transportable with two units in a Sikorsky CH-53G Sea Stallion helicopter, three units in a C-130 Hercules transport or four units in a C-160 Transall transport. A continuous evolution of the system concept and hardware configuration is being performed.

In 1993, the prototype equipped with GPS and thermal imaging fired four missiles and was used twice for German MoD Stinger acquisition trials.

In 1994, the prototype was modified to accept cueing data by a radio link from air defence radar.

In 1996, the prototype took part in very low temperature Stinger firings.

In 1998 it was again used in a comparative testing of various IR AD-seekers.

Description

The vehicle mounting the pedestal system is the Mercedes-Benz GD 250 (4 × 4) light all-terrain vehicle. The vehicle is fitted with a specially designed stabilised pedestal on its rear decking, which has direction controls for azimuth and elevation.

Other light vehicles such as the Peugeot P4 and Land Rover can also have the pedestal mount fitted on their cargo platform.

The pedestal can be fitted with a variety of sensors including either:

- A Forward-Looking InfraRed (FLIR) system for target acquisition (at night or in adverse weather)
- A Low-Light Level TV (LLLTV) device (for limited night or adverse weather capabilities)
- Or a laser range-finder mounted beside the primary target acquisition system to determine the distance to target-in-range decision.

MBDA Deutschland LFK GmbH European vehicle-mounted Low-Level Air Defence System (LLADS) on Mercedes-Benz GD 250 (4 × 4) all-terrain vehicle 0080368

The full LLADS sensor configuration characteristics available are shown in the table:

	LLLTV camera	FLIR camera	Laser range-finder	IFF and pre-assignment information
Limited night fighting capability:	Yes	No	Optional	Yes
Day/night fighting capability:	No	Yes	Optional	Yes

The gunner can use the control panel with monitor while seated in the cab or within the pedestal mount or remotely at a distance of up to 50 m away.

On either side of the mount are two vertically fixed standard Raytheon Systems Company Air-To-Air Stinger (ATAS) launcher modules, weighing 43.6 kg each and with a total of four ready-to-launch Stinger missiles. Full details of the Stinger missile versions are to be found in the Man-portable surface-to-air missile systems section of this book. The mount can also be used with LFK's combination launcher for VSHORAD missiles (see entry this section for launcher details).

The individual ATAS launcher module incorporates the launcher structure, a mechanical interface, an electronics package and a cooling system. It is operated by the fire-control system of the carrier via interface electronics and can be used with all the Stinger missile versions.

The field of fire is –°10 to +70° in elevation and 360° in traverse. Storage for up to four additional Stinger container-launcher tubes is available beneath the pedestal mount. The Stingers also retain their shoulder launch capability using the grip-stock assembly.

The vehicle also carries an antenna and communications system for pre-assignment tasking via the C3I network, a Northing Gyro to obtain a north reference and a Global Positioning System (GPS) unit to provide accurate position coordinates for defence networking the LLADS.

A generator set, operating remotely to the fire unit, supplies electrical power. For transportation, the set is carried on the vehicle's platform.

Variants

The standard Stinger ATAS module has also been adapted by MBDA Deutschland LFK GmbH for use with the 40 mm Saab Bofors Dynamics L/70 air defence gun. A boxed set of one ATAS module and target acquisition system is mounted on the top of the gun mount by the ammunition feed area.

Other examples of Stinger adaptations are the Tiger support helicopter, the BSH 1 escort helicopter and the 40-mm Navy Type 58 air defence gun.

Specifications

	LLADS
Dimensions and weights	
Length	
overall:	4.65 m
Width	
overall:	2.01 m
Height	
overall:	2.65 m
Weight	
combat:	2,850 kg
Mobility	
Configuration	
running gear:	wheeled
Firepower	
Armament:	8 (est.) estimated × Stinger SAM
Main armament traverse	
angle:	360°
speed:	70°/s
acceleration:	130 °/s²
Main armament elevation/depression	
armament front:	+70°/-10°
speed:	70°/s
acceleration:	110 °/s²

	FIM-92C Stinger
Dimensions and weights	
Length	
overall	1.52 m (4 ft 11¾ in)
Diameter	
body	70 mm (2.76 in)
Flight control surfaces	
span	0.091 m (3½ in) (wing)
Weight	
launch	10.1 kg (22 lb)
Performance	
Speed	
max Mach number	2.2
Range	
min	0.1 n miles (0.2 km; 0.1 miles)
max	2.4 n miles (4.5 km; 2.8 miles) (est.)
Altitude	
max	3,800 m (12,467 ft)
Ordnance components	
Warhead	1 kg (2.00 lb) HE fragmentation
Guidance	passive, IR (ultraviolet)
Propulsion	
type	solid propellant (boost and sustainer rocket motors, eject motor)

Status

Ready for production. Offered for export to approved countries.

Contractor

MBDA Deutschland LFK GmbH.

Standard Stinger Launching System (SLS)

Type

Tripod mounted man-portable Surface-to-Air Missile (SAM) system

Development

The Standard Stinger Launching System is a lightweight, self-contained unit that provides the structure, electronics and coolant to support the launch of two Standard Stinger missiles. It is compatible with helicopters, land and sea-based VSHORAD systems.

In 1999 it was shown on the Krauss-Maffei Wegmann Gepard 1A2 hybrid configuration upgrade.

The launcher is capable of performing repeated missile firing and has missile and coolant bottle reload capabilities without removal of the launcher from its parent-firing platform.

Description

Standard Stinger Launching System consists of the following sub-assemblies.

Launcher

The launcher sub-assembly is fitted with two ready-to-fire Raytheon or European (FIM-92C) produced Stinger missiles in their launcher tubes and rechargeable coolant bottle suitable for 40 engagements. A launcher adapter provides the mechanical interface between the launcher and the platform. It features launcher mounting and aligning fixtures.

Interface electronics

The interface electronics sub-assembly provides the interface between the launcher(s) and the launch platform fire-control system. Up to four launchers, that is eight ready-to-fire Stingers, can be controlled via the interface. The electronics are the centre for processing, controlling and distributing power, operator commands, signals and the responses between the various system elements: control panel or weapon fire-control computer, cyclic switches, intercom and launcher/missiles. It also provides signal buffering and scaling as required for the gunner's sight assembly. Two different interface outputs are provided - an analogue and/or a digital MIL-BUS-1553B.

Standard Stinger Launcher System with launcher for two missiles (top) and interface electronics (bottom)
0080407

Specifications

	Standard Stinger Launching System
Dimensions and weights	
Length	
overall	1.52 m (4 ft 11¾ in) (without adapter)
Width	
body	337 mm (13.27 in) (without adapter)
Height	
body	180 mm (7.09 in) (without adapter)
Weight	
loaded	43.6 kg (96 lb) (without adapter and complete with two missiles and full coolant gas bottle)

Status
In production. In service with the German Armed Forces (to equip the Tiger combat support helicopter). Offered for export to approved countries.

Contractor
MBDA - LFK GmbH (previously known as LFK-Lenkflugkörpersysteme GmbH).

Hungary

Hungarian - Igla-1E

Type
Self-propelled Surface-to-Air Missile (SAM) system

Development
The Hungarian Home Defence Force has developed a self-propelled Short-Range Air Defence (SHORAD) system to complement its Russian 9K31 Strela-1 (NATO SA-9 'Gaskin') low-altitude self-propelled surface-to-air missile systems.

The mobile SHORAD system was designed to be integrated into a low level air defence network with the Strela-1 and is based on the Russian Kolomna KBM 9K310 Igla-1E (NATO SA-16 'Gimlet') shoulder-launched, man-portable surface-to-air system (MANPADs). It has been assessed that the dual launcher for the Igla-1E is also capable of mounting the Igla missile or any other MANPADs.

Description
A two-round launcher assembly with sighting unit is mounted on the rear decking of a GAZ-630 4 × 4 truck chassis. This allows the system to have both cross-country mobility and the capability to defend units on the march. If required the system gunner can dismount from the truck and use the Igla-1E in the conventional man-portable role with the aid of a launching device (gripstock).

Status
Production as required. In service only with the Hungarian Army. Not believed to be offered for export. In early 1999 the Hungarian Army started to take delivery of nine mobile Mistral Coordination posts and 45 ATLAS twin MBDA Mistral SAM launchers, all of which are mounted on the rear of German

The truck-mounted system with the Igla-1E SAM in detail (P Beaver) 0509763

The truck-mounted system with the Igla-1E SAM in detail (P Beaver) 0509762

Mercedes-Benz Unimog (4 × 4) cross-country light truck chassis. These are believed to have initially supplemented the existing Igla-1E mobile system and will supplant or complement them in the longer term.

Contractor
Hungarian state factories.

India

Akash

Type
Self-propelled Surface-to-Air Missile (SAM) system

Development
Akash has been developed as part of the Integrated Guided Missile Development Programme (IGMDP) initiated by India in July 1983. The system is an all-weather medium range surface-to-air missile having a multidirectional, multitarget area defence capability. The weapon can simultaneously engage several air targets in a fully autonomous mode of operation.

The first trial firings were conducted in 1990 and as of December 2007 firings are still taking place.

In November 2007, firing trials took place at the Pokhran test facility in Rajasthan's Thar Desert. These test flights were carried out in near secrecy at Pokhran, which is 200 km from Pakistan's borders; previously at this test range India tested a string of nuclear weapons including a thermonuclear device.

The flight and ground elements of the weapon system are integrated in a plug-and-fight architecture. The hardware and software integration of various weapon system elements permits autonomous management of air defence functions such as programmable surveillance, target detection, target acquisition, tracking, identification, threat evaluation, prioritisation, assignment and engagement. Command-and-control nodes, communication links, self-propelled launchers and sensors are integrated with other air defence command-and-control networks through secure communication links. The system is also provided with advanced ECCM features at various levels. The weapon system is cost effective comparing to equivalent systems in the market. The weapon system has cross-country mobility and can be deployed by air, road and rail.

The first trial firings occurred in 1990, with the 10th stated test in September 1998. Since January 2005, the Akash system has been tested 16 times, including two crossing targets taken with live warheads. Akash has multiple-

Akash (T-72 Chassis) Army Version (DRDO) 1303196

3-D CAR System for use with Akash (DRDO) 0548763

targeting handling capacity, with digitally coded command guidance. Demonstration of simultaneous target interception capability against two live aerial targets, was successfully conducted in November 2005.

Description

The Akash Weapon System architecture is based on a Group Headquarters and a number of batteries. The system is customised on tracked (T-72 Main Battle Tank (MBT)) or wheeled chassis to provide area air defence against multivarious air threats to mobile, semi-mobile and static vulnerable forces and areas. The Akash air defence group covers a large volume of air space over the combat zone. The system can be operated either in the autonomous mode or in the Group Mode. The Akash Group consists of:

- Surveillance radars
- Control Centres
- Phased Array tracking and missile guidance radars
- Launchers
- Ground Support Equipment.

The Akash firing battery consists of:

- Four launchers (three missiles per launcher vehicle)
- Group Control Centre
- Central Acquisition Radar
- Battery Control Centre
- Battery Level Radar (a phased array tracking and guidance 3-D sensor)
- Battery Surveillance Radar (2-D sensor)
- Support Vehicles (power supply, missile transport and engineering support).

It is reported that a single battery can engage up to four targets simultaneously with eight missiles in flight.

Surveillance Sensor

The 3-D CAR is capable of detecting and tracking aerial targets up to a range of 150 km at an altitude of 18,000 m. It provides coordination in three dimensions of up to 200 targets to the Group Control Centre (GCC) through secure communication links. The data is used to cue the weapon control radar.

Group Control Centre (GCC)

Both the GCC and the Battery Control Centre (BCC) have ruggedised computers where real-time air picture from various sensors is integrated and data is processed. Decision support software carries out threat analysis and generates options for commanders. Automated target assignment and launch commands are generated for optional engagement in kill zone.

New Rajendra Radar on T-72 Chassis (DRDO) 1303198

Akash (Air Force) Version (DRDO) 1303195

C4I

The C4I software has been specially designed to meet Indian defence requirements, provide fusion of air pictures from various sensors, automatically track air targets, designation of track numbers to different targets, Identification of Friend or Foe (IFF), automatic assignment of target to GW batteries, automatic selection of launcher and decision support system for commander for launch and control of missiles. This feature also drastically reduces the requirement of manpower for operation of the system, as the complete operations from target detection to engagement are hands free. The advanced battle management software has been extensively field tested under realistic combat development conditions using multiple live targets. The system can also be integrated with legacy or futuristic radars and networks.

Multifunction Phased Array Radar

The multifunction phased array Rajendra radar variants have been configured on BMP and T-72 based tracked vehicles. The tracking and missile guidance radar configuration consists of a slewable phased array antenna of more than 4,000 elements, spectrally-pure TWT-based transmitter, two stage superheterodyne correlation receiver for three channels, high speed digital signal processor, real time management computer and a powerful radar data processor. The system has multiple target handling capability from any direction. Each radar can simultaneously engage four targets and guide eight missiles in a ripple mode. The radar has advanced ECCM features.

Missile

The 5.8 m long, 720 kg Akash utilises its integral Ram Rocket propulsion that provides all the way thrusting to a range of 25 -30 km with a velocity of 600-700 m/s from 1.5 km onwards. The Ram Jet system enables powered intercept, high manoeuvrability, much higher terminal velocities, high average speed, lesser reaction time, lower flight time and better engagements of receding targets vis-à-vis boost coast type of missiles. It also has a wider killing zone. The digital autopilot and guidance system are microprocessor based.

Launcher

Akash launcher carries three ready to fire Akash missiles. The Akash launcher for the army is based on a Russian T-72 MBT chassis manufactured under licence at the Medak Ordnance Factory, Tata Truck or Trailer mounted for the Air Force. The launcher is interfaced with BCC via line or radio, is fully automatic and remotely controllable, has a microprocessor controlled electromechanical servo system and is capable of check-out and auto launch of the missiles. It has its own built in GTE power source and can also be operated by DC power from housed DC batteries.

The Air Force trailer-mounted launcher uses hydraulics on each of the four corners of the launcher body that lifts the whole launcher vertically from the back of the trailer. This then leaves the trailer and the tractor towing vehicle free to move off and collect a further launcher with missiles, or recover a second launcher for replenishment of missiles. The launcher then become a static shelter type that is operated remotely. This static type system is used for defence of high value areas such as airfields, command centres and so on.

In March 2009 it was announced that the IAF has placed an order for 16 trailer mounted launchers. The launchers are to be to be delivered over a period of 33 months.

Ground Support Equipment (GSE)

Akash GSE system is designed for high reliability and maintainability. The system has built in diagnostics and check out systems with card level fault diagnostics and field replacement. Simulators for operator training are also built into the system.

Variants

DRDO is working on enhancement (Akash Mk2) of the capabilities of Akash Weapon System in its next version as pre-planned product improvement. During May and June of 2012, at least 10 firing's of the Akash missile took place from the Orissa Proving Range, these firing's have been reported as tests for the new surface-to-air Akash missile programme. Other reports have suggested that the tests included both Indian Army and Air Force missile systems flying against drone type targets over the sea.

Specifications

	Akash
Dimensions and weights	
Length	
overall	5.820 m (19 ft 1¼ in)
Diameter	
body	350 mm (13.78 in)
Flight control surfaces	
span	1.1 m (3 ft 7¼ in) (wing)
Weight	
launch	720 kg (1,587 lb)
Performance	
Speed	
max speed	1,361 kt (2,520 km/h; 1,566 mph; 700 m/s) (est.) (from 1.5 km)
max Mach number	2.05
Range	
min	1.6 n miles (3.0 km; 1.9 miles)
max	16.2 n miles (30.0 km; 18.6 miles)
Altitude	
min	30 m (98 ft)
max	18,000 m (59,055 ft)
Ordnance components	
Warhead	55 kg (121 lb) HE fragmentation
Fuze	proximity (radio with advanced signal processing features)
Propulsion	
type	ramjet

Status

The weapon system designs have been fine tuned through a number of development tests. Extensive field trials of the system were conducted by the services (both Army and Air Force) to include mobility and performance checks at Pokharan. In realistic combat conditions at Pokharan, the complete group of Akash Weapon System was fielded and its mobility and functionality assessed. The rigorous trials have established the ruggedness of various electronic and mechanical packages of the ground systems. Also the response of the Akash weapon system to various air threat scenarios has been assessed in detail. The tests have conclusively proved the combat worthiness of hardware and software integration of Akash weapon system. The immunity of the Akash weapon system to electronic countermeasure environment was separately tested and proved at Gwalior Air Force Base.

User trials to verify the consistency in performance of the total weapon system against low flying near-range target, long-range high-altitude target, crossing and approaching target and ripple firing of two missiles from the same launcher against a low altitude receding target, were conducted at ITR Chandipur during 14-21 December 2007. Akash missile successfully intercepted nine targets in successive launches. Fifth and last trial took place at 2:15 pm on 21 December at Chandipur in which the Akash missile destroyed an Unmanned Aerial Vehicle (UAV) (Lakshya) that was flying a trajectory simulating an air attack.

A decision to induct Akash Weapon System into the Indian Air Force has been taken by the Indian Government. During early January 2009, the Indian Air Force placed an order with Bharat Electronics Limited (BHL) for two squadrons of the Akash system.

The various subsystems have been sourced from private and public industry sources within India. The vendors have been chosen specifically for their mass production capabilities, manpower and quality management systems. This will ensure timely production deliveries. Cost competitiveness of the system has been a primary objective throughout the course of the programme.

During late March 2009, the Tata Group's defence arm received an order for supplying 16 indigenous Akash launchers for the Indian Air Force. The total contract for two regiments is worth an estimated INR1,200 crore, deliveries are to be completed within three years. At the same time, Bharat Electronics Limited (BEL) also received an order for the Rajendra phased array radar that is associated with the Akash system.

In June 2010, the Indian Army placed an order with BDL for the Akash system to be inducted into the Indian Army by 2011.

The first of the Akash squadrons will protect the Gwalior Air Base which is the home of the Mirage-2000 fighter aircraft. A second squadron due in December 2011 is expected to be deployed at the Lohegaon Air Base in Pune, a major base for the frontline Sukhoi-30MKI fighters. The remaining six are expected to be along the Sino-Indian border including Tuzpur, Bagdogra and Hasimara.

Contractor

Defence Research and Development Organisation (DRDO) (primary contractor).
DRDL in Hyderabad is responsible for system integration and missile development.
Bharat Dynamic Limited (BDL). Hyderabad has been identified as the main production partner for missile systems.
LRDE in Bangalore, for radar development and R and DE.
Bharat Electronics Limited (BEL), Bangalore has been identified as the Nodal production agency for radar and electronic support systems.
DRDO in Pune for the launcher.
Tata for the production of the Indian Air Force towed launcher vehicle.
ECIL, Hyderabad is identified as the main production partner.
L&T Mumbai identified as a main production partner.
TPCL Mumbai identified as a main production partner.
Chandipur-on-Sea test facility.
Wheelers Island test facility.
Pokhran test facility.
Kolar for functional evaluation.
Gwalior Air Force base for electronic countermeasure environment testing.

International

ASGLA

Type

Self-propelled Surface-to-Air Missile (SAM) system

Development

The German Company Rheinmetall working alongside the Ukrainian Company Arsenal have combined two separate air-defence systems into one to produce the ASGLA (ASRAD - IGLA). First identified at the Aviasvit Exhibition held in Kiev, Ukraine, September 2010. The system is formed from the Rheinmetall ASRAD-2 and the Ukrainian produced Igla-1M. Mounted on the chassis of an armoured troop carrier BTR-80, the system is intended for defence of pre-defined areas or mobile ground forces.

Description

The troop carrier BTR-80 with the added weight of the ASGLA weighs in at 1.3 tonnes and has a crew of three; commander, operator and driver. The vehicle is armed with four Igla-1M in position ready to fire whilst also boasting a 12.7 mm heavy machine gun for self-defence. Contained within the vehicle are eight resupply Igla-1M's.

The ASGLA is equipped with an electro-optical system on a platform that is stabilised in two planes. The electro-optics also include a night vision device, a daylight camera and a laser range-finder.

The complete platoon consists of:

- Command point acting as a control node
- Detection and fire-control post
- Up to eight launchers
- 3-D radar station.

The Rheinmetall X-Tar 3D radar fitted to the ASGLA self-propelled command post is equipped with a lattice phased array antenna with an IFF capability operating in the X-band. The phased array antenna is able to generate up to 12 simultaneous beams and incorporating an integrated Identification Friend-or-Foe antenna. The range of the antenna is 25-50 km with an update time of 1, 1.5 and 2.0 seconds.

Specifications

Not available.

Status

Development of the ASGLA system is complete and the system is now offered ready for export. As of March 2013 there have been no export orders placed for the system.

Contractor

Rheinmetall, Germany.
Arsenal, Ukraine.

ASGLA radar on BTR-80 (Rheinmetall) 1331702

ASRAD-R

Type
Self-propelled Surface-to-Air Missile (SAM) system

Development
Rheinmetall Defence Electronics (RDE) of Germany (former STN Atlas Elektronik of Germany) have developed the Advanced Short Range Air Defence System (ASRAD-R) as a private venture for the export market in cooperation with Saab Bofors Dynamics of Sweden (missile system).

Ericsson Microwave Systems of Sweden have assisted the companies with the HARD 3-D radar, as has Bodenseewerk GeräteTechnik (BGT) of Germany with the missile interface.

It builds on the experience obtained by RDE in the design and development of the original ASRAD system that has now been accepted by the German Army to meet its requirement for a low level air defence system called LeFlaSys (*Leichte Flugabwehrsystem*).

Full details of this system, based on a Rheinmetall Landsysteme Wiesel 2 armoured chassis, are given under Germany, in this section.

The first prototype of the ASRAD-R was completed late in 1998. This was followed by the first trials in Germany, and then the system went to Sweden early in 1999, where test launches of the RBS 70 Mk 2 took place later that year. Series production of 50 systems including 10 platoon command posts and seven command vehicles began in 2000. The German Army received its first system in June 2001 and deliveries were completed by the end of 2004.

ASRAD-R is aimed at existing users of the RBS 70-missile system as well as potential new customers. There are 13 known users of the RBS 70 missile system, most of which use the standard tripod version.

Description
ASRAD-R is a modular air defence concept and all-target missile system with high precision at long ranges down to ground level, for self defence also against bombers, fighter-bombers, modern threats also include stand-off weapons, UAVs and other small targets as well as armoured ground targets, well suited for protection of mobile units and defence of vital objects. The Bolide missile for ASRAD-R utilises the laser beam guidance of the RBS 70 and provides an intercept range of 8 km. ASRAD-R consists of an external pedestal system with four Missile Mk 2 in ready-to-launch position, on top of which is mounted the latest HARD 3-D radar system, although it is stated that the Missile Mk 1 can also be used with the system. However, the new BOLIDE missile from Saab Bofors Dynamics was scheduled to replace the missile Mk2 when it entered production in 2002.

The main advantage of the new ASRAD-R is that the weapons and sensors are now all on one platform and that the laser-guided RBS 70 missile is almost impossible to jam and has a greater effective range and altitude than the Stinger, Mistral and Igla.

Close-up of ASRAD-R system mounted on M113 series APC during prototype trials 0013967

ASRAD-R self-propelled air defence system on Mercedes Unimog 5000 in the ready-to-launch position 1112400

The current production Bolide missile has a maximum range exceeding 8,000 m against head-on targets, 4,000 m against crossing targets, with a speed of 250 m/s and a maximum altitude exceeding 5,000 m. The missile employs a booster and sustainer motor with low-smoke propellant that produces speeds greater than Mach 2 for short engagement time.

As well as launching the current production Bolide missile, ASRAD-R can also launch earlier production model missiles. According to Saab Bofors Dynamics, the laser beam guidance renders the system immune to jamming, gives high precision and delivers a head-on target engagement capability at long ranges.

So far, over 3,500 Missile Mk 2 have been built, with sales being made to eight countries. Total production of the RBS 70-missile system has now passed 15,000 missiles and 1,300 launchers.

The Bolide missile provides the RBS 70 and ASRAD-R systems with an all-target capability. It is fitted with a penetrating blast-fragmentation warhead combining a high explosive fill, 3,000 tungsten spheres and shaped charge, this is coupled to a dual impact/adaptive-proximity (three selectable modes) fuzing system. The missile is optimised for aerial targets, but in a self-defence role, surface targets can be engaged. When used to attack armoured surface or aerial targets, the warhead's shaped charge component can penetrate approximately 200 mm of armour. Other capabilities within the missile, such as the reprogrammable electronics suite, means that the Bolide missile can readily be upgraded with new or modified functions. The missile has been specifically designed for short time-of-flight with a high manoeuvrability.

Typical missions of the ASRAD-R include providing protection to static or mobile mechanised units against low flying aircraft and attack helicopters under day/night and bad weather conditions.

A SHOrt Range Air Defence (SHORAD) battery of nine ASRAD-R systems can defend troops on the move over a distance of 80 km and, when carrying out an area defence mission, the battery could engage targets within an area of 500 to 600 km^2.

The new pedestal on top of the hull is larger than that fitted to the original ASRAD. This contains: all electric azimuth and elevation drives; stabilised sensor suite, which includes infrared, daylight television and a laser range-finder for automatic target acquisition and tracking; and the standard laser beam transmitter unit, which is used to guide the RBS 70 to the target.

Elements of the stabilisation system are from the RDE fire-control systems originally developed for the Leopard 1 and Leopard 2 MBTs.

ASRAD-R self-propelled air defence system based on the widely deployed M113 series APC with four missile Mk 2 in the ready-to-launch position and HARD radar deployed 0036586

RBS 70 surface-to-air missile being launched from an ASRAD-R based on a M113 series chassis 0069496

Mounted on either side of the pedestal are two Missile Mk 2 with an additional six missiles carried inside the vehicle for manual reloading. According to RDE, four new Missile Mk 2 can be reloaded in less than 60 seconds.

Mounted on top of the pedestal is the Ericsson Microwave Systems Improved Helicopter and Aircraft Radar Detection (HARD) 3-D radar system which operates in the X-band and has a range of 20 km against aircraft, more than 12 km against hovering helicopters and can track 20 targets at once.

Compared with the original production HARD, the latest one has an 80 per cent increase in range performance, increased power output, lower noise figure, improved signal processing and Commercial Off-The-Shelf (COTS) data processing.

HARD was originally developed for the Swedish Army's RBS 90 SAM unit, with the Improved HARD version being developed for the German Army ASRAD programme.

In principle, this radar can be used when the vehicle is on the move, although it has not been tested in this role, and can be retracted for travelling purposes.

The system can exchange target information with other ASRAD-R systems or other SHORAD systems using radio datalinks.

The HARD 3-D radar with its integrated NATO standard Identification Friend or Foe (IFF) has been combined with the pedestal so that it can carry out a complete 360° target search as well as a 360° weapon deployment without a reciprocal interface. Offered with the IFF is an optional encrypted Mode 4 or National Secure Mode encoder.

ASRAD-R has a crew of four consisting of the commander, system and radar operator, a weapon operator and driver, all of whom are under full protection from small arms fire. The command-and-control system fitted also provides air picture data, threat analysis and built-in test equipment.

As an option, the system can be used by remote control, with the operator positioned up to 100 m away from the firing unit. Another option is the installation of the Pilkington Optronics ADAD on the ASRAD-R platform.

The ADAD is in service with the British Army in static and mobile versions and has been selected by the German Army for its ASRAD systems on the Wiesel 2 chassis.

Standard equipment includes a communications system based on VHF radios for voice and data transmission and a navigation system.

Although the prototype of the ASRAD-R is based on the widely deployed United Defense LP M113 series fully tracked armoured personnel carrier, RDE stress that it can be installed on a wide range of other chassis, tracked and wheeled. The series configuration of ASRAD-R for the Finnish Defence Forces will consist of the ASRAD-R platform, integrated in a shelter, which will be mounted on a Mercedes 5000 truck.

Specifications

Maximum altitude:	5,000 m
Range:	
(maximum)	8,000 m
(minimum)	200 m
Weight:	900 kg
Speed:	700 m/s
Warhead:	Blast Fragmentation with shaped charge

Variants

A naval variant was announced in February 2007, this is being offered by Rheinmetall Defence Electronics and Saab Bofors Dynamics.

Status

In production. Ordered by Finland in 2002 for 16 × ASRAD-R, orders were completed in June 2008.

Contractor

Rheinmetall Defence Electronics (RDE) GmbH.
Saab Bofors Dynamics AB.
Ericsson Microwave Systems AB.

Aster 15/30 (FSAF)

Type

Self-propelled Surface-to-Air Missile (SAM) system

Development

Feasibility studies for the French SYRINX programme (*système rapide interarmées à base d'engins et fonctionnant en bande X:* X-band fast tri-service missile system) were carried out from 1982-1984. Pre-development contracts were awarded in 1984 to Thomson-CSF (now Thales) for the Arabel radar and fire-control system and to Aerospatiale (now MBDA) for the Aster missile. The EUROSAM consortium was created in June 1989 by three major European aerospace companies (Aerospatiale, Alenia, and Thomson-CSF) known today as MBDA and Thales.

The EUROSAM Naval Surface-to-Air Anti-Missile (SAAM) systems are shipboard missile systems, which, according to the chosen configuration, range from self-defence and Local Area Protection (SAAM system with Aster 15 and Arabel or Empar radar) to Area Defence (SAAM AD system with Aster 30 and Aster 15 and Empar radar). The EUROSAM Naval Systems are designed to give the naval units protection against stand-off jammers, missile-armed attack aircraft, command-and-control aircraft and saturation attacks by sea-skimming and steep-diving anti-ship missiles even when they unmask at close range.

In August 1999, the Principal Anti-Air Missile System (PAAMS) contract was awarded to EUROPAAMS SAS as the prime contractor for the procurement of the system for the trinational air defence frigates programme. EUROSAM and UKAMS Ltd are main subcontractors and shareholders of EUROPAAMS.

The PAAMS is based on Aster 30 and Aster 15, and Empar or SAMPSON radar. It has been developed in parallel to the SAAM systems to equip the air defence frigates which are projected to enter the Royal Navy (with the Type 45 frigates), French Navy and Italian Navy (with the HORIZON frigates) from 2006 onwards.

The Surface-Air-Anti-Missile (SAAM) is now qualified in its French configuration and is installed on aircraft carrier Charles de Gaulle and on the SAWARI 2 frigates. The SAAM in its Italian configuration has also completed its qualification test run. Aircraft carrier Nuova Unita Maggiore (Andrea Doria) will be the first vessel to be equipped with the system.

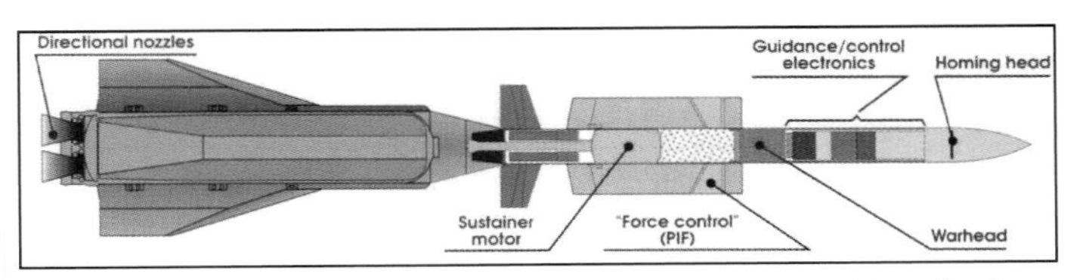

General arrangement drawing of Aster 30 missile showing position of main components 0080413

Close-up of the Arabel radar 0080414

Launch of Aster 15 missile during the trials programme (DGA/CEL) 0044108

Successful live firing trials of the Aster-15, from the French aircraft carrier, Charles De Gaulle, took place on 25 May 2005, followed on 1 June 2005 by an Aster-30 firing at the DGA Landes Test Centre (CELM) in south-western France. The Aster-15/30 missile is common to both the PAAMS naval and SAMP/T (*Sol-Air Moyenne Portée Terrestre*) land-based systems. The successful firing at sea would lead to land-based firings in the latter half of 2005.

On 26 July 2005, the first qualification firing of the Aster 30 SAMP/T took place at the Landes Test Centre. This qualification firing marks the first ever combined use of all the SAMP/T's system elements; the target engagement module, the Arabel radar and target identification module, the vertical launcher and the Aster 30 munition. The firing scenario was set up to demonstrate the capacity of the SAMP/T system to intercept a threat posed by a combat aircraft, in this case simulated by a C-22 target flying at an altitude of 7,000 m. The second qualification firing of the SAMP/T took place at the French DGA's (*Delegation Generale pour l'Armement*) CELM (*Centre d'Essai de Lancement des Missiles*) test facility on 20 December 2005. The final qualification firing of the Aster 30 took place on the range on the 14 November 2006. The objective was to be representative of the medium-range interception of a Self-Screening-Jamming (SSJ) aircraft.

The system will now be passed to the French and Italian armed forces for full operational evaluation.

The Land SAAM AD System (SAMP/T) (now known as Mamba in the French Air Force), is a land-based air-defence system based on the Aster 30 missile, developed under the leadership of MBDA, within the Franco-Italian Future Surface-to-Air Family (FSAF) programme. The system is a modern, cost-effective, advanced-design surface-to-air-defence system, incorporating highly innovative features. These features provide an outstanding anti-missile and anti-aircraft capability, as well as substantial growth potential for the future to maintain its destruction capability against the evolving threat and to protect the investment. It also provides, from user perspective, a great asset in terms of flexibility in system operation and mobility, making enemy counter-actions more difficult. The qualification campaign for the SAMP/T system began at the Poligono Interforze di Salto di Quirra (PISQ), the Italian defence test range in Sardinia. Firing trials by the Italian Army in May 2008 were undertaken firstly against a single target (Mirach 100) flying at an altitude of 150 m while being intercepted at a range of 15 km. A second test against a patrol of two Mirach 100 targets were also intercepted and engaged during evasive manoeuvres at 21 km.

The Aster SAMP/T has been designed to protect land forces and sensitive sites both from the conventional air threat and from the emerging threats posed by the new generation of stand-off weapons and cruise missiles, characterised by high speed, increased stealth and manoeuvrability.

The French Armament Procurement Agency (DGA) delivered in May 2009 the first mass-produced SAMP/T system to the Avord Air Force base in France for training purposes, these tests were completed in late 2010. On Monday 20 September 2010, the first operational weapon system was delivered to the

The EUROSAM Land system ground vertical launcher accommodates up to eight missile cells 0044107

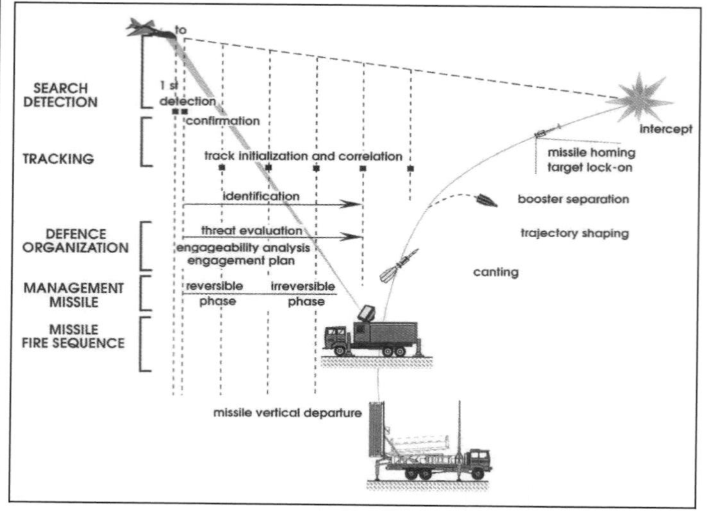

Typical target engagement sequence for the EUROSAM Land system surface-to-air missile system 0080415

air defence artillery squadron (EDSA) of French Air Force, Luxeuil Air Base, where the system is fully operational and is known as Mamba. It was also reported that the arrival of the SAMP-T Mamba coincided with the transfer of the tripod mounted Mistral to the Army and the decommissioning of Aspic and first generation Crotale missile systems. The remaining 10 systems on order by the French will gradually re-equip the Air Force five air-defence squadrons, based at:

- Mont-de-Marsan
- Saint-Dizier
- Avord
- Istres
- Luxeuil.

In October 2013 the Polish organisation Bumar announced "The Shield of Poland" programme whilst at the same time suggesting the use of the Aster-30 for the 1st Tier defence against conventional air-breathing targets. At the present time it is not known if Bumar were considering using the basic Aster-30 or the future Aster-30 Block 2. The Block 2 missile is being developed with ABM and ATBM in mind.

Description

SAMP/T

Aster SAMP/T fires the Aster 30 missile, a state-of-the-art, two-stage missile fitted with an active seeker with an interception range exceeding 100 km. The missile has a 15 kg warhead and an exceptional manoeuvrability of 50 *g*, completed by a transversal acceleration of 12 *g* in the end game. This gives

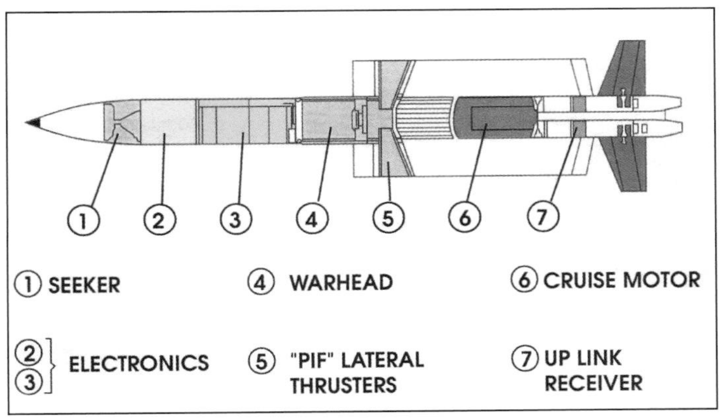

Main components of the Aster missile dart 0080417

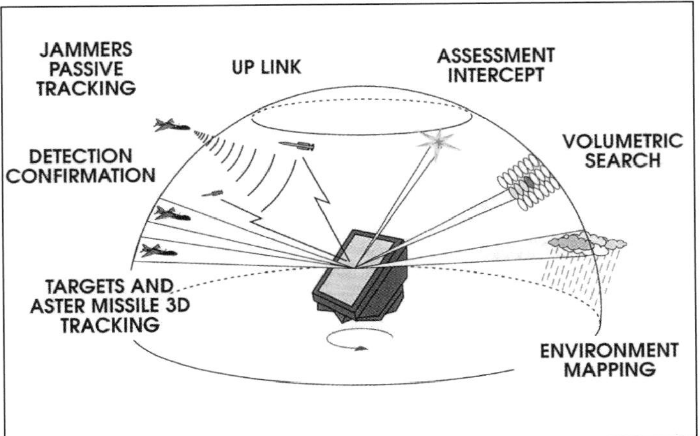

Main radar functions of the Arabel radar 0080416

Aster the capability to intercept all types of targets, including Theatre Ballistic Missiles (TBM).

The SAMP/T system comprises:

- The Aster 30 Block 1 is part of the Aster missile series designed and developed by MBDA. Aster missiles are based on vertical-launched two-stage missiles equipped with the 'PIF-PAF' manoeuvring system. The system combines aerodynamic control surfaces (PAF) with lateral thrusters (PIF) mounted at the missile's Centre of Gravity (CG). The PIF thrusters greatly increase the agility of missiles in end-game scenarios. A Command-and-Control (C2) module with integrated VHF and Link 16 communications.
- A fire-control system mostly based upon the Arabel Radar developed by Thales Air Systems. This multifunction radar has very accurate 360° surveillance, detection and tracking capabilities. The multifunction X-band Arabel radar also features an outstanding resistance to ECM and a refreshment rate of per second.
- A ground launcher system which includes four vertical launchers, with an option for up to six vertical launchers. Each launcher can carry eight Aster 30 missiles as well as two ground-based loading modules for quick rearming during operations. The vertical launchers are each capable of firing eight missiles in less than 10 seconds.

SAAM

The SAAM system provides the accuracy and the flexibility required to address a broad range of threats, such as air-breathing threat and tactical ballistic missiles.

Key Elements:

Multifunction radar:

- MultiFunction Radar (MFR radar) provides surveillance, target detection, tracking and up-link for target data transmission to the in-flight missiles.
- These functions are performed for a wide range of weather conditions and in case of ECM conditions.

Command-and-Control:

- The system's role is to engage targets whose characteristics - high velocity, low RCS, short unmask distance - require a short reaction time.
- The system provides the operators with all information about the tracks and the Firing Officer on the base of the system proposal to operate the Engagement Module (EM).
- The engagement is proposed through a ranking (priority of the engagement) based on different criteria (firing procedures, threatening track urgency, and interception rules) that have been defined during the preparation of the mission.
- The system allows the engagement of 10-12 targets from any direction.

Vertical launch:

- For Naval configurations, the Sylver is the vertical launcher of Aster 15 and Aster 30 missiles.
- Sylver ensures to the missiles a safe storage inside each of its eight cells and a nearly simultaneous delivery.
- For the SAMP/T the Launcher Module (LM) is mounted on a truck to achieve the required mobility. These launcher vehicles are to be fitted with the Lital LLN-93 Ring Laser Gyro Inertial Reference and Navigation Systems as part of the FSAF programme. It also contains eight Aster 30 ready-to-fire missiles.

Aster 30 missile fired from ground vertical launcher 0100549

Aster missile:

- The Aster is a two-stage missile to ensure maximum effectiveness of the interceptor stage.
- The solid-propellant booster stage is sized according to the mission. It separates from the Aster dart a few seconds after launch.
- The Aster dart is also equipped with a sustainer motor, a proximity fuze and a fragmentation warhead that is effective even against the most hardened targets.
- For maximum manoeuvrability, the terminal dart uses the 'PIF-PAF' system, as mentioned previously, which combines conventional aerodynamic control with a direct thrust control system that uses gas jets acting on the interceptor's centre of gravity.

Reaction time

The EUROSAM Land SAAM AD System role is to engage targets whose characteristics - high velocity and manoeuvrability, low RCS - require a short reaction time. Each EUROSAM Land SAAM AD System module contributes to a very short reaction time.

Fire power

Due to its ability to conduct 10 to 12 simultaneous engagements, with up to 16 to 24 missiles simultaneously in flight, the systems are able to face heavy attacks, which would saturate other conventional air-defence systems.

Flexibility

Land SAAM AD System flexibility in operation is characterised by a high level of:

- Interoperability and capability for integration with a distributed architecture
- Low manpower requirement
- Reliability, availability, maintainability owing to the fact that ILS policy has been applied since programme conception
- Robustness in battlefield environment.

The system is able to use the data provided by external sensors through the battalion operations centre, or directly when operating independently. When integrated within an Air Defence C3I, the EUROSAM Land SAAM AD System gets the opportunity to receive early warning data and so improves its interception capabilities.

Typical engagement sequence

During the surveillance mode, radar operations are coordinated under EM control to provide the most effective surveillance coverage. The Arabel MFR ensures the surveillance function in each fire unit. The radar will search its complete surveillance volume of 360° in azimuth and –5° to +90° in elevation. All track data can be sent to higher level of C2.

Two Fire Unit operational modes are possible:

(1) A centralised or coordinated mode, for which normally the decision to engage targets is taken by the higher level C2 facility.
(2) An autonomous mode, for which target engagements are decided by the Engagement Module.

Engagement phase

On receipt of the engagement command, the Engagement Module will send a missile activation message to the relevant launcher, which will respond by applying electrical power to the requested number of missiles (normally two).

Interior view of the Alenia-built Engagement Module for the EUROSAM Land System missile system 0080376

Each launcher can prepare up to two missiles simultaneously and launch them sequentially as a salvo. Each Fire Unit can have up to 16 missiles in the air at any one time. The engagement module will keep a record of how many missiles and firing channels are available.

The limited amount of performance data available can be summarised as follows:

- Employment: modules can be deployed up to 10 km apart
- Complete surveillance volume of 360° in azimuth and –5° to 90° in elevation
- Omni-directional engagement: 360° azimuth
- Multiple-tracking capability: more than 100 tracks
- Multiple target engagement capability: 10 targets simultaneously engaged
- Very short reaction time from target detection up to missile launching (less than 10 seconds)
- Very short time between two firings from the same launcher (less than one second)
- High salvo capability (eight missiles in 10 seconds from the same launcher)
- Possible integration within air defence C^3I
- Interception range up to 100 km (with cueing).

Variants

There are at least three variants of the system.

(1) Aster Block 1
(2) Aster Block 1+
(3) Aster Block 2.

A proposal to further develop the Aster Family of missiles has been suggested by MBDA, Safran and Thales. The proposal would include evolving the Aster Block 1 to an improved Block 1+ and a Block 2 model. The Block 1+ would have an intercept range in the region of 1,000 km, whereas the Block 2 would be expected to engage targets at 3,000 km. The new proposals are all based on the Aster 30 with the aim of ATBM and ABM, endo-atmospheric and exo-atmospheric. Work will be divided up on the new development as follows:

- MBDA will provide the missile
- Roxel would be responsible for the solid propulsion system
- Safran will provide seeker knowledge
- Thales will provide radar technology.

Specifications

	Aster 15	Aster 30
Dimensions and weights		
Length		
overall	2.7 m (8 ft 10¼ in) (dart)	5.2 m (17 ft 0¾ in) (combined)
	-	2.7 m (8 ft 10¼ in) (dart)
Diameter		
body	180 mm (7.09 in) (dart)	180 mm (7.09 in) (dart)
	-	380 mm (14.96 in) (booster)
Weight		
launch	310 kg (683 lb)	510 kg (1,124 lb)
Performance		
Speed		
max speed	1,944 kt (3,600 km/h; 2,237 mph; 1,000 m/s)	2,721 kt (5,040 km/h; 3,132 mph; 1,400 m/s)
Range		
min	0.9 n miles (1.7 km; 1.1 miles)	1.6 n miles (3.0 km; 1.9 miles)
max	16.2 n miles (30 km; 18.6 miles)	54.0 n miles (100 km; 62.1 miles)
Altitude		
max	15,000 m (49,213 ft)	20,000 m (65,617 ft)
Ordnance components		
Warhead	HE blast fragmentation	HE blast fragmentation
Fuze	proximity, time delay	proximity, time delay
Guidance	INS, mid-course update, radar (active Ku-band terminal seeker)	INS, mid-course update, radar (active Ku-band terminal seeker)
Propulsion		
type	solid propellant (rocket motor, 'PIF' direct thrust vector control)	two stage, solid propellant (rocket motor, 'PIF' direct thrust vector control)

Status

The SAMP/T Aster 30 qualification trials were concluded at the end of 2006. Operational evaluation started in 2007 and continued through 2008 with the last firing on 8 December 2008. The technical operation phase started in May 2009 with the first mass produced system being handed over to the Air Force Trial Centre. Production as required to complete orders. France and Italy have placed orders for 10 systems for the French Air Force and Italy for five.

On 6 March 2013, the SAMP/T system engaged and destroyed a Theatre Ballistic Missile (TBM) target in a live test firing at the Biscarrosse Firing Range in France. This live firing linked the actual SAMP/T system to the NATO BMD chain of command and was the final test of the system's capability and interoperability with NATO command-and-control ahead of the planned adding of the system to the inventory of NATO missile defence assets later in the year.

In mid-September 2013, the Singapore government announced that it is planning to acquire the Aster-30 for the Singapore Air Force as part of a multi-layered air defence system. It is assessed that the Aster-30 will be deployed to the Paya Lebar Air Base.

Contractor

EUROSAM (primary contractor).
MBDA (designer and main manufacturer).
Thales (seeker).
Lital SpA (LLN-93 Ring Laser Gyro Systems).

Barak 8MR

Type

Self-propelled Surface-to-Air Missile (SAM) system

Development

Information going back to July 2005 confirms that India and Israel agreed at that time to develop further the Israeli naval Barak system to create a new medium-range shipborne missile that will see service with both the Indian and Israeli navies.

On 25 March 2009, the Indian government signed another contract worth USD2 billion with the Israel Aerospace Industries (IAI) Ltd for the development of the MR-SAM (also referred to as Barak II and Barak-8MR) that is a follow-on to another contract dated February 2006.

This 2009 contract was for the land-based version of the missile system. It is understood that the new contract went before the Indian solicitor-general who felt that the government could go ahead with the deal if it considered the missile essential. At this time also, the joint Israeli and Indian Company 'Nova

Barak 8 on display (Israel Aerospace Industries (IAI)) 1331043

Barak 8 naval missile 1375611

Integrated Systems, a joint venture between IAI in Israel and Tata in India, signed a contract associated with the MR-SAM now known as Barak 8MR system.

Tata is the Indian state-owned company that deals with vehicles such as civilian cars, and military trucks, they have designed and developed the articulated vehicle for the Akash. It would not seem unreasonable therefore to expect this mobile SAM system to find itself mounted on a Tata truck with IAI technology used in the integration of the mobile mounted system.

At the time of the first documents signing in late 2005, the system requirements was for a missile that could intercept targets at distances up to 150 km, however, this range was soon reduced to 70 km. By August 2013 it became clear that there are in fact two variants of the Barak 8 weapon system, firstly Barak 8MR with a range of 70 km and Barak 8LR with a long range of 150 km. It is assessed that should this contract continue through to Initial Operational Capability (IOC), the system will replace not only the older S-125 (NATO SA-3 Pechora) but also the Akash. Manufacturing will take place entirely in India.

Description

The Barak 8MR was designed with the ability to engage multiple targets simultaneously for the protection of vital and strategic ground assets and area air defence. Taking into account the time already taken on the development of the naval system, it is estimated that the land-based variant should take no longer than four to five years. At this time India is expecting to deploy nine air-defence squadrons each with two Barak 8MR firing batteries. Each battery will consist of:

- Command-and-control centre
- Acquisition radar
- Guidance radar
- Three launchers with eight missiles each.

Variants

- Barak 8LR
- Barak NG

Specifications

	Barak 8MR
Performance	
Range	
max	37.8 n miles (70 km; 43.5 miles)
Ordnance components	
Guidance	INS, mid-course update (est)
Propulsion	
type	two stage, solid propellant (rocket motor)

Status

In late July 2013, the Israeli Navy began installing the Barak 8MR on their Sa'ar 5 warships. It is not known when the Indian Ground Forces will be receipt of the Barak 8MR but this can be expected sooner rather than later.

Contractor

Defence Research & Development Organisation (DRDO), India primary contractor.
Israel Aerospace Industries (IAI) Ltd, Israel is reported as a key partner providing a large proportion of the applicable technology.
Rafael, Israel producing the interceptor missiles.
IAI's ELTA Systems providing the radar.
Nova Integrated Systems Ltd.

Igla/Igla-1 missile launch unit

Type

Self-propelled Surface-to-Air Missile (SAM) system

Development

An Igla/Igla-1 Launch Unit has been developed by the Kolomna-based KBM (*Konstruktorskoye Byuro Mashynostroyeniaya*) and the then Thales Air Defence (now Thales France).

The system is a lightweight, self-contained unit that provides the structure, electronics and coolant to support the launch of two standard Igla/Igla-1 (normally) Man-Portable Surface-to-Air Missiles (MANPAD SAMs). The launcher is compatible with helicopters and land-based self-propelled Very Short-Range Air Defence (VSHORAD) systems such as the Blazer and Aspic.

The launcher is reloadable and therefore, is capable of performing repeated missile firing. Attached to the launcher are the missiles and seeker coolant bottles, while it is possible that larger coolant units can be fitted to the launcher vehicle, thus, allowing longer engagement times.

Description

The Igla/Igla-1 Launch Unit consists of the following sub-assemblies:

- Launcher module with ready-to-fire Igla/Igla-1 missiles in their launcher tubes. A launcher adapter provides the mechanical interface between the launcher module and the carrying platform
- Missile-aiming and launch-control equipment.

The module is capable of either single-shot or salvo firing (two missiles) against one target.

Specifications

	Igla/Igla-1 missile launch unit
Dimensions and weights	
Length	
overall	1.9 m (6 ft 2¾ in)
Width	
body	420 mm (16.54 in)
Height	
body	300 mm (11.81 in)
Weight	
loaded	65 kg (143 lb) (two missiles)

Status

Offered for export with production as required. The system is believed to be in service with an undisclosed country.

Contractor

Kolomna KBM (enquiries).
Thales France (enquiries).

IRIS-T SL

Type

Self-propelled Surface-to-Air Missile (SAM) system

Development

The IRIS-T SL [Surface-Launched] missile is a derivative of the air-to-air missile known simply as IRIS-T. Use of the missile in the surface launch role was not entirely new as ground launches of the IRIS-T had previously taken place at the test range, Salto di Quirra, Sardinia, as part of the development of the air-launched system.

The Definition of the programme commenced during mid-2004. Full development was planned to begin in November 2006 and culminate in live vertical-launched firing in 2009.

The initial system was to be the IRIS-T SLS [Surface Launched Short-range (Standard/SHORAD)], with the missiles launched at an angle, not vertically. However, it was announced in early April 2007 that another version of the SLS with longer-range and vertically-launched missiles was also in development. This model, the IRIS-T SL, was to act as a lower-tier complement to the International Medium Extended Air Defence System (MEADS). The effect of this was that a contract that should have been awarded early 2007 was delayed until summer of 2007.

IRIS-T SL missile firing (Diehl BGT) 1154418

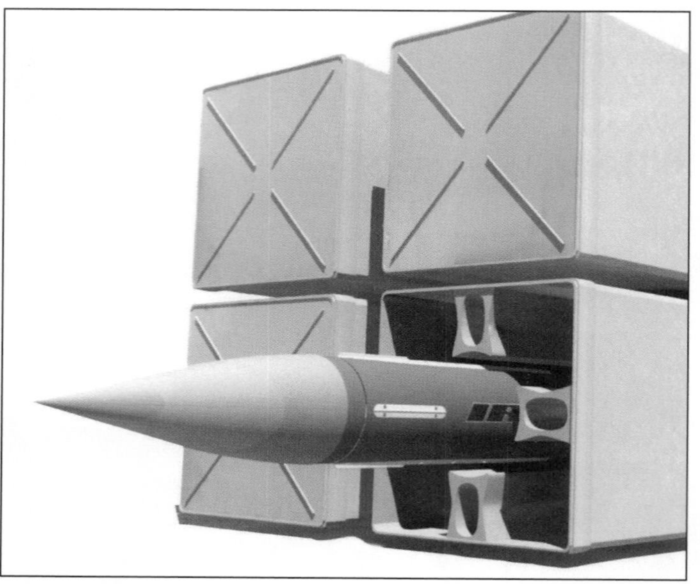

IRIS-T SL missile and launch canister (Diehl BGT) 1154419

Early in March 2008, Diehl BGT Defence demonstrated its IRIS-T SLS short-range system during live firings at the South African Overberg Test Range (*Overberg Toetsbaan* - OTB). The missile locked on to its target at a range in excess of 10 km and achieved a direct hit.

During the trials at Overberg, Diehl used the partner companies Saab, Giraffe AMB air-defence radar that provides surveillance of more than 100 km in range and 20 km in altitude whist also tracking up to 150 targets simultaneously.

In early September 2008, SENER and Diehl BGT Defence (BDB) signed an agreement programme framework, making SENER a subcontractor for the design of the IRIS-T SL missile control system. According to the signed agreement, SENER is in charge of the mechanical and electronic design and development of the actuators making up the control section.

In February 2009, the German government issued a request to MEADS International for the integration of the IRIS-T SL into its Medium Extended Air Defence System (MEADS). The requested work includes implementation of an inter-system plug-and-fight capability for IRIS-T SL, which will include software adaptation, but will not require the redesign of hardware.

During the early part of October 2009 the IRIS-T SL missile system, under the control of Diehl BGT Defence, carried out flight trials at the OTB (part of Denel) missile test range in Overberg, South African. It is understood that these firings primarily consisted of ballistic tests. The test also proved:

- All test objectives were achieved in the presence of representatives of the national customer;
- Safe missile launch from a carrier vehicle;
- Demonstration of the missile's flight-mechanical and aerodynamic characteristics;
- Operational performance of the newly developed rocket motor as well as controlled opening of the aerodynamic shroud.

Early in November 2013, at the Overberg Range, Diehl continued firing trials with the IRIS-T-SL as an ongoing part of the development programme. Both missiles fired destroyed their targets in excess of 20 km.

Description

Based on the European IRIS-T SRAAM there are three other applications of the IRIS-T missile:

- IRIS-T SL
- IRIS-T SLS
- IRIS-T-SLM

IRIS-T SL will be transported on a Mercedes-Benz Unimog 5000 TEL [Transporter Erector Launcher] that will be re-supplied by the same type Unimog 5000. The missiles will be equipped with a more powerful motor, Radio Frequency (RF) data up-link and a drag reducing nose cone to satisfy the German Air Force requirement for a future Medium-range air defence missile within the MEADS structure. Furthermore, the missile will be stored, transported and launched from a sealed container unit, produced in a fibre-reinforced plastic material that can be handled as a certified round. Each TEL is expected to carry up to eight ready-to-fire missiles. The crew for the TEL and system will comprise two persons.

The vertical-launched variant is equipped with a larger solid-propellant rocket motor to give the system an enhanced range of between 25 to 40 km.

The missile will be fitted with a drag-reducing nose cone that will be ejected prior to the end game, at which point the missile will lock-on-after-launch using its Imaging InfraRed (IIR) seeker to ensure a firm kill. The missile also employs a command data-link, as well as GPS-aided navigation.

A new three-dimensional (3D) radar is planned for use with the vertically launched system, the radar consists of six phased array antenna units mounted on a standard ISO pallet and carried on a MAN SX-45 four-axle truck.

Battle Management, Command, Control, Communications and Information (BMC4I) will be available to the system on a plug-in and fight principle similar to that of the MEADS system. It is intended that IRIS-T SL will form the short- and medium-range requirements of a layered defence that includes the MEADS for Germany only.

IRIS-T SL will be fully air transportable by the A400M transport aircraft with a drive on and off capability.

Variants

IRIS-T-SLM

The IRIS-T SLM is a vertical launch missile that provides comprehensive 360° protection against fixed wing aircraft, drones, helicopters, cruise missiles, guided weapons and rockets.

The seeker is locked onto the target to be engaged during missile flight with target data updated during the flight via the data link. The system is reported to be capable of engaging multiple targets at very short-range with short reaction times.

Specifications

	IRIS-T SL	IRIS-T SLS
Dimensions and weights		
Length		
overall	3.4 m (11 ft 1¾ in)	-
Diameter		
body	150 mm (5.91 in)	150 mm (5.91 in)
Performance		
Range		
max	16.2 n miles (30 km; 18.6 miles) (est.)	10.8 n miles (20 km; 12.4 miles) (est.)
Ordnance components		
Guidance	IIR (fire-and-forget seeker, command)	IIR (lock before launch, fire-and-forget seeker)

Status

Diehl Defence of Germany is to supply the IRIS-T SL system to the Swedish Armed Forces with deliveries commencing in 2016. This order was placed in March 2013 and will consist of the SLS [short-range system].

Contractor

Germany

Diehl BGT Defence GmbH & Co KG (primary contractor).

Mercedes-Benz (Unimog 5000 chassis Transporter-Erector-Launcher (TEL) and reload vehicle).

Greece

Hellenic Aerospace Industry AE.

Hellenic Defense Systems SA (EAS) (formerly the Greek Powder and Cartridge Co SA).

INTRACOM Defence Electronics.

Italy

Northrop Grumman Italia (formerly Lita).

MBDA Italia.

Norway

Nammo AS.

Spain

SENER Ingeniería y Sistemas SA.

Internacional de Composites Spain (ICSA).

Sweden

Saab Bofors Dynamics AB.

MADLS

Type

Self-propelled Surface-to-Air Missile (SAM) system

Development

Lenkflugkorpersysteme GmbH (LFK) has developed the Mobile Air Defence Launching System (MADLS) in response to existing requirements and inquiries from potential customers for a highly mobile and flexible light air-defence system for fast reaction/crisis intervention.

MADLS is based on the former Low-Level Air-Defence System (L-LADS) programme. Point defence on land and ship, as well as protection of units by a small dimension/low weight, highly mobile/easy transportable autonomous Very Short-Range Air-Defence (VSHORAD) system with high firepower, are the tasks and design parameters for MADLS.

MADLS is a combination launcher for the US Raytheon and/or the European-built FIM-92 Stinger. The ability to use this with other VSHORAD missiles such as the Kolomna KBM 9K310 Igla-1 (NATO SA-16 'Gimlet'), is the result of in-house development by LFK-Lenkflugkörpersysteme GmbH, based on experimental studies.

This is a land-based system designed to fire up to four fire-and-forget Surface-to-Air Missiles (SAM). The launcher is platform-mounted on a pallet that can be installed on the chassis of virtually any 4 × 4 vehicle, such as Land Rover, Jeep, Mercedes or other similar vehicles currently in use with armed forces worldwide. In its present configuration, the system can be transported easily by either aircraft or medium helicopters.

In 2000, the private venture programme for the development of MADLS started and since then it has been under evaluation by NATO countries in tactical analyses and in troop/firing trials.

A first testing campaign of the system was conducted in 2001 on a NATO partner's firing range. Trials and tests with components and subsystems were mainly performed during the L-LADS programme; additional new components and subsystems have already gone through extensive testing and/or operational use within other programmes/systems.

As of January 2004, firing trials of the newly developed Stinger platform MADLS were successfully concluded at the Danish test site of Oksbøl. The aim of these trials, which were conducted together with the Dutch Army, was assignment of the Stinger platform MADLS via radar, calculation of the target data and successful engagement of a small target drone with Stinger. A propeller-driven drone was used for target simulation. MADLS was mounted on a Wolf all-terrain vehicle and equipped with two Air-To-Air Stinger (ATAS) launchers for four Stinger missiles in total, a TV and IR camera and a laser range-finder.

The target acquisition radar, command connection between the radar and platform, support equipment and the Stinger missiles was provided by the Netherlands Armed Forces. EADS/LFK provided the radar and Stinger platform for the tests.

Description

MADLS is a mobile light air defence system, capable of firing various VSHORAD missiles. The standard configuration is designed for Stinger. MADLS can be used for both point defence of stationary objects and the protection of units on the march. The overall system essentially consists of a standardised pallet that can be loaded onto a vehicle, aircraft or ship, a battery director for the launcher and sensors, radio equipment, a fire-control unit, a remote control unit and interfaces that enable radar remote control and other functions.

The missile assembly is housed in an enclosed container that protects the weapons against mechanical damage and environmental stress. Each launcher assembly contains two missiles in their launch tubes, a pressurised gas coolant bottle and the control electronics interface. The missile launcher assemblies are mounted on a precision tracking director and the system is operated from a remote console that also displays the operating status information, which can be located 200 m away from the firing platform. The systems configuration is to support up to eight missiles, with additional missiles on board as reloads.

Pedestal-mounted MADLS (RADAMEC-LFK) 0595121

RADAMEC-LFK Mobile Air Defence Launching System (MADLS) mounted on Mercedes vehicle (RADAMEC-LFK) 0137897

The electro-optical suite includes the Radamec high-performance surveillance and tracking camera, an IR camera, and eye-safe, laser range-finder also mounted on the Radamec dual-axis director. The system has a video auto-tracker and is controlled via a serial data interface by the firing control computer. This further incorporates ruggedised Components-Off-The-Shelf (COTS) components running the German software similar to Panzer Haubitze 2000.

MADLS can be quickly unmounted from the vehicle chassis for remote firing when added safety or cover is required. The compact remote control panel is linked to the fire-control computer, which is mounted on the pallet by a single cable up to 100 m in length. The missile tracker can be raised hydraulically for combat and lowered for greater stability off-road or when a lower profile is needed.

MADLS features the following main items:

- Pallet with director included
- Tracker
- Launching equipment
- Sensors/communications/remote control
- Interfaces to command-and-control/pre-assignment/power supply
- Optional vehicle.

Pallet

The pallet is the supporting main mechanical element. The standard dimensions of the pallet structure are 2.50 × 1.84 × 0.40 m (L × W × H). The length and width can be changed to fit customers' vehicles requirements.

The pallet structure contains the following:

- Fire-control computer
- Compartment for the remote control panel
- Servo-electronics unit
- Power converter and distribution box
- Batteries
- Cable drums
- Communication system.

A compartment for the storage of the operator's equipment is also available. The power supply generator sits on top during transportation. The pallet structure carries the director with bridge on its hydraulic lifting device. The attachable adjustment and lifting system supports levelling and lifting of the pallet, including loading and unloading from vehicle.

Launching equipment

In the standard configuration, the missile launchers are installed left and right to the bridge on the director. The missiles are supplied and controlled from inside the launchers, which are coupled to the interface electronics. This unit receives the control-and-command signals from the fire-control computer and distributes those to the respective missile. It also acts as a continuous power supply. The interface electronics also provide status signals to the fire-control computer. It is connected to the fire-control computer via a MIL-Bus. The cryogenics system inside the launchers supplies argon cooling-gas from a special bottle to the missiles, whenever required.

Sensors, communication and remote control

The launcher and Sensor Bridge of the director carries the sensors, preferably a DLTV-camera, FLIR-system, and a laser range-finder. The major part of the communication systems, like amplifier, transceiver unit and the antenna, are integrated into the pallet. The remote control panel for the director can be installed in the car's cabin, in front of the assistant driver seat, or can be taken out for remote use, always together with the radios remote control, handset and headset. The maximum distance between remote control panel and fire-control computer is 100 m. The two are linked by glass fibre cable.

Interfaces to command-and-control, power supply

There are interfaces installed on the pallet, which connect to remote control/communication (such as the remote control panel including radio remote control/handset/headset), support pre-assignment (such as the Optical Targeting Device, Radar, IRST, CC-Systems and so on) and interface with the power supply generator or vehicle generator.

LFK-Lenkflugkörpersysteme GmbH combination launcher for VSHORAD missiles in Igla-1 version 0080369

Vehicle

The vehicle, which carries and/or transports the MADLS, is the choice of the customer. To mount the pallet on a cross-country vehicle, it needs a support structure, which is the vehicle's interface to the pallet. The pallet has defined standard ISO-corners, where the support structure of the vehicle fits in.

Operational aspects

The MADLS is able to operate in two main configurations:

- Firstly with the pallet on ground (ship deck)
- Secondly with the pallet on vehicle.

Autonomy and stand-along capability is given for both configurations. In the first case, it is a purely stationary operation. Power supply comes from the generator/batteries or mains. Pre-assignment is possible either by optical device, IRST or radar. Fire-control can be performed through a radio link with a command centre/higher CC-system. The MADLS-crew operates from a remote and protected location.

In the second case, it is a mobile (or quasi-stationary) operation. Power supply is from vehicle generator/batteries. Pre-assignment may be by optical device; fire-control is by radio link with a CC-system. The MADLS-crew operates from the driver's cabin.

Specifications

Launcher (without adapter unit)

Length:	1.589 m
Width:	0.371 m
Height:	0.207 m
Weight:	58 kg (approx. - complete with two Stinger missiles and full gas bottle)
Power supply:	24 V DC
Number of activations: (one gas bottle)	40 (approx.)

Pallet

Length:	2.50 m
Width:	1.84 m
Height:	
(pallet body only)	0.4 m
(transportation configuration)	1.28 m
(operational configuration)	1.74 m
Gross weight: (without vehicle)	1,250 kg
Levelling in terrain:	6°/3° slant

Director

Tracking limits:	
(azimuth)	n × 360°
(elevation)	−10° to +70°
Maximum payload mass:	300 kg
Slew speed:	100°/s
Minimum speed:	0.05°/s
Tracker:	in the servo-electronics unit of the director for automatic tracking in a video image

Launchers

Standard configuration:	2 × ATAS-launchers for 4 × Stinger missiles, including Interface Electronics Assembly (IEA)
Alternate launchers:	accommodate to missile type

Sensors

Standard configuration:	
(DLTV camera)	RADAMEC (Type 202-005)
(FLIR camera)	Zeiss 3rd generation
(laser range-finder)	Zeiss Molem

Fire-control computer

CPU:	Pentium III, 500 MHz+
Operating system:	Windows NT 4.0/2000
GPS:	GPS add-in board
Storage:	hard disk/CD-ROM drives
Casing:	hardened housing, class IP 65 protection
Self-test method:	BITE

Remote control panel

	15.1 in colour TFT display (1,024 × 768 px)
	touch screen
	three axes joystick
	compact case, class IP 65 protection
IFF:	via connection to radar, CC-system etc.

Target pre-assignment

optical target assignment device Leica VECTOR (binoculars with magnetic compass, inclinometer, laser range finder)
interfaced to remote control panel
RST systems for aerial targets
(V)SHORAD - radar systems
coupled through available interfaces on pallet to fire-control computer

Power supply

Power supply generator:	diesel combustion engine, 3.1 kW
	generator 2,000 W, 230 VAC, 50 Hz
	for 100 m remote operation, transportation on pallet
Power converter and distribution box:	installed in pallet for converting 230 VAC to 28 VDC and distribution
Battery pack:	installed in pallet, for silent operation of full system and buffer requirements
Alternate supplies:	vehicle generator
	mains supply (230 VAC)

Status

Following successful firing trials in Turkey, the Radamec-LFK MADLS has formally qualified for use with the US Raytheon-produced Stinger missile.

Development of the 9K310 Igla-1 version is complete.

The complete system is offered for export to approved countries.

Contractor

LFK-Lenkflugkörpersysteme GmbH, a subsidiary of Daimler-Benz Aerospace and RADAMEC UK.

NATO SHORADS/VSHORADS programme

Type

Self-propelled Surface-to-Air Missile (SAM) system

Development

The PG28/30 panel of NATO has been tasked to look at the next-generation Short Range Air-Defence System (SHORADS) missile programme to be used in army, navy and air force applications.

The study originally began under two separate NATO groups, which were tasked to look at both future Very Short-Range Air-Defence System (VSHORADS) and SHORADS requirements.

However, because of the time scales involved, it was considered appropriate to amalgamate the studies into one single programme, where the point and area defence capabilities could be met by the same weapon.

In 1992, NATO issued the Requests for Information to industry for this VSHORADS/SHORADS requirement. Neither upgrades nor off-the-shelf purchases of current weapons are likely to be featured in the programme because of the long time scale involved. A completely new weapon system is highly likely.

The UK MoD was selected to lead the programme. In January 1995, the then UK-based Management Office issued a tender invitation to conduct a feasibility study into future modular VSHORADS and SHORADS for the in-service time frame of 2010 and beyond. Two teams were formed to conduct the feasibility study of the requirement. The team leader for the first was the then Matra BAe Dynamics (with partners from France, Germany, Italy, the Netherlands, Sweden, Turkey and the US) with the then Thomson-CSF Airsys the leader for the second team (which includes partners from the Germany, the Netherlands, Norway, Turkey, the UK and the US). The study was to report after 24 months of evaluation.

UK sources still indicate that there are two programmes: SHORADS to meet Staff Target (Land) 4,058 with an in-service date of 2015; while VSHORADS is to meet Staff Target (Land) 4,059 with an projected in-service date of 2017.

Status

Feasibility stage. Until (or indeed if) the systems are moved into the developmental stage, very little information is expected to be made available to the open press.

RIM-7P (Sparrow)/Kub Launcher

Type
Self-propelled Surface-to-Air Missile (SAM) system

Development
During mid- to later-part of the 2000's, Wojskowe Zakłady Uzbrojenia (WZU) of Grudziadz, Poland in cooperation with the US Raytheon Company produced a technology demonstrator vehicle that improved the Kub launcher 2P25M2 to carry and launch the RIM-7P (Sparrow) missile. The integration of the missile on the launcher (2P25M2-P1) set enables an increase of tactical and combat parameters considerably whilst also providing NATO with a unique solution of inter-operational modernisation of air defence equipment. The prototype system containing three launch rails, two outer rails with RIM-7 missiles and the inner launch rail with a 3M9 missile, was first displayed at the MSPO defence exhibition in Kielce in September 2008, then again at Defendory 2008 in Athens, Greece.

Live firings of the system were first due to take place in 2007, however, the necessary US government permission and export licences were not granted in time and this caused a delay in the planned trials that lasted through to the last half of 2009.

Description
The modified boxed launcher system initially plans to mount two missiles on the outer two rails of the launcher with the third inner rail being used to carry an electro-optical based sensor. The vehicle shown in Kieice carried a mock-up of the SIC-12 originally developed for and used on the upgraded 9A33BM-P1 (Osa-AKM-P1) NATO SA-8 Gecko. This consists of a third-generation Sagem Iris TGS-100C thermal camera, a WZU-2 developed daylight camera, an eye safe laser range-finder, and a Raytheon AN/TPX-56 Mk XII compatible Identification Friend-or-Foe (IFF) interrogator set. Reports from WZU suggest that the new hardware, new software and hardware modifications were developed at WZU-2.

RIM-7P and 3M9 Missiles on Kub Launcher (Grzegorz Holdanowicz) 1290477

2P25M2 with RIM-7P and 3M9 (Grzegorz Holdanowicz)  1290476

RIM-7P missiles on 2P25M2 (Grzegorz Holdanowicz) 1331170

WZU have also suggested that the modified launcher would still be capable of launching the original 3M9 missiles for as long as stocks with adequate service life remained available.

Should the subsequent firing trials be successful and the system go into series production, Raytheon would deliver refurbished rounds from US stocks. These are expected to be equipped with new and probably upgraded rocket motors and the latest available software.

The RIM-7P missiles can be fitted with or without Jet Vane Control (JVC) units. These rotate and align the missile to the correct intercept path after clearing the launcher.

Variants
Two variants of the launcher and missile have been identified, the first with the missiles contained in their box type launcher and the second with the missiles mounted on the 2P25M2 launcher rails.

Specifications

	RIM-7P
Dimensions and weights	
Length	
overall	3.66 m (12 ft 0 in) (without JVC)
	3.85 m (12 ft 7½ in) (with JVC)
Diameter	
body	200 mm (7.87 in)
Flight control surfaces	
span	1 m (3 ft 3¼ in) (wing)
Weight	
launch	228 kg (502 lb) (without JVC)
	295 kg (650 lb) (with JVC)
Ordnance components	
Warhead	blast fragmentation, continuous rod
Fuze	proximity, impact
Propulsion	
type	solid propellant (Mk-58 boost-sustain motor with manual or remote safe and arm)

Status
As of December 2012, the system is still in development although proof of concept has been achieved. Missile firings were due to take place in the later months part of 2009 but these have never been confirmed.

Contractor
Wojskowe Zakłady Uzbrojenia S.A. (WZU) and Raytheon Company Missile Systems, Naval Weapon Systems.

Roland family

Type
Self-propelled Surface-to-Air Missile (SAM) system

Development
In 1964, Aérospatiale of France and Messerschmitt-Bölkow-Blohm (MBB) of Germany began design work on a low-altitude surface-to-air missile system, which eventually became known as the Roland. Aérospatiale had overall responsibility for the clear-weather version, called the Roland 1, and MBB (now EADS) overall responsibility for the Roland 2 all-weather version.

French Army upgraded Roland system during firing trials 0067655

At a later stage, a joint company called Euromissile was established to market this and other missiles produced by the two companies. The all-weather version (formerly called Roland 2) was offered together with the latest variant known as Roland 3.

The French Army had a requirement for 181 firing units based on the Giat Industries AMX-30 MBT chassis, designated for this purpose the AMX-30R. Of these, 181 were funded, with all 98 Roland 2 and 83 Roland 1 vehicles delivered. They were organised to provide the 51st, 53rd, 54th, 57th and 58th Roland regiments at corps level, each of three batteries with two troops of four fire units apiece.

Each Roland fire unit is accompanied in the field by a VAB (4 × 4) armoured personnel carrier mounting a 20 mm cannon in a Giat Industries T20-2 turret for ground and anti-helicopter over watch defence within the close-range dead zone of the Roland's engagement envelope. The first Roland fire units were delivered to the French Army in December 1977.

In the early 1990s, the Carole air-transportable version of Roland was procured for the French Reaction Force (FAR). (See entry in Shelter- and container-based surface-to-air missile systems section). The 20 converted AMX-30 units were used to re-equip the 54th Roland Regiment, which is part of FAR. The concept being that in the event of a force projection 4,000 km from France, a section of four Carole shelter systems could be deployed within 14 hours.

The first Roland firing trials unit was delivered to the German Army in 1978 as the replacement for the towed 40 mm L/70 Bofors guns.

In June 1981, the German Army officially took delivery of the first of 140 Roland SAM systems. The first operational units were in fact delivered to the anti-aircraft school at Rendsburg in 1980.

In July 1981, the 100th anti-aircraft rocket regiment of the German Army began re-equipment, followed by the 200th regiment in July 1982 and the 300th in July 1983. Each regiment has one Headquarters battery, three firing batteries (each with 12 fire units) and one support battery. In the German Army Thyssen Henschel (now Rheinmetall Landsysteme) bases the system on the chassis of the Marder 1 ICV manufactured. By May 2006, only thirty-two systems remain in active service with the German Army, the remainder have or are being decommissioned. All German Air Force assets were disposed of by mid-2006.

Brazil took delivery of four Roland 2 Marder fire units from Germany together with 50 missiles.

In 1984, the Spanish Defence Ministry selected the Roland for its mobile battlefield low-level air defence system. A contract worth Pta 29,000 million was placed for integration and co-production of the weapon system (nine Roland clear weather and nine Roland all-weather fire units on the Giat Industries AMX-30 MBT chassis with 414 missiles) in Spain.

There were six Spanish companies chosen to participate in the programme. Of these, INISEL was responsible for the electronic components (including the surveillance radars), fire units and production of the complete missile rounds.

Santa Bárbara Sistemas (now owned by General Dynamics of the US) was responsible for building the AMX-30 chassis, producing most of the missile components and total system integration.

CESELSA designed and built an indigenous IFF subsystem to meet national requirements.

German Army Roland 2 system on Marder 1 chassis 0509860

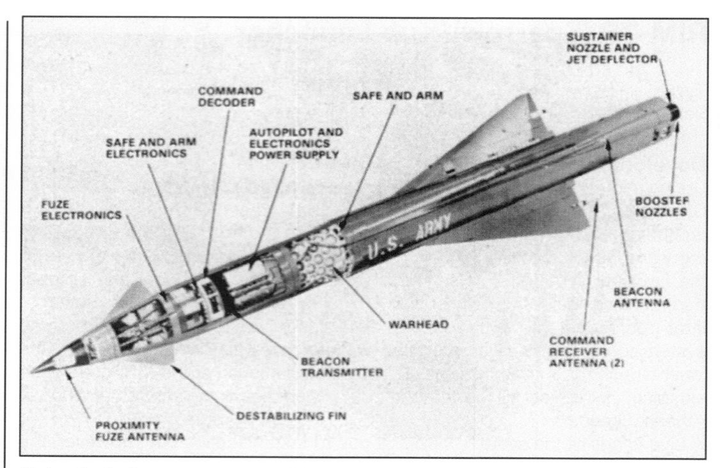

Roland missile common to all firing unit systems 0509462

To accommodate the system the Spanish Army has established the 350-man Roland Group under the auspices of the 71st Independent Air Defence Regiment. This comprises a Headquarters, a Headquarters battery, a service battery and two fire batteries each of two platoons with four fire units apiece.

Each platoon deploys two clear weather and two all-weather systems. Of the remaining two vehicles one is for use in crew and maintainer training at the Artillery School while the other is used by the Artillery Missile Weapons Maintenance Centre.

In reality the Roland Group is an administrative device for centralising personnel training and equipment maintenance to reduce overall costs and optimise use of available resources.

The fire batteries are attached operationally as *de facto* autonomous units to the 1st (Brunete) Armoured Division and the 2nd (Guzman el Bueno) Mechanised Division of the Immediate Intervention Forces.

Iraq has used its Roland systems in combat both during the Iran-Iraq Gulf War and the Gulf War against the Coalition Forces.

In February 1988, AEG delivered the first Roland air defence fire control and co-ordination command post (known as *Flugabwehr Gefechtsstand Roland*: FGR) to the Luftwaffe. Installed as a three-man operated NBC-protected shelter unit on a MAN (8 × 8) 10-ton truck. In total 21 were delivered by the end of October 1990.

The 2-D radar used is the TRM-L D-band frequency-agile model with integrated IFF system. This can distinguish between moving fixed-wing aircraft and helicopters as well as being capable of detecting Anti-Radiation Missiles (ARMs) and hovering helicopters.

Maximum elevation coverage is 60° within altitude limits of very low level to 6,000 m.

Detection range against a 1 m^2 target in high-intensity ECM and clutter is said to be 46 km with a maximum radar range of 60 km claimed.

The antenna is mounted on a hydraulically raised 12 m high mast assembly.

The whole system can be set up and made ready for operation within 15 minutes.

Two work stations are provided in the operator's section of the shelter, one for air situation processing and the other for operations control. The other two sections are the electronics bay and the protective systems bay with transmitter cooling, air conditioning and NBC units.

The FGR detects the targets (thus allowing the Roland fire units to shut down their surveillance radars so as to improve their own survivability), processes the target information and displays it on an air situation display with an indication of the nature of the threat.

The FGR commander decides on the most suitable air defence system, from up to 40 gun or missile systems he can control, to use for the engagement. The extensive onboard radio and cable-based communications suite is then used

Roland missile launch from an LVB air-transportable system on MAN (6 × 6) truck 0044109

Explosion of Roland missile during upgraded Roland trials 0067656

to inform the chosen weapon system of the target's parameters so that it can commence target acquisition and tracking. Radio or wire links while the voice links use either SEL SEM 80 and SEM 90 radios or field telephones carry out data transfer to and from the weapons systems. The system data renewal rate is two seconds.

For use with its Roland and Gepard twin 35 mm anti-aircraft systems the German Army has the HFlaAFüSys air defence warning network. As part of this, DASA produced and delivered the LÜR (*Luftraumüberwachungs radar*) which will be complemented by the TBR (*Tiefflugbereidsradar*) after the year 2008.

Roland as well as Gepard and the FlaSys will be integrated in the HFlaAFüSys network.

In November 1986, the Qatari Army placed an order for three batteries of three fire units each. One battery uses the AMX-30 chassis version while the other two use the shelter type. Deliveries and equipment training were completed in 1989.

In early 1991, the Roland (in both self-propelled and shelter form) was used by Iraq in the Gulf War against Coalition Forces. The Roland is believed to have been responsible for downing at least two Tornados.

With the reduction in tension in Europe, a number of Roland SAM systems used by France and Germany are surplus to requirements.

In mid-2001, Slovenia took delivery of six Roland 2 systems from Germany mounted on a MAN (8 × 8) truck chassis plus missiles, logistic support and training at an estimated cost of DM27 million.

A number of other countries in Europe have also expressed an interest in purchasing surplus Roland SAM systems.

Description

Roland 2

The Roland 2 SAM system has been designed to engage enemy aircraft flying at M1.2 or less between a minimum altitude of 10 m and a maximum altitude of 5,500 m and between a minimum effective range of 500 m and a maximum range of 6,300 m.

Radar

The system has two modes of operation, optical and radar, with possible switching from one to the other during an engagement. In both cases the target is first detected by the pulse-Doppler Siemens MPDR 16 D-band surveillance radar which rotates at 60 rpm and automatically suppresses fixed echoes.

The radar scanner, which can be operated while the vehicle is travelling, has an acquisition range of 1.5 to 16.5 km for a 1 m^2 target operating between speeds of 50 and 450 m/s and can be folded down behind the turret rear for transport. The radar is capable of detecting hovering helicopters.

Once the target has been detected, it is interrogated by either a Siemens MSR-400/5 (German vehicles) or an LMT NRAI-6A (French vehicles) IFF system, acquired and then either tracked by the tracking radar (in the radar

BKS Multifunctional displays of Euromissile Roland M3S SAM system 0080412

Roland F missile (Pierre Touzin) 0525733

mode) or by the operator using the optical sight (optical mode). In the radar mode at the tracking stage the radar beam is slaved to follow the target by misalignment voltage signals originating from the radar target-tracking channel.

The tracking radar, mounted on the front of the turret, is a two-channel, monopulse Doppler microwave Thales Domino 30 system; one channel tracks the target and the second locks in on a microwave source on the missile.

After launch, an infrared localiser on the antenna of the tracking radar is used to capture the missile within a distance of 500 to 700 m, at which range the missile has entered the pencil beam of the tracking radar. A second tracking channel follows the missile by means of a transponder carried on it. Missile deviation is calculated from the angular deviation between target/antenna and antenna/missile. The deviation information is supplied to the computer and from then the operation of the guidance loop is the same as that of the optical mode.

It is possible to switch from optical to radar guidance mode, or vice versa, in both the target pursuit and firing modes. This facility increases significantly the resistance of the Rolands to jamming in either mode.

Optical

In the optical mode the missile is slaved to the line of sight of the operator in the following manner. The sight measures the angular velocity of the target and the infrared localiser determines the deviation of the missile in relation to the line of sight. Using these data the computer calculates the required guidance commands which are then transmitted to the missile by a radio-command link. The signals received by the missile are then converted into jet-deflection orders.

Warhead

The two-stage Roxel (Fr) SD extruded double base solid propellant-powered missile itself has a launch weight of 66.5 kg, of which the multiple (hollow) projectile-charge warhead weighs 6.5 kg, including 3.3 kg of explosive that is detonated either by impact or a TRT electromagnetic continuous wave radar-type proximity fuze. Maximum lethal radius of the warhead's 65 projectile charges is approximately 8 m plus from the detonation point. The missile has an overall length of 2.4 m, wing span of 0.5 m and a body diameter of 0.16 m. The complete container (including the missile) weighs 85 kg, is 2.6 m long and has a diameter of 0.27 m.

Motor

The 1.7 second burn 1,600 kg thrust SNPE Roubaix boost rocket motor is of the extruded double-base solid propellant type and accelerates the missile to 500 m/s before the missile cruises at 544 m/s. The 200 kg thrust 13.2 seconds burn SNPE Lampyre sustainer rocket motor, located in front of the boost motor, then cuts in 0.3 seconds after booster burnout with its cast double-base solid propellant fuel to maintain speed until main motor burnout. Minimum flight time required by the weapon to arm itself is 2.2 seconds, with the maximum flight time around 13 to 15 seconds.

Launcher

Two missiles are carried ready to launch and another eight are carried in two revolver-type magazines each of which holds four rounds. It is delivered in its container, which also serves as a launcher tube.

Roland 3

An improved missile, Roland 3, with increased speed (570 m/s compared to 500 m/s) and range (8 km compared to 6.3 km) entered service in 1989.

Warhead

The new missile has a 9.2 kg warhead, which contains 5 kg of explosives and 84 projectile charges to increase its lethality without any change in its dimensions.

An improved proximity fuze coupled with a new 5,000 m/s maximum velocity fragmentation pattern (over 2.5 times the Roland 2 warhead fragment pattern maximum velocity) increases the lethal radius of the warhead. Maximum flight time is now approximately 16 seconds, missile weight is 75 kg, and complete container (including missile) 95 kg.

Motor

The uprated booster motor retains the missile's minimum effective engagement range of 500 m but allows an increase of 500 m in the maximum interception altitude limit to 6,000 m. It also allows targets taking evasive action at up to 9 *g* to be engaged out to the maximum limit of the missile's range.

Roland FR system (Pierre Touzin) 0525735

Response time

Response time for the first missile launch is six seconds and for the second missile, two to six seconds, depending on the target. Reloading the launcher from the magazines takes six seconds. A fresh batch of missiles can be reloaded in two to five minutes.

If required for defence of an airbase or other high-value target, up to eight Roland fire units can be co-ordinated and integrated into a defence network (as within the Federal Republic or Germany). Up to six Roland's can be auto-co-ordinated, (this is called 'netting'), amongst each other to avoid over or under-kill. Gun systems and fire-and-forget systems can make use of all target information by Roland in so-called cueing.

Roland upgrade programme

In 1988, the French and German MoDs decided to adopt an upgrade programme for their Roland fire units in order to maintain the system in service until 2015/20. The programme, known as 'VMV', involves the following areas of upgrading:

(1) Replacement of the current optical sight by the GLAIVE electro-optical integrated InfraRed (IR) sight assembly in order to provide a third operational mode for Roland - the IR mode. The GLAIVE sight is of modular design and is for use in heavy ECM environments in place of the radar sensors. It has an 8 to 12 micron wavelength IR Charge Coupled Device (CCD) thermal imager, a 16 km range Raman type eye-safe laser range-finder and EOCM-safe IR localiser. It is capable of automatic sector-based passive IR 8 to 12 micron wavelength surveillance in a 20 × 20° field of view, automatic and manual passive day/night 8 to 12 micron wavelength target passive mode tracking and 8 to 12 micron and 1 micron wavelength automatic missile tracking. Sighting in the visual spectrum will be performed using the direct optical sight or 0.9 micron wavelength TV camera.

(2) Simplification of the man/machine interfaces by use of a microprocessor-based turret and computer assembly for the fire unit known as the BKS system. This comprises three major subassemblies:

(3) LS control and guidance computer which is already integrated in existing Roland 3 fire units and optimises the guidance laws, increases overall system precision (especially for long-range and low-altitude engagements) and manages the various fire unit operating modes.

(4) KS co-ordination computer, which manages the fire unit tasks, allows input of data required for the co-ordination function (data exchanges concerning the air situation) and monitors availability of other fire unit equipment using built-in test equipment.

(5) BK commander's operations and fire unit control panel with ergonomically designed digital multifunction displays which maximise useful data as and when required during the surveillance, tracking and firing modes as well as supporting the system status functions.

All three sub-assemblies are linked through an MIL bus.

France intends to retrofit 53 AMX-30 and its 20 Carole shelter systems with this upgrade. The French upgrade will have the capability to launch the VT-1 missile from the Crotale NG. This capability will not be fitted to the German systems. The first order for eight upgrades for the French Army was placed in early 2000, with system qualification completed in 2001 and first units re-delivered to the French Army in mid-2002.

In October 2000 the German Federal Agency for Military Technology and Procurement (BWB) awarded a USD17.6 million contract to LFK to complete development work to extend the life of 140 German Army Roland systems to at least 2015.

Command Control and Surveillance System (HFLaAFüSys)

The Army Air Defence Command and Control System, Heeres Flugabwehr Aufklärungs-und Führungssystem (HFLaAFüSys), is a real-time operating short-range air-defence system that provides the integration of the three primary military functions of data acquisition and evaluation, command-and-control and weapon control.

HFLaAFüSys was designed to protect moving ground forces as well as important sites or defence areas against low- and/or medium-altitude aerial threats.

Advanced communication networks enable real-time and secure information exchange between sensor (data acquisition), command and control truck and the weapon systems. Thus, the integration of army air surveillance (FGR, LÜR) radar systems and the weapon systems Roland, Gepard with capabilities for automatic threat analysis and weapon co-ordination provide a common operational picture for effective army air defence.

With the low-level air picture interface a standard NATO data link connection enables co-operation with NATO partners which is currently realised with the US Army.

Missile alternatives

The Roland 3 system is currently being marketed with the Roland 3 missile described previously.

After the then Matra BAe Dynamics studies, which had been started in 1987, Daimler-Benz Aerospace together with Matra and Aérospatiale decided to develop a new missile, the RM5, with a maximum velocity of 1,600 m/s (M5) and improvements in range, maximum engagement altitude and manoeuvrability. This programme was totally funded by the partners and had to be shelved in July 1991 due to the lack of a launching customer.

VT1 SAM

In March 2001, following an international competition, Thales Air Defence Limited of Northern Ireland (previously Shorts Missile Systems Limited), was awarded a contract by the French element of Thales Air Defence (previously Thomson-CSF Airsys) to manufacture the VT1 SAM used with the Crotale New Generation SAM system.

The original VT1 SAM was designed and built, under contract to the now Thales Air Defence, by the now Lockheed Martin. Development of the VT1 commenced in the US in 1986 with first test firings being carried out in 1989.

A total of 1,000 VT1 missiles and 42 reusable four-round pods were subsequently built by Lockheed Martin with first deliveries being made in 1992. Production was completed some years ago and recent sales have been made from VT1 stocks held in France.

The first customer for the Crotale New Generation, or Crotale NG, as it is also referred to, was Finland who deploys it installed on the locally developed Patria Vehicles XA-180 (6 × 6) light armoured vehicle.

Other customers include the French Air Force (shelter mounted for airfield air defence) and Greece. There is also a naval version used by France and Oman.

Elements of the Crotale New Generation, but not the missile, are used in the South Korean Pegasus SAM system which is now in volume production for the South Korean Army. With modifications, the VT1 SAM can also be launched from the older Crotale SAM system which, as built, uses a R440 SAM designed and built by the now MBDA.

The existing VT1 SAM has a maximum speed of at least Mach 3.6 and a maximum effective range of over 11 km up to an altitude of at least 6 km and is fitted with a high explosive fragmentation warhead with a contact and proximity fuzing system.

Total value of the new VT1 contract to Thales Air Defence UK is about GBP50 million and covers research, development and production of an improved and more capable version of the VT1.

Thales Air Defence Limited will carry out work on the VT1 SAM at their main facility in Belfast as well as at their recently commissioned purpose built Multiple Ordnance Assembly and Test (MOAT) facility in County Down, Northern Ireland.

The MOAT facility currently carries out final assembly and testing on a wide range of missiles, including the Starburst and Starstreak SAMs and the Hellfire anti-tank guided missile launched by the Apache attack helicopter of the Army Air Corps.

French Army Roland SAM systems are to be modified to launch the VT1 SAM which has a longer range and greater altitude coverage.

Roland M3S

Since 1996, Euromissile has been developing the M3S configuration, which contains a new 3-D search radar and a new tracking radar in addition to the basic 'VMV' system.

One man can operate the ROLAND M3S, although two are required for sustained operations.

The operator can select one of the sensors, for example radar, TV or IR, to engage targets depending on the operational environment.

The Roland M3S system has been offered with the VT-1 to the Netherlands as a contender for the Netherlands Short-Range Air Defence System (NL-SHORADS) programme.

Roland air-transportable

In 1994, the family of Roland systems was completed with the first air-transportable version being produced. Designed for transport by C-160 Transall or C-130 Hercules cargo planes, a total of 30 fire units were produced, 20 during 1994 to 1995 for the FAR and 10 during 1998 and 1989 for the Reaction Forces of the German Air Force.

Specifications

	Marder 1 chassis
Dimensions and weights	
Crew:	3
Length	
overall:	6.915 m
Width	
overall:	3.24 m

	Marder 1 chassis
Height	
transport configuration:	2.92 m
Ground clearance	
overall:	0.44 m
Length of track on ground:	3.9 m
Weight	
combat:	32,500 kg
Ground pressure	
standard track:	0.93 kg/cm^2
Mobility	
Configuration	
running gear:	tracked
Power-to-weight ratio:	18.5 hp/t
Speed	
max speed:	70 km/h
Range	
main fuel supply:	520 km
Fuel capacity	
main:	652 litres
Fording	
without preparation:	1.5 m
Gradient:	60%
Vertical obstacle	
forwards:	1.15 m
Trench:	2.5 m
Electrical system	
vehicle:	24 V
Firepower	
Armament:	2 × hull mounted Roland Launcher SAM
Armament:	1 × hull mounted 7.62 mm (0.30) machine gun
Ammunition	
missile:	10

	Roland 2	Roland 3
Dimensions and weights		
Length		
overall	2.4 m (7 ft 10½ in)	2.4 m (7 ft 10½ in)
Diameter		
body	160 mm (6.30 in)	160 mm (6.30 in)
Flight control surfaces		
span	0.5 m (1 ft 7¾ in) (wing)	0.5 m (1 ft 7¾ in) (wing)
Weight		
launch	66.5 kg (146 lb)	75 kg (165 lb)
Performance		
Speed		
max speed	972 kt (1,800 km/h; 1,118 mph; 500 m/s)	1,108 kt (2,052 km/h; 1,275 mph; 570 m/s)
Range		
min	0.3 n miles (0.5 km; 0.3 miles)	0.3 n miles (0.5 km; 0.3 miles)
max	3.4 n miles (6.3 km; 3.9 miles)	4.3 n miles (8.0 km; 5.0 miles)
Altitude		
min	10 m (33 ft)	10 m (33 ft)
max	5,500 m (18,045 ft)	6,000 m (19,685 ft)
Ordnance components		
Warhead	6.5 kg (14.00 lb) HE fragmentation	9.2 kg (20.00 lb) HE fragmentation
Fuze	impact, proximity	impact, proximity
Propulsion		
type	solid propellant (Roxel (Fr), sustain stage)	solid propellant (Roxel (Fr), sustain stage)

Status

Production as required filling contracts with those countries that operate the Roland 3.

Full factory production commenced in 1988 and finished in 1993.

The Roland 3 became operational in 1989.

Roland 1 first mass production in 1978 became operational in 1977.

Roland 2 first mass production in 1981 became operational in 1984.

Roland systems are in service with Argentina, Brazil, France, Germany, Iraq, Nigeria, Qatar, Spain, Slovenia, the US and Venezuela.

In 2002, MBDA stated that a total production of the Roland SAM system comprised 650 firing posts and 26,000 missiles with sales being made to 11 countries.

By 1998, 3,800 missiles, including Roland 3, had been fired in trials, training and combat with a technical success rate of more than 98 per cent.

The Marder Roland units purchased by Brazil for the Army are now either in reserve or housed on display is local museums.

Contractor

MBDA (both in France and Germany).
Euromissile (marketing and sales).

Iran

Ra'ad

Type

Self-propelled Surface-to-Air Missile (SAM) system

Development

Iran has been working on what appears to be a copy of the Russian Buk 9M317 missile system. The system name is Ra'ad (Thunder), however in the western press there have been several transliterations such as Raad, a name that has already been allocated to the Iranian anti-tank missile. The renamed missile associated with the Ra'ad system is the Taer 2 (Bird 2).

The then Commander of the Aerospace Force of the Islamic Revolution Guards Corps (IRGC) Brigadier General Amir Ali Hajizadeh told reporters on 12 September 2012 that the system is the first completely indigenous system of the Sepah (IRGC) which has been designed and manufactured by committed Iranian technicians. The missile has been based on the Russian 9M317 but changes have been made within the Iranian Defence Industry to meet local theatre requirements.

Description

The Ra'ad surface-to-air missile system has been designed to engage targets such as fighter jets, cruise missiles, smart bombs, helicopters and drones (UAVs). From TV and photographic evidence, the system is assessed to consist of at least four launchers, two of which have what appears to be a radar mounted on the front of the system and have been called (for this description only) the master - the other two launchers are known as slaves. All four launchers mount three Taer-2 (lit Bird) missiles that fully resemble the Russian 9M317.

TV coverage shows at least two firing's at standard jet type targets both of which appeared to destroy the targets. Also noted in the TV coverage are four different launcher vehicles (two master and two slave), and at least three different types of missiles, some of which were undoubtedly range test missiles. On what appears to be live non-test missiles there have been observed apertures toward the front end of the missile just aft of the nose cone, assessed to be associated with some kind of fuzing for the warhead.

Variants

None known in country, however, the missile looks like the Russian 9M317 in every aspect, therefore, a variant could be the Russian Buk using the 9M317 missile.

Specifications

	Ra'ad
Performance	
Range	
min	0.5 n miles (1 km; 0.6 miles)
max	27.0 n miles (50 km; 31.1 miles)
Altitude	
max	22,860 m (75,000 ft) (est.) (other sources suggest 27,000 m)
Ordnance components	
Guidance	radar, mid-course update (est.)
Propulsion	
type	solid propellant

Status

As of late September 2012, possibly still in development although firing trials have taken place in Iran and have been released as a video to Iranian TV.

The Iranian Navy are known to have test fired the system in December 2012.

Contractor

Iranian Defence Industry Baseline (IRGC).

Rapier Project (Iran)

Type

Self-propelled Surface-to-Air Missile (SAM) system

Development

The Yamahdi Industries Group of the Iranian Aerospace Industries Organisation Rapier Project has succeeded in reverse-engineering and upgrading a version of the MBDA Rapier Mk 1 missile system. The Towed

Locally built (8 × 8) Behr cross-country truck chassis with Rapier system on rear decking 0048548

Rapier variant using the Mk 1 missile was originally supplied to Iran in 1970 and became operational in 1971.

A total of eight reverse-engineered weapons were fired during mid-1999, following a complete rebuild and upgrade of the design. It is thought that this upgrade and the test firings are the prelude to a more comprehensive programme to build a locally produced version of the Rapier.

Description

Apart from the reverse engineering of the missile itself, the Iranians have also deployed an indigenously designed self-propelled version. The Yamahdi Industries Group have taken the original towed launcher assembly and associated guidance units and mounted them on the rear decking of a locally built (8 × 8) Behr cross-country truck chassis to improve mobility and enhance survivability. It is assessed by most western analysts that even with these local modifications and upgrades the Rapier system deployed today in Iran will not have a credible anti-cruise missile or anti-UAV capability.

Status

Missile - In production. On order for the Iranian Armed Forces.

Self-propelled Rapier variant - believed to be in slow-rate production. In service with Iranian Armed Forces.

Offered for export at the Athens Defendory show 2002, no known confirmed customers as of January 2014.

Unconfirmed reports dated May 2008, have suggested that Iran has supplied Hezbollah with mobile Rapier 2s that have been reversed engineered in country. These weapons along with some Strela-2 missiles are reportedly in the hands of Hezbollah's first air-defence armed unit.

Contractor

Aerospace Industries Organization (AIO).
Yamahdi Missile Industries Group.

Ya-zahra (Crotale) Project

Type

Self-propelled Surface-to-Air Missile (SAM) system

Development

The Iranian Air Defence Industries Group of the Aerospace Industries Organisation Ya-zahra Project has succeeded in developing a reverse-engineered R440 missile version of the then French Thomson-CSF Airsys (now Thales France) Crotale system, parts of which were stated to have been captured from Iraq during the 1980 to 1988 Iran/Iraq War.

It is also believed that, as Iran has a number of Chinese-supplied FM-80 (Chinese Feimeng-80 - Flying Midge-80) systems and, possibly, follow-on FM-90 systems. Technology derived from those systems may have been used in the Ya-zahra Project.

Description

The main difference from the original Crotale system is that the Iranian version does not use a copy of the then Thomson-CSF Mirador IV radar, but is configured instead to operate with the Oerlikon Contraves Skyguard radar fire-control and guidance system. Improved countermeasures are fitted which allow the weapon to use a combination of radar/TV/InfraRed (IR) command to line-of-sight guidance against low-level targets and radar/IR guidance against medium-altitude targets.

The Single Shot Kill Probability using a single Ya-zahra missile (Shahab Thaqeb) is stated to be 0.8, which increases to 0.96 when a pair of missiles is used against a single target.

Variants

Ya Zahra 3 (YZ3), reported to be an indigenous Iranian design, developed and produced to suit the local theatre of operations. The system is based on the original French Crotale. The Ya Zahra 3 was first identified when used in Velayat 4, an Iranian air-defence exercise in mid-November 2012 played out against a backdrop of tension between Iran the West over Tehran's nuclear programme.

Shahab Thaqeb SAM being launched from its trailer-mounted platform and striking an aerial target. Inset: (bottom) missile out of launcher is similar to the French Crotale R440 missile and has similar capabilities 0522845

From photographic evidence, the missile canister/launch tube is mounted on a 2 × 2 platform not dissimilar to the Crotale vehicle but with a more modern look to it. A parabolic dish type antenna and what appears to be a TV camera is also mounted below the launcher. The missile is of the same configuration as a standard R-440 but will most probably have been updated considerably.

Information dated 27 Jan 13 confirmed that the Ya Zahra 3 went into serial production in January 2013. It is further reported that the air defence weapon is capable of identifying, intercepting and destroying aerial targets, including helicopters, UAVs and fixed wing aircraft.

Herz-9 (lit Talisman-9)

In mid-May 2013, reports on Iranian TV identified what was described as a new air defence missile in production aimed at low-level targets, from photographic and TV evidence, this system looks like a Ya Zahra (Crotale) mounted on a truck and also a towed trailer. Then again, it is just possible that this system is what was referred to earlier in the year (2013) as the YZ3 reported to have gone into production in January 2013.

Specifications

	Ya-zahra
Dimensions and weights	
Length	
overall	2.93 m (9 ft 7¼ in)
Diameter	
body	154 mm (6.06 in)
Weight	
launch	85.1 kg (187 lb)
Performance	
Speed	
max speed	1,458 kt (2,700 km/h; 1,678 mph; 750 m/s)
Range	
max	6.5 n miles (12 km; 7.5 miles) (400 m/s target)
	5.4 n miles (10 km; 6.2 miles) (300 m/s target)
	5.9 n miles (11 km; 6.8 miles) (rotary wing target)
Altitude	
max	5,500 m (18,045 ft)
Ordnance components	
Warhead	13.5 kg (29 lb) HE fragmentation
Fuze	IR, proximity
Guidance	radar, CLOS
	TV, CLOS
	IR, CLOS
Propulsion	
type	single stage, solid propellant

Status

In production. In service with the Iranian Armed Forces. Initially offered for export in Athens at the Defendory show 2002.

Contractor

Air Defence Industries Group of the Iranian Defence Industries Organization (DIO).

Israel

Air Defence Mobile System Weapon Station (ADMS)

Type

Self-propelled Surface-to-Air Missile (SAM) system

Rafael Air Defence Mobile System (ADMS) mounting four SA-7b 'Grail' man-portable SAMs and fitted to an Automotive Industries M240 Storm (4 × 4) light truck 0009327

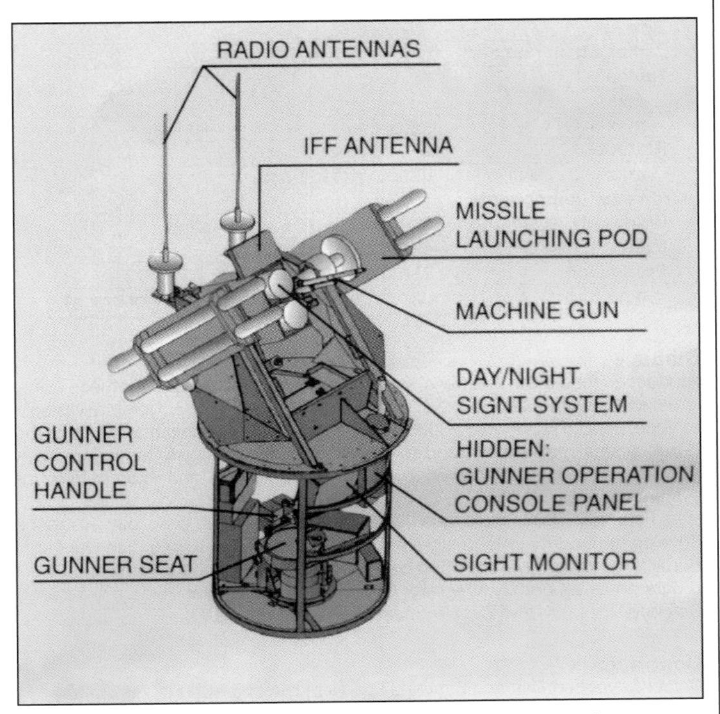

Cutaway showing various components of the ADMS weapon station 0099527

Development

The Rafael Advanced Defence Systems Ltd has developed the Air Defence Mobile Station (ADMS) Short Range Air Defence System (SHORADS) installation from its Overhead Weapon Station (OWS). This involves the attachment of a TD32 pedestal mount for four Russian Federation Kolomna KBM Strela-2M (NATO SA-7b 'Grail') SAMs or equivalents on a vehicle such as the Automotive Industries M240 Storm (4 × 4) light truck or on M113 Armoured Personnel Carrier (APC).

The system has been developed to meet the requirements of a foreign customer but can be modified to accommodate more modern and upgraded older man-portable missile systems if required.

Description

ADMS is an advanced self-propelled air-defence system that implements combat management techniques for integrated air-defence battery operation. The ADMS battery consists of a central Surveillance and Command Unit (SCU) and several Missile Launching Units (MLU). The SCU detects and clarifies the targets using surveillance radar and generates an Air Situation Picture (ASP). An advanced combat management system designates the targets to the MLU and automatically transmits the ASP and C2 to the MLU via radio datalink. The MLU engage the designated targets using their individual Electro-Optical (EO) day/night system. The MLU can also operate autonomously as a stand-alone fire unit.

The ADMS modular design enables easy upgrades and adaptation to the changing air-defence environment, which, in turn, enables the system to carry a full range of VSHORAD type missiles that can be integrated onto a variety of APCs.

Status

Production as required.

In service with an unidentified customer (believed to be Romania using CA-94 or CA-94M missile systems, see entry in this section).

Offered for export.

Contractor

Rafael Advanced Defence Systems Ltd, Platforms and Launchers Directorate (primary contractor).

Barak/ADAMS

Type

Self-propelled Surface-to-Air Missile (SAM) system

Development

In 1987, Rafael Advanced Defence Systems introduced the short-range surface-to-air missile Barak/ADAMS intended to protect ships and ground targets from aircraft and missile threats. Rafael simultaneously announced that trials had confirmed the missile's vertical launch capability and achieved a direct hit during the intercept of a TOW ATGW. The interception test was conducted with the TOW ATGW emulating an incoming skimmer missile.

In early 1998, it was revealed that the Barak was assessed for the application of boosters to increase its anti-air range and for a new warhead system to improve its lethality.

In early 1999 the Venezuelan Ministry of Defence awarded Holland Signaal Apparaten of the Netherlands a contract to supply a new air defence system to the Operational Air Defence Command of the Venezuelan Air Force. Signaal will deliver three systems each comprising a Signaal Flycatcher Mk 2 fire-control system with a Rafael Barak vertical launch missile assembly. The initial contract is worth some USD20 million to Rafael and the agreement includes the provision of an option to supply a further three systems at the same price. Delivery took place in 2001 through to early 2002. Under a 1997 contract Signaal is also upgrading the Venezuelan Air Forces Guardian air-defence system.

Description

The Air Defence Advanced Mobile System (ADAMS) vertical launch surface-to-air missile system is designed for vehicles such as the LAV-25, M2 Bradley IFV and MAN (8 × 8) truck as a shelter system, or simple ground-launcher configurations featuring eight, 12, 16 or more Barak missiles. Each lightweight canister-launcher occupies an area of only 0.1 m^2 per missile. The vertical launch configuration allows for an immediate 360° coverage around the launcher without the need for time-consuming mechanical aiming required by more traditional launch methods.

Artist's impression of self-contained ADAMS on an 8 × 8 truck chassis with one missile intercepting target and one missile being launched 0509465

Barak missile 0079445

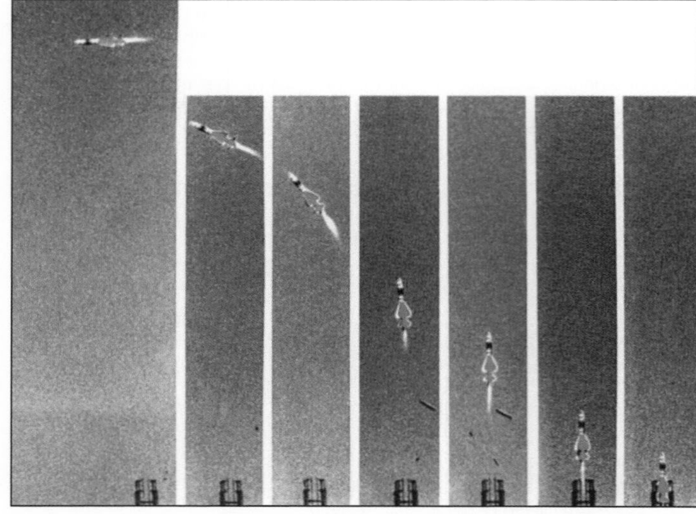

Sequence of photographs taken during Barak trials showing vertical launch and lean over manoeuvre 0080377

Pulse-Doppler search radar is fitted to the larger launchers to provide all the necessary data for the missile's guidance radar. The search radar can track up to 20 targets while still scanning for new contacts and has the ability to track targets operating within the speed range of M0.3 to 3. These are interrogated by an IFF system and handed over to the guidance radar if hostile.

The guidance radar is actually of the search-track-guidance type but has been optimised for use in the last two modes. Operating in the I/J and K-bands it provides target position and command to line-of-sight data for the Barak missile used in the system. Alternatively, for use in a heavy ECM environment electro-optical guidance can be fitted, although both can be used at the same time to guide missiles to two separate targets.

The Barak is stored in the 300 × 350 × 2,500 mm launcher tube with its wings and fins folded and are practically maintenance-free. A key feature of the missile is that, 0.6 seconds into the flight, it is able to vector over to as low as 25° below the horizon or as much as 85° above the horizon. It does this by using the rocket motor's virtually smokeless lower thrust setting and jettisonable thrust-vector control fins in the rocket efflux. In order to engage an incoming target, it uses high acceleration rocket boost mode, aerodynamic flight control fins and the motor in its final sustained thrust mode for high manoeuvrability.

Variants

ADAMS

ADAMS, the original name for the system.

Barak-1

Barak-1, the missile that went with the original system.

Barak-2

Barak-2, reported as the missile for Indian warships.

Barak-8

Barak-8MR (Medium-Range - 70 to 80 km), in development jointly with DRDO in India, this system will be truck mounted for the Indian Air Force but mounted on the Arjun Mk1 MBT hull for the Army.

The Israel Navy began fitting its Sa'ar-5 warships with the Barak-8MR in late July 2013. The missile which, is capable of speeds up to Mach 2.0 has an operational range of 70 km with a minimum intercept of 500 m.

Barak-8ER

Barak-8ER (Extended Range - 150 to 160 km), this variant is being developed with India for the Indian Navy and Air Force. The Air Force variant will be truck mounted.

Specifications

	Barak 1
Dimensions and weights	
Length	
overall	2.175 m (7 ft 1¾ in)
Diameter	
body	170 mm (6.69 in)
Flight control surfaces	
span	0.685 m (2 ft 3 in) (wing)
Weight	
launch	88 kg (194 lb)
Performance	
Speed	
max Mach number	2.17
Range	
min	0.3 n miles (0.5 km; 0.3 miles)
max	6.5 n miles (12 km; 7.5 miles)
Altitude	
max	10,000 m (32,808 ft)
Ordnance components	
Warhead	22 kg (48 lb) HE fragmentation
Fuze	laser, impact, proximity
Propulsion	
type	solid propellant (dual thrust, rocket motor)

Status

Production for export. The naval system has been sold to four countries - Chile, Israel and Singapore and six systems to India. The land-based version was sold to Venezuela. In November 2006, it was reported by Chilean authorities that the Barak system previously fitted to the ex-RN Country-class destroyers have been transferred to Talcahuano to act as on-shore anti-aircraft defence systems.

In February 2006, India and Israel signed their first ever joint weapons development contract to design and produce the Barak II for Indian warships. Barak II will be jointly developed by the Israel Aerospace Industries and Barak programmes secondary integrator Rafael, also by the Indian Hyderabad-based Defence Research and Development Laboratories (DRDL).

Contractor

Rafael Advanced Defence Systems Ltd, (enquiries).
Defence Research and Development Laboratories (DRDL).
Israel Aerospace Industries (IAI).

Iron Dome

Type

Self-propelled Surface-to-Air Missile (SAM) system

Development

An initial funding proposal from the then Rafael Armament Development Authority (now Rafael Advanced Defence System Ltd) to the Israeli Ministry of Defence to develop an anti-missile defence system to counter the Palestinian Qassam rocket attacks was made in January 2007. The system known as Iron Dome (Kippat Barzel) consists of an inexpensive kinetic interceptor based on Rafael's missile technology combined with an Elta Systems detection and fire-control radar. Initial Operational Capability (IOC) was 2009/10. Rafael claims the initial cost of each interceptor missile will be in the region of USD30,000 to 40,000 each, and the complete cost of the system is expected to be about USD600 million. The Israeli Air Force will oversee the development programme as the expected operators of the operational system.

The first launch of the Iron Dome missile, now known as Tamir, took place on 9 March 2008 with another successful firing taking place at the Ramon Air Force Base in July 2008.

Static tests of the highly sophisticated proximity fuzed warhead took place in January 2007 when the warhead successfully destroyed both a Qassam rocket and a 122 mm Grad-series round.

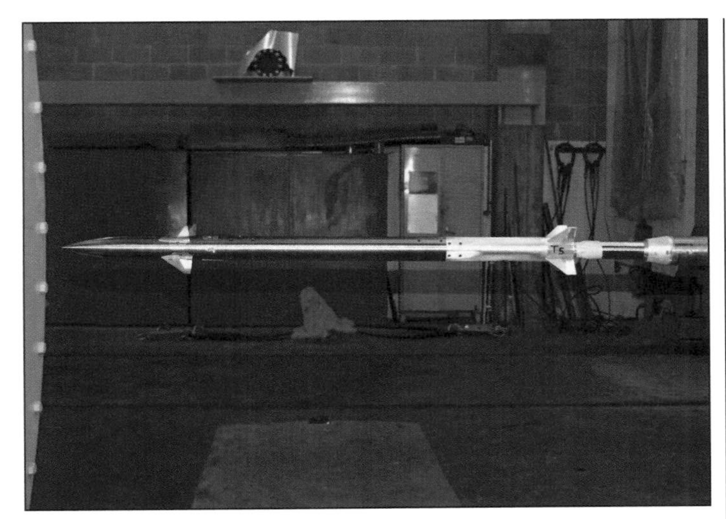

Iron Dome interceptor (Rafael) 1169756

Constituent parts of Iron Dome (Rafael) 1330593

At the end of March 2008, it was reported that the Israeli Air Force are updating their missile testing capabilities at the Palmachim Air Base that is situated just South of Tel Aviv; another base used for the developmental firings is the Ramon Air Force Base in the South of Israel. The upgraded facilities will be used for a wide range of programmes but most notably elements of the Iron Dome missile defence system.

In March 2009, the Ministry of Defence in Israel reported further critical testing of the Iron Dome system in which during the test, a number of rockets of the same type that were fired in recent years at Israel were launched as targets and the Iron Dome responded accordingly with successful intercepts. By July 2010, tests of the system were completed with the Iron Dome successfully intercepting the targets, all systems including radar, launcher and interceptors and control systems functioned well.

In July 2010, the Iron Dome passed its final series testing and is now fully operational.

Iron Dome Block 2 upgrade will include advances in interception capabilities and will respond to a wider range of threats. This Block 2 improvement package is expected to be completed within the next 12 months when a Block 3 will begin development. It is believed that the Block 3 will include improvements of the control capabilities of the system.

Description

Iron Dome is designed as an all-weather, multi-layered Counter Rocket, Artillery and Mortar (C-RAM) and Very Short Range Air Defence (VSHORAD) system that can defend an area of 150 km^2 and is to be part of a multi-layered defence system aimed at protecting Israel from both short-range missiles and rockets fired by militants in Gaza or Lebanon. The Iron Dome system is designed to detect incoming rockets and mortars within the ranges of 4-70 km and VSHORAD missile systems up to 10 km. The system utilises miniature missiles (Tamir) to engage and destroy in mid-air. The Iron Dome is designed to handle selective multiple threats simultaneously.

Iron Dome can distinguish between incoming threats whether targeted towards built-up areas or towards harmless areas, thus avoiding unwarranted interceptor launches and ensuring economy of effort.

The system makes use of the Israel Defence Forces (IDF) early warning radar system to identify short-range missiles and rockets in flight. The target acquisition process is performed in less than one second and is designed to launch the intercepting rocket about one second after the hostile rocket is detected on radar. The interceptor has a simple radio-frequency seeker covered with a plastic radome. This dome has a conical metallic shroud that protects the seeker in flight from aerodynamic heating and is ejected seconds before the interception, allowing the seeker Field-of-View (FoV) sight, identify and acquire the target and guide the missile to interception. The missile also boasts a proximity fuzed warhead details of which have not yet been released.

System components

The system battery will consist of:

- IAI Elta Systems EL/M-2084 multimission radar
- Battle management/fire-control centre
- Three missile launcher/firing units, each carrying 20 Tamir interceptors.

Launcher

Each mobile launcher will carry 20 missiles and the entire system consists of 50 trucks.

C4I/Radar

The Iron Dome uses two radar units, developed at IAI's Elta Systems, the radars are capable of tracking multiple targets and discriminate between benign threats and those threatening protected targets.

Variants

Iron Dome Block 2
Iron Dome Block 3

Specifications

	Tamir
Dimensions and weights	
Length	
overall	3 m (9 ft 10 in)
Diameter	
body	160 mm (6.30 in)
Weight	
launch	90 kg (198 lb)
Performance	
Range	
min	1.1 n miles (2 km; 1.2 miles)
max	21.6 n miles (40 km; 24.9 miles)
Ordnance components	
Warhead	kinetic
Fuze	proximity
Guidance	laser (RF homing head)
Propulsion	
type	solid propellant

Status

The system was supposed to be initially deployed in November 2010 in Sderot, in Southern Israel, however, in March 2011 the southern city of Beersheva received the first deployment to counter rockets fired into Southern Israel from the Gaza Strip. The Iron Dome is slated to defend Southern and Northern Israel from rocket attacks by Hamas and Hizbullah, furthermore, it is to become a key component in a multi-layered missile defence system that includes the Arrow anti-ballistic missile shield.

Reports from the Israeli Defence Force dated September 2009 confirmed that Iron Dome battery personnel had already begun exercising the battle management and weapon control system, simulating interception scenarios. This first deployment, is in the region of Sderot, comes after the successful completion of the July 2009 tests. The second battery was deployed in the Southern port city of Ashkelon on 21 August 2011 and a third on 31 August near Ashdod to bolster the south.

In view of heightened cross-border escalations, the fourth and fifth batteries were deployed in between September and November 2012. The production was increased as a result of the friction between Iran and Israel and the political situation in Syria. By 1 July 2013, six Iron Domes had been deployed, the sixth system to cover an area of the Gaza Strip.

According to the Israeli press, the envisaged objective force structure is to have 12 Iron Dome batteries by 2014. This will be an increment from the previously declared 9 batteries. The equipment for 10 more batteries has already been ordered.

Iron Dome is a key component in a multi-layered missile defence system that includes the Arrow anti-ballistic missile shield. The system constitutes the bottom tier, designed to intercept short-range missiles, rockets and mortar shells fired into northern Israel and in the south of the country.

The first overseas deployment was in Singapore in 2012 with the Singapore Air Force to counter allegedly Indonesian and Malaysian surface-to-surface threats.

Contractor

Rafael Advanced Defence Systems Ltd (primary contractor).
ELTA Systems Ltd (detection and fire-control radar).

Relampago

Type

Self-propelled Surface-to-Air Missile (SAM) system

Development

The Relampago was developed by Rafael Advanced Defence Systems Ltd (then known as Rafael Armament Development Authority), as a joint venture with the then Israel Aircraft Industries (now Israeli Aerospace Industries) for the export market, has developed the all-weather Relampago (Lightning) system.

Model of the Relampago (Lightning) mobile surface-to-air missile system with cell of 12 Barak-1 vertically launched SAMs at the rear 0080419

It is capable of engaging anti-radiation missiles, gliding bombs and cruise missiles as well as all current threats such as aircraft, helicopters and remotely piloted vehicles.

The system is based on the Rafael Barak vertically launched missile, which is already in production for naval and land-based applications. It can be deployed as a stand-alone unit, as part of a battery or be integrated into an air defence network to defend key strategic areas such as airfields, command centres or field units at brigade level or above. Each Relampago can defend an area of around 300 km^2.

As it is a modular system, elements can also be used to upgrade air defence gun batteries using their existing fire-control systems.

Description

A typical Relampago battery would comprise three fire units and a centralised command post, with the latter being netted into a higher command level to receive target information.

The system is mounted on a 6 × 6 or 8 × 8 cross country truck chassis. Tracking and surveillance radars are mounted on an elevated arm to the rear of the cab. Mounted at the very rear of the truck is the operator's cabin and the Barak missile cell with 12 ready to launch missiles in two rows of six. At the launch site two hydraulic stabilisers are lowered to the ground on either side of the truck.

The Barak missiles are stored in individual launch tubes with their wings and fins folded. In 1998, it was revealed that the Barak missile was being assessed for the application of boosters to increase its anti-air range and for a new warhead system to improve its lethality.

Variants

In 1993, plans were announced for a feasibility study to adapt the Barak for use on Russian air defence vehicles such as the ZSU-23-4 Shilka and SA-8 Gecko. Slovakia had a related programme known as Strop 2, which uses a similar concept to the Relampago system mounted on the same TATRA heavy truck chassis as the Strop 1 air defence gun vehicle.

Specifications

	Barak
Dimensions and weights	
Length	
overall	2.175 m (7 ft 1¾ in)
Diameter	
body	170 mm (6.69 in)
Flight control surfaces	
span	0.685 m (2 ft 3 in) (wing)
Weight	
launch	98 kg (216 lb)
Performance	
Speed	
max speed	1,400 kt (2,592 km/h; 1,611 mph; 720 m/s)
Range	
min	0.3 n miles (0.5 km; 0.3 miles)
max	6.5 n miles (12 km; 7.5 miles) (est.)
Altitude	
max	10,000 m (32,808 ft)
Ordnance components	
Warhead	22 kg (48 lb) HE fragmentation
Fuze	laser, impact, proximity
Propulsion	
type	solid propellant (dual thrust, rocket motor)

Status

If an order is received, the project will move to production. To date, no known order has been placed. Although a land-based missile system involving the Barak has been sold to Venezuela. Prime contractor for this is Thales Netherlands. In this application it is used with the Flycatcher Mk 2 fire-control radar.

As of November 2013, the system is no longer mentioned in Rafael or on-line files or brochures.

Contractor

Enquiries to Rafael Advanced Defence Systems Ltd.

Spyder

Type

Self-propelled Surface-to-Air Missile (SAM) system

Development

The then Rafael Armament Development Authority (now Rafael Advanced Defence Systems Ltd) of Israel, in collaboration with Israel Aerospace Industries (IAI) – MBT Missile Division and ELTA Radar Division, has developed and offered for sale a surface-to-air Python 5 and Derby Air-Defence System – Spyder ADS, also known as Spyder-SR [Short-Range] and Spyder-MR [Medium-Range].

All of the units associated with these systems can be air transported by a standard logistics support aircraft such as the C-130 or IL-76.

As a concept, the Spyder SAM is very similar to that of the US programme that combines the Raytheon Advanced Medium-Range Air-to-Air Missile (AMRAAM) with the US AM General High-Mobility Multipurpose Wheeled Vehicle (HMMWV), but with dual-type missiles and dual-type seekers.

According to Rafael, a proof-of-concept test launch has already been successfully carried out, and a number of undisclosed export customers have been briefed on the system. In particular, in 2005, during the Bangalore Air Show in India, the system was presented to the Indian Air Force, based on the conversion of Python 5 and Derby air-to-air missiles into anti-aircraft missiles.

A new and improved Spyder SAM, known as the Spyder-MR (Medium-Range), is near the end of its development. Rafael and IAI have introduced this new system into the Spyder family in order to combat a wide spectrum of threats, such as attack aircraft, bombers, cruise missiles, UAVs, UCAVs and stand-off weapons. In this project, Rafael is the prime contractor and IAI is the major sub-contractor.

This same Spyder team of Rafael, IAI-MBT and ELTA has successfully collaborated in the design, production and commissioning of the Barak Vertical Launch SAM system, that is operational within several Navies around the world, including the Israeli Navy.

Rafael/IAI SPYDER-SR air-defence system mounted on a 6 × 6 truck chassis (Rafael) 0563354

Spyder-SR Missile Firing Unit (MFU) (Rafael) 1043192

Spyder Transloader (Rafael) 0563355

Spyder-MR Launcher 1132763

Description

Basic Spyder (Spyder-SR) is a state-of-the-art, low-level, quick-reaction SAM system, designed to be capable of engaging aircraft, helicopters, UAVs, UVAVS and PGMs. It provides protection to high-value areas, as well as first-class defence for forces located in the combat area.

Propulsion

Both the Derby and Python missiles employ a solid fuel rocket motor, however, both missiles have been fitted with the all-new solid propellant booster motor that takes the missiles to maximum speed in a far quicker time and thus extends the range considerably.

Payload

11 kg with proximity fuse.

Guidance

The system incorporates Rafael's advanced air-to-air missiles, the Derby, an active radar missile, and the Python-5, a dual band Imaging InfraRed (IIR) Missile. Both missiles are the outcome of many years of research-and-development efforts utilising the most modern technologies.

Spyder missiles have full commonality with the air-to-air missile versions, and have the ability to engage a target with an RCS of 2 m^2. The system has been built on the modular structure that allows customer adaptation and future growth on a variety of platforms.

Spyder-MR is a medium-range air-defence system designed for the protection of urban areas as well as manoeuvring combat forces. This system, unlike its predecessor, is an all-weather, network-centric, multi-launch, quick-reactions ADS. The two missiles associated with the MR version are, again, the Derby active radar Beyond Visual Range (BVR) missile and the dual-band IIR Python-5.

Launcher

The basic launcher chassis is a Tatra or Mercedes-Benz or Actros. However for the Republic of Singapore Air Force MAN TGS is used.

The Spyder-SR launcher system is a electromechanical turret-based unit with infinite rotational capability (360° × n). The system can launch missiles in slant method with two modes of operation: Lock-On-Before Launch (LOBL)

Python test firing (Rafael) 1043193

Derby test firing (Rafael) 1043194

and Lock-On-After-Launch (LOAL) This launching method enables the LOBL feature that cannot be achieved in vertical-launch method. For this reason, many engagements can be carried out with the missile's seeker locked on the target before launch. A standard battery consists of one Command-and-Control Unit (CCU) plus radar and six Missile Firing Units (MFUs). The total number of missiles per battery consists, therefore, of 24, and of these, a maximum of 20 can be controlled in flight at any one time.

Each MFU carries on top of the cab two small wideband whip antennas, one for receiving, and the other a transmitter that passes data information to the four missiles. Address groups to those missiles are unique to that MFU.

Once the missiles have been launched and the engagement is complete, a transloading vehicle may refurbish the MFU with four new missiles that then have to shake hands with the MFU and identify themselves with the address groups of that particular transmitter. The whole reloading operation should take no longer than 15 minutes including the address-group identification.

Spyder-MR launcher is, again, mounted on a 6 × 6 truck, but this time with a vertical launcher offset slightly at 85°. This offset allows for a safety zone when launching the missiles in order that the booster motor will not drop on friendly forces or cause collateral damage to local civilian areas that the system is protecting. With the missiles ranged in vertical launchers, the reaction time to engage targets that have been declared hostile is reduced to two seconds, thus also allowing for a more pro-active engagement of pop-up targets such as terrain-hugging stand-off weapons.

C4I/Radar

The Spyder-SR system comprises a truck-mounted CCU with ELTA EL/M 2106NG ATAR 3D surveillance radar and truck-mounted Missile Firing Units (MFU) equipped with both IIR and RF missiles. The surveillance radar is a solid state L-band for ranges over 40 km and has advanced ECCM capabilities and can simultaneously track 60 targets with a 360° engagement day/night all-weather capability.

Spyder-MR utilises the flat-faced phased array radar ELTA MF STAR radar mounted on a 6 × 6 truck that is referred to as the Radar Sensor Unit (RSU). This RSU is linked to the CCU that is also linked to the Air-Defence network.

Other improvements over the Spyder-SR include an embedded training and debriefing unit and full maintenance capabilities, also available on Spyder-SR (Yh). The Spyder MFU carries any combination (IIR/RF) of four missiles on a rotation-able launcher assembly. Mounted on the cab top is TOPLITE, a highly stabilised multi-sensor, electro-optical, observation device, capable of operation during both day and night and severe weather conditions.

Wireless data link communication enables deployment of the MFUs at a distance of up to 10 km from the CCU. The system's high mobility allows quick deployment and operational agility. Spyder-MR with its vertical launcher carries up to eight missiles in any combination of the boosted Derby and

	Spyder-SR	Spyder-MR	Derby	Derby (upgraded)	Python 5	Python 5 (upgraded)
Dimensions and weights						
Length						
overall	-	-	3.621 m (11 ft 10½ in)	3.621 m (11 ft 10½ in)	3.0 m (9 ft 10 in)	3.0 m (9 ft 10 in)
Diameter						
body	-	-	160 mm (6.30 in)	160 mm (6.30 in)	160 mm (6.30 in)	160 mm (6.30 in)
Flight control surfaces						
span	-	-	0.64 m (2 ft 1¼ in) (wing)	0.64 m (2 ft 1¼ in) (wing)	0.35 m (1 ft 1¾ in) (wing)	0.35 m (1 ft 1¾ in) (wing)
Weight						
launch	-	-	118 kg (260 lb)	118 kg (260 lb)	105 kg (231 lb)	-
Performance						
Range						
min	0.5 n miles (1.0 km; 0.6 miles) (est.)	0.5 n miles (1.0 km; 0.6 miles)	-	-	-	-
max	8.1 n miles (15 km; 9.3 miles) (est.)	18.9 n miles (35 km; 21.7 miles) (est.)	-	-	-	-
Altitude						
min	20 m (66 ft)	20 m (66 ft)	-	-	-	-
max	9,000 m (29,528 ft)	16,000 m (52,493 ft)	-	-	-	-
Ordnance components						
Warhead	-	-	HE fragmentation	HE fragmentation	11 kg (24 lb) HE fragmentation	11 kg (24 lb) HE fragmentation
Fuze	laser, proximity	proximity, impact	laser, proximity	laser, proximity		
Guidance	INS, mid-course update	INS, mid-course update	active, IIR (dual band, LOBL and LOAL)	active, IIR (dual band, LOBL)	-	-
	active, radar (terminal phase)	active, radar (terminal phase)	-	-	-	-
Propulsion						
type	solid propellant (rocket motor)	solid propellant (rocket motor)	solid propellant (rocket motor)	solid propellant (rocket motor)	-	-
	-	solid propellant (booster)	-	solid propellant (booster)	-	-

Python-5. As yet there is no direct link between the long-range Arrow-2 batteries and the Spyder-MR, however, both systems will be reading the same general AD picture in order to utilise those missiles that are best equipped to deal with any incoming threat.

System Components

The Spyder system includes the following main components:
- Truck-mounted CCU with mounted radar
- IFF and communication
- Truck-mounted MFU and communication equipment
- Combination of Derby and Python-5 missiles
- Field Service Vehicle
- Missiles Supply Vehicle
- Ground support and special test equipment
- Built-in and add-on training equipment.

A standard Spyder missile battery has up to six missile-firing units per battery and a command-and-control unit. The Spyder-MR system includes the following main components:
- Truck-mounted CCU
- Separate truck-mounted phased array radar with communications equipment RSU
- Truck-mounted MFU with up to 8 × missiles per MFU
- Field service vehicle FSV
- Missiles Supply Vehicle MSV
- Integrated test equipment
- Integrated training equipment.

Specifications – *see table above*

Variants

Reports coming out of Israel (SIBAT) in April 2009 have indicated that a USD2.5 billion deal between Israel and India to jointly develop a new and advanced version of the Spyder SAM system.

Status

Development of the Spyder-SR is complete and the system has been offered for sale. Baseline missiles are in production for the air-to-air variants and are identical for the SAM role.

During February 2004, Rafael exhibited, for the first time, the basic Spyder (Spyder-SR) at the DEFEXPO Exhibition in New Delhi, India. This was followed a year later, in February 2005, with company officials reporting that the early stage of the Indian Air Forces (IAF) USD350 million air-defence tender is complete, beating like systems from Russia and South Africa.

In September 2006, India agreed to take delivery of up to four Spyder SAM systems. By December 2007, sources reported that the system had successfully completed its trials in India and that the contract was awaiting signature.

The contract for the Spyder system for India was signed on 1 September 2008 with deliveries of 18 systems beginning 2010 and continuing through early 2011 to August 2012.

Spyder-MR is near development completion. Firing trials have been completed in Finland, and the system is ready for deployment.

In May 2008, the basic Spyder system was offered to Sri Lanka as a means to defeat the Tamil Tiger Air Force, no decision was made as to whether or not to purchase.

During 2010, the Spyder system was delivered to Singapore for use with the Republic of Singapore Air Force (RSAF). Known in the RSAF as Ground-Based Air-Defence (GBAD) the system works together with the upgraded I-HAWK and RBS 70 air defence systems on a 24-hour a day basis to protect the skies over Singapore.

Contractor

Rafael Advanced Defence Systems Ltd, (primary contractor).
Israeli Aerospace Industry, (major subcontractor).
ELTA Radar Division, (radar production).
TATA Power, (maintenance for systems supplied to India only).

Japan

Type 93 Kin-SAM

Type

Self-propelled Surface-to-Air Missile (SAM) system

Development

In FY90 the Japanese Defence Agency (JDA) contracted Toshiba to develop a mobile version of its Type 91 Kin-SAM man-portable SAM system. The development of an InfraRed (IR) seeker began in 1979 for the Type 91.

Type 91 was deployed during 1991, at that time the Type 93 was in development. The result is the Type 93 system, which completed development in FY93, entered low-rate production and Initial Operational Capability (IOC) for the Japanese Ground Self-Defence Force in FY93.

Description

The Type 93 (also known as Closed Arrow) is based on the Kohkidohsha (4 × 4) 1,500 kg high-mobility vehicle chassis, the military version of the Toyota Mega Cruiser. The vehicle is fitted with a pedestal-mounted launcher/Electro-Optic (EO) system attached to the vehicle's rear decking area.

The launcher is comprised of two four-round Type 91 Kin-SAM missile pods (one pod on each side) and the focal plane (full imaging configuration) passive IR/visible EO system located between the pods. The launcher/EO system is free to traverse, and the pods are able elevate or depress; the exact minimum and maximum angular figures are unknown. Available details of the enhanced, third-generation, Type 91 Kin-SAM are given in the man-portable surface-to-air missile systems section.

Type 93 Self-Propelled SAM (Toshiba) 0577959

In most respects the operational use of the Toshiba Type 93 Kin-SAM is the same as that of the US developed Boeing Avenger SAM on the AM General HMMWV (4 × 4) chassis.

Specifications

	Type 91
Dimensions and weights	
Length	
overall	1.43 m (4 ft 8¼ in)
Diameter	
body	80 mm (3.15 in)
Weight	
launch	11.5 kg (25 lb)
Performance	
Speed	
cruise	1,258 kt (2,329 km/h; 1,447 mph; 647 m/s)
Range	
min	0.2 n miles (0.3 km; 0.2 miles)
max	2.7 n miles (5 km; 3.1 miles)
Ordnance components	
Warhead	HE fragmentation
Fuze	proximity, impact
Guidance	IR (two channels, 0.4 to 0.7 μm and 3.5 to 5.2 μm)
Propulsion	
type	solid propellant (sustain stage)

Status

IOC 1993. Low-rate production (10 vehicles FY94; figure for FY95 not available; eight vehicles for FY96, figure for FY97 not available).

In service with the Japanese Ground Self-Defence Force.

Not offered for export.

Contractor

Toshiba Corporation.

Korea (South)

Chun Ma (Pegasus)

Type

Self-propelled Surface-to-Air Missile (SAM) system

Development

Early in 1996, details were released of the Chun Ma (Pegasus-Winged Horse) Surface-to-Air Missile (SAM) system for which the prime contractor was the then Special Products Division of Daewoo Heavy Industries, now known as Doosan Infracore. The system was developed in combination with the state-run Agency for Defence Development and some 12 locally based subcontractors and the then Thomson-CSF of France.

The requirement for the Chun Ma's missile was first identified in 1987; this requirement was to meet the operational needs of the Republic of Korea Army (RoK) for a self-propelled all-weather SAM system to protect its mechanised forces. The development programme for the system began two years later in 1989.

The full-tracked chassis used for the Chun Ma SAM system is the latest in a long line of chassis developed by Daewoo to meet the operational requirements of the RoK Army.

So far the company has built over 2,000 Korean Infantry Fighting Vehicles (KIFV) and variants for the Republic of Korea Army and for export to Malaysia for deployment in Bosnia as part of United Nations forces.

Chun Ma (Pegasus) SAM being launched. The system carries a total of eight SAM in the ready-to-launch position 0080420

Scale model of the Chun Ma (Pegasus) SAM system with radars in operating configuration (Christopher F Foss) 0080422

The full tracked chassis used for the Chun Ma SAM system is much larger than that of the KIFV. This chassis is also being used for a number of other applications including an ammunition resupply vehicle and as the basis for the Flying Tiger twin 30 mm self-propelled air defence gun system now in production. Details of the Flying Tiger are given in the Self-propelled anti-aircraft gun section earlier in this volume.

By early 1996, two prototypes of the complete Chun Ma SAM system had been completed and firing trials began in 1997. Low rate pre-production of the system is believed to have started in late 1997 with some subsystems being imported. The first production of Pegasus began in early 1999 for operational deployment with the South Korean Army in late 1999. Cost of an individual Chun Ma is KRW15 billion and of a missile KRW280 million (at 1999 prices). December 2007 prices are USD16.2 million per unit, USD303.655 per missile.

Description

The Chun Ma (Pegasus/Flying Horse), also known as K-SAM is a short-range air-defence system designed to engage air threats flying at low altitudes. The system was designed for the protection of industrial facilities, infrastructures, mobile military units, ports and airports. The Chun Ma chassis is based on a modified K-200 Armoured Personnel Carrier (APC) and is of all-welded armour construction, which provides the occupants with protection from small arms fire and shell splinters. The driver is seated front left with the power pack to the right and this leaves the rear two-thirds of the vehicle clear for the missile system.

The power pack consists of a 520 hp Daewoo D2840L 10 V four-cycle, turbo-inter-cooled diesel engine coupled to a fully automatic transmission which gives a maximum road speed of 60 km/h and acceleration from 0 to 32 km/h in around 10 seconds. Total weight of the system is 26 tonnes. The crew is three.

It is believed that a 43 hp auxiliary power unit is fitted and standard equipment includes an NBC system and an automatic fire detection and suppression for the engine and crew compartments.

Mounted on top of the chassis is an electro-hydraulically power-operated unmanned turret with two banks of four SAM launchers each side. In the centre of the turret is the sensor package that consists of the E/F-band, solid-state, pulse-Doppler Surveillance radar accompanied with L-band IFF secondary radar with four operation modes – 1, 2, 3A and secure mode – with a range of 20 km. It has a track-while-scan capability and can track eight targets at once with automatic threat evaluation.

Mounted below the surveillance radar is the circular Ku-band TWT pulse-Doppler multi-operation mode, tracking radar with a tracking-range of 16 km and tracking altitude of 5 km. This has been designed to track hovering helicopters, fighters and other aircraft travelling at a maximum speed of M2.6 and has automatic ECCM capabilities. Both radar are of the frequency agility pulse compression type and, when travelling, the surveillance radar is lowered to the rear to reduce the overall height of the system.

Key system elements of the Chun Ma (Pegasus) SAM system 0080423

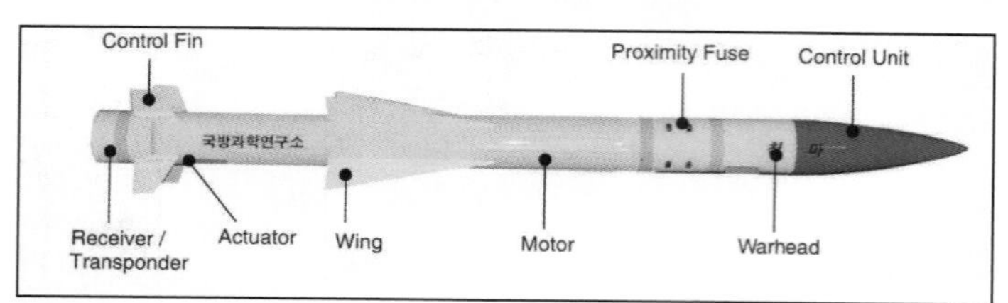

Chun Ma (Pegasus) SAM out of its launch/transport container showing key missile components 0080421

The accuracy of the missile system can be attributed to the Radar Guidance and Control System (RGCS) that utilises Forward Looking InfraRed (FLIR) and Command Line Of Sight (CLOS). The RGCS has a control and monitoring panel, a TV monitor, Synthetic Image generator, range finding key and a joystick. To the left of the tracking radar is the FLIR camera with two fields of view and a range of 15 km. To the right is the daylight TV camera with a range of 10 km, with the IR goniometer below used for initial gathering of the missile following launch. The latter has a 10° field of view.

According to Daewoo, the Chun Ma system can engage targets under day and night conditions regardless of battlefield clutter and hostile ECM conditions.

The whole turret, sensor package and the internal components were supplied by the then Thomson-CSF Airsys of France and are virtually identical to that used by the Crotale NG SAM which is in service with Finland (on a 6 × 6 Patria chassis) and France (shelter-mounted).

The missile used in the South Korean Chun Ma is of a different design to those used in the Crotale NG and has been developed in South Korea.

The 86.2 kg launch-weight-missile is in a sealed tube and is fitted with a laser proximity fuze and a 12 kg focused fragment warhead, which is claimed to give a high kill probability. It has four fixed wings two-thirds of the way from the nose and four control fins at the rear. Guidance is command-to-line of sight with maximum effective range quoted as 9 km, this being shorter than that of the French Crotale NG. Once expended, missiles are reloaded manually there being no provision for automatic loading of new missiles.

The missile operator is provided with a fire-control system with full colour multi-window display console and the software allows the Chun Ma to be integrated with other battlefield air defence assets and command systems.

Mounted at the front of the hull are two banks of four electrically operated smoke grenade launchers that fire over the frontal arc.

Variants

Black Fox

Doosan Infracore has offered for sale an 8 × 8 armoured wheeled variant mounted on, or probably known as Black Fox.

Specifications

	Chun Ma
Dimensions and weights	
Crew:	4
Length	
overall:	7.1 m
Width	
overall:	3.4 m
Height	
overall:	5.4 m
Weight	
combat:	26,000 kg
Mobility	
Configuration	
running gear:	wheeled
Power-to-weight ratio:	laden
Speed	
max speed:	100 km/h
Range	
main fuel supply:	500 km
Gradient:	60%
Side slope:	35%
Engine	Doosan D2840L, turbocharged, 4 stroke, diesel, 520 hp
Firepower	
Turret power control	
type:	electric
Main armament traverse	
angle:	360°
speed:	240°/s
Main armament elevation/depression	
armament front:	+65°/-1°

Missile specifications

Length:	2.17 m
Diameter:	0.15 m
Wing span:	unknown
Launch weight:	86.2 kg
Propulsion:	solid propellant rocket motor
Velocity:	M2.6 (885 m/s) other sources have suggest M2.7
Guidance:	Command to Line-Of-Sight (CLOS)
Warhead:	
(weight)	12 kg
(type)	HE focused fragmentation
Fuze:	active laser proximity and/or contact with 8 m lethal radius
Range:	
(maximum)	10.5 km
(maximum effective)	9 km
(minimum)	2 km
Altitude:	
(maximum)	6,000 m
(maximum effective)	5,000 m
(minimum)	unknown
Manoeuvrability:	30 *g*
Reaction time:	less than 10 s
Number of missiles per vehicle:	8

Status

Low-rate pre-production began in 1996/7 with six systems delivered by early 1999.

Entered operational service in December 1999.

During December 2007, the Chun Ma (Black Fox wheeled variant) system was offered to Saudi Arabia for a price of USD16.2 million per unit, no sale of any systems have been reported.

Contractor

Doosan Infracore Co Ltd, Changwon Plant Number 2, (primary contractor).
Goldstar (LG) Precision Instruments, (system missiles).
Daewoo (full-tracked chassis).
Samsung Electronics (now part of Thales - electronic components and radar development).

Iron Hawk II

Type

Self-propelled Surface-to-Air Missile (SAM) system

Development

Research into the Iron Hawk II (originally known as the M-SAM or KM-SAM) started back in 2001 after a decision was taken by the South Korean government to replace the US supplied HAWK that has been the backbone of the air defence for South Korea for many years. This was followed by a finalised agreement in 2006 and also in the same year a contract to develop and produce the missile for the system based on the Russian Almaz/Antei (Fakel) 9M96 that

Iron Hawk II missile (ADD) 1451706

FC radar (ADD) 1451708

can normally be found in the Russian S-400 system. The missile has been designed to take on targets such as the North Korean No Dong series of surface-to-surface (ballistic) weapon in the endo-atmosphere at around 15,000 to 18,000 m.

The following year in April 2007, another decision was taken to improve the performance of the basic Iron Hawk II missile particularly in altitude. This new missile will form part of the Phase 2 (Cheolmae 4-H) and will fly exo-atmospheric and is expected to be able to take targets in the ICBM role. The range of the new missile will be in the region of 150 km as opposed to the current 40 km. The Phase 2 missile is expected to become operational around the 2018 time.

Description

The new Iron Hawk II system was unveiled at a ceremony in the Agency for Defence Development (ADD) headquarters in the city of Daejeon on 15 December 2011. At that time the system was reported to have the capability to engage air-to-surface guided missiles, short-range ballistic missiles and multiple targets (up to six) with a single radar system. Also on the same day at the ADDs Research Institute in Yuseong, Daejeon, another briefing took place that gave a breakdown of the battery for the Iron Hawk II. The battery will consist of:

- Multifunctional Radar
- Combat Control Centre
- Launch Pad (Vehicles)
- Missiles.

The multifunction radar is reportedly able to detect target aircraft, distinguish between friendly and hostile (Identification Friend or Foe - IFF) and guide the missiles simultaneously.

The combat control centre has automated operational procedures incorporating the latest information and communications technology.

The launch pad uses a cold launch vertical ejection method, thereby reducing weight. Aligning the launch pad vertically eliminates the need to aim the pad in the direction of the launch, allowing faster launching.

The missile is based on the Russian 9M96 missile that was designed by Fakel a subsidiary of the Almaz/Antei Concern of Air Defence based in Moscow. The missile is cold launched and ignites the motor when clear of the launcher, then

Launcher for Iron Hawk II missile (ADD) 1451707

the missile flies using inertial navigation with mid-course updates from the radar before terminal engagement utilising the onboard nose cone radar. The missile has been designed as a Hit-To-Kill (HTK) type.

Variants

Phase 2 (Cheolmae 4-H)

Specifications

Iron Hawk II

Type:	medium-range, medium-altitude surface-to-air missile system
Length:	4.61 m
Diameter:	0.275 m
Guidance:	INS with mid-course updates then active homing
Propulsion:	Solid propellant rocket motor
Warhead:	Hit-To-Kill (HTK)
Maximum range:	40 km
Maximum altitude:	15-18,000 m
Number of targets:	6
Reaction time:	8-10 s
G load:	50 g

Status

First shown to the public in mid-December 2011, the system remains in development but was initially expected to become operational 2012/13. However, reports emanating from Seoul in September 2012 have now changed this date to 2015. Production of the missiles began in 2013 with the system now expecting to be fully operational by 2015 and complete by 2018/19.

Contractor

Agency for Defence and Development (ADD); Defence Agency for Technology and Quality (DATQ); LIG Nex 1; Samsung Thales; Doosan DST; Hanhwa; Kia Motors; and Lucky Goldstar.

Poland

2K12 Kub Polish Upgrade Programme

Type

Self-propelled Surface-to-Air Missile (SAM) system

Development

As part of a continuing improvement plan started in the early 2000's to their air-defence systems, the Polish Armed Forces incorporated new technology into the Polish Army's Russian-supplied 2K12 Kub self-propelled air-defence missile systems.

A contract of undisclosed value was signed in November 2005 with the MND's Wojskowe Zaklady Uzbrojenia Number 2 (WZU-2) depot in Grudziadz for the modernisation of the remaining 2K12 Kub (NATO SA-6 Gainful) surveillance and tracking radar 1S91M. The contract will bring the system up to the standard of the 1S91M1-P1 that is currently deployed in Hungary and Poland. Deliveries began in June 2006.

In 2007, Raytheon International unveiled a proposal to equip Polish Kub (NATO SA-6 'Gainful') with SL-AMRAAM missiles, based on the AIM-120C-5 standard. The proposal is connected to a Polish Kub modernisation programme conducted by WZU-2, Grudziadz, Poland. This modernised system will include an further improved 1S91M2-P1 fire-control radar and 9Sh33M optical tracking system fitted with a KT-1 CCD TV camera with continuous zoom. A thermal (InfraRed - IR) camera operating in the 8-12 µm band is also available; this has a range of 40-70 km in favourable working conditions.

Description

The military's WZU-2 workshop at Grudziadz is installing modern digital component electronic systems to replace existing vacuum-tube technology and enhance the ECCM capabilities and reduce system weight and InfraRed

(IR) signature. A PCO-supplied passive IR search sensor (to complement a previously fitted low-light TV tracker) is also being fitted together with a Radwar IFF interrogator assembly ISZ-01. The ISZ-01 is a medium-range IFF interrogator that is also being fitted to the mobile missile system Osa AK and Loara-G mobile gun system. This modernisation will ensure compatibility with NATO requirements and, following the delivery of a large quantity of 3M9M3E Kub missiles during 1980 and 1987, enable the system to remain in service until at least 2005 to 2007. Once all of the upgrades to the system have been completed the overall effect to the system will include:

- Increased resistance to passive interference and active jamming
- Increased detection of low RCS targets
- Passive day and night acquisition with long range IR and television cameras
- Application of IFF target identification system in the Mark XII mode 4 NATO standard
- Use of advanced spare parts allowing the supply of replacement spare parts necessary for normal operations
- Introduction of advanced methods and algorithms for digital data processing
- Enhanced radio electronic camouflage ECCM by application of radar sector blanking system
- Application of digital scaler for 2P25M launcher
- Elimination of adjustments and tuning for upgraded systems
- 2P25M launcher growth capability to launch state-of-the-art fire-and-forget missiles
- Integration of Dehumidification System
- Air conditioned crew cabin.

In July 2000, the Indian MoD placed a USD200 million contract with Poland's state-owned agency for defence imports and exports, Cenrex Trading Ltd, to modernise its existing inventory of OSA-AKM and Kvadrat self-propelled SAM systems. The work was completed by 2002 and involved upgrading transmissions by fitting new power packs, fitting new digital fire-control systems and replacing radars, signalling equipment and radio receivers. The Cenrex contractors I then integrated both systems into the Indian Army's field combat control centres.

1S91M radar for the Kub system 0099545

Tracked TEL Kub system 0080452

Status

In service with the Polish Army. However, there are plans already in process to replace all Kub systems in Poland with the NASAMS. By mid-2013 the planned upgrade to the Polish air-defence systems utilising the NASAMS was well underway, and it is expected that this will happen sometime in 2014/15.

Contractor

Polish Military Workshops.
Radwar Ltd (IFF system).

9K33 OSA-AK Polish Upgrade Programme

Type

Self-propelled Surface-to-Air Missile (SAM) system

Development

As part of its integration programme into NATO, the Polish Army initiated an upgrade programme for some of its 4th anti-aircraft regiment OSA-AK SAM systems, which are earmarked for use with NATO's Rapid Reaction Force. The upgrade is based on the fitting of Identification Friend or Foe (IFF) interrogator units, which are compatible with NATO air defence networks. The ISZ-01 IFF is a medium range system that is also being fitted to upgrades of the Polish Kub and Loara-G.

The upgraded system known as the SA-8P Sting will benefit from:

- Targets identification in Mark XII mode 4, (NATO) standard upgrade Mark XIIA mode 5, mode S
- Radio electronics camouflage by reducing radio location radiation to minimum as a result of application of passive opto-electronics target detection and tracking equipment (opto-electronics head, video tracker, laser range-finder)
- Increase of ECM by the application of the digital radar signal analysis system
- Application of advanced highly integrated parts allowing the supply of replacement parts necessary for normal operation
- Increase of the detection of low RCS targets
- Precise position determination by applying the inertial land navigation system supported by GPS
- Possibility of adapting the combat vehicle for launching a new generation of missiles of the type fire-and-forget (IR based)
- Introduction of advances methods and algorithms for digital processing and imaging of radio location signals and images from the thermal infrared cameras and TV cameras
- Elimination of adjustments and tuning.

Description

The OSA-AK was accepted into service with the Polish Army in 1980. It is designed for engaging air targets flying with speeds up to 1100 km/h at altitudes ranging from 25 through 5,000 m with a slant range from 1,500 to 10,000 m. The Polish system comprises of:

- 9A33BM2 combat vehicles
- 9M33M2 missiles
- 9V210BM3 technical service vehicles
- 9V242-1 control and measuring stations
- 9V914-aerodrome control tower with additional accessories.

In July 2000 the Indian MoD placed a USD200 million contract with Poland's state-owned agency for defence imports and exports, Cenrex Trading Ltd, to modernise its existing inventory of OSA-AKM and Kvadrat self-propelled SAM systems. The work was completed in 2002 and involves upgrading transmissions by fitting new power packs, fitting new digital fire-control systems and replacing radars, signalling equipment and radio receivers. The Cenrex contractors will then integrate both systems into the Indian Army's field combat control centres.

Polish upgraded Osa radar systems (IHS/Patrick Allen) 1149221

Rear view of Polish upgraded Osa system (IHS/Patrick Allen) 1149222

Status

Polish OSA-AK upgrade in service with the Polish Army. OSA-AKM upgrade undertaken for Indian Army completed in 2002. Poland is expected to replace all the Osa systems held in the country with the NASAMS.

According to assumptions of Poland's Land Force headquarters up to four battalions (20 batteries) of Poland's Kub systems should be modified to the Kub-M standard between 2013-2017, while four battalions of Osa-P are being upgraded under a programme due to end in 2018.

Contractor

Enquiries to Poland's state-owned agency for defence imports and exports Cenrex Trading Ltd.
Radwar Ltd (IFF system).

Poprad platform

Type

Self-propelled Surface-to-Air Missile (SAM) system

Development

Development of the Poprad anti-aircraft missile platform was driven by the need for a more mobile and controlled man-portable missile system. The Poprad is based on the Polish Grom family of missiles, which are themselves heavily influenced by the Russian Igla family of man-portable weapons. The Poprad pedestal can be mounted on a standard Iveco 4 × 4 truck and is completely capable of independent or coordinated defence of ground forces or areas of high value. The system is designed to take on targets such as low flying helicopters, Unmanned Aerial Vehicles (UAVs) and cruise missiles.

Description

Grom-2 missile

Production of the PZR Grom-2 (lit., Thunder) system is based entirely on Polish technology. The Gamrat Company located at Jaslo produces the propellant for the motor. The flight-control section is produced by the PZL Warszawall Company and a mobile, fully automated test unit, based on the all-terrain Tarpan Hunta chassis, has been developed by the Military Research Institute WITU of Zielonka and the WZU-2 Workshop at Grudziadz. The Grom-2 missile incorporates:

- New thermal batteries;
- New seeker unit with improved detector cooling;
- Miniature electronics (for example the fuze assembly has been reduced in size by some 30 per cent);
- Larger, more lethal warhead (due to a reduction in the fuel section size).

Poprad launcher (IHS/Patrick Allen) 1170683

Poprad

Poprad is a high-mobility, short-deployment and short-range anti-aircraft system capable of destroying air targets at low and middle altitudes, using homing guidance missiles of the GROM family or other MANPAD types. It is suited for inter-operation in an automated air defence command-and-control system, from which the data of targets to be engaged are acquired via digital links.

The tracking head makes it possible to engage fast and manoeuvring targets.

A Forward Looking Infrared (FLIR) camera has been employed for target acquisition and tracking, enabling effective use of the system in day and night conditions.

An IFF interrogator has been incorporated into the tracking head to minimise the possibility of mistakable engagement of friendly aircraft, which gives Poprad the possibility of operating more autonomously in the battlefield.

Small dimensions and light weight give the advantage of easy relocation of the unit at long distances with any means of transport. The launcher is based on an Iveco van chassis, its mechanical design makes it easily adaptable for installation on other types of vehicles.

The main components of the base carrier include:
Hydraulically erected tracking and aiming head, which is constituted by:

- Electro-optical sensors (thermal camera and laser range-finder);
- IFF interrogator;
- Four short-range missile launchers;
- Two-axis drive system.

Fire-control computer;

- Land navigation and north alignment system;
- Communication and data transmission equipment;
- Power supply system, including power generator.

Other features of the system include:

- High effectiveness of fire due to automatic target tracking (video tracker);
- Target engagement in day and night conditions;
- Capability of autonomous operation and integration in automated C2 systems;
- Low probability of intercept due to entirely passive target acquisition and tracking;
- Long time battery-powered operation, conducive to stealthy performance;
- Optional use of the IFF interrogator to avoid friendly fire casualties;
- High mobility;
- Short deployment time from the cruising trim;
- Erase of relocation with any transport means;
- Possible use of other type of missiles.

Variants

Kobra

In July 2005, Indonesia placed an order with the Polish Bumar Group for the mobile system known in Indonesia as Kobra, which is the Grom 2 missile, mounted in the same way as with the Poprad system on a mobile platform with four RADWAR Poprad anti-aircraft systems.

Kobra was developed by CNPEP RADWAR to protect static areas of high value such as command centres, air bases etc. while it can be also used at sea on small vessels in order to give air defence and a limited surface-to-surface capability. This surface-to-surface capability can also be used on land against soft skinned targets and light armoured vehicles.

The Kobra system is comprised of:

- 1 × lightweight S-band N-26B mobile Multi-beam Search Radar (MMSR);
- 2 × WD-95 Battery Command Vehicles (BCVs);
- 4 × Poprad mobile anti-aircraft systems;
- 14 × ZUR-23-2KG-I gun/missile anti-aircraft systems (each with a twin Grom 2 man portable SAM launcher);
- 1 × TR-23-2 simulator.

Each BCV can remotely control up to six ZUR-23-2KG-I and six Poprad platforms.

The N-26B radar has a range of up to 50 km and incorporated high electronic counter-countermeasures.

Zubr-P

In September 2008, CNPEP RADWAR showed for the first time the variant Zubr-P at the MSPO Defence exhibition in Kielce. The designation 'P' stands for Przeciwlotniczy which is translated as anti-aircraft.

Poprad and 4 × 4 Iveco truck (Grzegorz Holdanowicz) 0578259

The Zubr-P combines four Grom short-range man-portable missiles, TV and Forward Looking Infra-Red (FLIR) cameras with automatic target tracking, a laser range finder and the IKZ-02 short-range interrogator of the RADWAR's IFF Suprasl system.

Mounted on the AMZ-Kutno vehicle is the RADWAR N-26A/MMSR short-range E/F-band air surveillance radar. The system can engage targets less than 60 seconds from the march at a suitable firing position. Air-situation data can be received while on the move, by Radmor RRC-9500 or RRC-9311AP VHF radios, and analysed on a RADWAR REGA terminal that is mounted on the crew compartment next to the driver. This terminal can also be dismounted and it can control the system from a remote location several metres away from the base vehicle.

RADWAR N-26A/MMSR is already in service with the Indonesian Kobra systems and belongs to the family of radars also used by Polish shore-based naval units.

Specifications

	Iveco Grom-2 chassis
Dimensions and weights	
Crew:	2
Mobility	
Configuration	
running gear:	wheeled
Firepower	
Armament:	4 × Poprad Grom-2 SAM

	Grom-2
Dimensions and weights	
Length	
overall	1.596 m (5 ft 2¾ in)
Diameter	
body	72 mm (2.83 in)
Weight	
launch	10.5 kg (23 lb)
Performance	
Speed	
max speed	1,263 kt (2,340 km/h; 1,454 mph; 650 m/s)
Range	
min	0.3 n miles (0.5 km; 0.3 miles)
max	3.0 n miles (5.5 km; 3.4 miles)
Altitude	
min	10 m (33 ft)
max	3,500 m (11,483 ft)
Ordnance components	
Warhead	1.82 kg (4.00 lb) HE fragmentation
Guidance	passive, IR (2-band, lock-before-launch)
Propulsion	
type	solid propellant (sustain stage)

Status

Development complete and in use with the Polish and Indonesian armed forces. In 2009, it was released that Georgia has purchased four Poprad systems with 74 missiles.

Contractor

CNPEP RADWAR SA, Warsaw, Poland.
Zakłady Metalowe MESKO SA (missiles).

S-125 Neva-System Cyfrowy (Neva-SC)

Type

Self-propelled static Surface-to-Air Missile (SAM) system

Development

This modification to the old Soviet S-125 Pechora was developed by the Aviation and Armament Facility of the Military Academy of Technology Wojskowa Akademia Techniczna (WAT) in cooperation with Wojskowe Zakłady Elektroniczne (WZE), the S-125 Neva-SC (also known as the Pechora SC) is a locally modified highly digitised, mobile version of the original Russian-supplied S-125 Neva (Pechora) system supplied to Poland. The missile and overall system performance remain essentially the same (see entry in the Towed and static surface-to-air missile systems section). The first two squadrons (that is to say batteries) were deployed in early 1999 with the 1st and 26th Air Defence Brigades. These then become the backbone of the Polish ground-based air defence forces.

Description

Modification works is carried out by the Military Electronic Workshops Wojskowe Zakłady Elektroniczne (WZE) at Zielonka on behalf of the WAT. The original three 5P73 quadruple S-125 launchers of a Neva squadron have been reworked and installed on T-55 MBT family member chassis for mobility and survivability reasons. The first systems utilise either the BGT-67 armoured bridge-layer or WZT-1 armoured recovery vehicle chassis. Once these vehicle sources have been depleted then original T-55A MBT chassis will be used.

By switching to this mobile configuration, a Neva-SC squadron is said to be able to move from its firing position in considerably less than 30 minutes. Brochures issued by WZT refer to the system as the Pechora SC, Pechora being the export name designated to the system by the then Soviet Union, the original suppliers of the S-125 system.

The squadron's original SNR-125M (NATO 'Low Blow') tracking radar and operator's compartment have been installed on a heavily modified MAZ-543 8 × 8 chassis. These were formerly Polish Army R-17 (NATO SS-1 'Scud') tactical ballistic missile Tractor-Erector-Launcher (TEL) vehicles. The SNR-125M is equipped with a Radwar licence-built French Thomson-CSF IFF interrogator. The Neva-SC squadron has also been equipped with modern built-in training devices.

Digitisation and its associated electronics reduced a squadron's manpower requirements by 65 per cent and the number of vehicles required from 19 to 8. The latter is achieved by the loss of the P-18 3D-air surveillance radar (four trucks and trailers) and by switching the tracking radar to the truck-mounted configuration (where six trailers were originally needed). A Polish Neva-SC squadron receives its air defence picture from the integrated Polish air defence network, which uses modern PIT Institute 3-D radars.

Features as reported by WZT of the upgraded Pechora SC are:

- Digital technology and increased Mean Time Between Failure (MTBF)
- High hit probability
- Simple operation and maintenance
- High resistance against electronic warfare countermeasures
- High mobility (deployment and redeployment within minutes)
- Reduced operational crew (three officers)
- Reduced power consumption
- Command-and-control based on information of air situation from different sources
- Reduced number of components of system
- Possibility of operation within automated or autonomous command as well as the possibility of current exchange of information between radar systems of different Pechora Firing Units

Neva-SC launcher based on T-55 family chassis 0049389

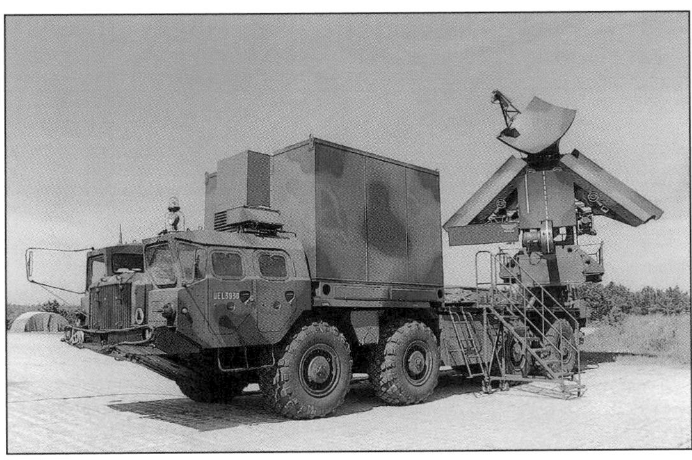

Hydraulically operated SNR-125M radar system and operator's cabin on modified MAZ-543 (8 × 8) cross-country truck chassis 0049390

- High Level of automation in command procedure
- Possibility of operation by un-leveled antennas and launchers
- Possibility of wide training scope
- Simple training of operation crew.

Upgrades to the basic S-125 system continue apace in Poland including incorporation of a new ZNO-X digital transmitter/receiver block developed by the Przemyslowy Instytut Telekomunikacji (PIT). The original magnetron hardware has been replaced using digital technology, including digital frequency synthesis. In 2009 it was revealed that the SNR-125 radar has now been mounted on an 8 × 8 truck and has been fully modernised with new hardware and software. This is now part of the system known in Poland as the S-125D. Additional equipment such as transporter/reloader vehicles, field generators, missile transporters and acquisition radars including new IFF are also part of the new S-125D.

Status
In production. In service with the Polish Air and Air Defence forces. Offered for export. It is known that the Czech Republic, Egypt and Hungary are interested in a similar modernisation of their S-125 Neva systems. Egypt has since chosen the RFAS (CIS) Pechora-2 upgrade package.

Poland is looking to replace all of the ageing ex-Soviet equipment in the not too distant future, most likely with equipment from western sources.

Contractor
Designed by the Aviation and Armament Facility of the Military Academy of Technology, Wojskowa Akademia Techniczna (WAT).
Modification work carried out by the Military Electronic Workshops, Wojskowe Zakłady Elektroniczne (WZE).

Romania

2K12 Kub/Kvadrat Maintenance and Repair Programme

Type
Overhaul and maintenance services of anti-aircraft systems

Development
As one of its major activities, RomArm has developed its own maintenance and repair facilities at its missile development and maintenance Electromechanical Factory in Ploiesti. The ex-Soviet 2K12 Kub (Kvadrat, export name) also known within NATO as the SA-6 Gainful, is known to have been repaired and maintained at Ploiesti. The Kub is a self-propelled tracked surface-to-air missile system designed and developed for the protection of the army at the Forward Edge of the Battlefield (FEBA).

Other systems that have been associated with this organisation include:
- CA-95 self-propelled air-defence system
- S-75 Volkhov air-defence system
- S-125 Neva air-defence system
- Osa-AKM self-propelled air-defence system.

Status
Refurbishment as required. Some refurbished vehicles are known to be in service with the Romanian Army. Offered for export.

Contractor
Electromecanica Ploiesti SA.

A-95

Type
Self-propelled Surface-to-Air Missile (SAM) system

Development
The Romanian Electromecanica Company has reverse engineered the Russian designed 9K31 Strela-1 (NATO SA-9 Gaskin) low-altitude surface-to-air missile system and has manufactured this on the locally developed TABC-79 (4 × 4) armoured personnel carrier chassis.

Description
Like the Strela-1, the Romanian system has a one-man turret with four missiles in the ready-to-fire position. The missiles are carried in launcher-containers, with two missiles on either side of the turret. Once these have been expended, the launcher-containers have to be discarded and new ones loaded manually.

When travelling, the missile containers are retracted flush with the top of the hull to reduce the overall height of the system. The vehicle is fully amphibious and is fitted with an NBC system.

The missile is designated A-95 and is similar to the 9M31M Russian missile. The data provided is from Romanian sources and may differ in some places from the information given for the original Russian missile.

The seeker used on the missile is built by Avionics Enterprise Aerofina and is designated 'co-ordinator A-95'. It is 127 mm in length, but when it is mounted in the missile, 13 mm is contained within the fuselage. Maximum diameter is 107 mm and unit weight is 1.45 kg.

The unit is an opto-mechanical-electrical system, operating in the 1.2 to 2.5 μm InfraRed (IR) waveband region. The optical system is mounted behind a 50 mm diameter hemispherical transparent dome and has a one degree field-of-view with a gimbal capability of ±38°. Seeker output takes the form of an electrical signal that indicates the angular speed of the missile-to-target sight line.

Specifications

	A-95
Dimensions and weights	
Length	
overall	1.803 m (5 ft 11 in)
Diameter	
body	120 mm (4.72 in)
Flight control surfaces	
span	0.36 m (1 ft 2¼ in) (wing)
Weight	
launch	30.5 kg (67 lb)
Performance	
Speed	
cruise	972 kt (1,800 km/h; 1,118 mph; 500 m/s)
Range	
min	0.3 n miles (0.5 km; 0.3 miles)
max	2.3 n miles (4.2 km; 2.6 miles)
Altitude	
min	30 m (98 ft)
max	3,500 m (11,483 ft)
Ordnance components	
Warhead	3 kg (6.00 lb) HE fragmentation
Guidance	passive, IR (operating in 1.2-2.5 μm waveband region)

Electromecanica A-95 low-altitude SAM system on a locally manufactured TABC-79 4 × 4 armoured vehicle chassis with conventional non-imaging IR sensor equipped missile above 0080425

Status
This weapon system is no longer in production, however limited numbers may still be held in reserve for service with the Romanian Army. To date, there have been no known export orders.

Contractor
Electromecanica Ploiesti SA.
Romtehnica (enquiries).

ADMS Weapon Station for CA-94 and CA-94M

Type
Self-propelled Surface-to-Air Missile (SAM) system

Development
In response to a Romanian Army request to provide a night engagement and improved target acquisition capability for the CA-94 and CA-94M man-portable SAM systems, the Israeli company Rafael Armaments Development Authority (now known as Rafael Advanced Defense Systems Ltd) in cooperation with Electromecanica SA that is a limited company and part of the National Company ROMARM, developed a mobile target-acquisition and launching system. This is known as the Air Defence Mobile System (ADMS) Weapon Station.

Description
A four-round pedestal launcher assembly is mounted on a light 4 × 4 vehicle chassis. Co-located with the launcher is the electro-optic all-weather and night-capable target acquisition sensor package that is based on a forward-looking infrared (FLIR) unit. The system was successfully trialled in September 1995, of the five trial rounds fired all hit their designated targets.

This system is sometimes known by ROMARM as the CA-95M, type ADMS integrated with SHORAR-TCP a self-propelled air defence system modernised version of the CA-95. The CA-95M has been designed to engage air targets flying at low latitudes, in direct visibility conditions and can be integrated into a mixed AA guns/missile system, the system is controlled by the SHORAR surveillance and acquisition radar.

A full description of the CA-94 (licence-built Russian Strela-2M) and CA-94M missile systems are given in the Man-portable surface-to-air missile systems section.

Status
The CA-94 is no longer in production, however, the CA-94M is probably produced on an as and when requirement basis. The CA-94M (CA-95M) is in service with Romanian Army, the CA-94 is not. To date, there have been no overseas sales of any of these weapon systems.

Contractor
Missile and self-propelled system
Electromecanica Ploiesti SA.

Enquiries
ROMARM S.A.

CA-95M

Type
Self-propelled Surface-to-Air Missile (SAM) system

Development
The CA-95M self-propelled air defence system is a modernisation of the CA-95 system both of which have been designed to destroy air targets flying at low altitudes in direct visibility conditions.

CA-95M launcher vehicle (Electromecanica) 1139163

SHORAD radar (Electromecanica) 0509831

The original CA-95 is itself a reverse engineered system based on the Soviet 9K31 Strela-1 (NATO SA-9 Gaskin) that was/is mounted on a BRDM-2 vehicle.

The CA-95M was mounted on a locally produced TABC-79 (4 × 4) armoured personnel carrier chassis.

In January 2008, the Romanian joint-stock company Electromecanica Ploiesti teamed up with the Israeli organisation Elbit System to develop a new mobile multirole launcher for the CA-95M. This new launcher known as Platforma Deschisa de Lansare a Rachetelor AA si Sol - Sol, is mounted on the ML-A95 launch vehicle with Romanian modified missiles designated as A95M-RC.

Description
The early CA-95M is mounted on a 4 × 4 TAB modified chassis that has the IL-A95M launching installation equipped with a special interface. The CA-95M can be integrated into a mixed AA guns/missiles system, controlled by the SHORAR surveillance/acquisition radar and Gun Star Night (GSN) fire-control units.

The modified CA-95M system consists of the ML-A95M launching vehicle with the A95M-RC missiles in their self contained launching canisters. The ML-A95M launch vehicle also boasts a surface-to-surface missile mounting for the STAR-80L guided rocket.

For maintenance and training purposes the manufacturer also offers the IV-A94M self-propelled checking installation and the IA-95M training installation.

CA-95M type ADMS integrated with SHORAR-TCP
Tracks and engages targets indicated by PC.
Modes:
- Broadcasting
- By indication
- Autonomous.

Assures:
- Calculation of the target coordinates to the ADMS position
- Self location on the tactical field
- Quick reaction time.

Tactical advantages:
- Permits the organisation of a mixed air-defence unit with Oerlikon GDF-103 guns
- Extend the efficiency air defence range of the unit
- Offers on move air coverage for AD unit.

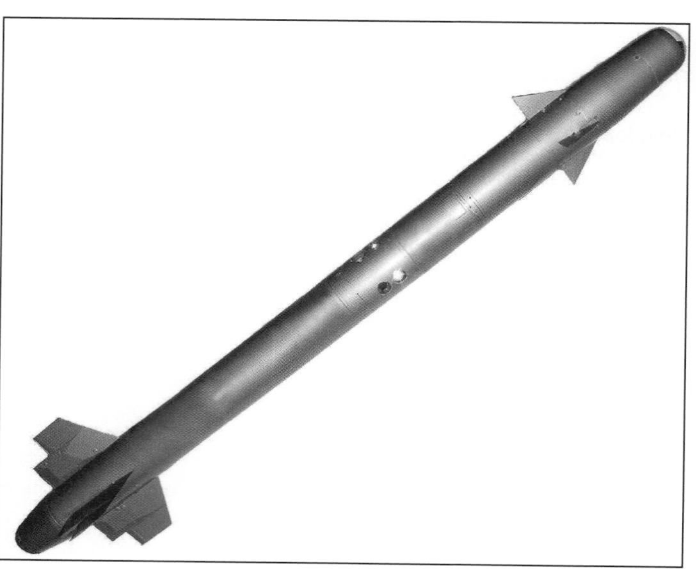

CA-95M missile (Electromecanica) 1139156

Specifications

	ML-A95M launching vehicle
Dimensions and weights	
Crew:	3
Mobility	
Configuration	
running gear:	wheeled
layout:	4 × 4
Firepower	
Turret power control	
type:	electric
Main armament traverse	
angle:	360°
speed:	36°/s
Main armament elevation/depression	
armament front:	+80°/-5°
speed:	18°/s

	A95M-RC
Dimensions and weights	
Length	
overall	1.803 m (5 ft 11 in)
Diameter	
body	120 mm (4.72 in)
Weight	
launch	30 kg (66 lb)
Performance	
Range	
min	0.4 n miles (0.8 km; 0.5 miles)
max	2.3 n miles (4.2 km; 2.6 miles)
Altitude	
min	50 m (164 ft)
max	3,500 m (11,483 ft)
Ordnance components	
Fuze	impact, proximity (2.5 m radius of action)
Guidance	IR (visible spectra homing with proportional guidance on 2 channels)
Propulsion	
type	two stage, solid propellant

Status
The early CA-95M system development is complete. This system has seen service with the Romanian Armed Forces. The system was to be phased out in favour of the new modified missile and launch vehicle, that was due to come on line in late 2009, however, no further information on the replacement vehicle was available at the time of writing.

Contractor
Marketing Agency
RomArm SA.

Manufacturer
Electromecanica Ploiesti SA.

Russian Federation

2K11 Krug

Type
Self-propelled Surface-to-Air Missile (SAM) system

Development
The 2K11 (US/NATO SA-4 'Ganef') medium to high-altitude surface-to-air missile system's overall development began in 1957/58 by the OKB-8 GKAT bureau under the direction of head designer LV Lyul'yev. The system was put into the Resolution of the USSR Council of Ministers on 27 March 1956. The requirement for the system included a flight range of 20 km at latitudes from 2 to 12-15 km at speeds up to 600 m/s. These figures were superseded in 1958 when the development of the 2K11 commenced, the new figures included speeds up to 600 m/s, altitudes between 3 and 25 km at ranges of up to 45 km. The supersonic ramjet engine was developed by OKB-670 GKAT. Initially the missile for the system was developed in two versions, both with different guidance systems. The 3M8 missile used a radio command system and the 3M10 used a combined radio-command mid-course guidance and terminal semi-active radar homing. The 3M8 version was finally selected as the missile for the system. The system made its first public appearance at the Moscow Parade in 1964.

The early version of the Krug entered service on 26 October 1964. The first operational deployment version, the Krug-A, entered service in 1967, with extensively modified versions, the Krug-M in 1971 and the Krug-M1 in 1974, which were developed to rectify problems discovered during army service.

A-4b 'Ganef' system with 3M8M2 missiles (Willis A Bullard) 0080426

Reload Ural-375 TZM (6 × 6) truck with integral crane carrying 3M8M1 variant 'Ganef' missile 0509757

Camouflaged 1S32 'Pat Hand' radar used with the SA-4 'Ganef' system from the rear with antenna traversed to rear and datalink raised 0509684

The system, known in the Russian Federation as the ZRD-SD (*Zenitniy Raketniy Kompleks - Srednoye Deistvie*: Anti-aircraft missile system - medium range), *Krug* (Russian for 'circle'), has been used by a number of armies. It is air-portable in the larger transport aircraft such as the An-22.

Polish Air Defence forces intend to use the system for several more years and continue actual launch training on the type. Modifications and upgrades to the Polish system are a continuing programme. The first Krug-M systems were delivered to Poland in 1976. The Polish Army uses the 3M8M3 missile.

The air defence elements for the military district (front) included several Krug brigades. The Krug brigade consisted of a brigade headquarters, three SAM battalions (each with nine ZU-23 twin 23 mm towed anti-aircraft gun systems), SSRTs (*Samokhodnaya Stantsiya Razvedki Tseleukazaniya:* mobile detection and designation radar 1S32 nicknamed 'Long Track' by NATO) and three Krug batteries and a technical battalion with one 1S32 and one N-41 height-finding 'Thin Skin' radar.

Each Krug battery in peacetime had one SSNR (*Samokhodnaya Stantsiya Navendeniya Raket:* mobile missile guidance station) 1S32 missile guidance station, nicknamed 'Pat Hand', three Krug SPU (*Samokhodnaya Puskovaya Ustanovka:* mobile launcher unit) launchers and four Ural-375 TZM

(*Transportno-Zaryazhayushchaya Mashina:* transporter-loader vehicle) reload vehicles each with an integral cradle crane to lift the single missile carried onto the SPU. Additional resupply missiles were carried by the missile technical battery singly on double-axle semi-trailers towed by ZIL-157V or ZIL-131V articulated tractors. On mobilisation for war a fourth launcher may be added to each battery.

The Krug system was normally deployed between 10 and 25 km behind the FEBA and forms part of an overall air defence system incorporating various man-portable, self-propelled and static SAMs with 23 mm, 30 mm and 57 mm anti-aircraft guns. The Krug has been replaced in front line Russian service by the 9K83 and 9K82 (NATO SA-12 'Gladiator/'Giant') brigades.

Description

Each TEL (Russian industrial index designation 2P24, 2P24M for the Krug-M version) consists of a tracked armoured chassis on top of which is mounted a hydraulically operated turntable carrying the two missiles. The chassis was at the time a new design, not based on an existing vehicle. The driver is seated at the front of the vehicle, on the left and has a single-hatch cover in front of which are two periscopes. At the front of the glacis plate is a splash board to stop water rushing up the glacis plate when the vehicle is fording. The engine is to the right of the driver with the remainder of the space in the vehicle taken up by the crew and electronics. There are hatches for the other crew members either side of the missile turntable.

The torsion bar suspension consists of seven dual rubber-tyre road wheels with the drive sprocket at the front and the idler at the rear, and four track-return rollers. Hydraulic shock absorbers are provided for the first and last road wheel stations. The vehicle has an air filtration and overpressure NBC system and an IR night vision system for the commander and driver but no amphibious capability.

The chassis has since been adopted for a number of other roles including the 152 mm 2S3 M-1973 self-propelled howitzer and minelaying. The latter model is called the GMZ and has the mines and minelaying equipment at the rear.

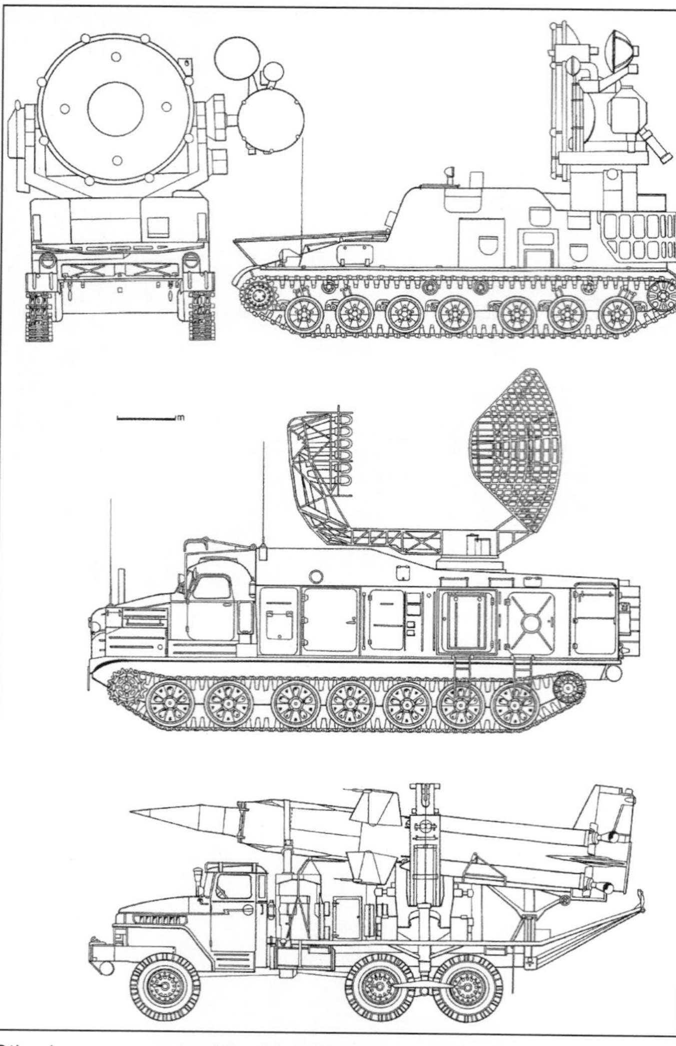

Other key components of the SA-4 'Ganef' air defence system (from top to bottom) include the 'Pat Hand' missile guidance radar, 'Long Track' surveillance radar and the TZM missile resupply vehicle on a Ural-375D (6 × 6) truck chassis. For rail transport, the antenna of the radar system is removed (Steven Zaloga) 0509435

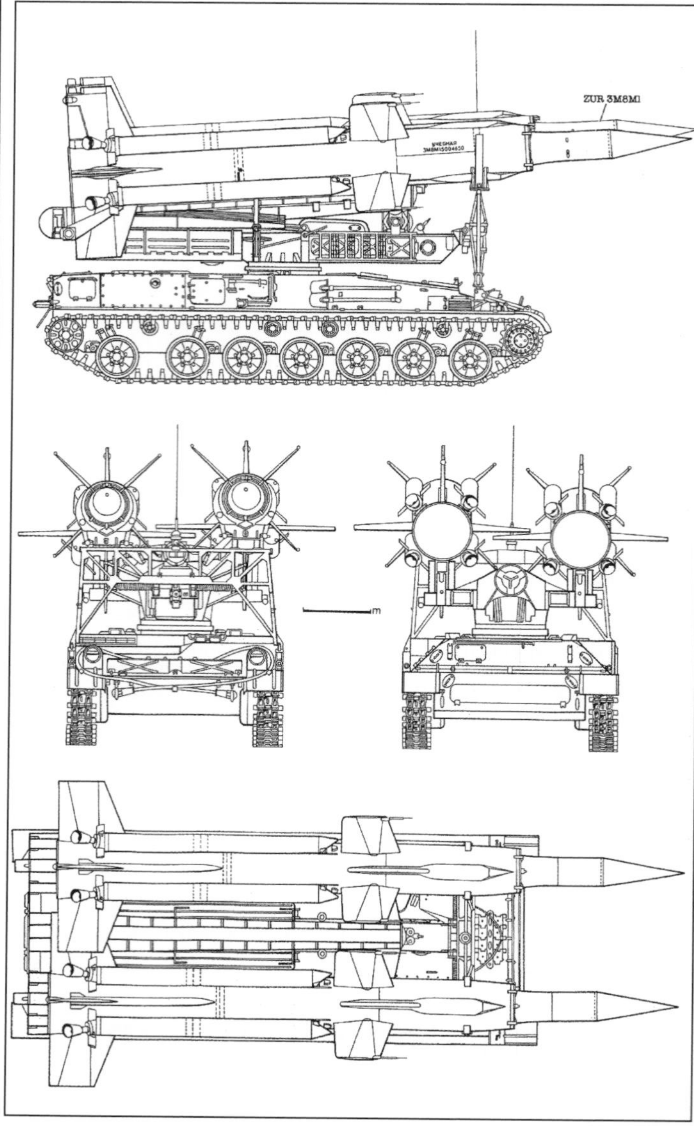

ZRK-SA (Krug) SA-4a 'Ganef' surface-to-air missile system with two-round launcher traversed to front (Steven Zaloga) 0509434

1S12 'Long Track' E-band surveillance radar on AT-T tracked chassis (Terry J Gander) 0080386

'Ganef' 3M8M2 missile on its depot handling trolley from rear clearly showing the four large boosters (Ted Dyke) 0080428

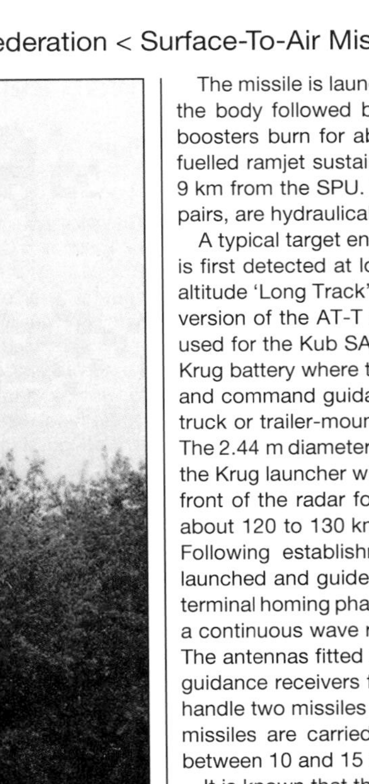

'Thin Skin' height finding radar which is used in a number of air defence systems including the SA-4 0080427

The launcher can be traversed through 360° and the missiles elevated up to an angle of 45° on their launcher arms for launching. Before the missiles can be launched, the rear vertical fins are replaced, the protective coverings of the ramjet air scoop and the various unit nozzles removed, the calliper clamps that hold each missile in the travelling position are released manually and the calliper frame folded forwards. The left missile is carried about 0.25 m above the right one. (Note: for transit purposes the rear vertical fins are removed.)

The missile is launched by four solid booster rockets mounted externally on the body followed by a kerosene fuelled ramjet sustainer. After launch the boosters burn for about 15 seconds and then fall away when the kerosene fuelled ramjet sustainer motor ignition speed of over M1 is attained at about 9 km from the SPU. The four fins are fixed and the four control wings, in two pairs, are hydraulically operated.

A typical target engagement is believed to take place as follows. The target is first detected at long range by a 1S12 175 km range, 30,000 m maximum altitude 'Long Track' early warning radar mounted on a lengthened ATT-426U version of the AT-T heavy artillery tractor with a large van body added (also used for the Kub SAM). 1S12 operates in the E-band and passes data to the Krug battery where the 1S32 H-band 'Pat Hand' continuous wave fire-control and command guidance radar takes over. Either H-band 240 km range N-41 truck or trailer-mounted height-finder radar also provides height information. The 2.44 m diameter 1S32 radar is mounted on essentially the same chassis as the Krug launcher with the whole assembly collapsed flat and a grille raised in front of the radar for road transit. This radar typically acquires the target at about 120 to 130 km and, when it is within 80 to 90 km, tracking can begin. Following establishment of a confirmed hostile track, a single missile is launched and guided to the target by the guidance beam with a semi-active terminal homing phase for the final flight stage. The missile is tracked in flight by a continuous wave radar transponder beacon attached to one of the tail fins. The antennas fitted to the edge of the four forward wings are the semi-active guidance receivers for the terminal homing system. If required, the 1S32 can handle two missiles per target in order to increase kill probability. The reserve missiles are carried on Ural-375 (6 × 6) trucks. Reloading the SPU takes between 10 and 15 minutes.

It is known that there have been as many as five sub-variants of the missile developed by the Lyul'yev design bureau at Sverdlovsk (now known as the Novator NPO, in Ekaterinburg) and believed to be designated 3M8, 3M8M, 3M8M1, 3M8M2 and 3M8M3. External differences between them were minimal, as any improvements were internal. The last two sub-variants remain the predominant types in service. The 3M8M1 is the 8.8 m long-nosed version (the SA-4a) with effective range limits of 9,000 to 50,000 m and effective altitude limits of 250 to 50,000 m. The 3M8M2 is the short-nosed 8.3 m version (SA-4b or 'Ganef' Mod 1). This has an improved close-range performance to reduce the dead zone above the SPU at the expense of losing some altitude. Both versions have a fuselage diameter of 0.86 m, a wing span of 2.3 m and a tail span of 2.73 m. The HE warhead weighs 150 kg and is detonated by a proximity fuze. The missile is armed 300 m from the launcher. The launch weight is stated to be 2,450 kg. A battery is likely to have one SPU fitted with the 3M8M2 and two SPUs with the 3M8M1 missile, although some SPUs were seen carrying one missile of each type. An electro-optical fire-control system is believed to be fitted for use in a heavy ECM environment.

A comparison of the Krug variants' performance capabilities is given in the accompanying table.

Krug Variants

Version	**Krug**	**Krug-A**	**Krug-M**	**Krug-M1**
IOC date:	1965	1967	1971	1974
Maximum target speed:	800 m/s	800 m/s	800 m/s	800-1,000 m/s
Engagement range:				
(minimum)	11,000 m	9,000 m	9,000 m	6-7,000 m
(maximum)	45,000 m	50,000 m	50,000 m	50,000 m
Target altitude:				
(minimum)	3,000 m	250 m	250 m	150 m
(maximum)	23,500 m	23,500 m	24,500 m	24,500 m
Missile launch weight:	2,450 kg	2,450 kg	2,450 kg	2,450 kg
Warhead weight:	150 kg	150 kg	150 kg	150 kg
Maximum missile speed:	800-1,000 m/s	800-1,000 m/s	800-1,000 m/s	800-1,000 m/s

	3M8M	**3M8M1**	**3M8M3**	**3M8M2**
Dimensions and weights				
Length				
overall	8.784 m (28 ft 9¾ in)	8.784 m (28 ft 9¾ in)	8.784 m (28 ft 9¾ in)	8.3 m (27 ft 2¾ in)
Diameter				
body	860 mm (33.86 in)	860 mm (33.86 in)	860 mm (33.86 in)	860 mm (33.86 in)
Flight control surfaces				
span	2.73 m (8 ft 11½ in) (wing)	2.73 m (8 ft 11½ in) (wing)	2.73 m (8 ft 11½ in) (wing)	2.73 m (8 ft 11½ in) (wing)
Weight				
launch	2,453 kg (5,407 lb)	2,453 kg (5,407 lb)	2,453 kg (5,407 lb)	2,453 kg (5,407 lb)
Performance				
Speed				
max speed	2,138 kt (3,960 km/h; 2,461 mph; 1,100 m/s) (est.)	2,138 kt (3,960 km/h; 2,461 mph; 1,100 m/s) (est.)	2,138 kt (3,960 km/h; 2,461 mph; 1,100 m/s) (est.)	2,138 kt (3,960 km/h; 2,461 mph; 1,100 m/s) (est.)
Ordnance components				
Warhead	150 kg (330 lb) HE fragmentation	150 kg (330 lb) HE fragmentation	150 kg (330 lb) HE fragmentation	150 kg (330 lb) HE fragmentation
Fuze	impact, proximity	impact, proximity	impact, proximity	impact, proximity
Guidance	semi-active, radar (command)	semi-active, radar (command)	semi-active, radar (command)	semi-active, radar (command)
Propulsion				
type	4 solid propellant	4 solid propellant	4 solid propellant	4 solid propellant
	kerosene, ramjet (sustain stage)	kerosene, ramjet (sustain stage)	kerosene, ramjet (sustain stage)	kerosene, ramjet (sustain stage)

Variants

If required, the Krug system and the 3M8 missile and its variants can be re-engineered into a supersonic missile target. A programming device is installed in the 1S32 'Pat Hand' to control Virazh and Virazh-M target missiles developed by the Novator design bureau. The manufacturers of Krug, the Kalinin Mechanical Engineering Plant, also offer a complete support and maintenance capability to its operators.

The Military Armament Works No2 in Grudziadz, Poland is offering an upgraded/modified version of the 2K11 Kurg system that includes a 1S32M2 fire-control radar and 2P24M2 launcher. The reported effects of the upgrade include:

- Increased resistance to passive and active interference
- Increased detection of low RCS targets
- Application of IFF target identification system in the Mark XII Mode 4 standard (NATO)]
- Application of advanced parts allowing the supply of replacement parts necessary for normal operations (digital systems of large scale of integration, integrated analogue systems, powerful semiconductor systems, etc.)
- Application of a modern long-range television imagery (consisting of a highly sensitive CCD camera and a computer-controlled LCD indicator), improving passive tracking of the target and radio electronic camouflage
- Reduced failure frequency of the equipment due to the application of highly reliable upgraded blocks and systems
- Elimination of time-consuming procedures of adjustments and tuning for all upgraded systems
- Introduction of advanced methods and algorithms for digital data processing.

Specifications

	2P24 SPU
Dimensions and weights	
Crew:	5
Length	
overall:	9.46 m
hull:	7.5 m
Width	
overall:	3.2 m
Height	
to top of roof mounted weapon:	4.472 m
Ground clearance	
overall:	0.44 m
Track	
vehicle:	2.66 m
Track width	
normal:	540 mm
Length of track on ground:	5 m
Weight	
combat:	28,200 kg
Mobility	
Configuration	
running gear:	tracked
Power-to-weight ratio:	17.33 hp/t
Speed	
max speed:	35 km/h
Range	
main fuel supply:	780 km
Fuel capacity	
main:	850 litres
Fording	
without preparation:	1.5 m
Gradient:	30%
Vertical obstacle	
forwards:	1 m
Trench:	2.5 m
Engine	V-59, V-12, water cooled, diesel, 520 hp
Firepower	
Armament:	2 × roof mounted SAM
Survivability	
Armour	
hull/body:	15 mm

Status

Production in Russia is complete. However, as of November 2009, Poland still has the system in its front-line inventory, but has planned to replace the system with a version of the US Patriot. The Polish Krug systems have been modified and upgraded in order to allow these to work with NATO forces, particularly with regards to NATO IFF standards. In August 2011 the last of the Krug systems attached to the 61st City of Skwierzyna Air Defence Regiment in Poland were retired. The system was first delivered to Poland in 1977.

Contractor

Designed by NPO Novator (Novator Design Bureau, OKB Novator, OKB Lyulev) that is now part of the Antei Scientific Industrial Organisation (now known as Almaz/Antei Concern of Air Defence).
Manufactured by the Kalinin Mechanical Engineering Plant (at Ekaterinburg).

2K12 Kub

Type

Self-propelled Surface-to-Air Missile (SAM) system

Development

I I Toropov's OKB-134 design bureau at Tushino began development of the Kub (NATO SA-6 'Gainful' Russian Industrial index number, 2K12) low-altitude surface-to-air missile system in 1959. It was completed by a design team under Yuriy N Figurovskiy.

Initial service entry was in 1967 and it was not seen in public until the 1967 Moscow Parade. The system for export, known as the *Kub* (Russian for 'cube') in Russia and the 2K12E *Kvadrat* (Russian for 'quadrant') entered full operational service in 1970, after a prolonged and troubled development and trial period. It is air-portable in the An-22 and Il-76 aircraft.

Series production of the 2K12 began at the Ulyanovsk Mechanical Plant in 1967 while first deliveries of the 2K12 Kvadrat to export customers began in 1971. The Kalinin Mechanical Engineering Plant also manufactured the 2K12 system.

The first known use in action was by Syria and Egypt during the 1973 Middle East war when the Kvadrat version proved highly effective against Israeli aircraft during the first few days of the war. The SA-6 forced aircraft to fly very low where they would encounter the ZSU-23-4 self-propelled anti-aircraft gun. It has been widely deployed, not only with the former Warsaw Pact, but also with many other countries that have received Russian aid.

Subsequent use of the Kub in combat has included the war between Iraq and Iran, the Syrian-Israeli missile crisis in Lebanon during 1981 and the Israeli invasion of Lebanon in 1982. Both Egypt and Libya have used it by both the Polisario Front and Algeria in border skirmishes with Morocco and during their 1977 seven-day border war. Libya used it during the 15 April 1986 bombing attack by the US and against French aircraft in the battles in northern Chad during 1986 and 1987. It has also been used by Angola against South African aircraft with negative results and was used extensively by Iraq in the two Gulf Wars and their aftermath. In 1994 and 1995 it was used by the Bosnian Serbs against NATO aircraft over flying Bosnia-Herzegovina shooting down a USAF Lockheed Martin F-16 jet.

Its latest usage was in the Kosovo/Serbian NATO campaign. Although some 70 odd 2K12 Kvadrat launchers were delivered to Yugoslavia in the 1970s, combat usage in the various local conflicts reduced these to some 25

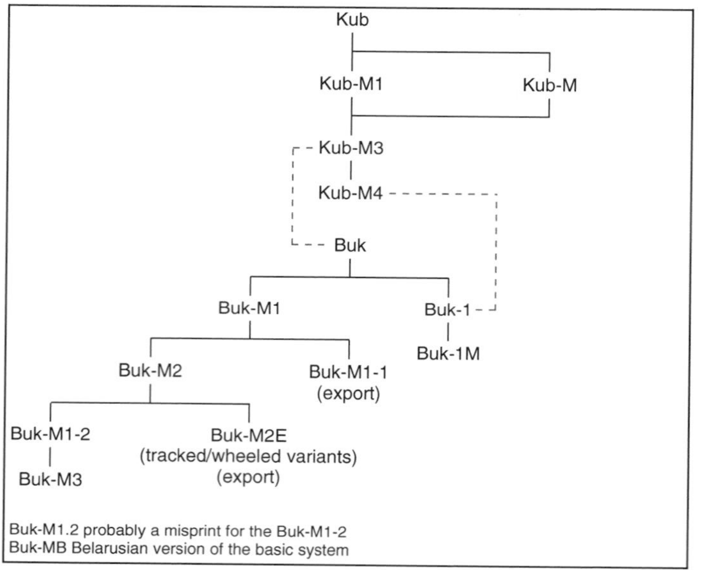

Kub family flow diagram (James C O'Halloran) 1040815

SA-6 'Gainful' SAM system with launcher traversed to front (Willis A Bullard) 0080452

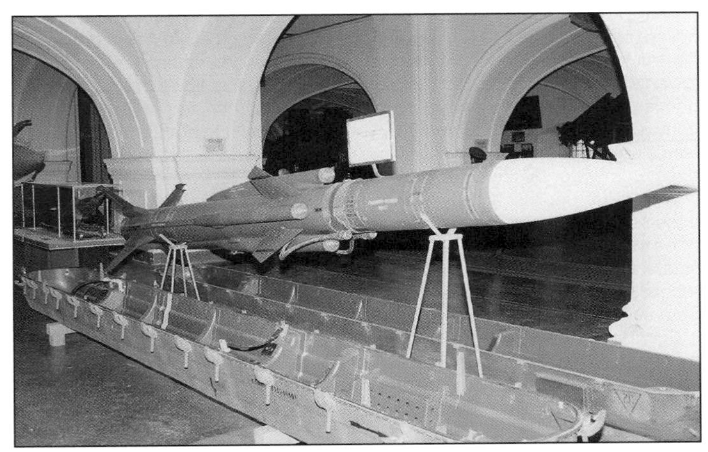

SA-6 'Gainful' surface-to-air missile with fins and wings attached (E Dyke)
0080393

distributed among four-air defence regiments of the field armies. Although some what elderly, with some probably having undergone the SDPR upgrade (see entry this section), the 2K12 systems still proved to be the most elusive of the Yugoslav air defence assets with few systems being destroyed during NATO Suppression of Enemy Air-Defence (SEAD) activities.

In army (corps) level air-defence units, the Kub is still sometimes found at divisional level, in the anti-aircraft regiment. Each SA-6 regiment consists of a regimental headquarters, a target acquisition battery (with one 'Thin Skin-B', a 'Score Board-A' IFF radar, one 'Spoon Rest' radar, one 'Flat Face' radar and one 'Long Track' radar) and five Kub batteries. Each battery in peacetime consists of one 1S91 'Straight Flush' radar vehicle, four Kub SPU launcher vehicles and four ZIL-131 TZM reload vehicles each with a large hydraulic crane centrally located on the tailboard and three reserve missiles. There are a further two Ural-375 crane trucks and 15 ZIL-131V or ZIL-157V articulated tractors towing double-axle semi-trailers carrying six resupply rounds each in the missile technical battery of the regiment. On mobilisation for war each battery would receive two further launchers, normally kept in storage. Other users normally have only four launchers per battery.

The improved 2K12M1 (Kub-M1) entered service in 1973, the 2K12M2 (Kub-M3) in 1976 and the 2K12M4 (Kub-M4) in 1978.

Description

Developed by the Astrov KB Design Bureau, the Kub SPU chassis is related to the ZSU-23-4 self-propelled air defence system and the ASU-85 self-propelled anti-tank gun. The SPU (Russian industrial index designation 2P25) is constructed of all-welded steel with the crew compartment at the front, missiles on the turntable immediately behind the crew compartment and the engine at the rear. The driver is seated at the front of the hull on the left side with the vehicle commander to the right. Both have a large windscreen to their front, which can be covered by a single-piece hatch cover hinged at the top, which opens on the outside. When the hatches are closed, forward observation is via periscopes mounted in the forward part of the crew compartment. On the glacis plate is a splash plate to stop water rushing up the front of the vehicle when fording.

The engine and transmission are at the rear of the hull. The torsion bar suspension system consists of six rubber-tyre road wheels with the drive sprocket at the rear and the idler at the front. There are no track-return rollers. Hydraulic shock absorbers are provided for the first and last road wheel stations. The SPU has a full PAZ/FVU air filtration and overpressure NBC system and infrared night vision equipment fitted as standard but the vehicle has no amphibious capability.

Three Kub missiles are carried on a turntable that can be traversed through a full 360° with the missile elevated on their launchers to a maximum of +85°. When travelling, the turntable is normally traversed to the rear and the missiles are horizontal to reduce the overall height of the vehicle.

SA-6 'Gainful' low-altitude surface-to-air launcher in firing position 0080392

1S91 'Straight Flush' radar system in close-up 0099543

It is estimated that the Kub (NATO designation 'Gainful' Mod 0) missile has a length of 5.8 m, body diameter of 0.335 m, wing span of 1.245 m, tail span of 1.524 m and has a launch weight of 630 kg with a 57 kg HE fragmentation warhead. The proximity and contact fuzes are armed after some 50 m of flight. The 3M9 missile family was developed by the Lyul'yev Design Bureau (now known as the Novator NPO).

The missile has an integral ramjet/rocket propulsion system. The latter accelerates the missile for 4.1 seconds generating 8,600 kg of thrust and a specific impulse of about 240 seconds after launch to a speed of about M1.5. The solid propellant booster rocket nozzle at the rear is then jettisoned and a solid-fuel ramjet with a much larger nozzle propels the missile. A solid-fuel gas generator with its own oxidising system feeds it. Air is introduced via the four ducts just in front of the centre body wings, maximum speed then reaches approximately M2.8 with a specific impulse of 1,200 seconds and maximum thrust of 1,540 kg. Total sustainer burn time is 22.5 seconds. The missile airframe can stand a maximum turning force of 15 *g* and a linear acceleration of 20 *g*. It is capable of 15 *g* maximum sustainable manoeuvres and is controlled by cruciform centre body wings with ailerons for roll control. Tail fins carry I-band mid-course command link receiver antennas and G/H-band beacon transmitter antennas. These antennas were not observed on the first version of the Kub seen in the 1960s (Russian industrial index designation for missile, 3M9). Later missile variants are believed to include the 3M9M, 3M9M1, 3M9M2, 3M9M3, 3M9M3E, 3M9M3A and 3M9MA. Terminal homing is of the semi-active radar type. Maximum target engagement speeds are 600 m/s for an approaching target and 300 m/s for a receding target.

The initial Kub version of the Kub missile had the following performance characteristics: minimum effective range 6 to 8,000 m, maximum effective range 22,000 m, minimum effective engagement height 100 to 200 m, maximum effective engagement height 7,000 m. In the latter system versions these figures were improved. The accompanying table gives an overview of the differences in system performances. Reload missiles are carried on modified ZIL-131 (6 × 6) trucks and are loaded manually onto the launcher by a crane carried on the rear of the loader vehicle. Reloading an SPU takes approximately 10 minutes.

The 'Straight Flush' (1S91) vehicle has a similar chassis to that of the Launcher. The radar has a range of 55 to 75 km and a 10,000 m altitude capability depending upon conditions and target size, and performs limited search, low-altitude detection/acquisition, pulse Doppler IFF interrogation, target tracking and illumination, missile radar command guidance and secondary radar missile tracking functions. The vehicle also carries the fire-control computers for the missile battery. Some modified 'Straight Flush' vehicles have been observed with a TV camera of 30 km range to enable the battery to remain in action even if the vehicle's radar are jammed or forced to shut down by the threat of anti-radiation missiles. 'Straight Flush' can also be

1S91 'Straight Flush' radar vehicle in travelling configuration 0099542

A Slovak missile crew lift a 3M9ME Kub round into the launcher (Miroslav Gyürösi) 1155620

linked to the launch vehicles by either a radio datalink or a 10 m long cable for direct data input to the launcher's systems. The datalink antenna is carried on the right forward hull corner of the SPU.

The upper foldable 'Straight Flush' 28 km range dish antenna is of the conical scanning type and is used for low-altitude H-band sector search scans, target tracking and target illumination. The lower parabolic antenna is of the G-band 55 to 75 km range medium-altitude target acquisition and early warning radar type, with the lower feed for medium- to high-altitude coverage and the upper feed for low-altitude coverage.

A typical engagement takes place as follows. The surveillance radar acquires potential targets at ranges of up to 275 km. The 'Thin Skin' radar and all the data are sent to the regimental operations centre either by landline or radio datalink confirm the target height. The operations van interprets the radar data and selects the appropriate battery to engage the targets. Data on the selected target are passed to a regimental signals van and then on to the firing battery command-and-control centre truck. This is usually located near to the battery's 'Straight Flush' engagement radar vehicle and connected to it by a landline. The data are then passed on to the 'Straight Flush'.

In most situations the 'Straight Flush' does not start transmitting until the target has been acquired and allocated by the regimental headquarters. Radar silence is maintained as long as possible to prevent compromise of the location of the battery to threat ELINT locating systems. The 'Straight Flush' can begin target acquisition at its maximum range of 75 km, and begin tracking and illumination at the 28 km mark. The 'Straight Flush' can only illuminate a single target and control three missiles at any one time, so Russian practice when a target track had been initiated was normally to order the launch of two and sometimes three missiles from one or more SPUs. These were tracked using their G/H-band tail fin tracking beacons to monitor the trajectories.

The 'Straight Flush' fire-control computer uses these signals to generate course corrections and each weapon is then guided on an intercept course via control transmission commands sent to the reference antenna located on the lower left tail fin. The missile's terminal attack phase is flown in a semi-active homing mode with the seeker homing in on the reflected energy from the continuous wave illumination radar. The course flown in both phases is of the lead pursuit type. At the intercept point the warhead is detonated either by direct contact or by the Doppler radar proximity fuzing system.

With radar up, reaction time from a dormant condition through the target acquisition, IFF interrogation and lock on phases to missile launch is about three minutes. If the radar vehicle is already active then the time taken for the sequence is reduced to between 15 to 30 seconds. A battery is able to become mobile and relocate to an alternative firing position in approximately 15 minutes from systems being shut down.

Egyptian Kub systems have had their auxiliary power unit turbines replaced by Garrett Turbine Engine Company GTP-30-150 APUs that supply up to 75 kW for system operations on the 'Straight Flush' SSNR (25 kW required) and SPU (35 kW required). The missiles themselves have also been renovated with Chinese and US assistance.

Variants

In 1977, the SA-6b (the NATO designation is 'Gainful' Mod 1), the 3M9M1/3M9M2/3M9M3 family of missiles plus the 9M38 entered service mounted on the 9A38 launch vehicle from the Buk system (SA-11 'Gadfly'). The SPU carried three SA-6b missiles and an associated 'Fire Dome' H/I-band missile guidance illuminator radar fitted on the front end of the launcher assembly. The SA-6b was initially deployed on the basis of one SPU per SA-6a battery, as an interim system until the complete SA-11 'Gadfly' system was fielded. By January 2007, this arrangement of Kub/Buk (now known as the Kub-M4) has been withdrawn from Russian service but is still offered for export upgrade by Rosoboroneksport.

The Ulyanovsk Mechanical Plant has been involved in a custom designed upgrade for one of the Kvadrat customers. This essentially improves the system's resistance to jamming and its overall efficiency. The modifications include improving the performance of the TV optical sight, the performance of the target illumination radar and the performance of the moving target selection system. It also introduces a lock-on capability for both current and older generation SA-6 missiles. Further upgrades developed by Ulyanovsk are detailed in the appropriate entry.

India contracted Poland to upgrade its 2K12E systems used by the Indian Army. The 2K12E systems are divided between its two Corps of Air Defence Artillery (CADA) SAM Groups, the 501st and 502nd, which each operate several regiments of Kub ('Gainful') and Osa ('Gecko') systems.

Poland, Romania and the former Yugoslavia have maintenance/upgrade programmes in place for the 2K12 system.

The Czech Republic Company Retia, has modernised the SAM systems held in their inventory and is now offering upgrades to the overseas customers. Testing by the Czech Armed Forces took place between 1 August and 19 October 2007 and by late November, the first of the systems was handed over to the 25th AA Missile Brigade in Strakonice. This modernisation programme includes:

- New solid-state RF units in the control radar SRC and tracking guidance radar SN (Straight Flush)
- Digital signal and data processing substitutes original analogue SRC unit
- Multi-function air-situation display
- Tactical communication (Link 1, Link 11B)
- Integrated IFF Mark 12 interrogator
- Integrated navigation GPS receiver.

This modernised system (also referred to as SURN reconnaissance and guidance radar post variant 1) also offers built-in on-line and off-line diagnostics, gathering and display, remote guidance of the SN allowing to set aside KPC vehicles.

Rosoboronexport is also offering what has been described as a First Stage Air Defence Missile System Modernisation for the Kvadrat, this consists of changes in the SURN reconnaissance and guidance radar post as follows:

- Replacement of the analogue moving target selection system with a digital one, featuring a suppression coefficient increased up to 28-30 dB
- Introduction of the tracked target classification system
- Extension of the illumination channel waveband from six to 12 lettered frequencies
- Replacement of electro-vacuum UHF amplifiers with solid-state ones, including substitution of their high-voltage power units for low-voltage ones, as well as introducing new electronic elements

Kub Variants

Version	Kub	Kub-M1	Kub-M3	Kub-M4
IOC date:	1967	1973	1976	1978
Maximum target speed:	600 m/s	600 m/s	600 km/s	600 m/s
Engagement:				
(minimum)	6-8,000 m	4,000 m	4,000 m	4,000 m
(maximum)	22,000 m	23,000 m	25,000 m	24,000 m
Target altitude:				
(minimum)	100 m	30 m	20 m	30 m
(maximum)	7,000 m	8,000 m	8,000 m	14,000 m
Missile launch weight:	630 kg	630 kg	630 kg	630 kg
Warhead weight:	57 kg	57 kg	57 kg	57 kg
Maximum missile speed:	600 m/s	600 m/s	600 m/s	600 m/s

- Replacement of the cathode-ray tubes of display system with colour LCDs, which notably increases the amount of life up to 10,000-15,000 hours, and also reduces power consumption and the number of operating adjustments.

Upgrades to the self-propelled launcher vehicles include:

- Introduction of testing and monitoring system providing real-time recording with subsequent playback of all data on operation of the ADMS major elements including radar, launcher and missiles.

The ADMS is equipped with a testing and measuring equipment set providing integral check-up of radar/launcher electronic equipment.

Czech Aspide Modification/Upgrade

In early March 2012, the Czech MoD announced that it has accepted the MBDA Aspide missile as the upgrade/modification to the 3M9M3 missile currently in use on the Czech Kub system. Furthermore, Retia a Czech Company in Pardubice has demonstrated a new missile guidance system solving the replacement of the original Soviet produced 3M9M3 missiles in support of the MBDA Aspide. These reports have suggested that the control system had to be completely changed for its integration into the Kub system. Apparently MBDA has cooperated fully with Retia on these changes since the conception of the upgrade in 2009.

There can appear to be little doubt that these control and missile changes will prolong the life span of the original Soviet Kub system and will be offered as upgrades to other countries that still have the original system within their inventories; by now any surviving Kub systems will most certainly require upgrades or modifications in order to bring the weapon into the 21st Century and capable of surviving on the modern day battlefield.

Specifications

	2P25TEL
Dimensions and weights	
Crew:	3
Length	
overall:	7.389 m
hull:	6.79 m
Width	
overall:	3.18 m
Height	
overall:	3.45 m
hull:	1.8 m
Ground clearance	
overall:	0.4 m
Track	
vehicle:	2.67 m
Track width	
normal:	360 mm
Length of track on ground:	3.8 m
Weight	
combat:	14,000 kg
Ground pressure	
standard track:	0.48 kg/cm^2
Mobility	
Configuration	
running gear:	wheeled
Power-to-weight ratio:	17.14 hp/t
Speed	
max speed:	44 km/h
Range	
main fuel supply:	260 km
Fuel consumption	
road:	0.96 litres/km
Fuel capacity	
main:	250 litres
Fording	
without preparation:	1.1 m
Gradient:	60%
Side slope:	30%
Vertical obstacle	
forwards:	1 m
Trench:	2.5 m
Engine	V-6R, in line-6, water cooled, diesel, 240 hp at 1,800 rpm
Gearbox	
type:	manual
forward gears:	5
reverse gears:	1
Steering:	clutch and brake
Suspension:	torsion bar
Electrical system	
vehicle:	24 V
Batteries:	2 × 12 V, 100 Ah
Firepower	
Armament:	3 × hull mounted SA-6 SAM
Survivability	
Armour	
hull/body:	9.4 mm

	Kub-M1
Dimensions and weights	
Length	
overall	5.8 m (19 ft 0¼ in)
Diameter	
body	335 mm (13.19 in)
Flight control surfaces	
span	1.245 m (4 ft 1 in) (wing)
Weight	
launch	630 kg (1,388 lb)
Performance	
Speed	
max speed	1,166 kt (2,160 km/h; 1,342 mph; 600 m/s)
Range	
min	2.2 n miles (4 km; 2.5 miles)
max	12.4 n miles (23 km; 14.3 miles)
Altitude	
min	100 m (328 ft) (radar mode)
	30 m (98 ft) (optical mode)
max	8,000 m (26,247 ft)
Ordnance components	
Warhead	57 kg (125 lb) HE fragmentation
Fuze	impact, proximity
Guidance	semi-active, radar
Propulsion	
type	ramjet (sustain stage)

Status

Production complete, (between 1967 and 1983, over 500 batteries were completed for use with the Soviet Union and the armed forces of 22 other client states).

Contractor

Design: Tikhomirov Instrument Research Institute.

Manufacture: Ulyanovsk Mechanical Plant and the Kalinin Mechanical Engineering Plant at Yekateninburg.

2K12 Kub M4 (M3/Kvadrat M3 system upgrade)

Type

Self-propelled Surface-to-Air Missile (SAM) system

Development

In 1995, the M4 upgrade of the 2K12 Kub (SA-6/'Gainful'), self-propelled system was revealed and offered for the export market. The export variant of the Kub is known as the Kvadrat. The original programme involved the integration of a Buk-M1 system (SA-11 'Gadfly') 9A310M1 self-propelled launcher unit, with attendant 9A39M1 launcher-loader into the basic 2K12-battery structure. The system with such an upgrade is known as the Kub-M4 and entered limited service with the Russian Army in 1998.

Description

The Kub-M4 upgrade doubles the engagement capability of the Kvadrat battery from one target to two and significantly complicates the defence suppression countermeasures required to negate the battery because of the different radar and missiles involved.

It also allows for the possibility of a missile 'ambush', where the attacker is misled into seeing only the Kub battery electronic characteristics and is then surprised by a Buk launcher engagement instead.

Russia is offering an upgrade package for the SA-6 'Gainful' SAM system that includes adding elements of the SA-11 'Gadfly' family 0080429

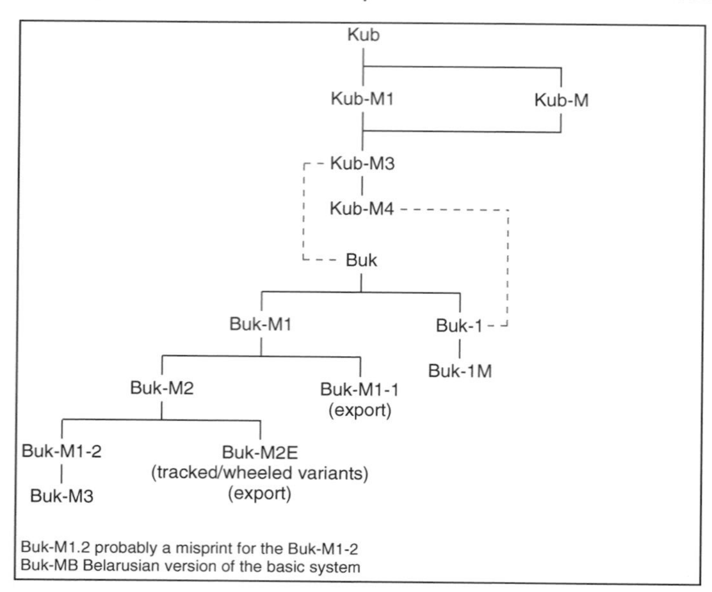

Kub family flow diagram (James C O'Halloran) 1040815

The enhanced combat performance characteristics are achieved by:

- Introduction of the Buk-M1-2 air defence missile system components into the Kvadrat missile system complement after their modification to interface with the Kvadrat
- Modernisation of the 1S91M2(M1) control, reconnaissance and guidance system by:
    - Introduction of digital moving target discrimination systems
    - Discrimination of target classes
    - Extension of the illumination channel frequency range
    - Replacement of microwave amplifiers built around travelling wave tubes by solid-state transistor amplifiers
    - Introduction of a coded telemetry channel to maintain communication with the command-and-control post.
- Replacement of command-and-control assets and an outdated pool of radar stations by modernised and new generation radar with the use of the 9S470M1-2 command post of the Buk-M1-2 missile system and the PORI-P2VM radar processing post.

The Kub-M3 modernisation to the Kvadrat missile system makes it possible for the weapon system to establish two options of interaction of missile battery combat assets and command-and-control posts:

(1) Six organic missile batteries with the modernised 1S91M1 (M2) can be controlled over new coded telemetry communications lines from the modernised 9S470M1-2 command post. The 9S470M1-2 receives radar data via the PORI-P2VM post and the modernised P-18 and P-19 radars, 9S18M1-1 radar and an organic PRV altimeter.

(2) Three organic missile batteries with modernised 1S91M1 (M2) are controlled over new coded telemetry communications lines from a modernised 9S470M1-2. These three batteries (each comprising one 9A310M1-2 and one 9A39M1) are controlled via standard telemetry communications lines (9S624 and 9S625) from the 9S470M1-2. The 9S470M1-2 receives radar data via the PORI-P2VM post from new-generation Nebo-SV and Kasta-2E2 radars or modernised P-18 and P-19 radars, 9S18M1 and an organic PRV altimeter.

(3) This third modernisation involved the integration into the Kub system of the 9A310M1-2 Transporter Erector Launcher And Radar (TELAR) with 9M317 missiles from the 9K37M1-2 Buk-M1-2 system, interfacing this with the 2P25M1 or M2. In its latest form, the integration of the 9A317 TELAR with 9M317 missiles from the 9K37M3 Buk-M2.

Variants

Kub (digitisation of analogue model)

Some years ago it was decided to introduce a digitizing programme into the Kub system to extend the operational life of the system and with a view to overseas sales to those many customers that were then holding the system in their inventory. The primary subsystems that were digitised were the radar and the radar consoles.

Kub/Aspide Czech MoD upgrade

In early March 2012, the Czech MoD announced that it has accepted the MBDA Aspide missile as the upgrade/modification to the 3M9M3 missile currently in use on the Czech Kub system. Furthermore, Retia a Czech Company in Pardubice has demonstrated a new missile guidance system solving the replacement of the original Soviet produced 3M9M3 missiles in support of the MBDA Aspide. These reports have suggested that the control system had to be completely changed for its integration into the Kub system. Apparently MBDA has cooperated fully with Retia on these changes since the conception of the upgrade in 2009.

There can appear to be little doubt that these control and missile changes will prolong the life span of the original Soviet Kub system and will be offered as upgrades to other countries that still have the original system within their inventories; by now any surviving Kub systems will most certainly require upgrades or modifications in order to bring the weapon into the 21st century and capable of surviving on the modern day battlefield.

Status

The original Kub system is no longer in production. The system has been withdrawn from Russian Air Defence service and replaced with the Buk. However, the Kvadrat is still offered as an export upgrade to the older Kub systems by Rosoboronexport.

In December 2008, sources from Moscow indicated that the Kvadrat (Kub/NATO SA-6 Gainful) system deployed in Iran is currently undergoing an upgrade/update from the Almaz Central Design Bureau. This upgrade includes deliveries of the 9M317E surface-to-air missile. This missile is associated with the Buk-M1-2 and it is therefore assessed that the upgrade to the Kvadrat is that of the Kub-M4.

Contractor

Enquiries: Rosoboronexport.
Design: Tikhomirov Scientific Research Institute of Instrument Design.
Manufacture: Ulyanovsk Mechanical Plant.

9K31 Strela-1

Type

Self-propelled Surface-to-Air Missile (SAM) system

Development

Developed by the then Nudelman OKB-16 design bureau (now the Moscow-based KB Tochmash Design Bureau of Precision Engineering) in parallel with the ZSU-23-4. The first Strela-1 (NATO SA-9/'Gaskin') launchers were produced in 1966 with the system attaining operational status in 1968. The Strela-1 low-altitude clear-weather surface-to-air missile system's first recorded combat use was in May 1981.

The system was issued to the anti-aircraft batteries of RFAS (CIS) motorised and tank regiments on the basis of four Transporter-Erector-Launcher (TEL) per battery to give a total of 16 per division. Replacement in the RFAS (CIS) was by the Strela-10 (SA-13/'Gopher') system.

The Strela-1 was also organic to the Russian Naval Infantry air defence batteries, with four launchers and four ZSU-23-4s assigned. These systems were used to supplement the air defences of amphibious warfare vessels by being deployed on deck in launch configuration. The Russian Naval Infantry has also replaced the Strela-1 with the Strela-10.

Although the system is no longer in use within the Russian Armed Forces, it is still widely used by many export nations as there is little or nothing that can go wrong with the system. As long as the missiles are regularly maintained there is little else that can go wrong except mechanical components on the BRDM amphibious launch vehicle.

Description

The system consists of a 9P31 BRDM-2 based Transporter-Erector-Launcher (TEL) with the normal turret and chain-driven belly wheels removed and replaced by a turret with four ready to launch SA-9 container-launcher boxes. These are normally lowered to the horizontal when travelling to reduce the overall height of the vehicle.

The original version of the Strela-1 missile was known as the 9M31 (NATO SA-9 'Gaskin' Mod 0) and used a 9E41 uncooled first-generation lead sulphide (PbS) infrared seeker operating in the 1 to 3 μm waveband region. This was supplemented by the 9M31M variant (Strela-1M/SA-9b/'Gaskin' Mod 1) which entered service in 1970 and has an improved 9E41 PbS seeker operating in the 1 to 5 μm waveband region to provide greater target sensitivity and lock on ability. When engaging a head-on target the system has a considerably reduced range. The SA-9 is fitted with a 3 kg HE fragmentation warhead and proximity fuze. The warhead has a lethal radius of 5 m and damage radius of 7.6 m.

SA-9 'Gaskin' SAM system based on BRDM-2 (4 × 4) amphibious chassis with four missiles ready to launch 0044113

BTR-60PU-12 command vehicle is used by a range of Russian air defence systems including the SA-9 0080430

SA-9 'Gaskin' surface-to-air missile out of its launcher box (E Dyke) 0080431

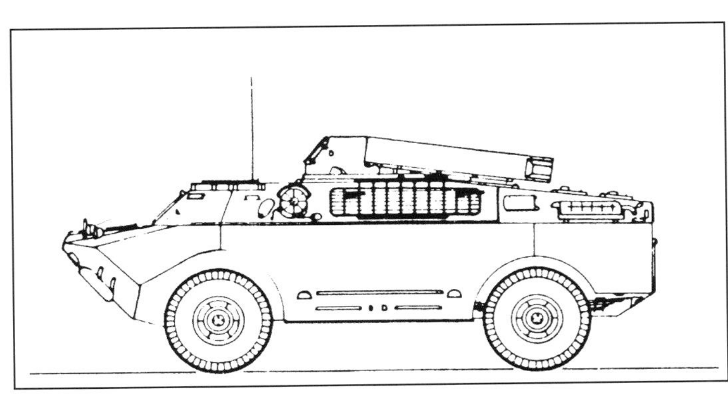

SA-9 'Gaskin' SAM system based on BRDM-2 chassis, with launching arms and missiles in travelling position 0509443

An overview of the different system performances is given in the accompanying table.

One SA-9 TEL, usually the battery commander, (SA-9 Mod A, BRDM-2A1 or SA-9A TEL) in each battery has been fitted with 9S16 Flat Box-A. The Flat Box A is a radar detection antenna, one is fitted either side of the hull above the front wheel housings, one under the left launch canisters pointing forward and one mounted on a small frame above the rear engine deck plate pointing rearwards. The azimuth scan sector for the system is 360°, the elevation capability 40° and the maximum detection range 30 km. The US Army refers to the TEL with no Flat Box-A system as the SA-9 Mod B, BRDM-2A2 or SA-9B. The vehicle crew of three includes the commander, driver and gunner. There is an infra-red system for the commander and driver to use at night. An air-filtration and overpressure NBC system is fitted as standard.

In the RFAS (CIS) armies the BTR-60PU-12 command vehicle of the Strela-1 unit is usually alerted by the Divisional Air Defence Regiment's command post as to a potential target's azimuth, range and altitude. This information is assessed with additional data from the unit's own visual observer network and any microwave transmissions picked up by the Strela-1 'Flat Box' TEL. The commander then instructs the unit as to which target should be engaged by which vehicle and orders the engagement(s) to commence.

The designated TEL moves to its predetermined launch site using its on board basic Inertial Navigation System (INS) located near to the vehicle commander's station. All the commander has to do to use the system is insert the co-ordinates of the SPU's 'hide' position, the co-ordinates of the pre-selected launch site and then read off the bearing and distance information calculated by the INS.

Once there, the gunner manually directs the one-man turret to the desired azimuth bearing and elevates the launcher assembly. He then selects the missile(s) to be used and acquires the target through an optical sight mounted behind a plexiglass window at the base of the turret. Once the target is acquired, the gunner presses the launch button to its initial stop position. This opens the designated missile's canister front door and unmasks the IR seeker. When lock on by the IR seeker is achieved, an aural signal sounds in the turret and the gunner depresses the launch button to its second and final stop position. This fires the missile's dual-thrust solid propellant rocket motor.

In combat the missiles are usually sequentially fired (two per target) to increase the kill probability with a time between rounds of about five seconds. Reloading is performed manually and takes about five minutes to accomplish. Operating in the visual and near IR regions the Strela-1 can home in on any air target from any direction in daytime since its tracking ability relies on a contrast (either positive or negative) with the background. At night the weapon homes in on the targets own radiation and are therefore limited to a tail chase engagement only.

SA-9 'Gaskin' SAM system in travelling configuration with missiles folded down 0509442

SA-9 'Gaskin' low-altitude SAM system showing 'Flat Box-A' passive radar detectors either side of hull front 0509441

Variants

The Romanians have licence-built the SA-9 'Gaskin' using the locally designed TABC-79 (4 × 4) APC chassis as the platform (see separate entry).

Specifications

	TEL
Dimensions and weights	
Crew:	3
Length	
overall:	5.8 m
Width	
overall:	2.4 m
Height	
transport configuration:	2.3 m
Ground clearance	
overall:	0.43 m
Track	
vehicle:	1.84 m
Wheelbase	3.1 m
Weight	
combat:	7,000 kg
Mobility	
Configuration	
running gear:	wheeled
layout:	4 × 4
Power-to-weight ratio:	20 hp/t
Speed	
max speed:	100 km/h
water:	10 km/h
Range	
main fuel supply:	750 km
Fuel capacity	
main:	290 litres
Amphibious:	yes

	TEL
Gradient:	60%
Vertical obstacle	
forwards:	0.4 m
Trench:	1.2 m
Engine	GAZ 41, V-8, water cooled, petrol, 140 hp at 3,400 rpm
Firepower	
Armament:	4 × hull mounted SA-9 SAM
Ammunition	
missile:	6
Main armament traverse	
angle:	360°
speed:	20°/s (est.)
Main armament elevation/depression	
armament front:	+80°/-20°
Survivability	
Armour	
hull/body:	14 mm

	Strela-1	Strela-1M
Dimensions and weights		
Length		
overall	1.803 m (5 ft 11 in)	1.803 m (5 ft 11 in)
Diameter		
body	120 mm (4.72 in)	120 mm (4.72 in)
Flight control surfaces		
span	0.36 m (1 ft 2¼ in) (wing)	0.36 m (1 ft 2¼ in) (wing)
Weight		
launch	30 kg (66 lb) (est.)	30.5 kg (67 lb)
Performance		
Speed		
max speed	972 kt (1,800 km/h; 1,118 mph; 500 m/s)	972 kt (1,800 km/h; 1,118 mph; 500 m/s)
Range		
min	0.5 n miles (1.0 km; 0.6 miles)	0.3 n miles (0.5 km; 0.3 miles)
max	2.3 n miles (4.2 km; 2.6 miles)	2.3 n miles (4.2 km; 2.6 miles)
Altitude		
min	50 m (164 ft)	30 m (98 ft)
max	3,000 m (9,843 ft)	3,500 m (11,483 ft)
Ordnance components		
Warhead	3 kg (6.00 lb) (est.) HE blast fragmentation	3 kg (6.00 lb) HE blast fragmentation
Fuze	radar, proximity	radar, proximity
Guidance	IR	IR
Propulsion		
type	single stage, solid propellant	single stage, solid propellant

Status
Production complete. Readers should be aware of the age of the system and take this into account when assessing the operational status of the weapon.

Contractor
Nudelman OKB-16 design bureau (now the KB Tochmash Design Bureau of Precision Engineering) (design).
Soviet state factories (manufacturer).

9K33 Osa

Type
Self-propelled Surface-to-Air Missile (SAM) system

Development
Development of the 9K33 Osa (NATO SA-8 Gecko) and Osa-M (Russian for 'wasp') short-range air defence missile system started in 1960 as a joint Navy-Ground Forces project. The Osa was the Ground Forces Project while the Osa-M was planned for the Navy. The missiles were developed according to unified tactical and technical requirements and there were no essential differences in their design. The lead developer for the acquisition and guidance system was NII-20 of the GKREh (State Committee for Radio Electronics).

The executive agency for the missile was the KB-82 (Moscow) while the ground components (including the launcher) were the responsibility of the SKB-203 (Central Ural SOVNARKhOZ). The naval mounting TsKB-34 (Leningrad), the sustainer propellant NII-9, the booster-propellant NII-6, the warhead NII-24, the tracking and power drive systems TsNII-173 (Kovrov) and the radio fuzing system NII-571.

The development programme was protracted with major redesigns of both the missile and launch platform required. Extensive range testing of the Osa for the Ground Forces was conducted at a test range in Kazakhstan in 1965 where many of the faults of the original system were discovered. The modified system

9T217BM2 resupply vehicle with forward part of roof covering folded to rear to show hydraulic crane that is used to lift three new missiles in their launch containers. Note that the stabilisers are also lowered to the ground 0509439

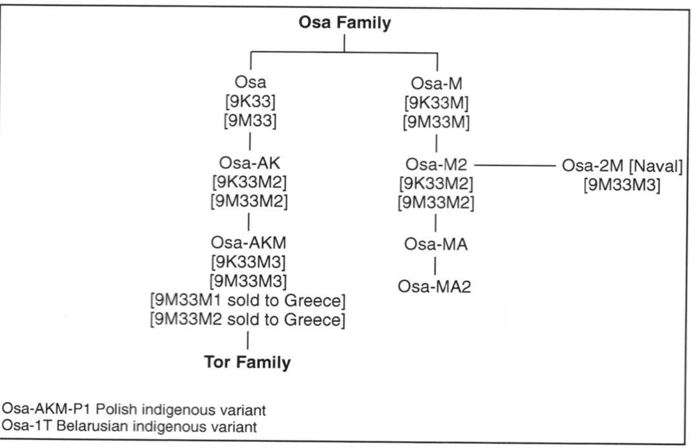

Osa Family flow diagram (James C O'Halloran) 1040814

was redesignated Osa-M with most of the individual systems having to be redesigned and also many of the developing agencies changed. For instance in 1964 the OKB-2 of the GKAT (State Committee for Aviation Technology) replaced KB-82 as the executive agency in charge of development, while the naval mount development was transferred to TsKB-7 (AO 'Arsenal' in Leningrad) in 1963. Another major change was the shift to the BAZ-5937 chassis.

The Russian Army accepted the OSA-M for service in 1972. Series production of the system began in 1971. Trials of the naval system Osa-M began in 1967 on board the Project 33 experimental vessel OS-24 (formerly the cruiser, *Voroshilov*). As with the Ground Forces system, a number of design faults were uncovered that necessitated further research and refinement of the system. In 1973, the OSA-M2 SAM system was accepted for service with the Navy, and was installed on the vessels of Projects 1135, 1134B, 1135-1, 1143, 1144, 1234 and others. Modernisation of the system now called OSA-MA began in 1975. The minimum altitude was lowered in the updated system from 60 m to 25 m.

Sea trials of the OSA-MA were held on a Project 1124 small ASW vessel (tactical number MPK-147) in the Black Sea. In 1979, the OSA-MA was accepted into service. Meanwhile, in the mid-1970s a limited programme was undertaken to enable Osa to engage not only helicopters in the hover but also helicopters on the ground with rotors running. This entered operational service in 1975 as the 9K33M2 OSA-AK.

In 1980, a further major modification entered service as the 9K33M3 OSA-AKM. This extended the engagement capability to all forms of low flying air targets such as aircraft, helicopters and remotely piloted vehicles. The OSA-AK and OSA-AKM have the US NATO designation SA-8b ('Gecko' Mod 1).

The OSA-AKM (and earlier variants) is issued on the scale of 20 fire units per divisional anti-aircraft regiment in place of the S-60 57 mm towed anti-aircraft gun.

The organisation of a SA-8 regiment is:

- A regimental headquarters
- A target acquisition and early warning battery with one 'Long Track', one 'Flat Face' and a 'Thin Skin-B' radar
- A missile support battery
- A transport company
- A maintenance company
- Five SA-8 batteries.

Each battery has four SA-8 SPU launcher vehicles and two 9T217BM2 TZM reload vehicles each with 12 reload rounds. The TZMs are supported by 24 ZIL-131 (6 × 6) cargo trucks used as missile transporter by the regimental Transport Company. The maintenance and repair unit uses the 9V242-1 automated mobile test station, the 9V914-adjustment tower, the 9V210M3-maintenance vehicle, the 9F372M3 SPTA-2 vehicle and the 9F632M electronic simulator for crew training. It has also replaced the SA-6 system in some units because of its greater mobility. It is air-portable in the Il-76 transport aircraft.

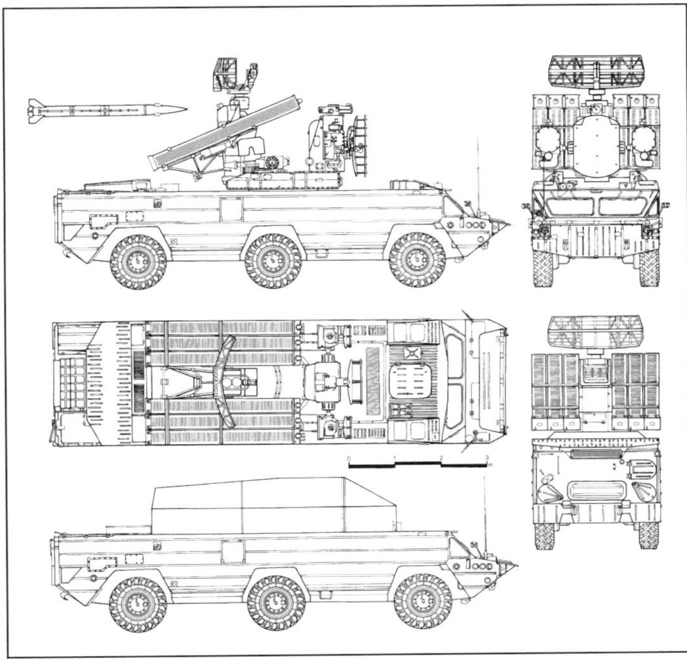

ZRK-SD Osa-AK (SA-8b 'Gecko' Mod 1) and its supporting TZM resupply vehicle (Steven Zaloga) 0509438

The system manufacturers are offering a restoration package for the OSA-AK system to ensure that the system maintains its operational capabilities. An upgrade package is also offered to convert the OSA-AK system to the same operational standard as the OSA-AKM.

In 1992, Greece acquired 12 Osa-AKM launchers from Germany. These form a complete artillery squadron of three batteries based in the eastern province of Thrace, near the border with Turkey. The systems use the 9M33M3-missile variant. In 1999, Greece ordered from Antei 16 further launchers. Greece also holds an option for a further four Osa-AKM systems.

India has asked Poland to upgrade its OSA-AKM units of the Indian Army. Poland has already performed a limited upgrade of its own Army's OSA-AK systems to ensure compatibility with NATO IFF requirements.

In September 2000, the South African Air Force revealed that many of the Angolan air defence vehicles captured intact during the mid 1980s were being used by the service to train combat pilots in air defence tactics. The other systems used include the ZSU-23-4 'Shilka,' the SA-9 'Gaskin' and the SA-13 'Gopher'.

The Czech Republic Company Retia is also offering a modernisation of the SAM system OSA-AKM, details of which were released in 2006.

Description

The Russian industrial index numbers for the Osa family system components are:

- 9A33BM3/9A33BM3-1 - launch vehicle (latest version)
- 9M33/9M33M3 - missile (latest version is 9M33M3)
- 9T217BM2 - transporter-loader vehicle
- 9V210M3 - technical maintenance vehicle
- 9F372M3/9F372M3-1 - vehicle with SPTA set
- 9F16M2 - ground support equipment for servicing the missiles
- 9F632M - electronic simulator for operators
- 9V914 - vehicle with adjusting equipment
- 9F16 - ground equipment set
- 9V242-1 - automatic mobile controls and tests station.

The Osa SPU is a six-wheeled design designated BAZ-5937. It is based on a number of earlier six-wheeled all-terrain vehicles developed by V A Grachev's design team at the Likhachev Automobile Plant (ZIL) in Moscow. The driver's compartment at the front of the vehicle has accommodation for two, the driver and commander, with access to it via a hatch in the roof. There are no other entrance/exit hatches apparent on the vehicle. The engine is at the rear.

Close-up of an SA-8b 'Gecko' launcher assembly 0099539

Russian Army OSA-AKM SA-8b 'Gecko' in travelling configuration with 'Land Role' surveillance antenna retracted 0099538

The vehicle is thought to have torsion bar suspension with steering on the front and rear axles and a central tyre pressure regulation system. Blast shields can be folded down over the windscreens to prevent damage when the missiles are launched. The vehicle is fully amphibious, being propelled in the water by two water-jets at the rear of the hull. Before entering the water, a trim vane, which is folded back onto the glacis plate when not in use, is erected at the front of the hull. The vehicle is fitted with an air filtration and overpressure NBC system together with IR systems for the commander and driver. Six command-guided missiles are carried ready-to-launch, three either side.

The main fire-control radar is at the rear of a one man gunner/radar operator position and folds back 90° to reduce the overall height of the vehicle for air transport and during high-speed road travel. It is known that the radar operates in the H-band with a 360° traverse and has a maximum range of 20 to 30 km (depending upon target type and altitude). Antenna rotation rate is 33 rpm. The complete conical-scan radar installation of the 'Gecko' has been assigned the NATO codename 'Land Role'.

In front of the radar is the guidance group comprising:

- A central monopulse target tracking radar with truncated sides, which operates in the J band and has a range of 25 km.
- Two monopulse missile guidance radar uplink transmitters, one either side of the tracking radar, which have truncated sides and limited traverse and operate in the J band.
- Two command link horns for missile gathering, one either side and below the missile guidance radars, which operate in the I band.
- Two rectangular devices which are believed to assist in tracking in an ECM environment, to the left and right of the missile guidance radars. Mounted on top of each missile guidance radar is a LLLTV/optical assist system for target tracking in low visibility and heavy ECM. Mounted on top of the tracking radar is what is thought to be a feed and below the tracking radar is a periscope, the exact role of which is uncertain. 'Land Role' is also known to have a short range target acquisition capability.

It is known that the two missile guidance radars operate on different frequencies, each controlling one missile in flight. This enables the system to engage a single target with a staggered two missile salvo four seconds apart that operate on different frequencies to avoid guidance problems and degrade the target's ECM capabilities.

The basic 9M33 (SA-8a/'Gecko' Mod 0) high-acceleration missile was designed by the Fakel MKB (formerly called the Grushin design bureau) is powered by a single-stage solid propellant rocket motor. The missile has a launch weight of about 126.3 kg, is 3.158 m long, 0.2096 m in diameter and has a wingspan of 0.65 m. The maximum speed is 500 m/s, minimum altitude is 50 m, maximum effective altitude 5,000 m. The minimum range is 2,000 m and the maximum range 9,000 m. Against an F-4 Phantom target the warhead's lethal radius at low altitude is 5 m and it is fitted with proximity and contact fuzes. The missile self-destructs after 25 to 28 seconds of flight following a signal generated by a clock mechanism.

In 1975, a modified missile the 9M33M (SA-8b or 'Gecko' Mod 1), was introduced into service. Contained in a rectangular launch box it has improved guidance and range characteristics. The booster section of the dual-thrust solid rocket has a two-second firing period while the sustainer section has a 15-second duration. The reloading time is five minutes. Performance was minimum altitude 25 m, maximum altitude 5,000 m, minimum range 1,500 m, maximum range 10,000 m. Missile speed remained the same.

Each battery also has two missile transloader based on the same chassis with a long coffin like blunt pointed tarp roofed structure covering the cargo space and crane. When operating, the blunt point area is raised and the tarped structure is slid to the rear. A total of 18 reloads in boxed sets of three is transferred to the SPUs by the hydraulic crane mounted centrally behind the vehicle cab. In the regiment's maintenance battery there is a single radar collimation vehicle using the same chassis. This has a collimation antenna, which lies on both sides of the vehicle and overhangs the rear during transit. In operation it is raised and mounted on each side of the hull directly behind the cab.

Combat deployment time is four minutes with system reaction 26 seconds. With the advent of the OSA-AKM the missile was further improved as the 9M33M3 to give the same basic performance characteristics but with enhanced flight characteristics to counter modern aerial target sets.

Close-up of 'Land Role' surveillance radar assembly 0099537

SA-8b 'Gecko' system with 'Land Role' surveillance system retracted for travelling and complete turret traversed to rear (Richard Stickland) 0069506

An overview of the different system performance characteristics is given in the accompanying table (Table 1).

Table 1 - Osa variants

Version	Osa-M	Osa-AK	Osa-AKM
IOC date:	1972	1975	1980
Maximum target speed:	420 m/s (approx.)	500 m/s (approx.)	500 m/s
Engagement range:			
(minimum)	1,500 m	1,500 m	1,500 m
(maximum)	9,000 m	10,000 m	10,000 m
Target altitude:			
(minimum)	50 m	25 m	25 m
(maximum)	5,000 m	5,000 m	5,000 m
Missile launch weight:	128 kg	128 kg	127 kg
Warhead weight:	15 kg	15 kg	15 kg
Maximum missile speed:	500 m/s	540 m/s	540 m/s

To train OSA-AKM crews, special training systems have been developed. These are the 2T269 integrated simulator, the 9F632M integrated simulator and the 2U448 radio frequency target and jamming simulator. The 9F632M are contained within a box body mounted on the rear of a ZIL-131 truck chassis. The system is used to train the 9A33 crews to accomplish their operational and fire missions without using real targets or expending actual missiles. The system is designed for use by army units, training centres and military colleges. The 9F632M vehicle has a crew of two, weighs 9,000 kg and is 7.45 m long, 2.4 m wide and 3.41 m high. It takes 30 minutes to deploy.

Antei is also marketing the 9F691 universal training target system. This is designed to enable air defence systems to track and engage small and fast targets flying at altitudes from 500 up to 5,000 m out to a maximum range of 20 km. The main modification is to the 9M33 missile, which has its warhead removed and a range self-destruct capability added. Other minor modifications are made to the missile and the 9A33BM2 launch vehicle. The 9F691 is based on the OSA-AK system variant.

In 1997, Antei revealed the 9F691 Saman universal training target system based on the OSA-AK launcher. The 9F691 is designed to provide a missile target for use with various air defence missile systems. The target is a 9M33 missile with its warhead removed and several internal modifications. It can simulate targets with radar cross-sections between 0.08 and 1.6 m^2, infrared signatures in the waveband range 3 to 5 μm and various single and twin target flight trajectories. The missile target can fly in the altitude range 50 to 5,000 m at distances up to 20 km. The launch vehicle, the 9A33BM2, is used to control the target.

An updated version is known as the Saman-M and is equipped with the additional path-generating device. A combat vehicle that has been updated to the target training system Saman-M can be used as an air defence weapon as part of the air defence missile system.

Variants

In 1991, SA-8b TELs were seen with an additional small-sized radar antenna fitted before the surveillance radar; this is the IFF antenna.

The first naval version developed was the 4K33 Osa-M. This was given the US/NATO designation SA-N-4a 'Gecko'. It entered operational service in 1973.

The system used the 9M33 missile, the 4S33 (ZIF-122) twin-rail launcher (with associated 20-round magazine) and the 4R33 'Bazan' fire-control radar assembly (NATO code name 'Pop Group'). The Osa-M was widely fitted to Russian naval ships of varying types.

The second version, the Osa-MA, was used only on the Project 1159 (NATO Koni-class) light frigates built for export. This used the 9M33M missile and entered operational service in 1979.

The final version, the Osa-2M system, is designated SA-N-4b 'Gecko' in the US/NATO series and uses the latest 9M33M3 missile. This system became operational in 1980.

The Belarussian company Tetraedr Scientific Industrial Unitary Enterprise offers an upgrade to the Osa-AK/AKM systems known as Osa-1T.

A new Osa-AKM variant was announced at the Defendory 2004 show in Athens, Greece. The Russian upgrade to the Osa-AKM self-propelled surface-to-air missile system has been developed by the joint-stock company OAO NIEMI (*Otkritoye Aktsionernoye Obschestvo Naucsno-Issledovatelskiy Elektromekhanicseskiy Institut* - Public joint-stock company Scientific-Research Electromechanical Institute) and Izhevskiy elektromekhanicseskiy zavod Kupol. Both companies are part of Almaz/Antei Concern of Air Defence, joint-stock company. Chief Designer, Iosif Matveyevics Drize and Deputy Chief Designer and Head of System Laboratory, Valentin Valentinovics lead the team which developed the upgrade and the original system.

The upgrade is being offered for Osa-AK and Osa-AKM systems, which use the 9A33BM2, 9A33BM3 and 9A33BM3-1 combat vehicles and 9M33M2 and 9M33M3 missiles. It will enhance the missile and modernise the vehicle.

For the missile, the team proposes modernisation of the onboard electronics, the use of a new solid propellant in the rocket motor and the installation of a new warhead. The latter will release fragments from much heavier alloy than was used in the original warhead. These are made from VNZh (*Volfram-Nikel-Zhelezo* - Tungsten-Nickel-Iron) alloy that includes 18 per cent tungsten. This alloy is relatively cheap and is made from a material created as otherwise-unwanted waste from tungsten manufacturing. Kill probability is 12 per cent higher than that of the original warhead.

Many of the changes to the combat vehicle update the electronics and electro-optics. A new day/night optronic device incorporating a laser range finder replaces the original 9Sh38-day camera. This offers an automatic target-tracking mode that has been used at ranges of up to 30 km.

The error signal separation unit and pulse-to-pulse cancellation unit now use state-of-the-art electronics, while new hardware increases the suppression coefficient of the moving-target selection system by up to 35dB.

A digital unit replaces the old analogue computer, flat-panel Liquid-Crystal Display (LCD) screens replace all three original cathode-ray tube displays, and outdated control consoles give way to more modern equivalents.

Replacement of the original LBV (*Lampa Begushey Volni* - travelling wave tube) microwave amplifiers in both all-round looking radar and guidance radar stations by new low-noise solid-state amplifiers improves the stability of both systems. The change involves replacing the existing OS93-25M and UV-75G sub-units by the new OS93-33 unit and Device 4; and replacing the OS93-30, OS93-29 and UV-67A sub-units with the new OS93-33 and M42152-11.

These changes reduce the radar noise factor by 4-6 dB and increase the dynamic range by 10-13 dB. This improves the detection of small targets and provides greater resistance to enemy jamming.

The old Identification Friend-or-Foe (IFF) interrogator is replaced by a NATO-compatible 44D3-01 unit, which operates in modes 1, 2 and 3/A. given additional hardware, it can simultaneously work in modes 3/A and 4. The 44D3-01 also has automated Built-In Test (BIT) capability.

A new communication (telecode) subsystem handles communications with higher command. This device has already been installed on the combat vehicles of the two Russian Army Osa-AKM air-defence regiments, having been fitted while the vehicles were undergoing general repair. It allows data and voice communication with newest patterns of Russian army mobile air-defence command post, such as the Ranzhir-M1 (9S737M1), Sborka-M1 (9S80M1) and PU-12M7 (9S482M7).

The 9S482M7 (PU-12M7) was upgraded and modernised during 2007-2008. The original version was based on a BTR-60 8 × 8 Armoured Personnel Carrier (APC) and relied on data from external radar sources. The latest variant uses a K1Sh1 wheeled vehicle based on the BTR-80 (GAZ 5903) 8 × 8 APC and incorporates a telescopic sensor mast. Within the telescopic mast are located the SRTR (*Stantsiya RadioTekhnicseskoy Razvedki*), passive ESM system and an electro-optical unit that is located at the masthead.

The mast also boasts four static antennas mounted halfway up the mast for short-range radar. Each antenna covers a sector, which when combined provides all-round coverage.

A trainable electro-optical unit mounted on top of the turret contains day and night cameras, providing hemispherical coverage. The PU-12M7 can accept target data from external surveillance radar, other fire-control systems and radar equipped air-defence weapons. Communications with assets up to 15 km away can be achieved by wire, with radio being used for ranges out to 40 km.

	9K33	9K33M3	9K33M
Dimensions and weights			
Length			
overall	3.13 m (10 ft 3¼ in)	3.158 m (10 ft 4¼ in)	3.158 m (10 ft 4¼ in)
Diameter			
body	208 mm (8.19 in)	210 mm (8.27 in)	210 mm (8.27 in)
Flight control surfaces			
span	0.65 m (2 ft 1½ in) (wing)	-	0.65 m (2 ft 1½ in) (wing)
Weight			
launch	127 kg (279 lb)	127 kg (279 lb)	126.3 kg (278 lb)
Performance			
Speed			
max speed	1,050 kt (1,944 km/h; 1,208 mph; 540 m/s)	-	1,050 kt (1,944 km/h; 1,208 mph; 540 m/s)
Range			
min	0.8 n miles (1.5 km; 0.9 miles) (target speed approx. 500 m/s)	0.8 n miles (1.5 km; 0.9 miles) (target speed approx. 500 m/s)	0.8 n miles (1.5 km; 0.9 miles) (target speed approx. 500 m/s)
	-	-	1.1 n miles (2 km; 1.2 miles) (target speed 0-360 km/h, using TV optical sight)
max	5.4 n miles (10 km; 6.2 miles) (target speed approx. 500 m/s)	5.4 n miles (10 km; 6.2 miles) (target speed approx. 500 m/s)	5.4 n miles (10 km; 6.2 miles) (target speed approx. 500 m/s)
	-	-	3.5 n miles (6.5 km; 4.0 miles) (target speed 0-360 km/h, using TV optical sight)
Altitude			
min	25 m (82 ft) (target speed approx. 500 m/s)	25 m (82 ft) (target speed approx. 500 m/s)	25 m (82 ft) (target speed approx. 500 m/s)
	-	-	10 m (33 ft) (est.) (target speed 0-360 km/h, using TV optical sight)
max	5,000 m (16,404 ft) (target speed approx. 500 m/s)	5,000 m (16,404 ft) (target speed approx. 500 m/s)	5,000 m (16,404 ft) (target speed approx. 500 m/s)
	-	-	5,000 m (16,404 ft) (target speed 0-360 km/h, using TV optical sight)
Ordnance components			
Warhead	14.5 kg (31 lb) HE fragmentation	14.5 kg (31 lb) HE fragmentation	15 kg (33 lb) HE fragmentation
Fuze	proximity	proximity	proximity, impact
Propulsion			
type	solid propellant	solid propellant	solid propellant

This then allows the PU-12M7 tactical control of an area out to 100 km, handling information for as many as 99 airborne targets. The operator can simultaneously track up to nine targets.

The vehicle navigation systems of both regiments have been improved by the addition of a CH-3001 combined Glonass/GPS satellite-navigation receiver. While the installation of more powerful air conditioners keeps the temperature within the vehicles at between +15°C and +30°C when the outside temperature is between +35°C and +50°C.

New monitoring and recording equipment can record the operation of the vehicle's on board systems, display data during missile firing and allow post-firing analysis of engagements. Analogue signals brought to sockets on the combat vehicle's communication terminal board can be converted to digital form.

To enhance the protection for the vehicle's crew and equipment protection, the upgrade offers extra protection for the sides of the front part of the vehicle body. This is achieved by a combination of quick-detachable add-on armour, plus permanently mounted armour elements inside the vehicle body. The total weight of these protection measures is no more than 500 kg.

One of the upgrade possibilities involves creating mixed Osa/Tor air-defence missile batteries - for example, one 9A331 Tor combat vehicle plus three 9A33BM3 Osa combat vehicles or a combination of two plus two. In such mixed batteries, the Osa combat vehicles would handle lower-tier threats at ranges of up to 5 km, with the Tor combat vehicles providing upper-tier coverage out to 10-12 km.

Retia (Czech Republic) OSA-AKM Modernisation includes:

- IFF Mark 12 interrogator integration including IFF antenna
- Modernisation of the surveillance radar SOC
- Digital signal and data processing
- Analogue and synthetic display (original IKO can be used)
- Installation of the Commander Terminal for complex air and tactical situation
- GPS navigation system integration
- Tactical transceiver for communication with superior command centre.

The Polish Military Armament Works Number 2 in Grudziadz is offering to either upgrade the basic OSA system based on digital technology and renaming the new weapon SA-8P Sting, or overhauling the OSA, guaranteeing inter-overhaul service period whilst restoring all the techno-operational parameters.

The effects of the upgrade will be:

- Targets IFF in Mark XII mode 4 (NATO) standard upgrade Mark XIIA mode 5, mode S
- Radio electronics camouflage by reducing radio location radiation to a minimum as a result of the application of passive opto-electronics target detection and tracking equipment (opto-electronics head, video tracker, laser range-finder)
- Increase of ECM by the application of the digital radar signal analysis system
- Application of advanced highly integrated parts allowing the supply of replacement parts necessary for normal operation
- Increase of the detection of low RCS targets
- Precise position determination by applying the inertial land navigation system supported by GPS
- Possibility of adapting the combat vehicle for launching the new generation missiles of the type fire-and-forget
- Introduction of advanced methods and algorithms for digital processing and imaging of radio location signals and images from thermal infrared cameras and TV cameras
- Elimination of adjustments and tuning.

Specifications

	OKA-AKM TEL
Dimensions and weights	
Crew:	5
Length	
overall:	9.14 m
Width	
overall:	2.8 m
Height	
hull:	1.845 m
transport configuration:	4.2 m
Ground clearance	
overall:	0.4 m
Wheelbase	3.075 m + 2.788 m
Weight	
combat:	18,800 kg
Mobility	
Configuration	
running gear:	wheeled
Speed	
max speed:	60 km/h
water:	10 km/h
Range	
main fuel supply:	500 km
Fuel capacity	
main:	350 litres
Amphibious:	yes
Vertical obstacle	
forwards:	0.5 m
Trench:	1.2 m
Engine	5D20 B-300, diesel, 300 hp at 2,000 rpm
Engine	gas turbine
Firepower	
Armament:	6 × SA-8 SAM

Status

Production within Russia of the Osa is complete. Final deliveries were made in 1998. The manufacturer however is currently offering modernised/upgraded systems to the Osa-AKM as standard for export.

India has signed a contract with the Polish firm of Centrex Trading Ltd for the upgrading of their SA-6 Kvadrat, SA-8 Osa-AK. This contract is expected to extend the life of these missiles beyond that of the normal 10-year storage life in addition to extended the maximum range of the Osa-AK from 10 to 15 km.

The Czech Republic in December 2006 retired the Osa-A-AKM from its inventory and replaced the system with the RBS 70 and the Bolide missile.

In 2010 Belarus offered an updated/upgrade to the basic Osa system. The upgrade includes the replacement of analogue components with digital ones, plus the replacement of previously installed missiles with more advanced models.

There is no question that the Osa system will continue for many years to come in the inventory of those countries that cannot afford to purchase newer and more modern air-defence, it is far more cost effective to upgrade these systems.

The British firm Babcock has been awarded a four year contract to overhaul and modernise former German Air Force Equipment including two 9K33 Osa radar systems. It is understood that these radar will then be used for electronic warfare training at RAF Spadeadam in Cumbria. It is also understood that under this contract, Babcock will fully refurbish and modernise both systems and provide spares, training and documentation and additional tasks.

Contractor

Almaz/Antei Concern of Air Defence.
Izhevsk Electromechanical Plant (manufacture).
Fakel (missiles).

9K35 Strela-10

Type

Self-propelled Surface-to-Air Missile (SAM) system

Development

The fully amphibious NBC-equipped Strela-10 (9K35 - NATO SA-13 'Gopher') mobile SAM system, with range-only radar is the successor of the 9K31 Strela-1 (NATO SA-9 Gaskin) entered operational service in 1975. In the then Soviet Army it replaced the far less capable Strela-1 system on a one-for-one basis to improve the mobility of the anti-aircraft batteries in the Motorised Rifle and Tank divisions. Development began with the Strela-10SV version that evolved into the first main production model, the Strela-10M.

Further development of the system has continued throughout the years and by September 2007 the 9K35M3-K Kolchan variant, mounted on a BTR-60 wheeled chassis, was displayed for the first time at the Moscow Air Show MAKS 2007.

A further development reported in June 2009 from the Belarusian company Belspetsvneshtekhnika offers export users an upgrade for the system known as Strizh, a night-vision optronic device.

By 2010, the KB Tochmash was reporting yet another variant this time known simply as the Strela-10M4. This latest version makes use of the more modern missile the 9M333 and also has the ability to use the 9M37M and 9M37MD weapons.

Description

The Russian industrial index numbers for the system components are:

- 9A35M2, 9A35M3-K - launcher vehicle with 9S16 (NATO 'Flat Box-B') passive radar detection system that gives a 360° azimuth and minimum 40° elevation coverage
- 9A34M2, 9A34M3-K - launcher vehicle
- 9M37 - missile, later versions are the 9M37M, 9M37MD and 9M333
- 9V839M - system check-out vehicle

'Dog Ear' early warning and target acquisition MT-LBU radar vehicle is used by SA-9 and SA-13 SAM units to provide target information 0509448

SA-13 'Gopher' with launcher assembly in firing position 0099540

SA-13 'Gopher' SAM system in travelling configuration, clearly showing 'Flat Box-B' passive antenna between two front hatches (Michael Jerchel) 0080442

- 9V915M, 9V915M-1 - technical maintenance vehicle
- 9F624/9F624M training simulator
- 9U111 - a 1,950 kg trailer-mounted 12 kW generator unit, designed to feed power to up to four 9A35M2, 9A35M3-K or 9A34M2, 9A34M3-K launcher vehicles at a distance of up to 30 m by cable while conducting maintenance or training operations
- 9S482M7 Control Post (see below)
- Drill and test equipment.

Like the MT-LB on which it is based, the SA-13 SAM system is fully amphibious. Before it enters the water, pontoons on either side of the hull are swung upward through 90° and locked into position.

The normal air-defence battalion for a regiment comprises four Strela-10, four ZSU-23-4, two or three BTR-60PU-12 command posts and five-man crew F/G-band MT-LBU (NATO 'Dog Ear') radar vehicle. The MT-LBU radar acquisition range is 80 km and the tracking range 35 km. A 9F624 will also be found which is the simulator/training device for the weapon systems crews. A modern and up-to-date version designated the 9M624M has also been developed for the Strela-10M system.

The Strela-10 is also organic to the Russian Naval Infantry air-defence batteries, having replaced the Strela-1 on a one-for-one basis.

There are two versions of the Strela-10 Transporter-Erector-Launcher And Radar (TELAR) variant of the MT-LBU vehicle in service, designated TELAR-1 and TELAR-2 by the US Army. Appraisal of both does not show any significant

SA-13 'Gopher' with launcher assembly being retracted to travelling configuration position 0099541

SA-13 'Gopher' SAM Strela-10M3 system with missiles retracted in the travelling position and showing 'Flat Box-B' passive radar detection system between commander's and driver's windscreen 0509451

SA-13 'Gopher' SAM Strela-10M3 system with missiles deployed in the firing position; note the 7.62 mm PKT machine gun at the front of vehicle and pontoons for amphibious operations retracted alongside the hull 0509450

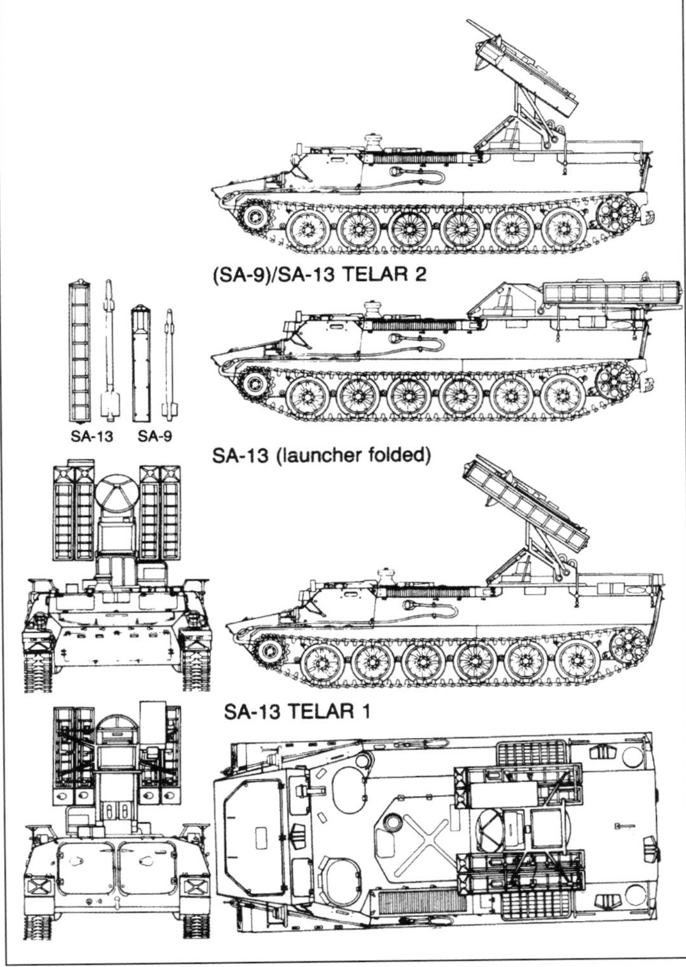

SA-13 'Gopher' mobile SAM system (Steven Zaloga) 0509449

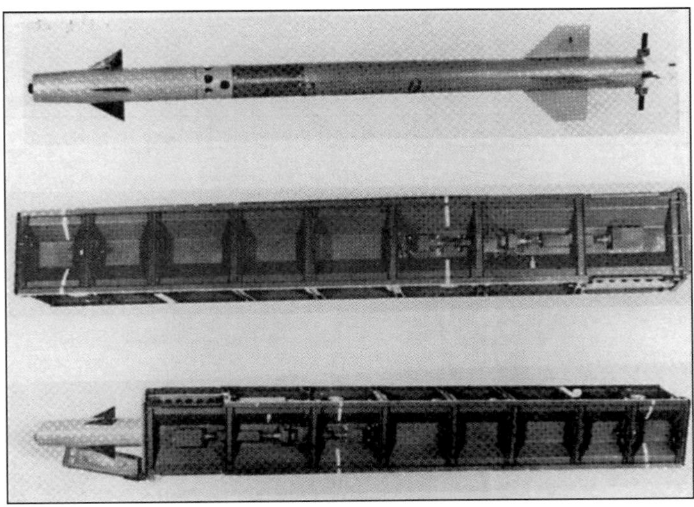

The 9M333 missile of the Strela-10M3 system out of its launcher (top), in launch container (centre) and emerging from launch container (bottom) 0509452

structural differences. However, it is known that the TELAR-1 (9A35) carries four 9S16 (NATO 'Flat Box-B') passive radar detection antenna units, one on either corner of the vehicle's rear deck, one facing after and one between the driver's vision ports at the front, whereas the TELAR-2 carries the 9S86 (NATO Snap Shot) range only radar.

The 9S16 system scan sector is 360°, 40° in elevation and range of up to 30 km. The Strela-10 battery commander uses the TELAR-1.

The original system is known as the ZRK-BD 9K35 Strela-10 in the then Soviet service. The basic 40 kg 9M37 missile is 2.19 m long, 0.12 m in diameter with a 0.4 m wing span and has a maximum speed of 517 m/s. It carries a three kg HE fragmentation warhead. The minimum range of the 9M37 is 500 m and the maximum effective range is 5,000 m, with altitude engagement limits of 25 to 3,000 m.

Depending upon the model, the missile utilises an improved passive lead sulphide (PbS) all-aspects Infra-Red (IR) seeker unit. The seeker operates in two individual frequency bands in the 1-5 μm wavelength region to give high discrimination against IR countermeasures such as flares and decoy pods, or a cryogenically-cooled passive indium antimonide (InSb) all-aspects IR seeker unit. As of 2004, a new and more capable homing head 9E47M was developed which replaced the original 9E47.

Normally the TELAR carries four ready-to-fire 9M37 missile container launchers and four reloads in the cargo compartment, but it has also been seen on numerous occasions with either 9M31 (NATO SA-9 'Gaskin') container-launcher boxes in their place or a mixture of the two. This enables the battlefield features of both missiles to be fully used by allowing the cheaper 9M31 to be deployed against the easier targets and the more expensive and sophisticated 9M37, against the difficult targets.

The missile mix also allows a choice of IR seeker types on the missiles for use against extremely low-altitude targets as well as in adverse weather conditions.

The 9S482M7 (PU-12M7) Command Post was upgraded and modernised during 2007/8. The original version was based on a BTR-60 8 × 8 Armoured Personnel Carrier (APC) and relied on data from external radar sources. The latest variant uses a K1Sh1 wheeled vehicle based on the BTR-80 (GAZ 5903) 8 × 8 APC and incorporates a telescopic sensor mast. Within the telescopic mast are located the SRTR (*Stantsiya Radiotekhnicseskoy Razvedki*), passive ESM system and an electro-optic unit that is located at the masthead. The mast also boasts four static antennas mounted halfway up the mast for short-range radar. Each antenna covers a sector, which when combined provides all-round coverage. A trainable electro-optical unit mounted on top of the turret contains day and night cameras, providing hemispherical coverage. The PU-12M7 can accept target data from external surveillance radar, other fire-control systems and radar equipped air-defence weapons. Communications with assets up to 15 km away can be achieved by wire, with radio being used for ranges out to 40 km. This then allows the PU-12M7 tactical control of an area out to 100 km, handling information for as many as 99 airborne targets. The operator can simultaneously track up to nine targets.

Some vehicles have a pintle-mounted PKT 7.62 mm machine gun in front of the forward hatch for local protection. Other vehicles have been seen with additional support railings for the system on the rear deck. The 450 to 10,000 m range circular parabolic 9S86 (NATO 'Snap Shot') radar antennas is located between the two pairs of missile canisters. This radar is a simple range-only set that aids computation of lead angle before a missile is launched, to prevent wastage of missiles outside the effective range of the system. The 1RL246-10-2 IFF system is also fitted; this determines target identification at ranges up to 12 km and altitudes of 100 to 5,000 m. Early versions of the Strela-10 IFF systems protruded beyond the leading edge of the launcher box mounting. This resulted in breakage and thus forced modifications to move the antenna back flush with the launch box mounting.

	9M37	9M37M	9M333
Dimensions and weights			
Length			
overall	2.19 m (7 ft 2¼ in)	2.19 m (7 ft 2¼ in)	2.223 m (7 ft 3½ in)
Diameter			
body	120 mm (4.72 in)	120 mm (4.72 in)	120 mm (4.72 in)
Flight control surfaces			
span	0.36 m (1 ft 2¼ in) (wing)	0.36 m (1 ft 2¼ in) (wing)	0.36 m (1 ft 2¼ in) (wing)
Weight			
launch	40 kg (88 lb)	40 kg (88 lb)	41 kg (90 lb)
Performance			
Speed			
max impulse speed	816 kt (1,512 km/h; 940 mph; 420 m/s)	-	-
max impulse Mach	1.2	-	-
max speed	1,005 kt (1,861 km/h; 1,156 mph; 517 m/s)	1,005 kt (1,861 km/h; 1,156 mph; 517 m/s)	1,069 kt (1,980 km/h; 1,230 mph; 550 m/s)
Range			
min	0.4 n miles (0.8 km; 0.5 miles)	0.4 n miles (0.8 km; 0.5 miles)	0.4 n miles (0.8 km; 0.5 miles)
max	2.7 n miles (5 km; 3.1 miles)	2.7 n miles (5 km; 3.1 miles)	2.7 n miles (5 km; 3.1 miles)
Altitude			
min	25 m (82 ft)	25 m (82 ft)	10 m (33 ft)
max	3,000 m (9,843 ft)	3,000 m (9,843 ft)	3,500 m (11,483 ft)
Ordnance components			
Warhead	3 kg (6.00 lb) HE fragmentation	3 kg (6.00 lb) HE fragmentation	5 kg (11.00 lb) HE fragmentation
Fuze	impact, proximity	impact, proximity	impact, proximity
Guidance	passive, IR	passive, IR	passive, IR (dual mode, photocontrast)
Propulsion			
type	single stage, solid propellant	single stage, solid propellant	single stage, solid propellant

SA-13 'Gopher' SAM system with 12 man-portable fire-and-forget SAM in the ready-to-launch position (IHS/Peter Felstead) 0069500

The first major model was the 9K35M Strela-10M in the 1970s. This was followed by the 9K35M2 Strela-10M2, with the 9M37M missile and finally the 9K37M3 with the 9M333 missile in the late 1990s.

From 2000 to 2001, the Strela-10M system was upgraded with the new Phoenix IR detection system. This upgrade included changes to the missile 9M37MD and computer changes within the firing vehicle. As a result of these changes, the missile range has also been extended. A low-light-level television system GEO-PZR2 for the night sighting has also been installed. The system now has the name Gyurza.

In June 2009, the Belspetsvneshtekhnika Company from Belarus offered an upgrade for export to users of the 9K35 system The offer was for a new Strizh night-vision optronic system that will allow operations in a passive mode by day and night and in bad weather conditions. The Strizh offers television and thermal channels and consists of:

- TV and thermal-imaging optronic systems installed on the vehicle's launcher and bore sighted to the existing 9Sh127 optical sight
- Video display and control console positioned within the turret at a location convenient to the operator
- Video information processing sub-system
- An interface with existing systems in the 9A34M or 9A35M combat vehicles
- Associated cables.

The monochrome TV channel operates at a wavelength of 0.6 to 1.0 μm and offers wide (19°) and narrow (1.5°) Fields of View (FoV). It has a built in thermal compensation and focusing system.

The thermal channel is based on an uncooled bolometer sensor operating in the 8-12 μm bands and has a 4.2° × 3.2° FoV. Both channels can be used at temperatures ranging from –40° up to +50°C.

These channels increase the range at which targets can be detected and tracked, whilst also improving the system's ability to engage small targets.

Variants

9K35M3-K - Kolchan

During MAKS 2007 in Moscow, a wheeled version of the Strela-10 system was displayed and is known as 9K35M3-K with the Russian name Kolchan. This version is mounted on a BTR-60 wheeled chassis. This much-improved version of the Strela-10 boasts the Trona-1 land navigation system, SOVI (*Sistema Obrabotki Videolzobrazheniy*) Okhotnik image-processing system, extended range and altitude of the missile. The chassis has also been improved with improved armour, a new diesel engine, new gearbox and improved one-disk clutch. The maximum road speed is increased to 80 km/h.

Strela-10M3

Apart from the Strela-10M and Strela-10M2, Russia developed a further version of the SA-13, known by the designation 9K35M3 Strela-10M3. This is designed for use in the mobile battle and to defend troops on the march from low-level attacks by aircraft and helicopters, precision-guided munitions and other flying vehicles such as reconnaissance RPVs. The subsystem designators have corresponding changes, for example 9A35M3-launcher vehicle. The 9V179-1 receiver and target designator and a 9V180 target designation execution system are fitted for centralised control of the vehicle from external mobile command-and-control posts, fitted with the ASPD-U data processing equipment.

The major change is the adoption of a dual-mode guidance system for the 9E425-missile seeker - optical 'photo-contrast' and dual-band passive IR. The missile, built by the Kovrovsky Mekhanichesky Zavod, accommodating this system is the 9M333. This weighs 41 kg at launch and, when in its container-launcher the 303 by 320 by 2,300 mm box-like canister, has a total mass of 72 kg. Target acquisition range using the optical 'photo-contrast' channel is between 2,000 and 8,000 m, while for the IR channel it is between 2,300 and 5,300 m. Altitude engagement limits are from 10 m up to 3,000 m at ranges of 800 to 5,000 m. Average missile speed is 550 m/s. Single-Shot Kill Probability (SSKP) is 0.6. The HE fragmentation rod warhead weighs 5 kg (including 2.6 kg of HE) and uses both contact and eight-laser pulsed active proximity fuzing systems. The actuation radius of the laser fuze is up to 4 m.

The dual-mode seeker ensures good IR decoy counter-countermeasures discrimination capability and optimum use of the system against diverse and extremely low-altitude targets, as well as in adverse weather conditions.

The 9V38M3 equipment for missile pre-launch preparation is also fitted. This allows for automatic external control, local manual control, emergency launch control and the possibility of crew training modes of platform operation. Both audible and visual signals are used during the system operation.

Strela-10M4

The latest non-export version of the Strela-10 system is designated 9K35M4 Strela-10M4. This is an upgrade package for the Strela-10M2, Strela-10M3 and earlier versions. The new Strela-10M4 battery will consist of one 9A35M4 battery command vehicle and three 9A34M4 firing vehicles. The external visible difference between the 9A35M4 and the earlier 9A35M3 is a new night-vision system and automatic target acquisition and tracking unit mounted on the right upper side of the launcher frame. Therefore the new system (upgrade) is intended to allow operation at night and under adverse weather conditions, plus the interception of smaller-sized targets such as UAVs and cruise missiles.

The spectral range of the thermal imaging channel is 3-5 μm and corresponds with the spectral range of the 9M333, 9M37M and 9M37MD missile seekers. The missile 9M333 has a multispectral seeker IR (3-4 μm) and visible channel. The system is equipped with a container with a thermal-imaging channel with the range, exceeded range of the missile seeker to provide 24 hour operation and tactical arrangement in adverse weather conditions. The display with a control panel is located in front of the operator in the combat vehicle.

A second version of the Strela-10M4 that is for export only is known as the SOSNA. This system is basically the Strela-10 chassis with a new launching system that carries:

	Strela-10SV	Strela-10M	Strela-10M2	Strela-10M3
Dimensions and weights				
Crew:	3	3	3	3
Length				
overall:	6.93 m	6.93 m	6.93 m	6.6 m
hull:	-	-	-	6.45 m
Width				
overall:	2.85 m	2.85 m	2.85 m	2.85 m
transport configuration:	-	-	-	2.3 m
deployed:	-	-	-	3.8 m
Height				
overall:	3.965 m	3.965 m	3.965 m	-
Weight				
combat:	12,080 kg	12,080 kg	12,080 kg	12,300 kg
Mobility				
Configuration				
running gear:	tracked	tracked	tracked	tracked
Speed				
max speed:	61.5 km/h	61.5 km/h	61.5 km/h	61.5 km/h
water:	6 km/h	6 km/h	6 km/h	6 km/h
Range				
main fuel supply:	500 km	500 km	500 km	500 km
Fuel capacity				
main:	-	-	-	450 litres
Amphibious:	yes	yes	yes	yes
Vertical obstacle				
forwards:	-	-	-	0.7 m
Trench:	-	-	-	2.7 m
Engine				YaMZ-238 V, diesel, 240 hp
Firepower				
Armament:	4 × SA-13 SAM	4 × SA-13 SAM	4 × SA-13 SAM	4 × SA-13 SAM
Ammunition				
missile:	8	8	8	8
Main weapon				
maximum range:	5.0 m	-	-	-
maximum vertical range:	3.5 m	-	-	-
minimum vertical range:	0.01 m	-	-	-
Main armament traverse				
angle:	360°	360°	360°	360°
speed:	100°/s	100°/s	100°/s	100°/s
Main armament elevation/depression				
armament front:	+80°/-5°	+80°/-5°	+80°/-5°	+80°/-5°
speed:	50°/s	50°/s	50°/s	50°/s
Survivability				
Armour				
hull/body:	-	-	-	7 mm

- A gyro stabilised platform TV system
- Thermal-imaging channel
- Laser rangefinder with a beam diverter
- Laser beam riding missile guidance equipment
- Thermal-imaging missile finder
- Built in adjustment device
- Climate control console
- Digital computer
- Control and display console
- Auto target tracking device
- Radar
- Analogue control unit
- Phase code converter unit
- Independent power source.

The Strela-10M4 Sosna carries a group of six missiles mounted either side of the turret, during the firing sequence, the missiles are guided for the first two seconds of flight and from then are laser beam riding.

Strela-10 with MANPADS

During 1999, a Strela-10 system was exhibited with the four missiles replaced with eight man-portable SAM.

Strela-10 missile with Palma

The Strela-10 missile family can be used with the Nudelman Design Bureau Palma naval air-defence system.

Serbian SAVA SAM system

A similar system, known as the SAVA, has been developed in the former Serbia and Montenegro. It is based on a modified BVP M80A mechanised infantry combat vehicle chassis. More recent information also suggests that the system (missiles in particular) are still in production at the Krusik Factor, near Valjevo in Serbia. Missiles bearing the logo of the Krusik Factory were seen at an arms exhibition along with other missiles of the Strela family.

Croatian Strijela 10 SAM system

R H Alan Torpedo of Croatia is producing a locally built version of the Strela-10 mounted on a modified armoured APC chassis, derived from the TAM (6 × 6) general purpose truck.

Czech upgraded Strela-10M systems

Following extensive trials with prototype systems, in 1996, the Czech Army started to receive a Strela-10M modernisation package developed by the RETIA company. Full details are given in this section.

Specifications – *see table on facing page and above*

Status

Production as required. Upgrade of older systems to Strela-10M4 standard offered for export.

Contractor

Design: KB Tochmash Design Bureau of Precision Engineering.
Production: Saratovskiy Zenit Machine Plant.
Mobile training simulators: Tulatocsmash.
9K35M3-K: Muromteplovoz Joint Stock Company.

9K37 Buk

Type

Self-propelled Surface-to-Air Missile (SAM) system

Development

The 9K37 Buk (NATO SA-11 'Gadfly') was developed in accordance with the Resolution of the CPSU Central Committee and USSR Council of Ministers dated 13 January 1972 using two self-propelled mounts, the 9A38 and the 9A310. The NIIP Research Institute of Instrument Design was assigned prime developer whilst the Sverdlovsk based Novator Design Bureau was tasked to develop the 9M38 missile. Developmental trials and firing took place at the Kapustin Yar Missile Test Range (KYMTR) starting in August 1975 and lasted until March 1979. The first Buk with the 9M38 missile became operational in 1978 associated with the Kub 3M9M3 missiles, this became known as the Kub-M4 system. However, the full configuration of Buk continued testing from November 1977 through 1979 and became operational in 1980. The system is

The Buk (SA-11 'Gadfly') system includes a Command Post (CP) with antenna erected (left) and Target Acquisition Radar (TAR) vehicle (right) with radar scanner erected (Christopher F Foss) 0080388

9S18M1 SA-11 radar vehicle in travelling configuration with radar retracted (Paul Jackson) 0080389

part of the replacement for the Krug (NATO SA-4 'Ganef') at the army (corps) level. The first Buk brigade became operational in 1980, although significant numbers were not deployed until the late 1980s.

The system is known in the Russian Federation as the 9K37 Buk (Russian for 'beech') with the complete system, including the radar and support equipment, having a Russian Industrial Index number, 9K37. The export version is known as the 'Gang', with the various sub-elements having the suffix, 'E' (for *Ehksportiynyi*; Russian for export) added to their designations, for example 9A310M1E. The first export order came from Syria in 1983, with deliveries beginning in 1986.

The former Yugoslavia ordered the system in the late 1980s with a small number of tracked launchers arriving before the Civil War broke out. It is believed that these were non-operational during the NATO Kosovo/Serbia bombing campaign. The principal sub-units are based on a full-tracked chassis developed and built by the Metrovagonmash Joint Stock Company from the GM-539 design. This provides protection from small arms fire as well as being sealed against NBC attack.

Due to a number of problems with the original Buk system, including the original surveillance radar Tube Arm, an improved system was already in development in 1980 to improve its combat capabilities, protection from countermeasures and anti-radiation missiles. Testing the new system was

9A310M1 launcher from the rear (Christopher F Foss) 0044117

conducted throughout 1982. Known as the Buk-M1, this entered service in 1983 and introduced the 9S18M1 'Snow Drift' surveillance radar into service. The Buk can also be used to engage tactical missiles and rockets; successful trials were undertaken in 1992 against simulated Pershing and MLRS type targets.

During June 2004, it was reported from Russia that work on upgrading the Buk SAM system's radar was continuing within the Almaz/Antei Concern of Air Defence and the development of a common surface-to-air missile system expected shortly. Furthermore, the concept was to make provision for a modular system able to engage targets in the short, medium and long ranges by re-configuring various units and modules.

Description

The original missile used by the Buk system is the 9M38, which was subsequently replaced with the 9M317. The 9M38 missiles employ the 9B-1103M (diameter 350 mm) seeker that has an acquisition range of 40 km for 5 m^2 RCS targets.

A total of six targets can be engaged simultaneously by a battery while they are flying on different bearings and at different altitudes and ranges. A typical battery comprises a Command Post (CP) vehicle, a Target Acquisition Radar (TAR) vehicle and six Self-Propelled Mounts (SPMs) that act as the launcher vehicles. A specialist Loader-Launcher (LL) vehicle that acts both as missile transloader and additional launch unit supports pair of launchers.

A Buk regiment comprises four such batteries and a Regimental Target Acquisition Battery with two long-range early-warning search radars.

The Buk self-propelled launcher is also offered as an upgrade to the 2K12 Kub/Kvadrat system. Known as the Kub-M4 system, it entered service in 1978 and involves a single dual-capable 9A310/9A38 SPM attached to each 2K12 battery to double the target engagement capabilities. A 9A39M1 loader-launcher supports the 9A310/9A38.

The Buk-M1-2 was shown at the MAKS-97 air show. This version is covered in a separate entry in this section: Tikhomirov Instrument Research Institute 9K37M1-2 Buk-M1-2 (SA-11 'Gadfly'/SA-17 'Grizzly') multipurpose low to high-altitude surface-to-air missile system.

The Buk-M1 anti-aircraft missile system battery is comprised of many individual elements, a overview of each follows.

9S470M1 Command Post (CP) vehicle

This provides control for the Buk battery and contains: the battery's data display and control system; digital fire-control solution computer, which automates target distribution, processing of target data, production of missile launch data and the guidance signals for the missiles in-flight; data transmission systems; navigation systems and command-and-control voice communications systems. It is also fitted with an auxiliary power supply unit.

Target Acquisition Radar (TAR, Russian designation 9S18M1/NATO designation 'Snow Drift')

This acquires potential aerial targets, interrogates them for Identification Friend or Foe (IFF) data and transmits their tracks and IFF status to the CP vehicle either by secure radio data or landline communications links.

Close up of the Buk (SA-11 'Gadfly) Self-Propelled Mount (SPM) with four-round launcher traversed to the rear (Christopher F Foss) 0044119

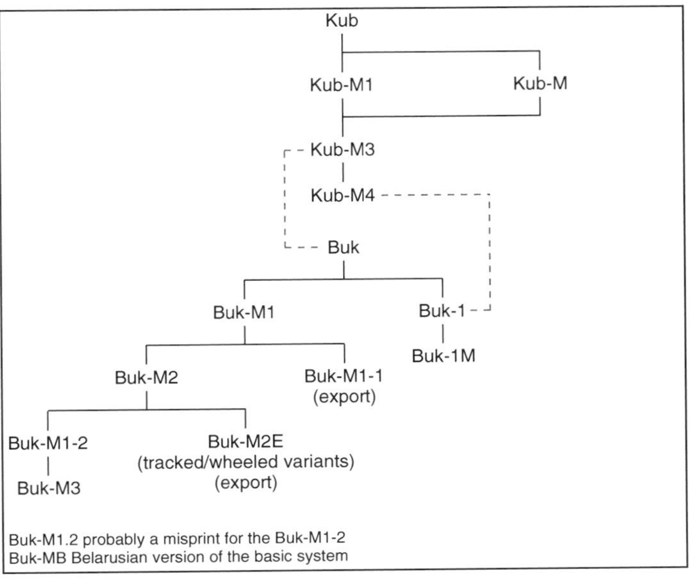

Buk family flow diagram (James C O'Halloran) 1040815

Apart from the radar assembly, which is carried in a lowered position for travel and hydraulically raised for operational use, the vehicle is also fitted with:

- A high-speed digital computer system together with associated control consoles for raw and processed radar displays
- An IFF interrogator system
- A TV tracker unit for use in heavy ECCM environments
- Land navigation equipment
- Voice and datalink communications systems
- An APU to power the different vehicle systems.

Sophisticated ECCM techniques and systems are employed with the 3-D circular scan radar to counter enemy jamming and clutter problems.

Self-Propelled Mount (SPM, Russian designation 9A310M1)

The SPM can operate either as part of the Buk-M1 system using the battery CP in the target designation mode or as an individual unit in its own right assigned to protect a specific sector and using its own onboard target illumination radar in a detection mode. The SPM may also work with its LL in an engagement with either vehicle acting as the missile launch platform. The SPM must, however, provide the entire target information and illumination to the LL during any such engagement.

The SPM can engage various types of aerial targets flying at altitudes between 15 and 22,000 m and has successfully demonstrated the capability to hit helicopters hovering at very low altitudes.

A total of four missiles are carried on the launch assembly platform, the fore part of which contains the H/I-band (NATO designation 'Fire Dome') monopulse guidance and tracking radar unit. For an engagement the four-round missile rail is raised to the desired launch angle. An IFF interrogator system is also carried, together with sophisticated ECCM subsystems for use with the radar.

The SPM can also be integrated into the Kub (SA-6 'Gainful') air defence network using that system's 1S91 ('Straight Flush') radar vehicle to provide all the target acquisition data in place of its own radar vehicle. This allows a hybrid Kub/Buk battery known as the Kub-M4 system (see 9A38 for further details) to engage two targets at a time, one by the Kub launchers and the other by the SPM.

Other items fitted to the SPM include:

- Various data processing and system control units
- A two-way data transmission system
- Land navigation devices
- A command-and-control voice communications system
- An APU to power the various onboard systems.

Buk (SA-11'Gadfly) specialist Loader-Launcher (LL) with four missiles in ready-to-launch position and four missiles for reloading (Christopher F Foss) 0080435

Self-propelled Mount (SPM, Russian designation 9A38)

This is the second SPM variant to be fielded with the Buk system and has a dual-capable launcher assembly suitable for use with both Kub and Buk missiles. A total of three ready-to-fire rounds are carried. The SPM carries the same communications and position orientation systems as used in the 2P25 SPM of the Kub. Kub systems are re-equipped with a replacement 9A38. In service with the Russian Army are designated Kub-M4.

Loader-Launcher (LL, Russian designation 9A39M1)

The LL is intended for the storing, transportation and, when required, the firing of Buk missiles. It is similar in appearance to the SPM but does not have the radar assembly mounted on the front of the launcher assembly. In addition to the four ready-to-launch missiles, a further four missiles are carried in pairs on either side of the launcher rail assembly in load carrying units. These can either be transferred to an SPM or used on the LL by a hydraulically operated collapsible crane unit fitted where the SPM is located.

The vehicle is also fitted with:

- An analogue computer system
- Land navigation devices
- Command-and-control voice communications system
- Secure inter-vehicle information and voice communications
- An APU to provide power to the various vehicle systems.

Specialist maintenance and transportation vehicles

To support the Buk-1M system in the field there is a series of specialist maintenance and transportation vehicles:

- 9B95M1 automated mobile test system for functionally testing the Buk missile. The test equipment is mounted on a Type ILPZ-738 two-wheel trailer and in a K-5 van bed carried on the rear of a ZIL-131 (6 × 6) truck chassis. Testing time for the missile is 15 minutes. The times needed for functional testing of the complete missile system components is one hour (in summertime) and three hours (in wintertime). The time needed to close down the automated mobile test unit is 24 minutes. Total weight is 13,000 kg.
- Maintenance and repair trucks utilising the Ural-43203-1012 (6 × 6) truck chassis and Type K2-4320 van body: RMT-1 (Russian designation 9B883M1 or MPTO-1) servicing vehicle for receiver and automatic azimuth/elevation control systems of the SPM as well as the secondary power sources of the SPM, CP and LL vehicles; RMT-2 (Russian designation 9B884M1 or MPTO-2) servicing vehicle for communications systems, electronics and associated systems on the CP, SPM and LL vehicles; RMT-3 (Russian designation 9B894M1 or MPTO-3) servicing vehicles for the TAR vehicle electronic subsystems. All the vehicles carry the tools, accessories and equipment to carry out the necessary repairs to their appropriate charges.

	9K37 Buk CP Vehicle	9K37 Buk TAR Vehicle	9K37 Buk SPM Vehicle	9K37 Buk LL Vehicle
Dimensions and weights				
Length				
overall:	8.371 m	-	9.3 m	-
transport configuration:	-	9.59 m	-	9.96 m
deployed:	-	9.58 m	-	-
Width				
overall:	3.25 m	3.25 m	-	-
transport configuration:	-	-	3.25 m	3.316 m
deployed:	-	-	9.03 m	-
Height				
transport configuration:	2.975 m	3.25 m	3.8 m	3.8 m
deployed:	8.785 m	8.02 m	7.72 m	-
Weight				
combat:	29,870 kg	35,000 kg	32,340 kg	35,000 kg
Mobility				
Configuration				
running gear:	wheeled	tracked	tracked	tracked
Speed				
max speed:	65 km/h	65 km/h	65 km/h	65 km/h
cross-country:	45 km/h	45 km/h	45 km/h	45 km/h
Range				
main fuel supply:	500 km (allowing 1h APU operation)	500 km (allowing 1 h APU operation)	-	500 km (including 1 h APU operation)
Firepower				
Armament:	**Armament:**	**Armament:**	4 × SA-11 'Gadfly' SAM	4 × SA-11 'Gadfly' SAM

Each vehicle also carries a set of spare parts. The vehicle set up and close down times are 15 minutes. Each vehicle can operate up to eight hours continuously. Total weight per vehicle is 13,450 kg.

- MT (or Russian designations 9B881M1 or MTO) general maintenance and repair vehicle for the CP, SPM and LL vehicles. This uses a Type K2-4320 van body on a Ural-43203-1012 (6 × 6) truck chassis and a four-wheel van body trailer. It carries all the necessary general-purpose test equipment, tools and accessories to maintain and repair these vehicles. Set up and close down times are 30 minutes. The vehicle can operate for 24 hours continuously. The total weight of the vehicle is 12,000 kg.
- MTO-ATG-M1 maintenance shop vehicle with a Type KM-131 van body on a ZIL-131 (6 × 6) truck chassis. It is fitted with a winch system and the necessary tools to undertake repairs to the running gear and suspensions of the SPM, TAR, CP and LL vehicles. The number of work places is three to six and the set up and close down times are five minutes.
- TV (Russian designations TM-9T229 or TM) transporter vehicle, which carries and acts as the field store for SA-11 reload rounds. The vehicle is based on KrAZ-255B (6 × 6) truck chassis and carries six Buk reload rounds in their containers or eight reload rounds without their containers. The missiles are loaded/unloaded from the vehicle by the 9T318 loading system. Total weight is 19,465 kg with eight missiles. Length is 10.4 m, width 3.065 m and height is 3.765 m with the missiles.
- 9T31M1 truck crane.
- UKS-400V compressor unit.

In operation, the Buk-M1 battery CP receives its target information from the TAR. Once the CP has assigned the target priority, it then assigns each confirmed hostile target to the most appropriately positioned SPM. This then commences its own radar scanning and, when locked on and the target is within engagement range, usually launches two missiles at it. Each SPM can engage a different target, allowing the battery to engage six separate targets flying in the same direction but at different altitudes. The single-shot kill probability against a manoeuvring target is said to be 0.6 to 0.7 and a non-manoeuvring target 0.8 to 0.9.

The single-stage Buk missile (Russian designation 9M38M1) is similar in many respects to the US Navy's Standard MR-1 RIM-66 weapon. The missile has an autopilot and is semi-active radar homing using a centimetric waveband. Propulsion is by a solid-fuel rocket motor. The steering is by a gas feed generator system that actuates the flight controls. The weapon has a 70 kg HE warhead and is fitted with radar proximity and contact fuzing systems. It has a kill zone of 17 to 20 m. There is also a further missile variant known as the 9M39M1. The original missile for the Buk system had the designation 9M38.

An overview of the different systems' performance capabilities is given in the accompanying table.

Variants

There is a naval version of the Buk-M1 known as the *Uragan* (Russian for 'hurricane') missile system. Developed by the Altair NPO design bureau, it was given the US/NATO designation SA-N-7 'Gadfly'. The export version is known as the *Shtil* (Russian for 'surf'). The latter has been exported to India for use on its Delhi-class destroyers (two systems per ship) and China.

A new variant/upgrade to the system from Ahut, a Minsk-based research and manufacturing enterprise, has been announced. The Buk-MB will be tested during the later part of 2005.

By early 2006 and again in 2010 and 2012, Russia had offered the Buk-M1-1 system to Malaysia. Although not confirmed by June 2010, it is assessed to be an export variant of the Buk-M1 and probably the Buk-M1-2.

In the early part of 2007, a booklet released to celebrate the 50th anniversary of the Kapustin Yar Missile Test Range (KYMTR) mentioned the Buk-M3 variant. This is the first indication of such a variant and the fact that trials have been carried out of the system at KYMTR.

Buk variants comparison

Version	Buk	Buk-M1	Buk-M1-2	Buk-M3
IOC date:	1980	1983	1998	<2006
Missile designator:	9M38	9M38M1	9M317	unknown
Maximum target speed:	800 m/s	800 m/s	1,200 m/s	unknown
Engagement range:				
(minimum)	5,000 m	3,000 m	3,000 m	unknown
(maximum)	25-30,000 m	32,000 m	42,000 m	unknown
Target altitude:				
(minimum)	25 m	15 m	15 m	unknown
(maximum)	18-20,000 m	22,000 m	25,000 m	unknown
Maximum ground target range:	n/app	n/app	15,000 m	unknown
Ballistic target range:				
(maximum)	n/app	n/app	20,000 m	unknown
(minimum)	n/app	n/app	3,000 m	unknown
Ballistic target altitude:				
(maximum)	n/app	n/app	16,000 m	unknown
(minimum)	n/app	n/app	2,000 m	unknown
Missile launch weight:	685 kg	685 kg	715 kg	unknown
Warhead weight:	70 kg	70 kg	70 kg	unknown
Maximum missile speed:	850 m/s	850 m/s	1,230 m/s	unknown
Missile length:	5.55 m	5.55 m	5.55 m	unknown
Missile diameter:	0.40 m	0.40 m	0.40 m	unknown
Wing span:	0.86 m	0.86 m	0.86 m	unknown
Maximum loading:	unknown	20 *g*	24*g*	unknown

Specifications

The command post accepts and transmits information to the brigade level command post vehicle, such as a Baikal or Polyana system.

	9M38M1
Dimensions and weights	
Length	
overall	5.55 m (18 ft 2½ in)
Diameter	
body	400 mm (15.75 in)
Flight control surfaces	
span	0.86 m (2 ft 9¾ in) (wing)
Weight	
launch	685 kg (1,510 lb)
Performance	
Speed	
max speed	1,652 kt (3,060 km/h; 1,901 mph; 850 m/s)
Range	
min	1.6 n miles (3.0 km; 1.9 miles)
max	17.3 n miles (32.0 km; 19.9 miles)

	9M38M1
Altitude	
min	15 m (49 ft)
max	22,000 m (72,178 ft)
Ordnance components	
Warhead	70 kg (154 lb) HE fragmentation
Fuze	radar, proximity, impact
Guidance	semi-active, radar
Propulsion	
type	single stage, solid propellant

Status

In production.

A future replacement for the Buk system and currently in development is the Vityaz. This system is reported as a medium-range (50 km) weapon and is in development at the new Almaz-Antei Concern of Air Defence. During June 2004, Yuri Bely, the Tikhomirov-NIIP's Director General, gave an interview in which he is quoted as saying that work on upgrading the Buk SAM system's radar was continuing within the Almaz/Antei Concern of Air Defence, with the development of a common surface-to-air missile system expected shortly. Furthermore, the concept is to make provision for a modular system able to engage targets in the short, medium and long ranges by reconfiguring various units and modules.

The Buk system passed to Finland in a weapons for debt repayment for the protection of Helsinki (capital), appears to have a major problem in so much as its vulnerability to jamming. This problem is rather serious and hence, the Finnish government has allocated EUR110 million for the appropriation of a replacement system. By January 2009, further information on the replacement of the Buk released into the open press by the Finish government suggests a formal decision has been made in favour of the NASAMS II system. It is understood that Finland may be looking to purchase the Aster15/30 system that has just become available. Estimates of the total cost of a new missile defensive system range between EUR400 and 500 million.

Contractor

Tikhomirov Instrument Research Institute (design).
Ulyanovsk Machine Plant(production).
Kalinin Mechanical Engineering Plant (production of 9A39M1 launcher-loader and launcher for the Buk-M1).
Phazotron-NIIR JSC (involved in the development of the phased array fire-control radar for the system).
Ekran, NII Priborostroeniya.
OAO Logicheskie Sistem.
Ryazan Plant.

9K37M1-2 Buk-M1-2

Type

Self-propelled Surface-to-Air Missile (SAM) system

Development

The Tikhomirov Scientific Research Institute of Instrument Design at the MAKS-97 air show, displayed the Buk-M1-2 missile system for the first time, although most Western analysts tended to believe that this was basically an upgrade of the Buk-M1 system. Evidence now suggests that, in effect, it is a completely new system, utilising the latest developments in missile technology and available Commercial Off-The-Shelf (COTS) components from the Buk-M1, for both air and coastal defence operations. The Buk-M1-2 is available both as a new build item and as a modernisation package for completely rebuilding existing Buk-M1 systems. This system is also advertised by the primary contractor Almaz/Antei Concern of Air Defence, in Moscow.

Development of the Buk-M1-2 system began in the early nineties between 1994 and 1997 and its design parameters were expanded to include the capability to engage radar detectable surface (both ground and naval) targets, 150 km range tactical ballistic missiles and air-launched weapons and their carriers. In order to capitalise on the existing Buk-1M infrastructure to meet these requirements and thus minimise overall costs, it was decided that the Buk-M1-2 systems would use a new missile type 9M317 but have minimal changes to the existing radar, electronics, software guidance algorithms and other associated equipment.

Extensive testing of the Buk-M1-2 system radar with live and telemetry fitted missiles occurred during the mid-nineties against 9K14 ballistic missile, 9M31 surface-to-air, Smerch MLRS rockets, 3M8 missile target simulators as well as ground (such as a Tu-16 bomber and a tactical ballistic missile launcher) and naval (minesweeper) targets. These proved to be extremely successful and in 1998 the Buk-M1-2 system officially entered service with the Russian Army.

Description

The missile chosen to replace the 9M38M1 weapon was the 9M317, designed and manufactured by the same Dolgoprudny Scientific and Production Enterprise (DNPP). The missile was again developed as a joint weapon for Russian Army and Navy air defence systems and has an extended engagement envelope when compared to the original Buk-M1 9M38M1 missile. It has a designed service life in the field of 10 years before it has to be refurbished and

9A310M1-2 launcher vehicle in travelling position with 9M317 missiles
0044118

can be treated as a round of ammunition without the need for checks and maintenance during its operational life. The weapon is also designed to be inherently safe in use with enhanced anti-explosion features, especially if involved in a fire-related accident.

The missile design allows for the missile control system and warhead fuzing system to be pre-programmed according to the type of engagement. The type of target engaged may be ballistic, aerodynamic, target size (for example, how small), surface target type or helicopter thus adjusting the Single-Shot Kill Probability to meet the engagement circumstances. Surface targets such as missile launchers and grounded aircraft/helicopters can be engaged with direct fire. The missile also comes as:

- 9M317 GMM full-scale dummy
- 9M317ME vertical-launch naval missile
- 9M317 UD training active missile
- 9M317 UR training sectional dummy.

The basic Buk-M1-2 system comprises the following elements:

(1) 9S470M1-2 command post - upgraded though incorporation of a new ballistic target selection and designation capability into its computer, including provision for ground/naval target characteristics and signature to ensure optimal attack angles to minimise the effect of ground or water clutter effects and to optimise the warhead contact fuzing parameters.

(2) 9S18M1 target acquisition radar - not needed to be updated as the anti-missile defence feature of a 120° azimuth and up to 55° in elevation scan sector mode already present to engage tactical ballistic and anti-radiation missiles is deemed adequate for the 9M317 missile. In trials the radar has shown the capability to successfully track destroyer sized targets at a range of 35 to 40 km.

(3) Up to six 9A310M1-2 self-propelled launchers - improved by incorporation of the 9M317 missile, the use of new missile guidance algorithms (including those used against surface targets) on its digital computer and the fitting of updated missile attachment equipment. These additions have also been integrated with other new and existing equipment such as a TV optical sight, laser range-finder, navigation and communications equipment, an IFF interrogator, a built-in simulator and documentation equipment. During the engagement the 9A310M1-2 is responsible for:
   - Detecting the target
   - Determining its IFF status
   - Automatically tracking it and identifying its type
   - Computing the engagement mission parameters
   - Assigning the launch to its own missile(s) or attached loader-launcher vehicle, launching the missile(s)

Transmitting radio correction commands to the weapon(s) in flight and evaluating the firing results.

(1) Ability to either engage the target(s) as part of a battery network, with the initial designation data transmitted from the battery's 9S470M1-2 command post or act independently within an assigned area of responsibility. The launcher is still capable of using the 9M38M1 missile.

(2) Up to six 9A39M1-2 launcher-loader vehicles, improved by missile attachment equipment and missile transport cradle re-positioning.

(3) Up to 72 9M317 missiles (when part of an air-defence battalion). If required, a customer can further request the incorporation of an Orion passive radar control system into their Buk-M1-2 system. Against naval targets the system has been tested without the use of the radar. An optical tracking mode with laser range finder has been successfully demonstrated to show system capabilities in a heavy ECM environment.

(4) Maintenance and support assets.

Variants

Certain elements of the Buk-M1-2 system can be incorporated into the 2K12 Kub or export 2K12E Kvadrat (SA-6 'Gainful') missile systems to enhance their capabilities. In the first instance, a 9A310M1-2 self-propelled launcher can be introduced into a 2K12 (or 2K12E) battery with command and control being exercised by the battery's existing 1S91 self-propelled surveillance and targeting vehicle. This increases the battery's simultaneous target engagement capability to two targets, one conducted by the 9A310M1-2 with an attached 2P25 launcher if required and the other to the 1S91 and its attendant 2P25 launchers. For a more comprehensive upgrade of the 2K12 (or 2K12E) battery the 1S91 vehicle can be replaced by the 9S470M1 Buk-M1-2 command post vehicle.

The 9M317 missile (9M317ME) can also be used with Russian Navy missile launchers associated with the "Yezh" systems (NATO SA-N-7/SA-N-12). Reports dating from April 2004 from the Almaz/Antei Concern of Air Defence (Naval Radio Electronics Scientific Research Institute Altair) confirm that a 9M317ME missile is a vertical-launch missile, fired from high-firing-rate cell-type shipborne launchers. This weapon is designed for installation in warships from 1,500-tonne displacement. It would seem likely, therefore, that a land-based vertical launch system is either in development or has been developed.

A further stage or modernisation involves the integration into the Kub system of the 9A310M1-2 Transporter Erector Launcher And Radar (TELAR) with 9M317 missiles from the 9K37M1-2 Buk-M1-2 system, interfacing this with the 2P25M1 and M2 self-propelled launchers from the Kub/Kvadrat system and, in its latest form, the integration of the 9A317 TELAR with 9M317 missiles from the 9K37M2 Buk-M2.

Further Buk-M1-2 System Upgrades

As part of an ongoing system improvement programme, the design team has developed the following items that can be incorporated into the Buk-M1-2 system:

(5) Modernised 9A317 launcher vehicle with new phased array radar antenna system to increase the number of simultaneous target engagements;
(6) Enhanced jamming immunity due to the specific adapting of the phased array radar beam to potential tactical and ECM operating environments;
(7) Enhancing radar effectiveness of the 9A317 by increasing the transmitter output power and incorporating a new SHF receiver to improve its sensitivity incorporating new high-speed digital computers and state-of-the-art digital signal processing.

Incorporating modernised 9A317 launcher vehicles into a Buk-M1-2 battery increases its simultaneous engagement capability from 6 to 10 or even 12 targets.

A future replacement for the Buk system and currently in development is the Vityaz. This system is a medium-range (50 km) weapon and is in development at the new Almaz/Antei Concern of Air Defence.

Specifications

	9M317	9M317ME
Dimensions and weights		
Length		
overall	5.55 m (18 ft 2½ in)	5.083 m (16 ft 8 in)
Diameter		
body	400 mm (15.75 in)	400 mm (15.75 in)
Flight control surfaces		
span	0.86 m (2 ft 9¾ in) (wing)	-
Weight		
launch	715 kg (1,576 lb)	581 kg (1,280 lb)
Performance		
Speed		
max speed	2,391 kt (4,428 km/h; 2,751 mph; 1,230 m/s)	-
Range		
min	1.6 n miles (3 km; 1.9 miles) (aerodynamic aerial target)	-
	1.6 n miles (3 km; 1.9 miles) (anti-missile target engagement)	-
	1.6 n miles (3 km; 1.9 miles) (helicopter target, flying V>50 m/s)	-
	1.9 n miles (3.5 km; 2.2 miles) (naval target)	-
	1.9 n miles (3.5 km; 2.2 miles) (detectable ground target)	-
max	22.7 n miles (42 km; 26.1 miles) (aerodynamic aerial target)	-
	10.8 n miles (20 km; 12.4 miles) (anti-missile target engagement)	-
	19.4 n miles (36 km; 22.4 miles) (helicopter target, flying V>50 m/s)	-
	5.4 n miles (10 km; 6.2 miles) (helicopter target, hovering V = 0 m/s)	-
	13.5 n miles (25 km; 15.5 miles) (naval target, destroyer sized)	-
	9.7 n miles (18 km; 11.2 miles) (naval target, missile boat sized)	-
	8.1 n miles (15 km; 9.3 miles) (est.) (detectable ground target)	-
Altitude		
min	15 m (49 ft) (aerodynamic target)	-
	100 m (328 ft) (anti-missile target)	-
max	25,000 m (82,021 ft) (aerodynamic target)	-
	16,000 m (52,493 ft) (anti-missile target)	-
Ordnance components		
Warhead	70 kg (154 lb) HE blast fragmentation	HE blast fragmentation
Fuze	impact (ground/waterborne)	-
	proximity (air)	
Guidance	semi-active, INS (with radio correction)	semi-active, INS (with radio connection)
	semi-active, radar (seeker)	-
Propulsion		
type	solid propellant (dual thrust)	solid propellant (dual thrust)

Status

In production and in service with the Russian Army. Existing Buk-M1 systems of the Belarussian Armed Forces are also being upgraded to this standard. Offered for export.

During the Langkawi (Malaysia) 11 December 2005 Defence Exhibition, the Buk-M1-2 was shown for the third time within a year to the Malaysian Armed Forces. By January 2006, a report that Russia, through Rosoboronexport (Russia's principal arms exporter), has offered a version designated as the Buk-M1-1 to Malaysia. The designation may well be a mistake for the Buk-M1-2 but then again it could also be a version specifically modified for the Malaysian area of operations. As of July 2013, no further information is available on this variant.

It has been suggested by the Russian Press, that the Buk-M1 and M1-2 variants have been directly procured by Cyprus, however, no confirmation to date has been available.

Contractor

System development: Tikhomirov Scientific Research Institute of Instrument Design.
Buk-M1-2 manufacture: Ulyanovsk Mechanical Plant.
9M317 missile - design and manufacture: Dolgoprudny Research Production Enterprise.
Primary contractor: Almaz/Antei Concern of Air Defence.
Ekran, NII Priborostroeniya.
OAO Logicheskie Sistem.
Ryazan Plant.

Antei-2500

Type

Self-propelled Surface-to-Air Missile (SAM) system

Development

The then Antei, (now known as Almaz/Antei Concern of Air Defence), designer of the S-300V, S-300VM, S-300V1, S-300V2 and S-300V4 missile systems, has developed for export the Antei-2500. The Antei-2500 (also translated as Antey) is the export version of the S-300VM (NATO SA-12 'Giant' and 'Gladiator') mobile SAM system from the earlier S-300V. There have been up to five batteries of the S-300VM deployed around Moscow. This provided protection against aircraft, tactical, ballistic and cruise missiles, and theatre ballistic missiles with launch ranges up to 2,500 km.

Description

Antei-2500 is a second-generation export system that is capable of simultaneously engaging 24 air-breathing targets, including stealth targets, or 16 ballistic targets. One of the new components is a sector scanning radar that operates at one revolution per second.

This, coupled with the improved characteristics of the radar information facilities and optimization of the radar signal processing, makes it possible to engage high-speed ballistic targets at velocities of up to 4,500 m/s and with radar cross-sections as low as 0.02 m^2. The S-300VM can engage targets with a radar cross-section as low as 0.01 m^2.

Radio command guidance signals correct the missile trajectory in the second stage of its flight, turning it 60° on the longitudinal axis to focus the warhead blast onto the target to optimise its blast-cone pattern.

9M82 and 9M83 missile launchers (Almaz/Antei) 0517567

The Antei-2500 uses 9M82M and 9M83M missiles with weight, size characteristics, guidance system concept and warhead effect remaining essentially the same as the S-300V/S-300V1/S-300V2 system missiles. What are enhanced, however, are the manoeuvring characteristics, so it can engage highly agile targets.

Maximum air-breathing target engagement range is 200,000 m at altitudes from 25 to 30,000 m. Ballistic targets can be destroyed at ranges up to 40,000 m at altitudes from 1,000 to 30,000 m.

Missile tests indicate that the system has a kill probability of 0.98 against ballistic missiles. In recent trials, successful firing of over 50 missiles have been made against missiles with terminal velocities of over 4,500 m/s while two Tu-16 bomber drone targets were destroyed at 200,000 m range.

The Antei Tor-M1 (now the Almaz/Antei Concern of Air Defence) is recommended to provide close-in protection for the Antei-2500 system.

A grouping consisting of several air-defence missile systems is controlled by the POLYANA-D4M1 Air Defence Brigade level automated command post.

Specifications

Antei-2500
(As for S-300V missiles with following exceptions:)

Maximum engagement range:	
(air breathing targets)	200,000 m
(ballistic missile targets)	40,000 m
(stealth air breathing targets)	25,000 m
Maximum altitude:	
(air-breathing targets)	30,000 m
(ballistic missile targets)	30,000 m
(stealth air breathing targets)	30,000 m
Minimum altitude:	
(air-breathing targets)	25 m
(ballistic missile targets)	1,000 m
(stealth air breathing targets)	25 m
Number of simultaneous targets:	24/48
Weight of warhead:	150 kg
Speed:	2,600 m/s

System Composition

Target detection and designation centre:

- up to four missile launching systems
- missile and technical support facilities
- training facilities.

The target detection and designation centre includes:

- 9S457ME command post
- 9S15ME target acquisition radar
- 9S19ME sector scanning radar

Each missile system includes:

- 1 × 9S32ME multi-channel missile guidance radar
- up to 6 × launcher vehicles with 4 × 9M83ME missile and 2 × 9M82ME missiles
- up to 6 × launcher/loader vehicles

Training facilities include:

- 9F681ME computer-based integrated training system
- missile training/combat and weight dimension mock-ups
- target detection radar 9S15ME and 9S19ME main facilities of air environment information that allow the system to conduct autonomous combat operations without using radar field and missile attack early warning system data.

Status

The Antei-2500 has been offered for export, with production from new but as of October 2008, no customers have been identified for the system. However, in September 2009, Venezuela using a USD2.2 billion loan from Russia signed a contract for the system, dates for delivery have not yet been released. Russian advisers suggest that the production of the S-300V is complete and should any orders be forthcoming the line will have to be completely restarted from new. However, should the numbers ordered for overseas sales be such that systems can be provided from obsolete or replaced (with S-400) from systems within Russia, it would probably be more economical for both parties if those coming from the Russian front line of defence or reserve stocks be prepared and made ready for export and then be sold than it would be to restart a complete production line.

Contractor

Almaz/Antei Concern of Air Defence.

Buk-M2 and M3

Type

Self-Propelled Surface-to-Air Missile (SAM) system

Development

The 9K40 Buk-M2 (NATO SA-17/'Grizzly') was developed and produced by the Tikhomirov Instrument Research Institute. The system is the evolutionary replacement system for the Buk-1 family (SA-11/'Gadfly') army level SAM and is similar in concept to that system.

Known as the Buk-M2, the complete system, including radar and support equipment, is code named as Ural (a Russian river). The system completed development in 1990 following trials and firings at the Kapustin Yar Missile Test Range (KYMTR).

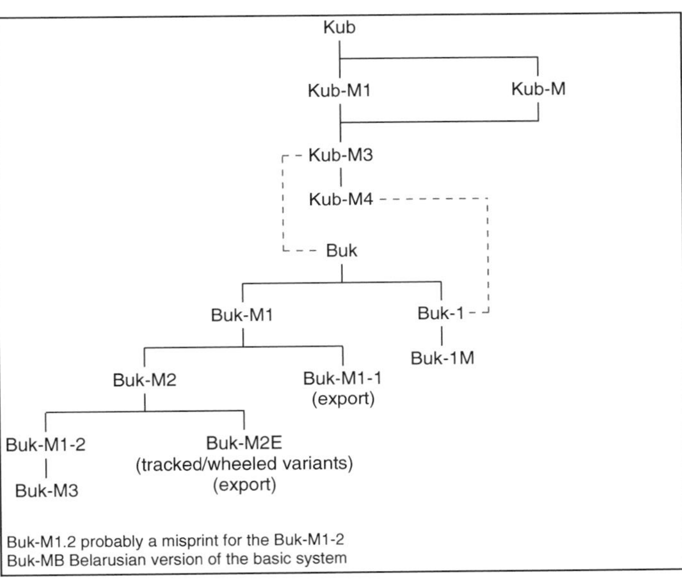

Buk family flow diagram (James C O'Halloran) 1040815

Command post vehicle used with SA-17 Buk-M2 SAM system (Steven Zaloga) 0080445

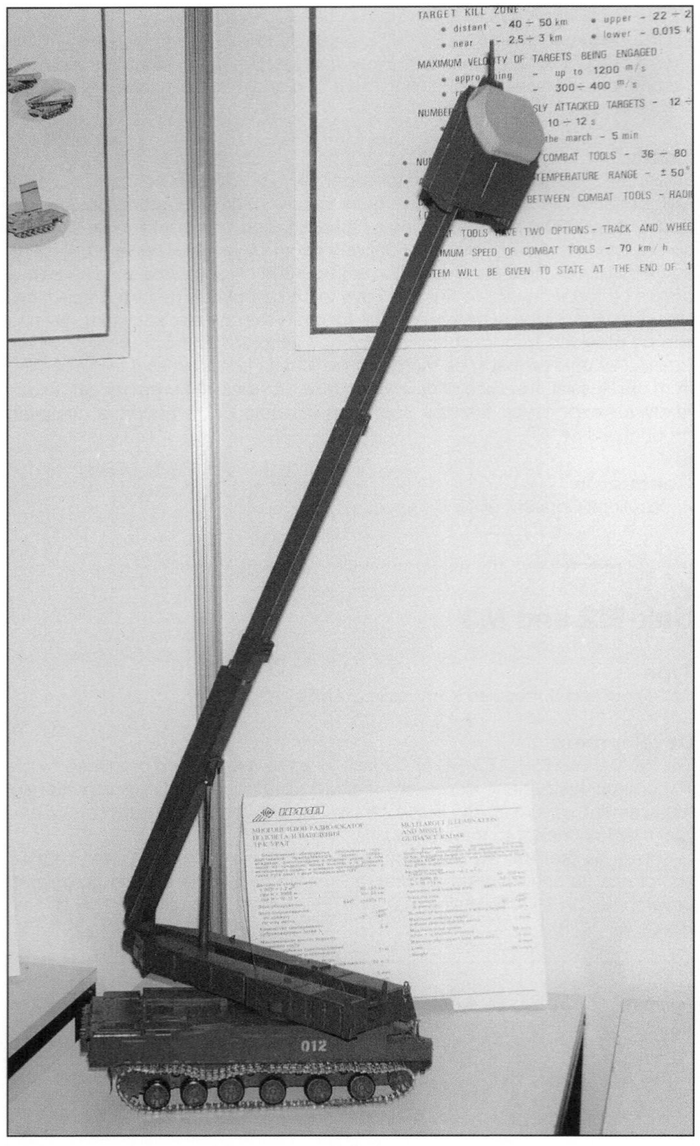

NATO codename 'Chair Back' illumination and missile guidance radar for SA-17 Buk-M2 deployed in operational configuration on top of 21 m telescopic mast (Steven Zaloga) 0080447

The Buk-M2 was formally accepted into service with the Russian Army during 1995, following troop acceptance trials during the early 1990s. However, according to a report in January 2008 by Colonel General Nikolai Frolov, the then Chief of the Army Air Defence Forces, the Buk-M2 was again reported as entering service with the Russian Army in 2008. There appears here to be some confusion between the Buk-M2 and the Buk-M3; the Buk-M2 was first reported as being accepted into service with the Russian Armed Forces in 1995, whilst the Buk-M3 was expected to enter service in 2008/09. Both of these dates emanated from Colonel General Frolov and therefore it is assessed that the M2 was indeed 1995 and that M3 was 2008/2009.

Reports emanating from Moscow in early September 2010, suggested that the Buk-M2 was successfully deployed to the Astrakhan Proving Grounds on the Lower Volga where test firings took place against manned and unmanned targets, including cruise missiles. Five targets were engaged and destroyed by the upgraded system that also boasts increased range plus the ability to lock onto as many as 24 incoming targets, as opposed to six with previous batteries.

The Buk-M2 is designed to engage strategic and tactical aircraft, tactical ballistic missiles, cruise missiles, tactical air-launched missiles, helicopters (including hovering targets), and RPV's at extremely low levels (around 10 m) up to 24,000 m altitude.

Information deriving from Moscow during August 2007 suggests that the Buk-M2E (export) variant can be integrated with independent parts of the S-300VM, a Buk-M3 is also known to exist.

An associated modernisation program involves the integration into the Kub system of the 9A310M1-2 transporter erector launcher and radar (TELAR) with 9M317 missiles from the 9K37M1-2 Buk-M1-2 system, interfacing this with the 2P25M1 or M2 self-propelled launchers from the Kub/Kvadrat system and, in its latest form, the integration of the 9A317 TELAR with 9M317 missiles from the 9K37M2 Buk-M2.

Description

The Buk-M2 system is available in two formats:

(1) A tracked vehicle version;
(2) A wheeled version, which utilises the following key elements of the system:
- Command post;
- Target acquisition radar;

Phased-array target acquisition radar used with SA-17 Buk-M2 SAM system (Steven Zaloga) 0080446

Self-propelled firing unit for SA-17 Buk-M2 showing launcher with four 9M317 missiles elevated (Steven Zaloga) 0080448

- Illumination and missile guidance radar;
- Launchers fitted to semi-trailers and towed by 6 × 6 tractor trucks.

The wheeled version is used mainly in static defence of high value areas, such as command centres, airfields and so on. It effectively replaces the S-125 system.

There is a Buk-M1-2 system that utilises the 9M317 missile. The Buk-M2 incorporates the best of design concepts and technical solutions as well as basic modes of operation used in the Buk-M1-2.

Command Post 9S510

The Command Post (CP) vehicle conducts analysis of the air situation around the launch unit, controls and monitors the various fire units and designates and assigns the targets to the individual launch vehicles. Communication is via extendable radio communications antenna and wire.

Target acquisition radar 9S18M1-3

This multi-function phased-array radar system vehicle provides the air situation picture to the Command Post by performing the target detection and IFF functions and target acquisition, tracking and classification. Communication is via extendable radio antenna and wire link. The radar is a modified variant of the 9S18M NATO codename 'Snow Drift' surveillance radar.

Illumination and missile guidance radar 9S36

This vehicle, known as Giraffe, utilises a high jam resistant electronically scanned phased-array engagement radar antenna mounted on top of a 21 m high telescopic arm mast. The radar, NATO code name 'Chair Back B', performs target acquisition and tracking (including those targets in a terrain-following flight profile), target classification and air situation analysis. In the system's firing mode it performs target illumination and transmits the course correction commands to the missile(s) in flight. It also controls the Loader-Launcher vehicles. Communication is via extendable radio antenna and wire link.

Self-propelled firing unit 9A317

This tracked vehicle can operate as: part of the SA-17 system using the battery CP in the target designation mode with a Loader-Launcher; a separate fire unit with a Loader-Launcher; or an individual unit in its own right assigned to protect an installation, using its onboard high jam resistant electronically controlled 'Chair Back A' phased-array radar. The latter can perform target acquisition, tracking and IFF functions, as well as providing target classification and air situation analysis. The vehicle's primary role, however, is target illumination, the transmission of missile course guidance corrections by radio and control of

Self-propelled loader launcher for SA-17 Buk-M2 which carries a total of eight 9M317 missiles (Steven Zaloga) 0080449

Wheeled version of Buk-M2 showing self-propelled loader launcher with four 9M317 missiles in the ready to launch position and a further four for rapid reloading (Steven Zaloga) 0080450

the Loader-Launcher. A total of four ready-to-fire missiles are carried on a turntable type launcher assembly mounted on the vehicle's rear decking. The launcher was designed and developed by the Start Scientific Production Enterprise, OAO/NPP in Yekaterinburg.

The phased-array radar is mounted at a raised angle at the head of the turntable assembly. Communication is via extendable radio antenna and wire link.

Self-propelled loader-launcher 9A316

This vehicle is intended for the storing, transportation and, when required, the launching of SA-17 missiles. It is similar in appearance to the self-propelled firing unit but does not have the radar assembly mounted on the front of the launcher turntable. Instead it has a further four missiles carried in pairs on either side of the turntable in load carrying units. This can either be used to reload itself or for reloading its parent self-propelled firing unit. A folding hydraulic crane is carried on the front top of the turntable assembly for transferring the missiles. Communication is via extendable radio antenna and wire link.

Missile 9M317

The 9M317 missile was developed by the Dolgoprudny Engineering Plant (DRPE JSC). It is a high-precision guided weapon integrated in the Army and Navy air defence systems and is intended to defeat modern as well as future aerodynamic, ballistic, surface and radio-controlled surface threats. It uses terminal, semi-active radar 9E240 for terminal homing with inertial mid-course guidance and radio datalink course correction updating. The propulsion system employs a two-phase solid propellant rocket motor to give the weapon a maximum velocity of 1,200 m/s, power aerodynamic controls laying the missile guidance trajectory with a minimum miss-in-the-point of its collision with the target. A second missile type with a delta-wing configuration is believed to be in development, the increased wing area provides for increased manoeuvrability when engaging a target. The 9M317 missile is used in the Army, in the air defence missile system Buk-M1-2 and the 9M317E missile is used in the Navy, in air defence missile system Shtil-1. The missiles can be integrated in the newly-developed systems providing them with multi-functional ability.

When engaging a hovering helicopter target, the 9M317 missile uses a newly developed low altitude-aiming algorithm.

Battery configuration

A typical Buk-M2 regiment comprises the following:

- A Command Post section with a Command Post vehicle and Target Acquisition vehicle;

Phased-array target acquisition radar (centre) used with wheeled version of Buk-M2 (Steven Zaloga) 0080451

- Two Type II battery units each with one Illumination and Missile Guidance radar vehicle and two Loader-Launcher vehicles;
- Four Type I battery units each with one self-propelled Fire Unit and one Loader-Launcher.

A regiment can typically engage between 12 and 24 targets simultaneously. Maximum speed on the march of a regiment is 70 km/h.

Variants

A Russian naval variant, the Yeozh is also in service with Project 965A Sarych-class destroyers.

Originally developed for the Buk-M2, the 9M317 is also used by the Shtil-1 naval system already sold to India and China. It is also being offered to potential customers as part of the upgraded Buk-M1-2 system.

A further variant using the 9M317ME missile is for vertical launch from warships and is also offered for sale.

Buk-M2E is the export variant and may be integrated at the customers request with independent parts of the Tor-M2E and S-300VM. The Buk-M2E variant has an increase in the engagement channels from six (6) in the M1 variant to 24 in the M2.

A wheeled variant of the Buk-M2E has been offered for export to Greece. At the Defendory 2008 exhibition held biannually in Athens, the wheeled variant with all of its associated components (see below):

- The KP command post;
- The SOTs target-acquisition radar;
- The SOU self-propelled fire unit;
- The SPN target illumination and guidance radar;
- The PZU launcher-loader unit.

The system was shown for the first time by the Almaz/Antei Concern for Air Defence and offered for sale. It is believed that work on the new system started in 2007 with mock-ups due for completion by the end of 2008. These vehicles are expected to be used to assess the mechanical characteristics of future planned vehicles.

Also at the MAKS 2007 exhibition in Moscow, a Buk-M3 was exhibited on what appeared to be a standard tracked vehicle, later reports have suggested that the Buk-M3 was expected to enter the Russia's Armed Forces inventory between 2008 and 2009.

Specifications

Battery configuration
(Has the following capabilities under autonomous operating conditions)

	Type I battery	**Type II battery**
Simultaneous target engagement capability:	up to 4 targets	up to 4 targets
Acceptance height of terrain in firing direction:	up to 2 m	up to 20 m
Reaction time:	4 s	8-10 s
Single firing unit launch rate:	1 missile/4 s	1 missile/4 s
Readiness times:		
(from march)	5 min	10-15 min
(after displacement)	20 s	n/avail

	Buk command post - wheeled	Buk command post - tracked	Buk target acquisition radar- wheeled	Buk target acquisition radar- tracked	Buk illumination and missile guidance radar- wheeled	Buk illumination and missile guidance radar- tracked	Buk Self-propelled firing unit	Buk Self-propelled Loader-Launcher - wheeled	Buk Self-propelled Loader-Launcher - tracked
Dimensions and weights									
Length									
overall:	21 m	8 m	21 m	8 m	21 m	8 m	8 m	21 m	8.0 m
Width									
overall:	3 m	3.3 m	3.3 m	3.3 m	3 m	3.3 m	3.3 m	3 m	3.3 m
Height									
overall:	3.8 m	3.8 m	3.8 m	3.8 m	3.8 m	3.8 m	3.8 m	3.8 m	3.8 m
Weight									
combat:	25,000 kg	30,000 kg	30,000 kg	35,000 kg	30,000 kg	36,000 kg	35,000 kg	35,000 kg	38,000 kg
Mobility									
Configuration									
running gear:	wheeled	tracked	wheeled	tracked	wheeled	tracked	wheeled	wheeled	tracked
Firepower									
Armament:	9M317 SAM	–	–	–	–	–	–	–	-

Missiles

	9M317	9M317ME	9M317E
Length:	5.550 m	5.083 m	5.550 m
Diameter:	0.400 m	0.400 m	0.400 m
Wing span:	0.860 m	0.860 m	0.860 m
Launch weight:	715 kg	581 kg	715 kg
Propulsion:	2-phase solid propellant rocket motor	solid	dual-mode solid propellant motor
Guidance:	multi-mode terminal semi-active radar homing with inertial mid-course and course correction radio datalink updating	multi-mode terminal semi-active radar homing with inertial mid-course and course correction radio datalink updating	multi-mode terminal semi-active radar homing with inertial mid-course and course correction radio datalink updating
Warhead:			
(weight)	50-70 kg	unknown	50-70 kg
(type)	HE fragmentation	HE with hitting elements	HE fragmentation
Fuze:	radar proximity and contact	unknown	radar proximity and contact
Maximum speed:	1,230 m/s	1,200 m/s	1,200 m/s
Maximum manoeuvrability:	24 *g*	24 *g*	24 *g*
Effective range:			
(maximum)			
(general)	42,000 m	45,000 m	n/avail
(Tactical Ballistic Missile - TBM)	n/avail	20,000 m	n/avail
(Cruise Missile-CM)	n/avail	15,000 m to 20,000 m	n/avail
(minimum)	3,000 m	3,000 m	n/avail
Effective altitude:			
(maximum)	25,000 m	25,000 m	n/avail
(minimum)	15 m	15 m	n/avail
Maximum target speed:	1,200 m/s	unknown	n/avail
(approaching)	≤1,200 m/s	unknown	n/avail
(receding)	300-400 m/s (depending on target type)	unknown	n/avail
Simultaneous target engagement:	unknown	24	n/avail

Status

Ready for series production on receipt of orders. In limited service with Russian armed forces (initial production systems only). The system probably entered full service with the Russian Army after testing in 2008. China and South Korea have also assessed the system.

In May 2008, it was announced that Saudi Arabia is to purchase twenty (20) Buk-M2E weapon systems, it is not known if these are to be the new wheeled version or tracked. No date for delivery has been set.

As announced by the Air Defence Force's chief Major General Alexander Leonov on 28 December 2013, the Russian Armed Forces will undergo a planned rearmament to replace the Buk-M1 and Buk-M2 systems with the advanced Buk-M3 commencing in 2016.

Contractor

V Tikhomirov Scientific Research Institute of Instrument Design.
Start Research and Production Enterprise.
Dolgoprudny Research Production Enterprise JSC.
Almaz Central Design Bureau JSC (Launcher development).

Kolchan

Type

Self-propelled Surface-to-Air Missile (SAM) system

Development

First shown at the MAKS 2007 show in Moscow, the Kolchan system is a block built self-propelled SAM that employs a standard BTR-60 vehicle and several types of missiles from other SAM systems. Missiles usable by the system includes the; 9M37, 9M37M, 9M37MD and 9M333. The system (9K35M3-K) is employed for the direct protection of troop units against low-flying targets in all types of combat situations; on the march or in defence of static installations.

Description

The Kolchan system is designed to engage fixed wing, rotary wing, UAV and cruise type targets in the presence of natural interference and/or optical InfraRed (IR) countermeasures. The system can be used by day or at night.

9M37 Launcher canisters 0517574

Carried in an upgraded 8 × 8 BTR-60 Armoured Personnel Carrier (APC), the system mounts a launcher with four launcher guides that hold the missiles in containerised boxes. Also mounted on the vehicle is:

- an aiming device;
- target acquisition and identification device;
- communication facilities;
- targeting information reception and distribution devices.

The primary purpose of the system's Electro-Optic Target Acquisition and Tracking Station (EOTATS) is to search for, detect and track air targets in the InfraRed (thermal) spectrum. This is based on a thermal-imaging module that provides round-the-clock search and acquisition of air targets.

Variants

The 9K35 Strela-10, or at least the 9M37 missiles from the system.

Specifications

Kolchan

Type:	self-propelled surface-to-air missile system
Range (slant):	
(maximum)	5,000 m
(minimum)	800 m
Altitude:	
(maximum)	3,500 m
(minimum)	1 10 m
Cross track parameter:	
(maximum)	3,000 m
(minimum)	0 m
Target speed:	
(maximum)	420 m/s
(minimum)	0 m/s
Missile launch condition:	head-on or tail chase
Missile load:	4 with 4 reserve

Status

Development complete and offered for sale as early as 2006/07. No known sales to date.

Contractor

Muromteplovoz JSC, (Murom, Vladimir Region, Russia).

Morfei

Type

Self-propelled Surface-to-Air Missile (SAM) system

Development

First identified in 1999 (NIOKR), the system went into OKR in 2007; the Morfei (also known as Morpheus) is a Very Short-Range Surface-to-Air (VSHORAD) missile system that has been designed for the point defence and protection of other long-range offensive and defensive strategic missile systems such as Inter-Continental Ballistic Missiles (ICBMs) and the S-400 and S-500. The system has also been described by the developers as Super Short-Range.

The missile is assessed to be the 9M100 (or a variant thereof) which in itself was initially a short-range air-to-air weapon, however, other sources have assessed the missile as the 9M338K22 with a range of up to 10 km. The launcher has been described as a vertical launcher with a high number of missiles, up to 36.

Description

The Morfei weapon system is an advanced VSHORAD system for the armed forces of Russia that will be capable of countering the latest generation of targets such as:

- Fixed wing aircraft
- Helicopters
- Cruise missiles
- Guided bombs
- Supersonic missiles
- Unmanned Aerial Vehicles (UAVs).

Radar

If the Morfei's development continues as originally conceived, the radar associated with the system will be unique. It will have an omnidirectional cupola-type and not a rotating one as is standard in other SAMs. Two types have been identified, the first is a multifunction dome type and the second is a triangular type with an option of an EO device mounted on top. Two designators identified from photography have so far been identified with the radar, these are 29Ya6 and 43Ya6. The multifunctional 29Ya6 was created under the Morfei programme and according to unconfirmed reports a circular phased array radar or APAA with dome lens. The hemispherical PAR consists of 2,048 transceiver modules.

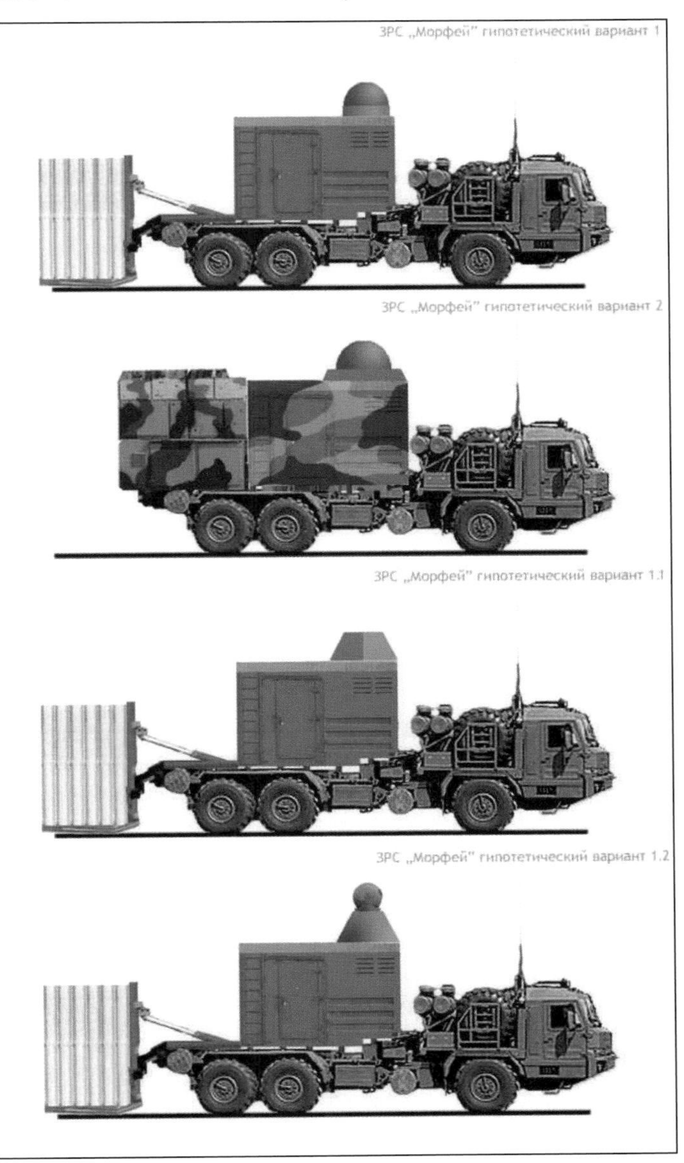

Morfei radars (RIA Novosti) 1445529

Variants

None known.

Specifications

Morfei

Type:	vertical launched Very Short-Range SAM (VSHORAD)
Length:	2.5 m
Diameter:	0.125 m
Weight:	70 kg
Warhead:	n/avail
Fuze:	Contact and remote
Propulsion:	solid propellant rocket
Guidance:	n/avail
Range:	
(maximum)	5 km (some unconfirmed sources report 10+ km)
(minimum)	n/avail
Altitude:	
(maximum)	<5,000 m
(minimum)	n/avail
Radar:	believed to be in the upper X-band or lower Ku-band supplemented with an Electro-Optics Suite (EOS)

Status

In development as of June 2013, not expected to become fully operational until 2015.

Contractor

Almaz/Antei Concern of Air Defence.

Pantsyr-S1-0

Type
Self-propelled Surface-to-Air Missile (SAM) system

Development
The Pantsyr-S1-0 was designed by KBP for air-defence of small-size military and industrial objects and areas against aircraft, helicopters, cruise missiles and high-precision weapons, as well as air-defence group coverage while repelling mass air threats.

Description
Unlike its big brother, the Pantsyr-S1, the S1-0 has no gunnery system attached but still uses the same 57E6 (57E6-E for export) surface-to-air missile. The 57E6-E missile shown at the FIDAE 2004 exhibition in Santiago, Chile, strongly resembles that of the Hermes missile previously exhibited and displayed within Russian magazines and publicity material.

The systems components are:
- Combat vehicle (up to six CVs in a battery)
- Battery command post
- Surface-to-air missiles
- Transloader vehicles (one vehicle for three CVs)
- Training facilities
- Maintenance facilities
- Common spare parts (SPTA) kit.

The Pantsyr-S1-0 features:
- Autonomous combat use
- Optical-electronic armament control system ensuring passive operation mode and high-precision guidance due to IR long wave band channel with logical signal processing and automatic target tracking
- Use of high-velocity and manoeuvrability SAM featuring high kill probability (P_k = 0.7 - 0.95) against all target types
- Round-the-clock combat use
- Operational coordination within a battery
- Modular design of the combat vehicle, enabling its versions on different carriers, wheeled, tracked and shelter variants.

Specifications
57E6 missiles

	57E6	57E6-E
Ammunition load:	8 in two banks of four	8 in two banks of four
Control system:	multiple-band radar-optical	multiple-band radar-optical and laser
Range:		
(maximum)	18,000 m	20,000 m
(minimum)	1,500 m	1,500 m
Altitude:		
(maximum)	10,000 m	10,000 m
(minimum)	5 m	15 m
Number of targets simultaneously engaged:	1	1
Crew of vehicle:	3	3

Status
Development complete. As of June 2013, no known foreign sales have been identified and it is assessed that they probably never will. The 57E6-E missile was offered for sale to South American countries at the FIDAE 2004.

Contractor
KBP Instrument Design Bureau, Tula, Russia.

57E6 missile for Pantsyr-S1-0 (KBP Instrument Design Bureau) 0563389

Pantsyr-S1-0 on 8 × 8 truck (KBP Instrument Design Bureau) 0137895

Phoenix

Type
Self-propelled Surface-to-Air Missile (SAM) system

Development
The Moscow-based Russian-Belarussian joint-venture group Industrial Company Defence Systems consortium (Oboronitelniye Sistemi) is marketing an InfraRed Search-and-Track (IRST) air-defence system called Phoenix (Feniks). The company has also been identified calling itself the Interstate Financial Industrial Group Defence Systems. Defence Systems reported that the Phoenix can be installed on both land and sea platforms to provide accurate detection, tracking and data classification of various target types. According to Russian sources, trials have taken place with the following air-defence systems:
- Tunguska
- Tor-M1
- Osa
- Pechora
- Strela-10
- Kub.

Description
The Phoenix is a panoramic passive InfraRed Surveillance System (IRSS) that is designed to automatically detect, identify and track air and surface targets, as well as measure their range, azimuth and elevation angles. It also provides detection, tracking and data classification, as well as accurate designation of various surface (tanks, fast patrol boats etc.) and aerial (subsonic and supersonic missiles, bombs, helicopters, UAVs etc.) targets. The system can discriminate between targets clustered in azimuth or elevation. It also minimises reaction time of combat systems while employed in conjunction with the radar.

The system can be deployed both independently and together with air defence weapon and radar systems.

Missile
The missile system advertised for use with the Phoenix is the Kolomna-based KBM Igla (NATO reporting name SA-18 'Grouse'), or Igla-1 (NATO reporting name SA-16 'Gimlet'). It is expected that these missiles, used with the Phoenix configuration, would draw their power and targeting information from within the self-propelled vehicle. Therefore, the weapons system can logically be expected to improve on time to target tracking and lock-before-launch capabilities. It has also been assessed that the Igla-S can also be used with the system bringing the Phoenix up to date with a very modern missile.

Computer module for the Phoenix air defence missile system 0116613

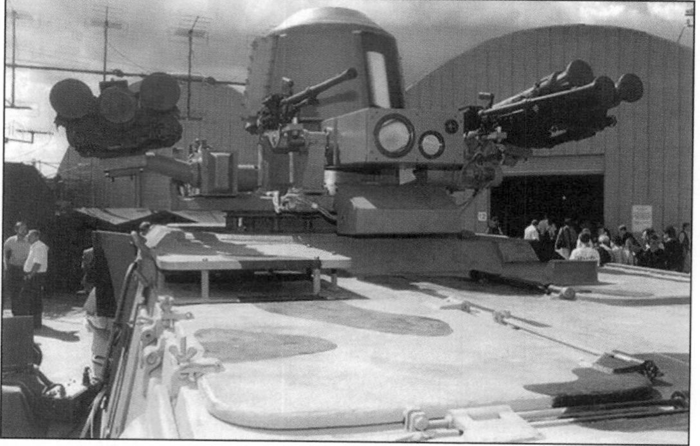

Infrared search and track device for the Phoenix 0116612

Vehicle/Station
The vehicle/station comprises:
- Electro-optic device
- Data display and control console with the control computer
- Special calculator
- Power supply unit
- The set of the connective cables and binders.

The following options are available for use with the Phoenix:
- Turret device
- Stabilised platform
- Laser range scope with the thermal imaging guidance channel.

Operation
The Phoenix, operates within the 8 to 12 micron Long-Wave InfraRed (LWIR) region and is claimed to be able to detect low-altitude targets at a range of 5 to 7 km, helicopters at 8 to 9 km, tactical aircraft at 15 to 18 km and cargo aircraft at over 20 km. It can track more than 50 land- and sea-based targets simultaneously.

The IRST includes the sensor pod, display and control console. It has full 360° coverage in azimuth and –10° to +40° in elevation. The IRST has been shown mounted on the roof of a GAZ-3937 with four fire-and-forget SAMs and one 7.62 mm PKT machine gun on both sides. The missile sighting system is situated between the pods of weapons and moves in elevation with these weapons.

Phoenix

	Phoenix	Improved Phoenix
Operating band:	8-12 μm band	8-12 μm band
Detection range:		
(low-altitude missiles)	5-7 km	5-7 km
(helicopters)	8-9 km	8-9 km
(aircraft (fighter-type))	15-18 km	15-18 km
(cargo planes)	>20 km	>20 km
Surface vessels: (ships)	unknown	15 km
Field of view: (in elevation)	≤18°	20°
Coverage:		
(in elevation)	–10° to +40°	–10° to +60°
(in azimuth)	360°	360°
Resolution:	0.5 mrad	0.5 mrad
System update rate:	0.5 Hz	unknown
Number of simultaneously tracked targets:	>50	100
Target updates:	unknown	2 × one second
Power consumption:	2 kVA	unknown
Weight: (opto-mechanical device)	250 kg	unknown

Status
As of August 2012, the system has completed development and is in limited production as required. It is believed to have been sold to Syria with the Igla missile system mounted on the Dzhgit dual rail launcher.

In late March 2012, the JSC announced that an improved version of the Phoenix is in development, JSC and the Optiko-Elektronniye Tekhnologii have formed a joint venture in this project.

Contractor
JSC Defence Systems.
Optiko-Elektronniye Tekhnologii (joint venture).

S-300P

Type
Self-propelled static Surface-to-Air Missile System (SAM)

Development
Development of the S-300P (NATO SA-10 'Grumble') air-defence system began in 1967 under the control of the Primary Developer now known as Almaz/Antei Concern of Air Defence, then known as the 'Almaz' Scientific Industrial Corporation (also known as the Almaz Central Design Bureau). The US/NATO codename SA-10 'Grumble' has been allocated to the system. Sub-contracting for the missiles was let to the Fakel missile design bureau, launchers to the MKB Start, with ground support equipment being worked in the Kalinin area at the Machine-Building Design Bureau (MKB).

The S-300P system was specifically designed as a semi-mobile all-weather strategic air-defence system to replace the S-25 Berkut and S-75 Desna/Dvina systems. The S-25 US/NATO codename is SA-1 'Guild' and the S-75 is 'Guideline'. The S-300P system and missile network was deployed in three concentric rings around Moscow for use against multi-target air-breathing threats and later in a limited Anti-Theatre Ballistic Missile (ATBM) capacity and cruise missiles scenario.

The system development was assigned to Boris V Bunkin, general designer for Almaz, at the time, while development of the accompanying missile was allocated to the 'Fakel MKB' (formerly the 'Grushin' Design Bureau) missile

64N6 radar vehicle of Almaz 86M6 Brigade command-and-control System (Paul Jackson) 0069507

SA-10 'Grumble' missiles being unloaded by a KrAZ-260 (6 × 6) truck using its onboard handling system (T J Gander) 0509758

design bureau. The missiles have a life expectancy of around 20 years in the launcher canisters, and this has been reflected in the announcement that older 5V55R missiles that have been withdrawn from service are to be converted into targets.

By 2000, there were some 1,900 plus S-300P launchers in various forms in service with the Russian air-defence force. These had not only replaced the S-25 system but had basically supplanted the S-75 and S-125 (US/NATO codename SA-3 'Goa') static SAM systems.

Since the initial conceptual design over 36 years ago, the S-300P family of systems has grown into what must be one of the largest weapons developments. The following flow diagram is believed to be as near to the family tree as is possible from the information that is available. Official Russian dates for Initial Operational Capability (IOC) of the various systems read as follows: S-300PT 1979; S-300-PT1 1981; S-300PS 1983; and S-300PMU1 1995.

Over the years, a number of S-300P developments/improvements/upgrades have been made which include the systems listed here.

Description
S-300PT (US/NATO codename SA-10a 'Grumble')
The S-300PT system (P = *podvizhnyi*, Russian for 'mobile') became operational in 1978/79 and used towed semi-trailer erector-launchers with four missiles mounted in pairs and erected towards the semi-trailer's front for launching. A set of outriggers at the front of the semi-trailer was erected to stabilise the platform before a launch could occur. Emplacement time was stated to be over 30 minutes.

The engagement radar used is the 5N63 (NATO codename 'Flap Lid') I/J-band phased array set. The usual battery configuration was three semi-trailer launchers and single 5N63 radar. The battery could simultaneously engage up to a maximum of six targets with 12 missiles under command guidance.

The radar was designed specifically for low altitude performance against air breathing targets. As a result, the antenna could be mounted and was often sighted on a mobile tower, which consisted of a 15 m long elevating mast on a

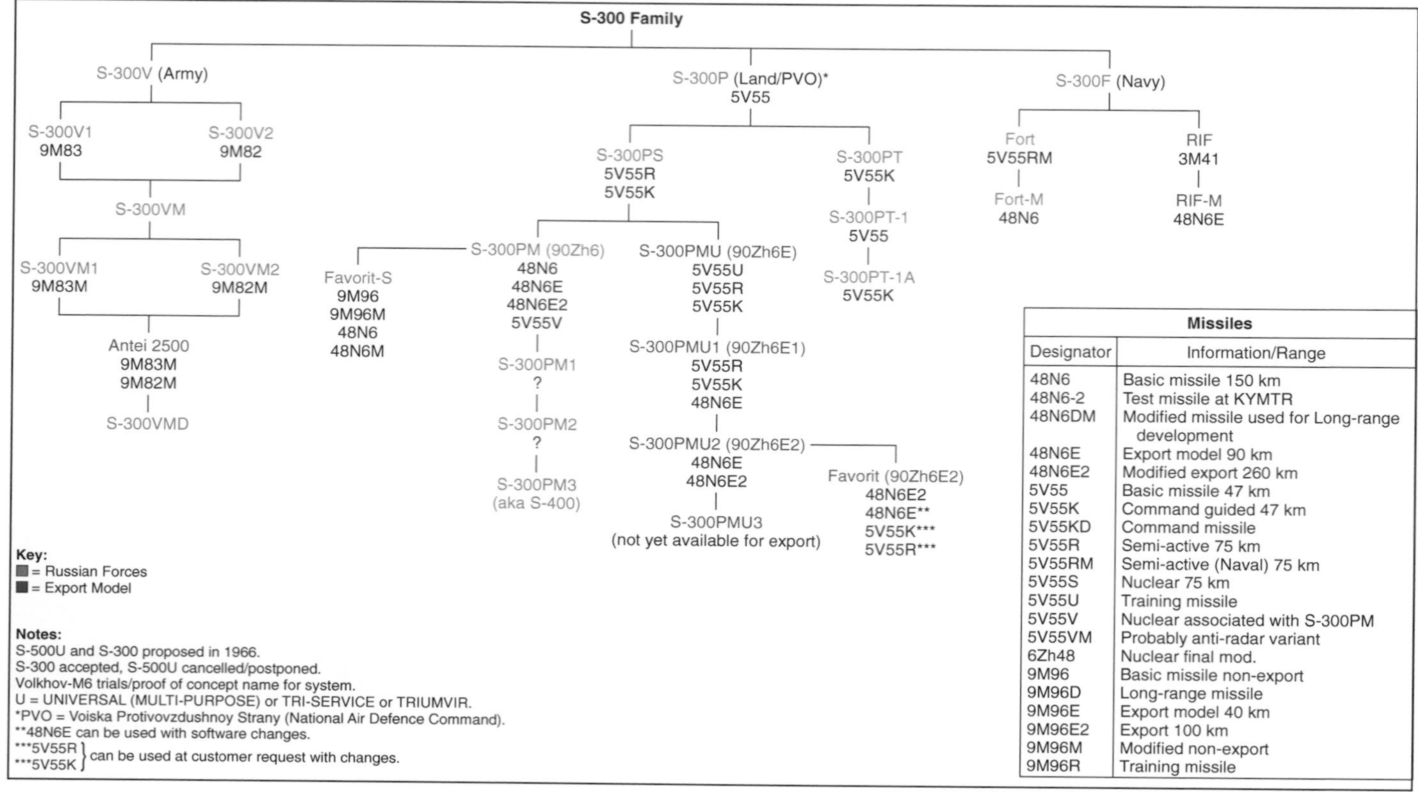

Missiles	
Designator	Information/Range
48N6	Basic missile 150 km
48N6-2	Test missile at KYMTR
48N6DM	Modified missile used for Long-range development
48N6E	Export model 90 km
48N6E2	Modified export 260 km
5V55	Basic missile 47 km
5V55K	Command guided 47 km
5V55KD	Command missile
5V55R	Semi-active 75 km
5V55RM	Semi-active (Naval) 75 km
5V55S	Nuclear 75 km
5V55U	Training missile
5V55V	Nuclear associated with S-300PM
5V55VM	Probably anti-radar variant
6Zh48	Nuclear final mod.
9M96	Basic missile non-export
9M96D	Long-range missile
9M96E	Export model 40 km
9M96E2	Export 100 km
9M96M	Modified non-export
9M96R	Training missile

S-300 Family flow diagram (James C O'Halloran)

0583822

ChMAP semi-trailer. The radar/engagement operators were located in an F-9 shelter unit that was located away from the radar mast.

The original missile used with the S-300P was the Fakel Design Bureau's 5V55K. This was the first Russian missile to incorporate a significant level of solid state electronics in its guidance system. The command guidance flight data was received from the battery's engagement radar. Maximum effective engagement range was 47 km with the single-stage missile using a solid-propellant rocket motor. The weapon was gas-ejected from its container-launcher to a height of 25 m before the main rocket motor fired.

At regimental and brigade level the 19Zh6 (NATO codename 'Tin Shield') 3-D S-band surveillance radar was used. This radar is manufactured at the PO Iskra radar plant. The Tin Shield radar was available in two versions, 19Zh6 and 36D6 (ST-68U and ST-68UM respectively. The 19Zh6 was used for early warning at the regimental and brigade levels, passing target data to the area command centre via the automated control system. The 36D6 could be trailer-mounted and used at the battery level and had the capability of interfacing directly with the 30N6-1 fire control/guidance radar bypassing the automated control system. The Tin Shield consisted of:

- The basic system mounted on a semi-trailer unit;
- An enhanced low altitude capability 40V6M1 tower assembly built specifically for use with the S-300P family.

Operational parameters are:

- Against a B-1 sized bomber with 1 m^2 radar cross-section at 100 m altitude - basic version detection range 45,000 m, mast version detection range 52,000 m;
- Against a cruise missile type target with 0.1 m^2 radar cross-section at 50 m altitude - basic version detection range 28,000 m, mast version detection range 38,000 m.

SA-10 mobile 'Grumble' missile resupply transporter-erector-launcher

0069508

The S-300P units, which entered service in 1980, were integrated into the national Russian PVO air-defence network fixed command-and-control system. Co-ordination of multiple static S-300P brigades is undertaken by the Proton NPO Universal-1E C3I system. However, due to inadequate range problems the S-300P system was rapidly upgraded to a later longer-range missile standard and supplemented by more modern mobile versions.

S-300PT-1 (US/NATO codename SA-10a 'Grumble')

In the mid-1980s, the S-300PT complex was modernised and its capabilities became analogous to those of the S-300PS complex. Its issue to the forces was carried out under the designation S-300PT-1. All of the containers for this variant of the complex are on vehicle semi-trailers in combat operations. The 5P851A launching installations used were developed with the aim of increasing reliability and ease of technical servicing of the 5P851 variant.

S-300PT-1A (US/NATO codename SA-10a 'Grumble')

Later the modified variant, S-300PT-1A was issued, and the previously issued S-300PT complexes underwent modification in the forces.

S-300PS (US/NATO codename SA-10b 'Grumble')

The self-propelled variant of the S-300P is the S-300PS, which was the standard production version and accepted for operational service around 1982/83. The S-300PS utilised the 5P85D-launching complex, along with the 5P85S. The 5P85S is equipped with a container F3S for preparation and control of the missile launch, a 5S18A autonomous power supply and up to two additional 5P85D launchers. The 5P85Ds are equipped with a 5S19A external power supply system, controlled via the F3S container on the 5P85S. All of the launching installations are mounted on the MAZ-543M all terrain four-axle heavy loader chassis and carry four transport-launch containers, with 5V55R missiles of different modifications. Later modifications and upgrades to the PS

SA-10 mobile 'Grumble' multifunction radar deployed in the operating configuration

0044114

SA-10 missiles elevated to fire 0080432

SA-10 mobile 'Grumble' transporter-erector-launcher vehicles deployed in the firing position with missiles elevated into vertical position 0044115

variant have taken place with Belarus receiving the system in 2006 with the very latest changes. During November 2006, four batteries of the new Belarus missile systems were deployed to the Polish border area within a day of Poland receiving the new F-16 fighter jets from the US. The F-16s arrived at Poland's Krzesiny air base near the western city of Poznan on the 9th November 2006 followed the day after on the Belarus side of the border by the S-300PS system. By January 2007, it was reported that Belarus had agreed to transfer some S-300PS systems to Iran in contravention of the UN Security Council approved sanctions. This was directly after a visit to Iran by the then Belarussian Defence Minister Leonid Maltsev and a meeting with his Iranian counterpart Mostafa Mohammad Najar.

By December 2008, the 2,566th Radioelectronic Weaponry Repair Plant in Borisov, Belarus, had displayed some components of the S-300PS for the then Prime Minister Alexander Lukashenko that had been repaired and updated by that organisation. Some of the equipment included:

- 5N63S;
- 5P85S;
- 5P85D.

Other equipment on display included an old Thin Skin (NATO Designation) height finding radar that had been painted the same light green colour as those vehicles of the S-300PS.

The S-300PS system is due to be replaced by the Vityaz in 2014 as the more modern system comes on-line and the outdated S-300PS become more obsolete. There is some question as to whether the Vityaz replacement is or could be known within Russia as the S-300V4.

S-300PM (US/NATO codename SA-10c 'Grumble')

This was the upgraded standard production version and accepted for operational service around 1990. Designated by the Russians, S-300PM (M = *modifikatsionniy,* Russian for 'modified'), this utilised the definitive 5P85T semi-trailer launcher with KrAZ-260 (6 × 6) heavy truck tractors. An export model of the 5P85T was given the designator 5P85TE. This erects its four paired missile container-launchers to the rear of the semi-trailer, thus avoiding

KrAZ-260V (6 × 6) tractor truck towing semi-trailer with SA-10a 'Grumble' missiles elevated for launch (Christopher F Foss) 0069501

SA-10 mobile 'Grumble' multifunction radar with antenna lowered for transport 0509444

the need to uncouple the tractor unit, hence considerably reducing the time needed to deploy the system. The 5P85TE launchers remain an option for use with the later S-300PMU1 and S-300PMU2 export models of the S-300PM. Reports dating from mid-2003 also suggested that a further export model was produced from the S-300PM, this is the FAVORIT-S. This export variant was produced from those S-300PM models that were redundant after the upgrade to the Moscow Air-Defence Ring. As far as is known, none have yet reached a customer.

New missiles were developed for the S-300PM system, the Fakel 48N6 and the 48N6E. Missile guidance was changed to the Track-Via-Missile (TVM) radar type using modified 'Flap Lid' engagement radar for control. An uprated rocket motor was used to increase the effective missile engagement range to a more acceptable 150 km.

Other versions of the S-300PM include:

- S-300PM1;
- S-300PM-1M;
- S-300PM2;
- S-300PM3 (this system has also been referred to as the S-400).

S-300PMU (US/NATO codename SA-10b 'Grumble')

The S-300PMU is the export variant of the S-300PS that entered service in 1982. It was specifically designed to improve system mobility and is basically the key element of the S-300P system, re-packaged to fit on a modified MAZ-543 (8 × 8) cross-country truck chassis. By employing these vehicles, emplacement time on a non-surveyed site is reduced to about five minutes.

The firing battery is preceded into its launch area by a 1T12-2M-survey vehicle, which prepares the site for battery occupation. New navigational equipment introduced during 2004, called NK Orientir, is designed to provide real-time positional information on an electronic map for the mobile SAM assets on which it is installed. It gives SAM unit commanders more accurate information on the position and orientation of the mobile vehicles that make up the firing battery than is normally available. The self-propelled 5P85 series launchers are accompanied by the self-propelled 5N63S-engagement radar, which is also mounted on the MAZ-543 chassis derivative. Supporting elements for the battery include 5T58 missile transport vehicles and 22T6 missile reloading vehicles. A full range of training facilities can also be provided.

Model of the 5V55/48N6 missile used in the SA-10 SAM system (Steven Zaloga) 0080433

The engagement radar has been modified for the S-300PMU system to permit it to control enlarged batteries of up to 12 launcher vehicles. This has increased the simultaneous engagement capability to six targets with up to two missiles per target allowed. As a result the S-300PMU brigade was reorganised to have six batteries, each with two three-launcher vehicles and an engagement radar. Each reorganised battery has one 5P85SU-launcher vehicle (identified by the presence of a command shelter located behind the vehicle cab) and one or two simplified 5P58DU-launcher vehicles (identified by the absence of the command shelter behind the cab).

The front level command element of S-300P mobile brigades was formed around either the Proton NPO Baikal-1 command vehicle system or the Belarus Agat NPO originated D4M Polyana C3I system, which coordinated the actions of the brigades and interfaced with higher PVO echelon and other missile assets.

The regimental/brigade level 19Zh6 surveillance radar is supplemented by the LEMZ 5N66M modified NATO (codename 'Clam Shell') low-altitude detection radar. The earlier version 5N66 was for use with the S-300PT. The 5N66M can be mounted on two versions of the Universal Tower. These are the 23.8 m high 40V6M mast assembly which takes one hour to deploy, or the 38.8 m high 40V6M2 mast assembly which takes two hours to deploy. The 5N66M radar has a detection range of 90,000 m against targets flying at 500 m altitude. It can also track up to a maximum of 180 targets. An export version of the 5N66M was produced and is known as the 76N6.

The missile used is the Fakel 5V55RUD (UD = *Usovershstvovanaya Dalnost*, Russian for 'improved range'). This missile had further rocket motor improvements to increase the maximum effective engagement range to 90,000 m. The minimum effective range is 5,000 m. Altitude limits are 10 to 27,000 m with target velocity up to 1,200 m/s. The system is capable of defeating tactical missiles with ranges of up to 300 km at distances of up to 30,000 m. The rate of fire is one missile every three to five seconds.

FAVORIT-S (US/NATO codename unknown)

During mid-2002, a statement by the head of the Russian Air Force mentioned that the S-300PM being made redundant from around the defensive rings of Moscow, would continue to be updated. The update would take the system to the same standard as that of the S-300PMU1 and PMU2, this system will be known as the FAVORIT-S.

No other details are available.

S-300PMU1 (90Zh6E) (US/NATO codename SA-10c 'Grumble')

The S-300PMU1 is an upgraded export system that has been developed from the S-300PM, which uses the 30N6E variant of the 30N6 (Flap Lid) radar. The S-300PMU1 was developed between 1985 and 1989, and was first shown at the 1992 Moscow Air Show. It differs from the earlier system in having more modern technology integrated into its various elements and by a major update of the software used in the high-speed computers.

SA-10 mobile 'Grumble' multifunction radar mounted on trailer and with unit elevated into vertical position for improved low-level coverage. Inset is the interior 0509445

The principal improvements over the S-300PMU include.

- The use of the 7.5 m long, 1,800 kg weight 48N6E (SA-10d) missile variant, with a 143 kg HE fragmentation warhead that increases the maximum effective intercept range of aircraft type targets to 150,000 m and reduces the minimum effective intercept altitude to 10 m. Minimum engagement range is 3,000 m. Missile diameter is increased to 0.515 m. Maximum missile velocity attained is 1,900 m/s within 12 seconds of launch. The weapon can withstand lateral *g*-loads amounting to 20 units. The missile in its container-launcher weighs 2,580 kg.
- The successful engagement of ballistic type targets launched at ranges of up to 1,000 km at distances up to 40,000 m using target designation from the 83M6E command-and-control system.
- Increasing the maximum target velocity capability from 1,167 m/s (4,200 km/h) to 2,788 m/s (10,000 km/h).
- Increasing the sector-scan radar coverage limits to improve the system's autonomous engagement capabilities.
- Adding extra crew training equipment to improve the level and standard of training.
- The use of an improved 45,500 kg vehicle mounted engagement radar, the 36N6E (also known by export designation 30N6E1). This has many improvements including a new-generation fire-control computer 40U6. There are three scan modes available: a 1° elevation × 90° azimuth for low altitude targets; a 13° elevation × 64° azimuth and 5° elevation × 64° azimuth for medium and high altitude targets; and 10° elevation × 32° azimuth for ballistic missile direction. Once the target is acquired the radar can be switched to either 4° elevation × 4° azimuth or 2° elevation × 2° azimuth sectors for automatic tracking and missile guidance. The radar is connected to the launchers by landlines, radio links or both. The radar antenna can also be mounted on the 40V6M tower if required.

The maximum number of targets engaged by the enhanced battery remains at six with up to a maximum of 12 missiles at any one time being guided simultaneously. The battery deployment time remains the same at five minutes, as does the firing rate of one missile every three seconds from a launcher. The maximum total of missiles available in the S-300PMU1 battery is 32 rounds.

The S-300 PMU1 system and its 86M6E command post can further be upgraded to the S-300 PMU2 capability through use of software upgrades as a field modification kit.

The S-300 PMU1 and later models of the system are capable of "silent engagements", where the battery radars are not turned on until the end of engagement and then only the 30N6-1. This is achieved by use of the Kolchuga-M passive electronic warfare equipment that was originally produced in the Ukraine but has now been reported as being produced in Russia. Trials using the Kolchuga-M are known to have been conducted at the Sary Shagan Missile Test Range from as early as September 2003.

S-300PMU2/FAVORIT (US/NATO codename SA-10d 'Grumble')

During MAKS-97 exhibition, the S-300PMU2/FAVORIT universal, mobile, multichannel anti-missile system was shown for the first time. The S-300PMU2 is a development of the S-300PM and PMU1.

S-300 systems sold to china

Date of order	Date of arrival	Battalions	Batteries	System type	No of TELs	Type of TELs	Type of missile	Total no of missiles	Cost
1991	1993	2	8	S-300PMU	32	5P85T 9KrAZ-260V)	5V55U	256 - 384 with 120 spare in 1994	USD220 million
1994	Late 1990s	2	8	S-300PMU1	32	5P85SE/DE	48N6E	196	USD400 million
2001	Unknown	2	8	S-300PMU1	32	Unknown (probably 5P85SE/DE)	48N6E	198	USD400 million
2003	2007/2008	4	16	S-300PMU2	64	5P85SE2/DE2	48N6E2	256	USD980 million

S-3OOF RIF systems sold to china

Date of order	Date of arrival	Quantity on ship	No ordered	System type	No of ships	Type of ship	Type of missile	Total no of missiles	Cost
2002	Unknown	6 revolvers each with 8 missiles	2	S-300F RIF	2	Type 051C Luzhou Class	5V55RM	96	USD200 million

Components and sub-components used with the system are:

- 48N6E2 missiles which have a new and improved warhead and reportedly have a fly-out range of up to 200 km but are shown as 150 km in the export variant. Note any combination of the following missiles may be used, but with varying control and ranges:
  - 48N6E - missile;
  - 5V55K - missile;
  - 5V55R - missile;
- 83M6E2 - control for the system consists of 54K6E2 Command Post (CP) and 64N6E2 Detection Radar (DR);
- 30N6E2 - illumination and guidance radar;
- 5P85SE2 - mobile launcher system;
- 5P85TE2 - transporter launcher system;
- 82Ts6E2 - technical support;
- 64N6E2 - early-warning radar.

Optional components include:

- 96L6E all altitude detection radar;
- 76N6 low altitude detection radar;
- 40V6M tower intended for lifting of the antenna post.

The Favorit SAM is intended for defence of important administrative-political, economical and military installations from air strikes, strategic cruise, aeroballistic, tactical and theatre ballistic missiles in heavy combat and electronic countermeasures environment.

Simulator

The Altek-300 training simulator for the combat crews of the S-300PMU2 was displayed on the Almaz stand at the Nizhny Novgorod defence exhibition in April 2002.

S-300PMU3 (US/NATO codename unknown)

Although mentioned several times at various defence exhibitions, nothing positive is known about what may constitute the S-300PMU3 system. At one stage, Western analysts believed that the S-300PMU2/FAVORIT, when modified to carry the four additional 9M96, 9M96E or 9M96E2, could be the S-300PMU3 variant. It has also been speculated that the S-400 export variant may be the S-300PMU3. However, as mentioned before, there is nothing positive and therefore the system must remain on the books as an unknown. One unconfirmed source has reported the S-300PM-1M as the Russian designator for the S-300PMU2 (export model).

96L6 3D low-altitude detection, surveillance and command post system

The Lianozovo electromechanical plant and the Lira KB Design Bureau displayed a model of their 96L6 radar vehicle at the IDEX '97 exhibition. The phased-array centimetric system is mounted on a MAZ-7930 (8 × 8) truck chassis and is intended for use with the S-300PMU2 system (the S-300PMU and S-300PMU1 can also make use of the radar) air-defence systems as an upgrade to replace several existing radars with a single, more capable system.

The 96L6 can function as low-altitude detection surveillance radar and as the battery command post. The maximum detection range is 300 km and it can track up to 100 targets simultaneously at speeds between 30 to 2,750 m/sec. If required, the radar antenna assembly can be fitted to a 40V6M tower unit for better low-altitude coverage. The 96L6 is considered an all-weather all-altitude radar system.

The cold launch common missile for the SA-10d and SA-N-6 systems is the 48N6E shown here in its launcher tube which has been cut away to show interior and missile with nose on left and tail on right (Christopher F Foss)
0080434

Almaz 83M6 Brigade command-and-control system

The command-and-control system used at brigade level with the S-300PMU missile battery system is designated the 83M6 and it is also designed to be used with the S-300PMU series, S-200DE and S-200VE air-defence systems. A typical Russian brigade has an 83M6 system with up to six batteries. Deployment time is said to be five minutes.

Two variants of the 83M6 are available:

- A mobile system mounted on MAZ-543 cross-country chassis;
- A semi-mobile system mounted on transportable shelters for use at static sites.

The six-man 83M6 comprises two elements:

(1) The 54K6 command post, which provides the command-and-control functions for a group of up to six launcher batteries. The control of the group is based on the data obtained from its associated radar and the airspace management information from the batteries under its control, the command-and-control systems of adjacent groups and from higher echelon air-defence command-and-control networks.The 54K6 automatically performs the following functions:
- Control of the associated 64N6 radar system.
- The acquisition, identification and tracking of up to 100 targets.
- Identification Friend or Foe (IFF) interrogation; prioritisation of target threats and selection of the most dangerous to hand-off to the individual batteries under its control.
- Command-and-control of group's ECCM subsystems in a heavy ECM environment; co-ordination of the batteries' autonomous actions.
- Coordinating the group's actions with adjacent and higher echelon command-and-control centres. The command post is fitted with operator consoles, a multiprocessor computer system and the various communications and monitoring systems to manage an air-defence battle with the group's available assets. Full crew training software and hardware is also fitted to train the command-post crew in autonomous group and combined battle management protocols.

(2) 64N6 (NATO designation Big Bird) 3-D long-range surveillance radar, which comprises a hydraulically raised antenna assembly and radar shelter mounted on a common semi-trailer. The phased-array antenna set has a double-sided antenna aperture. Its 3-D performance is obtained by rotation of the antenna once every 12 seconds and electronically scanning the antenna beam in both azimuth and elevation. Scan sector capability for detecting tactical ballistic missiles is also provided. This can detect aircraft and cruise missile type targets at ranges of up to 300 km and ballistic missiles with launching ranges of up to 1,000 km using the sector scan facility. An IFF transponder subsystem is fitted.The radar is used for target detection and tracking in normal, clutter and severe ECM environments. The data obtained are relayed to the command post for processing and assessment. The radar can detect targets with speeds of up to 2,788 m/s (10,000 km/h).The export designation of the system is 83M6E. The S-300PMU-2 system designation is 83M6E2.

Missile upgrade

Almaz in conjunction with the Fakel missile design bureau has developed a range missile upgrade packages suitable for the S-300P family, which includes:
5V55 missiles - 5V55, 5V55K, 5V55KD, 5V55R, 5V55RM, 5V55S, 5V55V and 5V55VM
48N6 missiles - 48N6, 48N6-2, 48N6DM, 48N6E and 48N6E2

Mobile repair base

Antei also market a mobile repair base facility comprising trailer mounted (4 × 4) workshops – a checkout and diagnostic workshop, a technological workshop, a machine workshop, a spare parts, tools and accessories facility,

	5V55K	5V55R	48N6
Dimensions and weights			
Length			
overall	7.25 m (23 ft 9½ in)	7.25 m (23 ft 9½ in)	7.50 m (24 ft 7¼ in)
Diameter			
body	514 mm (20.24 in)	514 mm (20.24 in)	515 mm (20.28 in)
Flight control surfaces			
span	1.133 m (3 ft 8½ in) (wing)	1.133 m (3 ft 8½ in) (wing)	1.133 m (3 ft 8½ in) (wing)
Weight			
launch	1,665 kg (3,670 lb)	1,665 kg (3,670 lb)	1,800 kg (3,968 lb)
Performance			
Speed			
max Mach number	6	-	-
Range			
min	2.7 n miles (5 km; 3.1 miles)	2.7 n miles (5 km; 3.1 miles)	2.7 n miles (5 km; 3.1 miles)
max	25.4 n miles (47 km; 29.2 miles) (target altitude 2,000 m plus)	40.5 n miles (75 km; 46.6 miles) (target altitude 2,000 m plus)	81.0 n miles (150 km; 93.2 miles) (target altitude 2,000 m plus)
	13.5 n miles (25 km; 15.5 miles) (target altitude 25 m and below)	-	-
Altitude			
min	25 m (82 ft)	25 m (82 ft)	10 m (33 ft)
max	25,000 m (82,021 ft)	25,000 m (82,021 ft)	27,000 m (88,583 ft)
Ordnance components			
Warhead	133 kg (293 lb) HE fragmentation	133 kg (293 lb)	143 kg (315 lb)
Fuze	proximity, impact	-	-
Propulsion			
type	solid propellant	solid propellant	solid propellant

a communications facilities checkout and diagnostics workshop and power generating units with their own tractor trucks. The checkout and diagnostics workshop uses the AS5-2 automated system to check and diagnose at least 90 per cent of the systems electronic components.

Variants

In 1969, the Altair Research and Development Corporation, as prime naval contractor, initiated a joint development programme with the Almaz Design Bureau for a naval version of the S-300 system called the S-300F Fort. This uses the 3M41 missile (alternative designation 5V55RM, a naval counterpart to the land-based 75 km range 5V55R missile), the 3S41 vertical launcher system with either 6-round B-203 or 8-round B-204 launcher units and the 3R41 'Volna' (NATO codename 'Top Dome') I/J band fire-control radar.

The system is designated SA-N-6 'Grumble' in the US/NATO series and is fitted to the first three Project 1144 Orlan (NATO Kirov class) nuclear-powered missile cruisers and the four Project 1164 Atlant (NATO Slava class) conventionally powered cruisers. The system was originally tested on the single Project 1134BE Berkut (NATO Kara class) cruiser named *Azov* over a six-year period from December 1977. The *Azov* was fitted with four B-204 vertical launcher assemblies and single 'Top Dome' radar. Final Russian naval service acceptance came in 1984.

The export version of the Fort system was first shown in 1993 and is named *Rif* (Russian for "coral reef").

The equivalent naval version to the S-300PMU-1 system is the S-300FM Fort-M system. This uses the latest 48N6 missile and a naval version of the 'Tombstone' fire-control radar. The Fort-M has been seen fitted to the fourth and final Project 1144 cruiser, the *Petr Vealiky*.

Slovakia has upgraded its S-300PMU radar early warning systems giving up to twice the range of the previous unmodified radars. The new configuration known locally as the ST-68MSK are upgrades of the basic ST-68U and ST-68UM that were part of the original purchase.

Specifications

	9P85S TEL
Dimensions and weights	
Crew:	4
Length	
overall:	9.4 m
Width	
overall:	3.1 m
Height	
overall:	3.7 m
Weight	
combat:	42,150 kg
Mobility	
Configuration	
running gear:	wheeled
layout:	8 × 8
Speed	
max speed:	60 km/h
Range	
main fuel supply:	650 km
Engine	D-12A-525, V-12, water cooled, diesel, 525 hp at 2,100 rpm
Firepower	
Armament:	4 × SA-10 SAM

Status

The S-300P family is no longer in production, export orders for customers will still be filled from refurbished systems taken from the front line when replaced with other systems such as Vityaz and/or S-400.

China has eight battalions of S-300PMU2 on order, four battalions of PMU1 delivered between 2003/2004, and eight battalions of S-300PMU delivered before 1999.

Cyprus had two battalions of S-300PMU1 deployed on the island of Crete and as of December 2007 these have been swopped with the Greek Armed Forces in return for Tor-M1's. Greece is the only NATO country to have in it's operational inventory the S-300 series missile systems, (some other NATO countries do possess these weapons but for analysis purposes only). On 13 December 2013 126th Combat Battalion of Hellenic Air Force (HAF) conducted the first Greek firing of the system, as part of Exercise 'White Eagle 2013', at the NATO Missile Firing Installation (NAMFI) located on the Akrotiri peninsula in Western Crete. A single missile was launched, which successfully engaged a jet-powered Scrab II subsonic target drone.

Kazakhstan has eight battalions of S-300PMU delivered in 2000.

Vietnam had a contract worth USD200 million for S-300PMU1s delivered during 2005/2006.

During a state visit to Moscow in December 2006, the President of Syria Bashar al-Assad showed particular interest in the S-300PMU-2 Favorit. According to reports from Israel, Assad signed a deal for an unknown quantity of S-300 units that would be delivered via Iran. The Iranian S-300 saga continues with Iran insisting that the system will be delivered eventually but the Russian authorities also insist that the system will only be delivered when other problems have been resolved. As of February 2013, there is still no sign of any delivery, indeed, Iran now claims to be able to produce their own version of the system.

Serbia and Montenegro is also reported to have some S-300 PM missiles and fire unit components. During the Kosovo conflict in 1999, these may have been interfaced with modified air-defence radars but not committed to battle.

The US DoD acquired 'Grumble' elements in 1994 from Belarus and in 1995 from the then Russian defence exports sales agency Rosvoorouzhenie (now Rosoboronexport). The US has bought the following:

- The command-and-control element of the S-300P system from Belarus
- Eight 5V55R
- Eight 5V55K
- Two 5V55RUD missiles from Ukraine
- 12 5P85 mobile launchers from an S-300PM unit (sold to East Germany before the 1989 re-unification and subsequently taken over by Germany)
- Several elements of the S-300P system from the Russian Federation.

All these systems have been tested by the US Army Missile Intelligence Command (USAMIC).

In early February 2014, it was announced by the Kazakhstan government that Russia will deliver free of charge 5 × battalions of S-300PS. Each battalion includes several different radar, command centres and six launch vehicles with 24 missiles. At the time of going to press it is not known the type of missile that is to be delivered but if they are the same as previously then the 5V55R variant is the favourite.

Contractor

Almaz/Antei Concern of Air Defence.

S-300V

Type
Self-propelled Surface-to-Air Missile (SAM) system

Development
The S-300V (NATO SA-12a 'Gladiator'/SA-12b 'Giant') is a multichannel all-weather Anti-Tactical Ballistic Missile (ATBM) and Surface-to-Air Missile (SAM) system. The Russians sometimes refer to the 9M82 'Giant' ATBM as the S-300V2, while the 9M83 'Gladiator' is known as the S-300V1.

The system achieved limited operational service with the Russian Federation in 1986 using the 9M83 missile. The 9M82 missile is believed to have achieved full operational capability in 1992. The complete system is classed as the world's first operational ATBM.

In 1997, the then Antei revealed that the S-300V system could, with a modest investment of money and within a limited time scale, be modernised to improve the missile range to 200,000 m against 2,500 to 3,000 km range ballistic missile targets. The system could also be integrated into customer-designated target data information, command-and-control systems.

Description
The S-300V is a fully mobile multichannel long-range air defence system designed to provide air defence of task forces and key military and state installations against: mass attacks by theatre and tactical ballistic, aero-ballistic and cruise missiles; strategic and tactical aircraft; as well as to engage loitering ECM aircraft and other air strike assets. The Russian designations so far identified as being associated with the S-300V (SA-12) family include:

S-300
Overall class designator for missile systems with ranges up to 150 km and engagement altitudes up to 25 to 30,000 m. Includes both the S-300P (NATO SA-10 'Grumble') and S-300V (NATO SA-12 Gladiator and Giant).

S-300V
The V (*Vysokopodvizhnyi* or 'high mobility') suffix denotes a high-mobility operational-tactical surface-to-air missile system with fast response times and

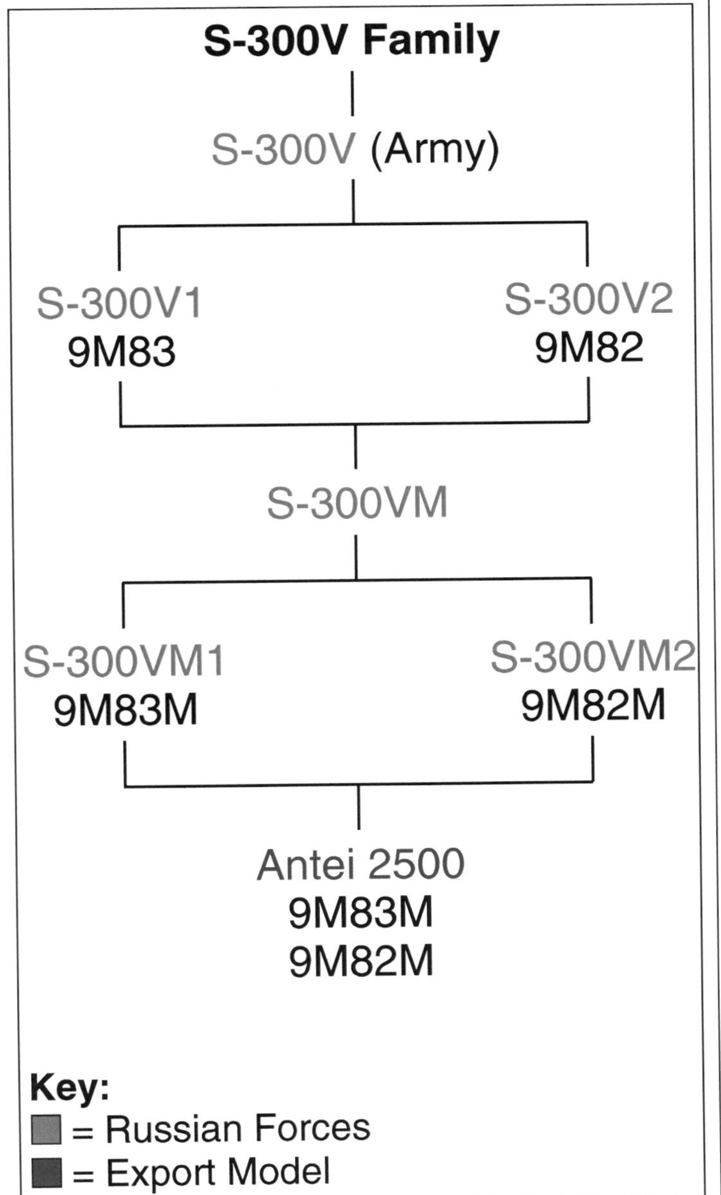

S-300V Family (James C O'Halloran) 0583824

SA-12 loader-launcher vehicle with crane traversed ready to load a missile (Christopher F Foss) 0080436

tactical mobility characteristics. The suffix 'V', has also been reported by Russian sources to indicate 'Voyskovoy', meaning tactical. The system uses the 75 km-range 9M83 and 100 km-range 9M82 missiles. The systems has been offered for export (see the Antei-2500 (S-300VM) entry below).

All the vehicles used in the system are based on a modern derivative of the full-tracked MT-T tractor chassis for increased mobility. All the individual units include their own land navigation system, communications systems and auxiliary power sources. The MT-T full-tracked vehicle series was originally designed and built at the Kharkov Morozov Machine Building Design Bureau in the Ukraine.

Full automation of deployment and out-of-action equipment operations allow the system to be operational within five minutes of arriving at a launch site and ready to travel five minutes after shutting down its subsystems.

A typical S-300V air defence missile brigade (ZRB: *Zeniitnaya Raketnaya Brigada*) comprises three elements:

(1) The headquarters, command and tracking battery with the 9S457/9S457-1 Command Post (CP) vehicle, a 9S15MT/9S15MV 360° surveillance radar vehicle (NATO designation 'Bill Board-A') and a sector scanning 9S19M2 radar (NATO designation 'High Screen') vehicle
(2) The four batteries, each of which has a multichannel 9S32/9S32-1 missile radar guidance station (NATO designation 'Grill Pan') and up to six Transporter-Erector-Launcher And Radar (TELAR) vehicles. The TELAR has a configuration to carry either two 9M82 or four 9M83 missiles
(3) Missile support and technical maintenance assets.

The normal battery make up, in theory, is a mixed one of four SA-12a TELARs and two SA-12b TELARs although, in reality, this has tended to be six SA-12a TELARs, with Russian sources indicating six SA-12b TELARs as an alternative. The SA-12b weapon acts primarily as an Anti-Tactical Ballistic Missile (ATBM) and the SA-12a as a dual-role anti-aircraft/anti-missile.

The total number of missiles per S-300V brigade varies from 96 to 192 rounds depending upon the number and types of TELARs in the batteries. For example, a solely SA-12a equipped brigade would have the maximum 192 (based on four rounds per TELAR and four reloads); a solely SA-12b equipped brigade would have 96 rounds (based on two rounds per TELAR and two reloads) and the normal mixed SA-12a/SA-12b brigade, 160 rounds.

The individual elements of the S-300V brigade are comprised of the following:

- 9S457/9S457-1 Command Post (CP) vehicle
- 9S15M/9S15MV Obzor-3 surveillance radar
- 9S19M2 Imbir sector scanning radar
- 9S32/9S32-1 missile guidance station
- 9A82 and 9A83 missile launcher vehicles

9A82 SA-12b 'Giant' SAM launcher with missiles elevated into vertical position and engagement radar at front of vehicle lowered for travelling (Christopher F Foss) 0080390

The 9S15MV surveillance radar has the NATO designation of 'Bill Board' and is shown here in operating position (Christopher F Foss) 0509446

9A83-1 launcher SA-12a 'Gladiator' with four missiles lowered into horizontal position (Steven Zaloga) 0080437

- 9A84 and 9A85 loader-launcher vehicles
- And the 9M82/9M83 missiles.

Data on each element is covered in detail in the following text:

9S457/9S457-1 Command Post (CP) vehicle

This tracked vehicle provides the command-and-control facilities for the brigade by monitoring up to a maximum of 70 initiated target tracks from the maximum of 200 identified targets that can be handed over from the 9S15 ('Bill Board-A') surveillance radar, allocating likely avenues of approach to the 9S19 ('High Screen') sector radar to search for high-speed missile targets, collating the resultant target data from all the available sources, including the 9S32 ('Grill Pan') systems and then assigning the threat priority. The 9S457 also receives information from external sources via the Polyana-D4 C3I system. The most dangerous of these (up to a maximum of 24) are then automatically designated to the most appropriate of the four 9S32 missile guidance stations of the batteries for engagement. Taking into account how dangerous the threat is, the degree of readiness of the missile batteries and the availability/status of the missiles performs the assignment. The whole engagement process is automated by the use of high-speed computer systems. Up to a maximum of 48 missiles can be simultaneously guided by the brigade assets at any one time against the maximum possible total of 24 targets engaged. Also, to ensure maximum engagement capability, highly sophisticated ECCM subsystems are fitted to all the S-300V radar elements that require them.

9S15M/9S15MV Obzor-3 surveillance radar (NATO designation 'Bill Board')

This provides long-range early warning and target acquisition data on up to 200 targets to the CP. For travelling, it folds down over the back of the vehicle. It has target coverage of 0° to +55° in elevation and completes a full rotation once

9S15MV 'Bill Board' surveillance radar of SA-12 air-defence system in retracted position for travelling (Christopher F Foss) 0044121

SA-12a 'Gladiator' SAM being carried on semi-trailer (Christopher F Foss) 0080438

SA-12b 'Giant' SAM being carried on semi-trailer (Christopher F Foss) 0044120

every six to 12 seconds. Range limits are said to be from 10,000 to 250,000 m, with a system accuracy of 30 to 35 minutes of arc in azimuth and 250 m in range.

9S19M2 Imbir sector scanning radar (NATO designation 'High Screen')

This is needed for the brigade's anti-missile role. The centre of the search sector and its parameters are defined by the CP. Regular scanning of the designated search sector at ±45° azimuth and up to 50° in elevation occurs. Automatic track initiation and transmission of the trajectory and track parameters follow each very high-speed target trajectory detected to the CP. The data are prioritised and the CP instructs the 9S19 station on which the very high-speed targets are to be continuously tracked. Range is stated as being between 20,000 and 175,000 m. The 9S19 can simultaneously follow up to 16 target tracks and identify up to six jamming sources (or 20 tracks and three jamming sources).

9S32/9S32-1 missile guidance station (NATO designation 'Grill Pan')

Each battery has a 9S32-1 missile guidance station attached to it. The station is manned by a crew of five and receives its targeting assignments from the CP. The phased-array radar on the top of the station's cabin is carried in the horizontal position for travel and hydraulically raised for operational use. It is used to provide the final track data of designated targets and to perform horizon searches in its assigned sector from where low-altitude targets are likely to appear. When searching for a target with a 2 m^2 radar cross-section, the maximum acquisition range is 150,000 m in its automatic mode, with pre-designation and 140,000 m in its manual mode. Accuracy is said to be seven to eight minutes of arc in azimuth, 0.7 to 1.4 m/s in velocity and 10 to 15 m in range. The antenna may be rotated to give 340° in azimuth coverage and surveys a 0° to 42° area in its normal operating mode. The vehicle is fitted with a prominent datalink antenna that is used to communicate with the

Ural-4320 (6 × 6) with scissor crane used for maintenance purposes with SA-12 SAM system (Christopher F Foss) 0080440

neighbouring launcher vehicles. The 9S32 are used to remotely control the target illumination engagement radars on the TELARs, by transmitting the data necessary for missile launching and guidance to them. It can track up to 12 targets and control up to six missiles against those targets simultaneously.

9A82 and 9A83 missile launcher vehicles

As stated, the basic three-man launcher vehicle can be configured to carry either two 9M82 'Giant' (the 9A82 TELAR) or four 9M83 'Gladiator' (the 9A83-1 TELAR) missile container-launcher tubes horizontally.

These tubes are raised to the vertical for missile launch and are carried horizontally for travel. When a TELAR receives the target data from its missile guidance station, it prepares either one or two missiles for launch. These can either be on their own vehicle or on an assigned LL vehicle. Once the missile(s) are fired, the TELAR works with the battery's 'Grill Pan' station by gathering the missile(s) in its radar beam and then transmitting the necessary missile guidance flight correction signals to intercept the target.

On the 9A83-1 TELAR, the engagement radar is mounted at the front of the launcher on a scissors-type mount which gives it a full 360° coverage in azimuth as well as full hemispheric coverage in elevation. On the 9A82 TELAR the radar is mounted in a semi-fixed position over the cab, giving 90° coverage to either side in azimuth or 110° coverage in elevation. The difference in the radar locations is because the 9A83-1 TELAR has a secondary anti-aircraft role and therefore needs a mast-mounted antenna to engage low-flying targets in any direction.

Typically, a target will be assigned either two missiles from one launcher or four missiles from two launchers. Pre-launch procedures take only 15 seconds and one missile can be fired every 1.5 seconds.

System deployment time is five minutes. The combat unit crews vary from three to five per vehicle and the combat weights from 35,000 to 48,000 kg.

9A84 and 9A85 loader-launcher vehicles

These are similar to their respective TELARs (9A84 for the 9A82 and 9A85 for the 9A83-1 TELAR), but have the radar installation replaced by a hydraulically-operated crane assembly for loading/unloading the missile containers. The loader-launcher has the primary role of replenishing the missile on the TELAR when necessary, but it can erect the missile tubes for its own launch sequence if the primary launcher load has been exhausted and there is insufficient time for reloading. However, it requires a TELAR to provide radar assistance for the engagement.

A full range of maintenance, missile reload semi-trailers, automatic test stations, servicing and repair vehicles are also included in the brigade to support the missiles, vehicles and associated equipment in the field.

Training facilities in the form of operator trainers, full-scale missile mock-ups and instructional training models of the missile types are also available.

9M82/9M83 missiles

The two-stage solid propellant-fuelled 9M82/'Giant' and 9M83/'Gladiator' missiles were developed by the Novator NPO design bureau and are designed for maximum commonality. The missiles are carried in recyclable container-

9A83-1 SA-12a 'Gladiator' launcher vehicle being carried on semi-trailer towed by MAZ-537 (8 × 8) tractor truck (Christopher F Foss) 0044122

Ural-4320 (6 × 6) with box body used in command post role for SA-12 SAM system (Christopher F Foss) 0080439

launcher tubes that have to be raised to the vertical before a launch can occur. The missile is then ejected from the container to a height of about 50 m and, once clear, ignites its propulsion system.

The main difference lies in the size of the first stage booster unit fitted and the performance it endows. The 9M82 'Giant' has a maximum velocity of 2,400 m/s, whereas the 9M83 'Gladiator' has a maximum velocity of 1,700 m/s. The warhead fitted is a 150 kg focused HE fragmentation type (with both heavy and light fragment sizes produced), coupled to a command-controlled detonation fuzing system that can be optimised in flight using command station, missile guidance and proximity fuzing data to suit the target type, flight trajectory and projected miss distance. A series of detonation cords in the directional warhead are detonated in a command-controlled sequence to ensure that the effectiveness of the main HE filling and subsequent fragmentation pattern is substantially enhanced. During system testing, over 60 ballistic-type targets were successfully intercepted. The 65 to 900 km range missiles either had their warheads destroyed or the flight trajectory changed from the projected impact point by up to 15 km by the S-300V warhead detonations.

The 9M82 'Giant' missile is designed to engage near-strategic ballistic missiles, tactical battlefield missiles and aircraft targets (including stand-off jamming platforms) at ranges of up to 100,000 m. Some American sources credit its maximum range capability as being 200,000 m. This may, however, reflect the weapon's capability against a non-manoeuvring subsonic target such as a Boeing E-3 AWACS aircraft.

The 9M83 'Gladiator' missile is designed to engage aircraft targets (including those performing manoeuvres of up to 7 to 8 *g*), tactical battlefield missiles, cruise missiles and some ballistic missile types.

A test firing at the Sary Shagan Missile Test Range (SSMTR) on 2 May 2001 was associated with the S-300V family. The resulting test report confirmed that a conventional warhead was used and that the aim of the trial was to improve missile defences in the Russian Federation. This could be another version of the S-300VM.

In November 2009, the Belarus firm, 'The Oboronitelniye Sistemi (Defence Systems)', released data on an improvement to the S-125 system that is now known as the S-125-2M. At the same time, the last paragraph of the press release included the following information. "A more radical third-phase upgrade to the S-125 would involve integrating the 9M83 vertically-launched missile used in the S-300V mobile air and missile-defence system. This would require the design and manufacture of new self-propelled launchers able to carry two 9Ya240 transporting and launching containers for 9M83 missiles, plus a folding mast carrying a target-illumination radar."

S-300V1

The V1 (*Vysokopodvizhnyi* or 'high mobility') suffix is used to indicate the S-300V high-mobility operational-tactical surface-to-air missile system using 9M83 missile types. The suffix 'V', has also been reported by Russian sources to indicate 'Voyskovoy', meaning tactical. Used by Belarussian, Russian and Ukrainian air defence units.

S-300V2

The V2 (*Vysokopodvizhnyi* or 'high mobility') suffix is used to indicate the S-300V high-mobility operational-tactical surface-to-air missile system using the 9M82 missile types. The suffix, 'V', has also been reported by Russian sources to indicate 'Voyskovoy', meaning tactical.

S-300V4

Since the beginning of 2011, mention of a S-300V4 system entering service with the Russian Ground Forces has been talked about in the open Russian Press. However, other sources have reported that information on the advanced S-300V4 remains classified. Again in January 2012, the system was mentioned in the same context as the Vityaz, that had previously been associated with the follow-on system to the Buk-M3, therefore the jury is still out as of February 2013 as to exactly what the S-300V4 is and what the system is based upon. What is known is that the V4 has a range of some 300 km and is to be deployed with Russian forces by 2015.

Variants

Antei-2500 (S-300VM)

The then Antei Concern developed the Antei-2500, which is an export variant of the S-300VM system. The S-300VM is a modification to the earlier S-300V series.

	9M83	9M82	S-300VM
Dimensions and weights			
Length			
overall	7.5 m (24 ft 7¼ in)	10 m (32 ft 9¾ in)	-
Diameter			
body	500 mm (19.69 in)	850 mm (33.46 in)	-
Weight			
launch	2,500 kg (5,511 lb)	4,600 kg (10,141 lb)	-
Performance			
Speed			
max speed	3,305 kt (6,120 km/h; 3,803 mph; 1,700 m/s)	4,665 kt (8,640 km/h; 5,369 mph; 2,400 m/s)	4,665 kt (8,640 km/h; 5,369 mph; 2,400 m/s)
Range			
min	3.2 n miles (6 km; 3.7 miles) (est.)	7.0 n miles (13 km; 8.1 miles)	7.0 n miles (13 km; 8.1 miles)
max	40.5 n miles (75 km; 46.6 miles) (aircraft type target)	54.0 n miles (100 km; 62.1 miles) (aircraft type target)	108.0 n miles (200 km; 124.3 miles) (aircraft type target)
	21.6 n miles (40 km; 24.9 miles) (ballistic target)	21.6 n miles (40 km; 24.9 miles) (ballistic target)	21.6 n miles (40 km; 24.9 miles) (ballistic target)
Altitude			
min	250 m (820 ft)	-	-
max	25,000 m (82,021 ft)	-	-
Ordnance components			
Warhead	150 kg (330 lb) HE fragmentation	150 kg (330 lb) HE fragmentation	150 kg (330 lb) HE fragmentation
Fuze	proximity (selectable)	proximity (selectable)	proximity (selectable)
Guidance	INS	INS	INS
	semi-active, radar	semi-active, radar	semi-active, radar
Propulsion			
type	two stage, solid propellant (sustain stage)	two stage, solid propellant (sustain stage)	two stage, solid propellant (sustain stage)

During the early part of 2007, a small reference publication celebrating the 50th anniversary of the Kapustin Yar Missile Test Range (KYMTR) made mention of yet another variant of the S-300V family, the designator used was S-300VMD, a previously unknown number. The developmental line starts with the Krug and follows through the S-300V, S-300VM and finally ends with the S-300VMD. The editor, therefore assumes that this is a follow-on type with the "D" probably identifying a long-range (Дальнегó) variant.

A further possibility for the VMD variant could be the VM that is integrated into the medium-range air-defence system Buk-M2E.

The term Eh or E is in a missile system suffix, for example Igla-1E, this means *Ehksportiynyi* or 'export model'. It is quite often found in the text of technical brochures or data sheets prepared for foreign nationals.

The S-300VM (Antei-2500) system is a long-range (200 km) multi-channel mobile air defence system that has been upgraded to protect task forces and vital national and military installations from mass attacks from: medium range ballistic, theatre and tactical ballistic missiles; aero-ballistic and cruise missiles; strategic and tactical aircraft; as well as to engage AWACS-type aircraft, reconnaissance and strike air systems; loitering ECM aircraft, and other air attack assets.

Specifications – *see table above*

Status

SA-12a and SA-12b - production as required as the production line is complete.

The S-300V system is offered for export. In 1997 it was reported that China was offered the S-300V system and the system had been reported to have been placed back in production to meet export orders, however Russian advisers have suggested that this, in fact, never happened and the production line is complete.

In March 2005, Russian sources indicated that the only true export of the S-300V system has been one to the US for evaluation purposes. However, by April 2013 elements of the system appeared in Venezuela and it is now known that the S-300VM (as described) has been purchased by Venezuela and is believed to be operational in theatre.

It was reported that the Indian Air Force had ordered six Antei-2500 fire units as an ATBM system but never materialised. Integration (according to Indian sources) was expected with the Akash and Trishul but with the cancellation of the Trishul this has now been abandoned.

Contractor

Almaz/Antei Concern of Air Defence.
Kalinin Mechanical Engineering Plant, (9M82 missile, the 9M83 missile, the 9A82 and 9A83-1 launchers and the 9A84 and 9A85 launcher-loaders).
Mariy El Machine-Building Plant, FGUP, Yoska Ola (production of the system).
Novator Experimental Design Bureau GUP/OKB Novator, (development and design of the missiles).
Machine-Building Plant, (Imeni M.I. Kalinina, OAO) (fuel pump systems for the missiles and full production of the system).
Start Scientific Production Enterprise, OAO/NPP Start (development of launchers for ground missile complexes).
State Scientific Research Institute of Instrument-Making FGUP/GNIIP, (systems for monitoring and preparation of launching the missiles).
Yegorshin Radio Plant, OAO, (multifunctional data transmission systems for the S-300V).
Istok State Science and Production Enterprise/GNPP Istok, (produces microwave and solid-state technology instruments for the S-300V).
Kirov Plant, OAO (performs the repairs for the chassis for the S-300V system).
Scientific Research Institute of Electronic Instruments FGUP/NIIEP, (develops and produces proximity detonation devices for the missile warheads).

S-400

Type

Self-propelled Surface-to-Air Missile (SAM) system

Development

The purpose of the development of the air defence system Triumph, is the high performance protection of political, economic and military highly valuable areas from air strikes, strategic cruise, tactic and operative tactical ballistic missiles, and ballistic medium-range missiles at conditions of combat and electronic counter-measures. A fourth-generation S-400 Triumf (Triumph) system (US/NATO designation SA-21 Growler) and sometimes referred to as the 40R6 or S-300PMU3, has been developed by the prime contractor Almaz/Antei Concern of Air Defence (PVO Konsern). First studies were conducted in the mid-1980s (some suggest 1986), till the early 1990s headed by the late Boris Bunkin from Almaz. The development commenced in 1993, when it was set out in a presidential decree that Russia needed to create a fourth-generation defence capability. However, constraints placed upon the

Launcher and associated S-400 missiles (IHS/Peter Felstead) 0105270

then Soviet defence industry dictated that new developments for these systems had to be restricted to 20-30 per cent with the remainder stemming from already deployed and tested systems. Hence, the main features of the S-300 were retained, including multiple containers and launchers, phased array radars and backwards compatibility with the older 5V55 and 48N6 missiles. At this time, it was also considered vital that Russia should be able to respond not only to attacks from the air but from space as well. Other Russian manufacturers include:

- The Fakel Machine Building Design Bureau (missiles)
- The Novosibirsk Research Institute of Measuring Instruments
- The St Petersburg Special Machine Design Bureau
- Kamyshin Crane Factory.
- Ashuluk Firing Test Range in Astrakhan Oblast
- Kapustin Yar Missile Test Range (KYMTR).

The S-400 was designed for use against envisaged existing and future air threat systems at that time including:

- Cruise missiles
- Tactical and strategic missiles
- Low-signature stealth aircraft
- AWACs type aircraft
- Standoff jammers.

The system employs multi-mode phased-array radar and signal processor systems, advanced highly automated crew stations and highly advanced target-engagement algorithms together with a variety of missile types to create a multi-layered defence. Aleksandr Lemanskiy, designer-general of the Almaz Scientific Industrial Corporation JSC, has said that the next step in the development of the S-400 system (post 2005/2006) will be active phased-array radars deployed with new high-speed missiles. He further quoted unidentified specialists as saying that the system is expected to remain in front line deployment into the second half of the 21st Century.

As of June 2006, Army General Aleksey Moskovskiy, the then Chief of the Russian Armed Forces Armaments Directorate and Deputy Minister of Defence, reported that the new S-400 would enter Russian service in 2007. By this time, however, non-missile equipment has been pre-deployed with the First Air Defence Corps, Special (formerly Moscow) Air Defence District, which was at that time using the older S-300 missiles. Personnel of the regiment were trained at Konstantin Zhukov Air Defence Academy at Tver, by defence industry experts.

During May 2007, the first battalion to be fully equipped with S-400 conducted final training exercises at a proving ground in Kazakhstan (Kapustin Yar Missile Test Range) prior to deployment into the Moscow Air-Defence Ring. The first fully operational site was located near the city of Elektrostal, 50 km east of Moscow. It is expected that up to 50 battalions will be finally equipped with the new system converting at a rate of one or two per year.

Full scale mock-up 9M96 (Christopher F Foss)

0048307

S-400 Launcher (IHS/Peter Felstead) 0048857

In mid-January 2008, Col. Gen. Alexander Zelin confirmed that the system would be deployed in the central region of the country additionally to Elektrostal as the system became available. This second system was deployed at Dmitrov at the end of 2008 after trials that began in August with the deployment being again in the Moscow region in what has been described as the European part. This second deployment of the S-400 will also tie in nicely with the strengthening of the Moscow air-defence in a major overhaul that commenced at the beginning of 2009. This upgrade/modernisation will be the 16th Air army of the Special Purpose Command, transforming into the Air Force and Air Defence army. The S-400 will be the primary theatre air-defence of Russia through to 2020 or even 2025. A third and fourth battalion have been formed in the Kaliningrad Oblast and the Far East. A fifth battalion is expected to be formed in the Southern Military District to cover the Southern Borders.

In early September 2011, reports emanating from Moscow detailed what has long been assessed by western analysts concerning problems encountered at the Avangard Mating Plant near Moscow. These reports confirmed the long-range missiles (40N6) had undergone at least two unsuccessful flights attributed to the nose-mounted seeker. The flight testing took place in December 2010 and March 2011. The 40N6 had originally been scheduled to complete state testing in 2010 but has been delayed even further into 2013. These continual delays have given rise to the speculation of changes in the planned deployment schedule. Reports suggest that a new long-term contract for S-400 production was due to begin in 2012, however these systems are not expected to be delivered until 2015. In July 2012, Russian Air Force Air Defence Units were reported to be in the process of taking delivery of the 40N6 long-range missile that had recently passed all state trials.

At least two new factories are being built in Russia for the S-400 production. However, it is more likely that these production factories will now be used for S-500 production; the S-500 will work in conjunction with the S-400 and not as a replacement.

Description

The S-400 Triumf is designed to be the foundation of Russia's advanced air defences until at least 2020 and probably even beyond 2025, the system is also designed to destroy stealth technology aircraft as well as missiles with speeds of 4.8 km per second from a range of up to 400 km. Trials held at the KYMTR and Ashuluk have confirmed the system has the capability to engage and destroy targets flying at speeds of 2,800 m/sec and at altitudes of 56 km. The type of missile used for these particular trials has not been verified.

Reports coming from the test ranges have indicated that the system may altogether employ seven types of missiles within three families four types of which are the 48N6, 48N6E, 48N6E-2 and 48N6E-3. A single missile designated 48N6DM has allegedly been specially modified for the S-400. For use with the system (sometimes also referred to as the S-300PMU3), the missile designer and manufacturer Fakel has developed the active radar seeker equipped 9M96, 9M96E and 9M96E2 family of medium-to-long range missiles. These missiles use a gas dynamic flight control system for improved manoeuvrability and were publicly revealed in 1998 (see entry in this section).

There is a further long-range family of missiles developed by the Fakel Design Bureau, which may have been deployed in the defence of Moscow since late 1999 or 2001 and even 2002. These missiles have an engagement range of up to 400 km and employ both semi-active and active engagement modes. The new long-range missile is reported to be much larger than its predecessors and deployed in pairs (two per TEL), rather than stacks of four as in previous S-300 systems. The missiles are reported to be part of the basic family designation 40N6 and probably the 40N6E with active and semi-active homing. The warheads each employ homing head seekers that can be switched to a search mode by ground command and home onto a target independently at ranges up to 400 km. At least one version will have an Over-The-Horizon (OTH) capability against jamming aircraft/airborne early warning aircraft.

The S-400 can also use the 48N6 group of missiles including the 48N6E2 and 48N6E3; these have been associated with the S-300PMU-1; and the S-300PMU-2 Favorit. New missiles such as the 48N6-2 and 48N6-3 will be able to handle targets as diverse as ballistic missile warheads travelling at speeds of approximately 5,000 m/s and small slow targets such as hang-gliders. It is further expected that the newer faster missiles that have yet to be developed, will employ Scramjet technology.

The naval variant of the system known as Rif-M and carried on board the cruiser Petr Veliky uses the same missiles. The Rif-M provides simultaneous engagement of six to eight targets and guidance of 12 to 16 missiles in flight.

The launchers contained within the ship are of various versions to accommodate the ammunition stock of 47 to 95 Mod.48N6E missiles in a barrel-type launcher, and four Mod.48N6E2 missiles and eight Mod.9M96E missiles in a modular type launcher. The target/missile channel capacity of the system is 6/12 for the 48N6E and 48N6E2 whilst the 9M96E is six to eight or 12 to 16 channels.

The S-400 (and the S-300PMU2) are the first Russian systems able to fire several types of missile from a standard launcher. As shown above, the S-400 currently has three missile types available to it, one of which is already operational; second has been tested and is in production and the third is being test fired. Unconfirmed reports on a fourth missile are being talked about but no open source reporting has yet taken place. Those missiles that are currently deployed with the S-400, are described by reliable Russian sources as surface-to-air missiles. What is needed today are missiles that can destroy targets in near space but these are still in development.

An S-400 system uses a central command-and-control vehicle with the multimode radar assembly and eight launcher units. A launcher can carry either four of the standard 48N6/48N6E2 missile container launchers (with one missile each), four of the 9M96 missile container launchers (with four missiles each) or a mixture of both. In all cases, a cold launch sequence ejects the weapon to a safe height, whereupon the main solid propellant rocket motor ignites. At the initial and middle flight phases, inertial guidance with radio command corrections is used. For the final phase the appropriate terminal guidance for the missile type is activated.

In mid-December 2012, information concerning missiles designated 77N6-N and 77N6-N1 became available, these missiles were reported to be the first Russian missiles with inert warheads, that can destroy nuclear warheads by direct impact. It was further reported that the two major production plants currently under construction in Russia have been specifically designated for the production of these missiles that will be used in both the S-400 and S-500. The S-400 as of December 2012 is using the 48N6 and 9M96 families of missiles.

55Zh6 Early Warning Radar

The 55Zh6 designator is the root for a new family of Early Warning Radars that have been designed and produced for modern air-defence systems under the auspices of the Almaz/Antei Concern of Air Defence in Moscow, the Nizhniy Novgorod Research Institute of Radio Engineering (NNIIRT) and AOA Nitel also in Nizhny Novgorod for radar production. The first indication of the 55Zh6 family occurred in 2001 when the system was still in development. The 55Zh6M is the first modification to the basic system, with the "M" traditionally being added after the first root designator 55Zh6 radar. This has undergone trials with the armed forces that have resulted in recommendations for modification to suit the intended air-defence system that it is required to work with. The export designator 55Zh6ME where the "E" is the export, would indicate that the air-defence system and the radar have been made available for export albeit with modifications assessed to be software and computer driven.

Furthermore, the appearance of the export designator would suggest that the S-400 system has found a home outside of Russia and is now available and will most likely be exported in the not too distant future.

Variants

S-500

It was reported in late 2002 that Russia was already working on an improvement to the S-400 known as Phase 2/Project Samoderzhets. From the lessons learnt from the Samoderzhets Project, these will be used to outline and govern the framework and interoperability of air defence units in Russia for the foreseeable future probably at least through to 2030.

During an interview in Moscow with Igor Ashburleie stated that two years after the first S-400 was handed over to the Russian Forces in 2007, the system will be modified to incorporate updates and upgrades before full series production. It may be at this point that the stationary (static) variant will be further developed. He also described the S-400 as an Element of an ABM system that has been assessed as the S-500.

Reportedly this system will consist of elements of the Antei-2500 for ABM defence as well as improved command-and-control and target handling capability, the S-500 is also believed to be related to the 45T6 anti-ballistic missile. It is further assessed that a new all encompassing system including ASAT, ABM, ATBM and AD will be probably deployed around 2015 and this may well be the so called Vlastelin Izmereniy.

S-400 Upgrade

Information released in the Almaz/Antei yearly report for 2010 confirmed the S-400 Upgrade giving years 2020/2021.

Specifications

S-400 Triumf

Number of simultaneous tracked targets:	not less than 300
Detection radar (azimuth × elevation):	
(aerodynamic targets)	360 × 14
(ballistic targets)	60 × 75
Range:	
(maximum)	
(aerodynamic targets)	250 km
(ballistic targets)	60 km
(minimum)	
(aerodynamic targets)	3 km
(ballistic targets)	5 km
Altitude:	
(maximum)	
(aerodynamic targets)	27,000 m
(ballistic targets)	27,000 m (trials at KYMTR and Ashuluk have proved destruction of targets at 56,000 m)
(minimum)	
(aerodynamic targets)	10 m
(ballistic targets)	2,000 m
Maximum target velocity:	4,800 m/s
Simultaneously engaged targets:	≥36
Simultaneously guided missiles:	≥72
Deployment time:	5 min

Status

Initial manufacturer's trials started in early 1999 and were completed in January 2000 at the Kasputin Yar Missile Testing Range (KYMTR) in Southern Russia. The 40N6 missile has been available since before June 2000, when it was suggested that the solid state rocket motor also be used for the testing and development of the Russian hypersonic scramjet. Pre-series production systems were delivered to the Russian Air Forces and air defence units in late 2000 for troop trials using missiles of the available S-300 systems. Operational deployment of the shorter range system began in the late 2001 and early 2002, when the new long-range target acquisition radars of 600 km were deployed operationally in the defence of Moscow. Further trials this time of a modified 48N6 missile with the designator 48N6DM, are known to have taken place at the KYMTR in April 2004. It is assessed that these trials were in fact associated with the S-400 and the missile is now in use with a standard battery. The developing company, Almaz/Antei has reported that this system was adopted for service in 2007 and that the first regiment equipped was deployed on combat duty in mid-2007.

As of the 6 August 2007, the first complete S-400 system became operational in the Moscow Air-Defence Ring. The Russian Air Force currently deploy more than 30 regiments equipped with S-300P missile complexes, all will be gradually replaced with S-400 systems.

During an interview with Colonel General Alexander Zelin on 5 April 2008, it was announced that the second S-400 regiment will become operational within the Moscow Air-Defence ring soon. This particular regiment is currently worked up and prepared for deployment. A second deployment was announced in March 2008 by the then Defence Minister Anatoly Serdyukov. It was also announced that 10 Pantsyr-S1 were to be purchased by the Russian Air Force and used in the close in protection role of the S-400. Subsequently in March/April 2010 these Panstsyr-S1 systems were delivered and became operational in the Moscow Ring. By March 2011, the Prime Minister Vladimir Putin during an interview with RIA Novosti announced that Russia was planning a revamp of all its air-defence network and that by 2020 all regiments will have received the new S-400 and the Pantsyr-S.

In early September 2009, it was announced that Saudi Arabia is in talks with Russia on the purchase of "several dozen" S-400 systems; each system will include at least eight launchers with 32 missiles and a mobile command post. Also interested in the S-400 are Turkey, Egypt and Iran.

The S-400 as of July 2012 has four regiments deployed including two in the Moscow region, one in Baltic Fleet and another in the Eastern Military District.

Sergey Chemezov, the first deputy general director of Rosoboronexport, has said that the United Arab Emirates and Iran are interested in equipping their armies with the latest types of Russian, air-defence weapons systems. Included in this, is the latest S-400 system, and talks over the next one or two years should be completed for the sale however, since then it has emerged that Iran ordered the S-300PMU-2 but was refused the sale by Russia. China is also expected to take delivery of the export variant early in the system's lifetime, especially as the Chinese government are allegedly reported to have paid USD500 million for a great deal of the trials.

In March 2007, the United Arab Emirates (UAE) and Russia were conducting technical consultations concerning the S-400 project and Russia was very confident that the sale will proceed either in 2007 or early 2008 however, by April 2013 nothing further has happened on this prospective sale.

In June 2007, the President of Venezuela embarked on a three country tour of Russia, Belarus and Iran with what is believed to be a shopping list for major arms purchases. Mobile air defence systems TOR-M1 and what can be described as an anti-missile system with a range of 250-300 km were included. This type of range would put such a purchase into the S-300 PMU2 class of SAM. However, the final purchase was of the S-300VM system (Antei-2500 export title)

During late October 2007, the Turkish Defence Industries Undersecretaries (SSM) released a request for information for a USD1.4 billion proposed programme for which Russia is expected to offer the S-400 Triumf SAM. The expected offer came in July 2008 when Russia replied that the S-400 was available for Turkey. On this occasion, Turkey gave the contract to China for the purchase of the HQ-9 a direct copy of the Russian S-300P.

In January 2008, Belarus asked Russia to supply the S-400 system with effect from 2010, it was understood at that time that the first export of the system will not be until late 2009 and possibly 2010, however, owing to unforeseen circumstances, exports of the complete system are to wait until after the system has been delivered to the Russian Armed Forces. In late February 2008, the then Russian Ambassador to Belarus confirmed that Belarus and Russia are to sign an agreement on the establishment of a single air-defence system based on the S-400. The agreement between Russia and Belarus was signed in April 2009 for 10 units.

Saudi Arabia (Riyadh) is planning to create a comprehensive air-defence system, which promises major contracts for the Russian defence industry. It is ready to buy nearly everything from Russia including, portable air-defence missile systems, Pantsyr-S1 and S-400 long-range SAMs.

Early October 2010, the then Deputy Minister of Industry and Trade in Russia, Yuri Borisov, told an Organisation of Collective Security military cooperation commission meeting that the S-400 system was not yet ready to be exported. Until the Russian Armed Forces receives the S-400 in the required numbers, any exports are out of consideration.

In February 2011, it was announced that Russia intends to deploy the second S-400 system into the Moscow defence ring along with the first of the systems, furthermore, the third system is to be deployed to Asia where it has been speculated that it will be installed on the southern most island of the Kuril chain in order to protect Russia's interests in the Far East and to counter any Japanese claims to the islands.

It was announced in August 2012 that the 4th S-400 regiment is to be deployed on the 16 August 2012 near the port city of Nakhodka in the Primorye Territory.

By late November 2013, Vladimir Putin the President of Russia announced that a further three S-400 regiments are to be equipped with the S-400 Triumf in 2014.

There are currently five regiments deployed, two are based around Moscow with the other three deployed near the city of Nakhodka in the Primorye Territory in Russia's Far East. Further deployments are expected in the Baltic region, also in the Southern Military District and in the Moscow defence ring.

Russia intends to deploy at least 28 regiments by 2020.

Contractor

Primary contractor: Almaz/Antei Concern of Air Defence.
Missile design and production: Fakel Machine-Building Design Bureau.
Sales and marketing: Rosoboronexport.
Ground support equipment: Kamyshin Crane Factory.
Radar development: NNIIRT Nizhny Novgorod Research Institute of Radio Engineering.
Radar production: AOA Nitel in Nizhny Novgorod.

S-500

Type

Self-propelled Surface-to-Air Missile (SAM) system

Development

The Russian Federation is working on a progressive Fifth Generation (S-500) air-defence, anti-space and anti-missile defence weapon system that advanced from the Experimental Research/Experimental Operational Construction and Theoretical Work (NIOKR) to the practical *Opitno Konstruktorsleya Rabota* (OKR - Operational Construction Work) stage in April 2007. In June 2007, General of the Army Vladimir Mikhailov, Commander-in-Chief of the Russian Air Force, announced that the next generation SAM system (the follow-on to the S-400) would be a mobile system and that Russia will only purchase mobile SAMs in the future, the days of the static systems are over. Mikhailov also reported that the fifth-generation system in development will undoubtedly inherit the best features of the S-300 and S-400 systems, but would be designed using the most advanced technology available at that time.

This information confirmed previous reports, some going as far back as 1999, that Russian designers have been working on a fifth-generation air-defence weapon system that will include air-defence, anti-ballistic missile and space defence capabilities. The technical requirements for the system will now be worked out including the programmes budget, several announcements by different high ranking persons have mentioned 2015 as the Initial Operational Capability (IOC) date, however, by December 2010, the earliest operational date for the full system was given as 2020 or even 2022.

During August 2007, the then Russian President Vladimir Putin visited what was described as a new Voronezh-type radar station in the village of Lekhtusiabout, 50 km north of St Petersburg. This radar and subsequent builds are part of a new Russian Air Defence system which is expected to be on-line in 2015, the same time as the new Fifth Generation S-500 is expected to start making its way into the Russian air and missile defence. Mr Putin further described the radar as "a first step in a large scale programme in this sphere that will be carried out to 2015".

It is understood from reports emanating from within the Russia Air Force, that as of 2009, the S-500 system was currently under design stage development at the Almaz/Antei Concern of Air Defence and is planned to be completed in 2012 or 2013. Furthermore, an inference was also drawn between the S-500 and a second version of the system called S-1000, but the difference between the two systems was not further explained.

Description

In October 1984, a report in the Financial Times on the then US Secretary of Defence William Cohen and Senator Curt Weldon's briefing on the S-500 by the Russians suggested that a new system beyond that of the S-400 was offered for joint development and cooperation, but that the system was still at the paper stage (NIOKR). The then Russian Deputy Defence Minister Nikolai Mikhailov, and several other top ranking Russian Generals, acknowledged that this new system will out perform the S-400 as well as the US Patriot PAC-3.

At a meeting that took place in Moscow on 27 February 2007, the Military Industrial Commission (which answers directly to the president) met to discuss the prospects and issues relating to the fifth generation system reporting that it must be an integrated system combining weapons, information and fire-control elements. Further reports dating from March 2007, also suggested that the Almaz/Antei will lead the effort at the industrial level and that the programme will likely involve the development of extended range missiles beyond that of the S-400.

Starting probably in 2010, the then older S-400, (which have been deployed in concentric rings in the defence of Moscow since July 2007), will start being modernised and upgraded with units and sub-units that will enhance the weapons characteristics in comparison with the current system. This type of practice is normal in the Russian Air Defence arena, when future systems are expected to be deployed, such as radar, missiles, launchers etc. before the new system become available, these sub-units are given to troops on the ground as a means of upgrading and modernisation. This also allows troops to train on future weapon units long before those systems are fully available and considered Initial Operational Capability (IOC). This practice was certainly used when the S-300 was to be replaced with the S-400 when in 1998, the 9M96 series of missiles was first introduced. Working on this assessment, it would therefore be natural to assume that sub-systems associated with the future S-500 are already being handed over to the air-defence troops in preparation for future deployment. Added to this, the fact that Russia has offered the export variant of the S-400 to the open arms market utilising the 48N6 series of missiles adds weight to the idea that something better is already deployed and that upgrades have begun that will finally lead to the IOC of a full S-500 deployment.

Further evidence of the S-400 being made available came to the fore in October 2011 with the release of the 55Zh6 family of new radars. The 55Zh6 designator is the root for a new family of Early Warning Radars that have been designed and produced for modern air-defence systems under the auspices of the Almaz/Antei Concern of Air Defence in Moscow, the Nizhniy Novgorod Research Institute of Radio Engineering (NNIIRT) and AOA Nitel also in Nizhny Novgorod for radar production. The first indication of the 55Zh6 family occurred in 2001 when the system was still in development. The 55Zh6M is the first modification to the basic system with the "M" traditionally being added after the first root designator 55Zh6 radar, This has undergone trials with the armed forces that have resulted in recommendations for modification to suit the intended air-defence system that it is required to work with. The export designator 55Zh6ME where the "E" is the export, would indicate that the air-defence system and the radar have been made available for export albeit with modifications assessed to be software and computer driven.

The new fifth-generation system that is now in OKR will then be slowly and methodically brought into the inventory of the Russian Air Defence and most likely culminating in IOC sometime around the reported date of 2015.

In late August 2009, Major General Oleg Barmin was quoted when reporting on a new-generation S-500 air-defence system that will consist not only of missiles capable of destroying various targets in the air, but also of high-altitude interceptor missiles intended to destroy targets in near-the-earth space.

The fifth-generation system will be designated *Yedinaya Sistema Zenitnova Raketnovo Oruzhiya Protivovozdushnoy Oboroni i Protivoraketnoy Oboroni* (EU SRO PWO-PRO) Air Defence And Missile Defence Missile Weaponry Joint System; it has been reported as the S-500 Prometheus.

The task of creating the new system is reported to have been one of the primary reasons for the amalgamation of the Almaz/Antei Concern of Air Defence with the subsequent combining of the facilities is the scientific designing schools of:

- PVO VVS (*Protivozdushnaya Oborona Voyenno-Vozdushnikh Sil*) Air Force Air Defence
- PVO VMF (*Protivozhushnaya Oborona Voyenno-Morskovo Flota*) Naval Air Defence
- PVO SV (*Protivozhushnaya Oborona Sukhoputnikh Voysk*) Army Air Defence
- ASU (*Avtomaticseskiye Sistemi Upravleniya*) Automated Control Systems
- PRO (*Protivoraketnaya Oborona*) Missile Defence.

In July 2007, the Agat Research Institute in Moscow reported it had tested the new fifth-generation digital active radar seekers it had developed. The seekers were tested in an surface-to-air role and as an air launched radar. Although the institute did not further identify the seekers used during these tests, or even a specific missile, other than SAM and air launched, the specific wording of the fifth-generation gives a clue as to where and how these seekers will be used in

9M82 missile in launch canister 0517568

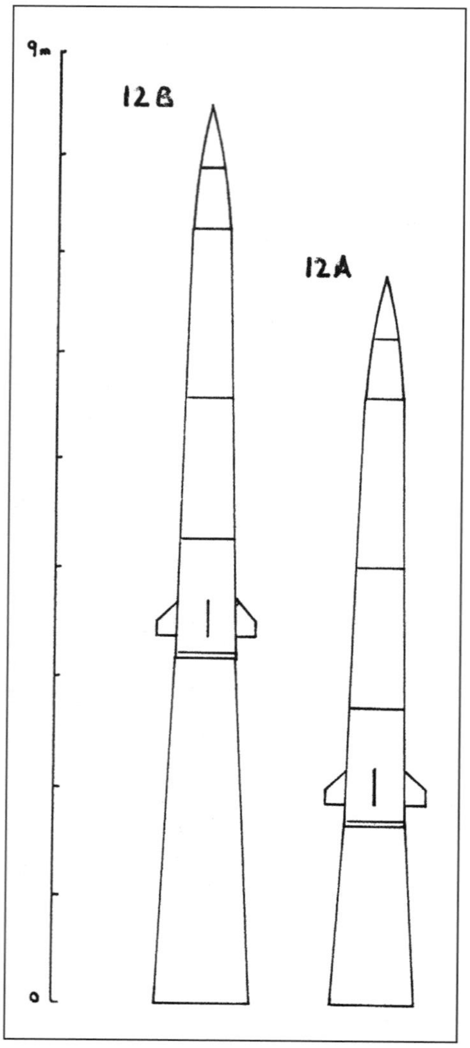

Line drawing of 9M82 and 9M83 0517569

the future. Up until this point, most of the development work had been carried out in ground-based laboratory facilities, supplemented by carry trials on aircraft. The seekers were described by the Agat General Designer, Professor Iosif Akopyan, as digital from the first intermediate frequency onwards. It is also understood that the seekers employ a reprogrammable onboard computer which allows the seeker to be adapted for whatever role may be required of the missile.

At the MAKS 2007, further information concerning a fifth-generation missile currently in development for the S-500 came to light, the missile known simply as Kh-96 was reported to have an engagement range of beyond 500 km. Although only a prototype, tests had proved the weapons capability to defeat enemy jamming and missile defences, operate in any theatre of operations and intercept ballistic and cruise missiles at any point of their trajectory.

Those regular readers of JLAD will now be questioning the Kh designation for a SAM, however, taking into account all of the above information, particularly the close involvement of Agat Research Institute in Moscow, and the PVO WS designation for the Air Force Air Defence will realise that what is in fact an air-launched designator may well be the basis for a surface-launched SAM. On the other hand, the Kh-96 could well be the designator for the missile that is launched from an air-defence aircraft to engage a ballistic target at some point of its trajectory and will indeed be part of an over arching system simply known as S-500. There are however some strong rumours that the missile for the land-launched S-500 will use a modernised version of the 9M82 (NATO SA-12 Giant). Indeed this is not the first time that these rumours have been heard, early trials going back as far as 2001/2 using the 9M82 missile reaching a maximum altitude of 150 km have been reported. If these rumours are correct, then a larger missile with greater acceleration can be expected.

In mid-December 2012, information concerning missiles designated 77N6-N and 77N6-N1 became available, these missiles were reported to be the first Russian missiles with inert warheads, that can destroy nuclear warheads by direct impact. It was further reported that the two major production plants currently under construction in Russia have been specifically designated for the production of these missiles that will be used in both the S-400 and S-500. The S-400 as of December 2012 is using the 48N6 and 9M96 families of missiles.

By August 2007, Russian Air Force Commander Colonel General Alexander Zelin mentioned that the fifth-generation air defence systems will be:

- more compact;
- more manoeuvrable; and
- will have superior technical characteristics.

Two years later in August 2009, General Zelin again released data on the S-500 to the press, this time he talked about Russia developing the S-500 surface-to-air missile system that would have ballistic missile tracking capabilities and aim at a targets that have a range of up to 3,500 km.

Based on the S-400, major enhancements to the electronics will allow the system to see further, higher and react quicker, significantly expanding its capability including repelling strikes from space.

As at late September 2009, further details about the systems capabilities were released giving a range of 600 km, simultaneously engage up to 10 targets with 10 independently guided sub-munitions from a 10-piece warhead that will make the system capable of destroying hypersonic ballistic targets. The new missile/system is now expected to be ready for field testing by 2012. The world's largest military training ground, Telemba located in Siberia, is currently being expanded as Russia readies to test its S-400 long-range missile defence system; it is also the most likely place where the S-500 will end up being tested. It has been reported by a source that additional areas of land to the training ground are being readied for these tests. Col. Sergei Kuryshkin, Telemba's chief, reported that the land is 160 km long and 70 km wide. Kuryshkin said the training ground would be used for S-400 long-range missile defence system tests as well as for other prospective missile defence systems; this is believed to be the S-500. The Telemba training ground, whose total area is more than 1.3 million hectares (approximately 5,020 square miles), was established in November 1960. The range is located in the Siberia's Buryat republic. Not only Russian but also Syrian, Ukrainian and Belarusian armed forces hold drills there. The field testing in Russia traditionally lasts for somewhere in the region of 12 to 15 months, followed some 30 months later by customer acceptance trials. Taking these figures into account, the assessment for the first limited deployment of the S-500 will be around late 2015 or early 2016.

By the 1 July 2013, General Valery Gerasimov, head of the General Staff in Moscow reported in a press statement that the first S-500 system should become available sometime in 2016. The speculation that management problems and troubles with cutting edge technology have largely been overcome and the 2017 date earlier announced has been brought forward by one year.

Variants

Information reaching the west in April 2010 suggested that a follow-on system to the S-500 is in the early design stage and has initially been called the S-600, however, by October 2010, another system (or variant of the system) was also identified as the S-1000. Additional to these is the S-550 that has been referred to. No further information on any of these systems is available.

Status

In development, advanced from theoretical (NIOKR) to practical (OKR) development since April 2007. A date of 2020 and 2022 have been mentioned but is as yet unconfirmed.

A statement released by the Air Force Commander at a press conference dedicated to the 100th anniversary of the Russian Air Force in June 2012 stated that the S-500 air defence missile systems could be with the Russian Armed Forces as early as 2013 however, this has not happened and only sub-components may have been allocated to the Armed Forces. All of the above dates have now been superseded by sometime in the year 2016.

Contractor

Almaz/Antei Concern of Air Defence, primary contractor.

Almaz Scientific Industrial Association, development of a fifth-generation missile Kh-96.

Fakel Design Bureau, primary design and developers of missiles particularly surface-to-air.

Kuntsevo Design Bureau, develops special equipment, trajectory measurement hardware, automated monitoring systems and electronic equipment.

Ulyanovsk Mechanical Plant, is responsible for the production of self-propelled SAM complexes and various types of radar, previously associated with the Kub and Buk systems.

Altair Research and Production Association, primarily develops and produces naval air-defence missile systems and naval artillery and is associated with the Rif system.

Agat R&D Institute, develops radar homing heads for missiles.

Avangard Mechanical Plant, specialises in the production of surface-to-air missiles and is associated with the S-300 and S-400 systems.

AVITEK Vyatka Machine-Building Enterprise is involved in the production of anti-aircraft guided missiles for ground troops, navy and PVO.

Rosoboroneksport offers the system for export.

Strelets Igla and Igla-S multiple launcher

Type

Pedestal mounted Surface-to-Air Missile (SAM) system

Development

The Strelets (Russian translation of 'Archer') multiple launcher unit was developed for use with the 9M39 Igla and Igla-S missiles. It provides an automatic remote launch capability in either single-round or salvo modes (from different launch modules) when mounted on various launch platforms; land-based, airborne and ship-borne.

Strelets launcher pack (Miroslav Gyürösi) 0578333

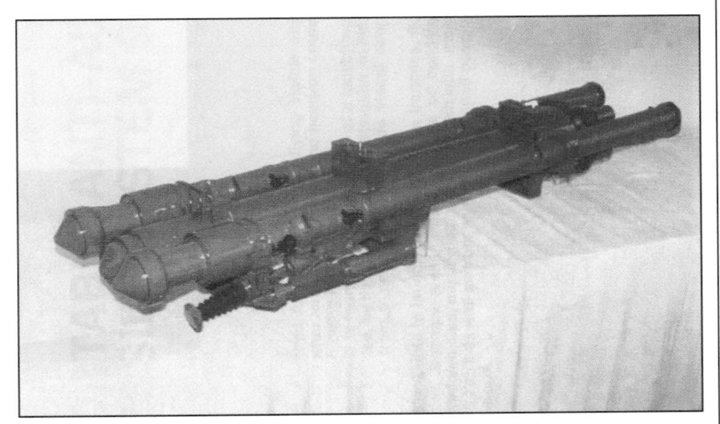
Kolomna KBM Strelets SEM multiple launcher for Igla and Igla-1 missiles 0024895

Description

The Strelets system comprises the following subsystems:

- A multipurpose launch module that can fire two to eight Igla family of missiles in their launch tubes and four ground supply units sufficient to activate the missiles up to four times.
- A control and communications system comprising a control unit (responsible for operational control of the modules and datalink with the launch platform's fire-control unit) and a power supply unit that converts the launch platform's power supply voltage to that required for operation of the Strelets component systems.
- An attachment set to secure the launcher modules onto the different carriers and to themselves (in two tiers).
- A test equipment set that can perform the periodic maintenance on the system equipment.

The Strelets turret modules can be mounted either on an armoured car, patrol boat or on the undercarriage of a helicopter gunship.

Variants

The Strelets can be mounted as part of the following systems.

- A mobile SHORAD system based on the Igla and Igla-1 weapons mounted on an Armoured Personnel Carrier (APC), Infantry Fighting Vehicle (IFV) or similar type vehicle.
- A medium-range air defence system to provide short-range cover.
- A dedicated self-propelled air defence system to extend kill envelopes and increase combat effectiveness.
- Combat helicopters armed with the Igla and Igla 1 as an air-to-air weapon.
- A ship-based SHORAD system known as the Ghibka (3M-47). This particular system has also been referred to as the Gubka and Gibka in the Western press. The missiles associated with the 3M-47 system are the Igla (9M39) and the Igla-S.
- Belarus has developed the Stalker 2T-reconnaissance vehicle, which mounts a dual Igla or Igla-1 missile launcher for self-defence purposes.
- Luchnik-E is an upgrade for the Strela-10M3 where the 9M35 missiles are removed and the Strelets launcher with Igla-S missiles is used.
- The Strelets SEM is another variant which is only a dual round launcher for use on UAVs.

Reports from the Machine Building Design Bureau (KBM) in Kolomna, Russia concerning a new Igla-Super missile has started reaching the West. This new missile is for use in the multiple launchers such as the Dzhigit and Strelets. The modifications to the missile include.

- A maximum range increased from 5.2 to 6.0 km.
- A laser proximity fuze.
- Improved control actuators.
- An improved warhead design that is the same weight as its predecessor, but allows for an adjustable (focused) fragmentation pattern.

Ring Sights Holding Company Ltd (UK) has developed and is producing, for an unknown South American customer, the LC-40-100-9K38 ring sight for the Igla missile system. The sight clamps on the Igla barrel at the position of the daytime only open sight, without mutual interference. Zeroing adjustments allow the sight to be aligned with the open sights. The system has a ring and bars graticule lit by Light Emitting Diodes (LEDs). It is adjustable in brightness to allow use with NVGs.

Specifications

Strelets Igla and Igla-S

Armament: (missile)	2 × Igla, Igla-S or training rounds
Dimensions: (loaded module)	1.77 × 0.43 × 0.25 m
Weight:	
(loaded module)	72 kg
(control and communications system)	≤35 kg
Missile max. continuous operation time from activation to firing:	60 s
Firing mode:	single or salvo
Max. number of activations of one missile (for 30 s):	4
Response time:	6.6 s
Reload time:	≤3 min (some reports suggest no more than 4 min)
Weight of control equipment:	24 kg
Types of fire:	one missile at a time, successive, salvo from various modules
Aiming techniques:	**a.** along line of sight **b.** by target designation at ±10° off homing head axis (for Igla-S only)
Mean time between failures:	no less than 500 hours

Test Equipment

9V5001 test equipment
9S846.51.00.000 single SPTA set
9S846.52.00.000 group SPTA set
9S846 equipment for the Strelets connection to the carrier and for adjustments and tests
9V866-2 mobile check point
9F719 test equipment set (stationary).

Status

In limited production and in service with the Russian Federation Army.

Offered for export in particular to Syria during January-February 2005. During the sales promotion to Syria, Igor Ivanov, the Russian Security Council Secretary, reported the need for a launch rail mounted on a vehicle, a patrol boat or a helicopter when describing the Strelets system that is on offer to Syria. The missiles will not work without this basic requirement. The offer was taken up and the launcher was mounted on the Phoenix passive air-defence vehicle.

In February 2011, a modified (upgraded) Strelets was offered for sale mounted in launcher boxes that can either contain four or eight Igla-S missiles. The boxes are reported to weigh either 80 kg for the four-missile launcher or 120 kg for the eight-missile launcher.

Contractor

Russian Federation state factories.
Konstruktorskoe Bjuro Machinostroenia (KBM) (developer).

Tor

Type

Self-propelled Surface-to-Air Missile (SAM) system

Development

The Almaz/Antei Concern of Air Defence (established 2002; previously, in particular, the Antei Concern) developed, through the 1980s, a mobile and highly automated integral SAM version of the Russian Navy's Kynshal (Klinok - export name) system, designed by the Altair NPO design bureau.

Designated the ZRK Tor (Russian for the Norse folklore god, Thor, NATO SA-15 'Gauntlet'), development started in accordance with the Resolution of the CPSU Central Committee and the USSR Council of Ministers of 4 February 1975, it entered limited service in March 1986 after joint tests of the system were conducted in 1983/84 and has gradually replaced ZRK Osa (NATO SA-8 'Gecko') systems at the division (brigade) level.

This system was developed by the Moscow Electromechanical Research Institute (NIEMI) and produced by the Izhevsk-based Kupol Electromechanical Plant. The Fakel Design Bureau headed by P.D. Grushin was responsible for the development and production of the missiles for the Tor system (9M330/9M331).

The Tor has been given the US/NATO designation SA-15 'Gauntlet' and is capable of engaging not only aircraft and helicopters but also RPVs, precision-guided weapons and various types of guided missiles. Since the original trial series vehicles, the Tor has undergone several substantial redesigns. The definite first production series model, under the designation Tor-M, was not fielded until 1991. The export model of the Tor-M is the Tor-M1.

At the 1998 Defendory International exhibition in Athens, the Greek DoD revealed it was ordering 21 Tor-M1 systems for its army under a USD560 million contract with a USD700 million option for another 29. The last of the 21 was delivered in August 2000. Greece has subsequently ordered a further 10 Tor-M1 in order to replace six units transferred to the Cypriot government and four units issued to the Greek Air Force to protect its

3D radar on Tor-M2 1529232

Tor (SA-15 'Gauntlet') system in operating configuration from the rear (Steven Zaloga) 0080391

Tor (SA-15 'Gauntlet') SAM system in operating configuration with all antennae erected 0509861

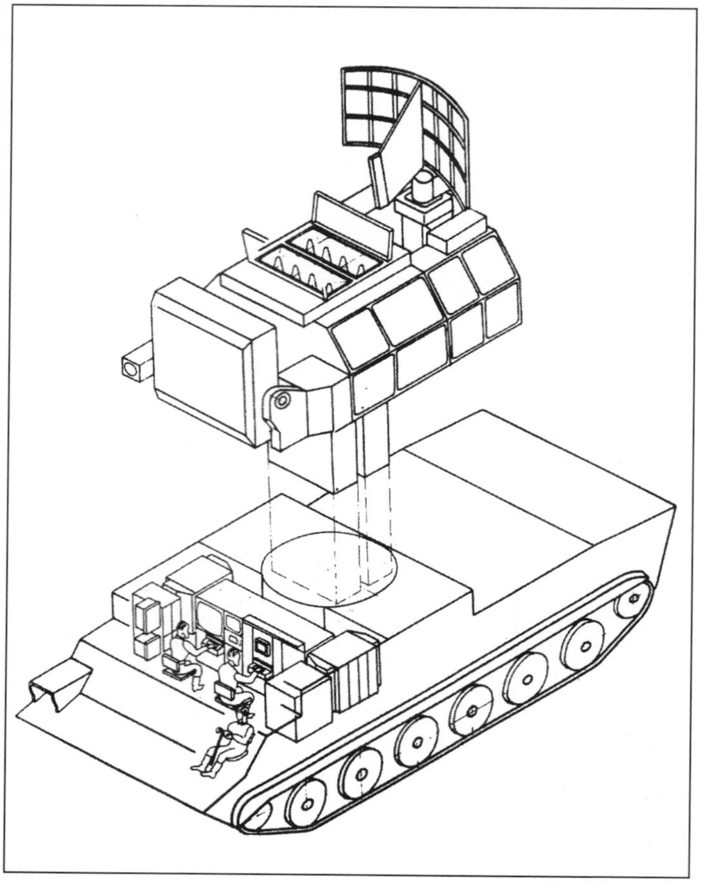

Position of main components of the Tor (SA-15 'Gauntlet') SAM system with magazine covers in open position to reveal vertically launched missiles 0509453

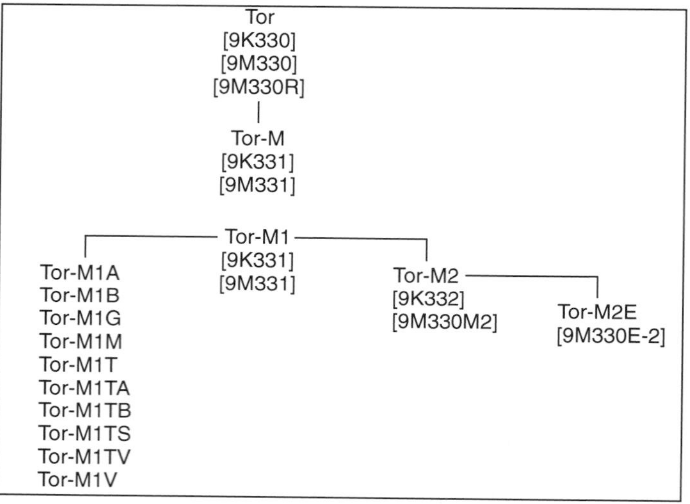

Tor Family flow diagram (James C O'Halloran) 1040813

S-300PMU-1 system, transferred from Cyprus. The only difference from a standard Tor-M1 is that the Greek units have an IFF system compatible with the NATO network.

Description

Tor system

The basic chassis used for the Tor system is the GM-355. The Tor-M uses the GM-355M, the export version of which, is the GM-5955. This is a slightly stretched vehicle in comparison with the GM-355 chassis. The Russian designation for the firing vehicle is 9A330 (9A331 for the Tor-M1). This vehicle weighs 34,000 kg and is produced by the Metrovagonmash Joint Stock Company.

The four-man crew comprises the vehicle commander, system operators and vehicle driver, all of who are seated in the front of the vehicle with the large box-like unmanned turret in the centre and the engine compartment at the rear. The driver is located on the left side of the vehicle front and has a windscreen immediately in front which, if required, may be covered by a hatch cover hinged at the top. This arrangement is similar to that used previously on the Kub (SA-6) and Shilka (ZSU-23-4) vehicles. The other crew members are seated to the driver's rear with the system PPI console on the right and the launch controls on the left.

The vehicle suspension consists of six dual rubber-tyre road wheels either side with the idler at the front, drive sprocket at the rear and three return rollers. An auxiliary gas turbine is fitted; this powers a 75 kW generator allowing the main diesel engine to be shut down when the system is deployed, so as to conserve fuel. Vehicle dimensions are 7.5 m long, 3.3 m wide and 5.1 m high (with radar erected). Maximum road speed is 65 km/h.

The Tor is not amphibious although it is air-portable. An NBC system is fitted as standard, as is a built-in training system.

On top of the turret rear is the mechanically steered H-band 25 km range surveillance radar antenna assembly, which is swung through 90° to the horizontal position for extended travelling purposes. The 3-D pulse-Doppler radar provides the range, azimuth, elevation and automatic threat evaluation data for up to 48 targets on the associated digital fire-control computer processing system. The surveillance radar is fitted with an IFF system.

Automatic track initiation can be performed on the 10 targets assessed as the most dangerous. These are then categorised and prioritised, in order of threat, by the computer for the engagement. All the operator has to do is to reconfirm the selection of the highest-priority target choice and track this selected target before pressing the 'fire' button.

Tor (SA-15 'Gauntlet') SAM system with 9M330 missile being vertically launched 0080444

Tor (SA-15 'Gauntlet') SAM system in travelling configuration with tarpaulin in position (Christopher F Foss) 0080443

Tor (SA-15 'Gauntlet') SAM system from the front clearly showing large, square, tracking radar (Christopher F Foss) 0044124

The Tor (SA-15 'Gauntlet') 9M330 missile out of its launcher with wings and fins in flight condition 0509454

Although the maximum radar range quoted is 25,000 m, the fast reaction time (five to eight seconds), which includes the fire-control computer assigning the target priority, suggests that it is probably greater than this.

At the front of the turret is the phased-array pulse-Doppler 25 km range K-band tracking radar, the beams of which are also electronically steered. This is capable of simultaneously tracking two targets with radar cross-sections down to 0.1 m^2 travelling at speeds of 0 to 700 m/s in all weather and at any time of day or night irrespective of threat ECM operations. The antenna assembly can also be folded down for travelling purposes. Mounted on the top left of this radar is a small vertical pointing antenna which, it is believed, gathers the missile after launch before handing it over to the main tracking/guidance system.

Lower down on the right side of the tracking radar is an autonomous automatic TV tracking system. This has a range of 20,000 m, which complements the tracking radar and enables the system to operate in battlefield clutter and heavy ECM environments.

Target radar surveillance is carried out on the move but the vehicle would normally come to a halt for missile launch. Although it is an autonomous system, it can be interfaced into an air defence network as it carries a specialist-coded datalink for such purposes. The Russians also state that its design is flexible enough to accommodate other manufacturers' command-and-control equipment if required by a customer. Command and a 9S737 Rangir armoured command vehicle, developed by the AGAT NPO in Minsk, Belarus, manage control of a Tor battery. Full details of the Rangir system are given in the Anti-aircraft control systems section.

Missile

The 9M330 and 9M331 short-range missile is intended to destroy various targets such as tactical and naval aircraft, helicopters, guided bombs and missiles. It is a single-stage missile employing a canard aerodynamic configuration with a freely-rotating after wing unit that unfolds after take off. The missile is launched vertically from its transporting container by means of a catapult mounted inside the container. The missile is launched without prior laying the launcher in the direction of the target.

The missile area comprises a box-like container that extends down below the level of the hull top and carries two four-pack 9M334-missile modules in the vertical position in the Tor-M and M1, whereas in the Tor, missiles are loaded and stored individually. Each Fakel SKB 9M330 (Tor) and 9M331 (Tor-M and M1) missile is set in its own sealed container-launcher box and requires no maintenance. A 9T244-specialist transportation/loader vehicle performs the reloading of the system.

On firing, a cold launch ejection system propels the missile upward to a height of 18 to 20 m, where a flight system turns the weapon on to the target bearing by means of a gas dynamic system. The main sustainer rocket motor then cuts in and the missile is command-guided to the intercept point where the radar proximity fuze is triggered. Flight control and guidance are accomplished via radio commands from the guidance radar mounted on the combat vehicle.

Structurally, the 9M330 missile consists of a number of sections housing radar fuze, missile control surface actuators, a gas dynamic missile deflection system, HE fragmentation warhead, onboard equipment units, dual-mode solid propellant rocket engine and control command receivers.

The modified version of the 9M330 is the 9M331. Effective range limits are from 1,500 to 12,000 m with target altitude limits between 10 and 6,000 m. The

Late production Tor-M2 (Tor-M2E) SAM system from the rear showing rectangular-shaped surveillance radar (Christopher F Foss) 0044125

maximum manoeuvring load factor limit on the weapon is 30 *g* against highly manoeuvrable (up to 12 *g*) small-sized, high-speed, low-flying targets.

A training split missile (9M330R) can be found on site and is used for operator familiarisation and training.

The Tor is backed up in the field by:

- The 9T244 loader-transporter wheeled vehicle (carrying two reload missile modules and a folding crane – a crew of two takes about 18 minutes to reload the Tor). The 9T244 is also used with the Tor-M2E.
- The 9V887 wheeled maintenance vehicle to maintain the 9A331 in the field (with a crew of three and able to service four 9A331s). A 9V887M2K is used at the battery level with the Tor-M2E. The 9V887-1M2K is used at the regimental level.
- The 9F399 wheeled SPTA system test vehicle (with a crew of two and able to service four 9A331s). The 9F399-1M2K is used with the Tor-M2E.
- The 9T245 wheeled transport vehicle with four missile packs on its flatbed and three others on a trailer (with a crew of two but not fitted with a crane). This vehicle is also used with the Tor-M2E.
- The 9S737 Rangir command post vehicle which coordinates the actions of a four vehicle Tor battery using their own radars or higher-echelon target data via secure telecommunication links (the crew is four). A specialised truck-mounted simulator, the 9F678, is also available. A variant known as 9S737MK is used with the Tor-M2E.
- The 9S482M1 (PU-12M7) can also be used with the Tor system. The 9S482M7 (PU-12M7) was upgraded and modernised during 2007/8. The original version was based on a BTR-60 8 × 8 Armoured Personnel Carrier (APC) and relied on data from external radar sources. The latest variant uses a K1Sh1 wheeled vehicle based on the BTR-80 (GAZ 5903) 8 × APC and incorporates a telescopic sensor mast. Within the telescopic mast are located the SRTR (*Stantsiya Radiotekhnicseskoy Razvedki*), passive ESM system and an electro-optic unit that is located at the masthead. The mast also boasts four static antennas mounted halfway up the mast for short-range radar. Each antenna covers a sector, which when combined provides all-round coverage. A trainable electro-optic unit mounted on top of the turret contains day and night cameras, providing hemispherical coverage. The PU-12M7 can accept target data from external surveillance radar, other fire-control systems and radar equipped air-defence weapons. Communications with assets up to 15 km away can be achieved by wire, with radio being used for ranges out to 40 km. This then allows the PU-12M7 tactical control of an area out to 100 km, handling information for as many as 99 airborne targets. The operator can simultaneously track up to nine targets.
- 9F116 used as a rigging equipment set and training simulator for the combat vehicle operators in the Tor-M2E.

Variants

The Russian Navy uses the Kynshal missile system. Designated SA-N-9 'Gauntlet' by the US/NATO, the Russian export version is known as 'Klinok' ('blade'), which uses the 9M330-2 missile.

Tor-M1

The Tor system underwent substantial redesign from its original trials batch until the production of the primary production version in 1991. At least three different surveillance radar configurations have been used. The initial production version used a mechanically scanned radar antenna while the latest version, the Tor-M1, uses an electronically scanned system on a noticeably more compact turret configuration. The radar is capable of detecting up to 48 targets at a distance of up to 27 km with the selection of up to 10 high threat targets with simultaneous engagement of two targets by two 9M331 missiles.

Ordered by Greece in 1998. The associated back-up vehicle modifications are designated 9V887M, 9V887-1M, 9F399-1M1 and the 9F116 rigging equipment set. Other vehicle designations remain the same.

The Tor-M1 was exported to China. In early 2000 the first units of a second batch of 20 Tor-M1s began to be delivered to China. A regiment in Chinese service has 16 Tor-M1s. Within China the Tor-M1 has been reverse engineered under the designation of HQ-17, however, no official information on the Chinese version of the system has been released by the Chinese government therefore assessments based on the original Russian version can only be made.

Tor (SA-15 'Gauntlet') SAM magazine cut away to reveal 9M330 missiles (Christopher F Foss) 0044126

A further upgraded version of the Tor-M1 has been cleared for export during 2005. At the MAKS 2005 exhibition held in Moscow, the Kupol Izhevsk Electromechanical Plant displayed a version that has been developed with four progressive stages of modernisation, allowing prospective customers to adopt the build standard that meets their operational needs and budget.

Tor-M1A (Tor-M1A)

The first is a software upgrade that will extend the maximum range against a crossing target to 8 km and raise the altitude coverage to 10,000 m. The upgraded vehicle is designated as 9A331-1A (Tor-M1A).

Tor-M1Б (Tor-M1B)

The second offers the same software upgrade but also adds what the manufacturers term as Zveno capability. This allows two Tor vehicles to operate as a team to improve their overall coverage. Information is exchanged between the two vehicles in real time by datalink. A single battery would consist of four vehicles working independently in two pairs. This version of the system is designated as 9A331-1B (Tor-M1B).

Tor-M1B (Tor-M1V)

The third upgrade designated 9A331-1V (Tor-M1V) combines the software and Zveno upgrades with increased resistance to radar jamming. This is provided mainly by new signal-processing algorithms and new operating modes. Other modifications to the M1V variant include:

- Extend the engagement envelope from 6-10 km n altitude and from 6-8 km in course parameter with target speeds of up to 200 m/s;
- Upgraded acquisition ability to guide the missiles onto targets diving from altitudes of 10-12 km at an angle of up to 85°;
- Increase the engagement range of cruise missiles, guided bombs and unmanned threats with a radar cross section of over 0.1 m^2 and speeds of up to 700 m/s from 5-7 km to 6-7.5 km;
- Integrate data from the acquisition radars of two Tor-M1 combat vehicles operating in different elevation sectors enabling simultaneous engagement of targets within the elevation section 0° to 64°;
- Increase the immunity of the guidance system's target channel against synchronous and asynchronous interferences luring away in speed and range, as well as against passive and active selective noise jamming in continuous and intermittent modes;
- Automatically restore target auto tracking in case of its disruption due to jamming.

Tor-M1Г (Tor-M1G)

The fourth and final modernisation exchanges the vehicle's existing 9Sh311 day only television sensor with a new 9Sh319 (GEO-PZR3) day and night camera. This is designated the 9A331-1G (Tor-M1G).

The first two modifications to the system have already been applied to many of the Russian Army systems. The modifications three and four have now completed trials and have been offered for export.

Table 1

Version	Tor	Tor-M	Tor-M1	Tor-M1T	Tor -M2/M2E	Tor-M2KM
IOC date:	1986	1988	1991	2000	2007-2008 est.	2008/09 est.
Chassis:	GM-355	GM-355M	GM-5955	Ural-5323	Wheeled Chain tracks Towed	Towed
Launcher designator:	9A330	9A331	9A331M	Unknown	9A331MU CV tracked version 9A331MK CV wheeled version	9A331 MK-1
System designator:	9K330	9K331	9K331M	Unknown	9K332	9K331 MK-1
Maximum target speed:	700 m/s	700 m/s	700 m/s	700 m/s	700 m/s	700 m/s
Engagement range (V target ≤300 m/s):						
(minimum)	1,500 m	1,500 m	1,000 m	-	1,000 m target RCS 1 m^2	1,000 m
(maximum)	12,000 m	12,000 m	12,000 m	-	12,000 m target RCS 1 m^2	15,000 m
Target altitude (V target ≤300 m/s):						
(minimum)	10 m	10 m	10 m	-	10 m target RCS 1 m^2	10 m
(maximum)	6,000 m	6,000 m	6,000 m	-	10,000 m target RCS 1 m^2	10,000 m
Cross range (V target ≤300 m/s):	-	-	8,000 m	-	8,000 m target RCS 1 m^2	8,000 m
Engagement range (V target ≤700 m/s):						
(minimum)	-	-	1,000 m	-	1,500 m	-
(maximum)	-	-	12,000 m	-	7,000 m	-
Target altitude (V target ≤700 m/s):						
(minimum)	-	-	10 m	-	50 m	-
(maximum)	-	-	7,000 m	-	6,000 m	-
Cross range (V target ≤700 m/s):	-	-	7,000 m	-	6,000 m	-
Missile launch weight:	165 kg	-	165 kg	-	-	-
Warhead weight:	15 kg	-	15 kg	-	-	-
Maximum missile speed:	850 m/s	-	860 m/s	-	-	-

An overview of the system variants' performances is given in the accompanying table (Table 1).

Tor-M2 and M2E

The Tor-M2 and M2E versions of the family of weapon systems were first displayed publicly at MAKS 2007; the Tor-M2 was inducted into the Russian Armed Forces after testing at the Ashuluk test ground was completed in late 2008 and early 2009. Further testing of an improved Tor-M2 continued until late 2013 and included the 9M338 missile, this system was inducted into the Russian Armed Forces in mid-November 2013.

The Tor-M2 was prompted by the need to considerably improve the performance of this class of weapon for use in short-range and quick reaction defence and is allegedly a 200 per cent improvement over the basic Tor-M1. Two versions were displayed and a third is assessed to be available with the first being mounted on a 6 × 6 truck with variable height restrictions. A second version is more standard in its approach and is mounted on a chain track.

The new and improved Tor-M2 (M2E export) variants have a new rectangular 3D target detection radar with a low radar cross-section capability down to 0.05 m^2 and a new thermal imaging device that is used for low thermal imaging signature detection. The 3D tracking radar tracks the targets and outgoing 9M338 missiles whilst sending guidance commands via datalink to the missile. This Tor-M2 system with the new radar has been positively identified with the Belarus Air-Defence forces, however, it is still not confirmed whether or not this is the Tor-M2 or Tor-M2E variant.

The optical tracking device does not need data from the tracking radar for missile guidance. The optical tracker locates the position of the target then tracks the target and outgoing missile whilst passing updated information to the missile via the datalink.

The number of channels has been increased from the M1 variant that has two to four and therefore can simultaneously take on four targets with four missiles in the air at any one time.

Other improvements in the system include:

- More channels and shorter reaction time;
- Larger sector for simultaneous engagement of targets;
- Improved performance of the acquisition radar including a new antennae;
- New advanced computing system;
- Modified TV optical sight;
- Wheeled basic cross-country chassis to improve operational comfort of the crew and reduce the cost. The new chassis is equipped with an automatic gearbox, automatic digital equipment for control and monitoring the system and a satellite navigation system.
- Battery Command Post of 9S737MK Ranzhir-MK.
- The 9M338 missile is smaller in size to its predecessor thus allowing the carrying capacity of the launcher to be doubled from eight to 16.

Integration with the S-300 (presumably the S-300V) system and the Antei-2500 is standard with the system engagement being controlled by either the S-300 or Antei-2500.

Crewing for new system has been reduced from three to two.

Tor-M2K

This is another variant of the Tor-M2 that is mounted on a wheel-type chassis.

Tor-M2KM

Yet another variant of the Tor-M2 that has been reported by Moscow as a modular build designated as 9K331MKM.

The system comprises:

- ACM (Autonomous Combat Module) 9A331MK-1
- SAMM (Surface-to-Air Missile Module) 9M334
- Four SAM 9M331
- BCP (Battery Command Post)
- TL (Transporter-Loader) 9T244K
- Maintenance vehicle 9B887M2KM
- Group SPTA set vehicle 9F399-1M2KM
- Rigging equipment set 9F116.

The ACM is designed so that it can be deployed independently or can be integrated with other air-defence systems. It houses target acquisition and tracking radar, missile guidance radar, test and control systems, life-support and independent power supply systems and eight SAMs. It is operated by a crew of two.

The target acquisition radar can process 48 targets simultaneously and can engage four different targets at one time. The radar are ECCM capable and can be integrated with other radar of the theatre for exchange of data. The acquisition range of the radar is 32 km and can acquire the target up to 32° from the horizon.

The SAMM is a unified construction designed to store, transport and launch guided SAMs. There are two SAMMs, which comprises a transporter/loader container with up to four SAMs loaded, as per operational requirement. The total missiles that can be carried are eight and the minimum missile loading time is 18 minutes.

The system weighs 15 tonnes. It can be installed on a truck chassis, trailers, semi-trailers and other platform types with appropriate carrying capacity and can also be transported by helicopter.

Tor-M2U

The Tor-M2U has been on the range at Kapustin Yar since August 2012, although no official photographs have yet been released of the system. It is understood that by the end of 2012 the system will be released to the Eastern Military District for operational use. The domestic Tor-M2U has been reported by various sources to have more 9M331 missiles in the launcher than the export Tor-M2 and also changes and modernisations in its electronics.

Tor variants

Shelter version Tor M1T

In 1995, it was revealed that a mobile trailer-mounted Tor-M1T system existed, the system being towed by a 6 × 6 lorry fitted with a van body.

The Tor-M1T is a derivative of the basic Tor-M1 and is manufactured in three versions:

- Wheeled (Tor-M1TA)
- Trailer (Tor-M1TB)
- Container-mounted Tor-M1TC).

The same mobile system is also used in conjunction with a four-wheel trailer fitted with a van body to provide the defence of civil and military rear area vital points. Both the lorry and the trailer provide power and other facilities to the launcher unit.

The final version revealed was a static shelter-mounted system with a second shelter used to provide the power and other facilities. Details of these versions can be found under the Shelter- and container-based surface-to-air missile systems section.

Specifications

	9M330	9M331
Dimensions and weights		
Length		
overall	2.895 m (9 ft 6 in)	2.895 m (9 ft 6 in)
Diameter		
body	230 mm (9.06 in)	230 mm (9.06 in)
Flight control surfaces		
span	0.65 m (2 ft 1½ in) (wing)	0.65 m (2 ft 1½ in) (wing)
Weight		
launch	165 kg (363 lb)	165 kg (363 lb)
Performance		
Speed		
max speed	1,555 kt (2,880 km/h; 1,790 mph; 800 m/s) (est.)	1,672 kt (3,096 km/h; 1,924 mph; 860 m/s)
cruise	1,263 kt (2,340 km/h; 1,454 mph; 650 m/s)	1,263 kt (2,340 km/h; 1,454 mph; 650 m/s)
Range		
min	0.8 n miles (1.5 km; 0.9 miles)	0.8 n miles (1.5 km; 0.9 miles)
max	6.5 n miles (12 km; 7.5 miles)	8.1 n miles (15 km; 9.3 miles)
Altitude		
min	10 m (33 ft)	10 m (33 ft)
max	6,000 m (19,685 ft)	6,000 m (19,685 ft)
Ordnance components		
Warhead	15 kg (33 lb) HE fragmentation	15 kg (33 lb) HE fragmentation
Fuze	radar, proximity	radar, proximity
Propulsion		
type	cold launched, two stage, solid propellant (sustain stage)	cold launched, two stage, solid propellant (sustain stage)

Status

In mid-2000, it was revealed that China was negotiating to licence build the Tor-M1. A total of 160 units was first mentioned, however two batches totalling 34 units with a third batch of more than 20 units were all that appeared. This was not sufficient to equip the 10 regiments as first planned. China is also believed to be reverse engineering the Tor-M1 under the programme designation of HQ-17 as a fallback measure if negotiations break down. Antei is thought to have received a contract to help develop such a system for the Chinese. China has also bought Tor-M1 simulators for training purposes.

By March 2007, it became clear that China had purchased a number of Tor-M1 and Tor-M1TA (wheeled and towed trailer version). These systems are garrisoned with the 38th Group Army near Beijing and the 31st Group Army near Taiwan. Both variants have been used during cruise missile air-defence exercises.

Again in March 2007, China has been making moves to end the procurement of any further Tor systems from Russia while there have been widespread but unconfirmed reports suggesting that the HQ-17 indigenous programme is now available in country with a Chinese improved version of the Tor-M1.

Contractor

Almaz/Antei Concern of Air Defence (primary contractor).
Izhevsk Electromechanical Plant (manufacturer).
NIEMI Research Institute (Tor-M1).
The Moscow Research Electro-Mechanical Institute (modifications and upgrades).

Serbia

2K12 Kub Maintenance and Modification Programme

Type

Self-propelled Surface-to-Air Missile (SAM) system

Radar associated with the Kub system (Michael Jerchel) 0099545

Kub Tel 0080452

Development

As part of its cost-effective approach to extend the shelf life of its air defence missile systems, while at the same time upgrading elderly ex-Soviet supplied Surface-to-Air Missile (SAM) systems, the former Yugoslavian SDPR (now Yugoimport - SDPR) developed an overhaul and modification package for the 2K12 Kub (NATO SA-6 'Gainful').

A second upgrade to the system was offered in October 2005 and has also been produced by Yugoimport.

In October 2012, yet another Serbian upgrade (number 3) to the old 2K12 Kub system has been offered for sale on the open arms market. The upgraded system is still undergoing tests but is expected to enter production very soon according to a recent report emanating from Belgrade.

Description

The 2K12 Kub is a tracked self-propelled SAM. The following upgrades and modifications have been made in order to extend the weapons ageing shelf-life:

- General system modifications designed to improve significantly the quality and reliability of the individual missile system components
- Installation of a solid-state low noise RF amplifier assembly
- Modification of the system including the 2P25 launcher vehicle to allow the additional capability of firing the 3M9M3 missile version
- Modification of the 2V8E control and test station (KIPS) to enable checking of 3M9M and 3M9M3 missile versions
- Installation of a digital MTI facility.

In addition to these upgrades, the following can also be performed to improve the 2K12 Kub/Kvadrat system:

- Overhaul of the 3M9M3/3M9M3E missile
- Overhaul of the 2V8E and 2V8EM1 control and test stations.

The second modification and upgrade to the system offered as of October 2005 includes:
Integration of electro-optic target acquisition, tracking and engagement system with the missile guidance station based on a thermal imaging camera, upgraded CCD TV camera and laser range-finder
T3-D cabin mounted three-dimensional automatic target tracking device
Upgrade of the 1S11 missile system surveillance radar through integration of new semi-conductor technology. This upgrade also includes, HF receivers with new low-noise HF amplifiers in target and missile channel instead of blocks 7Sh and 7Ash, logarithmic receiver with wide band technology integrated in place of the existing linear IF amplifier blocks 3M and digital moving target indicator instead of blocks 10M1 from cabinet 1M1
Upgrade of 1S31 missile system guidance radar through integration of new semi-conductor technology. This includes, HF receiver with new low noise HF amplifiers instead of blocks 7NSh and digital moving target indicator instead of blocks 10NM1 from cabinet 1M1.

Status

Production as required. In service with the former Yugoslavian Army (a number of the 2K12 systems operational with the then Yugoslavian Army during the Kosovo/Serbia conflict were believed modified to this standard. Very few of the operational vehicles were destroyed in combat). Offered for export but with no known takers.

Contractor

Yugoimport - SDPR.

SAVA

Type

Self-propelled Surface-to-Air Missile (SAM) system

Development

The fully amphibious SAVA mobile SAM system is an indigenously developed variant of the Russian 9K35 Strela-10 (NATO SA-13 'Gopher') system. It is intended to provide protection for highly mobile armoured units against attack by low-flying aircraft, helicopters and Remotely Piloted Vehicles (RPVs).

From October 2005, Yugoimport-SDPR has offered a complete missile system technical inspection and service life extension for the Strela-10 either in country or for those sold abroad by the Russian Federation. The technical inspection includes an overhaul of both the radar and the missiles.

The Yugoimport SDPR SAVA low-altitude surface-to-air missile is very similar in appearance to the Russian SA-13 'Gopher' system 0509479

Yugoimport SDPR SAVA low-altitude surface-to-air missile system in travelling configuration 0509478

Yugoimport SDPR SAVA low-altitude surface-to-air missile system launching a missile 0509476

Close-up of the nose of the Yugoimport SDPR SAVA low-altitude surface-to-air missile 0509477

Description

The four-round launch assembly is mounted on a modified locally designed and built BVP M80A APC chassis that acts as the Transporter-Erector-Launcher (TEL) vehicle. It is fitted with target acquisition and identification subsystems, electro-optic sight and launcher control systems.

The vehicle has a crew of three: a commander, gunner and driver. Apart from the four ready-to-fire missiles in their container-launchers, there are another six rounds stowed in container-launchers within the vehicle.

The missile used appears to be similar to an early Model 9M31M (Strela-10M) and is equipped with an improved all-aspects lead sulphide (PbS), passive infra-red homing seeker, operating at two individual frequency bands of the 1 to 5 μm near-IR spectrum region. A combined 3 kg HE fragmentation and expanding-rod warhead is fitted, together with an impact fuze and an active xenon lamp proximity fuzing system. Launch weight is 40 kg, missile length 2.2 m, diameter 0.12 m and wing span 0.4 m.

Altitude engagement limits are between 25 to 3,500 m. Maximum engagement range 5,000 m and maximum target detection range 10,000 m. Reaction time from initial target detection to missile launch is less than six seconds. The Single-Shot Kill Probability (SSKP) is said to be better than 0.6.

A typical SAVA battery comprises four to six TEL's and a radar detection vehicle. It is completed by a mobile trailer-mounted workshop for maintenance of the armoured vehicles, a mobile two-wheeled electric generator producing a 29 V 12 kW power supply, a lorry-mounted mobile automatic testing station for the missiles and a boxed gunner's simulation training set.

Variants

RH-ALAN Torpedo of Croatia has developed a similar system on a 6 × 6 armoured chassis. Available details are in this section under Croatia.

Specifications

	SAVA Tracked TEL
Dimensions and weights	
Crew:	3
Weight	
combat:	12,000 kg
Mobility	
Configuration	
running gear:	wheeled
Speed	
max speed:	60 km/h
water:	6 km/h
Range	
main fuel supply:	500 km
Amphibious:	yes

	SAVA Tracked TEL
Firepower	
Main armament traverse	
angle:	360°
speed:	100°/s (est.)
Main armament elevation/depression	
armament front:	+80°/-5°

	SAVA
Dimensions and weights	
Length	
overall	2.2 m (7 ft 2½ in)
Diameter	
body	120 mm (4.72 in)
Flight control surfaces	
span	0.4 m (1 ft 3¾ in) (wing)
Weight	
launch	40 kg (88 lb)
Performance	
Speed	
max speed	1,005 kt (1,861 km/h; 1,156 mph; 517 m/s)
Range	
min	0.4 n miles (0.8 km; 0.5 miles)
max	2.7 n miles (5 km; 3.1 miles)
Altitude	
min	25 m (82 ft)
max	3,500 m (11,483 ft)
Ordnance components	
Warhead	3 kg (6.00 lb) HE fragmentation, continuous rod
Fuze	proximity (active xenon flash lamp)
Propulsion	
type	solid propellant

Status
Believed to be built as a small prototype series. Probably non-operational or in very limited operation in Serbia. Apparently has not entered full series production and none are known to have been exported. By June 2013, the system has been removed from the Yugoimport website.

Contractor
Yugoimport - SDPR.

Singapore

Igla M113

Type
Self-propelled Surface-to-Air Missile (SAM) system

Development
The Singapore Ministry of Defence (MINDEF) Defence Science and Technology Agency (DSTA) has overseen a project that combines the locally produced Igla man-portable Surface-to-Air Missile (SAM) system (licensed produced from Russia) with the US produced M113 (A2 standard) tracked Armoured Personnel Carrier (APC) chassis.

The Igla M113 (sometimes known as Mechanised Igla) is an upgraded M113A1 that has been brought to the same standard as M113A2. The system is produced in two variants:

- Integrated Fire Unit (IFU)
- Weapon Fire Unit (WFU).

Both of these were developed at the DSTA - the MINDEF technology arm - and are intended to equip the Singapore Army's Combined Arms Divisions (CADs). The army has three CADs together with one rapid deployment division but it is not known how many Igla M113 are required.

The development programme was launched in the 1990s with the aim of providing the CADs with improved air-defence capabilities. Singapore officially purchased the Igla missile in 1997 and conducted live test firing at the South Africa's Overberg Toetsbaan Range in 1999.

Description
The Mechanised Igla is a SHOrt Range Air Defence (SHORAD) system that provides low-level air defence to the Army and its critical assets. It is operated by the Divisional Air Defence Group. The Integrated Fire Unit (IFU) comes with an integrated radar that is used to provide early warning of air threats to the Weapons Fire Unit (WFU). There is no difference between the IFU and WFU in terms of engagement capability. Both types have four ready-to-fire Igla with IRST optronics on the right and a self-protection 7.62 mm GPMG on the left. The two vehicles are wirelessly connected to form an air-defence umbrella for ground troops against low-flying aircraft and helicopters.

The M113 interior and roof have been modified to enable the installation of the launcher as well as to improve the overall inter-crew operation. Mounting the Igla missiles on the M113 chassis enhances mobility, protection and responsiveness. Human factor engineers were also involved to enhance the ergonomics of the hardware and man-machine interfaces.

A command-and-control system to link up various subsystems has also been developed. These subsystems include the radar and navigation system to offer the Igla M113 improved mobility, speed and accuracy. A basic platoon comprises 1 × IFU and 2 × WFU, the IFU with its integrated targeting radar, is able to cue itself and the 2 × WFU to engage the targets precisely.

The Igla missile produced in Singapore has reportedly been improved over its Russian variant with more modern electronics and front end technology.

Variants
There are none known.

Igla on M113 with integrated radar (Singaporean MINDEF) 1116422

M113 Singapore (Singaporean MINDEF) 1116423

Specifications

Mechanised Igla (Singapore)

Type:	short-range SAM
Length:	n/avail
Max effective range:	5 km
Max altitude:	3 km
Max missile speed:	570 m/s
Single shot kill probability:	85%
Radar detection range:	
(fixed wing)	14 km
(rotary wing)	8 km

Status
Development is complete, however it is not known how many are operational within the Singapore Army. As of December 2013, the system has not been offered for export.

Contractor
Defence Science and Technology Agency.

South Africa

ZA-HVM

Type
Self-propelled Surface-to-Air Missile (SAM) system

Development
The South African Army generated a requirement for a surface-to-air missile system with advanced cross-country mobility characteristics. The original concept was to work with the ZA-35 twin 35 mm self-propelled anti-aircraft gun system, however, development work on the ZA-35 has stopped.

The requirement called for the system to be capable of engaging aircraft and helicopters at ranges up to 10,000 m. In particular, aircraft using a short-range guided or smart weapon released from altitudes between 4,572 m and 6,096 m must be countered, as well as anti-armour helicopters with third-generation anti-tank guided weapons released from 5,000 to 6,000 m with very short exposure times.

ZA-HVM missile system on ZA-35 (8 × 8) chassis with surveillance antenna erected 0080455

SAHV-3 missile with transit container in background 0080456

The resulting system, developed by the then Kentron (now Denel Dynamics), was the ZA-HVM [High-Velocity Missile]. The system uses the basic structure of the ZA-35 gun, but with the ESD 110 acquisition radar modified to give a detection range of up to 25 km and a ceiling of 7,500 m. An Integrated Optical Radar Tracker (IORT) replaces the electro-optical sight. The IORT was also developed by ESD, which combines a K-band radar with TV and IR sensors, and uses the same optical auto tracker as the ZA-35 to track the target and the missile when used in the optical engagement mode.

Description
The ZA-HVM uses the SAHV-3 missile as developed for the Cactus 1000 upgrade programme. The missile is fitted with a K-band transponder, giving the option of radar or optical command guidance. This enables optimisation of the performance under adverse conditions of ECM, poor visibility or multi-path interference to be chosen.

There are four missiles in their container-launchers mounted on the launcher assembly in the ready to fire state. The launcher assembly can accommodate eight missiles if a wider chassis such as a tracked type is used.

	SAHV-3
Dimensions and weights	
Length	
overall	3.13 m (10 ft 3¼ in)
Diameter	
body	180 mm (7.09 in)
Flight control surfaces	
span	0.4 m (1 ft 3¾ in) (wing)
Weight	
launch	115 kg (253 lb)
Performance	
Speed	
max Mach number	3.5
Range	
min	0.4 n miles (0.8 km; 0.5 miles)
max	6.5 n miles (12 km; 7.5 miles)
Altitude	
min	30 m (98 ft) (aircraft)
	5 m (16 ft) (helicopter)
max	7,500 m (24,606 ft)
Ordnance components	
Warhead	20 kg (44 lb) HE fragmentation
Fuze	proximity (active RF)
Guidance	-
Propulsion	
type	solid propellant (smokeless)

Status
Prototype stage. As of February 2013, it was unlikely to be procured by the South African National Defence Force because of budget considerations.

As of February 2014, nothing further is available on this system.

Contractor
Denel Dynamics.

Sweden

RBS 70/M113

Type
Self-propelled Surface-to-Air Missile (SAM) system

Development
In March 1988, the then Bofors announced that it had test fired its latest vehicle-mounted application of the RBS 70 missile system, the RBS 70/M113 combination.

The system was designed to meet a Pakistan Army requirement for a mobile SAM system to protect mechanised units in the field. The conversion has gone into production in Pakistan, which also produces the RBS 70 system from supplied kits. Pakistan manufactures some parts of the RBS 70 missile under licence.

The missile system is transported in a folded-down state to present the M113 as a 'normal' APC and conceal its air defence role from overhead observers. Once in a combat situation and assigned a fire mission, the system is raised to its operating position.

In 1998, the then Bofors Missiles and STN ATLAS Elektronik (now Rheinmetall Defence Electronics) of Germany revealed a cooperative programme to integrate the Missile Mk 2-missile system with the Ericsson Microwave Systems HARD 3-D radar into the ASRAD.

Among the latest IFF systems developed by Thales, the TSA 1400 IFF interrogator has become the new standard for VSHORADS. The TSA 1400 is a miniaturised digital IFF interrogator suitable for all man portable and vehicular VSHORADS applications having a range from 250 m up to 100 km.

Saab Dynamics RBS 70/M113 SAM system in service with Pakistan 0044128

Given the designation ASRAD-R the development is designed for the export market and had been mounted on an M113A2 APC for prototype work. The ASRAD-R can be fitted to a wide variety of wheeled and tracked chassis. See entry on ASRAD under Germany for system details and the Man-portable SAM section for the missile description.

Description

The M113 variant chosen by Pakistan for the conversion is the locally assembled M113A2. The operating crew consists of four:
(1) The fire (and vehicle) commander;
(2) Missile operator;
(3) Loader/radio operator;
(4) Vehicle driver.
The vehicle driver sits at the front of the all-welded aluminium alloy hull on the left side and has a single-piece hatch cover that opens to the rear. To their right is the engine compartment. The fire commander is seated to the driver's rear, in the centre front of the troop compartment. Above the fire commander is a cupola, which can traverse through 360° and mounts a 12.7 mm calibre Browning M2 heavy machine gun. To their immediate left, on the crew compartment wall and above a bank of two radio transceiver sets, is the Target Data Receiver (TDR). This is connected to one of the radios and provides, from external surveillance radar, the combat control information and target data required for an engagement.

The loader/radio operator sits adjacent to these electronic units on a foldable seat with the missile operator alongside. Both face inwards and look directly at the RBS 70 missile system platform which is hinged to the compartment roof on its right side and held upright in the travelling position by two torsion springs. Behind this and on the vehicle's right wall, overhanging the track, is the missile store for six RBS 70, missile standard container-launcher tubes.

In normal mobile combat situations the fire commander receives a radio alert from the parent unit's Tactical Control Officer. The vehicle driver is then ordered to stop on a level piece of ground and the other crew members assume battle positions. The loader releases the crew compartment's rectangular two-piece foldable roof hatch and moves it to the open left position, the springs on the missile platform are released and, together with the missile operator, swings the unit upwards to act as the compartment's roof. The platform is automatically locked and secured in position.

The fire commander enters the vehicle's coordinates and references north into the TDR if they have not already been set before deployment. The missile operator leaves the vehicle via the rear power-operated ramp and climbs onto the missile platform where the lock of the missile hatch, located over the top of the missile store, is released and the RBS 70 launcher stand is assembled from its components.

In the meantime, the driver moves from the front of the vehicle to assist the loader who is preparing the first two missile containers inside the store for use. The driver removes the front end cap of the immediate ready to use round and the missile operator lifts this container up through the missile hatch, fits it onto the launcher stand and removes the rear end cap. The operators close the missile hatch and position themselves on the stand's seat where they connect up their intra-vehicle headphone communications unit and make the system ready for firing. Inside the vehicle, under the hatch cover, the loader prepares and positions the next missile container-launcher for use.

The fire commander and missile operator orientates the RBS 70-fire unit to the TDR and the former reports that the fire unit is ready for combat. When a target is assigned over the TDR the missile operator slew's the stand to the actual target bearing and starts a target engagement of the type described in the man-portable surface-to-air missile systems section, RBS 70 entry. The fire commander gives final permission to fire when the target is within effective range.

Once the engagement is concluded, the commander can order the missile operator and loader to reload the system. The operator locks the RBS 70-missile system sight in elevation discards the empty missile tube and repeats the loading sequence described above but without disconnecting the headphone unit. If disengagement or redeployment is required, the reverse of the operating sequence is followed to secure the vehicle for travelling.

The vehicle also carries a standard RBS 70 missile system field stand, which allows the fire unit to be deployed independently. For close-to-the-vehicle operations this must be within a 40 m radius and the fire commander decides upon the site to be used. The missile platform is deployed in the mobile engagement mode so that the missile operator can remove the sighting unit. The loader simultaneously removes the field stand from the vehicle and positions it at the designated site. The complete assembly is erected and a missile container attached. A signal cable is run from the vehicle to the field stand and the transceivers in the vehicle are connected to the TDR by twin cable. The fire commander then connects the field stand cable to the TDR and checks, with the missile operator that the cable link is functioning. If correct the missile operator reports that the missile is in the ready to fire position.

Locally designed and produced armoured (4 × 2) tractor of the Pakistan Army carrying a Saab Dynamics RBS 70 man-portable surface-to-air missile system 0080457

In the case of operations that are further than 40 m from the vehicle, because it has had to be abandoned, or it cannot be located in or near to the commander's designated site, then all of the portable RBS 70 missile system equipment is unloaded – namely the sight unit, field stand, the six missile containers, two radio transceivers, TDR with its signal cables and the system's accessories box. The deployment sequence then becomes the same as for the man-portable RBS 70 missile system.

Specifications

Full details of the Missile Mk 2 missile are given in the man-portable SAM systems section in this book.

	RBS 70/M113
Dimensions and weights	
Crew:	4
Length	
overall:	4.863 m
Width	
overall:	2.686 m
Height	
transport configuration:	2.04 m
deployed:	3.44 m
Weight	
combat:	11,600 kg
Mobility	
Configuration	
running gear:	tracked
Speed	
max speed:	67 km/h
cross-country:	23 km/h
Amphibious:	yes
Gradient:	60%
Vertical obstacle	
forwards:	0.61 m
Trench:	1.68 m
Engine	Detroit Diesel Model 6V-53T, 6 cylinders, water cooled, diesel, 275 hp at 2,800 rpm
Gearbox	
model:	Allison TX-100-1
forward gears:	3
reverse gears:	1
Suspension:	torsion bar
Electrical system	
vehicle:	24 V
Firepower	
Armament:	1 × RBS 70 SAM
Ammunition	
missile:	6
Survivability	
Armour	
hull/body:	aluminium 44 mm

Status

The first of the M113 chassis conversion for air-defence became operational in the early 1990s when the missiles were carried internally.

At January 2013, production as required and in service with the Pakistan Army (locally produced conversion on M113A2 APC).

The man-portable version is also used by the Pakistan Army, although these are mounted on armoured 4 × 2 tractors. The tractors also have two banks of electrically operated smoke grenade dischargers firing over the frontal arc. Similar vehicles are used for other weapon systems.

ASRAD-R. Prototype installation on M113 APC.

Contractor

RBS 70/M113 conversion package
Saab Dynamics AB.

RBS 70/ASRAD-R
Saab Dynamics AB.
Rheinmetall Defence Electronics GmbH.

M113 APC
BAE Systems Land & Armaments (formerly United Defence LP).
For some years Pakistan has been building the M113 series under licence.

Taiwan

Antelope

Type

Self-propelled Surface-to-Air Missile (SAM) system

Development

Development of the second-generation self-propelled Antelope (Tien Chien/Sky Sword) air defence system began in July 1995. Antelope is a tactical ground-based Surface-to-Air Missile (SAM) missile system that uses the InfraRed (IR) Tien Chien 1 Air-to-Air Missile (AAM) developed and produced from the Chung Shan Institute of Science and Technology (CSIST). The Tien Chien was developed as the successor to the AIM-9 Sidewinder and, subsequently, the AIM-9 can be used on the launcher, although this is not foreseen because the Tien Chien 1 missile allegedly outperforms the AIM-9.

A US supplied AM General 4 × 4 High Mobility Multi-purpose Wheeled Vehicle (HMMWV) chassis was used for the trials of the surface-to-air missile system.

Description

The Antelope can be used to intercept a variety of low-flying targets such as helicopters, fighter aircraft, attack aircraft and bombers. The production system comprises a six-tonne forward-control 4 × 4 truck with four ready-to-fire Tien Chien 1 (Sky Sword 1) passive InfraRed (IR) seeker-equipped SAMs. The missiles are mounted on a pedestal with an integrated radar and Forward-Looking InfraRed (FLIR) pod target detection system is also mounted on the top of the launcher. The radar is a small search type derived from the seeker of the Tien Kung 1 missile.

System components are:

- Target Acquisition System (TAS);
- Operation and control system;
- Four TC1 missiles;
- Communication system;
- Medium size of truck.

Two other vehicles provide communications and a multi-mode fire-control sensor facility. Two nearby air defence gun systems may also be integrated with an Antelope launcher. In the event of heavy jamming or the loss of the command vehicle, the Antelope can take over the role of the commander.

Chung Shan Institute of Science & Technology Antelope low-altitude surface-to-air missile deployed with four missiles ready to launch 0044129

The system can use any model of prime mover, or be dismounted for the defence of fixed sites. The Mission Control System (MCS), used to control a group of fire units, can be vehicle-mounted or in a building. The fire units are linked to the MCS either by radio or via fibre optic links. If the command facility is knocked out, one of the fire units can take over its functions to a limited degree.

The Antelope fire unit has a crew of two, which is comprised of a gunner and observer.

The vehicle has an electrical generator that can power the system at all times, allowing Antelope to engage targets while on the move, but most engagements would probably be conducted with the vehicle stationary. When the system is powered up, the fire-control computer checks the system via Built-In-Test (BIT) routines, then automatically reports the system status to the gunner via the control console. If repairs are necessary, line replaceable units can be replaced in the field. A short warm-up period follows, after which the gunner can input details of the area of the defended, plus preferred operation parameters.

If a firing command is received from the command centre, the system is armed, and will then search for targets in the predefined area. It will automatically lock on to any targets that do not give the correct IFF response. The integral Target Acquisition System (TAS) combines radar and a FLIR. When the TAS locates the target, the fire-control computer automatically slew's the turret and missiles to the correct bearing and elevation, allowing the missiles to be fired. The gunner can carry out the engagement from within the cabin, or can dismount and position a tripod-mounted control console up to 70 m from the vehicle.

The Antelope is capable of launch while on the move and has a brightness adjustable console for night or low-visibility operations. It operates successfully with both radar and FLIR systems and has its own self-contained simulator mode for training purposes. Typical engagement range is 9,000 m. The system can also be adapted for the Raytheon Missile Systems Stinger or MBDA Mistral man-portable systems. Remote operation up to 70 m away is possible.

The vehicle-mounted low-altitude air-defence fire-control system used is the track-while-scan, 30 km range CSIST CS/MPQ-78 360° coverage search radar with integral IFF system and associated 20 km range real-time tracking radar. An auxiliary passive electro-optical tracker with a built-in laser range-finder can be used at ranges of up to 10,000 m in clear weather.

Specifications

	Antelope
Dimensions and weights	
Length	
overall:	5 m (est.)
Height	
transport configuration:	3.2 m (est.)
deployed:	4.2 m (est.)
Weight	
combat:	5,000 kg (est.)
Mobility	
Configuration	
running gear:	wheeled
layout:	4 × 4
Speed	
max speed:	90 km/h

	Tien Chien 1
Dimensions and weights	
Length	
overall	2.87 m (9 ft 5 in)
Diameter	
body	127 mm (5.00 in)
Flight control surfaces	
span	0.64 m (2 ft 1¼ in) (wing)
Weight	
launch	90 kg (198 lb)
Performance	
Range	
max	4.9 n miles (9 km; 5.6 miles)
Altitude	
min	15 m (49 ft)
max	3,000 m (9,843 ft)
Ordnance components	
Warhead	HE blast fragmentation
Fuze	impact
	laser, proximity
Guidance	passive, IR (InSb)
Propulsion	
type	solid propellant (low-smoke)

Status

In low-rate production and in service with the Taiwan Air Force (one battalion ordered with a second projected) for air base defence. It has been confirmed that the Taiwanese Air Force was the first to field the Antelope. Offered for export but as of February 2013 no known customers.

Contractor

Chung Shan Institute of Science & Technology (CSIST).

Turkey

ATILGAN

Type

Pedestal mounted self-propelled Surface-to-Air Missile (SAM) system

Development

Under the 19-month contract Aselsan designed, developed and integrated the systems on two different vehicle types. The Pedestal Mounted Air Defence System (PMADS) is an indigenous product developed by Aselsan based on requirements of the Turkish Armed Forces (TAF), as a result of a cooperative research and development project supported by the Turkish Undersecratariat for Defence Industries (SSM). The ATILGAN variant is based on the M113A2 tracked Armoured Personnel Carrier (APC) chassis and was produced to meet the requirements of the Turkish Land Forces Command (TLFC). Aselsan, using inputs from the Turkish General Staff, Naval, Air Force and Land Forces Commands, chose the actual configuration.

The follow-on series production contract, awarded to Aselsan in December 2001, covers manufacturing and delivery of a total of 158 systems (70 ATILGAN and 88 ZIPKIN), with complementary maintenance and training components for the use of the TAF.

Description

The ATILGAN PMADS is a fully automated firing unit for a Very Short-Range Air Defence System (VSHORADS) armed with Stinger missiles. The system provides autonomous as well as co-ordinated operation with C3I systems and other air defence assets. The turret is currently integrated onto an M113A2 APC, though various other types of carrier vehicles may be used. The turret incorporates sub-systems of a Line Replaceable Unit (LRU) nature, making field-repairs or updates simpler.

Aselsan ATILGAN performing amphibious performance test 0099530

Aselsan ATILGAN performing a shoot-on-the move test 0099529

The main mission of ATILGAN is the low-level air defence of stationary and moving forward troops, convoys and tactical bases in the battlefield. The vehicle carries a three-man crew comprised of a commander, gunner and driver. All of the crew can be contained within the vehicle's armour protection when on station.

A gyrostabiliser one-man turret launcher assembly is fitted to the top of the M113A2 APC's rear decking; this carries eight ready-to-fire missiles and provides surveillance, acquisition and shoot-on-the-move capability. An additional eight reload rounds can be carried within the vehicle in their launcher-containers.

Although originally based on the Stinger missile, the infrastructure is suitable for the system to be reconfigured for different VSHORAD missiles.

The sensor suite incorporates a second-generation, two-field-of-view focal plane array thermal imager, and a daylight TV camera with zoom capability for passive day/night surveillance, target acquisition and tracking. A multipulse laser range finder is also integrated for target ranging.

Basic functional characteristics of the ATILGAN PMADS are as follows:

- Eight ready-to-fire Stinger missiles;
- One 12.7 mm Heavy Machine Gun (HMG) for self-defence;
- Target detection, tracking and missile firing at day/night conditions;
- Capability of firing on-the-move under armour protection;
- Identification of Friend/Foe (IFF);
- Capability of remote control of system functions 50 m away from the platform.

The ATILGAN PMADS fire-control computer provides automated functions such as:

- Remote control of all subsystems;
- Turret control and stabilisation;
- Automatic slewing of the turret to the target co-ordinates;
- Automatic target tracking;
- Automatic target type (fixed or rotary-wing) recognition;
- Target in range warning if the target is within the missile engagement envelope;
- Automatic super elevation and lead insertion before firing.

The fire-control computer's flexible hardware and software architecture allows it to accommodate continually evolving mission requirements. This flexible infrastructure also enables the integration of different SHORAD/VSHORAD missiles besides Stinger currently used. Additionally, it is possible to adapt the system to different mounting platforms, according to differing mission requirements.

All system functions are commanded from a unique control console that can also be dismounted from the vehicle. Remote operation at distances up to 50 m from the vehicle or platform can be carried out.

The system also incorporates a 12.7 mm HMG for self-defence and missile dead zone coverage. The HMG can be remotely commanded by use of the system control unit, and can be solenoid-fired in either short bursts, or in a continuous mode, both modes being selectable from the system control unit.

Aselsan ATILGAN performing a static firing trial 0099528

Aselsan ATILGAN PMADS mounted on an M113A2 series APC chassis (the 12.7 mm HMG is not fitted but would protrude from the oval-like hole) 1033334

Variants

Variants of the PMADS known as BORA carrying Russian Igla SAMs and suitable for use on naval platforms, are available.

A version of the PMADS armed with Russian Igla SAMs rather than the Raytheon Stingers used in the early PMADS was shown at the IDEF 2007 exhibition. Series production of the Igla version was expected during 2008, but this depended heavily on customer requirements. The Igla turret can carry four or eight missiles, while a central housing contains a second-generation thermal imaging system, a TV camera and a 20 km range laser rangefinder, plus a 12.7 mm HMG with 200 rounds of ready use ammunition.

Status

ATILGAN PMADS have been in service with the TAF since 2004/5. ATILGAN was expected to remain in production until 2009.

A modified version of ASELSAN's PMADS, as a drop-in unit to be integrated onto KMW manufactured Fennek, has been delivered for the use of the Netherlands' Armed Forces.

Contractor

Aselsan Electronics Industry Inc, (system).
Aselsan AS, (enquiries).

Stinger Weapon System Programme (SWP)

Type

Pedestal-mounted self-propelled Surface-to-Air Missile (SAM) system

Development

The Stinger Weapon System Programme (SWP) is also known as the Stinger Launching System (SLS). The SWP is based on the technical characteristics and capabilities of the Aselsan Pedestal Mounted Air Defence System (PMADS), which is fully capable of detecting, identifying, tracking, engaging and destroying targets such as fixed wing, rotary wing, UAVs and cruise missiles.

The contract awarded to Aselsan in September 2005 covers manufacturing and delivery of 18 SLS to be integrated with 4 × 4 Fennek vehicles manufactured by Krauss-Maffei Wegmann (KMW) and 18 systems assembled on a Mercedes-Benz G-VAGEN jeep, plus the Air-To-Air Launcher (ATAL) launcher manufactured by Raytheon for the use of the Royal Netherlands Army (RNLA).

Stinger Weapon Platform (SWP) (Aselsan) 1181655

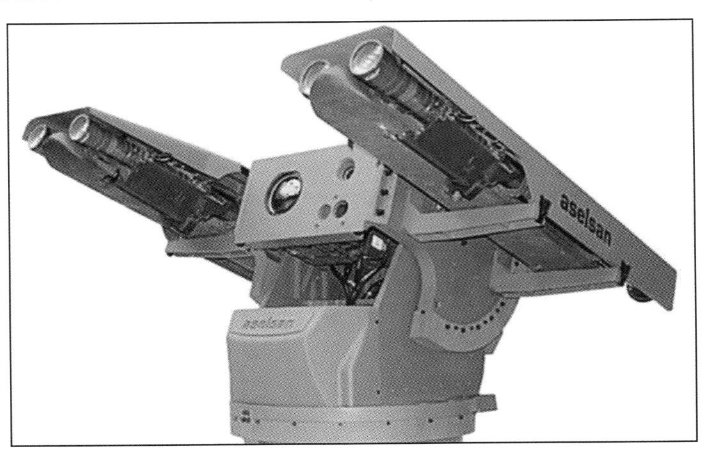

Stinger Launching System (SLS) (Aselsan) 1181654

Aselsan and Raytheon Missile Systems, the original designer and developer of the FIM-92 Stinger missiles, teamed up for the contract competition. Raytheon provided the SAMs for Aselsan to integrate onto Aselsan's launching systems mounted on the Fennek vehicles.

Description

The SLS, which is a major part of the SWP, is equipped with a digital Stinger ATAL. The system has been designed in order to protect high value assets in the brigade rear area, as well as mobile units under combat conditions.

SLS hardware and software were designed using the latest techniques and computer-aided design tools, and incorporate advanced features such as:

- Computer-controlled passive electro-optics sensor suite for surveillance, acquisition and tracking featuring Infrared (IR) and video imagers;
- Real-time assessment of target engagement range;
- Reduced missile firing time lines;
- Capability of information exchange and coordinated operation with Command, Control, Communication and Intelligence (C3I) operations centres.

The open hardware and software architecture used by the SLS makes it an easy and economical system to upgrade for improvements to the Stinger missile and other subsystems. The open architecture also makes the SLS flexible, allowing the integration of other VSHORAD missiles other than the Stinger.

The SWP provides the following basic functions:

- Lightweight, low-profile compact turret;
- Digital Stinger ATAL;
- Four ready-to-fire Stinger missiles under protective shields;
- Computer-controlled passive surveillance, acquisition and tracking features with Infra-Red (IR) and video imagers;
- Two-axis gyro-stabilised turret providing full target surveillance, detection, tracking in both stationary and on-the-move conditions;
- Fire-Control Computer (FCC) providing fully automated system functions:

(1) Turret slewing to targets assigned by air-defence command-control systems;
(2) Automatic target tracking using IR or daylight video imager;
(3) Target type (fixed/rotary wing) recognition;
(4) Automatic super elevation and lead angle insertion;
  - Computer controlled Laser Range-Finder (LRF) for target ranging;
  - Embedded trainer for hands-on training of the operator;
  - Adaptable to other VSHORAD missiles;
  - Low profile, allowing transportation by C-130/C-160 on different vehicles.

The system is capable of firing while the launcher vehicle is in motion, and was designed to fitted to wheeled or tracked vehicles, as well as small naval patrol boats. Configurations include ZIPKIN for light vehicles, ATILGAN for heavy vehicles and BORA for Naval platforms.

Variants

Igla and Mistral missiles may also be used by the launcher.

The Igla Launching System is a fully automated air defence weapon system making use of the Russian developed and produced Igla missiles. The system claims use against fixed and rotary wing targets, remotely piloted vehicles, unmanned aerial vehicles and cruise missiles.

Specifications

Stinger Weapon System

Type:	pedestal mounted self-propelled surface-to-air missile system
Armament:	Stinger FIM-92, Basic, RMP, Block-1
Elevation angle:	−10°/+72°
Traverse angle:	n × 360°
Elevation speed:	60°/s (maximum)
Traverse speed:	100°/s (maximum)

Status
In series production, deliveries were completed in 2008. The SWP Programme for the RNLA will establish an effective, reliable and state-of-the-art low-level air-defence infrastructure that has been evaluated as one of the most advanced air-defence systems currently on the market.

Contractor
Aselsan AS (system).
Krauss Maffei Wegmann (vehicle).
Raytheon (missile).

ZIPKIN

Type
Self-propelled Surface-to-Air Missile (SAM) system

Development
The ZIPKIN Pedestal-Mounted Air Defence System (PMADS) is an indigenous product developed by Aselsan based on the requirements of the Turkish Armed Forces (TAF), as a result of a co-operative R&D project supported by the Turkish Undersecretariat for Defense Industries (SSM).

The follow-on series production contract awarded to Aselsan in December 2001 covers manufacturing and delivery of a total of 158 systems (70 ATILGAN, 88 ZIPKIN) with maintenance and training components for the use of the TAF.

The ZIPKIN variant is mounted on the Land Rover Defender 130 chassis and was configured to meet the requirements of the Turkish Armed Forces.

Description
ZIPKIN PMADS is a fully-automated firing unit for Very Short-Range Air Defence System (VSHORADS) missiles such as the Stinger, providing autonomous, as well as coordinated operation with C3I systems and other air defence assets. Currently integrated onto a four-wheel drive vehicle, the modular turret incorporating LRU sub-systems, can also be integrated on various types of carrier vehicles.

The primary mission of the ZIPKIN is the low level air defence of fixed assets, like radar, air bases and harbours.

The vehicle carries a crew of two, gunner and driver. A pedestal-mounted launcher assembly is fitted to the rear decking, which carries four ready to fire missiles. Additional four reload rounds can be carried on the vehicle in their launcher-containers. The gunner is either seated in the vehicle cab or at a remote firing location positioned up to 50 m from the vehicle.

Although originally based on Stinger missile, the ZIPKIN PMADS can be configured to carry any one of a number of VSHORAD missile types.

The sensor suite incorporates a second-generation, two field of view focal plane array thermal imager, and a daylight TV camera with zoom capability for passive day/night surveillance, target acquisition and tracking. A multi-pulse laser range finder is also integrated for target ranging.

Basic functional characteristics of the ZIPKIN PMADS are as follows:
- Four ready-to-fire Stinger missiles
- 12.7 mm automatic machine gun for self-defence
- Target detection, tracking and missile firing at day/night conditions
- Identification of Friend/Foe (IFF)
- Capability of remote control of system functions 50 m away from the platform
- Passive surveillance, acquisition and tracking sensors, incorporating thermal and day TV cameras.

The ZIPKIN PMADS fire-control computer provides automated functions such as:
- Remote control of all subsystems
- Turret control and stabilisation
- Automatic slewing of the turret to the target co-ordinates

ZIPKIN system undergoing static firing trial 0099531

Rear view of ZIPKIN with PMADS trained to rear 0039468

The ZIPKIN Pedestal-Mounted Air Defence System (PMADS) mounted on a Land Rover Defender 130 chassis (Aselsan) 1379057

- Pre-assignment by the C3I system
- Automatic target tracking
- Automatic target type (fixed- or rotary-wing) recognition
- Automatic fire permission if the target is within the missile engagement envelope
- Automatic super elevation and lead insertion before firing.

The computer also has a flexible hardware and software architecture for evolving mission requirements.

All system functions are commanded from a unique control console that can also be dismounted from the vehicle for remote operation at distances up to 50 m from the platform. In remote control mode, the Optical Target Designator is used for the system to be directed automatically towards the target that the commander is tracking.

The system also has a secondary weapon integrated into the assembly, a 12.7 mm (0.5 in) calibre M3 heavy machine gun with 250 ready rounds mounted on an elastic cradle at the missile axes. The gun is used for both system self-defence and to cover the missile dead zones. The machine gun has full remote functions, controlled by the fire-control computer, passive surveillance, acquisition and tracking sensors incorporating thermal and daylight TV cameras. A multi-pulse laser range finder is also available for target ranging. The machine gun can be fired either in short burst or continuous mode selectable from the system control console.

ZIPKIN can easily be loaded into and transported by C-130 and C-160 aircraft.

Variants
Bora

A naval variant of PMADS has been developed called Bora.

Status
In service with the TAF since 2004.

Contractor
Aselsan A.S. Defence Systems Technologies Division, Ankara, Turkey.

United Kingdom

Common Anti-Air Modular Missile (CAMM)

Type
Developmental Surface-to-Air Missile (SAM) system

Development
The MBDA Soft Vertical Launch Technology (SVLT) programme at the MBDA Bedfordshire development facility is now known as the Common Anti-Air Modular Missile (CAMM), was first conceived during 1998 when the UK Ministry of Defence (MoD) set a conceptual contract with MBDA to study the stealthy launching of a Ground Based Air Defence (GBAD) Very Short Range Air Defence System/Short Range Air Defence System (VSHORADS/SHORADS).

The system is also known as the Future Local Area Air Defence System (FLAADS-Maritime).

Trials between 2002 and 2005 proved the principle of the launcher and the manoeuvrability of the missile with command after launch and the missile pitched over towards its intended target. Left and right movements have also been achieved during these trials.

By January 2003, MBDA was conducting tests using a real missile, with a real launcher. These trials included the ASRAAM-based hardware to prove the planned soft-launch scheme.

The system has been conceived as a joint service air defence missile and associated weapon system that could replace the Rapier air-defence system and the Royal Navy's Seawolf point defence missile system from 2015 onwards.

Further information released in January 2012 now talks of the system for the Royal Navy being known as the Sea Ceptor utilising what is now known as the CAMM-M.

The missile retained the existing ASRAAM mounting points, and was fitted into a container-launcher of square cross-section. The round sat on a piston below that was a gas generator that was fired to drive a (captive) piston up the length of the container, launching the missile vertically. This captive piston traps the gases produced by the gas generator, so removing the initial launch signature of the missile. As the missile left the tube, its fins unfolded and a series of four twin-nozzle thrusters mounted just aft of the fins fired to seer the round through a turnover manoeuvre. The main motor was then fired to begin powered flight.

MBDA has carried out captive carry trials of the new RF radar seeker developed to meet the requirements of the newly planned CAMM. These trials included a series of carry trials that took place early in 2011.

An announcement made on the 15 May 2013 confirmed that MBDA and Lockheed Martin Missiles have agreed to cooperate to jointly explore the naval market for the integration of the Mk-41 VLS with the MBDA CAMM-M Sea Ceptor missile.

Concept model of the CAMM (Doug Richardson)

1290481

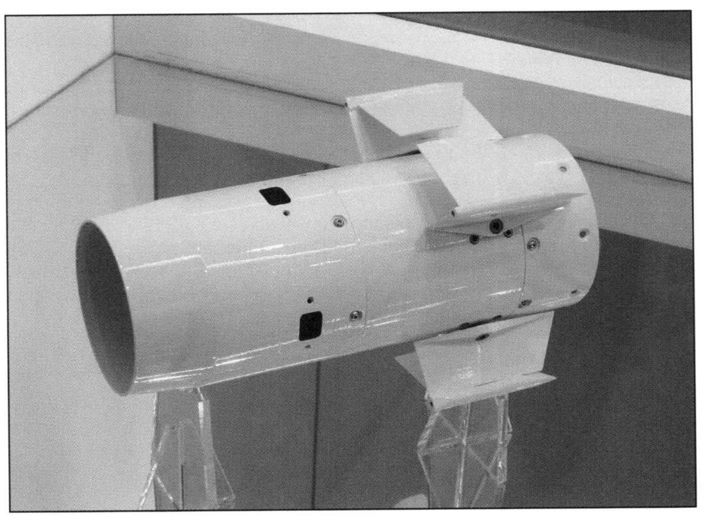

Aft folding fins thruster (Doug Richardson) 1290482

Description
Launcher

The launcher, which is made from composite materials, has been designed to be cheap to produce and may be a 'once only' use type. Basically, MBDA studied what was available at that time (1998) and applied this technology to its own designs based on the requirements of the MoD (UK) to produce a Hot Gas Charge (HGC) vertical launcher.

The container-launcher was approximately 3.25 m long and weighed about 45 kg.

It was confirmed by the British Ministry of Defence in January 2012 that the Sea Ceptor (for Naval applications) and the Ceptor (for Land applications) will both be soft launched from multiple launchers and attain speeds or around Mach 3 (1020 m/s). The naval CAMM-M will most probably use the Lockheed Martin Mk-41 VLS.

Missile

The missile currently used with the FLAADS-M is the Common Anti-air Modular Missile (CAMM), a high performance missile intended to deal with current and future threats. The demonstrator rounds were approximately 3 m in length, and retained the 166 mm diameter. Total weight was about 100 kg, slightly more than the 87 kg of the ASRAAM. Trials results proved the missile capable of flying up to 60 km from the launch point. MBDA has adopted many of the features of ASRAAM, including the solid propellant rocket motor, the laser proximity and impact fuze, and the blast fragmentation warhead. The major new components of CAMM are the active radar seeker, a dual-band weapon datalink that is undergoing trials in MBDA development facilities and an open architecture internal communications bus.

Artwork released by MBDA showed 12 CAMM rounds installed on a four-tonne truck in a configuration consisting of two six-packs each containing two horizontal banks of three missiles. Reports have also suggested that a naval version is being considered for the Future Surface Combatant (FSC) Type 26 ships or any other vessel selected to replace the UK Royal Navy's Type 22 and Type 23 frigates. This would probably consist of a quadruple package that would fit within the cells of the Sylver or Mk 41 vertical launchers.

Variants
FLAADS-M Future Local Area Air Defence System-Maritime also known as Sea Ceptor and CAMM-M.

FLAADS-L Future Local Area Air Defence System-Land also known as Ceptor.

Specifications

	CAMM
Dimensions and weights	
Length	
overall	3.0 m (9 ft 10 in)
Diameter	
body	166 mm (6.54 in)
Weight	
launch	87 kg (191 lb)
Performance	
Speed	
max speed	1,983 kt (3,672 km/h; 2,282 mph; 1,020 m/s)
max Mach number	3.0
Range	
max	13.5 n miles (25 km; 15.5 miles) (est.)
Ordnance components	
Warhead	HE fragmentation
Fuze	laser, proximity, impact
Guidance	active, radar (dual-band datalink)
Propulsion	
type	solid propellant

Status

The missile and system has completed its conceptual stage and now remains in the developmental.

The system is not intended to be in use before 2016 and possibly as far as 2020 it is therefore unlikely that updates on the basic information will be available on a regular basis.

As of January 2011, the missiles' component trials are expected to take at least another one or two years, including live round firings.

CAMM(M) (Maritime) also referred to as CAMM-M is planned to be the first CAMM to enter service, with a planned In Service Data (ISD) of 2016 for the Royal Navy Type 23 Frigates and later the Type 26 Global Combat Ships (GCS).

An announcement was made early in October 2013 that the Royal New Zealand Navy has purchased the Sea Ceptor variant of the CAMM weapon system for two frigates due for upgrade in the near future.

Contractor

MBDA.
BAE Systems.
EADS.
Finmeccanica.

THOR System with MMS firing Starstreak (Thales Air Systems Ltd) 1123818

Multi-Mission System (MMS)

Type

Self-propelled Surface-to-Air Missile (SAM) system

Development

Thales (UK) has developed and produced a modular Multi-Mission System (MMS) based on the Starstreak II that when mounted on the Pinzgauer vehicle was initially known as THOR. The MMS can be fitted to various types of mobile platforms and can be used in the air defence and surface defence/attack roles with missiles such as Hellfire and Ingwe. The MMS launcher turret weighs less than 500 kg and can be integrated into a network enabled force structure and be coordinated with Early Warning Command-and-Control systems or it can operate in an autonomous mode for specific missions. The MMS is completely air transportable.

Description

The MMS system is a highly automated system capable of delivering a rapid reaction response to threats from the air or the ground. The system has the capability to take on targets such as Armoured Personnel Carriers (APCs), static installations or terrorist platforms. The current MMS system is fitted to a standard British Army 6 × 6 Pinzgauer vehicle and consists of a modular design with five key elements:

(1) Open System architecture;
(2) Launcher system consisting of two banks of two missiles;
(3) Integrated guidance head with TV, IR, Laser range-finder and Laser Guidance Unit (LGU);
(4) Crew Station, two-man crew or one if operating in autonomous mode 1;
(5) Fire-control computer. An Identification Friend or Foe (IFF) is available as an optional extra.

The system currently uses the Thales Starstreak II missile but has been designed for multimission operations and can be configured quickly and easily to use any of the following:

(1) Four Starstreak II multirole missiles;
(2) Two Starstreak II multirole missiles and two fire-and-forget air defence missiles;
(3) Two Starstreak II multirole missiles and two anti-armour/anti-structure missiles;
(4) Four anti-armour/anti-structure missiles.

The turret that contains the launchers and targeting systems can traverse 360° in azimuth, and from –10° to +60° in elevation. The traverse rate is approximately 100°/s.

Firing Console inside THOR (Thales Air Systems Ltd) 1123819

The launcher system is a servo-driven automatic launcher with Integrated Guidance Head (IGH) for accurate target engagement.

The launcher has a full 360° in azimuth and in elevation –10° to +60°.

The launcher also contains 4 × missiles in the ready-to-fire mode. Within the IGH the IR camera is of the Focal Plane Array (FPA) type to provide day/night capability with a dual Field-of-View (FoV). The camera can be either a 3-5 microns version or 8-12 microns. The TV camera is a dual FOV camera (2°/8°).

The Laser Guidance Unit (LGU) transmits the coded laser grid used to guide the Starstreak missile. The Laser Range-Finder (LRF) provides information to the FC processor to ensure only targets within range are engaged at the optimum point.

A typical firing sequence would probably be for the target to be tracked by the Automatic Target Tracker (ATT) using the TV and/or IR detected signals. These signals provide accurate guidance and tracking information thereby freeing the operator from the task of manually tracking targets and thus increasing the kill probability and minimising operational training. The THOR could also make use of external data provided by the PAGE radar or ADAD systems.

Variants

Any of the following missiles may be mixed and matched depending on the scenario and mission of the vehicle:

- Igla and Stinger for fire-and-forget air defence
- Javelin
- Spike LR
- Hellfire
- Ingwe for armoured vehicle or bunker busting.

The then Thales Air Systems Limited announced in May 2003 their first multi-million pound export order for their Starstreak missile system. The order is to satisfy the Republic of South Africa's first phase of their Ground Based Air Defence (GBAD) programme – Local Warning Segment (LWS). This phase is intended to provide a fully integrated man-portable and air-transportable air-defence capability for South African troops deployed on peace-keeping operations. Armscor announced that they had placed a contract with Denel as the preferred prime contractor in December 2002. Subsequent to this, Thales has now successfully completed their negotiations with Kentron – the missile house subcontracted by Denel to provide the VSHORAD weapon. Discussions with the South Africans are on-going concerning future phases of their GBAD programme.

Future missiles that may be used with the system include Joint Common Missile (JCM) and Kinetic Energy Missiles such as HYKEM.

The military vehicles so far envisaged for use with the MMS include: Pinzgauer, Hummer, Duro/Adnam, Piranha and the M113 chassis.

Future integration of non-lethal and directed energy weapons is also being considered for future operations.

At September 2007, an MMS using and Starstreak II missile mounted on a Spanish Vehiculo de Alta Movilidad Tactico (VAMTAC) 4 × 4 high-mobility tactical vehicle is also available.

Specifications

Starstreak II

Type:	high velocity surface-to-air missile
Length:	1.397 m
Missile diameter:	0.127 m
Missile weight:	14 kg
Missile velocity:	>M3
Maximum range:	>7 km
Maximum altitude:	>5,000 m
Guidance:	laser beam guidance
Warhead:	three independent hit-to-kill KE/HE darts with armour piercing capability
Motor:	Titus cast double-based solid with a burn time of less than one second
g-load:	not available, classified information

Status
Sales are expected to the British Army also Spain, Malaysia, Middle East (Oman) and South Africa.

Contractor
Thales Air Systems Ltd.

Starstreak

Type
Self-propelled Surface-to-Air Missile (SAM) system

Development
In order to fulfil the British Army's General Staff Requirement (GSR) 3979, supplementing Rapier in the battlefield role of engaging late unmasking close support aircraft and ATGW-equipped hovering helicopters, the UK Ministry of Defence originally approached 11 different companies to provide a new High-Velocity Missile (HVM) design.

Of these, British Aerospace (now MBDA) and the then Short Brothers (now Thales Air Systems Ltd) were each awarded a 12-month project definition contract in 1984.

In late 1986, the latter was awarded a GBP356 million fixed-price contract to cover the system design, development, initial production and supply of the Starstreak HVM weapon.

In October 1995, the UK MoD initially accepted into service the Self-Propelled HVM (SP-HVM) having the commercial name of Starstreak. This has a turret with eight HVM in the ready-to-launch position, with additional missiles carried inside for manual loading. However, the in-service date of the first British Army Regiment, 12 Regiment, was not finally achieved until late 1997. Until then the only battery declared operational was 9 Battery, the first unit to be declared back in October 1995.

The current production SP-HVM is fitted with a roof-mounted passive Thales Optronics Air Defence Alerting Device (ADAD). However, the UK MoD has now endorsed a programme to increase the crew to four and provide an integrated night sight (thermal type) to give a full 24-hour capability.

One key advantage of the HVM over other air defence missiles is that its very high speed (MACH three plus), short flight time and laser guidance make it immune to all known countermeasures.

There have been 135 SP-HVM and over 7,000 missiles ordered in several tranches, with an additional 10 Alvis Stormers in the Troop Reconnaissance Vehicle (TRV) role. All of these have been delivered. A further order for 44 Stormer HVM may be placed in the future. A GBP200 million contract for additional HVM was awarded in December 1999.

The original in-service date of the SP-HVM was 1992 but there was some slippage owing to technical problems with both the Alvis Stormer chassis and the actual missile system. These problems have now been resolved. Stormer HVM is expected to stay in service well beyond the year 2015.

The major subcontractors to then Shorts Missile Systems (SMS) for the SP-HVM programme were:

Company	System
Alvis Vickers	Stormer chassis
Thales Optronics (Taunton)	Sighting system
Boxmore	Packaging
Raytheon Microelectronics	Electronic subassemblies
INSYS	Missile launch tubes
Nobel Explosives	Explosive for warhead
Thales Optronics	Air Defence Alerting Device
RO Defence	Rocket motor

Thales Lightweight Starstreak on Peugeot (4 × 4) chassis with six Starstreak SAM in ready-to-launch position 0099533

Thales Starstreak High-Velocity Missile (HVM) system mounted on a Stormer chassis with launcher traversed left 0099532

Control Technique Dynamics (CTD) Ltd servo motors are incorporated into the launcher and sighting system assembly.

In September 1998, the then SMS (now Thales Air Systems Ltd) announced that, following UK government clearance, it could offer armoured and lightweight self-propelled Starstreak systems for export. The armoured version can be installed on a wide variety of armoured wheeled and tracked chassis such as the MOWAG Piranha III (6 × 6) chassis, with eight ready-to-fire missiles and another 12 carried internally.

Initially, it was offered on the Stormer APC chassis, later the Lightweight Starstreak was made available. The latter is based on a Land Rover or Peugeot 4 × 4 unarmoured chassis with the Aspic fire-control system and six ready to fire missiles. Both variants have thermal sighting capability and automatic target tracking.

On 19 March 2008, a delegation from the UK Ministry of Defence witnessed the first demonstration of the latest variant of Thales Air Systems' self-propelled Starstreak missile system which incorporates automatic target tracking, new control consoles and the firing of the new standard missile (Starstreak II). The demonstration was carried out at the Manorbier Range in South Wales and was 100 per cent successful.

Description
The Stormer in the Starstreak launcher configuration carries a crew of three: driver, gunner and commander. It has eight ready-to-fire rounds in two armour-protected servo-controlled containers on the vehicle roof. Collocated with these is the Thales Optronics Mk 2 passive infra-red ADAD. Forward of the launcher is the gunner's surveillance, firing and target tracking turret which is fitted with an Thales Optronics (Taunton) servo-controlled target acquisition and tracking sight.

The ADAD Mk 2 provides for detection of targets, their prioritisation, operator alerting and automatic pointing of the Avimo weapon sight at the priority target in azimuth and elevation.

It is based on existing infrared system technology and offers an 8-12 micron waveband day-only operational capability for Starstreak which is totally independent of optical visibility, being able to see through battlefield smoke, haze and mist.

The system comprises three lightweight modules:

(1) Scanner Infra-red Assembly (SIA);
(2) Electronic Pack Processor Unit (EPPU);
(3) Electronic Pack Remote Display Unit (EPRDU).

A guidance beam transmitter is also housed in the sight unit and this is collimated to the target sight line. A total of 12 reload rounds are carried within the hull; these can be used to reload the missile containers.

Thales Starstreak HVM leaves its launcher which is mounted on a Stormer chassis 0069510

Thales Starstreak HVM showing method of manual reloading from the rear
0080458

Thales Lightweight Starstreak on Peugeot (4 × 4) chassis during firing trials
0069511

A new and upgraded launcher for the new Starstreak II missile is planned to enter service at the end of 2010. This new launcher also incorporates Automatic Target Tracking (ATT) and new control consoles.

A full description of the missile and method of operation is given in the Man-portable surface-to-air missile systems section.

Thales Air Systems has also been selected to act as the prime contractor for the Successor Identification Friend or Foe (SIFF) Mk 12 Mode 4 programme. The British Army will integrate this into the self-propelled Stormer HVM and the Lightweight Multiple Launcher (LML) systems used.

HVM with Thermal Imaging

Early in 2001, following a competition, the UK Defence Procurement Agency (DPA) awarded a GBP70 million contract to Thales Air Systems to provide the British Army's Self-Propelled High Velocity Missile system (SP-HVM) with a Thermal Sighting System (TSS).

Early in 1999, the Ground Based Air Defence Integrated Product Team (IPT) of the DPA issued an Invitation to Tender to the then SMS, to undertake the Phase I Feasibility Study (FS) to integrate a Thermal Sighting System (TSS) into the current SP-HVM. Five companies were selected by the then SMS to bid for the TSS.

The GBP70 million Phase II contract covers design, development and integration of the TSS into the SP-HVM and according to Thales Air Systems will create up to 25 high technology and skilled jobs in Belfast.

The TSS selected will use the Sensor Technology of Affordable Infra-Red Systems (STAIRS-C) technology developed by Thales Optronics (Glasgow and Staines) and the then Defence Evaluation and Research Agency (DERA) facility at Malvern. Thales Optronics (Taunton) and BAE Systems support this team.

For trials the Thales Starstreak HVM has been installed on this Swiss MOWAG Piranha (6 × 6) APC chassis
0525732

Thales Starstreak HVM of the British Army showing extensive on board stowage (Richard Stickland)
0525693

This is the first production application of STAIRS-C technology which is claimed to be the most advanced of its type developed in the UK and is also expected to be used for a number of other applications.

The current SP-HVM, of which 135 systems were delivered to the Royal Artillery, currently has a roof mounted Thales Optronics (Taunton) day sight. The installation of the TSS will provide the SP-HVM with a night-time capability equivalent to that of daytime. In addition, it will enhance the detection and recognition of targets in poor visibility.

For some years Thales Air Systems has been offering the export market two self-propelled versions of HVM fitted not only with a TSS but also an automatic target tracking capability. These are Lightweight Starstreak on a 4 × 4 chassis and Armoured Starstreak on an armoured platform such as the current tracked Stormer or the MOWAG Piranha (6 × 6).

This latest MoD contract brought the UK's recent investment in HVM up to GBP330 million. The earlier contracts were GBP60 million in September 2000 for additional LMLs (which has three missiles in the ready-to-launch position) and GBP200 million in December 1999 for additional HVM missiles.

Starstreak II

The Starstreak II missile has improved accuracy and performance which also adds to a much reduced training time for the operators and significantly reduces the workload of the system operator.

The enhancements to the missile itself mean that the range is extended beyond the standard missiles seven kilometre range, increased altitude and improved guidance precision and is optimised for use against small targets.

The all new Starstreak II is being offered on the open arms markets as a multirole missile that is not only capable of engaging aerial threats such as low-level Unmanned Air Vehicles (UAVs), but also attack helicopters and even light armoured vehicles and soft skinned warships.

Stormer chassis upgrade

The Defence Procurement Agency (DPA) has stated that it has a requirement to modify 135 SP-HVM and three driver training vehicles to include a fourth man seat. In addition, the 10 TRV Stormer vehicles will be fitted with machine gun mounts. Alvis Vickers (now BAE Systems Land & Armaments) will carry out this work.

Specifications

	Starstreak	Starstreak II
Dimensions and weights		
Length		
overall	1.397 m (4 ft 7 in)	1.397 m (4 ft 7 in)
Diameter		
body	127 mm (5.00 in)	127 mm (5.00 in)
Weight		
launch	16.82 kg (37 lb)	14 kg (30 lb)
Performance		
Speed		
max speed	1,983 kt (3,672 km/h; 2,282 mph; 1,020 m/s)	1,983 kt (3,672 km/h; 2,282 mph; 1,020 m/s)
Range		
min	0.2 n miles (0.3 km; 0.2 miles)	-
max	3.8 n miles (7 km; 4.3 miles)	3.8 n miles (7 km; 4.3 miles) (est.)
Altitude		
max	-	5,000 m (16,404 ft) (est.)
Ordnance components		
Warhead	3 kinetic, HE (dart-like sub-projectiles)	3 kinetic, HE (dart-like sub-projectiles)
Guidance	laser beam riding	laser beam riding
Propulsion		
type	two stage, solid propellant (sustain stage)	two stage, solid propellant (sustain stage)

Status

In production since 1993, and by late 1999, the 7,000 plus missiles and 135 Stormer SP-HVM had been delivered.

In December 1999, the UK MoD placed a contract with the then SMS worth GBP200 million, for additional HVM for delivery over a five-year period.

On 1 August 2004, another new contract worth USD326 million (GBP181+ million at USD1.8 to GBP1), was signed with Thales for the production and delivery of new missiles which began in 2007. Presumably these missiles will become part of the GBAD (UK) system.

The new launcher for the Starstreak II entered service at the end of 2010.

Contractor

Prime contractor

Thales Air Systems Ltd (prime contractor).
Control Technique Dynamics (CTD) Ltd, servo motors (sub-contractor).
BAE Systems Land & Armaments (formerly Alvis Vickers Ltd) (vehicle).

United States

Avenger AN/TWQ-1

Type

Self-propelled Surface-to-Air Missile (SAM) system

Development

In the early 1980s, the then Defence Systems Division of the Boeing Aerospace Company (now The Boeing Company) developed the Avenger air-defence system as a private venture. Total time from concept through to delivery to the US Army for trials was only 10 months.

The Avenger has been described by Boeing as an inexpensive, lightweight, shoot-on-the-move short-range air defence system that requires minimal operator training and has been designed for rapid firing.

The Boeing Avenger AM General High-Mobility Multipurpose Wheeled Vehicle (HMMWV) 0069513

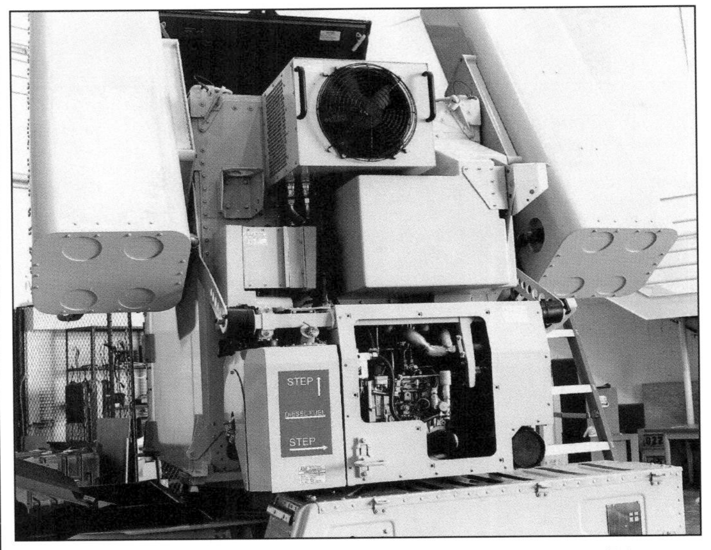

Avenger air-defence system fitted with Environmental Control Unit/Prime Power Unit (ECU/PPU) 0099535

The Avenger consists of an AM General 4 × 4 High-Mobility Multipurpose Wheeled Vehicle (HMMWV) with a turret mounted in the rear and eight missiles in the ready-to-launch position. The turret can also be deployed as a fixed stand-alone unit.

Target acquisition is either by direct vision using the optical sight or through the use of a Forward-Looking InfraRed (FLIR) system. Mounted either side of the turret are four Raytheon Stinger SAMs, identical to those used in the man-portable version.

During tests carried out in May 1984 by the US Army at the Yakima Washington Firing Centre, three live Stinger rounds were fired at ballistic aerial targets. The first shot was fired from the vehicle moving along an unprepared road about 32 km/h and scored a direct hit. The second shot was at night with the unit stationary and scored a direct hit, but for the third shot the vehicle was on the move and in the rain. It narrowly missed the target but was scored as a tactical kill as the missile passed within kill range of what would have been an attacking aircraft. Different gunners fired the three missiles who had never fired a missile before.

In August 1984 the US Army Air-Defence Board evaluated the Avenger system and during this evaluation 171 of the 178 fixed- and rotary-wing aircraft targets were successfully engaged by the system during day and night operations.

In 1986, the US Army issued a Request for Proposals (RfP) for a Pedestal-Mounted Stinger (PMS), or Line Of Sight - Rear (LOS-R) as one of the five key parts of the Forward Area Air-Defence System.

In August 1987, after evaluation of several competing designs, the then Defence Systems Division of Boeing Company was awarded a contract by the US Army Missile Command to commence production of the AN/TWQ-1 PMS air-defence system. The initial contract was for USD16.2 million for the first options buy of 20 systems. With options, the contract value reached USD232 million on 325 fire units over a five-year period together with associated logistic support. The second option covering 39 systems was exercised in 1988 (with deliveries from July 1989 through to June 1990), the third for 70 firing units in 1989, the fourth for 72 firing units in March 1990 and the fifth for 72 firing units in May 1991. A USD436 million multiyear production contract for 600 army systems and 79 marine corps systems over five years was approved with the 1991 budget. By June 1993, the US Army had ordered 1,004 Avenger systems.

First production Avenger systems were delivered in November 1988, the system becoming operational with the US Army in 1989, initially with the US Army 3rd Armoured Cavalry Regiment at Fort Bliss. By January 1997, 902 had been delivered. In FY92 the Avenger Pre-Planned Product Improvement (P3I) programme was initiated. This involved: the development and integration of the

Avenger Slew-to-Cue system on firing range 0034271

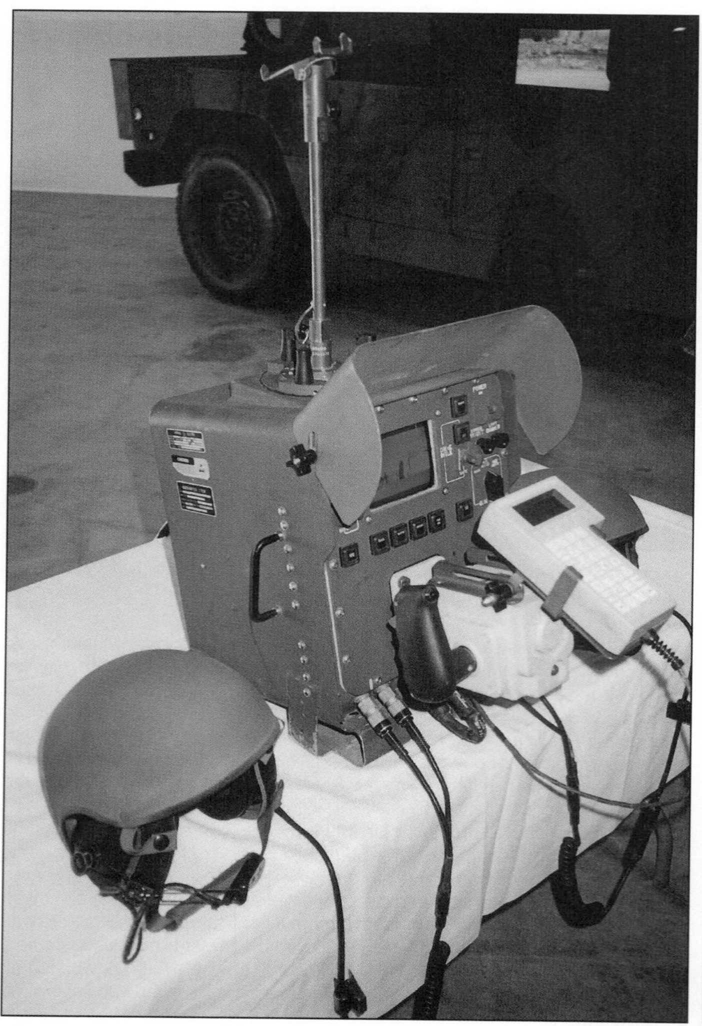

The Boeing Avenger air-defence system remote gunner's console showing hand controls (Scott R Gourley) 0080395

Environmental Control Unit/Prime Power Unit (ECU/PPU); interfacing the Avenger fire-control unit with the Forward Area Air Defence (FAAD) C2I system; and the development and the integration of the Fire Control System-1 (FCS-1). The last year of funding was FY94.

The Avenger was the first shoot-on-the-move air-defence weapon to enter production for the US Army. With the award of the production contract, the Avenger programme moved to Huntsville, Alabama, where the system is assembled, tested and delivered to the Army Missile Command at nearby Redstone Arsenal. Boeing's manufacturing facility at Oak Ridge, Tennessee makes the turret assembly, launcher mechanism and the base assembly that mounts the turret on the HMMWV.

The US Armed Forces' acquisition plan was for 1,004 systems with the US Army taking 767 and the US Marine Corps 237. From the total required by the US Army, a significant number will go to the reserve component, the Army National Guard. It is envisaged that approximately 450 more Avenger systems will be procured for National Guard beyond the current acquisition plan. The first Avenger firing by National Guard soldiers took place on 2 March 1996. As the Stinger has been sold to a number of foreign customers, Boeing believes that foreign military sales could eventually bring the total Avenger production figure to over 1,800 units. In addition to firing the original Stinger missile model, the Avenger is also able to launch follow-on models including the Stinger POST, Stinger Block 1 as well as the Mistral, Starburst and Starstreak from other missile manufacturers.

The first export sale of Avenger was through the US Army to Taiwan, which requested an initial buy of 1,299 Stinger-RMP missiles, 74 standard vehicle-mounted guided missile pedestal launchers, 96 HMMWV, 74 captive flight trainer Stinger-RMP missiles and other related equipment for its army. All of this was at an estimated cost of USD420 million in August 1996 and was accepted formally in October 1996. The 74 fire units were delivered in the beginning of 1999. South Korea, the Czech Republic, Egypt, Poland and Thailand have also expressed significant interest in obtaining Avenger systems.

In early 1994, the US Air Force officially identified a need for a point defence system to accompany rapidly deploying composite air wings to forward airfields in a major conflict.

Known as the Terminal/Point Air-Defence (TPAD) system the Mission Needs Statement identified the Avenger as a possible solution to the requirement. Each composite wing would have seven fire units. To date no operational needs document nor any funding has been earmarked for TPAD.

Main subcontractors are:

- General Dynamics Defense Systems - electric turret drive as used in the M2 Bradley Infantry Fighting Vehicle;
- CAI - CA-562 optical sight;
- DBA - auto-tracker;

Boeing Avenger air-defence system in travelling configuration 0099536

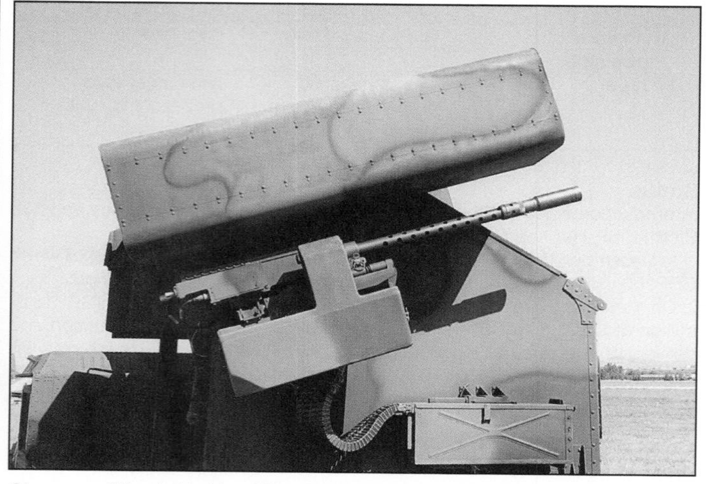

Close-up of the right side of The Boeing Avenger mount clearly showing the 12.7 mm M3P machine gun 0080462

- FN HERSTAL SA - 12.7 mm M3P machine gun;
- Raytheon - Forward-Looking Infra-Red (FLIR) System;
- KECO - heater and ventilator;
- Texstar - canopy;
- Raytheon Systems Company - CO_2 laser range-finder.

Raytheon supplies the IR-18 FLIR system, enabling Stinger to acquire targets at night and in bad weather. The Raytheon FLIR is a derivative of the IR-18 sensor developed by Thales Optronics, UK. The IR-18 FLIR is now produced at Raytheon.

The Avenger system can be installed on other types of chassis, tracked and wheeled, and is also fully air-transportable. During a demonstration at McChord air force base, Washington, it was shown that three HMMWV or five pallet-mounted systems and their crews could be carried in a C-130 Hercules transport aircraft while six HMMWV or 12 pallet-mounted systems and their crews could be transported in a C-141B Starlifter. The turret module can be carried by a UH-60 while a CH-47 Chinook, CH-46 Sea Knight or CH-53 Sea Stallion can airlift a complete PMS system.

For the US Army Avenger system, Stinger missiles are standard, but its design allows it to accommodate other sensors and other weapon systems including:

- Hellfire missiles;
- HYDRA 70/70 mm unguided rockets;
- Other infra-red seeking or laser-guided missiles.

Examples are the then Boeing/Matra BAe Dynamics Guardian. A mock-up installation of a pod of 36 Hyper-Velocity Rockets was also installed on the Avenger system.

In addition to the HMMWV and Bv 206 chassis, other potential chassis include the Commercial Utility Cargo Vehicle (CUCV), 2½ ton truck, M548 tracked cargo carrier and M113A3 APC.

The US Army planned a second Avenger upgrade programme to follow the Avenger P3I. Known as Avenger Block II this would have provided passive, on the move IR search and track, improved lethality against helicopters and 360° coverage. Funding for this upgrade programme is under review.

Description

The driver is seated on the left and, in addition to having all the controls required to drive and operate the HMMWV, he also has complete intercommunications with the gunner in the turret. All voice (intercom and radio) and system tones (IFF and missile) are provided.

The Boeing Avenger launching a Raytheon Stinger missile 0080464

Initial production AN/TWQ-1 Avengers were fitted with AN/PRC-77 and AN/VRC-47 radios and now accommodate the AN/VRC-91 SINCGARS radio system. FAAD C2I equipment is also incorporated as it is fielded with the gunner and driver communicating with each other via the AN/VIC-1 intercom system. The Stinger AN/PXX-3B interrogator operating in military Mode 4 provides IFF.

The driver also has access to the Remote-Control Unit (RCU) for a redundant control of the turret if required. This is fitted with the same system controls and displays as the turret and enables the Avenger crew to dismount and conduct engagements from remote positions up to 50 m from the fire unit. The RCU can be rotated by 180° to allow a crew member in the passenger seat to operate it, and it is fitted with training facilities.

Target engagement from the RCU is identical to engagement from within the turret through hand control switches and indicators on the gunner's console. Components connecting the RCU to the Avenger include a control console with FLIR display, driver's combat vehicle crewman helmet, cable connecting CVC helmet and built-in test terminal. The exception is that no adjustments to the FLIR can be made from the RCU and since the FLIR is susceptible to environmental and weather variations, it is critical to ensure proper adjustment of FLIR at all times to maintain effectiveness of the RCU.

The design of Avenger is modular so that it can accept advances in technology such as the replacement of Stinger by a laser beam rider missile, new sensors, advanced fire-control system, enhanced position locator, reporting system, and hand-held computer, HVR or a larger calibre weapon.

The gunner is seated in the electrically powered turret which can be traversed 360°. If required the complete fire unit can be removed from the HMMWV and used as a stand-alone system. The batteries in the base of the fire unit are interconnected in parallel to the HMMWV 24 V DC system to provide turret power. The Stinger pods, able to accommodate any Stinger model without modification, can be elevated from –10° to +70°. The gunner has a large transparent canopy for all-round observation; to aim the missiles, he looks through a sight glass on which he sees the projection of a driven reticule display. The reticule indicates the aiming point of the missile seeker, confirming to the gunner that the missile seeker is locked on the same target he is tracking and planning to engage.

Sensor package mount includes a CAI optical sight, Raytheon AN/VLR-1 (or IR-18) FLIR, DBA Automatic Video Tracker (AVT) and a Raytheon Systems Company CO_2 eye-safe laser range-finder. These enable the system to acquire and track targets under a wide range of operational conditions.

The FLIR with an electrically operated optics cover is mounted on the left launch arm beneath the missile pod. This is a self-contained system operating in the 8 to 12 μm wavelength region. It has dual-field capability and the gunner's foot pedal is used to select the field required. The gunner tracks the target either by direct vision or through the use of the FLIR system for night and poor-weather operation. The system is capable of auto tracking, lazing the distance to the target, seeing at night or in obscured weather and shooting either at rest or on-the-move at speeds of up to 35 km/h.

The AVT provides an automatic tracking mode. The FLIR video target-to-bore error signals determine the azimuth and elevation repositioning required by the turret drive system in order to maintain turret positioning on the target.

The laser range-finder is mounted on the left-hand launch arm behind the FLIR with target range being displayed on a hand-held display in the turret. Target range is processed by the Avenger control electronics for use in the automated fire permit and fire-control algorithms. The Avenger's FCS processes data from the LRF and displays an advisory fire permit symbol in the sight and FLIR display. The fire permit function maximises use of the Stinger's engagement boundaries. The electric turret drive is gyro-stabilised to maintain pod-aiming direction automatically regardless of the vehicle's movement.

The gunner has a hand controller on which the missile and gun controls are located. In addition, he can transfer tracking control to the automatic tracking systems, one of which uses signals from the uncaged missile seeker and the other data from the FLIR video auto-tracker, to track the target until the gunner is ready to fire. This allows the gunner to concentrate on target identification. The firing sequence is fully automated and the gunner has only to pull the fire trigger to initiate the launch sequence and immediately select and prepare the next missile for firing.

Interior of the four-round Stinger missile pod on The Boeing Avenger air-defence system without missiles installed (Scott R Gourley) 0080463

As a result of experience and the P3I programme in the Gulf War, US Army and Marine Corps Avengers have been equipped with Marvin Land Systems Environmental Control Unit/Prime Power Unit (ECU/PPU) systems. These provide the gunner with cooling and heating, protection from chemical and biological hazards, unobstructed fields of fire, as well as supplementary power to the turret battery system.

The unit can provide 3.66 kW of cooling or around 3.22 kW of heating. The power side produces approximately 6.0 kW at 28 V DC power, of which 1.5 kW powers the unit and the remainder, is fed back into the battery system for the Avenger turret. The unit weight is 190 kg, which includes fuel for eight hours of operating time.

For self-protection and for coverage of the Stinger dead zone, an M3P 12.7 mm machine gun is attached to the right-hand launch beam as supplementary armament. The M3P is an improved AN-M3 machine gun with a cyclic rate of fire of 1,100 rds/min, 5,000 mean rounds between failures, an IR/muzzle blast reducing flash hider and a 5 mil dispersion. There are 300 rounds of ammunition carried for ready use with additional rounds in reserve. Mounted either side of the turret is a pod of four Stinger low-altitude surface-to-air missiles. Full details of the Stinger missile are given in the Man-portable surface-to-air missile systems section.

In addition to the eight missiles in the ready to launch position, an additional eight Stingers are carried in reserve and a standard Stinger grip-stock is carried for use in the dismounted role. Reloading takes less than four minutes.

Variants

'Slew-To-Cue' (STC) Targeting System

During the Task Force XXI Advanced War fighting Experiment (AWE) held in March 1997, a platoon of six Avengers had a prototype AMCOM STC fitted. Following this, Boeing was awarded a USD5.45 million contract in March 1998 for further development of STC modification kit for the Avengers followed by production options for about 1100 kits. The STC upgrade permits the existing digital link between the FAAD C2 and the Avenger to automatically slew the turret so that it places the target in his field of view.

The upgrade comprises:

- New Avenger Fire-Control Computer (AFCC);
- Land Navigation System;
- Handheld Terminal Unit (HTU);
- Remote-Control Unit (RCU) upgrade;
- Simplified cab/turret layout;
- Slip ring.

Approval has been given to obtain STC Avenger capability under the War fighting Rapid Acquisition Program (RAP) to be fielded by 'select' Avenger units. A USD14.6 million option contract was placed in April 2000 for production and installation of an initial 100 modification kits. First installations started in spring 2000.

The programme is known as Avenger Block 1, with the first units to be used for training by mid-2000. Current planning indicates that 574 of 767 fielded US Army Avengers will receive the STC modification and the Avenger Fire-Control Computer (AFCC) but only 382 will receive upgraded Automatic Video Trackers (ANT) and 54 an upgraded FLIR due to funding constraints.

Turkish SP Stinger Systems

The Turkish company, Aselsan is offering a Pedestal-Mounted Air-Defence missile System (PMADS), similar in concept to the Avenger. There are two versions, a light vehicle-mounted system and an M113A2 APC-mounted version.

Laser Avenger System

The Laser Avenger System is a solid-state infrared laser that occupies one of the quadruple launcher box stations on the Avenger SAM system vehicle. The system is still in development (September 2011) but could be ready for IOC if funding becomes available from the Pentagon within a very short time.

Specifications

	Avenger Platform (M1097)
Dimensions and weights	
Crew:	2
Length	
overall:	4.953 m
Width	
overall:	2.184 m
Height	
overall:	2.59 m
Ground clearance	
overall:	0.406 m
Track	
vehicle:	1.81 m
Wheelbase	3.3 m
Weight	
combat:	3,900 kg
Mobility	
Configuration	
running gear:	wheeled
layout:	4 × 4
Speed	
max speed:	105 km/h
Range	
main fuel supply:	563 km
Fuel capacity	
main:	94 litres
Fording	
without preparation:	0.76 m
Angle of approach:	69°
Angle of departure:	45°
Gradient:	60%
Side slope:	40%
Vertical obstacle	
forwards:	0.56 m
Turning radius:	14.63 m
Engine	V-8, air cooled, diesel
Gearbox	
type:	automatic
forward gears:	3
reverse gears:	1
Transfer box:	2-speed
Brakes	
main:	hydraulic
parking:	disc
Tyres:	36 × 12.5 - 16.5
Electrical system	
vehicle:	24 V
Firepower	
Armament:	8 × hull mounted Stinger SAM
Armament:	1 × 12.7 mm (0.50) M3P machine gun
Main armament traverse	
angle:	360°
Main armament elevation/depression	
armament front:	+70°/-10°

	FIM-92C Stinger
Dimensions and weights	
Length	
overall	1.52 m (4 ft 11¾ in)
Diameter	
body	69 mm (2.72 in)
Flight control surfaces	
span	0.091 m (3½ in) (wing)
Weight	
launch	10.1 kg (22 lb)
Performance	
Speed	
max Mach number	2.2
Range	
min	0.1 n miles (0.2 km; 0.1 miles)
max	2.4 n miles (4.5 km; 2.8 miles) (est.)
Altitude	
min	200 m (656 ft)
max	3,800 m (12,467 ft)
Ordnance components	
Warhead	1 kg (2.00 lb) HE fragmentation
Guidance	passive, IR (ultraviolet homing seeker)
Propulsion	
type	solid propellant (sustain stage)

Status

The basic Avenger is in production (1,000 plus delivered in 2000), in service with Taiwan, the US Army, US Marine Corps and Army National Guard.

In June 2005, the Egyptian government has approached the US and requested 25 complete Avenger Fire Units in order to improve their air defence. This will provide Egypt with two additional short-range air-defence brigades of 12 Avenger fire units per brigade (six fire units per battalion).

The government of Egypt requested 164 Stinger Block 1 missiles configured for vehicle launch only, 12 fly-to-buy missiles, 25 Avengers, trainers, spares, engineering and technical assistance support, sentinel radars, SINCGAR radios, target/range/test support, containers, support equipment, spare and repair parts, publications and technical data, training, US government quality assurance teams services and other related elements of logistics support.

In May 2012, Boeing was awarded a USD83 million cost-plus-fixed-fee contract to provide support and engineering services to the weapon system. The execution deadline for the contract is March 2015.

Contractor

The Boeing Company/Boeing Integrated Defense Systems, (primary contractor).

The Boeing Company/Boeing Electronic Systems and Missile Defence, (missile).

Moog Components Group, (electro-mechanical and fibre optic products, including slip rings).

Wildwood Electronics, Huntsville, Alabama. (Provides electronic components, cable harness fabrication, environmental testing and quality assurance.)

Common Air Defense Launcher (CADL)

Type

Self-propelled Surface-to-Air Missile (SAM) system

Development

Raytheon and General Dynamics Armament Systems, have teamed together to produce an advanced concept development prototype of the Common Air Defence Launcher (CADL). Designed to fire several different missiles from the same platform.

Prototype of Common Air Defence Launcher based on HUMRAAM vehicle
0073282

CADL could be a potential replacement for the US Army and US Marine Corps Boeing Avenger air defence systems.

The concept has been developed to support US Army interest in a single air defence launcher that could fire several different types of missiles. As part of the US Army's projected transformation programme the service is looking to field a system that is capable of firing the Raytheon AIM-120 Advanced Medium-Range Air-to-Air Missile (AMRAAM). However, it has not yet been decided whether the HUMRAAM system itself will carry and fire several different missiles at the same time or be capable of firing a mix of missiles but carry only one type at a time.

At present the Common Air Defence Launcher system is some five to 10 years away from possible production and could, if required, be incorporated into a Pre-Planned Product Improvement (P3I) schedule to upgrade the HUMRAAM or the US Marine Corp's Complementary Low Altitude Weapon System (CLAWS).

In any event, the CADL is designed to be attractive to potential customers as a complement to both the HUMRAAM and the Avenger systems.

Description

The prototype uses a modified HUMRAAM launcher with a lightweight turret. The definitive production system will use an AM General M1097A2 High Mobility Multipurpose Wheeled Vehicle (HMMWV), which is capable of carrying and firing the:

(1) AMRAAM
(2) AIM-9X short-range air-to-air missile
(3) Stinger missiles × 4
(4) GAU-19 12.7 mm Gatling gun.

The Common Air Defence Launcher will also use Raytheon's AN/MPQ-64 Sentinel radar and a Kongsberg Defence and Aerospace Fire Detection Centre fire-control assembly.

Status

Advanced concept development with the initial prototype based on a modified HUMRAAM platform.

In March 2009, a proof-of-concept successful firing took place at the White Sands Missile Test Range in New Mexico when a High Mobility Artillery Rocket System (HIMARS) launcher fired two AMRAAM missiles in the surface-to-air mode. The HIMARS' onboard fire-control software and launcher system had been modified so it could successfully fire AMRAAM missiles.

After the test firings, a report confirmed that all of the test objectives were met. At that time it was believed that the firings proved the HIMARS as a feasible common launcher candidate, this was an advantageous proposition as the HIMARS was already in use by the US Army and US Marine Corps.

As of August 2011 no further data was available on this system.

Contractor

Raytheon Electronic Systems.
General Dynamics Armament & Technical Products.

Extended Area Protection and Survivability (EAPS) Programme

Type

Developmental Surface-to-Air Missile (SAM) system

Development

Northrop Grumman and Lockheed Martin were both awarded contracts in 2008 by the US Army Aviation and Missile Command/Research, Development and Engineering Centre at Redstone Arsenal in Huntsville Alabama, to design and demonstrate a prototype missile interceptor weapon system for the defence against rocket, artillery and mortar threats.

The contract calls for a system that will provide stationary and mobile forces with 360° hemispherical umbrella extending area protection from direct and indirect fire. The programme has a five-year time limit and will be based on a phased element.

The first phase will be the design and demonstration phase; included in the programme contract is a low-cost missile, launcher, fire-control radar and fire-control computer designed to defeat a wide range of Rocket, Artillery and Mortar (RAM) threats.

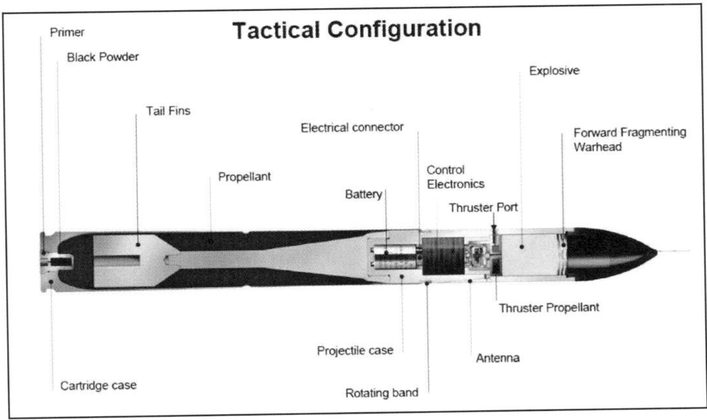

50 mm EAPS round (RDECOM-ARDEC) 1333015

Conceptual image of the EAPS counter-rocket and mortar system testbed vehicle (US Army) 1192741

Flight trials and demonstrations are to take place at the Yuma Proving Ground in Arizona.

In July 2010, the EAPS performed successfully in lethality tests and was also undergoing guidance testing in preparation for live field testing.

Late in 2010, both designs for the EAPS passed their critical design review elements clearing the way for flight testing to start.

During the first week of April 2012, EAPS was flight tested with a series of target tracking tests. This was followed by Controlled Test Vehicle flight tests in May 2012.

In addition to the missile programme there is also an Advanced Technology Demonstrator (ATO-D) using a gun-fire guided projectile. The programme is a joint US Army Research, Development and Engineering Command (RDECOM) and United States Army Armament Research, Development and Engineering Centre (ARDEC) project.

Description

The EAPS is a science and technology programme focused on developing and demonstrating critical technologies necessary to bridge the gap between the initial counter rocket, artillery and mortar capability and the enhanced protection envisioned in the future. The Lockheed Martin approach to the EAPS missile is for a small hit-to-kill kinetic energy interceptor that is just over 1 m in length and approximately 40 mm in diameter. The overall weight of the round is approximately 2 kg at launch. A Container Launching Unit (CLU) with up to 15 missile canisters - that will themselves contain four or six interceptors giving a maximum total of 90 - is also part of the system. Launch guidance will be provided by a Fire-Control Sensor (FCSen) system based on Lockheed Martin's Enhanced AN/TPQ-36 (EQ-36) counter-fire target acquisition radar, but fitted with a reconfigured antenna incorporating transceiver modules operating at a higher frequency. The EQ-36 is already deployed operationally in Afghanistan. The system leverages technology and algorithms developed for interceptors designed to defeat larger missile threats.

The final part of the system will be the computer driven battle management ruggedised laptop. Along with the Fire-Control System (FCS) this could simultaneously control multiple interceptors in the air at any given time. The exact number is not yet known, however, it is assessed to be less than 10 but more than two. The interceptors currently under development are fitted with a back-up data link to pass telemetry; this could be retained in the operational round to enable the missile to accept re-direction from a discrete emitter associated with the radar. Terminal guidance is affected via a radio-frequency seeker located in the nose of the missile that homes in on the radar's emissions reflected off the target. It is understood that a semi-active laser seeker is also in development. With one of the primary objectives of the Lockheed Martin EAPS programme being cost, it is understood that the missile should come in below the planned USD16,000 per shot.

Northrop Grumman approached the programme from a longer-range missile that will contain a dual fragmentation warhead that will be detonated with a proximity fuze. The missile itself is approximately 1.8 m in length with fairly simple canard flight controls; it is also fitted with thrust vector controls and two fragmentation warheads that create a cloud of debris to destroy incoming targets. Cost targets like its Lockheed Martin counterpart is expected to be below USD16,000 per shot.

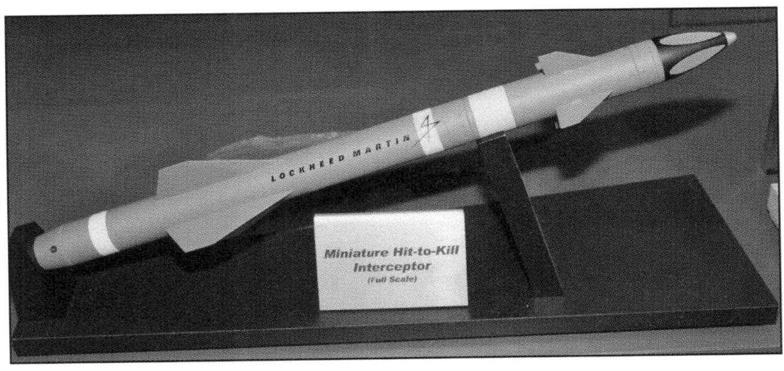

Lockheed Martin EAPS Interceptor (IHS/Patrick Allen) 1432807

The RDECOM-ARDEC programme has been in existence since at least 2005 (initial concept of requirements), with first non-classified reports becoming available in early 2007. Unlike the Lockheed Martin and Northrop Grumman missile programmes, this is to use a gun-launched guided projectile as its kill mechanism. Initial development studies involved a subcalibre (23.5 mm) dart-like guided projectile fired from a 64-calibre 60 mm smoothbore gun. The steel-cased finned dart was to contain guidance electronics in the rear and nose portions of the projectile. The kill mechanism was to be hit-to-kill, with a small tungsten insert, mounted between two electronics sections, to ensure hard-cased munition penetration. Later, many other calibres were assessed (20-82 mm) with research focused on a 50 mm course corrected guided projectile. The selected gun system was derived from the Bushmaster III 35.50 mm automatic cannon. The initial configuration of the projectile was an ogival hollow nose, cylindrical body with a long tail boom fitted with a fin stabilising unit, the current version uses an aero-spike instead of the ogival nose. Both versions use a forward projecting Multiple Explosively Formed Penetrator (MEFP) warhead as the kill mechanism. Course correction is via side-mounted thrusters. The potential system would use two turret-mounted Bushmaster III guns with a phased array detection and tracking radar mounted above the gun system. A conceptual computer generated image showed this mounted onto a MLRS tracked chassis.

Specifications

Lockheed Martin EAPS

Type:	kinetic energy hit-to-kill
Length:	1 m
Diameter:	0.4 m
Wing span:	unknown
Launch weight:	2 kg
Propulsion:	solid propellant
Guidance:	radar reflective with possible uplink
Warhead:	kinetic energy
Maximum speed:	unknown
Effective range:	
(maximum)	2.5 km
(minimum)	unknown
Effective altitude:	
(maximum)	unknown
(minimum)	unknown
Reload time:	unknown
Kill probability:	0.9

RDECOM-ARDEC EAPS

Type:	radar guided
Length:	unknown
Diameter:	50 mm
Wing span:	<50 mm
Launch weight:	unknown
Propulsion:	gun-fired
Guidance:	RF datalink controlling side-mounted thrusters
Warhead:	forward projecting MEFP
Fuze:	command
Maximum speed:	unknown
Effective range:	
(maximum)	unknown
(minimum)	unknown
Effective altitude:	
(maximum)	unknown
(minimum)	unknown
Reload time:	unknown
Kill probability:	unknown

Northrop Grumman EAPS

Type:	radar guided
Length:	1.8 m
Diameter:	unknown
Wingspan:	unknown
Launch weight:	unknown
Propulsion:	solid propellant
Guidance:	Ku-band radar command guidance with uplink
Warhead:	dual fragmentation
Maximum speed:	unknown
Effective range:	
(maximum)	>2.5 km
(minimum)	unknown
Effective altitude:	
(maximum)	unknown
(minimum)	unknown
Reload time:	unknown
Kill probability:	unknown

Status

All systems are in development with first firing trials having taken place in 2011/12.

Contractor

Lockheed Martin, Northrop Grumman Corporation and RDECOM-ARDEC.

M48 Chaparral/M48A1 Chaparral/M48A2 and M48A3 Improved Chaparral

Type

Self-propelled Surface-to-Air Missile (SAM) system

Development

The Chaparral low-altitude surface-to-air missile system from Lockheed Martin Missiles and Fire Control in Orlando, US, was initiated with the modification of the US Navy Sidewinder 1C (AIM-9D) air-to-air proportional navigation, infrared (IR) homing missile for ground-to-air launch. The launcher vehicle was based on the M113 family of chassis vehicles.

The study and evaluation programme for the Chaparral began in 1964 and was completed mid-to late 1965 at the Naval Weapons Centre, China Lake, California. The US Army Missile Command (MICOM) awarded a development contract the following year to Lockheed Martin.

The manufacturers, the Naval Weapons Centre and the White Sands Missile Range in New Mexico undertook development and testing. First production missiles were delivered to the US Army in 1967 with the systems being delivered and entered service in 1969. The manufacturer produced the M54 launch and control station, improved missile guidance sections and test equipment, as well as being responsible for overall system integration.

The Chaparral was one of the US Army air defence assets replaced in the active force component by the FAAD systems. In 1984 the US Army Air Defence Artillery School presented a plan to transfer Divisional Chaparral systems to the Army National Guard. The Chaparral systems have now been phased out of service with the National Guard and replaced by the Avenger system mounted on the 4 × 4 chassis.

A towed self-contained trailer system, designated M85, used the same launch and control system as the self-propelled M48 and designated the M54A1. The M85 weighs 5,250 kg empty and carries four ready to fire missiles when in the operational configuration with four stowed reloads and a power generator on an accompanying equipment carrying trailer. In early 1984, a USD10.6 million contract was placed for 13 systems of this type for use by the 9th Infantry Division on rapid deployment type exercises overseas. These systems are no longer operational with the US Army.

A version known as Sea Chaparral was also developed using a modified M54 launcher and is in use with the Taiwanese Navy aboard its larger warships. In 1996, Taiwan purchased 50 MIM-72J missiles for its naval vessels.

In 1983, the Pentagon announced a letter of offer to Egypt for the sale of 25 M48A2 Improved Chaparral self-propelled air defence systems, MIM-72F missiles and seven modified M577A2 command post vehicles at a total cost of USD112 million. This offer was accepted the following year and first deliveries were made early in 1988. At present Egypt has 50 Chaparral systems in its inventory. In 1999 an upgrade package was sold to Egypt for 30 of its Chaparral systems. In April 2000 it was revealed that Egypt was negotiating for a further 20 upgrades plus the possible transfer of an additional 72 fire units from US Army stocks. Associated with the Egyptian sale, Lockheed Martin Sanders provided 14 of its Tracked Search and Target Acquisition Radar Systems (TRACKSTARS), mounted on top of the modified M577A2 tracked command post vehicle chassis.

In 1986, Portugal ordered five Chaparral systems, 28 MIM-72F missiles, two AN/MPQ-54 Forward Area Alerting Radars (FAARs), spares and support equipment with a total value of USD45 million. These have now been delivered.

In 1986, Taiwan ordered 41 Chaparral fire units and associated spare parts with final deliveries made during 1989. Currently it has 45 Chaparral land systems and 22 Sea Chaparral systems. In 1996, the Taiwanese Marine Corps were in the process of acquiring eight Chaparral fire units with 148 MIM-72J missiles. The Taiwanese Army requested 148 MIM-72G missiles under the US DoD Excess Defence Articles (EDA) programme, which allows for surplus weapon systems to be acquired by friendly nations at greatly reduced costs. In April 2000, it was revealed that Taiwan was negotiating a transfer of technology to establish Chaparral missile maintenance and repair centre in the country, as well as, to upgrade its Navy's Sea Chaparral Systems.

Chaparral surface-to-air missile system showing main components, bows stowed at front of vehicle and anti-blast shield positioned over cab 0080396

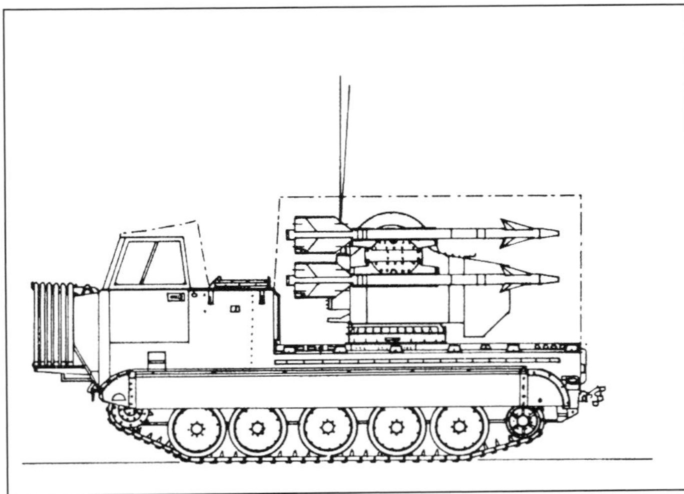

Chaparral surface-to-air missile system in travelling order without bows or tarpaulin cover in position 0509473

In 1996, it was revealed that Israel was due to take delivery of further Chaparral systems that included MIM-72J missiles. Tunisia was looking to acquire 200 MIM-72E missiles.

Other countries, especially Estonia, in the Baltic Republics area are interested in obtaining surplus Chaparral systems, Lockheed Martin has stated that a few hundred surplus Chaparral fire units are available. Costs for refurbishment are between USD600,000 to USD1 million per fire unit depending upon the scope of the task and to upgrade a fire unit the cost is up to USD220,000.

Lockheed Martin Orlando has developed a modular upgrade package, which includes:

- Expanding the ready-to-fire missiles to six
- Modifying the cab and fire-control systems
- Adding an improved FLIR sensor
- Netting each individual fire unit into an integrated air defence environment
- (And/or) Adding Hellfire missiles and the associated fire-control system.

All existing Chaparral operators have been briefed on the packages.

Description

A standard configuration Chaparral fire unit consists of two main elements, a tracked carrier and the M54 missile launch station. The carrier designated the M730 and based on the M548 tracked cargo carrier, uses components of the M113A1 armoured personnel carrier.

The crew of four comprises:

- A squad leader (who makes the target selections, identifications and issues the fire orders)
- Senior gunner (who operates the launch and control station and is system second in command)
- Vehicle driver (who can also function as an observer and/or radio operator)
- One back-up gunner (who primarily acts as target observer).

This allows 24 hour system operation/readiness.

The M730 has the engine and crew compartment at the front of the vehicle and supports the missile launch station behind the engine compartment. The crew compartment is equipped with front, sides, rear and top, which are removed before the missiles are launched. A torsion bar suspension and a drive system support the vehicle. This consists of tracks running on 5 × dual rubber-tyre road wheels on each side with drive sprockets at the front and idlers at the rear. There are no track-return rollers. The vehicle is fitted with infrared driving lights. When the flotation screen is erected around the rear of the hull the vehicle is fully amphibious, being propelled in the water by its tracks.

In 1982, a Product Improvement Programme (PIP) was approved to modify the M730 missile carrier for use with the RISE power train developed for the M113A1E1 APC. The M730 was in particular need of improvement to its drive train reliability and performance because of the Chaparral system combat weight. The first conversion to this standard took place in 1987 with the whole fleet of 596 modified by 1989. The vehicle is then known as the M730A2.

The launch and control station (designated M54), consists of the base structure and turret. The base structure provides mechanical support and contains essential auxiliary equipment including the engine/generator electrical power source, and storage for missiles, crew equipment and tools. The senior gunner's compartment, which is inside the turret, has filtered and conditioned air and an adjustable seat. All US Army systems have been fitted with NBC protection equipment. The missile control electronics operate in conjunction with the control panel switches and indicators to control the activation of the system, missile selection and sequencing, missile launch sequencing and test functions. Each missile launch station contains an IFF subsystem.

The Chaparral system uses a hydraulic turret drive that responds to rate commands from the gunner's hand control. The drive subsystem allows unrestricted movement in azimuth and movement in elevation from –9° to +90°. The air compressor is part of the missile air supply subsystem, supplying highly compressed air for cooling the infrared detectors. The subsystem accepts outside air, compresses, filters and purifies it and distributes it to each launcher assembly and the FLIR assembly.

TRACKSTAR integrated radar/command-and-control system based on an M577A2 tracked chassis 0080397

The main power unit, a petrol- (M48 and M48A1) or diesel- (M48A2 and M48A3) powered engine/generator set with associated power supply, provides the regulated power needed for all functions. If the power unit malfunctions the operator can continue for a limited period using the onboard storage batteries.

The missile itself is an in-line cruciform type with two pairs of canard control surfaces at the forward end and two pairs of fixed-wings at the rear. One pair of the rear wings is provided with rollerons to reduce roll rate. The missile is attached to the launch rail by hangers fixed to the rocket motor case. Four missiles are carried ready to launch and a further eight are carried in reserve.

The original missile was based on the AIM-9D Sidewinder, designated the MIM-72A, has a lead sulphide (PbS) seeker array, a launch weight of 86.9 kg, is 2.91 m long, has a diameter of 127 mm, a wing span of 0.715 m and is fitted with an 11.2 kg high-explosive warhead. Between 1970 and 1974 an improved missile, called the MIM-72C was developed, with all aspect seeker, and went into production in 1977, entering service in July 1978. It weighs 85.7 kg. The other dimensions remain the same and include an M817 Doppler radar proximity fuze developed by Harry Diamond Laboratories, a 12.6 kg M250 HE blast/fragmentation warhead developed by Picatinny Arsenal and an AN/DAW-1B all-aspect IR seeker with IRCCM developed by the manufacturer. Effective launch range is increased to over 9,000 m. Later versions of this missile, retrofitted with the M121 smokeless motor, are designated MIM-72E. Non-CCM versions (AN/DAW-1 seekers) are designated MIM-72F and MIM-72H. All versions are powered by a single-stage solid propellant rocket motor. The battlefield signature of the system has been reduced by the adoption of the Hercules Aerospace/Atlantic Research Corporation M121 smokeless motor. First installation of the M121 occurred in 1979 on the MIM-72C.

To provide a night and bad weather capability and to improve daylight performance in smoke and haze, the US Army retrofitted M48A2 launchers with a Forward-Looking Infra-Red (FLIR) thermal imaging device, with auto-track features. The TI FLIR contains 180 CMT 8 to 12 micron wavelength elements, has an 18 × 20° wide field of view and a 2 × 2.7° narrow field of view. It can operate in either a wide or narrow field of view to optimise the infrared detection capability of the receiver and improve the thermal image for the gunner. The optics on the receiver magnifies the image and the infrared target video obtained is presented on the video display located in the mount. At the moment of missile launch a small protective cover will briefly close over the FLIR optics to protect the sensitive optical elements. The system also contains a lightweight Mk XII IFF subsystem for target friend or foe identification. The FLIR is standard on all M48A3 launchers.

In addition, a new 30-hp diesel engine has been retrofitted to replace the 10-hp petrol engine used previously. This both increases the power available and introduces fuel commonality with the M730. With the RISE power train and FLIR, the designation changes to M48A3. With just the FLIR is M48A2 Improved Chaparral.

Chaparral of the Israel Defence Forces in its deployed configuration 0509474

Before the system can be used, the crew leaves the cab, folds down the windscreen, removes the cab cover and folds a six-piece hinged blast shield over the cab and engine compartment. An additional fixed blast shield protects the back of the engine compartment. The six bows and tarpaulin cover are then removed from the launcher area and the bows are stowed on the front of the hull. The launcher is then electro-mechanically raised into the firing position.

A typical daylight target engagement takes place as follows. Early warning is provided either by the AN/MPQ-49 Forward Area Alerting Radar or by a visual sighting. Once the gunner detects the target he moves the turret to acquire and maintain the aircraft in the centre of his sight or FLIR field of view. The turret can be traversed by 360°, the launch rails have an elevation of +90° and a depression of –9°. As the gunner tracks the target, an audio tone in his headset notifies him that the target is within infrared sensing range.

The gunner then launches a fire-and-forget missile, which operates under its own internal power. Proportional navigation guidance commands are generated from seeker tracking rates and used to control the missile flight path. Proximity fuzing ensures the warhead will detonate even without a direct hit. The fire-and-forget capability allows the gunner to begin to search for and attack another target immediately. The rate of fire of the basic system is four missiles/min with a full reload time of five minutes. The Single-Shot Kill Probability was assessed at 0.5 against targets with velocities between 0 and 300 m/s with the basic missile, but this has been substantially improved with later missile versions. The Chaparral surface-to-air missile has maximum range limits of 500 m to beyond 12,000 m and effective altitude limits of less than 15 m to greater than 3,000 m. The missile is armed after 180 to 340 m of flight.

A new guidance section called the Rosette Scan Seeker (RSS) and designated AN/DAW-2 was developed for the Chaparral missile under a contract placed by MICOM with the system manufacturer. Development started in 1982 and the seeker was type classified in August 1987. Missiles fitted with the RSS guidance section are known by the designation MIM-72G. The manufacturer was awarded a production contract for 441 RSS guidance sections plus depot test equipment for delivery in 1990 to 1992. The now Raytheon Electronic Systems was awarded a contract as a second source producer of 721 RSS guidance sections for delivery between 1990 and 1992. This second source production contract also contained options for an additional 422 sections; these were exercised to give Raytheon Electronics Systems contracts totalling USD52 million.

The RSS is based on software that can be reprogrammed to take into account evolving threats such as different heat signatures, flares and other infrared countermeasures. It is electronically reprogrammed through the umbilical connector. The AN/DAW-2 RSS provides two-colour spectral discrimination and scanning spatial discrimination. This helps distinguish between targets and offers a near total immunity to the countermeasures mentioned above.

An export version of the MIM-72G, the MIM-72J, has been developed. This offers the same 50 per cent increase in target acquisition range, the use of the AN/DAW-1 all-aspects seeker, the M121 smokeless rocket motor, M817 proximity fuze and M250 blast/fragmentation warhead. It will not have the same sophisticated infrared counter-countermeasure capability as the MIM-72G.

The MIM-72G has demonstrated launch, lock on and intercept of helicopter-type targets to beyond 9,000-m range. Launch and lock on of fixed-wing targets at 16,000 m range with final intercept occurring at 9,000 m and validation computer simulations against some tactical ballistic missile types of launch lock on at approximately 22,000 m with a projected intercept point at 9,000 m range.

According to the manufacturer, Chaparral achieved virtual immunity to all types of infrared countermeasures. Increased engagements and earlier launches were achieved by Chaparral with the Rosette Scan Seeker. Target acquisitions at 50 per cent longer range than the original guidance have been demonstrated. Flight tests including contact hits on a helicopter target beyond 8,000 m launch range and on fixed-wing targets beyond 12,000 m were successfully conducted.

Lockheed Martin Sanders TRACKSTAR Radar

Supporting the Chaparral Air Defence System for the Egyptian Army, the TRACKSTAR radar system is a self-contained 360°, D-band, 60 km range, pulse-Doppler AN/MPQ-49 derivative integrated radar/Command-and-Control (C2) system. It is used by the Egyptian Army with its Chaparral fire units and automatically broadcasts cueing, fire distribution and IFF data via its VHF radio or hard-wire data links to Integrated Weapon Display (IWD) operator control and processor units (total weight 8.2 kg) mounted in the Chaparral fire units.

Chaparral self-propelled air defence system used by Taiwan showing FLIR system between missiles on their launcher arms (L J Lamb) 0080468

The target cueing information is transmitted in terms of UTM coordinates and converted and orientated to each fire unit by the IWD processor. This has a target/display capacity for up to 32 targets and permits the input of operational parameters and selection of tactical modes. Command messages from the TRACKSTAR commander can also be added to the target information flow over the datalink and may be addressed to any or all the Chaparral fire units as required.

With IWD, the Chaparral gunner of the fire unit simply positions the boresight symbol over the designated target symbol, thereby pre-positioning the missile launch station to ensure that the target will appear in his field of view. This flexibility allows fire distribution tasks to be centrally controlled or delegated to gunners according to standard operating procedures.

Automatic netting of multiple TRACKSTAR systems is used to help uncover terrain-masked targets and provide additional advanced warning and cueing information.

Chaparral Chassis Service Life Extension Programme (CCSLEP)

The CCSLEP family is described in the 'Lockheed Martin Missiles and Fire Control - Orlando Chaparral Chassis Service Life Extension Programme' (CCSLEP) entry.

A USD1.4 million contract from Lockheed Martin to LaBarge Inc. has been placed as of November 2004 for the production of cable assemblies for the Chaparral Air Defence Missile system. The contract is expected to continue through to February 2005. LaBarge has produced the cable assemblies used in the missile launch system for more than a decade.

Specifications

	M48 Chaparral
Dimensions and weights	
Crew:	4
Length	
overall:	6.06 m
Width	
overall:	2.69 m
transport configuration:	2.54 m
Height	
overall:	2.68 m (with bows and tarpaulin cover)
Ground clearance	
overall:	0.4 m
Track	
vehicle:	2.159 m
Track width	
normal:	381 mm
Length of track on ground:	2.82 m
Weight	
standard:	6,425 kg
combat:	13,024 kg
Ground pressure	
standard track:	0.61 kg/cm²
Mobility	
Configuration	
running gear:	tracked
Power-to-weight ratio:	16.1 hp/t
Speed	
max speed:	67.2 km/h
water:	5.5 km/h
Range	
main fuel supply:	504 km
Fuel capacity	
main:	401 litres
Amphibious:	yes (with preparation)
Gradient:	60%
Side slope:	30%

	M48 Chaparral
Vertical obstacle	
forwards:	0.62 m
Trench:	1.68 m
Engine	Detroit Diesel Model 6V-53, 6 cylinders, water cooled, diesel, 210 hp at 2,800 rpm
Engine	Detroit Diesel Model 6V-53T, 6 cylinders, water cooled, diesel, 275 hp at 2,800 rpm
Gearbox	
model:	Allison TX-100
forward gears:	6
reverse gears:	2
Clutch	
type:	torque converter
Suspension:	torsion bar
Electrical system	
vehicle:	28 V
Firepower	
Armament:	4 × hull mounted Chaparral SAM
Ammunition	
missile:	8

	MIM-72G
Dimensions and weights	
Length	
overall	2.91 m (9 ft 6½ in)
Diameter	
body	127 mm (5.00 in)
Flight control surfaces	
span	0.63 m (2 ft 0¾ in) (wing)
Weight	
launch	86.2 kg (190 lb)
Performance	
Speed	
max Mach number	1.5 (est.)
Range	
min	0.3 n miles (0.5 km; 0.3 miles)
max	4.3 n miles (8 km; 5.0 miles) (helicopter target)
	4.9 n miles (9 km; 5.6 miles) (aircraft target)
Altitude	
min	15 m (49 ft)
max	3,000 m (9,843 ft) (est.)
Ordnance components	
Warhead	12.6 kg (27 lb) HE M250 blast fragmentation
Fuze	impact, proximity (radio-frequency Doppler, M817)
Propulsion	
type	solid propellant (Alliant Techsystems M121)

Status
Production complete (over 700 systems built with some 21,700 missiles produced). The system first entered service in 1969 but was phased out of the US Army between 1990 and 1998.

As of February 2007, a contract worth in excess of USD10 million for new Chaparral Rocket Motors has been placed by Madison Research, Huntsville, Alabama, with Alliant Techsystems (ATK). Production of these motors began April 2007 with the first deliveries in December 2007.

Contractor
Lockheed Martin Missiles and Fire Control.
The then United Defense LP now BAE Systems Land & Armaments (chassis).
The then Lockheed Martin Sanders now BAE Systems Electronics, Intelligence & Support (EI&S) (TRACKSTAR and Integrated Weapon Display (IWD).
LaBarge, Inc (production of cable assemblies).
Ford Motor Company.

SLAMRAAM

Type
Self-propelled Surface-to-Air Missile (SAM) system

Development
The US Army Aviation and Missile Command (AMCOM) Missile Research, Development and Engineering Centre (MRDEC) has completed the prototype of a medium-range self-propelled air-defence missile system known as HUMRAAM. It is based on the Raytheon Systems Company AIM-120 AMRAAM missile family and a modified AM General High-Mobility Multipurpose Wheeled Vehicle (HMMWV) 4 × 4 chassis.

Originally known as Project 559 by AMCOM and the HUMRAAM by Raytheon Systems Company, the programme was not funded and there was no recognised requirement as such. The system uses a 'creative repackaging' effort to use low-cost off-the-shelf missiles, launcher and vehicle components.

AIM-120 AMRAAM missile being launched from HUMRAAM prototype 0069514

US Army MICOM/Raytheon Systems Company HUMRAAM prototype fitted with five AIM-120 AMRAAM missiles 0009325

The total cost of the launcher system, including the vehicle but excluding the missiles, was only USD559,000, hence the title, Project 559, in order to produce a viable near term medium-range mobile SAM system.

The system will complement the Boeing Avenger air-defence system at the divisional level, for use against cruise missiles and UAVs, while also providing a rapidly deployable mobile air-defence system to accompany Rapid Reaction force units into the field.

The system is known by two names:

(1) Surface-Launched Advanced Medium Range Air-to-Air Missile (SLAMRAAM), as known by the US Army
(2) Complementary Low Altitude Weapon System (CLAWS), as known by the US Marine Corps.

In September 2005, the Raytheon Company was awarded a multimillion dollar contract for the fifth and sixth production systems for the CLAWS. The work was to be performed at Raytheon's plant at Andover, Massachusetts, and was expected to be completed by June 2006. By late 2007, the US Marine Corps had pulled out of the CLAWS contract and no longer supports the system.

In May 2006, the Boeing Company delivered the first Integrated Fire-Control Station (IFCS) to the US Army. This provides command-and-control for multiple SLAMRAAM fire units, enabling integrated fire-control capabilities. Altogether Boeing, as a subcontractor, is contracted to produce five of these IFCS systems.

During July 2006, SLAMRAAM achieved a third Critical Design Review (CDR) which confirmed that the system meets or exceeds the joint war fighters requirements providing a reliable and critical air-defence capability. The first CDR integrated the Fire-Control Shelter, and the second CDR was the Build 2 software.

As of 31 October 2006, CLAWS successfully completed the Joint Limited Technical Inspection that is part of the Production Representative Systems inspection at Raytheon Integrated Defence Systems Integrated Air Defence Centre, Andover, Massachusetts, United States. This clears the way for the US Marine Corps to accept into service the final two fire units.

In July 2008, the SLAMRAAM successfully completed system field integration testing at the White Sands Missile Test Range in New Mexico. The system demonstrated its ability to form a network of sensor elements tracking live targets and providing each battlefield element with a common air picture with fire quality data. Other units involved, included the Avenger with Stinger missiles and the Patriot SAM.

In late May 2009, Raytheon announced that the programme has received US Army approval for a long-lead acquisition, not to exceed USD30 million, for purchases leading to low rate initial production.

HUMRAAM firing AIM-120 AMRAAM missile 0049702

On 6 January 2011, the US Secretary of Defence announced the termination of the acquisition of the SLAMRAAM as part of a budget cutting effort.

Description

The Project 559 effort, emerged following the successful testing of a modified I-HAWK missile launcher to fire AMRAAM missiles during 'Safe Air 95' demonstration in Alaska.

It was discovered that AMRAAM was relatively insensitive to elevation and azimuth requirements, therefore, all that is required is to elevate an AMRAAM launcher and point it. The weapon can cope with azimuth angles up to ±60° to 70° off the launcher's centreline. A general sector of responsibility is assigned to the launcher, the system is cued for incoming targets and a missile is launched against a point in space where its terminal homing active seeker is activated.

The SLAMRAAM system is intended to replace all the US Army's short-range air defence weapon systems that employ the Stinger missile. The system is also intended to defend against threats from unmanned aerial vehicles and cruise missiles.

The prototype M1097 HMMWV launch vehicle is identical to that used in the Avenger system.

A simple aluminium mechanical structure interface that bolts into the same holes as the Avenger launcher assembly supports five USAF flight test LAU-128 launch rails taken from The Boeing Company F-15 Eagle fighter. The launch rails can accommodate and fire any of the three current AIM-120 missile types: the AIM-120A, AIM-120B and AIM-120C.

Missile

The AIM-120B is basically the AIM-120A with modified electronics that can be programmed in the field.

The AIM-120C introduced smaller, aerodynamic control surfaces and is significantly more agile than the AIM-120A/B and has a slightly longer range.

Unit Configuration

A typical HUMRAAM unit comprises a Fire Distribution Centre with advanced automated battle management system and operator situational awareness capability that can control up to eight HUMRAAM launchers. Communication is via Raytheon's Remote Terminal Unit (RTU) on each HUMRAAM.

Now known as SLAMRAAM, the launcher vehicles are located within the divisional area in such a way that each one is positioned along a primary target line for an attacker. This then becomes the launcher's primary sector of responsibility. At the desired position, the launcher assembly is elevated from its travelling position on the rear of the vehicle to the launch angle of approximately 30° and the crew vacates the vehicle to a remote firing position that can be up to 50 m away. A robotics package that will remotely control elevation and depression of the launcher assembly, as well as change the primary sector of responsibility direction, is under development.

The SLAMRAAM system consists of:

The launcher vehicle: The HMMWV-based launcher is the selected Joint launcher for the US Marine Corps CLAWS and the US Army SLAMRAAM programmes.

Forward Area Air-Defence Command, Control, Communications and Intelligence (FAAD C3I) vehicle: This vehicle was selected by the US Army as the baseline for its Integrated Fire-Control Stations and next-generation common AMD BMC4I. The Integrated Fire-Control Station (IFCS) is a vehicle-mounted shelter with two work stations, and is used to control the system. It is currently envisaged as mounted on Hummers, despite their mobility issues on difficult terrain.

Thales Raytheon Systems AN/MPQ-64 Sentinel Radar: Multitarget search and tracking is performed by the AN/MPQ-64 Sentinel radar. This radar is a modern 3-D pencil beam radar that features a large surveillance and track volume, phased-frequency electronic scanning antenna, X-band range gated Pulse Doppler operation and high survivability. It has a high scan rate (30 rpm), 75 km range, and is ECM resistant.

AMRAAM missiles: The Advanced Medium-Range Air-to-Air Missile (AMRAAM) is employed today by the armed forces of 18 nations. With both beyond-visual-range and non-line-of-sight capability, AMRAAM is an all-weather, day or night missile that provides operational flexibility with multiple simultaneous engagement capability. As of June 2007 (data released during the Paris Air Show), SLAMRAAM will be able to offer the added capabilities of the AIM-9X and SLAMRAAM-ER [Extended Range] missiles in addition to the already available AMRAAM missile.

The targets envisaged include manned aircraft, helicopters, UAVs and air breathing cruise missiles.

Fire Distribution Centre (FDC)

The Fire Distribution Centre (FDC) has been selected by the US Army as the baseline for its SLAMRAAM Integrated Fire-Control Station (IFCS) and next generation common AMD BMC4I.

Radar

The Sentinel Radar (AN/MPQ-64) has been in development for several years and was first delivered to the US Army in April 2006. It is the US premier air surveillance and target acquisition/tracking sensor for the US Army Short-Range Air Defence (SHORAD) programmes. This radar is a modern 3-D, pencil beam radar that features a large surveillance and track volume, a phase-frequency electronic scanning antenna, X-band range gated pulse Doppler operation and high survivability. Sentinel has a high scan rate (30 rpm), 75 km range, and is ECM resistant. Accurate, quick reacting and able to acquire targets far enough forward of friendly units, Sentinel allows air-defence weapons systems to engage hostile targets at optimum ranges. The radar automatically:

- Detects;
- Tracks;
- Identifies;
- Classifies;
- Reports airborne threats including:
    - Helicopters;
    - High-speed attack aircrafts;
    - Cruise missiles;
    - Unmanned Aerial Vehicles (UAVs).

In June 2010, Thales Raytheon Systems was awarded a USD21.8 million contract by the US Army to upgrade the AN/MP-64 Sentinel radar that involve the transmitters, receivers and exciters and also increase functional capabilities such as data processing and detection range for smaller targets.

Trials

Following successful firing trials at Eglin air force base in August 1996, the US Army Aviation and Missile Command constructed a second prototype. The US Marine Corps Systems Command subsequently conducted trials of the system in late August 1997 and September 1997. The latter tests using targeting information from an AEGIS cruiser and CWAR guidance achieved intercept ranges in excess of 15,000 m. The Marines refer to the system as the Complimentary Low-Altitude Weapon System (CLAWS).

Flight tests of the CLAWS system during January 2005, at the White Sands Missile Test Range, New Mexico, supported by elements of the 3rd Low Altitude Air-Defence Battalion (LAAD) and MACS 23 and 1 (Marine Air Control Squadrons 23 and 1), marked the completion of these flight tests and also marked the end of development testing. By late 2007, the US Marine Corps had withdrawn its support for the CLAWS system.

Specifications

HMMWV platform

Curb weight:	
(M1097A2)	2,681.8 kg
(M1113)	2,772.7 kg
Payload capacity:	
(M1097A2)	2,000 kg
(M1113)	2,454.5 kg
MIM-120 load weights:	
(6 missiles with launch rails)	1,363.6 kg
(5 missiles with launch rails)	1,136.4 kg
Length:	4.839 m
Height:	
(travelling to top of missiles)	2.440 m (approx.)
(travelling to top of vehicle cab)	1.829 m

AIM-120C-5

Length:	3.65 m
Diameter:	0.178 m
Wing span:	0.445 m
Fin span:	0.447 m
Weight:	161.5 kg
Motor:	solid propellent rocket motor
Maximum speed:	>1,360 m/s
Guidance:	active radar for final target approach plus the capability of 3rd party targeting
Warhead weight:	20.5 kg with a new self-destruct and warhead disarm feature
Fuze:	proximity and contact
Launcher:	rail and eject, air or land
Maximum range:	25,000 m
Maximum altitude:	4,000 m

Status

In February 1998, the US Marine Corps approved an Operational Requirements Document (ORD) calling for an IOC in FY02 and full operational capability with USMC Stinger armed LAADs having some 95 systems in service two years later.

In April 2000, the US DoD revealed that the US had approved a sale of the AMRAAM system to Egypt to replace its elderly 2K12 'Gainful' mobile SAM systems. Acquisition and deployment was scheduled for 2006 if funding was available. The size, configuration and associated spares, are still to be decided. The only US stipulation was that the AMRAAM missiles were to be configured for surface launch only.

In the FY01 budget requests, the USMC requested USD7.5 million for the programme. Total cost of the CLAWS was estimated to be USD88 million by FY06 with follow-on procurements past the date to raise it to USD150 million plus. The first of 20 CLAWS systems in the initial procurement would be delivered in FY05. Funding, however, still remains a continuing problem.

As of August 2004, CLAWS was advised by the US Ballistic Missile Defence Programme Office as the preferred weapon for short range terminal missile defence. A Notice For Information (NFI) issued by the US Marine Corps early in 2004 called for a replacement for the Stinger missile system with a requirement that the new weapon work alongside the CLAWS system.

During April 2005, the Raytheon Company made public its International SLAMRAAM programme during a roll-out in Washington. The International version is being marketed as an affordable and adaptable weapon system that can be fully integrated into other defence systems.

During April 2005, Raytheon proposed equipping the Polish Kub (NATO SA-6 'Gainful') systems with SLAMRAAM missiles, based on a standard AIM-120C-5 already purchased Polish F-16 aircraft.

The proposal is connected to a Polish Kub modernisation programme, conducted by Wojskowe Zakłady Uzbrojenia Number 2 (WZU-2, Grudziadz, Poland).

The SLAMRAAM system is operational and available, combining the fire-and-forget AMRAAM missile, the AN/MPQ-64 Sentinel surveillance radar, multiple launcher options, and a modern, advanced FDC that provides full control of all missile-launcher functions including the cueing of SHORAD units.

During October 2006, Raytheon in partnership with the Spanish Army and the US Air Force conducted three successful SLAMRAAM firings at the Swedish missile test range in Vidsel. These firings were to prove the new command detonation for the missile. Further firings were also conducted at the Andoya Rocket Range in Norway.

As of January 2007, Pakistan has signed a letter of offer and acceptance for the procurement of 500 AMRAAM and 200 AIM-9M missiles for air defence. It is believed that the system will be deployed with the Pakistan Air Force.

Field testing of the SLAMRAAM began in early 2008 at the White Sands Missile Test Range, New Mexico and was successfully completed in July 2008.

Low rate initial series production announced in May 2009.

On 6 January 2011, the US Secretary of Defence announced the termination of the acquisition of the SLAMRAAM as part of a budget cutting effort.

Contractor

US Army Aviation and Missile Command, (enquiries).
Raytheon Systems Company, (enquiries).
Boeing, (Integrated Fire-Control System production).
Wildwood Electronics, Huntsville, Alabama (provides electronic components, cable harness fabrication, environmental testing and quality assurance).
Thales-Raytheon Systems Co. LLC joint venture in Fullerton, CA for the AN/MPQ-64 Sentinel air-defence radar.

Sword

Type

Self-propelled Surface-to-Air Missile (SAM) system

Development

The Sword short-range missile defence concept consists of fire-control radar with sufficient accuracy to command guide interceptors to hit and kill sophisticated cruise and short-range ballistic missiles.

In addition, the weapon system also has the ability to negate anti-radiation missiles, short- and long-range rockets, UAVs, aircraft, helicopters and helicopter-launched air-to-ground missiles.

As of November 2013, no further information concerning this system is available.

Description

The objective fire-control system will be capable of all-weather 360° on-the-move search, detect and track; non-cooperative threat identification; and simultaneous engagement of multiple threats.

The objective Sword interceptor will be a low-cost round fired from retrofitted existing field launchers such as the current in-service Lockheed Martin Vought Systems Multiple Launch Rocket System (MLRS) which is currently used in the surface-to-surface application.

To provide sufficient fire power to counter high volume saturation threats, the MLRS launcher can be modified and equipped with 108 Sword rounds.

The Sword tracked vehicles will provide continuous protection of forward area manoeuvre forces while the wheeled units can be strategically positioned to defend fixed-site assets and convoys.

The Sword system operates either autonomously or with the existing command, control, communications and intelligence theatre missile defence and air defence architecture.

When the proposed technology demonstration is complete, it could provide a limited operational capability.

Status

Technology Demonstrator status. The Air Defence Battle Laboratory at Fort Worth and the Weapons Directorate of the US Army Space and Missile Defence Command are working together to support a near term demonstration of the Sword systems ability. This is being accomplished by constructing a trailer-mounted demonstration fire unit and modifying one of several missiles to accept guidance commands from the radar. The interceptor used will be retrofitted with a transceiver for bi-directional communication, divert thrusters and a guidance package that will allow smart steering, given a command update from the fire-control system.

Contractor

US Army Space and Missile Defence Command.
Enquiries to Lockheed Martin Missiles and Fire Control, previously know as Lockheed Martin Vought Systems.

Towed Anti-Aircraft Guns

Bulgaria

ZU-23-2

Type
Towed anti-aircraft gun

Development
The original ZU-23-2 was developed in the late 1950s in order to engage low-flying targets at a range of 2.5 km as well as armoured vehicles at a range of 2 km. A further development of the basic weapon lead to the ZSU-23-4 Shilka. Since 1983, the Bulgarian Arsenal JSC has also been manufacturing the Russian ZU-23-2 twin 23 mm towed anti-aircraft gun as well as its associated ammunition. Full details of the standard ZU-23-2 are given in this section, under Russian Federation.

According to Arsenal it has produced over 5,000 of these weapons for the home and export markets.

As far as is known, ZU-23-2 is almost identical to that manufactured in the former Soviet Union, although there may be minor manufacturing differences. The information supplied is from Bulgarian sources.

Description
The ZU-23-2 can be used to engage an aerial target out to a maximum slant range of 2,500 m. Maximum altitude 1,500 to 2,000 m. Ground targets out to a range of 2,000 m.

Aircraft targets are engaged using an optical/mechanical sight designated the ZAP-23. Ground targets are engaged by using a T-3 telescopic direct laying sight.

The 2A14 type cannon normally fires two types of fixed 23 × 153 B ammunition; High Explosive Incendiary Tracer (HEI-T) and Armour-Piercing Incendiary Tracer (API-T). Each barrel is fed by a belt containing 50 rounds with the full combat load being 100 rounds. Muzzle velocity is 970 m/s, cyclic rate of fire is 1,600 to 2,000 rounds/min though the practical rate of fire is up to 400 rounds/min.

Ammunition for the 23 mm ZU-23-2 anti-aircraft gun is also produced in Bulgaria by the Arsenal Joint Stock Company (JSC). Natures presently available are; HEI-T, API-T, blank, HEI-T inert and Frangible Armour-Piercing Discarding-Sabot Tracer (FAPDS-T).

The actual ZU-23 cannon is designated the 2A14 while the cannon carrier is designated the 2A13.

Variants
This is a clear weather system, with no provision for off carriage fire-control. Bulgaria also produces a special naval mount for the ZU-23-2 that is called the ZU-23-2F.

ZSU-23-2 Upgrade
Details of this upgrade are given in a separate entry under International.

ZU-23-2M
There are three versions of the basic ZU-23-2M:

Close-up of Bulgarian-built ZU-23-2 light anti-aircraft gun showing sighting system
0509764

Bulgarian-built ZU-23-2 light anti-aircraft gun in travelling configuration
0509765

ZU-23-2M1
Semi-automatic Mount Control - this version has a reduction of personnel and facilitation of the operation by initiation of electric actuating mechanisms and increase of the Mount possibility to destroy fast moving targets. This first version also consists of a gun computer, actuating mechanisms and sensors and auxiliary equipment.

ZU-23-2M2
This upgrade controls the system fire by increasing its efficiency, enlarging the zone of destruction, reducing the number of attending personnel and facilitating their operation. The operator performs the target acquisition by manual or semi-automatic setting of the target into the vision range of the multipurpose aiming unit, then the electronic system performs the tracking of the target and by electric actuating mechanisms also performs the aiming. The second upgrade consists of the ZU-23-2M system, control computer, system for visual observation and designation, power supply system, gun computer, actuating mechanisms and sensors and auxiliary equipment.

ZU-23-2M3
In this version the fire-control system is designed to control the fire of up to 24 guns from one battery commander console. The battery fire-control system consists of up to 24 ZU-23-2M systems, battery commander console, radar tracker, acoustic devices for target detection (optional), radio-relay system for data transmission, system for visual observation and designation, electronic control system of ZU-23-2, vehicle for installation of radar tracker and radio-relay system for data transmission and auxiliary equipment.

Specifications

	ZU-23-2
Dimensions and weights	
Crew:	5
Length	
transport configuration:	4.57 m
Width	
transport configuration:	1.83 m
Height	
transport configuration:	1.87 m
Ground clearance	
overall:	0.56 m
Track	
vehicle:	1.67 m
Mobility	
Carriage:	trailer mounted two wheeled
Firepower	
Armament:	2 × 23 mm cannon
Ammunition	
main total:	100
Main weapon	
length of barrel:	2.01 m
calibre length:	23 calibres
breech mechanism:	vertical sliding block
maximum range:	2,500 m

	ZU-23-2
maximum vertical range:	1,500 m
operation:	gas automatic
cartridge:	23 mm × 153 mm B
Rate of fire	
cyclic:	2,000 rds/min
practical:	1,600 rds/min
Main armament traverse	
angle:	360°
Main armament elevation/depression	
armament front:	+90°/-10°

23 mm ammunition

Type	HEI-T	API-T
Fuze:	MG-25/V19-UK	nil
Primer:	KV-3	KV-3
Material:		
(projectile body)	steel	steel
(cartridge case)	steel	steel
(driving band)	copper	copper
Weight:		
(projectile)	188.5 g	190 g
(cartridge case)	172 g	172 g
Length:		
(projectile)	108.2 mm	99.3 mm
(cartridge case)	151.5 mm	151.5 mm
(overall)	235 mm	235 mm
Muzzle velocity:	970 m/s	970 m/s
Chamber pressure:	286.8 MPa	286.8 MPa

Status
Production as required. In service with Bulgaria and many other undisclosed countries.

Contractor
Arsenal JSC.

Chile

Browning M2

Type
Towed anti-aircraft gun

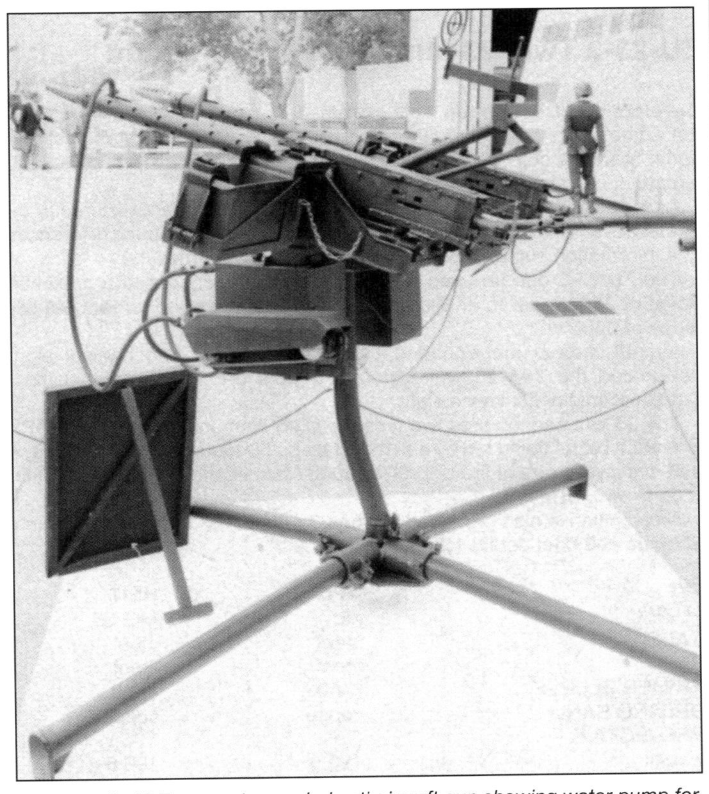

Chilean twin 12.7 mm water-cooled anti-aircraft gun showing water pump for circulating water through barrel cooling liners; two 12 V batteries used to power water pump can be seen either side of the central pedestal tube (T J Gander) 0509481

Description
This twin 12.7 mm (.50 cal.) Browning M2 anti-aircraft gun uses water-cooled barrels that have been converted from the normal air-cooled barrels. Weapon elevation and mount traverse is manual.

The water-cooling barrel liners are contained in perforated jackets around each barrel and the barrels are fitted with new multi-baffle muzzle brakes. Water is circulated via flexible hoses through the barrel liners by a single water pump, powered by two 12 V batteries. The complete system is mounted on a modified M63 pedestal mounting. A cartwheel sight is mounted above and between the two M2 machine guns for laying the weapons onto the target.

The 12.7 mm water-cooled Browning M2 machine guns are converted from aircraft guns where the receiver mechanism and ammunition feed remain unchanged.

There are indications that the modifications have been introduced by the Chilean Air Force.

This is a clear weather system with no provision for off carriage fire control.

Status
Production complete. Reported to be in service with the Chilean Air Force. As far as it is known this system has not been exported.

Contractor
Browning.

FAM-2M

Type
Towed anti-aircraft gun

Development
The first Chilean 20 mm anti-aircraft guns in the FAMIL series were produced under the name Sogeco; FAMIL is now a subsidiary of Sogeco. The FAMIL first guns in the series used Hispano-Suiza HS.820 cannon, these taken from redundant De Havilland Vampire T.11 aircraft. The latest guns in the series use 20 mm Oerlikon Contraves (now Rheinmetall Air Defence AG) KAD B16 and B17 cannon with a muzzle velocity of 1,040 m/s. The cannon were renamed from HS.820 to KAD after Oerlikon Contraves purchased Hispano-Suiza's armament division in the 1970s. These guns are mainly used for airfield defence, but they can also be used against ground targets. The difference between the two versions is the feed mechanisms - one is fed from the left whilst the other is fed from the right.

Description
The carriages on all the guns produced so far are visually similar, they are supported in the firing position on four folding outriggers for stabilisation. For transport, the gun is carried on a removable twin-wheeled yoke carriage. A small petrol engine on the right-hand side of the carriage supplies power for the mounting. The engine drives a hydraulic system which is controlled by the gunner using a single joystick control. The final production models also have a 24 V electrical system for the gun sight supplied from two 12 V batteries. A shield is installed to provide some protection from shell splinters.

Early models use a simple cartwheel sight, but the latest models have a BAE Systems gyroscopic reflector sight that can be used together with a BAE Systems Avionics target injection training system. In service now is a central control system that controls up to four guns from a single control box mounting a single joystick to act as a fire director system for a battery.

The Hispano-Suiza 820 cannon versions use drum magazines each holding 120 rounds of ready use ammunition. The Oerlikon Contraves KAD cannon are belt-fed from box magazines, each box holding a belt of 200 rounds.

This is a clear weather air defence system with no provision for off carriage fire-control.

FAMIL FAM-2M twin barrelled anti-aircraft gun fitted with HS.820 20 mm cannon (T J Gander) 0509480

Specifications

	FAM-2M
Dimensions and weights	
Weight	
combat:	1,700 kg
Mobility	
Carriage:	trailer mounted two wheeled with outriggers
Firepower	
Armament:	2 × 20 mm cannon
Main weapon	
maximum range:	1,500 m
feed mechanism:	200 round boxed (per barrel)
operation:	gas automatic
cartridge:	20 mm × 139 mm
Rate of fire	
cyclic:	2,100 rds/min

Status
Production as required. In service with the Chilean Air Force. As far as it is known, this system has not been exported. As of January 2013, no further information on this system is available.

Contractor
FAMIL SA.

China

12.7 mm Type 54

Type
Towed anti-aircraft gun

Development
The China North Industries Corporation (NORINCO) 12.7 mm anti-aircraft machine gun Type 54 is essentially the Soviet 12.7 mm Degtyarev Krupnokalibernyi Model 38/46 (DShKM) heavy machine gun, produced in China under the designation of the Type 54.

It has been designed to engage both air and ground targets. In the first application the tripod is almost vertical, while in the second application it is almost horizontal. The 12.7 mm Type 54 is also used on a number of Chinese armoured fighting vehicles as an air defence weapon.

Description
The Type 54 is a 12.7 mm, gas-operated reloading, belt-fed, crew-served, automatic weapon that is effective against low flying aircraft, soft-skinned ground targets and personnel. The ammunition is fed from the left in boxes of 70 rounds, two types of 12.7 × 108 mm ammunition can be fired:

NORINCO 12.7 mm Type 54 anti-aircraft machine gun being used in air defence role 0509482

- Armour-Piercing Incendiary (API)
- Armour-Piercing Incendiary Tracer (API-T).

Both types of ammunition have a muzzle velocity of 820 m/s (other sources put the muzzle velocity at 850 m/s). The API round will penetrate 10 mm of conventional steel armour at a range of 800 m. The foresight is a pillar that slides up and down, while the rear sight has twin vertical pillars with a U-back sight between them. In addition, there is a special anti-aircraft sight that requires two people to use it to its best advantage. Two of the three tripod legs are fixed with the third leg being fully adjustable.

Variants
Type 54-1
The Type 54-1 is an improved version of the Type 54 weapon and weighs 45 kg, with elevation in the air defence role being from –26° to +72°.

Type 59
The Type 59 anti-aircraft machine gun was developed from the Type 54 and has similar design and performance characteristics.

Specifications
12.7 mm Type 54

Calibre:	12.7 mm
Number of barrels:	1
Practical rate of fire:	600 rds/min
Cartridge:	12.7 × 108
Muzzle velocity:	820 m/s (other sources 850 m/s)
Operation:	gas, automatic
Method of locking:	projecting lugs
Feed:	belt with 50 rounds
Weight:	
(gun)	35.7 kg (other sources suggest 34 kg)
(barrel)	12.7 kg
Effective range against air targets:	1,600 m (other sources suggest 2000 m)
Effective range against ground targets:	1,500 to 2,000 m (other sources suggest maximum range 2500 m)
Elevation in air defence role:	–34° to +78°
Traverse in air defence role:	360°
Traverse in ground defence role:	120°

Status
Production as required. In service with the People's Liberation Army, Bangladesh Army and the Pakistan Army.

Contractor
China North Industries Corporation (NORINCO).
This is also made in Pakistan by the Pakistan Ordnance Factories (POF).

12.7 mm Type 77

Type
Towed anti-aircraft gun

Development
The China North Industries Corporation (NORINCO) 12.7 mm Type 77 anti-aircraft machine gun has been designed to engage both air and ground targets. In the first application the tripod is almost vertical, while in the second it is almost horizontal. It is assessed that the system can be pintle-mounted, vehicle-mounted or tripod-mounted dependant on the mission expectancy.

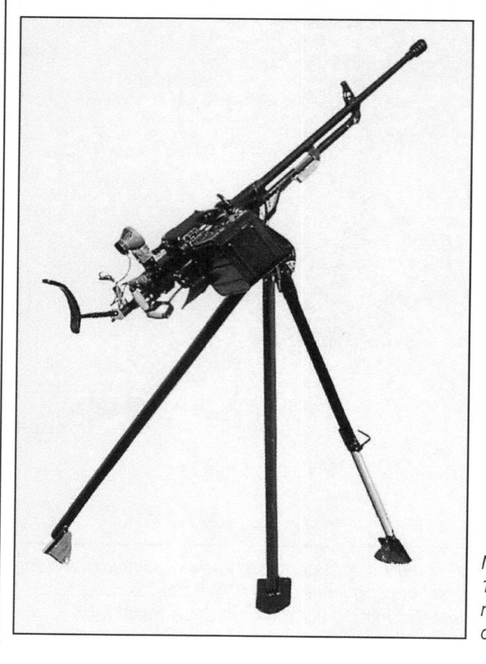

NORINCO Type 77 12.7 mm anti-aircraft machine gun in air defence role 0509873

Description

This system is similar in many ways to the Soviet DShK and was probably designed with the DShK in mind. The weapon appears to be recoil operated and has both ground and anti-aircraft sights. A distinctive pepper pot muzzle brake is provided. The operator is provided with a shoulder-type stock and twin hand grips.

Ammunition is fed from the left in boxes of 60 rounds. Two basic types of 12.7 × 108 mm ammunition can be fired:

- Armour-Piercing Incendiary (API)
- Armour-Piercing Incendiary Tracer (API-T).

Both types of ammunition have a muzzle velocity of 800 m/s.

More recently the Type 54-1 Armour-Piercing Discarding Sabot (APDS) round with a tungsten alloy penetrator for engaging armoured ground targets has been developed. This can also be fired by the other NORINCO 12.7 mm machine guns.

Two of the tripod legs are fixed with the third one being adjustable.

Specifications

12.7 mm Type 77

Calibre:	12.7 mm
Cartridge:	12.7 × 108 mm
Number of barrels:	1
Rate of fire: (cyclic)	650-700 rds/min (also reported as 650-750 rds/min)
Weight: (with empty ammunition box)	56.1 kg
Length in ground role:	2.15 m
Width in ground role:	3.10 m
Axis of fire:	
(ground role)	0.36 m
(air defence role)	1.3 m
Effective range:	
(air targets)	1,600 m
(ground targets)	1,500 m
(armoured ground targets)	800 m
Elevation in air defence role:	–15° to +80°
Traverse:	
(air defence role)	360°
(ground role)	120°

Status

Production as required. In service with the People's Liberation Army (PLA), Cambodia and Pakistan. Photographic evidence obtained in May 2006 also suggests that the Tamil Tigers had access to these weapons, it is not sure if they came via Pakistan or direct from China.

Contractor

China North Industries Corporation (NORINCO).

12.7 mm Type 85

Type

Towed anti-aircraft gun

Development

The China North Industries Corporation (NORINCO) 12.7 mm Type 85 anti-aircraft machine gun, also known as Type W-85, has been developed as a multipurpose weapon for use against both air and ground targets.

In the first application the tripod is almost vertical, while in the second it is almost horizontal.

Compared with the older Type 54 12.7 mm anti-aircraft machine gun, it is some 58 per cent lighter. It can be disassembled easily into three parts, the heaviest of which does not exceed 20 kg; these are the weapon, mount and 12.7 mm ammunition box. In the past this weapon has also been referred to as the Type W-85 by NORINCO.

Description

The Type 85 gun is gas operated, belt fed and exceptionally light, with an all-up weight some 58 per cent lighter than the obsolescent Type 54, the Chinese version of the Soviet DShK. The 12.7 mm Type 85 machine gun can be stripped and reassembled quickly without tools and the barrel can be changed quickly in the field. According to NORINCO, the machine gun mechanism has a malfunction rate not exceeding 0.2 per cent. The barrel is certified for 3,500 rounds while the remainder of the weapon exceeds 7,000 rounds.

The 12.7 mm Type 85 machine gun is fitted with a rear mounted optical telescope enabling it to engage a wide range of battlefield targets under both day and night conditions. This can be replaced quickly by a much larger special anti-aircraft sight.

The weapon can fire the following types of 12.7 × 108 mm ammunition, all with a muzzle velocity of 800 m/s:

- Type 54 Armour-Piercing (AP).
- Type 54 Armour-Piercing Incendiary (API).
- Type 54 Armour-Piercing Incendiary Tracer (API-T).

It can also fire the Type 54-1 Armour-Piercing Discarding Sabot (APDS) with a tungsten alloy cored subprojectile at a muzzle velocity of 1,150 m/s. Ammunition feed is from the left, with each metal box holding 60 rounds and two of the tripod legs are fixed with the third one being adjustable.

NORINCO Type 85 12.7 mm anti-aircraft machine gun in air defence configuration 0509874

Specifications

12.7 mm Type 85

Calibre:	12.7 mm
Cartridge:	12.7 × 108 mm
Number of barrels:	1
Rate of fire:	
(practical)	80-100 rds/min
(cyclic)	650-750 rds/min
Weight:	
(system)	39 kg
(gun only)	18.5 kg
(gun mount)	15.5 kg
Length: (in horizontal position)	2.050 m
Width: (in horizontal position)	1.160 m
Axis of fire:	
(weapon in AA role)	1.224 m
(weapon in ground role)	320 mm
Effective range:	
(air targets)	1,600 m
(ground targets)	1,500 m
(armoured targets)	800 m
Elevation in horizontal position:	–15° to +25°
Maximum elevation:	–10° to +80°
Traverse:	
(horizontal position)	120°
(air defence role)	360°

Status

Production as required. In service with the People's Liberation Army. Probably in service with other countries such as Pakistan.

Contractor

China North Industries Corporation (NORINCO).

14.5 mm QJG02 (Type 02)

Type

Towed anti-aircraft gun

Development

Developed by the NORINCO 356 Factory, the QJG02 (also known as Type 02) is a single barrel 14.5 mm anti-aircraft gun that has been deployed with the Peoples Liberation Army (PLA) ground forces from 2007 onwards although it is believed production may have started as early as 2004/05.

As with all heavy calibre machine guns the weapon has dual capability; primarily the gun is designed with air defence in mind, however a secondary role in the ground defence is also an option. The gun may also be found on light PLA(N) ships for short-range defence against low flying targets and light armoured surface boats.

Unconfirmed reports have also suggested the system may also be found deployed with the PLAs coastal defence forces and mountain units stationed in Southern China.

Description

The system has been designed to replace the Type 56 copy of the Russian ZPU-1. The QJG02 (Type 02) is of a gas-operated design with a quick removable barrel that is mounted on a tripod and operated by a five-man crew. The system normally travels on a two-wheel single axle carriage that can be towed by a truck or APC.

The gun is belt-fed from a 50-round cartridge case.

Variants

QJG02G is an improved variant that with increased weight of 110 kg that has a soft mount utilising an annular floating spring limiting recoil and improving accuracy.

Specifications

	QJG02
Dimensions and weights	
Weight	
combat:	73 kg
Mobility	
Firepower	
Armament:	1 × 14.5 mm
Main weapon	
maximum range:	2,000 m
feed mechanism:	50 round belt
Rate of fire	
cyclic:	600 rds/min
practical:	100 rds/min

Status

In production and in use with the PLA and PLA(N).

Contractor

China North Industries Corporation (NORINCO), Factory 356.

14.5 mm Type 56/58/75/80

Type

Towed anti-aircraft gun

Description

Type 56 LAAG

The China North Industries Corporation (NORINCO) 14.5 mm anti-aircraft guns all use the Chinese version of the Soviet 14.5 mm KPV heavy machine gun. The 'base' model as far as Chinese production is concerned is the Type 56, a direct copy of the Soviet ZPU-4. The Type 56 fires three types of 14.5 × 114 mm ammunition: API, API-T and Incendiary. These are all identical to their then Soviet counterparts.

NORINCO 14.5 mm Type 75-1 anti-aircraft machine gun in firing position
0509492

NORINCO 14.5 mm Type 56 anti-aircraft machine gun in firing position
0509871

The Type 56 has been exported widely and is also produced in North Korea.

North Korea fitted with the locally manufactured version of the Type 56 has developed at least two full tracked self-propelled anti-aircraft gun systems. These are called the M1984 and VTT 323 and available details are given in the Self-propelled anti-aircraft gun section earlier in this volume. As far as it is known these are only in service with North Korea.

All of these are clear weather systems and are suitable for engaging slow targets flying at low level.

This gun is being replaced in the PLA by the QJG02 anti-aircraft system.

	Type 56 LAAG	Type 58 LAAG	Type 75 LAAG	Type 75-1 LAAG	Type 80 LAAG
Dimensions and weights					
Crew:	5	-	-	-	-
Length					
overall:	4.54 m	3.9 m	2.93 m	2.93 m	2.5 m
Width					
overall:	1.86 m	1.66 m	1.62 m	1.62 m	1.8 m
Height					
overall:	2.34 m	1.1 m	1.07 m	1.07 m	2.2 m
Weight					
standard:	1,000 kg	-	-	-	-
combat:	2,100 kg	560 kg	140 kg	140 kg	214 kg
Mobility					
Carriage:	trailer mounted four wheeled	trailer mounted two wheeled	-	trailer mounted two wheeled	trailer mounted two wheeled
Firepower					
Armament:	4 × 14.5 mm machine gun	2 × 14.5 mm machine gun	1 × 14.5 mm machine gun	1 × 14.5 mm machine gun	1 × 14.5 mm machine gun
Main weapon					
length of barrel:	1.348 m	1.348 m	1.348 m	1.348 m	1.348 m
maximum range:	2,000 m (air target)	2,000 m (air target)	2,000 m (air target)	2,000 m (air target)	2,000 m (air target)
maximum range:	1,000 m (ground target)	1,000 m (ground target)	1,000 m (ground target)	1,000 m (ground target)	1,000 m (ground target)
feed mechanism:	150 round boxed	150 round boxed	80 round boxed	80 round boxed	80 round boxed
Rate of fire					
cyclic:	2,200 rds/min	1,100 rds/min	550 rds/min	550 rds/min	600 rds/min (est.)
Main armament traverse					
angle:	360°	360°	360°	360°	360°
Main armament elevation/depression					
armament front:	+90°/-7°	+90°/-15°	+85°/-10°	+85°/-10°	+85°/-15°

NORINCO 14.5 mm Type 58 anti-aircraft machine gun in firing position 0509872

NORINCO 14.5 mm Type 80 anti-aircraft machine gun in travelling position 0509493

NORINCO 14.5 mm Type 58 anti-aircraft machine gun deployed in the direct fire role 0509692

Type 58 LAAG
The NORINCO Type 58 is a direct copy of the Soviet ZPU-2 and resembles the late production version of the original. For engaging ground targets, a direct view sight with a magnification of ×3.5 is fitted.

Type 75 LAAG
The NORINCO Type 75 is the Chinese version of the ZPU-1 and is designed to be pack transported if necessary. The Type 75-1 has a more complex optical sight, which may have image intensification characteristics.

NORINCO 14.5 mm Type 58 anti-aircraft machine gun showing computing sight in detail 0509693

Type 80 LAAG
The NORINCO Type 80 is a version of the ZPU-1, re-engineered to Chinese standards and with a redesigned carriage enabling it to be fired from the prone position against ground targets. It has a more robust carriage than either of the Type 75s and may be towed behind a light (4 × 4) vehicle.

The Types 58, 75, 75-1 and 80 14.5 mm light anti-aircraft guns all fire the same 14.5 mm fixed ammunition as the Type 56.

Specifications – *see table on facing page*

Status
Production as required. In service with the Chinese and other armed forces including Cameroon (18 Type 58), Nepal (30 Type 56), Pakistan (various models) and other undisclosed countries. This system was observed in December 2006/January 2007 during a Chinese land-based air-defence gunnery exercise.

Contractor
China North Industries Corporation (NORINCO).

23 mm Type 80

Type
Self-Propelled Light Anti-Aircraft Gun (LAAG) system

Development
The China North Industries Corporation (NORINCO) twin 23 mm Light Anti-Aircraft Gun (LAAG) Type 80 has been in production for many years. It is a reverse-engineered copy of the Soviet ZU-23-2 with several minor differences.

The system has been offered for export and is known as the Giant Bow Twin 23 mm AA Gun Air Defence System consisting of eight twin 23 mm AA guns used in conjunction with one Battery Optical-electronic Commander Vehicle.

From December 2005, NORINCO is also offering an upgraded version of the Giant Bow which is known as the Giant Bow II. The upgraded version utilises the TY-90 missiles that were originally designed for use on helicopters as short-range infrared air-to-air missiles.

Along with the quadruple launchers, the new system will also have a 3-D search radar that operates in conjunction with the BCV.

Description
The Type 80 LAAG is operated by a crew of two on the mount and three PLA off; the off mounted crew serve as ammunition carriers and loaders for the two metallic linked 50-round ammunition belts in the box magazines.

In all operational respects, the gun is similar to the Russian weapon and it is only effective against low-flying targets at slant ranges up to 2,500 m and an altitude of 1,500 m. It can also be used to engage ground targets out to 2,000 m effective range.

The cannon fires two types of 23 × 152 mm Belted (23 × 152 B) fixed ammunition. Both are copies of the standard ammunition used with the ZU-23-2.

The 23 mm HEI-T round is fitted with a nose-mounted fuze that contains a self-destruct device. The projectile is filled with Hexal based explosive (RDX/Al/Wax) with a tracer in its rear end. The fuze delaying mechanism ensures that the projectile only explodes after penetrating the target. If the projectile fails to hit the target then it will automatically explode in 5 to 11 seconds under the action of the self-destruct device.

The 23 mm API-T round penetrates the target first, behind armour damage is caused by the projectile, armour spall, plus the incendiary charge built into the projectile.

Both rounds have the same ballistics, and HEI-T and API-T rounds are normally incorporated into belts in a ratio of 3:1.

NORINCO 23 mm light anti-aircraft gun Type 80 from the rear in the firing position 0509690

NORINCO 23 mm light anti-aircraft gun Type 80 from the side in the firing position 0509691

Variants

Type 80 air defence system

The Type 80 Air Defence System (ADS) is comprised of a battery of six 23 mm Type 80 LAAGs used in conjunction with the Type 800 trailer-mounted laser course director.

Giant Bow

The Giant Bow Twin 23 mm AA Gun Air Defence System comprises eight twin barrelled Type 80 23 mm LAAGs used in conjunction with one Battery Optical-electronic Commander Vehicle. A crew of five operates the Giant Bow. Two of which are seated one each side of the mount at the rear, whilst the other three are off the carriage. These three serve as ammunition carriers and loaders for the two metallic linked 50-round ammunition belts in a steel magazine.

The Operating bands of the Commander Vehicle are:

- TV camera: 0.4-0.75 μm
- Thermal imaging camera: 8-12 μm
- Laser range-finder: 1.06 μm.

The 23 mm Type 80 LAAG has two sighting systems, one for engaging airborne targets and the other for ground targets. The ground target sight is known as the T3. The T3 has a magnification of ×3.5 with a field of view of 4° 30'. The range setting device being from 0 to 2,000 m.

The Type 80 takes 15 seconds to come into action from the travelling position and takes 35 seconds to come out of action to the travelling configuration.

Giant Bow II

The Giant Bow II has additional anti-aircraft missiles in the inventory. The system comprises:

- Giant Bow II Battery Command Vehicle (BCV)
- AS901A 3-D post target designation radar
- SAM launchers using the TY-90 SHORAD missiles
- 23 mm towed Giant Bow AA guns.

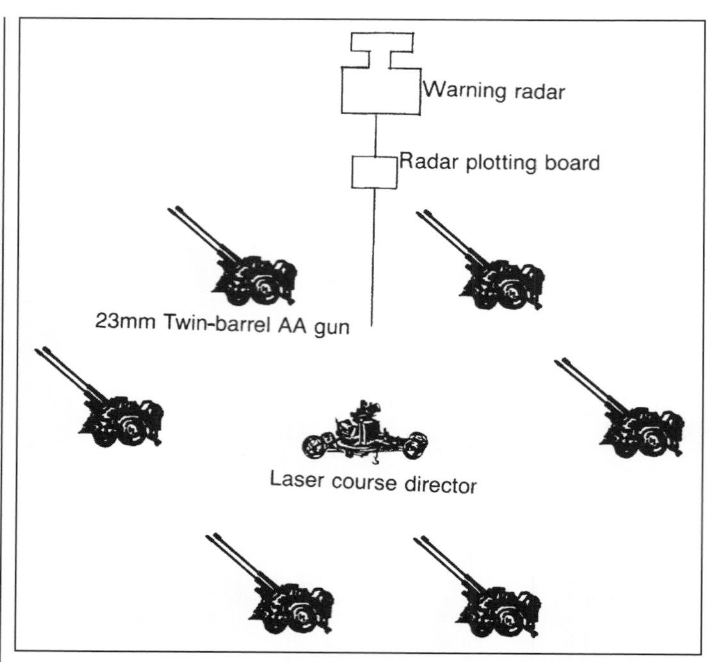

Block diagram of 23 mm Type 80 air defence system which consists of six twin 23 mm Type 80 light anti-aircraft guns and a Type 800 laser course director 0509870

The Giant Bow II air defence system can be deployed in AA gun only format or as the AA gun and missile configuration. The AS901A radar can be deployed either with the gun or gun-missile or if used independently with cooperative engagements the radar can be left out of the battery all together.

Missile

The TY-90 missile has been described by NORINCO as an IR guided missile that employs advanced multi-element InSb seeker that is capable of both head and tail-chase attack capability. The missile also boasts high manoeuvrability, anti-jamming capability and lethal warhead with a proximity fuze. The targets the system has been designed against include:

- Low-flying aircraft
- Attack helicopters
- Unmanned Aerial Vehicles (UAVs)
- Stand-off precision guided weapons.

Radar

The search radar provides the system with autonomous tactical air situation awareness, together with multi-target and multi-layer engagement capability. It is described as a three-dimensional (3-D) planar-array tactical search radar with Identification Friend or Foe (IFF) capabilities. The system can automatically detect, identify, track and report multiple airborne threats to the battery command vehicle. The radar operates in L-band and has a detection range of 30 km.

Battery Command Vehicle

The Battery Command Vehicle (BCV) is operated by a crew of six. It incorporates:

- Batter Command, Control, Communication and Intelligence (BC3I) capabilities
- Fire-control functions
- Electro-Optical (EO) sensors
- Manportable optical command telescope.

The EO sensors include thermal imaging, television camera and laser range-finder.

The Giant Bow II can be deployed in AA-gun only or gun-missile integrated configurations. Both configurations comprise three twin AA guns and/or quadruple SAM launchers, together with one BCV, one 3-D search radar and combat support vehicles.

Specifications

	23 mm Type 80	23 mm Type 80 (Upgraded)
Dimensions and weights		
Crew:	5	5
Length		
transport configuration:	4.57 m	4.57 m
deployed:	4.57 m	-
Width		
transport configuration:	1.83 m	1.83 m
deployed:	2.88 m	-
Height		
transport configuration:	1.87 m	1.87 m
deployed:	1.82 m	-
Weight		
standard:	950 kg	-
combat:	950 kg	-

	23 mm Type 80	23 mm Type 80 (Upgraded)
Mobility		
Speed		
towed:	70 km/h	-
Firepower		
Armament:	2 × carriage mounted 23 mm cannon	2 × carriage mounted 23 mm cannon
Main weapon		
length of barrel:	2.01 m	-
breech mechanism:	vertical sliding block	-
maximum range:	7,000 m (effective)	-
maximum vertical range:	5,100 m (effective)	-
minimum vertical range:	2,500 m (effective)	-
feed mechanism:	50 round belt	-
operation:	gas automatic	gas automatic
Rate of fire		
cyclic:	2,000 rds/min (est.)	2,000 rds/min (est.)
practical:	400 rds/min (est.)	400 rds/min (est.)
Main armament traverse		
angle:	360°	360°
Main armament elevation/depression		
armament front:	+90°/-10°	+90°/-10°

Ammunition (All systems)

Calibre:	23 mm
Cartridge:	23 × 152 B
Muzzle velocity:	970 m/s
Maximum chamber pressure:	3,100 kg/cm^2
Length: (complete round)	236 mm
Weight:	
(complete round)	450 g
(HEI-T projectile)	188.5 g
(API-T projectile)	190 g
(HEI-T explosive)	13 g
(API-T incendiary charge)	4.3 g
Tracer burning time:	5 s
Fuze self-destruct:	5-11 s

Status

As of January 2014, there is only one known export sales of this Type 80-air defence system and that is to Indonesia. Production as required. In service with the Chinese Armed Forces and Indonesian Air Force.

The Giant Bow has been offered for export in two variants, Giant Bow and Giant Bow II.

Contractor

China North Industries Corporation (NORINCO).

25 mm Type 85

Type

Towed anti-aircraft gun

NORINCO twin 25 mm Type 85 light anti-aircraft gun in firing position 0509491

Development

At Asiandex '86, a new twin 25 mm calibre towed low-altitude defence anti-aircraft gun was unveiled. It is believed to have the designation WA709. Other sources have stated that this is designated the WA309, with the People's Liberation Army designation being Type 85. The Type 85 gun was developed in 1985 exclusively for the export market.

Description

This is very similar to the Russian ZU-23-2 with distinctive parallel flash eliminators and a handle on each barrel to facilitate a quick barrel change.

It has a crew of three with the elevation operator on the left, the commander in the centre and the traverse operator on the right. The cannon's elevation and traverse controls are manual. The fixed ammunition types used are HE-T, AP-T and HEAP-T, all with a muzzle velocity of 1,050 m/s. The ammunition used is thought to be based on the NATO 25 × 137 mm cartridge.

Two accessories are now available for use with this weapon, a communications system and a so-called laser air route device.

Self-propelled 25 mm system

In late 1999, the People's Liberation Army displayed a new self-propelled air defence system armed with four 25 mm cannon and four fire-and-forget SAMs based on a tracked chassis. This is believed to use the same 25 mm cannon as the Type 85 towed system. This system was originally known as the Type 90-2, but has since been referred to as the Type 95. Details are given in a separate entry in the Self-propelled anti-aircraft guns/surface-to-air missiles section.

Specifications

	Type 85
Dimensions and weights	
Crew:	3 (plus off-mount loaders)
Length	
transport configuration:	4.68 m
Width	
transport configuration:	2.04 m
Height	
transport configuration:	2.08 m
Weight	
standard:	1,500 kg
Mobility	
Carriage:	trailer mounted two wheeled
Speed	
towed:	60 km/h
Firepower	
Armament:	2 × 25 mm cannon
Main weapon	
maximum vertical range:	5,000 m
operation:	gas automatic
Rate of fire	
cyclic:	1,200 rds/min (est.) (all barrels)
Main armament traverse	
angle:	360°
Main armament elevation/depression	
armament front:	+90°/-10°

Status

Development complete, ready for production on receipt of firm orders. In service with the People's Liberation Army under the designation of the Type 85. As of April 2013 no further information was available on this system.

Contractor

China North Industries Corporation (NORINCO).

35 mm Type 90

Type

Towed anti-aircraft gun

Development

Since the mid-1990s China North Industries Corporation (NORINCO) has been marketing a twin 35 mm (Type 90) towed anti-aircraft gun system that is a licensed copy of the then Swiss Oerlikon Contraves GDF series covered in detail later in a separate entry in this section.

As of early 2009, Oerlikon Contraves is now known as Rheinmetall Air Defence.

Late in 1999 during a major parade in Beijing, the People's Liberation Army (PLA) exhibited a number of these systems towed by 6 × 6 trucks. A locally manufactured fire-control system, very similar to the Rheinmetall Air Defence trailer-mounted Skyguard, was also shown.

Details of the original Rheinmetall Air Defence version are given in a separate entry under Anti-aircraft control systems.

Late in November 2012, NORINCO released photographic evidence of the Type 90 gun mounted on the Wanshan WS-2400 chassis vehicle. The Type 90 was only previously available as a towed anti-aircraft gun.

NORINCO twin 35 mm towed anti-aircraft gun system in travelling configuration
0509868

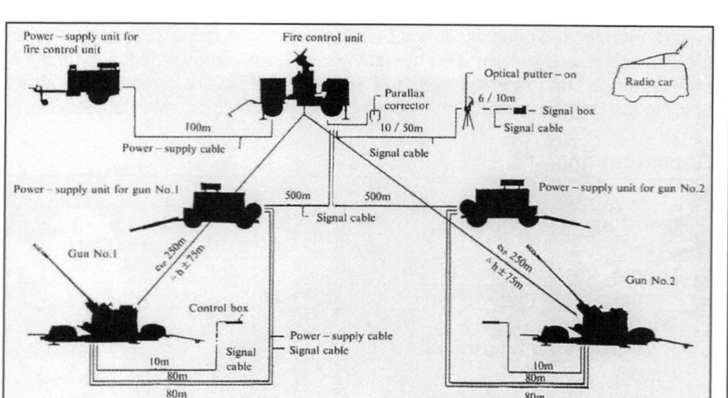

Block diagram of NORINCO twin 35 mm towed anti-aircraft gun system with one fire-control unit controlling two weapons
0509869

Description

The Type 90 anti-aircraft gun system has been designed specifically to engage low flying targets such as helicopters, cruise missiles and UAVs. As far as is known, technical details of the Chinese twin 35 mm towed anti-aircraft gun system are identical to early production Swiss GDF weapons, although not the latest production GDF-005 with the Rheinmetall Air Defence Gunking computerised fire-control system.

A fire-control system designed by the Chinese North Automatic Control Technology Institute is the Sky Shield Short-Range Optronic Fire-Control System for Air Defence. This fire-control system integrates together into a single combat system the following:

- Battalion-scaled command posts
- Search radar
- Optical director
- Thermal imaging
- Laser range-finder
- Fire-control computer
- Automatic guns and missiles.

The system allows the commander to direct each combat unit via automatic control and communication devices.

Four types of Sky Shields have been reported, these are:

- Expanded Configuration 1 - to control three new double-barrel 37 mm air defence guns and one PL-9C missile launcher
- Expanded-Configuration 2 - to control two new double-barrel 35 mm air defence guns and one PL-9C launcher
- Extended Configuration 1 - controls six 74-Type double-barrel 37 mm and three new double-barrel 37 mm air defence guns and one PL-9C missile launcher.

Extended Configuration 2 - controls six 74-Type double-barrel 37 mm and two double-barrel 35 mm air defence guns and one PL-9C missile launcher.

NORINCO is also marketing 35 mm ammunition for this weapon including:

- HEI
- HEI-T
- TP-T.

Details on these ammunition natures are given in the specifications table titled 'Ammunition'.

NORINCO does not manufacture any of the more advanced natures of 35 mm Rheinmetall Air Defence 35 mm ammunition such as Advanced Hit Efficiency And Destruction (AHEAD).

Typically, two NORINCO twin 35 mm towed anti-aircraft guns would be controlled by a single fire-control radar system of Chinese or foreign origin.

The twin 35 mm system is mounted on a four-wheeled carriage and, when deployed in the firing position, the wheels are raised off the ground and the carriage is supported at four points, one at either end of the carriage and one either side on outriggers.

Each 35 mm cannon is fed from a magazine situated on the outside of the weapon with reserve ammunition being carried on the rear of the mount ready for rapid reloading. Each barrel is provided with 56 rounds of ready use ammunition with another 63 rounds in reserve.

The mount is electrically powered with power supply from an off-carriage generator and manual controls are provided for emergency use.

Open press literature from NORINCO has aligned this gun system with the PL-9C air-to-air missiles used in the land-based air-defence role, the AF902 fire-control radar and offers the complete integrated air defence system package.

Specifications

Ammunition

	HEI	HEI-T	TP-T
Length: (round)	387 mm	387 mm	387 mm
Weight:			
(round)	1.58 kg	1.565 kg	1.58 kg
(projectile)	0.550 kg	0.535 kg	0.550 kg
(propellant)	0.330 kg	0.330 kg	0.330 kg
(explosive)	0.112 kg	0.098 kg	n/app
Muzzle velocity:	1,175 m/s	1,175 m/s	1,175 m/s
Fuze:	KZD338	KD338	n/app
Time of flight:			
(1,000 m)	0.96 s	0.96 s	0.96 s
(2,000 m)	2.18 s	2.18 s	2.18 s
(3,000 m)	3.8 s	3.8 s	3.8 s

Variants

Self-propelled (WS-2400) Type 90.

Status

In production.

In service with the People's Liberation Army. The Type 90 is assessed to be replacing the Type 65 and Type 74 twin 37 mm guns in the ground forces.

Contractor

China North Industries Corporation (NORINCO).

37 mm Types 55, 65, 74, 74SD and P793

Type

Towed anti-aircraft gun

Description

37 mm Type 55 LAAG

The most commonly encountered model of the Chinese China North Industries Corporation (NORINCO) 37 mm anti-aircraft gun is the Type 55, a direct copy of the Soviet 37 mm M1939 from which it differs in few details. In their standard production configuration, the 37 mm guns have hydraulic recoil buffers and spring recuperators. The breechblocks are of the rising block type and all of these weapons are clear-weather only.

37 mm Type 65 LAAG

The 37 mm Type 65 is a direct copy of the Soviet twin-barrelled version of the M1939 while the Type 74 is a model with revised detail engineering to suit Chinese manufacturing methods and can operate with radar fire-control at a slightly increased rate of fire compared to the Type 65. The Type 65 and Type 74 use the same ammunition as the Type 55.

37 mm Type 74 LAAG

According to NORINCO the 37 mm twin-barrel Type 74 can be used to engage air targets within a slant range of 3,500 m with a self destruct range of 4,000-4,700 m; if required the gun can engage ground targets.

The weapon is usually deployed in batteries of six 37 mm guns with each battery being provided with a radar and fire director so that the gun can fire not only at visible targets but also targets above clouds and at night.

A fire-control system designed by the Chinese North Automatic Control Technology Institute is the Sky Shield Short-Range Optronic Fire-Control System for Air Defence. This system integrates battalion-scaled command posts, search radar, optical director, thermal imaging, laser range-finder,

NORINCO twin 37 mm LAAGs fitted with shields and captured in Kuwait
0509485

NORINCO twin 37 mm anti-aircraft gun Type 65 in firing position 0509486

China National Electronics Import & Export Corporation Type 311A fire-control radar system 0509487

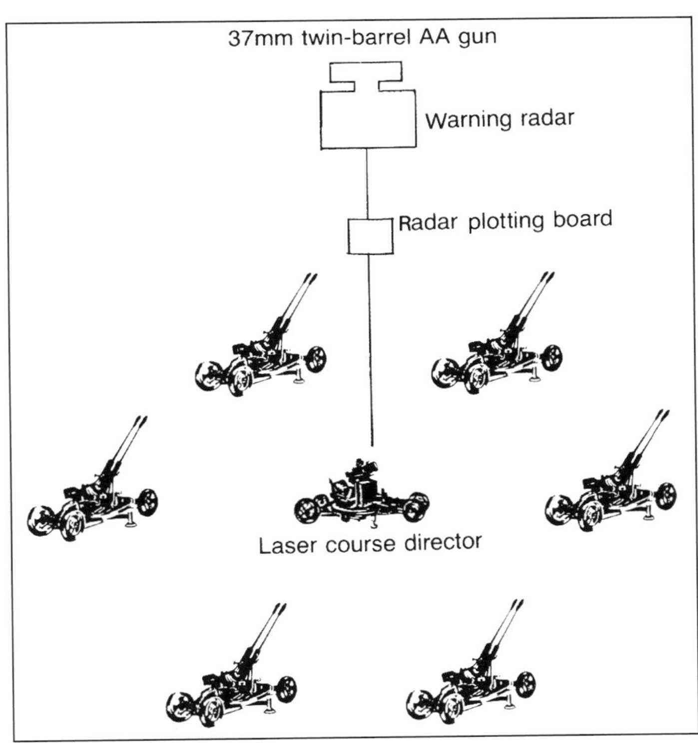

Key components of the NORINCO Type 74SD twin 37 mm air defence system 0509866

fire-control computer, automatic guns and missiles together into a combat system. The system allows the commander to direct each combat unit via automatic control-and-communication devices. Four types of Sky Shields have been reported, these are: Expanded Configuration 1 - to control three new double-barrel 37 mm air defence guns and one PL-9C missile launcher; Expanded-Configuration 2 - to control two new double-barrel 35 mm air defence guns and one PL-9C launcher; Extended Configuration 1 - controls six

Type 702 fire-control radar deployed with stabilisers lowered 0509490

NORINCO 37 mm Type 74 anti-aircraft gun in firing position 0509488

Type 801 director in position with outriggers extended 0509489

74-Type double-barrel 37 mm and three new double-barrel 37 mm air defence guns and one PL-9C missile launcher; and Extended Configuration 2 - controls six 74-Type double-barrel 37 mm and two double-barrel 35 mm air defence guns and one PL-9C missile launcher.

Also associated with the Type 74 twin 37 mm gun is the Type 702A fire-control radar. Both radar and gun have been offered for sale since January 2007.

The Type 74 is fitted with an automatic vector sight that allows it to fire at air or ground targets by direct aiming.

The system can engage aerial targets with a maximum speed of 300 m/s at 220-240 rounds per minute, per barrel.

37 mm Type 74SD LAAG

This is essentially the 37 mm Type 74 with the servo-system and electric firing device removed and an elevating and traversing mechanism and indicator dial being used in its place. It has an effective slant range of 3,000 m. This forms a

NORINCO 37 mm Type P793 in firing position with outriggers extended 0509865

key part of a complete weapon system which consists of six twin 37 mm Type 74SD towed anti-aircraft guns and a Type 800 laser course director system with additional target information coming from a warning radar via a radar plotting board.

A fire-control system designed by the Chinese North Automatic Control Technology Institute is the Sky Shield Short Range Optronic Fire-Control System for Air Defence. This system integrates all the factors like every battalion-scaled command posts, search radar, optical director, thermal imaging, laser range-finder, fire-control computer, automatic guns and missiles together into a combat system. The system allows the commander to direct each combat unit via automatic control-and-communication devices. Four types of Sky Shields have been reported, these are: Expanded Configuration 1 - to control three new double-barrel 37 mm air defence guns and one PL-9C missile launcher; Expanded-Configuration 2 - to control two new double-barrel 35 mm air defence guns and one PL-9C launcher; Extended Configuration 1 - controls six 74-Type double-barrel 37 mm and three new double-barrel 37 mm air defence guns and one PL-9C missile launcher; and Extended Configuration 2 - controls six 74-Type double-barrel 37 mm and two double-barrel 35 mm air defence guns and one PL-9C missile launcher.

37 mm Type P793 LAAG

The Type P793 is a twin-barrelled 37 mm anti-aircraft gun based on the Type 74 but with many new features. The twin guns themselves have a higher rate of fire and are available with two types of barrel and rate of fire. The Type A gun has a muzzle velocity of 880 m/s. The Type B has a muzzle velocity of 1,000 m/s. The rate of fire of a single P793 Type A barrel is 220 to 240 rds/min, while that of a Type B barrel is 270 to 300 rds/min. Both types of barrel continue to use Type 55 ammunition. The effective slant range for a Type A is 3,500 m while for the Type B it is 4,000 m.

The P793 carriage has its own generator powered by an engine developing 7 hp at 3,750 rpm. The generator powers the carriage electrical controls and also the electro-optical sight system. The sight is known as the JM 831 (which has also been referred to as the C335) and appears to be of the Italian Officine Galileo pattern having a ×5 magnification and a 12° field of view.

The P56 is stated to be a self-powered electro-optical aiming sight that takes three to four seconds to be ready for firing once a target has been detected. The sight can cope with target velocities of 60 to 350 m/s. Maximum lead angle is 25°. Target ranges are between 100 and 1,500 m, while ground target maximum range is 3,000 m. Maximum barrel angular acceleration for traverse is 95°/s^2, while that for elevation is 80°/sec^2. Powered elevation is from –5° to +87° while manual elevation is from 0° to +82.5°.

The P793 has 182 37 mm rounds stored on the carriage and has a crew of five or six. It can be towed at speeds of up to 60 km/h on roads or 25 km/h on dirt roads. The weight of the complete P793 is 3,100 kg.

37 mm Ammunition

The Type 74, Type 74 SD and the Type P793 all use fixed ammunition called the Type 55 which are listed in the specifications table labelled as 'Ammunition'.

All rounds have a temperature range of –40 to +50°C. The AP-T round will penetrate 40 mm of conventional steel armour at an angle of 30° and the AP/HE round will penetrate 25 mm of conventional steel armour at an angle of 30°.

NORINCO Type 390 integrated gun/missile air defence system with PL-9 SAM launcher on left and Type 702 fire-control radar on right 0509867

Pakistan 37 mm upgrade

Pakistan has around 700 Chinese-supplied twin 37 mm light anti-aircraft guns fitted with their original manual controls and optical sights.

There were three Western companies each supplied with a Chinese-built twin 37 mm system which were then upgraded and tested by the respective manufacturers before being shipped to Pakistan for extensive tracking and firing trials. These were completed in mid-1989.

Contraves Italiana fitted its Gunking sight, that is already in volume production, for a number of weapons including the Oerlikon Contraves twin 35 mm GDF series of towed anti-aircraft guns. Saab Tech Vetronics has fitted its Universal Tank and Anti-Aircraft System (UTAAS) fire-control system which is installed in a version of the new Swedish Army Combat Vehicle 90 while Officine Galileo has fitted its Vanth/MB sight which is a further development of the P75. Details of these sights are given in the Towed anti-aircraft gun sights section. In the end Pakistan adopted none of these upgrade packages and more recently Pakistan has developed its own upgrade package and details of this are given in a separate entry under Pakistan.

Fire-control systems

All the 37 mm towed systems can be coupled to a fire-control radar. Known as the Type 311A, this uses a four-wheel trailer van with an I/J-band search, acquisition and tracking radar antenna on a roof-mounted pedestal. The set operates on three pre-selected frequencies that can be manually switched without any adjustments. Minimum and maximum ranges are 500 m and 30,000 m respectively, against low radar cross-section fighter-sized targets with speeds of less than 550 m/s. An eight-tonne truck is needed to tow the four-tonne trailer and this carries the ancillary equipment and the power supply generator. The radar is also used with the 57 mm Type 59 towed systems. Further development of the Type 311 has led to the Type 311B and Type 311C models. The Type 311B introduces an integral IFF system, increased frequency coverage and a maximum detection range of 35,000 m using a new antenna design. The Type 311C goes one stage further and has frequency-agile radar with a maximum range of 40,000 m. The minimum range stays the same in both cases.

Type 80 weapon system

This system essentially consists of six twin 37 mm Type 74 light anti-aircraft guns coupled to a central distribution box, power supply units and a Type 702 fire-control radar. The system has been designed to engage air targets with a slant range of 3,500 m but can also be used to engage ground and water targets.

The trailer-mounted Type 702 fire-control system searches and tracks the targets, calculates the firing data and then passes this information on to the guns with the fire-control radar deciding when to open fire.

The system has four modes of operation: automatic, semi-automatic, manual alignment of indicator and manual aiming.

The Type 702 system comprises the roof-mounted fire-control radar, digital fire-control computer, TV monitor, main power supply unit and an auxiliary power supply unit.

The fire-control radar has a maximum target detection range and target tracking range of 40 km. Its range tracking error is 15 m with range tracking speed being a maximum of 600 m/s. The digital fire-control computer installed is an Intel 8086.

Type 80 air defence network

This system consists of three six-gun batteries of twin 37 mm Type 74 light anti-aircraft guns, a Type 801 director, a Type 702 fire-control radar and a Type 703 battalion Command, Control, Communication and Information (C3I) system.

Type 801 director

This comprises an optical sight, laser range-finder, laser data processor, single joystick tracker, a microcomputer and a power supply unit. The system is mounted on a four-wheeled trailer.

Ammunition

	HE-T	AP-T	HE	APHE
Length:				
(round)	381.88-384.97 mm	382.49-386.22 mm	386 mm	n/avail
Weight:				
(round)	1.417 kg	1.444 kg	1.417 kg	1.44 kg
(projectile)	0.732 kg	0.758 kg	0.732 kg	0.755 kg
Muzzle velocity:	866 m/s	880 m/s	866 m/s	868 m/s
Maximum chamber pressure:	274.6 MPa	284.4 MPa	274.6 MPa	284.4 MPa
Tracer duration:	6 s	6 s	nil	6 s
Fuze:	ML-1	n/app	ML-1	n/avail
Maximum slant range:	8,500 m	8,500 m	8,500 m	8,500 m
Maximum firing altitude:	6,700 m	6,700 m	6,700 m	5,000 m
Effective firing altitude:	3,000 m	3,000 m	3,000 m	1,500 m

Its optical sight has a magnification of ×10 and a 7° field of view with a viewing range of 15 km. The laser range-finder has a maximum range of 7,000 m and a ranging error of ±5 m. The system also incorporates an Intel 8086 microprocessor and a power generating system.

Type 390 integrated gun/missile air defence system
This system includes twin 37 mm light anti-aircraft guns, Type 702 fire-control systems, Type 801 director, command systems and the Ibis low-altitude search radar together with a ground-launched version of a PL-9 air-to-air missile. Details of the NORINCO PL-9 surface-to-air missile are given in the entry for this system in the Self-propelled surface-to-air missiles section. Production as required (since at least 1991). The system is in service with the PLA and in Pakistan as the DK-9.

North Korean 37 mm SPAAG
It is known that North Korea has developed a twin 37 mm self-propelled ant-aircraft gun system called the M1992. It has similar ballistic capabilities to the Chinese twin 37 mm towed weapons. Available details are given in the Self-propelled anti-aircraft gun section under Korea, North.

Specifications

	Type 65 LAAG	Type 74 LAAG
Dimensions and weights		
Length		
transport configuration:	6.36 m	6.205 m
Width		
transport configuration:	1.80 m	1.816 m
Height		
transport configuration:	2.25 m	2.28 m
Weight		
combat:	2,700 kg	2,835 kg
Mobility		
Carriage:	trailer mounted four wheeled	trailer mounted four wheeled
Speed		
towed:	60 km/h (est.)	35 km/h
Firepower		
Armament:	2 × 37 mm cannon	2 × 37 mm cannon
Main weapon		
length of barrel:	2.73 m (est.)	2.73 m (est.)
breech mechanism:	vertical sliding block	vertical sliding block
maximum range:	8,500 m	8,500 m
maximum range:	3,500 m (slant)	3,500 m (slant)
maximum vertical range:	6,700 m	6,700 m
feed mechanism:	5 round clip	5 round clip
operation:	recoil automatic (buffer and spring recuperator)	recoil automatic (buffer and spring recuperator)
Rate of fire		
cyclic:	360 rds/min (est.)	480 rds/min (est.)
Main armament traverse		
angle:	360°	360°
speed:	-	50°/s
Main armament elevation/ depression		
armament front:	+85°/-10° (manual drive)	+87°/-5° (manual drive)
speed:	-	30°/s

Status
Production as required.

Contractor
China North Industries Corporation (NORINCO).

57 mm Type 59

Type
Towed anti-aircraft gun

Development
The China North Industries Corporation (NORINCO) 57 mm Type 59 anti-aircraft gun is a close copy of the Soviet 57 mm S-60 and differs in few details from the original production. If necessary the Type 59 could be used in the surface role to engage land and water targets.

NORINCO 57 mm Type 59 anti-aircraft gun in travelling configuration 0509864

NORINCO 57 mm Type 59 anti-aircraft gun towed by Type 59 artillery tractor 0509483

NORINCO Type GW-03 anti-aircraft fire director in travelling configuration 0509484

Description
The 57 mm Type 59 fires two types of fixed ammunition:
- APC-T
- HE-T.

Ammunition is fed from the left in clips of four rounds.

The HE-T has a muzzle velocity of 1,012 m/s and is produced with a steel cartridge case (total weight 6.31 kg) or a copper cartridge case (total weight of 6.48 kg). This shell uses a DRD34 proximity fuze that has been developed by GNGC. The total length of the round is 538.48 mm, and has a range of 12,000 m with a maximum altitude of 8,800 m. The maximum effective range is however only 6,000 m with an average effective radius of the fuze 5 to 8 m.

Time into action of the Type 59 is stated to be one minute and time to prepare for travelling is two minutes. A twin-barrel 57 mm naval version, known as the Type 66, has also been developed and is in service with the People's Liberation Army (Navy) (PLA(N)). This uses a water-cooling system for the barrels and a chain-type hoist system for the ammunition.

All the towed 57 mm systems use the China North Industries Corporation (NORINCO) Type GW-03 anti-aircraft fire-control director with the LLP12 fire-control computer system and test unit. This four-wheel trailer-mounted system utilises a 3 m wide stereoscopic optical range finder with onboard computation facilities to control a complete battery of 57 mm Type 59 anti-aircraft guns firing against either airborne or surface targets. The GW-03 can engage targets between 780 and 31,600 m. It can be coupled to fire-control radar if required. The Type 80 twin 57 mm self-propelled anti-aircraft gun system, fully described in the Self-propelled anti-aircraft gun section, uses the same gun and ammunition as the 57 mm Type 59 anti-aircraft gun.

In late 2006 and early 2007, this gun was observed during a Chinese land-based air-defence exercise operating with a fire-control radar similar to the Type 702.

Variants

Details of the NORINCO Type 80 twin 57 mm self-propelled anti-aircraft gun on a full tracked chassis are given in a separate entry under the Self-propelled anti-aircraft gun systems section.

Specifications

	Type 59
Dimensions and weights	
Crew:	8 (est.)
Length	
transport configuration:	8.6 m
Width	
transport configuration:	2.07 m
Height	
transport configuration:	2.46 m
Weight	
combat:	4,780 kg
Mobility	
Carriage:	trailer mounted four wheeled
Speed	
towed:	35 km/h
Firepower	
Armament:	1 × 57 mm cannon
Main weapon	
length of barrel:	4.39 m
muzzle brake:	multibaffle
maximum range:	6,000 m (effective)
feed mechanism:	4 round clip
operation:	recoil automatic
Rate of fire	
cyclic:	120 rds/min (est.)
Main armament traverse	
angle:	360°
speed:	24°/s
Main armament elevation/depression	
armament front:	+87°/-5°
speed:	15°/s

Status

Production complete. In service with Bangladesh, Burma (Myanmar) (12), China, Guinea (12), Pakistan, Sudan and Thailand (24). As far as it is known there has been no recent (April 2013) production of this weapon in China.

Early in 1997, it was reported that most of the 57 mm Type 59 anti-aircraft guns used by Thailand were non-operational due to lack of spare parts and logistical support.

Contractor

China North Industries Corporation (NORINCO).
Proximity fuzes - GNGC.

Czech Republic

12.7 mm M53

Type

Towed anti-aircraft gun

Development

The quad 12.7 mm M53 anti-aircraft gun was developed in Czechoslovakia in the late 1950s and is a two-wheeled carriage fitted with four Soviet 12.7 mm M1938/46 DShKM Heavy Machine Guns (HMGs).

Description

The M53 Czech Republic anti-aircraft system comprises four ex-Soviet DShKM 12.7 mm HMGs in a 2 × 2 arrangement on a two-wheeled carriage, each fed with a 50-round drum magazine. In the firing position, the two rubber-tyre road wheels are removed and the weapon is supported on three outriggers.

The M53 fires an API projectile weighing 49.5 g, with a muzzle velocity of 840 m/s. This projectile can penetrate 20 mm of conventional steel armour, at an incidence of 0° at a range of 500 m. Each 12.7 mm machine gun is fed from a drum type magazine that holds 50 rounds of ammunition.

The M53 is no longer in front-line service with the Czech Republic and Slovakia. The Egyptian Army had a number of BTR-152 (6 × 6) armoured personnel carriers fitted with the Quad 12.7 mm M53 anti-aircraft machine gun system in the rear and having full 360° traverse. These are no longer in front-line service with Egypt, but some were encountered in Afghanistan. There is also a Czech twin-barrelled version mounted on a UAZ-469 (4 × 4) light vehicle, but its exact status is uncertain.

Quad 12.7 mm M53 anti-aircraft machine gun, captured by US forces, with stabilisers lowered and showing drums each holding 50 rounds (US Army)
0509769

The 12.7 mm M53 anti-aircraft machine gun has an effective slant range of 1,006 m. Effective altitude limit with an elevation of +45° is 671 m, while effective altitude limit with an elevation of +65° is 914 m.

Specifications

	12.7 mm M53
Dimensions and weights	
Crew:	6
Length	
transport configuration:	2.9 m
Width	
transport configuration:	1.57 m
Height	
transport configuration:	1.78 m
Track	
vehicle:	1.5 m
Weight	
standard:	2,830 kg
combat:	628 kg
Mobility	
Carriage:	trailer mounted two wheeled with outriggers
Tyres:	5.00 × 16
Firepower	
Armament:	4 × 12.7 mm (0.50) machine gun
Main weapon	
length of barrel:	0.967 m (1.588 m with muzzle brake)
maximum range:	6,500 m
maximum vertical range:	5,600 m
feed mechanism:	50 round drum
operation:	gas automatic
cartridge:	12.7 mm × 108 mm
Rate of fire	
cyclic:	2,400 rds/min (est.)
practical:	320 rds/min (est.)
Main armament traverse	
angle:	360°
Main armament elevation/depression	
armament front:	+90°/-7°

Variants

BTR-152 - The Egyptian Army is known to have fitted some of their BTRs with a quad 12.7 mm M53 that came direct from the Czech factory. It is understood (assessed) that this variant is no longer in front line use but held in reserve.

Status

Production of the system is complete, and it is no longer marketed. The system is believed to have seen service with the Cuban (Army, reserve), Egyptian (Army, reserve), Vietnamese (Army, reserve) and other countries in the Middle East and Southeast Asia. The M53 has also been observed in Afghanistan mounted in the rear of BTR-152 (6 × 6) APCs.

Contractor

Former Czech state factories.

30 mm M53

Type
Towed anti-aircraft gun

Development
The Czech Republic and Slovakia use the same 30 mm automatic anti-aircraft gun M53 in place of the Soviet 23 mm ZU-23-2. The system entered service in the late 1950s. The M53 weapon is heavier than the ZU-23-2 and has a slower rate of fire, but its effective anti-aircraft range is 3,000 m compared with the ZU-23-2's 2,500 m.

This system is also known in the Czech Republic as the M53/59 Praga V3S, where the Praga is a 6 × 6 wheeled vehicle that can often be seen with the M53 mounted on the rear.

Description
In action, the four wheels are raised off the ground and the carriage is supported on four jacks, one at the front, one at the rear and one each side on outriggers. The guns are gas operated, fully automatic and are fitted with quick-change 30 mm barrels. Ammunition is fed horizontally in clips of 10 rounds. The following types of fixed ammunition can be fired:

- API with the projectile weighing 0.45 kg and a muzzle velocity of 1,000 m/s, which will penetrate 55 mm of steel armour at an incidence of 0° at a range of 500 m.
- HEI with the projectile weighing 0.45 kg and a muzzle velocity of 1,000 m/s.

The twin 30 mm anti-aircraft gun M53 is a clear weather-only system with no provision for radar control. There is also a self-propelled model of the M53 called the M53/59 based on an armoured Praga V3S (6 × 6) truck which is described in the Self-propelled anti-aircraft gun section. The guns of the M53/59 have a higher rate-of-fire as the vertical magazines. Each hold 50 rounds of fixed ammunition in clips of 10 rounds.

The basic system has a secondary role as a ground support weapon against unarmoured or lightly armoured targets.

Variants
The Cuban Army has modified a number of BTR-60P (8 × 8) armoured personnel carriers to carry the M53 system. The Czech Republic has removed the twin 30 mm cannon and fitted a 73 mm SPG-9 recoilless gun on a few of its 30 mm automatic anti-aircraft gun M53 systems for training purposes.

Specifications

	30 mm M53
Dimensions and weights	
Crew:	4
Length	
transport configuration:	7.587 m
Width	
transport configuration:	1.758 m
Height	
transport configuration:	1.575 m
axis of fire:	0.86 m
Ground clearance	
overall:	0.3 m
Track	
vehicle:	1.575 m
Weight	
standard:	2,100 kg
combat:	1,750 kg
Mobility	
Carriage:	trailer mounted four wheeled with outriggers
Firepower	
Armament:	2 × 30 mm cannon
Main weapon	
length of barrel:	2.429 m
muzzle brake:	multibaffle
maximum range:	9,700 m
maximum vertical range:	6,300 m
feed mechanism:	10 round clip
cartridge:	30 mm × 120 mm
Rate of fire	
cyclic:	1,000 rds/min (est.)
practical:	200 rds/min
Main armament traverse	
angle:	360°
Main armament elevation/depression	
armament front:	+85°/-10°

30 mm automatic anti-aircraft guns M53 in travelling order 0509503

30 mm automatic anti-aircraft gun M53 defending 'Bar Lock' radar installation 0509504

Status
Production complete. No longer marketed.

Contractor
Czech state factories.

Egypt

ZU-23M

Type
Towed anti-aircraft gun

Development
The ZU-23M is essentially the ex-Soviet 23 mm ZU-23-2 produced under licence at Heliopolis, in Egypt by Abu Zaabal Engineering Industries (Factory 100). It differs only slightly from the original and fires the same 23 mm ammunition. It has also been referred to as the SH-23M. The original ZU-23-2 was developed in the late 1950s in order to engage low-flying targets at a range of 2.5 km as well as armoured vehicles at a range of 2 km. A further development of the basic weapon lead to the ZSU-23-4 Shilka.

Description
The ZU-23M has a five-man detachment, two of whom are on the weapon, one on either side of the mount and the other three off the carriage, with two of these acting as ammunition members. Each barrel is fed from a 50-round belt of ammunition in a steel magazine and a new magazine can be loaded in 10 seconds. The ammunition feeding system has been designed to prevent the firing of the last round and the moving parts stay to the rear, ready to receive the first round of the new belt so that firing can start immediately.

The ZU-23M twin 23 mm towed anti-aircraft gun can engage aircraft and helicopters at altitudes up to 1,500 m and ranges up to 2,500 m. Lightly armoured ground targets can be engaged out to a maximum range of 2,500 m.

The weapon is fitted with a sighting device to calculate the future position of targets up to a speed of 300 m/s. The sighting device is also provided with a direct-fire telescope to engage both stationary and moving ground targets.

The Maadi Co for Engineering Industries (also called Factory 54) manufactures the twin 23 mm cannon (2A14) used in the system. Ammunition for this weapon is also manufactured in Egypt. These are made in API and HEI both of which have a muzzle velocity of 970 m/s.

Variants
Ramadan 23

In 1987, Factory 100 and Oerlikon Contraves fitted two ZU-23Ms with the Oerlikon Contraves Gunking laser/computer sighting system, which is already in volume production for a number of other applications including the Oerlikon Contraves twin 35 mm GDF towed anti-aircraft gun system.

The Egyptians call the ZU-23M/Gunking combination the Ramadan and firing trials were completed in late 1987. As far as is known, the Egyptian Army has not yet procured this version of the ZU-23M.

Egyptian-built 23 mm ZU-23M in firing position (Christopher F Foss) 0509505

Ramadan 23 has just one man on the mount, with full power control from an onboard electric motor, powered by an off-carriage generator. One man can control two ZU-23Ms with not only greater response, but with a greatly enhanced kill probability. Full details of the Oerlikon Contraves Gunking sight are given in the Towed anti-aircraft gun sights section.

Self-propelled twin 23 mm system (Egypt/France)
This uses the basic 23 mm ZU-23-2 cannon and is covered in the Self-propelled anti-aircraft gun section under Egypt. This system uses the United Defence LP full-tracked M113A2 Series APC chassis. It is believed that between 40 and 50 of these systems have been completed. Production of this system is now complete.

Specifications

	ZU-23M
Dimensions and weights	
Crew:	5
Mobility	
Speed	
towed:	70 km/h
Firepower	
Operation	
stop to fire first round time:	20 s (est.)
out of action time:	35 s (est.)

Status
Production as required. In service with the Egyptian and Sudanese armed forces.

Contractor
Abu Zaabal Engineering Industries Company, Cairo, Egypt. This factory is affiliated with the Ministry of Military Production.

Finland

Finnish upgraded ZU-23-2

Type
Towed anti-aircraft gun

Development
The two Finnish companies, Instrumentointi Oy (now known as the Insta Group Oy) and Patria Engineering have joined forces to develop and market an upgrade package for the widely deployed Russian twin 23 mm ZU-23-2 light anti-aircraft gun, which is covered in detail in a separate entry. The system is known in Finland as the 23 ItK 95.

Following user trials with three prototype systems, volume production is understood to have begun in 2000 for the Finnish Armed Forces.

According to both companies, the upgrade package improves the effectiveness and hit probability of the weapon with the addition of a gyro-stabilised gun, an APU and laser range-finder.

Description
The complete ZU-23-2 upgrade package consists of:
- The Air Defence Gun Sight (ADGS)
- Thermal imaging system including monitor
- Coarse optical sight
- 24 V DC all electric laying system
- Resolvers

Upgraded twin 23 mm ZU-23-2 light anti-aircraft gun deployed in the firing position 0007659

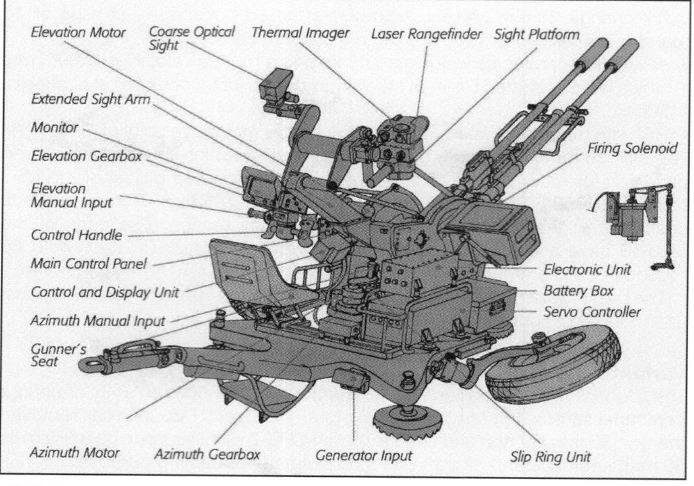

ZSU-23-2 light anti-aircraft gun system upgraded by Insta Group Oy and Patria Engineering and showing positions of main improvement 0100680

- Control handle
- Electric firing system
- Slip ring unit
- Buffer batteries
- Portable power unit
- Welded steel subsystems to allow new components to be installed on the gun and cables for electric integration of various components.

The ZU-23-2 is normally operated by a crew of two, seated one either side at the rear. The upgraded weapon is operated by one person, seated on the left side with the space on the right side being occupied by part of the upgrade package.

On the upgraded ZU-23-2 the original sight and old manual computing system have been replaced by the new ADGS.

Key parts of the ADGS are the sight platform, laser range finder, electronic unit as well as the control and display unit. By integrating a thermal imager with the ADGS full night operational capabilities have been received.

The ADGS provides accurate gun orders based upon precise target tracking and laser ranging combined with modern prediction techniques. Inaccurate estimates of the target range and speed are no longer required.

The ADGS allows the operator to track the target even when aim-off is being applied and updated so re-tracking or re-acquiring the target is unnecessary. In this way, sudden disturbances to the sight and the operator, which could cause a loss of aim at the crucial point of the engagement, are avoided.

All the operator needs to do is align the reticule of the monitor on the target by means of the control panel and press a certain button on it. As soon as the computer has calculated the aim-off, the barrels are automatically driven to the aim-off position without disturbing the aim.

The system indicates to the operator when the aim-off are accurately applied as well as the range to the target. Pressing a trigger on the control handle does firing of the gun.

The upgraded gun is powered by maintenance-free batteries, which are in a special box located on the forward right side of the mount. The batteries are charged via a slip ring unit with a portable generator. To facilitate unlimited rotation of the upper mount a slip ring unit is used.

If required, an external parallax computer can be connected to the upgraded ZU-23-2 to enable targets to be designated from an external source such as radar or an optical target designator. One computer system and radar could control up to four upgraded ZU-23-2 light anti-aircraft guns.

Insta Group Oy has also developed a complete range of simulators and target designation systems for the ZU-23-2 and other light anti-aircraft guns.

Status
In production, and in limited service with the Finnish Defence Force and the Finnish Navy from 2000.

Contractor
Insta Group Oy.
Patria Engineering Oy.

France

Cerbere 76T2

Type
Towed anti-aircraft gun

Development
The then Giat Industries Cerbere (now Nexter Systems) twin 20 mm automatic anti-aircraft gun is, essentially, the German Rheinmetall (now called Rheinmetall DeTec AG) twin 20 mm system, with the original Rh-202 cannon replaced by the French Giat Industries M693 (F2). A full description of the mount is given in the entry for the original German system in this section.

Description
The French designation for the Cerbere is the 76T2. It was adopted by the French Air Force for close-range protection of airfields with Thales France (formerly Thales Defence Systems) Crotale SAM batteries. A total of 299 systems were delivered to the French Air Force between 1980 and 1985.

In a more advanced version, the mounting can be integrated into a defence system and directed automatically at the target, either by surveillance radar or a director equipped with a DALDO target indicator helmet. As far as is known, users of the system have not yet adopted this version.

Specifications

	Cerbere
Dimensions and weights	
Crew:	3
Length	
transport configuration:	5.05 m
deployed:	4.02 m
Width	
transport configuration:	2.39 m
deployed:	2.06 m
Height	
transport configuration:	2.075 m
axis of fire:	0.765 m
deployed:	1.67 m
Weight	
standard:	2,150 kg
combat:	1,600 kg
Mobility	
Carriage:	trailer mounted two wheeled with outriggers
Firepower	
Armament:	2 × 20 mm cannon
Main weapon	
feed mechanism:	270 round belt (per barrel)
operation:	delayed blowback (with locked breech)
cartridge:	20 mm × 139 mm
Rate of fire	
cyclic:	1,800 rds/min (est.)
Main armament traverse	
angle:	360°
Main armament elevation/depression	
armament front:	+83°/-5° (manual)

Status
Production for the system is complete but can be resumed if further orders are placed. In service with the French and Somali Army. The system was first offered for export in 2007.

Contractor
Nexter Systems.

Nexter Systems' Cerbere 76T2 twin 20 mm automatic anti-aircraft gun in firing position 0509506

Tarasque 53T2

Type
Towed anti-aircraft gun

Development
The then Giat Industries (now Nexter Systems) Tarasque 20 mm automatic anti-aircraft gun, official designation 53T2, was selected as the standard weapon in its class for the French Army and first production weapons were delivered in 1982.

The system is armed with a single 20 mm French former Giat Industries M693 (F2) cannon which is also installed in the Cerbere twin 20 mm automatic anti-aircraft gun. Certain versions of the former Giat Industries AMX-30 MBT, the former Giat Industries AMX-10P IFV and the South African Ratel 20 IFV. In South Africa, LIW, a division of Denel, has also manufactured the 20 mm M693 cannon under licence.

The former Giat Industries Tarasque 53T2 20 mm automatic anti-aircraft gun is a clear weather only system and the French Army does not use this in conjunction with any all-weather fire-control system.

Nexter Systems Tarasque 53T2 20 mm automatic anti-aircraft gun in firing position (Pierre Touzin) 0509507

Renault Trucks Defence TRM 2000 (4 × 4) truck which is used by the French Army to tow the former Giat Industries Tarasque 53T2 20 mm LAAG as well as carrying its crew and ammunition 0509508

Renault Trucks Defence TRM 2000 (4 × 4) truck with former Giat Industries 53T2 automatic anti-aircraft gun in rear (Pierre Touzin) 0509698

Description

Although designed primarily for anti-aircraft use, Tarasque can also be used against personnel and light armoured vehicles. Its low weight makes it easily transportable across rough country and it can also be carried slung under a helicopter.

A heat motor driven rotary pump charges an oil receiver supplying the traversing and elevation hydraulic motors and the hydraulic re-cocking mechanism. If the motor is unserviceable a hand-operated pump can charge the receiver. Maximum powered traverse speed is 40°/s. Hand wheels are also fitted for emergency use; one turn of one hand wheel gives 10° of traverse and one turn of the other gives 6.5° in elevation.

The gunner can select either single shots or fully automatic, with the hydraulic firing mechanism actuated by the gunner depressing his right foot. Safety controls, ammunition selector and hydraulic reloading controls are to the right of the gunner's position. On the forward part of the mount on the left side are two discs, adjustable to within 25 mils, to prevent the gun being fired in predetermined zones.

The M348 day sight includes a reticule sight for anti-aircraft and a 5.2 telescopic sight with a magnification of ×1 for use against ground targets.

The 20 mm M693 (F2) cannon has dual feeds and the weapon has 90 rounds of HEI/HEI-T ammunition and 40 rounds of APDS ammunition. Standard HSS 820 ammunition can be fired, including:

- APDS (French designation OPT-SOC) with a muzzle velocity of 1,300 m/s;
- HEI (French designation OEI) with a muzzle velocity of 1,050 m/s;
- HEI-T (French designation OEIT) with a muzzle velocity of 1,050 m/s;
- Practice Tracer (French designation OXT);
- Practice Inert (French designation OX).

The APDS projectile will penetrate 20 mm of conventional steel armour at an incidence of 0° at a range of 1,000 m.

The Tarasque is carried on a two-wheeled carriage that can be towed by a (4 × 4) light vehicle and takes only 15 seconds to bring into action. A normal crew would consist of one on the gun and two ammunition members.

The system can also be mounted in the rear of trucks such as the Renault TRM 2000 (4 × 4) which is the standard vehicle in its class in the French Army.

Specifications

	53T2
Dimensions and weights	
Crew:	3
Width	
transport configuration:	1.9 m
Track	
vehicle:	1.72 m
Weight	
standard:	840 kg
combat:	650 kg
Mobility	
Carriage:	trailer mounted two wheeled
Firepower	
Armament:	1 × 20 mm
Armament:	1 (surface action role)
Main weapon	
length of barrel:	2.065 m
operation:	delayed blowback
cartridge:	20 mm × 139 mm
Rate of fire	
cyclic:	900 rds/min
Main armament traverse	
angle:	360°
speed:	80°/s
Main armament elevation/depression	
armament front:	+83°/-8°
speed:	40°/s

Status

Production of the system is complete, but can be resumed if additional orders are placed.

Contractor

Nexter Systems.

Germany

20 mm AAG

Type

Towed anti-aircraft gun

Development

Rheinmetall (now Rheinmetall Air Defence) developed this anti-aircraft gun system under contract to the German Department for Ordnance Technology and Procurement. By 2000, over 1,800-twin 20 mm systems had been delivered to German and other forces. As far as it is known there has been no production of this complete weapon in recent years.

Rheinmetall 20 mm AA twin gun air defence system deployed in firing position 0509770

Rheinmetall 20 mm AA twin gun air defence system in travelling order 0509515

The main advantages of the system are:

- A high rate of fire;
- Accurate laying of the guns by the computerised optical fire-control system;
- Large ammunition supply;
- Rapid elevation and traverse due to hydraulic servo-drive;
- No requirement for an external power source;
- Suitability for operations under a wide range of climates and a very short training time for operators.

Description

The 20 mm AA twin gun air defence system consists of the following main components; two Rheinmetall 20 mm MK 20 Rh 202 cannon, ammunition supply, fire-control equipment, laying mechanism, cradle, upper carriage, lower carriage and the two-wheeled trailer.

Table 1

Ammunition nature	APDS-T	API-T	HEI	HEI-T	TP/TP-T	Break Up
German designation(s):	DM63	DM43A1	DM101A1	DM51A2 and DM81/A1	DM48, DM48A1, DM98, DM98A1/A2/A3 and DM88A1	DM78A1 and DM78A2
Weight of projectile:	111 g	111 g	120 g	120 g	120 g	120 g
Muzzle velocity:	1,150 m/s	1,100 m/s	1,055 m/s	1,055 m/s	1,045 m/s	1,095 m/s

Rheinmetall 20 mm AA twin gun air defence system in firing position 0509699

The 20 mm Rheinmetall MK 20 Rh 202 is a gas-operated, fully-automatic cannon firing 20 × 139 mm ammunition. The cannon has also been installed on the following:
- The Marder 1 IFV;
- Luchs (8 × 8) reconnaissance vehicle;
- FIAT/Otobreda Type 6616 armoured car;
- Wiesel 1 light armoured vehicle;
- Rheinmetall Landsysteme (previously Henschel Wehrtechnik) Condor (4 × 4) armoured personnel carrier;
- VCTP infantry combat vehicle (Argentina);
- The Norwegian 20 mm automatic anti-aircraft gun FK 20-2 (which is described in this section);
- German Navy ships, when used in the naval role it is designated as the S20 Naval Gun Mount.

The weapon has a low recoil force and can be stripped into the main assemblies without any tools. Using the belt feeder Type 3 allows the cartridges can be fed selectively from two sides, the left gun from the left and above and the right gun from the right and above.

The ammunition feed comprises two ammunition boxes, two flexible belt clips and two belt-centring mechanisms. The two ammunition boxes are fixed to the upper carriage, one right and one left of the twin guns. Each box contains 270 rounds of ammunition and there are another 10 rounds in the feed mechanism. Flexible cartridge belt guide channels connect the ammunition boxes to the belt-centring devices, which centre the cartridges in the links of disintegrating belts and introduce belted ammunition to the feed mechanism.

The laying mechanism is mounted over the rear bearings of the upper carriage and carries the laying system. Maximum powered traverse speed is 80 degrees per second and maximum powered elevation speed is 48 degrees per second.

Weapons are fired by a foot-operated mechanism fitted with a safety device. Interlocks that can be overridden stop the gun before the last round is fired. A selector lever enables the gunner to choose single shots or sustained fire, with either electric or mechanical firing from one or both barrels. The types of fixed ammunition that can be fired is listed in the table at the top of the page.

The Italian Galileo Avionica P56 computing sight (can be replaced by the P75 fire-control system with a laser range-finder) has the following main components:
- Monocular optical sight with a magnification of ×5 and a 12° field-of-view with a swivelling objective prism for laying the gun against air and ground targets;
- Electronic analogue computer for calculating lead values;
- Joystick with two degrees of freedom for the speed control of the line of sight;
- Input panel for target speed and crossing point distance of air targets and target distance of ground targets.

The cradle is made of a low-density high-strength alloy and its trunnions engage in the trunnion bearings of the upper carriage and rotate the cradle in elevation. During firing the cradle acts as the recoil bed of the guns. At the forward end of the cradle are two hinged covers to enable the weapons to be removed and at the top it is closed by two lockable hinged frames, which also hold the belt, feed mechanisms and the ejectors. The trigger mechanism and the cocking device are also housed on the cradle.

The upper carriage is also made of a low-density high-strength alloy with the lying mechanism fitted to the sidewalls. It also carries the gunner's seat and at the top of the upper carriage are the trunnion bearings in which the cradle fits. The centre of gravity of the cradle system, when fitted with the cannon is off centre, with static weight equilibrium achieved by two balances. Two multi-stage elevating gears are flanged to the cheeks of the upper carriage. On the left side of the upper carriage is a firing pedal and the hand lever for locking the guns is on the right.

The lower carriage has three outriggers, two of which are adjustable for levelling. On the lower carriage is a brake lever for locking the upper carriage and at the end of each outrigger is a socket for securing the mounting on the trailer.

The two-wheeled trailer is used to transport the mounting with the lower carriage secured to the trailer by a swivel ring at the front and two lockable devices at the rear. The towing bracket of the trailer is adjustable to fit the height of the coupling of the vehicle and couplings and wiring for 12 or 24 V lighting are standard. When the mounting is put on or taken off the trailer, the wheels are secured by wedges, this being the only task, which requires a crew of three.

Optional equipment includes:
- Taboo facility which can be programmed by the gunner to prevent the guns firing at specific objects;
- S11 laying exercise kit for the simulation of programmed flight paths with the sight for training purposes;
- Video training system and radio equipment to link the gun with the battery commander or command post.

Variants

The French Air Force uses an identical mount fitted with French Giat Industries 20 mm M693 (F2) cannon under the designation 76T2, or Cerbere. A description of this model is given in the French section.

Specifications

	MK 20 Rh 202 AAG
Dimensions and weights	
Crew:	4
Length	
overall:	2.61 m
transport configuration:	5.035 m
deployed:	4.05 m
Width	
transport configuration:	2.36 m
deployed:	2.3 m
Height	
transport configuration:	2.075 m
axis of fire:	0.735 m
deployed:	1.67 m
Weight	
standard:	2,160 kg
combat:	1,640 kg
Mobility	
Carriage:	trailer mounted two wheeled with outriggers
Firepower	
Armament:	2 × 20 mm MK 20 Rh 202 cannon
Main weapon	
length of barrel:	1.84 m (excluding muzzle brake)
maximum range:	2,500 m
maximum vertical range:	1,600 m
operation:	gas automatic
cartridge:	20 mm × 139 mm
Rate of fire	
cyclic:	2,000 rds/min (est.)
Main armament traverse	
angle:	360°
Main armament elevation/depression	
armament front:	+83.5°/-5.5° (manual)

Status

Production as required.

Contractor

Rheinmetall Waffe Munition GmbH.

Greece

Artemis

Type
Towed anti-aircraft gun

Development
The Artemis twin 30 mm air defence system was designed in the mid-1980's by the then Hellenic Arms Industry SA (now Hellenic Defence Systems) to meet the operational requirements of the Greek Armed Forces for low-level air defence. This Greek system is related to the German Mauser/Rheinmetall Landsysteme (previously KUKA) Mk 30 Model F air defence system Arrow.

The system is mounted on a four wheeled towed carriage with a total of 500 rounds ready to use.

Both the Twin Cannon Carriage (TCC) and the associated family of 30 mm ammunition have been proven for field use.

The targeting capability of the system is dependent on the requirements of the customer and can be a choice from various 3-D fire-control systems on board the unit or can be remotely controlled by a distant fire-control system. In the standard production system Artemis is supplied with a gyroscopic 2-D tracking device for local operation.

In the course of development of the Artemis system, Hellenic Arms Industry has collaborated with various manufacturers of fire-control systems so proving the TCC's integration ability and performance within a low and very low-level air defence system.

For the Greek Armed Forces, Bofors and Siemens jointly developed the fire-control system and battery coordination posts with the acquisition radar.

A typical battery would have consisted of one battery co-ordination post, two trailer-mounted fire-control systems and eight twin 30 mm guns.

For a number of reasons this configuration has never been fully operational with the Greek Armed Forces and in 1996, a successful trial took place in Crete in which the Oerlikon Contraves Skyguard computerised fire-control system was used to control two Artemis twin 30 mm light anti-aircraft guns.

The Greek Air Force already operates the Skyguard computerised fire-control system with its Sparrow surface-to-air missile systems, covered in a separate entry.

Description
The Artemis 30 fire unit is a twin 30 mm cannon system towed on a twin-axle split-type carriage. The axle nearest the towing arm carries the generator which powers the cannon system in action.

When deployed, this axle/generator assembly is removed and placed some distance away from the mount. The deployed mount is lowered and levelled by means of three hydraulically operated pads, two on outrigger arms, while the rear axle is power-retracted upwards.

The weapons are mounted on each side of a horizontal drum assembly, which elevates the cannon. This assembly is placed on a central support on a turntable effecting the weapon traverse movement. The circular ammunition hoppers holding 500 linked rounds (250 for each cannon) are also on the central support. Each cannon receiver has a protective housing which covers the cannon to the base of the barrel. Each of these boxes can be opened to reveal the mechanism to clear jams or for routine servicing. By removing the gun barrel, the receivers can be swung outwards for more involved repairs and maintenance. The linked ammunition is fed from the hoppers upward through the central drum and into the feeders from the inside. Spent links and cases are ejected through slots in the cannon's housing.

The weapons are the Mauser 30 mm Mk 30 Model F cannons, with EBO-produced cold-forged 30 mm barrels with a constant rifling pitch. The twin-cannon upper mount assembly has an unlimited 360° arc in traverse, while the elevation arc is from –5° to +85°.

The mount has three distinct modes of operation:

1. operation via a remote fire-control system;

2. operation by a gunner seated directly behind the central mount support. In this mode the gunner is supplied with all necessary controls including a periscope for ground targets and a gyroscopic angle predicting sight for air target engagement;

3. emergency operation by the gunner (no power supplied to the mount).

Weapon aiming is accomplished via hand wheels, firing via a foot trigger. The weapons use the well-known 30 × 173 mm family of ammunition including:

- Armour-Piercing Discarding Sabot (APDS)
- Armour-Piercing Discarding Sabot - Tracer (APDS-T)
- High-Explosive Incendiary Self Destruct (HEI-SD)
- High-Explosive Incendiary Self Destruct - Tracer (HEI-SD-T)
- Semi-Armour-Piercing High-Explosive Incendiary (SAPHEI)
- Semi-Armour-Piercing High-Explosive Incendiary - Tracer (SAPHEI-T)
- Target Practice (TP)
- Target Practice - Tracer (TP-T).

The basic gun sight fitted to the Artemis system, which is also referred to as the A-30 twin cannon carriage, is a day sighting system. If required, a Thales Optronics lightweight infra-red camera could be fitted for enhanced operational effectiveness.

Variants
An enhanced hybrid version of the TCC with four Stinger fire-and-forget missiles was revealed in late 1996 with trials yet to take place as of late 2000. This is a collaborative programme between Hellenic Arms Industry and LFK of Germany. The Stinger SAM is already in service with the Greek armed forces. The hybrid system performance depends on the customer's choice for an onboard optro-electronic fire-control system. The inclusion of an infra-red

Artemis 30 in travelling configuration 0509516

Artemis 30 deployed in firing position 0509517

Oerlikon Contraves Skyguard fire-control system controlling Artemis 30 anti-aircraft gun system 0509879

Artemis 30 anti-aircraft gun system fitted with two pods each containing two Stinger SAMs 0509880

camera on such a fire-control system and its coupling with external search radar would allow for maximum operational use of the hybrid system. The hybrid system has full remote-control capability via the remote fire-control system.

In a typical target engagement, the Stinger SAMs would be used to engage the targets at longer range with the 30 mm cannon being used to engage close in targets.

Although the prototype uses Stinger SAMs it could also be used with other fire-and-forget SAMs such as the Russian SA-16 and SA-18.

A naval version of this system is currently under study.

A version of this system covering its integration onto the chassis of the locally manufactured Austrian Steyr-Daimler-Puch 4K 7FA full-tracked armoured personnel carrier has been proposed.

Specifications

	Artemis
Dimensions and weights	
Length	
transport configuration:	7.95 m
Width	
transport configuration:	2.375 m
Height	
transport configuration:	2.25 m
Ground clearance	
overall:	0.26 m
Track	
vehicle:	1.8 m
Wheelbase	4.35 m
Weight	
standard:	6,900 kg
combat:	5,900 kg
Mobility	
Carriage:	trailer mounted four wheeled (double wheeled cruciform)
Speed	
towed:	80 km/h
Firepower	
Armament:	2 × 30 mm cannon
Ammunition	
main total:	500 (ready to use)
Main weapon	
length of barrel:	2.59 m (with muzzle brake)
maximum range:	8,400 m
operation:	gas automatic
cartridge:	30 mm × 173 mm
Rate of fire	
cyclic:	1,600 rds/min
Main armament traverse	
angle:	360°
speed:	90°/s
acceleration:	185 °/s²
Main armament elevation/depression	
armament front:	+85°/-5°
speed:	70°/s
acceleration:	140 °/s²

Status

Production complete. In service with all three branches of the Greek Armed Forces. No known exports.

Contractor

Hellenic Defence Systems SA.

International

ZU-23-2 Upgraded

Type

Towed anti-aircraft gun

Development

The original ZU-23-2 was developed in the late 1950s in order to engage low-flying targets at a range of 2.5 km as well as armoured vehicles at a range of 2 km. A further development of the basic weapon lead to the ZSU-23-4 Shilka. In 2000, the Arsenal company of Bulgaria, which had been producing the Russian twin 23 mm ZU-23-2 light anti-aircraft gun for many years (see Bulgaria), announced that they had developed an enhanced version of this weapon which was being marketed to Greece.

It is believed that Greece has 506 of these twin 23 mm ZU-23-2 light anti-aircraft guns in service, which were supplied by Germany from former East German Army stocks.

This upgrade is being promoted to the Greek Army in association with EBO - Hellenic Arms Industry, and ARSCO - a consortium of Bulgaria Arsenal Company and Scorpion of Greece.

The first prototype of this upgraded twin 23 mm ZU-23-2 light anti-aircraft gun was completed and commenced its trials in Bulgaria late in 2000.

Description

The upgrades carried out have been designed to address the main drawback of the twin 23 mm ZU-23-2 light anti-aircraft gun. These include poor range, manual elevation and traverse and its limited adverse weather capabilities.

The upgrade includes:

- the installation of a new electro-optical sighting system to enable targets to be engaged in poor weather conditions and at night, and
- an increase in detection range from 2,000 to 6,000 m.

The original twin 23 mm ZU-23-2 light anti-aircraft has manual elevation and traverse controls while the upgraded system has a joystick controlled electro-hydraulic laying mechanism.

The upgraded system therefore has a faster response time and is also claimed to have a tighter shot dispersion pattern.

According to the consortium, the upgraded system is now capable of attacking aerial targets up to a maximum speed of 250 m/s, this is claimed to be a 500 per cent improvement over the original system.

The consortium marketing this twin 23 mm ZU-23-2 upgrade package claim the comparative figures for the system set out in the specifications:

Specifications

Model	Standard ZU-23-2	Upgraded ZU-23-2
Aiming speed:		
(traverse)	30, 60°/s	90°/s
(elevation)	40°/s	90°/s
Range of sight system operation:	2,000 m	6,000 m
Target speed:	0 to 300 m/s	0 to 400 m/s
Speed of target that can be hit:	0 to 50 m/s	0 to 200 m/s
Fire effectiveness with target speed of 200 m/s:	0.023	0.40
Night operation:	not possible	possible
Crew:	5	4

Some sources have indicated that the Greek Army may elect to go for a phased upgrade programme for cost reasons. The first phase would add powered controls and the second phase the enhanced sighting system.

Also being marketed is a command post, which could control a number of upgraded twin 23 mm ZU-23-2 light anti-aircraft guns.

Status

Prototype trials started in 2000, further information on this has not been made available to date.

Contractor

Arsenal Company, (see Development).

Iran

23 mm LAAG

Type

Towed anti-aircraft gun

Development

For some years, the Armament Industries Division of the Aerospace Industries Organisation (AIO) in Iran and the SANAM Industry Group has been building a local version of the Russian twin 23 mm automatic anti-aircraft gun ZU-23-2, which is covered in detail later in this section.

Description

As far as is known, the Iranian-produced ZU-23-2 is an identical copy of the Russian-designed weapon, brought up-to-date. AIO has provided a brief specification on this system.

The weapon system relies on its crew for alignment to its target, but can be layed-on by radar. The system is relatively effective against slow and low-flying targets, but generally ineffective against fast moving targets.

The Ammunition Industries Division of the AIO manufactures a range of 23 mm ammunition including HEI-T and API-T natures for the ZU-23-2 LAAG. This ammunition is identical to the OFZT and BZT natures manufactured in the Russian Federation.

The AIO also manufactures a 12.7 mm tripod-mounted anti-aircraft gun called the Dooshka which is similar to the Chinese NORINCO 12.7 mm Type 77 (see China) as well as the German Rheinmetall 7.62 mm MG3 machine gun. The latter is widely used in the ground and low-level air defence roles.

Variants

Mesbah-1 a towed anti-aircraft system with eight 23 mm cannons (4 × ZU-23-2).

Iranian-built ZU-23-2 twin 23 mm light anti-aircraft gun deployed in the firing position
0007661

Specifications

	ZU-23-2
Dimensions and weights	
Weight	
standard:	950 kg
Mobility	
Firepower	
Armament:	2 × 23 mm cannon
Main weapon	
length of barrel:	1.88 m (excluding flash suppressers)
maximum range:	2,500 m (slant)
maximum range:	2,000 m (horizontal)
maximum vertical range:	1,500 m
cartridge:	23 mm × 153 mm B
Rate of fire	
cyclic:	1,000 rds/min (est.)

Status
In production and in service with various branches of the Iranian Armed Forces. As far as it is known, Iran has not exported any of these ZU-23-2 light anti-aircraft guns however, it is possible that some may have found there way into central Africa.

A naval single barrel variant is also known to exist. This consists of a single barrel anti-aircraft machine gun/standing type and is a suitable weapon against targets at a distance of 2,500 m within an altitude of 1,500 m. This type of gun has been designed for mounting on light fast patrol boats and in coastguard deployment.

Contractor
Defence Industries Organization (DIO).
SANAM Industry Group manufacturers.

Saeer

Type
Towed anti-aircraft gun

Development
Iran has allegedly developed and produced its own 100 mm medium-range anti-aircraft gun for use with the Islamic Revolutionary Guards Corps (IRGC). First identified in late November 2011 when it was reported by Iranian News that a battery of the new system had been delivered to IRGC personnel.

It is just possible (some may say more than probable) that the Saeer gun is an upgrade to the old Russian 100 mm air defence weapon KS-19 or the Chinese equivalent Type-59. What is known is that Iran has a considerable stockpile of these older weapons and may have modified them with automatic positioning and engagement radar.

Description
The Saeer system is described by Iranian press as a medium-range gun that operates with high precision. Reported to be fully automatic without crew intervention, the system is assisted by radar and optical devices for interception of targets. The gun reportedly fires at a rate of 12 to 15 shells per minute.

Variants
None known.

Specifications
Saeer

Type:	100 mm towed anti-aircraft gun
Rounds per min:	12-15 shells

Status
Identified in late November 2011 when first issued to the Islamic Revolution Guards Corps (IRGC).

Contractor
Iranian Arms Industry.

Samavat

Type
Towed anti-aircraft gun

Development
Due to the imposed embargo on armaments, Iran has indigenously developed (some reports suggest copied) the Oerlikon 35 mm twin anti-aircraft gun. Very little information is available on this system but from photographic evidence it appears to be a copy of the original Oerlikon (now Rheinmetall Air Defence) system.

Description
The 35 mm rapid-fire cannon is a short-range defence system capable of targeting various aircraft and missiles. Iran originally purchased the GDF-001 gun with its accompanying Super Fledermaus radar for the defence of high value areas such as airfields etc. The Samavat is equipped with an optical fire-controlled system and a new advanced locally produced radar system that can target aircraft, Unmanned Aerial Vehicles (UAVs), helicopters and cruise missiles.

The twin cannons are mounted on a twin axle, four-wheeled towed chassis with what appears to mount an automatic shell feeder on the left side of the system. The chassis is fitted with at least three retractable stabilisers.

Variants
None known.

Specifications
Samavat

Type:	twin towed anti-aircraft gun
Maximum effective range:	4,000 m
Maximum effective altitude:	3,000 m
Rate of fire:	100 rounds per sec (reported)
Elevation:	–5° to +92°
Traverse:	360°
Crew:	5

Status
Development complete with series production having commenced in September 2008.

Contractor
Aeronautics Industries Organization (AIO), Iran.

Italy

40L70

Type
Towed anti-aircraft gun

Development
Oto Melara produced, as a private venture, a field-mounting version of its well-established twin 40 mm naval anti-aircraft gun system for use against low-flying aircraft and missile targets. The gun is fully automatic and no personnel are required on the mounting when it is in action. It can be used in conjunction with a wide range of fire-control systems such as the Thales Nederland BV Flycatcher.

The combination of two Oto Melara Twin 40L70s and a Flycatcher is marketed as a low-level air defence system known as Guardian. Venezuela acquired a total of 36 mounts in the early 1980s as part of a Guardian system, which are operated by the army.

Typical roles of the Guardian air defence system, according to Oto Melara, include the protection of high-value targets in the rear area including airfields, missile sites, ammunition storage areas, headquarters and high-value civilian targets of strategic importance such as oilfields.

Oto Melara Twin 40L70 Field Mounting in travelling configuration with outriggers and stabilisers retracted 0509519

Oto Melara Twin 40L70 Field Mounting deployed in the firing position and firing twin 40 mm L/70 guns 0509520

During trials at the Italian School of Anti-Aircraft Artillery, inert 40 mm rounds were fired against a towed sleeve target. A total of 218 rounds were fired of which 78 per cent (171 rounds) came within 4 m of the target. A further 12.9 per cent (28 rounds) came within 8 m. Using proximity fuzed 40 mm ammunition, the sleeve target was shot down in all cases with the first rounds of the burst.

There has been no recent production of this system. Marketing is now being concentrated on the Oto Melara Twin Fast Forty Field Mounting, which is covered in detail in the previous entry.

The Flycatcher systems used by Venezuela are being upgraded with new ECCM kits as well as the supply of a number of Reporter surveillance radars and Mirador optronic fire-control systems. First deliveries were made in mid-1999. Details of the Mirador are given in the Anti-aircraft control systems section later in this volume.

Description

The guns used with the Oto Melara Twin 40L70 are two Bofors Defence 40 mm L/70 guns, joined together to form a twin elevating mass by a specially designed twin cradle. The barrels are set 300 mm apart to reduce the influence of recoil forces of the elevating and training mass and to reduce weight.

Rounds are fed into the magazine from four-round clips. The magazine holds a total of 444 or 736 rounds of ammunition dependant on type and is housed on the traversing platform and consists of four horizontal layers that move the rounds towards a vertical ammunition elevator.

The magazine is divided into two independent sections, each of which delivers ammunition to the corresponding gun to continue fire even when either gun is out of operation. The magazine trains with the mounting and the supply system is arranged so that if one gun is out of action the other can continue to fire. The forward section of the magazine supplies the left-hand gun and the rear section supplies the right-hand gun.

A 400 V-60 Hz motor which supplies either fast or slow drive, drives the system. The slow drive operates the magazine conveyors and the scuttle transferring the rounds to the lower chain hoist. The fast drive operates at speeds in excess of 300 rpm to drive the ammunition chain hoists and the scuttle at the top of the hoist. From this top scuttle the fast drive also feeds the rounds into fan-shaped shifters which move the rounds through 90° into the gun feeders. Empty cases are ejected down a chute to the front of the gun. The chain hoists take four rounds from the magazine every 0.7 seconds. A series of brakes and slipping clutches detect any misfeeds. Rounds are fed into the lower magazine manually via the side hatches and, when fully loaded, each mounting can hold 444 rounds.

Ammunition details are given in the entry for the Bofors Defence 40 mm L/70 gun under Sweden.

Specifications

	40L70
Dimensions and weights	
Length	
transport configuration:	8.2 m
Width	
transport configuration:	3.155 m
Height	
transport configuration:	3.47 m (est.)
Ground clearance	
overall:	0.3 m (est.)
Track	
vehicle:	2.5 m
Wheelbase	5 m
Weight	
standard:	10,400 kg
Mobility	
Carriage:	trailer mounted four wheeled with outriggers
Firepower	
Armament:	2 × 40 mm cannon
Ammunition	
main total:	736
Main weapon	
length of barrel:	2.8 m
calibre length:	70 calibres
breech mechanism:	vertical sliding block
maximum range:	12,500 m
maximum vertical range:	8,700 m
feed mechanism:	444 round clip
cartridge:	40 mm × 365 mm R
Rate of fire	
cyclic:	600 rds/min (est.)
Main armament traverse	
angle:	360°
speed:	100°/s
acceleration:	130 °/s²
Main armament elevation/depression	
armament front:	+85°/-8°
speed:	70°/s
acceleration:	150 °/s²

Status

Production as required. In service with Venezuelan Army who ordered 36, however this figure cannot be verified.

Contractor

Oto Melara SpA.

40 mm AAG

Type

Towed anti-aircraft gun

Development

Oto Melara (previously Breda) manufactured the Bofors Defence 40 mm L/70 anti-aircraft gun under license, as well as a number of naval mounts of both its own and Bofors Defence design. The first Oto Melara 40 mm L/70 guns were produced for the Italian Army in 1969.

Description

Oto Melara has also developed an Automatic Feeding Device (AFD), which can be fitted to single versions of the 40 mm L/70 anti-aircraft gun, as well as naval weapons of this calibre. The AFD can be provided by Oto Melara in kit form, installed on new production guns or fitted when the guns are returned to Oto Melara for overhaul.

The conversion consists of substituting parts of the elevating mass to increase the cyclic rate of fire from 240 to 300 rds/min - where this has not already been done - and installing the ammunition feeding complex on the platform. The performance of the gun is not affected by the modification and loading and firing rates remain constant through all the elevation range.

Main advantages of the system are the higher rates of fire and the reduced manning requirements as only one person is required on the mount when the optical fire-control system is being used.

Loading is simple and quick and only one loader is required. The magazine, which comprises three layers, rests on the traversing platform and four-round clips of 40 mm ammunition are fed into the magazines via prepared ramps. The magazine holds 144 rounds of ammunition.

When firing, the rounds are automatically fed along the layer and taken up in threes by the elevation chain to a fan-shaped shifter at trunnion axis level. This, through differentials, adjusts to barrel elevation and conveys the rounds to the feeder. The power to carry out this operation is provided by an electric motor. Oto Melara has also developed additional improvements for this system, such as a new digital servo system using buffer batteries for emergency silent operations.

Oto Melara 40 mm L/70 anti-aircraft gun in travelling order with automatic feeding device 0509521

Oto Melara-built 40 mm L/70 anti-aircraft gun in firing position with automatic feeding device on rear part of traversing platform 0509522

The Oto Melara 40 mm L/70 anti-aircraft gun can fire all natures of 40 mm ammunition, with the exception of the latest Bofors Defence developed 3P round.

Specifications

	40 L/70
Dimensions and weights	
Crew:	1
Length	
transport configuration:	7.28 m
Width	
transport configuration:	2.289 m
Height	
transport configuration:	2.655 m
Wheelbase	4.025 m
Weight	
standard:	4,800 kg
Mobility	
Firepower	
Rate of fire	
cyclic:	300 rds/min
Main armament traverse	
angle:	360°
speed:	45°/s
Main armament elevation/depression	
armament front:	+85°/-5°
speed:	85°/s

Status
Production complete. It is understood that 252 × 40 mm L/70 guns were built for the Italian Army and 50 for the Hellenic Army. As of May 2013 there has been no recent production of this weapon system.

Contractor
Oto Melara SpA.

Fast Forty

Type
Towed anti-aircraft gun

Development
The Twin Fast Forty Field Mounting is a further development of the Oto Melara Twin 40L70 Field Mounting covered in the following entry.

By late 2002, this system developed as a private venture for the export market was at the prototype stage. There are reports that the Italian Navy has taken delivery of the Fast Forty.

Description
Fitted with Oto Melara-designed recoiling masses, the cyclic rate of fire has been increased from 300 to 450 rds/barrel giving the system a cyclic rate of fire of 900 rds/min.

To get the gun into action the carriage is emplaced in the firing position by swivelling the outriggers, lowered by means of a hydraulic device which rotates the axles and levelled using the six outriggers' jacks.

Standard equipment on both this and the previous model is digital control, which allows for:

- The easy and fast introduction of prohibited zones by means of a keyboard.
- Rapid alignment of the gun and fire-control system.
- Automatic correction of line of fire in relation to platform inclination.
- Visualisation of diagnostic messages on the local control panel and control of the gun by means of serial transmission of gun orders through a two-wire cable.

A new concept autoloader capable of being re-cocked by a shorter recoil, carries out the cartridge ramming and case ejection in a faster way without additional stress for the round itself, thanks to light materials (titanium) and progressive way of accelerating and decelerating parts.

In addition, the ramming route is shorter in respect to the current 40L70. Hydropneumatic motors for moving the shift tongue and the breechblock substitute more effectively than mechanical springs.

Electrical DC motors driving epi-cyclic gearbox train the gun.

The interior of the watertight cupola and magazine are supplied with an air conditioning system.

The Oto Melara Twin Fast Forty Field Mounting can accommodate all forms of Bofors Defence 40 mm L/70 ammunition, including proximity fused rounds. Ammunition details are given for the Bofors Defence 40 mm L/70 gun under Sweden. The Oto Melara Twin Fast Forty Field Mount can fire the latest Bofors Defence 3P advanced ammunition.

The field mounting consists of a 360° traverse training platform supported on a wire race bearing. The platform has parallel vertical lightweight aluminium alloy trunnion supports and holds the servo and ammunition feed motors, together with the upper ammunition feed mechanism and system junction box.

The magazine holds a total of 444 rounds of ammunition and is housed on the traversing platform and consists of four horizontal layers that move the rounds towards a vertical ammunition elevator.

The magazine is divided into two independent sections, each of which delivers ammunition to the corresponding gun to continue fire even when either guns is out of operation. The upper section of the mounting is completely enclosed in a watertight, reinforced glass fibre cupola fitted with three servicing hatches, one at the rear and one in either side.

Under the mounting platform inside the carriage is the magazine. The carriage is normally on four wheels and two axles. Due to a frame controlling the barrels during run, the Fast Forty, notwithstanding its increased rate of fire, maintains a very high accuracy (SD less than 1 mrad).

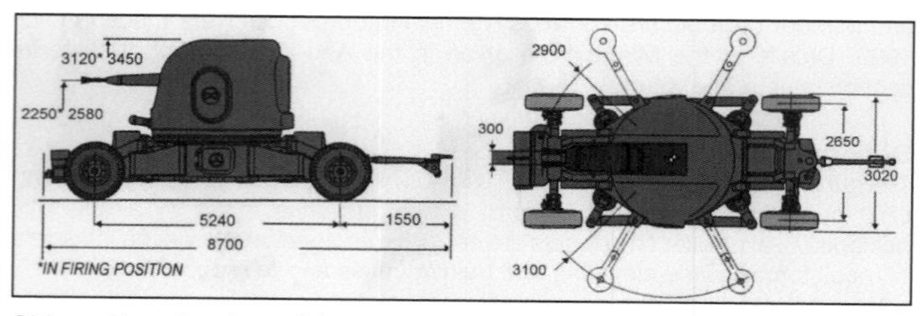

Side and top drawings of Oto Melara Twin Fast Forty Field Mounting in travelling configuration 0100681

Oto Melara Twin Fast Forty Field Mounting deployed in the firing position 0509771

Once emplaced, pickets are driven into the ground from the outrigger feet and connections to the gun from external generator and fire-control systems are made via junction boxes. The external generator supplies 200 V AC 400 Hz.

Digital technology allows for the quick and easy introduction of firing limit zones, obstacle contouring, tilt measurement for automatic correction of the line of fire and BITE with messages. A control panel, at the rear of the carriage, provides for the direct operation of the mounting.

The off carriage fire-control system can be positioned up to 1,000 m from the Oto Melara Twin Fast Forty Field Mounting, with one of these typically controlling two systems.

Specifications

	Fast Forty
Dimensions and weights	
Length	
transport configuration:	8.2 m
Width	
transport configuration:	3.155 m
Height	
transport configuration:	3.47 m (est.)
Ground clearance	
overall:	0.3 m
Track	
vehicle:	2.5 m
Wheelbase	5 m
Weight	
standard:	12,150 kg
Mobility	
Carriage:	trailer mounted four wheeled with outriggers
Firepower	
Armament:	2 × 40 mm cannon
Main weapon	
length of barrel:	2.8 m
calibre length:	70 calibres
breech mechanism:	vertical sliding block
maximum range:	12,500 m (horizontal)
maximum vertical range:	8,700 m
feed mechanism:	736 round
cartridge:	40 mm × 365 mm R
Rate of fire	
cyclic:	900 rds/min
Main armament traverse	
angle:	360°
speed:	100°/s
acceleration:	130 °/s²
Main armament elevation/depression	
armament front:	+85°/-8°
speed:	70°/s
acceleration:	150 °/s²

Status
Development complete.

Contractor
Oto Melara SpA.

Sentinel

Type
Towed anti-aircraft gun

Development
The Oto Melara Twin 30 mm anti-aircraft gun has been developed as a private venture and it uses two 30 mm Mauser Models F guns mounted side-by-side on a mobile field mounting. The Twin 30 mm is intended to operate as an independent unit and has its own power source and electro-optic fire-control unit.

Description
On tow, the Twin 30 mm has its barrels pointing to the rear and is carried on a pair of two-wheeled axles. To bring the weapon into action, the wheels are raised on swivels until they are clear of the ground. The gun is then supported on four levelling jacks, one each at the front and rear and the two side jacks on outward-folding outrigger arms. The gun traverse features the mounting on a turntable, with the aimer, who is provided with a shield, seated to the left of the barrels. According to Oto Melara, the total time taken for the system to come into action, for example, from coming to a halt until it is ready to fire, is 20 seconds.

The second prototype has hydraulic wheel-drives for limited self-propulsion, an automatic outrigger extension and levelling and electro-optic improvements. Other improvements include a new gunner's position, new shield and a different generator at carriage rear.

Second prototype of the Oto Melara Sentinel Twin 30 mm anti-aircraft gun system which has a hydraulic road wheel drive system 0509523

The aimer is provided with a Galileo Avionica Vanth (or the M75) electro-optic fire-control system. This comprises:

- A control panel
- A laser range-finder
- An optical aiming device
- A computer combined in one unit.

To track a target, the target line of sight is held by operating the system joystick to enable the lead angle to be automatically applied to the guns while at the same time the laser range-finder also provides fire data. Passive infra-red night aiming equipment is optional and at all times target acquisition data from external sources such as radar may be fed into the system.

Power for the mounting and the fire-control system is provided by a power supply unit carried over the rear axle. The main power source is an HATZ 3L 40C four-stroke diesel engine. This air-cooled engine has three cylinders and a capacity of 2.5 litres. Using a direct injection system, the engine has a maximum speed of 3,000 revolutions per minute. SACCARDO GS 132 M/16 three-phase brushless alternators supply electrical power.

The Twin 30 mm Mauser Model F cannons are mounted side by side and are provided with 500 rounds of ammunition. Each barrel has a rate of fire of 800 round per minute (cyclic) and the ammunition feed uses flexible chutes to guided belted rounds to supply the two 30 millimetre cannons. The ammunition used is of the GAU-8/A type, originally developed for the A-10 ground attacks aircraft used only by the US Air Force and includes:

- Armour Piercing Discarding Sabot (APDS)
- High-Explosive Incendiary (HEI)
- Armour-Piercing Incendiary (API)
- High Explosive Incendiary - Self Destruct (HEI-SD)
- Target Practice (TP).

It can also fire the latest Frangible Armour Piercing Discarding Sabot - Tracer (FAPDS-T) which has a higher muzzle velocity and increases the effective range of the system.

Specifications

	Sentinel
Dimensions and weights	
Length	
transport configuration:	6.46 m
Width	
transport configuration:	1.76 m
Height	
transport configuration:	1.94 m
Ground clearance	
overall:	0.43 m
Track	
vehicle:	1.76 m
Wheelbase	3.5 m
Weight	
combat:	5,000 kg
Mobility	
Carriage:	trailer mounted four wheeled with outriggers
Firepower	
Armament:	2 × 30 mm cannon
Main weapon	
length of barrel:	2.458 m
calibre length:	82 calibres
feed mechanism:	500 round
cartridge:	30 mm × 173 mm
Rate of fire	
cyclic:	1,600 rds/min
Main armament traverse	
angle:	360°

	Sentinel
speed:	120°/s
acceleration:	150 °/s^2
Main armament elevation/depression	
armament front:	+85°/-5°
speed:	80°/s
acceleration:	120 °/s^2

Status
Development complete. Ready for production, no reported customers as of June 2013.

Contractor
Oto Melara SpA.

Korea (South)

M167 Vulcan

Type
Towed anti-aircraft gun

Development
For some years the then Daewoo Defence Products of the Republic of Korea, now known as Doosan Infracore, manufactured under licence, the US, now General Dynamics Armament and Technical Products, (originally known as General Electric Armament Systems and, for a short time, Lockheed Martin Armament Systems) 20 mm M167 Vulcan anti-aircraft gun system.

It is believed that at least 150 systems were built in South Korea with production now complete.

Description
Details of a self-propelled version on the Korean Infantry Fighting Vehicle chassis are given a separate entry.

The M167 is a towed short-range six-barrelled gun based on the M61 Vulcan 20 mm rapid-firing Gatling gun, originally designed for the US Army. Its role was to protect forward area combat elements and rear area critical assets.

The system is no longer in operational use within the US forces, having been replaced over time by the Avenger and Phalanx close in weapon systems.

As far as it is known, the system manufactured in South Korea is identical to the original US version, although some indigenous differences in the specifications are believed to exist. Ammunition for this weapon is made in South Korea.

South Korean-built 20 mm M167 Vulcan air defence system with outriggers deployed 0509524

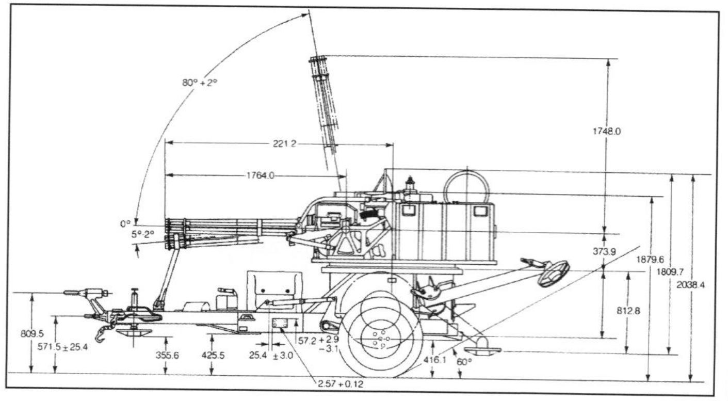

Side elevation drawing of Doosan 20 mm M167 Vulcan anti-aircraft gun system showing main dimensions 0509700

20 mm Vulcan showing range only radar and generator on forward part of carriage 0509540

Specifications

	M167
Dimensions and weights	
Crew:	1
Length	
transport configuration:	4.724 m
deployed:	3.861 m
Width	
transport configuration:	2.007 m
deployed:	3.708 m
Height	
transport configuration:	2.057 m
deployed:	1.829 m
Ground clearance	
overall:	0.444 m
Weight	
standard:	1,429 kg
Mobility	
Carriage:	trailer mounted two wheeled with outriggers
Speed	
towed:	72 km/h
Fording	
without preparation:	0.991 m
Firepower	
Armament:	6 × 20 mm cannon
Main weapon	
length of barrel:	1.524 m
operation:	electrically powered
cartridge:	20 mm × 102 mm
Rate of fire	
cyclic:	3,000 rds/min
Main armament traverse	
angle:	360°
Main armament elevation/depression	
armament front:	+80°/-5°

Status
Production as required and in service with the Republic of Korea Army.

As far as it is known, there have been no export sales of this system from Korea.

The system was in service with US forces from 1970 through 1994.

Contractor
Doosan Infracore Co Ltd.

Norway

20 mm FK 20-2

Type
Towed anti-aircraft gun

Development
The 20 mm automatic anti-aircraft gun FK 20-2 was a joint development between Hispano-Suiza of Switzerland (then part of Oerlikon Contraves now Rheinmetall Air Defence), Kongsberg Defence & Aerospace of Norway, Rheinmetall Industrie AG of Germany and Kern Company AG.

20 mm automatic anti-aircraft gun FK 20-2 deployed in firing position in direct fire role (Michael Jerchel) 0509701

20 mm automatic anti-aircraft gun FK 20-2 in travelling configuration being towed by Unimog (4 × 4) truck (Michael Jerchel) 0509702

Description

The system, which was developed in the late 1960s, basically consists of:

- Modified HSS 669 mount fitted with a Rheinmetall 20 mm MK 20 Rh 202 automatic cannon (as installed in the Marder 1 IFV, Luchs 8 × 8 reconnaissance and recovery vehicle)
- FIAT/Otobreda Type 6616 armoured car
- Wiesel 1 light airborne armoured vehicle
- VCTP infantry combat vehicle
- Single naval gun S 20 SV 20
- Rheinmetall Landsysteme Condor (4 × 4) armoured personnel carrier
- Twin 20 mm automatic anti-aircraft gun, with the cradle, ammunition cases and flexible feed channels designed by Kongsberg Defence & Aerospace
- Optical sight designed by Kern.

The FK 20-2 can be used against both ground and air targets and can be quickly dismantled without special tools into loads suitable for carrying over short distances. The FK 20-2 weighs only 620 kg in travelling order compared with the twin 20 mm MK 20 Rh 202 automatic anti-aircraft gun which weighs 2,160 kg in travelling order.

The FK 20-2 consists of five main components:

- The cannon
- Cradle
- Upper mounting with seat
- Lower mount
- Carriage.

The Rheinmetall 20 mm MK 20 Rh 202 cannon is gas operated and fully automatic. Of 160 rounds of 20 mm ammunition carried, 75 are in each of the side magazines and 10 rounds (normally APDS-T or API-T) are in the magazine on top of the cannon. The types of fixed ammunition (20 × 139 mm) which can be fired are detailed in the entry under Germany for the Rheinmetall 20 mm anti-aircraft twin gun system.

More recently Raufoss (today Nammo AS) has developed the 20 mm NM75 (now NM75 F2) multipurpose round to meet the requirements of the German and Norwegian armies. This round does not detonate until it is inside the target and also has an improved incendiary effect.

The lower mount is of the tripod type and the two shorter trails may be disconnected for transport. The two-wheeled carriage is horseshoe shaped and the tow bar may be set at different positions to compensate for different towing vehicles.

The cradle, which carries the gun and guides its recoil, is made of cast aluminium alloy and pivots on the upper mounting. It contains the recoil brakes, components forming part of the fire selector and trigger mechanism and a hinged frame assembly, which contains the feeder mechanism and belt guides. The frame assembly can be latched semi-raised for quick inspection of the gun during action and fully raised to remove the top feeder mechanism. The cocking crank is on top of the cradle.

The upper mounting holds the cradle and contains the elevation and traverse mechanisms, which include adjustable twin gears to prevent play in the transmission. Operation of the quick-release locking device enables the upper mounting to be removed from the lower mounting. Fire selection, triggering and main locking controls are on the elevation and traverse hand wheels. The optical sight is mounted between the hand wheels. The gunner's seat can be folded forward to provide the gunner with a prone support position when the gun is being used against ground targets. The mount has a one-piece shield, which can be removed.

The optical sight has a magnification of ×5 for use against ground targets and ×1.5 for use against air targets. The partially combined light paths for the air target sight give individual sighting images and optical values for both targets. Lead angle curves in the sight simplify firing against air targets at various speeds. The sighting image of the ground target sight comprises simple cross hair and range prediction lines for firing at targets at ranges of 500, 1,000 and 1,500 m.

Specifications

	FK 20-2
Dimensions and weights	
Crew:	3
Length	
transport configuration:	4 m
deployed:	3.72 m
Width	
transport configuration:	1.86 m
deployed:	1.8 m
Height	
transport configuration:	2.2 m
axis of fire:	0.58 m
deployed:	1.2 m
Ground clearance	
overall:	0.38 m
Track	
vehicle:	1.62 m
Weight	
standard:	620 kg
combat:	440 kg
Mobility	
Carriage:	trailer mounted two wheeled with outriggers
Firepower	
Armament:	1 × 20 mm cannon
Main weapon	
length of barrel:	2.61 m (including muzzle brake)
maximum range:	2,000 m
operation:	gas automatic
cartridge:	20 mm × 139 mm
Rate of fire	
cyclic:	1,030 rds/min (est.) (has also been reported as up to 1,000 rds/min)
Main armament traverse	
angle:	360°
Main armament elevation/depression	
armament front:	+80°/-8°

Status

Production complete. In service with Germany (Army) and Norway (216, Army). No longer being marketed.

Contractor

Kongsberg Defence & Aerospace.

Pakistan

37 mm AAG (Upgraded)

Type

Towed anti-aircraft gun

Development

Some years ago Pakistan carried out extensive tests of three China North Industries Corporation (NORINCO) twin 37 mm Light Anti-Aircraft Guns (LAAG) fitted with a Western computerised fire-control system to improve their first round hit probability against low flying aircraft and helicopters.

These computerised fire-control systems were supplied by Contraves Italiana (Gunking), the then CelsiusTech (Universal Tank and Anti-Aircraft Sight) and Galileo Avionica, the then Officine Galileo (Vanth/MB). However, Pakistan adopted none of these sighting systems for cost reasons.

In 1995, the College of Electrical and Mechanical Engineering in Rawalpindi carried out a training exercise to improve the efficiency of the twin 37 mm LAAG by adding powered controls and a new fire-control system.

Trials with this were so successful that in 2000 another four systems were converted in association with the then Dr A Q Khan Research Laboratories (now known as 'The Institute of Industrial Control and Systems') and the Pakistani Army from late 2000 evaluated these. As a consequence the Laser Aided Automatic Fire Control System (also known as the Laser Aiming Sight LAS786P) produced by IICS was adopted.

Close-up of the off carriage aiming system of the upgraded twin 37 mm LAAG clearly showing the laser range-finder mounted on top (Christopher F Foss) 0100682

Upgraded twin 37 mm LAAG upgraded with a new all electric gun control equipment and computerised fire-control system (Christopher F Foss) 0100683

The LAS786P provides azimuth, elevation and range of a target up to 10 km, whilst the computer TBC786P receives data from LAS786P, computes trajectory of a moving target and predicts future position of the target. This information enables calculation of optimum lead and super elevation of the gun.

The Directorate General Munitions Production, Defence Production Division, is carrying out marketing of the upgraded twin 37 mm system.

Description

Mounted on a tripod offset from the twin 37 mm LAAG is the LAS786P or LRH786Q laser range-finder under which is the DGM768P digital goniometer. This is pointed at the target manually and provides accurate ranging information.

The laser range-finder fires at the rate of one pulse per second and the digital goniometer gives the azimuth and elevation angles of the target with an accuracy of +1 mils.

The information is sent to the TBC786P ballistic computer that is mounted on the carriage of the twin 37 mm LAAG. The ballistic computer also calculates the advance and super elevation.

The tracking and ballistic computer is built around a high-performance, state-of-the-art, high-speed, large memory single chip micro controller. This instructs the control computer to position the gun at the required angle to hit the target.

In addition, new all electric traverse and elevation units have been installed, although manual controls are retained for emergency use. There is also a joystick that allows one person to lay the weapon onto the target and then open fire using the analogue computing sight. An off carriage generator is provided. Electrical and mechanical firing is possible.

One off carriage fire-control system would typically be used to control the fire of four-twin 37 mm LAAG. At the present time this is a clear weather system with no enhancements being made of the old 37 mm ammunition which is fed to the weapon in clips of five rounds.

According to the DGMP, the sophistication of the system can be gauged from the fact that the whole system is based on three computers running in

Close-up of TBC786P ballistic computer installed on the carriage of the upgraded twin 37 mm LAAG (Christopher F Foss) 0100684

parallel, communicating with each other and the gun is responding to the commands of one of the computers, for example the control computer.

Status

As of June 2013, it is understood that the system whilst still undergoing trials has not yet been offered for sale on the open arms market.

Contractor

Marketing enquiries to Directorate General Munitions Production (DGMP); Institute of Industrial Controls and Systems (IICS), for Laser Aided Automatic Fire Control System.

Type 54

Type

Towed anti-aircraft gun

Description

This is an automatic weapon (some reports suggest a license production of the DShK) produced in the Pakistan Ordnance Factories employing the 12.7 × 108 mm Armour-Piercing Incendiary (API), Armour-Piercing Incendiary Tracer (APIT) and Armour-Piercing Hard-Core (APHC) ammunition. The system is effective against low-flying aircraft and ground targets, especially light armoured vehicles etc.

The gun is supplied complete with tripod, anti-aircraft sighting system and chromium-plated spare barrel.

This gun can also be mounted on tanks such as the new Al Khalid (MBT 2000) and armoured personnel carriers, both in the anti-aircraft mode and in ground modes.

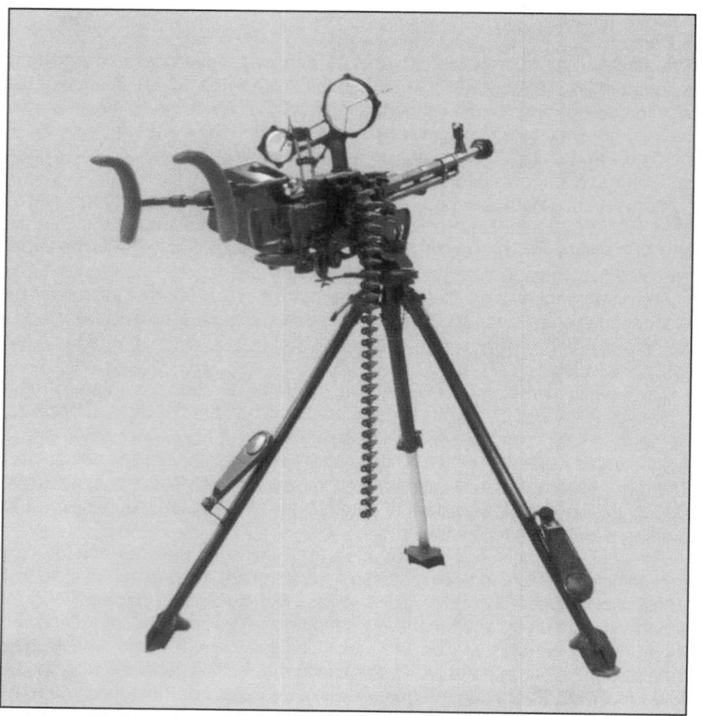

Type 54 AAG (Pakistan Ordnance Factories) 1333357

Specifications

Type 54

Weight:	
(gun complete with tripod)	92 kg
(gun body)	33 kg
(gun mount)	53 kg
(anti-aircraft sight)	2.2 kg
(ammunition box, 70 round linked-belt)	14 kg
Traverse:	
(vertical)	–26° to 78°
(horizontal)	
(air targets)	360°
(ground targets)	120°
Effective range:	
(air targets)	1,600 m
(ground targets)	1,500 m
Armour penetration:	
(API at 500 m)	15 mm
(API at 800 m)	10 mm
Muzzle velocity:	818-833 m/s (some reports 850 m/s)
Cyclic rate of fire:	600 rds/min
Packaging:	
(contents)	2 gun bodies with 2AA sights, spare parts and accessories packed in a wooden box
(dimensions)	126 × 43 × 47 cm
(weight)	131 kg

Status

In production and in use with the Pakistan Army. As of June 2013, it is not known if or how many of these weapons have been sold.

Contractor

Pakistan Ordnance Factories (POF).

Poland

ZUR-23-2KG

Type

Towed anti-aircraft gun

Development

This system is basically the Soviet ZU-23-2 twin 23 mm towed anti-aircraft gun system, manufactured under licence in Poland for some years, fitted with two locally produced Grom man-portable surface-to-air missiles. Full details of the Grom are given later in the Man-portable surface-to-air missile systems section.

The Tarnow Mechanical Works completed the first prototype of this system in 1985 and it entered service with the Polish Army the following year.

According to the prime contractor, the 23 mm cannon allow aerial targets to be engaged out to a maximum range of 2,000 m and up to a maximum altitude of 1,500 m.

The Grom fire-and-forget surface-to-air missiles can be used to engage aircraft approaching up to a maximum altitude of 3,500 m and from the rear up to an altitude of 5,500 m down to a minimum altitude of 10 m.

Twin 23 mm automatic anti-aircraft gun and missile system ZUR-23-2KG deployed in the firing position from the front 0509881

Twin 23 mm automatic anti-aircraft gun and missile system ZUR-23-2KG deployed in the firing position from the rear 0509525

Description

The 23 mm anti-aircraft mobile gun and missile system ZUR-23-2KG mounts a collimator sight and short-range surface-to-air missiles.

The main components of the system are:

- Two 23 mm 2A14 cannon
- Two Grom short-range surface-to-air missiles
- CKE-2 Collimator sight, which also includes an emergency ring-shaped sight.

The CKE-2 sight is electrically operated with the battery supplying sufficient power for 10 hours of operation. When the system is being towed, for example by a UAZ-469 (4 × 4) light vehicle, or STAR-266 (6 × 6) truck, provision is made for the battery to be charged.

The installation of the CKE-2 sight enables the rapid introduction of the most probable range to the target as a range-finder transmitter is installed and brightness of the sight can be adjusted to suit the operating environment.

Bringing the system from the travelling to firing configuration takes 15 to 20 seconds, while it takes 35 to 45 seconds to bring the system from firing configuration to the travelling configuration.

The crew of four consists of the commander, sight operator and two gunners.

The two 2A14 23 mm cannons have a combined cyclic rate of fire between 1,800 and 2,170 rds/min, although the combined practical rate of fire is 400 rds/min. Muzzle velocity of the 23 × 152 mm B ammunition is 970 m/s.

The ZUR-23-2KG-1's simulator, the TR-23-2, is designed to train personnel to efficiently operate the 23 mm gun/missile system. A constituent part of the simulator is an instructor's station, equipped with a PC workstation. An instructor can simulate different targets and several scenarios, allowing the instructor to give an objective evaluation of the training. The simulator is adapted for transportation by the haulage vehicle.

Types of ammunition used:

- Armour-Piercing Incendiary with Tracer API (BZT)
- High-Explosive Incendiary with Tracer HEI (OFZT)
- Armour-Piercing Discarding Sabot with Tracer APDST.

Variants

ZU-23-2MR Wrobel

This was developed for naval applications and also consists of two 23 mm cannon and two launchers for the Strela-2M SAM. However, this is a turret system rather than an unprotected mount, as is the case with the ZUR-23-2S Jod. The sighting system GP-02MR is mounted with the ZU-23-2MR.

ZU-23-2K

Yet another modification to the basic ZU-23-2, this time with the system using the CKE-1 collimator sight.

Igla/Igla-1

Due of the flexibility of the system, various man-portable SAM can be fitted, it would be reasonable to expect to see the Igla-1 and/or Igla system replacing the Grom missiles currently fitted.

Specifications

	ZUR-23-2KG
Dimensions and weights	
Crew:	4
Length	
transport configuration:	4.68 m
Width	
transport configuration:	1.9 m
Height	
transport configuration:	2 m
Ground clearance	
overall:	0.36 m
Track	
vehicle:	1.73 m
Weight	
standard:	1,120 kg

	ZUR-23-2KG
Mobility	
Carriage:	trailer mounted two wheeled
Batteries:	1 × 12 V
Firepower	
Armament:	2 × 23 mm cannon
Main weapon	
length of barrel:	2.5 m (including flash suppressors)
breech mechanism:	vertical sliding block
maximum range:	5200 m
maximum vertical range:	3500 m
minimum vertical range:	0 m
feed mechanism:	50 round belt
operation:	gas automatic
cartridge:	23 mm × 152 mm B
Rate of fire	
cyclic:	2,170 rds/min
practical:	400 rds/min
Main armament traverse	
angle:	360°
speed:	60°/s
Main armament elevation/depression	
armament front:	+75°/-10°
speed:	40°/s

Status
Production as required. In service with the Polish Army, offered for export. As of June 2012, there was only one other known export of this system and that was to Indonesia.

Contractor
Tarnow Mechanical Works (ZUR-23-2S).
Mesko Metal Works(rocket system).
Prexer Ltd(sighting systems).
FTE Cenzin Co Ltd (marketing).

Romania

12.7 mm AAG (Romanian)

Type
Towed anti-aircraft gun

Development
This Romanian 12.7 mm light anti-aircraft machine gun is similar to the Russian 12.7 mm DShKM machine gun.

Mounted either on a tripod or on a two-wheeled carriage the gun can be supplied by SN ROMARM S.A.

Description
It has been designed to engage both air and ground targets. In the first application the tripod is almost vertical while in the second application it is almost horizontal.

The foresight is a pillar that slides up and down, while the rear sight has two vertical pillars with a U-back sight between them. In addition, there is a special anti-aircraft sight which requires two personnel to use it to its best advantage.

Ammunition is fed from the left, in boxes of 50 rounds and three types of 12.7 × 108 mm ammunition, plus blanks, can be fired (see Table 1 for details).

The B-32 API round will penetrate 20 mm of steel armour at a range of 100 m while the BZT will penetrate 15 mm of armour at a range of 100 m.

Table 1

Type	HEI	API	API-T
Cartridge designation:	MDZ	B-32	BZT
Weight:			
(bullet)	44.6 g	49 g	44.6 g
(cartridge)	122 g	126 g	122 g
Length:			
(bullet)	64.6 mm	64.6 mm	64.5 mm
(cartridge)	145.5 mm	147 mm	147 mm
Bullet:			
(core material)	steel	steel	steel
(jacket material)	steel	brass	brass
Muzzle velocity:	817 m/s	817 m/s	817 m/s

SN ROMARM S.A. 12.7 mm light anti-aircraft machine gun on tripod 0044165

Specifications

	12.7 mm AAG (Romanian)
Dimensions and weights	
Length	
transport configuration:	1.65 m
deployed:	1.65 m
Width	
transport configuration:	0.78 m
deployed:	0.78 m
Height	
transport configuration:	0.5 m
deployed:	0.9 m
Weight	
standard:	157 kg
Mobility	
Firepower	
Armament:	1 × 12.7 mm (0.50) machine gun
Ammunition	
main total:	50
Main weapon	
length of barrel:	1.065 m
maximum range:	1,500 m
feed mechanism:	50 round belt
operation:	gas automatic
cartridge:	12.7 mm × 108 mm
Rate of fire	
cyclic:	600 rds/min (est.)
practical:	100 rds/min (est.)
Main armament traverse	
angle:	360°
Main armament elevation/depression	
armament front:	+80°/-20°

System in ground mode

Elevation/depression:	+10°/–0°
Traverse:	90°

Status
Production complete. As far as is known the system is in service with Romanian Armed Forces. As of March 2011, this system is no longer marketed by SN ROMARM S.A.

Contractor
SN ROMARM S.A.

14.5 mm ZU-2/MR-4 (Romania)

Type
Towed anti-aircraft gun

Development
The Romanian Company SN ROMARM S.A. manufactures two types of 14.5 mm towed light anti-aircraft machine gun.

Both of these are armed with the 14.5 mm KPVI heavy machine gun that were developed in Russia and are also manufactured by SN ROMARM S.A. Both use a two-wheeled light carriage designed and built in Romania.

Description
The twin 14.5 mm model is referred to as the ZU-2 and is similar to the Russian ZPU-2 covered in detail later in this section, while the quad 14.5 mm model is referred to as the MR-4. The latter model is a local design of the Russian equivalent ZPU-4 and is mounted on a heavier four-wheeled carriage.

Table 1

Type	API	API-T	HEI
Designation:	B-32	BZT	MDZ
Muzzle velocity:	980/995 m/s	995/1,015 m/s	1,000/1,100 m/s
Weight (bullet):	64 g	60.3 g	58 g
Chamber pressure:	324 MPa	324 MPa	324 MPa
Cartridge case:	lacquered steel	lacquered steel	lacquered steel
Bullet construction:	jacketed/steel core/incendiary	jacketed/steel core/incendiary/tracer	steel shell/HE filled/compression fuze
Primer:	Berdan	Berdan	Berdan
Armour penetration:	38 mm/100 m	20 mm/100 m	nil

Romanian twin 14.5 mm ZU-2 light anti-aircraft gun deployed in the firing position 0064563

Romanian quad 14.5 mm MR-4 light anti-aircraft gun deployed in the firing position 0064562

Both weapons can fire three different types of 14.5 mm ammunition, which are stowed in two box-type magazines:

- HE-I (MDZ)
- API (B-32)
- API-T (BZT).

Brief details of this 14.5 mm ammunition are listed in Table 1 (see top of page).

Specifications

	ZU-2	MR-4
Dimensions and weights		
Crew:	3	5
Length		
transport configuration:	3.9 m	4.3 m
deployed:	3.9 m	2.7 m
Width		
transport configuration:	1.665 m	1.86 m
deployed:	1.665 m	1.90 m
Height		
transport configuration:	1.5 m	1.81 m
deployed:	1.1 m	1.47 m
Ground clearance		
overall:	0.27 m	0.30 m
Track		
vehicle:	1.48 m	1.64 m
Weight		
standard:	660 kg	1,360 kg
Mobility		
Carriage:	trailer mounted two wheeled	trailer mounted two wheeled
Speed		
towed:	60 km/h	60 km/h
Firepower		
Armament:	2 × 14.5 mm machine gun	4 × 14.5 mm machine gun
Main weapon		
length of barrel:	1.342 m	-
cartridge:	14.5 mm × 114 mm	14.5 mm × 114 mm
Rate of fire		
cyclic:	600 rds/min	600 rds/min
practical:	300 rds/min	-
Main armament traverse		
angle:	360°	360°
Main armament elevation/depression		
armament front:	+88°/-15°	+88°/-10°

Status
Production complete. In service with the Romanian Armed Forces. As of June 2013 there are no known export sales of these weapons and the system is no longer marketed by ROMARM.

Contractor
SN ROMARM S.A.

30 mm AAG (Romanian)

Type
Towed anti-aircraft gun

Development
The Romanian-developed twin 30 mm towed anti-aircraft gun (30 mm AAG) is armed with two 30 mm A 436 automatic cannons which are on a mount, installed on a four-wheeled carriage.

Description
The A 436 is a Romanian RATMIL design and is said to be similar to the Russian 30 mm NN 30 cannon. This cannon is air-cooled and gas operated and has a maximum cyclic rate of fire of 600 rounds per minute. The barrels are provided with flash suppressers.

Belted 30 × 210 B ammunition is fed from a box-type magazine holding 30 rounds of ammunition. Three different natures of 30 mm ammunition are available:

- HEI
- HE-T
- AP.

Muzzle velocity for all natures is 1,050 m/s.

Romanian twin 30 mm anti-aircraft gun system deployed in firing position with radar scanner on left side of mount 0509526

Romanian twin 30 mm anti-aircraft gun system deployed in the firing position from the rear (Paul Beaver) 0044150

Another two boxes of ammunition are normally kept for immediate use. The rate of fire can be selected from a command device, for example, single shot, 100, 250 or 500 rounds per minute, per barrel.

The on board electrical supply provides power subsystems with the 12 V 45 Ah battery being mounted on the carriage. Mount traverse and weapon elevation is accomplished by a two-speed manual gear mechanism.

This anti-aircraft gun can be used against air targets with a maximum speed up to 300 m/s and flying at an altitude of between 50 and 3,500 m. Slant range is 4,100 m.

The system can also be used against ground targets such as static and moving light armoured vehicles out to a maximum range of 1,000 m. Against unarmoured targets at a maximum range of 2,000 m.

When deployed in the firing position the carriage is supported at four points, each being adjustable; one of these is at either end of the carriage and the other one either side on outriggers.

Romanian twin 30 mm anti-aircraft gun deployed in the firing position 0064561

In Romanian Army service this system is normally towed by a Romanian DAC 665 T (6 × 6) truck which also carries the gun crew, ready use ammunition and gun stores.

Variants

This twin 30 mm anti-aircraft gun system can also be supplied fitted with the ACT-30 firing device mounted on the left side of the mount.

The ACT-30 includes radar with a maximum range of 15 km, which determines the motion parameters of the target, computes the firing data and automatically actuates the gun to move into the correct elevation in order to engage the target. This action takes two seconds.

Fire-control system

Details of the fire-control system for this and the Oerlikon Contraves GDF series twin 35 mm towed anti-aircraft gun systems are given in the Anti-aircraft control systems section under International.

Specifications

	30 mm AAG (Romanian)
Dimensions and weights	
Length	
transport configuration:	6.40 m
deployed:	6.40 m
Width	
transport configuration:	2.10 m
deployed:	2.10 m
Height	
transport configuration:	2.20 m
deployed:	1.80 m
Weight	
standard:	3,460 kg
Mobility	
Carriage:	trailer mounted four wheeled
Firepower	
Armament:	2 × 30 mm cannon
Main weapon	
feed mechanism:	30 round belt (per barrel)
operation:	gas automatic
cartridge:	30 mm × 210 mm B
Rate of fire	
cyclic:	1,000 rds/min (est.)
Main armament traverse	
angle:	360°
Main armament elevation/depression	
armament front:	+85°/-5°

Status

Production as required. Believed to be still in service with Romanian Army as of June 2013.

Contractor

SN ROMARM S.A.

Russian Federation

KS-19

Type

Towed anti-aircraft gun

Development

The KS-19, 100 mm towed anti-aircraft gun was introduced in the late 1940s as the replacement for the 85 mm M1939 and M1944 anti-aircraft guns, which are covered in detail in the following sections.

Table 1

Ammunition nature	AP-T[1]	APC-T	HE	HE-FRAG	FRAG
Projectile designation:	BR-412B	BR-412D	F-412	OF-412	O-415
Fuze model:	MD-8	DBR-2	RGM	V-429	VM-30/VL-30L
Weight:					
(projectile)	15.89 kg	16 kg	15.91 kg	15.61 kg	15.44 kg
(bursting charge)	0.56 kg	0.63 kg	2.159 kg	1.46 kg	1.58 kg
Type of bursting charge:	RDX/Al	RDX/Al	TNT	TNT	n/avail
Muzzle velocity:	1,000 m/s	900 m/s	900 m/s	900 m/s	900 m/s

[1] will penetrate 185 mm of steel armour at incidence of 0° at range of 1,000 m

Late production model of KS-19, designated KS-19M2, in travelling order with outriggers retracted and ordnance in travelling lock 0509494

100 mm KS-19 anti-aircraft gun deployed in the firing position 0509766

The KS-19 is no longer in service with the Russian Army, having been replaced by surface-to-air missiles, but it is still in very limited use by some countries. The KS-19 has also been manufactured in China as the Type 59.

The effectiveness of the 100 mm anti-aircraft gun KS-19 against modern aircraft is very limited.

Description

When travelling, the mount is traversed to the rear and the 100 mm ordnance is held in position by a travelling lock at the rear of the carriage. In the firing position the wheels are raised off the ground and the carriage is supported on four screw jacks, one at each end of the carriage and one either side on outriggers.

The KS-19 has a power rammer, automatic fuze setter and a single-round loading tray. The KS-19 has on-carriage fire-control equipment but is normally used in conjunction with the PUAZO-6/19 director and the SON-9/SON-9A (NATO codename 'Fire Can' and operated in the A/B-band) fire-control radar. It is also reported that the KS-19 is used in conjunction with the PUAZO-7 director and the SON-4 (NATO codename 'Whiff') fire-control radar.

Ammunition is of the fixed type and is fed from the left. The KS-19M2 AAA may also be employed in the surface support role and this is reflected by the various types can be fired (see Table 1 at the top of the page).

Specifications

	KS-19
Dimensions and weights	
Crew:	15
Length	
transport configuration:	9.45 m
Width	
transport configuration:	2.35 m
Height	
transport configuration:	2.201 m
axis of fire:	1.682 m
Ground clearance	
overall:	0.33 m
Track	
vehicle:	2.165 m
Wheelbase	4.65 m
Weight	
standard:	9,550 kg
Mobility	
Carriage:	trailer mounted four wheeled
Firepower	
Armament:	1 × 100 mm cannon
Main weapon	
length of barrel:	5.742 m
breech mechanism:	horizontal sliding block (semi-automatic)
muzzle brake:	multibaffle
maximum range:	21,000 m (horizontal)
maximum vertical range:	15,000 m (proximity fuzed)
feed mechanism:	100 round
Rate of fire	
cyclic:	15 rds/min
Main armament traverse	
angle:	360°
Main armament elevation/depression	
armament front:	+85°/-3°

Status

Production complete.

Iran, although not mentioned as one of the recipient countries for export of the KS-19 has developed its own variant of a 100 mm AA gun known as Sa'eer. It is not for certain if this gun was developed from the stock of obsolete Soviet equipment that found its way into Iran or of a Chinese copy of a Type-59.

Contractor

Russian state factories.

M1939

Type

Towed anti-aircraft gun

Development

In the late 1930s, the then Factory No. 8 at Kaliningrad, near Moscow, developed a 45 mm automatic anti-aircraft gun under the designation of the 49-K. This was similar in operation and design to the Swedish Bofors L/60 weapon.

Following extensive trials, the 49-K was accepted for service with the Russian Army as the 45 mm Model 1939 Automatic Anti-aircraft Gun.

This was not placed in production.

In January 1938, the Artillery Directorate requested that the then Artillery Factory No 8 develop a 37 mm version of the 49-K.

This was rapidly built under the designation of the ZIK-37, subsequently renamed the 61-K. Firing trials commenced in October 1938 with the weapon installed on a four-wheeled ZU-7 carriage which used the wheels and tyres from the GAZ-AA truck.

Twin barrel version of 37 mm automatic anti-aircraft gun M1939 in service with Egyptian Army (Egyptian Army) 0509496

Table 1

	FRAG-T	FRAG-T	AP-T
Projectile designation:	OR-167	OR-167N	BR-167
Fuze model:	MG-8	B-37	n/app
Weight:			
(projectile)	0.732 kg	0.708 kg	0.77 kg
(bursting charge)	0.035 kg	0.036 kg	n/app
Filling:	RDX/aluminium	A-IX-2 (RDX/aluminium/wax)	nil
Muzzle velocity:	880 m/s	880 m/s	880 m/s
Armour penetration:			
(500 m at 0° obliquity)	n/app	n/app	47 mm
(1,000 m at 0° obliquity)	n/app	n/app	37 mm

37 mm automatic anti-aircraft gun M1939 fitted with shield and with outriggers in position (Franz Kosar) 0509497

37 mm automatic anti-aircraft gun M1939 in travelling position but with barrel not in travelling lock (Steven Zaloga) 0509695

The 61-K was soon accepted for service under the designation of the 37 mm Automatic Anti-aircraft Cannon M1939 and the first order placed for 900 units. It is believed that total production amounted to almost 20,000 weapons with final deliveries made in 1945.

The M1939 has been built in China as the Type 55. A twin version built for export is known to be in service with Algeria and Egypt and was manufactured in China as the Type 65. The Russian Navy in both single (70-K) and twin liquid-cooled mounts (V-11M) has also used the 37 mm gun.

The effectiveness of the 37 mm automatic anti-aircraft gun M1939 against modern aircraft is very limited and it has no poor weather or all-weather capability. As far as is known, there have been no recent programmes to improve this weapon or its associated ammunition.

Description

In the firing position, the wheels are raised off the ground and it is supported by four screw jacks, one at the front, one at the rear of the carriage and one either side on outriggers. When travelling, the barrel is pointed to the rear and is held in position by a lock hinged at the rear of the carriage.

Most countries have removed the shield, which weighs about 100 kg. The M1939 is a clear weather system only, with no provision for radar fire-control.

The M1939 can fire the types of fixed 37 mm ammunition, which is fed in clips of five rounds, listed in Table 1 at top of page.

The High Velocity Armour Piercing (HVAP) ammunition nature is no longer used. It was capable of penetrating 57 mm of steel armour at 1,000 m.

The 37 mm M1939 anti-aircraft gun has an effective slant range of 2,499 m. Effective altitude limit with an elevation of +45° is 1,768 m and with an elevation of +65° is 2,865 m. Self-destruct range is 4,389 m.

Specifications

	M1939
Dimensions and weights	
Crew:	8
Length	
transport configuration:	5.50 m
Width	
transport configuration:	1.785 m (without shield)
Height	
transport configuration:	2.105 m (without shield)
axis of fire:	1.1 m
Ground clearance	
overall:	0.36 m (travelling)
Track	
vehicle:	1.545 m
Wheelbase	3.259 m
Weight	
combat:	2,100 kg (without shield)
Mobility	
Carriage:	trailer mounted four wheeled
Tyres:	6.50 × 20
Firepower	
Armament:	1 × 37 mm cannon
Main weapon	
length of barrel:	2.729 m
breech mechanism:	vertical sliding block
maximum range:	9,500 m (horizontal)
maximum vertical range:	6,700 m
feed mechanism:	200 round
operation:	recoil automatic
cartridge:	37 mm × 250 mm R
Rate of fire	
cyclic:	170 rds/min (est.)
practical:	80 rds/min
Main armament traverse	
angle:	360°
Main armament elevation/depression	
armament front:	+85°/-5°

Status

Production complete. This system has seen service with a number of countries, but due to age, plus the lack of ready spares, it is likely that other guns have been broken down and used as spares. Therefore, the operational capability within these countries cannot be confirmed.

There is a separate entry for the Chinese 37 mm weapons. These are marketed by China North Industries Corporation (NORINCO) and include the Type 55, Type 65, Type 74, Type 74SD and Type P793. North Korea has developed self-propelled versions of the Chinese twin 37 mm system mounted on a tracked chassis. Available details of this are given in self-propelled anti-aircraft gun section under Korea, North.

Contractor

Russian state factories.

Artillery Plant No. 8, Kaliningrad; No. 4, Krasnoyarsk; No. 586, Kolomna.

Also Chinese state factories and Polish state factories.

M1939 and M1944

Type

Towed anti-aircraft gun

Development

The 85 mm anti-aircraft gun M1939 (KS-12) was designed by M N Loginov and G D Dorokhin at the then Artillery Plant No 8, Kaliningrad near Moscow and was introduced into the Red Army shortly before the start of the Second World War as the replacement for the 76.2 mm M1938.

The ordnance of the M1939 was later adopted for use in the SU-85 assault gun and later still the T-34/85 tank. The M1939 is no longer in use within Russia and has been replaced by surface-to-air missile systems.

China North Industries Corporation (NORINCO) has also manufactured this weapon under the designation of the Type 56. It is understood that production of this was completed in China many years ago.

85 mm anti-aircraft gun M1939 in travelling order with outriggers retracted and ordnance in travelling lock (Steven Zaloga) 0509694

85 mm anti-aircraft guns M1944 deployed in the firing position and showing different muzzle brake from earlier M1939 0509495

Description

When travelling, the mount together with 85 mm ordnance is traversed to the rear and the ordnance is held in position by a travelling lock at the rear of the carriage. In the firing position the carriage is supported on four screw jacks, one at either end of the carriage and one either side on outriggers.

The M1939 was often seen without the shield. The original KS-12 had built-up ordnance whereas the later KS-12A had monobloc ordnance.

The M1939 is provided with on-carriage fire-control equipment but is normally used in conjunction with the PUAZO-6/12 director and the SON-9/SON-9A (NATO designation 'Fire Can') A/B-band fire-control radar. This fire-control system is also used with the 85 mm anti-aircraft guns, M1939 and M1944.

The ammunition fired by the M1939 is of the fixed type, some of which is interchangeable with that fired by 85 mm assault guns, field and tank guns. In addition to firing all the rounds of the Soviet 85 mm field guns, the M1939 85 mm anti-aircraft gun has special cartridges with time-fused projectiles for firing at aircraft, details are given in Table 1:

Table 1

Ammunition type	FRAG	FRAG
Projectile designation:	O-365[1]	O-365M
Fuze model:	T-5[2]	VM-2[3]
Weight:		
(projectile)	9.2 kg	9.24 kg
(bursting charge)	0.64 kg	0.776 kg
Type of bursting charge:	TNT	TNT
Muzzle velocity:	792 m/s	792 m/s

[1] There are variants in O-365 projectiles giving different bursting charge weights as well as projectile weights
[2] Powder train time fuze can be set from 1.6 to 32 s
[3] Mechanical clockwork fuze can be set from 0.8 to 30 s

Other projectiles include:

- AP-T BR-365 (fitted with MD-5 fuze)
- AP-T BR-365K (fitted with MD-8 fuze)
- HVAP-T BR-365P (no fuze)
- HVAP BR-365-PK (no fuze).

The effective slant range of the 85 mm M1939 is 8,382 m. Effective altitude limit with an elevation of +45° is 5,944 m. Effective altitude limit with an elevation of +65° is 7,620 m while self-destruct range is 10,516 m.

It was succeeded in production in 1944 by the M1944, which was designed by G D Dorokhin and had a number of modifications including a longer ordnance with a T-shaped muzzle brake. The M1944 fired the same ammunition as the M1939 except that the HE round had a more powerful propellant charge that increased its muzzle velocity to 900 m/s compared with the 792 m/s of the basic round. The complete HE round weighed 15.9 kg compared with the 15.1 kg of the standard round. The Soviet Army in large numbers did not use the M1944 as it was soon replaced by the 100 mm KS-19. Production of the M1944 was undertaken in Czechoslovakia after the Second World War.

Specifications

	M1939	M1944
Dimensions and weights		
Crew:	7	7
Length		
transport configuration:	7.049 m	8.2 m
Width		
transport configuration:	2.15 m	2.15 m
Height		
transport configuration:	2.25 m	2.25 m
axis of fire:	1.55 m	1.55 m
Ground clearance		
overall:	0.4 m	0.4 m
Track		
vehicle:	1.8 m	1.8 m
Weight		
standard:	4,300 kg	5,000 kg
combat:	4,300 kg	5,000 kg
Mobility		
Carriage:	trailer mounted four wheeled	trailer mounted four wheeled
Tyres:	34.00 × 7	34.00 × 7
Firepower		
Armament:	1 × 85 mm cannon	1 × 85 mm cannon
Main weapon		
length of barrel:	4.693 m	5.743 m
breech mechanism:	vertical sliding block	vertical sliding block
muzzle brake:	multibaffle	double T type
maximum range:	15,650 m	18,000 m
maximum vertical range:	10,500 m	11,600 m
feed mechanism:	150 round	150 round
Rate of fire		
cyclic:	20 rds/min (est.)	20 rds/min (est.)
Main armament traverse		
angle:	360°	360°
Main armament elevation/depression		
armament front:	+82°/-3°	+82°/-3°

Status

Production complete.

Contractor

Czech state factories and Soviet state factories (former).
Artillery Plant No 8, Kaliningrad, near Moscow and evacuation site in Sverdlovsk. China North Industries Corporation (NORINCO) has built this weapon under the designation of the Type 56 AAG.

NSV

Type

Towed anti-aircraft gun

Development

The 12.7 mm NSV heavy machine gun (also known by the Russian name Utjos - the original code name for the project) was designed in the late 1960s by a three-man team consisting of G I Nikitin, J M Sokolov and V I Volkov. The NSV was seen as a replacement for the older, widely deployed 12.7 mm DShK and DShKM weapons.

The first application was for the NSV vehicle turret-mounted version which is standard fit for the T-64, T-72, T-80, T-84 and T-90 MBTs.

The second version to enter service was the 6T7 tripod system which has a manual traverse of 360° and manual elevation. Targets are engaged using an SPP K10-T collimating optical day sight with iron sights provided for back up.

The third model to enter service is a specialised air defence version known as the 6U6 which consists of a tripod, weapon mount for the 12.7 mm NSV heavy machine gun, folding seat for the gunner, VK-4 reflex anti-aircraft day sight and a PU telescopic day sight for engaging ground targets.

Description

The NSV is a gas operated, air cooled, belt fed, automatic machine gun that fires from an open bolt. When engaging air targets the gunner operates from the sitting position with the seat folded up. The portable mount weighs 55 kg and 92.5 kg with the 12.7 mm NSV machine gun and a box of 70 rounds of ammunition.

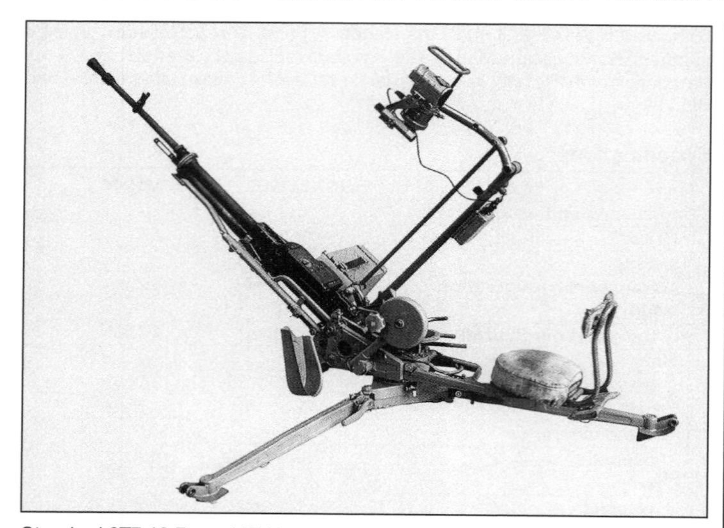

Standard 6T7 12.7 mm NSV heavy anti-aircraft machine gun system deployed in firing position 0509878

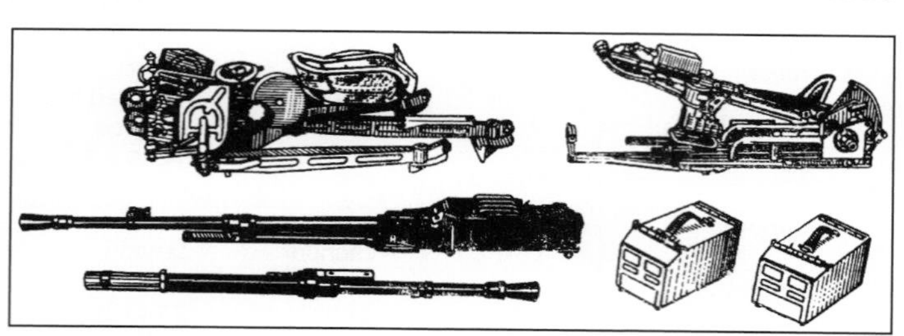

For transportation across rough terrain the 6T7 12.7 mm NSV heavy anti-aircraft machine gun system can be disassembled into five manpack loads 0509697

The 12.7 mm weapon has a cyclic rate of fire of 680 to 800 rds/min with the barrel capable of being changed quickly after about 1,000 rounds have been fired.

The NSV normally fires API ammunition with every fourth round being an API-T. The API will penetrate 20 mm of conventional steel armour at a range of 500 m and 13.2 mm of conventional steel armour at a range of 1,000 m. In addition to API and API-T rounds, ball types can also be fired.

Effective range against anti-aircraft targets is 1,500 m while against ground targets it is around 1,500 to 2,000 m.

This is a clear weather system only with no provision for off-carriage fire-control.

Variants

Anti-aircraft, machine gun/grenade launcher mount

For use during convoy escort work in Afghanistan, the 30 mm AGS-17 grenade launcher was attached to the standard 12.7 mm 6U6 anti-aircraft machine gun mount. This combination is known as the anti-aircraft, machine gun/grenade launcher mount.

Front and rear frames are attached to the machine gun cradle on this mount. The standard cradle for the 30 mm AGS-17 is attached at the rear hinge of the machine gun cradle with its forward part being clamped into the front frame with the help of a spring-loaded clamp. This is used if it is necessary to flip the grenade launcher so that the operator has direct access to the barrel chamber and feed mechanism of the machine gun.

Turning an adjustment screw in the controlling section of the mechanism increases the regulator on the elevation balancing hand wheel mechanism.

There are two separate organic sights on the mount, the OP-81 optical ground sight and the OP-80 collimator anti-aircraft sight.

UTIOS

The 12.7 mm NSV machine gun and its associated 6T7 mount is also manufactured in Poland by Zakłady Mechaniczne Tarnow with marketing being carried out at the Cenzin Ltd foreign trade enterprise. The Polish name for this system is the UTIOS.

NSVT

The NSVT (Tank) is a modification that has been made to allow the gun to be fitted to most Russian Main Battle Tanks (MBTs).

Specifications

12.7 mm NSV HMG

Calibre:	12.7 mm
Cartridge:	12.7 × 108 mm
Muzzle velocity:	845 m/s
Number of barrels:	1
Operation:	gas, automatic
Method of locking:	rotating bolt
Width:	154 mm
Height:	166 mm
Rate of fire:	
(cyclic)	680-800 rds/min
(practical)	80-100 rds/min
(on 6T7 mount)	270 rds/min
Sights:	
(aircraft targets)	1,500 m
(ground targets)	2,000 m
Maximum range:	6,000 m
Effective range:	1,500-2,000 m

6T7 Mount

Ammunition supply:	70 rounds
Rate of laying: (at 2 revolutions of hand wheel per second)	
(elevation, 1st speed (fine lay))	4.1°/s
(elevation, 2nd speed)	30°/s
(traverse, 1st speed)	9°/s
(traverse, 2nd speed)	55°/s
Sight magnification:	×3.5
Weight:	
(mount, ammunition box and 70 rnds)	92.5 kg
(gun)	25 kg
(loaded ammunition box)	12.5 kg
Length:	
(deployed, no elevation)	2.52 m
(deployed, +85° elevation)	1.65 m
Height:	
(deployed, no elevation)	0.765 m
(deployed, +85° elevation)	1.885 m
Width:	
(deployed, no elevation)	1.185 m
(deployed, +85° elevation)	1.68 m

Status

The 12.7 mm NSV is widely used by countries that use the T-72 MBT.

As far as is known the anti-aircraft version is only used by the Russian Federation.

It is currently (January 2014) offered on the export market and it is possible that some although not proven, systems have been exported.

Contractor

Molot Vyatskiye Polyany Machine Building Plant JSC (gun).
KBP Instrument Design Bureau (gun).
Novosibirsk Instrument-Building Plant (sight), Production Association, GUP/PO NPZ.
This machine gun is also manufactured under licence in a number of other countries including Poland which also produces the 6T7 mount.

S-60

Type

Towed anti-aircraft gun

Development

The 57 mm automatic anti-aircraft gun (AZP) S-60 was designed by Central Design Bureau and introduced in 1950 as the replacement for the older 37 mm M1939 anti-aircraft gun. Main improvements over the latter include greater range and the facility to use an off-carriage fire-control system.

In the Soviet Army, the S-60 was issued on the scale of 24 per tank division and 24 per motorised rifle division. Each of them had an anti-aircraft regiment, which had four batteries each of six guns, with each battery having two

Ammunition type	**FRAG-T**	**FRAG-T**	**APHE-T**
Projectile designation:	OR-281	OR-281U	BR-281 and BR-281U
Fuze model:	MG-57	MGZ-57	MD-10
Weight:			
(projectile)	2.81 kg	2.85 kg	2.82 kg
(bursting charge)	0.168 kg	0.154 kg	0.018 kg
Type of bursting charge:	A-IX-2 (RDX/Al/Wax)	A-IX-2 (RDX/Al/Wax)	A-IX-2 (RDX/Al/Wax)
Muzzle velocity:	1,000 m/s	1,000 m/s	1,000 m/s
Armour penetration:			
(500 m)	n/avail	n/avail	106 mm/0° obliquity
(1,000 m)	n/avail	n/avail	96 mm/0° obliquity

57 mm automatic anti-aircraft gun S-60 in travelling configuration of the Polish Army (Jaroslaw Cislak) 0044151

57 mm S-60 automatic anti-aircraft gun installed on MT-LB multipurpose armoured vehicle for use in ground role 0044152

three-gun platoons. The headquarters battery had two P-15 ('Flat Face') target acquisition radars and each six-gun battery had a single SON-9 ('Fire Can') fire-control radar, later replaced or supplemented by the SON-9A. The S-60 57 mm automatic anti-aircraft gun has been replaced in all Russian units by the OSA (SA-8) mobile (6 × 6) SAM system.

In turn, the OSA (SA-8) has been replaced by the Almaz/Antei Concern of air defence Tor system (SA-15) which has eight vertically launched surface-to-air missiles in the ready-to-launch position.

The 57 mm S-60 was first used in combat during the Korean War and a total of 5,725 were built between 1950 and 1957, when production is understood to have been completed.

As late as September 2010, the S-60 gun is still in service with many countries and is still being upgraded and modified with the latest modification being handled by the Polish Armed Forces where the system is to be modified for remote control.

Description

In the firing position, the wheels are raised off the ground and the carriage is supported on four screw jacks, one at the front and one at the rear of the carriage and one either side of the carriage on outriggers. The gun can be fired from its wheels in an emergency. Fire-control equipment consists of a reflex sight for anti-aircraft use and a telescope sight for ground use. There are four modes of operation:

- Manual with the hand wheels operated by the crew
- Assisted, with the hand wheels operated by the crew assisted by a servo-motor
- Automatic, remotely controlled by a director
- Zero indicator, remotely controlled by radar.

When originally introduced, it was used in conjunction with the PUAZO-5 director and the SON-9 radar. Today, it is used in conjunction with the PUAZO-6/60 director and the SON-9 or SON-9A A/B-band radar, but in recent years improved director and radar combinations have entered service. Russian sources have also mentioned that the S-60 could be used in conjunction with the Vaza RPK-1 radar fire-control system that laid the guns onto the target by remote control.

Photographs of Soviet-built 'Flap Wheel' I-band anti-aircraft radars associated with the 57 mm S-60 anti-aircraft gun in Iraqi service, show that the radar has been modernised. A low-light television camera, similar to that seen on the 'Land Roll' radar on SA-8 vehicles and some 'Low Blow' radars associated with the Neva system, has been mounted on top of the 'Flap Wheel' antenna. 'Flap Wheels' with long cables have also been observed, enabling them to be placed 200 m away from the firing position. These modifications will increase the effectiveness of the S-60, especially when confronted by chaff or electronic jamming.

The top of each side of the shield folds forward through 180°. The ammunition, which is fed to the gun in four-round clips, is not interchangeable with that used by the 57 mm ASU-57 self-propelled anti-tank gun or the 57 mm

Close-up of ammunition feed tray on 57 mm S-60 automatic anti-aircraft gun (Christopher F Foss) 0509875

PUAZO series director which is used with the 57 mm automatic anti-aircraft gun S-60 (Steven Zaloga) 0509767

towed anti-tank guns due to a different configuration. A horizontal feed tray on the left side of the mounting holds one clip of four 57 mm rounds and a second clip can be placed on an upright stand that rotates with the mounting.

The following types of fixed ammunition can be fired by the S-60, see table at the bottom of the opposite page.

Effective slant range using onboard optical sights is 3,993 m, which increases to 6,005 m with radar. Effective altitude limit with an elevation of +45° is 2,835 m with onboard optical sights and 3,627 m with radar, effective altitude limit with an elevation of +65° is 4,237 m with onboard optical sights and 5,425 m with radar. Self-destruct range of the FRAG-T projectile is quoted as 7,224 m.

Russian sources have stated that after firing 50 rounds of ammunition, firing has to stop to allow the barrel to cool down.

Variants

Type 59

The S-60 was manufactured in China by China North Industries Corporation (NORINCO) as the Type 59. Details of the Type 59 and its associated fire-control systems are given earlier in this section under China.

ZSU-57-2

The ZSU-57-2 twin 57 mm self-propelled anti-aircraft gun has the same ballistic performance as the S-60 and uses the same ammunition.

Type 80

The Chinese equivalent of the Russian ZSU-57-2.

Naval S-60

The 57 mm gun is also used in a number of naval applications.

MT-LB with S-60

The MT-LB multipurpose full-tracked armoured personnel carrier has been observed fitted with a 57 mm S-60 for use in the ground role.

Specifications

	S-60
Dimensions and weights	
Crew:	7
Length	
transport configuration:	8.6 m
Width	
transport configuration:	2.054 m

	S-60
Height	
transport configuration:	2.46 m
axis of fire:	1.3 m
Ground clearance	
overall:	0.38 m
Weight	
standard:	4,875 kg
combat:	4,775 kg
Mobility	
Carriage:	trailer mounted four wheeled
Speed	
towed:	60 km/h
Tyres:	34.00 × 7
Firepower	
Armament:	1 × 57 mm cannon
Main weapon	
length of barrel:	4.390 m (with muzzle brake)
muzzle brake:	multibaffle
maximum range:	12,000 m
maximum vertical range:	6,000 m (with off-carriage fire control)
feed mechanism:	4 round clip
operation:	recoil automatic
cartridge:	57 mm × 347 mm SR
Rate of fire	
cyclic:	120 rds/min (est.)
practical:	70 rds/min
Main armament traverse	
angle:	360°
Main armament elevation/depression	
armament front:	+87°/-2°

Status
Production complete.

As of October 2012, this system is still being upgraded and modified by local factories and institutions.

Contractor
Gun system

Aleksinskiy Khimicheskiy Kombinat.

The S60 has also been manufactured in Hungary under the designation of SZ-60 and in China with the local designation being the Type 59.

Ammunition

Aleksinskiy Khimicheskiy Kombinat, Gosudarstvennoye Proizvodstvennoye Obedinesiye (GPO Aleksinskiy Khimkombinat), produces rounds for S-60.

SOSNA

Type
Towed anti-aircraft gun

Development
In 1997, it was revealed that the KB Tochmash Design Bureau, located in Moscow (which has considerable experience in the development of surface-to-air missiles), was developing a new trailer-mounted 30 mm air defence system called SOSNA for the export market. By 2004/5 the system made its first public appearance with a confirmation that this is a Tochmash private venture.

Description
The SOSNA system is intended to engage threats that include precision-guided weapons, cruise missiles and other air breathing targets. SOSNA is armed with the 2A38M 30 mm twin-barrelled automatic cannon, which is already used in the 2K22M Tunguska self-propelled anti-aircraft gun/Surface-to-Air Missile (SAM) system, which has been in service with the Russian Army for some years and has been exported to India. Details of the 2K22M are given in the Self-propelled anti-aircraft guns/surface-to-air missile section.

Elements of this are also used in the more recent 96K6 Pantsyr-S1 self-propelled air defence system on an 8 × 8 chassis which has recently been ordered by the United Arab Emirates.

The 2A38M 30 mm twin-barrelled automatic cannon is water-cooled and has a cyclic rate of fire up to 2,400 rounds per minute. The effective range in the air defence role is 3,000 m in altitude with a maximum slant range of 4,000 m.

The weapon uses the 30 × 165 mm cartridge and fires High Explosive - Tracer (HE-T) and High Explosive Incendiary (HEI) natures of ammunition. The 2A38M 30 mm cannon is manufactured by the Tulamashzavod facility in Tula.

SOSNA is based on a two-axle trailer which, when deployed in the firing position, is supported on four outriggers, two either side.

The power-operated weapon mount is positioned in the centre of the trailer with the operator seated in a fully enclosed cabin on the left side and the 30 mm 2A38M cannon on the right. A total of 600 rounds of ready use, 30 mm

SOSNA-A 30 mm low-level air defence system showing electro-optics pod in front of gunner's cabin 0010000

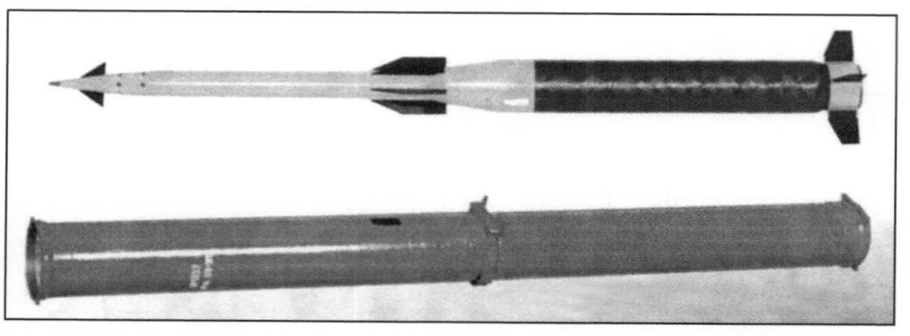

SOSNA-R surface-to-air missile out of its launcher tube as in flight (top) and missile in launcher canister (bottom) 0045791

SOSNA 1022688

ammunition is carried and the carriage is provided with a generator. An IFF system can also be added if required by the customer. The 30 mm magazine is mounted to the right of the 30 mm 2A38M cannon.

Mounted in front of the operator's cabin is the ball-type electro-optic device (optronic control system) that is similar in appearance to that used on the MBDA Rapier Field Standard C (or Jernas for the export market) SAMs system used by the British Army and RAF Regiment. This modular electro-optic fire-control package has a day/night sighting system with laser range-finder and digital computer. The Urals Optical Mechanical Plant Production Association has developed the electro-optical device.

The gunner is provided with a control panel with all of the necessary controls to lay the weapon onto the target. He also has an optical collimator sight with a wide field of view.

The fire-control system is said to ensure high effectiveness against both ground and air targets.

The automatic optical module has been designed for independent or automatic target detection (with target information being received from the command post or a radar system). It can track targets in automatic or semi-automatic modes and measures the target's line-of-sight angle range. This information is transferred to the digital computer, which computes the required information to lay the weapon onto the target.

Although the first example of SOSNA is trailer-mounted, the prime contractor stresses that this system is of modular design so it can also be mounted on a variety of tracked and wheeled chassis, which can take a payload of 3,500 kg.

To further increase the effectiveness of the SOSNA system, it can also be provided with a thermal tracking and surveillance system as well as being integrated with other search and target designation systems.

If the system is fitted with the passive, automatic, infrared air target detector, the target can be detected and automatically tracked without the assistance of the gunner. All the latter has to do is press the finding button when the target is in range of the weapons.

To further enhance its air defence capability, the SOSNA system can be fitted with SAMs. In a typical target engagement, the missiles would be used to engage targets at long range (up to 10 km in range and up to 5 km in altitude) and then with the 30 mm 2A38M gun being used to engage close-in targets which cannot be engaged by the missiles.

It was originally believed that standard Kolomna KBM Igla (NATO SA-18 'Grouse') or older Igla-1 (NATO SA-16 'Gimlet') missiles would be used, but it has been revealed that a new longer-range laser-guided missile called SOSNA-R will now be employed.

The new missile is designated as the 9M337, is very similar in appearance to 9M311/9M311M used by the 2K22M and the 9M335 used by the Pantsyr-S1 truck mounted air-defence gun/missile system. The Nudelman Precision Engineering Design Bureau has developed the missile.

The SOSNA-R, however, is a high speed maintenance free smaller missile with a maximum speed of 875 m/s with an average reported flight speed of 565 m/s. The missile itself is 2.317 m in length and 72 mm in diameter with the booster motor having a diameter of 140 mm. The electro-optic package has been modified to accept the laser guidance system for the SOSNA-R, which is claimed to be highly effective and difficult to jam.

The missile takes 11 seconds to reach a range of 8 km and is claimed to have an effective altitude of up to 3.5 km and an effective range from 1.3 to 8 km. In addition to being used with the SOSNA air defence system, the laser-guided SOSNA-R can also be used with the naval Palma air defence system.

The missile weighs 29.5 kg with the 5 kg warhead of advanced design being fitted with a 12-channel laser proximity fuze. The missile is transported and launched from a sealed container, which weighs 42 kg and is 2.4 m long and 153 mm in diameter.

Variants

Current variants of the SOSNA at present are the:

- SOSNA
- SOSNA-A
- SOSNA-R.

Specifications

Model	SOSNA-A Gun system	SOSNA Gun + missile system
Crew:	2	2
Effective range:		
(maximum)	4,000 m	10,000 m
(minimum)	100 m	1,300 m
Effective altitude:		
(maximum)	3,000 m	5,000 m
(minimum)	n/avail	2 m
Cyclic rate of fire:	2,400 rpm	2,400 rpm
Muzzle velocity:	960 m/s	n/avail
Ammunition:		
(30 mm)	300 rounds	300 rounds
(missiles)	n/avail	4
Target speed:	0 - 300 m/s	0 - 500 m/s
Kill probability:	>0.6	>0.95
Reaction time:	3 to 5 sec	3 to 5 sec

Missile (SOSNA-R)

Length:	2,317 mm
Diameter:	
(missile)	0.072 m
(booster motor)	0.140 m
Weight:	
(in canister)	42 kg
(launch weight)	30 kg
Guidance:	combined missile guidance system with a smoke resistant radio command system used in the launch phase (first second) and a precision countermeasures-immune laser coded beam riding system after engine separation.
Warhead:	
(type)	two section blast continuous-rod fragmentation with piercing effect
(weight)	5 kg
Fuze:	12-channel laser proximity and contact with a circular sensitivity diagram compensating guidance errors and command for armour piercing mode
Propulsion:	solid propellant motor with a progressive burning grain
Speed:	875 m/s
Duration of flight to 8,000 m:	11.5 s

Mount

Crew:	3
Armament:	30 mm 2A38M cannon; 4 × SAM (optional)
Calibre: (cannon)	30 mm
Cartridge:	30 × 165 mm
Ammunition:	600 rounds (ready use)
Length:	6.8 m
Height:	2.7 m
Axis of fire:	1.8 m
Ground clearance:	0.350 m
Weight:	6,300 kg
Elevation/depression:	+85°/–5°
Traverse:	360°
Towing vehicles:	Ural-432 or KAMAZ-4310 (typical)

Status

Development complete, offered for sale on the arms open market.

Contractor

KB Tochmash Design Bureau of Precision Engineering (primary contractor).
Tulamashzavod facility in Tula (2A38M 30 mm cannon).
Urals Optical Mechanical Plant Production Association (development of the electro-optic device).
Nudelman Precision Engineering Design Bureau (development of missile).

ZPU-1, ZPU-2 and ZPU-4

Type

Towed anti-aircraft gun

Development

ZPU-1

The ZPU-1 14.5 mm light anti-aircraft gun was introduced into Soviet Army service immediately after the Second World War. Like the ZPU-2 and ZPU-4, the ZPU-1 uses the 14.5 mm Vladimirova KPV (*Krubnokalibernyj Pulemet Vladimirova*) Heavy Machine Gun (HMG), which has a quick-change barrel. The 14.5 mm KPV series is also fitted to a number of AFVs including:

- BRDM-2 reconnaissance vehicle
- BTR-60PB (8 × 8) APC
- BTR-70 (8 × 8) APC
- BTR-80 (8 × 8) APC
- Czech OT-64 (8 × 8) APCs models:
  - OT-64C(1)
  - OT-64C(2)
  - OT-62C tracked APC.

It was reported that at least one insurgent faction in Afghanistan had managed to obtain approximately two ZPU-1s. The faction attempted to install these on the rear of a flatbed truck for air-defence, both were destroyed within 12 hours by coalition forces mid-April 2009. China has produced its own version of the KPV which is known as the Type 56 heavy machine gun.

ZPU-2

The first model of the ZPU-2 entered service in 1949, had larger mudguards and a double tubular tow bar. In the firing position, the wheels are removed and the weapon rests on a three-point platform, each point having a screw jack for levelling.

The late production model ZPU-2 is lighter and lower, has narrower mudguards and a lighter single tow bar. In the firing position the wheels are raised clear off the ground but not removed.

ZPU-4

The ZPU-4 entered service with the Soviet Army in 1949 and although no longer in front-line service with the Russian Federation Army, like the ZPU-2, it may be found in second-line units defending airfields and other high-priority targets.

Description

All of the ZPU series of 14.5 mm anti-aircraft guns are clear weather systems with no provision for off-carriage fire-control.

All weapons in this series fire the following fixed ammunition:

- API (BS 41) projectile weighing 64.4 g, with a muzzle velocity of 1,000 m/s, which will penetrate 32 mm of conventional steel armour at an incidence of 0° at a range of 500 m
- API-T (BZT) projectile weighing 59.56 g
- I-T (ZP) projectile weighing 59.68 g.

The 14.5 mm ZPU series of anti-aircraft machine guns has an effective slant range of 1,402 m, effective limit at +45° elevation is 975 m while effective altitude limit at +65° is 1,280 m.

ZPU-1

The two-wheeled carriage of the ZPU-1 was designed by Vodop'yanov and Rachinskiy. For transport in rough terrain, it can be dismantled into units weighing about 80 kg each. The quick-change barrel is air cooled, with the ammunition box on the right side.

BTR-152A twin 14.5 mm SPAAG, late production vehicle with central tyre pressure regulation system 0509876

BTR-40A twin 14.5 mm SPAAG 0509877

14.5 mm ZPU-1 anti-aircraft gun in firing position in Zimbabwe (UK Land Forces) 0509500

Late production model ZPU-2 (Franz Kosar) 0509501

Early model ZPU-2 in travelling order (US Marine Corps) 0509502

ZPU-2 LAAG of the South African Army deployed in firing position (T J Gander) 0007662

ZPU-4 anti-aircraft machine gun deployed in firing position and showing outriggers either side (Christopher F Foss) 0509768

ZPU-2

The ZPU-2 is no longer in front-line service with the Russian Federation as it has been replaced by the similar 23 mm ZU-23. It is distinguishable by its ammunition boxes, which are horizontal rather than vertical and its parallel flash hiders. The ZPU-2 is manufactured in China by China North Industries Corporation (NORINCO) as the Type 58 (see China). It has been manufactured in Romania under the local designation of the ZU-2. Additional details of this weapon and their version of the ZPU-4 are given in this section under Romania.

ZPU-4

The ZPU-4 has a four-wheel carriage designed by Leshchinskiy. In the firing position, the wheels are raised off the ground and the carriage is supported at four points, screw jacks at each end of the carriage and on outriggers on each side of the carriage that are also provided with screw jacks. The weapon can be brought into action in 15 to 20 seconds but can, if required, be fired with the wheels in the travelling position. The ZPU-4 is manufactured in China by China North Industries Corporation (NORINCO) and the Type 56 and has been produced in North Korea, as is the ZPU-2.

Variants

Various vehicles which are fitted with variants of the ZPU-series mountings have been produced.

BTR-40A SPAAG

The BTR-40A SPAAG entered service in 1950 and is essentially a BTR-40 (4 × 4) armoured personnel carrier with a twin 14.5 mm ZTPU-2 mount in the rear.

The mount has manual traverse through a full 360° with weapon elevation from –5° to +80°. This was phased out of service with the Soviet Army many years ago but can still be found in service in the developing nations.

BTR-152A SPAAG

The BTR-152A SPAAG entered service in 1952 and is essentially a BTR-152 (6 × 6) armoured personnel carrier with a twin 14.5 mm ZTPU-2 mount in the rear.

The mount has manual traverse through a full 360° with weapon elevation from –5° to +80°. It was phased out of service with the Russian Federation Army many years ago but can still be found in service in the developing nations.

	ZPU-1	ZPU-2	ZPU-2 (late)	ZPU-4
Dimensions and weights				
Crew:	4	5	5	5
Length				
transport configuration:	3.44 m	3.536 m	3.871 m	4.53 m
Width				
transport configuration:	1.62 m	1.92 m	1.372 m	1.72 m
Height				
transport configuration:	1.34 m	1.83 m	1.097 m	2.13 m
axis of fire:	0.635 m	0.8 m	-	1.02 m
Ground clearance				
overall:	0.28 m	-	0.27 m	0.458 m
Track				
vehicle:	1.384 m	-	1.1 m	1.641 m
Weight				
standard:	413 kg	994 kg	649 kg	1,810 kg
combat:	413 kg	639 kg	621 kg	1,810 kg
Mobility				
Carriage:	trailer mounted two wheeled	trailer mounted two wheeled	trailer mounted two wheeled	trailer mounted four wheeled
Tyres:	4.50 × 16	6.50 × 20	6.50 × 20	6.50 × 20
Firepower				
Armament:	1 × 14.5 mm cannon	2 × 14.5 mm cannon	2 × 14.5 mm cannon	4 × 14.5 mm cannon
Main weapon				
length of barrel:	1.348 m	1.348 m	1.348 m	1.348 m
maximum range:	8,000 m	8,000 m	8,000 m	8,000 m
maximum vertical range:	5,000 m	5,000 m	5,000 m	5,000 m
feed mechanism:	50 round (est.) drum	50 round (est.) drum	50 round (est.) drum	50 round (est.) drum
cartridge:	14.5 mm × 114 mm	14.5 mm × 114 mm	14.5 mm × 114 mm	14.5 mm × 114 mm
Rate of fire				
cyclic:	600 rds/min	1,200 rds/min	1,200 rds/min	2,400 rds/min
practical:	150 rds/min	300 rds/min	300 rds/min	600 rds/min
Main armament traverse				
angle:	360°	360°	360°	360°
Main armament elevation/depression				
armament front:	+88°/-8°	+90°/-7°	+85°/-15°	+90°/-10°

ZPU-2 anti-aircraft gun deployed off its carriage with trailer to rear (Michael Green) 0509696

Specifications – *see table above*

Status
Production complete.

Contractor
Chinese, North Korean, Romanian and Russian state factories.

ZU-23M

Type
Towed anti-aircraft gun

Development
The Russian twin 23 mm ZU-23-2 towed Light Anti-Aircraft Gun (LAAG) is the most widely used weapon of its type in the world and full details are given in a separate entry.

Over 60 countries still use the twin 23 mm ZU-23-2 and it has also been made under licence, or copied, by a number of other countries including Bulgaria, China, Egypt, Iran and Poland.

The main drawback of the twin 23 mm ZU-23-2 is that it has manual elevation and traverses and is essentially a clear weather air-defence system, these problems have effectively been solved with the updated and modified ZU-23M. Effective range is 200 to 2,500 m.

Russian upgraded twin 23 mm ZU-23M light anti-aircraft gun deployed in firing position 0044153

The twin 23 mm ZU-23-2 LAAG is normally operated by a crew of five of which two are on the actual mount. Each 23 mm 2A14 cannon has a cyclic rate of fire per barrel of 800-1,000 rounds/minute while the effective rate of fire per barrel is 200 rounds/minute. Each barrel is provided with a box type magazine that holds 50 rounds of ready-use ammunition.

In 1999 the ZU-23-2 fitted with two Kolomna KBM 9M39 Igla (NATO SA-18 'Grouse') fire-and-forget surface-to-air missiles mounted above the 23 mm cannon was shown for the first time. In order to accomplish this modification (which could also handle the 9M342 Igla-s) the system is fitted with the PM-23M1-1 launch module tohandle the missiles. Details are given in the entry for the ZU-23-2 system.

Description
The Russian K B Tochmash Design Bureau of Precision Engineering based in Moscow, with the assistance of the Podolsky Elektromechanichesky Zavod Stock Corporation, has developed an upgrade package for the twin 23 mm ZU-23-2 LAAG. Brief details of this were released in late 1998. As far as it is known this system has been developed mainly for the export market.

According to K B Tochmash, the upgrade package automates the aiming of the weapons, allows targets to be engaged under day and night conditions and gives a higher kill probability against faster targets. This upgraded system is called the ZU-23M.

Power for the system is initially drawn from two batteries mounted in a box below the gun operators seat, these batteries have up to two hours life before either an external power supply is required either from portable generators or local mains.

To allow for faster elevation and traverse capabilities, all electric drives have been fitted which allow the weapons to be traversed at a maximum speed of 90°/second and elevated at a similar speed.

Mounted above the right ammunition box is an automatic optronic electronic search and tracking package. This comprises a day TV camera, infrared camera and a laser range-finder. A collimating sight with built-in viewer is provided for the gunner, as only one person is now required to be on the mount.

According to the K B Tochmash Design Bureau, the target kill effectiveness of a standard twin 23 mm ZU-23-2 is just 0.023 while that of the upgraded ZU-23M is increased to 0.4.

Accuracy of tracking was up to 30 minutes and when upgraded is up to three minutes. Targets could only be engaged up to a maximum speed of 50 m/s while the upgraded system can engage targets up to 200 m/s.

Variants

ZU-23M1

To further improve the air-defence capability of the twin 23 mm ZU-23M system, the Open Joint Stock Company Podolsky Electromechanical Plant of Special Engineering (PEMZ Spetsmash) and the Closed Joint Stock Company NTC Elins have mounted a pod of two fire-and-forget SAMs (same as for Igla MANPADs) above the left 23 mm cannon.

These could be the Kolomna KBM Igla (SA-18 'Grail'). When fitted with these SAMs, targets can be engaged with a speed up to 360 metres/second with target effectiveness up to 0.5.

In a typical target engagement the missiles would be used to engage targets at long range while the guns would be used to engage close-in targets. When fitted with the pod of missiles the upgraded twin 23 mm ZU-23M is called the ZU-23M1.

A number of further enhancements are also possible if required by the user. These include the installation of an Identification Friend or Foe (IFF) system, target designation data receiver, acquisition radar and a portable target designation system.

Status

Developments complete. A variant of the ZU-23M is also produced in the Czech Republic.

Contractor

K B Tochmash Design Bureau of Precision Engineering - also known as Nudelman Precision Engineering Design Bureau.
Open Joint Stock Company Podolsky Electromechanical Plant of Special Engineering (PEMZ0 Spetsmash) (ZU-23M1 variant).
Closed Joint Stock Company NTC Elins (ZU-23M1 variant).

ZU-23-2

Type

Towed anti-aircraft gun

Development

In 1954, the Russian Main Artillery Directorate issued a requirement for single, twin and quadruple 23 mm cannon as the replacement for the then current ZPU-1, ZPU-2 and ZPU-4 14.5 mm weapons (see separate entry for details).

In 1955, the now KBP (then TsKB-14) presented its concept for a ZU-1 single 23 mm, ZU-14 twin 23 mm mount and the 2A14 automatic cannon. The first prototype of the ZU-1 was completed in 1956 with the ZU-2 being completed the previous year. The ZU-1 was similar to the 14.5 mm ZGU-1 mountain weapon and its mount was designated the 6U3.

Trials with the ZU-1 and ZU-2 were completed late in 1956 but the ZU-1 was dropped and never placed in production. Prototypes of the ZU-2 had two and four wheeled carriages but in the end the two-wheeled version was selected.

Extensive testing of the ZU-2 was completed in early 1960 and it was finally accepted for service under the designation of the ZU-23 (or ZSU-23-2 as it is also referred to). The complete weapon has the industrial index of the 2A13 with the cannon being the 2A14. Production was undertaken by Factory No 535.

Twin 23 mm automatic anti-aircraft gun ZU-23-2 of the Polish Army deployed in the firing position (IHS Jane's Intelligence Review) 0509498

Twin 23 mm automatic anti-aircraft gun ZU-23-2 on its air drop platform (Steven Zaloga) 0044154

Twin 23 mm automatic anti-aircraft gun ZU-23-2 with wheels removed and mounted in the rear of a truck (Israel Defence Forces) 0509499

In the Russian Army it was issued on the scale of 24 per airborne division, 12 in the divisional artillery element and four in each of the three airborne rifle regiments. The airborne divisions have now been restructured as airborne rifle divisions and TOEs show that they no longer have ZU-23-2 towed guns as part of their equipment. Instead they have 48 Strela-1 (SA-9 'Gaskin') SAM systems, 12 in each of the three airborne rifle regiments and 12 in the divisional artillery element.

Description

In the firing position the wheels of the carriage are raised off the ground and the weapon is supported on its triangular platform which has three screw-type levelling jacks. The quick-change barrels have flash suppressers. The 23 mm guns used in the ZU-23-2 are also used in the quad ZSU-23-4 self-propelled anti-aircraft gun system, but in this latter application the barrels are water-cooled.

The ZU-23-2 fires fixed ammunition with the same cartridge case dimension as the obsolete 23 mm automatic aircraft gun model VYa ammunition (23 × 152 B). The ZU-23-2 fires two types of 23 mm ammunition:

- API-T (BZT)
- HEI-T (MG25).

The API-T projectile weighs 0.189 kg, has a muzzle velocity of 970 m/s and will penetrate 25 mm of conventional steel armour, at an incidence of 0° at a range of 500 m. The HEI-T projectile weighs 0.19 kg and has a muzzle velocity of 970 m/s. The ZU-23-2 has an effective slant range of 2,012 m with ammunition having a self-destruct range of 3,780 m.

The ZU-23-2 is a clear weather only system with no provision for radar fire-control. Care must be taken not to confuse the 14.5 mm ZPU-2 (late model) with the ZU-23-2, which has different flash suppressers and horizontal rather than vertical ammunition boxes on either side.

There are separate entries for the Bulgarian, Chinese, Egyptian, Iranian and Polish-built models of the ZU-23-2 light anti-aircraft gun systems.

Variants

2A14 upgrade

Tulamashzavod Joint Stock Company is now offering to upgrade the 23 mm 2A14 guns used in the ZU-23-2 to the latest 2A14M standard.

The original weapon has a barrel life of 8,000 rounds with the latest 2A14M having a barrel life of 10,000 rounds. Under normal operational conditions, once each barrel has fired 100 rounds it becomes too hot and is removed and then placed in a water container. The second barrel is then quickly fitted.

Each ZU-23-2 weapon is normally provided with two replacement barrels as part of its standard equipment.

Upgraded ZU-23M

Full details of this and the ZU-23M1 with missiles, are given in a separate entry.

ZSU-23-2 with two Igla SAM in the ready-to-fire position (IHS/Peter Felstead)
0064564

ZSU-23-2M1 (with Igla SAM)
During a defence exhibition held in Moscow in 1999, a ZSU-23-2M1 LAAG was shown fitted with two Kolomna KBM Igla (SA-18 'Grouse') fire and forget surface-to-air missiles in the ready-to-launch position. These were mounted above the two 23 mm cannon.

New 23 mm PMP ammunition
Pretoria Metal Pressings of South Africa (now PMP Denel) have developed a 23 × 152 B family of ammunition to add to their product range.

This features a significant flash reduction additive that reduced the flash signature to a practical zero while remaining completely compatible in handing and ballistic terms with other 23 mm rounds.

PMP produce 23 mm HEI and API-T rounds, the rounds form part of the PMP New Generation family of ammunition.

New Oerlikon Contraves 23 mm ammunition
Oerlikon Contraves Pyrotech of Switzerland have developed a 23 mm Frangible Armour Piercing Discarding Sabot - Tracer (FADS-T) round called the PMA 276.

This is the subcalibre frangible armour-piercing discarding sabot projectile with tracer. The tungsten alloy penetrator, complete with integrally mounted ballistic cap, is held in a moulded plastic discarding sabot and weighs 150 g.

Muzzle velocity of the FAPDS-T PMA 276 is 1,180 m/s compared with 1,000 m/s for the conventional 23 × 152 B round. The additional velocity reduces the projectile time of flight and enhances the chances of hit as well as extending the effective combat range.

In addition, on-target performance against aerial and lightly-armoured targets is said to be high.

Finland
This country has developed and placed in production an upgrade package for the ZU-23-2. Details are given earlier in this section under Finland. This entered service with the Finnish Defence Forces in 2000 following extensive trials with three upgraded systems.

Bulgarian upgrade
Details of the Bulgarian 23 mm ZU-23-2 upgrade developed by Arsenal of Bulgaria and EBO (Hellenic Arms Industry of Greece) are given earlier in this section under International.

ZU-23-2 on BTR-152
It is known that a number of BTR-152 (6 × 6) armoured personnel carriers were fitted with the twin 23 mm automatic anti-aircraft gun ZU-23-2 in place of the 14.5 mm ZPU-2 anti-aircraft machine guns. These were supplied to a number of countries including the former East Germany. The Israel Defence Force also captured a quantity in the Lebanon.

ZU-23-2 on Namibian truck
Namibia has a number of upgraded ZU-23-2 systems mounted on the rear of locally manufactured (4 × 4) mine-protected armoured vehicles.

ZU-23-2 on ZIL-135
During the war in Afghanistan, a number of ZIL-135 (8 × 8) trucks used by the Soviet Army were fitted with ZU-23-2 weapons in the rear cargo area for convoy protection duties.

ZU-23-2 on BTR-D AFV
The Russian airborne forces are known to have mounted the twin 23 mm ZU-23-2 automatic anti-aircraft gun on the roof of the BTR-D air assault transporter, which is a member of the BMD-1 airborne combat vehicle family. This was used in combat during the Russian occupation of Afghanistan.

ZUMLAC system
The South African Army has fitted captured ZU-23-2 weapons on the rear of specially modified SAMIL 100 (6 × 6) mine-protected chassis and details of this system are given in the Self-propelled anti-aircraft guns section under South Africa.

Within the South African National Defence Force the ZU-23-2 is designated the GA6.

Specifications

	ZU-23-2
Dimensions and weights	
Crew:	5
Length	
transport configuration:	4.57 m
Height	
transport configuration:	1.87 m
axis of fire:	0.62 m
Ground clearance	
overall:	0.36 m
Track	
vehicle:	1.67 m
Weight	
standard:	950 kg
combat:	950 kg
Mobility	
Carriage:	trailer mounted two wheeled
Tyres:	6.00 × 16
Firepower	
Armament:	2 × 23 mm cannon
Main weapon	
length of barrel:	2.01 m
breech mechanism:	vertical sliding block
feed mechanism:	50 round belt
operation:	gas automatic
cartridge:	23 mm × 152 mm B
Rate of fire	
cyclic:	2,000 rds/min (est.)
practical:	400 rds/min (est.)
Main armament traverse	
angle:	360°
Main armament elevation/depression	
armament front:	+90°/-10°

Status
Production complete.

Contractor
Russian state factories.
Also licence-produced in Bulgaria, China, Czech Republic, Egypt, Iran and Poland, which have separate entries under this section. In some cases they have been modified to meet local operational requirements.

Serbia

20/1 mm M75

Type
Towed anti-aircraft gun

Development
The 20/1 mm M75 anti-aircraft gun is a simple lightweight weapon mounting a single licence-produced HSS-804 20 mm L/70 cannon and is intended for infantry use against both air and ground targets.

This weapon has been widely used in the ground role in the former Yugoslavia.

This is a clear weather system only with no provision for off-carriage fire-control. The list of users should be treated with caution, as it is difficult to determine exactly which model of this 20 mm system some countries use.

20/1 mm M75 anti-aircraft gun in travelling position 0509547

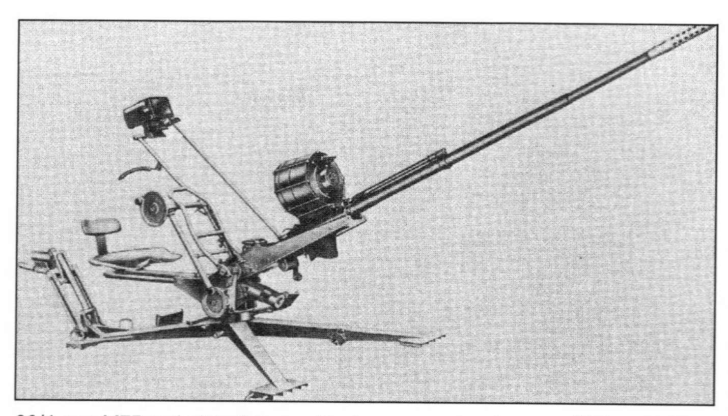

20/1 mm M75 anti-aircraft gun, with drum magazine, in normal firing position 0509548

Description

The M75 is transported on a two-wheeled single axle bogie which also carries ammunition and sight boxes. In action, the M75 can be emplaced by a two man crew and when the bogie is removed the gun rests on a tripod mounting with the actual carriage height varied by the mounting feet angles; the rear tripod leg foot angle adjustment doubles as a variable height towing eye. For traverse the gunner's feet push the gun round but for elevation a small hand wheel is provided. The M75 is light enough to be carried in or on vehicles but can be stripped down in only 60 seconds for pack carriage on four animals. As the heaviest sub-assemblies weigh no more than 55 kg, man pack carriage is possible. Assembly and disassembly require no tools. The M75 may be fired direct from the bogie if necessary.

The M75 can be fitted with a simple cartwheel sight for anti-aircraft use but the more usual sight is the M73, a day reflex sighting device with a tritium source providing scale illumination. The M73 is normally calibrated for an average target range of 800 m and target speeds of 50, 100, 150, 200, 250 and 300 m/s; it uses an elliptical grid. For use against ground targets at ranges up to 1,000 m a ×3.8 magnification day sighting telescope is used.

The normal ammunition feed used on the M75 is a 60-round drum magazine but when using armour-piercing ammunition a 10-round box magazine may be used. Ammunition is the same as that used on the M55 (20 × 110).

The Swiss Hispano-Suiza Company, on whose weapon the 20/1 mm M75 anti-aircraft gun is based, was taken over by Oerlikon Contraves many years ago. This company was taken over by Rheinmetall DeTec AG late in 1999. Oerlikon Contraves no longer manufactures 20 mm light anti-aircraft guns.

Specifications

	20/1 mm M75
Dimensions and weights	
Crew:	4
Width	
overall:	1.51 m
Ground clearance	
adjustable:	0.215 m to 0.315 m
Track	
vehicle:	1.215 m
Weight	
combat:	260 kg
Mobility	
Carriage:	trailer mounted two wheeled
Firepower	
Armament:	1 × 20 mm cannon
Main weapon	
length of barrel:	1.956 m
maximum range:	5,500 m
maximum vertical range:	4,000 m
maximum altitude:	2,500 m
feed mechanism:	60 round drum
feed mechanism:	10 round boxed
cartridge:	20 mm × 110 mm
Rate of fire	
cyclic:	700 rds/min
Main armament traverse	
angle:	360°
Main armament elevation/depression	
armament front:	+83°/-10°

Status

Production as required. This system has seen service in Bosnia Herzegovina, Croatia, Macedonia, Serbia, Montenegro, Slovenia and Zambia. The M75 was inducted into the Former Yugoslavian National Army in 1979.

Contractor

Yugoimport - SDPR.

20/3 mm M55 A2

Type

Towed anti-aircraft gun

Development

The 20/3 mm M55 A2 anti-aircraft gun forms the basic component of a family of three similar weapons all based on the same triple-gun mounting; the other two weapons are the M55 A3 B1 and the M55 A4 B1 (see separate entries). The first 20/3 mm M55 guns were produced in 1955. This weapon was widely used in the ground role in the conflict in the former Yugoslavia.

This is a clear weather system only with no provision for off-carriage fire control. The list of users should be treated with caution, as it is difficult to determine exactly which model of this 20 mm system some countries use.

The Swiss Hispano-Suiza Company, on whose weapons the 20/3 mm M55 A2 is based, was taken over by Oerlikon Contraves many years ago. Late in 1999 Oerlikon Contraves was taken over by Rheinmetall DeTec AG of Germany. This company no longer markets or manufactures 20 mm light anti-aircraft guns.

Description

Although many local modifications have been introduced since 1955, the 20/3 mm M55 A2 is basically a licence-built weapon comprising three Hispano-Suiza HSS-804 20 mm L/70 cannon mounted on the HSS 630-3 towed carriage. The three guns are arranged on the mounting horizontally with the central gun positioned slightly to the rear of the two outboard guns; this allows the three 60-round drum magazines to be positioned close together for loading, and concentrates the three barrels. In action, the carriage rests on three outrigger legs that are folded outwards into place, with the two carriage wheels raised off the ground. In an emergency the guns can be fired directly from the towed carriage. A spring suspension is used on the carriage, which is equipped with a hydraulic system for the parking and automatic towing brakes. The weapon can be towed at speeds of up to 80 km/h by a light (4 × 4) truck.

The gunner is positioned behind the gun barrels, aims manually with elevation and traverse hand wheels and fires via a foot pedal. Two equilibrates are provided. For fire-control the M55 A2 is fitted with a PANS-20/3 optical-

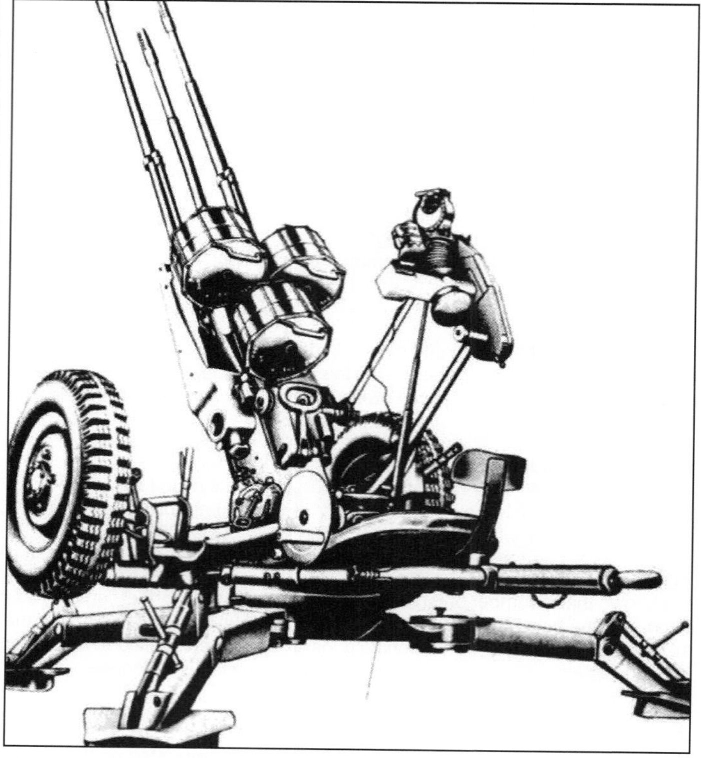

20/3 mm M55 A2 anti-aircraft gun in firing position 0509543

Ammunition natures

Type	HEI-T	HE-T	HEI	HE	API	API-T	AP-T
Designation:	M57	M57	M57	M57	M60	M60	M60
Projectile weight:	137 g	137 g	132 g	132 g	142 g	142 g	142 g
Bursting charge:	TNT/Incendiary or RDX/Al	TNT or RDX	TNT/Incendiary or RDX/Al	TNT or RDX	nil	nil	nil
Total weight of round:	261 g	261 g	257 g	257 g	274 g	274 g	274 g
Muzzle velocity:	850 m/s	850 m/s	850 m/s	850 m/s	840 m/s	840 m/s	840 m/s

mechanical automatic sight. The gunner inserts the target information which includes target range, course angle, angles of dive and climb, and target speed which is inserted in increments of 50 m/s up to 300 m/s. With the target information inserted, the gunner can then track the target directly using the sight graticule. Targets flying at speeds up to 1,000 km/h can be engaged at ranges up to 1,500 m although for slower targets the effective range is 2,000 m.

The (20 × 110) ammunition used with the M55 gun family is detailed in the accompanying table at top of page.

Additional rounds include TP-T (M57), TP-T (M79), TP (M57), Blank (M77) and drill. All of these have brass cartridge cases 110 mm long with the propellant being a single-base nitrocellulose (NC)-based powder.

Variants

El Salvador SPAAG

El Salvador has a small number of self-propelled versions of this system mounted on a much modified US-supplied M1114 full-tracked armoured reconnaissance vehicle chassis. Like the basic 20/3 mm system, these are clear weather systems only.

Specifications

	20/3 mm M55 A2
Dimensions and weights	
Crew:	6
Length	
transport configuration:	4.3 m
Width	
transport configuration:	1.27 m
Height	
transport configuration:	1.47 m
Ground clearance	
overall:	0.23 m
Weight	
standard:	970 kg
combat:	1,100 kg
Mobility	
Carriage:	trailer mounted two wheeled
Firepower	
Armament:	3 × 20 mm cannon
Main weapon	
length of barrel:	1.956 m
maximum range:	5,500 m
maximum vertical range:	4,000 m
maximum altitude:	2,500 m
feed mechanism:	60 round drum (per barrel)
cartridge:	20 mm × 110 mm
Rate of fire	
cyclic:	2,100 rds/min
Main armament traverse	
angle:	360°
Main armament elevation/depression	
armament front:	+83°/-5°

Status

Production of the system is complete.

Contractor

Yugoimport - SDPR.

20/3 mm M55 A3 B1

Type

Towed anti-aircraft gun

Development

The 20/3 mm M55 A3 B1 anti-aircraft gun is a Yugoslav-derived variant of the basic M55 A2 weapon. It uses the same locally produced Hispano-Suiza HSS-804 20 mm L/70 guns in a triple arrangement and the carriage is derived from the single axle HSS 630-3. The main change on the M55 A3 B1 is that a small Wankel engine has been added to the right of the gunner's position to provide power for both elevation and traverse.

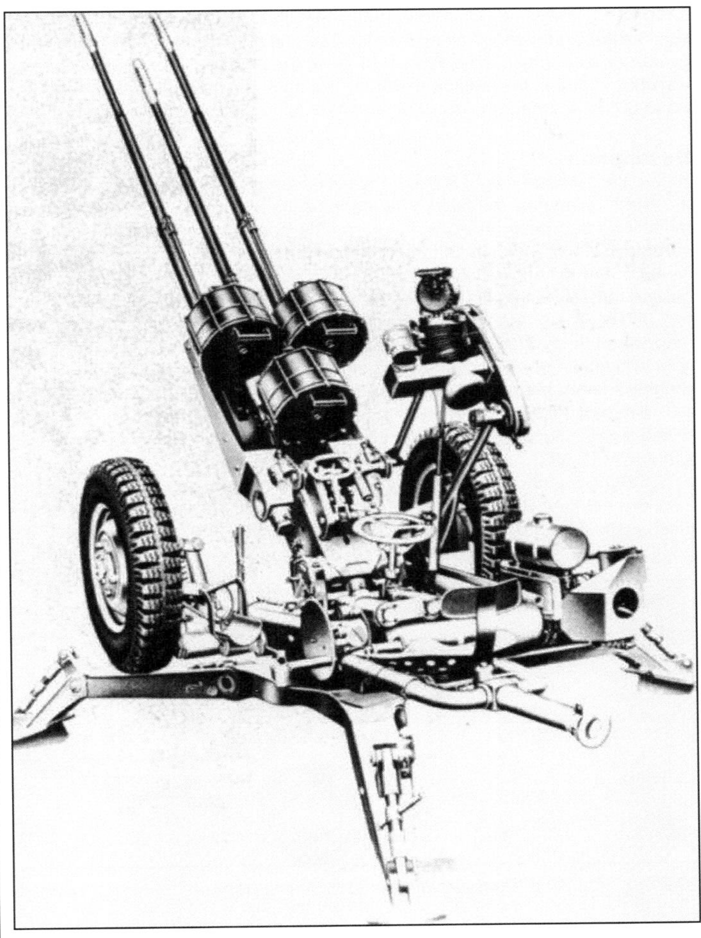

20/3 mm M55 A3 B1 anti-aircraft gun in firing position 0509544

The Wankel engine has a power output of 8 hp. Each chamber has a capacity of 160 cm³ and consumes, on medium load, 2.4 litres per hour of fuel (the petrol/oil mix is 50:1). The power unit has three main components:

- The control unit, which the gunner operates via a single joystick for traverse and elevation
- The transmission system
- The Wankel engine.

In traverse the power unit can move the carriage at rates between 0.3° and 70°/s. In elevation the barrel can be moved at rates between 0.3° and 50°/s. Manual control is available.

The 20 mm M55 weapon is produced in four basic variants; M55A2B1, M55A3B1, M55A4B1 and M55A4M1.

Description

In all other respects the M55 A3 B1 is the same as the M55 A2 and uses the same PANS-20/3 sighting system and ammunition types with a firing rate of 700 rounds per minute. The only change is that the weight without the ammunition drums is 1,150 kg and with all three loaded drums fitted the weight is 1,236 kg.

This is a clear weather system only with no provision for off-carriage fire-control. The list of users should be treated with caution, as it is difficult to determine exactly which model of this 20 mm system some countries use.

The Swiss Hispano-Suiza Company, on whose weapons the 20/3 mm M55 A3 B1 anti-aircraft gun is based, was taken over by Oerlikon Contraves many years ago. Oerlikon Contraves was taken over by Rheinmetall DeTec of Germany late in 1999. This company no longer manufactures 20 mm light anti-aircraft guns.

Status

Production complete. The system has seen service in Bosnia Herzegovina, Croatia, Macedonia, Slovenia and Serbia and Montenegro.

Contractor

Yugoimport - SDPR.

20/3 mm M55 A4 B1

Type
Towed anti-aircraft gun

Development
With the 20/3 mm M55 A4 B1 the Yugoslav weapon designers have combined the established M55 triple-gun mounting with a local adaptation of the carriage and sighting system of the Hispano-Suiza HSS 666A (now the Oerlikon Contraves twin 20 mm GAI-D01 automatic anti-aircraft gun). While the three licence-built HSS-804 20 mm barrels have been retained, the M55 carriage has been much revised to accommodate a Wankel engine under the gunner's seat, a licence-built version of the Italian Officine Galileo P56 computer sight, a small shield and a compressed air system for firing the guns. The engine provides the power for a hydraulic drive system to move the carriage in both elevation and traverse. In action the maximum rate of motion of the carriage in both modes is up to 80°/s. Acceleration in traverse is 120°/s^2 and in elevation 60°/s^2.

Description
Before the computer sight is used, the target range and speed are inserted into the sight computer, the latter in speeds up to 350 m/s, while the maximum range that can be inserted is 1,200 m. The computer sight will then lay off automatically for traverse and angle of lead (up to 21°). The gunner controls the weapon aim by a joystick mounted on the console under the sight unit. For ground targets, the sight unit uses a separate sight with range divisions of 100 to 2,500 m. The ground sight has a magnification of ×4 (the aircraft sight has a magnification of ×1.1). For rough alignment of the barrels with a target the gunner uses an open horizontal grid sight over the sight unit. Once the gun is roughly aligned the optical system of the computer sight is then used.

The maximum towing speed of the M55 A4 B1 is 70 km/h.

The ammunition types used with the M55 A4 B1 are the same as those used with the M55 A2.

The BOV-3 triple 20 mm self-propelled anti-aircraft gun system uses the 20/3 mm M55 A4 B1 system on a (4 × 4) armoured chassis. Details of this are given in the Self-propelled anti-aircraft gun section.

20/3 mm M55 A4 B1 anti-aircraft gun in travelling position 0509545

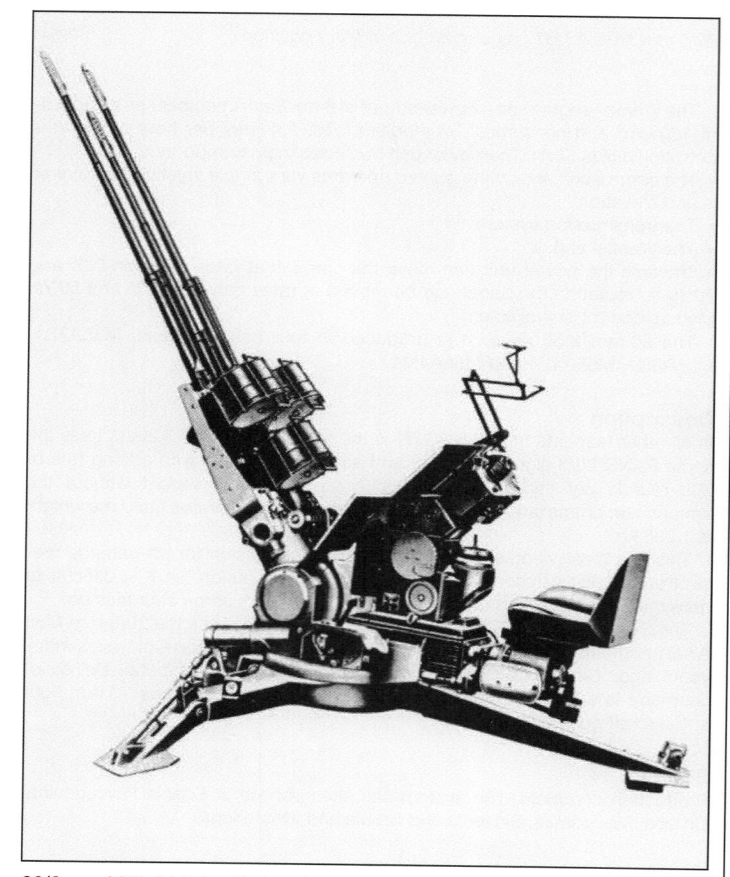

20/3 mm M55 A4 B1 anti-aircraft gun in firing position 0509546

This weapon has been widely used in the ground role in the recent conflict in the former Yugoslavia.

This is a clear weather system only with no provision for off-carriage fire-control. The list of users should be treated with caution, as it is difficult to determine exactly which model of this 20 mm system some countries use.

The Swiss Hispano-Suiza Company, on whose weapons the 20/3 mm M55 A4 B1 anti-aircraft gun is based, was taken over by Oerlikon Contraves many years ago. Oerlikon Contraves was taken over by Rheinmetall DeTec AG late in 1999. This company no longer manufactures 20 mm light anti-aircraft guns.

Specifications

	M55 A4 B1
Dimensions and weights	
Crew:	6
Weight	
standard:	1,350 kg
combat:	1,096 kg
Mobility	
Carriage:	trailer mounted two wheeled
Firepower	
Armament:	3 × 20 mm cannon
Main weapon	
length of barrel:	1.956 m
maximum range:	5,500 m
maximum vertical range:	4,000 m (under 80°)
maximum altitude:	2,500 m
feed mechanism:	60 round drum
cartridge:	20 mm × 110 mm
Rate of fire	
cyclic:	700 rds/min
practical:	60 rds/min
Main armament traverse	
angle:	360°
Main armament elevation/depression	
armament front:	+83°/-5°

Status
Production complete. The 20 mm M55 is produced in four basic variants: M55A2B1, M55A3B1, M55A4B1 and M55A4M1. The system has seen service in Bosnia Herzegovina, Croatia, Macedonia, Slovenia and (former) Serbia and Montenegro. As of April 2011, the state of maintenance of the systems is uncertain.

Contractor
Yugoimport - SDPR.

Singapore

40 mm L70 FADM

Type
Towed anti-aircraft gun

Development
Allied Ordnance of Singapore PTE Ltd (AOS), a subsidiary of the now Singapore Technologies Kinetics (STK), has extensive experience in 40 mm air defence systems for both land and sea applications. Singapore Technologies Kinetics is itself the land systems arm of Singapore Technologies Engineering (ST

Allied Ordnance of Singapore 40 mm L70 Field Air Defence Mount (FADM) deployed in the firing position 0509703

Engineering). A further subsidiary company of ST Kinetics - Advanced Material Engineering Pte Ltd (AME) - designs and manufactures conventional and smart munitions advanced protective materials as well as guided system components.

In 1994, the company disclosed that it was developing, as a private venture, the 40 mm L70 Field Air Defence Mount (FADM). This is an improved version of the standard combat-proven Saab Bofors Dynamics 40 mm L70 towed anti-aircraft gun covered later in this section.

The 40 mm Field Air Defence Mount (FADM) was successfully proof fired and tested in 1996 before being placed into initial production. Serial production has now commenced for the Singapore Armed Forces.

Description

The 40 mm L70 FADM is an improved field version of the NATO standard 40 mm L70 Air-Defence Weapon System. It has been reported to be capable of intercepting high speed aerial and surface targets. It is equipped with an electric drive driven aiming system for high tracking and aiming accuracies in quick time. Today's air defence threats, with their higher speeds and greater manoeuvrability, make more demands on the air defence weapon. Modern air defence weapon systems must offer high-hit probabilities, fast response times, autonomy and round-the-clock operation.

FADM has been designed to meet these requirements and is said to be capable of intercepting all kinds of high-speed aerial and surface targets. To carry out this mission it has the following advanced features:

- Electric drive aiming system using brushless motors to attain high tracking and aiming accuracy in very quick time.
- Battery powered, providing immediate and non-interruptible source of power supply on demand.
- 101-round dual-compartment 40 mm magazine to allow for ample engagements without reloading and immediate selection of two types of ammunition providing the unique capability of defending against aerial and surface targets simultaneously and instantly.
- Fires the highly destructive 40 mm L70 Pre-Fragmented High-Explosive (PFHE) at a cyclic rate-of-fire of 330 rds/min to provide an effective and sustained defence against modern air targets.
- On-mount Optronic Fire-Control System (OFCS) with FLIR for autonomous, round-the-clock operations and ECM immunity.
- New rigid one-piece carriage providing a stable firing platform on land.
- An automatic levelling device to allow for very quick deployment in less than three minutes with an accuracy of ±1 mrad. Immediate deployment is also possible under emergency situations, for example, the gun can be lowered to the ground with its outriggers spread for stability and firing can commence almost immediately.

Although the AOS 40 mm L70 FADM is fully autonomous, it can be interfaced easily with any remote fire-control system. The weapon system is complemented by a full range of 40 mm L70 ammunition including PFHE, HE-T and PT. Production of this ammunition is undertaken in Singapore.

Variants

Naval Air Defence Mount (NADM)

A naval variant has been advertised by the Singapore production facility, this is known simply as the NADM. Other than the naval version's copula mounting, there are two main differences between the land and the naval system, these are; a weight reduction from 6,000 to 3,300 kg and an increase of the elevation's lower limit from –4° to –8°.

Specifications

	L70 FADM
Dimensions and weights	
Length	
overall:	6.60 m
Width	
overall:	2.25 m
Height	
transport configuration:	2.42 m
Ground clearance	
overall:	0.39 m
Track	
vehicle:	2 m
Wheelbase	4.05 m
Weight	
standard:	6,000 kg
Mobility	
Carriage:	trailer mounted four wheeled
Electrical system	
vehicle:	415 V
Batteries:	48 V, 100 Ah
Firepower	
Armament:	1 × 40 mm cannon
Main weapon	
length of barrel:	2.8 m
rifling:	16 m
breech mechanism:	vertical sliding block
maximum vertical range:	4,000 m
feed mechanism:	101 round
cartridge:	40 mm × 365 mm R
Rate of fire	
cyclic:	330 rds/min
Fire-control	
system:	RS-422 or MIL-STD-1553B
Main armament traverse	
angle:	360°
speed:	100°/s
acceleration:	152 °/s²
Main armament elevation/depression	
armament front:	+80°/-4°
speed:	60°/s
acceleration:	152 °/s²

Electro-optic fire-control system

Laser range-finder	
Wavelength:	1.54 µm (eye safe)
Pulse repetition rate:	continuous, 3 Hz
Accuracy:	7 m
Range:	6,000 m
Technology:	Nd:YAG and Raman shift cell (standard atmosphere and NATO standard)

FLIR

Wavelength:	8-12 µm (TCM detectors) (options for 3-5 µm)
Cooler:	integral stirling cycle
Cool-down time:	3 min
Field-of-view:	
(NFoV)	2 × 3°
(WFoV)	4 × 6° (pupil 7.6 mm)
Detection:	10 km on aircraft
Tracking:	7 km on aircraft
CCIR output:	25 images/s
Resolution:	0.21 mrad (narrow field of view)
Maximum resolution:	200 pts/line × 575 lines

Fire-control computer

Ballistic computation with 3 integrated ballistic switchable
Future target calculation
Firing sequence and auto tracking
Overall aiming accuracy of less than 2 mrad
Access for parameters such as wind, ammunition mean point of impact, muzzle velocity and air conditions

Reflex sight

Magnification:	×1
Field of view:	
(horizontal)	20° (min)
(vertical)	16° (min)
Diameter of collimator:	78 mm

Status

In production for the Singapore Armed Forces.

Contractor

Allied Ordnance of Singapore (Pte) Ltd.

Sweden

40 mm L/70 modernisation package

Type

Towed anti-aircraft gun

Development

The then Bofors Defence (now BAE Systems Bofors) 40 mm L/70 Renovation and Modernisation (REMO) programme comprises a basic package, mainly for extending the gun's technical life span and several upgrading kits for increasing the effectiveness of the weapon. The REMO package upgrades the older L/70 40 mm gun that first entered service with the Swedish Army in 1951.

The REMO programme is modular and renders step-by-step upgrading possible as funding becomes available or the threat evolves. The REMO work can be carried out by Bofors Defence in Sweden or in the user's own facilities.

Description

The programme consists of function packages that are designed to extend the life, improve the availability and performance of the gun that includes a higher rate of fire, bringing it into line with today's systems, as well as reduce LCC, the basic package comprises:

- A mechanical renovation
- Installation of a new 48 V DC electrical laying system with digital servo-amplifier

Upgraded 40 mm L/70 light anti-aircraft gun deployed in the firing position
0509530

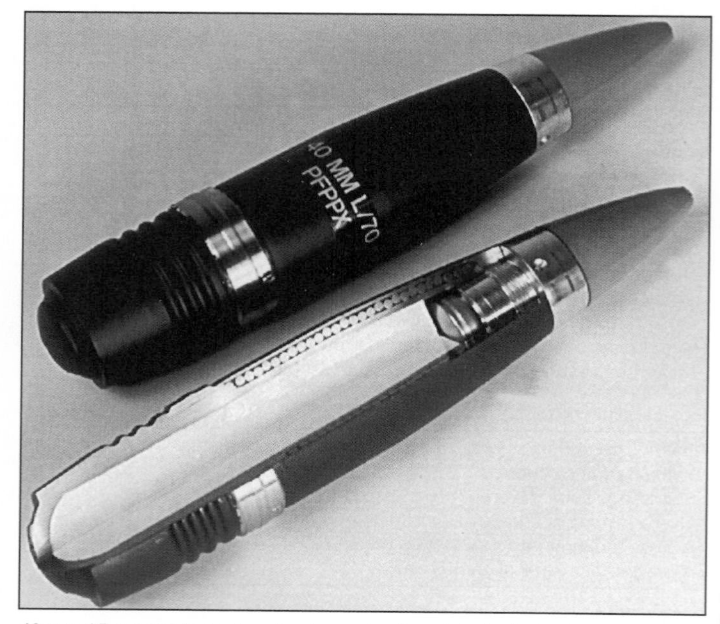

40 mm 3P projectile (top) and cutaway to show interior (below) 0509882

- A new electrical installation.

With the new electrical laying system, the hydraulic system and most of the original laying transmission system are discarded.

The main power supply for the laying system consists of batteries (5 kW) that are located in the lower mounting. These batteries are dimensioned for several hours of operation without charging. A smaller 220 V, 50 Hz power supply unit or mains, can be used for compensating charge of the batteries. As an alternative to the new electrical laying system the original electro-hydraulic laying system can be modernised and retained.

With the basic package, all difficult to obtain spare parts are replaced by easily accessible ones, so granting a considerable extension to the weapon's technical life span. The basic package design can be furnished easily with the upgrading modules.

The major upgrading module comprises:

- An integrated fire-control system
- A family of 40 mm L70 ammunition
- 3P programmer
- A kit for increased rate-of-fire and kits for increased firing endurance.

The integrated fire-control system, which converts the gun to an autonomous firing unit, consists of the Saab Tech Vetronics Universal Tank and Anti-Aircraft Sight (UTAAS) sight and fire-control computer. Details of the UTAAS fire-control system are given in the Towed anti-aircraft gun sights section. This sight is already in volume production for the CV 9040 (Sweden) and CV 9030 (Norway) armoured vehicles. In this application the sight is fitted in a two-man power-operated turret.

The 3P programmer enables current 40 mm L/70 light anti-aircraft guns to fire the new, programmable 3P air burst ammunition. It consists of the operator's panel, programming electronics and the antenna unit located in the feed device front guide.

The ammunition function mode is from the operator's panel. Programming data is fed from the programmer via the antenna unit to the fuze immediately before ramming. To use all 3P modes the necessary data input to the programmer is range to estimated intercept point by the integrated fire-control system or connection to an external fire-control system is necessary. Some 3P modes can, however, be used by the installation of a laser range-finder.

The high rate-of-fire kit, which comprises:

- Conversion kits for the original feed device and automatic loader
- New extractors
- New operating cam and a replacement recoil buffer, increases the cyclic rate-of-fire from 260 to 300 rds/min.

The kit for increased firing endurance comprises a 26-round magazine and a 96-round ammunition rack. The 26-round magazine enables three to four engagements to be carried out without reloading. It is fitted to the breech casing and the feed device is loaded with clipped ammunition. Each clip of four 40 mm rounds is connected to the next clip by a further clip on the opposite side. As the rounds are fed into the feed device the clips are automatically ejected. During transportation of the gun the magazine is kept in a folded-down position.

The 96-round ammunition rack increases the combat load of ammunition. It consists of two compartments, each containing 48 rounds of easily accessible clipped ammunition and replaces the original 48-round rack. Personnel standing on the ground can refill from behind the ammunition rack.

Further upgrading options are available including a muzzle velocity radar, optical target designator for enhanced target acquisition, integrated power supply unit and a device for automatic conversion from the travelling to the firing position and vice versa.

Status

Production as required. Elements of this upgrade were included in the 40 mm L/70 systems upgraded for Netherlands some years ago. For trials purposes a Swedish Navy Weapon Systems 40 mm L/70 has been fitted with an all-electric gun control system and this is the model installed on the 40 mm TRIDON (6 × 6) self-propelled system.

An announcement made in September 2007 confirmed that the Indian Ordnance Factories Board (OFB) and the then BAE Systems SWS of Sweden are to start development of a non-government funded upgrade programme for the 2,000 or so L-70 anti-aircraft guns in the Indian inventory. As of January 2014, no further information on this upgrade is available.

Contractor

BAE Systems Bofors.
NAMMO AS, produces the HE-T and TP-T for the L/70 and fuzes for 40 mm ammunition.
As from June 2007, Ordnance Factory Board (OFB) (Pakistan) supplies 40 mm ammunition to Turkey for their L/70 towed AAGs.

L/70

Type

Towed anti-aircraft gun

Development

The Bofors Defence 40 mm L/70 automatic anti aircraft gun entered service with the Swedish Army in 1951 as the successor to the Bofors 40 mm L/60.

Some L/60s still exist in Argentina but are held in reserve stocks. Six of these L/60 40 mm guns were handed over to Paraguay in November 2008.

The L/70 is still being marketed, there has been no recent production of the Bofors 40 mm L/70 weapon system. Upgrade packages continue to be marketed by the company.

In 1998, it was stated that Sweden was to pass, free of charge, one battalion of 18 Bofors 40 mm L/70 anti aircraft guns, plus support equipment, to Estonia, Latvia and Lithuania. It has been confirmed that at least one of these countries, Estonia, declined to take delivery of these weapons.

Description

There were two basic models of the L/70, Type A and B. Type A is fed from an external power source, whereas the Type B has a three phase 220 V 50 Hz Auxiliary Power Unit (APU) mounted on the rear of the carriage.

Upgraded Bofors Defence 40 mm/L70 gun of the Royal Netherlands Army showing the APU mounted on the forward part of the carriage
(Directorate of Material, Royal Netherlands Army, Communications Office)
0509772

Table 1

Ammunition type	PFHE Mk 2	HCHE	HE-T	APC-T	TP -T
Weight of complete round:	2.4 kg	2.4 kg	2.5 kg	2.5 kg	2.5 kg
Weight of shell:	0.88 kg	0.87 kg	0.96 kg	0.92 kg	0.96 kg
Weight of explosive:	0.12 kg	0.165 kg	0.103 kg	none	none
Type of explosive:	octal	octal	hexotonal	n/app	n/app
Round length:	534 mm	534 mm	534 mm	534 mm	534 mm
Muzzle velocity:	1,025 m/s	1,030 m/s	1,005 m/s	1,010 m/s	1,005 m/s

Bofors Defence 40 mm L/70 automatic anti-aircraft gun Type B in firing position 0509704

Bofors Defence 40 mm L/70 automatic anti-aircraft gun system from rear with BOFI equipment, empty cartridge cases being ejected forward. This particular 40 mm L/70 has the ammunition stay fitted on top of the loading device (T J Gander) 0509528

Bofors Defence 40 mm L/70 automatic anti-aircraft gun with fair weather BOFI equipment 0509529

The high rate of fire of 240 rds/min is obtained by ramming the rounds during the run out, with the empty cartridge cases ejected towards the end of recoil. The empty cartridge case is deflected into a chute at the front of the mounting.

Ammunition is fed in four round clips to the feed guides by the automatic loading device by two loaders positioned one either side of the gun. Waist high supports protect the loaders against falling during the high training acceleration of the mounting. An ammunition stay can be placed on top of the automatic loading device to serve as a magazine, permitting 26 rounds to be fired from the unmanned gun. Two ready use ammunition racks, holding 48 rounds, are fitted at the rear end of the gun platform. These racks are fed from the outside by an ammunition supply party and are emptied from the inside by the two loaders.

Bofors Defence 40 mm L/70 BOFI all-weather system showing radar mounted above gun barrel 0007663

The monobloc barrel is provided with a flash suppresser. The recuperator spring encircles the rear part of the barrel and this, together with the recuperator spring, forms an easily exchangeable unit. The recoil buffer is hydraulic and the breech mechanism has a vertically sliding breechblock, which opens and closes automatically.

There are two close range sights, model NIFE SRS 5, fitted on a sight bracket on the breech casing of the gun, one for the elevation layer and one for the traversing layer. Elevation and traverse are electro hydraulic with maximum elevation speed 45° per second and maximum traverse speed 85° per second. There are manual controls for emergency use.

In the remote control mode the power operation devices for elevating and traversing are controlled by the input signals received from a fire-control system connected to the gun by a cable.

The Bofors Defence precision remote control system with transistorised amplifiers is used. In remote control the data from the fire-control equipment is transmitted with one cable and with Type B there is also a cable for connecting the gun with the power supply unit.

In local control one person on the left side of the platform operates the gun. This joystick is used in combination with the close range sight if a central fire-control system is not being used, is out of order, or if the gun is being operated as an independent unit.

A firing-limiting gear for the electrical firing is provided and is set by pushing stop bolts, one for every 10° of traverse and setting the highest limited elevation for any of the zones limited in traverse.

The types of 40 mm ammunition now produced by Bofors Defence and NAMMO AS are detailed in Table 1 at top of page.

The Pre-Fragmented High-Explosive (PFHE) Mk 2 pre-fragmented round has a proximity fuze with an effective range up to 6.5 m against aircraft and 4.5 m against missiles. Flight time to 1,000 m is 1.1 seconds, to 2,000 m is 2.44 seconds and to 3,000 m is 4.44 seconds. The PFHE projectile is made up of high grade steel, which, together with the explosive charge, gives a large number of fragments. To increase the projectile's effectiveness, 650 tungsten carbide pellets, with a penetration capability of about 14 mm of duraluminium, are contained in the walls of the forward section. The 'boat tail' of the projectile improves large fragment dispersion.

The new High Capacity High-Explosive (HCHE) shell is a multipurpose projectile designed for use against all types of target from light aircraft to vessels and armoured personnel carriers. It can, to a certain extent, replace conventional types of HE and AP ammunition. The casing of the HCHE shell is manufactured from special steel, which is sufficiently strong to allow penetration of armour plate without breaking up. The shell also protects the post impact delay fuze, which means it will explode only when inside the armoured target. The shell has a filling of 165 g of octal, a powerful high-explosive.

More recently Bofors Defence has developed two new rounds for 40 mm L/70 guns, these are the Pre-Fragmented Programmable Proximity Fused 3P (PFPPX) and the Armour-Piercing Fin-Stabilised Discarding Sabot (APFSDS). The first production application for these two new rounds is the Saab Bofors Dynamics/Hägglunds Vehicle CV 9040 full tracked infantry fighting vehicle now in production for the Swedish Army. The 3P round cannot be fired from a standard 40 mm L/70 weapon.

The Bofors Defence 40 mm L/70 anti aircraft gun is normally used in conjunction with central fire-control systems such as the Flycatcher designed by Hollandse Signaalapparaten and used by India and the Netherlands (army and air force) and the Swiss Skyguard and Super Fledermaus systems.

During 1980, the Ericsson Microwave Systems' Giraffe search radar, designed for use with the RBS 70 surface to air missile, was successfully used in conjunction with the 40 mm BOFI gun to shoot down an attacking missile.

Late production versions of the Bofors Defence 40 mm L/70 include:

- The electrical power unit mounted on the gun
- Rate of fire increased to 300 rds/min (cyclic)
- Hydraulic reversion to towing position
- Proximity fuze paralysing device
- As an option, a Doppler radar mounted on the right side of the weapon for measuring muzzle velocity.

BOFI fair weather gun system

The Bofors Defence 40 mm Bofors Optronic Fire-control Instrument (BOFI) gun system consists of a modified version of the basic 40 mm L/70 Type B, BOFI electro-optic fire-control system and proximity fused ammunition. The FCE is integrated with the gun and is based on a computer, which calculates the angles of aim off to the target. The range to the target is continuously provided by a laser range-finder. The operator, continuously keeping the gun aimed at the target, measures the movements of the target. Once tracking has been established the computer automatically takes over the aiming of the gun and the operator has only to make minor corrections to obtain accurate tracking. The operator observes the target through a sighting device, which is a combination day-and-night sight with a light amplifier.

An optical target indicator makes target designation or central search radar linked to the BOFI gun via a Target Data Receiver (TDR) located beside the gun. Target data are transmitted to the TDR by wire or radio. By 1983 approximately 200 fair weather BOFI guns had been delivered for service in Europe and Asia, and the former Yugoslavia is known to have taken delivery of a quantity of these systems.

BOFI all weather gun system

This is one of the more recent developments. Over 100 systems have been sold and the first customer was Malaysia. This version has a multisensor fire-control system using a J band pulse Doppler radar as the main sensor. The radar gives the system an all-weather operation and automatic acquisition and tracking capabilities. It operates on the MTI mode at target acquisition and switches over automatically to frequency agility in the tracking mode, which radically improves tracking accuracy. The radar sensor can be backed up by the electro optics for tracking supervision or noise tracking using radar for angular tracking and laser for ranging. All the sensors can be used in different combinations for maximum flexibility and jamming resistance. A total of 22 ready use and 96 reload rounds is carried on each BOFI mount with the gun normally firing two second bursts of 10 rounds against a target using the PFHE Mk 2 ammunition (which was introduced in 1982) to an effective range of over 4,000 m.

Netherlands 40 mm L/70 upgrade

Late in 1987 the then Bofors Defence was awarded a contract from the Royal Netherlands Army for the modernisation of 60 Bofors Defence 40 mm L/70 anti aircraft guns, which were originally built under licence in Netherlands in the 1950s. The then Bofors Defence was awarded the contract, worth SKR220 million after the Royal Netherlands Army considered modernising the 40 mm L/70 weapons or purchasing new Oerlikon Contraves twin 35 mm GDF series systems.

The contract covered the production of the modernisation kits and delivery of proximity fused ammunition. Dutch industry was involved in the programme as the first six weapons were modernised by the then Bofors Defence and delivered to Netherlands in August 1989 with the remaining 54 weapons being modernised in Netherlands by RDM Technology.

These modernised 40 mm L/70 guns were used by three units of the Royal Netherlands Army, one regular and two reserves with the first becoming operational in 1991. One Flycatcher fire-control system will control two 40 mm L/70 anti aircraft guns.

The modernisation included the installation of a new servo system, new amplifiers, increased rate-of-fire kit (now 300 rds/min), ammunition racks and a diesel power unit.

As a result of defence cutbacks some of these weapons are now for sale.

Spanish Bofors Weapon Systems upgrade

In 1987, the Spanish Army took delivery of its first Felis electro optic automatic tracking system for the Bofors Defence 40 mm L/70 anti aircraft guns.

Felis consists of a high definition TV set with automatic tracking coupled to a telemetry laser, portable target designator and a radar interface. Spain uses the Contraves LPD 20 radar although other types can be used.

The system has three modes of operation:

- The first of these is radar acquisition which sends initial information to start the automatic tracking optical sequence
- The second mode uses the portable autonomous visual designator which starts the same sequence
- The third option uses a predetermined TV scanning pattern until acquisition is achieved.

Inisel is the prime contractor for the system previously called Linca and it was developed by Santa Bárbara.

By the end of 1987, one tracking system plus eight 40 mm L/70 weapons fitted with the kits were being used for trials purposes with a total of 100 systems delivered to the Spanish Army from 1988 to 1993.

Bofors Defence 40 mm L/70 modernisation package

Details of this are given in a separate entry under Sweden, Bofors Defence 40 mm L/70 modernisation package.

Electrical power supply unit

This has been developed to meet the requirements of the Royal Netherlands Army and is an on mount self contained, retro fit, diesel powered electrical generator set. It provides the necessary power enabling the 40 mm L/70 to be fully autonomous and is fitted with an electric start motor and a fuel tank, which has 20 litres of fuel, this being sufficient for eight hours of continuous use.

LVS upgrade

Details of the Bofors Defence LVS air defence sight are given in the Towed anti-aircraft gun sights section.

40 mm AOS L70 FADM

Details of the Allied Ordnance of Singapore 40 mm L70 Field Air Defence Mount (FADM), currently in production, are given under the entry for Singapore, 40 mm L70 Field Air Defence Mount.

Indian OFB 2007 Upgrade Project

An announcement made in September 2007 confirmed that the Indian Ordnance Factories Board (OFB) and the then BAE Systems SWS of Sweden are to start development of a non-government funded upgrade programme for the 2,000 or so L-70 anti-aircraft guns in the Indian inventory.

Specifications

	L/70 Type A	L/70 Type B
Dimensions and weights		
Crew:	6	6
Length		
transport configuration:	7.29 m	7.29 m
Width		
transport configuration:	2.225 m	2.225 m
Height		
transport configuration:	2.349 m	2.349 m
axis of fire:	1.335 m	1.335 m
Ground clearance		
overall:	0.39 m (travelling)	0.39 m (travelling)
Track		
vehicle:	1.8 m	1.8 m
Wheelbase	4.025 m	4.025 m
Weight		
standard:	4,800 kg	5,150 kg
Mobility		
Carriage:	trailer mounted four wheeled with outriggers	trailer mounted four wheeled with outriggers
Firepower		
Armament:	1 × 40 mm cannon	1 × 40 mm cannon
Ammunition		
main total:	96 (in racks)	96 (in racks)
Main weapon		
length of barrel:	2.8 m	2.8 m
calibre length:	70 calibres	70 calibres
breech mechanism:	vertical sliding block	vertical sliding block
maximum vertical range:	4,000 m (est.)	4,000 m (est.)
feed mechanism:	4 round clip	4 round clip
cartridge:	40 mm × 365 mm R	40 mm × 365 mm R
Rate of fire		
cyclic:	260 rds/min	260 rds/min
Main armament traverse		
angle:	360°	360°
Main armament elevation/depression		
armament front:	+90°/-4°	+90°/-4°

Status

Production as required.

Contractor

Bofors Defence AB.

It is, or has been, produced under licence in India, Italy, Norway, Singapore, Spain (by SA Placencia de las Armas) and the UK.

As of June 2007, Ordnance Factory Board (Pakistan) is supplying 40 mm ammunition to Turkey for their L70 towed AAGs.

Indian Ordnance Factory Board (OFB) upgrade programme.

The then BAE Systems SWS of Sweden upgrade programme.

Allied Ordnance of Singapore Pte Ltd produce a local version of the L70.

NAMMO AS, produces the HE-T and TP-T for the L/70 and fuzes for 40 mm ammunition.

Switzerland

20 mm GAI-B01

Type
Towed anti-aircraft gun

Development
The Oerlikon Contraves (now Rheinmetall Air Defence) 20 mm GAI-B01 automatic anti-aircraft cannon (gun) (formerly known as 10ILa/5TG) is fitted with a Rheinmetall Air Defence 20 mm cannon model KAB-001 (former designation 5TG) and was the lightest of the extensive Rheinmetall Air Defence range of 20 mm anti-aircraft weapons. It can be used in both anti-aircraft and ground defence and is normally operated by a detachment of three but can also be two, with one on the mount and the other one/two acting as ammunition handlers.

Description
The mount is normally carried on a two-wheeled carriage that can be towed by most light vehicles but it can also be mounted on the rear of a (4 × 4) or a (6 × 6) truck chassis. If required, the mount and carriage can be dismantled into individual loads.

The 20 mm GAI-B01 automatic anti-aircraft gun has four main components: the 20 mm cannon, mount, aiming equipment and the carriage.

The Rheinmetall Air Defence 20 mm automatic cannon KAB-001 is a positively locked, gas-operated weapon with mechanical ignition and can fire single shots or fully automatically. The breech is cocked using the manual cocking device and is held open automatically after firing the last round. The lower part of the weapon recoils in the cradle together with the barrel, whereas the trigger housing is locked firmly with the cradle. Ammunition is fed from a magazine, which can be changed in three seconds. Three types of magazine are available:

- 50-round drum weighing 41.5 kg loaded and 24.5 kg empty
- 20-round drum weighing 23.5 kg loaded and 17 kg empty
- 8-round box weighing 8 kg loaded and 4.5 kg empty.

The KAB cannon uses the less common 20 × 128 mm cartridge. Performance is almost identical to the 20 × 139 mm and the projectiles models used in original Oerlikon ammunition are the same. Technical data on the 20 × 139 mm can be found in the entry for Rheinmetall Air Defence 20 mm GAI-D01 anti-aircraft gun.

The mount is the standard base for the cannon and consists of:

- The cradle with weight compensatory and trunnions
- Pivot with elevation drive and trigger to support the cradle and allow elevation and traverse movement of the cannon
- Sight bracket to support the sight
- Tripod as a firm base when firing without the wheels in position
- The chassis with slide plate for transport and rapid change of position.

The gunner controls elevation with the upper hand wheel, one revolution of which gives 10° of elevation. Traverse is free with the gunner using his feet.

For engaging air targets an ellipse day sight or a Delta IV day sight can be fitted. The former can be used against aircraft and also ground targets after fitting a dioptre. It has a glass fibre or metal graticule with the appropriate lead marks. The Delta IV sight can be used against aircraft and ground targets with the lead ellipse being controlled by a mechanical attachment. For engaging ground targets, a telescope with a magnification of ×3.7 is secured to the cradle on a separate bracket. Effective range against low-flying aircraft is 1,500 m while against helicopters it is 2,000 m.

The two-wheeled carriage has a ground clearance of 0.34 m which can be reduced to 0.2 m by repositioning the wheels. A slide plate allows the gun to slide over obstacles.

When used as an anti-aircraft gun, the tripod legs are horizontal and the gunner is seated at the rear of the mount, the gun has an elevation of +85°, a depression of –5° and a total traverse of 360°.

Rheinmetall Air Defence 20 mm GAI-B01 automatic anti-aircraft gun in firing position with 50-round drum magazine and optical sight 0509535

Rheinmetall Air Defence 20 mm GAI-B01 light anti-aircraft gun mounted on rear of Steyr-Daimler-Puch 6 × 6 vehicle. Behind the truck a carriage is towed enabling the weapon to be deployed away from vehicle if required, by tactical situation. Spare magazines are also carried on the vehicle 0509536

When used against ground targets, the tripod legs remain horizontal but the gunner lies prone and uses the lower elevating gearing and the lower trigger. In this role the gun has an elevation of +25°, a depression of –5° and a total traverse of 60°.

The mount can also be set up in a higher position for engaging ground targets. The side spades are folded downwards and packing pieces are inserted. The towing hook, which is used as a trail spade, is pivoted downwards until the cannon is horizontal. When used in this mode, the gun has an elevation of +35°, a depression of –5° and a total traverse of 80°. In addition the gun can also be fired with the wheels in position.

Two crew can bring the gun into action from the travelling position in about 20 seconds. The chassis is pulled to the rear and the gun is placed on its tripod support.

This is a clear weather air defence system with no provision for onboard fire-control. Target information can be supplied from the Rheinmetall Air Defence LPD-20 search radar, an illustration of which appears in the entry for the Rheinmetall Air Defence twin 35 mm GDF series anti-aircraft guns.

Specifications

	20 mm GAI-B01
Dimensions and weights	
Crew:	3
Length	
transport configuration:	3.85 m
deployed:	4.71 m
Width	
transport configuration:	1.55 m
deployed:	1.55 m
Height	
transport configuration:	2.5 m
axis of fire:	0.425 m
deployed:	1.2 m
Ground clearance	
adjustable:	0.2 m to 0.34 m
Weight	
standard:	547 kg
combat:	405 kg
Mobility	
Firepower	
Armament:	1 × 20 mm cannon
Main weapon	
length of barrel:	2.4 m
maximum vertical range:	2,000 m (est.) (effective)
operation:	gas automatic
cartridge:	20 mm × 128 mm
Rate of fire	
cyclic:	1,000 rds/min
Main armament traverse	
angle:	360°
Main armament elevation/depression	
armament front:	+85°/-5°

Status
Production complete. No longer marketed. In service with many countries including Austria (560 army, 36 air force), Singapore (30), South Africa, Spain (329) and Switzerland (1,700). Around 40 Rheinmetall Air Defence 20 mm LAAG of unspecified type were supplied via Italy to the Mujahideen guerrillas in Afghanistan.

Contractor
Rheinmetall Air Defence.

20 mm GAI-C01 and GAI-C04

Type
Towed anti-aircraft gun

Development
The GAI-C01 and GAI-C04 automatic anti-aircraft cannon (guns) (formerly known as HS 639-B3.1 and HSS 639-B5 respectively), both from Oerlikon Contraves (now Rheinmetall Air Defence), are fitted with a gas-operated system. These guns can be used as both anti-aircraft and grounds weapons and are normally manned by a crew of three; one on the mount and the other two acting as ammunition handlers. The mount is normally carried on a two-wheeled carriage that can be towed by most light vehicles and, if required, the mount and carriage can be dismantled into individual loads.

Description
The GAI-C01 has an Rheinmetall Air Defence cannon type KAD-B13-3 (former designation HS 820-SL7° A3-3) with single feed from the right and the GAI-C04 has an Rheinmetall Air Defence cannon type KAD-B14 (former designation HS 820-SL7° A4) with dual feed. Elevation and traverse are manual with one revolution of the hand wheel giving 8° of elevation. Traverse is by using a pedal and the gunner can also unclutch the traverse mechanism to give free traverse.

Rheinmetall Air Defence 20 mm GAI-C01 automatic anti-aircraft gun in firing position 0509533

Rheinmetall Air Defence 20 mm GAI-C04 automatic anti-aircraft gun in firing position 0509534

The gunner can select either single shots or full automatic and has a Delta IV reflector day sight with a magnification of ×1 for engaging aerial targets plus a telescopic day sight with a magnification of ×2.5 that can be swung into position for engaging ground targets. The aircraft sight has two graticules, the first for aircraft flying at high speed (up to 900 km/h) and the other for slower flying aircraft or helicopters (up to 200 km/h). The graticule can be illuminated by two 4.5 V batteries.

Mounted on the right side of the GAI-C01 is a 75-round belt magazine that weighs 44 kg when loaded, while the GAI-C04 has two 75-round box magazines, one each side of the mount. Ammunition details are as for the Rheinmetall Air Defence 20 mm GAI-D01 anti-aircraft gun.

This is a clear weather 20 mm air defence system with no provision for off-carriage fire-control. Target information can be supplied by an Rheinmetall Air Defence LPD 20 trailer mounted search radar, an illustration of which appears in the entry for the Rheinmetall Air Defence twin 35 mm GDF series towed anti-aircraft gun system.

Variants
The South African National Defence Force (SANDF) has a locally developed, self-propelled version of this weapon on a day (4 × 4) truck. Details are given in the Self-propelled anti-aircraft guns section under South Africa. A total of 84 of these systems are currently in service with the South African National Defence Force.

Specifications

	GAI-C01	GAI-C04
Dimensions and weights		
Crew:	3	3
Length		
deployed:	3.87 m	3.87 m
Width		
deployed:	1.7 m	1.7 m
Height		
axis of fire:	0.5 m	0.5 m
deployed:	1.45 m	1.45 m
Weight		
standard:	512 kg	535 kg
combat:	370 kg	435 kg
Mobility		
Carriage:	trailer mounted two wheeled	trailer mounted two wheeled
Firepower		
Armament:	1 × 20 mm cannon	1 × 20 mm cannon
Main weapon		
length of barrel:	1.84 m	1.906 m
maximum vertical range:	2,000 m (est.) (effective)	2,000 m (effective)
operation:	gas automatic	gas automatic
cartridge:	20 mm × 139 mm	20 mm × 139 mm
Rate of fire		
cyclic:	1,050 rds/min	1,050 rds/min
Main armament traverse		
angle:	360°	360°
Main armament elevation/depression		
armament front:	+83°/-7°	+83°/-7°

Status
Production complete, no longer marketed. The system has seen service with Chile (GAI-C01), Indonesia (10), Nicaragua, South Africa (GAI-C01) and other undisclosed countries.

Contractor
Rheinmetall Air Defence.

20 mm GAI-C03

Type
Towed anti-aircraft gun

Development
The Oerlikon Contraves (now Rheinmetall Air Defence) 20 mm GAI-C03 automatic anti-aircraft gun (formerly known as HS 639-B4.1) is fitted with an Rheinmetall Air Defence gas-operated 20 mm cannon model KAD-A01 (former designation HS 820 SAA1). The carriage and mount are identical to those used for the 20 mm GAI-C01 and GAI-C04 automatic anti-aircraft guns. Full details of these are given in a separate entry. The weapon was and can be used in both the anti-aircraft and ground roles and was/is normally crewed by a detachment of three, one on the mount and the other two acting as ammunition handlers. The mount is normally carried on a two-wheeled carriage, which can be towed by most light vehicles, and, if required, the mount and carriage can be dismantled into individual loads.

Rheinmetall Air Defence 20 mm GAI-C03 automatic anti-aircraft gun in firing position 0509537

Description

This weapon uses a crew of three with a two minute set up time. Elevation and traverse are manual with one revolution of the hand wheel giving 8° of elevation. A pedal controls traverse and the gunner can release the clutch mechanism to give free traverse.

The gunner can select either single shots or full automatic and has a Delta IV reflector day sight with a magnification of ×1 for engaging aerial targets plus a telescopic day sight with a magnification of ×2.5 that can be swung into position for engaging ground targets. The anti-aircraft sight has two graticules, one for aircraft flying at high speed (up to 900 km/h) and the other for slower-flying aircraft and helicopters (up to 200 km/h). The graticule can be illuminated by two 4.5 V batteries.

Mounted over the cradle of the KAD-A01 cannon is a drum-type magazine that holds 50 rounds of ammunition and weighs 36 kg loaded and 20 kg empty. Ammunition details were given in the Rheinmetall Air Defence 20 mm GAI-D01 anti-aircraft gun entry.

This is a clear weather air defence system with no provision for on board fire-control. Target information can be provided by an Rheinmetall Air Defence LPD-20 trailer mounted search radar, an illustration of this is given in the entry for the Rheinmetall Air Defence GDF series twin 35 mm anti-aircraft guns.

Specifications

	GAI-C03
Dimensions and weights	
Crew:	3
Length	
deployed:	4.27 m
Width	
deployed:	1.7 m
Height	
axis of fire:	0.5 m
Weight	
standard:	495 kg
combat:	342 kg
Mobility	
Carriage:	trailer mounted two wheeled
Firepower	
Armament:	1 × 20 mm cannon
Main weapon	
length of barrel:	2.24 m
maximum vertical range:	2,000 m (effective)
operation:	gas automatic
Rate of fire	
cyclic:	1,050 rds/min
Main armament traverse	
angle:	360°
Main armament elevation/depression	
armament front:	+83°/-7°

Status

Production complete. No longer marketed. The system has seen service with many undisclosed countries. No further information is available on this system.

Contractor

Rheinmetall Air Defence.

20 mm GAI-D01

Type

Towed anti-aircraft gun

Development

The Oerlikon Contraves (now Rheinmetall Air Defence) twin 20 mm GAI-D01 automatic anti-aircraft gun (formerly known as HSS-666A) was designed to fill the gap between single manual 20 mm anti-aircraft guns such as the GAI-C01, GAI-C03 and the GAI-C04 and, the much more sophisticated and effective Rheinmetall Air Defence twin 35 mm anti-aircraft gun Type GDF-002.

It can be used as an anti-aircraft and ground weapon and is normally operated by a detachment of five, with only one on the mount. The prototype was produced during 1976 and the first production example appeared in 1978.

Although production was completed in Switzerland some time ago it has been manufactured under license in Turkey by MKEK for the Turkish Land Forces.

Description

The system normally requires a crew of five with a set up time of one minute or less. The GAI-D01 is fitted with two 20 mm KAD cannons, a left-feed model designated the KAD-B16 and a right-feed model designated the KAD-B17. The gunner can select single shot, rapid single shot, and automatic fire limited and automatic fire unlimited. Each cannon is provided with 120 rounds of ready use ammunition, with each full magazine weighing 68 kg. The types of linked ammunition that can be fired are detailed in Table 1 at the bottom of the page.

The upper mounting consists of the cradle with belt damper unit, counterweight for the elevating assemblies, Galileo P56 (covered in detail separately) sighting and aiming unit, Wankel engine and an adjustable gunner's seat. Two containers on each side of the mounting provide compressed air for the electro-pneumatic operation of the trigger mechanism. The lower mounting consists of the tripod support and attachment points for the travelling carriage. The carriage is a tubular construction and has a towing eye, which is adjustable for height, independently sprung wheels, and a hydraulically operated over-running brake, which can also be operated by hand. The GAI-D01 can be brought into action in 60 seconds from the travelling position by a team of five.

Elevation and traverse are hydraulic, with maximum traverse speed of 80°/sec and maximum elevation speed of 48°/sec. Manual controls are provided for emergency use.

The Italian Galileo Avionica P56 sight is used to engage low-flying aircraft to a range of 1,500 m and surface targets to 2,000 m. The target is initially engaged using the mechanical auxiliary sight and then located and tracked using the optical sight, which has a magnification of ×5 and a 12° field of view. In the 'air target' mode (automatic/hydraulic) the unit calculates the overall lead on a continuous basis. Surface targets may also be engaged using this mode or a manual/mechanical mode.

The P56 sight consists of an optical sight, an electro-mechanically operated servo-drive unit for the view prism and a computer, plus a gun laying system consisting of a control-stick assembly, hydraulic unit, mechanical transfer gearboxes and manual controls.

Facilities are provided for integrating the GAI-D01 automatic anti-aircraft gun with an early warning radar such as the Contraves LPD-20 and, connecting the gun to a remote input equipment specifying target speed, crossover point range, fire release and also incorporating a two-way intercom. An alternative

Rheinmetall Air Defence twin 20 mm GAI-D01 automatic anti-aircraft gun in firing position clearly showing Wankel engine under gunner's seat 0509885

Table 1

NATO designation	API-T	AP-T	HEI-T	HEI	HEI (BF)	SAPHEI-T	SAPHEI	TP-T/TP
Weight of projectile:	128 g	110 g	125 g	125 g	128 g	125 g	125 g	125 g
Weight of explosive:	n/avail	n/avail	5.6 g	10 g	8 g	4.7 g	4.7 g	n/avail
Weight of propellant:	53 g	53 g	53 g	53 g	53 g	53 g	53 g	53 g
Weight of complete round:	325 g	322 g	337 g	337 g	325 g	337 g	337 g	337 g
Muzzle velocity:	1,040 m/s	1,100 m/s	1,040 m/s	1,040 m/s	1,040 m/s	1,040 m/s	1,040 m/s	1,040 m/s

sight with a radar data decoder was available. An illustration of the Rheinmetall Air Defence LPD-20 radar system, which is no longer manufactured, is given in a separate entry for the Rheinmetall Air Defence twin 35 mm GDF series of towed anti-aircraft guns which are still being manufactured and marketed.

Specifications

	GAI-D01
Dimensions and weights	
Crew:	5
Length	
transport configuration:	4.59 m
deployed:	4.555 m
Width	
transport configuration:	1.86 m
deployed:	1.81 m
Height	
transport configuration:	2.34 m
axis of fire:	0.6 m
deployed:	1.3 m (at 0° elevation)
Track	
vehicle:	1.86 m
Weight	
combat:	1,330 kg
Mobility	
Carriage:	trailer mounted two wheeled
Firepower	
Armament:	2 × 20 mm cannon
Main weapon	
length of barrel:	1.906 m
maximum range:	2,000 m (effective)
operation:	gas automatic
Rate of fire	
cyclic:	2,000 rds/min
Main armament traverse	
angle:	360°
Main armament elevation/depression	
armament front:	+81°/-3°

Status
Production completed in Switzerland; however, serial production may still take place in Turkey by MKEK for the Turkish Army. No longer offered for sale. The system has seen service with Bolivia, Guatemala, Turkey (400) and other undisclosed countries.

Contractor
Rheinmetall Air Defence.

25 mm GBI-A01

Type
Towed anti-aircraft gun

Development
The GBI-A01 cannon began development in the mid-to-late 1950s as a basic low-level direct-fire anti-aircraft gun. This single barrel gun uses manual handwheels to lay-on the AA target but also has a limited capability against surface light skinned targets.

Description
The Oerlikon Contraves (now Rheinmetall Air Defence) 25 mm GBI-A01 automatic anti-aircraft gun is fitted with a gas-operated Rheinmetall Air Defence KBA-C cannon and can also be used to engage light AFVs and other battlefield targets. The weapon is normally manned by a crew of three, one on the mount and the other two acting as ammunition handlers.

The 25 mm GBI-A01 is a clear-weather air defence system although target information can be provided from other radar and sensor systems. Typical radar would be the Rheinmetall Air Defence LPD-20, an illustration of which appears in a separate entry. There is no provision for automatic lying of the system onto the target.

The weapon has three firing positions:

- Normal firing position mounted on its tripod with an elevation of +70°, depression of –10° and a traverse of 360°

Rheinmetall Air Defence 25 mm GBI-A01 automatic anti-aircraft gun in firing position 0509884

- Mounted on its travelling carriage with the rear supported by the towing eye, with an elevation of +70°, no depression and a traverse of 360°
- Emergency mode with the carriage still coupled to the towing vehicle, with an elevation of +70°, depression of -10° and a traverse of 45° left and 45° right.

Elevation and traverse are manual with one revolution of the elevation hand wheel gives 4° of elevation and one revolution of the traversing hand wheel gives 10° of traverse. The gunner can unclutch the traverse mechanism if required to give free traverse. The mount and travelling gear can be dismantled into individual loads.

The gunner can select either single-shot fire or full automatic and has a binocular sight for engaging ground targets and a Delta sight for air targets. Mounted either side of the dual-feed KBA-C cannon is a 40-round box magazine each of which weighs 33 kg when loaded. The types of ammunition that can be fired are detailed in the Table 1 at the bottom of the page.

Specifications

	GBI-A01
Dimensions and weights	
Crew:	3
Length	
transport configuration:	4.72 m
deployed:	4.17 m
Width	
transport configuration:	1.8 m
deployed:	1.79 m
Height	
transport configuration:	1.65 m
axis of fire:	0.5 m
deployed:	1.45 m
Ground clearance	
overall:	0.4 m (travelling)
Weight	
standard:	666 kg
Mobility	
Carriage:	trailer mounted two wheeled
Firepower	
Armament:	1 × 25 mm KBA cannon
Ammunition	
main total:	40
cannon:	40
Main weapon	
length of barrel:	2.182 m
muzzle brake:	multibaffle
maximum vertical range:	2,500 m (est.)
operation:	gas automatic
Rate of fire	
cyclic:	570 rds/min
Main armament traverse	
angle:	360°
Main armament elevation/depression	
armament front:	+70°/-10°

Table 1

NATO designation	APFSDS-T	APDS-T	FAPDS-T	TPDS-T	HEI-T/HEI	SAPHEI-T/SAPHEI	TP-T/TP
Weight of projectile:	130 g	150 g	150 g	150 g	180 g	180 g	180 g
Weight of propellant:	105 g	105 g	105 g	105 g	91 g	91 g	91 g
Weight of complete round:	440 g	480 g	460 g	486 g	500 g	500 g	500 g
Muzzle velocity:	1,405 m/s	1,335 m/s	1,310 m/s	1,335 m/s	1,100 m/s	1,100 m/s	1,100 m/s

Status
Production complete. In service with undisclosed countries. Although this weapon is no longer marketed, it can be found in many parts of the world.

Contractor
Rheinmetall Air Defence.

Skyguard/35 mm AHEAD

Type
Towed anti-aircraft gun

Development
Oerlikon Contraves (now Rheinmetall Air Defence) of Switzerland has developed, as a private venture, the AHEAD ammunition to expand the defensive capability of 35 mm medium calibre air defence guns against the modern threats of precision-guided munitions, cruise missiles, anti-radiation missiles, remotely piloted vehicles and other high-technology weapons.

AHEAD stands for Advanced Hit Efficiency And Destruction. Successful demonstrations of the system from a twin 35 mm GDF-005, automatic anti-aircraft gun system used in conjunction with a Skyguard fire-control system were carried out in Austria in 1993.

By late 2000, at least three countries, including Canada and Oman, had ordered AHEAD for their twin 35 mm GDF series anti-aircraft guns and volume production was under way.

South Africa has selected the 35 mm AHEAD ammunition for use with its new LIW twin 35 mm Dual-Purpose Gun (DPG) naval mount, which will be installed on the four corvettes ordered by the South African Navy. It is also fired by the BAE Systems, RO Defence Millennium 35/1000 Naval Gun System.

Description
The new-generation projectile of the AHEAD ammunition consists of an outer shell enclosing the heavy metal payload (spin-stabilised subprojectiles), the payload ejection charge and the programmable electronic fuze. Programming of the timer is achieved at the multipurpose AHEAD muzzle base as the projectile leaves the muzzle.

The muzzle base consists of three coils. While the projectile passes through the first two coils, its exact velocity is determined and processed with target information as supplied by the fire-control computer. The exact projectile flying time is calculated and imparted to the fuze by electro-induction as the projectile passes through the third coil.

As a result, the time fuze will accordingly detonate the payload ejection charge at a given distance in front of the target, forming a cone of the heavy metal spin-stabilised subprojectiles and directing it into the target. Through their high-kinetic energy, these subprojectiles are capable of defeating cruise missiles, drones, remotely piloted vehicles and other precision-guided munitions, whatever their front-end armour protection.

The upgrading of existing Rheinmetall Air Defence 35 mm/Skyguard air defence systems to cope with the new threat is possible as part of a retrofit programme.

The AHEAD system has considerable development potential assuring multiple applications for the future, according to the manufacturer.

AHEAD is usable with all 35 mm Rheinmetall Air Defence cannons, towed and self-propelled, provided that the AHEAD specific gun computer and the AHEAD muzzle base are fitted.

A typical target engagement takes place as follows:

- The target is first detected by the search radar of the fire-control radar and then displayed on the PPI.
- The tracking radar then locks on to the target at a range of about 4 km and then continually feeds all relevant information to the computer, which passes information to the actual weapons.

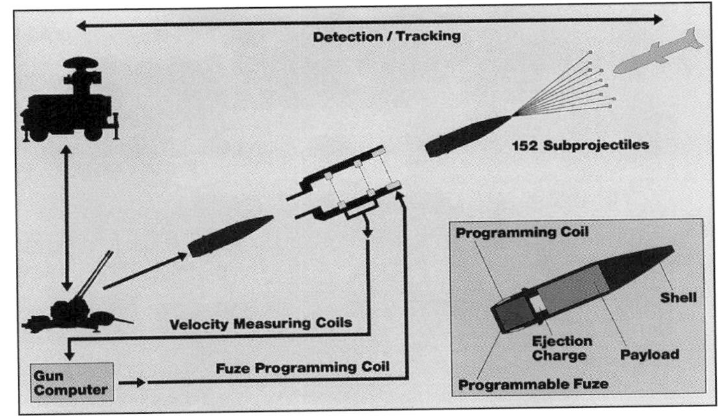

AHEAD schematic showing the method of target engagement. Although this is shown for the twin 35 mm GDF towed anti-aircraft gun system with the Skyguard fire-control system, AHEAD is also applicable to twin 35 mm self-propelled air defence systems 0509776

- As soon as the target comes within range, the 35 mm cannon opens fire. As each projectile passes through the triple coil, muzzle velocity gauge, its exact velocity is determined between coils one and two. Within microseconds the gun-mounted electronic processor determines the time of flight and imparts this through electrical induction via coil three to the timer in the projectile's base fuze.
- While the projectile is in flight, the onboard timer counts down the thousandths of a second until the zero mark is reached. At this time, the time fuze detonates the payload ejection charge so forming a cone of heavy metal spin-stabilised sub-projectiles that are directed into the target.

During trials in the Middle East early in 1995, a Skyguard/35 mm fire unit, fielded by the Abu Dhabi (United Arab Emirates) Armed Forces was AHEAD upgraded in country.

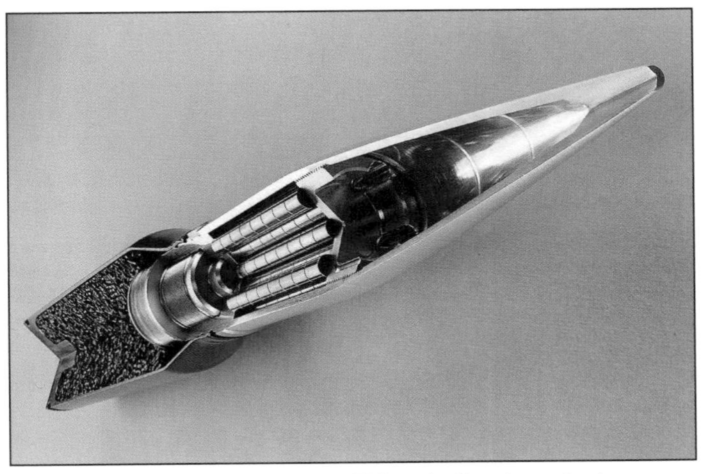

This photograph shows the 35 mm AHEAD projectile cutaway to show interior. The payload consists of 152 tungsten cylinders each of which weighs 3.3 g 0064566

An AHEAD modified Skyguard fire-control/twin 35 mm gun air defence battery in action showing AHEAD ammunition impact in lower right inset 0044158

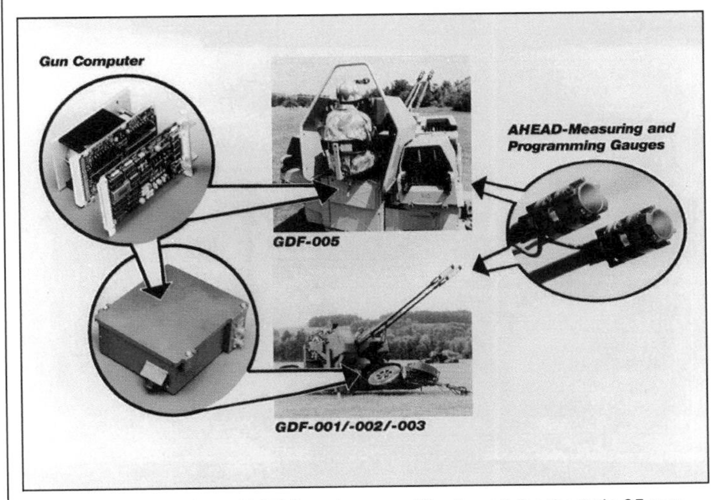

Key components of the AHEAD system modification kit for the twin 35 mm GDF series anti-aircraft gun. Photograph shows the computer installation depending on the standard of build (GDF-001/002/003) as a separate unit or GDF-005 in the existing cabinet. Each 35 mm barrel is fitted with the new AHEAD measuring and programming gauges 0509777

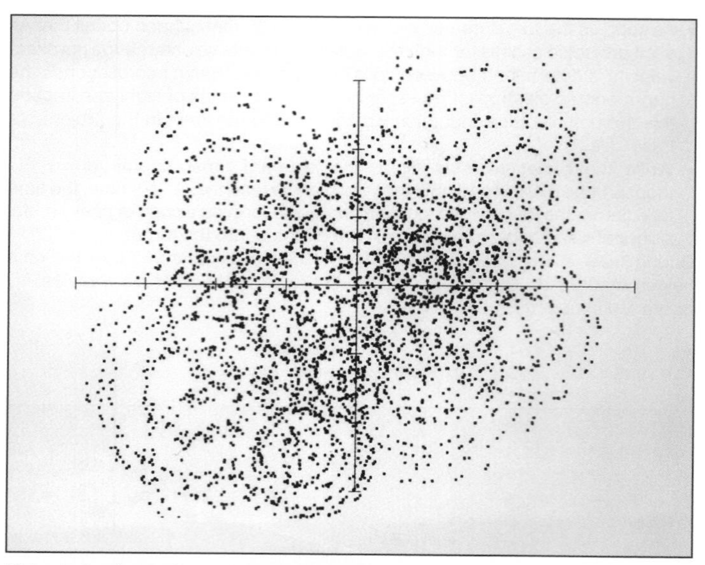

This photo shows the typical distribution pattern of sub-projectiles of a short burst of AHEAD rounds 0100687

The live firings, conducted in part under poor weather conditions, were against Banshee targets. On each of the four successful firings, a burst of about 24 AHEAD rounds was fired, which in each case resulted in numerous impacts of spin-stabilised sub-projectiles in the flying target.

Status

In production and in service with at least three countries; Canada, Oman and United Arab Emirates (UAE).

Contractor

Rheinmetall Air Defence AG.

Skyguard GDF-002/003 and GDF-005

Type

Towed anti-aircraft gun

Development

In the late 1950s, the then Oerlikon-Bührle (now Rheinmetall Air Defence) started development of a twin 35 mm towed automatic anti-aircraft gun. The first prototype of this was completed in 1959 under the designation 1 ZLA/353 MK. This entered production as the 2 ZLA/353 MK but was subsequently redesignated the GDF-001.

In 1980, the GDF-002 model was introduced. This advanced version has a MBDA (former designation Ferranti) sight instead of an Xaba sight and digital data transmission. By 2000, over 2,000 GDF twin 35 mm systems had been produced with sales made to around 30 countries.

The Rheinmetall Air Defence twin 35 mm GDF-002 automatic anti-aircraft gun is used primarily as an anti-aircraft weapon but can also be applied against ground targets. It can be used on its own with its onboard optical sight, but is normally used in conjunction with an off-carriage fire-control system. A typical battery would consist of two GDF series anti-aircraft guns, each with a power supply unit and fire-control unit. The fire-control unit was originally the Contraves Super Fledermaus, now replaced in production in Switzerland by the Rheinmetall Air Defence Skyguard fire-control system that is much more effective. Details of Skyguard are given in the Anti-aircraft control systems section.

GDF-001 anti-aircraft gun fitted with XABA auxiliary sight with lead computer 0100688

In May 1985, the GDF-005 was introduced. This is an overall improvement of the GDF-001/2/3 and the earlier models can be modified to the GDF-005 standard by the use of combat improvement kits supplied by Rheinmetall Air Defence. The GDF-005 features a new autonomous gun sighting system, an onboard power supply system, an automatic reload and other improvements. The GDF-001 can be upgraded to any of the other gun standards by using kits. The NDF-B is particularly useful as it introduces automatic ammunition replenishment from auxiliary magazines to the main magazines.

Late in 1999, during a parade held in China, the People's Liberation Army displayed a very similar twin 35 mm towed anti-aircraft gun together with a radar fire-control system similar to Skyguard. Details of this gun are in this section under China. As far as it is known China has not exported any of these weapons.

Description

The Rheinmetall Air Defence twin 35 mm GDF-002 automatic anti-aircraft gun consists of the following main components: two KDB (former designation 353 MK) cannon, cradle, two automatic ammunition feed mechanisms, upper mount, lower mount and the sighting system.

The Rheinmetall Air Defence KDB cannon is a positively locked gas-operated fully-automatic weapon. The weapon housing, together with the barrel, slides in the cradle during recoil. The cannon cover contains the ammunition feed mechanism and does not move during recoil. The manual cocking device is also mounted on the cannon cover. The barrels have progressive twist rifling, are fitted with muzzle brakes and, if required, can also be fitted with muzzle velocity measuring equipment. The cyclic rate of fire per barrel is 550 rounds per minute, the total combined rate of fire is 1,100 rounds per minute.

Rheinmetall Air Defence twin 35 mm GDF-005 automatic anti-aircraft gun in firing position with Skyguard fire-control system in right background 0509532

Rheinmetall Air Defence twin 35 mm GDF-002 automatic anti-aircraft gun in travelling configuration with twin 35 mm cannon in travelling lock 0044160

Rheinmetall Air Defence twin 35 mm GDF-005 deployed and showing generator mounted on rear of carriage 0509883

Table 1

NATO designation	HEI-T	HEI	HEI (BF)	SAPHEI-T	FAPDS	TP-T/TP	AHEAD
Weight:							
(projectile)	535 g	550 g	550 g	550 g	375 g	550 g	750 g
(explosive)	98 g	112 g	70 g	22 g	0 g	0 g	1.0 g
(propellant)	330 g	330 g	330 g	330 g	330 g	330 g	330 g
(complete round)	1.565 kg	1.580 kg	1.580 kg	1.552 kg	1.440 kg	1.580 kg	1.78 kg
Muzzle velocity:	1,175 m/s	1,175 m/s	1,175 m/s	1,175 m/s	1,440 m/s	1,175 m/s	1,050 m/s

Oerlikon-Contraves GDF-003 twin 35 mm anti-aircraft gun fitted with MBDA Type GSA Mk 3 auxiliary sight and built-in automatic breech block lubrication system 0100689

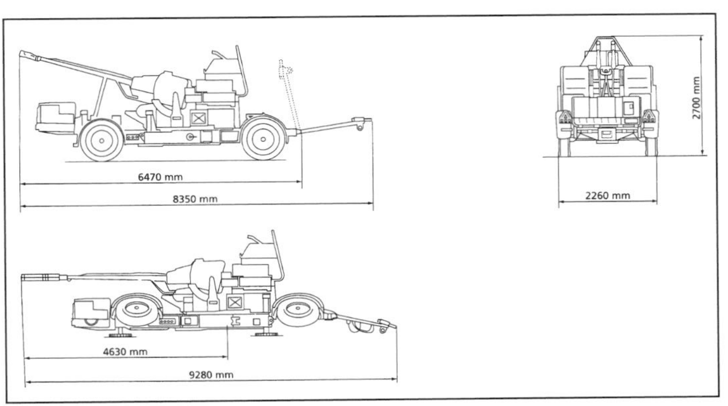

Outline drawings of Rheinmetall Air Defence GDF-005 twin 35 mm anti-aircraft gun in travelling and deployed configuration 0100690

Rheinmetall Air Defence twin 35 mm GDF-005 anti-aircraft guns with camouflage covers in position 0007668

The cradle is designed to carry both guns and is on the elevation axis. It contains the hydromechanical recoil mechanism, which absorbs the recoil forces. The ammunition containers are on each side of the cradle and rotate with it. Each fully loaded ammunition container holds 56 rounds. The ammunition is reloaded in seven-round clips from the reloading container and passed through the upper mount trunnions to the cannon. The drive for the feed is independent of the cannons and uses electric spring motors. Rewinding the spring motors, which is normally automatic, can also be done manually. Power supply to the system is on a separate two-wheel trailer.

The upper mount determines traverse of the guns, which is in the pivot bearing of the lower mount. The upper mount platform supports the auxiliary aiming equipment, the seats for the crew and both 63-round reloading containers. The guns have a maximum elevation speed of 56°/s and a maximum traverse speed of 112°/s.

The lower mount forms the stable base of the gun. It comprises the two-axle chassis and the outriggers with the levelling spindles for three-point support in the firing positions. Raising and lowering the levelling spindles and raising the wheels are done electro-hydraulically or manually in the case of power failure.

Some countries use the Rheinmetall Air Defence LPD-20 surveillance radar to pass on target information to the Skyguard fire-control system and subsequent target engagement by the twin 35 mm GDF series anti-aircraft guns 0509705

A crew of three can bring the weapon into the firing position in 2 minutes 30 seconds, 1 minute 30 seconds by one. A hand pump is also fitted and when it is used the weapon can be brought into action in five minutes.

The sighting equipment consists of a MBDA sight Type GSA Mark 3, a ground target sight mounted on the MBDA sight housing and an optical alignment sight. The target range is the only parameter to be adjusted in action on the MBDA sight.

Ammunition

The air defence types of 35 mm ammunition that can be fired are listed in Table 1 at the top of the page.

Modification packages

Rheinmetall Air Defence is now offering a number of modification packages for the twin 35 mm GDF series of towed anti-aircraft guns and brief details of these are in Table 2:

Table 2

Combat improvement kit	NDF-A	NDF-C
MBDA sight:	yes	no
Weapon optimisation:	yes	yes
Weapon lubrication:	yes	yes
Camouflage:	yes	yes
Automatic re-loaders:	no	yes
Cab for gunners:	no	no
Integrated power supply:	no	yes

The NDF-A kit comprises several changes, notably the addition of the MBDA Type GSA Mark 3 sight for engaging air and ground targets, a quick-erect camouflage assembly, a transistorised power supply unit and the KDC-02 gun modification package. The NDF-C kit is used to bring any of the GDF-001/2/3 series up to the GDF-005 standard. In this version the gunner does not have a completely enclosed cab and the power unit is mounted on the rear of the trailer and lowered to the ground while firing, so as not to impede the guns. A Rheinmetall Air Defence Gunking sight system is fitted. Full details of the Gunking sight system, which can be fitted onto other towed anti-aircraft guns, for example the Russian twin 23 mm ZU-23-2, are given separately.

By 2000, over 350 NDF-C kits had been sold, with customers including Austria, Finland, Switzerland (108 upgraded to this standard) and Saudi Arabia.

The Austrian Rheinmetall Air Defence Skyguard fire-control radars for the twin 35 mm GDF series guns are being modernised in Switzerland. By 2004, they will have:

- A laser range-finder
- Updated combat and simulation software
- Automated threat analysis
- Electronic warfare improvements.

GDF-005

Introduced in May 1985, the 35 mm Rheinmetall Air Defence GDF-005 has several overall improvements that can be retrofitted to existing GDF versions using modification kits.

One of the main improvements is the fitting of the Gunking 3-D autonomous computer-controlled optronic sighting system, which eliminates the need for the gunner to estimate target parameters. A built-in microcomputer processes all target data such as target range obtained via a laser range-finder, muzzle velocity and meteorological data, to generate lead data for the gun control system. The Gunking system allows engagements out to a possible range of 4,000 m. The addition of a fully automatic 35 mm re-loader reduces the number of crew on the gun from three to one (the layer) and at the same time the number of rounds on each gun has been increased to 280 rounds, enough for 10 combat bursts. The re-loaders are powered by a hydraulic system which also supplies power for the automatic weapon and breech block lubrication system (under a new gun cover) and power for the hydraulic emergency trigger systems. The 35 mm cannon used are the type KDC with a breech recoil brake, rate of fire attenuation and a firing pin lock. A new onboard integrated power supply unit supplies not only the gun control systems but also provides power for emplacing the gun. When on site an electro-hydraulic circuit is used to extend the outriggers, operate the jacks and pivot the carriage wheels to their inclined position. The power supply unit is also lowered to the ground. Levelling is carried out using a push-button control and, for use in an emergency, a hand-operated pump is provided. Permanently attached camouflage material is optional.

The weight of a fully loaded GDF-005 is 7,700 kg, unloaded weight is 7,250 kg.

Known customers for the GDF-005 include Canada, Cyprus and Malaysia.

Variants

35 mm AHEAD ammunition system

With the Oerlikon Ahead Air Burst Munition, a cloud of sub-projectiles is thrown towards the attacking target. A short burst of Ahead rounds produces a high density of sub-projectiles in front of the attacking target, so that even the smallest targets is hit by a sufficient number of impacts to achieve a mission abort kill. Further details of this are given in a separate entry. Canada, Oman and one undisclosed country have adopted the 35 mm AHEAD system for their GDF 007 series anti-aircraft gun systems.

Skyguard Gun missile system

Each battery consists of one Rheinmetall Air Defence Skyguard fire-control system, two missile launchers each with four Sparrow surface-to-air missiles in the ready to launch position and two twin Rheinmetall Air Defence 35 mm GDF-002/003 or 005 cannon. Details of this system are given in a separate entry in the Static and towed surface-to-air missile systems section.

Skyguard Gun/Kentron SAM systems

Details of this system, announced for the first time in November 1994, are given in a separate entry, under Static and towed surface-to-air missile systems section, International, Kentron SAHV-IR/Rheinmetall Air Defence 35 mm Skyguard air defence system.

Marksman SPAAG

The then British BAE Systems, RO Defence, Marksman twin 35 mm self-propelled air defence system uses KD series 35 mm cannon and the same family of ammunition. Details of this system are given in a separate entry, under Self-propelled anti-aircraft guns, United Kingdom, RO Defence Marksman twin 35 mm anti-aircraft turret. Marksman is in service in Finland on T-55 MBT chassis.

Gepard SPAAG

The German Gepard twin 35 mm self-propelled air defence system uses KD series 35 mm cannon and the same family of ammunition. Details of this are given in a separate entry, under the Self-propelled anti-aircraft guns section, International, Krauss-Maffei Wegmann/Rheinmetall Air Defence Gepard and CA 1 twin 35 mm self-propelled anti-aircraft gun systems. Part of the German and Netherlands Gepard fleet is now being upgraded with first deliveries late in 1998. Romania has also taken delivery of a quantity of these systems from the German Army.

Type 87 SPAAG

The Japanese Type 87 twin 35 mm self-propelled air defence system uses KD series 35 mm cannon and the same family of ammunition. Details of this are given in a separate entry under the Self-propelled anti-aircraft guns section, Japan, Type 87 twin 35 mm self-propelled anti-aircraft gun system.

Specifications

	Skyguard GDF-003
Dimensions and weights	
Crew:	3 (local mode)
Length	
transport configuration:	7.8 m
deployed:	8.83 m
Width	
transport configuration:	2.26 m
deployed:	4.49 m
Height	
transport configuration:	2.6 m
axis of fire:	1.28 m
deployed:	1.72 m
Ground clearance	
overall:	0.33 m
Track	
vehicle:	1.9 m
Wheelbase	3.8 m
Weight	
standard:	6,300 kg
Mobility	
Carriage:	trailer mounted four wheeled with outriggers
Firepower	
Armament:	2 × 35 mm cannon
Main weapon	
length of barrel:	3.15 m
maximum vertical range:	4,000 m (effective)
feed mechanism:	238 round (total on gun)
cartridge:	35 mm × 228 mm
Rate of fire	
cyclic:	1,100 rds/min (est.)
Main armament traverse	
angle:	360°
Main armament elevation/depression	
armament front:	+92°/-5°

Status

GDF-001 and GDF-002 production complete, no longer offered. GDF-003 and GDF-005 in production. By the year 2000, over 2,000 GDF series weapons had been manufactured with sales being made to over 30 countries.

It was announced at the end of October 2013 that Nigeria had fabricated spare parts to refurbish its 35 mm anti-aircraft guns that were originally supplied to the country in 1979 from Switzerland. These guns are now deployed with the 31st Artillery Brigade in Nigeria.

Contractor

Rheinmetall Air Defence AG.

Skyshield 35 AHEAD

Type

Towed anti-aircraft gun

Development

In June 1995, Oerlikon Contraves of Switzerland (now Rheinmetall Air Defence) unveiled its Skyshield 35 Advanced Hit Efficiency And Destruction (AHEAD) Air Defence System (ADS) which had been developed by the company as a private venture.

The system had been developed to meet the market requirement for a twin 35 mm air defence system with increased capabilities over the current in-service Rheinmetall Air Defence twin 35 mm GDF series which is covered in detail in a separate entry.

Development of the Skyshield 35 AHEAD ADS was completed late in 1995 and extensive system trials took place in June 1996.

These were carried out at the Nettuno firing range in Rome, where the Skyshield 35 AHEAD hit a Hayes TRX target being towed by a Gates Learjet and then switched and engaged a ground target represented by a static Falcon missile.

The ability of the new Skyshield 35 fire-control unit to control the older Rheinmetall Air Defence GDF-002 twin 35 mm towed anti-aircraft gun was also successfully demonstrated when a 6 m long towed sleeve target was shot down.

Marketing of the new system is now under way to existing users of the current twin 35 mm GDF towed anti-aircraft gun system and to potential new customers.

As the new system is lighter and more compact than the older twin 35 mm GDF system, production costs are expected to be lower and local production would also be easier.

In late 1997, the Skyshield 35 AHEAD ADS was successfully demonstrated in Spain when it engaged towed sleeve and Banshee drone targets.

A MOOTW/C-RAM 35 mm gun module (Rheinmetall) 1332662

A MOOTW/C-RAM sensor module (Rheinmetall) 1346809

The unmanned Skyshield 35 AHEAD Revolver Gun Mount (RGM) is armed with a single 35/1000 revolver cannon for which 228 rounds of ready-use ammunition are carried 0100686

Rheinmetall Air Defence Skyshield 35 AHEAD Air Defence System showing the unmanned Sensor Unit (right) and the AHEAD Revolver Gun Mount (RGM) (left) 0100685

To provide extended range coverage, the Skyshield 35 AHEAD ADS can also be integrated with the pallet-mounted ADATS surface-to-air missile system with eight missiles in the ready-to-launch position 0509773

A further development of the Skyshield is the Oerlikon Skyshield Military Operations Other Than War (MOOTW)/Counter-Rockets, Artillery and Mortar (C-RAM) system. Also known in Germany as the NBS (*Nachstbereichs-Schutzsystem*) Modular Automatic and Network-capable Targeting and Interception System (MANTIS), the system started life in 2007 with a German Bundesamt fur Wehrtechnik und Beschaffung (BWB) requirement to develop a system for the protection of German forward operating bases. The system went through trials in 2009 culminating in the German Army order for the system to protect troops on the battlefield.

Description

The Skyshield ADS features a modular design, flexible capabilities and a high level of automation, as well as being easy to transport. A typical Skyshield 35 AHEAD ADS would consist of two 35 mm Revolver Gun Mounts (RGM), each armed with the new Rheinmetall Air Defence 35/1000 rapid fire revolver cannon, first revealed late in 1994, and a new modular Fire-Control Unit (FCU).

The new system has been designed to engage not only aircraft and helicopters but also a wide range of other battlefield targets including cruise missiles, remotely piloted vehicles and precision guided munitions.

The unmanned RGM armed with the Rheinmetall Air Defence 35/1000 gas-operated revolver cannon is provided with 228 rounds of ready-use 35 mm ammunition which is fed from the right via a linkless ammunition feed system. Cyclic rate-of-fire is 1,000 rds/min. The AHEAD anti-missile system involves firing a burst of approximately 24 rounds toward the target intercept point. At a precise moment in time – programmed into each round at the muzzle of the barrel and corresponding to a point 10 or 40 m ahead of the target – 152 high density tungsten sub-projectiles are released from each AHEAD round, forming a compact and destructive cloud.

This ammunition load is sufficient for about 20 target engagements. The entire gun mount is battery operated with an integrated motor/generator being fitted for recharging. The mount also has automatic electronic tilt correction.

The FCU consists of two parts, the unmanned Sensor Unit (SU) and the detached Command Post (CP) housing the two operators for complete command and control. One of the operators is responsible for command and engagement and the other for target tracking.

The shelter-based CP also contains the generator, computer for data processing and the datalink. If required, these sub-units can be removed for positioning inside buildings.

The splitting of the FCU into two parts offers a number of advantages including:

- Improved crew survivability against anti-radiation missiles;
- Allows the SU to be positioned well away from the CP with maximum distance between the two being 500 m.

The SU is claimed to be highly ECM resistant and has X-band pulse Doppler search (top) and tracking (lower) radars. Target detection alarm with track-while-scan and threat evaluation is automatic and followed by hand over to the tracking sensors which are to the right of the tracking radar.

The multi-sensor tracking unit combines the radar and a TV/laser/FLIR electro-optical module, which has high sensor dynamics for interference-free precision tracking. The sensor unit has manually operated coarse levelling and automatic electronic tilt correction.

According to Rheinmetall Air Defence the Skyguard 35 ADS has a very short reaction time and time taken from target detection to opening fire is 4.5 seconds.

One of the key features of the Skyshield 35 ADS is that it is lightweight and easy to transport. All main units have pallet-concept or ISO container standard base frames (2.44 × 3.00 m) and are fitted with corners and fasteners for lifting sling nets.

This allows the individual units to be loaded by a forklift onto trucks, railway wagons and into transport aircraft with maximum unit-weight of 3,000 kg. The individual units can be quickly positioned on the tops of buildings, oil rigs and other high-value strategic targets, by helicopter.

Two (6 × 6) trucks, each fitted with a hydraulic crane, can carry a complete Skyshield 35 AHEAD air defence system quickly over long distances.

The Skyshield 35/1000 fires all standard Rheinmetall Air Defence 35 mm ammunition plus the new AHEAD ammunition, development of which has now been successfully completed and demonstrated. Additional details of AHEAD ammunition are given in a separate entry. Canada, Oman and one other undisclosed country for existing 35 mm GDF systems have already ordered the AHEAD round.

Using the AHEAD concept, the target is destroyed by multiple impacts of high density metal spin-stabilised sub-projectiles. Each 35 mm AHEAD projectile contains a payload of 152 sub-projectiles which are released by a small explosive charge initiated by the precision programmable time fuze, the high spin rate rapidly dispersing the sub-projectiles. The release is timed to be just ahead of the attacking target.

The current Rheinmetall Air Defence twin 35 mm GDF-series of towed anti-aircraft guns remains in production, with sales now past 2,000 units. These are normally used in conjunction with the Rheinmetall Air Defence Skyguard fire-control system.

Rheinmetall Air Defence Skyshield 35 Gun/Missile integrated Air Defence System with two 35 mm AHEAD Gun Mounts on the left. Command Post Sensor Unit and ADATS SAM 0509774

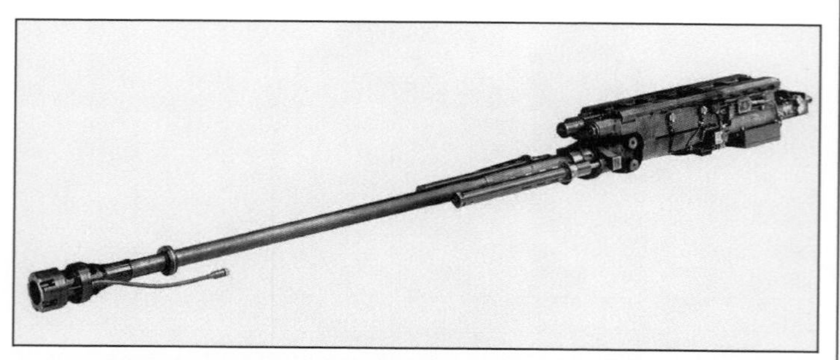

The Rheinmetall Air Defence 35/1000 gas-operated revolver cannon showing muzzle brake and muzzle velocity measuring equipment 0509775

The GDF-series of guns can also be modified to fire the new AHEAD ammunition, which has a significant increase in hit and kill probability over current types of 35 mm ammunition. South Africa has also selected the 35 mm AHEAD ammunition for use with its LIW (a division of Denel) Naval Air Defence System 35 mm Light Gun Mount (NADS 35 LGM), previously known as the twin 35 mm Dual-Purpose Gun (DPG), with 14 mountings having been produced for the South African Navy.

Variants

In addition to the pallet-mounted Skyshield 35 AHEAD ADS, other variants are also being studied, these include a naval mount, Skyguard III/35 mm twin-gun modernisation programme and a self-propelled mount.

Typically the latter would be based on a (6 × 6) cross-country truck on the rear of which would be mounted the 35 mm cannon and its associated fire-control system.

To enable targets to be engaged at longer ranges, Rheinmetall Air Defence Inc. Canada, has developed an unmanned missile module along similar lines, which has a total of eight (four either side) ADATS missiles in the ready-to-launch position with control being carried out from the command post.

During system operation, the fire-control unit Track-While-Scan (TWS) X-band search radar with integrated IFF first acquires the target. Targets at long range, after threat evaluation, will be automatically assigned to the missile control console.

After target acquisition, 2-D or 3-D fire-control unit radar tracking information is transferred to the EO operator. Passive target tracking is initiated by using the thermal imager sensor for night and adverse weather, or the TV sensor for daytime and clear weather.

The missile is then launched and automatically guided to the target by the encoded laser beam. The fire-control unit, line of sight and the tracker are always free to control the two 35 mm revolver cannon for target engagement at shorter ranges. This allows one fire unit to be able to engage two targets at the same time and change to the next target within seconds.

The ADATS missile system, is used by Canada (self-propelled) and Thailand (shelter) and has a maximum speed of over M3 and an effective range of up to 10 km. Details of the ADATS system are given in the Self-propelled and Shelter- and container-based surface-to-air missiles sections.

The Skyguard III/35 mm twin-gun modernisation programme has been designed to cope with 21st Century air threats and includes all the fire-control functions necessary for successful AHEAD engagement of small, fast and steep attacking missile targets with respect both to elevation coverage and detection range.

Small, Compact and modular, the main components of the new Skyshield 35 – AHEAD fire units offer maximum crew protection. These include two unmanned, remotely controlled 35/1000 revolver gun mounts, complemented with a fire-control unit consisting of an unmanned sensor unit with a detached command post housing the two operators, assuring complete command and control of the fire unit.

The self-propelled battlefield air defence system is composed of separate self-propelled modules like the 35 mm AHEAD gun turret with the 35/1000 revolver cannon and the tactical command post equipped with the new X-TAR 3-D radar providing an instrumented range of 25 or 50 km. This new system is also suitable for engaging ground targets.

The MOOTW/C-RAM, is a further development of the Skyshield and know as NBS MANTIS when in service with the German Army. Trials in 2009 culminated in the C-RAM version of the system for sales to overseas customers and a contract with the German Army for the MANTIS system.

In January 2011, the German Air Force (Luftwaffe) took control of the MANTIS system and responsibility for all air defence. A standard MANTIS system consists of six 35 mm automatic guns modules that are each capable of firing up to 1,000 rounds per minute, a ground control unit and two sensor units.

Specifications

Revolver Gun Mount

Weight:	
(with ammunition)	3,600 kg
(without ammunition)	3,200 kg
Length:	
(with 35 mm cannon)	5.428 m
(mount)	2.991 m
Width:	2.438 m
Height:	1.60 m
Armament:	1 × 35 mm 35/1000 revolver cannon
Ammunition:	224 rounds
Power supply:	8 × 12 V batteries; separate generator supplies built-in battery charger

Sensor unit

Weight:	3,200 kg
Length:	2.991 m
Width:	2.438 m
Height:	
(top of tracker)	2.220 m
(top of surveillance radar)	2.970 m

Command post

Weight:	2,500 kg
Length:	2.991 m
Width:	2.438 m
Height:	2.220 m

Status

Development complete.

In April 2007, the German BWB ordered two Skyshield 35 AHEAD air defence systems optimised for the CRAM role. Under the terms of this contract Rheinmetall Air Defence completed development of the CRAM version ready for qualification trials that took place in 2009. One of these systems, now known as NBS C-RAM/MANTIS, has now been deployed with the German Luftwaffe at the 1st air defence squadron (*Flugabwehrraketengeschwader* 1) based in Husum, Germany, the second system is protection for the German Army assets in Kunduz, Afghanistan. The German Army took formal delivery of the MANTIS system on 26 November 2012 at a ceremony in Husum, Germany, home of the Air Defence Missile Squadron 1 Schleswig-Holstein.

The latest upgrade to the Rheinmetall 35 mm gun systems is the GDF-009.

Contractor

Rheinmetall Air Defence AG.

Taiwan

40 mm L70/T92

Type

Towed anti-aircraft gun

Development

The Republic of China (Taiwan) 202nd Arsenal Materiel Production Centre has re-invented a towed low-level Anti-Aircraft Gun (AAG) system based on a modified Swedish Bofors Defence 40 mm L/AAG system.

Reports suggest that trials of the first prototype weapons have been completed in Taiwan and that consideration is being given to manufacture 76 × series production systems.

For the development programme, a number of prototypes of the system are believed to have been completed. At least one of these used the same four-wheeled carriage as the Swiss Oerlikon Contraves twin 35 mm GDF towed AAGs that are also in the inventory of Taiwan.

A self-propelled version for more mobile air defence is under consideration. Meanwhile, those that are produced will be issued to two battalions for use in airfield defence.

L70 T92 self-propelled anti-aircraft gun 1132343

Description

The system, designated as 40 mm L70/T92 and is armed with the Bofors Defence 40 mm L70 barrel that is normally fed with clips of four rounds of ammunition with the empty cartridge cases ejected from the forward part of the mount. Each mount is also provided with what appears to be a French Sagem day/night optical sight. Effective range of the system is reported to be around 4,000 m.

When fully deployed, it is assessed that two systems would be controlled by one fire-control system, equipped with tracking and surveillance radar.

The Taiwanese system is fitted with a large magazine originally developed for naval applications, and holds 101 rounds of ready-use 40 mm ammunition. This feature enables a significant number of targets to be engaged (dependant on number of rounds expended per target) before reloading is required.

A generator is installed on the rear of the carriage to provide power for elevation and traverse, although manual controls are also provided for emergency use.

The automatic levelling system enables the system to be deployed in less than five minutes, according to the developer.

The system is operated by a crew of three.

Ammunition

40 mm ammunition for the L/70 weapon has been made in Taiwan by the Ji-Ning Machinery Corporation for some years. A more recent TC77 Pre-Formed High-Explosive (PFHE) is being optimised for the air-defence role. In addition to the HE content, the TC77 PFHE contains 640 tungsten pellets, capable of neutralising a range of aerial targets, particularly missiles and soft-skinned helicopters.

Other ammunition available includes:

- XTC96 Armour-Piercing Discarding Sabot (APDS)
- XTC97 Armour-Piercing Fin Stabilised Discarding Sabot (APFSDS)
- TC74 High-Explosive Tracer (HE-T)
- TC93 Armour-Piercing Capped Tracer (APC-T).

Variants

A self-propelled version for more mobile air defence is under consideration.

Specifications

	L70/T92
Mobility	
Carriage:	trailer mounted four wheeled
Firepower	
Armament:	1 × 40 mm cannon
Main weapon	
maximum range:	4,000 m (est.)

Status

Prototype development is complete. The L70/T92 may now go into series production for 76 pieces to equip at least two battalions for airfield defence.

As of December 2012, the T92 the gun system has still not gone into full scale (series) production due to development and fabrication problems encountered with both the gun and its associated T'ien Yung (CS/MPQ-561) radar.

Contractor

202nd Arsenal Materiel Production Centre (primary contractor).
Ji-Ning Machinery Corporation (ammunition).

Turkey

20 mm/35 mm Cansas

Type

Towed anti-aircraft gun

Development

The production facility of MKEK Turkey, (*Makina ve Kimya Endüstrisi Kurumu* or 'Mechanical and Chemical Industries Corporation')-Cansas *(Cankiri Silah Sanayi ve Ticaret AS)* was originally established to manufacture, under licence, the Swiss Oerlikon Contraves GAI-DO1 twin 20 mm light anti-aircraft gun system (covered in detail in this section) which entered production in late 1985.

Turkish MKEK-Cansas-built twin 20 mm GAI-DO1 light anti-aircraft gun system deployed in the firing position 0509778

Turkish MKEK-Cansas-built twin 35 mm GDF-003 towed anti-aircraft gun system deployed in firing position with weapons at maximum elevation 0509779

No details of quantities of these weapons built in Turkey have been released although it is estimated that by late 2000 the total was around 400 plus.

This was followed by licensed production of the Oerlikon Contraves GDF-003 twin 35 mm towed anti-aircraft gun system (covered in detail in this section).

No details of quantities of the 35 mm weapon system have been released although by late 2000 it was estimated that around 120 had been produced.

There was a competition to supply radar fire-control systems for the Turkish-built twin 35 mm GDF-003 anti-aircraft guns, details of which are given in the Inventory section.

Today the MKEK-Cansas facility can manufacture weapons of up to 64 mm in calibre. The main shareholder in Cansas is MKEK, which is the largest state-owned company in the Turkish defence industry. The other important shareholder in the company is the Turkish Armed Forces Foundation.

In addition to manufacturing, the complete 20 mm and 35 mm GAI-DO1/GDF-003 anti-aircraft gun systems, the company has manufactured under licence the Giat Industries 25 mm M811 cannon. This is being installed in locally manufactured Giat Industries Dragar turrets for installation in the FNSS Defence Systems (formerly FMC-Nurol) Armored Infantry Fighting Vehicles.

Muzzle Velocity Radar (MVR)

Aselsan released details of the 7941 MVR System which is designed to measure a wide variety of field, anti-aircraft and naval artillery shells and provides accurate muzzle velocity data to enhance hit probability.

Reports suggested that the gun types using the 7941 are all standard weapons. Functionally is to measure muzzle velocity and calibrate its host weapon on the basis of the measurement of the velocities of multiple rounds. The 7941 MVR operates in the old X-band frequency range (8-12.5 GHz) and covers the muzzle velocity range of 50-2000 m/s.

Status

As of May 2013, the 20 mm and 35 mm towed anti-aircraft guns are still in production, offered for sale and in service with the Turkish Armed Forces.

Contractor

MKEK - Cansas.

United States

M16 half-track modernised by Giat Industries with four 12.7 mm machine guns replaced by two Giat Industries 20 mm cannon 0007669

12.7 mm M55

Type

Towed anti-aircraft gun

Development

Development of the 12.7 mm (0.50) (Quad) anti-aircraft machine gun M55 began in 1942 by the then Kimberly-Clark Corporation it entered US Army service the following year. About 10,000 M55s were delivered to the US Army between 1943 and 1953, with the final contractor being the Bowen and McLaughlin Corporation (now part of United Defence LP). It was replaced in US Army service by the 20 mm M167 Vulcan anti-aircraft gun system. Both systems have now been phased out of US Army service and all US Army air defence is now surface-to-air missiles, for example, Raytheon Systems Company Stinger MANPADS and Stinger self-propelled.

The 12.7 mm (0.50) (Quad) anti-aircraft machine gun M55 is a clear weather system with no provision for off-carriage fire-control.

M16 half-track armed with four 12.7 mm anti-aircraft machine guns. When deployed for firing, the upper parts of the hull sides fold downwards to give the weapons all-round traverse (Kensuke Ebata) 0509542

Description

The full designation of the M55 is the Mount, Gun, Trailer, Multiple Calibre .50 Machine Gun, M55 and it is composed of two main parts: the M45C Mount and the M20 Trailer.

The M45C Mount is a power-driven semi-armoured gun mount with a self-contained power unit. A power charger produces electrical current to be stored in two 6 V storage batteries on which the electrical system operates. The mount has four 12.7 mm (0.50) M2 HB machine guns, two each side. The early models have two M2 ammunition chests each side, but later mounts have the chests replaced by ammunition box trays.

Power is directed by a pair of control handles immediately in front of the operator's seat on the mount. The machine guns are fired by a solenoid and will continue to fire and load automatically as long as the gunner applies pressure to both the triggers on the control handles.

The gunner aims the weapons using an M18 reflex sight which projects a graticule image, focused at infinity, on an inclined glass plate. The graticule image consists of four concentric circles, corresponding to various aircraft speeds, and three dots on a vertical line in the centre of the field of view are used to determine line of sight and to compensate for gravity pull on the projectile. As the gunner looks through the plate the target can be seen superimposed on the graticule image. Turret traverse and gun elevation speeds are 60° per sec.

The machine guns fire various natures of 12.7 × 99 fixed ammunition, details of which are in Table 1 at the bottom of the page.

In the firing position the wheels are removed by using the three lifting jacks (one at the front and two at the rear of the carriage) and the complete unit is then lowered to the ground to provide a stable-firing platform. The two-wheeled trailer M20, can be towed by a light (4 × 4) Jeep-type vehicle, but its small diameter tyres allow it to be towed on roads at a maximum speed of only 16 km/h. Normally the trailer is carried in the rear of a 21/2 tonne (6 × 6) truck, which is equipped with special loading and unloading equipment.

12.7 mm (Quad) anti-aircraft machine gun M55 in firing position. Note that only two of the four box magazines are fitted to this system (US Army) 0509541

Variants

Second World War variants

During the Second World War, the M45 Mount was mounted on Trailer Mount M17, which in turn was mounted on the M51 four-wheel carriage, and could be towed by a (4 × 4) or (6 × 6) truck. Two self-propelled models were also built: the M16, which was based on the M3 half-track, and the M17 based on the M5 half-track.

Colombian self-propelled M55

The Colombian Army has fitted a small number of its M8 (6 × 6) armoured cars, manufactured during the Second World War, with the M55 anti-aircraft gun system. The primary role of these is the defence of high value strategic targets from ground attack. Up to 20 of these M8/M55 combinations may be in service.

Israeli TCM-20 system

The MBT Division of Israel Aircraft Industries has developed a new model armed with twin 20 mm cannon called the TCM-20, details of which are given in the entry under Towed anti-aircraft guns, Israel, RAMTA Structures and Systems TCM-20 twin 20 mm anti-aircraft gun system.

Brazilian M55 upgrade

Brazil has developed a twin 20 mm modernised version of the M55.

Table 1

Type	Designation	Projectile weight	Muzzle velocity
AP	M2	45.88 or 46.53 g	885 m/s
API	M8	42.06 g	888 m/s
API-T	M20	39.66 g	888 m/s
Ball	M2	46.1 or 46.79 g	858 m/s
Ball	M33	42.9 g	888 m/s
Incendiary	M1	41.02 g	901 m/s
Incendiary	M23	33.18 g	1,036 m/s
Training	M10	41.67 g	873 m/s
Training	M17	14.67 g	873 m/s
Training	M21	45.3 g	867 m/s

The Taiwanese T75 is armed with two 20 mm cannon and is essentially an upgraded M55 system (Raymond Cheung) 0509780

Giat Industries M55 upgrade

In 1989, Giat Industries of France announced that it had, as a private venture, modernised an M16 half-track self-propelled anti-aircraft gun system. Two Giat Industries 20 mm cannon with the gun control and sighting systems, also being improved, have replaced the four 12.7 mm machine guns. Trials of the prototype system were carried out in 1989, and at late 2000, none had been sold of this update package.

Thai upgraded M55

Details of this upgrade were last given in 1995/1996, the current status of which is unknown.

Taiwanese Upgraded M55

It is known that Taiwan has upgraded an M55 system to a new configuration known as the T75 which is armed with two 20 mm cannons each fitted with a muzzle brake. A new all-electric gun control system has been installed and new 20 mm magazines mounted externally either side of the mount.

Turkish self-propelled M55

The Turkish Land Forces have completed a prototype of a self-propelled model of the M55 based on a locally modified United Defence M113 chassis. Details are given in the entry under Self-propelled anti-aircraft guns, Turkey, MKEK - CANSAS 35 mm and 20 mm anti-aircraft gun production.

Specifications

	M55
Dimensions and weights	
Crew:	4
Length	
transport configuration:	2.89 m
Width	
transport configuration:	2.09 m
Height	
transport configuration:	1.606 m
deployed:	1.428 m
Ground clearance	
overall:	0.178 m
Track	
vehicle:	1.524 m
Weight	
standard:	1,338 kg (travelling order with trailer)
combat:	975 kg (M45C only)
Mobility	
Carriage:	trailer mounted two wheeled
Fording	
without preparation:	0.457 m
Angle of approach:	10°
Angle of departure:	20°
Firepower	
Armament:	4 × 12.7 mm (0.50) Browning machine gun
Main weapon	
length of barrel:	1.143 m
maximum range:	1,500 m (effective)
maximum vertical range:	1,000 m (effective)
feed mechanism:	210 round belt (per barrel)
operation:	recoil automatic
cartridge:	12.7 mm × 99 mm
Rate of fire	
cyclic:	2,200 rds/min
practical:	600 rds/min
Main armament traverse	
angle:	360°
Main armament elevation/depression	
armament front:	+90°/-10°

Status

Production complete.

Contractor

United Defence LP.

20 mm M167 Vulcan

Type

Towed anti-aircraft gun

Development

Development of the 20 mm Vulcan Air Defence system began under the direction of the United States Army Weapons Command at Rock Island Arsenal, Illinois, in 1964. Two versions of the Vulcan were developed a self-propelled model called the M163 (development designation XM163) and a towed model called the M167 (development designation XM167).

Prime contractor for both models is now General Dynamics Armament and Technical Products of Burlington, Vermont.

After trials, carried out by the US Army Air Defence Board at Fort Bliss, Texas, and at Aberdeen Proving Ground, Maryland, in 1965, the system was accepted for service as the replacement for the 12.7 mm (quad) M55 anti-aircraft gun system. First production M167s were delivered to the US Army in 1967. The final service model is the M167A1, which incorporates no fundamental changes from the M167. The M167 was phased out of US Army service in 1994 at which time some 63 systems remained in service.

The manufacturer is no longer marketing the M167 system although surplus weapons may be available from US Army stocks. Total production of the 20 mm M167 system was 626 units.

Description

The M167 was a towed version of the 20 mm Vulcan Air Defence system that was mounted on the M42A1 two-wheeled carriage and is a lightweight version of the M163 self-propelled system which is fully described in the Self-propelled anti-aircraft guns section. The M167 has the advantage of being helicopter transportable and can therefore be used in tactical situations where the tracked version cannot be employed.

The M167 consists of a 20 mm M168 Vulcan gun, linked ammunition feed system and a fire-control system with a radar tracking capability of 5 kms, all mounted in an electrically powered turret. The towed M42A1 carriage has its own power generator for re-charging, mounted on the forward part of the carriage.

The six-barrelled 20 mm M168 cannon has two rates of fire, 1,000 rds/min, normally used against ground targets and 3,000 rds/min normally used against aircraft. Maximum effective anti-aircraft range is 1,200 m and maximum effective ground range is 2,200 m (some sources have suggested 3,000 m). The gunner can select 10, 30, 60 or 100-round bursts (in the high rate of fire only). The dispersion pattern can be increased, by fitting a special muzzle spread, which results in an increased hit probability. The ammunition container is on the left side and holds 300 or 500 rounds of ammunition. The 20 mm M168 cannon can fire the following types of 20 × 102 mm fixed ammunition:

- M53 (APT) with the projectile weighing 0.100 kg and a muzzle velocity of 1,030 m/s
- M54 (HPT) with the projectile weighing 0.127 kg, muzzle velocity dependant on required pressure test
- M55A2 (TP) with the projectile weighing 0.098 kg and a muzzle velocity of 1,030 m/s
- M56A3 (HEI) with the projectile weighing 0.103 kg and a muzzle velocity of 1,030 m/s
- M220 (TPT) with the projectile weighing 0.097 kg and a muzzle velocity of 1,030 m/s
- M242 (HEIT) with the projectile weighing 0.094 kg and a muzzle velocity of 1,030 m/s.

20 mm Vulcan with dual wheels deployed in firing position 0509539

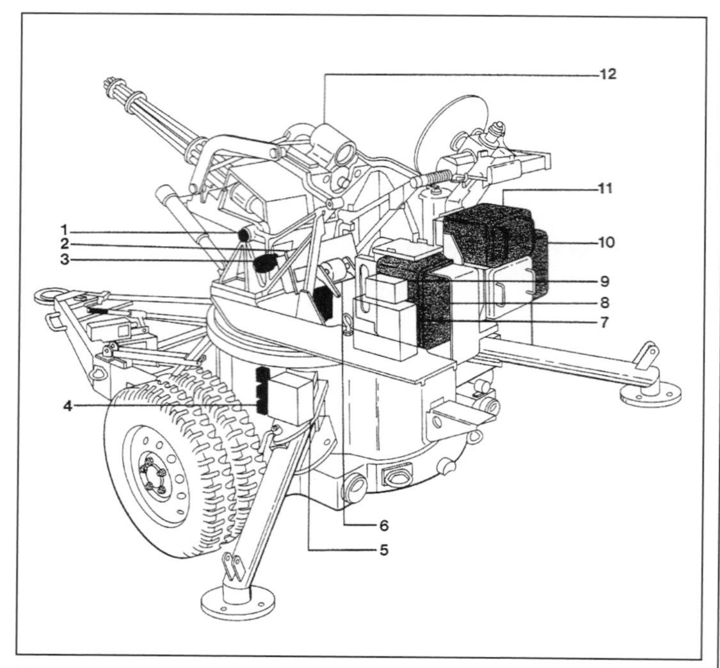

Key components of Lockheed Canada Product Improved 20 mm Vulcan Air Defense System (PIVADS) are (1) Elevation synchro (2) Control panel (3) Elevation drive (4) Servo amplifiers (three) (5) Distribution box (6) Azimuth drive (7) Electronics unit (8) Radar power supply (9) Voltage converter (10) Radar unit 2 (11) Radar unit 4 (12) Director gun sight 0509538

20 mm Vulcan showing range-only radar and generator on forward part of carriage (Michael Jerchel) 0509540

The system has five modes of operation: radar, manual, ground, external or test. The first of these is the normal operational mode with the 5,000 m range radar determining the target range and range rate. The fire-control computer calculates the predicted target flight path, positions the cannon in super elevation and azimuth to achieve target impact, takes in rate aid to take over track from the gunner and stabilises the sight graticule in the target line of sight.

The ready to fire indicator lamp on the sight lights up when the fire-control computer determines that all conditions are satisfied. These conditions are predicted impact is within ammunition range, lead angle is less than 25°, elevation angle is less than 80°, radar is locked on and rate aid starts.

When deployed in the firing position, the M167 is stabilised by three jacks, one at the front and one either side on outriggers.

Variants

PIVADS

Lockheed Canada Inc are currently building and marketing an improved fire-control system for the Vulcan system, which includes a new optical sight, a digital processor, and harmonic drives for azimuth and elevation. A director-type optical sight is provided as the angle-tracking device, which permits rapid acquisition and accurate tracking when turret disturbances are present. The sight contains integrating rate gyros for azimuth and elevation tracking, in conjunction with proportional handgrip controls. A new digital processor replaces the existing analogue computer to improve weapon tracking and operational capabilities. The system also includes built-in test equipment and permits multiple ballistics selection via the operator control panel. System response and point accuracies are also improved with the new Lockheed Canada system. Harmonic drives replace older azimuth and elevation gear trains to improve stiffness, backlash and power consumption characteristics. A test prototype, designated Product Improved Vulcan Air Defence System (PIVADS) was tested extensively by the US Army at Fort Bliss in June 1979.

Late in 1982, the US Army awarded Lockheed Electronics a USD19 million contract for the supply of 285 PIVADS kits for the 20 mm Vulcan Air Defence System in both towed and self-propelled configurations. Of the 285 systems, 122 were for the towed M167 series and the remaining 163 for the M167 self-propelled systems.

In 1990, all manufacturing for this system was transferred to Lockheed Canada Inc, in Ottawa, Ontario. The company, a part of the Lockheed Electronics Systems Group, was then awarded its first PIVADS contract for USD4.2 million to supply spares to the US Army.

Basic Vulcan

This is a simplified version of the M167 Vulcan produced for commercial export sales. The basic changes are to the fire-control system, which does not use the range-only radar of the M167 but a range update computer, lead computing gun sight and a control panel which includes burst length, estimated range and target speed settings, mode selection and controls for positioning the electrically powered turret.

Four operating modes are available to the gunner: range update, external range, fixed mode and ground.

With range update the gunner identifies the target visually, estimates the target speed and selects the burst length and predicted range to open fire. These settings are made on the control panel. Using the hand controls, the gunner moves the turret to acquire the target in the sight, keeping the sight gyro caged by using a button on the left-hand grip. On acquisition, releases the cage button and tracks the target using the centre of the sight graticule. After tracking the target to the pre-selected range the gunner opens fire. The computer then updates the range data automatically and supplies the system with the lead angles and super elevation throughout the engagement.

In the external range mode an extra crew member signals the predicted range and signals the gunner to open fire. The extra crew member then updates the range setting via a hand-held external range unit.

In the fixed mode the lead computing gun sight is fixed with pre-set range and target speed.

In the ground mode the lead computing sight is mechanically caged so that the system is set for use at 1,000 m range.

The rest of the Basic Vulcan is the same as the normal M167. The only data change is that the combat weight is reduced to 1,406 kg.

Dual-Wheel Vulcan

All US Army M167 Vulcan's were modified by the addition of a second road wheel on each side of the carriage. This increases cross-country performance. The extra wheel is available as a modification kit and, when fitted, the combat weight of the M167A1 is increased to 1,732 kg. The outside wheel track is then 2.271 m and the inside track 1.76 m.

Specifications

	M167
Dimensions and weights	
Crew:	1
Length	
transport configuration:	4.906 m
Width	
transport configuration:	1.98 m
Height	
transport configuration:	2.038 m
deployed:	1.651 m
Track	
vehicle:	1.752 m
Weight	
standard:	1,588 kg
combat:	1,565 kg
Mobility	
Carriage:	trailer mounted two wheeled with outriggers
Firepower	
Armament:	1 × 20 mm cannon
Main weapon	
length of barrel:	1.524 m
maximum range:	2,200 m
maximum vertical range:	1,200 m (effective)
operation:	electrically powered
cartridge:	20 mm × 102 mm
Rate of fire	
cyclic:	3,000 rds/min
Main armament traverse	
angle:	360°
Main armament elevation/depression	
armament front:	+80°/-5°

Status

Production of the Vulcan system is complete. The M167 Vulcan is no longer in use with the US Army and no longer marketed.

Contractor

General Dynamics Armament and Technical Products.

This weapon has also been built under licence in South Korea by the then Special Products Division of Daewoo Heavy Industries. Details are given in the South Korea entry under Towed anti-aircraft guns.

Centurion Phalanx Block 1B

Type
Towed anti-aircraft gun

Development
The US Army and Raytheon Missile Systems have developed and deployed a mobile, ground-based variant of the Phalanx (Mk 15) Block 1B (Mod 29) 20 mm Close-In Weapon System (CIWS) for Counter-Rockets, Artillery and Mortars (C-RAM) defence.

The Land-Based Phalanx (now known as the Centurion system) was first tested for possible Land-Based Phalanx Weapon System (LPWS) missions in November 2004 and successfully proved its ability to provide unparalleled land-based protection against rockets, artillery and mortar rounds.

As of June 2006, at least two systems had been deployed to Iraq and were operational at Camp Anaconda near Balad. The system was selected as a near-term solution in response to an operational submission to provide theatre forces with protection against rocket, artillery and mortar attacks in a counter-insurgency operational environment, in particular the forward operating basis.

Initial testing of this land-based variant was reported to have been conducted during the latter part of 2004 and early 2005 at the Yuma Proving Ground, Arizona, against 60 mm, 81 mm and 120 mm mortar bombs at low and high elevations. Following the successful pre-action qualification trials at Yuma, the first two LPWS systems arrived in Iraq in May 2005. The first battery using the new system has been augmented by USN personnel familiar with the Phalanx.

Continuing upgrades to the Phalanx system are ongoing; in late June 2006, the Raytheon Company from Tucson, Arizona, were awarded a continuance contract for Phalanx close-in weapons system and associated spares. It is not known if the modifications in this contract will effect those systems deployed in Iraq but the upgrades will include:
- Five thermal imagers
- Three block 1B Aegis upgrade kits
- Two conversions
- Required spares and support equipment.

Again, in August 2006, Raytheon was awarded another contract, this time for ancillary equipment for army land-based Phalanx weapon system. The ancillary equipment was not specified in the contract, but work will be carried out in Louisville, Kentucky, and is expected to be completed by April 2007.

In September 2008, Raytheon was awarded a contract for modification, engineering and technical services in support of the Mk15 Phalanx Close In Weapon System (CIWS).

As with the September 2008 contract, a further January 2009 contract was set by Raytheon Missile Systems with General Dynamics Armament and Technical Products worth USD18 million to upgrade the Block 1B with enhanced fire-control capability, optimised gun barrels and integrated Forward Looking InfraRed (FLIR) system. A further contract for upgrades and conversions, system overhauls and associated hardware was set with Raytheon in May 2009. This upgrade is aimed specifically at the Block 1B for both the US Navy and the Land-Based system that is deployed to Iraq.

Description
Introduced to the US Navy service in the late 1970s for last-ditch anti-ship missile defence, the Phalanx CIWS has been the subject of substantial product improvements over the past 25 years.

The latest Block 1B variant, now being fielded by the USN and a number of allied navies, incorporates:
- An 8-12 µm-band FLIR sensor suite. The Block 1B FLIR provides the capability to search, track and engage threats while simultaneously providing a detect, prioritise and kill assessment feature that is effective in both daylight and night-time environments.
- An automatic acquisition video tracker.
- Improved Ku-band search-and-track radar with closed loop spotting technology to detect threats early in their flight and then hand over to the track mode only when those targets are determined to threaten the area protected by the system.
- New local and remote-control stations.
- An Optimised Gun Barrel set (fixed in a stiffened cluster) for the M61A1 six-barrel 20 mm Gatling gun firing M-246 or M-940 self-destruct rounds at a selectable rate of 3,000 or 4,500 shots per minute.
- New Mk 244 Enhanced Lethality Cartridge (ELC) ammunition.

The gun assembly itself is mounted on a flat-bed trailer along with:
- Generator
- Power supply unit
- Control cabin.

Raytheon engineers have reportedly modified the operational software programme to reflect the changed operating environment and target set. Another change is the use of M-246/M-940 High-Explosive Incendiary Tracer round Self-Destruct (HEI-T-SD) ammunition rather than the Mk149/Mk244 Armour-Piercing Discarding Sabot (APDS) tungsten-penetrator rounds associated with the naval version.

This Army Phalanx System is presumably integrated into the Forward Area Air-Defence Command-and-Control system, therefore allowing the system to communicate freely with existing air defence sensors and other army battle command systems.

Variants
Shipborne anti-missile systems
An upgrade to the shipborne and most likely the towed anti-aircraft variant has been contracted with the addition of the All Light Marine TeleVision (ALMTV) camera programme. Full rate production of the camera began in 2009. The all-light-level camera includes a 18:1 visible/near IR zoom lens coupled to a low light Electron Multiplication Charge Coupled Device (EMCCD) sensor and interface control module housed in a ship compatible with environmental housing.

Centurion on Low Boy Trailer (Raytheon) 1192379

Phalanx - shipborne version (IHS/Harry M Steele) 0588633

Centurion at Basra Air Station (IHS/Patrick Allen) 1368579

Centurion future configuration (Raytheon) 1192946

New hardware consisting of a reflecting-telescope optics for a high-powered laser, and a side-mounted laser illuminator operating at a much lower power level, have been fitted to the Centurion anti-aircraft gun system. Trials carried out in Albuquerque, New Mexico, have shown that the laser mounted on the side of the Centurion can destroy a 60 mm mortar at a range in excess of 450 m. Further tests are scheduled with increased range for the laser out to 15 km. Raytheon have reported that should the laser system get the permission, it can be deployed within 12 months.

At AUSA in November 2008, Raytheon and Oshkosh teamed up to show a prototype Centurion defensive system by placing a rapid fire, area defence Phalanx gun on the back of a diesel-electric, 13-tonne Heavy Expanded Mobility Tactical Truck (HEMTT). The diesel-electric HEMTT is a stabilised wheeled platform designed to protect ground forces also saves around 20 per cent more fuel; a diesel engine drives a generator, which then drives electric drive motors to propel the vehicle. A significant amount of power is necessary in order to drive the Phalanx gun system, so not only will the HEMTT vehicle provide this, it also brings 120 kW of military grade exportable power to the battlefield.

In November 2008, the US Army had 22 Phalanx guns protecting its troops and foreword bases in Iraq.

Specifications

Phalanx

Type:	20 mm close-in weapon system
Gun:	M61A1 20 mm cannon
Calibre:	20 mm
Cartridge:	20 × 102 mm
Barrel length:	53 calibres
Gun drive:	pneumatic
Muzzle velocity:	1,030 m/sec
Rate of fire:	dual-rate, 3,000 or 4,500 rounds/minute
Magazine:	1,500 rounds of HEI-T-SD
Range: (horizontal)	1,470 m
Fire-control:	fully integrated closed loop with manual override and advanced radar and computer technology
Mount drive:	electric
Traverse:	310°
Elevation:	–25° to +85°
Elevation/training speed:	115°/s
Elevation/training acceleration:	458°/s^2
Power requirements:	440 V, 60 Hz, three phase, 19 kW search, 70 kW transient
Search radar:	Ku-band, digital MTI
Track radar:	Ku-band, pulse Doppler monopulse
E/O sensor:	FLIR imaging system with automatic acquisition tracker
Weight:	
(gun and mounting)	6.6 tonnes
(trailer mounted)	24.0 tonnes

Status

Development of the systems is complete. These were deployed with the US and UK forces in Iraq but have since been redeployed to Afghanistan.

Open press reporting in early June 2007 confirmed that up to six systems have been purchased by the UK and are deployed in Afghanistan.

In May 2008, the producers of the Land-Based Phalanx (also known as C-RAM), was awarded a contract for the Mk15 CIWS from the US Army for ordnance alteration kits, spares and associated hardware. The US Army has procured the CIWS to support the war on terrorism.

As of September 2008, Phalanx is installed on approximately 187 USN ships and is in use in 20 foreign navies.

Contractor

Raytheon Missile Systems (primary contractor).

General Dynamics Armament and Technical Products, Saco, Maine (gun systems production).

Ball Aerospace and Technologies Corp (for the ALMTV upgrade).

Burlington, Vt-based Technology Centre (programme management).

Ethan Allen Firing Range in Jericho, Vt (testing of the system).

Man-Portable Surface-To-Air Missile Systems

Bulgaria

Igla-1E

Type
Man-portable surface-to-air missile system (MANPADs)

Development
The Bulgarian Company, Vazovski Machinostroitelni Zavodi (VZM) was producing, under licence at the time, the Russian-improved second-generation Kolomna KBM 9K310E Igla-1E (NATO SA-16 Gimlet) version of the Igla-1 variant of the Igla man-portable low-altitude surface-to-air missile family. Its Russian industrial index number for the complete export system is 9K310E. A decision was taken during the early part of 1971, by the then Soviet Union, to discontinue improvements in the Strela-2/2M/3 systems and to concentrate on a new class and family of missiles known as Igla ('Needle'). Using the experience gained from previous MANPADs and in particular the cooled seeker technology; the first (Igla-1) of the new type of missiles (9M313) used the Strela-3 seeker with an added aerospike-like device. The motor for the 9M313 missile was new, as were most of the electronics package and certain parts of the warhead, namely the detonators.

VZM discontinued production of the Igla-1E some years ago now, therefore no further upgrades or modifications to this particular system can be expected.

Description
The Igla-1E is designed to engage low-flying manoeuvrable, non-manoeuvrable and stationary hovering targets by detecting hot metal surfaces and engine plume exhaust using FM tracking logic. During approaching and receding engagements the system can be used against head-on targets with speeds of up to 360 to 400 m/s. Tail-chase targets with speeds of up to 320 m/s can also be engaged. Apart from being able to launch the missile from stationary vehicles, or ground sites, it is also possible to fire it from an armoured vehicle travelling on even ground at a speed not greater than 20 km/h.

The Igla-1E weapon system consists of the following elements: combat equipment; target indication and reception aids; maintenance facilities and teaching and training facilities.

Combat equipment
The system's combat equipment consists of:

(1) The missile, in its launch tube, which comprises five sections:
- Nose section - a passive IR seeker (operating in the 3.5 to 5 micron wavelength region) incorporating a cooled target homing device and electronics unit that creates the guidance commands which are sent to the missile's control section. Just before impact, the seeker's FM tracking logic system shifts the aiming point from the engine exhaust region towards the central fuselage area at the junction with the wings. An aerospike-like device, supported by three arms, is fitted to reduce nose drag and kinetic heating of the sensor dome
- Control section - the missile control vanes, control actuators and systems, accommodate all the elements needed to fly the missile to its target
- Warhead section - a 1.27 kg high explosive fragmentation warhead and an impact/proximity fuzing circuit. The Igla-1E does not detonate any residual solid propellant left over at the point of engagement, unlike the basic Igla-1
- Dual-thrust solid propellant sustainer motor section - ignites at a safe distance from the gunner once the missile, has left the tube. The sustainer is designed to accelerate the missile to its cruising speed in its first mode of operation and then maintain its flight in the second operating mode. The sustainer also carries the wing unit which is fitted with four folding fins which open at launch to stabilise the missile aerodynamically and rotate it about its longitudinal axis
- Booster section - imparts initial velocity to the missile to enable it to leave the launch tube and spins the missile up about its longitudinal axis.

(2) A 2.8 kg weight glass fibre launch tube, this is used to transport, aim and launch the missile. The launch tube mounts:
- A sighting device with fore and rear sights and a light indicator
- A set of electrical coils to spin up the seeker's homing device and a means for electrical and mechanical interface to the missile
- Grip-stock launching mechanism
- Ground power supply unit which incorporates the thermal battery providing the system with power in the pre-launch period and a high-pressure compressed air bottle for seeker head cooling.

Target indication and reception aids
Target indication and reception aids include an electronic plotting board system with integrated R-255PP radio that can display the tactical air situation within a 12.5 km radius. Up to four targets at a time can be plotted and the commander of the launch team can determine the greatest threat track. The Igla-1E is fitted with an IFF of the customer's own choice.

Maintenance facilities
Maintenance facilities include an equipment set designed to prove functionality of the tactical and training systems deployed in the field. The equipment set can either be carried on a truck chassis or be statically installed at a base to provide technical checks of 12 missiles or 12 launcher mechanisms per hour.

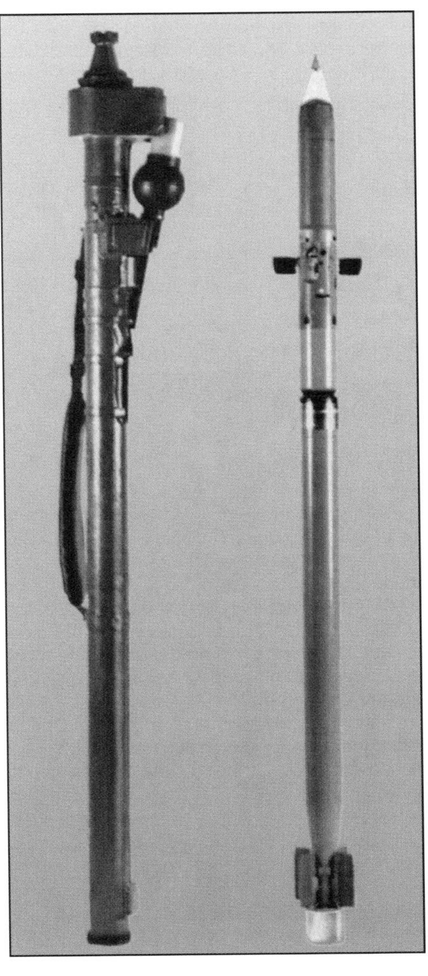

SA-16 (Igla-1E) low-altitude surface-to-air missile system
0024886

Teaching and training facilities
Teaching and training facilities consist of a field trainer, a training set, a launch control set and a full-scale mock-up. The training set is designed to provide for instructor training of gunners. A total of 70 electronic simulated firings is considered sufficient to train an individual gunner. The field trainer includes an IR target simulator on a stand with aircraft models and IR emission source, three inert training missiles, power unit and instructor's console.

Variants
As far as it is known, Bulgaria does not manufacture any of the specialised models of the Igla-1E (SA-16) system.

Since the system is a standard Russian design, it can be fitted with the LOMO MOWGLI-2 night sight.

Specifications

	Igla-1E
Dimensions and weights	
Length	
overall	1.673 m (5 ft 5¾ in)
Diameter	
body	72 mm (2.83 in)
Weight	
launch	10.8 kg (23 lb)
Performance	
Speed	
cruise	1,108 kt (2,052 km/h; 1,275 mph; 570 m/s)
Range	
min	0.3 n miles (0.5 km; 0.3 miles)
max	2.8 n miles (5.2 km; 3.2 miles) (receding target)
	2.4 n miles (4.5 km; 2.8 miles) (approaching target)
Altitude	
min	10 m (33 ft)
max	3,500 m (11,483 ft)
Ordnance components	
Warhead	1.27 kg (2.00 lb) HE fragmentation
Fuze	impact
Guidance	passive, IR (single channel 3.5-5 μm wavelength, FM tracking logic seeker)
Propulsion	
type	solid propellant (sustain stage)

Status
Vazovski Machinostroiteini Zavod has ceased all production of all man-portable surface-to-air missile systems in Bulgaria. However, a limited number of missiles/systems produced before cessation of production are believed/assessed to be available for service from reserve stocks with the Bulgarian armed forces. Offered for export. The UN Arms Register indicates Bulgaria sold Igla-1 (SA-16) weapons to Peru in 1994 and 1995 (the 1994 buy was for 56 launchers and 190 missiles and the 1995 buy for 21 systems, the latter was for the Peruvian Air Force).

Contractor
Vazovski Machinostroitelni Zavodi.

Strela-2M

Type
Man-portable surface-to-air missile system (MANPADs)

Development
The Russian first-generation KBM Mashynostroeniya Strela-2M, Russian industrial index number 9K32M, man-portable missile system was at the time licence-produced by the Bulgarian company, Vazovski Mashinostroitelni Zavodi (VMZ). It was specifically designed to destroy visually acquired aerial targets, such as helicopters and aircraft, by detecting solar reflections and hot metal surfaces using AM tracking logic.

Originally produced at the KBM Mashynostroeniya in Kolomna, Russia, the basis behind the Strela-2M (NATO SA-7b Grail) missile system is the clear passive single channel infra-red optical detector mounted at the front of the missile. These detectors were originally developed at the Leningrad Sovnarkhoz, which later became the Leningrad (now St Petersburg) Optical Mechanical Plant (LOMO), in Russia. Also involved was the State Optical Institute (GOI) which was headed at that time by G.Goryachkin.

As of October 2006, the missile was produced at the Arsenal Armament, Ammunition Company, in Kazanlak, Bulgaria.

Description
The Strela-2M missile system consists of the following elements (described here with their Russian industrial index numbers):

(1) 9M32M missile. This is comprised of four sections:
- Nose section - with a single-channel passive infra-red transparent AM tracking logic seeker unit (designated in Russian TGS: *Teplovaya Golovka Samonavedeniya*) and auto-pilot that receives the target signal from the TGS and creates the guidance commands that are sent to the missile's control section.
- Control section - with the missile control vanes; the angular speed sensor (designated in Russian, DUS; *Datchik Uglovykh Skorostej*), which senses the angular speed of the missile's airframe oscillations and translates them into signals used to stabilise the missile on its desired flight trajectory; a solid propellant gas generator and an onboard power supply, which powers the control vanes and features a turbine driven by diversion of exhaust gases from a burning propellant charge located in the solid propellant gas generator.
- Warhead section - with a 1.17 kg HE fragmentation warhead (0.37 kg of HE) and detonating system. The latter comprises a remote arming device, contact and grazing fuzing circuit and a self-destruct mechanism and can only be activated following missile launch.
- Propulsion section - with a launch booster that provides the missile with its initial 28 m/s velocity to drive it out of the launch tube with an angular rotation speed of 28 ±5°/s. Once clear of the gunner (approximately 5 to 6 m away) and with the four rear fins deployed, the first grain of the sustainer motor fires and starts to accelerate the missile up to its maximum flight speed. Between 80 and 250 m downrange the second propellant grain ignites. A total of 4.2 kg of solid propellant fuel is carried.

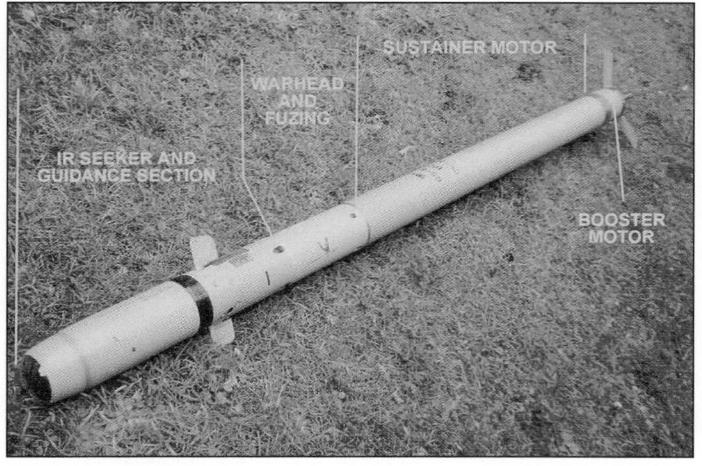

Strela-2M (Colin King) 0074265

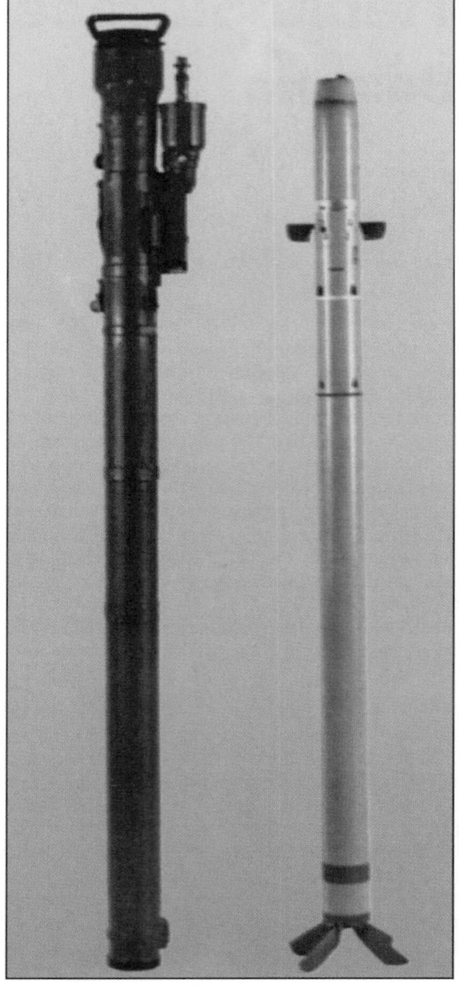
Licence-built SA-7b (Strela-2M) low-altitude surface-to-air missile system 0024888

(2) 9P54M 3 kg weight disposable glass fibre launch tube; this is used to transport, aim and launch the missile.
- 9P58 grip-stock launching mechanism.
- 9B17 disposable thermal battery power supply unit; this is fitted to the grip-stock and supplies both 22 V and 40 V DC to the appropriate grip-stock electronics, missile electronics and missile seeker head. Following a 1 to 1.3 second activation period, the battery will operate for a further 40 seconds before it has to be replaced.

The best engagement aspect against jet aircraft moving at speeds of up to 260 m/s, is a pursuit flight profile, while for propeller-driven aircraft and helicopters at speeds of up to 150 m/s, a pursuit or head-on engagement is possible.

The system can also engage and destroy hovering targets, provided they are within the launch envelope and emitting sufficient heat energy for the seeker to lock on. The effectiveness of the system increases if a remote target acquisition system, such as a radar and target movement indicator, is used. If required, it is also possible to launch the weapon from an armoured vehicle travelling on even ground at a speed not greater than 20 km/h.

The complete missile system is capable of operating at a relative air humidity of 98 per cent between –40°C and +50°C. It should be noted that some of the information given in the specification table (from Bulgarian sources) differs from that given in Russian sources.

Variants
The then RFAS developed a number of variants of this missile for home and export markets but as far as it is known none of these was produced in Bulgaria.

Since the system is a standard Russian design, it could be fitted with the Leningrad Optical Mechanical Organisation (LOMO) MOWGLI-2 night sight.

Specifications

	9M32M
Dimensions and weights	
Length	
overall	1.438 m (4 ft 8½ in) (fins folded)
Diameter	
body	72 mm (2.83 in)
Weight	
launch	985 kg (2,171 lb)
Performance	
Speed	
cruise	972 kt (1,800 km/h; 1,118 mph; 500 m/s)
Range	
min	0.4 n miles (0.8 km; 0.5 miles)

	9M32M
max	1.5 n miles (2.8 km; 1.7 miles) (approaching target)
	2.3 n miles (4.2 km; 2.6 miles) (rededing target)
Altitude	
min	15 m (49 ft) (est.) (missile may be affected by horizon and ground reflective heat)
max	2,300 m (7,546 ft)
Ordnance components	
Warhead	1.17 kg (2.00 lb) HE fragmentation
Fuze	impact
Guidance	passive, IR (single channel 0.7-1.5 μm wavelength, AM logic tracking seeker)
Propulsion	
type	solid propellant (sustain stage)

Status
Vazovski Mashinostroitelni Zavodi has ceased all production of man-portable surface-to-air missile systems in Bulgaria. However, a number of missiles and systems produced before cessation of production are believed/assessed to be available for service from reserve stocks with the Bulgarian Armed Forces. Some transportation boxes marked with the VMZ logo have been found at Libyan abandoned storage sites, the missiles themselves were very old and probably not in pristine condition.

Contractor
Vazovski Mashinostroitelni Zavodi.

Strela-3

Type
Man-portable surface-to-air missile system (MANPADs)

Development
The Bulgarian Company Vazovski Mashinostroitelni Zavodi (VMZ) produced at the time for export the second-generation Russian KOLOMNA KBM Strela-3 (Russian industrial index number for complete system 9K34) man-portable missile system under license. The Strela-3 (NATO SA-14 Gremlin) is designed to engage low-flying manoeuvrable, non-manoeuvrable and stationary hovering targets by detecting hot metal surfaces and engine plume exhaust using Frequency Modulation (FM) tracking logic.

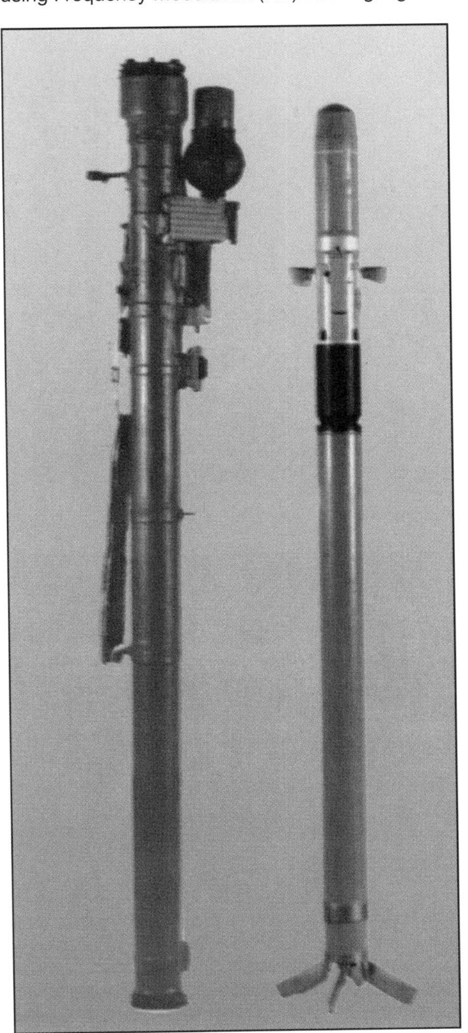

Licence-built SA-14 (Strela-3) low-altitude surface-to-air missile system 0024887

Modifications and improvements to the design of both the Strela-2 and the Strela-2M took place between 1970 and 1973, resulting in the Strela-3 system. The Strela-3 has the ability to cool the passive InfraRed (IR) seeker by means of a nitrogen gas bottle before detection and lock-on of the target. This gives the seeker a full head-on engagement capability.

Description
The Strela-3 missile system consists of the following elements (described here with their Russian industrial index numbers where known); combat equipment, maintenance facilities, and teaching and training facilities.

Combat equipment
The Strela-3 combat equipment comprises the following:

(1) 9M36 (the 9M36-1 is the export variant), missile in its launch tube. This comprises four sections:
- Nose section - a two-channel cooled passive infrared opaque seeker head operating in the 3.5 to 5 micron wavelength region (designated in Russian TGS: *Teplovaya Golovka Samonavedeniya)* and auto-pilot that receives the target signal from the TGS and creates the guidance commands that are sent to the missile's control section
- Control section - the missile control vanes, control actuators and systems, accommodates all the elements needed to fly the missile to its target
- Warhead section - a 1 kg high explosive fragmentation warhead and a detonating system. The latter comprises a remote arming device, contact and grazing fuze circuits and a self-destruct mechanism which operates after 14 to 17 seconds of flight if the missile has not hit the target
- Propulsion section - a launch booster that provides the missile with its initial velocity to drive it out of the launch tube. Approximately 0.3 second into the flight and some 5.5 m from the gunner, with the rear fins deployed, the dual-thrust sustainer rocket motor ignites and accelerates the missile in its first operating mode to its maximum flight speed. In its second operating mode it sustains the missile velocity.

(2) 9P59 2.95 kg weight glass fibre launch tube. (This is used to transport, aim and launch the missile. The launch tube mounts a sight device with fore and rear sights and a light indicator, a set of electrical coils to spin up the seeker's homing device, a means of mechanical and electrical interface to the missile, a grip-stock and a ground power supply unit. The launch tube can be used up to five times.)

(3) 9P58 grip-stock launching mechanism.

(4) Ground power supply unit (incorporating the thermal battery that provides the system with power in the prelaunch period and a high-pressure compressed air bottle for seeker heads cooling).

Maintenance facilities
Maintenance facilities include an equipment set that proves the functionality of the operational and training systems in the field. The equipment set can be either carried on a truck chassis or be emplaced in a static firing position. The set provides for technical checks of 12 missiles or 12 launcher mechanisms per hour.

Teaching and training facilities
Teaching and training facilities consist of a 9F730-field trainer designed to provide for instructor training of gunners. A total of 70 electronic simulated firings is considered sufficient to train an individual gunner. The 9F730 equipment includes a 9F635 IR target simulator on a stand with aircraft models and IR emission source, a 9K310 training missile and launcher. This is the same equipment used in the Igla system.

The complete Strela-3 missile system is capable of operating in high humidity, between temperatures of –38°C to +50°C.

In head-on engagements, the system can be used against targets with speeds of up to 310 m/s, while, against receding targets, the maximum target speed is reduced to 260 m/s. A competent gunner should be ready to engage a second target within 35 seconds. If required, the missile may also be fired from an armoured vehicle travelling on even ground at a speed not greater than 20 km/h.

Variants
RFAS developed a number of variants of the Strela-3 but, as far as it is known, none of these were produced in Bulgaria.

Since the system is a standard Russian design, it could be fitted with the Leningrad Optical Mechanical Organisation (LOMO) MOWGLI-2 night sight.

Specifications

	9M36
Dimensions and weights	
Length	
overall	1.42 m (4 ft 8 in) (fins folded)
Diameter	
body	72 mm (2.83 in)
Weight	
launch	10.3 kg (22 lb)
Performance	
Speed	
cruise	914 kt (1,692 km/h; 1,051 mph; 470 m/s)
Range	
min	0.3 n miles (0.6 km; 0.4 miles) (est.) (approaching/receding target)

	9M36
max	1.1 n miles (2 km; 1.2 miles) (approaching target, jets)
	2.4 n miles (4.5 km; 2.8 miles) (approaching target, helicopters and piston-engine aircraft)
	2.2 n miles (4 km; 2.5 miles) (receding target, jets)
	2.4 n miles (4.5 km; 2.8 miles) (receding target, helicopters and piston-engine aircraft)
Altitude	
min	15 m (49 ft) (est.)
max	1,500 m (4,921 ft) (approaching target, jets)
	3,000 m (9,843 ft) (approaching/receding target, helicopters and piston-engine aircraft)
	1,800 m (5,906 ft) (receding target, jets)
Ordnance components	
Warhead	1 kg (2.00 lb) HE fragmentation
Fuze	impact
Guidance	passive, IR (single channel 3.5-5 µm wavelength, FM tracking logic seeker)
Seeker view angle	
vertical	40°/–40°
Propulsion	
type	solid propellant (dual thrust, sustain stage)

Status
Vazovski Mashinostroitelni Zavodi has ceased production of all man-portable surface-to-air missile systems in Bulgaria. However, a limited number of missiles/systems produced before cessation of production are believed/assessed to be available for service from reserve stocks with the Bulgarian Armed Forces. The Strela-3 has been replaced by the Igla family of missiles.

Contractor
Vazovski Mashinostroitelni Zavodi.

China

FN-6 (HY-6)

Type
Man-portable surface-to-air missile system (MANPADs)

Development
The China National Precision Machinery Import & Export Corporation (CNPMIEC) is marketing a man-portable passive infrared homing missile system designated for export as the FN-6 (Fei Nu-6). The system is designed for use with an optional clip-on optical sight and IFF units to enhance its tactical engagement capabilities. The all aspects weapon is specifically designed for use against low and very low altitude targets such as fighters, fighter-bombers and helicopters. A single-shot-kill probability is claimed to be around 0.7 per cent. Targets can be engaged while they are manoeuvring at up to 4 *g*.

FN-6 man-portable surface-to-air missile in ready-to-fire position and fitted with optical sight and IFF subsystem 0073251

Description
The FN-6 (export designation) is known by the Chinese as the HY-6 (Hong Ying-6). The missile has an overall length of 1.495 m, diameter of 71 mm and a complete system weight (including gripstock and missile in its launch tube) of 17 kg.

Gas/coolant bottle
The combined battery/coolant gas bottle unit is mounted under the forward part of the launcher, which gives the impression of an imbalance at the systems centre of gravity which is similar to the configuration used on the QW-1 Vanguard. When questioned on this subject, Chinese sources claim that modern digitised electronics allows for a much lighter and smaller grip stock, thus the ability to mount this firing unit much further forward than normal.

Propulsion
The solid fuel booster motor ignites inside the launch tube to eject the missile and falls away when the weapon is clear of the tube. The main sustainer rocket motor then ignites to accelerate the missile to its maximum velocity in excess of 600 m/s.

The missile has four fins at the rear with four control surfaces near the front. The four forward control fins are permanently deployed, unlike a normal man-portable SAM, in that the fins are stored within the main missile body and deployed once the missile has left the launcher.

Guidance
The passive infrared pyramidal nose-mounted seeker has a four-cell mid-range infrared detector unit. Full-digital technology is adopted by the infrared seeker so that the information sampling, signal processing, anti-jamming logic judgement, target adaptive guidance are all carried out by the computer. In appearance, the seeker is not dissimilar to that fitted to the MBDA Mistral. The seeker technology is designed to counter both infrared modulated jamming and artificially generated decoy jamming systems, as well as having an inherent resistance to natural background interferences such as bright clouds and hot ground objects.

Variants
The FN-6 is available in a number of configurations for use on vehicles, helicopters and small naval combatants. One of the latter systems seen is a four-round (in pairs) pedestal launcher assembly mounted immediately behind and above the open bridge of a small naval craft with the system operator located in the bridge well.

FN-6 system without optical sight 0033937

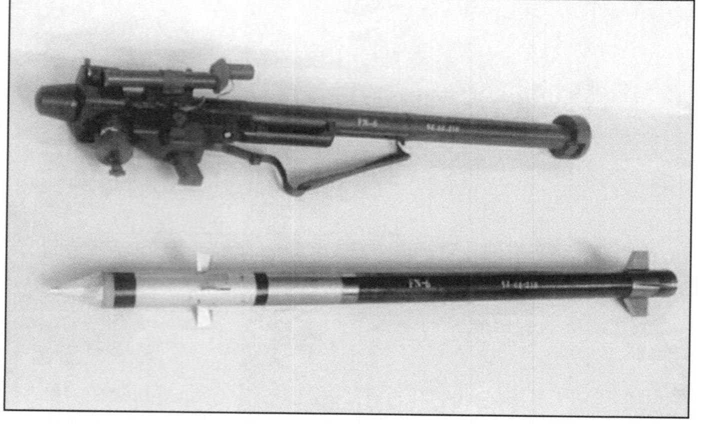

FN-6 system with launcher assembly (above) and missile out of its launch tube (below) 0033939

FN-16
In November 2008 the Zhuhai Air Show, held bi-annually in China, displayed the FN-16. Currently no information on the system has been confirmed, this will be reported as soon as it becomes available.

Specifications

	FN-6
Dimensions and weights	
Length	
overall	1.495 m (4 ft 10¾ in)
Diameter	
body	71 mm (2.80 in)
Weight	
launch	10.77 kg (23 lb)
Performance	
Speed	
cruise	1,166 kt (2,160 km/h; 1,342 mph; 600 m/s) (est.)
Range	
min	0.3 n miles (0.5 km; 0.3 miles)
max	3.0 n miles (5.5 km; 3.4 miles) (est.)
Altitude	
min	15 m (49 ft)
max	3,800 m (12,467 ft) (est.)
Ordnance components	
Guidance	passive, IR (pyramidal nose seeker)
Propulsion	
type	solid propellant (sustain stage)

Status
In production and in service with the Chinese Armed Forces where it is known as the HY-6 (Hong Ying).

Offered for export. Reports suggesting that the system was offered to Malaysia have subsequently turned out to be true, in October 2008, the Malaysian Army took delivery of a small quantity of the systems that included 16 launchers. It is expected that the FN-6 will eventually replace the British Starburst missiles held in the Malaysian inventory, therefore it is likely that further deliveries of the FN-6 will occur. Malaysia is known to have had up to 70 Starburst at one time in its inventory.

At the end of March 2008, it was revealed that Sudan purchased a number of these missile systems; they may also be found in Cambodia and Peru.

In mid-February 2013, photographic evidence of the FN-6 was obtained in the hands of Syrian rebels, it is not known how many or where these weapons came from.

Contractor
China National Precision Machinery Import and Export Corporation (CNPMIEC).
Poly Technologies Inc.

FN-16

Type
Man-portable surface-to-air missile system (MANPADs)

Development
The FN-16 (Feinu (Fei NU)) or Flying Crossbow, also known as the HY-6 within the PLA) is a new third generation, man-portable weapon system from the Chinese company, Poly Technologies Inc.

The FN-16 has been designed mainly for battlefield air defence to intercept low latitude and ultra-low altitude air targets.

This system was first identified at the IDEX 2009 defence exhibition; the system has completed development and is reported by Chinese sources at the exhibition to be in use in the PLA and with an unidentified export customer. The system has been exported to Malaysia, Cambodia, Sudan and Peru. FN-16 is the export designation for the system.

Description
The FN-16 is an improvement over the original FN-6 in particular with an extended range of 6,000 m. In addition to a high energy rocket motor, the system adopts advanced technologies such as Imaging InfraRed and Ultra-Violet (IIR/UV) two-colour rosette-scan quasi-imaging seeker, fully digitised design, laser proximity fuse to achieve strong anti-interference capability.

The FN-16 missile system features:

- IIR/UV two-colour rosette-scan quasi-imaging seeker
- Laser proximity and contact fuses
- Solid propellant high-energy motor
- An IFF antennae with a clip-on optical sight may also be fitted to the missile launcher.

Variants
FB-6A
The FB-6A is a mobile version which is mounted on a high-mobility cross-country chassis. This system consists of a one-person power-operated turret with a bank of four FN-16 in the ready-to-launch position either side.

Additional missiles are carried for manual reloading.

Specifications

	FN-16
Dimensions and weights	
Length	
overall	1.6 m (5 ft 3 in)
Diameter	
body	72 mm (2.83 in)
Weight	
launch	11.5 kg (25 lb)
Performance	
Speed	
cruise	1,166 kt (2,160 km/h; 1,342 mph; 600 m/s)
Range	
min	0.3 n miles (0.5 km; 0.3 miles)
max	3.2 n miles (6.0 km; 3.7 miles) (est.)
Altitude	
min	10 m (33 ft)
max	4,000 m (13,123 ft)
Accuracy (CEP)	1.5 m (4 ft 11 in) (helicopter target)
Ordnance components	
Fuze	laser, proximity, impact
Guidance	passive, IR (dual-colour rosette-scan, fire-and-forget)
Propulsion	
type	solid propellant (high energy motor)

Contractor
Poly Technologies Inc.
China National Precision Machinery Import and Export Corporation (CNPMIEC).

HN-5

Type
Man-portable surface-to-air missile system (MANPADs)

Development
The China National Precision Machinery Import-Export Corporation (CNPMIEC), HN-5 (Hong Nu or Red Cherry) is a product-improved version of the first-generation Russian Strela-2 (SA-7 'Grail') man-portable surface-to-air missile system that is fully described later in this section.

Development of the HN-5, which has also been reported on occasion as the Hong Ying - 5 (Red Tassel - 5), was started in 1975 under the auspices of the Chinese State Council and the Military Commission of the Central Committee of the Communist Party of China (CMC). The first Strela-2 systems were apparently obtained from Egypt in 1974, as a technology gift, following the help China provided in the 1973 Yom Kippur War.

The work, which was carried out by the Office of Industry for National Defence (OIND) Liaoning Province branch in four stages, started in March 1975 and finished in April 1985 when the weapon's design certification was finally approved.

In January 1979, at the same time as the HN-5 development was under way, the Shanghai No 2 Bureau of Machinery and Electronics was given clearance to develop the Hong Nu-5A (or HN-5A) missile system. Preliminary work had been started in 1975 and the final design certification was given in November 1986. The HN-5A was accepted into PLA service in the early 1990s.

The HN-5A is capable of making tail-on engagements against jet aircraft or head-on engagements against propeller-driven aircraft and helicopters under visual aiming conditions. The weapon has seen combat use in a number of regional conflicts.

The main improvements of the HN-5A over the original HN-5 include: a greater detection range of the IR homing seeker; reduced susceptibility to IR background sources, such as bright clouds, by the incorporation of a background noise rejection device into the IR seeker; and a larger HE warhead.

Chinese gunner taking aim with Hong Nu-5 (HN-5) man-portable SAM system 0509829

The last man-portable version of the system was the HN-5B, which has detail improvements over the preceding models. These include thermoelectric cooling of the IR seeker.

A version, the HN-5C, for use with self-propelled platforms, also exists (see entry in Self-propelled surface-to-air missile section).

Transfer of HN-5 component technology has apparently been made to Pakistan for use in the production of the then Dr A Q Khan Research Laboratories, now the Institute of Industrial Control Systems (IICS), Anza Mk I man-portable SAM system.

Description

The HN-5 system comprises:

- A launch tube that serves as an aiming device and launcher, as well as a carrying case (this launch tube can be reused on a very limited number of occasions);
- A grip stock firing unit (designated SK-5A) mounted under the forward part of the launcher, which provides launch information and ensures correct firing of the missile. The gripstock can be reused any number of times;
- A thermal battery mounted on the forward part of the grip stock to provide power.

The 13 kg missile itself is composed of four sections: the infrared seeker section fitted with background noise rejection devices; the control actuator which contains a gas generator; the warhead and fuze; and the rocket motor with rear fins attached. The infrared seeker is designed to detect the thermal radiation emitted from the target and converts this into missile steering commands to guide the missile by proportional navigation to an intercept point.

To complement the HN family, the China National Precision Machinery Import-Export Corporation (CNPMIEC) also offers the CH-3A integral test and measuring vehicles based on a 6 × 6 cross-country chassis, as well as a training system.

The CH-3A integral test and measuring vehicle has a target simulation table with an IR source attached, an integral tester and a self-contained power unit.

The training system includes: a launcher mounted dummy missile and firing unit for use in operating, aiming and firing practice; a reusable battery; a monitor score recorder; and a moving target simulator.

Variants

A mobile version is designated the HN-5C (see entry in Self-propelled SAM section).

Reports from Pakistan in November 2011 suggest a mobile variant of the HN-5 with a quadruple launcher mounted on an M113A2 chassis. The HN-5s are described as locally built, however, this would be somewhat odd and against a well known trend of Pakistan producing its own man-portable SAMs known as Anza. It is assessed that the system reported probably contained the Anza missiles that are similar in all respects to the HN-5 series.

Reports from March 2001 suggest that Myanmar has purchased an HN-5N variant. This is the first (and last) known report of the (N) missile.

Specifications

	HN-5A	HN-5B
Dimensions and weights		
Length		
overall	1.44 m (4 ft 8¾ in)	1.44 m (4 ft 8¾ in)
Diameter		
body	72 mm (2.83 in)	72 mm (2.83 in)
Weight		
launch	13 kg (28 lb)	13 kg (28 lb)
Performance		
Range		
min	0.4 n miles (0.8 km; 0.5 miles)	0.4 n miles (0.8 km; 0.5 miles)
max	2.4 n miles (4.4 km; 2.7 miles)	2.4 n miles (4.4 km; 2.7 miles)
Altitude		
min	50 m (164 ft)	50 m (164 ft)
max	2,300 m (7,546 ft)	5,000 m (16,404 ft) (est.)
Ordnance components		
Warhead	0.5 kg (1.00 lb) HE fragmentation	0.5 kg (1.00 lb) HE fragmentation
Guidance	passive, IR (fire and forget tail chasing only)	passive, IR (fire and forget tail chasing only)
Propulsion		
type	solid propellant	solid propellant

Status

Production of the HN-5 family is complete. The HN-5 was in service with the Chinese Armed Forces and North Korea, however, it has probably been replaced by more up-to-date equipment such as the QW series of man-portable systems. The HN-5A has also seen service in Afghanistan, Bangladesh, People's Liberation Army, Iran, Iraq (current status uncertain), North Korea (licence-built), Myanmar (Burma), Pakistan (locally built derivative Anza Mk I) and Thailand.

In January 1997, it was revealed that Thailand was negotiating with China for a USD5.5 million contract covering 30 launchers, 90 missiles and four testing units. Marketing of the HN-5A was being undertaken by CNPMIEC and the weapons were due to join around 230 HN-5A systems already in service with the Royal Thai Army.

All 36 × HN-5s sold to Bolivia in 1996 have now been destroyed with US Government aid.

Contractor

Export marketing

China National Precision Machinery Import-Export Corporation (CNPMIEC).

Qian Wei (QW) family of manportable weapon systems

Type

Man-portable surface-to-air missile systems

Development

The China Aerospace Science and Industry Corporation (CASIC) is the primary contractor for the Qian Wei (QW) family of manportable Surface-to-Air Missile systems (SAMs). The front end production has been subcontracted to the Liuzhou Changhong Machinery Manufacturing Corporation (also known as the Liuzhou Changhong Jiqi Zhizao Gongsi), the development of which was the responsibility of the Luoyang Optoelectro Technology Development Centre.

Listed here allows comparison between the various weapons at-a-glance.

The China National Precision Machinery Import and Export Corporation (CNPMIEC) is responsible for the export marketing of these weapons.

As of February 2014 the very latest information on this family suggests that the QW-11 and QW-18 would logically be the next generation of weapon systems, however, the designators of the Gripstock and Batteries indicates some form of commonality between those systems these include SK-18, SK-18Z on the Gripstock and DL-18 on the batteries.

Specifications

	QW-1	QW-1A	QW-1G	QW-1M	QW-2	QW-3 (IR)	QW-3 (Laser)	QW-4	QW-11	QW-18
Dimensions and weights										
Length										
overall	1.526 m (5 ft 0 in)	1.532 m (5 ft 0¼ in)	-	1.526 m (5 ft 0 in)	1.59 m (5 ft 2½ in)	2.1 m (6 ft 10¾ in)	2.1 m (6 ft 10¾ in)	-	1.526 m (5 ft 0 in)	1.52 m (4 ft 11¾ in)
Diameter										
body	71 mm (2.80 in)	-	-	71 mm (2.80 in)	72 mm (2.83 in)	-	-	-	71 mm (2.80 in)	71 mm (2.80 in)
Weight										
launch	10.36 kg (22 lb)	-	-	11.46 kg (25 lb)	11.32 kg (24 lb)	-	-	-	10.86 kg (23 lb)	10.76 kg (23 lb)
Performance										
Speed										
max speed	1,166 kt (2,160 km/h; 1,342 mph; 600 m/s)	-	-	-	1,166 kt (2,160 km/h; 1,342 mph; 600 m/s)	1,458 kt (2,700 km/h; 1,678 mph; 750 m/s) (est.)	1,458 kt (2,700 km/h; 1,678 mph; 750 m/s)	1,244 kt (2,304 km/h; 1,432 mph; 640 m/s)	1,166 kt (2,160 km/h; 1,342 mph; 600 m/s)	-
Range										
min	0.3 n miles (0.5 km; 0.3 miles)	-	-	0.3 n miles (0.5 km; 0.3 miles)	0.3 n miles (0.5 km; 0.3 miles)	0.3 n miles (0.5 km; 0.3 miles)	0.4 n miles (0.8 km; 0.5 miles)	0.3 n miles (0.5 km; 0.3 miles)	0.3 n miles (0.5 km; 0.3 miles)	0.3 n miles (0.5 km; 0.3 miles)
max	2.7 n miles (5 km; 3.1 miles)	-	-	2.7 n miles (5 km; 3.1 miles)	3.2 n miles (6 km; 3.7 miles)	3.2 n miles (6 km; 3.7 miles)	4.3 n miles (8 km; 5.0 miles)	3.2 n miles (6 km; 3.7 miles)	2.7 n miles (5 km; 3.1 miles)	2.7 n miles (5 km; 3.1 miles)

	QW-1	QW-1A	QW-1G	QW-1M	QW-2	QW-3 (IR)	QW-3 (Laser)	QW-4	QW-11	QW-18
Altitude										
min	30 m (98 ft)	-	-	15 m (49 ft)	10 m (33 ft)	10 m (33 ft) (est.)	4 m (13 ft)	4 m (13 ft)	30 m (98 ft)	10 m (33 ft)
max	4,000 m (13,123 ft)	-	-	4,000 m (13,123 ft)	3,500 m (11,483 ft) (est.)	4,000 m (13,123 ft)	5,000 m (16,404 ft)	4,000 m (13,123 ft)	4,000 m (13,123 ft)	4,000 m (13,123 ft)
Ordnance components										
Warhead	1.5 kg (3.00 lb) HE	-	-	-	1.42 kg (3.00 lb) HE	HE	HE continuous rod	HE fragmentation	1.42 kg (3.00 lb) HE	1.42 kg (3.00 lb)
Fuze	proximity	proximity								
Guidance	IR	IR (with enhanced counter-measures)	passive, IR (dual band)	IR	semi-active, laser	IIR	IR	IIR (enhanced dual band)		
Propulsion										
type	single stage, solid propellant	solid propellant	solid propellant	single stage, solid propellant	solid propellant (sustain stage)	solid propellant (sustain stage)	solid propellant (sustain stage)	single stage, solid propellant	solid propellant (sustain stage)	solid propellant (sustain stage)

Status
All are believed to be operational including the training round that is a life size and weight round.

Contractor
China Aerospace Science and Industry Corporation (CASIC).
China National Precision Machinery Import and Export Corporation (CNPIEC).
Louzhou Changhong Machinery Manufacturing Corporation.
Luoyang Optoelectro Technology Development Centre.
Shenyang Aerospace Company at Xinle.
Xi'an Tianwei Electronic System Engineering.
119 Factory.

QW-1

Type
Man-portable surface-to-air missile system (MANPADs)

Development
China National Precision Machinery Import and Export Corporation (CNPMIEC) is marketing the second-generation all-aspect Qian Wei (QW-1) ('Advanced Guard' or 'Vanguard') man-portable system for use against high-speed jets, propeller-driven aircraft and helicopters at low and very low altitudes. The China Aerospace Science and Industry Corporation (COSIC or CASIC) factory at Xinle, the Shenyang Aerospace Company has developed the QW-1 family of weapon systems as a follow-on to the older HN series of weapons. The system was unveiled at the 1994 Farnborough Air Show under the export name QW-1 Vanguard.

Compared to its HN-5 predecessor, the QW-1 has major improvements in terms of speed of response, target aspect capabilities, guidance seeker sensitivity and propulsion system. In many respects the technology involved is similar to that used in the Raytheon Missile Systems FIM-92 Stinger and Kolomna KBM Igla systems, that is the seeker elements are used to detect hot metal surfaces and engine plume exhaust using FM tracking logic.

Description
The QW-1 Vanguard missile system consists of the following elements:
- Missile - which comprises the following sections:
  (1) passive InfraRed (IR) seeker nose section and the circuitry that receives the target signal from the seeker, creating the proportional navigation guidance commands that are sent to the missile's control section;
  (2) control section, which contains the two missile control vanes and all the elements needed to fly the missile to the target;
  (3) warhead section with the HE chemical energy fragmentation warhead and fuzing system;
  (4) propulsion section with a jettisonable booster motor and dual-thrust solid propellant sustainer rocket motor;
  (5) wing unit that has four flip-out folding tail fins to provide stability to the missile in flight.
- Disposable glass fibre launch tube; this is used to transport, aim and launch the missile
- Grip-stock launching mechanism
- Disposable cylindrical thermal battery and high-pressure gas bottle for cooling the seeker.

Normally operated as part of a two-man team the gunners can, on the march, carry the system on their backs by means of a harness.

The system has been designed to engage targets like:
- Jet aircraft
- Propeller driven aircraft
- Low flying helicopters
- Unmanned Aerial Vehicles (UAVs).

The missile can be deployed with AAA guns to form a multi-layer air defence system or on board warships for very short-range last ditch defence. A helicopter mounted version is also believed to exist.

Close-up of the CNPMIEC QW-1 Vanguard low-altitude SAM system showing sights and trigger mechanism (Christopher F Foss) 0509664

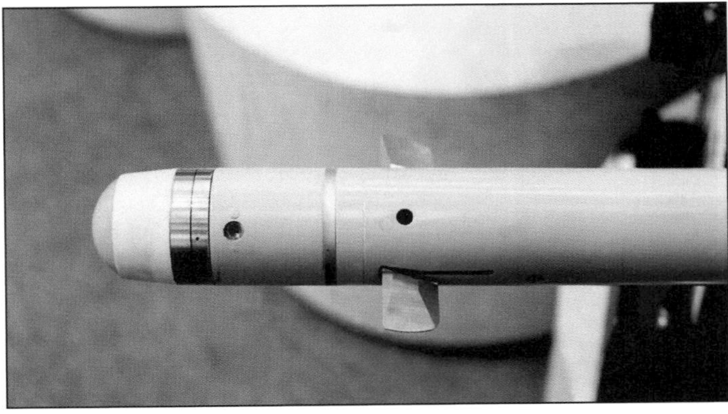
Close-up of the nose section of the CNPMIEC QW-1 Vanguard SAM 0509827

The operating sequence is as follows. The gunner is alerted by visual detection of a target. The assistant then selects a firing site and removes the front and rear covers of the launcher tube and raises it to the shoulder firing position. If identified as hostile, the firing trigger is depressed part way to activate the electric battery and coolant bottle to provide power and gas to the missile's systems.

The gunner then aims at and tracks the target until continuous audio and visual signals are received. These indicate seeker lock on. The gunner then sets the lead angle and depresses the firing trigger all the way. Between 0.3 and 0.8 seconds later, the booster motor sends the missile out of the firing tube to a safe point where it is jettisoned and the dual-thrust sustainer motor kicks in to power the missile in flight.

The missile is guided by proportional navigation to the impact point with the target, whereupon a target adaptive homing circuit cuts in just before impact to ensure maximum damage is caused. Once the missile is launched, the gunner is free to seek protective shelter or make ready for the next engagement.

Variants
Other versions known to have been developed include air-to-air, self-propelled, an octuple naval mount and a twin-round mount for naval applications. The helicopter air-to-air missile was first shown in 1998 at the Zhuhai air show and is designated TY-90. Range is said to be around 10,000 m.

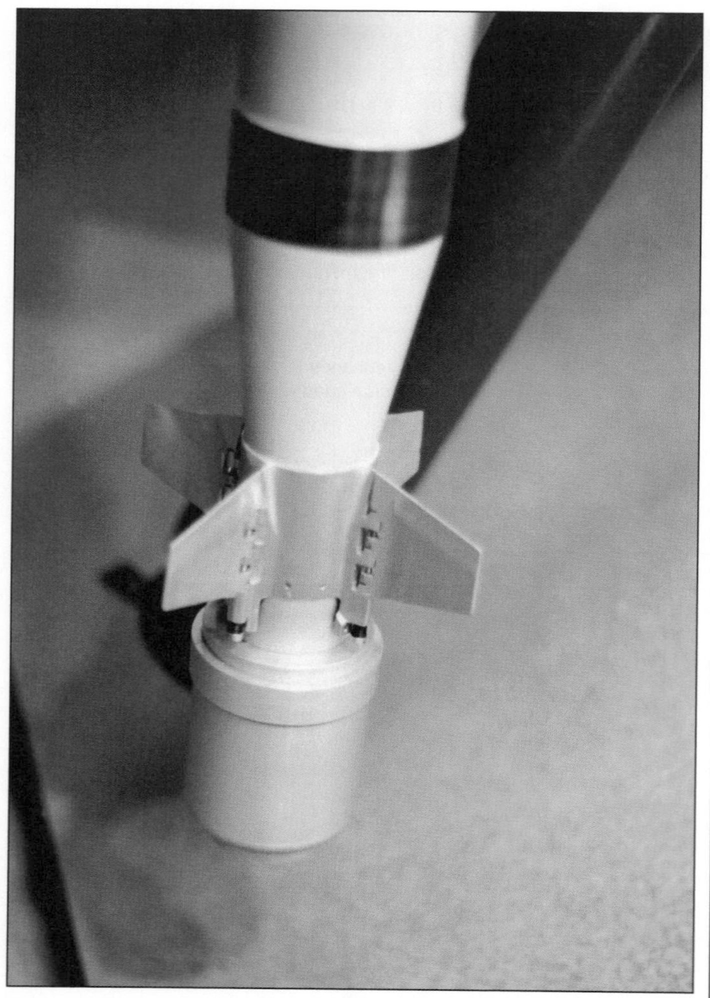

Close-up of the tail section of the CNPMIEC QW-1 Vanguard SAM 0509828

CNPMIEC QW-1 Vanguard missile out of its launcher with launcher in the background (Christopher F Foss) 0509665

The QW-1 and its technology was exported to Pakistan, where it was produced as the Anza Mk II. It is assessed that, for Pakistan to produce this weapon, transfer of component technology from the QW-1 programme took place.

Shown for the first time at the Zhuhai Air Show in 2006, the QW-1G variant was with two other systems QW-11 and QW-18. The QW-1G uses a new firing mechanism (NATO Gripstock) that is smaller and squarer in shape than the older mechanism and is designated SK-11. The battery is similar in shape to the QW-1 and is designated DL-11. Other variants of the QW-1 family include:

- QW-1
- QW-1A
- QW-1G
- QW-1M
- QW-1T2.
- QW-01M

The QW-1A (sometimes referred to in Chinese open press as 'Warm Winds') has been displayed as a small regional security system that entails the use of a tripod mounted radar, datalink communications and up to four man-portable SAMs.

Early in February 2013, a cargo of weapons that included manportable SAMs gave rise to the suggestion that the Iranian Misagh-2 could be based on the Chinese QW-1M. Previous thoughts and assessments on this associated the Misagh-2 with the QW-2.

Very early on in March 2013, photographic evidence from another cargo of missiles from a captured Iranian Dhow included five delivery cases of QW family of missiles. Inside each box was the obligatory 2 × missiles, 4 × batteries. The box was marked as QW-01M (QW-Zero One M) as were the missiles. This is the first evidence of the QW-01M missile.

Specifications

	QW-1
Dimensions and weights	
Length	
overall	1.477 m (4 ft 10¼ in) (est.)
Diameter	
body	71 mm (2.80 in)
Weight	
launch	10.36 kg (22 lb)
Performance	
Speed	
cruise	1,166 kt (2,160 km/h; 1,342 mph; 600 m/s)
Range	
min	0.3 n miles (0.5 km; 0.3 miles)
max	2.7 n miles (5 km; 3.1 miles)
Altitude	
min	30 m (98 ft)
max	4,000 m (13,123 ft)
Ordnance components	
Warhead	1.5 kg (3.00 lb) HE fragmentation
Guidance	passive, IR (cooled InSb seeker)
Propulsion	
type	solid propellant (dual thrust)
	solid propellant (sustain stage)

Status

In production. In service with the People's Liberation Army and Marines. Offered for export under the western name of Vanguard.

Contractor

China Aerospace Science and Industry Corporation (CASIC) (primary contractor).
Space Electronic Systems Engineering Company Limited, (data communications - QW-1A).
China National Precision Machinery Import and Export Corporation (CNPMIEC) (export marketing).
119 Factory (also known as Shenyang Hangtian Xinle Ltd) a subsidiary of COSIC, missile development.
Xi'an Tianwei Electronic System Engineering (system development).

QW-2

Type

Man-portable surface-to-air missile system (MANPADs)

Development

The China National Precision Machinery Import and Export Corporation (CNPMIEC) is promoting the third-generation Qianwei-2 ('Vanguard' - export name) (QW-2) man-portable surface-to-air missile system in its series of passive infrared homing fire-and-forget low-altitude SAMs. QianWei can also be translated as 'Advanced Guard'.

The InfraRed (IR) guided fire-and-forget missile was developed at the Shenyang based CASIC 119 Factory (also known as Shenyang Hangtian Xinle).

In many respects the all-aspect proportional guidance QW-2 is similar to the Russian Kolomna KBM Igla-1 (NATO SA-16 'Gimlet') system.

It has the same diameter as the Igla-1 (0.072 m) but it is slightly shorter and lighter. The QW-2 missile also has a different seeker design, which is stated to be highly resistant to a variety of countermeasures. The seeker elements are used to detect hot metal surfaces and engine plume exhaust using FM tracking logic, with the addition that the latter can recognise and distinguish chemical flare countermeasures and filter them from the guidance picture.

It is assessed that both Iran and Pakistan are using the QW series of weapons on which to base their own indigenous man-portable SAMs.

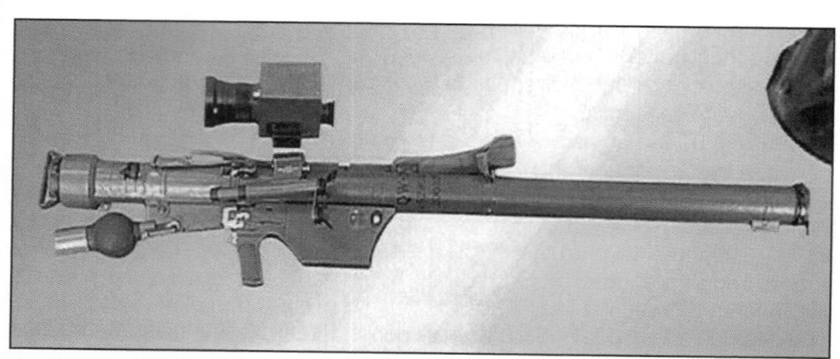

QW-2 system (CASIC) 0536855

TV camera for the QW-2 system (James C O'Halloran) 0536857

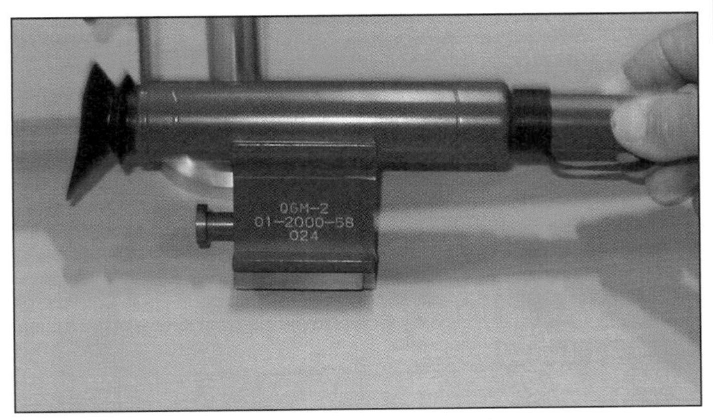

New mechanical sighting device QGM-2 for the QW-2 system (James C O'Halloran) 0536856

The QW-2 system probably entered IOC with the PLA sometime between 1998 through 2002.

Description

First identified at the Farnborough Air Show in the UK, although some other sources like to record the 1998 Paris Air Show. The QW-2 (export identification - Vanguard-2) is a third generation man-portable SAM developed for use with the PLA. The system effective altitude limits differ depending on source used, but are commonly stated from Chinese sources to be from 10 to 3,500 m through to 4,000 m with a minimum and maximum slant range of 500 and 6,000 m.

Propulsion

It takes less than five seconds to react with the solid fuel booster rocket motor ejecting the weapon from the launcher tube. At a safe distance the main solid rocket sustainer motor accelerates the missile to its maximum flight velocity of 600 m/s. The missile itself weighs 11.32 kg at launch, the warhead 1.42 kg and the complete system 18.4 kg in the firing position.

Guidance

The missile features a cooled two colour IR seeker that locks on to its target before launch and is reported to contain IR countermeasures.

A TV night vision only system can be fitted to the system, which is removed during daylight operations. This TV system does not have a range capability.

Variants

The QW family of missiles keeps expanding. To date (February 2014), known systems include: QW-1, QW-1A, QW-1G (now known to be the original designator for the QW-11), QW-1M, QW-1T2, QW-2, CQW-2, QW-3 and QW-4, QW-11 and QW-18.

Of these systems, modified variants also exist.

The QW-2 is reported to be licensed produced in Pakistan as the Anza Mk III.

The CQW-2 is the vehicle mounted variant of the QW-2. The missile is interchangeable with the shoulder-launched version.

Specifications

	QW-2
Dimensions and weights	
Length	
overall	1.59 m (5 ft 2½ in)
Diameter	
body	72 mm (2.83 in)
Weight	
launch	11.32 kg (24 lb)
Performance	
Speed	
max speed	1,166 kt (2,160 km/h; 1,342 mph; 600 m/s)
Range	
min	0.3 n miles (0.5 km; 0.3 miles)
max	3.2 n miles (6 km; 3.7 miles)
Altitude	
min	10 m (33 ft)
max	4,000 m (13,123 ft) (est.)
Ordnance components	
Warhead	1.42 kg (3.00 lb) HE fragmentation
Fuze	impact, proximity
Guidance	passive, IR (dual-band, proportional navigation techniques)
Propulsion	
type	single stage, solid propellant (sustain stage)

Status

This system has been in series production since October 2002.

The QW-2 is reported to be operational with the PLA's Type 95 SPAAG self-propelled anti-aircraft gun system, therefore it is to be assumed that it is in use with the PLA ground forces.

It is offered for export. Bangladesh and Pakistan are the only known users of the QW-2 missile system to date. However, information gained at the Asian Aerospace exhibition in Seoul, ROK, suggested that Pakistan has purchased the production line for the system complete from China and is producing under the name of Anza Mk III. It is expected that Bangladesh will replace the extremely old HN-5 series of man-portable SAMs with the QW-2.

In early February 2013, a shipment of arms from Iran was seized in Yemen, amongst this shipment were manportable SAMs. Reports from Yemen after inspecting the manpads have suggested that the Iranian Misagh-2 previously assessed to be developed on the Chinese QW-2 may indeed have been developed from the QW-1M.

The cost per round on the open arms market ranges from USD99,500 through USD1.25 million.

Contractor

Liuzhou Changhong Machinery Manufacturing for production.

CASIC 119 Factory (also known as Shenyang Hangtian Xinle), development.

China National Precision Machinery Import and Export Corporation (CNPMIEC) for promotion and sales.

QW-3

Type

Man-portable surface-to-air missile system (MANPADs)

Development

The Liuzhou Changhong Machinery Manufacturing Corporation (Liuzhou Changhong Jiqi Zhizao Gongsi), a manufacturing enterprise directly controlled by the China Aerospace Science and Industry Corporation (CASIC) produces the Qian Wei family of weapon systems. According to company officials at the Zhuhai Air Show in October 2004, the QW-3 (Qian Wei-3, or Vanguard 3) is a third-generation plus system that entered serial production two or three years previously, that would place the Initial Operational Capability (IOC) around the 2001/2002 era. It is reasonable, therefore, to assess that development commenced on this version of the QW family sometime in the early to mid-1990s. The Luoyang Optoelectro Technology Development Centre is also involved in the development of the front-end technology for the missile.

Open-source Chinese literature has suggested that at least two, and possibly three versions of the QW-3 exist, and that the system can be deployed with the PLA ground forces and PLA(N); in the future an airborne version is expected to be deployed on rotary wing assets.

Description

The missile is a two-stage, boost-and-sustain weapon that weighs approximately 23 kg at point of launch. The InfraRed (IR) version of the QW family of missiles is said to be similar to the Russian Kolomna KBM Igla-1 (9K310-1).

The QW-3 missiles are capable of laser semi-active homing guidance, IR homing, and possibly, as suggested by Western intelligence sources, a combined laser and IR guidance. The laser guidance missile in comparison with the IR has a slightly slower rate of chasing, 15° per second. This missile when used against very low-flying targets compensates for this slow rate, as targets at very low altitudes cannot manoeuvre as quickly as those at higher altitudes.

The system has been designed to deal with targets at low and very low altitudes, such as helicopters, cruise missiles and other air-to-ground munitions. The system has inbuilt IR anti-jamming circuits with an omni-directional attack capability.

As with its predecessors, the QW-3 will most likely be found protecting troops at the Forward Edge of the Battle Area (FEBA), and in the rear for the protection of high-value targets, command centres and political high-risk areas.

Variants

QW-1
QW-2
QW-4

A report coming from the 2002 Zhuhai Air Show generated by CASIC officials suggested that the QW-3 is the preferred missile mounted on the FL-2000 self-propelled SAM system. However, the system can also mount the QW-1 and QW-2.

Specifications

	QW-3
Dimensions and weights	
Length	
overall	2.1 m (6 ft 10¾ in)
Weight	
launch	23 kg (50 lb)
Performance	
Speed	
max speed	1,458 kt (2,700 km/h; 1,678 mph; 750 m/s)
Range	
min	0.4 n miles (0.8 km; 0.5 miles)
max	4.3 n miles (8 km; 5.0 miles)
Altitude	
min	4 m (13 ft)
max	5,000 m (16,404 ft)
Ordnance components	
Warhead	HE fragmentation, continuous rod
Fuze	laser, proximity, impact
Guidance	semi-active, IR
	semi-active, laser, IR (dual mode)
Propulsion	
type	two stage, solid propellant

Status

Development is complete. The first foreign customer for the system is reported to be Indonesia who have taken delivery of the IR version of the system. These missiles have been inducted into the Indonesian Air Force as of September 2006, and were reported to be with the then peace mission in Lebanon.

Contractor

China Aerospace Science and Industry Corporation (CASIC) (lead organisation).
Liuzhou Changhong Machinery Manufacturing Corporation (Liuzhou Changhong Jiqi Zhizao Gongsi) (production of the front end).
Luoyang Optoelectro Technology Development Centre (development of front end).

QW-4

Type

Man-portable surface-to-air missile system (MANPADs)

Development

It was reported late in 2001/2002 that the China Aerospace Science and Industry Corporation (CASIC or COSIC as is sometimes reported) is leading the development of the new third-generation (++) Qiang Wei-4 (QW-4) SHOrt-Range Air-Defence (SHORAD) system. The requirements of the system include the use of Imaging InfraRed (IIR) technology for the front end of the missile in particular the detector array that is being developed by the Luoyang Optoelectro Technology Development Centre and the Liuzhou Changhong Machinery Manufacturing Corporation (Liuzhou Changhong Jiqi Zhizao Gongsi), who are responsible to the CASIC for the production of the system. The overall system is probably assembled at their premises.

Although the system was originally reported as a man-portable (MANPAD) type, it is entirely possible that the missile will be used in other than the MANPAD scenario, such as mounted on land vehicles, warships and possibly helicopters.

Description

The QW-4 (Qian Wei-4, or Vanguard 4) is a new generation of SAM systems that has been designed primarily as a MANPAD for the defence of high-value units and areas of political importance.

The weapon is required to intercept cruise missiles, rotary-wing and fixed-wing aircraft, and Unmanned Aerial Vehicles (UAVs). Reports coming from the CASIC have also suggested that the system is required to have a long-range detection, a large kill zone and omni-directional targeting. This would then suggest that the stand-alone MANPAD, as has been traditionally deployed, will now have some form of command-and-control added to the system. This may come as a portable radar, radio or high speed data link or even all three combined.

The incorporation of an IIR seeker will give the missile immunity initially (dependant on time taken to bring the missile system to IOC) to flares and other heat-based defence systems as may be deployed by aircraft or helicopters.

Other third party sources have suggested that this new missile system has new fully electric control surfaces that provide a smoother flight path curve and thus greatly improved accuracy.

Variants

QW-1
QW-2
QW-3

Specifications

	QW-4
Performance	
Speed	
max speed	1,244 kt (2,304 km/h; 1,432 mph; 640 m/s)
Range	
min	0.3 n miles (0.5 km; 0.3 miles)
max	3.2 n miles (6 km; 3.7 miles)
Altitude	
min	4 m (13 ft)
	2 m (7 ft) (deployed at sea)
max	4,000 m (13,123 ft)
Ordnance components	
Warhead	HE fragmentation
Fuze	laser, proximity
Guidance	IIR
Propulsion	
type	single stage, solid propellant

Status

As of May 2013, this system is probably still in development, which is believed to have started either late 2001 or early 2002. Due to the closed nature of the Chinese missile industry, the exact status of the system is unknown.

Contractor

China Aerospace Science and Industry Corporation (CASIC) is the lead organisation with the Luoyang Optoelectro Technology Development Centre, which is responsible for the development of front end. The Liuzhou Changhong Machinery Manufacturing Corporation is responsible for the production of the front end. China National Precision Machinery Import and Export Corporation (CNPMIEC) will take responsibility for the advertising and export of the system when this is available.

QW-11

Type

Man-portable surface-to-air missile system (MANPADs)

Development

This weapon was displayed at the China Zhuhai Air Show, advertising documentation described the system as having been designed to take on specific targets, particularly the cruise missile threat.

Description

The Qian Wei-11 (QW-11) has been reported in Chinese literature as originally designated the QW-1G (Qian Wei-1G), and was specifically designed to take on terrain-hugging cruise missiles while still retaining the capability to engage low-flying aircraft. The missile (a development of the Qian Wei 1 variant) employs an advanced impact and laser proximity fuze with digital information processing. The missile on display in Zhuhai showed a red band just forward of the warhead that contained 8 × IR/Laser lenses probably for warhead fuzing. Compared to impact-only fuzes deployed on earlier models, the laser proximity provides for improved ECCM to counter the cruise missile threat.

The missile itself is similar in all aspects to earlier models with the only distinguishing features being identified in the firing unit or gripstock as it is known within NATO. The gripstock has the designator SK-11 and is a new shape, more square with the appearance of micro-electronic circuitry with technological advances in comparison to the older gripstocks.

The power source (battery) however continues with the older shape that is similar to that of the FIM-92A Stinger, this is mounted in the upright position forward of the gripstock and is designated DL-11.

At the front end of the missile is an opaque convex dome that has no aero-spike, this then suggests that the seeker operates in the 3 to 5 micron band. Chinese sources have also reported this seeker of having the further improved variant featuring an enhanced dual band infrared seeker developed, in order that the target is not only tracked via the exhaust heat but also by the temperature difference of the surface of the target.

Variants

QW-1G.

Specifications

	QW-11
Dimensions and weights	
Length	
overall	1.47 m (4 ft 9¾ in)
Diameter	
body	71 mm (2.80 in)

	QW-11
Weight	
launch	10.68 kg (23 lb)
Performance	
Speed	
max speed	1,166 kt (2,160 km/h; 1,342 mph; 600 m/s)
Range	
min	0.3 n miles (0.5 km; 0.3 miles)
max	2.7 n miles (5 km; 3.1 miles)
Altitude	
min	30 m (98 ft)
max	4,000 m (13,123 ft)
Ordnance components	
Warhead	0.57 kg (1.00 lb) HE fragmentation
Fuze	impact, proximity
Guidance	passive, IR (cooled InSb, dual band)
Propulsion	
type	single stage, solid propellant (sustain stage)

Status
Development is complete as the system has been offered for sale at the Zhuhai Air Show 2006 and again at Farnborough 2008. As of March 2014 no customers have been identified.

Contractor
Shenyang Aerospace Company in Xinle, China. A subsidiary of the China Aerospace Science and Industry Corporation.

QW-18

Type
Man-portable surface-to-air missile system (MANPADs)

Development
The QW-18 (Qian Wei-18) was shown/identified for the first time at the Zhuhai Air Show in China during November 2006. The China Aerospace Science and Industry Corporation (CASIC) subsidiary, the Shenyang Aerospace Company in Xinle, developed this new man-portable missile system to engage targets such as:

- Helicopters;
- Propeller-driven aircraft;
- Terrain-hugging cruise missiles;
- Jet aircraft.

The system has been described in Chinese literature as a follow-on to the earlier Qian Wei (QW) family of missile systems designed with specific targets in mind.

Description
The HQ-18 is a further improved/upgraded missile that employs has a highly sensitive seeker that makes use of dual-band InfraRed (IR) technology which looks at the target and initially seeks the hottest point, normally the tail or engine but also the temperature difference caused by in flight friction of the skin effect of the target. During flight it will then update the target information to the surface hot points and attack those more vulnerable sections of the target.

The seeker has a conventional convex window with no aerospike, it is opaque to visible light and has been assessed to use the 1.2-2.5 μm and 3-5 μm IR bands.

The missile also employs strong anti-jamming techniques that are reported to be resistant to jammers and multiple flares launched in a sequential series dependant on heat and thus IR frequency as well as jamming from complex backgrounds and sunlight reflected from aluminium foil. These counter jamming techniques are enabled throughout the course both before and during flight.

QW-18 System (IHS/Patrick Allen) 1322084

The systems firing unit (known in NATO as the gripstock) is of the older type and designated SK-18.

The battery is mounted horizontally similar to the FN-6.

Variants
QW-18A

The QW-18A first displayed at Airshow China in November 2010; the system is reported to have new electric servo control actuators for improved guidance and flight characteristics. Furthermore, a laser proximity fuze not seen on the earlier QW-18 and Imaging-InfraRed (IIR) seeker.

Specifications

	QW-18
Dimensions and weights	
Length	
overall	1.526 m (5 ft 0 in)
Diameter	
body	71 mm (2.80 in)
Weight	
launch	10.76 kg (23 lb)
Performance	
Range	
min	0.3 n miles (0.5 km; 0.3 miles)
max	2.7 n miles (5 km; 3.1 miles)
Altitude	
min	10 m (33 ft)
max	4,000 m (13,123 ft)
Ordnance components	
Warhead	1.42 kg (3.00 lb)
Guidance	IR (enhanced dual-band tracing temperature difference of the skin of the target)
Propulsion	
type	solid propellant (sustain stage)

Status
Development is complete. The system was offered for sale at the Zhuhai Air Show 2006 but as of March 2014 no export customers have been identified.

Contractor
Shenyang Aerospace Company in Xinle, China. A subsidiary of the China Aerospace Science and Industry Corporation (CASIC).

Egypt

Sakr Eye

Type
Man-portable surface-to-air missile system (MANPADs)

Development
The Egyptian Army has had in it's inventory the Soviet first-generation designed Strela-2/Strela-2M (SA-7a/SA-7b) 'Grail' man-portable low-altitude surface-to-air missile for many years.

In the late 1970s, an alleged reverse engineered and improved product version of the Strela-2M (SA-7b) was placed in production and subsequently called Ayn-al-Saqr, the Sakr Eye (Eye of the Falcon). The then Sakr Factory for Developed Industries at Almaza manufactured it.

Qualification occurred in 1982 and it was first shown in public in late 1984 with pre-series production commencing in March 1985. Full production started in 1986 following extensive Egyptian Army trials, with Initial Operational Capability (IOC) being achieved in 1987 to 1988.

Sinai 23 0509412

Sakr Eye SAM system fitted with night sight and IFF system 0509396

Egyptian sources indicated that it cost USD180 million to develop Sakr Eye with all funding coming from the Egyptian Defence Ministry.

Sakr Eye was deployed with the Egyptian units assigned to the Coalition forces in the Gulf War of 1991.

Egypt apparently supplied a small quantity of the Strela-2 system to China and North Korea in 1974, as a technology gift, for help given during the 1973 Yom Kippur War. North Korea subsequently reverse engineered their own model. China produced their HN-5 system based on the Strela-2.

Description

The system is a lightweight, short-range anti-aircraft missile with infrared passive homing guidance. The Sakr Eye system comprises three main components: the missile itself, thermal battery and the grip stock, (sometimes called the launching mechanism). The missile is transported and launched from an expendable launch tube, which is made of glass fibre and is fitted with two aiming sights positioned on the left forward side, an acquisition indicator lamp and a thermal battery mounted under the front, which has sufficient power for 40 seconds. The thermal battery can be replaced in the field if required.

The missile is a fire-and-forget type and consists of:

- An InfraRed (IR) homing head
- Guidance and control section
- Warhead
- Solid propellant propulsion section.

The grip stock or launching mechanism combines the firing mechanism and the logic circuits and, when attached to the launch tube, carries out the firing sequence and authorises ejection of the missile in either the manual or automatic modes. Once a target has been engaged, the launcher tube is discarded and the grip stock attached to a new launcher. The original Soviet Strela-2 launcher can be reused up to five times, it is assumed that the Egyptian version will have similar capabilities.

In the absence of an integrated Identification Friend or Foe (IFF) system, a typical target engagement sequence is as follows:

The gunner acquires the target and aligns the weapon using the open sight and then selects either manual firing mode; audio and visual cues are given to the gunner to indicate when the target is within the weapon's engagement envelope. The trigger is then squeezed according to the short delay; the booster accelerates the missile until the sustainer's motor is ignited to propel it to the target. If a target is not encountered within 16 seconds, Sakr Eye destroys itself. Homing is by means of a passive IR seeker that was developed in conjunction with Thomson-CSF and is more sensitive than the original. It is fitted with a background IR noise rejection filter to avoid the seeker being led astray by extraneous IR sources such as clouds. A contact fuze that also provides for graze functionality insures initiation of the warhead's HE fill.

The Sakr Eye ammunition container has two missiles fitted with thermal batteries plus two spare batteries, while the grip stock container has one grip stock. It takes 10 seconds to be prepared for action and can engage aircraft travelling at a maximum speed of 280 m/s in a pursuit or flying at a maximum speed of 150 m/s head on. Successful trials have also been carried out with Sakr Eye fitted with the US CA-563 optical sight. This incorporates increased magnification (×3 with a 22° field of view) with advanced sighting techniques that include a direct computer sight digital interface with the missile, so that positive acquisition and confirmation of specific target lock on can occur.

The optical magnification is incorporated to improve identification at extended ranges. Light Emitting Diode (LED) cueing is provided to indicate system status such as uncaged seeker or an out-of-tolerance condition. A night vision image intensification module can also be incorporated into the CA-563 optical sight. Optional equipment includes a Thomson-CSF PS-340 IFF unit, the antenna of which is attached to the right side of the launcher, with the interrogator electronics unit hanging on the operator's belt and a nigh vision sight.

The Egyptian Army Sakr Eye is normally operated by a three-man team comprising the commander, gunner and wireless operator, with a long wheelbase jeep (4 × 4) light vehicle carrying two missile teams.

Variants

Pedestal-mounted Sakr Eye

In addition to the standard man-portable version, the prototype of a single-round ring pedestal-mounted version has been built for installation on a light vehicle or ship. A helicopter-mounted version has also been proposed.

M113A2-mounted Sakr Eye

The Dassault Electronique Sinai 23 twin 23 mm self-propelled air defence system mounted on a United Defence LP M113A2 chassis, is also armed with a total of six (three each side) Sakr Eye SAMs, details of which are given in the self-propelled anti-aircraft guns section.

Specifications

	Sakr Eye
Dimensions and weights	
Length	
overall	1.4 m (4 ft 7 in)
Diameter	
body	72 mm (2.83 in)
Weight	
launch	9.9 kg (21.00 lb)
Performance	
Range	
max	2.3 n miles (4.2 km; 2.6 miles)
Altitude	
min	30 m (98 ft) (est.)
	150 m (492 ft) (from helicopter)
max	2,300 m (7,546 ft)
Ordnance components	
Warhead	1 kg (2.00 lb) HE fragmentation
Fuze	impact
Propulsion	
type	solid propellant (sustain stage)

Status

Production as required. The system has seen service with the various Afghanistan factions, Egyptian Army and, it is believed several other undisclosed countries.

Reports from Middle Eastern intelligence agencies have confirmed that the Hamas terrorist organisation and Palestinian terrorist organisations have access to the Sakr Eye.

As of August 2010, the Sakr Eye system is now referred to by its production factory as a legacy system, and is probably no longer in production.

Contractor

Sakr Factory for Developed Industries.

France

Mistral-1

Type

Tripod mounted man-portable surface-to-air missile system (MANPADs)

Development

In 1977, a technology group formed by the French Joint Chiefs of Staff and the *Délégation Générale pour l'Armement* (DGA, French Weapons Procurement Authority) began a study of different short-range gun and missile point defence surface-to-air systems to meet a tri-service requirement for a Very Low-Level Air Defence (VLLAD) system.

In 1979 the study narrowed to the procurement of a new third-generation missile system to be known as the *Sol-Air-Très Courte Portée* (SATCP, surface-to-air very short range).

An operational programme index was established that year by the *Direction des Engins/Délégation Générale pour l'Armement* (Missile Division of the French Weapons Procurement Authority) to develop a weapon common to all three services. Subsequently, the SATCP has become known as the Mistral.

In March 1980, an evaluation trial was held to examine the proposals put forward by five competing companies. These were quickly narrowed down to the projects proposed by the then Matra (now MBDA), Aerospatiale and TDA (formerly Thomson Brandt). In September that year, following further technical and feasibility studies, Matra was named as the company responsible for developing and producing the Mistral.

On 1 December 1980, the development contract for the basic man-portable tripod-launched version (MANPADS or Man-Portable Air Defence System) was placed, naming the following as the main subcontractors:

- Matra as the prime contractor
- *Société Anonyme de Télécommunications* (SAT) for the homing head
- *Société Européenne de Propulsion* (SEP) for the rocket motor
- *Société Nationale des Poudres et Explosifs* (SNPE) for the solid propellants
- SAFT for the thermal batteries
- Manurhin for the warhead, safety arming device and missile container-launcher tube.

Sigma with three Mistral SAMs (MSI-Defence Systems) 1033984

Test firings were started in 1983 and the last of the 37 scheduled launches was completed in March 1988. The first production systems were delivered and became fully operational within the French Army and French Air Force during late 1988. After Operational Valuation (OPVAL) trials conducted by the French forces in 1989, Mistral achieved initial operational capability with the three French services in January 1990.

Current French Army deployment is to group the Mistral MANPADs launchers in corps-level air defence support regiments, which are part of an Artillery Brigade. These will each deploy a number of batteries that will comprise four to six sections, each of six launchers and an alert system. The French Air Force uses its MANPADs launchers for defence of its air bases.

During the 1991 Gulf War, the air defence of the French 'Daguet' Division units on the move was assigned to Mistral MANPADS.

In September 1996, it was revealed that the New Zealand Army had purchased the MANPADs system to meet a Very Low-Level Air Defence System (VLLADS) requirement. Deliveries were made towards the end of 1997. The contract, worth USD22.75 million, included 12 launchers, 23 missiles, training simulators and logistic support. The unit equipped is G troop, 43 Air Defence Battery with five launchers. The remainders are in reserve.

In 1998, an unnamed South American country and Brunei, the latter being the 23rd country to order Mistral systems, placed contracts.

In mid-2000, it was revealed, by MBDA, that production had switched to the Mistral 2 missile, with improved aerodynamics and other enhancements to improve its performance. The Mistral 2 is in service with the French Army, Navy and Air Force and several other countries.

Description

MANPADs

The Mistral MANPADs system comprises:

- The missile in its container-launcher tube
- The vertical tripod stand
- A pre-launch electronics box
- Daytime sighting system
- Battery/coolant unit.

A thermal sight of SAGEM (Ratis) or Thales Optronics (RATS 3) for night-time firing and an IFF interrogation may be added.

The basic assembly can be broken down into two 20 kg loads - the containerised missile and the pedestal mount with its associated equipment for carriage by the missile team commander and the gunner respectively. In operational use, the system will normally be transported in a light vehicle to the deployment area where it will be man packed to the firing site by the team.

The Mistral-1 is a slim, two-stage cylindrical type with a booster motor to eject it from the launch tube and a SNECMA SPS composite solid rocket sustainer motor to accelerate it to its maximum speed of M2.5. Flight control is exercised by movable canard control surfaces located in the weapon's front region. The 3 kg HE fragmentation warhead with 1 kg of HE uses 1,600 embedded tungsten spheres to achieve increased penetration of the target surface and is fitted with both contact and proximity fuzes to ensure detonation. The MBDA proximity fuze is an active laser type.

The cooled narrow field-of-view (to avoid decoys), passive InfraRed (IR) seeker is derived from technology used on the MBDA Magic 2 air-to-air missile programme. It has a multi-element cruciform array sensor with digital processing that allows a full, head-on, non-after burning jet combat aircraft to be acquired at ranges of about 6,000 m and light combat helicopters with reduced IR signatures at ranges of 4,000 m or more. The two-colour, all-aspects seeker waveband operating regions are believed to be in the 2 to 4 and 3.5 to 5 μm. The seeker can apparently tilt up or down by 38°.

An IR transparent magnesium fluoride pyramidal-shaped seeker cover was used to reduce the drag factor appreciably. This increases the Mistral's manoeuvring capabilities considerably during the terminal phase of the flight.

In autonomous mode, the team commander is in charge of liaison with the section fire-control centre, identifying the target and ordering the engagement. The gunner then carries out the firing sequence of target acquisition, system lock on and firing.

However, in the French forces, Mistral will be used in a coordinated manner, with the firing units linked by radio, through a weapon terminal, to a platoon coordinating centre.

The French Army is using the Thomson-CSF Samantha alerting system, while MBDA is proposing for export, associated with the AIDA terminal (or the new generation BADO), a Mistral Coordination Post (MCP) based on the Oerlikon Contraves SHORAR radar. This allows the user to optimise coordination and control of a platoon's firing units.

At a launch site, the firing post is erected first, using its adjustable legs on uneven terrain to level both the attached height-adjustable gunner's seat and the two-handed firing grip. The latter is fitted with:

- A safety lever to avoid accidental operation of the seeker activation lever.
- A homing head unlocking button to release the seeker from a target which is not to be engaged.
- A seeker activation lever to initiate battery power.
- Detector cooling and missile gyroscope spin-up.
- A firing trigger.

The missile in its container tube is placed on the tripod, linked with the battery/coolant unit, the sighting device and the pre-launch electronics box. The battery/coolant unit supplies the electrical power required by the missile before launch and supplies the coolant necessary to cool the detector cells of the seeker head for lock on. Once initiated, the unit operates for a period of 45 seconds and then has to be replaced, which takes a short time to do. It takes approximately 60 seconds to assemble the Mistral system in the ready-to-fire state at a firing site.

The target can be designated in one of three ways:

1. By the team commander, using information passed over a radio net from an off-site observation network.

2. Visually, by the team commander, using binoculars.

3. By the gunner using the AIDA or BADO cuing terminal-bearing designation transmitted by radio from a coordination post.

Once a target is designated in azimuth, the gunner acquires it in elevation and begins tracking it. The aiming reticule is displayed luminously and continuously, which allows the gunner to follow the pre-launch sequence.

MBDA Mistral MANPADs system deployed in the field 0093170

ALBI launcher assembly mounted on Panhard VBL (4 × 4) armoured vehicle 0093171

MBDA Mistral Surface-to-Air Missile being launched from the MANPADs system 0024891

Atlas twin launcher system with two MBDA Mistral SAMs on ACMAT (4 × 4) truck 0056296

When the gunner depresses his firing trigger, this causes the booster motor to ignite and thus accelerate the missile to a muzzle velocity of around 40 m/s. Before the missile emerges, the motor burns out in order to protect the gunner from the rocket motor's efflux effects. Once free of the launch tube and at 15 m from the launcher, the booster motor falls away and the 2.5 second burn composite propellant-fuelled sustainer motor fires to accelerate the missile to its maximum speed. The weapon is then guided to the target's exhaust by the proportional navigation on board passive IR homing system using a gyro as reference. Control is achieved by the canard surfaces.

Maximum flight time possible is 14 seconds. As soon as a round is fired, the expended launch tube is discarded and a new one fitted in approximately 10 seconds. Total engagement time from firing sequence initiation to weapon launch is less than five seconds without early warning of a target and around three seconds if a warning is provided. The single-shot kill probability is claimed to be very high.

Variants

Atlas lightweight twin-round launcher system

The Mistral pedestal-mounted twin-round ATLAS (*Affût Terrestre Léger Anti-Saturation:* Advanced Twin Launcher Anti-air Strikes) is designed for use as either a ground or vehicle mount to protect vital points or troops on the move.

It comprises a portable launcher operated by one gunner, with two ready-to-fire rounds.

When used on a vehicle it is possible to dismount the system and quickly configured it for the ground role. The Atlas system has been sold to Abu Dhabi, Belgium, Cyprus and Hungary.

In April 1997, a USD100 million contract was signed between the Hungarian Ministry of Defence and the then Matra BAE Dynamics (now MBDA) for Mistral missiles, 45 Atlas firing posts mounted on German Mercedes-Benz Unimog (4 × 4) light trucks and nine MCP systems.

Air-To-Air Mistral (ATAM)

For air-to-air use the airborne system is designated the ATAM or *Air-Air Très Courte Portée* (AATCP, air-to-air very short range) which uses 70 kg two-round groups of missiles on weapons pylons with internal electronics boxes to arm helicopters. The French Army has bought it for its Gazelle armed helicopters using the TH-200 gyro stabilised roof-mounted gun sight for target designation.

The ATAM system entered French Army service in June 1996. Three squadrons of ATAM-equipped Gazelles are operational.

Hungarian Army Mistral Atlas twin launcher system deployed in the field with MCP on the left and launcher on the right 0062754

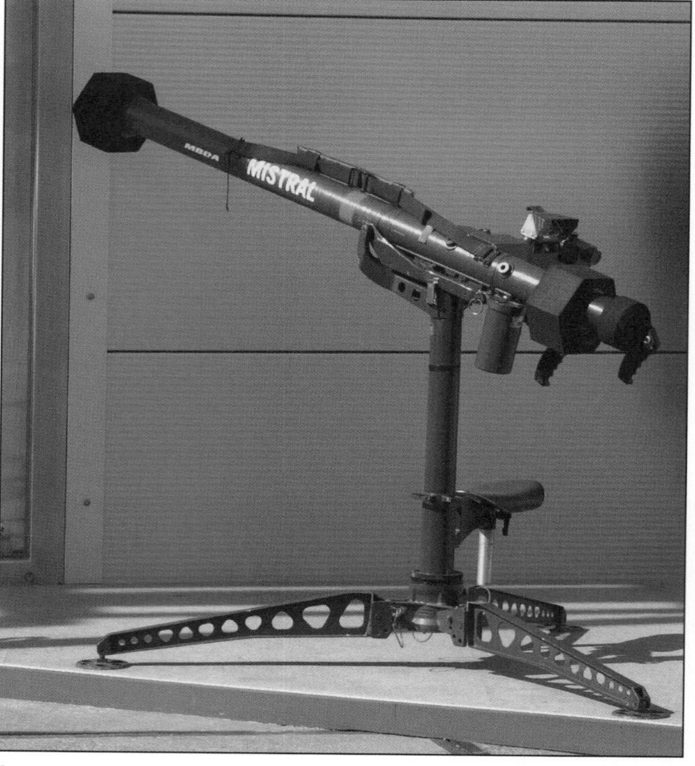

Sigma/Mistral (MSI Defence) 1109214

The launch envelope in the air-to-air role is said to be from ground level up to 5,000 m at ranges from 600 to 5,500 m. The launch sequence, from target acquisition, seeker lock on and firing, takes just over four seconds. The laser proximity fuze is used to engage targets flying at altitude, whilst the impact fuze can be selected against low-altitude targets where other obstructions, for example trees, could set off the proximity fuze. The pilot selects the fuzing option before launch. The ATAM can engage high-speed aircraft as well as helicopters.

The system can be operated at speeds to in excess of 150 kt.

In January 2007, it was announced that following the signing of the contract dated July 2006, MBDA will supply the ATAM version of Mistral to India's Hindustan Aeronautics Ltd (HAL) for a new combat version of HAL's Dhruv light utility helicopter. This is the first time Mistral will be deployed operationally in India. The Dhruv became fully operational in late 2009.

In September 2007, it was announced that the Spanish Government has agreed to purchase the air-to-air version of the Mistral for the Tiger helicopters. Exact numbers were not released but Spain has an acquisition contract for 24 Tiger combat helicopters in the Spanish HAD (Helicoptero de Apoya y Destruccion, attack and destruction helicopter) version.

Sadral naval system

For naval use, it becomes the 1,500 kg six-round *Système d'AutoDéfense Rapprochée Anti-aérienne Légère* (SADRAL, light short-range Anti-aircraft self-defence system) for use on surface ships of all sizes. The navies of Abu Dhabi (using a derivative version), France, Qatar and Thailand have adopted this version.

Simbad lightweight twin-round naval system

This is a naval lightweight twin-round launcher designed primarily for installation on various types of smaller vessels, logistic vessels and support ships to provide them with a degree of autonomous Anti-aircraft self-protection.

The Simbad system can be fitted to any type of 20 mm cannon mounting and has been ordered by Brazil, Cyprus, France, Indonesia, Norway, Singapore and an unnamed South American country. The UK Royal Navy trial of the Simbad system on the Royal Fleet Auxiliary Ship Black Rover took place during March 2005.

Simbad-RC

The Simbad-RC is the latest (February 2013) system in the Simbad family and is a remote-controlled short-range naval air defence system. Prototypes are in production and system qualification is expected in 2014. The system has already been sold by MBDA to an un-named customer with deliveries due to begin in 2015.

The Simbad-RC gyro-stabilised automatic launcher turret holds two Mistral 2 fire-and-forget missiles and can be fitted with a thermal and large field of view camera. The SMU-RC operator terminal located inside the ship can manage two turrets and can be integrated with the ship's combat system, radar and electro-optical sensors.

With a weight of 350 kg the turret has a bearing of ±160° and an elevation –15° to +65° and can be fired in less than five seconds after preparation.

Sigma combined gun/missile stabilised naval mount

Sigma is a multirole weapon system of modular concept with options for carrying a variety of close-range surface-to-air missiles such as the Mistral (3) or Igla (4) and 25 or 30 mm cannon.

Tetral naval four round launcher

This system was developed to use the Mistral 2 missile and has been adapted for use on all types of surface vessels.

Aspic automated firing unit

Mistral has been adapted for use on flatbed 4 × 4 vehicles with the Aspic Automated Firing Unit, a full description of which appears in the Self-propelled surface-to-air missiles section of this book. The system was decomissioned from the French Air Force in September 2010.

ALBI

In 1999, the Royal Army of Oman ordered a new launching system originally developed for the French Army known as ALBI. ALBI is a derivative of the Atlas twin-launcher but fitted on a turret which can be integrated on armoured vehicles. The version selected by the RAO is for their Panhard VBL (Light Armoured Vehicle) already in service with their reconnaissance units. The system can be deployed less than five seconds after the vehicle has stopped. In addition to the two missiles mounted on the turret, the vehicle transports four complementary missiles.

Mistral Coordination Post (MCP)

The MCP has been designed to be used with Mistral MANPADs or ATLAS firing units to defend the rear areas of infantry, artillery or Vital Point units. It maximises the attrition rate of attacking aircraft, even during saturation raids, thereby considerably reducing the overkill. The MCP also avoids the firing of weapons at out-of-range targets and maximises the safety of friendly aircraft. It is essential for night operation.

MBDA has developed several land, naval and airborne weapon systems based on the Mistral with a Mistral Coordination Post (MCP). In March 2011, the French Army and MBDA the missile manufacturer chose the Matis MP3 optronic aiming sight offered by Sagem, which is now part of the Safran group, to modernise the French Army's 186 Mistral firing posts. The MP3 is a day and night sight that enhances the Mistral missile's capabilities in several key areas:

Extended detection and identification ranges

Greater operability under different weather conditions

Integration of day and night channels in the same system.

The system is based on existing equipment and has a 2-D Shorar surveillance radar from Oerlikon Contraves Italiana with IFF that can be used either in an autonomous mode or linked to an overall C3I system and receives tracks from an external source. The radar has a detection capability of 17 km in upper elevation and 28 km in lower elevation. Up to 200 targets can be tracked simultaneously with the TDMA (distribution time sharing) mode of the fire management console allowing, after target classification, transmission of firing data to the various launch platforms. It also comprises:

- A command post station for two operators
- A coordination terminal for each MANPADs or Atlas firing unit (AIDA)
- A radio datalink network to transmit two-way voice and data in TDMA (distributed time sharing) mode
- A command system to communicate with the next higher command echelon.

The system (with the exception of the coordination terminals) is built into a small, highly mobile air transportable-wheeled vehicle Unimog 1350 (4 × 4) and can coordinate up to 11 firing posts.

The MCP has been sold to Hungary. Brunei has also chosen it but the actual contract has yet to be placed.

Mistral Simulator

DoDaam Systems (South Korea) has developed a Mistral Simulator that uses virtual reality based on the computer network and 3-D graphic-oriented technology. The simulator is portable and can be used for tracking and fire, IFF, identification of targets, day and night training. The system consists of:

- Trainee computer
- Instructor computer
- Mistral mockup
- Visual equipment.

Specifications

	Mistral-1
Dimensions and weights	
Length	
overall	1.86 m (6 ft 1¼ in)
Diameter	
body	92.5 mm (3.64 in)
Flight control surfaces	
span	0.18 m (7 in) (wing)
Weight	
launch	18.7 kg (41 lb)
Performance	
Speed	
max speed	1,711 kt (3,168 km/h; 1,969 mph; 880 m/s)
Range	
min	0.2 n miles (0.3 km; 0.2 miles)
max	3.2 n miles (6 km; 3.7 miles) (est.)
Altitude	
min	5 m (16 ft)
max	3,000 m (9,843 ft)
Ordnance components	
Warhead	3 kg (6.00 lb) HE 1,600 fragmentation tungsten spheres
Fuze	laser, impact, proximity
Guidance	passive, IR
Propulsion	
type	solid propellant (eject motor)
	solid propellant (sustain stage)

Status

Mistral-1 - production completed in the late 1990s. Mistral 2 - in production from 2000.

15,000+ rounds ordered for the Mistral-1, since series production started in 1990, by 36 different armed forces in 25 countries: eight in Europe, eight in Asia, three in the Middle East and five in South America. Several other countries in Europe, Asia and the Middle East were also assessing it.

Known to have seen service with several countries.

Contractor

MBDA.

DoDaam (development and production of the Mistral Simulator).

PT Pindad (Indonesia), development with MBDA for use on ATLAS system with Pindad Light Military Vehicle.

Mistral-2

Type

Man-portable surface-to-air missile system (MANPADs)

Development

In 2000, it was revealed by MBDA, that production had switched to the Mistral-2 missile, with improved aerodynamics and other enhancements to improve its performance. The Mistral-2 is in service with the French Army, Navy and Air Force and several other countries.

In December 2006, it was announced by MBDA that it would join with Indonesia's PT Pindad Persero to form an industrial partnership to market an Indonesian variant of the ATLAS (Mistral-2) short-range air defence system-based on Pindad's Light Military Vehicle. MBDA are hopeful that this will fill the Indonesian requirement for a mobile, quick reaction, vehicle-mounted missile system for protection against fast and highly manoeuvrable aerial targets.

During September 2007, it was announced that an Evolved Mistral has been developed with the primary aim for inclusion with the Multi-Mission System (MMS) mounted on vehicles such as the THOR and Spanish VAMTAC. This Evolved Mistral has been designed with the low-level, low-signature airborne threats in mind such as Unmanned Aerial Vehicles (UAVs) and cruise missiles. The evolved Mistral also has an extended range, altitude and precision.

In February 2011, MBDA signed a contract with an unidentified customer (possibly Saudi Arabia) for the integration of the Mistral missile with a Multi-Purpose Combat Vehicle (MPCV). First deliveries of the MPCV were made in November 2013. The MPCV was developed by MBDA in cooperation with Rheinmetall Defence Electronics (RDE) in Germany.

It is understood that the unidentified customer will be receiving MPCV kits for completion and installation of a final assembly in country.

Description

The Mistral-2 is a short-range (up to 6 km), man-portable, fully digital, heat-seeking surface-to-air missile that is capable of intercepting a wide variety of aerial targets, even those having a low or very low InfraRed (IR) signature. It is characterised by an outstanding success rate of 96 per cent from more than 3,000 live firings; a high effectiveness against manoeuvring targets and has demonstrated a high effectiveness against wing aircraft, nap-of-the-earth helicopters and Unmanned Aerial Vehicles (UAVs) as well as cruise missiles. Several land, naval and airborne weapon systems based on the Mistral-2 have been developed.

MPCV with Mistral (MBDA) 1517328

Omani ALBI system with Mistral-2 0059324

The Mistral-2 missile is fully compatible with all existing Mistral 1 firing systems. It uses the very latest technologies in electronics and signal processing, combined with aerodynamic improvements to increase its firing envelope. It is fitted with a new booster, an enhanced IR seeker unit and new fin-and-control surfaces. Apart from increasing its firing envelope these changes also significantly improve its interception capabilities against helicopter and anti-ship missiles.

The maximum speed has been increased to greater than Mach 2.6.

MBDA has developed several land, naval and airborne weapon systems based on the Mistral-2 with a Mistral Coordination Post (MCP). In March 2011, the French Army and MBDA the missile manufacturer chose the Matis MP3 optronic aiming sight offered by Sagem, which is now part of the Safran group, to modernise the French Army's 186 Mistral firing posts. The MP3 is a day and night sight that enhances the Mistral missile's capabilities in several key areas:
Extended detection and identification ranges
Greater operability under different weather conditions
Integration of day and night channels in the same system.

Variants

Mistral 1

ATLAS/Pindad: ATLAS fires the Mistral-2 short-range (up to 6.5 km) SAM capable of intercepting a wide variety of aerial target including those with a low IR signature.

Pindad's vehicle is a four-wheeled light military vehicle with a gross weight of less than four tonnes, measuring 4.5 m in length, 1.6 m in width and 1.8 m in height. It is powered by a four-cylinder diesel engine and equipped with power steering.

Simbad-RC

The Simbad-RC is the latest system in the Simbad family and is a remote-controlled short-range naval air defence system. Prototypes are in production and system qualification is expected in 2014. The system has already been sold by MBDA to an un-named customer with deliveries due to begin in 2015.

The Simbad-RC gyro-stabilised automatic launcher turret holds two Mistral-2 fire-and-forget missiles and can be fitted with a thermal and large field of view camera. The SMU-RC operator terminal located inside the ship can manage two turrets and can be integrated with the ship's combat system, radar and electro-optical sensors.

With a weight of 350 kg the turret has a bearing of ±160° and an elevation –15° to +65° and can be fired in less than five seconds after preparation.

Specifications

	Mistral-2
Dimensions and weights	
Length	
overall	1.86 m (6 ft 1¼ in)
Diameter	
body	0.093 mm (0.00 in)
Weight	
launch	18.7 kg (41 lb)
Performance	
Speed	
max speed	1,711 kt (3,168 km/h; 1,969 mph; 880 m/s) (est.)
Range	
min	0.3 n miles (0.6 km; 0.4 miles)
max	3.5 n miles (6.5 km; 4.0 miles) (est.)
Altitude	
max	3,000 m (9,843 ft)
Ordnance components	
Warhead	3 kg (6.00 lb) HE fragmentation
Fuze	impact
	laser, proximity
Guidance	passive, IR (with cooled multi-cell seeker, proportional navigation)
Propulsion	
type	solid propellant (SNECMA SPS, eject motor)

Status

In production and known to be in service with Estonia, France, Indonesia, Saudi Arabia and the United Arab Emirates (UAE).

Contractor

MBDA (main).
DoDaam (Mistral Simulator).
PT Pindad Persero (Indonesia - Development with MBDA for use on ATLAS system with Pindad Light Military Vehicle).

Gaza and the West Bank

Al-Quds Martyrs Brigade

Type

Man-portable surface-to-air missile system (MANPADS)

Development

A spokesman and leading member of the military wing of the Al-Quds Martyrs Brigade confirmed that the organisation is attempting to develop Surface-to-Air Missiles (SAMs). Late in 2004, it was also confirmed by the then Hamas leader Nizar Rayyan that "Hamas engineers are trying to develop anti-aircraft missiles". The only details as yet known are that Adnan al-Ghul (the Hamas weapon designer - killed by the Israelis during an air strike on 21 October 2004), was working on a new SAM and Anti-Tank Guided Missile (ATGM) system.

In mid-December 2007, reports from Israel suggested that Hizbullah had shot down an Israeli helicopter near the village of Yater. Reports of the weapon used were confusing; some claimed that a missile was used, others a rocket, but Hizbullah insisted that the helicopter was shot down with a new missile known simply as the 'Waad' (Arabic for Promise). Another report dated 20 December 2007, involved the Hamas group which claimed that Hamas had fired SAMs at Israeli aircraft and helicopters which were operating over the Gaza Strip. It is understood that if any or all of these incidents involved locally produced SAMs.

It was reported that several Palestinian Islamic Jihad (PIJ) Al-Quds Brigade operatives were injured in February 2006 while attempting to use a home-made SAM to shoot down an Israeli helicopter.

As of June 2010, verification of the existence of the Waad missile systems is still not available. Other reports concerning the use of man-portable SAMs supplied from China via Iran are still filtering through to the west.

In May 2011, sources close to *IHS Jane's* expressed as doubtful that the Al-Quds Brigade still has the capability to work on development of a home

Strela-2M complete system 0509835

Cutaway training model of the Misagh-1 0059325

made man-portable SAM. With the amount of ammunition that is currently stored and readily available there would be no need to go through the experience of such a programme.

Description

Based on past knowledge of non-government organisation SAM systems, plus the backing from Iran for such organisations, a reverse-engineered missile in the class of the ex-Soviet Strela-2/2M or the Iranian Misagh-1 are the most likely candidates for such a weapon. Reports from Israeli intelligence sources assess that at least 10 Strela-2 (Strela-2M) shoulder-launched anti-aircraft missiles have been smuggled into the Gaza Strip in recent years. It is probable, although not certain, that all or some of these weapons have been exploited and reverse engineered as part of the programme to indigenously develop man-portable SAMs.

Iran has further agreed to supply other Russian-made SAMs in addition to the indigenously produced Misagh to replace those lost during confrontation with Israel in 2006. Included in this offer are:

- Strela-2/2M ex-Soviet man-portable SAM
- Strela-3 ex-Soviet man-portable SAM
- Igla-1E ex-Russian man-portable SAM
- QW-1 ex-Chinese man-portable SAM.

Variants

- Strela-2
- Strela-2M
- Misagh-1

Status

Reported to be in development since March 2006. As of August 2013, nothing further has been detailed about this system.

Contractor

Al-Quds Martyrs Brigade.

India

Igla-1

Type

Man-portable surface-to-air missile system (MANPADs)

Development

According to Indian press reports, the Indian Army and the Indian Electrical and Mechanical Engineers (IEME) have used their own resources to overhaul life-expired, Russian-designed Surface-to-Air Missiles (SAMs). Systems that have been reworked and most probably upgraded to reflect modern technology include the Igla-1 (SA-16 Gimlet) and SA-08 (Osa-AK).

Description

Guidance

No change is expected in the guidance, which is passive infraRed (IR), fire-and-forget.

Test Equipment

The Laser Science and Technology Centre (LASTEC) based in Delhi, has developed a portable opto-electronic test equipment for man-portable SAMs whose seeker technology operates in the 3-5 µm band. This test equipment, simply known as IRGM, was developed under Project Pareekshak, is battery operated and simulates the target's IR signature. Specifications are:

Spectral Output: 3-5 µm band

Power Density: 500 nW/cm^2 and 5 µW/cm^2.

Specifications

	Igla-1
Dimensions and weights	
Length	
overall	1.673 m (5 ft 5¾ in)
Diameter	
body	72 mm (2.83 in)
Weight	
launch	10.8 kg (23 lb)
Performance	
Speed	
max speed	1,108 kt (2,052 km/h; 1,275 mph; 570 m/s)
Range	
min	0.3 n miles (0.5 km; 0.3 miles)
max	2.8 n miles (5.2 km; 3.2 miles)
Altitude	
min	10 m (33 ft)
max	3,500 m (11,483 ft)
Ordnance components	
Warhead	1.27 kg (2.00 lb) HE fragmentation
Fuze	impact
Guidance	passive, IR (fire-and-forget 3.5-5.0 µm band)
Propulsion	
type	solid propellant (eject motor)
	solid propellant (sustain stage)

Status

Reports emanating from internal sources in India in April 2008, have suggested that the Indian Army has launched global tenders worth INR3,800 crore to acquire a new range of quick reaction SAMs to provide air cover to its rapid formations.

The new missiles seek to replace the army's ageing shoulder-fired Strela-2/2M missiles as well as the Soviet acquired ZU-23 rapid firing guns.

It is assumed that the Igla-1 missiles that are overhauled with a life increase will fit this bill until other systems can be procured.

Contractor

NPO MBDB.
Indian Electrical and Mechanical Engineers (IEME).
Laser Science and Technology Centre (LASTEC).

International

European Stinger-RMP

Type

Man-portable surface-to-air missile system (MANPADs)

Development

In 1983, the Federal Republic of Germany and the US signed an MoU on the production of the (Raytheon Electronic Systems) Stinger man-portable missile system. The implementation agreement for this was signed in 1986, to allow Germany series production of the Stinger-Reprogrammable Microprocessor Mod 3 (RMP) (FIM-92D) weapon and, with Germany as lead country, in cooperation with any interested European NATO country.

The latter subsequently became the subject of a European MoU, signed by Germany, Greece, Netherlands and Turkey. The cost sharing, in accordance with the respective hardware take-off was:

- Turkey - 40 per cent (launch and flight motors at Roketsan)
- Germany - 36 per cent
- Netherlands - 15 per cent
- Greece - 9 per cent.

German soldier using Stinger man-portable SAM in field 0509397

The European prime contractor for the system was LFK-Lenkflugkörpersysteme GmbH, a subsidiary of EADS Deutschland. On the government side, the collaboration was organised under the Stinger Project Group (SPG) with the German Federal Agency for Military Technology and Procurement (BWB) acting as the responsible authority on the customer side. More than 20 Greek, Turkish, Dutch and German companies were involved in the programme, which continued until 2002.

Description

The first pilot production batch was completed in 1992 to 1993, with full production starting in 1993. The first production systems were delivered to the four participating countries in October 1993. A total of 12,000 Stinger-RMP missiles with a smaller number of grip-stocks plus training and support equipment were manufactured by the four countries during 2000. The country totals were:

- Germany - 4,500
- Turkey - 4,800
- Greece - 1,100
- Netherlands - 1,600.

During September 1997, production was switched to the Stinger Block 1 (FIM-92E) version with the first deliveries made in late 1998.

The IFF antenna and interrogator normally associated with the Stinger system are not part of the licence production.

A full description of the Raytheon Electronic Systems Stinger missile system is given later in this section.

Status

In production. In service with Germany, Greece, Netherlands and Turkey. Greece and Turkey are considering placing additional orders. Italy plans to procure European source missiles as well. No further information was available on the Italian purchase at the time of updating this file.

Contractor

EADS Deutschland GmbH (prime European contractor).

Iran

Misagh-1

Type

Man-portable surface-to-air missile system (MANPADs)

Development

The Misagh-1 was developed by the Shahid Shah Abadi Industrial Complex in Tehran that is now part of the Aerospace Industrial Organisation of Iran (some western sources report that the development is by the Shahid Kazemi Industrial Complex, also in Tehran, with only the manufacture being undertaken by the Shahid Shah Abadi Industrial Complex). The Misagh-1 is an all-aspect, passive infrared homing, Man-portable surface-to-air missile system (MANPADs). This system is comprised of; the missile in its disposable launch tube, a grip-stock (firing unit) and a replaceable battery/coolant unit which is mounted towards the front of the launcher all of which is contained in its transportable packing case.

Description

In appearance, the weapon looks very similar to the China National Precision Machinery Import and Export Corporation (CNPMIEC) QW-1 Vanguard (see QW-1 Vanguard entry, under China, in this section).

The Misagh-1 is a second-generation missile that has a 1.42 kg HE warhead and probably an impact type fuzing system, although a proximity type would be more logical. Its length is 1.477 m, diameter 0.071 m and it has a launch weight of 10.86 kg. The complete system weighs 16.9 kg.

At launch the missile is boosted out of its tube by a booster motor at the rear of the weapon which drops away to allow the sustainer rocket motor to fire. The missile is then accelerated to a maximum speed of 600 m/s. Four stabilisation fins are fitted at the missile rear with two canard surfaces located to the rear of the seeker head.

Variants

Other launcher configurations such as twin-round systems, shipboard mounts and vehicle mounts using the Misagh-1 are believed to be in development.

Cutaway training model of the Misagh-1 0059325

A Misagh-2 variant (probably based on the Chinese QW-2) has been developed as a follow-on to the Misagh-1.

During an interview with the then Defence Minister for Iran, mention was made of the Misagh-3 being in development for the Iranian Navy.

Specifications

	Misagh-1
Dimensions and weights	
Length	
overall	1.477 m (4 ft 10¼ in)
Diameter	
body	71 mm (2.80 in)
Weight	
launch	10.86 kg (23 lb)
Performance	
Speed	
max speed	1,166 kt (2,160 km/h; 1,342 mph; 600 m/s)
Range	
min	0.3 n miles (0.5 km; 0.3 miles)
max	2.7 n miles (5 km; 3.1 miles)
Altitude	
min	30 m (98 ft)
max	4,000 m (13,123 ft)
Ordnance components	
Warhead	1.42 kg (3.00 lb) HE fragmentation
Fuze	impact
Propulsion	
type	solid propellant
	solid propellant (sustain stage)

Status

The Misagh-2 is in production, the Misagh-1 is now reported as production complete. The Misagh-2 is in service with the Iranian armed forces whereas the Misagh-1 has been reported by US officials as the missile that is being smuggled into Iraq for use by insurgent groups to down coalition forces fixed- and rotary-wing assets. Both systems have at some stage been offered for export.

Contractor

Shahad Shah Abadi Industrial Complex.

Misagh-2

Type

Man-portable surface-to-air missile system (MANPADs)

Development

The Misagh-2 was developed as a follow-on to the original Misagh-1. Misagh-2, also known as the Mithaq-2 (probably a Western misspelling) is an advanced version of the Misagh-1. This new missile, like its predecessor, is a very short-range passive infrared-guided missile developed/produced by the Iranian Defence Ministry's Shahid Kazemi Industrial Complex in Tehran.

On 6 February 2006, the then Defence Minister General Mustafa Mohammed-Najjar, inaugurated the production line for the Misagh-2 and commented that the system is capable of defeating IRCM electronic warfare systems. Given its similar appearance to the Chinese QW-2 (itself reportedly a copy of the Russian Igla family of missiles) and according to reports that the Misagh-1 is based on the Chinese QW-1, this would indicate that the Misagh-2 is probably in the same class of weapon as the Chinese QW-2 (Russian Igla). However, in early February 2013, other evidence gained from a seized cargo of weapons shipped from Iran via Yemen indicated that the Misagh-2 could be based upon the QW-1M. No further details have yet been released by Iran, however, it is expected that the speed and range of the missile will have been increased over the Misagh-1.

In September 2007, the Iranian Ministry of Defence confirmed the production of simulators for the Misagh-2 man-portable SAM. Iranian news reports identified this as one of the 70 types of indigenous simulators in use by the Iranian Armed Forces, including the Revolutionary Guard Corps and the Basij militia.

Description

The Misagh-2 is reported to be a license production of the Chinese QW-2 series of man-portable surface-to-air missile systems.

Variants

Misagh-1

Specifications

	Misagh-2
Dimensions and weights	
Diameter	
body	71 mm (2.80 in)
Weight	
launch	12.74 kg (28 lb)

	Misagh-2
Performance	
Speed	
max speed	1,361 kt (2,520 km/h; 1,566 mph; 700 m/s) (est.)
Range	
max	2.7 n miles (5 km; 3.1 miles)
Altitude	
max	3,500 m (11,483 ft)
Ordnance components	
Warhead	1.42 kg (3.00 lb) HE
Guidance	active, IR (proportional navigation)
Propulsion	
type	solid propellant (sustain stage)

Status

The Misagh-2 is in production, the Misagh-1 is now reported as production complete. The Misagh-2 is in service with the Iranian Armed Forces whereas the Misagh-1 has been reported by US officials as the missile that was being smuggled into Iraq for use by insurgent groups to down fixed- and rotary-wing targets. Both systems have at some stage been offered for export.

Contractor

Shahad Shah Abadi Industrial Complex.

Israel

Red Sky-1

Type

Man-portable surface-to-air missile system (MANPADs)

Development

Israel Military Industries (IMI) have developed the Red Sky Very-Short-Range Air-Defence (VSHORAD) system that is based on a modular design, which enables the system to interface with different types of man portable Surface-to-Air Missiles (SAMs) with minimal adaptation changes. Development of the system was in response to customer demands and the need for VSHORAD including command-and-control on the battlefield. At present, the system has been developed around the Russian Strela and Igla missile with two rounds mounted on the launching tripod. It is possible however to mount other man-portable SAMs without any major changes to the system.

Description

The Red Sky provides automatic simultaneous multiple target detection and tracking for the protection against low-flying aircraft. The system comprises:
1. Infra-Red Search (IRS) scanner; mounted on a two-axis gimballed tripod and has a Wide Field of View (WFOV) Forward-Looking Infra-Red (FLIR) camera, enabling continuous passive scan and day and night electro-optical coverage of a WFOV. The FLIR camera also produces real-time video images that are processed by the C3 unit's advanced target detection algorithm. The horizontal field of regard is ±160° whilst in the vertical the field of regard is ±30°. The IRS sectors in the horizontal are 30° to 180° and in the vertical 10° and 20°. The IRS FoV is 10°. The scanning rate of the IRS is 40°/s.
2. Track-and-Launch (TL) unit; supports two missiles, laser range-finder and high resolution, dual WFoV of 5° and NFoV 1° FLIR camera. The TL also tracks the target and provides target coordinates. Upon target recognition and user decision a missile is launched from the TL. The field of regard in the horizontal is ±160° whilst in the vertical –10° to +70° is standard.
3. Command, Control and Communication (C3) is a PC-based system for processing, detection and fire-control, processing incoming video signals, and automatically detecting and engaging aerial targets. An optional video recorder enables the C3 unit to be used for post engagement analysis and/or training purposes.
Red Sky provides the following features and capabilities:

- Autonomous search, detect, identify and fire capability
- Compatibility with most fire-and-forget SAMs
- Passive Day/Night Search While Track
- Multiple targets automatic detection
- Automatic calculation of firing envelope
- Short reaction time
- Minimum signature
- Light weight
- User friendly man machine interface
- Built-in simulator and recording
- Highly cost-effective solution.

Red Sky has been created for single-soldier operation, while only two additional persons are required to carry and set up the equipment. The system's modularity enables helicopter/naval/light ground vehicle as well as manual transportation. The overall weight of the system (three units) is 98 kg. The power supply required to operate at full efficiency is 110/220 V AC or 28 V DC.

Missile

Igla and Strela but is compatible with any fire-and-forget type missiles.

Variants

Red Sky-2 is a modular integrated sensor and man-portable system follow on to the Red Sky-1.

Specifications

Red Sky-1

Detection range	n/avail
Scanner FOR:	
(horizontal)	160°
(vertical)	30°
Scanner FoV:	10°
Scanning Rate:	40°/s
Tracker FOR:	
(horizontal)	160°
(vertical)	–10° to +70°
Tracker WFOV:	
(WFoV)	5°
(NFoV)	1°
Battery:	110/220 V AC or 28 V DC
System weight:	98 kg

Status

Development is complete, it is understood that the Red Sky-2 has superseded the Red Sky-1 variant.

Contractor

Israel Military Industries (IMI) Ltd, Advanced Systems Division.

Red Sky-2

Type

Man-portable surface-to-air missile system (MANPADs)

Development

This is a Short-Range Air-Defence (SHORAD) modular system, similar to the basic Red Sky. Improvements have been made with emphasis on the Field Of Regard (FOR) detection system in the vertical.

The system is designed to be effective against low flying surface skimming targets and also has been designed to integrate with different types of missile including the US Stinger and Russian Strela and Igla families, Polish Grom furthermore, the system can handle the Chinese QW systems.

Description

The Red Sky-2 is a fast deployment, cost effective mobile and modular system for point Defence of strategic assets, essential air corridors and rapid deployment forces. The system provides automatic simultaneous multiple target detection and tracking for the protection against low-flying aircraft. The system comprises:
1. The Infra-Red Search (IRS) unit is mounted on a two-axis gimballed tripod and has a Wide Field of view (WFoV) Forward-Looking Infra-Red (FLIR) camera, enabling continuous passive scan and day-and-night electro-optical coverage of a wide Field of View (FoV). The FLIR camera also produces real-time video images that are processed by the C3 unit's advanced target detection algorithm. The horizontal field of regard is ±160° whilst in the vertical, the field of regard is ±30°. The IRS sectors in the horizontal are 30° to 180° and in the vertical 10° and 20°. The IRS FOV is 10°. The scanning rate of the IRS is 40°/s.
2. The Track-and-Launch (TL) unit supports the customers two missiles, laser range-finder and high resolution, dual WFOV of 5° and NFOV 1° FLIR camera. The TL also tracks the target and provides target coordinates and identification. Upon target recognition and user decision, a missile is launched from the TL. The field of regard in the horizontal is ±160° whilst in the vertical –10° to +70° is standard.
3. The Command, Control and Communication (C3) is a PC-based system for processing, detection and fire-control, processing incoming video signals while automatically detecting and engaging aerial targets. An optional video recorder enables the C3 unit to be used for post engagement analysis and/or training purposes.
Red Sky-2 provides the following features and capabilities:

- Autonomous search, detect, identify and fire capability
- Compatibility with most fire-and-forget SAMs
- Passive Day/Night Search While Track
- Multiple targets automatic detection
- Automatic calculation of firing envelope
- Short reaction time
- Minimum signature
- Light weight
- User friendly man machine interface
- Built-in simulator and recording
- Highly cost-effective solution.

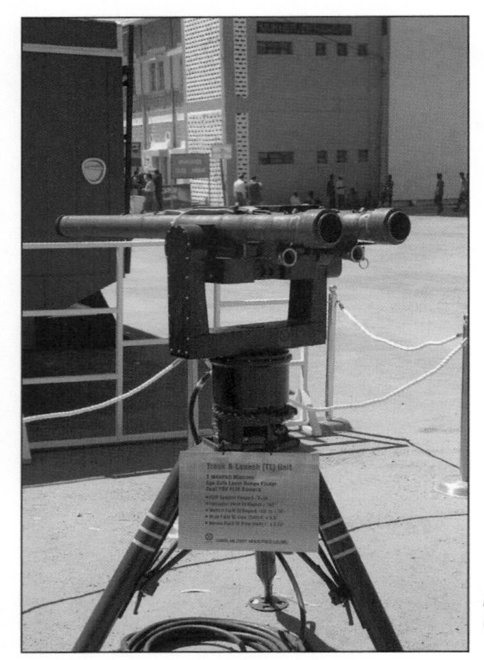

Red Sky-2
(IHS/Patrick Allen)
1136861

Infra-Red Search (IRS) Unit (IHS/Patrick Allen)
1136862

Red Sky-2 has been created for single-soldier operation, while only two additional persons are required to carry and set up the equipment. The system's modularity enables helicopter/naval/light ground vehicle as well as manual transportation. The overall weight of the system (three units) is 98 kg. The power supply required to operate at full efficiency is 110/220 V AC or 28 V DC.

Variants

Red Sky-1

More details of this variant can be found in a separate entry.

Specifications

Red Sky-2

Detection range:	15 km
IRS Scanner FOR:	
(horizontal FOR)	360°
(vertical FOR)	±25°
(horizontal scanning sectors)	30° to 180°
(vertical scanning sectors)	11° and 22°
TL field of regard (FOR):	–10° to +70°
Scanner FOV:	10°
Scanning rate:	36°/s
Tracker FOR:	
(horizontal)	160°
(vertical)	–10° to +70°
Tracker WFOV:	
(WFoV)	5°
(NFoV)	1°
Battery:	110/220 V AC or 28 V DC
Weight:	98 kg (system)

Status

Development is complete, it is believed that the system has found several export customers but this information has not yet been released.

Contractor

Israel Military Industries (IMI) Ltd, Advanced Systems Division.
Controp Precision Technology, for IRS and EO systems.

Japan

Kin-SAM Type 91

Type

Man-portable surface-to-air missile system (MANPADs)

Development

The Japanese Defence Agency (JDA) has been funding advanced Infrared (IR) seeker technology at its Technical Research and Development Institute (TRDI) since 1979.

In 1982, the Japanese Self-Defence Force started to look for an indigenous enhanced third-generation man-portable system to succeed/complement the American Raytheon Electronic Systems Stinger systems, used by the Ground Self-Defence Force (GSDF) as its standard divisional Short-Range Air-Defence (SHORAD) weapon.

Known initially as the Keiko (or SAM-X/91-shiki Keitai Chitaikû Yûdôdan), it is also known within the Japanese Self Defence Forces (JSDF) as Hand Arrow. Development was initially deferred until 1987 when the JDA funded the final development phase. In Fiscal Year-88 (FY88), the programme was transferred from the TRDI to the Toshiba Corporation for engineering development.

This was completed in FY90 and the system entered low-level production in 1991 as the Type 91 Kin-SAM portable surface-to-air missile system. First deployment was in FY94 with the Japanese Ground Self-Defence Force.

Description

The missile is of standard configuration, using separate current technology solid propellant booster and sustainer rocket motors.

The homing system, however, is of a dual-mode, focal plane, full imaging configuration type that uses both Imaging-IR (IIR) and optical regions for guidance. These are believed to be the 0.4 to 0.7 micron visible and 3.5 to 5.2 micron IR regions respectively. All the operator has to do is to lock the head on to the target whereupon a high-resolution Charge Coupled Device (CCD) 'memorises' its appearance and causes the weapon to follow an all-aspects attack flight profile which is extremely resistant to any defensive countermeasures that may be employed. This guidance technology also has considerable potential for resistance to laser countermeasures, although it is not known to what extent, if any, this capability is exploited.

Missile length is approximately 1.43 m, diameter 0.080 m and launch weight 11.5 kg. Minimum engagement range is said to be effectively some 300 m and maximum range 5,000 m.

Variants

Type 93 Kin-SAM

In FY90 the JDA contracted Toshiba to develop a mobile version of the Type 91 Kin-SAM system based on the Kohkidohsha (4 × 4) 1,500 kg high-mobility vehicle chassis and fitted with an eight-round boxed pedestal launcher mount together with associated IR and optical fire-control sighting systems. The designation is the Type 93 surface-to-air missile launcher (Type 93 Kin-SAM), more details of which can be found in the entry in Self-propelled SAM section.

Type 91 Kai

In 2007, the Type 91 Kai became operational with the Japanese Defence Forces, this is an improved version of the original Type 91 that has both day and night capability. It is understood that the Type 91 Kai will be used on the Type 93 mobile variant.

Specifications

	Type 91
Dimensions and weights	
Length	
overall	1.43 m (4 ft 8¼ in)
Diameter	
body	80 mm (3.15 in)
Weight	
launch	11.5 kg (25 lb)
Performance	
Speed	
cruise	1,258 kt (2,329 km/h; 1,447 mph; 647 m/s)
Range	
min	0.2 n miles (0.3 km; 0.2 miles)
max	2.7 n miles (5 km; 3.1 miles)
Ordnance components	
Warhead	HE fragmentation

	Type 91
Fuze	proximity, impact
Guidance	IIR (two channels, lock-before-launch, 0.4 to 0.7 and 3.5 to 5.2 μm)
Propulsion	
type	solid propellant (sustain stage)

Status
Low-rate production (the system is procured in sets, which are sufficient launchers and associated equipment to furnish a platoon - FY91 (13), FY92 (25), FY93 (25), FY94 (data not available), FY95 (data not available), FY96 (18), FY97 onwards (data not available). Believed to be in service with the Japanese Ground Self-Defence Force but may also be held in reserve. Not offered for export.

Contractor
Toshiba Corporation.

Korea (North)

North Korean man-portable surface-to-air missile systems

Type
Man-portable surface-to-air missile system (MANPADs)

Development
The first man-portable Surface-to-Air (SAM) systems used in the development of the North Korean MANPADs were apparently a small number of Strela-2 (NATO SA-7 Grail), obtained from Egypt in 1974 as part of a technology gift for the help that Korea had given Egypt during the 1973 Yom Kippur War. Subsequently, North Korea reverse engineered the system and produced its own local copy, known locally as the Hwasung Chong, from the late 1970s onwards.

The Strela-2 clone was followed into service during the mid-1980s by quantities of the People's Republic of China (PRC) HN-5A system. This was then licence-built by North Korea from the late 1980s. Also, in the late 1980s, the Russians provided both the Strela-3 (NATO SA-14 Gremlin) and Igla-1 (NATO SA-16 Gimlet) systems. Both of which were subsequently produced under license. The latter is believed to have entered production in North Korean weapons plants around the 1993 time frame. Small quantities of both these license built North Korean weapons have been exported, including batches of both types to Cuba.

The latest North Korean man-portable SAM project, which started in the mid-1990s, following the clandestine acquisition of several American FIM-92A Stinger systems, is the reverse engineering of a Stinger clone. By the latter part of 1999, US intelligence was indicating that a small quantity of Stinger look-a-like weapons were in service with the North Korean Armed Forces. It is also likely, now both Pakistan and Iran are producing missiles which bear a close resemblance to the Chinese QW-1 Vanguard man-portable SAM, that North Korea will probably have its own production plans for this weapon.

As of July 2012, there has been no confirmation that North Korean Factories are reverse engineering or cloning the Russian Igla-S.

Status
Strela-2 copy - production complete. In service with North Korean Armed Forces.
HN-5A copy - production complete. In service with North Korean Armed Forces.
Strela-3 copy - in production. In service with North Korean Armed Forces and Cuba.
Igla - in production. In service with North Korean Armed Forces.
Igla-1 copy - in production. In service with North Korean Armed Forces and Cuba.
Stinger copy - in limited production. In service with North Korean Armed Forces.

Contractor
North Korean State Arsenals.

Korea (South)

Chiron (Singung)

Type
Man-portable surface-to-air missile system (MANPADs)

Development
Development of the Chiron (Singung) (also reported by some western sources as Shingung or Shin-Gung or Shin-Kung) man-portable surface-to-air missiles system (MANPADS SAM) by the NEX1 Future Company Ltd. is believed to have commenced during 1995 when the system was then known as the KP-SAM. The system requirements included the ability to take on targets such as:
- fixed-wing aircraft
- rotary-wing helicopters
- UAVs
- cruise missiles.

Development of the missile system began in 1998 and continued through 2005. Series production was expected to commence during 2004, with deployment and IOC during the same year, however, extended trials meant that the system was not ready for series production and deployment before September 2005. Trials during the research and development phase were carried out using both fixed and rotary wing targets.

The programme was originally aimed at developing a new man-portable SAM for the protection of troops in the foreword area.

During discussions with South Korean engineers, they admitted that the seeker technology was Russian (Leningrad Optical Mechanical Association - LOMO) but that the control section, warhead and motor were of a South Korean design.

Description
The Singung is a shoulder-fired MANPAD system that is reportedly cheaper to produce than the Stinger and Mistral supplied by the US and France respectively. It is estimated that at the cost of USD153,000 per round, the South Korean government will save themselves millions over a period of time.

The Chiron (Singung) is a two-man, fire-and-forget system that can also be fired in the same way as the Mistral, which is mounted on a tripod. The system boasts the following:
- man-portable SAM
- operational in day and night time conditions
- night vision sight and IFF
- high terminal velocity with dual thrust propulsion
- able to cope with fixed wing aircraft, helicopters, UAV and cruise missiles
- fire-and-forget missile
- Infra-Red Counter-Counter Measures (IRCCM) with a two colour seeker IR and UV
- can be linked with alerting system by radio.

Although the Chiron (Singung) SAM is a man-portable system it would normally be deployed with other elements with target information such as height, range and speed being provided by the truck-mounted low-altitude surveillance radar TPS-830KE. This system would typically be deployed at platoon level.

Other platforms that can be utilised as launchers for the system include:
- vehicular
- shipborne
- helicopters.

Simulator
Nex1 Future has developed a simulator to train man-portable air-defence system operators. The Chiron (Singung) short-range portable SAM simulator was developed under the KP-SAM programme. The simulator is based on hardware from the operational weapon, but incorporates a monitor used to display the simulated engagement to the trainee.

Portable surface-to-air missile Chiron (Singung) E (Nex1 Future Co Ltd) 0526700

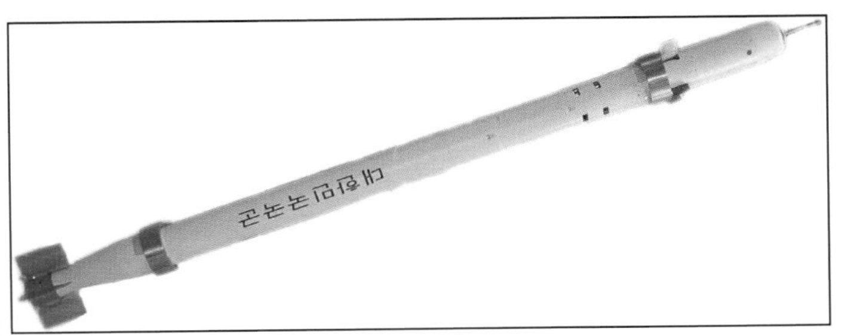

Chiron missile (Robert Karniol) 1116764

Variants

During its trials period, the system was sometimes dubbed the KP-SAM.

Specifications

	Chiron
Dimensions and weights	
Length	
overall	1.68 m (5 ft 6¼ in)
Diameter	
body	800 mm (31.50 in)
Weight	
launch	14 kg (30 lb) (est.)
Performance	
Speed	
cruise	1,361 kt (2,520 km/h; 1,566 mph; 700 m/s) (est.)
cruise Mach number	2.1 (est.)
Range	
max	3.8 n miles (7 km; 4.3 miles)
Altitude	
max	3,500 m (11,483 ft)
Ordnance components	
Warhead	2.5 kg (5.00 lb) HE blast fragmentation
Fuze	impact, proximity (at 1.5 m)
Guidance	IR (2-colour)
Propulsion	
type	solid propellant (dual pulse)

Status

Research and development are complete; the system went into series production in September 2005, with deployment in March 2006 although other sources have quoted entry into service with the ROK Army 2005.

There has been limited information that India plans to acquire the missile system from Nex1 sometime prior to 2014.

Contractor

Agency for Defence and Development, South Korea.
Nex1 Future Co Ltd., South Korea.

Pakistan

Anza Mk II

Type

Man-portable surface-to-air missile system (MANPADs)

Development

The Institute of Industrial Control Systems (IICS) (formerly known as the Dr A. Q. Khan Research Laboratories) at Kahuta, Pakistan has developed the Anza (Lance) Mk II passive InfraRed (IR) homing man-portable surface-to-air missile system, for the Pakistan Armed Forces. This system is a further development of the Anza Mk I and Anza Mk IB.

Description

The Anza Mk II is comparable, in the technical specifications provided, with other types of man-portable SAM systems of its era (second- and third-generation). In general appearance and performance it resembles the Chinese QW-1 Vanguard system, but with various minor differences between the two.

Anza Mk II low-altitude SAM system 0132699

Anza Mk II low-altitude SAM system deployed in the kneeling firing position 0509670

The system has a high sustained velocity with high manoeuvrability and an all aspect engagement capability with effective engagement limits stated to be 500 to 5,000 m. The passive IR seeker is InSb (cooled) has been considerably improved over the Anza Mk I version to provide an all-aspects engagement capability.

Transition time from movement to ready for operation is less than 10 seconds. In flight the missile has a self destruction time of 14-18 seconds. The system has been used in the border skirmishes with India during the late 1990s.

Training Simulator for Anza Mk-II (ATS-II)

The ATS-II training simulator has been designed, developed and produced for the operators of Anza Mk-II man-portable SAM system. The simulator can effectively impart training to the operators in carrying out a sequence of steps that lead to accurate launching of the weapon system. The ATS-II consists of:

- Four Mk-II training missiles
- Firing units
- Simulated ground batteries
- PC based control, monitoring and scoring unit operated by an instructor
- Cable interconnections
- Target simulator, consisting of an electric bulb (IR source) moving along an overhead wire. This can simulate speed, course and IR characteristics of all typical targets.

The training missiles are connected to the control, monitoring and scoring unit by cables. The operators aim, capture and track a simulated target which emits IR radiation. The simulator monitors and records separately, all operational steps carried out by each trainee operator and assesses launch accuracy. The recorded data is displayed on the screen and can be printed out for subsequent analysis.

Specifications (Technical)

Number of gunners that can be trained at one time:	4
Target visual angular velocity:	≤15°/s
Target tracking mode:	Tail-on/head-on
Target acquisition course:	left or right
Types of targets that can be simulated:	Jet aircraft, propeller aircraft, helicopter
Height of target:	
(minimum)	300 m
(maximum)	5,000 m
Power source:	220 V ±10%, 50 Hz single phase

Variants

Anza Mk III

Pakistan is developing under license, a modified version of the Chinese Qian Wei-2 man portable SAM, known locally as the Anza Mk III.

Specifications

	Anza Mk II
Dimensions and weights	
Length	
overall	1.447 m (4 ft 9 in)
Diameter	
body	72 mm (2.83 in)
Weight	
launch	10.5 kg (23 lb)
Performance	
Speed	
cruise	1,166 kt (2,160 km/h; 1,342 mph; 600 m/s)
impact velocity	583 kt (1,080 km/h; 671 mph; 300 m/s)
Range	
min	0.3 n miles (0.5 km; 0.3 miles)
max	2.7 n miles (5 km; 3.1 miles)
Altitude	
min	30 m (98 ft)
max	4,000 m (13,123 ft)
Ordnance components	
Warhead	0.55 kg (1.00 lb) HE fragmentation
Fuze	impact (electromechanical)
Guidance	passive, IR (cooled InSb, proportional and terminal adaptive guidance)
Propulsion	
type	solid propellant (sustain stage)

Status

In production (full-scale production started 1994). In service with the Pakistan Army since September 1994. Also used by the Pakistan Air Force.

Malaysia received the Anza Mk II for use with the Malaysian Army. This is part of an order placed with the Pakistan government for the Baktar-Shikan anti-tank missiles and Anza Mk II.

It was reported in January 2008, that Pakistan has developed the Anza Mk III; this weapon is based on the Chinese Qian Wei-2 man portable SAM, full details can be found in the Anza Mk III file.

Contractor

Institute of Industrial Control Systems (IICS).

Anza Mk III

Type

Man-portable surface-to-air missile system (MANPADs)

Development

Pakistan is producing under license the Anza Mk III low-altitude Surface-to-Air Missile (SAM) system based on the Chinese Qian Wei-2 (QW-2) man portable SAM. Chinese technology on thermal-imaging and infrared night-vision capabilities have been transferred to the IICS and will be used in the Pakistan further development of the system. Local modifications to the license production models are likely to have been added to suit the requirements of the Pakistan Armed Forces (PAF) and also local conditions for which the system is intended for deployment. It is not known yet, whether either of the models will be available for export. However if one is to be exported, it would most likely be the one of the license production version.

Description

The Anza Mk III can be used either from a mobile or fixed launcher and/or man-portable. The system was primarily designed as a short-range, man-portable, shoulder-fired missile that has drawn its ongoing development from an original idea developed in Russia known as the Strela-2. There is naturally some controversy between the US with the FIM-43 Red Eye and the Russian Strela as to who was the first to develop these weapons. As it turns out neither is correct - the basic concept was first worked on by the Germans during the Second World War.

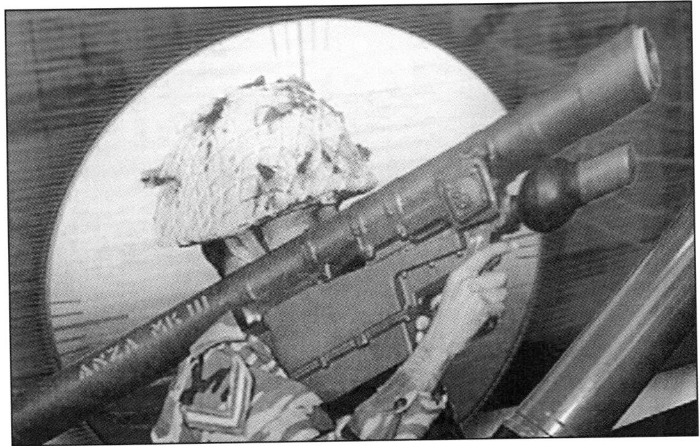

Anza Mk III publicity photograph (IICS) 1132639

The Russian Strela family of missiles has been sold extensively throughout the world with copies and license production leading to many variants on the theme. The Strela was finally replaced in Russia with the Igla system and again like its predecessor, it has been copied and licensed produced by such countries as:

- Bulgaria
- China
- North Korea
- Poland
- Singapore
- Vietnam.

Launcher/Gripstock (firing mechanism)

The Launcher reportedly bears a striking resemblance to the Russian 9P322-1 used with the 9K313-1 Igla-1 (NATO SA-16 Gimlet) and is expected to have the capability to load either the Anza Mk III or the Igla-1. If this is so then, the Pakistan missile will also have to employ identical electrical connections for the Igla and for both the missile and the Gripstock. Like the launcher, the Gripstock appears to be based on the Russian unit 9P519 with the 9B238 battery to provide the power for launch and flight. Again, if this is so, the system will almost certainly contain infrared countermeasures, Identification Friend or Foe (IFF) and probably links to a passive or active radar/targeting source.

Variants

Anza Mk I No longer in production.
Anza Mk II The current system in used with the Pakistan Armed Forces (PAF).

Specifications

Anza Mk III/QW-2

Type:	shoulder-launched man portable surface-to-air missile
Missile dimensions:	
(length)	1.59 m
(diameter)	n/avail
(weight)	11.32 kg
Weight including launcher:	18 kg
Propulsion:	solid propellant boost and sustain
Guidance:	dual-band infrared with proportional and infrared terminal homing
Warhead:	
(weight)	0.55 kg
(type)	HE shaped charge (assessed). Some reports suggest HE fragmentation with contact and graze fusing
Average velocity:	600 m/s
Range:	
(maximum)	6,000 m
(minimum)	500 m
Altitude:	
(maximum)	3,500 m
(minimum)	10 m
Launcher:	fixed, mobile or shoulder-launched

Status

The Anza Mk III went into series production for the Pakistan Army reportedly in 2006.

Contractor

Primary contractor, Institute of Industrial Control Systems (IICS).

Poland

Anti-Helicopter Mine

Type

Man-portable anti-helicopter mine

Development

Research and development of the anti-helicopter mine was conducted under the specific targeted Project Number 4867 and was partly funded by the Ministry of Science and Information Society Technologies. The system is a requirement from the Polish Armed Forces for defence against low flying helicopters both manned and unmanned for the defence of areas of high value.

Description

The system is designed to engage rotary type targets (both manned and unmanned) that come within a range of up to 150 m at speeds up to 300 km/h. The mine is remotely armed and disarmed by an optical signal therefore it may be used many times over. It can be partially buried which facilitates camouflage and significantly reduces the dispersion area.

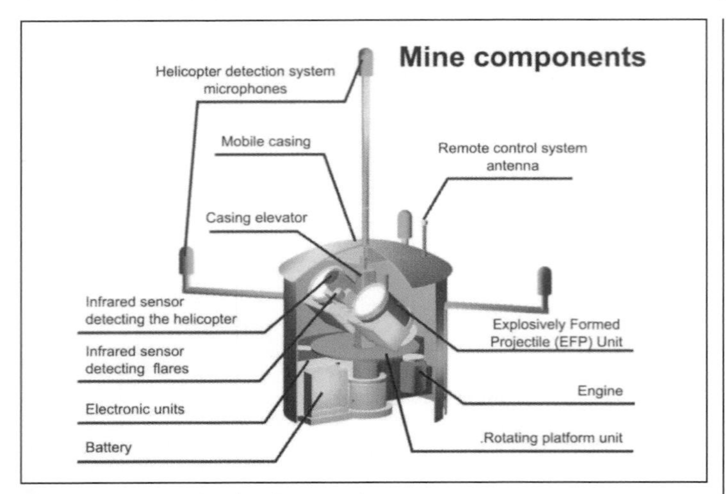

Main components of anti-helicopter mine 1335224

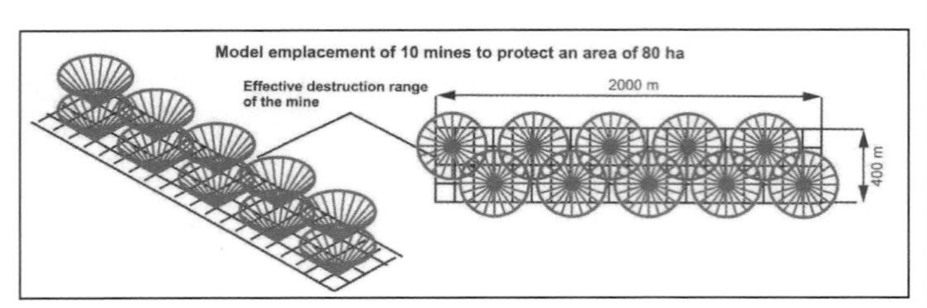

Anti-helicopter defended area layout 1335225

A smart targeting and command centre, using digital analysis, detects the helicopter and then decides whether or not to fire the two Explosively Formed Projectiles (EFP) at the target.

As the mine is only equipped with passive sensors, acoustic and Infra-Red (IR), it is therefore undetectable from the helicopter target. This arrangement of sensors also provide a low probability of false classification and identification of the target.

When the acoustic sensors detect a helicopter, the casing on the mine is lifted. The lifting activates the fuze. The EFPs are launched when the target is acquired by the IR sensor and comes within the mine range. The mine is provided with a flare detector and is resistant to these types of disturbances.

The mines can be employed individually or in patterns sown up to and including 10 set at 45° to each other in order to protect an extended area for example: 2,000 m × 400 m.

Variants

None known.

Specifications

Anti-Helicopter Mine

Type:	man-portable anti-helicopter mine
Diameter:	100 mm
Weight:	22 kg
Maximum effective range:	150 m
Minimum range:	10 m
Maximum velocity:	2,600 m/s
Maximum speed of target:	300 km/h
Number of projectiles:	2
Armour piercing capacity:	35 mm to 60 mm
Number of passive sensors:	5
Mine operating time:	35 min
Target acoustic detection:	800 m to 1,000 m
Target recognition:	≤500 m
Angle of rotation in azimuth:	360°
Angular resolution:	1°
Temperature range:	–40°C to +70°C

Status

Development complete, offered for sale on the open arms market. As of April 2012 no known customers have been identified for this system.

Contractor

Wojskowe Zakłady Uzbrojenia S.A. (WZU).

Grom

Type

Man-portable surface-to-air missile system (MANPADs)

Development

PZR Grom, (Przeciwiotniczy Zestaw Rakietowy - anti-air rocket propelled system) was designed to target low-flying aircraft such as helicopters. Development of the initial version of the system, the Grom-1 'Thunderbolt', began in 1992 after Poland left the old Soviet bloc and license was declined for the production of the Russian Federation's Igla system.

The first examples entered Polish Army service in 1995.

There is much unconfirmed speculation that the Polish intelligence service managed to purchase the Igla design plans and thus the design and development of the Grom was kept very short.

From 1996, the weapon was deployed with selected units of the Polish Army, such as the 6th Air Assault Brigade and the 10th Armoured Cavalry Brigade. The Grom-1 was in production until 1999.

In 1999, final tests were conducted of the definitive production Grom-2 system. Series production of the Grom-2 began in 2000 and is continuing today.

Grom is a short-range surface-to-air anti-aircraft system, capable of destroying air targets at low and middle altitudes, using homing guidance missiles. It is suited for inter-operation in an automated air-defence command-and-control system, from which the data of targets to be engaged are acquired via digital links.

In December 2007, RADWAR signed a contract to deliver four sets of Blenda target acquisition systems to the Naval anti-aircraft battalions as part of an upgrade that is to include the S-60 gun and the Grom-2 man portable SAM.

Description

Grom-1

The external appearance of the Grom-1 is similar to the Russian Igla-1 (NATO SA-16 'Gimlet'). Some reports have suggested that the Grom-1 is in fact a licence production of the Igla-1, which would explain the striking similarity. However, this is denied by ZM Mesko, who note that the basic difference is that the Grom-1 missile is immune to the thermal decoys dispensed from military aircraft as a countermeasure to the missile seeker. ZM Mesko also note that the Russian-produced Igla-1 can be decoyed away from its intended target.

Other differences noted between the Igla and the Grom-1 include:

- Highly integrated electronics with significant reduction in size
- Warhead lethality improved by over 50 per cent
- Programmable, digital gripstock enabling control and recording of the pre-launch parameters.

It is also possible to reprogramme the gripstock-firing unit according to varying battlefield conditions.

Grom-2

Production of the Grom-2 system is based entirely on Polish technology. The Gamrat Company located at Jaslo produces the propellant for the motor. The flight-control section is produced by the PZL Warszawall Company and a mobile, fully automated test unit, based on the all-terrain Tarpan Hunta chassis, has been developed by the Military Research Institute WITU of Zielonka and the WZU-2 Workshop at Grudziadz. The Grom-2 missile incorporates:

- New thermal batteries
- New seeker unit with improved detector cooling
- Miniaturised electronics (for example the fuze assembly has been reduced in size by some 30 per cent)
- Larger, more lethal warhead (due to a reduction in the fuel section size).

IKZ-02 IFF System

Associated with the GROM missile systems is the IKZ-02 short-range Identification Friend or Foe (IFF) system. The IKZ-02 was developed and produced by the Scientific Industrial Centre of Professional Electronics RADWAR SA, which is part of the Bumar Group.

Grom-2 (ZM Mesko) 1037387

Grom missiles deployed with 23 mm gun (Miroslav Gyürösi) 0533190

Truck-mounted Grom (RADWAR) 0567288

The IKZ-02 is an IFF interrogator set that is compatible with the Mk 10A and Mk 12 systems. It is mainly dedicated for integration with short-range SAM systems. The primary features of the IKZ-02 include:

- Small and light-weight
- Built-in battery operation and antennae
- Suitable for man-portable SAMs
- Suitable for short-range gun/missile systems
- Modes: 1, 2, 3/A, 4 (or national secure mode)
- Voice messages, including distance to replying objects
- RS-422 serial communication interface
- Optional fire-disable signal.

The IFF result can be sent to the fire-control system to automatically prevent engaging targets identified as friendly.

The IKZ-02 has also been associated with the ZGS-158 Electro-Optical Tracking Head that is mounted on a Polish Army 6 × 6 truck for air target tracking and also the Poprad Anti-Aircraft Mobile Missiles Launcher.

Variants

Poprad

The Grom was originally designed for and is primarily a man-portable SAM system. A truck-mounted version known locally as the Poprad system exists that sees four missiles mounted in two up-and-over pairs on a pedestal launcher on an Iveco 4 × 4 small truck. The main components are:

- Base carrier
- Hydraulically erected tracking and aiming head, including:
  (1) Electro-optical sensors (thermal camera and laser range-finder)
  (2) IFF interrogator (IKZ-02)
  (3) Four short-range missile launchers
  (4) Two-axis drive system.
- Fire-control computer
- Land navigation and north alignment system
- Communication and data transmission equipment
- Power supply system, including power generator.

A Forward-Looking Infra-Red (FLIR) camera has been employed for target acquisition and tracking, enabling effective use of the system in day and night conditions.

Kobra

When used in the battery format, the Poprad air-defence system is also known as the Kobra system. Kobra comprises:

- Mobile multibeam search radar with a range of 40 km
- 4 × Poprad mobile anti-aircraft systems
- 2 × WD-95 battery command vehicles
- 4 × ZUR-23-2KG JODEK-G 23 mm gun/missile anti-aircraft systems.

Kobra is reported to have been acquired by Indonesia late 2005 or early 2006.

ZUR-23-2

The Grom SAM is also deployed on various Polish, ZUR-23-2 gun missile systems.

Launcher pedestal of Grom (RADWAR) 0567289

Tracked (tank) mounted version

A tracked (tank) mounted version has also been identified.

Naval variant

A Naval variant, mounted on small vessels, is also known to exist. This consists of two missiles, mounted individually either side of a pedestal. The power for the systems is drawn from the ships' internal supplies.

Piorun

Also known as the Improved Grom, this system is for use with the Kusza twin launcher that is vehicle mounted on the Polaris Ranger 6 × 6. Full details are available in this section under the title Piorun.

Specifications

	Grom-2
Dimensions and weights	
Length	
overall	1.596 m (5 ft 2¾ in)
Diameter	
body	72 mm (2.83 in)
Weight	
launch	10.5 kg (23 lb)
Performance	
Speed	
max speed	1,263 kt (2,340 km/h; 1,454 mph; 650 m/s)
Range	
min	0.3 n miles (0.5 km; 0.3 miles) (est.)
max	3.0 n miles (5.5 km; 3.4 miles)
Altitude	
min	10 m (33 ft)
max	3,500 m (11,483 ft)
Ordnance components	
Warhead	1.82 kg (4.00 lb) HE fragmentation
Guidance	passive, IR (2-band, lock-before-launch)
Propulsion	
type	solid propellant (sustain stage)

Status

Grom-1 - production complete 1999. In service with selected Polish Army units.

Grom-2 - reported as the latest version of the family of missiles, in production and in use with Polish Army. Some reports have suggested that between 200 and 300 missiles/launchers were ordered for different applications, including installation on ZUR-23-2KG, ZSU-23-4MP and Project 877EM (Kilo-class submarine Orzel). Further development of the Grom is underway to extend the range and new proximity fuze and improve the electronic counter-countermeasures.

Reports from Russian forces operating in Chechnya have suggested that at least two Polish produced Grom man-portable missiles have been found and positively identified. However, according to Polish officials, no Grom systems have ever been exported from Poland to a non-government end user and these two Groms may have been part of a batch delivered in 2007 or early in 2008 to Georgia's Ministry of Defence. Russian forces operating in South Ossetia had previously reported capturing Grom missile systems in Georgia.

Information released to the open press in late March 2014 confirmed the talks between Lithuania and Poland on the purchase of the Grom missile system. As to the exact variant and its use in Lithuania, by 24 March 2014 this information had not been released by the Lithuanian National Defence.

Contractor

Main: Zakłady Metalowe Mesko SA. This manufacturer is often referred to as ZMT (*Zaklady Mechaniczne Tarnow*), but this is not correct. These two companies are independent manufacturers of completely different products. ZM Mesko is the company located at Skarzysko-Kamienna, whereas Zaklady Mechaniczne Tarnow is the company located in Tarnow.

Flight control section: PZL Warszawall Company.
Fully automated test unit: Military Research Institute, WITU of Zielonka and WZU-2 Workshop at Grudziadz.
Motor propellant: Gamrat Company.
KZ-02 IFF system: CNPEP RADWAR SA.

Romania

CA-94

Type

Man-portable surface-to-air missile system (MANPADs)

Development

The Russian Kolomna-based KBM Strela-2M (Russian industrial index number 9K32M) man-portable missile system has been produced in Romania, after reverse engineering, by the Electromecanica, Industrial Group of the Army (GIA) State factory. The CA-94 system was specifically designed to destroy the following:

- Visually acquired aerial targets;
- Helicopters;
- Aircraft;
- RPVs.

Engagement aspect for targets with speeds of 260 m/s is a pursuit flight profile, while for propeller-driven aircraft and helicopters at speeds of up to 150 m/s a head-on engagement is also possible. The system can also engage and destroy hovering targets provided they are within the launch envelope and emitting sufficient heat energy for the seeker to lock on.

The effectiveness of the system increases if a remote targets acquisition system, for example radar and targets movement indicator, is used. If required, it is also possible to launch the weapon from a moving armoured vehicle, but only if travelling on even ground at a speed not greater than 20 km/h.

Description

The CA-94 missile system consists of the following elements (shown here with their Romanian industrial index numbers):

- A-94 missile;
- IL-02 disposable launch tube;
- ML-94 grip-stock launching mechanism;
- BT-A94 disposable thermal battery power supplies unit.

The A-94 missile comprises four separate sections:

1. Nose section, with a single channel passive infrared transparent seeker unit (designated TGS: *Teplovaya Golovka Samonavedeniya*) and autopilot that receives the target signal from the TGS and creates the guidance commands that are sent to the missile's control section.

2. Control section, consisting of: the missile control vanes; the angular speed sensor (designated DUS: *Datchik Uglovykh Skorostej*), which senses the angular speed of the missile's airframe oscillations and translates them into signals used to stabilise the missile on its desired flight trajectory; a solid propellant gas generator; and an on board power supply which powers the control vanes (this comprises a turbine driven by diversion of exhaust gases from a burning propellant charge located in the solid propellant gas generator).

3. The complete 3 kg warhead section a 1.15 kg HE chemical energy (0.37 kg HE charge) smooth-cased fragmentation warhead and detonating system. The latter comprises a remote arming device, contact and grazing fuzing circuits and a self-destruct mechanism which can only be activated following missile launch.

4. Propulsion section, with a launch booster that provides the missile with the thrust required to drive it out of the launch tube with an initial velocity of 28 m/s and an angular rotation rate of 28 ±5°/s. Once clear of the gunner (approximately 5 to 6 m away) and with the four rear fins deployed, the first grain of the sustainer motor ignites and starts to accelerate the missile to its maximum flight speed. Between 80 and 250 m further away the second propellant grain ignites. A total of 4.2 kg of solid propellant fuel is carried.

The IL-02 glass fibre launch tube is used to transport, aim and launch the missile. An ML-94 grip-stock is attached to it to act as the launching mechanism. Fitted to this is the disposable BT-A94 thermal battery power unit, which supplies both 22 V and 40 V DC to the appropriate grip-stock electronics, missile electronics and missile seeker head. Following a 1 to 1.3 second activation period the battery will operate for a further 40 seconds before it has to be replaced.

The complete missile system is capable of operating at a relative air humidity of 98 per cent between –40 and +50°C.

The associated training system is designated IA-94 and the mobile or fixed check out system IV-94.

The specification table is from Romanian sources and it should be noted that some of the information differs from that given in Russian sources.

Variants

Arsenalul Armatei Regia Autonoma developed an updated version of the CA94, known as the CA94M, at its Electromechanical Factory Ploiesti facility.

The CA-94 underwent trials with the Rafael Air Defence Mobile System (ADMS).

Utilising the IL-A94 launcher, the Romanian Army mounted the CA-94/CA-94M missile on the Gepard self-propelled air-defence system with four missiles either side of the main turret. The upgrade has been offered by Karuss-Maffei Wegmann and Electromecanica Ploiesti.

Specifications

	A-94
Dimensions and weights	
Length	
overall	1.438 m (4 ft 8½ in) (fins folded)
Diameter	
body	72 mm (2.83 in)
Weight	
launch	9.8 kg (21.00 lb)
Performance	
Speed	
cruise	972 kt (1,800 km/h; 1,118 mph; 500 m/s)
cruise Mach number	1.4
Range	
min	0.4 n miles (0.8 km; 0.5 miles)
max	1.5 n miles (2.8 km; 1.7 miles) (approaching target)
	2.3 n miles (4.2 km; 2.6 miles) (receding target)
Altitude	
min	30 m (98 ft) (est.)
max	2,300 m (7,546 ft)
Ordnance components	
Warhead	3 kg (6.00 lb) HE blast fragmentation
Fuze	impact
Guidance	passive, IR (single channel)
Propulsion	
type	solid propellant (dual thrust, sustain stage)

Status

Production of the CA-94 is complete, the system is no longer offered for export. This product was prototype tested in joint development with Rafael but is not in production with the company. Financing for the CA-94 and CA-94M has stopped.

Contractor

Romtehnica (exporter).
Electromecanica Ploiesti SA, Industrial Group of the Army.
Rafael Advanced Defense Systems Ltd.

CA-94M

Type

Man-portable surface-to-air missile system (MANPADs)

Development

The CA-94M is an indigenously developed and produced modernisation of the first-generation CA-94, itself a reversed engineered variant of the Russian Kolomna/KBM Strela-2M man-portable missile system. The CA-94M is designed to destroy aerial targets flying at low and very low altitude in all directions. Qualification testing of the system occurred during 1998.

The CA-94M system differs from the CA-94 in having the following:

- An updated missile the A-94M with 1990s hybrid technology electronics;
- A 0.5 m activation radius proximity fuzing system;
- A preformed HE-fragmentation warhead;
- An automatic lead angle indicator system;
- A 1.3 second short reaction time thermal battery power supply unit.

These features allow for a direct line-of-sight day/night firing capability, enhanced systems reliability and a higher kill probability against hovering helicopters, head-on and pursuit targets.

Description

The CA-94M programme was aimed at improving the performance of the Russian system by increasing the target hit probability and by introducing a proximity fuze with a activation radius of 0.5 m. The destructive capability of the missile was further improved with a new warhead with an improved explosive charge and higher efficiency fragmentation effect.

The CA-94M system consists of the following elements:

- A-94M missile;
- Disposable launch tube;
- Grip-stock launching mechanism;
- Disposable thermal battery power supplies unit.

Field Training Units

The IAC-A94 is the associated field training system and IV-A94M is the mobile or fixed testing system.

Variants

The system can be used with the Rafael Air Defence Mobile System (ADMS) that was believed to be in service with the Romanian Armed Forces (see entry in Self-propelled surface-to-air missiles section).

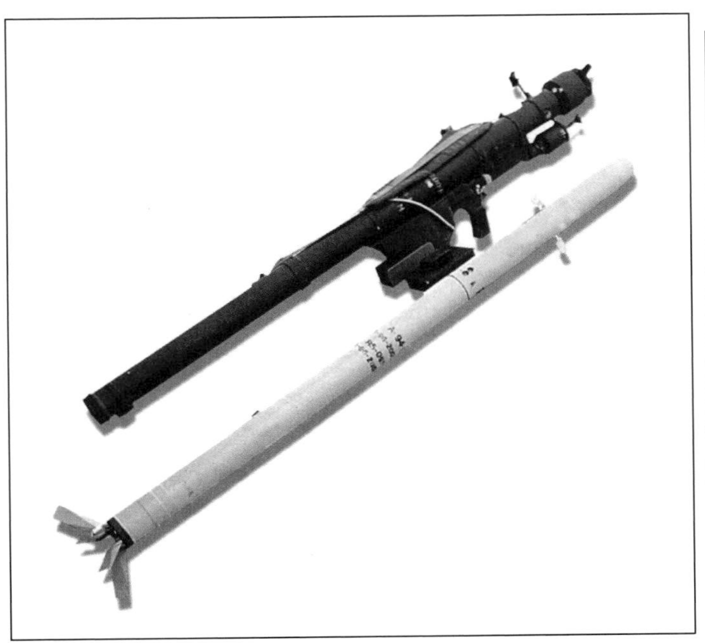

CA-94M low-altitude surface-to-air missile system with launcher above and proximity fuzed missile as in flight below 0024894

Krauss-Maffei Wegmann and Electromecanica Ploiesti have also integrated the IL-A94 type launcher on the weaponry of the Gepard Self-Propelled Air-Defence system for the Romanian Army. The missiles are mounted either side of the main turret in quad packs and can be targeted by means of the Gepard radar. The system is fully capable of day and night searching and firing even in bad weather conditions.

Specifications

	A-94M
Dimensions and weights	
Weight	
launch	15 kg (33 lb)
Performance	
Speed	
max speed	972 kt (1,800 km/h; 1,118 mph; 500 m/s)
max Mach number	1.4
Range	
min	0.3 n miles (0.5 km; 0.3 miles) (approaching target)
	0.3 n miles (0.6 km; 0.4 miles) (receding target)
max	1.8 n miles (3.3 km; 2.1 miles) (approaching target)
	2.5 n miles (4.6 km; 2.9 miles) (receding target)
Altitude	
min	30 m (98 ft)
max	2,300 m (7,546 ft)
Ordnance components	
Warhead	HE fragmentation
Fuze	proximity (0.5 m)
Guidance	passive, IR (self guiding, proportional navigation)

Status
Production of the system is complete. The CA-94M had been in service with the Romanian Army, but is now either held in reserve or stocks destroyed. The CA-94M was offered for export, but with no known clients.

Contractor
Romtehnica.
Electromecanica Ploiesti SA.

Russian Federation

Anti-helicopter mine

Type
Man-portable anti-helicopter mine

FKP GkNIPAS anti-helicopter mine (vehicle placed version) (FKP GkNIPAS) 0533544

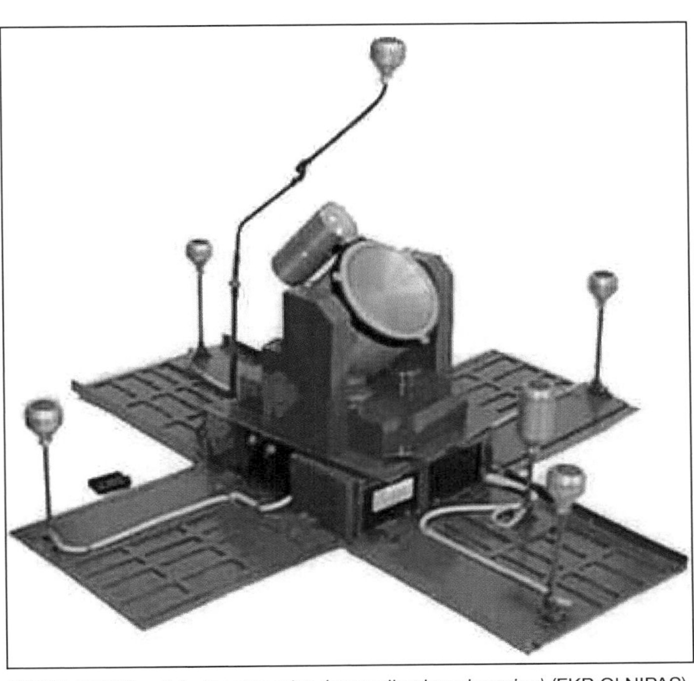

FKP GkNIPAS anti-helicopter mine (manually placed version) (FKP GkNIPAS) 0533545

Development
The anti-helicopter mine has been in development since at least 1998 by the Federal Governmental Enterprise "State Governmental Scientific-Testing Area of Aircraft Systems", Russia. Prototypes and mock-ups of the anti-helicopter mine systems have been in preliminary testing since 1999, but lack of funds at the time held up the rate of development. The development and testing of prototypes using real helicopters has been carried out. If the necessary financing is provided the development may be completed in a short time.

Several dozen anti-helicopter mines are required to enable the system to receive full certification and patents with the Russian authorities. The system has been described as an autonomous air-defence complex.

In July 2007, the anti-helicopter mine was also noted in the context of an anti-UAV device, optimised to respond to hostile UAV signatures.

Description
The anti-helicopter/UAV mine is designed to defeating low flying targets with a high-speed Explosively-Formed Penetrator (EFP) at a range between 150-180 m. The mine is aimed at defending high value areas from helicopters, to block runways of enemy airfields and to psychologically influence helicopter pilots to make them flight at higher altitudes.

At least two versions of the mine are known to exist:
- Mine version for manual placement;
- Mine version for placement from delivery vehicles.

The mine is composed of a warhead and a proximity fuze.

During the Defendory Defence Exhibition held in Athens in 2002, four variants of the system were exhibited on a display panel and one version of the mine's (hardware) was also displayed.

The mine can detect a target by an acoustic system at ranges between 800 to 1,000 m and also a UHF sensor. It then aims the warhead in the direction of the target, and scanning by means of a multi-channel infrared sensor, determines the true direction of the target and the instant of the warhead detonation.

Anti-Helicopter Mine (GkNIPAS) 1499177

The warhead consists an explosive charge, detonation system, casing and liner. Upon detonation the liner formed into one or several EFPs, with a total mass of about 500 g, this projected towards the target at an initial velocity of 2,500 m/s.

The anti-helicopter mine can be remotely controlled via a radio channel.

Specifications

Anti-helicopter mine

Length:	
(missile)	n/avail
Diameter:	
(missile)	n/avail
Wing span:	n/avail
Weight:	
(missile)	13 kg
Propulsion:	1,000 grams of explosive charge
Guidance:	on acoustic direction-finding data
Warhead:	
(weight, liner)	500 g
(type)	explosively-formed projectile charge
Fuze:	proximity, 1 on the base of the IR sensor
Maximum velocity:	2,500 m/s
Sustainer burn time:	n/avail
Detection range:	≥1,000 m
Target speed:	≥100 m/s
Range:	
(maximum)	150-180 m (other reports suggest up to 400 m)
(minimum)	n/avail
Altitude:	
(maximum)	100 m
(minimum)	n/avail
Penetration at 100 m:	12 mm
Working temperature:	–30° to +50°

Status

Uncertain. This weapon bears a striking resemblance to the US Wide Area Munition that is currently in use with the US Army. However, it would appear that the Russians on this occasion have beaten the US by several years.

A very similar system had been offered on the open arms market during the Defendory 2008 biannual exhibition held in Athens, Greece, this time the system was produced in Poland. It is now known for sure that Russia has progressed with the final development and production of these types of weapons, and that other countries are (or have) continued this trend and have already some in production.

Bulgaria is also known to have developed anti-helicopter mines.

Contractor

FKP GkNIPAS and State Scientific and Research Proving Ground of Aviation Systems.

Igla

Type

Man-portable surface-to-air missile system (MANPADs)

Development

Known by the Russian GRAU designation 9K38, Igla (Russian for 'needle'), the NATO-designated SA-18 'Grouse' (9K38) was developed by the Konstruktorskoye Byuro Mashinostroenia (KBM) and is produced by the V A Degtyarev Plant in the Vladimir region of the Russian Federation. This is a third-generation, man-portable, surface-to-air missile system and it is given the Russian industrial index number 9K38, for the complete system. It entered service in 1983. Development began in 1971 but it was preceded into service by the simplified export variant Igla-1. This happened because problems were encountered with development of several of the Igla's major components.

The Igla is a further improvement to the KBM/Degtyarev family of low-altitude MANPADs and incorporates technology innovations from the Igla-1 and Strela-3 systems. It features a dual-channel IR seeker and highly sophisticated FM-tracking logic target discrimination selection unit to defeat sophisticated (pyrotechnically, blinking and modulated) IR decoys. The seeker is believed to operate in the 1.5 to 2.5 and 3.0 to 5.0 μm infrared waveband regions. Commonality remains with the Igla-1 in terms of the rocket motor body, on board and ground power supply units and the equipment used in servicing the missile systems and personnel training. The latter can also be used for the Igla-1 system.

A contract has been agreed with Vietnam for the local production of the Igla missile system. Initially the Russian Federation will supply components for assembly in Vietnam. However, it is understood that local production will continue thereafter. This contract has been brokered by the Leningrad Optical Mechanical Association (LOMO) – the developers and producers of the Igla seeker.

The Igla was apparently used in the Kosovo conflict by Serbian forces.

Description

Basic 9K38 version

The Igla man-portable air defence system is designed to engage visible turbo-jet, turbo-prop and piston-engined aircraft, as well as helicopters and remotely piloted vehicles, both approaching and receding, under conditions of natural cluster and manmade infrared interferences.

In head-on engagements, the system can be used against targets with speeds of up to 360 to 400 m/s, while against receding targets, the maximum target speed is reduced to 320 m/s.

The basic 9K38 Igla missile system consists of the following elements:

- Combat equipment
- Target indication and reception aids
- Maintenance facilities
- Teaching and training facilities.

The system's combat equipment includes:

9M39 missile in the 9P39 launch tube, comprising five sections:

1. Nose section, with a passive dual-channel LOMO JSC (Optical and Mechanical Joint-Stock Company (St Petersburg)) 9E410 seeker head that incorporates a cooled target homing device and an electronics unit. This creates the guidance commands that are sent to the missile's control section. Immediately before impact, the seeker's FM-tracking logic system shifts the aiming point from the engine exhaust region towards the central fuselage area at its junction with the wings. An aerodynamic spike is fitted to reduce wave drag and kinetic heating of the sensor dome.

2. Control section, with the missile control vanes and which accommodates all the elements needed to fly the missile to its target.

3. Warhead section, with a 1.27 kg HE chemical energy fragmentation warhead (0.405 kg of explosive plus, depending upon target aspect, about 0.6 to 1.3 kg of remaining solid propellant fuel that is detonated by the warhead), an impact fuze and explosion generator to detonate the remaining propellant.

4. A dual-thrust, high impulse, shorter-burning rocket solid propellant sustainer section, the sustainer ignites at a safe distance from the gunner once the missile has left the tube. The sustainer is designed to accelerate the missile to its cruising speed in its first mode and then sustain it in flight in the second mode. Any remaining un-burnt propellant is detonated when the missile warhead is initiated, the propellant acting as secondary charge, so boosting the overall explosive effects. The sustainer carries the wing stabiliser unit, which comprises four folding fins that open to stabilise the missile aerodynamically and rotate it about its longitudinal axis.

5. Booster section, which imparts the initial velocity to the missile to leave the launch tube and which spins the missile up about its longitudinal axis.

9P39 glass fibre launch tube, which is used to transport, aim and fire the missile. The launch tube mounts a sighting unit with fore and rear sight and light indicator, a set of electrical coils to spin up the seeker's homing device, a means for a mechanical and electrical interface to the missile, a grip-stock and a ground power supply unit. The launch tube can be used up to five times.

9P516 grip-stock or launching mechanism.

Igla Missile System 1195007

Multiple vehicle mounted Igla 1195008

Electronic plotting board for use with the SA-18 'Grouse' (Igla) man-portable surface-to-air missile system IL15-1 0509389

SA-18 'Grouse' (Igla) missile out of its launcher (front), with complete launcher to rear (Christopher F Foss) 0509738

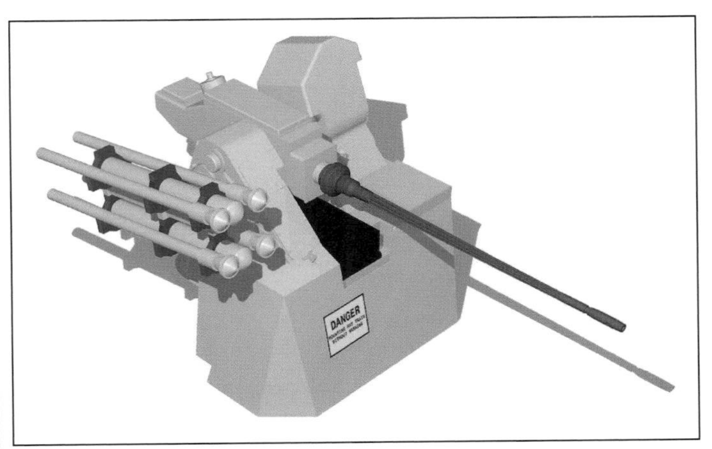

Sigma/Igla (MSI Defence) 1129546

Ground power supply unit that incorporates the thermal battery (9B238) which provides the system with power in the pre-launch period and a high-pressure compressed air bottle for seeker head cooling.

Target indication and reception aids include a man-portable electronic plotting board with integrated R-255PP radio that can display the location, direction and identity of aerial targets within a 12.5 km radius. Up to four targets at a time can be plotted and the team commander can determine the greatest threat track. The system can be interfaced with an IFF interrogator. As of November 2005, The Russian Federal State Unitary Enterprise (FGUP) and The Federal Research and Production Centre (FNPTs) Radioelektronika of Kazan in Tatarstan has developed an improved version of the NATO standard 1L229D IFF interrogator for use with man-portable SAM systems. The 1L229D is intended to identify aircraft fitted with Mk 10 and Mk 10a IFF transponders to prevent missiles being launched against friendly aircraft.

The 9S482M7 can also be used as a Control Post for target indication.

The 9S482M7 (PU-12M7) was upgraded and modernised during 2007/8. The original version was based on a BTR-60 8 × 8 Armoured Personnel Carrier (APC) and relied on data from external radar sources. The latest variant uses a K1Sh1 wheeled vehicle based on the BTR-80 (GAZ 5903) 8 × 8 APC and incorporates a telescopic sensor mast. Within the telescopic mast are located the SRTR (*Stantsiya Radiotekhnicseskoy Razvedki*), passive ESM system and an electro-optical unit that is located at the masthead.

The mast also boasts four static antennas mounted halfway up the mast for short-range radar. Each antenna covers a sector, which when combined provides all-round coverage.

A trainable electro-optical unit mounted on top of the turret contains day and night cameras, providing hemispherical coverage. The PU-12M7 can accept target data from external surveillance radar, other fire-control systems and radar equipped air-defence weapons. Communications with assets up to 15 km away can be achieved by wire, with radio being used for ranges out to 40 km.

This then allows the PU-12M7 tactical control of an area out to 100 km, handling information for as many as 99 airborne targets. The operator can simultaneously track up to nine targets.

Maintenance facilities include an equipment set to prove the functionality of the operational and training equipment in the field. The equipment set can either be carried on a truck chassis or be statically installed at military bases and provides technical checks of 12 missiles or 12 launcher mechanisms per hour.

Teaching and training facilities comprise:

(1) The 9F635 field trainer which is designed to provide instructional training for gunners. The equipment includes: a target simulator on a stand with aircraft models and infrared emission source; three 9F727 inert training missiles; power unit; and instructor's console displaying information on the individual gunner's operations.
(2) The 9K38 Maket full-scale mock-up intended to train the gunner in handling and operating the system.
(3) The 9F633 training set (multipurpose trainer) designed for instructor training of gunners, to prepare them psychologically for the actual missile launch and to conduct practice firings under an instructor's control. The set consists of an inert training missile, a launch training mechanism, a practice missile in a launch tube. It is fitted with a booster unit, instructor's console, power unit (thermal battery with charger) and a stand-supported target simulator. Operating experience suggests that 70 electronic launches at simulated air targets and one live weapon launch using the full size and weight training unit, is sufficient for a gunner to attain the level of a qualified operator.

The missile system is capable of operating in high humidity at temperatures between −44°C and +50°C. It also remains safe if hit by small arms fire, and remains combat ready if dropped from a height of 1 m.

When live firing the Igla missile, the gunner receives air situation data either from the electronic plotter display with target trajectories illuminated and identified by colours, or by direct visual observance in the given task sector. The gunner must then visually detect the target, activate the ground power supply unit, aim at the target via the fore and rear sights and pull the trigger back full against its stops. Subsequently, the entire missile launch procedure, IFF interrogation, activation of the onboard missile systems and missile launch are carried out automatically. The gunner has only to follow the missile's target through the fore and rear sights before the missile launch commences. It takes 1.7 seconds from trigger pull to missile launch.

Once the missile leaves the launch tube and the tail fins are deployed, the guidance unit is programmed to pitch over the missile to the lead point, flying it to an altitude that eliminates premature ground contact when firing against targets at 10 m or less altitude. At the same time, the first fuze safety mechanism is disabled. The second safety mechanism is unlocked by longitudinal acceleration produced by the sustainer at a distance of no less than 120 m from the gunner. At this point, the fuzing mechanism is fully armed.

The Igla can also be equipped with a built-in IFF interrogator that automatically prohibits firing at a friendly target.

Guidance to the target is by proportional convergence logic, using the dual-channel passive IR seeker to generate the missile guidance commands. The seeker has a target selection logic unit which is sufficiently sensitive to detect different IR decoys being launched from supersonic targets with emission levels higher than that of the target itself and having launch period intervals as short as 0.3 seconds. It is also suitable for use in areas where localised fires are found or when small calibre AA guns are in use.

At the very last moment before hitting the target, the programmed identification logic unit orders a shift of the missile aiming point from the engine exhaust area towards the central section of the fuselage. The missile impacts the target and a delay action fuzing circuit is initiated which detonates both the high-explosive warhead and the remaining solid sustainer rocket motor propellant after penetration of the airframe has been achieved. This ensures maximum damage to the target.

Variants

Komar (NATO SA-N-10)

A naval variant known as the Komar (NATO SA-N-10) completed its Russian state trials and has started to be fitted to Russian naval ships. The system consists of:

- An integral mount
- An operator station
- A power supply unit.

The Komar is produced by the RATEP Open Joint Stock Company that is part of the Almaz/Antei Concern of Air Defence, and has headquarters at Serpukhov near Moscow.

Gibkha 3M-47

The Gibkha is a lightweight, short-range missile system that is based around the basic Igla 9M39 missile. The system will make use of the parent ships radar and other sensors. First installed on the Buyan- and Buyan-M-class corvettes it is also to be installed on the new Vladivostok-class amphibious ships that themselves are variants of the French Mistral-class Landing Platform Helicopter (LPH).

Specifications

	9M39
Dimensions and weights	
Length	
overall	1.7 m (5 ft 7 in)
Diameter	
body	72.2 mm (2.84 in)
Weight	
launch	10.6 kg (23 lb)
Performance	
Range	
min	0.3 n miles (0.5 km; 0.3 miles) (approaching target)
	0.4 n miles (0.8 km; 0.5 miles) (receding target)
max	2.4 n miles (4.5 km; 2.8 miles) (approaching target)
	2.8 n miles (5.2 km; 3.2 miles) (receding target)
Altitude	
min	10 m (33 ft)
max	2,000 m (6,562 ft) (approaching target, jet)
	3,000 m (9,843 ft) (approaching target, piston engined aircraft or helicopter)
	2,500 m (8,202 ft) (receding target, jet)
	3,500 m (11,483 ft) (receding target, piston engined aircraft or helicopter)
Ordnance components	
Warhead	1.27 kg (2.00 lb) HE fragmentation
Fuze	impact, time delay
Guidance	passive, IR (dual-channel, believed 1.5-2.5 and 3.0-5.0 µm, FM-tracking logic seeker with countermeasures discrimination unit)
Propulsion	
type	solid propellant (dual thrust, sustain stage)

Status

In production at the V A Degtyarev Plant, Vladimir region, Russian Federation.

An announcement was made early February 2013 that Brazil was interested in purchasing two batteries of Igla missiles and three batteries of Pantsyr-S1. By early April 2013, the first of these systems had arrived in country.

It was reported in early March 2014, during the Russian occupation of the Crimea, that several dozen Igla manportable missile systems had gone missing from Ukrainian armouries. The exact number and type of systems has not been confirmed.

Contractor

Kolomna KBM Engineering Design Bureau, (development).
VA Degtyarev Plant, (production).
VNIIEF, (warhead initiation detonators).
Almaz-Antei Concern of Air Defence, (marketing the Gibka).
Kursk Kristall Joint-Stock Company, (produces sub-assemblies for the Igla missile system's guidance).

Igla-1

Type

Man-portable surface-to-air missile system (MANPADs)

Development

Russian designation, Igla-1 ('Igla' lit. Russian for 'needle'), (NATO SA-16 'Gimlet') was developed by the Konstruktorskoe Bjuro Mashinostroenia (KBM) in Kolomna and is produced by the VA Degtyarev Plant in the Vladimir region of Russia. This is an improved second-generation man-portable surface-to-air missile system given the Russian industrial index number 9K310 for the complete system.

The 9K310 Igla-1 missile system is the predecessor of the 9K38 Igla MANPADS family and differs from it by not being resistant to thermal decoys and having a lower combat effectiveness. The other characteristics of the two systems are the same.

The Igla grip-stock (firing assembly) can be attached to the Igla-1 missile system. The power supply unit, teaching and training systems of the two weapons are standardised. The maintenance equipment of the Igla-1 can, with modifications, be used with the Igla system.

The Igla-1 has seen combat use in Angola and with Iraq during the 1991 Gulf War. In the latter case, it proved to be the most potent of the Iraqi man-portable SAM systems, downing a number of coalition fixed-wing aircraft. It has also been used in the Ecuador-Peruvian border conflict, the resurgence of the Angola civil war, the Kosovo conflict and the ongoing Russian Federation border conflicts.

Description

9K310 version

The Igla-1 entered service with the then Soviet Armed Forces in March 1981 and is designed to engage low-flying manoeuvrable and non-manoeuvrable targets, including stationary hovering ones, by detecting hot metal surfaces and engine plume exhaust using FM tracking logic. In head-on engagements, the system can be used against targets with speeds of up to 360 to 400 m/s. Against receding targets, the maximum target speed is reduced to 320 m/s.

The 9K310 Igla-1 weapon system consists of the following elements: combat equipment; target indication and reception aids; maintenance facilities; and teaching and training facilities.

The system's combat equipment comprises:

(1) 9M313 missile in its launch tube and comprising five sections:
- Nose section, with a passive IR seeker (operating in the 3.5 to 5 µm wavelength region) and incorporating a cooled target homing device and electronics unit which creates the guidance commands which are sent to the missile's control section. Just before impact, the seeker's FM tracking logic system shifts the aiming point from the engine exhaust region towards the central fuselage area at the junction with the wings. A three-pronged aerodynamic spike is fitted to reduce wave drag and kinetic heating of the sensor dome.
- Control section, with the missile control vanes and which accommodates all the elements needed to fly the missile to its target.
- Warhead section, with a 1.27 kg HE chemical energy fragmentation warhead and an impact fuzing circuit.
- Dual-thrust solid propellant sustainer motor section; the sustainer ignites at a safe distance from the gunner once the missile has left the tube. The sustainer is designed to accelerate the missile to its cruising speed in its first mode of operation and then maintain its flight in the second operating mode. The sustainer also carries the wing unit, which is fitted with four folding fins which open at launch to stabilise the missile aerodynamically and rotate it about its longitudinal axis.
- Booster section, which imparts initial velocity to the missile in order to leave the launch tube and spins the missile up about its longitudinal axis.

(2) 9P322 glass fibre launch tube; this is used to transport, aim and launch the missile. The launch tube mounts: a sighting device with fore and rear sights and a light indicator; a set of electrical coils to spin up the seeker's homing device and a means for electrical and mechanical interface to the missile; a grip-stock; and a ground power supply unit. The launch tube can be used up to five times.

(3) 9P519 grip-stock launching mechanism.

(4) 9B238 ground power supply unit which incorporates the thermal battery that provides the system with power in the pre-launch period and a high-pressure compressed air bottle for seeker head cooling.

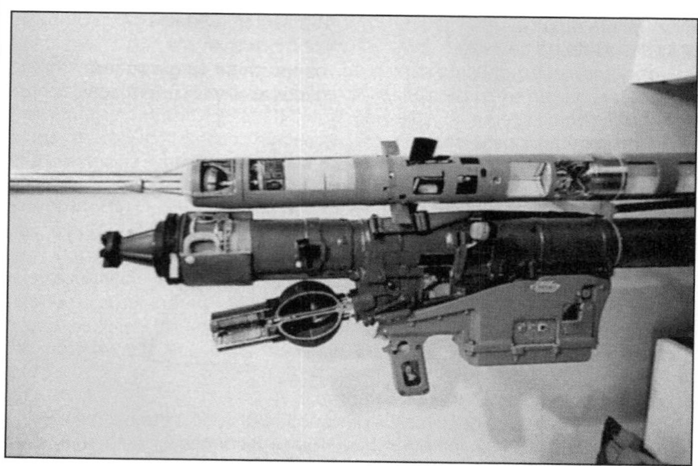

Cutaway of the forward end of the SA-16 'Gimlet' (Igla-1) surface-to-air missile system
0509390

SA-16 'Gimlet' (Igla-1) surface-to-air missile system deployed in the field 0509391

Igla-1 Front End 0041454

Target indication and reception aids include an electronic plotting board system with integrated radio that can display the tactical air situation within a 12.5 km radius. Up to four targets at a time can be plotted and the commander of the launch team can determine the greatest threat track Maintenance facilities include an equipment set designed to prove functionality of the tactical and training systems deployed in the field. The equipment set can either be carried on a truck chassis or be statically installed at a base to provide technical checks of 12 missiles or 12 launcher mechanisms per hour. Teaching and training facilities consist of a field trainer, a training set, a launch control set and a 9K310 Maket full-scale mock-up. The training set is designed to provide for instructor training of gunners. A total of 70 electronic simulated firings is considered sufficient to train an individual gunner. The 9F635 field trainer includes an IR target simulator on a stand with aircraft models and IR emission source, three 9F729 inert training missiles, power unit and instructor's console.

Variants

There have been three Igla-1 variants developed: the Igla-1E (for the export market), the Igla-1M and the Igla-1 S. They differ from the basic Igla-1 in the following respects:

- Their fuel remnants are not detonated
- The Igla-1E system is equipped with an IFF interrogator, tailored to the customer's own requirements
- The Igla-1M is configured without an IFF interrogator
- The Igla-1S, a modern variant that is substantially improved (more below).

The Igla-1 system can also be used with the twin-round Dzhigit (Russian for 'a skilful horseman in the Caucasus') mount launcher.

The Igla-1 system can also be used with the Strelets (Russian for 'archer') multiple launcher system.

In September 2007, shown for the first time at MAKS 2007, a twin-mount container Igla missile was mounted on an MT-LBM with the designator 6M1B5.

Belarus has developed the Stalker 2T-reconnaissance vehicle that mounts a dual Igla or Igla-1 missile launcher for self-defence purposes.

During the Coalition invasion of Iraq, in a factory just north of Baghdad, an unusual Igla-1 missile was found that had a red seeker window covering the IR sensor. The normal seeker windows are either clear, as in the older Strela-2 systems, or opaque, as in the Strela-3 and Igla systems. Previous models with opaque seeker windows employed some form of sensor cooling and thus had an effective head-on target capability.

It has been assessed that use of a red seeker window, over that of the otherwise normally clear one, is to shift the seekers IR operating band, reducing the missiles overall sensitivity, but increasing its kill capability due to the lack of countermeasures employed against this particular seeker type by the coalition forces. Of those seekers so far identified, both had cyrillic markings, and to date the only factory known that works within the this spectrum is the Nalchik Electrovacuum Plant in Kabardino-Balkar Republic. The Nalchik organisation advertises amongst other items, Vidicons sensitive in this region of the spectrum for observation systems in transport and electronic guidance systems.

Igla-1 S

The Igla-1 S was developed by the Machine Building Design Bureau (KBM) in Kolomna, Russia. In Russia it is also referred to as the Igla-Super. This missile was produced by VA Degtyarev Plant in the Vladimir region of Russia and is for use in the multiple launchers, such as the Dzhigit and Strelets. The modifications to the missile include:

- Maximum range increased from 5.2 to 6.0 km
- Maximum and minimum altitude 10 - 3500 m
- A laser proximity fuze
- Improved control actuators
- An improved warhead design that is the same weight as its predecessor, but uses an adjustable focused fragmentation pattern.
- It is also assessed that this variant will have improved electronic warfare capabilities.

Further details can be found in the separate entry within this section.

LOMO MOWGLI-2 night sight

The electro-optical design house, LOMO, is offering its MOWGLI-2 night sight for use with the Igla-1. The 1.9 kg unit measures 308 × 96 × 76 mm and incorporates an optical system with 66 mm fl 1:1 objective lens, giving a 24° circular field-of-view and a magnification of ×2.5. A 25 mm second-generation image intensifier tube is fitted and two AA size alkaline batteries power the unit for more than 18 hours. Detection range against fixed or rotary wing aircraft in starlight conditions is said to be more than 4,500 m.

Arsenal Central Design Bureau Igla-1M upgrade

This upgraded system is covered in a separate entry.

Specifications

	Igla-1
Dimensions and weights	
Length	
overall	1.673 m (5 ft 5¾ in)
Diameter	
body	72 mm (2.83 in)
Weight	
launch	10.8 kg (23 lb)
Performance	
Speed	
cruise	1,108 kt (2,052 km/h; 1,275 mph; 570 m/s)
Range	
max	2.8 n miles (5.2 km; 3.2 miles) (receding target)
	2.4 n miles (4.5 km; 2.8 miles) (approaching target)
Altitude	
min	10 m (33 ft)
max	2,000 m (6,562 ft) (approaching target, jets)
	3,000 m (9,843 ft) (approaching target, helicopters and piston-engined aircraft)
	2,500 m (8,202 ft) (receding target, jets)
	3,500 m (11,483 ft) (receding target, helicopters and piston-engined aircraft)
Ordnance components	
Warhead	1.27 kg (2.00 lb) HE, chemical warfare agent fragmentation
Fuze	impact
Seeker view angle	
vertical	40°/–40°
Propulsion	
type	solid propellant (sustain stage)

Status

In production.

Contractor

Development: Konstruktorskoe Bjuro Mashinostroenia (KBM).

Licence-built by Bulgaria (Igla-1E), North Korea and other unspecified countries.

Igla-S

Type
Man-portable surface-to-air missile system (MANPADs)

Development
The Igla-S (sometimes known as the Igla-Super or Special) is the latest generation system coming from Russia that is referred to within NATO/USA as SA-24 Grinch. It has further improvement over the basic Igla and the Igla-1 family of missile systems, outperforming them in effectiveness, reliability, service life and survivability. In addition, the system retains all merits of the previous Russian MANPADs; shoulder firing by a single gunner, "fire-and-forget" concept, high resistance to background clutter and thermal countermeasures, easy aiming and launching, easy maintenance and training, high covertness of use, retained operability in extreme operational environments.

Description
The Igla-S (Igla-Super or Special) has been designed to engage front line aircraft, helicopter, cruise missiles and Unmanned Aerial Vehicles (UAVs) under direct visibility conditions both day and night. The system can take a target from both the head-on and tail chase in background clutter and thermal countermeasures environments. Head on targets with a velocity of 400 m/s and a tail chase target of 320 m/s can be engaged.

The new Igla-S system performance surpasses the level of the previous systems, the basic Igla and the Igla-1. Changes and improvements to the system result in:

- Considerable increase in warhead lethality against all target types (increased HE quantity by 1.5 times as compared to Igla)
- Variable-depth detonation of the missile warhead in impact mode, detonation depth adjustment dependant on impact speed
- Proximity detonation of the missile's warhead when a body-to-body (impact) does not occur, with optimization of the detonation point
- Detonation of the sustainer's remaining solid propellant at impact
- Advanced missile accuracy characteristics
- Extended firing range up to 6 km as compared with 5.2 km for Igla
- Night firing capability due to application of a night vision device with holder
- Easy installation on various platforms.

The Russian designations for the Igla-S are 9M342 for the missile and 9K338 for the system. The battery and coolant are 9B238; two missiles and two batteries can be found in the 9Ya710 packaging box.

Mowgli-2M Night Sight
A new night sight has been developed for use with the Igla-S; the 1PN97M Mowgli-2M is a further development of the basic Mowgli-2. The night sight is loosely based on the French uncooled micro-bolometer sensor operating in the 8 to 12 micron band, but the optics and electronics are produced in Russia. The Field of View (FoV) is 20° × 15°, with a detection range against a fighter target of 7.7 km by day or night. For a helicopter, the range decreases to 6 km by day but increases to 8 km at night. The Mowgli-2M runs on a 12 V DC supply from a battery that allows for up to six hours of continuous operation. For operational reasons a spare battery can be found with the system thus increasing the use of the sight to 12 hours.

1L228D IFF
The Dzigit dual launcher can be additionally fitted with a night sight, an IFF interrogator using the same frequency band and operating modes as the Mk 10 and Mk 12 systems.

Dzhigit with 1L228D IFF (Rosoboronexport) 1484152

Igla-S missile 1195006

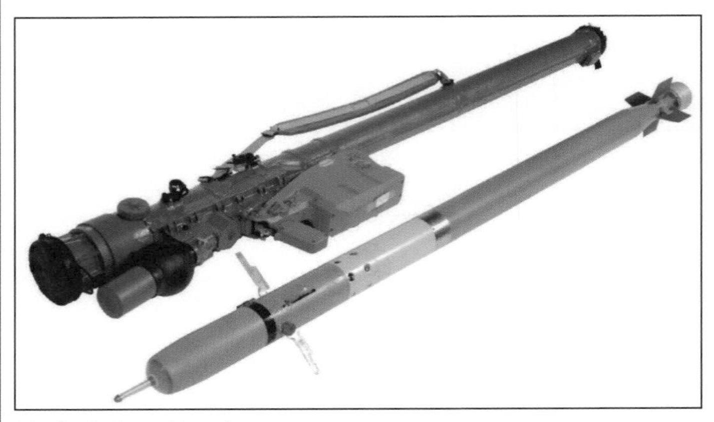

Igla-S missile and launcher 1110312

Variants
Igla
Igla-1

Specifications

	Igla-S
Dimensions and weights	
Length	
overall	1.635 m (5 ft 4¼ in)
Diameter	
body	72.2 mm (2.84 in)
Weight	
launch	11.7 kg (25 lb)
Performance	
Range	
max	3.2 n miles (6 km; 3.7 miles) (receding target)
Altitude	
min	10 m (33 ft)
max	3,500 m (11,483 ft) (receding target, piston engined aircraft or helicopter)
Ordnance components	
Warhead	2.5 kg (5.00 lb) HE fragmentation
Fuze	impact, proximity
Propulsion	
type	solid propellant (dual thrust, with a 10 second sustain burn time, sustain stage)

Status
Development is complete and the system has been offered for sale on the open arms market.

It was announced in November 2008 that LOMO in St. Petersburg through Rosoboronexsport, has sold the Igla-S system to Venezuela. Further to this, at the same time, another contract with Vietnam was also announced.

During October 2008, it was announced that Thailand spent the equivalent of USD3.8 million purchasing seven launchers and 36 missiles of the Igla-S system.

The Igla-S system has also been sold to Syria and Libya, both of whom received the missiles for mounting on multiple launchers such as the Strelets. However, it has been confirmed by the developer and producer that these particular missiles cannot be used as a man-portable, shoulder-launched variant. Furthermore, when fired from the Syrian and Libyan multiple launchers, the missiles cannot be fired in salvo as the launch tubes are mounted to close together.

The Igla-S missile system can also be mounted on the Dzhigit dual rail launcher system along with the 1L228D IFF system.

Late in 2013, on parade in Burkina Faso, the dual rail launchers for the Igla-S were identified mounted on the back of Toyota Land Cruiser trucks. At least 6 could be seen a various times. It is therefore assessed that the Igla-S missile may well be in country either purchased direct from Russia or maybe these weapons have been liberated from Libya.

Contractor

Primary contractor: KB Mashynostroeniya (KBM).
Import/export of military equipment: Rosoboroneksport.
Production: V.A. Degtyarev Plant (JSC).
Seeker technology: LOMO PLC.
RFNC Institute of Experimental Physics.

Strela-2

Type

Man-portable surface-to-air missile system (MANPADs)

Development

In mid-1960, several Russian defence industry enterprises were tasked with undertaking feasibility studies on the use of a low-level air defence system. Following an evaluation of these studies, the Konstruktorskoye Byuro Mashinostroenia (KBM) in Kolomna was assigned the programme to develop a first-generation low-level man-portable air defence system. The system was subsequently given the designation 9K32 and the name, 'Strela-2' ('Strela' being Russian for 'arrow'). The US designation assigned was SA-7a and the NATO name 'Grail' Mod 0.

Flight tests began in 1964 and development was completed in 1967. The SA-7a (Russian industrial index number for missile 9M32) entered Soviet Army service in 1968. Due to its fairly primitive uncooled 0.2 to 1.5 micron wavelength lead sulphide (PbS) seeker head with a 1.9° field of view and 9°/second tracking rate, it was only able to engage a target effectively when it was fired from directly behind at the very hot metal exhaust area. This tail-chase situation resulted in it only being able to engage aircraft flying at up to 220 m/s.

This early type of uncooled seeker was also easily saturated by false targets as it did not have any filter system to screen out spurious heat sources.

The acquisition range varied, depending upon aircraft type and background, from 600 to 2,100 m. It could also be saturated by solar reflection from clouds and go wildly off course. The grip-stock design was simplified because the gunner himself had to decide at what instant the missile should be launched.

Following tests in August 1969, series production for the Soviet Army began in 1970.

The first recorded combat use of the SA-7a was in 1969, during the 1968 to 1970 Egyptian-Israeli War of Attrition. At the same time, the Strela-2 was given to the North Vietnamese Army to counter the American mass use of helicopters. During the war in Vietnam from 1972 to 1975, 589 Strela-2 and Strela-2M missiles were launched against Vietnamese aircraft/helicopter targets, resulting in 204 hits where the target was either shot down or damaged.

The Strela-2 and Strela-2M systems were widely deployed in the various wars between the Arabs and Israelis. For example, from June 1968 until June 1970, the Egyptian Air Defence Forces fired 99 missiles at Israeli targets that resulted in 36 aircraft being damaged or shot down. In 1974, the Syrian Air Defence forces shot down 11 Israeli aerial targets within two months with these missiles. Egypt apparently gave small quantities of the Strela-2 variants to the People's Republic of China and the Democratic Republic of Korea in 1974 as a technology gift for their aid during the 1973 Yom Kippur War.

The Strela-2 was used in combat during:

- The 1982 Falklands conflict with Argentina (no kills);
- The 1982 invasion of Lebanon (by Syria);
- The 1980 to 1988 Gulf War (by both Iraq and Iran);
- Angola (by the MPLA and Cuban troops against South Africa);
- Nicaragua (by both the Sandinistas and the Contras);
- South Yemen during the January 1986 civil war;
- Thai-Laotian and Thai-Kampuchea borders (by Laotian, Kampuchea and Vietnamese forces);
- North/South Yemen border by North Yemen-supported rebels;
- 1978 Ugandan-Tanzanian War;
- Chad region by Libya and Libyan supported guerrillas against French aircraft.

The relative simplicity of the SA-7 has also resulted in widespread distribution to various guerrilla and terrorist groups throughout the world. The Tamil Tigers used the weapon against the Sri Lanka Air Force.

Despite its faults, the SA-7 achieved its design aim of forcing enemy pilots to fly above the minimum radar detection altitude of RFAS-type radars, making them vulnerable to higher-echelon air defence systems. In several conflicts it has also had the added effect of causing enemy pilots to adopt new higher-altitude weapons delivery tactics and this has resulted in a significant degrading of their bombing accuracy and their capacity to aid ground forces.

Description

The system consists of the missile in its green (for operational round) re-loadable launch container canister (Russian industrial index number 9P54), a grip-stock (Russian industrial index number 9P53 with a 24 pin connector between it and the canister) and a can-like thermal battery.

When operating the Strela-2 system, the gunner detects and identifies the target either visually or by means of a separate IFF interrogator (the interrogator was not available for export). They then start the ground power supply unit to energise the seeker and the electronics unit of the grip-stock. These actions and procedures take between four and six seconds. The gunner pulls the trigger back to its full stop thus activating the automatic target lock on and launch mechanism which estimates, within 0.7 seconds, sufficiency of the target's emissive power and permissibility of the angular tracking rate. If these parameters are within the specification limits, the on board power supply unit is activated and 0.8 seconds later the missile ignites. The continuous light of the red pilot lamp on the sight and the monotonous sound of the buzzer indicate correct actions of the gunner and that target parameters are within the allowable limits.

Should the launch conditions be violated, the seeker is automatically caged and disrupted light and sound signals alert the gunner to the necessity to take aim again. After hearing the specific sound of the on board power supply activation, the gunner must enter a lead angle in the flight direction of the target. The missile starts off with ignition of the booster which accelerates the missile within 0.05 seconds to 27 to 31 m/s and makes the missile rotate clockwise at a rate of up to 19 to 21 rev/s. The booster operation is terminated in the tube. After leaving the tube the missile opens its tail and control fins, the first safety mechanism is disabled and the self-destruct timer is enabled in case the missile misses its target. Approximately 0.3 seconds after the missile leaves the tube and at a distance of over 5.5 m from the gunner, the sustainer cuts in, accelerating the missile within the first 1.7 seconds of flight to 430 m/s and then maintaining it. During this acceleration phase (under the action of longitudinal acceleration) the second safety mechanism is disabled at a distance of no less than 120 m from the gunner and the fuze is now fully armed.

Throughout its flight, the seeker head continually determines the target's angular velocity and the on board AM tracking logic guidance system produces flight data to resolve the difference between the direction that the head is

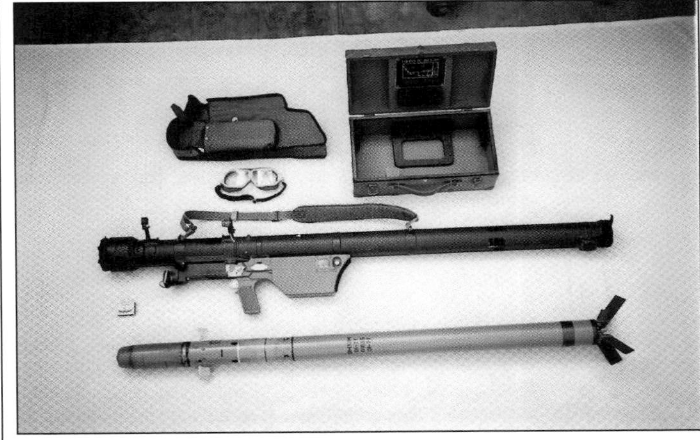

Key component parts of the SA-7 'Grail' (Strela-2M) man-portable surface-to-air missile system, including launch tube and missile, grip-stock firing mechanism carrying case, protective cover bag and goggles 0509835

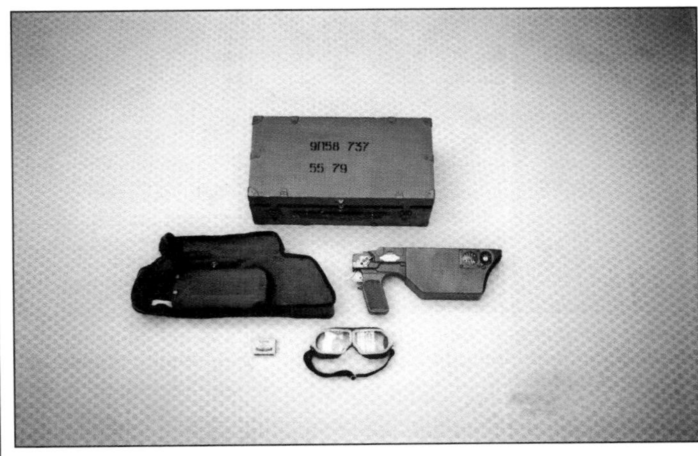

SA-7 'Grail' (Strela-2M) grip-stock firing mechanism with protective cover bag, carrying case and goggles 0509836

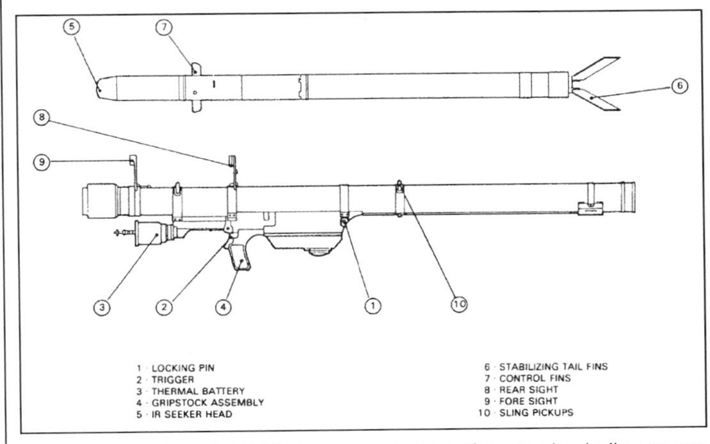

Main components of SA-7 'Grail' man-portable surface-to-air missile 0509395

Close-up of Polish SA-7 'Grail' (Strela-2M) man-portable surface-to-air missile system with Pelengator (Russian name for an RF direction-finder) on the gunner's helmet
0509393

pointing and the weapon's trajectory. At target impact the fuze detonator operates and the 0.37 kg of explosive within the warhead is initiated. The missile self-destructs after 14 to 17 seconds of flight if it misses the target.

Variants

In 1981, a helicopter self-defence mounting was seen for the first time on Mi-24 'Hind' gun ship helicopters. This involved the fitting of quadruple Strela-2M launcher arrangements on the helicopter's weapon carriers. Subsequently a two-round Strela-2M version was seen mounted on a Polish Air Force Mil Mi-2 'Hoplite' variant on either side of the fuselage. The Serbian Air Force use a single-round Strela-2M launcher on each side of the Type NNH weapon pylons attached either side of its GAMA gun ship variants of the licence-built Eurocopter SA 342L Gazelle helicopter. Air-to-air range of the Strela-2M remains at 4,200 m.

There are also several multiple-round naval launcher assemblies that have been produced for the Strela-2/3 family of missiles.

Egypt has reverse engineered the Strela-2M under the designation Sakr Eye and China has also reverse engineered under the HN-5 designation. The basic HN-5 (equivalent to the Strela-2) was followed by the HN-5A, which is equivalent to the Strela-2M with some further improvements, including seeker cooling to enhance sensitivity. The system was further modified to produce the HN-5B and HN-5C versions.

The former Federal Republic of Yugoslavia (Serbia and Montenegro now (as of June/July 2006) Serbia) licence-built the system at the Krusik Factory near Valjevo. The designation was Strela-2M/A. The Strela-2M system has also been produced in the Czech Republic, Slovakia, Bulgaria, Poland and Romania under licence. As recently as June/July 2006, new missiles bearing the Krusik company logo have been identified (along with other missiles of the Strela family) at arms exhibitions.

Specifications

	Strela-2
Dimensions and weights	
Length	
overall	1.438 m (4 ft 8½ in)
Diameter	
body	72 mm (2.83 in)
Weight	
launch	9.15 kg (20.00 lb)
Performance	
Speed	
max speed	836 kt (1,548 km/h; 962 mph; 430 m/s)
Range	
min	0.4 n miles (0.8 km; 0.5 miles)
max	1.8 n miles (3.4 km; 2.1 miles)
Altitude	
min	50 m (164 ft)
max	1,500 m (4,921 ft)
Ordnance components	
Warhead	1.17 kg (2.00 lb) HE fragmentation
Fuze	impact
Guidance	passive, IR
Propulsion	
type	solid propellant (sustain stage)

Status

Production complete. However, the system can be re-manufactured to a later standard using the technology described in the Variants section.

It is also in widespread service with various guerrilla/terrorist/insurgent groups throughout the world.

Cambodia destroyed all 233 Strela-2/2M stocks during the last week of March 2004 under an agreement with the US funding programme.

The United States and Hungary agreed on 27 September 2005 to cooperate in the destruction of 1,540 of Hungary's Strela-2 missile systems and all related equipment in the Hungarian inventory.

In December 2008, it was announced that Belarus has destroyed all remaining stocks of Strela-2/Strela-2M systems.

Known to have been in service with several countries, however many have now been removed from inventories, placed in reserve, or destroyed with the aid of the US Funding Programme.

Contractor

Development

Konstruktorskoe Bjuro Mashinostroenia (KBM).

Licence-built by Bulgarian, Czech Republic, Slovak, Polish, Romanian (including an upgraded version) and former Yugoslav (including an upgraded version) state factories. It has been reverse engineered. Egypt, North Korea and China have all produced improved copies.

Strela-2M

Type

Man-portable surface-to-air missile system (MANPADs)

Development

The Strela-2M (Russian industrial index number 9K32M, US designation SA-7b, NATO designation 'Grail' Mod 1), with the 9P58 grip-stock, was specifically designed to automatically solve the problem of achieving the optimum time for the missile launch. The missile (9M32M) was still primarily a tail-chase weapon, but now has a limited engagement capability against approaching helicopters and piston-engine aircraft.

The Strela-2M can also be used with a small passive Radio Frequency (RF) direction finder fixed to the front of the operator's helmet. This picks up the emissions from aircraft radars and radar altimeters.

Description

The system consists of the missile in its green (operational round) re-loadable launch container canister (Russian industrial index number 9P54M), a grip-stock (Russian industrial index number 9P58 and a 28-pin connector between it and the canister) and a can-like thermal battery.

In a joint venture with the LOMO JSC (Optical and Mechanical Joint Stock Company (St Petersburg)), InfraRed (IR) seeker design bureau, Kolomna KBM, is offering a new seeker head to extend the life of the Strela-2M. Designated as the 9E16M seeker, the assembly replaces the existing 9E46 single-waveband channel passive IR seeker with a dual channel unit. The upgraded weapon has been given the designation Strela-2M2.

Strela-2 connected to improvised power pack (Bashaer al-Nasr Brigade)
1513282

Dual mounted Strela-2M (Miroslav Gyürösi) 1290452

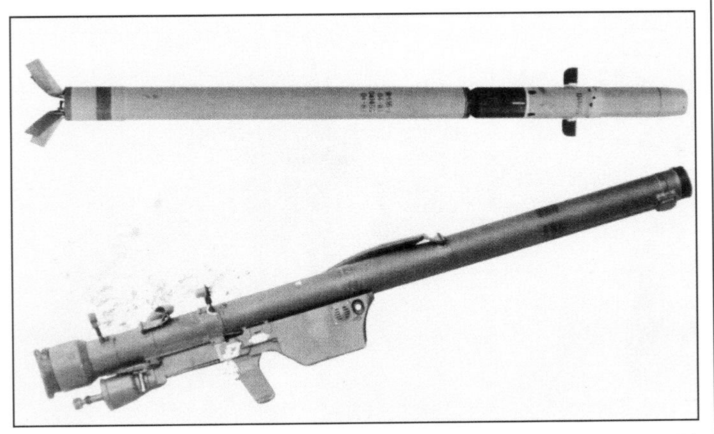

Strela-2M missile and launcher 0509398

In 2004, an upgraded seeker/homing head was identified as the RD-93, again this seeker was for the Strela-2M. It is not known how many of these seekers were put into full service or how wide the distribution was.

LOMO is also offering its MOWGLI-2 night sight for use with the Strela-2M. The 1.9 kg unit measures 308 × 96 × 76 mm and incorporates an optical system with 66 mm f/1.1 objective lens, giving a 24° circular field of view and a magnification of ×2.5. A 25 mm second generation image intensifier tube is fitted and two AA size alkaline batteries power the unit for more than 18 hours. Detection range against fixed or rotary wing aircraft in starlight conditions is said to be more than 4,500 m.

Another night-sight GEO PZR1 is also available for the man-portable SAM systems, this may however be the designator for the MOWGLI-2.

During late 2013, video obtained from the Free Syrian Army showed the system being used with the aid of a cluster of motorcycle batteries to extend the life of the missile during engagements against Syrian Army helicopters. This is not the first time evidence of such a modification has been obtained, earlier reports also suggest that Hamaz in the Gaza Strip and Hizbullah in the Lebanon have also attempted this modification.

Variants

In 1981, a Strela helicopter self-defence mounting was seen for the first time on Mi-24 'Hind' gunship helicopters. This involved the fitting of quadruple Strela-2M launcher arrangements on the helicopter's weapon carriers. Subsequently a two-round Strela-2M version was seen mounted on a Polish Air Force Mil Mi-2 'Hoplite' variant on either side of the fuselage. The Serbian Air Force use a single-round Strela-2M launcher on each side of the Type NNH weapon pylons attached either side of its GAMA gun ship variants of the licence-built Eurocopter SA 342L Gazelle helicopter. Air-to-air range of the Strela-2M remains at 4,200 m.

Egypt has reverse engineered the Strela-2M under the designation Sakr Eye and China has also reverse engineered under the HN-5 designation. The basic HN-5 (equivalent to the Strela-2) was followed by the HN-5A, which is equivalent to the Strela-2M with some further improvements, including seeker cooling to enhance sensitivity. The system was further modified to produce the HN-5B and HN-5C versions.

The former Federal Republic of Yugoslavia (former Serbia and Montenegro (as of June/July 2006) Serbia) licence-built the system at the Krusik Factory near Valjevo. The designation was Strela-2M/A. The Strela-2M system has also been produced in the Czech Republic, Slovakia, Bulgaria, Poland and Romania under licence. In July 2006, a new missiles was identified bearing the Krusik company logo along with other missiles of the Strela family at arms exhibitions.

Specifications

	Strela-2M
Dimensions and weights	
Length	
overall	1.438 m (4 ft 8½ in)
Diameter	
body	72 mm (2.83 in)
Weight	
launch	9.6 kg (21.00 lb)
Performance	
Speed	
max speed	836 kt (1,548 km/h; 962 mph; 430 m/s)
Range	
min	0.4 n miles (0.8 km; 0.5 miles)
max	2.3 n miles (4.2 km; 2.6 miles)
Altitude	
min	30 m (98 ft)
max	2,300 m (7,546 ft)
Ordnance components	
Warhead	1.17 kg (2.00 lb) HE fragmentation
Fuze	impact
Guidance	passive, IR
Propulsion	
type	solid propellant (sustain stage)

Status

Production is complete. The system has however been license manufactured in many countries throughout the world, and also copied in others. The system because of its relative cheapness and ease of operation is a favourite of terrorist organisations and is held in the inventory of many.

Cambodia destroyed all 233 Strela-2/2M stocks during the last week of March 2004 under an agreement for US funding programme. It was announced in October 2008 that Cyprus has also withdrawn its stock of 324 systems and will be destroying them in the very near future.

In December 2008, it was announced that Belarus has destroyed all remaining stocks of Strela-2/Strela-2M systems.

Contractor

Konstruktorskoe Biuro Mashinostroenia (KBM).

Strela-2/Strela-3 Multiple round mounts

Type

Tripod mounted man-portable Surface-to-Air Missile (SAM) system

Development

Over the years a number of multiple-round launcher assemblies for the Strela-2 and Strela-3 family of man-portable SAM have been developed. Available details for some of those systems are given below:

Description

MTU-4S four-round launcher system

The MTU-4S simple pedestal mount was specifically developed for use with Strela-2 and Strela-3 weapons aboard naval ships. The gunner has to aim the assembly at the target manually, turn the missile's power supply on and, after target lock on, launches the weapon. Elevation limits are from –8° to +64°. Launcher weights empty 229.5 kg, with four Strela-2 systems 289.5 kg and with four Strela-3 weapons 295.5 kg. The US DoD designation with the Strela-2 was SA-N-5 and with the Strela-3, SA-N-8.

MTU-4US four-round launcher system

The MTU-4US is an improved version of the MTU-4S pedestal mount for use with the Strela-2M and Strela-3 missile systems. Fitted to Tarantul-1-class missile boats of the Polish Navy and has a light operated target warning system for the gunner.

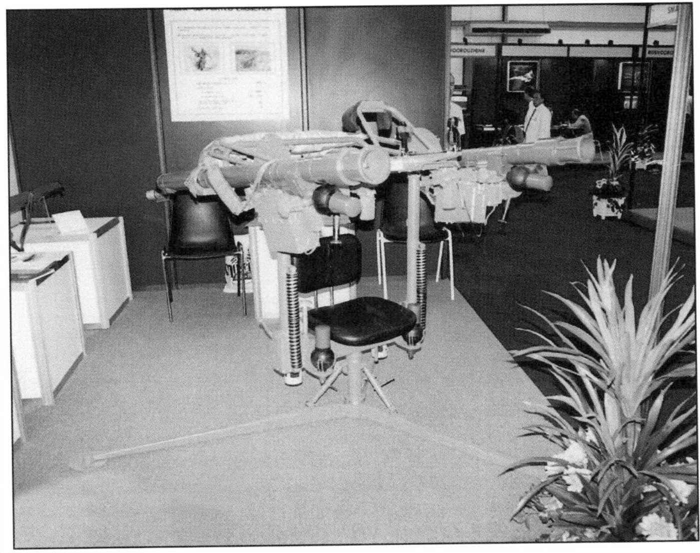

Dzhigit dual rail missile launcher 0509739

FASTA-4 four-round launcher system
The FASTA-4 system was a former East German improved version of the MTU-4S four-round pedestal mount for naval and land-based uses. The former East German Army fielded the system in 1979 mounted on the rear of Robur L-1801A (4 × 4) 1,800-kg trucks to defend rear area targets such as airfields. As far as is known, there are no longer any of these land-based systems operational in any country. Reports have also been made of a similar system being used by Egypt in the 1973 Yom Kippur War. There was also an improved version designated the FASTA-4M, which was an improved version of the MTU-4SU system.

FAM-14 four-round launcher assembly
This was produced by the former East Germany and is a specialised naval mount. It was sold to a number of countries, including Poland.

Status
Production of each system is complete. Most of the versions remain in service with the Russian Federation, former Warsaw Pact and former Soviet client state navies. In due course these launchers are being replaced by the Dzhigit dual rail launchers.

As of September 2012, no further information is available on these launcher systems.

Contractor
Altair.

Strela-3

Type
Man-portable surface-to-air missile system (MANPADs)

Development
The (*Perenosniy Zenitniy Raketniy Kompleks* - PZRK: portable air defence system) Strela-3 (Russian for arrow) was developed by the Konstruktorskoe Bjuro Mashinostroenia (KBM) in Kolomna as a second-generation man-portable surface-to-air missile system and given the Russian industrial index number for the complete system of 9K34.

The Strela-3 (a GRAU designation) was given the NATO designation SA-14, nicknamed 'Gremlin' when it entered operational service in 1974. Development started in 1968, with factory testing starting in 1970 and completed in 1972.

It replaced, in some instances while in others complemented, the earlier Strela-2 and Strela-2M (SA-7) series weapons and is designed to engage low-flying manoeuvrable and non-manoeuvrable targets, including stationary hovering ones.

Compared with its predecessors, the Strela-3 missile system solved the problem of firing a compact target-seeking missile against fast approaching jets. Moreover, the use of a cooled seeker detector ensured high immunity against background infrared sources. Sun-illuminated clouds and the horizon line do not impose any practical restrictions on firing as the seeker is designed to use FM tracking logic to detect hot metal surfaces and engine plume exhaust.

In head-on engagements, the system can be used against approaching targets at speeds of up to 310 m/s. Against receding targets, the maximum target speed is 260 m/s.

The Strela-3 system has seen combat in Angola, El Salvador, various RFAS ethnic struggles and with Iraq during the 1991 Gulf War.

In August 2007, a Strela-3 using the 9M36-1 missile was captured from the Tamil Tiger terrorist group in Sri Lanka and again in 2010, many were recovered in Libya after a very short and bloody campaign with insurgents and coalition forces fighting together to topple the Ghadaffi regime.

Description
The Strela-3 missile system consists of the following elements (described here with their Russian industrial index numbers) combat equipment; maintenance facilities and teaching and training facilities.

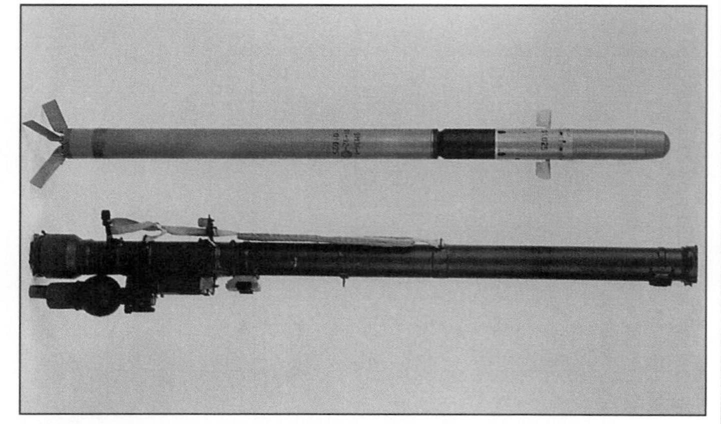

Strela-3 (SA-14 'Gremlin') low-altitude surface-to-air missile (top) and launcher (bottom). Note the missile is facing the wrong way for the actual launcher 0007251

The LOMO MOWGLI-2 slip-on night sight 0039292

Strela-3 with African Forces 0509392

The Strela-3 combat equipment comprises the following:

- Missile in its launch tube comprising four separate sections:
    - Nose section, with a single-channel cooled passive Frequency Modulated (FM) infrared transparent seeker head operating in the 3.5 to 5 micron wavelength region, designated in Russian, *Teplovaya Golovka Samonavedeniya* (TGS) and autopilot that receives the target signal from the TGS and creates the guidance commands that are sent to the missile's control section.
    - Control section, with the missile control vanes and which accommodates all the elements needed to fly the missile to its target.
    - Warhead section, with a 1.15 kg HE chemical energy fragmentation warhead and a detonating system. The latter comprises a remote arming device, contact and grazing fuze circuits and a self-destruct mechanism which operates after 14 to 17 seconds of flight if the missile has not hit the target.
    - Propulsion section, with a launch booster that provides the missile with its initial velocity to drive it out of the launch tube. Approximately 0.3 second into the flight and some 5.5 m from the gunner, with the rear fins deployed, the dual-thrust sustainer rocket motor ignites and accelerates the missile in its first operating mode to its maximum flight velocity and then sustains the speed in the second mode.
- 9P59 glass fibre launch tube. (This is used to transport, aim and launch the missile. The launch tube mounts a sight device with fore and rear sights and light indicator, a set of electrical coils to spin up the seeker's homing device, a means of mechanical and electrical interface to the missile, a grip-stock and a ground power supply unit. The launch tube can be used up to five times.)
- 9P58 grip-stock launching mechanism.
- Ground power supply unit (incorporating the thermal battery that provides the system with power in the pre-launch period and a high-pressure compressed air bottle for seeker head cooling).

Maintenance facilities include an equipment set that provides the functionality of the operational and training systems in the field. The equipment set can be either carried on a truck chassis or be statistically installed at military bases. The set provides for technical checks of 12 missiles or 12 launcher mechanisms per hour.

Teaching and training facilities consist of a 9F730-field trainer designed to provide for instructor training of gunners. A total of 70 electronic simulated firings is considered sufficient to train an individual gunner. The 9F730 equipment includes a 9F635 IR target simulator on a stand with aircraft models and IR emission source, a 9K310 training missile and launcher.

The missile system is capable of operating in high humidity between temperatures –38 to +50°C.

LOMO MOWGLI-2 night sight
The electro-optical design house LOMO is offering its MOWGLI-2 night sight for use with the Strela-3. The 1.9 kg unit measures 308 × 96 × 76 mm and incorporates an optical system with 66 mm fl 1:1 objective lens giving a 24° circular field-of-view and a magnification of times 2.5. A 25 mm second-

generation image intensifier tube is fitted and two AA size alkaline batteries power the unit. These power the system for more than 18 hours. Detection range against fixed or rotary wing aircraft in starlight conditions is said to be more than 4,500 m.

Specifications

	Strela-3
Dimensions and weights	
Length	
overall	1.42 m (4 ft 8 in) (fins folded)
Diameter	
body	72 mm (2.83 in)
Weight	
launch	10.3 kg (22 lb)
Performance	
Speed	
cruise	914 kt (1,692 km/h; 1,051 mph; 470 m/s)
impact Mach number	1.4
Range	
min	0.3 n miles (0.5 km; 0.3 miles) (est.) (approaching target)
	0.3 n miles (0.6 km; 0.4 miles) (est.) (receding target)
max	1.1 n miles (2 km; 1.2 miles) (approaching target, jets)
	2.4 n miles (4.5 km; 2.8 miles) (approaching target, helicopters and piston-engine aircraft)
	2.2 n miles (4 km; 2.5 miles) (receding target, jets)
	2.4 n miles (4.5 km; 2.8 miles) (receding target, helicopters and piston-engine aircraft)
Altitude	
min	15 m (49 ft) (est.)
max	1,500 m (4,921 ft) (approaching target, jets)
	3,000 m (9,843 ft) (approaching target, helicopters and piston-engine aircraft)
	1,800 m (5,906 ft) (receding target, jets)
	3,000 m (9,843 ft) (receding target, helicopters and piston-engine aircraft)
Ordnance components	
Warhead	1.15 kg (2.00 lb) HE blast fragmentation
Fuze	impact
Guidance	passive, IR (single channel 3.5-5 μm wavelength)
Seeker view angle	
vertical	40°/–40°
Propulsion	
type	solid propellant (sustain stage)

Status
Production complete; this system was used by non-regular forces (terrorists/insurgents) in Iraq to attack a commercial aircraft during take-off from Baghdad airport in November 2003. The aircraft received a hit on its port wing engine but still managed to land safely.

Contractor
Konstruktorskoe Bjuro Mashinostroenia (KBM) Kolomna (development and primary contractor).
Licence-built by Bulgaria (see entry this section), North Korea and Belarus.
Beltechexport, offered for sale.
SAMEL-90 plc, electronics production for Bulgarian missile.

Verba

Type
Man-portable surface-to-air missile system (MANPADs)

Development
The Verba (lit Willow) man-portable Surface-to-Air Missile system (SAM) was first identified back in 2007 when it was in the flight/design and testing stage. Further trials took place between 2009 and 2010 at the 726 Army Training Centre in the Krasnodar Region Yeisk. It is understood that the system entered service in 2011 after the completion of trials and is currently (March 2014) with the land forces troops.

During an interview for Krasnaya Zvezda, the then Ground Troops CinC General-Colonel Aleksandr Postnikov indicated that the Ground Troops Air Defence will receive modernised S-300V4 systems, Buk-M3, Tor-M2U and manportable Igla-S and Verba SAMs.

The Verba system has not yet been offered for sale but it is understood that it has already been shown to potential customers in Asia.

It is also understood that Verba can be used in conjunction with the multiple launcher known as Strelets, a system that has already been sold extensively abroad.

Description
The Verba MANPAD is reported to employ a 3 × waveband optical homing head and is to replace all predecessor Strela and Igla missile systems.

Variants
None known.

Specifications
Verba

Type:	man-portable SAM
Length:	not available
Diameter:	unknown
Weight:	unknown
Guidance:	3 × band infrared
Warhead:	
(type)	HE fragmentation
(weight)	1.5 kg plus
Fuze:	probably impact and proximity
Motor:	probably black powder solid propellant boost with a solid sustain (SPR TTX)
Velocity:	500 m/s
Range:	
(maximum)	6,400 m
(minimum)	50 m
Altitude:	
(maximum)	4,500 m
(minimum)	10 m
Reaction time:	8 sec

Status
Believed to have entered service with the Russian Armed Forces in 2011 but has not yet been offered for export.

Contractor
Engineering Design Bureau in Kolomna.726th Army Training Centre, Krasnodar Region, Yeisk, Russia.

Serbia

Strela-2M/A

Type
Man-portable surface-to-air missile system (MANPADs)

Development
The Strela-2M/A (Yugoslav military designation S-2M/A) is a locally built derivative of the Russian Strela-2M (see SA-7b 'Grail' Mod 1 entry this section) man-portable surface-to-air missile system. License production of the missile in the former Yugoslavia began in 1984.

As of August 2012, further improvements have been reported by Yugoimport. These include the introduction of head-on engagement capability and extended power supply thus enabling an extended target acquisition and engagement time. Other improvements include a new warhead, electronics and heat-seeking guidance system.

In July of 2013, a report emanating from Serbia suggested that Serbian engineers were working to adapt the 9K32M (Strela-2M) for the air-to-surface role, they are also reported to be planning to resume production of the missile with a new composite rocket motor and passive seeker.

Description
The S-2M/A varies from the Russian model in that the electronic systems in the single channel passive infrared seeker have been miniaturised, allowing the warhead section to be enlarged and increased in weight by 20 per cent.

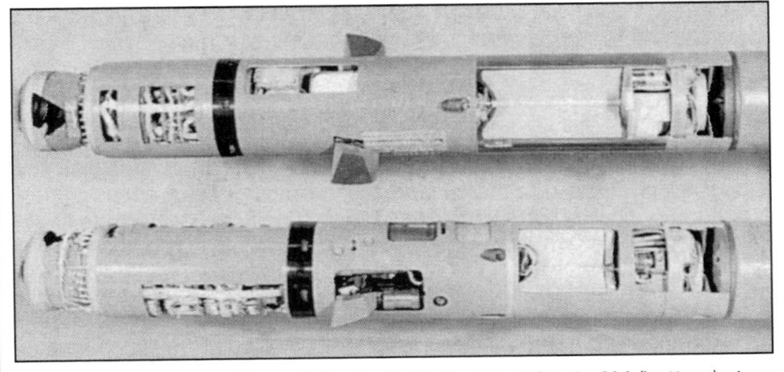

Comparison cross-sections of Strela-2M/A (top) and Strela-2M (bottom) steer and control units with warhead size 0509407

Yugoslav-built Strela-2M surface-to-air missile system with operator standing
0509841

Yugoslav-built Strela-2M surface-to-air missile system with operator kneeling
0509842

Advantages when compared to the standard Strela-2M warhead are:

- A 40 per cent increase in the weight of the explosive charge carried (0.518 kg versus 0.370 kg)
- A 40 per cent increase in the blast effects
- A 30 per cent increase in the overall fragmentation effect
- A 30 per cent increase in the total warhead efficiency
- A 0.2 per cent increase in the single-shot kill probability figure.

In addition to the above the basic system remains similar in appearance and capability to the original ex-Soviet Strela-2M, but with two other upgrades.

- The system is reported to have been upgraded with a new seeker that is capable of head-on engagements. This then suggests that the new seeker is cooled in some way, either by means of a nitrogen bottle or possibly (but not probable) the Chinese tried-and-failed way of thermoelectric cooling.
- In conjunction with the probable nitrogen cooling, an extended power supply set, enabling extended target acquisition and engagement time, has also been added. This would then suggest the appearance of the 9B238 or 9P51 battery and cooling bottle seen on the Igla and Strela-3 systems respectively.

All the factions involved in the internal unrest in the former Yugoslavia used the Strela-2M/A. It is known that it shot down a number of fixed- and rotary-wing aircraft including, it is believed, the Royal Navy Sea Harrier FRS Mk 1 lost in April 1994 over Bosnia.

During the 1999 Kosovo conflict it saw widespread use against NATO air assets and is believed to have been responsible for the majority of the Unmanned Aerial Vehicles (UAVs) lost on operations over Kosovo.

Its effectiveness against NATO fixed-wing and rotary-wing aircraft protected by sophisticated countermeasures was minimal.

Specifications

	Strela-2M/A
Dimensions and weights	
Length	
overall	1.44 m (4 ft 8¾ in)
Diameter	
body	72 mm (2.83 in)
Weight	
launch	1.32 kg (2.00 lb)
Performance	
Altitude	
max	2,300 m (7,546 ft)
Ordnance components	
Warhead	1.32 kg (2.00 lb) HE fragmentation
Fuze	impact
Guidance	passive, IR (single-channel)
Propulsion	
type	solid propellant (sustain stage)

Status

Production is reported to have recommenced with a new composite rocket motor and passive seeker as of July 2013.

Contractor

Yugoimport - SDPR.

Sweden

Bolide

Type

Man-portable surface-to-air missile system (MANPADs)

Development

Bolide is the new-generation missile for the laser-beam-guided air defence missile systems RBS 70 and ASRAD-R in their various applications both for existing and new systems.

The Bolide missile is based on the RBS 70 Missile Mk 2, but further developed, with the latest generation of computers and state-of-the-art electronics, providing many new features, increasing the effectiveness of the missile and the speed from 1.6 Mach in the basic RBS 70 to 2.0 in the Bolide.

Bolide is a modern air-defence missile in the VSHORAD/SHORAD-class and is reported to be effective against fast small targets, targets taking evasive manoeuvres and targets with enhanced protection. Production started during 2002.

Besides the new Bolide missile, the latest generation of RBS 70 (RBS 70NG) utilises a third-generation Thermal Imager - BORC, a new digital IFF Interrogator, a new External Power Supply and a new Training Simulator using commercially available PCs.

Description

The beam-riding Bolide is a fourth-generation missile that provides the RBS 70 and ASRAD-R systems with a wider target capability due to an adaptable proximity fuze function and the combined preformed fragmentation (3,000 tungsten spheres) and shaped charge warhead. The missile is optimised for aerial targets, but in a self-defence role, surface targets can be engaged. Other capabilities within the missile, such as the re-programmable electronics suite, means that the Bolide can readily be upgraded with new or modified functions. The missile has been specifically designed for short time-of-flight with a high manoeuvrability.

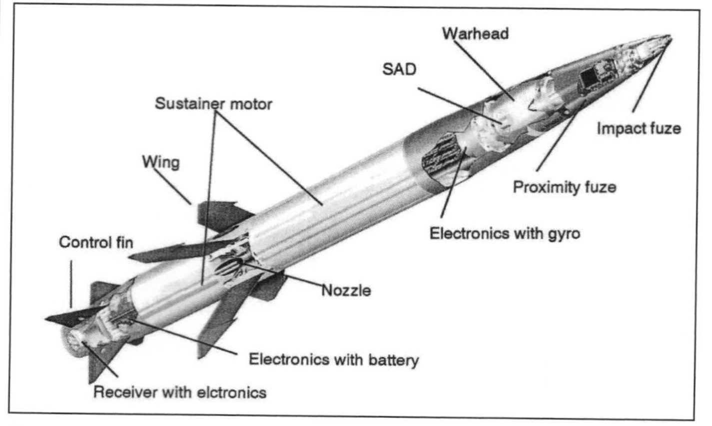

Saab Bofors Bolide low-altitude surface-to-air missile system 0095040

Guidance

The laser beam guidance allows the missile to target low altitude threats operating with terrain background, such as combat helicopters. Furthermore, the missile's reported unjammable laser beam guidance provides for a short reaction time and head-on engagement capability, with high accuracy and high kill probabilities at long range down to the lowest possible altitudes.

Fuze

The Bolide missile proximity fuze provides greater effect against small and dark targets as a result of:
- Laser beam lobes directing forward
- Higher laser pulse repetition frequency
- Adaptive proximity fuze delay
- Increased number of laser lobes
- Smart algorithms.

This proximity fuze uses electro-optical techniques and is built up with a transmitter channel and a receiver channel. The transmitter channel consists of laser diodes sequentially transmitting laser pulses in narrow lobes equally spaced and forward directed. The receiver channel consists of one detector, which sequentially receives information from outwards and rearwards pointing lobes so that the transmitter lobes cross the receiver lobes within the required area. These implemented techniques allow for strictly limited areas and ranges within which the proximity fuze is sensitive and extreme high resistance against jamming. When operating with the proximity fuze activated, a wing tip, rotor or the nose of an aircraft can initiate the proximity fuze. The proximity fuze can be disconnected prior to launch; the warhead is initiated then only by impact.

As a result of the above, the proximity fuze can be used in three different modes:

(1) Normal - against aircraft and helicopters
(2) Small target - against, for example, cruise missiles
(3) Off - where only the impact fuze is activated.

Warhead

The warhead and Safety and Arming Fuze (SAF) consists of an octol (HMX/TNT) based high-explosive filling, cast in a convex-surfaced steel casing with 3,000 embedded tungsten pellets. This gives an effective dispersion of the fragments, both forward and to the rear. Mounted in the front section of the warhead is a shaped charge, effective against armoured targets. The SAF initiates the warhead when a signal from the proximity or the impact fuze is received, but only after the SAF has been armed. The warhead is armed 250 m from the firing unit; the minimum intercept range.

Motor

The new larger motor (by six inches) for the Bolide missile is a double-base high-energy motor that has a longer burn than its predecessor. This also gives the missile an extended range and higher speed.

Possible future upgrade

Discussions with Saab Bofors Dynamics revealed the following are in line for future upgrades or modifications to the RBS 70/Bolide system that will be known as the RBS 70 NG (New Generation):
- Selective fusing by the operator prior to launch
- Semi-active laser
- Modulated Uplink commands
- Integrated sighting system
- Enhanced gunner aids.

Specifications

	Bolide
Dimensions and weights	
Length	
overall	1.32 m (4 ft 4 in)
Diameter	
body	106 mm (4.17 in)
Flight control surfaces	
span	0.320 m (1 ft 0½ in) (wing)
Performance	
Speed	
launch	97 kt (180 km/h; 112 mph; 50 m/s)
max speed	1,322 kt (2,448 km/h; 1,521 mph; 680 m/s) (est.)
Range	
min	0.1 n miles (0.25 km; 0.2 miles)
max	4.3 n miles (8 km; 5.0 miles) (est.)
Altitude	
max	5,000 m (16,404 ft) (est.)
Ordnance components	
Warhead	multipurpose fragmentation, shaped charge 3000 tungsten spheres
Fuze	laser, proximity
Guidance	laser beam riding (using pulsed, low-energy laser (GaAs) with a wavelength of 0.9 μm)
Propulsion	
type	solid propellant

Status

In production. In August 2002, Finland placed an order for the system, deliveries began during 2004. Five years later, in August 2007, Finland also placed an order for RBS 70 with deliveries expected in 2008.

Several orders were placed by Australia in 2003. All of these systems have been delivered.

The system can also be found in Taiwan, mounted on the AV7 Amphibious vehicles.

In mid-December 2013, the Lithuanian Ministry of Defence announced that the current stock of RBS 70 systems held in country would be upgraded during 2014/15 to use the Bolide missile. Part of the upgrade will also make the system able to use night vision devices.

Contractor

Saab Bofors Dynamics AB.

RBS 70

Type

Man-portable surface-to-air missile system (MANPADs)

Development

In late 1967, the Commander-in-Chief of the Swedish Army commissioned a special air defence committee to review air defence requirements and asked the committee to recommend what equipment should be developed or bought to meet requirements.

On the given economic and strategic grounds, the committee chose a combination of Saab JA-37 Viggen interceptors and short-range missile systems. The latter had to be cheap enough to be procured in large numbers yet still be able to operate under adverse Electronic CounterMeasures (ECM) conditions. The committee also recommended that the chosen system should replace the 20 mm cannon and General Dynamics Redeye (known locally as the Rb69) shoulder-launched SAM at brigade level and the Bofors 40 mm and 57 mm anti-aircraft guns at divisional level.

After reviewing all the alternatives available, a development contract for the RBS 70 (Robotsystem 70) missile system was placed with Bofors in mid-1969. It was intended at this stage to procure only the missile in its container-launcher tube, the control system and the sight and stand, with target detection being carried out visually.

However, studies carried out by the Commander-in-Chief of the army, showed that a more effective system would be produced if a search radar and Identification Friend or Foe (IFF) system were included. Therefore, in mid-1972, development contracts were also placed with SATT Elektronik AB for the PI-69 IFF system and for the RBS 70 and with LM Ericsson for its PS-70/R Basic Giraffe search radar (now known as the Ericsson Microwave Systems AB PS-70/R Giraffe 40 radar).

The first delivery of RBS 70 missile systems for trial purposes were made in late 1973 with user trials conducted between 1974 and 1975. In a three-phase evaluation programme the Swedish Army fired more than 100 complete test rounds fitted with telemetry heads. In 1975, the programme was completed satisfactorily and, in June of that year, the first production orders were placed for the RBS 70 missiles, sights, stands and PI-69 IFF sets. The first order for production of Basic Giraffe radar sets was not placed until 1978. The first production day-only RBS 70 missile system sets were delivered to Swedish Army training units in 1976, with the first operational units being formed the following year. The first production radar sets were delivered in 1979.

In the late seventies, Bofors completed development of the RBS 70 missile system by introducing a wider observing angle for the missile laser receiver that enlarges the engagement envelope by between 30 and 50 per cent depending upon the tactical situation.

In September 1986, the Swedish Defence Matériel Administration awarded the contract to FFV Aerotech, part of the Celsius group and a number of other companies, for the REnovation and MOdification (REMO) of the RBS 70 missile system. Total value is SEK250 million of which SEK140 million will go to a consortium consisting of FFV Aerotech and Bofors. The work continued until the year 2000.

To complete the RBS 70 missile system family, Bofors was awarded a development contract in 1984 by the Swedish Defence Matériel Administration (FMV) for the RBS 70M (M=*Mörker*, Swedish for 'dark') day/night missile system. Contracts were also placed with Ericsson Microwave Systems AB for complementary search and tracking radars, Giraffe 75 and a thermal imaging system for the weapon and to Hägglunds Vehicle for the conversion of its Bv 206 articulated tracked vehicle as the fire unit.

Now known by the designation RBS 90, the first production systems were delivered late in 1991 (see entry in Static and towed surface-to-air missile systems).

Only Sweden uses the RBS 90 SAM, although it is being marketed overseas.

In mid-1999, the Pakistan Army deployed RBS 70 missile systems in the vicinity of the Pakistan Navy Atlantic aircraft crash site following that aircraft's shooting down by Indian Air Force Mig-21 fighters.

As of October 2007, FLIR Systems Inc, announced that it had received a USD9.5 million order from Saab Bofors Dynamics in Karlskoga, Sweden, for custom designed thermal imaging systems and components that will be integrated into the RBS 70 system for delivery to an unknown export customer. Deliveries began in 2008 and extended into 2010.

In December 2013, the Ministry of National Defence in Lithuania reported that a contract had been signed by the Lithuanian Army and the Swedish Company SaaB Dynamics AB for the modernisation in 2014/15 of the RBS 70 missile system.

The modernisation will include night vision devices and also allow the new-generation missile Bolide to be used with the system. RBS 70 is expected to remain within the inventory until at least 2019.

Saab Bofors Dynamics RBS 70 missile system deployed in firing position and with IFF system fitted to lower part of tripod 0509837

Saab Bofors Dynamics RBS 70 VLM based on Land Rover (4 × 4) chassis with RBS 70 SAM leaving the launcher 0024896

Norwegian Army Low-Level Air Defence System
A modified version of the Giraffe has been developed under a contract placed in 1983 for the Norwegian Army. Known as the Norwegian Army Low-Level Air Defence System (NALLADS) project, the updated Giraffe 50 All-Terrain (AT) radar contains a specially designed data processing unit and is known under the Norwegian designation NO/MPY-1 radar system. It is installed on a Bv 206 tracked vehicle with the power generator and communications equipment in the front car, the radar and command-and-control units in the trailer car. The G-band radar and back-to-back IFF antenna can be raised to a height of 7 m on an extendable arm. The frequency agile radar provides automatic location, pop-up handling, identification and evaluation of threats and can handle up to 20 targets simultaneously with automatic track initiation on its two operator stations (each of which is equipped with a 19 inch (483 mm) full colour raster scan display). The derived data can be sent to terminals at each one of up to 20 RBS 70 launch stations through a specially developed communications system. Up to three NALLADS can be linked together to defend an extremely large area. Ericsson developed the software for the system in conjunction with Ericsson Radar Norge, while the weapon-control terminals come from Siemens (Norway). The first prototype system was delivered to the Norwegian Army in 1987 with a SEK615 million contracts placed in 1989. Deliveries started in 1992.

Target detection performances are: 70 km range/7,000 m altitude against a 10 m^2 Radar Cross-Section (RCS) target; 50 km range/5,200 m altitude against a 3 m^2 RCS target; 37.5 km range/4,000 m altitude against a 1 m^2 RCS target and 22.5 km range/2,200 m altitude against a 0.1 m^2 RCS target.

Additional details of the Giraffe 50AT search radar are given under Sweden in the anti-aircraft control systems section.

Description

Missile Mk 0

The basic 15 kg Missile Mk 0 (Robot70, Rb70) is a two-stage, Roxel solid propellant (Epictete casted double base), rocket-motor powered type. It is never removed from its container-launcher tube when in the field but, once fired, the tube is discarded. Both impact and active laser proximity fuzes are fitted to the missile's warhead, a dual-mode HE shaped charge pre-fragmented device which incorporates some 2,000 to 3,000 tungsten pellets, each with an optimised diameter of 3 mm. The proximity fuze is set for short-range

Saab Bofors Dynamics RBS 70 missile system installed on a Cadillac Gage V-200 (4 × 4) APC chassis as used by the Singapore Air Force (David Boey) 0024897

The Clip-On Night Device (COND) for the RBS 70 SAM system 0007254

activation to avoid premature detonation during operations close to reflecting surfaces such as ice, snow and water. For very low-altitude targets, like helicopters flying nap-of-the-earth profiles or behind natural obstacles (trees), the proximity fuze can be disabled before launch by a switch on the gunner's left hand aiming grip, warhead functioning now is only by direct impact with the target. The shaped charge can penetrate all aerial armoured targets as well as lightly-armoured vehicles and surface targets.

In addition to the essentially smokeless Imperial Metal Industry's sustainer rocket motor, the missile body also houses:

- A receiver unit which senses deviation from the laser line of sight
- A small computer that converts these deviations signals into guidance pulses that command the missile automatically to follow the centre of the laser beam.

Maximum engagement altitude is around 4,000 m while the minimum altitude is effectively ground level.

Missile Mk 1

The slightly heavier 16.5 kg Missile Mk 1 (Rb70 ös) is essentially the same as the basic Missile Mk 0 but uses a laser guidance sensor unit which increases the rearward field of view from 40° to 57°. This considerably enlarges the available engagement envelope.

Missile Mk 2

The Missile Mk 2 (Rb90) is different internally but remains the same in overall size and weight. The electronics have been considerably miniaturised to fit a larger sustainer and new combined warhead with over 3,000 tungsten pellets and shaped charge. The new sustainer increases both the missile velocity

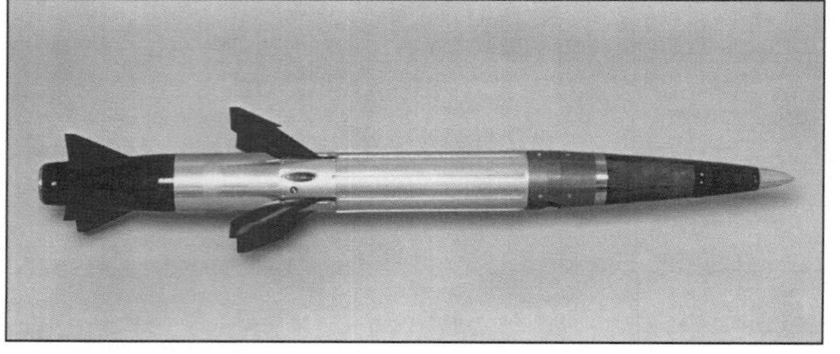

Saab Bofors Dynamics Missile Mk 2 used with the RBS 70 missile system 0062755

Ericsson Microwave Systems PS-70 (Giraffe 40) radar deployed with 12 m antenna erected
0007252

profile and the maximum range and altitude (up to 7,000 m as opposed to 6,000 m of the other round types against slow-moving targets and from 3,000 to 4,000 m altitude). A slightly higher maximum velocity is reached and, once in the coasting phase, the velocity decreases more slowly, so that when the missile reaches its maximum range it is flying slightly faster than the earlier round types. The missile guidance control method is described as the Linear Quadratic Method, based on the Kalman Theory.

Fire Unit

The basic fire unit comprises a stand and the sight as the two major parts. Each constitutes a one-person pack with a third team member carrying missiles in container-launcher tubes. The IFF equipment, if used, forms another portable pack, as does the target data receiver terminal, which is used in conjunction with the Giraffe search radars.

For an engagement, the tubular stand assembly is removed from its carrying harness and the three legs unfolded and roughly levelled by adjustments to one of them. The operator's seat is then unfolded from the vertical central tube and the gyrostabiliser sight, battery power supply, IFF unit and container-launcher missile tube attached. All the electrical connections are made to these units, the operator's headset and the data receiver terminal if present. A well-trained crew will take only 30 seconds to complete all these procedures.

Giraffe Radar

When the Giraffe 40 PS-70/R Tgb 40 (6 × 6), weighing 16,160 kg including the off-road truck-mounted G-band pulse Doppler radar, is networked with the launcher unit before operations can begin, a prismatic compass is set up behind the vehicle pointing at a mirror located on the edge of the radar's cabin roof. The angle between true north and the radar reference direction is read from the compass and set into the radar by means of a switch on the control panel to ensure that the Position Plan Indicator (PPI) scan is orientated to the north. A point at least 20 km to the south and west is then chosen on the map of the area, the radar's position is read and its co-ordinates, relative to the chosen point, are inserted on the control panel by a thumb wheel. Each fire unit attached to the radar is then also orientated to true north by prismatic compasses at each site and the relative positions of the radar and fire units established on a common grid for command-and-control reasons. For defence of an area containing a number of high-value targets, the firing units are deployed around 4,000 m apart thus allowing a company to defend an area of approximately 250 km^2, well overlapping each other's effective area.

The radar has MTI facilities and is fitted with an antenna that is elevated on a hydraulically operated mast to a height of 12 m. The operating cab houses three operators who detect and manually track targets on the digital PPI with another man plotting the air situation as updated by higher echelons on a map and this is used by the commander as the basis for his orders.

System operation

A radar search is initiated on receipt of a radio report from the higher air defence echelons, which indicates a potential target is approaching. The maximum detection ranges of 3 m^2 and 0.1 m^2 RCS targets, at speeds between 30-1,800 m/s, are 28 km and 12 km respectively. Maximum-instrumented search range is 40 km. Up to three targets can be handled simultaneously between the radar's altitude limits from very low level up to 10,000 m. The target's speed, course and direction are then passed up to a maximum of nine fire units by radio or cable link to the target data receivers as required by the radar vehicle commander. With the search radar's aid in its tracking mode (maximum range 20 km) the fire unit can engage to the maximum range of the missile.

The target data receiver unit computer takes the information sent, applies a parallax correction, displays the required angle of traverse and range to the target on a small screen and transmits an acoustic signal to the gunner who slews the sight and launcher assembly using two aiming handles until they hear a pulse tone on their headset. This indicates the sight is aligned with the approaching target in azimuth and the gunner then starts to search in elevation until they acquire it. Once the target is in missile range, indicated again by the target data receiver, the gunner fires the weapon by depressing a button with their left thumb. The latest target data receiver has a built-in threat evaluation providing priority of the assigned targets and also presenting first and last time to fire.

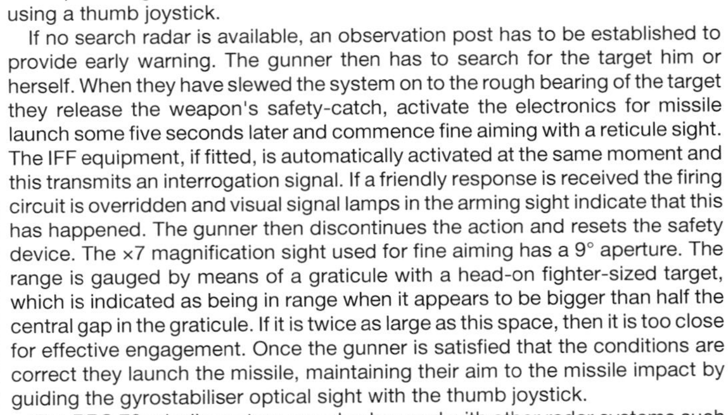

The laser guidance unit is activated and a Bofors booster motor on the missile is ignited to propel it out of the tube. For operator safety the motor cuts out before the end of the missile leaves the tube. The booster motor is jettisoned at a point several metres from the muzzle and the round's four centre-body fins and four rear cruciform control surfaces unfold. The sustainer motor ignites and the guidance receiver on the missile starts to sense the modulated laser guidance beam. This missile is now stabilised and flies in a cross-wing configuration. The onboard computer then translates these received signals into commands for moving the electrically operated control surfaces. Once at maximum velocity the sustainer cuts out and the missile continues on course in its coasting mode. To ensure a hit, the gunner has only to keep the target in the middle of the cross hairs of the gyrostabiliser sight by using a thumb joystick.

If no search radar is available, an observation post has to be established to provide early warning. The gunner then has to search for the target him or herself. When they have slewed the system on to the rough bearing of the target they release the weapon's safety-catch, activate the electronics for missile launch some five seconds later and commence fine aiming with a reticule sight. The IFF equipment, if fitted, is automatically activated at the same moment and this transmits an interrogation signal. If a friendly response is received the firing circuit is overridden and visual signal lamps in the arming sight indicate that this has happened. The gunner then discontinues the action and resets the safety device. The ×7 magnification sight used for fine aiming has a 9° aperture. The range is gauged by means of a graticule with a head-on fighter-sized target, which is indicated as being in range when it appears to be bigger than half the central gap in the graticule. If it is twice as large as this space, then it is too close for effective engagement. Once the gunner is satisfied that the conditions are correct they launch the missile, maintaining their aim to the missile impact by guiding the gyrostabiliser optical sight with the thumb joystick.

The RBS 70 missile system can also be used with other radar systems such as the Lockheed Martin PSTAR and the Ericsson Microwave Systems HARD.

Reloading

Reloading takes less than five seconds and the empty tube is discarded. The padded end caps of a new container-launcher are removed; this is then hooked on to the stand and secured by a lever which also connects the electrical contacts for the pre-launch power supply to the missile and the firing signal from a battery in the sight unit.

Simulator

An RBS 70 missile system simulator is available with a Target Path Unit based on a personal computer solution. Simulator training of the operator takes about 15 to 20 hours, typically spread over 10 to 13 days.

RBS 70 Trainer

Kongsberg Defence and Aerospace (KDA) began series production of 23 upgraded trainers for the Saab Bofors Dynamics RBS 70 short-range SAM systems of the Royal Norwegian Army in April 2002. The RBS 70 trainer upgrade includes the replacement of expensive and maintenance intensive original equipment and other obsolete components by Commercial Off-The-Shelf (COTS) technology. The gyro and mirror are removed and realistically simulated, and the CRT display to represent the RBS 70's coarse and telescope sights is replaced by high-resolution flat panel LCD technology. The Clip-On Night Device (COND) for night targeting is also simulated. The RBS 70 training system does not use a dome with laser projection of targets, typical of systems developed during the 1980s, as this is regarded an obsolete technology by the Norwegian company. The simulation software and the instructor software run on a standard Windows 2000 PC workstation.

KDA is now proposing a solution based on the Norwegian upgrade for the Australian Defence Forces planned air-defence training centre (Project Land 19), for which the invitations to tender were issued in mid-2002. Sweden is also looked upon as a potential customer.

Variants

RBS 70 VLM

There is a vehicle-launched version of the RBS 70 missile system available under the designation RBS 70 VLM. The RBS 70 VLM can be mounted on almost any wheeled or tracked vehicle. A target data receiver and a radio set normally form part of the unit. An ordinary field stand in the fire unit equipment ensures that the fire unit can be used independently of the vehicle. The mounted unit has one missile in the ready to fire position and six to eight others in reserve. A typical chassis for the RBS 70 VLM system is the British Land Rover (4 × 4).

RBS 70/M113A2 vehicle-mounted system

In response to a mid-1980s requirement from the Pakistan Army for an RBS 70 installation to defend mechanised units, Bofors adapted its VLM system for use with the M113A2 tracked Armoured Personnel Carrier (APC). A full description of the system is given in the Self-propelled surface-to-air missiles section under Pakistan.

ASRAD-R RBS 70 Installation

Saab Bofors Dynamics AB and STN ATLAS Elektronik have developed a version of the latter's ASRAD system that is fitted with the RBS 70 Missile Mk 2 and Ericsson Microwave Systems 3D HARD radar for the export market (see ASRAD entry in Self-propelled SAM section).

RBS 70 Clip-On Night Device (COND)

To increase the RBS 70 missile system operational capacity beyond its daylight engagement, a COND for attachment to the front of the day sight unit has been developed and is in series production. This weighs 24 kg (including an integral battery power pack and coolant bottle) and uses a 23-detector element, 8 to 12 µm waveband and infra-red scanner unit for target imaging. Thales Optronics is supplying the scanner and electronic control unit, similar to the Class I common modules supplied for the Bofors RBS 56 BILL ATGW. The thermal picture generated from the unity magnification equipment of the sight is injected directly into the front end of the day sight allowing the laser beam to pass through. No alignment is necessary. The detection range against aircraft is better than 10,000 m and against helicopters better than 6,000 m. By 1998 approximately half of the 13 customers for the RBS 70 system had placed orders for the COND. These include Australia, Indonesia, Ireland, Norway and Venezuela. The Swedish Army is also expected to place an order.

RBS 90 SAM system

This system was only used by the Swedish Army, production is now complete. The system has been removed from the Swedish inventory.

RBS 70/REPORTER

The day-only RBS 70 missile system has also been interfaced with the Dutch Hollandse Signaalapparaten REPORTER radar system in place of the Giraffe set. This uses a portable operations room shelter-mounted on the rear of any 1½ ton military vehicle, a generator unit and a 5 m extendable radar antenna mounted on a two-wheel trailer towed by the radar truck.

The 2-D I/J-band radar is fitted with an integrated IFF system and has a 40 km detection range on targets flying between 15 and 5,000 m. An MTI facility is fitted and the system can track up to 12 targets automatically from a 20 km range. The information derived is then routed automatically via a one-way data transmission system to all the firing units in the field to ensure their positions are not revealed. An unlimited number of such units can receive these data. The total time from target detection by the radar operator to the actual reception of an alarm at a firing unit can be as little as four seconds.

RBS 70 NG

The new RBS 70 NG is the latest air defence system presented by Saab. Now with integrated 24/7 all-target capability Saab's new RBS 70 NG VSHORAD system uses the Bolide missile that has been developed for any combat situation. The system boasts integrated sighting solution, enhanced missile operator aids, unmatched range and unjammable laser guidance. The system has been designed with the view to take on small Radar Cross Section (RCS) targets such as cruise missiles and UAVs out to a range of 8 km and altitude of up to 5,000 m.

Specifications

	RBS 70 Mk 2
Dimensions and weights	
Length	
overall	1.318 m (4 ft 4 in)
Diameter	
body	106 mm (4.17 in)
Flight control surfaces	
span	0.32 m (1 ft 0½ in) (wing)
Weight	
launch	16.5 kg (36 lb)
Performance	
Speed	
launch	97 kt (180 km/h; 112 mph; 50 m/s) (est.)
max speed	1,127 kt (2,088 km/h; 1,297 mph; 580 m/s) (est.)
Range	
min	0.1 n miles (0.2 km; 0.1 miles) (est.)
Altitude	
max	4,000 m (13,123 ft)
Ordnance components	
Warhead	1.1 kg (2.00 lb) HE fragmentation, shaped charge
Fuze	impact
	laser, proximity (3 modes; off, normal, small target)
Guidance	laser beam riding (modulated)
Propulsion	
type	solid propellant (Roxel Epictete cast double-base, booster and sustainer rocket motor)

Status

In production (Missile Mk 2 missile only). By 1999, total production of the RBS 70-missile system amounted to well over 1,300 launchers and 15,000 missiles. In 2002, the RBS 70 missile was replaced by the new Bolide missile with enhanced coverage and small target capability. (See details of the Bolide missile.) The Brazilian Army signed a procurement contract with Saab for a number of RBS 70 air-defence systems including simulators, night vision equipment and a test set with maintenance tools and spares. In service with the following countries:

Argentina

A small quantity have been delivered to the Marine Corps.

Australia

Early in 1985 the Australian Army selected the RBS 70 missile system as its Very Low-Level Air Defence System (VLLADS) and a contract was placed with Bofors for the supply of 60 launchers and support items. A Memorandum of Understanding (MoU) was also drawn up between Australia and Sweden, which provides for assurances on non-interrupted product support for the system.

The first contract was valued at SEK87 million with additional orders subsequently being placed. Australia has bought the Clip-On Night Device (COND).

The RBS 70, when used at the Battery level in support of the Rapier, would normally consist of three troops, each with five RBS 70s. The complete Battery would then include three Rapier launchers and three Blindfire radars (one of each per troop) and fifteen RBS 70s (five per troop).

Australia has also bought five Lockheed Sanders PSTAR alerting radars for cueing its RBS 70 missile systems. These were delivered by July 1998. In 2001, Australia announced two new procurements of RBS 70 systems including radars.

In June 2003, Saab Bofors Dynamics won a AUD42.5 million contract from the Australian Army for new air defence equipment. The deal includes Saab Bofors Dynamics-built RBS 70 laser-guided missiles, Lockheed Martin portable search and target acquisition radars and a tactical command-and-control system. Deliveries to the 16th Air Defence Regiment of the Royal Australian Artillery was completed in 2006.

Bahrain

The Bahrain Defence Force purchased RBS 70 missile systems in 1979 with the first deliveries being made in 1980.

Czech Republic

The Czech military will receive 16 new RBS 70 mobile anti-aircraft missile systems by 2007 and Omnipol will be the firm to act as an intermediary in the transaction worth about KCS1 billion. The RBS 70 will replace the older Soviet made Strela-2M. The final delivery of the systems was made in October 2007; the 25th Anti-Aircraft Brigade took delivery of the final 16 fully equipped launchers complete with IFF subsystems and BORC third-generation clip-on thermal imagers, two training simulators and almost 100 missiles.

Finland

In 2002, Finland placed an order for 128 × RBS 70 Mk 2 missiles, the deal was worth USD30 million. Part of this deal worth USD120 million was for the ASRAD-R systems. In August 2007, Finland placed a second order worth SEK600 million for the complete system, including missiles and maintenance equipment.

Indonesia

A number of RBS 70 missile systems have been delivered for use by the Indonesian Army. The Giraffe radar system is also in service with the system. It is known that the Reporter radar is also used. Indonesia has also bought the Clip-On Night Device (COND).

Ireland

The Irish Army purchased four RBS 70 missile systems in the late 1970s.

An order for a Giraffe 40 radar mounted on a 4 × 4 truck chassis was placed in 1985. This was delivered in 1986 and uses AN/VEC-46 radios for data transmission and AN/PRC-77 radios to receive the target data at the missile post positions. The alternative to the radio net is a standard telephone cable link. Ireland has bought the Clip-On Night Device (COND).

Latvia

First delivered to Latvia in 2006/7 with three Giraffe-40 surveillance radars.

Lithuania

The Lithuanian Air Force accepted the RBS 70 from Sweden during 2005. Between 5 and 6 October the same year, troops of the Air Defence Battalion of the Lithuanian Air Force, together with their Latvian counterparts, conducted the first ever exercise of live firing using the RBS 70 missile system.

Malaysia

Towards the end of November 2008, it was reported by sources in Malaysia that the RBS 70 was deployed as the MANPAD variant within the Malay inventory.

Norway

Following a Norwegian Army competitive evaluation in the late 1970s for a MANPADS missile system, a contract worth SEK400 million was placed for 110 RBS 70 missile systems and 27 Giraffe 40 radars. These were used from 1981 onwards to reorganise the air defence batteries of six field brigades with three Giraffe 40 radars, 18 RBS 70 launchers and 12 20 mm light anti-aircraft guns each. The first radars were delivered in 1982.

In 1984, a second MANPADS evaluation contest was held for a system to re-equip the rest of the field brigades. The RBS 70 missile system again won and a SEK700 million contract was placed with deliveries taking place between 1987 and 1990. The contract also involved a long-term maintenance agreement between Bofors, the Norwegian Ministry of Defence and participation by Norsk Forsvarsteknologi (the then Hughes and Kongsberg) in manufacturing some of the electronic components of the system and undertaking the final assembly and testing of the sights.

The contract also contained an option, which has been taken up for additional systems for use by the Norwegian Coastal Defence Force. Most of these systems have already been delivered.

In April 1988, Norway awarded a further SEK500 million contract to Bofors for the Missile Mk 2 as the third phase of its RBS 70 missile system procurement programme for its army and coastal defence force. The missiles were delivered between 1990 and 1992 with the contract including options for additional missile batches as required.

In September 1989, Norway awarded Bofors an SEK800 million plus contract for RBS 70 missile systems and Missile Mk 2 to replace the 40 mm L/60 Bofors anti-aircraft guns used as the main component of the Norwegian Air Force air defence network around its airbases. Delivery began in 1991 and continued until 1994.

Production of the RBS 70 missile system field sight started at NFT in 1986/1987 with final deliveries in late 1993. In January 1997, it was revealed that Bofors had been awarded a contract worth SEK190 million from the Norwegian military procurement agency for the Clip-On Night Device (COND). First deliveries were made in 1998. In April 1999, a SEK19 million contract was placed for COND spares, support equipment and simulator units.

Pakistan

In 1986, Pakistan ordered missiles and RBS 70 missile systems with Giraffe 40 radar systems. These were delivered from 1986 onwards and include both ground and various vehicle type mounts including M113A2 APC installations. Pakistan is now making certain parts of the RBS 70 missile system and assembling the system from complete knock down kits.

Singapore

The Singapore Air Defence Systems Division (a merger of the Singapore Air Defence Artillery and the Air Forces Systems Command in 1995) operates several RBS 70 missile systems that are carried on Cadillac Gage V-200 (4 × 4) APCs. The Singapore Air Defence Systems Division has three battalions, the 3rd Defence Artillery Battalion, 6th Defence Artillery Battalion and 9th Defence Artillery Battalion equipped with RBS 70, Igla and Giraffe 40 radars. Their roles are to provide air defence to the Army Division and vital installations within Singapore. A fourth battalion, the 18th Defence Artillery Battalion, uses the Mistral MANPADS.

The Giraffe 40 radars of the Army Air Defence Artillery have been locally mounted on modified Mercedes-Benz (6 × 6) truck chassis. Singapore Automotive Engineering (SAE) carried out the conversion work.

Sweden

The Swedish Army deploys a large number as described in the development section of this entry. In December 1988, the Swedish Armed Forces placed an SEK200 million contract for Missile Mk 2.

In January 2007, Sweden purchased from Kongsberg in Norway the Ground Based Air Defence Operation Centre (GBADOC) for use with the Nordic Battlegroup. This makes Sweden the 4th country to adopt the system, and their GBADOC will be upgraded to include Swedish national data links for integration of the Saab Ericsson Giraffe AMB radar and the RBS 70 as well as higher echelon units. The GBADOC was delivered in 2007 and is expected to be fully operational in 2008.

Thailand

In October 1996, it was revealed that the Royal Thai Air Force had selected the RBS 70 missile system with the Missile Mk 2 for airbase defence. A total of three launchers and 15 missiles were delivered during 1997 together with several Giraffe 40 radars.

Tunisia

In 1979, the Tunisian Army ordered 60 RBS 70 missile systems and 12 Basic Giraffe radar systems, which were delivered in 1980 to 1981.

United Arab Emirates

In 1979, Dubai placed an order for a number of RBS 70 missile systems and missiles, which were delivered from 1980 onwards.

Venezuela

A number of RBS 70 missile systems have been delivered for use by the Venezuelan Army. Venezuela has bought the Clip-On Night Device (COND) sight.

In 1998 the Venezuelan Army ordered four truck-mounted Ericsson Microwave Systems Giraffe 75 surveillance radar and combat control centres for use with its RBS 70 missile systems. The C-band radar has an instrumented range of 75 km and features automatic threat evaluation, with a hovering helicopter detection function integrated into the radar. The systems are mounted on the rear of German MAN LX90 6 × 6 cross-country vehicles.

In February 1999 an order was placed for the Missile Mk 2 and a limited amount of missile-related equipment.

Contractor

Saab Bofors Dynamics AB (RBS 70 missile system).
Ericsson Microwave Systems AB, Ground Systems Division (Giraffe series radars).

Ukraine

Igla-1M Upgrade Package

Type

Man-portable surface-to-air missile (MANPADs) system

Development

The Arsenal Central Design Bureau in the Ukraine was founded in 1954 as a structural sub-division of the Arsenal Industrial Plant in Kiev, and is now offering an upgrade for the 9K310 Igla-1 (NATO SA-16 Gimlet) man-portable SAM system to improve its low-altitude capabilities against targets such as turbo-props, jets, propeller driven aircraft, cruise missiles and helicopters in head-on and rear aspect engagements where thermal background and InfraRed (IR) jamming environments are present.

Description

The Igla-1M upgrade comprises:

- Replacement of the existing 9E418 seeker by a new model with enhanced performance capabilities against IR jamming and thermal background phenomena. A new seeker head that is designated I1-2000 is offered by the Arsenal Central Design Bureau for portable air-defence missile systems for use against aircraft, helicopters, cruise missiles and drones. (Technical specifications are listed in Table 1 below.)
- Modification of several missile subsystems to support the new seeker head model.

Table 1

Guidance method:	proportional navigation
Target lock-on range:	5-7 km
Minimum FoV:	40°
Field of vision:	±40°
Maximum angular target auto tracking speed:	15° per second
Maximum readiness time:	5 s
Minimum autonomous operation time:	14 s
Maximum power consumption:	35 W
Weight:	1 kg

Specifications

Igla-1M upgrade

Maximum firing range:	
(approaching target)	4-5,000 m
(receding target)	≤4,000 m
Altitude:	
(maximum)	3,500 m
(minimum)	10 m
Lethality:	0.4-0.6
Average probability of jamming rejection for delay-to-target signal ratio of up to 30×:	0.7

Status

Ready for production. Offered for export.

In 2012, a further upgrade to the Igla-1M was offered by Arsenal, this is the UA-424, which is reported to provide enhanced capabilities against IR countermeasures along with improved single-launch kill probability. The UA-424 also offers extended range whilst engaging its targets head-on.

As of January 2013, no further information on this upgrade was available, however, according to the Russian Defence Ministry, the Ukrainian Arsenal plant delivered, or planned to deliver, 50 9K310 and 400 9M313 missiles with missile seekers upgraded by the Arsenal plant.

Contractor

Arsenal Central Design Bureau (I1-2000 seeker head).
Scientific Industrial Complex Progress (9E418 seeker head).

Ukrainian seeker assemblies

Type

Overhaul and maintenance services of anti-aircraft systems

Development

At the IDEX '97 military show, the then Ukrainian firms Arsenal Central Design Bureau and the 'Progress' Nizhi Scientific Complex displayed a variety of passive infrared seekers suitable for use with the Russian Kolomna KBM families of man-portable Surface-to-Air Missile (SAM) systems.

Ukrainian weapons designers are known to have upgraded air-defence systems such as the Strela-2/2M, the Igla and Igla-1. Unfortunately, the information derived from various defence exhibitions is not always constant.

Description

The seeker types shown for use with the Russian Kolomna KBM man-portable family of missile systems at the show and available for export included:

- The UA24 for the Strela-2M (SA-7b 'Grail') system
- The UA424 for the Igla (SA-18 'Grouse') system
- The UA418 for the Igla-1 (SA-16 'Gimlet') system.

The Arsenal Central Design Bureau is also offering the Igla-1M upgrade of the Igla-1 system utilising the I1-2000 seeker head (see entry within this section).

Later in 2012, the same Company (Arsenal Central Design Bureau) was also offering Optical seekers 36-45 for the Strela-2/2M and UA-424 for the Igla-1M.

The 36-45 optical seeker for the Strela-2/2M is reported to enable the system to defeat IR decoy flares and interference both natural and electronic.

The UA-424 for the Igla-1M provides enhanced IR countermeasures and improved single launch kill probability with a reported increase in range.

Another upgrade is the 336-24 that is for the Igla system. As with all of these upgrades the 336-24 offers improved IR countermeasures but additionally a range increase to 5,200 m.

Status

Production as required. Offered for export in Ukrainian weapons brochures and manuals.

Contractor

Arsenal Central Design Bureau.
Scientific Industrial Complex Progress.

United Kingdom

Javelin

Type

Man-portable surface-to-air missile system (MANPADs)

Development

The Thales UK Javelin close-range air defence weapon was developed under contract to the British MoD from 1979 as a follow-on to the Blowpipe system. It was first revealed in September 1983, by which time initial firing trials had already been completed.

The Javelin was designed to counter a wide range of low level air defence targets and it employs Semi-Automatic Command to Line-Of-Sight (SACLOS) guidance, rather than infrared detection, to engage its target. The Javelin can also be used against helicopters and has a secondary limited surface-to-surface capability.

In comparison with the earlier Blowpipe, the Javelin has a new warhead, a more powerful second stage motor for increased range and it has also benefited from advances in video micro-processing techniques. The updated guidance system, which has been incorporated in a new aiming unit, makes the operator's task much easier. The aiming unit of Javelin has been designed so that it is compatible with the current Blowpipe system.

Thales UK stresses that the Javelin does not rely on the target's infrared signature and is therefore almost impossible to counter by decoys such as hot flares. Compared with the earlier Blowpipe, training time is much reduced with the Javelin system. First production Javelins were completed in 1984.

In June 1984, the then Shorts Missile Systems announced a second order worth GBP35 million from the UK MoD for production of Javelin. This increased home and export sales to GBP120 million. In the middle of the same year, it was announced that the Javelin had been selected by the Royal Navy to provide special protection against kamikaze-type attacks on naval vessels, especially those operating in the Middle East. In January 1985, the then Shorts Missile Systems announced that it had received a third production contract valued at GBP25 million for the Javelin, which brought the total Javelin order book to over GBP160 million.

According to the manufacturers, the effectiveness of the Javelin close-range SAM was demonstrated in British Army practice camps where so many Skeet targets were destroyed, that at least one camp had to be delayed through lack of targets. One aimer of 10 (Assaye) Air Defence Battery had a 100 per cent success rate in 1985 when eight Skeet targets in eight engagements were destroyed.

During the 1986 British Army Equipment Exhibition, the manufacturers revealed that it was acting as Project Manager for all aspects of Close-Air Defence Weapon Systems (CADWS) on behalf of the MoD. This included Blowpipe, Javelin and Starstreak, but also the system integration for enhancements to UK CADWS, the then Thales Optronics Air Defence Alerting Device (ADAD), the IFF equipment, the thermal imaging night sight and the Air Defence Command and Information System (ADCIS). By mid-1993, the Javelin had been replaced by Starburst in front line and reserve unit operational use. The Javelin stocks remaining have been reserved for training use only. Production of the Javelin missile has been completed.

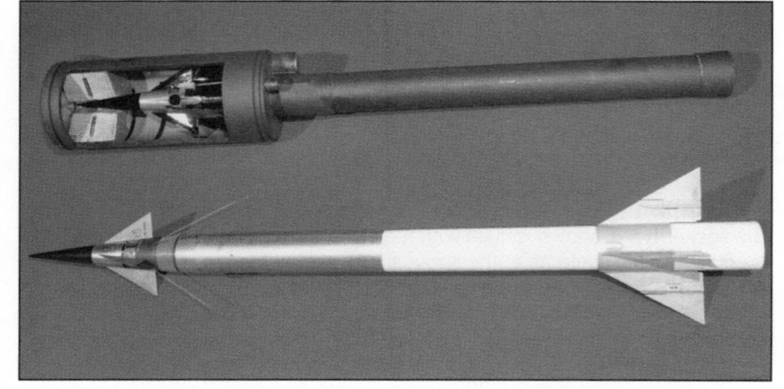

Thales UK Javelin missile out of its launcher tube (bottom) with missile in its launcher tube (top) 0509401

Standard man-portable Thales UK Javelin SAM system deployed with Thales Optronics Air Defence Alert Device (ADAD) in the foreground 0509672

Description

The Javelin SAM system consists of two main components: the missile sealed within its launching canister and the aiming unit; lightweight carrying cases are provided for both.

The canister in which the missile is factory-sealed is a lightweight environmental container, designed to act as a recoilless launcher. It houses the guidance aerial, the electrical connections and the thermal battery to power the aiming unit after missile launch.

The front cap is blown off by gas pressure when the missile gyro is fired and the laminated rear closure is ejected at launch.

The missile is a 1.4 m long slender tube with the fuzes in the tip and the warhead in the centre. The guidance equipment is in the forward part of the body and the rocket motors are in the rear. There are four delta shaped aerofoils in the nose for aerodynamic control and four at the tail for ballistic stability. A self-destruct facility is incorporated. The nose section is free to rotate independently of the main body, to which it is attached by a low-friction bearing. Twist and steer commands to the control fins guide the missile with a high response rate.

The aiming unit is a self-contained firing and control pack with a pistol-grip firing handle on the right side. It contains a stabilised sighting system, which provides manual target tracking and automatic missile guidance through a solid state TV camera.

Digital commands from the camera are fed to a microprocessor and the resultant guidance demands are transmitted to the missile by radio. The simple controls on the handle include: the firing trigger, thumb-controlled joystick and system, fuze mode and superelevation switches. Other controls include channel selector switches for the transmitter and an automatic crosswind correction switch.

The then Marconi Avionics was awarded its first production contract, worth over GBP5 million, for the advanced television guidance system used in the Javelin in May 1984. The automatic gather-and-guidance system comprises: a miniature solid state Charge-Coupled Device (CCD), television camera and zoom lens, sophisticated signal processing electronics and a two-axis sub-miniature gyro assembly. The company's Electro-Optical Products Division, Basildon and the gyro assembly produced the camera unit and associated data extraction equipment by the Gyro Division, Rochester. The complete electro-optical and gyro subsystem is contained within the operator's lightweight aiming unit. This was made possible by the use of a CCD imaging array (a light-sensitive microchip) to form the TV picture and by high-density electronic packaging involving multi-layer hybrid microcircuits.

The Javelin is made ready for action by clipping the aiming unit on to the canister, which takes less than five seconds.

Information of an imminent attack can be received over the radio net or by the team scanning the horizon visually. The aimer acquires the target in the monocular sight and switches on the system, selecting the frequency of the guidance transmitter and the mode of the fuze (proximity or impact). This activates the tracking electronics and projects an illuminated stabilised aiming mark (a red circular reticule) into their field of view. The target is tracked briefly

Thales UK man-portable Javelin SAM being launched from a surface ship of the Royal Navy 0509402

Lightweight Multiple Launcher deployed in the field 0509741

with the aiming mark to establish a lead angle, the safety catch is released and the trigger pressed. Range is indicated stadiametrically in the aimer's eyepiece, which has a magnification of ×6, compared with ×5 of a Blowpipe.

The firing trigger activates two thermal batteries, one of which supplies the power to the aiming unit while the other supplies power to the missile.

The gyro of the missile is run up by the action of the smokeless propellant gas generator, initiated by the thermal battery. The gas overpressure blows off the canister sealing cover and the missile is boosted from the canister by the first stage motor (the same type as used in the Blowpipe) which burns out 0.2 second before the missile emerges from the tube. Then, at a safe distance from the aimer, the weapon is accelerated to its supersonic burnout speed, by the second stage sustainer motor.

The Javelin's wing assembly comprises four wings mounted on a central tube and is housed at the forward end of the canister until the round is launched. The wing tips are folded in this stowed position to reduce the diameter of the canister. While the missile is being launched, the main body of the missile passes through the wing assembly, which is arrested by a band of tape around the rear body. When the missile is clear of the launch canister its wing tips are unfolded by the roll action that the booster motor applies to the rear body. A slight cant on the wings then rolls the missile continuously throughout flight in order to maintain aerodynamic stability. The missile is not armed until it is at a safe distance from the aimer and if guidance signals are lost it self-destructs. Javelin retains the twist and steer control method of Blowpipe with the control surfaces being mounted on the nose section, which is free to rotate.

The camera detects the missile flares and, using digital techniques, transmits guidance demands to the missile. The TV guidance datum line is collimated with the aiming mark, which is maintained on target by the gunner using the thumb joystick. In the event of sight failure, the integral TV camera system tracks the missile flares and sends an error signal via the command link to adjust its flight trajectory as needed.

The warhead is detonated either by the pre-set impact or proximity fuze.

Variants

Lightweight Multiple Launcher (LML)

The LML has been developed to provide the Javelin with a multi-engagement capability. All LML systems use three standard Javelin canister missiles and a standard shoulder-launched aiming unit as clip-on equipment.

In the free-standing application, the support tube is held vertically by the tripod legs, which pivot off an eccentric support collar. Screw jacks positioned between the top of the sleeve and the legs are used to level the launcher.

When deployed in a trench, the support collar can be slid down the support tube partially and then clamped at the appropriate height. This means one leg hangs vertically while the other legs are used to provide lateral support to the launcher.

To deploy the LML, the tripod stand is erected by unfolding the legs and sliding the leg support collar to the bottom of the support tube, which is then locked in position. The LML head is then lifted on to a spigot on top of the tripod stand and the sight arm is released from its stowed position and unfolded into its operational state. After fitting the aiming unit on to the sight arm saddle and loading the three missiles in position, the LML is ready for action and the aimer now follows normal Javelin operational procedures. The LML is in service with UK and overseas forces.

Lightweight Multiple Launcher (Vehicle) (LML(V))

The LML(V) is suitable for mounting on many types of armoured personnel carriers and has been mounted on the Shorland S53 (4 × 4) Air Defence Vehicle, variant of the Shorland S52 Armoured Patrol Car. It can also be mounted on unarmoured vehicles such as the US AM General HMMWV (4 × 4).

The turret ring is fitted over a hatch opening and is provided with its own integral hatch cover. The ring carries a pintle for mounting the vehicle variant of the traverse head. Turret traverse relative to the turret ring is plus or minus 40 degrees. Six missiles in their container launchers are stowed in racks either side at the rear of the S53 vehicle.

The ring is provided with a handgrip and frictional brake, enabling the aimer to slew it to approximate target bearing and then track the target. The LML(V) is in service overseas.

Lightweight Multiple Launcher (Naval) (LML(N))

The LML(N) was developed for naval applications.

Specifications

	Javelin
Dimensions and weights	
Length	
overall	1.39 m (4 ft 6¾ in)
Diameter	
body	76 mm (2.99 in)
Flight control surfaces	
span	0.275 m (10¾ in) (wing)
Weight	
launch	12.7 kg (27 lb)
Performance	
Range	
min	0.2 n miles (0.3 km; 0.2 miles) (est.)
max	3.0 n miles (5.5 km; 3.4 miles) (est.) (approaching target, against helicopters)
	2.4 n miles (4.5 km; 2.8 miles) (est.) (approaching target, against jet aircraft)
Altitude	
min	10 m (33 ft) (est.)
max	3,000 m (9,843 ft)
Ordnance components	
Warhead	2.74 kg (6.00 lb) HE fragmentation
Fuze	impact, proximity
Guidance	SACLOS
Propulsion	
type	two stage, solid propellant

Status

Production of the Javelin missile system is complete, (over 16,000 missiles were produced). It is no longer marketed. The system has seen service with Botswana, Chile, South Korea, Malaysia, Oman, Peru, and the United Kingdom.

Contractor

Thales UK (primary contractor).

Moog Components Group Ltd (electro-mechanical and fibre optic products, including slips rings, etc).

Starburst

Type

Man-portable surface-to-air missile system (MANPADs)

Development

The Starburst close air defence missile system (in house project designation Javelin S15) was developed from the middle 1980s onwards, as an advanced unjammable variant of the Javelin man-portable low-altitude SAM system.

Thales UK Starburst Lightweight Multiple Launcher (LML) firing a Starburst 0062756

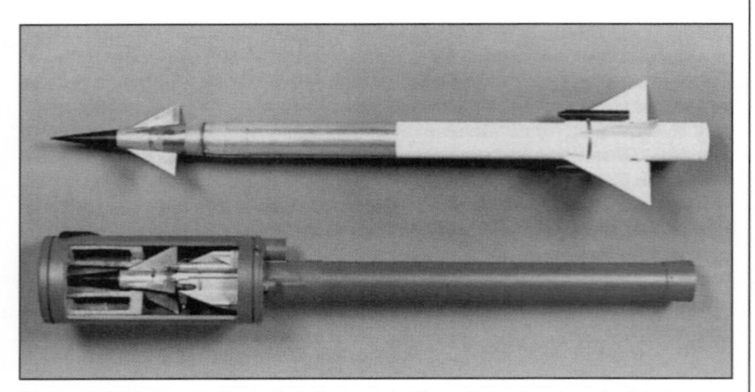

Thales UK Starburst close air defence missile system showing the Starburst missile (top) and cutaway missile canister with Starburst missile in its ready-to-launch position (bottom) 0509400

Thales UK Starburst Multiple Launcher launching a Starburst SAM 0007256

Designed to meet an MoD requirement placed on behalf of the British Army, the Starburst maintains all the proven characteristics of the parent system's airframe and aiming unit but incorporates the laser optical command guidance technology of the high-velocity follow on, Starstreak, to significantly increase the weapon's single shot hit probability.

The first shoulder-launched firing took place in 1986 and development was completed in 1989. This was also the year that the British Army formally accepted the weapon for service, with the first deliveries being made.

Operational deployment began in early 1990. Starburst has now replaced all Javelins for operational use by both regular and reserve army units.

Starburst was deployed during Operation Desert Storm by 10 (Assaye) Battery, 40th Field Regiment Royal Artillery, where it maintained an operational availability of 100 per cent for all battlefield days during the war.

In its simplest form the Starburst system consists of two units - a single missile in its container launcher and a clip on aiming unit. However, its prime application is a three round system which utilises the then Shorts Lightweight Multiple Launcher (LML), providing both increased firepower and enhanced tracking capability. Depending upon configuration of the LML, the Starburst can be used for ground, trench, vehicle or naval applications.

In October 1994, it was announced that the Kuwait Ministry of Defence had confirmed its 1993 order for the Starburst missile system. Included in the GBP50 million package was the Thales Starlite Thermal Imager, which gives the Starburst system a 24 hour capability.

The Kuwait Air Force is believed to have received 48 Starburst Lightweight Multiple Launchers and 250 missiles, although other sources now indicate that the number of missiles was in fact possibly greater. Training of the Kuwait Air Defence Brigade (Kuwait Air Force) personnel on the Starburst system began in early 1995, by which time over 10,000 missiles had been built.

In November 1996, an MoU was signed with the government of Qatar for the provision of a package of defence equipment valued at GBP500 million from a number of UK firms. Included in the package was the Starburst close air defence missile system. The system probably ceased production around 2001.

Description

The Starburst missile consists of a two-stage Royal Ordnance Westcott motor, pre-fragmented blast warhead and dual mode impact/proximity fuze. Twist and steer commands are sent to the forward-mounted steering control surfaces. Ballistic stability is provided by the rear fins, which house the two interconnected laser transceiver guidance units. The latter act as the relays between the laser guidance unit and the missile's forward-positioned electronics and control section.

Each of the transceiver units incorporates a laser receiver, a signal processor and a transmitter in a small cylindrical pod. The reasons for two electrically interconnected pods being fitted are system redundancy and the prevention of any possible screening effects acting upon the guidance signals.

The transmitter is mounted in the nose of the pod and relays the command uplink data to the missile's forward-mounted electronics. The optical data signals are detected by small pop-up antennas connected to the control unit, which, apart from software changes, is essentially unchanged from that used on the Javelin.

The missile canister is a sealed lightweight environmental container, which acts as a recoilless launcher tube and is discarded after use. It houses an electrical interface connector to pass firing signals from the aiming unit to the missile. At launch, the front cap of the canister is blown off by the gas pressure when the missile gyro is fired.

The reusable 408 × 342 × 203 mm Avimo Ltd laser guidance unit consists of a laser guidance head and a control unit. The former, which requires no alignment procedure, has an optical stabilisation system, a guidance transmitter, an aiming mark injector and a ×6 magnification monocular sight - all housed in an environmentally sealed light-alloy casting.

The control unit consists of a lightweight moulded case with an environmentally sealed compartment (containing the wind offset switch and electronic assemblies for processing and control), an externally mounted battery box and an attached control handle assembly. The control handle contains the joystick, trigger, system on/off switch, fuze selection switch and superelevation button. The SB14 IFF unit (described under Starburst LML below) can also be fitted.

In combat, when the gunner receives a target indication they acquire it in the monocular sight and select 'system on'. They then track the target by moving the weapon system so that coincidence is maintained with the aiming mark. This action automatically generates lead angles, in both azimuth and elevation. The gunner then operates the trigger mechanism.

Thales UK Starburst Lightweight Multiple Launcher (LML) vehicle on Land Rover (4 × 4) of the Royal Malaysian Army 0509838

Thales UK Starburst close air defence missile system deployed in standing configuration with missile leaving the launcher 0007255

The missile is launched from the canister by the first stage motor and, at a safe distance from the gunner, is boosted to supersonic velocity by the second-stage rocket motor. The first stage rocket motor is completely burnt within the length of the canister and is jettisoned upon emergence from the canister. The gunner continues to track the target by keeping the aiming mark superimposed over it by use of the thumb-operated joystick. The missile guidance system lock on is automatically maintained on the centre of the aiming ring. On reaching the target, the missile's warhead is detonated either by impact or the proximity fuzing circuit. If, after launch, it is realised that the target is in fact a friendly aircraft, the gunner has the facility to command the missile to self-destruct.

A Trainer Set, suitable for both initial and continuation training, is also available for both the shoulder-launched and LML Starburst applications.

Variants

Starburst Lightweight Multiple Launcher (LML)

The Starburst LML is similar to the LML described in the later Javelin entry, except that it uses three standard Starburst canistered missiles and a standard Starburst Aiming Unit as the clip-on equipment. If required, both a thermal imager (for 24 hour operation) and an IFF system can be fitted. The latter system weighs 3.5 kg (including battery) and its dimensions are 254 × 180 × 110.5 mm. Designated SB14, the system is capable of Selective Identification Feature (SIF) 1, 2, 3/A and secure operation with Mark XII compatibility. A removable memory module is mounted on the side of the interrogator, which can be programmed with Secure Mode and one SIF or two SIF modes. Memory capacity provides up to four days of SIF and crypto secure operations and takes account of code transition period conflicts. Range is out to 10,000 m and time to identify is 0.2 s. BITE and Friend status indications are visually displayed with optional audio synthesised voice.

Dimensions of the LML head assembly are 697 × 533 × 514 mm in field use, the tripod stand 1,930 × 305 mm diameter in field use. Weights are: head assembly field use 18.2 kg, operational use 15.6 kg; tripod stand field use 16 kg, operational use 14.7 kg. Time into action with three missiles and the standard laser guidance unit is less than two minutes.

Starburst Vehicle Multiple Launcher (VML)

The Starburst VML is similar to the turret ring mounted strengthened head VML described in the Javelin entry, except that it uses three standard Starburst canistered missiles and a standard Starburst Aiming Unit as the clip-on equipment. Its weight is 44 kg empty; its dimensions with missiles loaded are 0.793 × 0.927 × 1.441 m and the associated hatch ring is 78 kg. The three canistered missiles weigh 45.6 kg and the laser guidance unit 8.5 kg.

Starburst Naval Multiple Launcher (NML)

The Starburst NML consists of a lightweight tubular turret supporting power-assisted azimuth and elevation drives for two missile panniers, each containing four ready-to-fire Starburst missiles. The elevation arm supports the guidance and fire-control unit, incorporating a thermal imaging acquisition and tracking sensor for 24 hour operation. Cueing is by the ship's target acquisition systems or the weapon operator.

Starburst Starlite Thermal Imager

The Starlite TI is designed as a clip-on to be fitted to a Starburst laser guidance unit. It operates in the 8 to 12 μm waveband region, providing detection of head-on attacking fighters at 9,000 m and helicopters at 7,000 m. It operates independently of the laser guidance unit and has an 8 × 6° field of view. Weight is 6 kg and dimensions 25 × 20 × 15 mm. Its battery pack has a life of up to 80 hours.

Starburst SR2000

As a result of a collaborative venture between Radamec Defence Systems and Thales UK the shipborne Starburst SR2000 system has been developed. This combines the Starburst air defence missile on a six round fully stabilised launch platform with Radamec's 2400 Electro-Optical Tracking System. This forms an integrated air defence guided missile system. The Radamec 2400 can track targets in excess of 12,000 m, allowing the Starburst missiles to engage strike aircraft and helicopters at well beyond weapon release range. The system also has an anti-missile capability and can be used effectively against surface vessels.

Starburst upgrades

Since Starburst's service introduction, Thales UK has introduced a number of system upgrades. These include:

(1) The substitution of an MBDA-CSF radar proximity fuze to increase the proximity fuzing distance by around 400 per cent over the current system, giving greater effectiveness to the pre-fragmented blast warhead's lethal radius. The fuzing modification was tested during 1991 and has been available from the beginning of 1992.

(2) The introduction of a non-rechargeable Crompton Vidor lithium sulphur dioxide battery pack as an alternative to the current three rechargeable Ni/Cd batteries, obviating the need to carry many spare batteries and a recharging unit. The lithium battery is effective for more than 300 engagements.

(3) The adoption of a Simrad KN200 Image Intensifier clip-on sight unit for operations during periods of low light and/or bad visibility. The unit weighs 1.4 kg and has dimensions of 210 × 116 × 140 mm (excluding the optical relay). Power requirements are two batteries providing 3V. The intensifier image is projected into the monocular of the laser guidance unit, with no change to the operator's eye position for day or dusk use. The lightweight nature allows the unit to be used for both the shoulder-launch and the multiple launcher applications.

Specifications

	Starburst
Dimensions and weights	
Length	
overall	1.394 m (4 ft 7 in)
Flight control surfaces	
span	0.197 m (7¾ in) (wing)
Weight	
launch	8.5 kg (18.00 lb)
Performance	
Range	
min	1.6 n miles (3 km; 1.9 miles)
max	3.8 n miles (7 km; 4.3 miles) (est.)
Ordnance components	
Warhead	2.74 kg (6.00 lb) HE blast fragmentation
Fuze	impact, proximity
Guidance	laser beam riding
Propulsion	
type	two stage, solid propellant

Status

This system is no longer in production, however, Starburst has seen service with Canada, Jordan, Kuwait, Malaysia and the UK.

Contractor

Thales UK.

Starstreak I

Type

Man-portable surface-to-air missile system (MANPADs)

Thales Starstreak High-Velocity Missile shortly after launch, showing booster falling away. The Lightweight Multiple Launcher has three Starstreak HVM in the ready-to-launch position 0509740

Thales Starstreak Lightweight Multiple Launchers on the production line
0024898

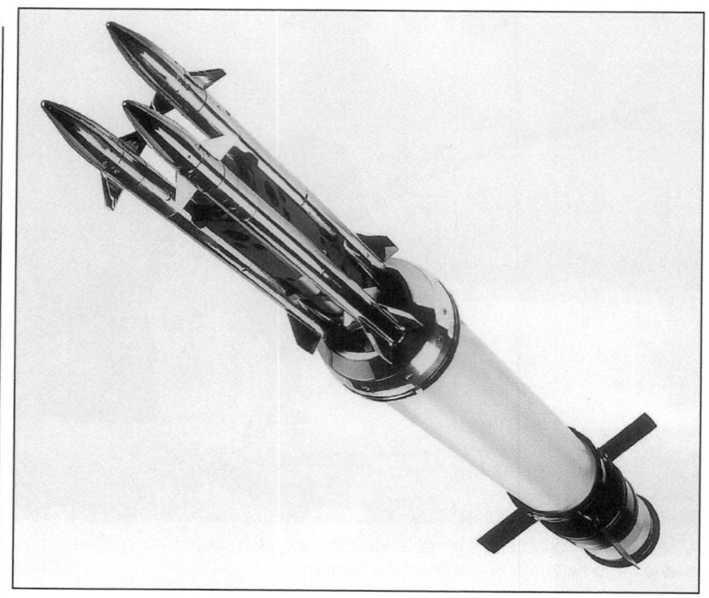

Thales Starstreak High-Velocity Missile out of its canister and showing three manoeuvrable darts each containing a high-explosive warhead 0056298

Development

General Staff Requirement (GSR) 3979 was drawn up by the British MoD, following requests from the then British Army of the Rhine, for an air defence system to supplement the tracked Rapier SAM system then in service, especially in the Forward Edge Battlefield Area (FEBA). The then Royal Armament Research and Development Establishment had already carried out a detailed study that showed that a high-velocity missile system rather than a gun or a gun/missile mix was the best solution. GSR 3979 required not only a self-propelled version of the High-Velocity Missile (HVM), but also a three-round lightweight launcher and a single round man-portable launcher.

Originally, 11 companies showed an interest in the project, but in the end this was narrowed down to three competitors: British Aerospace, Marconi Command and Control Systems and the then Shorts Missile Systems.

Late in 1984, the British MoD awarded both the then British Aerospace (now MBDA) and Shorts Missile Systems (SMS - now Thales Air Systems Ltd, part of the Thales Group) a one year project definition contract valued at GBP3 million each for the HVM, although each company invested some of its own money as well. Both companies submitted their proposals and detailed costings in October 1985.

In June 1986, the MoD decided to go ahead with the Starstreak and, in November 1996, awarded a GBP356 million contract for the system design and development, and initial production and supply of 4,100 Tranche 1 Starstreak high-velocity missiles.

Thales had, however, already started development of an HVM under the company project number S14, following a very detailed analysis of current and future air threats. They showed that the major threat of the future would be very fast attacking fixed-wing aircraft and the late unmasking attack helicopter. An HVM was the only type of missile, which would enable targets to be defeated before they released their weapons.

By the time they were awarded the contract by the MoD, Thales had already carried out over 100 test firings of the HVM since 1982 as part of a technology demonstrator programme aimed at minimising risk during the full development phase. The Starstreak HVM is complementary to the MBDA Rapier SAM and the former is deployed more forward than the latter.

According to plans at that time, the British Army were to have two regular close air defence regiments – 12 Regiment (of three 36 SP HVM batteries) and 47 Regiment (SP HVM and man-portable role during late 2000). In addition, three Territorial Army reserve man-portable regiments (of battery strength) will also re-equip in the next three to four years while the newly created 106 Regiment war establishment reinforcement unit will have two Rapier and two Starstreak batteries.

In October 1995, the UK MoD accepted the Self-Propelled High-Velocity Missile system (SP HVM), which has the commercial name of Starstreak. This has a turret with eight HVMs in the ready-to-launch position with additional missiles being carried inside for manual loading. The officially released in-service date for the UK Army was September 1997.

In addition, the SP HVM carries the Lightweight Multiple Launcher (LML) version of HVM, which has three missiles in the ready-to-launch position as well as the man-portable version of the HVM. It is understood that the three-round LML version is not yet fully operational with the British Army.

The current production SP HVM is fitted with a roof-mounted passive Thales Optronics Air Defence Alerting Device (ADAD), but the UK MoD is expected shortly to endorse a programme to increase the crew to four and provide an integrated night sight of the thermal type to give a full 24-hour capability.

One key advantage of the HVM over other air defence missiles is its very high speed (of greater than Mach 3). Its short flight time and laser guidance made it immune to all known countermeasures at that time.

A total of 135 SP HVMs have now been completed. In addition there are 10 Alvis Stormers in the Troop Reconnaissance Vehicle (TRV) role, which have been delivered. The TRV is essentially a SP HVM without the weapons fit. In principle, if a SP HVM becomes disabled with its HVM system still operational, the disabled vehicle's HVM assembly can be transferred to it by use of a fitter's vehicle with a crane.

The original in-service date of the SP HVM was 1992 but there was some slippage because of technical problems with both the Alvis Stormer chassis and the actual missile system. These problems have now been overcome.

Thales has already marketed the Starstreak HVM to a number of selected NATO countries. In 1998, the UK government gave clearance for Starstreak to be offered on the world export market.

In September 1999, SMS arranged, as part of its DSEi show presentation, a trial demonstration for potential customers at Shoeburyness, Essex. Part of the trial was Starstreak's use against an FV432 APC chassis, demonstrating the weapon's anti-armour capability – a point that Thales was then emphasising. At the DSEi exhibition in September 2005, a new mobile version of the Starstreak family, based on a Multi Mission System (MMS) called the Thor, was exhibited. This Thor (MMS) system was mounted on a standard 6 × 6 Pinzgauer vehicle along with TV (day/night), IR and laser range-finder. An additional laser designator could also be fitted if the missiles are mixed and matched with the Hellfire for surface action.

By September 2007, and following trials at the Castle Martin Range in Pembrokeshire the MMS, on which the Thor was designed, has now passed the final design configuration stage and has been shown on a Spanish Vehiculo de Alta Movilidad Tactico (VAMTAC) 4 × 4 high-mobility tactical vehicle.

Firing trials at the South African Overberg Test Range near Cape Town were conducted late October through to early November 2007. The South African Army has ordered the Starstreak missile for its GBAD future system and will also be using the Starstreak in its original role as a man portable SAM.

In January 2009, a GBP200 million Air Defence Availability Project (ADAPT) contract with Thales Air Systems Ltd in Belfast, was set to enhance the systems through-life support and address the mid-life update programme. This allowed Starstreak to remain in operational use through 2013.

Description

All three versions of the Starstreak use the same basic missile. There is a separate entry for the Armoured Starstreak variant on the Alvis Stormer chassis in the Self-propelled surface-to-air missiles section. This version also carries a Lightweight Multiple Launcher (LML) for dismounted use.

The Starstreak missile is sealed in an environmental container that also acts as the launcher unit. It requires no field testing, as the only launch preparation required is the connection of the reusable aiming unit.

The missile itself consists of a two stage, solid propellant rocket motor assembly. An initial first stage blip motor to eject the weapon from the tube and a second stage boost motor. The total burn time for both is a fraction of a second with a payload separation system mounted on the front end of the second stage motor. This payload bus supports three winged darts which each have guidance and control circuitry, a high-density penetrating explosive warhead and delay action fuzing.

The Avimo aiming unit contains all the systems required for the engagement cycle and comprises two separate and detachable assemblies.

- A light alloy casting hermetically sealed optical head with an optics stabilisation system, aiming mark injector unit and aimer's monocular sight. All three of these are used to acquire and track the target.
- A hermetically sealed control unit in a lightweight moulded case which contains the power supply unit (with one lithium sulphur dioxide battery pack) and the electronics units required for processing and control. An attached control handle contains the joystick controller, trigger assembly, system switch, wind offset switch and super elevation button.

In combat, the aimer acquires the target in the monocular sight and selects 'system on' which energises the aiming unit battery supply. A space stabilised aiming mark is injected into the centre of the field of view of the aimer who then tracks the chosen target by moving the launcher assembly to maintain the target in coincidence with the aiming mark. This permits lead angles in both azimuth and elevation to be generated and ensures that the missile is brought on to the target at the end of its boost phase.

After this pre-launch tracking phase is completed, the aimer presses the firing trigger; this causes a pulse of power to be transmitted from the aiming system power supply to the missile booster unit, where it causes the first-stage Royal Ordnance Brambling (now Roxel) SD extruded double-based propellant blip rocket motor to ignite. The Starstreak is ejected from its launch tube by this motor and is completely burns out within the length of the container in order to safeguard the operator. The motor accelerates the missile to a high exit velocity, while its canted exhaust nozzles impart sufficient roll on the weapon to create a centrifugal force that unfolds a set of flight stabilising fins. The first-stage motor then separates from the main missile body and falls away as the Starstreak emerges from the canister.

At a safe distance from the gunner, the main Royal Ordnance (Roxel) Titus CDB propellant second-stage rocket motor cuts in to accelerate the missile to an end-of-boost velocity which is in the region of M3.5. Approximately 400 m down range the motor burns out. The attenuation in thrust triggers the automatic payload separation of the three hittile darts, which are held in place by three plastic sabot clips. Upon clearing the missile body, the projectiles are independently guided in a fixed formation with a radial separation of some 1.5 m. The projectiles individual onboard guidance systems using the launcher's laser beam and a 2-D laser generated grid matrix (or lower power laser information shield at it is sometimes described) are laid on the target.

The individual darts ride the laser beam projected by the aiming unit which incorporates two laser diodes, one of which is scanned horizontally and the other vertically to produce the required 2-D matrix. Each dart then uses its on board guidance package of a laser receiver in its tail with its internal processing logic to control a set of two broad-chord canard delta wings mounted on the spinning fore body so as to hold its flight formation within this matrix. Upon receipt of the guidance signal, a clutch actuator unit within the dart momentarily inhibits the spinning fore body to allow the dart to be directed back on track. This is achieved by the canard wings, which steer it back towards the centre of the grid on line to the target. The laser itself does not illuminate the target. Four broad chord fins with swept leading edges are fitted to the dart's rear assembly. Separation of the darts also initiates arming of the warheads.

The darts, each 0.396 m long, 0.022 m in diameter, have a dense tungsten alloy housing encasing the impact/delay action fuzed warhead. The darts rely primarily on their kinetic energy for target penetration, with the inertial impact forces generated activating the delay action fuze mechanism. The explosive component contained within the centre of the tungsten warhead (0.5 kg total warhead weight) is detonated by the delay fuze when the dart is within the confines of the target. The resulting detonation of the HE charge fragments the tungsten alloy warhead housing, the many hundreds of splinters created ensuring maximum target disruption.

All the operator has to do after the launch is to continue to track the target and maintain the sight aiming mark on it using the joystick. Once the engagement is over, the operator discards the empty launch tube and connects a fresh one to the aiming unit. Maximum effective range is around 7,000 m, which is the maximum distance at which the darts can retain sufficient manoeuvrability and energy to catch and penetrate a modern 9 *g* manoeuvring target.

As Starstreak does not rely on a heat source for guidance, it can engage targets from all angles including head on. A Single-Shot Kill Probability (SSKP) of 0.96 has been mentioned in connection with the system.

The LML version consists of three standard Starstreak rounds in a 'traffic light' configuration and a man-portable aiming unit mounted on a traverse head that can be slewed quickly through a full 360°. The system can stand above ground or be sited in a trench.

The basic Starstreak HVM is a clear weather system only, but target information can come from a number of other sensors such as the Thales Optronics Air Defence Alerting Device (ADAD) which was ordered in its original Mk 1 version by the British Army in 1987. This is operational with the Starburst missile system and is also mounted as the Mk 2 version on the Stormer Starstreak vehicle.

Missile improvements to enhance performance past current requirements were tested in late 1998.

The UK MoD placed a GBP28 million order for 1,000 Tranche 1A Starstreak missiles in June 1995 and followed this with a further GBP37 million contract for 1,000 Tranche 1B missiles in September 1995. In June 1997, a third contract for 1,000 Tranche 1C missiles was signed that brought the total number of missiles on order for the British Army to around 7,000. Late in 1999, a further order worth GBP200 million for additional HVM missiles was placed by the UK MoD.

The Tranche 1C and later missiles are to the Starstreak Improvement Programme (SIP) configuration, the result of a complete lethality redesign. Starstreak is envisaged to remain in operational service until at least 2020.

SMS has also entered into a collaboration programme with the Lockheed Martin Corporation to market Starstreak for US customers.

Variants

Air-To-Air-Starstreak (ATASK)

In September 1988, the then Shorts Missile Systems teamed with McDonnell Douglas Helicopters and Lockheed Martin Electronics and Missiles for a weapon integration programme on the AH-64 Apache helicopter for close range air-to-air engagements. The system, known as ATASK, consists of one or more two-round missile panniers (weighing approximately 50 kg each and similar to that fitted to the Boeing Avenger air defence vehicle) and a guidance transmitter. The pannier to aircraft attachment is by a standard 14 in electronic release unit. In July 1995, a 17-month USD6 million contract was awarded to the then Shorts Missile Systems as prime contractor to undertake the integration and clearance process on Apache for the US Army Aviation Directorate. The first phase, demonstrating that the weapon could be successfully fired from the helicopter, was conducted in 1995 to 1996. The second phase, under a further USD13 million contract, included full integration of the guidance system with the Apache's sighting device and took place during 1997 and 1998. The third phase is planned to be a shoot off and is due to take place in the near future. It is expected to last some two years and will be conducted on an AH-64D Apache Longbow. ATASK is also applicable to other helicopters such as the PAH-2. Effective range is over 5,000 m. Other variants being studied are a shipborne system and the use of technology in the US DoD Strategic Defence Initiative (SDI) concept.

Seastreak

During 1989, the then Shorts Missile Systems (now Thales Air Systems Ltd) exhibited a ship borne version of Starstreak, known as Seastreak, which is intended to provide a close in defence against low-level strike aircraft and anti-ship missiles.

In 1991, Shorts displayed the Starstreak pannier for use on its Naval Multiple Launcher system. Two panniers, each containing three missiles, can be fitted to the one-man mount.

South Africa Ground-Based Air Defence (GBAD)

The Starstreak missile was delivered to South Africa during the first quarter of 2006 as part of a man-portable and armoured vehicle mountable system to form part of the South African Ground-Based Air Defence (GBAD). Trials at the Denel Overberg Test Range began late October through November 2007.

In late November 2010, further trials at the Overberg Test Range were undertaken to improve the training and experience of South African operators in firing live rounds. The missiles were fired against the locally-manufactured Locats aerial target system. The Locats vehicle simulates small-sized conventional air threats and has a maximum speed of 310 km/h. It is not known what variant (Mk1 or Mk2) of Starstreak missiles fired during the tests but it is assessed that probably the initial trials used the Mk 1 and later trials made use of the upgraded Mk 2.

Multi Mission System (MMS)

The Multi Mission System (MMS) comprises a servo-driven turret and the weapon operator's human machine interface. The turret houses a Stabilised Integrated Guidance Head (IGH) containing a daylight and low light Charge Coupled Device (CCD) cameras and a thermal imaging system, integrated with an Automatic Target Tracker (ATT), UHF alert radar (optional), Starstreak laser-guidance unit and optional laser range-finder.

The new MMS system also employs an Evolved Starstreak HVM that extends the range, altitude and precision with a view to enhancing its small target, low signature engagement capability.

Starstreak II

In September 2007, Thales announced the development of Starstreak II. This system has increased range beyond the 7 km of the original system as well as increased altitude and improved precision guidance. The first known overseas customer for this system will be Thailand. The Thai government signed the contract for Starstreak II and the Lightweight Multiple Launchers (LML) in mid-November 2012.

Rapid Ranger

The new Rapid Ranger comprises multiple weapon launchers, electro-optical observation and guidance system. The Rapid Ranger has also been associated with the Spanish Vamtac vehicle for the Indonesian purchase of the Force Shield purchase in January 2014.

Specifications

	Starstreak I
Dimensions and weights	
Length	
overall	1.397 m (4 ft 7 in)
Diameter	
body	127 mm (5.00 in)
Weight	
launch	16.82 kg (37 lb)
Performance	
Range	
min	0.2 n miles (0.3 km; 0.2 miles)
max	3.8 n miles (7 km; 4.3 miles)
Ordnance components	
Guidance	laser beam riding (probably in 3-5 μm waveband region)
Propulsion	
type	two stage, solid propellant (booster sustainer)

Status

In production with over 7,000 ordered. The missile has been in service with the British Army in vehicle-mounted configuration and MANPADs versions since 1995.

In December 1999, the UK MoD placed a contract with SMS worth GBP200 million, for additional HVM for delivery over a five-year period.

On 1 August 2004, another new contract worth USD326 million was signed with Thales for the production and delivery of new missiles which started in 2007. Presumably these missiles will become part of the GBAD (UK) system.

A mid-life update/upgrade is expected to take place in the near future which will probably include changes within the three tungsten darts.

Contractor

Thales Air Systems Ltd.

Starstreak II

Type
Man-portable surface-to-air missile system (MANPADs)

Development
The Starstreak missile system has been designed for use against aircraft and lightly armoured vehicles. The main sensor is based on optronics rather than radar to detect, track and engage the target. Thales as the primary contractor has largely funded development of the new Mk II variant version of the HVM missile from its own organisation. In June 1997, a third contract for 1,000 Starstreak Tranche 1C missiles was signed with the UK Ministry of Defence (MoD) that brought the total number of missiles on order for the British Army to around 7,000. Late in 1999, a further order worth GBP200 millions for additional HVM missiles was placed by the UK MoD.

Missile improvements to enhance performance according to past and current requirements, were tested in late 1998.

The Tranche 1C and later missiles, are currently in the Starstreak Improvement Programme (SIP) known as Starstreak II configuration which is the result of a complete lethality redesign. Starstreak is envisaged to remain in operational service until at least 2020.

In September 2007, Thales Defence first showed the Starstreak II, a much improved successor to the basic Starstreak missile. The new missile is improved in range and targeting and also has the ability to operate at higher altitudes.

A mid-life update/upgrade is expected to take place in the near future which will probably include changes within the three tungsten darts. This update was announced in January 2009 with a GBP 200 million contract that extended the life of the system through to 2013.

Late in November 2010, trials at the South African Overberg Test Range took place with the South African Army operators developing experience using live rounds against a locally manufactured Locats aerial target system. It is not known however if the missiles used were of the Starstreak I or II variant but it is assessed that probably the Mk I was used in the early trials with the Mk II used in the later trials.

Description
All versions of the Starstreak use the same basic missile. The missile is sealed in an environmental container that also acts as the launcher unit. It requires no field testing, as the only launch preparation required is the connection of the reusable aiming unit.

The missile itself consists of a two-stage, solid propellant rocket motor assembly, an initial stage blip motor to eject the weapon from the tube and a second stage boost motor. This payload bus supports three winged darts, each dart has its own guidance and control circuitry, high-density penetrating explosive warhead and delay action fuzing system. The mid-life update for the system includes new darts that each contain their own exclusive warheads that are designed to destroy the target more efficiently.

The missile is guided by two laser beams (gathering beam and riding beam) projected into a two dimensional matrix by the aiming unit. The three darts fly in a formation approximately 1.5 m in radius and have enough kinetic energy on impact with the target to penetrate and destroy it. However, each dart also contains enough HE to cause the dart to break up into small fragments, thus enhancing the kill probability.

Associated with the Starstreak II is the Thales Multi Mission System (MMS). This is a lightweight vehicle mounted turret system that can be equipped with the Starstreak and/or other missiles of a similar configuration. A further type of launcher for the Starstreak is the new Rapidranger that comprises multiple weapon launchers, electro-optical observation and guidance system.

Variants
Starstreak I

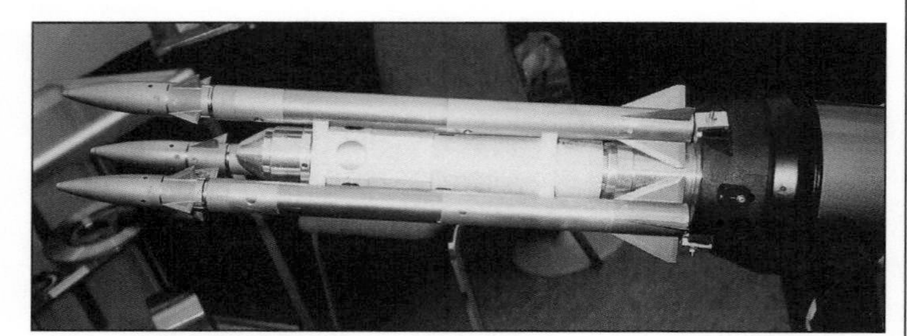

Three dart Starstreak 0062633

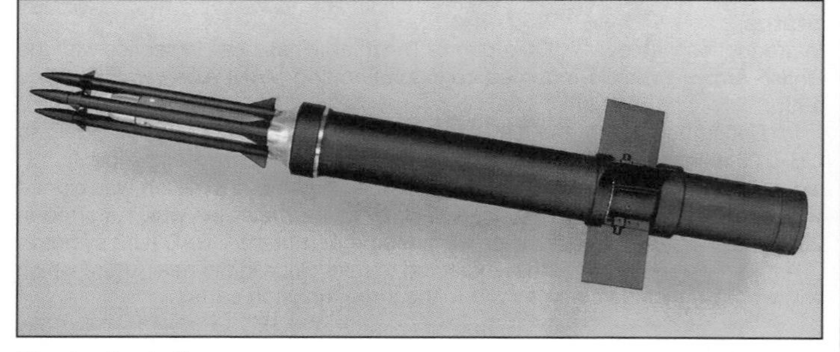

Starstreak missile 0053937

Specifications

	Starstreak II
Dimensions and weights	
Length	
overall	1.397 m (4 ft 7 in)
Diameter	
body	127 mm (5.00 in)
Weight	
launch	14.0 kg (30 lb)
Performance	
Speed	
max speed	1,983 kt (3,672 km/h; 2,282 mph; 1,020 m/s) (est.)
Range	
max	3.8 n miles (7 km; 4.3 miles) (est.)
Altitude	
max	5,000 m (16,404 ft) (est.)
Ordnance components	
Warhead	kinetic, HE 3 dart-like sub-projectiles
Guidance	laser beam riding (probably in 3-5 µm waveband region)
Propulsion	
type	two stage, solid propellant (sustain stage)

Status
Developmental trials and testing will continue although the Starstreak II has been offered to certain customers in Asia. In mid-November 2012, Thailand signed a contract with Thales for the supply of the Starstreak (presumably the II variant) and the Lightweight Multiple Launchers (LML), no dates for delivery have yet been announced.

In mid-October 2013, the UK Ministry of Defence (MoD) placed a further order for 200 Starstreak missile in order to increase the stocks currently held for the UK Armed Forces.

Indonesia in January 2014 announced the purchase of Starstreak missiles from Thales UK for integration into the Force Shield system, also associated with this purchase is the Control Master 200 air-defence radar.

Contractor
Thales Air Systems Ltd and PT LEN Industries, Indonesia, for integration of part of the Force Shield System.

United States

FIM-92 Stinger

Type
Man-portable surface-to-air missile system (MANPADs)

Development
The Stinger has been developed ever since the early designs of the FIM-92A, there are now seven members of the FIM-92 ranging from the 92A through 92G:
FIM-92A the first batch of Stingers dating from 1978
FIM-92B POST improvement over 92A, production 1983

Stinger missile shortly after launch with booster falling away 0509839

FIM-92C an upgrade dating from 1987
FIM-92D countermeasures upgrades 1990
FIM-92E upgraded software suite and sensor, production 1995
FIM-92F another software suite upgrade from 2001
FIM-92G upgrades to earlier 92D model.

In May 1992, the then Hughes Aircraft company and General Dynamics Corporation announced that Hughes had agreed to acquire the General Dynamics missile business. This acquisition included the General Dynamics (then Air Defense Systems) division, whose primary manufacturing facilities were in Pomona and Rancho Cucamonga, California. The unmanned strike systems portion of the Convair Divisions primary plants was in San Diego and Sycamore Canyon, California.

In September 1992, Hughes Missile Systems Company announced that it was consolidating missile-manufacturing activities from several locations into its plant in Tucson, Arizona. This meant the end of major production lines in San Diego, Pomona and Rancho Cucamonga, California and Camden, Arkansas. The consolidations were completed by 1994. In early 1998 Raytheon took over Hughes Missile Systems and the company is now known as Raytheon Electronic Systems.

The General Dynamics FIM-43A Redeye system achieved its initial operational capability in 1967. Even as it did so, a joint work programme between the US Army and the General Dynamics Pomona Division was in the second year of studying new design concepts and initiating the testing of components for a Redeye II weapon system with an all-aspects target engagement capability.

In 1972, this Advanced Seeker Development Programme (ASDP) eventually gave rise to the second-generation man-portable XFIM-92A. This Stinger design has a more sensitive seeker head and an improved kinematic performance, compared with its predecessor, with the addition of a forward aspect engagement capability to its flight envelope and an integral Identification Friend or Foe (IFF) system.

However, the first guided tests in 1974, at the White Sands Missile Range resulted in a number of problems being found with the Stinger. This caused the US Army Missile Command (MICOM) to request that the Ford Aerospace Aeronutronic Division develop what was designated the Stinger Alternate system, using a reusable laser beam device attached to the launcher assembly as the guidance system.

To cure problems found with the system and to reduce the continually rising costs, General Dynamics initiated a design review, which resulted in a 15 per cent reduction in the total number of electronic parts used and the introduction of a separate grip-stock assembly. These changes made a considerable improvement in the test results obtained in the 1975 firing and, by February 1976, the US Department of Defense was satisfied that the early guidance difficulties had been overcome. They were so convinced that in 1977, the funding used for the development of the Stinger Alternate was stopped. In 1978, following an engineering development programme, which had needed only 130 test round firings to validate the design, Stinger was finally released for production. This started in 1979 with the first production systems being delivered the same year and the first military units achieving initial operational capability status in February 1981 with the basic FIM-92A Stinger version.

In mid-1977, after a four-year advanced development programme and just before the basic Stinger was released for production, General Dynamics was awarded a full-scale engineering development contract for the next generation of Stinger. This involved the fitting of a microprocessor-controlled Passive Optical Seeker Technique (POST) homing head which uses a dual InfraRed (IR) and Ultra-Violet (UV) rosette-pattern image scanning guidance technique to enhance the missile's target detection capabilities. The use of the different seeker only involves a modular change to the weapon and allows it to discriminate effectively between a target, any deployed IR decoy flares and background clutter, when they lie within detectable range, thus preventing a false launch.

Limited procurement of this FIM-92B Stinger-POST version began in 1983 alongside the earlier variant with the production of both ending in 1987. Operational deployment of Stinger-POST systems to the US Army began in July 1987. A total of 15,669 FIM-92A (Basic Stinger) and just fewer than 600 FIM-92B (Stinger-POST) missiles were produced. The last Stinger-POST rounds were produced by August 1987.

As a further increase to the effectiveness of Stinger, General Dynamics began development, in September 1984, of what is essentially a fourth-generation man-portable SAM system. Known as the Stinger-Reprogrammable Micro-Processor (RMP) system, the change allows the onboard digital micro-processor to be periodically updated with new software to counter any new threat technology, instead of having to go through a missile redesign each time. Production of this FIM-92C model began in November 1987 at the General Dynamics Valley Systems Division Stinger plant in California. The export version of the Stinger-RMP does not have the reprogrammable module but contains embedded InfraRed CounterMeasures (IRCM) to defeat all known NATO threats. In March 1988, Valley Systems Division was awarded a USD695 million multiyear Stinger production contract to produce over 20,000 rounds through to 1991. A final total of 29,108 Stinger-RMP missiles is expected for the US Army. Additional rounds are being procured for foreign military sales and the other US Armed Forces.

Raytheon Electronic Systems continues production of the Stinger-RMP at its Tucson facility at over 1,000 missiles a year.

Before General Dynamics' production of the Stinger-RMP, on 2 September 1987, the US Army MICOM selected the then Raytheon Missile Systems Division as the second source contractor for production of this version. The initial USD24.6 million contract was for 400 Stingers with a USD54.4 million option for an additional 1,500 missiles, which was exercised in 1989. Raytheon was allowed to compete with the Hughes-acquired General Dynamics from 1990 onwards for the annual production contracts. This was done in order to keep the overall acquisition costs down for the US armed services. Raytheon was declared qualified for bidding purposes for the FY91 contract bid. In April 1990, Raytheon received a USD45.1 million contract to produce 1,383 missiles. In the following year General Dynamics reverted to the sole source supplier. The US Army requirement was for 29,108 FIM-92C Stinger-RMP rounds with last funding for procurement being provided in FY92.

Stinger man-portable SAM system
(Directorate of Materiel Royal Netherlands Army, Communication Office)
0509742

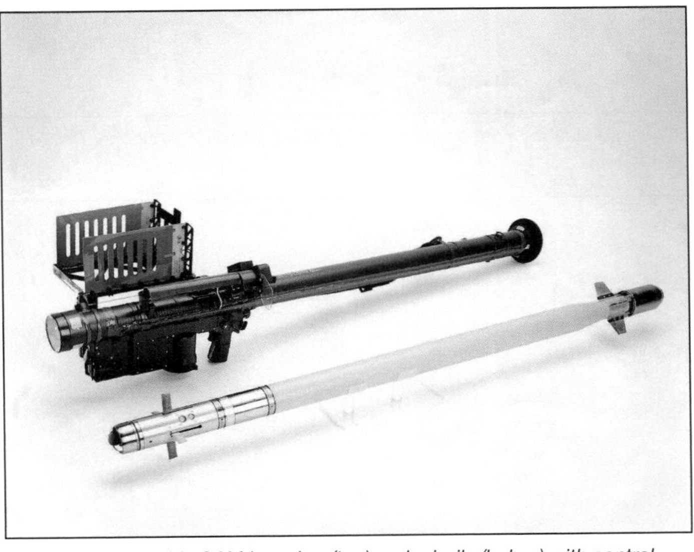

Stinger man-portable SAM launcher (top) and missile (below) with control surfaces and fins in the unfolded position as they would be in flight 0509843

In FY92, an upgrade contract was placed to improve the FIM-92A/B/C performance against the latest countermeasures. Known as the FIM-92E Block 1 rounds, modifications were made to the RMP software to see its low-signature targets such as UAVs, cruise missiles and light helicopters in even more cluttered countermeasures environments. Block 1 variants use a roll frequency sensor to provide trajectory shaping. A ring-laser gyro roll sensor and a lithium battery are also fitted. First production deliveries were made of the Stinger Block 1 rounds in 1995. The programme will involve upgrading all the remaining FIM-92A and FIM-92B missiles in the inventory to the standard of FIM-92D, involving up to 8,500 rounds plus; it is reported to have been finished in FY99. The Stinger Block 1 is also made by the European Stinger Project Group, which switched production to this version. Stinger Block 1 is in production as new build and retrofit. Block 1 missiles are deployed with front echelons Bradley Linebacker, Avenger and aviation units as a helicopter Air-To-Air missile.

In mid-2000, Italy chose the Stinger Block 1 to arm its A129 Mangusta attack helicopter.

It has been reported that plans to develop and produce the Advanced Stinger (Block 2) from the year 2001 onwards may have been put on hold. The Block 2 was originally an evolutionary technology insertion which would have provided for the replacement of the current FIM-92E seeker with an advanced technology Focal Plane Array (FPA) imaging IR seeker, known as the Small Diameter Imaging Seeker (SDIS), to increase the detection range. It is estimated that a total of up to 5,000 missiles would have been retrofitted for use in both the forward air defence and Air-To-Air role against helicopters in clutter, unmanned aerial vehicles, cruise missiles and stealthy modern fixed-wing aircraft.

The European Stinger-RMP Production Programme is covered earlier in this section under International projects.

In German service it is known as the Fliegerfaust-2 (FLF-2) and is deployed with the army, navy and air force. A series of launch platforms was developed by a subsidiary of Daimler-Benz Aerospace for use on ships, wheeled and tracked vehicles.

Stinger SAM system fitted with the AN/PAS-18 Wide-Angle Stinger Pointer (WASP) 0509673

Stinger man-portable surface-to-air missile system from the side (US Army) 0509840

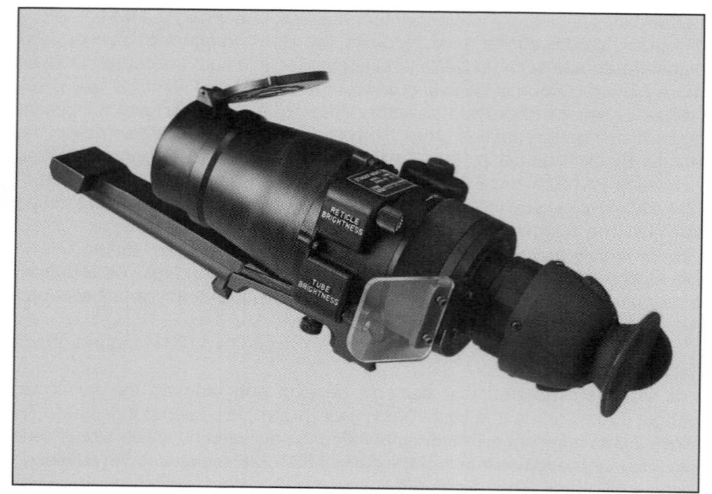

ITT Stinger F4960 Night Sight which uses a third-generation intensification system 0509674

Stinger man-portable surface-to-air missile system from the front 0056299

Both Germany and the Netherlands have also undertaken trials on their man-portable Stinger systems with an early warning radar system to enhance its performance. The Royal Netherlands Army used the Hollandse Signaalapparaten Radar Equipment Providing Omni-directional Reporting of Targets at Extended Ranges (REPORTER) mobile trailer-mounted I/J-band radar system with an integral IFF system in a highly successful series of tests in late 1985. The radar provided early warning of targets up to 40 km away and flying between 15 and 4,000 m altitude which were then handed over to a Stinger launch team for engagement.

The Basic Stinger received worldwide attention during the Soviet invasion of Afghanistan, when over 250 Russian fixed-wing aircraft and helicopters were destroyed by Mujahideen guerrillas, using US-supplied Stingers. Despite limited training, the Mujahideen achieved over 80 per cent combat success with the Stinger missile. To date, the Basic Stinger has been responsible for 270 confirmed kills against both fixed- and rotary-wing aircraft types. Those systems provided by the US to the Mujahideen are still being found in and around Afghanistan today. During the latter part of September 2005, the Pakistan paramilitary Frontier Corps seized six US-produced Stinger missiles and a huge cache of arms and ammunition deposited in a compound in Mohammad Gath.

Other nations which have received varying numbers of Stinger systems include Bahrain, Chad, France, Iran, Israel, Japan, South Korea, Pakistan, Qatar, Saudi Arabia and the United Kingdom. Of these, France and Chad have used limited numbers successfully against Libyan aircraft during the 1986 to 1987 border skirmishes. The British Special Air Service (SAS) used a small number of FIM-92A Stingers during the 1982 Falklands (Malvinas) conflict, where they destroyed an Argentine Air Force FMA IA 58A Pucara twin-propeller close-support aircraft during the 21 May San Carlos amphibious landings.

In September 1988, Switzerland chose the export variant of the Stinger-RMP for its man-portable air defence system. A maximum of 2,500 will be procured at a cost of USD315 million.

In US service all four armed services use the weapon and the US Air Force has small detachments trained to defend airfields and Vital Points (VPs), especially in Asia at the South Korean airbases used by its units. It has also been revealed that the American President's residences in Washington and elsewhere are protected by specialist Stinger teams in case of an aerial attack by terrorist organisations.

Each of the US Army's armoured, mechanised, light infantry, airborne and air assault divisional Air Defense Artillery (ADA) battalions have a Stinger Platoon (of four sections) with each of its four batteries. For the airborne and air assault divisions, three of the sections have five two-man teams each, while the fourth has only three teams to give a divisional total of four two-man section HQs and 72 firing teams. For the division 86s mechanised and armoured units, the number of Stinger teams is reduced to 60, whereas the light infantry divisions only have 40. A team is normally equipped with an M998 series (4 × 4) HMMWV light vehicle, a GSQ-137 Target Alert Data Display Set (TADDS) comprising a 6 kg portable unit with a display, audio warning and VHF receiver, two AN/PPX-3 IFF interrogators and a basic load of six Stingers. The TADDS warns the team of an approaching aircraft, provides a tentative identification and gives approximate range and azimuth to the target. The datalink between the team and the radar can handle 49 friendly and 49 unknown (that is, hostile) targets. Each Patriot fire-control platoon also carries one Stinger team set (less the TADDS) as part of its normal equipment allowance. During a heavy attack, both team members would shoulder a launcher, providing two independent ready-to-launch weapons with four extra weapons available.

Early warning is provided by the eight 1 to 15 km (on a 0.2 m^2 radar cross-section target) range pulse Doppler D-band MPQ-49 Forward Area Alerting Radars (FAARs), with integral AN/TPX-50 (Mk XII) IFF systems, held in the ADA headquarters radar platoon, which transmit target position data to the TADDS by radio link. Increasingly, however, the FAARs are being used without TADDS and the Stinger teams are being cued on to a target by a voice communication VHF radio link direct to the radar operator.

The Ground-Based Sensors (GBS) are replacing the older FAARs in the division air defence alerting radar role. The GBS is a pulse Doppler, 3-D phased array X-band radar that provides Low-Altitude target data at altitudes from 0 to 4,000 m and at ranges of up to 40 km. It can automatically detect, track, classify, identify and report a target in 360° azimuth coverage.

In the US Marine Corps the Stinger system is assigned to the Low-Altitude Air Defence (LAAD) battalions which have two firing batteries each of three platoons. A platoon has three sections each of five Stinger teams that are each equipped with an HMMWV light vehicle and four Stinger missiles.

A Marine Expeditionary Force (MEF) is assigned a fully automated Tactical Air Command centre, two Tactical Air Operations Centres, a Light Anti-Aircraft Missile (LAAM) battalion and a complete LAAD battalion (of 90 Stinger teams). A Marine Expeditionary Brigade (MEB) is assigned a Tactical Air Command Centre, a Tactical Air Operations Centre, an LAAM battalion and an LAAD battery (equating to four I-HAWK batteries and 45 Stinger teams), whilst a Marine Expeditionary Unit (MEU) has a single LAAD platoon (of 15 Stinger teams) attached.

The US Navy uses Stinger teams to supplement warship and support vessel close-range air defences in high-threat areas. A team of two is normally employed with the gunner located within a circular pedestal-type open mount. The other team member acts as a target locator using information sent over the vessel's internal communications net. Stinger is also the principal air weapon on the US Navy's fleet of special operations patrol boats.

In 1991 to 1992, the US Army deployed the Bradley SHORAD vehicle to replace Vulcan self-propelled guns in its Heavy Mechanised Divisions. This was followed by the M6 Linebacker vehicle (see entry in self-propelled surface-to-air missile section).

To date, Stinger is integrated or deployed on over 20 vehicle and helicopter platforms in 19 countries and over 40 services.

During the 1999 Kashmir conflict between India and Pakistan, a US-made Stinger was credited with bringing down one of the three Indian aircraft lost in combat operations.

Stinger weapon systems in Operations Desert Storm and Desert Shield

During the 1991 Gulf War, Stinger systems were deployed extensively by a number of the Coalition forces, but as far as is known, did not engage any Iraqi targets. Post-war debriefing of Iraqi command staff credited Stinger with deterring the use of the 500-strong Iraqi fleet of helicopters that included numerous Hind gunships.

The then latest version of the Stinger, Stinger-RMP, equipped with the latest MOD IV software, was deployed with at least three of the four US armed services in support of the Gulf operations. Basic Stinger and export variant Stinger-RMP systems were also deployed with the armed forces of several of the other nations of the Coalition forces.

Virtually every US Army ground combat unit had the Stinger either in the MANPADS role or as the principal armament of the wheeled Avenger PMS fire unit.

Since initial deployment Stinger has, in combat, defeated more modern fixed-wing aircraft and helicopters than any other fighter, helicopter or missile system worldwide since the Second World War.

The US Marine Corps used the same two-man MANPADS team as the US Army, with the team (gunner and team chief) carried in an HMMWV with four ready-to-fire Stingers and two reload missile rounds.

The US Navy used MANPADS teams aboard its ships, primarily for close in defence against small aircraft and very small surface craft. The US Army used the ATAS variant on its OH-58C helicopter in the Air-To-Air role, whilst the US Air Force may have deployed some of its specially trained Air Police in the MANPADS role to guard some of the air bases it used in the Arabian Peninsula region.

Future man-portable SAMs

Between May and August 1994, the US Marine Corps issued a Request For Information (RFI) concerning the hardware replacement and a possible expansion of existing capabilities to incorporate surface threats for the replacement of the current man portable SAM systems held in the inventory. IOC is required for FY08/09 and as of August 20121, nothing further has been reported on any such system.

Description

The FIM-92 Stinger is a personal portable InfraRed (IR) homing Surface-to-Air Missile (SAM) that can be adapted to engage targets from ground vehicles, helicopters or shipborne. The most up-to-date and modern Stinger is a lightweight fire-and-forget, two colours, short-range air defence missile designed for maximum lethality and survivability. The advanced passive two colours IR/UV detector and advanced algorithms optimise target acquisition and ensure effectiveness against all known countermeasures. Stinger has embedded software to handle all known threats. In addition, the seeker combined with proportional navigation guidance allows the missile to acquire, track and engage targets at any aspect.

The basic Stinger system comprises the launcher assembly with a missile, a grip-stock, an IFF interrogator and an argon gas Battery Coolant Unit (BCU) (which consists of the squib activated argon gas coolant unit and electrical generating chemical battery).

The launcher assembly consists of a glass fibre launch tube with frangible end covers, a sight, desiccant, coolant line, gyro-boresight coil and a carrying sling. A detachable grip-stock that has a receptacle for the BCU is fitted with an IFF connector. The grip-stock is also fitted with an impulse generator (BCU energised), a seeker head un-cage bar, a weapon launch trigger, an AN/PPX-1 IFF interrogator switch and a foldable antenna and control electronics for the missile gyro.

The missile has a two-stage, three-phase rocket motor. A separable launch motor ejects the missile followed by an advanced 'boost-sustain' motor, which provides high supersonic speed and agility out to maximum range. In its FIM-92A version it is fitted with a second-generation cooled passive IR conical scan reticule seeker head with discrete electronic components to provide signal processing. They process the IR energy received from the target in the 4.1 to 4.4 µm wavelength region to determine its relative angle and then, by using a proportional navigation guidance technique, continually predict an intercept point.

In the FIM-92B version, the reticule seeker unit is replaced by one that uses an optical processing system. This has two detector materials, one sensitive to IR (in the waveband region 3.5 to 5.0 µm) and the other responsive to UV energy (in the waveband region 0.3 to 0.4 µm), together with two microprocessors which are integrated into microelectronic circuitry for the signal processing phase. The latest Stinger-RMP takes this one stage further by introducing a microprocessor reprogramming facility into the circuitry to allow for new threat characteristics and guidance tailoring. The logic allows for recognition of countermeasures and their filtering out from the seeker's guidance picture.

In all cases, the seeker output is sent as steering data to the guidance assembly, which converts it into guidance signal format for the control electronics. This module then commands the two movable (of four) forward control surfaces to manoeuvre the weapon on to the required intercept course. The control concept used is known as the single channel rolling airframe type and, as such, considerably reduces both the missile weight and manufacturing costs. As the weapon nears its target, the seeker head activates its Target Adaptive Guidance (TAG) circuit within one second of impact to modify its trajectory away from the exhaust plume towards the critical area of the IR target itself. The fuzing system allows for both contact activation as well as missile self-destruction after 20 seconds of flight time following the launch. The Picatinny Arsenal 3 kg warhead carried has a smooth titanium-alloy fragmentation casing to ensure that the desired penetrating blast/fragmentation and incendiary effects are achieved. The titanium-alloy case gives the required strength to weight ratio and additional pyrophoric effects. The original explosive fill is reported to be the aluminised high-blast explosive HTA-3 (HMX 49 per cent, TNT 29 per cent and aluminium flake powder 22 per cent), this reported to weight between 0.8 to 1.0 kg (1.75 to 2.25 lbs).

A typical tactical engagement follows this sequence of events. Once alerted to a target the gunner shoulders the system, inserts the BCU into its grip-stock receptacle and unfolds the IFF antenna. The gunner then removes the front protective cover of the launcher tube to reveal the IR or IR/UV transparent frangible disc, raises the open sight assembly and connects the belt pack IFF interrogator unit via a cable to the grip-stock. The gunner is now ready to acquire the target visually. The gunner does this by using the sight and estimating its range with the estimation facility of the system. If required, the gunner now interrogates the target using the AN/PPX-1 system. This can be done without having to activate the weapon. The azimuth coverage of the 10 km range IFF system is essentially the same as that of the optical sight enabling the gunner to associate responses with the particular aircraft that is in view. An audio signal 0.7-second after the IFF challenge switch is depressed provides the gunner with the cue as to whether the target is friendly or an unknown for possible engagement.

If it is decided that the target is unfriendly, the gunner continues to track the aircraft and activates the weapon system by depressing the impulse generator switch. This causes the impulse generator to energise the BCU which then releases its 6,000 PSI pressurised argon gas coolant to the IR detector and generates a dual-polarity ±20 V DC output for at least 45 seconds. The cooling takes three to five seconds. It provides all the pre-launch electrical power required for the seeker coolant system, gyro spin-up, launcher acquisition electronics, guidance electronics, activation of the missile's onboard thermal battery and ignition of the ejector motor.

The seeker is allowed to look at the target through the IR or IR/UV transparent front launcher disc, and when the detector for acquisition to have occurred receives sufficient energy, an audio signal is sent to the gunner. Total time required for tracking and missile activation is about six seconds. The gunner then depresses the seeker un-cage bar and, using the open sight, inserts the superelevation and lead data.

The newer versions of Stinger (Block 1 and Block 2) preclude the need to super elevate the missile.

Once this is accomplished, the gunner depresses the firing trigger that activates the missile battery. This powers all the missile functions after launch and operates for around 19 seconds until the dual-polarity ±20 V DC output drops below the required minimum for use. A brief time delay operates, following which the umbilical connector to the grip-stock is retracted and a pulse is sent to ignite the ejector motor. Total time to motor ignition from depression of the firing trigger is only 1.7 seconds. Upon ignition, the initial thrust generated imparts roll to the missile airframe and starts the fuze timer system. The missile and its exhaust then break through the frangible discs at either end of the launcher tube.

Before the missile completely clears the end of the tube, the ejector motor burns out in order to protect the gunner from the rocket blast, and two movable control surfaces spring out. Once it clears the tube, the two fixed and the four fixed and folded tail fins open out and the ejector motor is jettisoned. The missile then coasts to a predetermined safe distance from the gunner where the fuze timer ignites the dual-thrust Atlantic Research Mk 27 solid propellant rocket motor. When the correct acceleration rate is reached after one second of flight, the Magnavox M934 time delayed impact fuzing circuit for the warhead is armed and the self-destruct timer started.

The seeker continues to track the target throughout the flight with the electronics processing the received signals to eliminate or reduce the line of sight pointing angle to the target. The weapon flies a proportional navigation path to the interception point near to which the TAG circuit is activated and a signal is generated within the seeker head to add bias to the steering signal causing the missile airframe to guide itself into a vulnerable part of the target. Even if the target is using 8 *g* manoeuvres, the missile is still capable of engaging it.

Once the gunner has depressed the trigger and the missile has left the launch tube, they are free, either to get another weapon round, to assemble another missile round for a further engagement (which takes less than 10 seconds), take cover or move to another location.

US Army training needs indicate that 136 hours of instruction is required on the Stinger system before weapon qualification is given. The M60 field handling and M134 tracker head training versions are used for instruction.

Variants

Air-To-Air Stinger (ATAS)

In the late seventies the then General Dynamics began development of an Air-To-Air Stinger (ATAS) system which completed its full-scale engineering development phase in late 1986. First production deliveries were made to the US Army in mid-1988 and the Flight Structure Division of the Western Gear Corporation was the subcontractor responsible for building the launcher structural assembly.

Weighing 55.9 kg fully loaded, the lightweight two-round launcher unit is available for use on the following:

- US Army Bell OH-58A/C Kiowa (one launcher)
- Bell AH-58D Warrior (two)
- Sikorsky UH-60 Black Hawk (one)
- Boeing Company AH-64 Apache (two)
- Boeing Company AH-6F/G (one) helicopter
- Bell AH-1 Cobra (one) helicopter.

The Stinger ATAS system is also being qualified for use on the Eurocopter Tiger combat helicopter. fixed-wing aircraft such as the Boeing V-22 Osprey can also be outfitted with ATAS launchers on their underwing pylons. Operational deployment of ATAS by the US Army occurred in 1990 on the OH-58C, and is continuing on the OH-58D and special operations helicopters. Depending upon funding, the US Army will also fit ATAS to a portion of its Apache fleet. The RAH-66 Comanche will have Stinger-RMP fitted in a specially developed Stinger Universal Launcher. The upgrade of Stinger to the Block 1 configuration is primarily to enhance ATAM capabilities.

The next generation upgrade of the ATAS system is known as the Stinger Universal Launcher (SUL), which is under development to accommodate missile improvements, ground platforms, helicopters (like Longbow Apache and Comanche) and to be compatible with MIL-STD-1760A avionics databus interfaces. Italy has chosen the Stinger to arm its attack helicopters.

Bradley Stinger Fighting Vehicle (BSFV)

The self-propelled Vulcan gun system was replaced in the air defence battalions of heavy mechanised divisions by a SHORAD version of the Bradley Infantry Fighting Vehicle (IFV). The Bradley is manned by a three-man crew and carries a two-man Stinger team, which is responsible for engaging low-level enemy aerial targets. The SHORAD crews monitor the air defence early warning networks of AWACS aircraft, Patriot missile batteries. The vehicle itself has had several modifications to improve stowage and ammunition supplies. Each battalion had 30 Bradley vehicles.

The first battalion to be equipped was the 4th Battalion, 3rd Air Defense Artillery in Germany, which supported the 3rd Infantry Division.

M6 Linebacker upgrade

The M6 Bradley Linebacker upgrade is described in the self-propelled surface-to-air missiles section.

Stinger hybrid systems

A number of programmes have been implemented to upgrade older Vulcan and Chaparral systems with more capable Stinger missiles. The Israeli self-propelled Stinger programme is known as the Mahbet (see entry in the self-propelled surface-to-air missiles section) and provides a four-pod Stinger launcher with a Vulcan Gatling gun turret mounted on an M113 Armoured Personnel Carrier (APC). Raytheon Electronic Systems is also promoting 'combined arms' systems which include Stinger and TOW mounted on each side of most standard IFV turrets.

Stinger MANPADS Alerting and Cueing System (MACS)

Raytheon Electronic Systems has produced the MACS device for the MANPADS Stinger team. It is designed to provide the team with the capability to automatically cue the gunner to acquire designated targets via accurate 3-D track data from the Forward Area Air Defence (FAAD) Command, Control Intelligence (C2I) ground-based sensor radar network. The track information is transmitted to the MACS vehicle subsystem communications controller via radio link and is then forwarded to the team chief's weapon terminal over a wireless RF digital link.

The system displays the data to the team leader as a computerised graphical situation display on the hand-held weapon terminal, so that the air threat can be quickly evaluated and a target selected for engagement by the gunner.

Once a target is designated it is automatically sent to the gunner subsystem over a digital wireless RF Local Area Network. The gunner is quickly cued to align to the target through a display of flashing light LED direction indicators on the weapon sight. Red LEDs guide them in azimuth and elevation to align the weapon's field of view on the target so that it can be acquired by the seeker head. When the target is computed to be within the Stinger's engagement envelope the gunner is informed by green LED indicators. The MACS can also be used to control and co-ordinate multiple Stinger gunners. The system can also be used by a platoon leader to give a status roll-up capability whereby the position and status of each MANPADS team can be graphically displayed. This allows platoon leaders to position their Stinger teams to best defend on-the-move assets and provide real-time information on which targets are being engaged by each unit and their weapon status and condition.

Tripod-mounted Stinger

Raytheon Electronic Systems also privately developed the tripod-mounted Stinger system. This has four ready-to-fire missiles mounted on two ATAS launcher shoes at 90° to the vertical and is fitted with an integrated high-magnification optical sight. A Marconi Forward Looking InfraRed (FLIR) tracking system is also fitted to allow Stinger launches at night and in bad weather. The one-man system weighs less than 136.4 kg and can be mounted on the rear of a vehicle if required.

It can also be interfaced with a higher-echelon command-and-control network using positive gunner cueing and is fitted with automatic missile sequencing and seeker uncaging. Traverse capability is a full 360° and the elevation limits are −10° to +50°.

In late 1987, the tripod-mounted Stinger system was tested in South Korea for use in the airfield defence role. Growth potential includes the fitting of a laser rangefinder and a go/no go fire-control computer.

Pedestal-mounted Stinger

This is fully described in the self-propelled surface-to-air missiles section of this book. This is usually referred to as the Avenger.

Aselsan pedestal-mounted Air Defence Missile System (PMADS)

The Turkish company, Aselsan, has developed two versions of the pedestal-mounted Stinger for its armed forces, the Land Rover-mounted Zipkin and the M113A2-mounted Atilgan. The Turkish Armed Forces received the first 10 of 148 pedestal mounted stinger missiles on 26 November 2004. Aselsan will ultimately supply 70 Atilgan and 78 Zipkin for use with the Turkish forces.

Dual-Mounted Stinger (DMS)

Per Udsen of Denmark and Raytheon have developed a lightweight, pedestal variant of Stinger (see entry this section). The total weight of the system with missiles is 95 kg, with the elevation assembly weighing 25 kg, the mast assembly 38 kg, the gunner's seat 5 kg and the removable terminal shelf 2 kg. The system is capable of being emplaced by two to three people in less than 90 seconds.

In September 1998 the US Government revealed a Foreign Military Sales (FMS) package to Taiwan. Part of the package was a USD180 million sale of 61 DMS systems, support and associated test equipment, spares and repair parts. These are used by Army and Marines.

In May 2005, eight HMMWVs on their way to Lithuania (out of an order for a total of 60 HMMWVs), were fitted with the Dual-Mounted Stinger launchers and the RMP/Block 1 International missile (54 in total). These HMMWVs are expected to form a battery, working with the Raytheon Thales AN/MPQ-64 Sentinel radar.

LFK-Lenkflugkörpersysteme GmbH (now EADS/LFK)Tripod adapted Stinger

The then LFK-Lenkflugkörpersysteme GmbH, a subsidiary of Daimler-Benz Aerospace, developed a relatively simple adaptation of two Stinger launcher assemblies on a lightweight tripod.

Raytheon Wide-Angle Stinger Pointer (WASP) AN/PAS-18

Raytheon Electronic Systems has developed a lightweight night sight attachment known as WASP which has been manufactured in small quantities against special orders. The sight was adopted by the US Marine Corps as its definitive Stinger night sight. Three units were delivered for trials in August 1991. A USD10.15 million fixed price contract was placed in November 1993 for the sight with completion of production by January 1996. In April 1994, a USD4.5 million contract was placed by the Danish MoD for AN/PAS-18 sights, spares and depot equipment to be used by the Danish Army and Navy. The sight is also used by the Netherlands.

Approximately 1,200 sights have been delivered with the German Federal Agency for Weapons Technology and Procurement announcing a new contract for 60 systems for use by the German Air Force. Potential value of this contract, which includes options on additional systems for the German Army and Navy, was USD5 million. In Germany the AN/PAS-18 is also used on the SA-16 Igla-1 MANPADS. Adaptation of the SA-16 to accommodate the sight was undertaken by Euroatlas GmbH.

Based on the company's short-range thermal sight technology, modifications have been made to the scanner and objective lens to accommodate the Stinger launch envelope. An RS-170 TV format output is provided for remote viewing and/or videotaping for training and post engagement analysis. A quick-release mounting bracket allows the WASP to be mated to the next round in under 10 seconds.

Total weight (including a BA 5847/U disposable lithium battery) is about 2.27 kg with dimensions of 292.1 × 101.6 × 152.4 mm. The field of view is 20° horizontal by 12° vertical with the viewing dome using a fixed illumination reticle and the 3 to 5 μm spectral waveband region. Battery life is 10 hours. With a Ni/Cd battery for training, the life is seven hours.

	FIM-92A	FIM-92B	FIM-92C
Dimensions and weights			
Length			
overall	1.47 m (4 ft 9¾ in)	1.47 m (4 ft 9¾ in)	1.47 m (4 ft 9¾ in)
Diameter			
body	690 mm (27.17 in) (est.)	690 mm (27.17 in) (est.)	690 mm (27.17 in) (est.)
Flight control surfaces			
span	0.091 m (3½ in) (wing)	0.091 m (3½ in) (wing)	0.091 m (3½ in) (wing)
Weight			
launch	10.4 kg (22 lb)	10.4 kg (22 lb)	10.4 kg (22 lb)
Performance			
Speed			
max Mach number	2.2	2.2	2.2
Range			
min	0.1 n miles (0.2 km; 0.1 miles)	0.1 n miles (0.2 km; 0.1 miles)	0.1 n miles (0.2 km; 0.1 miles)
max	4.3 n miles (8 km; 5.0 miles)	4.3 n miles (8 km; 5.0 miles)	4.3 n miles (8 km; 5.0 miles)
	2.2 n miles (4 km; 2.5 miles) (est.) (effective)	2.6 n miles (4.8 km; 3.0 miles) (est.) (effective)	2.6 n miles (4.8 km; 3.0 miles) (est.) (effective)
Altitude			
max	3,500 m (11,483 ft)	3,800 m (12,467 ft)	3,800 m (12,467 ft)
Ordnance components			
Warhead	3 kg (6.00 lb) HE, incendiary blast fragmentation	3 kg (6.00 lb) HE, incendiary blast fragmentation	3 kg (6.00 lb) HE, incendiary blast fragmentation
Fuze	impact, time delay	impact, time delay	impact, time delay
Guidance	passive, IR	passive, IR	passive, IR
Propulsion			
type	solid propellant (sustain stage)	solid propellant (sustain stage)	solid propellant (sustain stage)

Raytheon Thermal Weapon Sight (TWS)
Raytheon Electronic Systems has adapted its TWS unit for use on a number of weapon systems including the Stinger MANPADS. This is a thermo-electrically-cooled passive forward-looking infrared focal plane array sensor with an easily exchangeable telescope and graticule device for night-time use.

ITT Defense F4960 Stinger night sight
The Stinger night sight is a third-generation image intensifier system based on the AN/PVS-4 weapon sight technology. It incorporates a 60 mm, f1.2 objective lens (which provides a ×2.26 magnification and a 23.5° circular field of view) with a 25 mm first- and third-generation F4844 image intensifier tube. The spectral response region is from 600 to 900 nm.

Total length is 312.4 mm (386.1 mm with the mounting bracket attached), diameter 104.1 mm and weight 1.91 kg (2.27 kg with the bracket). The sight is powered by two AA size alkaline batteries, which provide on average up to 30 hours of usage.

The illuminated tracking/aiming reticle has a fully adjustable brightness control and is similar to the existing Stinger reticle. The combination of the lens and image intensifier tube allows acquisition of targets at ranges of up to 7,000 m and identification at 4,500 to 5,000 m even under starlight conditions.

A total of 150 examples of the sight have been supplied to the US Marine Corps as the interim Stinger night sight system. These were deployed in the 1991 Gulf War.

LAV-air defence gun/missile system
This is described in the self-propelled anti-aircraft guns section.

LFK-Lenkflugkörpersysteme GmbH (now EADS/LFK) self-propelled Stinger
Details of this system developed by the German company LFK-Lenkflugkörpersysteme GmbH, are given in the self-propelled surface-to-air missiles section.

Krauss-Maffei Wegmann Gepard 1A2 hybrid configuration
Krauss-Maffei Wegmann are offering the Gepard 1A2 hybrid configuration self-propelled anti-aircraft gun system. Two twin-round clip-on Stinger (or Igla) are fitted to the externally mounted 35 mm cannon and use the vehicle's tracking and surveillance sensors to ensure target identification and lock-on.

Extended Range Stinger
In 1990, an improved motor was undergoing preliminary research and development by Atlantic Research Corporation. The motor is intended to provide significant improvements in the effective range and ceiling for the FIM-92C. The US Army is conducting trials as to whether to include the Extended Range Motor into the Stinger Block II upgrade. This upgrade supports ranges of up to 10 km.

North Korean Stinger
In testimony to a US Senate panel in October 1997, a North Korean defector revealed that the North Koreans had obtained Stinger missiles for reverse engineering production. It is believed that a Stinger clone-building programme has started, with US intelligence indicating small numbers in service by the end of 1999.

Naval Stinger
In late 1998, it was revealed that Raytheon and Radamec Defence Systems had teamed up to develop a fully autonomous lightweight air defence system known as Naval Stinger for patrol to frigate-class vessels. The US Special Operations Command is developing a naval launcher for the US patrol craft. The naval DMS is currently deployed on Danish patrol ships.

SM Swiss Munition Enterprise Stinger Tracking and Launch System (STLS)
SM are marketing the STLS training system to simulate a Stinger launch in terms of the initial launch sequence and target acquisition procedures. The STLS uses a modified missile fitted with an STLS boost motor that mimics all the launch characteristics of a real weapon but only flies some 125 m down range on a ballistic trajectory. The launch velocity (38 m/s), noise (168 dB) and recoil impulse (<3.6 N.s) are the same as produced during an actual launch, the weight of the system is identical and an optional smoke assembly can also be fitted. The whole system allows the trainee to experience a live firing at a fraction of the cost of a real launch.

Specifications *See table top of page*

Status
The basic Stinger production is complete with 15,669 rounds built. Stinger-POST, production took place 1983 to 1987, with just under 600 rounds built. Stinger-RMP, in production since 1987 and by 1997 over 44,000 Stinger-RMPs had been manufactured for the home and export markets with production running at around 700 per month. In early 1998 the United States Defense Security Agency (DSA) released data on US International missile sales to the end of FY96 (30 September 1996). For Stinger, total deliveries through FY96 were 5,510, total orders through FY96 were 6,584, with missiles yet to be delivered 1,434. Production of Stinger Block 1 (FIM-92D) began in 1995 with up to 8,500 plus earlier model missiles being retrofitted to this standard. Stinger Block 2, although in the advanced development with production stage, is probably going to be cancelled. Parallel production of an initial 12,500 Stinger-RMPs at the European Stinger Production Group, with the participation of Germany, Greece, the Netherlands and Turkey, is under way. Between 1998 and 2004 Germany took delivery of FIM-92C Stinger missile systems. These are known in the country as Fleigerfaust-2. Stinger-RMP and Block 1 are in rate production in Tucson, Arizona for the US (all services) and 18 other nations.

In May 2003, the Redstone Arsenal in Huntsville, Alabama ordered 139 Stinger Block 1 missiles from Raytheon for delivery between January 2005 and April 2005. The contract is worth USD16.4 million.

In July 2009, Raytheon received an order from the US Army on behalf of Foreign Military Sales for 171 Stingers for Taiwan and 178 Stingers for Egypt and Turkey.

Stinger Missile Replacement Programme
The US Naval Surface Warfare Centre Crane Division is seeking technical and cost information to assist the US Marine Corps in planning its proposed Stinger Missile Replacement (SMR) effort. The service may buy up to 2,000 units to succeed its Stinger systems. Candidate technologies of interest include kinetic- and directed-energy weapons in addition to conventional missile approaches.

The weapon would be operated primarily against air targets, including cruise missiles and ultra-light aircraft, with a secondary ground-to-ground capability against trucks, shelters, generator houses and lightly armoured vehicles.

The use of multiple warhead option for different effects is desired. The ability to fire from within enclosures is also required, and ease of use is a key consideration.

Tentative requirements for the SMR include a weight below 11.4 kg and a length of less than 1.5 m. It should have a range of at least as good as that of Stinger, being available for firing within 10 seconds of initial turn-on (desired) or 90 seconds (required).

Stinger (STLS)
Engineers at Schweiz Unternehmung have developed the Stinger Simulation Unit Stinger Tracking and Launch Simulator (STLS). A dummy missile, with a maximum range of 150 m, exposes the operators to smoke, noise and recoil. Tracking may be practised as well against actual targets.

The Royal Netherlands Army is to buy the United Industrial Subsidiary AAI Corporation Advanced Moving Target Simulator (AMTS) system for air-defence training for the Netherlands Armed Forces involved with the Stinger weapon system. The simulator, designated Stinger Trainer by the RNLA, will be installed at the Joint Air Defence School at De Peel Air Forces Base. As of January 2009, the new mobile Stinger Weapon Platforms (Fennek and Mercedes-Benz G-Vagen jeep) are fully operational with 11 and 13 air-defence batteries.

In September 2013, trials took place in Finland on the FIM-92 RMP Block 1 variant with a view to purchase from the United States. Announcements made in February 2014 confirmed the order.

Contractor

Raytheon Electronic Systems.

LFK-Lenkflugkörpersysteme GmbH (prime European contractor; now EADS/LFK).

Leica (Swiss co-production).

Junghans Feinwerktechnik (PDSD M934E6 point detonating fuze w. SD production).

DME Corporation (produces a simulation and training package for Stinger operators).

InterSense (produces the inertial tracking technology used in the DME Corporations simulator and training package).

Static And Towed Surface-To-Air Missile Systems

Belarus

S-75M3

Type
Static Surface-to-Air Missile (SAM) system

Development
Designed to protect vital administrative, industrial and military installations, the S-75M3 upgrade/modernisation by the Belarussian Company Belvneshpromservice is offered for sale with the 5Ya23 missile.

Description
The S-75M3 upgrade is designed to engage targets with a radar cross section of 0.2 m^2 such as reconnaissance aircraft, strategic bombers and fighter-bombers. The system comprises the following:

- RSN-75V tracking and guidance radar
- Six launchers with one 5Ya23 missile each
- Power supply means
- An additional P-18 radar is available at the customers request.

Radar
The RSN-75V radar is arranged on three trailers consisting of a transmitter-receiving post, control post and equipment cabin. The radar also has TV tracking channels operating the automatic and manual modes. The air defence missile complex assets are accommodated and transported in trailers and semi-trailers interconnected by cables. The power source can either be by means of a mobile diesel generator or industrial mains.

Missile
The 5Ya23 missile is a two-stage weapon that employs the normal aerodynamic configuration. The jettisonable booster stage is stabiliser-finned throttleable solid-propellant rocket motor, while the second stage mounts a liquid-propellant turbopump-fed sustainer motor. Structurally, the sustainer stage comprises a number of sections housing a radio fuze, High-Explosive (HE) fragmentation warhead, on board equipment, propellant tanks, liquid-propellant motor, control surface actuators and steering command receivers.

Fan Song Radar (Miroslav Gyürösi) 0560698

S-75 missile on launcher (Miroslav Gyürösi) 0560697

Launcher
The missile is fired from an oblique launcher laid in azimuth and elevation. The missile is controlled and guided to its target by radio commands received from the ground guidance radar, and the warhead is initiated by commands from the radio fuze or the ground guidance radar after the missile approaching the target has reached the effective range.

Variants
A version of the S-75 missile family upgraded in Romania is also known as S-75M3.

Specifications

S-75M3
(5Ya23)

Type:	low to medium altitude SAM
Length:	not available
Range:	
(maximum)	60 km
(minimum)	unknown
Altitude:	
(maximum)	30,000 m
(minimum)	100 m
Kill probability: (P_K)	
(up to 50 km range)	0.4-0.97
(50-60 km range)	unknown
Booster motor:	solid propellant
Sustainer motor:	liquid-propellant (turbopump-fed)
Guidance:	radio commands received from the ground guidance radar
Warhead:	HE fragmentation
Fuze:	radio command
Radar cross section targets:	0.2 m^2

Status
Offered for sale/upgrade/modernisation.

Contractor
Belvneshpromservice.

S-125-2T (Pechora-2T)

Type
Static and towed Surface-to-Air Missile (SAM) system

Development
The Belarussian company, Tetraedr, is offering a modification/upgrade of the Soviet S-125M (M1) (NATO SA-3 'Goa') medium-range Surface-To-Air Missile (SAM) system that will then be known as the Pechora-2T.

The purpose of the upgrade is an increase in the combat efficiency and operating performance of the S-125 Pechora. This is achieved with the use of new guidance methods for the anti-aircraft guided missiles, substitution of the existing equipment to digital circuitry base and new instrumentation introduced into the missile system on the basis of the latest developments in the fields of signal processing, digital engineering and high technologies.

Firing trials of the Pechora-2T were carried out at the Sary Shagan Missile Test Range (SSMTR) in Kazakhstan during April 2003; November 2004 and October 2007. The purpose of these trials were to prove the concept of engaging small size targets with high speeds and low altitudes. The system is also able to perform under complex jamming conditions.

The later trials in 2007 were to demonstrate the performance of the modified system beyond the original systems official capabilities.

Several firings took place at the SSMTR against targets PM-6 (parachute retarded), MR-9 (rocket target) and Balaban a derivative of the 5Ya23/V-759 (SA-2). The missile used by the S-125-2T is the 5V27D at ranges out to 32.2 km at heights of 12,000 m. The reported flight time of the 5V27D was 49.3 seconds close to the time at which the round would have self-destructed.

Description
The Pechora-2T is mounted on wheeled vehicles for increased mobility and survivability.

The Pechora-2T modified system allows for the engagement of two targets simultaneously, with a salvo of one missile per target. There are limitations to the control system for the missile firings, in that the cluster of two targets has to be within a very limited area (box) that has been defined as 7 × 7 beam of the guidance radar.

UNV-2T Fire-Control Radar (Tetraedr Scientific Industrial Unitary Enterprise)
1147406

Pechora-2T control panel (Tetraedr Scientific Industrial Unitary Enterprise)
1036516

The engagement of two targets is possible by means of a new UVK-125-2T command generating system, installed in the radar vehicle in place of the older UK-80 console. This uses two new guidance methods that have been identified as KDC and MTP.

The architecture of the Radio Frequency (RF) system has been revised and the High-Frequency (HF) receiving subsystem now uses solid-state electronics. The solid-state Ultra-High Frequency (UHF) amplifiers installed in the guidance radar reception channels feature a low-noise coefficient and a wide dynamic band which, combined with the correlation-and-filtering signal-processing modes, accounts for the tangible enhanced jamming immunity of the SAM system. Jamming immunity of the Pechora-2T SAM system stands at 2,700 W/MHz, against active jamming from an equivalent distance of 100 km. With the original Pechora system, this stood at 24 W/MHz. Up to 50 per cent of the instrumentation of Pechora-2T, as compared to the basic Pechora version, has been converted to new circuitry. Reliability and the life cycle of the SAM system equipment have been considerably stepped up, the time required for maintenance shortened, and the nomenclature of spare parts required narrowed down.

Changes to the system include:

(1) Enhancing the system combat effectiveness by:
- Installing the UVK-125-2T command generation device, incorporating the innovative KDC and MTP SAM guidance methods
- Installing the fire operator's unit UK-31-2T
- Installing the new APP-125-2T automated launch device
- Introducing the new 100 km and 50 km range scales of the missile guidance radar
- Transmitting device upgrade.

Pechora-2T (Tetraedr Scientific Industrial Unitary Enterprise) 1036515

Fire Control Point for UNV-2T
(Tetraedr Scientific Industrial Unitary Enterprise) 1147408

Missile launch, S-125-2T (Tetraedr Scientific Industrial Unitary Enterprise)
1147409

(2) Stepping-up the SAM system jamming protection by:
- Upgrading the UHF receive channel in UV-40 cabinet
- Correlation and filtering signal processing
- Widening the dynamic band of the target channel receiving device
- Introduction of new mixed (manual tracking - automatic tracking) target tracking mode.

(3) Enhancing the system combat effectiveness by:
- Installing the new UK-40-2T training simulator
- Launch control system upgrade
- Upgrading the UnV antenna post and 5P73 launchers power tracking drives. Additional options
- Replacing the 9Sh38 Karat tele-optical sight with OES-2T electro-optical system
- Reducing the SAM system in/out of action time to 25 minutes (using the vehicle chassis supplied by the customer).

Variants

A Pechora-2TM is also known to exist that has been reported to use the 5V27 (V601) missile by Tetraedr.

On 1 December 2008, Tetraedr received its first order to upgrade 27 × S-125 systems for an unidentified customer to the Pechora-2TM standard. In this case the letter T in the designation represents the Tetraedr Company whilst the M stands for mobilniy (mobile).

The S-125-2TM combat assets include:

- SNR-125-2TM missile guidance radar
- 5P73-2TM launcher with four guide rails
- 5V27 SAMs
- APSS-2TM autonomous power supply system

The TRV-2TM technical support, transportation and reloading vehicle is also available as is the P-18T radar target acquisition and target designation system.

Specifications

	PECHORA-2T	PECHORA-2TM
Travelling mode to deployment:	25 min	25 min
Channel capacity, target:	2	unknown
Target detection range:	100 km	unknown
Maximum target speed:		
(head on)	1,100 m/s	900 m/s
(tail on)	300 m/s	300 m/s
Altitude:		
(minimum)	20 m	20 m
(maximum)	25,000 m	24,848 m (25,000 m)
Range:		
(minimum)	3,500 m	3,500 m
(maximum)	35,400 m	35,400 m (36,000 m)
Jamming immunity: (spectrum density at a distance of 100 km)	2,700 W/MHz	2,700 W/MHz
Guidance methods:	KDC and MTP	unknown
Minimal RCS target:	0.02 m^2	0.02 m^2

Single missile kill probability

	5V27D	5V27
Tactical fighter:	0.85-0.97	0.85-0.92 at ranges up to 20 km; 0.89-0.90 at ranges of 25-30 km
Helicopter:	0.75-0.95	unknown
Cruise missile:	0.45-0.95	unknown
Manoeuvring target:	0.55-0.97	unknown

Status

Proving trials were completed at SSMTR. Between 2002 and 2005, the UE Tetraedr has performed several export contracts related with upgrade of S-125 Pechora Air-Defence Missile System (ADMS) to S-125-2T Pechora-2T level. The S-125-2T Pechora-2T ADMS is adopted in three countries by the Commonwealth of Independent States (CIS) Joint Air Defence Group (combined with S-300/SA-10 Grumble ADMS Air-Defence Group on the territory of the Republic of Kazakhstan). The system, as reported by Tetraedr, was also sold to an unidentified African country in April 2005.

Contractor

Tetraedr Scientific Industrial Unitary Enterprise.

China

FT-2000

Type

Mobile Surface-to-Air Missile (SAM) system used on a static site

Development

First reported back in July 2004, the Hong Qui-9 (HQ-9) also known by its export designation FT-2000 was reported to be based on the Russian S-300PMU that had been licensed produced in China.

During the Farnborough International Air Show in 1998, the Chinese National Precision Machinery Import and Export Corporation (CPMIEC) displayed and offered the FT-2000 Surface-to-Air Missile (SAM) system. As expected, the external configuration strongly suggests the Russian S-300PMU.

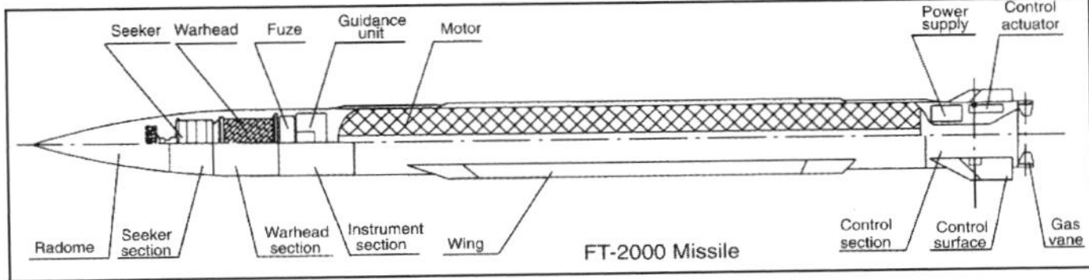

Line drawing of FT-2000 missile (CPMIEC) 0132432

FT-2000 showing four vertical launch canisters (CPMIEC) 0132084

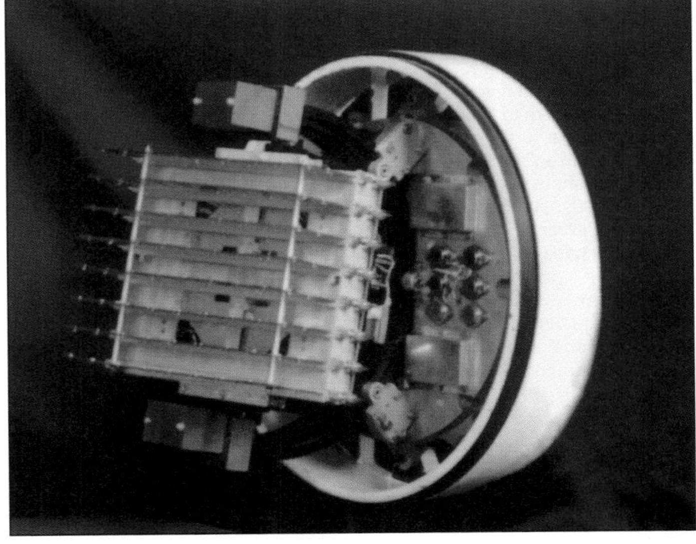

Passive Seeker for the FT-2000 missile (CPMIEC) 0132433

By late 2008 or early 2009, the naval variant has also been offered for sale, this bears the PLAN designation HHQ-9 (Hai Hong Qui - 9) and the export FD-2000.

In early January 2010, the HQ-9 system underwent a test firing in China when a salvo of two missiles were launched at a ballistic target; reports suggest that the target was destroyed and the firing was a complete success.

Description

The FT-2000 is a surface-to-air anti-radiation missile system with a passive detection and guidance system, which is designed to engage targets like the stand-off jammers, radiation sources and AWACS. The weapon system has wide-band passive radar stations, missiles and launching vehicles. The system is capable of full integration with other Chinese SAMs, in particular, those of Russian origin.

Drawings of the missile released through CPMIEC brochures show a missile similar to the US Patriot and the Russian 5V55. Although similar overall, there appears to be two/four added strakes (labelled as 'wings' in the diagram) running the length of the body, these clearly shown and labelled. Either CPMIEC has got its facts and drawings wrong, or this may be a Chinese puzzle. Photography of the HHQ-9 at lift-off clearly shows there are no strakes on the missile, neither are there strakes on the Russian 5V55.

Reports coming from within China indicate that the FT-2000 has been deployed with the PLAAF air-defence forces based in the Fujian Province near the Taiwan Straits. Co-located within this area are the Russian produced S-300PMU1 SAMs.

The FT-2000 makes use of the HT-233 phased array fire-control radar that is able to track 50 targets simultaneously. What is not known, is how this radar and the FT-2000 system fit in with the Russian S-300PMU1 command-and-control systems.

Radar

The FT-2000/HQ-9 has associated radar, HT-233 phased array engagement radar that is believed to be modelled on the Russian 30N6 Flap Lid, the Type 305A (also known as the K/LLQ-305A) that is modelled on the Thales GM400 AESA and is the equivalent of the Russian 64N6 Big Bird and the Type 120 (also known as K/LLQ-120) low altitude acquisition radar that is similar in use to the Russian 76N6 Clam Shell.

Variants

FT-2000A - (see separate entry).
FD-2000 - also known by its PLAN number of HHQ-9 is the air-defence version of the system.

Specifications

	HQ-9
Dimensions and weights	
Length	
overall	6.8 m (22 ft 3¾ in)
Diameter	
body	466 mm (18.35 in)
Weight	
launch	1,300 kg (2,866 lb)
Performance	
Range	
min	6.5 n miles (12 km; 7.5 miles)
max	54.0 n miles (100 km; 62.1 miles)
Altitude	
min	3,000 m (9,843 ft)
max	20,000 m (65,617 ft)
Ordnance components	
Guidance	passive (anti radiation seeker 2–18 GHz)
Propulsion	
type	solid propellant

Status
Development is complete. It was reported in March 2008, that China has offered this system to Turkey as part of the long-range acquisition requirement laid down by the Turkish Undersecretariat for the Defence Industry (SSM). Test firings have taken place in China during early January 2010 against a ballistic target, both shots were reported as successful.

By June 2010, it was reported that Pakistan has ordered the system and is expecting delivery in the very near future, however, by May 2013 this system has still not been delivered and there is no sign of it being delivered.

Contractor
Chinese state factories.
Marketing and exports: China National Precision Machinery Import and Export Corporation (CPMIEC)

FT-2000A

Type
Static and towed Surface-to-Air Missile (SAM) system

Development
According to non-proven sources within China, the FT-2000A is based on a heavily modified Hong Qi-2 (HQ-2, Red Flag-2) missile body. However, the overall appearance of the FT-2000 gives the impression of a copied or reverse engineered ex-Russian S-300PMU system. Described as an AWACS killer, this SAM system has been designed for export.

Description
The FT-2000A has been equipped with a passive radio frequency (RF) homing seeker operating in the 2 to 6 GHz waveband region covering the NATO E, F and G bands. Also fitted, as new systems, are a 60 kg HE-fragmentation warhead, a millimetric-frequency proximity fuze, guidance and control section and a cut-off valve unit for thrust adjustment in flight. The latter extends the engagement range of the missile to 60,000 m. Stated maximum engagement altitude is 18,000 m.

The FT-2000A system can be used either as part of a composite fire unit or as a stand-alone battery. In the former case, a specialised fire-control unit for the battery and several FT-2000A launchers are required to allow one-for-one

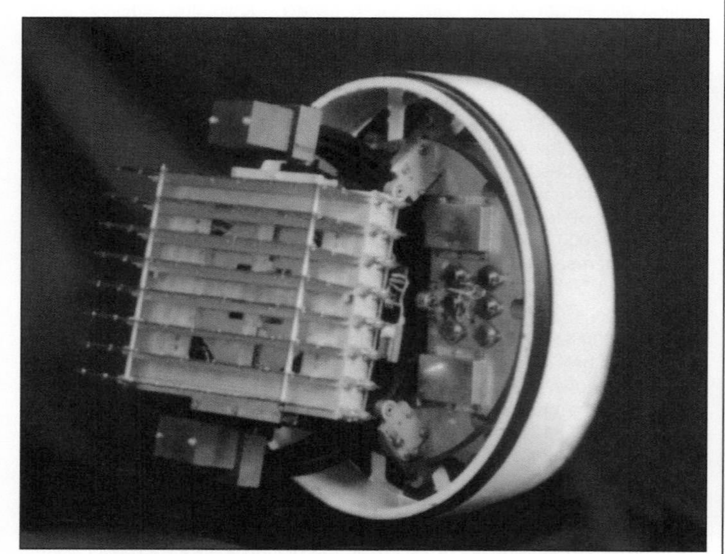

Passive RF seeker for FT-2000 (CPMIEC) 0132433

FT2000 Missile (IHS/Patrick Allen) 1024283

replacement of the conventional SAMs. In the second case, a stand-alone battery consists of a central command-and-control station with 12 single-rail launchers. The central station has a master work together using a triangulation technique to determine the angle and range of targets emitting in the 2 to 6 GHz waveband region.

Before launch, the designated engagement missile seeker is loaded with the target's emitted RF signature gained from the sensors and once launched, relies on this for the in-flight tracking and intercept phases. Thus the whole engagement cycle is totally passive, relying completely on the RF emitted from the target.

Variants
FD-2000, this is believed to be the air-defence variant of the FT-2000.

Status
The system is designed for the export market as an update for Chinese and Russian Federation (CIS) HQ-2/SA-2 'Guideline' systems.

Contractor
Chinese state factories.
China National Precision Machinery Import & Export Corporation (CPMIEC).

Hongqi-2

Type
Static and towed Surface-to-Air Missile (SAM) system

Development
The China National Precision Machinery Import and Export Corporation (CPMIEC) Hong Qi-2 (HQ-2, Red Flag-2) was the result of a redesign of the HQ-1 system. The basic HQ-1 system was purchased from the then Soviet Union in the mid to late 1950s, and was known as the System-75 Dvina. It consisted of:
- Guidance Station RSN-75
- Launchers SM-63
- V-750 missiles
- Ground Support Equipment.

On arrival in China, the system was renamed HQ-1. Shortly after arrival of the weapon and system components, China and the USSR diplomatically fell out. China was therefore forced to reproduce spare parts for its operational systems. Furthermore, copying and reverse engineering resulted in a modified HQ-1 incorporating anti-jamming techniques this system became known as the HQ-2.

Hongqi-2B variant mobile surface-to-air missile system with missile elevated (Eric Ditchfield) 0509564

Hongqi-2B variant being launched from its tracked transporter/launch platform 0509709

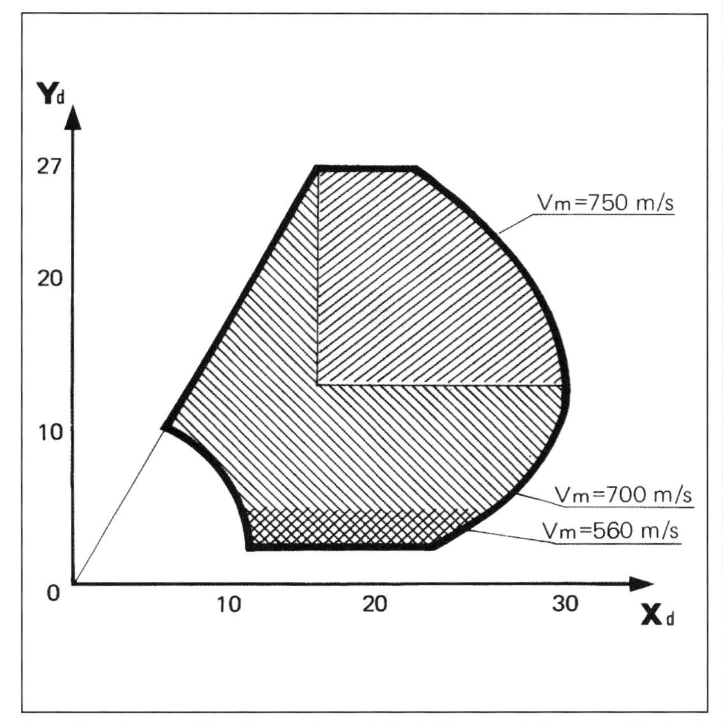

Operational zone of the Chinese Hongqi-2B SAM extends from an altitude of 1 km up to 27 km with maximum slant range being 35 km. The V_m on chart is speed of target 0509565

In April 1965, coordination of the system development was assigned to the No 2 Research Academy (RA) of the 7th Ministry of Machine Building. An accelerated development programme then followed which resulted in the weapon passing its type certification by the end of 1966. However some reports suggest that the then 2nd Space Academy (now China Academy of Defence Technology) became responsible for the design along with the 139 Factory and 786 Factory producing the missile and ground stations respectively. This SAM system was then designated the HQ-2. In July 1967, the complete system received its design certificate approval for production. It was then used operationally to shoot down a Lockheed U-2 in September 1967.

Export marketing of the missile system is the responsibility of the China National Precision Machinery Import and Export Corporation (CPMIEC).

The HQ-2 saw combat service with the People's Liberation Army Air Force in the late 1960s against Taiwanese-flown U-2s; the missiles claimed five U-2s, one in each of the years 1967, 1968 and 1970 and two in 1969. Iran also used an HQ-2 during its war with Iraq. The system was also used against US Air Force SR-71 reconnaissance aircraft by North Korea. The latter country has also imported HQ-2 technology from China in order to maintain its own stocks of the weapon and Russian-supplied System-75s.

Over the 30 years of HQ-2 production a series of variants has been produced to form a complete family; these are the HQ-2, HQ-2A, HQ-2B, HQ-2F, HQ-2J and HQ-2P. The missiles have been matched by improvements to the missile guidance radars. The HQ-2 family members also have a secondary ground and anti-shipping surface-to-surface attack capability.

Hongqi-2B missile being prepared for launch on its tracked transporter/launch platform 0509566

'Gin Sling-B' radar for Hongqi-2B surface-to-air missile system deployed 0509567

HQ-2 technology packages have been passed to Egypt, Iran and North Korea. North Korea has developed its own extensive missile development and upgrade programmes based on HQ-2 designs.

Description

HQ-2J model

The HQ-2J is also known as the Chinese M7 system. This has a surface-to-surface capability with a range of 120 km and a CEP of 0.5 km. The HQ-2J launcher battalion comprises the following:

- Launching installation with a fire-control vehicle
- Six semi-fixed trainable single rail launchers (with 360° traverse and elevation limits of +11 to +65°)
- 24 missiles (six on the launcher rails and the rest either at the battalion reload storage facility in canisters or carried ready for loading on the transloader vehicles)
- Six transloader vehicles with semi-trailers carrying single ready to load missiles
- Associated technical support vehicles including those for missile check and maintenance/liquid propellant filling/solid propellant charge loading
- Power station sub-unit with three power generating vehicles and one power distribution vehicle
- Radar sub-unit with a missile guidance station comprising a radar van, a radar control vehicle, a battalion command post vehicle and a maintenance/spare parts vehicle.

The two-stage HQ-2J missile is similar in configuration to the Russian SA-2 being 10.842-m long and weighing 2,326 kg at launch. The diameter of the solid-propellant rocket booster motor stage is 0.654 m. The high-thrust liquid-fuelled sustainer rocket-propelled main body stage is 0.5 m.

SJ-202 guidance station for Hongqi-2 SAM system deployed 0509568

Internally, the weapon has been upgraded in terms of electronic systems and fitted with an improved design warhead with a wide multiple fragmentation scattering pattern. The target speed at high altitude can be 750 m/s and at low altitude 420 m/s. The missile can be used in head-on and tail-chase engagements.

The basic all-weather target acquisition/tracking and missile guidance radar used with the HQ-2J has the NATO designation 'Gin Sling-A'. In appearance it is similar to the Russian NATO designation 'Fan Song' system and has a similar performance, with the ability to track a single target and guide up to three missiles to it.

HQ-2B model

The HQ-2B battalion is a mobile version of the HQ-2J unit and comprises the following components:

- Six trainable single-rail launchers carried on the rear decking of lengthened NORINCO Type 63 light tank chassis, all the missile/launcher firing controls, power supply systems and necessary electronics are located in the hull area with access provided by latched side panels
- 24 missiles (six on the launcher vehicles with the remainder at the battalion reload storage site or on transloader vehicles)
- Variable number of transloader vehicles and a ZD-2(B) missile guidance station.

The HQ-2B missile is similar to the HQ-2J described previously but is fitted with a moderate dispersion angle HE fragmentation warhead, an anti-jam digital code radio-command guidance system, an integrated circuit autopilot and an FM phase comparison proximity fuzing circuit. The weapon has head-on, crossing target and tail-chase engagement capabilities.

The Single-Shot Kill Probability (SSKP) of a single round against a target with a velocity of less than or equal to 565 m/s is 0.92 and against a target with a velocity of less than or equal to 750 m/s is 0.73. Targets up to a maximum velocity of 1,000 m/s can be engaged.

If required, the HQ-2B can be fired from the semi-fixed single rail launcher assembly used for the other HQ-2 family members.

The ZD-2 (B), missile guidance station comprises the following subsystems:

- 2FA(B) radar receiver-transmitter trailer (believed to have the NATO codename 'Gin Sling-B') which is fitted inside its cabin with the following:

(1) The main E/F-band radar transmitter unit
(2) The main E/F-band radar receiver unit
(3) An auxiliary I/J-band range measuring radar transmitter
(4) A monopulse radar receiver unit
(5) An antenna feeder system
(6) An antenna servo-system.

On the trailer's cabin roof are:

(1) The main radar's elevation scan antenna
(2) The main radar's azimuth scan antenna
(3) The target illuminator radar antenna
(4) The D-band missile command guidance transmitter antenna
(5) The secondary telescope and video TV camera tracker head assembly
(6) The auxiliary range measuring radar antenna receiver box.

The missiles own tracking transponder system operates in the D-band.

- 2X(B) truck-mounted display shelter which acts as the combat command centre for the battalion. This contains the IFF display, the video TV tracker and large PPI screen displays, the radar and launcher control systems, the battalion system status displays, main battalion and higher echelon communications and a training engagement simulator system.
- 2M(B) truck-mounted command communications shelter is equipped with the battalion and higher echelon signals equipment including the cryptography units for encoding and decoding of signal traffic.

The guidance station can operate in normal main radar target acquisition scan, selected target track and illumination, monopulse (as an ECCM technique) and TV tracking modes. The co-ordinate difference between the target and the missile is obtained by using the relative co-ordinate system in the 2Z(B)-shelter truck. The ranges and angle co-ordinates of the target and the missile are also obtainable with the secondary digital tracking system.

Maximum detection range of the main radar system is in excess of 100 km, with a stable automatic tracking range of over 75 km. The maximum auxiliary range measuring capability is in excess of 60 km and the secondary TV tracking range for use in heavy ECM environments is around 45 km.

- The 2Z(B) truck-mounted co-ordinate shelter which carries the TV video and IFF electronics, the co-ordinate tracking system for the targets and missiles, and a microcomputer for firing solution computations and generation of missile command-and-control instructions
- The 2P(B) truck-mounted power distribution centre which carries the distribution board system for the output of the generator trucks. It supplies by cable the power outputs to all the guidance unit subsystems

Hongqi-2J variant surface-to-air missile being carried on a semi-trailer towed by a 6 × 6 truck 0509890

- Three 2D(B) truck-mounted generator systems (one of which acts as a spare). Each system has one shelter-mounted 50 Hz diesel generator and converter unit to provide three-phase 50 Hz and three-phase 400 Hz AC electrical supplies to the power distribution centre truck.

Variants

HQ-2A

The concept of a new HQ-2 version was proposed in early 1973 to counter low-altitude targets and electronic technology observed during the Vietnam War.

In August 1973, the development of what became known as the HQ-2A began with the main performance, aerodynamic and basic system characteristics remaining unchanged.

Only the ECCM features and the weapon system's low-altitude engagement capabilities were to be altered. Between 1978 and 1982, the system's developmental and design certification firing trials were undertaken. Final design certification approval for production was given in June 1984. The modifications made to the HQ-2A allowed the system to increase the horizontal firing angle from the original ±55° to ±75°. The speed of the missile was also improved to 1,200 m/s with improvement in the *g* capability from 1 to 1.5 *g*. A TV camera was also observed for the first time improving the electronic countermeasure capability of the system.

HQ-2B

The concept of the HQ-2B (as described previously) was first considered in July 1978 as a further improvement on the HQ-2A. In June 1979, its development was assigned to the same team, which designed the HQ-2A. The system was modified to improve mobility, missile guidance and missile ECCM, fuzing and warhead lethality. The fuze is of the FM phase comparison type. Operational testing and design certification trials took place from 1980 to 1986.

HQ-2 systems with the SJ-202 radar set

For use with both the HQ-2 family members and the SA-2 'Guideline' system CPMIEC has produced the SJ-202 phased-array target acquisition/tracking and missile guidance station comprising a highly automated radar van and a command-and-control truck. The station is easily integrated with the power generation units of the missile unit to which it is attached.

Maximum detection range is 115 km and tracking range 80 km. Up to four missiles can be guided at any one time to attack either one or two targets simultaneously. Response time is quoted as eight seconds. This radar may also be used with the CPMIEC KS-1 SAM.

8610 (CSS-8) SSM (HQ-2J) (M7)

Design of the export 8610 programme by the Second Academy began in 1986. It is based on the HQ-2 system, the first stage being a solid propellant booster whilst the engine on the second stage uses storable liquid propellants. The main difference is the control system, which differs completely from that on the SAM. Iran has procured the 8610 in some numbers. The CSS-8 has a range of 120 km. In addition, standard HQ-2s can be used in the surface-to-surface role, especially using airbursts against hostile radars. The CSS-8 has also been referred to as the M-7, B-610 and BG-510 tactical ballistic missile by the Chinese and is launched from a similar tracked chassis to that used for the Hongqi-2B surface-to-air missile system which is illustrated in this entry. In Iran it has been given the local name 'Tamdar' or 'Tandar'.

Specifications

	HQ-2J	HQ-2B
Dimensions and weights		
Length		
overall	10.842 m (35 ft 6¾ in)	10.842 m (35 ft 6¾ in)
Diameter		
body	1,700 mm (66.93 in)	1,700 mm (66.93 in)
	2,500 mm (98.43 in) (booster stage)	2,500 mm (98.43 in) (booster stage)
Weight		
launch	2,326 kg (5,127 lb)	2,322 kg (5,119 lb)

	HQ-2J	HQ-2B
Performance		
Speed		
max speed	2,333 kt (4,320 km/h; 2,684 mph; 1,200 m/s)	2,430 kt (4,500 km/h; 2,796 mph; 1,250 m/s)
Range		
min	3.8 n miles (7 km; 4.3 miles)	3.8 n miles (7 km; 4.3 miles)
max	18.4 n miles (34 km; 21.1 miles)	18.9 n miles (35 km; 21.7 miles)
Altitude		
min	500 m (1,640 ft)	1,000 m (3,281 ft)
max	27,000 m (88,583 ft)	27,000 m (88,583 ft)
Ordnance components		
Warhead	HE fragmentation	HE fragmentation
Fuze	proximity	proximity
	uplink command	uplink command
Propulsion		
type	solid propellant (with storable liquid fuel, sustain stage)	solid propellant (with storable liquid fuel, sustain stage)

Status
Production for the original system is complete. Various members of the HQ-2 family have seen service with Albania, Pakistan, People's Liberation Army, Iran (eight battalions purchased in early 1989), technology used in its own SA-2-type production programme. North Korea has its own extensive HQ-2/SA-2 development and upgrade programme.

Contractor
Chinese state factories.
China National Precision Machinery Import and Export Corporation (CNPMIEC) (export marketing).

Hongqi-61A

Type
Static and towed Surface-to-Air Missile (SAM) system

Development
The China National Precision Machinery Import and Export Corporation (CNPMIEC) Hong Qi-61A (also known as Red Flag-61A, HQ-61A, RF-61A or SD-1A) system was developed from the navy of the People's Liberation Army HQ-61 (SD-1) shipboard SAM system under the auspices of the Shanghai No 2 Bureau of Machinery and Electronics (now the Shanghai Academy of Spaceflight Technology - SAST). The original concept for this land-based low- to medium-altitude HQ-61 version was decided upon in 1976 by the Chinese State Council and the Military Commission of the Central Commission of the Communist Party of China.

In 1979, the project definition stage was finalised under the leadership of the Shanghai No 2 Bureau under the designation of HQ-41 but later renumbered to HQ-61 and, in November 1984, the first trial firings were successfully undertaken to prove the various components.

Between April and June 1986, the system certification firing trials were undertaken, with the final design certification being awarded in November 1989 before production commenced.

The army version was first seen in public during the November 1986 Beijing defence exhibition to meet an operational date of 1989.

The system was designed to engage targets flying at low- to medium-altitudes. A minimum range of 2.5 km and a maximum range of 12.5 km in the horizontal plane, extending in an arc to an altitude of around 10,000 m.

As far as is known the Hong Qi-61A missile has not been exported by China and is no longer marketed by CNPMIEC. The system is believed to have entered service with the People's Liberation Army in only a limited capacity.

Description
The missile itself is single stage with a solid propellant rocket motor. The missile has a chain-type high-explosive warhead, dual-fuze system, homing head using a continuous wave semi-active homing guidance system which is said to have an anti-jamming feature, autopilot, airborne power supply using a combustion turbine generator to supply AC and DC power for all missile components and four hydraulic servo-operated canard fins.

A typical army battery of RF-61A launchers consists of four Hong Yan CQ-261 series (6 × 6) trucks each with two rail-mounted missiles on a turntable at the rear of the vehicle, mobile generators, command post vehicle, tracking and illuminating radar vehicle and a target indicating radar vehicle. A full support unit is also provided for technical back up of the battery in the field. The target indicating radar vehicle is also based on a (6 × 6) chassis which is similar in appearance to the Russian-designed 'Flat Face' target acquisition radar used with the S-125 Pechora SAM system.

The launcher vehicle has a fully enclosed forward control cab with an anti-blast shield to its immediate rear. When in the firing position, the launcher is supported on four stabilisers, two each side. Reload missiles are brought up by a missile resupply truck and loaded with the aid of a crane.

A typical target engagement would take place as follows. The target is first detected by the target indicating radar vehicle and after being confirmed as hostile, is tracked and illuminated by the Continuous Wave (CW) tracking and illuminating radar vehicle. When the target is within range a missile is launched. It appears that the system can engage only one target at a time with typically one missile from each of two to three launchers.

Hong Qi-61A SAM system with stabilisers lowered and launcher traversed to left 0509560

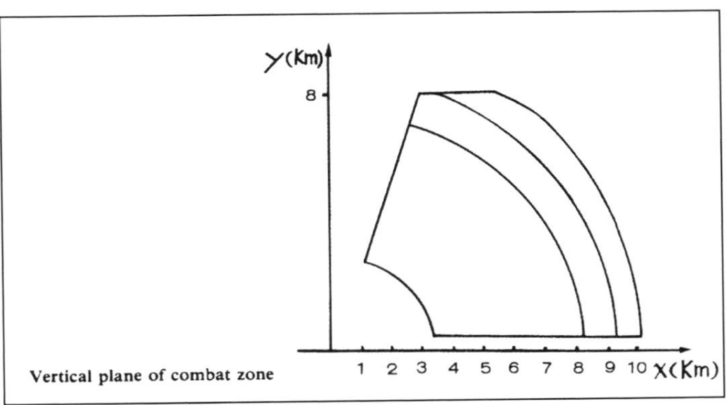

Typical target envelope of Hong Qi-61A SAM system in vertical plane 0509561

Hong Qi-61A missile being launched during trials 0509562

CNEIEC Model 571 C-band radar deployed in field showing two elliptic parabolic net reflectors one above the other. A similar vehicle carries the generator 0509563

No details of the tracking and illuminating radar system have been disclosed, although photographic imagery examined shows a dish-type antenna with a TV camera mounted coaxially to the right for use in an ECM environment, or passive operations during clear weather engagements.

The target indicating radar vehicle is based on a different (6 × 6) 2,500 kg cross-country truck model, as is the power supply truck.

The truck-mounted C-band radar system uses the Chinese designation Model 571 and has two elliptic parabolic net-type reflectors. The speed of rotation is 3 or 6 rpm. Other features include moving target indication and frequency-hopping agility. According to the Chinese, the Model 571 radar has been designed specifically for low-altitude warning and displays both the slant range and azimuth of aircraft targets detected.

Specifications

	HQ-61A
Dimensions and weights	
Length	
overall	3.99 m (13 ft 1 in)
Diameter	
body	286 mm (11.26 in)
Flight control surfaces	
span	1.116 m (3 ft 8 in) (wing)
Weight	
launch	320 kg (705 lb) (est.)
Performance	
Speed	
max Mach number	3.0
Range	
min	1.3 n miles (2.5 km; 1.6 miles)
max	6.5 n miles (12 km; 7.5 miles)
Altitude	
max	10,000 m (32,808 ft)
Ordnance components	
Warhead	40 kg (88 lb) HE fragmentation
Fuze	acoustic, proximity, impact
Guidance	semi-active, radar
Propulsion	
type	solid propellant (sustain stage)

Status

Production complete. Possibly in very limited service with the People's Liberation Army, but more likely in reserve. The HQ-61 was not been exported and it is no longer marketed.

Contractor

Chinese state factories.

China National Precision Machinery Import and Export Corporation (CNPMIEC) (export marketing).

Model 571 radar: China National Electronics Import and Export Corporation (CNEIEC) (marketing).

KS-1/KS-1A (HQ-12)

Type

Mobile Surface-to-Air Missile (SAM) system used on a static site

Development

The China National Precision Machinery Import and Export Corporation (CNPMIEC) first offered the KaiShan-1 (KS-1) at the Paris Air Show in 1991, an improved version known as the KS-1A was later developed in the late 1990s.

The KS1 is designated by the People's Liberation Army (PLA) as the Hong Qi 12 (HQ-12). The KS-1 is reported to be complementary to, and also reported by US and Chinese sources as the replacement for, the earlier HQ-2 systems.

Development began in the early 1980s with the first missile firing taking place in 1989. The developmental programme was completed in 1994.

A derivative of the second (upper) stage of the HQ-2 missile, the KS-1 was designed to engage aircraft, helicopters and Remotely Piloted Vehicle (RPV) systems at altitudes of 500 m to a maximum of 25,000 m. The system also has a limited engagement capability against air-launched tactical missiles. The KS-1 system has multi-tracking and multi-engagement capabilities.

In the late 1990s the system underwent an upgrade primarily from the single arm fixed launcher that was based on the HQ-2 launcher to a dual rail truck mounted (6 × 6) version with a new phased array radar. This was first offered for sale in 2001. The KS-1A is an upgraded version of the KS-1 and offers all-weather air defence against targets in the medium and high altitudes at medium range. Both systems were developed by the China Jiangnan Space Industry Company who are also known as the Base 061 (061 Base).

Description

KS-1

The basic KS-1 battery configuration makes use of four-leg pedestal mounts for the launchers and a towed SJ-202 radar antenna.

Seen for the first time at the Zhuhai Air Show 2002 was a new dual-arm launcher mounted on a 6 × 6 new mobile launch platform that is associated with the KS-1A upgraded version of the KS-1.

KS-1A

The KS-1A boasts a new dual-rail launcher mounted on the rear of a Chinese produced truck. The reported leading features include:

- High overload capability;
- Large killing zone;
- Multi-target engagement capability (guiding six missiles to intercept three targets);
- High jamming immunity in severe ECM environment;
- Effective operation in high automation level, mobility and shorter response time than KS-1;
- Missile miniaturisation and less ground technical support.

The KS-1A system comprises:

- An operational unit that includes 24 missiles
- One HT-233 3D C-band phased array guidance radar station
- Six dual-launcher vehicles
- Six transport-loading vehicles
- Two power supply vehicles
- One frequency conversion power distribution vehicle

The technical support required for the system includes:

- One missile test vehicle
- Three missile transport vehicles
- Two tool vehicles
- One power supply vehicle
- One set of technical support equipment

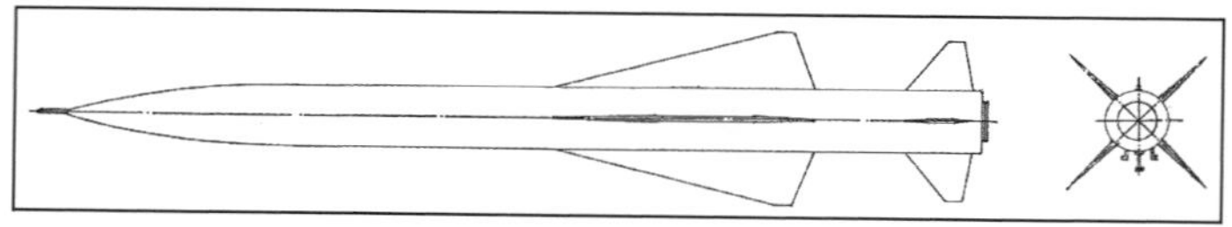

Side and front drawing of the KS-1 SAM 0509557

Scale model of the two-round KS-1 launcher with launcher traversed to right (Christopher F Foss) 0509558

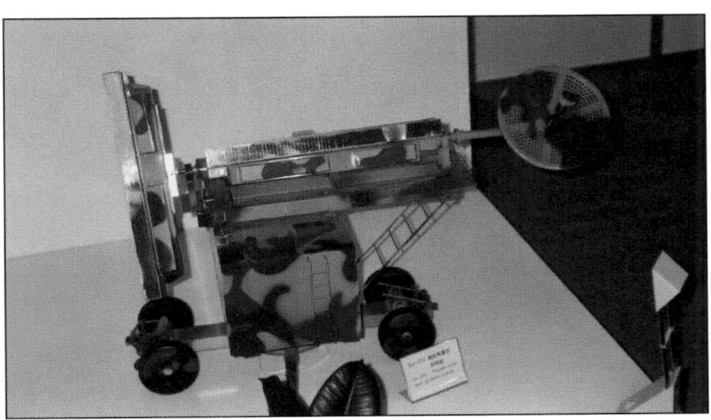

The phased-array radar of the KS-1 SAM system appears to be similar to that used with the HQ-2 SAM system (Christopher F Foss) 0509559

Chinese KS-1 SAM on its ground handling trolley 0509785

KS-1A new single arm launcher (R Karniol) 0531426

- One electronics maintenance vehicle
- Two spare parts vehicles
- One missile testing and metrology vehicle

Yet another truck mounted version is known to exist which has the missile mounted on the rear of a 6 × 6 truck but is enclosed within a box launcher.

There have been reports that another guidance radar known as the H-200 is available for the system.

Variants

The following variants are known to exist:

- Single-rail fixed launcher;
- Dual-rail fixed launcher;
- Dual-rail truck-mounted launcher;
- Dual-box truck-mounted launcher.

Specifications

	KS-1	KS-1A
Dimensions and weights		
Length		
overall	5.6 m (18 ft 4½ in)	5.644 m (18 ft 6¼ in)
Diameter		
body	400 mm (15.75 in)	400 mm (15.75 in)
Flight control surfaces		
span	1.2 m (3 ft 11¼ in) (est.) (wing)	1.2 m (3 ft 11¼ in) (est.) (wing)
Weight		
launch	886 kg (1,953 lb)	886 kg (1,953 lb)
Performance		
Speed		
max speed	2,333 kt (4,320 km/h; 2,684 mph; 1,200 m/s)	2,333 kt (4,320 km/h; 2,684 mph; 1,200 m/s)
Range		
min	3.8 n miles (7 km; 4.3 miles)	2.7 n miles (5 km; 3.1 miles)
max	22.7 n miles (42 km; 26.1 miles)	20.5 n miles (38 km; 23.6 miles) (target speed 720 m/s)
	-	22.7 n miles (42 km; 26.1 miles) (target speed 420 m/s)
	-	27.0 n miles (50 km; 31.1 miles) (target speed 320 m/s)
Altitude		
min	500 m (1,640 ft)	300 m (984 ft)
max	25,000 m (82,021 ft)	27,000 m (88,583 ft)
Ordnance components		
Warhead	100 kg (220 lb) HE fragmentation	100 kg (220 lb) HE fragmentation
Fuze	proximity	proximity
Propulsion		
type	solid propellant (single chamber dual-thrust)	single stage, solid propellant (dual-thrust)

Status

Probably in low-rate production with only a few systems delivered to the PLA for trial and evaluation. There is no evidence that the basic KS-1 system is in service with PLA although some western analysts believe a date of 1996 for IOC.

KS-1 offered for export, but with only one known customer Malaysia.

The PLA units that have this system may well be training/demonstration organisations for export purposes.

The KS-1A has reportedly been the subject of a contract between the Chinese government and Malaysia, although it is hard to accept that a technologically advanced country such as Malaysia would prefer this system when other Western systems with a proven capability are available.

The fact that the KS-1A was given an HQ number in December 2007, would associate the system with full operational capability within the PLA.

Contractor

Main

China Jiangnan Space Industry Company (also known as Base 061 [061 Base]).

Enquiries

China National Precision Machinery Import and Export Corporation (CNPMIEC).

Lieying-60 (HQ-64/LY-60)

Type

Mobile Surface-to-Air Missile (SAM) system used on a static site

Development

The China National Precision Machinery Import and Export Corporation (CNPMIEC) is offering on behalf of the Shanghai 2nd Mechanical-Electronic Bureau (now known as the Shanghai Academy of Space Flight Technology (SAST)), the Lieying-60, or LY-60 (Leiying translates as 'falcon') all-weather day- and night-capable system that has been developed by SAST for both land and sea applications against low- and medium-altitude fast jet targets, slow low-level helicopters and sea-skimming missiles. It can also be used in mixed missile and anti-aircraft gun batteries. The LY-60 known as the HQ-64 by the People's Liberation Army (PLA) was developed during the late 1980s and early 1990s as the replacement for the HQ-61 system and was first deployed with the PLA in October 1994 although other sources/agencies have suggested IOC was early in 1994.

The system is an indigenous development from the Aspide missile, that in itself was developed from the US AIM-7 Sparrow air-to-air missile.

Description

The LY-60 is a short- to medium-range, low- to- medium-altitude SAM system. The system can detect up to 40 targets and track up 12 whilst engaging three targets.

A typical land battery fire unit would comprise one (4 × 4) truck-mounted surveillance radar, three (4 × 4) truck-mounted target tracking/illumination radars, one emergency power supply vehicle and six (6 × 6) Hanyang (HY473) truck-mounted transporter-launcher platforms. Each of the platforms has five missiles that are ready to launch from individual sealed containers. In the

Tracking/illuminating radar vehicle for LY-60 SAM 0509786

Surveillance radar vehicle of the LY-60 0100712

LY-60D variant the launcher vehicle has four missiles mounted in a 2 × 2 structure.

The fire unit is complemented by a technical support unit, which comprises:

- A transportation and reloading vehicle;
- A test vehicle;
- An electronic maintenance vehicle;
- An electromechanical maintenance vehicle;
- A tools support vehicle;
- A spares and meter vehicle;
- A power supply vehicle.

The surveillance radar detects the target aircraft and then conveys it to the appropriate tracker/illuminator radar vehicle for the engagement. The system uses continuous wave semi-active homing guidance principles and, with the allocated assets, the battery can process up to 40 targets, track 12 and engage three simultaneously. System reaction time is nine seconds, with the maximum permissible launcher elevation angle being 70°. The Single-Shot Kill Probability (SSKP) is 0.6 to 0.8 depending upon target type. The use of the moving target tracking system and frequency agile technology ensures that the system has a good anti-jamming capability.

The LY-60 missile is similar in appearance to the Italian Alenia (now Alenia Aeronautica) Aspide used in that company's Spada air defence system.

It weighs 220 kg at launch and has a maximum speed of M3.0. Two pairs of fully movable wings and four stabilising tail fins are fitted.

Italy concluded a deal with China in the 1988 to 1989 time period that involved provision of the air-launched Aspide to the air force of the People's Liberation Army. Several small batches were provided that was due to become the basis of a locally developed variant. The Italian government following the 1990 Tiananmen Square problems blocked subsequent co-operation.

The missile body comprises four separate sections:

- **Guidance section** - located at the missile nose and contains the radome, semi-active radar seeker, fuzing circuit and secondary power source.
- **Warhead section** - includes the firing circuit, safety and arming circuits and the HE-prefabricated steel ball-type fragmentation warhead.
- **Control section** - contains the autopilot, electro-hydraulic power unit, control surface servo system and rear receiver mixer of the seeker head. The four movable wings, an umbilical connector and the front missile suspension unit fit on to the section body.
- **Motor section** - contains a single-stage solid propellant rocket motor. The four fixed tailfins, a motor ignition connector and the rear missile suspension unit fit on to this section.

Variants

LY-60N

The naval version is designated the LY-60N and is capable of engaging sea-skimming missiles.

Model of LY-60 launch vehicle (Duncan Lennox) 0100713

A further variant known as the LY-60D has been offered for sale by the Aerospace Long-March International Trade Co., Ltd.

Air-to-air variant

The LY-60 airframe has been given the designations PL-11 or FD-60 for use in the air-to-air role.

Specifications

	LY-60D	LY-60
Dimensions and weights		
Length		
overall	-	3.69 m (12 ft 1¼ in)
Diameter		
body	-	203 mm (7.99 in) (forward)
	-	208 mm (8.19 in) (aft)
Flight control surfaces		
span	-	0.68 m (2 ft 2¾ in) (wing)
Weight		
launch	-	220 kg (485 lb)
Performance		
Speed		
max Mach number	-	3.0
Range		
min	0.5 n miles (1 km; 0.6 miles)	0.5 n miles (1 km; 0.6 miles)
	0.8 n miles (1.5 km; 0.9 miles) (effective against cruise missile)	-
max	9.7 n miles (18 km; 11.2 miles)	9.7 n miles (18 km; 11.2 miles) (est.)
	5.4 n miles (10 km; 6.2 miles) (effective against cruise missile)	-
Altitude		
min	30 m (98 ft)	30 m (98 ft)
max	12,000 m (39,370 ft)	12,000 m (39,370 ft) (est.)
Ordnance components		
Warhead	-	33 kg (72 lb) HE fragmentation prefabricated steel balls
Fuze	impact, proximity	
Guidance	semi-active (CW)	semi-active, radar (autopilot with CW)
Propulsion		
type	solid propellant	single stage, solid propellant

Status

In production. In very limited service (probably for trials and testing only) with the People's Liberation Army (in air launched, land-based and naval forms) and also with the Peoples Liberation Army Air Force (PLAAF). The system has also been exported to Pakistan.

Contractor
Shanghai Academy of Space Flight Technology.
China Aerospace Science and Technology Corporation.
Aerospace Long-March International Trade Co., Ltd.

Enquiries
China National Precision Machinery Import and Export Corporation (CNPMIEC).

Germany

Tripod Adapted Stinger (TAS)

Type
Tripod mounted man-portable Surface-to-Air Missile (SAM) system

Development
The Tripod Adapted Stinger (TAS) is a relatively simple adaptation of two Raytheon Electronic Systems or European-built Stinger missile launcher assemblies on a lightweight tripod mounting for all-weather, day and night use. The then LFK GmbH proposed a modular configuration to meet the requirements of the German Air Force.

The TAS can be transported by any vehicle or hand carried in three-man transportable loads. The electrical power to operate the TAS is supplied by either a battery or a mobile external power supply.

Description
The TAS consists of:
- Two stinger man portable missiles in their launcher tubes with grip stocks (launching mechanisms) and battery coolant units;
- A tripod with attached seat and elevation axis with attached hand grips (option: without seat);
- Electrical equipment with micro-controller, azimuth encoder and headset (option: with radio connected target data display);
- Eight to 12 micron wavelength thermal imaging sight assembly with 3 × 4° and 9 × 12° fields of view.

The TAS can be operated autonomously or in a pre-assignment mode based on any passive InfraRed (IR) search-and-track device or other sensors, including radar and other SHORAD systems. The required cueing data of a target are its position, flight direction and speed.

When installed at the firing position, the gunner sets the coordinates and the Northing reference. Cueing data are received via a radio set and a warning tone is heard within the gunner's headset or target data are shown on the display.

MBDA Deutschland LFK Tripod Adapted Stinger (TAS) low-altitude surface-to-air missile system deployed in firing position 0509832

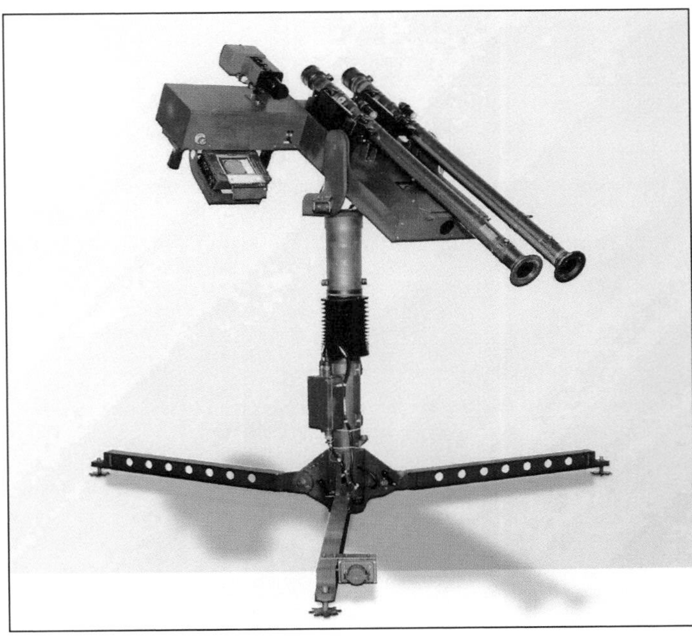

Tripod variant without seat for standing operator, with target display, mini-FLIR and BCU holder 0093174

The gunner activates a Stinger missile by means of the hand grips. The gunner turns the tripod on to the given direction and attempts to acquires the target. After Stinger lock on, the seeker's acquisition tone is heard. Looking into the imaging sight, the gunner is able to identify whether the missile has locked on to the correct target. Once satisfied, the target is then engaged.

Specifications
Tripod Adapted Stinger (TAS)

Crew:	1 (gunner)
Dimensions:	
(height)	1.3 m
(diameter)	1.6 m
Weight:	
(tripod complete)	47 kg
(electrical equipment without battery)	23 kg
(thermal imager)	6 kg
(2 Stingers, each with grip stock and 3 BCUs)	37 kg
Missiles:	2 ready-to-fire Stinger
Weapon reaction time:	<5 s
Reload time:	<1 min (2 missiles using 2 men)

Variants
MBDA Deutschland LFK GmbH has a number of adaptations for ship and land-based guns. These involve the fitting of Stinger to standard gun mounts, for example 40 mm L/70 Bofors.

Status
Ready for production, up to 4,500 missiles have been delivered to the German Air Force but the number of Tripod Launchers is unknown. An Igla family variant has also been developed.

Contractor
MBDA Deutschland LFK GmbH.

International

Barak extended - range

Type
Developmental Surface-to-Air Missile (SAM) system

Development
In January 2007, Israel and India agreed to a cooperative programme to extend the range of the Barak missile system, the new system has been named as Barak-NG and/or Barak-8, (Barak-8 is the Israeli Naval Terminology, Barak-NG is the Indian Naval name). This new system is to be a land-based variant with a range at least twice that of the Barak-8/Barak-NG's 70 to 80 km, therefore the new missile is expected to fly at least 150 km.

Israel Aerospace Industries (IAI) representing Israel and the Defence Research Development Organisation (DRDO) representing India signed a memorandum of agreement early in 2008. This was followed in April 2009 by a full contract with IAI and DDRO as the two primary contractors, delivery dates are estimated as 2017/18.

Barak-8 ER and Barak-8 missiles (IHS/Patrick Allen)
1379147

Barak-8 launcher (IHS/Patrick Allen) 1380387

Description

Source's in India suggest that the two countries have agreed to extend their ongoing Barak-8/Barak-NG project with an extended range missile that should be capable of intercepting targets out to ranges of at least 150 km. The system will probably also make use of a vertical launcher similar to if not the Mk 41 VLS system.

It is understood that the new programme will have a limited Anti-Tactical Ballistic Missile (ATBM) capability and will most likely complement the ongoing other DRDO programme based on the Prithvi II missile known as AXO/PAD. Linkage between the two systems was first identified in March 2006 and again in July 2007.

Indian sources have revealed that the new land-based air-defence system could be the replacement for the Indian Air Force Soviet-developed and produced Pechora (NATO SA-3 Goa) system that has been deployed in India for far too many years. This new system is expected to have a full 360° interception coverage making use of its VLS. IAI has reported that the new system will be more sophisticated than the Patriot PAC-3 and more capable, hence India's lack of further interest in the purchase of the PAC-3.

Variants

ADAMS/Barak

ADAMS/Barak is the basic version of the family of missiles with an engagement range out to 12 km.

Barak-8

Barak-8 is the Israeli naval name for the medium-range missile that can fly 70 - 80 km. The Barak-8 uses a new 360° phased array radar system. The fourth-generation missiles and radar are designed to work under the toughest operational scenarios. Barak-8 is not a simple upgrade of the Barak system, but has been specifically built for high performance, performance which is not available elsewhere on the market.

The Barak-8 has commenced fitting into Sa'ar-5 class warships of the Israeli Navy as from late July 2013.

Barak-NG

Barak-NG is the Indian naval nomenclature for the medium-range missile that also flies 70-80 km.

Barak-ER

Barak-ER is the extended-range land-based and shipborne version that is expected to fly out too or beyond 150 km.

Specifications

Barak extended - range

Type:	land-based static or towed extended-range version of the Barak family
Length:	not available
Maximum range:	150 km (estimate)

Status

A memorandum of agreement has been signed and the Barak-8ER is in full development.

Contractor

Israeli Aerospace Industries (IAI), joint primary contractor.
Defence Research and Development Organisation (DRDO), joint primary contractor.

Defender

Type

Static and towed Surface-to-Air Missile (SAM) system

Development

The Defender air-defence missile system from Rafael Advanced Defence Systems and the then Thales Nederland (now just Thales), was first identified late in 2001 and again early 2002, when the system was offered to the North Atlantic Treaty Organisation (NATO) using the Barak-1 missile, however at this time it is understood that NATO refused this offer as the system was not NATO compatible.

Development of the system has continued over the years with integration testing and non-firing launch exercises. A live successful intercept, tracking and firing using the Barak missile has taken place at the Shdema Test Range in southern Israel in February 2006. The target (drone) was detected by the Flycatcher Mk 2 radar at a range of 20 km and was then continuously tracked until at a range of 7 km, when the Barak missile was launched. Intercept of the target took place at a range of 6 km.

The system modular architecture facilitates an independent upgrading of its subsystems, when and if required, thus allowing the manufacturer of the Defender system ease of implementation of all eventual future software and/or hardware upgrades to those potential and operational customers.

Description

The Defender is an all-weather, day-and-night short-range air-defence system for the protection of high value-military and civilian areas. The system is designed to counter all conventional and modern air threats, including fixed- and rotary-wing aircraft, Unmanned Aerial Vehicles (UAVs) of all classes, standoff Precision-Guided Munitions (PGMs) and cruise missiles.

The Defender system has en suite for interface with Very Short Range Air-Defence (VSHORAD) weapon systems, such as Anti-Aircraft (AA) guns and MAN-Portable Air-Defence (MANPAD) systems.

The Defender can optionally control a mix of up to three medium-calibre AA guns simultaneously. Furthermore, the Defender can provide MANPADs with early warning and cueing data by broadcasting the air picture via radio/data-link. Target data is displayed at the MANPAD fire unit.

Intercept envelopes extend the range to 12 km against helicopters, 10 km against aircraft and 8 km against missiles, enabling an effective air defence covering an area exceeding 300 km^2. The intercept of crossing targets aimed towards neighbouring sites is effective within crossing ranges of several kilometres (1 to over 5 km).

Launcher for the Defender Missile System 1154421

Computer-generated image of Defender launch 1154420

Missile

The Barak missile was initially designed as a naval round and to be a relatively low-cost point defence weapon, primarily used for ship borne defence. The Defender system uses the Barak-1 missile for Ground-Based Air Defence (GBAD).

Externally the missile is of a conventional configuration, with the aerodynamic surfaces arranged in a cruciform configuration around the missile body. This consists of clipped delta wings with long and short chord clipped delta tail surfaces.

The Barak missile is employed in the Israel Navy and is also in use in a number of navies around the world.

The Barak weapon boasts a high manoeuvring capability of the missile at the end-game phase that is considered an additional important and desirable asset of the system, ensuring its effectiveness in a wide variety of combat scenarios. Missile rate-of-turn capability of almost 40°/s throughout the whole flight, this provides the missile with high agility throughout the entire velocity range. Accurate guidance from the Flycatcher Mk 2 radar determines a very low miss-distance of the missile-to-target ratio, resulting in an extremely high kill probability.

The laser proximity fuze initiates the high explosive fragmentation 22 kg warhead that enhances the lethal effect of the missile. The maintenance-free Barak missile is contained within a sealed launching canister that protects the missile not only in the operation, but also in storage and handling, contributing to readiness and reliability of the weapon, confirmed by an automatically executed Built-In self-Test (BIT), prior to launch.

Launcher

Vertical 8-cell launcher with a full 360° coverage, thus enabling multiple target engagement capability achieved by both short reaction time and short engagement cycles. This enhances the efficiency of the system and the launcher to defend an area in the event of saturation attacks. Furthermore, due to the vertical-launch capability of the system, time is not wasted on launcher slewing and elevation movements and therefore reaction times are significantly reduced.

In Command-to-Line-Of-Sight (CLOS) engagements, each engagement must be completed before re-targeting and launching of subsequent missiles. Due to the short-range capability and short reaction times, sequential engagement of three-to-four Mach 1.0 targets on one bearing are possible.

Flycatcher MK 2 - Hybrid Weapon Control Centre

Flycatcher Mk 2 - Hybrid Weapon Control Centre is the core of a hybrid, integrated layered air-defence concept developed by Thales. The Flycatcher Mk 2 (an upgrade of the Flycatcher Mk 1) is able to control both guns and missiles simultaneously.

The radar, operating in Ka-band with a large RF dynamic range, combined with a very narrow main guidance beam, provides a high level of missile guidance immunity to Electronic CounterMeasure (ECM) interference.

A directional command link, operating within the very narrow main lobe during the guidance phase ensures communication immunity in an Electronic Warfare (EW) saturated environment.

The surveillance radar is equipped with an Identification Friend or Foe (IFF) system that allows the Defender to rapidly classify incoming targets as either hostile or friendly. The Search Radar detection range exceeds 25 km in the air-defence mode, and 50 km in the command centre mode. A stand-alone Defender System naturally employs the air-defence mode.

When two or more Defender units are employed together with optional gap-filler surveillance radar, one of the Defender units operates as a central command site in the command centre mode.

Target engagement sequence

The engagement sequence associated with the target commences with the detection of the target by the Search Radar and is followed by an IFF interrogation and threat evaluation, facilitating the establishment of engagement priority.

After detection the target is designated to the fire-control radar, which slews to the target direction and executes an automatic acquisition process, providing target data to the Guidance Computer.

Target position and missile optimal trajectory are continuously calculated to ensure the optimum/correct firing moment. When the predicted intercept range is within the operational envelope of the system, the launch sequence is initiated. The sequence ends with the rocket motor ignition.

Since missile velocity in the initial phase of its flight is still insufficient for aerodynamic control the Thrust Vector Control (TVC) unit performs the steering at this stage. The unit deflects the motor exhaust gases in the desired direction, as commanded by the autopilot, to bring the missile into the capture beams - wide and narrow - facilitating the entrance of the missile into the main guidance beam.

After completion of the inertial lean-over manoeuvre, when the missile acquires sufficient velocity for aerodynamic control, the TVC unit is jettisoned. Upon the entrance of the missile into the tracking/main guidance beam, the missile's communication assembly generates a beacon signal and receives guidance commands, transmitted by the uplink antenna.

In the guidance phase, the Guidance Computer receives target and missile measurements from the tracking radar. These measurements enable the calculation of the guidance commands that guide the missile in the main antenna beam towards target intercept. The guidance commands are transmitted to the missile by the fire-control radar.

Shortly before the estimated intercept of a target, the GC initiates the activation of the proximity fuze. When the proximity fuze detects the target it initiates the warhead detonation at an optimal time delay calculated to inflict on the target the highest possible degree of damage.

Target destruction is assessed instantly after the intercept thus enabling immediate determination of the follow-on engagements.

System Components

- Barak-1 missiles in their individual launcher canisters
- Eight cell launcher
- Flycatcher Mk 2 Hybrid weapon control centre
- AAA guns (optional)
- Man-portable SAMs (optional).

Variants

Barak naval weapon system
Barak MR
Barak ER
Barak 8.

	Defender	Barak-1
Dimensions and weights		
Length		
overall	-	2.175 m (7 ft 1¾ in)
Diameter		
body	-	170 mm (6.69 in)
	-	685 mm (26.97 in) (forward wings)
Weight		
launch	-	88 kg (194 lb) (est.)
Performance		
Speed		
max Mach number	-	2
Range		
min	0.3 n miles (0.5 km; 0.3 miles)	-
max	6.5 n miles (12 km; 7.5 miles)	-
Altitude		
max	10,000 m (32,808 ft)	-
Ordnance components		
Warhead	-	22 kg (48 lb) HE fragmentation
Fuze	laser, proximity, impact	
Guidance	CLOS	
Propulsion		
type	solid propellant, two stage	

Status

The system has been in development since the early 2000's, firing trials have taken place in Israel early in 2006 but since then very little information has been released about the system, although it has been reported that Venezuela ordered three Defender systems with the Flycatcher Mk 2 in 1998.

Contractor

Thales (Flycatcher radar, primary contractor).
Rafael Advanced Defence Systems Ltd.

HAWK XXI and Surface Launched AMRAAM

Type

Static and towed Surface-to-Air Missile (SAM) system

Kongsberg Defence & Aerospace-developed Fire Distribution Centre (FDC) mounted here on an HMMWV (4 × 4) chassis
(Kongsberg Defence and Aerospace) 1108058

Operator workstations of the Kongsberg Defence & Aerospace developed Fire Distribution Centre (FDC) showing display positions
(Kongsberg Defence and Aerospace) 1108061

Development

The US Raytheon Company has teamed with Kongsberg Defence & Aerospace of Norway for the international marketing and sales of HAWK XXI and Surface Launched - AMRAAM (SL-AMRAAM). This weapon is an all-weather day/night, low to medium-altitude air defence missile system that offers the enhanced range and Anti-Tactical Ballistic Missile (ATBM) intercept capability. The HAWK XXI and SL-AMRAAM air defence systems by Raytheon and Kongsberg are designed with Commercial Off-the-Shelf (COTS) software equipment and based on an open architecture. They are modular and flexible to procure and customers can choose a range of variants of the system:

- HAWK XXI: HAWK with 3D sensor and state-of-the-art BMC4I;
- SL-AMRAAM: Surface-Launched AMRAAM;
- HAWK-AMRAAM: a combination of HAWK and AMRAAM.

Description

HAWK XXI

The HAWK XX1 system is a more advanced and compact version of the HAWK PIP-3. The system has been upgraded with a AN/MPQ-64 Sentinel 3D surveillance radar and a command post - Fire Distribution Centre (FDC) provided by Kongsberg of Norway, the same FDC as used in Kongsber's NASAMS. The combat efficiency, readiness times and survival is increased, while manpower and maintenance/support are reduced by approximately 30 per cent. The HAWK XXI is interoperable with higher echelon and controls subordinate Very Short-Range Air Defence Systems/Short-Range Air Defence Systems (VSHORADS/SHORADS) systems that are complementary to the HAWK. The HAWK XXI is prepared for SL-AMRAAM, which will increase the

AMRAAM firing from HMMWV vehicle (Kongsberg Defence and Aerospace) 1108060

number of fire channels and total firepower. The FDC open system architecture allows the customer to tailor the system to specific technical and operational requirements and other government factors unique to existing geopolitical, logistics and economic situations.

The missiles used with the system are the MIM-23K standard with an improved high-explosive blast fragmentation warhead that increases the kill zone of the weapon.

SL-AMRAAM

This is available with two different off-the-shelf launcher alternatives, facilitating tailoring to customer needs and missions. The CLAWS launcher - based on the high mobility HMMWV vehicle, carries four to six AMRAAM missiles. The NASAMS canister launcher - carrying 6 × AMRAAM missiles. The FDC command post is available for containerised solutions such as ISO shipping containers, armoured wheeled or tracked vehicles, HMMWV-class and other light vehicles. SL-AMRAAM has beyond-visual-range capability and multiple simultaneous engagement capability. Other capabilities are:

- Autonomous missile guidance and non-line-of-sight capability;
- A large number of parallel firing channels;
- Supports the flexibility to deploy force packages with the number and type of sensors and missiles tailored to the mission;
- Capable of connecting to different sensors, 3D, 2D, EO and IR of different type and manufacture;
- Capable of operating even without employing its own ground based sensors;
- Capable of operating with a passive sensor network;
- System with a minimum number of operators, only one or two to conduct a battle with a large number of firing channels and a high firing rate;
- Networking or combining multiple air defence units doubling and tripling the defended area relative to the range of the individual missile;
- Maintains combat ability even after loss of elements (fire-control, sensors and launchers with missiles) through real time netting of units;
- Exploits the capabilities of different missiles of different type and manufacture;
- AMRAAM missiles are used both from air and GBAD platforms thus reducing logistics cost and inventory diversity.

HAWK-AMRAAM

Combines the two combats proven missiles in one system. The two missiles are complementary to each other with the range and altitude of the HAWK with the firing rate and beyond-visual-range capability of the AMRAAM.

Fire Distribution Centre (FDC)

The FDC was designed based on open architecture. Sensors and launchers can be dispersed up to 25 km (terrain dependent) connected via field wire, radio, cable or a combination. A number of batteries (fighting unit), each with an FDC, sensors (up to eight 2D/3D) and launchers (up to 12 AMRAAM/HAWK, normally three or four), can be netted together in hard real time and sharing a local air picture.

This distributed architecture allows any launcher and any sensor to be connected to any FDC in a network centric manner, increasing the overall survival rate and jamming resistance. Loss of modules (sensors, communication, launchers or FDC) has limited consequences on the fighting capability.

The system can be fully integrated with the MIM-104 Patriot system and, utilising the Patriot radar or other sensor information, can provide a short-range TBM intercept capability.

The active RF seeker of the AMRAAM provides for parallel fire channels. Netted sensors and a common non-redundant air picture allow for engagement from launchers without line-of-sight to the target. The lofted (over-and-above) trajectory of AMRAAM allows engagement of targets exploiting terrain screening such as Unmanned Combat Aerial Vehicles (UCAVs), Unmanned Aerial Vehicles (UAVs), and combat helicopters.

Status

Ready for production on receipt of an order.

This system has been chosen by several countries but in different configurations, some of those countries include:

- The Royal Norwegian Air Force was the first to introduce SL-AMRAAM in the NASAMS programme;
- The Hellenic Army (HAWK);
- Spanish Army (NASAMS);
- Turkish Air Force (HAWK;
- US (SL-AMRAAM).

Contractor

Raytheon Company.
Kongsberg Defence & Aerospace AS.

SAHV-IR/35 mm Skyguard

Type

Static and towed Surface-to-Air Missile (SAM) system

Development

The South African company Kentron (now Denel Dynamics) combined with the then international company Oerlikon Contraves (now Rheinmetall Air Defence AG) to offer an integrated missile/anti-aircraft gun air-defence system that extends the coverage of a Skyguard AA gun system with the minimum of added complexity or equipment.

The system is based on a battery comprising:

- A Skyguard fire-control radar unit
- Two twin-barrelled Rheinmetall Air Defence 35 mm GDF-005 anti-aircraft guns
- One (with the option of a second) eight-round SAHV-IR (see entry in this section) containerised missile launcher based on a modified GDF-005 gun platform to ease logistic support.

Description

The missile is able to engage both high-speed aircraft targets at altitudes from below 31 m to 8,000 m. It is also capable of countering attacks from anti-tank helicopters equipped with short-range anti-tank guided weapons.

The missile launcher is aimed at the predicted intercept point beyond the attacking aircraft's weapon release range by the Skyguard fire-control radar so that the SAHV-IR can be launched. Once this happens the radar is free to acquire a second target.

The SAHV-IR missile flies under inertial control of its digital autopilot while its passive Infra-Red (IR) seeker scans a volume of airspace around the calculated target position. When the two-colour, narrow band IR seeker detects the target it locks on automatically and the missile enters its proportional terminal homing mode. The scan pattern for this Lock On After Launch seeker mode at an intercept range of up to 7,700 m makes allowances for any target manoeuvres and the seeker type ensures rejection of IR countermeasures. For very close-range engagements at targets at 1,536 m altitude and up to 3,000 m range and at very low targets at up to 2,000 m range, the seeker's Lock-On-Before-Launch (LOBL) mode is automatically activated. Against distant targets with an intercept range in excess of 8,000 m the missile uses guidance during its motor burn phase via an IR tracker and a low probability of intercept RF link, to provide launch error correction. Known as gather and forget the missile then enters the lock on after launch fire-and-forget mode for the terminal phase.

The weapon uses an active proximity fuze and a 20 kg pre-fragmented warhead. Missile flight control is achieved by a set of linear electromechanical servos driving independent tail-mounted control fins. Propulsion is by a high-energy, low-smoke solid composite propellant rocket motor. A comprehensive built-in test function provides for low-maintenance operation.

Artist's impression of Kentron SAHV-IR/Rheinmetall Air Defence 35 mm Skyguard air-defence system deployed in firing position 0044361

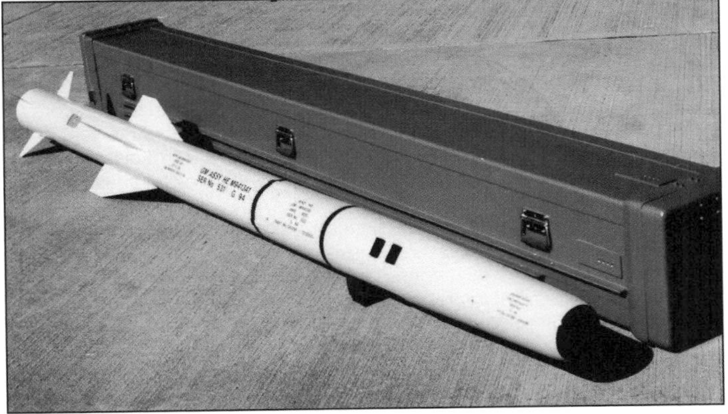

Kentron SAHV-IR missile beside its launcher-container box 0054271

The carriage of the eight-round SAHV-IR missile launcher is the same as that used for the Rheinmetall Air Defence GDF twin 35 mm anti-aircraft gun system (Christopher F Foss) 0509895

Specifications

	SAHV-IR
Dimensions and weights	
Length	
overall	3.25 m (10 ft 8 in)
Diameter	
body	180 mm (7.09 in)
Flight control surfaces	
span	0.5 m (1 ft 7¾ in) (wing)
Weight	
launch	132 kg (291 lb)
Performance	
Speed	
max Mach number	2.0
Range	
min	0.5 n miles (1 km; 0.6 miles) (est.)
max	6.5 n miles (12 km; 7.5 miles)
Altitude	
min	30 m (98 ft) (fixed-wing aircraft)
	5 m (16 ft) (helicopter)
max	8,000 m (26,247 ft) (est.)
Ordnance components	
Warhead	23 kg (50 lb) HE fragmentation
Fuze	laser, proximity
Guidance	INS (mid-course phase)
	passive, IR (two colour, with proportional navigation)
Propulsion	
type	solid propellant (smokeless composite, sustain stage)

Status

Ready for production. Initial firing trials took place in 1997. Offered for export. To date, there have been no known orders.

Contractor

Rheinmetall Air Defence AG.
Denel Dynamics.

Skyguard/Sparrow

Type

Static and towed Surface-to-Air Missile (SAM) system

Development

This air-defence system is a joint development between Oerlikon Contraves of Switzerland (now known as Rheinmetall Air Defence) and Raytheon Company of the United States. It combines two proven systems, the Skyguard

Typical Skyguard/Sparrow/twin 35 mm air defence battery with the then Oerlikon Contraves Skyguard fire-control system, one twin four-round Sparrow SAM launchers and two twin GDF series 35 mm anti-aircraft gun systems 0044362

Rear view of four-round launcher for Sparrow missiles. Carriage is identical to that employed in twin 35 mm anti-aircraft gun which is also deployed in the battery 0509578

Close-up of launcher showing two Sparrow missiles in their transport/launcher containers each side of the operator's position with illuminator antenna mounted forward and below his position 0509577

Taiwanese Skyguard/Sparrow four-round launcher in travelling configuration (L J Lamb) 0509710

fire-control system, normally used to control Oerlikon Contraves twin 35 mm GDF series towed anti-aircraft guns and the Raytheon AIM-7E/AIM-7F/AIM-7M Sparrow missile which, in its normal application, is air-launched.

Typically, the missile would be used to engage targets at longer ranges with the guns used at ranges of 4,000 m or less.

The US Navy in the Surface-to-Air Missile (SAM) role as part of its Basic Point Defence System uses a variant of the AIM-7E, designated RIM-7E-5. Another variant, the RIM-7H-5, is used in the NATO Sea Sparrow Surface Missile System. More recently, the US Navy has adopted the RIM-7M with the RIM-7P as the follow-on weapon.

In 1982, the Egyptian Air Defence Command ordered 18 battery sets of the Skyguard/Sparrow air defence system from Contraves (Italy) to equip three autonomous air defence brigades. The first deliveries were made in December 1984 and final deliveries were in 1987. Egypt calls the system Amoun and a typical section consisted of one Skyguard fire-control system, two Oerlikon Contraves GDF-003 twin 35 mm towed anti-aircraft guns and two four-round Sparrow SAM launchers. One section can engage three targets at once, two with missiles and one with guns. Reaction times are 4.5 seconds for the guns and 8 seconds for the Sparrows.

The system delivered to Egypt has 16 modifications, including a new search antenna unit to reduce the effects of clutter, a new computer and software and a revision of the operator's console to incorporate three operators. Optical target detection range is 15 km and radar range 20 km.

Egypt is understood to have a requirement for additional Amoun sections so that its remaining obsolete Russian systems can be phased out of service. Further orders would involve a more significant proportion of local production, which was minimal in the initial order.

Kuwait took delivery of six Amoun batteries for use by its air force but promptly lost them to Iraq during the invasion.

Greece selected the system under the name, 'Velos' (Arrow) in 1983 using RIM-7M missiles. The first was delivered in October 1984.

In October 1985 Spain ordered 13 Italian Alenia Marconi Systems Spada launchers and 200 Aspide missiles for integration with existing Skyguard fire-control systems and Oerlikon Contraves twin 35 mm GDF-005 towed anti-aircraft gun batteries. The Spanish Navy on its frigates already used the Aspide missiles. Total value of this contract was USD220 million and included Spanish co-production.

Oerlikon Contraves delivered the first battery with the remaining six assembled at the Bazán factory in Spain. Marconi Spain was responsible for modification kits and the launcher electronic groups and EISA manufactured the electrical cables. Bazán produced the missile container and the launcher's electrical and mechanical components.

The Spanish name for this system is 'Toledo'. By the end of 1989, all systems had been delivered. As the Spanish Skyguard fire-control systems were delivered in the late 1970s, the contract also covered the supply of seven modification packages to bring these up to the latest configuration.

The 1st Group of the 73rd Air Defence Regiment is equipped with Toledo. The unit is deployed in Southwest Spain and is assigned the air defence of the army's 3rd Mechanised Division and the Spanish Navy's base at Cartagena. The 1st Group has three batteries with a total of six Super Skyguard fire-control systems, 12 GDF twin 35 mm anti-aircraft guns and 12 Toledo four-round SAM launchers. The remaining Skyguard fire-control system and Toledo launcher are assigned to the Spanish Army's Artillery School, with two Roland SAM systems for training duties.

Tactical firing trials of the Skyguard/Sparrow were carried out in October 1980 at the Naval Weapons Centre, China Lake, California. Three missiles, one AIM-7E and two AIM-7Fs, were fired against remotely controlled aircraft targets, Northrop QT-38 Talon for the AIM-7E and North American QF-86 Sabre for the AIM-7F.

The AIM-7E was not fitted with a warhead, but intercepted the target and passed well within the lethal radius of the warhead. The first AIM-7F was fitted with a telemetry pack in place of the warhead, but scored a direct hit on the right under wing fuel tank. The second AIM-7F hit the front fuselage of the target causing extensive damage to the air intake and cockpit. The drone aircraft then went out of control and subsequently crashed.

In 1980, a demonstration of the Skyguard/Sparrow took place at the NATO Missile Firing Range (NAMFI) in northern Crete using production equipment that had already been delivered to a customer.

A remotely piloted Chukka target drone, which has roughly the dimensions and speed of some cruise missiles, was used as a target. The drone approached the weapon position at an angle from the front, flying 700 m over the weapon site at a speed of more than 200 m/s. The first launch scored a physical hit at more than 12 km range sending the drone into the sea. In the second intercept, the missile flew past the new drone at a 1 m distance; since the Sparrow missile would have been triggered by its proximity fuze in a real engagement, it was considered a kill.

In mid-1991, Oerlikon Contraves stated that an AIM-7F missile, launched under Skyguard control from an Oerlikon Contraves SAM launcher for a customer force in May 1991, scored a direct hit when it shot down a target drone flying at a speed of approximately 200 m/s at a range in excess of 10 km.

The success was achieved within a customer-specified system acceptance test programme, which involved one dry rehearsal and one hot firing against a target drone flying at high-subsonic speed. The weather conditions prevailing at the time of the test were described as marginal with reduced visibility, rain and heavy gusts of wind.

Sparrow SAM being launched under control of the Skyguard fire-control system 0509579

Description

The Skyguard carries out search and identification using a 20 km I/J-band radar, tracks the selected target by radar using a pulse Doppler K-band set or TV, computes intercept to facilitate the engagement and then aims the launcher at the target via digital data cable transmission.

Launcher

The four-round launcher is mounted on the carriage of the Oerlikon Contraves twin 35 mm GDF towed anti-aircraft gun system and the missiles are fired through the covers. The folded wings deploy after separation. Two rounds are mounted each side of the operator's position with the illuminator antenna mounted forward and below his position.

Radar

The I-band illuminator provides continuous wave illumination of the target out to 13 km (20 km in the case of Egyptian systems) and the missile's seeker homes on the reflected signal.

Variants

Skyguard/Aspide

This is similar to the Skyguard/Sparrow but uses the four-round Alenia Marconi Systems Aspide launcher used in the Spada air defence system in service with Italy and Thailand. The system was successfully demonstrated in 1981 and subsequently adopted by Spain under the name, 'Toledo'. Spain already has the Skyguard and twin 35 mm guns in service. Cyprus has also bought the Skyguard/Aspide system with 12 systems in service. A further purchase of 12 systems is the subject of protracted negotiations with the Italian government. The system in Cypriot service has the name Othello. Under the name Amoun, the first batteries for firing trials in Kuwait were delivered in September 2007 from Rheinmetall, Italy. The first trials were two Amoun batteries each with a Skyguard fire-control system and two 35 mm anti-aircraft guns.

Amoun

Amoun is the local Skyguard/Aspide system's name when deployed in Egypt.

Toledo

Toledo is the local Skyguard/Aspide system's name when deployed in Spain.

Othello

Othello is the local Skyguard/Aspide system's name when deployed in Cyprus.

Velos

The name of the system used in Greece. The Velos system is the air defence backbone of the Hellenic Air Force. In 2006, Rheinmetall Defence, subsidiary Oerlikon Contraves, finalised a contract to overhaul and upgrade the system in order to assure logistical support for the system in the future. This also included training of troops with advanced technology and simulators. The 35 mm gun section of the system uses the AHEAD ammunition to improve the capability against small targets such as cruise missiles, UAVs and high-speed anti-radiation missiles.

The radar's detection capabilities have been greatly improved, as has the complete human-machine interface technology. Additionally, Oerlikon Contraves fitted the missile module with a new set of digital electronic pods.

The Velos fire unit consists of:

- Skyguard fire-control system that controls two Oerlikon 35 mm guns and two missile launchers equipped with Sparrow missiles
- Long-range SUGI radar to coordinate the systems
- Link-11B for full integration into the Greek air defence system.

Velos has been fully qualified on completion of live firing events at the NAMFI Firing Range.

Specifications

	AIM-7F
Dimensions and weights	
Length	
overall	3.66 m (12 ft 0 in)
Diameter	
body	0.203 mm (0.01 in)
Flight control surfaces	
span	1.02 m (3 ft 4¼ in) (wing)
Weight	
launch	234 kg (515 lb)
Performance	
Speed	
max Mach number	2.5
Range	
max	10.8 n miles (20 km; 12.4 miles)
Altitude	
min	15 m (49 ft)
max	5,000 m (16,404 ft)
Ordnance components	
Warhead	39 kg (85 lb) HE fragmentation
Fuze	impact, proximity
Guidance	semi-active, radar
Propulsion	
type	single stage, solid propellant

Status

Production as required. In mid 1999, it was announced that the Greek Ministry of Defence (MoD) was to upgrade its Velos systems to allow a 15 to 20 year extension of life. This involved the installation of the NDF-C kit to bring the 35 mm guns up to the GDF-005 standard, the fitting of a Ka-band radar to allow up to 20 targets to be tracked (rather than the current eight) and an optional infrared camera system. The IFF will be upgraded to Mk XII standard, a Kalman filter installed and three C-2003 computers with plasma screens fitted.

Contractor

Rheinmetall Air Defence, previously know as Oerlikon Contraves AG.
Raytheon Company.

David's Sling

Type

Static and towed Surface-to-Air Missile (SAM) system

Development

Announced on 24 May 2006, the Israeli Missile Defence Organisation (IMDO) in cooperation with US Authorities, has selected the Rafael Advanced Defence Systems Ltd and Raytheon, Tucson, Arizona to develop a new terminal missile defence interceptor with the biblical name known as David's Sling (Kela David) using the Raytheon Stunner missile. David's Sling is also known as the new Magic Wand system.

Raytheon will take the lead on the David's Sling missile whereas Rafael will lead on a complementary system known as Iron Dome.

David's Sling is an element of the Medium- and Short-Range Missile Defence (SRMD) programme and is considered by both Israel and the US to have a critical and immediate need for a medium-range, highly effective and affordable weapon in this class.

Announced early November 2007, out of a total US aid budget of USD155 million for FY08, USD37 million was allocated for the continuing development of the Stunner/Kela David. The first systems are still expected to be available around the time frame of 2014 with full operational capability by 2015.

Late in December 2007, the US Congress further announced USD205 million for US/Israeli defence projects. This was simultaneous with another announcement regarding the US missile defence system and its integration with the Israeli Arrow.

Stunner was formally released/introduced to the Israeli press in January 2008, in the presence of the then Israeli Prime Minister Ehud Olmert whilst visiting the Rafael Advanced Defence Systems organisation. The second showing occurred during the Paris Air Show in 2013 when the stunner missile was displayed and press information was released.

Prototype Stunner (Rafael) 1169330

Stunner Launcher (Rafael) 1193013

In December 2008, the Stunner missile completed its high-angle of attack supersonic wind tunnel tests held in the newly refurbished Test Tunnel A at the Arnold Engineering Development Centre (AEDC) in St. Louis, Missouri.

In late December 2011, the US agreed to double the special aid it gives to Israel for the development and implementation of anti-missile system David's Sling and Arrow 3. The amount of aid increased from USD129 million to USD235.7 million.

Firing trials are known to have taken place on the Southern Range in Israel starting in November 2011 and have continued with other firings taking place in January 2013. The system has been reported by Rafael to be fully operational by 2015.

A further firing took place in mid-November 2013 against a ballistic missile target, this test firing turned out to be very successful.

Description

Known as a Medium-Range Anti-Ballistic Missile Defence Program (MRABMDP), the system is primarily designed to attend to the burgeoning Medium-Range Ballistic Missile (MRBM) threat coming from Syria, Lebanon and Iran. The missile has been designed to intercept long-range rocket threats with ranges of between 40 and 250 km.

Raytheon also wants to ensure that the system is multi-mission capable and that it can be inserted into existing and future US terminal missile and air-defence systems. This approach will additionally provide the US Army with an all-weather, hit-to-kill, low-cost extended air-defence option for the future.

The interceptor for the Israeli version will be based on next-generation Rafael Python dual-wave Imaging InfraRed (IIR) missile technology and advanced low-cost Raytheon tactical missile technology. It is assessed that a US version based on the next-generation AMRAAM may also be in development in the near future. David's Sling has two targeting and guidance systems, a radar and an electro-optical sensor, installed in its nose-tip.

In November 2006, the Raytheon Company selected the Alliant Techsystems Inc (ATK) Tactical Systems to develop a composite booster motor for the Stunner interceptor. ATK will perform work on the motor at Rocket City, West Virginia and Luka, Mississippi.

It has been stressed that the requirement for the system is not to take on targets such as the Hizbullah 122 mm Katyusha, instead the perceived targets are long-range rockets launched within the 40 to 250 km zone such as the Iranian produced Zilzal and Fajr missiles. David's Sling is to be integrated into an Israeli layered defence system consisting of:

- Arrow-2 (Arrow-3 in development) (also known as Hetz in Israel);
- Patriot PAC-2 (Yahalom);
- David's Sling (Kela David);
- Iron Dome (Cap).

Launcher

It is expected that the launcher will be fully mobile with a complement of numerous canister/boxed-launched missiles. It would also make sense for the missile rounds to be launched vertically given the time and short distance the missiles are expected to travel before interception.

C4I/radar

Taking into account the requirement of the system for terminal missile defence against MRBMs at very short-range within extremely short time frames, the command-and-control will be required to link in with the IMDO Arrow-2 with the Green Pine radar system and its follow-on system (Arrow-3) along with the Patriot existing and future control systems. The radar is developed by Israel Aircraft Industries (IAI) Elta Systems. In August 2008, some reports suggested that the US production X-band radar tracking equipment currently used in the US-BMD programme, has been identified as the radar most likely to be associated with David's Sling.

System components

- Launcher;
- missiles;
- radar;
- communications;
- ground support equipment.

Variants

The Israeli government is looking to further develop the David's Sling system into an air-to-air variant. Rafael has been examining this possibility as a low-cost means to developing this system.

Specifications

	Stunner
Dimensions and weights	
Length	
overall	5 m (16 ft 4¾ in) (est.)
Weight	
launch	100 kg (220 lb)
Performance	
Speed	
cruise	1,944 kt (3,600 km/h; 2,237 mph; 1,000 m/s)
Range	
max	135.0 n miles (250 km; 155.3 miles) (est.)
Ordnance components	
Warhead	kinetic kill vehicle
Guidance	mid-course update (command)
	passive, IR
Propulsion	
type	solid propellant, two stage (high propellant fill, three pulse motor, sustain stage)

Status

A new unit has been formed in Israel to man the first David's Sling battery that will be based in central Israel sometime in 2014/15. There is some evidence that suggests certain items, such as the launcher box and pre-production missiles, may have been produced as early as 2004.

As of June 2013 the system IOC is expected in 2014/15. Meanwhile, in late December 2011, the US agreed to double the special aid it gives to Israel for the development and implementation of anti-missile system David's Sling.

Some sources have reported that the system was originally required to replace the MIM-23 HAWK and MIM-104 Patriot in the Israeli arsenal.

Contractor

Rafael Advanced Defence Systems Ltd.
Raytheon Missile Systems.
IAI Elta Systems.
Alliant Techsystems Inc (ATK) Tactical Systems, (booster motor development).
Israel Missile Defence Organisation (IMDO), (management for the new system).

Iran

Mersad/Shahin

Type

Static and towed Surface-to-Air Missile (SAM) system

Development

First displayed to Iranian journalists on 6 June 2009, the Shahin (lit., Falcon) missile is part of the Mersad system that has been designed to engage targets such as fixed wing, rotary wing aircraft and allegedly ballistic missiles. The research and production phases of the system included missile, missile interceptors, hardware and software networks and launch facilities carried out by Iranian industry experts. It is understood that the system is heavily based on the US HAWK (MIM-23) originally supplied to Iran from the USA with additional I-HAWK missiles from Israel.

Description

Very little information on the system has been released by Iran except the range that is quoted at 40+ km and the fact that there are probably three types of missile; the Shahin as the basic (probably reverse engineered MIM-23), the

A side view showing two Shahin missiles shown to the public in June 2009 (Iranian Student's News Agency (ISNA)) 1379844

A side view of the Iranian Shahin missile shown to the public in June 2009 (Iranian State TV (IRINN)) 1379843

A rear view of the Shahin missile shown to the public in June 2009 (Iranian Student's News Agency (ISNA)) 1379845

Shahin 1 with the first locally developed and produced modification and upgrade, and finally the Shalamche missile that is reported to have increased range and contain electronic countermeasures (on a smart basis).

It is assessed that the Shalamche is probably based on the Improved version of the I-HAWK that was passed to Iran by Israel in return for Israeli hostages.

Variants

Mersad-1 (Ambush)

Reports coming from inside Iran during April 2010 indicated for the first time that the Mersad-1 Medium-Range Air-Defence System utilises the Shahin-1 missile. Serial production of the Mersad began in early 2010 with the aim of supplying Iran's Defence Forces by the end of the year.

The system is reportedly equipped with Iranian produced radar signal processing technology. The system is aimed at engaging targets at low and medium-ranges and contains home-developed and -produced electronic equipment for guidance and target acquisition.

Mersad-2

Flight trials of this system utilising the Shalamche missile suggest this is a longer range system with added countermeasures to enable to system/missile to operate in an Electronic Warfare hostile environment. The Shalmche is reported to fly at Mach 3.0.

Three Shahin missiles shown during a military parade in September 2008 (Iranian Student's News Agency (ISNA)) 1379846

Specifications

Shahin

Type:	medium-range Surface-to-Air Missile (SAM)
Length:	unknown
Diameter:	unknown
Propulsion:	solid propellant rocket engine
Speed:	900+ m/s
Range:	
(maximum)	
(Shahin I)	30 km
(Shahin II)	60 km
(minimum)	unknown
Altitude:	
(maximum)	unknown
(minimum)	unknown
Guidance:	semi-active radar homing
Launcher load:	3

Status

Entered series production in June 2009, probably in service with the Iranian Armed Forces.

Contractor

Iranian Arms Industry.

Sayad-1

Type

Static Surface-to-Air Missile (SAM) system

Development

The Aerospace Industries Organisation (AIO) Sayad-1 Programme (also reported as the Sayyad-1, Hunter-1) has succeeded in reverse-engineering a derivative of the ex-Soviet Union S-75 Dvina/Volkhov (NATO SA-2 Guideline) missile system family, Chinese and Russian Federation versions of which have been in service with the Iranian Armed Forces since at least 1985.

China began supplying Iran in 1985 with their version of the S-75 known as the HQ-2B. Using the HQ-2 as a baseline for further development and modernisation the AIO in Iran was able to incorporate modern Russian technology into the Iranian version, now known as the Sayad-1.

Description

The Sayad-1 is a two stage guided missile, with a large solid propellant booster stage fitted with four very large delta fins.

The major upgrades within the weapon are that the missile has been completely rebuilt internally to incorporate modern electronics. In fact according to Iranian sources, apart from the missile's physical characteristics and overhauled rocket motor propulsion system, the Sayad-1 weapon has little else in common with the original Russian system design.

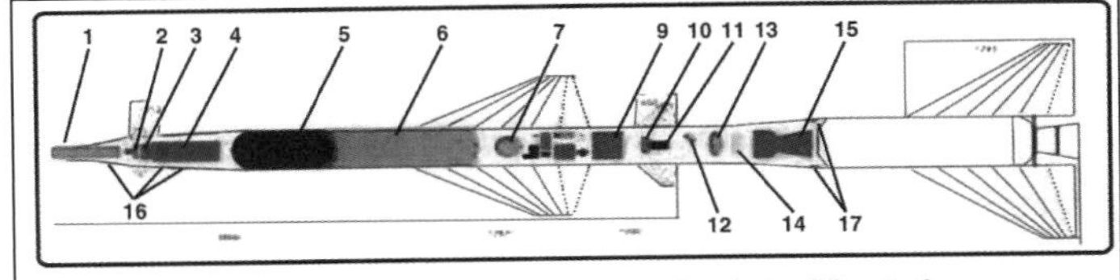

Sayad-1 (A) breakdown (Aerospace Industries Organization) 0526364

The Sayad-1 is deployed in conjunction with Russian Federation-supplied S-200 Angara (NATO SA-5 'Gammon') systems for use against medium-to-high altitude targets. It is possible that North Korea has had an input to the technology required in order to produce this weapon.

Reports from the then Iranian Defence Minister Admiral Ali Shamkhani following initial discussions on 1 October 2001 with Russia's Minister of Defence Sergei Ivanov confirm that Iran has expressed an interest in the procurement of the S-300PMU2 Favorit surface-to-air missile system in order to protect its strategic facilities. By mid-2012 it became evident that the Russian produced S-300 was not going to be exported to Iran and probably never will be.

Furthermore, the Iranian Army has placed a large purchase order for the 9K37 Buk-M1 and 9K331 Tor-M1 mobile SAM systems. This purchase will enhance Iran's defensive missile capability many fold. These systems will be available to Iran within the next few years, at which time the phasing out of the Sayad-1 project would probably take place.

Variants

Sayad-1 (A)

Sayad-1 (A) an improved version of the basic Sayad-1.

Sayad-2 is a further development of the Sayad-1 series that are themselves developments from the Chinese HQ-2 and before the HQ systems the S-75 from the old Soviet Union. It is further assessed that there is or has been a large input of technology from North Korea in the ongoing development and upgrade of this older type of system. Which ever way forward is taken, Iran, China and North Korea still have a 1950s designed and developed missile that was heavily influenced from the German world war-2 target requirement and has deservedly found its way into the history books.

Sayad-2

This system has been reported as an upgraded version of the Sayad-1 with increased precision, range and warhead power.

Status

As of December 2012, in production and in service with Iranian Armed Forces.

Contractor

Defence Industries Organization (DIO).

Sayad-1 (A)

Type

Static Surface-to-Air Missile (SAM) system

Development

The Sayad-1 (A) static Surface-to-Air Missile (SAM) system is a further development of the basic Sayad-1, which in itself is a development of the Russian S-75 (NATO SA-2 Guideline) and Chinese HQ-2 systems. The basic Sayad-1 was also heavily influenced by the Chinese HQ-2 and may also have benefited from North Korean technology input. Other reports that have not yet been confirmed have suggested that the Sayad was also strongly influenced by the US produced HAWK and Standard missiles, those reports are largely from US sources.

Description

The Sayad-1 (A) (named after Lt. Gen. Ali Sayyad-Shirazi) is an improved version of the Sayad-1 missile that can be used against low- to high-altitude and medium-range targets. The priority mission of the system is defined as protection of economic, political and social centres and the destruction of bombers and reconnaissance targets.

The missile employs an inertial navigation system with command guidance. Mounted on the nose of the missile is a passive infrared seeker that acquires a heat source from the target and allows the missile to lock-on at end game.

Variants

Sayed-1 is a development of the S-75 produced by the Aerospace Industries Organisation of Iran.

Sayed-2 is a further development of the Sayed-1 series that are themselves developments from the Chinese HQ-2 and before the HQ systems the S-75 from the old Soviet Union. It is further assessed that there is or has been a large input of technology from North Korea in the ongoing development and upgrade of this older type of system. But which ever way forward is taken, Iran, China and North Korea still have a 1950s designed and developed missile that was heavily influenced from the German world war-2 target requirement and has deservedly found its way into the history books.

This weapon system was allegedly first deployed in April 2011 and is now reported to have been deployed in all air-defence units across Iran.

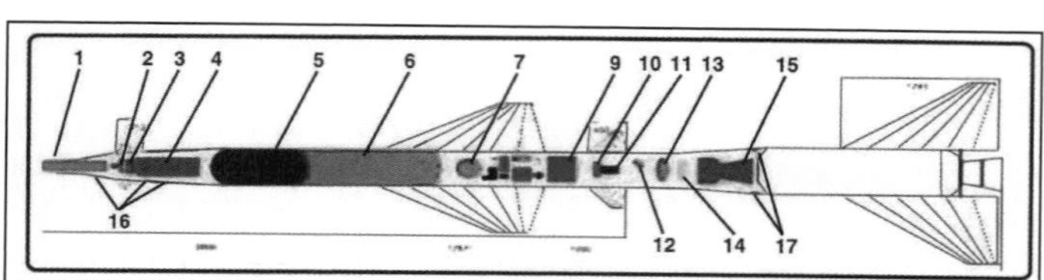

Sayad-1 (A) Surface-to-Air Missile (SAM) breakdown (AIO) 0526364

Specifications

	Sayad-1 (A)
Dimensions and weights	
Length	
overall	10.842 m (35 ft 6¾ in)
Diameter	
body	500 mm (19.69 in)
Flight control surfaces	
span	2.5 m (8 ft 2½ in) (booster, wing)
	1.7 m (5 ft 7 in) (missile, wing)
Weight	
launch	2,320 kg (5,114 lb)
Performance	
Speed	
max speed	2,333 kt (4,320 km/h; 2,684 mph; 1,200 m/s)
Range	
max	10.8 n miles (20 km; 12.4 miles)
Ordnance components	
Warhead	195 kg (429 lb) HE
Guidance	INS (command)
	passive, IR
Propulsion	
type	solid propellant (booster)
	liquid propellant (sustain stage)

Status

As of October 2010, in production and in service with the Iranian Armed Forces.

Contractor

Aerospace Industries Organization (AIO) of Iran.

Sayad-2

Type

Static and towed Surface-to-Air Missile (SAM) system

Development

The Sayad-2 is yet another further development/upgrade to the old Sayad-1 and Sayad-1A. The Sayad family of missile systems found in Iran are all derivatives from the Chinese HQ-2 which, in itself is a copy/reverse engineered ex-Soviet S-75 missile.

It has been assumed/assessed that the Sayad-2 development has been heavily influenced with North Korean technology, however, no matter what, Iran, China and North Korea still have a very old missile that was first designed and developed way back in the early 1950s from German world war-2 target requirements.

By mid-November 2013, another weapon under the Sayad family was released to the open press from Iran; this was named Sayyad-2. The eagle-eyed amongst the readers will have noticed that this weapon has been spelt as Sayyad and not Sayad. The question here is has the spelling been a bad translation on the part of western analysts or is this a new weapon?

Open press reports (western) on this new missile have associated the new one with and reports of an upgrade to the Sayad-1 and Sayad-1A, so once again what is Sayyad, is this the same missile system as the previously reported Sayad-1 and Sayad-1A, have the western press jumped to the wrong conclusion without looking at the facts or is the Sayyad a completely new missile?

Firstly, the Sayad-1 and 1A data has come direct from source Aerospace Industries Organization (AIO) and also in brochures collected during various defence shows, this information and photographic evidence clearly shows reverse engineered HQ-2/S-75 systems and was admitted as such. When the first Sayad-2 was identified, the same sources admitted that it was indeed an upgrade from the original Sayad programme and again brochures were issued showing a large liquid propellant missile with solid propellant boosters as in the HQ-2/S-75 systems.

However, the Sayyad-2 new release from Iran in November 2013 shows a much smaller missile associated with the Talash system. This new missile has a solid propellant motor and strongly resembles a reverse engineered SM-1 missile that is known to be in the inventory of Iran for some years. Indeed, the only part of the missile that bears any resemblance to the HQ-2/S-75 are the fuze strips next to the warhead section.

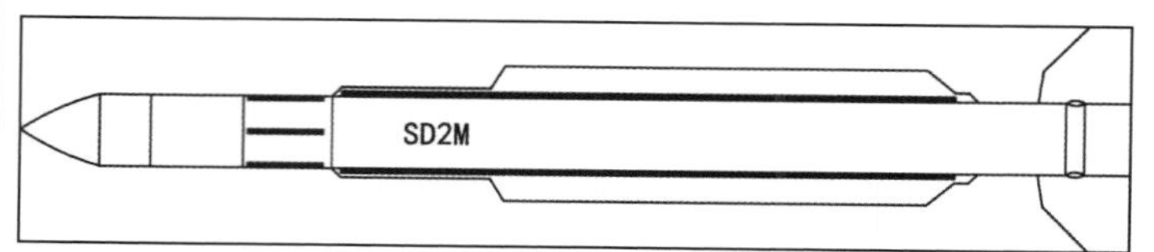

SD-2 Sayyad-2 Missile (Sam O'Halloran) 1521779

Sayad-2 in Iran (Aerospace Industries Organization (AIO)) 1461801

Finally, the Sayyad-2 has been shown with a new four box launcher that is carried on a flatbed truck. A battery of the system has been reported to consist of six launcher trucks, radar and a command and control truck, other ancillary equipment will also accompany any such battery.

Description

Reports have suggested that the Sayad-2 was first unveiled in April 2011 when the system was first deployed to the Khatamol-Anbia Air Defence Base and subsequently displayed to the Iranian press. The then Commander of the Air Base Brigadier General Farzad Esmayeeli further reported that the system would or was now being deployed extensively over Iran in air-defence units.

Launcher

The launcher for the system is identical to that of the other Sayad family and ex-Soviet S-75 (NATO SA-2 Guideline). Based on a semi-mobile trailer with a single ramp launcher that can be towed by a Mercedes-Benz 2624 6×6 truck.

Missile

The missile is as would be expected a standard two-stage liquid fuelled rocket with a large solid propellant booster that has four large delta shaped wings. The booster is ejected some seven to 10 seconds into flight.

The rocket will then be command guided to its target with command fuzing at end game. Reports coming from inside Iran also suggest the system does include Electronic Counter-Counter Measures (ECCM) features.

Furthermore, some reports have suggested that the system can cope with targets with low Radar Cross Section (RCS), but no figures for such a capability have been released. If this is so, then much work/improvement to the basic radar of the system would have to have been done for the system to cope with such.

Variants

Sayad-1
Sayad-1A
Sayyad-2

Specifications

	Sayad-2
Dimensions and weights	
Length	
overall	10.84 m (35 ft 6¾ in)
Performance	
Speed	
max speed	2,333 kt (4,320 km/h; 2,684 mph; 1,200 m/s)
Range	
max	54.0 n miles (100 km; 62.1 miles)
Ordnance components	
Warhead	200 kg (440 lb) HE fragmentation
Guidance	CLOS
Propulsion	
type	liquid propellant, solid propellant, two stage

Status

Reported to be in service in Iran from 2011.

Contractor

Aerospace Industries Organization (AIO) of Iran.

Israel

Arrow Weapon System (AWS)

Type

Static and towed Surface-to-Air Missile (SAM) system

The IAI Elta Great Pine (Super Green Pine) radar system (Israel Aerospace Industries) 1411974

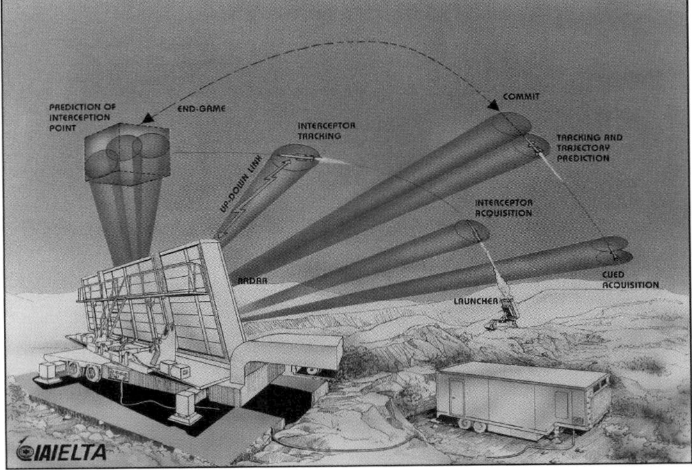

Key components of the Arrow Weapon System (AWS) 0509794

Development

As part of the US Department of Defence (DoD) Strategic Defence Initiative (SDI) research and development effort, an initial demonstrator contract was placed with the then Israel Aircraft Industries (IAI) Electronic Division (now Israel Aerospace Industries) in July 1988. This contract was to develop, manufacture, integrate and flight test four prototypes of the 7 m long, single-stage solid propellant Chetz (Arrow)-1 ATBM together with its early warning, guidance and proximity fuzed warhead subsystems.

Total cost of the contract for the project was USD158 million, with the US DoD funding 80 per cent and the Israeli MoD, in the form of the Israel Missile Defence Organisation (IMDO), the other 20 per cent. The system was initially designed to counter air-breathing threats and SRBMs of up to 1,000 km. Arrow is also capable of intercepting single or multiple targets in their final stage of flight at altitudes up to 40,000 m and at ranges up to 100,000 m. With a Rafael high-explosive focused fragmentation warhead, the directional capability built in should be able to destroy any target within a 50 m radius.

Elta Electronics Industries Ltd, of IAI's Electronic Group, has developed a combined solid-state phased-array EL/M-2080 L-band early warning and fire-control radar. The radar, codename 'Green Pine' and developed from Elta's 'Music' phased-array radar, uses an active array which is electronically scanned in elevation and azimuth to provide search, detection, alert, track and guidance operations simultaneously. Detection range is approximately 500 km. It can track and illuminate missile targets with speeds in excess of 3,000 m/s and can guide an Arrow to within 4 m of the target. The radar system comprises radar, power unit, cooling system and control centre transportable units fitted to semi-trailer chassis.

In January 1989, it was announced that Lockheed Martin Missiles and Space Company had signed an agreement with IAI to involve itself in the follow-up full-scale development and production phases following a successful completion of the demonstrator test programme.

The first demonstration flight using the Arrow-1 developmental missile was flown at an Israeli Mediterranean Sea test range in August 1990. The second flight took place in March 1991, the third in October 1991 and the fourth in September 1992.

The second stage of development, known as the Arrow Continuation Experiments (ACES) or Arrow-2, lasted approximately 45 months and cost USD350 million. The Arrow-2 is a two-stage weapon with solid propellant rocket motor booster and sustainer. It is approximately 700 kg less than the weight of the basic Arrow (Arrow-1). The US DoD paid 72 per cent and 28 per cent by the Israel Missile Defence Organisation (IMDO) that was formed in 1991 by the Israeli MoD. A total of 11 test flights were planned. However, due to problems experienced during the second and third full-scale test flights the programme was delayed by 12 months. The fifth flight took place in February

Elta L-band EL/M-2080 'Green Pine' radar system which is the heart of the Arrow Weapon System (AWS) 0509795

Launch of Israel Aircraft Industries Arrow-2 missile in July 1995 0509796

Inside view of 'Citron Tree' fire-control system (Tadiran) 0039167

Arrow missile and six-round launcher (Doug Richardson) 0039166

1993. The missile flew at M9.0 and met nearly all its test objectives. The sixth test occurred in July 1993 but the Arrow-1 was not launched due to the target missile going wild and having to be destroyed. The repeat sixth launch was deemed a failure as, although the Arrow-1 successfully acquired and tracked a second Arrow-1 acting as a surrogate 'Scud', the warhead failed to detonate. A seventh test launch was made on 12 June 1994, with all facets of the engagement being successful in all respects. The Arrow-1 warhead fragments hit the target missile.

A test firing of the Arrow (reported as Arrow Mk 3) took place off the West Coast of the US against a Scud target that was launched from a vessel at sea, in an experimental field by the US Navy. The trial was a complete success, although the target was a basic Scud and no effort as far as is known was made to confuse or decoy the Arrow during the trial. A further trial was planned for late 2004, this trial was more vigorous than the last and included components from the Boeing organisation that have been designed for the future Arrow System Improvement Programme (ASIP). It was expected that this trial used the Arrow Mk 4 missile; a good indication therefore that something new will be included such as decoy countermeasures. What was reported as a routine development ASIP test, took place on 5 December 2005. This was also reported as the 14th interceptor test and the ninth test of the complete weapon system. The test objectives were to demonstrate the system's improved performance in an operational scenario against an incoming target that represented the threat for the State of Israel.

In August 2005, reports from Israel suggested that a new version of the Arrow (known locally as Mini-Arrow) was to be developed to tackle the problem of cross-border rocket attacks. This system is known as Stunner or David's Sling and as of May 2010 is ready for deployment following extensive trials. Full details of the Stunner can be found in the International section of the Static and Towed missile part of this title.

It has been reported that the Israeli Air Force are updating the missile testing capabilities at the Palmachim Air Base that is situated just South of Tel Aviv. These facilities will be used for a wide range of programmes including elements of the Arrow 2. Simultaneously with the above announcement, the Israeli Aerospace Industry was awarded a contract by the Israeli MoD for follow-on production of an undisclosed number of additional Arrow 2 interceptors. Whether these two items are coincidental or related to each other is still not known.

A full system test on the Arrow 2 system took place on 7 April 2009 against a target that was simulating an Iranian Shahab-3 surface-to-surface missile, the test was reported to be successful. Complimentary to the April 2009 trial, in February 2011, a test shot of the Arrow system that included a new test missile and new elements that had recently been added took place on the West Coast of the US. This trial identified, tracked and engaged a target that was fired from a platform in the Pacific Ocean simulating a target that could likely be used against Israel at some time in the future.

On 10 February 2012, the Arrow Block-4 was tested using the Oren Adir (Great/Magnificent Pine) radar system against a target simulating a potential threat from and Iranian surface-to-surface missile. Block 4 is currently being installed and deployed across the country. This particular upgrade includes technologies for synchronizing with US systems that allow closer working conditions and cooperation with US weapons. An Arrow Link 16 upgrade was developed as part of the M4 upgrade to enable the AWS to cooperate with US MIM-104 Patriot and PAC-3 batteries, and to include US Aegis ships with SM-3 missiles. There are or have been several Block numbers allocated to the Arrow 2 system. These are:

- Block-2
- Block-3
- Block-4
- Block-5.

The first test of the Block-2 took place in January 2003. The first test of the Arrow 2 Block-3 took place on 11 February 2007 when the missile successfully intercepted and destroyed a Black Sparrow simulated ballistic missile target. The Block-4 was tested during a launch on 15 April 2008 also against a Black

Sparrow target. It is understood that the Block-4 upgrades improves the process of discrimination and the transmission of target data for better situation control of the missile. The Block-5 has been defined as an upgrade to the complete Arrow system to merge the lower-tier Arrow and the exo-atmospheric Arrow into a single national missile defence system. Block-5 will also optimise the existing Oren Adir (Great/Magnificent Pine) radar to operate with the AN/TPY-2 radar.

Description

The first Arrow 2 launch occurred on 30 July 1995. A number of tests of Arrow-2 have been undertaken through 1996 to 1999. In November 1999, the first system-wide test of the AWS was successfully undertaken. A simplified TM-91 Arrow missile target was destroyed.

By September 2000, a total of seven out of eight Arrow 2 launch tests and three out of three system tests had been declared successful.

Operational deployment of the AWS, codename 'Homa' (Fence), was originally stated to be early 1998. This was shifted to before the end of 1999. Two batteries will be deployed that will be capable of protecting up to 85 per cent of Israel's population. The first will be placed near Tel Aviv and the second to the south of Haifa. Due to missile proliferation in the area a third battery has been ordered. This will provide additional coverage to civilian population centres.

Each battery is estimated to cost USD170 million and will have:

- Four six-round launcher series trailers;
- A truck-mounted Launch Control Centre (LCC);
- A truck-mounted Communications Centre (CC);
- A semi-trailer-mounted Fire-Management Centre (Etrog Zahav) (FMC);
- A semi-trailer-mounted Elta L-band 'Green Pine' (Oren Yarok)-based radar system;
- A semi-trailer-mounted Radar Control Centre (RCC);
- A semi-trailer-mounted radar power generator unit and a semi-trailer-mounted radar cooling unit.

The FMC is codenamed 'Citron Tree'.

The Arrow 2 missile is hot-launched vertically from its container launcher with manoeuvring by flexible nozzles. The missile flies a 'double kick' flight profile of initial burn and secondary burn towards the target. From 1997 through to 2001, the US provided USD40 million a year to Israel to fund continued development and deployment of the Arrow 2. The seeker used is dual mode with a passive IR system to acquire and track TBMs and an active radar unit to combat air-breathing targets amid heavy countermeasures at lower altitudes.

Variants

Mini-Arrow (Stunner/David's Sling)

This system has a requirement to take on targets such as large-calibre rockets that have a range of up to 90 km, such as the Iranian-produced Fajr 5 thought to be in service with Hizbullah in Lebanon. It is feasible therefore that the Mini-Arrow is to be used to protect high-value targets and populated civilian areas. Funding for the new programme has been requested from the US in advance of Fiscal Year 2006 (FY06) and would be overseen by the Homa Directorate of the Israeli Ministry of Defence. Should the Mini-Arrow be developed, this would mark a change in policy for Israel, which had previously looked to directed energy weapons, most notable of which is the THEL and MTHEL both of which are considered obsolete.

The stated technical specification for the new system, which is required to counter the threat from Hizbullah El Fagar and Zazal short-range missiles, is to intercept at shorter ranges with a much improved reaction time. The current missiles deployed within Israel, the Arrow-II and Patriot PAC-3, are considered too large to counter this new threat.

ASIP

The AWS is to be extended beyond the Arrow-2. In March 1997 the Israeli government announced that if it can get additional funding, then AWS will be extended to cover a follow-on programme informally known as 'Arrow Beyond 2000' now known as the Arrow System Improvement Programme (ASIP). This began in 1998 and was extended through 2008. Total AWS costs budgeted until 2005, were stated to be USD1.6 billion, although this figure was exceeded. In July 2005, the IMDO approached the US Senate Appropriations Committee for USD16 million to support the new ASIP programme. This new missile system was the subject of a competition between IAI and Rafael with their respective US partners.

The 15th test for the interceptor missile and the 10th for the complete system took place on 2 February 2007. The Israeli MoD has reported that the test was part of the Arrow System Improvement Programme (ASIP) and was carried out jointly by Israel and the US. The missile used for the intercept was reported as Arrow 2M4 (hardware version) Block 3 (software) missile produced jointly by Boeing and IAI/MLM, a newer interceptor version, the M5, which is capable of increased manoeuvrability were tested in April 2007. The objectives of the test were many and included:

- To examine improved performance including widening of the interception envelope;
- To test the system in an integrated operational configuration engaging two batteries located at a distance from each other;
- To check implementation of lessons from previous tests.

On 27 March 2007, an upgraded Arrow ABM was tested as part of the ongoing Arrow System Improvement Programme (ASIP), the missile was fired from an upgraded launcher. Israeli information/sources report that the test was fully successful.

In May 2007, the MDA additionally allocated approximately USD80 million each year from 2010 through 2013 for post ASIP activities, details for which will be determined by the two countries over the coming years.

Arrow 3 (Hetz 3)

In November 2007, after a visit to Washington by the then Defence Minister for Israel Ehud Barak, it was announced that USD98 million were to be allocated for FY08 for the Arrow system; emphasis was placed on the Arrow 3 having the capability to intercept ballistic missiles in space.

Funded in a multiyear programme entitled 'Tefen', the Arrow 3 is a significant upgrade to the currently deployed Arrow 2 weapon system. The Arrow 3 is reported to be faster and more accurate than it's predecessor and capable of exo-atmospheric intercepts well beyond the range of the Arrow 2 and will use the upgraded EL/M 2084 Great Pine. The Green Pine (Arrow 2 radar) is being replaced by the Oren Adir (Great/Magnificent Pine) that will support both Arrow 2 and Arrow 3, the Great Pine is already deployed at a few sites within Israel. The new radar has both improved range and resolution by up to 50 per cent. Video evidence released by the Israeli Defence Ministry shows that the Arrow 3 interceptor has two powered stages; the first boost stage is reported to have an extremely high burnout velocity and is steered by Thrust Velocity Control (TVC). The Kill Vehicle (KV) has a dual-pulse motor that coasts the KV toward the target using the first pulse for initial trajectory correction and, as the IR sensor acquires the target, the second pulse is used for the final homing. The KV is equipped with a hemispheric, gimballed IR homing sensor; there is no explosive warhead on the Arrow 3 interceptor.

Further congressional funding for the Arrow 3 programme was allocated to the then Israeli Defence Minister Ehud Barak. Officials say the Arrow 3 system is an advanced model of an Arrow interception missile and that the missile will help Israel intercept ballistic missiles at higher altitudes and at greater distances than the current model.

In May 2009, the US finally agreed to provide the full funding (some reports suggest USD100 million) for the development and production of the Arrow 3 anti-missile system.

On or about 26 July 2010, Israel and the USD signed an agreement to develop the Arrow 3 ballistic missile defence system capable of shooting down long range missiles outside the Earth's atmosphere.

Reports emanating from the Israeli Ministry of Defence suggest that the Arrow 3 will join the Iron Dome and the Magic Wand systems to provide a multi-layered defence to the threats faced by Israel. All three systems will be operated by the Israel Air Force (IAF) who are responsible for the Arrow system as it stands today.

There are a number of logistically characteristics of the Arrow 3 system; these are:

- Arrow 3 can re-target whilst in flight, for instance if the first of a dual-missile launch kills the intended target, the second missile can divert to a second target;
- Arrow 3 uses the same resources as Arrow 2 such as launchers and battle management centres;
- Arrow 3 is fully interoperable with the Arrow 2.

Late in December 2011, The US House and Senate Committees on Appropriations approved the doubling of the special aid it gives Israel for the development and implementation of anti-missile systems; these included the Arrow 3 and David's Sling.

In mid-February 2013, a joint exercise with the US forces allowed Israel to test the Arrow 3 in an exercise designed to defend Israel from the threat of an Iranian strike. The Israeli ministry reported later that the test was a major milestone in the development of the Arrow 3 Weapon System.

The Arrow 3 is expected to take up to two more years before the system will become fully operational.

Specifications

	Arrow-2
Dimensions and weights	
Length	
overall	7 m (22 ft 11½ in)
Diameter	
body	800 mm (31.50 in)
Weight	
launch	1,300 kg (2,866 lb)
Performance	
Speed	
max Mach number	9.0
Range	
max	48.6 n miles (90 km; 55.9 miles)
	37.8 n miles (70 km; 43.5 miles) (effective)
Altitude	
min	8,000 m (26,247 ft)
max	50,000 m (164,042 ft)
Ordnance components	
Warhead	HE fragmentation

Status

In production. Initial deployment has taken place with final development and integration taking place simultaneously. A total of two batteries have been procured and deployed about 100 km apart; one in Palmachim Air Force Base, south of Tel Aviv, and the other in Ein Shemer, about halfway between Tel Aviv and Haifa. However, Israel is set to acquire a third battery foreseeing a growing threat from surface-to-surface missiles from Iran and Syria.

In 1997, Turkey began preliminary enquiries about the possibility of obtaining an AWS network for its defence. Negotiations concerning the purchase of several systems were believed to have taken place in 2000 with both Israel and the US.

India has taken delivery (2003) of the first of two Green Pine radar systems, presumably for use with its Russian Federation-(CIS) supplied anti-ballistic missile capable systems. However, after the air raid by the Tamil Tiger Air Force (LTTE) late March 2007, the Indian Defence Minister George Fernandes urgently applied through the Indian-Israeli liaison officer for a third Green Pine system, even on a year's loan, for deployment south of New Delhi.

Israeli officials hope to obtain elements of the Patriot PAC-3 to improve a missile defence command composed of the PAC-2 and Arrow-2 TBM. This will create a two-tiered system; the Arrow will handle the upper tier and Patriot will handle the lower tier. At the same time, Israel planned to deploy two Arrow-2 batteries during the latter part of 2002 and early 2003. The first and central Arrow-2 facility is at an air force base south of Tel Aviv, the second is being prepared at Ein Shemer, east of the Hadera.

A new EL/M-2083 phased array radar, developed and manufactured by Israel Aircraft Industries (now Israel Aerospace Industries), Elta Systems Group (IAI/Elta) has allegedly been incorporated into the Extended Air Defence Aerostat System. The EL/M-2083 early warning and control radar has been designed to detect hostile approaching targets from long ranges, with a particular emphasis on low altitudes. Data is then fed to a central air defence command-and-control centre where it is used to maintain an extended air situation picture. This picture will in turn be available to the Arrow, Patriot and shorter-range air defence systems for layered defence in depth.

As of February 2008, the Arrow 3 was in development with plans to test the anti-ballistic missile system for the first time in late 2008, with the Arrow 4 being at the design stage. Further mention of the Arrow-3 system was made in a press interview in mid-December 2007 regarding the five layered defence system that included the Arrow 2, MIM-104 Patriot PAC-2 (and upgraded Yahalom missile), Arrow 3 (for upper tier layer defence), Davids Sling and Iron Dome.

Late in December 2011, The US House and Senate Committees on Appropriations approved the doubling of the special aid it gives Israel for the development and implementation of anti-missile systems; these included the Arrow 3 and David's Sling.

The IDF is to deploy a fourth Arrow system. Currently, three Arrow sites have been identified, Palmachim Base, Ein Shemer and a third in the southern district. The fourth location has not yet been officially released by the IDF but is almost certainly deployed to counter any threat coming from Iran.

Late January 2012, an announcement from Tel Aviv confirmed the sale of Arrow 2 to an unidentified Asian country (not India) but most likely either Singapore, Taiwan, Japan or South Korea.

A second test flight of the Arrow 3 took place early in January 2014 from the Palmahim Air Base south of Tel Aviv. The missile is reported to have carried out manoeuvres in response to a virtual incoming threat missile. This flight trial was also the first time the system's radar tracked the target and sent commands to the interceptor. The test also involved the Arrow 3 Command and Control system.

Contractor

Israel

MLM Division of Israel Aerospace Industries (IAI) (prime contractor: for AWS, Arrow-2, Arrow-3, launcher and launcher control).
Elta Electronics Industries Ltd of Israel Aircraft Industries (fire-control radar).
TADIRAN Electronics Ltd (Battle Management, Command, Control and Communications (BM/C3) fire-control centre).
Rafael Advanced Defence Systems Ltd (warhead). As of January 2011, Rafael is set to purchase Israeli Military Industries Limited (IMI) who currently build the missile motors at their Givon factory.

United States

American companies contributing to AWS include Lockheed Martin, Marietta, the then Loral Fairchild and Raytheon's Amber Engineering.
Boeing has become a second supplier for Arrow components and Arrow development, in order to boost production rates for the missile.
Alliant Techsystems received a USD17.6 million contract to prepare production of the following:

- Filament wound first- and second-stage rocket motor cases;
- Electrical cabling;
- First-stage nozzle components and propellant materials;
- A metallic skirt;
- Interstage parts.

Wildwood Electronics, Huntsville, Alabama (provides electronic components, cable harness fabrication, environmental testing and quality assurance).

Italy

Aramis/Aspide

Type

Static and towed Surface-to-Air Missile (SAM) system

Development

The Aramis air defence system from MBDA is designed for all weather multiple intercept area defence at the brigade level. In its simplest form it comprises the following elements:

MBDA Aramis/Aspide area multiple intercept system with six-round missile launcher in foreground and Battery Command Post (BCP) and radar in background 0007966

(1) Battery Control Post (BCP)
(2) Aspide Launching Unit (LU).

The BCP integrates an Nuclear, Biological and Chemical (NBC) - and ElectroMagnetic Protection (EMP)-protected two-man command post module with a search radar assembly on the rear decking of a (4 × 4) cross-country lorry chassis.

The antenna of the 50 simultaneous track-capable 2-D frequency-agile radar is mounted on a 10 m extendable mast and provides detection ranges of up to 45 km on tactical air targets.

An Identification Friend or Foe (IFF) subsystem is incorporated into the primary radar to provide automatic identification of the tracks. The radar also has a built-in jamming location capability.

The operations centre module is fitted behind the mast assembly and monitors the output from the radar, launchers and higher echelon authorities to provide a coordinated battery mission plan response to any threat on the post's digital mapping and high-definition colour graphics computing display facilities.

A Global Position System (GPS) navigation system ensures accurate position information. Tracks are assessed for threat, prioritised and assigned to one of the four Aspide LUs which the BCP controls. If required, the threat can be designated to all other anti-air defence weapons in the air defence network via a real-time automatic broadcast or on an optional weapon terminal radio link for up to 10 small anti-air weapons deployed within a minimum 10 km range.

The Aspide LU is an unmanned, self-powered six-round launcher unit that is under the direct supervision of the BCP during an engagement. The target designation is sent either via a 1,000 m long telephone cable or a maximum 5,000 m range coded data radio link.

Description

The two rows of three, ready-to-fire missiles are carried in their container-launcher boxes on an elevatable frame assembly with the target illumination radar fitted directly below them on the frame's front edge.

Upon receiving the target designation data, the LU autonomously performs the acquisition and tracking of the target and all engagement functions until the target is destroyed. Each LU can perform its own engagement under the control of the BCP so that up to four separate targets can be engaged simultaneously by the battery. The system takes about eight minutes to deploy. The Aramis/Aspide system can use either the Aspide Mk1 or the Aspide 2000 missile. The Aspide Mk1 is similar to the US AIM-7E Sparrow missile with a Selenia monopulse semi-active seeker and SNIA-Viscosa solid-propellant rocket motor. The latter (Aspide 2000) has a new single-stage solid propellant rocket motor, improved ECCM features, higher velocity, higher lateral acceleration and a 30 to 40 per cent range increase.

Both Aspide missiles are described in the specifications.

Specifications

Aspide Mk 1 missile

Type:	single-stage, low to medium altitude
Length:	3.7 m
Diameter:	0.203 m
Wing span:	0.68 m
Launch weight:	220 kg
Propulsion:	SNIA Viscosa single-stage solid propellant rocket motor
Guidance:	semi-active radar homing
Warhead type:	HE-fragmentation
Fuze:	contact and proximity
Single shot kill probability:	80%
Speed:	M2.0 (approx.) Some sources suggest 650 m/s

Maximum range:	>15,000 m
Altitude:	
(maximum)	>6,000 m
(minimum)	10 m (approx.)
Launcher:	towed unmanned 6-round trainable

Aspide 2000 missile

Type:	single-stage, low to medium-altitude
Length:	3.7 m
Diameter:	0.203 m
Body:	0.203 m
Motor:	0.234 m
Wing span:	0.68 m
Launch weight:	241 kg
Propulsion:	single-stage solid propellant rocket motor
Guidance:	semi-active radar homing
Warhead type:	HE-fragmentation
Fuze:	contact and proximity
Speed:	M2.5+
Maximum range:	24,000 m
Altitude:	
(maximum)	>8,000 m
(minimum)	10 m (approx.)
Launcher:	towed unmanned 6-round trainable

Status
Ready for production. Being marketed in South America, the Middle East and Asia.

Contractor
MBDA.

Spada

Type
Static Surface-to-Air Missile (SAM) system

Development
The Spada point defence missile system was developed in the late 1970s to meet an urgent Italian Air Force requirement to defend air bases and other key areas.

To reduce both development time and costs it was decided to use components already proven in other applications. For example, the Aspide Mk 1 missile is identical to that used in the naval air-defence role, the TIR pulse Doppler radar has been developed from the naval Alenia Marconi Systems (now MBDA) Orion 30X radar tracker incorporating a J-band illuminator which has been in widespread service for some years.

Trials with the first prototype system were completed in 1977, with full technical and operational evaluations undertaken by the Italian Air Force during 1982 and 1983.

By 1986, a total of 12 systems had been ordered by the Italian Air Force, with the first battery having become operational in 1983. Additional four systems were ordered in 1991 to complete the total procurement requirement of four battalions.

In 1986, the Royal Thai Air Force ordered one complete Spada battery of four six-round launchers, which was delivered in 1988. The Royal Thai Navy already uses the Aspide missile in its Albatross naval point defence system.

In 1996, an upgrade programme was started that consists mainly of enhancing the command-and-control architecture in terms of its hardness and software capabilities.

Key features of the Spada point defence system have been summarised by MBDA as large area of cover with high single-shot kill probability. Flexibility in the system configuration, deployment and sighting, good co-ordination, reliable target identification capability, high resistance to enemy ECM and the possibility of being integrated into a national air-defence system.

Heart of the Spada low-level air-defence system is the detection centre with its Pluto SIR which is a modified Alenia Marconi Systems Pluto E/F-band radar 0509581

Discussions are underway with the Italian Air Force concerning the upgrading of its Spada systems to a standard close to that of the latest Spada 2000 standard. As Aspide was designed to have a 24-year life with a planned maintenance and refurbishment after 12 years then the half-life maintenance period can be used to convert the system.

During December 2005, the Italian Air Force successfully carried out a series of live firing trials of its upgraded Spada air-defence system to qualify for the Mid-Life Update (MLU) programme. These trials also tested the quality of the work by MBDA Italy together with the Raggruppamento Temporaneo di Imprese (RTI) - a temporary industrial consortium within this technology upgrade programme. All firing met the standards required by the Air Force.

The MLU mainly involved upgrades to the Command-and-Control (C2) system and its air-transportability. The upgrades have served to further enhance the Spada system's operational capability. Technology improvements focused particularly on the hardware and software within Spada's C2, adding important functions, such as mission planning and training. These improvements also brought Spada into line with anti-pollution regulations regarding its cooling gases. Hardware and software within Spada's launchers have also been upgraded. Servo-control management has been achieved with new software that now incorporates highly sophisticated algorithms to improve significantly the launcher's mobility.

The electrical systems generators have also been upgraded to better facilitate maintenance and to comply with current safety regulations.

Spada 2000 is an advanced, ground-based missile system capable of operating in dense ECM environments to provide all weather, day and night area defence against combat aircraft and incoming missiles. It is a development from the original Skyguard/Aspide system and now benefits from the increased engagement range of the powerful Aspide 2000 missile as well as the increased detection range and track management capabilities of the new Detection Centre (DC).

Spada 2000 is integrated within a shelter system, allowing for both tactical and strategic mobility, including air-transportability from a C-130 transport aircraft.

Description
A typical Spada battery has two main components; the Detection Centre (DC) and the Firing Sections (FSs).

The detection centre consists of the Search and Interrogation Radar (SIR), which comprises the SIR antenna pedestal, its equipment shelter and the Operational Control Centre (OCC) shelter, plus a generator.

Six-round Aspide missile launcher in travelling configuration being towed by a truck fitted with a hydraulic crane for missile resupply 0509582

Alenia Marconi Systems TIR used at each firing section with TV sensor to its left 0509583

The MBDA Aspide Mk 1 multirole missile 0509584

Alenia Marconi Systems Spada operational control centre 0509585

One DC controls up to four firing sections, each of which comprises the following:

(1) Two Missile Launchers (MLs) with six missiles in the ready-to-fire position
(2) The Fire-Control Centre (FCC) consisting of the Tracking and Illuminating Radar (TIR) antenna pedestal with its TV sensor
(3) Control Unit (CU) shelter for the TIR
(4) Firing-control equipment and a generator.

The Firing Sections can be deployed up to 5 km away from the Detection Centre.

The largest size battery can cover an area of up to 800 km^2 with a maximum of 72 missiles in the ready-to-launch position, which can be fired singly or in salvos of two.

The task of the DC is to search for targets, identify, evaluate and designate them to the FSs. Through the SIR, targets in the area covered are searched for, detected, plot extracted and selectively IFF interrogated. Tactical control of the battery is achieved by means of the associated OCC. Plot initiation, track-while-scan and data updating of a number of tracks, their identification, threat evaluation and target designation to the FSs for engagement are carried out.

All the operations listed above are carried out automatically by the system.

The SIR consists of low-altitude search radar, particularly suitable for dense clutter environments, advanced ECM and interrogation radar with decoding capability on IFF answers.

The OCC consists of three operational consoles, a data processing system and a number of digital and phonic operational communication links.

The task of the FSs, which represent the reaction centre of the system, is to acquire and destroy assigned targets. At the FCC, TIR performs target acquisition, tracking and illumination for missile guidance. As an additional mode of operation, the TIR provides for target search (all-round or sector scan), detection and self-designation. The TV set is used both as a back up to the radar and as an aid for target identification and discrimination and for kill assessment. The associated CU supervises the aforementioned functions. Automatic or manual control of the firing action by means of displays and communications equipment is also monitored.

The MLs provide for missile storing, aiming, selection and setting to fire, as well as for automatic launching sequence of the six missiles on each launcher.

The TIR consists of tracking pulse radar, a CW transmitter for illumination and a common antenna with its own pedestal.

The CU consists of an operational console, a data processing system and a number of digital and phonic operational communication links.

The ML consists of a slew able stand, which supports a frame-type structure on which two rows of three missiles canisters each are located. This structure can be moved in elevation, both for positioning to the firing angle and for canister loading/unloading operations.

The launcher has six rounds in the ready-to-launch position and, when in the firing position, is supported on four outriggers that can be adjusted for height. The launcher can be traversed through 360° at a slew rate of 50°/s with elevation up to 30°.

Reload rounds in their containers are loaded with a hydraulic crane carried on the missile resupply vehicle.

The main components of the Spada point defence system, such as the shelters and fire units, are coupled to a mobiliser wheeled system with the units then towed behind trucks. The mobiliser basically consists of two pairs of wheels that are connected to the ends of the shelters as well as both the SIR antenna and the missile launchers.

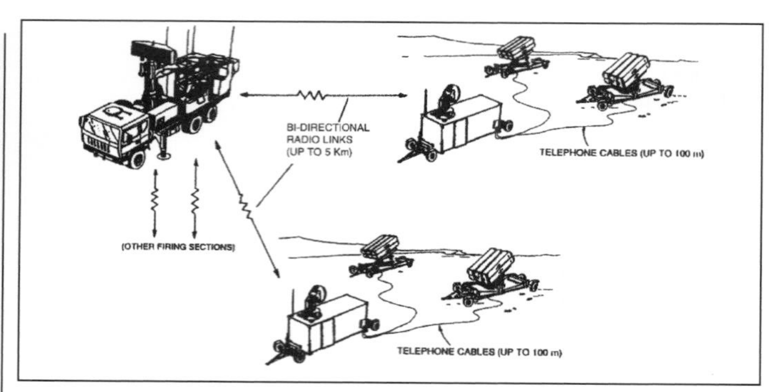

Typical Spada 2000 air-defence system with Detection Centre (left), two Fire-Control Centres and four six-round launchers 0007967

Latest configuration of Aspide six cell launching clearly showing six missile launch tubes and their antenna 0007968

In addition to the shelters a SIR antenna platform can be autonomously loaded/unloaded from a truck by means of a positioning device consisting of four lifting legs which are fitted as standard.

The Aspide Mk 1 missile is a J-band semi-active homing weapon able to guide itself onto the target by sensing the CW electromagnetic energy, either reflected by the illuminating target or emitted by a self-screening jammer. It is propelled by a high-thrust single-stage, solid propellant rocket motor, the missile uses proportional navigation to direct itself by collision course towards the target.

According to Alenia Marconi Systems the missile assures a high Single-Shot Kill Potential (SSKP), even in the presence of a dense and sophisticated ECM environment and at very low altitude. This is a consequence of the missile's guidance accuracy (due to the adoption of the monopulse inverse-type radar receiver), active radar proximity fuze and the sizeable pre-fragmented-type warhead.

The electrical and hydraulic power generator of the Aspide missile has been developed by Microtecnica and is powered by a solid propellant gas generator. It consists of a high-speed gas turbine, reduction gears, an alternator and a hydraulic pump. The hydraulic system is a closed loop, ensuring the availability of full hydraulic power for the entire duration of guided flight.

The missile is housed in and fired from a canister that is also used for transportation and stowing.

In addition to engaging aircraft and helicopters, Spada has a capability against RPVs and air-to-surface missiles.

The air-to-air version is hypersonic and the surface-to-air version is supersonic. They have many common components although their fore and aft fins are different, the former being movable and the latter fixed.

Variants

Skyguard/Aspide

The Skyguard/Aspide point defence system is an integration of the then Oerlikon Contraves (now Rheinmetall Air Defence) Skyguard fire-control system, Rheinmetall Air Defence twin GDF 35 mm towed anti-aircraft guns and the MBDA Aspide missile. The then Alenia Marconi Systems developed the tracker and illuminator radar installed on the launcher.

Al Amoun, obtained by Kuwait in 2005 is based on the original Skyguard system and is meant for the defence of Kuwait from cruise missiles.

Spada 2000

The Spada 2000 was developed by MBDA (formerly Alenia Marconi Systems) in 1994 and is an upgraded version of the Spada system with significant operational and technological improvements. These include:

(1) Tactical and strategic mobility through equipment integration onto sheltered units, suitable for air transporting by Lockheed Martin C-130 aircraft;

(2) Increased firepower and engagement range through the adoption of the Aspide 2000 missile. The Aspide 2000 has a new single-stage solid propellant rocket motor, improved ECCM features, higher velocity, higher lateral acceleration and a 30 to 40 per cent range increase to 24,000 m;
(3) Capability of co-ordinating up to 10 additional small anti-air weapons including MANPADS systems integrated into the local area air-defence network.

The basic configuration of the SPADA 2000 consists of; the DC; and two FS' that may be expanded up to four each with two missile launchers, each missile launcher houses six ready-to-fire Aspide 2000 missiles.

The Detection Centre embodies into a single sheltered unit the Thales Defence Systems Ltd's RAC 3-D search and interrogation radar and the Operational Centre. Radar coverage is greater than 45 km.

The Firing Section is based upon the Fire-Control Centre (including the tracking and target illumination radar and the Control Unit) and the six-round ready-to-fire missile launchers.

Both the Detection Centre and the Firing Sections can be deployed in a flexible way around the objective to be defended to maximise the air-defence protective level required. Two radio links of up to 5 km between the Detection Centre and Firing Sections and 100 m between launchers and the Fire-Control Centre by telephone cable.

Additional features include:

- Provision for integration into a national air-defence network
- Built-in simulation programme for operator training
- Mission planning capability.

The operational performance of the Aspide 2000 missile allows it to engage air threats at ranges well beyond many other Very SHOrt RAnge Defence (VSHORAD) missiles while engaging targets before release of their standoff air-to-surface weapon systems. This upgrade of the original Aspide missile embodies the latest electronic features to improve its guidance characteristics and effectiveness in heavy-clutter and Electronic Counter-Measure (ECM) environments.

Aspide 2000 is compatible with all systems currently using the Aspide multirole missile by the simple application of a cost-effective modification kit.

Similarly, all existing Aspide multirole missiles can be retrofitted to the Aspide 2000 configuration with only marginal costs.

In September 2007, the Pakistan Air Force chose the MBDA Spada 2000 air defence system for its low to medium air-defence requirements. It is likely that this system will replace the ageing French Crotale including those that are Chinese copies.

Specifications

	Aspide 2000	Aspide Mk 1
Dimensions and weights		
Length		
overall	3.7 m (12 ft 1¾ in)	3.7 m (12 ft 1¾ in)
Diameter		
body	203 mm (7.99 in)	203 mm (7.99 in)
	234 mm (9.21 in) (motor)	-
Flight control surfaces		
span	0.68 m (2 ft 2¾ in) (wing)	0.68 m (2 ft 2¾ in) (wing)
Weight		
launch	241 kg (531 lb)	220 kg (485 lb)
Performance		
Speed		
max Mach number	2.5	2.0 (est.)
Range		
max	13.0 n miles (24 km; 14.9 miles)	8.1 n miles (15 km; 9.3 miles) (est.)
Altitude		
min	10 m (33 ft) (est.)	10 m (33 ft) (est.)
max	8,000 m (26,247 ft) (est.)	6,000 m (19,685 ft) (est.)
Ordnance components		
Warhead	HE fragmentation	HE fragmentation
Fuze	impact, proximity	impact, proximity
Guidance	semi-active, radar	semi-active, radar
Propulsion		
type	single stage, solid propellant	single stage, solid propellant

Status

The system is in production and in service with the Italian Air Force. Batteries based on three fire-control centres with tracking and illumination radars and six launchers.

Royal Thai Air Force (with battery based on two fire-control centres).

Spanish Air Force, (Spada 2000 with two batteries ordered with each based on two fire-control centres - first battery delivery October 1998). The Spanish systems use the Thales RAC 3-D radar, Aerospatiale Matra command-and-control systems and can control up to four vehicle-mounted MBDA Mistral MANPADS.

Kuwait upgraded its Aspide missile systems to the SPADA 2000 configuration with an order placed in 2007.

Contractor

MBDA.

Japan

Chu-SAM

Type

Static and towed Surface-to-Air Missile (SAM) system

Development

In 1990, it was revealed that the Japanese Defence Agency (JDA) Technical Research and Development Institute (TRDI), and the Gijutsu-Kenkyu-Honbu, had started research and development work on system components with the Mitsubishi Electronics Corporation (MELCO) for an indigenous Medium Surface-to-Air Missile (M-SAM). This M-SAM was to replace the Japanese Defence Ground Force's eight groups of Raytheon I-HAWKs.

The demonstration and validation phases of the Chu-SAM (also known as Type 03 and SAM-4) were funded through FY89 to FY91 with the project engineering development phase due to start in FY93. However, this was put back by at least one year for the TRDI to confirm that Japanese companies had

A Chu-SAM Type 03 Transporter Erector Launcher (TEL) in travelling mode
1417092

The rear portion of a Chu-SAM Type 03 Transporter Erector Launcher (TEL) showing missile canisters in firing position
1416928

A Chu-SAM Type 03 missile resupply vehicle 1416929

A Chu-SAM Type 03 fire-control radar vehicle 1416927

A Chu-SAM Type 03 power supply vehicle 1416926

the technological base to complete the project without any outside assistance. The development was scheduled to take place between FY94 and FY97 with operational testing due in 1999, initial production around 2000 to 2001 and initial deployment in 2002, however, with the designation Type 03 this would also suggest the Japanese consider 2003 for Initial Operational Capability (IOC). This original time scale has changed due to evolving system requirements and funding constraints on the Japanese defence budget. The milestones are now believed to have moved two to three years or so in the future.

In December 2000, it was revealed that the Chu-SAM was due to have its first live firing tests at the White Sands Missile Range, New Mexico in Japanese FY01. Procurement funding is intended to start in Japanese FY03 with first operational deliveries one to two years later.

On 20 November 2006, at the MacGregor Range of the White Sands missile test range in New Mexico, a Japanese Air Self Defence Force (JASDF) training exercise took place that included a live firing of the Chu-SAM. Previous firings of the Chu-SAM have taken place at White Sands (up to six), these have been attributed to research and development firing, however the training exercise that took place in November 2006 indicates that the Chu-SAM is indeed the replacement of the I-Hawk system that had been used previously in these JASDF exercises. It was further announced that similar training is scheduled to take place using the Chu-SAM yearly.

Description

The weapon is primarily for use against aircraft but will also have both anti-tactical missile and anti-cruise missile capabilities. The missile has four clipped-tip delta fins at the nose and four similar sized fins at the rear. The Chu-SAM system is envisioned as having a 50 km range, depending upon the target type, and a maximum engagement altitude of up to 10,000 m. It will also have an over-the-horizon mode to allow it to be used in the mountainous regions of Japan.

The system has a simultaneous multi-target engagement capability using a multi-functional phased array, separate target acquisition and detection radar. The phased array radar is reported to be capable of tracking up to 100 targets simultaneously, whilst engaging 12. The missile is to be vertically launched from a truck-mounted launcher, with each fire unit having four launch vehicles; a six-round Vertical Launch System (VLS) assembly, a multi-beam, multi-function truck-mounted phased array search and fire-control radar, a truck-mounted fire-control post and a truck-mounted communications station. A Missile Group will have four or five fire units.

The Chu-SAM itself is likely to be a single-stage thrust vector-controlled weapon with a solid propellant rocket motor and fitted with state-of-the-art warhead, proximity and contact fuzing and guidance system technologies. The latter is to include initial command guidance with active radar, terminal homing radar.

Variants

It is highly probable that an extended range two-stage version may also be developed with a 40-second coasting period between stage ignitions to allow high *g* terminal manoeuvring against low-level cruise missiles.

Specifications

Missile length:	4.9 m
Diameter:	0.32 m
Launch weight:	570 kg
Warhead:	High Explosive Fragmentation
Guidance:	Inertial with command updates, and active radar terminal seeker
Propulsion:	Solid propellant
Speed:	M2.0 to M3.0
Range:	
(maximum)	50 km
(minimum)	Unknown
Altitude:	
(maximum)	10,000 m
(minimum)	Unknown

Status

The development phase came to an end during January 2007, after which the system then went into production and finally deployment with the Japanese defence forces. Those deployments were:

- Anti-aircraft Artillery Training Unit 2003
- 2nd Anti-aircraft Artillery Group 2007
- 8th Anti-aircraft Artillery Group 2008.

Contractor

Mitsubishi Heavy Industries (MHI) Ltd, the parent company of Mitsubishi Electronics Co.

Technical Research and Development Institute, Japanese Defence Agency (project coordination).

Type 81 Tan-SAM

Type

Self-propelled static Surface-to-Air Missile (SAM) system

Development

The requirement for the Tan-SAM missile system, (*Tan* is the Japanese word for 'short'), was generated in 1966/1967 when a replacement for the US-supplied divisional level static 75 mm M51 Skysweeper and self-propelled M15A1 37 mm/12.7 mm anti-aircraft guns (now phased out of service) was requested by the then Japanese Ground Self-Defence Force (JGSDF), now known simply as the Japanese Defence Force.

System research and fabrication of basic prototype systems were undertaken by Toshiba in 1967 to 1968 with actual construction of experimental units taking place in 1969 to 1970. The first complete prototype systems were built between 1971 and 1976 with the technical testing of the individual system components taking place during 1972 to 1977. In 1978 to 1979, the operational tests of the complete system were undertaken and upon their successful conclusion the JGSDF standardised the Tan-SAM as the Type 81 short-range surface-to-air missile system. Initial Operational Capability (IOC) began in 1981 as did yearly production contracts with its manufacturer Tokyo Shibaura Denki Company (Toshiba Electric).

As part of the current JGSDF plans, each of its divisional groups is assigned four fire units of Tan-SAM systems with the first deployments made to the four divisions located on the northern island of Hokkaido.

A fire unit comprises one Fire-Control System (FCS) vehicle, two quadruple-round launcher vehicles and a few support vehicles with a total team of 15.

During the mid-1970s, the then Japanese Air Self-Defence Force (JASDF) began a project to increase the survivability of its airfields and radar bases in terms of both active and passive defences. The Tan-SAM was chosen to provide the outer defence ring against enemy aircraft that had 'leaked' through

Close-up of four-round missile launcher of Type 81 Tan-SAM low-altitude surface-to-air missile system in travelling configuration (K Nogi) 0509586

Type 81 Tan-SAM low-altitude surface-to-air missile launcher with missiles on upper arm only and stabilisers lowered (Kensuke Ebata) 0509587

the forward interceptor and area defence SAM barriers. Whilst the Raytheon Systems Company FIM-92A Stinger and the Nippon Steel Works Company licence-built 20 mm M167 Vulcan Air Defence System provided the progressively shorter-ranged second- and third-line point defence rings.

The JASDF deployment is in composite SAM-AAA battalions; for an airbase this usually comprises three batteries:

1. Airbase defence battery with:
   - Two Type 81
   - 24 Stingers
   - 16 M167 20 mm towed Vulcan cannons.
2. Radar station defence battery with:
   - Two Type 81
   - 24 Raytheon Systems Company Stingers
   - Six M167 Vulcan cannons.
3. SAM battery of:
   - Raytheon Systems Company Patriots
   - 24 Raytheon Systems Company Stingers
   - Six M167 Vulcan cannons.

A total of six composite SAM-AAA battalions will be raised.

Both the JGSDF and the JASDF have also taken delivery of two further fire units that have been exclusively assigned to the training role.

The Japanese Maritime Defence Force also generated a requirement for the Type 81 Tan-SAM and bought two fire units in each of the FY89, 90 and 92 budgets for the protection of naval bases.

Total procurement of the three services was 57 fire units for the Ground Defence Force, 30 fire units for the Air Defence Force and six for the Maritime Defence Force.

Description

The missile system is a short-range fully mobile weapon that is of the single-stage, fire-and-forget type with four clipped cruciform centre body wings and four movable tail-mounted cruciform control fins.

Propulsion

Propulsion is by a Nissan Motor solid fuel rocket that exhausts an excessive amount of white smoke that both visually marks the launch site and allows an observant target to evade the oncoming weapon.

Payload

The missile contains an HE fragmentation warhead with contact or radar proximity fuze that is activated to detonate the explosive. The lethal radius of the warhead is 5 to 15 m depending on the target type. A self-destruct circuit has not been fitted.

Guidance

The guidance system uses a Kawasaki Heavy Industries inertial autopilot for the first part of the flight and then switches to a Toshiba Infra-Red (IR) passive homing seeker in the missile nose compartment.

Fire-control system of Type 81 Tan-SAM low-altitude surface-to-air missile in operational configuration with stabilisers deployed (Kensuke Ebata) 0509588

Type 81 Tan-SAM-kai missile 0517754

Before launch, the scanning angle (in degrees) of the seeker head is pre-programmed by the FCS computer to avoid the missile tracking the sun. This data is calculated from the continuously updated information on the target position. The FCS also controls the Fuji Heavy Industries aluminium launcher assembly movements so that a round cannot be fired directly at the sun by accident.

Once the missile is in flight and has reached the point at which the seeker is activated, the IR head starts to scan the pre-programmed area of the sky to find the target. The guidance unit then locks on and the missile continue to follow the shortest course to intercept.

Although adverse weather conditions affect the missile performance, it is still said to be comparable to that of the Euromissile Roland under the same conditions. The intercept capabilities in cloud remain good. However, although no IR filters are fitted, targets employing IR decoy systems or manoeuvring within or near to the sun's disc, as viewed from the missile, stand a good chance of defeating its seeker by simply overwhelming it with spurious heat sources. Some protection is provided by the special electronic precautions incorporated into the design of the search pattern scanning programme (Intermediate Frequency - IF).

For operations against extremely low-altitude targets in a heavy electronic jamming environment or targets from the rear of the launcher where radar is of marginal effect, control can be switched from the FCS vehicle to a tripod-mounted off-vehicle optical sight/control unit which is carried by each of the launchers. Once this is activated, the electronic link between the FCS and the launcher unit is automatically disengaged and the module is slaved to the sight.

At a fire unit position, the personnel responsible for an engagement consist of the commander, radar operator and two launch operators. Each vehicle has to be levelled and stabilised by its hydraulically operated jacks and interconnected by electric cables and the field telephone/datalink system. The normal distance between the FCS and the two launchers is approximately 300 m. The total time for these preparations is about 30 minutes. Once these have been completed, the surveillance radar on the FCS commences searching its assigned defence area. If only a single target is found, the engagement is relatively simple but, for multiple targets, priorities have to be assigned. When a designated target comes into effective range of the chosen missile on its launcher a visual signal is lit in the FCS and a missile is launched using the guidance technique described.

The in-flight seeker lock-on feature allows a Tan-SAM fire unit to launch either two missiles simultaneously or successively while the first is still homing on to the target. Thus, theoretically, up to four targets can be engaged by a single fire unit, however, in practice this is doubtful as a single missile cannot be guaranteed a 100 per cent hit probability. The actual hit probability of a Tan-SAM missile is officially stated to be 75 per cent, even in cloud.

If required, the FCS module, the two launcher modules and their associated generators can be removed from their vehicles and used either for a fixed static site or as a helicopter-transported unit by JGSDF Kawasaki/Vertol CH-47J Chinook for distances up to 100 km and over. In either case a small dozer tractor is required to set up the launch site.

Fire-Control System (FCS)

The FCS module weighing approximately 3,054 kg is mounted on the rear of a modified Isuzu Motors Ltd Type 73 (6 × 6) 3,000 kg truck and consists of a 30 kW generator unit immediately aft of the driving cab with the system control cabin to the rear. On top of the cabin roof is a 1-m wide, 1.2 m high 3-D phased-array pulse Doppler radar antenna that is mechanically steered in azimuth and elevation. The vehicle is stabilised on site by three hydraulically operated outriggers. No armour or NBC protection is provided. The radar search range is around 30,000 m and an integral IFF interrogation facility is fitted. The antenna rotates at 10 rpm and sweeps 360° in azimuth and 15° in elevation during a full rotation. In a sector search it automatically sweeps 110° in azimuth and 20° in elevation. The system has a multitarget capability with each threat being assigned its own number by the FCS computer. The future position and elevation of an individual threat is calculated, then all the information on the various targets is displayed on the three CRT scopes below the antenna in the form of a target symbol, range, altitude and direction data. The unit commander will assign target priorities and indicates which he intends to engage to the radar operator. The radar operator then moves a cursor on to each of the designated targets and presses a button to select it for precise tracking by the radar. Up to six targets can be tracked in this manner at any one time, each one being displayed on a CRT together with its continuously updated evaluation data. The launch data is fed into the selected weapon's onboard guidance system by the FCS computer that also directs the appropriate launcher to turn and elevate according to the position of the target. Once in range, the missile is fired.

Launcher

The launcher/generator system weighs approximately the same as the FCS and is mounted on the same type of truck chassis. It has no armour or NBC protection but has four hydraulically operated stabiliser jacks. The module is rectangular with two arms that have launcher rails on their upper and lower sides. On the front ends of each are two IR seeker covers that can be rotated through 180° to protect the missile heads during transit.

Each missile is loaded onto the launcher by hydraulically operated loading platforms mounted on each side of the vehicle. The missile, which arrives in a simple box-like container, is manually picked up and placed on the loading platform by the fire unit's crew after removing the container's cover. The loader lifts the round hydraulically into the loading position where the aft end of the launcher rail is slid back to reveal two latches that are then inserted into a slit on the missile body to hold it in place. The rail is moved back to its original position that automatically advances the round into the launch position. This process is repeated four times until all four launch rails are loaded. The total time taken for this action is approximately three minutes. The integral generator is rated at 30 kW and provides power when the vehicle engine is switched off.

Variants

Type 81 Tan-SAM-kai

In 1983, following initial funding by the JDA, Toshiba began development of the improved Type 81 Tan-SAM-kai (kai being a Kaizo or modification symbol) missile. During the remainder of the 1980s, the funding remained low level until FY89 when the JDA placed a contract with Toshiba for a six-year development programme. This effectively started in 1990 with the programme completed with FY94 budget money. Production of the missile started with the FY96 budget although the system was designated SAM-1C in 1995.

The major improvements include the following:

- An active phased-array radar seeker head.
- A smokeless rocket motor with increased thrust to increase speed and engagement range (to 14,000 m).
- A mid-course update link in the missile to allow the fire-control system to send flight corrections in order that manoeuvring aircraft can be engaged successfully.
- The fitting of a thermal imaging optical guidance capability to the fire-control system in place of the current tracker so as to improve effectiveness in ECM conditions.

Type 81 Tan-SAM-kai 2

It is understood that further development of the system, known locally as the Tan-SAM-kai 2 is already underway but it is not known if this work is for a complete new system or merely an upgrade based on the Tan SAM-kai. Work was supposed to have started as early as 2005.

Specifications

	Type 81 Tan-SAM
Dimensions and weights	
Length	
overall	2.7 m (8 ft 10¼ in)
Diameter	
body	160 mm (6.30 in)
Flight control surfaces	
span	0.60 m (1 ft 11½ in) (wing)
Weight	
launch	100 kg (220 lb)
Performance	
Speed	
max speed	1,586 kt (2,938 km/h; 1,825 mph; 816 m/s)
max Mach number	2.4
Range	
min	0.3 n miles (0.5 km; 0.3 miles) (est.)
max	3.8 n miles (7 km; 4.3 miles)
Altitude	
min	15 m (49 ft)
max	3,000 m (9,843 ft) (est.)
Ordnance components	
Warhead	9.2 kg (20.00 lb) HE fragmentation
Fuze	impact
	radar, proximity
Guidance	passive, IR (pre-programmed autopilot)
Propulsion	
type	solid propellant (sustain stage)

Status

Type 81 Tan-SAM

Production of fire units and missiles complete (1,800 Type 81 missiles built by 1995, to maintain a stockpile requirement of 16 missiles per fire unit - or two missiles per launcher rail). In service with the Japanese Ground Defence Force (57 fire units), Japanese Air Defence Force (30 fire units) and Japanese Maritime Defence Force (six fire units). As of August 2013, the system is still not offered for export.

Type 81 Tan-SAM-kai

In production. Initial production of missile and upgrading of Type 81 Tan-SAM fire units started with the FY96 budget. Not offered for export.

Tan-SAM-kai 2

In development.

Contractor

Toshiba (Tokyo Shibaura Electric) Company Limited (prime contractor and system integration).

Main subcontractors

Isuzu Motors Limited (vehicles).
Former Nissan Motor Company Limited now part of IHI Aerospace Co LtdCompany, Aerospace Division (rocket motor).
Fuji Heavy Industries (FHI) Ltd (launcher assembly, less electronics).
Kawasaki Heavy Industries (KHI) Ltd (autopilot and missile body).

Korea (North)

North Korean static surface-to-air missiles systems

Type

Static Surface-to-Air Missile (SAM) system

Development

In 1961, the then Soviet Union and the Democratic People's Republic of Korea (DPRK) signed a treaty on mutual aid and defence. Under this treaty, the Korean People's Air Force (KPAF) received a small number of surface-to-air missile systems (type unknown) for the defence of Pyongyang.

This delivery was followed in 1962/1963 by the S-75 Dvina (NATO SA-2 Guideline) weapons, including the setting up of a basic in-country capability to assemble, test and maintain the systems.

In the late 1960s, as the Soviet-DPRK relationship deteriorated with yet another arms embargo being implemented by the then Soviet Union. As a result of this embargo, the DPRK turned to the People's Republic of China (PRC) for it's defensive missile systems. The result of this was that from 1971 onwards, the Chinese helped reorganising the various Soviet-established missile support programmes. Shortly thereafter, the PRC also started deliveries of its Hong Qi-2 (HQ-2 NATO CSA-1) SAMs.

Some earlier reports have suggested that the Chinese exported the HQ-1 SAM to North Korea, however, this is probably untrue as the HQ-1 designator was allocated to the then Soviet Union supplied S-75 system which consisted of the guidance station RSN-75, launchers, missiles V-750 and ground support equipment.

As with North Korea, the Soviet Union also diplomatically fell out with China shortly after deliveries of the S-75 commenced to China. This then forced the Chinese to copy and re-engineer the few S-75's (HQ-1) that were held by them, and renaming the system HQ-2. No doubt several Chinese requirement modifications were made to the system when copying, thus enabling the Chinese to claim the HQ-2 as their own. It is assessed that these were the systems that China subsequently sold to North Korea.

S-200 site similar to North Korean 0509575

Description

In the mid-1970s, the PRC expanded its missile delivery programme to include the technology package necessary to enable the DPRK to assemble, maintain, upgrade and finally, in the late 1970s allow local production of the HQ-2 using PRC-supplied critical sub-assemblies.

As part of its overall missile production requirement, in 1975 the DPRK began a dedicated ballistic missile programme based on a surface-to-Surface Missile (SSM) variant of the HQ-2 to supplement the secondary SSM capability of the of the S-75. This programme lasted until 1979, when the resources being used were switched to the indigenous production and improvement of the HQ-2/S-75 type SAMs. This also happens to coincide with the Chinese improvements and upgrades to the HQ-2 system, to enable the minimum engagement altitude to be improved.

In the mid-1980s, the DPRK began exporting its home built HQ-2/S-75 missiles and technology to both Iran and Egypt.

In 1984, following the visit of Kim-Il-Sung to Moscow, the DPRK took delivery of a variety of air defence weapons, including the Igla-1, S-125 Pechora, S-200 Vega, 2K12 Kvadrat, ZSU-23-4 Shilka and Strela-10. As part of these new arms deliveries it is probable that maintenance, repair and upgrade packages were included.

The distribution and usage of the static air defence weapons are subordinate to the North Korean Air Force and is given in the following paragraphs.

In late 1987, North Korea took delivery of S-200 (SA-5) systems for use in the Korean People's Air Force (KPAF) Air Defence Command.

The first S-200 regiment of two six-launcher battalions was deployed to the Sarlwon-Pyonsan area, 40 to 100 km north of the DMZ, with the second regiment located on the east coast in the Wonsang-Hamhung region. Both units can reach into the northern sectors of South Korea's airspace, threatening both AWACS aircraft and strategic stand-off reconnaissance assets.

The regiments are integrated with the advanced 1980s technology Ukrainian Iskra 36D6 (NATO 'Tin Shield') 3-D surveillance trailer-mounted radar system optimised for low-level coverage. The radars also serve the four HQ-2/S-75 'Guideline' brigades (with 15 regiments) and the two independent S-125 (SA-3b 'Goa') regiments already in service with the KPAF.

The S-band radar has an instrumented range of 180 km and elevation limits of –0.5° to +30° with accuracy stated to be 0.3° in azimuth, 250 m in range and 400 m in height. The system can track up to 120 targets simultaneously and is usable against helicopters, cruise missiles and other small radar cross-section targets.

The entire country comprises a single air defence district with all operations controlled from the Combat Command Post (CCP) located at the KPAF headquarters, Pyongyang. The district is subdivided into three Sector Commands:

- Northwest
- Northeast
- Southern which includes Pyongyang subsector.

Each sector consists of a HQ, an Air Direction Control Centre (ADCC), EW radar regiment(s), air defence fighter division(s), SAM regiment(s), an AAA division and other independent AAA units.

When a target is detected, fighters are alerted and launched while the SAM and AAA unit's initiate tracking. Any subsequent SAM or AAA engagement is then coordinated with the fighter division's headquarters and the CCP. The idea is that, if the aircraft are unable to engage or are unsuccessful in their interception then the SAMs deployed in 'belts' will either destroy the target or force it, by virtue of making it take evasive action, into a position where covering AAA guns in 'ambush' positions are able to effect its destruction.

The HQ-2/S-75 regiments of three six-launcher battalions are each deployed mainly in a west coast 'belt' that runs along the Kaesong-Sarlwon-Pyongyang-Pakch'on-Sinuljus axis and two east coast 'belts' that lie along the Wonsan-Hamhung-Sinp'o and Chongjin-Najin lines. A number of independent HQ-2/S-75 regiments are located in 'dead' areas and around strategic areas between these 'belts'.

The S-200 regiments are used to support these 'belts' as are the two S-125 units made up of four batteries, each with four quadruple-missile launchers, one in the west and the other in the east. It is assessed that North Korea is in possession of a total 10 batteries of S-125 systems, each battery having four quadruple launchers and one fire control radar. Seven batteries are reported to be positioned to defend the capital of Pyongyang, the other three are deployed to defend the Yongbyon nuclear research centre. It is understood that 8 batteries were delivered to North Korea between 1984 and 1986, these were added to the two that were already in country.

The SAM force, based on the Russian Federation military practice of three ready missiles per launcher, disposes of over 1,200 missiles; 810 HQ-2/S-75, 384 S-125 and 72 S-200. With the strategic war storage stock added, the grand total is increased to the 2,500 to 3,000 round level.

The individual HQ-2/S-75/S-125/S-200 sites generally follow the Russian patterns except that the majority of their EW, target acquisition and GCI radars are located either in large underground concrete-reinforced NBC-protected bunker complexes or dug into mountains. The sites comprise a tunnel, control room, crew quarters and steel blast proof doors. An elevator raises the radar to the surface when required. There are also many dummy radar and SAM sites together with genuine alternative site positions for the SAM launchers.

Status

Production as required. As of August 2010, some HQ-2/S-75 derivatives in service with Iran and North Korea.

Contractor

North Korean state arsenals.

Norway

NASAMS

Type

Static and towed Surface-to-Air Missile (SAM) system

Development

In January 1989, the Royal Norwegian Air Force (RNoAF) awarded a contract for the National Advanced Surface-to-Air Missile System (NASAMS) to a joint venture company formed by Kongsberg Defence & Aerospace and Raytheon Company.

The contract provided for the development of a medium-range mobile ground-launched version of the Raytheon AIM-120 Advanced Medium-Range Air-to-Air Missile (AMRAAM) under the designation NASAMS. Two batteries were fielded at the end of 1989, initially to replace the four obsolete Nike Hercules SAM batteries that were deployed in southeast Norway to defend the Gardemoen and Rygge airbases near Oslo, and the other two in 1992, disbanded two of the Nike Hercules batteries.

Ultimately, by the end of the 1990s, all six of the Norwegian Adapted HAWK (NOAH) batteries that were then in service will also have been upgraded to the NASAMS configuration. This includes the TPQ-36A (of which 24 are currently in Norwegian service for use with the NOAH batteries) and the Fire Distribution Centre (FDC, of which Kongsberg has produced 25 for the NOAH batteries).

The development of NASAMS was completed in June 1993, with live fire testing at the US Naval Air Warfare Centre at San Nicolas Island, California. In addition to nearly 100 controlled target exercises, four live rounds were launched with a 100 per cent intercept rate. One of the tests involved the dual engagement of two widely separated targets by a single launcher. The missiles killed both targets at almost the same instant.

The production phase started in 1992 with delivery of the first two batteries to the RNoAF, for fielding in late 1994.

In 1994, a contract was signed for conversion of the NOAH batteries to NASAMS. This also included updating the FDCs and replacing the I-HAWK missile with the fire-and-forget AMRAAM. This gives the RNoAF a dramatic increase in its SAM firepower and reaction time. The replacement programme was completed in the late 1990s.

Raytheon AMRAAM during a test of the NASAMS system 0509592

Raytheon AN/TPQ-36A air defence radar deployed in operating conditions
0509593

Launcher for the NASAMS air defence system is mounted on the rear of a truck and is deployed on the ground when required 0509594

AMRAAM being loaded into the ground launcher for the NASAMS air defence system 0509595

In October 1996, it was revealed that the Norwegian Army would be running a mid-life upgrade to the system with expected delivery to the armed forces in 2006 and full IOC by 2007. The modified NASAMS system is known as the NASAMS II.

In September 1998, Norway declared its NASAMS operational after a series of live fire tests held at the White Sands test range in the US.

In September 2000, a USD34.7 million contract from Kongsberg Defence & Aerospace was awarded to Raytheon for the supply and delivery of AN/MPQ-64 3-D radars and missile in surface electronics for four NASAMS fire units to be delivered to the Spanish Air Force. Under the EUR86.6 million contracts from Spain, Kongsberg Defence & Aerospace will deliver four FDC and eight launchers based on the NASAMS used by the Norwegian Air Force. Deliveries were due to end in September 2003. These systems are now believed operational.

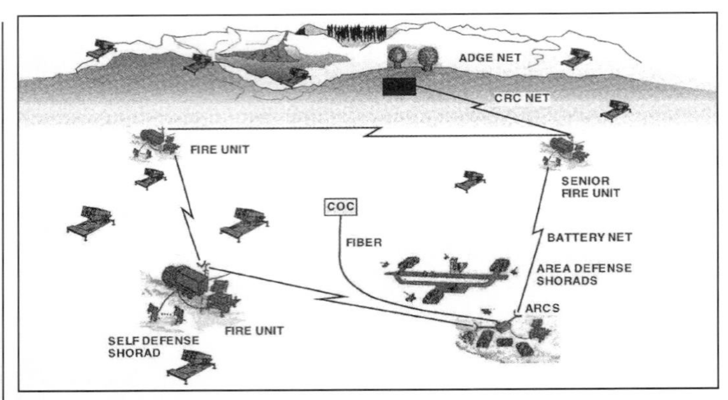

Key components of the NASAMS system 0509713

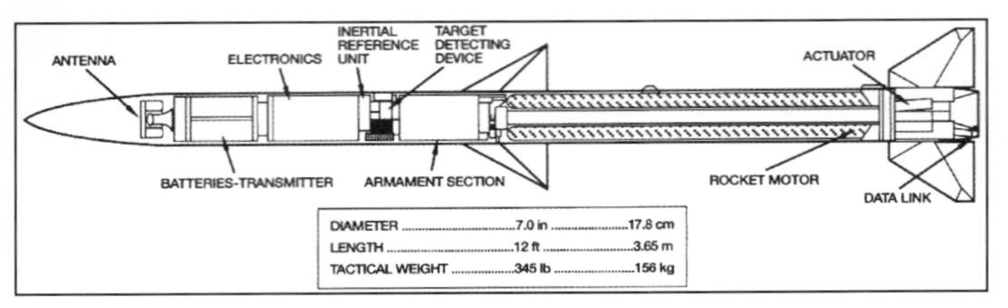

Cross-section of the Raytheon AMRAAM series missile used with the NASAMS 0509714

In September 2011, Kongsberg announced a probable future upgrade to the NASAMS system utilising the AIM-9X air-to-air missile and the Evolved Sea Sparrow Missile (ESSM). Future trials were planned and carried out in October 2011 with the AIM-9X on the back of those trials that have already taken place in mid-2011 when the IRIS-T was launched from the surface launcher.

In mid-June 2012 at the Andoya Rocket Range in Norway, Raytheon and Kongsberg carried out firing trials of what is now known as Hawk XXI using Raytheon produced ground launched Evolved Sea Sparrow Missile (ESSM). The aim of the test was to demonstrate the proven capabilities of the Kongsberg Fire Distribution Centre (FDC), the NASAMS canister launcher, the Thales/Raytheon Systems AN/MPQ-64F1 Sentinel radar and Raytheon's Hawk High Power Illuminator (HPI) as the semi-active missile guidance source.

Integration of the NASAMS launcher with Hawk XXI provided an opportunity to use other missiles apart from the AIM-120 Advanced Medium-Range Air-to-Air Missile (AMRAAM).

Hawk XXI is the latest configuration of the Hawk defence system and is designed to engage and destroy fixed-wing aircraft, helicopters, UAVs, cruise missiles and can provide Anti-Tactical Ballistic Missile (ATBM) capability against short-range ballistic missiles.

Description

A NASAMS Fire Unit (FU) consists of three truck-mobile six-round launcher platform subsystems, an FDC and 3-D AN/TPQ-36A radar. The FU composition allows for up to 60 targets to be engaged independently.

Up to four FDCs can be linked together in a command network to form the battery, so that if required, any of the four radars can provide targeting data to any of the nine to 27 available launcher platforms. One FDC is dedicated to the SHORAD role in the battery configuration.

The gimballed missile launcher is pallet-mounted on a truck and offloaded at a launch site where it is levelled by adjusting a hydraulic lifting system. The missile pallet is raised to point the missile at the fixed launch angle and slewed in azimuth to the desired launch bearing. The launcher is trainable in 360° for quick reaction against targets from any direction.

The six missiles are attached to standard LAU-129 aircraft launcher rails and protected by canisters with all-weather covers. The missiles are standard solid-state AIM-120B/C series all-weather all-aspect air-to-air rounds, which use a Northrop strap down inertial guidance unit for the initial flight phase and a Raytheon I/J-band active radar all-aspect look-down seeker for terminal homing. It also has a home-on-jam capability. Flight control is by cruciform rear fins. The HE blast/fragmentation warhead and fuze are activated in the final approach to the target and are detonated either by the proximity delay or contact fuzing circuits.

In operation the missiles will be maintained in a powered down condition until a firing command is initiated for the selected weapon at the FDC. Once this is received at the launcher, the missile start sequence is activated and boost motor ignition starts the launch 1.4 seconds later.

Initial position data are fed into the AIM-120 series inertial guidance unit before launch by an indigenously designed and built navigation system mounted on the launcher. This guides the weapon towards the predicted target location during the early part of the flight trajectory.

Upon approaching the location, the frequency-agile active radar seeker is switched on in a high-Pulse Repetition Frequency (PRF) mode to acquire and track the target. The missile then changes to its medium-PRF mode for the terminal acquisition and track attack phase.

The AN/TPQ-36A low-altitude I/J-band range gated pulse Doppler frequency-agile search and track radar uses a 2 × 1.8° 3-D pencil beam and phase beam scanning antenna to provide accurate target range, bearing and elevation track data to the manned field mobile FDC shelter. The radar antenna rotates mechanically in azimuth to provide 360° coverage with an elevation coverage between limits of –10° to +55°. Beam scheduling is computer-controlled and adaptive to mission requirements. It can track up to 60+ targets at ranges up to 75 km and is fitted with an integral Mk XII IFF subsystem. Acquisition range on a low-radar cross-section fighter-sized target is said to be 40 km. The system has a two-second update rate.

The FDC contains:

- An ADC 2000 Command-and-Control workstation with Commercial Off-The-Shelf (COTS) computers including Battle Management Command, Control, Communications and Information (BMC3I) software (in Ada)
- A user-friendly Man/Machine Interface (MMI) providing all the necessary functions for airspace control
- Track management (including Track Correlation and Jam Strobe Triangulation)
- Target identification
- Threat evaluation and weapon assignment.

The FDC also contains the Battery and Launcher communications equipment and interfaces to the radar control computer.

The NASAMS battery has an autonomous launch-and-leave capability allowing engagements that could have more than 50 independently targeted AMRAAMs in the air at any one time from a fire unit.

Also, as NASAMS utilises the Raytheon MPQ-64F2 a network enabled capability distributed architecture that does not require launchers/missiles to be co-located with the rest of the system then, as the launchers can be placed up to 25 km away, this has the effect of providing a large area coverage.

These factors, plus the weapon's very short reaction time, considerably increase the enemy's perceived problems in trying to overwhelm the RNoAF air defence network. The RNoAF carried out a mid-life update on the NASAMS called NASAMS II, with the upgraded version handed over in mid-2006. The system also has an up and over trajectory capability to engage very low-flying targets behind rough terrain features.

Variants

- NASAMS II;
- NASMAS upgrade to a mobile system.

Specifications

	AIM-120
Dimensions and weights	
Length	
overall	3.655 m (12 ft 0 in)
Diameter	
body	178 mm (7.01 in)
Flight control surfaces	
span	0.64 m (2 ft 1¼ in) (wing)
Weight	
launch	156 kg (343 lb)
Performance	
Speed	
max Mach number	3.0 (est.)
Range	
max	10.8 n miles (20 km; 12.4 miles) (est.)
Ordnance components	
Warhead	20.4 kg (44 lb) HE fragmentation
Fuze	proximity, impact
Guidance	INS, mid-course update
	active, radar
Propulsion	
type	solid propellant (sustain stage)

Status

In service with the Royal Norwegian Air Force (eight batteries), Spain and Finland. Offered for export. The Spanish Air Force took delivery of four NASAMS fire units and eight launchers in 2003. In late 2008 at the Medano del Oro firing range in Mazagon, Spain, these systems were tested and proved to be in full operational condition.

NASAMS is now complemented with the NASAMS II system, which is replacing the basic system as more units become operational.

An announcement was made in December 2011 that Norway is to spend USD67 million to upgrade the NASAMS system to a self-propelled launcher vehicle this is to include upgrades to the fire control system and radar.

In mid-June 2013, Raytheon confirmed the system is now known as the National Advanced Surface-to-Air System and not as previously Norwegian Advanced Surface-to-Air System.

An announcement was made in January 2014 that NASAMS has been sold to the Sultanate of Oman for a value of USD1.28 billion. This sale will include ground support equipment, a full training package and technical assistance.

Contractor

Kongsberg Defence & Aerospace.
Raytheon Missile Systems.

NASAMS II

Type

Static and towed Surface-to-Air Missile (SAM) system

Development

On 27 January 2003, Thales Raytheon Systems, an equally owned transatlantic joint venture between Raytheon Company and Thales Group, announced a fixed price contract to upgrade the National Advanced Surface-to-Air Missile System II (NASAMS II) TPQ-36A surveillance radar to a similar standard to that of the AN/MPQ-64. All Thales Raytheon Systems elements of the NASAMS II programme were to be accomplished over a 42-month period.

A further development of the NASAMS II system, this time for the Netherlands Army will use the TRML-3D mobile surveillance radars.

During October 2006, the NASAMS II was successfully tested as part of the NATO Electronic Warfare Force Integration Programme. This included countermeasures to intruder aircraft that attempted to disturb the radar system using advanced jamming techniques. This was the second test of the system's electronic countermeasure capability. In August 2006, three NASAMS batteries from the Spanish Land Force (Ejercito) also test fired the NASAMS systems on the Swedish Vidsel Test Range. Targets used during these trials included cruise missiles, UAVs and drone shots using telemetry rounds.

In mid-2011, at the Andoya Rocket Range in Northern Norway, the NASAMS underwent a live firing exercise to prove the systems capabilities. The exercise known as Silver Arrow is an annual event intended to test the weapons operational status and in addition, as a proof to war fighters that NASAMS can detect, identify and destroy targets.

In September 2011, Kongsberg announced a probable future upgrade to the NASAMS system utilising the AIM-9X air-to-air missile and the Evolved Sea Sparrow Missile (ESSM). Trials have taken place with the AIM-9X on the back of those trials that have already taken place in mid-2011 when the IRIS-T was launched from the surface launcher.

Early January 2013, Kongsberg signed a contract worth NOK300 million to deliver an upgrade to the Royal Norwegian Air Force NASAMS II air defence system. The contract comprises upgrading of existing missile launchers with new electronics and software for increased performance and lifetime extension.

Description

There will be no new systems procured for NASAMS II. The programme will consist of hardware upgrades to the missile, launcher, and Fire Direction Control (FDC). The system will be more mobile and as the battery is part of an air defence network, target information can be obtained from external sources. If needed, the equipment can be airlifted into position. Most of the battery positions are also likely to be prepared in advance. NASAMS II will have an Over-The-Horizon (OTH) targeting capability.

A dedicated mobile Ground-Based Air Defence Operation Centre (GBADOC) is part of the NASAMS II programme. This is designed to link (using Link 16) the battery and any associated Royal Norwegian Air Force (RNoAF) NASAMS units into the upper echelon air defence network so that all the participating NASAMS units share the same air picture. The GBADOC has the same hardware configuration as a NASAMS FDC but uses a different set of software. If the GBADOC is hit during combat then any other NASAMS FDC can take over simply by loading the GBADOC software.

As of January 2007, Kongsberg has sold the GBADOC to Sweden for use by the EU's new Swedish-led Nordic battle group. The GBADOC will be modified and upgraded to include Swedish national data links for the integration of the Saab Ericsson Giraffe AMB radar and the RBS 70 Very Short Range Air-Defence (VSHORAD) system as well as higher echelon units.

The GBADOC was delivered to Sweden during 2007 and is currently fully operational.

In mid-June 2012 at the Andoya Rocket Range in Norway, Raytheon and Kongsberg carried out firing trials of what is now known as Hawk XXI using Raytheon produced ground launched Evolved Sea Sparrow Missile (ESSM). The aim of the test was to demonstrate the proven capabilities of the Kongsberg Fire Distribution Centre (FDC), the NASAMS canister launcher, the Thales/Raytheon Systems AN/MPQ-64F1 Sentinel radar and Raytheon's Hawk High Power Illuminator (HPI) as the semi-active missile guidance source.

Integration of the NASAMS launcher with Hawk XXI provided an opportunity to use other missiles apart from the AIM-120 Advanced Medium-Range Air-to-Air Missile (AMRAAM).

Hawk XXI is the latest configuration of the Hawk defence system and is designed to engage and destroy fixed-wing aircraft, helicopters, UAVs, cruise missiles and can provide Anti-Tactical Ballistic Missile (ATBM) capability against short-range ballistic missiles.

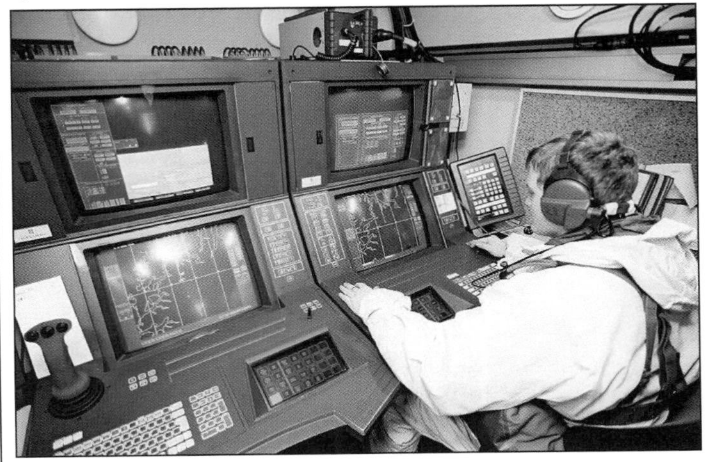

Fire Distribution Centre (FDC) 0578781

Variants
- NASAMS
- NASAMS upgrade to mobile system.

Status
Contracted in 2003, deliveries started 2006 with full IOC in 2007.

The US Air Force Rapid Capabilities Office acquired, modified, tested and deployed the NASAMS to the National Capital Region at the direction of the Chairman of the Joint Chiefs of Staff in time to support the Presidential Inauguration day.

The system is reported to be operational in Finland, Norway, Spain, the Netherlands and the US.

In December 2006, the Netherlands Army placed a firm contract for their Future Ground-Based Air Defence (FGBAD) with EADS for the following equipment:
- TRML-3D mobile surveillance radars - supplied by EADS;
- Mobile command-and-control operation centres - provided by EADS and the then Oerlikon;
- Digital radio communication network - provided by the then Oerlikon;
- Six NASAMS II - supplied by Kongsberg.

It was announced in mid-December 2011 that Norway is to spend USD67 million to upgrade its NASAMS to include a self-propelled launcher vehicle with upgrades also to the fire-control system and radars.

In mid-January 2013, the first of 12 Ground Master 400 (GM 400) long-range air defence radar systems was handed over to the Finnish Air Force to provide extended surveillance of the nation's air space.

In mid-June 2013, it was confirmed by Raytheon that the NASAMS now stands for National Advanced Surface-to-Air System rather than previously known as Norwegian Advanced Surface-to-Air System.

Contractor
Kongsberg Defence Systems.
Raytheon Company.
ThalesRaytheonSystems.

Poland

S-200C Wega-C Upgrade Programme

Type
Static and towed Surface-to-Air Missile (SAM) system

Development
As part of the continuing improvements to their air defence systems, the Polish Armed Forces are incorporating the technology developed for the Newa-SC (S-125 Neva-SC, NATO SA-3 'Goa') self-propelled modernisation into the Polish Air and Air Defence Force's (WLOP) two battalions of the Russian Federation-supplied S-200s.

These were fully merged within NATO's Integrated Air Defence System in 2005.

The designator for the upgraded system is S-200C Wega-C. The first battalion of the upgraded S-200C Wega-C version of the Mrzezyno-based 78th Independent Air Defence Missile Regiment became fully operational in 2003. The second battalion was upgraded in the 2003 to 2004 period. The upgrade programme was launched in the Autumn of 1998.

Description
The S-200C was originally designed in the 1950s at the Petr Grushin Design Bureau what is now part of the Almaz/Antei Concern of Air Defence and entered into service with the Russian Armed Forces in 1967, the present upgrade was developed by a Warsaw Military Academy of Technology team and is being manufactured by the WZU-2 depot in Grudziadz. The upgrade has basically

A map display within the modernised K2 command cabin (Miroslav Gyürösi)
0532202

The updated consoles in the modernised K2 command cabin
(Miroslav Gyürösi) 0532201

turned the previous analogue system into a more modern digital one, it also allows for a reduction of crewing levels, with the direct tactical team reduced from twelve to only three people. The K-9 commander's cabin was completely eliminated and now the commander uses one of three work stations in the upgraded K-2C cabin.

Other objectives of the upgrade include:
- To increase the efficiency of the command, fire-control and combat-group teamwork;
- To provide a greater level of automation as evident in the reduction from 12 to 3 operators;
- To improve the technical performance and reliability of the system.

Modernisation of the K1V antenna post has included removal of the interrogator associated with the Soviet-era Kremniy-2 Identification Friend of Foe (IFF) system and updating of the modulator by adding modern solid-state electronics.

The redesign of the K2V cabin is extensive, eliminating approximately 85 per cent of the existing consoles, which used analogue electronics based on vacuum tubes and ferrite memory. A new digital command console and three upgraded original analogue consoles handle all the functions formerly carried out by the unmodified K2 cabin and the K9M. New technology is used in the target-tracking channel in order to make the processes of target search and the modernised system is able to engage targets flying at speeds ranging from 500-1,500 m/sec. A new method for measuring target range allows the modernised system to measure target ranges out to 450 km, making this measurement continuously without degrading the other functions of the target-illumination radar.

The S-200C currently works within the automated air-defence network and receives air picture data from external sources, including command and reporting centres. Other sources include the NATO E-3 AWACS (to be improved after implementation of Link 11B/Link 16 terminal) and networked 3-D air-defence radars.

The S-200C has no Anti-Tactical Ballistic Missile (ATBM), or Anti-Ballistic Missile (ABM) capabilities.

Status
Modernisation programme is being implemented. In service with the Polish Air and Air Defence Force (two squadrons).

Contractor
WZU-2 Wojskowe Zaklady Uzbrojenia (Number 2 Military Armament Works).

Romania

S-75M Volkhov maintenance and repair programme

Type
Overhaul and maintenance services of anti-aircraft systems

Development
As one of its major activities, Arsenalul Armatei has developed its own maintenance and repair facilities at its missile development and maintenance Electromechanical Factory, located at Ploesti, for the Russian-built S-75M Volkhov-M (SA-2 'Guideline') static surface-to-air missile system.

Status
Refurbishment as required. Some refurbished systems are in service with the Romanian Army. Offered for export. The facility is believed to also undertake refurbishment of the RD-75 upgraded S-75M3 Volkhov-M3 system.

Contractor
Arsenalul Armatei Regie Autonoma.

S-75M3 Volkhov-M3

Type
Static Surface-to-Air Missile (SAM) system

Development
In the early to mid-1970s, the then Soviet Union developed an upgrade for its S-75 Volkhov-M missile system based on the operational experience gained in a radar jamming environment with the S-75 in both the Middle East and Vietnam. Part of the upgrade was the addition of the RD-75 radar range-finder assembly. Subsequently this upgrade was adopted by, apart from the Soviet Union itself, only the Romanian Air Force within the then Warsaw Pact under the designation S-75M3 Volkhov-M3.

Description
The RD-75 radar range finder consists of a single 4 × 4 trailer-mounted cabin with a relatively small all-round coverage antenna reflector. The antenna is slaved to the main RSN-75 ('Fan Song-E') missile guidance radar and follows the same assigned target.

In normal operating mode, the RSN operates in its active mode measuring the three main target parameters - two angles that indicate its horizontal and vertical positions and its range. If intense jamming is experienced, the two units, the RD-75 and RSN are coordinated. The RSN operates in its passive mode, tracking the target via its onboard jammer and measuring its horizontal and vertical position from these emissions. The RD-75 takes over the third parameter acquisition requirement - that of range - by operating in its active mode.

The dual frequency RD-75 has two widely separated centimetric operating wavelengths, each with a different polarisation. Further incorporated Electronic Counter-Counter Measures (ECCM) systems allow it to function successfully in the jamming environment.

Variants
The S-75M3 as a designation also exists in a modification/upgrade from the Belarussian Company Belvneshpromservice.

RD-75 trailer-mounted radar range-finder unit 0033943

The RD-75 antenna has two separate waveguide feeds 0033944

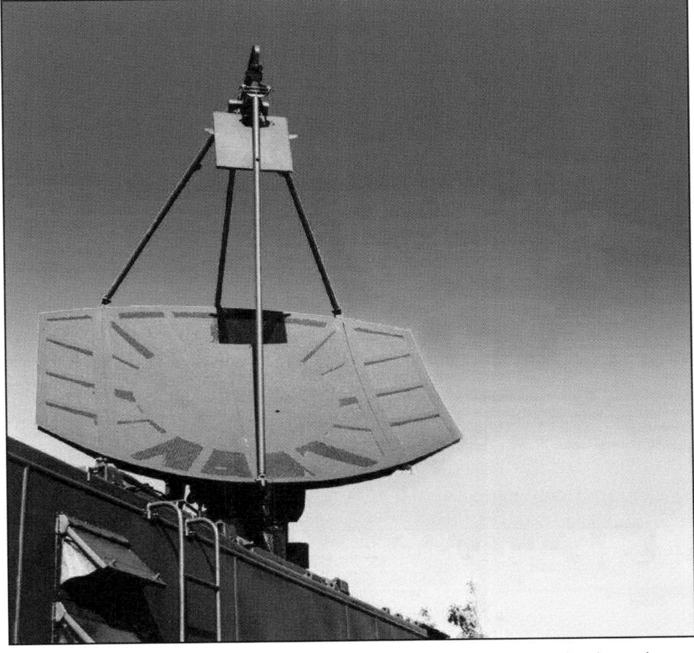
Close-up view of RD-75 antenna showing twin feeds and polarisation mirror 0033945

Status
Production complete. Believed to be in service with the Romanian Air Force, but being replaced by the upgraded MIM-23 HAWK.

The 5Ya23 missile used in the S-75M3 has been renovated and is now used as a ballistic missile target.

Contractor
Maintained and overhauled in Romania by Arsenalul Armatei Regie Autonoma.

S-125 Neva maintenance and repair programme

Type
Overhaul and maintenance services of anti-aircraft systems

Development
As one of its major activities, Arsenalul Armatei has developed its own maintenance and repair facilities at its missile development and maintenance Electromechanical Factory, located at Ploesti, for the Russian-built S-125 Neva (SA-3 'Goa') static surface-to-air missile system.

Status
Refurbishment as required. Refurbished systems are currently in service with the Romanian Army. Offered for export.

Contractor
Arsenalul Armatei Regie Autonoma.

Russian Federation

Dzhigit 203-OPU

Type
Tripod mounted man-portable Surface-to-Air Missile (SAM) system

Development
The Dzhigit (Russian for 'a skilful horseman in the Caucasus', also transliterated as Djhigit) has been designed for use with the Igla-S, Igla and Igla-1 family of man-portable SAM systems.

The one-man pedestal single and/or salvo launcher ensures either simultaneous or sequential salvo firing of two missiles against the same target to improve the kill probability by up to 1.5 times. The launcher can be installed on a variety of mount types.

Description
The gunner has all the controls required to sight and fire the weapons within easy reach of his seat. These include two functionally interconnected handle units. The controls are used to:

- Initiate supply of power and cooling agent to the individual missiles
- Select the firing mode (ripple or single) against approaching or receding targets
- Engage targets in the missile(s) pursuit flight trajectory envelope(s).

Launcher showing 1L228D IFF (IHS/Patrick Allen) 1138406

Static defence Dzhigit launcher with two Igla (SA-18 'Grouse') man-portable missiles showing sighting system between missiles. The launcher can also be used with Igla-1 missiles 0509388

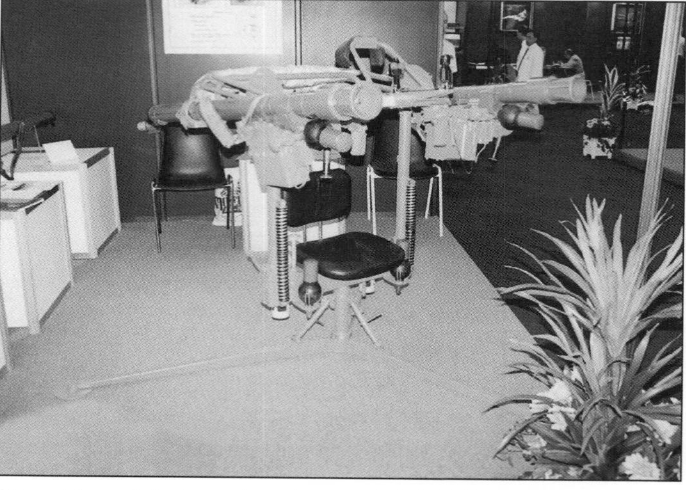

Static defence Dzhigit launcher with two Igla (SA-18 'Grouse') man-portable missiles. Note the two replacement ground power supply units below gunner's seat 0509739

To aim and track a target, the gunner has a sight unit that produces both audio and visual signals on achieving target lock. A removable headrest is provided to fix the gunner's eye relative to the sight line. A removable overhead shade is also provided to protect the gunner from excessive sun in hot climates. Maximum temperature range limits are from −44 to +50°C.

Additionally, the Dzhigit launcher can be fitted with a night sight and an IFF interrogator 1L228D using the same frequency band and operating modes as the Mk 10/Mk 12 system, as well as the target designation assets. A test vehicle and test-and-monitoring equipment are used to provide functional testing of the launcher and its components and localise malfunctions, as well as to provide maintenance of the launcher.

Variants
The Dzhigit system can be mounted as:

- A mobile vehicle platform. When fitted on a light (4 × 4) vehicle a support platform is provided that houses six reload rounds in storage compartments
- A ship-based SHORAD system
- Static ground mount.

The more modern 9K338 Igla-S (NATO SA- 24 Grinch) also developed by the KBM in Kolomna will use this style of launcher. This new missile is for use in multiple launchers such as the Dzhigit and Strelets. Modifications to the missile include:

- Maximum range increased from 5.2 to 6.0 km
- Maximum and minimum altitude 10 - 3500 m
- A laser proximity fuze
- Improved control actuators
- An improved warhead design that is the same weight as its predecessor, but uses an adjustable focused fragmentation pattern.

Specifications
Dzhigit 203-OPU

Crew:	1
Missiles:	2 Igla-S, Igla or Igla-1 family of missiles
Dimensions:	
vehicle mounted:	
(length)	2.165 m
(width)	1.585 m
(height (firing position with shade))	1.925 m
(height (travelling position))	0.5 m
ground mounted:	
(diameter across supports)	2.315 m
(width (of elevatable section))	1.2 m
(height (with shade))	1.52 m
(folded)	0.88 m
Weight:	
(without missiles on vehicle)	175 kg
(without missiles on ground)	80 kg
(with missiles on ground)	105 kg
Elevation limits:	−10° to +70°
Traverse:	360°
Reload time:	
(on vehicle)	3 min
(on ground)	2.5 min
Reloads:	6 (on vehicle)

Status
Limited production. In service with Russian Army. Offered for export and known to have been sold to Jordan, Syria, UAE and Singapore, 12 in total.

Contractor
Development: Konstruktorskoe Bjuro Mashinostroenia (KBM).
Production: Russian state factory.

S-75 family (SA-2 'Guideline')

Type
Static Surface-to-Air Missile (SAM) system

Development
Development of the S-75 Dvina (Russian river name; US/NATO designations SA-2 'Guideline') system began in the mid-fifties under the direction of the 2nd Main Directorate of the USSR Council of Ministers and the Ministry of Aviation's KB-2 (A Raspletin, P Grushin and B Korobov).

The system was designed as a mobile, medium to high-altitude SAM, for use against non-manoeuvring targets such as medium to high-altitude bombers.

It was also designed to be more suitable than the S-25, R-113 (NATO SA-1 'Guild') SAM for nationwide deployment. The designation for the system changed as it evolved through its operational development period.

In 1955, the first operational PVO-Strany (*Voyska Protivovozduchnoy Oborony Strany* - Troops of the National Air Defence) S-75 system with the V-750 missile regiments of three 1 × 6 rail launcher battalions were formed with one of the initial deployment near the strategically important city of Sverdlovsk.

On 1 May 1960, the Sverdlovsk units fired a total of 14 V-750 missiles against a Lockheed U-2 high-altitude reconnaissance aircraft flown by Gary Powers of the CIA. The subsequent detonation of the missiles at high altitude not only forced the U-2 to crash, thereby precipitating an international crisis, but also destroyed a PVO-Strany MiG-19 interceptor. The result of the incident was that the US ceased all further U-2 over flights of Russian territory, losing a valuable strategic intelligence source.

The next incident involving the S-75 was when Chinese People's Liberation Army Air Defence Missile units shot down a Taiwanese-based Lockheed U-2 over Nanching in September 1962. The S-75 and its locally built copy, the HQ-2, were subsequently used on many occasions in the 1960s against further Taiwanese-based U-2s and US Ryan pilotless reconnaissance drones, scoring at least another eight U-2s by 1970.

The initial Chinese incident was rapidly followed by the 1962 Cuban missile crisis during which a US Air Force U-2 was lost on 27 October to a V-750 missile while flying over the Cuban naval base at Banes.

In mid-1965, the S-75 system was introduced into the North Vietnamese Air Defence Network, claiming its first victim, a US Air Force McDonnell Douglas F-4C Phantom, on 24 July of that year. The system was subsequently used throughout the various bombing campaigns against the North Vietnamese military infrastructure.

Surprisingly, despite the initial uses of the S-75, it was not until the Vietnam War that the US gained the necessary raw intelligence data on the system. The intelligence concerning the weapon's proximity fuzing, its terminal phase guidance signals and the warhead's overpressure characteristics at detonation enabled the US to design a suitable ECM package to counter the weapon.

This was obtained on 13 February 1966 during a flight over North Vietnam by a specially radar-enhanced high-altitude Ryan 147E ELINT pilotless drone which relayed the information back to a monitoring station until a V-750 destroyed it.

This was followed on 22 July 1966 by another special flight involving a Ryan 147F drone protected by onboard ECM equipment. A total of 11 V-750s were fired at the drone before one managed to defeat the ECM coverage and destroy it.

In the same year that the S-75 entered the Vietnam War, the Indian Air Force used it operationally during the 1965 Indo-Pakistan War. Obtained in 1963, the first examples of an eventual 25 battery force were deployed around New Delhi and several of the key airfields in that area. The only confirmed kill was near Delhi on 6 September when an Indian Air Force Antonov An-12 Cub transport was shot down, mistaken for a Pakistan Air Force Lockheed C-130 Hercules. However, during the latter stages of the war, the Pakistan Air Force's sole high-altitude Martin RB-57F reconnaissance aircraft was bracketed by two V-750s at about 15,850 m altitude causing sufficient damage for it to crash land on return to its base.

In December 1965, a USAF RB-57F was destroyed by a V-750 on a flight over the Black Sea, near the Russian coastline.

Egypt began receiving the S-75 at the same time as India and had 18 batteries in service by the time the June 1967 war with Israel started. According to US Corona satellite reconnaissance assets, Egypt had 35 known SAM sites and six known SAM support facilities available for use. These fired only 22 missiles, destroying two Mirage IIICJ fighters (on 7th and 8th June respectively) and one complete battalion, including radars, was captured by the Israel Defence Force with another eight battalions destroyed by the Israeli Air Force.

SA-2 'Guideline' SAM launcher without missile in former Yugoslavia (Richard Stickland) 0044360

During the following 1968 to 1970 War of Attrition, hundreds of V-750s were supplied to Egypt and the weapon scored its first kill in this war on 9 March 1969 when an Israeli Piper Cub observation aircraft was destroyed. Between then and the 1973 war the number of kills it had made increased to about 10.

On 19 March 1993, Georgian Air Defence units around Sukhumi shot down a Russian Air Force Sukhoi Su-27 'Flanker' fighter by using V-750 missiles. This is one of several Russian Air Force aircraft and helicopters shot down during several localised conflicts in the Caucasus during the early 1990s.

A more recent combat use was in November 1994 against NATO aircraft operating over western Bosnia.

The Bosnian and Krajinian Serbs have also used the S-75 in the ground-to-ground role against Bosnian Muslim and Croatian targets. At least 18 'Guidelines' were fired during November and December 1994 against ground targets using modified fuze systems for ground contact or very low-level airburst. It was also the subject of NATO air defence suppression raids during the 1995 bombing of Bosnian Serb targets.

Whenever the performance of the weapon is analysed, a high number of launches per kill is found. However, the missile has proved its worth by reducing the accuracy and effectiveness of the enemy's air power. This is done by diverting valuable effort to SAM suppression missions, restricting the use of reconnaissance assets and, most importantly, forcing enemy aircraft to adopt tactics or fly lower where other air defence systems such as guns, interceptors or different missile types can prey on their increased vulnerability.

By mid-1993, the number of S-75 launchers in Russian air defence regiments had declined from a peak of over 4,600 in the late 1960s through approximately 2,400 in mid-1988 to approximately 150 in 1996. The system was effectively withdrawn from service by the year 2000.

During its long life, the weapon has been subjected to numerous modifications, both internally and to its guidance systems. Most of these were prompted by operational experience and the need to rectify problems found in combat.

The original V-750 missile used in the first S-75 Dvina SAM system was assigned the US/NATO codename SA-2a/'Guideline' Mod 0. The system entered operational service in November 1957. The associated E-band missile guidance radar was assigned the NATO codename 'Fan Song-A'.

A further missile variant used with this system was the V-750V, which was given the same SA-2a 'Guideline' Mod 0 codename. The system was capable of engaging targets flying at up to 1,100 km/h at altitudes between 3,000 and 22,000 m.

The S-75 was quickly superseded by the SA-75 Desna system which entered operational service in 1959. This was assigned the missile codename SA-2b/'Guideline' Mod 1 and used the V-750VK and V-750VN missile variants. The associated radar was assigned the codename 'Fan Song-B'. The main

S-75 on launcher (Miroslav Gyürösi) 0560697

ZIL-131V (6 × 6) tractor truck towing semi-trailer carrying SA-2 SAM 0007969

SNR-75 (NATO Fan Song) Radar (Miroslav Gyürösi) 0560698

improvements were in the missiles themselves and on the radar with the deletion of the original upper parabolic antenna fitted to the vertical orthogonal antenna. The system could engage targets at altitudes between 500 and 30,000 m and ranges up to 34,000 m. Target speed could be up to 1,500 m/s.

The next system entered service in 1961 and was designated the S-75M Volkhov. The associated missile was the V-750M and was assigned the designation SA-2c/'Guideline' Mod 2. The radar was the G-band 'Fan Song-C' set. The V-750M was identical to the V-750VK/V-750VN weapons but with improved performance. The target engagement altitudes were now 400 to 30,000 m at ranges up to 43,000 m. The maximum target speed was increased to 680-m/s.

During the early to mid-1960s, two further S-75M variants were fielded. The S-75M Volkhov system with the V-750SM missile assigned the codename SA-2d/'Guideline' Mod 3 and the S-75M Volkhov system with the V-750AK missile assigned the codename SA-2e/'Guideline' Mod 4. The G-band radar associated with both variants has the codename 'Fan Song-E'. A G-band 'Fan Song-D' was also developed but never actually entered operational service.

The V-750SM differed significantly from the original V-750 in having four enlarged dielectric uplink guidance receiver strip antennas under prominent covers on the forward side of the missile instead of the usual two sets of four. It also has a longer barometric nose probe and several other differences associated with the sustainer motor casing. The V-750AK is essentially the same as the V-750SM externally but has upgraded internal components and a larger, more bulbous warhead section for either a conventional HE fragmentation or command-detonated nuclear warhead. Unlike the V-750SM the V-750AK was not exported and has remained a solely Russian-operated weapon.

The 'Fan Song-E' radar has two parabolic antennas added above the horizontal orthogonal antennas to provide Lobe-On-Receiver-Only (LORO) Electronic Counter-Counter Measures (ECCM).

TsKB Almaz (the S-75 design bureau) undertook a crash programme to improve the ECCM capabilities and engagement envelope of the system as a result of the combat experience gained by the US forces on S-75 in Vietnam. The capture of a battalion with at least 12 S-75 system missiles, and associated RSN-75 radar equipment by Israel in the 1967 Six Day War, further added to this importance.

Using technology drawn from the more advanced S-125 Neva, the bureau produced the S-75M Volkhov version with the V-755 missile (US/NATO codename SA-2f/'Guideline' Mod 5 missile) and RSN-75 ('Fan Song-F') radar. Prototype trials were undertaken in 1968, with the first production battalions being operationally deployed in late 1968. A number of these systems were rushed to Egypt, for use along the Suez Canal, in the latter part of the War of Attrition. Further deliveries were made to Vietnam between 1970 and 1971, for incorporation into its air defence network. The missile could engage targets with velocities up to 1,200 m/s head-on or up to 300 m/s in tail-chase engagements.

The major changes were in the RSN-75 radar. This reverted back to the original E-band but with a higher output, scintillation suppression and manual plus mixed mode tracking. The model is readily distinguished from the earlier models, having a small distinctive box-like housing centrally located over the horizontal orthogonal antenna that replaces the RSN-75's pair of LORO parabolic dish antennas.

The 'cab' contains the necessary electro-optical tracking and guidance equipment for a two-man team to track a target in a severe ECM environment where the normal automatic electronic tracking mode has been jammed. The optical systems allow target acquisition and missile guidance using the C-band UHF command link at altitudes down to approximately 100 m.

The other significant change in the SA-2f is in the missile guidance package, which now has a capability to home on targets using strobe jamming. The S-75 guidance system was the responsibility of the Aleksandr Raspletin design bureau. Target speed was now up to 3,700 m/s.

In the early to mid-1970s, following further operational experience in the Middle East and Vietnam, the Soviets introduced the S-75M3 Volkhov-M3 variant into service using the RD-75 radar range-finder. Only Romania and the Soviet Union used it.

Front view of SA-2 SAM being carried on a semi-trailer 0509790

In the mid- to late 1990s the S-75 Volga-M and S-75 Volga-2 upgrade packages were developed.

The S-75 family was used by Iraq during the 1991 Gulf War and in the subsequent clashes with US/UK forces. It was also used in the NATO Kosovo/Yugoslavia air campaign.

The S-300P 'Grumble' system has effectively replaced the SA-2 'Guideline' in Russian service.

Description

The Petr Grushin design bureau developed the V-750 missile which is a two-stage weapon with a large solid propellant booster stage fitted with four very large delta fins. The missile itself has a storable, liquid fuel, sustainer rocket motor, which uses an inhibited red fuming nitric acid/kerosene fuel mix. Towards the mid-section is a set of four cropped delta-shaped wings with a second in-line set of small fixed fins at the nose and a third in-line set of slightly larger powered control fins at the tail.

Warhead

The warhead of the SA-2a/b/c/e/f models is fitted forward of the main fins and behind the nose-mounted guidance assembly. Maximum fragmentation radius against a high-altitude target such as a U-2 is around 244 m due to the rarefied atmosphere. At medium to low levels against fighter-sized targets the kill radius is about 65 m and the radius for severe damage is 100 to 120 m. The weapon has a CEP figure of 75 m with the large effective radius compensating for any system inaccuracies. The warhead of the SA-2a/b/c/d/f weighs 195 kg (90 kg of which is HE) and is a HE internally grooved controlled-fragmentation type with proximity, contact and command-type fuzing available. The 295-kg nuclear warhead for the SA-2e variant is believed to have a yield of 15 kT. The conventional warhead for the e-model weighs the same.

Launcher

The whole S-75 system, including the launcher, is designed to be simple and easy to operate with the minimum of specialised training. In practically all user countries the pattern of a battalion site is as follows: six semi-fixed trainable single rail launchers are deployed in a hexagonal arrangement, about 60 to 100 m apart. They can either be dug into pits, left at ground level or hardened by being dug in and surrounded by concrete revetments.

Radar

In the centre of the launchers is the battery command post with the fire-control team and its computer, the RSN-75 ('Fan Song') missile control radar, the P-12 (NATO designation 'Spoon Rest-A' truck-mounted or 'Spoon Rest-B') early warning radar and usually six reload rounds on their articulated trailers. Again, the fire-control team, its equipment and the radars can either be van-mounted above ground, simply dug in or located underground in hardened concrete bunkers. Camouflage is used as required.

The battalion's early warning and target acquisition P-12/'Spoon Rest' A-band radar has a range of 275 km using a large Yagi antenna array.

At regimental HQ there is a fourth P-12/'Spoon Rest', a van-mounted P-15 (NATO codename 'Flat Face') 250 km range C-band search and tracking radar with two elliptical parabolic reflectors and a PRV-11 (NATO codename 'Side Net') 180 km range E-band nodding height-finder radar mounted on a box-bodied trailer. There is also a radar control truck and a 'Mercury Grass' truck-mounted command communications system for linking the HQ to the three battalions.

Once the P-12 and P-15 radars detect a target entering the regiment's assigned defence zone, it is interrogated by the HQ's IFF system (NATO codename 'Score Board'). If designated hostile, the regimental HQ identifies which of its three battalions is the most suitable one for the engagement and transmits the contact details in the form of range, altitude and bearing either by radio or landline from the 'Mercury Grass' command station to the battalion's radar elements.

The 'A' to 'E' models of the RSN-75/Fan Song is normally operated by a four-man crew (the 'F' variant has six crew): a range tracking officer and three enlisted men that serve as angle track operators.

The RSN-75/'Fan Songs' operate in two basic modes (except the 'E' and 'F' systems that have a third intermediate mode for low-altitude search and tracking): target acquisition and automatic tracking.

In the target acquisition mode the radar searches for the target to which it has been alerted by the P-12 battalion set. The track-while-scan capability allows it to transfer the data on one target's bearing, altitude and velocity to the fire-control computer whilst continuing the scanning to acquire other targets for follow-up tracking. The RSN-75/'Fan Song-E' is able to track up to six targets simultaneously. Once there is sufficient data for the engagement the radar is switched over to its automatic tracking mode and then to its missile guidance mode.

Maximum radar range of the E-band RSN-75/'Fan Song-A/B/F' models varies between 60 and 120 km depending upon target type, altitude and operating conditions. The G-band RSN-75/'Fan Song-D/E' maximum range is extended to between 75 and 145 km under the same parameters.

The main element of an RSN-75/'Fan Song' is a pair of orthogonal trough antennas, one horizontal and one vertical, which emit two 'flapping' fan-shaped radar beams in their respective planes. The separate azimuth and elevation beams of the early RSN-75/'Fan Song' radars sweep through the target aircraft and a pair of enlisted men keep the target in the centre of the scan pattern by using manually operated controls.

As already stated the RSN-75/'Fan Song-E' set has an additional LORO ECCM facility known as the GSh mod built in, while the RSN-75/'Fan Song-F' has had this replaced by an electro-optical guidance mode option for use in heavy ECM environments.

In some countries which only deploy early versions of the S-75, the elderly ground-mounted P-8 Dolphin (NATO codename 'Knife Rest-A') or truck-mounted P-10 (NATO codename 'Knife Rest-B/C') radars may be used in lieu of P-12/'Spoon Rest'. They are A-band sets and have an operating range in the order of 150 to 200 km.

As soon as the computer has a firing solution on a target, a launcher is brought to bear, elevated to between +20 and +80° and blast deflectors erected. Up to three missiles can be fired and controlled against a single target. Launch interval is six seconds.

Once a missile is fired, its solid fuel booster is ignited and this burns for three to five seconds to take the weapon away from the launch site; 0.5 seconds later the burnout booster unit is jettisoned. After the launch the fire-control computer continues to receive target data from the RSN-75/'Fan Song' which is now tracking the missile as well. The computer continually generates commands to guide the missile to the target and these are transmitted over a C-band UHF radio beam uplink to four (on the S-75/SA-2d/e) or eight (on the S-75/SA-2b/c) strip antennas mounted forward and aft of the missile's centre body wings. The onboard guidance unit accepts these and adjusts the missile's trajectory using the movable control fins aft. A V-750 must pick up its narrow UHF line of sight guidance beam within six seconds of launch otherwise it goes ballistic and does not guide. The 3,500 kg total thrust liquid fuel sustainer burns for a total of 22 seconds with the V-750 attaining its maximum velocity only when it reaches around 7,630 m altitude. The latter version of the missile has a sustainer burn time of 47 to 63 seconds.

This means that the missile has limited capability and manoeuvrability when engaging tactical aircraft. Once guided to the vicinity of its target, the weapon's fuzing system is command activated by the fire-control computer and this detonates the warhead either by proximity to the target or by receipt of a command signal. A self-destruct unit detonates the warhead after 60 seconds of unguided flight time following launch or after 115 seconds, if closure with the target during guided flight has not been made. Reloading a launcher takes about 12 minutes using the articulated reload trailer and its (6 × 6) tractor.

A North Vietnamese tactic developed for the S-75 system was the use of a 'light battalion' organisation. This used only one or two launchers, fire-control radar and no acquisition and tracking radar or supporting sub-units. The time taken to come out of action and be ready to move was thus reduced to approximately 15 minutes, some 30 minutes less than the Russian norm of 45 minutes for a full battalion outfit. These 'light battalions' were used extensively by the Vietnamese operating ambush tactics whereby several such units would occupy previously prepared positions and would wait, studying the air position in the region before attacking a single target.

Over the years, various defence techniques have been developed to counter the S-75 models. These include ECM systems, the deployment of large quantities of chaff to confuse the RSN-75/'Fan Song' guidance radar, and actually outmanoeuvring the missile in flight. However, the best to date has proved to be the use of specialist aircraft to suppress the SAM system by electronically and physically attacking it.

Variants

There are many variants of the original S-75/SA-2 theme. The least talked about are the 17D, 18D and 22D. These missiles were developed from the V-755 missile and were Ramjet Rockets. Although never operationally deployed, they played a large part in the future development of Ramjet missiles in Russia.

	SA-2a	SA-2b	SA-2c	SA-2d	SA-2e	SA-2f
Dimensions and weights						
Length						
overall	10.6 m (34 ft 9¼ in)	10.8 m (35 ft 5¼ in)	10.8 m (35 ft 5¼ in)	11.2 m (36 ft 9 in)	10.8 m (35 ft 5¼ in)	10.8 m (35 ft 5¼ in)
Diameter						
body	650 mm (25.59 in) (booster)	650 mm (25.59 in) (booster)	650 mm (25.59 in) (booster)	650 mm (25.59 in) (booster)	650 mm (25.59 in) (booster)	650 mm (25.59 in) (booster)
	500 mm (19.69 in) (missile)	500 mm (19.69 in) (missile)	500 mm (19.69 in) (missile)	500 mm (19.69 in) (missile)	500 mm (19.69 in) (missile)	500 mm (19.69 in) (missile)
Flight control surfaces						
span	2.5 m (8 ft 2½ in) (booster, wing)	2.5 m (8 ft 2½ in) (booster, wing)	2.5 m (8 ft 2½ in) (booster, wing)	2.5 m (8 ft 2½ in) (booster, wing)	2.5 m (8 ft 2½ in) (booster, wing)	2.5 m (8 ft 2½ in) (booster, wing)
	1.7 m (5 ft 7 in) (missile, wing)	1.7 m (5 ft 7 in) (missile, wing)	1.7 m (5 ft 7 in) (missile, wing)	1.7 m (5 ft 7 in) (missile, wing)	1.7 m (5 ft 7 in) (missile, wing)	1.7 m (5 ft 7 in) (missile, wing)
Weight						
launch	2,287 kg (5,041 lb)	2,287 kg (5,041 lb)	2,287 kg (5,041 lb)	2,450 kg (5,401 lb)	2,450 kg (5,401 lb)	2,395 kg (5,280 lb)
Performance						
Speed						
max Mach number	3.5	3.5	3.5	3.5	3.5	3.5
Range						
min	4.3 n miles (8 km; 5.0 miles)	5.4 n miles (10 km; 6.2 miles)	5.0 n miles (9.3 km; 5.8 miles)	3.8 n miles (7 km; 4.3 miles)	3.8 n miles (7 km; 4.3 miles)	3.2 n miles (6 km; 3.7 miles)
max	16.2 n miles (30 km; 18.6 miles)	18.4 n miles (34 km; 21.1 miles)	23.2 n miles (43 km; 26.7 miles)	23.2 n miles (43 km; 26.7 miles)	23.2 n miles (43 km; 26.7 miles)	31.3 n miles (58 km; 36.0 miles)
Altitude						
min	3,000 m (9,843 ft)	500 m (1,640 ft)	400 m (1,312 ft)	400 m (1,312 ft)	400 m (1,312 ft)	100 m (328 ft)
max	22,000 m (72,178 ft)	30,000 m (98,425 ft)	30,000 m (98,425 ft)	30,000 m (98,425 ft)	30,000 m (98,425 ft)	30,000 m (98,425 ft)
Ordnance components						
Warhead	195 kg (429 lb) HE fragmentation	195 kg (429 lb) HE fragmentation	195 kg (429 lb) HE fragmentation	195 kg (429 lb) HE fragmentation	295 kg (650 lb) HE fragmentation	195 kg (429 lb) HE fragmentation
Fuze	proximity (command)	proximity (command)	proximity (command)	proximity (command)	proximity (command)	proximity (command)
Propulsion						
type	solid propellant (sustain stage)	solid propellant (sustain stage)	solid propellant (sustain stage)	solid propellant (sustain stage)	solid propellant (sustain stage)	solid propellant (sustain stage)

China has developed its own modified version of the S-75 under the designation HQ-2.

China has also developed the FT-2000A modification for its HQ-2 missile. The original licence-built version was the HQ-1. China has also produced a surface-to-surface missile version of the HQ-2, known as the 8610 system.

Iran has reverse engineered the S-75 under the name Sayyed-1 Project.

North Korea has its own HQ-2/S-75 development, modification and local production programme.

Arab British Dynamics reverse-engineered the V-750 'Guideline' to meet the requirements of the Egyptian Air Defence Command using technology obtained from North Korea, but it was not placed in production. It had the local name Tayir as Sabah ('Early Bird').

A naval version, the V-753 missile with the system named the M-2 Volkhov-M, was tried on the Project 70E 'Sverdlov' class cruiser conversion *Dzerzhinski*. Converted during 1960/1962 with a twin-rail launcher in place of X-turret and an eight round magazine, the Heard on video as being the radar associated with the new Tor system project was not a success.

Iraq has modified some of its SA-2 stockpile to accept an infrared homing seeker. As far as is known this has not seen any combat use in the recent Middle East conflicts. In late 1999, the US DoD stated that Iraq had developed, manufactured and deployed an improved S-75 system with hybrid electronics to improve its engagement range. These upgrades have apparently been met in combat during the numerous Anglo-American strikes against the Iraqi air defence network from around September 1999 onwards.

It is also reported that American ELINT operations have detected similar system upgrades in Bosnia and North Korea. Modern radars, laser range-finders, digital fire control systems and electro-optical trackers from a variety of West European, Chinese, Russian, Israeli and South African companies are believed to be involved.

In 1993, the Volga-M modernisation programme under the management of the Almaz NPO (the successor to the original S-75 design bureau) was revealed. A further development has been the Volga-2.

The (former) Yugoslavian concern, SDPR is offering an upgrade package for the S-75 Volga. Romania is also offering a maintenance and repair programme for the S-75M Volkhov system.

The Belarussian Company Belvneshpromservice is offering a S-75M3 variant utilising the 5Ya23 missile.

As more of the S-75 missiles become obsolete, some are being converted into targets for training air-defence troops on the test ranges. The latest (as of June 2005) is known as the RM-75, which is due to go to test ranges in the third quarter of 2005 for testing and also to be used as a target for the new S-400 programme. The RM-75 target missiles have both high and low altitude modifications, the RM-75VU for high altitude and the RM-75VU1 for low altitudes.

Specifications *See table at foot of previous page*

Status
Production complete. The S-75 system has been in service with several countries in some form or another, however, as of November 2013 most of those countries may well have the system in reserve.

Contractor
Almaz/Antei Concern of Air Defence.

S-75 Volga-2

Type
Static Surface-to-Air Missile (SAM) system

Development
The S-75 Volga-2 (NATO SA-2 'Guideline') is an upgrade package offered by Almaz/Antei Concern of Air Defence (PVO Concern) for the S-75 Volga missile system to extend its cost-effective service life.

Based on the use of advanced digital and solid-state electronics originally developed for the S-300PMU-1 (SA-10 'Grumble'), a total of 78 analogue units originally used for signal processing, target and missile tracking, generation of flight-control commands and combat crew training have been replaced by 12 new digital systems. This cuts overall maintenance time by a factor of 2 to 2.5 and the use of digital algorithms for data processing and flight control commands significantly improves the operational capabilities of the system to a standard approaching that of the latest SAM generations.

Description
A comparison of the S-75 Volga-2 performance characteristics with the standard S-75 Volga system is given in the table at the foot of the page:

Status
Ready for production and offered for export. It is not known if or how many of the upgraded systems have been exported.

Contractor
Almaz/Antei Concern of Air Defence.

Enquires to
Rosoboronexport.

S-125 Neva/Pechora (SA-3 'Goa')

Type
Static and towed Surface-to-Air Missile (SAM) system

Development
The S-125 Neva (US/NATO designations SA-3 'Goa') system began development in 1956 as a low- to medium-altitude complement to the larger S-25/R-113 (SA-1), S-75 (SA-2) and S-200 (SA-5) medium- to high-altitude surface-to-air missile systems. The name, 'Neva' (a Russian river) was assigned to the system.

Prototype state trials began in 1959. Initial service entry deployment began in 1961 on trainable twin launchers at static sites. The S-125 using the 5V24 (V-600) missile entered operational service with the PVO-Strany (*Voyska Protivovozduchnoy Oborony Strany* - Troops of the National Air Defence) missile units for use on airfield defence, low-level air defence around long-range SAM systems.

The rear area protection of army and fronts was the responsibility of the S-75 (SA-2) in conjunction with the S-125 system. In this last role, the missile units were not subordinate to the ground forces, but integrated into the overall army or front air defence network.

At the same time, the S-125 Neva entered naval service as the 4K90 (M1 Volna), again with the V-600 missile. The naval system was also known as the M-1 Volna (NATO SA-N-1) system, and achieved operational status with the Russian Navy aboard cruisers and destroyers.

In 1973, the 5P73 quadruple rail launcher was introduced into service with air defence forces to supplement the original 5P71 twin launcher in areas of strategic importance.

In 1964, the improved S-125M Neva-M system entered service. This used the V-601 missile and later the 5V27 and then 5V27V missiles. The naval upgrade was known as the 4K91 system. The export version of the system was designated S-125 Pechora.

The S-125 Neva is usually formed into regiments of four battalions.

The first recorded combat use of the S-125 was in 1970. PVO-Strany units including several S-125 missile regiments, were deployed to Egypt with twin launchers to form a joint Egyptian-Russian air defence network to cover the Suez Canal Zone during the last phase of the 1968 to 1970 Egyptian-Israeli War of Attrition.

Although a large number of men and much equipment was lost to Israeli air attacks, the PVO-Strany units shot down a total of five McDonnell Douglas F-4E Phantoms with the S-125 and were instrumental in forcing on the Israelis the August 1970 UN ceasefire.

	Volga	**Volga-2**
Altitude range:	100-30,000 m	100-30,000 m
Maximum range:		
(500 m altitude)	24,000 m	27,000 m
(5,000-25,000 m altitude)	40,000-55,000 m	45,000-60,000 m
(30,000 m)	50,000 m	55,000 m
Maximum crossing target range:		
(500 m altitude)	22,000 m	26,000 m
(5,000-25,000 m altitude)	38,000-50,000 m	40,000-57,000 m
(30,000 m altitude)	34,000 m	45,000 m
Resolution:		
(range)	250 m	250 m
(azimuth)	1.5°	1.5°
Single shot kill probability:		
50,000 m range	0.40-0.97 (depending upon target type)	0.56-0.98 (depending upon target type)
50,000-60,000 m range	n/avail	0.41-0.98 (depending upon target type)
Update rate:	10 s	5 s
Number of tracked targets/min:	250	25

ZIL-131 (6 × 6) truck carrying two SA-3 'Goa' SAMs 0044358

SA-3 'Goa' SAM 5P73 four-round launcher from the rear (Christopher F Foss) 0509893

The next combat use came in late 1972 when the North Vietnamese used small numbers of S-125s against US aircraft during the Linebacker raids. Their first and only recorded kill was also against an F-4 Phantom of the US Marine Corps.

Its major combat test, however, came during the 1973 Yom Kippur War when both Syria and Egypt used large numbers against the Israeli Air Force. The Egyptians started the war with an air defence network of 146 SAM batteries, approximately a third of which used S-125s. The Syrians deployed a total of 34 SAM batteries, 15 (including three S-125s) being located between Damascus and the Golan Heights.

The Arab Air Defence networks of both fronts fired around 2,100 missiles of the AS-75/S-125 and Kub types to destroy a total of 46 Israeli aircraft as confirmed kills, six of which were assessed as S-125 victims. However, in return, the Egyptians lost a number of S-75, S-125 and Kub batteries, which were captured intact by the Israelis and subsequently made available to US intelligence experts.

Since 1973, the S-125 has been used in combat by:

- Iraq (during the 1980 to 1988 Iraq/Iran war, and the Gulf Wars of 1990 to 1991 and 2003);
- Syria (two batteries in the 1982 air-to-ground Bekaa Valley air defence war);
- Libya (during the April 1986 US Navy and Air Force raids);
- Angola (against South African aircraft during various battles in southern Angola).

The system has also seen service with Serbia against NATO aircraft.

By late 1989 there were still over 300 S-125 battalion sites in operation, using either four semi-mobile twin 5P71 or fixed 5P73 quadruple rail launchers. This had been reduced in the Russian Federation to some 200 sites by mid-1993 and to very few launchers by 1997/1998. A number of other battalion sites still remain in other member states.

During September 2004, the old Soviet Union's strategic war stockpiles of S-125 missiles in the Leningrad Region were destroyed. The system, which was placed into reserve in 1994, was considered too volatile to keep any longer and was therefore destroyed.

Description

The initial V-600 or 5V24 missile has a large 2.6 second burn jettisonable 14-element solid fuel booster section fitted with rectangular fins that rotate through 90° at launch. The smaller missile body has an 18.7 second burn, single-element solid fuel, and sustainer rocket and is fitted with four fixed fins aft and four movable control surfaces forward. After booster jettison the second stage is captured in the radar beam and guidance signals are sent via antennas on the rear fins to place the missile on an intercept trajectory.

In the initial 1961 5V24 Neva version (US designation SA-3a, NATO designation 'Goa' Mod 0), guidance is by command throughout the flight.

SA-3 5P71 'Goa' SAM twin-round launcher 0509572

Close-up of SA-3 5P73 'Goa' four-round launcher 0509573

The Neva-M, V-601 or 5V27 missile version, introduced into service in 1964 (US designation SA-3b, NATO designation 'Goa' Mod 1), has been improved. A further improved missile, the 5V27V variant, appeared later and was assigned the same codename.

Long-range early warning and target acquisition is usually handled by a van-mounted P-15 (NATO designation 'Flat Face') C-band 210 km range two co-ordinate (azimuth and positional angle) radar with two stacked elliptical parabolic antennas utilising a 5° vertical and 2° horizontal beamwidth. In many S-125 battalions the P-15 has been replaced by the P-15M set (NATO designation 'Squat Eye') which has a similar performance but has had its antenna mounted on a 20 to 30 m mast to improve the low-altitude coverage. A PRV-11 (NATO designation 'Side Net') 180 km range 32,000 m altitude E-band height-finder radar is also used.

All target data generated are passed on to the S-125 battalion's organic trailer-mounted fire-control radar SNR-125 (NATO designation 'Low Blow'). Wherever possible, the four launchers at an S-125 site are positioned in a hand-shaped pattern with the thumb consisting of the revetted long-range search and the palm the SNR-125 radar. The latter is controlled from a van or bunker and is optimised for low- to medium-target monitoring using an unusual antenna configuration of two electro-mechanically scanned parabolic dishes set above each other and optimised to pick objects out of the ground clutter. Maximum acquisition range is 110 km and tracking range of the I-band system is between 40 and 85 km depending on the target size, altitude and operational conditions. Radiated beam width is 12° in the fan and 1.5° in the direction of scan. It can track six aircraft simultaneously and guide one or two missiles at once.

For operating in a heavy ECM environment, upgraded S-125 system SNR-125 radars have been fitted with 25 km range TV cameras to give the fire-control team the same data as from the emitting radar and allow a passive mode only command guidance interception to be performed.

ZIL-131 (6 × 6) truck from the rear, carrying two SA-3 'Goa' SAMs 0044359

Trailer-mounted 'Low Blow' fire-control radar which is used with the SA-3 'Goa' SAM system 0509791

Against F-4 sized targets at low level the missiles HE fragmentation warhead has a lethal burst radius of 12.5 m. The warhead is armed after the missile has travelled 50 m with the Doppler radar fuze being activated by command signal when the weapon is 300 m from the launcher. The proximity radar fuze has an effective operating distance of 10 to 50 m depending upon target type. If the missile fails to intercept, another signal is sent to either change the trajectory or self-destruct.

The trainable launchers are ground-mounted but can be relocated.

Missiles are normally transported in pairs from battalion storage areas on modified ZIL-131 (6 × 6) or ZIL-157 (6 × 6) trucks and loaded onto the launchers with the aid of a conveyor. It takes only a minute to load the missiles onto the rails, but the duration between missile launches is about 50 minutes due to missile preparation, truck transit and other reloading procedures.

The missile's ability to dive also allows it to be used against surface targets and naval vessels.

Variants

Work on a naval S-125 system began in 1956 with the first prototype being sent to sea on the Project 56K (NATO 'Kotlin') class destroyer *Bravyy* during 1960 to 1962. Known as the M-1 Volna system (US/NATO codename SA-N-1A 'Goa') it used ZIF-101 twin-rail launchers, V-600 (4K90) missiles and the Yatagan command and Parus radar director.

This was followed by the improved M-1M Volna-M system with the V-601 (4K91, US/NATO codename SA-N-1b 'Goa') missile and the ZIF-102 twin-rail launcher. A further variant, the M-1P Volna-P was developed between 1974 and 1976 with improved guidance up-links and a television sight (9Sh33) added to the radar director to improve low-level engagements. This was used only with the ZIF-101 launcher and had a Yatagan-P control system.

The NATO codename for the Yatagan was 'Peel Group'. The system was further upgraded in the 1980s to the Volna-N standard to improve interception capabilities against sea-skimming targets.

Apart from the naval Volna systems already mentioned, the Almaz NPO design bureau revealed in 1993 that it had developed a modernised export system, the Pechora-M. This included the new Yatagan engagement radar.

In 1993, proposals were made for updating exported S-125s with terminal IR homing, in a joint update involving both Russian Federation and Western programmes.

The latest upgrade offered by the Almaz/Antei Concern is the Pechora-2 system (see Russian Federation, Static and Towed Systems, Almaz/Antei Concern of Air Defence S-125 Pechora-2).

Yugoslav upgrade

The (former) Yugoslavian concern SDPR is offering an upgrade package for the S-125 Pechora system.

Polish upgrade

Poland has repackaged the S-125 system as a self-propelled unit to enhance mobility and survivability.

Romanian maintenance programme

Romania is offering a maintenance and repair programme for the S-125 Neva system.

Georgian reconstruction programme

During 2004 the Georgian S-125 air defence missile system that was inherited from the USSR army was completely reconstructed and upgraded. Although information is scant, presumably the upgrades included the standard analogue to digital electronics package.

Cambodian destruction programme

Reports coming from Cambodia during November 2005 confirmed the destruction of the old Pechora (SA-3) missiles that were in country. The stockpile of old missiles was destroyed at Srok Phnom Sruoch, about 65 km west of the capital, Phnom Penh. Cambodia originally acquired 36 such systems from the former Soviet Union in the 1980s.

Ukraine S-125-2D

Late in 2010, the Ukrainian Company Aerotechnica released information on modifications it had made to the original Pechora that used the 5V24 and 5V27 missiles. These modifications/upgrades effected only the supporting radar and command-and-control systems, leaving the missiles in there original state. It is reported that every component in the UNV-2 mobile radar, with the exception of the frequency magnetron has been replaced by up-to-date modern technology parts. The improvements made in the operational capability of the system include, reduced crew to half of its original number. It is understood that the radar now has built in countermeasures plus, increases in the speed and reliability of the system.

At least four of these system upgrades have been delivered to an unidentified Southern African country with a second contract signed during September 2010's African Aerospace and Defence show in Cape Town. Again this second customer has not been named but it is strongly suspected that Angola will be the host country.

Specifications

	5V24	5V27
Dimensions and weights		
Length		
overall	5.88 m (19 ft 3½ in)	5.95 m (19 ft 6¼ in)
Diameter		
body	550 mm (21.65 in) (booster)	550 mm (21.65 in) (booster)
	390 mm (15.35 in) (missile)	390 mm (15.35 in) (missile)
Flight control surfaces		
span	1.22 m (4 ft 0 in) (wing)	1.22 m (4 ft 0 in) (wing)
Weight		
launch	933 kg (2,056 lb)	953 kg (2,101 lb)
Performance		
Speed		
max Mach number	3.5	3.5
Range		
min	2.2 n miles (4 km; 2.5 miles)	2.2 n miles (4 km; 2.5 miles)
max	8.1 n miles (15 km; 9.3 miles)	13.5 n miles (25 km; 15.5 miles)
Altitude		
min	100 m (328 ft)	100 m (328 ft)
max	10,000 m (32,808 ft)	14,000 m (45,932 ft)
Ordnance components		
Warhead	60 kg (132 lb) HE 4G90 3,500 fragmentation	18 kg (39 lb) HE 5B 4,500 fragmentation
Fuze	radar, proximity, impact	radar, proximity, impact
Propulsion		
type	solid propellant (sustain stage)	solid propellant (sustain stage)

Status

Production of the original system is complete, however, upgrades and modifications for export still continue to date.

The S-125 system has been taken out of service in Russia, some of which have been destroyed.

Upgrades and modifications to export systems are available. There are a large number of S-125 sites operational within India especially in the north west on the border between India and Pakistan. However, it is understood that India is looking to replace all the S-125s probably with an Israeli SAM.

Contractor

Almaz/Antei Concern of Air Defence.

S-125 Pechora-M (SA-3 'Goa')

Type

Static and towed Surface-to-Air Missile (SAM) system

Development

The former Almaz Antei Concern of Air Defence (now known as the Almaz Central Design Bureau JSC) has developed with the Fakel Missile Design Bureau (responsible for the design, development and production of the missiles) an upgrade package for the Russian S-125 Pechora (SA-3 'Goa') SAM system. Known as the S-125 Pechora-M.

The Pechora is the export variant of the S-125 Neva system. The main areas of improvement for the system include:

- Replacement of the analogue subsystems with digital units in the areas of the system's missile guidance control command generator, launch command unit and crew training equipment;
- Improvement of the missile's guidance algorithm;
- Introduction of additional electronic subsystems to provide overall increase in system resistance to active jamming;
- Automation of the control system's verification subsystems;
- Reduction in time spent on technical servicing requirements;
- Incorporation of automatic target acquisition and tracking modes in the TV guidance channel;
- Incorporation of an infrared TV channel;
- Incorporation of a built-in IFF system;
- Upgrading of the missile.

A Pechora four-rail launcher, photographed in Moscow 1999 (IHS/Peter Felstead) 0104269

Description

A comparison of the Pechora-M with the basic Pechora system is given in the table at the foot of the page:

The modernisation package is delivered in the form of ready to use assemblies for installation at the missile battery site.

Status

Production as required. First systems delivered in 1994. Offered for export. It is known that India has upgraded its S-125 systems and it is believed this may be related to the Pechora-M system standard.

Contractor

Almaz/Antei Concern of Air Defence.

S-125 Pechora-2 Upgrade Package

Type

Static and towed Surface-to-Air Missile (SAM) system

Development

The S-125 Pechora-2 is an upgrade package offered by the Almaz/Antei Concern for Air Defence (PVO Concern) for the S-125 missile system. Pechora is the export name given to the Neva (S-125) system. It is based on advanced digital and solid-state electronics and subsystems originally developed for the S-300 (SA-10 'Grumble'). The then Almaz Central Design Bureau joined with the following companies.

- The Fakel Mechanical Engineering Design Bureau.
- The Moscow Radio Engineering Plant.
- The Design Office of Special Purpose Designing.
- Several other Russian and Belarussian companies (Oboronitelniye Sistemy) to significantly increase the mobility, range and lethality of the missile system especially in the modern ECM environment.

The most significant change is the system set up time which has been reduced from 90 to 30 minutes. The system is said to be capable of engaging attacking missiles and cruise missiles, as well as aircraft, helicopters or UAVs. Detection range against stealth type aircraft is increased from 16 km to more than 30 km whilst the kill probability against a cruise missile is up to 55 per cent for a single missile launch and up to 80 per cent for a two-round salvo. Through-life costs are said to be considerably reduced and maintenance time is cut by a factor of between eight and 10. The number of spare parts required has also been considerably reduced, from 3,000 to just 300.

Description

Each Pechora-2 battery can consist of the following components:

- Up to two platoons each with four 5P73-2 vehicle-mounted launchers on KAMAZ 6 × 6 heavy truck chassis, (the 5P73-2 launcher is a two-rail modified version of the original 5P73 four-rail static launcher).
- One or two UNV missile control radar (depending upon the number of platoons in the battery).

The UNV model 1999 truck-mounted fire-control radar is a derivative of the original SNR-125 ('Low Blow') radar and can track up to 16 targets simultaneously but can guide only a single missile. Thus a single unit with four launchers and two UNV radars can only engage two targets simultaneously. An automatic TV channel optical drift indicator and a night vision device are available as options. These allow the system to become all-weather capable by passively tracking targets by day or night in adverse weather conditions and visibility. They also function as a second target-tracking channel allowing a UNV radar tracking system to engage two targets simultaneously. Thus potentially doubling the battery's simultaneous engagement capability.

The control post vehicle, UNK commands post vehicle and the Casta-2E2 radar vehicles are all based on the KAMAZ-4320 6 × 6 truck chassis. All three vehicles are interconnected with the maximum distance between the UNV radar and UNK command post being 150 m. However, the control post vehicle

Parameter	Pechora	Pechora-M
Engagement area:		
(altitude limits)	50-18,000 m	20-20,000 m
(maximum range)	25,000 m	32,000 m
(single-shot kill probability)	0.5-0.9	0.72-0.99
(detection range against stealth type target)	10-16 km	≥ 30 km
ECCM capabilities:		
(angular errors of tracking in angle deception jamming)	5-10 min	1-1.5 min
(angular errors of target tracking in clutter and passive (ECM environment))	3-5 min	1-2 min
(prolongation of target tracking in a clutter environment)	no	yes
(angular errors of tracking a low altitude radar target)	5 min	2-3 min
(angular errors of tracking using TV optical system)	1 min (approx.)	0.2 min
Maintenance:		
(relative number of test parameters required)	100%	50%
(relative reduction in routine maintenance time)	n/avail	2-2.5 times
Launch capabilities:		
(mode of launch operation)	semi-automatic	automatic
(accuracy of missile impact point co-ordinate estimate)	1.5-3 km	0.3 km
(indication of guaranteed engagement area)	no	yes

may be located up to 10,000 m away. Communication equipment, a navigation system and suitable interfaces allow the command-and-control systems to receive radar information from higher echelon sources.

The Casta-2E2 (or 39N6E) UHF band radar consists of a command-and-control communications truck, antenna truck and a mobile generator truck-mounted unit. A remote operator post can be installed at a distance of up to 300 m from the radar. The Casta-2E2 incorporates an IFF subsystem and provides automatic target detection, range/azimuth/altitude measurements and tracking functions, as well as track details of targets at low and very low altitudes. Its mast-mounted antenna can be positioned just above tree top height or up to a maximum height of 50 m. Operational range is just below 150 km with a resolution of 360 m in range and 5.5° in azimuth. Up to 50 targets can be tracked simultaneously.

The Casta-2E2 incorporates several ECCM features, which includes:

- The radar, which is frequency-agile and can be switched manually or automatically over 10 carrier frequencies
- Phase modulation is applied to the transmitted pulses
- A minimal false alarm rate is maintained by special sort criteria based processing of received signals combined with changes in pulse repetition frequency.

A special device also suppresses wide-band reflections from discrete meteorological formations without compromising target detection probability. The basic radar functions are automated and monitored by Built-In Test Equipment (BITE), which allows the system to operate continuously for up to a minimum of 20 days.

The missile used is an upgraded version of the 5V27 (5V27DE), with a new rocket motor giving an increased range of 28,000 m (compared to 25,000 m for a standard 5V27) and a new HE-fragmentation warhead with a radio frequency proximity fuze. The 5V27DE has been designed to more effectively engage incoming targets than its older counterpart the 5V27.

Variants

Pechora-2M

Russia has been testing a new protection device for the new Pechora-2M system. This device, which substantially enhances the combat survivability of the Pechora-2M, has been developed by Oboronitelnyye Sistemy (Defence Systems) over the course of two years. Development has been a joint effort, with a number of leading Russian design bureaux and enterprises namely:

- The Second Research Institute of the Russian Defence Ministry
- The Kuntsevo Design Bureau
- Plant of Radio Engineering Equipment in St. Petersburg
- Pulsar Plant in Moscow
- Moscow Plant of Radio Engineering.

Exercises held during the early part of 2005 successfully defeated an incoming missile up to 400 m away from the antenna post which forms part of the overall system, thereby preserving the combat capacity of the Pechora-2M. These decoys mimic the signal from the radar being defended, radiating at a higher power level than that created by the sidelobes of the radar's antenna. This KRTZ device is probably the same, or at the very least, associated with the devices found on the Almaz/Antei Concern Medium Range Air Defence Missile System (MRADS).

In early May 2013, claimed US sources associated the Pechora-2M with the NATO designator SA-26, however it is not known what name has also been applied to the system or even if this is true.

Pechora-2A

Pechora-2A is a new generation of the S-125M1 Pechora SAM. Trials showed the following increase in the system's combat capabilities:

- Enlargement of kill zone
- Greater efficiency in engaging modern targets
- Enhanced ECCM capabilities
- Automatic tracking in optical/TV channel.

It has been reported that up to 50 per cent of the system has had the older analogue equipment replaced with new digital equipment based on modern technological services. This has resulted in:

- Longer service life
- Better tactical and operating characteristics
- Reduced volume and time required for maintenance
- Power sharing.

S-125-2TM

S-125-2TM variant is a modification from the basic S-125 Pechora found in Vietnam. During December 2011, live firing of the system on the Vietnamese missile range TB1 in Luc Ngan district, Bac Giang province, took place along with the S-300, ZSU-23 and 57 mm gun systems.

Specifications

5V27 Missile Variants

	Upgraded 5V27	5V27DE
Effective range:		
(minimum)	3.5 km	unknown
(maximum)	28 km	32 km
Heading parameter:	up to 24	unknown
Maximum target speed:	700 m/s	700 m/s
Target acquisition time:	2.5-3 s	2.5-3 s
Altitude:		
(maximum)	20,000 m (some sources report 18,000 m)	unknown, although believed to be increased
(minimum)	20 m	unknown
ECCM capacity:	2,000 W/MHz	enhanced
Minimum target radar cross section:	0.3 m^2	unknown
Number of channels during tracking of target/missile:	1/2	1/2
Number of missiles on one launcher:	4	2
In/out deployment time:	180 min	30 min

Status

The initial missile-firing and system integration tests were started in 1999.

Ordered by Egypt to update its S-125 Pechora inventory (contract awarded to upgrade 50 systems). In April 2006, the Russian/Belarus company Oboronitelniye Sistemy (Defence Systems) started implementing the second stage of the contract with Egypt. Under this contract, some 30 battalions of S-125 systems will be upgraded in three stages. In addition to the upgrade, Egypt has received an Electronic-Warfare (EW) package that protects the Pechora against anti-radar missiles.

Negotiations are also underway with several other countries with at least one further contract reported as being close to signature.

Tajikistan has taken delivery of the new S-125-2M (Pechora-2M) in January 2009. The system uses the upgraded 5V27DE missiles that extend the range to 30 km.

Early July 2012, Kyrgyzstan announced a contract with Russia for the upgrade of their S-125 systems to the standard of S-125-2M (Pechora-2M).

Contractor

Almaz/Antei Concern of Air Defence.
Enquiries to: Rosoboronexport.

S-200 Angara/Vega (SA-5 'Gammon')

Type

Static Surface-to-Air Missile (SAM) system

Development

Development of the S-200 Angara (Russian river), US/NATO designations SA-5 'Gammon', began in the early 1950s by the Aleksandr Raspletin (SB-1) (guidance) and Petr Grushin (missile) design bureaux. Almaz/Antei Concern of Air Defence (PVO Concern) now markets the system. The S-200 was designed to meet a requirement for a long-range low to high-altitude surface-to-air missile to complement the S-25 (NATO SA-1) and S-75 (NATO SA-2) weapons. This was due to neither system being considered sufficient to deal with the projected new generation of American high-speed high-altitude penetrating strategic bombers. The final target set envisaged for the system included the F-4, B-52, F-111, SR-71 and active jamming standoff aircraft.

Initial deployment of an S-200 trials unit using the original 5V21 missile took place from 1963 to 1964 on the outskirts of Tallinn in Estonia. The first operational regiments were deployed in 1966 with 18 sites and 342 launchers in service by the end of the year. By 1967, when the S-200 was officially accepted into service, the site total had risen to 22, by 1968 to 40, and by 1969 to 60. The growth in numbers then gradually increased throughout the 1970s and early 1980s, following the reorganisation of the PVO-Strany into the V-PVO (*Voyska Protivovozduchnoy Oborony* - Troops of the Air Defence) in 1981, until the peak of 130 sites and 1,950 launchers was reached in 1985-86. The numbers deployed had declined to some 200 launchers by 1997-98. Replacement by the S-300P (SA-10 'Grumble') family is essentially complete. Redundant systems are sold off to provide much needed revenue for modernisation programmes.

The S-25 (NATO SA-1), S-75 (NATO SA-2), S-125 (NATO SA-3), S-200 (NATO SA-5) and later missile systems remained under the control of the ZRV (*Zenitayye Raketnyye Voyshky* - Anti-Aircraft Rocket Troops) component with the addition of all mobile surface-to-air systems from the former troops of Troop Air Defence of the Ground Forces.

The initial version, the S-200 (US designation SA-5A) fielded in 1963, used the 150 km range 5V21 command-guided missile with a conventional warhead. The second version, the S-200V Angara (US designation SA-5B) fielded in

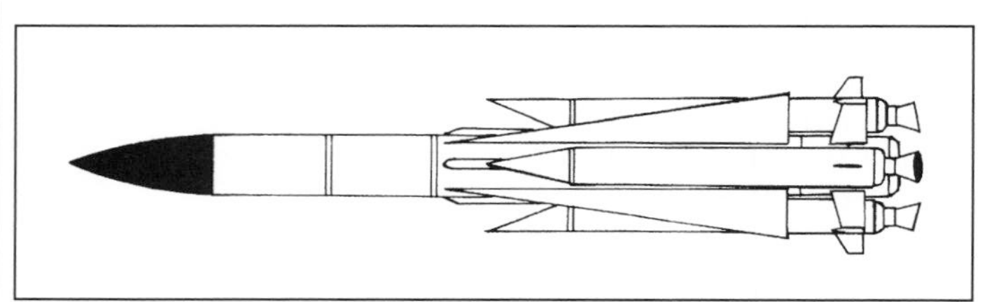

SA-5 Provisional drawing of the SA-5 'Gammon' SAM 0010166

SA-5 'Gammon' SAM deployed (Christopher F Foss) 0517864

SA-5 'Gammon' SAM on its reload cart, this is used to move the missile from its environmental shelter via a light railway to the launcher 0509575

'Square Pair' missile guidance radar used with the SA-5 'Gammon' SAM 0509576

SA-5 'Gammon' SAM of the Czech Republic on its semi-trailer resupply vehicle which is towed by a KrAZ-255V (6 × 6) tractor truck 0118083

The 'Tin Shield' early warning/Ground Control Intercept (GCI) radar which forms part of the North Korean overall SA-5 'Gammon' SAM system 0509789

1971, used the longer-range 250 km 5V28 missile. This was a dual-capable weapon using either a command detonated or proximity fused conventional 217 kg HE-fragmentation warhead or a command detonated 25 kT yield nuclear warhead with a semi-active radar terminal homing seeker. The only export version with a conventional warhead was designated S-200VE (Vega).

The final version was the S-200D Angara (US designation SA-5c), with an IOC date of 1975 and which used the 5V28V missile with enhanced manoeuvrability, an improved radar guidance system using upgraded ECCM capabilities and a maximum range of 300 km. The conventional warhead only export version was then designated S-200DE Vega. There is also a possibility of an anti-radar homing, passive seeker missile version having been developed in the 1970s for use against standoff jammers and early variants of AWACS type aircraft.

In 1983, the Russians began to establish 'S-200' sites in Warsaw Pact territories with battalions positioned in East Germany (near Rostock and Rudolstadt), Czechoslovakia (near Plzen) and western Hungary (near Szombathely). Before this, the only known site outside official borders was in Mongolia. The former East German S-200s were withdrawn from service by 1993, in part because the fuel in the external boosters was judged to pose an unacceptable environmental hazard.

In the wake of the Syrian Air Defence debacle in the Lebanese Bekaa Valley during 1982, four six-launcher battalions were sent to Syria in late 1982 and early 1983. Two sites were activated: Dumayr (40 km east of the capital Damascus) and Sharsar (in north-east Syria to the south-east of Homs). Initially manned by V-PVO troops, they were turned over to the Syrian Air Defence Command in 1985 to initially form two, then four two-battalion Independent Air Defence Missile Regiments with the additional sites located at As Suwayda and Mesken. According to Russian reports, the S-200 has been successfully used in action against the Israelis. In mid-May 2013, Hassan Nasrallah the then Hizbullah group leader revealed in a press conference that Syria had supplied the Pantsyr-S1E and the S-200 air defence systems to the group for impending use against Israel. It is easy to understand the Pantsyr system with Hizbullah but a lot more difficult to accept the S-200.

Following the Syrian success in obtaining S-200s, Libya began negotiations in 1984 with the first of three brigades delivered the following year. Operated by the Libyan Arab Air Defence Command, each brigade comprises two six-launcher S-200 battalions and two four-quadruple launcher S-125 battalions. On 24 March 1986, the S-200 brigade at Sirte on the Gulf of Sidra launched at least five missiles at US Navy aircraft operating in international airspace. All missed and the brigade's radars were subsequently attacked by US Navy aircraft firing Anti-Radiation Missiles (ARMs) on at least two occasions during the 24-25 March. The other two sites are at Benghazi and Oka Ben Nafi.

In the Russian Federation, the S-200 was organised into Air Defence Rocket Brigades. These consisted of two or three regiments with two five S-200 battalions of six launchers each. These were arranged in circles around the battalion's 5N62 ('Square Pair') radar and two or three battalions of S-125s,

	S-200 Angara	S-200V Angara	S-200D Angara
Dimensions and weights			
Length			
overall	10.5 m (34 ft 5½ in)	10.72 m (35 ft 2 in)	10.72 m (35 ft 2 in)
Diameter			
body	860 mm (33.86 in)	860 mm (33.86 in)	860 mm (33.86 in)
Flight control surfaces			
span	2.85 m (9 ft 4¼ in) (wing)	2.85 m (9 ft 4¼ in) (wing)	2.85 m (9 ft 4¼ in) (wing)
Weight			
launch	-	7,018 kg (15,472 lb)	7,100 kg (15,652 lb)
Performance			
Speed			
max speed	4,860 kt (9,000 km/h; 5,592 mph; 2,500 m/s)	4,860 kt (9,000 km/h; 5,592 mph; 2,500 m/s)	4,860 kt (9,000 km/h; 5,592 mph; 2,500 m/s)
Range			
min	3.8 n miles (7 km; 4.3 miles)	3.8 n miles (7 km; 4.3 miles)	3.8 n miles (7 km; 4.3 miles)
max	81.0 n miles (150 km; 93.2 miles)	135.0 n miles (250 km; 155.3 miles)	162.0 n miles (300 km; 186.4 miles)
Altitude			
min	300 m (984 ft)	300 m (984 ft)	300 m (984 ft)
max	20,000 m (65,617 ft)	29,000 m (95,144 ft)	40,000 m (131,234 ft)
Ordnance components			
Warhead	217 kg (478 lb) HE fragmentation	217 kg (478 lb) HE fragmentation	217 kg (478 lb) HE fragmentation
Fuze	uplink command, proximity	uplink command, proximity	uplink command, proximity
	active, radar	active, radar	active, radar
Propulsion			
type	4 solid propellant (strap-on boosters with liquid dual-thrust rocket motors)	4 solid propellant (strap-on boosters with liquid dual-thrust rocket motors)	4 solid propellant (strap-on boosters with liquid dual-thrust rocket motors)

each with four quadruple launchers to provide low altitude cover for both the installations being guarded and the S-200 battalions. In addition, each of the SAM battalions had a sub-unit of Automatic Anti-Aircraft (AAA) guns of either light (23 mm ZU-23) or medium calibre (57 mm S-60).

Description

The headquarters unit of an S-200 regiment has a radar section with a single D-band (NATO designation 'Big Back') 500 km plus range early warning radar.
Each missile battalion has:

- One P-35M (NATO designation 'Bar Lock-B') E/F-band 320 km range target search and acquisition radar with an integral D-band IFF system;
- One H-band 5N62 (NATO designation 'Square Pair') 270 km plus range missile guidance radar;
- Six trainable semi-fixed single rail 5P72 launchers with supporting loader units;
- Pre-launch preparation cabins and diesel-powered electricity generator stations.

The missile itself is of the single-stage type with four jettisonable, wraparound, solid fuel booster packs. Each booster is 4.9 m long, 0.48 m in diameter and has a single fin span length of 0.35 m from the booster body. The basic 5V21 is 10.5 m long overall and has a maximum wing span of 2.85 m. The main body is 0.86 m in diameter and has a single stage liquid fuel 10,000 kg total thrust, 51-150 second duration rocket engine sustainer.

The minimum range of around 7 km is due to the three-second to five-second, 160,000 kg thrust total duration booster burn time and jettison requirements which limit it to engagements against relatively large non-manoeuvring targets at ranges up to the maximum of the missile model. Guidance beyond the 60 km booster jettison point is by course correction command signals from the 5N62 ('Square Pair') radar with the missiles own active radar terminal homing seeker head being activated near to the projected intercept point for the final guidance phase. The large HE warhead is detonated either by a command signal or the onboard proximity fuzing system.

PVO Concern upgrade package for S-200VE Vega system

In 2000, the then Almaz Design Bureau revealed an upgrade package for the S-200VE system. This involved the following enhancements:

- Addition of the 5V28M missile version to the system to increase its engagement range to 300 km, the same as the later S-200D/S-200DE;
- Increasing the individual battery deployment locations of a battalion up to a maximum of 100 km from the battalion headquarters central target data source;
- Augmenting the target illumination radar with an aural indicator to give the target type;
- Improvements to the guidance system to ensure that low radial speed targets can be successfully intercepted.

The final improvement package can be revised according to customer specified requirements.

Variants

It is reported that Azerbaijan modified a few S-200 missiles it had for use in the surface-to-surface role.

Poland is upgrading its S-200 systems locally to a mobile configuration known as S-200C.

The former Yugoslavia is offering an overhaul and modification package for the S-200 Vega system.

The S-225, also known as ABM-2 in the west was developed between 1968 and 1978 and tested at Sary Shagan. The A-135 ABM system came into service before the S-225.

Specifications *See table top of page*

Status

Production complete. Has been seen in service with several countries but may now be considered obsolete or held in reserve by others. Iran is known to have updated and upgraded it stock of S-200 systems from 2007, with test firings having taken place in November 2010.

An unidentified Middle Eastern country made advances to Rosoboronexport during the Dubai 2005 Air Show for the purchase of the S-200 Angara system. It is understood that an order was placed for this redundant system.

Contractor

Almaz/Antei Concern of Air Defence (design).
Rosoboronexport (enquiries).

SOSNA-R/9M337

Type

Static Surface-to-Air Missile (SAM) system

Development

Designed by AE Nudelman's Moscow-based KB Tochmash Design Bureau, the private venture 9M337 missile is very similar in appearance to the 9M311 and 9M335 weapons used in the 2K22 Tunguska and Pantsyr-1 systems respectively. Nudelman reported that the mobile system that was nearing completion in early April 2009 is now complete and will be installed on an upgraded MT-LB multipurpose tracked armoured fighting vehicle.

Description

The light hypersonic SOSNA-R is, however, smaller and faster, having a maximum velocity of 875 m/s. The 9M337 has a body diameter of 72 mm with the booster rocket motor being 140 mm in diameter. Guidance is by 'In Laser Beam' with the electro-optical package of the platform modified to accept the guidance system. The missile if fitted with what has been described as a spread warhead and impact/non-impact laser fuse with continuous circular beam

SOSNA twin 30 mm low-level air defence system showing electro-optical pod in front of gunner's cabin 0010000

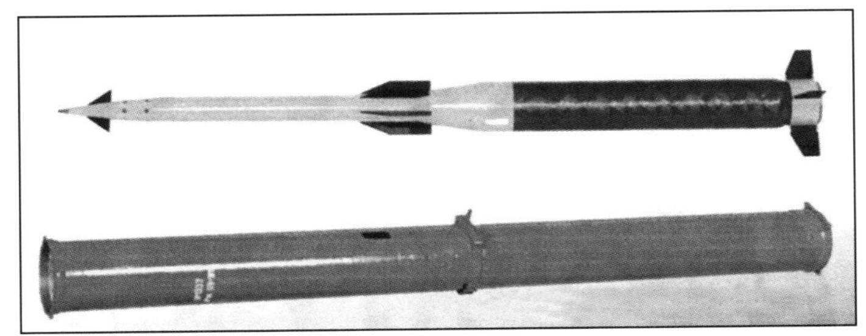

SOSNA-R surface-to-air missile out of its launcher tube as in flight (top) and in missile launcher canister (below) 0045791

pattern and adaptive burst time. The automated optical module has a thermal day or night vision channel and laser channels for range-finding and missile guidance. The accuracy of the laser range finder is ±5 m with an operating range against an aircraft or helicopter of at least 12,000 m.

The missile is used with the private venture trailer mounted SOSNA twin-barrelled 2A38M 30 mm anti-aircraft gun. It takes 11.5 seconds to reach its maximum effective range of10,000 m and has a minimum effective range of 1,300 m. Altitude engagement limits are from 3 up to 5,000 m.

Missile weight at launch is 30 kg, with the advanced design HE-fragmentation warhead weighing 5 kg. A 12-channel laser proximity and contact fuzing system is fitted. The missile is carried in a 2.4 m long, 153 mm diameter cylindrical container-launcher tube which weighs 42.5 kg fully loaded.

The launcher is the 2A38.

Although the primary role of the system is anti-air, the missile can also be used against ground light armoured vehicles.

Variants

A naval system called the Palma also uses the 9M337 missile. This is a joint venture between KB Tochmash (missile) and Tulamashzavod (guns and mount) for the export market. Two 30 mm calibre AO-18KD cannons are integrated with a dedicated electro-optic sensor and eight ready-to-fire short-range anti-aircraft missiles such as the SOSNA-R, Strela-10, Stinger or Mistral. The Palma has already been sold to an undisclosed customer. The first of three mounts was delivered in January 1999.

SOSNA is the self-propelled fully tracked AFV variant that employs a turret supporting a total of 12 × 9M337 laser-guided missiles in the ready-to-launch position. Sited between the two banks of missiles are two advanced electro-optical packages for acquisition, tracking and laser guidance and are claimed to be usable under almost all weather conditions. An automatic target tracker is also fitted.

Specifications

	SOSNA-R
Dimensions and weights	
Length	
overall	2.317 m (7 ft 7¼ in)
Diameter	
body	72 mm (2.83 in)
	130 mm (5.12 in) (booster)
Weight	
launch	30 kg (66 lb)
	42.5 kg (93 lb) (in canister-launcher tube)
Performance	
Speed	
max speed	1,701 kt (3,150 km/h; 1,957 mph; 875 m/s)
max Mach number	2.5
impact velocity	972 kt (1,800 km/h; 1,118 mph; 500 m/s) (target engagement)
Range	
min	0.7 n miles (1.3 km; 0.8 miles)
max	5.4 n miles (10 km; 6.2 miles)
Altitude	
min	2 m (7 ft)
max	5,000 m (16,404 ft)
Ordnance components	
Warhead	5 kg (11.00 lb) blast fragmentation, armour piercing
Fuze	laser, proximity, impact (12-channel, optional command for armour-piercing)
Guidance	laser beam riding
Propulsion	
type	solid propellant (sustain stage)

Status

Ready for production. Offered for export. As of June 2013 no customers have been identified.

Contractor

Design: Nudelman Precision Engineering Design Bureau.
KB Tochmash Design Bureau.
Podolsky Elektromechanichesky Zavod.
Rosoboronexport.

Volga-M (SA-2 'Guideline')

Type

Static Surface-to-Air Missile (SAM) system

Development

In 1993, the then Almaz/Antei Concern of Air-Defence (now part of the Almaz Central Design Bureau) revealed an upgrade package for the S-75 Volga variant of the S-75 (SA-2 'Guideline') SAM family.

Known as the Volga-M, the areas of improvement/upgrade features guidance technology, computer processing techniques and other modern electronics including:

- Replacement of the analogue subsystems by digital units in the areas of the system's MTI system, missile guidance control command generator, launch command unit and crew training equipment
- Improvement of the missile's guidance algorithm
- Introduction of digital electronic subsystems to provide increased resistance to active pulse-modulated repeat jammer units
- Automation of the control system's verification subsystems
- Reduction in time spent on technical servicing requirements.

A comparison of the Volga-M to an unmodified Volga system is given in the table at the foot of the page.

Description

The modernisation package is delivered in the form of ready to use assemblies for installation at the missile battery site.

Status

Production as required.

First systems delivered to Russian air-defence forces in 1994 but have now all been superseded.

Offered for export. Several countries in the Middle East and Asia may have purchased the system as an upgrade package rather than purchasing new SAM systems.

A further development is the Volga-2 programme.

Contractor

Almaz/Antei Concern of Air Defence.

Model	Volga	Volga-M
Engagement area:		
(altitude limits)	100-30,000 m	100-30,000 m
(maximum range)	55,000 m	67,000 m
(single-shot kill probability)	0.4-0.97	0.56-0.98
ECCM capabilities:		
(angular errors of tracking in angle deception jamming)	6-12 min	1.5-2 min
(angular errors of target tracking in clutter and passive ECM environment)	3-5 min	1-2 min
(prolongation of target tracking in a clutter environment)	no	yes
Maintenance:		
(relative number of test parameters required)	100%	40%
(relative reduction in routine maintenance time)	n/avail	2-2.5 times
Launch capabilities:		
(mode of launch operation)	semi-automatic	automatic
(accuracy of missile impact point co-ordinate estimate)	2-5 km	0.5 km
(indication of guaranteed engagement area)	no	yes

Serbia

S-125 Pechora maintenance and modification programme

Type
Static and towed Surface-to-Air Missile (SAM) system

Development
As part of its cost-effective approach to upgrading elderly Russian supplied surface-to-air missile systems, the then Yugoslavian (now Serbia) SDPR developed an overhaul and modification package for the S-125 Pechora (SA-3 'Goa') static surface-to-air missile system. The first modification/upgrade being the integration of a thermal imaging laser system into the missile's radar guidance station for terminal homing.

Description
At present the existing daylight TV system is used for acquisition, tracking and firing at targets operating in an intense radar-jamming environment aimed specifically against the weapon's normal radar guidance operating mode. In TV guidance mode there is no reliable information against target range, therefore a three-guidance method is applied, which enables targets operating at speeds of up to 300 m/s to be engaged.

Also, there is little or no capability against a target operating in poor visibility, at night or directly in the direction of the sun. The latter problem is overcome by the addition of a thermal imaging system. The advantages of which in comparison to the existing daylight TV mode and radar mode guidance systems is shown in the accompanying table.

For the S-125 Pechora system the target acquisition ranges in conditions of good visibility for its daylight TV system is 40,000 m for a bomber-sized target and 27,000 m for a fighter-sized target. The target distance is obtained by the addition of a laser rangefinder module, which is aligned with either the axis of the TV system or the thermal imaging system. This allows data to be obtained on one or more targets in intense ECM environments when there is no other way to obtain the range: the ability for the modified system to engage in the semi-automatic radar homing mode of operation using the radar so that targets over the TV cameras limiting 300 m/s and up to the maximum permissible 700 m/s velocity can be engaged.

The command activation of the weapons fuzing system at the optimum distance of 60 m between the target and the missile thereby preventing either a deliberate malfunction or premature detonation of the weapon's radio frequency fuze due to ECM.

Other modifications/upgrades are:
- Protection against ARM weapons by use of decoys
- Modernisation of the weapon control recording equipment
- Installation of a digital MTI system
- Installation of a logarithmic receiver system.

In addition to these upgrades the following can also be performed to improve the S-125 Pechora system (note the Serbia Montenegro designations may differ from the original Russian designations).
- Overhaul of the 5V27U Pechora missile.
- Overhaul and modification of P-14 'Tall King' metric wavelength early warning radar. This involves interfacing the P-12 'Spoon Rest' UHF-band early warning radar operations shelter with the antenna assembly of the P-14 'Tall King' radar. The installation of K-14 equipment to reduce the vulnerability of the radar against ARMs and the fitting of solid-state HF amplifier technology to increase the sensitivity of the radar receiver.
- Overhaul and modification of the P-12 'Spoon Rest' UHF-band early warning radar. This is aimed at improving the performance characteristics so that they match those of the P-18 radar system. The improvements include the installation of a solid-state HF amplifier and logarithmic amplifier to increase the receiver sensitivity and enhance the anti-jamming features; and the replacement of mechanical pre-adjustment controls by electronic sub-systems; modification of the transmitter to increase power output; modification of the antenna assembly to reduce the radiating pattern beam formed from a 12° sector to an 8° sector; decreasing the antenna lower limiting tilt angle to -5° so as to improve identification of low-altitude targets; the installation of an MTI capability; and the removal of unnecessary sub-systems such as the original direction-finder, altitude indicator, Y-transformer, AP-3 automatic system and unit assemblies Nos 59 and 75.
- Overhaul and modification of the P-15 'Flat Face' 2D UHF-band early warning radar.

Status
Production as required. In service with the Serbian Army. Offered for export. The system is believed to have been used in the Kosovo conflict of 1999.

Contractor
Yugoimport - SDPR.

SDPR S-75 Volga maintenance and modification programme

Type
Static and towed Surface-to-Air Missile (SAM) system

Description
As part of its cost-effective approach to upgrading elderly Russian supplied surface-to-air missile systems, the Yugoslavian SDPR has developed the following overhaul and modification package for the S-75 Volga (SA-2 'Guideline') static surface-to-air missile system:
- Integration of a thermal imaging laser system into the missile's radar guidance station. At present, the existing daylight TV system is used for acquisition, tracking and firing at targets operating in an intense radar jamming environment aimed specifically against the weapon's normal radar guidance operating mode. In the TV guidance mode there is no reliable information against target range, therefore a three-guidance method is applied, which enables targets operating at speeds of up to 300 m/s to be engaged. Also, there is little or no capability against a target operating in poor visibility, at night or directly in the direction of the sun.The latter problem is overcome by the addition of a thermal imaging system. The advantages of which in comparison to the existing daylight TV mode and radar mode guidance systems is shown in the accompanying table.The target distance is obtained by the addition of a laser rangefinder module, which is aligned with either the axis of the TV system or the thermal imaging system. This allows data to be obtained on one or more targets in intense ECM environments when there is no other way to obtain the range. The ability for the modified system to engage in the semi-automatic radar homing mode of operation using the radar so that targets over the TV cameras target limiting 300 m/s velocity and up to the maximum permissible 700 m/s velocity can be engaged. The command activation of the weapons fuzing system at the optimum distance of 60 m between the target and the missile thereby preventing either a deliberate malfunction or premature detonation of the weapon's radio frequency fuze due to ECM.
- Protection against ARM weapons by use of decoys.
- Modernisation/improved control of the weapon control recording equipment.
- Installation of a digital MTI system.
- Installation of a logarithmic receiver system.

In addition to these upgrades the following can also be performed to improve the S-75 Volga system (note the Yugoslavian designations may differ from the original Russian designators):
- Overhaul of the V775 Volga missile.
- Overhaul and modification of P-14 'Tall King' metric wavelength early warning radar. This involves interfacing the P-12 'Spoon Rest' UHF band early warning radar operations shelter with the antenna assembly of the P-14 'Tall King' radar. The installation of K-14 equipment to reduce the vulnerability of the radar against ARMs and the fitting of a solid-state HF amplifier technology to increase the sensitivity of the radar receiver.
- Overhaul and modification of the P-12 'Spoon rest' UHF-band early warning radar. This is aimed at improving the performance characteristics so that they match those of the P-18 radar system. The improvements include the installation of a solid-state HF amplifier and logarithmic amplifier to increase the receiver sensitivity and enhance the anti-jamming features; and the replacement of mechanical pre-adjustment controls by electronic subsystems; modification of the transmitter to increase power output; modification of the antenna assembly to reduce the radiating pattern beam formed from a 12° sector to an 8° sector; decreasing the antenna lower limiting tilt angle to –5° so as to improve identification of low altitude targets; the installation of an MTI capability; and the removal of unnecessary sub-systems such as the original direction-finder, altitude indicator, Y-transformer, AP-3 automatic system and unit assemblies Nos 59 and 75.
- overhaul and modification of the P-15 'Flat Face' UHF-band early warning radar.

Status
As of September 2012, this particular upgrade has ceased to form any part of the SDPR advertising for upgrades of old ex-Soviet equipment.

The S-125 Pechora (NATO SA-3 Goa) is still advertised as is a general statement on air-defence radars.

Therefore it is assessed that should a customer with a specific request be found production could possibly be activated, however, if this is not the case the system modifications will no longer be produced.

Advantages and disadvantages of S-125 upgrade with thermal imaging system mode in comparison to existing daylight TV and radar guidance modes

Characteristic	Daylight TV mode	Thermal imaging mode	Radar mode
Rain	disadvantage	disadvantage	advantage
Mist	disadvantage	advantage	advantage
Smoke	disadvantage	advantage	advantage
24-hour operation	disadvantage	advantage	advantage
ECM resistance	advantage	advantage	disadvantage
Passive operation	advantage	advantage	disadvantage
Acquisition of close-range targets	advantage	advantage	disadvantage
Multitarget capability	advantage	advantage	disadvantage

Advantages and disadvantages of S-75 upgrade with thermal imaging system mode in comparison to existing daylight TV and radar guidance modes

Characteristic	Daylight TV mode	Thermal imaging mode	Radar mode
Rain	disadvantage	disadvantage	advantage
Mist	disadvantage	advantage	advantage
Smoke	disadvantage	advantage	advantage
24-hour operation	disadvantage	advantage	advantage
ECM resistance	advantage	advantage	disadvantage
Passive operation	advantage	advantage	disadvantage
Acquisition of close-range targets	advantage	advantage	disadvantage
Multitarget capability	advantage	advantage	disadvantage

In service with the former Serbia and Montenegro Army. Offered for export. The system is believed to have been used in the Kosovo conflict of 1999.

Contractor

Yugoimport - SDPR.
Fotona dd (TV trackers. This company now based in the Republic of Slovenia).

South Africa

SAHV-IR

Type

Static and towed Surface-to-Air Missile (SAM) system

Development

In order to simplify the ground equipment of the SAHV-3 (Umkhonto) missile system (South African Crotale upgrade in the Self-propelled surface-to-air missiles section), Kentron has conducted trials with the two-colour seeker of its Darter air-to-air missile on the SAHV-3 missile body for use in a Lock On After Launch (LOAL) mode.

The missile can be launched from a variety of trainable launchers, a typical example being the towed system illustrated which carries a pack of four container-launchers. The system can equally well be mounted on a variety of self-propelled launchers.

Description

The Umkhonto-IR missile is a vertically launched, high-velocity, infrared homing missile specifically designed for all-round defence against fixed and rotary wing targets, these targets also include missiles. The system is supplied as a missile group for easy integration into naval combat suites or ground-based air defence units. The missile requires a ground tracker to aim the launcher and to establish the course for the initial flight path. This may take the form of a remote Radar Fire-Control Unit (FCU) or accurate 3-D acquisition radar, or, in the simplest form, a TV or Forward Looking InfraRed (FLIR) tracker mounted on the launcher itself to generate its own target vector with target designation coming from a 2-D acquisition radar.

The InfraRed (IR) seeker is a fully developed item, which is in service with the South African Air Force. The remaining missile subsystems are identical to those used for the SAHV-IR missile except that the missile wings are moved forward on the motor case to compensate for the increased body lift caused by the longer nose section.

The Darter's two-colour IR seeker has excellent flare rejection capability, enabling the weapon to be used in the presence of optical/IR countermeasures. Its other significant feature is the use of a separate inertial reference unit, which is electrically coupled to the seeker telescope. This allows the seeker head to scan a cone of ±15° in less than 1.5 seconds.

The missile is launched with a small lead angle and flies under onboard inertial control, with the seeker scanning the target space until the target is detected. The seeker locks on and the missile enters its proportional navigation terminal homing flight phase.

This allows a target approaching in a direct head-on aspect to be engaged at much greater range than can be obtained with a conventional IR missile which must wait for target detection before launch. For short-range engagements, the missile may be used in the conventional Lock On Before Launch (LOBL) mode. In either case, the launcher is free to engage a subsequent target as soon as the missile is fired. More details of the weapons operational capabilities are given in the Kentron SAHV Oerlikon Contraves 35 mm Skyguard air defence system entry earlier in this section under International.

Kentron SAHV-IR missile being launched from a towed launcher 0509596

Parachute target engaged by a Kentron SAHV-IR missile 0509597

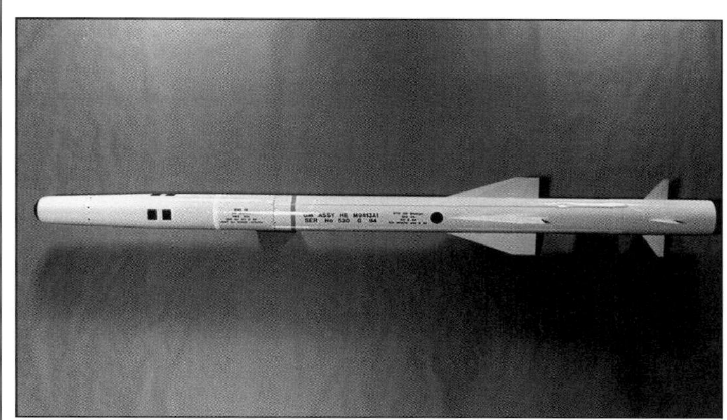

Kentron SAHV-IR missile 0509799

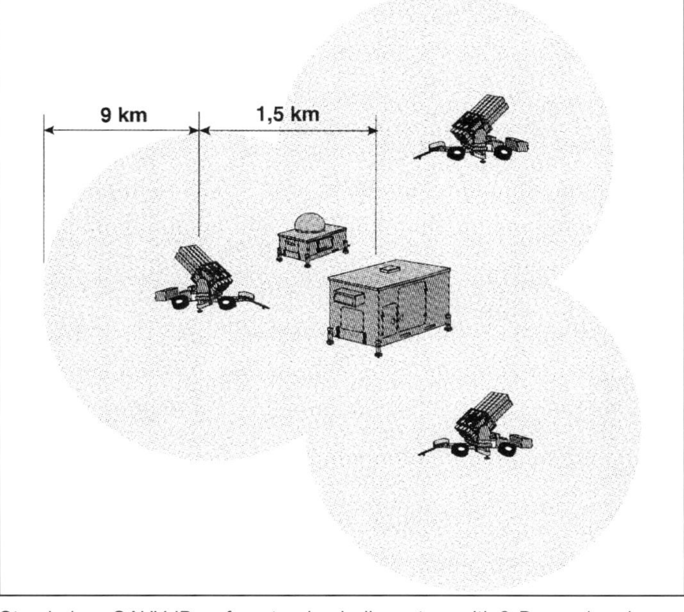

Stand-alone SAHV-IR surface-to-air missile system with 3-D search radar designation 0509896

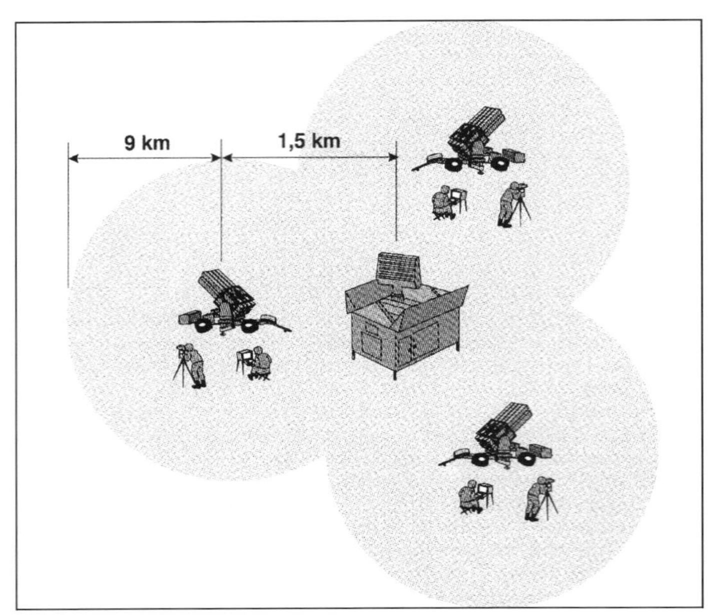

Stand-alone SAHV-IR surface-to-air missile system with 2-D search radar designation 0509897

The missile uses the same smokeless solid fuel rocket motor as the SAHV-3, together with the 20 kg HE blast/fragmentation warhead and active laser proximity fuze system. The velocity is reduced as compared to the SAHV-3 because of the drag of the hemispherical IR seeker nose dome. A peak velocity of around Mach 3.0 is reached with the missile attaining its maximum range of 8,000 m in 12-13 seconds. Kentron is currently investigating the use of a Mach-breaker spike to reduce the excessive drag so that the SAHV-IR performance more closely approaches that of the SAHV-3.

Variants
A vertical launch variant, originally known as the SAHV-IRN, now known as the Umkhonto-IR (Umkhonto the native South African for, 'spear') has been developed for the South African Navy and the Umkhonto Ground Based Launcher (GBL). A planned upgrade will involve an active radar seeker and a larger rocket motor to give full all-weather capability and a greater range.

Specifications

	Umkhonto-IR	SAHV-IR
Dimensions and weights		
Length		
overall	3.32 m (10 ft 10¾ in)	3.25 m (10 ft 8 in)
Diameter		
body	180 mm (7.09 in)	180 mm (7.09 in)
Flight control surfaces		
span	0.5 m (1 ft 7¾ in) (wing)	0.4 m (1 ft 3¾ in) (wing)
Weight		
launch	135 kg (297 lb)	132 kg (291 lb)
	-	178 kg (392 lb) (in container-launcher)
Performance		
Speed		
max Mach number	2.0	3.0
Range		
min	-	0.5 n miles (1 km; 0.6 miles)
max	6.5 n miles (12 km; 7.5 miles)	4.6 n miles (8.5 km; 5.3 miles)
Altitude		
min	-	30 m (98 ft) (fixed-wing aircraft)
	-	5 m (16 ft) (rotary-wing aircraft)
max	8,000 m (26,247 ft)	7,315 m (23,999 ft) (est.)
Ordnance components		
Warhead	23 kg (50 lb) HE blast fragmentation	20 kg (44 lb) HE blast fragmentation
Fuze	laser, proximity	
Guidance	INS, mid-course update	INS (mid-course phase)
	passive, IR (two-colour with proportional navigation (terminal))	passive, IR (two-colour with proportional navigation (terminal))
Propulsion		
type	solid propellant (smokeless, sustain stage)	solid propellant (smokeless, sustain stage)

Status
Development is complete.

Contractor
Denel Dynamics (formerly known as Kentron).

Umkhonto GBL

Type
Static and towed Surface-to-Air Missile (SAM) system

Development
The Umkhonto Ground-Based Launcher (GBL) missile system was designed to engage and defeat targets used in simultaneous air attacks of fixed and/or rotary wing types including missiles. The GBL design is based on the already proven South African naval Block 1 and Finnish naval system that utilises the Umkhonto-IR Block 2 missile rounds. The Block 2 has a range of 15 km and is fitted to the Finnish Hameenmaa-class mine layers and Hamina-class corvettes.

There are unconfirmed reports that the Umkhonto has been selected by Algeria for use on its Meko frigates.

Firing trials of the GBL have taken place at the Overberg Test Range, where the missile destroyed 2 × targets at 15 km and 1 × target at 20 km, the targets used were the Low-Cost Aerial Target Systems (LOCATS). All three targets were engaged with the missiles using lock-on-after-launch mode.

Description
The GBL can be deployed either on a vehicle or as a stand alone system that is connected to the command-and-control centre via radio, hard-wired or fibre-optic links.

The Umkhonto can also be used by the SA Army Phase 2 of its Ground-based Air Defence System (GBADS) along with the Reutech Radar Systems new RSR-320 dual-band 3D radar. The reaction time for the system to engage a target is 2.5 seconds.

Variants
- Umkhonto
- Umkhonto-IR
- Umkhonto extended range beyond 20 km
- Umkhonto radar guided variant.

Umkhonto GBL (Denel) 1332056

Mobile GBL (OMC) 1330982

Specifications

	Umkhonto GBL
Dimensions and weights	
Length	
overall	3.32 m (10 ft 10¾ in)
Diameter	
body	180 mm (7.09 in)
Flight control surfaces	
span	0.5 m (1 ft 7¾ in) (wing)
Performance	
Speed	
max speed	1,322 kt (2,448 km/h; 1,521 mph; 680 m/s)
max Mach number	2.0
Range	
max	10.8 n miles (20 km; 12.4 miles)
Altitude	
max	8,000 m (26,247 ft)
Ordnance components	
Warhead	23 kg (50 lb) HE
Guidance	IR, mid-course update

Status
Production of the Umkhonto Block 1 has ceased, only the Block 2 is now in production at Denel.

Contractor
Denel Pty Ltd.

Sweden

RBS 23 BAMSE

Type
Static and towed Surface-to-Air Missile (SAM) system

Development
Following the 1989 definition contract, which was completed in 1991, the then Bofors Missiles and the Swedish Defence Materiel Administration (FMV) signed a USD200 million contract in May 1993 to develop the RBS 23 Bofors Advanced Medium-Range Surface-to-Air Evaluation (BAMSE) missile system for the Swedish Defence Forces.

The programme is a joint project between the then Bofors Missiles and Ericsson Microwave Systems AB. Saab Bofors Dynamics AB developed the missile and the Missile Control Centre (MCC) and retains overall responsibility for system integration. Under a separate USD80 million contract from FMV, Ericsson Microwave Systems is to develop the new-generation Giraffe 3-D surveillance radar and the fire-control radar; the Command, Control, Communications and Intelligence (C3I) cabin-based system which is splinter proof and has NBC protection. Ericsson Microwave Systems delivered its prototypes in 1997 with Saab Bofors Dynamics finishing full-scale development of BAMSE in 1998. In late 1999, a report was accepted by the Swedish Supreme Commander that procurement of three BAMSE batteries, which would comprise a total of three central radar units and nine firing units plus missiles, would take place after 2003. An order worth SEK55 million was placed in late 1999 by the FMV for the RBS 23 BAMSE System with Saab Bofors Dynamics AB.

Saab Bofors Dynamics BAMSE missile being fired from the launcher 1048692

In Swedish service, BAMSE is an all-weather system that fills the gap above the altitude coverage of VSHORADS and SHORADS systems. Its range of well over 15 km and 15,000 m altitude capability makes the system suitable for the protection of vital positions, such as air bases, naval bases and HQs, as the missile range exceeds most attack aircraft's electro-optical controlled weapons' stand-off engagement distances. If the weapon is launched outside the range of the BAMSE system, it is still capable of defeating the attacking missile. The system is capable of destroying:

- Fixed wing aircraft
- Rotary wing aircraft
- Bombers and stand off weapons such as cruise missiles
- Anti-radiation missiles
- Guided bombs
- Unmanned Ground Vehicles (UAVs).

The system has been offered for export.

BAMSE is also suitable for the protection of mobile army units because of its high mobility and the short reaction times required for redeployment and preparation to fire. The system is air-transportable.

Description
The BAMSE air defence missile system is an Automatic Command-to-Line-Of-Sight (ACLOS) missile system that has been developed as a dedicated ground-based air-defence weapon. It has an all-weather capability and a range exceeding the stand-off range of electro-optically controlled weapons.

A BAMSE battery will normally comprise one Surveillance Co-ordination Centre (SCC) and three MCCs (maybe up to six). The SCC comprises Giraffe AMB 3-D-surveillance radar that operated in the 5.4 to 5.9 GHz band (NATO G-band) with a characteristic 8 or 13 m antenna mast (to allow it to be positioned behind terrain obstacles) and an integrated BM/C4I function. The SCC co-ordinates and controls other type of air defence systems and includes such features as automatic instrument tracking at ranges 30, 60 and 120 km plus of 150 targets with IFF and continuous threat evaluation and engagement planning. Full elevation coverage is up to 70° with altitude detection ceiling of 20,000 m. Up to six MCCs can be individually co-ordinated by an SCC. The MCCs communicate with the SCC by cable, fibre optic cable or different types of radio. To ensure a reliable link, the Swedish Army's standard TS9000 tactical radio system has been chosen with its high degree of jamming resistance. The distance to the SCC can vary, but 15 to 20 kilometres is regarded as normal. The MCC can carry out-group and split tracking, track correlation, jammer triangulation and full radar netting with other Giraffes.

The MCC contains all the essential elements for the target engagement. It is towed by a cross-country vehicle, which also carries the missile re-supply transloader. A deployment and fire preparation takes 10 minutes from on-the-

Saab Bofors Dynamics BAMSE six-missile launcher (Saab Bofors Dynamics) 1048693

Ericsson Microwave Systems Giraffe AMB (SCC) on Scania (6 × 6) truck with antenna erected into operating configuration 0509718

march position, with reload less than five minutes, and six seconds from target detection to launch of missile. On launch, the missile is gathered in a wide beam and then uses the narrow beam for guidance to target.

An elevatable mast on the roof of the MCC raises the fire-control radar, Thermal Imaging System (TIS) working in the 8 to 12 micron-band and IFF sensor assembly. Guidance of the missile in flight is by automatic command-to-line of sight and is performed via the K_a-band (34 to 35 GHz) fire-control pulse-Doppler monopulse radar, which is a further development of the Ericsson Microwave Systems Eagle tracking radar.

The capability of the MCC to acquire and track low-flying targets is enhanced by the radar's ability to 'look over' terrain obstacles. Target acquisition and tracking are initiated from the SCC by target assignment. The missile fire-control system presents the relevant information to the two operators housed in the NBC and splinter protected MCC. The missiles are fired from an elevatable stand assembly, holding six SPX ready-to-fire missiles, located on the roof of the MCC. Reloading the system takes less than four minutes.

The missile is of the high-acceleration, high-velocity type, with a maximum speed of about Mach 3.0 and is fitted with a pre-fragmented HE warhead with proximity and impact fuzes. The weapon is stowed in a cylindrical container-launcher tube and will be capable of engaging all types of aerial targets, from small high-speed cross-sectional area missiles, such as cruise missiles, anti-radiation missiles and Unmanned Aerial Vehicles (UAVs), to large slow cross-sectional area targets such as transport aircraft.

The MCC and SCC each have Built-In Test Equipment (BITE), integrated simulators and recording equipment. The last two items will allow operators to practice combat system training without the need to fire any missiles.

BAMSE will be able to provide protection for air bases and other strategic installations such as population centres, as well as mobile units, under all weather conditions. It will be fully air-portable in the Lockheed Martin Hercules C-130 transport aircraft in service with the Swedish Air Force.

Specifications

RBS 23 BAMSE Missile

Length:	2.6 m
Diameter:	
(missile)	0.32 m
(wing)	0.60 m
Weight:	85 kg
Warhead:	
(type)	pre-fragmented HE with shaped charge function
Fuze:	adaptive proximity and impact
Guidance:	Automated Command-to-line-of-Sight (ACLOS)
Motor:	solid propellant
Velocity:	>M3.0
Range:	
(maximum)	>20 km
(minimum)	<1 km
Altitude:	
(maximum)	>15,000 m
(minimum)	n/avail
Shelf life:	15 years

RBS 23 system

FCR instrumented range:	30 km
Altitude coverage:	15,000 m
Coverage area:	>1,500 km^2
Missiles:	6 ready-to-fire missiles
Deployment time:	MCC <10 min Battery <15 min
Reloading time:	<5 min for all 6 missiles

Status

In series production for the Swedish government, however, hardware delivered will be placed in storage pending a review of the future role of the Swedish Armed Forces.

Cutbacks announced on 21 September 2007 proposed by the Swedish government for the Fiscal Year 2008 (FY08) lead to this decision.

In January 2014, SaaB announced that they had joined with an Indian transport company Ashok Leyland and would compete for the Indian Army Short-Range SAM programme by offering the BAMSE mounted onto the Ashok Leyland high-mobility vehicles.

Contractor

Saab Bofors Dynamics AB.

Switzerland

Skyguard

Type

Static anti-aircraft control system

Development

The Skyguard all-weather fire-control radar was originally developed in the 1970s by Oerlikon Contraves (now Rheinmetall Air Defence) of Switzerland, with the assistance of L M Ericsson of Sweden and Siemens-Albis of Switzerland. It was designed to control the Rheinmetall Air Defence GDF series twin 35 mm towed anti-aircraft guns.

Since then, well over 600 Skyguard all-weather fire-control radar systems have been built for the home and export markets and the system has been continuously updated. In addition to controlling anti-aircraft guns, it can also control surface-to-air missiles or a combination of towed anti-aircraft guns and surface-to-air missiles.

Although most commonly used with Rheinmetall Air Defence twin 35 mm GDF series anti-aircraft guns, in 1996, the system underwent a successful demonstration in Crete with the Greek Artemis 30 twin 30 mm towed anti-aircraft gun system. Greece already uses Skyguard with its Skyguard/Sparrow air defence missile system.

Description

The Fire Control Unit (FCU) Skyguard controls short-range air defence weapons for use against manned and unmanned aerial targets such as; fighter aircraft, fighter bombers, helicopters, drones, cruise missiles, guided missiles and guided bombs. The FCU is trailer-mounted and fully air conditioned, with the cabin comprised of fire-resistant reinforced glass fibre polyester.

Rheinmetall Air Defence Skyguard fire-control system of the United Arab Emirates deployed in the desert (Christopher F Foss) 0509914

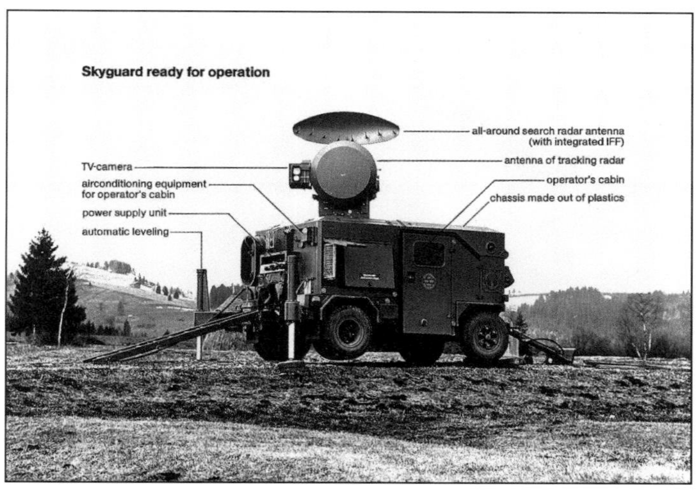

Main components of the Rheinmetall Air Defence Skyguard fire-control system 0509726

The antenna head, containing the search radar, tracking radar and TV camera, folds away within the cabin while travelling and, when required, is erected under full power control. When deployed in the operational position, Skyguard is supported on three outriggers that have fine levelling. As well as being mounted on a trailer, it can also be fitted on a variety of other vehicle types.

The pulse Doppler search radar has a maximum range of 20 km and key features include high detection probability against low-flying air threats, high information rate, fine resolution cell, high clutter suppression, automatic target alarm, automated acquisition procedure for tracking radar, high resistance to ECM, elaborate video presentation and optional IFF, according to customer requirements.

The pulse Doppler tracking radar is mounted below the search radar and features monopulse evaluation, fine resolution, accurate tracking, automated target exchange under computer control, air-to-surface missile detection and alarm, resistance to ECM and second tracking system when tracking missiles.

The TV tracking system is mounted to the right of the tracking radar and allows high-precision target tracking using a video processor, computer-aided tracking using joystick Vidicon (with or without light intensifier) or SEC-Vidicon camera.

The heart of the system is the CORA II - MB digital computer which carries out a number of tasks including threat evaluation based on search radar information, optimal hitting point data for the air defence guns, calculation of burst duration for optimal hit probability, calculations for guided missile deployment, calculations of target tactical data and monitoring of all Skyguard subsystems.

In addition, the computer carries out full automatic functional and performance checks of the entire system as well as simulation combat situations for purposes of realistic training.

The Skyguard control console can be operated by one or two crew and has a PPI that displays Doppler and 'raw' video as well as symbols and markers for the presentation of the tactical situation. A scope is provided to check target tracking and to judge the ECM situation.

There is also a tactical display with numerical readout, monitoring of the TV tracking system, roller ball to control markers, joystick for the manual control of the tracker, matrix panel for data input and output and an intercommunications system.

The power supply for the Skyguard system, a four-stroke petrol or diesel engine, is mounted internally but can be removed for external operation.

Information from the Skyguard fire-control system is sent to the weapons via standard two-wire field telephone connection.

Chinese Skyguard

During a parade held in China late in 1999, the People's Liberation Army displayed a new air defence system that included twin 35 mm towed anti-aircraft guns (similar to the Rheinmetall Air Defence GDF series) and a trailer-mounted fire-control system which is similar to the Skyguard.

Greek Skyguard

Greece is reported to be considering upgrading its Skyguard air defence system, which includes 12 fire-control systems, 20 quadruple RIM-7M missile launchers and 44 GDF-003 twin 35 mm, towed anti-aircraft guns. Early warning and target detection is carried out by four Ericsson Microwave Systems Super Giraffe radars.

Swiss Skyguard upgrade

In 1999, it was revealed that the Swiss Army had awarded a contract to Rheinmetall Air Defence to upgrade a batch of 66 Skyguard all-weather fire-control systems.

The first upgraded Skyguard system was completed in the third quarter of 1999 with final deliveries made during late 2001.

The upgrade includes improved search capabilities, more resilient Electronic Counter CounterMeasures (ECCM), better target detection and acquisition, enhanced data processing, more efficient operator performance aids, improved availability and maintainability.

Increased ECCM has been achieved by using an automated frequency jump and an improved staggered pulse-repetition frequency. A Track While Scan (TWS) system is installed with automatic target detection and selection facilities.

The installation of a high-performance laser range-finder provides full 3-D electro-optical tracking, while an improved tracker control, with power boost and Kalman filtering, reduces lock on times and improves tracking accuracy.

The fire-control system is fitted with a new C2003 computer with new software, offering a better-hit probability (using prediction algorithms with adaptive Kalman filters and much improved processing time). Provision will also be used for using the new Rheinmetall Air Defence 35 mm AHEAD ammunition in the future.

Built-in test equipment and modern modular components offer better service availability and repair when down.

In addition, operators will benefit from an automated engagement sequence with two-button operation, engage-and-fire and a new air conditioning system in the operator's cabin.

Variants

Details of the Skyguard/Sparrow air defence system are given in the Static and towed surface-to-air missile section under International. Details of the Skyguard/SAHV-3 air defence system, first announced in November 1994, are given in the same section.

Ericsson Microwave Systems of Sweden has developed the new Eagle add-on radar sensor for the Skyguard fire-control system.

There are separate entries under Brazil for the AVIBRAS FILA and under Taiwan for the CS/MPQ-78, both of which are similar to the Rheinmetall Air Defence Skyguard.

Status

The basic Skyguard is complemented with the Skyguard II and III upgrades.

In service with many countries including Argentina, Austria, Canada (10), Cyprus, Egypt, Greece (12), Malaysia, Oman, Saudi Arabia, Spain, Switzerland, Taiwan, Thailand (with twin 30 mm guns) and the United Arab Emirates.

Contractor

Rheinmetall Air Defence.

Skyguard III

Type

Static anti-aircraft control system

Description

Together with suitable weapons, the Skyguard III Fire Control Unit (FCU) provides protection of high-valued objects in a stand-alone or networked configuration. The system controls anti-aircraft guns and missile launchers for use against manned aircraft such as fighters, bombers and helicopters. It is also designed for use against unmanned aerial targets, in particular fast, small missiles in a steep diving attack. Skyguard can perform the following tasks:

- Air space surveillance over the complete elevation range with each antenna revolution
- High detection and acquisition performance also of extremely small and steeply attacking aerial targets
- Target tracking, engagement and fast, automatic target change
- Control of anti-aircraft guns, and missile launchers against:

(1) Manned aircraft such as fighters, bombers and helicopters
(2) Unmanned aerial targets, in particular fast, small missiles in a steep diving attack
(3) Mission adaptable local or remote operation for maximum operator safety.

Variants

Skyguard III has two variants, the original basic Skyguard and the basic Skyguard upgraded to the Skyguard III standard.

Rheinmetall Air Defence Skyguard III FC radar (Rheinmetall Air Defence) 0548208

Skyguard III (Rheinmetall Air Defence) 0127332

Status
Production as required.

Contractor
Rheinmetall Air Defence.

Skyguard Modernisation (upgrading to Skyguard III)

Type
Static anti-aircraft control system

Development
Oerlikon Contraves (now Rheinmetall Air Defence) initiated a modernisation programme for the Skyguard Fire Control Units (FCUs). The modernisation takes into account experience gained from the already operational air defence systems Skyguard, Skyguard II and Skyshield and is based on state-of-the-art Skyshield technology. It meets all requirements and demands ensuring the successful protection of vital points and sensitive areas.

Description
Together with suitable weapons, the Skyguard III Fire Control Unit (FCU) provides protection of high-valued objects in a stand-alone or networked configuration. It performs the following tasks:

- Air space surveillance over the complete elevation range with each antenna revolution
- High detection and acquisition performance also of extremely small and steeply attacking aerial targets
- Target tracking, engagement and fast, automatic target change
- Control of anti-aircraft guns and missile launchers against: manned aircraft such as fighters, bombers and helicopters; and unmanned aerial targets, in particular fast, small missiles in a steep diving attack
- Mission adaptable local or remote operation for maximum operator safety.

Variants
There is the basic Skyguard, Skyguard III and the modernisation programme.

Skyguard FCU Modernisation (Rheinmetall Air Defence) 0548211

Status
Production as required (modernisation).

Contractor
Rheinmetall Air Defence.

Skyshield 35

Type
Static anti-aircraft control system

Development
Skyshield 35 Ahead Air Defence System has been developed and produced as a private venture by Oerlikon Contraves (now Rheinmetall Air Defence) meet the requirements of present and future low-level air defence. To counter attacking air targets at ranges of up to 10 km, Rheinmetall Air Defence provides a gun missile weapon mix system, integrating the missile launcher into the Skyshield 35 AHEAD Fire Control Unit (FCU). This gun/missile-integrated system is a powerful combination layered air-defence system which provides optimal protection of vital points, it is based on a single FCU for managing both guns and surface-to-air missiles.

Description
The Skyshield 35 AHEAD FCU offers maximum crew protection and is of small, compact, lightweight and modular design. It consists of an unmanned sensor unit and a detached command post housing two operators (three operators if integrated with a missile launcher).

Skyshield 35 AHEAD FCU is suitable for controlling the 35 mm twin field air defence guns, types GDF-003 and GDF-005. If the GDF-003/05 guns are upgraded to fire the AHEAD ammunition, the user has available the full Ahead capability. Such AHEAD upgraded 35 mm GDF guns are still able to fire the conventional types of 35 mm ammunition.

Rheinmetall Air Defence Skyshield 35 FC radar (Rheinmetall Air Defence) 0548209

Together with suitable weapons, the Skyshield 35 FCU provides protection of high-valued objects in a stand-alone or netted configuration. It performs the following tasks:

- Air space surveillance over the complete elevation range with each antenna revolution
- High detection and acquisition performance also of extremely small and steeply attacking aerial targets
- Target tracking, engagement and fast, automatic target change
- Control of anti-aircraft guns and missile launchers against: manned aircraft such as fighters, bombers and helicopters; and unmanned aerial targets, in particular fast, small missiles in a steep diving attack.

Status

Development is completed and the Skyshield 35 FCU is ready for production. It is being offered as an upgrade to the Skyguard system.

Contractor

Rheinmetall Air Defence.

Taiwan

Sky Sword I

Type

Static Surface-to-Air Missile (SAM) system

Development

In mid-1988, the Taiwanese Cabinet, in its report to the Taiwanese Legislature, revealed that the Ministry of National Defence had successfully undertaken modification of the Sky Sword (Tien Chien) at the Chung Shan Institute of Science and Technology (CSIST). Sky Sword I is an infrared passive homing air-to-air missile converted for mobile, naval and static ground-launch applications. The latest configuration is the Antelope Self-Propelled system (see separate entry in the Self-Propelled surface-to-air missile section) which has been ordered by the Taiwanese Air Force.

The Sky Sword I has supplemented the US-supplied MIM-72 Chaparral/Sea Chaparral weapons.

Description

The all-aspect Sky Sword I is similar in physical appearance to the AIM-9G/H Sidewinder. Flight control is by cruciform canard fins. Maximum range is estimated to be 9,000 m within altitude limits of 15 to 6,000 m.

Specifications

	Sky Sword I
Dimensions and weights	
Length	
overall	2.87 m (9 ft 5 in) (est.)
Diameter	
body	127 mm (5.00 in) (est.)
Weight	
launch	90 kg (198 lb) (est.)
Performance	
Speed	
max Mach number	1.0 (est.)
Range	
max	4.9 n miles (9 km; 5.6 miles) (est.)
Altitude	
min	15 m (49 ft) (est.)
max	6,000 m (19,685 ft) (est.)
Ordnance components	
Warhead	HE blast fragmentation
Guidance	passive, IR
Propulsion	
type	solid propellant

Sky Sword 1 0022864

Status

In production and in service with the Taiwanese Army and Air Force, as of May 2012 this system has still not been offered for export.

Contractor

Chung Shan Institute of Science and Technology (CSIST).

Tien Kung I

Type

Static and towed Surface-to-Air Missile (SAM) system

Development

In the mid- to late 1970s, the Sun-Yat-sen (Chung Shan Institute of Science and Technology (CSIST)), began work on a new surface-to-air missile system for the Taiwanese Army-proposed 1990s Sky Net air defence system network that was based on the US-supplied MIM-103B I-HAWK missile.

A higher-thrust rocket motor was to be installed to increase both the speed and maximum engagement altitude. The airframe and control surfaces were redesigned in order to cope with the higher velocities, increase the weapon's manoeuvring capabilities while a new guidance system was fitted. This combined mid-course command guidance with a terminal Semi-Active Radar (SAR) seeker head which allowed the missile to fly an energy-efficient flight path to the vicinity of the target where the seeker head would take over for the final attack. The theory was that the actual radar illumination of the target which produces the reflected electromagnetic energy signals for the seeker to home on would be initiated only during the last seconds of the engagement in order to give the target the minimum amount of time either to evade or commence ECM. Ultimately, this project produced a missile design that resembled a scaled-up Raytheon AIM-54 Phoenix.

Due to the acquisition of more sophisticated missile aerodynamic technology, in the form of an 85 per cent transfer from Raytheon and the US government, mainly those of the MIM-104 Patriot design, the Chung Shan Institute was directed to stop working on the above. In 1981, it had started the development of the Tien Kung I (TK-1/Sky Bow I) missile system based on the Patriot.

Various combinations of different components and aerodynamic configurations to suit local production needs were tried. Using computer-aided design techniques coupled with an exhaustive scale model wind tunnel test programme. In the final stages of the work, full-scale test firings that included one missile configuration, fitted with much larger tail surfaces than Patriot. Firing trials were completed in 1985/1986.

In 1993, formal discussions began with Raytheon on the mobile Modified Air Defence System (MADS which is also known as Patriot T) which complements the Sky Bow programmes. In 1994, the MADS programme, which is based on the Raytheon Patriot, was unfavourably viewed because of the costs involved - estimated to total some USD1.3 billion.

In September 1993, the first fully operational Sky Bow I system was deployed to replace a HAWK unit in northern Taiwan. By 1997, four out of a planned six batteries with 500 missiles had been deployed. The sites are located at Keeling (in northern Taiwan), Kaohsiung (in southern Taiwan), Peng-hu (in the Pescadores Islands) and Tung Yin Island (off the Chinese mainland). The original deployment plan called for nine batteries with 700 missiles.

Tien Kung I SAM quad-box launcher (DTM) 0509599

Tien Kung I SAM being launched (DTM) 0509600

CS/MPG-25 CW illumination radar used with the Tien Kung I SAM system (DTM) 0509601

Following the critical cost review of the MADS, the Taiwanese Defence Ministry placed an order in mid-1994 with Raytheon, worth around USD600 million, for three fire units and some 200 missiles (to PAC-2 standard). The first unit was delivered in January 1997 to replace a Nike Hercules battalion currently in reserve status. The unit is deployed in the north of Taiwan to provide ATBM coverage for the capital Taipei. It is believed that the systems would be upgraded with PAC-3 missiles at a later stage. Late in 1996, it was stated that Taiwan had phased all Nike Hercules SAMs out of service.

Description

The Tien Kung I physical appearance and basic operational parameters remain similar to the Patriot. For the target acquisition, tracking and mid-course missile guidance requirements the army has deployed the CSIST/GE ADAR-1. This is a semi-trailer mounted 450 km range Chang Bei ('Long White') multi-function phased array F-band (old S-band) radar. The associated fire-control computer system and the CS/MPG-25 continuous wave dish antenna illuminator radar that are tied into the main phased-array radar, on a similar time-share basis to that employed by the US Navy's ship borne RCA AEGIS air defence system to allow multiple target engagement capability.

Each Tien Kung I battery (Fire Unit) is said to have one Chang Bei and two J-band 200 km range CS/MPG-25 illuminator radars servicing three or four, four-round missile launchers.

The Chung Shan Institute developed the Chang Bei radar with technological assistance from General Electric Company's RCA Electronic Systems department.

Tien Kung I active radar seeker (DTM) 0509602

Tien Kung I missiles being carried on (4 × 2) trucks for display purposes (L J Lamb) 0509720

The CS/MPG-25 is solely a Chung Shan Institute development and was derived from the I-HAWK AN/MPQ-46 HPI radar but is estimated to be something like 60 per cent more powerful in output. Improved EW, ECM and IFF capabilities were also introduced.

Before these Chung Shan Institute developments, operational Tien Kung I batteries were fitted with Tien He (Sky Union) data interface electronics to operate with the fire-control radars of I-HAWK batteries. Post-deployment of the radars, the Tien He (interface) system was introduced for both systems to co-ordinate their actions during an engagement.

There is no Track-Via-Missile (TVM) homing capability, as this technology was not included as part of the package released to Taiwan. Despite this, the basic SAR seeker-equipped version is being supplemented in service by a variant fitted with a passive all-aspect liquid nitrogen-cooled infra-red indium antimonide seeker. This provides the battery with the option of firing more than one missile type during a single or multiple target engagement. This variant was tested successfully in April 1985 against a HAWK missile target.

The trailer-mounted launcher station is almost identical to the one used for the Patriot, with an exception of minor details such as the frangible covers on the four container-launcher boxes. There is an underground silo launch version also.

Variants

A ground based version housed in underground shelters exists, this is designed to survive Suppression of Enemy Air Defence (SEAD) attacks.
Tien Kung II.
Tien Kung III.

Specifications

	Tien Kung I
Dimensions and weights	
Length	
overall	5.3 m (17 ft 4¾ in)
Diameter	
body	410 mm (16.14 in)
Weight	
launch	915 kg (2,017 lb)
Performance	
Speed	
max speed	2,644 kt (4,896 km/h; 3,042 mph; 1,360 m/s)
max Mach number	4.0
Range	
max	37.8 n miles (70 km; 43.5 miles) (est.)
Altitude	
min	100 m (328 ft)
max	24,000 m (78,740 ft)

	Tien Kung I
Ordnance components	
Warhead	90 kg (198 lb) HE fragmentation
Fuze	impact, proximity
Guidance	INS (command)
	semi-active, radar
	passive, IR (homing seeker)
Propulsion	
type	single stage, solid propellant (dual-thrust, sustain stage)

Status

In production and in service with the Taiwanese Army. As of November 2013, this system has not been offered for export. Known to be deployed at the Sanchih base in suburban Taipei. TK-1 has also been deployed by the ROC Army on the outlying islands of Penghu Island and Dong Ying island, bringing all of the Taiwan Strait, and parts of PRC's Fujian Province and Zhejiang and Guangdong Provinces within range.

Reports dating from August 2011 have suggested that the ROC intend to upgrade the system to the TK-2 standard and retire completely the TK-1.

Contractor

Chung Shan Institute of Science and Technology (CSIST).

Tien Kung II

Type

Static Surface-to-Air Missile (SAM) system

Development

Using the technology and experience gained from its previous surface-to-air missile development projects (Tien Kung I - TK-1/Sky Bow I), the Chung Shan Institute of Science and Technology (CSIST) developed the replacement for the Taiwanese Army's MIM-14 Nike Hercules systems.

Known as the Tien Kung II (TK-2/Sky Bow II), the first test firings apparently took place in the mid-1980s with limited deployment occurring in September 1998. The Sky Bow II officially replaced the Nike Hercules in October 1996 when the latter system was retired from Taiwanese usage after 37 years of active service. The associated search radar range is 300 km.

Tien Kung II vertical box launcher mounted on a locally produced launch ramp (DTM) 0509603

Description

The Tien Kung II is essentially the Tien Kung I round equipped with a second-stage solid propellant rocket booster motor section.

Two launch systems are used:

- A locally built single rail ramp
- Underground clusters of vertical launch systems.

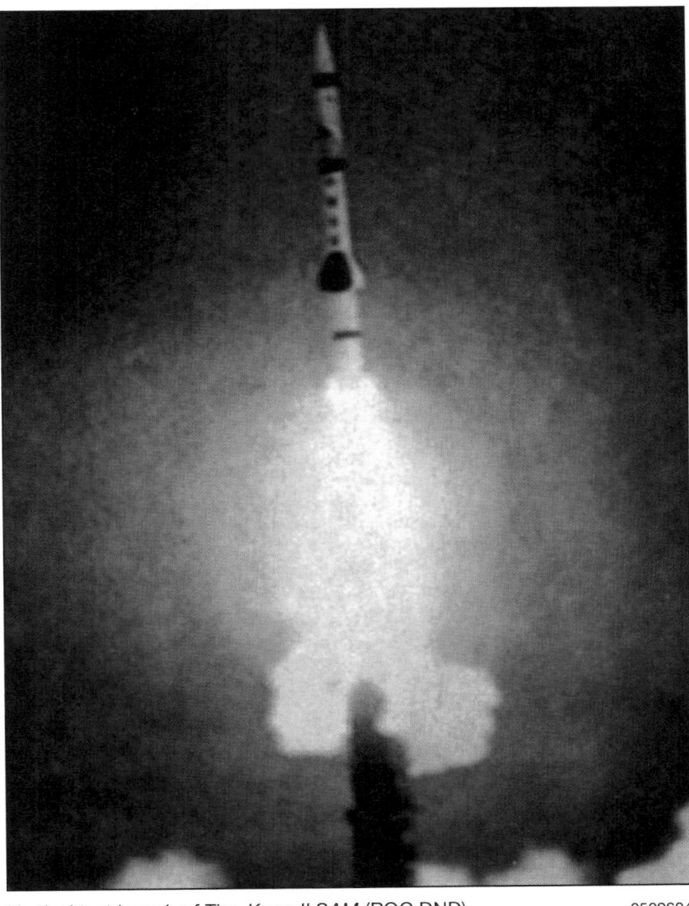

Vertical test launch of Tien Kung II SAM (ROC DND) 0509604

Underground launcher for Tien Kung II SAM system (DTM) 0509605

In 1994, it was revealed that the Sky Bow II would have an anti-tactical ballistic missile capability. Improvements were made to the fire-control, phased array, multiple-target tracking system and the weapon's manoeuvrability. In 1997, the anti-missile weapon capability was dropped following the acquisition of the Patriot system.

The Tien Kung II has an active radar homing seeker system that operates in the 8-12 GHz frequency range, sources have indicated that the radar technology for the seeker was purchased from the USA by CSIST in the mid 1980s.

There has been speculation from unproven sources that the system has also been modified into a surface-to-surface weapon, however, as most SAMs have this very limited capability it would not be surprising if a secondary mode of operation for this type of engagement does exist. It is known that the TK-2 has been used as a sounding rocket to perform upper atmospheric research for the ROKs civilian space programme.

Variants

Naval Tien Kung II

The vertical launch capability is to be used on a planned series of Kwang Hua I Batch 2 frigates.

Tien Kung III

The Tien Kung III SAM is based on the earlier Tien Kung 1 and 2 and is similar in configuration. This missile makes use of the Chang Bai (Long White) mobile phased-array radar and is said to have a performance similar to that of the Patriot PAC-2. The system was expected to begin flight testing mid-2008, however this was never confirmed, as of September 2010 no further information is available on this system or the trials.

Tien Chi

The Tien Chi (Sky Halberd) is a surface-to-surface missile derivative of the Tien Kung II. It was test fired during 1997. The system is apparently funded as a response to the People's Republic of China ballistic missile network aimed at Taiwan. The Tien Chi was reported as being ready for initial operational capability during December 2001.

Specifications

	Tien Kung II
Dimensions and weights	
Length	
overall	8.1 m (26 ft 7 in)
Diameter	
body	410 mm (16.14 in) (first stage)
	570 mm (22.44 in) (second stage)
Weight	
launch	1,115 kg (2,458 lb)
Performance	
Speed	
max Mach number	4.5
Range	
max	81.0 n miles (150 km; 93.2 miles)
Altitude	
max	30,000 m (98,425 ft)
Ordnance components	
Warhead	90 kg (198 lb) HE fragmentation
Fuze	impact, proximity
Guidance	INS, mid-course update
	active, radar
Propulsion	
type	solid propellant (sustain stage)

Status

Tien Kung II

In production and in service with the Taiwanese Army only. The system can be found at the Lung-tan Missile Test Site and is deployed at Peng-hu, Tungyin and Sanchih.

Tien Kung III

Development phase. This system has not been offered for export.

Contractor

Chung Shan Institute of Science and Technology (CSIST).

United Kingdom

Jernas

Type

Static and towed Surface-to-Air Missile (SAM) system

Development

The MBDA Jernas (Arabic word for a falcon that has just come to maturity and is at the peak of its predatory capability) air-defence system was first shown at the IDEX '93 Exhibition in Abu Dhabi.

Jernas SAM system 0509899

Jernas missiles on launcher (IHS/Patrick Allen) 1066707

The private venture development of Jernas started in the mid 1990s, retaining all the capability of the Rapier FSC system but with additional netting features which increase its operational flexibility and ensure that it can be easily configured to meet customer requirements.

Jernas can be configured to customer requirements ranging from a passive day/night availability obtained from the launcher operating in isolation, through to the full all-weather, dual-engagement capability of a complete system. The latter is capable of fully automatic engagements against a variety of target types, including ground attack aircraft, multiple pop-up helicopter attack, Anti-Radiation Missiles (ARMs), Remotely Piloted Vehicles (RPVs) and cruise missiles.

Description

Jernas utilises the eight-round launcher, the monopulse tracking radar and the 3-D Dagger surveillance radar of Rapier FSC. Together with the option of a new air conditioned two-person Tactical Operations Cab (TOC) that can be mounted on the rear of a suitable vehicle such as a (4 × 4) Land Rover Defender, tracked Swedish all-terrain BAE Systems, Global Combat Systems Bv206. At one stage a modified MOWAG Piranha (8 × 8) armoured personnel carrier was also being marketed but at July 2007 this was not the case and it is no longer available.

The TOC houses the Operator Control Unit (OCU) used by the operator for controlling the engagement and the Tactical Control Unit (TCU) which is used by the unit commander for combat management. As the controls are duplicated at each station, either person can take full control at any time.

The system has been designed for worldwide operation to meet UK MoD specifications. These include:

- The capability to operate in intense electronic and other countermeasures environments

- The meeting of the latest chemical and nuclear hardening requirements to operate in NBC conditions
- The ability to function in temperatures of up to +50°C.

Jernas can fire the Rapier Mk 1 and Mk 2 missile variants. If required by the user, further options are available for Jernas, including a helmet pointing sight as an aid to visual target acquisition. The complete Jernas system can engage two different targets at once, one with the electro-optical tracker and the other with the radar tracker.

The frequency-agile J-band pulse Doppler Dagger surveillance radar has processing capacity for over 75 targets per second to be detected and tracked simultaneously. It also automatically prioritises the primary threat tracks. The Dagger has a scan speed of 60 rpm and can also drive a number of launcher units to increase operational flexibility.

A typical engagement for the complete Jernas fire unit occurs as follows. The fire unit monitors all potentially hostile targets in its assigned airspace using its Dagger surveillance radar. The highest priority target is automatically acquired and tracked by both the radar and electro-optical trackers. To engage this target the operator simply presses the fire button and the missile is launched. It is then automatically commanded to intercept using the differential monopulse frequency-agile all-weather F-band radar tracker.

Simultaneously, the system will automatically acquire and track the second priority target using the passive IR electro-optical tracker. Thus, once the first missile is in flight, this second target is engaged. This operation is controlled by the operator who tracks the target whilst the missile is automatically guided to intercept.

The Dagger uses the Mk 12 IFF system to aid in its threat assessment. A high elevation guard beam automatically stops the Dagger's transmissions when the presence of an anti-radiation missile is detected.

A networking capability has been added for maximum flexibility.

Specifications

	Jernas Launcher
Dimensions and weights	
Length	
overall:	4.1 m
Width	
overall:	2.2 m
Height	
overall:	2.6 m
Weight	
combat:	2,400 kg
Mobility	
Firepower	
Armament:	8 × turret mounted Rapier Mk2 SAM
Rate of fire	
practical:	7 rds/min
Main armament traverse	
angle:	360°
Main armament elevation/depression	
armament front:	+60°/-10°

Jernas system

Maximum detection range:	>15,000 m
Maximum target altitude coverage:	5,000 m
Azimuth target coverage:	360°
Maximum target acquisition range:	>15,000 m
Maximum target missile engagement range:	>8,000 m
Guidance:	Automatic Command-to-Line-Of-Sight (ACLOS)
Automatic reaction time:	<5 s
Second target engagement time:	<3 s
Automatic threat assessment:	>75 targets/s
Operation:	all-weather day and night
Built-In Test (BIT):	automatic, with continuous system monitoring

Rapier Mk 2 missile

Type:	single-stage, low-altitude
Length:	2.24 m
Diameter:	0.166 m
Wing span:	0.381 m
Launch weight:	43 kg
Propulsion:	Roxel (UK) Epicete casted double base two-stage boost/coast solid propellant rocket motor
Guidance:	see text
Warhead type:	dual-mode SAP HE fragmentation
Fuze:	active multimode laser and contact
Maximum speed:	>M2.5
Manoeuvrability:	>35 *g* throughout dynamic range
Maximum engagement:	
(range)	>8,000 m
(altitude)	5,000 m
Launcher:	mobile trainable eight-round trailer-mounted

Status

In production. Malaysia signed a contract for the Jernas as part of its Air Defence modernisation programme in 2002, with deliveries having begun in late 2005. The system, comprising nine launchers, three tracking radars and three surveillance radars, plus the latest Rapier Mk 2 SAM.

In 2003, MMC Defence Sdn Bhd was appointed by MBDA, UK to design, R&D, build, test and commission Jernas support vehicles for the Malaysian Army. The company has since designed the Towing Vehicle Installation Kit (TVIK) and Combat Repair Vehicle Installation Kit (CRVIK) according to MBDA's technical specifications.

According to the then Malaysian Armed Forces Chief, Admiral Mohd Anwar Mohd Noor, the Jernas with the Mk 2 missile became operational with the Malaysian forces during early 2006.

Contractor

MBDA (primary contractor).
MMC Defence Sdn Bhd (support vehicles).

Rapier

Type

Static and towed Surface-to-Air Missile (SAM) system

Development

The Rapier low-level surface-to-air missile system was developed by the former British Aircraft Corporation (now MBDA), Guided Weapons Division from the early 1960s onwards to meet the requirements of the British Army and Royal Air Force Regiment for a missile system to replace the towed 40 mm L/70 Bofors air defence guns then in use. The operational requirements laid down for Rapier can be summarised as:

- Short reaction time with the ability to be taken into and out of action quickly
- Compactness and low weight
- High rate-of-fire and -kill potential
- Good defence coverage with maximum/minimum range performance
- The ability to engage targets with speeds from zero to M1.5 from ground level up to at least 3,000 m.

Before receiving MoD funding for the ET-316, the then British Aircraft Corporation invested its own money in a simpler weapon, then called Sightline, which eventually became ET-316, or Rapier as it was officially named in January 1967.

Design studies began in 1963 and Rapier was announced for the first time in September 1964 with the first unguided firing trials carried out in 1965.

In April 1967, the first successful guided engagement of a live target took place at the RAE range at Aberporth. This was a drone Meteor representing a crossing target and was flying at an altitude of 914.4 m at a range of 3,048 m from the firing site. The target was hit and destroyed. Joint Services evaluation trials of the Rapier at the Woomera range commenced in 1968.

The first production order was placed by the British MoD in June 1967 and the first production units were delivered in July 1970. The system achieved initial operational capability with the British Army and Royal Air Force in 1973. In addition to extensive trials in the UK, cold weather trials were carried out at Cold Lake in Alberta, Canada, and hot and wet trials were carried out in Australia.

9 (Plassey) Light Air Defence Battery, Royal Artillery, based at Kirton-in-Lindsey carried out troop trials of Towed Rapier, which also included a flight from No 63 Squadron, Royal Air Force Regiment. Kirton-in-Lindsey was the base of the Royal Artillery's first Rapier regiment, 12 Light Air Defence Regiment.

Rapier deployed in the field with tracker radar and generator on the right and four-round missile launcher on left with Rapier Mk 1 missile leaving launcher
0044356

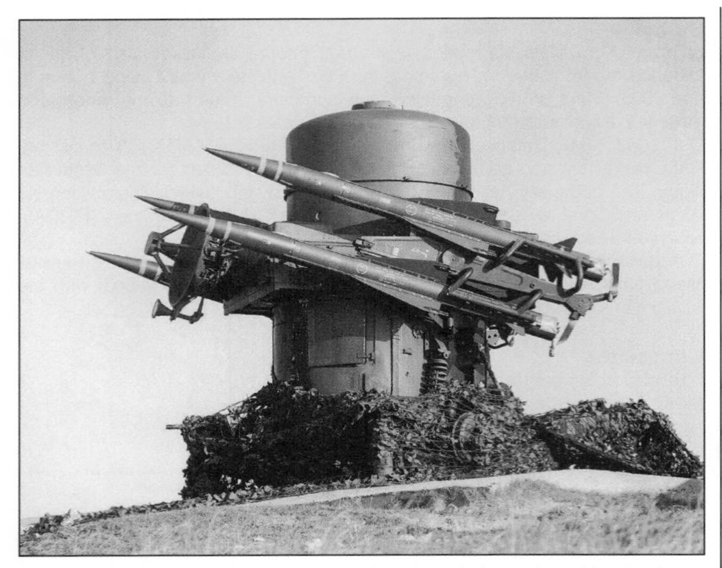

Four-round Rapier missile launcher deployed ready for action with wheels removed and supported on four jacks 0044355

Rapier SAM leaves its four-round launcher 0044354

Under the British MoD's Options for Change programme, the Royal Artillery has four air defence regiments, two with Towed Rapier and two with the Thales UK Starstreak High-Velocity Missile (HVM) system.

In the British Army, Rapier is used as the second and third line of battlefield air defence. The first is the Starburst man-portable system which is used as a gap filler for the Rapiers, mostly as a Vital Point (VP) defence system and a route defence system. The second line is provided by the Towed Rapier and Thales UK Starstreak HVM vehicles and the third by the Towed Rapier Systems.

The Thales UK HVM and Rapiers are employed in three main roles. VP defence to give localised protection to small-sized target locations such as HQs, vehicle choke points and logistic dumps. Each VP is allocated either four or six launchers. They are deployed in such a manner as to give mutual support and to try to destroy an attacker before it releases weapons. The Battery Command Post (BCP) normally coordinates the VP defences, but if a troop is allocated an independent VP task, command-and-control devolves to the troop commander.

The second mission role is route defence of main supply or unit withdrawal routes. A battery can protect up to 30 km, although this depends greatly on the local terrain. Two fire units are normally located at each end of the route to be defended in 'blocking' positions, and the other eight are deployed between 1 to 2 km off the line of march in alternate positions either side of the route.

It is the third mission role, however, that is used most often as it covers the largest area of battlefield. The Air Defended Area (ADA) can be used to cover both VPs and routes with a battery's Tactical Area Of Responsibility (TAOR). Once assigned this task, the troop commanders and their staff reconnoitre likely firing positions within their TAORs and send the details back to the BCP either by burst transmission radio message or courier.

The Battery Commander and his Command Post officer feed the information into the BCP processor, which selects the best 12 sites and a number of reserve sites according to firing arcs and radar coverage.

When this is finished the fire units are moved to their new locations. For a Thales UK Starstreak HVM or battery, the same three mission roles are assigned but because of increased mobility the individual fire unit commanders have much more latitude and can move on their own initiative when the position they are occupying becomes indefensible. For support of armoured units in direct combat with the enemy, the Starstreak HVM or battery elements will be located a number of kilometres behind, in order to remain out of range of enemy support fire weapons.

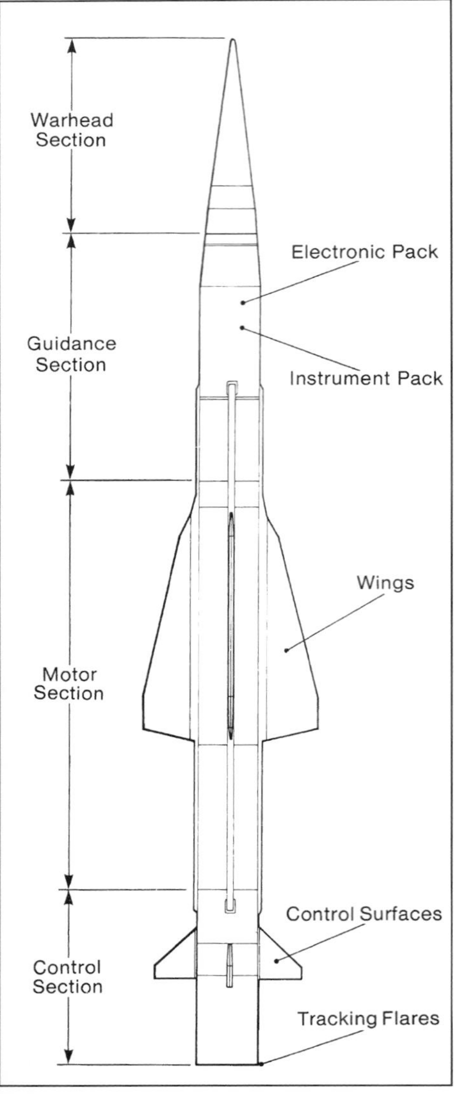

Drawing of the Rapier Mk 1 SAM showing position of key components 0509607

'T' Battery of 12 Air Defence Regiment, Royal Artillery used rapier, during the 1982 Falklands campaign. A total of 12 launchers were deployed with reduced logistic support and no tracker radars.

At the time of the Falklands conflict, the UK Rapier force was undergoing a systems upgrade to improve its performance, but the Rapier unit deployed to the Falklands was the original Field Standard A model.

'T' Battery of 12 Air Defence Regiment was followed by four tracker radar's from the Royal School of Artillery, eight Rapiers and tracker radars from 63 Squadron, Royal Air Force Regiment and finally 9 Battery, 12 Air Defence Regiment. The tracker radar was never used in action.

Rapier was, however, first used in combat by Iran in the second half of 1972 when it shot down an Iraqi Air Force Tu-22 'Blinder' supersonic bomber attacking Kurdish rebels in the Iran/Iraq border region. During the Gulf War, Rapier was responsible for a number of Iraqi Air Force losses.

A typical fire unit consists of a Land Rover carrying the optical tracker and towing the four-round launcher and generator power supply, a Land Rover towing the tracker radar (if used) and generator power supply and a Land Rover carrying nine reserve missiles.

Each battery also has a battery HQ and a repair section which has a Land Rover equipped with diagnostic and performance test gear and tows a trailer carrying a quantity of ready use spares. There is also a battery repair team with two vehicles, one for optical and hydraulic repairs and the other for major electronic repairs, and an ordnance spares vehicle.

The first Royal Air Force Regiment unit to receive Rapier was No 63 Squadron based at North Luffenham, UK, and this was subsequently deployed at Gutersloh, Germany, in mid-1974.

The second unit to be equipped was No 58 Squadron, which was deployed to RAF Laarbruch, Germany. No 27 Squadron was the first to receive the tracker radar in late 1977.

A Royal Air Force Ground-Based Air Defence (GBAD) Force Rapier squadron comprises an HQ flight, an engineering flight of four forward repair teams (of which two are for Rapier) and a second echelon maintenance section and two Rapier flights, each with three fire units. The fire units comprise eight crew, a Towed Rapier launcher, a radar tracker, two one-tonne (4 × 4) light vehicles and an LWB Land Rover with missile resupply trailer. A basic load of 17 missiles for the unit is carried.

Since it was introduced into the British Army and Royal Air Force Regiment in the early 1970s, there has been a series of phased developments not only to improve the performance of the system but also to keep the system ahead of the threat, which is constantly changing.

In mid-1987, for example, the British MoD placed an order with the then British Aerospace worth over GBP5 million for the supply of new digital computers for the Rapier air defence systems to replace older computers. By

Tracking radar deployed in the field. Kits were developed to give better target acquisition, ECCM and missile guidance performance 0509608

1987, more than 50 per cent of the launcher's major assemblies comprised 1980s technology including the key areas of computing and radars. This has had a significant effect on Rapier system reliability as well as increasing system automation.

These modifications are incorporated into current production systems and are offered to current Rapier users as part of a programme of modifications.

By 1998, the total signed orders for Towed Rapier and Tracked Rapier exceeded 700 fire units and 25,000 missiles, of which more than 12,000 missiles have been fired during development, training and combat. By 2000, production of Rapier in its original towed form was complete.

Individual parts of the Towed Rapier air defence system can be carried slung under medium- or heavy-lift helicopters such as the SA 330 Puma and CH-47 Chinook. A C-130 Hercules transport can carry one complete system with radar tracker or two optical systems with vehicles.

For the export market, Rapier was tailored to meet the specific requirements of each customer, but the British Armed Forces use designations to avoid confusion.

The basic Rapier, together with its tracker radar, was called Field Standard A (FSA). This was followed in 1979 to 1980 by Field Standard B1 (FSB1) that has a free-standing tripod-mounted 'pointing stick' which, when pointed at a target and activated by pressing a trigger, aligns the optical tracker automatically. There are also improvements to both ECCM and ECM, including an improved planar array antenna and an automatic code changer for the IFF. The Royal Air Force upgraded its FSB1 systems to the FSB1(M) (Modified) standard, the reliability of which is some 50 per cent higher than the predecessor system. These were subsequently replaced by FSC systems.

In mid-1994, an invitation to tender was issued to the then BAE Dynamics as prime contractor for the design, development and supply of a tactical control man/machine interface known as the Tactical Display Unit (TDU) for the FSB1(M). The TDU replaces the SEZ, SLICOS and RDU units and is a single, two-man-operated portable system that utilises processed target data from the surveillance radar and tactical data sent over the Rapier communication net. The TDU presents information to the tactical controller simultaneously over two displays: the Primary Position Indicator Display (PPID) and one or more secondary information displays. An initial quantity of 31 TDUs was required with up to 40 more on option.

The original FSB1(M) programme was divided into three areas: launcher modifications (B1); mid-life improvements (B1MLI); and tracking radar modifications (TR1LBI - Tracker Radar Mid-Life Improvements). The programme began in 1989 and was completed in 1991.

In 1985, troop trials began of the Field Standard B2 (FSB2) electro-optic system, known as Rapier Darkfire or FSB2 (EO). This involved the introduction of an infra-red tracker which replaced the optical unit, a new six-round launcher incorporating an improved planar array Racal/Northern Telecon (formerly STC) pulse Doppler surveillance radar, an automatic code changer for IFF and the Console Tactical Control system for the fire unit commander. First deliveries

Land Rover 127 of the Royal Air Force Regiment towing a tracker radar 0509609

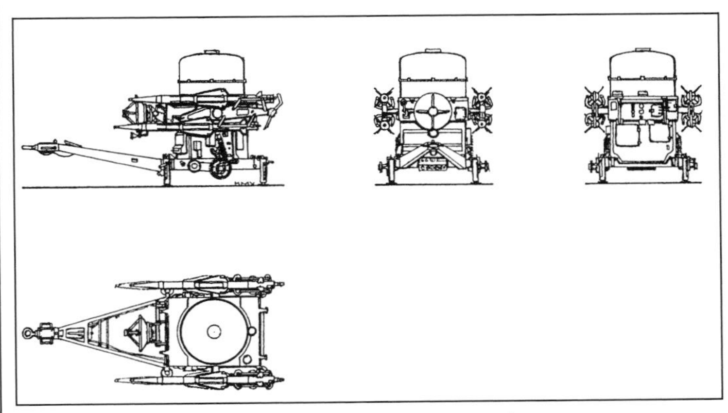

Four-view drawing of Rapier fire unit (Henry Morshead) 0509611

were made in 1988. The completion of programmed deliveries, which total four batteries of FSB2 systems plus support, spares and maintenance, was achieved by the end of 1992. By July 2007 the Darkfire was no longer in service.

The Rapier FSC and Jernas systems are described under separate entries.

Description

The main component of the Towed Rapier system is the launcher, which is carried on a two-wheeled A-frame trailer. When in the firing position, the wheels are removed and it is stabilised on four adjustable legs. The engagement envelope is from –5° to +60° in elevation and 360° in traverse. The F-band surveillance radar (range 15 km, altitude 3 km) is located in the launcher base. The aerial is housed under this radome and rotates approximately once a second. Northern Telecon made the planar array aerial.

The radar, together with the IFF, provides for the early detection and warning of approaching aircraft and helicopters. The launcher turntable rotates through 360° and carries four missiles in the ready-to-launch position, two on each side. The command transmitter and aerial, which provide the link between the computer in the base of the launcher and the missiles in flight, is located between the two banks of two missiles.

The optical tracker stands on a tripod which has a levelling jack on each leg and consists of a static column with a rotating head providing 360° coverage in azimuth. Movable prisms in the rotating head provide between –10° to +60° elevation coverage. The wide field of view is 20° and the narrow field of view 4.8°. The computer normally controls operation with operator override at any time to select the narrow option.

The manual tracking system is optical, with the operator using a joystick-operated servo-driven unit to track the target. The guidance system uses a fully automated TV system which gathers the missile in flight after launch using an 11° field of view, then automatically switches over to a 0.55° field of view to guide it to interception with the target.

The operator is provided with a binocular sight for target tracking and has a few simple controls to operate the system. A second monocular sight is provided at the rear of the static column, which an instructor can use to monitor students' performances during field training.

The Tactical Control Unit (TCU) provides tactical control facilities and is connected by cable between the launcher and optical tracker. It is divided into 32 sectors in azimuth, each sector covering 11.25°. By operating sector switches, blind arcs can be built up as required to provide 'safe' channels for friendly aircraft or to set in priority arcs of fire.

The Rapier missile has a streamlined monocoque body of circular cross-section and consists of four main sections: warhead, guidance, propulsion unit and control.

The 1.4 kg warhead section contains the semi-armour-piercing warhead, with 0.4 kg of explosive, a safety and arming unit and crush fuze. A collapsible plastic nosecone is fitted to the penetration head to provide optimum aerodynamic shape.

The guidance section is in two parts: the electronic pack and the instrument pack. The Imperial Metal Industries, Troy propulsion unit is an integral dual-thrust two-stage rocket motor and gives the missile a maximum speed of around 650 m/s. The control section contains the hot gas-driven control

From left to right are the pointer, Tactical Control Unit (TCU) and the Optical Tracker (OT) 0509806

surface actuation mechanism, which controls the missile in flight, and pyrotechnic flares to facilitate TV gathering and tracking. The same Mk 1 missile was used for both the optical and tracker radar guidance modes and for Tracked Rapier. A missile flying to maximum range takes 13 seconds with the minimum time of flight (required to arm) being three seconds. The Single-Shot Kill Probability (SSKP) is stated to be over 70 per cent.

Rapier is manufactured as a round of ammunition and requires no maintenance testing or servicing once it has left the ordnance depot except for routine changing of desiccators. When stored in controlled conditions the missile has a shelf-life of 10 years.

In May 1988, the hybrid Rapier Mk 1E missile design was trialled against RPV-sized battlefield targets. The Mk 1E uses the propulsion system of the existing Mk 1 weapons with an advanced 'intelligent' passive infrared proximity and graze fuze system and a revised high explosive pre-formed fragmentation warhead. It was envisaged that the missile would complement rather than replace the Mk 1 to enhance the system's capabilities against smaller-sized aerial targets. Production started in 1989.

The Mk 2 Rapier missile, which is now the only version in production, is fully compatible with all versions of the Rapier missile system including Optical Rapier, tracker Rapier, Rapier FSC and Jernas.

In mid-1992, the then BAE (Dynamics) was awarded a GBP11 million contract to modify all Towed Rapier systems currently in service with the British Army and Royal Air Force Regiment to fire the Mk 2 missile. The programme included changes to the existing launcher computer software, as well as the Tactical Control Unit, the Marconi Radar Tracker system and all associated maintenance support software packages. Deliveries commenced in late 1995.

The 43-kg Mk 2 missile exists in two versions and has a 15 to 20 per cent increase in range. One version retains the semi-armour-piercing warhead of the original; the other has a combined fragmentation and armour-piercing warhead and dual crush and proximity fuzes.

The then Thomson-Thorn Missile Electronics produced a small pre-production order of the active Infra-Red (IR) fuzes for the initial Rapier Mk 2A missile in 1990. Full production of the fuze began in mid-1991 as a milestone objective in the GBP10 million developments and initial production contract awarded to the company in 1986.

The fuze system comprises an IR laser transmitter and four quadrant receiver optics units, coupled with intelligent signal processing to determine the optimum range to target position at which detonation of the fragmentation warhead is to be initiated.

The DN181 K-band 10-km range tracker radar was originally developed by the then Marconi to enable Rapier to cope with the night and all-weather threat. The non-coherent frequency-agile radar was designed to meet the same characteristics of size, lightweight and mobility as the rest of the system. It is therefore trailer-mounted and can be plugged simply into any Rapier system.

In operation, the monopulse radar employs differential tracking of both the missile and target using a very narrow pencil beam to achieve the accuracy required.

The development of the tracker radar commenced under contract to the MoD in 1968 with first prototypes completed in 1970. The first pre-production system was handed over in 1973 and the main production order was placed the same year. First production units entered service with the MoD in 1979. Iran placed its first order in 1974. The tracker radar was also successfully tested with the US Chaparral SAM system during trials in the US. Over 350 were built for the home and export markets before production ceased.

The tracker radar is mounted on trailer chassis, which is similar to the launcher. When deployed for action, the wheels are removed and it is supported on four adjustable legs. The base housing contains the electrical, electronic and hydraulic power assemblies and is static during operation.

The upper assembly carries the main reflector and sub-reflector assembly, the TV gathering unit with its power supply and the RF unit. This assembly rotates in azimuth and the aerial system in elevation.

The addition of the tracker radar allows Rapier to engage targets successfully in darkness and poor visibility. As it is autonomous the target tracking and acquisition process decreases system reaction time, increases kill probability and reduces human operations to a monitoring role. Once the surveillance radar has picked up a target, the azimuth bearing is passed automatically to the tracking radar, which slew rapidly onto this bearing. A high-speed search pattern is carried out and the precise position of the target is established.

As soon as the tracking radar is locked on to the target, the operator is informed by an audio tone that the radar is tracking and he then switches to the radar mode to allow the tracking radar to control the Rapier system. Immediately the target comes within range, a lamp indicator tells him he is free to launch a missile. All the operator has to do is to press the launch button.

Once the Rapier missile is launched, the radar tracks both the target and the missile. Error signals are derived automatically within the radar system and passed to the command guidance unit, which uses encoded signals to direct the missile flight path to reduce the error angle to zero.

The tracker radar uses a standard Rapier generator and requires only the cable connection to the launcher unit and to the lamp indicator system control switch on the optical tracking unit.

The system control switch enables the operator to select either radar or optical guidance for the engagement at any point up to the moment of firing.

A mobile optical fire unit consists of two Land Rovers (or similar vehicles), the launcher and a light trailer.

The first Land Rover carries the optical tracker and four missiles in their travelling boxes, and tows the launcher with its generator set rear-mounted. The second Land Rover carries stores and supplies for the fire unit and tows the missile resupply trailer, carrying a further nine missiles in their travelling boxes. The fire unit can be brought into and out of action in under 20 minutes and can be manned by three men, although five are normally used for sustained operations. Rapier can also be integrated into an overall air defence system.

An optical fire unit can be converted into a tracker radar unit simply by the addition of a tracker radar towed by a third Land Rover, which is similar to that which tows the launcher. It carries four missiles in their travelling boxes and tows the tracker radar unit with its generator set rear-mounted.

A fully mobile tracker radar fire unit thus consists of three light vehicles, three light trailers, with a total of 17 ready-use/reload missiles. In terms of capabilities, it has its own surveillance radar, IFF system, guidance computer, day, night and all-weather tracking systems, ready-use and resupply missiles and power supply units. It can thus operate autonomously or be integrated, with the necessary communications system, as part of an air defence network.

A typical Rapier target engagement takes place as follows:

The surveillance radar is continuously rotating through 360° looking for aircraft or helicopters, which come within range. When detected, a target is automatically interrogated by the Cossor IFF system. If no friendly reply is received, the operator is alerted by an audible signal in the headphones. At the same time, the rotating head on the optical tracker and radar tracker automatically lines up with the target in azimuth followed by the launcher. The radar tracker then begins its automatic elevation search while the operator undertakes a manual visual search. If a radar tracker is available it will be used as the prime means of target engagement and the procedure will be as described. However, an alternative manual engagement procedure can be used. The operator then undertakes an elevation search to acquire the target, after which he switches to the track mode and begins to track the aircraft visually using a joystick control. He is then able to identify the aircraft visually. Information from the optical tracker and the surveillance radar is fed into the system computer in the launcher and these data are used to calculate whether or not the aircraft is within effective range of the system.

When the aircraft comes within firing range a visual signal appears in the operator's field of view and he immediately presses the firing button to initiate a missile launch.

The computer calculates toe-in and turns the launcher towards the optical line-of-sight. The missile is automatically gathered along the sight line by the TV system until impact. During the missile flight the only task of the operator is to keep tracking the target.

The original Mk 1 missile has a semi-armour-piercing warhead, which penetrates the aircraft skin. The crush fuze, after a short delay, detonates the high explosive inside the aircraft, causing the target to be destroyed. This contact technique has caused Rapier to be known as a 'hittile' system.

Once the target has been engaged, the operator can switch back to search so that another engagement sequence can be started immediately, or a second missile may be launched at either the same target or another target in the operator's field of view.

A trained crew in less than 2.5 minutes can reload four replacement missiles. The surveillance radar can detect low-flying targets in the presence of heavy ground clutter out to a range of more than 15 km and the missile itself can engage targets at 7,000 m, giving an overall intercept coverage of 150 km^2 per fire unit. Numerous firings have demonstrated that the system reaction time (from when the target is detected until a missile is launched) is about six seconds.

Towing vehicle

In the British Army, the Rapier was normally towed by a one-tonne (4 × 4) Land Rover. Following trials in 1990, the Royal Air Force Regiment placed an order with Land Rover for 214 long wheel base 127 V-8 powered Land Rovers to tow Rapier systems in both the United Kingdom and overseas.

For Operation Granby, (Gulf War 1990/91), a number of SUPACAT (6 × 6) all-terrain vehicles were modified for use as Towed Rapier Support Vehicles and deployed to the Middle East.

Variants

Rapier II or (Rapier B1X)

This is the current export name of the basic Towed Rapier and is based on the requirement laid down by the British MoD for use by the Royal Artillery and

Royal Air Force air defence units. The Rapier II (or Rapier B1X upgrade or equivalent) has been sold to Australia, Oman, Singapore, Switzerland, Turkey and the US.

Rapier FSC
This is now in operational service with the British Army.

Tracked Rapier
With the introduction of the Rapier FSC and the Thales UK Starstreak HVM, the Tracked Rapier system was withdrawn from British Army use in 1997.

Specifications

	Rapier Launcher	Tracker radar
Dimensions and weights		
Length		
overall:	4.064 m	4.14 m
Width		
overall:	1.765 m	1.753 m
Height		
overall:	2.134 m	-
transport configuration:	-	2.032 m
deployed:	-	3.378 m
Weight		
combat:	1,227 kg	1,186 kg
Mobility		

	Rapier Mk 1
Dimensions and weights	
Length	
overall	2.24 m (7 ft 4¼ in)
Diameter	
body	133 mm (5.24 in)
Flight control surfaces	
span	0.381 m (1 ft 3 in) (wing)
Weight	
launch	42.6 kg (93 lb)
Performance	
Speed	
max speed	1,263 kt (2,340 km/h; 1,454 mph; 650 m/s)
Range	
min	0.3 n miles (0.5 km; 0.3 miles) (optical/tracker Rapier)
max	3.8 n miles (7 km; 4.3 miles) (est.)
Altitude	
min	15 m (49 ft) (est.)
max	3,000 m (9,843 ft) (est.)
Ordnance components	
Warhead	1.4 kg (3.00 lb) HE semi-armour piercing, blast fragmentation
Fuze	time delay, impact
Guidance	semi-active, CLOS (optical, thermal or automatic, command)
Propulsion	
type	two stage, solid propellant

Status

Production complete. A considerable number of ex-UK Armed Forces Towed Rapier and Tracked Rapier launcher assemblies are available for refurbishment and resale. Rapier has been sold to the following countries:

Abu Dhabi
The Rapier systems were bought back by MBDA as part of a Hawk aircraft acquisition programme in 1989.

Australia
A contract for 20 optical Rapier systems was signed in December 1975, with the first optical Rapier fire unit accepted in October 1978. The contract included test equipment and a base repair facility. In January 1977, an order was placed through the then British Aerospace worth GBP13 million for 10 tracker radars to give the system an all-weather capability. Rapier became operational with the 16 Air Defence Regiment, (integrated with ready reserves) Australian Army, in 1980.

The 16 Air Defence Regiment deploys 12 of the launchers with two at the RAEME School at Albury, four at the School of Artillery in Manly, Sydney and the remaining two in operational reserve. In 1984, a contract was placed for FSB1 upgrade kits. These were subsequently embodied in Australia.

In early 1995, a GBP2 million contract was placed with Matra BAE Dynamics to incorporate the Cossor IFF 892 Mk 12 system into Rapier.

Brunei
The government of Brunei announced its intention to purchase Rapier tracker radar in late 1978 with the contract, worth over GBP30 million, being signed in 1979 and first units delivered in 1983. Matra BAE Dynamics was also awarded a contract worth more than GBP3 million by Brunei for construction of a missile firing range on its coast. It is understood that a total of 12 Rapier fire units and four tracker radars were delivered.

Indonesia
Late in 1984, Indonesia signed a contract worth GBP100 million for optical Rapier air defence systems for use by the army. A second order, valued at over GBP100 million, was announced in December 1985. The third order, worth GBP40 million, was signed in December 1986, to bring the total to 51 optical systems. In early 1987, Indonesia became the first export customer for the then British Aerospace Battery Command Post Processor System for Rapier. Total value of the contract, for five systems plus spares, was GBP500,000.

The Battery Command Post Processor, which is already used by the British armed forces, links all elements of the battery by radio. Data can be passed using tactical data entry devices coupled to existing combat radio, which permits the automatic transmission of data to and from the command post. Messages can also be passed in voice to the command post radio operator who enters the data, using a keyboard, into the battery command post processor. All relevant data can be received, stored, processed and displayed on request. By greatly speeding up management and computation tasks, which have previously been carried out, manually, the assessment times for deployment are much reduced.

Iran
The first order, placed in June 1970 and worth GBP47 million, covered technical maintenance and support training. It originally deployed an optical version but subsequently deployed tracker radars. In Iran, Rapiers are manned by the air force and a total of 45 launchers to equip five batteries was originally procured, although most are now non-operational due to non-availability of spare parts owing to the long-standing British government arms embargo. Iran is now reverse engineering Rapier technology (see entry in self-propelled SAMs section).

Oman
An order was placed in mid-1974 for 28 fire units valued, with spares and training, at GBP47 million. These were delivered in 1977. Oman purchased 12-tracker radars in 1980. Oman has contracted for modification upgrade kits to raise its Rapiers to the latest export standard. The GBP40 million plus contract involved shipping the systems back to the UK where MBDA has brought them up to the locally designated SAHAM standard. This is equivalent to the B1X configuration. The upgrade involves ECM and ECCM improvements, the enhanced planar array antenna, the Rapier Mk 2 missile and an automatic IFF code changer. Oman has also received 12 Rapier launchers and six-tracker radar's from Qatar. Oman has three Rapier squadrons: numbers 10, 12 and 22.

The first firing of an upgraded SAHAM Rapier system using the Mk 2 missile took place at the Royal Air Force of Oman's (RAFO) firing range in the south of Oman during late September/early October 1995.

Qatar
The army ordered Rapier in 1981 and took delivery of 12 launchers and six tracker radar's but these have since been passed to Oman.

Singapore
The air force deploys with the 165th Singapore Air Defence Artillery Squadron 12 Rapier launchers, which were ordered in 1981 at a cost of GBP60 million. They are used in conjunction with six-tracker radars.

Singapore is one of the five customers that has upgraded to the Rapier B1X (or equivalent) upgrade standard.

Switzerland
The Swiss government formally proposed the purchase of 60 Rapier units in June 1980 at an initial cost of CHF1.2 billion (GBP315 million) which included GBP50 million for tracker radar's which are issued on the scale of one per Rapier launcher. The order was finally placed with the then British Aerospace in December 1980 with a proportion of the equipment built in Switzerland shared between government factories and private industry.

Firing trials of the first Swiss Rapier units were completed in the UK in 1982. The Swiss use Pinzgauer (6 × 6) 2,000-kg light vehicles to tow their Rapier systems.

The Swiss Rapier systems have a number of modifications to meet their unique operational environment, including an improved acquisition and tracking capability in mountainous terrain, improved ECCM capability, PPI tactical display and Swiss IFF system.

The target display-and-control system allows the operator to select automatic or manual target designation, threat assessment and effective countermeasures against electronic jamming, as well as fast identification of defects in the system.

MBDA also supplied a ‘special to Swiss’ requirements prototype Rapier Classroom Acquisition and Tracking Trainer (RCATT) for evaluation. This was followed by a contract in 1991 for eight production standard versions. 1993 completed deliveries.

In mid-1994, a GBP8 million, two-and-a-half-year development contract was awarded to the then Matra BAE Dynamics for the upgrading of the training equipment associated with the Rapier system. Included in the work was the development of a new engagement simulator for training the commander. In addition, several mid-life improvements will be made to enhance the performance and maintainability of the surveillance radar and the tracker radar. The RCATTs will be modified to link to the commander's engagement simulator so that both the commander and the operator can be trained together. The development contract will be followed by a separate production contract.

In May 1995, a GBP6 million contract was awarded by the Swiss Defence Technology Procurement Agency to the then Matra BAE Dynamics for the development of the Swiss Mid-Life Improvements (SWIMLI) Phase 2. The contract covers the development of improvements to the surveillance radar, tracker radar and launcher. The contract also contains an option for production. Development work on the SWIMLI programme started in April 1994. The system upgrade is equivalent to the B1X package but tailored to

Swiss needs. In February 1999, Matra BAE Dynamics was awarded a GBP50 million plus contract to implement the production phase of the SWIMLI programme. The programme is known by the local designation of KAWEST II.

Each Swiss Army mechanised division has one battalion with 20 mm LAAGs and a second with Rapier SAM systems. The latter has an HQ battery and two Rapier batteries. A total of three Rapier militia battalions have been formed. The 60th Rapier system was handed over on 1 May 1986.

Turkey

In August 1983, Turkey placed a GBP146 million order for 36 Rapier launchers and 12 tracker radars. These were delivered between November 1983 and late 1985 and used by the air force. In mid-1985, a second contract for 36 launchers and 15-tracker radars was placed but, due to financial problems, was not implemented until December 1985. Additional 14 launchers and 13-tracker radars were obtained from USAF Europe.

In February 1996, the then British Aerospace Defence Ltd announced that it had been awarded a GBP70 million plus contract to upgrade Turkish Air Force Rapier systems to the Rapier II configuration. The Rapier II is very similar to the Rapier B1X and introduces digital technology to not only give an increased performance but significant improvements in Reliability, Availability and Maintainability (RAM). The programme was completed in May 1999.

In addition to modifications to the actual Rapier launcher and the radar tracker, a commander's display unit has also been introduced. The work was carried out over a three-year period in the UK with the then GEC-Marconi (radar tracker) and Racal-Thorn Defence (surveillance radar) being the major subcontractors.

In late 1998, talks were underway between the UK and Turkey about co-producing Mk 2 Rapier missiles for Turkey over a 10-year period. An additional batch would be produced for the UK over the same period. In mid-1999, the contract was signed for the joint production of the Rapier Mk 2. A total of 840 Mk 2 missiles for Turkey and 1,800 for the UK will be built, with first deliveries in 2002.

United Kingdom

Army - Following recent changes in the UK Armed Forces the British Army now fields one Regiment FSC version (three batteries each of two troops with four fire units apiece).

The Resident Rapier unit also provides the short-range missile defence of the Falkland Islands. A four-launcher flight detachment deployed to Belize was withdrawn in early 1994.

United States

(Turkey) - In October 1985, it was announced that the US government had placed an order under the US European Air Defence Initiative programme for 14 Rapier launchers with 11 tracker radar systems to defend two USAF bases in Turkey which are manned by Turkish military personnel. The bases, at Incirlik and Izmir had six launchers apiece in a Turkish Short-Range Air Defence (SHORAD) squadron. These systems together with 515 missiles were handed over in June 1995 to the Turkish Air Force under the Excess Defence Articles (EDA) programme.

(UK) - In 1981, the USAF and British government reached an agreement for the supply of 32 Rapier launchers with attendant tracker radar systems to defend US bases in the UK. Manned by the Royal Air Force Regiment the three squadrons were formed between 1983 and 1985. Total value of the contract was approximately GBP327.5 million with 28 of the systems being used operationally and four for training.

These Rapier systems have now been returned to the United States where they are used for range training work. In 1996, the USAF turned over 312 Mk 1 and 306 Mk 2 missiles to the UK under the EDA programme.

Zambia

This country was supplied with 12 optical Rapiers in 1971 to defend the country against the then Rhodesian air attack, but they rapidly became unserviceable after the withdrawal of the BAE support team in 1977.

Contractor

MBDA.

MBDA Dagger surveillance and target acquisition radar for Rapier FSC SAM system. It has an advanced planar array and uses the same trailer as other parts of the Rapier FSC system 0509900

Trials firing of the Rapier FSC low-level air defence system 0044357

Rapier Field Standard C (FSC)

Type

Static and towed Surface-to-Air Missile (SAM) system

Development

In November 1986, the then British Aerospace Stevenage (now MBDA) was awarded a contract worth some GBP1.6 billion for the design, development and initial production of the Rapier FSC air defence weapon for delivery to the British Army and Royal Air Force Regiment in the mid-1990s and beyond.

Development of Rapier FSC actually started in 1983 as part of the continuing Rapier improvement programme contract for the UK Ministry of Defence.

While the earlier Rapier system is considered to be effective against the current low-level air threat, a major improvement programme was thought necessary if the system was to remain effective in the 1990s and beyond.

Main improvements of Rapier FSC over the existing system can be summarised as:

- Higher rate of fire
- Greater operational flexibility
- New and more effective missiles
- Extensive built-in test capability
- Enhanced reliability
- Increased robustness
- A significant improvement in resistance to ECM
- Has protection against severe battlefield environments, including NBC.

Rapier FSC has been designed to counter low and ultra-low-level air threats, including fast ground attack aircraft, pop-up helicopters, RPVs and cruise missiles, under all weather conditions and in an ECM and NBC environment.

For this contract, the then British Aerospace agreed with the UK MoD an incentive pricing arrangement bound by a maximum price for the whole package. The contract, as announced in late 1986, covered a production order sufficient to re-equip one of the Royal Artillery's Air Defence Regiments with three troops of four fire units each. Plus, four Royal Air Force Regiment Ground-Based Air Defence (GBAD) Force squadrons and a training/evaluation unit (Numbers 15, 16, 26, 37 squadrons and the Rapier JRTU training unit). Each squadron has two flights with three fire units.

In November 1993, the then British Aerospace Defence, Dynamics Division, stated that Rapier FSC was in full-scale production and that the certificate of design for the system had been agreed by the UK MoD. The total cost of the programme is now estimated to be GBP1.85 billion over a 10-year period.

It was originally expected that 205 Rapier FSC fire units would be purchased by the UK but this was cut back to a total of 57.

The first British unit to receive the FSC was No 26 Squadron, RAF Regiment at Laarbruch in Germany. The squadron first took delivery of FSC in February 1995 but the equipment was not officially purchased until 1 April 1996. The intervening period involved the unit in a series of trials and development of the equipment. On 1 April 1996, the squadron was officially declared to NATO as being operational.

Under earlier orders, Matra BAE Dynamics received an order for 57 Rapier FSC systems of which 14 were not originally to have a tracking radar.

Late in 1995, the then Dynamics Division of British Aerospace Defence Limited was awarded a contract by the UK MoD worth over GBP100 million for the production of a further quantity of Rapier FSC Surface-to-Air Missile (SAM) ground equipment.

This contract, Tranch 1A of the Rapier FSC contract, covers the supply of an additional 14 radar trackers, spares and support equipment.

The latter includes three second-line maintenance trucks, SEERT, TERSE and SEGRT which are used to repair faulty Line Replaceable Units (LRUs) from the three Rapier prime units.

MBDA Rapier FSC fire unit deployed with eight Rapier Mk 2 missiles in the ready-to-launch position 0509804

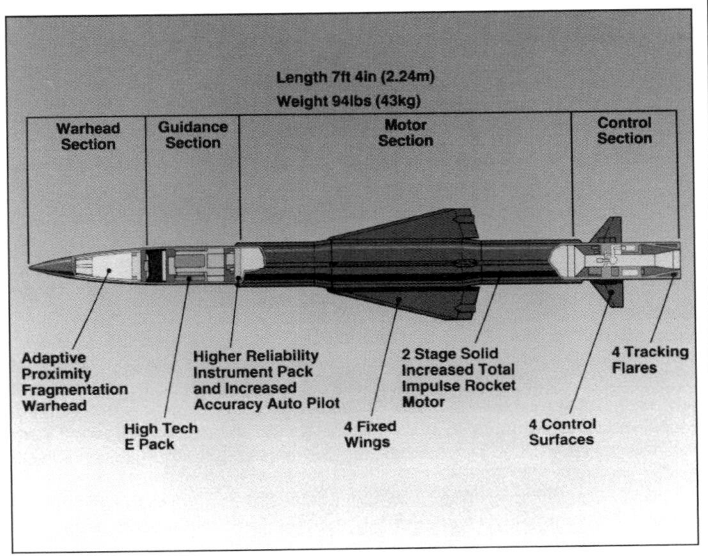

Main components of the Rapier Mk 2 missile which can be fired by the Rapier FSC system 0509606

The additional radar tracker order will fulfil the British Armed Forces operational requirement for all Rapier FSC units to have a radar tracking capability.

Rapier FSC was produced at the then Matra BAE Dynamics (now MBDA) manufacturing facility at Lostock with engineering support provided by Bristol and Stevenage.

Key subcontractors to MBDA for the Rapier FSC SAM systems were:

- BAE Systems payload for the Tracker Radar Trailer. This is a monopulse, all-weather tracking radar and for missile gather/guidance
- BAE Systems (formerly Siemens Plessey), Cowes, IOW, 3-D surveillance radar payload
- Cossor, Harlow, Mk 10A IFF system (replaced by Mk12, Mode 4 IFF in 2005)
- Thales, command transmitter which is fitted to both the launcher and tracker radar trailers for dual-fire capability
- Commercial Hydraulics (Keelavite), Warwick, integral diesel generator and hydraulic supply unit
- RADAMEC, Chertsey, electro-optical tracker yoke and servo system for each launcher
- Marshalls of Cambridge, towing vehicle installation kit (stowage for missiles and ancillary equipment), and containers for support equipment
- ROXEL, Rapier missile warheads
- ROXEL, missile motor
- Thales, missile fuze.

Rapier FSC deployed in the field with surveillance radar (left), tracker radar (centre) and fire unit (right) 0509901

Camouflaged Rapier FSC launcher with eight missiles in ready-to-fire position (left) and tracker radar (right) 0069516

The Rapier FSC has been offered as a contender for Brunei Armed Forces requirement and for the Netherlands SHORAD requirement of eight to 12 systems for its army and six to eight systems for its air force.

Description

A typical Rapier FSC fire unit consists of three elements, each of which is towed by a Leyland 4 ton (4 × 4) truck that also carries 15 missiles and associated stores.

Rapier FSC can engage targets through a full 360° out to a range of 8,000 m plus. The engagement altitude is reported as at least 5,000 m.

The fire unit, or launcher trailer, has eight Rapier missiles in the ready-to-launch position with automatic infra-red tracking. A manual acquisition and computerised tactical control facilities can provide an engagement capability by day and night.

By day, an optical acquisition facility can be used to acquire and designate targets for engagement. The IR tracker, mounted between the two banks of four missiles, has a passive scanning mode, which can be employed to search for, acquire and track targets by day or night.

A planar array transmitter mounted on the turntable sends secure guidance commands to the missile in flight. This provides the operator with a remote viewing system that allows him to work from a protected position.

The MBDA surveillance radar trailer can be added to provide a fully automatic engagement capability and a radar tracker trailer to provide all-weather operations.

The MBDA 3-D surveillance radar, also known as the Dagger, has a multibeam planar array antenna scanning at either 60 or 30 rpm providing extremely low sidelobes and good multitarget discrimination. It has a compact high-power transmitter employing Travelling Wave Tube (TWT) technology, wideband receiver unit and high-speed digital processing. It rejects clutter, is resistant to ECM and protects against anti-radiation missiles. It has integral IFF equipment.

Dagger tracks and simultaneously processes over 75 targets per second while carrying out automatic identification. It can detect hovering helicopters and very small targets such as RPVs. By using modern filtering techniques it can detect small targets against heavy ground clutter. Excellent range resolution and the variable PRF help to eliminate range ambiguities and mutual interference.

The frequency-agile J-band pulse Doppler Dagger can detect multiple targets with their range, bearing and elevation to aid rapid acquisition by the trackers. The most important tracks are displayed on an associated tactical control unit. A high elevation guard beam automatically stops its transmissions when the presence of an anti-radiation missile is detected.

The new-generation tracking radar gives Rapier FSC an all-weather capability by day and night and incorporates its own missile command link with frequency management techniques being used to evade hostile ECM. This gives Rapier FSC a dual-fire capability.

The differential monopulse frequency-agile all-weather F-band tracking radar has very narrow beams and low sidelobes to provide high-accuracy target and missile tracking with multitarget discrimination and ultra-low-level tracking capability. It has a high-power TWT transmitter, high-speed real-time distributed processing and standardised processing hardware, integral command link transmitter for system dual capability (same as on fire unit) and an integral radar missile gathering unit providing all-weather operation.

In addition, there is the Operator's Control Unit (OCU), Tactical Control Unit (TCU) and a Manual Acquisition Facility (MAF).

The OCU is used by the operator to control engagements, while the unit commander, for combat management of the weapon system, uses the TCU. They have controls and displays for controlling and optimising the weapon system performance in a multiple threat environment.

Both the OCU and the TCU continuously monitor the combat readiness of the system and allow access to automatic Built-In Test (BIT) and fault diagnosis. As all controls are duplicated at both units, one man from either unit can operate the complete system.

The MAF allows targets to be acquired passively when radar transmissions are not desirable. It can also be used with the fire unit alone to provide an air defence capability during initial system deployment or in those situations where only the fire unit can be deployed.

All three elements of Rapier FSC use a Common Trailer Base (CTB), which is not only much stronger, but is also easier to decontaminate thanks to its smoother shape. The CTB is supplied by MBDA to British Aerospace Defence Systems and GEC-Marconi who then install their respective systems into the trailer.

Each trailer also has its own integrated diesel-electric power supply, air conditioning unit and liquid-cooled electronics unit.

The trailer interconnections incorporate fibre optics and nuclear-hardened microchips that offer increased resistance to radiation and Electro-Magnetic Pulse (EMP) effects. The CTB also has a low-thermal/acoustic signature for survival in ground suppression environments.

Extensive standardisation of equipment, including standard PECs, power supplies and components ease the test, repair and logistic overheads. Second-line maintenance includes the use of three truck-mounted systems: SEERT, TERSE and SEGRT which are used to repair faulty Line Replaceable Units (LRUs) from the three Rapier FSC prime units.

Rapier FSC fires the Mk 2 missile that replaced the earlier Mk 1 in production in 1990. Externally it is very similar to the earlier Mk 1 but all-major components have been replaced or upgraded. New pyrotechnics provide a 10-year shelf-life without the requirement for interim maintenance.

There were originally two versions of the Mk 2; the Mk 2A and the Mk 2B but all current production is designated just the Mk 2. The Mk 2A retained the semi-armour-piercing warhead of the Mk 1 and was fitted with an impact fuze.

The latter Mk 2B has a dual-purpose semi-armour-piercing fragmentation warhead which is triggered by a contact or active multimode laser proximity fuze developed by the then Thomson-Thorn Missile Electronics. The Mk 2B has been designed to engage small airborne targets such as cruise missiles and RPVs.

The Mk 2B missile fuze employs an infra-red laser transmitter and four quadrant receiver optics, coupled with intelligent signal processing. The 'point of closest approach' algorithms determine the optimum range to the target at which to detonate the warhead, or holds off triggering until impact.

The Mk 2 missile can be fired by Rapier FSC and most of the earlier Rapier air defence systems. It also has increased range due to the new ROXEL Thermopylae two-stage elastomer-modified double-base propellant motor that provides a Mach 2 plus boost coast profile.

A typical Rapier FSC target engagement would take place as follows. The surveillance radar would first detect, interrogate and form tracks on hostile aircraft.

Information describing target tracks is passed to a threat assessment algorithm, which allocates the target to be engaged according to threat priorities that are defined by the customer.

Once the target has been allocated, the operator has the choice of selecting either a radar tracker or an electro-optical tracker engagement.

In the case of the former, the radar tracker is directed to acquire the highest-priority target and after acquisition the target is automatically tracked. The operator then activates the fire button and the missile is launched. Once launched, the radar-gathering sensor guides the missile into the tracker beams where it is tracked differentially until target impact. The missile guidance commands are transmitted from the radar tracker command transmitter.

The missile is guided towards the target-using Automatic Command-to-Line Of Sight (ACLOS), with the target being destroyed either by a direct hit or proximity-fuzed fragmentation warhead.

With the first missile in flight, the fire unit is directed to acquire the next highest-priority threat and on acquisition a second missile is launched and guided towards this target using guidance commands from the command transmitter on the fire unit.

The ability of Rapier FSC to engage two targets was demonstrated early in 1991. The first missile was launched and guided by the radar tracker against a Jindivik target and seconds later, while the first target engagement was still in progress, a second missile was launched under control of Rapier FSC's electro-optical tracker against a second target that was simulating a ground attack aircraft.

Specifications

	Rapier FSC launcher
Dimensions and weights	
Length	
overall:	4.1 m
Width	
overall:	2.2 m
Height	
overall:	2.6 m
Weight	
combat:	2,400 kg
Mobility	
Firepower	
Armament:	8 × turret mounted SAM
Rate of fire	
practical:	7 rds/min
Main armament traverse	
angle:	360°
Main armament elevation/depression	
armament front:	+60°/-10°

	Rapier Mk 2
Dimensions and weights	
Length	
overall	2.24 m (7 ft 4¼ in)
Diameter	
body	133 mm (5.24 in)
Flight control surfaces	
span	0.381 m (1 ft 3 in) (wing)
Weight	
launch	43 kg (94 lb)
Performance	
Speed	
max Mach number	2.5 (est.)
Range	
max	4.3 n miles (8 km; 5.0 miles) (est.)
Altitude	
max	5,000 m (16,404 ft)
Ordnance components	
Warhead	HE shaped charge, fragmentation (preformed tungsten cubes)
Fuze	IR, laser, impact, proximity (active multimode)
Propulsion	
type	two stage, solid propellant

Status

Production complete - could be restarted upon receipt of an order. In service with the British Army where the system is considered to be far superior to the original basic Rapier. Late in 1999, the UK MoD placed an additional order for around 1,800 Mk 2 missiles for the British Army. Under a 10-year programme, these will be co-produced in Turkey, at the Roketsan factory in Elmadag (near Ankara), and the UK. Turkey will take an additional 840 rounds. Initial delivery started in 2002.

RAF Air Command (ex-RAF Strike Command) has confirmed that the out-of-service date for the Royal Air Force Regiment Rapier FSC was 2008; at this time all air defence became the responsibility of the British Army.

Contractor

MBDA.

United States

MIM-14 Nike-Hercules

Type

Static Surface-to-Air Missile (SAM) system

Development

Development of the MIM-14 Nike-Hercules SAM system commenced in 1954 with the then prime contractor being Western Electric Company of Burlington, North Carolina.

The Nike-Hercules was developed as the replacement for the older Nike-Ajax (MIM-3) which was range limited and only had a HE warhead. A total of 15,000 Nike-Ajax missiles were produced but none remain in service. The Nike-Ajax was also supplied to Belgium, Denmark, France, Germany, Greece, Italy, Japan, the Netherlands, Norway, Taiwan and Turkey.

Nike-Hercules on its M94 mobile launcher, ready for launch (US Army) 0509625

Nike-Hercules surface-to-air missile being launched 0509626

Delivery of production Nike-Hercules systems to the US Army commenced in January 1958 and was completed in March 1964, although production for export continued after this date. Three different models of the Nike-Hercules were produced: the MIM-14A, MIM-14B and MIM-14C, the MIM-14B being the most common. Total production amounted to over 25,500 missiles of which 2,650 were exported under the Foreign Military Sales programme and 1,764 under the Military Aid Program (MAP). Production of the missile was undertaken at the US Army Ordnance Missile Plant at Charlotte, North Carolina, run by the then Douglas Aircraft company.

Nike-Hercules missile showing four boosters at rear (US Army) 0509627

The Nike-Hercules was deployed by the US Army in the static role in the United States (including Alaska), Okinawa, Taiwan, Hawaii, Thule AFB and Greenland. US Army semi-mobile units were deployed in Miami (at the Homestead AFB), Germany and South Korea.

In the Continental United States (CONUS) static versions were employed to provide defence of critical installations and urban population centres while semi-mobile units were deployed to protect field armies and theatres of operation.

Each air defence battalion of the US Army consisted of a headquarters battery and four firing batteries. Each battery can operate as part of an air defence network or as an autonomous unit and is capable of multiple launches during a single engagement.

The system was designed to engage aircraft flying at altitudes in excess of 30,480 m and at ranges reaching 154 km. During trials it successfully intercepted short-range ballistic missiles such as Corporal and other Nike-Hercules missiles. During its service life, three versions of the system were deployed: the basic Nike-Hercules (IOC date 1958), Improved Nike-Hercules (IOC 1960) and the Nike-Hercules Anti-Tactical Ballistic Missile (ATBM) system (this was deployed to Germany, IOC date believed to be 1961). The missile was also capable of being used in the ground-to-ground role with a contact fused nuclear warhead.

This missile system was supplied to a number of NATO countries and Mitsubishi Heavy Industries produced a non-nuclear version under licence for the then Japanese Air Self-Defence Force. This is known as the Nike-Hercules-J.

Deployment of the Nike-Hercules in the US Army reached its peak in 1963 when no fewer than 134 batteries were operational in the CONUS as well as Alaska, Okinawa, Taiwan and Germany. By 1974, they had all been disbanded in the CONUS apart from those retained for training. The successor to Nike-Hercules is the Patriot system.

Description

The Nike-Hercules missile is a two-stage weapon consisting of a solid propellant, computer-controlled missile body and a cluster of four solid propellant booster rockets. The missile airframe, wings and booster clustering hardware are made of aluminium and the booster cases are steel.

The missile body is sharply tapered at the nose and is flared back to a maximum diameter of 800 mm. The missile has four delta-shaped wings, with elevons to control roll and steering, and four small linearisation fins are attached forward of the wings.

The booster cluster is composed of four individual booster rockets and has a cross-sectional width of 876 mm with the four trapezoidal fins attached to the aft end of the cluster.

The Nike-Hercules is launched by remote control, normally at an angle of approximately 85°, and when the booster is jettisoned the guidance system is activated, programming the missile to roll toward the target and dive into the intercept plane. Steering orders direct the missile to the optimum burst point. The warhead is either the high-explosive or nuclear type. The US has long since removed all nuclear warheads for this SAM system and only HE warheads remained in service.

Battery of four Nike-Hercules missiles deployed in static role (Michael Green)
0509906

Nike-Hercules missile on its static launcher showing missile reloading rails on left (Michael Green)
0509907

In addition to the missile itself, key components of Nike-Hercules are a 180 km range low-power acquisition radar, high-power acquisition radar, target tracking radar, missile tracking radar, electronic data processing equipment and remote-controlled launchers.

After the system had been in service for some years a later development, the 315-km range High-Power Acquisition Radar (HIPAR) was added as part of the improved Nike-Hercules programme. This improved the target detection capability against targets up to M2.0. It is also capable of detecting air breathing missiles and rockets with speeds up to M3.0.

The mobile HIPAR variant used five semi-trailers, vans housing radar transmitter and receiver and control equipment. Before the introduction of HIPAR some 20 vehicles were required to move the radar system.

In operation the target is first detected by the acquisition radar and is then interrogated by the associated AN/TPX-46 IFF Mk XII interrogator; if confirmed as hostile, its location is transferred to the target tracking radar which pinpoints it for intercept purposes. When the target is within range a missile is launched and the missile tracking radar issues guidance and orders to the missile until it reaches the target.

The system operators are located in battery, tracking and launcher control trailers.

In 1981, contracts were placed for a number of improvements to the system with McDonnell Douglas refurbishing and modifying NATO Nike-Hercules missiles.

Norden Systems provided the Digital Computer System, which is based on the PDP-11/34M mini-computer. This receives missile and target position inputs from tracking radars, solves the intercept problem and issues guidance commands to the missile. In addition it performs various routines such as fault diagnosis.

Variants

South Korea is known to have produced the NHK-I/II version optimised for the ground-to-ground role. These have an enlarged booster motor section. At least three battalions with 12 launchers in total have been raised to use this configuration.

Specifications

	MIM-14B
Dimensions and weights	
Length	
overall	12.141 m (39 ft 10 in)
	4.34 m (14 ft 2¾ in) (first stage booster)
	8.19 m (26 ft 10½ in) (second stage)
Diameter	
body	879 mm (34.61 in) (first stage booster)
	800 mm (31.50 in) (second stage)
Flight control surfaces	
span	3.510 m (11 ft 6¼ in) (first stage booster, fin)
	2.286 m (7 ft 6 in) (second stage, fin)
Weight	
launch	4,868.6 kg (10,733 lb)
	2,354.5 kg (5,190 lb) (first stage booster)
	2,514.1 kg (5,542 lb) (second stage)
Performance	
Speed	
max Mach number	3.65
Range	
max	83.7 n miles (155 km; 96.3 miles) (surface-to-air)
	98.8 n miles (182.9 km; 113.6 miles) (surface-to-surface)
Altitude	
min	1,000 m (3,281 ft)
max	30,480 m (100,000 ft)
Ordnance components	
Warhead	502.7 kg (1,108 lb) HE M17 and M135 (272.7 kg HBX-6 explosive) blast fragmentation
Propulsion	
type	solid propellant

Status

The programme to replace the South Korean Nike Hercules was initially designated M-SAM however by 2012 and after press releases it is now known as Iron Hawk II. A variety of systems are competing for an order for a 60 km plus range weapon. A total of 48 launchers for two battalions are required with dual anti-missile and air breathing target engagement capability.

Production complete.

Contractor

Western Electric Company (now AT & T Technologies).
AAI Incorporated (major subcontractor).
Bell Telephone Laboratories (major subcontractor for guidance systems).
General Electric Company (major subcontractor).
McDonnell Douglas Astronautics (major subcontractor for the airframe).
Raytheon Company (major subcontractor).

MIM-23 HAWK

Type

Static Surface-to-Air Missile (SAM) system

Development

The Homing All the Way Killer (HAWK) semi-active radar seeking medium-range SAM system commenced development in 1952 with the US Army awarding a full-scale development contract to the Raytheon Company for the missile in July 1954. Northrop was to provide the launcher and loader, radars and fire-control.

The first guided test firing took place in June 1956 with the development phase completed in July 1957. Initial Operational Capability (IOC) of the Basic HAWK, MIM-23A, took place in August 1960 when the first US Army battalion was activated. In 1959, a NATO Memorandum of Understanding (MoU) was signed for NATO HAWK between France, Italy, the Netherlands, Belgium, Germany and the US for co-production of the system in Europe. In addition, special grant aid arrangements were made to deliver European-built systems to Spain, Greece and Denmark and direct sale arrangements of US-built systems were made with Japan, Israel and Sweden. The Japanese sale soon expanded into a country-to-country co-production agreement with production initiated in 1968. In the same region, the US also made grant aid deliveries of the HAWK to Taiwan and South Korea.

Radar types and ranges (km)

Target radar	AN/MPQ-50		AN/MPQ-48		AN/MPQ-46		AN/MPQ-51
Cross-section	PRF setting		CWAR mode		HPI box search		
Area (m2)	High	Low	CW	FM	High	Low	
1	79	72	52	48	75	72	63
2.4	98	90	65	60	93	89	78
3	104	96	69	63	99	93	83

I-HAWK missile leaves its three-round M192 launcher during US test on Patriot/HAWK interoperability trials 0509612

However, in 1964, in order to counter advanced threats, especially at low altitude, the US Army initiated a modernisation programme known as the HAWK Improvement Program (HAWK/HIP). This involved a number of changes to the basic system. The most important of which were:

- The addition of a central information coordinator with a digital automatic data processor at the battery headquarters for target processing threat ordering and target intercept evaluation
- Updating of the missile to the Improved-HAWK MIM-23B configuration with a larger warhead (75 kg versus 54 kg)
- An improved small guidance package and a higher-performance rocket motor.

The system modifications allowed both the missile and its Continuous Wave (CW) illuminating radar to discriminate a target from ground clutter by using its velocity, ensuring low-altitude coverage. Type-classified in 1971, all US Army and Marine Corps HAWK battalions were subsequently retrofitted to the I-HAWK standard by 1978. In 1974, the enlarged NATO HAWK Production and Logistics Organisation (NHPLO), awarded Raytheon a contract for joint production of I-HAWK components in Europe under the auspices of the HAWK European Limited Improvement Programme. This improvement programme was carried out from 1974 to 1982. Japan followed suit with an I-HAWK joint production agreement in 1977. NHPLO included Greece, which joined in 1972, but excluded Norway (which joined in 1986), Denmark (which joined in 1976) and Belgium (which eventually rejoined in 1979).

In 1973, the US Armed Forces started a second modernisation effort under the designation HAWK-PIP (for Product Improvement Program). This involved three phases:

Phase I

PIP Phase I fielded with the US forces in 1979 to 1981, which included:

- An Acquisition Radar (CWAR) transmitter to double the output power and increase detection range
- The addition of digital Moving Target Indication (MTI) to the Pulse Acquisition Radar (PAR)
- The inclusion of Army Tactical Data Link (ATDL) communications within the system.

Phase II

Phase II upgrade modification developments started in 1978 and were fielded in 1983 to 1986. These greatly improved the reliability of the High-Powered Illumination (HPI) radar by replacing vacuum-tube circuits with modern solid-state technology and added the Tracking Adjunct System (TAS) optical tracking system for operation in an ECM environment to the HPI, the Battery Control Central (BCC) and the Platoon Command Post (PCP).

The HAWK Phase II system is deployed in two basic configurations: a battery with two fire sections and an Assault Fire Platoon (AFP) with one fire section. A third configuration is the TRIAD battery, which is a combination of the two, a battery with one fire section and two AFPs.

The standard Phase II battery consists of:

- BCC (TSW-12)
- ICC (MSQ-110)
- PAR (MPQ-50)
- CWAR (MPQ-55)
- ROR (MPQ-51)
- Two fire sections, each consisting of an HPI (MPQ-57) and three launchers (M192).

An AFP consists of:

- A PCP (MSW-18)
- CWAR (MPQ-55)
- HPI (MPQ-57)
- three launchers (M192).

The US Army has reconfigured all its HAWK units to AFPs. However, a number of other countries still retain a battery configuration.

Belgium, Denmark, France, Italy, Greece, Netherlands and Germany have implemented Phase I and Phase II. Other countries are following suit.

French Army M192 launcher being towed by a Berliet (6 × 6) truck (Pierre Touzin) 0069517

French Army HPI radars in travelling configuration (Pierre Touzin) 0069518

The Netherlands and Germany have introduced into service the STN Atlas Elektronik GmbH HAWK Electro-Optical Sensor (HEOS). A total of 93 systems have been ordered (83 for Germany and 10 for Netherlands). The HEOS is positioned with HPI radar between its two antennas. It consists of a thermal imaging sight with the thermal sensor operating in the 8 to 12 μm infrared range. It has a day and night capability and two fields of view. Tracking range is believed to be over 100 km. The HEOS can be fitted onto any I-HAWK system that can accept the Northrop Tracking Adjunct System (see later in entry).

Due to the reorganisation in Germany, the projected Luftwaffe SAM structure for the 1990s is six mixed wings (with three wings of one group of Patriot, Roland and I-HAWK each, and three wings each of one group of Patriot and one group of I-HAWK); five in the former West Germany and one in the former East Germany. A total of 24 I-HAWK squadrons are operational with a further 12 placed in reserve status.

In 1996, DASA delivered the first seven of a total of 25 mobile antenna towers to the German Procurement Agency (BWB) for use with the Luftwaffe I-HAWK battery sites. Similar to the 80 in-service Patriot systems, it is possible to interchange them within the mixed SAM group. The I-HAWK antenna towers feature four transmit/receive parabolic dish antennas each 1 m in diameter. These antennas can be individually adjusted in the vertical and horizontal direction and extended (within a few minutes) to a maximum height of 34 m. Two people can ready the tower in approximately 20 minutes. Even at high wind speeds the tower does not deviate more than 2° in order to maintain precise antenna alignment for the radio relay links. Maximum permissible wind speed for the I-HAWK antenna mast is 37 m/s. The Danish I-HAWKs are used by the *Luftvårnsgruppen* (Air Defence Group of Tactical Air Command) which has 10 units:

- Eskadrille (three squadron) ESK 531-534
- ESK 541-544 deployed on Sjoelland (three squadrons protecting Copenhagen)
- Fyn (two squadrons of leased US Army I-HAWK systems)
- At the airbases of Karup (one squadron)
- Skrydstrup (one squadron).

In 1991, the Danes deployed an indigenous video tracking system for its I-HAWKs. The system utilises two optical sensors mounted with the fire-control system for long-range (40,000 m) tracking and short-range (20,000 m) search functions. Depending upon the tactical situation, the system is designed to switch between the optical and radar modes or use a combination. Thus the HPI can be left off until immediately before firing as the search-and-track phase can be totally passive.

In August 1998, the Danish Air Materièl Command awarded INFOCOM Systems A/S and Thomson-CSF Communications a contract to implement a new communications network for the I-HAWK systems. This was part of the Royal Danish Air Force's Danish Enhanced Hawk (DEHAWK) project.

Interior of Kongsberg Defence & Aerospace Fire Direction Centre as supplied to Greek Armed Forces for its HAWK upgrade 0069519

Functional check and delivery of the first Daimler Chrysler Aerospace antenna mast tower for Improved HAWK systems 0509908

Phase III

The Phase III upgrade programme started development in 1981 and is complete for the US Armed Forces. Phase III makes major modifications to many of the system's major items. The Range Only Radar (ROR) and Information Co-ordination Central (ICC) have been deleted from the system and the BCC replaced by a Battery Command Post (BCP). Major electronic modifications, which included the addition of distributed microcomputers and greatly improved computer software, were made to the BCP, PCP, CWAR and HPI. The TPQ-29 Field Trainer was replaced with the Integrated Operator Trainer (IOT).

However, the major system operational change made by Phase III is the addition of a single scan target detection capability and a Low-Altitude Simultaneous HAWK Engagement (LASHE) system to the HPI, by employing a fan beam antenna, to provide a wide-angle, low-altitude illumination pattern, in order to allow multiple engagements against saturation raids.

The US government conducted a Missile Reliability Restoration (MRR) programme between 1982 and 1984.

Three-round I-HAWK M192 missile launcher deployed (US Army)

0007985

Tracking Adjunct System (TAS) which is mounted on top of the HAWK's HPI radar system (of which a second unit is seen behind the gunner) 0509613

Concurrent with the MRR programme, ECCM improvements were incorporated into MIM-23C and MIM-23E updated versions of the HAWK missile to counter specific ECM threats. A subsequent modification to improve performance for low-altitude engagements in high clutter was incorporated into the MIM-23G model in 1990.

The most recent change to the missile modified the warhead and fuzing circuit to improve performance against Tactical Ballistic Missiles (TBMs). The modifications are believed to include the use of a higher-energy explosive, altering the fuze system to react against both TBM and air-breathing targets and replacing the current-generation 30 grain fragment warhead with a 540 grain one. This missile, which is believed to be designated MIM-23J/K, has been approved for production by the US Army. By early 1996, 300 plus HAWK rounds had been upgraded to this standard. A warhead kill of a TBM target was achieved during test firings at the White Sands Missile Range during May 1991, after being cued by Patriot acquisition data. Two missiles were fired during the test, the first achieving a direct hit and the second providing a warhead kill. These were passed onto the US Marine Corps for their I-HAWK systems but, by late 1998, all of these HAWK missile systems had been completely phased out of US Army and Marine Corps service.

Raytheon completed development of the HAWK Advanced Training Simulators (HATS) in 1991. The HATS system is designed as a cost-effective alternative to in-field training with actual tactical hardware and provides realistic hands-on training on up to 90 per cent of the critical tasks associated with the HAWK system.

The HATS provides three-dimensional computer-aided mock-ups of the PCP, HPI and CWAR, so as to furnish training for HAWK operators, officers and maintenance personnel. The simulators closely duplicate the operational 'feel' of the tactical hardware and include all controls, displays, test points and adjustments that are required for training of the operator and maintenance

US Army personnel setting up an AN/MPQ-50 Improved Pulse Acquisition Radar set 0509614

M35 (6 × 6) truck towing M192 HAWK launchers (Peter Siebert) 0509721

Raytheon designed and produced Multiple Role Survivable Radar (MRSR) and HAWK missile 0509615

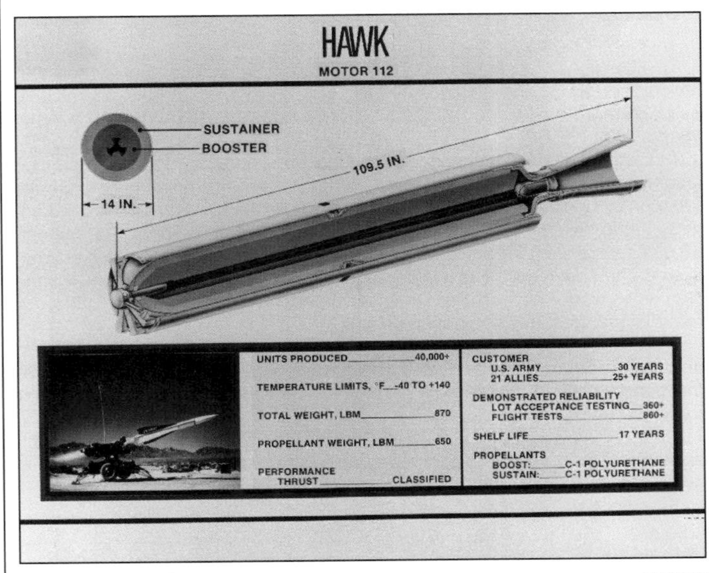

HAWK motor 112 0509616

personnel. Maintenance training capability includes adjustments, various test procedures, troubleshooting and module removal and replacement. The operator training utilises authentic air defence scenarios, which replicate the Integral Operator Trainer built into the PCP.

The Phase III HAWK System is fielded in three configurations: an Assault Fire Platoon (AFP), an Assault Fire Platoon Plus (AFP+) and a Battery. The AFP will consist of a CWAR, PCP and HPI and four launchers. The AFP+ has the same complement as the AFP with a PAR added. The Battery has a CWAR, PAR, BCP, two HPIs and six launchers.

Several foreign allies, including the Netherlands, have ordered Phase III up-graders to keep their HAWK units current. On 1 August 1995, Saudi Arabia placed a contract worth USD118 million for Phase III kits. Egypt also placed a contract, worth an estimated USD303 million, for HAWK missile system ground equipment and upgrade kits for its HAWK batteries. In January 1997, a USD206 million contract was signed with the US Army MICOM on behalf of the government of Egypt for the remanufacture of eight I-HAWK fire units provided by the US government as Excess Defence Articles. Under the contract, Raytheon will produce modification kits to upgrade the fire units to the latest configuration and remanufacture the remaining components to a 'like new' condition. These fire units will join the previous 24 Phase III fire units currently in service. In 1998, at the Defendory Exhibition, the Greek MoD indicated that Raytheon will be upgrading its I-HAWK batteries to the Phase III configuration with a new Fire Direction Centre. This will allow interoperability with the Patriot fire units that Greece is acquiring. The contract was signed in 1999. During Operations Desert Storm/Desert Shield, US forces deployed 24 HAWK fire units to the Gulf region.

Norway developed its own HAWK upgrade scheme known as the Norwegian Adapted HAWK (NOAH - see entry under Norway in this section). This involved the lease of I-HAWK launchers, HPI radars and missile loaders from the US and their integration with Hughes-Kongsberg Vaapenfabrik (HKV) Acquisition Radar and Control Systems (ARCS) in place of the normal search, acquisition and ranging radars. The ARCS is a combination of the Raytheon 3-D AN/TPQ-36A Low-Altitude Surveillance Radar (LASR) and the Kongsberg Defence & Aerospace Fire Direction Centre. An early version of the HEOS passive infra-red acquisition and tracking system is also in use. A total of six batteries were procured from 1987 onwards for airfield defence at Bodø, Andøya, Bardufoss, Evenes, Ørland and Vårnes. These were replaced in the late 1990s by the NASAMS system.

During the early 1970s to early 1980s, the I-HAWK system was also sold to a number of countries in the Middle East and Asia. To maintain their HAWK system's viability, the Israelis have upgraded it to the PIP Phase II standard with the addition, in the mid-1970s, of a Super Eye electro-optical TV system for detection of aircraft at 30 to 40 km and identification at 17 to 25 km. Israel has also modified its I-HAWK systems for use at altitudes of up to 24,384 m. One of these was used in southern Lebanon to shoot down a Syrian Air Force MiG-25R 'Foxbat-B' photo-reconnaissance aircraft flying at M2.5 on a high-level 21,336 m sortie near Jouniehj, north of Beirut.

Not surprisingly, Israel was also the first country to use the Basic HAWK in combat when, during the 1967 June war, it had destroyed one of its own Dassault Mystère IVA fighters with an incapacitated pilot on board to prevent it crashing into the nuclear weapons facility at Dimona. They followed this by the first true combat launch on 24 May 1969, when an Egyptian MiG-21 flying near Kantara, over the Suez Canal, was hit at about 6,700 m altitude. By the end of the War of Attrition in August 1970, Basic HAWK had accounted for 12 Egyptian aircraft (one Ilyushin Il-28 'Beagle', four Sukhoi Su-7 'Fitters', four MiG-17 'Frescoes' and three MiG-21 'Fishbeds').

During the 1973 Yom Kippur War, around 75 HAWK rounds were used against Syrian, Iraqi, Libyan and Egyptian aircraft and destroyed four MiG-17s, one MiG-21, three Sukhoi Su-7s, one Hawker Hunter, one Dassault Mirage V and two Mil Mi-8 helicopters. Included among the kills were several multiples using just one missile.

Its next use by the Israelis was in its modified I-HAWK configuration during the June 1982 Peace for Galilee war when a Syrian MiG-23 was destroyed. This was followed by the MiG-25 incident.

By March 1989, Israeli Air Force missile units had shot down 42 Arab aircraft using Basic HAWK, Improved HAWK and Chaparral SAMs.

However, before the last two Israeli uses, Iran's Armed Forces used I-HAWKs against the Iraqi Air Force on several occasions. In 1974, Iran supported the Kurds in a rebellion against Iraq using HAWKs to shoot down most of the 18 planes claimed by the Kurds and then, on 14 and 15 December 1974, destroyed two Iraqi fighters on armed reconnaissance missions over Iran. Following the 1980 invasion and up to the end of the war, Iran is believed to have shot down at least 40 aircraft with the weapon.

In the mid-nineties the Iranians modified the I-HAWK missile to act as an air-to-air missile on its F-14A Tomcats.

Singapore operates the I-HAWK with its Air Force's 163rd Singapore Air Defence Artillery Squadron. In May 1990, two batteries of this six-battery squadron were reassigned to the Integrated Air Defence System (IADS) of the Five Power Defence Agreement (FPDA) with Australia, Malaysia, New Zealand and the United Kingdom to replace Singapore's Bloodhound squadron.

France became a combat user of I-HAWK when it deployed one battery to Chad to defend the capital, N'Djamena. On 7 September 1987, it shot down a Libyan Air Force Tupolev Tu-22 'Blinder' bomber, which was trying to attack the airport.

HAWK tracked loader, reloading a HAWK launcher 0007986

STN Atlas Elektronik GmbH HAWK Electro-Optical Sensor (HEOS) mounted between antennas of HPI radar system 0509617

Kuwaiti I-HAWK batteries were used extensively against Iraqi aircraft and helicopters during the August 1990 invasion. A total of 23 Iraqi aircraft and helicopters were destroyed by the three (of five available) I-HAWK batteries that were able to engage.

Among the HAWK items designed and developed by Northrop are the loader/transporter, crane attachment, ramps, winter kits, hoisting beam, hoist adaptor, track cleats, wings/elevons, actuator and the tracking adjunct system.

By 2000, some 750 loader/transporters, 1,700 three-round launchers, 38,000 sets of wings/elevons, 38,000 actuators and over 500 tracking adjunct systems had been built. The withdrawal of the HAWK from US Army and Marine Corps service has resulted in many systems becoming available for refurbishment and resale or used as aid.

In late 1999, the US Army awarded Raytheon a USD9.7 million contract to develop the Advanced Infra-Red Tracking Adjunct System (AIRTAS) to improve the performance of the I-HAWK systems. The AIRTAS will be available for foreign I-HAWK improvements.

As of 17 February 2006, Raytheon Company's Integrated Defence Systems appointed Merex Inc, as its exclusive licensee for Hawk Phase II and Phase III ground system components, spares and repairs. This should now provide the Hawk customers with continued, reliable, whole-life system support.

MIM HAWK Missiles

1959 - MIM-23A
1971 - MIM-23B
1982 - MIM-23C and MIM-23D
1990 - MIM-23E and MIM-23F
1994 - MIM-23J and MIM-23K
1995 - MIM-23G and MIM-23H
1995 - MIM-23L and MIM-23M

Patriot/HAWK Interoperability

Patriot/HAWK Interoperability has been developed and fielded to gain the benefits of effective air-defence co-ordination when these two systems occupy the same defence area. With the introduction of Patriot's Post Deployment Build-2 (PDB-2) software, an enhanced interoperability function was provided which allowed air-defence commanders the ability to organise HAWK and Patriot fire units under the control of the Patriot Information Co-ordination Central (ICC). The interoperability functions of the ICC software have been further expanded in PDB-3 to provide improved capabilities in terms of Patriot/HAWK operations as well as Master Battalion enhancements. Overall, the ICC software has been automated to the maximum extent to complete HAWK-related tasks of target identification, threat assessment and target assignments. The PDB-3 software also accommodates the HAWK Phase III improvements of passive target identification increased update rates and improved HPI search patterns.

A major improvement of the PDB-3 software is the new Master Battalion capability. This enhancement provides the Patriot ICC with the ability to function as a Brigade and to control HAWK battalions as well as subordinate Patriot battalions. The ICC software has been programmed to prioritise threat targets and assign the most cost-effective missile system to engage the target. During the US Army's May 1991 Patriot/Hawk testing the Patriot ICC demonstrated the capability to designate TBM targets to the HAWK system for engagement.

US Marine Corps TPS-59/HAWK anti-TBM capability

Also during the May 1991 testing, the US Marine Corps successfully investigated the use of a modified Martin Marietta (formerly General Electric, now Lockheed Martin) D-band AN/TPS-59 tactical long-range radar system to perform the anti-TBM search and track functions for a HAWK fire-control unit. No actual firing took place.

Martin Marietta modified the radar and Sensis Corporation to allow it to track TBM targets. This was achieved by expanding its 3-D surveillance coverage to a higher elevation (from 19° against airborne targets to 65° against TBMs), extending its range and adding a Sensis multi-scan correlator to reduce clutter and eliminate false targets. An unmodified AN/TPS-59 has a maximum altitude limit of 30,480 m and a range of 556 km (at 0 dBsm) and 240 km (at –10 dBsm) against airborne targets. With the modifications, the limits against TBMs are increased to 742 km (at 0 dBsm), 556 km (at –10 dBsm) and 240 km altitude. The HAWK missile was also upgraded by modifying its fuzing system and warhead to the MIM-23J/K standard.

The US Marine Corps intends the anti-TBM capability in its HAWK batteries to form part of its Integrated Air-Defence programme. The weapon is intended to engage FROGs, SS-21 'Scarabs' and, in certain cases, 'Scud' missiles. The HAWK ATBM capability involves reinforcing the pedestal of the AN/TPS-59 radar so that it can be pointed at a higher angle relative to the horizon. The introduction of an Air-Defence Communications Platform (ADCP) in order that the TPS-59 can send data direct to the HAWK fire units, rather than the Tactical Air Operations Centre. This is to speed hand-off, modifying the HAWK launcher's hydraulic system so that the missile can be elevated into a superelevation position and kept there ready for firing. The HAWK missile used is the MIM-23K Improved Lethality Missile configuration developed for the US Army.

Part of a US Marine Corps initiative jointly funded with BMDO (although there has been substantial foreign interest), the upgraded radar has a TMD mode allowing detection of TBMs at up to a 400 n mile range and 500,000 ft altitude and provides cuing via a JTIDS link. The battery command post would be modified to accept cuing data.

The HAWK surface-to-air missile systems used by the US Marine Corps have now been withdrawn from service. Before this happened it was reported that the US Marine Corps had some 1,000 MIM-24K equivalent missiles and 11 modified AN/TPS-59 radars available.

HAWK Mobility, Survivability and Enhancement (HMSE) programme

For further development and exploitation of the HAWK system and its evolution into a system for the early part of the 21st Century, the first step foreseen by the manufacturer is the USD40 million HMSE programme. The goal of this 1995 deployment date programme by the US Army/Marine Corps and the Netherlands is to improve the system's overall mobility and reduce both the emplacement and march order times. Full-scale development started in 1989 and was completed in late 1992. A USD12.9 million contract was placed with Raytheon in July 1993 for 43 HAWK Mobility Kits. The US Marine Corps received the HMSE from late 1993 for all its HAWK systems. Initial operational capability for the US Marine Corps HMSE was late 1994.

There are four major features of this upgrade:

(1) The launcher will be modified to remove its remaining vacuum-tube circuits and replace them with a modern digital laptop microcomputer. This change permits improved computation of missile launch obstruction avoidance and provides full-duplex serial datalink communications between each launcher and the PCP. This last change replaces the current large, heavy, multi-conductor data cables with a telephone-type field wire inter connector to the PCP.
(2) Mechanical modifications will be made to the launcher to allow it to move during road marches, with three ready-to-fire HAWK missiles mounted in place. The change greatly reduces launcher emplacement and march order times by eliminating the requirement to load missiles during emplacement and remove missiles during march order.
(3) An upgrade to launcher hydraulics will be made to minimise electrical power requirements and convert the hydraulics from a fully active motor-pump design to a stored energy system employing rechargeable accumulators. The modifications will improve reliability by eliminating or replacing existing high-failure rated items and the hydraulic heat exchangers. The elimination of the present motor-pump and the incorporation of current-generation low-leakage servo valves allowed for the removal of the heat exchanger and both elevation pressure reduction circuits providing a reduction in required prime power by a factor of six. Acoustic and IR signatures are also minimised with the hydraulic modifications.
(4) The incorporation of a North Finding System (NFS) into the system. This device contains a north-seeking gyroscope and a digital computer. When mounted on a HAWK radar or launcher, the NFS can provide rapid determination of the unit's azimuth alignment with respect to the true north

reference. With this device, each unit of the HAWK system can be independently and accurately aligned day or night in all weather conditions without using the existing optical telescope and the need for line of sight between units.

Operational benefits of the mobility enhancement programme are said to be:

- A 50 per cent reduction in HAWK prime movers and towed loads (that is, from 14 to seven per AFP)
- A 50 per cent reduction in pre-siting time (from 60 to 30 minutes)
- A 61 per cent reduction in emplacement time (from 40 to 15.5 minutes)
- A 67 per cent reduction in march order time (from 30 to 10 minutes)
- A 120 per cent increase in launcher electronics reliability (from 411 to 906 hours MTBF)
- A 67 per cent reduction in launcher electronics maintenance (18 to six hours per year per launcher)
- A 28 per cent reduction in C-141B Starlifter sorties per AFP (from 15 to 11).

Also the digital electronics include a small missile get-ready capability in preparation for the possible later addition of either the Sparrow or AMRAAM missile to the HAWK launcher.

The installation of a Phase III digital computer to the launcher electronics also allows a greater dispersion of the major AFP items from 110 m apart to 2,000 m, thus significantly enhancing survivability.

Agile Continuous Wave Acquisition Radar (ACWAR)/Multiple Role Survivable Radar (MRSR)

The next change was expected to be the introduction of the Agile Continuous Wave Acquisition Radar (ACWAR).

The ACWAR is an evolution of the HAWK Continuous Wave (CW) radar technology. It performs full 3-D target acquisition over a 360° azimuth sector and large elevation angles.

The ACWAR programme was initiated to meet increasingly severe tactical air-defence requirements, and the equipment is designed for operation in the 1990s and the 21st Century. ACWAR replaces two current HAWK radars to reduce vehicle, manpower and logistics needs. At the same time, the system provides full 3-D track information on a large number of targets to accuracy's sufficient for cuing and control of other remote weapons as well as data netting. ACWAR is mobile, helicopter and C-130 transportable and can be vehicle- or trailer-mounted.

ACWAR consists of an all-solid-state exciter/transmitter, all-digital radar control, row board transmit and receive antenna construction for precise side lobe control, mechanical steering in azimuth and electronic steering in elevation. ACWAR uses a CW frequency-agile, phase-coded waveform. Digitised target data are sent from the radar to the HAWK PCP or equivalent. The antenna assembly is mast-mounted with a hydraulic arm elevation system.

ACWAR technologies are being used in the US Army/Raytheon Multiple Role Survivable Radar (MRSR) programme. Major purposes of this advanced development programme include survivability against Anti-Radiation Missiles (ARMs) without decoys or EMCON and robust ECM performance against the increasing threat, while continuously providing an accurate 3-D air picture. It has an MTBF of over 350 hours and an MTTR of 0.6 hours resulting in an availability of 97 per cent. It has the processing capability to track over 100 targets simultaneously.

The RBS 97 (FU 97) systems in use in Sweden have a modified illumination radar for use in Sweden only.

Latest version of the Sparrow HAWK has eight missiles (the first design had nine), which allows the system to be loaded onto a C-130 aircraft (Scott Gourley) 0509618

Description

The I-HAWK missile is a certified round requiring no field maintenance or testing. The certification is maintained through periodic batch acceptance testing, annual service firing and periodic batch sampling at special maintenance facilities operated by the contractor.

It is of the single-stage cruciform configuration with a dual-thrust Aerojet M112 solid propellant rocket motor; flight control is achieved by elevons located on the trailing edges of the four rear fins.

A typical I-HAWK battery engagement occurs in the following sequence.

Detection of the target happens when the acquisition radar beam returns match the required parameters of the automatic electronic data processor in the battery information co-ordination central unit. The target then becomes eligible for automatic threat ordering.

For medium- to high-altitude target detection the radar used is the AN/MPQ-50 pulse-acquisition set. This equipment also has several advanced Electronic Counter-Counter Measures (ECCM) receivers to overcome specific types of jammers. An off-the-air tuning capability also permits frequency changes to avoid enemy jamming.

For low-altitude targets, coverage is provided by an AN/MPQ-55 or AN/MPQ-62 (Phase I or Phase II) improved continuous wave radar operating in alternate continuous wave and frequency modulation/continuous wave modes in order to provide range rate and range data. The two acquisition radars are synchronised at a scan rate of 20 rpm with the frequency modulation applied at a rate asynchronous to the scan rate so as to prevent any possibility of a dead coverage situation occurring on successive scans.

The data processor correlates with the IFF transponder returns to identify the accepted target as either friendly or hostile. If it is the latter, the processor directs one of the battery's available fire units to engage the target.

The unit's associated high-power illuminator is slewed to the correct azimuth and automatically commanded to actively seek a designated elevation sector for the target. Once detected, the radar locks on to the reflected electromagnetic energy from the aircraft and begins to track in radial speed and angular position.

The illuminator has both continuous wave low-altitude and sector search capabilities and tracks the target throughout the air-defence engagement cycle. If the illuminator is not able to obtain the target's range because of enemy countermeasures then the AN/MPQ-51 range-only radar is activated. This is maintained in an operational state during the engagement by being placed in a receive-only mode with its control system continually scanning its tuneable bandwidth region for the presence of jamming. It can be called into operation either automatically by the illuminator or by manual means and will transmit on a non-jammed area of its bandwidth to obtain the required target range data.

The target information is transmitted to the selected launcher assembly for the engagement. This slews to the same azimuth and elevation angle used by the illuminator and provides the power to activate the missile gyros, electronics and hydraulic systems. Once this sequence happens and the gyros are up to speed, a lead angle command is sent from the data processor via the illuminator to the launcher. The missile seeker is then space stabilised and the launcher unit realigns itself to take into account the lead angle value. On completion of the realignment the motor ignition sequence is started and the missile is fired.

The missile uses a proportional navigation collision course trajectory, with in-flight guidance commands generated by the semi-active radar homing inverse monopulse solid-state seeker head fitted.

At the terminal interception point a radio frequency proximity fuze with impact override is used to detonate the HE directed fragmentation pattern warhead. This warhead type is used specifically to increase the Single-Shot Kill Probability (SSKP) in multiple target situations, a facility, which the Israelis exploited on several occasions during the 1973 Yom Kippur war.

HAWK Platoon Command Post with stabilisers lowered (Michael Jerchel) 0509619

Immediately after warhead detonation, a target intercept evaluation is performed by the battery Tactical Control Officer using the illuminator Doppler data to determine whether or not the missile has destroyed the target, or if further missile launches are required.

The SSKP of the MIM-23B is around 0.85, while that of the original MIM-23A is approximately 0.56.

A fuller description of the various ground equipment components of the HAWK systems are as follows.

AN/MPQ-50 Pulse Acquisition Radar (PAR)

The PAR is the primary source of high- to medium-altitude aircraft detection for the battery. The C-band frequency allows the radar to perform in an all-weather environment. The radar incorporates a digital MTI to provide sensitive target detection in high-clutter areas and a staggered pulse repetition rate to minimise the effects of blind speeds. The PAR also includes several ECCM features and uses off-the-air tuning of the transmitter. In the Phase III configuration the PAR is not modified.

CW Acquisition Radar (CWAR)

Aircraft detection at the lowest altitudes, in the presence of heavy clutter, is the primary feature the CWAR brings to HAWK. The CWAR and PAR are synchronised in azimuth for ease of target data correlation. Other features include FM ranging, Built-In Test Equipment (BITE) and band frequencies. FM is applied on alternate scans of the CWAR to obtain target range information. During the CW scan, range rate minus range is obtained. The Automatic Data Processor (ADP) in the ICC processes this information to derive target range and range rate. This feature provides the necessary data for threat ordering of low-altitude targets detected by the CWAR.

The Phase III programme makes some major modifications to the CWAR. The basic function of the CWAR, as the system's low-altitude acquisition sensor, remains unchanged, however the transmitted waveform was changed to permit the radar to determine both target range and range rate on a single scan. A Digital Signal Processor (DSP) using a Fast Fourier Transform (FFT) was added to digitally process target Doppler into detected target data. The DSP provides this digital data to a new microcomputer located in the CWAR. The microcomputer performs much of the CWAR target processing, formerly done by the ADP in the ICC, and transmits the processed target track data to the PCP/BCP in serial digital format over a field telephone wire interconnection. The full-duplex digital link eliminates the need for a large, heavy, multiconductor cable between the CWAR and PCP/BCP.

Battery Control Central (BCC)

The BCC provides the facilities for the man/machine interface. The Tactical Control Officer (TCO) is in command of all the BCC operations and maintains tactical control over all engagement sequences. The TCO monitors all functions and has the authority and facilities to enable or pre-empt any engagement or change established priorities. The tactical-control assistant assists the TCO in detection, identification, evaluation and co-ordination with higher commands. The tactical control console gives these two operators the necessary target and battery status information and controls required.

The azimuth-speed operator has the sole mission of obtaining the earliest possible detection of low-altitude targets. The azimuth-speed indicator console and separate radar B-scope display, provide ICWAR target data for this purpose. Targets selected for manual engagements are assigned to one of the two fire-control operators. Each operator uses the fire-control console displays and controls for rapid HPI target lock, target track, missile launch and target intercept evaluation.

In the Phase III configuration, the BCC is removed from the system and replaced by the BCP described here.

Information Co-ordination Central (ICC)

The ICC is the fire-control data processing and operational communications centre for the battery. It provides rapid and consistent reaction to critical targets. Automatic detection, threat ordering, the IFF followed by automatic target assignment and launch functions are provided by the ICC. The ICC contains an ADP, IFF, and battery terminal equipment and communications equipment.

The ADP comprises an Electronic Data Processor (EDP) and a Data Take-Off unit (DTO). The DTO forms the interface between the other system equipment and EDP. With the exception of inputs from a solid-state reader and outputs to a printer, all communications with the ECP are through the DTO. The EDP is a militarised, general-purpose digital computer especially adapted to this role.

Phase III eliminates the ICC and transfers its data processing and communications functions to the Phase III PCP and BCP described here.

Platoon Command Post (PCP)

The PCP is used as the fire-control centre and command post for the AFU. It can also be used to replace an ICC. The PCP provides manual and automatic target processing, IFF, intra-unit, intra-battery and army air-defence command post communications and the displays and fire-control equipment for the three-man crew. It is essentially an ICC with a tactical display and engagement control console, a central communications unit, status indicator panel and an automatic data processor. The tactical display and engagement control console provides the man/machine interface for the AFP. The interior of the shelter is divided into two compartments: the tactical officer, radar operator and communications operator occupy the forward compartment with the display console, status panel, power distribution panel and communications equipment; the rear compartment contains the ADP, air conditioning unit and IFF equipment.

Phase III Platoon Command Post (PCP)

The Phase III PCP performs most of the same functions for Phase Assault Fire Platoon (AFP) as the PCP performed for the AFU. The new PCP uses the same shelter as the original PCP and is operated by a crew of three, consisting of Tactical Officer (TO), Radar Operator (RO) and Communications Operator. Some of the original equipment is retained in the new PCP, however a large proportion is replaced by the recently designed Phase III equipment.

The interior layout of the shelter is considerably changed with the communications operator, radios, computers, IFF set and air conditioning equipment both relocated and changed. The entire shelter interior is air conditioned with the larger, relocated air conditioner supplying cooling for the shelter as well as all the electronic equipment. The addition of a Nuclear, Biological and Chemical (NBC) Gas-Particulate Filter Unit (GPFU) and an entryway airlock provides positive air pressurisation of the shelter for protection of the crew and equipment from NBC effects.

Phase III electronic equipment modifications replaced the ADP with modern high-speed microcomputers and more densely packed memories, substituted the TO's, RO's and TAS display for two new computer-driven display systems and provided full-duplex serial digital datalink communications from the PCP to both the CWAR and HPI.

The PCP features an Integral Operator Trainer (IOT) which provides an on-site capability for HAWK operational training. This trainer, housed within the Automatic Data Processor, provides the realistic target simulation necessary to enable all the fire-control capabilities inherent to a fire unit to be exercised. The IOT's software memory contains 25 simulated target scenarios, including multiple, manoeuvring and ECM targets.

When the AFP is configured as an AFP+, which includes a PAR, the PCP used is the same as before with the exception of an additional microcomputer placed into the digital computer rack. This additional computer capacity is used to process PAR data and interfaces the PAR with the PCP.

Battery Command Post (BCP)

In Phase III battery configuration both the BCC and ICC are replaced by a single BCP. This reduces the operating crew from six personnel to the following four operators: a TO, two ROs and a Communications Operator. The physical configuration of the BCP is identical to that of the Phase III PCP described above with the exception of a Second Radar Operator's (SRO's) console placed in the left corner adjacent to the Tactical Display and Engagement Control Console (TDECC) and additional microcomputers placed into the digital computer rack.

High-Power Illuminator (HPI)

HPI automatically acquires and tracks designated targets in azimuth, elevation and range rate. It serves as the interface with unit-supplying azimuth and elevation launch angles computed by the ADP for up to three launchers. The HAWK missile for guidance also receives the HPI J-band energy reflected off the target. A missile reference signal is transmitted directly to the missile by the HPI. Target track is continued throughout the missile flights and, after intercept, HPI Doppler data are used for kill evaluation. The HPI receives target designations from the BCC and automatically searches a given sector for rapid target lock on. The HPI incorporates ECCM and BITE.

The Phase II programme includes two major modifications to HPI. One is the addition of a wide-beam transmitting antenna which is used to illuminate a much larger volume for missile guidance during use of the Low-Altitude Simultaneous HAWK Engagement (LASHE) mode of operation against multiple target attacks. The second is the addition of a digital microcomputer which processes HPI target data and provides full-duplex serial digital communications between the HPI and the PCP.

The Northrop Tracking Adjunct System (TAS), used in the HAWK-PIP Phase II upgrade for the HPI radar, was derived from the US Air Force's Target Identification System, Electro-Optical (TISEO) device and provides a passive tracking capability with remote real-time video presentation.

The day-only TAS is designed to complement the illuminator and can be used either coincidentally with, or independent of, the radar line of sight. Manual or automatic acquisition and tracking modes, rate memory and preferential illumination are the key features of the system.

Northrop Tracking Adjunct System (TAS)

Northrop Corporation Electronic Systems Division TAS is the video target identification system installed on the HPI radar. It enhances the HAWK missile system survivability by allowing for increased passive operations in providing preferential target illumination and performing the raid count, recognition and classification as well as damage assessment tasks.

It comprises a two field of view closed-circuit TV camera system, which is mounted on a gyro-stabilised platform and enhanced by a ×10 magnification telescope.

It is currently a day-only system that has been upgraded to:

- Improve its daytime performance (in terms of increased range and haze penetration capabilities)
- Add an automatic target search capability
- Add an infra-red focal plane array for day/night usage.

The fully functional day/night system is then designated Improved TAS (ITAS). Final development of the ITAS ended in 1991 with the field demonstration and trials phase in early 1992.

Production of TAS devices for the US Marine Corps began in 1980 and exports have been made to seven overseas I-HAWK users. By early 1999 over 500 had been produced.

AN/MPQ-51 Range-Only Radar (ROR)

This is a K-band pulse radar that provides quick response range measurement when the other radars are denied range data by enemy countermeasures. During a tactical engagement, the radar is designated to obtain ranging information, which is used in the computation of the fire command. The ROR reduces its vulnerability to jamming by transmitting only when designated. The ROR is not retained in the Phase III system.

M192 Launcher (LCHR)

LCHR supports up to three ready-to-fire missiles and is activated only on the initiation of the fire cycle. When the fire button is activated in the BCC or PCP, several launcher functions occur simultaneously: the launcher slews to designated azimuth and elevation angles, power is supplied to activate the missile gyros, electronic and hydraulic systems, and the launcher activates the missile motor and launches the missile. The launcher is equipped with electronic cut outs and sensing circuits that allow firing in all emplacement situations.

Raytheon has completed a proof-of-principle test, which demonstrated significant improvements in launcher performance, field support and survivability and field life expectation.

Based on launcher hydraulic and power levelling system modifications the following enhancements have been achieved:

- Prime power requirement reduction from 21.1 to 3.4 kW
- Elimination of the hydraulic heat exchanger and the launcher battle ALERT mode
- A considerable reduction in the launcher's battlefield infra-red and acoustic signatures
- A significant reduction in the time required to both raise and level the launcher as well as raise the superstructure hatch.

Variants

Self-propelled HAWK

To increase the mobility of some of its Basic HAWK batteries, the US Army at one stage fielded several self-propelled HAWK platoons.

These consisted of three tracked M727 vehicles, based on the M548 tracked cargo carrier, each carrying three ready-to-fire missiles and towing one piece of ground equipment. They have now been withdrawn from service. Israel is believed to have a number of these launchers still in service.

Sparrow HAWK demonstration programme (now known as Hawk XXI)

Another tri-service HAWK project is a system called Sparrow HAWK. It combines elements of both these Raytheon-produced missile systems. The standard M192 launcher is modified to allow eight Sparrow (Sea Sparrow) missiles to be placed on the same launcher, in lieu of the three I-HAWKs.

In January 1985 at the Naval Weapons Centre, China Lake, California, a modified missile launcher was used for field demonstration tests. A HAWK fire unit manned by a US Marine Corps team successfully fired Sparrow missiles at two unmanned aircraft targets. Earlier tests on the mobility of the system were carried out at the US Marine Corps Air Station, Yuma, Arizona.

A typical reduced-manpower Sparrow HAWK fire unit would comprise an IPCP, an ICWAR, an HPI, two M192 launchers with I-HAWK missiles and one M192 launcher with Sparrow missiles. The fire unit is also able to convert a launcher in the field for either I-HAWK or Sparrow compatibility by using the solid-state digital electronic system changes in the launcher assembly.

The two missile types can be deployed concurrently within the fire unit either in the HPI antenna pencil beam or simultaneous target engagement modes of operation.

The HAWK missile loader and missile pallets are eliminated and replaced by truck transportation and crane loading. Two four-round Sparrow or one three-round I-HAWK clip is used to reduce the loading time.

If carried by a Lockheed Martin C-130 Hercules or helicopter the M192 launcher can be loaded with either two I-HAWK or eight Sparrow missiles in the operationally ready-for-use condition. This considerably cuts down on the time required to go into action at the landing site.

In mid-June 2012 at the Andoya Rocket Range in Norway, Raytheon and Kongsberg carried out firing trials of what is now known as Hawk XXI using Raytheon produced ground launched Evolved Sea Sparrow Missile (ESSM). The aim of the test was to demonstrate the proven capabilities of the Kongsberg Fire Distribution Centre (FDC), the NASAMS canister launcher, the Thales/Raytheon Systems AN/MPQ-64F1 Sentinel radar and Raytheon's Hawk High Power Illuminator (HPI) as the semi-active missile guidance source.

Integration of the NASAMS launcher with Hawk XXI provided an opportunity to use other missiles apart from the AIM-120 Advanced Medium-Range Air-to-Air Missile (AMRAAM).

Hawk XXI is the latest configuration of the Hawk defence system and is designed to engage and destroy fixed-wing aircraft, helicopters, UAVs, cruise missiles and can provide Anti-Tactical Ballistic Missile (ATBM) capability against short-range ballistic missiles.

AMRAAM HAWK demonstration programme

At 'Safe Air 95' the US Missile Command demonstrated the firing of the AMRAAM missile from a modified M192 HAWK missile launcher using standard battery radar systems. The launcher has two four-round AMRAAM clips fitted in lieu of the three-round I-HAWK clip. Visually it is similar to the Sparrow HAWK configuration. An international consortium of Raytheon and Kongsberg is offering a HAWK-AMRAAM hybrid system of this type (see separate entry in this section). Greece is to upgrade its I-HAWK fire units to the Phase III using the Fire Direction Centre of the HAWK-AMRAAM system.

Danish Enhanced HAWK (DEHAWK) Ground Based Air-Defence (GBAD) system

The preliminary design review of DEHAWK began in January 1998. Prototype system acceptance checks, followed by first operators training, began in February 1999, with the first deliveries of the upgraded units made to the Royal Danish Air Force in January 2001. Developed by TERMA Elektronik AS of Denmark in partnership with Thomson-CSF Airsys of France the DEHAWK programme transforms the existing eight I-HAWK squadrons into 16 autonomous Firing Units (FU), two per squadron. The first DEHAWK live firing took place in September 2000. It is envisaged that the current I-HAWK missile and its associated guidance radar will be replaced in the DEHAWK GBAD system by a new longer-range missile type by the year 2007.

Each FU has its own Thomson-CSF Airsys RAC-3D search radar (range 100 km) with integrated IFF as well as a TERMA developed Operations Centre, the existing MPQ-46 HIPIR and three modernised I-HAWK launcher assemblies. All the FU systems are interconnected via 1,000 m long fibre-optic cable links to allow maximum flexibility in tactical deployment.

The individual FUs and two new mobile SAM Wing Operations Centres (SAMWOC) will operate under the TERMA's air defence Command, Control, Communications and Intelligence (C3I) system via the DEHAWK SAM Area Communication System (SACS). The latter was ordered from the Danish firm INFOCOM Systems A/S and the French company Thomson-CSF Communications in 1998. The SACS network includes 10 SAM Communications Units (SCUs) in shelters with a 20 m high electrical-hydraulically-operated antenna mast on a trailer. The mobile SCUs are interconnected by a meshed network of tactical line-of-sight radio links, which provide high capacity, internet protocol based multimedia communications using Asynchronous Transfer Mode Technology.

The first SAMWOC was delivered in February 2001 and the second in December 2001. The last fire unit will also be delivered in December 2001. Each fire unit will also have a section with six Stingers MANPADS for local air defence.

As of November 2004, Terma A/S has developed a C3 system as part of the DEHAWK system. Housed in a box-body, it consists of three identical workstations and is connected to sensors and weapons systems in a wide area network through radio access points. All sensors deliver track information to feed a real-time air picture. Operators positively control the weapons systems. A battle management system is also provided to support planning, logistics and other combat support activities. The system can be adapted to a variety of weapons.

The I-HAWK PIP III systems that were sold by the Netherlands to Romania have since been upgraded and modernised to a Romanian standard and are now known as HAWK XXI. The upgrades were carried out in country by Electromecanica Ploesti in cooperation with Raytheon Missiles and Kongsberg Defence and Aerospace. Electromecanica was responsible for upgrading the missiles with PIP III kits, whilst Raytheon worked on the MPQ-64 Sentinel radar and Kongsberg the fire-control centres similar to those used in NASAMS II.

Swedish RBS 97 (FU 97)

In Sweden the HAWK system is known as the RBS 97 or FU 97, the system has an intercept range of 40 km with a maximum altitude of 18 km. A new communications system the TS 9000 is in use in Sweden only whilst the illumination radar has been modified to Swedish standards.

Mersad in Iran

Announced in March/April 2010, Iran has started series production of a 40 km range system known locally as Mersad. Photography and video released by the Iranian authorities would indicate that the Mersad is directly descended from the US produced HAWK sold to Iran in the 1970s. It is assessed (although not proven) that part of the improvement to the original system includes taking the electronics from the analogue to digital age.

Specifications

	MIM-23A	MIM-23B
Dimensions and weights		
Length		
overall	5.08 m (16 ft 8 in)	5.08 m (16 ft 8 in)
Diameter		
body	370 mm (14.57 in)	370 mm (14.57 in)
Flight control surfaces		
span	1.19 m (3 ft 10¾ in) (wing)	1.19 m (3 ft 10¾ in) (wing)
Weight		
launch	584 kg (1,287 lb)	627 kg (1,382 lb)
Performance		
Speed		
max Mach number	2.7 (est.)	2.7 (est.)
Range		
min	1.1 n miles (2 km; 1.2 miles) (high-altitude target)	0.8 n miles (1.5 km; 0.9 miles) (high-altitude target)
	1.9 n miles (3.5 km; 2.2 miles) (low-altitude target)	1.3 n miles (2.5 km; 1.6 miles) (low-altitude target)
max	17.3 n miles (32 km; 19.9 miles) (high-altitude target)	21.6 n miles (40 km; 24.9 miles) (high-altitude target)
	8.6 n miles (16 km; 9.9 miles) (low-altitude target)	10.8 n miles (20 km; 12.4 miles) (low-altitude target)
Altitude		
min	60 m (197 ft)	60 m (197 ft)
max	13,700 m (44,948 ft)	17,700 m (58,071 ft)

	MIM-23A	MIM-23B
Ordnance components		
Warhead	54 kg (119 lb) HE blast fragmentation	75 kg (165 lb) HE blast fragmentation
Fuze	impact, proximity	impact, proximity
Guidance	semi-active, radar (with proportional navigation)	semi-active, radar (with proportional navigation)
Propulsion		
type	solid propellant (dual-thrust, sustain stage)	solid propellant (dual-thrust, sustain stage)

Status
HAWK and I-HAWK production is complete and was replaced in service in 1994 by the MIM-104 Patriot.

Contracts
For European manufacture of the Improved HAWK air-defence system, Raytheon was systems contractor under the direction of the NATO HAWK Management Office (NHMO) in Paris, France. In the US, Raytheon produced the missile guidance and control units and limited quantities of ground equipment, certain missile parts, final missile assembly and overhaul and conversion of the Basic HAWK equipment. A Raytheon subsidiary, Raytheon European Management and Systems Company (REMSCO), oversaw major European industrial activities. NATO states participating in the programmes are Belgium, Denmark, France, Germany, Greece, Italy, Netherlands and Norway.

In late 1995, all but France and Italy withdrew from the NATO HAWK Production and Logistics Organisation (NHPLO) that was set up to plan and introduce HAWK modification and maintenance improvements and complete system development and US Production.

Contractor
Raytheon Company, (enquiries).
Merex Inc., (Hawk Phase II and Phase III ground system components, spares and repairs).
Intracom Defence, Greece (integrated tactical communications).

MIM-104 Patriot

Type
Static and towed Surface-to-Air Missile (SAM) system

Development
The concept of a mobile all-weather air-defence missile was started in 1961 by the US Army Missile Command (MICOM) Research and Development Directorate as the Field Army Ballistic Missile Defence System (FABMDS), then became the Army Air-Defence System 1970 (AADS-70).

By 1965, the design had been specified and the missile assigned the designation XMIM-104 Surface-to-Air Missile-Development (SAM-D) before project management was placed under MICOM direction. Shortly after this, in

Table 1

Year	Features improved/added
1986	Fuze processor; Guidance ECCM; Strobe jammer engagement mode; Jammer engagement (using correlation technique) mode; Out-of-sector launch capability (±80°)
1987	Radar enhancement Phase I; IFF upgrade
1988	PAC-1 deployment; Interoperability Block 1; Clutter canceller; Maintenance improvements; Pulse Doppler discrimination capability
1989	Transmitter maintenance upgrade; DLU upgrade
1990	PAC-2 deployment
1991	Standoff jammer counter capability; Pulse Doppler search/track capability; Interoperability Block 2; Guidance enhancement; Optical disc
1992	Out-of-sector launch capability (360°)
1994	Radar enhancement Phase II; EMP enhancement; VHSIC WCC; ARM decoy deployment
1995	PAC-3 Configuration 1 deployment
1996	Block I Improvements; Missile; PAC-3 Configuration 2 deployment; Extended range; Launcher; Remote-control launcher; Radar enhancement Phase III; Dual TWT transmitter; Improved S/T receiver; Command, Control and Communications (C3); JTIDS interface; SINCGARS/HCS interface; HIMAD C IMPS; BDE ICC; Identification; Mk XV IFF interface; NCTR
1998	Block II Improvements; Radar enhancement Phase IV; 360° intercept capability; Battle management; Communications; Discrete target identification Phase II
2000	PAC-3 Configuration 3 deployment Adds the DWT tube unit, a new radio-frequency exciter to radar to improve performance further and detection of small targets in cluttered environments; a Classification, Discrimination, Identification phase III; extends remote launch capability from CRP distance of 10 to over 30 km; and plans to equip each US Army fire unit with six PAC-2 and two PAC-3 launchers.
2000	Programme to upgrade the PAC-2 missiles to the GEM+ standard inception. Raytheon Company received a contract for 770 missiles. By April 2006 515 had been delivered. The remainder were to be delivered during 2006/2007.
2002	GEM+ Missile: Guidance Enhanced Missile Plus (GEM+) adds a low noise oscillator for improved acquisition and tracking performance in clutter against lower cross section targets. The GEM+ missile provides an upgraded capability to defeat air-breathing, cruise missile, and ballistic missile threats, and integrated with the PAC-3 system of radars.
2005/6	Patriot Lower Tier Project using developed Patriot PAC 2 GEM system Post Deployment Build-6 (PDB-6) software. Also known as Configuration-3 and GEM-T, PDB-6 system's capability is to search, detect, track, engage and kill two surrogate full-body tactical ballistic missile targets.
2006	The German Patriot PAC-2 system is being updated and modernised in the communications and data processing package for the five air-defence missile battalions. This package is to improve the interoperability with other air-defence systems as part of overall upgrade to the system.
2006	Raytheon Southeast Asia Systems Company in Andover, US, has been awarded a firm fixed price contract for the technical assistance for the Kuwaiti Patriot missile system. Work is to be performed in Kuwait and is expected to be completed by January 2011. By this time Kuwait is expected to have taken delivery of the Patriot PAC-3.
2007	In January 2007, two contracts were let to the Raytheon Company for modification and fore body to GEM+ frequency generator upgrades. Both contracts to continue through April 2009.
2007	8 October 2007, Raytheon received a contract to begin the Patriot 'Pure Fleet' modernisation programme which was followed in December with a further contract to include tactical assets. In October 2008, a contract for add-on items was also let to be finished in February 2010. On 23 February 2009, a USD9,244,000 firm fixed price contract for PATRIOT Pure Fleet 12 Lot Add on Items. Work to be performed at Andover, Massachusetts, with a completion date of 28 February 2010. The US Army Contracting Command, Aviation & Missile Command Contracting Centre, Redstone Arsenal, Alabama, is the contracting activity (W31P4Q-07-C-0151).
2007	9 November 2007, Tecro, a US based company, obtained a contract to work on the Patriot Configuration 2 Ground Systems Upgrade.
2008	July 2008, Kuwait issued a contract to Raytheon to upgrade all Patriot's in country to Config 3 standard, this will make them consistent with the US Army's Pure Fleet initiative.
2009	January 2009, Taiwan Patriot is to be upgraded with kits for radar and command-and-control components to the Config-3 standard. The upgrade will allow the Patriot to fire the latest version of the missile, the Patriot PAC-3.
2009	Raytheon Company has been awarded a new contract to continue "Pure Fleet" upgrades of US Army Patriot Air and Missile Defense Systems. The US Army Aviation and Missile Command contract provides USD115 million to upgrade four additional US Army Patriot Air and Missile Defense Systems to Configuration-3 status. The upgrades involve enhancements to Patriot system ground components, particularly the radar. This contract supports a "Grow the Army" initiative.
2011	In April 2011, Raytheon was awarded a contract for Patriot GEM-T missiles improvement from the MIM-104 Patriot PAC-2. This is a follow-on contract as part of AMCOMs Patriot missile continuous technology refreshment programme initiated in 2000. In June 2011, the Kingdom of Saudi Arabia awarded Raytheon a contract to upgrade/update its PAC-2 systems to the Config-3 standard, this improvement will be carried out with Mercury Computer Systems of Massachusetts and will include radar signal processing and Applications Ready Subsystems (ARS). Taiwan will also receive the same upgrade, this will then allow the system to make use of the Lockheed Martin PAC-3 missile.
2011	In November 2011, Israel is to upgrade its Patriot PAC-2 launchers that currently fire four missiles up to 16. This is both a software and hardware upgrade.

M901 Patriot launcher (Hennie Keeris) 0560888

Patriot AN/MPQ-53 radar set undergoing company tests before delivery 0007983

April 1966, the US Department of Defense (DoD) issued contract definition awards to the then Raytheon, Hughes and RCA with Raytheon receiving an advanced missile development contract in May 1967.

The first test launch occurred in February 1970. The engineering development programme commenced in 1972. This consisted of two phases, the first being a nine-round flight series in 1973 to test the missile systems, then a 14-round series in 1974-75, to evaluate the Track-Via-Missile (TVM) guidance system concept.

The success of these tests led in 1976 to the DoD initiating the Full-Scale Engineering Development phase of what was then called the MIM-104 Patriot missile system.

By 1981, all the development and operational evaluation trials were completed and limited production by Raytheon had commenced. The first production systems were delivered to the US Army in June 1982 and a follow-on test and evaluation programme was conducted in 1983. This, however, uncovered a number of hardware and training problems, which were subsequently resolved with Patriot undergoing a further series of firings in 1984. Following the successful completion of these firings, the US Army declared Patriot to be fully operational.

In order to keep Patriot performance effective against a changing threat, the US Army has a continuous Patriot Improvement Research and Development programme under way.

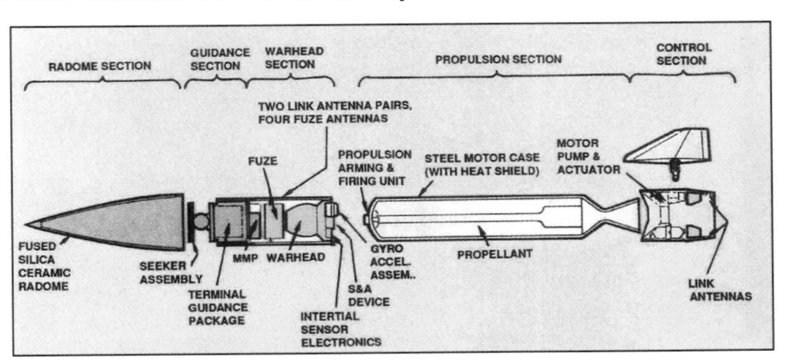

Main components of the Patriot surface-to-air missile 0022190

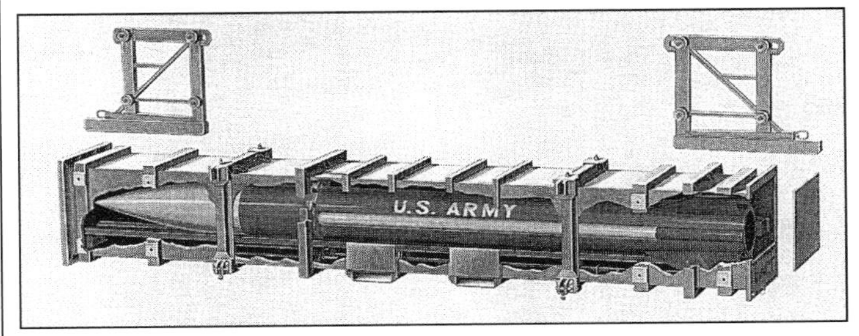

Cutaway view of the Patriot missile sealed in its container which serves as the launch tube for the missile when it is fired. The missile's airframe and its canister are manufactured by Lockheed Martin under contract to Raytheon, which (as prime contractor) manufactures the missile's guidance and control electronics and all of the ground equipment for the missile system 0509621

For the full Patriot Product Improvement Programme schedule, please see Table 1 on the previous page:

Of the features added and/or improved, the following information is known:

- The first, known as Phase 1B, consists of a radar enhancement design and software development for jammer engagements, guidance ECCM, a 16-missile launch capability and better battalion resource management. This was completed in 1986;
- The second, Phase 1A, started in 1986 and was a radar enhancement software development by Raytheon for a Patriot out-of-sector launch capability. It was completed in 1989;
- The last is the Anti-Tactical Missile (ATM) programme, which consists of two parts.

The Patriot Level-1 ATM (or PAC-1), which requires only software changes to the ground radar itself (that is, it does not require destruction of the target's warhead) for it to achieve a mission kill capability against short-range ballistic missiles by giving it an upward trajectory and a high-angle sector search capability to track them in flight. The idea is to protect the Patriot system itself. This ability was successfully demonstrated on 11 September 1986 when a modified production-line Patriot intercepted a Lance missile at the White Sands Missile Range and knocked it off its intended course. The Patriot Level-2 ATM (or PAC-2) was also successfully tested at White Sands on 4 November 1987 when a suitably modified Patriot destroyed another Patriot in flight, together with its warhead, which was simulating an SRBM. The PAC-2 modification involves further software changes, a new missile warhead casing, with enhanced explosives and a fuzing system with a second set of forward-angled beams designed to optimise warhead detonation against targets with a very high closing rate so as to increase sensitivity and reduce the system's reaction time. This then allows the larger 45.4 g size of the fragmentation pattern splinters produced in the explosion to perform a catastrophic kill of the target by destroying both the missile body and its warhead and is an overall improvement to the system's area defence role.

The Level-1 ATM capability was deployed with the US Army in Europe in July 1988. Level-2 ATM capability was deployed in August 1990 for the Desert Shield Gulf deployment. The deployment was transitioned to Desert Storm with the outbreak of hostilities in January 1991. On 18 January, Patriot conducted a historic first interception of a hostile TBM threat. In all, 88 modified 'Scud' TBMs were fired into Israeli and Saudi Arabian airspace by Iraq. There were 53 of these fired into Patriot-defended areas and of those, Patriot intercepted 51. A total of 157 Patriot missiles were expended in engagements.

Upgrades to the Patriot PAC-2 include the congressionally mandated Quick Reaction Program (QRP). This includes improvements to the radar intended to increase range and performance by distinguishing missile debris from the actual warhead. The incorporation of NAVSTAR Global Positioning System (GPS) receivers to improve the navigation component, a remote launch capability of up to 10 km for the system and faster warhead fuze responses to allow engagements against faster re-entry vehicles. QRP improvements were introduced into service, with two of the US Army's Patriot battalions upgraded by early 1994. A companion programme, the Guidance Enhanced Missile (GEM) includes engineering improvements to the PAC-2 missile by Hercules Aerospace following a two-year development subcontract from Lockheed Martin in 1992. These include a lightweight composite case and advanced propellant that increases the effective range of the missile by 30 to 40 per cent.

As an adjunct to the Patriot system MICOM awarded a USD2.7 million contract in May 1983 to the General Instrument Corporation's Government Systems Division to develop and build a decoy unit that could entice attacking anti-radiation missiles from Patriot air-defence missile sites. The technological challenge in the programme was the size of the decoy which apparently

Patriot AN/MPQ-53 radar set which is built by Raytheon 0044350

The four-round M901 Patriot launcher used by the German Air Force mounted on the rear of a MAN (8 × 8) truck for improved cross-country mobility. The vehicle is known by the designation Startstation LS 0509904

Raytheon Patriot surface-to-air missile system deployed 0044349

The German Air Force 34 m telescopic antenna mast system for Patriot has been designed by DASA's Dornier-led Information and Communications Systems Division and is carried on the rear of a MAN (6 × 6) truck. The vehicle is known by the designation Antennenmastanlage AMA. A total of 80 units was produced between 1989 and 1992 0044348

became a device portable by two personnel. In December 1987, MICOM started seeking bids from industry to undertake a two-year Full-Scale Engineering Development (FSED) programme involving the development, production and qualification of five decoy units.

In February 1989, the USD4.715 million FSED contract was awarded to Brunswick Corporation's Defence Division. The system is known as an Anti-Radiation Missile Decoy (ARM-D) and will transmit an RF signal of similar frequency and amplitude to the AN/MPQ-53 radar in order to divert the track point of the incoming threat weapon.

The US Army procurement objective for Patriot was for 10 battalions of 60 Fire Units (seven battalions with 42 Fire Units in Germany and three battalions with 18 Fire Units in the Continental US for training and so on). By mid-1989, all 10 battalions had been activated with seven deployed to Germany as part of the 32nd Air-Defence Missile Command. Currently the West European battalions have three Fire Units each instead of the six in their TOE. Each 600-man Patriot battalion comprises six batteries each of two firing platoons and four launchers. Each battery headquarters and firing platoon also includes an FIM-92A Stinger man-portable SAMs two-man team set of equipment for close-range low-level air defence. At battalion level there is also a (6 × 6) truck-mounted Information Co-ordination Centre (ICC) which provides battalion command-and-control and the interface with other air-defence assets. This is performed by (6 × 6) truck-mounted communications relay units and attendant (6 × 6) truck-mounted Antenna Mast Groups (AMGs).

In December 1995, the PAC-3 Configuration 1 was fielded. This has a number of improvements over the Patriot QRP, especially in battle management, Command, Control, Communications and Intelligence (C3I) and incorporates the Patriot PAC-2/GEM high *g*, tail control, blast fragmentation warhead missile.

In 1996, the first PAC-3 Configuration 2-equipped unit was fielded with improvements and modifications to the radar, communications and other systems. One of the most prominent changes is a significant improvement in target Classification, Discrimination and Identification (CDI) ability.

In March 1997, at the Kwajalein Atoll, the upgraded guidance-enhanced missile version of Patriot successfully intercepted a Scud missile using a blast fragmentation warhead at close range. The test, conducted by the BMDO and the US Army Space and Strategic Defence Command, was the second successful test against a Scud missile at the same missile range test location.

In 2000, the definitive PAC-3 Configuration 3 system was fielded with the Lockheed Martin Missiles and Fire Control - Dallas PAC-3 missile in conjunction with the GEM and earlier missiles used in Configuration 2. A minimum of two launchers per battery are PAC-3 capable.

Further development possibilities include expansion of the datalink capability to include US Navy and MEADS waveband regions.

The US DoD promoted Patriot as a NATO follow-on system for the MIM-14 Hercules and some MIM-23 HAWK systems. In February 1979, the US, Belgium, Denmark, France, Germany, Greece and the Netherlands signed a NATO Memorandum of Understanding (MoU) for a two-year study on the most practical and economical ways to acquire and produce the Patriot system. This was itself a follow-on study to Project Successor, a joint US-German analysis, concluded in 1978, to evaluate Patriot's suitability for a European air-defence role.

By early 1988, only two of the MoU signatories had actually procured the system. In 1984, the Netherlands ordered four squadrons totalling 20 launchers and 160 missiles. The first of these, the 502nd Anti-aircraft Squadron of the Royal Netherlands Air Force 5th Guided Missile Group, became operational in April 1987 with a second squadron of the Group attaining operational status in July 1988. By 1990, the remaining two squadrons were operational with the 3rd Guided Missile Group. A follow-on order for four more fire units, 32 launchers and 256 missiles was approved by the Netherlands government.

In Europe, collocation of the NATO SAM units with Army Corps areas is designed to enable better integration with the short-range air defences of the ground forces. This is being facilitated by the introduction in the early 1990s of the Mobile SAM Operation Centre (SAMOC) which will replace the current air-defence liaison teams.

The SAMOC will also provide the links between the Patriot ICC and the CRCs, Sector Operation Centre and, ultimately, the relevant Allied Tactical Air Force's air-defence operations centre.

The German Air Force AN/MPQ-53 radar set for Patriot is mounted on the rear of a MAN (8 × 8) truck and is shown here in the operating configuration. The vehicle is known by the designation Radargerät RS 0509622

Patriot ECS installed on the rear of a MAN (6 × 6) truck used by the German Air Force. This is called the 'Feuerleitanlage' ECS 0509623

In place of US (6 × 6) trucks the Netherlands uses the DAF YTZ 2300 (6 × 6) truck to tow the four-round Patriot launcher and the AN/MPQ-53 radar set, and YAZ 2300 (6 × 6) 10,000 kg trucks carry the engagement control station.

In the same year that the Netherlands ordered its systems, Germany agreed to buy the system and in 1985/1986 ordered the first squadrons it was going to purchase. It was accepted, through a supplementary compensation agreement with the US, that an additional 12 squadrons were to be supplied from US Army stocks and manned by German personnel. The 36 Luftwaffe squadrons were used to form six wings (*Geschwader*), each of six air-defence squadrons (*Flugabwehr-raketenstaffeln*), one headquarters squadron, a direct support unit and four radio relay groups. The major difference between this Patriot equipment and the US Army systems is that the prime German contractor, SI Sicherungstechnik GmbH & Co KG, is using German MAN (8 × 8) vehicles and accessories wherever possible. The first Luftwaffe firing trials were completed in November 1987 at the White Sands Missile Range using its own training equipment. The first operational units were handed over in August 1989. In 1998, it was stated that the US systems operated by the Germans were being formally handed over to Germany. As of August 2006, the Luftwaffe use the DT-35 subsonic aerial target for training of Patriot SAM missile crews.

Due to the reorganisation in Germany, the projected Luftwaffe SAM structure is six mixed wings (each of one group of Patriot and one group of I-HAWK), five in the former West Germany and one in the former East Germany.

On 19 December 2006, at the Federal Office for Military Technology and Procurement (BWB), a contract was signed for the improvement of the combat effectiveness of the Patriot air-defence missile system. This contract covers the modernisation of the communications and data processing system for five air-defence missile battalions. With the system improvement, the fire-control and communication system is being modernised. In addition, this will improve the interoperability with other air-defence systems as part of the integrated air defence. The prime contractor for the present contract is LFK, Lenkflugkorpersysteme GmbH, Unterschleißheim. The subcontractors for this modernisation programme are EADS Deutschland GmbH, DATUS AG, Sinus Electronic GmbH, Aschenbrenner Gerätebau GmbH, TE-SYSTEMS GmbH and Elnic GmbH. The modernisation of the communications system is, after the increase in radar range and the launcher upgrade, the third substantial product improvement package for the Patriot weapon system.

In May 1999, it was revealed that Germany was considering an advanced upgrade for its Patriot systems due to the MEADS programme problems. The upgrade would involve conversion of 210+ (out of the existing 320 launcher five units) to the PAC-2+ missile configuration (German designation for missiles with PAC-2 hardware and PAC-3 software) and the remaining launchers to the full PAC-3 level, starting in 2001. Also envisaged is the integration of an SEL Defence Systems pulse-Doppler, pulse compression UHF/metre wave cuing radar designated MCR-21.

In 1984, the Japanese Defence Agency selected Patriot for the Nike Hercules-J as its long-term replacement. In 1986, Mitsubishi started producing the missile under a licensed agreement from Raytheon. An initial buy of two batteries (one in knockdown form for local assembly) was made in 1985 to form training units. A total of six Japanese Air Self-Defence Force missile groups were re-equipped at the rate of one per year between 1986 and 1991 with four batteries, each of five launchers, and a total of 1,000 missiles (980 of which were produced locally). The PAC-1 systems were upgraded to PAC-2 standard between the fiscal years 1994 and 1999. In April 2006, the US Army Aviation and Missile Command, from the Redstone Arsenal in Alabama, awarded a contract to Raytheon for the upgrade of the M818E2 fuzes to the M818E3A configuration for Japan. This fuze is used on the PAC-2 missile system.

The 1990 to 1991 Gulf War resulted in 21 US Army batteries with 132 launchers being deployed to Saudi Arabia, seven batteries in Israel (two Israeli, one Dutch and four US Army) and four batteries in Turkey (two US Army and two Dutch). Of the foreign systems in Israel, the Israeli Self-Defence Air Force has subsequently received two US Army batteries and one German battery (supplied after the war) as grant aid and purchased a third US Army battery through the FMS programme. All the systems include PAC-2 missile capability and have received the QRP upgrade under the US Army 1993 contract with Raytheon.

An initial 1990 USD513 million contract for seven (six operational and one training) Patriot Fire Units, 48 launchers, six AN/MPQ-53 radars and six engagement control centres with 384 PAC-2 missiles, was agreed for Saudi Arabia. All the systems were supplied as ex-US Army stock from 1993 onwards. A further USD1.03 billion contract for 13 Fire Units and 761 PAC-2 missiles was placed in December 1992. In January 1993, Kuwait placed a USD327 million contract 'via the US Army' for the supply of five Patriot Fire Units and 210 PAC-2 missiles.

In mid-1994, the Taiwanese Ministry of Defence placed a USD380 million plus commercial contract with Raytheon for three MADS fire units and some 200 missiles. The first fire unit was delivered in January 1997. As of September 2004, the US Air Force planned to build (locate) an early warning radar system at Leshan military base on Leshan Island. Approval for the sale of the early warning radar from the US was given in April 2004. The 32 m high and 30 m high radar can be linked with the three Patriot PAC-2 air-defence missile systems currently deployed in northern Taiwan. In March 2006, the Ministry of National Defence (MND) asked for funding to upgrade the three PAC-2 missile batteries and at the same time asked for the purchase of Patriot PAC-3 batteries in the annual budget, beginning March 2007.

At Defendory in 1998, the Greek MoD stated that it was ordering the Patriot. In an order worth some USD1.1 billion, the Greeks acquired four fire units with an option on two more to the latest PAC-3 configuration with GEM missiles. The first was delivered in 2001. In October 1999, three leased Patriot PAC-2 systems were delivered to Greece at a cost of USD8.2 million for a three-year period. These fire units have been made operationally ready and deployed at three separate sites, two forward deployed in the Aegean and the third in Greece since February 2000. All three were upgraded to the PAC-3 in 2001.

In early 1999, it was revealed that the Patriot Air-Defence system had been released for sale to Egypt. Any ensuing contract is expected to be comparable to at least that of the Greek one and would probably involve PAC-3 configuration fire unit systems.

In June 1999, Turkey made informal inquiries in Washington concerning the price and technical specifications of the Patriot.

In 2001, PDB-5 was deployed to Turkey with the Netherlands Air Defence Group (Patriot PAC-2 GEM missiles) to protect high-value areas, in case of attack during the Iraq/Coalition war. PDB-5 also meant that, if necessary, the Dutch systems could be reinforced with PAC-3 missiles, with no further changes necessary.

In 2002, Raytheon completed the upgrade of the Patriot PAC-2 missile that became known as the Guidance Enhanced Missile Plus (PAC-2 GEM+). GEM+ missiles are essentially PAC-2 systems that still use the larger PAC-2 missiles but are refurbished, modernised and integrated with the PAC-3 system of radars, communications and so on.

In 2004, three PAC-3 missile operational tests were conducted at White Sands Missile Range, New Mexico. All three tests were against threat representative targets and employed operational doctrine, with each test increasing in complexity. All the tests were successfully conducted, achieving intercept of all targets. The PAC-3 missile has intercepted 17 targets in 19 opportunities during its testing phase, which stretches back several years and represents the most successful testing of any air and missile defence interceptor of this complexity.

Between May and July 2006, further trials of PDB-6 – composed of user-requested improvements, planned performance improvements that resulted from lessons learned in operations during the Iraq war – were demonstrated at the White Sands Missile Test Range. These successful trials further demonstrated the PDB-6 software upgrade works with all variations of the Patriot.

As of February 2007, tests were completed on the Patriot Configuration 3 system composed of a Patriot Guidance Enhanced Missile-T (GEM-T) that has been designed to engage land-attack cruise missiles and a range of ballistic missile threats through applying technical enhancements and modifications. The Configuration 3 missile system uses the newly developed Post Deployment Build-6 (PDB-6) system software.

In March 2007, Raytheon received a contract with the intent of the Pure Fleet effort to upgrade Patriot fire units for the Army's worldwide requirements to

Communications vehicle of the German Air Force Patriot system uses a MAN (4 × 4) truck and is called the 'Richfunklage' 0509905

German Air Force Patriot missile resupply vehicle which carries four missiles and tows a trailer carrying a further four missiles 0509624

provide units with configuration-3 capability following the Army's February 2006 decision to upgrade additional tactical Patriot fire units. The initial work includes software and hardware upgrades, top Patriot test stations and engineering to address obsolescence in the factory and key suppliers. This is the first step to achieving Pure Fleet capabilities. Raytheon IDS is the prime contractor for the Patriot system and the system integrator for the PAC-3 configuration system including the affordable Guidance Enhanced Missile-T. GEM-T missiles are PAC-2 missiles that are refurbished and modernised at Raytheon's Integrated Air Defence Centre in Andover, United States. Through the upgrade process, older components are replaced, new technology is inserted and reliability is increased. In December 2007, a further contract was issued to Raytheon for the Pure Fleet tactical assets.

In November 2007, the US Company Tecro was awarded a contract for Patriot Configuration 2 Ground Systems Upgrade as part of a Foreign Military Sales programme to Taiwan. The Taipei Economic and Cultural Representative Office in the US has requested an upgrade and refurbishment of Taiwan's existing Patriot fire units to the latest Army Configuration 3 Ground Support Equipment.

In March 2008, Raytheon was again awarded a contract valued at USD115 million for maintenance support of the Patriot system including those sold overseas to foreign clients. In addition to this contract, it was announced that the Patriot system is to be sold to South Korea to form the SAM-X programme. Up to 48 fire systems purchased from Germany under a contract dated March 2005/7 have been integrated into the South Korean national command-and-control structure. In July 2008, it was announced that these missiles were upgraded to GEM-T standard.

Late November 2008, Raytheon received another contract valued at USD77.4 million to upgrade the system in order that the missiles can detect and engage tactical ballistic missiles.

Description

The MIM-104A Patriot missile is a certified round which is shipped, stored and fired from its Lockheed Martin rectangular box-like container-launcher. Each container-launcher box is 6.1 m long, 1.09 m wide and 0.99 m high. Weight empty is 794 kg and loaded 1,696 kg. The missile requires no testing or maintenance in the field, periodic lot sampling of missiles on launchers and in storage provides assurance of the weapon's capability. In configuration it is a single-stage missile with four sections. At the front is the Raytheon guidance and radome compartment containing the autopilot controls, guidance electronics and the monopulse seeker unit, with its steerable 30.5 cm diameter antenna. Next is the warhead section, made by Picatinny Arsenal, which contains four flush-mounted guidance antennas, inertial sensors and the fixed, strip line E/F-band fuzing (M818E1/M818E2), arming and blast/fragmentation warhead devices. In April 2006, Raytheon received a contract to upgrade the M818E2 fuze to the M818E3A configuration for the Japanese missiles. This fuze is used on the PAC-2 system. This is followed by the high-strength steel propulsion section, which contains a Thiokol 11.5 second burn 10,909 kg thrust TX-486 solid propellant rocket motor. At the tail is the Lockheed Martin control section, which supports the control actuation system, four aerodynamic control surfaces and two further flush-mounted guidance antennas.

Velocity at motor burnout is 1,700 m/s with the missile able to undertake 20 *g* continuous manoeuvres and 30 *g* short-term manoeuvres. This allows it to cope with targets performing continuous 6 *g* evasive manoeuvres.

Maximum flight time is 170 seconds and minimum flight time (that is, time required to arm) is 8.3 seconds.

In the field the Patriot battery consists of:

- AN/MPQ-53 phased-array radar mounted on a two-axle M860 semi-trailer with a 5 ton (6 × 6) M818 truck as the tractor;
- (6 × 6) truck-mounted AN/MSQ-104 Engagement Control Station (ECS);
- AN/MSQ-24 (6 × 6) power plant truck with two 150 kW diesel-powered gas-turbine AC generator units;
- Two firing platoons, each with four four-tube M901 launching stations on M860A1 trailers with their own individual 15 kW generators and secure VHF data links to the ECS and M818 tractors.

Ground Support Equipment (GSE)

There is also support equipment in the form of missile reload trailer transporters and their tractors, a maintenance centre truck and trailer, a battery replaceable unit small truck transporter, and a large battery replaceable unit semi-trailer transporter with a maintenance vehicle tractor. There can also be a (6 × 6) truck with a GTE/Sylvania extendable AMG for communications with battalion headquarters, other units and higher echelons.

Communications

Communications within the battalion are via voice and digital data. Six operational nets, two data and four voice, are used with at least 50 km between the Battalion's Command and Co-ordination ICC and up to six ECSs, 40 km between each ECS and at least 1 km between the ECS and M901s in a typical deployment pattern. An announcement made during August 2005, confirmed that the US Army was to adopt High Capacity Line-Of-Sight (HCLOS) radios in the Patriot system. Ultra Electronics Tactical Communication Systems division of Montreal, Canada, won a contract to supply the HCLOS radios to the US Army for use with the Patriot missile system. The AN/GRC-245(V) HCLOS radio will replace the AN/GRC-103(V) for point-to-point voice and data communications. This radio system provides full-duplex operation in Band 1 (225-400 MHz) or – as used by the Patriot system – Band 3+ (1,350-2,690 MHz).

AN/MSQ-104 Engagement Control Station (ECS)

This is the only manned station in the Patriot battery and requires three operators. Inside are two operator console positions, one communications station, the digital weapons control computer, the VHF datalink terminal, three radio relay terminals and the battery status panel with a hardcopy unit beneath. In operation, it sequences the battery through all tactical engagement procedures, monitors the operational status of the various systems, conducts automatic fault-finding and location as required, and provides the human control part of the man/machine interface for the battery.

A typical engagement involves the radar being assigned its search sector then automatically adapting itself via the ECS's No. 2 operating station with its environmental control panel to both the natural and hostile electronic environments it finds. It then modifies its operational functions as required. The radar search is carried out by the surveillance and detection beam with the radar informing the ECS when a detection occurs. The ECS then verifies the track by looking at several returns and, at the appropriate time, initiates IFF interrogation using the target track and illumination beam. The ECS then orders all tracks, establishes their engagement priorities and schedules the engagement. When an engagement decision is made, either in the manual, semi-automatic or fully automatic mode, the ECS selects the launcher to be used and sends any pre-launch data to the chosen missile through the VHF datalink. It also notifies the radar, at the time of launch, where to look for the missile. The initial course turn executed by the weapon is either commanded by the simple, self-contained guidance system aboard or by the pre-set launch instructions from the ECS. Once in the air, the missile is acquired by the radar and this initiates a missile track and command uplink beam to monitor its flight and command it to follow, using instructions from the ECS computer, an efficient energy-saving trajectory to the vicinity of the designated target. At this point, the TVM terminal homing technique, described below, using the missile's onboard TVM track and down link systems, is initiated by the ECS. Just before the missile's closest approach to the target, the warhead is detonated, producing a fragmentation pattern of 1.94 g splinters, by the E/F-band proximity fuzing device.

The Patriot ECS can also display the air picture from an E-3 Sentry AWACS aircraft via an automatic datalink through the appropriate NATO/USAF sector operations centre and command, control and communications centre. The Tactical Control Officer can then pass any relevant track information to other air defence by FM radio.

As part of the PAC-3 improvements the ECS was equipped with an optical disk data recorder from 1994. Designed to replace the current tape recorder system, it will take data from the weapon control computer, including radar tracking and missile flight path data, and record them to help determine whether the missile successfully counters a target.

The Patriot Antenna Mast Group (AMG) and Communications Relay Group (CRG) in deployed configuration (Raytheon) 0022191

AN/MPQ-53 multifunction phased-array radar

The AN/MPQ-53 G-band frequency-agile phased-array radar is automatically controlled from the ECS by the digital weapons control computer. Mounted on a trailer it has a 5161-HAWK platoon element array for the search and detection, target track and illumination and missile command and uplink beams. At any one time the system is able to handle between 100 target tracks and support up to nine missiles in their final moments of engagement using TVM terminal homing. This technique involves the missile's passive monopulse seeker array being directed by the ECS to look in the direction of the target which then begins to intercept increasingly precise returns from the reflected electromagnetic energy signals. This, in turn, triggers the G/H-band onboard downward datalink, which is offset in frequency from the target track and illumination beam and which transmits target data from the missile guidance package to the ECS computer, via the circular 251-element TVM receive-only array, at the lower right of the antenna group. The ECS uses this information to calculate guidance instructions, which are passed to the missile by the radar's G/H-band command and uplink beam. The phase-coded information is received on the missile by the two sets of guidance antennas, these transmit it to the guidance electronics which in turn use it to move the control surfaces. This procedure is repeated until the point of closest approach when the warhead is detonated. At no time in the engagement are any data actually processed on the missile itself.

Radar interrogation of a target is carried out by an AN/TPX-46 (V) 7 IFF system using a linear antenna array set below the main array position. There are also five diamond-shaped 51-element arrays: two individual ones, above the IFF array set at the bottom corners of the main array and a set of three, centred below the level of the TVM receiver array, near the lower edge of the front face of the planar radar housing. These are sidelobes cancellers, used to reduce the effects of enemy jamming. In a jamming environment the TVM technique is still usable because the system can measure range difference and, as it already has the angular measurement of the target, it can determine the difference in path length without loss of range resolution. It has the same range resolution as a non-jamming target engagement.

The 3 km to 170 km range radar performs its surveillance, tracking, guidance and ECCM functions in a time-shared manner by using the weapon's computer to generate 'action-cycles' that last in milliseconds. Up to 32 different radar configurations can be called up with the beams tailored for long-range, short-range, horizon and clutter, guidance and ECCM functions in terms of their power, waveform and physical dimensions. The data rate for each function can also be independently selected to give 54 different operational modes, so that, for example, a long-range search can be conducted over a longer time period than a horizon search for low-altitude pop-up targets. None of the functions require any given time interval which therefore allows a random sequence of radar actions at any one time, considerably adding to an attacker's ECM problem. The search sector is 90° and the track capability 120°.

When in place the radar is connected to both the ECS and generator vehicles by cabling.

In July 1993, Raytheon was awarded a USD62 million contract by the US Army Missile Command for further upgrades to the Patriot's radar system. The funding, for the Patriot's Classification, Discrimination and Identification (CDI) Phase III programme, allows the radar to distinguish tactical ballistic missile warheads from other parts breaking off from the missile, as it re-enters the Earth's atmosphere.

AN/MSQ-24 electrical power plant

This comprises two turbine-driven Deko Products 150 kW AC generator units with power cable reels, control panel and fuel tank transported on the rear of a standard US Army 5 ton (6 × 6) truck. Either generator can supply the required power for both the ECS and radar.

M901 launcher station

This is a remotely operated traversable four-round launcher station, mounted on the rear of an M860 two-axle semi-trailer with its own 15 kW generator, datalink terminal and electronics pack.

Patriot Battery Command Post based on HMMWV vehicle 0039514

Time required to reload the full basic 20 missile complement of a five-launcher battery is 60 minutes.

The US Army is now looking at a launcher system that can fire THAAD and Patriot missiles. The launcher is known as the Common Launcher and is the subject of a development contract awarded to Lockheed Martin Missiles and Space Company.

Training

Live missile firing and target tracking exercises are currently using the EADS DT-35 subsonic aerial target with a typical flight speed of 360 kts. For increased performance the DT-35-200, which operates at speeds in excess of 400 kts are used. These targets are in production at a rate of roughly 200 per year for NATO and NATO countries.

Variants

Patriot Multi-Mode Seeker (MMS) programme

The Patriot Multi-Mode Seeker (MMS) Programme was the subject of a US/German component demonstration project to provide a low-risk but much improved tactical air defence against SRBMs, low-observable cruise missiles and aircraft and Stand Off Jammers (SOJs). The initial Patriot work used funds from the original US/German Roland-Patriot agreement and involved Raytheon, AEG, Lockheed Martin and Telefunken SystemTechnik.

The US DoD and German MoD Memorandum of Agreement defined Phase 2 Patriot as part of an Extended Air-Defence project. This resulted in prototype flight tests in 1992.

The PAC-3 programme consisted of a major upgrade of the Patriot radar and a new missile, which could be either the multimode Patriot (with a fragmentation warhead) or the Lockheed Martin Vought ERINT (using hit-to-kill). The plan was to increase defended area and lethality. The radar improvements were aimed to increase detection range, target identification capability, capability against low radar cross-section targets, extend range and increase firepower and survivability.

The main changes involved were improvements to the radar by introducing both a dual travelling-wave tube transmitter and hardware design changes to reduce the system's internal noise while modifying the missile.

The radar changes were to enhance the effective radiated power level and reduce the signal-to-noise ratio components of the radar range equation in order to offset the reduction in the anticipated target's radar cross-section component.

The seeker programme was the subject of a 1989 NATO award for a two-year research and development contract to complete the integration of the unit.

The modified Patriot missile, initially known as PAC-3, had a 0.76 m long rocket motor extension, a new multivariable autopilot (to make the missile more responsive to acceleration commands) and a Ka-band active radar seeker in addition to the normal TVM guidance system. Under certain circumstances the new rocket motor (designed and developed by Hercules Aerospace Company under a March 1992 contract) was expected to double the range of the weapon so it can engage AWACS aircraft flying over territory deep in the rear of an Army's sphere of operations. In the ATM role the motor would also allow the weapon to increase its engagement altitude so as to provide more time in engaging a multiple incoming SRBM raid. The PAC-3 could engage TBMs at twice the altitude and protect 16 times the ground area of the PAC-2 missiles deployed in Operation Desert Storm.

The combination of the improvements would also give Patriot fire units the capability of driving SOJ aircraft such as the Antonov An-12PP 'Cub-C', Mil Mi-8PPA, Mi-8MTI, Mi-8MTSh and Mi-8MTPB Hip-J/K over the horizon, thus improving all the divisional and corps level communications assets and the effective range of all radar-based systems within its vicinity, which the jammers would otherwise have seriously degraded had they remained in line-of-sight to them.

The seeker itself used a 0.406 m aperture and was designed to give considerable improvements in the missile's engagement envelope against observable targets and allow the system to engage 'stealth'-type cruise missiles. It would also permit the weapon to attack low-altitude threats even if they re-mask themselves behind terrain features in trying to escape a missile. The active seeker is used in the last few tenths of a second of the flight as the fuze system.

One other area of improvement included incorporating the ability to allow launchers of a fire unit to be placed under the control of a neighbouring ECS while their own ECS and radar are being moved to another location.

In February 1994, the Patriot PAC-3 upgrade had the Lockheed Martin Vought missile chosen as its designated missile. The Patriot MMS programme would continue as a technology demonstrator and as a back-up should the PAC-3 encounter difficulties during its engineering development programme.

In October 1996, Germany's DASA LFK-LenkflugKörpersysteme signed an agreement with Lockheed Martin Missiles and Fire Control - Dallas to develop a PAC-3 system for the German Luftwaffe Patriot system. In 1997, the Luftwaffe for the programme allocated funds. The planned upgrades include a much better capability to detect targets in clutter and jamming environments, improved datalink communications and C2 for faster data traffic, the capability to offset launches up to 15 km away and the capability to remote launch from adjacent engagement centres.

As of July 2005, defence officials in Germany and South Korea held talks concerning the prospective purchase by South Korea of redundant German Patriot PAC-2 missiles. It is believed that Seoul is looking to purchase about 48 PAC-2 missiles as part of its SAM-X programme. Further reports have suggested that this sale will allow the Germans to reduce their number of Patriot batteries from 32 to 24, in accordance with German military requirements for the future. The missiles were delivered in November 2008 and were updated to GEM-T standard before deployment in 2011/12.

With the introduction of the PAC-3 missile, the radar power will be significantly enhanced and further expansion of the remote launching capability to 30 km will be made. This will allow 1,000 km range TBMs to be engaged.

Reconfigured Patriot subsystems

At the Association of the United States Army (AUSA) annual meeting in 2000, Raytheon displayed several newly configured Patriot subsystems to allow the system to remain in the field until 2025. These new subsystems include:

- Information and Co-ordination Central/Engagement Control Station (ICC/ECS) which controls the engagement. All the equipment from the ICC/ECS has been re-packaged into a Common Operating Shelter onto the rear of a C-130 transportable HMMWV as the Standardised Integrated Command Post System (SICPS) for either Patriot or HUMRAAM systems;
- Light Antenna Mast Group using a Common Communications Shelter design vehicle and a Light Antenna Mast Group vehicle on HMMWV chassis;
- Battery Command Post (BCP) based on HMMWV vehicles. The first five were to be fielded in September 2001 to allow the air picture from higher echelons to be presented at battery level.

Patriot-HAWK Phase III interoperability

With the introduction of the HAWK Phase III it will become possible for the Patriot ICC to supply target data to HAWK Assault Fire Units. This allows the HAWK to engage targets faster without the need to search for them in elevation and to share the more sophisticated IFF of the Patriot to clarify their identities. They can also be used together in the ATM role.

In an April 1988 demonstration programme test, using Patriot radar to cue the High-Powered Illuminator (HPI) radar of a HAWK III system onto a Patriot missile simulating an SRBM target, the HAWK successfully destroyed it at 8,000 m altitude and 8,000 m down range.

In November 1990, the US Army validated the HAWK anti-ballistic missile software with a live fire test and, in June 1991, tactical ballistic missile targeting data were passed from a Patriot air-defence system via a secure digital communications link to a HAWK unit. This then successfully fired two missiles at the target using both HAWK and Patriot system software to control the interception. One HAWK achieved a direct hit and the other provided a warhead kill.

The HAWK/Patriot interoperability is already part of US Army air-defence doctrine, entering service in 1988. The new capability was provided by the Patriot software change known as Post Deployment Build II (PDB II). The first operational test was by the 69th ADA Brigade using three batteries of the 8-43rd ADA (Patriot) Battalion and two AFPs of B/3-60th ADA (I-HAWK) Battalion during Reforger 88. This was followed by the October 1988 Exercise Hammer when a composite I-HAWK/Patriot defence, using the PDB II, successfully engaged a mass raid of approximately 100 aircraft.

The Block 1 enhancement (with the Patriot PDB II changes) allows the integration of I-HAWK fire units into a Patriot battalion with automatic target identification and a complete air picture compilation (of all I-HAWK and Patriot tracks) as Battalion Track Data Records being made in the Patriot ICC. Automatic engagement recommendations are then made with target allocation to the appropriate Patriot/I-HAWK fire unit. In the latter case the allocation has to be made by manual means.

The complete data compilation record can also be provided directly by the ICC to an adjacent battalions AN/TQA-73 missile minder via the Army Tactical Data Link 1 network if that unit has lost its own link to the parent ADA Brigade headquarters. However, it will not pass remote tracks received from a higher echelon nor will it process adjacent I-HAWK battalion tracks that do not correlate with its own air picture.

The follow-on Block 2 enhancement (with the Patriot PDB III changes) brings the Brigade Operations Centre ICC concept into the picture as one of the Patriot Battalion ICCs assumes the role of a centralised battle management Brigade/Master Battalion ICC, it combines information and then distributes engagement assignments to other subordinate battalions which can either be grouped in an ADA brigade or cluster of several SAM battalions.

This allows up to 12 I-HAWK fire units, six Patriot fire units or a mixed battalion force up to a total of 10 fire units (in a ratio of two I-HAWK to one Patriot fire unit) to be integrated into a single command-and-control structure using appropriate Army Tactical Data Link 1 and Patriot Digital Information Link networks.

The Master ICC is able to control both AN/TSQ-38 and AN/TSQ-73 systems as well as another ICC. It is also able to implement remote track reception from higher echelons and process adjacent I-HAWK battalion tracks to correlate with its own air picture. Additionally it provides a full-track management facility and can perform automatic threat ordering, identification processing of I-HAWK Phase II/Phase III and Patriot fire unit data and target allocation tasking to all its subordinate Patriot and/or I-HAWK battalions but not to individual fire units themselves.

When the recipient battalion assigns the target to one of its fire units for engagement the other fire units in the battalion are notified by messages appropriate to their systems.

The programme to provide these capabilities is being pursued jointly by the US, the Netherlands and Germany.

As all HAWK missiles have now been phased out of US Army and Marine Corps service, this is no longer an active US Army programme.

Specifications

Missile
(PAC-1 if not stated)

Type:	single-stage, low to high-altitude
Length:	5.18 m
Diameter:	0.41 m
Wing span:	0.92 m
Launch weight:	700 kg
Propulsion:	single-stage solid propellant rocket motor
Guidance:	command with TVM semi-active homing
Warhead:	
PAC-1:	
(weight)	91 kg
(type)	M248 Composition B HE blast/fragmentation with two layers of pre-formed fragments or Octol 75/25 HE blast/fragmentation
PAC-2:	
(weight)	84 kg
(type)	blast/fragmentation with a single layer fragmentation case with larger (20 × 20 × 20 mm) and heavier preformed fragments
Fuze:	
(PAC-1)	M818E1 proximity
(PAC-2)	M818E2 dual-mode
(PAC-2 GEM)	improved system
Maximum speed:	M5.0
Range:	
(maximum)	160,000 m
(minimum)	3,000 m
Altitude:	
(maximum)	24,240 m
(minimum)	60 m
Launcher:	mobile trainable four-round semi-trailer

Status

In production – PAC-2 missile entered production in 1989. By 2000, over 172 fire units and 8,600 missiles had been assembled by US manufacturers.

The four Patriot and four Improved HAWK squadrons formed part of the 3rd and 5th Netherlands Missile Groups headquartered at Blomberg and Stolzenau as part of the NATO SAM belt in Germany, and moved back to the Netherlands between 1994 and 1996.

The two groups merged into one 900-strong group at De Peel comprising four TRIple Air-Defence (TRIAD) squadrons. TRIAD integrates Patriot and I-HAWK radar and fire-control assets to form a squadron with one Patriot and two Improved HAWK fire platoons and one information coordination centre.

In mid-1995, Netherlands awarded a contract to upgrade its Patriot systems to the PAC-3 standard. The upgrade maintained the Dutch Patriot's commonality with the US Army and German Luftwaffe systems.

The US Army announced in November 2001 that it is considering a surface-to-surface Patriot. Modifications of the original MIM-104 missile began in 2003.

Raytheon announced in December 2001 its proposal to add a new front end to older Patriots in order to give them a hit-to-kill capability, similar to the PAC-3.

Initial delivery of the Patriot GEM+ fore bodies to the US Army was completed during October 2002. GEM+ missiles are PAC-2 missiles that have been refurbished and modernised at Raytheon's Andover, Massachusetts, manufacturing facility. This upgrade includes the following:

- Older components replaced;
- New technology inserted and reliability increased;
- A modernised fuze is added that provides significant performance improvements against tactical ballistic missile targets;
- A new low noise front end is inserted to increase the seeker's sensitivity and improve acquisition and track performance against smaller radar cross-section targets.

During February 2004, Raytheon was awarded a contract for the continuous technology refreshment of Patriot PAC-2 fore bodies with the GEM+ frequency generator upgrade. This upgrade is carried out in Bedford, Massachusetts, and was completed by 31 July 2006. A further contract, released on 31 March 2005, announced that Raytheon, working with the US Army, is to upgrade 90 PAC-2 missiles to the GEM-T configuration, while at the same time upgrading an additional 73 PAC-2 GEM missiles to the GEM-C configuration.

On 31 January 2005, Lockheed Martin received a USD532 million contract for 156 PAC-3 missiles for the US Army, the Netherlands and Japan. This represents the first international sales of the PAC-3 missiles. Of the 156 missiles being produced, 32 will go to the Netherlands and 16 missiles to Japan.

In May 2007, another contract was awarded to both the US Army and FMS sales for the Patriot GEM-T variant; 230 missiles are to be produced including upgrades and spare parts.

In August 2008, it was revealed that the Israeli Patriots had been fitted with an Israeli developed Sniper system that enables identification of a target whether missile or aircraft, while remaining several hundreds of kilometres away from the Israeli borders. The IAF stressed that this system is currently reserved for Israel's exclusive use; Patriot batteries elsewhere in the world are not equipped with Sniper.

In February 2009, the Greek Company Intracom renewed its contract with NAMSA a subsidiary of NATO for the maintenance of its European arsenal of Patriot missiles, this contract covers the maintenance in Germany, Greece, Netherlands and Spain.

An announcement made in Seoul, South Korea in late October 2012 confirmed that the government wants to purchase the software upgrade for the PAC-2 system to bring it up to PAC-3 standard. No date has yet been set for this upgrade.

During early 2013, some Patriot systems were moved primarily from Europe to the border between Turkey and Syria for the defence of Turkey should Syria pose a threat. The systems which number six in total are manned by Netherlands, German, Turkish and US forces.

By 18 November 2013, US and Netherlands had both extended the deployment of the Patriot systems for one more year.

In November 2013, the Japanese Air Self-Defence Force (JASDF), completed test firing of Patriot Air and Missile Defence Systems at McGregor Range, N.M. as part of annual service practice that included target engagement and firing in battlefield tactical mission scenarios.

Contractor

Raytheon Systems Company (prime system contractor for the Patriot Air and Missile Defence System).
Lockheed Martin Missiles and Fire Control (principal subcontractor and Prime Contractor on the PAC-3 Segment upgrade to the Patriot Air Defence System).
Siemens AG (principal German Patriot contractor).
Mitsubishi Heavy Industries (principal Japanese Patriot contractor (licence builder).
Brunswick Corporation, Brunswick Defence Division (ARM-D System).
Moog Components Group (electro-mechanical and fibre-optic products, including slip rings).
EADS North America (DT-35 family of targets for live firing and target tracking exercises).
Walton Construction Co, Kansas City (engineering services).
Intracom Defence Electronics, Greece (responsible for the control and repair of the actuators of the Patriot launchers based in Europe).
Aselsan of Turkey, co-development with Raytheon of the antenna mast group for Config-3 for UAE.
Mercury Computer Systems of Massachusetts, radar signal processing and applications ready subsystems to bring radars up to Config 3 standard.

PAC-3 being launched from Patriot launcher during trials 0100707

PAC-3 intercept of a low-level target 0100706

Patriot PAC-3

Type

Static Surface-to-Air Missile (SAM) system

Development

The MIM-104F Patriot Advanced Capability-3 (PAC-3 - formerly ERINT) design programme began in 1983, but flight testing was not funded until 1987, when the prime contractor, Lockheed Martin Missiles and Fire Control - Dallas, was awarded a contract worth USD80 million. Originally conceived as a follow-on from the earlier Flexible Lightweight Agile Guided Experiment (FLAGE) for intercepting Tactical Ballistic Missiles (TBMs) at altitudes in excess of 15,000 m, the programme has been extended to provide an intercept capability against both TBMs and air-breathing missiles.

ERINT competed with Patriot Multi-Mode Seeker (MMS) for the missile to be part of the PAC-3 system. In early 1994, the ERINT was chosen for the PAC-3 programme with the Patriot radar system to be modified accordingly.

PAC-3 Configuration 1: 1995 - a new pulse Doppler processor that significantly improved radar performance was added; also, Engagement Control Station (ECS) and Information Coordination Central (ICC) upgrades; improved weapons-control computer throughput, memory and reliability. ECS and ICC upgrades also added optical disk and embedded data recording equipment to decrease computer access times, improve reliability and provide a tactical data recording capability. Patriot Guidance Enhanced Missile (GEM) included a faster warhead fuze to improve kill probability against tactical ballistic missiles and a new low-noise missile seeker section to expand the missile's engagement area.

PAC-3 Configuration 2: 1996 - improved the Communications Processor and added the Joint Tactical Information Distribution System (JTIDS) to improve communications and interfaces with joint forces. Fielded Post Deployment Build (PDB) 4 software which improves multi-function radar performance, detects small radar cross section targets and improves system detection, identification, and engagement of anti-radiation missiles and aircraft carrying those missiles.

PAC-3 Configuration 3 Ground Equipment: 2000 - addition of dual travelling wave tube units and a new radio frequency exciter to the radar to further improve radar multi-function performance and detection of small targets in cluttered environments. A Classification, Discrimination Identification Phase III programme which significantly improves radar range performance to discriminate and identify a tactical ballistic missile warhead from other target debris or objects.

Post Deployment Build 5 software improves radar multi-function performance, determines tactical ballistic missile impact and launch points, and provides interfaces with the Theatre High Altitude Area Defence System (THAAD). Remote Launch improvements increased the location of launchers from the ECS from a distance of 10 km to 30 km. This dramatically increased the Patriot-defended area. PDB-5 was deployed to Turkey with the Netherlands Air Defence Group (Patriot PAC-2 GEM missiles) to protect high-value areas in case of attack during the Iraq/Coalition war. PDB-5 also meant that, if necessary, the Dutch systems could be reinforced with PAC-3 missiles with no further changes necessary. A further PDB-6, composed of user-requested improvements, planned performance improvements and improvements that

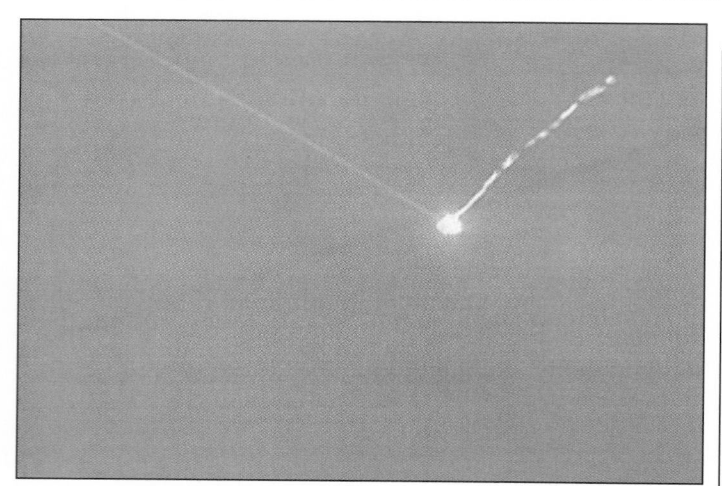

PAC-3 intercept of a high-level target 0100708

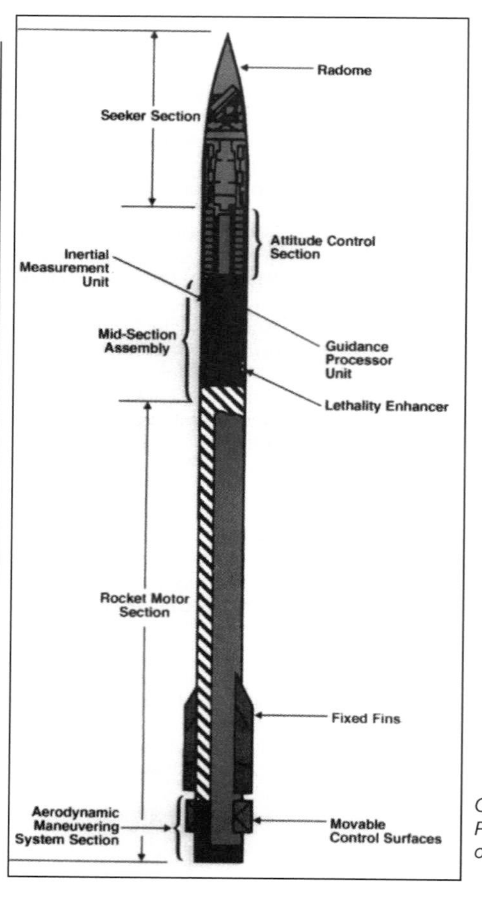

Outline drawing of the PAC-3 showing position of main components 0007981

resulted from lessons learned in operations during the Iraq war, was demonstrated at the White Sands Missile Test Range, New Mexico, from May to July 2006. The successful trials further demonstrated the PDB-6 software upgrade works with all variations of the Patriot system.

The PAC-3 uses hit-to-kill as its prime kill mechanism. It also employs a ring-type warhead of small, explosive-propelled, high-density tungsten pellets, which is considered only as a lethality enhancer against air-breathing targets, such as cruise missiles and aircraft.

PAC-3 is one of the world's most sophisticated technologies. The PAC-3 Missile boasts 12 successes out of 13 flights over the past three years, with nine intercepts in 10 attempts, an overall success rate of 92 per cent for the flight test programme.

The PAC-3 Missile intercept successes are:

- 15 March 1999 - successful intercept of TBM
- 16 September 1999 - successful intercept of TBM
- 5 February 2000 - successful intercept of TBM
- 22 July 2000 - successful intercept of low-flying cruise missile
- 28 July 2000 - successful intercept of low-flying cruise missile
- 14 October 2000 - successful intercept of TBM
- In 2000, Lockheed Martin won the contract to include the Patriot PAC-3 missile in the Medium Extended Air Defence System (MEADS). After winning the contract, Lockheed replaced the blast fragmentation warhead in PAC-3 with a hit-to-kill millimetre wave seeker as part of the Missile Segment Enhancement (MSE) upgrade programme
- 31 March 2001 - first tactical ripple mode test - successful intercept of TBM by first PAC-3 Missile and successful tactical self-destruct of second PAC-3 Missile
- 9 July 2001 - successful intercept of an F-4 remotely piloted aircraft by PAC-3 Missile
- 19 October 2001 - successful intercept of advanced cruise missile
- Early 2003 - deployed to Middle East to support Gulf War II
- July 2003 - start of 51 month software update by Lockheed Martin. Flight trials expected September 2006
- 4 March 2004 - ripple fired testing at White Sands, against a PAAT (Patriot-As-A-Target) simulating a Scud type missile
- 15 September 2005 - software changes to the system and other improvements lead to a successful test firing at White Sands, against a full-body aerodynamic TBM target
- May 2006 - Lockheed Martin Missiles and Fire Control was contracted to a PAC-3 Missile Segment Enhancement (MSE) upgrade; this comprises the PAC-3 Missile - a highly agile hit-to-kill interceptor, the PAC-3 Missile canisters (in four packs), a Fire Solution Computer and an Enhanced Launcher Electronics System. The PAC-3 MSE programme also includes flight software, flight-testing, modification and qualification of subsystems, production planning and tooling, and support for full Patriot system integration. The MSE programme will span 57 months, flight testing began in September 2006. Flight test programmes include one controlled test flight and two guided intercept tests against threat-representative Tactical Ballistic Missiles (TBMs). All testing will be conducted at White Sands Missile Range. In January 2008, Lockheed Martin received a contract worth USD66 million to incorporate the PAC-3 MSE missile into the MEADS programme. The MSE missile includes a more powerful warhead with hit-to-kill capability, new motor for longer-range and more responsive control surfaces.
- Information released from Dallas, Texas, early April 2013 reported that a multiple PAC-3 MSE test took place at the White Sands Missile Range, N.M. The missiles were ripple fired to intercept a Tactical Ballistic Missile with the first PAC-3 MSE destroying the target and the second self destructing as ordered.

In addition to the 12 successful PAC-3 Missile flight tests, the PAC-3's predecessor, the Extended-Range Interceptor, hit three times in a row during the demonstration/validation programme in 1994. Two of those tests involved TBM targets and one air-breathing target (simulating a cruise missile or aircraft).

First low-rate initial production began in late 1999. This was followed by further testing in February and July 2000. Selection for the MEADS programme also took place in May 2000. Customer acceptance flights and trials started in 2001 and culminated in the 4th low-rate production in December 2002. Elements of the command-and-control were then shipped to Israel in late 2002 with the complete system being sent to the Gulf to protect US troops during the Gulf War II in early 2003.

The programme requirement was for 1,500 missiles, 180 modified Patriot launchers and 74 modified radar. Total cost at 1994 prices was USD3 billion, with Lockheed Martin Missiles and Fire Control - Dallas receiving a contract for the missiles and Raytheon for PAC-3 integration into the Patriot launchers and the associated fire-control equipment. The final totals may well be short of this.

The Lockheed Martin Corporation, Grand Prairie, Texas, was awarded a modification to a cost-plus-incentive-fee contract of USD7,500,000 in September 2003 for initial production facility additional tools and equipment for the PAC-3 missile programme. Work on this contract was carried out at Grand Prairie and was completed by September 2004.

An additional contract in February 2004, for USD214,277,004 for 133 PAC-3 missile four packs and ground support equipment has also been allocated. This contract was completed by April 2006. As with the September 2003 contract, all work was carried out at Grand Prairie, Texas.

Further development contracts associated with the Patriot PAC-3 issued 13 April 2006 to Lockheed Martin were for:

- Missiles
- Launcher mod kits
- Parts library
- Storage and ageing
- Missile mid-section audits
- Interim contractor depot support
- PALS FSC
- Shorting plugs
- Test set cables
- Concurrent spares
- Replenishment parts.

The work on these contracts were completed by July 2008.

Patriot PAC-3 has been chosen as the weapon for the Medium Extended Air Defence System (MEADS) development programme being undertaken by Lockheed Martin, Alenia Marconi Systems and Daimler Chrysler Aerospace. In 2004, the Combined Aggregate Programme (CAP) was first identified with the start of the development of the first fire unit. The CAP is the process by which the Patriot PAC-3 system transitions to the MEADS. The MEADS mission is to provide low-to-medium altitude air and missile defence with the capability to counter, defeat, or destroy tactical ballistic missiles, air-breathing threats (including cruise missiles), Unmanned Aerial Vehicles (UAVs), tactical air-to-surface missiles and anti-radiation missiles. MEADS will be interoperable with other airborne, ground-based and sea-based sensors, early trials for which, using the Patriot system, have been completed. The CAP programme plans called for a system design review in 2009 and the start of production in the first quarter of fiscal year 2013. The US Army expect MEADS to achieve initial operating capability in 2017 with four units.

Patriot PAC-3 Cost Reduction Initiative (CRI) missiles have been sold to Germany to supplement the existing fielded Patriot systems.

Patriot PAC-3 at Kadens AFB in Okinawa (Itsuo Inouye) 1187258

Lockheed Martin confirmed in January 2007 that it received a contract from the US Missile Defence Agency to continue studies into a possible air-launched variant of the Patriot PAC-3 known as the Air-Launched Hit-To-Kill (ALHTK) initiative. The AAM is expected to engage ballistic missiles and cruise type targets during their boost phase (ballistic missile) and early part of flight. The system is further expected to draw target data information gathered by systems such as the JLENS or satellite.

This contract follows the Risk Assessment (RA) contract that concluded in April 2006 that it had studied and identified the feasibility for such a system.

During March 2007, the 500th PAC-3 missile seeker was handed over to the US Army in Huntsville, Alabama. Additionally, at this time Lockheed received a further contract for the production of 112 Patriot PAC-3 missiles.

In May 2008, the US Army conducted a controlled flight test of the Patriot PAC-3 Missile Segment Enhancement (MSE) interceptor at the White Sands Missile Test Range in New Mexico. This test demonstrated launch canister hardware, missile functionality, interfaces and integration with the system and missile fly-out functions.

In February 2010 at the White Sands Missile Test Range, the PAC-3 MSE variant was test fired against a target that was simulating a tactical ballistic missile. The test was deemed a success.

Further testing of the basic Patriot PAC-3 missile also reported as the Cost Reduced Initiative (CRI) took place at White Sands in November 2011 with ripple firing against a tactical ballistic target.

In mid-June 2012, it was announced that the Turkish production organisation Roketsan is to produce the new single canister launch system to replace existing four-pack canisters, this should enable more flexibility of the PAC-3 system when deployed in theatre.

Description

The PAC-3 is designed to utilise the Patriot launcher unit. It is 5 m long, has a maximum body diameter of 0.255 m and weighs 312.4 kg at launch. A PAC-3 launch canister is compatible with the Patriot launcher, but will hold four PAC-3 missiles instead of one Patriot missile thus effectively quadrupling the ready missile inventory per launcher without increasing the force structure.

The missile has four fixed fins and four aerodynamic control surfaces located just aft of its centre and 180 mini solid-propellant altitude-control thruster motors mounted in a special attitude control section located in its fore body section, to effect flight-control manoeuvres. A high-performance HTPB solid propellant rocket motor with lightweight composite case is used as the propulsion system.

The missile uses an inertial guidance package for the fly-out guidance phase to fly to the predicted intercept point. Initial target data are acquired by the system's fire-control radar, which pre-programmes the onboard guidance package before the launch.

If required, the weapon's trajectory can be updated during the flight using the fire-control radar system. In the last few seconds of the flight, a Boeing Company Electronic Systems and Missile Defence radome-covered, nose-mounted, gimballed Ka-band pulse Doppler range gated radar seeker antenna assembly is activated to terminally guide the missile to the target. The high-power coherent radar also doubles as the proximity fuzing system to detonate the lethality enhancer, which uses rings of tungsten high density pellets to enhance its effectiveness against air-breathing targets such as aircraft or cruise missiles.

The export model of the Patriot PAC-3 consists of:

- M901 launchers
- AN/MPQ-53 fire-control radars
- AN/MSQ-104 engagement control stations
- OA-9054(V) 41G antennae mast groups
- MIM-104D or GEM missiles.

The PAC-3 is designed to intercept short-range TBMs over a wide range of closing velocities. At impact, the PAC-3 weighs 142 kg.

JLENS

The Joint Land Attack Cruise Missile Defence Elevated Netted Sensor System (JLENS) consists of the aerostat, mobile mooring station, power and fibre optic data transfer tethers, and ground support equipment and will in the future provide surveillance and engagement support to the PAC-3 and MEADS programmes. The aerostat has undergone an increase in size from 71 m to 74 m in order to accommodate the necessary lift increase to 7,000 lb of total payload. JLENS is primarily an integration effort based on relatively mature technologies from other programmes and is designed to provide over-the-horizon detection and tracking of land-attack cruise missiles. In addition to providing a significant cruise missile defence capability, the JLENS system will also be capable of tracking surface moving targets and tactical ballistic missiles during their boost phase, and passing target data to various weapon systems and platforms across the military service. JLENS will provide a long-duration, wide-area cruise missile defence capability while also providing elevated communications capabilities. Each JLENS consists of a long-range surveillance radar and a high-performance fire-control radar, each integrated onto a large aerostat connected via tether to a ground-based processing station. System testing was scheduled to begin late 2009 through 2010 with programme completion in 2012.

As at April 2008, JLENS passed its Orbit Preliminary Design Review (PDR), this took four days and involved a comprehensive assessment of the design maturity.

Variants

Air-launched variant from the F-16, F-22 and F-35 has been in development since at least 2006. A risk assessment contract concluded in April 2006 and was followed by a Feasibility Study that was completed in January 2007. Lockheed has received a USD3 million contract from the Missile Defence Agency (MDA) to continue the Air-Launched Hit-To-Kill (ALHTK) initiative, that would enable fighter aircraft to carry and launch Patriot PAC-3 missiles to intercept hostile ballistic and cruise missiles.

PAC-3 MSE

Under the Patriot PAC-3 MSE [Missile Segment Enhancement] programme, the company (Lockheed Martin Missiles and Fire Control) have incorporated a larger, more powerful motor into the missile for added thrust, along with larger fins and other structural modifications for more agility. The modifications have extended the missile's reach by up to 50 per cent. The larger fins, which will collapse to allow the missile to fit into the current PAC-3 launcher, although a heavier missile than the previous PAC-3, will reduce the number that can be carried on a launcher from 16 to 12. This will provide the interceptor with increased capability and manoeuvrability against faster and more sophisticated ballistic, cruise missiles and air-breathing threats.

These enhancements are the natural, pre-planned evolution of a system that was base lined in 1994.

A further software improvement to the system known as the PDB-6 (Post Deployment Build), will allow Config 3 units to discriminate targets such as anti-radiation missiles, helicopter, UAVs and cruise missiles.

The PAC-3 MSE variant was selected as the primary interceptor for the multinational MEADS system in September 2006.

Upgrades incorporated into the PAC-3 MSE also include:

- The solid rocket motor has a second pulse and is larger in diameter.
- The thermal batteries have remained the same in size but have been improved with increased performance and as such longer mission time.
- Aerodynamic surfaces are larger and the span of the after control surfaces is greater to accommodate the increased performance envelope.
- The missile is packaged in a single canister that stacks to provide flexibility for the Patriot or MEADS launcher.

Specifications

	Patriot PAC-3	Patriot PAC-3 MSE
Dimensions and weights		
Length		
overall	5.205 m (17 ft 1 in)	5.205 m (17 ft 1 in)
Diameter		
body	255 mm (10.04 in)	255 mm (10.04 in)
Weight		
launch	315 kg (694 lb)	312 kg (687 lb)
Performance		
Speed		
max speed	3,305 kt (6,120 km/h; 3,803 mph; 1,700 m/s)	-

	Patriot PAC-3	Patriot PAC-3 MSE
Range		
max	8.1 n miles (15 km; 9.3 miles)	18.9 n miles (35 km; 21.7 miles)
Altitude		
max	-	36,000 m (118,110 ft)
Ordnance components		
Warhead	kinetic (back-up HE fragmentation enhanced warhead)	kinetic
Guidance	INS, mid-course update	INS, mid-course update
	active, radar (Ka-band)	active, radar (Ka-band)
Propulsion		
type	solid propellant (Atlantic Research composite rocket motor with special attitude control section for in-flight manoeuvring)	solid propellant

Status

The US Army announced, in November 2001, that it was considering a surface-to-surface Patriot. Modifications of the original MIM-104 missile began in September 2003.

In February 2006, Lockheed Martin delivered the first PAC-3 Stockpile Reliability Test (SRT) Missiles to the US Army in Camden, AR.

The Netherlands received their first system during early 2007. On 30 March 2007, the first anti-missile PAC-3 missiles were transferred to the Guided Missiles Group in a short ceremony at De Peel air force base in the Netherlands.

On 7 September 2006, PAC-3 Cost Reduction Initiative (CRI) missiles as well as associated equipment and services, were requested by Germany with notification to Congress as a Foreign Military Sale. Up to 72 PAC-3 CRI missiles, 12 each Missile Round Trainers, support equipment, modification kits, publications, spare and repair parts, US Government and contractor technical assistance and other related elements of logistics support.

Raytheon announced, in December 2001, that it proposed adding a new front end to older Patriots in order to give them a hit-to-kill capability, similar to the PAC-3.

The Patriot PAC-3 missile system was first deployed in small numbers only to Suwon Air Base in South Korea during 2003. The Suwon deployment consists of six firing units attached to the 1st Battalion 43rd Air Defence Artillery organisation.

Japan received the Patriot PAC-3 during 2005. License production of the system by Mitsubishi Heavy Industries has been agreed between the US and Japanese governments.

The Japanese government had requested (7 September 2004), from the US, a possible sale of 20 × PAC-3 systems including:
- 120 Guided missiles
- Support equipment
- Modification kits
- Fire-control computer
- Publications
- Personnel training
- Spare parts and repair kits
- Logistics support.

The whole deal is worth approximately USD79 million. In addition to this, in June 2006, three PAC-3 batteries were immediately purchased and deployed to the Island of Okinawa with up to 600 US troops. This was in direct response to the threat from North Korea and an imminent Taepodong-2 MRBM launch. On the 30 March 2007, the first of the PAC-3 launchers were installed at an air base to the north of Tokyo. The Japanese Self Defence Forces have set up a ring of Patriot PAC-3 defences around the city of Tokyo. By January 2008, trials had begun in order to tune and test these systems while the third site was activated in February 2008 at the Takeyama base in Yokosuka, near the US seventh fleet naval base. A fourth site has been completed at Kasumigaura in Ibaraki to the northeast of Tokyo. The Tokyo defence ring with this last installation is now considered complete.

Saudi Arabia is also likely to upgrade their missile systems to the PAC-3 standard. Both Saudi Arabia and Kuwait have been fully briefed on the PAC-3 system and contracts were originally expected late 2005 or early 2006. However, in July 2008, Kuwait issued a contract to Raytheon to upgrade their PAC-2's to the Config 3 status which includes upgrades and support consistent with the US Army's Pure Fleet initiative. The PAC-3 is already operational in Israel, Germany, the Netherlands, Belgium, Japan and Spain and was also offered to India in July 2008.

Lockheed Martin has received a USD532 million contract for 156 PAC-3 missiles for the US Army, Netherlands and Japan.

During a visit to the US by the Indian Defence Minister, Shri Pranab Mukherjee, the US Department of Defense made a verbal offer to provide technical briefings on the PAC-3 anti-missile system.

In January 2007, Lockheed Martin confirmed preliminary talks for a potential sale of the Patriot PAC-3 to Turkey.

In August 2007, it was announced from Jerusalem that the Israeli Air Force is to buy the Patriot PAC-3 to replace the PAC-2 variant currently in use. No date has yet been set for the first delivery.

In December 2007, the United Arab Emirates placed an order with the US for the following:
- 288 × Patriot PAC-3 missiles
- 216 × Patriot GEM-T missiles
- 9 × Launching Stations (4 per fire unit)
- 8 × Antenna Mast Groups on trailers
- 8 × Antenna Mast Groups for Tower Mounts
- AN/GRC-245 Radios
- Single Channel Ground and Airborne Radio Systems (SINCGARS, Export)
- Multifunctional Information Distribution System/Low Volume Terminals
- Generators
- Electrical Power Units
- Trailers
- Communication and support equipment
- Publications
- Spare and repair parts
- Repair and return parts
- US government and contractor technical assistance
- Other related elements of logistics support.

By late 2008, it became apparent that the PAC-3 for use in the UAE would have to be capable of interfacing with the THAAD system, that has also been purchased by the UAE.

On 2 September 2008, Germany announced the IOC of their Patriot PAC-3 Config 3 missiles system having completed trials and other upgrades. Firing trials of the system were carried out in October 2008 at the White Sands Missile Test Range in the US. The upgrades to the system include:
- PAC-3 Missile Segment Launcher Electronics
- Fire Solution Computer upgrade
- PAC-3 missiles
- Interface with other Patriot systems.

In May 2009, a new contract set with Lockheed Martin was awarded for the PAC-3 Guidance Processor Unit Redesign. The estimated date for completion of the redesign was Sept 2011 it is assessed that the redesign was completed during this time frame.

In November 2012, Qatar became the latest Gulf State to ask for the Patriot Configuration-3 system. The package is reported to be:
- 11 × Patriot Configuration-3 Modernised Fire Units
- 11 × AN/MPQ-65 radar
- 11 × AN/MSQ-132 engagement control systems
- 30 × antenna mast groups
- 44 × M902 launching stations
- 246 × Patriot MIM-104E GEM-TBM with canisters
- 2 × Patriot MIM-104E GEM-T test missiles
- 768 × Patriot PAC-3 missiles with canisters
- 10 × Patriot PAC-3 test missiles with canisters.

In the notification to Congress package requested, estimated cost would be USD9.9 billion.

In mid-August 2013, the US agreed a contract worth USD308 million with Kuwait for the supply of 244 PAC-3 missiles and 72 launcher modification kits and associated tooling, including programme management.

In October 2013, the Japanese Air Self-Defense Force (JASDF) completed its annual Patriot missile test firings at McGregor Range, New Mexico depicting tactical mission scenarios. Patriot Advanced Capability-3 (PAC-3) missiles successfully engaged tactical ballistic missile representative targets during test firings conducted at White Sands Missile Range, New Mexico, Raytheon announced on 20 November as part of the Field Surveillance Program (FSP) intended to demonstrate the viability and performance of existing fielded PAC-3 stocks. The sample missiles are selected at random to be used in annual test firings that allow an independent assessment of their production batch.

Contractor

Lockheed Martin Missiles and Fire Control (prime contractor).

Raytheon Missile Systems (JLENS programme and prime contractor for Systems Integration).

Atlantic Research Corp (solid propellant composite motors).

Mitsubishi Heavy Industries (MHI) Ltd (Japanese PAC-3).

Moog Components Group (electro-mechanical and fibre optic products including slip rings).

Roketsan, Turkey, (single canister launch system).

THAAD

Type

Static Surface-to-Air Missile (SAM) system

Development

The US Army Terminal High Altitude Area Defence (THAAD) (previously known as the Theatre High Altitude Defence) programme is being managed within the DoD by a Programme Executive Officer (PEO). Oversight is provided by the Strategic Defence Initiative (SDI) office and the Missile Defence Agency.

The THAAD system is a long-range, land-based theatre defence system that acts as the upper tier of a basic two-tiered defence against ballistic threats. The THAAD system is an easily transportable battery of weapons, designed to hit-to-kill with incoming tactical and theatre ballistic missiles at endo-atmospheric and exo-atmospheric heights with ranges of up to 195 km. This

Second flight of the Terminal High-Altitude Area Defence (THAAD) missile in July 1995 0509807

Terminal High-Altitude Area Defence (THAAD) radar deployed in the field. This X-band radar has a surveillance range of 1,000 km against tactical ballistic missiles 0044353

would then allow the current air-defence systems to preserve their primary mission of anti-aircraft defence.

US Army predictions indicate that THAAD is capable of addressing 80 per cent of the projected US Army Common Target Set of Theatre Ballistic Missile (TBM) threat inventories. Combined with PAC-3 the two systems together address 76 per cent of the threat inventory. The latter means that there are a large number of TBMs against which a near-leak-proof level of effectiveness is possible. The two-tier concept is based on the notion that a perfect missile system cannot be built.

The THAAD system includes truck-mounted launchers, interceptor missiles, radar, and fire-control and communications management based on air-transportable trucks. It would be cued either by space-based sensor satellites such as Brilliant Eyes or 9.2 m^2 array-sized I/J-band Ground-Based Radar (GBR) which provides surveillance, threat identification and classification functions.

In September 1992, the US Army awarded the contractor team, led by the then Lockheed Martin Missiles and Space Company, a USD689 million contract to develop the THAAD system. The subcontractors who work with Lockheed Martin on the programme include:

- United Technologies Corporation;
- Chemical Systems Division, for the THAAD solid propellant rocket motor;
- Aerojet, for the boost motor;
- Rockwell International Rocketdyne Division;
- Lockheed Sanders;
- Honeywell Space Systems Group;
- Litton Data Systems Division for computer software and the Tactical Operations Centre development;
- Loral Infra-Red & Imaging Systems Group (for the THAAD infra-red terminal guidance two-axis gimballed staring seeker assembly);
- Westinghouse Marine Division;
- Dornier GmbH;
- BAE Systems provides the front end seeker;
- Arnold Engineering Development Centre (AEDC), pitch motor and trials.

The initial demonstration/validation contract for the GBR was also let, in September 1992, to a team led by Raytheon Systems. The terms of the contract included the building of three radars for Theatre Missile Defence. Two of these were operational evaluation systems with a 9.2 m^2 aperture and are used with the THAAD system. They are Lockheed Martin C-130 compatible and road transportable and operate using full field of view phased-array antennas to acquire theatre missile threats at a range of up to 1,000 km. These radar use 25,344 I/J-band solid-state transmit/receive modules in the antenna phased array making them lighter and easier to maintain in the field. The complete radar comprises:

- Antenna;
- Electronics unit;
- Cooling unit;
- Operator control unit;
- Diesel generator.

Plans at the time indicated that 1,422 missiles, 99 launchers and 14 GBR radar would be procured. Total costs were estimated at USD9 billion for the THAAD and USD5.4 billion for the Raytheon GBR radar. There is a distinct possibility that Japan may well become involved due to the 1998 North Korean missile developments. However, it became evident from later purchases that Japan would use the Patriot PAC-3 and the shipborne Aegis system.

The main subcontractors at the time were:

- TRW Incorporated (for software);
- Texas Instruments Incorporated (for the solid-state modules);
- Digital Equipment Corporation (for signal and data processors);
- EBCO (for the National Missile Defence variant radar turrets);
- Datatape (for tape recorders);
- the then Hughes Aircraft Company (for the National Missile Defence variant travelling-wave tubes).

Transition of the system to the Missile Defence Agency (MDA) happened in October 2001.

During July 2005, the Missile Defence Agency announced a successful integration trial of the THAAD radar and command-and-control system against live targets at White Sands Missile Test Range (WSMR), New Mexico. During this testing, the THAAD radar acquired, tracked and classified in-bound expended booster and separate re-entry vehicles through a single-stage event. Using the track data from the radar, the command, control, battle management and communications system conducted threat assessment, object tasking and engagement planning. Furthermore, it also launched a simulated interceptor, providing acquisition and intercept data to the radar.

By late 2005, the producers of the missile, Lockheed Martin (Troy, Alabama), were preparing for the first flight of the second-generation missile. This missile is reported to be close to the operational weapon with the added telemetry requirements for the flight test. This is to be a missile-only flight as there will be no target in the air, but the seeker developed by BAE Systems was installed on the missile and was part of the trial. After completion of these test flights (15 in all), the system was moved from the WSMR test area to the Pacific Missile Range Facility (PMRF) in Hawaii sometime around the end of 2006. It has been reported that the trials in Hawaii took place due to range constraints at WSMR.

A further trial at WSMR took place on 22 November 2005 involving Raytheon Company's THAAD radar. The phased array radar successfully acquired, tracked and communicated with the THAAD missile during these trials. Radar trials also took place during April/May 2006 at WSMR, where the tests conducted demonstrated the first fully integrated radar, launcher, fire-control and missile operations and engagements against a simulated target.

Further testing of the THAAD radar, this time against live targets, took place on 12 July 2006, again at the WSMR. This test did include a successful intercept of a HERA target, even though interception was not a primary objective. A further reason for the tests was to conduct seeker characterisation including uplinks, downlinks and target acquisition and tracking by the interceptor's seeker.

The final flight test of THAAD at WSMR was a missile fly-out with no target involved, during late 2006.

As of 18 October 2006, THAAD equipment started to arrive at the Pacific Missile Range Facility (PMRF). Testing at PMRF allows for more robust test scenarios against missiles launched from sea-based platforms and the location in the Pacific Ocean means there will no longer be a need for the now-characteristic THAAD Energy Management System (TEMS) manoeuvre, or corkscrew, that the missile has made as it comes out of the launcher. That manoeuvre was required strictly because of space limitations at WSMR.

In December 2006, Lockheed received a contract for fire unit fielding, support equipment and initial spares. This contract is for the first two complete fire units including 48 interceptors, six launchers and two fire-control and communications units.

The first of the Hawaiian live missile firing tests took place on 25 January 2007 at the Pacific Missile Range Facility. This test was conducted by the US Missile Defence Agency with the system engaging and intercepting unitary target. The test was also to prove the BAE Systems infrared missile seeker performance in detecting the target and passing the data to the missile computer for use in guidance and control to its target. This was the second time the seeker trials have been concluded successfully, the first was at the White Sands Missile Test Range in New Mexico in July 2006. A second flight test on 6 April 2007 was also a success with the target intercepted in the mid-endo-atmosphere.

In late October 2007, a firing from the Pacific Missile Range Facility to test the exo-atmospheric capabilities of the system was carried out. This firing also tested the integration of the radar, launcher, Fire Control and Communications and interceptor. Following the successful tests on 19th October 2007, at least two more tests were carried out in 2008. The first in the Spring and will be the first intercept of a separating target, the second later in the year was a two-shot salvo test.

In June 2008, a successful interception of a separating target in the mid-endo atmosphere took place at the Pacific Missile Range Facility, in Kauai.

The Battle Management Command, Control and Intelligence components of THAAD are mounted in hardened shelters carried on HMMWV (4 × 4) chassis
0044352

THAAD launcher on Oshkosh PLS (10 × 10) truck chassis with launcher elevated
0044351

Late in August 2008, the Boost Motor passed its final qualification testing at Aerojet's Sacramento headquarters at the low-end of the operating temperature demonstrating its reliability in extreme cold weather conditions. Earlier in 2008, the Motor also passed its uppermost operating temperature trials. These tests cleared the way for Aerojet to deliver qualified boost motors to the US Army for fielding in 2009.

Arnold Engineering Development Centre (AEDC) tested and verified the pitch motor ignition and performance in September 2008. The motor is used to pitch the THAAD kill vehicle to a predetermined angle prior to impacting simulated threat targets.

On 18 March 2009, Lockheed Martin and the MDA conducted the sixth test of the THAAD weapon system at the Kauai test range. This flight demonstrated the systems ability to detect, track and intercept a separating target inside the Earths atmosphere. This was the first salvo firing, with two THAAD interceptors launched against a single separating target that is a tactical option for the system. In this trial the first interceptor destroyed the target whilst the second was destroyed by range safety officers.

The seventh test took place on 28 June 2010, the test involved the intercept of a short-range endo-atmospheric unitary target. The system continues to undergo development and testing against ballistic missiles of all ranges in all phases of flight.

Description

The THAAD system must also be able to cue other weapon systems and interface with other air-defence data information networks to allow battle management and command, control and communications tasking of highly complicated attack scenarios in a distributed manner.

The THAAD system is intended to provide the upper tier of a layered defence of high-value targets such as airfields with complementary weapon systems such as the Raytheon Patriot PAC-3.

Missile

The 6.17 m long, 900 kg launch weight THAAD missile is a single-stage, solid-fuelled weapon that is the first missile defence system with both exo-atmospheric and endo-atmospheric intercepts designed to provide upper-tier defence. While on its launcher, primary power will be provided by lead acid batteries automatically recharged by a tactically quiet generator. The THAAD missile will employ thrust vector technology for manoeuvring and a high performance liquid Pratt and Whitney Rocketdyne (a United Technologies Corporation, Company), Divert-and-Attitude-Control System (DACS) for terminal manoeuvring of the forecone, which separates from the missile body before impact. A target object map and predicted intercept point will be provided to the missile before launch, though it will be able to receive in-flight updates via uplinks. Terminal homing of the forecone is by a gimbal-mounted infra-red seeker unit, with the seeker looking through an uncooled side window with a shroud which will separate before terminal homing. THAAD will Hit-To-Kill (HTK).

The extremely fast, 2,800 m/s, THAAD missiles are expected to engage targets out to 200 km plus, and intercept missiles as high as 150,000 m in altitude. The latter is especially needed in order to engage weapons with nuclear, biological or chemical warheads safely.

In a scenario known as 'shoot-look-shoot', THAAD weapons engaging two possible targets should have a single-shot kill probability of 0.9, the terminal phase of the attack employing onboard indium antimonide focal plane array infrared sensors to close in on the target missile.

The initial part of the 48-month Demonstration/Validation (Dem/Val) phase involved the building of two US Army Palletised Loading System-based truck-mounted launchers, two Tactical Operations Centre (TOC) shelters and 20 missiles. The missiles were used at the White Sands Missile Range for flight tests that started in September 1994.

In October 1993, the US Army re-configured the THAAD missile to ensure maximum HTK capability against incoming ballistic missile threats. The missile diameter has been increased with more propellant added to produce a heavier missile that still flies at the required speeds.

Trials THAAD GBR complex with (front left) antenna, (middle left) electronics unit, (upper left) the operator control centre vehicle, (middle right) the unit's cooling assembly and (far right) the power supply unit
0022201

In November 1996, the seventh flight test of THAAD was delayed to early 1997. This followed continuing preflight reviews by the BMDO and Lockheed Martin that uncovered problems with the missile's inertial measurement unit software. These problems were fixed and the missile was recertified. In March 1997, the delayed launch was undertaken but failed to destroy a simulated ballistic target; the fourth failure in four attempts since December 1995. Additional testing of components was reported to be required during and after the missile assembly process along with high-fidelity simulations of both the missile and its components. Two independent review teams, established after the fourth failure, made these recommendations while validating the THAAD concept as sound.

The THAAD research and development costs are said to have risen to USD7.7 billion from the previous USD6.3 billion as a result of the programme changes. The additional money will cover extra flight tests, an extended flight-tests schedule and the risk reduction activities for component engineering and manufacturing development. Lockheed Martin is re-qualifying all missile components.

In May 1998, the eighth test was conducted. Control was lost just after launch and it crashed about two miles from the launch point. Although this was the fifth consecutive failure, the eight flights had in fact achieved 28 of the planned 30 flight test objectives. The next set of flight tests during 1999 gave mixed results.

In June 2000, in order to begin the Engineering and Manufacturing Development (EMD) phase of the THAAD programme, the US Army Space and Missile Defence Command awarded Lockheed Martin approximately USD4 billion in a contract. Lockheed Martin supplied seven launchers, six command-and-control stations and 30 missiles for EMD testing.

In August 2000, Lockheed Martin awarded Raytheon Company a USD1.4 billion plus contract to design, develop and manufacture three EMD radars for the THAAD programme. The contract also included hardware design, development and manufacturing of six Battle Management/Command Control, Communications and Information Tactical/Shelter Groups.

In October 2000, the Lockheed Martin Company Sanders received an eight-year USD225 million contract from Lockheed Martin in support of the THAAD EMD phase. Sanders will deliver 49 flight and flight representative seekers and provide the Missile Check out Console (MCC) ground equipment used in missile fabrication and flight tests. The first EMD flight test took place in 2004.

In July 2007, BAE Systems in Nashua, New Hampshire, received a contract from Lockheed Martin to begin full production of the missile infrared seeker.

Radar

On 2 September 2004, the THAAD X-band radar (AN/TPY-2) successfully tracked a tactical ballistic missile during a successful test of a Patriot missile interceptor missile at White Sands Missile Range, New Mexico. The Patriot missile successfully intercepted a ballistic missile and a cruise missile target during the test. The target missile was flown in a low endo-atmospheric trajectory as part of the Patriot PAC-3 test programme. The THAAD radar utilised production hardware and software to correctly classify the object as a threatening TBM, tracked the object to intercept, and gathered hit assessment data for cost/mission analysis.

The AN/TPY-2 is a high-power, phased array, transportable X-Band radar designed to detect, track and discriminate ballistic missile threats. The AN/TPY-2 was developed by Raytheon Integrated Defence System (IDS) and the software jointly developed by Raytheon and Lockheed Martin. The Raytheon Company, Woburn, Massachusetts, was awarded a contract in February 2007 for the manufacture, delivery and integration support of the radar components for the system. The contract was completed in May 2010.

However, during the last few days of October 2010, the Missile Defence Agency (MDA) awarded Raytheon Company a USD190 million fixed price contract to construct, integrate and test a new AN/TPY-2 radar.

In June 2007, Raytheon completed all factory acceptance testing on its second THAAD radar and shipped it off to the White Sands Missile Test Range for final testing and acceptance by the MDA.

Raytheon's AN/TPY-2 provides a common mission capability: in terminal-based mode with the THAAD Fire-Control Center in support of the THAAD weapon system, and in a forward-based mode with Command, Control, Battle Management and Communications, enabling MDA's Ballistic Missile Defense System. It is a phased array radar, capable of search, threat detection, classification, discrimination and precision tracking at extremely long ranges.

Radar Simulation - SOLD

Following the seventh test in Hawaii on 28 June 2010, test personnel used the Simulation-Over-Live-Driver (SOLD) software system to inject multiple simulated threat scenarios into the THAAD radar. This exercised THAAD's capability to track and engage a mass raid of enemy ballistic missiles.

Battery

Each operational THAAD battery will comprise nine M1075 truck-mounted launchers, weighing 40,000 kg fully loaded, (with eight missile container-launchers two TOCs and a GBR). The launcher vehicle is 12 m long and 3.25 m high. Reloading takes 30 minutes. During October 2002, Oshkosh Truck Corporation delivered two Heavy Expanded Mobility Tactical Trucks with Load Handling Systems, modified for use as launch platforms for the THAAD weapon system. The Battle Management/Command, Control, Communication and Intelligence (BMC3I) components are mounted in hardened shelters carried on High Mobility MultiWheeled Vehicles (HMMWVs).

Variants

Lockheed Martin is developing the concepts of a navy THAAD and an International THAAD (or ITHAAD) for export. It has been reported that Israel has had discussions with the US late in 2006 for the THAAD system to replace the ageing Arrow currently deployed in theatre. No confirmation has been forthcoming from Israel.

Specifications

	THAAD
Dimensions and weights	
Length	
overall	6.17 m (20 ft 3 in)
Diameter	
body	340 mm (13.39 in) (propulsion unit)
	370 mm (14.57 in) (kill vehicle)
Weight	
launch	800 kg (1,763 lb)
Performance	
Speed	
max speed	5,443 kt (10,080 km/h; 6,263 mph; 2,800 m/s)
Range	
max	108.0 n miles (200 km; 124.3 miles) (est.)
Ordnance components	
Warhead	kinetic
Guidance	INS, GPS (command)
	passive, IR
Propulsion	
type	solid propellant
	liquid propellant (kill vehicle)

Status

EMD phase. Low rate missile production (final integration, assembly and testing) began at the Lockheed Martin Factory in Troy, Alabama on 26 May 2004. This low rate production will support the missiles that were used for flight testing which began during late 2004. The first THAAD missile manufactured in Troy will be a pathfinder missile, used to demonstrate and validate test processes and procedures. Immediately following the pathfinder missile, the first of 16 developmental flight test missiles will be manufactured.

The US Army has revealed that the Theatre High-Altitude Area Defence (THAAD) system will be able to track and intercept Inter Continental Ballistic Missiles (ICBM). THAAD is to be a layer of the US Missile Defence Agency's (MDA) strategy to defeat incoming ICBMs.

In 2005 the THAAD system was passed to Lockheed Martin Missiles and Fire Control in Dallas for continued development. In December 2006, Lockheed Martin Space Systems Company, Sunnyvale, California, was awarded a contract for fire unit fielding, support equipment and initial spares. This contract ran until 2011.

The first of the series production started in March 2007 with Aerojet boost motors for the new missile. A letter contract was signed in Sacramento, California on 5 March that provided production through to 2009.

The Initial Operational Capability (IOC) of one fire unit planned by the MDA was handed over for limited operational use in fiscal year 2009. A second system became available during 2010.

Other reports dating from December 2008 have suggested that the UAE will acquire 147 missiles and for the THAAD to work hand in glove with the Patriot PAC-3's that have already been purchased. The missile number was reduced to 96 in January 2012.

In late October 2009, it was announced by the US Army that a second THAAD battery was activated with the US Army at Fort Bliss, Texas. The first battery was handed over to the army in May 2008.

During April 2013, a THAAD battery was moved to the Pacific island territory of Guam in response to threats from North Korea that the US base on Guam is now within range of the North Korean ballistic missiles. The 95-person crew required to man this system was taken from Fort Bliss, Texas.

In late September 2013, an announcement was made that the THAAD missile system has been sold to the United Arab Emirates (UAE) for a total of what is assessed as in excess of USD4 billion. The system is not expected to arrive in country before 2015 when the contract states 96 missiles with nine launcher including spare parts and training data.

It is understood that the US is also in talks with Japan, Qatar, Saudi Arabia and South Korea for the sale or deployment of the THAAD system.

Contractor

Lockheed Martin Missiles and Fire Control (prime contractor for THAAD).

Lockheed Martin Space Systems Company, Sunnyvale, California (contracted for fire unit fielding, support equipment and initial spares).

Raytheon Systems Company (prime contractor for the Theatre Missile Defence - Ground-Based X-band Radar).

Concurrent (provides the RedHawk real-time Linux for hardware in the loop simulation during testing).

Hamilton Sundstrand (Thrust Vector Actuation (TVA) devices).

Aerojet, Sacramento, California, (THAAD boost motor).

BAE Systems (infra-red seeker development and production).

Arnold Engineering Development Centre (AEDC) (for pitch motors and testing).

Shelter- And Container-Based Surface-To-Air Missile Systems

Canada

ADATS

Type
Shelter- and container-based Surface-to-Air Missile (SAM) system

Development
The ADATS shelter-mounted system is one of the available configurations in the ADATS family of missile systems. It is designed for the static defence of high-value installations and can be acquired with or without radar. In the latter case it receives its target information from an external data source.

Trailer or flatbed truck can easily relocate the shelter system. The system retains its anti-armour capability.

Description
In 1993, the Royal Thai Air Force purchased one shelter-mounted ADATS fire unit for integration into its existing air-defence system for defending air bases. The ADATS fire unit receives information on its target(s) from a Skyguard fire-control station.

The Royal Thailand Air Force has conducted operational firings of ADATS in a coastal (over water) environment during hot and humid conditions on a regular basis.

In 1999 and 2000, ADATS scored a one-shot hit against a small sub-scale Snipe drone flying at a range of 6.7 km. The RTAF ADATS unit has had a success rate of 100 per cent to date.

A full description of the ADATS system, the ADATS missile and the system's operational capabilities is given in the Self-propelled surface-to-air missiles section.

Specifications

ADATS missile

Length:	2.05 m
Diameter:	0.125 m
Launch weight:	51 kg
Propulsion:	smokeless solid propellant rocket motor
Guidance:	laser beam-rider using digitally coded CW CO_2 laser
Warhead:	
(weight)	12.5 kg
(type)	dual-purpose HE fragmentation shaped charge
Fuze:	range-gated laser proximity with variable time delay and nose-mounted impact/crush
Maximum speed:	>M3
Maximum effective range:	
(air and ground targets)	10,000 m
(high-speed manoeuvring aerial targets)	8,000 m
Engagement altitude:	
(maximum)	7,000 m
(minimum)	ground level

Shelter-mounted ADATS, as selected by Thailand 0044229

Shelter-mounted ADATS launches a missile during operational test firings in Thailand (Rheinmetall Canada) 1419785

Status
Production as required, the latest standard offered is the ADATS 401.

In service in Thailand, one fire unit with the Royal Thai Air Force using Skyguard fire-control system as the target information source.

The Canadian government announced in 2012 that ADATS is to be phased out of Canadian service.

Contractor
Rheinmetall Canada Inc.

China

HQ-7 (FM-80)

Type
Shelter- and container-based Surface-to-Air Missile (SAM) system

Development
At the 1989 Dubai Aerospace Show, the China National Precision Machinery Import and Export Corporation (CNPMIEC) displayed, for the first time, the HQ-7 (also known by its export name of FM-80; FeiMing-80, which translates in English to Flying Midge 80) all-weather low- and very-low-altitude surface-to-air missile system.

The system was developed using as a basis the French Crotale (originally developed for South Africa under the name Cactus) from 1979 onwards to meet a requirement of the People's Liberation Army General Staff. The then 2nd Research Academy, with Institutes No 23 and No 206 also involved, managed the programme. The former for the radar fire-control system and the latter for the ground equipment.

Test missiles were produced in 1983 with first firings in 1985. Design certification was undertaken from July 1986 to June 1988. Production began shortly afterwards for use as a divisional air defence.

HQ-7 SAM launcher deployed in the firing position from the rear 0044230

Shelter-mounted surveillance radar used with HQ-7 surface-to-air missile system 0044231

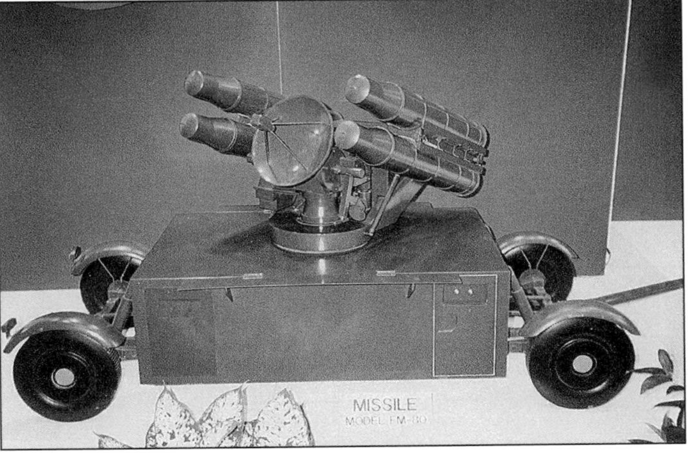

Model of CNPMIEC HQ-7 SAM launcher system showing four missiles in ready-to-launch position and tracking radar (Christopher F Foss) 0509707

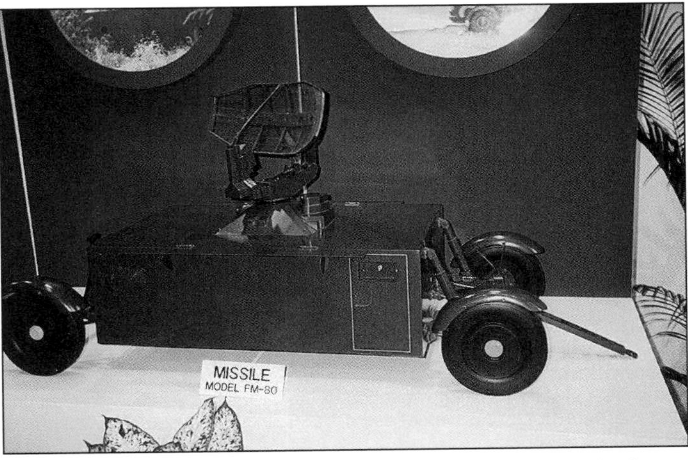

Model of CNPMIEC HQ-7 shelter-mounted surveillance radar showing four wheels (Christopher F Foss) 0509708

The development of the HQ-7 appears to have been aided considerably by a technology transfer package from France involving the then Thomson-CSF Crotale SAM (R-440) system. The resultant HQ-7 system is very similar in physical and technical characteristics to both the self-propelled and shelter versions of Crotale.

A typical shelter-mounted operational battalion comprises three operational sections (batteries) and a technical support section (battery), with direct support comprising:

- An electronic maintenance vehicle;
- Mechanical maintenance vehicle;
- Electronics spares shelter;
- Collimation mast trailer;
- Missile testing vehicle;
- Missile launcher testing vehicle;
- Mobile 12 kW generator vehicle;
- Missile transporter-loader vehicle;
- Missile transporting vehicle;
- 50 kW generator vehicle;
- Indirect support maintenance groups (with various special test benches and standard test equipment).

Each battery comprises:

- Search unit (SS);
- Three firing units (FS);
- Three optical aiming systems;
- Four 50 kW generators.

In 1998, an enhanced (export) system, the FM-90 was revealed (full details of which can be found in the Self-propelled surface-to-air missile section).

In 1999 it was revealed that Iran had reverse-engineered some components and subsystems of the Crotale system. This programme, designated as the Ya-zahra Project, may well have incorporated technology from the FM-80 and FM-90 systems.

Description
(Shelter-mounted version)

The FM-80 (export designation) is an all-weather, low and very low altitude air defence missile weapon system, which can effectively cope with various threat vehicles. It is mainly used for field air defence to protect key military installations and important facilities from air threats.

The operational section has a multitarget interception capability as follows:

- The three firing units can simultaneously deal with three targets either coming from one direction or from different directions.
- A single firing unit can engage targets from the same approaching group four times.
- A single firing unit can engage a single target from four different target groups attacking from four different directions.

With the components it comprises, the system can operate in one of three-missile guidance operating modes. The exact mode is chosen by the operator to fit the operational scenario. The modes are:

- IR - radar, where the target is tracked by the radar, the missile is launched and gathered by the IR localiser and the radar is used to track and measure the missile's angular deviation so as to generate the command radio guidance system signals and guide the weapon to the intercept point.
- TV-IR - radar, where the target is tracked by the passive TV system, the missile is launched and gathered by the IR localiser and its range and angular deviation determined by the radar. The command radio guidance system then generates the control orders to guide it to the target intercept point.
- Manual operation, used where the tracking radar's target channel is subjected to jamming so the operator tracks the target using the handgrip of the TV tracker. The radar measures the range and angular deviation of the target and the missile is guided by the command radio guidance system.

The shelter-mounted HQ-7 unit is used to search, identify, evaluate and classify the targets. It then designates the most dangerous targets and distributes the information to the allocated firing unit(s). If its radar is jammed then the optical aiming units of the firing units can be used to acquire and designate targets.

The search unit comprises the following subsystems:

- A 4.15 m long, 1.33 m high and 2.6 m wide trailer-mounted shelter unit weighing 8,700 kg when fully loaded, and towing vehicle.
- An E/F-band pulse Doppler search radar which has an operational range of 3,200 to 18,400 m, a beamwidth in elevation (single beam) of 0° to 27° (or switchable in 0° to 7°) and in azimuth of 3.6°. Target designation accuracy is 80 m in range and 7 mrad in azimuth.
- A data processing unit which is able to process 30 targets and in conjunction with the radar system track 12 simultaneously, classify the incoming targets, evaluate the threat aspect and then allocate the three most dangerous targets to the designated firing unit(s).
- A wire or radio datalink network between itself and the firing units.
- An IFF system.
- A radio station.

As already stated, if the search unit is subjected to hostile jamming or fails to work because of other reasons the firing unit's optical aiming device can be used to visually track the target and send data to the firing unit. This device has a 7° field of view and a maximum operational range (in fair visibility) of more than 12,000 m.

The firing unit function receives the target designation data (from either the search unit or the optical aiming device) enabling the monopulse tracking radar to search in elevation. When the radar or passive TV tracking system acquires the target and goes to automatic tracking, the firing unit computes the firing solution. Once the engagement parameters are met, the designated missile is fired and the IR localiser determines the weapon's angular deviation and it is gathered through the radio remote-control commands.

Once the missile has entered into the main radar beam, the IR localiser hands over the guidance to the radar. This then measures the relative angular deviation of the missile and the target and flight-control commands are generated to guide the missile to the intercept point with the target. If the firing unit radar is jammed then an alternative guidance mode is chosen.

The firing unit comprises the following subsystems:

- A 4.15 m long, 1.33 m high, 2.6 m wide trailer-mounted shelter unit weighing 11,000 kg when fully loaded, and a towing vehicle.
- A monopulse J-band tracking radar with an operational range up to 17,000 m, rotational limits in azimuth of 360° and in elevation of –5° to +70° and relative accuracy of 0.1 mrad. The missile control transmitter operates in the I/J-bands using a 10° antenna beamwidth and an operational range of 12,000 m plus.
- A TV tracking system with an operational range, in fair weather, better than 15,000 m, a 3° field of view and a tracking accuracy of 0.3 mrad.
- An infrared (IR) localiser with a 10° × 10° wide field of view and 4° × 5° narrow field of view.
- A data processing unit.
- A wire or radio datalink network between itself, the search unit and other firing units.
- A four-tube launcher, turret and missile sequencer assembly. The missiles can be launched either singly or in a ripple mode with three seconds between firings.

The 40 kW trailer-mounted generators are used individually to supply power to the search and firing units.

The HQ-7 missile is 3 m long and weighs 84.5 kg at launch. The missile is in a sealed cylindrical container, which also acts as the launch tube. Propulsion is by a single-stage solid propellant rocket motor, that gives the weapon a maximum speed of 750 m/s. Flight control is by cruciform canard fins. The warhead is of the focused HE fragmentation type and is fitted with a passive proximity fuzing system. Its lethal radius is 8 m. The minimum operational range is 500 m and maximum operational range is 8,600 m against a 400 m/s target, 10,000 m against a 300 m/s target and approximately 12,000 m against helicopters. Operational altitude limits are from 30 to 5,000 m.

The single-shot kill probability of a single round is greater than 0.8 while for two rounds and against the same target it is greater than 0.96.

Variants

FM-80

The FM-80 is the export model of the HQ-7.

HHQ-7

The HHQ-7 (Hai Hong Qui) is the naval variant of the HQ-7.

FM-90

The FM-90 is export model of the HQ-9, which is an upgraded FM-80 model.

FM-90(N)

The FM-90(N) is an export model for naval applications.

Specifications

HQ-7 missile

Length:	2.89 m
Diameter:	0.156 m
Wing span:	0.54 m
Launch weight:	84.5 kg
Propulsion:	solid propellant rocket motor
Guidance:	command-to-line-of-sight
Warhead:	
(weight)	n/avail
(type)	HE focused-fragmentation
Fuze:	passive proximity
Velocity:	750 m/s
Effective range:	
(maximum)	
(400 m/s target)	8,600 m
(300 m/s target)	10,000 m
(helicopter)	12,000 m
(minimum)	500 m
Effective altitude:	
(maximum)	5,000 m
(minimum)	30 m
Kill probability:	>80%
Reaction time:	6 s
Maximum detection range of target:	18.4 km
Maximum tracking range of target:	17 km

Status

Production as required. The HQ-7 as a system is in service with People's Liberation Army, People's Liberation Air Force and People's Liberation Navy.

There are unconfirmed reports that the system may have been exported to Iran. Another possible purchaser is Pakistan that already uses the then French Thomson-CSF Crotale system.

Contractor

Chinese state factories (prime contractor).

2nd Research Academy, with Institutes No 23 and No 206 (subcontractors).

Enquiries

Chinese National Precision Machinery Import and Export Corporation (CNPMIEC).

France

Crotale

Type

Shelter- and container-based Surface-to-Air Missile (SAM) system

Development

The shelter version of the then Thomson-CSF Matra (now Thales France) Crotale low-altitude all-weather surface-to-air missile system was developed as part of the Crotale system family for use in the static defence role.

Crotale shelter firing unit deployed in firing position 0509457

Crotale system deployed (Thales) 0096398

Description

The shelter system involved the fitting of the Crotale acquisition and fire unit systems onto truck and two-wheeled trailer units. The complete system is air-transportable.

Originally the R-440 missile associated with the Crotale system was developed by the then Thomson-Houston and Mistral in France for South Africa where it was named the Cactus; a full description of the Crotale system and the R440 missile, including the system's operational capabilities is provided in the Self-propelled surface-to-air missiles section.

Specifications

R440 missile

Length:	2.89 m
Diameter:	0.15 m
Wing span:	0.54 m
Launch weight:	84 kg
Propulsion:	solid propellant rocket motor
Guidance:	command-guided
Warhead:	
(weight)	15 kg
(type)	HE focused-fragmentation
Fuze:	contact and proximity
Maximum speed:	750 m/s
Effective range (250 m/s target velocity):	
(maximum)	
(head-on target)	9,500 m
(crossing target)	5,500 m

(minimum)	2
(head-on target)	500 m
(crossing target)	2,000 m
Effective altitude (250 m/s target velocity):	
(maximum)	4,500 m
(minimum)	1 15 m

Status
Production complete. In service with the UAE, (four acquisition plus eight fire units). As far as is known, there have been no upgrades or updates to the system within the time frame 2010/2014.

Contractor
Thales France.
MBDA (missile).

Crotale NG

Type
Shelter- and container-based Surface-to-Air Missile (SAM) system

Development
The shelter version of the then Thomson-CSF Crotale NG (New Generation) low-altitude all-weather surface-to-air missile system was developed as part of the Crotale NG system family for use in the static defence role (trailer configuration) and protection of mobile forces (vehicle-mounted configuration).

The shelter system involves the fitting of the Crotale NG combined acquisition, tracking, firing and computer fire-control system onto a shelter unit that can be static-mounted or carried by a variety of vehicles Including a semi-trailer or a United Defence LP FVS987 tracked vehicle. The complete system is air-transportable.

The first prototype shelter-mounted system was used in 1988 for tracking trials while the first complete shelter system trials took place in 1990 with the Mach 3.5 VT-1 missile.

By mid-1991, the French Air Force had selected the shelter form of Crotale NG to protect its air bases. This variant is fitted on a completely self-contained air-transportable shelter suitable for rapid overseas deployment operations. An initial batch of 12 fire units were ordered with the first becoming operational in 1995.

Shelter-mounted version of Crotale New Generation on trailer unit with outriggers deployed 0007989

Crotale New Generation on trailer of the French Air Force being towed by TRM 10 000 (6 × 6) truck (Pierre Touzin) 0069520

At the 1998 Defendory International Exhibition, the Greek MoD revealed it had ordered the Crotale NG system for the Hellenic Air Force and Hellenic Navy. It is believed that a total of 11 systems; nine for the Air Force and two for the Navy in shelter form with VT-1 missiles, are involved. The contract is worth an estimated USD200 million.

Description
A full description of the Crotale NG system, the VT-1 missile and the system's operational capabilities is provided in the Self-propelled surface-to-air missiles section.

Specifications
VT-1 missile

Length:	2.29 m
Diameter:	0.165 m
Wing span:	n/avail
Launch weight:	75 kg
Propulsion:	solid propellant rocket motor
Guidance:	command-guided
Warhead:	13 kg HE fragmentation with contact and proximity fuzing
Warhead kill zone:	8 m
Maximum speed:	M3.5
Effective range:	
(maximum)	11,000 m
(minimum)	500 m
Effective altitude:	
(maximum)	6,000 m
(minimum)	very low
Reload time:	10 min

Status
In production. In service with the French Air Force (initial batch of 12 air-portable shelters delivered). Ordered by Greece in 1998 (11 systems, nine for the air force, two for the navy). These were delivered in 2001. Before these were delivered, Greece leased four Crotale systems from France to train crews. These were delivered to Andhravidha Air Base, in July 2000, at no cost.

A variant known as Crotale Mk 3 has flown in test firings at the CELM test centre in Biscarrosse.

Contractor
Radar and prime contractor system integrator: Thales Defence Systems.
Missile: Lockheed Martin Vought.
Integrated tactical communications: Intracom Defence, Greece.

Shahine (ATTS)

Type
Shelter- and container-based Surface-to-Air Missile (SAM) system

Development
The Air-Transportable Towed System (ATTS) version of the Shahine low-altitude all-weather surface-to-air missile system was developed as part of the Shahine 2 contract awarded in 1984 to the then Thomson-CSF (now Thales France) by the Saudi Arabian government.

Shahine-ATTS firing unit with six missiles in ready-to-launch position and four stabilisers lowered to ground 0509456

The contract with Saudi Arabia involved the outfitting of the improved Shahine 2 modular acquisition and fire unit systems on to towed 4 × 4 trailer units.

The complete system is air-transportable in transport aircraft such as the Lockheed Martin C-130 Hercules.

Description

A full description of the Shahine system, the R460 missile and the system's operational capabilities is provided in the Self-propelled surface-to-air missiles section of this publication.

Specifications

R460 missile

Length:	3.12 m
Diameter:	0.156 m
Wing span:	0.59 m
Launch weight:	100 kg
Propulsion:	cast double-base solid propellant rocket motor
Guidance:	command-guided
Warhead:	
(weight)	15 kg
(type)	HE focused fragmentation
Fuze:	contact and proximity
Maximum speed:	M2.8
Maximum effective range (250 m/s target velocity):	
(head-on target)	11,800 m
(crossing target)	8,000 m
Minimum effective range (250 m/s target velocity):	
(head-on target)	500 m
(crossing target)	2,000 m
Maximum effective altitude (250 m/s target velocity):	6,000 m
Minimum effective altitude (250 m/s target velocity):	15 m

Status

Production is complete.

In service only with the Royal Saudi Land and Air Defence Forces (10 acquisition and 19 fire units).

As of February 2014, there is no further information on this system.

Contractor

Thales France.
MBDA (missile).
Protac SA(now part of Bayern-Chemie) (motor).

International

Roland

Type

Shelter- and container-based Surface-to-Air Missile (SAM) system

Development

Roland 1 entered service with the French Army in 1977, Roland 2 in 1981 and Roland 3 in 1988. The first Euromissile Roland shelter version was developed to meet export customer static defence requirements. Argentina procured four, one of which was used operationally during the 1982 Falklands War and subsequently captured by the British. Of the eight missiles it fired, one shot down a Royal Navy BAE Sea Harrier FRS 1 and two others apparently hit 454 kg (1,000 lb) free-fall bombs.

Iraq also bought 100 shelter versions for use in the 1980-88 Gulf War against Iran, during which the systems shot down a number of Iranian aircraft, which were attacking strategic Iraqi targets.

The fire units were subsequently used in the 1990 to 1991 Gulf War against the Western-led Coalition forces, a conflict in which they again proved to be effective systems, being rated one of the most important Iraqi air defence weapons to be avoided. The system is believed to have been responsible for downing at least two Tornado fighter-bombers.

In 1983, the shelter version was chosen by NATO to protect German and US airbases. A total of 27 fire units to defend three US bases, 60 for 12 German airbases and eight for training use was procured together with an additional 20 for use by the German Navy to protect three of its airbases. Deliveries were completed over the periods 1986-90 to the German Air Force, 1987-89 to the United States Air Force and 1988-90 to the German Navy. Apart from the German Navy systems, all the others are manned by the German Air Force.

The systems are installed on MAN SX90 (8 × 8) high-mobility, cross-country truck chassis. The Roland 3 missile was also adopted for the system, which is known as the FlaRakRad.

Euromissile Roland FlaRakRad (8 × 8) heavy shelter version in travelling configuration with radars retracted 0509920

Air-transportable Roland (Carol version FAR) SAM system being unloaded from a tactical transport aircraft 0007990

Other countries which bought the initial heavy version of the Roland shelter system are Qatar (six shelter versions on MAN SX90 (8 × 8) chassis with Roland 2 missile and three shelter versions on their own) and Venezuela (six shelter versions with Roland 2 missiles).

Since 1996, Euromissile has developed the M3S configuration, which contains new 3-D search radar, a new tracking radar and the basic VMV Roland upgrade components.

The operator can select one of the sensors, for example radar, TV or optical, to engage targets, depending upon the operational environment.

In 1994, the family of Roland systems was complete when the first Carol aluminium structure air-transportable version was finished. Designed for transport by C-160 Transall or C-130 Hercules-type cargo aircraft, a total of 20 fire units was produced by converting AMX-30 systems during 1994 and 1995 for the French Reaction Force (FAR). The system weighs 8,400 kg without missiles, is 4.5 m long, 2.4 m wide and, in addition to the two Roland 3 missiles in the ready to fire position, has another eight in two vertical magazines for automatic reloading.

The French Army uses an ACMAT (6 × 6) tractor truck to tow a Lohr semi-trailer on to which the shelter unit can slide for transportation purposes. The shelter can also be transported by rail if required. The German Air Force took delivery of 10 heavy shelter conversions of the air-transportable, form-mounted systems in 1998 and 1999 for its rapid-reaction force, the Krisen Reactions Kräfte (KRK). The air-transportable shelter retains nearly all the components of the Roland 2 system and is fitted to receive the BKS control and monitoring system.

Description

A full description of the Roland system, the Roland 2/3 missiles and the system's operational capabilities is given in the Self-propelled surface-to-air missiles section.

Command Control and Surveillance System (HFLaAFüSys)

The Army Air Defence Command and Control System, Heeres Flugabwehr Aufklärungs-und Führungssystem (HFLaAFüSys), is a real-time operating short-range air-defence system that provides the integration of the three primary military functions of data acquisition and evaluation, command-and-control and weapon control.

HFLaAFüSys was designed to protect moving ground forces as well as important sites or defence areas against low-, and/or medium-altitude aerial threats.

Advanced communication networks enable secure, real-time information exchanges between sensor (data acquisition), command-and-control truck and the weapon systems. Thus, the integration of army air surveillance (FGR,

Euromissile Roland FlaRakRad (8 × 8) heavy shelter version in operational configuration on MAN SX90 (8 × 8) cross-country truck chassis with both tracking and surveillance antennas erected 0509826

Air-transportable Roland (Carol version FAR) SAM in travelling configuration being towed by ACMAT (6 × 6) truck as used by the French Army 0509921

LÜR) radar systems and the weapon systems Roland, Gepard with capabilities for automatic threat analysis and weapon co-ordination provide a common operational picture for effective army air defence.

With the low-level air picture interface, a standard NATO datalink connection enables cooperation with NATO partners which is currently realised with the US Army.

Specifications

	Roland
Dimensions and weights	
Length	
overall	2.4 m (7 ft 10½ in)
Diameter	
body	160 mm (6.30 in)
Flight control surfaces	
span	0.5 m (1 ft 7¾ in) (wing)
Weight	
launch	75 kg (165 lb)
Performance	
Speed	
max speed	1,108 kt (2,052 km/h; 1,275 mph; 570 m/s)
Range	
min	0.3 n miles (0.5 km; 0.3 miles)
max	4.3 n miles (8 km; 5.0 miles)
Altitude	
min	10 m (33 ft)
max	6,000 m (19,685 ft)
Ordnance components	
Warhead	9.2 kg (20.00 lb) HE fragmentation (multihollow)
Fuze	impact, proximity
Guidance	CLOS
Propulsion	
type	solid propellant (sustain stage)

Status
Production as required. As of March 2013, no further information on this system has been made available.

Contractor
Marketing and sales enquiries
Euromissile.

Rocket motor and propellant
Roxel France.

Russian Federation

Tor-M1 (SA-15 'Gauntlet')

Type
Shelter- and container-based Surface-to-Air Missile (SAM) system

Development
A series of shelter-mounted low- to medium-altitude surface-to-air missile systems has been developed by the then Antei Concern (now Almaz/Antei Concern of Air Defence) from its self-propelled Tor-M1 system. Available details are as follows:

- Tor - M1TA paved road mobile system - with the large box-like, unmanned, eight-round, vertical-launch fire unit of the tracked Tor-M1 system mounted on a 4 × 4 Ch MZAP - 8335.5 trailer which is towed by a URAL - 5323 (6 × 6) truck housing the fire unit's control cabin. Hydraulic jacks are lowered from the trailer when the system is in its firing position. Full 360° traverse of the launcher assembly is available.
- Tor - M1TB towed version for paved roads - with both the unmanned fire unit and the control cabin mounted on towed (4 × 4) trailer units. The cabin on an SMZ - 792B trailer and the fire unit on a Ch MZAP - 8335.5 trailer. This can be used to defend both rear area military and civilian vital points.
- Tor - M1TS stationary version - whereby the fire unit is mounted directly on to the ground via an elevated platform assembly that allows full 360° traverse. The control cabin is located separately in a KP4 shelter and can be at a similar or lower height.

In all three cases the control cabin is connected to the fire unit by cable links of up to 50 m in length. If required, these systems can be integrated into a customer's regional air defence network.

Description
A full description of the Tor system, the Fakel 9M330 and 9M331 missile with its 9M334 four-round missile launcher modules and the Tor's operational capabilities is given in the Self-propelled surface-to-air missiles section.

Variants
Tor-M2U, is the new domestic variant of the Tor family of systems. A description of the Tor-M2U can be found in the Self-Propelled section under Russia.

Specifications

	9M330	9M331
Length:	3.5 m	3.5 m
Diameter:	0.35 m	0.35 m
Wing span:	not available	not available
Launch weight:	165 kg	165 kg
Propulsion:	thruster jets with two-stage eject and solid-propellant rocket motor	thruster jets with two-stage eject and solid-propellant rocket motor
Guidance:	radio command guidance	radio command guidance
Warhead:		
(weight)	15 kg	15 kg
(type)	HE fragmentation	HE fragmentation
Fuze:	radar frequency proximity	radar frequency proximity

Transportable fire unit of Tor-M1 mounted on 4 × 4 ChMZAP - 8335.5 trailer with radars erected and covers open to show missiles (Christopher F Foss) 0007991

Maximum missile speed:	700-850 m/s	860 m/s
Average missile speed:	650 m/s	650 m/s
Target speed:		
(maximum)	700 m/s	700 m/s
(minimum)	0 m/s	0 m/s
Effective range:		
(maximum)	12,000 m	15,000 m
(minimum)	1,500 m	1,500 m
Effective altitude:		
(maximum)	6,000 m	6,000 m
(minimum)	15 m	10 m
Reaction time for system after target detection:	5-8 s	5-8 s

System model	Tor - M1TA	Tor - M1TB	Tor - M1TS
Weight:			
(control cab)	1,000 kg	9,000 kg	5,000 kg
(fire unit)	14,000 kg	14,000 kg	10,000 kg
Time to set up:	8 mins	10 mins	n/avail
Time to pack up:	12 mins	15 mins	n/avail

Status

Production on receipt of orders. Offered for export.

Contractor

Almaz/Antei Concern of Air Defence (primary contractor).
Antei Scientific Industrial Organisation (design).
Izhevsk Electromechanical Plant (system manufacturer).
Fakel (missile manufacture).

Universal Multipurpose Land/Sea/Air Missiles

India

Maitri

Type
Universal (land-/sea-/ground-based) missile system

Development
Development of the Maitri (lit., Friendship) system was announced by the Defence Research and Development Organisation (DRDO) in March 2007 although MBDA first approached the DRDO about this project in November 2005. The Defence Research and Development Laboratory (DRDL) an organisation within the DRDO, will jointly develop this new generation Low-Level, Quick-Reaction Missile (LLQRM) for the Indian Navy, Air Force and Ground Forces whilst the Bharat Dynamics Limited (BDL) is expected to be the primary production factory.

It is also understood that another DRDO agency, the Electronics and Radar Development Establishment, in Bangalore will develop two indigenous radar for the project. It is also understood that the new radar are to be three-dimensional (3D) and capable of handling up to 150 targets with ranges up to 200 km.

Two other code names have been associated with the Maitri project – Revati and Rohini – allegedly for use with the Indian Navy and Air Force respectively.

The missile onboard seeker and guidance assembly will probably be developed in France whilst the platform integration will take place in India.

Description
The Maitri system will come in two packages, both of which are expected to be vertical launched. The Navy will get their system in a vertical launcher that will probably be fitted in the forward part of the ship, whereas the Army (Ground Forces) and the Air Force will come on mobile wheeled and tracked vehicles.

The missile shown during a press day is a very slim weapon with four fins running from a quarter of the way down from the front end to just before the rear thruster area where the fins are cut off strait across. On the rear are four triangular shaped fins used for stabilisation.

Variants
None known.

Specifications

	LLQRM
Performance	
Range	
max	8.1 n miles (15 km; 9.3 miles) (est.) (low)
	10.8 n miles (20 km; 12.4 miles) (est.) (high)
Ordnance components	
Propulsion	
type	solid propellant

Status
As of July 2013, the system is in development.

Contractor
Defence Research and Development Organisation (DRDO); Defence Research and Development Laboratory (DRDL); MBDA; and Electronic and Radar Development Establishment (LRDE).

Russian Federation

Hermes

Type
Self-propelled universal (multipurpose) missile system

Development
The Hermes Universal long-, medium-, and short-range anti-tank/air-defence missile has completed tests and is ready for series production, according to Russia's KBP (*Konstruktorskoye Byuro Priborostroyeniya*) Instrument Design Bureau.

Development of the Hermes missile system was first identified in early 1999, when KBP announced the new family of 'universal' missiles, intended for use from land, sea and air. The system is primarily designed for short, medium and long-range destruction of tanks, armoured vehicles, static high value areas such as command bunkers, small ships, helicopters and low-velocity air targets. Presumably, low-velocity air targets include such items as cruise missiles and UAVs. Other primary factors taken into account for the development of this family of missiles include ranges exceeding that of the defensive weapons of a typical target, a warhead sufficient to penetrate 1,000 mm of armour and an all weather capability.

Hermes mounted on KAMAZ truck (KBP Instrument Design Bureau) 0563387

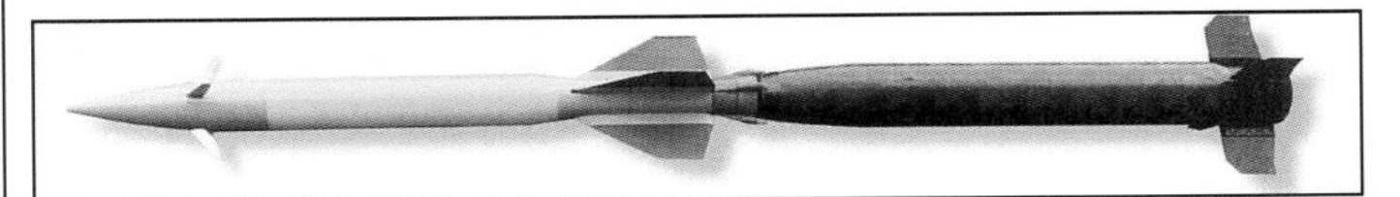
Hermes missile (KBP Instrument Design Bureau) 0563389

Hermes small/patrol ship fit (KBP Instrument Design Bureau) 0563388

Description
At least three versions of the Hermes are believed to exist:
- Hermes (land-based)
- Hermes-A (*Aviatsionnyi* - Airborne)
- Hermes-K (*Korabelnyi* - Shipborne).

Basic Hermes
The basic Hermes is mounted on a KAMAZ type truck with two launchers, each containing twelve missiles. Early artist concepts of the system showed the missiles mounted on a BMP-3 tracked chassis, it is not known whether or not this mounting has been pursued. The basic Hermes can also be shelter- or container-based on static sites for shore side defence. As far as has been reported by the KBP, this type of installation will contain the medium-range 40 km missile only. Command vehicles and target identification and tracking radars will also be integrated with the truck and static installation.

The basic sub-systems of the Hermes consists of:
- Combat vehicle
- Control vehicle
- Air-reconnaissance vehicle
- Command vehicle
- Transport-load vehicle
- Maintenance vehicle.

All vehicles are equipped with communication, navigation, power supply and life-support facilities.

	Hermes-A	Hermes-K	Hermes-K (ER)	Hermes
Dimensions and weights				
Diameter				
body	170 mm (6.69 in) (booster section)	170 mm (6.69 in) (booster section)	210 mm (8.27 in) (booster section)	210 mm (8.27 in) (booster section)
	130 mm (5.12 in) (sustainer section)	130 mm (5.12 in) (sustainer section)	-	130 mm (5.12 in) (sustainer section)
Weight				
launch	110 kg (242 lb)	110 kg (242 lb)	130 kg (286 lb)	130 kg (286 lb)
Performance				
Speed				
max speed	1,944 kt (3,600 km/h; 2,237 mph; 1,000 m/s)	1,944 kt (3,600 km/h; 2,237 mph; 1,000 m/s)	2,527 kt (4,680 km/h; 2,908 mph; 1,300 m/s)	2,527 kt (4,680 km/h; 2,908 mph; 1,300 m/s)
Range				
max	10.8 n miles (20 km; 12.4 miles) (est.)	10.8 n miles (20 km; 12.4 miles)	54.0 n miles (100 km; 62.1 miles)	54.0 n miles (100 km; 62.1 miles)
Ordnance components				
Warhead	28 kg (61 lb) HE fragmentation	28 kg (61 lb) HE fragmentation	30 kg (66 lb) HE fragmentation	28 kg (61 lb) HE fragmentation
Guidance	IR (terminal)	IR (terminal)	IR (terminal)	IR (terminal)
	INS, one-way datalink (mid-course)	INS, one-way datalink (mid-course)	INS, one-way datalink (mid-course)	INS, one-way datalink (mid-course)
Propulsion				
type	solid propellant	solid propellant	solid propellant	solid propellant

Hermes-A

Hermes-A is the helicopter mounted version, trials of which have been carried out using the Ka-52 attack helicopter. In November 2007, KBP released information concerning additional carriers, including combat fixed-wing aircraft.

The system has been developed on the basis of modern scientific and technical backgrounds such as TV/thermal means of target detection with ranges up to 15 - 25 km; it is also reported that it is possible to integrate the surveillance channels to make target detection more reliable under various conditions.

The Hermes-A weapon system consists of:

Optronic system

This includes:

- Two-channel laser designator-rangefinder, which performs measurement of the range to targets and their automatic illumination
- Target auto-tracker, which also stabilises the line-of-sight of the TV and thermal channels and of the target illuminating channels
- Video monitor
- Joystick.

Armament

This consists of supersonic missiles located on the launchers on the pylons of the helicopter. The missile carries a multipurpose high-explosive fragmentation warhead and a high-precision automatic guidance system, which ensures direct hit on a small-size target.

Fire-control system

This comprises:

- Computing system for control of the weapon system, which realises algorithms of missile guidance and the logic of the weapon system operation in all functional states and modes
- Multifunction control console which links the operator with the firmware of the weapon system
- Automatics unit, which realises the missile launch cyclogram observing safety precautions
- Data-exchange pathway
- Software.

Hermes-K

Hermes-K is the shipborne (40 km range) version that is primarily mounted on the AK-306 gun mount modified for missile application found on small vessels and patrol boats. The larger 100 km range missiles can be fitted to larger classes of ships when a mix of missiles can be expected.

The "K" variant has been designed to combat:

- Ground targets
- Surface targets
- Low-velocity air targets (helicopters).

The Hermes-K components consist of:

- Hermes guided missiles in launch tubes
- System control equipment
- Communication and data link facilities
- Maintenance facilities.

Missile

The missile design and layout is compatible with the ground launcher of the Pantsyr-S1 with the large solid booster at the rear of a rounder and stubbier dart. Although in this case, the missile/dart back-end is nothing like those previously seen coming from Russian Design Bureaux. The tail fins employ a sloping leading edge design with a less pronounced slope following a flat surface. These tail fins are much larger than the Pantsyr design, giving the impression of a conventional looking missile round. Just aft of the nose of the missile are what appears to be four standard KBP-designed small rectangle control fins.

There has been at least two independent sources reporting that the Pantsyr missile and the Hermes missiles can be interchanged giving both systems a great deal of flexibility. In one of these reports it was also suggested that a more modern long-range missile is in development.

At least three types of seeker can be employed with the system and possibly even four. All versions of the missile are reported to have a semi-active, two-channel laser for terminal homing with laser designation from a third party. However, it is unclear from the literature whether the missile's seeker can also laser designate the target in an active form. Two other seekers that are in development are the passive InfraRed (IR) fire-and-forget type and an active millimetric wave-band seeker. Reports coming from KBP suggest that when using the semi-active/active laser seekers, a salvo of two missiles per target can be fired, but when using the IR or active radar, up to 12 missiles can be airborne at one time.

The solid rocket booster comes in two sizes, the standard with a diameter of 0.17 m and the long-range version that has a diameter of 0.21 m. The Tunguska and Pantsyr booster motor also employs the same 0.17 m diameter system.

Specifications *See table top of page*

Status

Completed research, development, and testing and offered for export, available for series production. As of July 2013, there have been no known customers.

Contractor

KBP Instrument Design Bureau.

Khrizantema

Type

Self-propelled Universal (multipurpose) missile system

Development

Design and development at the KB Mashynostroyenia of the Khrizantema system began during the 1980s, with the first prototype completed in 1990. The KBM Engineering Design Bureau produced the first of this new class of weapon and installed it on the BMP-3 chassis. The chassis/launcher is the 9P157 (9P157-2) that is supplied by the Aggregate Plant in Saratov, with the design authority for the automotive section at the Kurgan Machine Construction Plant. There have also been some spurious reports of the chassis designation 9M157-2, however this would not fit into the Russian designation system as 'M' is normally associated with 'Missile' whereas the 'P' is normally the launcher.

The system that has undergone trials and testing throughout the 1990s was first shown to the Open Press in 1996. Trials series production was completed in 2003 and full series production was expected to commence in 2004. Originally, production was reported to begin in 1998, but this later turned out to be for the trial series production missiles only.

Description

The Khrizantema is a supersonic new generation of self-propelled, multipurpose guided missile system that has been designed as an all weather multi-purpose missile that can engage various targets such as:

- Modern main battle tanks that are equipped with Explosive Reaction Armour (ERA)
- Light armoured combat vehicles

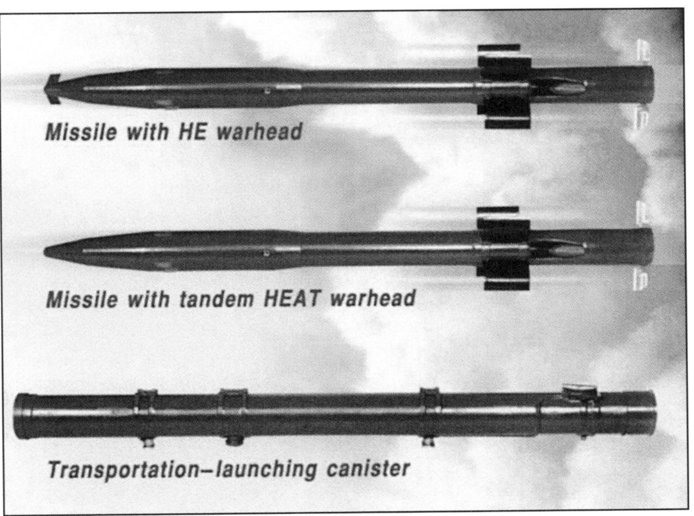

Khrizantema weapon family (KBM-KB Mashynostroeniya) 0529825

Khrizantema-S self-propelled anti-tank system has completed service-acceptance trials (KBM Engineering Design Bureau) 0531751

- Fortified structures
- Pill-boxes
- Small seagoing vessels
- Slow, low-flying piston engine aircraft
- Helicopters.

The Khrizantema system may be installed upon tracked chassis vehicles, small ships or helicopters. The land-based army system comprises:

- A combat vehicle with guidance equipment and ready-to-fire guided missiles
- Ready-to-fire 9M123 (9M123-2) HEAT missiles and 9M123F (9M123F-2) HE missiles
- Radar for the detection of targets
- Stationary and mobile test equipment
- Training aids.

The 9M123 missiles are armed with various types of warheads, which will be used depending on the target. The BMP-3 chassis carries up to 15 missiles in hermetically sealed expendable containers that are available for automatic reload. Within these 15 missiles, selections of differing types of warheads are available depending on the threat and action scenario. The system is supplied with its own mobile test equipment and training aids. The crew of a BMP-3 will normally consist of two men, commander/gunner and driver, both of whom are seated in the front of the combat vehicle under full armour protection.

Variants

Khrizantema-S

Rosoboronexport in Moscow offers Khrizantema-S, mounted on the 9P157-2 BMP-3 ICV chassis for export. An air-launched variant is still in the developmental stage was not expected to be ready until late 2004 or 2005, to date (September 2006), as of August 2010, no further information has been made available on this variant.

The purpose and design of the system is to engage Main Battle Tanks (MBTs), sea targets, low-level aerial targets, field fortifications, manpower in shelters and at open sites. This system is capable of engaging targets day or night and in adverse weather conditions.

Known features of the Khrizantema-S are:

- Round-the-clock all weather combat capability
- Simultaneous guidance of two missiles to two targets
- Ability to automatically track the target and guide the missile (fire-and-forget principle)
- Electronic warfare proof capability
- High manoeuvrability and survivability
- Missiles in TLC (9M123 tandem shaped-charge, 9M123F HE).

The Khrizantema-S was finally accepted into service with the Russian Armed Forces in mid-July 2013.

Specifications

9M123 missiles

Length:	2.057 m
Diameter:	
(missile)	0.15 m
(wingspan)	0.3 m
Weight:	
(missile)	46 kg
(missile in canister)	54 kg
(9M123/9M123-2 warhead)	8.0 kg
Propulsion:	single solid propellant
Guidance:	
(mode 1)	automatic millimetric-waveband (100-150 GHz) radar
(mode 2)	semi-automatic laser beam riding
Warhead:	
(9M123/9M123-2)	tandem shaped charge HEAT warhead used in the anti-tank role
(9M123F/9M123F-2)	HE Fragmentation warhead used in the anti-air role
(thermobaric)	a thermobaric warhead has been associated with the Khrizantema system, this is not mentioned on any export brochures and is expected therefore not to be available except to Russian forces
Armour penetration:	
(9M123/9M123-2)	1,100-1,200 mm (path-length) behind ERA
Maximum speed:	400 m/s
Effective range:	
(maximum)	5,000-6,000 m
(minimum)	400 m
Effective altitude:	
(maximum)	n/avail
(minimum)	n/avail
Maximum target speed:	n/avail

Status

Development complete, trial series production took place during 2003. Full series production of the basic Khrizantema commenced during 2004 with the system assuming Initial Operational Capability (IOC) status with the Russian Armed Forces. The Khrizantema-S was not accepted into service until mid-July 2013. At the Moscow Air Show 2006, a system with the 9M123 and 9M123M missiles were offered for sale.

Contractor

KBM Mashynostroeniyan Khrizantema (primary contractor).
Saratov Equipment Plant GUP/SAZ (develops and supplies the 9P157 chassis).
Kurgan Machine Construction Plant (supplies the automotive section).
All Russian Scientific Research Institute of Experimental Physics, Russian Federal Nuclear Center/VNIIEF (development of the fragmentation warhead and trials).
Signal, All Russian Scientific and Research Institute, GUP/VNII Signal (developed the AC and DC electric and hydraulic-electric system).

Kvartet

Type

Self-propelled universal (multipurpose) missile system

Development

The Kvartet system using the Kornet-E missiles is a KBP Instrument Design Bureau third-generation primary anti-tank missile system that also ensures engagement of ERA-protected current and future tanks, fortifications, lightly

Kliver - Naval version of the Kvartet (Christopher F Foss) 0022561

armoured low-velocity air (hovering helicopters) and surface targets by day and night, under adverse weather and radar/optical jamming. Kvartet is a development of the Kornet anti-tank system with a multiple launcher that can be mounted on any self-propelled chassis or a single tripod man portable launcher.

Description

Kvartet vehicle mounted launcher variant utilises the automated 9P163-2 launcher that has four rails and electro-mechanical gears installed on light infantry vehicles. The laser guidance channel of the stabilised sight enhances the whole system lethality.

The 9P163-2 launcher consists of (note the Kvarted-M uses the 9P163-2M launcher):

- Turret with four rails and missiles mounted
- 1P45M-1 sighting-and-guidance unit
- 1PN79M-1 thermal imager
- Electronic unit and gunner's seat.

Ammunition stowage is placed separately. Up to four missiles can be launched without reloading. Target search is simplified with track automatic processing due to electromechanical gearing. Two missiles may be launched in salvo against one target within the same beam.

The single tripod mounted launcher comprises of combat assets, maintenance facilities and training aids. These assets include:

- 9P163-1 launcher
- 1P45-1 sight tracker
- 1PN79-1 thermal sight
- 9M133-1 and 9M133F-1 missiles.

Variants

- Kvarted-M
- Kornet
- Kornet-E
- Kliver

Specifications

Kvartet

Missile:	9M133-1/9M133F-1
Length:	1.2 m
Diameter:	0.152 m
Wingspan:	0.46 m
Weight:	
(portable launcher)	26 kg (some reports say 27 kg)
(4 × rail vehicle launcher)	600 kg
Warhead:	
(9M133-1)	tandem HEAT
(9M133F-1)	fuel-air-explosive based thermobaric (10 kg TNT equivalence)
Motor:	solid propellant rocket motor
Range:	
(maximum)	5.5 km
(minimum)	100 m
Range with thermal imager:	
(maximum)	1,200 m
(minimum)	1,000 m
Speed:	240 m/s
Ammunition:	
(full load)	8 missiles (some later KBP reports suggest up to 9 missiles)
(ready-to-fire)	4 missiles
(reserve)	5 missiles
Guidance:	semi-automatic, laser beam riding
(in traverse)	±180°
(in elevation)	−10° to +15°
(at rolling)	±15°
(at pitching)	±5°
Rate of Fire:	1-2 rounds/min
Reload time:	30 s

Status

Reports emanating from Israel have suggested that a very small number of these missiles may be in the hands of Hamas and Hizbullah.

Contractor

KBP Instrument Design Bureau.

RVV-AE

Type

Towed universal (multipurpose) missile system

Development

The RVV-AE Ground-Based Air Defence (GBAD) (also known is the West as AMRAAMski) missile system from the Russian Company Vympel is based on the previous R-77 (NATO AA-12 Adder) Air-to-Air Missile (AAM) system, development of which began back in the early 1980's, probably as early as 1982. The RVV-AE missile is being offered for sale as a multipurpose (Universal) ground-based, air-launched and naval weapon. The RVV-AE version of the missile is the export variant of the R-77. The RVV-AE missile, intended for combat employment from land-based launchers, is stored in a transport-launch container.

Vympel RVV-AE in launch canister (P Butoswki) 1116479

RVV-AE on S-60 launcher (Vympel Design Bureau) 1127961

During 1993, one year before the R-77 AAM version was declared IOC, vertical launch trials of the missile for proof of concept were believed to have been carried out at a test range in Russia. Furthermore, ship borne simulating trials were also carried out with a larger diameter motor incorporating Thrust Vector Control (TVC) folding fins and wings. By 1995, one year after the R-77 IOC, the Vympel Design Bureau was actively studying several enhanced versions of the basic AAM missile. These studies included an IR variant (R-77E or TE), an increased range variant with a ramjet engine (R-77M-PD), the first modification programme with a heavier missile (R-77M) and the ground-launched variant (R-77-ZRK).

In 1996, at the Athens Defendory Exhibition in Greece, Vympel showed the RVV-AE missile with a solid rocket booster for air defence. At that time, no mention was made of the launcher or even when the system would be ready (if ever) for sale.

RVV-AE launch canister (Robert Hewson) 1127962

During MAKS-2005 in Zhukovsky, Moscow, Vympel finally revealed the missile, launch canister and launcher. The system uses four missile canisters which are mounted one a simple S-60 (57 mm anti-aircraft gun) carriage, with a pair of canisters on each side of the mounting.

Description

The missile is cued to its target by a remote command-and-control unit attached to a ground-based air defence system, typically the 9K33 Osa (NATO SA-8). Once the R-77 seeker is tracking the target, the missile is launched under its own guidance. Engagement range is equivalent to the 9B-1348 seeker range; about 12 km. The minimum range is reported as 1.2 km. Vympel is promoting the new variant as an affordable air-defence system, claiming it is at least 25 per cent cheaper than the US equivalent AIM-120 Advanced Medium-Range Air-to-Air Missile (AMRAAM) currently used in the Surface-Launched (SLAMRAAM) variant.

Russian open-press sources such as Voyenno-Promyshlenny Kuryer have suggested that up to six of the RVV-AE missile launcher/containers can be mounted on a Kub (2P25) tracked vehicle (NATO SA-6 'Gainful') as a means of upgrading the Kub system with an export potential.

Variants

R-77 air-to-air missile with the export designator RVV-AE.

Specifications

	RVV-AE
Dimensions and weights	
Length	
overall	3.6 m (11 ft 9¾ in)
Diameter	
body	200 mm (7.87 in)
Flight control surfaces	
span	0.4 m (1 ft 3¾ in) (wing)
Weight	
launch	175 kg (385 lb)
Performance	
Range	
min	0.6 n miles (1.2 km; 0.7 miles)
max	6.5 n miles (12 km; 7.5 miles)
Altitude	
min	20 m (66 ft)
max	9,000 m (29,528 ft)
Ordnance components	
Warhead	22.5 kg (49 lb) HE fragmentation
Fuze	laser, proximity

Status

As of July 2013, the status of the system is uncertain. Development and proof of concept is probably complete; the missile on the S-60 launch platform was offered for sale at MAKS-2005 in Zhukovsky, Moscow.

RVV-AE missile (Robert Hewson) 1127963

Contractor

Vympel Design Bureau (primary contractor).
Aleksinskiy Khimicheskiy Kombinat, producer of the S-60 platform.

South Africa

GBADS South Africa

Type

Universal (Land/Sea/Ground-based) missile system

Development

The South African National Defence Force (SANDF), Ground Based Air Defence System (GBADS) moved a step nearer fruition in September 2002. An announcement that the government procurement agency Armscor is in the process of negotiating a contract with Denel as the preferred prime contractor for the GBADS Local Warning Segment (LWS), was made during the African Aerospace and Defence show at Waterkloof airbase near Pretoria.

A contract was announced in May 2003 for the Thales Starstreak missile system to satisfy South Africa's first phase of the GBAD programme that has also been referred to as the Project Sable GBADS System.

Project Sable uses a building block approach to integrate all current and future South African Army Air-Defence Artillery (ADA) assets into a single system of systems. This is to include the thirty GDF 05 L90 twin 35 mm towed cannon with its four Reutech Radar Systems (RRS) ESR220 Thutlwa (also known as Kameelperd or Giraffe) Mobile Battery Fire-Control Post Systems. Additionally, eight Thales Starstreak Lightweight Multiple Launchers and 100 Starstreak high velocity missiles acquired under another Project Guardian. It is also assessed that in the future a new Denel developed missile and Continuous-Wave (CW) 3-D radar (an upgrade to the Thutlhwa) will also become part of the Sable system.

In March/April 2006, during Exercise 'New Horizon 8', the South African Air Force demonstrated the capability of integrating civilian radars, civilian flight plan data, strobe data, shipping information, GBAD and Naval air-defence

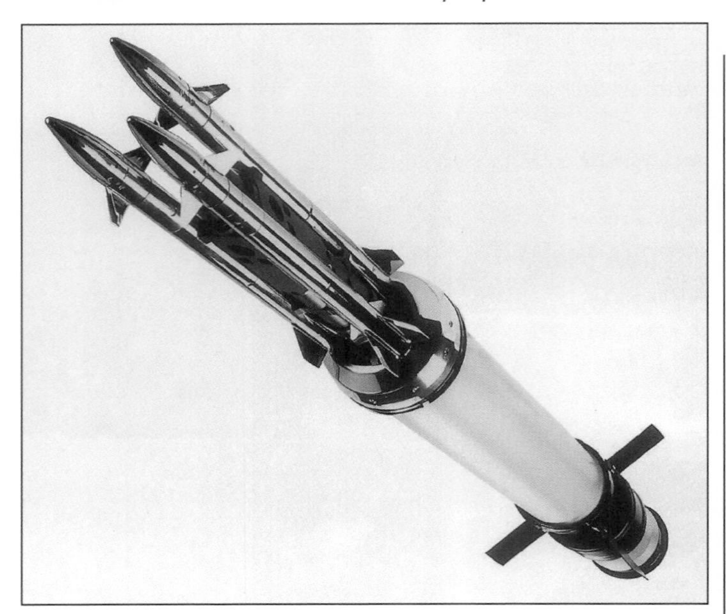

Starstreak missile (Unknown) 0056298

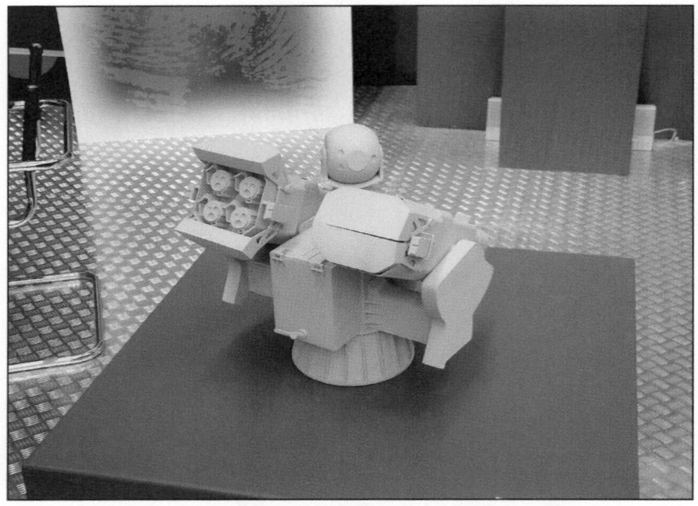

Model of proposed remote missile turret equipped with Starstreak 0523879

information at a central node. All of this information was subsequently used to enable GBAD assets to cover airspace used by friendly combat aircraft.

The exercise formed part of a continuing technology project to address interoperability of weapons, sensors and information systems in the joint air-defence domain. Information was passed from various radar sources direct to the SAAF Mobile Sector Control Centre (MSCC). From there the SANDF coordinator passed weapon control data to an Army Battery Fire-Control Post using the Link ZA data message standard. No voice communications were used at any time during the exercise. The exercised/demonstration went a long way towards proving the concept of semi-positive control of GBAD in South Africa.

The radar/sources used during the demonstration included:

- A transportable Umlindi (Sentinel) long-range surveillance radar (upgraded Plessey AR3D belonging to SAAF)
- Marconi S-711 Tactical Mobile Radar (SAAF)
- Alenia ATCR-33 military air traffic control radar at Overberg Air Base near Overberg Missile Test Range (SAAF)
- Thales Star 2000 ATCR at Cape Town International Airport
- Thales RSM-9701 radars at De Aar and Blesberg
- Thales MRR radar from new MEKO-class patrol corvette, SAS Amatola operating off shore at the OTR range
- SAN coast watch radar system
- South African Army RRS ESR-220 Kameelperd gap-filler radar and battery command post.

In March 2007, details released by the South African Department of Defence (DoD) suggested that the project to acquire a new GBAD system has fallen critically behind schedule. A five-month delay occurred because of financing problems. A joint Armscor/army project team and Denel have now developed an approach to minimise further delays, while the group itself has put a new management team in place for this project. Armscor is happy that all foreseeable risks have been identified and detailed risk mitigation plans have been developed.

The first missile firing by an Army air defence crew was expected to take place in October 2007, however, this was delayed by one month and took place at the Overberg Test Range near Cape Town late October early November. The Overberg Test Range is part of the Denel organisation. Reports have suggested

Kentron containerised Umkhonto SHORADS installation 0523878

that up to eight missiles were fired at Low Cost Aerial Targets systems (LOCATs) after each of the eight operators had proved their capability by at least 1,000 hours on simulators. It is presumed that each operator had one live round to fire at the LOCATs.

In April 2008 at the DSA Defence Exhibition in Malaysia, Reutech Radar Systems displayed the new RSR 940 and RSR 942 Spider radars that were at this time undergoing trials in South Africa with a view to future deployment with a GBADS system. The RSR 940 is a mobile land-based, rapid deployment, continuous scanning surveillance radar that is capable of detecting and tracking low-level air targets as well as surface targets. On-board power generation and communications allows for autonomous operation of the system. The instrumentation range of the radar is 45 km.

The RSR 942 is an electronic sector scanning surveillance radar based on the Spider RSR 940 and is also highly mobile. The radar offers a man-portable configuration with similar characteristics to the RSR 942 and is ideally suited for the detection and recognition of moving personnel, helicopters and small boats in predefined sectors that may extend over land, sea and air.

Late in July 2010, a small contract for risk reduction studies for a ground based launcher was awarded to Denel Dynamics under the Project Protector. Protector is a technology programme being funded by the South African Army ADA through the Department of Defence - Defence Secretariats' Defence Matériel Division's Directorate Technology that is headed by Rear Admiral Derek Dewey.

In March 2012, it was reported that Denel Integrated Systems Solutions (DISS) had completed the contractual hardware delivery for the GBADS and that the system is to become operational.

Description

Phase 1 (also known as the LWS phase) is a fully integrated air defence system for support of SANDF in theatre. Phase 1 was planned to be implemented within six years, and expected to include a battery of eight to 12 dismounted Starstreak Very short-Range Air Defence Systems (VSHORADS) launchers with a 6 km engagement range.

Phase 2 was planned between 2009 to 2014. Representative of this phase is Kentron's Umkhonto mobile missile system.

Phase 3 is planned between 2015 to 2022. The procurement of two Mechanised Air Defence Systems (MECADS) batteries, one equipped with self-propelled VSHORADS, both missile and gun based, and the other with additional Short-Range Air Defence Systems (SHORADS) missiles.

Command-and-Control

Originally known by the name "Project Legend", the C2 system for the South African GBADs was initiated in the period 2001/2002 and was sent to industry in 2006. Saab Systems South Africa was allocated the contract in August 2008 and is to deliver the system in 18 months (2010). The software is known by the brand name 'Chaka' and is entirely South African designed and produced. Chaka was developed according to some years of experience with Brigade-level war-gaming (simulation) software, branded by the company as BattleTek.

Variants

GBADS 2

GBADS 2, also known as Project Protector, will use the same missile that is currently deployed with the SA Navy's Umkhonto point defence system. The system has also been given the acronym All-Weather Surface-to-Air Missile (AWSAM). In July 2011, the system moved into the study and testing phase on the launcher.

Status

In development. In November 2007, the system was on the Overberg Range tor trials firings. Handover of the complete Starstreak system was expected during a period covering 2009 to 2010; by July 2011 this delivery had not been confirmed.

The South African Army Air Defence Artillery's ESR220 Thutlwa mobile GBAD fire-control post that is also associated with the GDF 05 L90 was deployed along with GDF 05 to the South Sudan airfield of Juba in July 2011 to support the SAAF in providing radar coverage of the airfield for Airspace Management/Air Traffic control for SA government deployed military contingents in support of visiting heads of state and other VIP's at the airport during independence day celebrations.

In March 2012, it was reported that Denel Integrated Systems Solutions (DISS) had completed the contractual hardware delivery for the GBADS and that the system is to become operational.

Contractor

GBADS Industrial Team, which is comprised of:
Armscor (government procurement agency).
Denel (Pty) Ltd (prime contractor).
Denel Dynamics (previously known as Kentron).
Denel Integrated Systems Solutions (DISS) (hardware contractor).
African Defence Systems (Pty) Ltd (ADS) (part of the Thales group).
BAE Systems.
Reutech Radar Systems.
Saab Systems South Africa (C2 system).
Thales UK (Belfast) (Starstreak missile system).

United Kingdom

LMM

Type

Universal (land/sea/ground-based) missile system

Development

Thales Air-Defence Systems Limited, Belfast, is developing a new concept Lightweight Multirole Missile (LMM), the cost effective Short-Range Air-Defence (SHORAD) missile system based on the Starburst. It is understood that the system is a private venture for the company and that since early 2008, up to GBP2 million have been invested.

LMM is the latest in a long line of pedigree missile systems designed and developed in Belfast at what was Shorts Brothers and is now Thales UK. Going back to the 1950s and 1960s, the Very Short-Range Air-Defence (VSHORAD) Seacat missile system was developed. This saw action in the Falklands War and its land-based variant, the Tigercat, was used by the Iranians during the Iraq/Iran War.

During the 1970s Blowpipe was introduced, originally designed and developed, like the LMM, as a multirole missile system. During the 1980s, the Javelin and Starburst were introduced and at the end of the 1990s the 30 kg Starstreak VSHORAD missile was deployed which is still the only multiple warhead (three tungsten darts) missile system anywhere in the world.

Leading into the 21st Century, Thales gained orders to produce the VT1 SHORAD and by 2002, the Next Generation Light Anti-Armour Weapon (NLAW).

The LMM missile concept is believed to have commenced early in 2006 resulting in ground-to-ground proof of concept firing trials in November of the same year. Over a period of time, the LMM missile is envisaged to replace the Starburst whilst complementing the extremely capable Starstreak. As the system was to be a privately funded development, several constraints were placed upon the LMM namely:

- Costs had to be kept to a minimum (value engineering).
- Controlled flight; the system had to be capable of precision strike.
- Multirole; available for all branches of the armed forces.
- Export; essential if the system was to be successful.

In June 2008, the pre-production missile had been test-fired against ground targets, UAVs with an 11 ft wingspan and drone targets with 'V' shaped 13 ft wingspan.

Description

System

The system has been designed to address targets on the land, sea and air. These targets are best described as:

Sea Targets

- Fast Attack Craft/Fast Infiltration Attack Craft (FAC/FIAC)
- Landing Craft
- Surfaced Submarines.

Land Targets

- Soft-skinned wheeled vehicles
- Tracked vehicles (not Main Battle Tanks - MBTs)
- High value static locations, for example command centres, shore batteries and so on
- Light Vehicles.

Air Targets

- UAVs
- Helicopters
- Light Aircraft
- Cruise Missiles (a capability will probably be added during a mid-life update; as of July 2013, the missile has not been modelled against these targets).

In order to address the cost problem, components for the new LMM were either acquired Off-The-Shelf (OTS) or utilised from other Thales systems. Re-engineering of sub-components has been kept to a minimum, although some of these, such as the seeker gyro, has already taken place.

Motor

The basic rocket shell and motor (the motor being used for trial purposes only) have been acquired from the Starburst missile. This uses the Epictete motor containing Cast Double-Base (CDB) propellant. The trial motor is to be replaced by a new motor which is under development by Roxel. The motor is expected to cost up to one-third the price of the current Starstreak motor. The motor is to be a two-stage design, the first stage launches the missile from the launcher, the second stage accelerates the LMM to a speed of greater than Mach 1.5. Roxel is an Anglo-French Company formed in February 2003 by the merging of CELERG of France and Royal Ordnance Rocket Motors of the United Kingdom.

Warhead

The LMM warhead is a mixture from two different Thales weapons. With minimum re-working of both previous warheads, in particular the 3 kg blast fragmentation taken from Starburst and the shaped charge from the Blowpipe. The missile is provided with maximum lethal effect against lightly-armoured land and air-defence targets, performing better than both its predecessor systems.

UAV/Naval/Helicopter Applications for LMM (Thales UK) 1304029

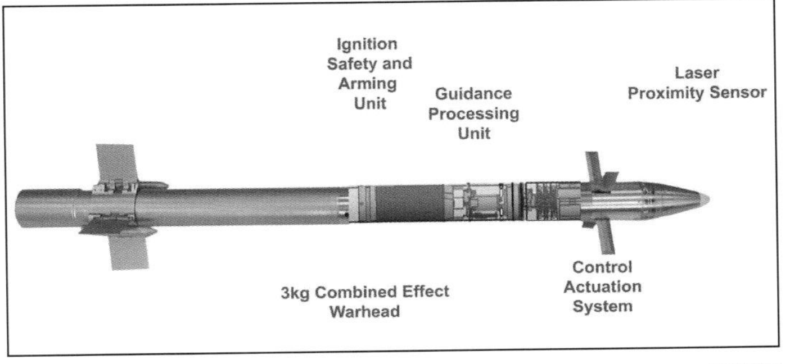

LMM missile (Thales UK) 1304030

The Thales multi-aperture laser proximity fuze unit on its own and fitted to the LMM (Thales UK) 1304031

Fuze
The Thales Missile Electronics multi-aperture laser proximity fuze employs beam-cutting technology at close range. It is derived from an already qualified fuze fitted to the Starburst missile.

Control/Actuating System
The current control and actuating system, which is undergoing further miniaturisation, employs three DC motors and four control surfaces. One motor controls both the top and bottom surfaces, the other two motors control the left and right surfaces individually, the combined system controlling the missile's roll, pitch and yaw.

Guidance
LMM guidance is provided via an Electro-Optical (EO) tracker system fitted to the launch platform. Contained within an optically stabilised mount, this typically comprises a Charge Coupled Device (CCD), TV and thermal cameras with an automatic target tracker and the missile laser guidance unit.

A variety of alternative guidance options are being developed, with the initial focus on Laser Beam Riding (LBR) and a Semi-Active Laser (SAL) seeker. The currently envisaged guidance modes are as follows:

- **Mode a:** LBR and SAL.
- **Mode b:** Infra-Red (IR) terminal homing.
- **Mode c:** Imaging Infra-Red (IIR) terminal homing (a future mid-life upgrade, Thales group in France is looking at low cost seekers possibly in the 14 micron band).
- **Mode d:** INS/GPS.

For laser beam riding the current missile is fitted with two laser receivers mounted on the tail assembly. These receivers have been developed from those used on the darts of the Starstreak multi-dart missile system. To capture the missile at launch the laser guidance system is initially in a wide-angle Field-Of-View (FOV) mode (a slight misnomer as the FOV is still only a few degrees), once captured the beam collapses to a narrow angle to guide the missile accurately to its target.

Announced in early June 2012, Goodrich Corporation (Goodrich Sensors, Plymouth, UK) has been awarded the contract to its SiIMU02 inertial measurement units for the LMM missile. The SiIMU02 will be used for guidance within the missile and will assist to increase the delivery accuracy of the weapon.

Launcher
As the system has been designed for multiplatforms, the following can be expected:

Naval Platforms
- MSI Defence System Naval Gun.

Air Platforms
- Helicopters, particularly the Westland type
- UAV, trials have already been carried out using the Schiebel Camcopter S-100 with no detriment to the UAV on launch
- Naval trials (not yet associated with the LMM) using Camcopter S-100 have been conducted with the Spanish and Pakistan Navies, it may therefore be logical to expect at the very least offers to supply and fit the LMM, should these Navies go ahead with S-100 purchases.

Land Platforms
- Thales MultiMission System (MMS)
- Lightweight Weapon Station
- Remote Weapon Station in conjunction with Kongsberg
- Shoulder-launched (MANPADs).

The launcher loads will be as many as six - seven on the naval platforms and also on the helicopter. The Schiebel Camcopter has been flown with one missile mounted either side of the cab. The land launchers will be as required depending on the mission of the vehicle. Man-portable launchers will be single for shoulder-launch or a multiple one, if mounted on a tripod.

Specifications

	LMM
Dimensions and weights	
Length	
overall	1.3 m (4 ft 3¼ in) (est.)
Diameter	
body	760 mm (29.92 in) (est.)
Flight control surfaces	
span	0.26 m (10¼ in) (est.) (wing)
Weight	
launch	13 kg (28 lb)
Performance	
Speed	
max speed	1,069 kt (1,980 km/h; 1,230 mph; 550 m/s) (est.)
Range	
min	0.2 n miles (0.40 km; 0.2 miles) (est.)
max	4.3 n miles (8 km; 5.0 miles) (est.)
Ordnance components	
Warhead	3.9 kg (8.00 lb) HE fragmentation, shaped charge
Fuze	laser, proximity (multi-apeture)
Guidance	semi-active, laser, laser beam riding (mode a)
	IR (mode b)
	IIR (mode c)
	INS, GPS (mode d)
Propulsion	
type	two stage, solid propellant

Status
As of July 2013, the missile and system are reported to be on time and within budget.

Contractor
Thales UK.
Roxel SAS, Goodrich Sensors, Plymouth, UK.

United States

AERAM

Type
Universal (land/sea/ground-based) missile system

Development
The US Army and Raytheon are developing a supersonic Surface-to-Air Missile (SAM). This system known as the Advanced Extended Range Attack Missile (AERAM) that will have a range of more than 100 km and will fill part of the army's future air defence requirement.

AERAM has been designed to provide a cost-effective solution against subsonic Cruise Missiles, Unmanned Aerial Systems and other subsonic airborne threats.

There have already been suggestions that the missile, with its loitering capability, will be an excellent submarine launched weapon when used in the clandestine sea-base scenario.

During late 2004, Northrop Grumman Corporation successfully completed launching trials from an Ohio class submarine USS *Georgia* whilst dived off the coast of San Diego of what has become known as SACS [Stealthy Affordable Capsule System]. This is the fourth such trial of the SACS system, the first going back as far as December 2002, when tests were performed in test pools at the SRI Corral Hollow Experiment Site in Tracy, California. Two more tests, the first in March 2003 and another, in July 2004 preceded this trial. These tests were performed under an agreement from Naval Sea Systems Command for the Submarine Payloads and Sensors programme.

During Exercise Silent Hammer, further testing and launching from the Georgia took place. This was to evaluate the universal encapsulation that entails the development of methods to enable submarines to affordable launch a variety of weapons and organic unmanned off-board systems took place. The module was installed in a D-5 missile tube.

Initially the AERAM is expected to be a complementary munition launched from a SLAMRAAM type launcher.

Description
AERAM is a research and development programme to demonstrate a cost effective solution with the challenge to provide a near-term demonstration, leveraging existing systems and components, of a capability that will provide the US military with the ability to engage at extended ranges, cruise missiles, UAVs and other symmetrical or asymmetrical airborne threats. Primarily, the missile is being designed to fit into the army's next-generation air defence system, the Surface-Launched Advanced Medium Range Air-to-Air Missile (SLAMRAAM). It will use the same launcher as the SLAMRAAM, easing the

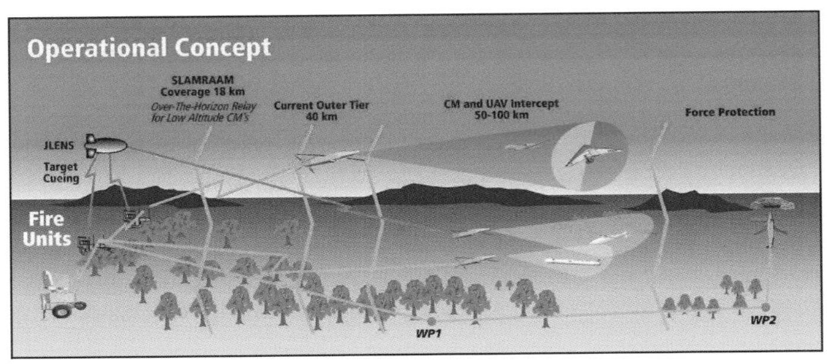

Artist concept of operation (Raytheon Company) 0585513

logistic and training burden. Currently SL-AMRAAM is optimised for supersonic threats, whilst AERAM will be more suited for longer-range, slower (subsonic) moving targets like cruise missiles, Unmanned Aerial Vehicles (UAVs) and helicopters.

A secondary surface-to-surface capability will also be available to the AERAM system.

Missile

The missile is expected to use unique turbofan propulsion with an afterburner and a variable geometry design. On launch the wings will be deployed fully with the throttle of the engine back giving a loitering speed of Mach 0.3. However, with the wings fully swept back and the after burner on the missile will achieve speeds in excess of Mach 1.2. This changeover from loiter to afterburner will naturally be accomplished during flight by data link transmission.

Data Link

The data link that is used to provide positive control and guidance of the missile to the target area at extended ranges. The link is also used to pass data and target imagery back to the Command Centre.

Warhead

The missile is also being designed to accept a variety of warheads, under consideration are warheads from the AIM-9 Sidewinder and the blast fragmentation and Metal Augmented Charge (MAC) versions of the AGM-114 Hellfire air-to-surface missiles. It is expected that the warhead used in the missile will be dependent upon the systems requirements at the time of firing, that being fully conversant with the threat.

Seeker

Under consideration are seekers from the Stinger Block 1 (SAM), the Navy Rolling Airframe Missile (RAM) and the AIM-9L, M and X variants of the Sidewinder air-to-air missile. The developers are also considering the use of a RF seeker, the Advanced Electronic Scanning Array.

As with the warheads, the seekers used on the operational rounds are expected to be dependent on the threat. If not interchangeable, then at the very least an assortment of missiles may be carried with various warheads and seekers. This capability allows systems like AERAM to operate in a densely populated airspace consisting of commercial, private, and military aircraft.

Status

In development; the programme is currently in the Science and Technology Phase, but army officials had hoped to transition it to system design and development by 2009 however, it is unclear if this happened by the target date. Assuming that all goes well with the development of the system, then operational capability could be achieved around 2015/17 time frame.

Until then, Raytheon is under a five-year contract to develop the technology. The goal is to build seven missiles and test them before the system design phase.

Further exercises with the SACS launched from a submerged submarine are believed to have taken place during 2006. These exercises involved the use of the AIM-9X missile in the air-defence role.

Contractor

Raytheon Company, Missile Systems division.

APKWS

Type

Universal (land/sea/ground-based) missile system

Development

The original contract for the basic Block I Advanced Precision Kill Weapon System (APKWS) system was let with General Dynamics Armament and Technical Products, a business unit of General Dynamics. The USD53.8 million contract was let by the US Army Aviation and Missile Command, Redstone Arsenal, Huntsville, Alabama. The contract period of performance initially extended through 31 July 2005.

The Low Cost Precision Kill (LCPK) Advanced Technology Demonstration (ATD) programmes was completed in October 2002. The technology that was demonstrated in this ATD became the APKWS in 2003.

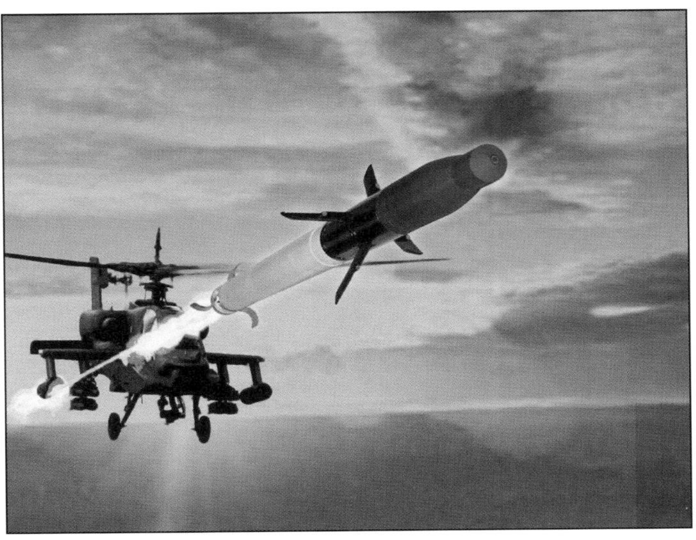

Advanced Precision Kill Weapon System (APKWS) (BAE Systems) 0526320

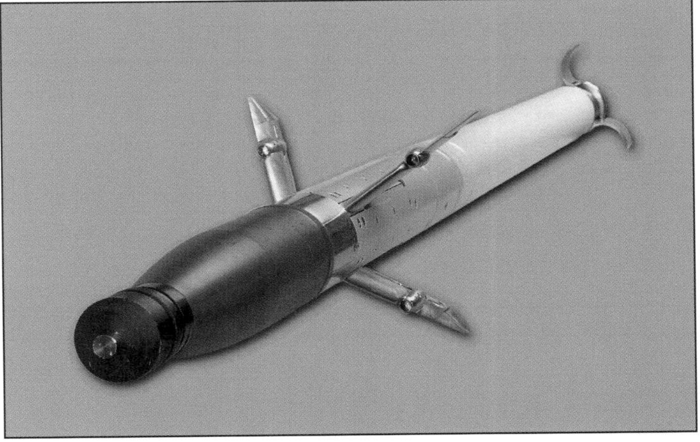

APKWS II round showing fin-mounted laser seekers (BAE Systems) 1155219

Flight testing of the APKWS began at Eglin Air Force base in Florida in 2002, culminating in the fifth successful firing from an M260 rocket launcher during September 2004 at a stationary target 5,000 m being illuminated by a third party laser. Earlier tests were conducted at the Army's Yuma Proving Grounds at targets ranging from 1,500 to 5,500 m.

Due to changing customer requirements, the APKWS programme was curtailed in January 2005 with the decision being reversed by May of the same year. Production of the Hydra-70 Rocket, on which the APKWS system was based, still continues.

A further US government requirement of the of the APKWS programme is in an advanced stage of development and is known as the APKWS II. BAE Systems Electronic & Integrated Solutions, Nashua, has been selected as prime contractor for this programme, and the company has received an initial USD96.1 million for development, demonstration qualification and limited production of the laser-guided rockets.

As of 19 March 2007, BAE Systems reported that the APKWS II was on cost and on track, but added that funding has been zeroed out in the FY08 budget request. In response to this, BAE are putting the programme on hold.

In November 2009, BAE announced that the APKWS had entered its final phase of testing intending to confirm both production capability and reliable accuracy. Previous trials of the system, both air and ground launched, had proved successful on at least 18 firings since 2002. Qualification testing and demonstration firings were due to take place in 2010.

In July 2010, the APKWS programme finally entered low-rate production.

By March 2012, the APKWS II system achieved IOC and was sent to Afghanistan with the Marine Corps.

In July 2012, a full rate production contract was awarded by the US Navy to BAE Systems with first deliveries in October the same year.

Description

The APKWS programme integrated an advanced seeker on a previously unguided 2.75 inch (69.85 mm) Hydra-70 rocket.

Block I of the APKWS programme is to integrate a HE warhead (M151/MK-152). The required range is 5 km with an expected goal of 6 km or nearly 3.8 miles.

Future Block upgrades to the APKWS programme includes an improved rocket motor to extend the range beyond 6 km and other warheads.

The US Army and Special Operations Forces (SOF) are thought to have desired an adaptation to the APKWS programme, the adaptation being the use of a flechette based warhead (M255A1/E1) over HE. The flechette warhead expels approximately 1,200 (3.9 g) steel flechettes in an expanding pattern, akin to a shotgun. The principle was also used in the US APERS class of artillery projectiles (M494, M546, M580 and M581). This type of warhead is ideal as an Air Defence Weapon System against rotary and fixed wing aircraft, UAVs and other low-flying objects but is currently touted as an anti-personnel weapon.

BAE Systems of Nashua, New Hampshire, US, has been competitively selected as the primary contractor and for the development and provision of the mid-body guidance system. The APKWS allows for the use of flechette and HE warheads without interference to the warhead effectiveness.

BAE Systems uses a mid-body guidance approach that employs its Distributed Aperture Semi-Active Laser Seeker (DASALS) with the seeker mounted in the fins rather than in the head of the missile, this mid-body seeker provides an exponential increase in effectiveness by guiding the APKWS to a precise point in space and expelling the Flechette at the optimum distances for target coverage. During environmental testing, the wings and optics that are sealed within the mid-body guidance section were also protected from sand, dust vibration, ice and other environmental hazards that are likely to be found in combat scenarios.

APKWS II is a 2.75 inch laser-guided rocket that will provide crews of the US Army Apache and Marine Corps Cobra attack helicopters and other platforms with precision-strike capability against targets that do not require a seven inch Hellfire missile, an option that is not presently available. The missile is a low-cost alternative that is designed to destroy non-armoured targets that are close to civilian assets and/or friendly forces.

Specifications

	APKWS	APKWS II
Dimensions and weights		
Diameter		
body	-	70 mm (2.76 in)
Performance		
Range		
max	3.2 n miles (6 km; 3.7 miles)	3.0 n miles (5.5 km; 3.4 miles) (est.)
Ordnance components		
Warhead	HE flechette	-
Guidance	semi-active, laser	-
Propulsion		
type	solid propellant	solid propellant (Mk 66 (Mod 4) motor)

Status

The fifth in the series of firing trials took place in September 2004, further flight trials continued through 2005 before moving into the demonstration phase.

Further flight testing with the missile in operational configuration took place in May 2007, these tests were held in partnership with the US Navy programme office and were considered successful when the missile struck its target with the 2 m requirement.

In November 2009, in final phase of testing with qualification trials 2010. Low level production contract started in July 2010.

Contractor

BAE Systems Electronic Solutions Nashua, New Hampshire, US (primary contractor plus guidance and control section).

General Dynamics (rocket motor and warhead).

Laser Weapon Systems

Germany

Rheinmetall Laser Air Defence System (RLADS)

Type
Laser weapon systems

Development
During late November 2011, Rheinmetall demonstrated the operational potential of combining a 10 kW laser weapon installed on a Skyshield gun turret. During this demonstration in which a Unmanned Aerial Vehicle (UAV) was engaged and brought down, the laser utilised the Oerlikon Skyguard 3 Fire Control Unit for target tracking and laying on. The cost involved in the development for this project is primarily from Rheinmetall with some funding from the German Bundesamt fur Wehrtechnik und Beschaffung (BWB).

Description
The RADLS has been designed with Counter Rocket Artillery and Mortar (CRAM) in mind in order to counter the threat from incoming rockets in an air defence scenario. The system when complete is expected to be containerised to allow it to be rapidly transported by truck or helicopter. Rheinmetall Weapons and Munitions are responsible for the power supply, cooling systems, solid state laser and the beam-forming element. The RADLS is considered to be a precision weapon that can offer graduated levels of escalation that range from non-lethal to lethal and also a very low risk of collateral damage. Rheinmetall expects a high-energy laser weapon system with an output of 100 kW to be available within the next three to five years. Further trials were expected between 2012 and 2016.

Variants
A 1-kW laser weapon module mounted on a TM 170 type vehicle demonstrated for use against IEDs.

Status
As of February 2013, the system is in development at the full scale demonstrator stage.

Rheinmetall Laser Air Defence System (RADLS) (Rheinmetall) 1445527

Rheinmetall Laser system truck mounted for counter IEDs (Rheinmetall) 1445528

Contractor
Rheinmetall Air Defence of Switzerland.
Fraunhofer Institute for Applied Optics and Precision Engineering (IOF).
Rheinmetall Weapon and Munitions and Air Defence Division of Germany.
Ochsenboden Proving Ground, Switzerland.

India

India Laser Weapon Systems

Type
Laser weapon systems

Development
India is planning to develop laser weapon systems that can be Submarine, Surface Ship, Airborne and Land-Based to intercept an incoming missile launched from a distance of up to 2,000 km away. This development will be headed by the Defence Research and Development Organisation (DRDO) working with The Laser and Science Technology Centre (LASTEC).

Initially the first system is expected to be associated with India's Anti-Ballistic Missile systems that are currently in development, the first laser will take at least 10-15 years to come on-line.

It is assessed that the first of these systems will be designed to destroy ballistic missile targets carrying nuclear or conventional warheads in their boost phase. During an interview with the LASTEC Director Anil Kumar Maini in late August 2010, it was confirmed that India intends to use laser (Directed Energy Weapons - DEW) to engage an enemy missile in its boost and/or terminal phase.

A project known simply as KALI [Kilo Ampere Linear Injector] is a linear electron accelerator that was first identified as far back as 1985; this was a DRDO/Bhabha Atomic Research Centre (BARC) programme that finally started life in 1989. Although not in itself a laser weapon system as we know it today, the potential for the system to fulfil the role as a military DEW system is still there. It is assessed that if weaponised, it is very likely that KALI would be integrated into the Ballistic Missile Defence programme PAD/AAD.

India is conducting tests on a developmental 25 kW laser system to hit a missile during its terminal phase at a distance of 5-7 km. The laser is designed to heat up the skin of the target to between 200°C to 300°C, with a resultant warhead detonation or fuel-tank structural failure.

LASTEC is also working on a vehicle-mounted gas dynamic laser based DEW system under the project name of Aditya. This project should be complete within three years and will be used as a technology demonstrator. Ultimately India has designs on developing solid-state lasers for use by all three of the armed forces.

A further air-defence system known simply as Dazzler has already been tested; the Dazzler can engage aircraft and helicopters at a range of 10 km and will be ready for induction within the next two years according to DRDO scientists. Taking into account the two year induction from the time of reporting, then this would have been 2012/13, however, there has been no evidence of any laser weapon system coming from India in this time period.

Variants
None known.

Specifications

Aditya Weapon System

Type:	laser defence system
Range:	
(boost phase)	2,000 km
(terminal phase)	5-7 km

Dazzler Weapon System

Type:	laser defence system
Range:	>10 km

Status
As of February 2013, the laser weapon systems are still at the very early stages of research and development.

Contractor
Defence Research and Development Organisation (DRDO).
The Laser and Science Technology Centre (LASTEC).
Bhabha Atomic Research Centre (BARC).

Israel

Iron Beam

Type
Laser weapon system

Development
Israel has been working on a High Energy Laser (HEL) weapon system with the US for many years, originally with the Tactical High Energy Laser (THEL) and then the MTHEL, a mobile variant of the THEL. It is logical therefore to expect that the Iron Beam has been born on the back of this knowledge but with a more up-to-date capability.

Description
The Iron Beam is an HEL weapon system that has been designed to engage targets at very short-range of 4,500 m or less. The system will deal with threats that fly outside the scope of the Iron Dome radar-guided rocket system.

Iron Beam uses a pair of multi-kilowatt solid-state lasers to engage incoming projectiles at very short range. First pictures of the system show the HEL mounted in two ISO containers but Rafael admit that the system could be produced to customers requirements and easily be mounted on armoured vehicles or in some other configuration.

The system comprises:
- air defence radar
- a command-and-control unit
- two HEL systems.

A typical engagement would consist of the Iron Beam's air defence radar acquiring the target, then a thermal camera takes over the tracking until it is engaged simultaneously by both HELs who focus the beams on the target on an area about the size of a coin.

Variants
None known.

Specifications

Type:	High Energy Laser (HEL) air-defence weapon
Maximum range:	4,500 m

Status
January 2014, still in development but is expected to become operational as part of the layered defence of Israel sometime in 2015.

Contractor
Rafael Advanced Defence Systems Ltd.

Artists drawing Iron Beam (Rafael) 1529220

United States

HORNET

Type
Laser weapon system

Development
Northrop Grumman are working on a new Hazardous Ordnance Engagement Toolkit (HORNET) quick-reaction directed-energy laser weapon system that has been designed to defeat small, supersonic missiles targeted at aircraft landing and taking off from military and civilian airfields.

HORNET has been conceived from the company's Mobile Tactical High Energy Laser (MTHEL) (MTHEL is the mobile follow-on to the THEL demonstrator system that is now complete) that is a US/Israeli programme designed to defeat incoming tactical munitions such as MLRS. The HORNET has been described in open press as a slightly different, upgraded MTHEL configuration that can protect an airport against a full range of man portable missile threats.

Like the MTHEL, the HORNET is to be based on a megawatt-class deuterium-fluoride chemical laser that can fire a lethal energy beam at a target at the speed of light.

It has been reported by Northrop Grumman that considerable analysis of the concept has been undertaken and that this has been briefed to the US Air Force and homeland security representatives. It is expected that one strategically placed HORNET could protect medium-to-large airfields, shielding aircraft during their most vulnerable landing and take-off times.

Description
The system will most likely be mounted on a wheeled transporter for mobility, while at the same time having the capability of static installation. Technology to drive the system for both the military and civilian use, plus the hardware involved will probably be identical, although it must be expected that the military version will have to be militarised for operational use within a theatre of conflict, such as Afghanistan.

Variants
The HORNET system has been derived from the MTHEL. For more information please check the record on the system in this section.

Specifications
HORNET

Type:	directed energy laser weapon system
Class:	megawatt-class deuterium-fluoride chemical laser

Status
In development as of August 2012 and believed to be still at the conceptual stage.

Contractor
Northrop Grumman.

Laser Area Defence System (LADS)

Type
Mobile ground-based laser weapon system

Development
The Tucson-based Raytheon Missile Systems has developed - without government backing - a new laser air-defence beam weapon that has been designed to engage and destroy rockets, mortars and small missiles with bursts of laser light. The system is known today as the Laser Area Defence System (LADS). The system has been described by Raytheon in that it is intended to provide a short-range point-defence weapon as a replacement for the Phalanx Close in Weapon System (CIWS), utilising existing Phalanx hardware and systems.

In early January 2007, Raytheon reported that it had developed and tested an early prototype solid-state laser adapted to an existing, Raytheon-produced ship-defence system: Phalanx. This test involved the destruction of a static 60 mm HE mortar round at 715 m - almost three quarters of a kilometre.

Tests carried out during 2007 against a moving target at tactical ranges in excess of 500 m, proved the concept of operations and the lethality of the system.

In January 2008, further tests proved the laser capable of taking on targets such as mortars. Another demonstration of the system's capability took place during the Autumn of 2012.

The test allegedly proved that solid-state lasers used in everyday civil applications could be made powerful enough for use in advanced weapon systems.

The LADS system is powered by a commercially-available generator or grid supplies and has reportedly been developed over a period of just six months using off-the-shelf solid-state lasers, coupled with commercially available optics technology.

Shown publicly for the first time at the Farnborough Air Show in July 2010, Raytheon reported that the 50 kW beam produced by the LADS system has destroyed several (at least four) Unmanned Aerial Vehicles (UAVs) in recent tests conducted at San Nicolas Island off the coast of California.

Description
LADS uses current solid-state laser technology that is available Off-The-Shelf (OTS) for tackling the threat of mortars and Katyusha-class rockets used by non-regular forces in Gaza and the West Bank.

Laser Area Defence System (LADS) (Raytheon) 1308135

The laser has been mounted on the carriage of a Phalanx Close-In-Weapon-System (CIWS). The Phalanx CIWS is a land- and ship-based defence system designed to automatically track incoming missiles with radar and InfraRed (IR) technology and destroy them with a hail of projectiles from it's electrically powered 20 mm six-barrelled Gatling gun.

The Phalanx CIWS, like the LADS, is a point defence system and therefore it would seem reasonable to assume that the radar associated with the Phalanx CIWS could also be used for target tracking and laying-on the LADS.

After completion of development, the LADS system is expected to be fully mobile and have the operational capability to simultaneously engage multiple targets at increased tactical ranges.

Raytheon's laser system is not without its inherent problems however, at sea in the moist oceanic air-conditions or when in mist/fog, much of the laser's energy may be absorbed or scattered before it reaches its target, therefore making it less effective (reducing its range), or ineffective.

There are also some materials that can absorb the laser light, that if used as a surface covering on such targets, will also defeat the system.

Variants

None at present are known of.

Specifications

Type:	solid-state laser
Length:	not available
Range:	>500 m (trials)

Status

In developmental trials, the first prototype complete.

By July 2008, trials proved the system could engage and destroy targets such as mortar rounds and MLRS, and by the 29 of July 2008, the Jerusalem Post was reporting that the system was deployed near the Gaza Strip in Israel. It has also been reported that the US military in Iraq has ordered LADS.

As of November 2009, the LADS was still officially in development but is not expected to become operational in the near future.

Video footage of live testing conducted at San Nicolas Island off the coast of California was shown for the first time at the Farnborough Air Show in July 2010, the video footage covered the shooting down of at least four UAVs at significant distances.

Contractor

Raytheon Missile Systems, Tucson, Arizona, US.

Laser Avenger

Type

Laser weapon systems

Development

Boeing Missile Defence Systems, Arlington, Virginia, has funded an indigenously developed Laser Avenger weapon system that could be at the Initial Operational Capability (IOC) within 12 months if funded by the Pentagon. The Boeing Laser Avenger is mounted on the AN/TWQ-1 Avenger combat vehicle developed by Boeing in Huntsville Alabama.

First proof of concept and prototype low power exercise firings took place in Huntsville Alabama, at the Redstone Arsenal during September 2007 against seven types of static targets including Improvised Explosive Devices (IED), Unexploded Ordnance (UXO) and Unmanned Aerial Vehicles (UAVs). These test firings were followed by further testing at the White Sands Missile Test Range in New Mexico in December 2008.

Since the 2007 firings, Boeing has doubled the laser power and reduced the development cost by 35 per cent and increased the ruggedness and ease of manufacture. Furthermore, Boeing has also added Target Tracking (TT) software aimed at the counter-UAV capability demonstration.

By late January 2009, information was released that further testing against live UAV targets had recently taken place at the White Sands Missile Range in New Mexico, where three UAVs were detected then tracked by the system and then one engaged and destroyed. The reported range was described as "an operationally relevant range".

Developmental trials were carried out in April 2009 utilising the Avenger Laser to heat a small airframe target such as a UAV, in order to improve the on board Stinger missile's lock-on capability have been observed on the range. A Stinger round then successfully engaged the UAV target at a range of about 1,000 m.

Description

The Laser Avenger is an InfraRed (IR) solid state laser that is designed to engage and destroy the smaller variety of UAV that has proved so difficult for the standard Surface-to-Air Missile (SAM) in the past. The prototype is mounted on the standard HMMWV chassis used for the air-defence Avenger normally equipped with a 0.5 inch, Heavy Machine Gun (HMG) and two quadruple-boxed launchers containing FIM-92 Stinger mounted either side of the turret.

The laser is mounted as a replacement for one of the quadruple launchers. The other launcher remains, as does the HMG.

The system is capable of multiple target acquisition against a complex background no details of the detection equipment have yet been released, then tracking and engaging at the point when burn through could logically be expected (depending on target) with a single shot.

Boeing Laser Avenger (Boeing) 1332650

The laser system is completely integrated to the Avenger turret and works independently of on-board wide field of view camera/Forward Looking InfraRed (FLIR) systems. The laser system can work together with on-board wide field of view camera/FLIR systems and slew-to-cue fire-control system of the Avenger turret.

Once a target has been sighted, the laser system will 'zoom in' on the target and use advanced UAV tracking algorithms to acquire and track the target even in heavy background clutter. Only after positive identification will it fire the high energy laser.

Variants

The Mk 38 Tactical Laser System (TLS). Boeing is working with BAE Systems and the US Navy on the Mk 38 TLS that can protect a ship against a small boat or other vessels. The beam director is derived from the Boeing mount on the Avenger air-defence system currently being built in Huntsville, Alabama. The shipborne laser system reportedly fits into the circular base of the Mk 38 gun mount.

Specifications

Laser Avenger

Type:	solid-state, infrared, laser weapon system
Power:	kilowatt-class laser powered by the Avenger's engine alternator (no further details have been released)
Range:	operationally relevant
Number of shots available:	100% duty cycle

Status

In development with IOC expected within 12 months if funded by the Pentagon.

Contractor

Boeing Missile Defence Systems, Arlington, Virginia.
Boeing Combat Systems, St. Louis, Missouri.

Mobile Tactical High Energy Laser (MTHEL)

Type

Ground-based mobile air-defence laser weapon system

Development

Mobile Tactical High-Energy Laser (MTHEL) programme is the follow-on to the Tactical High-Energy Laser (THEL) advanced concept technology demonstrator that was completed in 2000. It was reported that in December 2001, Israel and the US were about to begin talks that would determine the type of MTHEL system they would develop jointly to replace the THEL. The MTHEL testbed is located at the US Army Space and Missile Command's (SMDC) High Energy Laser Systems Test Facility on White Sands Missile Range in New Mexico. In early 2003, MTHEL made the transition from SMDC to the Programme Executive Office for Air and Missile Defence.

The aim of the MTHEL programme is to develop, demonstrate, test, and field the first mobile Directed Energy Weapon (DEW) capable of detecting, tracking, engaging and destroying rockets, artillery and mortar, unmanned aerial vehicles, cruise missiles, short-range ballistic missiles and air-to-ground munitions.

The US Army was expecting to award a contract in FY04 to begin development of the selected MTHEL approach. Brigadier-General Yair Dori, Commander of the Israel Air Defence Force (2003), says that there is a place for MTHEL with the IDF in the future, but at the moment it is too big: "We cannot afford to deploy the MTHEL system in the stage of development that it is in now - principally for the survivability of the system. We need a smaller more mobile solution. Therefore one of the principal goals of the MTHEL programme is to reduce the size by a factor of 5:1 in comparison with the testbed currently at White Sands". No contract is known to have been awarded.

During live firing trials in November and December 2002, MTHEL successfully destroyed 152 mm artillery shells in flight. Engagement of mortar bombs during 2003 has been programmed.

Further tests of the MTHEL took place during May 2004, when the joint US-Israeli mobile laser gun, designed to knock down rockets in flight, successfully tracked a live target in a test in New Mexico. Operators decided against performing a second part of the test and did not try to shoot down the rocket with the laser.

According to the Defence Ministry, the Mobile Tactical High-Energy Laser (MTHEL) test took place at the White Sands Missile range. The primary objective was to test if the mobile system could track a large-calibre rocket. If it did, then operators were to attempt to intercept it with the stationary Tactical High Energy Laser (THEL) as the MTHEL went through the motions.

"Preliminary data indicates a track was established but the secondary objective of destroying the target was not attempted", a ministry statement said. "This was the first attempt to establish a track on this type of target."

Description

MTHEL is a mobile, stand-alone, ground based, directed energy weapon system that is expected to be deployed on a tractor towing vehicle that will contain all the necessary ancillary equipment.

MTHEL is the natural follow-on to the Tactical High Energy Laser (THEL) that is currently being considered for deployment with US forces to counter mortar and rocket attacks.

Status

In development at the White Sands Missile Test Range. MTHEL was not expected to be available for deployment before 2006-2008. At one point, Israel was expected to show an interest in procuring MTHEL units for use on its Northern border, but as of November 2009, there were no positive moves towards this goal. It is probable that any future decision on MTHEL will be delayed even further.

Reports dated January 2006 from the Pentagon have suggested that the US government has terminated the Nautilus programme. One of the reasons given was the system's problems in meeting the US Army's mobility requirement. However, events in July/August 2006 such as the 31-day war against Hizbullah in Lebanon have brought the THEL to the fore again. The Israeli government re-examined the development of the laser-based anti-rocket system with a view to deploy up to six static sites in the North of Israel by the year 2008/9 however, these deployments have not occurred.

Completing the development of the THEL system is estimated to cost USD200 million. As of August 2011, no further information on the MTHEL system is available.

The MTHEL system will remain in JAAD for a further 12 months as a tactical explanation of how the laser air defence systems have got to where they are in 2013/14; after that the file will be archived.

Contractor

TRW.
US Army Space and Missile Command.

Skyguard (Laser Air Defence)

Type

Laser weapon system

Development

Northrop Grumman Space Technology (NGST) Redondo Beach, California announced in July 2006 that it was developing the Skyguard land-based air-defence system for US government agencies and allies that require near-term defence against Short-Range Ballistic Missiles (SRBM), short- and long-range artillery rockets, artillery shells, mortars, Unmanned Aerial Vehicles (UAVs) and cruise missiles.

Development is taking place at the White Sands Missile Test Range, New Mexico, US where the predecessor (THEL) was also developed.

The system was offered to Israel in December 2006 for protection against the 122 mm Katyusha MLRS along the northern borders and Gaza Strip, however probably owing to the now deployment of the Iron Dome CRAM system it will be very unlikely that the Skyguard will ever be accepted into service in Israel. The contract offer reported that the system could be fully operational within six months (June 2007), therefore the editor assumes that the Skyguard system has completed development and is awaiting sale. As of November 2009, no such order has been placed and there are no prospects for any such order in the pipeline.

In October 2006, the US Department of Homeland Security selected the Skyguard as a possible protection for civilian aircraft and commercial airports from the threat of man-portable surface-to-air missiles systems being launched from the ground at either landing or taking off aircraft.

Description

Skyguard is a high-energy laser defensive system based on a chemical laser generator. It has been developed using technical knowledge gained from previous laser defence systems such as THEL. However Skyguard has an increased power ratio and a larger beam making it a much more capable system to engage those targets such as SRBMs and UAVs. The design of the system makes it ideal for the defence of not only troops deployed in theatre but also large military installations, civilian population and/or industrial areas.

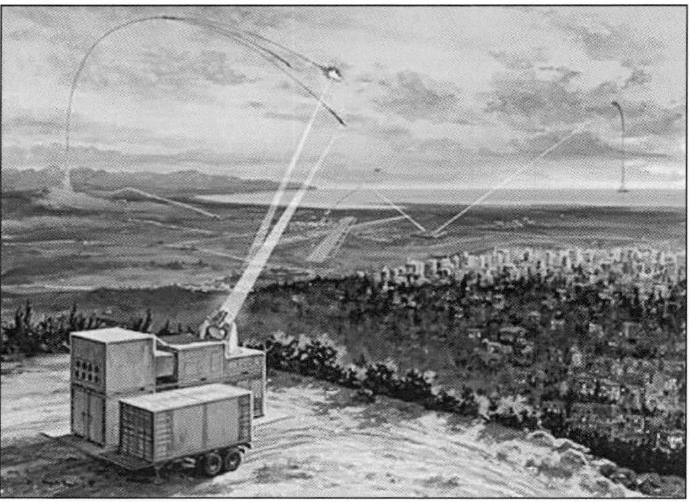

Skyguard land-based laser air-defence system (Northrop Grumman) 1162234

With its modular construction and compact transportable laser weapon, Skyguard is a multi-mission soldier-operated system that has been fully designed for field deployment and operations.

It has been reported that a single Skyguard system can defend deployed forces, large military installations, civilian population or industrial areas, it is also reported that only one Skyguard can form a protective shield of about 10 km.

Variants
Skyguard is derived from the Tactical High Energy Laser (THEL).

Specifications

Type:	land-based air-defence system
Maximum range:	10 km

Status
Developed at White Sands Missile Test Range, New Mexico, US.

In December 2006, this system was offered to Israel for protection of its northern borders and also along the Gaza Strip. The first Skyguard could be operational in Israel within 18 months of the first order being placed.

During 1997 the Skyguard system was rejected by Israel in favour of the Iron Dome system from Rafael, however, by late December 2007, the then Prime Minister Ehud Barak requested a review of the laser weapon that was to be complete by 2008. No further details have been released on this system.

Contractor
Northrop Grumman Space Technology (NGST).
RADA Electronic Industries Ltd., Netanya, Israel, (subcontractor).

Talon

Type
Laser-based air-defence system

Development
A collaborative project between Northrop Grumman and the then United Defense (now BAE Systems Land & Armaments) was undertaken to provide a vehicle-mounted laser air-defence system under the concept name of Talon. The Talon concept is a vehicle-based Counter Rocket And Mortar (CRAM) system designed to utilise a 100 kW solid-state laser to shoot down rockets, mortars, artillery and enemy Unmanned Aerial Vehicles (UAVs) and Land Attack Cruise Missiles (LACM's). The vehicle uses an all terrain prototype 8 × 8 wheeled vehicle that United Defense was at that time developing. The crew of the vehicle will consist of a driver, laser operator and possibly a third to rotate in and out with the other crew members on long missions.

In late May 2005, the US Army's Space and Missile Defense Command released a formal request for proposals for a 100 kW laser with a high-beam quality, as well as concepts for integrating the laser onto a ground vehicle or a UAV. The UAV would have a mission of air-to-ground precision strike.

In July/August 2006, Northrop Grumman commenced construction of a dedicated facility at their Directed Energy Production Plant at the Space Technology sector's Space Park campus for low-rate production of high-energy lasers for military weapon systems. One of the first applications to benefit from this facility will be the Joint High-Power Solid-State Laser (JHPSSL) Phase programme, where a laser power of 100 kW (the same as the requirement for the Talon) will be demonstrated. Other lasers with a greater output than 100 kW will also be tested here along with electric lasers for various military platforms, such as armoured combat vehicles. The JHPSSL system is designed to accelerate solid-state laser technology for military uses. This includes force protection and precision strike missions for air-, sea- and ground-based platforms.

The JHPSSL development programme's timeline follows:

- Phase 1 (2002): Northrop Grumman addresses risk reduction of the technologies necessary to obtain high power and beam quality simultaneously and is awarded Phase.
- Phase 2 (2005): Northrop Grumman scaled Phase 1 technologies to greater than 25 kW using two chains, showing further scalability to 100 kW and beyond. The company is awarded Phase 3.
- Phase 3 (February 2007): first demonstration milestone: JHPSSL team enters integration and test phase after exceeding all demonstration requirements for the first gain module; it produced a power level of more than 3.9 kW, operated for 500 seconds at 20.6 per cent efficiency. There are four gain modules per each laser chain.
- Phase 3 (December 2007): second demonstration milestone: First laser chain successful demonstration, exceeding all target requirements.

Description
The Talon is a highly mobile, ground-based, solid-state laser system capable of providing mobile forces with a multirole defence against rocket, artillery, mortar and other aerial attacks such as Unmanned Aerial Vehicles (UAVs).

The laser weapon is to be mounted on a manned fully mobile ground vehicle similar to those being produced by then United Defense for the US Army's now cancelled Future Combat System (FCS) programme. The most difficult engineering challenges were then, according to officials from United Defense, the power management in the ground vehicle, which provides power to the all the subsystems on the vehicle, including the hybrid electric drive and laser, and the dissipation of the heat created by the laser when operating.

Artists impression of laser air-defence system (Northrop Grumman) 1116215

Future Combat System Vehicle (United Defence) 1116214

In addition to providing a protective shield against incoming threats, Talon's high resolution optics will provide launch point prediction with high accuracy, this will allow for decisive counter-battery fire and will improve the freedom of manoeuvre for the joint force.

Variants
A similar system developed by ATK for the US Navy specifically designed against anti-ship cruise missile attacks was delivered in May 2006 to the Naval Research Laboratories for testing. This particular directed energy weapon uses a pulse chemical laser in a cylindrical package.

Status
In the prototype stage. As of January 2014, no further data has been released on this system.

Contractor
Northrop Grumman Corporation (laser).
The then United Defense, now BAE Systems Land & Armaments (vehicle).

Vesta

Type
Laser weapon system

Development
Northrop Grumman from Redondo Beach, California, have announced the development of the Vesta, Compact, High-Power, Solid-State Laser (CHPSSL) for multiple military missions including the protection of fixed sites from precision strike by cruise missiles, UAVs and manned aircraft. The Vesta system could also be used for the protection of Naval and airborne assets.

Vesta has been developed by Northrop Grumman Space Technology using internal company funds and without a government contract. The continuing work on the Vesta represents 'leap-ahead' technology and is expected to greatly shorten the timeline for lasers to go from the laboratory onto the battlefield. Furthermore, the Vesta requires only a mains electricity power source to run, this will ensure significant reduction in the size and weight of the laser and will enable the system greater mobility, something that is lacking on the battlefield today, whilst retaining beam time and quality that can be focused onto the target.

Tests at the White Sands Missile Range, New Mexico during early 2009, proved that Northrop's laser can generate a beam of laser-light with an average power output of 15 kW for more than 20 minutes. Vesta has been designed to operate at this power and beam quality level indefinitely. This laser is powerful enough to engage and shoot down enemy missiles, rocket artillery, drones and mortar bombs. With the power of the laser required to engage these targets, two other problems are still left to be resolved. The first of these is heat dissipation, the conversion of electrical power to laser-light is not 100 per cent efficient so a lot of waste 'heat' is generated. The second is size; the laser needs to be miniaturised to make it small enough to be fitted to a wide range of practical vehicle types.

Description

As of March 2007, the Vesta has only been shown publicly as a laboratory mock-up mounted on a breadboard.

Laboratory testing of the system have proved excellent beam quality at consistent power levels whilst continuously running. The Vesta has demonstrated more than 20 minutes of continuous operation with no degradation.

The Vesta laser uses gain modules to produce the 15 kW output power by combining modules to form amplifier chains, these chains can be used as building blocks for higher power lasers. Individual gain modules have been tested at greater than 4 kW continuous output. The system could be used on platforms such as B-2, C-130, UAVs, Army tactical vehicles, warships and even unmanned warships.

Under the Joint High Power Solid State Laser (JHPSSL) Phase 3 programme that is funded by the US Navy, Army and Air Force, the amplifier chains will be used to reach the 100 kW required for the development of a future solid-state system.

Vesta's critical features include:

- High power;
- Beam quality;
- Long run time;
- Compact laser;
- Low weight.

Variants

Vesta II

The Vesta II is a compact, stand-alone, 15 kW high-power, solid-state laser developed for the US Air Force Research Laboratory, at Kirtland Air Force Base, NM.

The system is a compact, high-power laser available for laboratory use as a testing device for:

- Lethality testing;
- Atmospheric propagation testing;
- Long-range imaging;
- Laser weapon applications.

Vesta II is now part of the Joint High Power Solid State Laser (JHPSSL) programme in which the US military has set a power level goal of 100 kW to shoot down rockets, missiles, artillery shells and mortar bombs.

Firestrike

Based on a laser beam combining architecture validated by Northrop Grumman over several years with the Joint High Power Solid State Laser (JHPSSL) programme, the Firestrike laser offers war fighters a 15 kW fieldable laser as well as a combinable Line Replacement Unit (LRU) building block for much higher power.

Specifications

Vesta

Type:	compact, high-power, solid-state laser
Power source:	electrical
Power level:	greater than 12.5 kW, 15 kW, indefinitely operating in steady state, continuous wave
Run time:	as required by mission[1]
Beam quality:	1.5 tight beam focusing[2]

Note:
[1] Run time measures how long the laser can fire.
[2] Demonstrated beam quality of less than 1.3 times the diffraction limit at 15 kW power level and indefinite run time.

Status

As of January 2014, the Vesta remains in development at the laboratory stage and also proving capabilities at the White Sands Missile Range.

Contractor

Northrop Grumman Space Technology, primary contractor.

Artillery and Air Defence In Service

Inventory

Artillery and Air Defence In Service

Introduction
Listings by country of equipment in service, held in reserve or due to enter service some time in 2014/2015.

Artillery: covers towed guns and howitzers, towed anti-tank guns and self-propelled artillery systems, tracked and wheeled.

Air Defence: covers static missiles, self-propelled gun and missile weapons and gun systems going back to the 1960s and 70s. These systems have been incorporated in order that customers can follow trends in Air Defence for interested countries. This information should also help in forecasting future trends in Air Defence.

This data is based on the latest available information at the time of going to the press and in some cases updates that are given in the main text.

ARS - artillery rocket system
ATG - anti-tank gun
FG - field gun
FH - field howitzer
GH - gun howitzer
Ma - marines
MRS - multiple rocket system
NG - national guard
R - reserve
SPG - self-propelled gun
SPGH - self-propelled gun howitzer
SPH - self-propelled howitzer
UG - upgraded

Afghanistan

Artillery
122 mm D-30 howitzer
130 mm M-46 FG
152 mm D1 howitzer
122 mm BM-21 MRS
220 mm Urgan MRS

Air Defence
S-75 Dvina
S-125 Pechora-M
37 mm (Single) M1939
57 mm (Single) S-60
85 mm (Single) KS-12
100 mm (Single) KS-19
9K32 Strela-2
9K32M Strela-2M
HN-5A
9K35 Strela-10
9K36 Strela-3
9K310 Igla-1
9K310 Igla-1E
FIM-92A Stinger
12.7 mm M53
14.5 mm (Single) ZPU-1
14.5 mm (Twin) ZPU-2
14.5 mm (Quad) ZPU-4
23 mm (Quad) Shilka
23 mm (Twin) ZU-23-2

Albania

Artillery
152 mm Type 66 GH

Air Defence
HQ-2A
9K32M Strela-2M
HN-5A
23 mm (Single) ZPU-1
23 mm (Twin) ZU-23-2
37 mm (Single) M1939
57 mm (Single) S-60
85 mm (Single) KS-12
100 mm (Single) KS-19

Algeria

Artillery
SU-100 SPG
122 mm 2S1 SPH
152 mm 2S3 SPGH
57 mm M1943 ATG
85 mm D-44 FG
100 mm T-12 ATG
122 mm M1931/7 howitzer
122 mm M1938 howitzer
122 mm D-30 howitzer
122 mm D-74 FG
130 mm M-46 FG
152 mm M1931/7 GH
122 mm BM-21 MRS
140 mm BM-14-16 MRS
240 mm BM-24 MRS
300 mm BA 9A52 Smerch MRS

Air Defence
S-75 Dvina
S-125 Pechora-M
S-125 Pechora-M (5V27)
S-300PMU-2
2K12 Kvadrat
2K12 Kvadrat (3M9)
Pantsyr-S1
Pantsyr-S1 (9M311)
9K33 Osa-AK
9K33 Osa-M (9M33)
9K33 Osa-AK (9M33)
9K32 Strela-2
9K32M Strela-2M
2K12 Kub
9K31 Strela-1
9K31 Strela-1 (9M31)
14.5 mm (Twin) ZPU-2
14.5 mm (Quad) ZPU-4
20 mm (Various) Unknown types
23 mm (Twin) ZU-23-2
23 mm (Quad) ZSU-23-4
37 mm (Single) M1939
57 mm (Single) S-60
85 mm (Single) KS-12
100 mm (Single) KS-19
130 mm (Single) KS-30

Angola

Artillery
122 mm 2S1 SPH
152 mm 2S3 SPH
76 mm M1942 FG
85 mm D-44 FG
122 mm D-30 howitzer
130 mm M-46 FG
152 mm D-20 GH
122 mm BM-21 MRS
140 mm BM-14-16 MRS
122 mm RM-70 MRS
240 mm BM-24 MRS

Air Defence
S-75 Dvina
S-125M Pechora-M
S-125M Pechora-M (5V27)
2K12 Kvadrat
2K12 Kvadrat (3M9)
9K33 Osa
9K33 Osa (9M33)
9K31 Strela-1
9K31 Strela-1 (9M31)
9K35 Strela-10
9K35 Strela-10 (9M35)
23 mm (Twin) ZU-23-2
57 mm (Twin) ZSU-57-2
9K32 Strela-2
9K32M Strela-2M
9K36 Strela-3
9K310 Igla-1
14.5 mm (Single) ZPU-1
14.5 mm (Twin) ZPU-2
14.5 mm (Quad) ZPU-4
23 mm (Twin) ZU-23-2
23 mm (Quad) ZSU-23-4
37 mm (Single) M1939
57 mm (Single) S-60

Argentina

Artillery
155 mm Mk F3 SPG
105 mm Model 56 P howitzer (and Ma)
105 mm M101 howitzer (and Ma)
155 mm M114 howitzer
155 mm CITEFA Mod 77 howitzer
155 mm CITEFA Mod 81 howitzer
155 mm CITEFA L 45 CALA 30/2 howitzer
127 mm CP3 MRS
127 mm SAPBA MRS
105 mm Pampero MRS (and Ma)
160 mm LAR-160 MRS

Air Defence
9K32 Strela-2
9K32M Strela-2M
20 mm (Twin) TCM-20
20 mm (Twin) GAI-D01
35 mm (Twin) GDF-2
9K32 Strela-2
9K32M Strela-2M
Blowpipe
Roland
Roland 2
Tigercat
RBS 70
20 mm (Twin) Rh-202
30 mm HS 831
40 mm L/60
40 mm L/70
90 mm M117
30 mm HS 816

Armenia

Artillery
122 mm 2S1 SPH
152 mm 2S3 SPGH
85 mm D-44 gun
100 mm T-12 ATG
122 mm D-30 howitzer
152 mm D-1 howitzer
152 mm D-20 GH
152 mm 2A36 gun
122 mm BM-21 MRS
273 mm WM-80 ARS

Air Defence
S-75M Volkhov
S-125 Pechora-M
S-300PT
S-300PM
S-300PM (5V55R)
S-300V1
2K11 Krug
2K11 Krug (3M8M1)
9K32 Strela-2
9K32M Strela-2M
9K33 Osa
9K33 Osa (9M33)
9K31 Strela-1
9K35 Strela-10
9K36 Strela-3
9K38 Igla
9K38 Igla launchers

Australia

Artillery
105 mm Hamel Light Gun
155 mm M198 howitzer
155 mm M777 howitzer

Air Defence
FIM-43C Redeye
RBS 70
RBS 70 Mk 2
Rapier B1X
Rapier-1
Rapier B-1M

Austria
Artillery
155 mm M109/M109A2 SPH
155 mm M109A5Oe SPH
Air Defence
20 mm (Single) GAI-B01
35 mm (Twin) GDF-002
35 mm (Twin) GDF-005
40 mm (Single) L/70
40 mm (Twin) M42
Mistral

Azerbaijan
Artillery
85 mm D-44 FG
100 mm T-12 ATG
120 mm 2S9 SPM/H
122 mm 2S1 SPH
122 mm BM-21 MRS
122 mm D-30 howitzer
130 mm M-46
152 mm 2A36 gun
152 mm 2S3 SPH
152 mm D-20 gun howitzer
203 mm 2S7 SPG
Air Defence
S-75M Volkhov
S-125 Pechora-M
S-200 Angara
S-300PMU-2
2K11 Krug
9K32 Strela-2
9K32M Strela-2M
9K34 Strela-3
9K310 Igla-1
9K33 Osa
9K31 Strela-1
9K35 Strela-10

Bahrain
Artillery
203 mm M110A2 SPH
155 mm M109A5 SPH
105 mm Light Gun
155 mm M198 howitzer
227 mm MLRS
Air Defence
35 mm (Twin) Generic
40 mm (Single) L/70
Crotale
Crotale (R-440)
FIM-92A Stinger
I-HAWK
I-HAWK (MIM-23B)
RBS 70

Bangladesh
Artillery
6 pounder ATG
76 mm Type 54 FG
105 mm Model 56 P howitzer
122 mm Type 86 (D-30) howitzer
122 mm Type 54 howitzer
122 mm Type 56 FG
130 mm Type 59-1 FG
155 mm NORA SPG
Chinese MRS
Air Defence
HN-5A
FB-6A
QW-2
14.5 mm Generic
20 mm HS-804
37 mm (Twin) Type 55
57 mm (Twin) Type 59
FM-90
LY-80
Buk-M1-2
FM-90N
FM-90N (R-440)

Belarus
Artillery
100 mm T-12 ATG
120 mm 2S9 SPM/H
122 mm 2S1 SPH
152 mm 2S3 SPH
152 mm 2S5 SPG
152 mm 2S19 SPG
203 mm 2S7 SPG
122 mm D-30 howitzer
152 mm D-20 GH
152 mm D-1 howitzer
152 mm 2A65 howitzer
152 mm 2A36 gun
122 mm BM-21 MRS
220 mm BM 9P140 MRS
300 mm BM 9A52 Smerch MRS
Air Defence
S-75M Volkhov
S-125 Pechora 2T
S-200 Angara
S-300PM
S-300PM (5V55R)
9K32 Strela-2
9K32M Strela-2M
9K33M Osa-AKM
9K37 Buk
S-300V1
9K35M Strela-10M
9K34 Strela-3
9K331 Tor-M1
9K332 Tor-M2 possibly Tor-M2E variant
9K310 Igla-1
9K38 Igla
Stalker-2T
23 mm (Quad) ZSU-23-4M5

Belgium
Artillery
105 mm LG 1 Mk II gun
Air Defence
I-HAWK
I-HAWK (MIM-23B)
Mistral 1
Mistral-1 (ATLAS launchers)
20 mm M167
35 mm Gepard
20 mm HS-804

Benin
Artillery
105 mm Light Gun
105 mm M2 howitzer
122 mm D-30 howitzer
Air Defence
9K32 Strela-2
9K32M Strela-2M
9K31 Strela-1
14.5 mm (Quad) ZPU-4

Bolivia
Artillery
75 mm Bofors M1935 FG
75 mm M116 P howitzer
105 mm M101 howitzer
105 mm FH-18 howitzer
122 mm Type 54-1 howitzer
Air Defence
HN-5A
20 mm Oerlikon
37 mm (Single) Type 65

Bosnia-Herzegovina
Artillery
76 mm M48 mountain gun
100 mm T-12 ATG
105 mm Light Gun
105 mm M101 howitzer
105 mm M56 howitzer
105 mm Model 56 pack howitzer
120 mm 2S9 SPM/H
122 mm 2S1 SPH
122 mm D-30 howitzer (+ Bosnian Serbs)
122 mm D-30M howitzer
130 mm M-46 FG (+ Bosnian Serbs)
130 mm M-59-M FG
130 mm Model 62 FG
152 mm M-84 GH
152 mm D-20 GH
155 mm M114A2 howitzer
122 mm BM-21 MRS
122 mm (40-round) APRA MRS
128 mm Plamen-S MRS
128 mm M-63 MRS
262 mm M-87 Okran MRS
Air Defence
9K310 Igla-1
9K36 Strela-3
9K32M Strela-2M
9K31 Strela-1
S-75M Volkhov
9K31 Strela-1
23 mm (Twin) ZU-23-2
30 mm (Twin) M53/59
23 mm (Quad) ZSU-23-4
57 mm (Twin) ZSU-57-2

Botswana
Artillery
105 mm Light Gun
105 mm Model 56 P howitzer
155 mm Soltam howitzer
122 mm APRA MRS
Air Defence
9K32 Strela-2
9K32M Strela-2M
9K310 Igla-1
Javelin
20 mm M167

Brazil
Artillery
105 mm M108 SPH
105 mm M7 SPH
155 mm M109A2 SPH
105 mm M101 howitzer (and Ma)
105 mm M102 howitzer
105 mm Model 56 P howitzer
105 mm Light Gun (and Ma)
155 mm M114 howitzer
108m mm FGT 108R MRS (and Ma)
ASTROS II MRS
Air Defence
9K38 Igla
9K38 Igla (9M39)
9K310 Igla-1
9K338 Igla-S
Mistral
Roland
Roland 1
Roland 2
RBS 70
12.7 mm M55
35 mm (Twin) GDF-001
40 mm M1
40 mm (Single) L/70

Brunei
Artillery
N/A
Air Defence
Rapier
Rapier-1
Mistral

Bulgaria
Artillery
SU-100 SPG
122 mm 2S1 SPH
85 mm D-44 FG
85 mm SD-44 FG
100 mm M1944 FG
100 mm T-12 ATG
122 mm M1931/7 gun
122 mm M1938 howitzer
152 mm D-20 GH
152 mm M1937 GH
122 mm BM-21 MRS
132 mm BM-13 MRS

Air Defence
S-75 Volkhov
S-125 Pechora
S-125 Pechora (5V27)
S-300P
S-300P (5V55K)
S-200 Angara
9K32 Strela-2
9K32M Strela-2M
9K36 Strela-3
9K310 Igla-1
2K11 Krug
2K11 Krug (3M8M1)
2K12 Kub
2K12 Kub (3M9)
9K31 Strela-1
9K31 Strela-1 (9M31)
9K35 Strela-1
9K35 Strela-1 (9M35)
9K33 Osa
9K33 Osa (9M33)
9K33 Osa (9M33M3)
9M33 Osa-M
14.5 mm (Twin) ZPU-2
23 mm (Twin) ZU-23-2
23 mm (Twin) ZU-23-2F
23 mm (Quad) ZSU-23-4
57 mm (Single) S-60
85 mm (Single) KS-12
100 mm (Single) KS-19

Burkina Faso
Artillery
105 mm M101 howitzer
107 mm Type 63 MRS
Air Defence
9K32 Strela-2
9K32M Strela-2M
14.5 mm ZPU series
20 mm (Twin) TCM-20

Burundi
Artillery
122 mm D-30 howitzer
122 mm BM-21 MRS
Air Defence
Strela-2M
14.5 mm (Quad) ZPU-4

Cambodia
Artillery
76 mm M1942 FG
100 mm T-12 ATG
122 mm D-30 howitzer
122 mm M1938 howitzer
130 mm Type 59-1 FG
122 mm Type 60 FG
107 mm Type 63 MRS
122 mm BM-21 MRS
132 mm BM-13 MRS
140 mm BM-14 MRS
Air Defence
S-125 Pechora
S-125 Pechora (5V27)
FN-6
HN-5
HN-5A
9K32 Strela-2
12.7 mm (Single) Type 77
14.5 mm (Single) ZPU-1
14.5 mm (Twin) ZPU-2
14.5 mm (Quad) ZPU-4
37 mm (Single) M1939
57 mm (Single) S-60

Cameroon
Artillery
85 mm Type 56 FG
105 mm M101 howitzer
130 mm Type 59-1 FG
130 mm M-46 FG
130 mm Model 1962 FG
155 mm Soltam ATMOS SPG
122 mm BM-21 MRS
122 mm (40-round) APRA MRS
Air Defence
14.5 mm (Twin) Type 58
20 mm (Twin) TCM-20
35 mm (Twin) GDF-001
37 mm (Single) M1939

Canada
Artillery
155 mm M109A1/M109A2 SPH (R)
105 mm LG1 Mk II
105 mm C1 howitzer
105 mm C1 howitzer (UG)
155 mm M777 howitzer
Air Defence
Starburst
Blowpipe
Javelin
ADATS LOS-FH
35 mm (Twin) GDF-005
40 mm (Single) L/60

Cape Verde
Artillery
75 mm gun
76 mm gun
Air Defence
9K32 Strela-2
9K32M Strela-2M
14.5 mm (Single) ZPU-1
23 mm (Twin) ZU-23-2

Chad
Artillery
105 mm M2 howitzer
122 mm 2S1 SPG
130 mm BM-13 MRS
Air Defence
9K32 Strela-2
9K32M Strela-2M
FIM-92A Stinger
FIM-43 Redeye
14.5 mm (Single) ZPU-1
14.5 mm (Twin) ZPU-2
14.5 mm (Quad) ZPU-4
23 mm (Twin) ZU-23-2

Chile
Artillery
105 KH178 howitzer (marines)
155 mm Mk F3 SPG
105 mm Model 56 P howitzer
105 mm M101 howitzer
155 mm M68/M68/M71 M71 GH
155 mm G5 GH (army and marines)
160 mm LARS MRS
M108 command post vehicle
Air Defence
Javelin
Mistral 1/Mygale
Crotale
Crotale (R-440)
12.7 mm (Twin) Generic
12.7 mm (Quad) M55
20 mm (Twin) FAM-2M
35 mm (Twin) GDF-005
Aspic
NASAMS
Blowpipe
Avenger
FIM-92C Stinger RMP Block 1
20 mm (Single) GAI-C04
20 mm (Twin) Rh-202
20 mm (Twin) TCM-20
40 mm (Single) L/70
M163 Vulcan
M167 Vulcan
35 mm Gepard 1A

China
Artillery
120 mm 2S23 SP/GM
122 mm Type 54-1 SPH
122 mm Type 83 SPH
152 mm Type 83 SPH
120 mm Type 89 TD
155 mm PLZ45 SPH
122 mm PLZ-07 SPH
152 mm PLZ-05 (PLZ52) SPH
122 mm Type 85 SPH
122 mm Type 85
122 mm YW531C SPH
122 mm WZ551 SPH
57 mm Type 56 ATG
76 mm Type 54 ATG
85 mm Type 56 FG
100 mm Type 73 ATG
100 mm Type 86 ATG
100 mm Type 59 FG
122 mm Type 60 FG
122 mm Type 54 howitzer
122 mm Type 83 howitzer
122 mm D-30 howitzer
122 mm M1931 gun
130 mm Type 59 FG
130 mm Type 59-1 FG
152 mm Type 66 GH
152 mm Type 54 howitzer
152 mm Type 83 FG
152 mm Type 86 gun
155 mm Type WAC-21 GH
107 mm Type 63 MRS
107 mm Type 63-1 MRS
107 mm Type 81 MRS
122 mm Type 81 MRS
122 mm Type 84 MRS
122 mm Type 89 MRS
122 mm Type 90 MRS
130 mm Type 63/Type 63-1 MRS
130 mm Type 70 MRS
132 mm BM-13-16 MRS
140 mm BM-14-16 MRS
180 mm Type 71 MRS
252 mm Type 85 MRS
253 mm Type 81 MRS
253 mm Type 81-II MRS
273 mm Type 83 MRS
284 mm Type 74 MRS
300 mm A100 ARS
300 mm A120 ARS
305 mm Type 79 ARS
320 mm WS-1 ARS
350 mm M-1B MRS
425 mm Type 762 MRS
Air Defence
S-300PMU
S-300PMU-1
S-300PMU-1 (5V55R)
S-300PMU-1 (48N6)
S-300PMU-2 Favorit
S-300PMU-2 Favorit (48N6E2)
S-300 (type unidentified)
HQ-2B
HQ-2J
HQ-9 (FT-2000)
HQ-9 (FT-2000A)
23 mm (Twin) Generic
57 mm (Single) Type 59
85 mm (Single) Type 56
100 mm (Single) Type 59
QW-1
QW-2
FN-6
HN-5A
HN-5B
HN-5C
9K32M Strela-2
9K32M Strela-2M
HQ-61A
FM-80
Crotale (R-440/R-440N)
FM-90
PL-9C
HQ-17/9K331 Tor-M1
9K331 Tor-M1
9K331 Tor-M1 (9M331)
HQ-16
LY-60
KS-1A HQ-12
Type 95-2
2K22M Tunguska
TY-90
14.5 mm (Quad) Type 56
14.5 mm (Twin) Type 58
14.5 mm (Single) Type 75
14.5 mm (Twin) Type 75-1
14.5 mm (Single) Type 80

23 mm (Twin) Type 80
25 mm (Twin) Type 85
35 mm (Twin) Generic
37 mm (Twin) Type 55
37 mm (Twin) Type 65
37 mm (Twin) Type 74
37 mm (Twin) Type P793
57 mm (Single) Type 59
57 mm (Twin) Type 80
57 mm (Twin) ZSU-57-2
85 mm (Single) Type 56
100 mm (Single) Type 59

Colombia

Artillery

105 mm LG1 Mk III
105 mm M101 howitzer
155 mm/52 calibre (from Spain)

Air Defence

M48 Chaparral
M48 Chaparral (AIM-7F)
12.7 mm (Quad) M55
40 mm (Single) L/70

Congo-Brazzaville

Artillery

57 mm M1943 ATG
75 mm M116 P howitzer
76 mm M1942 FG
85 mm Type 56 FG
100 mm M1944 FG
100 mm M1938 howitzer
122 mm D-30 howitzer
122 mm 2S1 SPH
130 mm M-46 FG
152 mm D-20 GH
122 mm BM-21 MRS
140 mm BM-14 MRS

Air Defence

14.5 mm (Twin) ZPU-2
14.5 mm (Quad) ZPU-4
23 mm (Quad) ZSU-23-4
37 mm (Single) M1939
57 mm (Single) S-60
100 mm (Single) KS-19

Congo, Democratic Republic of

Artillery

122 mm 2S1 SPG
57 mm M1943 ATG
75 mm M116 P howitzer
85 mm Type 56 FG
122 mm M1938 howitzer
122 mm 2S1 SPH
122 mm D-30 howitzer
122 mm Type 60 FG
130 mm Type 59-1 FG
107 mm Type 63 MRS
122 mm BM-21 MRS
130 mm (30-round) Type 82 MRS
152 mm 2S3 SPGH

Air Defence

9K32 Strela-2
9K32M Strela-2M
14.5 mm (Quad) ZPU-4
20 mm TCM-20
37 mm M1939/Type 63
40 mm L/60

Côte d'Ivoire

Artillery

122 mm BM-21 MRS
105 mm M2 howitzer
105 mm M-1950 How

Air Defence

20 mm Generic
40 mm (Single) L/60
20 mm (Twin) M3 VDAA

Croatia

Artillery

122 mm 2S1 SPH
76 mm ZIS-3 FG
76 mm M48 mountain gun
85 mm D-44 FG
100 mm T-12 ATG
105 mm M56 howitzer
105 mm M101 howitzer
120 mm 2S9 SPM/H
122 mm M1938 howitzer
122 mm D-30 howitzer
130 mm M-46 FG
152 mm D-20 GH
152 mm M-84 GH
155 mm L33 CITEFA Mod 77 howitzer
203 mm M115 howitzer
70 mm Heron MRS
122 mm BM-21 MRS
122 mm LOV RAK 124/128 MRS
122 mm M96 Typhoon MRS
128 mm Plamen-S MRS
128 mm M-63 MRS
262 mm M-87 Okran MRS

Air Defence

S-300PS
9K32 Strela-2
9K32M Strela-2M
9K36 Strela-3
9K310 Igla-1
9K31 Strela-1
9K37 Buk
9K35 Strela-10
FIM-92A Stinger
Strijela
14.5 mm (Twin) ZPU-2
14.5 mm (Quad) ZPU-4
20 mm M30/38
20 mm M75
20 mm M55
30 mm BOV-3
30 mm (Twin) M53/59
57 mm (Twin) ZSU-57-2

Cuba

Artillery

SU-100 SPG
122 mm 2S1 SPH
122 mm 2S3 SPGH
76 mm M1942 FG
85 mm D-44 FG
85 mm SD44 FG
85 mm D-48 ATG
100 mm T-12 ATG
122 mm D-30 howitzer
122 mm M1938 howitzer
130 mm M-46 FG
152 mm D-20 GH
122 mm BM-21 MRS
140 mm BM-14 MRS

Air Defence

S-75M Volkhov
S-125 Pechora-M
S-125 Pechora-M (5V27)
9K32 Strela-2
9K32M Strela-2M
9K36 Strela-3
9K310 Igla-1
2K12 Kvadrat
2K12 Kvadram (3M9)
9M33 Osa-M
9K31 Strela-1
9K35 Strela-10
9K35 Strela-10 (9M35)
14.5 mm (Single) ZPU-1
14.5 mm (Twin) ZPU-2
14.5 mm (Quad) ZPU-4
23 mm (Twin) ZU-23-2
23 mm (Quad) ZSU-23-4
30 mm (Twin) M53
30 mm (Twin) M53 Local mod
37 mm (Single) M1939
57 mm (Single) S-60
57 mm (Twin) ZSU-57-2
85 mm (Single) KS-12 M1939
100 mm (Single) KS-19

Cyprus

Artillery

155 mm Mk F3 SPH
155 mm/45 calibre Zuzana SPG
75 mm M116 P howitzer
25 pounder FG (R)
100 mm M1944 FG
105 mm M56 howitzer
128 mm M-63 MRS
155 mm TR gun
122 mm BM-21 MRS

Air Defence

9K32 Strela-2
9K32 Strela-2M
Mistral
9K331 Tor
9K331 Tor (9M331)
Aspide
20/3 mm M55 A2
35 mm (Twin) GDF-005
40 mm Mk 1

Czech Republic

Artillery

152 mm Dana SPG
122 mm RM-70 MRS

Air Defence

S-75M Volkhov
S-125 Pechora-M
S-125 Pechora-M (5V27)
S-200 Angara
S-200 Angara (5V28)
S-300PMU
S-300PMU (5V55R)
2K12 Kvadrat
2K12 Kvadrat (3M9)
30 mm (Twin) M53
2K11 Krug
2K11 Krug (3M8M1)
9K33 Osa
9K33 Osa (9M33)
9K31 Strela-1
9K31 Strela-1 (9M31)
9K35M Strela-10M
9K35M Strela-10M (9M35M)
9K32 Strela-2
9K36 Strela-3
9K310 Igla-1
RBS 70
RBS 70 with Bolide
30 mm (Twin) M53
30 mm (Twin) M53/59
57 mm (Single) S-60

Denmark

Artillery

155 mm M109A3 SPG

Air Defence

I-HAWK
I-HAWK (MIM-23B)
40 mm L/70
FIM-43C Redeye
FIM-92A Stinger
12.7 mm (Quad) M55

Djibouti

Artillery

122 mm D-30 howitzer

Air Defence

20 mm (Twin) Tarasque 53T2
23 mm (Twin) ZU-23-2
40 mm (Single) L/70

Dominican Republic

Artillery

105 mm M101 howitzer

Air Defence

20 mm Generic
0.5 in M2HB
M1 40 mm

Ecuador

Artillery

155 mm MK F3 SPG
105 mm M101 howitzer
105 mm Model 56 P howitzer
155 mm M198 howitzer
155 mm M114 howitzer
122 mm RM-75 MRS
122 mm BM-21 MRS

Air Defence

9K310 Igla-1
9K310 Igla-1E

9K38 Igla
9K32 Strela-2M
Mistral
Blowpipe
9K33 Osa
9K33 Osa (9M33)
20 mm GAI-C01
14.5 mm (Single) ZPU-1
14.5 mm (Twin) ZPU-2
20 mm M167 Vulcan
20 mm M163 Vulcan
23 mm (Quad) ZSU-23-4
35 mm (Twin) GDF-002
40 mm (Single) L/70
40 mm (Single) M1A1

Egypt

Artillery

122 mm SP122 SPH
122 mm T-55 chassis SPH
155 mm M109A2 SPH
85 mm D-44 FG
100 mm M1944 FG
122 mm D-30 howitzer
122 mm Type 60 FG
130 mm M-46 FG
130 mm Type 59-1M FG
152 mm D-20 GH
152 mm D-1 howitzer
155 mm EH52
122 mm BM-21 MRS
122 mm Sakr-18 MRS
122 mm Sakr-36 MRS
227 mm MLRS
FROG-7 ARS

Air Defence

2K12 Kvadrat
2K12 Kvadrat (3M9)
I-HAWK
I-HAWK (MIM-23B)
S-75 Dvina
S-125 Pechora-M
S-125 Pechora (5V27)
S-125-2M Pechora-2M
Crotale
Crotale (R-440)
35 mm Amoun
20 mm M53
20 mm (Single) M57
20 mm Vulcan
23 mm (Twin) ZU-23-2
35 mm GDF-003
37 mm (Single) M1939
57 mm (Single) S-60
85 mm (Single) KS-12
100 mm (Single) KS-19
FIM-92A Stinger
FIM-92C Stinger RMP
Sakr Eye
9K32 Strela-2
9K32 Strela-2M
9K38 Igla
9K31 Strela-1
M48 Chaparral
M48 Chaparral (MIM-72C)
Avenger
SA-17
12.7 mm (Quad) M53
14.5 mm (Twin) ZPU-2
14.5 mm (Quad) ZPU-4
20 mm M53
20 mm M57
20 mm M167 Vulcan
23 mm (Twin) ZU-23-2
23 mm (Quad) ZSU-23-4
37 mm Generic
57 mm (Single) S-60
57 mm (Twin) ZSU-57-2

El Salvador

Artillery

105 mm M56 howitzer
105 mm M101 howitzer
105 mm M102 howitzer

Air Defence

FIM-43 Redeye
9K32 Strela-2
9K32M Strela-2M
9K36 Strela-3
20/3 mm M55 A2
20 mm TCM-20

Eritrea

Artillery

152 mm 2S5 SPG
122 mm 2S1 SPH
85 mm D-44 FG
122 mm D-30 howitzer
130 mm M-46 FG
122 mm BM-21 MRS
220 mm BM 9P140 MRS

Air Defence

9K32 Strela-2
9K32M Strela-2M
9K38 Igla
S-125 Pechora
23 mm (Quad) ZSU-23-4
57 mm (Single) S-60

Estonia

Artillery

105 mm m61/37 FH
122 mm D-30 howitzer
155 mm FH-70

Air Defence

Mistral 2
23 mm (Twin) ZU-23-2

Ethiopia

(Status of all equipment is uncertain)

Artillery

122 mm 2S1 SPH
85 mm D-44 FG
85 mm D-48 ATG
105 mm M101 howitzer
122 mm D-30 howitzer
122 mm M1938 howitzer
130 mm M-46 FG
152 mm D-1 howitzer
152 mm 2S19 SPG
155 mm M114 howitzer
122 mm BM-21 MRS
200 mm BMD-20 MRS

Air Defence

S-75 Dvina
S-125 Pechora-M
9K32 Strela-2
9K32M Strela-2M
9K31 Strela-1
14.5 mm (Single) ZPU-1
14.5 mm (Twin) ZPU-2
14.5 mm (Quad) ZPU-4
23 mm (Twin) ZU-23-2
23 mm (Quad) ZSU-23-4
37 mm (Single) M1939
57 mm (Single) S-60

Finland

Artillery

122 mm 2S1 SPH
152 mm 2S5 SPH
105 mm m61/37 FH
122 mm D-30 howitzer (H 63)
122 mm m/38 FH
130 mm m/54 FG (K 54)
152 mm 2A36 (15 K 89) gun
152 mm H 37 A GH
152 mm H 88-40GH
152 mm K 89 GH
155 mm 155 GH 52 APU
122 mm RM-70 MRS
227 mm MLRS

Air Defence

9K32 Strela-2
9K32 Strela-2 (9M32)
9K32M Strela-2M
9K32M Strela-2M (9M32M)
9K36 Strela-3
9K310 Igla-1
9K310 Igla-1 (9M313)
9K38 Igla
FIM-92 (RMP) Block 1
S-125M1 Pechora-M1
Crotale NG
9K37 Buk-M1
ASRAD-R
RBS 70 Mk II
Mistral
NASAMS II
23 mm (Twin) ZU-23-2
35 mm (Twin) GDF-005
35 mm Marksman
40 mm (Single) L/60
40 mm (Single) L/70
57 mm (Twin) ZSU-57-2

France

Artillery

155 mm GCT SPG
155 mm CAESAR SPG
155 mm TR gun
227 mm MLRS

Air Defence

Mistral
Crotale
Crotale NG
Crotale NG (VT-1)
SAMP/T
20 mm (Twin) Cerbere
FIM-92A Stinger
Mistral
Roland 2
HAWK
HAWK (MIM-23A)
I-HAWK
I-HAWK (MIM-23B)
20 mm (Twin) Tarasque
20 mm (Single) 53T1
30 mm generic

Gabon

Artillery

105 mm M2 howitzer
140 mm Teruel MRS
130 mm FG Type 59-1
107 mm MRS
122 mm MRS

Air Defence

Mygal
Mistral
20 mm Kriss
23 mm (Twin) ZU-23-2
37 mm (Single) M1939
40 mm (Single) L/70

Georgia

Artillery

152 mm 2S3 SPGH
207 mm 2S7 SPG
85 mm D-44 gun
100 mm T-12 ATG
122 mm D-30 howitzer
122 mm Dana SPG
152 mm 2A65 howitzer
152 mm 2S19 SPG
122 mm BM-21 MRS
122 mm RM-70 MRS
160 mm LAROM MRS

Air Defence

S-75M Volkhov
S-125-M Pechora
S-200 Angara
9K32 Strela-2
9K32M Strela-2M
9K36 Strela-3
9K310 Igla-1
Grom
9K31 Strela-1
9K37 Buk-M1
9K37 Buk-M1 (9M38)
9K33 Osa
9K33 Osa (9M33)
Poprad
Poprad (Grom)
100 mm (Single) KS-19
23 mm (Twin) ZU-23-2
23 mm (Quad) ZU-23-4

Gaza and the West Bank

Artillery
N/A

Air Defence
9K33 Osa
9K32 Strela-2
9K32M Strela-2M
9K36 Strela-3
9K338 Igla-S
Rapier 2
14.5 mm KPV

Germany

Artillery
155 mm PzH 2000 SPG
227 mm MLRS

Air Defence
FIM 92C Stinger RMP
I-HAWK
I-HAWK (MIM-23B)
Patriot (MIM-104)
Patriot
Patriot PAC-3 (MIM-104)
20 mm (Twin) Generic
9K310 Igla-1
9K38 Igla
FIM-43 Redeye
FIM-92A Stinger
RBS 70
Roland
ASRAD
20 mm (Single) FK 20-2
35 mm Skyshield
35 mm (Twin) Gepard

Greece

Artillery
155 mm M109/M109A1B/
M109A2/M109A5 SPH
155 mm M109A3GE
155 mm/52-calibre PzH 2000 SPH
203 mm M110 SPH
105 mm M101 howitzer
105 mm Model 56 P howitzer
155 mm M114 howitzer
122 mm RM-70/85 MRS
227 mm MLRS

Air Defence
Skyguard/Sparrow
Nike-Hercules
Patriot
Patriot PAC-2
Patriot PAC-2 (MIM-104)
Crotale NG
Crotale NG (VT-1)
9K331 Tor-M1
9K331 Tor-M1 (9M331)
S-300PMU-1
S-300PMU-1 (5V55R)
ASRAD-R
FIM-43 Redeye
FIM-92A Stinger
FIM 92C Stinger RMP
MIM-23A HAWK
9K33 Osa
9K33 Osa (9M33)
20 mm (Twin) Generic
23 mm (Twin) ZU-23-2
23 mm (Quad) ZSU-23-4
30 mm (Twin) Artemis
35 mm (Twin) Generic
40 mm (Single) L/70
40 mm (Single) L/60
40 mm (Twin) M42A1

Ghana

Artillery
N/A

Air Defence
9K32 Strela-2
9K32M Strela-2M
14.5 mm (Twin) ZPU-23-2
14.5 mm (Quad) ZPU-23-4
23 mm (Twin) ZU-23-2

Guatemala

Artillery
75 mm M116 howitzer
105 mm M101 howitzer
105 mm M102 howitzer
105 mm M56 howitzer

Air Defence
20 mm M55
40 mm (Twin) M42
20 mm (Twin) TCM-20
GAD-D01

Guinea

Artillery
76 mm M1942 FG
85 mm D-44 FG
122 mm M1931/37 gun
130 mm M-46 FG
130 mm Model 1982 gun
220 mm BM 9P140 (16-round) MRS

Air Defence
9K32 Strela-2
9K32M Strela-2M
14.5 mm (Quad) ZPU-4
30 mm (Twin) M53
37 mm (Single) M1939
57 mm (Single) S-60
100 mm (Single) KS-19

Guinea-Bissau

Artillery
85 mm D-44 FG
122 mm M1938 howitzer
122 mm D-30 howitzer

Air Defence
9K32 Strela-2
9K32M Strela-2M
14.5 mm (Quad) ZPU-4
23 mm (Twin) ZU-23-2
37 mm (Single) M1939
57 mm (Single) S-60

Guyana

Artillery
130 mm M-46 FG

Air Defence
9K32 Strela-2
9K32M Strela-2M

Honduras

Artillery
105 mm M101 howitzer
105 mm M102 howitzer
155 mm M198 howitzer

Air Defence
20 mm TCM-20
20 mm M197
20/3 mm M55 A2

Hungary

Artillery
100 mm T-12 ATG
152 mm GH D-20

Air Defence
S-75M Volkhov
S-125 Pechora-M
S-125 Pechora-M (5V27)
S-200 Angara
9K31 Strela-1
9K31 Strela-1 (9M31)
23 mm (Twin) ZU-23-2
9K32 Strela-2
9K32M Strela-2M
9K36 Strela-3
9K310 Igla-1
9K38 Igla
2K11 Krug
2K11 Krug (3M8M1)
2K12 Kvadrat
2K12 Kvadrat (3M9)
9K35 Strela-10
Mistral (Atlas)
Mistral
23 mm (Twin) ZU-23-2
23 mm (Quad) ZSU-23-4
57 mm (Single) S-60

India

Artillery
130 mm Vijayanta SPG
152 mm 2S19 SPG (unconfirmed)
76 mm mountain howitzer
76 mm M48 mountain gun (P)
105 mm FG
130 mm M-46 FG
155 mm M-46 FG (upgrade)
105 mm Model 56 P howitzer
155 mm FH-77B
122 mm D-30 howitzer
122 mm BM-21 MRS
122 mm LRAR MRS
214 mm Pinacha MRS
300 mm BM 9A52 Smerch MRS

Air Defence
S-75M Dvina
S-125 Pechora (5V27)
S-125 Pechora
S-200 Angara
Akash
Spyder
Spyder-MR
9K310 Igla-1
Tigercat
Mistral
2K12 Kvadrat
2K12 Kvadrat (3M9)
9K32 Strela-2
9K32M Strela-2M
9K36 Strela-3
9K38 Igla
9K310 Igla-1
9K33 Osa
9K33 Osa-AK
9K33M Osa-AKM
9K33M Osa-AKM (9M33)
9K33 Osa (9M33)
9K31 Strela-1
9K31 Strela-1 (9M31)
9K37 Buk
9K37 Buk (9M38M1)
9M317
9M317 (9M317ME)
9M38
9K35M3 Strela-10M3
9K35 Strela-10
9K331 Tor
(Twin) 2K22 Tunguska
(Twin) 2K22M1 Tunguska-M1
(Twin) 2K22M1 Tunguska-M1 (9M311)
9M311
2K22M Tunguska
2K22M Tunguska (9M311)
9M311
Barak-1
Barak-8
Barak-8 (aka Barak-NG and Barak-2)
Barak-8ER
20 mm Generic
23 mm (Twin) ZU-23-2
23 mm (Quad) ZSU-23-4
23 mm (Quad) ZU-23-4
30 mm (Twin) 2K22M Tunguska
40 mm (Single) L/70
40 mm (Single) L/60
Lora

Indonesia

Artillery
76 mm M48 mountain gun
105 mm Light Gun Mk II
105 mm M56 howitzer
105 mm M101 howitzer
122 mm M1938 howitzer (and Ma)
155 mm FH-88 FH
70 mm NDL-40 MRS
122 mm RM-70 MRS

Air Defence
Mistral
Rapier
Rapier-1
RBS 70
9K310 Igla-1

Kobra
Grom-2
TD-2000B
Giant Bow-2
QW-3
20 mm (Twin) Generic
23 mm (Twin) ZU-23-2
40 mm (Single) L/70
57 mm (Single) S-60

Iran

Artillery

122 mm Raad-1 SPG
122 mm 2S1 SPH
155 mm M109/M109A1 SPH
155 mm Raad-2 SPG
175 mm M107 SPG
170 mm M1978 SPG
203 mm M110 SPG
85 mm D-44 FG
105 mm M101 howitzer
122 mm D-30 (SHAFIE) howitzer
122 mm HM40 howitzer
122 mm Type 60 FG
122 mm Type 54 howitzer
130 mm Type 59-1 FG
152 mm D-20 GH
155 mm M114 howitzer
155 mm HM41 howitzer
155 mm WAC-021 GH
155 mm G5 GH
155 mm GH N-45 GH
203 mm M115 howitzer
107 mm Type 63 MRS
122 mm Type 81 MRS
122 mm BM-21 MRS
122 mm BM-11 MRS
240 mm M1985 MRS
240 mm Falaq-1 MRS
333 mm Falaq-2 ARS
NAZEAT MRS
SHAHIN 1 ARS
SHAHIN 2 ARS
OGHAB ARS
HADID ARS
ARASH ARS
NOOR ARS
FADGR ARS
HASEB ARS
ZELZAL ARS
FADJR 5 ARS
FADJR 3 ARS
FADJR 6 ARS

Air Defence

Tigercat
Rapier
Rapier (Rapier-1)
I-HAWK
I-HAWK (MIM-23B)
HQ-2
HQ-7/FM-80
S-200 Angara
2K12 Kub
9K331 Tor-M1
9K331 Tor-M1 (9M338)
9K331T Tor-M1T
Pantsyr-S1
23 mm (Twin) ZU-23-2
40 mm M1
FIM-92A Stinger
Igla-S
9K32 Strela-2
9K32M Strela-2M
9K36 Strela-3
QW-1 (Misagh-1)
QW-2 (Misagh-2)
FM-80 (R-440)
FM-90 Ya-zaha
FN-90
RBS 70
HN-5A
14.5 mm (Twin) ZPU-2
14.5 mm (Quad) ZPU-4
23 mm (Twin) ZU-23-2
23 mm (Quad) ZSU-23-4
35 mm (Twin) Generic
37 mm (Single) M1939/Type 55
40 mm (Single) L/70
57 mm (Single) S-60
57 mm (Twin) ZSU-57-2

Ireland

Artillery

25-pounder gun
105 mm Light Gun

Air Defence

RBS 70
40 mm (Single) L/70
40 mm (Single) L/60

Israel

Artillery

155 mm M109/M109A1 SPH
155 mm L33 SPH (R)
155 mm Sherman SPH (R)
175 mm M107 SPG
203 mm M110A2 SPH
105 mm M101 howitzer
122 mm D-30 howitzer
130 mm M-46 field gun
155 mm M114 howitzer
155 mm M-71 GH
155 mm 845/839P GH
122 mm BM-21 MRS (R)
240 mm BM-24 MRS (R)
290 mm MAR ARS
350 mm MAR ARS
227 mm MLRS

Air Defence

FIM-43C Redeye
FIM-92A Stinger
FIM-92C Stinger RMP
Chetz
MIM-48 Chaparral
MIM-48 Chaparral (MIM-72C)
MIM-23A HAWK
MIM-23B I-HAWK
MIM-104 Patriot PAC-2
Patriot PAC-2 MIM-104
Patriot PAC-2 GEM+
Spyder
Iron Dome
20 mm HS-804
20 mm (Twin) TCM-20
20 mm M167 VADS
20 mm M163
23 mm (Twin) ZU-23-2
23 mm (Quad) ZSU-23-4
37 mm (Single) M1939
40 mm (Single) L/70

Italy

Artillery

155 mm PzH 2000 SPG
155 mm M109L SPH
155 mm FH-70
227 mm MLRS

Air Defence

Spada
Nike-Hercules
SAMP/T
Skyshield
Mistral
FIM-92A Stinger
FIM 92C Stinger RMP
HAWK
I-HAWK (MIM-23B)
Skyguard
FIM-92A Stinger
12.7 mm (Quad) M55
25 mm SIDAM
40 mm (Single) L/70

Japan

Artillery

155 mm Type 75 SPH
155 mm Type 99 SPH
203 mm M110A2 SPH
155 mm FH-70
227 mm MLRS

Air Defence

FIM-92A Stinger
HAWK
HAWK (MIM-23A)
HAWK (MIM-23B)
I-HAWK
Nike-Hercules
Nike-Hercules (MIM-14)
Patriot
Patriot (MIM-104)
Patriot PAC-3
Type 81 Tan-SAM
Type 91 Kin SAM
Type 93 Kin SAM
20 mm M167 Vulcan
35 mm Type 87
35 mm (Twin) GDF-001

Jordan

Artillery

105 mm M52 SPH
155 mm M109/M109A2 SPG
155 mm M44 SPH
203 mm M110A1/M110A2 SPH
105 mm M101 howitzer
105 mm M102 howitzer
155 mm M59 gun
155 mm M114 howitzer
203 mm M115 howitzer
227 mm HIMARS

Air Defence

9K33M Osa-AKM
9K33M Osa-AKM (9M33)
9K35 Strela-10
9K35 Strela-10 (9M35)
23 mm (Quad) ZSU-23-4
9K32 Strela-2
9K32M Strela-2M
9K36 Strela-3
FIM-43 Redeye
Tigercat
Mistral
I-HAWK
I-HAWK (MIM-23B)
Patriot PAC-3
Pantsyr-S1
Starburst
Dzhigit
9K38 Igla
9K38 Igla
12.7 mm (Quad) M55
20 mm VADS
20 mm M163 VADS
40 mm (Twin) M42A1 Duster

Kazakhstan

Artillery

122 mm 2S1 SPG
152 mm 2S3 SPGH
100 mm T-12 ATG
122 mm D-30 howitzer
122 mm Senser SPG
120 mm 2S9 SPH/M
152 mm D-20 GH
152 mm 2A65 howitzer
152 mm 2A36 gun
122 mm BM-21 MRS
220 mm BM 9P140 MRS
220 mm (30-round) TOS-1 MRS

Air Defence

S-75 Dvina
S-125-2T
S-200 Angara
S-300 PMU
S-300 PMU (5V55R)
9K32 Strela-2
9K32M Strela-2M
9K36 Strela-3
9K310 Igla-1
ZSU-23-4
Various

Kenya

Artillery

105 mm Light Gun
105 mm Model 56 P howitzer
122 mm BM-21 MRS

Air Defence

Tigercat
Mistral
20 mm (Twin) TCM-20
20 mm Generic
40 mm (Single) L/70

Korea, North

Artillery

122 mm M1977 SPH
122 mm M1981 SPG
122 mm M1991 SPG
122 mm M1985 SPG
130 mm M1975 SPG
130 mm M1981 SPG
130 mm M1991 SPG
130 mm M1992 SPH
152 mm M1974 SPH
152 mm M1977 SPH
85 mm D-44 FG
85 mm Type 56 FG
85 mm D-48 ATG
100 mm M1944 FG
122 mm M1931/7 gun
122 mm M1938 howitzer
122 mm D-74 FG
122 mm Type 60 FG
122 mm Type 54 howitzer
122 mm D-30 howitzer
130 mm M-46 FG
130 mm Type 59 FG
152 mm M1937 GH
152 mm M1938 howitzer
152 mm Type 66 GH
107 mm Type 63 MRS
122 mm BM-11 MRS
122 mm BM-21 MRS
122 mm M1977 MRS
122 mm M1985 MRS
130 mm Type 63 MRS
140 mm RPU-14-16 MRS
200 mm BMD-20 MRS
240 mm BM-24 MRS
240 mm M1991 MRS
FROG-7 ARS

Air Defence

S-75 Dvina
S-125 Pechora-M
S-125 Pechora-M (5V27)
S-200 Angara
HQ-9 or S-300
HN-5A
2K12 Kub
9K32 Strela-2
9K32M Strela-2M
9K35 Strela-10
9K36 Strela-3
9K310 Igla-1
9K38 Igla
FIM-92A Stinger
14.5 mm (Single) ZPU-1
14.5 mm (Twin) ZPU-2
14.5 mm (Quad) ZPU-4
14.5 mm M1983
14.5 mm (Quad) M1984
14.5 mm (Single) M38/46 DShK
23 mm (Twin) ZU-23-2
23 mm (Quad) ZSU-23-4
23 mm (Twin) M1992
30 mm M1990
30 mm M1992
37 mm (Twin) M1992
37 mm (Single) M1939
37 mm (Twin) Type-65
37 mm (Twin) Type-74
57 mm M1985
57 mm (Twin) ZSU-57-2
57 mm (Single) S-60
85 mm (Single) KS-12
100 mm (Single) KS-19

Korea, South

Artillery

155 mm M109A2 SPH
155 mm Thunder K9 SPG
175 mm M107 SPG
203 mm M110 SPH
105 mm M101 howitzer
105 mm KH-178 howitzer
155 mm KH-179 howitzer
155 mm M114 howitzer
155 mm M59 gun
175 mm M107 SPG
130 mm KM809A1 MRS
227 mm MLRS

Air Defence

Mistral
Javelin
FIM-92A
FIM-92A Stinger
HAWK (MIM-23A)
I-HAWK
I-HAWK (MIM-23B)
9K310 Igla-1E
9K38 Igla
Nike-Hercules
Nike-Hercules (MIM-14)
Patriot
Crotale-NG
Chun Ma
Chiron
12.7 mm (Quad) M55
20 mm VADS
30 mm (Twin) Flying Tiger
35 mm (Twin) GDF-003
40 mm M1
40 mm (Single) L/70

Kuwait

Artillery

155 mm M109A1B SPH
155 mm GCT SPG (R)
155 mm PLZ45 SPG
300 mm BM 9A52 Smerch

Air Defence

FIM-92A Stinger
Starburst
Aspide-2000
HAWK III
I-HAWK
I-HAWK (MIM-23B)
Patriot PAC-2
Patriot PAC-2 (MIM-104)
Patriot PAC-2 Configuration-3
Patriot PAC-2 GEM-T Configuration-3
9K33 Osa
9K33 Osa (9M33)
35 mm Amoun
Sakr Eye
9K32M Strela-2M
9K36 Strela-3
2K12 Kub

Kyrgyzstan

Artillery

100 mm T-12 ATG
120 mm 2S9 SPM/H
122 mm 2S1 SPH
100 mm M1944 ATG
122 mm D-30 howitzer
122 mm M198 howitzer
152 mm D-1 howitzer
122 mm BM-21 MRS
220 mm (16-round) Urugan MRS

Air Defence

S-75 Dvina
S-125 Pechora-M
9K32 Strela-2
9K32 Strela-2M
9K36 Strela-3
9K310 Igla-1
57 mm (Single) S-60
23 mm (Quad) ZSU-23-4

Laos

Artillery

75 mm M116 P howitzer
85 mm D-44 FG
105 mm M101 howitzer
122 mm M1938 howitzer
122 mm D-30 howitzer
130 mm M-46 FG
155 mm M114 howitzer

Air Defence

S-125 Pechora-M
S-125 Pechora-M (5V27)
9K32 Strela-2
9K32M Strela-2M
9K310 Igla-1
14.5 mm (Single) ZPU-1
14.5 mm (Twin) ZPU-2
14.5 mm (Quad) ZPU-4
23 mm (Twin) ZU-23-2
23 mm (Quad) ZSU-23-4
37 mm (Single) M1939
57 mm (Single) S-60

Latvia

Artillery

100 mm M53 FG

Air Defence

RBS 70
14.5 mm (Quad) ZPU-4
40 mm (Single) L/70

Lebanon

Artillery

105 mm M101 howitzer
105 mm M102 howitzer
122 mm M1938 howitzer
122 mm D-30 howitzer
130 mm M-46 FG
155 mm Model 50 howitzer
155 mm M114 howitzer
155 mm M198 howitzer
122 mm BM-21 MRS
122 mm BM-11 MRS

Air Defence

9K32 Strela-2
9K32M Strela-2M
9K36 Strela-3
9K310 Igla-1
14.5 mm (Quad) ZPU-4
20/3 mm M55 A2
23 mm (Twin) ZU-23-2
23 mm (Quad) ZSU-23-4
40 mm (Twin) M42

Liberia

Artillery

75 mm M116 P howitzer

Air Defence

23 mm (Quad) ZU-23-4

Libya

(Status of all equipment is uncertain)

Artillery

122 mm 2S1 SPH
152 mm 2S3 SPGH
152 mm Dana SPG
155 mm Palmaria SPG
155 mm M109 SPH
105 mm M101 howitzer
122 mm D-30 howitzer
122 mm D-74 FG
130 mm M-46 FG
107 mm Type 63 MRS
122 mm BM-11 MRS
122 mm BM-21 MRS
122 mm RM-70 MRS
130 mm M51 MRS
FROG-7 ARS

Air Defence

S-75 Dvina
S-125 Pechora
S-125 Pechora (5V27)
S-200 Angara
Crotale
Crotale (R-440)
9K33 Osa
9K33 Osa (9M33)
23 mm (Twin) ZU-23-2
2K12 Kvadrat
2K12 Kvadrat (3M9)
9K32 Strela-2
9K32M Strela-2M
9K338 Igla-S
9K31M Strela-1
9K31 Strela-1 (9M31M)
9K35 Strela-10
9K35 Strela-10 (9M35)
14.5 mm (Twin) ZPU-2
14.5 mm (Quad) ZPU-4
23 mm (Twin) ZU-23-2
23 mm (Quad) ZSU-23-4
30 mm (Twin) M53/59

40 mm (Single) L/70
57 mm (Single) S-60
100 mm (Single)

Lithuania

Artillery

105 mm M101 howitzer

Air Defence

9K32 Strela-2
FIM-92C RMP
RBS 70 Mk1
40 mm (Single) L/70

Macedonia

Artillery

76 mm M48 mountain gun
76 mm M1942 gun
105 mm M101 gun
105 mm M56 howitzer
122 mm D-30 howitzer
122 mm M1938 howitzer
122 mm M-63 MRS
122 mm BM-21 MRS

Air Defence

9K35M Strela-10M
9K35M Strela-10M (9M35)
9K310 Igla-1
FIM-92A Stinger
9K32M Strela-2M
30 mm RH-20
20 mm M75

Madagascar

Artillery

105 mm M2 howitzer
105 mm M101 howitzer
122 mm D-30 howitzer

Air Defence

14.5 mm (Quad) ZPU-4
37 mm Type 55

Malawi

Artillery

105 mm Light Gun

Air Defence

Blowpipe
14.5 mm (Quad) ZPU-4

Malaysia

Artillery

105 mm Model 56 P howitzer
155 mm FH-70
155 mm G5 GH
ASTROS II MRS

Air Defence

9K38 Igla
Anza Mk II
KS-1A
Javelin
Rapier Jernas
Rapier-2
Starburst
FN-6
40 mm (Single) L/70
35 mm GDF-005

Mali

Artillery

85 mm D-44 FG
100 mm M1944 FG
122 mm D-30 howitzer
130 mm M-46 FG
122 mm BM-21 MRS

Air Defence

S-125 Pechora
9K32 Strela-2
14.5 mm (Quad) ZPU-4
37 mm (Single) M1939
57 mm (Single) S-60

Malta

Artillery

N/A

Air Defence

14.5 mm (Quad) ZPU-4
40 mm L/70

Mauritania

Artillery

122 mm D-30 howitzer
105 mm M101 howitzer
122 mm D-74 FG

Air Defence

9K32 Strela-2
9K32M Strela-2M
9K31 Strela-1
14.5 mm (Single) ZPU-1
14.5 mm (Twin) ZPU-2
14.5 mm (Quad) ZPU-4
23 mm (Twin) ZU-23-2
37 mm (Single) M1939
57 mm (Single) S-60
100 mm (Single) KS-19

Mauritius

Artillery

N/A

Air Defence

9K32 Strela-2
9K32M Strela-2M

Mexico

Artillery

75 mm M8 SPH
105 mm M7 SPH
75 mm M116 P howitzer
105 mm M56 howitzer
105 mm How (China)
FIROS 6 MRS

Air Defence

RBS 70
9K38 Igla
9K338 Igla-S
20 mm Generic

Moldova

Artillery

120 mm 2S9 SPM/H
100 mm MT-12 ATG
122 mm M1938 howitzer
152 mm 2A36 gun
152 mm D-20 GH
220 mm BM 9P140 MRS

Air Defence

S-125 Pechora
S-200 Angara
9K32 Strela-2
9K32M Strela-2M
9K36 Strela-3
9K310 Igla-1
23 mm (Twin) ZU-23-2
57 mm (Single) S-60

Mongolia

Artillery

85 mm D-44 ATG
100 mm T-12 ATG
100 mm M1944 FG
122 mm D-30 howitzer
122 mm M1938 howitzer
130 mm M-46 FG
152 mm D-1 howitzer
122 mm BM-21 MRS

Air Defence

9K32 Strela-2
S-75 Dvina
S-125M Pechora-M
S-125M Pechora-M (5V27DE)
14.5 mm (Twin) ZPU-2
23 mm (Twin) ZU-23-2
23 mm (Quad) ZSU-23-4
37 mm (Single) M1939
57 mm (Single) S-60

Montenegro

Artillery

105 mm M7 SPH
122 mm 2S1 SPH
76 mm M48 mountain gun
76 mm M1942 FG
85 mm D-44 FG
100 mm T-12 ATG
100 mm M87 ATG
105 mm M56 howitzer
105 mm M101 howitzer
105 mm M18 FH
105 mm M18 (M) FH
105 mm M18/40 FH
122 mm D-30 howitzer
122 mm M1938 howitzer
130 mm M-46 FG
152 mm M1937 GH
152 mm D-20 GH
152 mm M84 GH
155 mm M65 howitzer
155 mm M59 gun
155 mm M114 howitzer
122 mm BM-21 MRS
128 mm M-63 MRS
128 mm M-77 MRS
128 mm M-85 MRS
128 mm Plamen-S MRS
262 mm M-87 MRS
FROG-7 ARS

Air Defence

9K32 Strela-2
9K32M Strela-2M
9K36 Strela-3
9K310 Igla-1
20/3 mm M55 A2
20/3 mm M55 A4 B1
20/1 mm M75
30 mm M53 BOV-3
30 mm M53/59

Morocco

Artillery

SU-100 SPG
155 mm M109A1B SPH
155 mm Mk F3 SPG
203 mm M110A2 SPH
81 mm M125A1 SPM
76 mm M1942 FG
85 mm D-44 FG
105 mm Model 56 P howitzer
105 mm M101 howitzer
105 mm Light Gun
122 mm D-30 howitzer
130 mm M-46 FG
155 mm M-50 howitzer
155 mm M198 howitzer
155 mm M114 howitzer
155 mm FH-70
122 mm BM-21 MRS

Air Defence

9K32 Strela-2
9K32M Strela-2M
M48 Chaparral
M48 Chaparral (MIM-72C)
2K22M Tunguska
2K22M Tunguska (9M311)
14.5 mm (Twin) ZPU-2
14.5 mm (Quad) ZPU-4
20 mm M-167 VADS
20 mm M-163 VADS
23 mm (Twin) ZU-23-2
37 mm (Single) M1939
57 mm (Single) S-60
100 mm (Single) KS-19

Mozambique

Artillery

76 mm M1942 FG
85 mm D-44 FG
85 mm D-48 ATG
100 mm M1944 FG
122 mm M1938 howitzer
122 mm D-30 howitzer
130 mm M-46 FG
152 mm D-1 howitzer
122 mm BM-21 MRS

Air Defence

9K32 Strela-2
9K32M Strela-2M
S-75 Dvina
S-125M Pechora-M

S-125M Pechora-M (5V27)
9K31 Strela-1
9K31 Strela-1(9M31)
14.5 mm (Single) ZPU-1
14.5 mm (Twin) ZPU-2
14.5 mm (Quad) ZPU-4
23 mm (Twin) ZU-23-2
37 mm (Single) M1939
57 mm (Twin) ZSU-57-2
57 mm (Single) S-60

Myanmar

Artillery

6-pounder ATG
76 mm M48 mountain gun
17-pounder ATG
25-pounder FG
105 mm M101 howitzer
122 mm BM-21 MRS
130 mm M-46/Type 59 FG
155 mm towed artillery
5.5 in gun
SH1 155 mm (6 × 6) SPG
155 mm (8 × 8) SPG
107 mm Type 63 MRS
105 mm M-56 howitzer
240 mm M1991 MRL

Air Defence

HN-5A
9K38 Igla
9K310 Igla-1
Bloodhound Mk 2
2K12 Kvadrat
S-125 Pechora-2M
HQ-2
S-200
37 mm Type 74
40 mm M1
57 mm Type 59

Namibia

Artillery

122 mm D-74 FG
122 mm BM-21 MRS

Air Defence

9K32 Strela-2
9K32M Strela-2M
14.5 mm (Quad) ZPU-4
23 mm (Twin) ZU-23-2

Nepal

Artillery

3.7 in mountain howitzer
105 mm Model 56 P howitzer
105 mm Light Gun

Air Defence

40 mm (Single) L/60
14.5 mm Type 56
37 mm Generic

Netherlands

Artillery

155 mm PzH 2000 SPG

Air Defence

Patriot
Patriot (MIM-104)
Patriot PAC-2
Patriot PAC-2 (MIM-104)
Patriot PAC-3
40 mm (single) L/70
FIM-92A Stinger
FIM 92C Stinger RMP
NASAMS II
Stinger Weapon Platforms
35 mm (twin) Gepard
40 mm (single) L/70

New Zealand

Artillery

105 mm Light Gun

Air Defence

Mistral
Mistral 1

Nicaragua

Artillery

76 mm ZIS-3 ATG
100 mm M1944 ATG
105 mm M101 howitzer
122 mm D-30 howitzer
152 mm D-20 GH
122 mm BM-21 MRS

Air Defence

9K31 Strela-1
9K32 Strela-2
9K32M Strela-2M
9K36 Strela-3
9K33 Osa
9K33 Osa (9M33)
9K310 Igla-1
14.5 mm (Single) ZPU-1
14.5 mm (Twin) ZPU-2
14.5 mm (Quad) ZPU-4
20 mm GAI-C01
23 mm (Twin) ZU-23-2
37 mm (Single) M1939
57 mm (Single) S-60
100 mm (Single) KS-19
57 mm ZSU-57-2

Niger

Artillery

N/A

Air Defence

20 mm (Twin) M-3 VDA
20 mm Generic

Nigeria

Artillery

155 mm Palmaria SPG
105 mm Model 56 P howitzer
122 mm D-74 FG
122 mm D-30 howitzer
130 mm M-46 FG
130 mm Model 1962 FG
122 mm BM-21 MRS
122 mm APR 40 MRS

Air Defence

9K32 Strela-2
9K32M Strela-2M
Blowpipe
Roland
Roland 2
23 mm (Quad) ZSU-23-4
23 mm (Twin) ZU-23-2
40 mm (Single) L/60

Norway

Artillery

155 mm M109A3GN SPH
227 mm MLRS (P)

Air Defence

I-HAWK NOAH
I-HAWK (MIM-23B)
40 mm (Single) L/70
RBS 70
RBS 70 Mk 2
Mistral
NASAMS
NASAMS (AIM-120A)
20 mm (Single) FK 20-2

Oman

Artillery

105 mm Light Gun
122 mm D-30 howitzer
130 mm M-46 FG
130 mm Type 59-1M FG
122 mm Type 90 (40-round) MRL
155 mm FH-70

Air Defence

Rapier
Rapier-1
Rapier-2
Avenger
SLAMRAAM
SLAMRAAM (AIM-120C-7)
9K32 Strela-2
9K32M Strela-2M
Blowpipe
Javelin
Sakr Eye
FIM-92 (RMP) Block 1
Mistral-2/ALBI
Vadar
20 mm GAI-D01
20 mm (Twin) TA20
23 mm (Twin) ZU-23-2
35 mm (Twin) GDF-005
40 mm (Single) L/60

Pakistan

Artillery

105 mm M7 SPH
155 mm M109/M109A2 SPH
203 mm M110A2 SPH
85 mm Type 56 FG
100 mm Type 59 FG
105 mm M101 howitzer
105 mm Model 56 P howitzer
122 mm Type 83 (D-30) howitzer
122 mm Type 54 howitzer
122 mm Type 60 FG
130 mm Type 59-1 FG
130 mm M-46 FG
155 mm M59 gun
155 mm M114 howitzer
155 mm/52 cal Panter
155 mm M198 howitzer
203 mm M115 howitzer
122 mm MRS
122 mm BM-11 MRL

Air Defence

Crotale-2000
Crotale-2000 (R-440)
Crotale-4000
HQ-2
HQ-2B
HQ-9 (FT-2000)
Mistral 1
SLAMRAAM
Anza Mk I
Anza Mk II
Anza Mk III
Anza SP
HN-5A
QW-1
FIM-92A Stinger
FIM-43 Redeye
RBS 70
RBS 23 BAMSE
LY-60
Spada-2000
Aspide-2000
12.7 mm Type 54
12.7 mm (Quad) M55
14.5 mm Anza/14.5 mm
14.5 mm (Twin) ZPU-2
14.5 mm (Quad) ZPU-4
23 mm (Twin) ZU-23-2
35 mm (Twin) GDF 002
35 mm (Twin) GDF 005
37 mm (Twin) Type 55/65
40 mm M1
57 mm Type 59
FN-6

Paraguay

Artillery

75 mm Bofors M1935 FG
105 mm M101 howitzer

Air Defence

40 mm M1A1
40 mm L/60
40 mm Generic

Peru

Artillery

122 mm 2S1 SPH
155 mm Mk F3 SPG
155 mm M109A2 SPG ?
105 mm M101 howitzer
105 mm Model 56 P howitzer
122 mm D-30 howitzer
130 mm M-46 FG
155 mm M114 howitzer
155 mm Model 50 howitzer

107 mm RO 107 MRS
122 mm BM-21 MRS

Air Defence
S-125 Pechora
S-125 Pechora (5V27)
S-75 Dvina
9K310 Igla-1
9K32 Strela-2
9K32M Strela-2M
9K36 Strela-3
9K38 Igla
9K331 Tor-M1
Javelin
QW-11
FN-6
23 mm (Twin) ZU-23-2
23 mm (Quad) ZSU-23-4
40 mm M1
40 mm (Single) L/60
40 mm (Single) L/70

Philippines
Artillery
105 mm Model 56 P howitzer
105 mm M101 howitzer (and Ma)
155 mm M114 howitzer
155 mm M68 GH

Air Defence
N/A

Poland
Artillery
122 mm 2S1 SPH
152 mm Dana SPG
155 mm/52 calibre Krab SPG*
122 mm BM-21 MRS
122 mm RM-70 MRS

Air Defence
S-75 Dvina
S-125 Newa
S-125 Newa (5V27)
S-200 Angara
Aster-30
VL Mica
Patriot
23 mm (Twin) ZU-23-2
Newa Neva-SC
Grom I
9K32 Strela-2
9K32M Strela-2M
9K36 Strela-3
2K11 Krug
2K11 Krug (3M8M1)
2K12 Kub-M
2K12 Kub-M (3M9M)
9K33 Osa-AK
9K33 Osa-AK (9M33)
9K31 Strela-1
9K31 Strela-1 (9M31)
9K35M Strela-10M
9K35M Strela-10M (9M35)
23 mm (Twin) ZU-23-2KG
23 mm (Twin) ZUR-23-2S Jod
23 mm (Quad) ZSU-23-4
57 mm (Single) S-60

Portugal
Artillery
105 mm Light Gun
155 mm M109A5 SPH
105 mm M101 howitzer
105 mm Model 56 pack howitzer
155 mm M114 howitzer

Air Defence
MIM-23B I-HAWK
FIM-92E Stinger
Blowpipe
M48 Chaparral
M48 Chaparral (MIM-72C)
M48 Chaparral (MIM-72G)
12.7 mm (Quad) M55
20 mm M163 VADS
20 mm (Twin) Generic
40 mm L/60

Qatar
Artillery
155 mm Mk F3 SPG
ASTROS MRS

Air Defence
9K32 Strela-2
9K32M Strela-2M
Rapier
Rapier-1
Blowpipe
FIM-92A Stinger
Mistral 1
Roland 2
Sea Cat (Tiger Cat)

Romania
Artillery
122 mm 2S1 SPH
122 mm M1989 SPH
57 mm M1943 ATG
76 mm M48 mountain gun
76.2 mm mountain gun
85 mm D-48 ATG
98 mm mountain howitzer
100 mm M1977 ATG
122 mm M1931/7 gun
130 mm M1982 gun
152 mm 1981 GH
152 mm M1984 howitzer
152 mm M1985 GH
152 mm M1938 howitzer
122 mm APR-40/MRS
160 mm LAROM 160 MRS

Air Defence
S-75 Dvina
S-125 Pechora
HAWK XXI
I-HAWK (MIM-23B)
VL Mica
Mistral
9K32 Strela-2
9K32M Strela-2M
2K12 Kvadrat
2K12 Kvadrat (3M9)
CA-94
CA-94M
9K33 Osa
9K31 Strela-1
CA-95
CA-95M
14.5 mm (Twin) ZPU-2
14.5 mm (Quad) ZPU-4
23 mm (Quad) ZSU-23-4
30 mm Generic
30 mm (Twin) M53
35 mm (Twin) GDF-003
35 mm (Twin) Gepard
37 mm (Single) M1939
57 mm (Single) S-60
57 mm (Twin) ZSU-57-2
85 mm (Single) M1939
100 mm (Single) KS-19

Russian Federation
Artillery
120 mm 2S9 SPM/H
120 mm 2S23 SP/GM and NI
122 mm 2S1 SPH (and NI)
152 mm 2S3 SPH
152 mm 2S5 SPGH
152 mm 2S19 SPG
203 mm 2S7 SPG
240 mm 2S4 SPM
76 mm M1966 mountain gun
85 mm D-44 field gun (R and NI)
100 mm T-12 ATG (and NI)
122 mm D-30 howitzer
125 mm 2A45M ATG
130 mm M-46 FG (R)
152 mm D-1 howitzer
152 mm D-20 GH
152 mm 2A36
152mm 2A65
152 mm 2A61
122 mm BM-21 MRS (and NI)
122 mm M1975 MRS
122 mm M1976 MRS
220 mm Splav BM 9P 140 MRS
300 mm BM 9A52 Smerch MRS
TOS-1 MRS

Air Defence
S-200 Angara/Vega
S-300PM
S-300PM-1M
S-300PS
S-300PT
S-300PT-1
S-300PT-1A
S-300V1/2
S-300VM1/2
S-400 Triumph
2K11 Krug
2K12 Kub
2K22 Tunguska
9K31 Strela-1
9K32 Strela-2
9K32M Strela-2M
9K33 Osa
9K33M Osa-AKM
9K35 Strela-10
9K35M Strela-10M
9K35M3-K Kolchan
9K36 Strela-3
9K37 Buk
9K38 Igla
9K310 Igla-1
9K331 Tor
Pantsyr-S1
6M1B5 MT-LBM
14.5 mm (Single) ZPU-1
14.5 mm (Twin) ZPU-2
14.5 mm (Quad) ZPU-4
23 mm (Twin) ZU-23-2
23 mm (Quad) ZSU-23-4
37 mm (Single) M1939
57 mm (Single) S-60
57 mm (Twin) ZSU-57-2
85 mm (Single) KS-12
100 mm (Single) KS-19
130 mm (Single) KS-30

Rwanda
Artillery
105 mm M101 howitzer
122 mm Type 54-1 howitzer
122 mm D-30 howitzer
152 mm Type 59 howitzer
122 mm RM-70 MRS

Air Defence
9K310 Igla-1

São Tomé and Príncipe
Artillery
N/A

Air Defence
14.5 mm (Quad) ZPU-4

Saudi Arabia
Artillery
155 mm PLZ45 SPH
155 mm M109A1B/M109A2 SPH
155 mm/52 cal CAESAR SPG
155 mm GCT SPG
105 mm M101 howitzer
105 mm M102 howitzer (NG)
120 mm NEMO SPM (NG)
155 mm M114 howitzer
155 mm FH-70
155 mm M198 howitzer
ASTROS II MRS

Air Defence
Crotale-4000
Crotale (R-440N)
Crotale (R-440)
Crotale
I-HAWK
I-HAWK (MIM-23B)
I-HAWK III
Patriot PAC-2 (MIM-104)
Patriot PAC-2
Patriot PAC-2 upgraded to PAC-3 standard
20 mm M163 Vulcan Air Defence System (VADS)
30 mm (Twin) AMX-30 DCA
35 mm GDF
40 mm (Single) L/70

9K310 Igla-1
Dzhit
FIM-92A Stinger
FIM-43 Redeye
Mistral
Mistral 2
Shahine
Shahine (R-460)

Senegal

Artillery

75 mm M116 P howitzer
105 mm M101 howitzer
155 mm Model 50 howitzer

Air Defence

20 mm (Single) 53T2
40 mm (Single) L/60

Serbia

Artillery

105 mm M7 SPH
122 mm 2S1 SPH
76 mm M48 mountain gun
76 mm M1942 FG
85 mm D-44 FG
100 mm T-12 ATG
100 mm M87 ATG
105 mm M56 howitzer
105 mm M101 howitzer
105 mm M18 FH
105 mm M18 (M) FH
105 mm M18/40 FH
120 mm 2S9 SPH/M
122 mm D-30 howitzer
130 mm M-46 FG
152 mm M1937 GH
152 mm D-20 GH
152 mm M84 GH
155 mm M65 howitzer
155 mm M59 gun
155 mm M114 howitzer
122 mm BM-21 MRS
128 mm M-63 MRS
128 mm M-77 MRS
128 mm M-85 MRS
128 mm Plamen-S MRS
262 mm M-87 MRS

Air Defence

2K12 Kub
9K37 Gang
9K37 Gang (9M38)
9K31 Strela-1
S-125 Neva-M1T
S-125 Pechora (5V27)
2K12 Kvadrat (3M9)
9K31 Strela-1
9K31 Strela-1 (9M31)
9K35 Strela-10
9K35 Strela-10 (9M35)
9K36 Strela-3
9K310 Igla-1
20/3 mm M55 A2
20/3 mm M55 A3 B1
20/3 mm M55 A4 B1
20/1 mm M75
20 mm BOV-3
23 mm (Quad) ZSU-23-4
30 mm (Twin) M53
30 mm M53/59
30 mm (Twin) BOV
37 mm (Single) M1939
40 mm (Single) L/70
40 mm Mk 1
57 mm (Single) S-60
57 mm (Twin) ZSU-57-2
85 mm (Single) KS-12
100 mm (Single) KS-19
9K33 Osa (9M33)
ZIF-22

Seychelles

Artillery

122 mm D-30 howitzer
122 mm BM-21 MRS

Air Defence

9K32 Strela-2
9K32M Strela-2M
14.5 mm (Quad) ZPU-4
57 mm (Single) S-60

Sierra Leone

Artillery

25-pounder FG

Air Defence

9K32 Strela-2
9K32M Strela-2M
9K36 Strela-3
Blowpipe
12.7 mm Generic
14.5 mm ZPU-2/4
23 mm ZU-23-2

Singapore

Artillery

120 mm SRAMS (on Bronco)
105 mm LG1 Light Gun (reserve)
155 mm M68 GH (R)
155 mm M71 GH (R)
155 mm FH-77 GH
155 mm FH-2000
155 mm SPG Primus
155 mm Pegasus howitzer
227 mm HIMARS

Air Defence

9K38 Igla
FIM-92A Stinger
I-HAWK
I-HAWK (MIM-23B)
Mistral 1
Mistral 2
Rapier B1X
Rapier-1
Rapier-2000
Rapier-2
Spyder
RBS 70
20 mm GAI-C01
35 mm (Twin) GDF-001
35 mm (Twin) GDF-002
40 mm (Single) L/70
Aster-15
Barak-1

Slovakia

Artillery

122 mm M-30 FH (R)
122 mm 2S1 SPG
152 mm Dana SPG
155 mm Zuzana SPG
122 mm D-30 howitzer
122 mm RM70 MRS

Air Defence

2K12M2 Kub-M2
S-75 Dvina
S-125 Pechora
S-200 Angara
S-300PMU
30 mm (Twin) M53
9K31 Strela-1
9K32 Strela-2
9K32M Strela-2M
9K35 Strela-10
9K36 Strela-3
9K310 Igla-1
9K38 Igla
30 mm (Twin) M53
30 mm (Twin) M53/59
57 mm (Single) S-60

Slovenia

Artillery

M-84 MBT
T-55 MBT
120 mm NEMO SPM
105 mm M101 howitzer
155 mm Model 845 gun
128 mm M-63 MRS

Air Defence

9K31 Strela-1
9K32M Strela-2M
9K310 Igla-1
Roland-2
Pantsyr-S1
20 mm M75
30 mm M53/59
57 mm (Twin) ZSU-57-2

Somalia

(Status of all equipment is uncertain)

Artillery

76 mm M1942 FG
85 mm D-44 FG
85 mm D-48 ATG
100 mm M1944 FG
105 mm Model 56 howitzer
122 mm Type 60 F
122 mm D-30 howitzer
122 mm M1938 howitzer
130 mm Type 59-1 FG
130 mm M-46 FG
152 mm D-1 howitzer
155 mm M114A1 howitzer
155 mm M198 howitzer
180 mm S-23 gun
122 mm BM-21 MRS
132 mm BM-13 MRS
140 mm BM-14-16 MRS
140 mm BM-14-17 MRS
240 mm BM-24 MRS

Air Defence

9K38 Igla
9K310 Igla-1
9K32 Strela-2
9K32M Strela-2M
FIM-43 Redeye
2K12 Kub
23 mm (Twin) ZU-23-2

South Africa

Artillery

155 mm G6 SPG
155 mm G4 GH (R)
155 mm G5 GH
127 mm Valkiri MRS (R)
127 mm Bataleur MRS

Air Defence

Crotale
Crotale (R-440)
9K32 Strela-2
9K32M Strela-2M
9K36 Strela-3
Locally developed MANPADS
SAHV-3
Starstreak
Tigercat
14.5 mm (Single) ZPU-1
14.5 mm (Twin) ZPU-2
20 mm (Single) Ystervark
20 mm (Single) GAI-C04
20 mm (Single) GAI-C01
20 mm (Single) GAI-B01
35 mm (Twin) GDF-002
35 mm (Twin) GDF-005
23 mm (Twin) ZU-23-2
35 mm (Twin) GDF

Spain

Artillery

155 mm M109/M109A1/
M109A1B SPH
105 mm Model 56 P howitzer (and Ma)
105 mm Light Gun
155 mm M114 howitzer
155 mm 155/52 APU SBT-1

Air Defence

Spada-2000
Aspide-2000
NASAMS
Mistral
FIM-92A Stinger
HAWK
HAWK (MIM-23A)
I-HAWK
I-HAWK (MIM-23B)
Patriot PAC-2
Patriot PAC-2 (MIM-104)
Mistral
Roland-2
35 mm Toledo
12.7 mm (Quad) M55

20 mm (Single) GAI-B01
35 mm (Twin) GDF-002
40 mm (Single) L/70

Sri Lanka

Artillery

76 mm M48 mountain gun
85 mm Type 56 FG
122 mm Type 54-1 howitzer
152 mm GH Type 66 (D-20)
25 pounder FG
130 mm Type 59-1 FG
122 mm RM-70 (40-round) MRL
122 mm (30-round) MRL

Air Defence

9K38 Igla
9K310 Igla-1
12.7 mm Type 56
20 mm TCM-20
23 mm ZU-23-2
40 mm (Single) L/60
40 mm (Single) L/70

Sudan

Artillery

155 mm Mk F3 SPG
76 mm Type 54 ATG
85 mm D-44 FG
85 mm D-48 ATG
85 mm M1944 FG
100 mm M1944 ATG
100 mm M1944 FG
105 mm M101 howitzer
105 mm M1938 howitzer
122 mm 2S1 SPH
122 mm D-30 howitzer
122 mm Type 54 howitzer
122 mm Type 60 GH
122 mm D-74 FG
130 mm Type 59-1 FG
130 mm M-46 FG
155 mm M114 howitzer
107 mm Type 63 MRS
122 mm Sakr MRS
122 mm BM-21 MRS

Air Defence

9K32 Strela-2
9K32M Strela-2M
FIM-43 Redeye
FN-6
S-75 Dvina
9K38 Igla
14.5 mm (Twin) ZPU-2
14.5 mm (Quad) ZPU-4
20 mm M163 VADS
20 mm M167 VADS
23 mm (Twin) ZU-23-2
37 mm (Single) M1939
40 mm (Single) L/60
85 mm (Single) KS-12
100 mm (Single) KS-19

Sweden

Artillery

155 mm FH-77B
155 mm Archer SPG

Air Defence

I-HAWK
I-HAWK (MIM-23B)
RBS 23 BAMSE
FIM-43 Redeye
RBS 70
RBS 90
Bolide
40 mm (Single) L/70
40 mm CV 90AD

Switzerland

Artillery

M109 SPH (UG)

Air Defence

Bloodhound
FIM-92D RMP Stinger
Rapier
Rapier Mk 1
Rapier Mk 2
35 mm (Twin) Skyguard

Syria

Artillery

100 mm M1944 T-34 SPG
122 mm 2S1 SPG
122 mm D-30 T-34 SPG
152 mm 2S3 SPGH
76 mm M48 mountain gun
85 mm D-44 FG
100 mm M1944 FG
122 mm M1938 howitzer
122 mm M1931/7 gun
122 mm D-74 FG
122 mm D-30 howitzer
130 mm M-46 FG
152 mm D-20 GH
152 mm M1937 GH
152 mm 2A36 gun
180 mm S-23 gun
240 mm M-240 mortar
107 mm Type 63 MRS
122 mm BM-21 MRS
FROG-7 ARS

Air Defence

2K12 Kub
2K12 Kub (3M9)
S-300P
S-75 Dvina
S-125 Pechora
S-125 Pechora (5V27)
S-200E Angara
9K38 Igla
23 mm (Twin) ZU-23-2
130 mm (Single) KS-30
2K12 Kvadrat
2K12 Kvadrat (3M9)
9K31 Strela-1 (9M31)
9K31 Strela-1
9K32 Strela-2
9K32M Strela-2M
9K33 Osa
9K33 Osa (9M33)
9K35 Strela-10
9K35 Strela-10 (9M35)
9K36 Strela-3
9K37 Gang
9K37 Gang (9M38)
9K40 Buk
9K40 (9M317)
Strelets
9K38 Igla
Pantsyr-S1
Pantsyr-S1 (9M311)
12.7 mm M53
14.5 mm (Twin) ZPU-2
14.5 mm (Quad) ZPU-4
20 mm (Single) Hispano-Suiza
20 mm (Triple) Hispano-Suiza
23 mm (Twin) ZU-23-2
23 mm (Quad) ZSU-23-4
30 mm HS-661
37 mm (Single) M1939
57 mm (Twin) ZSU-57-2
57 mm (Single) S-60
85 mm KS-12
100 mm (Single) KS-19

Taiwan

Artillery

105 mm M108 SPH
155 mm M44 SPH
155 mm XT-69 SPH
155 mm XT-69 SPG
155 mm M109/M109A2 SPH (and Ma)
155 mm M109A5 SPH
203 mm M110A2 SPH (and Ma)
75 mm M116 P howitzer
105 mm M101 howitzer
105 mm T64 howitzer
155 mm T65 howitzer
155 mm M114 howitzer (and Ma)
155 mm M59 gun
203 mm M115 howitzer (and Ma)
240 mm howitzer
117 mm Kung Feng VI MRS
126 mm Kung Feng III MRS
126 mm Kung Feng IV MRS (and Ma)
RT2000 ARS

Air Defence

Antelope
Avenger
RBS 70
FIM-92A Stinger
FIM-92 Block-1
Nike Hercules (MIM-14)
Patriot PAC-2
Patriot PAC-2 (MIM-104)
Patriot PAC-3
Patriot PAC-3 (MIM-104)
Tien Kung I
Tien Kung II
Sky Sword I
Mistral 1
I-HAWK (MIM-23B)
I-HAWK
M48 Chaparral
M48 Chaparral (MIM-72C)
Skyguard
12.7 mm (Quad) M55
35 mm (Twin) GDF-001 (Tien Tun)
40 mm M1
40 mm (Single) L/70
40 mm (Single) L70/T92
40 mm (Twin) M42A1

Tajikistan

Artillery

122 mm D-30 howitzer
122 mm M1938 howitzer
122 mm BM-21 MRS

Air Defence

S-75 Dvina
S-125 Pechora-2M
S-125-2M Phase II
9K33 Osa
9K32 Strela-2
9K32M Strela-2M
9K36 Strela-3
9K310 Igla-1

Tanzania

Artillery

76 mm M1942 FG
85 mm Type 56 FG
122 mm D-30 howitzer
122 mm Type 54 howitzer
130 mm Type 59-1 FG
130 mm M-46 FG
122 mm BM-21 MRS

Air Defence

9K32 Strela-2
9K32M Strela-2M
S-125 Pechora
S-125 Pechora (5V27)
2K12 Kvadrat
2K12 Kvadrat (3M9)
9K31 Strela-1
14.5 mm (Single) ZPU-1
14.5 mm (Twin) ZPU-2
14.5 mm (Quad) ZPU-4
23 mm (Twin) ZU-23-2
37 mm Type 55

Thailand

Artillery

155 mm M109A5 SPH
105 mm LG1 Mk II
105 mm Model 56 PH
105 mm M425 howitzer
105 mm M101 howitzer (UG)
105 mm M101 howitzer
105 mm M102 howitzer
105 mm L119 Light Gun
130 mm Type 59-1 FG
155 mm M71 GH
155 mm M114 howitzer
155 mm M68 GH (Ma)
155 mm GC 45 GH (Ma)
155 mm M198 howitzer
155 mm GH H-45 GH
155 mm CAESAR SPG (6 × 6)
70 mm MRS
130 mm MRS

122 mm Type 83 MRS
130 mm Type 85 MRS

Air Defence

RBS 70 Mk 2
RBS 70
Blowpipe
Spada
Aspide Mk-1
ADATS LOS-FH
HAWK
HAWK (MIM-23A)
20 mm (Twin/Triple) M39
30 mm (Twin) Arrow
37 mm (Twin) Type 74
40 mm M1
40 mm (Single) L/70
40 mm (Twin) M42
57 mm Type 59
FIM-43 Redeye
HN-5A
Mistral
Mistral (SADRAL)
9K38 Igla
12.7 mm (Quad) M55
20 mm M167A1 VADS
20 mm (Single) M163 VADS

Togo

Artillery

105 mm M102 howitzer
122 mm 2S1 SPH

Air Defence

14.5 mm (Quad) ZPU-4
37 mm M1939

Tunisia

Artillery

155 mm M109A2 SPH
105 mm M101 howitzer
155 mm M114 howitzer
155 mm M198 howitzer

Air Defence

RBS 70
M48 Chaparral (MIM-72C)
M48 Chaparral
M48 Chaparral (MIM-72E)
9K32 Strela-2
9K32M Strela-2M
20 mm M55
37 mm Type 55
40 mm M42A1 Duster

Turkey

Artillery

105 mm M52/M52A1 SPH
155 mm M52T SPG
105 mm M108 SPH
155 mm M44T SPG
155 mm M109 SPH
175 mm M107 SPG
203 mm M110 SPH
155 mm Firtina SPG
75 mm M116 P howitzer
105 mm M101 howitzer
155 mm M59 gun
155 mm M114 howitzer
155 mm/52 calibre Panter howitzer
203 mm M115 howitzer
107 mm MRS
122 mm T-122 MRS
227 mm MRS
300 mm T-300 MRS

Air Defence

Rapier
Rapier-1
Rapier-2000
Rapier-2
Nike-Hercules
I-HAWK
I-HAWK (MIM-23B)
FIM-43 Redeye
FIM-92A Stinger
FIM 92C Stinger RMP
ZIPKIN
ATILGAN
12.7 mm (Quad) M55
20 mm (Twin) GAI-D01
35 mm (Twin) GDF 003
20 mm (Twin) Rh 202
40 mm M1
40 mm (Single) L/70
40 mm (Twin) M42A1 Duster
RIM-116A RAM
12.7 mm (Single) STAMP

Turkmenistan

Artillery

120 mm 2S9 SPM/H
122 mm 2S1 SPH
152 mm 2S3 SPGH
85 mm D-44 gun
100 mm MT-12 ATG
122 mm D-30 howitzer
122 mm M1938 howitzer
152 mm D-1 howitzer
122 mm BM-21 MRS
220 mm (16 round) Urugan MRS
300 mm BM 9A42 Smerch MRS

Air Defence

S-75 Dvina
S-125 Pechora
S-200 Angara
9K32 Strela-2
9K32M Strela-2M
9K36 Strela-3
9K310 Igla-1
9K35 Strela-10
9K33 Osa
96K6 Pantsyr-S1
23 mm (Quad) ZSU-23-4
57 mm (Single) S-60

Uganda

Artillery

76 mm M1942 FG
122 mm D-30 howitzer
122 mm M1938 howitzer
155 mm Soltam ATMOS SPG
107 mm (12 round) Type 63 MRS

Air Defence

9K32 Strela-2
9K32M Strela-2M
14.5 mm (Single) ZPU-1
14.5 mm (Twin) ZPU-2
14.5 mm (Quad) ZPU-4
23 mm (Twin) ZU-23-2
37 mm (Single) M1939

Ukraine

Artillery

120 mm 2S9 SPM/H
122 mm 2S1 SPH
152 mm 2S3 SPGH
155 mm 2S5 SPG
152 mm 2S19 SPG
203 mm 2S7 SPG
100 mm T-12 ATG
122 mm D-30 howitzer
152 mm D-20 GH
152 mm 2A65 howitzer
152 mm 2A36 gun
122 mm BM-21 MRS
132 mm BM-13 MRS
220 mm BM 9P140 MRS
300 mm BM 9A52 Smerch MRS
FROG-7 ARS

Air Defence

S-75 Volkhov
S-125 Pechora
S-125M Pechora (Ukraine Mod)
S-200D
S-200V
9K37M1 Buk-M1
S-300V1
S-300PS
S-300PT
2K11 Krug
2K12 Kub
9K32 Strela-2
9K32M Strela-2M
9K33 Osa AKM
9K31 Strela-1
9K36 Strela-3
9K35 Strela-10
9K37M Buk-M
9K38 Igla
9K310 Igla-1
9K331 Tor
S-300V1
S-300V2
23 mm (Quad) ZU-23-4
9K22M Tunguska

United Arab Emirates

Artillery

155 mm M109L47
155 mm Mk F3 SPG
155 mm G6 SPG
105 mm Light Gun
130 mm Type 59-1 FG (R)
70 mm (40-round) LAU (and RL [Sharjar])
120 mm AGRAB Mk 2 SPM
122 mm FIROS-30 MRS
122 mm Type 90 MRS(?)
227 mm HIMARS
300 mm BM 9A52 Smerch MRS

Air Defence

I-HAWK
I-HAWK (MIM-23B)
Rapier
Rapier-1
Patriot PAC-2 GEM-T
Patriot PAC-3
THAAD
9K36 Strela-3
9K38 Igla
9K310 Igla-1
Blowpipe
Crotale
Crotale (R-440N)
Crotale (R-440)
Mistral
Mistral 2
RBS 70
Pantsyr-S1
Pantsyr-S1 (9M311)
20 mm (Twin) Panhard M3 VDA
35 mm (Twin) GDF-002series
20 mm M55A2

United Kingdom

Artillery

155 mm AS90 SPG
105 mm Light Gun
227 mm MLRS

Air Defence

Bloodhound-2
Centurion
FIM-92A Stinger
FIM 92C Stinger RMP
Javelin
Rapier
Rapier FSC
Rapier-1
Starstreak
Starburst
9K310 Igla-1

United States

Artillery

155 mm M109 series SPH
105 mm M101 howitzer
105 mm M102 howitzer (R)
105 mm M119A1/M119A2 Light Gun
155 mm M198 howitzer (R)
155 mm M777/M777A1/M777A2 howitzer
227 mm MLRS
227 mm HIMARS

Air Defence

FIM-92A Stinger
FIM-92B Stinger
FIM 92C Stinger RMP
FIM-92E Stinger
9K38 Igla
9K36 Strela-3
9K310 Igla-1
M6 Bradley Linebacker
Patriot
Patriot PAC-3
NASAMS II
SLAMRAAM

Roland
Roland-2
ADATS
Rapier
Rapier-1
FIM 92C Stinger RMP
25 mm LAV with Stinger

Uruguay

Artillery

75 mm Bofors M1935 FG
105 mm M101 howitzer
105 mm M102 howitzer
155 mm M114 howitzer
122 mm 2S1 SPH
122 mm RM-70 MRS

Air Defence

20 mm M167 VADS
20 mm TCM-20
40 mm L/60

Uzbekistan

Artillery

120 mm 2S9 SPM/H
122 mm 2S1 SPH
152 mm 2S3 SPGH
203 mm 2S7 SPG
100 mm T-12 ATG
122 mm D-30 howitzer
152 mm D-20 GH
152 mm 2A36 gun
122 mm BM-21 MRS

Air Defence

S-75 Dvina
S-125 Pechora
S-200 Angara
9K32 Strela-2
9K32M Strela-2M
9K36 Strela-3
9K310 Igla-1

Venezuela

Artillery

155 mm Mk F3 SPG
155 mm M109 SPH
75 mm M116 P howitzer
105 mm Model 56 P howitzer (and Ma)
105 mm M101 howitzer
155 mm M114 howitzer
160 mm LARS MRS

Air Defence

Barak
Roland
Roland 2
Antei-2500
S-300P
Buk-M2
S-125-2M Pechora Phase II
20 mm TCM-20
20 mm (Twin) S530
23 mm ZU-23-2
40 mm (Twin) 40L70
40 mm (Single) L/70
40 mm (Single) L/60
40 mm (Twin) M42
RBS 70
Mistral
9K338 Igla-S
9K331 Tor-M1

Vietnam

Artillery

SU-100 SPG
120 mm 2S9 SPM/H
152 mm 2S3 SPG
175 mm M107 SPG ?
76 mm M1942 FG
85 mm Type 56 FG
85 mm D-44 FG
85 mm D-48 ATG
100 mm M1944 FG
100 mm T-12 ATG
105 mm M101 howitzer
122 mm D-30 howitzer
122 mm D-74 FG
122 mm M1938 howitzer
122 mm Type 54 howitzer
122 mm Type 60 FG
130 mm M-46 FG
130 mm Type 59 FG
130 mm Type 59-1 FG
152 mm Type 66 GH
152 mm D-20 GH
152 mm M1937 howitzer
152 mm D-1 howitzer
107 mm Type 63 MRS
122 mm BM-21 MRS
130 mm Type 63 MRS
140 mm BM-14-16 MRS

Air Defence

S-75 Desna
S-75M Volga
S-125 Pechora
S-125 Pechora (5V27)
S-125-2TM
S-300 PMU-1
S-300 PMU-1 (48N6)
2K12 Kvadrat
2K12 Kvadrat (3M9)
9K31 Strela-1
9K32 Strela-2
9K32M Strela-2M
9K32 Strela-2
9K32M Strela-2M
9K36 Strela-3
9K38 Igla
9K310 Igla-1
Kashtan (9M311)
12.7 mm (Quad) M53
14.5 mm (Single) ZPU-1
14.5 mm (Twin) ZPU-2
14.5 mm (Quad) ZPU-4
23 mm (Twin) ZU-23-2
23 mm (Quad) ZSU-23-4
30 mm (Twin) M53
37 mm (Single) M1939
57 mm (Twin) ZSU-57-2
57 mm (Single) S-60
85 mm (Single) KS-12
100 mm (Single) KS-19
130 mm (Single) KS-30

Yemen

Artillery

76 mm M1942 FG
85 mm D-44 FG
100 mm M1944 ATG
105 mm M101 howitzer
122 mm 2S1 SPH
122 mm D-30 howitzer
122 mm M1931/7 gun
122 mm M1938 howitzer
122 mm M-46 FG
152 mm D-20 GH
152 mm M114 howitzer
122 mm BM-21 MRS
140 mm BM-14 MRS
220 mm BM 9140 (16-round) MRL
FROG-7 ARS
SU-100 SPG

Air Defence

20 mm M167 VADS
20 mm (Single) M163 VADS
23 mm (Twin) ZU-23-2
23 mm (Quad) ZSU-23-4
37 mm (Single) M1939
57 mm (Single) S-60
85 mm (Single) KS-12
2K12 Kub
9K31 Strela-1
9K32 Strela-2
9K32M Strela-2M
S-75 Dvina
S-125 Pechora

Zambia

Artillery

25-pounder FG
76 mm M1942 FG
122 mm BM-21 MRS
122 mm D-30 howitzer
130 mm M-46 FG

Air Defence

S-125 Pechora
S-125 Pechora (5V27)
9K32 Strela-2
9K32M Strela-2M
Rapier
Rapier-1
Tigercat
14.5 mm (Quad) ZPU-4
20 mm M55
37 mm M1939
57 mm (Single) S-60
85 mm (Single) KS-12

Zimbabwe

Artillery

107 mm Type 63 MRS
122 mm D-30 howitzer
122 mm RM-70 MRS (Mod 70/85)
122 mm Type 60 FG

Air Defence

14.5 mm (Single) ZPU-1
14.5 mm (Twin) ZPU-2
14.5 mm (Quad) ZPU-4
23 mm (Twin) ZU-23-2
37 mm (Single) M1939
9K32 Strela-2
9K32M Strela-2M

Contractors

Argentina

Dirección General de Fabricaciones Militares
Av Cabildo 65, C1426AAA, Buenos Aires, Argentina
Tel: (+54 11) 47 79 34 00
Fax: (+54 11) 45 76 55 43
e-mail: info@ffmm.gov.ar
Web: www.fab-militares.gov.ar

Belarus

Belvneshpromservice
Rue Kazinets 2, 220099, Minsk, Belarus
Tel: (+375 17) 219 07 08
Fax: (+375 17) 278 24 08
e-mail: reception@bvpservice.com
Web: www.bvpservice.com

Brazil

Avibras Indústria Aeroespacial SA
Rodovia dos Tamoios km 14, Jacareí,
PO Box 278 12315-020, São Paulo, Brazil
Tel: (+55 12) 39 55 60 00
Fax: (+55 12) 39 51 62 77
e-mail: gspd@avibras.com.br
Web: www.avibras.com.br

Bulgaria

Arsenal 2000 JSCo
100 Rozova Dolina Street, 6100, Kazanlak, Bulgaria
Tel: (+359 431) 637 40
(+359 431) 500 00
Fax: (+359 431) 637 83
(+359 431) 500 01
e-mail: arsenal2000@arsenal2000.com
marketing@arsenal2000.com
Web: www.arsenal-bg.com
www.arsenal2000.com

Beta Industry Corp JSC
Industrial zone, 5980, Cherven Briag, Bulgaria
Tel: (+359 659) 922 17
Fax: (+359 659) 939 66
e-mail: beta@beta.bg
Web: www.beta.bg

Vazovski Mashinostroitelni Zavodi AD (VMZ AD)
1 Ivan Vazov Boulevard, 4330, Sopot, Bulgaria
Tel: (+359 3134) 98 48
(+359 2) 980 55 04
Fax: (+359 3134) 60 25
(+359 3134) 98 10
e-mail: office@vmz.bg
Web: www.vmz.bg

Canada

General Dynamics Land Systems - Canada (GDLS-C)
(a subsidiary of General Dynamics Land Systems, US)
2035 Oxford Street East, Building 10, London, Ontario, N5V 2Z7, Canada
Tel: (+1 519) 964 59 00
Fax: (+1 519) 964 54 88
e-mail: gdlscanada@gdls.com
Web: www.gdlscanada.com
www.gdls.com
www.generaldynamics.com

Rheinmetall Canada Inc
(a subsidiary of Rheinmetall Defence, Germany)
225 Boulevard du Seminaire Sud,
Saint-Jean-Sur-Richelieu, Québec, J3B 8E9, Canada
Tel: (+1 450) 358 20 00
Fax: (+1 450) 358 17 44
e-mail: info@rheinmetall.ca
Web: www.rheinmetall.ca

China

Aerospace Long-March International Trade Co Ltd (ALIT)
No 7 Building, Section 15, ABP Beijing, No 188, Nansihuan Xilu, Fengtai, Beijing, China
Tel: (+86 10) 56 53 37 00
(+86 10) 56 53 37 01
Fax: (+86 10) 56 53 37 77
(+86 10) 56 53 38 88
(+86 10) 56 53 39 99
e-mail: alitchina@alitchina.com
Web: www.alitchina.com

China North Industries Corporation (NORINCO)
Headquarters
12A Guang An Men Nan Jie,
PO Box 2932 100053, Beijing, China
Tel: (+86 10) 63 52 99 88
Fax: (+86 10) 63 54 03 98
e-mail: norinco@norinco.cn
Web: www.norinco.com

Poly Technologies Inc
27/F, New Poly Plaza, No.1 Chaoyangmen Beidajie, 100010, Beijing, Dongcheng, China
Tel: (+86 10) 64 08 22 88
Fax: (+86 10) 64 08 29 88
Web: www.poly.com.cn

Croatia

Agencija Alan doo
Nike Grskovca 15, 10000, Zagreb, Croatia
Tel: (+385 1) 378 08 06 (export)
(+385 1) 378 08 05 (import)
(+385 1) 378 08 02 (gm office)
Fax: (+385 1) 378 08 40 (export)
(+385 1) 378 08 32 (import)
(+385 1) 378 08 38 (gm office)
e-mail: export.dep1@aalan.hr
Web: www.aalan.hr

Egypt

Sakr Factory for Developed Industries
(a subsidiary of Arab Organisation for Industrialisation, Egypt)
4.5km Cairo - Suez Road, Sakr Factory,
PO Box 2 Heliopolis, Cairo, Egypt
Tel: (+20 2) 22 69 21 55
(+20 2) 22 69 33 14
Fax: (+20 2) 22 69 12 10
(+20 2) 22 69 19 78
e-mail: sakrmarketing@aoi.com.eg
Web: www.aoi.com.eg

France

Eurosam GIE
(a subsidiary of MBDA, UK)
Centre d'affaires La Boursidière, Bâtiment Kerguelen, Le Plessia- Robinson, F-92357, Cedex, France
Tel: (+33 1) 41 87 14 16
Fax: (+33 1) 41 87 14 42
Web: www.eurosam.com

Nexter
Headquarters
13 route de la Minière, F-78034, Versailles, Cedex, France
Tel: (+33 1) 30 97 37 37
(+33 1) 30 97 36 41
Fax: (+33 1) 30 97 39 78
(+33 1) 39 24 16 93
(+33 1) 30 97 39 00
Web: www.nexter-group.fr
www.nexter-group.com

Thales
Corporate Headquarters
45 rue de Villiers, F-92526, Neuilly sur Seine, Cedex, France
Tel: (+33 1) 57 77 80 00
Fax: (+33 1) 57 77 83 00
e-mail: press.office@thalesgroup.com
info@thalesgroup.com
dtri@thalesgroup.com
Web: www.thalesgroup.com

Germany

Diehl BGT Defence GmbH & Co KG
PO Box 10 11 55 D-88641, Ueberlingen, Germany
Alte Nussdorfer Strasse 13, D-88662, Ueberlingen, Germany
Tel: (+49 7551) 89 01
Fax: (+49 7551) 89 28 22
e-mail: info@diehl-bgt-defence.de
Web: www.diehl.com

GLS Gesellschaft für Logistischen Service mbH
(a subsidiary of Krauss-Maffei Wegmann GmbH & Co KG, Germany)
PO Box 500231 D-80972, München, Germany
Krauss-Maffel Strasse 11, D-80997, München, Germany
Tel: (+49 89) 81 40 50
Fax: (+49 89) 812 63 80
e-mail: info@kmweg.de
Web: www.kmweg.de/

Krauss-Maffei Wegmann GmbH & Co KG (KMW)
Headquarters
Krauss-Maffei-Strasse 11, D-80997, München, Germany
Tel: (+49 89) 81 40 54 98
Fax: (+49 89) 81 40 49 77
e-mail: info@kmweg.de
kmw-press-service@kmweg.de
Web: www.kmweg.com
www.kmweg.de

MBDA
Headquarters
(a joint venture between BAE Systems, UK (37.5 per cent), Airbus Group, Netherlands (37.5 per cent) and Finmeccanica, Italy (25 per cent))
PO Box 1340 D-86529, Schrobenhausen, Germany
Hagenauer Forst 27, D-86529, Schrobenhausen, Germany
Tel: (+49 8252) 990
Fax: (+49 8252) 99 61 20
Web: www.mbda-systems.com
www.mbda.net

MBDA Deutschland LFK GmbH
Hagenauer Forst 27, D-86529, Schrobenhausen, Germany
Tel: (+49 8252) 99 25 49
Fax: (+49 8252) 99 38 71
e-mail: communications@mbda-systems.de
Web: www.mbda-systems.com

MTU Friedrichshafen GmbH
Maybachplatz 1, D-88040, Friedrichshafen, Germany
Tel: (+49 7541) 900
(+49 7541) 90 70 40
(+49 7541) 90 47 41
Fax: (+49 7541) 90 50 00
(+49 7541) 90 70 84
(+49 7541) 90 70 81
e-mail: info@mtu-online.com
Web: www.mtu-online.com

Rheinmetall Defence Electronics GmbH
Headquarters
(a subsidiary of Rheinmetall AG, Germany)
Brueggeweg 54, D-28309, Bremen, Germany
Tel: (+49 421) 457 01
Fax: (+49 421) 457 29 00
e-mail: info-rde@rheinmetall.com
Web: www.rheinmetall-defence.de

Rheinmetall Waffe Munition GmbH (RWM)
Headquarters
Rheinmetall W & M GmbH
(a subsidiary of Rheinmetall AG, Germany)
Heinrich-Ehrhardt-Strasse 2, D-29345, Unterluess, Germany
Tel: (+49 58) 27 80 02
Fax: (+49 58) 27 10 90
e-mail: info@rheinmetall-defence.com
Web: www.rheinmetall-detec.de

Greece

Hellenic Aerospace Industry SA (HAI)
PO Box 23 GR-320 09, Schimatari, Greece
Tel: (+30 22620) 520 00
Fax: (+30 22620) 521 70
e-mail: contact@haicorp.com (public relations)
Web: www.haicorp.com

Hellenic Defence Systems SA
(a subsidiary of EBO Group of Companies, Greece)
1 Ilioupoleos Avenue, Athinai, Greece
Tel: (+30 210) 979 90 00
(+30 210) 979 91 62
e-mail: info@eas.gr
Web: www.eas.gr

India

Ordnance Factory Board (OFB)
Ayudh Bhawan, 10-A S K Bose Road, Kolkata (Calcutta), 700001, India
Tel: (+91 33) 22 43 90 27
(+91 33) 22 43 04 72
(+91 33) 22 43 21 03
(+91 33) 22 48 50 77
(+91 33) 22 48 50 80
(+91 33) 22 48 91 21
(+91 33) 22 48 91 28
Fax: (+91 33) 22 48 17 48
e-mail: ofbtrade@dataone.in
export.ofb@nic.in
Web: ofbindia.gov.in
ofbindia.nic.in

Iran

Defence Industries Organisation (DIO)
Pasdaran Street, Entrance of Babaie Highway, PO Box 19585-777 Tehran, Iran
Tel: (+98 21) 22 56 28 83
(+98 21) 22 59 57 57
(+98 21) 22 54 20 59
Fax: (+98 21) 22 55 19 61
(+98 21) 22 55 19 60
e-mail: marketing@diol.org
Web: www.diomil.ir

Israel

ELTA Systems Ltd (IAI/ELTA)
(a subsidiary of Israel Aerospace Industries Ltd, Israel)
100 Yitzhak, Hanasi Boulevard, PO Box 330 IL-77102, Ashdod, Israel
Tel: (+972 8) 857 23 33
(+972 8) 857 23 12
(+972 8) 857 24 10
Fax: (+972 8) 856 18 72
e-mail: market@elta.co.il
Web: www.iai.co.il
www.elta-iai.com

Israel Aerospace Industries Ltd (IAI)
Headquarters
Ben Gurion International Airport, IL-70100, Tel-Aviv, Israel
Tel: (+972 3) 935 33 43
(+972 3) 935 31 11
(+972 3) 935 00 00
Fax: (+972 3) 935 85 16
(+972 3) 935 82 78
(+972 3) 935 31 31
(+972 3) 935 41 62
e-mail: corpmkg@iai.co.il
helpdesk@iai.co.il
Web: www.iai.co.il

Israel Aerospace Industries Ltd (IAI)
MBT Systems Missiles and Space Group
Yehud Industrial Zone, PO Box 105 IL-56000, Beer Yacov, Israel
Web: www.iai.co.il

Israel Military Industries (IMI)
PO Box 1044 IL-4711001, Ramat Hasharon, Israel
Tel: (+972 3) 548 52 22
Fax: (+972 3) 548 61 25
e-mail: imimrktg@imi-israel.com
Web: www.imi-israel.com

Israel Military Industries Ltd (IMI)
Rocket Systems (RSD)
PO Box 1044 IL-47100, Ramat Hasharon, Israel
Tel: (+972 8) 924 26 84
(+972 8) 927 74 47
(+972 8) 927 74 49
Fax: (+972 8) 925 28 96
e-mail: iweinreb@imi-israel.com
Web: www.imi-israel.com

Rafael Advanced Defense Systems Ltd
Corporate Headquarters
Rafael Manor
PO Box 2250 IL-3102102, Haifa, Israel
Tel: (+972 4) 879 44 44
(+972 4) 879 47 14
Fax: (+972 4) 879 46 57
e-mail: intl-mkt@rafael.co.il
customersupport@rafael.co.il
Web: www.rafael.co.il

Rafael Advanced Defense Systems Ltd
Rafael Systems (RSD)
PO Box 2250 IL-31021, Haifa, Israel
Tel: (+972 4) 879 47 17
Fax: (+972 4) 879 46 57
e-mail: intl-mkt@rafael.co.il
Web: www.rafael.co.il

Italy

MBDA
Headquarters
(a joint venture between BAE Systems, UK (37.5 per cent), Airbus Group, Netherlands (37.5 per cent) and Finmeccanica, Italy (25 per cent))
Via Carciano 4-50, 60-70, I-00131, Roma, Italy
Tel: (+39 06) 877 11
Web: www.mbda-systems.com

Oto Melara SpA
(a subsidiary of Finmeccanica SpA, Italy)
Via Valdilocchi, 15, I-19136, La Spezia, Italy
Tel: (+39 0187) 58 11 11
Fax: (+39 0187) 58 26 69
e-mail: press-office@otomelara.it
communication@otomelara.it
Web: www.otomelara.it

Selex ES
Head Office
(a subsidiary of Finmeccanica SpA, Italy)
Via Tiburtina Km 12,400, I-00131, Roma, Italy
Tel: (+39 06) 415 01
Fax: (+39 06) 413 14 36
(+39 06) 413 11 33
e-mail: info@selex-si.com
Web: www.selex-si.com

Japan

Japan Steel Works Ltd (JSW)
Head Office
Gate City Ohsaki West Tower, 11-1, Osaki 1-chome, Shinagawa-ku, Tokyo, 141-0032, Japan
Tel: (+81 3) 57 45 20 01
Fax: (+81 3) 57 45 20 25
Web: www.jsw.co.jp

Mitsubishi Heavy Industries Ltd (MHI)
Head Office
Mitsubishi Jukogyo Kabushiki Kaisha
16-5 Konan, 2-Chome, Minato-ku, Tokyo, 108-8215, Japan
Tel: (+81 3) 67 16 31 11
Fax: (+81 3) 67 16 58 00
Web: www.mhi.co.jp

Toshiba Corporation
Principal Office
1-1 Shibaura, 1-chome, Minato-ku, Tokyo, 105-8001, Japan
Tel: (+81 3) 34 57 45 11
Fax: (+81 3) 54 44 92 84
Web: www.toshiba.co.jp

Jordan

King Abdullah II Design and Development Bureau (KADDB)
PO Box 928125 11190, Amman, Jordan
Tel: (+962 6) 625 60 24
(+962 6) 460 32 30
Fax: (+962 6) 460 32 35
(+962 6) 562 72 03
(+962 6) 460 32 43
e-mail: info@kaddb.com
sales@kaddb.com
Web: www.kaddb.mil.jo
www.kaddbinvest.com

Korea, South

LIG Nex1 Co Ltd
Hapjeong-dong, Hapjeong Building, 19, Yanghwa-ro, Mapo-gu, Seoul, 121-885, Korea, South
Tel: (+82 2) 69 42 53 00
(+82 2) 16 44 20 05
Fax: (+82 2) 69 42 54 54
Web: www.lignex1.com

Netherlands

Thales Nederland BV
Headquarters
(a subsidiary of Thales, France)
PO Box 42 NL-7550 GD, Hengelo, Netherlands
Haaksbergerstraat 49, NL-7554 PA, Hengelo, Netherlands
Tel: (+31 74) 248 81 11
Fax: (+31 74) 242 59 36
e-mail: info@nl.thalesgroup.com
Web: www.thales-nederland.nl

Pakistan

Institute of Industrial Control Systems (IICS)
(a subsidiary of Global Industrial & Defence Solution (GIDS), Pakistan)
Dhoke Nusah, Dakhli Gangal, Near Chatri Chowk, PO Box 1398 46000, Rawalpindi, Pakistan
Tel: (+92 51) 447 00 70
Fax: (+92 51) 447 00 76
e-mail: info@iics.com.pk
Web: www.iics.com.pk

Pakistan Ordnance Factories (POF)
Exports
(a subsidiary of Defence Export Promotion Organisation, Pakistan)
Board Office Secretariat, Wah Cantonment, 47040, Rawalpindi, Pakistan
Tel: (+92 51) 931 41 01
Fax: (+92 51) 931 40 58
(+92 51) 927 14 00 (industrial and commercial relations)
(+92 596) 931 41 00 (industrial and commercial relations)
(+92 596) 931 40 80
e-mail: export@pof.gov.pk
exports@pof.gov.pk
Web: www.pof.gov.pk

Poland

Bumar Electronics SA
ul. Poligonwa 30, PL-04 051, Warszawa, Poland
Tel: (+48 22) 540 21 01
Fax: (+48 22) 673 78 78
e-mail: office_dbe@bumar.com
Web: www.bumar.com

Cnpep Radwar SA
Marketing Department, 30 Poligonowa Street, PL-04-051, Warszawa, Poland
Tel: (+48 22) 540 22 00
Fax: (+48 22) 813 48 84
e-mail: office_dbe@bumar.com
Web: www.bumar.com

Huta Stalowa Wola SA (HSW SA)
Military Production Centre (CPW)
Ulica Kwiatkowskiego 1, PL-37-450, Stalowa Wola, Poland
Tel: (+48 15) 813 77 77
(+48 15) 813 48 30
Fax: (+48 15) 813 49 84
e-mail: cpw@hsw.pl
Web: www.hsw.pl

Huta Stalowa Wola SA (HSW SA)
Technical-Trading Office
ul Leszno 14, PL-01-192, Warszawa, Poland
Tel: (+48 15) 813 48 30
Fax: (+48 15) 813 49 84
Web: www.hsw.pl

Military Armament Plant No 2 (WZU)
Wojskowe Zaklady Uzbrojenia nr 2
ulitsa Parkowa 42, PL-86-300, Grudziadz, Poland
Tel: (+48 56) 644 62 00
Fax: (+48 56) 462 37 83
e-mail: wzu@wzu.pl
Web: www.wzu.pl

Romania

Aerostar SA
9 Condorilor Street, R-600302, Bacau, Romania
Tel: (+40 234) 57 50 70
e-mail: aerostar@aerostar.ro
Web: www.aerostar.ro

Russian Federation

Almaz CMDB JSC
Warshavskaya 50, 196128, St Petersburg, Russian Federation
Tel: (+7 812) 369 59 25
e-mail: office@almaz-kb.sp.ru
Web: almaz-kb.com

Almaz-Antey Concern of Air Defence
ul. Vereyskaya, 41, 121471, Moskva, Russian Federation
Tel: (+7 495) 276 29 80
Fax: (+7 495) 276 29 81
e-mail: antey@almaz-antey.ru
Web: www.almaz-antey.ru

JSC Defence Systems
(a subsidiary of Oboronprom United Industrial Corporation, Russia)
29 Vereiskaya Street, 121357, Moskva, Russian Federation
Tel: (+7 95) 440 09 12
Fax: (+7 95) 440 06 87
e-mail: defensys@defensys.ru
Web: www.defensys.ru

JSC Motovilikhinskiye Zavody
Motovilikha Works
35 1905 Goda Street, 614014, Perm, Russian Federation
Tel: (+7 342) 260 73 03
(+7 342) 260 73 00
Web: mz.perm.ru

JSC V Tikhomirov Scientific Research Institute of Instrument Design (JSC NIIP)
3 Gagarin Street, 140180, Zhukovsky, Moscow Region, Russian Federation
Tel: (+7 495) 556 23 48
Fax: (+7 495) 721 37 85
(+7 495) 721 35 59
e-mail: niip@niip.ru
corp@niip.ru
Web: www.niip.ru

KBP Instrument Design Bureau
59 Shcheglovskaya Zaseka Street, 300001, Tula, Russian Federation
Tel: (+7 4872) 41 02 10
(+7 4872) 41 00 68
Fax: (+7 4872) 42 61 39
(+7 4872) 46 98 61
e-mail: kbkedr@tula.net
Web: www.kbptula.ru

KBP Instrument Design Bureau
Kievskaya street 7, gate 7, level 5, Moskva, Russian Federation
Tel: (+7 495) 544 56 00
Fax: (+7 495) 544 56 01
e-mail: office@kbpmoscow.ru
Web: www.kbptula.ru

Konstruktorskoye Byuro Mashinostroyeniya (KBM)
(a subsidiary of Open Joint-Stock Company V A Degtyarev Plant, Russia)
42 Oksky Avenue, 140402, Kolomna, Moscow Region, Russian Federation
Tel: (+7 496) 616 31 50
Fax: (+7 496) 615 50 04
e-mail: kbm-kbm@mail.ru
Web: www.kbm.ru

Kurganmashzavod Joint Stock Company
17 Mashinostroitley Avenue, 640027, Kurgan, Russian Federation
Tel: (+7 3522) 53 22 44
(+7 3522) 57 49 76
Fax: (+7 3522) 23 20 71
e-mail: root@kurganmash.ru
Web: www.kurganmash.ru
www.kmz.ru

Novosibirsk Instrument Making Plant (NPZ)
D Kovalchuk 179/2, 630049, Novosibirsk, Russian Federation
Tel: (+7 383) 216 08 33
(+7 383) 228 52 59
Fax: (+7 383) 226 15 94
e-mail: info@npzoptics.ru
sales@npzoptics.ru
Web: www.npzoptics.com

Public Joint Stock Company Plant No 9
Artillery Plant No.9
(a subsidiary of Uralvagonzavod, Russia)
pl. First Five, 620012, Ekaterinburg, Russian Federation
Tel: (+7 343) 327 59 00
Fax: (+7 343) 336 66 84
e-mail: zavod9@r66.ru
Web: www.zavod9.com

Splav Scientific Production Enterprise
(a subsidiary of State Research and Production Enterprise Bazalt, Russia)
Sheglovskaya Zaseka 33, 300004, Tula, Russian Federation
Tel: (+7 4872) 46 45 86
(+7 4872) 46 46 47
Fax: (+7 4872) 41 14 74
(+7 4872) 31 52 87
e-mail: market@splav.org
mail@splav.org
foreign@splav.org
Web: www.splav.org

State Governmental Scientific-Testing Area of Aircraft Systems (FKP GkNIPAS)
140250, Belozersky, Voskresensky District, Russian Federation
Tel: (+7 495) 556 07 09
Fax: (+7 495) 556 07 40
e-mail: info@fkpgknipas.ru
gknipas@trancom.ru
Web: www.fkpgknipas.ru

Ural Plant of Transport Engineering - JSC Uraltransmash
(a subsidiary of Uralvagonzavod, Russia)
ul.Frontovyh brigades, 29, 620027, Ekaterinburg, Russian Federation
Tel: (+7 343) 336 71 11
(+7 343) 336 70 71
e-mail: utranspost@etel.ru
Web: www.uraltransmash.com

Serbia

Yugoimport SDPR
2 Bulevar Umetnosti, 11150, Beograd, Serbia
Tel: (+381 11) 222 44 44
Fax: (+381 11) 222 45 99
e-mail: fdsp@eunet.rs
office@yugoimport.com
Web: www.yugoimport.com

Singapore

Allied Ordnance of Singapore (Pte) Ltd (AOS)
(a subsidiary of Singapore Technologies Kinetics Ltd, Singapore)
249 Jalan Boon Lay, Singapore, 619523, Singapore
Tel: (+65) 64 68 00 41
Fax: (+65) 64 68 71 43
Web: www.stengg.com

South Africa

Denel Dynamics
PO Box 7412 0046, Centurion, South Africa
Tel: (+27 12) 671 19 11 (switchboard)
(+27 12) 671 10 01 (marketing)
Fax: (+27 12) 671 17 79
e-mail: market@deneldynamics.co.za
Web: www.deneldynamics.co.za

Rheinmetall Denel Munition Pty Ltd
Reeb Road, Firgrove, 7130, Somerset West, South Africa
Tel: (+27 21) 850 29 11
Fax: (+27 21) 850 20 11
e-mail: marketing@rheinmetall-denelmunition.com
Web: www.rheinmetall-defence.com

Thales Defence Systems (ADS)
(a subsidiary of Thales Air Defence SA, France)
55 Richards Road, Halfway House, 1685, Gauteng, Midrand, South Africa
Tel: (+27 11) 313 91 23
Fax: (+27 11) 313 91 68
Web: www.thalesgroup.com

Spain

EXPAL
Avenida del Partenon 16, Campo de las Naciones, E-28042, Madrid, Spain
Tel: (+34 917) 22 02 35
(+34 917) 22 01 00
Fax: (+34 917) 22 01 01
(+34 917) 22 20 95
e-mail: contact.es@maxam.net
expal@expal.biz
Web: www.expal.biz

General Dynamics European Land Systems-Santa Bárbara Sistemas
Headquarters
(a part of General Dynamics European Land Systems, Austria)
Parque Empresarial Cristalia, Edificio 7/8, Vía de los Poblados 3, E-28033, Madrid, Spain
Tel: (+34 91) 585 02 40
Fax: (+34 91) 585 02 18
e-mail: info.sbs@gdels.com
sales.sbs@gdels.com
Web: www.generaldynamics.com
www.gdels.com

Sweden

BAE Systems Land & Armaments
Global Combat Systems Weapons
SE-691 80, Karlskoga, Sweden
Tel: (+46 0586) 73 30 00
Fax: (+46 0596) 73 30 12
Web: www.baesystems.com

Saab Electronic Defence Systems
Solhusgatan 10, Kallebaecks Teknikpark, SE-412 89, Göteborg, Sweden
Tel: (+46 31) 794 90 00
Fax: (+46 31) 794 90 02
Web: www.saabgroup.com

Switzerland

RUAG Defence Ltd
(a subsidiary of RUAG Holding Ltd, Switzerland)
Allmendstrasse 86, CH-3602, Thun, Switzerland
Tel: (+41 332) 28 21 11
Fax: (+41 332) 28 20 47
e-mail: info.landsystems@ruag.com
marketing.landsystems@ruag.com
info.defence@ruag.com
info@ruag.com
Web: www.ruag.com

Rheinmetall Air Defence AG
(a subsidiary of the Rheinmetall Defence Group, Germany)
Birchstrasse 155, CH-8050, Zürich, Switzerland
Tel: (+41 44) 316 22 11
Fax: (+41 44) 311 31 54
e-mail: info@rheinmetall-ad.com
Web: www.rheinmetall-defence.com

Taiwan

Chung Shan Institute of Science and Technology (CSIST)
No 15, Shi Qi Zi, Gaoping Village, Longtan Township, Taoyuan, Taiwan
Tel: (+886 3) 411 21 17
Fax: (+886 3) 411 54 60
e-mail: csist@csistdup.org.tw
Web: www.csistdup.org.tw

Turkey

Aselsan AS
Head Office
PK 1 TR-06172, Cankaya, Ankara, Turkey
Mehmet Akif Ersoy Mahallesi 296, Cadde No 16, TR-06370, Macunköy, Ankara, Turkey
Tel: (+90 312) 592 10 00
Fax: (+90 312) 354 13 02
(+90 312) 354 26 69
e-mail: marketing@aselsan.com.tr (international marketing)
Web: www.aselsan.com.tr

Makine ve Kimya Endüstrisi Kurumu (MKEK)
General Directorate
Tandogan, TR-06330, Ankara, Turkey
Tel: (+90 312) 296 10 00
(+90 312) 296 10 09
Fax: (+90 312) 213 13 62
e-mail: mkek@mkek.gov.tr
Web: www.mkek.gov.tr

Roketsan
Roket Sanayii Ticaret AS
Roketsan Roket Sanayii ve Ticaret AS
Kemalpasa Mah. Sehit Yuzbasi, Adem Kutlu Sk. No.21, TR-06780, Elmadag, Ankara, Turkey
Tel: (+90 312) 860 55 00
Fax: (+90 312) 863 42 08
e-mail: marketing@roketsan.com.tr
Web: www.roketsan.com.tr

Ukraine

State Kyiv Design Bureau 'Luch' (SKDB 'Luch')
2 Melnikova Street, 04050, Kiyev, Ukraine
Tel: (+380 044) 483 07 45
(+380 044) 484 18 98
Fax: (+380 044) 483 13 94
e-mail: kb@luch.kiev.ua
Web: www.luch.kiev.ua

United Arab Emirates

Emirates Defense Technology
Airport Road, Bin Jabr Building,
PO Box 46711 Abu Dhabi, United Arab Emirates
Tel: (+971 2) 641 96 60
Fax: (+971 2) 641 81 84
e-mail: inquiries@emiratesdefense.com
Web: emiratesdefense.com

International Golden Group (IGG)
Baniyas Tower,
PO Box 43999 Abu Dhabi, United Arab Emirates
Tel: (+971 2) 626 66 61
Fax: (+971 2) 626 70 70
e-mail: igg@iggroup.ae
Web: www.iggroup.ae

Jobaria Defense Systems
(a member of The Tawazun Group)
PO Box 2175 Abu Dhabi, United Arab Emirates
Tel: (+971 2) 555 75 44
Fax: (+971 2) 555 76 27
e-mail: info@jobaria.ae
Web: www.jobaria.ae

United Kingdom

BAE Systems Land & Armaments
Global Combat Systems Vehicles
Armstrong Works, Scotswood Road,
Newcastle-upon-Tyne, NE99 1BX, United Kingdom
Tel: (+44 191) 273 88 88
Fax: (+44 191) 273 23 24
Web: www.baesystems.com

MBDA
UK Headquarters
(a joint venture between BAE Systems, UK (37.5 per cent), Airbus Group, Netherlands (37.5 per cent) and Finmeccanica, Italy (25 per cent))
11 Strand, London, WC2N 5RJ, United Kingdom
Tel: (+44 207) 451 60 00
e-mail: salesenquiries@mbda-systems.com
Web: www.mbda-systems.com

Moog Components Group Ltd
(a subsidiary of Kaydon Corporation, US)
30 Suttons Park Avenue, Suttons Business Park, Reading, Berkshire, RG6 1AW, United Kingdom
Tel: (+44 118) 966 60 44
Fax: (+44 118) 966 65 24
e-mail: mcg@moog.com
Web: www.moog.com

Thales Holding UK Ltd
UK Headquarters
(a subsidiary of Thales, France)
2 Dashwood Lang Road, The Bourne Business Park, Addlestone, Weybridge, Surrey, KT12 2NX, United Kingdom
Tel: (+44 1932) 82 48 00
Fax: (+44 1932) 82 49 48
Web: www.thalesgroup.com

United States

Airbus Group North America
(a subsidiary of Airbus Group, Netherlands)
2550 Wasser Terrace, Suite 9000, Herndon, Virginia, 20171, United States
Tel: (+1 703) 466 56 00
Fax: (+1 703) 466 56 01
Web: www.northamerica.airbus-group.com

BAE Systems Land & Armaments
Headquarters
1101 Wilson Boulevard, Suite 2000, Arlington, Virginia, 22209-2444, United States
Tel: (+1 703) 312 61 00
Fax: (+1 703) 312 61 11
Web: www.baesystems.com/landarmaments

BAE Systems Land & Armaments
US Combat Systems, Headquarters
(a subsidiary of BAE Systems Land & Armaments, US)
1300 Wilson Boulevard, Suite 700, Arlington, Virginia, 22209, United States
Tel: (+1 703) 907 82 00
Fax: (+1 703) 894 35 44
Web: www.baesystems.com/landandarmaments

Browning
Parts or Product Service
One Browning Place, Arnold, 63010, United States
Tel: (+1 800) 322 46 26 (ext. 2860)
Web: www.browning.com

General Dynamics Armament & Technical Products
Headquarters
(a subsidiary of General Dynamics Corp, US)
Four Lake Pointe Plaza, 2118 Water Ridge Parkway, Charlotte, North Carolina, 28217, United States
Tel: (+1 704) 714 80 00
Fax: (+1 704) 714 80 04
e-mail: gdbusdev@gdatp.com
Web: www.gdatp.com
www.generaldynamics.com

General Dynamics Armament & Technical Products
Technology Center
128 Lakeside Avenue, Burlington, Vermont, 05401-4985, United States
Tel: (+1 802) 657 61 00
Fax: (+1 802) 657 62 92
e-mail: gdbusdev@gdatp.com
Web: www.gdatp.com

General Dynamics Land Systems (GDLS)
Headquarters & Sterling Heights Complex
(a subsidiary of General Dynamics Corporation, US)
PO Box 2074 Warren, Michigan, 48090-2074, United States
38500 Mound Road, Sterling Heights, Michigan, 48310-3200, United States
Tel: (+1 586) 825 40 00
Fax: (+1 586) 825 40 13
e-mail: info@gdls.com
Web: www.gdls.com
www.generaldynamics.com

General Motors Corporation (GMC)
Global Headquarters
Renaissance Center,
PO Box 300 Detroit, Michigan, 48243, United States
Tel: (+1 313) 556 50 00
Web: www.gm.com

Lockheed Martin Corporation
Global Headquarters
6801 Rockledge Drive, Bethesda, Maryland, 20817, United States
Tel: (+1 301) 897 60 00
Fax: (+1 301) 897 62 52
Web: www.lockheedmartin.com

Lockheed Martin Maritime Systems and Sensors
(a subsidiary of Lockheed Martin Mission Systems & Sensors, US)
1210 Massillon Road, Akron, Ohio, 44315-0001, United States
Tel: (+1 330) 796 22 00
Web: www.lockheedmartin.com/ms2

Raytheon Company
Global Headquarters
870 Winter Street, Waltham, Massachusetts, 02451-1449, United States
Tel: (+1 781) 522 30 00
e-mail: info@raytheon.com
Web: www.raytheon.com

Raytheon Integrated Defense Systems (IDS)
Headquarters
(a subsidiary of Raytheon Company, US)
50 Apple Hill Drive, Tewksbury, Massachusetts, 01876, United States
Tel: (+1 978) 858 42 16
Fax: (+1 978) 858 91 94
e-mail: ids_communications@raytheon.com
Web: www.raytheon.com

Raytheon Missile Systems (RMS)
Headquarters
(a subsidiary of Raytheon Company, US)
1151 East Hermans Road, Tucson, Arizona, 85706, United States
Tel: (+1 520) 794 30 00
Fax: (+1 520) 794 28 99
Web: www.raytheon.com

Rock Island Arsenal
HQ AMCCOM
Bulding 90, 1 Rock Island Arsenal, Rock Island, Illinois, 61299-5000, United States
Tel: (+1 309) 782 60 01
e-mail: usarmy.ria.imcom-central.mbx.usag-ria-pa@mail.mil
Web: www.ria.army.mil

ThalesRaytheonSystems Co LCC
(a joint veture between Raytheon, US (50 per cent) and Thales SA, France (50 per cent))
1801 Hughes Drive,
PO Box 34055 Fullerton, California, 92834-9455, United States
Tel: (+1 714) 446 32 32
Fax: (+1 714) 446 27 75
Web: www.thalesraytheon.com

The Boeing Company
Defense, Space and Security
(a subsidiary of The Boeing Company, US)
PO Box 516 St Louis, Missouri, 63166, United States
Tel: (+1 314) 232 02 32
(+1 562) 797 20 20
Web: www.boeing.com

The Boeing Company
Strategic Missiles and Defense Systems
(a part of Boeing Integrated Defense Systems, US)
499 Boeing Boulevard, Building 48-18,
PO Box 516 Huntsville, Alabama, 35824, United States
Tel: (+1 256) 461 28 03
(+1 256) 461 21 21
Fax: (+1 256) 461 22 52
Web: www.boeing.com

US Army Space & Missile Defense Command (USASMDC)
Public Affairs
PO Box 1500 Huntsville, Alabama, 35807-3801, United States
Tel: (+1 256) 955 38 87
(+1 719) 554 19 83
Web: www.smdc.army.mil

Indexes

Alphabetical Index

A

B

C

D

E

F

G

H

N

O

P

Q

R

S

T

U

V

W

X

Y

Z

Manufacturers' Index

C

China Precision Machinery Import - Export Corporation (CPMIEC)

Chinese state factories

Chung-Shan Institute of Science and Technology CSIST)

Closed Joint Stock Company NTC Elins

CNPEP RADWAR S.A.

Combined Service Forces

Concurrent

Control Technique Dynamics (CTD) Ltd

D

Daewoo

Daimler-Benz Aerospace

Defence Agency for Technology and Quality (DATQ)

Defence Industries Organization (DIO)

Defence Products Industries

F

G

H

I

J

K

N

R

T

U

X

Y

Z

NOTES

NOTES